KU-730-374

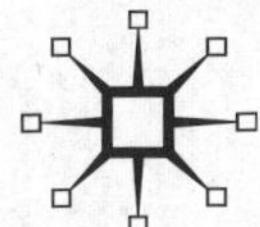

The editor of ***The Grants Register*** cannot undertake any correspondence in relation to grants listed in this volume.

While every care has been taken in compiling the information contained in this publication, the publisher and editor do not accept any responsibility for any errors or omissions therein.

First published 2013 by
PALGRAVE MACMILLAN

Palgrave Macmillan in the UK is an imprint of Macmillan Publishers Limited, registered in England, company number 785998, of Houndmills, Basingstoke, Hampshire RG21 6XS.

Palgrave Macmillan in the US is a division of St Martin's Press LLC,
175 Fifth Avenue, New York, NY 10010

Palgrave Macmillan is the global academic imprint of the above companies and has companies and representatives throughout the world.

Palgrave® and Macmillan® are registered trademarks in the United States, the United Kingdom, Europe and other countries.

ISBN: 978-0-230-39285-4
ISSN: 0072-5471

This book is printed on paper suitable for recycling and made from fully managed and sustained forest sources. Logging, pulping and manufacturing processes are expected to conform to the environmental regulations of the country of origin.

A catalogue record for this book is available from the British Library.

A catalog record for this book is available from the Library of Congress.

The Grants Register

2014

Thirty-Second Edition

palgrave
macmillan

LIST OF CONTENTS

PREFACE

The thirty-second edition of *The Grants Register* provides a detailed, accurate and comprehensive survey of awards intended for students at or above the postgraduate level, or those who require further professional or advanced vocational training.

Student numbers around the world continue to grow rapidly, and overseas study is now the first choice for many of these students. *The Grants Register* provides comprehensive, up-to-date information about the availability of, and eligibility for, non-refundable postgraduate and professional awards worldwide.

We remain grateful to the institutions which have supplied information for inclusion in this edition, and would also like to thank the International Association of Universities for continued permission to use their subject index within our Subject and Eligibility Guide to Awards.

The Grants Register database is updated continually in order to ensure that the information provided is the most current available. Therefore, if your details have changed or you would like to be included for the first time, please contact the Assistant Editor, at the address below. If you wish to obtain further information relating to specific application procedures, please contact the relevant grant-awarding institution, rather than the publisher.

Susan Povey
Assistant Editor
The Grants Register
Palgrave Macmillan
Houndmills,
Basingstoke,
RG21 6XS,
United Kingdom
Tel: 144 (0)1256 302683
Fax: 144 (0)1256 353774
Email: grants.editor@adi-mps.com

HOW TO USE *THE GRANTS REGISTER*

For ease of use, *The Grants Register 2014* is divided into five sections:

- *The Grants Register*
- Subject and Eligibility Guide to Awards
- Index of Awards
- Index of Discontinued Awards
- Index of Awarding Organisations

The Grants Register

Information in this section is supplied directly by the awarding organisations. Entries are arranged alphabetically by name of organisation, and awards are listed alphabetically within the awarding organisation. This section includes details on subject area, eligibility, purpose, type, numbers offered, frequency, value, length of study, study establishment, country of study, and application procedure. Full contact details appear with each awarding organisation and also appended to individual awards where additional addresses are given.

Subject and Eligibility Guide to Awards

Awards can be located through the Subject and Eligibility Guide to Awards. This section allows the user to find an award within a specific subject area. *The Grants Register* uses a list of subjects endorsed by the International Association of Universities (IAU), the information centre on higher education, located at UNESCO, Paris (please see pp. 873–876 for the complete subject list). It is further subdivided into eligibility by nationality. Thereafter, awards are listed alphabetically within their designated category, along with a page reference where full details of the award can be found.

ACADIA UNIVERSITY

Room 214B, Horton Hall, 18 University Avenue, Wolfville, NS, B4P 2R6, Canada
Tel: (1) 902 585 1914
Fax: (1) 902 585 1096
Email: theresa.starratt@acadiau.ca
Website: www.acadiau.ca
Contact: Ms Theresa Starratt, Graduate Studies Officer-Research and Graduate Studies

Acadia University is an institution that is commited to providing a liberal education based on the highest standards. The University houses a scholarly community that aims to ensure a broadening life experience for students, faculty and staff.

Acadia Graduate Awards

Subjects: English, political science, sociology, biology, chemistry, computer science, geology, psychology, education and recreation management, mathematics and statistics.
Purpose: To financially support students.
Eligibility: Open to registered full-time graduate students at Acadia University. In order to be eligible for an award students must have a GPA of not less than 3.00 in their major field in each of their last 2 years of undergraduate study.
Level of Study: Postgraduate
Type: Award
Value: Up to Canadian $9,000
Length of Study: 1–2 years

ENGINEERING

GENERAL

Any Country

A*STAR Graduate Scholarship (Overseas) 11
A*STAR International Fellowship 12
Association for Women in Science Educational Foundation Predoctoral Awards 125
AUC Assistantships 102
AUC Laboratory Instruction Graduate Fellowships 102
AUC University Fellowships 103
Berthold Leibinger Innovation Prize 155
BITS HP Labs India PhD Fellowship 159
BRI PhD Scholarships 162

Index of Awards

All awards are indexed alphabetically with a page reference.

Index of Discontinued Awards

This Index lists awards previously included within *The Grants Register* which are no longer being offered, have been replaced by another programme, or are no longer relevant for inclusion in the publication.

Index of Awarding Organisations

A complete list of all awarding organisations, with country name and page reference.

THE GRANTS REGISTER

A-T CHILDREN'S PROJECT

5300 W. Hillsboro Blvd. #105, Coconut Creek, FL, 33073, United States of America
Tel: (1) 954 481 6611
Fax: (1) 954 725 1153
Email: info@atcp.org
Website: www.atcp.org
Contact: Dr Cynthia Rothblum-Oviatt, Science Co-ordinator

The A-T Children's Project is a non-profit organization that raises funds to support and co-ordinate biomedical research projects, scientific conferences and a clinical centre aimed at finding a cure for ataxia-telangiectasia, a lethal genetic disease that attacks children, causing progressive loss of muscle control, as well as cancer and immune system problems.

Ataxia-Telangiectasia Childrens Project Research Grant

Subjects: Although basic science grant proposals are welcome, our funding efforts are focused on translational (bench to beside) research, clinical studies and proposals that apply innovative and novel strategies for suggesting, developing evaluating specific disease-modifying and symptomatic interventions.
Purpose: To accelerate first-rate, international scientific research on Ataxia-Telangiectasia to help find life-improving therapies or a cure for children A-T.
Eligibility: Open to applicants of all ages and nationalities.
Level of Study: Research
Type: Grant
Value: Up to US$75,000
Length of Study: 1–2 years
Frequency: Annual
Application Procedure: Application procedures can be found online at www.atcp.org/grantguidelines
Closing Date: September 1st
Funding: Private

Ataxia-Telangiectasia Doctoral Fellowship Award

Subjects: Although basic science grant proposals are welcome, our funding efforts are focused on translational (bench to bedside) research, clinical studies and proposals that apply innovative and novel strategies for suggesting, developing and evaluating specific disease-modifying and symptomatic interventions.
Purpose: To accelerate first rate, international research on ataxia-telangiectasia to help find life-improving therapies or a cure for children with A-T.
Eligibility: Open to Post Docs with academic excellence and with one year experience or less post degree. Applicants must be nominated for this award by their principal investigator (PI). Any interested PI who would like to nominate a new post doc for the A-T Post Doctoral Fellowship Award must send a Letter of Intent to the A-T Children's Project. This letter (not to exceed two pages) should include a brief abstract describing the proposed research, specific aims and an estimated budget.
Level of Study: Postdoctorate
Type: Funding support
Value: US$30,000–40,000
Length of Study: 1–2 years
Frequency: Annual
Application Procedure: Application procedures can be found online at www.atcp.org/grantguidelines
Closing Date: September 1st and March 1st
Funding: Private

AARON SISKIND FOUNDATION

C/o School of Visual Arts, MFA Photography, 209 East 23rd Street, New York, 10010, United States of America
Tel: (1) 212 592 2363
Fax: (1) 212 592 2366
Email: info@aaronsiskind.org
Website: www.aaronsiskind.org

The Foundation works to preserve and protect Aaron Siskind's artistic legacy, and foster knowledge of and appreciation of his art.

Individual Photographer's Fellowship

Subjects: Photography-based art. Eligible work must be based on the idea of the lens-based still image, but grant recipients work in forms as diverse as digital imagery, video, installations, documentary projects and photo-generated print media.
Purpose: To stimulate excellence and the promise of future achievement in the photographic field.
Eligibility: Open to citizens or permanent residents of the United States of America. Applications sent from outside the United States of America will not be accepted.
Level of Study: Postgraduate
Type: Fellowship
Value: Up to $10,000 each
Country of Study: United States of America
No. of awards offered: 5
Application Procedure: Candidates must send an application form, 10 x 35 mm slides of their work, a slide list, stamped addressed return envelope together with their curriculum vitae and statement of plans for intended work to main address.
Closing Date: May 18th
Funding: Private
Additional Information: The application procedure is explicit and candidates are advised to check the website.

ABBEY AWARDS

1 St. Lukes Court, 136 Falcon Road, London, SWII 2LP, United Kingdom
Email: contact@abbey.org.uk
Website: www.abbey.org.uk
Contact: Faith Clark, Administrator

The Abbey Awards offer all-expenses-paid residencies for painters in excellent studio apartments at the British School in Rome. The 9-month Abbey Scholarship is awarded to an earlycareer artist while the 3-month Abbey Fellowships are for mid-career painters with an established record of achievement. Please see The Britsh School at Rome.

Abbey Awards

Subjects: Applicants must be fine artists specializing in painting.
Purpose: To enable painters to take up residencies at the British School at Rome.
Eligibility: Open to painters only, from UK or USA or from other countries provided they have lived for at least 5 years in UK or USA.
Level of Study: Unrestricted
Type: Scholarships and fellowships
Value: Scholarship - £16,000, Fellowships - £6,500
Length of Study: Scholarships - 9 months, Fellowships - 3 months
Frequency: Annual
Study Establishment: British School at Rome
Country of Study: Italy
No. of awards offered: 4
Application Procedure: Application forms are available on website www.abbey.org.uk and must be accompanied by a CD showing not more than 8 JPEG images of recent work, alongwith a cheque for £25/US$50.
Closing Date: January 13th (18.00 hrs Monday)
Funding: Private
Contributor: The Incorporated Edwin Austin Abbey Memorial Scholarships Trust (in US)
No. of awards given last year: 4
No. of applicants last year: 146

Abbey Harris Mural Fund

Subjects: Mural painting.
Purpose: To provide grants to artists who have been commissioned to create murals in public places or in charitable institutions in the United Kingdom.
Eligibility: Awards are for mural painters and the work must be carried out in the United Kingdom. There are other restrictions.
Level of Study: Unrestricted
Type: Grant
Value: Approx. UK£3,000
Country of Study: United Kingdom

No. of awards offered: 1 or 2
Application Procedure: Applicants request for application forms can be made by email information at www.abbey.org.uk.
Closing Date: Applications are accepted at any time. Decisions may be made at twice-yearly meetings of the Trustees in May and November
Funding: Trusts
Contributor: E A Abbey Memorial Trust Fund for mural painting in Great Britain and E Vincent Harris Fund for mural decoration
No. of awards given last year: 1
No. of applicants last year: 3

THE ABDUS SALAM INTERNATIONAL CENTRE FOR THEORETICAL PHYSICS (ICTP)

Strada Costiera 11, Trieste, 34151, Italy
Tel: (39) 040 224 0111
Fax: (39) 040 224 163
Email: sci_info@ictp.it
Website: www.ictp.it
Contact: May Ann Williams, Public Information Officer

The Abdus Salam International Centre for Theoretical Physics (ICTP) is an institution for research and high-level training in physics and mathematics, mainly for scientists from developing countries. It also maintains a network of associate members and federated institutes.

Abdus Salam ICTP Fellowships

Subjects: Physics and mathematics.
Purpose: To enable qualified applicants pursue research in the fields of condensed matter physics, mathematics and high-energy physics.
Eligibility: Open to qualified applicants of any nationality who have a PhD in physics or mathematics.
Level of Study: Postdoctorate
Type: Fellowship
Value: Monthly stipend and round-trip expenses where applicable, and allowances according to the length of the visit
Length of Study: Up to 1 year
Frequency: Dependent on funds available
Study Establishment: ICTP
Country of Study: Italy
No. of awards offered: Varies
Application Procedure: Applicants must visit the ICTP website.
Closing Date: Applications are accepted at any time
Funding: Government
Contributor: The Italian government, IAEA and UNESCO

ABERYSTWYTH UNIVERSITY

Postgraduate Admissions Office, Studen Welcome Centre, Aberystwyth University, Penglais, Aberystwyth, SY23 3FB, United Kingdom
Tel: (44) 19 7062 2270
Fax: (44) 19 7062 2921
Email: pg-admissions@aber.ac.uk
Website: www.aber.ac.uk

Located in beautiful surroundings, the Aberystwyth University provides an ideal learning environment. Its information services are among the best in the United Kingdom, and students also enjoy free access to one of the six copyright libraries in the United Kingdom, the National Library of Wales, which is adjacent to the University.

Aberystwyth International Excellence Scholarships

Subjects: Studies leading to the award of a Master degree in one of the relevant academic departments. The university offers a wide range of Taught Master's courses within the faculties of science, social sciences and arts.
Purpose: To enable Non-EU students to undertake full-time Master's study at Aberystwyth University.
Eligibility: Open to Non-EU candidates who have obtained at least an Upper Second Class (Honours) Degree or equivalent in their degree examination and who wish to study full-time. To be eligible you must: (i) be a non-eu fee payer. (ii) have or expect to obtain at least a Second Upper Class Honours Degree. (iii) apply and be offered a place for full time taught Master's study at Aberystwyth University.
Level of Study: Graduate, MBA, Postgraduate
Type: Scholarship
Value: £2,000 and £3,000 depending on academic achievement
Length of Study: 1 year
Frequency: Annual
Study Establishment: Aberystwyth University
Country of Study: United Kingdom
No. of awards offered: Approx 50 per year
Application Procedure: Applicants must complete an application form, available from the Postgraduate Admissions Office or from the website. Selection is based on the application for admission including a statement of purpose and references.
Closing Date: July 27th
Funding: Private
Contributor: Aberystwyth University
No. of awards given last year: 59
No. of applicants last year: 120
Additional Information: For all enquiries, please contact the Postgraduate Admissions Office: pg-overseas@aber.ac.uk.

Doctoral Career Developement Scholarship (DCDS) Scheme

Subjects: Studies leading to the award of a PhD in any of the University's 17 academic departments. The university offers a wide range of research opportunities within the faculties of science, social sciences and arts.
Purpose: To enable United Kingdom and European Union and International students to undertake full-time doctoral study at Aberystwyth University.
Eligibility: Open to United Kingdom, European Union and International candidates who have obtained at least an Upper Second Class (Honours) Degree or equivalent in their degree examination and who wish to study full-time. To be eligible you must: (i) be a uk/eu fee payer. (ii) have or expect to obtain at least a second upper class Honours degree. (iii) apply and be offered a place for Doctoral study at Aberystwyth University. International candidates must demonstrate their English language ability through the achievement of 7.0 in IELTS or a recognized equivalent.
Level of Study: Doctorate, Postgraduate, Research
Type: Team
Value: United Kingdom fees, annual stipend based on UK research council rates and travel bursary.
Length of Study: 1 year, in the first instance, but usually renewable for up to 3 years subject to satisfactory academic progress
Frequency: Annual
Study Establishment: Aberystwyth University
Country of Study: United Kingdom
No. of awards offered: Usually 12 per year
Application Procedure: Applicants must apply for the PhD programme using the UKPASS online system or on an application form from the university website www.aber.ac.uk/en/postgrad/how to apply/ There is no separate application form for the DCDS. Selection is based on the course application for admission including the research proposal and two references.
Closing Date: March 1st. Applicants for courses in the Department of International Politics or Institute of Geography & Earth Sciences should submit their applications by February 1st.
Funding: Private
Contributor: Aberystwyth University
No. of awards given last year: 12
No. of applicants last year: 200
Additional Information: For all enquiries, please contact the Postgraduate Admissions Office: pg-admissions@aber.ac.uk

International Masters Scholarships

Purpose: To enable international students process new masters courses in the Institute of Biological, Environmental, Rural Sciences.
Eligibility: Open to international (non-EU) applicants for the degree of food plus water security (masters), animal science (masters), livestock science (masters), managing the environment (masters) and equine science (masters).
Level of Study: Graduate, Postgraduate

Type: Scholarship
Value: £5,000 or £3,000 for masters scholarships and £3,000 for specialist masters scholarships
Length of Study: 1 year
Frequency: Dependent on funds available
Study Establishment: Aberystwyth University
Country of Study: United Kingdom
No. of awards offered: 17 (up to 11 masters scholarships in food plus water security, animal sciences or livestock science and up to 6 specialist masters scholarships in managing the environment or equine science)
Application Procedure: Please submit a course application using the UKPass online application tool or complete and submit an application form downloaded from website: www.aber.ac.uk/en/postgrad/how-toapply. A separate application for the scholarship is not required. However, a separate scholarship letter must be submitted with the course application. Full details are available on our website at www.aber.ac.uk/en/postgrad/funding-fees or email at pg-admissions@aber.ac.uk.
Closing Date: April 1st
Funding: Private
Contributor: Institute of Biological, Environmental and Rural Science - Aberystwyth University
Additional Information: For enquiries please contact the Postgraduate Admissions Office at pg-admissions@aber.ac.uk.

Law and Criminology Masters Scholarships

Subjects: Full-time studies leading to the award of LLM.
Purpose: To enable UK/EU and International students to undertake masters courses in the Department of Law plus Criminology.
Eligibility: Open to applicants from the UK/EU and International on the basis of academic merit.
Level of Study: Graduate, Postgraduate
Type: Scholarship
Value: £5,000
Length of Study: 1 year
Frequency: Dependent on funds available
Study Establishment: Aberystwyth University
Country of Study: United Kingdom
No. of awards offered: 5
Application Procedure: Please submit a course application using the UKPass online application tool or complete and submit an application form downloaded from our website: www.aber.ac.uk/en/postgrad/howtoapply. A separate application form for the scholarship is not required.
Closing Date: March 1st
Funding: Private
Contributor: Department of Law plus Criminology, Aberystwyth University
Additional Information: For enquiries please contact the Postgraduate Admissions Office at pg-admissions@aber.ac.uk.

Law and Criminology Research Scholarships

Subjects: Full-time studies leading to the award of PhD in the Department of Law and Criminology.
Purpose: To enable UK/EU and International students to undertake doctoral studies in the Department of Law and Criminology.
Eligibility: Open to UK/EU and International candidates with academic excellence (at least an Upper Second Class Honours degree or equivalent) and wish to study full time. To be eligible you must a) have or expect to have obtained at least a Second Upper Class Honours degree or equivalent and b) apply and have been offered a place for doctoral study in the Department of Law and Criminology at Aberystwyth University.
Level of Study: Doctorate, Postgraduate
Type: Scholarship
Value: £2,000 per year
Length of Study: 3 years
Frequency: Dependent on funds available
Study Establishment: Aberystwyth University
Country of Study: United Kingdom
No. of awards offered: 2
Application Procedure: Candidates must apply for the PhD programme using the UKPass online system or on an application form from the university's website at www.aber.ac.uk/en/postgrad/how-toapply. There is no separate form for the Law and Criminology scholarships. Selection is based on the course application form, the research proposal and references.
Closing Date: March 1st
Funding: Private
Contributor: Department of Law and Criminology, Aberystwyth University
No. of awards given last year: 1
No. of applicants last year: 10
Additional Information: For all enquiries please contact the Postgraduate Admissions Office at pg-admissions@aber.ac.uk.

School of Management and Business Masters Scholarships

Purpose: To enable International, UK and EU students to undertake full-time master's study in the School of Management and Business at Aberystwyth University.
Eligibility: Open to International, UK and EU applicants based on academic and personal/professional criteria.
Level of Study: Graduate, Postgraduate
Type: Scholarship
Value: £2,000–3,000
Length of Study: 1 year
Frequency: Annual
Study Establishment: Aberystwyth University
Country of Study: United Kingdom
No. of awards offered: Up to 20 MBA Ambassador masters scholarships; up to 10 management master's scholarships (including International Business Management, Management and Marketing Masters, Management Masters, Management and Finance Masters); up to 10 Specialist Masters Scholarships (including Accounting and Finance Masters, International Finance Masters and Advanced Marketing Masters)
Application Procedure: Applicants must first apply for their chosen course in the School of Management and Business using the UKPass online application system or an application form downloaded from our website www.aber.ac.uk/en/postgrad/howtoapply. Applicants are also required to submit a 1,000 word essay, the details of which may be found at www.aber.ac.uk/en/smb/masters-scholarships.
Closing Date: April 30th
Funding: Private
Contributor: School of Management and Business, Aberystwyth University
No. of awards given last year: 40
No. of applicants last year: 120
Additional Information: For all enquiries, please contact the Postgraduate Admissions Office at pg-admissions@aber.ac.uk.

ABMRF/THE FOUNDATION FOR ALCOHOL RESEARCH

1122 Kenilworth Drive, Suite 407, Baltimore, MD, 21204, United States of America
Tel: (1) 410 821 7066
Fax: (1) 410 821 7065
Email: grantinfo@abmrf.org
Website: www.abmrf.org
Contact: Erin L Teigen, Grants program Administrator

ABMRF/The Foundation for Alcohol Research is a non-profit independent research organization that provides support for scientific studies on the use of alcoholic beverages. It awards grants to study changes in drinking patterns, effects of moderate use of alcohol on health and well being and the mechanisms underlying the behavioural and biomedical effects of alcohol.

ABMRF/The Foundation for Alcohol Research Project Grant

Subjects: Medical and behavioural sciences.
Purpose: To support new knowledge in order to understand the effects of alcohol on health and behaviour.
Level of Study: Doctorate
Type: Project grant
Value: Up to US$50,000 per year

Length of Study: Up to 2 years
Frequency: Biannual
Study Establishment: Non-profit universities and research institutions
Country of Study: United States of America or Canada
No. of awards offered: 30–40
Application Procedure: Applicants must complete an application form, available on request or from the website.
Closing Date: February 15th
Funding: Private
No. of awards given last year: 40
No. of applicants last year: 150
Additional Information: South Africa is also the country of study.

THE ACADEMY OF NATURAL SCIENCES

1900 Benjamin Franklin Parkway, Philadelphia, PA, 19103-1195, United States of America
Tel: (1) 215 299 1000
Fax: (1) 215 299 1028
Email: webmaster@ansp.org
Website: www.ansp.org
Contact: Dr Lois Kuter, Volunteer Coordinator

Founded in 1812, the Academy of Natural Sciences is the oldest continually operating museum of its kind in the Western Hemisphere. The Academy's mission is to create the basis for a healthy and sustainable planet through exploration, research, and education.

Böhlke Memorial Endowment Fund

Subjects: Ichthyology.
Purpose: To support graduate students and postdoctoral researchers to work with ichthyology collection and library at the academy.
Eligibility: Open to graduate students and recent postdoctoral researchers.
Level of Study: Postdoctorate, Graduate
Type: Funding support
Value: Less than $500
Frequency: Dependent on funds available
Study Establishment: The Academy of Natural Sciences
Country of Study: United States of America
Application Procedure: A letter of application outlining proposed research and tentative budget should be sent.
Additional Information: Contact: Dr John Lundberg, The Academy of Natural Sciences

The Don and Virginia Eckelberry Fellowship

Purpose: Eckelberry endowment are used to grant fellowships to artists whose work might be enhanced by an opportunity to have access to museum collections and to engage in field studies.
Eligibility: Open to painters, sculptors and graphic artists, regardless of age, sex or nationality.
Type: Fellowship
Value: Offer funding for research, travel and field studies
Frequency: Annual
Application Procedure: A letter of intent (no more than two pages) in which the applicant should briefly outline his or her career and explain the reasons for seeking the fellowship. If a specific project is to be involved, it should be described and/or supported through supplementary information. Most important, the letter should explain how the fellowship would advance the applicant's artistic aspirations and goals. A current curriculum vitae. Two letters of recommendations from professionals familiar with your work. 35mm slides or a CD with untouched digital images of a representative sampling of your work (no more than 20 pieces) and a self-addresed, stamped envelope for their return (if desired).
Closing Date: October 15th

Jessup and McHenry Awards

Subjects: The Jessup award is for zoology and the McHenry award is for botany.
Purpose: To assist predoctoral and postdoctoral students working with biological collections at the Academy of Natural Sciences in Philadelphia.
Eligibility: Jessup funds are awarded competitively to students wishing to conduct studies at the postgraduate, doctoral and postdoctoral levels under the supervision or sponsorship of a member of the curatorial staff of the Academy. The awards are not available for undergraduate study. These awards are intended to assist predoctoral and postdoctoral students within several years of receiving their PhDs. Students commuting within the Philadelphia area are ineligible.
Level of Study: Doctorate, Postdoctorate, Postgraduate, Predoctorate
Type: Fellowship
Value: The stipend for subsistence is US$375 per week. Fellowships may include round trip travel costs of up to a total of US$500 for North American applicants, including Mexico and the Caribbean, and US $1,000 for applicants from other parts of the world. This is not guaranteed.
Length of Study: 2–16 weeks
Frequency: Annual
Study Establishment: The Academy of Natural Sciences
Country of Study: United States of America
Application Procedure: Applicants must send queries, requests for information and supporting information to the given address.
Closing Date: March 1st or October 1st
Funding: Private
No. of awards given last year: 3
Additional Information: The provision of scientific supplies and equipment is the responsibility of the student and the sponsoring curator. Contact Dr Ted Daeschler, Jessup McHenry Fund Committee for further information.

For further information contact:

Jessup-McHenry Fund Commitee
Contact: Dr Edward Daeschler

John J. & Anna H. Gallagher Fellowship

Purpose: To offer an opportunity for the study of rarer microscopic, multicellular invertebrate animals.
Eligibility: Open to candidates involved in original postdoctoral or sabbatical research on the systematics of microscopic invertebrates with priority for the study of rotifers.
Level of Study: Postdoctorate
Type: Fellowship
Value: US$36,000 plus benefits
Length of Study: 1 year
Frequency: Dependent on funds available
Study Establishment: The Academy of Natural Sciences
Country of Study: United States of America
Application Procedure: Application materials may be sent by email or post.
Closing Date: December 31st
No. of awards given last year: 1

For further information contact:

Chair, Gallagher Fellowship Committee, The Academy of Natural Sciences
Contact: Dr Ted Daeschler

ACADEMY OF TELEVISION ARTS & SCIENCES

5220 Lankershim Boulevard, North Hollywood, CA, 91601-3109, United States of America
Tel: (1) 818 754 2800
Fax: (1) 818 761 2827
Email: websupport@emmys.org
Website: www.emmys.org
Contact: Michele Fowble

The Academy of Television Arts & Sciences promotes creativity, diversity, innovation and excellence through recognition, education and leadership in the advancement of the telecommunications arts and sciences. The Academy has become both a place for serious discussion and a place to celebrate the industries finest achievements with its annual Emmy Awards ceremonies.

Fred Rogers Memorial Scholarship

Subjects: Children's media.
Purpose: To support and encourage an aspiring upper division or graduate student to pursue a career in children's media that furthers the values and principles of Fred Rogers work.
Eligibility: Open to upper division graduate students of accredited colleges or universities. Candidates must have the ultimate goal of working in the field of children's media.
Level of Study: Doctorate, Postgraduate
Type: Scholarship
Value: US$10,000
Frequency: Annual
Country of Study: United States of America
No. of awards offered: 4
Application Procedure: Candidates must complete an extensive form that includes background information about the themselves and plan for the use of the scholarship money. The application must also include recommendations from 2 people, either faculty members or professionals from the children's media industry who have worked with the applicant. Application forms are available online.
Closing Date: February 08th
Contributor: The Academy of Television Arts & Sciences in association with Ernst & Young LLP
Additional Information: In addition to the monetary award, successful applicants will work with a mentor from the Children's Programming Peer Group during the academic year.

ACADIA UNIVERSITY

Room 214 Horton Hall, 18 University Avenue, Wolfville, NS, B4P 2R6, Canada
Tel: (1) 902 585 1914
Fax: (1) 902 585 1096
Email: theresa.starratt@acadiau.ca
Website: www.acadiau.ca
Contact: Ms Theresa Starratt, Graduate Studies Officer-Research and Graduate Studies

Acadia University is an institution that is committed to providing a liberal education based on the highest standards. The University houses a scholarly community that aims to ensure a broadening life experience for students, faculty and staff.

Acadia Graduate Awards

Subjects: English, political science, sociology, biology, chemistry, computer science, geology, psychology, education and recreation management, mathematics, applied geomatics and statistics, social and political thought.
Purpose: To financially support students.
Eligibility: Open to registered full-time graduate students at Acadia University. In order to be eligible for an award, students must have a GPA of not less than 3.00 in their major field in each of their last 2 years of undergraduate study.
Level of Study: Postgraduate
Type: Award
Value: Up to Canadian $9,000
Length of Study: 1–2 years
Frequency: Annual
Study Establishment: The Division of Research and Graduate Studies at Acadia University
Country of Study: Canada
No. of awards offered: Limited
Application Procedure: Applicants must write for details. In almost all cases, to be automatically considered for funding, applicants need to apply by February 1st of each year.
Closing Date: When all spots have been filled
Additional Information: Recipients of an Acadia Graduate Award should expect to undertake certain duties during the academic year (up to a minimum of 10 hours per week and a maximum of 120 hours per semester) as a condition of tenure of the award. The specific duties will be established by agreement at the beginning of each academic year with your department/school coordinator.

THE ACOUSTICAL SOCIETY OF AMERICA (ASA)

Suite 1NO1, 2 Huntington Quadrangle, Melville, NY, 11747-4502, United States of America
Tel: (1) 516 576 2360
Fax: (1) 516 576 2377
Email: asa@aip.org
Website: http://asa.aip.org
Contact: Elaine Moran, ASA Office Manager

The Acoustical Society of America (ASA) is primarily a voluntary organization and attracts the interest, commitment and service of a large number of professionals. Since its inception in 1929, the Society has grown steadily in membership and stature.

ASA Frederick V. Hunt Postdoctoral Research Fellowship

Subjects: Acoustics.
Purpose: To carry out Professor Hunt's wish that his estate be used to further the science of, and education in, acoustics.
Eligibility: Open to candidates who have obtained a PhD and are members of the ASA.
Level of Study: Postdoctorate, Research
Type: Fellowship
Value: US$55,000
Length of Study: 1 year
Frequency: Annual
Country of Study: United States of America
No. of awards offered: 1
Application Procedure: Applicants can download the form from the website.
Closing Date: September 1st

ACTION CANADA

#513 – 1489 Marine Drive, West Vancouver, BC, V7T 1B8, Canada
Tel: (1) 778 881 7961
Fax: (1) 604 569 5697
Email: actioncanada@actioncanada.ca
Website: www.actioncanada.ca

Action Canada is a national organization committed to building leadership for the future of Canada through an innovative fellowship program.

Action Canada Fellowships

Subjects: All subjects.
Purpose: To support applicants in the early years of their careers.
Eligibility: Open to applicants who are citizens of Canada.
Level of Study: Postgraduate
Type: Fellowship
Value: Canadian $20,000
Length of Study: 11 months
Frequency: Annual
Country of Study: Canada
No. of awards offered: Up to 20
Application Procedure: A completed application form along with a curriculum vitae and proof of Canadian citizenship must be submitted. See the website for details.
Closing Date: February 17th
Contributor: Action Canada
No. of awards given last year: 16
No. of applicants last year: 199

ACTION MEDICAL RESEARCH

Vincent House, North Parade, Horsham, West Sussex, RH12 2DP, England
Tel: (44) 14 0321 0406
Fax: (44) 14 0321 0541
Email: info@action.org.uk
Website: www.action.org.uk

Action Medical Research is dedicated to preventing and treating disease disability by funding vital medical research in UK-based hospitals and universities. The remit focuses on child health with an emphasis onclinical research or research at the clinical/basic interface. Research applications are judged by rigorous peer review.

Action Medical Research Project Grants

Subjects: A broad spectrum of research with the objective of preventing and treating disease and disability and alleviating physical disability. The remit focuses on child health to include problems affecting pregnancy, childbirth, babies, children and adolescents.
Purpose: To support one precisely formulated line of research.
Eligibility: Open to researchers based in the United Kingdom. Grants are not awarded to MRC units, other charities or for higher education.
Level of Study: Unrestricted
Type: Grant
Value: Varies
Length of Study: Up to 3 years, assessed annually
Frequency: Dependent on funds available, see the website
Study Establishment: Hospitals, universities and recognised research establishments in the UK
Country of Study: United Kingdom
No. of awards offered: Varies
Application Procedure: Applicants must submit a one-page outline of the project before an application form can be issued. Full details and outline proposals are available on the website.
Closing Date: March 29th or November 22nd. See the website for details.
Funding: Private, trusts
Contributor: Voluntary income
No. of awards given last year: 20
No. of applicants last year: 159

Action Medical Research Training Fellowship

Subjects: A broad spectrum of research with the objective of preventing and treating disease and disability and alleviating physical disability. The remit focuses on child health to include problems affecting pregnancy, childbirth, babies, children and adolescents.
Purpose: To enable the training of young medical and non-medical graduates in research techniques and methodology in areas of interest to Action Medical Research.
Eligibility: Open to medical and non-medical graduates. Although it is not limited to United Kingdom citizens, those who do not hold United Kingdom citizenship must be able to show that they have all the required statutory documentation, e.g. work permits, to cover the period of the fellowship. Preference will be given to candidates resident in the UK. No grants are made purely for higher education. Further guidelines are available on the organisation's website.
Level of Study: Doctorate, Postdoctorate, Postgraduate
Type: Fellowship
Value: Varies
Length of Study: Up to 3 years
Frequency: Dependent on funds available
Study Establishment: A hospital, university department or recognised research institute in the UK
Country of Study: United Kingdom
No. of awards offered: Varies
Application Procedure: Applicants must submit a one-page outline of the project before an application form can be issued. Full details and outline proposal forms are available on the website.
Closing Date: August 31st. See the website for details.
Funding: Private, trusts
Contributor: Voluntary income
No. of awards given last year: 3
No. of applicants last year: 46
Additional Information: Fellowships are advertised separately each year in June/July.

ADELPHI UNIVERSITY

1 South Avenue, PO BOX 701, Garden City, NY, 11530-0701, United States of America
Tel: (1) 516 877 3412/3080
Fax: (1) 516 877 3424
Email: ucinfo@adelphi.edu
Website: www.adelphi.edu/

Adelphi University is the oldest institution of higher education for liberal arts and sciences on Long Island.

Adelphi University Athletic Grants

Subjects: Athletics.
Purpose: To support students who demonstrate exceptional ability in the area of athletics.
Eligibility: The student's athletic performance/record is the initial criterion for consideration of this award.
Level of Study: Postgraduate
Type: Scholarship
Value: Covers full tuition, fees, room, and board
Length of Study: 1 year
Frequency: Annual
Study Establishment: Adelphi University
Application Procedure: Students must file an admissions application.
Closing Date: February 15th

Adelphi University Dean's Award

Subjects: All subjects.
Purpose: To reward academic achievement and cocurricular activities.
Eligibility: Deans Awards are awarded to entering full-time freshmen with very good academic performances. Ordinarily, recipients have minimum SAT scores ranging from 1580 to 1790 and rank in the top 25% of their high school class.
Level of Study: Postgraduate
Type: Scholarship
Value: $8,000–11,500
Length of Study: 1 year
Frequency: Annual
Study Establishment: Adelphi University
Application Procedure: Students must file an admissions application.
Closing Date: February 15th
Funding: Private

Adelphi University Presidential Scholarship

Subjects: All subjects.
Purpose: To reward exceptional academic achievement and cocurricular activities.
Eligibility: Open to Presidential scholars typically having combined critical reading and mathematics Scholastic Achievement Test (SAT) scores of at least 1,300 and rank in the top 10 per cent of their high school class. The performance on the writing section of the SAT will be carefully evaluated and will be considered in the determination of the final scholarship award.
Level of Study: Postgraduate
Type: Scholarship
Value: $15,000–16,000
Length of Study: 1 year
Frequency: Annual
Study Establishment: Adelphi University
Application Procedure: Students must file an admissions application.
Closing Date: February 15th
Funding: Private

Adelphi University Provost Scholarship

Subjects: All subjects.
Purpose: To reward students with excellent academic performance and co curricular activities.
Eligibility: Open to Provost scholars having combined SAT scores minimum 1800 (critical reading, math and writing) and rank in the top 15% of their high school class or have a minimum transfer GPA of 3.5.
Level of Study: Postgraduate
Type: Scholarship
Value: For Freshmen: $12,000–14,500; for Transfers: $7,500–11,000
Length of Study: 1 year
Frequency: Annual
Study Establishment: Adelphi University
Application Procedure: Students must file an admissions application.
Closing Date: February 15th
Funding: Private

Adelphi University Talent Awards

Subjects: Theater, dance, art and music.
Purpose: To support students who demonstrate exceptional talent in the areas of theater, dance, art, or music.
Eligibility: Open to candidates who demonstrate exceptional talent in the areas of theater, dance, art or music. Candidates must declare a major and will be required to initially submit a portfolio or audition in their area of concentration.
Level of Study: Postgraduate
Type: Award
Value: $4,000–9,500
Length of Study: 1 year
Frequency: Annual
Study Establishment: Adelphi University
Application Procedure: Students must file an admissions application.
Closing Date: February 15th
Funding: Private

Adelphi University Trustee Scholarship

Subjects: All subjects.
Purpose: To support freshmen with the most outstanding academic achievement and cocurricular activities.
Eligibility: Open to trustee scholars having combined critical reading and math SAT scores exceeding 1,350 and rank in the top 10% of their high school class.
Level of Study: Postgraduate
Type: Scholarship
Value: $16,500–24,000
Length of Study: 1 year
Frequency: Annual
Application Procedure: Students must file an admissions application.
Closing Date: February 15th
Funding: Private
Contributor: Adelphi University

THE ADOLPH AND ESTHER GOTTLIEB FOUNDATION, INC.

380 West Broadway, New York, NY, 10012, United States of America
Tel: (1) 212 226 0581
Fax: (1) 212 226 0584
Email: sross@gottliebfoundation.org
Website: www.gottliebfoundation.org
Contact: Sheila Ross, Grants Manager

The Adolph and Esther Gottlieb Foundation is a non-profit corporation registered with the state of New York. It was established to award financial aid to mature creative painters, sculptors and printmakers.

Gottlieb Foundation Emergency Assistance Grants

Subjects: Painting, sculpture and printmaking.
Purpose: To provide interim financial assistance to creative visual artists whose need is the result of unforeseen catastrophic incident.
Eligibility: Open to painters, sculptors and printmakers who can demonstrate a minimum of 10 years of involvement in a mature phase of their work and who do not have the resources to meet the costs incurred by a catastrophic event e.g. fire, flood or emergency medical expenses. The disciplines of film, photography or related forms are not eligible unless the work involves directly, or can be interpreted as, painting or sculpture.
Level of Study: Unrestricted
Type: Grant
Value: Up to US$10,000. US$4,000 is typical on a one time basis only
Frequency: Dependent on funds available
Country of Study: Any country
No. of awards offered: Varies
Application Procedure: Applicants must complete and submit an application form which is available from the Foundation throughout the year and may be requested by telephone. Second party requests are honoured only when the applicant is physically unable to communicate with the Foundation.
Closing Date: Please write for details
Funding: Private
Contributor: An endowment
Additional Information: Maturity is based on the level of technical, intellectual and creative development of the artist. The programme does not cover general indebtedness, dental work, unemployment, capital improvements, long-term disabilities or project funding. Review procedures for completed applications begin as soon as they are received. Full review generally takes about four weeks from the time an application is complete. Situations with imminent deadlines will receive priority. When a situation warrants it, reviews can be completed within 24 to 48 hours.

Gottlieb Foundation Individual Support Grants

Subjects: Painting, sculpture and printmaking.
Purpose: To recognize and support serious, fully committed painters, sculptors and printmakers who are in financial need.
Eligibility: Open to creative painters, sculptors and printmakers who have been in a mature phase of their work for at least 20 years and require financial assistance to continue this work. United States residency is not required. The Gottlieb Foundation does not provide funding for organizations, projects of any type, educational institutions, students, graphic artists or those working in crafts. The disciplines of photography, film, video or related forms are not eligible unless the work directly involves, or can be interpreted as, painting or sculpture.
Level of Study: Unrestricted
Type: Grant
Value: Varies
Length of Study: 1 year
Frequency: Annual
Country of Study: Any country
No. of awards offered: 12 every year
Application Procedure: Applicants must include a current application form, available from the foundation, and a small group of slides of the artist's work that illustrates the progressive development of the art for at least a 20-year period. These slides must be properly labelled and dated. Applicants must also include a written statement in narrative form. This statement should include outside jobs which have helped support the artist's career, changes in artistic approach that have occurred, and other facts which can aid the review panel in forming an accurate picture. All aspects of artistic history, i.e. education, exhibitions, etc., should be described, and dates must be provided for all information. Financial disclosure, which entails completing a disclosure page and submitting a copy of a federal tax return for the past year, is necessary in the determination of financial need. A stamped addressed envelope for the return of supplementary materials must also be included.
Closing Date: December 15th
Funding: Private
Contributor: An endowment
Additional Information: Artists who have been awarded a grant must allow one year to elapse before reapplication. Only first person written requests for application forms will be honoured.

AFRICA EDUCATIONAL TRUST

Africa Educational Trust, 18 Hand Court, London, WC1V 6JF, United Kingdom
Tel: (44) 020 7831 3283
Fax: (44) 020 7242 3265
Email: m.omona@africaeducationaltrust.org
Website: www.africaeducationaltrust.org
Contact: Ms May Omona, Programme Manager

Andrzejewski Memorial Fund

Subjects: Language, literature and education.
Purpose: To support individual study or field work on the Horn Region or to assist the study, production or provision of materials which will support language, literature, education or broadcasting in the region.
Level of Study: Postgraduate
Type: Grant
Application Procedure: Successful applicants will be required to submit a report of approximately 2,000 words on the use of their award, no later than 3 months after the completion of their course or project. Applicants are asked to ensure that they meet the above conditions as regards to region and area of their study. A letter of application for this award should be sent to Africa Educational Trust.

Funding: Trusts
Additional Information: Please check the website for more information.

Kenneth Kirkwood Fund

Purpose: This fund was established in 1998 in honour of the memory of Kenneth Kirkwood, the first Professor of Race Relations and Co-ordinator of African Studies at St Antony's College, Oxford.
Eligibility: Priority will be given to those students who are studying subjects which are relevant to the development of their home countries. It is expected that first consideration will be given to students studying at St Antony's College, Oxford. However, students from other colleges and universities will also be eligible for awards. Applicants should be reaching the final months of their period of study.
Level of Study: Postgraduate
Type: Grant
Value: £500
Frequency: Ongoing
Country of Study: United Kingdom
Application Procedure: Applicants should write a short letter of application and send it to Africa Educational Trust..
Funding: Trusts
Additional Information: The fund provides small grants for maintenance, fees or for emergency payments for students from Africa, particularly Southern Africa. Both undergraduate and post graduate students are eligible to apply to the Kenneth Kirkwood Fund.

Small Emergency Grants

Purpose: The fund is administered for African students who have run into unexpected financial difficulties during the final months of their course. The goal of the Small Grants Programme is to enable African students to complete their course and to return home.
Eligibility: Applicants must come from Africa and be studying in the UK on a student's visa. The financial difficulty must be unexpected. The student must be in the final 4 months of his/her study. All applications must be supported in writing by the student's supervisor, tutor or head.
Level of Study: Postgraduate
Frequency: Ongoing
Country of Study: United Kingdom
Application Procedure: If you meet all of the above criteria, and would like to apply, please write a short letter marked Small Emergency Grants Programme and send to Africa Educational Trust..
Funding: Trusts
Additional Information: If eligible, you will be sent a Small Emergency Grants Form to fill in. Please note that you can only apply during the final 4 months of your course of study and that grants are available for both undergraduate and postgraduate study.

UCL-AET Undergraduate International Outreach Bursaries

Purpose: To provide help to AET-nominated students who have accepted offers of places at University College London.
Eligibility: Be a national of any African country (including Madagascar), currently living in an African country, and have one or both parents living in an African country, or are orphaned; be currently attending school or have recently completed school in an African country. See the details in website.
Level of Study: Postdoctorate
Country of Study: United Kingdom
Application Procedure: See the website http://www.africaeducationaltrust.org/userfiles/file/AET%20GrantScholarshipLeaflet%20Nov2012.pdf.
Closing Date: To be confrmed
Funding: Trusts

AFRICAN MATHEMATICS MILLENNIUM SCIENTIFIC INITIATIVE

School of Mathematics, University of Nairobi, PO Box 30197, Nairobi, GPO 00100, Kenya
Tel: (254) 20 445 0934
Email: ammsi@uonbi.ac.ke
Website: www.ammsi.org
Contact: Professor Wandera Ogana, AMMSI Programme Director

The African Mathematics Millennium Science Initiative (AMMSI) is a distributed network of mathematics research, training and promotion throughout sub-Saharan Africa. It has five regional offices located in Botswana, Cameroon, Kenya, Nigeria and Senegal. It is a project established by the Millenium Science Initiative (MSI), administered by the Science Initiative Group (SIG). The primary goal of the MSI is to create and nuture world-class science and scientific talent in the developing world by strengthening S&T capacity through integrated programmes of research and training, planned and driven by local scientists in the field of mathematics.

Research/Visiting Scientist Fellowships

Subjects: Mathematics.
Purpose: To encourage research and postgraduate teaching in mathematics.
Eligibility: Candidates should undertake research and postgraduate teaching in mathematics at any university in sub-Saharan Africa, should be a staff member at a university and hold at least a Master's degree, and should obtain an official invitation from the host institution.
Level of Study: Postgraduate
Type: Fellowships
Value: US$5,000
Length of Study: 1–12 months
No. of awards offered: 5
Application Procedure: Application forms should be filled and submitted online. In exceptional circumstances, hard copy application forms may be obtained from the nearest AMMSI Regional Coordinator whose contact details can be found in the website.
Closing Date: September 15th (check with website)

For further information contact:

School of Mathematics, University of Nairobi, PO Box 30197, Nairobi, 30197/GPO 00100, Kenya
Tel: (254) (20) 235 8569/(20) 445 0934
Fax: (254) (20) 445 0934
Email: progoffice@ammsi.org/ammsi@uonbi.ac.ke
Website: http://www.ammsi.org/?q=node/6
Contact: Wandera Ogana, AMMSI Programme Director

AFRICAN NETWORK OF SCIENTIFIC AND TECHNOLOGICAL INSTITUTIONS (ANSTI)

UNESCO Nairobi Office, UN Gigiri Complex, UN Avenue, PO Box 30592, Nairobi, Kenya
Tel: (254) 2 7622 619/20
Fax: (254) 2 7622 538/750
Email: info@ansti.org
Website: www.ansti.org
Contact: ANSTI Coordinator

The African Network of Scientific and Technological Institutions (ANSTI) is an organ of co-operation that embraces institutions engaged in the fields of science and technology. To date it has 126 member institutions in 33 countries in Sub-Saharan Africa.

ANSTI/DAAD Postgraduate Fellowships

Subjects: Basic and engineering sciences.
Purpose: To enable students to pursue Master's and PhD courses in the basic and engineering sciences.
Eligibility: Open only to African nationals who are staff members of ANSTI member institutions. Applicants must possess a good Bachelor's degree and must be below 36 years of age.
Level of Study: Doctorate, Postgraduate

Type: Fellowship
Value: Varies (approx. US$12,000)
Length of Study: Varies (approx. 1.5 years)
Frequency: Annual
Study Establishment: ANSTI member institutions
Application Procedure: Applicants must complete an application form, available by contacting ANSTI by mail, fax or email, or by visiting our website.
Closing Date: May 31st
Funding: Government
Contributor: DAAD/UNESCO
No. of awards given last year: 5
No. of applicants last year: 52
Additional Information: The applicant is responsible for gaining admission into the university of his or her choice. Preference is given to graduates with a few years of experience.

AFRICAN WILDLIFE FOUNDATION

African Wildlife Foundation, Ngong Road, Karen, PO Box 310, 00502, Nairobi, Kenya
Tel: (254) 20 2765000
Fax: (254) 20 2765030
Email: africanwildlife@awfke.org/charlottefellowship@awfke.org
Website: http://www.awf.org/
Contact: AWF Charlotte Fellowship Program

The Charlotte Conservation Fellows Program

Subjects: Conservation.
Purpose: To provide support for African nationals pursuing Master's degrees or doctoral research. The program was launched in tribute to the late Charlotte Kidder Ramsay, a long-time conservationist.
Eligibility: Applicants must:
1. Be nationals of eligible countries: Benin, Botswana, Burkina Faso, the Democratic Republic of the Congo, Mozambique, Namibia, Niger, South Africa, Zambia, or Zimbabwe.
2. Be 21–40 years of age.
3. Have secured a place at an appropriate university and be ready to commence studies within one year of the award of the fellowship. Students applying to study in African universities will be given special consideration.
4. Have exemplary work experience of not less than five years that demonstrates a strong commitment and outstanding motivation to the conservation of Africa's natural heritage.
5. Have a field research plan, which has a direct link or is relevant to AWF and host country conservation needs.
6. Demonstrate that they will remain working in conservation in home country in order to use their skills to make a contribution to conservation. Applicants should submit a letter of recommendation or support from their employer to show that they are engaged in conservation in home country. If they are in the diaspora, they should get a letter of recommendation from a relevant government ministry to demonstrate their willingness to return to home country to work in conservation.
7. Qualified women candidates are especially encouraged to apply.

Level of Study: Doctorate, Postgraduate
Type: Fellowship
Value: Up to $25,000
Frequency: Annual
No. of awards offered: 3–6
Application Procedure: See the website for details.
Closing Date: See the website for details
Funding: Foundation

For further information contact:

Website: www.awf.org/section/people/education/charlotte

THE AGA KHAN FOUNDATION

The Aga Khan Development Network, PO Box 2049, 1-3 Avenue de la Paix, Geneva 2, 1211, Switzerland
Tel: (41) 22 909 7200
Fax: (41) 22 909 7292
Email: info@akdn.org
Website: www.akdn.org

The Aga Khan Foundation is a non-denominational, international development agency established in 1967 by His Highness the Aga Khan. Its mission is to develop and promote creative solutions to problems that impede social development, primarily in Asia and East Africa. Created as a private, non-profit foundation under Swiss law, it has branches and independent affiliates in 15 countries. It is a modern vehicle for traditional philanthropy in the Ismaili Muslim community under the leadership of the Aga Khan.

Aga Khan Foundation International Scholarship Programme

Subjects: All subjects.
Purpose: To financially support outstanding students from developing countries who have no other means of financing their studies.
Eligibility: Open to candidates under the age of 30 with excellent academic records and genuine financial need. Candidates are expected to have some years of work experience in their field of interest.
Level of Study: Doctorate, Postgraduate
Type: Scholarship
Value: The Foundation assists students with tuition fees and living expenses only.
Length of Study: 1–2 years
Frequency: Annual
Application Procedure: Application forms are available on January 1st each year from AKF offices or Aga Khan Education Services/Boards in applicants countries of current residence. Completed applications should be returned to the agency or to the address indicated on the front of the form.
Closing Date: March 31st
Funding: Foundation, government, private
Contributor: Aga Khan Foundation
Additional Information: The foundation accepts applications from countries where it has branches, affiliates or other AKDN agencies which can help with processing applications and interviewing applicants. At present, these are Bangladesh, India, Pakistan, Afghanistan, Tajikistan, Syria, Kenya, Tanzania, Uganda, Mozambique, Madagascar, France, Portugal, UK, USA and Canada. Please see the website for further details http://www.akdn.org/akf_scholarships.asp.

AGENCY FOR SCIENCE, TECHNOLOGY AND RESEARCH (A*STAR)

1 Fusionopolis Way, #20-10 Connexis North Tower, 138632, Singapore
Tel: (65) 6826 6111
Fax: (65) 6777 1711
Email: contact@a-star.edu.sg
Website: www.a-star.edu.sg/astar

A*STAR comprises of the Biomedical Research Council (BMRC), the Science and Engineering Research Council (SERC), Exploit Technologies Private Ltd (ETPL), the A*STAR Graduate Academy (AGA) and the Corporate Planning and Administration Division (CPAD). Both BMRC and SERC promote, support and oversee the public sector R&D research activities in Singapore.

*A*STAR Graduate Scholarship (Overseas)*

Subjects: Biomedical sciences, physical sciences and engineering.
Purpose: To provide financial assistance for a research intensive training programme at PhD level and to support Singapore's knowledge-based economy.
Eligibility: Open to Singapore applicants.
Level of Study: Doctorate, Research, PhD
Type: Scholarship
Value: Successful candidates will be given support for up to 4 years of academic pursuit leading to a PhD. Additional two years of post-doctoral training for selected partner universities will also be considered for those who qualify. Full tuition fees, monthly sustenance allowance, annual book allowance, computer allowance, thesis allowance
Length of Study: 4 years
Frequency: Annual
Study Establishment: A*STAR

Country of Study: Singapore
Application Procedure: Please refer to the website for more information www.a-star.edu.sg/ags. Applicants can apply at https://scholarships.a-star.edu.sg/.
Closing Date: Open all year round
Funding: Government
Additional Information: The AGS (overseas)is tenable at the following partner universities: Imperial college of London, University of Illinois at Urbana-Champaign, Karolinska Institutet, University of Dundee, Carnegie Mellon University,University of Cambridge and University of Oxford.

*A*STAR International Fellowship*

Subjects: Biomedical sciences, physical sciences and engineering.
Purpose: For post-graduate students to undertake post-doctoral studies at the top overseas laboratories to further enrich their research careers.
Eligibility: Open to Singapore applicants.
Level of Study: Postdoctorate, Research
Type: Fellowship
Value: Full tuition fees, monthly overseas living allowance, monthly sustenance allowance, settling-in allowance, return airfares, conference allowance, health insurance coverage
Length of Study: 2 years
Frequency: Annual
Study Establishment: A*STAR
Country of Study: Any country
Application Procedure: Download AIF application form from www.a-star.edu.sg/aif.
Closing Date: Open all year round
Funding: Government

National Science PhD Scholarship

Subjects: Science, engineering and biomedical science disciplines.
Purpose: To offer support for PhD training in various Science, engineering and biomedical science disciplines. In return, scholars will return to make their mark on the R & D landscape in Singapore.
Eligibility: Open to Singapore applicants.
Level of Study: Research, PhD
Type: Scholarship
Value: All rates or allowances vary according to country and university
Length of Study: Varies
Frequency: Annual
No. of awards offered: Varies
Application Procedure: Please refer to the website for more information.
Closing Date: April 1st
Additional Information: For further information please visit the website www.a-star.edu.sg/nss_phd

Singapore International Graduate Award

Subjects: Research areas in Biomedical sciences and Physical sciences and Engineering.
Purpose: The scholarship is offered to international students to do a PhD programme in an A*STAR research institute.
Eligibility: Open to all international students who are graduates with a passion for research and excellent academic results, good skills in written and spoken English and good reports from academic referees.
Level of Study: Doctorate
Type: Award
Value: Tuition fees for 4 years of PhD studies; Stipend of US$24,000 per year, one time air fare grant of $1,500 and one time settling-in allowaance of $1,000
Length of Study: 4 years
Frequency: Biannual
Country of Study: Singapore
Application Procedure: Please apply online at www.singa.a-star.edu.sg.
Closing Date: August Intake - January 1st, January Intake - June 1st
Contributor: The Singapore International Graduate Award (SINGA) is a collaboration between the Agency for Science, Technology & Research (A*STAR), the National University of Singapore (NUS) and the Nanyang Technological University (NTU).
Additional Information: Please visit website for details on award www.singa.a-star.edu.sg.

Singapore International Pre-graduate Award

Subjects: Science, engineering and biomedical science disciplines.
Purpose: To provide a unique opportunity for top international students to experience the vibrant scientific environment in Singapore and A*STAR.
Eligibility: Open to international students in science and engineering related disciplines with good skills in written and spoken English and with excellent academic results.
Level of Study: Doctorate, Research
Type: Award
Value: Monthly stipend of $1,500
Length of Study: 2–6 months
No. of awards offered: Varies
Application Procedure: Applicants must refer to the website www.a-star.edu.sg/sipga.
Closing Date: See website
Additional Information: There are 4 application cycles per year. The nomination closing dates are February 1, May 1, August 1, and November 1 every year. For more information, please contact Ms Joanna Ng at joanna_ng@a-star.edu.sg.

AGRICULTURAL HISTORY SOCIETY

MSU History Department, PO Box H, Mississippi State, MS, 39762, United States of America
Tel: (1) 501 569 8782/(662) 268 2247
Fax: (1) 501 569 3059
Email: CSTROM@Rollins.edu
Website: www.aghistorysociety.org
Contact: Claire Strom, Editor, Agricultural History

The Agricultural History Society recognizes the roles of agriculture and agri-business in shaping the political, economic, social and historical profiles of different countries worldwide. Since 1927, the Society's publication, *Agricultural History*, has been the international journal for the field and publishes innovative research on agricultural and rural history.

Everett E Edwards Awards in Agricultural History

Subjects: Agricultural and rural history.
Purpose: To encourage and reward scholarly work in the field. The award was established in 1953 in memory and recognition of the outstanding services of Everett Eugene Edwards, a long-time agricultural historian and editor of *Agricultural History* from 1931 to 1951.
Eligibility: Open to any graduate or doctoral student submitting an article to *Agricultural History* during the calendar year.
Level of Study: Graduate
Type: Award
Value: US$200 plus publication in the Journal
Frequency: Annual
Country of Study: Any country
No. of awards offered: 1
Application Procedure: Applicants must submit three copies of their manuscript, prepared in accordance with the latest edition of the *Chicago Manual of Style*, to the editor.
Closing Date: December 31st
Additional Information: The award is presented annually to the author of the winning article at the Agricultural History Society's Presidential Luncheon. In addition, the winning submission is published in the fall issue of *Agricultural History*. Further information is available on request.

Gilbert C Fite Dissertation Award

Subjects: Agricultural and rural history.
Purpose: To reward the best dissertation.
Eligibility: Open to any doctoral student who has completed a PhD dissertation.
Level of Study: Doctorate
Type: Award
Value: US$300
Frequency: Annual
Country of Study: Any country
No. of awards offered: 1

Application Procedure: Applicants must complete forms and send three copies to the editor.
Closing Date: 31 December
Additional Information: Further information is available on request.

Theodore Saloutos Award

Subjects: Agricultural and rural history in the United States of America.
Purpose: To reward the best book published annually in the United States of America on the subject of agricultural history.
Eligibility: Books must be based on substantial primary research and should represent a major new scholarly interpretation or reinterpretation of agricultural history scholarship.
Level of Study: Unrestricted
Type: Prize
Value: US$500
Frequency: Annual
Country of Study: Any country
Application Procedure: Applicants must send four copies of the book to the editor. Books may be nominated by their authors, the publisher, a member of the award committee, or a member of the Society.
Closing Date: 31 December.
Additional Information: Further information is available on request.

Vernon Carstensen Memorial Award in Agricultural History

Subjects: Agricultural and rural history.
Purpose: To promote research and publication. The award was established in 1980 to recognize Vernon Carstensen's services to agricultural history by his former students.
Eligibility: Open to any author who has been published in the quarterly journal of *Agricultural History* during the calendar year.
Level of Study: Doctorate, Postdoctorate
Type: Award
Value: US$200
Frequency: Annual
Country of Study: Any country
No. of awards offered: 1
Application Procedure: All published articles per issue and year are considered.
Closing Date: The Autumn issue of the Journal each year
Additional Information: Vernon Carstensen served as editor of Agricultural History from 1953 to 1957 and as president of the *Agricultural History* Society from 1957 to 1958. Further information is available on request.

AIDS UNITED

Fellowship Enquiries, 1424 K Street, N.W., Suite 200, Washington, DC, 20005, United States of America
Tel: (1) 202 530 8030, 202 408 4848 ext. 248
Fax: (1) 202 530 8031
Email: information@aidsaction.com/Zamora@aidsunited.org
Website: www.aidsaction.org
Contact: Pedro Zamora, Coordinator Fellowship Program

AIDS Action, founded in 1984, is the National AIDS Organization dedicated to the development, analysis, cultivation, and encouragement of sound policies and programs in response to the HIV epidemic.

The Pedro Zamora Public Policy Fellowship

Subjects: Research into a variety of public health and civil rights issues related to HIV prevention, treatment and care.
Purpose: To fund postgraduate students seeking experience in public policy and government affairs focussed on HIV/AIDS issues.
Eligibility: Open to graduate and undergraduate students and young professionals. Candidates need strong research, writing, and organizational skills and a willingness to work in a professional office. Ability to work independently in a fast-paced environment is critical. Familiarity with HIV-related issues and the legislative processes is preferred. Fellows must commit to working a minimum of 30 hours per week for 8 weeks. People of color, women, gay, lesbian, bisexual, and transgender individuals, and HIV positive individuals are encouraged to apply.
Level of Study: Graduate, Postgraduate
Type: Fellowship
Value: Stipend plus expenses
Length of Study: Up to 26 weeks
Frequency: 3 times per year
Country of Study: United States of America
Application Procedure: Applicants must apply with covering letter, curriculum vitae, writing sample and essay. Check website for up-to-date details.
Closing Date: November 1st, March 15th, July 15th
Funding: Private

THE AIREY NEAVE TRUST

The Airey Neave Trust, PO Bo 111, Leominster, HR6 6BP, England
Tel: (44) 20 7833 4440
Email: info@aveyneevetrust.org.uk/aireyneavetrust@gmail.com
Website: www.aireyneavetrust.org.uk
Contact: Hannah Scott, The Trustees

Initiated in 1979, the Airey Neave Trust sponsors research into the protection of personal freedom under the law against the threat of terrorism, political violence and torture. It also provides financial support for any person who is a refugee, with particular emphasis on postgraduates and those retraining in their professions.

The Airey Neave Research Fellowships

Subjects: Research into the protection of personal freedom under the law against the threat of terrorism, political violence and torture.
Purpose: To support serious research connected with national and international law and human freedom.
Eligibility: Open to academics wishing to undertake research in the field of human freedom.
Level of Study: Postdoctorate, Research, Fellowships
Type: Fellowship
Value: £5,000 available for seminars and workshops, with certain guidelines
Length of Study: Up to 2 years
Study Establishment: An institution attached to a particular university in the United Kingdom
Country of Study: United Kingdom
No. of awards offered: 1–3 per year
Application Procedure: Contact the organisation for more details. No application form required but copy of an upto date curriculum vitae, certificate of mental fitness, 2 referees research plan up to 500 words. Applications should be sent to: aireyneavetrust@gmail.com.
Closing Date: October 31st
Funding: Trusts
No. of awards given last year: 4
No. of applicants last year: 6
Additional Information: The trust also supports work in the following areas: (i) research into transnational networks and their implications for international security. The study will investigate the use of networks by criminal and terrorist organizations for activities such as money-laundering and arms sales (Centre of International Studies, Cambridge University); and (ii) research assistance to the Metropolitan Police in preventing, deterring, disrupting and detecting terrorist activity, particularly in its preparatory stages.

For further information contact:

Contact: Hannah Scott, Administrator

AKADEMIE SCHLOSS SOLITUDE

Solitude 3, Stuttgart, 70197, Germany
Tel: (49) 711 99 61 90
Fax: (49) 711 99 61 95 0
Email: mail@akademie-solitude.de
Website: www.akademie-solitude.de
Contact: Angela Butterstein, PR/Press

A public foundation located on the grounds of the boroque castle Schloss Solitude near Stuttgart, the Akademie Schloss Solitude operates an international artist program awarding live/work fellowships to artists (architecture, the visual arts, the performing arts, design, literature, music/sound and video/film/new media), emerging scholars, scientists, and professionals. Hundreds of fellows from more than 70 countries have used the Akademie to realize and advance their work and projects.

Akademie Schloss Solitude Fellowships

Subjects: Architecture, visual arts, performing arts, design, literature, music/sound, video/film/new media, natural sciences, humanities, economics, and chess.
Purpose: To promote young gifted artists, scholars, scientists and professionals.
Eligibility: Fellowships are awarded to artists, emerging scholars, scientists and professionals, who are not older than 35 or who have finished their basic studies not more than 5 years before applying to Akademie Schloss Solitude. Currently enrolled university or college students (at the time of application) will not be considered for selection. Several fellowships are also awarded regardless of the applicant's age.
Level of Study: Postgraduate, Postdoctorate, Professional development
Type: Residential fellowships
Value: €1,100 per month plus other benefits, ready furnished studio
Length of Study: 3–12 months
Frequency: Every 2 years
Country of Study: Germany
No. of awards offered: 50–70
Application Procedure: The independent jury consists of a jury chairperson and generally ten specialist jurors who independently allocate the fellowships for their respective disciplines. New jurors are nominated every 24 months. The next application round will take place between July and October for a fellowship. Applications sent outside of the application round cannot be considered and the materials will be disposed.
Closing Date: October 31st
Funding: Foundation
Contributor: Lottery of the State of Baden-Wuerttemberg
No. of awards given last year: 67
No. of applicants last year: 1,800
Additional Information: The Akademie offers 45 studios to fellows and guests. It operates in the intermediary space between private and public; where art is reflected upon and produced but where it also finds a connection to the public. The Akademie places particular value on offering its guests another quality of time – one that is better than the artists would experience in their daily lives. A residence in a place like Solitude should therefore be understood as an investment in the future; an investment that, for both the participating artists and the institutions, may bear fruit much later.The wide-reaching program of events (performances, readings, concerts and exhibitions and symposias) organized each year by the Akademie (approx. 80–100 events, most of them public) is devoted exclusively to the works of the Akademie's guests and fellows.

ALBERT ELLIS INSTITUTE

45 East 65th Street, New York, NY, 10065, United States of America
Tel: (1) 212 535 0822
Fax: (1) 212 249 3582
Email: info@albertellis.org
Website: www.albertellisinstitute.org
Contact: Fellowships Office

The Albert Ellis Institute is a non-profit training and therapy institute chartered by the regents of the University of the State of New York, specializing in cognitive behaviour therapy and rational emotive behaviour therapy.

Albert Ellis Institute Clinical Fellowship

Subjects: Psychology and counselling.
Purpose: To provide in-depth, hands-on training in cognitive behavioural therapy and rational emotive behaviour therapy.
Eligibility: Candidates must be in a doctoral programme, hold a PhD, MSW, MD or RN and be licence-eligible in their place of practice. There are no other restrictions.
Level of Study: Postdoctorate, Postgraduate, Predoctorate
Type: Fellowship
Value: US$9,000
Length of Study: 1–2 years
Frequency: Annual
Country of Study: United States of America
No. of awards offered: Varies
Application Procedure: Candidates must obtain applications and further information by writing to the Institute.
Closing Date: February 15th
Funding: Private
No. of awards given last year: 3
No. of applicants last year: 30
Additional Information: The programme begins in mid-July.

ALBERTA INNOVATES HEALTH SOLUTIONS

Suite 1500, Bell Tower, 10104-103 Avenue, Edmonton, AB, T5J 4A7, Canada
Tel: (1) 780 423 5727
Fax: (1) 780 429 3509
Email: pamela.valentine@albertainnovates.ca
Website: www.albertainnovates.ca; www.ahfmr.ab.ca
Contact: Dr Pamela Valentine, Interim Vice-President

Alberta Innovates Health Solutions supports a community of researchers who generate knowledge that improves the health and quality of life of Albertans and people throughout the world. The long-term commitment is to fund basic patient and health research based on international standards of excellence and carried out by new and established investigators and researchers in training.

AIHS Clinician Fellowships

Subjects: Medical research.
Purpose: To provide an opportunity for research training to candidates who have completed clinical sub-speciality training requirements.
Eligibility: Open to candidates: hold or be completing an MD/PhD, health professional degree or health professional degree-PhD and should have completed all Canadian clinical requirements, be licensable in Alberta, and must have a practice permit and/or license to work as a health professional in Alberta.
Level of Study: Postgraduate
Type: Fellowship
Value: Canadian $5,000 research allowance plus a stipend of $70,000
Length of Study: 1 years, with a possibility of renewal for a maximum of 3 years
Frequency: Annual
Study Establishment: An appropriate institution, usually in Alberta
Country of Study: Canada
No. of awards offered: Varies
Application Procedure: Applicants must complete an application form. Candidates must submit, in full, the original application to the AIHS office by the deadline.
Closing Date: April 1st
Funding: Government
No. of awards given last year: 20
No. of applicants last year: 42

For further information contact:

Email: grants.health@albertainnovates.ca

AIHS Graduate Studentship

Subjects: Medical research.
Purpose: To provide opportunities for support for individuals undertaking health-related research areas in pursuit of a Master's or PhD.
Eligibility: Applicants must be currently enrolled in a graduate program at an Alberta University undertaking health-related research training leading to thesis-based graduate degree.
Level of Study: Doctorate, Graduate
Type: Studentship
Value: Canadian $2,000 research allowance plus a stipend of $30,000
Length of Study: 1 years with the possibility of renewal for a maximum of 5 years of support; 2 years maximum at the masters level

Frequency: Annual
Study Establishment: A university in Alberta
Country of Study: Canada
No. of awards offered: Varies
Application Procedure: Candidates must submit, in full, the original application to the AIHS offices by the deadlines.
Closing Date: April 1st
Funding: Government
No. of awards given last year: 49
No. of applicants last year: 200
Additional Information: Please contact at grants.health@albertainnovates.ca

AIHS MD/PhD Studentship

Subjects: Medical research.
Purpose: To provide opportunities for support to individuals undertaking combined MD-PhD degrees at an Alberta University.
Eligibility: Applicants must be currently enrolled at an Alberta University in a combined MD-PhD program. Support is available to applicants who have completed their first year of medical school or graduate training.
Level of Study: Graduate
Type: Studentship
Value: Canadian $2,000 research allowance plus a $30,000 stipend
Length of Study: Tenable for maximum of 6 years
Frequency: Annual
Study Establishment: A University in Alberta
Country of Study: Canada
No. of awards offered: Varies
Application Procedure: Submit completed application to AIHS by the deadline.
Closing Date: April 1st
Funding: Government
No. of awards given last year: 6
No. of applicants last year: 9
Additional Information: Please contact to grants.health@albertainnovates.ca.

AIHS Postgraduate Fellowships

Subjects: Medical research.
Purpose: To provide opportunities for individuals to pursue postgraduate health-related research at an Alberta University.
Eligibility: Open to candidates with a PhD, MD, DDS, DVM or DPharm degree. Normally, support will not be provided beyond 5 years after receipt of the PhD degree, or beyond 8 years after receipt of the MD, DDS, DVM or DPharm degrees.
Level of Study: Postdoctorate
Type: Fellowship
Value: Canadian $5,000 research allowance plus a stipend of $50,000
Length of Study: 1 year with the possibility of renewal for a maximum of 3 years of support
Frequency: Annual
Study Establishment: Usually at a university in Alberta
Country of Study: Canada
No. of awards offered: Varies
Application Procedure: Applicants must complete an application form. Candidates must submit, in full, the original application to the AIHS office by the deadline.
Closing Date: March 1st and October 1st
Funding: Government
No. of awards given last year: 39
No. of applicants last year: 156
Additional Information: Please contact at grants.health@albertainnovates.ca.

ALEXANDER GRAHAM BELL ASSOCIATION FOR THE DEAF AND HARD OF HEARING

3417 Volta Place North West, Washington, DC, 20007, United States of America
Tel: (1) 202 337 5220
Fax: (1) 202 337 8314
Email: financialaid@agbell.org
Website: www.agbell.org
Contact: Wendy F Will, Youth and Family Programs Manager

The Alexander Graham Bell Association for the Deaf and Hard of Hearing helps families, health care providers and education professionals understand childhood hearing loss and the importance of early diagnosis and intervention. Though advocacy, education, research and financial aid, AG Bell helps to ensure that every child and adult with hearing loss has the opportunity to listen, talk and thrive in mainstream society. With chapters located in the United States and a network of international affiliates, AG Bell supports its mission: Advocating Independence through Listening and Talking!

AG Bell College Scholarship Program

Subjects: All subjects.
Purpose: Merit-based scholarship for full-time students who are deaf and hard of hearing, use listening and spoken language, and who attend a mainstream college.
Eligibility: Eligibility requirements are subject to change. Please visit the AG Bell website (http://nc.agbell.org/page.aspx?pid=493) for the most current information.
Level of Study: Unrestricted
Type: Scholarship
Value: Varies from year to year, which ranged from $1,000–10,000 last year.
Frequency: Annual
Country of Study: Any country
No. of awards offered: Varies
Application Procedure: Applicants should visit the website for the most uploaded information. Inquiries may be sent to financialaid@agbell.org.
Closing Date: Varies. Please visit the website.
Funding: Private
Contributor: Members and donors
No. of awards given last year: 18
No. of applicants last year: Approx. 150

ALEXANDER S ONASSIS PUBLIC BENEFIT FOUNDATION

7, Aeschinou Street, Athens, GR-105 58, Greece
Tel: (30) 210 371 3000
Fax: (30) 210 371 3013
Email: ffp@onassis.gr
Website: www.onassis.gr
Contact: Deputy Director, Human Resource Manager

The Alexander S Onassis Public Benefit Foundation establishes and supports public benefit projects, offers services and makes contributions to other public benefit institutions for medical care, education, literature, religion, science, research, journalism, art, cultural matters, history, archaeology and sport. It also awards prizes, grants and scholarships to both Greeks and foreigners.

Onassis Foreigners' Fellowship Programme Educational Scholarships Category B

Subjects: Greek language, Greek literature, Greek history and civilization.
Purpose: To render possible the acquaintance, collaboration and exchange of information between the scholarship recipients and their Greek colleagues in Greek schools, education or other relevant departments of Greek universities.
Eligibility: Open to active elementary or high school foreign teachers who teach the Greek language, modern or ancient, Greek literature, Greek history and Greek civilization. Only persons other than Greek

nationals are eligible. However, persons of Greek descent, second generation and on, are also eligible providing they are permanently residing and working abroad or currently studying in foreign universities.
Type: Fellowship
Value: A monthly allowance of €1,200 plus hotel accommodation and a round trip air ticket
Length of Study: Up to 6 months
Frequency: Annual
Country of Study: Greece
No. of awards offered: 5
Application Procedure: Copies of the announcement and the relevant nomination and application forms are available daily at the Foundation's Secretariat or from the website.
Closing Date: January 31st
Additional Information: The programme presupposes that the scholarship recipients will continue offering their services to their country of origin after they have completed their training in Greece.

Onassis Foreigners' Fellowship Programme Educational Scholarships Category C

Subjects: All subjects.
Purpose: To render possible the acquaintance, collaboration and exchange of information between the scholarship recipients and their Greek colleagues in Greek schools, education or other relevant departments of Greek universities.
Eligibility: Open to foreign postgraduate students and PhD candidates up to 40 years of age who pursue studies in universities, scholarly or research centres or fine art schools either outside Greece or in Greece.
Level of Study: Doctorate, Postgraduate
Type: Fellowship
Value: A monthly allowance of €850 plus hotel accommodation and a round trip air ticket
Length of Study: 5–10 months
Frequency: Annual
Country of Study: Greece
No. of awards offered: 10
Application Procedure: Copies of the Announcement and the relevant nomination and application forms are available daily at the Foundation's Secretariat or from the website.
Closing Date: February 29th
Additional Information: The programme presupposes that the scholarship recipients will continue offering their services to their country of origin, after they have completed their training in Greece.

Onassis Foreigners' Fellowships Programme Research Grants Category AI

Subjects: Humanistic sciences, political sciences and the arts.
Purpose: To enable full members of national academies and full university professors whose scholarly or artistic work has been widely acclaimed and who wish to visit Greece in order to conduct scholarly research or to collaborate with educational institutions, research institutions or organizations.
Eligibility: Only people of non-Greek nationality are eligible. However, persons of Greek descent, second generation and on, are also eligible providing they are permanently residing and working abroad or currently studying in foreign universities. Candidates must have had a professional academic career of at least 10 years.
Level of Study: Postdoctorate
Type: Research grant
Value: A monthly allowance of €1,000 plus a round trip air ticket and hotel accommodation.
Length of Study: 3–6 months
Frequency: Annual
Country of Study: Greece
No. of awards offered: 10
Application Procedure: Copies of the Announcement and the relevant nomination and application forms are available daily at the Foundation's Secretariat or from the website.
Closing Date: February 29th
Additional Information: Any candidate wishing to apply under this programme should specify the category in which they want to be considered in order to receive the relevant nomination and application form. Only one application form for one of the categories can be submitted.

Onassis Foreigners' Fellowships Programme Research Grants Category AII

Subjects: All subjects.
Purpose: To enable university or equivalent institutions' faculty, researchers, PhD holders, artists and musicians, and translators of Greek literature who wish to come to Greece either for scholarly research co-operation with a Greek university, research centre or institute or for their artistic creation or translation.
Eligibility: Only candidates who are not Greek nationals are eligible. However, candidates of Greek descent, second generation and on, are also eligible providing they are permanently residing and working abroad or currently studying in foreign universities. Candidates must have had a professional academic career of at least 10 years.
Level of Study: Postdoctorate
Type: Fellowship
Value: A monthly allowance of €1,000 plus hotel accommodation and a round trip air ticket
Length of Study: Up to 3–6 months
Frequency: Annual
Country of Study: Greece
No. of awards offered: 15
Application Procedure: Copies of the Announcement and the relevant nomination and application forms are available daily at the Foundation's Secretariat or from the website.
Closing Date: February 29th
Additional Information: Any candidate wishing to apply under this programme should specify the category in which they want to be considered in order to receive the relevant nomination and application form. Only one application form for one of the categories can be submitted.

ALEXANDER VON HUMBOLDT FOUNDATION

Press, Communications and Marketing, Bonn, 53173, Germany
Tel: (49) 228 833 455
Fax: (49) 228 833 441
Email: regine.laroche@avh.de
Website: www.humboldt-foundation.de
Contact: Ms Regine Laroche

The Alexander von Humboldt Foundation is a non-profit foundation established by the Federal Republic of Germany for the promotion of international research co-operation. It enables highly qualified scientists and scholars not resident in Germany to spend extended periods of research in Germany and promotes the ensuing academic contacts. The Humboldt Foundation promotes an active world-wide network of researchers.

Alexander Von Humboldt Professorship

Subjects: All subjects.
Purpose: To recruit on a long-term basis top scientists and scholars from abroad for research in Germany and to give German universities–also in cooperation with non-university research institutions–the opportunity to establish or upgrade internationally visible research focus areas, thus raising their profile.
Eligibility: Academics of all disciplines from abroad, who are internationally recognized as leaders in their field.
Level of Study: Research
Type: Professorship
Value: Up to €180,000 per year
Length of Study: 5 years
Frequency: Annual, 2 selection rounds each year
Study Establishment: German universities and research institutions
Country of Study: Germany
No. of awards offered: Up to 10
Application Procedure: Applications may be made by German universities. Non-university research institutions may also submit nominations jointly with a German university.
Closing Date: April 15th and October 15th
Funding: Government

Anneliese Maier Research Award

Subjects: Humanities, social science, cultural science, law, and economics.
Purpose: To promote research collaboration between outstanding researchers from abroad and specialist colleagues in Germany, contributing towards the further internationalisation of the humanities and social sciences in Germany.
Eligibility: Researchers from abroad who already number among the established leaders in their subject as well as researchers who are not yet so advanced in their scientific careers but who are already internationally established researchers.
Level of Study: Research
Type: Prize
Value: €250,000 for annually
Frequency: Annual
Study Establishment: Universities and Research Institutions
Country of Study: Germany
No. of awards offered: Up to 8
Application Procedure: Nominations may be submitted by established academics in Germany. Direct applications are not accepted.
Closing Date: April 30th
Funding: Government

Feodor Lynen Research Fellowships for Experienced Researchers

Subjects: All subjects.
Purpose: To enable researchers from Germany to carry out a long-term research project of their own choice in cooperation with an academic host at a research institution abroad. The host must be an academic working abroad who has already been sponsored by the Humboldt Foundation.
Eligibility: This programme targets outstanding academics from Germany who completed their doctorates less than twelve years ago and whose work demonstrates an independent academic profile. Typically applicants should be working at least at the level of an assistant professor or junior research group leader or as an independent researcher in a comparable position.
Level of Study: Research, Postdoctorate, Professional development
Type: Fellowship
Value: The provisions of the fellowship comprise a basic monthly sum and a monthly foreign allowance which may differ according to location and marital status.
Length of Study: 6–18 months; The fellowship is flexible and can be divided into as many as three stays within three years
Frequency: Annual, 3 selection rounds each year
Study Establishment: Research institutions or Universities
No. of awards offered: Up to 150 per year (for postdoctoral and experienced researchers together)
Application Procedure: Information regarding the application procedure is available at the website of the Humboldt Foundation.
Closing Date: Applications can be made at any time
Funding: Government

Feodor Lynen Research Fellowships for Postdoctoral Researchers

Subjects: All subjects.
Purpose: To enable researchers from Germany to carry out a long-term research project of their own choice in cooperation with an academic host at a research institution abroad. The host must be an academic working abroad who has already been sponsored by the Humboldt Foundation
Eligibility: Open to researchers from Germany with above average qualifications, at the beginning of their academic career, who have completed their doctorate or comparable academic degree (PhD, CSc or equivalent) less than four years prior to the application. Please check the website for further details.
Level of Study: Postdoctorate
Type: Fellowship
Value: A basic monthly sum and a monthly foreign allowance which may differ according to location and marital status. In addition marital and child allowances may be paid.
Length of Study: 6–24 months
Frequency: Annual, 3 selection rounds each year
Study Establishment: Research institutions or universities
No. of awards offered: Up to 150 (for postdoctoral and experienced researchers together)
Application Procedure: Information regarding the application procedure is available at the website of the Humboldt Foundation.
Closing Date: Applications can be made at any time
Funding: Government

Friedrich Wilhelm Bessel Research Award

Subjects: All subjects.
Purpose: To support scientists and scholars, internationally renowned in their field, who are expected to continue producing cutting-edge achievements which will have a seminal influence on their discipline beyond their immediate field of work.
Eligibility: The nominee must be recognised internationally as an outstanding researcher in his/her field and has a doctorate completed less than 18 years ago.
Level of Study: Research
Type: Prize
Value: €45,000
Length of Study: 6–12 months
Frequency: Annual, Two selection rounds each year
Study Establishment: Universities and research institutions
Country of Study: Germany
No. of awards offered: Up to 25
Application Procedure: Nominations may be submitted by established academics in Germany. Direct applications are not accepted.
Closing Date: Nominations are accepted throughout the year
Funding: Government
Additional Information: Selection committee meetings are held twice a year in Spring and Autumn.

Georg Forster Fellowships for Experienced Researchers

Subjects: The research proposal must address issues of significant relevance to the future development of the candidate's country of origin, and, in this context, promise to facilitate the transfer of knowledge and methods to developing and emerging countries.
Purpose: To enable researchers from developing and emerging countries to carry out long-term research project of their own choice with an academic host at a research institution in Germany.
Eligibility: This programme targets outstanding academics from developing and emerging countries (excluding the Republic of China and India) who completed their doctorates or comparable academic degrees (PhD, CSc or equivalent) less than 12 years ago and whose work demonstrates an independent academic profile. Typically, applicants should be working at least at the level of an assistant professor or junior research group leader, or as an independent researcher in a comparable position.
Level of Study: Professional development, Research, Postdoctorate
Type: Fellowship
Value: €3,150 per month (includes a mobility lump sum and a contribution towards health and liability insurance as well as additional benefits)
Length of Study: 6–18 months; the fellowship is flexible and may be divided up into as many as three stays in Germany within three years; plus 12 months return fellowship
Frequency: Annual, 3 selection rounds each year
Study Establishment: Universities or research institution
Country of Study: Germany
No. of awards offered: Up to 60 per year (for postdoctoral and experienced researchers together)
Application Procedure: The complete application should be submitted to the Humboldt Foundation at least four to seven months ahead of the prospective selection date. Information regarding the application procedure is available at http://www.humboldt-foundation.de/web/georg-forster-fellowship-experienced.html.
Funding: Government
Additional Information: Applications should be sent directly to the Foundation. If this is not possible, applications may be submitted via the branch offices of the DAAD or the German embassy or consulate which will then forward them to the Humboldt Foundation.

Georg Forster Research Fellowships for Postdoctoral Researchers

Subjects: The research proposal must address issues of significant relevance to the future development of the candidate's country of

origin, and in this context, promise to facilitate the transfer of knowledge and methods to developing and threshold countries.
Purpose: To enable researchers from developing and emerging countries to carry out long-term research projects of their own choice with an academic host at a research institution in Germany.
Eligibility: Open to researchers from developing and emerging countries (excluding Republic of China and India) with above average qualifications, at the beginning of their academic career who have only completed their doctorate or comparable academic degree (PhD, CSc or equivalent) in the last 4 years. Please check the website for further details.
Level of Study: Postdoctorate
Type: Fellowship
Value: €2,650 per month (including a mobility lump sum and a contribution towards health and liability insurance as well as additional benefits.)
Length of Study: 6–24 months, plus 12 months return fellowship
Frequency: Annual, 3 selection rounds each year
Study Establishment: Universities or research institutions
Country of Study: Germany
No. of awards offered: Up to 60 (for postdoctoral and experienced researchers together)
Application Procedure: Information regarding the application procedure is available at the website of the Humboldt Foundation.
Closing Date: Applications are accepted at any time
Funding: Government
Additional Information: Applications should be sent directly to the Foundation. If this is not possible they may be submitted via the branch offices of the DAAD or the German Embassy or the Consulate, which will then forward them to the Humboldt Foundation.

German Chancellor Fellowships for Prospective Leaders

Subjects: Candidates from all professions and disciplines, but especially from the humanities, law, social sciences and economics.
Purpose: To enable prospective leaders from the USA, the Russian Federation or the People's Republic of China to carry out an individual project tailored to their professional development and goals in cooperation with a German host they have selected themselves.
Eligibility: Applicants must hold at least a Bachelor's or comparable degree completed less than 12 years prior to the start of the fellowship (September 1st of the year following the application) and be citizens of the United States, the Russian Federation, or the Republic of China. Candidates must have gained work experience and have already shown outstanding leadership potential in their careers.
Level of Study: Postgraduate, Professional development, requires at least a Bachelor's degree
Type: Fellowship
Value: €2,150–2,750 monthly (in exceptional cases up to €3,650) including a mobility lump sum and a contribution to health and liability insurance
Length of Study: 1 year (beginning in September), preceded by a mandatory German language course for Fellows with little or no German language skills
Frequency: Annual
Study Establishment: Any institution/organization in Germany
Country of Study: Germany
No. of awards offered: 10 to US citizens, 10 to Russian citizens and 10 to Chinese citizens
Application Procedure: Information regarding the application procedure is available at the website of the Humboldt Foundation.
Closing Date: Varies, once per year. Please see our website for further information
Funding: Government
No. of awards given last year: 10 to US, 10 to Russian and 10 to Chinese citizens
Additional Information: The fellowship begins with a 4-week introductory seminar in Berlin and Bonn.

Hezekiah Wardwell Fellowships

Subjects: Music.
Purpose: To allow young, highly-gifted musicians or musicologists from Spain to pursue further training or advanced studies at a college of music or conservatoire in Germany mentored by an established music teacher.
Eligibility: The Fellowship is aimed at candidates from Spain in transition to their professional careers. Therefore the Fellowship period applied for cannot start later than 3 years after the end of the graduation as a musician or musicologist. The Humboldt Foundation expects candidates to establish contact with supervisors prior to submitting their applications and to obtain binding clarification on whether they can be accepted as students and be given academic supervision.
Level of Study: Graduate, Postgraduate, Professional development
Type: Fellowships
Value: €800 per month and a sum of €250 on arrival for travel expenses
Length of Study: 10 months
Frequency: Annual
Study Establishment: College of Music, Conservatoire, University in Germany
Country of Study: Germany
No. of awards offered: Up to 10
Application Procedure: Information regarding the application procedure is available at the website of the Humboldt Foundation.
Closing Date: January 10th
Funding: Foundation

Humboldt Research Award

Subjects: All subjects.
Purpose: To support outstanding scientists and scholars from all disciplines from abroad whose fundamental discoveries, new theories, or insights have had a significant impact on their own discipline and who are expected to continue producing cutting-edge achievements in future.
Eligibility: Eminent foreign researchers at the peak of their academic careers and in leading positions, such as full professors or directors of institutes, may be nominated.
Level of Study: Research
Type: Prize
Value: €60,000
Length of Study: Up to 1 year
Frequency: Annual, 2 selection rounds each year
Study Establishment: Universities and research institutions
Country of Study: Germany
No. of awards offered: Up to 100
Application Procedure: Applicants are nominated by eminent German scientists and scholars. Direct applications are not accepted.
Closing Date: Nominations are accepted throughout the year
Funding: Government
Additional Information: Selection committee meetings are held twice a year in Spring and Autumn.

Humboldt Research Fellowships for Experienced Researchers

Subjects: All subjects.
Purpose: To enable researchers from abroad to carry out long-term research project of their own choice in cooperation with an academic host at a research institution in Germany.
Eligibility: This programme targets outstanding academics who completed their doctorates less than 12 years ago and whose work demonstrates an independent academic profile. Typically, applicants should be working at least at the level of an assistant professor or junior research group leader, or as an independent researcher in a comparable position.
Level of Study: Postdoctorate
Type: Fellowship
Value: €3,150 per month (includes a mobility lump sum and a contribution towards health and liability insurance as well as additional benefits)
Length of Study: 6–18 months; The fellowship is flexible and may be divided into as many as three stays in Germany within three years
Frequency: Annual, 3 selection rounds each year
Study Establishment: Universities or research institutions
Country of Study: Germany
No. of awards offered: Up to 600 per year (for postdoctoral and experienced researchers together)
Application Procedure: Information regarding the application procedure is available at the website.
Closing Date: Applications are accepted at any time

Funding: Government
Additional Information: Applications should be sent directly to the Foundation or through diplomatic or consular offices of the federal Republic of germany in the candiadtes respective countries.

Humboldt Research Fellowships for Postdoctoral Researchers

Subjects: All subjects.
Purpose: To enable researchers from abroad to carry out a long-term research project of their own choice in cooperation with an academic host at a research instituition in Germany.
Eligibility: Open to researchers from abroad with above average qualifications, at the beginning of their academic career who have only completed their doctorate or comparable academic degree (PhD, CSc or equivalent) in the last 4 years. Please check the website for further details.
Level of Study: Postdoctorate
Type: Fellowship
Value: €2,650 per month (including a mobility lump sum and a contribution towards health and liability insurance, plus additional benefits)
Length of Study: 6–24 months
Frequency: Annual, 3 selection rounds each year
Study Establishment: Universities or research institutions
Country of Study: Germany
No. of awards offered: Up to 600 per year (For Postdoctoral and Experienced researchers together)
Application Procedure: Information regarding the application procedure is available at the website.
Closing Date: Applications are accepted at any time
Funding: Government
Additional Information: Applications should be forwarded directly to the Foundation or through diplomatic or consular offices of the Federal Republic of Germany in the candidates' respective countries.

JSPS Research Fellowships for Postdoctoral Researchers

Subjects: All subjects.
Purpose: To enable highly qualified postdoctoral researchers from Germany to carry out research projects of their own choice in cooperation with an academic at a selected national research institution in Japan.
Eligibility: Open to highly qualified post-doctoral researchers from Germany, who have completed their doctorate in the last 6 years.
Level of Study: Postdoctorate
Type: Fellowship
Value: ¥362,000 per month plus additional allowances
Length of Study: 12–24 months
Frequency: Annual, 3 selection rounds each year
Study Establishment: A university or other research institution
Country of Study: Japan
No. of awards offered: Up to 20
Application Procedure: Information regarding the application procedure is available at the website of the Humboldt Foundation.
Closing Date: Applications can be made at any time
Funding: Government

Konrad Adenauer Research Award for Canadian Scholars

Subjects: Humanities and social sciences.
Purpose: To promote academic collaboration between Canada and the Federal Republic of Germany.
Eligibility: Open to highly qualified Canadian scholars, whose research work in the humanities or the social sciences has earned international recognition and who are among the group of leading scholars in their respective area of specialization.
Level of Study: Research
Type: Prize
Value: Up to €50,000
Length of Study: Up to 1 year
Frequency: Annual
Study Establishment: Universities and research institutions
Country of Study: Germany
No. of awards offered: 1 per year
Application Procedure: Canadian universities and research institutions may nominate Canadian scholars of international renown in the humanities and social sciences for the award. Direct applications are not accepted. Nomination forms and further information can be obtained from the awards coordinator of the Royal Society of Canada.
Closing Date: Please see the website.
Funding: Government
Additional Information: Pre-selection recommendations will be made jointly by the Royal Society of Canada and the University of Toronto and submitted to the Humboldt Foundation.

For further information contact:

Awards Coordinator: The Royal Society of Canada, 170 Waller Street, Ottawa, ON, K1N 9B9, Canada

Max Planck Research Award

Subjects: On an annually alternating basis, the call for nominations addresses areas within the natural and engineering sciences, the life sciences, and the humanities.
Purpose: To enable excellent scientists and scholars of all nationalities who are expected to continue producing outstanding academic achievements in international collaboration to pursue research of their own choosing.
Eligibility: Open to scientists and scholars of all nationalities who are recognized internationally as outstandingly qualified academics.
Level of Study: Research
Type: Prize
Value: €750,000
Length of Study: 3–5 years
Frequency: Annual
Study Establishment: Universities and research institutions
Country of Study: Germany or abroad
No. of awards offered: 2 per year, 1 researcher working in Germany and 1 researcher working abroad.
Application Procedure: The presidents/vice chancellors of universities and the heads of research institutions in Germany are eligible to make nominations (a detailed list of persons can be found on our website). Direct applications are not accepted.
Closing Date: January 31st
Funding: Government
Additional Information: Selection occurs once per year.

Sofja Kovalevskaja Award

Subjects: All subjects.
Purpose: To enable successful top ranking junior researchers from abroad to spend five years building up working groups and working on a high-profile, innovative research project of their own choice at a research institution of their own choice in Germany.
Eligibility: Open to highly qualified scientists and scholars who have a doctorate or comparable academic degree (PhD, CSc or equivalent), completed with distinction less than 6 years prior to the date of application.
Level of Study: Research
Type: Prize
Value: Up to €1.65 million
Length of Study: 5 years
Frequency: Every 2 years
Study Establishment: Universities and research institutions
Country of Study: Germany
No. of awards offered: Up to 8
Application Procedure: Applicants may apply directly to the Humboldt Foundation.
Funding: Government
Additional Information: Selection committee meetings are held once every two years. Please check our website for further details.

THE ALFRED L AND CONSTANCE C WOLF AVIATION FUND

2060 State Highway 595, Gavilan Community, New Mexico, Lindrith, 87029, United States of America
Tel: (1) 575 774 0029
Email: mail@wolf-aviation.org
Website: www.wolf-aviation.org
Contact: Rol Murrow, Executive Director

The Wolf Aviation Fund was established in the wills of Alfred L. and Constance C. Wolf. The Wolf Foundation hopes to help people and projects that benefit general aviation by identifying talented, worthy individuals - often working in collaboration with others - and worthwhile projects and providing them support.

Wolf Aviation Fund Grants Program

Subjects: Aviation.
Purpose: To promote and support the advancement of personal air transportation by seeking and funding the most promising individuals and worthy projects which advance the field of general aviation.
Level of Study: Postgraduate
Type: Grant
Value: Varies
Frequency: Annual
No. of awards offered: Varies
Application Procedure: For more details please visit the website.
Closing Date: December 15th
Additional Information: In preparing proposals the foundation strongly encourages applicants to browse the Resources section and learn how to find out what kinds of grants might be available from any source and to learn in general how to properly prepare grant requests.

ALFRED P SLOAN FOUNDATION

630, Fifth Avenue, Suite 2550, New York, NY, 10111, United States of America
Tel: (1) 212 649 1649
Fax: (1) 212 757 5117
Email: stella@sloan.org; researchfellows@sloan.org
Website: www.sloan.org; www.sloan.org/fellowships
Contact: Erica Stella, Fellowship Administrator

The Alfred P Sloan Foundation is a philanthropic non-profit institution which was established in 1934 by Alfred Pritchard Sloan. The Foundation makes grants primarily to support original research and broad-based education related to science, technology, economic performance and the quality of American life.

Sloan Industry Studies Fellowships

Subjects: Economics, management, engineering, political science, sociology and other related fields.
Purpose: To enhance the careers of the very best young faculty members in the interdisciplinary field of industry studies and to support the development of research in industry studies.
Eligibility: Open to candidates who hold a PhD or an equivalent degree and must be members of the regular faculty.
Level of Study: Postgraduate
Type: Fellowship
Value: US$45,000
Length of Study: 2 years
Frequency: Annual
Country of Study: United States of America
No. of awards offered: 5
Closing Date: October 15th
Funding: Foundation
No. of awards given last year: 5
No. of applicants last year: 33

Sloan Research Fellowships

Subjects: Chemistry, computational and evolutionary molecular biology, computer science, economics, mathematics, neuroscience and physics.
Purpose: To enhance the careers of the very best young faculty members in specified fields of science.
Eligibility: A candidate must hold a PhD (or equivalent) in chemistry, computational or evolutionary molecular biology, computer science, economics, mathematics, neuroscience, ocean sciences, physics, or a related field; be members of the regular teaching faculty (i.e. tenure track) of a college, university, or other degree-granting institution in the United States or Canada; normally, be no more than six years from completion of their most recent PhD or equivalent, as of the year of their nomination.
Level of Study: Postdoctorate
Type: Fellowship
Value: $50,000
Length of Study: 2 years
Frequency: Annual
No. of awards offered: 118
Application Procedure: Applicants must check the website for detailed procedure.
Closing Date: September 15th
Funding: Foundation
No. of awards given last year: 118

ALFRED TOEPFER FOUNDATION

Georgsplatz 10, Hamburg, 20099, Germany
Tel: (49) 04033 4020
Fax: (49) 04033 5860
Email: mail@toepfer-fvs.de
Website: www.toepfer-fvs.de

Alfred Toepfer Scholarships

Subjects: Humanities, social sciences.
Purpose: To provide scholarships for doctoral candidates from the humanities and the social sciences who are in the final stages of research on European issues.
Eligibility: Candidates should be no older than 30 years and should be from Central and Eastern Europe: Albania, Armenia, Azerbaijan, Belarus, Bosnia-Herzegovina, Bulgaria, Croatia, Czech Republic, Estonia, Georgia, Hungary, Kosovo, Latvia, Lithuania, Macedonia, Moldavia, Poland, Romania, Russia, Slovakia, Slovenia, Ukraine. Applicants should have a good knowledge of German.
Level of Study: Doctorate
Type: Scholarship
Value: €920 per month
Length of Study: Up to 1 year
Frequency: Annual
Country of Study: Germany
No. of awards offered: 30–50
Application Procedure: Applications need not be submitted on a special form.
Closing Date: November 30th
Contributor: Alfred Toepfer Foundation
Additional Information: Available on www.toepfer-fvs.de

For further information contact:

Email: ericke(at)toepfer-fvs.de
Contact: Hélène Ericke, Scholarship Programme Administrator

Max Brauer Award

Purpose: To honour personalities and institutions in the City of Hamburg for their services to the city's cultural, scientific, or intellectual life and for extraordinary impulses for the preservation of its architecture and architectural monuments, its city and landscape and renewal of the city, as well as its tradition and its customs.
Type: Award
Value: €15,000
Frequency: Annual
Contributor: Alfred Toepfer Foundation in association with European school
Additional Information: The selection of the Preisträgerin and/or the winner is decided by the Kuratorium Max Brauer award.

For further information contact:

Tel: 49 (0)40/33 402 - 16
Email: luthe(at)toepfer-fvs.de
Contact: Ricarda Luthe

ALICIA PATTERSON FOUNDATION

1090 Vermont Avenue, NW, Suite 1000, Washington, DC, 20005, United States of America
Tel: (1) 202 393 5995
Fax: (1) 301 951 8512
Email: info@aliciapatterson.org
Website: www.aliciapatterson.org
Contact: Ms Margaret Engel, Director

The Alicia Patterson Foundation gives grants to professional print reporters and photojournalists to investigate a subject of their choice. Their reports are published in a quarterly magazine, the *Alicia Patterson Foundation Reporter*, and on the Foundation's website.

Alicia Patterson Journalism Fellowships

Subjects: Journalism.
Purpose: To give working print journalists, a chance to spend a year researching and writing on a topic of their choosing.
Eligibility: Open to print journalists, e.g. reporters, editors, photographers with at least 5 years of full-time professional experience and who are citizens of the United States of the America.
Level of Study: Professional development
Type: Fellowships
Value: The fellowship stipend is $40,000 for 12 months and $20,000 for 6 months and must cover your travel and research costs
Length of Study: 1 year
Frequency: Annual
Country of Study: Any country
No. of awards offered: 5–9
Application Procedure: Candidates must use the Alicia Patterson Foundation application form and are also required to submit a three-page proposal, a two-page autobiographical essay, three clips, four letters of reference and a budget.
Closing Date: Postmarked October 1st
Funding: Private
No. of awards given last year: 7
No. of applicants last year: 197

ALL INDIA COUNCIL FOR TECHNICAL EDUCATION (AICTE)

NBCC Place, 4th Floor, Eastern Tower, Bhishma Pitamah Marg, Pragati Vihar, Lodhi Road, New Delhi, 110 003, India
Tel: (91) 011 24369619-9622
Fax: (91) 011 24369633
Email: rid@aicte.ernet.in
Website: www.aicte-india.org
Contact: Advisor (Quality Assurance)

All India Council for Technical Education (AICTE) was set-up in November 1945 as a national-level Apex Advisory Body to conduct survey on the facilities on technical education and to promote development in the country in a coordinated and integrated manner. The statutory AICTE was established on May 12, 1988 with a view to proper planning and coordinated development of technical education system throughout the country, the promotion of qualitative improvement of such education in relation to planned quantitative growth and the regulation and proper maintenance of norms and standards in the technical education system. The purview of AICTE (the Council) covers programmes of technical education including training and research in engineering, technology, architecture, town planning, management, pharmacy, applied arts and crafts, hotel management, and catering technology etc. at different levels.

Career Awards for Young Teachers

Subjects: All subjects.
Purpose: To support young teachers.
Eligibility: Open to candidates below 35 years (relaxation of 5 years in age limit for women candidates), who are regular teachers in an AICTE-approved technical institutions/university departments and holds at least a postgraduate degree with consistent good academic career and an aptitude for research.
Level of Study: Postgraduate
Type: Award
Value: Total amount of grant is 10,50,000 for 3 years. 4,60,000 (1st year), 3,10,000 (2nd year), 2,80,000 (3rd year) as per the norms of the scheme.
Length of Study: 3 years
Frequency: Annual
Study Establishment: AICTE approved technical institutions/university departments
No. of awards offered: 20
Application Procedure: Candidates must apply to AICTE in prescribed format through the Head of the Institution to which the applicant is attached.
Closing Date: August 31st

ALL SAINTS EDUCATIONAL TRUST

Suite 8C, First Floor, VSC Charity Centre, Royal London House, 22-25 Finsbury Square, London, EC2A 1DX, United Kingdom
Tel: (44) 020 7256 9360
Fax: (44) 20 7621 9758
Email: aset@aset.org.uk
Website: www.aset.org.uk
Contact: Mr Stephen Harrow FKC, Clerk to the Trustees

The Trust makes grants to help with the costs of the formal training or better qualification of individual teachers, and also funds educational advance by other means, e.g. research. There is particular emphasis on religious education, home economics or related subjects, and multi-cultural endeavour linked to those areas. (UK/EU and Commonwealth of Nations only.)

All Saints Educational Trust Corporate Awards

Subjects: Religious education, home economics and kindred subjects, as well as multicultural and interfaith education.
Purpose: To offer assistance to individuals and institutions within certain specified terms of reference.
Level of Study: Unrestricted
Type: Award
Value: Varies. UK£125,000 over 3 years is the maximum awarded
Length of Study: Up to 5 years
Country of Study: United Kingdom
No. of awards offered: Subject to the availability of funds
Application Procedure: Applicants must complete an application form, available on request from the Clerk to the Trust.
Closing Date: April 2nd
Funding: Private
No. of awards given last year: 8
No. of applicants last year: 15
Additional Information: The award must be used or applied for in the United Kingdom. Further information is available on request or from the website.

All Saints Educational Trust Personal Scholarships

Subjects: Religious education, home economics and multicultural education.
Purpose: To give support to persons who work in certain capacities associated with education.
Eligibility: Open to individuals over 18 years of age who are, or who intend to become, teachers. Grants for research are open to individuals. The award must be used with United Kingdom Awards to commonwealth citizens and will be for full-time postgraduate study at Marker's level only.
Level of Study: Foundation programme, Graduate, Postgraduate, Professional development
Type: Grant
Value: Varies, usually UK£500–10,000 but occasionally more (Commonwealth Scholars)
Length of Study: 1–3 years
Study Establishment: Recognized educational institutions in the United Kingdom
Country of Study: United Kingdom
No. of awards offered: Dependent on availability of funds
Application Procedure: Applicants must complete an application form, available on request from the Clerk to the Trust or from the website.
Closing Date: March 1st for UK/EU applications. April 1st for Commonwealth applications
Funding: Private
No. of awards given last year: 10 UK/EU, 8 Commonwealth
No. of applicants last year: 26 UK/EU, 41 Commonwealth
Additional Information: Enquiries should not be delayed until the offer of a place on a course of study has been confirmed. Further information is available on request or from the website

THE ALLEN FOUNDATION, INC.

PO Box 1606, Midland, MI, 48641-1606, United States of America
Tel: (1) 517 832 5678
Fax: (1) 517 832 8842
Email: dbaum@allenfoundation.org
Website: www.allenfoundation.org
Contact: Dale Baum, Secretary

Established in 1975 by agricultural chemist William Webster Allen, the Allen Foundation makes grants to projects that benefit human nutrition in the areas of education, training and research.

Allen Foundation Grants

Subjects: Human nutrition in the areas of health, education, training and research.
Purpose: To assist in the field of human nutrition, to fund relevant nutritional research and to encourage the dissemination of information regarding healthful nutritional practices and habits.
Eligibility: Open to non-profit organizations that are able to provide a copy of their federal Internal Revenue Service certification of 501(c)3 tax-exempt status. If applying from outside the United States of America, applicants must send their country's counterpart or equivalent of the tax-exempt form. Individuals, non-profit organizations without a current exempt status, conferences, seminars, symposia, sponsorship events, fund-raising events and religious organizations without a secular community designation are not eligible for the award. If a grant proposal involves primarily academic research, the grant should be conducted under the leadership of a full-time, principal investigator who is a regular faculty member with tenure or on a tenure track.
Level of Study: Research
Type: Grant
Country of Study: United States of America
No. of awards offered: 1
Application Procedure: Applicants must submit an application form via email only. Application forms and further information can be obtained from the website.
Closing Date: December 31st
Additional Information: Any applications received after this deadline will be reviewed for the following year's applications. The Board of Trustees will announce their decision for successful applicants in June. Because of the number of proposals received and the limited resources of the Foundation, applicants should never view possible declinations to fund their proposals or delays in reviewing their proposals as judgements on the actual merits of their proposals. The Foundation does not directly administer the programmes it funds. For further information, visit the website or contact the Allen Foundation Inc.

ALPHA KAPPA ALPHA EDUCATIONAL ADVANCEMENT FOUNDATION, INC. (AKA-EAF)

5656 South Stony Island Avenue, Chicago, IL, 60637, United States of America
Tel: (1) 773 947 0026
Fax: (1) 773 947 0277
Email: akaeaf@akaeaf.net
Website: www.akaeaf.org

AKA-EAF Financial Need Scholarship

Subjects: All subjects.
Purpose: To promote lifelong learning by securing charitable contributions, gifts and endowed funds to award scholarships, fellowships and community assistance awards.
Eligibility: Applicants must be a full-time and currently enrolled student at an accredited campus based degree-granting institution, must plan to continue their academic pursuits in the fall of the grant year, demonstrated exceptional academic achievement/or financial need, leadership, volunteer, civic and academic services
Level of Study: Graduate, Postgraduate
Type: Scholarship
Value: US$750–2,500
Frequency: Annual
Application Procedure: The only means to obtain an application is through the website
Closing Date: April 15th
Funding: Foundation
No. of awards given last year: Undergraduate 42 and Graduate 38
No. of applicants last year: Undergraduate 217 and Graduate 71

AKA-EAF Merit Scholarships

Subjects: All subjects.
Purpose: To promote lifelong learning by securing charitable contributions, gifts and endowed funds to award scholarships, fellowships and community assistance awards.
Eligibility: Applicants must be a full-time and currently enrolled student at an accredited campus based degree-granting institution, must plan to continue their academic pursuits in the fall of the grant year, demonstrated exceptional academic achievement/or financial need, leadership, volunteer, civic and academic services
Level of Study: Graduate, Postgraduate
Type: Scholarship
Value: US$750–2,500
Frequency: Annual
Application Procedure: The only means to obtain an application is through the website
Closing Date: April 15th
Funding: Foundation
No. of awards given last year: Undergraduate 91 and Graduate 36
No. of applicants last year: Undergraduate 263 and Graduate 158

ALZHEIMER'S AUSTRALIA

1 Frewin Place, Scullin, Australian Capital Territory, 2614, Australia
Tel: (61) 02 6254 4233
Fax: (61) 02 6278 7225
Email: secretariat@alzheimers.org.au; nat.admin@alzheimers.org.au
Website: www.alzheimers.org.au
Contact: Dinusha Fernando, Research Development Manager

Alzheimer's Australia Research Ltd (AAR) was established as the research arm of Alzheimer's Australia to provide funds and disseminate into Alzheimer's disease and other forms of dementia. AAR provides annual research grants and a key priority is to support emerging researchers and to encourage the next generation of dementia researchers.

AAR Postdoctoral Fellowship in Dementia

Subjects: Fellowship can be for biological or psychosocial research, but the research proposal must be judged to relate to research in dementia.
Purpose: To support a PhD graduate undertaking research in an area related to dementia.
Eligibility: Open to citizens of Australia and permanent residents.
Level of Study: Postdoctorate
Type: Fellowship
Value: $50,000 per year
Length of Study: 2 years
Frequency: Dependent on funds available
Country of Study: Australia
No. of awards offered: To be confirmed
Application Procedure: Please check the website for details.
Closing Date: Mid-April
Contributor: Alzheimer's Australia Research Limited (AAR)
No. of awards given last year: 2
No. of applicants last year: 14

AAR Rosemary Foundation Travel Grant

Subjects: Allow a researcher to travel overseas to learn new techniques in a variety of dementia-related fields and/or network with international research teams conducting dementia-related research.
Purpose: To enable an Australian researcher to travel overseas in order to learn new techniques and/or network with international dementia research teams.
Eligibility: Open to citizens of Australia or permanent residents.
Level of Study: Postgraduate, Research
Type: Grant
Value: $15,000

Length of Study: 1 month
Frequency: Dependent on funds available
No. of awards offered: 1
Application Procedure: Please check the website for details.
Closing Date: Mid-April
Contributor: The Rosemary Foundation for Memory Support Inc.
No. of awards given last year: 1
No. of applicants last year: 5

Dementia Grants Program

Subjects: Behavioural and cognitive sciences, biological sciences, or medical and health sciences.
Purpose: To fund research into all aspects of dementia.
Eligibility: Open to new investigators who are citizens of Australia or permanent residents.
Level of Study: Postgraduate
Type: Grant
Value: $30,000 per year (full Phd scholarship up to 3 years) and $7,500 per year (top-up PhD Scholarships up to 2 years)
Frequency: Annual
Country of Study: Australia
Application Procedure: Check website for further details.
Closing Date: April 5th
Contributor: Alzheimer's Australia Research Limited (AAR)
No. of awards given last year: 5
No. of applicants last year: 23
Additional Information: Please see the website for further details http://www.fightdementia.org.au/research-publications/dementia-grants-program.aspx.

Hazel Hawke Research Grant in Dementia Care

Subjects: Suitable projects might include research in carer support, best quality care practices, activities and non-pharmaceutical therapies for people with dementia, or any other aspect of dementia care research.
Purpose: To fund research in dementia care.
Eligibility: Open to citizens of Australia or permanent residents.
Level of Study: Postdoctorate, Postgraduate, Research
Type: Grant
Value: $30,000
Frequency: Annual
Country of Study: Australia
No. of awards offered: To be confirmed
Application Procedure: Please check the website for details.
Closing Date: Mid-April
Funding: Foundation
Contributor: Hazel Hawke Alzheimer's Research and Care Fund
No. of awards given last year: 2
No. of applicants last year: 24

ALZHEIMER'S DRUG DISCOVERY FOUNDATION (ADDF)

57 W 57th Street, Suit 904, New York, NY, 10019, United States of America
Tel: (1) 212 901 8000
Fax: (1) 212 901 8010
Email: nthakker@alzdiscovery.org
Website: www.alzdiscovery.org
Contact: Mr Niyati Thakker, Grants Associate

The ADDF is an affiliated public charity of the Institute for the Study of Aging (ISOA), a private foundation founded by the Estèe Lauder family in 1998. The charity was established in 2004 to enable the public to work in advancing the common mission of supporting scientists and to rapidly accelerate the discovery and development of drugs to prevent, treat, and cure Alzheimer's disease, related dementias and cognitive aging.

ADDF Grants Program

Subjects: Early identification, prevention and treatment of Alzheimer's disease and cognitive decline.
Purpose: To promote the research and development of technology and therapies to identify, treat and prevent cognitive decline, Alzheimer's disease and related dementias.
Eligibility: There are no eligibility restrictions.
Level of Study: Unrestricted
Type: Grant
Value: Negotiable
Length of Study: 1–3 years
Frequency: Dependent on funds available, There is no funding cycle
Study Establishment: A non-profit public foundation
Country of Study: Any country
Application Procedure: Candidates must submit a letter of intent through our online submission system at www.alzdiscovery.org
Closing Date: Letters of intent are accepted at any time. There are quarterly deadlines for full proposals.
Funding: Foundation, government, individuals, private
No. of awards given last year: 29
Additional Information: In addition to funding research activities, the foundation sponsors and/or co-sponsors conferences, scientific and medical workshops to advance knowledge on issues related to Alzheimer's disease and cognitive vitality.

For further information contact:

Email: hfillit@aging-institute.org
Website: http://www.alzdiscovery.org/index.php/research-programs/grant-opportunities
Contact: Howard Fillit, Executive Director
Contact: Niyati Thakker, Grants Associate

ALZHEIMER'S RESEARCH TRUST

The Stables, Station Road, Great Shelford, Cambridge, CB22 5LR, England
Tel: (44) 1223 843 899
Fax: (44) 1223 843 325
Email: enquiries@alzheimers-research.org.uk
Website: www.alzheimers-research.org.uk
Contact: Tegwen Ecclestone, Research Grants Officer

Alzheimer's Research Trust is the leading United Kingdom research charity for dementia. ART funds work in any area of research that promises to further the understanding of the basic disease process in Alzheimer's and related dementias, or that is directed to early detection, identifying risk factors, or progress towards effective treatments.

Alzheimer's Research Trust, Clinical Research Fellowship

Subjects: The basic disease process, symptoms and treatments in Alzheimer's disease and related dementias.
Purpose: To support clinical research by a medically qualified applicant in the field of dementia.
Eligibility: Applicants are required to have secured the sponsorship of a senior established investigator in the institution where the Fellowship is to be held. The Fellowship must be based in the UK with the lead supervisor.
Level of Study: Postdoctorate
Type: Fellowship
Value: Full salary plus contribution towards research and travel costs of upto £10,000 per year
Length of Study: Up to 3 years
Frequency: Annual
Country of Study: United Kingdom
No. of awards offered: 1
Application Procedure: Applicants must apply online by visiting https://a-r.org.uk and submitting 20 paper copies of the application form.
Closing Date: January 25th
Funding: Individuals, trusts
Contributor: Charitable sources
No. of applicants last year: 1

Alzheimer's Research Trust, Emergency Support Grant

Subjects: The basic disease processes, symptoms and treatments in Alzheimer's disease and related dementias.
Purpose: To bridge funding shortfalls in research.
Eligibility: Members of the Alzheimer's Research Trust network only.
Level of Study: Research
Type: Grant

Value: Up to £30,000
Length of Study: A few weeks–1 year
Frequency: Annual, Anytime
Country of Study: United Kingdom
No. of awards offered: As necessary
Application Procedure: Applicants must apply online by visiting https://a-r.org.uk and submitting 20 paper copies of the application form.
Closing Date: November 23rd
Contributor: Charitable sources
No. of awards given last year: 6
No. of applicants last year: 7

Alzheimer's Research Trust, Equipment Grant

Subjects: The basic disease process in Alzheimer's disease and related dementias.
Purpose: To speed up and increase the accuracy and efficiency of research.
Eligibility: Must be UK-based.
Level of Study: Research
Type: Grant
Value: UK£10,000–100,000
Frequency: Annual, Twice a year
Country of Study: United Kingdom
No. of awards offered: 1 or more
Application Procedure: Applicants must apply online by visiting https://a-r.org.uk and submitting 20 paper copies of the application form.
Closing Date: November 23rd
Funding: Individuals, trusts
Contributor: Charitable sources
No. of awards given last year: 2
No. of applicants last year: 10
Additional Information: Applicatons for joint funding for larger pieces of equipment are accepted.

Alzheimer's Research Trust, Major Project or Programme

Subjects: The basic disease process in Alzheimer's disease and related dementias.
Purpose: To support imaginative and high-quality research.
Eligibility: Collaborations with researchers outside the UK will be considered providing the lead applicant is UK-based.
Level of Study: Research
Type: Grant
Value: UK£150,000–1,000,000
Length of Study: 2–5 years
Frequency: Annual
Country of Study: United Kingdom
No. of awards offered: 1 or more
Application Procedure: Applicants must apply online by visiting https://a-r.org.uk and submitting 20 paper copies of the application form.
Closing Date: January 25th
Funding: Individuals, trusts
Contributor: Charitable sources
No. of awards given last year: 5
No. of applicants last year: 25

Alzheimer's Research Trust, PhD Scholarship

Subjects: The basic disease processes in Alzheimer's disease and related dementias.
Purpose: To contribute towards research in the field, and to help ensure that bright young graduates are inducted into this area.
Eligibility: Must be UK-based supervisors to apply for grant.
Level of Study: Postgraduate
Type: Scholarship
Value: Full fees and stipend of UK£15,000 or £16,000 in London and research cost of UK£10,000. Fees paid at home/EU rate
Length of Study: 3 years
Frequency: Annual
Country of Study: United Kingdom
No. of awards offered: Up to 5
Application Procedure: Applicants must apply online by visiting https://a-r.org.uk and submitting 20 paper copies of the application form.
Closing Date: October 26th
Funding: Individuals, trusts
Contributor: Charitable sources
No. of awards given last year: 5
No. of applicants last year: 27

Alzheimer's Research Trust, Pilot Project Grant

Subjects: The basic disease process in Alzheimer's disease and related dementias.
Purpose: To fund innovative research projects and pilot studies.
Eligibility: Lead applicant must be based in the UK.
Level of Study: Research
Type: Grant
Value: Up to UK£30,000
Length of Study: Up to 2 years
Frequency: Annual, Twice a year
Country of Study: United Kingdom
No. of awards offered: 1 or more
Application Procedure: Applicants must apply online by visiting https://a-r.org.uk and submitting 20 paper copies of the application form.
Closing Date: October 26th
Funding: Individuals, trusts
Contributor: Charitable sources
No. of awards given last year: 9
No. of applicants last year: 33
Additional Information: For projects involving inherently expensive techniques, such as imaging, it may be possible to apply for up to £50,000.

Alzheimer's Research Trust, Research Fellowships

Subjects: The basic disease process in Alzheimer's disease and related dementias.
Purpose: To allow junior postdoctoral researchers of demonstrated ability and high potential to carry out further research.
Eligibility: Sponsor must be UK-based.
Level of Study: Postdoctorate
Type: Fellowship
Value: Full salary plus contribution towards research and travel costs of up to £10,000 per year
Length of Study: Up to 3 years
Frequency: Annual
Country of Study: United Kingdom
No. of awards offered: Up to 3
Application Procedure: Applicants must apply online by visiting https://a-r.org.uk and submitting 20 paper copies of the application form.
Closing Date: January 25th
Funding: Individuals, trusts
Contributor: Charitable sources
No. of awards given last year: 2
No. of applicants last year: 10

Alzheimer's Research Trust, Preparatory Clinical Research Fellowship

Subjects: Alzheimer's disease and related dementias.
Purpose: To support clinical research by a medically qualified applicant in the field of dementia.
Eligibility: Open to clinically qualified persons with an honorary clinical contract. Applicants are required to have secured a supervisor, a senior established investigator in the institution where the fellowship is to be held.
Level of Study: Postdoctorate
Type: Fellowship
Value: Full salary plus contribution towards research and travel costs (up to £10,000)
Length of Study: 1 year
Frequency: Annual, Biannual
Country of Study: United Kingdom
No. of awards offered: 1 or 2
Application Procedure: Applicants must apply online by visiting https://a-r.org.uk/ and submitting 20 paper copies of the application form.
Closing Date: November 23rd
Funding: Individuals, trusts
Contributor: Charitable sources
No. of applicants last year: 2

Sabbatical/Secondment

Subjects: Alzheimer's disease and related dementias.
Purpose: To allow tenure/tenure-track researchers to enrich their research programmes and establish collaborations.
Eligibility: Open to tenure/tenure-track researchers. The lead applicant and point of contact must be based in a UK academic/ research institution.
Level of Study: Research
Value: Full salary and research and travel costs (up to £50,000)
Length of Study: 6–12 months
Frequency: Annual, Biannual
Country of Study: United Kingdom
No. of awards offered: 1
Application Procedure: Applicants must apply online by visiting https://a-r.org.uk/ and submitting 20 paper copies of the application form.
Closing Date: November 23rd
Funding: Individuals, trusts
Contributor: Charitable sources
No. of awards given last year: 1
No. of applicants last year: 1

Senior Research Fellowship

Subjects: Alzheimer's disease and related dementias.
Purpose: To allow experienced postdoctoral researchers of demonstrated ability and high potential to carry out further research.
Eligibility: Open to outstanding researchers who have completed their terminal degree within the last 3–10 years. Applicants are required to have secured a sponsor, a senior established investigator in the institution where the fellowship is to be held.
Level of Study: Postdoctorate
Type: Fellowship
Value: Full salary, support staff and running costs (up to £330,000)
Length of Study: Up to 3 years
Frequency: Annual
Country of Study: United Kingdom
No. of awards offered: 1
Application Procedure: Applicants must apply online by visiting https://a-r.org.uk/ and submitting 20 paper copies of the application form.
Closing Date: January 25th
Funding: Individuals, trusts
Contributor: Charitable sources
No. of awards given last year: 3
No. of applicants last year: 6

Travelling Research Fellowship

Subjects: Alzheimer's disease and related dementias.
Purpose: To provide fellows with an opportunity to travel abroad to develop collaborations, learn new techniques and complete projects to advance their career.
Eligibility: Open to outstanding researchers who have completed their terminal degree within the last 10 years. Applicants are required to have secured two supervisors, both senior established investigators in the institutions (UK and abroad) where the fellowship is to be held.
Level of Study: Postdoctorate
Type: Fellowship
Value: Full salary and contribution towards research and travel costs (up to £55,000)
Length of Study: Up to 3 years
Frequency: Annual
Country of Study: United Kingdom
No. of awards offered: 1
Application Procedure: Applicants must apply online by visiting https://a-r.org.uk/ and submitting 20 paper copies of the application form.
Closing Date: January 25th
Funding: Individuals, trusts
Contributor: Charitable sources
Additional Information: The final 6 months must be spent in the UK.

Travelling Research Fellowship US

Subjects: Alzheimer's disease and related dementias.
Purpose: To provide fellows with an opportunity to travel to the US to develop collaborations, learn new techniques and complete projects to advance their career.
Eligibility: Open to outstanding researchers who have completed their terminal degree within the last 10 years. Applicants are required to have secured two supervisors, both senior established investigators in the institution where the fellowship is to be held.
Level of Study: Postdoctorate
Type: Fellowship
Value: Full salary and contribution towards research and travel costs (up to £55,000)
Length of Study: Up to 3 years
Frequency: Annual
Country of Study: United Kingdom
No. of awards offered: 1
Application Procedure: Applicants must apply online by visiting https://a-r.org.uk/ and submitting 20 paper copies of the application form.
Closing Date: January 25th
Funding: Individuals, trusts
Contributor: Charitable sources
Additional Information: The final 6 months of the fellowship must be spent in the UK.

ALZHEIMER'S SOCIETY

Devon House, 58 St Katharine's Way, London, E1W 1LB, England
Tel: (44) 020 7423 3500
Fax: (44) 020 7423 3501
Email: enquiries@alzheimers.org.uk
Website: www.alzheimers.org.uk
Contact: Dr Richard Harvey, Director of Research

The Alzheimer's Society is the leading care and research charity for people with all forms of dementia, their families and carers.

Alzheimer's Society Research Grants

Subjects: All forms of dementia, particularly Alzheimer's disease and vascular dementia.
Purpose: To support research into the cause, cure and care of dementia.
Eligibility: Awards may only be held by United Kingdom institutions. Non-United Kingdom-based researchers may be subcontractors.
Level of Study: Postdoctorate, Research
Value: UK£1,000,000 per year is committed to research. Fellowship Grants are approx. UK£200,000. Project Grants are approx. UK£350,000
Length of Study: Up to 5 years
Country of Study: United Kingdom
No. of awards offered: Varies
Application Procedure: Applicants must complete an application form available from the website.
Closing Date: Fellowship Grants - October 28th. Project Grants - Febuary 25th
Funding: Commercial, government, private, trusts
No. of awards given last year: 7
No. of applicants last year: 71

AMERICA–ISRAEL CULTURAL FOUNDATION (AICF)

1140 Broadway, Suite #304, New York, NY, 10001, United States of America
Tel: (1) 212 557 1600
Fax: (1) 212 557 1611
Email: info@aicf.co.il
Website: www.aicf.org
Contact: Mr Gideon Paz, Executive Director

The America–Israel Cultural Foundation (AICF) has been promoting and supporting the arts in Israel for over 60 years. Through its Sharett Scholarship Program, the AICF grants hundreds of study scholarships each year to Israeli students of the arts, music, dance, visual arts, film, television and theatre, mainly for studies in Israel. The AICF also provides short-term fellowships to artists and art teachers and financially supports various projects in art schools, workshops, master classes, etc.

AICF Sharett Scholarship Program

Subjects: Performing arts, visual arts, design, film or television.
Purpose: To respond to Israel's ever-evolving artistic life and the needs of her artists and institutions.
Eligibility: Open to Israeli citizens only.
Level of Study: Unrestricted
Type: Scholarship
Value: US$750–2,000
Length of Study: Varies
Frequency: Annual
Country of Study: Any country
No. of awards offered: More then 1,400 scholarships, fellowships and grants, mostly for students in Israel
Application Procedure: Applicants must complete and submit an application form with recommendations and prerequired repertoire. Application forms are available from February 1st of each year.
Closing Date: End of February
Funding: Private
Contributor: America–Israel Cultural Foundation
No. of awards given last year: 1,110
No. of applicants last year: 2,300
Additional Information: The programme is revised on an annual basis. For more detailed information, please contact the Foundation after February 1st.

THE AMERICAN ACADEMY IN BERLIN

Hans Arnhold Center, Am Sandwerder 17-19, Wannsee, Berlin, D-14109, Germany
Tel: (49) 30 804 83 118
Fax: (49) 30 804 83 111
Email: applications@americanacademy.de
Website: www.americanacademy.de
Contact: Manager of Fellows Section

The American Academy in Berlin provides a unique bridge between Germany and America – a bridge created through the scholarship and creativity of distinguished individuals involved in cultural, academic and public affairs.

American Academy in Berlin Prize Fellowships

Subjects: Public affairs, culture, humanities, journalism, law, fiction and non-fiction writing, poetry, history, sociology and literature.
Purpose: To provide residential fellowship opportunities to scholars and professionals.
Eligibility: Open to scholars, writers, and professionals who are permanent residents of the United States. US expatriates are not eligible. Candidates in academic fields must have completed their doctorates at the time of application. The Academy does not accept applications from visual artists.
Level of Study: Postgraduate
Type: Fellowship
Value: A stipend of $5,000 per month of residency, round-trip airfare to Berlin, and apartment with partial board
Length of Study: Academic semester
Frequency: Annual
Study Establishment: American Academy in Berlin
Country of Study: Germany
Application Procedure: For application guidelines, restrictions, and forms please visit our website: www.american academy.de/home/fellows/applications
Closing Date: October 1st
Funding: Private
No. of awards given last year: 24
No. of applicants last year: 250
Additional Information: The academy especially encourages people on sabbatical or study leave from their home institutions and organizations.

THE AMERICAN ACADEMY IN ROME

7 East 60 Street, New York, NY, 10022-1001, United States of America
Tel: (1) 212 751 7200
Fax: (1) 212 751 7220
Email: info@aarome.org
Website: www.aarome.org
Contact: Grants Management Officer

The American Academy in Rome is the only American overseas centre for independent study and advanced research in the fine arts and the humanities. It provides a unique opportunity for interaction between artists and scholars working in up to 18 different disciplines. The Academy offers a number of fellowships, residencies and grants for visiting artists and scholars, as well as organizing a variety of events, including concerts, readings, symposia and exhibitions.

American Academy in Rome Fellowships in Design Art

Subjects: Interior design, industrial design, architecture, landscape architecture, set design, urban design, urban planning, conservation and historic preservation and graphic design.
Purpose: To financially support students of design art.
Eligibility: Open to citizens of the United States who have working experience in the related field for at least 7 years and are also currently working in the same field.
Level of Study: Professional development
Type: Fellowship
Value: US$15,000 for 6 months fellowship and US$27,000 for 11 months fellowship
Length of Study: 6 or 11 months
Frequency: Annual
Country of Study: Italy
No. of awards offered: 30
Application Procedure: Applicants can complete an online application posted on the Academy's website (www.aarome.org)
Closing Date: November 1st (extended deadline of November 15th for an additional fee)
No. of awards given last year: 30
No. of applicants last year: 692

AMERICAN ACADEMY OF CHILD AND ADOLESCENT PSYCHIATRY

3615 Wisconsin Avenue North West, Washington, DC, 20016-3007, United States of America
Tel: (1) 202 966 7300
Fax: (1) 202 966 2891
Email: research@aacap.org
Website: www.aacap.org
Contact: Deputy Director of Research & Training

The American Academy of Child and Adolescent Psychiatry is a national, professional medical association established in 1953 as a non-profit organization to support and improve the quality of life for children, adolescents and families affected by mental illnesses.

AACAP Educational Outreach Program for Child and Adolescent Psychiatry Residents (former Travel Grant Program)

Subjects: Child and adolescent psychiatry.
Purpose: To help defray the cost of attending the AACAP's Annual Meeting in Boston, MA.
Eligibility: Candidates must be child and adolescent psychiatry residents at the time of the AACAP Annual Meeting and must be currently enrolled in a residency programme in the United States of America. All awardees must attend the Young Leaders Awards Luncheon and other stated events, and serve as a monitor for 1 day at the Annual Meeting.
Level of Study: Postdoctorate
Type: Travel grant
Value: Up to US$1,000 for travel expenses to the AACAP/CACAP Joint Annual Meeting in Toronto, Ontario, Canada, participation in various AACAP/CACAP Joint Annual Meeting events.
Frequency: Annual

Country of Study: United States of America
No. of awards offered: 40
Application Procedure: For updated application guidelines, applicants must visit the AACAP website at www.aacap.org/awards/pfizertravel.htm
Closing Date: July 12th
Contributor: ACCAP
No. of awards given last year: 50

Jeanne Spurlock Minority Medical Student Clinical Fellowship in Child and Adolescent Psychiatry

Subjects: Psychiatry and mental health.
Purpose: To support work during the summer with a child and adolescent psychiatrist mentor.
Eligibility: Applications are accepted from African American, Asian American, Native American, Alaskan Native, Mexican American, Hispanic and Pacific Islander students in accredited United States of America medical schools.
Level of Study: Graduate
Type: Fellowship
Value: Up to $3,500
Length of Study: 12 weeks
Frequency: Annual
Country of Study: United States of America
No. of awards offered: Up to 14
Application Procedure: For updated application information, applicants must visit the AACAP website at www.aacap.org/awards/index.htm
Closing Date: February 15th
Funding: Government
Contributor: CMHS
No. of awards given last year: 12

Jeanne Spurlock Research Fellowship in Drug Abuse and Addiction for Minority Medical Students

Subjects: Psychiatry and mental health.
Purpose: To support work during the Summer with a child and adolescent psychiatrist research mentor.
Eligibility: Applications are accepted from African American, Asian American, Native American, Alaskan Native, Mexican American, Hispanic and Pacific Islander students in accredited United States of America medical schools. All applications must relate to substance abuse research.
Level of Study: Graduate
Type: Fellowship
Value: Up to US$4,000
Length of Study: 12 weeks
Frequency: Annual
Country of Study: United States of America
No. of awards offered: Up to 5
Application Procedure: For updated application information on the Spurlock Fellowships, please visit website.
Closing Date: February 15th
Funding: Government
Contributor: NIDA
No. of awards given last year: 1

THE AMERICAN ACADEMY OF FACIAL PLASTIC AND RECONSTRUCTIVE SURGERY (AAFPRS)

310 S. Henry Street, Alexandria, VA, 22314, United States of America
Tel: (1) 703 299 9291
Fax: (1) 703 299 8898
Email: info@aafprs.org
Website: www.aafprs.org
Contact: Research Programme

The American Academy of Facial Plastic and Reconstructive Surgery (AAFPRS) Foundation represents 2,700 facial plastic and reconstructive surgeons throughout the world. Its main mission is to promote the highest quality facial plastic surgery through education, the dissemination of professional information and the establishment of professional standards. The AAFPRS was created to address the medical and scientific issues confronting facial plastic surgeons.

Leslie Bernstein Grant

Subjects: Facial plastic and reconstructive surgery.
Purpose: To encourage original research projects that will advance facial plastic and reconstructive surgery.
Eligibility: Open to all AAFPRS members.
Level of Study: Professional development
Type: Grant
Value: US$25,000
Length of Study: 3 years
Study Establishment: The recipient's practice or institution
Country of Study: Any country
No. of awards offered: 1
Application Procedure: Applicants must submit an application form and other documentation including a curriculum vitae and research proposal. Application forms and guidelines are available on the web at www.entlink.net//research/grant/foundation.funding-opportunities.cfm.
Closing Date: January 15th
Funding: Private
Contributor: Dr Leslie Bernstein
No. of awards given last year: 1
No. of applicants last year: 3
Additional Information: All applications must be submitted through the Centralized Otolaryngology Research Efforts (CORE) programme. Please refer to the website for details.

Leslie Bernstein Investigator Development Grant

Subjects: Facial plastic surgery or clinical or laboratory research.
Purpose: To support the work of a young faculty member in facial plastic surgery conducting significant clinical or laboratory research, as well as the training of resident surgeons in research.
Eligibility: Open to AAFPRS members who are involved in the training of resident surgeons.
Level of Study: Postgraduate
Type: Research grant
Value: US$15,000
Length of Study: 3 years
Frequency: Annual
Study Establishment: The recipient's institution
Country of Study: United States of America
No. of awards offered: 1
Application Procedure: Applicants must submit an application form and other documentation including a curriculum vitae and research proposal. Application forms and guidelines are available on the web at www.entlink.net//research/grant/foundation-funding-opportunities.cfm.
Closing Date: January 15th
Funding: Private
Contributor: Dr Leslie Bernstein
No. of awards given last year: 1
No. of applicants last year: 2
Additional Information: All applications must be submitted through the Centralized Otolaryngology Research Efforts (CORE) programme (see the website listed above for information and application).

Leslie Bernstein Resident Research Grants

Subjects: Facial plastic surgery.
Purpose: To stimulate resident research in projects that are well conceived and scientifically valid.
Eligibility: Open to AAFPRS members. Residents at any level may apply even if the research work will be done during their fellowship year. All applicants are required to have the sponsorship and oversight of the department chair or an AAFPRS member as mentor.
Level of Study: Postgraduate
Type: Research grant
Value: US$5,000
Length of Study: 2 years
Frequency: Annual
Study Establishment: The recipient's institution
Country of Study: United States of America
No. of awards offered: Up to 2
Application Procedure: Applicants must submit an application form and other documentation including a curriculum vitae and research

proposal. Application forms and guidelines are available on the web www.entlink.net//research/grant/foundation-funding-opportunities.cfm.
Closing Date: January 15th
Funding: Private
Contributor: Dr Leslie Bernstein
No. of awards given last year: 1
No. of applicants last year: 4
Additional Information: Residents are encouraged to enter early in their training so that their applications may be revised and resubmitted if not accepted the first time. All applications must be submitted through the Centralized Otolaryngology Research Efforts (CORE) programme (see the website listed above for information and application).

THE AMERICAN ALPINE CLUB (AAC)

710 Tenth Street, Suite 100, Golden, CO 80401, United States of America
Tel: (1) 303 384 0110
Fax: (1) 303 384 0111
Email: getinfo@americanalpineclub.org
Website: www.americanalpineclub.org
Contact: Janet Miller, Grants Administrator

The American Alpine Club (AAC) is a national non-profit organization that has represented mountaineers and rock climbers for almost a century. Since its inception in 1902, the AAC has been the only national climbers' organization devoted to the exploration and scientific study of high mountain elevations and polar regions of the world, and the promotion and dissemination of knowledge about the mountains and mountaineering through its meetings, publications and libraries. It is also dedicated to the conservation and preservation of mountain regions and other climbing areas and the representation of the interests and concerns of the American climbing community.

AAC Mountaineering Fellowship Fund Grants

Subjects: Rock climbing.
Purpose: To encourage young American climbers to visit remote areas and seek out climbs more technically demanding than they would normally undertake.
Eligibility: Applicants must be 25 years of age or under, citizens of the United States of America and experienced climbers. Membership of the American Alpine Club is a prerequisite. Members of a single expedition may apply individually, but organized groups or expeditions are ineligible. Grants are not available for the purpose of climbing instruction.
Level of Study: Unrestricted
Type: Grant
Value: US$300–800
Frequency: Annual
Country of Study: United States of America
No. of awards offered: 5–16
Application Procedure: Applicants must download application forms from the website and submit them online.
Closing Date: April 1st and November 1st
Funding: Private
No. of awards given last year: 13
No. of applicants last year: 17
Additional Information: Grants will be based on the excellence of the proposed project and evidence of mountaineering experience. A report must be written upon project completion.

AAC Research Grants

Subjects: Scientific research focusing on mountain and polar areas.
Purpose: To recognize a specific contribution to scientific endeavour germane to mountain regions and alpine research projects.
Eligibility: There are no restrictions on eligibility, but grants will not be awarded for academic tuition. Applications are considered in terms of their scientific or technical quality and the purposes for which the funds and the AAC are established.
Level of Study: Postgraduate
Type: Research grant
Value: US$200–1000
Frequency: Annual
Country of Study: Any country
No. of awards offered: Varies
Application Procedure: Applicants must call or write for application forms, which are also available from the website.
Closing Date: March 1st
Funding: Private
Contributor: The Arthur K. Gilkey Memorial Research Fund, the R.L. Putnam Research Fund, and the Bedayn Research Fund
Additional Information: A report must be submitted upon completion of the project.

AMERICAN ANTIQUARIAN SOCIETY (AAS)

185 Salisbury Street, Worcester, MA, 01609-1634, United States of America
Tel: (1) 508 755 5221
Fax: (1) 508 753 3311
Email: academicfellowships@mwa.org
Website: www.americanantiquarian.org
Contact: Paul Erickson, Director of Academic Programs

The American Antiquarian Society (AAS) is a learned society that was founded in 1812 in Worcester, MA. The Society maintains a research library of American history and culture up to 1876 in order to collect, preserve and make available for study the printed records of the United States of America.

AAS American Society for 18th Century Studies Fellowships

Subjects: American 18th century studies.
Eligibility: Open to candidates holding the PhD or equivalent degree at the time of the application.
Level of Study: Postdoctorate
Type: Fellowships
Value: $1,850 per month or $1,350 per month including housing
Length of Study: 1 month
Frequency: Annual
Study Establishment: The Society's Library in Worcester, Massachusetts
Country of Study: United States of America
No. of awards offered: 1
Application Procedure: All application material is available from our website: www.americanantiquarian.org
Closing Date: January 15th
Funding: Private
Contributor: The American Society for 18th Century Studies and the AAS
No. of awards given last year: 1

AAS Joyce Tracy Fellowship

Subjects: Early American history and culture.
Purpose: To support research on newspapers or magazines for projects using these resources as primary documentation.
Eligibility: Doctoral candidates may apply.
Level of Study: Doctorate, Postdoctorate
Type: Fellowship
Value: $1,850 per month or $1,350 per month including housing
Length of Study: 1 month
Frequency: Annual
Study Establishment: The Society's Library in Worcester, Massachusetts
Country of Study: United States of America
No. of awards offered: 1
Application Procedure: All application material is available from our website: www.americanantiquarian.org
Closing Date: January 15th
Funding: Private
Contributor: An endowment established in memory of Joyce Tracy
No. of awards given last year: 1

AAS Kate B and Hall J Peterson Fellowships

Subjects: Early American history to 1876.
Purpose: To enable persons, who might not otherwise be able to do so, to travel to the Society in order to make use of its research facilities.
Eligibility: Doctoral candidates may apply
Level of Study: Doctorate, Postdoctorate
Type: Fellowship
Value: $1,850 per month or $1,350 per month including housing
Length of Study: 1–3 months
Frequency: Annual
Study Establishment: The Society's Library in Worcester, Massachusetts
Country of Study: United States of America
No. of awards offered: 10
Application Procedure: All application material is available from our website: www.americanantiquarian.org
Closing Date: January 15th
Funding: Private
Contributor: The late Hall J Peterson and his wife Kate B Peterson

AAS Reese Fellowship

Subjects: American bibliography and the history of the book in America to 1876.
Purpose: To support research.
Eligibility: Doctoral candidates may apply
Level of Study: Doctorate, Postdoctorate
Type: Fellowship
Value: $1,850 per month or $1,350 per month including housing
Length of Study: 1 month
Frequency: Annual
Study Establishment: The Society's Library in Worcester, Massachusetts
Country of Study: United States of America
Application Procedure: All application material is available from our website: www.americanantiquarian.org
Closing Date: January 15th
Contributor: The William Reese Company, New Haven, CT
No. of awards given last year: 2

AAS-National Endowment for the Humanities Visiting Fellowships

Subjects: Early American history and culture.
Purpose: To make the Society's research facilities more readily available to qualified scholars.
Eligibility: Fellowships may not be awarded to degree candidates or for study leading to advanced degrees, nor may they be granted to foreign nationals unless they have been resident in the United States of America for at least 3 years immediately prior to receiving the award.
Level of Study: Postdoctorate
Type: Fellowship
Value: The maximum stipend available is US$50,400
Length of Study: 4–12 months
Frequency: Annual
Country of Study: United States of America
No. of awards offered: 3
Application Procedure: All application material is available from our website: www.americanantiquarian.org
Closing Date: January 15th
Funding: Government
Contributor: NEH
No. of awards given last year: 3
Additional Information: Fellows may not accept teaching assignments or undertake any other major activities during the tenure of the award. Other major fellowships may be held concurrently.

AAS-North East Modern Language Association Fellowship

Subjects: American literary studies.
Purpose: To support research.
Eligibility: Doctoral candidates may not apply.
Level of Study: Postdoctorate
Type: Fellowship
Value: $1,850 per month or $1,350 per month including housing
Length of Study: 1–3 months
Frequency: Annual
Study Establishment: The Society's Library in Worcester, Massachusetts
No. of awards offered: 1
Closing Date: January 15th
Funding: Private
Contributor: Jointly funded by NEMLA and AAS
No. of awards given last year: 1

ACLS Frederick Burkhardt Fellowship

Subjects: All subjects supported by the AAS library.
Purpose: To support research.
Eligibility: Candidates must be recently tenured humanists selected on the basis of their scholarly qualifications, the scholarly significance of the project and the appropriateness of the proposed study to the Society's collections.
Level of Study: Postdoctorate
Type: Fellowship
Value: A maximum stipend of US$75,000
Length of Study: 1 year
Frequency: Annual
Study Establishment: The Society's Library in Worcester, Massachusetts
Country of Study: United States of America
No. of awards offered: 11
Application Procedure: All application material is available from ACLS website: www.acls.org/burkguid.htm
Closing Date: October 2nd
Funding: Private
Contributor: The Andrew W Mellon Foundation and ACLS
No. of awards given last year: 8, 1 at AAS

American Historical Print Collectors Society Fellowship

Subjects: American prints of the 18th and 19th centuries.
Purpose: To support research or projects using prints as primary documentation.
Eligibility: Doctoral candidates may apply.
Level of Study: Doctorate, Postdoctorate
Type: Fellowship
Value: $1,850 per month or $1,350 per month including housing
Length of Study: 1 month
Frequency: Annual
Study Establishment: The Society's Library in Worcester, Massachusetts
Country of Study: United States of America
Application Procedure: All application material is available from our website: www.americanantiquarian.org
Closing Date: January 15th
Funding: Private
Contributor: The American Historical Print Collectors Society and the AAS
No. of awards given last year: 1

Barbara L. Packer Fellowship

Subjects: The Barbara Packer Fellowship supports research on the Transcendentalists in general, and particularly Ralph Waldo Emerson, Margaret Fuller, and Henry David Thoreau.
Purpose: To support research.
Eligibility: Doctoral candidates may apply.
Level of Study: Doctorate, Postdoctorate
Type: Fellowship
Value: $1,850 per month or $1,350 per month including housing
Length of Study: 1 month
Frequency: Annual
Study Establishment: The Society's Library in Worcester MD
Country of Study: United States of America
No. of awards offered: 1
Application Procedure: All application material is available at www.americanantiquarian.org.
Closing Date: January 15th
Funding: Private
Contributor: The Ralph Waldo Emerson Society

Drawn to Art Fellowship

Subjects: American art, visual culture or other projects that will make substantial use of graphic materials as primary sources.
Purpose: To support research.
Eligibility: Doctoral candidates may apply.
Level of Study: Doctorate, Postdoctorate
Type: Fellowship
Value: $1,850 per month or $1,350 per month including housing
Length of Study: 1 month
Frequency: Annual
Study Establishment: The Society's Library in Worcester, Massachusetts
Country of Study: United States of America
No. of awards offered: 1
Application Procedure: All application material is available from our website: www.americanantiquarian.org
Closing Date: January 15th
Funding: Private
Contributor: Diana Korzenik
No. of awards given last year: 1

Jay and Deborah Last Fellowship in American History Visual Culture

Subjects: American art, visual culture, or other projects that will make substancial use of graphic materials as primary sources.
Purpose: To support students who wish to carry out research in the related fields.
Eligibility: Doctoral candidates may apply.
Level of Study: Doctorate, Postdoctorate
Type: Fellowship
Value: $1,850 per month or $1,300 per month including housing
Length of Study: 1–3 months
Frequency: Annual
Study Establishment: Study at the Society's library in Worcester, Massachusetts.
No. of awards offered: 10
Application Procedure: You can find more information and downloadable form at www.americanantiquarian.org
Closing Date: January 15th
Funding: Private
Contributor: Jay and Deborah Last

Justin G. Schiller Fellowship

Subjects: The Schiller Fellowship supports research from any disciplinary perspective on the production, distribution, literary content or historical context of American children's books to 1876.
Purpose: To support research.
Eligibility: Both postdoctoral scholars and doctoral candidates from any disciplinary perspective may apply.
Level of Study: Doctorate, Postdoctorate
Type: Fellowships
Value: $1,850 per month or $1,350 per month including housing
Length of Study: 1 month
Frequency: Annual
Study Establishment: The Society's Library in Worcester, MD
Country of Study: United States of America
Application Procedure: Application materials are available on the Society's website at www.americanantiquarian.org.
Closing Date: January 15th
Funding: Private

Linda F. and Julian L. Lapides Fellowship

Subjects: The Lapides Fellowship supports research on printed and manuscript material produced in America through 1865 for (or by) children and youth.
Purpose: To support research.
Eligibility: Doctoral candidates may apply.
Level of Study: Doctorate, Postdoctorate
Type: Fellowship
Value: $1,850 per month or $1,350 per month including housing
Length of Study: 1 month
Frequency: Annual
Study Establishment: The Society's Library in Worcester, MD
Country of Study: United States of America
No. of awards offered: 1
Application Procedure: Application materials are available on the Society's website at www.americanantiquarian.org.
Closing Date: January 15th
Funding: Private
Contributor: Linda F. and Julian L. Lapides

Stephen Botein Fellowship

Subjects: The history of the book in American culture to 1876.
Purpose: To support research.
Eligibility: Doctoral candidates may apply.
Level of Study: Doctorate, Postdoctorate
Type: Fellowship
Value: $1,850 per month or $1,350 per month including housing
Length of Study: Up to 2 months
Frequency: Annual
Study Establishment: The Society's Library in Worcester, Massachusetts
Country of Study: United States of America
No. of awards offered: 1–2
Application Procedure: All application material is available from our website: www.americanantiquarian.org
Closing Date: January 15th
Funding: Private
Contributor: An endowment established by the family and friends of the late Mr Botein
No. of awards given last year: 2

AMERICAN ASSOCIATION FOR CANCER RESEARCH (AACR)

615 Chestnut Street, 17th Floor, Philadelphia, PA, 19106-4404, United States of America
Tel: (1) 215 440 9300
Fax: (1) 215 440 9313
Email: aacr@aacr.org
Website: www.aacr.org
Contact: Ms Sheri Ozard, Program Co-ordinator

The American Association for Cancer Research (AACR) is a scientific society of over 17,000 laboratory and clinical cancer researchers. It was founded in 1907 to facilitate communication and dissemination of knowledge among scientists and others dedicated to the cancer problem, and to foster research in cancer and related biomedical sciences. It is also dedicated to encouraging the presentation and discussion of new and important observations in the field, fostering public education, science education and training, and advancing the understanding of cancer etiology, prevention, diagnosis and treatment throughout the world.

AACR Career Development Awards

Subjects: Translational or clinical cancer research.
Purpose: To support cancer research by junior faculty.
Eligibility: Applicants must have a doctoral degree (including PhD, M. D., D.O., D.C., N.D., D.D.S., D.V.M., Sc.D., D.N.S., Pharm.D., or equivalent doctoral degree, or a combined clinical and research doctoral degree) in a related field and not currently be a candidate for a further doctoral or professional degree.
Level of Study: Research, Postdoctorate
Type: Award
Value: US$50,000 per year
Length of Study: 2 years
Frequency: Annual
Study Establishment: Universities or research institutions
Country of Study: Any country
No. of awards offered: Varies
Application Procedure: Candidates must be nominated by a member of AACR and must be an AACR member or apply for membership by the time the application is submitted. Associate members may not be nominators. The online application is available at the AACR website. Please see the website for further details regarding eligibility http://www.aacr.org/Uploads/DocumentRepository/Grants/2012_FCC_CDA_PG.rev.pdf.
Closing Date: February 1st at 12:00 noon, Eastern Time
Funding: Private

Contributor: The Cancer Research and Prevention Foundation, the Susan G Komen Breast Cancer Foundation, Genentech Inc., the Pancreatic Cancer Action Network
No. of awards given last year: 6
No. of applicants last year: 75

For further information contact:

Tel: (215) 446 7191
Fax: (215) 440 9372
Email: grants@aacr.org
Website: http://www.aacr.org/Uploads/DocumentRepository/Grants/2012_FCC_CDA_PG.rev.pdf
Contact: Hanna Hopfinger, Program Associate

AACR Fellowship

Subjects: Translational or clinical research.
Purpose: To foster meritorious cancer research.
Eligibility: Applicants must have a doctoral degree (including PhD, MD, DO, DC, ND, DDS, DVM, ScD, DNS, PharmD, or equivalent doctoral degree, or a combined clinical and research doctoral degree) in a related field and not currently be a candidate for a further doctoral or professional degree. Please see the website for further details regarding eligibility http://www.aacr.org/home/scientists/aacr-research-funding/current-funding-opportunities-for-postdoctoral-or-clinical-research-fellows.aspx#FCC.
Level of Study: Postdoctorate
Type: Fellowships
Value: One-year grant of $55,000 to support the salary and benefits
Length of Study: 1–3 years
Frequency: Annual
Study Establishment: Universities or research institutions
Country of Study: Any country
No. of awards offered: Varies
Application Procedure: Candidates must be nominated by a member of the AACR. Candidates must be an AACR member or apply for membership by the time the application is submitted. The online application is available from the AACR website.
Closing Date: January 24th noon, Eastern Time
Funding: Private
Contributor: Amgen, Inc., AstraZeneca, Bristol-Myers Squibb Oncology, the Cancer Research and Prevention Foundation and Genentech BioOncology, Inc, MedImmune, National Brain Tumor Foundation
No. of awards given last year: 7
No. of applicants last year: 125

For further information contact:

Contact: Hanna Hopfinger, Program Associate

AACR Gertrude B. Elion Cancer Research Award

Subjects: Cancer research.
Purpose: To foster meritorious basic, clinical or translational cancer research.
Eligibility: Applicants must have a doctoral degree (including PhD, MD, DO, DC, ND, DDS, DVM, ScD, DNS, PharmD, or equivalent doctoral degree, or a combined clinical and research doctoral degree) in a related field and not currently be a candidate for a further doctoral or professional degree. Please see the website for further details regarding eligibility http://www.aacr.org/home/scientists/aacr-research-funding/current-funding-opportunities-for-junior-faculty.aspx#elion.
Level of Study: Postdoctorate, Research
Type: Research award
Value: $150,000 over two years ($75,000 per year) for salary and benefits, laboratory supplies and limited travel for the grant recipient)
Length of Study: 1 year
Frequency: Annual
Study Establishment: Universities or research institutions
Country of Study: Any country
No. of awards offered: 1
Application Procedure: Candidates must be nominated by a member of the AACR. The online application is available at the AACR website.
Closing Date: January 23rd at noon, Eastern Time
Funding: Private
Contributor: GlaxoSmithKline
No. of awards given last year: 1
No. of applicants last year: 20

For further information contact:

Tel: (215) 446-7191
Fax: (215) 440-9372
Email: grants@aacr.org
Website: http://www.cals.wisc.edu/Research/funding/DisplayOpp.php?OppID=306#
Contact: Ms Hanna Hopfinger, Program Associate

AACR Scholar-in-Training Awards

Subjects: Cancer research.
Purpose: To allow individuals to attend the AACR Annual Meeting and Special Conferences.
Eligibility: All graduate and medical students, postdoctoral fellows and physicians-in-training who submit abstracts may be considered for a Scholar-in-Training Award. Please see the website for further details regarding eligibility http://www.aacr.org/home/scientists/travel-grants/aacr-scholar-in-training-awards-other-conferences-and-meetings.aspx.
Level of Study: Doctorate, Graduate, Postdoctorate, Postgraduate, Predoctorate
Type: Award
Value: US$400–2,000
Frequency: Annual
No. of awards offered: Varies
Application Procedure: No application is needed. Qualified persons who want to be considered should follow the instructions included in the abstract submission materials for the AACR Annual Meetings or Special Conference. If a candidate is eligible based on the above criteria, a certification form confirming his or her status will be requested at a later date.
Closing Date: Varies, please contact AACR for details
Funding: Corporation, individuals, private
Contributor: AFLAC Inc., AstraZeneca, Aventis, Bristol-Myers Squibb Oncology, Genentech, GlaxoSmithKline, ILEX, ITO EN Limited, Novartis Pharmaceuticals, the Avon Foundation, Susan G. Komen Breast Cancer Foundation and Pezcoller Foundation
No. of awards given last year: 300
No. of applicants last year: Approx. 2,000

AMERICAN ASSOCIATION FOR RESPIRATORY CARE

9425 N. MacArthur Boulevard, Suite 100, Irving, TX, 75063-4706, United States of America
Tel: (1) 972 243 2272
Fax: (1) 972 484 2720
Email: info@aarc.org
Website: www.aarc.org
Contact: Administrative Assistant

The American Respiratory Care Foundation is dedicated to the art, science, quality and technology of respiratory care. It is a non-profit organization formed for the purpose of supporting research, education and charitable activities and to promote prevention, quality treatment and management of respiratory-related diseases.

NBRC/AMP Gareth B Gish, MS RRT Memorial Postgraduate Recognition Award

Subjects: Respiratory care and prevention.
Purpose: To assist qualified individuals in the pursuit of training leading to an advanced degree.
Eligibility: Must be a respiratory therapist who has at least a Baccalaureate degree with a 3.0 cumulative GPA or better on a 4.0 scale, or the equivalent. If no Baccalaureate degree, then a letter of matriculation from a graduate program and an undergraduate transcript will suffice. Please see the website for further details regarding eligibility http://www.arcfoundation.org/awards/postgraduate/gish.cfm.
Level of Study: Postgraduate, Professional development
Type: Award

Value: Up to US$1,500 plus airfare, a certificate of recognition, one night's lodging and registration for the AARC International Respiratory Congress. May also include a cash award up to $1,500.
Frequency: Annual
Country of Study: United States of America
No. of awards offered: 1
Application Procedure: Candidates must return a completed, signed and notarized application form, provide three letters of reference attesting to the applicant's character, academic ability and professional commitment and supply an essay of at least 1,200 words. This must describe how the award will assist the applicant in reaching the objective of an advanced degree and the candidate's ultimate goals of leadership in healthcare. Application forms can be downloaded and printed out from the website. Please see the website for further details http://www.arcfoundation.org/awards/postgraduate/gish.cfm.
Closing Date: June 15th

Parker B Francis Respiratory Research Grant

Subjects: Respiratory care and related topics.
Purpose: To provide financial assistance for research programmes.
Eligibility: Open to qualified investigators in the field of respiratory care. The principal investigator may be a physician or respiratory therapist. However, a respiratory therapist must be the co-principal investigator if a physician is the principal applicant for the award.
Level of Study: Professional development
Type: Research grant
Value: The amount and frequency of this award is at the discretion of the Board of Trustees of the American Respiratory Care Foundation and dependent upon the quality of the proposals.
Frequency: Annual
Country of Study: United States of America
Application Procedure: Candidates must apply directly to the Foundation Executive Office. Complete details can be found in the Application for Research Grant packet available from the Foundation. Please see the website for further details http://www.arcfoundation.org/awards/research_grants/francis.cfm.
Closing Date: 1st September to 31st December
Funding: Private
Contributor: Parker B Francis Foundation
Additional Information: In 1993, the Parker B Francis Foundation provided an endowment to the American Respiratory Care Foundation to make funds available to provide financial assistance for research programmes.

William F Miller, MD Postgraduate Education Recognition Award

Subjects: Respiratory care and prevention.
Purpose: To assist a professional therapist pursuing postgraduate education which will lead to an advanced degree.
Eligibility: Must be a respiratory therapist who has at least a Baccalaureate degree with a 3.0 cumulative GPA or better on a 4.0 scale, or the equivalent. If no Baccalaureate degree, then a letter of matriculation from a graduate program and an undergraduate transcript will suffice. Please see the website for further details regarding eligibility http://www.arcfoundation.org/awards/postgraduate/miller.cfm.
Level of Study: Postgraduate, Professional development
Type: Award
Value: Up to US$1,500 plus airfare, certificate of recognition, one night's lodging and registration for the AARC International Respiratory Congress.
Frequency: Annual
Country of Study: United States of America
No. of awards offered: 1
Application Procedure: Candidates must return a completed, signed and notarized application form, provide three letters of reference attesting to the candidate's character, academic ability and professional commitment and supply an essay of at least 1,200 words. This must describe how the award will assist the candidate in reaching the objective of an advanced degree and the candidate's ultimate goals of leadership in healthcare. Please see the webiste for further details http://www.arcfoundation.org/awards/postgraduate/miller.cfm.
Closing Date: June 15th
Funding: Private

AMERICAN ASSOCIATION FOR WOMEN RADIOLOGISTS (AAWR)

1891 Preston White Drive, Reston, VA, 20191, United States of America
Tel: (1) 713 965 0566
Fax: (1) 713 960 0488
Email: admin@aawr.org
Website: www.aawr.org

Alice Ettinger Distinguished Achievement Award

Subjects: Radiology.
Purpose: To recognize long-term contribution to radiology and to the American Association for Women Radiologists.
Eligibility: Open to AAWR members only.
Level of Study: Unrestricted
Type: Award
Value: Plaque
Frequency: Annual
Country of Study: Any country
No. of awards offered: 1
Application Procedure: Candidates must submit a current curriculum vitae and letters of support.
Closing Date: June 30th
Contributor: Membership dues

Eleanor Montague Distinguished Resident Award in Radiation Oncology

Subjects: Radiation oncology.
Purpose: To honor a resident radiation oncologist on the basis of outstanding contributions to clinical care, teaching, research and/or public service.
Eligibility: Open to candidates in the field who are members of the AAWR as of January 1st of the year of the award.
Level of Study: Unrestricted
Type: Award
Value: Plaque
Frequency: Annual
Country of Study: Any country
No. of awards offered: 1
Application Procedure: Candidates must submit an application including a letter of nomination, a letter of concurrence and a curriculum vitae.
Closing Date: June 30th
Contributor: Membership dues
Additional Information: Nominees will be evaluated on the basis of outstanding contributions in clinical care, teaching, research, or public service.

Lucy Frank Squire Distinguished Resident Award in Diagnostic Radiology

Subjects: Radiology.
Purpose: To honor a resident diagnostic radiologist on the basis of outstanding contributions to clinical care, teaching, research and/or public service.
Eligibility: Open to candidates in the field of diagnostic radiology who are members of the AAWR as of January 1st of the year of the award.
Level of Study: Unrestricted
Type: Award
Value: Plaque
Frequency: Annual
Country of Study: Any country
No. of awards offered: 1
Application Procedure: Candidates must submit an application including a curriculum vitae, a letter of nomination and a letter of concurrence.
Closing Date: June 30th
Contributor: Membership dues
Additional Information: Nominees will be evaluated on the basis of outstanding contributions in clinical care, teaching, research, or public service.

Marie Sklodowska-Curie Award

Subjects: Radiology.
Purpose: To honor an individual who has made an outstanding contribution to the field of Radiology.
Eligibility: There are no nationality restrictions and nominees must be members of the AAWR.
Level of Study: Unrestricted
Type: Award
Value: Plaque
Frequency: Annual
Country of Study: Any country
No. of awards offered: 1
Application Procedure: Complete curriculum vitae should accompany the nomination, as well as letter(s) of support addressing the unique role the nominee has undertaken in clinical care, teaching, and/or research and the accomplishments that set her/him apart.
Closing Date: June 30th
Contributor: Membership dues

AMERICAN ASSOCIATION OF FAMILY AND CONSUMER SCIENCES (AAFCS)

400 N. Columbus Street, Suite 202, Alexandria, VA, 22314-2752, United States of America
Tel: (1) 703 706 4600
Fax: (1) 703 706 4663
Email: cislamd@aafcs.org
Website: http://www.aafcs.org
Contact: Ms Amy Campbell, Grants Management Officer

Founded in 1909 as the American Home Economics Association, the American Association of Family and Consumer Sciences (AAFCS) is an organisation of members dedicated to improving the quality of individual and family life through programs that educate, influence public policy, disseminate information and publish research findings. Representing nearly 16,000 professionals in the family and consumer sciences, AAFCS members include elementary, secondary and post secondary educators and administrators, co-operative extension agents and other professionals in government, business and non-profit sectors.

Moselio Schaechter Distinguished Service Award

Purpose: This award, named in honor of Professor Moselio Schaechter, former ASM President, honors an ASM member who has shown exemplary leadership and commitment towards the substantial furthering of the profession of microbiology in research, education or technology in the developing world.
Eligibility: Individuals (for example: microbiologists who have been instrumental in setting up properly functioning clinical microbiology laboratories or successful biotechnology services based on microbiology; academicians who have developed high quality undergraduate or graduate training programs; researchers who have demonstrated leadership in the context of the region) from the upper-middle, lower-middle, and low-income countries as determined per World Banks's classification. The nominee must be a national or a permanent resident of a qualifying country and have a full-time professional appointment in the microbiological sciences or a related field for at least ten years in a country or region of the developing world. The nominees may not be currently serving on any ASM Board or Committee and can not be an ASM Ambassador or Country Liaison at the time of the nomination deadline. The nominee must be an ASM member at the time of nomination.
Type: Award
Value: The award consists of a $4,000 cash prize to defray expenses associated with traveling to and attending the ASM General Meeting; an engraved plaque to be presented during the International Reception at the ASM General Meeting; publication of the awardee profile in the International Affairs section of Microbe.
Length of Study: 3 years
Closing Date: July 1st
Funding: Commercial
Contributor: GlaxoSmithKline
Additional Information: Please see the website for further details http://www.asm.org/index.php/grants-fellowships/moselio-schaechter-distinguished.

AMERICAN ASSOCIATION OF LAW LIBRARIES (AALL)

105 W. Adams Street, Suite 3300, Chicago, IL, 60603, United States of America
Tel: (1) 312 939 4764
Fax: (1) 312 431 1097
Email: scholarships@aall.org
Website: www.aallnet.org

The American Association of Law Libraries (AALL) was founded in 1906 to promote and enhance the value of law libraries to legal and public communities, to foster the profession of law librarianship and to provide leadership in the field of legal information. Today, the AALL represents law librarians and related professionals who are affiliated with a wide range of institutions including law firms, law schools, corporate legal departments and courts, and local, state and federal government agencies.

AALL and Thomson West-George A Strait Minority Scholarship Endowment

Subjects: Law librarianship.
Eligibility: Open to degree candidates in an accredited library or law school. Preference is given to individuals with previous service to, or interest in, law librarianship and who intend to pursue a career in law librarianship. Applicants must be members of a minority group as defined by the current United States of America government guidelines.
Level of Study: Graduate
Type: Scholarship
Value: Up to US$3,500 for tuition and school-related expenses
Frequency: Annual
Study Establishment: Accredited library schools or accredited law schools
Country of Study: Any country
No. of awards offered: Varies
Application Procedure: Applicants must write for details or download an application form from the website.
Closing Date: April 1st

AALL James F Connolly LexisNexis Academic and Library Solutions Scholarship

Subjects: Law librarianship.
Eligibility: Awarded to library school graduates with law library experience who are presently attending an accredited law school with the intention of pursuing a career as a law librarian. Preference will be given to individuals who have demonstrated an interest in government documents.
Level of Study: Graduate
Type: Scholarship
Value: Up to US$3,000 for tuition and school-related expenses
Frequency: Annual
Study Establishment: ABA-accredited Law Schools
Country of Study: Any country
No. of awards offered: Varies
Application Procedure: Applicants must write for details or download an application form from the website.
Closing Date: April 1st

For further information contact:

53 West Jackson Boulevard, Suite 940, 53 West Jackson Boulevard, Suite 940Chicago, IL, 60604, United States of America

AALL LexisNexis/John R Johnson Memorial Scholarship Endowment

Subjects: Law librarianship.
Eligibility: Candidates who apply for AALL educational scholarships, types I–IV, become automatically eligible to receive the LexisNexis/John R Johnson Memorial Scholarship.
Level of Study: Graduate
Type: Scholarship
Value: Up to US$2,000 for tuition and school-related expenses
Frequency: Annual
Study Establishment: ALA-accredited library schools or ABA-Accredited Law Schools

Country of Study: Any country
No. of awards offered: Varies
Application Procedure: Applicants must write for details or download an application form from the website.
Closing Date: April 1st

AMERICAN ASSOCIATION OF NEUROLOGICAL SURGEONS (AANS)

5550 Meadowbrook Drive, Rolling Meadows, IL, 60008-3852, United States of America
Tel: (1) 847 378 0500
Fax: (1) 847 378 0600
Email: info@aans.org
Website: www.aans.org
Contact: Julie Qattrocchi, Development Coordinator

Founded in 1931 as the Harvey Cushing Society, the American Association of Neurological Surgeons (AANS) is a scientific and educational association with more than 6,500 members worldwide. The AANS is dedicated to advancing the speciality of neurological surgery in order to provide the highest quality of neurosurgical care to the public. All active members of the AANS are certified by the American Board of Neurological Surgery, The Royal College of Physicians and Surgeons (Neurosurgery) of Canada or the Mexican Council of Neurological Surgery, AC. Neurological surgery is the medical speciality concerned with the prevention, diagnosis, treatment and rehabilitation of disorders that affect the entire nervous system including the spinal column, spinal cord, brain and peripheral nerves.

NREF Research Fellowship

Subjects: Any field of neurosurgery.
Purpose: To provide training for neurosurgeons who are preparing for academic careers as clinician investigators.
Eligibility: Open to MDs who have been accepted into, or who are in, an approved residency training programme in neurological surgery in North America.
Level of Study: Postdoctorate
Type: Fellowship
Value: US$40,000 for a 1-year fellowship
Length of Study: 1–2 years
Frequency: Annual
Country of Study: Other
Application Procedure: Applicants must send a completed application, sponsor statement, programme director comments and letters of recommendation. Responses to questions 1–9, a curriculum vitae and photographic images must also be submitted. Applications are available at the website www.aans.org
Closing Date: October 31st
Funding: Private
Contributor: Corporations and membership
No. of awards given last year: 5
No. of applicants last year: 25
Additional Information: Notification of awards will be made by February 28th. After notification of the award, the applicant must indicate acceptance, in writing, no later than April 1st. If unwilling to accept the award by that date, funds will be awarded to the first runner-up. A report of findings and accounting of funds will be expected at the halfway point and upon completion of the fellowship. Normally, no more than one award per year will be made to any one institution. Individuals who accept a grant from another source, NIH or private, for the same research project will become ineligible for the award. A budget must be prepared by the applicant and the sponsor indicating how the grant funds will be expended. It is the policy of the NREF to fund only direct costs involved with the research awards. This means no fringe benefits, publication costs or travel expenses. The signature representing the applicant's institution's financial officer on page four should be that of their chief financial officer or grants and contracts manager. The award will be made payable to the institution and disbursed by it according to its institutional policy.

For further information contact:

Email: nref@aans.org

NREF Young Clinician Investigator Award

Subjects: Any field of neurosurgery.
Purpose: To fund pilot studies that provide preliminary data used to strengthen applications for more permanent funding from other sources.
Eligibility: Candidates must be neurosurgeons who are full-time faculty in teaching institutions in North America and in the early years of their careers.
Level of Study: Postdoctorate
Type: Award
Value: One-year research project grant of US$40,000
Length of Study: 1 year
Frequency: Annual
No. of awards offered: 1
Application Procedure: Candidates must send a completed application, sponsor statement, programme director comments and letters of recommendation. Responses to questions 1–9, a curriculum vitae and photographic images must also be submitted. Applications are available at the website www.aans.org
Closing Date: October 31st
Funding: Private
Contributor: Corporations and membership
No. of awards given last year: 3
No. of applicants last year: 20
Additional Information: Notification of awards will be made by February 28th. After notification of the award, the applicant must indicate acceptance, in writing, no later than April 1st. If unwilling to accept the award by that date, funds will be awarded to the first runner-up. A summary report and an accounting of funds will be expected upon completion of the award. Normally, no more than one award per year will be made to any one institution. Individuals who accept a grant from another source, NIH or private, for the same research project will become ineligible for the award. The award is for those budget items necessary to pursue proper research. It may be used entirely, or in part, for stipend. A budget must be prepared by the applicant and sponsor indicating how the award funds will be expended. It is the policy of the NREF to fund only direct costs involved with the research awards. This means no fringe benefits, publication costs or travel expenses.

For further information contact:

Email: NREF@aans.org

William P. Van Wagenen Fellowship

Subjects: Any field of neurosurgery.
Purpose: To fund quality research in which the plan for a period abroad has been designed.
Eligibility: All senior neurological residents in approved neurosurgery residency programs.
Level of Study: Postdoctorate
Type: Travelling fellowship
Value: $120,000 stipend for living and travel expenses to a foreign country for a period of 12 months. A family travel and living allowance of $6,000 is available if spouse and/or children are accompanying the Fellow. In addition, $15,000 of research support is available to the University, hospital or laboratory, which has agreed to sponsor the Van Wagenen Fellow. If needed, $5,000 is available for medical insurance.
Length of Study: 6–12 months
Frequency: Annual
Country of Study: Country of study must be different than the country of residence
No. of awards offered: 2
Application Procedure: Application should be submitted with letters of reference, including one from the applicant's Program Director. A letter from the proposed sponsor and documentation of intent to pursue an academic career, while not required, will strengthen the application.
Closing Date: October 1st
Funding: Private
Contributor: William P Van Wagenen
No. of awards given last year: 1
No. of applicants last year: 6
Additional Information: By December 31st, the Chairman of the Van Wagenen Selection Committee will notify the winning applicant, who

will be expected to implement the fellowship within 6 months following notification. A formal announcement of the award will be made at the Annual Meeting of the AANS. Applications and additional information regarding the William P Van Wagenen Fellowship can be located at www.aans.org

For further information contact:

Tel: 847-378-0500
Email: nref@aans.org

AMERICAN ASSOCIATION OF OCCUPATIONAL HEALTH NURSES FOUNDATION (AAOHN)

AAOHN National Office, 7794 Grow Drive, Pensacola, FL, 32514, United States of America
Tel: (1) 850 474 6963; 800 241 8014
Fax: (1) 850 484 8762
Email: aaohn@aaohn.org; aaohngov@dancyamc.com
Website: www.aaohn.org

Securing the future by improving the health and safety of the Nation's workers.

AAOHN Professional Development Scholarship

Subjects: Occupational and Environmental Health.
Eligibility: Candidates must be employed in the field of occupational and environmental health nursing.
Level of Study: Postgraduate, Professional development
Value: $3,000Academic Study, UPS Foundation ($2,500 each – two awards); Continuing Education, Moore Medical ($1,500 each – two awards), Georgia State Chapter ($1,000 each – two awards).
Length of Study: 1
Frequency: Annual
No. of awards offered: 6
Application Procedure: Submit an application form, a narrative of 500 words or less describing career goals and how the scholarship will enable continued education activity, and also supply a letter of support.
Closing Date: January 6th

THE AMERICAN ASSOCIATION OF PETROLEUM GEOLOGISTS (AAPG) FOUNDATION

PO Box 979, Tulsa, OK, 74101 0979, United States of America
Tel: (1) 918 560 2664
Fax: (1) 918 560 2642
Email: jterry@aapg.org
Website: http://foundation.aapg.org/
Contact: Mrs Jane Terry

Established by the American Association of Petroleum Geologists (AAPG) in 1967, the AAPG Foundation is a public foundation, qualified to receive gifts that are tax-deductible for United States of America taxpayers, in support of worthwhile educational and scientific programmes or projects related to the geosciences.

American Association of Petroleum Geologists Foundation Grants-in-Aid

Subjects: Earth and geological sciences.
Purpose: To support graduate (Master's or PhD) students whose research can be applied to the search for, and development of, petroleum and energy-minerals resources, and to related environmental geology issues.
Eligibility: Open to graduate and doctorate students of any nationality.
Level of Study: Doctorate, Graduate
Type: Grant
Value: A maximum of US$3,000
Country of Study: Any country
No. of awards offered: Varies
Application Procedure: Applicants must complete an application form and submit certified college academic transcripts and signed statements from professors commenting on the applicant's academic credentials and endorsements. Applicants must apply online the website being http://aapg.gia.confex.com/aapg_gia/2009/index.html.
Closing Date: January 31st
Funding: Foundation
No. of awards given last year: 84
No. of applicants last year: 300
Additional Information: Grants are to be applied to expenses directly related to the student's thesis work, such as summer fieldwork, analytical analyses, etc. Funds are not to be used to purchase capital equipment, or to pay salaries, tuition or room and board during the school year. Students are eligible to win twice.

AMERICAN ASSOCIATION OF UNIVERSITY WOMEN EDUCATIONAL FOUNDATION (AAUW)

AAUW, 1111 Sixteenth Street, NW, Washington, DC, 20036, United States of America
Tel: (1) 202 785 7700
Fax: (1) 202 872 1425
Email: connect@aauw.org
Website: www.aauw.org
Contact: Gloria Blackwell, Director of Fellowships, Grants and International Programs

AAUW has a long and distinguished history of advancing educational and professional opportunities for women in the United States of America and around the globe. One of the world's largest sources of funding for graduate women, AAUW is providing more than 3 million in funding for more than 200 fellowships and grants to outstanding women and non-profit organizations.

AAUW American Fellowships

Subjects: All subjects.
Purpose: To encourage and support women conducting Doctoral/Postdoctoral research in related fields.
Eligibility: American Fellowships support women doctoral candidates completing dissertations or scholars seeking funds for postdoctoral research leave from accredited institutions. Candidates must be U.S. citizens or permanent residents. Candidates are evaluated on the basis of scholarly excellence; the quality and originality of project design; and active commitment to helping women and girls through service in their communities, professions, or fields of research.
Level of Study: Doctorate, Postdoctorate, Research
Type: Fellowship
Value: Dissertation Fellowship: $20,000; Postdoctoral Research Leave Fellowship: $30,000; Summer/Short-Term Research Publication Grant: $6,000
Frequency: Annual
Study Establishment: Any accredited institution
Country of Study: Any country
Closing Date: November 15th
Funding: Foundation
No. of awards given last year: 66
No. of applicants last year: 1175
Additional Information: Please note that materials sent to the Washington, D.C. office will be disqualified and will not be reviewed. Please see the website for further details http://www.aauw.org/learn/fellowships_grants/american.cfm.

For further information contact:

AAUW American Fellowships, PO Box 4030, Iowa City, Iowa, 52243 4030, United States of America

AAUW Career Development Grants

Purpose: To support women who hold a bachelor's degree and are preparing to advance their careers, change careers, or re-enter the work force.
Eligibility: Primary consideration is given to women of color and women pursuing their first advanced degree or credentials in nontraditional fields.

Level of Study: Graduate, Postgraduate
Type: Grant
Value: $2,000–12,000 (Funds are available for tuition, fees, books, supplies, local transportation, and dependent care.)
Length of Study: 1 year
Frequency: Annual
No. of awards offered: Varies
Closing Date: December 15th
Additional Information: Materials sent to the Washington, DC office will be disqualified and will not be reviewed.

For further information contact:

C/O ACT, Inc., 101 ACT Dr., Iowa City, IA, 52243 4030; 52243 9000, United States of America
Tel: (1) 319/337-1716 ext. 60
Email: aauw@act.org
Website: http://www.aauw.org/learn/fellowships_grants/career_development.cfm

AAUW Case Support Travel Grants

Subjects: All subjects.
Purpose: To enable Legal Advocacy Fund-supported plaintiffs, their lawyers, and related experts to speak at state meetings or conventions about LAF-supported cases, sex discrimination issues in the workplace and higher education, and the work of LAF.
Eligibility: Please see website for conditions.
Level of Study: Postgraduate, Research
Type: Grant
Value: The grant covers the speaker's travel, lodging, and meal expenses
Frequency: Annual
Application Procedure: Apply online.
Closing Date: October 15th
Additional Information: Please note that materials sent to the Washington, DC office will be disqualified and will not be reviewed.

For further information contact:

Fax: 202 463 7169
Email: laf@aauw.org
Website: http://www.aauw.org/act/laf/travelgrant.cfm

AAUW Community Action Grants

Subjects: All subjects.
Purpose: To provide seed money to individual women, local community-based non-profit organizations, AAUW branches and AAUW state organizations for innovative programmes or non-degree research projects that engage girls in mathematics, science and technology.
Eligibility: Applicants must be women who are citizens or permanent residents of the United States of America. Special consideration will be given to AAUW members and AAUW branch and state applicants who seek partners for collaborative projects. Collaborators can include local schools or school districts, businesses and other community-based organizations. 2-year grants are restricted to projects focused on girls' achievement in mathematics, science or technology. Projects must involve community and school collaboration. The fund supports planning and coalition-building activities during the 1st year and implementation and evaluation the following year.
Type: Grant
Value: $2,000-7,000 (One-year grant); $5,000-10,000 (two-year grant)
Length of Study: 1 or 2 years
Frequency: Annual
No. of awards offered: 5
Application Procedure: Applicants must write for an application form, which is also available from the website.
Closing Date: January 15th
Funding: Private
Additional Information: Please note that materials sent to the Washington, DC office will be disqualified and will not be reviewed. Two types of grant are available. 1-year grants are for short-term projects. Topic areas are unrestricted but should have a clearly defined educational activity. 2-year grants are for longer term programmes and are restricted to projects focused on K-12 girls achievement in mathematics, science and/or technology. Funds support planning activities and coalition-building during the 1st year and implementation and evaluation the following year.

For further information contact:

C/O ACT, Inc., 101 ACT Dr., Iowa City, IA, 52243-9000, United States of America
Tel: (1) 319/337 1716 ext. 60
Email: aauw@act.org

AAUW Eleanor Roosevelt Fund Award

Subjects: Equity and education.
Purpose: To remove barriers to women's and girls' participation in education; to promote the value of diversity and cross-cultural communication; and to develop greater understanding of the ways women learn, think, work, and play.
Eligibility: To be eligible for the award, projects or activities must take place within the U.S. and recipients or organizational representatives must reside in the U.S. at the time the award is given. AAUA programs are not eligible for this award.
Level of Study: Professional development
Type: Award
Value: $5,000 plus travel expenses to attend the AAUW National Convention.
Frequency: Every 2 years
No. of awards offered: 1
Application Procedure: Please see the website for details.
Closing Date: August 1st to December 1st
Funding: Private
No. of awards given last year: 1

For further information contact:

Tel: 202/728 3300
Email: fellowships@aauw.org
Website: http://www.aauw.org/learn/awards/erfund.cfm

AAUW International Fellowships

Subjects: All subjects
Purpose: To support full-time study or research to women who are not US citizen or permanent residents.
Eligibility: Open to women who are not citizens of the United States of America or permanent residents, who hold a United States of America Bachelor's degree or equivalent. Applicants must be planning to return to their home country upon completion of degree and/or research. English proficiency is required.
Level of Study: Doctorate, MBA, Postdoctorate, Postgraduate, Professional development
Type: Fellowship
Value: Master's/First Professional Degree Fellowship: $18,000;Doctoral Fellowship: $20,000;Postdoctoral Fellowship: $30,000
Length of Study: 1 year
Frequency: Annual
Study Establishment: Any accredited institution
Country of Study: United States of America
Application Procedure: Applicants must complete an application for each year applying. Applications must be obtained from the customer service centre or the AAUW website between August 1st and December 15th. Three letters of recommendation, transcripts and a minimum score of 550 on the Test of English as a Foreign Language (213 computer-based) are also required. Order for brochure at http://www.act.org/aauw/brochurerequest.html.
Closing Date: December 1st
Funding: Foundation
No. of awards given last year: 36
No. of applicants last year: 1,194
Additional Information: Please note that materials sent to the Washington DC office will be disqualified and will not be reviewed.

For further information contact:

C/O ACT, Inc., 101 ACT Dr., Iowa City, IA, 52243-9000, United States of America
Tel: (1) 319 337 1716 ext. 60
Email: aauw@act.org
Website: http://www.aauw.org/learn/fellowships_grants/international.cfm

AAUW Selected Professions Fellowships

Subjects: Architecture, computer/information sciences, engineering, mathematics/statistics, business administration, law and medicine.
Purpose: To support women who intend to pursue a full-time course of study at accredited institutions during the fellowship year in one of the designated degree programs where women's participation has been low.
Eligibility: Women candidates who intend to pursue a full-time course of study at accredited US institutions during the fellowship year in one of the designated degree programs where women's participation traditionally has been low (see list below). Applicants must be US citizens or permanent residents. Please check the website for further details of eligibility.
Level of Study: Doctorate, Postdoctorate, Postgraduate
Type: Fellowships
Value: $5,000–18,000
Length of Study: 1 year
Frequency: Annual
Study Establishment: Any accredited institution
Application Procedure: Check website for details.
Closing Date: January 10th
Additional Information: Materials sent to the Washington DC office will be disqualified and will not be reviewed.

For further information contact:

C/O ACT, Inc., P.O. Box 4030, Iowa City, IA, 52243 4030, United States of America
Tel: (1) 319 337 1716 ext. 60
Email: aauw@act.org
Website: http://www.aauw.org/learn/fellowships_grants/selected.cfm

AMERICAN AUSTRALIAN ASSOCIATION (AAA)

50 Broadway, Suite 2003, New York, NY, 10004, United States of America
Tel: (1) 212 338 6860
Fax: (1) 212 338 6864
Email: information@aaanyc.org
Website: www.americanaustralian.org
Contact: Diane Sinclair, Director of Education

The American Australian Association (AAA), founded in 1948, is the largest non-profit organization in the United States devoted to relations between the United States, Australia and New Zealand, with operations throughout the tri-state and the New England regions. Its goal is to encourage stronger ties across the Pacific, particularly in the private sector.

AAA Education Fund Program

Subjects: Science, engineering, mining, medicine.
Purpose: To assist students doing research in the field of science, engineering, mining medicine and other related fields.
Eligibility: Open to applicants who are American citizens or permanent residents of the USA doing research or study at the graduate level. Applicants must also show proof of the arrangements made at an Australian university or institution where the research or study will be conducted for the next academic year.
Level of Study: Research
Value: Up to US$30,000 each year to Australians and Americans for advanced research and study in the United States and Australia respectively
Frequency: Annual
No. of awards offered: Varies
Closing Date: October 31st

THE AMERICAN CENTER OF ORIENTAL RESEARCH (ACOR)

656 Beacon Street, 5th Floor, Boston, MA, 02215-2010, United States of America
Tel: (1) 617 353 6571
Fax: (1) 617 353 6575
Email: acor@bu.edu
Website: www.bu.edu/acor
Contact: Donald R Keller, Associate Director

The American Center of Oriental Research (ACOR) in Amman, Jordan, is a private, non-profit academic institution dedicated to promoting research and publication in the fields of archaeology, anthropology, history, languages, biblical studies, Arabic, Islamic studies and other aspects of near eastern studies.

ACOR Jordanian Graduate Student Scholarships

Subjects: Jordan cultural heritage.
Purpose: To assist Jordanian graduate students with the annual costs of their academic programs.
Eligibility: Open to Jordanian citizens and currently enrolled in either a Master's or doctoral program in a Jordanian university. Applicants who demonstrate excellent progress in their programs will be eligible to apply in consecutive years.
Level of Study: Predoctorate
Type: Scholarship
Value: $3,000
Frequency: Annual
Country of Study: Jordan
No. of awards offered: 4
Closing Date: February 1st
Funding: Corporation, private
Additional Information: Eligible to nationals of Jordan only.

ACOR Jordanian Travel Scholarship for ASOR Annual Meeting

Purpose: To assist Jordanian scholars in participating in and delivering a paper at the American Schools of Oriental Research Annual Meeting in mid-November in the United States.
Eligibility: Only those applicants whose papers have been accepted to be delivered in the academic program of ASOR's Annual Meeting will be eligible to be considered for this scholarship.
Level of Study: Predoctorate, Postdoctorate
Type: Travel grant
Value: $3,500 each
Frequency: Annual
No. of awards offered: 2
Application Procedure: Two travel scholarships of $3,500 each to assist Jordanians participating and delivering a paper at the ASOR Annual meeting in mid-November in the United States. Academic papers should be submitted through the ASOR's website (www.asor.org/am) by February 1st. Final award selection will be determined by the ASOR program committee.
Closing Date: February 1st
Funding: Private
Contributor: ACOR
Additional Information: Eligible to the nationals of Jordan only.

ACOR-CAORC Fellowship

Subjects: Humanities, natural and social sciences.
Purpose: To provide fellowships for Master's and predoctoral students.
Eligibility: Open to masters and doctoral students candidates of US.
Level of Study: Postgraduate, Predoctorate
Type: Fellowship
Value: $20,200
Length of Study: 2–6 months
Country of Study: Jordan
No. of awards offered: 3 or more
Closing Date: February 1st
Funding: Government
Additional Information: Topics should contribute to scholarship in Near Eastern studies.

ACOR-CAORC Postgraduate Fellowships

Subjects: Natural and social sciences and humanities.
Purpose: To financially support students undertaking research in Jordan.
Eligibility: Open to post-doctoral scholars and scholars with a terminal degree in their field, pursuing research or publication projects in the natural and social sciences, humanities and associated disciplines relating to the Near East. U.S. citizenship required.
Level of Study: Postdoctorate
Type: Fellowships
Value: Maximum award is $31,800. Awards may be subject to funding.
Length of Study: 2–6 months
Frequency: Annual
Country of Study: Jordan
No. of awards offered: 2 or more
Application Procedure: Candidates need to submit letters of recommendation, health insurance and waiver forms along with the application packet.
Closing Date: February 1st
Funding: Government
Additional Information: A prior recipient of an ACOR-CAORC Fellowship is not eligible for this award for a period of 2 years.

Bert and Sally de Vries Fellowship

Subjects: Archaeology.
Purpose: To support a student for participation on an archaeological project.
Eligibility: Open to enrolled undergraduate or graduate students of any nationality.
Level of Study: Graduate, Predoctorate, Undergraduate
Type: Fellowship
Value: $1,200
Frequency: Annual
Country of Study: Jordan
No. of awards offered: 1
Closing Date: February 1st
Funding: Private
Contributor: ACOR
Additional Information: Senior project staff whose expenses are being borne largely by the project are ineligible. Eligible to any nationality except Jordanian.

Frederick-Wenger Jordanian Educational Fellowship

Subjects: Related to Jordan's cultural heritage is preferred.
Purpose: To assist Jordanian student with the cost of their education.
Eligibility: Open to enrolled undergraduate or graduate students with Jordanian citizenship.
Level of Study: Graduate, Undergraduate
Type: Fellowship
Value: US$1,500
Frequency: Annual
Country of Study: Jordan
No. of awards offered: 1
Closing Date: February 1st
Funding: Private
Contributor: ACOR
Additional Information: Eligibility is not limited to a specific field of study, but preference will be given to study related to Jordan's cultural heritage. Eligible to nationals of Jordan only.

Harrell Family Fellowship

Subjects: Archaeology.
Purpose: To support a graduate student for participation on an archaeological project.
Eligibility: Open to enrolled graduate students of any nationality.
Level of Study: Graduate, Predoctorate
Type: Fellowship
Value: $1,800
Frequency: Annual
Country of Study: Jordan
No. of awards offered: 1
Closing Date: February 1st
Funding: Private
Contributor: ACOR
Additional Information: Senior project staff whose expenses are being borne largely by the project are ineligible. Eligible to any nationality except Jordanian.

James A. Sauer Fellowship

Subjects: Archaeology.
Purpose: To assist a Jordanian graduate student in cost of their academic program.
Eligibility: Open to US or Canadian citizens.
Level of Study: Graduate
Type: Residential fellowships
Value: Provides 1 month residency at ACOR and a stipend of $400.
Length of Study: Academic year
Frequency: Every 2 years
Country of Study: Jordan
No. of awards offered: 1
Closing Date: February 1st
Funding: Private

Jennifer C. Groot Fellowship

Subjects: Archaeology.
Purpose: To support beginners in archaeological fieldwork who have been accepted as staff members on archaeological projects with ASOR/CAP affiliation in Jordan.
Eligibility: Open to a graduate student with US or Canadian citizenship.
Level of Study: Graduate, Undergraduate
Type: Fellowship
Value: $1,800 each
Frequency: Annual
Country of Study: Jordan
No. of awards offered: 2 or more
Closing Date: February 1st
Funding: Private

Kenneth W. Russell Fellowship

Subjects: Archaeology, anthropology, conservation or related areas.
Purpose: To support a graduate student for participation in an ACOR-approved archaeological research project, which has passed an academic review process. Senior project staff members whose expenses are being borne largely by the project are ineligible.
Eligibility: Please check website for details
Level of Study: Graduate
Type: Fellowship
Value: $1,800
Frequency: Every 2 years
No. of awards offered: 1
Closing Date: February 1st
Funding: Private
Additional Information: For this cycle the competition is closed to Jordanian students, but open to enrolled graduate students of all other nationalities.

MacDonald/Sampson Fellowship

Subjects: Ancient Near Eastern languages and history, archaeology, Bible studies and comparative religion.
Purpose: To support Canadian students.
Eligibility: Open to enrolled undergraduate or graduate students with Canadian citizenship or landed immigrant status.
Level of Study: Graduate, Undergraduate
Type: Fellowship
Value: The ACOR residency fellowship option includes room and board at ACOR and a stipend of $600. The travel grant option provides a single payment of $1,800 to help with any project related expenses.
Length of Study: 6 weeks
Frequency: Annual
Country of Study: Jordan
No. of awards offered: 1
Closing Date: February 1st
Funding: Private
Additional Information: ACOR for research in the fields of Ancient Near Eastern languages and history, archaeology, Bible studies, or comparative religion, or a travel grant to assist with participation in an archaeological field project in Jordan. Both options are open to

enrolled undergraduate or graduate students of Canadian citizenship or landed immigrant status.

National Endowment for the Humanities (NEH) Fellowship

Subjects: Modern and classical languages, linguistics, history, jurisprudence, philosophy, archaeology, comparative religion, ethics, and the history, criticism, and the theory of the arts.
Eligibility: One to two awards of four to six months for scholars who have a PhD or have completed their professional training. Fields of research include: modern and classical languages, linguistics, literature, history, jurisprudence, philosophy, archaeology, comparative religion, ethics, and the history, criticism, and theory of the arts. Social and political scientists are encouraged to apply. Applicants must be U.S. citizens or foreign nationals living in the U.S. three years immediately preceding the application deadline.
Level of Study: Doctorate, Postdoctorate
Type: Fellowship
Value: $25,200
Length of Study: 6 months
Frequency: Annual
Country of Study: Jordan
No. of awards offered: 1
Closing Date: February 1st
Funding: Government
Additional Information: Awards must be used between May 15th and December 31st

Pierre and Patricia Bikai Fellowship

Subjects: Archaeology.
Purpose: 1–2 month residency at Acor in Amman.
Eligibility: Open to enrolled graduate students of any nationality participating on an archaeological project or research in Jordan.
Level of Study: Graduate
Type: Fellowship
Value: The fellowship includes room and board at ACOR and a monthly stipend of $600.
Length of Study: 1–2 months residential
Frequency: Annual
Country of Study: Jordan
No. of awards offered: 1
Closing Date: February 1st
Funding: Private
Additional Information: Eligible to any nationality except Jordanian.

THE AMERICAN CHEMICAL SOCIETY (ACS)

1155 16th Street, NW, Washington, DC, 20036, United States of America
Tel: (1) 202 872 4600
Fax: (1) 202 776 8008
Email: help@acs.org; awards@acs.org
Website: http://portal.acs.org

The American Chemical Society is a self-governed individual membership organization that consists of more than 160,000 members at all degree levels and in all fields of chemistry. The organization provides a broad range of opportunities for peer interaction and career development, regardless of professional or scientific interests. The program and activities conducted by ACS today are the products of a tradition of excellence in meeting member needs that dates from the Society's founding in 1876.

ACS Ahmed Zewail Award in Ultrafast Science and Technology

Subjects: Physics, chemistry, biology, or related fields.
Purpose: To recognize outstanding and creative contributions to fundamental discoveries or inventions in ultrafast science and technology.
Eligibility: The award will be for outstanding and creative contributions by a nominee to fundamental discoveries or inventions in ultrafast science and technology in the areas of physics, chemistry, biology, or related fields. Please see the website for further details regarding eligibility.
Level of Study: Postgraduate
Type: Award
Value: The award will consist of $5,000 and a certificate. Up to $2,500 for travel expenses to the meeting at which the award will be presented will be reimbursed
Frequency: Annual
Application Procedure: See website for details
Closing Date: November 1st
Funding: Trusts
Contributor: Ahmed Zewail Endowment Fund

For further information contact:

Tel: Phone: 202 872 4575
Fax: Fax: 202 776 8008
Email: awards@acs.org

ACS Award for Achievement in Research for the Teaching and Learning of Chemistry

Subjects: Chemical sciences.
Purpose: To recognize outstanding contributions to experimental research that have increased our understanding of chemical pedagogy.
Eligibility: This award recognizes contributions to experimental research that have increased our understanding of chemical pedagogy and led to the improved teaching and learning of chemistry.
Level of Study: Postgraduate
Type: Award
Value: The award consists of $5,000 and a certificate. Up to $2,500 for travel expenses to the meeting at which the award will be presented will be reimbursed
Length of Study: 1 year
Frequency: Annual
No. of awards offered: 1
Application Procedure: See website for details.
Closing Date: November 1st
Funding: Commercial
Contributor: Pearson Education
Additional Information: Experimental research should include one or more recognized techniques and designed such as: control-group designs with random assignments of subjects or the use of in-tact class sections; factorial designs, bi-variate or multivariate correlation studies, etc.

ACS Award for Creative Advances in Environmental Science and Technology

Subjects: Environmental science and technology.
Purpose: To encourage creativity in research and technology.
Eligibility: The award will be granted regardless of race, gender, age, religion, ethnicity, nationality, sexual orientation, gender expression, gender identity, presence of disabilities, and educational background.
Level of Study: Postgraduate
Type: Award
Value: The award consists of $5,000 and a certificate. Up to $2,500 for travel expenses to the meeting at which the award will be presented will be reimbursed
Frequency: Annual
Application Procedure: See the website for more details.
Closing Date: November 1st
Contributor: ACS Division of Environmental Chemistry
Additional Information: Air Products and Chemicals, Inc., in memory of Joseph J. Breen, established the award in 1978 and sponsored the award until 2010.

For further information contact:

Tel: 202 872 4575
Fax: 202 776 8008
Email: awards@acs.org

ACS Award for Creative Research and Applications of Iodine Chemistry

Subjects: Chemistry.

Purpose: To support, promote, and motivate global research of iodine chemistry and develop its use and knowledge through applications.
Eligibility: A nominee must have performed outstanding and creative research related to iodine chemistry or its applications. Please see the website for further details regarding eligibility.
Level of Study: Postgraduate
Type: Award
Value: The award consists of $10,000 and a certificate. Up to $1,000 for travel expenses to the meeting at which the award will be presented will be reimbursed. The award is presented biennially in odd-numbered years
Frequency: Biennially in odd-numbered years
Application Procedure: A completed nomination form and curriculum vitae must be sent.
Closing Date: November 1st
Contributor: Sociedad Quimica y Minera de chile S.A. (SQM S.A.)

For further information contact:

Tel: 202 872 4575
Fax: 202 776 8008
Email: awards@acs.org

ACS Award for Creative Work in Synthetic Organic Chemistry

Subjects: Organic chemistry.
Purpose: To recognize and encourage creative work in synthetic organic chemistry.
Eligibility: A nominee must have accomplished outstanding creative work in synthetic organic chemistry that has been published. Please see website for further details regarding eligibility.
Level of Study: Postgraduate
Type: Award
Value: The award consists of $5,000 and a certificate. Up to $1,000 for travel expenses to the meeting at which the award will be presented will be reimbursed
Frequency: Annual
Application Procedure: See the website for more details
Closing Date: November 1st
Contributor: The Aldrich Chemical Company, Inc.

For further information contact:

Tel: 202 872 4575
Fax: 202 776 8008
Email: awards@acs.org

ACS Award for Encouraging Disadvantaged Students into Careers in the Chemical Sciences

Subjects: Chemistry.
Purpose: To recognize significant accomplishments by individuals in stimulating students, underrepresented in the profession, to elect careers in the chemical sciences and engineering.
Eligibility: Nominees for the award may come from any professional setting: academia, industry, government or other independent facility. Please see the website for further details regarding eligibility.
Level of Study: Postgraduate
Type: Award
Value: The award consists of $5,000 and a certificate. A grant of $10,000 will be made to an academic institution, designated by the recipient, to strengthen its activities in meeting the objectives of the award. Up to $1,500 for travel expenses to the meeting at which the award will be presented will be reimbursed.
Frequency: Annual
Country of Study: United States of America
Application Procedure: Completed nomination and optional support forms must be submitted to the awards office.
Closing Date: November 1st
Contributor: The Camille and Henry Dreyfus Foundation, Inc.

For further information contact:

Tel: 202 872 4575
Fax: 202 776 8008
Email: awards@acs.org

ACS Award for Encouraging Women into Careers in Chemical Sciences

Subjects: Chemistry.
Purpose: To recognize individuals who have significantly stimulated the interests of women in chemistry.
Eligibility: Open to candidates of all nationalities. Nominees for the award may come from any professional setting: academia, industry, government, or other independent facility.
Level of Study: Professional development
Type: Award
Value: The award consists of $5,000 and a certificate. A grant of $10,000 will be made to an academic institution, designated by the recipient, to strengthen its activities in meeting the objectives of the award. Up to $1,500 for travel expenses to the meeting at which the award will be presented will be reimbursed.
Frequency: Annual
No. of awards offered: 1
Application Procedure: For more details visit the website.
Closing Date: November 1st
Contributor: The Camille and Henry Dreyfus Foundation, Inc.

For further information contact:

Tel: 202 872 4575
Fax: 202 776 8008
Email: awards@acs.org

ACS Award in Chromatography

Subjects: Chemistry.
Purpose: To recognize outstanding contributions to the fields of chromatography.
Eligibility: A nominee must have made an outstanding contribution to the fields of chromatography with particular consideration given to developments of new methods. Please see the website for further details regarding eligibility.
Level of Study: Postgraduate
Type: Award
Value: The award consists of $5,000 and a certificate. Up to $2,500 for travel expenses to the meeting at which the award will be presented will be reimbursed.
Frequency: Annual
Application Procedure: See the website for more details.
Closing Date: November 1st
Contributor: SUPELCO, Inc.

For further information contact:

The Awards Office
Tel: 202 872 4575
Fax: 202 776 8008
Email: awards@acs.org

ACS Award in Colloid and Surface Chemistry

Subjects: Chemistry.
Purpose: To recognize outstanding scientific contributions to colloid and/or surface chemistry in North America.
Eligibility: The nominee must be a resident of North America and must have made outstanding scientific contributions to colloid and/or surface chemistry. Please see the website for further details regarding eligibility.
Level of Study: Postgraduate
Type: Award
Value: The award consists of $5,000 and a certificate. Up to $2,500 for travel expenses to the meeting at which the award will be presented will be reimbursed.
Frequency: Annual
Application Procedure: Completed nomination and optional support forms must be submitted.
Closing Date: November 1st
Contributor: Procter & Gamble Company

For further information contact:

Tel: 202 872 4575
Fax: 202 776 8008
Email: awards@acs.org

ACS Award in Colloid and Surface Chemistry

Subjects: Chemistry.
Purpose: To support outstanding graduate students during their final 2 years of doctoral thesis research.
Eligibility: Open to candidates who are enrolled in a full-time graduate programme leading to a PhD degree at an accredited university within the United States.
Level of Study: Research
Type: Award
Value: US$20,000
Length of Study: 2 years
Frequency: Annual
Study Establishment: Any accredited university
Country of Study: United States of America
Application Procedure: See the website www.membership.acs.org
Closing Date: November 1st
Funding: Corporation
Contributor: The Cognis Corporation
Additional Information: Contact at Division of Colloid and Surface Chemistry.

ACS National Awards

Subjects: Chemical sciences.
Purpose: To recognize premier chemical professionals in extra-ordinary ways.
Level of Study: Postgraduate
Type: Grant
Length of Study: 2 years
Frequency: Annual
No. of awards offered: 57
Application Procedure: See website for details.
Closing Date: November 1st
Additional Information: Inquiries concerning awards should be directed to the office of the National Awards office awards@acs.org.

ACS Petroleum Research Fund

Subjects: Petroleum-related chemistry science.
Purpose: The ACS Petroleum Research Fund will support innovative fundamental research, advanced scientific education, and the careers of scientists, to aid in significantly increasing the world's energy options
Eligibility: Open to candidates in the field of petroleum-related chemistry science.
Type: Funding support
Value: US$90,000
Length of Study: 1–3 years
Frequency: Annual
Application Procedure: See website for details.
Closing Date: December 9th
Funding: Trusts
Contributor: Petroleum Research Fund

ACS PRF Scientific Education (Type SE) Grants

Subjects: Scientific education and fundamental research in the petroleum field.
Purpose: To provide partial funding for foreign speakers at major symposia.
Eligibility: Open to non-profit institutions throughout the United States of America and worldwide for speakers coming to conferences in the United States of America, Canada and Mexico and speaking within the PRF Trust.
Level of Study: Unrestricted
Type: Grant
Value: Up to $1,500 per foreign speaker with a maximum of $4,500 per symposium (three or more visiting speakers)
Frequency: Annual
Country of Study: United States of America
No. of awards offered: Varies
Application Procedure: For more information regarding the ACS Awards program contact awards@acs.org or 202-872-4575.
Closing Date: Applications are accepted at any time
Funding: Private
Contributor: A private trust

ACS Roger Adams Award in Organic Chemistry

Subjects: Organic chemistry.
Purpose: To recognize and encourage outstanding contributions to research in organic chemistry.
Eligibility: The award will be granted regardless of race, gender, age, religion, ethnicity, nationality, sexual orientation, gender expression, gender identity, presence of disabilities, and educational background.
Type: Award
Value: The award consists of a medallion and a replica, a certificate and $25,000
Frequency: Biennially in odd-numbered years
Application Procedure: A completed nomination form must be sent as an email attachment. See the website www.chemistry.org for further details.
Closing Date: November 1st
Contributor: Organic Reactions, Inc. and Organic Synthesis, Inc.

For further information contact:

Tel: 202 872 4575
Fax: 202 776 8008
Email: awards@acs.org

ACS Stanley C. Israel Regional Award

Subjects: Chemical sciences.
Purpose: To recognize individuals who have advanced diversity in the chemical sciences and significantly stimulated or fostered activities that promote inclusiveness within the region.
Eligibility: Nominees may come from academia, industry, government, or independent entities, and may also be organizations, including ACS Local Sections and Divisions. The nominee must have created and fostered ongoing programs or activities that result in increased numbers of persons from diverse and underrepresented minority groups, persons with disabilities, or women who participate in the chemical enterprise.
Level of Study: Postgraduate
Type: Award
Value: The award consists of a medal and a $1,000 grant to support and further the activities for which the award was made. The award also will include funding to cover the recipient's travel expenses to the ACS regional meeting at which the award will be presented.
Frequency: Annual
Application Procedure: See website for details.
Closing Date: April 2nd, August 3rd
Contributor: ACS Committee on Minority Affairs
Additional Information: First deadline is for Middle Atlantic (MARM), Central (CERM), and Northwest (NORM) and second deadline is for Midwest (MWRM), Northeast (NERM), Southeastern (SERMACS), Southwest (SWRM), and Rocky Mountain (RMRM).

For further information contact:

Fax: 202 776 8003
Email: Diversity@acs.org

Alfred Burger Award in Medicinal Chemistry

Subjects: Medicinal chemistry.
Purpose: To acknowledge outstanding contributions to research in medicinal chemistry.
Eligibility: The award will be granted for outstanding contributions in the field of medicinal chemistry regardless of race, gender, age, religion, ethnicity, nationality, sexual orientation, gender expression, gender identity, presence of disabilities, and educational background.
Level of Study: Postgraduate
Type: Award
Value: The award consists of $3,000 and a certificate. Up to $2,500 for travel expenses to the meeting at which the award will be presented will be reimbursed.
Frequency: Biennically in even-numbered years
Application Procedure: A completed nomination form available on the website www.chemistry.org must be sent as an email attachment.
Closing Date: November 1st
Contributor: Glaxo SmithKline

For further information contact:

Tel: 202 872 4575
Fax: 202 776 8008
Email: awards@acs.org

Anselme Payen Award

Subjects: Chemistry.
Purpose: To encourage outstanding professional contributions to the science and chemical technologies of cellulose and its allied products.
Eligibility: Open to all scientists conducting research in the field of cellulose. Please see the website for further details regarding eligibility http://cell.sites.acs.org/anselmepayenaward.htm.
Level of Study: Postgraduate
Type: Award
Value: US$3,000 and a bronze medal
Frequency: Annual
Application Procedure: A completed nomination form available on the website www.membership.acs.org must be sent.
Closing Date: December 1st
Contributor: Cellulose, paper and textile division, ACS

For further information contact:

11215 N. Jason Drive, RM 1107, Peoria, IL, 61604, United States of America
Tel: (1) 309 681 6338
Fax: (1) 309 681 6691
Contact: Dr Gordon Selling, Awards Chair

Arthur C. Cope Scholar Awards

Subjects: Organic chemistry.
Purpose: To recognize outstanding achievement in the field of organic chemistry.
Eligibility: The award will be granted regardless of race, gender, religion, ethnicity, nationality, sexual orientation, gender expression, gender identity, presence of disabilities, and educational background. Please see the webiste for further details regarding eligibility.
Level of Study: Postgraduate
Type: Award
Value: The award consists of $5,000, a certificate, and a $40,000 unrestricted research grant to be assigned by the recipient to any university or nonprofit institution. Up to $2,500 for travel expenses to the fall national meeting will be reimbursed.
Frequency: Annual
No. of awards offered: 10
Application Procedure: See the website.
Closing Date: November 1st
Contributor: The Arthur C. Cope Fund

For further information contact:

1155 16th Street NW, Washington, DC, 20036 4801
Tel: 202 872 4575
Fax: 202 776 8008
Email: awards@acs.org

Earle B. Barnes Award for Leadership in Chemical Research Management

Subjects: Chemistry and chemical engineering.
Purpose: To recognize outstanding achievements in chemical research management.
Eligibility: Open to candidates who are citizens of the United States. The award is intended to recognize individuals who have demonstrated outstanding leadership and creativity in promoting the sciences of chemistry and chemical engineering in research management.
Level of Study: Postgraduate
Type: Award
Value: US$5,000 and a certificate. Up to $2,500 for travel expenses to the meeting at which the award will be presented will be reimbursed.
Frequency: Annual
Application Procedure: A completed nomination form must be submitted to awards@acs.org as an email attachment.
Closing Date: November 1st
Contributor: The Dow Chemical Company Foundation

Ernest Guenther Award in the Chemistry of Natural Products

Subjects: Organic chemistry.
Purpose: To recognize and encourage outstanding achievements in analysis, structure elucidation and chemical synthesis of natural products.
Eligibility: Open to all applicants for their accomplished outstanding work.
Type: Grant
Value: US$6,000, a medallion and a certificate. Up to $2,500 for travel expenses to the meeting at which the award will be presented will be reimbursed.
Frequency: Annual
Application Procedure: A completed nomination form and optional support forms must be mailed to awards@acs.org
Closing Date: November 1st
Contributor: Givaudan

For further information contact:

1155 16th Street NW, Washington, DC, 20036 4801, United States of America
Tel: (1) 202 872 4575
Fax: (1) 202 776 8008
Email: awards@acs.org

F. Albert Cotton Award in Synthetic Inorganic Chemistry

Subjects: Inorganic chemistry.
Purpose: To recognize distinguished work in synthetic inorganic chemistry.
Eligibility: The award recognizes outstanding synthetic accomplishment in the field of inorganic chemistry. The award will be granted regardless of race, gender, age, religion, ethnicity, nationality, sexual orientation, gender expression, gender identity, presence of disabilities and educational background.
Level of Study: Postgraduate
Type: Award
Value: US$5,000 and a certificate. Up to $2,500 for travel expenses to the meeting at which the award will be presented will be reimbursed.
Frequency: Annual
Application Procedure: A completed application form to be submitted as an email attachment to awards@acs.org
Closing Date: November 1st
Funding: Private
Contributor: F. Albert Cotton Endowment Fund

For further information contact:

1155 16th Street NW, Washington, DC, 20036 4801, United States of America
Tel: (1) 202 872 4575
Fax: (1) 202 776 8008
Email: awards@acs.org

Frederic Stanley Kipping Award in Silicon Chemistry

Subjects: Chemistry.
Purpose: To recognize distinguished contributions to the field of silicon chemistry.
Eligibility: Open to all candidates who have contributed to the field of silicon chemistry. There are no limits on age or on nationality. A nominee must have made distinguished contributions in the field of silicon chemistry during the 10 years preceding the current nomination
Level of Study: Postgraduate
Type: Award
Value: US$5,000 and a certificate. Up to $2,500 for travel expenses will be reimbursed to the spring national meeting at which the award will be presented and to the U.S. based Silicon Symposium to deliver an award address.
Frequency: Biennically in even-numbered years
Application Procedure: A completed nomination form to be mailed to awards@acs.org
Closing Date: November 1st
Funding: Corporation
Contributor: Dow Corning Corporation

For further information contact:

1155 16th Street NW, Washington, DC, 20036 4801, United States of America
Tel: (1) 202 872 4575
Fax: (1) 202 776 8008
Email: awards@acs.org

Glenn T. Seaborg Award for Nuclear Chemistry

Subjects: Chemistry.
Purpose: To recognize and encourage research in nuclear and radiochemistry or their applications.
Eligibility: nominee must have made outstanding contributions to nuclear or radiochemistry or to their applications. The award will be granted regardless of race, gender, age, religion, ethnicity, nationality, sexual orientation, gender expression, gender identity, presence of disabilities and educational background.
Level of Study: Postgraduate
Type: Award
Value: US$5,000 and a certificate. Up to $2,500 for expenses to the meeting at which the award will be presented will be reimbursed.
Frequency: Annual
Application Procedure: See the website.
Closing Date: November 1st
Contributor: ACS Division of Nuclear Chemistry and Technology

For further information contact:

1155 16th Street, NW, Washington, DC, 20036 4801, United States of America
Tel: (1) 202 8721 4575
Fax: (1) 202 776 8008
Email: awards@acs.org

Herbert C. Brown Award for Creative Research in Synthetic Methods

Subjects: Chemistry.
Purpose: To recognize and encourage outstanding and creative contributions to research in synthetic methods.
Eligibility: nominee must have accomplished outstanding and creative research that involved the discovery and development of novel and useful methods for chemical synthesis. Please see the website for further details regarding eligibility.
Level of Study: Postgraduate
Type: Award
Value: US$5,000 a medallion with a presentation box and a certificate. Up to $2,500 for travel expenses to the meeting at which the award will be presented will be reimbursed.
Frequency: Annual
Application Procedure: A completed application form available on the website must be sent as an email attachment to awards@acs.org
Closing Date: November 1st
Contributor: Purdue Borane Research Fund and Herbert C. Brown Award Endowment

For further information contact:

1155 16th Street NW, Washington, DC, 20036 4801
Tel: 202 872 4575

Ipatieff Prize

Subjects: Chemistry.
Purpose: To recognize outstanding chemical experimental work in the field of catalysis.
Eligibility: Open to candidates who should not have passed his or her 40th birthday on April 30 of the year in which the award is presented, and must have done outstanding chemical experimental work in the field of catalysis or high pressure. Special weight will be given to independence of thought and originality. The award may be made for investigations carried out in any country and without consideration of nationality.
Level of Study: Postgraduate
Type: Prize
Value: The award will consist of the income from a trust fund and a certificate. The financial value of the prize may vary, but it is expected that it will be approximately $5,000 and that it will be awarded every three years. Travel expenses related to conferment of the award will be reimbursed.
Frequency: Every 3 years, triennial review
Country of Study: Any country
Application Procedure: Contact the awards office.
Closing Date: November 1st
Contributor: Ipatieff Trust Fund
Additional Information: Preference will be given to American chemists.

For further information contact:

1155 16th Street, NW, Washington, DC, 20036 4801, United States of America
Tel: (1) 202 872 4575
Email: awards@acs.org

Irving Langmuir Award in Chemical Physics

Subjects: Chemistry and physics.
Purpose: To encourage research in chemistry and physics.
Eligibility: A candidate must have made an outstanding contribution to chemical physics or physical chemistry within the 10 years preceding the year in which the award is made. The award will be granted without restriction, except that the recipient must be a resident of the United States and the monetary prize must be used in the United States or its possessions.
Level of Study: Postgraduate
Type: Award
Value: US$10,000 and a certificate. Up to $2,500 for travel expenses to the meeting at which the award will be presented will be reimbursed.
Frequency: Every 2 years, Biennically in even-numbered years
Country of Study: United States of America
Application Procedure: Contact the Awards office.
Closing Date: November 1st
Funding: Foundation
Contributor: GE Global Research and The American Chemical Society Division of Physical Chemistry

For further information contact:

1155 16th Street NW, Washington, DC, 20036 4408, United States of America
Tel: (1) 202 872 4575
Email: awards@acs.org

James Bryant Conant Award in High School Chemistry Teaching

Subjects: Teaching.
Purpose: To recognize outstanding teachers of high school chemistry in the United States.
Eligibility: Open to candidates who are actively engaged in the teaching of chemistry in high school. The nominee must be actively engaged in the teaching of chemistry in a high school (grades 9–12). Please see the website for further details regarding eligibility.
Level of Study: Postgraduate
Type: Award
Value: US$5,000 and a certificate. Up to $2,500 for travel expenses to the meeting at which the award will be presented will be reimbursed.
Frequency: Annual
Country of Study: United States of America
Application Procedure: A completed nomination form must be submitted as an email attachment.
Closing Date: November 1st
Contributor: The Thermo Fisher Scientific, Inc.
Additional Information: A certificate will also be provided to the recipient's institution for display.

For further information contact:

1155 16th Street, NW, Washington, DC, 20036 4801
Tel: 202 872 4575
Email: awards@acs.org

Peter Debye Award in Physical Chemistry

Subjects: Physical chemistry.
Purpose: To encourage and reward outstanding research in physical chemistry.
Eligibility: Open to all candidates without regard to age or nationality. Please see the website for further details regarding eligibilty.
Level of Study: Postgraduate
Type: Award
Value: US$5,000 and a certificate. Up to $2,500 for travel expenses to the meeting at which the award will be presented will be reimbursed.
Frequency: Annual
Application Procedure: A completed nominations form to be sent as an email attachment to awards@acs.org

Closing Date: November 1st
Contributor: E.I. du Pont de Nemours and Company

For further information contact:

1155 16th Street, NW, Washington, DC, 20036 4801
Tel: 202 872 4575
Email: awards@acs.org

Ronald Breslow Award for Achievement in Biomimetic Chemistry

Subjects: Chemistry.
Purpose: To recognize outstanding contributions to the field of biomimetic chemistry.
Eligibility: The award will be granted regardless of race, gender, age, religion, ethnicity, nationality, sexual orientation, gender expression, gender identity, presence of disabilities and educational background for outstanding contributions to the field of biomimetic chemistry.
Level of Study: Postgraduate
Type: Award
Value: US$5,000 and a certificate. Up to $2,500 for travel expenses to the meeting at which the award will be presented will be reimbursed.
Frequency: Annual
Application Procedure: A completed application form along with a curriculum vitae must be submitted as an email attachment to awards@acs.org
Closing Date: November 1st
Funding: Trusts
Contributor: The Ronald Breslow Endowment

For further information contact:

1155 16th Street, NW, Washington, DC, 20036 4801
Tel: 202 872 4575

AMERICAN COLLEGE OF RHEUMATOLOGY

2200 Lake Boulevard NE, Atlanta, GA, 30319, United States of America
Tel: (1) 404 633 3777
Fax: (1) 404 633 1870
Email: ref@rheumatology.com
Website: www.rheumatology.org
Contact: Damian Smalls, Senior Specialist, Awards and Grants

The American College of Rheumatology (ACR) is the professional organization of rheumatologists and associated health professionals who share a dedication to healing, preventing disability and curing more than 100 types of arthritis and related disabling and sometimes fatal disorders of the joints, muscles and bones.

Amgen Fellowship Training Award

Subjects: Rheumatic diseases.
Purpose: To help ensure that a highly trained workforce is available to provide competent clinical care to those affected by rheumatic diseases.
Eligibility: Only training directors at ACGME-accredited institutions in good standing may apply. The rheumatology fellowship training programme director at the institution will be responsible for the selection and appointment of trainees. Award applicant must be a citizen or non-citizen national of the United States of America, or be in lawful possession of a permanent resident card. Individuals on temporary (J1, H1) or student visas are not eligible.
Level of Study: Doctorate
Type: Fellowship
Value: Recipients will receive $25,000 to support the salary and fringe of one trainee. The award is paid directly to the sponsoring institution and payments are disbursed in two equal installments of $12,500.
Length of Study: 1 year
Frequency: Annual
Country of Study: United States of America
Application Procedure: Application forms are available on website.
Closing Date: August 1st
Funding: Corporation, foundation
Contributor: Amgen, Inc.
No. of awards given last year: 31
No. of applicants last year: 37

Amgen Pediatric Research Award

Subjects: Rheumatology.
Purpose: To recognize and promote scholarship in the field of pediatric rheumatology.
Eligibility: Candidates must be trainees enrolled in a recognized ACGME-accredited pediatric rheumatology programme or a related research laboratory. Trainees must be preparing for a career in pediatric rheumatology. In addition, candidates must submit an abstract to the upcoming ACR Annual Scientific Meeting. There are no citizenship requirements for this award.
Level of Study: Doctorate, Postdoctorate, Research
Type: Award
Value: US$1,000 cash award, waiver of registration fees and $1,000 to cover travel and hotel expenses for the ACR Annual Scientific Meeting
Length of Study: 1 year
Frequency: Annual
Country of Study: United States of America
Application Procedure: Application forms are available online at www.rheumatology.org
Closing Date: August 1st
Funding: Corporation, foundation
Contributor: Amgen Inc.
No. of awards given last year: 1
No. of applicants last year: 5
Additional Information: Please check website for further information http://www.rheumatology.org/foundation/index.asp

Career Development in Geriatric Medicine Award

Subjects: Geriatrics and rheumatology.
Purpose: To support career development for junior faculty in the early stages of their research career.
Eligibility: To be eligible for the award, the candidate must: be a member of the ACR; have completed a rheumatology fellowship leading to certification by the ABIM and be within the first 3 years of his/her faculty appointment; and possess a faculty appointment at the time of the award. Award applicant must be a citizen or non-citizen national of the United States of America, or be in lawful possessions of a permanent resident card. Individuals on temporary (J1, H1) or student visas are not eligible.
Level of Study: Doctorate, Professional development, Research
Type: Research award
Value: US$75,000 per year plus US$3,000 in travel grants
Length of Study: 2 years
Frequency: Annual
Country of Study: United States of America
No. of awards offered: 2
Application Procedure: Application forms are available on website http://www.rheumatology.org/foundation/index.asp
Closing Date: August 1st
Funding: Foundation
Contributor: Association of Subspecialty Professors
No. of awards given last year: 1
No. of applicants last year: 1

Clinician Scholars Educator Award

Subjects: Rheumatology.
Purpose: To recognize and support rheumatologists dedicated to providing high-quality clinical educational experience to trainees.
Eligibility: Candidates must be ACR members with experience in the training of medical students, residents and fellows in rheumatology. Candidates must be affiliated with an LCCME-accredited school or an ACGME-accredited training programme in internal medicine, pediatrics or rheumatology. A faculty appointment is not essential. If the candidate is not a citizen or non-citizen national of the United States of America, the candidate must provide evidence that he or she is eligible to remain in the United States of America throughout the period of the award.
Level of Study: Doctorate, Professional development
Type: Award
Value: US$50,000 per year
Length of Study: 3 years

Frequency: Annual
Country of Study: United States of America
Application Procedure: Application forms are available on website http://www.rheumatology.org/foundation/index.asp
Closing Date: August 1st
Funding: Foundation
Contributor: Rheumatology Research Foundation
No. of awards given last year: 3
No. of applicants last year: 8
Additional Information: This is a Career Development Award.

Ephraim P. Engleman Endowed Resident Research Preceptorship

Subjects: Rheumatology.
Purpose: To introduce residents to the speciality of rheumatology by supporting a full-time research experience.
Eligibility: Candidates currently enrolled in ACGME-accredited training programmes in internal medicine, paediatrics, or med/paeds are eligible. Candidates may apply during any year of their residency; however preference is given to 1st and 2nd year residents. Preceptors are responsible for approving the research plan and must be members of the ACR. Candidates must be citizen or non-citizen national of the United States of America or be in lawful possession of a permanent resident card. Individuals on temporary (J1, H1) or student visas are not eligible.
Level of Study: Research, Doctorate, Postdoctorate
Type: Scholarship
Value: Up to US$15,000
Length of Study: 3 months
Frequency: Annual
Country of Study: United States of America
No. of awards offered: 1
Application Procedure: Application forms are available on website.
Closing Date: 4 cycles – February 1st, May 2nd, August 1st, November 1st
Funding: Foundation, individuals
Contributor: Funding for this award is made possible through an endowment from Dr Ephraim P. Engleman.
No. of awards given last year: 1
No. of applicants last year: 3

Health Professional Research Preceptorship

Subjects: Rheumatic diseases.
Purpose: To introduce students to rheumatology-related healthcare by supporting full-time research by a graduate student in the broad area of rheumatic diseases.
Eligibility: Only students enrolled in graduate school are eligible. Preceptors are responsible for selecting student applicants and must be members of the ARHP.
Level of Study: Graduate
Type: Scholarship
Value: US$3,500 student stipend, US$1,000 for laboratory expenses, US$1,000 stipend for the mentor and up to US$1,000 for travel funds for the student to attend the ACR/ARHP annual scientific meeting.
Length of Study: 1 academic year
Frequency: Annual
Country of Study: United States of America, Canada or Mexico
No. of awards offered: Up to 10
Application Procedure: Only complete applications submitted online by the deadline will be accepted.
Closing Date: 4 cycles – February 1st, May 2nd, August 1st, November 1st
Funding: Corporation, foundation
Contributor: Abbott Endowment for Rheumotology Development
No. of awards given last year: 5
No. of applicants last year: 6

Investigator Award

Subjects: Rheumatic diseases.
Purpose: To provide support for basic science and clinical investigators engaged in research relevant to the rheumatic diseases and to support junior investigators during the period that they are developing a project that will be competitive for NIH funding.
Eligibility: Applicants must be members of the ACR or ARHP, have a doctoral degree and must be clinician scientists.
Level of Study: Professional development, Doctorate, Postdoctorate, Research
Type: Research award
Value: US$75,000 annual salary plus $50,000 per year.
Length of Study: 3 years
Frequency: Annual
Country of Study: United States of America
Application Procedure: Application forms are available on website.
Closing Date: August 1st
Funding: Foundation
Contributor: Rheumatology Research Foundation
No. of awards given last year: 2
No. of applicants last year: 6
Additional Information: For any questions regarding eligibility, please contact the Rheumatology Research Foundation.

Lawren H. Daltroy Fellowship in Patient-Clinician Communication Award

Subjects: Rheumatology.
Purpose: To improve patient-clinician interactions through the development of more qualified and trained clinicians and investigators in the field of patient–clinician communication.
Eligibility: Eligible candidates must be ARHP members. A candidate must have either a doctoral degree or a clinical degree and apply in partnership with a mentor possessing a doctoral degree. The award is not intended for physician members of the ACR or ARHP.
Level of Study: Doctorate, Professional development
Type: Award
Value: Up to US$7,000 per year
Length of Study: 1 year
Frequency: Annual
Country of Study: United States of America
Application Procedure: Application forms are available on website.
Closing Date: August 1st
Funding: Foundation, private
Contributor: Rheuminations, Inc.
No. of awards given last year: 1
No. of applicants last year: 2

Marshall J. Schiff, MD, Memorial Fellow Research Award

Subjects: Rheumatology.
Purpose: The purpose is to recognize outstanding scholarship in the field of rheumatology and provide fellows-in-training who are authors or co-authors of abstracts submitted to the ACR/ARHP Annual Scientific Meeting an opportunity to attend the meeting to present their abstract.
Eligibility: Candidates must be a rheumatology fellows-in-training enrolled in an adult or pediatric rheumatology training program at an ACGME-accredited institution. Candidates must also be the author co-author of a submitted abstract of the current year to the abstract submission portal.
Level of Study: Doctorate, Research
Type: Research award
Value: Recipients will receive an award of $1,500 plus reimbursement up to $1,000 travel expenses (including air fare, hotel and meals) to attend the upcoming ACR/ARHP Annual Scientific Meeting. Recipients will also receive complimentary registration for the meeting. This is a one-year award
Length of Study: 1 year
Frequency: Annual
Country of Study: United States of America
No. of awards offered: 2
Application Procedure: Submit online application by the due date.
Closing Date: August 1st
Funding: Individuals, private
Contributor: Marshall J. Schiff Family
No. of awards given last year: 2
No. of applicants last year: 7
Additional Information: Please check at http://www.rheumatology.org/foundation/index.asp.

Medical and Pediatric Resident Research Award

Subjects: Rheumatology.
Purpose: To motivate outstanding residents to pursue subspecialty training in rheumatology by providing an opportunity to attend the ACR annual scientific meeting.

Eligibility: A candidate must be a resident enrolled in an ACGME-accredited pediatric, medicine or combined medicine/pediatric residency programme who is interested in rheumatology. The candidate must be an author or co-author of an abstract submitted to the upcoming ACR annual meeting. He must be a citizen or non-citizen national of the United States of America, or be in lawful possession of a permanent resident card. Individuals on temporary (J1, H1) or student visas are not eligible.
Level of Study: Doctorate, Research
Type: Research award
Value: US$750 cash prize plus US$1,000 to cover travel expenses and hotel accommodations for the ACR annual scientific meeting. Registration fees will be waived
Length of Study: 1 year
Frequency: Annual
Country of Study: United States of America
Application Procedure: Only online applications submitted through the REF Web site will be accepted.
Closing Date: August 1st
Funding: Corporation, foundation
Contributor: Abbott Endowment for Rheumatology
No. of awards given last year: 2
No. of applicants last year: 7
Additional Information: Please check website for further information http://www.rheumatology.org/foundation/index.asp

Medical Student Clinical Preceptorship

Subjects: Rheumatology
Purpose: To introduce students to the specialty of rheumatology by supporting a full-time clinical experience.
Eligibility: Only students enrolled in LCME or AOA COCA accredited medical schools or undergraduate students who have been accepted into medical school are eligible.
Level of Study: Graduate, Predoctorate
Type: Scholarship
Value: US$1,500 student stipend, US$500 stipend for mentor and US $1,000 for travel funds for the student to attend the ACR/ARHP Annual Scientific Meeting
Frequency: Annual
Country of Study: United States of America, Canada or Mexico
No. of awards offered: Up to 10
Application Procedure: Only complete applications submitted online by the deadline will be accepted.
Closing Date: 4 cycles – February 1st, May 2nd, August 1st, November 1st
Funding: Corporation, foundation
Contributor: Abbott Endowment for Rheumatology Development
Additional Information: Eligible to the nationals of Mexico also.

Medical Student Research Preceptorship

Subjects: Rheumatology
Purpose: To introduce students to the specialty of rheumatology by supporting a full-time research experience.
Eligibility: Only students enrolled in LCME or AOA COCA accredited medical schools undergraduate students who have been accepted into medical school are eligible.
Level of Study: Graduate, Postdoctorate, Research
Type: Scholarship
Value: US$3,000 student stipend, US$1,000 for the mentor and US $1,000 for travel expenses for the student to attend the ACR/ARHP Annual Scientific Meeting
Frequency: Annual
Country of Study: United States of America, Canada or Mexico
No. of awards offered: Up to 10
Application Procedure: Only complete applications submitted online by the deadline will be accepted.
Closing Date: 4 cycles – February 1st, May 2nd, August 1st, November 1st
Funding: Corporation, foundation
Contributor: Abbott Endowment for Rheumatology Development
Additional Information: Eligible to nationals of Mexico also.

Paula De Merieux Rheumatology Fellowship Award

Subjects: Rheumatic diseases.
Purpose: To help ensure that a diverse and highly trained workforce is available to provide competent clinical care to those affected by the rheumatic diseases.
Eligibility: Only training directors at ACGME-accredited institutions in good standing may apply. The trainee must be an underrepresented minority or a woman, i.e., either Black American, Native American (American Indian, Alaska Native, Native Hawaiian), Mexican American, Puerto Rican or any other minority category. Award applicant must be a citizen or non-citizen national of the United States of America, or be in lawful possession of a permanent resident card. Individuals on temporary (J1, H1) or student visas are not eligible.
Level of Study: Doctorate
Type: Training award
Value: US$25,000
Length of Study: 1 year
Frequency: Annual
Country of Study: United States of America
Application Procedure: Applications forms are available on website.
Closing Date: August 1st
Funding: Foundation, private
Contributor: The Dr Paula de Merieux estate
No. of awards given last year: 1
No. of applicants last year: 6

Scientist Development Award

Subjects: Rheumatology.
Purpose: To provide a training programme to rheumatology fellows or rheumatologists in the early stages of their career on aspects of clinical investigations through a structured, formal training programme.
Eligibility: Candidates must be an ACR or ARHP member, have a doctoral level degree, must be clinician scientists.
Level of Study: Professional development, Doctorate, Postdoctorate, Research
Type: Award
Value: US$50,000 for first year, $75,000 for second year and $100,000 for third year.
Length of Study: Up to 2 years
Frequency: Annual
Country of Study: United States of America
Application Procedure: Application forms are available online at www.rheumatology.org
Closing Date: August 1st
Funding: Foundation
Contributor: Rheumatology Research Foundation
No. of awards given last year: 5
No. of applicants last year: 11
Additional Information: This ia a Career Development Award. For any questions regarding eligibility, contact the Rheumatology Research Foundation.

Student Achievement Award

Subjects: Rheumatology.
Purpose: To recognize outstanding medical and graduate students for significant work in the field of rheumatology and provide an opportunity to attend the ACR annual scientific meeting.
Eligibility: Medical student candidates must be enrolled in an LCME-accredited medical school; graduate student candidates must be enrolled in an accredited institution. In addition, students must submit an abstract to the annual scientific meeting. The student must have made a significant contribution to the work submitted in order to be considered for the award. Award applicant must be a citizen or non-citizen national of the United States of America, or be in lawful possession of a permanent resident card. Individuals on temporary (J1, H1) or student visas are not eligible.
Level of Study: Doctorate, Graduate, Postgraduate
Type: Award
Value: US$750 cash award, waiver of registration fees and up to $1000 to cover hotel and travel expenses for the ACR Annual Scientific Meeting
Length of Study: 1 year
Frequency: Annual

Application Procedure: Application forms are available on website http://www.rheumatology.org/foundation/index.asp
Closing Date: August 1st
Funding: Corporation, foundation
Contributor: Abbott Endowment for Rheumatology Development
No. of awards given last year: 16
No. of applicants last year: 22

Training Development Award

Subjects: Rheumatology
Purpose: To help ensure that a diverse and highly trained workforce is available to provide competent clinical care to those affected by rheumatc diseases.
Eligibility: Only training directors at ACGME-accredited institutions in good standing may apply.
Level of Study: Doctorate
Type: Fellowship
Value: US$50,000 to support the salary and fringe of one trainee. Other trainee costs (e.g. fees, health insurance and other educational expenses) are to be incurred by the recipient's institutional program. Funds to attend the ACR/ARHP Annual Scientific Meeting will be administered through the ACR Fellows-in-Training (FIT) Travel Scholarship
Frequency: Annual
Country of Study: United States of America
Closing Date: August 1st
Funding: Foundation
Contributor: Rheumatology Research Foundation
No. of awards given last year: 2
No. of applicants last year: 10

AMERICAN COLLEGE OF SPORTS MEDICINE (ACSM)

401 West Michigan Street, Indianapolis, IN, 46202-3233, United States of America
Tel: (1) 317 637 9200
Fax: (1) 317 634 7817
Email: mwayne@acsm.org
Website: www.acsm.org
Contact: Megan Wayne, Coordinator

American College of Sports Medicine (ACSM) was founded in 1954. Since that time, they have applied their knowledge, training and dedication in sports medicine and exercise science to promote healthier lifestyles for people around the globe. Working in a wide range of medical specialities, allied health professions and scientific disciplines, their members are committed to the diagnosis, treatment and prevention of sports-related injuries and the advancement of the science of exercise.

ACSM Carl V. Gisolfi Memorial Fund

Subjects: Thermoregulation, exercise or hydration.
Purpose: To honor Carl V. Gisolfi's contributions to ACSM and the exercise science field and to encourage research in thermoregulation, exercise, and hydration.
Level of Study: Research
Type: Funding support
Value: $5,000
Frequency: Annual
No. of awards offered: 1
Application Procedure: Check website for details http://www.acsm.org/find-continuing-education/awards-grants/research-grants/2011/08/15/carl-v.-gisolfi-memorial-fund.
Closing Date: January 18th
Contributor: ACSM
Additional Information: Contact the Research Administration and Programs Department regarding this grant at Tel: (317) 637 9200, ext. 143 or email questions to Jane Senior at jsenior@acsm.org or Michael F. Dell at mdell@acsm.org.

ACSM Clinical Sports Medicine Fund

Subjects: Sports medicine.
Purpose: To stimulate clinical research in sports medicine and research in clinical sports medicine.
Eligibility: Open to MDs, DOs, PTs, ATCs, and other medical professionals (ACSM members only) involved in the conduct of patient-based clinical research.
Level of Study: Research
Type: Funding support
Value: $5,000
Frequency: Annual
No. of awards offered: 1
Application Procedure: Check website for details.
Closing Date: January 18th
Funding: Foundation
Contributor: Clinical Sports Medicine Endowment
Additional Information: Contact the Research Administration and Programs Department regarding this grant at Tel: (317) 637-9200, ext. 143 or email questions to Jane Senior at jsenior@acsm.org or Michael F. Dell at mdell@acsm.org.

ACSM International Student Award

Purpose: To support students fees and personal expenses.
Eligibility: Applicants must be a student at the time of the 2013 ACSM Annual Meeting and must not hold a completed doctoral degree of any type (PhD, Ed.D., M.D., DPH, etc.) at the time of application. Previous International Student Award winners are not eligible to apply. Please see the website for further details regarding eligibility http://www.acsm.org/find-continuing-education/awards-grants/international-awards/2011/08/16/international-student-award.
Level of Study: Research
Type: Award
Value: $1,000
Frequency: Annual
No. of awards offered: 3 (Maximum)
Application Procedure: Check website for details.
Closing Date: February 1st

ACSM Raymond and Rosalee Weiss Endowment

Subjects: Physical, mental and emotional benefits of physical activity.
Purpose: To provide on-going financial support to sponsor research on the subject of the physical, mental and emotional benefits of physical activity.
Eligibility: One project will be funded for applied rather than for basic research, with the intent of applying the results to programs involving physical activity and sports.
Level of Study: Research
Type: Funding support
Value: US$1,500
Frequency: Annual
No. of awards offered: 1
Application Procedure: Check website for details http://www.acsm.org/find-continuing-education/awards-grants/research-grants/2011/08/16/raymond-and-rosalee-weiss-research-endowment.
Closing Date: January 18th
Funding: Foundation
Additional Information: Contact the Research Administration and Programs Department regarding this grant at Tel: (317) 637-9200, ext. 143 or email questions to Jane Senior at jsenior@acsm.org or Michael F. Dell at mdell@acsm.org.

ACSM Research Endowment

Subjects: Basic and applied research in science.
Purpose: The Research Endowment is dedicated to the advancement of basic and applied research in exercise science.
Eligibility: Funding is primarily targeted for new or junior investigators, within 7 years of attaining a terminal degree (e.g. PhD, EdD). Only one application per person is allowed.
Level of Study: Research
Type: Research
Value: US$10,000
Frequency: Annual
No. of awards offered: 1–2
Closing Date: January 18th
Funding: Foundation
Additional Information: Contact the Research Administration and Programs Department regarding this grant at Tel: (317) 637 9200 ext. 143 or email questions to Jane Senior at jsenior@acsm.org or Michael F. Dell at mdell@acsm.org.

Doctoral Student Research Grant
Subjects: All subjects.
Purpose: To assist doctoral students.
Eligibility: Doctoral students enrolled in full-time programs.
Level of Study: Postgraduate, Research, Doctorate
Type: Grant
Value: Up to US$5,000 for experimental subjects, supplies and small equipment needs
Length of Study: 1 year
Frequency: Annual
No. of awards offered: 10
Application Procedure: Check website for further details.
Closing Date: January 18th
Additional Information: Contact the Research Administration and Programs Department regarding this grant at Tel: (317) 637 9200, ext. 143 or email questions to Jane Senior at jsenior@acsm.org or Michael F. Dell at mdell@acsm.org.

NASA Space Physiology Research Grant
Subjects: Exercise, weightlessness and musculoskeletal physiology.
Eligibility: Open to US residents who are doctoral students enrolled in full-time programs. Doctoral students enrolled in full-time programs are eligible to apply. Grants can range up to $5,000 and are available for a one-year period.
Level of Study: Doctorate, MBA, Postgraduate, Research
Type: Research grant
Value: US$5,000
Length of Study: 1 year
Frequency: Annual
No. of awards offered: 1–2
Application Procedure: Check website for further details http://www.acsm.org/find-continuing-education/awards-grants/research-grants/2011/08/16/nasa-space-physiology-research-grants.
Closing Date: January 18th
Contributor: National Aeronautics and Space Administration (NASA)
Additional Information: Funds are available after October 1st. Contact the Research Administration and Programs Department regarding this grant at Tel: (317) 637-9200, ext. 143 or email questions to Jane Senior at jsenior@acsm.org or Michael F. Dell at mdell@acsm.org.

Paffenbarger-Blair Fund for Epidemiological Research on Physical Activity.
Subjects: Sports medicine
Purpose: To encourage researchers early in their career to become involved with physical activity epidemiology.
Eligibility: Open to ACSM member at the time of application submission. Please see the website for details regarding eligibility http://www.acsm.org/find-continuing-education/awards-grants/research-grants/2011/08/18/paffenbarger-blair-fund-for-epidemiological-research-on-physical-activity.
Level of Study: Postdoctorate, Postgraduate
Type: Funding support
Value: US$10,000
Length of Study: 1 year
Frequency: Annual
Country of Study: United States of America
No. of awards offered: 1
Application Procedure: Applications can be downloaded from the ACSM website, applicants are expected to apply within 2 years of receiving a postgraduate degree or completion of clinical training
Closing Date: January 18th
Funding: Foundation
Additional Information: Contact the Research Administration and Programs Department regarding this grant at Tel: (317) 637-9200, ext. 143 or email questions to Jane Senior at jsenior@acsm.org or Michael F. Dell at mdell@acsm.org.

AMERICAN CONGRESS OF OBSTETRICIANS AND GYNECOLOGISTS (ACOG)

PO Box 70620, Washington, DC, 20024-9998, United States of America
Tel: (1) 800 673 8444, 202 638 5577
Fax: (1) 202 863 4992, 202 638 5577
Email: grants@acog.org
Website: www.acog.org
Contact: Mrs Lee Cummings, Director of Corporate Relations

The American College of Obstetricians and Gynecologists (ACOG) is a membership organization of obstetricians and gynecologists dedicated to the advancement of women's health through education, advocacy, practice and research.

ACOG Abbott Nutrition Research Award on Nutrition in Pregnancy
Subjects: Obstetrics and gynecology with a focus on nutrition in pregnancy with a special interest in the areas of obesity, diabetes and women age 20–39.
Purpose: To provide seed grant funds to a junior investigator for clinical research in the area nutrition in pregnancy.
Eligibility: Applicants must be ACOG Junior Fellows or Fellows at the time of application.
Level of Study: Postdoctorate
Type: Research grant
Value: US$25,000 plus $1,000 travel stipend to attend the ACOG Annual Clinical Meeting
Length of Study: 1 year
Frequency: Annual
Country of Study: United States of America or Canada
No. of awards offered: 1
Application Procedure: Applicants must sumit six copies of a proposal consisting of a hypothesis, objectives, specific aims, background and significance and experimental design, and references. One page budget is required, applicants's curriculum vitae and a letter of support from the program director, departmental chair or laboratory director.
Closing Date: January 4th
Funding: Commercial
Contributor: Abbott Nutrition
No. of awards given last year: 1
No. of applicants last year: 6
Additional Information: Further information can be found on the member side of the website www.acog.org

ACOG Bayer Health Care Pharmaceticals Research Award in Long Term Contraception
Subjects: Obstetrics and gynecology.
Purpose: To provide seed grant funds to junior investigators for clinical research in the area of non-daily contraception, such as proper use and administration, factors affecting patient acceptance and compliance, and non-contraceptive benefits.
Eligibility: Candidates must be an ACOG Junior Fellow or Fellow and in an approved obstetrics-gynecology residency program or within 3 years post-residency.
Level of Study: Postgraduate
Type: Research grant
Value: US$25,000 plus $1,000 travel stipend to attend the ACOG annual clinical meeting.
Length of Study: 1 year
Frequency: Annual
Country of Study: United States of America or Canada
No. of awards offered: 1
Application Procedure: Candidates must submit six copies of a research proposal, written in eight pages or less, and should include the following: hypothesis, objectives, specific aims, background and significance, experimental design, and references. A one-page budget, curriculum vitae and letter of support from the program director, departmental chair or laboratory director must also be submitted.
Closing Date: October 1st
Funding: Commercial

Contributor: Bayer Healthcare Pharmaceuticals
No. of awards given last year: 1
No. of applicants last year: 4
Additional Information: Further information can be found on the member side of the website www.acog.org

ACOG/Bayer Health Care Pharmaceuticals Research Award in Contraceptive Counseling

Subjects: Obstetrics and gynecology.
Purpose: To provide seed grant funds to junior investigators for clinical research in the area of contraceptive counselling, such as education to parents to recognize the options available to them and the risks, benefits and total costs associated with each option, patient preference for contraception options and patient selection criteria for different forms of contraception.
Eligibility: Candidates must be ACOG Junior Fellows or Fellows who are in an approved obstetrics or gynecology residency programme or within 3 years of postresidency.
Level of Study: Postgraduate
Type: Research award
Value: US$25,000 plus a $1,000 travel stipend to attend the ACOG Annual Clinical Meeting
Length of Study: 1 year
Frequency: Annual
Country of Study: United States of America or Canada
No. of awards offered: 1
Application Procedure: Candidates must submit six copies of a proposal consisting of a hypothesis, objectives, specific aims, background and significance, experimental design and methods. These must not exceed six typewritten pages in total. A curriculum vitae, letter of support from the programme director, departmental chair or laboratory director, references and a one-page budget must also be submitted.
Closing Date: January 4th
Funding: Commercial
Contributor: Bayer Healthcare Pharmaceuticals
No. of awards given last year: 1
No. of applicants last year: 4
Additional Information: Further information can be found on the member side of the website www.acog.org

ACOG/Eli Lilly and Company Research Award for the Prevention and Treatment of Osteoporosis

Subjects: Obstetrics and gynecology with a focus on osteoporosis.
Purpose: To provide seed grant funds to a junior investigator for clinical research in the area of prevention and treatment for osteoporosis.
Eligibility: Candidates must be ACOG Junior Fellows or Fellows who are in an approved obstetrics/gynecology residency program, or within 3 years post-residency.
Level of Study: Postdoctorate
Type: Research award
Value: US$$15,000 and includes a $1,000 travel stipend for attendance at the ACM.
Length of Study: 1 year
Frequency: Annual
Country of Study: United States of America or Canada
No. of awards offered: 1
Application Procedure: Candidates must submit six copies of a proposal consisting of a hypothesis, objectives, specific aims, background and significance, and experimental design and references. A one page budget is required, applicant's curriculum vitae and a letter of support from the program director, departmental chair or laboratory director.
Closing Date: January 4th
Funding: Commercial
Contributor: Eli Lilly and Company
No. of awards given last year: 1
No. of applicants last year: 5
Additional Information: Further information can be found on the member side of the website www.acog.org.

ACOG/Hologic Research Award for the Prevention of Cervical Cancer

Purpose: To provide seed grant funds to a junior investigator for clinical research in the area cervical cancer prevention and to provide opportunity to advance knowledge on issues related to cervical cancer.
Eligibility: Applicants must be ACOG Junior Fellows or Fellows who are in an approved obstetrics/gynecology residency program, or within three years post-residency.
Level of Study: Postdoctorate
Type: Research award
Value: US$25,000 plus a US$1,000 travel stipend to attend the ACOG Annual Clinical Meeting
Length of Study: 1 year
Frequency: Annual
Country of Study: United States of America & Canada
No. of awards offered: 1
Application Procedure: Applicants must submit six copies of a proposal consisting of a hypothesis; objectives, specific aims, background and significance, and experimental design, and references. A one-page budget is required, applicant's curriculum vitae and a letter of support from the program director, departmental chair, or laboratory director.
Closing Date: January 4th
Funding: Commercial
Contributor: Hologic
No. of awards given last year: 1
No. of applicants last year: 5
Additional Information: Please check the website www.acog.org for more details.

ACOG/Kenneth Gottesfeld-Charles Hohler Memorial Foundation Research Award in Ultrasound

Subjects: To provide grant funds to a junior investigator to support research or advanced training that is ultrasound specific and dedicated to a practical clinical use in a new or unique approach.
Purpose: To provide grant funds to support work in ultrasound and its application to obstetrics and gynecology.
Eligibility: Applicants must be ACOG Junior Fellows or Fellows who are in an approved obstetrics/gynecology residency programme, or within 5 years of completion of their residency or fellowship.
Level of Study: Postdoctorate
Value: One grant of US$10,000 or two grants of US$5,000 will be provided at the discretion of the review committee plus US$1,000 travel stipend to attend the ACOG Annual Clinical Meeting
Length of Study: 1 year
Frequency: Annual
Country of Study: United States of America or Canada
No. of awards offered: 1 or 2
Application Procedure: Applicants must submit six copies of a proposal or eight pages or less consisting of a hypothesis; objectives, specific aims, background and significance, and experimental design, and references. An advanced training proposal must be eight pages or less and include the site, the dates, the proposed curriculum and the individuals responsible for the training. A one-page budget is required and applicant's curriculum vitae.
Closing Date: January 4th
Funding: Foundation
Contributor: Kenneth Gottesfeld-Charles Hohler Memorial Foundation
No. of awards given last year: 1
No. of applicants last year: 8
Additional Information: Further information can be found on the member side of the website www.acog.org

ACOG/Merck and Company Research Award on Adolescent Health Preventive Services

Purpose: To provide seed grant funds to a junior investigator for research in the area of prevention in adolescent health and to provide opportunity to advance knowledge on issues related to adolescent health, specifically to develop and test strategies: to incorporate young adolescents into ob/gyn practice for the ACOG recommended initial adolescent reproductive health visit (ages 13–15) and on-going preventive care at all ages; to provide clinical preventive services such as immunization and screening; and, to provide health guidance and counseling to both adolescents and parents on issues such as sexual behavior, tobacco, alcohol and other drugs, safety, weight and exercise.
Eligibility: Applicants must be ACOG Junior Fellows or Fellows.

Level of Study: Postdoctorate
Type: Research grant
Value: US$15,000 plus a US$1,000 travel stipend to attend the ACOG Annual Clinical Meeting
Length of Study: 1 year
Frequency: Annual
Country of Study: United States of America & Canada
No. of awards offered: 1
Application Procedure: Applicants must submit 6 copies of a proposal consisting of a hypothesis; objectives, specific aims, background and significance, and experimental design, and references. A one-page budget is required, applicant's curriculum vitae and a letter of support from the program director, departmental chair, or laboratory director.
Closing Date: January 4th
Funding: Commercial
Contributor: Merck and Company
No. of awards given last year: 1
No. of applicants last year: 6
Additional Information: Please check the website www.acog.org for further details.

ACOG/Merck and Company Research Award on Immunization

Purpose: To provide seed grant funds to an ACOG Junior Fellow or Fellow for research in the area of immunization and to provide opportunity to advance knowledge on issues related to immunization in the ob/gyn practice. Topics of interest include, but are not limited to: immunization history of patients; frequency of immunization of patients by an ob/gyn; incidence of infectious diseases preventable by immunization; prohibitions to immunizations in ob/gyn patients; pregnancy and immunization; ob/gyn patients interest in immunization; and, any other related area.
Eligibility: Applicants must be ACOG Junior Fellows or Fellows.
Level of Study: Postdoctorate
Type: Research grant
Value: US$15,000 plus a US$1,000 travel stipend to attend the ACOG Annual Clinical Meeting
Length of Study: 1 year
Frequency: Annual
Country of Study: United States of America or Canada
No. of awards offered: 1
Application Procedure: Applicants must submit six copies of a proposal consisting of a hypothesis; objectives, specific aims, background and significance, and experimental design, and references. A one-page budget is required, applicant's curriculum vitae and a letter of support from the program director, departmental chair, or laboratory director.
Closing Date: January 4th
Funding: Commercial
Contributor: Merck and Company
No. of awards given last year: 1
No. of applicants last year: 7
Additional Information: Please check the website www.acog.org for further details.

ACOG/Ortho Women's Health and Urology Academic Training Fellowships in Obstetrics and Gynecology

Subjects: Obstetrics and gynecology.
Purpose: To provide opportunities for especially qualified residents or Fellows to spend an extra year involved in responsibilities that will train them for academic positions in the speciality.
Eligibility: Open to ACOG Junior Fellows or Fellows who have completed at least 1 year of training, and are considered by the director of their residency programme to be especially fitted for a career in medical education or academic obstetrics and gynecology.
Level of Study: Postgraduate
Type: Research grant
Value: US$30,000 stipend plus a $1,000 travel stipend to attend the ACOG Annual Clinical Meeting
Length of Study: 1 year
Frequency: Annual
Country of Study: United States of America or Canada
No. of awards offered: 2
Application Procedure: Applicants must submit six copies of a proposal consisting of a hypothesis, objectives, specific aims, background and significance, experimental design and references. These must not exceed eight typewritten pages in total. A curriculum vitae, letter of support from the programme director, departmental chair or laboratory director are also required.
Closing Date: January 4th
Funding: Commercial
Contributor: Ortho-Women's Health and Urology
No. of awards given last year: 2
No. of applicants last year: 8
Additional Information: Further information can be found on the member side of the website www.acog.org

Warren H Pearse/Wyeth Pharmaceuticals Women's Health Policy Research Award

Subjects: Obstetrics and gynecology.
Purpose: To provide funds to support research in the area of health care policy.
Eligibility: The principal or co-principal investigator must be an ACOG Junior Fellow or Fellow. Proposals will be considered with regard to innovation, potential utility of the research, ability to generalize results and demonstrated capability of the investigator.
Level of Study: Postgraduate
Type: Research award
Value: US$15,000 plus $1,000 travel stipend to attend the ACOG Annual Clinical Meeting.
Length of Study: 1 year
Frequency: Annual
Country of Study: United States of America or Canada
No. of awards offered: 1
Application Procedure: Candidates must submit six copies of a proposal consisting of a hypothesis, objectives, specific aims, background and significance, experimental design and methods. These must not exceed eight typewritten pages in total. A curriculum vitae, letter of support from the programme director, departmental chair or laboratory director, references and a one-page budget must also be submitted.
Closing Date: January 4th
Funding: Commercial
Contributor: Wyeth Pharmaceuticals
No. of awards given last year: 1
No. of applicants last year: 7
Additional Information: Further information can be found on the member side of the website www.acog.org.

AMERICAN CONGRESS ON SURVEYING AND MAPPING (ACSM)

6, Montgomery Village Avenue, Suite 403 Gaithersburg, Maryland, 20879, United States of America
Tel: (1) (240) 632 9716 ext. 109
Fax: (1) (240) 632 1321
Email: ilse.genovese@acsm.net
Website: www.acsm.net

The ACSM is a nonprofit association dedicated to serving the public interest and advancing the profession of surveying and mapping.

AAGS Graduate Fellowship Award

Subjects: Geodetic surveying.
Purpose: To support students in the field of geodetic surveying or geodesy.
Eligibility: Open to students enrolled in a programme in geodetic surveying. Preference will be given to applicants having at least two years of employment experience in the surveying profession.
Level of Study: Graduate
Type: Award
Value: $2,000
Length of Study: 2–4 years
Frequency: Annual
Application Procedure: See the website.
Closing Date: February 17th
Funding: Foundation
Contributor: American Association for Geodetic Surveying (AAGS)

No. of awards given last year: 1
No. of applicants last year: 5

AAGS Joseph F. Dracup Scholarship Award

Subjects: Surveying.
Purpose: To offer better opportunities to students in geodetic science programmes.
Eligibility: Open to students enrolled in 4-year programme in surveying. Preference will be given to applicants from programs with a significant focus on geodetic surveying.
Level of Study: Graduate, Postgraduate
Value: US$2,000
Length of Study: 4 years plus
Frequency: Annual
Application Procedure: A completed application form must be submitted.
Closing Date: February 17th
Funding: Foundation
No. of awards given last year: 1
No. of applicants last year: 5

ACSM Fellows Scholarship

Subjects: Any ACSM disciplines.
Purpose: To encourage, recognize and support exceptional surveying and mapping students.
Eligibility: Open to students with a junior or higher degree and enrolled in four-year degree programs in surveying or in closely related programs such as geomatics or surveying engineering.
Level of Study: Graduate, Postgraduate
Type: Scholarship
Value: US$2,000
Length of Study: 4 years
Frequency: Annual
Application Procedure: Applicants must submit a completed application form along with proof of membership in ACSM, a complete statement indicating educational objectives, future plans of study or research, professional activities and financial need, three letters of recommendation and a complete original official transcript.
Closing Date: February 17th
Funding: Corporation
Additional Information: Prior scholarship winners are eligible to apply in succeeding years providing all appropriate criteria are satisfied.

Berntsen International Scholarship In Surveying

Subjects: Surveying.
Purpose: To encourage, recognize and support exceptional surveying and mapping students.
Eligibility: Open to students enrolled in four-year degree programs in surveying or in closely related programs such as geomatics or surveying engineering.
Level of Study: Graduate
Type: Scholarship
Value: US$1,500
Length of Study: 2 years
Application Procedure: Applicants must submit a completed application form along with proof of membership in ACSM, a complete statement indicating educational objectives, future plans of study or research, professional activities and financial need, three letters of recommendation and a complete original official transcript.
Closing Date: February 17th
Funding: Commercial
Contributor: Berntsen International Inc., of Madison, Wisconsin
Additional Information: Prior scholarship winners are eligible to apply in succeeding years providing all appropriate criteria are satisfied.

Berntsen International Scholarship in Surveying Technology

Subjects: Surveying technology.
Purpose: To encourage exceptional students in surveying and mapping.
Eligibility: Open to students who are registered in a 2-year programme in surveying technology.
Level of Study: Graduate
Type: Scholarship
Value: US$500
Length of Study: 2 years
Frequency: Annual
No. of awards offered: 1
Application Procedure: A completed application form must be submitted.
Closing Date: February 17th
Funding: Commercial
Contributor: Bernsten International Inc., of Madison, Wisconsin
No. of awards given last year: 1
No. of applicants last year: 24

The Cady McDonnell Memorial Scholarship

Purpose: To recognize a women student enrolled in the field of surveying.
Eligibility: Open to female candidates who are residents of one of the following western states: Alaska, Arizona, California, Colorado, Hawaii, Idaho, Montana, Nevada, New Mexico, Oregon, Utah, Washington, Wyoming.
Level of Study: Postgraduate
Type: Scholarship
Value: US$1,000
Length of Study: 2–4 years
Frequency: Annual
No. of awards offered: 1
Application Procedure: A completed application along with proof of legal home residence and ACSM membership website form must be submitted. See the website for further information.
Closing Date: February 17th
Funding: Foundation
Contributor: National Society of Professional Surveyors
No. of awards given last year: 1
No. of applicants last year: 24

The Schonstedt Scholarships in Surveying

Subjects: Surveying.
Purpose: To provide opportunities for students in surveying.
Eligibility: Open to students enrolled in a four-year degree programme in surveying. Preference will be given to applicants with junior or senior standing.
Level of Study: Postgraduate
Type: Scholarship
Value: US$1,500
Length of Study: 4 years
Frequency: Annual
No. of awards offered: 2
Application Procedure: See the website.
Closing Date: February 17th
Funding: Commercial
Contributor: Schonstedt Instrument Company of Kearneysville, West Virginia
No. of awards given last year: 1
No. of applicants last year: 7
Additional Information: Schonstedt donates a magnetic locator to the surveying program at each recipient's school.

Tri State Surveying and Photogrammetry Kris M. Kunze Memorial Scholarship

Subjects: Surveying.
Purpose: To provide financial assistance.
Eligibility: Open to candidates who are citizens of United States of America. First priority candidates are licensed Professional Land Surveyors or Certified Photogrammetrists pursuing college level courses in Business Administration or Business Management. Second priority candidates are certified Land Survey Interns pursuing college level courses in Business Administration or Business Management. Third priority candidates are full-time students enrolled in a two or four year degree program in Surveying and Mapping pursuing a course study including Business Administration or Business Management.
Level of Study: Postgraduate
Type: Scholarship
Value: US$1,000
Length of Study: 2–4 years
Frequency: Annual

Application Procedure: A completed application form along with proof of ACSM membership must be submitted. See the website for further information.
Closing Date: February 17th
Funding: Foundation
Contributor: Kris M. Kunze Memorial Scholarship Fund
No. of awards given last year: 1
No. of applicants last year: 12

AMERICAN COUNCIL OF LEARNED SOCIETIES (ACLS)

633 3rd Avenue, 8th Floor between 40th & 41st Streets, New York, NY, 10017 6795, United States of America
Tel: (1) 212 697 1505
Fax: (1) 212 949 8058
Email: sfisher@acls.org
Website: www.acls.org

The American Council of Learned Societies (ACLS) is a private non-profit federation of 68 national scholarly organizations. The mission of the ACLS, as set forth in its Constitution is the advancement of humanistic studies in all fields of learning in the humanities and the social sciences and the maintenance and strengthening of relations among the national societies devoted to such studies.

ACLS American Research in the Humanities in China

Subjects: Humanities.
Purpose: To enable scholars to carry out research in the People's Republic of China.
Eligibility: This program is open to scholars in the humanities and related social sciences who have received a PhD or its equivalent by the time of application. Please see the website for further details regarding eligibility http://www.acls.org/programs/arhc/.
Level of Study: Postdoctorate
Type: Research grant
Value: Up to $50,400
Length of Study: 4–12 months
Frequency: Annual
Study Establishment: A university or research institute
Country of Study: China
No. of awards offered: Approx. 5
Application Procedure: Applicants must write for details.
Closing Date: October 2nd, 9 p.m. Eastern Daylight Time
Contributor: The National Endowment for the Humanities

ACLS Charles A. Ryskamp Research Fellowship

Subjects: Humanities and related social sciences
Purpose: To support advanced assistant professors and untenured associate professors in the humanities and related social sciences whose scholarly contributions have advanced their fields and who have well designed and carefully developed plans for new research and to provide time and resources to enable research under optional conditions.
Eligibility: Open to candidates holding a PhD or equivalent and employed in tenure-track positions at degree-granting academic institutions in the United States, remaining so for the duration of the fellowship. Please see the website for further details regarding eligibility http://www.acls.org/programs/ryskamp/.
Level of Study: Postdoctorate
Type: Fellowship
Value: $64,000, plus $2,500 for research and travel, and the possibility of an additional summer's support
Length of Study: One academic year, plus one summer if justified by a persuasive case
Frequency: Annual
Country of Study: United States of America
No. of awards offered: 12
Application Procedure: Apply online (ofa.acls.org).
Closing Date: October 2nd, 9 p.m. Eastern Daylight Time
Funding: Foundation
Contributor: Andrew W. Mellon Foundation

ACLS Digital Innovation Fellowships

Subjects: Humanities and related social sciences and digital humanities.
Purpose: To support an academic year dedicated to work on a major scholarly project that takes a digital form.
Eligibility: This program is open to scholars in all fields of the humanities and the humanistic social sciences. Candidates must have a PhD degree conferred prior to the application deadline. (An established scholar who can demonstrate the equivalent of the PhD in publications and professional experience may also qualify.) U.S. citizenship or permanent resident status is required as of the application deadline.
Level of Study: Postdoctorate
Type: Fellowship
Value: US$60,000 stipend (for academic year's leave from teaching), plus up to US$25,000 in project costs like access to tools, personnel for digital production, collaborative work and dissemination and preservation of digital projects created or enhanced under the Fellowship.
Length of Study: One year
Frequency: Annual
Country of Study: United States of America
No. of awards offered: 8
Application Procedure: Completed applications must be submitted through the ACLS Online Fellowship Application system.
Closing Date: October 2nd, 9 p.m. Eastern Daylight Time
Funding: Foundation
Contributor: Andrew W. Mellon Foundation
Additional Information: Check website for further details at http://www.acls.org/programs/digital/.

ACLS Humanities Program in Belarus, Russia and Ukraine

Subjects: History, archeology, literature, linguistics, film studies, art history.
Purpose: To sustain individuals doing exemplary work, so as to assure continued future leadership in the humanities.
Eligibility: Applicants should be involved in studies related to performing arts, ethnographic and cultural studies, gender studies, philosophy or religious studies. ACLS organizes annual regional meetings for advisers and grant recipients.
Level of Study: Research
Type: Grant
Value: Varies
Frequency: Annual
No. of awards offered: Varies
Application Procedure: Application forms can be obtained by writing to hp@acls.org. Please see the website for further details http://www.acls.org/programs/hp/.
Closing Date: November 18th
Funding: Foundation
Additional Information: In 2007, with the help of the Carnegie Corporation of New York and the American Council of Learned Societies, the International Association for the Humanities (IAH) was founded as an independent association of humanities scholars primarily in Belarus, Russia, and Ukraine to help represent the post-Soviet region in the international scholarly community. Starting in the academic year 2010-11, IAH organizes a competition for short-term grants in the humanities. For further details and currect competition application form, see the IAH website.

ACLS/Chiang Ching-kuo Foundation (CCK) Comparative Perspectives on Chinese Culture and Society

Subjects: Humanities and related social science.
Purpose: To provide support for all conferences and publications on new perspectives on Chinese culture and society.
Eligibility: Open to candidates who are affiliated with a university or research institution and who hold a PhD degree. There are no restrictions as to citizenship of participants or location of the project.
Level of Study: Doctorate, Postgraduate
Type: Scholarship
Value: Up to $25,000 for conferences; $10,000–$15,000 for workshops and seminars; up to $6,000 for planning meetings
Frequency: Annual
Country of Study: Any country

Application Procedure: Applications should be submitted by mail to fellowships@acls.org or by courier. Please see the website for further details http://www.acls.org/programs/chinese-culture/.
Closing Date: September 28th
Funding: Foundation
Contributor: Chiang Ching-kuo Foundation for International Scholarly Exchange

ACLS/New York Public Library Fellowship

Subjects: Humanities and humanistic social sciences
Purpose: The ACLS and the NYPL offer a collaborative programme to provide up to 5 residential fellowships at the Library's Dorothy and Lewis B. Cullman Center for Scholars and Writers. The center provides opportunities for up to 15 fellows to explore and use the collections of NYPL Humanities and Social sciences Library.
Eligibility: Application for an ACLS/NYPL residential fellowship has the same eligibility requirements, application form, and schedule as the ACLS fellowship programme.
Level of Study: Postdoctorate
Type: Fellowship
Value: Stipend for the NYPL residential fellowships will be $60,000.
Length of Study: 6–12 months
Frequency: Annual
Study Establishment: New York Public Library
Country of Study: United States of America
No. of awards offered: Up to 15
Application Procedure: Please see the website for further details http://www.acls.org/programs/acls/.
Closing Date: September 30th
Additional Information: Because this is a collaborative fellowship, applicants for the ACLS/NYPL residential fellowships must also apply to the Dorothy and Lewis B. Cullman Center for Scholars and Writers at the NYPL. An NYPL application form is available at http://www.nypl.org/csw may be requested from csw@nypl.org.

ACLS/SSRC/NEH International and Area Studies Fellowships

Subjects: The societies and cultures of Asia, Africa, the Near and Middle East, Latin America and the Caribbean, Eastern Europe and the former Soviet Union.
Purpose: To encourage humanistic research in area studies.
Eligibility: Applicants must be citizens or permanent residents of the United States as of the application deadline date, and hold a PhD degree. However, an established Scholar who can demonstrate the equivalent of a PhD in publications and professional experience may also qualify. Scholars pursuing research and writing on the societies and cultures of Asia, Africa, the Near and Middle East, Latin America and the Caribbean, East Europe and the Former Soviet Union are eligible. Scholars currently enrolled for any degree are not eligible.
Level of Study: Postdoctorate
Type: Fellowship
Value: Up to US$65,000 for full professor and equivalent, $45,000 for associate professor and equivalent, and $35,000 for assistant professor and.
Length of Study: 6–12 months
No. of awards offered: Approx. 10
Application Procedure: Applications must be made to the ACLS Fellowship Programme and all requirements and provisions of that programme must be met. The Fellow must submit a final report to both NEH and ACLS. Note that applications must also be made to the competition for residential fellowships administered separately by the NYPL Center for Scholars and Writers.
Closing Date: September 28th
Contributor: American Council of Learned Societies (ACLS), Social Science Research Council (SSRC), National Endowment for the Humanities (NEH)
Additional Information: Please see the website for further details http://www.acls.org/programs/acls/.

Andrew W. Mellon Foundation/ACLS Early Career Fellowships Program Dissertation Completion Fellowships

Subjects: Humanities and related social sciences.
Purpose: To assist graduates in the last year of their PhD dissertation writing.
Eligibility: Open to a PhD candidate in a humanities or social science department in an American University. Applicant should not be in the degree programme for more than 6 years and the successful candidates cannot hold this fellowship after the 7th year.
Level of Study: Doctorate
Type: Fellowships
Value: US$25,000, plus funds for research costs of up to $3,000 and for university fees of up to $5,000.
Length of Study: 1 year
Frequency: Annual
Country of Study: United States of America
No. of awards offered: 65
Application Procedure: Applicants can apply online (ofa.acls.org). Please see the website for further website http://www.acls.org/programs/dcf/.
Closing Date: November 9th
Contributor: The Andrew W. Mellon Foundation

For further information contact:

Email: grants@acls.org

Andrew W. Mellon/ACLS Recent Doctoral Recipients Fellowships

Subjects: Humanities and social science.
Purpose: To provide support and assistance for young scholars to complete their dissertation and later to advance their research after being awarded the PhD.
Eligibility: Open to scholars who hold the Dissertation Completion Fellowships of any nation. Candidates must be in the final year of dissertation completion at the time of application. Those who have completed the dissertation are not eligible. Candidates for this program must be PhD. candidates in a humanities or social science department in the United States. Candidates from other departments may be eligible if their project is in the humanities or related social sciences, and their principal dissertation advisor holds an appointment in a humanities or related social science field.
Level of Study: Doctorate, Postgraduate
Type: Fellowships
Value: US$35,000
Length of Study: 1 year
Frequency: Annual
Country of Study: United States of America
No. of awards offered: 25
Application Procedure: Applicants will have to fill an online application form (ofa.acls.org). Please see the website for further details http://www.acls.org/grants/Default.aspx?id=514.
Closing Date: December 9th
Funding: Foundation

Early Career Postdoctoral Fellowships in East European Studies

Subjects: Humanities and social sciences.
Purpose: To offer support for postdoctoral research and a written account of the work in East European studies.
Eligibility: Open to candidates who are citizens or permanent residents of the United States. A candidate must hold a PhD degree conferred prior to the application deadline; however, an established scholar who can demonstrate the equivalent of the PhD in publications and professional experience may also qualify. A candidate must be at an early career stage; tenured faculty are not eligible.
Level of Study: Postdoctorate, Research
Type: Fellowships
Value: US$25,000
Length of Study: 6–12 consecutive months
Frequency: Annual
Study Establishment: American Council of Learned Societies
Country of Study: United States of America or Germany
Application Procedure: Completed applications must be submitted through the ACLS Online Fellowship Application system (ofa.acls.org). Please see the website for further details http://www.acls.org/grants/Default.aspx?id=534.
Closing Date: November 9th
Additional Information: These fellowships are to be used for work outside East Europe, although short visits to the area may be proposed as part of a coherent programme primarily based elsewhere.

Project must clearly demonstrate the necessity to conduct research at German universities or research institutes on countries of East and Central Europe or on Germany's relations with those countries.

Frederick Burkhardt Residential Fellowships for Recently Tenured Scholars

Subjects: Humanities.
Purpose: To encourage more adventurous, more wide-ranging and longer-term patterns of research; the specific goal of the fellowship year is a major piece of scholarly work.
Eligibility: Candidates must be employed in a tenured position at a degree-granting academic institution in the United States, remaining so for the duration of the fellowship.
Level of Study: Postdoctorate
Type: Fellowships
Value: US$75,000 stipend
Length of Study: 1 year
Frequency: Annual
Study Establishment: A participating national research centre
Country of Study: United States of America
No. of awards offered: 9
Application Procedure: Completed applications must be submitted through the ACLS Online Fellowship Application system (ofa.acls.org). Please see the website for further details http://www.acls.org/programs/burkhardt/.
Closing Date: September 28th
Funding: Foundation
Contributor: Rockefeller Foundation

Henry Luce Foundation/ACLS Dissertation Fellowships in American Art

Subjects: Art history, focusing on a topic in the history of the visual arts of the United States.
Purpose: To assist students at any stage of PhD dissertation research or writing.
Eligibility: Applicants must be United States citizens and have completed all requirements for a PhD except the dissertation before beginning tenure. They must also be in a department of art history. A student whose degree will be granted by another department may be eligible if the principal dissertation advisor is in a department of the history of art. In all cases the dissertation topic should be object orientated. Students preparing theses for the Master's of Fine Arts degree are not eligible.
Level of Study: Graduate, Predoctorate
Type: Fellowship
Value: US$25,000, plus up to $2,000 as a travel allowance.
Length of Study: 1 year, non-renewable
Country of Study: United States of America
No. of awards offered: 10
Application Procedure: Apply online (ofa.acls.org). Please see the website for further details http://www.acls.org/programs/american-art/.
Closing Date: November 9th
Contributor: The Henry Luce Foundation

AMERICAN COUNCIL ON RURAL SPECIAL EDUCATION (ACRES)

West Virginia University, 509 Allen Hall, PO Box 6122, Morgantown, WV, 26506 6122, United States of America
Tel: (1) 304 293 3450
Fax: (1) 435 797 3572
Email: acres-sped@mail.wvu.edu
Website: http://acres-sped.org/
Contact: David Forbush, Headquarters Co-ordinator

ACRES Scholarship

Subjects: Special education in the areas of the handicapped, those with specific learning disabilities and the socially disadvantaged.
Purpose: To give a rural teacher an opportunity to pursue education and training not otherwise affordable within his or her district.
Eligibility: Applicants must be citizens of the United States of America, currently employed by a rural school district as a certified teacher in regular or special education, working with students with disabilities or with regular education students and retraining to a special education career. Please see the website for further information regarding eligibility http://acres-sped.org/scholarships.
Level of Study: Graduate
Type: Scholarship
Value: Up to US$1,000
Length of Study: 1 year
Frequency: Annual
Country of Study: United States of America
No. of awards offered: 1
Application Procedure: Applicants must complete and submit an application form with an essay and two letters of recommendation. Applicants should access application materials online.
Closing Date: February 15th
Funding: Private
Additional Information: The award will be announced at the March ACRES conference.

For further information contact:

United States of America
Tel: (1) 304 293 4384
Email: acres-sped@mail.wvu.edu

AMERICAN COUNCILS FOR INTERNATIONAL EDUCATION ACTR/ ACCELS

1828 L Street N.W., Suite 1200, Washington, DC, 20036, United States of America
Tel: (1) 202 833 7522
Fax: (1) 202 833 7523
Email: outbound@americancouncils.org
Website: www.americancouncils.org
Contact: American Councils Headquarters Program Officer, Russian and Eurasian Outbound Programs

American Councils for International Education advances scholarly research and cross border learning through the design and implementation of educational programs that are well grounded in key world languages, cultures and regions. We contribute to the creation of new knowledge, broader professional perspectives, and personal and intellectual growth through international training, academic exchange, collaboration in educational development, and public diplomacy. With a presence in the U.S., Russia and Eurasia for nearly four decades, in addition to representation in over thirty countries across Asia, the Middle East and Southeastern Europe, American councils strives to expand dialog among students, scholars, educators and professionals for the advancement of learning and mutual respect in the diverse communities and societies in which we work.

Collaborative Research Grants in the Humanities

Subjects: Humanities including such disciplines as anthropology, modern and classical languages, history, linguistics, literature, jurisprudence, philosophy, political science, archaeology, comparative religion, sociology and ethics.
Purpose: To provide fellowships for humanities research in Eastern Europe and Eurasia.
Eligibility: Scholars who hold a PhD degree or other relevant terminal degree. Please see the website for further details regarding eligibility http://www.americancouncils.org/programDetail.php?program_id=NTc.
Level of Study: Postdoctorate, Research
Type: Grant
Value: $50,400 in NEH funds for a grant period of from six to twelve months nor more than $25,200 in NEH funds for a grant period of four to five months
Frequency: Annual
Country of Study: Eastern Europe and Eurasia
Application Procedure: Proposals must include plans to work with at least one collaborator in the field. Especially encouraged are applications with a strong regional focus and the potential to broaden and strengthen international academic linkages beyond the traditional centres such as Moscow, St Petersburg, Warsaw and Prague.
Closing Date: February 15th
Funding: Foundation

Contributor: National Endowment for the Humanities (NEH)
Additional Information: All applications will receive consideration without regard to any non-merit factor such as race, colour, religion, sex, sexual orientation, national origin, marital status, age, political affiliation or disability.

For further information contact:

1776 Massachusetts Avenue, NW, Suite 700, Washington, DC, 20036, United States of America

Title VIII Combined Research and Language Training Program

Subjects: Humanities and social sciences.
Purpose: The American Councils Combined Research and Language Training (CRLT) Program serves graduate students, postdoctoral scholars and faculty who, in addition to support for research in Eurasia, require supplemental language instruction.
Eligibility: Applicants must be US citizens or permanent residents. All competitions for funding are open and merit based. Applicants must be scholars in humanities and social sciences who have attained at least an intermediate level of proficiency in Russian or their proposed host-country language, typically students at relatively early stages of their dissertation research. However, participants may be at more advanced stages in their careers and applications from established scholars seeking to develop their proficiency in new languages are welcome.
Level of Study: Doctorate, Graduate, Postdoctorate, Postgraduate, Predoctorate, Research
Type: Fellowship
Value: $5,000–25,000.
Length of Study: 3–9 months
Frequency: Annual
Application Procedure: Applicants must submit a 2 to 3 page research proposal and bibliography, curriculum vitae, archive lists (if relevant), 1 page research synopsis in the host-country language, application form, copy of the inside page of their passport and 2 letters of recommendation from colleagues, professors or other qualified persons who are familiar with the applicant's work. At least one letter of recommendation must directly address the applicant's language skills and ability to conduct research in the host country. Please see the website for further details http://researchfellowships.americancouncils.org/researchscholar.
Closing Date: October 1st
Funding: Government
Contributor: Programme for the Study of Eastern Europe and the Independent States of the former Soviet Union (Title VIII), US Department of State
Additional Information: Programmes are available in Central Asia, Russia, South Caucasus, Ukraine and Moldova. A wide range of topics receive support each year, all funded research must contribute to a body of knowledge enabling the US to better understand the region and formulate effective policies within it. All applicants should clearly describe the policy-relevance of their work, be it in anthropology, literature, history, international relations, political science or some other field. Applications sent through fax or email will not be considered.

Title VIII Research Scholar Program

Subjects: Humanities and social sciences.
Purpose: To provide full support for graduate students, independent scholars and faculty seeking to conduct research for 3 to 9 months in Central Asia, Russia, South Caucasus, Ukraine and Moldova.
Eligibility: Applicants must be US citizens or permanent residents. All competitions for funding are open and merit based.
Level of Study: Graduate, Postdoctorate, Postgraduate, Predoctorate, Research
Type: Fellowship
Value: $5,000–25,000
Length of Study: 3–9 months
Frequency: Annual
Application Procedure: Applicants must submit a 2 to 3 page research proposal and bibliography, curriculum vitae, archive lists (if relevant), 1 page research synopsis in the host-country language, application form, copy of the inside page of their passport and 2 letters of recommendation from colleagues, professors or other qualified persons who are familiar with the applicant's work. At least 1 letter of recommendation must directly address the applicant's language skills and ability to conduct research in the host country. Please see the website for further details http://researchfellowships.americancouncils.org/researchscholar.
Closing Date: October 1st
Funding: Government
Contributor: Programme for the Study of Eastern Europe and the Independent States of the former Soviet Union (Title VIII), US Department of State
Additional Information: A wide range of topics receive support each year, all funded research must contribute to a body of knowledge enabling the US to better understand the region and formulate effective policies within it. All applicants should clearly describe the policy-relevance of their work, be it in anthropology, literature, history, international relations, political science, or some other field. Applications sent through fax or email will not be considered. Title VIII South East European Research Program is part of this program.

Title VIII Southeast European Language Training Program

Subjects: Literature (linguistics) and cultural studies
Purpose: The American Councils Southeast European Language training program offers academic year, semester and summer programs for independent language study in Albania, Bosnia-Herzegovina, Bulgaria, Croatia, Macedonia, Montenegro, Romania, Kosovo and Serbia.
Eligibility: Open to students at the MA and PhD level, as well as postdoctoral scholars and faculty who have at least elementary language skills. Applicants must be US citizens or permanent residents.
Level of Study: Graduate, Research, Doctorate, Postdoctorate, Postgraduate, Predoctorate
Type: Fellowship
Value: Varies
Length of Study: 1–9 months
Frequency: Annual
Application Procedure: Applicants must plan to study for at least one month in the region. Study trips for periods of 4–9 months are particularly encouraged. Applicants should explain how their plans for language-study support their overall research goals. A wide range of interests and research goals have received support each year, all funded research must contribute to a body of knowledge enabling the US policy makers to better understand the region.
Closing Date: October 1st
Funding: Government
Contributor: Programme for the Study of Eastern Europe and the Independent States of the former Soviet Union (Title VIII), US Department of State
Additional Information: Fellowships provide: full tuition at major university or educational institution in Southeastern Europe, international round trip airfare from the fellow's home city to host city, a monthly living and housing stipend, health insurance of upto $50,000 per accident or illness, Visa support as necessary, graduate-level academic credit through Bryn Mawr College for programmes providing 7 weeks or more of full-time instruction, ongoing logistical support from American Councils offices throughout the region. Applications sent through fax or email will not be considered.

Title VIII Special Initiatives Research Fellowship program

Subjects: Policy-relevant research in Armenia, Azerbaijan, Georgia, Kazakhstan, Kyrgyzstan, Tajikistan and Turkmenistan.
Purpose: To financially support innovative programmes in scholarly research.
Eligibility: Open to permanent residents or citizens of the United States who hold a PhD in a policy-relevant field. Please see the website for further details regarding eligibility http://www.americancouncils.org/programDetail.php?program_id = NzA = .
Level of Study: Doctorate, Postdoctorate, Research
Type: Fellowships
Value: US$35,000
Length of Study: 4–9 months overseas
Frequency: Annual
Application Procedure: Applicants can download the application form from the website. The completed application form must be

submitted along with a research proposal, research bibliography, a curriculum vitae, copy of the inside page of passport and 2 letters of recommendation.
Closing Date: September 30th
Funding: Government
Contributor: Programme for the Study of Eastern Europe and the Independent States of the former Soviet Union (Title VIII), US Department of State
Additional Information: Applications sent through fax or email will not be considered.

THE AMERICAN DENTAL ASSOCIATION FOUNDATION (ADA)

211, East Chicago Avenue, Chicago, IL, 60611-2678, United States of America
Tel: (1) 312 440 2500
Fax: (1) 312 440 3526
Email: adaf@ada.org
Website: www.ada.org
Contact: The Director

The purpose of the ADA Foundation Charitable Assistance Programs is to provide a measure of financial assistance to individuals who have financial hardship, whether due to educational needs, chemical dependency, disability or disaster.

ADA Foundation Allied Dental Student Scholarships

Subjects: Dental hygiene, dental assisting and dental laboratory technology.
Purpose: To defray study expenses including tuition fees, books and living expenses.
Eligibility: Open to citizens of the United States of America only. Applicants must either be entering their 1st year (dental assisting) or final year (dental laboratory technology and dental hygiene). Applicants must have a minimum grade point average of 3.0 based on a 4.0 scale and show financial need of at least US$1,000.
Level of Study: Professional development
Type: Scholarship
Value: Up to $135,000 total funding.
Frequency: Annual
Country of Study: United States of America
No. of awards offered: Approx. 54 scholarships
Application Procedure: Application forms are available from the dental hygiene, dental laboratory technology and dental assisting programme directors, and are distributed by school officials. Application forms must be original, typed, completed and signed with the assistance of school officials. Applicants must submit a completed application form, including the Academic Achievement Record Form and Financial Needs Assessment Form signed by school officials, two typed reference forms sealed and noted on the back of the envelopes by the referees and a typed biographical sketch.
Closing Date: October 10th
Funding: Private
Contributor: The ADA Foundation
No. of awards given last year: 15 for dental hygiene, 10 for dental assisting and 5 for dental laboratory technology
Additional Information: This scholarship is not renewable. Please see the website for further details http://www.ada.org/applyforassistance.aspx#adaf.

AMERICAN DIABETES ASSOCIATION (ADA)

1701 North Beauregard Street, Alexandria, VA, 22311, United States of America
Tel: (1) 703 549 1500, ext. 2362
Fax: (1) 703 549 1715
Email: grantquestions@diabetes.org
Website: http://professional.diabetes.org/grants

The American Diabetes Association (ADA) is the nation's leading non-profit health organization providing diabetes research information and advocacy. The mission of the organization is to prevent and cure diabetes, and to improve the lives of all people affected by diabetes. To fulfil this mission, the ADA funds research, publishes scientific findings and provides information and other services to people with diabetes, their families, healthcare professionals and the public.

ADA Career Development Awards

Subjects: Diabetes-related research.
Purpose: To allow exceptionally promising new investigators to conduct research.
Eligibility: Career Development Award applicants must be four to seven years out of their post-doctoral or clinical fellowship in order to apply. Applicants must hold an assistant professorship within his/her institution, or will be promoted to this position upon receipt of the award. Applicant must have demonstrated the ability to conduct research independently of their former mentor by appearing as the senior or corresponding author on at least one previous publication relevant to the grant topic, preferably without mentor as co-author.
Level of Study: Postdoctorate, Professional development
Type: Research grant
Value: Awards are up to $150,000 per year for up to five years, plus 15% allowable indirect costs. Additionally, applicants may request a $25,000 stipend for additional equipment for each of the first two years.
Length of Study: 5 years, non-renewable
Frequency: Bi-annual
Country of Study: United States of America
No. of awards offered: Varies, depending on funds available
Application Procedure: Applicants must write for details. All applications must be submitted online via the website www.diabetes.org/research.
Closing Date: January 15th
Funding: Foundation, individuals, private
Additional Information: Each year of funding, after the first, is contingent upon approval by the ADA of the recipient's research progress report, and the availability of funds.

ADA Junior Faculty Award

Subjects: Diabetes.
Purpose: To support investigators who are establishing their independence as diabetes researchers.
Eligibility: At the time of application, Junior Faculty Award applicants must be either: 1) senior post-doctoral or clinical fellows (more than 3 years of research experience since doctoral degree) and will receive their first full-time faculty/staff position by the start date of the award, or 2) junior faculty holding any level of faculty appointment up to and including Assistant Professor. Applicants with more than 10 years research experience beyond conferral of their doctoral degree are not eligible for this award.
Level of Study: Postdoctorate, Professional development
Type: Research grant
Value: Up to US$120,000 per year, and up to 15 per cent for indirect costs plus up to US$10,000 per year towards repayment of the principal on loans for a doctoral degree such as the MD or PhD.
Length of Study: 3 years
Frequency: Bi-annual
Country of Study: United States of America
No. of awards offered: Varies, depending on funds available
Application Procedure: Applicants must write for details. All applications must be submitted online via the website at www.diabetes.org/research. Please see the website for details http://professional.diabetes.org/Diabetes_Research.aspx?cid=89696.
Closing Date: January 15th
Funding: Foundation, individuals, private

ADA Mentor-Based Postdoctoral Fellowship Program

Subjects: Diabetes.
Purpose: To support the training of scientists in an environment most conducive to beginning a career in research.
Eligibility: There are no citizenship requirements for the Fellow. However, the investigator must be a citizen of the United States of America or a permanent resident, and must also hold an appointment at a United States of America research institution and have sufficient research support to provide an appropriate training environment for the Fellow. The Fellow selected by the investigator must hold an MD or PhD degree and must not be serving an internship or residency during the fellowship period. The Fellow must not have more than 3

years of postdoctoral research experience in the field of diabetes or endocrinology at the commencement of this fellowship.
Level of Study: Postdoctorate
Type: Fellowship
Value: Up to US$45,000 per year
Length of Study: Up to 4 years
Frequency: Annual
Country of Study: United States of America
No. of awards offered: Varies, depending on funds available
Application Procedure: Applicants must complete an application form. All applications must be submitted online via the website www.diabetes.org/research. Contact at grantquestions@diabetes.org for submitting applications.
Closing Date: January 15th for July 1st funding
Funding: Foundation, individuals, private

ADA Research Awards

Subjects: Aetiology and pathophysiology of diabetes.
Purpose: To assist investigators, new or established, who have a particularly novel and exciting idea for which they need support.
Eligibility: Open to citizens of the United States of America and permanent residents or those who have applied for permanent resident status, who have MD or PhD degrees, or, in the case of other health professions, an appropriate health or science-related degree. Applicants must hold full-time faculty positions or the equivalent at university-affiliated institutions within the United States of America and its possessions.
Level of Study: Research
Type: Research grant
Value: US$20,000–100,000 per year, for a maximum of 3 years, of which a maximum of US$20,000 can be used for principal investigator salary support, and up to 15 per cent for indirect costs. Each year of funding after the first is contingent upon approval by the ADA of the recipient's research progress report, and the availability of funds
Length of Study: Up to 3 years
Frequency: Bi-annual
Country of Study: United States of America
No. of awards offered: Varies, depending on funds available
Application Procedure: All applications must be submitted online via the website at www.diabetes.org/research
Closing Date: January 15th for July 1st funding and July 15th for January 1st funding
Funding: Foundation, individuals, private

Clinical Scholars Program

Subjects: Diabetes.
Purpose: To produce leaders in the fields of research, teaching and patient care by supplying clinicians-in-training the opportunity to contribute to the process of discovery in diabetes research laboratories/clinics. The Clinical Scholars Program will supply a unique opportunity to effectively integrate medical students into the process of discovery with an emphasis on patient-oriented research experiences.
Eligibility: Open to institutions within the United States of America and its possessions. The application must be initiated by the student, and the student must have a qualified sponsor. The student must have completed at least 1 year of medical school and the sponsor must hold a faculty position within an accredited medical school in the United States of America and be a citizen of the United States of America or a permanent resident.
Level of Study: Doctorate, Postgraduate
Type: Scholarship
Value: $30,000 per year. Applicants may request up to $20,000 for the student's stipend and up to $10,000 for lab expenses.
Length of Study: 1 year
Frequency: Annual
Country of Study: United States of America
No. of awards offered: Varies
Application Procedure: Applicants must write for details. All applications must be submitted online via the website at www.diabetes.org/research.
Closing Date: January 15th for July 1st funding
Funding: Foundation, individuals, private

AMERICAN FEDERATION FOR AGING RESEARCH (AFAR)

55 West 39th Street, 16th Floor, New York, NY, 10018, United States of America
Tel: (1) 212 703 9977
Fax: (1) 212 997 0330
Email: grants@afar.org or info@afar.org
Website: www.afar.org
Contact: Director, Grant Programs

The American Federation for Aging Research (AFAR) is a leading non-profit organization supporting biomedical aging research. Since its founding in 1981, AFAR has provided approximately US$124 million to more than 2,600 new investigators and students conducting cutting-edge biomedical research on the aging process and age-related diseases. The important work AFAR supports leads to a better understanding of the aging process and to improvements in the health of all Americans as they age.

AFAR Research Grants

Subjects: Biomedical and clinical topics. Basic mechanisms of aging.
Purpose: To help junior faculty to carry out research that will serve as the basis for longer term research efforts.
Eligibility: Open to junior faculty with an MD or PhD degree.
Level of Study: Postdoctorate, Research
Type: Research grant
Value: US$100,000
Length of Study: 1–2 years
Frequency: Annual
Country of Study: United States of America
No. of awards offered: Approx. 15
Application Procedure: Applicants must complete and return the application by the annual deadline. These are available from the website.
Closing Date: December 17th
Funding: Foundation, private
Contributor: AFAR

The Cart Fund, Inc.

Subjects: Alzheimers disease.
Purpose: To encourage exploratory and developmental Alzheimers disease research projects.
Eligibility: Open to applicants from within the United States whose projects have the potential to advance biomedical research.
Level of Study: Research
Value: US$250,000
Length of Study: Up to 2 years
Frequency: Annual
No. of awards offered: 1
Application Procedure: A letter-of-intent that includes sufficient details of the study must be submitted.
Closing Date: Jaunary 18st
Funding: Private
Contributor: American Federation of Aging Research and The Rotary CART Fund

Ellison Medical Foundation/AFAR Postdoctoral Fellows in Aging Research Program

Subjects: The fundamental mechanisms of aging.
Purpose: The program addresses the current concerns about an adequate funding base for postdoctoral fellows (both MDs and PhDs) who conduct research in the fundamental mechanisms of aging.
Eligibility: Postdoctoral fellows at all levels of training are eligible.
Level of Study: Postdoctorate, Research
Type: Fellowship
Value: US$47,114–55,670
Length of Study: 1 year
Frequency: Annual
No. of awards offered: Up to 15
Closing Date: December 27th. Deadline for the LOI is October 17th
Funding: Foundation
Contributor: The Ellison Medical Foundation

Additional Information: Candidates who submitted a Letter-of-Intent by the deadline date will be invited to submit a full application by November 1st.

The Glenn/AFAR Breakthroughs in Gerontology Awards

Subjects: Projects that focus on genetic controls of aging and longevity, on delay of aging by pharmacological agents or dietary means, or which elucidate the mechanisms by which alterations in hormones, antioxidant defenses, or repair processes promote longevity are all well within the intended scope of this competition. Projects that focus instead on specific diseases or on assessment of health care strategies will receive much lower priority, unless the research plan makes clear and direct connections to fundamental issues in the biology of aging. Studies of invertebrates, mice, human clinical materials or cell lines are all potentially eligible for funding. Although preliminary data are always helpful for evaluating the feasibility of the experiments proposed, the emphasis in review will be on creativity and the likelihood that the findings will open new vistas and approaches to aging research that might merit intensive follow up studies.
Purpose: To provide timely support to a pilot research program
Eligibility: To be eligible, applicants must at the time they submit their proposal be full time faculty members at the rank of Assistant Professor or higher. A strong record of independent publication beyond the postdoctoral level is a requirement. Applications from individuals not previously engaged in aging research are particularly encouraged, as long as the research proposals show high promise for leading to important new discoveries in biological gerontology. Applicants who are employees in the NIH Intramural programme are not eligible. The proposed research must be conducted at any type of non-profit setting in the United States.
Level of Study: Research
Value: Up to US$200,000
Length of Study: 2 years
Frequency: Dependent on funds available
No. of awards offered: 2
Closing Date: December 17th
Funding: Private
Contributor: The Glenn Foundation for Medical Research
No. of awards given last year: 2

The Julie Martin Mid-Career Award in Aging Research

Subjects: Basic aging research.
Purpose: To encourage outstanding mid-career scientists.
Eligibility: Open to mid-career (Associate Professors) scientists whose research could lead to novel approaches to aging, and also whose research is high risk.
Level of Study: Research
Type: Award
Value: US$500,000 (at the level of $125,000 per year), in addition up to US$50,000 may be requested for indirect costs.
Length of Study: 4 years
Frequency: Annual
No. of awards offered: 2
Closing Date: December 17th
Funding: Foundation
Contributor: Ellison Medical Foundation

Medical Student Training in Aging Research (MSTAR) Program

Subjects: Medical science, geriatrics
Purpose: This programme provides medical students, early in their training, with an enriching experience in aging-related research and geriatrics.
Eligibility: Any allopathic or osteopathic medical student in good standing, who will have successfully completed 1 year of medical school at a US institution by June. Evidence of such likelihood must be provided at the time of application. Applicants must be citizens or non-citizen nationals of the United States, or must have been lawfully admitted for permanent residence (i.e., in procession of a currently valid Alien Registration Receipt Card I-551, or some other legal verification of such status.) Individuals on temporary or student visas and individuals holding PhD, MD, DVM, or equivalent doctoral degrees in the health sciences are not eligible. The NIA and other sponsoring organizations have a strong interest in continuing to diversify the research workforce committed to advancing the fields of aging and geriatric research. Therefore, students who are members of ethnic or racial groups underrepresented in these fields, students with disabilities, or students whose background and experience are likely to diversify the research or medical questions being addressed, are encouraged to apply.
Level of Study: Graduate, Medical student
Type: Scholarship
Value: A stipend of US$1,748 per month
Length of Study: 8-12 week
Frequency: Annual
No. of awards offered: 130
Application Procedure: Applications can be completed through www.afar.org
Closing Date: January 31st
Funding: Government, private

Paul Beeson Career Development Award in Aging Research

Subjects: Medical sciences.
Purpose: To bolster the current severe shortage of academic physicians who have the combination of medical, academic and scientific training relative to caring for other people.
Eligibility: Applicants must be citizens of the United States of America or permanent residents, be full-time faculty members with clear potential for long-term faculty appointments. To be eligible a candidate must (1) have clinical doctoral degree (e.g. MD, DO, DDS) or its equivalent and have completed clinical training; (2) commit at least 75 per cent of his/her full-time professional effort to the goals of this award; (3) be a US citizen or non-citizen national of the United States or a permanent resident alien; (4) be at a for-profit or non-profit organization, public or private institution (such as universities, colleges, hospitals and laboratories), units of state and local governments or eligible agencies of the federal government provided the demonstrated environment has a commitment to the geriatric population and capacity to support the scholar's career development.
Level of Study: Professional development, Research
Type: Grant
Value: US$600,000–800,000
Length of Study: 3–5 years
Frequency: Annual
Country of Study: United States of America
No. of awards offered: 5–7
Application Procedure: Applicants must complete and return the application by the annual deadline. Application forms can be found at www.beeson.org.
Closing Date: December 6th
Funding: Government, private
Contributor: The National Institute on Aging, John A Hartford Foundation, Commonwealth Fund, Atlantic Philanthropies and Starr Foundation

For further information contact:

Website: www.beeson.org

The Rosalinde and Arthur Gilbert Foundation/AFAR New Investigator Awards in Alzheimer's Disease

Subjects: Neurosciences, research related to Alzheimer's disease.
Purpose: To support important research and encourage junior investigators in the United States and Israel to pursue research and academic careers in the neurosciences, and Alzheimer's disease in particular.
Eligibility: Open to applicants in their four years of a junior faculty appointment and establishing independent research activities.
Level of Study: Research
Type: Grant
Value: US$100,000 each
Length of Study: 1–2
Frequency: Annual
No. of awards offered: 5
Application Procedure: Check the website for instruction sheet and application.
Closing Date: December 15th
Funding: Private
Contributor: The Rosalinde and Arthur Gilbert Foundation

For further information contact:

Website: www.afar.org

AMERICAN FOUNDATION FOR AGING RESEARCH (AFAR)

Dept. of Biological Sciences, University at Albany, 1400 Washington Avenue, Albany, NY, 12222, United States of America
Tel: (1) 518 437 4448
Fax: (1) 919 515 2047
Email: afar@agingresearchfoundation.org
Website: www.ncsu.edu/project/afar/
Contact: Dr Paul F Agris, President

The American Foundation for Aging Research (AFAR) promotes and supports researchand education that will elucidate the basic processes involved in the biology of aging and age associated disease, by awarding scholarships and fellowships to young, motivated scientists.

Dr Vincent Cristofalo Memorial Fund, Cecille Gould Memorial Fund Award in Cancer Research, Richard Shepherd Fellowship, Agris-Rokaw Fellowship Award

Subjects: Aging and cancer research.
Purpose: To encourage young people to pursue research in age-related health problems and the biology of aging.
Eligibility: Open to graduates enrolled in degree programmes such as MS, PhD, MD or DDS at institutions within the United States of America. Must be conducting cellular, molecular or genetic research on aging or age-related illnesses such as cancer, diabetes or Alzheimer's. Sociological and psychological research is not accepted in these programmes.
Level of Study: Doctorate, Graduate, Postgraduate
Type: Fellowship
Value: US$1,000 per semester or summer
Length of Study: Between 4 months and 1 year
Frequency: Annual
Study Establishment: Educational institutions
Country of Study: United States of America
No. of awards offered: 5–10
Application Procedure: Applicants must undertake the two levels of review: a pre-application form to determine eligibility, and a full application. Applicants should submit a request for a pre-application. There is no charge for the submission of the pre-application or the full application.
Closing Date: There is no deadline
Funding: Private
No. of awards given last year: 15
No. of applicants last year: 115
Additional Information: Pre-applications are submitted through the AFAR website. AFAR is a national, tax-exempt, non-profit, educational and scientific charity but not affiliated with North Carolina State University, or any other institution.

AMERICAN FOUNDATION FOR PHARMACEUTICAL EDUCATION (AFPE)

2107 Wilson Boulevard, Suite 700, Arlington, VA, 22201 3042, United States of America
Tel: (1) 703 875 3095
Email: info@afpenet.org
Website: www.afpenet.org
Contact: Administrative Assistant

The mission of the AFPE is to advance and support pharmaceutical sciences education at US schools and colleges of pharmacy.

AAPS/AFPE Gateway to Research Scholarships

Subjects: Pharmaceutics.
Purpose: To encourage graduates from any discipline to pursue a PhD in a pharmacy graduate programme.
Eligibility: Open to students who are enrolled in the last 3 years of a Bachelor of science or PharmD programme at a United States school or college of pharmacy, or Baccalaureate degree programme in a related field of scientific study at any college and have completed at least one year of the degree program. Candidates must have a demonstrated interest in, and potential for, a career in any of the pharmaceutical sciences and be enrolled for at least one full academic year following the award of the scholarship. United States citizenship or permanent resident status is not required.
Level of Study: Postgraduate, Professional development
Type: Scholarship
Value: US$5,000. No less than $4,000 is provided as a student stipend for a full calendar year, $500 is provided to attend an AAPS Annual Meeting, and up to $500 may be used by the sponsoring faculty member in direct support of the research effort.
Frequency: Annual
Country of Study: United States of America
No. of awards offered: 3
Application Procedure: Applicants must write for details.
Closing Date: January 23rd
Contributor: American Association of Pharmaceutical Scientists, American Society of Health-System Pharmacists, ASHP Foundation, AstraZeneca Pharmaceuticals, GlaxoSmithKline, Johnson & Johnson Pharmaceutical Research & Development, Novartis Pharmaceuticals, Ortho-McNeil Janssen Scientific Affairs, Pfizer, Inc., Procter & Gamble, United States Pharmacopeia (USP), Wyeth

AFPE Clinical Pharmacy Post-PharmD Fellowships in the Biomedical Research Sciences

Subjects: Pharmacology including topics such as cost benefit and cost effectiveness of pharmaceuticals, the impact of current or future legislation on drug innovation and healthcare in the nation, the economics of healthcare and the quality of life in changing patterns of healthcare delivery systems, the contribution of the pharmaceutical industry, the economic impact of research and new drugs, and healthcare cost containment issues.
Purpose: Outstanding Pharm.D. graduates who have completed one or more post-doctoral residencies or fellowships.To obtain advanced education and training in relevant areas of the biomedical and related basic sciences in order to become competent clinical scientists.
Eligibility: Open to all pharmacy faculty members who have a strong record of research.
Level of Study: Doctorate, Postdoctorate
Type: Fellowship
Value: US$27,500 per year
Length of Study: 1–2 years
Frequency: Annual
Study Establishment: An Institute of Higher Education
Country of Study: United States of America
No. of awards offered: 1
Application Procedure: Applicants must complete an application form and should write for details.
Closing Date: February 15th

AFPE Gateway to Research Scholarship Program

Subjects: Pharmacology.
Purpose: To encourage individuals in a pharmacy college to pursue a PhD within a pharmacy college.
Eligibility: Open to students who are enrolled in the last 3 years of a Bachelor of Science or PharmD programme at a United States school or college of pharmacy, or Baccalaureate degree programme in a related field of scientific study at any college. Candidates must have a demonstrated interest in, and potential for, a career in any of the pharmaceutical sciences and will be enrolled for at least one full academic year following the award of the scholarship. United States citizenship or permanent resident status is required.
Level of Study: Postgraduate, Professional development
Type: Scholarship
Value: Up to US$5,000. No less than $4,000 is provided as a student stipend for a full calendar year. No more than $1,000 may be used by the sponsoring faculty member in direct support of the research effort.
Frequency: Annual
Study Establishment: An approved college of pharmacy
Country of Study: United States of America
No. of awards offered: Up to 11
Application Procedure: Applicants must write for details.
Closing Date: January 23rd

AFPE Predoctoral Fellowships

Subjects: Any of the pharmaceutical sciences, including pharmaceutics, pharmacology, manufacturing pharmacy and medicinal chemistry.
Purpose: To offer fellowship support leading to a PhD degree.
Eligibility: Open to students who have completed at least three semesters of graduate study and who have no more than three years remaining to obtain a PhD degree in a graduate programme in the pharmaceutical sciences administered by, or affiliated with, a United States school or college of pharmacy. The award is also open to students enrolled in joint PharmD and PhDs, if a PhD degree will be awarded within three additional years. Applicants must be United States citizens or permanent residents.
Level of Study: Doctorate, Postdoctorate, Postgraduate
Type: Fellowships
Value: Predoctoral Fellowship in Pharmaceutical Science awards at $11,000 per year; new and renewal Predoctoral Fellowship in Pharmaceutical Science awards at $6,500 per year and Clinical Pharmaceutical Science Fellowships at $6,500 per year
Length of Study: 1 year, renewable for 2 additional years
Frequency: Annual
Study Establishment: An appropriate university
Country of Study: United States of America
No. of awards offered: Up to 58 (4 new and 44 renewal Predoctoral Fellowships in pharmaceutical science and 10 new Clinical Pharmaceutical Science Fellowships)
Application Procedure: Applicants must write for details.
Closing Date: March 1st
Additional Information: Succesful applicants are eligible to apply for one renewal for another full year (12 months) if you will be enrolled full-time in your PhD program for another full academic year after AFPE Fellowship ends.

AMERICAN FOUNDATION FOR SUICIDE PREVENTION (AFSP)

120 Wall Street, 22nd Floor, New York, NY 10005, United States of America
Tel: (1) 212 363 3500
Fax: (1) 212 363 6237
Email: inquiry@afsp.org
Website: www.afsp.org
Contact: Jonathan Dozier-Ezell, Grants Manager

The American Foundation for Suicide Prevention (AFSP) is the only national non-profit organization exclusively dedicated to understanding and preventing suicide through research and education, and to reaching out to people with mood disorders and those affected by suicide.

AFSP Distinguished Investigator Awards

Subjects: The clinical, biological or psychosocial aspects of suicide.
Purpose: Awarded to investigators at the level of associate professor or higher with a proven history of research in the area of suicide.
Level of Study: Research
Type: Grant
Value: Up to $50,000 per year for a one or two-year period
Length of Study: 1–2 years
Frequency: Annual
Country of Study: Worldwide
No. of awards offered: Varies
Application Procedure: Applicants should consult the website or contact the organization for full details. Application form must be completed. Applications accepted via email only at grants@afsp.org.
Closing Date: November15th
Funding: Private
No. of applicants last year: 7
Additional Information: Decisions regarding awards are made in May and funding begins in October.

AFSP Pilot Grants

Subjects: The clinical, biological or psychosocial aspects of suicide.
Level of Study: Research
Type: Grant
Value: Up to $15,000 per year for a two-year period, or $30,000 for one year.
Length of Study: 1–2 years
Frequency: Annual
Country of Study: Worldwide
No. of awards offered: Varies
Application Procedure: Applicants should consult the website or contact the organization for full details. Application form must be completed. Applications accepted via email only at grants@afsp.org.
Closing Date: November 15th
Funding: Private
No. of awards given last year: 4
No. of applicants last year: 31
Additional Information: Decisions regarding awards are made in May and funding begins in October.

AFSP Postdoctoral Research Fellowships

Subjects: The clinical, biological or psychosocial aspects of suicide.
Purpose: Postdoctoral Research Fellowships are training grants designed to enable young investigators toqualify for independent careers in suicide research.
Eligibility: Applicants must have received a PhD 3 years prior to application for the fellowship.
Level of Study: Postdoctorate, Research
Type: Fellowship
Value: US$104,000
Length of Study: 2 years
Frequency: Annual
Country of Study: Worldwide
No. of awards offered: Varies
Application Procedure: Applicants should consult the website or contact the organization. Application form must be completed. Applications accepted via email only at grants@afs.org.
Closing Date: November 15th
Funding: Private
No. of awards given last year: 2
No. of applicants last year: 11
Additional Information: Decisions regarding awards are made in May and funding begins in July.

AFSP Standard Research Grants

Subjects: The clinical, biological or psychosocial aspects of suicide.
Purpose: Awarded to individual investigators at any level.
Level of Study: Research, Postdoctorate
Type: Grant
Value: Up to $45,000 per year for a one or two-year period
Length of Study: 2 years
Frequency: Annual
Country of Study: Worldwide
No. of awards offered: Varies
Application Procedure: Applicants should consult the website or contact the organization. Application form must be completed. Applications accepted via email only at grants@afsp.org.
Closing Date: November 15th
Funding: Private
No. of awards given last year: 9
No. of applicants last year: 67
Additional Information: Decisions regarding awards are made in May and funding begins in October.

AFSP Young Investigator Award

Subjects: The clinical, biological or psychosocial aspects of suicide.
Purpose: Awarded to those at the level of assistant professor or lower.
Level of Study: Research
Type: Award
Value: Up to $37,500 per year for a one or two-year period. Plus a mentor's fee of $5,000 per year.
Length of Study: 2 years
Frequency: Annual
Country of Study: Worldwide
No. of awards offered: Varies
Application Procedure: Applicants should consult the website or contact the organization. Application form must be completed. Applications accepted via email only at grants@afsp.org.

Closing Date: November 15th
Funding: Private
No. of awards given last year: 5
No. of applicants last year: 28
Additional Information: Decisions regarding awards are made in May and funding begins in October. Investigators should be at the level of assistant professor or lower.

AMERICAN FOUNDATION FOR THE BLIND (AFB)

1000 Fifth Ave., Suite 350, Huntington, WV, 25701, United States of America
Tel: (1) 304 523 8651
Fax: (1) 800 232 5463
Email: afbinfo@afb.net
Website: www.afb.org; www.afb.org/scholarships.asp

The American Foundation for the Blind (AFB) is a national non-profit organisation that expands possibilities for people with vision loss. AFB's priorities include broadening access to technology; elevating the quality of information and tools for the professionals who serve people with vision loss; and promoting independent and healthy living for people with vision loss by providing them and their families with relevant and timely resources.

Karen D. Carsel Memorial Scholarship

Subjects: Literature or Music
Purpose: To provide financial help to blind students who want to further a career.
Level of Study: Postgraduate
Type: Scholarship
Value: US$500
Frequency: Annual
No. of awards offered: 1
Application Procedure: The candidate will have to submit the evidence of legal blindness, official transcripts, proof of acceptance into a programme, evidence of economic need, 3 letters of recommendation, and typewritten statement describing educational and personal goals, work experience, extra-curricular activities and how scholarship monies will be used.
Closing Date: March 31st

AMERICAN GEOPHYSICAL UNION (AGU)

2000 Florida Avenue, N.W., Washington, DC, 20009-1277, United States of America
Tel: (1) 202 462 6900
Fax: (1) 202 328 0566
Email: service@agu.org
Website: www.agu.org
Contact: Director, Outreach and Research Support

The American Geophysical Union (AGU) is an international scientific society with more than 45,000 members, primarily research scientists, dedicated to advancing the understanding of Earth and space and making the results of the AGU's research available to the public.

F.L. Scarf Award

Subjects: Solar-planetary science.
Purpose: To award outstanding dissertation research that contributes directly to solar-planetary science.
Eligibility: Open to all candidates with a PhD (or equivalent) degree.
Level of Study: Doctorate
Value: US$1,000, a complimentary ticket for the SPA dinner, and a certificate
Frequency: Annual
No. of awards offered: 1
Application Procedure: Nominations to be sent to outreach administrator at AGU.
Closing Date: March 23rd
Contributor: The Space Physics and Aeronomy section of AGU
Additional Information: Awardee will have the opportunity to deliver an invited paper on the dissertation topic at appropriate SPA session at the upcoming AGU Fall Meeting.

For further information contact:

Tel: 1 202 777 7502
Email: leadership@agu.org; mgl.scarf.award@nrl.navy.mil; dwilliams@agu.org

Horton (Hydrology) Research Grant

Subjects: Hydrology including its physical, chemical or biological aspects, life sciences, physical sciences, social sciences, school of public affairs, school of law.
Purpose: To support research in hydrology and water resources.
Eligibility: There are no eligibility restrictions.
Level of Study: Postdoctorate
Type: Grant
Value: US$10,000 and related travel expenses ($500 U.S. /$1,000 international).
Length of Study: 1 year
Frequency: Annual
No. of awards offered: Varies
Application Procedure: Applicants must submit four copies of the application form, an executive summary, a statement of purpose, a detailed budget and two letters of recommendation. Applicants should contact the AGU for further details.
Closing Date: May 1st
No. of awards given last year: 2
No. of applicants last year: 27

The Mineral and Rock Physics Graduate Research Award

Subjects: Mineral and rock physics.
Purpose: To recognize outstanding contributions by young scientists.
Eligibility: Open to students who have completed their PhD.
Level of Study: Doctorate
Type: Award
Value: US$500, a certificate and public recognition at the annual Mineral and Rock Physics Reception at the AGU fall meeting
Frequency: Annual
No. of awards offered: Varies
Application Procedure: A letter of nomination along with a curriculum vitae, two supporting letters and 3 reprints or preprints of the nominee's work should be sent.
Closing Date: May 1st
Contributor: Mineral and Rock Physics community at AGU

For further information contact:

Department of Geology and Environmental Geoscience, Northern Illinois University, Davis Hall 312, Normal Road, DeKalb, IL, 60115, United States of America
Email: hwatson@niu.edu
Contact: Dr Heather C. Watson, Assistant Professor

AMERICAN HEAD AND NECK SOCIETY (AHNS)

AHNS, 11300 W. Olympic Boulevard, Suite 600, Los Angeles, CA, 90064, United States of America
Tel: (1) 310 437 0559
Fax: (1) 310 437 0585
Email: admin@ahns.info
Website: www.headandneckcancer.org
Contact: Joyce Hasper, Research Grants Enquiries

The purpose of the American Head and Neck Society (AHNS) is to promote and advance the knowledge of prevention, diagnosis, treatment and rehabilitation of neoplasms and other diseases of the head and neck.

AHNS Pilot Research Grant

Subjects: Diseases of the head and neck.
Purpose: To support students who wish to try a pilot project in head and neck-related research.
Eligibility: Open to residents and fellows in the junior faculty.

Level of Study: Doctorate, Postgraduate
Type: Award
Value: US$10,000
Length of Study: 1 year
Frequency: Annual
Study Establishment: A university in the United States of America
Country of Study: United States of America
No. of awards offered: 2
Closing Date: Letter of Intent Due December 15th, Applications due on January 16th
Funding: Private

AHNS Surgeon Scientist Career Development Award (with AAOHNS)

Subjects: Cancer and other diseases of the head and neck.
Purpose: To support research in the pathogenesis, pathophysiology, diagnosis, prevention or treatment of head and neck neoplastic disease.
Eligibility: Open to surgeons beginning a clinician-scientist career.
Level of Study: Postdoctorate
Type: Award
Value: US$35,000 per year (non-renewable)
Length of Study: 2 years
Frequency: Annual, Available only during odd numbered years
Study Establishment: A university in the United States of America
Country of Study: United States of America
No. of awards offered: 1
Application Procedure: The grants are reviewed through the Academy CORE (combined otolaryngologic research evaluation) process.
Closing Date: December 15th (letter of intent), January 16th (application submission).
Funding: Private
Additional Information: Submit your LOI early to gain advanced access to the full application. Forms are available through American Academy of Otolaryngology CORE

AHNS Young Investigator Award (with AAOHNS)

Subjects: Cancer and other diseases of the head and neck.
Purpose: To support research in neoplastic disease of the head and neck.
Eligibility: Candidate must be a member of AHNS.
Level of Study: Doctorate
Type: Award
Value: US$20,000 per year
Length of Study: Up to 2 years
Frequency: Annual
Study Establishment: A university in the United States of America
Country of Study: United States of America
No. of awards offered: 1
Application Procedure: The grants are reviewed through the Academy CORE (combined otolaryngologic research evaluation) process.
Closing Date: December 17th (letter of intent), application due on January 15th.
Funding: Private
Additional Information: Submit your LOI early to gain advanced access to the full application. Forms are available through American Academy of Otolaryngology CORE.

AHNS-ACS Career Development Award

Subjects: Diseases of the head and neck.
Purpose: To facilitate research in connection with career development.
Eligibility: Applicants must be a member or candidate member of ACS and AHNS. Applicants must be within 5 years of completion of training, and be full-time faculty member.
Level of Study: Postgraduate
Type: Award
Value: US$40,000 per year (non-renewable)
Length of Study: 2 years
Frequency: Annual
Study Establishment: A university in the United States of America
Country of Study: United States of America
Funding: Private

AMERICAN HEALTH ASSISTANCE FOUNDATION (AHAF)

22512 Gateway Center Drive, Clarksburg, MD, 20871, United States of America
Tel: (1) 800 437 2423
Fax: (1) 301 258 9454
Email: info@ahaf.org
Website: www.ahaf.org
Contact: Dr Kara Summers, Grants Coordinator

The American Health Assistance Foundation (AHAF) is a non-profit charitable organization that funds research and public education on age related and degenerative diseases including: Alzheimer's disease, macular degeneration, glaucoma and heart and stroke diseases. The organization also provides emergency financial assistance to Alzheimer's disease patients and their care givers.

AHAF Alzheimer's Disease Research Grant

Subjects: Neurology, biomedicine, biochemistry, biophysics, molecular biology and pharmacology.
Purpose: To enable basic research on the causes of and treatments for Alzheimer's disease.
Eligibility: The principal investigator must hold the rank of assistant professor or equivalent, or higher.
Level of Study: Postdoctorate, Doctorate, Postgraduate
Type: Grant
Value: $400,000 for standard awards; $150,000 for pilot awards, $100,000 for postdoctoral fellowship awards
Length of Study: 1–3 years
Frequency: Annual
Study Establishment: Non-profit institutions and organizations
Country of Study: Any country
No. of awards offered: Varies
Application Procedure: Applicants must complete an online application form. The current application form should be requested for each year or can be downloaded from the website.
Closing Date: October 19th
Funding: Private
No. of awards given last year: 14
No. of applicants last year: 89
Additional Information: The ADR program offers three types of awards: Standard Awards, Pilot Awards and Postdoctoral Fellowship Awards. Check website for closing date information.

AHAF Macular Degeneration Research

Subjects: Ophthalmology, biomedicine, biochemistry, biophysics, genetics, molecular biology and pharmacology.
Purpose: To enable basic research on the causes of, or the treatment for, macular degeneration.
Eligibility: The principal investigator must hold a tenure track or tenured position and the rank of assistant professor or higher.
Level of Study: Research
Type: Grant
Value: Up to US$100,000. Grants may be renewed on a competitive peer review basis.
Length of Study: 2 year
Frequency: Annual
Study Establishment: Non-profit institutions and organizations
Country of Study: Any country
No. of awards offered: Varies
Application Procedure: Applicants must complete an application form. The current application form should be requested for each year or can be downloaded from the website.
Closing Date: Letters of intent due July 11th of each year. Application due in November
Funding: Private
No. of awards given last year: 7
No. of applicants last year: 31

AHAF National Glaucoma Research

Subjects: Ophthalmology, biomedicine and pharmacology.
Purpose: To enable basic research on the causes of or treatments for glaucoma.

Eligibility: The principal investigator must hold the rank of assistant professor or equivalent, or higher.
Level of Study: Doctorate, Research
Type: Grant
Value: Up to US$100,000
Length of Study: 1–2 years
Frequency: Annual
Study Establishment: Non-profit institutions and organizations
Country of Study: Any country
No. of awards offered: Varies
Application Procedure: Applicants must complete an application form. The current application form should be requested for each year and can also be downloaded from the website.
Closing Date: October 26th
Funding: Private
No. of awards given last year: 9
No. of applicants last year: 28

AMERICAN HEART ASSOCIATION (AHA)

National Center, 7272 Greenville Avenue, Dallas, TX, 75231, United States of America
Tel: (1) 800 242 8721, 888 242 2453
Fax: (1) 214 706 1341
Email: Review.personal.info@heart.org; sessionsadmin@heart.org
Website: www.americanheart.org
Contact: Ms Juanita Morales, Manager

The American Heart Association (AHA) is a non-profit, voluntary health organization funded by private contributions. Its mission is to reduce disability and death from cardiovascular diseases and stroke. To support this goal, the Association has given more than US$2 billion to heart and blood vessel research since 1949.

AHA Fellowships

Subjects: Cardiovascular diseases, stroke, basic science, clinical, bioengineering/biotechnology and public health problems.
Purpose: To help students initiate careers in cardiovascular research by providing assistance and training in research activities broadly related to cardiovascular function and diseases.
Eligibility: Open to permanent residents or citizens of the United States. Also for exchange visitors under J-1, temporary worker under H-1, H-1B, O-1 visas. Canadian or Mexican citizens engaged in professional activities or student visa are also eligible.
Level of Study: Research, Predoctorate
Type: Fellowships
Value: Stipend of US$22,000 per year. Fringe benefit: $3,000.
Length of Study: 1 or 2 years (You may submit an application to compete for a third year of funding, if eligible.)
Frequency: Annual
Country of Study: United States of America
No. of awards offered: Varies
Closing Date: January 10th, next closing date is July 1st
Additional Information: Participation by women and minorities is encouraged. Visit our Web site in mid-March for information.

AMERICAN HERPES FOUNDATION

433 Hackensack Avenue, 9th Floor, Hackensack, NJ, 07601, United States of America
Tel: (1) 201 883 5852
Fax: (1) 201 342 7555
Email: IHMF@hbase.com
Website: www.herpes-foundation.org
Contact: Loretla A. Ponesse

American Herpes Foundation is a non-profit organization dedicated to improving the management of herpes virus infections. Our initiatives focus primarily on clinician education and awareness.

American Herpes Foundation Stephen L Sacks Investigator Award

Subjects: Research areas include HSV1 and 2, VZV, EBV, CMV, and HHV6 and 8.
Purpose: To recognize and encourage newer researchers who have completed significant research in the herpes virus area.
Eligibility: Residents, fellows or junior faculty up to the 5th year of faculty appointment are eligible.
Level of Study: Physicians-in-training or researchers-in-training
Type: Cash prize
Value: US$5,000 cash prize
No. of awards offered: 2
Application Procedure: Candidates must submit a letter of nomination, completed application form, curriculum vitae, biographical sketch and documentation of research.
Closing Date: Call for details
Funding: Private

AMERICAN HISTORICAL ASSOCIATION

400 A Street, S.E., Washington, DC, 20003-3889, United States of America
Tel: (1) 202 544 2422
Fax: (1) 202 544 8307
Email: info@historians.org
Website: www.historians.org
Contact: Matthew Keough, Executive Office Assistant

The American Historial Association (AHA) is a non-profit membership organization founded in 1884 and was incorporated by Congress in 1889 for the promotion of historical studies, the preservation of historial documents and artefacts and the dissemination of historical research.

Albert J Beveridge Grant

Subjects: The history of the United States of America, Latin America or Canada.
Purpose: To promote and honor outstanding historical writing and to support research in the history of the Western hemisphere.
Eligibility: Open to American Historical Association members only.
Level of Study: Doctorate, Postdoctorate, Postgraduate
Type: Grant
Value: A maximum of US$1,000
Frequency: Annual
Country of Study: Any country
No. of awards offered: Varies
Application Procedure: Applicants must apply online at www.historians.org/prizes/beveridgegrantinfo.htm
Closing Date: February 15th
Funding: Private
No. of awards given last year: 9
No. of applicants last year: 114

Bernadotte E Schmitt Grants

Subjects: The history of Europe, Asia and Africa.
Purpose: To support research in the history of Europe, Africa and Asia, and to further research in progress.
Eligibility: Open to American Historical Association members only.
Level of Study: Doctorate, Postdoctorate, Postgraduate
Type: Grant
Value: Up to US$1,000
Frequency: Annual
Country of Study: Any country
No. of awards offered: Varies
Application Procedure: Applicants must apply online at www.historians.org/prizes/schmittgrantinfo.htm
Closing Date: February 15th
Funding: Private
No. of awards given last year: 16
No. of applicants last year: 96

J Franklin Jameson Fellowship

Subjects: The collections of the Library of Congress.
Purpose: To support significant scholarly research for one semester in the collections of the Library of Congress by scholars at an early stage in their careers in history.
Eligibility: Applicants must hold a PhD degree or equivalent, must have received this degree within the past 7 years, and must not have published or had accepted for publication a book-length historical

work. The fellowship will not be awarded to complete a doctoral dissertation.
Level of Study: Postdoctorate
Type: Fellowship
Value: US$5,000
Length of Study: 1 semester
Frequency: Annual
Country of Study: United States of America
No. of awards offered: 1
Application Procedure: Applicants must refer to the website www.historians.org/prizes/jameson_fellowship.htm for instructions.
Closing Date: March 15th
Funding: Government, private

Littleton-Griswold Research Grant

Subjects: American legal history, law and society.
Purpose: To further research in progress.
Eligibility: Open to American Historical Association members only.
Level of Study: Doctorate, Postdoctorate, Postgraduate
Type: Research grant
Value: Up to US$1,000
Frequency: Annual
Country of Study: Any country
No. of awards offered: Varies
Application Procedure: Applicants must apply online at www.historians.org/prizes/littleton-griswaldgrantinfo.htm
Closing Date: February 15th
Funding: Private
No. of awards given last year: 30
No. of applicants last year: 6

Michael Kraus Research Grant

Purpose: To support research in colonial American history, especially the inter-cultural aspects of American and European interaction.
Eligibility: Open to American Historical Association members only.
Level of Study: Doctorate, Postdoctorate
Type: Grant
Value: Up to $1,000
Frequency: Annual
Application Procedure: Please visit www.historians.org/prizes/krausgrantinfo. Apply through website only.
Closing Date: February 15th
Funding: Private
No. of awards given last year: 3
No. of applicants last year: 20

AMERICAN INDIAN GRADUATE CENTER (AIGC)

3701, San Mateo NE, #200, Albuquerque, NM, 87110, United States of America
Tel: (1) 505 881 4584
Fax: (1) 505 884 0427
Email: web@aigcs.org
Website: www.aigc.com

American Indian Graduate Center (AIGC) is the only national non-profit organization dedicated to aiding Indian graduate students in all fields of study. As a non-profit organization, AIGC prides itself on maintaining a very low administrative cost. In fact 90 per cent of all contributions goes directly to student services. AIGC will help plan and produce the social, economic and political changes needed to ensure the long-term positive development of the communities by providing extraordinary numbers of talented, highly skilled and exceptionally trained Indian professionals.

AIGC Accenture American Indian Scholarship Fund

Subjects: Fields of study: business, high technology, medicine, law and engineering fields
Purpose: To financially support American Indian and Alaskan Native students seeking higher education.
Eligibility: Open to candidates who are members of a U.S. federally recognized American Indian or Alaska Native Group and who are able to demonstrate involvement with Native American activities or affairs. Must be a full-time student. Undergraduates must have a 3.25 GPA Cumulative at their 7th semester of high school. Graduates must have a 3.00 Cumulative GPA in their undergraduate years.
Level of Study: Graduate, Doctorate, MBA
Type: Scholarships
Value: Varies
Frequency: Annual
Country of Study: United States of America
No. of awards offered: 7
Application Procedure: Application forms can be downloaded from the AIGC website.
Closing Date: May 4th
Contributor: Accenture LLP

AMERICAN INNS OF COURT

1229 King Street, 2nd Floor, Alexandria, VA, 22314, United States of America
Tel: (1) 703 684 3590, 571 319 4703
Fax: (1) 703 684 3607
Email: info@innsofcourt.org; cdennis@innsofcourt.org
Website: www.innsofcourt.org
Contact: Cindy Dennis, Awards and Scholarships Coordinator

American Inns of Court is designed to improve the skills professionalism and ethics of the bench and bar. An American Inn of Court is an amalgam of judges, lawyers, and in some cases law professors and law students. In short it is our mission to foster excellence in professionalism, ethics, civility and legal skills.

Pegasus Scholarship Trust Program for Young Lawyers

Subjects: English legal system.
Purpose: To support talented young American lawyers travel to London, England.
Eligibility: Refer to website.
Level of Study: Graduate
Type: Scholarship
Value: All transportation costs to and from the United States, accommodation and a stipend sufficient for meals and public transport
Length of Study: 6 weeks
Frequency: Annual
Country of Study: United Kingdom
No. of awards offered: 2
Application Procedure: Complete available online application.
Closing Date: October 30th

Warren E. Burger Prize

Subjects: Any topic that addresses issues of legal excellence, civility, ethics, and professionalism.
Purpose: To encourage outstanding scholarship that promotes the ideals of excellence, civility, ethics, and professionalism within the legal profession.
Eligibility: Open to judges, lawyers, professors, students, scholars, and other authors.
Type: Prize
Value: US$5,000
Application Procedure: Check website for further details.
Closing Date: June 1st
Funding: Trusts
Additional Information: The winning essay will be published in the *South Carolina Law Review* and the Warren E. Burger Prize will be presented to the author at the American Inns of Court annual Celebration of Excellence at the United States Supreme Court on October 20th.

AMERICAN INSTITUTE FOR ECONOMIC RESEARCH (AIER)

250 Division St, PO Box 1000, Great Barrington, MA, 01230, United States of America
Tel: (1) 888 528 1216
Fax: (1) 413 528 0103
Email: info@aier.org; fellowships@aier.org
Website: www.aier.org

The American Institute for Economic Research (AIER), founded in 1933, is an independent scientific educational organization. The Institute conducts scientific enquiry into general economics with a focus on monetary issues. Attention is also given to business cycle analysis and forecasting as well as monetary economics.

AIER Summer Fellowship

Subjects: Scientific procedures of enquiry, monetary economics, business cycle analysis and forecasting.
Purpose: To further the development of economic scientists.
Eligibility: Open to graduating seniors who are entering doctoral programmes in economics, or those enrolled in doctoral programmes in economics for no longer than 2 years. The programme is not designed for those enrolling into business school.
Level of Study: Postgraduate
Type: Fellowship
Value: US$500 weekly stipend plus room and full board
Length of Study: two 2-week sessions
Frequency: Annual
Study Establishment: AIER
Country of Study: United States of America
No. of awards offered: 10–12
Application Procedure: Applicants must submit a completed application form, curriculum vitae, personal statement, writing sample, an outline of the proposed course of study and official transcripts. Scholastic references should be sent directly to the director from the referees.
Closing Date: March 23rd
Funding: Private
No. of awards given last year: 17
No. of applicants last year: 47
Additional Information: There are three core seminars in property rights, scientific procedures of inquiry, and sound money. Each seminar meets two to three hours per week over two weeks.

AMERICAN INSTITUTE FOR SRI LANKAN STUDIES (AISLS)

155 Pine Street, Belmont, MA 02478, United States of America
Email: rogersjohnd@aol.com
Website: www.aisls.org
Contact: John Rogers

The American Institute for Sri Lankan Studies (AISLS) was established in 1995, to foster excellence in American research and teaching on Sri Lanka, and to promote the exchange of scholars and scholarly information between the US and Sri Lanka. The Institute serves as the professional association for US-based scholars and other professionals who are interested in Sri Lanka.

AISLS Dissertation Planning Grant

Subjects: Social sciences and humanities with a focus on Sri Lanka.
Purpose: To fund students enrolled at an American university who intend to do dissertation research in Sri Lanka, to be able to make a predissertation visit to Sri Lanka and investigate the feasibility of their topic, sharpen their research design or make other practical arrangements for future research.
Eligibility: Open to candidates enrolled in a PhD programme (or equivalent) in a university in the US.
Level of Study: Doctorate
Type: Grant
Value: US$390 per week and reimbursement for roundtrip airfare and any visa fees paid to the Sri Lankan government
Length of Study: 8 weeks
Frequency: Annual
Application Procedure: Applicants can download the application cover sheet from the website. The completed application cover sheet along with curriculam vitae, a copy of graduate transcript, a project narrative, one-page project bibliography and confidential letter of recommendation is to be sent.
Closing Date: December 1st
Contributor: American Institute for Sri Lankan Studies
Additional Information: Submissions by fax or email will not be accepted.The country of study is Sri Lanka.

AISLS Fellowship Program

Subjects: Social sciences and humanities with a focus on Sri Lanka.
Purpose: To encourage research and a better understanding of Sri Lanka.
Eligibility: Open to citizens of the United States who hold a PhD or equivalent academic degree.
Level of Study: Research, Postdoctorate
Type: Fellowship
Value: US$3,200 per month, reimbursement for roundtrip airfare and research expenses
Length of Study: 2–9 months
Frequency: Annual
Application Procedure: Applicants can download the application cover sheet from the website. The completed application cover sheet along with a curriculum vitae and a description of the proposed study should be sent.
Closing Date: December 1st
Contributor: American Institute for Sri Lankan Studies
Additional Information: Submissions by fax or email will not be accepted. Country of study is Sri Lanka.

THE AMERICAN INSTITUTE OF BAKING (AIB)

1213 Bakers Way, PO Box 3999, Manhattan, Kansas, 66505-3999, United States of America
Tel: (1) 785 537 4750
Fax: (1) 785 537 1493
Email: kembers@aibonline.org
Website: www.aibonline.org
Contact: Mr Ken Embers, Registrar

The AIB is a non-project corporation, founded by the North American wholesale and retail baking industries in 1991. AIB's staff includes experts in the fields of baking production, research related to experimental baking, cereal science and nutrition; food safety and hygiene; occupational safety and maintenance engineering.

Baking Industry Scholarship

Subjects: Food processing.
Purpose: To support students who are planning to seek employment in the baking or food processing industry.
Eligibility: Open to applicants who are enrolled at the American Institute of Baking in the 16 week Baking Science and Technology class.
Level of Study: Postgraduate
Type: Scholarship
Value: US$500 to the cost of the full tuition
Frequency: Annual
Country of Study: United States of America
No. of awards offered: Varies
Application Procedure: AIB awards scholarships annually to self-sponsored* students attending AIB's 16-week Baking Science and Technology (BS&T) course. See the website for further details.
Closing Date: May 1st
Funding: Commercial
No. of awards given last year: 20
No. of applicants last year: 30

AMERICAN INSTITUTE OF BANGLADESH STUDIES (AIBS)

203 Ingraham Hall, 1155 Ovservatory Drive, Madison, WI, NY, 53706, United States of America
Tel: (1) 608 261 3062
Fax: (1) 608 265 3062
Email: aibsinfo@aibs.net
Website: www.aibs.net
Contact: Laura Hammond, Administrative Program Manager

The American Institute of Bangladesh Studies (AIBS) is a consortium of US universities and colleges involved in research on Bangladesh. It strives to improve the scholarly understanding of Bangladesh culture and society in US and to promote educational exchange between the two countries.

AIBS Junior Fellowships

Eligibility: Open to an individual member of AIBS; currently in the ABD phase of your PhD program; is in the data collection and writing stage of the dissertation, and must be United States citizen or permanent resident.
Level of Study: Research
Type: Fellowship
Value: Fellowships will be equivalent to $920 per month, plus research and dependents' allowances and economy Round-Trip Air Transportation will be provided via the most direct route and using the Bangladesh carrier whenever possible
Length of Study: 6–12 months
Frequency: Annual
Application Procedure: Applications must be received by September 1st with consideration for study beginning in the spring or summer of the following year.
Closing Date: September 15th
Contributor: U.S. Department of State Bureau of Educational and Cultural Affairs through the Council of American Overseas Research Centers
Additional Information: Please contact AIBS for more information.

AIBS Pre-Dissertation Fellowships

Purpose: The American Institute of Bangladesh Studies offers short-term grants to graduate students pursuing studies of Bangladesh funded by the U.S. Department of State Bureau of Educational and Cultural Affairs through the Council of American Overseas Research Centers.
Eligibility: Open to an individual member of AIBS and must have completed at least one year of graduate study in a recognized PhD granting institution and must be a US citizen.
Type: Fellowship
Value: $3,000 plus round-trip air fare
Frequency: Annual
Application Procedure: Check at https://www.aorcapp.wisc.edu/ for application download.
Closing Date: September 15th
Additional Information: Please note that all applications require a $25 processing fee. Awarding a grant to a student from a US institution that is a member of the AIBS will incur no additional fees but for submissions from non-represented institutions there is a $250 membership fee to be paid before an award can be dispersed.

AIBS Senior Fellowship

Subjects: Area and cultural studies.
Purpose: To improve the scholarly understanding of Bangladesh culture and society in the United States.
Eligibility: Open to citizens of the United States. Senior Fellowships are available to applicants who have obtained a PhD. Junior Fellowships are available to applicants who are ABD. Pre-dissertation Fellowships are available to PhD students who have completed at least one year of graduate study.
Level of Study: Postdoctorate, Predoctorate, Research, Doctorate
Type: Fellowships
Value: US$1,000 per month and other benefits
Length of Study: 4–12 months
Frequency: Annual
Country of Study: Bangladesh
Application Procedure: Applicants can download the application form from the website.
Closing Date: September 15th
No. of awards given last year: Varies
No. of applicants last year: Varies
Additional Information: See AIBS Website for additional information (aibs.net)

AIBS Travel Grants

Subjects: All subjects.
Purpose: AIBS can provide funding for conference travel for the presentation of papers or organization of panels that include topics relevant to Bangladesh Studies at scholarly conferences.
Level of Study: Graduate
Type: Travel grant
Value: Up to $600 for travel to conferences within the United States. International travel grants have been suspended until further notice due to budgetary constraints
Application Procedure: Students who wish to avail of this scholarship grant should submit the following requirements to the Fellowship Coordinator: a duly accomplished scholarship application form, updated resume, letter of support from a previous or current faculty mentor, proof of acceptance of your paper, conference travel proposal that does not exceed three pages, and a summary of budget including travel costs, conference registration and hotel costs. Applicants may send their application to rmcdermo@barnard.edu.
Closing Date: Any time as needed
Additional Information: Please write to rmcdermo@barnard.edu for more information on Travel Grant.

AMERICAN INSTITUTE OF CERTIFIED PUBLIC ACCOUNTANTS (AICPA)

1211 Avenue of the Americas, New York, NY, 10036-8775, United States of America
Tel: (1) 212 596 6200
Fax: (1) 212 596 6213
Email: service@aicpa.org
Website: www.aicpa.org

The American Institute of Certified Public Accountants (AICPA) is a national, professional organization for all Certified Public Accountants. Its mission is to provide members with the resources, information and leadership that enable them to provide valuable services in the highest professional manner to benefit the public as well as employers and clients. In fulfilling its mission, the AICPA works with state CPA organizations and gives priority to those areas where public reliance on CPA skills is most significant.

AICPA Fellowship for Minority Doctoral Students

Subjects: Accounting or taxation.
Purpose: To ensure that professors of ethnically diverse backgrounds are represented in college and university accounting classrooms.
Eligibility: Open to African American, Hispanic American or Native American who have applied to or been accepted into a doctoral program with a concentration in accounting or taxation. Applicants must be CPAs or plan to pursue the CPA credential.
Level of Study: Doctorate
Type: Fellowship
Value: Up to US$12,000
Length of Study: Up to 5 years
Frequency: Annual
Country of Study: United States of America
No. of awards offered: 21
Application Procedure: Applicants must submit their application form, transcript and reference letters. See website for details.
Closing Date: April 1st
Funding: Foundation
No. of awards given last year: 20

For further information contact:

American Institute of CPAs - Team 331, 220 Leigh Farm Road, Durham, NC, 27707
Email: scholarships@aicpa.org
Website: www.aicpa.org

AICPA John L. Carey Scholarship

Subjects: Accounting.
Purpose: To provide financial assistance to liberal arts degree holders pursuing graduate studies in accounting and the CPA designation. These awards are intended to encourage liberal arts students to consider professional accounting careers.
Eligibility: Open to US applicants who have obtained a liberal arts degree prior to enrolling in a graduate accounting programme. Applicants must visit website for additional eligibility information.
Level of Study: Graduate
Type: Scholarship
Value: US$5,000 for 1 year (non-renewable)
Length of Study: Full-time for entire academic year
Frequency: Annual

Country of Study: United States of America
No. of awards offered: 10
Application Procedure: Applications can be downloaded from www.aicpa.org
Closing Date: April 1st
Funding: Foundation
Contributor: AICPA Foundation

For further information contact:

American Institute of CPAs - Team 331, 220 Leigh Farm Road, Durham, NC 27707

AICPA Scholarship for Minority Accounting Students

Subjects: Accounting, finance, taxation or other related program.
Purpose: To provide financial awards to accounting students of ethnically diverse backgrounds.
Eligibility: Applicants must be U.S. citizens or permanent residents, minority students and full-time graduate students. Applicants must be declared accounting, tax, finance or other related majors who plan to pursue the CPA credential.
Level of Study: Graduate
Type: Award
Value: US$3,000 per academic year
Length of Study: Up to 3 years
Frequency: Annual
Country of Study: United States of America
No. of awards offered: 80
Application Procedure: Applicants must submit a completed application form along with official transcripts, one letter of recommendation and a brief essay.
Closing Date: April 1st
Funding: Foundation
Contributor: AICPA foundation
No. of awards given last year: 78

AICPA/Accountemps Student Scholarship

Subjects: Accounting, finance or information systems.
Purpose: To provide financial assistance to students who are majoring accounting, finance, or information systems.
Eligibility: Open to full-time master's level student at an accredited college or university in the United States maintained an overall GPA and major GPA of at least 3.0. Applicants must be US citizen or permanent resident and AICPA student affiliate member.
Level of Study: Graduate, Postgraduate
Type: Scholarships
Value: US$2,500
Length of Study: 1 year
Frequency: Annual
Country of Study: United States of America
No. of awards offered: 10
Application Procedure: Applications can be downloaded from www.aicpa.org.
Closing Date: April 1st
Funding: Corporation
Contributor: RHI/Accountemps

AMERICAN INSTITUTE OF INDIAN STUDIES (AIIS)

1130 East 59th Street, Chicago, Illinois, IL 60637, United States of America
Tel: (1) 773 702 8638
Email: aiis@uchicago.edu
Website: www.indiastudies.org
Contact: Dr Elise Auerbach, US Director

The American Institute of Indian Studies (AIIS) is a consortium of American colleges and universities that supports the understanding of India, its people and cultures. AIIS offers a range of fellowships for research in India. It also supports individuals studying the performing arts, operates language programmes in India and offers research facilities to scholars in India.

AIIS Junior Research Fellowships

Subjects: India, its people and culture.
Purpose: To support the advancement of knowledge and understanding.
Eligibility: Open to doctoral candidates at United States of America colleges and universities.
Level of Study: Doctorate
Type: Fellowship
Length of Study: Up to 11 months
Frequency: Annual
Study Establishment: An Indian university
Country of Study: India
Application Procedure: Applicants must write for further information.
Closing Date: July 1st

AIIS Senior Performing and Creative Arts Fellowships

Subjects: Performing and creative arts.
Eligibility: Open to accomplished practitioners of the performing arts of India and creative artists who demonstrate that study in India would enhance their skills, develop their capabilities to teach or perform in the United States of America, enhance American involvement with India's artistic traditions and strengthen their links with peers in India.
Level of Study: Unrestricted
Type: Fellowship
Frequency: Annual
Country of Study: India
Application Procedure: Applicants must write for further information.
Closing Date: July 1st

AIIS Senior Research Fellowships

Subjects: South Asian studies.
Purpose: To enable scholars to pursue further research in India.
Eligibility: Open to scholars who hold a PhD or its equivalent and are either citizens of the United States of America or resident aliens teaching full-time at United States of America colleges and universities.
Level of Study: Postdoctorate
Type: Fellowship
Length of Study: Up to 9 months
Frequency: Annual
Country of Study: India
Application Procedure: Applicants must write for further information.
Closing Date: July 1st

AIIS Senior Scholarly/Professional Development Fellowships

Subjects: India, its people and culture.
Purpose: To support the advancement of knowledge and understanding.
Eligibility: Open to established scholars who have not previously specialized in Indian studies and to established professionals who have not previously worked or studied in India.
Level of Study: Professional development
Type: Fellowship
Length of Study: 6–9 months
Frequency: Annual
Country of Study: India
No. of awards offered: Varies
Application Procedure: Applicants must write for further information.
Closing Date: July 1st

AMERICAN LIBRARY ASSOCIATION (ALA)

50 E. Huron street, Chicago, IL, 60611, United States of America
Tel: (1) 800 545 2433 ext. 4274
Fax: (1) 312 440 9374
Email: ala@ala.org
Website: www.ala.org
Contact: Ms Melissa Jacobsen, Manager, Prof. Dev.

Each year the American Library Association (ALA) and its member units sponsor awards to honour distinguished service and foster professional growth.

ALA 3M/NMRT Professional Development Grant

Subjects: Library studies.
Purpose: To encourage professional development and participation by new ALA members in national ALA and NMRT activities.
Eligibility: Open to members of the ALA and the New Members Round Table (NMRT) who are working within the territorial United States.
Level of Study: Professional development
Type: Grant
Value: Round trip airfare, lodging, conference registrations fees and some incidental expenses
Frequency: Annual
Country of Study: Any country
No. of awards offered: 3
Application Procedure: Applicants must submit nominations to the NMRT Professional Development Grant at ALA.
Closing Date: December 16th
Funding: Commercial
Contributor: 3M Library Systems

ALA AASL Frances Henne Award

Subjects: Library media.
Purpose: To enable an individual to attend an AASL national conference or ALA Annual Conference for the first time.
Eligibility: Open to school librarians specialists with less than 5 years in the profession.
Level of Study: Unrestricted
Type: Grant
Value: US$1,250
Frequency: Annual
Country of Study: Any country
No. of awards offered: 1
Application Procedure: Visit www.ala.org/aasl/aaslawards/aaslawards to apply.
Closing Date: February 1st
Funding: Commercial
Contributor: ABC-CLIO
No. of awards given last year: 1

ALA AASL Information Technology Pathfinder Award

Subjects: Library media.
Purpose: To recognize and honour a school librarian for demonstrating vision and leadership through the use of information technology to build lifelong learners.
Eligibility: Open to school librarians, supervisors or educators.
Level of Study: Professional development
Type: Scholarship
Value: US$1,000 to the specialist and US$500 to the library.
Frequency: Annual
Country of Study: Any country
No. of awards offered: 1
Application Procedure: Applicants must write for details. Visit www.ala.org/aasl/aaslawards/aaslawards to apply.
Closing Date: February 1st
Funding: Commercial
Contributor: Follett Software Company
No. of awards given last year: 2

ALA AASL Research Grant

Subjects: Library science.
Purpose: To enable an individual to conduct innovative research aimed at measuring and evaluating the impact of school library programmes on learning and education.
Eligibility: Open to qualified researchers of any nationality.
Level of Study: Professional development
Type: Research grant
Value: Up to US$2,500
Frequency: Annual
Country of Study: Any country
No. of awards offered: 1
Application Procedure: Visit www.ala.org/aasl/aaslawards/aaslawards to apply.
Closing Date: February 1st
Funding: Commercial
Contributor: The Highsmith Company
No. of awards given last year: 2

ALA Beta Phi Mu Award

Subjects: Education for librarianship.
Purpose: To recognize distinguished service.
Eligibility: Open to library school faculty members or others in the library profession.
Level of Study: Professional development
Value: US$1000 and a citation
Frequency: Annual
No. of awards offered: 1
Application Procedure: Applicants must submit 6 copies of nominations to the ALA Awards Programme Office.
Closing Date: December 1st
Funding: Private
Contributor: The Beta Phi Mu International Library Science Honorary Society
No. of awards given last year: 1

ALA Bogle Pratt International Library Travel Fund

Subjects: Library science.
Purpose: To enable ALA members to attend their first international conference.
Eligibility: Open to ALA members.
Level of Study: Professional development
Type: Travel grant
Value: US$1,000
Length of Study: To attend their first international conference.
Frequency: Annual
Country of Study: Any country
No. of awards offered: 1
Application Procedure: Applicants must write for details.
Closing Date: January 1st
Funding: Private
Contributor: Bogle Memorial Fund and the Pratt Institute School of Information and Library Science
No. of awards given last year: 1

ALA Bound to Stay Bound Book Scholarships

Subjects: Library science.
Purpose: To support study in the field of library service to children in an ALA-accredited programme.
Eligibility: Applicants must be citizens of the U.S. or Canada
Level of Study: Graduate, Postgraduate
Type: Scholarship
Value: US$7,000 each
Frequency: Annual
Country of Study: United States of America or Canada
No. of awards offered: 4
Application Procedure: Applicants must write or email for details. The scholarship process is open annually from October - March. The scholarship application and the reference forms are both online submissions.
Closing Date: March 1st
Funding: Commercial
Contributor: Bound to Stay Bound Books, Inc.
Additional Information: Money will be paid directly to the school.

ALA Carroll Preston Baber Research Grant

Subjects: Library service.
Purpose: To encourage innovative research that could lead to an improvement in library services to any specified group or groups of people.
Eligibility: Any ALA member may apply. The Jury would welcome projects that involve both a practicing librarian and a researcher.
Level of Study: Unrestricted
Type: Research grant
Value: Up to US$3,000
Length of Study: Up to 18 months
Frequency: Annual
Country of Study: Any country
No. of awards offered: 1
Application Procedure: Applicants must submit an application including a research proposal.
Closing Date: January 3rd
Funding: Private
Contributor: Eric R Baber

No. of awards given last year: 1
No. of applicants last year: 5
Additional Information: The project should aim to answer a question that is of vital importance to the library community and the researchers should plan to provide documentation of the results of their work. The jury would welcome proposals that involve innovative uses of technology and proposals that involve co-operation between libraries and other agencies, or between librarians and persons in other disciplines.

ALA Christopher J. Hoy/ERT Scholarship

Subjects: Library and information studies.
Purpose: To allow individuals to attend an ALA-accredited programme of library and information studies.
Eligibility: Open to U.S./Canadian citizen or permanent resident who will be attending an ALA-accredited programme of library and information studies leading to a Master's degree.
Level of Study: Postgraduate
Type: Scholarship
Value: US$5,000
Frequency: Annual
Country of Study: Any country
No. of awards offered: 1
Application Procedure: Applicants must write for details. Please contact the ALA Scholarship Clearinghouse at 1-800-545-2433 ext. 4279 or scholarships@ala.org.
Closing Date: March 1st
Funding: Private
Contributor: The family of Christopher J Hoy

ALA David H Clift Scholarship

Subjects: Library science.
Purpose: To enable a worthy candidate to begin a Master's degree.
Eligibility: Open to qualified citizens of Canada or the United States of America pursuing a Master's degree in library science in an ALA-accredited programme.
Level of Study: Postgraduate
Type: Scholarship
Value: US$3,000 in two installments
Frequency: Annual
Country of Study: Any country
No. of awards offered: 1
Application Procedure: Applicants must write for details.
Closing Date: March 1st

ALA Eli M. Oboler Memorial Award

Subjects: Intellectual freedom and freedom to read.
Purpose: To award the best published work in the field.
Eligibility: There are no eligibility restrictions.
Level of Study: Unrestricted
Type: Award
Value: US$500 and a Certificate
Frequency: Every 2 years
Country of Study: Any country
No. of awards offered: 1
Application Procedure: Applicants must submit the nominated documents with nominating form.
Closing Date: December 1st, of odd-numbered years
Funding: Private
Contributor: Intellectual Freedom Round Table (IFRT) of the American Library Association (ALA).

ALA Elizabeth Futas Catalyst for Change Award

Subjects: Library science.
Purpose: To recognize and honour a librarian who invests time and talent to make positive changes in the profession of librarianship by taking risks to further the cause, helping new librarians grow and achieve, working for change within the ALA or other library organizations and inspiring colleagues to excel or make the impossible possible.
Type: Award
Value: US$1,000 and a citation
No. of awards offered: 1
Application Procedure: Applicants must submit 6 copies of the application.
Closing Date: December 1st
Contributor: An endowment administered by the ALA
No. of awards given last year: 1

ALA Equality Award

Subjects: Pay equity, affirmative action, legislative work and non-sexist education.
Purpose: To recognize an outstanding contribution towards the promotion of equality in the library profession. The contribution may be either a sustained one or a single outstanding accomplishment.
Eligibility: Open to members of the library profession.
Level of Study: Professional development
Value: US$1,000 plus a citation
Frequency: Annual
No. of awards offered: 1
Application Procedure: Applicants must submit 6 copies of nominations to the ALA Awards Programme Office.
Closing Date: December 1st
Funding: Commercial
Contributor: The Scarecrow Press
No. of awards given last year: 1

ALA Facts on File Grant

Subjects: Library science.
Purpose: To award a library for imaginative programming that would make current affairs more meaningful to an adult audience. Programmes, bibliographies, pamphlets, and innovative approaches of all types and in all media are eligible.
Eligibility: Open to adult librarians.
Level of Study: Professional development
Type: Grant
Value: US$2,000
Frequency: Annual
Country of Study: Any country
No. of awards offered: 1
Application Procedure: Applicants must submit a proposal accompanied by a statement of objective, identification of the current issues, the target audience and the extent of community involvement planned, an outline of planned activities for conducting and promoting the project, a budget summary and details of how the project will be evaluated.
Funding: Commercial
Contributor: Facts On File, Inc.
No. of awards given last year: 1

ALA Frances Henne/YALSA/VOYA Research Grant

Subjects: Library science.
Purpose: To provide seed money to an individual, institution or group for a project to encourage research on library service to young adults.
Eligibility: Open to applicants of any nationality. Applicants must be personal members of YALSA
Level of Study: Unrestricted
Type: Research grant
Value: US$1,000
Frequency: Annual
Country of Study: Any country
No. of awards offered: 1
Application Procedure: Applicants must write for details.
Closing Date: December 1st
Contributor: Voice of Youth Advocates/Scarecrow Press.

ALA Frederic G Melcher Scholarship

Subjects: Library Science.
Purpose: To provide financial assistance for the professional education of men and women who intend to pursue children's librarianship in an ALA-accredited program.
Eligibility: Open to qualified young persons who have been accepted for admission to an appropriate school.
Level of Study: Graduate, Postgraduate
Type: Scholarship
Value: US$6,000
Frequency: Annual
Study Establishment: An ALA-accredited school
Country of Study: United States of America or Canada
No. of awards offered: 2

Application Procedure: Applicants must write or email for details. The scholarship process is open annually from October to March. Only online applications and online references will be accepted.
Closing Date: March 1st
Funding: Private
Additional Information: Money will be paid in two equal amounts at the beginning of the first two semesters or quarters in which the recipient is enrolled.

ALA H W Wilson Library Staff Development Grant

Subjects: Library science.
Purpose: To award a library organization whose application demonstrates greatest merit for a programme of staff development designed to further goals and objectives of the library organization.
Eligibility: A library organization is defined as:individual library library system group of cooperating libraries state governmental agency local, state, or regional association Staff development is defined as:"a program of learning activities that is developed by the library organization and develops the on-the-job staff capability and improves the abilities of personnel to contribute to the overall effectiveness of the library organization."
Type: Grant
Value: US$3,500 and a citation
Frequency: Annual
No. of awards offered: 1
Application Procedure: Applicants must submit 6 copies of the application and documentation to the ALA Awards Programme Office.
Closing Date: December 1st
Contributor: The H W Wilson Company
No. of awards given last year: 1

ALA Jesse H. Shera Award for Distinguished Published Research

Subjects: Library science.
Purpose: To honour an outstanding and original paper reporting the results of research related to libraries.
Eligibility: Authors of nominated articles need not be Library Research Round Table (LRRT) members but the nominations must be made by LRRT members. All entries must be research articles published in English during the calendar year previous to the competition. All nominated articles must relate in at least a general way to library and information studies.
Level of Study: Unrestricted
Type: Prize
Value: US$500
Frequency: Annual
Country of Study: Any country
No. of awards offered: 1
Application Procedure: Applicants wishing to nominate research articles for this award should send three copies of each article together with a covering letter stating that they are a current member of LRRT or that they are acting in their role as journal editor.
Closing Date: March 29th
Funding: Private
No. of awards given last year: 1
No. of applicants last year: 12

ALA John Phillip Immroth Memorial Award

Subjects: Intellectual freedom.
Purpose: To recognize a notable contribution to intellectual freedom fuelled by personal courage.
Eligibility: Open to intellectual freedom fighters. Individuals, a group of individuals or an organization are eligible for the award.
Level of Study: Unrestricted
Type: Award
Value: US$500 plus a citation
Frequency: Annual
Country of Study: Any country
No. of awards offered: 1
Application Procedure: Applicants must submit a detailed statement explaining why the nominator believes that the nominee should receive the award. Nominations should be submitted to IFRT Staff Liaison at the ALA.
Closing Date: February 17th
Funding: Private
Contributor: Intellectual Freedom Round Table (IFRT) of the American Library Association
For further information contact:

Tel: 312 280 4220/800 545 2433, ext. 4220
Fax: 312 280 4227
Email: bcampbell@ala.org
Contact: Bryan Campbell

ALA Joseph W Lippincott Award

Subjects: Library work with professional library associations.
Purpose: To recognize distinguished service in the profession of librarianship, including outstanding participation in professional library activities, notable published professional writing or other significant activities.
Eligibility: Open to librarians.
Level of Study: Professional development
Type: Award
Value: US$1,000 plus a citation
Frequency: Annual
No. of awards offered: 1
Application Procedure: Applicants must submit 6 copies of nominations to the ALA Awards Programme Office.
Closing Date: December 1st
Funding: Private
Contributor: The late Joseph W Lippincott
No. of awards given last year: 1

ALA Ken Haycock Award for Promoting Librarianship

Subjects: Library science.
Purpose: Honours an individual for contributing significantly to the public recognition and appreciation of librarianship through professional performance, teaching or writing.
Type: Award
Value: US$1,000 and a citation
Frequency: Annual
No. of awards offered: 1
Closing Date: December 1st
Funding: Private
Contributor: Kenneth Haycock, PhD

ALA Lexis/Nexis/GODORT/ALA "Documents to the People" Award

Subjects: Library science.
Purpose: To provide funding for research in the field of documents librarianship or in a related area that would benefit the individual's performance as a documents librarian or make a contribution to the field.
Eligibility: Open to individuals and libraries, organizations and other appropriate non-commercial groups.
Level of Study: Unrestricted
Type: Award
Value: US$3,000
Frequency: Annual
Country of Study: Any country
No. of awards offered: 1
Application Procedure: Applicants must submit nominations to the GODORT Staff Liaison at the ALA.
Closing Date: December 1st
Funding: Commercial
Contributor: Lexis/Nexis

ALA Loleta D. Fyan Grant

Subjects: Library service.
Purpose: To facilitate the development and improvement of public libraries and the services they provide.
Eligibility: Applicants can include but are not limited to local, regional or state libraries, associations or organizations including units of the ALA, library schools or individuals.
Level of Study: Unrestricted
Type: Research grant
Value: Up to US$5,000
Frequency: Annual
Country of Study: Any country
No. of awards offered: 1 or more

Application Procedure: Applicants must submit an application form in addition to a proposal and budget to the ALA Staff Liaison. E-mail subject line should read "2011 Fyan Award Proposal". Please do not fax or mail.
Closing Date: December 21st
Funding: Private
No. of awards given last year: 1
No. of applicants last year: 10
Additional Information: The project must result in the development and improvement of public libraries and the services they provide, have the potential for broader impact and application beyond meeting a specific local need, should be designed to effect changes in public library services that are innovative and responsive to the future and should be capable of completion within 1 year.

For further information contact:

American Library Association
Tel: 312 280 3217
Email: cbourdon@ala.org
Contact: Cathleen Bourdon, Associate Executive Director, Communications & Member Relations

ALA Marshall Cavendish Excellence in Library Programming

Subjects: Library science.
Purpose: To recognize a school or public library for programmes that have community impact and respond to community needs.
Eligibility: Eligible programs or particular interest for consideration includes: support of educational programs, library programs for children and adults, reading and literature programs for children, library programs for young adults, programming for multi-ethnic groups, community outreach, literacy programs and providing programs and services for persons with disabilities.
Type: Award
Value: US$2,000 and a citation
Frequency: Annual
No. of awards offered: 1
Application Procedure: Applicants must submit 6 copies of the application and supporting material.
Closing Date: December 1st
Funding: Corporation
Contributor: Marshall Cavendish Corporation

ALA Mary V Gaver Scholarship

Subjects: Library science.
Purpose: To assist library support staff specializing in youth services.
Eligibility: Open to library support staff who are citizens of the United States of America or Canada who are pursuing a Master's degree in library science.
Level of Study: Unrestricted
Type: Scholarship
Value: US$3,000
Frequency: Annual
Country of Study: Any country
No. of awards offered: 1
Application Procedure: Applicants must write for details.
Closing Date: March 1st
No. of awards given last year: 1
Additional Information: Apply online between October and March at www.ala.org/scholarships.

ALA Melvil Dewey Medal

Subjects: Library science.
Purpose: To award an individual or group for recent creative professional achievement in library management, training, cataloguing, classification and the tools and techniques of librarianship.
Type: Award
Value: Medal and citation
Frequency: Annual
No. of awards offered: 1
Application Procedure: Applications can be downloaded from the ALA website. Applicants must submit 6 copies of nomination form.
Closing Date: December 1st
Contributor: The OCLC Forest Press
No. of awards given last year: 1

ALA Miriam L Hornback Scholarship

Subjects: Library science.
Purpose: To assist an individual pursuing a Master's degree.
Eligibility: Open to ALA or library support staff who are pursuing a Master's degree in library science and who are citizens of the United States of America or Canada.
Level of Study: Postgraduate
Type: Scholarship
Value: US$3,000
Frequency: Annual
No. of awards offered: 1
Application Procedure: Applicants must write for details.
Closing Date: March 1st
No. of awards given last year: 2

For further information contact:

Email: klredd@ala.org
Contact: Kimberly L. Redd, Program Officer

ALA NMRT/EBSCO Scholarship

Subjects: Library science.
Purpose: To enable an individual to begin an MLS degree in an ALA-accredited programme.
Eligibility: Open to citizens of the United States of America and Canada.
Level of Study: Postgraduate
Type: Scholarship
Value: US$1,000
Frequency: Annual
Country of Study: Any country
No. of awards offered: 1
Application Procedure: Applicants must write for details.
Closing Date: April 1st

ALA Penguin Young Readers Group Award

Subjects: Library science.
Purpose: To allow children's librarians to attend the Annual Conference of the ALA.
Eligibility: Open to members of the Association for Library Service to Children with between 1 and 10 years of experience who have never attended an ALA Annual Conference.
Level of Study: Professional development
Type: Award
Value: US$600
Frequency: Annual
Country of Study: Any country
No. of awards offered: 4
Application Procedure: Applicants must telephone or email for details. Contact the chair of the ALSC Grant Administration committee, Nancy Baumann, with questions at horsepwr_2000@yahoo.com.
Closing Date: December 21st
Funding: Commercial
Contributor: Penguin Young Readers Group
No. of awards given last year: 4

ALA Schneider Family Book Award

Subjects: Library science.
Purpose: The Schneider Family Book Awards honour an author or illustrator for a book that embodies an artistic expression of the disability experience for child and adolescent audiences.The book must emphasize the artistic expression of the disability experience for children and or adolescent audiences. The book must portray some aspect of living with a disability or that of a friend or family member, whether the disability is physical, mental or emotional.
Eligibility: The person with the disability may be the protagonist or a secondary character. Definition of disability. Dr. Schneider has intentionally allowed for a broad interpretation by her wording, the book "must portray some aspect of living with a disability, whether the disability is physical, mental, or emotional." This allows each committee to decide on the qualifications of particular titles. Books with death as the main theme are generally disqualified. The books must be published in English. The award may be given posthumously. Term of eligibility extends to publications from the preceding two years, e.g. 2007 awards given to titles published in 2006 and 2005.

This may be changed to one year when the award is well established. Books previously discussed and voted on are not eligible again.
Type: Award
Value: US$5,000 and a citation for each winner
Frequency: Annual
No. of awards offered: 3
Application Procedure: Applicants must submit 8 copies of the application.
Closing Date: December 1st
Funding: Private
Contributor: Katherine Schneider

ALA Shirley Olofson Memorial Awards

Subjects: Library science.
Purpose: To allow individuals to attend ALA conferences.
Eligibility: Open to members of the ALA who are also current or potential members of the New Members Round Table. Applicants should not have attended any more than five conferences.
Level of Study: Unrestricted
Type: Award
Value: US$1,000
Frequency: Annual
Country of Study: Any country
No. of awards offered: Varies
Application Procedure: Applicants must write for details. Fill out the application online.
Closing Date: December 10th
Contributor: The New Members Round Table (NMRT) and the Shirley Olofson Memorial Award Committee

ALA Spectrum Initiative Scholarship Program

Subjects: Library and information studies.
Purpose: To encourage admission to an ALA recognized Master's degree programme by the four largest underrepresented minority groups.
Eligibility: Open to citizens of the United States of America or Canada only, from one of the largest underrepresented groups. These are African American or African Canadian, Asian or Pacific Islander, Latino or Hispanic and native people of the United States of America or Canada.
Level of Study: Postgraduate
Type: Scholarship
Value: US$5,000
Frequency: Annual
Country of Study: United States of America or Canada
No. of awards offered: 25–50
Application Procedure: Applicants must request details via fax or visit the website.
Closing Date: March 1st
Additional Information: Applications accepted from mid-October to March 1st each year.

For further information contact:

Tel: 800 545 2433 ext 5048
Email: spectrum@ala.org

ALA Sullivan Award for Public Library Administrators Supporting Services to Children Award

Subjects: Library science.
Purpose: To an individual who has shown exceptional understanding and support of public library service to children while having general management/supervisory/administrative responsibility that has included public library service to children in its scope.
Eligibility: Please use a separate form to submit a statement explaining this nominee's contribution, which will include the following: brief career summary - title, library and dates (Chronological order); educational background; membership/participation in professional organizations - functions, dates; publications, productions and presentations and other significant contributions.
Type: Award
Value: Citation and commemorative gift
Frequency: Annual
No. of awards offered: 1
Application Procedure: Applicants must submit six copies of the application.
Closing Date: February 15th
Funding: Private
Contributor: Peggy Sullivan, PhD

ALA W. David Rozkuszka Scholarship

Subjects: Library science.
Purpose: To provide financial assistance to an individual who is currently working with government documents in a library.
Eligibility: Open to applicants currently completing a Master's programme in library science.
Level of Study: Postgraduate
Type: Scholarship
Value: US$3,000
Frequency: Annual
Country of Study: Any country
No. of awards offered: 1
Application Procedure: Applicants must write for details.
Closing Date: December 31st

For further information contact:

PO Box 515, Gleneden Beach, OR, 97388, United States of America
Tel: (1) 541 992 5461
Email: asevetson@hotmail.com
Contact: Andrea Sevetson

ALA W.Y. Boyd Literary Award for Excellence in Military Fiction

Subjects: The writing and publishing of outstanding war-related fiction.
Purpose: To award an author who has written a military novel that honours the service of American veterans and military personnel during a time of war: 1861–1865, 1914–1918 or 1939–1945.
Eligibility: Novel must have been published during the year prior to the award; incidents of war can consitute the main plot of the story or merely provide the setting and young adult and adult novels only.
Type: Award
Value: US$5,000 and a citation
Frequency: Annual
No. of awards offered: 1
Application Procedure: Applicants must submit six copies of the application.
Closing Date: December 1st
Funding: Private
Contributor: William Young Boyd II

ALA YALSA/Baker and Taylor Conference Grant

Subjects: Library science.
Purpose: To allow young adult librarians who work directly with young adults in either a public library or a school library, to attend the Annual Conference of the ALA.
Eligibility: Open to members of the Young Adult Library Services Association with between 1 and 10 years of library experience who have never attended an ALA Annual Conference.
Level of Study: Professional development
Type: Grant
Value: US$1,000 each
Frequency: Annual
Country of Study: Any country
No. of awards offered: 2
Application Procedure: Applicants must submit applications to the Young Adult Library Services Association, ALA, by e-mail to Nichole Gilbert at ngilbert@ala.org. Attachments must be named the applicants last name, underscore, the name of the grant, e.g., Gilbert_BakerTaylor.
Closing Date: December 1st

ALA/Information Today, Inc. Library of the Future Award

Subjects: Library science.
Purpose: To honour an individual library, library consortium, group of librarians or support organization for innovative planning for applications of, or development of, patron training programmes about information technology in a library setting.
Type: Award
Value: US$1,500 and a citation

Frequency: Annual
No. of awards offered: 1
Application Procedure: Applicants must submit six copies of the application.
Closing Date: December 1st
Funding: Private
Contributor: Information Today, Inc.
Additional Information: The American Library Association offers a number of other awards in various fields related to library science, including the following medals and citations with no cash prizes: the Randolph Caldecott Medal, the James Bennett Childs Award, the Dartmouth Medal, the John Newberry Medal, the Laura Ingalls Wilder Medal, the ASCLA Exceptional Service Award, the Armed Forces Librarians Achievement Citation, the Francis Joseph Campbell Citation, the Margaret Mann Citation, the Isadore Gilbert Mudge Citation, the Esther J Piercy Award, the Distinguished Library Service Award for School Administrators and the Trustees Citations. A full list of awards is available from the ALA.

For further information contact:

Contact: Cheryl Malden

DEMCO New Leaders Travel Grant

Subjects: Library science.
Purpose: To enhance professional development and improve the expertise of public librarians new to the field by making possible their attendance at major PLA professional development activities.
Eligibility: Open to qualified public librarians, MLS, PLA member.
Level of Study: Professional development
Type: Travel grant
Value: Plaque and travel grant of up to US$1,500 per awardee
Frequency: Annual
Country of Study: Any country
No. of awards offered: 1
Application Procedure: Visit website www.pla.org.
Closing Date: December 1st
Funding: Corporation
No. of awards given last year: 3

For further information contact:

Email: jkloeppel@ala.org
Contact: Julianna Kloeppel

EBSCO ALA Annual Conference Sponsorship

Subjects: Library science.
Purpose: To allow librarians to attend the ALA Annual Conference.
Eligibility: Applicants must be ALA members and must not supervise another professional librarian (MLS).
Level of Study: Professional development
Type: Travel grant
Value: Up to US$1,000 for expenses
Frequency: Annual
Country of Study: Any country
No. of awards offered: 7
Application Procedure: Applicants must submit 6 copies of the application and essay.
Closing Date: December 1st
Funding: Commercial
Contributor: EBSCO Subscription Services

Scholastic Library Publishing Award

Subjects: Library work with children and young people to high school age.
Purpose: To recognize a librarian whose unusual contribution to the stimulation and guidance of reading by children and young people exemplifies outstanding achievement in the profession. The award is given either for outstanding continuing service, or in recognition of one particular contribution of lasting value.
Eligibility: Open to community and school librarians.
Level of Study: Professional development
Type: Award
Value: US$1,000 plus a citation
Frequency: Annual
No. of awards offered: 1
Application Procedure: Applicants must submit six copies of the application to the ALA Awards Programme Office.
Closing Date: December 1st
Funding: Commercial
Contributor: Gale, a part of Cengage learning
No. of awards given last year: 1

AMERICAN LUNG ASSOCIATION

1301 Pennsylvania Ave., NW, Suite 800, Washington, DC, 20004, United States of America
Tel: (1) 202 785 3355
Fax: (1) 202 452 1805
Email: info@lungusa.org
Website: www.lungusa.org
Contact: Ms Evita Mendoza

The American Lung Association is the oldest voluntary health organization in the United States, with a National Office and constituent and affiliate associations around the country. Founded in 1904 to fight tuberculosis, the American Lung Association today fights lung disease in all its forms, with special emphasis on asthma, tobacco control and environmental health.

Lung Health (LH) Research Dissertation Grants

Subjects: Psychosocial, behavioral, health services, health policy, epidemiological, biostatistical and educational matters related to lung disease.
Purpose: To provide financial assistance for Doctoral research training for dissertation research on issues relevant to lung disease.
Eligibility: Open to citizens or permanent residents of the United States.
Level of Study: Doctorate
Type: Research grant
Value: US$21,000 per year
Length of Study: 1–2 years
Frequency: Annual
Country of Study: United States of America
No. of awards offered: 1
Closing Date: October 20th
Funding: Corporation, foundation, government
Contributor: American Lung Association
Additional Information: Up to $16,000 of the award may be used for a student stipend. Funds may not be used for tuition. The award will terminate at the time the awardee is granted a doctoral degree.

For further information contact:

Research and Program Services, American Lung Association, 14 Wall Street, 8th Floor, New York, NY, 10005, United States of America

AMERICAN METEOROLOGICAL SOCIETY (AMS)

45 Beacon Street, Boston, MA, 02108-3693, United States of America
Tel: (1) 617 227 2425
Fax: (1) 617 742 8718
Email: amsinfo@ametsoc.org
Website: www.ametsoc.org

The American Meteorological Society (AMS) promotes the development and dissemination of information and education on the atmospheric and related oceanic and hydrologic sciences and the advancement of their professional applications. Founded in 1919, AMS has a membership of more than 11,000 professionals, professors, students and weather enthusiasts.

AMS Graduate Fellowship in the History of Science

Subjects: History of the atmospheric and related oceanic and hydrologic sciences.
Purpose: To provide financial support to students who wish to complete a dissertation in the related fields.
Eligibility: Open to graduate students in good standing who propose to complete the dissertation in the subject mentioned above.
Level of Study: Research, Graduate
Type: Fellowship

Value: US$15,000 stipend
Length of Study: 1 year
Frequency: Annual
Country of Study: United States of America
No. of awards offered: 1
Application Procedure: Applicants must submit a complete application form along with a cover letter with curriculum vitae, official transcripts from undergraduate and graduate institutions, a typewritten, detailed description of the dissertation topic and proposed research plan (maximum 10 pages) and 3 letters of recommendation. See website for details www.ametsoc.org.
Closing Date: February 8th
Funding: Foundation
Contributor: Member donations
No. of awards given last year: 1
Additional Information: Any questions regarding the fellowship opportunity may be directed to Donna Fernandez, 617 227 2426 ext 246 or Stephanie Armstrong, 617 227 2426 ext 235. Please see the website for further details http://www.ametsoc.org/amsstudentinfo/scholfeldocs/gradfellowshipscience.html.

AMS Graduate Fellowships

Subjects: Atmospheric and related oceanic and hydrologic sciences.
Purpose: To attract promising young scientists to prepare for careers in the atmospheric and related oceanic and hydrologic fields.
Eligibility: Applicants must be US citizens or hold permanent resident status entering their first year of graduate school and provide evidence of acceptance as a full-time student at an accredited US institution at the time of the award. Applicants must have a minimum grade point average of 3.25 on a 4.0-point scale.
Level of Study: Graduate
Type: Fellowship
Value: $24,000 to each recipient for a 9-month period
Length of Study: 9 months
Frequency: Annual
Country of Study: United States of America
No. of awards offered: 14
Application Procedure: Application form should be completed and written references, official transcripts, and GRE score reports, may be sent under separate cover. Applications and written references can be sent via email to dFernandez@ametsoc.org. Also refer to website www.ametsoc.org.
Closing Date: January 18th
Funding: Corporation, foundation, government
Contributor: Industry leaders and government agencies
No. of awards given last year: 13
Additional Information: The evaluation of applicants will be based on applicant's performance as an undergraduate student, including academic records, recommendations and GRE scores. Please see the website for further details http://www.ametsoc.org/amsstudentinfo/scholfeldocs/.

AMERICAN MUSEUM OF NATURAL HISTORY (AMNH)

Central Park West, 79th Street, New York, NY, 10024-5192, United States of America
Tel: (1) 212 769 5606
Fax: (1) 212 769 5427
Email: yna@amnh.org
Website: www.amnh.org
Contact: Ms Maria Dixon, Office of Grants & Fellowships

For 125 years, the American Museum of Natural History (AMNH) has been one of the world's pre-eminent science and research institutions, renowned for its colletions and exhibitions that illuminate millions of years of the Earth's evolution.

AMNH Annette Kade Graduate Student Fellowship Program

Subjects: Vertebrate zoology, invertebrate zoology, paleontology, physical sciences and anthropology.
Purpose: To partner with French and German institutions and to permit an exchange of graduate students.
Eligibility: Open to students who are engaged in full-time research towards a Master's or PhD degree.
Level of Study: Postgraduate
Type: Fellowship
Value: US$2,500 monthly stipend for housing and food for a 3 month stay. Travel will be provided in the form of roundtrip airfare of up to US $1,500.
Length of Study: 3 months
Frequency: Annual
Country of Study: United States of America and Europe
No. of awards offered: 4
Application Procedure: Applicants must submit their curriculum vitae, project description, reference letters and transcripts along with their application form. There are three (3) parts to the application process that must be completed by deadline time online.
Closing Date: March 1st
Funding: Foundation
Additional Information: Each student will pursue a predetermined research project with a science mentor. This mentor will also ensure that the student is exposed to the other disciplines in the Museum. Research interest must be in one of the subjects mentioned above and it must be in keeping with the Museum's research interests. Note that this program is by invitation only. If you are interested in participating, an AMNH curator must nominate you.

AMNH Research Fellowships

Subjects: Vertebrate zoology, invertebrate zoology, paleozoology, anthropology, astrophysics and Earth and planetary sciences.
Purpose: To provide financial support to recent postdoctoral investigators and established scientists to carry out a specific project within a limited time period.
Eligibility: Open to students with a Doctoral degree or an equivalent degree.
Level of Study: Postdoctorate, Research
Type: Research fellowship
Value: Varies
Length of Study: 2 years
Frequency: Annual
Study Establishment: American Museum of Natural History
Country of Study: United States of America
Application Procedure: Applicants can obtain the application form online or from the office of grants and fellowships. The application requires a project description with bibliography, budget, curriculum vitae including list of publications and letters of recommendation.
Closing Date: November 15th

AMERICAN MUSIC CENTER

322 8th Avenue, Suite 1401, New York, NY, 10001, United States of America
Tel: (1) 212 366 5260
Fax: (1) 212 366 5265
Email: jclarke@amc.net
Website: www.amc.net
Contact: Jenny Clarke, Manager of Grantmaking Programmes

The American Music Center is a non-profit membership and service organization. The Center's mission is to build a national community for new American music.

American Music Center Composer Assistance Program

Subjects: Music.
Purpose: To support individual composers to realize their music in performance.
Eligibility: American composers in good standing with the American Music Center.
Level of Study: Professional development
Type: Fellowship
Value: Total of approximately $80,000 annually. The maximum one-time grant is up to US$5,000.
Length of Study: Variable
Frequency: Annual
Country of Study: United States of America
No. of awards offered: Varies

Application Procedure: Applicants must download guidelines from website.
Closing Date: October 15th
Funding: Private
Contributor: The Helen F Whitaker Fund

For further information contact:

American Music Center 30 West 26th Street, Suite 1001, New York, NY, 10010-2011, United States of America
Contact: Jennifer Clarke, Grants Manager

AMERICAN MUSICOLOGICAL SOCIETY (AMS)

6010 College Station, Brunswick ME, 04011-8451, United States of America
Tel: (1) 207 798 4243
Fax: (1) 207 798 4254
Email: ams@ams-net.org
Website: www.ams-net.org
Contact: A L Hipkins, Office Manager

The American Musicological Society (AMS) was founded in 1934 as a non-profit organization, with the aim of advancing research in the various fields of music as a branch of learning and scholarship. In 1951, the Society became a constituent member of the American Council of Learned Societies.

Alfred Einstein Award

Subjects: Musicology.
Purpose: To honour a musicological article of exceptional merit by a scholar in the early stages of his or her career.
Eligibility: Open to citizens or permanent residents of Canada or the United States of America.
Level of Study: Professional development
Type: Prize
Value: Varies
Frequency: Annual
No. of awards offered: 1
Application Procedure: Applicants must be nominated. The committee will entertain articles from any individual, including eligible authors who are encouraged to nominate their own articles. Nominations should include the name of the author, the title of the article and the name and year of the periodical or other collection in which it was published. A curriculum vitae is also required.
Closing Date: May 1st
Funding: Private
No. of awards given last year: 1
No. of applicants last year: 26

Alvin H Johnson AMS 50 Dissertation One Year Fellowships

Subjects: Any field of musical research.
Purpose: To encourage research in the various fields of music as a branch of learning and scholarship.
Eligibility: Open to full-time students registered for a doctorate at a North American university, who have completed all formal degree requirements except the dissertation at the time of full application. Open to all students without regard to nationality, race, religion or gender.
Level of Study: Doctorate, Postgraduate
Type: Fellowship
Value: US$20,000 stipend
Length of Study: 1 year
Frequency: Annual
Country of Study: United States of America or Canada
No. of awards offered: 3
Application Procedure: Application forms will be sent via the Directors of Graduate Study at all doctorate-granting institutions in North America. They will also be available directly from the Society and the website. Applications must include a curriculum vitae, certification of enrolment and degree completed and two supporting letters from faculty members, one of whom must be the principal adviser of the dissertation. A detailed dissertation prospectus and a completed chapter or comparable written work on the dissertation should accompany the full application. All documents should be submitted in triplicate.
Closing Date: December 17th
Funding: Private
No. of awards given last year: 4
No. of applicants last year: 67
Additional Information: Any submission for a doctoral degree in which the emphasis is on musical scholarship is eligible. The award is not intended for support of early stages of research and it is expected that a recipient's dissertation will be completed within the fellowship year. An equivalent major award from another source may not normally be held concurrently unless the AMS award is accepted on an honorary basis.

AMS Subventions for Publications

Subjects: Musicology.
Purpose: To help individuals with expenses involved in the publication of works of musical scholarship, including books, articles and works in non-print media.
Eligibility: Open to younger scholars and scholars in the early stages of their careers. Proposals for projects that make use of newer technologies are welcomed.
Level of Study: Professional development
Type: Grant
Value: US$500–2,000 with a maximum of US$2,500 available
Application Procedure: Applicants must submit a short, written abstract of up to 1,000 words describing the project and its contribution to musical scholarship, a copy of the article or other equivalent sample, a copy of a contract or letter of agreement from the journal editor or publisher indicating final acceptance for publication, and a detailed budget and explanation of the expenses to which the subvention would be applied. Wherever possible expenses should be itemized. If the publication is a book a representative chapter should be submitted.
Closing Date: February 15th to August 15th
Funding: Private
No. of awards given last year: 16
No. of applicants last year: 53
Additional Information: No individual can receive a subvention more than once in a 3-year period.

Howard Mayer Brown Fellowship

Subjects: Musicology.
Purpose: To increase the presence of minority scholars and teachers in musicology.
Eligibility: Open to candidates who have completed at least 1 year of academic work at an institution with a graduate programme in musicology and who intend to complete a PhD in the field. Applicants must be members of a group historically underrepresented in the discipline, including African Americans, Native Americans, Hispanic Americans and Asian Americans. Candidates will normally be citizens or permanent residents of the United States of America or Canada. There are no restrictions on age or gender.
Level of Study: Postgraduate
Type: Fellowship
Value: US$20,000 for 12 months
Length of Study: 1 year
Frequency: Annual
Study Establishment: An institution which offers a graduate programme in musicology
Country of Study: United States of America or Canada
Application Procedure: Applicants must be nominated. Nominations may come from a faculty member of the institution at which the student is enrolled, from a member of the AMS at another institution, or directly from the student. Supporting documents must include a letter summarizing the candidate's academic background, letters of support from three faculty members and samples of the applicant's work such as term papers or any published material.
Closing Date: December 17th of the year in which the fellowship is awarded
Funding: Private
No. of awards given last year: 2
No. of applicants last year: 22

Additional Information: The AMS encourages the institution at which the recipient is pursuing his or her degree to offer continuing financial support. Further information is available on request.

Noah Greenberg Award

Subjects: Musicology.
Purpose: To provide a grant-in-aid to stimulate active co-operation between scholars and performers by recognizing and fostering outstanding contributions to historical performing practices.
Eligibility: Both scholars and performers may apply. Applicants need not be members of the Society.
Level of Study: Professional development
Type: Award
Value: $2000 and a certificate
Frequency: Annual
No. of awards offered: 1–2
Application Procedure: Applicants must submit three copies of a description of the project, a detailed budget and supporting materials such as articles or tapes of performances which are relevant to the project. Applications must be sent to the chair of the Noah Greenberg Award Committee.
Closing Date: August 15th
Funding: Private
No. of awards given last year: 1
No. of applicants last year: 27

Otto Kinkeldey Award

Subjects: Musicology.
Purpose: To award the work of musicological scholarship such as a major book, edition or other piece of scholarship that best exemplifies the highest quality of originality, interpretation, logic, clarity of thought and communication.
Eligibility: The work must have been published during the previous year in any language and in any country by a scholar who is a citizen or permanent resident of Canada or the United States of America.
Level of Study: Professional development
Type: Prize
Value: Varies
Frequency: Annual
No. of awards offered: 1
Application Procedure: Applicants must write for details.
Closing Date: May 1st
Funding: Private
No. of awards given last year: 1
No. of applicants last year: 34
Additional Information: Further information is available on request.

Paul A Pisk Prize

Subjects: Musicology.
Purpose: To encourage scholarship.
Eligibility: Open to graduate students whose abstracts have been submitted to the Programme Committee of the Society and papers accepted for inclusion in the Annual Meeting. Open to all students without regard to nationality, race, religion or gender.
Level of Study: Graduate
Type: Prize
Value: Varies
Frequency: Annual
Country of Study: United States of America or Canada
No. of awards offered: 1
Application Procedure: Applicants must submit three copies of the complete text paper to the chair of the Pisk Prize Committee. The submission must be accompanied by a statement from the student's academic adviser affirming the graduate student status of the applicant.
Closing Date: October 1st
Funding: Private
No. of awards given last year: 1
No. of applicants last year: 29
Additional Information: Further information is available on request.

AMERICAN NUCLEAR SOCIETY (ANS)

555 North Kensington Avenue, La Grange Park, IL, 60526, United States of America
Tel: (1) 708 352 6611
Fax: (1) 708 352 0499
Email: outreach@ans.org
Website: www.ans.org
Contact: Scholarship Programme

The American Nuclear Society (ANS) is a non-profit, international, scientific and educational organization. It was established by a group of individuals who recognized the need to unify the professional activities within the diverse fields of nuclear science and technology.

Alan F Henry/Paul A Greebler Scholarship

Subjects: Reactor physics.
Purpose: To aid students pursuing studies in the field of nuclear science.
Eligibility: Open to full-time graduate students at a North American university engaged in Master's or PhD research in the area of nuclear reactor physics or radiation transport. Applicants may be of any nationality.
Level of Study: Graduate, Postdoctorate
Type: Scholarship
Value: US$3,500
Length of Study: Varies
Frequency: Annual
Study Establishment: An accredited institution
Country of Study: United States of America
No. of awards offered: 1
Application Procedure: Applicants must complete an application form available from the organization. An official grade transcript and three completed confidential reference forms are also required. Applications on line at www.ans.org/honors/scholarships
Closing Date: February 1st
No. of awards given last year: 1
Additional Information: Further information is available either on request or from the website.

ANS Undergraduate/Graduate Pittsbugh Local Section Scholarship

Subjects: Nuclear science and technology.
Eligibility: Two Pittsburgh Local Section Scholarships – one for a graduate student (studying nuclear science and technology) and one for an undergraduate student (studying nuclear science and technology) who either have some affiliation with Western Pennsylvania or who attend school at a nearby university within the region. Applicants must check the Pittsburgh Local Section box on the scholarship form.
Value: US$2000–3,500; undergraduate/graduate US$3,000
Frequency: Annual
Study Establishment: An accredited institution
Country of Study: United States of America
Application Procedure: Applications on line at www.ans.org/honors/scholarships
Closing Date: February 1st

Delayed Education Scholarship for Women

Subjects: Must be a mature woman whose undergraduate studies in nuclear science, nuclear engineering or a nuclear-related field have been delayed for at least one year.
Value: US$4,000
Frequency: Annual
Study Establishment: An accredited institution
Country of Study: United States of America
No. of awards offered: 1
Application Procedure: Applications on line at www.ans.org/honors/scholarships
Closing Date: February 1st

Everitt P Blizard Scholarship

Subjects: Radiation protection and shielding.
Purpose: To aid students pursuing studies in the field of radiation protection and shielding.

Eligibility: Open to full-time graduate students in a programme leading to an advanced degree in nuclear science, nuclear engineering or a nuclear-related field. Applicants must be citizens of the United States of America or permanent residents and be enrolled in an accredited institution in the United States of America.
Level of Study: Graduate
Type: Scholarship
Value: $3,000
Length of Study: Varies
Frequency: Annual
Study Establishment: An accredited institution
Country of Study: United States of America
No. of awards offered: 1
Application Procedure: Applicants must complete an application form available from the organization. An official grade transcript and three completed confidential reference forms are also required. Applications on line at www.ans.org/honors/scholarships
Closing Date: February 1st
No. of awards given last year: 1
Additional Information: Further information is available either on request or from the website.

James F Schumar Scholarship

Subjects: Materials science and technology for nuclear applications.
Eligibility: Open to citizens of the United States of America or holders of a permanent resident visa who are full-time graduate students enrolled in a programme leading to an advanced degree
Level of Study: Graduate
Type: Scholarship
Value: $3,000
Frequency: Annual
Study Establishment: An accredited institution
Country of Study: United States of America
No. of awards offered: 1
Application Procedure: Applicants must submit a request for an application form that includes the name of the university the candidate will be attending, the year the candidate will be in during the autumn of the award, the major course of study and a stamped addressed envelope. Completed applications must include a grade transcript and three confidential reference forms. Candidates must be sponsored by an ANS section, division, student branch, committee, member or organization member. The applicant should indicate on the nomination form that he or she is applying for the MSTD Scholarship. Applications on line at www.ans.org/honors/scholarships
Closing Date: February 1st
No. of awards given last year: 1
Additional Information: Further information is available either on request or from the website.

John and Muriel Landis Scholarship Awards

Subjects: Nuclear physics and engineering.
Purpose: To help students who have greater than average financial need.
Eligibility: Candidates should be planning to pursue a career in nuclear engineering or a nuclear-related field. Candidates must have greater than average financial need, and consideration will be given to conditions or experiences that render the student disadvantaged. Applicants need not be citizens of the United States of America.
Level of Study: Graduate
Type: Scholarship
Value: US$4,000
Frequency: Annual
Study Establishment: An accredited institution
Country of Study: United States of America
No. of awards offered: Up to 8
Application Procedure: Applicants must request an application form and include the name and a letter of commitment from the university the candidate will be attending, the year the candidate will be in the Autumn of the award, the major course of study and a stamped addressed envelope. Completed applications must include a grade transcript and three confidential reference forms. Candidates must be sponsored by an ANS section, division, student branch, committee, member or organization member. Applications on line at www.ans.org/honors/scholarships
Closing Date: February 1st

John R. Lamarsh Scholarship

Subjects: Nuclear science and technology.
Eligibility: US and non-US applicants must be ANS student members enrolled in and attending an accredited institution in the United States. Academic accomplishments must be confirmed by transcript.
Value: $2,000
Frequency: Annual
Study Establishment: An accredited institution
Country of Study: United States of America
Application Procedure: Applications on line at www.ans.org/honors/scholarships
Closing Date: February 1st
No. of awards given last year: 1

Operations and Power Division Scholarship Award

Subjects: Nuclear science and technology.
Eligibility: US and non-US applicants must be ANS student members enrolled in and attending an accredited institution in the United States. Academic accomplishments must be confirmed by transcript. The OPD Scholarship is intended for an undergraduate or graduate student. Applicants: must be enrolled in a course of study leading to a degree in nuclear science or engineering at an accredited institution in the United States; must have completed a minimum of two complete academic years in a four-year nuclear science or engineering program; must be U.S. citizens or possess a permanent resident visa; must have intentions of working in the nuclear power industry.
Value: $2,500
Frequency: Annual
Study Establishment: An accredited institution
Country of Study: United States of America
No. of awards offered: 1
Application Procedure: Applications on line at www.ans.org/honors/scholarships.
Closing Date: February 1st

Robert A Dannels Memorial Scholarship

Subjects: Nuclear science or nuclear engineering.
Eligibility: Open to citizens of the United States of America or holders of a permanent resident visa who are full-time graduate students enrolled in a programme leading to an advanced degree or in a graduate-level course of study leading towards a degree in mathematics and computation. Handicapped persons are encouraged to apply.
Level of Study: Graduate
Type: Scholarship
Value: $3,500
Frequency: Annual
Study Establishment: An accredited institution
Country of Study: United States of America
No. of awards offered: 1
Application Procedure: Applicants must submit a request for an application form that includes the name of the university the candidate will be attending, the year the candidate will be in during the Autumn of the award, the major course of study and a stamped addressed envelope. Completed applications must include a grade transcript and three confidential reference forms. Candidates must be sponsored by an ANS section, division, student branch, committee, member or organization member. Applications on line at www.ans.org/honors/scholarships
Closing Date: February 1st

Verne R Dapp Memorial Scholarship

Subjects: Nuclear science or nuclear engineering.
Eligibility: Open to citizens of the United States of America or holders of a permanent resident visa who are full-time graduate students enrolled in a programme leading to an advanced degree.
Level of Study: Graduate
Type: Scholarship
Value: $3,000
Frequency: Every 2 years
Study Establishment: An accredited institution
Country of Study: United States of America
No. of awards offered: 1
Application Procedure: Applicants must submit a request for an application form that includes the name of the university the candidate

will be attending, the year the candidate will be in during the Autumn of the award, the major course of study and a stamped addressed envelope. Completed applications must include a grade transcript and three confidential reference forms. Candidates must be sponsored by an ANS section, division, student branch, committee, member or organization member. Applications on line at www.ans.org/honors/scholarships
Closing Date: February 1st

Vogt Radiochemistry Scholarship Award

Subjects: Student must be enrolled in or proposing to undertake research in radioanalytical chemistry, analytical chemistry or analytical applications of nuclear science.
Eligibility: US and non-US applicants must be ANS student members enrolled in and attending an accredited institution in the United States. Academic accomplishments must be confirmed by transcript.
Level of Study: Graduate, or undergraduate
Type: Scholarship
Value: US$2,000 if awarded to an undergraduate (Junior/Senior) US $3,000 if awarded to a graduate.
Frequency: Annual
Country of Study: United States of America
No. of awards offered: 1 per year
Application Procedure: Applications on line at www.ans.org/honors/scholarships
Closing Date: February 1st
No. of awards given last year: 1

Walter Meyer Scholarship

Subjects: Nuclear physics and engineering.
Eligibility: Open to full-time graduate students in a programme leading to an advanced degree in nuclear science, nuclear engineering or a nuclear-related field. Applicants must be citizens of the United States of America or permanent residents and be enrolled in an accredited institution in the United States of America.
Type: Scholarship
Value: $3,500
Length of Study: Varies
Frequency: Every 2 years
Study Establishment: An accredited institution
Country of Study: United States of America
No. of awards offered: 1
Application Procedure: Applicants must complete an application form available from the organization. An official grade transcript and three completed confidential reference forms are also required. Applications on line at www.ans.org/honors/scholarships.
Closing Date: February 1st
Additional Information: Further information is available either on request or from the website.

AMERICAN NUMISMATIC SOCIETY (ANS)

75 Varick Street, floor 11, New York, NY, 10013, United States of America
Tel: (1) 212 571 4470
Fax: (1) 212 571 4479
Email: wartenberg@numismatics.org
Website: www.numismatics.org
Contact: Dr Ute Wartenberg Kagan, Executive Director

The mission of the American Numismatic Society (ANS) is to be the preeminent national institution advancing the study and appreciation of coins, medals and related objects of all cultures as historical and artistic documents. It aims to do this by maintaining the foremost numismatic collection and library, supporting scholarly research and publications, and sponsoring educational and interpretative programmes for diverse audiences.

Grants for ANS Summer Seminar in Numismatics

Subjects: Numismatics.
Purpose: To provide a selected number of graduate students with a deeper understanding of the contribution that this subject makes to other fields.
Eligibility: Open to applicants who have had at least 1 year's graduate study at a university in the United States of America or Canada and who are students of classical studies, history, near eastern studies or other humanistic fields.
Level of Study: Postgraduate
Type: Grant
Value: $4,000
Length of Study: 9 weeks during the Summer
Frequency: Annual
Study Establishment: Museum of the American Numismatic Society
Country of Study: United States of America
No. of awards offered: Approx. 5
Application Procedure: Applicants must write well in advance for details of the application process.
Closing Date: February 15th
Funding: Private
No. of awards given last year: 12
No. of applicants last year: 21
Additional Information: One or two students from overseas are usually accepted to the seminar but will not receive a grant.

THE AMERICAN OCCUPATIONAL THERAPY FOUNDATION (AOTF)

4720 Montgomery Lane, PO Box 31220, Bethesda, MD, 20814, United States of America
Tel: (1) 301 652 6611
Fax: (1) 301 656 3620
Email: aotf@aotf.org
Website: www.aotf.org

The AOTF is a charitable, non-profit organization created in 1965 to advance the science of occupational therapy and increase public understanding of its value.

The A. Jean Ayres Award

Subjects: Occupational therapy.
Purpose: To recognize occupational therapy clinicians, educators and researchers who have made significant contributions to their profession.
Eligibility: Be initially certified by the National Board for Certification in Occupational Therapy, formerly the American Occupational Therapy Certification Board. Be a member in good standing of the American Occupational Therapy Association. Not be serving as a member of the American Occupational Therapy Foundation Board of Directors.
Level of Study: Doctorate
Type: Award
Value: US$500, a plaque and acknowledgement at the Annual conference of the American Occupational Therapy Association
Frequency: Annual
No. of awards offered: 2
Application Procedure: Five copies of the nomination package must be sent to the chairman of the AOTF Awards of Recognition Committee.
Closing Date: Deadline for receipt of 2013 nominations for awards to be given at the AOTA 2013 Annual Conference and Exhibition is October 1, 2012.
Funding: Foundation
Contributor: American Occupational Therapy Foundation
No. of awards given last year: 1

AOTF Certificate of Appreciation

Subjects: Occupational therapy.
Purpose: To recognize outstanding service toward the Foundation.
Eligibility: Any individual, agency, business or other institution. May not be a current voting member of the AOTF Board of Directors.
Level of Study: Doctorate, Postgraduate
Type: Award
Value: A certificate of appreciation and acknowledgement at the annual conference
Frequency: Annual
Application Procedure: Four copies of a letter of nomination and candidate's curriculum vitae must be sent.
Closing Date: November 1st
Funding: Foundation
Contributor: American Occupational Therapy Foundation

AOTF Scholarship

Eligibility: Open to Masters-level students with academic excellence.
Type: Scholarship
Value: Multiple scholarships available raging from US$375–5,000
Frequency: Annual
No. of awards offered: 40
Application Procedure: Online application can be found on AOTF website: www.aotf.org.
Closing Date: Varies – generally November or December
Funding: Foundation
No. of awards given last year: Varies over 50 available
No. of applicants last year: Varies

AMERICAN ORCHID SOCIETY

Fairchild Tropical Botanic Garden, 10901 Old Cutler Road, Coral Gables, FL, 33156, United States of America
Tel: (1) 305 740 2010
Fax: (1) 305 740 2011
Email: theaos@aos.org
Website: www.aos.org
Contact: Ms Pamela S Giust, Awards Registrar

Grants for Orchid Research

Subjects: Orchid research.
Purpose: To advance scientific study of orchids in every respect and to assist in the publication of scholarly and popular scientific literature on orchids.
Eligibility: There are no eligibility restrictions.
Level of Study: Postgraduate
Type: Grant
Value: $500–12,000
Length of Study: Up to 3 years
Frequency: Annual
Country of Study: Any country
No. of awards offered: Varies
Application Procedure: Applicants must write for guidelines.
Closing Date: January 1st

For further information contact:

Tel: 561 404 2000
Email: theaos@aos.org
Website: www.aos.org
Contact: Executive Director

AMERICAN ORIENTAL SOCIETY

Hatcher Graduate Library, University of Michigan, Ann Arbor, MI, 48109-1190, United States of America
Tel: (1) 734 647 4760
Fax: (1) 734 763 6743
Email: jrodgers@umich.edu
Website: www.umich.edu/˜aos/
Contact: Grants Management Officer

The American Oriental Society is primarily concerned with the encouragement of basic research in the languages and literatures of Asia.

Louise Wallace Hackney Fellowship for the Study of Chinese Art

Subjects: Chinese art, with special relation to painting, and the translation into English of works on the subject.
Purpose: To remind scholars that Chinese art, like all art, is not a disembodied creation, but the outgrowth of the life and culture from which it has sprung. It is requested that scholars give special attention to this approach in their study.
Eligibility: Open to United States citizens who are doctoral or postdoctoral students and have successfully completed at least 3 years of Chinese language study at a recognized university, and have some knowledge or training in art. In no case shall a fellowship be awarded to Scholars of well-recognized standing, but shall be given to either men or women who show aptitude or promise in the said field of learning.
Level of Study: Postdoctorate
Type: Fellowship
Value: $8,000
Length of Study: 1 year
Frequency: Annual
Study Establishment: Any institution where paintings and adequate language guidance is available
Country of Study: Any country
No. of awards offered: 1
Application Procedure: Applicants must submit the following materials in duplicate: a transcript of their undergraduate and graduate course work, a statement of personal finances, a four page summary of the proposed project to be undertaken including details of expense, and no less than three letters of recommendation.
Closing Date: March 1st
Funding: Private
Additional Information: It is possible to apply for a renewal of the fellowship, but this may not be done in consecutive years.

For further information contact:

Hackney Fellowship, American Oriental Society, Hatcher Graduate Library, Ann Arbor, MI, 48109-1205

THE AMERICAN OTOLOGICAL SOCIETY (AOS)

3096 Riverdale Road, The Villages, FL, 32162, United States of America
Tel: (1) 352 751 0932
Fax: (1) 352 751 0696
Email: segossard@aol.com
Website: www.americanotologicalsociety.org
Contact: Ms Shirley Gossard, Administrator

The AOS is a society focused upon 'aural' medicine. The society's mission is to advance and promote medical and surgical otology, encouraging research in the related disciplines.

American Otological Society Research Grants

Subjects: All aspects of otosclerosis, Ménière's disease and related disorders.
Purpose: To encourage and support academic research in sciences related to the investigation of Otosclerosis or Ménière's Disease.
Eligibility: Open to physicians and doctorate level investigators in the United States and Canada only.
Level of Study: Postdoctorate, Postgraduate
Type: Research grant
Value: Up to US$55,000 per year. No funding is provided for the investigator's salary
Length of Study: 1 year, renewable
Frequency: Annual
Country of Study: United States of America or Canada
No. of awards offered: Varies
Application Procedure: A research proposal will qualify for review when it involves anatomical, physiological, biochemical, pharmacological, physical, genetic, environmental, psychological, pathological or audiological investigations, or other studies which are related in some way to Otosclerosis or Ménière's disease. Applications are reviewed by the Board of Trustees of the AOS Research Fund.
Closing Date: January 31st
Funding: Private
Contributor: American Otological Society, Inc.
No. of awards given last year: 3
No. of applicants last year: 10
Additional Information: The grants and reference letters are to be submitted via email to Shirley Gossard, Administrator for the American Otological Society Research Fund, segossard@aol.com and to Dr. John Carey, Executive Secretary of the American Otological Society Research Fund, jcarey@jhmi.edu.

For further information contact:

Research Fund of the American Otological Society, Inc., Johns Hopkins University, School of Medicine, Department of Otolaryngol-

ogy-Head & Neck Surgery, 601 N. Caroline Street, JHOC 6210, Baltimore, MD, 21287-0910
Tel: 410 955 7381
Fax: 410 955 0035
Email: jcarey@jhmi.edu
Website: http://www.americanotologicalsociety.org/information.html
Contact: John P. Carey, MD, Executive Secretary

American Otological Society Research Training Fellowships

Subjects: All aspects of otosclerosis, Ménière's disease and related disorders.
Purpose: To support research.
Eligibility: Open to physicians, residents and medical students in the United States of America and Canada.
Level of Study: Postgraduate
Type: Fellowship
Value: Up to US$40,000 depending on position and institutional norms ($35,000 for stipend, $5,000 for supplies)
Length of Study: 1–2 years
Frequency: Annual
Country of Study: United States of America or Canada
No. of awards offered: Varies
Closing Date: January 31st
Contributor: American Otological Society, Inc.
No. of awards given last year: 1
No. of applicants last year: 1
Additional Information: The organization requires institutional documentation that facilities and faculty are appropriate for the requested research.

For further information contact:

Research Fund of the American Otological Society, Inc., Johns Hopkins University, School of Medicine, Department of Otolaryngology-Head & Neck Surgery, 601 N. Caroline Street, JHOC 6210, Baltimore, MD, 21287-0910, United States of America
Tel: (1) 410 955 7381
Fax: (1) 410 955 0035
Email: jcarey@jhmi.edu
Contact: John P. Carey, MD, Executive Secretary

AMERICAN PHILOSOPHICAL SOCIETY

104 South Fifth Street, Philadelphia, PA, 19106-3387, United States of America
Tel: (1) 215 440 3429
Fax: (1) 215 440 3450
Email: lmusumeci@amphilsoc.org
Website: www.amphilsoc.org
Contact: Linda Musumeci, Director of Grants and Fellowships

The American Philosophical Society is an eminent scholarly organization of international reputation and promotes useful knowledge in the sciences and humanities through excellence in scholarly research, professional meetings, publications, library resources and community outreach.

Daland Fellowships in Clinical Investigation

Subjects: Internal medicine, neurology, pediatrics, psychiatry and surgery.
Purpose: To award a limited number of fellowships for research in clinical medicine including the fields of internal medicine, neurology, pediatrics, psychiatry and surgery. For the purposes of this award, the committee emphasizes patient-orientated research.
Eligibility: Candidates are expected to have held the MD degree for less than 8 years. The fellowship is intended to be the first postclinical fellowship, but each case will be decided on its merits. Preference is given to candidates who have less than 2 years of postdoctoral training. Applicants must expect to perform their research at an institution in the United States of America, under the supervision of a scientific adviser.
Level of Study: Research, Post-MD
Type: Fellowship
Value: US$40,000 each for the first and second year
Length of Study: 1 year, with renewal for a further year if satisfactory progress is demonstrated.
Frequency: Annual
Country of Study: United States of America
Application Procedure: Applicants must complete an application form. Information and forms are available from the website. Candidates must be nominated by their department chairman in a letter providing assurance that the nominee will work with the guidance of a scientific adviser of established reputation who has guaranteed adequate space, supplies, etc. for the Fellow. The adviser need not be a member of the department nominating the Fellow, nor need the activities of the Fellow be limited to the nominating department. As a general rule, no more than one fellowship will be awarded to a given institution in the same year of competition. Application forms must be sent to the Daland Fellowship Committee along with letters of support from the scientific adviser and another expert.
Closing Date: September 17th
Funding: Private
No. of awards given last year: 1
No. of applicants last year: 20

Franklin Research Grant Program

Subjects: Scholarly research: as the term is used here, covers most kinds of scholarly inquiry by individuals leading to publication. It does not include journalistic or other writing for general readership, the preparation of textbooks, case books, anthologies or other materials for use by students or the work of creative and performing artists.
Purpose: To contribute towards the cost of scholarly research in all areas of knowledge, except those in which support by government or corporate enterprise is more appropriate.
Eligibility: Applicants are normally expected to have a doctorate, but applications are considered from persons whose publications display equivalent scholarly achievement. Grants are never made for predoctoral study or research. It is the Society's longstanding practice to encourage younger scholars. The Committee will seldom approve more than two grants to the same person within any 5 year period. Applicants may be residents of the United States of America, citizens of the United States of America on the staffs of foreign institutions or foreign nationals whose research can only be carried out in the United States of America. Institutions are not eligible to apply. Applicants expecting to conduct interviews in a foreign language must possess sufficient competence in that language and must be able to read and translate all source materials.
Level of Study: Postdoctorate, Research
Type: Grant
Value: The maximum grant is US$6,000. The budget year corresponds to the calendar year, not the academic year. If an applicant receives an award for the same project from another granting institution, the Society will consider limiting its award to costs that are not covered by the other grant.
Length of Study: Two programs of 1–2 and 2–4 months
Frequency: Annual
Country of Study: Any country
No. of awards offered: Varies
Application Procedure: Applicants must complete an online application. Information and access to the online application portal are available through the website.
Closing Date: December 17th
Funding: Private
No. of awards given last year: 62
No. of applicants last year: 409
Additional Information: If an award is made and accepted, the recipient is required to provide the Society with a one-page report on the research accomplished during the tenure of the grant, and a one-page financial statement.
Franklin grants are taxable income.

The Lewis and Clark Fund for Exploration and Field Research

Subjects: Archeology, anthropology, biology, ecology, geography, geology, linguistics and paleontology.
Purpose: To encourage exploratory field studies for the collection of specimens and data and to provide the imaginative stimulus that accompanies direct observation.

Eligibility: Grants are available to doctoral students, the competition is open to US residents wishing to carry out research anywhere in the world. Foreign applicants must either be based at a US Institution or plan to carry out their work in the United States. Applicants should ask their academic advisor to write one of the two letters of recommendation, specifying the student's qualifications to carry out the proposed work and the educational content of the trip.
Level of Study: Doctorate, Research
Type: Grant
Value: Up to $5,000
Frequency: Annual
Country of Study: Any country
No. of awards offered: Varies
Application Procedure: Applicants must complete an online application. Information and access to the online application portal are available through the website.
Closing Date: February 1st
Funding: Private
No. of awards given last year: 28
No. of applicants last year: 260
Additional Information: Lewis and Clark Fund grants are taxable income, but the Society is not required to report payments.

Library Resident Research Fellowships

Subjects: Library collections research. Fields include early American history and culture, Atlantic history, history of science, technology and medicine, history of eugenics and genetics, history of physics especially quantum physics, history of natural history in the 18th and 19th centuries, Native American history, culture and languages, caribbean and slavery studies.
Purpose: To support research in the American Philosophical Society library's collections.
Eligibility: Scholars who reside beyond a 75-mile radius of Philadelphia will be given some preference. The fellowships are open to both citizens of the United States of America and foreign nationals who are holders of a PhD or equivalent, PhD candidates who have passed their preliminary exams and independent scholars. Applicants in any relevant field of scholarship may apply.
Level of Study: Doctorate, Postdoctorate, Predoctorate, Research
Type: Fellowship
Value: $2,500 per month
Length of Study: 1–3 months
Frequency: Annual
Country of Study: United States of America
No. of awards offered: Varies
Application Procedure: Applicants must complete an online application. Information and access to the online application portal are available through the website.
Closing Date: March 1st
Funding: Private
No. of awards given last year: 22
No. of applicants last year: 51
Additional Information: Comprehensive, researchable guides and finding aids to the society's collection are available online at http://www.amphilsoc.org/library/search (check catalogs and guides). Applicants are strongly encouraged to consult the library staff by mail or phone regarding the collections. A list of these guides and further information can be found on the website or by contacting the American Philosophical Society Library.
Awards are taxable income, but the Society is not required to report payments.

Phillips Fund for Native American Research

Subjects: Native American linguistics and ethnohistory, and the history of the study of Native Americans in the continental United States of America and Canada.
Purpose: To financially support research in archives or in the field.
Eligibility: Open to graduate students who have passed their qualifying examinations for either the Master's or doctoral degrees. Postdoctoral applicants are eligible. Applicants may be residents of the United States of America or Canada or Foreign Nationals Planning to carry out work in the United States of America or Canada.
Level of Study: Doctorate, Predoctorate, Postdoctorate, Postgraduate, Research
Value: The average award is approx. $2,500 and grants do not exceed $3,500. This is to cover travel, tapes and informants' fees and is not for general maintenance or the purchase of permanent equipment
Length of Study: 1 year
Frequency: Annual
Country of Study: United States of America & Canada
No. of awards offered: Varies
Application Procedure: Applicants must complete an online application form. Information and access to the online application portal are available through the website. A complete application includes all information requested on the form and two letters of support.
Closing Date: March 1st
Funding: Private
No. of awards given last year: 17
No. of applicants last year: 52
Additional Information: If an award is made and accepted, the recipient is required to provide the Society's Library with a brief formal report and copies of any tape recordings, transcriptions, microfilms, etc., that may be acquired in the process of the grant-funded research as well as a release for scholarly use.
Grants are taxable income, but the Society is not required to report payments.

THE AMERICAN PHYSIOLOGICAL SOCIETY (APS)

9650 Rockville Pike, Bethesda, MD, 20814-3991, United States of America
Tel: (1) 301 634 7164
Fax: (1) 301 634 7241
Email: webmaster@the-aps.org
Website: www.the-aps.org
Contact: Ms Linda Jean Dresser, Executive Assistant

The American Physiological Society (APS) is a non-profit scientific society devoted to fostering education, scientific research and the dissemination of information in the physiological sciences. The Society strives to play a role in the progress of science and the advancement of knowledge.

APS Conference Student Award

Subjects: Biology and physiology.
Purpose: To encourage the participation of young scientists in training at the APS conferences.
Eligibility: Open to graduate students wishing to present a contributed paper at an APS conference.
Level of Study: Graduate
Type: Award
Value: Cash award of $500 and complimentary conference registration
Length of Study: The duration of the conference
Frequency: Dependent upon meetings scheduled
Study Establishment: Any APS conference
Country of Study: United States of America
No. of awards offered: Varies
Application Procedure: Applicants must submit an abstract to the APS. Candidates must indicate on the abstract page a desire to be considered for the award and should contact the APS for further details.
Closing Date: Please write for details
No. of awards given last year: Varies, depending on meeting/conference dates. Please check with the individual meeting websites.

APS Mass Media Science and Engineering Fellowship

Subjects: Physiology or any related subject.
Purpose: To enable promising young scientists to work in the newsroom of a newspaper, magazine, radio or television station, sharpening their ability to communicate complex scientific issues to non-scientists and helping to improve public understanding of science.
Eligibility: Open to graduate or postgraduate students of physiology, or a related subject, preferably with a background in scientific writing.
Level of Study: Graduate, Postgraduate
Type: Studentship
Value: $4,500 stipend
Length of Study: 10 weeks
Frequency: Annual

Study Establishment: The newsroom of a newspaper, magazine or radio or television station.
Country of Study: United States of America
No. of awards offered: 1
Application Procedure: Applicants must complete an application form, available from Alice Ra'anan, Public Affairs Office, American Physiological Society.
Closing Date: January 5th
No. of awards given last year: 1

APS Minority Travel Fellowship Awards

Subjects: Biology and physiology.
Purpose: To increase the participation of predoctoral and postdoctoral minority students in the physiological sciences.
Eligibility: Open to advanced predoctoral and postdoctoral students. Students in the APS Porter Physiology Development programme are also eligible. Minority faculty members at MBRS and MARC eligible institutions may also submit applications.
Level of Study: Postdoctorate, Predoctorate
Type: Travel grant
Value: Funds for travel to attend either the Experimental Biology meeting or one of the APS conferences.
Length of Study: The duration of the conference or meeting
Country of Study: United States of America
No. of awards offered: Varies
Application Procedure: Applicants must contact the Education Office of the APS for further details.
Closing Date: January 15th. Deadline is 1 week after the EB abstract deadline or APS conference abstract deadline.
Contributor: NIDDK and NIGMS

For further information contact:

Email: education@the-aps.org
Contact: Dr Marsha Matyas

Caroline tum Suden Professional Opportunity Awards

Subjects: Biology and physiology.
Purpose: To provide funds for junior physiologists to attend and fully participate in the Experimental Biology meeting.
Eligibility: Open to graduate students or postdoctoral Fellows who are APS members or sponsored by an APS member.
Level of Study: Graduate, Postdoctorate
Type: Award
Value: $500 per award, complimentary registration for the meeting.
Length of Study: The duration of the conference
Frequency: Annual
Study Establishment: An APS Experimental Biology meeting
Country of Study: United States of America
No. of awards offered: Up to 41 awards. APS offers four abstract-based awards for graduate students and postdoctoral fellows: Caroline tum Suden/Frances Hellebrandt Professional Opportunity Awards (36); Fleur L. Strand Professional Opportunity Award (1); Steven M. Horvath Professional Opportunity Awards (2); Gabor Kaley Professional Opportunity Awards (2).
Application Procedure: Applicants must submit an abstract to APS and should contact the Education Office for further details.
Closing Date: Coincides with EB meeting abstract deadline.
No. of awards given last year: 36
Additional Information: Recipients are obliged to attend the Experimental Biology meeting and present a paper.

Porter Physiology Fellowships for Minorities

Subjects: Biology and physiology.
Purpose: To support the training of talented students entering careers in physiology by providing predoctoral fellowships for underrepresented students (African Americans, Hispanics, Native Americans, Native Alaskans and Native Pacific Islanders).
Eligibility: Open to underrepresented ethnic minority applicants, i.e. African Americans, Hispanics, Native Americans, Native Alaskans or Native Pacific Islanders who are citizens or permanent residents of the United States of America or its territories.
Level of Study: Graduate, Predoctorate, Postdoctorate
Type: Fellowship
Value: US$23,500 stipend
Length of Study: 1–2 years
Frequency: Annual
Study Establishment: Universities or research establishments
Country of Study: United States of America
No. of awards offered: Varies
Application Procedure: Applicants must contact the Education Office of the APS for further details.
Closing Date: January 15th

Procter and Gamble Professional Opportunity Awards

Subjects: Biology and physiology.
Purpose: To provide funds to predoctoral students allowing them to fully participate in the Experimental Biology meeting.
Eligibility: Open to predoctoral students who are within 1–1.5 years of completing a PhD degree and wish to present a paper at the meeting. Applicants must be student members of the APS or have an adviser or a supporting sponsor who is an APS member.
Level of Study: Predoctorate
Type: Award
Value: $500 per award and complementary registration for the Experimental Biology meeting
Length of Study: The duration of the conference
Frequency: Annual
Study Establishment: The APS Experimental Biology meeting
Country of Study: United States of America
No. of awards offered: Varies
Application Procedure: Applicants must submit an abstract to APS and should contact the Education Office for further details.
Closing Date: Please write for details
No. of awards given last year: 9

AMERICAN PSYCHOLOGICAL ASSOCIATION MINORITY FELLOWSHIP PROGRAM (APA/MFP)

Minority Fellowships Program (MFP), 750 First Street, N.E., Washington, DC, 20002 4242, United States of America
Tel: (1) 800 374 2721, 202 336 5500
Fax: (1) 202 336 6012
Email: mfp@apa.org
Website: www.apa.org/pi/mfp
Contact: Administrative Assistant

The American Psychological Association's (APA) Minority Fellowship Program (MFP) is an innovative, comprehensive and coordinated training and career development program that promotes psychological and behavioural outcomes of ethnic minority communities. MFP is committed to increasing the number of ethnic minority professionals in the field and enhancing our understanding of the life experiences of ethnic minority communities.

MFP Mental Health and Substance Abuse Services Doctoral Fellowship

Subjects: Clinical, counselling and school psychology. Predoctorate and Postdoctorate
Purpose: To promote culturally competent behavioural health services and policy for ethnic minority populations and increase the number of ethnic minority psychologists providing behavioural health services and developing policy for ethnic minority populations.
Eligibility: Applicants must be citizens or permanent residents of the United States of America enrolled full-time in an APA-accredited doctoral programme at the time the fellowship is awarded. Postdoctoral applicants will be considered for up to five years after graduation. An additional factor among the many considered is the applicant's ethnic minority group including, but not limited to, Blacks or African Americans, Alaskan Natives, American Indians, Asian Americans, Hispanics or Latinos and Pacific Islanders, and/or those who can demonstrate commitment to a career in psychology related to ethnic minority health.
Level of Study: Doctorate
Type: Fellowship
Value: Predoctorate $21,600 and Postdoctorate up to $43,476
Length of Study: 1 year, renewable for up to 3 years
Frequency: Annual
Country of Study: Any country

No. of awards offered: Approximately 25 Predoctorate and 2 Postdoctorate
Application Procedure: Applicants must submit a completed application, cover letter, addictions training plan, curriculum vitae, recommendations essay, references, transcripts and Graduate Record Examination scores. Further information and application forms are available on request.
Closing Date: January 15th
Funding: Government
Contributor: The Substance Abuse and Mental Health Administration
No. of awards given last year: 17 Predoctorate and 1 Postdoctorate
No. of applicants last year: 201 Predoctorate and 6 Postdoctorate
Additional Information: Please see the website for further details http://www.apa.org/pi/mfp/index.aspx.

For further information contact:

Email: www.apa.org/pi/mfp

THE AMERICAN PSYCHOLOGICAL FOUNDATION (APF)

750 First Street, NE, Washington, DC, 20002 4242, United States of America
Tel: (1) 202 336 5843
Fax: (1) 202 336 5812
Email: foundation@apa.org
Website: www.apa.org/apf
Contact: Parie Kadir, Program Officer

The American Psychological Foundation (APF) is a non-profit, philanthropic organization that advances the science and practice of psychology as a means of understanding behaviour and promoting health, education and human welfare. APF seeks to advance psychology and its impact on improving the human condition. The work of the Foundation is ongoing, sometimes urgent, always important and always growing.

Alexander Gralnick Research Investigator Prize

Subjects: Serious mental illness.
Purpose: To support exceptional research and mentoring accomplishments in the area of serious mental illness.
Eligibility: Applicants must have a doctoral degree (PhD, PsyD or MD) and must have a record of significant research productivity (for at least 8 years) and provide evidence of continuing creativity in the area of serious mental illness research. Applicants must have an affiliation with an accredited college, university or other research/treatment institution.
Level of Study: Doctorate, Research
Type: Prize
Value: US$20,000
Length of Study: 5 years
Frequency: Every 2 years, Biennial
No. of awards offered: 1
Application Procedure: Please see the website for further details http://www.apa.org/apf/funding/gralnick.aspx. Submit a completed application online at http://forms.apa.org/apf/grants/.
Closing Date: April 15th
Funding: Foundation
No. of awards given last year: 1

APF Division 29 Early Career Award

Subjects: Psychotherapy and psychology.
Purpose: To recognise an early career psychologist for promising contributions to psychotherapy, psychology and the division of psychotherapy.
Eligibility: Open to members of Division 29 who are within 7 years of receiving their Doctorate. Applicants must have demonstrated achievement related to psychotherapy theory, practice, research or training.
Level of Study: Doctorate
Type: Award
Value: US$2,500
Frequency: Annual
No. of awards offered: 1
Application Procedure: Check website for further details http://www.apa.org/apf/funding/div-29.aspx. Submit a completed application online at http://forms.apa.org/apf/grants.
Closing Date: January 1st
Funding: Foundation
Contributor: APA Division of Psychotherapy (Division 29)
No. of awards given last year: 1

APF Kenneth B and Marnie P. Clark Early Career Grant

Subjects: Child self-identity, academic achievement.
Purpose: To stimulate the line of inquiry that the Clarks pioneered regarding the impact of race and power on the personal development of children in the USA. Also to encourage early career psychologists to promote the understanding of the relationship between self-identify and academic achievement with an emphasis on children in grade levels K-8.
Eligibility: Applicant must be a Psychologist with an EdD., PsyD., or PhD from an accredited university candidate must be no more than 5 years postdoctoral. Familiarity with the Clarks' work is essential.
Level of Study: Doctorate, Postdoctorate, Research
Type: Fellowships
Value: $10,000
Length of Study: 1 year
Frequency: Every 2 years
No. of awards offered: 1
Application Procedure: Check website for further details www.apa.org/apf. Submit a completed application online at http://forms.apa.org/apf/grants.
Closing Date: June 15th
Funding: Foundation

APF Pre-college Psycology Grant Program

Subjects: Psychology, secondary education.
Purpose: To provide financial support for efforts aimed at improving the quality of education in psychological science and its application in the secondary schools.
Eligibility: Applicants must be educational institutions or a 501(c)(3) non-profit organizations or affiliated with such an organization. Proposals for programs must focus on supporting the education of talented high school students. IRB approval is required for any research project involving human participants.
Level of Study: Research, Unrestricted
Type: Grant
Value: Up to $20,000
Length of Study: 1 year
Frequency: Annual
No. of awards offered: Varies by year
Application Procedure: Check website for details http://www.apa.org/apf/funding/pre-college.aspx. Submit a completed application online at http://forms.apa.org/apf/grants/.
Closing Date: May 1st
Funding: Foundation
No. of awards given last year: 2

APF/COGDOP Graduate Research Scholarships

Subjects: Graduate-level psychological research.
Purpose: To assist graduate students of psychology with research costs associated with the Master's thesis or doctoral dissertation.
Eligibility: Graduate students enrolled in an interim master's program or doctoral program are eligible to apply. If a student is currently enrolled in a terminal master's program, the student must intend to enroll in a PhD program. Students at any stage of graduate study are encouraged to apply. The purpose of the scholarship program is to assist graduate students of psychology with research costs associated with the master's thesis or doctoral dissertation. The American Psychological Association Science Directorate administers the granting of the scholarships. Please see the website for details regarding eligibility http://www.apa.org/apf/funding/cogdop.aspx.
Level of Study: Graduate, Research
Type: Scholarships
Value: The $5,000 Harry and Miriam Levinson Scholarship, the $5,000 William and Dorothy Bevan Scholarship, the $3,000 Ruth G. and Joseph D. Matarazzo Scholarship, the $2,000 Clarence J. Rosecrans Scholarship, the $1,000 William C. Howell Scholarship, the

$1,000 Peter and Malina James and Dr. Louis P. James Legacy Scholarship
Length of Study: 1 year
Frequency: Annual
No. of awards offered: 13
Application Procedure: Please check website http://apa.org/apf/funding/cogdop.aspx for detailed application instructions.
Closing Date: June 30th
Funding: Foundation
No. of awards given last year: 13

For further information contact:

APA Science Directorate, 750 First Street, NE, Washington, DC, 20002-4242, United States of America

Benton Meier Neuropsychology Scholarships

Subjects: Neuropsychology and psychology.
Purpose: To funnd scholarships for promising graduate students enrolled in neuropsychology programs.
Eligibility: Applicants must have completed their doctoral candidacy and demonstrated research competence with strong area commitment. IRB approval is required for any research project involving human participants.
Level of Study: Graduate
Type: Scholarships
Value: Two annual $2,500 scholarships
Length of Study: 1 year
Frequency: Annual
Country of Study: United States of America
No. of awards offered: 2
Application Procedure: Submit a completed application online at http://forms.apa.org/apf/grants/. Check the website for further details http://www.apa.org/apf/funding/benton-meier.aspx.
Closing Date: June 1st
Funding: Foundation
No. of awards given last year: 2
Additional Information: Candidates must submit a letter that documents their scholarly or research accomplishments, explains their financial need and describes for what purpose the financial award will be used.

Charles L. Brewer Distinguished Teaching of Psychology Award

Subjects: Teaching of psychology.
Purpose: To recognize an outstanding career contribution to the teaching of psychology.
Eligibility: Open to those who have a proven track record as an exceptional teacher of psychology.
Level of Study: Doctorate, Research
Type: Award
Value: $2,000 award and all-expense paid round trip and plaque presented at the APA Convention. Awardees are invited to give a special address at the APA Convention.
Frequency: Annual
No. of awards offered: 1
Application Procedure: Submit a completed application online at http://forms.apa.org/apf/grants/ or mail to the American Psychological Foundation. Check website for further details http://www.apa.org/apf/funding/brewer.aspx.
Closing Date: December 1st
Funding: Foundation
No. of awards given last year: 1

Division 17 Counseling Psychology Grant

Subjects: Counseling psychology.
Purpose: To support activities for the advancement of counseling psychology.
Eligibility: Membership in Society of Counseling Psychology, educational Institution or 501(c)(3) nonprofit organization or affiliation therewith, IRB approval must be received from host institution before funding can be awarded if human participants are involved, early career psychologists (no more than seven years post-doctoral) are encouraged to apply.
Level of Study: Research
Type: Grant
Value: Up to US$5,000
Length of Study: 1 year
Frequency: Annual
No. of awards offered: Varies by year
Application Procedure: Check website for further details – www.apa.org/apf/. Submit a completed application online at http://forms.apa.org/apf/grants/.
Closing Date: April 1st
Funding: Foundation
Contributor: APA Society for Counseling Psychology (Division 17)
No. of awards given last year: 3

Elizabeth Munsterberg Koppitz Child Psychology Graduate Fellowships

Subjects: Child psychology.
Purpose: To provide fellowships and scholarships for graduate student research in the area of child psychology.
Eligibility: Open to applicants who have academically progressed through the qualifying exams for Doctoral study, with a demonstrated research competence and area commitment. IRB approval is required for any research project involving human participants.
Level of Study: Graduate, Predoctorate
Type: Fellowships
Value: Up to four research awards of up to $25,000 each; up to two $5,000 scholarships for runners-up.
Length of Study: 1 year
Frequency: Annual
Country of Study: United States of America
No. of awards offered: Varies by year
Application Procedure: Submit a completed application online at http://forms.apa.org/apf/grants/. Check website for details – www.apa.org/apf/.
Closing Date: November 15th
Funding: Foundation, trusts
No. of awards given last year: 6 scholarships and 3 runner-up scholarships
Additional Information: Consideration will be given to psychological research that creates significant new understandings that facilitate the development and functioning of children and youth.

Esther Katz Rosen Graduate Student Fellowships

Subjects: Psychological understanding of gifted and talented children and adolescents.
Purpose: To support activities related to the psychological understanding of gifted and talented children and adolescents.
Eligibility: Applicants must be graduate students who have achieved doctoral candidacy. Applicants must be in good academic standing at an accredited university in the US or Canada and be enrolled in a graduate program during the fellowship year. The applicant's home institution must provide a tuition waiver. IRB approval is required for any research project involving human participants.
Level of Study: Graduate, Research
Type: Grant
Value: $20,000
Length of Study: 1 year
No. of awards offered: Up to 3
Application Procedure: Check website for details http://www.apa.org/apf/funding/rosen.aspx. Submit a completed application online at http://forms.apa.org/apf/grants/.
Closing Date: March 1st
Funding: Foundation
No. of awards given last year: 1

F J McGuigan Young Investigator Research Prize on Understanding the Human Mind

Subjects: Psychology, psychophysiological perspective and the human mind.
Purpose: To support an early-career psychologist engaged in research that seeks to explicate the concept of the human mind from a primarily psychophysical perspective (physiological and behavioral research may also qualify.)
Eligibility: Earned a doctoral degree in psychology or in a related field and be no more than seven years post-doctoral degree at the nomination deadline, an affiliation with an accredited college, university or other research institution. Please see the website for

further details regarding eligibility http://www.apa.org/apf/funding/mcguigan-prize.aspx.
Level of Study: Research
Type: Research prize
Value: $25,000
Length of Study: 2 years
Frequency: Every 2 years
No. of awards offered: 1
Application Procedure: Check website for details – www.apa.org/apf/. Submit a completed application online at http://forms.apa.org/apf/grants/ or send materials by mail.
Closing Date: March 1st
Funding: Foundation
No. of awards given last year: 1

For further information contact:

APF McGuigan Young Early Career Prize, American Psychological Association, Science Directorate, 750 First Street, NE, Washington, DC, 20002-4242

Henry P. David Grants for Research and International Travel in Human Reproductive Behavior and Population Studies

Subjects: Population studies or human reproductive behavior.
Purpose: To support young professionals with a demonstrated interest in behavioural aspects of human reproductive behaviour or an area related to population concerns.
Eligibility: Applicants must be graduate students conducting dissertation research or early career researchers with not more than seven years' postgraduate experience with a demonstrated interest in human reproductive behaviour or relate population concerns. Open to applicants in all relevant disciplines who have a demonstrated psychological approach to their work, with preference given to psychologists. IRB approval is required for any research project involving human participants.
Level of Study: Doctorate, Graduate, Postgraduate, Research
Type: Grant
Value: Up to US$1,500 for research, up to US$1,500 for International travel
Length of Study: 1 year
Frequency: Annual
No. of awards offered: Varies by year
Application Procedure: Check website for further details http://www.apa.org/apf/funding/david.aspx. Submit a completed application online at http://forms.apa.org/apf/grants/.
Closing Date: December 1st and February 15th
Funding: Foundation
No. of awards given last year: 4 (2 – research and 2 – travel)
Additional Information: Applicants may apply for one or both grants.

Lizette Peterson-Homer Injury Prevention Grant Award

Subjects: Research related to the prevention of injuries in children.
Purpose: To support university-based research into the psychological and behavioural aspects of injury prevention for children and adolescents.
Eligibility: Applicants must be either a student or faculty at an accredited university. Applicants must have a demonstrated research competence and area commitment. IRB approval is required for any research project involving human participants.
Level of Study: Research
Type: Research grant
Value: Up to US$5,000
Length of Study: 1 year
Frequency: Annual
Country of Study: United States of America
No. of awards offered: 1
Application Procedure: Check website for further details http://www.apa.org/apf/funding/peterson-homer.aspx. Submit a completed application to Paul Robins, PhD at robinsp@email.chop.edu (phone 215 590 7594).
Closing Date: October 1st
Funding: Foundation
Contributor: APA Society of Pediatric Psychology (Division 54)
No. of awards given last year: 1

Additional Information: For further information please contact at sharon.berry@childrensmn.org.

Paul E. Henkin School Psychology Travel Grant

Subjects: Travel grant to psychological convention.
Purpose: To provide support to defer the costs of registration, lodging, and travel for student members of APA Division 16 to attend the APA Annual Convention.
Eligibility: Student membership in APA Division 16, demonstrated commitment to pursuit of a school psychology career, those receiving any APA travel reimbursement for convention attendance are ineligible.
Level of Study: Graduate
Type: Travel grant
Value: Up to $1,000 to defer the costs of registration, lodging and travel
Frequency: Annual
No. of awards offered: 1
Application Procedure: Application form should include letter of recommendation, 500-word essay and curriculum vitae. Submit a completed application online at http://forms.apa.org/apf/grants/. Please see the website for further details http://www.apa.org/apf/funding/henkin.aspx.
Closing Date: April 15th
Funding: Foundation
No. of awards given last year: 1
Additional Information: The grant monies may not be used for food, drink or any materials that are not included in the registration fee.

Randy Gerson Memorial Grant

Subjects: Understanding of couple, family dynamics or multi-generational processes.
Purpose: To provide grants for graduate student projects in family and/or couple dynamics and/or multi-generational processes.
Eligibility: Applicants must be graduate students in psychology enrolled full-time and in good standing at an accredited university. Applicants must have a demonstrated competence in area of the proposed work. IRB approval is required for any research project involving human participants.
Level of Study: Graduate, Research
Type: Grant
Value: $6,000
Length of Study: 1 year
Frequency: Annual
No. of awards offered: 1 (usually)
Application Procedure: Check website for details http://www.apa.org/apf/funding/gerson.aspx. Submit a completed application online at http://forms.apa.org/apf/grants.
Closing Date: February 1st
Funding: Foundation
No. of awards given last year: 1

Roy Scrivner Memorial Research Grants

Subjects: Psychology, family psychotherapy and LGBT issues.
Purpose: To support graduate student research on LGBT family psychotherapy, particularly research leading to dissertations.
Eligibility: Applicants must be advanced graduate students, in good standing, endorsed by supervising professor, with a demonstrated commitment to LGBT family issues. IRB approval is required for any research project involving human participants.
Level of Study: Graduate, Research
Type: Grant
Value: Up to $12,000
Length of Study: 1 year
Frequency: Annual
No. of awards offered: 1
Application Procedure: Check website for details http://www.apa.org/apf/funding/scrivner.aspx. Submit a completed application online at http://forms.apa.org/apf/grants/.
Closing Date: December 31st
Funding: Foundation
No. of awards given last year: 1

Theodore Millon Award in Personality Psychology

Subjects: Science of personality psychology including the areas of personology, personality theory, personality disorders and personality measurement.
Purpose: This award honors an outstanding psychologist engaged in advancing the science of personality psychology including the areas of personology, personality theory, personality disorders and personality measurement.
Eligibility: APF encourages nominations for individuals who represent diversity in race, ethnicity, gender, age, disability and sexual orientation. Nominees should be no less than 8 years and no more than 20 years post doctoral degree. Please see the website for further details regarding eligibility http://www.apa.org/apf/funding/millon.aspx.
Level of Study: Research
Type: Award
Value: $1,000 and a plaque at the APA convention
Frequency: Annual
No. of awards offered: 1
Application Procedure: Applications should include a cover letter outlining the nominee's contributions to science of personality psychology, a copy of an abbreviated curriculum vitae and up to two letters of recommendation. Self-nomination is permitted. Check website for more details – www.apa.org/apf/. Please send completed applications to the below address or by email to the Awards Committee Chair at div12apa@comcast.net.
Closing Date: November 1st
Funding: Foundation
Contributor: APA Society of Clinical Psychology (Division 12)
No. of awards given last year: 1

For further information contact:

PO Box 1082, Niwot, CO, 80544-1082, United States of America
Contact: Division of Clinical Psychology

Timothy Jeffrey Memorial Award in Clinical Health Psychology

Subjects: Clinical health psychology.
Purpose: To recognize the outstanding commitment to clinical health psychology by a full-time provider of direct clinical services.
Eligibility: Applicants must be fully licensed clinical health psychologists and members of APA and APA Division 38 (Health Psychology). Applicants should typically spend a minimum of fifteen to twenty hours weekly in direct, face-to-face patient care, in assessment or therapy, in individual or group settings.
Level of Study: Doctorate
Type: Award
Value: $3,000
Frequency: Annual
No. of awards offered: 1
Application Procedure: Nominations must be accompanied by a current curriculum vitae, at least one letter of support from a non-psychologist, professional colleague (letters will not be accepted from students or supervisees), and one letter from a psychologist colleague. Nomination letters should describe the nominee's practice, professional activities and commitment to the field.
Closing Date: May 1st
Funding: Foundation
Contributor: APA Health Psychology (Division 38)
No. of awards given last year: 1
Additional Information: Please see the website for further details http://www.apa.org/apf/funding/jeffrey.aspx.

For further information contact:

PO Box 1838, Ashland, VA, 23005
Website: www.health-psych.org
Contact: APA Division 38 Awards Committee

Violence Prevention and Intervention Grant

Subjects: Psychology and violence prevention.
Purpose: To encourage the transfer of psychological science to the prevention of violence in our society and facilitate the implementation of innovative community programmes aimed at interventions to reduce violence.
Eligibility: Applicants should be a 501(c)(3) non-profit organisation or educational institution or affiliated with such an organisation. Applicants must have a demonstrated capability for research or intervention in the violence prevention area (and, where relevant, community support). IRB approval is required for any research project involving human participants.
Level of Study: Doctorate, Research
Type: Grant
Value: Up to $20,000
Length of Study: 1 year
Frequency: Annual
No. of awards offered: 1
Application Procedure: Check website for details – www.apa.org/apf/. Submit a completed application online at http://forms.apa.org/apf/grants/.
Closing Date: June 1st
Funding: Foundation
No. of awards given last year: 1

Visionary and the Drs Rosalee G and Raymond A Weiss Research and Program Innovation Grants

Subjects: Psychology.
Purpose: Grants seek to seed innovation through supporting research, education and intervention projects and programs that use psychology to solve social problems.
Eligibility: Applicants can be educational institutions, 501(c)(3) non-profit organisations, or affiliated with such organisations. Demonstrated competence and capacity to execute the proposed work is required. IRB approval is required for any research project involving human participants. Please see the website for further details regarding eligibilty http://www.apa.org/apf/funding/vision-weiss.aspx.
Level of Study: Doctorate, Unrestricted
Type: Grant
Value: Annual 'Visionary' grants are available in amounts ranging from $2,500 to $20,000. The annual 'Weiss' grant is available for up to $2,500.
Length of Study: 1 year
Frequency: Annual
No. of awards offered: Annual
Application Procedure: Check website for details – www.apa.org/apf/. Submit a completed application online at http://forms.apa.org/apf/grants/.
Closing Date: March 1st
Funding: Foundation
Contributor: The American Psychological Foundation
No. of awards given last year: 1 visionary grant and 1 weiss grant
Additional Information: Multi-year grants are no longer available.

Wayne F Placek Grants

Subjects: Behavioural and social sciences.
Purpose: To encourage scientific research to increase the general public's understanding of homosexuality and to alleviate the stress that gay men and lesbians experience in this and future civilizations.
Eligibility: Must be either a doctoral]level researcher or graduate student affiliated with an educational institution of a 501(c)(3) nonprofit research organization. Graduate students and early career researchers are encouraged to apply.
Level of Study: Doctorate, Research
Type: Grant
Value: $15,000 in research support
Frequency: Annual
No. of awards offered: 2
Application Procedure: Please see the website for further details http://www.apa.org/apf/funding/vision-weiss.aspx. Application forms can be downloaded from the website.
Closing Date: March 1st
Funding: Foundation
Contributor: Placek Fund

AMERICAN PUBLIC POWER ASSOCIATION (APPA)

1875 Connecticut Avenue, NW, Suite 1200, Washington DC, 20009 5715, United States of America
Tel: (1) 202 467 2900
Fax: (1) 202 467 2992
Email: DEED@PublicPower.org
Website: www.PublicPower.org/DEED
Contact: Ms Michele Suddleson, DEED Program Manager

APPA is a service organization for the nation's more than 2,000 community-owned electric utilities that serve more than 43 million Americans. Its purpose is to advance the public policy interests of its members and their consumers, and provide member services to ensure adequate, reliable electricity at a reasonable price with the proper protection of the environment.

DEED (Demonstration of Energy-Efficient Developments) Student Research Grant and Internship

Subjects: Engineering, mathematics or computer science.
Purpose: To promote the involvement of students studying in energy-related disciplines in the public power industry and to increase awareness of career opportunities in public power.
Eligibility: Open to students studying in energy-related disciplines. Applicants must be enrolled in an accredited university in the United States or Canada and must be sponsored by a DEED member utility (check APPA's website for instruction).
Level of Study: Graduate, Doctorate, Postdoctorate, Postgraduate
Type: Grant
Value: US$4,000
Frequency: Annual
Country of Study: United States of America
No. of awards offered: 10
Application Procedure: Applications must be completed in full as per the instructions, and submitted with all required signatures and an academic transcript, by the deadline. Please see the website for further details.
Closing Date: February 15th and October 15th
Funding: Private
No. of awards given last year: 10
Additional Information: Currently only the United States of America has DEED members. Applicants should visit the website for a listing of members and for additional scholarship information and tips.

For further information contact:

Tel: 202 467 2960
Email: DEED@APPAnet.org
Website: http://www.publicpower.org/research/index.cfm?ItemNumber=16257

THE AMERICAN RESEARCH CENTER IN EGYPT (ARCE)

8700 Crownhill Blvd. Suite 507, San Antonio, TX, 78209-1130, United States of America
Tel: (1) 210 821 7000
Email: info@arce.org
Website: www.arce.org
Contact: Dina Aboul Saad

The American Research Center in Egypt (ARCE) is the professional society in the United States of America for specialists on all periods of Egypt's cultural history. It is also a consortium of universities and museums that supports archaeological and academic research in Egypt via fellowships, and whose membership is open to the public.

ARCE Fellowships

Subjects: Arts and humanities, Near East studies and humanistic social sciences.
Purpose: To support research in Egypt.
Eligibility: Open to citizens of the United States of America who are predoctoral candidates. Postdoctoral candidates should be nationals of the United States of America or foreign nationals who have been teaching at an American university for 3 years or more.
Level of Study: Doctorate, Postdoctorate, Museum curators
Type: Fellowship
Value: Varies
Length of Study: 3 months–1 year
Frequency: Annual
Study Establishment: ARCE
Country of Study: Egypt
No. of awards offered: 10–17
Application Procedure: Please visit the Fellowship Application Instructions page and the Fellowship Application Forms page to view instructions and download all materials.
Closing Date: January 15th
Funding: Government
No. of awards given last year: 12
No. of applicants last year: 31
Additional Information: Please see the website for details http://www.arce.org/grants/fellowships/funded.

AMERICAN RESEARCH INSTITUTE IN TURKEY (ARIT)

University of Pennsylvania Museum, 3260 South Street, Philadelphia, PA, 19104-6324, United States of America
Tel: (1) 215 898 3474
Fax: (1) 215 898 0657
Email: leinwand@sas.upenn.edu
Website: http://ccat.sas.upenn.edu/ARIT
Contact: Nancy Leinwand, Executive Director

The American Research Institute in Turkey's (ARIT) aim is to support U.S based scholarly research in all fields of the humanities and social sciences in Turkey through administering fellowship programmes at the doctoral and postdoctoral level and through maintaining research centres in Ankara and Istanbul.

ARIT Fellowship Program

Subjects: Research on ancient, medieval, or modern times in Turkey, in any field of the humanities and social sciences are eligible.
Purpose: To enable scholars and advanced graduate students interested in research in the field of humanities and social science.
Eligibility: Turkish law requires foreign scholars to obtain formal permission to carry out research at institutions in Turkey.
Level of Study: Predoctorate, Doctorate, Postdoctorate, Postgraduate, Research
Type: Fellowship
Value: US$4,000–16,000
Length of Study: 1 year
Frequency: Annual
Study Establishment: Turkey and its branch centres in Istanbul and Ankara
Country of Study: Turkey
No. of awards offered: Varies
Application Procedure: In order to be considered, applicants must provide complete information in their applications. Applications and three letters of recommendation must be received by the due date.
Closing Date: November 1st
Funding: Government, private
Contributor: United States Department of State, Bureau of Educational and Cultural Affairs; the National Endowment for the Humanities
No. of awards given last year: 8
No. of applicants last year: Approx. 80
Additional Information: ARIT fellowships applicants are responsible for obtaining their research permissions and visas. In general, researchers should seek permission to carry out research from the Directors of the Institutions in which they intend to work.

ARIT Mellon Advanced Fellowships in Turkey

Subjects: Humanities and social sciences.
Purpose: To bring Eastern and Central European scholars to Turkey to carry out research.
Eligibility: Open to nationals of Bulgaria, Czech Republic, Slovakia, Poland, Hungary and Romania as well as Estonia, Latvia, Lithuania. Preference will be given to scholars in the early stages of their careers who have not had an opportunity for extensive travel.
Level of Study: Postdoctorate

Type: Fellowship
Value: Up to US$20,000
Length of Study: 3–4 months
Frequency: Annual
Study Establishment: Either of ARIT's two research establishments in Ankara or Istanbul
Country of Study: Turkey
No. of awards offered: 3–4
Application Procedure: Applicants must submit an application form, project statement and references. For further information contact ARIT. Tel.: (215) 898-3474; fax: (215) 898-0657; e-mail: ARIT executive director [leinwand at sas.upenn.edu].
Closing Date: November 1st
Funding: Private
Contributor: Andrew W. Mellon Foundation and the Council of American Overseas Research Centers
Additional Information: Further information is available on request or from the website www.mellon.org

Fellowships for Intensive Advanced Turkish Language Study in Istanbul, Turkey

Subjects: Turkish language
Purpose: To provide full travel and fellowship to students and scholars for participation in the summer program in advanced Turkish language at Bogazici University in Istanbul.
Eligibility: Applicant must be a citizen, national or permanent resident of the United States.
Level of Study: Graduate, Predoctorate
Type: Fellowship
Value: Fellowship includes round-trip airfare to Istanbul, application and tuition fees, and a maintenance stipend.
Length of Study: 8 weeks
Frequency: Annual
Study Establishment: Bogazici University
Country of Study: Turkey
No. of awards offered: Approx. 15
Application Procedure: Application forms are available at the ARIT website http://ccat.sas.upenn.edu/ARIT. For application material contact: Director, Erika H.Gilson, 110 Jones Hall, Princeton University, Princeton, NJ 08544-1008, USA. e-mail: ehgilson at princeton.edu. Or Nancy Leinwand at ARIT leinwand at sas.upenn.edu.
Closing Date: November 1st
Funding: Government
Contributor: US Department of Education, Fulbright-Hays Group Projects Abroad Programme
No. of awards given last year: 17
No. of applicants last year: 65

For further information contact:

Summer Program in Turkish Language and Culture, Bogaziçi University, Istanbul, Bebek, 34342, Turkey
Tel: (90) 90 212 257 5039
Fax: (90) 90 212 265 7131
Email: tlcp@boun.edu.tr
Website: http://www.boun.edu.tr/special/web.html
Contact: Erika H Gilson, The Language Center

Kenan T. Erim Fellowship for Archaeological Research at Aphrodisias

Subjects: Excavation and/or research in the field of art history and archaeology to be carried out at the site of Aphrodisias in Turkey.
Purpose: To enable scholars to carry out research in the field of art history and archaeology to be carried out at the site of Aphrodisias in Turkey.
Eligibility: Scholars or advanced graduate students engaged in excavation at the site of Aphrodisias or research on material from that site are eligible to apply.
Level of Study: Doctorate, Predoctorate, Research
Type: Fellowships
Value: US$2,375
Frequency: Annual
Country of Study: Turkey
No. of awards offered: Varies
Application Procedure: Applicants must provide a completed application in order to be considered. The application and a letter of acceptance from the director of excavations at Aphrodisias, in addition to two letters of reference, must be received by the due date. Student applications must include a transcript. For further information please call (215) 898-3474, fax (215) 898-0657, or email the ARIT executive director [leinwand at sas.upenn.edu].
Closing Date: November 1st
Funding: Private
Contributor: American friends of Aphrodisias
No. of awards given last year: 1
No. of applicants last year: 3
Additional Information: Projects should be included within the Aphrodisias excavation permit.

KRESS/ARIT Pre-Doctoral Fellowship in the History of Art and Archaeology

Subjects: The history of art and architecture from antiquity to the present, and archaeology.
Eligibility: Open to advanced graduate students engaged in research in Turkey. Applicants must have fulfilled all preliminary requirements for the doctorate except the dissertation by June 2007, and before beginning any ARIT-sponsored research. Candidacy is open to US citizens and non-US applicants who matriculated at US or Canadian institutes. Pre-doctoral applicants may also qualify for ARIT Fellowship in the Humanities and Social Sciences.
Level of Study: Doctorate, Predoctorate, Research
Type: Fellowship
Value: Up to US$17,000
Length of Study: 1 academic year. Awards for shorter periods of time are also available
Frequency: Annual
Study Establishment: Either of ARIT's two research establishments in Ankara or Istanbul
Country of Study: Turkey
No. of awards offered: 2–4
Application Procedure: Applicants must submit an application form accompanied by three letters of recommendation.
Closing Date: November 1st
Funding: Private
Contributor: The Samuel H. Kress Foundation
No. of awards given last year: 2
No. of applicants last year: 14

NEH ARIT-National Endowment for the Humanities Fellowships for Research in Turkey

Subjects: All subjects of the humanities and interdisciplinary approaches to social sciences, prehistory, history, art, archaeology, language and literature.
Purpose: To support research on ancient, medieval or modern times.
Eligibility: Open to scholars who have completed their formal training and plan to carry out research in Turkey may apply. They may be US citizens or three year residents of the US. Please consult ARIT headquarters on questions of eligibility. Advanced scholars also may apply for ARIT Fellowships in the Humanities and Social Sciences.
Level of Study: Postdoctorate, Professional development, Research
Type: Fellowship
Value: US$16,800–50,400
Length of Study: 4–12 months
Frequency: Annual
Study Establishment: Either of ARIT's two research establishments in Ankara or Istanbul
Country of Study: Turkey
No. of awards offered: 2–3
Application Procedure: Applicants must submit an application form, project statement and references. For further information call (215) 898-3474, fax (215) 898-0657, or e-mail ARIT executive director [leinwand@sas.upenn.edu].
Closing Date: November 1st
Funding: Government
Contributor: National Endowment for the Humanities (NEH)
No. of awards given last year: 3
No. of applicants last year: 16
Additional Information: The hostel, research and study facilities are available at ARIT's branch centers in Istanbul and Ankara

For further information contact:

Contact: Nancy Leinwand, Director

W.D.E. Coulson and Toni M. Cross Aegean Exchange Program

Subjects: Any field of humanities and social sciences from prehistoric to modern times to conduct research in Greece.
Purpose: To provide an opportunity for Turkish scholars to meet with their Greek colleagues and to pursue research interests in the museum, archives and library collections and at the sites and monuments of Greece.
Eligibility: Applicants must be Turkish citizens and must have their primary academic affiliation with a university in Turkey. They must have completed all PhD cousework and passed all qualifying examinations for the degree before entering the tenure of the fellowship.
Level of Study: Doctorate, Postdoctorate
Type: Fellowship
Value: Up to $7,500
Length of Study: minimum of 1 month
Frequency: Annual
Study Establishment: American School of Classical Studies of Athens
Country of Study: Greece
No. of awards offered: Up to 3
Application Procedure: Applicants should submit a 5-page statement of purpose, along with application and at least 2 letters of recommendation. Five copies of the applications (except for the recommendation letters) should be submitted.
Closing Date: November 1st
Funding: Government
Contributor: U.S. Department of State Educational and Cultural Affairs, Council of Overseas Research Centers
No. of applicants last year: 10

For further information contact:

Arit-Ankara Temsilcilii, Sehit Ersan Caddesi 24/9, Cankaya, Ankara, 06680, Turkey
Tel: (90) 312 427 2222
Fax: (90) 312 427 4979

THE AMERICAN SCHOOL OF CLASSICAL STUDIES AT ATHENS (ASCSA)

6-8 Charlton Street, Princeton, NJ, 08540-5232, United States of America
Tel: (1) 609 683 0800
Fax: (1) 609 924 0578
Email: ascsa@ascsa.org
Website: www.ascsa.edu.gr
Contact: Ms Mary E Darlington, Executive Associate

Established in 1881, the American School of Classical Studies at Athens (ASCSA) offers both graduate students and scholars the opportunity to study Greek civilization, first hand, in Greece. The ASCSA supports and encourages the teaching of the archaeology, art, history, language and literature of Greece from early times to the present.

A.G. Leventis Foundation Scholarships

Subjects: Graduate students and postdoctoral scholars in Byzantine studies from any university worldwide to attend month-long program in intermediate level Medieval Greek language and philology at the Gennadius Library of the ASCSA. Site and museum trips.
Purpose: Medieval Greek summer session at the Gennadius Library.
Eligibility: Graduate students in field of late antique, postantique, Byzantine or Medieval studies enrolled in any university in the world. Minimum two years of college level Classical Greek required. College professors worldwide, if space permits.
Level of Study: Postgraduate, Graduate, Postdoctorate, Predoctorate
Value: Covers the costs to attend the month-long session in Athens. (Tuition and housing provided.) Participant pays for international travel to and from Greece, meals, local transportation, incidentals
Length of Study: 1 month (late June to late July)
Frequency: Every 2 years
No. of awards offered: 12
Application Procedure: Applicants must complete online application. For guidelines and application visit www.ascsa.edu.gr after September.
Closing Date: January 15th
Funding: Foundation, private
No. of awards given last year: 12
No. of applicants last year: 38

ASCSA Advanced Fellowships

Subjects: Classical art history, history of architecture and study of pottery.
Eligibility: Open to students enrolled in the United States of America or Canadian institutions who have completed one year as a regular or student associate member of the ASCSA.
Level of Study: Predoctorate, Postgraduate
Type: Fellowship
Value: A stipend of $11,500 plus room, board, and waiver of School fees are available to students who have completed the Regular Program or one year as a Student Associate Member and plan to return to the School to pursue independent research, usually for their PhD.
Length of Study: 1 academic year
Frequency: Annual
Study Establishment: ASCSA
Country of Study: Greece
No. of awards offered: 7
Application Procedure: Applicants must complete online applications. For guidelines and application visit www.ascsa.org.
Closing Date: February 19th
Funding: Private
No. of applicants last year: 20
Additional Information: The fellowships include: the Edward Capps, the Doreen C Spitzer and the Eugene Vanderpool Fellowships (subject unrestricted); the Samuel H Kress Fellowships in art history; the Gorham P Stevens Fellowship in the history of architecture; and the Homer A and Dorothy B Thompson Fellowship in the study of pottery. Ione Mylonas Shear in Mycenaean Archaeology or Athenian architecture.

ASCSA Fellowships

Subjects: Classical philology and archaeology, post-classical Greek studies or a related field.
Eligibility: Open to students who hold a Bachelor of Art degree but not a PhD, and who are preparing for an advanced degree in classical studies or a related field. Applicants must be affiliated with a college or university in the United States of America or Canada.
Level of Study: Predoctorate, Graduate
Type: Fellowship
Value: US$11,500 stipend plus fees, room and partial board
Length of Study: 1 academic year
Frequency: Annual
Study Establishment: ASCSA
Country of Study: Greece
No. of awards offered: 13
Application Procedure: Applicants must complete online applications. For guidelines and application visit www.ascsa.edu.gr.
Closing Date: January 15th
Funding: Private
No. of awards given last year: 13
No. of applicants last year: 19

ASCSA Research Fellowship in Environmental Studies

Subjects: Earth sciences, geological sciences and archaeological sciences.
Purpose: To support research on studies from archaeological contexts in Greece.
Eligibility: Doctoral candidates working on their dissertation and postdoctoral scholars with well-defined projects that can be completed during the academic year of the fellowship.
Level of Study: Graduate, Postgraduate, Predoctorate, Postdoctorate, Doctorate
Type: Fellowship
Value: US$15,500–27,000 stipend depending on seniority and experience

Length of Study: 1 academic year
Frequency: Annual
Study Establishment: The Malcolm H Wiener Research Laboratory for Archaeological Science, ASCSA
Country of Study: Greece
No. of awards offered: 1
Application Procedure: Applicants must complete online applications. For guidelines and application visit www.ascsa.edu.gr.
Closing Date: January 15th
Funding: Private
No. of awards given last year: 1
No. of applicants last year: 8

ASCSA Research Fellowship in Faunal Studies

Subjects: Biological sciences, life sciences and archaeological sciences.
Purpose: To study faunal remains from archaeological contexts in Greece.
Eligibility: Doctoral candidates working on their dissertation and postdoctoral scholars with well-defined projects that can be completed during the academic year of the fellowship. There is no citizenship requirement.
Level of Study: Doctorate, Graduate, Postdoctorate, Postgraduate, Predoctorate
Type: Fellowship
Value: US$15,500–27,000 stipend depending on seniority and experience
Length of Study: 1 academic year
Frequency: Annual
Study Establishment: The Malcolm H Wiener Research Laboratory for Archaeological Science, ASCSA
Country of Study: Greece
No. of awards offered: 1
Application Procedure: Applicants must complete online applications. For guidelines and application visit www.ascsa.edu.gr.
Closing Date: January 15th
Funding: Private
No. of awards given last year: 1
No. of applicants last year: 8

ASCSA Research Fellowship in Geoarchaeology

Subjects: Earth sciences, geological sciences and archaeological sciences.
Purpose: To support research on a geoarchaeological topic in Greece.
Eligibility: Doctoral candidates working on their dissertation and postdoctoral scholars with well-defined projects that can be completed during the academic year of the fellowship. There is no citizenship requirement.
Level of Study: Doctorate, Graduate, Postdoctorate, Postgraduate, Predoctorate
Type: Fellowship
Value: US$15,500–27,000 stipend depending on seniority and experience
Length of Study: 1 academic year
Frequency: Annual
Study Establishment: The Malcolm H Wiener Research Laboratory for Archaeological Science, ASCSA
Country of Study: Greece
No. of awards offered: 1
Application Procedure: Applicants must complete online applications. For guidelines and application visit www.ascsa.edu.gr.
Closing Date: January 15th
Funding: Private
No. of awards given last year: 1
No. of applicants last year: 8

Coulson-Cross Aegean Exchange

Subjects: Graduate students/scholars in ancient studies, archaeology, classical studies, anthropology, Byzantine, post-Byzantine, Ottoman studies and modern Greek or Turkey studies to pursue research in Ankara or Istanbul at ARIT.
Purpose: To provide Greek scholars the opportunity to pursue research in Turkey.
Eligibility: Greek nationals including the staff of the Ministry of Culture, doctoral candidates and faculty members of Greek institutions of higher education.
Level of Study: Graduate, Postdoctorate, Postgraduate, Predoctorate
Type: Fellowship
Value: US$250 per week plus airfare
Length of Study: 2 weeks to 2 months
No. of awards offered: 2–4
Application Procedure: Submit Associate Membership application should be submitted online at the ASCSA website at http://www.ascsa.edu.gr/index.php/admission-membership/student-associate-membership.The application should include a curriculum vitae, statement of the project to bepursued during the period of grant (up to three pages in length), two letters of reference fromscholars in the field commenting on the value and feasibility of the project.
Closing Date: March 15th
Funding: Government
No. of awards given last year: 2
No. of applicants last year: 3
Additional Information: Eligible to the nationals of Greece. The awards will be announced later in the spring.

George Papioannou Fellowship

Subjects: Modern Greek studies.
Purpose: Study of George Papaioannou papers at the Gennadius Library of the ASCSA for a study of the Greek Civil War.
Eligibility: Open to all nationalities.
Level of Study: Postdoctorate, Postgraduate, Predoctorate
Type: Fellowship
Value: €1,000
Length of Study: Up to 2 months
Application Procedure: Applicants must complete online application. For guidelines and application, please visit www.ascsa.edu.gr.
Closing Date: January 15th
Funding: Private
No. of awards given last year: 1
No. of applicants last year: 1

Harry Bikakis Fellowship

Subjects: North American or Greek graduate student researching ancient Greek law or Greek graduate student working on a American School of Classical Studies archaeological excavation.
Eligibility: Field of study is ancient Greek law for North American or Greek graduate students or Greek graduate students working on an ASCSA archaeological excavation.
Level of Study: Graduate, Predoctorate
Type: Fellowship
Value: US$1,875
Length of Study: Varies
Frequency: Annual
Study Establishment: ASCSA
Country of Study: Greece
Application Procedure: Applicants must complete online applications. For guidelines and application visit www.ascsa.edu.gr.
Closing Date: January 15th
Funding: Private
No. of awards given last year: 1
No. of applicants last year: 1
Additional Information: Eligible to the nationals of Greece.

J Lawrence Angel Fellowship in Human Skeletal Studies

Subjects: Biological sciences, life sciences and archaeological sciences.
Purpose: To study human skeletal remains from archaeological contexts in Greece.
Eligibility: Doctoral candidates working on dissertations and scholars holding a PhD or equivalent degree.
Level of Study: Doctorate, Graduate, Postdoctorate, Postgraduate, Predoctorate
Type: Fellowship
Value: US$15,500–27,000 stipend, depending on seniority and experience
Length of Study: 1 academic year
Frequency: Annual

Study Establishment: The Malcolm H Wiener Research Laboratory for Archaeological Science, ASCSA
Country of Study: Greece
No. of awards offered: 1
Application Procedure: Applicants must complete online applications. For guidelines and application visit www.ascsa.edu.gr.
Closing Date: January 15th
Funding: Private
Contributor: The Malcolm H Wiener Research Laboratory for Archaeological Sciences at the American School at Athens
No. of awards given last year: 1
No. of applicants last year: 8

Jacob Hirsch Fellowship

Subjects: Pre-classical, classical or post-classical archaeology.
Purpose: To support individuals completing a project that requires a lengthy residence in Greece.
Eligibility: Open to graduate students of American or Israeli institutions who are writing a dissertation and to recent PhD graduates completing a project in Greece graduate students, citizens of Israel, at such as a dissertation in archaeology for publication. Applications will be judged on the basis of appropriate credentials including referees.
Level of Study: Predoctorate, Postdoctorate, Postgraduate
Type: Fellowship
Value: US$11,500 stipend plus room, board and waiver of school fees
Length of Study: 1 academic year, non-renewable
Frequency: Annual
Study Establishment: ASCSA
Country of Study: Greece
No. of awards offered: 1
Application Procedure: Applicants must complete online applications. For guidelines and application visit www.ascsa.edu.gr.
Closing Date: January 15th
Funding: Private
No. of awards given last year: 1
No. of applicants last year: 14

M Alison Frantz Fellowship

Subjects: Post-classical studies in late antiquity, Byzantine studies, post-Byzantine studies and modern Greek studies.
Eligibility: Open to PhD candidates enrolled in institutions in the United States of America or Canada must be recent PhD candidates and all candidates must show a need to use the Gennadius Library.
Level of Study: Predoctorate, Doctorate, Postdoctorate
Type: Fellowship
Value: US$11,500 stipend plus room, board and waiver of fees
Length of Study: 1 academic year
Frequency: Annual
Study Establishment: The Gennadius Library
Country of Study: Greece
No. of awards offered: 1
Application Procedure: Applicants must complete online application. For guidelines and application visit www.ascsa.edu.gr.
Closing Date: January 15th
Funding: Private
No. of awards given last year: 1
No. of applicants last year: 6
Additional Information: This fellowship was formerly known as the Gennadeion Fellowship.

NEH Fellowships

Subjects: Ancient, classical and post-classical studies, including but not limited to the history, philosophy, language, art and archaeology of Greece and the Greek world, art history, literature, philology, architecture, archaeology, anthropology, metallurgy and environmental studies from prehistoric times to the present.
Eligibility: Open to doctoral and postdoctoral scholars who are citizens of the United States of America or foreign nationals with 3 years residency in the United States of America immediately preceding the application deadline.
Level of Study: Doctorate, Postdoctorate
Type: Fellowship
Value: A maximum stipend of US$21,000 for a 5-month project and US$42,000 for a 10-month project
Length of Study: 5–10 months
Frequency: Annual
Study Establishment: ASCSA
Country of Study: Greece
No. of awards offered: 2–4
Application Procedure: Applicants must complete online applications. For guidelines and application visit www.ascsa.edu.gr.
Closing Date: November 9th
Funding: Government
No. of awards given last year: 1 academic year, 2 partial year
No. of applicants last year: 16

Oscar Broneer Traveling Fellowship

Subjects: PhD candidate or recent PhD past member of the ASCSA or American Academy in Rome using the resource center as a base for research in Greece or Italy in alternate years.
Eligibility: Past member of the ASCSA or AAR. Approved dissertation if PhD candidate or recent postdoctoral scholar without tenure. Projects for any period of study in the humanities; preference for research in classical antiquity.
Level of Study: Postdoctorate, Postgraduate, Predoctorate
Type: Fellowship
Value: Up to $30,000
Length of Study: 3–6 months
No. of awards offered: 1
Application Procedure: Applicants must complete online applications, including budget. For guidelines and application visit www.ascsa.edu.gr.
Closing Date: January 15th
Funding: Private
No. of awards given last year: 1
No. of applicants last year: 9

AMERICAN SCHOOLS OF ORIENTAL RESEARCH (ASOR)

656 Beacon Street, 5th Floor, Boston, MA, 02215 2010, United States of America
Tel: (1) 617 353 6570
Fax: (1) 617 353 6575
Email: asor@bu.edu
Website: www.asor.org
Contact: Britta Abeln, Office Coordinator

The American Schools of Oriental Research's (ASOR) mission is to initiate, encourage and support research into, and public understanding of the people and cultures of the near East from the earliest times by fostering original research, archaeological excavations and explorations, by encouraging scholarship in the basic languages, cultural histories and traditions of the near Eastern world.

ASOR Mesopotamian Fellowship

Subjects: Social sciences, history, archaeology in Middle East studies.
Purpose: To financially support field research in ancient Mesopotamian civilization carried out in Middle East.
Eligibility: Open to applicants affiliated with an institution that is a corporate member of ASOR or who have an individual membership. See website for further details.
Level of Study: Research
Type: Fellowship
Value: US$9,000. $1,000 of this Fellowship amount will be allocated for registration and travel support to the ASOR Annual Meeting.
Length of Study: 3–12 months
Frequency: Annual
No. of awards offered: 1
Application Procedure: Applicants need to submit cover sheet (with contact information, ASOR membership information, title of project, and brief abstract) and a short proposal. Applicants currently in graduate degree programs should provide three recommendations.
Closing Date: March 1st
Additional Information: This fellowship is primarily intended to support field/research projects on ancient Mesopotamian civilization carried out in the Middle East, but other research projects such as museum or archival research related to Mesopotamian studies may also be considered.

For further information contact:

Tel: 617 353 6570
Email: asor@bu.edu
Website: http://www.asor.org/fellowships/mesopotamia.shtml

ASOR W.F. Albright Institute of Archaeological Research/ National Endowment of the Humanities Fellowships

Subjects: Archaeology, history, religion/theology, art history, literature/english/writing, social sciences, anthropology, geography, near and Middle East studies.
Purpose: To financially support scholars holding a PhD or equivalent degree with a research project.
Eligibility: Open to citizens of the United States or alien residents residing in the United States for the last 3 years. Please see the website for further details regarding eligibility http://www.neh.gov/divisions/research/fellowship/albright-institute-archaeological-research-jerusalem.
Level of Study: Doctorate
Type: Fellowships
Value: Up to $50,400 for 12 months and $18,900 for 4.5 months. Stipend varies with the duration of the fellowship.
Length of Study: 4–12 months
Frequency: Annual
Country of Study: United States of America
No. of awards offered: 6
Application Procedure: A completed application form must be sent.
Closing Date: October 1st
Contributor: National Endowment of the Humanities
Additional Information: Residence at the Institute in Jerusalem is preferred.

For further information contact:

Albright Fellowship Committee, Department of Art and Art History, Providence College, Providence, RI, 02918, United States of America
Tel: (1) 401 865 1789
Fax: (1) 401 865 2410
Email: jbranham@providence.edu
Website: www.aiar.org
Contact: Professor Joan R. Branham, Chair

AMERICAN SOCIETY FOR ENGINEERING EDUCATION (ASEE)

1818 North Street NW, Suite 600, Washington, DC, 20036, United States of America
Tel: (1) 202 331 3500/3525/202 649 3834
Fax: (1) 202 265 8504
Email: sttp@asee.org
Website: www.asee.org
Contact: Mr Michael More, Projects Department

The American Society for Engineering Education (ASEE) is committed to furthering education in engineering and engineering technology by promoting excellence in instruction, research, public service and practice, exercising worldwide leadership, fostering the technological education of society and providing quality products and services to members.

Air Force Summer Faculty Fellowship Program

Subjects: Engineering, science and mathematics.
Purpose: To stimulate professional relationships among SFFP participants, the scientist and engineers at Air Force Research facilities.
Eligibility: Applicants to the Air Force Summer Faculty Fellowship Program (SFFP) must be citizens or legal permanent residents of the United States. Applicants must hold a full-time appointment at a US college or university. Participants are expected to conduct research at an Air Force Research Laboratory Directorate, US Air Force Academy, or the Air Force Institute of Technology. Please see the website for further details regarding eligibility http://sffp.asee.org/.
Level of Study: Doctorate, Postdoctorate, Postgraduate, Professional development, Research
Type: Fellowship
Value: Weekly stipend based on the status of the participant's application. The levels for the AF SFFP faculty are Assistant Professor: $1,300; Associate Professor: $1,500 and Professor: $1,700. The levels for the AF SFFP graduate students are: pursuing a master's degree: $884; pursuing a PhD: $1,037. A daily expense allowance of $50 per workday for those with temporary residence.
Length of Study: 8–12 weeks
Frequency: Annual
Country of Study: United States of America
No. of awards offered: Varies
Application Procedure: Apply online at www.asee.org/sffp.
Closing Date: December 14th at 5:00 p.m. EST
Funding: Government
Contributor: Air Force Office of Scientific Research, US Air Force Academy and the Air Force Institute of Technology
No. of awards given last year: 90
No. of applicants last year: 250
Additional Information: Each AF SFFP award is for one summer; however, previous AF SFFP awardees may apply for renewal of their award for a 2nd and 3rd year, providing they wish to renew at the same AF SFFP laboratory and will work with the same AF SFFP research advisor.

Naval Research Laboratory Post Doctoral Fellowship Program

Subjects: Computer science, artificial intelligence, plasma physics, acoustics, radar, fluid dynamics, chemistry, materials, science and many more specialist fields.
Purpose: To increase the involvement of creative and highly trained scientists to scientific and technical areas of interest and relevance to the US Navy.
Eligibility: US citizens and permanent residents.
Level of Study: Doctorate, Postdoctorate
Type: Fellowship
Value: Up to a maximum of $75,000 and travel and relocation allowance
Length of Study: 1 year, renewable for a second and third year
Frequency: Annual
Study Establishment: Naval Research Laboratory
Country of Study: United States of America
No. of awards offered: 40
Application Procedure: Apply online at www.asee.org/nrl or http://www.asee.org/fellowships/nrl/apply.cfm.
Closing Date: No deadline. Applications are accepted and processed on an on-going basis.
Funding: Government
Contributor: US Navy
No. of awards given last year: 35
Additional Information: A group health insurance program is provided for participants (paid for by the fellowship) and Optional for dependents (paid for by participant).

For further information contact:

Email: postdocs@asee.org
Website: http://nrl.asee.org

NDSEG Fellowship Program

Subjects: Aeronautical/astronautical engineering, biosciences, chemical engineering, chemistry, civil engineering, cognitive, neural and behavioral science, computer/computational sciences, electrical engineering, geosciences, materials science and engineering, mathematics, mechanical engineering, naval architecture and ocean engineering, oceanogrphy, physics.
Purpose: To increase the number of US citizens and nationals trained in science and engineering disciplines of military importance.
Eligibility: Must be a US citizen or national. Applicants must be at or near beginning of graduate studies in one of the above-named fields. Applicants must be either enrolled in their final year of undergraduate studies or have completed no more than the equivalent of 2 year's of full-time graduate study in the field in which they are applying. Exceptional circumstances may qualify other applicants as being at the early stages of their graduate studies.
Level of Study: Doctorate
Type: Fellowship

Value: NDSEG Fellowships last for three years and pay for full tuition and all mandatory fees, a monthly stipend, and up to $1,000 a year in medical insurance.
Length of Study: 3 years
Frequency: Annual
Country of Study: United States of America
No. of awards offered: 200
Application Procedure: Apply online at www.asee.org/ndseg.
Closing Date: December 16th
Funding: Government
Contributor: US Department of Defense
No. of awards given last year: 200
No. of applicants last year: 2,000
For further information contact:

Email: ndseg@asee.org
Contact: Rachel Kline

ONR Summer Faculty Research Program

Subjects: Science technology, engineering and mathematics.
Purpose: To allow university faculty members to collaborate with the Navy Scientist on issues of mutual interest.
Eligibility: US citizen or permanent resident, must hold teaching or research appointment at US college or university.
Level of Study: Postdoctorate
Type: Fellowship
Value: $1,400–1,900 ($1,400 per week at the Summer Faculty Fellow level, $1,650 per week at the Senior Summer Faculty Fellow level, and $1,900 per week at the Distinguished Summer Faculty Fellow level).
Length of Study: 10 weeks
Frequency: Annual
Country of Study: United States of America
No. of awards offered: Varies
Application Procedure: Apply online at www.asee.org/summer.
Closing Date: December 17th through February 16th
Funding: Government
No. of awards given last year: 73
No. of applicants last year: 544
Additional Information: There are three levels of appointment: Summer Faculty Fellow, Senior Summer Faculty Fellow, and Distinguished Summer Faculty Fellow. Each fellow will be reimbursed for expenses incurred on an optional pre-program visit to the sponsoring laboratory and one round-trip encompassing travel to the sponsoring laboratory at the beginning of the program and travel back to their home residence at the end of the program.

For further information contact:

Email: onrsummer@asee.org
Website: http://www.onr.navy.mil/Education-Outreach/Summer-Faculty-Research-Sabbatical.aspx
Contact: Artis Hicks

SMART Scholarship for Service Program

Subjects: Science, technology, mathematics and engineering.
Purpose: To support the education of future scientists and engineers.
Eligibility: Applicants must be a US citizen or national, pursuing a degree in science, technology, engineering or mathematics, at least 18 years old, have a minimum of 3.0 average on a 4.0 scale and must be enrolled in a US college or university if applying for undergraduate funding.
Level of Study: Doctorate, Graduate
Type: Scholarship
Value: Full tuition and education related fees (does not include items such as meal plans, housing, or parking), stipend paid at a rate of $25,000–38,000 depending on degree pursuing (may be prorated depending on award length), paid summer internships, health Insurance allowance up to $1,200 per calendar year, book allowance of $1,000 per academic year, employment placement after graduation.
Length of Study: Varies
Frequency: Annual
Country of Study: United States of America
No. of awards offered: Varies
Application Procedure: Apply online at www.asee.org/smart
Closing Date: December 17th
Funding: Government
Contributor: US Department of Defense
No. of awards given last year: 200
No. of applicants last year: 1800

AMERICAN SOCIETY FOR MICROBIOLOGY (ASM)

1752 N Street North West, Washington, DC, 20036, United States of America
Tel: (1) 202 942 9323
Fax: (1) 202 942 9353
Email: awards@asmusa.org
Website: www.asm.org/awards
Contact: Ms Leah Gibbons, Program Assistant

The American Society for Microbiology (ASM) is the oldest and largest single life science membership organization in the world. With 43,000 members throughout the world. The ASM represents all disciplines of microbiological specialization including microbiology education. The ASM's mission is to promote research and research training in the microbiological sciences and to assist communication between scientists, policymakers and the public to improve health, the environment and economic well-being.

Abbott Award in Clinical and Diagnostic Immunology

Subjects: Clinical or diagnostic immunology.
Purpose: To honour a distinguished scientist in the field of clinical or diagnostic immunology.
Eligibility: Nominees must demonstrate significant contributions to the understanding of the functioning of the host immune system in human disease, clinical approaches to diseases involving the immune system or development, or clinical application of immunodiagnostic procedures.
Level of Study: Unrestricted
Type: Award
Value: US$2,500 cash prize, a commemorative medal and domestic travel to the ASM General Meeting. The winner will also be asked to present the Abbott Award in Clinical and Diagnostic Immunology lecture.
Frequency: Annual
Country of Study: Any country
No. of awards offered: 1
Application Procedure: Self-nominations will not be accepted. Nominations must consist of a nomination form that includes a specific description of the nominee's contributions, a curriculum vitae including a list of the nominee's publications and two additional supporting forms. Only one of the three individuals involved in the nomination may be employed at the nominee's institution. Forms can be found at www.asm.org/awards.
Closing Date: July 1st
Funding: Commercial
Contributor: Abbott Laboratories, Diagnostic Division
No. of awards given last year: 1
Additional Information: ASM awards are granted at the discretion of award selection committees and may not be awarded every year.

Abbott-ASM Lifetime Achievement Award

Subjects: Microbiology.
Purpose: To honour a distinguished scientist for a lifetime of outstanding contributions to the microbiological sciences.
Eligibility: Open to mature scientists, both active and retired, from all relevant areas of microbiology.
Level of Study: Unrestricted
Type: Award
Value: US$20,000 cash prize, a commemorative medal and travel to the ASM General Meeting where the Laureate delivers the Abbott-ASM Lifetime Achievement Award lecture.
Frequency: Annual
Country of Study: Any country
No. of awards offered: 1
Application Procedure: Self-nominations will not be accepted. Nominations must consist of a nomination form that includes a description of the nominee's outstanding research accomplishments, a curriculum vitae including a list of nominee's publications and two additional forms of support. Only one of the three individuals involved

in the nomination may be employed at the nominee's institution. Forms can be found at www.asm.org/awards.
Closing Date: July 1st
Funding: Commercial
Contributor: Abbott Laboratories
No. of awards given last year: 1
Additional Information: ASM awards are granted at the discretion of the award selection committees and may not be awarded every year.

ABMM/ABMLI Professional Recognition Award

Purpose: Recognizes a Diplomate of the American Board of Medical Microbiology (ABMM) or the Americal Board of Medical Laboratory Immunology (ABMLI) for outstanding contributions to the professional recognition of certified microbiologists and/or immunologists and the work they do.
Eligibility: Primary consideration will be given to ABMM or ABMLI Diplomates who have made significant contributions to the advancement and public recognition of the profession over and above scientific achievements or board-related activites.
Type: Award
Value: A commemorative piece and travel to the ASM General Meeting where the winner will present the ABMM/ABMLI Professional Recognition Award lecture
Application Procedure: Self nominatins will not be accepted. Nominations must consist of a nomination form, a curriculum vitae, including a list of publications and two supporting forms. Only one of the three individuals involved in the nomination may be employed at the nominee's institution. Forms can be found at www.asm.org/awards.
Closing Date: July 1st

ASM Founders Distinguished Service Award

Purpose: Honors a member of the ASM for outstanding contributions and commitment to the ASM as a volunteer at the national level.
Eligibility: Selection is based on commitment to furthering the goals of the ASM, ability to inspire commitment from others, and signifance of contributions to the membership of ASM and its various audiences. The nominee must be an ASM member in good standing who has served in a volunteer capacity for ASM at the national level (e.g. as a member of committees or editorial boards or as a workshop leader) for a minimum of five years, and has not held office as ASM President, Secretary, Treasurer, or Chair of a CPC Board or the American Academy of Microbiology.
Level of Study: Unrestricted
Type: Award
Value: A commemorative piece and travel to the ASM General Meeting
Frequency: Annual
Application Procedure: Self-nominations and more than one nomination per nominee will not be accepted. Only one nominating form and two supporting forms are accepted per nomination. The two supporters must be persons other than the nominator who are familiar with the nominees qualifications and accomplishments only one of the three individuals involved in the nomination may be employed at the nominees institution. The nominator and supporters must not share employers. Nominations must have a curriculum vitae, nomination form and supporting forms. Forms can be found at www.asm.org/awards.
Closing Date: July 1st
Funding: Private
Contributor: American Society for Microbiology

ASM Graduate Microbiology Teaching Award

Subjects: Microbiology.
Purpose: To recognize an individual for exemplary teaching of microbiology and mentoring of students at the graduate and postgraduate level and for encouraging them to subsequent achievement.
Eligibility: Nominees must be currently teaching microbiology in a recognized college or university, have devoted a substantial portion of their time during the past 5 years to teaching graduate students in microbiology and have a minimum of 10 years of total teaching experience. Nominees may have engaged in research or other concerns, provided that teaching graduate students remained a substantial activity.
Level of Study: Unrestricted
Type: Award
Value: US$2,500 cash prize, a commemorative piece and travel to the ASM General Meeting
Frequency: Annual
Country of Study: Any country
No. of awards offered: 1
Application Procedure: Self-nominations will not be accepted. Nominations must consist of a nomination form that specifically addresses how the nominee fulfils the award eligibility, including a record of teaching responsibilities, manifests of distinguished teaching, innovations, publications, special awards or other pertinent information, a curriculum vitae including a list of the nominee's publications and two additional supporting form. Only one of the three individuals involved in the nomination may be employed at the nominee's institution. Forms can be found at www.asm.org/awards.
Closing Date: July 1st
Funding: Private
Contributor: American Society for Microbiology
No. of awards given last year: 1
Additional Information: ASM awards are granted at the discretion of award selection committees and may not be awarded every year.

BD Award for Research in Clinical Microbiology

Subjects: Clinical microbiology.
Purpose: To honour a distinguished clinical microbiologist for outstanding research accomplishments leading to or forming the foundation for important applications in clinical microbiology.
Eligibility: Open to clinical microbiologists.
Level of Study: Unrestricted
Type: Award
Value: US$2,000 cash prize, a commemorative piece and travel expenses to the ASM General Meeting where the winner will present the BD Award lecture
Frequency: Annual
Country of Study: Any country
No. of awards offered: 1
Application Procedure: Self-nominations will not be accepted. Nominations must consist of a nomination form that describes the nominee's activities and accomplishments pertinent to the award, a curriculum vitae including a list of publications and two additional supporting forms. Only one of the three individuals involved in the nomination may be employed at the nominee's institution. Please visit www.asm.org/awards to review the forms.
Closing Date: July 1st
Funding: Commercial
Contributor: BD Diagnostic Systems
No. of awards given last year: 1
Additional Information: ASM awards are granted at the discretion of award selection committees and may not be awarded every year.

bioMérieux Sonnenwirth Award for Leadership in Clinical Microbiology

Subjects: Clinical microbiology.
Purpose: Recognizes a distinguished microbiologist for the promotion of innovation in clinical laboratory science, dedication to ASM and the advancement of clinical microbiology as a profession.
Eligibility: Open to distinguished microbiologists, who are identified with clinical microbiology.
Level of Study: Unrestricted
Value: US$2,000 cash prize, commemorative piece and travel to the ASM general meeting where the winner will be asked to present the bioMérieux Sonnenwirth Award lecture.
Frequency: Annual
Country of Study: Any country
No. of awards offered: 1
Application Procedure: Self-nominations will not be accepted. Nominations must consist of a nomination form that describes the nominee's activities and accomplishments pertinent to the award, a curriculum vitae including a list of publications and two additional supporting forms. Only one of the three individuals involved in the nomination may be employed at the nominee's institution. Forms can be found at www.asm.org/awards.
Closing Date: July 1st
Funding: Commercial

Contributor: BioMérieux Inc.
No. of awards given last year: 1
Additional Information: ASM awards are granted at the discretion of award selection committees and may not be awarded every year.

Carski Foundation Distinguished Undergraduate Teaching Award

Subjects: Microbiology education.
Purpose: To recognize a mature individual for distinguished teaching of microbiology to undergraduate (pre-baccalaureate) students and who has encouraged them to subsequent achievements.
Eligibility: Nominees must be currently teaching microbiology in a recognized college or university. A substantial portion of his or her time during the past 5 years must have been devoted to teaching undergraduate students in microbiology and a minimum of 10 years total teaching experience is required. Nominees may have engaged in research or other concerns, provided that teaching undergraduates remained a substantial activity.
Level of Study: Unrestricted
Type: Award
Value: US$2,500 cash prize, commemorative piece and travel to the ASM General Meeting where the winner will present the Carski Award lecture
Frequency: Annual
Country of Study: Any country
No. of awards offered: 1
Application Procedure: Self-nominations will not be accepted. Nominations must consist of a nomination form, a nominating letter detailing teaching responsibilities, manifests of distinguished teaching, innovations, publications and special awards, a curriculum vitae and two additional supporting forms. Only one of the three individuals involved in the nomination may be employed at the nominee's institution. Please visit www.asm.org/awards to view the forms.
Closing Date: July 1st
Funding: Foundation, private
Contributor: The Carski Foundation
No. of awards given last year: 1
Additional Information: ASM awards are granted at the discretion of award selection committees and may not be awarded every year.

Cubist - ICAAC Award

Subjects: Microbiology.
Purpose: To stimulate research in antimicrobial chemotherapy and honour outstanding sustained achievement.
Eligibility: Nominees must be actively engaged in research involving development of new agents, investigation of antimicrobial action or resistance to antimicrobial agents and/or the pharmacology, toxicology or clinical use of those agents. They must not have served on an ICAAC Program Committee within the past 2 years.
Level of Study: Unrestricted
Type: Award
Value: US$10,000 cash prize, a commemorative piece and travel expenses paid to attend the ICAAC where the winner will deliver the Cubist - ICAAC Award lecture.
Frequency: Annual
Country of Study: Any country
No. of awards offered: 1
Application Procedure: Self-nominations will not be accepted. Nominations must consist of a nomination form that includes a specific description of the research on which the nomination is based, a curriculum vitae including a list of publications and two additional supporting forms. Only one of the three individuals involved in the nomination may be employed at the nominee's institution. Forms can be found at www.asm/awards.
Closing Date: April 1st
Funding: Commercial
Contributor: Cubist, Pharmaceuticals
No. of awards given last year: 1
Additional Information: ASM awards are granted at the discretion of award selection committees and may not be awarded every year.

D.C. White Research and Mentoring Award

Purpose: Recognizing distinguished accomplishments in interdisciplinary research and mentoring in microbiology, this award honors D.C. White, who was known for his interdisciplinary scientific approach and for being a dedicated and inspiring mentor.
Eligibility: Consideration will be given to the breadth of the nominee's contributions, as well as their originality and overall impact. There are no age restrictions, but the nominee must have a distinguished record of accomplishments in microbiological research. Nominees in all areas of microbiology will be considered.
Level of Study: Unrestricted
Type: Award
Value: A cash prize of US$5,000, a commemorative piece, and travel to the ASM General Meeting, where the laureate delivers the D.C. White Research and Mentoring Award Lecture
Application Procedure: Self-nominations and more than one nomination per nominee will not be accepted. Only one nominator and two suuporters are accepted per nomination. Two supporters must be persons who are familiar with the nominee's qualifications and accomplishments. One must be someone who can comment specifically on the nominee's mentoring. Only one of the three individuals involved in the nomination may be employed at the nominee's institution. The nominator and supporters must not share employers. Nominations must include a curriculum vitae, including a list of publications, list of those mentored, nominating form and supporting forms. Forms can be found at www.asm.org/awards.
Closing Date: July 1st
Contributor: David C. White's Family and Friends

Eli Lilly and Company-Elanco Research Award

Subjects: Microbiology and immunology.
Purpose: To reward fundamental research of unusual merit in microbiology or immunology by an individual on the threshold of his/her career, who has not reached his/her 45th birthday.
Eligibility: Nominees must be working in the United States of America or Canada at the time of application and must be actively involved in the line of research for which the award is to be made. They must not have reached their 45th birthday by April 30th of the year the award is given.
Level of Study: Unrestricted
Type: Award
Value: US$5,000 cash prize, a commemorative piece and travel expenses to the ASM General Meeting where the winner will present the Eli Lilly and Company-Elanco Research Award lecture.
Frequency: Annual
Country of Study: Any country
No. of awards offered: 1
Application Procedure: Self-nominations will not be accepted. Nominations must consist of a nominating form that includes a specific description of the research on which the nomination is based, verification of the date of birth, i.e. a photocopy of driver's licence, passport or birth certificate, a curriculum vitae including a list of publications and two additional supporting forms. Only one of the three individuals involved in the nomination may be employed at the nominee's institution. Forms are located at www.asm.org/awards.
Closing Date: July 1st
Funding: Commercial
Contributor: Eli Lilly and Company-Elanco
No. of awards given last year: 1
Additional Information: ASM awards are granted at the discretion of award selection committees and may not be awarded every year.

EMD Millipore Alice C. Evans Award

Subjects: Microbiology.
Purpose: To recognize contributions toward the full participation and advancement of women in microbiology.
Eligibility: Nominees can be any member of ASM who has made major contributions toward fostering the inclusion, development and advancement of women in careers in microbiology. Nominees must demonstrate commitment to women in science through mentorship and advocacy and by setting an example through scientific and professional achievement.
Level of Study: Unrestricted
Value: A commemorative piece and travel to the ASM general meeting
Frequency: Annual
Country of Study: Any country
No. of awards offered: 1

Application Procedure: Self nominations will not be accepted. Nominations must consist of a nomination form that describes the nominees major contributions toward fostering the inclusion, development and advancement of women in careers in microbiology, a curriculum vitae including a list of publications and two additional supporting forms. Only one of the three individuals involved in the nomination may be employed at the nominee's institution. Forms can be found at www.asm.org/awards.
Closing Date: July 1st
Funding: Corporation
Contributor: EMD Millipore
No. of awards given last year: 1
Additional Information: ASM awards are granted at the discretion of award selection committees and may not be awarded every year.

Gen-Probe Joseph Public Health Award

Subjects: Microbiology.
Purpose: To honour a distinguished microbiologist who has exhibited exemplary leadership and service in the field of public health.
Eligibility: Nominees must be a microbiologist identified with public health.
Level of Study: Unrestricted
Value: A cash prize of $2,500 a commemorative piece and a travel stipend to attend the ASM general meeting where the winner will present the Gen-Probe Joseph Public Health Award lecture
Frequency: Annual
Country of Study: Any country
No. of awards offered: 1
Application Procedure: Self nominations will not be accepted. Nominations must consist of a nomination form that describes the nominee's leadership and service in the field of public health, a curriculum vitae including a list of publications and two additional supporting forms. Only one of the three individuals involved in the nomination may be employed at the nominee's institution. Forms are located at www.asm.org/awards.
Closing Date: July 1st
Funding: Commercial
Contributor: Gen-Probe
No. of awards given last year: 1
Additional Information: ASM awards are granted at the discretion of award selection committees and may not be awarded every year.

GlaxoSmithKline International Member of the Year Award

Purpose: Honors a distinguished microbiologist who exhibited exemplary leadership in the international microbiological community. It recognizes an international ASM member for education, communication, research, and advancement of the profession to the international microbiology community while demonstrating a commitment to the ASM.
Eligibility: The nominee can be any international member of ASM who has made major contributions toward the advancement of the microbiological sciences within the international community through education, research, communication, and leadership. The nominee must not have served on the International Board, the International Education Committee or the International Membership Committee within the past two years.
Type: Award
Value: A commemorative piece and travel to the ASM General Meeting where the laureate delivers the GSK Award lecture
Application Procedure: Self-nominations and more than one nomination per nominee will not be accepted. Only one nominating form and two supporting forms are accepted per nomination. The two supporters must be persons other than the nominator who are familiar with the nominee's qualifications and accomplishments. Only one of the three individuals involved in the nominating process may be employed at the nominee's institution. The nominator and supporters must not share emplyers. Nominations must have a curriculum vitae, including list of publications, nominating form and supporting forms. Forms are at www.asm.org/awards.
Closing Date: July 1st
Funding: Commercial
Contributor: GlaxoSmithKline

ICAAC Young Investigator Award

Subjects: Microbiology, including the discovery and application of chemotherapeutic agents and other sciences associated with infectious diseases.
Purpose: To recognize and reward young investigators for research excellence and potential in microbiology and infectious diseases.
Eligibility: Nominees must have completed postdoctoral research training in microbiology or infectious diseases no more than 3 years prior to presentation of the award. One of the awards is earmarked for a researcher working in the area of HIV who resides and works in North America.
Level of Study: Doctorate, Postdoctorate
Type: Award
Value: US$3,000 cash prize to support travel to the ICAAC and commemorative piece
Country of Study: Any country
No. of awards offered: Up to 5
Application Procedure: Self-nominations will not be accepted. Nominations must consist of a nomination form including a specific description of research, a curriculum vitae including a list of publications and two additional supporting forms. Only one of the three individuals involved in the nomination may be employed at the nominee's institution. Forms are located at www.asm.org/awards.
Closing Date: April 1st
Funding: Commercial
Contributor: The Human Health Division of Merck USA and the American Society for Microbiology
No. of awards given last year: 5
Additional Information: ASM awards are granted at the discretion of award selection committees and may not be awarded every year.

Maurice Hilleman/Merck Award

Purpose: ASM's premier award for major contributions to pathogenesis, vaccine discovery, vaccine development, and/or control of vaccine-preventable diseases. The award is presented in memory of Maurice R. Hilleman, whose work in the development of vaccines has saved the lives of many throughout the world.
Eligibility: The nominee must have made outstanding achievements in pathogenesis, vaccine discovery, vaccine development, and/or control of vaccine-preventable diseases.
Level of Study: Unrestricted
Type: Award
Value: A cash prize of US$20,000, a commemorative piece, and travel to the ASM General Meeting where the laureate delivers the Maurice Hilleman/Merck Award Lecture
Application Procedure: Self applications will not be accepted. Nominations must consist of a nomination form, a curriculum vitae, including a list of publications, a nominating form and two supporting forms. Only one of the three individuals involved in the nomination may be employed at the nominee's institution. Forms are located at www.asm.org/awards.
Closing Date: July 1st
Funding: Commercial
Contributor: Merck & Co., Inc

Merck Irving S. Sigal Memorial Award

Subjects: Microbiology.
Purpose: To recognize excellence in basic research in medical microbiology and infectious diseases.
Eligibility: Nominees must be no more than 5 years beyond completion of postdoctoral research training in microbiology or infectious diseases at the time of the nomination deadline.
Level of Study: Doctorate, Postdoctorate
Type: Award
Value: A commemorative piece and a cash prize of $2,500
Frequency: Annual
Country of Study: Any country
No. of awards offered: Up to 2
Application Procedure: Self nominations will not be accepted. Nominations must consist of a nomination form that describes the nominee's excellence in basic research in medical microbiology and infectious diseases, a curriculum vitae including a list of publications and two additional supporting forms. Only one of the three individuals involved in the nomination may be employed at the nominee's institution. Forms can be found at www.asm.org/awards.

Closing Date: July 1st
Funding: Commercial
Contributor: Merck Research Laboratories
No. of awards given last year: 2
Additional Information: ASM awards are granted at the discretion of award selection committees and may not be awarded every year.

Moselio Schaechter Distinguished Service Award

Subjects: Microbiology.
Purpose: This award, named in honor of Professor Moselio Schaechter, former ASM President, honors an ASM member who has shown exemplary leadership and commitment towards the substantial furthering of the profession of microbiology in research, education or technology in the developing world.
Eligibility: Individuals (e.g. microbiologists who have been instrumental in setting up properly functioning clinical microbiology laboratories or successful biotechnology services based on microbiology; academicians who have developed high quality undergraduate or graduate training programs; researchers who have demonstrated leadership in the context of the region) from the upper-middle, lower-middle and low-income countries as determined per World Bank's classification. The nominee must be a national or a permanent resident of a qualifying country and have a full-time professional appointment in the microbiological sciences or a related field for at least ten years in a country or region of the developing world. The nominees may not be currently serving on any ASM Board or Committee and can not be an ASM Ambassador or Country Liaison at the time of the nomination deadline. The nominee must be an active ASM member at the time of nomination deadline.
Level of Study: Unrestricted
Type: Award
Value: £4,000 cash prize to defray expenses assosiated with travelling to and attending the ASM General Meeting an engraved plaque to be presented during the International Reception at the ASM General Meeting; publication of the awardee profile in the International Affairs section of Microbe
Frequency: Annual
Application Procedure: Nominations must consist of the following: curriculum vitae, including a list of publications, emailed to awards@asmusa.org, nominating form and supporting form. Only one of the three individuals involved in the nomination may be employed at the nominee's institution. Forms can be found at asm.org/awards.
Closing Date: July 1st
Funding: Commercial
Contributor: GlaxoSmithKline

Procter & Gamble Award in Applied and Environmental Microbiology

Subjects: Environmental microbiology, applied microbiology.
Purpose: To recognize distinguished achievement in research and development in applied (non-clinical) and environmental microbiology.
Eligibility: Nominees must show outstanding accomplishment in research or development in the appropriate field. They must be actively engaged in research or development at the time that the award is presented.
Level of Study: Unrestricted
Type: Award
Value: US$2,000 cash prize, a commemorative medal and travel to the ASM General Meeting where the winner the present the Procter & Gamble Award lecture
Frequency: Annual
Country of Study: Any country
No. of awards offered: 1
Application Procedure: Self-nominations will not be accepted. Nominations must consist of a nomination form that describes the work that has stimulated the nomination, a curriculum vitae including a list of publications and awards and two additional supporting forms. Only one of the three individuals involved in the nomination may be employed at the nominee's institution. Forms can be found at www.asm.org/awards.
Closing Date: July 1st
Funding: Commercial
Contributor: Procter & Gamble
No. of awards given last year: 1
Additional Information: ASM awards are granted at the discretion of award selection committees and may not be awarded every year.

Promega Biotechnology Research Award

Subjects: Biotechnology.
Purpose: To honour outstanding contributions to the application of biotechnology through fundamental research, developmental research or reduction to practice.
Eligibility: An outstanding contribution can be a single exceptionally significant achievement or the aggregate of a number of exemplary achievements.
Level of Study: Unrestricted
Type: Award
Value: US$5,000 cash prize, a commemorative piece and travel to the ASM General Meeting where the winner will present the Promega Biotechnology Research Award lecture
Frequency: Annual
Country of Study: Any country
No. of awards offered: 1
Application Procedure: Self-nominations will not be accepted. Nominations must consist of a nominating form that includes a description of the nominee's research, a curriculum vitae including a list of publications and two additional supporting forms. Only one of the three individuals involved in the nomination may be employed at the nominee's institution. Forms may be found at www.asm.org/awards.
Closing Date: July 1st
Funding: Commercial
Contributor: The Promega Corporation
No. of awards given last year: 1
Additional Information: ASM awards are granted at the discretion of the selection committee and may not be awarded every year.

Raymond W. Sarber Award

Subjects: Microbiology.
Purpose: To recognize students at the undergraduate or predoctoral levels for research excellence and potential.
Eligibility: Nominees must be at the undergraduate or predoctoral level, attending an accredited institution in United States, in an academic programme involving microbiology.
Level of Study: Predoctorate
Value: a cash prize of $2,000, a commemorative piece
Frequency: Annual
Country of Study: Any country
No. of awards offered: Up to 2
Application Procedure: Self nominations will not be accepted. Nominations must consist of a nomination form that describes the nominee's research excellence and potential, a personal statement from the student, a curriculum vitae, two additional supporting letters. One form must be from a supervisor or mentor, while the other form may be from an academic advisor or colleague. Only one of the three individuals involved in the nomination may be employed at the nominee's institution. Forms can be viewed at www.asm.org/awards.
Closing Date: July 1st
Funding: Private
Contributor: American Society for Microbiology
No. of awards given last year: 1
Additional Information: ASM awards are granted at the discretion of award selection committees and may not be awarded every year.

Scherago-Rubin Award

Subjects: Clinical microbiology.
Purpose: To recognize an outstanding, bench-level clinical microbiologist.
Eligibility: Nominees must be a non-doctoral level clinical microbiologists involved primarily in routine diagonostic work, rather than in research, who has distinguished himself or herself with excellent performance in clinical laboratory.
Level of Study: Graduate, Predoctorate
Value: A commemorative piece and $2,000 cash prize
Frequency: Annual
Country of Study: Any country
No. of awards offered: 1
Application Procedure: Self nominations will not be accepted. Nominations must consist of a nomination form from supervisor that

describes the nominee's routine disgnostic work that has distinguished him or her in the clinical laboratory, a curriculum vitae including a list of publications and two additional supporting forms. Only one of the three individuals involved in the nomination may be employed at the nominee's institution. Forms are located at www.asm.org/awards.
Closing Date: July 1st
Funding: Private
Contributor: American Society for Microbiology
No. of awards given last year: 1
Additional Information: ASM awards are granted at the discretion of award selection committees and may not be awarded every year.

Siemens Healthcare Diagnostics Young Investigator Award

Subjects: Microbiology.
Purpose: To recognize research excellence and potential and to further the educational or research objectives of an outstanding young clinical scientist.
Eligibility: The nominee must be no more than 5 years beyond completion of postdoctoral training
Level of Study: Postdoctorate
Type: Award
Value: US$2,500 cash prize, a commemorative piece and travel to the ASM General Meeting
Frequency: Annual
Country of Study: Any country
No. of awards offered: 1
Application Procedure: Self-nominations will not be accepted. Nominations must consist of a nomination form, a curriculum vitae including a list of publications, abstracts and manuscripts in preparation and two additional supporting forms documenting the nominee's research excellence and anticipated impact of the award on achievement of the nominee's career objectives. Only one of the three individuals involved in the nomination may be employed at the nominee's institution. Forms are located at www.asm.org/awards.
Closing Date: July 1st
Funding: Commercial
Contributor: Siemens Healthcare Diagnostics
No. of awards given last year: 1
Additional Information: ASM awards are granted at the discretion of award selection committees and may not be awarded every year.

USFCC/J. Roger Porter Award

Subjects: Microbiology.
Purpose: To recognize outstanding efforts by a scientist who has demonstrated the importance of microbial biodiversity through sustained curatorial or stewardship activities for a major resource used by the scientific community.
Eligibility: Nominees will have greatly aided other scientists by demonstrating the fundamentals of culture collections and related resources and the rich biodiversity that such collections preserve.
Level of Study: Unrestricted
Value: A cash prize of $2,000, a commemorative piece and travel to the ASM general meeting where the winner will present the USFCC/J. Roger Porter Award lecture
Frequency: Annual
Country of Study: Any country
No. of awards offered: 1
Application Procedure: Self nominations will not be accepted. Nominations must consist of a nomination form that describes the nominee's outstanding efforts demostrating the importance of microbial biodiversity through sustained curatorial or stewardship activities for a major resource used by the scientific community, a curriculum vitae including a list of publications and two additional supporting forms. Only one of the three individuals involved in the nomination may be employed at the nominee's institution. Forms are located at www.asm.org/awards.
Closing Date: July 1st
Funding: Commercial, private
Contributor: The United States Ferderation for Cultural Collections and The American Society for Microbiology
No. of awards given last year: 1
Additional Information: ASM awards are granted at the discretion of award selection committees and may not be awarded every year.

William A Hinton Research Training Award

Subjects: Microbiology.
Purpose: To honour an individual who has made outstanding significant contributions towards fostering the research training of underrepresented minorities in microbiology.
Eligibility: Nominees must have contributed to the research training of undergraduate students, graduate students, postdoctoral Fellows or health professional students. Their efforts must have led to the increased participation of underrepresented minorities in microbiology.
Level of Study: Unrestricted
Type: Award
Value: US$2,500 cash prize, a commemorative piece and travel to the ASM General Meeting
Frequency: Annual
Country of Study: Any country
No. of awards offered: 1
Application Procedure: Self-nominations will not be accepted. Nominations must consist of a nominating form highlighting the nominee's activities and accomplishments pertinent to the award, a curriculum vitae including a list of publications and two additional supporting forms. Only one of the three individuals involved in the nomination may be employed at the nominee's institution. Forms are found at www.asm.org/awards.
Closing Date: July 1st
Funding: Private
Contributor: American Society for Microbiology
No. of awards given last year: 1
Additional Information: ASM awards are granted at the discretion of award selection committees and may not be awarded every year.

THE AMERICAN SOCIETY FOR PHOTOGRAMMETRY & REMOTE SENSING (ASPRS)

The Imaging and Geospatial Information Society 5410 Grosvenor Lane, Suite 210, Bethesda, MD, 20814 2160, United States of America
Tel: (1) 301 493 0290
Fax: (1) 301 493 0208
Email: asprs@asprs.org; scholarships@asprs.org
Website: www.asprs.org

The American Society for Photogrammetry and Remote Sensing (ASPRS) was founded in 1934. It is a scientific association serving over 7,000 professional members around the world whose mission is to advance knowledge and improve understanding of mapping sciences to promote the responsible applications of photogrammetry, remote sensing, geographic information systems and supporting technologies.

Abraham Anson Memorial Scholarship

Subjects: Geospatial science, photogrammetry, remote sensing, surveying and mapping.
Purpose: To encourage students who have an exceptional interest in pursuing scientific research or education in geospatial science or technology related to photogrammetry, remote sensing, surveying and mapping to enter a professional field where they can use the knowledge of their discipline to excel in their profession.
Eligibility: Open to students currently enrolled or intending to enroll in a college or university in the United States of America for the purpose of pursuing a program of study to enter a profession in which education in geospatial science or technology related to photogrammetry, remote sensing, surveying and mapping will advance the value of those disciplines within that profession.
Level of Study: Research
Type: Scholarship
Value: Certificate, US$1,000 and one year student membership in the society
Frequency: Annual
Country of Study: United States of America
No. of awards offered: 1
Application Procedure: Applicants must submit listing of courses, internships and special projects taken, a transcript of all courses completed, grades obtained, two letters of recommendation and a personal statement. Applicants must visit organization website for

complete details http://www.asprs.org/Awards-and-Scholarships/Abraham-Anson-Memorial-Scholarship.html.
Closing Date: October 17th (Check with website)
Funding: Foundation, individuals
Contributor: Anson Bequest

ASPRS Robert N. Colwell Memorial Fellowship

Subjects: Remote sensing and geospatial information technologies.
Purpose: To encourage and commend college/university students or postdoctoral researchers who display exceptional interest, desire, ability and aptitude in the specified field and who have a special interest in developing practical uses of these technologies.
Eligibility: Open to student enrolled or intending to enrol in a college or university in the United States or Canada, or a recently graduated postdoctoral researcher who is pursuing a programme of study aimed at starting a professional career.
Level of Study: Doctorate, Postgraduate
Type: Fellowship
Value: US$5,000, a certificate and a one-year student or associate membership (new or renewal) in ASPRS.
Frequency: Annual
Country of Study: United States of America or Canada
No. of awards offered: 1
Application Procedure: Applicants must include a listing of courses, transcripts, listing of internship, 3 letters of recommendation and statement of purpose along with a completed application form.
Closing Date: October 17th
Contributor: ASPRS Foundation
Additional Information: Please see the website for further details http://www.asprs.org/Awards-and-Scholarships/Robert-N-Colwell-Memorial-Fellowship.html.

ASPRS Ta Liang Memorial Award

Subjects: Remote sensing.
Purpose: To facilitate research-related travel by outstanding graduate students.
Eligibility: Graduate Student members of ASPRS.
Level of Study: Graduate
Type: Grant
Value: US$2,000 and a hand-engrossed certificate
Frequency: Annual
No. of awards offered: Varies
Application Procedure: Applicants must submit an application form and recommendation letters. In addition to that they must also submit a 2 page statement detailing the plan for research-related travel, transcripts and Graduate Record Examination scores.
Closing Date: October 17th (Check with website)
Funding: Individuals, corporation
Contributor: ASPRS Foundation
Additional Information: The recipient is obligated to provide ASPRS and Ta Liang's family, a report of his/her accomplishments during the travel for which the award is granted. The selection committee will evaluate each application, applying necessary weights as indicated below: 40% on the basis of scholastic record; 30% on the basis of the research travel plan; 20% on the basis of letters of recommendation; 10% on the basis of community service activities. Please see the website for details http://www.asprs.org/Awards-and-Scholarships/Ta-Liang-Memorial-Award.html.

ASPRS William A. Fischer Memorial Scholarship

Subjects: Remote sensing data/techniques.
Purpose: To facilitate graduate-level studies and career goals directed towards new and innovative uses of remote sensing data/techniques that relate to the natural, cultural or agricultural resources of the Earth.
Eligibility: Open to current or prospective graduate student.
Level of Study: Graduate
Type: Scholarship
Value: US$2,000 and a certificate
Length of Study: 1 year
Frequency: Annual
Country of Study: United States of America
No. of awards offered: Varies
Application Procedure: Applicants must submit a 2-page statement detailing applicant's educational and career plans for continuing studies in remote sensing applications, letters of recommendation and transcripts.
Closing Date: October 17th (Check with website)
Funding: Corporation, individuals
Contributor: ASPRS Foundation
Additional Information: Please see the website for details http://www.asprs.org/Awards-and-Scholarships/William-A-Fischer-Memorial-Scholarship.html.

BAE Systems Award

Subjects: Photogrammetry, remote sensing.
Purpose: To reward top quality research and publication by young students (under age 35, as of the application deadline) at master's or doctoral level and to encourage researchers to use the ASPRS annual conference as a vehicle to publish and present their findings.
Eligibility: Applicants must be a student enrolled in a Master's or doctoral programme at a recognised institution, must be the principal of the paper accepted for publication, must be a current student member of ASPRS.
Level of Study: Doctorate, Postgraduate
Type: Grant
Value: US$2,000 equally divided among the recipients, if more than one is selected and a certificate.
Frequency: Annual
Country of Study: Any country
No. of awards offered: Varies
Application Procedure: Applicants must submit an abstract and once it is accepted for presentation, the full paper should be submitted along with the application for the award.
Closing Date: October 17th (Check with website)
Funding: Private
Contributor: BAE Systems
Additional Information: Both the paper and the application must be submitted on or before the required deadline for receipt of the paper for the conference proceedings. Please see the website for details http://www.asprs.org/Awards-and-Scholarships/BAE-Systems-Award.html.

Francis H. Moffitt Memorial Scholarship

Subjects: Surveying, photogrammetry and geospatial mapping.
Purpose: To encourage upper-division undergraduate and graduate-level college students to pursue a course of study in surveying and photogrammetry leading to a career in the geospatial mapping profession.
Eligibility: Open to students currently enrolled or intending to enroll in a college or university in the United States of America or Canada, who are pursuing a program of study in surveying or photogrammetry leading to a career in the geospatial mapping profession.
Level of Study: Research
Type: Scholarship
Value: US$4,500, a certificate and a one year student or associate membership in ASPRS.
Frequency: Annual
Country of Study: United States of America or Canada
No. of awards offered: 1
Application Procedure: Applicants must submit listing of courses undertaken, internships and special projects taken, a transcript of all courses completed, grades obtained, two letters of recommendation and a personal statement. Applicants must visit organization website for complete details.
Closing Date: October 17th
Funding: Foundation, individuals
Contributor: ASPRS, MAPPS and ACSM
Additional Information: The committee will assign additional points to the final scores for those applicants having attributes that reflect Professor Moffitt's career and contributions. Please see the website for details http://www.asprs.org/Awards-and-Scholarships/Francis-H-Moffitt-Memorial-Scholarship.html.

GeoEye Foundation Award for the Application of High-Resolution Digital Satellite Imagery

Subjects: Remote sensing, image processing.
Purpose: To support remote sensing education and stimulate the development of applications high-resolution digital satellite imagery for applied research by graduate students.

Eligibility: Applicant should be full-time undergraduate or graduate student at an accredited college or university in the United States or Canada with image processing facilities appropriate for conducting the proposed work. Applicant should be a Student member of ASPRS.
Level of Study: Graduate
Value: A grant of satellite imagery data up to 500 square kilometers (potential value of up to $20,000), and a certificate inscribed with the recipient's name.
Frequency: Annual
Study Establishment: Accredited college or university in the United States or Canada
Country of Study: United States of America or Canada
No. of awards offered: 1
Application Procedure: Applicants must submit the application form along with letters of recommendation and a brief 2-page proposal.
Closing Date: October 17th
Funding: Private
Contributor: GeoEye
Additional Information: The imagery award is one 500 sq. km scene from the GeoEye archive. Request for more square kilometers will be considered on a case-by-case basis. Please see the website for details.

John O. Behrens Institute for Land Information (ILI) Memorial Scholarship

Subjects: Geospatial science or technology or land information systems/records.
Purpose: To encourage those who have an exceptional interest in pursuing scientific research or education in geospatial science or technology or land information systems/records to enter a professional field where they can use the knowledge of this discipline to excel in their profession.
Eligibility: Open to students currently enrolled or intending to enroll in a college or university in the United States of America for the purpose of pursuing a program of study that prepares him/her to enter a profession in which education in geospatial science or land information disciplines will advance the value of those disciplines within that profession.
Level of Study: Research
Type: Scholarship
Value: Certificate, US$1,000, and one year student membership in the society
Frequency: Annual
Country of Study: United States of America
No. of awards offered: 1
Application Procedure: Applicants must submit listing of courses undertaken, a transcript of all courses completed, grades obtained, two letters of recommendation and a personal statement. Applicants must visit organization website for complete details.
Closing Date: October 17th (Check with website)
Funding: Foundation, individuals
Contributor: Institute for Land Information

Paul R. Wolf Memorial Scholarship

Subjects: Surveying, mapping, photogrammetry.
Purpose: To encourage and commend college students who display exceptional interest, desire, ability and aptitude to enter the profession of teaching surveying, mapping or photogrammetry.
Eligibility: Applicants must be a graduate student currently enrolled or intending to enroll in a college or university in the Unite'd States, who is pursuing a programme of study in preparation for entering the teaching profession in the general area of surveying, mapping or photogrammetry.
Level of Study: Graduate
Type: Scholarship
Value: US$3,000 and a certificate.
Frequency: Annual
Country of Study: United States of America
No. of awards offered: 1
Application Procedure: Applicants must submit the application forms along with list of courses taken, academic grades, a transcript of all college and university level courses completed, two letters of recommendation from faculty members, papers, reports, other items produced by the applicant, a statement of teaching experience and a 2 page statement detailing applicant's plans for continuing studies to become an educational professional in surveying, mapping or photogrammetry.
Closing Date: October 17th (Check with website)
Funding: Individuals
Contributor: Friends and colleagues of Paul R. Wolf.
Additional Information: Please see the website for details http://www.asprs.org/Awards-and-Scholarships/Paul-R-Wolf-Memorial-Scholarship.html.

Robert E. Altenhofen Memorial Scholarship

Subjects: Theoretical photogrammetry.
Purpose: To encourage college students who display exceptional interest and ability in the theoretical aspects of photogrammetry.
Eligibility: Undergraduate or graduate student members of ASPRS.
Level of Study: Graduate
Type: Award
Value: US$2,000 and a certificate
Frequency: Annual
Country of Study: Any country
No. of awards offered: 1
Application Procedure: Applicants should submit application form, letters of recommendation, 2-page statement regarding plans for continuing studies, papers, reports or other items and academic transcripts.
Closing Date: October 17th (Check with website)
Funding: Foundation
Contributor: ASPRS Foundation and the estate of Mrs. Helen Altenhofen.
Additional Information: Please see the website for details http://www.asprs.org/Awards-and-Scholarships/Robert-E-Altenhofen-Memorial-Scholarship.html.

Z/I Imaging Scholarsip

Subjects: Signal processing, image processing, photogrammetry.
Purpose: To facilitate graduate-level studies and career goals adjudged to address new and innovative uses of signal processing, image processing techniques and the application of photogrammetry to real-world techniques within the earthimaging industry.
Eligibility: Applicant should be a member of ASPRS. Should be a student currently pursuing graduate-level studies or who plans to enroll for graduate studies.
Level of Study: Graduate
Type: Scholarship
Value: US$2,000 as one-year Scholarship and a certificate.
Length of Study: 1 year
Frequency: Annual
Study Establishment: Recognised college or university
Country of Study: Any country
No. of awards offered: 1
Application Procedure: Applicants should submit the application forms along with 2-page statement detailing educational and career plans and two reference forms from faculty members who have knowledge of the applicant's capabilities.
Closing Date: October17 (Check with website)
Funding: Private
Contributor: Intergraph
Additional Information: Please see the website for details http://www.asprs.org/Awards-and-Scholarships/Z-I-Imaging-Scholarship.html.

AMERICAN SOCIETY FOR QUALITY (ASQ)

600 North Plankinton Avenue, Milwaukee, WI, 53203, United States of America
Tel: (1) 414 272 8575
Fax: (1) 414 272 1734
Email: help@asq.org
Website: www.asq.org

The American Society for Quality (ASQ) is the world's leading authority on quality. With more than 100,000 individual and organizational members, this professional association advances learning, quality improvement and knowledge exchange to improve business results and to create better workplaces and communities worldwide.

Ellis R. Ott Scholarship for Applied Statistics and Quality Management

Subjects: Statistics.
Purpose: To encourage students to pursue a career in a field related to statistics and/or quality management.
Eligibility: Open to candidates who are planning to enroll or are enrolled in a Master's degree or higher level programme in the United States or Canada.
Level of Study: Doctorate, Postgraduate
Type: Scholarships
Value: US$5,000
Length of Study: 1 year
Frequency: Annual
Country of Study: United States of America
No. of awards offered: 6
Application Procedure: Applicants can download the application form from the website. The completed application form along with curriculum vitae, academic transcripts and 2 letters of recommendation are to be submitted.
Closing Date: April 1st

For further information contact:

55 Buckskin Path, Plymouth, MA, 02360, United States of America
Tel: (1) 774 413 5268
Email: lynne.hare@comcast.net
Contact: Dr Lynne B Hare

The Richard A. Freund International Scholarship

Subjects: It covers the engineering, statistical, managerial and behavioral foundations of theory and application of quality control, quality assurance, quality improvement, and total quality management.
Purpose: This scholarship honours the memory of Richard A. Freund, a past president of ASQ. It is for graduate study of the theory and application of quality control, quality assurance, quality improvement and total quality management.
Eligibility: Applicants do not need to be ASQ members. Applicants should have a GPA of 3.25 or higher (undergraduate and graduate). Concentration must be in quality control, quality assurance, quality improvement, total quality management or similar quality emphasis.
Level of Study: Postgraduate
Type: Scholarship
Value: US$5,000
Frequency: Annual
Study Establishment: Study may take place in one's own country or in another country.
No. of awards offered: Varies
Closing Date: April 1st
Funding: Foundation
Additional Information: Please see the website for further details http://asq.org/about-asq/awards/freundscholar.html.

AMERICAN SOCIETY OF HEATING, REFRIGERATING AND AIR CONDITIONING ENGINEERS, INC. (ASHRAE)

1791 Tullie Circle, North East, Atlanta, GA, 30329-2305, United States of America
Tel: (1) 404 636 8400
Fax: (1) 678 539 2112
Email: mvaughn@ashrae.org
Website: www.ashrae.org
Contact: Mr Michael R Vaughn, Manager of Research and Technical Services

The American Society of Heating, Refrigerating and Air Conditioning Engineers, Inc. (ASHRAE) is an international organization of 50,000 people with chapters all over the world. The Society is organized for the sole purpose of advancing the arts and sciences of heating, ventilation, air conditioning and refrigerating for public's benefit through research, standards writing, continuing education and publications.

ASHRAE Grants-in-Aid for Graduate Students

Subjects: Heating, refrigeration, air conditioning and ventilation.
Purpose: To stimulate interest through the encouragement of original research.
Eligibility: Open to graduate engineering students capable of undertaking appropriate and scholarly research.
Level of Study: Doctorate, Postgraduate
Type: Grant
Value: Up to US$10,000 depending upon the needs and nature of request.
Length of Study: Usually for 1 year or less, non-renewable.
Frequency: Annual
Study Establishment: The grantee's institution
Country of Study: Any country
No. of awards offered: 10–25
Application Procedure: Applicants must complete an application form, available from the website. An application form must also be submitted by the faculty advisor.
Closing Date: December 15th
Funding: Private
No. of awards given last year: 18
No. of applicants last year: 48

AMERICAN SOCIETY OF HEMATOLOGY (ASH)

2021 L Street NW, Suite 900, Washington, DC, 20036, United States of America
Tel: (1) 202 776 0544
Fax: (1) 202 776 0545
Email: ash@hematology.org
Website: www.hematology.org
Contact: Administrative Assistant

The mission of the American Society of Hematology (ASH) is to further the understanding, diagnosis, treatment and prevention of disorders affecting the blood, bone marrow and the immunologic, hemostatic and vascular systems, by promoting research, clinical care, education, training and advocacy in hematology.

American Society of Hematology Abstract Achievement Awards and Outstanding Abstract Achievement Awards

Subjects: Haematology.
Purpose: To help individuals to defray annual meeting expenses.
Eligibility: Applicants must be medical students graduate students, resident physicians or postdoctoral Fellows who are both first author and present of an abstract.
Level of Study: Professional development
Type: Travel award
Value: US$500 and annual meeting travel reimbursement for Outstanding Abstract Achievement Award winners.
Frequency: Annual
No. of awards offered: Varies
Application Procedure: Applicants must submit an abstract for the annual meeting and identify themselves as 'travel/merit award' applicants. Applicants must also include a letter from their Training Programme Director requesting travel support and indentifying need.
Closing Date: Early August
Funding: Commercial, foundation
No. of awards given last year: 195
Additional Information: Visit ASH's Acknowledgement of Commercial Support page for more information on support for the Abstract Achievement and Outstanding Abstract Achievement Awards. Please see the website for further details http://www.hematology.org/Awards/Abstract-Achievement/2630.aspx.

American Society of Hematology Minority Medical Student Award Program

Subjects: Haematology.

Purpose: To provide support for a summer research programme of 8–12 weeks and for travel to the Society's annual meeting.
Eligibility: Applicants must be minority medical students enrolled in either MD, MD/PhD or equivalent DO programmes and must be citizens or permanent residents of the United States of America or Canada.
Level of Study: Professional development, Research
Type: Grant
Value: Stipend of US$5,000 and $2,000 allowance for travel, complimentary subscriptions to Blood, guidance of a research and career-development mentor.
Length of Study: 8–12 weeks
Frequency: Annual
Study Establishment: Depends on participant, US or Canadian university
Country of Study: United States of America or Canada
No. of awards offered: 10
Application Procedure: All applicants must complete the Minority Medical Student Award Program application available from the website http://www.hematology.org/Awards/MMSAP/2624.aspx.
Closing Date: March 8th
Funding: Commercial
No. of awards given last year: 9
No. of applicants last year: 11
Additional Information: For additional information, please contact Courtney Krier, Award Programme Coordinator, at ckrier@hematology.org or by phone at 202 776 0544, ext. 1168. Notification of awards will be by May 1st.

American Society of Hematology Scholar Awards

Subjects: Haematology.
Purpose: To encourage haematologists to begin a career in research by providing partial salary or other support.
Eligibility: To be eligible for the Junior Faculty Scholar Award, applicants must be within the first 2 years of their initial faculty appointment as an assistant professor, and for the Fellow Scholar Award, applicants must have more than 2 years, but less than 6 years of postdoctoral research training. Applicants must work in a United States of America or Canadian institution.
Level of Study: Research
Type: Award
Value: US$1,00,000 for Fellow Scholars and US$1,50,000 for Junior Faculty Scholars
Length of Study: 2–3 years
Frequency: Annual
No. of awards offered: Varies
Application Procedure: A letter of intent must be submitted by early May and it should include a signed cover letter, abstract of the proposed project (350 words or less), applicant's curriculum vitae and should identify which award category the applicant is applying for.
Closing Date: June 3rd (letter of intent); August 2nd (application)
Funding: Commercial, foundation
Contributor: American Society of Hematology
No. of awards given last year: 17
No. of applicants last year: 83
Additional Information: For detailed information, applicants must visit the website or contact Courtney Krier at the American Society of Hematology. Please see the website for more details http://www.hematology.org/Awards/Scholar/2407.aspx.

American Society of Hematology Trainee Research Award Program

Subjects: Haematology.
Purpose: To provide support for a research project of 3 months and for travel to the Society's annual meeting.
Eligibility: Applicants must be medical students or residents and selected undergraduates only. The programme is open to ACGME-accredited institutions in the United States of America, Mexico and Canada that have a training programme director in haematology or a related area. Please see the website for further details regarding eligibility http://www.hematology.org/Awards/Next-Generation-Research-Scientists-Award/2627.aspx#b.
Level of Study: Professional development
Value: a $5,000 stipend and $1,000 each year for two years to support attendance at the ASH Annual Meeting.
Length of Study: either a short hematological research project for a minimum of three months, or a long hematological research project between three and 12 months
Frequency: Annual
No. of awards offered: 1
Application Procedure: All applicants and institutions must complete the Trainee Award application form available from the website. Applications MUST be submitted electronically to training@hematology.org.
Closing Date: October 1st (Short project) and August 1st (Long project)
Funding: Commercial
Contributor: American Society of Hematology
Additional Information: For any additional information regarding the programme, please contact Joe Basso, Training Manager, at 202 776 0544.

American Society of Hematology Visitor Training Program

Subjects: Haematology.
Purpose: To provide scientists and haematologists in developing countries an opportunity to gain valuable clinical experience, technology training or laboratory experience.
Eligibility: Applicants must be scientists and haematologists from developing countries as defined by the American Society of Hematology.
Level of Study: Professional development
Type: Grant
Value: ASH will fund approved costs for travel and living expenses (applicants must submit a budget).
Length of Study: Up to 12 weeks
Frequency: Annual
No. of awards offered: Varies
Application Procedure: Applicants must complete the visiting trainee programme application and submit it with a letter of recommendation from the proposed host institution. They will need to identify a site and host for their proposed short-term clinical or laboratory experience and give a clear statement of the topic or goal of the training programme. Application forms are available from the website.
Closing Date: May 2nd
Funding: Commercial
Contributor: ASH
Additional Information: For any additional information regarding the programme, please contact Clare Kelley, International Programs Specialist, at pckelley@hematology.org or 202-776-0544, ext. 4902. Please see the website for further details http://www.hematology.org/Awards/Visitor-Training-Program/2196.aspx#Benefits.

ASH-AMFDP Award

Subjects: Heamatology.
Purpose: To increase the number of underrepresented minority scholars in the field of heamatology with academic and research appointments.
Eligibility: Applicants must be from historically disadvantaged backgrounds, US citizens or permanent residents at the time of application deadline, completing their formal clinical training.
Value: $75,000 and an annual grant of $30,000
Length of Study: 4 years
Frequency: Annual
No. of awards offered: Varies
Application Procedure: Applications are available online at http://www.hematology.org/Awards/ASH-AMFDP/2224.aspx#a.
Closing Date: March 14th
Funding: Commercial, foundation
Contributor: American Society of Hematology and the Harold Amos Medical Faculty Development Program (AMFDP)
Additional Information: The Scholar will also have the opportunity to attend the ASH annual meeting for the four years of the award, and gets a subscription to the ASH journal, Blood. The costs of economy travel, four-night hotel stay, and $100 per day for meals will be covered by ASH.

For further information contact:

8701 Georgia Ave. Suite 411, Silver Spring, MD, 20910-3713
Contact: Harold Amos, Medical Faculty Development Programme

AMERICAN SOCIETY OF INTERIOR DESIGNERS (ASID) EDUCATIONAL FOUNDATION, INC.

608 Massachusetts Avenue North East, Washington, DC, 20002-6006, United States of America
Tel: (1) 202 546 3480
Fax: (1) 202 546 3240
Email: education@asid.org
Website: www.asidfoundation.org
Contact: Education Department

The American Society of Interior Designers (ASID) Educational Foundation represents the interests of more than 30,500 members including interior design practitioners, students and industry and retail partners. ASID's mission is to be the definitive resource for professional education and knowledge sharing, advocacy of interior designers' right to practice and expansion of interior design markets.

ASID/Joel Polsky Academic Achievement Award

Subjects: Interior design.
Purpose: To recognise an outstanding student's interior design research or thesis project.
Eligibility: Open to applicants of any nationality. Research papers or doctoral and Master's theses should address such interior design topics as educational research, behavioural science, business practice, design process, theory or other technical subjects.
Level of Study: Postgraduate
Type: Prize
Value: US$5,000
Frequency: Annual
Country of Study: Any country
No. of awards offered: 1
Application Procedure: Applicants must write for details. Please see the website for details http://www.asidfoundation.org/ASID_FOUNDATION/SCHOLARSHIPS_and_AWARDS.html.
Closing Date: March 12th
Additional Information: Entries will be judged on actual content, breadth of material, comprehensive coverage of topic, innovative subject matter and bibliography or references.

ASID/Joel Polsky Prize

Subjects: Interior design.
Purpose: To recognise outstanding academic contributions to the discipline of interior design through literature or visual communication.
Eligibility: Entries should address the needs of the public, designers and students on topics such as educational research, behavioural science, business practice, design process, theory or other technical subjects.
Level of Study: Unrestricted
Type: Prize
Value: US$5,000
Frequency: Annual
Country of Study: Any country
No. of awards offered: 1
Application Procedure: Applicants must write for details.
Closing Date: March 12th
Additional Information: Material will be judged on innovative subject matter, comprehensive coverage of topic, organization, graphic presentation and bibliography or references.

ASID/Mabelle Wilhelmina Boldt Memorial Scholarship

Subjects: Interior design.
Purpose: To encourage talented practicing interior designers to advance their professional developmentthrough graduate study and research.
Eligibility: Applicants must have been practising designers for a period of at least five years prior to returning to graduate level. Preference will be given to those with a focus on design research. The scholarship will be awarded on the basis of academic or creative accomplishment, as demonstrated by school transcripts and a letter of recommendation.
Level of Study: Graduate
Type: Scholarship
Value: US$2,000
Frequency: Annual
Study Establishment: A degree granting institution
Country of Study: Any country
No. of awards offered: 1
Application Procedure: Applicants must write for details. Please see website for details http://www.asidfoundation.org/NR/rdonlyres/4FAD2C3B-F880-4067-8901-9180B0E428F6/0/Scholarship_BOLDT-Web.pdf.
Closing Date: March 1st
Contributor: American Society of Interior Designers (ASID) Educational Foundation, Inc.

AMERICAN SOCIETY OF MECHANICAL ENGINEERS (ASME INTERNATIONAL)

Two Park Avenue, New York, NY, 10016-5990, United States of America
Tel: (1) 800 843 2763
Fax: (1) 212 591 7143, 212 591 7856
Email: CustomerCare@asme.org
Website: www.asme.org/education/enged/aid
Contact: Theresa Oluwanifise, Coordinator Educational Operations

Founded in 1880 as the American Society of Mechanical Engineers (ASME International), today ASME International is a non-profit educational and technical organization serving a worldwide membership.

ASME Graduate Teaching Fellowship Program

Subjects: Mechanical engineering.
Purpose: To encourage outstanding students, especially women and minorities, to pursue a doctorate in mechanical engineering teaching and to encourage the engineering education as a profession.
Eligibility: Applicant must be a PhD student in mechanical engineering, with a demonstrated interest in an academic career. A master's degree or passage of qualifying exam is required. A lecture-responsibility teaching assistantship commitment from the applicant's department is required. Please see the website for details regarding eligibility http://www.asme.org/about-asme/scholarship-and-loans/graduate-teaching-fellowships.
Level of Study: Postgraduate, Doctorate
Type: Fellowship
Value: US$5,000 per year
Length of Study: 2 years
Frequency: Annual
Country of Study: United States of America
No. of awards offered: 4
Application Procedure: Applicants must submit an undergraduate grade point average, Graduate Record Examination scores, two letters of recommendation from faculty or their MS committee, a graduate transcript, transcripts of all academic work, a statement about faculty career and a current curriculum vitae.
Closing Date: February 15th
No. of awards given last year: 4
No. of applicants last year: 10
Additional Information: In the terms of the fellowship, the awardee must teach at least one lecture course. The applicant's department head must certify, prior to the award or continuation notice, the commitment of a teaching assistantship and the lecture assignment anticipated.

Elisabeth M and Winchell M Parsons Scholarship

Subjects: Mechanical engineering.
Purpose: To assist ASME student members working towards a doctoral degree.
Eligibility: Selection is based on academic performance, character, need and ASME participation. Applicants must be citizens of the United States of America and be enrolled in a United States of America school in an ABET-accredited mechanical engineering department. No student may receive more than one auxiliary scholarship or loan in the same academic year.
Level of Study: Doctorate
Type: Grant
Value: US$2,000
Frequency: Annual

Country of Study: United States of America
No. of awards offered: Approx. 2
Application Procedure: Application forms are available from the website.
Closing Date: March 15th

For further information contact:

5025 Iroquois Avenue, Lakewood, CA, 90713, United States of America
Tel: (1) 562 920 3653
Email: cindipool@gmail.com
Website: http://www.asme.org/about-asme/scholarship-and-loans/about-asme-scholarships
Contact: Cynthia Pool

Marjorie Roy Rothermel Scholarship

Subjects: Mechanical engineering.
Purpose: To assist students working towards a Master's degree.
Eligibility: Selection is based on academic performance, character, need and ASME participation. Applicants must be citizens of the United States of America and must be enrolled in a United States of America school in an ABET-accredited mechanical engineering department. No student may receive more than one auxiliary scholarship or loan in the same academic year.
Level of Study: Graduate
Type: Scholarship
Value: US$2,000
Frequency: Annual
Country of Study: United States of America
No. of awards offered: 6–8
Application Procedure: Application forms are available from the website.
Closing Date: March 15th

For further information contact:

332 Valencia Street, Gulf Breeze, FL, 32561, United States of America
Tel: (1) 850 932 3698
Email: eprocha340@aol.com
Website: http://www.asme.org/about-asme/scholarship-and-loans/about-asme-scholarships
Contact: Mrs Otto Prochaska

Rice-Cullimore Scholarship

Subjects: Mechanical engineering.
Purpose: To aid a foreign student pursuing graduate work for a Master's or doctoral degree in the United States of America.
Eligibility: Open to candidates from any country except the United States of America. Selection is based on academic performance, character, need and ASME participation. No student may receive more than one auxiliary scholarship or loan in the same academic year.
Level of Study: Doctorate
Type: Scholarship
Value: US$2,000
Length of Study: 1 year
Frequency: Annual
Country of Study: United States of America
Application Procedure: Applicants must apply in their home country through the local institute of International Education Embassy (IEE) or Education Offices at the United States of America Embassy. Only applications received from the IEE will be considered.
Closing Date: Please contact the organization.
Additional Information: Contact Ella Baldwin-Viereck, Chair, e-mail: ellabv@earthlink.net. Please see the website for further details http://www.asme.org/about-asme/scholarship-and-loans/about-asme-scholarships.

AMERICAN SOCIETY OF NEPHROLOGY (ASN)

1510 H Street, NW, Suite 800, Washington, DC, 20005, United States of America
Tel: (1) 202 659 0599, 202 640 4660
Fax: (1) 202 659 0709, 202 637 9793
Email: email@asn-online.org
Website: www.asn-online.org
Contact: Grants Co-ordinator

The American Society of Nephrology (ASN) was founded in 1967 as a non-profit corporation to enhance and assist the study and practice of nephrology, to provide a forum for the promulgation of research and to meet the professional and continuing education needs of its members.

ASN-ASP Junior Development Grant in Geriatric Nephrology

Subjects: Geriatric and gerontologic aspects of nephrology.
Purpose: To support developing academic subspecialists interested in careers in the field.
Eligibility: Open to individuals who are within the first 3 years of a faculty appointment. Candidates must have completed a subspecialty internal medicine fellowship leading to a certification in nephrology by the American Board of Internal Medicine. All candidates must have United States of America citizenship or permanent resident status classification in the United States of America, and must be active ASN members at the time of application. Please see the website for further details http://www.forschen-foerdern.org/ausschreibungen/asn-asp-nia-junior-development-grant-in-geriatric-nephrology/en.
Type: Grant
Value: US$25,000 per year and one-time travel grant of US$2,500
Length of Study: 2 years
Frequency: Annual
No. of awards offered: 1
Application Procedure: Applicants must submit four copies of the grant application form, available online, which must include the department chairman's letter, division director's letter (if applicable) and three letters of reference. To access the application: http://grants.nih.gov/grants/guide/rfa-files/RFA-AG-12-002.html.
Closing Date: December 14th at 4:00 p.m. EST
Contributor: ASP, ASN
No. of awards given last year: 2
No. of applicants last year: 7
Additional Information: Applicants to the ASN-ASP-NIA Junior Development Grant in Geriatric Nephrology must first apply for the GEMSSTAR award through the NIA. The GEMSSTAR application closes on Monday, October 1, 2012.

ASN-AST John Merrill Grant in Transplantation

Subjects: Biomedical research related to transplantation.
Purpose: To foster the independent careers of young investigators in biomedical research related to transplantation.
Eligibility: Applicant must be an active member of the ASN and hold an MD or PhD or equivalent degree. At the time of submission the applicant's membership must be current and their dues paid. Appointment to full-time faculty must be confirmed in writing by the department chair.
Level of Study: Postdoctorate, Postgraduate
Value: US$100,000
Length of Study: 2 years
Frequency: Annual
Study Establishment: ASN and AST
Country of Study: United States of America
No. of awards offered: Varies
Application Procedure: Applicants must submit four copies of the grants application form, letters from the chairman and division director, a curriculum vitae and a research proposal (no longer than 10 pages). Preference will be given to an investigator performing biomedical research related to transplantation.
Closing Date: January 27th
Additional Information: For more information contact Benjamin Schuster by email at bschuster@asn-online.org or grants@asn-online.org.

Carl W Gottschalk Research Scholar Grant

Subjects: Nephrology.
Purpose: To provide funding for young faculty to foster evolution to an independent research career and a successful application a National Institutes of Health (NIH) R01 grant or equivalent.
Eligibility: Applicants must an active member of the ASN and hold an MD or PhD or equivalent degree. At the time of submission the applicant's membership must be current and their dues paid. Appointment to full-time faculty must be conformed in writing by the department chair.
Level of Study: Postgraduate, Postdoctorate
Type: Grant
Value: US$100,000 for 2 years. A maximum of 10 per cent of the whole amount can be used to cover indirect costs at the candidate's sponsoring institution.
Length of Study: 2 years
Frequency: Annual
Country of Study: United States of America
No. of awards offered: Varies
Application Procedure: Applicants must submit four copies of the grant application form, the letters from the chairman and division director, a curriculum vitae and a research proposal (no longer than 10 pages).
Closing Date: January 27th
Contributor: Co-sponsored by the Kidney and Urology Foundation of America
Additional Information: Please visit the website for further details.

For further information contact:

Tel: 202 640 4660
Email: grants@asn-online.org or bschuster@asn-online.org
Contact: Mr Benjamin Schuster

AMERICAN SOCIETY OF TRAVEL AGENTS (ASTA) FOUNDATION, INC.

1101 King ST, Suite 200, Alexandria, VA, 22314, United States of America
Tel: (1) 703 739 2782
Fax: (1) 703 684 8319
Email: askasta@asta.org
Website: www.astanet.com

ASTA is the world's largest association of travel professionals. Its mission is to enhance the professionalism and profitability of member agents through effective representation in industry and government affairs, education and training, and by identifying and meeting the needs of the traveling public.

Alaska Airlines Scholarship

Subjects: Travel and tourism.
Purpose: To encourage students pursuing a career in the field of travel and tourism.
Eligibility: Open to applicants who are enrolled in a 4 year travel and tourism programme at a university.
Level of Study: Postgraduate
Type: Scholarship
Value: Up to US$2,000
Frequency: Annual
No. of awards offered: 1
Application Procedure: Application form can be downloaded from the website http://www.scholarships4students.com/alaska_airlines_scholarship.htm.
Closing Date: August 29
Funding: Foundation
Contributor: ASTA Foundation
Additional Information: All inquiries and requests should be directed to the Tourism Cares Scholarship Department at scholarship@astahq.com

ASTA Arizona Chapter Professional Development Scholarship

Subjects: Travel and tourism.
Purpose: To encourage serious academic study in the field of travel and tourism.
Eligibility: applicants must be permanentU. S. residents, with a minimum of two years of experience in the travel industry, who have successfully completed the program, or attended the conference. They must also be individual members of the ASTA Arizona Chapter, or be employed at an office where at least one travel agent is a member of the chapter.
Level of Study: Postgraduate
Type: Scholarship
Value: Up to $500 (aggregate annual amount)
Length of Study: 4 years
Frequency: Annual
Country of Study: United States of America
No. of awards offered: 1
Application Procedure: Application form available on the website http://www.asta.org/Education/content.cfm?ItemNumber=2552&navItemNumber=614.
Closing Date: June 1st at 5pm Eastern Time.
Funding: Foundation
Contributor: ASTA Foundation
Additional Information: All inquiries and requests should be directed to the Tourism Cares Scholarship Department at scholarships@tourismcares.org.

Healy Scholarship

Subjects: Travel and tourism.
Eligibility: Open to candidates enrolled in a 4 year college/university course of study.
Level of Study: Postgraduate
Type: Scholarship
Value: US$2,000
Length of Study: 4 years
No. of awards offered: 1
Application Procedure: See the website.
Closing Date: August 30th
Funding: Foundation
Contributor: ASTA Foundation
Additional Information: All inquiries and requests should be directed to the Tourism Cares Scholarship Department at scholarships@tourismcares.org.

For further information contact:

Tel: 703-739-8721
Contact: Verlette Mitchell, Manager

Southern California Chapter/Pleasant Hawaiian

Subjects: Travel and tourism.
Purpose: To encourage people to take up travel and tourism business as their profession.
Eligibility: Citizen or permanent resident of the United States, permanent resident of California, full-time or part-time undergraduate student enrolled in a travel-and-tourism- or hospitality-related program of study, enrolled at a college or university in California, cumulative 3.0 grade point average (GPA), or greater, on a U.S. 4.0 scale, enrolled at an accredited two-year college, and will be entering second year in the fall of the calendar year of application, and will have completed a minimum of 30 credit hours by the end of this semester or term; or enrolled at an accredited four-year college or university, and will be entering junior year in the fall of the calendar year of application, and will have completed a minimum of 60 credit hours by the end of this semester or term; or enrolled at an accredited four-year college or university, and will be entering senior year in the fall of the calendar year of application and will have completed a minimum of 90 credit hours by the end of this semester or term.
Level of Study: Postgraduate
Type: Scholarship
Value: US$2,000
Length of Study: 4 years
Frequency: Annual
No. of awards offered: 2
Application Procedure: A completed form and general application requirements must be submitted.
Closing Date: April 9th
Contributor: ASTA Foundation

Additional Information: All inquiries and requests should be directed to the Tourism Cares Scholarship Department at scholarships@tourismcares.org.

AMERICAN SOCIOLOGICAL ASSOCIATION (ASA)

1430 K Street, NW, Suite 600, Washington, DC, 20005, United States of America
Tel: (1) 202 383 9005
Fax: (1) 202 638 0882
Email: minority.affairs@asanet.org
Website: www.asanet.org
Contact: Karina Havrilla, Minority and Student Affairs Coordinator

The American Sociological Association (ASA), founded in 1905, is a non-profit membership association dedicated to advancing sociology as a scientific discipline and profession serving the public good. With over 13,200 members, the ASA encompasses sociologists who are faculty members at colleges and universities, researchers, practitioners and students. About 20 per cent of the members work in government, business or non-profit organizations.

ASA Minority Fellowship Program

Subjects: Sociological research in any subarea of sociology.
Purpose: To support the development and training of minority sociologists, to attract talented minority students interested in any subarea of sociological research and to facilitate success in their respective graduate programs.
Eligibility: Applicants must be enrolled in (and have completed one full academic year) in a program that grants the Phd in sociology. Please see the website for details.
Level of Study: Predoctorate, Graduate
Type: Fellowship
Value: US$18,000
Length of Study: 1 year, renewable
Frequency: Annual
Study Establishment: Varies
Country of Study: United States of America
No. of awards offered: Varies
Application Procedure: Applicants must submit their complete application package to the Minority Fellowship Program in one package. The complete application package consists of a fellowship application, essays, three letters of recommendation, official transcripts and a curriculum vitae.
Closing Date: January 31st
Funding: Private
Contributor: NIMH
No. of awards given last year: 4
No. of applicants last year: 60

AMERICAN STATISTICAL ASSOCIATION (ASA)

732 North Washington Street, Alexandria, VA, 22314-1943, United States of America
Tel: (1) 703 684 1221
Fax: (1) 703 684 2037
Email: asainfo@amstat.org
Website: www.amstat.org

The American Statistical Association (ASA) is a scientific and educational society, founded in 1839, to promote excellence in the application of statistical science across the wealth of human endeavour.

Gertrude M. Cox Scholarship

Subjects: Statistics.
Purpose: To provide financial assistance to students and encourage more women to enter statistically orientated professions.
Eligibility: Application is limited to women who are citizens or permanent residents of the United States or Canada and who are admitted to full-time study in a graduate statistics program by July 1 of the award year. Women in or entering the early stages of graduate training (MS or PhD) are especially encouraged to apply.
Level of Study: Doctorate
Type: Scholarship
Value: US$2,000 and a certificate
Frequency: Annual
No. of awards offered: Varies
Application Procedure: Applicants can download the application form from the website.
Closing Date: April 1st
Contributor: ASA Committee on Women in Statistics and the Caucus for Women in Statistics
Additional Information: Please see the website for more details.
For further information contact:

Contact: Chair, Gertrude Cox Scholarship

AMERICAN TINNITUS ASSOCIATION (ATA)

522 S.W. Fifth Avenue, Suite 825, Portland, OR, 97207-0005, United States of America
Tel: (1) 503 248 9985 x218
Fax: (1) 503 248 0024
Email: amy@ata.org
Website: www.ata.org
Contact: Amy Harris, Director of Research

The American Tinnitus Association's (ATA) mission is to cure tinnitus through the development of resources that advance tinnitus research.

American Tinnitus Association Scientific Research Grants

Subjects: Tinnitus.
Purpose: To identify the mechanisms of tinnitus, to improve treatments and to identify a cure.
Eligibility: Those scientists and doctors, worldwide, who are seeking tinnitus-related research funding and are affiliated with non-profit institutions.
Level of Study: Doctorate, Graduate, Postdoctorate, Postgraduate, Predoctorate, Research
Value: ATA awards a maximum of $150,000 at $50,000 per year over three years.
Length of Study: 1–3 years
Frequency: Annual
Country of Study: Any country
No. of awards offered: Varies
Application Procedure: Applicants must write for grant application policies and a procedures brochure. These documents and also the applications can be downloaded from the ATA website.
Closing Date: November 1st
Funding: Individuals, private
Contributor: Sufferers of tinnitus
No. of awards given last year: 7
No. of applicants last year: 30
Additional Information: Please see the website for further details http://www.ata.org/research/grant-program.

THE AMERICAN UNIVERSITY IN CAIRO (AUC)

PO Box 2511, 113 Kasr El Aini Street, Cairo, 11511, Egypt
Tel: (20) 2794 2964
Fax: (20) 2795 7565
Email: ocm@aucegypt.edu; aucegypt@aucnyo.edu; ouc@aucegypt.edu
Website: http://www.aucegypt.edu/Pages/default.aspx
Contact: Mrs Sawsan Mardini, Director of Graduate Students Services

The American University in Cairo (AUC) provides quality higher and continuing education for students from Egypt and the surrounding region. The University is an independent, non-profit, apolitical, non-sectarian and equal opportunity institution. English is the primary language of instruction. The University is accredited in the United States of America by the Commission of Higher Education of the Middle States Association of Colleges and Schools.

AUC African Graduate Fellowship

Subjects: Arts, humanities, business administration, engineering or information science.
Purpose: To enable outstanding young men and women from Africa to study for a Master's degree.
Eligibility: Open to African nationals, not including Egyptians, with Bachelor's degrees, an academic record of not less than 'Very Good' and an overall grade point average of 3.0 on a 4.0 scale or the equivalent. Candidates must also show proficiency in the English language by either submitting a Test of English as a Foreign Language with TWE score of 550 or above, or taking the AUC's ELPET exam.
Level of Study: Graduate
Type: Fellowship
Value: A waiver of tuition fees. A monthly stipend of LE 1,200. Student services and activities fee. Graduation fees. Medical service and health insurance fees. A monthly housing allowance of LE 1250. In support of their professional training fellows are assigned 12 hours per week of related academic or administrative work.
Length of Study: 2 academic years and the intervening Summer session
Frequency: Annual
Study Establishment: AUC only
Country of Study: Egypt
No. of awards offered: 5
Application Procedure: Applicants must complete an application form available from the Office of Graduate Studies and Research.
Closing Date: December 15th
Funding: Private
No. of awards given last year: 3
No. of applicants last year: 45
Additional Information: For inquiries please contact grad@aucegypt.edu. Please see the website for further details http://www.aucegypt.edu/admissions/grad/finsup/pages/african.aspx.

AUC Arabic Language Fellowships

Subjects: All subjects.
Purpose: To support fully admitted international graduate students who need to satisfy their degree requirement.
Eligibility: International graduate students. Full admission to a graduate program that requires Arabic language proficiency to satisfy degree requirement. A minimum overall GPA of 3.2 on a 4.0 scale, or equivalent.
Level of Study: Graduate, Postgraduate
Type: Fellowship
Value: A waiver of 50 percent of the tuition for the ALI intensive Arabic fall, spring or summer program or 50% off ALING courses.
Length of Study: One semester or one summer session
Frequency: Annual, every semester
Study Establishment: AUC
Country of Study: Egypt
No. of awards offered: 5
Application Procedure: Applicants must submit fellowship application online to grad@aucegypt.edu.
Closing Date: April 1st for Fall, September 15th for Spring, February 1st for Summer
Additional Information: Fellows are assigned 5 hours per week of related academic or administrative work. Please see the website for further details http://www.aucegypt.edu/admissions/grad/finsup/Pages/ArabicLanguageFellowships.aspx.

AUC Assistantships

Subjects: Arts, humanities, business administration, engineering and information science.
Purpose: To support graduate-level teaching or research assistants who do not receive tuition waivers.
Eligibility: Fully accepted graduate students enrolled in two or more courses or actively engaged in thesis work are given preference over those not enrolled in the graduate programme. Applicants who have completed their MA or MS, are preparing for a PhD, and have or are receiving academic degree training may also receive assistantships as postmaster's assistants.
Level of Study: Graduate
Type: Award
Value: Hourly rate for Master's degree is LE 25. Hourly rate for bachelor's degree holders is LE 20.
Length of Study: 1 semester, renewable
Study Establishment: AUC
Country of Study: Egypt
No. of awards offered: Varies
Application Procedure: Applications must be made to the relevant department. Please see the website for further details AUC Assistantships.
Closing Date: September 1st for the Fall, January 1st for the Spring, and June 1st for the Summer

AUC Graduate Merit Fellowships

Subjects: Business, communication, computer science, social and behavioural sciences.
Purpose: To recognize and reward outstanding new or continuing graduate students who wish to pursue full-time study in one of the graduate programmes.
Eligibility: Open to students who are fully admissible to one of the graduate programmes at AUC and who have a Bachelor's degree with a minimum overall grade point average of 3.4 on a 4.0 scale and a minimum of 3.5 in their major. Students who are already enrolled in one of AUC's graduate programmes and have a minimum grade point average of 3.7 in their graduate courses are also eligible to apply.
Level of Study: Graduate
Type: Fellowship
Value: A tuition waiver of one, two, or three, courses per semester
Length of Study: 1 year, renewable.
Frequency: Annual
Study Establishment: AUC
Country of Study: Egypt
No. of awards offered: 18
Application Procedure: Applications should be submitted online to grad@aucegypt.edu. Please see the website for further details http://www.aucegypt.edu/admissions/grad/finsup/Pages/GraduateMeritFellowship.aspx.
Closing Date: May 1st
Contributor: AUC
No. of awards given last year: 18
No. of applicants last year: 100
Additional Information: Please note that the Merit Fellowship provides a partial tution waiver to international students.In support of their professional training, fellows are assigned 18 hours per week of related academic or administrative work.

AUC International Graduate Fellowships in Arabic Studies, Middle East Studies and Sociology/ Anthropology

Subjects: Arabic studies, Middle East studies, sociology or anthropology.
Purpose: To recognize and award outstanding new international graduate students who wish to pursue full-time study.
Eligibility: Candidates must have completed an appropriate undergraduate degree with a minimum overall grade point average of 3.4 on a 4.0 scale or equivalent.
Level of Study: Graduate, Postgraduate
Type: Fellowship
Value: A waiver of tuition fees, a monthly stipend of LE 1200 and a monthly housing allowance of LE 1250 or accommodation in the university dormitory. Medical service and health insurance fees.
Length of Study: 2 academic years and the intervening summer session.
Frequency: Annual
Study Establishment: AUC
Country of Study: Egypt
No. of awards offered: 2
Application Procedure: Applicants must complete an application form, available from the website and submit online to grad@aucegypt.edu.
Closing Date: February 1st
No. of awards given last year: 2

No. of applicants last year: 30
Additional Information: Fellows are assigned 18 hours per week of related academic or administrative work. Please see the website for further details http://www.aucegypt.edu/admissions/grad/finsup/Pages/InternationalGraduateFellowships.aspx.

AUC Laboratory Instruction Graduate Fellowships

Subjects: Computer science, engineering, chemistry and physics.
Purpose: To recognize and support outstanding graduate students who wish to pursue full-time study in engineering, computer science, chemistry or physics.
Eligibility: Full admission to a graduate program in the school of sciences and engineering. Bachelor of Science degree with a minimum overall GPA of 3.2 on a 4.0 scale, or its equivalent. Students who are already enrolled in one of these graduate programs and have a minimum GPA of 3.2 in their graduate courses. Please see the website for further details regarding eligibility http://www.aucegypt.edu/admissions/grad/finsup/Pages/LaboratoryInstructionGraduate-Fellowships.aspx.
Level of Study: Graduate
Type: Fellowship
Value: A tuition waiver of one, two or three courses per semester. Student services and activities fee. A monthly stipend of up to LE 1,500.
Length of Study: One semester, renewable for a maximum period of two years
Frequency: Annual
Study Establishment: AUC
Country of Study: Egypt
No. of awards offered: 13
Application Procedure: Applications and supporting documents must be submitted online to grad@aucegypt.edu.
Closing Date: June 1st for the fall semester and November 8th for the spring semester
Additional Information: In support of their professional training, fellows are assigned 24 hours per week of laboratory instruction work.

AUC Nadia Niazi Mostafa Fellowship in Islamic Art and Architecture

Subjects: Islamic art and architecture.
Purpose: To recognize and award outstanding Egyptian graduate students who wish to pursue full-time study in the programme. The award is for 2nd-year Egyptian students already enrolled in the programme.
Eligibility: Full admission to the graduate program in Arabic studies with a specialization in Islamic art and architecture. Minimum qualifications of an overall GPA of 3.2 on a 4.0 scale.
Level of Study: Postgraduate
Type: Fellowship
Value: A tuition waiver. Student services and activities fee. A monthly stipend of up to LE 750 for 10 months.
Length of Study: Two semesters of full time graduate study
Frequency: Annual
Study Establishment: AUC
Country of Study: Egypt
No. of awards offered: 1
Application Procedure: Applications should be submitted online to grad@aucegypt.edu. Please see the website for further details http://www.aucegypt.edu/admissions/grad/finsup/Pages/NadiaNiaziMostafaFellowship.aspx.
Closing Date: May 15th
Funding: Private
No. of awards given last year: 1
Additional Information: Fellows are assigned 12 hours per week of related academic or administrative work.

AUC Ryoichi Sasakawa Young Leaders Graduate Scholarship

Subjects: Humanities and social sciences.
Purpose: To educate outstanding young men and women who have demonstrated a high potential for future leadership in international affairs, public life and private endeavour.
Eligibility: Full admission into one of AUC's M.A. programs in the humanities and social sciences. A Bachelor's degree with a minimum overall GPA of 3.2 on a 4.0 scale, or equivalent. Actively participated in extra curricular activities. Both new and continuing graduate students may apply but preference is given to those who require four semesters to complete their degree.
Level of Study: Graduate
Type: Scholarship
Value: A waiver of tuition fees (nine credits per semester or as required by the program). Student services and activities fee. Medical service and health insurance fees (for international students).
Length of Study: 2 years
Frequency: Annual
Study Establishment: AUC only
Country of Study: Egypt
No. of awards offered: 3
Application Procedure: Applicants must complete the online application form. For inquiries grad@aucegypt.edu. Please see the website for further details http://www.aucegypt.edu/admissions/grad/finsup/Pages/RyoichiSasakawa.aspx.
Closing Date: February 1st
Funding: Private
Contributor: The Tokyo Foundation
No. of awards given last year: 3
No. of applicants last year: 25

AUC Teaching Arabic as a Foreign Language Fellowships

Subjects: Arabic, education and teacher training.
Purpose: To acquire language teaching skills.
Eligibility: A Bachelor's degree with a minimum overall GPA of 3.0 on a 4.0 scale. Special consideration in selection is given to those with previous TAFL experience and/or excellent qualifications in the Arabic language.
Level of Study: Graduate
Type: Fellowship
Value: A tuition waiver, a monthly stipend of LE 850 and medical service and health insurance fees (for international fellows)
Length of Study: Two academic years and the intervening summer session.
Frequency: Annual
Study Establishment: AUC
Country of Study: Egypt
No. of awards offered: 3
Application Procedure: Applicants must complete the online application form and submit to grad@aucegypt.edu. Please see the website for further details http://www.aucegypt.edu/admissions/grad/finsup/Pages/TAFL.aspx.
Closing Date: February 1st
Contributor: AUC
No. of awards given last year: 3
No. of applicants last year: 10

AUC Teaching English as a Foreign Language Fellowships

Subjects: Education.
Purpose: To acquire language teaching experience.
Eligibility: A BA with a minimum overall GPA of 3.2 or "Very Good". Native or near native proficiency in English. Special consideration in selection is given to those with: TEFL/TESL experience in the Middle East; knowledge of Arabic or other languages; BA degree or course work in linguistics, English or a related field.
Level of Study: Graduate
Type: Fellowship
Value: A waiver of tuition fees and a monthly stipend of LE 850 and medical insurance. Non-residents of Egypt are provided with accommodation in the university dormitory or with a monthly housing allowance of LE 1250, medical service and health insurance fees and one-way home travel.
Length of Study: 2 academic years and the intervening summer session
Frequency: Annual
Study Establishment: AUC
Country of Study: Egypt
No. of awards offered: Varies
Application Procedure: Applicants must complete an online application form and submit to grad@aucegypt.edu. Please see the website for further details http://www.aucegypt.edu/admissions/grad/finsup/Pages/TEFL.aspx.

Closing Date: February 1st
Contributor: AUC
No. of awards given last year: 10
No. of applicants last year: 50

AUC University Fellowships

Subjects: Art and humanities, business administration and management, engineering, mass communication and information, mathematics and computer science, social and behavioural sciences.
Purpose: To assist new and continuing graduate students who display superior performance in their academic endeavors and who wish to pursue full-time study in the graduate program at AUC.
Eligibility: A Bachelor's degree with a minimum overall GPA of 3.2 on a 4.0 scale, or equivalent. Students already enrolled in one of the graduate programs and have a minimum GPA of 3.2 in their graduate courses.
Level of Study: Graduate
Type: Fellowship
Value: A waiver of tuition fee; student services and activities fees; a monthly stipend of LE 600 for 10 months.
Length of Study: One semester and may be renewed for a maximum period of 2 years. The fellowship may cover a summer session.
Frequency: Annual
Study Establishment: AUC
Country of Study: Egypt
No. of awards offered: Varies
Application Procedure: Applicants must complete an online application form and submit to grad@aucegypt.edu. Please see the website for further details http://www.aucegypt.edu/admissions/grad/finsup/Pages/University.aspx.
Closing Date: May 1st for the fall semester and November 8th for the spring semester.
Contributor: AUC
No. of awards given last year: 18
No. of applicants last year: 123
Additional Information: Fellows are assigned 10–12 hours per week of work with faculty members in teaching and research activities.

AUC Writing Center Graduate Fellowships

Subjects: English, grammar, education and native language, literacy education, teaching and learning.
Purpose: To provide outstanding students with valuable teaching, academic experience and to involve them as tutors in AUC's Writing Center.
Eligibility: BA degree with a minimum overall grade point average of 3.2 on a 4.0 scale, or its equivalent. Students already enrolled in a HUSS graduate program and who have achieved a minimum overall GPA of 3.4 in their graduate courses.
Level of Study: Graduate
Type: Fellowship
Value: A waiver of tuition fee; student services and activities fee; a monthly stipend of LE 600 for 10 months; medical service and health insurance fees.
Length of Study: One semester, renewable for a maximum period of two years.
Frequency: Annual
Study Establishment: AUC
Country of Study: Egypt
No. of awards offered: Varies
Application Procedure: Applicants must complete an online application form and submit to grad@aucegypt.edu. Please see the website for further details http://www.aucegypt.edu/admissions/grad/finsup/Pages/WritingCenter.aspx.
Closing Date: April 1st for the fall semester and November 8th for the spring semester
Contributor: AUC
No. of awards given last year: 1
No. of applicants last year: 5
Additional Information: As part of their fellowship and in support of their professional training, fellows are assigned 10 hours of work per week in the Writing Center.

AMERICAN UROLOGICAL ASSOCIATION (AUA) FOUNDATION

1000 Corporate Boulevard, Linthicum, MD, 21090, United States of America
Tel: (1) 410 689 3700
Fax: (1) 410 689 3800
Email: grants@auafoundation.org
Website: http://www.auanet.org/content/contact/contact.cfm

The association founded by Ramon Guiteras in 1902 takes care for the advancement of urologic patient care. Its wide range of services including publications, research, the annual meeting, continuing medical education (CME), and the formulation of health policy fosters the highest standards of urological care. Its main aim is to network and collaborate physicians to increase educational opportunities.

MD Post-Resident Fellowship in Urology

Subjects: Urology.
Purpose: To learn scientific techniques, to provide evidence of current and/or prior interest/accomplishments in research.
Eligibility: Open to applicants who are willing to spend at least 80% of their time on their research project, should prepare the scholar to become an independent investigator capable of obtaining independent research grant support, must have an interest in the delivery of health services, economics, and policy also are encouraged to submit applications for consideration.
Level of Study: Predoctorate
Type: Fellowship
Value: US$60,000 per year
Length of Study: 2 years
Frequency: Annual
No. of awards offered: 1–10
Application Procedure: Check website for further details.
Closing Date: July 1st
Funding: Foundation
Contributor: An accredited medical education/research institution or department within such an institution must sponsor the Applicant.

MD/PhD One Year Fellowship in Urology

Subjects: Urology.
Purpose: To learn scientific techniques.
Eligibility: Open to candidates who are either a trained urologist or basic scientist within 5 years of completing residency or doctorate, must be willing to spend at least 80% of their time on their research project. PhD applicants must spend 100% of their time on their research project to prepare the scholar to subsequently become an independent investigator capable of obtaining independent research grant support.
Level of Study: Doctorate, Predoctorate
Type: Fellowship
Value: US$30,000
Length of Study: 1 year
Frequency: Annual
No. of awards offered: 5–10
Application Procedure: Check website for further details.
Closing Date: July 1st
Funding: Foundation

PhD Post-Doctoral Fellowship in Urology

Subjects: Urology.
Purpose: To support researcher urologic diseases and conditions and/or related diseases and dysfunctions.
Eligibility: Open to candidates who must be a basic scientist within 5 years of earning a PhD with a research interest in urologic diseases.
Level of Study: Doctorate, Postdoctorate
Type: Fellowship
Value: US$30,000
Length of Study: 2 years
Frequency: Every 2 years
No. of awards offered: 5–20
Application Procedure: Check website for further details.
Funding: Foundation
Contributor: An accredited medical education/research institution or department within such an institution

Urology Research Scholarship Program

Subjects: Urologic research.
Purpose: To support young scientists to begin a career in urologic research.
Eligibility: Open to young men and women who are interested in pursuing a career in urologic research.
Level of Study: Doctorate, Postgraduate
Type: Scholarship
Length of Study: 1–2 years
Application Procedure: Check website for further details.
Closing Date: August 24th

AMERICAN VENOUS FORUM FOUNDATION

555 E. Wells Street Suite 1100, Milwaukee, WI, 53202, United States of America
Tel: (1) 414 918 9880
Fax: (1) 414 276 3349
Email: venous-info@administrare.com; info@veinforum.org
Website: www.venous-info.com

The American Venous Forum Foundation grants research awards and prizes that stimulate and recognize excellence in published (science) writing on laboratory and clinical research in the study of venous disease.

The BSN-Jobst Inc. Research Award

Subjects: Venous disease.
Purpose: To award a grant to a research fellow or a resident in an ACGME programme who has a specific interest in the diagnosis and treatment of venous disease.
Eligibility: Residents and fellows in a vascular training program, as well as physicianswho have completed their training within the past five years can apply.
Level of Study: Doctorate, Graduate, Postdoctorate
Type: Award
Value: US$50,000
Length of Study: 1 year
Frequency: Annual
Country of Study: United States of America
No. of awards offered: 1
Application Procedure: Candidates must visit the website for details http://www.veinforum.org/About-Us/Awards-and-Recognition/BSNJobst-Research-Fellowship4.aspx.
Closing Date: September 17th at 11:59 a.m. CST
Contributor: BSN-Jobst Inc. and the American Venous Forum Foundation
No. of awards given last year: 1
No. of applicants last year: 4

Servier Traveling Fellowship

Subjects: Venous disease.
Purpose: To enable young physicians to travel throughout the United States of America and abroad to visit centres of excellence in the management of venous disease.
Level of Study: Doctorate, Graduate, Postdoctorate
Type: Travelling fellowship
Value: US$3,000
Length of Study: Travel to the Annual Meeting of the AVF
Frequency: Annual
Country of Study: United States of America and abroad
No. of awards offered: 2
Application Procedure: Applicants must visit the website for details http://www.veinforum.org/about-us/awards-and-recognition/servier-traveling-fellowship.aspx.
Closing Date: September 6th
Contributor: Sigvaris Inc. and the American Venous Forum Foundation
No. of awards given last year: 1

THE AMERICAN-SCANDINAVIAN FOUNDATION (ASF)

58 Park Avenue at 38th Street, New York, NY, 10016, United States of America
Tel: (1) 212 779 3587
Fax: (1) 212 249 3444
Email: grants@amscan.org
Website: www.amscan.org
Contact: Director of Fellowships and Grants

The ASF is a publicity supported, non-profit organization that promotes international understanding through educational and cultural exchange between the US and the Nordic countries

American-Scandinavian Foundation Award for Study in Scandinavia

Subjects: All subjects.
Purpose: To pursue research or study in one or more Scandinavian country.
Eligibility: Applicants must be US citizens or permanent residents and must have a well-defined research or study project that makes a stay in Scandinavia essential. The applicants must have completed their undergraduate education by the start of their project in Scandinavia. Team projects are eligible, but each member must apply as an individual, submitting a separate, fully-documented application.
Level of Study: Postgraduate, Professional development, Research
Type: Fellowship
Value: $23,000 (fellowship) or $5,000 (grant)
Length of Study: 1 year (fellowship) or 1–3 months (grant)
Frequency: Annual
No. of awards offered: Varies
Application Procedure: Application forms are available online at www.amscan.org.
Closing Date: November 1st
Funding: Foundation
Contributor: American-Scandinavian Foundation (ASF)
No. of awards given last year: 25
No. of applicants last year: 138
Additional Information: For projects that require a command of one or more Scandinavian (or other) languages, candidates should defer application until they have the necessary proficiency.

The American-Scandinavian Foundation–Visiting Lectureship

Subjects: Public policy, conflict resolution, enviromental studies, multiculturalism and healthcare.
Eligibility: Applicants must be Norwegian or Swedish scholars or experts who shall go to the United States for up to six months. They must teach one course at an American institution on public policy, conflict resolution, environmental studies, healthcare or multiculturalism. The scholars must participate at a NORTANA conference, SASS conference, and/or a Swedish Teachers' Conference.
Level of Study: Postdoctorate
Type: Grant
Value: $20,000 plus $5,000 in-country travel allowance and J 1 Visa sponsorship as a scholar
Length of Study: 6 months
Application Procedure: Check website for further details.
Closing Date: November 4th and February 6th
Funding: Private
No. of awards given last year: 2
No. of applicants last year: 10

Awards for Scandinavians

Subjects: All subjects.
Purpose: To fund Scandinavians to undertake study or research programs in the United States.
Eligibility: Applicants must be citizens of Denmark, Finland, Iceland, Norway or Sweden. Check website for further details.
Level of Study: Professional development, Research, Graduate, Postgraduate
Type: Award
Value: $500,000
Length of Study: 1 year

Frequency: Annual
Country of Study: United States of America
No. of awards offered: Varies
Application Procedure: Check the website for further details.
Closing Date: Varies
Funding: Foundation
No. of awards given last year: 55
Additional Information: The number and size of awards granted annually varies widely between countries. Contact the ASF's cooperating organizations for specific information regarding eligibility, award size and application deadlines.

Visiting Lectureships

Subjects: Public policy, conflict resolution, environmental studies, multiculturalism and health care.
Purpose: To support American Universities and colleges to host Norwegian and Swedish lecturers.
Eligibility: Open to all American colleges and universities. The award is appropriate not just for Scandinavian studies departments, but for any department or inter-disciplinary program with an interest in incorporating a Scandinavian focus into its course offerings.
Level of Study: Postdoctorate
Type: Lectureship/Prize
Value: $20,000 as research/teaching stipend, $5,000 in-country travel stipend for lecture appearances outside home institution and J–1 visa sponsorship as a short-term scholar (up to 6 months) through the ASF Visitor Exchange Program.
Length of Study: 6 months
Frequency: Annual
No. of awards offered: 1
Application Procedure: Applicants are required to submit a pre-proposal, from which finalists will be selected. Finalists will then be required to submit more detailed and specific information, including confirmation of the lecturer and institutional support, for a later deadline.
Closing Date: November 7th (letter of intent); and February 4th (final proposal)
Funding: Foundation
No. of awards given last year: 2
No. of applicants last year: 10
Additional Information: Lectureships should be in the area of contemporary studies with an emphasis on one of the five areas: Public Policy; Conflict Resolution; Environmental Studies; Multiculturalism; Healthcare.

THE AMITY SCIENCE & TECHNOLOGY FOUNDATION (ASTF)

Amity University Campus, Sector 125, Noida, India
Tel: (91) 120 2445252
Fax: (91) 120 2432200
Email: ssaran@amity.edu
Website: www.amity.edu
Contact: Professor Sunil

The Amity Science & Technology Foundation (ASTF) was established with the aim of helping India become a global leader in the field of science and technology.

ASTF PhD Fellowships

Subjects: Applied sciences, biotechnology, engineering, forestry and environment, herbal and microbial studies, high vacuum technology, Indian heritage crops and their products, nanotechnology, organic agriculture, telecommunication and other areas in science, engineering and technology.
Purpose: To enable brilliant young researchers and scientists who possess the competence and motivation to carry out cutting edge research in thrust areas that will impact the development of the Indian Nation.
Eligibility: Throughout first class academic record up to Master's Degree.
Level of Study: Doctorate
Type: Fellowships
Value: Indian Rupees 2,70,700 per year. The fellowship includes academic fees, subsidy towards cost of on-campus or off-campus accommodation, mediclaim for hospitalisation and a stipend of Indian Rupees 1,000 per month.
Length of Study: 3 years
Frequency: Annual
Study Establishment: Amity Science & Technology Foundation
Country of Study: India
No. of awards offered: 100
Closing Date: July 28th
Additional Information: Please see the website for further information www.amity.edu/astf/fellowship.htm.

ASTF Postdoctoral Research Associate Fellowships

Subjects: Applied sciences, biotechnology, engineering, forestry and environment, herbal and microbial studies, high vacuum technology, Indian heritage crops and their products, nanotechnology, organic agriculture, telecommunication and other areas in science, engineering and technology.
Purpose: To enable brilliant young researchers and scientists who possess the competence and motivation for carrying out cutting-edge research in thrust areas that will impact the development of the Indian nation.
Eligibility: Open to candidates with PhD in relevant area with evidence of sustained high quality research and throughout first class academic record.
Level of Study: Postdoctorate
Type: Fellowships
Value: Indian Rupees 1,87,900 per year. The fellowship includes a stipend of Indian Rupees 12,000 per month plus HRA, and mediclaim for hospitalization
Length of Study: 3 years
Frequency: Annual
Study Establishment: Amity Science & Technology Foundation
Country of Study: India
No. of awards offered: 100
Application Procedure: Please visit the website www.amity.edu/astf
Closing Date: July 28th
Funding: Foundation
Additional Information: For further information, please see the website.

ANGLIA RUSKIN UNIVERSITY

Cambridge Campus, East Road, Cambridge, Cambridgeshire, CB1 1PT, United Kingdom
Tel: (44) 845 271 3333
Email: angliaruskin@enquiries.uk.com
Website: www.anglia.ac.uk

The International Merit Scholarship

Purpose: The International Merit Scholarship scheme provides awards to well-qualified students applying for any full-time Bachelor's or Master's course.
Eligibility: Awards are made on the basis of academic merit and level of competence in English language.
Level of Study: Doctorate, Postgraduate
Type: Scholarship
Value: £500 or £1,000 (in some countries a laptop alternative will be offered)
Frequency: Annual
No. of awards offered: Varies
Application Procedure: For details of other scholarships which may be available for you from outside of Anglia Ruskin, you should refer to the British Council website.
Closing Date: August 31st
Additional Information: Once you apply for a course at Anglia Ruskin, you will be automatically considered for a scholarship, and details of any award given will be included in your offer letter. Please see the website for further details http://www.anglia.ac.uk/ruskin/en/home/study/international/fees.html.

ANGLO-AUSTRIAN MUSIC SOCIETY

Richard Tauber Prize for Singers, 158 Rosendale Road, London, SE21 8LG, England
Tel: (44) 20 8761 0444
Fax: (44) 20 8766 6151
Email: info@aams.org.uk
Website: www.aams.org.uk
Contact: Jane Avery, Secretary

The Anglo-Austrian Music Society promotes lectures and concerts and is closely associated with its parent organization, the Anglo-Austrian Society, which was founded in 1944 to promote friendship and understanding between the people of the United Kingdom and Austria through personal contacts, educational programmes and cultural exchanges. AAMS awards the Richard Tauber prize for singers.

Richard Tauber Prize for Singers

Subjects: Vocal musical performance.
Eligibility: Open to singers of any country resident in the United Kingdom or Austria. Applicants must be aged over 21 years and not older than 30 years.
Level of Study: Postgraduate
Type: Prize
Value: First prize: UK£5,000 plus public recital in London; second prize: UK£2,500; other prizes totalling UK£3,500
Frequency: Every 2 years
No. of awards offered: 5
Application Procedure: Applicants must complete an application form. The registration fee is £30
Closing Date: March 12th
Funding: Private
Additional Information: Preliminary auditions are held in London and Vienna in April. Applicants must attend these auditions at their own expense. A final public audition is held at Wigmore Hall London in June. Additional prizes are 1. Adele Leigh Memorial Prize – £2,500, 2. Ferdinand Rauter Memorial prize for Accompanists – £1,000, 3. Schubert Society Lied Prize – £500 plus recital for the Schubert Society of Britain.

THE ANGLO-DANISH SOCIETY

43 Maresfield Gardens, London, NW3 5TF, England
Tel: (44) 1332 513 932
Email: scholarships@anglo-danishsociety.org.uk
Website: www.anglo-danishsociety.org.uk
Contact: Mrs Margit Staehr, Administrator

The Anglo-Danish Society exists to promote closer understanding between the United Kingdom and Denmark by arranging lectures, outings, social gatherings and other events of interest for its members and their guests. The society administers scholarship funds which help Danish students visit the United Kingdom or British students visit Denmark, for the purpose of advanced or postgraduate studies.

Anglo-Danish Society / Ove Arup Foundation Scholarships

Subjects: Research concerning the Built Environment leading to publication.
Purpose: Foundation scholarships, to promote Anglo-Danish relations.
Level of Study: Doctorate, Postdoctorate, Postgraduate, Professional development, Research
Type: Scholarship
Value: UK£1,500 per award (Min.)
Frequency: Dependent on funds available
Study Establishment: Universities
No. of awards offered: 1
Application Procedure: Application forms can be obtained by e-mailing at scholarships@anglo-danishsociety.org.uk. Completed applications may be submitted from 1st January 2013 onwards.
Closing Date: March 1st
Funding: Foundation
Contributor: Ove Arup Foundation
No. of awards given last year: 1
No. of applicants last year: 12
Additional Information: Country of Study: Denmark (for the British); UK (for Danes). Please see the website for details http://www.anglo-danishsociety.org.uk/artman/publish/scholarships.shtml.

The Anglo-Danish Society Scholarships

Subjects: Anglo-Danish cultural and scientific interests.
Purpose: To promote Anglo-Danish relations.
Eligibility: Open to graduates of Danish and British nationality. The country of study is Denmark (for the British) and the United Kingdom (for Danes).
Level of Study: Postgraduate, Professional development, Doctorate, Postdoctorate
Type: Scholarship
Value: Maximum UK£2,000 per award
Length of Study: Up to 6 months
Frequency: Dependent on funds available
Study Establishment: Universities
No. of awards offered: 4–6
Application Procedure: Application forms can be obtained by sending an e-mail to scholarships@anglo-danishsociety.org.uk. Completed applications may be submitted from 1st January 2013 onwards. Applicants should submit their applications by e-mail or by post to the Administrator's address. Please see the website for further details http://www.anglo-danishsociety.org.uk/artman/publish/scholarships.shtml.
Closing Date: March 1st
Funding: Commercial, private
No. of awards given last year: 6
No. of applicants last year: 80
Additional Information: Country of Study: Denmark (for the British); UK (for Danes)

THE ANGLO-JEWISH ASSOCIATION

Suite 21, 58 Acacia Road, London, NW8 6AG, England
Tel: (44) 20 7449 0909
Email: info@anglojewish.org.uk
Website: www.anglojewish.org.uk
Contact: Julia Samuel, Chairman, Education Committee

The Anglo-Jewish Association gives grants to Jewish students in financial need at universities in the United Kingdom.

Anglo-Jewish Association Bursary

Subjects: All subjects.
Purpose: To assist Jewish students in full-time higher education who are in financial need.
Eligibility: Open to Jewish students of any nationality in financial need up to the age of 35.
Level of Study: Postgraduate, Doctorate, Graduate, Undergraduate
Type: Bursary
Value: Up to UK£2,000 per year
Frequency: Annual
Study Establishment: University
Country of Study: United Kingdom
No. of awards offered: 20
Application Procedure: Application form must be completed and returned with all enclosures. This form can be downloaded from the website. Please email us for advice about the scholarships and assistance with the completion of the application forms.
Closing Date: April 30th
Funding: Trusts
No. of awards given last year: 9
Additional Information: Please see the website for further details http://www.anglojewish.org.uk/index.php/scholarships.

APPRAISAL INSTITUTE EDUCATION TRUST

200 W. Madison St., Suite 1500, Chicago, IL, 60606, United States of America
Tel: (1) 312 335 4133
Fax: (1) 312 335 4134
Email: educationtrust@appraisalinstitute.org
Website: www.aiedtrust.org

The Appraisal Institute is an international membership association of professional real estate appraisers, with more than 21,000 members and 99 chapters throughout the United States of America, Canada and abroad. Its mission is to support and advance its members as the choice for real estate solutions and uphold professional credentials, standards of professional practice and ethics consistent with the public good.

Appraisal Institute Education Trust Designation Scholarship

Subjects: Real estate appraisal.
Purpose: To provide financial assistance to take appraisal institute courses leading to the MAI or SRA designation.
Eligibility: Open to outstanding Appraisal Institute member working toward MAI or SRA designation, have paid Appraisal Institute membership in full and must be current with standards and ethics requirement for associate members.
Level of Study: Postdoctorate, Postgraduate, Professional development
Type: Scholarship
Value: Covers the cost of one Appraisal Institute advanced education course
Length of Study: 1 year
Frequency: Annual
Study Establishment: Appraisal Institute
Country of Study: United States of America
No. of awards offered: Varies
Application Procedure: Applicant must submit an application form. Please see the website for further details http://www.appraisalinstitute.org/education/scholarship.aspx.
Closing Date: January 1st, April 1st, July 1st, and October 1st
Funding: Private

Appraisal Institute Education Trust Graduate Scholarship

Subjects: Real estate appraisal, land economics, real estate or allied fields.
Purpose: To finance the educational endeavours of individuals concentrating in real estate or allied fields.
Eligibility: Applicants must be masters or doctoral candidate, full-time or part-time student at a US degree college or university with a strong academic record and must demonstrate financial need.
Level of Study: Doctorate, Graduate, MBA, Postgraduate
Type: Scholarship
Value: $2,000
Length of Study: 1 year
Frequency: Annual
Country of Study: United States of America
No. of awards offered: Varies
Application Procedure: Applicants must visit the organization website for application procedures at http://www.appraisalinstitute.org/education/scholarship.aspx.
Closing Date: April 15th
Funding: Private

Appraisal Institute Education Trust Minorities and Women Designation Scholarship

Subjects: Real estate appraisal.
Purpose: To support minorities and women associate members who are active in appraising and need financial assistance to take appraisal institute courses.
Eligibility: Open to outstanding appraisal institute member working toward MAI or SRA designation, have paid appraisal institute membership in full and must be current with standards and ethics requirement for associate members. Applicants must be minority as identified by the US census bureau: black, Asian, Pacific Islander, Hispanic, American Indian or Alaskan native.
Level of Study: Postdoctorate, Postgraduate, Professional development
Type: Scholarship
Value: Covers the cost of one appraisal institute advanced education course.
Length of Study: 1 year
Frequency: Annual
Study Establishment: Appraisal Institute
Country of Study: United States of America
No. of awards offered: Varies
Application Procedure: Applicant must submit an application form. Please see the website for further details http://www.appraisalinstitute.org/education/scholarship.aspx.
Closing Date: January 1st, April 1st, July 1st, and October 1st
Funding: Private

Appraisal Institute Education Trust Minorities and Women Educational Scholarship

Subjects: Real estate appraisal or related fields.
Purpose: The Minority and Women Educational Scholarship is geared towards college students working towards a degree in real estate appraisal or a related field. The scholarship is to help offset the cost of tuition.
Eligibility: Applicant must be a member of a racial, ethnic or gender group underrepresented in the appraisal profession and full- or part-time student enrolled-in real estate related courses at a degree-granting college/university or junior college/university. Individuals must have proof a cumulative grade point average of no less than 2.5 on 4.0 scale and have demonstrated financial need.Scholarship award must be used in the same calendar year as awarded by the committee.
Level of Study: Graduate, Postgraduate
Type: Scholarship
Value: US$1,000 per person
Frequency: Annual
Country of Study: United States of America
No. of awards offered: Varies
Application Procedure: An official student transcript for all college work completed to date. A 500-word written essay stating why applicant should be awarded the scholarship. Two letters of recommendation from previous employers and/or college professors. An attestation that the scholarship will be applied toward tuition/books expense as stated in the application. Optional: Applicants are asked to include a head and shoulders photograph as scholarship recipients may be profiled in Appraisal Institute newsletter/news releases.
Closing Date: April 15th
Funding: Private
Additional Information: Applicants must visit the website www.aiedtrust.org for further information.

ARAB-BRITISH CHAMBER CHARITABLE FOUNDATION (ABCCF)

43 Upper Grosvenor Street, London, W1K 2NJ, United Kingdom
Tel: (44) 020 7235 4363
Fax: (44) 020 7245 6688
Email: abccf@abcc.org.uk
Website: www.abcc.org.uk
Contact: Mr

The Arab-British Chamber Charitable Foundation (ABCCF) provides funding for Arab postgraduate students studying at British universities.

ABCCF Student Grant

Subjects: Agriculture, forestry and fishery, architecture and town planning, business administration and management, education and teacher training, engineering, mathematics and computer science, mass communication and information science, social sciences or transport.
Purpose: To assist Arab nationals in financial need, while they are at United Kingdom universities, to undertake studies in subjects of potential value to the Arab world.

Eligibility: Open to nationals of an Arab League State. Applicants must have United Kingdom student visa status. Applicants must show a commitment to return to the Arab world on completion of the postgraduate programme. Applicants should hold a first class or higher second class honours degree, or its equivalent grade of 'very good' or GPA 3+, from a recognised university. They should also meet the minimum standard in the English Language that is required by the British university.
Level of Study: MBA, Research, Doctorate, Postgraduate
Type: Grant
Value: Up to UK£2,000 per academic year
Length of Study: 3–4 years
Frequency: Annual
Study Establishment: A university in the United Kingdom
Country of Study: United Kingdom
No. of awards offered: Up to 30
Application Procedure: Applicants must complete an application form that is sent only to applicants who have confirmed that their circumstances meet with the ABCCF's criteria. Other supporting documentation is required, e.g. transcripts of degrees, academic references, proof of citizenship and visa status, university acceptance or registration and a written undertaking to return to the Arab world after graduation.
Closing Date: None
Funding: Commercial
Contributor: The Arab-British Chamber of Commerce, London
No. of awards given last year: 30
No. of applicants last year: 200

THE ARC OF THE UNITED STATES

11825 K Street, NW, Suite 1200, Washington, DC, 20006, United States of America
Tel: (1) 800 433 5255
Fax: (1) 202 534-3731
Email: info@thearc.org
Website: www.thearc.org
Contact: Sue Swenson, Executive Director

The Arc of the United States advocates for the rights and full participation of all children and adults with intellectual and developmental disabilities. Together with our network of members and affiliation chapters, we improve systems of supports and services, connect families, inspire communities and influence public policy.

Distinguished Research Award

Subjects: The prevention or improvement of mental retardation.
Purpose: To reward an individual or individuals whose research has had a significant impact on the prevention or improvement of mental retardation.
Type: Award
Value: The recipient of the award will receive a plaque, US$1,000 and a trip to speak at the Research and Prevention Luncheon of The Arc's National Convention
Frequency: Annual
No. of awards offered: 1
Application Procedure: Applicants must send the original and five copies of a nomination to The Arc.
Closing Date: April 4th
Funding: Private
No. of awards given last year: 1
No. of applicants last year: 5

ARCTIC INSTITUTE OF NORTH AMERICA (AINA)

The University of Calgary, 2500 University Drive North West, Calgary, AB, T2N 1N4, Canada
Tel: (1) 403 220 7515
Fax: (1) 403 282 4609
Email: arctic@ucalgary.ca
Website: www.arctic.ucalgary.ca
Contact: Executive Director

Created in 1945, the Arctic Institute of North America (AINA) is a non-profit membership organization and a multidisciplinary research institute for the University of Calgary.

Jennifer Robinson Memorial Scholarship

Subjects: Northern biology.
Purpose: To award a graduate who best exemplifies the qualities of scholarship that the late Jennifer Robinson brought to her studies at the Kluane Lake Research Station. The scholarship committee looks for evidence of Northern relevance and a commitment to field orientated research.
Eligibility: Applicants should contact the organization for eligibility details and guidelines.
Level of Study: Graduate
Type: Scholarship
Value: Canadian $5,000
Length of Study: 1 year, with the possibility of renewal
Frequency: Annual
Country of Study: Canada
No. of awards offered: 1
Application Procedure: There is no application form. Applicants must submit a brief description, of 2–3 pages, of the proposed research, including a clear hypothesis, relevance, title and statement of the purpose of the research, the area and type of study, and the methodology and plan for the evaluation of findings. Any collaborative relationship or work should be briefly identified. Three academic reference letters, a complete curriculum vitae with copies and a separate sheet of paper listing current sources and amounts of research funding including scholarships, grants and bursaries should also be submitted. Applicants are requested to include their email address, if they have one, upon submitting applications.
Closing Date: January 10th
Funding: Private
No. of awards given last year: 1
Additional Information: The winning applicant will be notified by the selection committee in February. Information is also available on our website www.arctic.ucalgary.ca

Jim Bourque Scholarship

Subjects: Education, environmental studies or traditional knowledge of telecommunications.
Purpose: To financially support those in post-secondary training.
Eligibility: Open to Canadian Aboriginal mature or matriculating students who are enrolled in postsecondary training in the relevant subject areas.
Type: Scholarship
Value: Canadian $1,000
Length of Study: 1 year
Frequency: Annual
Country of Study: Canada
No. of awards offered: 1
Application Procedure: There is no application form. Applicants must submit, in 500 words or less, a description of their intended programme of study and the reasons for their choice. In addition, applicants must include a copy of their most recent college or university transcript, a signed letter of recommendation from a community leader e.g. Town or Band Council, Chamber of Commerce, Metis Local, a statement of financial need indicating funding already received or expected and a proof of enrolment into, or application for, a post-secondary institution. Applications are evaluated based on need, relevance of study, achievements, return of investment and overall presentation of the application.
Closing Date: July 15th
Funding: Private
Additional Information: The recepient of the award will be announced in August. Information is also available on our website www.arctic.ucalgary.ca

Lorraine Allison Scholarship

Subjects: Canadian issues.
Purpose: To promote the study of Northern issues.
Eligibility: Open to any student enrolled at a Canadian university in a programme of graduate study related to Northern issues, whose application best addresses academic excellence, a demonstrated commitment to Northern research and a desire for research results to

be beneficial to Northerners, especially Native Northerners. Candidates in biological science fields will be preferred, but a social science topic will also be considered. Scholars from Yukon, the North West Territories and Nunavut are encouraged to apply.
Level of Study: Graduate
Type: Scholarship
Value: Canadian $3,000
Length of Study: 1 year with the possibility of renewal following receipt of a satisfactory progress report and reapplication
Frequency: Annual
Country of Study: Canada
No. of awards offered: 1
Application Procedure: Applicants must submit a two-page description of the Northern studies programme and relevant projects being undertaken, three letters of reference from the applicant's current or past professors, a complete curriculum vitae with academic transcripts and a separate sheet of paper listing current sources and amounts of research funding, including scholarships, grants and bursaries. There is no application form.
Closing Date: January 10th
Funding: International Office, trusts
Additional Information: The selection committee will notify the winning applicant in February. Information is also available on website http://www.arctic.ucalgary.ca/scholarships.

ARD INTERNATIONAL MUSIC COMPETITION

Bayerischer Rundfunk, Munich, 80335, Germany
Tel: (49) 89 5900 2471
Fax: (49) 89 5900 3573
Email: ard.musikwettbewerb@brnet.de
Website: www.ard-musikwettbewerb.de

The ARD International Music Competition is an event organized by the Association of Public Broadcasting Station in Germany. The Competition, held annually in September in Munich, Germany, covers various categories and is open to all nationalities.

ARD International Music Competition Munich

Subjects: Music, categories vary annually.
Purpose: To support and reward a selection of young musicians who are at concert standard.
Eligibility: Open to musicians of any nationality. Age restrictions apply.
Level of Study: Graduate
Type: Competition
Value: 1st prize - €10,000; 2nd prize - €7,500; 3rd prize - €5,000. Audience prizes of € 1,500 and other prizes and stipends
Frequency: Annual
Study Establishment: Conservatories, university schools of music and music academies, but also includes advanced private studies
Country of Study: Any country
No. of awards offered: The competition includes 4 categories. For each category 3 prizes are offered
Application Procedure: Applicants must complete and submit an application form, an application fee and a CD. Please see the website for further details.
Closing Date: March 31st
Funding: Corporation
Contributor: Public radio stations in Germany
No. of awards given last year: 13
No. of applicants last year: 446

For further information contact:

Email: ard.musikwettbewerb@brnet.de
Website: www.ard-musikwettbewerb.de

ARKANSAS SINGLE PARENT SCHOLARSHIP FUND (ASPSF)

614 East Emma Avenue, Suite 119, Springdale, AR, 72764, United States of America
Tel: (1) 479 927 1402
Fax: (1) 479 927 0755
Email: rnesson@jtlshop.jonesnet.org
Website: www.aspsf.org
Contact: Christina Womack, IT Director

The ASPSF is a private, non-profit corporation, which was established in 1990 in recognition of the severe impoverishment of single-parent families in Arkansas.

Arkansas Single Parent Scholarships

Subjects: All subjects.
Purpose: To provide financial assistance to Arkansas residents who are single parents and pursuing a course of instruction that will improve their income-earning potential.
Eligibility: Open to applicants who are single parents living in Arkansas who are considered economically disadvantaged.
Level of Study: Postgraduate, Professional development, (MAT only)
Type: Scholarship
Value: Varies
Frequency: Semesterly
Country of Study: United States of America
Application Procedure: See the website.
Closing Date: Varies by country
Funding: Commercial, corporation, foundation, government, individuals, private, trusts
Contributor: Varies
Additional Information: Eligible - Any country (those with visas making them eligible to study in US). To apply for a scholarship or to get involved, contact the affiliate SPSF serving the county you live in.

ARTHRITIS NATIONAL RESEARCH FOUNDATION (ANRF)

200 Oceangate, Suite 830, Long Beach, CA, 90802, United States of America
Tel: (1) 800 588 2873
Fax: (1) 562 983 1410
Email: anrf@ix.netcom.com, hbelisle@curearthritis.org
Website: www.curearthritis.org
Contact: Ms Helene Belisle, Executive Director

The Arthritis National Research Foundation (ANRF) provides funding for highly qualified postdoctoral researchers associated with major research institutes, universities and hospitals seeking to discover new knowledge for the prevention, treatment and cure of arthritis and related rheumatic diseases. The ANRF fills a much-needed niche in the field of rheumatic disease research by providing support for young postdoctoral investigators, often providing the first major funding in their research careers.

ANRF Research Grants

Subjects: Arthritis, rheumatic diseases and related immune disorders.
Purpose: To support research focusing on arthritic diseases, such as osteoarthritis, rheumatoid arthritis, lupus and related rheumatic and autoimmune diseases.
Eligibility: Applicants must hold an MD or PhD degree. Applicants need not be citizens of the United States of America, but must conduct their research at United States of America institutions. Applications will be accepted from postdoctorates and faculty members, with priority going to those scientists who do not already hold awards from the NIH or the Arthritis Foundation.
Level of Study: Postdoctorate, Research
Type: Research grant
Value: US$20,000–75,000
Length of Study: 1 year, with potential for renewal
Frequency: Annual
Study Establishment: Qualifying non-profit institutions in the United States of America

Country of Study: United States of America
No. of awards offered: 8–15
Application Procedure: Applicants must visit the website for further information. A copy of the grant guidelines may be obtained via telephone or email. Please see the website for further details http://curearthritis.org/.
Closing Date: January 18th at 5:00 p.m. PST
Funding: Foundation, individuals, private
Contributor: Individuals
No. of awards given last year: 13
No. of applicants last year: 70

ARTHRITIS RESEARCH UK

Copeman House, St Mary's Court, St Mary's Gate, Chesterfield, Derbyshire, S41 7TD, England
Tel: (44) 0 300 790 0400
Fax: (44) 0 300 790 0401
Email: info@arthritisresearchuk.org
Website: www.arthritisresearchuk.org
Contact: Mr Michael Patnick, Head of Research & Education

Arthritis Research UK is the fourth largest medical research charity in the United Kingdom, and the only charity in the country dedicated to finding the cause of and cure for arthritis, relying entirely upon voluntary donations to sustain its wide-ranging research and educational programmes.

Career Development Fellowships

Subjects: Arthritis and related musculoskeletal diseases.
Purpose: To attract and retain talented postdoctoral scientists, nurses and allied health professionals in research relevant to arthritis and related musculoskeletal diseases.
Eligibility: Open to candidates working in institutions in the United Kingdom who should normally have at least 3 years of postdoctoral research experience.
Level of Study: Postdoctorate, Professional development, Research
Type: Fellowship
Value: Salary plus reasonable running costs and small items of equipment
Length of Study: Up to 5 years
Frequency: Annual
Study Establishment: A university department or similar research institute preferably within a multidisciplinary research group
Country of Study: United Kingdom
No. of awards offered: Varies
Application Procedure: Applications for funding are available via an online system accessible from the website
Closing Date: June 19th
Funding: Private
Contributor: Voluntary charitable contributions
No. of awards given last year: 5
No. of applicants last year: 28

Clinical PhD Studentships (funding for institutional departments)

Subjects: Arthritis and related musculoskeletal diseases.
Purpose: To provide training for medically qualified clinicians in a high-quality research environment leading to a PhD and allow institutions to recruit candidates of the highest calibre.
Eligibility: Open to university departments who can provide an appropriate scientific/clinical training environment and a scientifically robust project relevant to the aims of the Arthritis Research UK.
Level of Study: Doctorate, Postgraduate, Professional development, Research
Type: Studentship
Value: A clinical salary over 3 years at the appropriate level for the appointed candidate (usually at SpR level) with running expenses and essential equipment
Length of Study: 3 years
Frequency: Annual
Study Establishment: A university hospital or recognized research institute
Country of Study: United Kingdom
No. of awards offered: 2
Application Procedure: Applications for funding are available via an online system accessible from the website http://www.arthritisresearchuk.org/research/grants-you-can-apply-for/types-of-grant/phd-studentships.aspx.
Closing Date: August 7th
Funding: Private
Contributor: Voluntary charitable donations
No. of applicants last year: 6

Clinical Research Fellowships

Subjects: Arthritis and related musculoskeletal diseases.
Purpose: Aim to provide an opportunity for training in clinical and/or laboratory research techniques in a project that demonstrates clear relevance to the aims of Arthritis Research UK in a centre of excellence in the UK.
Eligibility: Open to medical graduates (including orthopaedic surgeons), usually during speciality training, who are expected to register for a higher degree, usually a PhD.
Level of Study: Professional development, Research, Doctorate
Type: Fellowship
Value: Salaries will be according to age and experience on the appropriate clinical salary scale. Applications may also be made for reasonable running costs although tuition fees will not generally be provided.
Length of Study: 3 years
Frequency: Annual
Study Establishment: A university, hospital or recognized research institute
Country of Study: United Kingdom
No. of awards offered: Varies
Application Procedure: Applications for funding are available via an online system accessible from the website.
Closing Date: January
Funding: Private
Contributor: Voluntary charitable contributions
No. of awards given last year: 5
No. of applicants last year: 16

Clinician Scientist Fellowship

Subjects: Arthritis and related musculoskeletal diseases.
Purpose: To provide a combination of clinical training with a period of postdoctoral research in any discipline relevant to arthritis and related musculoskeletal diseases.
Eligibility: Open to medical/surgical graduates who have completed their first period of research training (in most cases having obtained a PhD) and who are committed to a career in academic medicine.
Level of Study: Postdoctorate, Professional development, Research
Type: Fellowship
Value: Fellows will be paid on the Clinical Lecturer Scale with the first year's salary reflecting their current specialist registrar or equivalent salary. The salary will rise annually in line with the Clinical Lecturer Scale which will continue to apply even if the applicant obtains their CCST during the course of the fellowship.
Length of Study: 3–5 years
Frequency: Annual
Study Establishment: A university, hospital or recognized research institute
Country of Study: United Kingdom
No. of awards offered: Varies
Application Procedure: Applications are available online accessible from the website.
Closing Date: October 10th
Funding: Private
Contributor: Voluntary charitable donations
No. of awards given last year: 1
No. of applicants last year: 3
Additional Information: Candidates will be expected to submit a timetable for their clinical training that has been approved by their postgraduate training committee. Please see the website for further details http://www.arthritisresearchuk.org/research/grants-you-can-apply-for/types-of-grant/fellowships/clinician-scientist-fellowship.aspx.

Educational Project Grant

Subjects: Arthritis and related musculoskeletal diseases.

Purpose: To fund a number of projects aimed at investigating innovative approaches to enhance the education of healthcare professionals and the public on musculoskeletal disorders. Calls are in relation to specific research questions and further details can be found on our website.
Level of Study: Research
Type: Project grant
Value: Varies – typically up to a maximum of £250,000 per project
Length of Study: Up to 2 years
Frequency: Dependent on funds available
Study Establishment: A university, hospital or recongnized research institute
Country of Study: United Kingdom
Application Procedure: Please see the website for details.
Closing Date: September 19th for submission of an intent form and April 17th for submission of a full application form
Funding: Private
Contributor: Voluntary charitable contributions
Additional Information: Those interested in submitting an intent form are advised to speak beforehand to Keir Windsor, Education Manager, on 01246 558033. Technical queries relating to the application process or timing should be directed to Rowan Roberts, Education Awards Administrator, on 01246 558033 or r.roberts@arthritisre-searchuk.org.

Educational Research Fellowships

Subjects: Arthritis and related musculoskeletal diseases.
Purpose: To provide an opportunity for training and research in an educational project relevant to arthritis and related musculoskeletal diseases.
Eligibility: Open to doctors (usually at specialty training level in rheumatology, orthopaedics or general practice), nurses, allied health professionals and educationalists. Candidates will be required to register for a higher degree (MD, MSc, PhD or equivalent).
Level of Study: Doctorate, Postgraduate, Professional development, Research
Type: Fellowship
Value: Salaries will be according to age and experience on the appropriate clinical or scientific scale. Reasonable running costs are also included. The total budget available for this call is £150,000.
Length of Study: 1, 2 or 3 years (full-time or part-time equivalent)
Frequency: Annual
Study Establishment: A university, hospital or recognized research institute
Country of Study: United Kingdom
No. of awards offered: Varies
Application Procedure: Applications for funding are available via an online system accessible from the website.
Closing Date: January 9th
Funding: Private
Contributor: Voluntary charitable contributions
Additional Information: Please email Rowan Roberts at r.robert-s@arthritisresearchuk.org if you would like any further information about the application process. Please read our information for applicants carefully before applying for a grant.

Equipment Grants

Subjects: Arthritis and related musculoskeletal diseases.
Purpose: To fund major items of equipment costing in excess of UK£30,000 that will facilitate multiple projects and make a lasting impact on rheumatological research over a period of many years.
Eligibility: Open to established units with a track record of research in arthritis and musculoskeletal disease.
Level of Study: Research
Type: Grant
Value: Varies
Length of Study: Up to 3 years
Frequency: Biannual
Study Establishment: A university, hospital or recognized research institute
Country of Study: United Kingdom
No. of awards offered: Varies
Application Procedure: Applications for funding are available via an online system accessible from the website.
Closing Date: May 29th
Funding: Private
Contributor: Voluntary charitable contributions
No. of awards given last year: 3
Additional Information: Applications are considered twice yearly by the Research Sub-Committee, with a decision approximately 6 months after the deadline date.

Foundation Fellowships

Subjects: Arthritis and related musculoskeletal diseases.
Purpose: To retain the best PhD students as postdoctoral research-ers and provide an opportunity for Fellows to develop independent research ideas at an early stage in their career.
Eligibility: Open to PhD students in the final year of training, or within 1 year of completing their PhD at the time of application. Fellowships cannot be taken up until the PhD has been awarded.
Level of Study: Postdoctorate, Professional development, Research
Type: Fellowship
Value: Salary, plus an additional fixed salary supplement for non-clinical fellows, plus reasonable running costs and small items of equipment
Length of Study: 3 years
Frequency: Annual
Study Establishment: A university, hospital or recognized research institute
Country of Study: United Kingdom
No. of awards offered: Varies
Application Procedure: Applications for funding are available via an online system accessible from the website.
Closing Date: March 6th
Funding: Private
Contributor: Voluntary charitable donations
No. of awards given last year: 3
No. of applicants last year: 12
Additional Information: Applications should be made by the applicant and their sponsor(s) (who should not be the applicant's previous PhD supervisor). Preference will be given to applicants who wish to move to another laboratory in the UK or, if remaining in the same institution, spend a period of time in another laboratory, preferably abroad.

Nurse and Allied Health Professional Educational Training Bursaries

Subjects: Arthritis and related musculoskeletal diseases.
Purpose: To promote awareness and understanding of arthritis and related musculoskeletal diseases among nurses and allied health professionals through research, practical experience, presentation of research or formal education in rheumatology.
Eligibility: Open to nurses and allied health professionals who are, or are eligible to be, registered with the Health Professionals Council (HPC), Nursing and Midwifery Council (NMC) or other appropriate regulatory body. Applicants should have at least 3 years relevant post-registration work experience and at least 1 year's experience in rheumatology/arthritis. Please see the website for details regarding eligibility.
Level of Study: Postgraduate, Professional development, Unrest-ricted
Type: Bursary
Value: £1,000 for a short training/diploma course or modules or £5,000 for a full MSc course or taught/professional doctorate.
Length of Study: Full MSc course or short training/diploma course
Frequency: Biannual
Study Establishment: A recognized training establishment in the UK
Country of Study: United Kingdom
No. of awards offered: Varies
Application Procedure: Applications for funding are available via an online system accessible from the website.
Closing Date: As from November 1st a rolling deadline applies. This means you can apply at any time (as long as this is at least 2 months before the start of the course or course module).
Funding: Private
Contributor: Voluntary charitable donations

Nurse and Allied Health Professional Educational Travel Awards

Subjects: Arthritis and musculoskeletal diseases.

Purpose: To enable nurses and allied health professionals to attend a national or international meeting in order to present a paper or poster, chair a session or lead a discussion, provided the meeting has a clear relevance to the aims and interests of Arthritis Research UK or to visit wards/units in their speciality for the purpose of studying other methods of care of patients with arthritis and related musculoskeletal diseases.
Eligibility: Open to nurses and allied health professionals who are, or are eligible to be, registered with the Health Professions Council (HPC), Nursing and Midwifery Council (NMC) or other appropriate regulatory body. Applicants should have at least 3 years relevant post-registration work experience and at least 1 year's experience in rheumatology/arthritis. Please see the website for details regarding eligibility.
Level of Study: Postgraduate, Professional development, Unrestricted
Type: Travel award
Value: Two categories of funding: Up to a maximum of £500 is offered towards the costs of attending a national or international meeting or Up to a maximum of £2,000 is offered to enable nurses or allied health professionals.
Frequency: Any time
Study Establishment: A recognized training establishment, a national or an international meeting
Country of Study: United Kingdom
No. of awards offered: Varies
Application Procedure: Applications for funding are available via an online system accessible from the website.
Closing Date: A rolling deadline so one can apply at any time (up to 2 months before the planned meeting or visit).
Funding: Private
Contributor: Voluntary charitable donations

Nurses and Allied Health Professional Training Fellowships

Subjects: Arthritis and related musculoskeletal diseases.
Purpose: To enable nurses and allied health professionals to undertake training in clinical or basic science research at a UK institution leading to a PhD or other appropriate higher degree.
Eligibility: Open to nurses and allied health professionals who are, or are eligible to be, registered with the Health Professionals Council (HPC), Nursing and Midwifery Council (NMC) or other appropriate regulatory body. Applicants should have at least 3 years relevant post-registration work experience and be committed to the care of patients with arthritis and related musculoskeletal diseases.
Level of Study: Professional development, Postgraduate, Research
Type: Fellowship
Value: Salary commensurate with the applicant's current salary at the time of appointment on the appropriate clinical salary scale with reasonable running costs, although tuition fees will not generally be provided.
Length of Study: Full-time over 3 years or part-time over 5 years
Frequency: Annual
Study Establishment: A university, hospital or recognized research institute
Country of Study: United Kingdom
No. of awards offered: Varies
Application Procedure: Applications for funding are available via an ONLINE system accessible from the website.
Closing Date: February 13th
Funding: Private
Contributor: Voluntary charitable donations

Orthopaedic Clinical Research Fellowship

Subjects: Arthritis and related musculoskeletal diseases.
Purpose: To encourage trainee orthopaedic surgeons to undertake research relevant to arthritis and related musculoskeletal diseases.
Eligibility: Open to fellows or members of the Royal College of Surgeons of Edinburgh and/or the Royal College of Surgeons of England and in good standing. Successful candidates will be expected to register for a higher degree (usually a PhD).
Level of Study: Professional development, Research, Doctorate
Type: Fellowship
Value: Varies
Length of Study: Up to 3 years
Frequency: Annual
Study Establishment: A university, hospital or recognized research institute
Country of Study: United Kingdom
No. of awards offered: 1
Application Procedure: Applications for funding are available via an ONLINE system accessible from arc's website.
Closing Date: June 9th
Funding: Private
Contributor: Voluntary charitable contributions
No. of awards given last year: 1
No. of applicants last year: 3

PhD Studentships (funding for institution departments)

Subjects: Arthritis and related musculoskeletal diseases.
Purpose: To encourage the best science graduates to embark on a research career in any discipline relevant to arthritis and related musculoskeletal diseases.
Eligibility: Open to university departments allied to rheumatology for projects that have clear relevance to the aims of Arthritis Research UK and provide training in research in a multidisciplinary environment.
Level of Study: Professional development, Research, Doctorate, Postgraduate
Type: Studentship
Value: Incremental stipend, United Kingdom tuition fees and limited running costs
Length of Study: 3 years
Frequency: Annual
Study Establishment: A university, hospital or recognized research institute
Country of Study: United Kingdom
No. of awards offered: Varies
Application Procedure: Applications for funding are available via an ONLINE system accessible from the website.
Closing Date: August 7th
Funding: Private
Contributor: Voluntary charitable donations
No. of awards given last year: 6
No. of applicants last year: 33
Additional Information: Proposals may be submitted for collaborative studentships by universities with an industrial sponsor/supervisor.

Programme Grants

Subjects: Arthritis and related musculoskeletal diseases.
Purpose: Programme Grants provide longer-term support where the aim is to answer an interrelated set of questions on a broader front than would be feasible with projects and, at the same time, should provide an opportunity for innovation and creativity.
Eligibility: Established groups undertaking research relevance to the strategic goals of Arthritis Research UK and which have a substantial research track record already supported by project grants from Arthritis Research UK, research councils or other funding bodies operating peer review.
Level of Study: Research
Type: Grant
Value: Up to £1,200,000
Length of Study: Up to 5 years
Frequency: Annual
Study Establishment: A university, hospital or recognized research institute
Country of Study: United Kingdom
No. of awards offered: Varies
Application Procedure: Applications for funding are available via an online system accessible from the website.
Closing Date: December 11th
Funding: Private
Contributor: Voluntary charitable contributions
No. of awards given last year: 3
No. of applicants last year: 22
Additional Information: Applicants should note that up to one year should be allowed for the full process of the programme grant evaluation to take place.

Project Grants

Subjects: Arthritis and related musculoskeletal diseases.

Purpose: To provide support for projects designed to seek an answer to a single question or group of related questions, providing they offer promise of advancement in the understanding of arthritis and related musculoskeletal diseases.
Eligibility: Open to applicants working at institutions in the United Kingdom who must have previous investigation and research experience.
Level of Study: Professional development, Research
Type: Project grant
Value: Up to £300,000
Length of Study: Up to 3 years
Frequency: Biannual
Study Establishment: A university, hospital or recognized research institute
Country of Study: United Kingdom
No. of awards offered: Varies
Application Procedure: Applications for funding are available via an online system accessible from the website.
Closing Date: May 29th
Funding: Private
Contributor: Voluntary charitable contributions
No. of awards given last year: 23
Additional Information: Grants are made in support of specific research projects. Individuals with proposals of potentially significant strategic importance to Arthritis Research UK in excess of given value of award limit may approach the medical director for further discussions by emailing research@arthritisresearchuk.org.

Senior Research Fellowships

Purpose: To support outstanding medical or scientific graduates committed to a research career in any discipline relevant to arthritis and related musculoskeletal diseases and with proven ability in establishing an independent research programme.
Eligibility: Open to outstanding medical or scientific graduates with 6 to 12 years postdoctoral research experience who do not hold an established academic post. Applicants will be expected to secure a commitment from their host institution for continued funding at the end of Arthritis Research UK support.
Level of Study: Research
Type: Fellowships
Value: Salaries will be according to age and experience on the appropriate scientific or clinical salary scale. Application may also be made for a supporting technician, running costs and essential equipment.
Length of Study: 5 years
Frequency: Annual
Study Establishment: A university, hospital or recognized research institute
Country of Study: United Kingdom
No. of awards offered: Varies
Application Procedure: Applications are available online accessible from the website.
Closing Date: October 9th
Funding: Private
Contributor: Voluntary charitable contributions
No. of awards given last year: 1
No. of applicants last year: 7

THE ARTHRITIS SOCIETY

393 University Avenue, Suite 1700, Toronto, ON, M5G 1E6, Canada
Tel: (1) 416 979 7228 ext. 393
Fax: (1) 416 979 1149, 416 979 8366
Email: scara@arthritis.ca/info@on.arthritis.ca
Website: www.arthritis.ca
Contact: Ms Julie Wysocki, Manager, Reserach and career Development Program

The Arthritis Society is Canada's principal charity devoted solely to funding and promoting arthritis research and care.

Arthritis Society Research Fellowships

Subjects: Arthritis.
Purpose: To provide financial support so that candidates can pursue full-time research.
Eligibility: Open to highly qualified candidates with preference given to candidates intending to embark on a research career in Canada. Candidates must hold a PhD, MD, DDS, DVM, PharmD or the equivalent.
Level of Study: Postdoctorate
Type: Fellowship
Value: Based on institution scales
Length of Study: 2 years, usually beginning on July 1st with a possibility of renewal
Frequency: Dependent on funds available
Study Establishment: Ordinarily, universities. Out-of-country training may be arranged in order to obtain specific expertise
Country of Study: Canada
No. of awards offered: Varies
Application Procedure: Applicants must complete and submit an application form with further documentation as outlined in the regulations.
Closing Date: December 1st
Funding: Private, individuals
Contributor: Public donors
No. of awards given last year: 7
No. of applicants last year: 13
Additional Information: Fellowships are awarded by the Society on the advice of the Review Panel. The Society reserves the right to approve or decline any application without stating its reasons.

Geoff Carr Lupus Fellowship

Subjects: Lupus.
Purpose: To provide advanced training to a rheumatologist.
Eligibility: Open to nationals of any country specializing in lupus at an Ontario lupus clinic, with preference given to Canadians.
Level of Study: Postdoctorate
Type: Fellowship
Value: Canadian $65,000, a travel allowance up to Canadian $1,000.
Length of Study: 1 year
Frequency: Annual
Study Establishment: An approved Ontario lupus clinic
Country of Study: Canada
No. of awards offered: 1
Application Procedure: Applicants must submit an application with three letters of recommendation and a letter of acceptance from a proposed supervisor. The letter of acceptance must include an outline proposed training programme and a certified transcript of their undergraduate record.
Closing Date: December 15th
Funding: Private
Contributor: Lupus Ontario
No. of awards given last year: 1
Additional Information: Awardees may be subject to personal income tax in Canada or in accordance with the applicable law of their country of origin.
The Fellowship is administered by Lupus Ontario, 2900 John Street, Suite 301, Markham, Ontario, L3R 5G3, Tel. 905-415-1099 or 1-877-240-1099, fax: 905-415-9874, email: admin@lupusontario.org.
Awards are made on the recommendation of The Arthritis Society's Medical Advisory Committee.

Metro A Ogryzlo International Fellowship

Subjects: Clinical rheumatology.
Purpose: To provide advanced training to individuals from a developing country.
Eligibility: The successful candidate will have completed his or her training in general medicine and have a substantial prospect of returning to an academic position in his or her own country. Canadian citizens or landed immigrants are not eligible.
Level of Study: Postdoctorate
Type: Fellowship
Value: Up to a maximum of Canadian $31,000 per year
Length of Study: 1 year, non-renewable
Frequency: Dependent on funds available
Study Establishment: A rheumatic disease unit or arthritis centre
Country of Study: Canada
No. of awards offered: 1
Application Procedure: Applicants must submit an application including letters of recommendation from three sponsors, letter of

acceptance from the proposed supervisor, an outline of the proposed training programme and a certified transcript of their undergraduate record.
Closing Date: December 1st
Funding: Private
Contributor: Public donors
Additional Information: Fellows may not receive remuneration for any other work or hold a second major scholarship, except that, with the approval of their supervisors, they may engage in and accept remuneration for such departmental activities as are conducive to their development as clinicians, teachers or investigators. Ordinarily, a Fellow who is not a graduate of a medical school in the United States of America, the United Kingdom, Republic of Ireland, Australia, New Zealand or South Africa must take the Medical Council of Canada evaluating examination to obtain the Medical Council of Canada certificate before an education licence can be issued.

ARTHUR RUBINSTEIN INTERNATIONAL MUSIC SOCIETY

12 Huberman Street, Tel Aviv, 64075, Israel
Tel: (972) 3 685 6684
Fax: (972) 3 685 4924
Email: competition@arims.org.il
Website: www.arims.org.il
Contact: Ms Idith Zui, Director

The Arthur Rubinstein International Music Society was founded by Jan Jacob Bistritzky in 1980 in tribute to the artistry of Arthur Rubinstein (1887–1982) and to maintain his spiritual and artistic heritage in the art of the piano. The Society organizes and finances the Arthur Rubinstein International Piano Master Competition and the Hommage à Rubinstein worldwide concert series and festivals, awards scholarships, runs music courses and master classes, organizes lectures and memorial festivals and issues publications and recordings.

Arthur Rubinstein International Piano Master Competition

Subjects: Piano.
Purpose: To reward talented pianists with the capacity for multi-faceted creative interpretation of composers, ranging from the pre-classic to the contemporary era.
Eligibility: The competition is open to pianists of all nationalities. Candidates must be 18–32 years of age. Candidates who study with a Jury member (not including Master Classes) shall not be allowed to compete.
Level of Study: Professional development
Type: Prize
Value: The first prize is a competition gold medal plus US$25,000, the second prize is a competition silver medal plus US$15,000 and the third prize is a competition bronze medal plus US$10,000. The fourth, fifth, and sixth prizes are US$3,000 each. $3,000 will be granted for outstanding performance of Chamber Music. A special grant of $500 who completes in Stage II. US$500 travel reimbursement.
Frequency: Every 3 years
Country of Study: Any country
No. of awards offered: 10
Application Procedure: Applicants must complete an application form according to the rules stipulated in the prospectus of the Arthur Rubinstein International Piano Master Competition. Details are available from the organization's website http://www.arims.org.il/index.php/the-competition/next-competition.
Closing Date: March 1st
Funding: Government, private
No. of applicants last year: 185
Additional Information: Next competition will take place May 13–29.

ARTIST TRUST

1835, 12th Ave, Seattle, WA, 98122, United States of America
Tel: (1) 206 467 8734
Fax: (1) 206 467 9633
Email: info@artisttrust.org
Website: www.artisttrust.org
Contact: Taura Hoppenjans, Programs Assistant

Artist Trust is a non-profit organization whose sole mission is to support and encourage individual artists working in all disciplines in order to enhance community life throughout Washington state.

Artist Trust Fellowships

Subjects: Art.
Purpose: To reward practicing professional artists of exceptional talent and demonstrated ability. The fellowship is a merit-based, not a project-based award.
Eligibility: Candidates must be 18 years or older, and should not be a matriculated student. Only Washington State residents may apply.
Level of Study: Unrestricted, Students may not apply
Type: Fellowship
Value: US$7,500
Frequency: Annual, Awarded in two-year cycles: music, media, literature and crafts disciplines are awarded in even-numbered years. Disciplines are awarded in odd-numbered years. Emerging fields and cross-disciplinary performing, visual, traditional and folk artists.
No. of awards offered: 16
Application Procedure: Candidates must complete an application form (available online), submit work samples, proof of Washington State residency, a curriculum vitae and a work sample description.
Closing Date: June 12th / February 24th
Funding: Commercial, corporation, foundation, government, individuals, private
Contributor: Washington State Arts Commission
No. of awards given last year: 16
No. of applicants last year: 400–500 applicants in visual arts, dance, design, theatre

Arts Innovator Awards

Subjects: Art.
Purpose: To reward artists of all disciplines who are originating new work, experimenting with new ideas, taking risks and pushing boundaries in their respective fields.
Eligibility: Candidates must be 18 years or older and should not be a matriculated student only. Washington State residents only may apply.
Type: Award
Value: $25,000
Frequency: Annual
No. of awards offered: 4
Application Procedure: Candidates must complete an application form (available online), submit work samples, proof of Washington State residency, a curriculum vitae, essay and letter of recommendation.
Funding: Foundation, government, individuals, private, commercial, corporation
Contributor: The Dale and Leslie Chihuly Foundation
No. of awards given last year: 4
No. of applicants last year: Approx. 200

Grants for Artist Projects (GAP) Program

Subjects: Art.
Purpose: To provide support for artist-generated projects, which can include (but are not limited to) the development, completion or presentation of new work.
Eligibility: All candidates must be 18 years or older, and Washington State residents. All disciplines and interdisciplinary projects are eligible.
Level of Study: Unrestricted, students may not apply
Type: Grant
Value: A maximum of US$1,500 for projects
Frequency: Annual
No. of awards offered: 60
Application Procedure: Candidates must submit an application form (available online), work sample, proof of Washington State residency, a curriculum vitae and a work sample description.
Closing Date: May 10th
Funding: Commercial, corporation, foundation, government, individuals, private
No. of awards given last year: 60
No. of applicants last year: Approx. 675

Irving and Yvonne Twining Humber Award for Lifetime Artistic Achievement

Subjects: Visual arts.
Purpose: To reward a female visual artist over the age of 60 from Washington State.
Eligibility: Artists must be nominated. Nominees must be female, over the age of 60, a Washington State resident and a visual artist who has been practicing for 25 years or more.
Level of Study: Unrestricted
Type: Award
Value: US$10,000
Frequency: Annual
No. of awards offered: 1
Application Procedure: Nomination forms are available by mail or online.
Closing Date: December 15th
Funding: Commercial, corporation, foundation, government, individuals, private
Contributor: Mrs Twining Humber (deceased)
No. of awards given last year: 1
No. of applicants last year: Approx. 50

ARTS AND HUMANITIES RESEARCH COUNCIL (AHRC)

AHRC Polaris House, North Star Avenue, Swindon, SN2 1FL, England
Tel: (44) 01793 41 6000
Fax: (44) 117 987 6600
Email: e.wakelin@ahrc.ac.uk
Website: www.ahrc.ac.uk
Contact: Dr Emma Wakelin, Associate Director of Programmes

The Arts and Humanities Research Council (AHRC) funds postgraduate study and research within the United Kingdom's Institutions of Higher Education. The AHRC supports Master's courses and doctoral research within a huge subject domain ranging from history, modern languages and English literature, to music and the creative and performing arts. The AHRC makes awards on the basis of academic excellence.

Collaborative Doctoral Awards

Subjects: Employment-related skills and training.
Purpose: To encourage and develop collaboration and partnerships between Higher Education Institution (HEI) departments and non-academic organisations and businesses.
Eligibility: Eligibility to apply to the partnership route is based on the number of awards held in partnership with HEIs across the previous 5 years. Organisations, or consortia, which together or separately have been awarded a total of 5 or more awards under the AHRC CDA schemes are eligible to apply to the partnership route.
Level of Study: Doctorate
Type: Studentship
Length of Study: 3 years
Frequency: Annual
Study Establishment: Any approved Institute of Higher Education
Country of Study: United Kingdom
No. of awards offered: 60
Application Procedure: Applications to become Collaborative Doctoral Partners with AHRC will be submitted "off JeS" via e-mail using word document attachments. Please refer to the following guidance carefully using the headings and word limits stipulated. Due to the tight timetable for assessment applications that do not meet the specified criteria may not be given the opportunity to make amendments and may be rejected.
Closing Date: October 15th at 4pm (off JeS)
Funding: Government
Additional Information: Please see the website for further details or email at culturesandheritage@ahrc.ac.uk.

Professional Preparation Master's Scheme

Subjects: Art and design, interpreting and translation, librarianship, archives and information management, museum studies and heritage management, creative writing, archaeology, classics and literature, linguistics, modern languages, music, drama, dance and performing arts and medieval and modern history, conservation.
Purpose: To provide funding to allow students to undertake Master's or postgraduate diploma courses that focus on developing high-level skills and competencies for professional practice.
Eligibility: Applicants must be resident in the United Kingdom or the European Union and be graduates of a recognized Institute of Higher Education or be expecting to graduate by July 31st preceding the start of the course. Applicants should refer to the AHRC guide for full details.
Level of Study: Postgraduate
Type: Studentship
Value: For full-time, UK£10,000 in London, UK£8,000 elsewhere, plus tuition fees of up to UK£3,168. These amounts are subject to change. Please consult the organization. For part-time, maintenance grant of UK£250 per year and tuition fees of UK£1,584 per year.
Length of Study: One year full-time or two year part-time. Two year full-time or four year part-time. Up to a maximum of three years.
Frequency: Annual
Study Establishment: An Institute of Higher Education
Country of Study: United Kingdom
No. of awards offered: Varies
Application Procedure: Applicants must download and complete an application form available on the website.
Funding: Government
No. of awards given last year: 351
No. of applicants last year: 1,521

Research Preparation Master's Scheme

Subjects: Archaeology, classics and ancient history, cultural and media studies, English language and literature, history of art and architecture, law, linguistics, modern languages, music, drama, dance and performing arts, philosophy, religious studies, art and design, medieval and modern history, museum studies, librarianship, and information studies.
Purpose: To support students undertaking Master's courses that focus on advanced study and research training explicitly intended to provide a foundation for further research at doctoral level.
Eligibility: Applicants must be resident in the United Kingdom or the European Union, be graduates of a recognized Institute of Higher Education or be expecting to graduate by July 31st preceding the start of the course. Applicants should refer to the AHRC guide for full details.
Level of Study: Postgraduate
Type: Studentship
Value: For full-time, UK£10,600 per year maintenance grant for London-based students and UK£8,600 per year for students based elsewhere plus tuition fees up to UK£3,168 per year. For part-time, maintenance grant of UK£275 per year and tuition fees UK£1,584 per year
Length of Study: One year full-time or two year part-time Master's course. Two year full-time or four year part-time will exceptionally be considered.
Frequency: Annual
Study Establishment: An approved Institute of Higher Education
Country of Study: United Kingdom
No. of awards offered: Varies
Application Procedure: Applicants must download and complete an application form available on the website. Students should talk to the institution at which they plan to study in order to obtain information on their organisational selection processes.
Funding: Government
No. of awards given last year: 466
No. of applicants last year: 1,779

THE ARTS COUNCIL OF WALES

Bute Place, Cardiff, CF10 5AL, Wales
Tel: (44) 0845 8734 900
Fax: (44) 0292 0441 400
Email: info@artswales.org.uk
Website: www.artswales.org.uk
Contact: Tracy Shellard

The Arts Council of Wales is the national organisation with specific responsibility for the funding and development of the arts in Wales. Most of its funds come from the National Assembly for Wales, but it also distributes National Lottery funds to the arts in Wales.

Arts Council of Wales Creative Wales Award

Subjects: Art and design, humanities, music and performing arts.
Purpose: To develop their creative practice.
Eligibility: Applicants must be aged 18 or above and not in full-time education. Must be able to demonstrate originality, excellence and purpose in their work. Applicants must live in Wales and demonstrate a commitment to Wales.
Type: Award
Value: Up to £25,000
Frequency: Annual
Country of Study: Wales
No. of awards offered: Varies
Application Procedure: Read the general guide to Arts Council of Wales funding for individuals and follow steps of the application process.
Closing Date: Published deadlines.
Contributor: Arts Council of Wales
Additional Information: The award might include the creation of new, experimental and innovative work that takes forward the art form and artistic practice. Please see the website for further details http://www.artswales.org.uk/what-we-do/funding/creative-wales.

Arts Council of Wales Major Creative Wales Award

Subjects: Art and design, humanities, music and performing arts
Purpose: To develop their creative practice.
Eligibility: Applicants must be aged 18 or above and not in full-time education. Must be able to demonstrate originality, excellence and purpose in their work. applicants must live in Wales and demonstrate a commitment to Wales.
Type: Award
Value: £20,000–25,000
Length of Study: Every 3 years
Frequency: Annual
Country of Study: Wales
No. of awards offered: Varies
Application Procedure: Potential applicants should read the Arts Council of Wales general guide to funding for individuals and follow the step-by-step guide to the application process.
Closing Date: Published deadlines. See website for exact details.
Contributor: Arts Council of Wales
Additional Information: Artists should demonstrate a consistent level of achievement and contribution within the area of professional practice in Wales.

Arts Council of Wales Project Grants

Subjects: Art and design, humanities, music and performing arts.
Purpose: To allow individuals and/or organisations to explore project ideas or to build their creative, artistic and professional capability over time.
Eligibility: Individuals - Applicants must be aged 18 or above and not in full-time education. Must be able to demonstrate originality, excellence and purpose in their work. Applicants must live in Wales and demonstrate a commitment to Wales. Applicants will need to demonatrate their ability to manage public funds effectively.Organizations - Applicants must have provision of artistic activity in their memorundum or articles of association. Have been in existence for more than two financial years and have two years' accounts, during which time it must have a track record of delivering arts activity. Have appropriate Equal Opportunities and Child Protection Policies in place. Have a bank or building society bank account with at least two signatories.
Type: Grant
Value: £250–5,000 for individuals and £250–30,000 for organisations
Frequency: Annual
Country of Study: Wales
No. of awards offered: Varies
Application Procedure: Applicants must submit application forms. A step-by-step process is noted within the General Guide to Arts Council of Wales Funding for organizations/individuals.
Closing Date: Published deadlines. See website for exact details.
Contributor: Arts Council of Wales
Additional Information: Projects must be undertaken in Wales.

Beyond Borders

Subjects: Music.
Purpose: To increase range of innovative and experimental work produced and to increase the number of people attending performances of new music.
Eligibility: Open to Wales individuals whose application is fully supported by a partner organisations. Open to organisations based in Wales or intend to present work in Wales.
Type: Grant
Value: £5,000–15,000
Frequency: Annual
Country of Study: Wales
No. of awards offered: Varies
Application Procedure: Contact the senior music officer at regional ACW office for specifc guidelines.
Closing Date: February 10th. See website for exact details.
Contributor: The PRS for Music Foundation, Creative Scotland, Arts Council Wales, Colwinston Charitable Trust and Arts Council of Northern Ireland. The scheme is administered by PRSF.

For further information contact:

PRSF, 29-33 Berners street, London, W1T 3AB, United Kingdom
Website: www.prsfoundation.co.uk

Wales Arts International

Subjects: Art and design, music and performing arts.
Purpose: To support professional arts practitioners and arts organisations based in Wales to explore international partnerships and projects, outside the UK.
Eligibility: Applicants must be aged 18 or above and not in full-time education. Must be able to demonstrate originality, excellence and purpose in their work. applicants must live in Wales and demonstrate a commitment to Wales.
Type: Funding support
Value: Up to £3,000.
Frequency: Monthly
No. of awards offered: Varies
Application Procedure: Contact Wales Arts International to discuss your project. Applicants must submit an application form.
Closing Date: Published deadlines.
Contributor: Arts Council of Wales
Additional Information: Please see the website for details http://www.wai.org.uk/funding.

ARTS NSW

Level 9, St James Centre, 111 Elizabeth Street, Sydney, NSW, 2000, Australia
Tel: (61) 1800 358 594
Fax: (61) 0292 284 722
Email: mail.artsnsw@communities.nsw.gov.au
Website: www.arts.nsw.gov.au

Arts NSW is part of the NSW Department of the Arts, Sport and Recreation. Arts NSW advises the Minister for the Arts on all aspects of the arts and cultural activity. Arts NSW works closely with the state's 8 major cultural institutions, providing policy advice to Government on their operations. The institutions manage significant cultural heritage collections and provide services and programmes throughout the state and beyond.

Asialink Residency Program

Subjects: Arts management, literature, performing arts, and visual arts/crafts.
Purpose: To promote cultural understanding, information exchange and artistic endeavour between Australia and Asian countries.
Eligibility: Open to the Australian citizens or permanent residents who have at least 3 years professional experience in their field.
Level of Study: Unrestricted
Type: Grant
Value: Up to Australian $12,000
Length of Study: 3 months

No. of awards offered: 30
Application Procedure: Check website for further details.
Closing Date: September 7th
Contributor: Arts NSW, in association with the Asialink Centre.

For further information contact:

Arts Program, The Asialink Centre, The University of Melbourne, Sidney Myer Asia Centre, Vic 3010, Australia
Tel: (61) 03 8344 4800
Fax: (61) 03 9347 1768
Email: arts@asialink.unimelb.edu.au
Website: www.asialink.unimelb.edu.au/arts

Helen Lempriere Travelling Art Scholarship
Subjects: Arts.
Purpose: To enable visual artists who are at the beginning of their career to undertake a programme of study or training overseas.
Eligibility: Open to applicants who are within the first 8 years of their professional practice as artists and are Australian citizens.
Level of Study: Postgraduate
Type: Scholarship
Value: Australian $60,000 per year
Length of Study: 1–2 years
Frequency: Annual
Country of Study: Australia
No. of awards offered: 1
Application Procedure: Applicants must submit a completed application form, including a summary of their qualifications and professional record and an outline of the programme of study to be undertaken.
Closing Date: Check website for details.
Funding: Government, trusts
Contributor: Helen Lempriere Bequest and Art NSW
Additional Information: The judging committee will make its decision on the basis of both artistic ability and potential, and the suitability of the proposed programme of study. The scholarship is not intended for established artists.

NSW Premier's History Awards
Subjects: History.
Purpose: To establish values and standards in historical research and publication and promote the excellence in the interpretation of history.
Eligibility: Open only to the citizens of Australia.
Level of Study: Unrestricted
Type: Award
Value: Australian $15,000 each for 5 categories
Frequency: Annual
No. of awards offered: 6
Application Procedure: Check website for further details.
Closing Date: April 11th
Additional Information: Nominees may enter a published book or ebook in only one of the following categories: Australian History Prize; General History Prize; New South Wales Community and Regional History Prize; Young People's History Prize. Nominees may enter a non-print media work such as a film, television, or radio program, a DVD or website in one or both of the following categories (if appropriate): Young People's History Prize; Multimedia History Prize. (Note: If entering into both categories, only one nomination form needs to be completed.). All works must have been first published, produced, performed or made publicly available between April 1st and March 31st.

For further information contact:

Literature and History Program Staff
Tel: 02 9228 5533/1800 358 594
Email: arts.funding@communities.nsw.gov.au

NSW Premier's Literary Awards
Subjects: Literature.
Purpose: To provide appropriate recognition of the best writers from Australia.
Eligibility: Open only to the residents of Australia.
Level of Study: Unrestricted
Type: Award
Value: Australian $315,000 in prizes (13 categories)
Frequency: Annual
No. of awards offered: 13
Application Procedure: Check website for further details. Contact for details: Gregory Kenny, Senior Policy Officer, Arts NSW, gregory.kenny@arts.nsw.gov.au.
Closing Date: October 18th

NSW Premier's Translation Prize
Subjects: Poetry, stage and radio plays, and fiction and non-fiction works of literary merit.
Purpose: To acknowledge the contribution made to literary culture by Australian translators.
Eligibility: Open to the Australian translators who translate literary works into English from other languages and whose body of literary translation has been published or performed in recent years.
Level of Study: Unrestricted
Type: Prize
Value: Australian $30,000 and a commemorative medallion.
Frequency: Every 2 years
No. of awards offered: 1
Application Procedure: Check website for further details.
Closing Date: December 5th
Contributor: NSW Government through Arts NSW, and the Community Relations Commission for a Multicultural NSW in association with Sydney PEN.

For further information contact:

Literature and History Program Staff
Tel: 02 9228 5533

THE ASCAP FOUNDATION

One Lincoln Plaza, New York, NY, 10023 7142, United States of America
Tel: (1) 212 621 6219
Fax: (1) 212 595 3342
Email: mspudic@ascap.com
Website: www.ascapfoundation.org
Contact: Michael Spudic

The American Society of Composers, Authors and Publishers (ASCAP) is a membership association of over 260,000 composers, songwriters, lyricists and music publishers. It is dedicated to nurturing the music talent of tomorrow, preserving the legacy of the past and sustaining the creative incentive for today's creators through a variety of educational, professional and humanitarian programmes and activities which serve the entire music community. ASCAP's function is to protect the rights of its members by licensing and paying royalties for the public performances of their copyrighted works.

ASCAP Foundation Morton Gould Young Composer Awards
Subjects: Music composition.
Purpose: To encourage talented young composers by providing recognition, appreciation and monetary awards.
Eligibility: Open to United States citizens or permanent residents who have not reached their 30th birthday by January 1st in the year of competition. Original concert music of any style will be considered. However, works which have previously earned awards or prizes in any other national competition are ineligible. Arrangements are also ineligible.
Level of Study: Unrestricted
Type: Award
Value: $30,000
Frequency: Annual
No. of awards offered: Varies from year to year
Application Procedure: Applicants must complete an application form and other materials. For more information write to organization.
Closing Date: February 1st
Funding: Private, foundation
Contributor: The ASCAP Foundation's Jack and Amy Norworth Memorial Fund, the Leo Kaplan Fund and the Frank & Lydia Bergen Foundation
No. of awards given last year: 39

Additional Information: Each year the top award winner receives an additional cash prize, The ASCAP Foundation Leo Kaplan Award.

For further information contact:

c/o ASCAP, New York, NY, 10023, United States of America
Email: concertmusic@ascap.com
Website: http://www.ascapfoundation.org/programs/awards/young-composer-awards.aspx
Contact: Cia Toscanini

The ASCAP Foundation Rudolf Nissim Prize

Subjects: Music composition.
Purpose: To encourage talented composers of concert music by providing recognition, appreciation and a monetary award to the composer of the winning score.
Eligibility: Open to all living concert composer members of ASCAP. Prior winners of this Prize are ineligible. The bound score (copy, not original manuscript), of one published or unpublished original concert work (no arrangements) requiring a conductor scored for full orchestra, chamber orchestra, or large wind/brass ensemble (with or without soloists and/or chorus) will be considered. The work must not have been previously premiered by paid professionals.
Level of Study: Unrestricted
Type: Award
Value: US$5,000
Frequency: Annual
No. of awards offered: 1
Application Procedure: Applicants must complete an application form and other materials. For further details mail to mspudic@ascap.com.
Closing Date: November 15th
Contributor: The ASCAP Foundation
Additional Information: For guidelines and application: http://www.ascap.com/~/media/files/pdf/career-development/nissimapp.pdf.

The Herb Alpert Young Jazz Composer Awards

Subjects: Music composition.
Purpose: To encourage talented young jazz composers by providing recognition, appreciation and monetary awards.
Eligibility: Open to United States citizens or permanent residents who have not reached their 30th birthday by December 31st in the year of the competition. Only completely original music will be considered. Arrangements are not eligible. Compositions which have previously earned awards or prizes in any other national competition are ineligible.
Level of Study: Unrestricted
Type: Award
Value: US$25,000
Frequency: Annual
No. of awards offered: Varies from year to year
Application Procedure: Applicants must complete an application form and other materials.
Closing Date: December 1st
Contributor: The Gibson Foundation and The ASCAP Foundation Bart Howard Fund
Additional Information: For guidelines and application: http://www.ascapfoundation.org/programs/awards/herb-alpert-composer.aspx.

ASHRIDGE MANAGEMENT COLLEGE

Berkhamsted, Hertfordshire, HP4 1NS, England
Tel: (44) 1442 843491
Fax: (44) 1442 841209
Email: mba@ashridge.org.uk
Website: www.ashridge.org.uk
Contact: Ms Jane Tobin, MBA Admissions Manager

Ashridge Business School's expertise, built up through many years of experience as a provider of executive development, has deliberately shaped their mission to help practicing and experienced managers become even more effective as leaders, and in so doing, fulfil their individual potential and that of the organization.

Ashridge Management College Full-Time MBA and Executive MBA Scholarships

Subjects: MBA studies.
Purpose: To fund students pursuing a full-time MBA.
Eligibility: Applicant must be accepted in a full-time MBA programme. Preferably, the applicant should be an individual applying from a non-profit organization, a female applicant or a need-based applicant. Applicants must have 3–5 years business or managerial experience.
Level of Study: MBA
Type: Scholarship
Value: UK£6,000
Length of Study: 1 year
Frequency: Annual
Study Establishment: Ashridge Business School
Country of Study: United Kingdom
No. of awards offered: 5
Application Procedure: Applicants must submit a form, with two references and either a Graduate Management Admission Test score or an Ashridge test score. English language skill is required. All candidates are asked to come for an interview.
Closing Date: December 1st
Funding: Commercial
Contributor: Natwest Bank, the Bank of Scotland and the Association of MBAs
No. of awards given last year: 5
No. of applicants last year: 7
Additional Information: Ashridge also awards two bursaries per year to suitable applicants who are employed in the charity sector, provided that the applicants carry out their project work for their employing charitable organization.

ASIA RESEARCH INSTITUTE(ARI) NATIONAL UNIVERSITY OF SINGAPORE

The Institute Manager, 469A Tower Block, 10-01 Bukit Timah Road, 259770, Singapore
Tel: (65) 6516 3810
Fax: (65) 6779 1428
Email: arisec@nus.edu.sg
Website: www.ari.nus.edu.sg

The Asia Research Institute (ARI) was formed in July 2001 as one of the strategic initiatives of the National University of Singapore (NUS). The mission of the Institute is to provide a world class focus and resource for research on the Asian region, located at one of its communication hubs.

ARI (Senior) Visiting Research Fellowships

Subjects: Social sciences and humanities (areas of particular interest to the institute).
Purpose: Intended for outstanding active researchers from both the Asian region and the world, to complete an important programme of research in social sciences and humanities.
Eligibility: Open to active researchers from all over the world.
Level of Study: Postdoctorate, Research
Type: Fellowship
Value: Up to Singaporean $2,250 per month
Length of Study: 1 year or less
Frequency: Annual
Country of Study: Asia
Application Procedure: Applicants must send their curriculum vitae, a synopsis of the research project and at least 1 sample of their published work.
Closing Date: September 1st (Check with website)
Additional Information: Applicants may send their applications through email, facsimile or mail. Please see the website for further details http://www.ari.nus.edu.sg/article_view.asp?id=1461.

ARI Postdoctoral Fellowships

Subjects: Social sciences and humanities (areas of particular interest to the institute).
Purpose: Intended for outstanding active researchers from both the Asian region and the World, to complete an important programme of research in social sciences and humanities.
Eligibility: Candidates must have fulfilled all requirements of the PhD within the last 2 years. If you are a PhD candidate at the point of application, you may also apply provided that you are confirmed for graduation between June/July or December/January. A letter from

your university will be required to confirm your graduation before your proposed start date.
Level of Study: Postdoctorate, Research
Type: Fellowship
Value: Fixed monthly salary of S$5,000 and a monthly housing allowance of S$500. Support for research and fieldwork, and conference attendance (on application and subject to approval).
Length of Study: Contract is tenable for a period of one year in the first instance with a possibility of extension to two years.
Frequency: Annual
Country of Study: Asia
Application Procedure: Applicants must send their curriculum vitae, a synopsis of the research project and at least 1 sample of their published work.
Closing Date: September 30th (Check with website)
Additional Information: Applicants can email, facsimile or mail their applications. Please see the website for further details http://www.ari.nus.edu.sg/article_view.asp?id=1461.

Asian Graduate Students Fellowship

Subjects: Social sciences and humanities (areas of particular interest to the Institute).
Purpose: To enable students working in the social sciences and humanities on Asian topics to be based at the National University of Singapore, and make use of the wide range of library resources.
Eligibility: Open to Asian citizens enrolled for a fulltime advanced degree at a university in an Asian country except Singapore.
Level of Study: Graduate, Research
Type: Fellowship
Value: A monthly allowance of SGD1,220, a monthly housing allowance of SGD250, a settling-in allowance of SGD150, a sum of $100 on a reimbursement basis for miscellaneous expenses, a one-time round trip travel subsidy by the most economical and direct route on a reimbursement basis upon being accepted for the fellowship.
Length of Study: Period of 2.5 months
Frequency: Annual
Country of Study: Asia
Application Procedure: Applicants must send their curriculum vitae, a synopsis of the research project and at least 2 letters of reference. See the website for application details.
Closing Date: November 15th
Contributor: ARI
Additional Information: The applicants may email, facsimile or mail their applications. Please see the website for details http://www.ari.nus.edu.sg/article_view.asp?id=273.

One-Year Visiting (Senior) Research Fellowship

Level of Study: Doctorate, Postdoctorate
Type: Fellowship
Value: Monthly honorarium of $2,250, plus complimentary university housing and travel assistance
Length of Study: 12 months
Additional Information: Applications for One-Year Visiting (Senior) Research Fellowship are also invited for commencement between April, July or October. The appointment will have a normative tenure of one year, though shorter periods may be negotiated. Interested applicants should have at least a PhD with a few years of postdoctoral research experience.

THE ASIALINK CENTRE

Level 4, Sidney Myer Asia Centre, The University of Melbourne, Parkville, VIC, 3010, Australia
Tel: (61) 3 8344 4800
Fax: (61) 3 9347 1768
Email: j.shaw@asialink.unimelb.edu.au
Website: www.asialink.unimelb.edu.au
Contact: Jacyl Shaw, Chief Executive Officer

Asialink is Australia's leading centre for the promotion of public understanding of the countries of Asia and Australia's role in the region. Headquartered at the University of Melbourne, it engages the corporate, media, arts, education, health and community sectors in Australia and Asia through numerous, diverse programmes and initiatives.

Dunlop Asia Fellowships

Subjects: Social service, local community development, regional organisation building, peace-keeping, public health/welfare, appropriate technology, environment/resource management, arts/culture and sport.
Purpose: To provide opportunities for young Australians who are committed towards making a lasting contribution to Australia–Asia relations.
Eligibility: Open to candidates who are between 21 and 40 years of age, are Australian citizens and are able to demonstrate commitment to a career with a regional focus.
Level of Study: Professional development
Type: Fellowships
Value: Up to Australian $15,000
Length of Study: 3–12 months
Frequency: Annual
Country of Study: Asia
No. of awards offered: 1–3 per year
Application Procedure: Applicants can download the form from the website. E-mail application to dunlop@asialink.unimelb.edu.au.
Closing Date: November 11th (Check with website)
Funding: Trusts
Contributor: Asialink is supported by the Myer Foundation and the University of Melbourne
No. of awards given last year: 2
No. of applicants last year: 20
Additional Information: Please see the website for further details http://www2.asialink.unimelb.edu.au/cpp/exchanges/dunlopmedal.html.

ASIAN CULTURAL COUNCIL (ACC)

6 West 48th Street, 12th floor, New York, NY, 10036, United States of America
Tel: (1) 212 843 0403
Fax: (1) 212 843 0343
Email: acc@accny.org
Website: www.asianculturalcouncil.org
Contact: Jennifer Goodall, Executive Director

The Asian Cultural Council (ACC) is a foundation that supports cultural exchange in the visual and performing arts between the United States and the countries of Asia. The emphasis of the ACC's programme is on providing individual fellowships to artists, scholars and specialists from Asia undertaking research, study and creative work in the United States. Grants are also made to United States citizens pursuing similar work in Asia.

ACC Fellowship Grants Program

Subjects: Visual and performing arts.
Purpose: To provide fellowship opportunities for research, training, travel and creative work.
Eligibility: The Asian Cultural Council grants are open to citizens and permanent residents of the United States. In Asia, the council's grants are only open to citizens and permanent residents of the countries of Asia eastward through Japan and Indonesia.
Level of Study: Doctorate, Postdoctorate, Postgraduate, Professional development
Value: Varies
Length of Study: 1–12 months
Frequency: Annual
No. of awards offered: Approx. 150
Application Procedure: Applicants must send a brief project description to the Council by October 15th. If the proposal falls within the Council's guidelines, application forms will be forwarded to individual candidates or more detailed information will be requested from institutional applicants.
Closing Date: November 16th at 5p.m. Eastern Standard Time
Funding: Corporation, foundation, individuals, private
No. of awards given last year: 158
No. of applicants last year: 281
Additional Information: Please see the website for further details http://www.asianculturalcouncil.org.hk/en/app/information_and_deadline.

ACC Humanities Fellowship Program

Subjects: Archaeology, conservation, museology and the theory, history and criticism of architecture, art, dance, film, music, photography and theater.
Purpose: To assist American scholars, graduate students and specialists in the humanities to undertake research, training and study in Asia.
Eligibility: Open to Asian individuals who are seeking grant assistance to conduct research, study, receive specialized training, undertake observation tours or pursue creative activity in the United States in the visual and performing arts. Americans seeking aid to undertake activities in Asia are also eligible to apply.
Level of Study: Professional development, Unrestricted, Research, Graduate, Postdoctorate, Postgraduate, Predoctorate
Type: Fellowship or Grant
Value: Varies
Length of Study: 1–9 months
Frequency: Annual
Country of Study: United States of America or other countries if appropriate
No. of awards offered: Varies
Application Procedure: Applicants should send a brief description of the activity for which assistance is being sought to the Council.
Closing Date: November 16th at 5 p.m. Eastern Standard Time
Funding: Foundation, government, individuals, private
Contributor: The JDR 3rd Fund and the Andrew W. Mellon Foundation
No. of awards given last year: 11 (Humanities Program only)
No. of applicants last year: 49 (Humanities Program only)
Additional Information: The programme also supports American and Asian scholars participating in international conferences, exhibitions, visiting professorships and similar projects. Please see the website for further details http://www.asianculturalcouncil.org.hk/en/app/information_and_deadline.

ASIAN DEVELOPMENT BANK (ADB)

6 ADB Avenue, Mandaluyong City 1550, Manila, 0980, Philippines
Tel: (63) 632 4444
Fax: (63) 636 2444
Email: information@adb.org
Website: www.adb.org

The Asian development bank is a multilateral development financial institution. Its work is aimed at improving the welfare of the people in Asia and the Pacific.

ADB Fully Funded Internships

Subjects: Finance.
Purpose: To provide promising students with practical professional experience in the broad field of development finance.
Eligibility: Open to candidates enrolled in a Master's or PhD-level programme at a recognized academic institution.
Level of Study: Postgraduate
Type: Internship
Value: A daily stipend, round trip airfare and suitable accommodation, limited medical insurance and a certificate of completion to successful participants.
Frequency: Annual
No. of awards offered: 6
Application Procedure: See the website http://www.gsid.nagoya-u.ac.jp/sotsubo/Internship_ADB.pdf.
Closing Date: January 31st (Check with website)

ADB Research Fellowships

Subjects: All subjects.
Purpose: To support research.
Eligibility: Open to nationals of one of ADB's member countries.
Type: Fellowship
Value: All costs covered by the fellow's academic institution, grant agency or research fellow
Frequency: Annual
No. of awards offered: Up to 5
Application Procedure: A completed application form along with curriculum vitae must be submitted.
Additional Information: Please see the website for details.

ADB-Japan Scholarship Program

Subjects: Economics, management, science and technology.
Purpose: To provide an opportunity for further studies.
Eligibility: Open to applicants who are not more than 35 years of age.
Level of Study: Postgraduate
Type: Scholarship
Value: Full tuition fees, a monthly subsistence and housing allowance, travel expenses and an excellence for books
Length of Study: 1–2 years
Frequency: Annual
No. of awards offered: 300
Closing Date: See website for exact details.
Additional Information: Please see the website for details http://www.adb.org/site/careers/japan-scholarship-program/main.

ASIAN MUSLIM ACTION NETWORK (AMAN)

House 1562/113, Soi 1/1 Mooban Pibul, Pracharaj Road, Bangsue, Bangkok, 10800, Thailand
Tel: (66) 662 913 0196
Fax: (66) 662 913 0197
Email: aman@arf-asia.org
Website: www.arf-asia.org/aman

Asian Muslim Action Network (AMAN) was established in 2002. t is a network of progressive Muslims in Asia that promotes human dignity and social justice for all by encouraging intercultural and inter religious dialogue and co-operation.

AMAN Research Fellowship Programme for Young Muslim Scholars

Subjects: Popular Islam, globalization and identity politics, Islam and changing gender realities and Islam values, economic activities and social responsibilities.
Purpose: To encourage innovative research on issues concerning economic, socio-political and cultural changes taking place in the diverse Muslim communities of Southeast Asia.
Eligibility: Open to nationals of Southeast Asian countries who are not more than 40 years of age.
Level of Study: Postgraduate, Research
Type: Grant
Value: US$5,000
Length of Study: 6 months
Frequency: Annual
Application Procedure: Applicants can download the application form from the website and submit the completed form along with a complete research proposal.
Closing Date: November 30th
Additional Information: Women are strongly encouraged to apply. Please see the website for further details http://www.scholarshipnet.info/postgraduate/asia-research-fellowship-programme-for-young-muslim-scholars/.

ASIAN SCHOLARSHIP FOUNDATION (ASF)

4th Floor, Vanissa Building, Chidlom, Ploenchit Road, Pathumwan, Bangkok, 10330, Thailand
Tel: (66) 2655 1615/6/7
Fax: (66) 2655 7977
Email: info@asianscholarship.org
Website: www.asianscholarship.org

The Asian Scholarship Foundation (ASF) is an Asian led, non-profit organization funded by a grant from the Ford Foundation that is mandated to strengthen regional capacity to produce scholarly research on Asian societies, create a network of Asian specialists on Asian studies in Asia and develop a regional perspective among scholars working in the field of Asian studies.

ASF Asia Fellow Awards

Subjects: Arts, culture, humanities and social sciences.

Purpose: To increase the overall awareness of intellectual resources in the countries of Northeast, South and Southeast Asia and to contribute to the growth of long-range capabilities for cross-regional knowledge sharing.
Eligibility: Open to citizens and residents of Bangladesh, Bhutan, Brunei, Cambodia, the People's Republic of China, Hong Kong, India, Indonesia, Japan, Laos, Malaysia, the Republic of Maldives, Myanmar, Nepal, Pakistan, the Philippines, Singapore, South Korea, Sri Lanka, Taiwan, Thailand, and Vietnam. Applicants must not be more than 45 years of age. However, those up to 50 years old, proposing to do research in the field of humanities may be given special consideration.
Level of Study: Postgraduate, Research
Type: Fellowship
Value: Covers international travel allowance, monthly living allowance, limited accident and health insurance, field trip and language training allowance, research allowance, excess baggage/shipping allowance
Length of Study: 6–9 months
Frequency: Annual
Application Procedure: Applicants must download the application form from the website and send the completed application materials to the ASF office or affiliate offices in the country or region of their citizenship.
Funding: Foundation
Additional Information: Those who are currently enrolled in a degree programme, or have just completed a degree programme for less than 1–2 years will not be eligible to apply. Those who were recipients of a Ford Foundation fellowship grant within the last 2 years prior to the application are also ineligible. All fellows are required to attend the Orientation Program in conjunction with the Annual Conference. Please see the website for details http://www.asianscholarship.org/asf/General%20information_Grant%20Information.php.

ASSOCIATED BOARD OF THE ROYAL SCHOOLS OF MUSIC

24 Portland Place, London, W1B 1LU, England
Tel: (44) 20 7636 5400
Fax: (44) 20 7637 0234
Email: abrsm@abrsm.ac.uk
Website: www.abrsm.org
Contact: Director of Finance & Administration

The Associated Board of the Royal Schools of Music is the world's leading provider of graded music examinations with over 500,000 candidates each year in over 80 countries. It is also a major music publisher and a provider of professional development courses and seminars for music teachers.

Associated Board of the Royal Schools of Music Scholarships

Subjects: Instrumental and vocal performance.
Purpose: To enable exceptionally talented young musicians to study at one of the four Royal Schools of Music.
Eligibility: For undergraduate study you should normally be at least 17 years of age by 31 January in the year of entry to the course. Postgraduates should normally be at least 21 years of age by 31 January in the year of entry. You must be a national of a country that is not a member of the European Union.
Level of Study: Postgraduate, Professional development
Type: Scholarship
Value: Tuition fees, grant of £5,000 per year towards living expenses and a flight ticket home upon satisfactory completion of the course.
Length of Study: 1–4 years, according to designated course
Frequency: Annual
Study Establishment: The Royal Academy of Music, the Royal College of Music, the Royal Northern College of Music or the Royal Scottish Academy of Music and Drama
Country of Study: United Kingdom
No. of awards offered: 8
Application Procedure: Applicants must submit an application form, health certificate, examination marks, forms, testimonials, and an authenticated cassette tape of recent performance. Candidates should apply to the Board's representative in their own country or directly to the Board in London.
Closing Date: January 31st
Funding: Private
Additional Information: Please see the website for details http://www.advance-africa.com/Associated-Board-International-Scholarships.html.

ASSOCIATED GENERAL CONTRACTORS OF AMERICA (AGC)

2300 Wilson Boulevard, Suite 400, Arlington, VA, 22201, United States of America
Tel: (1) 548 3118, 800 242 1767
Fax: (1) 548 3119
Email: info@agc.org
Website: www.agc.org

The Associated General Contractors of America (AGC), the voice of the construction industry, is an organization of qualified construction contractors and industry-related companies dedicated to skill, integrity and responsibility. Operating in partnership with its chapters, the association provides a full range of services satisfying the needs and concerns of its members, thereby improving the quality of construction and protecting the public interest.

AGC The Saul Horowitz, Jr. Memorial Graduate Award

Subjects: Construction or civil engineering.
Purpose: To provide financial assistance to students who wish to pursue higher studies.
Eligibility: Open to applicants enrolled or planning to enroll, in a Master's or Doctoral level construction or civil engineering programme as a full-time student.
Level of Study: Doctorate, Postgraduate
Type: Scholarship
Value: US$7,500. Paid in 2 installments of US$3,750
Length of Study: Up to 5 year
Frequency: Annual
Country of Study: United States of America
No. of awards offered: Varies
Application Procedure: Applications will only be accepted online. Apply online at scholarship.agc.org. Applicants must send an email to foundation@agc.org.
Closing Date: November 3rd

ASSOCIATION FOR INTERNATIONAL PRACTICAL TRAINING (AIPT)

10400 Little Patuxent Parkway, Suite 250, Columbia, MD, 21044-3519, United States of America
Tel: (1) 410 997 2200
Fax: (1) 410 992 3924
Email: aipt@aipt.org
Website: www.aipt.org

The AIPT is a non-profit organization that promotes international understanding through cross-cultural, on-the-job, practical training exchanges for students and professionals.

Jessica King Scholarship Fund

Subjects: Hotel management.
Purpose: To inspire young hospitality students and professionals.
Eligibility: Open to young Americans who wish to succeed in the international hospitality field.
Level of Study: Postgraduate
Type: Scholarship
Value: US$2,000 each (awarded in two parts: $1500 upon approval of program and $500 at the end of 12 months or the end of the program) and travel and related program costs.
Length of Study: 1 year
Frequency: Annual
No. of awards offered: Varies

Application Procedure: Applications are to be submitted using the AIPT Americans Abroad Exchange Program application. Contact AIPT for further details.
Closing Date: Applications are accepted year-round.
Contributor: The King Family

ASSOCIATION FOR LIBRARY SERVICE TO CHILDREN

American Library Association, 50 East Huron Street, Chicago, IL, 60611-2795, United States of America
Tel: (1) 800 545 2433 ext. 2163
Fax: (1) 312 280 5271
Email: alsc@ala.org
Website: www.ala.org

The Association for Library Service to Children develops and supports the profession of children's librarianship by enabling and encouraging its practitioners to provide the best library service to our nation's children.

The Bound to Stay Bound Books Scholarship

Subjects: Library science.
Purpose: For the education of men and women who intend to pursue an MLS or advanced degree and who plan to work in the area of library service to children.
Eligibility: See the website.
Level of Study: Postgraduate
Type: Scholarship
Value: US$6,500 each
Frequency: Annual
Country of Study: United States of America
No. of awards offered: 4
Closing Date: March 1 (for three references details)

The Frederic G. Melcher Scholarship

Subjects: Library science.
Purpose: To provide financial assistance to professionals who plan to work in children's librarianship.
Eligibility: Open to candidates with academic excellence, leadership qualities, and desire to work with children.
Level of Study: Postgraduate
Type: Scholarship
Value: Two US$6,000
Frequency: Annual
Country of Study: United States of America
No. of awards offered: 2
Application Procedure: See the website.

ASSOCIATION FOR SPINA BIFIDA AND HYDROCEPHALUS (ASBAH)

ASBAH House, 42 Park Road, Peterborough, Cambridgeshire, PE1 2UQ, England
Tel: (44) 0845 450 7755
Fax: (44) 0173 355 5985
Email: info@asbah.org
Website: www.asbah.org
Contact: Mrs Lyn Rylance, Secretary to the Directorate

The Association for Spina Bifida and Hydrocephalus (ASBAH) is a voluntary organization that works for people with spina bifida and hydrocephalus. The charity lobbies for improvements in legislation and provides advisory and support services to clients and their families or carers, in addition to supplying information to professionals and sponsoring medical, social and educational research.

ASBAH Bursary Fund

Subjects: Any subject that will improve the chances of employment for people with spina bifida, hydrocephalus or both.
Purpose: To help with expenses of further or higher education courses approved by, but not organized by, ASBAH.
Eligibility: Open to individuals with spina bifida and hydrocephalus. Applicants must be resident in the United Kingdom.
Level of Study: Unrestricted
Type: Studentships and bursaries
Value: Course fees and other expenses
Length of Study: Varies
Frequency: Dependent on funds available
Study Establishment: Varies
Country of Study: England, Wales and Northern Ireland only
No. of awards offered: Varies
Application Procedure: Applicants must complete an application form, available from Mrs L. Rylance, Secretary to the Directorate.
Closing Date: Applications are accepted at any time (Check with website)
Funding: Private
Contributor: Charitable donations
No. of awards given last year: 2
No. of applicants last year: 2
Additional Information: Applicants are normally visited by an ASBAH Area Adviser prior to an award being considered.

ASBAH Research Grant

Subjects: Medical sciences, natural sciences, education and teacher training, recreation, welfare and protective services.
Purpose: To support research in an area directly related to spina bifida and/or hydrocephalus, and to explore ways of improving the quality of life for people with these conditions through medical, scientific, educational and social research.
Eligibility: Applicants must be resident in the United Kingdom.
Level of Study: Postgraduate
Value: Varies
Length of Study: Varies
Frequency: Dependent on funds available
Study Establishment: Varies
Country of Study: United Kingdom
No. of awards offered: Varies
Application Procedure: Applicants must make an initial enquiry to the Chief Executive. If the proposed research is considered to be interesting, the applicant will be asked to complete an application form. Applications must be submitted on time to the committees, which meet in February and September to October.
Closing Date: January 1st and August 1st
Funding: Individuals, private, trusts
Contributor: Charitable donations
No. of awards given last year: 1
No. of applicants last year: 1
Additional Information: Award subjects: biology and life sciences; economics; medicine and surgery; social sciences; teacher training and education; theology and religious studies.

ASSOCIATION FOR WOMEN IN SCIENCE EDUCATIONAL FOUNDATION

1321 Duke Street, Street 210, Alexendria, VA, 22314, United States of America
Tel: (1) 703 894 4490
Email: awis@awis.org
Website: www.awis.org
Contact: Dr Barbara Filner, President

The Association for Women in Science Educational Foundation provides fellowships to assist women students studying the sciences.

Association for Women in Science Educational Foundation Predoctoral Awards

Subjects: Life, physical, behavioural or social science and engineering.
Purpose: To promote the participation of women in engineering and the sciences.
Eligibility: Open to female students enrolled in any life, physical, behavioural or social science or engineering programme leading to a PhD degree. Applicant must have passed the departmental qualifying exam and except to complete the PhD within two-and-one-half years, at the time of application.
Level of Study: Doctorate, Predoctorate
Value: US$1,000
Length of Study: 2 years

Frequency: Annual
Study Establishment: An Institute of Higher Education
Country of Study: Anywhere for United States of America citizens; in the United States of America for others
No. of awards offered: 5–10
Application Procedure: Applicants must submit an application including a basic form, a five-page summary of the candidate's dissertation research, two recommendation report forms and official transcripts of all coursework conducted at postsecondary institutions. Forms are available on the website.
Closing Date: January 29th
Funding: Foundation
No. of awards given last year: 10
No. of applicants last year: 100
Additional Information: Winners are notified by email in June and announced publicly in the Autumn issue of the AWIS Magazine.

ASSOCIATION OF AMERICAN GEOGRAPHERS (AAG)

1710 Sixteenth Street, North West, Washington, DC, 20009-3198, United States of America
Tel: (1) 202 234 1450
Fax: (1) 202 234 2744
Email: psolis@aag.org; gaia@aag.org
Website: www.aag.org
Contact: Dr Patricia Solis, Director of Outreach and Strategic Initiatives

The Association of American Geographers (AAG) is a non-profit organization founded in 1904 to advance professional studies in geography and to encourage the application of geographic research in business, education and government. The AAG was amalgamated with the American Society of Professional Geographers (ASPG) in 1948.

AAG Dissertation Research Grants

Subjects: Geography.
Purpose: To support dissertation research.
Eligibility: Open to candidates without a doctorate at the time of the award, who have been AAG members for at least 1 year at the time of application and who have completed all PhD requirements except the dissertation by the end of the semester or term following the approval of the award. The candidates' dissertation supervisor must certify eligibility and proposals should demonstrate high standards of scholarship.
Level of Study: Postdoctorate
Type: Research grant
Value: Up to US$500
Frequency: Dependent on funds available
Country of Study: United States of America
No. of awards offered: 3
Application Procedure: Digital submissions are required. All information must be entered on our online application form. Also, applicants must submit seven copies of a dissertation proposal of no more than 1,000 words and seven copies of the completed forms. The proposal should describe the problem that is to be solved, outline the methods and data to be used and summarize the results expected to be found. Budget items should also be included within the body of the proposal. For further detials mail to grantsawards@aag.org.
Closing Date: December 31st
Funding: Private
Contributor: Members
No. of awards given last year: 6
No. of applicants last year: 8
Additional Information: By accepting an AAG dissertation grant, awardees agree to submit a copy of the dissertation, and a report that documents expenses charged to the grant, to the AAG Executive Director. AAG support must also be acknowledged in presentations and publications. The awards include the Robert D Hodgson Memorial PhD Dissertation Fund the Paul Vouras Fund and the Otis Paul Starkey Fund. Please visit the website for further details http://www.aag.org/cs/grants/dissertation#applications.

AAG NSF International Geographical Union Conference Travel Grants

Subjects: Geography.
Purpose: To provide Travel Grants to the IGU conference.
Eligibility: All scientists employed by US agencies, firms and academic institutions may apply for support. All grantees must: be citizens or hold permanent residency in the United States of America, be registered for the main international congress or main regional conference even if their presentations are scheduled for a symposium or study group, and travel via a US carrier in accordance with US government regulations.
Level of Study: Professional development
Type: Grant
Value: $1,250 each to junior scholars, including graduate students and $1,000 each to senior scholars.
Frequency: Annual
Country of Study: As applicable
No. of awards offered: Total 30: 15 (Junior scholars), 15 (Senior scholars)
Application Procedure: Applications must be submitted digitally using the online application form provided in website.
Closing Date: February 4th (to submit abstract); March 1st (to attend conference)
Contributor: National Science Foundation
Additional Information: Please see the website for further details http://www.aag.org/cs/grantsawards/igutravel.

AAG Research Grants

Subjects: Geography.
Purpose: To support research and field work expenses.
Eligibility: Open to candidates who have been AAG members for at least 2 years at the time of application. Proposals that, in the opinion of the committee, offer the prospect of obtaining substantial subsequent support from private foundations or federal agencies and that address questions of major import to the discipline will be given preference.
Level of Study: Postdoctorate, Professional development
Type: Research grant
Value: $1,000 (Maximum)
Country of Study: Any country
No. of awards offered: Varies
Application Procedure: Digital submissions are required. All information must be entered on our online application form.
Closing Date: December 31st
Funding: Private
Contributor: AAG members
No. of awards given last year: 5
No. of applicants last year: 8
Additional Information: No awards are made if proposals are not suitable or for Master's and doctoral dissertation research. Guidelines are printed in the AAG Newsletter. Please see the website for further details http://www.aag.org/cs/grants/research.

Anne U White Fund

Subjects: Geochemistry.
Purpose: To enable people, regardless of any formal training in geography, to engage in useful field studies and to have the joy of working alongside their partners.
Eligibility: Open to candidates who have been AAG members for at least 2 years at the time of application. Proposals that, in the opinion of the committee, best meet the purposes for which Anne and Gilbert White set up the funds will be given preference.
Level of Study: Professional development
Type: Funding support
Value: US$1,500 each
Frequency: Annual
No. of awards offered: 1–2
Application Procedure: Candidates must complete seven application forms, available from the website or by request from the executive assistant, Ehsan M Khater. Successful candidates will be announced on or about March 31st. For further details mail to grantsawards@aag.org.
Closing Date: December 31st
Funding: Private
No. of awards given last year: 2
No. of applicants last year: 9

Additional Information: By accepting the Anne U White grant, awardees agree to submit a two-page report that summarizes results and documents expenses underwritten by the grant to the AAG Executive Director. In 1989, Gilbert and Anne White donated a sum of money to the Association of American Geographers to establish the Anne U White Fund. Gilbert White and other donors have subsequently added substantially to the original gift. Please see the website for further details http://www.aag.org/cs/about_aag/grants_and_awards/aag_research_grants/anne_white_fund.

The George and Viola Hoffman Fund

Subjects: Historical, contemporary, systematic and regional geographic studies.
Purpose: To provide financial support towards a Master's thesis or doctoral dissertation on a geographical subject in Eastern Europe.
Level of Study: Doctorate, Postgraduate
Type: Grant
Value: US$350–500
Length of Study: 1 year
Frequency: Annual
No. of awards offered: 1
Application Procedure: Applications should be limited to 2,500 words and should include a statement of the problem to be pursued; methods to be employed, including field study; schedule for the work; competence in the language of the area; and a bibliography of pertinent literature, accompanied by a letter stating the professional achievements and the goals of the individual and a letter of support from a sponsoring faculty member. Digital submissions are strongly encouraged. Please submit your application through our online application form. In exceptional cases, paper submissions may be accepted, but only with prior approval (requests should be sent to grantsawards@aag.org).
Closing Date: December 31st
Funding: Private
Additional Information: Please see the website for further details http://www.aag.org/cs/grants/hoffman.

J. Warren Nystrom Award

Subjects: Geography.
Purpose: To support a paper based upon recent dissertations in geography.
Eligibility: This competition is restricted to AAG members who have received the PhD degree since April 1st and who have been members of the AAG for last 3 years. See the website for more details.
Level of Study: Postdoctorate, Doctorate
Value: Approximately $1,000
Frequency: Annual
Country of Study: Any country
No. of awards offered: Varies
Application Procedure: Digital submission is required. Applications should be sent online to meeting@aag.org.
Closing Date: September 22nd
Funding: Private
Contributor: AAG members
No. of awards given last year: 1
No. of applicants last year: 15
Additional Information: Awards are made for papers presented at the annual meeting of the Association. Please see the website for further details http://www.aag.org/cs/nystrom.

Visiting Geographical Scientist Program

Subjects: Geography.
Purpose: To stimulate interest in geography.
Eligibility: To qualify for the program, at least one-half of the institutions hosting a visiting scientist during the academic year must have active chapters of Gamma Theta Upsilon (an active chapter is one that has reported initiates in the past two years).
Level of Study: Professional development
Type: Grant
Value: US$100 per institutional visit to each visiting scientist and will reimburse the visitor up to US$600 for travel costs to and from the area of the institutions visited.
Frequency: Annual
Study Establishment: Institution with an active chapter of Gamma Theta Upsilon
Country of Study: United States of America
Application Procedure: Applicants must write to Oscar Laron, VGSP Coordinator, at the main organization address at olarson@aag.org.
Contributor: Gamma Theta Upsilon (GTU), the International Geographical Honor Society
Additional Information: The VGSP requires one scientist to visit two or more institutions over a one to three-day period. Please see the website for further details http://www.aag.org/cs/vgsp.

ASSOCIATION OF CALIFORNIA WATER AGENCIES (ACWA)

910 K Street, Suit 100, Sacramento, CA, 95814-3577, United States of America
Tel: (1) 916 441 4545
Fax: (1) 916 325 2316
Email: acwabox@acwa.com
Website: www.acwa.com

The Association of California Water Agencies has been a leader in California water issues since 1910. Its primary mission is to assist its members in promoting the development, management and reasonable beneficial use of water in an environmentally balanced manner.

ACWA Scholarship

Subjects: Water resources.
Purpose: To support students in water resources-related fields.
Eligibility: Eligible students must be California residents attending a four-year, publicly funded college or university in California full-time as a junior or senior during the year the scholarship is awarded.
Level of Study: Postgraduate
Type: Scholarship
Value: US$3,000 each
Length of Study: 4 years
Frequency: Annual
No. of awards offered: 2
Application Procedure: Eligible candidates must submit a completed application form and two or three recommendations and e-mail them to awards@acwa.com. See the website.
Closing Date: April 1st
Contributor: ACWA
Additional Information: Please see the website for details http://www.acwa.com/content/scholarship/acwa-scholarship-application-and-guidelines.

Clair A. Hill Scholarship

Subjects: Water resources.
Purpose: To provide financial assistance to students in the water-related fields.
Eligibility: Eligible students must be California residents attending California colleges or universities full-time as a junior or senior during the year the scholarship is awarded.
Level of Study: Postgraduate
Type: Scholarship
Value: US$5,000
Length of Study: Academic year
Frequency: Annual
Country of Study: United States of America
No. of awards offered: 1
Application Procedure: See the website.
Closing Date: February 1st
Contributor: Association of California Water Agencies
Additional Information: The top three candidates will be chosen by the East Bay Municipal Utility District scholarship committee upon receipt of a completed application form and at least two, but not more than three, recommendations for each candidate. Please see the website for details http://www.acwa.com/content/clair-hill-scholarship-0.

ASSOCIATION OF CLINICAL PATHOLOGISTS

189 Dyke Road, Hove, East Sussex, BN3 1TL, England
Tel: (44) 1273 775700
Fax: (44) 1273 773303
Email: info@pathologists.org.uk
Website: www.pathologists.org.uk
Contact: Administrative Assistant

The Association of Clinical Pathologists promotes the practice of clinical pathology by running postgraduate education courses and national scientific meetings and has a membership of 2,000 worldwide.

Student Research Fund

Subjects: Research projects within laboratory medicine (undergraduate) or to support students undertaking BSc/BMED Sci.
Purpose: To encourage undergraduates to undertake some research within laboratory medicine, to raise the profile of laboratory medicine within the minds of undergraduates and to aid in recruitment of young graduates.
Type: Scholarship
Value: Up to £150 per week for a maximum of six weeks (Funding of a small project), £5,000 maximum (Financial sponsorship to help support living expenses during their extra undergraduate year), Up to £1000 (Funding for the cost of consumables). See the website for details.
Frequency: Annual
No. of awards offered: Up to 5
Application Procedure: Applicants should complete the relevant application form, including a brief statement of not more than 400 words outlining their interest in laboratory medicine, and include a full CV. Applications should include details of the work to be undertaken as part of the project or during the BSc and be supported in writing by the project supervisor, or by the Head of Department in which the student will be placed.
Closing Date: Applications for BSc support must be received by June 29th. For project funding applications are considered throughout the year
Contributor: Association of Clinical Pathologists
No. of awards given last year: 3
No. of applicants last year: 5
Additional Information: Please return completed form by post to ACP Postgraduate Education Secretary,and also by e-mail to rachel@pathologists.org.uk.

THE ASSOCIATION OF COMMONWEALTH UNIVERSITIES

Woburn House, 20-24 Tavistock Square, London, WC1H 9HF, United Kingdom
Tel: (44) 20 7380 6700
Fax: (44) 20 7387 2655
Email: info@acu.ac.uk
Website: www.acu.ac.uk
Contact: Ms Natasha Lokhun, Communications Officer

ACU Titular Fellowship

Subjects: Preference will be given to applicants in the following priority subject areas: agriculture, forestry and food sciences; biotechnology; development strategies; earth and marine sciences; engineering; health and related social sciences; information technology; management for change; professional education and training; social and cultural development; university development and management.
Purpose: The ACU Titular Fellowships provide opportunities for staff from member universities and employees working in industry, commerce or public service in a Commonwealth country to spend periods of time at other member universities or relevant institutions outside their own country. Founded by particular sponsors, each of the ACU Titular Fellowships is dedicated to a specific purpose.
Eligibility: Applicants must: be on the staff of, or a nominee of, an ACU member university or a Commonwealth inter-university organisation, or be working in industry, commerce or public service in a Commonwealth country, and be of proven high ability. ACU Titular Fellowships are not intended for degree courses or for immediately postdoctoral programmes, and cannot be held in the country in which the applicant currently works or concurrently with other awards.
Level of Study: Professional development, Research
Type: Fellowship
Value: Up to £5,000 according to the actual programme, and intended to cover: one international return air fare, in-country ground travel, medical insurance, board and lodging, fees (where the approved programme includes a formal training programme)
Length of Study: Up to a maximum of 6 months
Frequency: Annual
Study Establishment: Except where specific institutions are stipulated, Titular Fellowships are tenable at either a university in ACU membership or in industry, commerce or public service
No. of awards offered: 8 +
Application Procedure: Complete the application form and submit by email to acuawards@acu.ac.uk by the stated deadline (TBC). Applications must be approved by the executive head (vice-chancellor, president or rector) of a university in ACU membership, or the head of the candidate's firm/company. The ACU will also consider applications approved by the chief executive officer of a Commonwealth inter-university organisation. Applications must include all relevant signatures and be converted to (and submitted in) PDF format. The ACU will accept scanned copies of the relevant signature pages. For more information, visit http://www.acu.ac.uk/member_-services/fellowships_mobility/acu_titular_fellowships.
Closing Date: Check with website
Funding: Private
Contributor: The Association of Commonwealth Universities
Additional Information: Except where specific countries of tenure are stipulated, Titular Fellowships are tenable in any Commonwealth country (other than the applicant's home country). Please see the website for details https://www.acu.ac.uk/membership/titular-fellowships/.

Canada Memorial Foundation Scholarship

Subjects: Any subject.
Purpose: Awards are offered for postgraduate study leading to a university degree.
Eligibility: Candidates must be UK citizens and hold, or be in their final year of undergraduate study and expect to attain, a minimum of an upper second class undergraduate degree.
Level of Study: Postgraduate
Type: Scholarship
Value: Maintenance allowance, return airfare, approved tuition fees, book, thesis, travel and health insurance allowance.
Length of Study: 1 year
Frequency: Annual
Study Establishment: Universities/higher education institutions
Country of Study: Canada
Application Procedure: Complete the application form and print and post five hard copies of it to the ACU by the stated deadline (TBC), accompanied by a document giving proof of UK citizenship (in top copy of application only) and a certified transcript for your undergraduate degree (or partial transcript if final year of studies in progress) (in each application).
Closing Date: Please check with website
Funding: Foundation
Contributor: Canada Memorial Foundation
Additional Information: Please see the website for details https://www.acu.ac.uk/focus-areas/scholarship-administration/canada-memorial-foundation-scholarships.

ASSOCIATION OF MANAGEMENT DEVELOPMENT INSTITUTIONS IN SOUTH ASIA (AMDISA)

Secretariat, University of Hyderabad Campus, Central University Post Office, Hyderabad, 500 046, India
Tel: (91) 40 64545226, 40 64543774
Fax: (91) 40 23013346
Email: amdisa@amdisa.org
Website: www.amdisa.org
Contact: Executive Director

Association of Management Development Institutions in South Asia (AMDISA) was established in 1988, with the initiative of leading management development institutions in the SAARC region. It is the only association that networks management development centres across 8 nations and promotes partnership between business schools, business leaders and policy administrators for enhancing the quality and effectiveness of management education in South Asia.

Commonwealth AMDISA Regional Doctoral and Post Doctoral Fellowship

Subjects: Management and related social science and other disciplines.
Purpose: To provide financial and academic institutional assistance to PhD scholars and younger faculty in the South Asian countries, and to contribute to the developement of South Asian academic perspectives networks and communities in management and related areas.
Eligibility: Open to resident citizens of South Asian Commonwealth countries to undertake comparative studies in their home country or at least one of the countries listed. For doctoral fellowships, applicants must be registered PhD scholars in a recognized institution. For postdoctoral fellowships, applicants must be below 50 years of age, must have a doctorate degree and must be employed as a full-time teacher/researcher. See the website for details.
Level of Study: Postdoctorate, Doctorate
Type: Fellowship
Value: UK£1,500–4,500
Length of Study: 1 year
Frequency: Dependent on funds available
Study Establishment: Recognized institution where the fellow is a registered PhD scholar or full-time teacher-researcher in the home country.
Country of Study: Home Country
No. of awards offered: 8
Application Procedure: Through the Fellowship Announcement a detailed outline of the research proposal and other supporting documents to be submitted in hard and soft copies is communicated to eligible applicants. The applications are evaluated and fellowships are awarded by a 5-member AMDISA Regional Fellowship Committee.
Closing Date: December 31st
Funding: International Office
Contributor: Commonwealth Fund for Technical Cooperation (CFTC)
No. of awards given last year: 8
No. of applicants last year: 15
Additional Information: Countries in which fellows can undertake comparative studies are Bangladesh, India, Malaysia, Maldives, Pakistan, Singapore and SriLanka. Please see the website for details http://www.amdisa.org/FellowshipAnnouncement.pdf.

ASSOCIATION OF MOVING IMAGE ARCHIVISTS (AMIA)

1313 North Vine Street, Hollywood, CA, 90028, United States of America
Tel: (1) 323 463 1500
Fax: (1) 323 463 1506
Email: amia@amianet.org; kkersels@amianet.org
Website: www.amianet.org

The Association of Moving Image Archivists (AMIA) is a non-profit professional association established to advance the field of moving image archiving by fostering co-operation among individuals and organizations concerned with the acquisition, preservation, exhibition and use of moving image materials.

AMIA Kodak Fellowship in Film Preservation

Subjects: Filmmaking, video, historic preservation and conservation.
Purpose: To foster the education and training of the next generation of moving image archivists.
Eligibility: Open to applicants who are not below 21 years of age and are full time students with a grade point average of at least 3.0 in their recent academic programmes.
Level of Study: Postgraduate
Type: Fellowship
Value: A $4,000 scholarship for the upcoming academic year. Complementary registration to the AMIA Conference. A four-week summer internship at Kodak and other film restoration facilities in Los Angeles.
Length of Study: 1 year
Frequency: Annual
Country of Study: United States of America
No. of awards offered: 1
Application Procedure: Applicants must submit an application form, official transcript, essay of no more than 1,000 words describing the applicant's interest and 2 reference letters.
Closing Date: May 1st
Funding: Foundation
Contributor: Eastman Kodak Company
Additional Information: Transportation and housing will be provided at no cost, and the student will be paid an hourly wage to offset other living expenses. Please see the website for further details http://www.amianet.org/events/kodak.php.

AMIA Scholarship Program

Subjects: Filmmaking/video, historic preservation and conservation.
Purpose: To financially support students who wish to pursue careers in moving image archiving.
Eligibility: Open to students who have a minimum grade point average of 3.0. Applicants must be full-time students.
Level of Study: Postgraduate
Type: Scholarships
Value: Minimum of US$4,000 each
Length of Study: One academic year
Frequency: Annual
Country of Study: United States of America
No. of awards offered: 4
Application Procedure: Applicants must submit an application form, transcript, essay and reference letters. Applicants need only submit one application form and one set of supporting documents to be eligible for all four awards.
Closing Date: May 1st
Funding: Foundation
Contributor: Mary Pickford Foundation, Sony Pictures Entertainment, The Rick Chace Foundation, and Universal Studios for Film and Television Heritage
Additional Information: The program awards four annual scholarships - the Mary Pickford Scholarship, the Sony Pictures Scholarship, The Rick Chace Foundation Scholarship, and the Universal Studios Preservation Scholarship. For additional information concerning the AMIA Scholarship Program, contact the AMIA office or email amia@amianet.org.

For further information contact:

Contact: Janice Simpson, AMIA Managing Director

The AMIA/Rockfeller Archive Center Visiting Archivist Fellowship

Purpose: To provide assistance and first hand experience, the AMIA/Rockfeller Archive Center Visiting Archivist Fellowship is awarded each year to a professional archivist from the developing world interested in improving their skills and knowledge through a study at the Centre.
Eligibility: The applicant must be employed as archivists or in a closely related field, the applicant must also have a BA or equivalent, professional archival training, and 2–5 years experience as an archivist.
Type: Fellowship
Value: US$2,500 (stipend), US$3,500 (travel expenses), and complementary registration to the AMIA Annual Conference.
Length of Study: 2 week
No. of awards offered: Varies
Application Procedure: The applicant must submit a duly filled application form, resumes or vitae, proof of employment, an essay of no more than 1,000 words describing the applicant's goals, and two letters of recommendation.
Closing Date: June 30th
Contributor: Rockfeller Archive Center in cooperation with the Association of Moving Image Archivists

Additional Information: Please see the website for further details http://www.amianet.org/events/rockefeller.php.

Image Permanence Institute Internship

Subjects: Imaging science and technology.
Purpose: To give a student of merit who is committed to the preservation of moving images the opportunity to acquire practical experience in preservation research.
Eligibility: The applicant must be enrolled for the next moving image programme academic year and must have a grade point average of atleast 3.0.
Level of Study: Postgraduate
Type: Internship
Value: US$5,000 stipend to be used for living expenses and reimbursement of travel fare
Length of Study: 3 months
No. of awards offered: Varies
Application Procedure: Applicant should submit a duly filled internship application form, an essay of no more than 1,000 words and current resume.
Closing Date: May 1st
Contributor: Rochester Institute of Technology (RTI) and the Society for Imaging Science and Technology (IS&T)
Additional Information: Please see the website for further details http://www.amianet.org/events/ipi.php.

ASSOCIATION OF RHODES SCHOLARS IN AUSTRALIA

University of Melbourne, Victoria, 3010, Australia
Tel: (61) 3 8344 4000
Fax: (61) 3 8344 5104
Email: d.cookson@unimelb.edu.au
Website: www.research.unimelb.edu.au
Contact: Dr David Cookson, Vice-Principal (Research)

Rhodes Scholarships were created under the will of Cecil John Rhodes, the British colonial pioneer and statesman, who died in 1902. Rhodes hoped to provide future leaders of the English-speaking world with an education, which would broaden their views and develop their abilities. Rhodes hoped that those who gained these benefits from Oxford and his scholarships would go on to improve the lot of humanity, and work towards maintaining peace between nations. The Rhodes Trust, which administers the scholarships, is based in Oxford. Its Secretary, also Warden of Rhodes House, assists, guides and advises Rhodes Scholars in residence, and the Warden's office administers financial and other aspects of the scholarships. The Secretary to the Trust is also responsible for the operation of all the Selection Committees in the various constituencies, and is assisted in this in most of the larger constituencies by national secretaries. The Australian Secretary of the Trust coordinates the work of the Selection Committees and generally represents the Trust in Australia.

Association of Rhodes Scholars in Australia Scholarship

Subjects: All subjects.
Purpose: To enable an overseas Commonwealth student to undertake research in Australia.
Eligibility: Open to graduates of a Commonwealth university approved by the committee administering the bursary. Graduates must currently be enrolled as higher degree research students at their home university, be Commonwealth citizens and may not be graduates of an Australian or New Zealand university. Awards are not normally made to students who at the time of their visit to Australia will be in the first 10 months or the final 6 months of candidature.
Level of Study: Postgraduate
Type: Scholarship
Value: Australian $20,000 including travel expenses
Length of Study: 6–12 months
Frequency: Every 2 years
Study Establishment: A university
Country of Study: Australia
No. of awards offered: 1
Application Procedure: Applicants must apply for information and application forms, available through the website.
Closing Date: July 31st
Funding: Private
Contributor: Charitable donations from former Australian Rhodes Scholars
Additional Information: Please see the website for further information http://www.research.unimelb.edu.au/rgc/grants/find/schemes/uom/rhodes_scholars or contact Dr David Cookson (8344 2058).

ASSOCIATION OF SURGEONS OF GREAT BRITAIN AND IRELAND

Association of Surgeons of Great Britain and Ireland, 35-43 Lincoln's Inn Fields, London, WC2A 3PE, England
Tel: (44) 20 7973 0300
Fax: (44) 20 7430 9235
Email: admin@asgbi.org.uk
Website: www.asgbi.org.uk
Contact: Mrs Laura Andrews, Administrative Assistant

The founding objectives of the Association of Surgeons of Great Britain and Ireland, in 1920, were the advancement of the science and art of surgery and the promotion of friendship among surgeons. As other surgical specialities developed, the Association came to represent general surgery, encompassing breast, colorectal, endocrine, laparoscopic, transplant, upper gastrointestinal and vascular surgery.

Moynihan Travelling Fellowship

Subjects: General surgery.
Purpose: To enable specialist registrars or consultants to broaden their education, and to present and discuss their contribution to British or Irish surgery overseas.
Eligibility: Open to either specialist registrars approaching the end of their higher surgical training or consultants in general surgery within 5 years of appointment after the closing date for applications. Candidates must be nationals of and residents of the United Kingdom or the Republic of Ireland, but need not be Fellows or affiliate Fellows of the Association. They may be engaged in general surgery or a subspeciality thereof.
Level of Study: Postdoctorate
Type: Fellowship
Value: Up to UK£5,000
Frequency: Annual
Country of Study: Any country
No. of awards offered: 1
Application Procedure: Applicants must submit 12 copies of an application, which must include a full curriculum vitae giving details of past and present appointments and publications, a detailed account of the proposed programme of travel, costs involved and the object to be achieved. Applications must be addressed to the Honorary Secretary at the Association of Surgeons. For further details contact Bhavnita Borkhatria at bhavnita@asgbi.org.uk.
Closing Date: October 5th
Funding: Private
Contributor: Charitable association funds
No. of awards given last year: 1
No. of applicants last year: 6
Additional Information: Shortlisted candidates will be interviewed by the Scientific Committee of the Association, which will pay particular attention to the originality, scope and feasibility of the proposed itinerary. The successful candidate will be expected to act as an ambassador for British and Irish surgery and should therefore be fully acquainted with the aims and objectives of the Association of Surgeons and its role in surgery. After the completion of the fellowship, the successful candidate will be asked to address the Association at its annual general meeting and to provide a written report for inclusion in the Executive Newsletter. A critical appraisal of the centres visited should form the basis of the report. Please see the website for further details http://www.asgbi.org.uk/en/awards_fellowships/moynihan_travelling_fellowship.cfm.

ASSOCIATION OF UNIVERSITIES AND COLLEGES OF CANADA (AUCC)

Corporate Services and Scholarships Division, 350 Albert Street, Suite 600, Ottawa, ON, K1R 1B1, Canada
Tel: (1) 613 563 1236
Fax: (1) 613 563 9745
Email: awards@aucc.ca
Website: www.aucc.ca
Contact: Mr Luc Poulin

The AUCC is a non-profit, non-governmental association that represents Canadian universities at home and abroad. The Association's mandate is to foster and promote the interests of higher education in the firm belief that strong universities are vital to the prosperity and wellbeing of Canada.

Canadian Agri-food Policy Institute (CAPI) Award

Subjects: Agriculture.
Purpose: To elicit the views of graduate students on the future government policies or models that will help Canada thrive in the emerging agri-food world.
Eligibility: Open to Canadian citizens or permanent residents enrolled in graduate studies in one of the eligible institutions at the time of submitting application. Applicants must submit two signed reference letters from university professors, executive summary, position paper and proof of registration from the educational institution.
Type: Award
Value: One for $10,000 and two for $5,000 each. This marks the third and final year of the program, which will have awarded a total of $40,000.
No. of awards offered: 3
Application Procedure: Applicants must apply online via the organization website.
Closing Date: January 31st
Contributor: The Canadian Agri-Food Policy Institute (CAPI) and Farm Credit Canada (FCC).

Frank Knox Memorial Fellowships

Subjects: Arts and sciences including engineering, business administration, design, divinity studies, education, law, public administration at the John F. Kennedy School of Government, medicine, dental medicine and public health.
Purpose: To offer an opportunity to students from Canada who wish to graduate from the Harvard University.
Eligibility: Open to Canadian citizens or permanent residents who have recently graduated or who are about to graduate from an institution in Canada which is a member or affiliated to a member of the AUCC. Applications from students presently studying in the United States of America will not be considered, although applications will be considered from recent graduates who are working in the United States of America and will be applying to the MBA programme.
Level of Study: MBA, Postgraduate
Type: Fellowship
Value: Full Harvard tuition and mandatory health insurance fees and a stipend for the living expenses.
Length of Study: 10-month academic year. Up to two years of study at Harvard for students in degree programs requiring more than one year of study.
Frequency: Annual
Study Establishment: Harvard University
Country of Study: United States of America
No. of awards offered: Up to 3
Application Procedure: Applicants must apply directly to the graduate school of their choice. Applicants are responsible for gaining admission to Harvard University by the deadline set by the various faculties. Further information and application forms are available on request or from the website.
Closing Date: November 30th
Additional Information: Holders of this award may not accept any other grant for the period of this fellowship unless approved by the Committee on General Scholarships and the Sheldon Fund of Harvard University.

L'ORÉAL Canada For Women in Science Fellowships, With Support of the Canadian Commission for UNESCO

Subjects: Engineering/pure and applied sciences and life sciences.
Purpose: To support research by Canadian female scientists.
Eligibility: Open to female Canadian citizens or permanent residents who are already engaged in pursuing research at the postdoctoral level.
Level of Study: Doctorate, Postdoctorate, Postgraduate
Type: Fellowship
Value: Canadian $20,000 each
Length of Study: 1 academic year
Frequency: Annual
Country of Study: Canada
No. of awards offered: 2
Application Procedure: Further information and application forms are available on request or from the website.
Closing Date: April 2nd
Funding: Private
Contributor: L'ORÉAL Canada and Canadian Commission for UNESCO
No. of awards given last year: 2
Additional Information: Fellowships will alternate each year between the life sciences and engineering/pure and applied sciences.

Public Safety Canada Research Fellowship Program in Honour of Stuart Nesbitt White

Subjects: Research in Emergency Management. Preferred disciplines are regional planning, engineering, environmental studies, computer science, geography, sociology, economics and/or areas such as risk modeling and system science.
Purpose: To encourage PhD research in two areas of the PSC's mission, which is to enhance the safety and security of Canadians in their physical and cyber environments.
Eligibility: Open to Canadian citizens or permanent residents of Canada, who have completed their coursework (first year of PhD study) or applying to a PhD program not requiring coursework as they begin their research program. Candidates must intend to use the fellowship to assist them in completing a graduate degree (preferably a doctorate) that includes a thesis on a topic related to the fields of study, work must be carried out at an accredited university in Canada or abroad.
Level of Study: Doctorate
Type: Fellowship
Value: Canadian $19,250. Transportation and board will be paid in addition to the value of the award.
Length of Study: One year; however, award holders may re-apply annually by submitting a new application form.
Frequency: Annual
Study Establishment: Any accredited university in Canada or abroad
Country of Study: Canada
No. of awards offered: Up to 8
Application Procedure: Application can be downloaded from the website.
Closing Date: March 31st. Check website for exact information.
Funding: Government
Additional Information: Further information is available on request or from the website. Resources permitting, the award winners will attend a one day symposium in Ottawa in late spring/early summer to present information on the progress/results of their work. Attendees will include fellowship winners for the year, PSEPC representatives and other invited guests. Students cannot accept both this award and financial awards from other federal government programs (e.g. NSERC, SSHRC, or CIHR).

THE ASTHMA FOUNDATION OF NEW SOUTH WALES (AFNSW)

Level 3, 486 Pacific Highway, St Leonards, NSW, 2065, Australia
Tel: (61) 02 9906 3233
Fax: (61) 02 9906 4493
Email: training@asthmafoundation.org.au
Website: www.asthmansw.org.au
Contact: Executive Director

The vision of the Asthma Foundation of New South Wales (AFNSW) is to eliminate asthma as a major cause of illness and disruption within the New South Wales community. Fundraising efforts assist with the promotion and funding of research activities that are aimed at helping the Foundation to achieve this vision. The Foundation has been supporting and helping people with asthma since 1961. It is a registered charity committed to helping people with asthma, their families and careers.

Asthma Research Postgraduate Scholarships

Subjects: Medical, scientific, and clinical research into asthma, its causes, triggers, and impact.
Purpose: To expand the body of knowledge towards the causes of asthma and its possible cure.
Level of Study: Postgraduate, Postdoctorate
Type: Scholarship
Value: Australian $22,500
Length of Study: 1 year
Frequency: Annual, (dependent on funds available)
Country of Study: Australia
No. of awards offered: Varies
Application Procedure: Applicants must complete application forms. Short-listed candidates will be interviewed.
Closing Date: November 19th
Funding: Private
Contributor: Private donors
No. of awards given last year: 2
No. of applicants last year: 3

Asthma Research Project Grants

Subjects: Medical, scientific and clinical research into asthma, its causes, triggers and impact.
Purpose: To expand the body of knowledge towards the causes of asthma and its possible cure.
Level of Study: Postdoctorate, Postgraduate
Type: Project grant
Value: AU$50,000
Length of Study: 1 year
Frequency: Annual, (dependent on funds available)
Country of Study: Australia
No. of awards offered: Varies
Application Procedure: Applicants must complete application forms. Short-listed candidates will be interviewed.
Closing Date: November 13th
Funding: Private
Contributor: Private donors
No. of awards given last year: 4
No. of applicants last year: 11

ASTON UNIVERSITY

Aston University, Aston Triangle, Birmingham, B4 7ET, United Kingdom
Tel: (44) 0121 204 3000
Fax: (44) 0121 204 3696
Email: a.levey@aston.ac.uk
Website: www1.aston.ac.uk/homepage
Contact: Alison Levey, Academic Registrar

Aston university is a long-established research – CED university, known for its world-class teaching quality, strong links to industry, government and commerce, and its friendly and safe campus environment. Aston is currently ranking 1st outside London for graduate employment and is consistently featured in the top30 universities in the country.

ABS (Aston Business School) Home & EU Scholarships

Purpose: To assist students with tuition fees. Please note this is what has been offered in 2011 and we cannot confirm what will be available for 2013/14. Students should check our website: aston.ac.uk/abs
Eligibility: Scholarships are only available for students who are self-funded (i.e. not sponsored by an organisation). Applicable to students who are applying for full-time MSc programmes.
Level of Study: Postgraduate
Type: Scholarship
Value: Up to £3,000
Frequency: Annual
Study Establishment: Aston University
Country of Study: United Kingdom
Application Procedure: Essay in no more than 1,000 words, explaining why you think intercultural awareness and competence is important, and how you will make a positive contribution to creating a dynamic and inclusive student community in the context described above. Draw on theory and/or personal experience as appropriate. You must have applied for a full-time MSc with ABS – include student number with essay/
Closing Date: May 31st
No. of awards given last year: 8
No. of applicants last year: Unlimited
Additional Information: The Aston Business School Scholarship Panel will make their decision by June 30th. all successful applicants will be informed shortly afterwards. – The decision of the panel is final, – Aston Business School reserves the right to conduct a phone interview with suitable candidates.

For further information contact:

Email: abs-grad@aston.ac.uk

LHS (School of Life and Health Sciences) Postgraduate Masters Scholarships – Commonwealth Shared Scholarship Scheme

Subjects: MSc molecular toxicology and MSc psychology of health and illness.
Purpose: These scholarships are for students from developing Commonwealth countries who would not otherwise be able to study in the United Kingdom.
Eligibility: Applicants must hold an offer for MSc in molecular toxicology or MSc in psychology of health and illness.They must hold a undergraduate Bachelor's degree at either First/Upper Second class or equivalent (work experience cannot be accepted as an alternative). They must be a national of an eligible Commonwealth country and permanently living in that country. They must have the minimum English language requirement for the programme as no funding will be given for pre-sessional language programmes.They must not have studied for one year or more in a developed country previously.They must not be employed by a national government or an organisation owned or part-owned by the government (parastatal organisation) – higher education institutions are exempted from this restriction.
Level of Study: Postgraduate
Type: Scholarship
Value: These are fully funded scholarships and include living stipend
Frequency: Annual
Study Establishment: Aston University
No. of awards offered: 2
Application Procedure: Application form is available on the website.
Closing Date: May 1st
No. of awards given last year: 2
No. of applicants last year: 16

For further information contact:

School of Life and Health Sciences, Aston University, Aston Triangle, Birmingham, B4 7ET, United Kingdom
Email: c.m.hoban@aston.ac.uk
Contact: Postgraduate Admissions

LSS (School of Languages & Social Studies) Bursaries for MA Students (Home & EU)

Purpose: Cover cost of tuition fees.
Eligibility: Applicant should be a native speaker with competence in English. You are not eligible to apply for these scholarships if you already hold a Commonwealth Scholarship.
Level of Study: Doctorate
Type: Bursary
Value: Covering half of MA tuition fees.
Frequency: Annual
No. of awards offered: 3
Application Procedure: If you wish to be considered for the bursary, you must submit an essay of approximately 1,000 words along with your application on the following:

What do you consider to be your major strengths and weaknesses? How would you be able to support international students on programmes in the School of Languages and Social Sciences? Email your completed essay to lss_pgadmissions@aston.ac.uk.
Closing Date: June 24th
Additional Information: In return, the students will complete up to six hours a week pastoral work for the School.

LSS (School of Languages & Social Studies) Bursaries for MA Students (International)

Purpose: Contribution to cost of tuition fees.
Eligibility: You will be considered for this bursary if you:
Meet the entry requirements for your programme of study (Honours Degree and IELTS score – refer to the relevant course outline on website for full details); and submit a 2,000 word essay on the topic: How would the receipt of a scholarship contribute to your programme of study?
Level of Study: Doctorate
Type: Bursary
Value: £1,500 each
Frequency: Annual
No. of awards offered: 3
Application Procedure: Email your completed essay to lss_pgadmissions@aston.ac.uk.
Closing Date: April 29th
Additional Information: You are not eligible to apply for these scholarships if you already hold a Commonwealth Scholarship.

LSS (School of Languages & Social Studies) Postgraduate Scholarship for African & South American Students

Subjects: TESOL studies, translation studies, TESOL and translation studies and applied linguistics.
Purpose: To improve educational opportunities for students who would not otherwise be able to consider study in the UK.
Eligibility: The scholarships are available to students applying to on campus programmes offered in the School of Languages and Social Sciences.
Level of Study: Postgraduate
Value: £7,000 towards the cost of MA course fees
Frequency: Annual
Study Establishment: Aston University
Country of Study: United Kingdom
No. of awards offered: 9
Application Procedure: Write a two-page letter stating why you feel you deserve the scholarship, how you hope to benefit from the opportunity and the contribution you hope to make in your home country following the degree.
Submit a 2,000–3,000 word essay on one of the following topics:
What do you consider to be your major strengths and weaknesses? How might an Aston MA in TESOL, translation studies or applied linguistics help you address them and enhance your prospects?
What would you be able to contribute to other students in your group?
Describe a significant success or failure from your professional experience and explain what you discovered as a result.
Send confirmation of sponsorship or maintenance funds for your fees and living expenses.
You may send a confirmation letter from your sponsor or proof of available funds to cover your expenses whilst studying at Aston University.
Submit both pieces of work to lss_pgadmissions@aston.ac.uk.
Closing Date: May 1st
Contributor: Ferguson Foundation

School of Engineering and Applied Science Postgraduate International Scholarship Scheme

Subjects: Data communication networks, engineering management, industrial enterprise management, IT project management, mechanical engineering (modelling), product design enterprise, product design innovation, supply chain management telecommunications technology, software engineering and sensors and sensing systems.
Purpose: To assist with the cost of tuition fees.
Eligibility: The scholarships are merit-based and to be eligible an applicant should have or expect to achieve a First Class (Honours) Degree or equivalent, hold the required English Language qualification, and be able to demonstrate the potential for outstanding achievement.
Level of Study: Postgraduate
Type: Scholarship
Value: £5,000
Length of Study: 1 year
Frequency: Annual
Study Establishment: Aston University
Country of Study: United Kingdom
Application Procedure: To be considered for a scholarship, the applicant must have submitted a postgraduate application form and hold either a conditional or unconditional offer for one of the MSc courses listed above.
Closing Date: July 13th
Contributor: School of Engineering and Applied Science, Aston University

The Su Youn Scholarship (Korea)

Subjects: This award is for a student from South Korea taking a Postgraduate Taught Program. There is no subject restriction on which Postgraduate Taught Program is studied.
Purpose: To support students from South Korea studying at Aston University.
Eligibility: Applicants must be considered full overseas fee payers and nationals of South Korea who are coming directly from South Korea to study at Aston University; hold a conditional or unconditional offer for any postgraduate taught programme of study at Aston University; agree to work with Aston University's International Office to share with future students their experiences of their time at Aston and to assist with the promotion of Aston University's programmes where requested.
Level of Study: Postgraduate
Type: Scholarship
Value: £4,500 towards tuition fees
Length of Study: 1 year
Frequency: Dependent on funds available
Study Establishment: Aston University
Country of Study: United Kingdom
No. of awards offered: 1
Application Procedure: Suitable applicants must submit an application in the form of a 500-word essay which outlines why they would benefit from the scholarship, the difference it would make to them and explain the reasons why they would be a suitable person to share their experiences at the University and help promote Aston to future students.
Applications should be sent directly to the International Office by email to IntScholarships@aston.ac.uk with the subject title "Su Youn Scholarship". An email acknowledgement of receipt will be sent. For further information on scholarships, visit www.aston.ac.uk/international-students/finance.
Closing Date: July 29th
Funding: Individuals
No. of awards given last year: 1
Additional Information: Open to applicants from South Korea.

ATAXIA UK

Lincoln House, Kennington Park, 1-3 Brixton Road, London, SW9 6DE, England
Tel: (44) 20 7820 3900
Fax: (44) 20 7582 9444
Email: research@ataxia.org.uk
Website: www.ataxia.org.uk
Contact: Mrs Julie Greenfield, Research Projects Manager

Ataxia UK is the leading charity in the United Kingdom working with and for people with ataxia. It will support research projects and related activities in order to enhance scientific understanding of ataxia, develop and evaluate therapeutic and supportive strategies and encourage wider involvement with ataxia research.

Ataxia UK PhD Studentship

Subjects: Any aspect of both inherited and sporadic progressive ataxias including Friedreich's ataxia and other cerebellar ataxias.

Purpose: To further research into causes and treatments for prevention of progressive ataxias.
Eligibility: Proposals are accepted from academic institutions, private sector research companies and suitably qualified individuals. There are no restrictions on age, nationality or residency.
Level of Study: Predoctorate
Type: Project grant
Value: Covers stipend, tuition fees and contribution to consumables for project.
Length of Study: Varies
Frequency: Dependent on funds available
Country of Study: Any country
No. of awards offered: Varies
Application Procedure: Applicants must complete an application form available from Ataxia UK Research Projects Manager at research@ataxia.org.uk
Closing Date: Normally three times a year, refer to the website for details.
Funding: Commercial, individuals, private, trusts
Additional Information: There are a number of priority areas of research and these can be obtained from the Research Projects Manager.

Ataxia UK Research Grant

Subjects: Any aspect of both inherited and sporadic progressive ataxias including Friedreich's ataxia and other cerebellar ataxias.
Purpose: To further research into causes of and treatments for progressive ataxias.
Eligibility: Proposals are accepted from academic institutions, private sector research companies and suitably qualified individuals. There are no restrictions on age, nationality or residency.
Level of Study: Unrestricted
Type: Project grant
Value: Varies
Length of Study: Varies
Frequency: Dependent on funds available
Country of Study: Any country
No. of awards offered: Varies
Application Procedure: Applicants must complete an application form available from Ataxia's Research Projects Manager at research@ataxia.org.uk
Closing Date: Normally three times a year, refer the website for details
Funding: Private, commercial, individuals, trusts
No. of awards given last year: 11
Additional Information: There are a number of priority areas of research and these can be obtained from the Research Projects Manager.

Ataxia UK Travel Award

Subjects: Biology and life sciences, medicine and surgery. Any aspect of both inherited and sporadic progressive ataxias including Friedreich's ataxia and other cerebellar ataxias.
Purpose: To enable researchers to present their ataxia research at national and international conferences.
Eligibility: Proposals are accepted from academic institutions, private sector research companies and suitably qualified individuals. There are no restrictions on age, nationality or residency, although preference will be given to events which could potentially benefit patients and researchers in the UK or Europe.
Level of Study: Unrestricted
Type: Travel grant
Value: Dependent on the conference
Frequency: Dependent on funds available
Country of Study: Any country
No. of awards offered: Varies
Application Procedure: Applicants must complete an application form available from Ataxia's Research Projects Manager at research@ataxia.org.uk.
Closing Date: Normally three times a year. Contact Research Projects Manager for precise dates.
Funding: Commercial, individuals, private, trusts
No. of awards given last year: 1
No. of applicants last year: 3
Additional Information: There are a number of priority areas of research and these can be obtained from the Research Projects Manager.

ATHENAEUM INTERNATIONAL CULTURAL CENTRE

3 Adrianou Street, Athens, GR, 105 55, Greece
Tel: (30) 210 321 1987
Fax: (30) 210 321 1196
Email: contact@athenaeum.com.gr
Website: www.athenaeum.com.gr
Contact: Mrs Irene Mega, Executive Secretary

The Athenaeum International Cultural Centre is a non-profit association dedicated to preserving the memory of Maria Callas. The organization was founded in 1974 by a group of inspired artists who wanted to contribute to the development and evolution of musical education and culture in Greece.

Maria Callas Grand Prix International Music Competition

Subjects: Musical performance, although disciplines vary with each competition. singing such as opera, oratorio-lied and the piano.
Purpose: To recognize outstanding artists.
Eligibility: Open to musicians of all nationalities. Age restrictions: up to 30 years for women and 32 years for men.
Level of Study: Unrestricted, Postgraduate
Type: Competition
Value: Please contact the organization
Frequency: Annual
Country of Study: Any country
No. of awards offered: Varies according to the category
Application Procedure: Applicants must request for full details and application procedures or visit the website. There is a registration fee of €120.
Closing Date: December 15th
Funding: Government, private
Contributor: The Ministry of Culture and The Cultural Organization of the Municipality of Athens
No. of awards given last year: 6 in opera and 3 in piano
No. of applicants last year: 70 for singing and 52 for the piano competition
Additional Information: Concert appearances are arranged for the winners.

Paolo Montarsolo Special Prize

Subjects: Bass singing by males in the 2011 Maria Callas Grand Prix.
Purpose: To honour the best young bass singer of the 2011 Maria Callas Grand Prix.
Eligibility: Applicants should not be older than 32 years.
Type: Prize
Value: €2,000
Frequency: Every 2 years
No. of awards offered: 1
Application Procedure: Application form must be filled in.
Closing Date: December 15th
Funding: Government, private
Contributor: The Ministry of Culture and the Cultural Organization of the Municipality of Athens
Additional Information: Please refer to the website www.athenaeum.com.gr/english/grand for further details.

For further information contact:

Athenaeum, International Cultural Centre 3, Adrianou Str., Thission Athens, 105 55, Greece

ATLANTIC SALMON FEDERATION (ASF)

PO Box 5200, St Andrews, NB, E5B 3S8, Canada
Tel: (1) 506 529 4581
Fax: (1) 506 529 4438
Email: emerrill@asf.ca, savesalmon@asf.ca
Website: www.asf.ca/index.php
Contact: Ms Ellen Merrill, Executive Assistant

The Atlantic Salmon Federation (ASF) is an international, non-profit organization that promotes the conservation and management of the Atlantic salmon and its environment. ASF has a network of seven regional councils, a membership of over 150 river associations and 40,000 volunteers. Regional offices cover the salmon's freshwater range in Canada and the United States of America.

Affiliate of the year Award

Purpose: The Affiliate of the Year award recognizes outstanding leadership and achievement in wild Atlantic salmon conservation within the federation's affiliate structure.
Eligibility: Only regional council presidents, in consultation with their board of directors, are eligible to submit nominations.
Type: Award
Value: The winning organization receives an engraved plaque and a cheque for $500 to be used toward their conservation programs
Frequency: Annual
Closing Date: Contact organisation

ASF Roll of Honor

Purpose: The ASF Roll of Honor is presented annually to individuals who exhibit outstanding commitment to wild Atlantic salmon conservation at the grass-roots level. The award was initiated to provide a means of acknowledging the hard work that is performed on behalf of the Atlantic salmon by the many volunteers within the federation's affiliate network - the so-called unsung heroes who are the life blood of the salmon conservation movement.
Eligibility: Nominations are solicited from the federation's regional councils, affiliates and its membership at large. Candidates must be ASF members in good standing.
Frequency: Annual

Lee Wulff Conservation Award

Purpose: To award an individual who has made outstanding long-term contributions to Aild Atlantic salmon conservation. The award reflects efforts on a regional or national level rather than a local or river specific contribution.
Type: Award
Frequency: Annual
Application Procedure: Nominations are welcome from ASF directors and membership at large.

Olin Fellowship

Subjects: Salmon biology, management and conservation.
Purpose: To support individuals seeking to improve their knowledge or skills in advanced fields, while looking for solutions to current problems in Atlantic salmon biology, management and conservation.
Eligibility: Open to citizens and legal residents of the United States of America or Canada. Applicants need not be enrolled in a degree programme to be eligible.
Level of Study: Unrestricted
Type: Fellowship
Value: Canadian $1,000–3,000
Frequency: Annual
Study Establishment: Any accredited university, research laboratory or active management programme
Country of Study: United States of America or Canada
No. of awards offered: Varies
Application Procedure: Applicants must complete an application form, available on request or from the website. www.asf.ca/olin-fellowships.ps.html
Closing Date: March 15th
Funding: Private
Contributor: Memberships and foundation grants
No. of awards given last year: 2
No. of applicants last year: 2

T. B. (Happy) Fraser Award

Subjects: Practical conservation of the Atlantic salmon.
Purpose: The T. B. "Happy" Fraser Award was inaugurated in 1975 to honor Happy Fraser. Past President and General Manager of The Atlantic Salmon Association ASF Canada), in recognition of his many important contributions to Atlantic salmon conservation during his lifetime.
Type: Award
Frequency: Annual
No. of awards offered: 1

ATSUMI INTERNATIONAL SCHOLARSHIP FOUNDATION (AISF)

3-5-8 Sekiguchi, Bunkyo-ku, Tokyo, 112-0014, Japan
Tel: (81) 03 3943 7612
Fax: (81) 03 3943 1512
Email: office@aisf.or.jp
Website: www.aisf.or.jp/index.html

The Atsumi International Scholarship Foundation (AISF) was established to help promote a greater sense of Japan's role in the process of internationalization. Its goal is to provide financial assistance to foreign students with superior academic skills to further their studies at graduate schools in Japan.

AISF Scholarship

Subjects: All subjects with a focus on Japan.
Purpose: To help scholars to visit Japan and experience the diverse culture by not only developing a deeper academic knowledge of Japan but by cultivating a greater appreciation for its social richness as well.
Eligibility: Open to candidates of non-Japanese nationality who are PhD candidates enrolled in a graduate school in Kanto area.
Level of Study: Doctorate
Type: Scholarships
Value: ¥200,000 per month
Length of Study: 1 year
Frequency: Annual
Country of Study: Asia
No. of awards offered: 12
Application Procedure: Applicants can avail the application forms from the appropriate foreign students scholarship office school or from the AISF office.
Closing Date: September 30th

THE AUSTRALIA COUNCIL FOR THE ARTS

PO Box 788, Strawberry Hills, NSW, 2012, Australia
Tel: (61) 2 9215 9000
Fax: (61) 2 9215 9111
Email: mail@australiacouncil.gov.au
Website: www.australiacouncil.gov.au

The Australia Council for the Arts is the Australian Government's arts funding and advisory body. Each year, the council delivers more than $160 million in funding for arts organisations and individual artists across the country. Individuals, groups and organisations can apply to the Australia Council for funding. Individuals must be Australian citizens or have permanent resident status in Australia. All amounts are in Australian dollars.

Aboriginal and Torres Strait Islander Arts Fellowship

Subjects: Craft art, literature, performing arts, new media.
Purpose: These grants provide financial support for two years to Aboriginal and Torres Strait Islander artists to enable them to undertake a major creative project or program in their artform.
Eligibility: Open to practicing Aboriginal and Torres Strait Islander artists who are able to demonstrate at least 10 years experience as a practicing professional artist.
Level of Study: Postgraduate
Type: Fellowship
Value: Australian $45,000 per year
Length of Study: 2 years
Frequency: Annual
Country of Study: Australia
No. of awards offered: Varies
Application Procedure: Apply online. For further details mail to atsia@australiacouncil.gov.au.
Closing Date: November 19th
Funding: Government

Aboriginal and Torres Strait Islander Arts Presentation and Promotion

Purpose: Presentation and Promotion grants support projects that promote Aboriginal and Torres Strait Islander artists and their work regionally, nationally and internationally through publications, recordings, performances, exhibitions and international export.
Eligibility: Open to Aboriginal and Torres Strait Islander artists and community organisations and Aboriginal and Torres Strait Islander and non-indigenous arts organisations (including publishers).
Value: Up to a maximum of $10,000 for CD/DVD recording projects involving writing, recording, production, manufacture, distribution and promotion.
Frequency: Annual, Biannual
Application Procedure: Apply online. Please mail to atsia@australiacouncil.gov.au for further information.
Closing Date: November 19th
Additional Information: Overseas applicants for international projects must provide written evidence of co-funding from the host country or organisation. All applicants are encouraged to seek further funding support for the project from other sources.

Aboriginal and Torres Strait Islander Arts Skills and Arts Development

Purpose: These grants support Aboriginal and Torres Strait Islander artists, groups, organisations and accredited non-Indigenous organisations to develop their ideas and skills such as: mentorship programs, arts workshops, professional development programs, conferences, seminars or planning and development programs.
Eligibility: Applicants should be Aboriginal and Torres Strait Islander individuals, organisations or groups.
Value: Varies
Frequency: Biannual
Application Procedure: Apply online.
Closing Date: November 19th
Additional Information: Applicants are encouraged to seek funding from a number of sources. Please mail to atsia@australiacouncil.gov.au for further information.

Aboriginal and Torres Strait Islander New Work Grant

Subjects: Theatre production, writing, and music.
Purpose: These grants support Aboriginal and Torres Strait Islander artists, groups, organisations and accredited non-Indigenous organisations to create new work with an expected public outcome.
Eligibility: Open to Aboriginal and Torres Strait Islander artists who demonstrate artistic merit and innovation.
Level of Study: Professional development
Type: Grant
Value: Varies
Length of Study: Up to 12 months
Frequency: Annual
Country of Study: Australia
Application Procedure: Apply online. For further information mail to atsia@australiacouncil.gov.au.
Closing Date: November 19th
Funding: Government
Additional Information: All applicants are encouraged to seek further funding support for the project from other sources.

Art Fare: Australian Art Export

Purpose: These grants support the Australian contemporary visual arts sector, including public and private galleries and individuals, to attend international visual art, design and craft fairs.
Value: $4,000 (individuals) or up to $20,000
Application Procedure: Apply online.
Closing Date: August 30th
Additional Information: Please mail to artsdevelopment@australiacouncil.gov.au for further information.

Booked: Travel Fund for Publishers and Literary Agents

Purpose: This grants program provides international travel support for Australian publishers and literary agents to attend book fairs to promote Australian works and authors and generate international demand for Australian literature.
Type: Travel grant
Value: Up to $10,000
Application Procedure: Apply online.
Closing Date: No closing date – ongoing
Additional Information: Please mail to artsdevelopment@australiacouncil.gov.au for further information.

Career Pathways – Professional Development

Purpose: These grants provide funding for individual artists and arts workers to build their professional capacity as community arts and cultural development workers.
Value: Up to $12,000
Application Procedure: Apply online.
Closing Date: September 6th
Additional Information: Please mail to cp@australiacouncil.gov.au for further information.

Career Pathways – Structured Mentorship

Purpose: These new grants support tailored mentorship arrangements that build the skills and capacities of practitioners at different stages of their career in identified areas that contribute to the sustainability and quality of community arts and cultural development practice.
Value: $10,000–30,000
Application Procedure: Apply online.
Closing Date: September 6th
Additional Information: Please mail to cp@australiacouncil.gov.au for further information.

Chosen Cultural Apprenticeships or Residencies – Aboriginal and Torres Strait Islander Arts

Purpose: 'Chosen' allows Indigenous communities to take control and plan for how they will nurture early career people from their community/sector in the arts and/or culture.
Value: $10,000–20,000
Application Procedure: Apply online.
Closing Date: December 19th
Additional Information: Please mail to atsia@australiacouncil.gov.au for further information.

Cité Residency

Purpose: A three-month residency at the Australia Council's residential studio at the Cité Internationale des Arts in Paris is available for a community arts and cultural development artist or arts worker from March 24th to June 23rd.
Value: $10,000
Application Procedure: Apply online.
Closing Date: September 6th
Additional Information: Potential applicants are encouraged to discuss their application. For information and any questions relating to this grant, please contact to cp@australiacouncil.gov.au.

Community Partnerships - Projects

Subjects: Geographic, demographic, and social contexts.
Purpose: These grants provide funding for individuals, groups and organisations to develop and implement community arts and cultural development projects with a range of partners. These projects may or may not have a public outcome. Consideration of an evaluation strategy is recommended.
Eligibility: Open to individuals, groups and organisations.
Value: Up to $20,000
Length of Study: Varies
Frequency: Annual
No. of awards offered: Varies
Application Procedure: Apply online.
Closing Date: September 6th
Additional Information: Please mail to cp@australiacouncil.gov.au for further information.

For further information contact:

Tel: 02 9215 9034
Email: cp@australiacouncil.gov.au
Contact: Community Partnerships

Creative Australia – New Art (Creative Development)

Purpose: This initiative supports artists to develop ambitious experimental art projects and partnerships that intersect with broader culture, leading to the production of major new works.
Value: $30,000
Application Procedure: Apply online.
Closing Date: August 1st
Additional Information: Please mail to inter-arts@australiacouncil.gov.au for further information.

Creative Australia – New Work

Purpose: Creative Australia new work grants are for dance artists wishing to explore innovative ideas and collaborations to make major new dance works.
Value: Up to $50,000
Application Procedure: Apply online.
Closing Date: August 16th
Additional Information: Please mail to dance@australiacouncil.gov.au for further information.

Dance Artform Development

Subjects: Dance.
Purpose: To provide organisations with funding for programs and services that benefit a range of dance artists.
Eligibility: This category is open to Organisations only. Open to young and emerging artists and will be assessed as a discrete group.
Type: Grant
Length of Study: Up to 1 year
Frequency: Biannual
Application Procedure: Apply online.
Closing Date: August 16th
Additional Information: Please mail to dance@australiacouncil.gov.au for further information.

Don Banks Music Award

Purpose: This award honours a senior individual's outstanding contribution to music in Australia.
Type: Award
Value: $60,000
Application Procedure: Apply online.
Closing Date: September 17th
Additional Information: Please mail to music@australiacouncil.gov.au for further information.

The Dreaming Award

Purpose: This award supports a young artist aged 18–26 years to create a major body of work through mentoring and partnerships, either nationally or internationally.
Value: $20,000
Application Procedure: Apply online.
Closing Date: November 19th
Additional Information: Please mail to atsia@australiacouncil.gov.au for further information.

Experimental Art Grants

Purpose: These grants support artists, groups and organisations investigating experimental arts.
Eligibility: To be eligible, you must meet the general eligibility requirements and the specific eligibility requirements given below. This category is open to eligible artists, groups and organisations.
Length of Study: Up to 12 months
Application Procedure: Apply online. An assessment panel will assess applications according to the selection criteria above. The outcomes will be published in an Assessment Meeting Report.
Closing Date: October 4th
Additional Information: To keep this program as open as possible, Inter-Arts will consider any proposal from artists proposing to explore emerging or experimental arts. Please mail to inter-arts@australiacouncil.gov.au for further information.

Festivals Australia

Purpose: These grants support regional, remote and community festivals to present quality arts projects which have not been presented before, and would not be possible without financial support.
Eligibility: In order to apply you must be, or must apply through, a registered legal entity (with an ABN) or an incorporated organisation, which is able to produce an annual audited financial statement. Check website for complete details.
Application Procedure: Apply online.
Closing Date: September 6th
Additional Information: Please mail to artsdevelopment@australiacouncil.gov.au for further information. Potential applicants are encouraged to discuss their application. For information and any questions relating to this grants program.

For further information contact:

Tel: 02 9215 9176
Email: T.Kita@australiacouncil.gov.au
Contact: Tara Kita, Program Officer, Market Development

Going Global

Purpose: Going Global is a quick-response fund to support international touring of contemporary performing arts. It provides support for presentations, tours, and 'go-see' funding for international presenters to see Going Global supported works on tour.
Eligibility: To be eligible, you must meet the general eligibility requirements and the specific eligibility requirements given at the website. This category is open to individuals, groups and organisations. If an application is deemed ineligible, it will not be assessed.
Value: Up to $20,000 for a one-off presentation or up to $50,000 for a tour
Application Procedure: Apply online.
Closing Date: Applications must be received eight weeks prior to the project commencement date
Additional Information: Please mail to artsdevelopment@australiacouncil.gov.au for further information.

International Performing Arts Market (IPAMS) Travel Fund

Purpose: This grants program supports Australian participation at key international performing arts markets to build international audiences for Australian arts.
Eligibility: To be eligible, you must meet the general eligibility requirements and the specific eligibility requirements given at the website. This category is open to individuals, organisations and groups.
Value: $4,000
Application Procedure: You are encouraged to apply online for this grant. To begin an online application, use the 'Apply online' button in the right-hand column. Please select 'Market Development' when prompted to choose a board for your application. For other ways to apply, please see How to apply in Your application.
Closing Date: November 29th
Additional Information: Please mail to artsdevelopment@australiacouncil.gov.au for further information.

Languages Other Than English – Publishing Initiative

Purpose: This initiative supports the translation and publication of works by living Australian authors writing in languages other than English, including the publication of Indigenous titles in language.
Eligibility: To be eligible, you must meet the general eligibility requirements and the specific eligibility requirements given at the website.
Value: Up to $15,000
Application Procedure: Apply online.
Closing Date: October 1st
Additional Information: Please mail to literature@australiacouncil.gov.au for further information.

Literature Program

Subjects: Literature.
Purpose: These grants provide funding to established Australian organisations that support Australia's literary infrastructure.
Eligibility: Open to Australian organisations which meet the eligibility requirements and provide the necessary support materials.
Type: Programme

Value: Covers production, program and/or operational costs.
Length of Study: 1 year
Frequency: Annual
No. of awards offered: Varies
Application Procedure: Apply Online. Check website for further details. Applicant must provide evidence of his eligibility and all required support material by the application closing date. For further details contact the Program Officer.
Closing Date: October 1st
Additional Information: Please mail to literature@australiacouncil.gov.au for further information.

For further information contact:

Tel: 02 9215 9057
Email: l.byrne@australiacouncil.gov.au
Contact: Lucy Byrne, Program Officer, Literature

Music Skills and Development

Subjects: Music
Purpose: To support skills development for professional artists and arts workers.
Eligibility: Open to individuals and organisations which meet the eligibility requirements and provide the necessary support materials.
Value: Up to $10,000
Frequency: Biannual
No. of awards offered: Varies
Application Procedure: Apply online.
Closing Date: September 17th
Additional Information: There are two subcategories: individuals and groups (established and emerging), and organisations (legally constituted). Please mail to music@australiacouncil.gov.au for further information.

For further information contact:

Tel: 02 9215 9108
Email: p.keogh@australiacouncil.gov.au
Contact: Peter Keogh, Program Officer

New Work – Theatre

Purpose: New Work grants support one-off projects that create new theatre work. Creative development grants cover the stages of creating a new work, such as research, workshops or commissioning a writer. Production grants are for public performances of a new work and the stages leading up to it, where a presenting partner has already committed to a project. New Work productions can be new pieces, or new productions of existing works. Young artist initiative grants are available to artists aged 30 or under to create new work.
Eligibility: To be eligible, you must meet the general eligibility requirements and the specific eligibility requirements given at the website. This category is open to individuals, groups and organisations.
Value: Up to $35,000 for creative development, up to $15,000 for young artists and no set amount for productions
Application Procedure: Apply online.
Closing Date: August 5th
Additional Information: Please mail to theatre@australiacouncil.gov.au for further information.

New Work – Writing and Recording

Purpose: These grants are designed to support one-off projects that involve the creation or recording of music of all types. They support one-off projects that involve the creative development and recording of new Australian music, including sound art and media art.
Eligibility: To be eligible, you must meet the general eligibility requirements and the specific eligibility requirements given at the website. Only composers, songwriters, sound/media artists and those involved with the creation of the new artistic work for which funding is being requested can apply to this category. If you are collaborating with others to create new work for a group, you may nominate one of the collaborators to apply on behalf of the group.
Value: Up to $20,000
Application Procedure: Apply online.
Closing Date: November 18th
Additional Information: Please mail to music@australiacouncil.gov.au for further information.

OZCO Dance Fellowship

Subjects: Dance.
Purpose: This grant is designed to support an established dance artist to undertake creative or professional development.
Eligibility: This category is only open to individuals who are practising artists or arts workers. Applicant must meet the general eligibility requirements.
Level of Study: Professional development, Postdoctorate
Type: Fellowship
Value: Australian $50,000 per year
Length of Study: 2 years
Frequency: Annual
Country of Study: Australia
No. of awards offered: Varies
Application Procedure: Applicants are encouraged to apply online for this category.
Closing Date: August 16th
Funding: Government
Additional Information: Please contact to dance@australiacouncil.gov.au for further information.

For further information contact:

Tel: 02 9215 9164
Email: a.burnett@australiacouncil.gov.au
Contact: Adrian Burnett, Program Manager, Dance

OZCO Music Fellowship

Subjects: Music.
Purpose: These two-year fellowship support outstanding, established music artists to produce new work and/or undertake professional development.
Eligibility: This category is open to individuals who meet the general eligibility requirements. Fellowship recipients may not apply for a Project Fellowship where the start date of the Project Fellowship is less than five years after the end date of their Fellowship.
Level of Study: Postdoctorate, Professional development
Type: Fellowship
Value: Australian $100,000 ($50,000 per year for two years)
Length of Study: 2 years
Frequency: Annual
Country of Study: Australia
No. of awards offered: Varies
Application Procedure: Applicants are encouraged to apply online for this category.
Closing Date: September 17th
Funding: Government
Additional Information: Fellowship recipients may apply for funding from other categories during the term of the Fellowship, with the exception of the project Fellowships initiative. Fellowships are granted only once in an artist's lifetime.

For further information contact:

Tel: 02 9215 9115
Email: music@australiacouncil.gov.au
Contact: Andy Rantzen, Program Officer, Music

OZCO Music Project Fellowship

Subjects: Music.
Purpose: These grants support mid-career and established artists to develop significant creative and/or developmental projects over a period of up to 12 months.
Eligibility: Music artists working in music theatre and indigenous music artists are particularly encouraged to apply.
Level of Study: Postgraduate, Professional development
Type: Fellowship
Value: Australian $30,000
Length of Study: 1 year
Frequency: Annual
Country of Study: Australia
No. of awards offered: Varies
Application Procedure: Apply online.
Closing Date: September 17th
Funding: Government
Additional Information: Please mail to music@australiacouncil.gov.au for further information.

For further information contact:

Tel: 02 9215 9108
Email: music@ozco.gov.au; p.keogh@australiacouncil.gov.au
Contact: Peter Keogh, Program Officer, Music

Playing Australia
Purpose: These grants assist the touring of professionally produced performing arts across Australia, including regional and remote areas, where there is a demonstrated public demand and tours are otherwise not commercially viable.
Eligibility: This category is open to organisations only.
Application Procedure: Apply online. You are encouraged to apply online for this grants program. To begin an online application, use the 'Apply online' button in the right-hand column. Please select 'Market Development' when prompted to choose a board.
Closing Date: December 2nd
Additional Information: Please mail to artsdevelopment@australiacouncil.gov.au for further information.

For further information contact:

Tel: 02 9215 9176
Email: T.Kita@australiacouncil.gov.au
Contact: Tara Kita, Program Officer, Market Development

Projects – Creative Development
Purpose: These grants provide support for the research and creative development of new dance works.
Eligibility: To be eligible, you must meet the general eligibility requirements and the specific eligibility requirements given below. This category is open to individuals, groups and organisations. Dance Key Organisations are not eligible to apply to this category.
Application Procedure: You are encouraged to apply online for this grant. To begin an online application, use the 'Apply online' button in the right-hand column. Please select 'Dance' when prompted to choose a board for your application. For other ways to apply, please see How to apply in Your application.
Closing Date: August 16th
Additional Information: Please mail to dance@australiacouncil.gov.au for further information.

For further information contact:

Tel: 02 9215 9179
Email: k.morcombe@australiacouncil.gov.au
Contact: Kiri Morcombe, Program Officer, Dance

Projects – Interconnections
Purpose: As the result of research undertaken into the sector, the Major Performing Arts Board has established a fund of $120,000 for 2012/13 to support a program of co-commissions and innovative collaborations between Major Performing Arts organisations and Key Organisations, or Major Performing Arts organisations and small to medium arts organisations.
Eligibility: This category is open by inivation only. Only Major Performing Arts organisations are eligible to apply for collaborations between Major Performing Arts organisations and Key Organisations, or Major Performing Arts organisations and small to medium arts organisations.
Value: A grant for up to $30,000 to forge new relationships
Application Procedure: You must submit your application via the Australia Council online applications page.
Closing Date: No closing date
Additional Information: Please mail to artsorganisations@australiacouncil.gov.au for further information.

Projects – Presentation
Subjects: Dance.
Purpose: The purpose of this grant is to provide support for dance works with a presentation outcome. This can include final stage creative development and presentation/s and remounts of dance works.
Eligibility: To be eligible, you must meet the general eligibility requirements and the specific eligibility requirements given at the website. This category is open to individuals, groups and organisations. Key Organisations are not eligible to apply to this category.
Application Procedure: You are encouraged to apply online for this grant. To begin an online application, use the 'Apply online' button in the right-hand column. Please select 'Dance' when prompted to choose a board for your application. For other ways to apply, please see How to apply in Your application.
Closing Date: August 16th
Additional Information: Please mail to dance@australiacouncil.gov.au for further information.

For further information contact:

Tel: 02 9215 9179
Email: k.morcombe@australiacouncil.gov.au
Contact: Kiri Morcombe, Program Officer, Dance

Projects with Public Outcomes
Purpose: These grants provide funding for community arts and cultural development projects that have strong public presentation outcomes and involve both arts and non-arts partners. Individuals, groups and organisations are welcome to apply provided they have sought financial support and/or co-funding from other sources.
Eligibility: To be eligible, you must meet the general eligibility requirements and the specific eligibility requirements given at the website. This category is open to individuals, groups and organisations.
Value: $20,000–35,000
Application Procedure: You are encouraged to apply online for this grant. To begin an online application, use the 'Apply online' button in the right-hand column. Please select 'Community partnerships' when prompted to choose a board for your application. For other ways to apply, please see How to apply in Your application.
Closing Date: September 6th
Additional Information: Please mail to cp@australiacouncil.gov.au for further information.

Promotion – Literature
Purpose: This grant supports the promotion of Australian literature and activities that foster greater awareness and appreciation of Australian creative writing.
Eligibility: These grants are for organisations only. If the activities are taking place overseas, the overseas organisation must submit the application.
Application Procedure: You are encouraged to apply online for this grant. To begin an online application, use the 'Apply online' button in the right-hand column. Please select 'Literature' when prompted to choose a board for your application. For other ways to apply, please see How to apply in Your application.
Closing Date: October 1st
Additional Information: Please mail to literature@australiacouncil.gov.au for further information.

The Red Ochre Award
Purpose: The Aboriginal and the Torres Strait Islander Arts Board established The Red Ochre Award in 1993 to pay tribute to an Aboriginal or Torres Strait Islander artist who, throughout their lifetime, has made outstanding contributions to the recognition of Aboriginal and Torres Strait Islander arts, both nationally and internationally.
Eligibility: Nominations will be accepted from arts and community organisations and individuals. Nominations may only be made for living artists and individuals cannot nominate themselves.This award is not project based and, therefore is not given to assist any particular project, program or intended activity.
Value: $50,000
Application Procedure: Apply online. Contact organisation for details.
Closing Date: November 19th
Additional Information: Please mail to atsia@australiacouncil.gov.au for further information.

For further information contact:

Tel: 02 9215 9067
Email: f.trotman-golden@australiacouncil.gov.au
Contact: Frank Trotman-Golden, Program Officer

Skills and Arts Development – Residencies

Purpose: Skills and Arts Development – Residency grants enable professional development opportunities for craftspeople, designers, media artists and visual artists. The residencies are available in set blocks of time and are located in Barcelona, Berlin, Helsinki, Liverpool, London, New York, Paris, Rome and Tokyo.
Eligibility: To be eligible, you must meet the general eligibility requirements and the specific eligibility requirements given at the website. This category is open to individuals and groups of artists only.
Value: $10,000, $25,000 or $35,000
Application Procedure: You are encouraged to apply online for this grant. To begin an online application, use the 'Apply online' button in the right-hand column. Please select 'Visual arts' when prompted to choose a board for your application. For other ways to apply, please see How to apply in Your application. Check more details at website.
Closing Date: August 19th
Additional Information: Please mail to visualarts@australiacouncil.gov.au for further information.

Visions of Australia

Purpose: Visions of Australia supports the development and touring of major public exhibitions of Australian cultural material throughout Australia, particularly into regional and remote areas.
Eligibility: To be eligible, you must meet the general eligibility requirements and the specific eligibility requirements given at the website. This category is open to organisations only.
Application Procedure: You are encouraged to apply online for this grant. To begin an online application, use the 'Apply online' button in the right-hand column (http://www.australiacouncil.gov.au/grants/2013/visions-of-australia). Please select 'Market Development' when prompted to choose a board for your application. For other ways to apply, please see How to apply in Your application.
Closing Date: December 2nd
Additional Information: Please mail to artsdevelopment@australiacouncil.gov.au for further information. Potential applicants are encouraged to discuss their application. For information and any questions relating to this grants program, please contact Tata Kita.

For further information contact:

Tel: 02 9215 9176
Email: t.kita@australiacouncil.gov.au
Contact: Tara Kita, Program Officer, Market Development

Visual Arts Skills and Arts Development

Subjects: Visual arts.
Purpose: To enable professional development opportunities for craftspeople, designers, media artists, visual artists, arts writers and curators.
Eligibility: Open to individuals and groups which meet the eligibility requirements and provide the necessary support materials.
Value: Supports artists to undertake professional development activities in Australia or overseas. $10,000 for Barcelona, Helsinki, Liverpool, London, New York, Paris, Rome or Tokyo; $25,000 for New York; $35,000 for Berlin residencies.
Length of Study: More than 1 year (General). 3-month residency in Barcelona, Helsinki, Liverpool, London, New York, Paris, Rome or Tokyo; 6-month residency in New York; 12-month residency in Berlin.
Frequency: Annual
No. of awards offered: Varies
Application Procedure: Apply online.
Closing Date: August 21st
Additional Information: Please mail to visualarts@australiacouncil.gov.au for further information.

For further information contact:

Tel: 02 9215 9336
Email: r.petersen@australiacouncil.gov.au
Contact: Romany Petersen, Program Officer, Visual Arts

AUSTRALIAN ACADEMY OF THE HUMANITIES (AAH)

GPO Box 93, Canberra, ACT, 2601, Australia
Tel: (61) 2 6125 9860
Fax: (61) 2 6248 6287
Email: enquiries@humanities.org.au
Website: www.humanities.org.au
Contact: Administration Officer

The Australian Academy of the Humanities (AAH) was established under Royal Charter in 1969 for the advancement of the scholarship, interest in and understanding of the humanities. Humanities disciplines include, but are not limited to, history, classics, English, European languages and cultures, Asian studies, philosophy, the arts, linguistics, prehistory and archaeology and cultural and communications studies.

AAH Humanities Travelling Fellowships

Subjects: Humanities disciplines as per the Academy's charter.
Purpose: To enable short-term study abroad.
Eligibility: Open to scholars resident in Australia and who are working in the field of humanities. Fellows of the Academy are ineligible for awards. Preference shall be given to scholars in the earlier stages of their careers, and who are not as well placed to receive funding from other sources. They should have a project going forward that requires a short visit overseas for its completion or advancement. The proposed work should not form part of the requirement for a higher degree. Funds are not given for conference attendance. Country of study is any country except Australia.
Level of Study: Postdoctorate, Doctorate
Type: Fellowship
Value: Australian $4,000 each
Length of Study: At least 2 weeks
Frequency: Annual
Study Establishment: An appropriate research centre
Country of Study: Any country
No. of awards offered: 10
Application Procedure: Applicants can download application form and guidelines from the academy website www.humanities.org.au
Closing Date: End of July
Funding: Government
No. of awards given last year: 11
No. of applicants last year: 26
Additional Information: Please visit the Academy's website, www.humanities.org.au for detailed information and application procedures. Source of funding: Bequests.

AAH Visiting Scholar Programmes

Subjects: Arts and humanities.
Purpose: To encourage scholarly contact with scholars from both Russia/the former USSR and Indonesia/South–East Asia and to assist scholars from those countries to obtain access to research materials held in Australia.
Eligibility: Applicants must be identified as being appropriate representatives at Australia-based conferences.
Level of Study: Doctorate, Postdoctorate
Type: Award
Value: $7,000 (for 2 scholars from Russia and the Former USSR) and $4,000 (for 2 scholars from Indonesia and South–East Asia)
Frequency: Annual
Country of Study: Australia
No. of awards offered: 4
Application Procedure: Applicant (Australian host scholar) must send the Secretariat a brief explanation of the reason for the visit, a copy of the visiting scholar's curriculum vitae and a list of their most significant publications (in English), a provisional itinerary listing speaking engagements, potential contact with Australian scholars and research institutions to be visited and a provisional budget for the expenditure of the funds.
Closing Date: July 31st
Funding: Government
No. of awards given last year: 2
No. of applicants last year: 7
Additional Information: Eligible to nationals of: Russia and SE Asia.

For further information contact:

Tel: 02 6125 8950
Email: grants@humanities.org.au
Contact: Jorge Salavert

Bilateral Exchange Programme

Subjects: Humanities.
Purpose: In 2011, the Australian Academy of the Humanities and the Accademia Nazionale dei Lincei signed a new agreement to foster research collaboration between humanities scholars from both countries.
Type: Programme
Value: Up to $3,000 towards an economy class airfare to Italy and the Accademia Nazionale dei Lincei will disburse an all-inclusive allowance of €1,500 for the month of the visit
Application Procedure: The successful recipient will be selected from each country on alternating years, with Australia selecting the first exchange scholar. Please read the Rules of Award carefully before submitting your application.
Closing Date: Please check website
Contributor: The Australian Academy of the Humanities

The Crawford Medal

Subjects: Humanities.
Purpose: The Max Crawford Medal recognises outstanding achievement in the humanities by young Australian scholars currently engaged in research, and whose publications contribute towards an understanding of their discipline by the general public.
Type: Award
Frequency: Every 2 years
No. of awards offered: 1
Application Procedure: The award of the Max Crawford Medal is based on a nomination process. The Rules of the Award should be consulted before nominations are prepared. Please consult the Rules of Award, available to download at http://www.humanities.org.au/Grants/CrawfordMedal.aspx, before submitting a nomination.
Closing Date: July 31st
No. of awards given last year: 1

The McCredie Musicological Award

Purpose: The McCredie Musicological Award is funded through a bequest to the Academy from Professor Andrew McCredie FAHA.
Eligibility: The award is granted on the basis of work which leads to outstanding contribution in musicology. Recipients need to be Australia-based and their work must have been conducted in an Australian university. The scholar must be under forty years of age at the closing date of the call for nominations in the year when the award is made.
Type: Award
Application Procedure: The award of the McCredie Medal is based on a nomination process. Please consult the Rules of the Award (available for download at www.humanities.org.au/Grants/McCredie-MusicologicalAward.aspx) before submitting a nomination.
Closing Date: July 31st
Additional Information: For further information on the McCredie Medal or to subscribe to the Grants Mailing List, please contact the Programme Manager at grants@humanities.org.au or call the Academy on (02) 6125 9860.

Publication Subsidy Scheme

Subjects: Humanities.
Purpose: The scheme is designed to assist humanities scholars based in Australia. Both independent scholars and those working within an institution are eligible to apply.
Eligibility: Applications are assessed by the Academy's Awards Committee and rated according to academic merit and demonstrated need for a subsidy. Priority is given to works that require a subsidy for their viability as a publishing venture, or for the inclusion of essential items such as illustrations, photographs or maps. Applicants need to prove that they have limited access to funds to complete publication of the work, or that without the subsidy the work would go either unpublished or be published in a diminished capacity. Please note that a Publication Subsidy cannot be awarded to books that are published before the Awards Committee makes its decision.
Value: Up to $3,000
Application Procedure: Applicants must consult the Rules of the Award before applying. These will be available at website when the programme opens. Applications must include all of the following: a cover sheet, signed and dated by the applicant; a completed Application Form; a synopsis of the project (3–5 pages); an assessment of the work by a recognised scholar in the field; and correspondence from the publisher demonstrating support for the work; a publisher's reader's report.
Additional Information: For further information on the Publication Subsidy Scheme, or to subscribe to the Grants Mailing List, please contact the Programme Manager at grants@humanities.org.au or call the Academy on (02) 6125 9860.

AUSTRALIAN BIO SECURITY-CRC (AB-CRC)

Brisbane, OLD Building 76, Molecular Biosciences, The University of Queens land, St Lucia, QLD 4072, Australia
Tel: (61) (0) 7 3346 8866
Fax: (61) (0) 7 3346 8862
Email: corinna.lange@abcrc.org.au
Website: www1.abcrc.org.au
Contact: Mrs Corinna Lange, Communications Manager

The mission of the ABCRC is to protect Australia's public health, livestock, wildlife and economic resources through research and education that strengthens the national capability to detect, diagnose, identify, monitor, assess, predict and to respond to emerging infectious disease threats.

AB-CRC Honours Scholarships

Subjects: Biosecurity and emerging infectious diseases.
Purpose: To encourage students of high academic ability to take the first step in their career path as a researcher. To build research capacity in high priority areas related to biosecurity.
Eligibility: Scholarships will be awarded preferentially to Australian residents and students from the Asia-Pacific region.
Level of Study: Postgraduate
Type: Scholarship
Value: Australian $5,000 per year (full-time) or $2,500 per year (part-time).
Length of Study: 1 year or 2 years for a part-time scholarship
Frequency: Annual
Study Establishment: AB-CRC participating university
Country of Study: Australia
Application Procedure: Contact the scholarships Administrator officer.
Closing Date: October 31st
For further information contact:

Tel: 08 9266 1634
Email: debra.gendle@abcrc.org.au
Contact: Debra Gendle

AB-CRC PhD and Masters-by-Research Scholarships

Subjects: Bio security and emerging infectious diseases.
Purpose: To expand training opportunities.
Level of Study: Postgraduate, Doctorate, Postdoctorate
Type: Scholarship
Value: An annual stipend of up to $25,000 and a development allowance of $2,000. Masters scholarships of up to $22,500 per year.
Length of Study: 1–3 years
Frequency: Annual
Study Establishment: AB-CRC participating university
Country of Study: Australia
No. of awards offered: Varies
Application Procedure: Request application form.
Closing Date: October 24th
Contributor: Curtin University of Technology, James Cook University, Murdoch University, The University of Queensland and The University of Sydney.

AB-CRC Professional Development Scholarships

Subjects: Bio Security and emerging infectious diseases.
Purpose: To enhance linkages with research projects of relevance to the AB-CRC. To expand our capability to support the training of specialists. To provide students with access to the AB-CRC network and enhanced learning opportunities.
Eligibility: PhD students enrolled at AB-CRC partner organisations.
Level of Study: Professional development
Type: Scholarship
Value: Australian $2,000 per year.
Length of Study: Varies
Frequency: Annual
Study Establishment: AB-CRC participating university
Country of Study: Australia
No. of awards offered: 5
Application Procedure: Students must submit a Professional Development Plan. Contact the Scholarships Administration officer.
Closing Date: November 16th
Additional Information: Candidates will be required to sign a confidentially agreement. Funding awarded in a Professional Development Scholarship will be on a sliding scale depending upon the student's enrolment date.

For further information contact:

Email: debra.gendle@abcrc.org.uk
Contact: Debra Gendle

AUSTRALIAN CATHOLIC UNIVERSITY (ACU)

PO Box 456, Virginia, Brisbane, QLD, 4014, Australia
Tel: (61) 07 3623 7100
Email: futurestudents@acu.edu.au
Website: www.acu.edu.au

ACU National shares with universities worldwide a commitment to quality in teaching, research, and community engagement and to provide an excellent higher education experience for the entire student body, through upholding the principles of free enquiry, and academic integrity. The University strives to create an environment of support and challenge, where intellectual engagement and cognitive development occur as part of a deeper concern to help people learn intellectually, spiritually, culturally, and socially. It provides excellent higher education for its entire diversified and dispersed student body. Through fostering and advancing knowledge in education, health, commerce, the humanities, the sciences and technologies, and the creative arts, Australian Catholic University seeks to make a specific contribution to its local, national, and international communities.

The ACU Business Foundation Scholarship for Postgraduate Studies

Eligibility: Eligibility will be on the grounds of academic achievement, professional experience and a commitment to making a positive contribution to self and society.
Type: Scholarship
Value: Tuition fee to the value of Australian $200 per unit of study. The total value of the scholarship is Australian $2,400
Length of Study: 12 units
Application Procedure: After reviewing the Terms and Conditions of the scholarship at http://www.acu.edu.au/about_acu/faculties,_institutes_and_centres/business/for_prospective_students/scholarships/business_foundation_scholarship, click on the Apply Now link at the bottom left of the page to access the online application form.Select the ACU Business Foundation Scholarship for Postgraduate Studies from the drop-down menu and complete all required fields. If you have questions or difficulties with your application, please email at acub.postgrad@acu.edu.au and include your preferred number for telephone contact. You may also send the required documentation to this.

ACU National Equity Bursaries

Purpose: These bursaries are intended to financially assist students from low socio-economic backgrounds and are a once only bursary valued at $2,000.
Type: Bursary
Frequency: Annual
No. of awards offered: 20

Anne Lyons Memorial Travel Scholarship (Postgraduate)

Eligibility: Open to any postgraduate student in the final year of study on any campus of the University. The Scholarship may be in the form of either a travelling fellowship for one semester to carry out relevant research into an area of social justice, or a visiting fellowship for a suitable local or overseas scholar to attend Australian Catholic University for one semester.
Type: Travel award
Value: Up to $6,000
Frequency: Annual
No. of awards offered: 1
Application Procedure: Applications should include details of the applicant's research fellowship/placement in the area of social justice, including evidence that the applicant has been accepted into this placement; a budget estimate outlining airfares and accommodation expenses; a copy of the applicant's academic transcript; and two referee reports, one academic from the Head of School or Course Coordinator and one personal.
No. of awards given last year: 1

Archdiocese of Brisbane Theology Scholarships

Subjects: Theology.
Purpose: To provide support and encouragement for postgraduate students undertaking higher degree studies in Theology at the University's Brisbane campus.
Level of Study: Postgraduate
Type: Scholarship
Value: Up to Australian $2,500
Frequency: Annual
Study Establishment: Australian Catholic University's Brisbane campus
Application Procedure: Candidates can obtain further information from the Course Co-ordinator. Online applications only.
Closing Date: March 17th
Contributor: Archdiocese of Brisbane

For further information contact:

Website: www.acu.edu.au/scholarships

Australian Catholic Superannuation Retirement Fund - Honours and Postgraduate Scholarship

Eligibility: Open to students who have completed undergraduate studies through any Faculty at Australian Catholic University (ACU) and who have secured a place in an honours program or a postgraduate course in the Faculty of Education at ACU.
Type: Scholarship
Value: This scholarship has a total value of $5,000, which will be paid in two instalments
Frequency: Annual
No. of awards offered: Up to 4
Application Procedure: Applications must include evidence of academic merit, by having achieved a minimum average of Distinction in the studies on which the applicant's eligibility for the scholarship is based; and a report from an academic and/or professional supervisor commenting on the quality and originality of the applicant's academic and/or professional work.
Additional Information: Scholarship recipients must be enrolled in the course at each semester census date to be eligible for the scholarship payment.

Australian Catholic University Foundation - Social Work Rural Placement Scholarship

Purpose: This scholarship established in 2006 to provide financial support for social work students, towards the costs of undertaking a rural or international placement.
Eligibility: Open to students currently enrolled in the Bachelor or Master of Social Work at the Canberra campus (Signadou) of the University who wish to undertake a rural or international placement, with priority being given to rural placements.
Type: Scholarship
Value: Up to $1,500

Frequency: Annual
No. of awards offered: 1 or more
Application Procedure: Applications should include evidence of academic merit; and an explanation of how the Scholarship could financially support their intended program of study.

Australian Catholic University International Student Scholarship

Eligibility: Applicants must be a citizen of a country other than Australia or New Zealand; have gained and formally accepted a place in an undergraduate or postgraduate program at Australian Catholic University (ACU); attend at the relevant ACU campus as a full-time student; not be in receipt of another major scholarship or award; not be attending ACU as a Study Abroad or Exchange student.
Level of Study: Postgraduate, Research, Undergraduate
Type: Scholarship
Frequency: Annual
No. of awards offered: Up to 22
Application Procedure: Applications should include details of the applicant's course of study at ACU; evidence of academic merit including achievement of an average of at least 80 per cent across the applicant's secondary study (for undergraduate students) or undergraduate study (for postgraduate students), whether at ACU or elsewhere.
Additional Information: Two Higher Degree by Research scholarships equivalent to the full-time fees and stipend for the recipient's enrolling in a Masters by Research or Doctoral course of study; and 20 undergraduate and postgraduate scholarships equivalent to one-half of the full-time fees for the recipient's course of study. Each scholarship will be awarded for the duration of the recipient's course, subject to satisfactory academic progress.

Bob and Margaret Frater Travel Scholarship

Purpose: To recognise and reward those teachers within Catholic primary schools who display leadership qualities and commitment and who will contribute to the continuing development and enhancement of the school system through participation in an international experience.
Eligibility: Open to persons who are enrolled in, or have completed, studies at Australian Catholic University (or one of its predecessor colleges) and who are working in the teaching profession in a New South Wales or Australian Capital Territory Catholic primary school.
Type: Scholarship
Value: Up to $10,000
Frequency: Annual
No. of awards offered: 1
Application Procedure: Applications should include a proposal of approximately 1,200 words identifying a professional development activity (i.e. conference or other activity) for which the applicant seeks support and evidence of acceptance to participate in that event; a budget outline of estimated expenses relating to the applicants attendance at his/her chosen conference or activity; evidence of postgraduate study (either completed or currently being undertaken); and demonstrated commitment to using their qualifications to work for the betterment of the educational community.
Additional Information: Shortlisted applicants will be required to attend an interview prior to a final decision being made.

Co-op Bookshop Scholarship

Eligibility: Open to students from rural or regional areas enrolled in any course at any campus at Australian Catholic University who are Australian citizens or permanent residents.
Type: Scholarship
Value: The scholarship has a value of $1,000 in a gift-in-kind, Co-op Bookshop Book Voucher which has an 18-month expiry date
Frequency: Annual
No. of awards offered: 1

Commonwealth Accommodation Scholarship

Type: Scholarship
Value: $4,754
Frequency: Annual
Application Procedure: Applicants must be of Australian Aboriginal or Torres Strait Islander descent and identify as an Australian Aboriginal or Torres Strait Islander; be enrolled or about to commence study in an undergraduate course or a graduate diploma (or equivalent postgraduate course of study) in an area of National Priority; be enrolled as a Commonwealth supported student for that course of study; must not have completed the requirements of a course of study equivalent to or higher than an Australian bachelors award unless such an award is a prerequisite to your current undergraduate course of study or your current course is in the area of National Priority; demonstrate financial hardship; and be enrolled, or about to commence study as a full time student (exceptional circumstances may be considered for part time study).
Additional Information: You may apply for a Commonwealth Accommodation Scholarship if you are enrolled in one of the following course types: all Bachelor and Honours degrees; Graduate Diploma in Education (Secondary); Graduate Diploma in Midwifery; Master of Teaching (Primary) and Master of Teaching (Secondary). For further details on the selection criteria and to apply visit at http://www.uac.edu.au/equity.

Commonwealth Education Costs Scholarship

Eligibility: Applicants must be of Australian Aboriginal or Torres Strait Islander descent and identify as an Australian Aboriginal or Torres Strait Islander; be enrolled or about to commence study in an undergraduate course or a graduate diploma (or equivalent postgraduate course of study) in an area of National Priority; be enrolled as a Commonwealth supported student for that course of study; must not have completed the requirements of a course of study equivalent to or higher than an Australian bachelors award unless such an award is a prerequisite to your current undergraduate course of study or your current course is in the area of National Priority; demonstrate financial hardship; and be enrolled, or about to commence study as a full time student (exceptional circumstances may be considered for part time study).
Type: Scholarship
Value: $2,377 per year
Frequency: Annual
Additional Information: You may apply for a Commonwealth Education Costs Scholarship if you are enrolled in one of the following course types: all Bachelor and Honours degrees; Graduate Diploma in Education (Secondary); Graduate Diploma in Midwifery; Master of Teaching (Primary) and Master of Teaching (Secondary). For further details on the selection criteria and to apply visit at http://www.uac.edu.au/equity/.

Council of Catholic School Parents (NSW) Indigenous Postgraduate Scholarship (IES)

Subjects: Education.
Purpose: To encourage involvement of parents and the community in education.
Eligibility: Open to Indigenous students enrolled in a postgraduate course within the faculty of education at the Strathfield campus, and have a particular focus, interest, or understanding of the importance of parent and community involvement in education.
Type: Scholarship
Value: Australian $1,000 and a certificate
Frequency: Annual
Study Establishment: Strathfield campus
Country of Study: Australia
Application Procedure: Candidates can obtain further information from Yalbalinga Indigenous Unit, Strathfield campus. Online applications only.
Closing Date: March 18th
Contributor: Council of Catholic School Parents (NSW)

For further information contact:

Website: www.acu.edu.au/scholarships

Coursework scholarship: Padnendadlu Postgraduate Scholarship

Eligibility: Applicants must be Aboriginal and/or Torres Strait Islander women; must be enrolled in a Masters or Doctorate by coursework at a South Australian university; must not be employed full-time, or on fully-paid study leave; must not have received an AFUW-SA Scholarship in this category in 2012.
Type: Scholarship
Value: Up to Australian $4,000 each

Frequency: Annual
No. of awards offered: At least one
Application Procedure: Selection of winners is based primarily on academic merit, but also on financial need, the importance of the purpose for which the bursary will be used to the progress or completion of the degree, community activities and other interests. Your application must reach the Trust Fund's secretarial service by deadline.
Closing Date: March 1st
Funding: Individuals, private
Additional Information: Receipt of all applications will be acknowledged by letter. In the normal course of events the successful applicants will be notified by the last day of June in the year of the award. Subsequently, the list of winners will be posted on this website in early July. The decision of the selection committee will be final and no correspondence will be entered into.

Dooleys Lidcombe Catholic Club Postgraduate Scholarships

Eligibility: Open to commencing students enrolled in any postgraduate course at the Strathfield or North Sydney campuses of the University with preference given to students from western Sydney.
Type: Scholarship
Value: $4,000
Frequency: Annual
No. of awards offered: 1
Application Procedure: Applications should include a written statement of approximately 200 words that outlines how the Scholarship could financially support the applicant's intended program of study; evidence of the successful completion of his/her undergraduate studies with a minimum Credit average.
Contributor: Dooleys Lidcombe Catholic Club Ltd
Additional Information: Please check website for further information.

Faculty of Business Executive Dean's Scholarship

Purpose: The Faculty of Business offers scholarships for international students.
Eligibility: You must be an international student; enrol full-time in an ACU Faculty of Business Bachelor Degree or postgraduate coursework program; commence studies in the eligible program in 2012 or 2012.
Type: Scholarship
Value: The Scholarship will subsidise tuition fees by Australian $1,000 in the first semester of study. Upon successful completion of all enrolled units, tuition fees in the next semester will also be subsidised by Australian $1,000. The Scholarship will continue for all semesters in the program, providing successful completion of units is maintained
Frequency: Annual
Application Procedure: ACU will consider all eligible students for the Scholarship at the time of application for admission. No further application is required. Recipients will receive confirmation of the Scholarship in the Letter of Offer issued by ACU.
Additional Information: Please check website for more information.

Francis Carroll Scholarship

Purpose: Established by public appeal in 2006, in honour of Archbishop Francis Carrolls contribution to the Archdiocese of Canberra and Goulburn, on the occasion of his retirement as Archbishop.
Eligibility: Open to commencing students who relocate from rural or regional areas of the Archdiocese of Canberra/Goulburn or the Diocese of Wagga Wagga to undertake an Education course at the Canberra campus of the University.
Type: Scholarship
Value: Up to $2,500
Frequency: Annual
No. of awards offered: 1
Application Procedure: Applications should include evidence of academic merit (based on UAI score or entry rank); evidence of the applicants contribution to and/or leadership in community service; an explanation on how the Scholarship could financially support their intended program of study; evidence of rural or regional background; and one referees report from the applicants school Principal which addresses the above selection criteria.

George Alexander Foundation Bursaries

Purpose: To assist with professional experience placement costs associated with Education and Nursing courses at the Ballarat campus. Preference will be given to applicants undertaking rural or remote placements.
Eligibility: open to all students undertaking any year of an Education or Nursing course at the Ballarat campus of the University.
Type: Bursary
Value: $1,000
Frequency: Annual
No. of awards offered: 10
Application Procedure: An applicant must provide an explanation on how the Bursary could financially support his/her intended completion of the professional experience requirements of the applicants course. In the case of students undertaking rural or remote placements, this may include factors such as additional travel and accommodation costs, and/or loss of income from part-time work.

Indigenous Access Scholarship

Type: Scholarship
Value: $4,485
Frequency: Annual
Application Procedure: Applicants must: be of Australian Aboriginal or Torres Strait Islander descent and identify as an Australian Aboriginal or Torres Strait Islander; be enrolled or about to commence study in an undergraduate course or a graduate diploma (or equivalent postgraduate course of study) in an area of National Priority; be enrolled as a Commonwealth supported student for that course of study; must not have completed the requirements of a course of study equivalent to or higher than an Australian bachelors award unless such an award is a prerequisite to your current undergraduate course of study or your current course is in the area of National Priority; demonstrate financial hardship and be enrolled, or about to commence study as a full time student (exceptional circumstances may be considered for part time study).
Additional Information: You may apply for an Indigenous Access Scholarship if you are enrolled in one of the following course types: all Bachelor and Honours degrees; Graduate Diploma in Education (Secondary); Graduate Diploma in Midwifery; Master of Teaching (Primary) and Master of Teaching (Secondary).

Michael Myers Theology Scholarship

Eligibility: Open to students who are living in rural or regional Victoria and who are enrolled in any undergraduate or postgraduate theology course at or through the Ballarat campus of the University.
Type: Scholarship
Value: Up to $5,000
Frequency: Annual
Application Procedure: Applications should include evidence of academic merit; evidence of the applicants contribution to and/or leadership in service to the rural or regional Church community; and an explanation of how the Scholarship could financially support the applicant's intended program of study and/or other evidence of the applicant's circumstances for the purpose of assessment of financial need.
Additional Information: Please check website for more information.

Nano Nagle Scholarship (IES)

Subjects: All subjects.
Purpose: To commemorate and honour the work of their founder Nano Nagle who established the Congregation.
Eligibility: Open to Indigenous students undertaking postgraduate studies at ACU National's Brisbane Campus in areas of concern to the Presentation Sisters, with a preference given to study in education.
Level of Study: Graduate
Type: Scholarship
Value: Australian $1,250 per year
Frequency: Annual
Study Establishment: Australian Cahtolic University National's Brisbane Campus
Country of Study: Australia
Application Procedure: Candidates can obtain further information from Weemala Indigenous Unit, Brisbane campus. Online applications only.
Closing Date: March 18th

Contributor: Presentation Sisters, Queensland Congregation
For further information contact:

Website: www.acu.edu.au/scholarships

Pratt Foundation Bursary (IES)

Subjects: All subjects.
Purpose: To make available a bursary to a suitably qualified Aboriginal and Torres Strait Islander student undertaking postgraduate study at ACU
Eligibility: Open to suitably qualified Aboriginal and Torres Strait Islander student undertaking postgraduate study at ACU National.
Level of Study: Graduate
Type: Bursary
Value: Australian $2,500
Frequency: Annual
Study Establishment: Australian Catholic University (ACU)
Country of Study: Australia
No. of awards offered: 1
Application Procedure: Candidates can obtain further information from Weemala Indigenous Unit, Brisbane campus. Online applications only.
Closing Date: March 18th
Funding: Foundation
Contributor: The Pratt Foundation
For further information contact:

Website: www.acu.edu.au/scholarships

QIEC Super Scholarship

Purpose: To provide financial support to enable students enrolled in the School of Education teacher programs to experience a rural and or remote Queensland education setting thereby; supporting rural and remote communities through provision of student teachers, providing a unique professional development experience for students, encouraging graduating students of the School of Education (Brisbane QLD) to consider working in rural and remote education settings in QLD and building an awareness of QIEC Super and its brand amongst students of the University.
Eligibility: Open to students enrolled in any course offered by the School of Education (Qld) at the Brisbane campus of the University who have previously undertaken a minimum of one practicum placement, and provide a report pertaining to the success of that placement; are prepared to undertake a four week or longer practicum placement in a rural, regional or remote area based in Queensland consistent with QIEC Super's business reach (grid); have not previously been a recipient of this Scholarship; and can demonstrate that they or their family live at least 100kms from the designated practicum placement.
Type: Scholarship
Value: $1,000 ($500 pre placement payment [subject to confirmation of practical placement]and $500 post placement payment [subject to successful completion of practicum])
Frequency: Annual
No. of awards offered: Up to 5
Application Procedure: Applications must include a written statement of 200 words or less explaining how a practicum placement in a rural, regional, or remote school could benefit the applicant's learning and practice as a future teacher; and a copy of the applicant's academic transcript to demonstrate capacity to undertake practicum.

Sophia Scholarship

Purpose: Established in 2011, the Sophia Scholarship was initiated by the Deputy Vice-Chancellor (Students Learning and Teaching) to reward students who are participating in extraordinary activities for the common good.
Eligibility: Open to full-time or part-time students from any campus at either undergraduate or postgraduate level. Applications are also invited from ACU Alumni.
Type: Scholarship
Value: Up to $5,000
No. of awards offered: More than one
Application Procedure: Applications must include a written statement of no more than 250 words detailing applicant's involvement and impact on an extraordinary activity or project that is for the benefit of the common good. Statement should also detail how financial support can help progress or have an impact on the activity or project.
Additional Information: Recipients will be selected by the Deputy Vice-Chancellor (Students, Learning and Teaching) and one or more Associate Vice-Chancellors.

Victoria India Doctoral Scholarships

Purpose: The Victorian International Research Scholarship (VIRS) is offered in partnership between the Victorian Government and Victorian Universities. The Scholarship supports high-calibre international PhD scholars to undertake research in Victoria.
Eligibility: Open to an international student; who accepted into a doctoral programme at ACU; intend to complete the majority of work related to the doctorate in Victoria; not have completed a degree equivalent to an Australian doctorate; willingness to act as an ambassador for the (VIRS) program.
Type: Scholarship
Value: Australian $90,000 over three years
Closing Date: November 30th
Additional Information: For further information, visit Study Melbourne, or email your application (MS Word document, 2.8 MB) to VIC.cand@acu.edu.au. Please note that you will also be asked to submit an application for candidature at ACU.

AUSTRALIAN CENTRE FOR BLOOD DISEASES (ACBD)

6th Floor, Burnet Tower, 89 Commercial Road, Melbourne, VIC, 3004, Australia
Tel: (61) 3 990 30122
Fax: (61) 3 990 30228
Email: acbd@med.monash.edu.au
Website: www.acbd.monash.org

The Australian Centre for Blood Diseases (ACBD) brings together the skills and facilities of separate yet complementary organizations to enhance understanding of blood and its diseases. Its aim is to provide excellence in the diagnosis and treatment of blood conditions as well as play a leading role in the advancement of knowledge in this increasingly important area of medicine.

Firkin PhD Scholarship

Subjects: Cardiovascular disciplines.
Purpose: To undertake a PhD programme at the ACBD or affiliated institutes comprising AMREP.
Eligibility: Open to students interested in pursuing doctorate studies in cardiovascular disciplines, and who have the appropriate graduate qualifications.
Level of Study: Graduate
Type: Scholarships
Value: Australian $22,500 per year
Length of Study: 3 years
Frequency: Annual
No. of awards offered: Up to 4
Application Procedure: For further information, please contact Dr Robert Medcalf.
Closing Date: See website for exact details.

For further information contact:

Australian Centre for Blood Diseases, Monash University, 6th Floor, Burnet Building, AMREP, Commercial Road, Melbourne, Victoria 3004, Australia
Email: Robert.Medcalf@med.monash.edu.au
Contact: Dr Robert Medcalf, Associate Professor

AUSTRALIAN FEDERATION OF UNIVERSITY WOMEN (AFUW)

School of Education, University of Ballarat, PO Box 663, Victoria, Ballarat, 3353, Australia
Tel: (61) 3 9557 2556
Email: AFGW.Fellowships@gmail.com
Website: www.afuw.org.au
Contact: Dr Jacqueline Wilson, AFUW Vic Membership Secretary

Australian Federation of University Women (AFUW) Victoria was formed in 1922 as part of the international network of women Graduates for the benefit of women and society. AFUW Victoria is a member association of AFUW and serves a number of benefits both at personal and societal level in providing women with opportunities.

AFUW William and Elizabeth Fisher Scholarship

Subjects: All subjects.
Purpose: To promote the advancement of graduate women worldwide and their equality of opportunity through initiatives in education, friendship and peace.
Eligibility: Open to female students who are enrolled in postgraduate research degree at a Victorian University.
Level of Study: Postgraduate
Type: Scholarship
Value: Australian $6,000 and two bursaries of $3,000 each.
Length of Study: 1 year
Frequency: Annual
Study Establishment: Victorian University
Country of Study: Australia
No. of awards offered: 2
Application Procedure: A complete application form must be sent across by post.
Closing Date: March 31st

For further information contact:

AFUW Fellowships Convenor, PO Box 4066, Bay Village, NSW, Australia

THE AUSTRALIAN FEDERATION OF UNIVERSITY WOMEN, SOUTH AUSTRALIA, INC. TRUST FUND (AFUW-SA, INC.)

18 Humphries Terrace, Kilkenny, SA, 5009, Australia
Website: www.afuwsa-bursaries.com.au
Contact: Ms Heather Latz, Bursaries Trustee

Graduate Women South Australia (GWSA) advocates at state, national and international level for the protection of human rights and the education, health and status of women and girls. GWSA assists women and men in tertiary education in South Australia via bursaries.

Dentistry Scholarships: Winifred E. Preedy Postgraduate Scholarship

Eligibility: Applicants must be women of any nationality; must be a past or present student of the University of Adelaide Dental School; must be studying a postgraduate degree in dentistry or an allied field at an Australian university; must have completed one year of their postgraduate degree; must not be employed full-time, or on fully-paid study leave; must not have previously won the Winifred E Preedy Postgraduate Bursary.
Type: Scholarship
Value: Up to Australian $4,000
Frequency: Annual
No. of awards offered: 1
Application Procedure: Selection of winners is based primarily on academic merit, but also on financial need, the importance of the purpose for which the bursary will be used to the progress or completion of the degree, community activities and other interests. Your application must reach the Trust Fund's secretarial service by deadline.
Closing Date: March 1st
Contributor: The bequest of Winifred E. Preedy BDS (1901-1989)
Additional Information: Receipt of all applications will be acknowledged by letter. In the normal course of events the successful applicants will be notified by the last day of June in the year of the award. Subsequently, the list of winners will be posted on this website in early July. The decision of the selection committee will be final and no correspondence will be entered into.

Diamond Jubilee Bursary

Subjects: All subjects.
Purpose: To assist in the completion of a coursework postgraduate degree.
Eligibility: Open to men or women of any nationality, enrolled in a postgraduate degree by coursework at South Australian Universities. Applicants must not be in full-time paid employment or on fully paid study leave during the tenure of the bursary.
Level of Study: Doctorate, MBA, Postgraduate
Type: Bursary
Value: Australian $2,500
Length of Study: The bursary must be used within 1 year of the date of the award
Frequency: Annual
Study Establishment: A South Australian university
Country of Study: Australia
No. of awards offered: 1
Application Procedure: Candidates should complete an application form and submit it with evidence of enrolment at the institution at which the qualification will be obtained, as well as copies of official transcripts and a curriculum vitae. Application forms can be downloaded from the website and sent by post.
Closing Date: March 1st
Funding: Private
No. of awards given last year: 2
No. of applicants last year: 7

Doreen McCarthy, Barbara Crase, Cathy Candler and Brenda Nettle Bursaries

Subjects: All subjects.
Purpose: To assist in the completion of a Master's or PhD research degree.
Eligibility: Open to men and women of any nationality. Applicants must have completed 1 year of postgraduate research, and must be enrolled at a university in South Australia. Applicants must not be in full-time paid employment or on fully paid study leave.
Level of Study: Postgraduate, Doctorate
Type: Bursary
Value: Australian $2,500
Length of Study: The bursary must be used within 1 year of the date of the award
Frequency: Annual
Study Establishment: A South Australian university
Country of Study: Australia
No. of awards offered: 4
Application Procedure: Candidates should complete an application form and submit it with evidence of enrolment at the institution at which the qualification will be obtained, as well as copies of official transcripts, a curriculum vitae and a list of publications. Application forms can be downloaded from the website and sent by post.
Closing Date: March 1st
Funding: Private
No. of awards given last year: 9
No. of applicants last year: 26

Padnendadlu Graduate Bursary

Subjects: All subjects.
Purpose: To assist in the completion of a graduate diploma or graduate certificate.
Eligibility: Applicants must be Australian Aboriginal or Torres Strait Islander women undertaking a graduate diploma or graduate certificate at South Australian universities. Applicants must not be in full-time paid employment or on fully paid study leave.
Level of Study: Graduate
Type: Bursary
Value: Australian $1,000
Frequency: Annual
Study Establishment: A South Australian University
Country of Study: Australia
No. of awards offered: 1+
Application Procedure: Candidates should complete an application form and submit it with evidence of enrolment at the institution at which the qualification will be obtained, as well as copies of official transcripts and a curriculum vitae. An application form can be downloaded from the website and sent by post.
Closing Date: March 30th
Funding: Private

Padnendadlu Postgraduate Bursary

Subjects: All subjects.
Purpose: To assist in the completion of a postgraduate degree.
Eligibility: Applicants must be Australian indigenous women undertaking postgraduate degrees at South Australian universities. Applicants must have completed 1 year postgraduate research and must not be in full-time paid employment or on fully paid study leave during the tenure of the bursary.
Level of Study: Postgraduate, Doctorate
Type: Bursary
Value: Australian $3,500
Frequency: Annual
Study Establishment: A South Australian university
Country of Study: Australia
No. of awards offered: At least 1
Application Procedure: Candidates should complete an application form and submit it with evidence of enrolment at the institution at which the qualification will be obtained, as well as copies of official transcripts and a curriculum vitae. Application forms can be downloaded from the website and sent by post.
Closing Date: March 1st
Funding: Private

Winifred E Preedy Postgraduate Bursary

Subjects: Dentistry or a related field.
Purpose: To assist women in the completion of a higher degree.
Eligibility: Open to women of any nationality who are students of the University of Adelaide's Dental School in the past or present, and are enrolled as postgraduate students in dentistry or an allied field. The applicant must have completed 1 year of the postgraduate degree, and must not be in full-time paid employment or on fully paid study leave, and must not have previously won this bursary.
Level of Study: Doctorate, Postgraduate
Type: Bursary
Value: Australian $4,000
Length of Study: The bursary must be used within 1 year of the date of the award
Frequency: Annual
Study Establishment: Anywhere, if the applicant is a past student at the University of Adelaide's Dental Faculty in Australia. Otherwise, the applicant must be a student at the University of Adelaide's Dental Faculty
Country of Study: Australia
No. of awards offered: 1
Application Procedure: Applicants must complete an application form and submit it with evidence of enrolment at the institution at which the qualification will be obtained, as well as copies of official transcripts, a curriculum vitae and a list of publications. Application forms can be downloaded from the website and sent by post.
Closing Date: March 1st
Funding: Private

THE AUSTRALIAN MUSIC FOUNDATION

Goar Lodge, Smith's Green, Takeley, Herts, East Sussex, CM22 6NS, United Kingdom
Tel: (44) 1279 871114
Email: info@amf-aus.org
Website: www.amf-uk.com
Contact: Mr Michael Letchford, Executive Assistant AMF

The Australian Music Foundation was established in 1980 and exists to offer financial support to outstanding young Australian musicians who wish to pursue post-graduate music courses in leading music education institutions in Europe. The Foundation aims to support and enhance that facet of modern Australian society, offering opportunities to outstanding young instrumentalists and singers to further their careers, studying with top teachers and coaches in the well-established music colleges and academies of Europe.

Australian Music Foundation Award

Subjects: Music.
Purpose: To offer financial support to outstanding young Australian musicians who wish to pursue post-graduate music courses in leading music education institutions in Europe.
Eligibility: Open to outstanding young Australian musicians who wish to pursue post-graduate music courses in leading music education institutions in Europe.
Level of Study: Graduate
Type: Award
Value: Up to $35,000
Frequency: Annual
Study Establishment: Music colleges in Europe
Country of Study: European Union
No. of awards offered: Varies
Application Procedure: Check the website for further information.
Closing Date: April 30th
Funding: Foundation

AUSTRALIAN NATIONAL UNIVERSITY (ANU)

Fees and Scholarships Office, Building X-005, Canberra, ACT 0200, Australia
Tel: (61) 2 6125 8124
Fax: (61) 2 6125 7535
Email: research.scholarships@anu.edu.au
Website: www.anu.edu.au
Contact: Manager Fees and Scholarships

The Australian National University (ANU) was founded by the Australian Government in 1946 as Australia's only completely research-orientated university. It comprises of seven colleges and many research schools.

Alex Rodgers Travelling Scholarship: ANU College of Physical & Mathematical Sciences

Subjects: Astronomy and astrophysics.
Eligibility: Full-time current RSAA PhD student. A successful applicant is ineligible to apply again. Australian and New Zealand citizens; Permanent residents of Australia; International students.
Type: Scholarship
Value: Up to Australian $6,000, depending on available funds. The committee may decide to divide an award amongst a number of deserving candidates
No. of awards offered: Up to 2
Application Procedure: General information and application forms are available online at: www.mso.anu.edu.au/education.
Closing Date: February 28th each year
Additional Information: Please contact to graduate.enquiries@mso.anu.edu.au for more information.

Allan White Scholarship: ANU College of Physical and Mathematical Sciences

Subjects: Earth sciences.
Eligibility: Open to Australian and New Zealand citizens; permanent residents of Australia; international students.
Type: Scholarship
Value: Australian $6,500 per year tax free, in fortnightly payments
Length of Study: 3 years full-time, with a possible extension of 6 months
Closing Date: August 31st (international applicants) or October 31st (domestic applicants)
Additional Information: Please contact to student.admin.rses@anu.edu.au for more information.

ANU Alumni Association PhD Scholarships

Subjects: All subjects.
Purpose: To assist international students with study in Australia.
Eligibility: Open to nationals of Japan, Malaysia, Thailand and Singapore. Applicants must meet university admission requirements. To be eligible for scholarsip, applicants must have a bachelor's degree with first class or upper second class honours or a research master's degree from a recognized university.
Level of Study: Doctorate
Type: Scholarship
Value: Approx. Australian $23,728 (2012 rate) (tax-free), payment of the tuition fees for the duration of the stipend, an additional allowance for dependent children of married scholars, travel to Canberra and a

grant for the reimbursement of some removal expenses. A thesis reimbursement allowance is also available
Length of Study: Normally tenable for 3 years, renewable for 6 months
Country of Study: Australia
No. of awards offered: 1 each for nationals of eligible countries
Application Procedure: Applicants must complete an application form, available on request or from the website www.anu.edu.au/sas/scholarships/research/scholarships_international_students/index.
Closing Date: Published through the Alumni Association of each country
Contributor: The Australian National University

ANU Doctoral Fellowships

Subjects: All subjects.
Purpose: To help Doctoral students pursue research and dissertation writing.
Eligibility: Open to candidates who have obtained their graduate degree from a university located in the United States.
Level of Study: Doctorate
Type: Fellowships
Value: Australian $1,600 per month, one round-trip economy class airfare and some funding for research
Length of Study: 3–9 months
Frequency: Annual
Study Establishment: Australian National University
Country of Study: Australia
Application Procedure: Applicants can download the application form from the website. The completed application form can be sent by post or electronically.
Closing Date: January 30th
Additional Information: Fellows accompanied by their family and staying for 6 months or more at the ANU will be entitled to family support of up to Australian $5,000, depending on the number of family members and other circumstances.

For further information contact:

ANU Southeast Asian Studies Fellowship Selection Committee, HR Academic Research School of Pacific and Asian Studies, The Australian National University, Canberra, ACT, 0200, Australia
Email: hr.rspas@anu.edu.au

ANU Graduate Access Scholarship

Subjects: All subjects.
Eligibility: Applicants must be current enrolled students in a graduate program of study at the ANU; must supply supporting evidence of a permanent disability; and demonstrate financial hardship.
Type: Scholarship
Value: $4,000
Length of Study: Up to 4 years
Frequency: Annual
No. of awards offered: 1
Application Procedure: Complete the application form, available on website.
Closing Date: March 9th
Additional Information: Please contact to coursework.scholarships@anu.edu.au for further more information.

ANU Indigenous Australian Reconciliation PhD Scholarship

Subjects: All subjects.
Purpose: To assist an Indigenous student to undertake a graduate study.
Eligibility: Open to Indigenous Australians.
Level of Study: Doctorate
Type: Scholarship
Value: Australian $23,728 (last year rate) in fortnightly payments and travel to Canberra from within Australia; reimbursement of some removal expenses; thesis reimbursement allowance.
Length of Study: Dependent upon the programme for which the scholarship is awarded
Frequency: Annual
Study Establishment: Open to Indigenous Australians
Country of Study: Australia
No. of awards offered: 1
Application Procedure: Applicants must complete an application form available on-line at http://students.anu.edu.au/apply_online.asp.
Closing Date: October 31st
Contributor: The Australian National University
Additional Information: Open to Indigenous Australians only.

For further information contact:

Fees and Scholarships Office
Contact: Manager

ANU Indigenous Graduate Scholarship

Subjects: All subjects.
Purpose: To assist Indigenous students to undertake a graduate study.
Eligibility: Open to Indigenous Australians.
Level of Study: Doctorate, Graduate, Postgraduate
Type: Scholarship
Value: Australian $23,728 (2012 rate) in fortnightly payments and travel to Canberra from within Australia; reimbursement of some removal expenses; thesis reimbursement allowance.
Length of Study: Dependent upon the programme for which the scholarship is awarded
Frequency: Annual
Country of Study: Australia
No. of awards offered: 1
Application Procedure: Applicants must complete an application form available on-line at http://students.anu.edu.au/apply_online.asp.
Closing Date: October 31st
Contributor: The Australian National University
Additional Information: Open to Indigenous Australians only.

ANU PhD Scholarships

Subjects: All subjects.
Purpose: To assist international and Australian students to undertake a PhD or professional doctorate by research.
Eligibility: Applicants must meet university admission requirements. To be eligible to be considered applicants must have completed a bachelor's degree with at least an upper second class honours or a master's degree with a research component or equivalent from a recognized university.
Level of Study: Doctorate
Type: Scholarship
Value: A stipend of Australian $23,728 (last year rate) per year tax free, and if applicable, an additional allowance for dependent children of international scholars plus economy travel to Canberra and a grant for the reimbursement of some removal expenses and a thesis reimbursement allowance.
Length of Study: 3 years, renewable for 6 months
Frequency: Annual
Country of Study: Australia
No. of awards offered: Varies
Application Procedure: Applicants must complete an application form available on-line at http://students.anu.edu.au/apply_online.asp
Closing Date: October 31st for Australian citizens, permanent residents, and New Zealand citizens and August 31st for international applicants
Contributor: The Australian National University

ANU University Research Scholarship

Subjects: All subjects.
Purpose: To assist both international and domestic students to undertake a doctorate philosophy or professional doctorate degree by research.
Eligibility: Open to candidates with who have obtained a bachelor's degree with first class honours, or a research master's degree from a recognized university. Candidates must fulfil the university admission requirements and be admitted to programme.
Level of Study: Doctorate
Type: Scholarship
Value: A stipend of Australian $23,728 (last year rate) per year tax free, travel to Canberra and a grant for the reimbursement of some removal expenses is available. A thesis allowance is also available.
Length of Study: 3 years in the first instance, with a possible extension of 6 months
Frequency: Annual

Country of Study: Australia
Application Procedure: Applicants must complete an application form available on-line at http://students.anu.edu.au/apply_online.asp.
Closing Date: October 31st for Australian citizens, permanent residents, and New Zealand students and August 31st for International students
Contributor: The Australian National University

CECS: College Postgraduate International Award

Eligibility: The Scholarships are offered on the basis of offer of a place in the Master Degree by Coursework programs offered by the College of Engineering and Computer Science and other criteria as set out from time to time by the Program Authority. Eligible students will be automatically considered for the Scholarship on academic merit. The Scholarship is offered on the condition that the recipient is admitted to and continues to pursue a full-time postgraduate program of study at this University and in a program offered by the College of Engineering and Computer Science.
Type: Award
Value: Stipend is $5,000, payable in two equal instalments of $2,500 at the beginning of each semester for one year. Scholars are required to meet all costs associated with their studies including travel, accommodation, books and incidental expenses. The scholar is responsible for paying International Tuition fees for the duration of the program.
Frequency: Annual
No. of awards offered: Up to 5
Additional Information: A scholar may not hold concurrently another scholarship awarded by the University or another University College. A scholar must obtain permission from the College Coursework Scholarships Committee to hold any other scholarship or award concurrently with a College of Engineering and Computer Science Postgraduate Scholarship.

For further information contact:

ANU College of Engineering and Computer Science
Tel: 02 6125 0677
Email: student.services@cecs.anu.edu.au
Contact: Professor Chris Johnson

CECS: College Postgraduate International Honours Award

Type: Award
Value: Stipend is $5,000, payable in two equal instalments of $2,500 at the beginning of each semester for one year
Frequency: Annual
No. of awards offered: Up to 5
Additional Information: The Scholarships are offered on the basis of offer of a place in the Master Degree by Coursework programs offered by the College of Engineering and Computer Science and other criteria as set out from time to time by the Program Authority.

For further information contact:

Associate Dean (Education), ANU College of Engineering and Computer Science
Tel: 02 6125 0677
Email: student.services@cecs.anu.edu.au
Contact: Professor Chris Johnson

Chris Heyde Scholarship: ANU College of Physical and Mathematical Sciences

Eligibility: Open to Australian and New Zealand citizens; permanent residents of Australia; international students.
Type: Scholarship
Value: Australian $5,000 to assist with expenses associated with travel with direct relevance to thesis work, including conference attendance and undertaking research at another location
Application Procedure: Online applications can be made at http://www.maths.anu.edu.au/study/graduate/forms/application.php.
Additional Information: Please check website for more information.

CMBE and CPMS: Indigenous Australian Graduate Scholarship

Eligibility: Applications are only open to Indigenous Australians enrolling in or enrolled in a graduate program of study at ANU. Applicants for a coursework scholarship must hold at least a Bachelors degree. Applicants for a research scholarship must hold a Bachelors degree with at least upper second-class honours, or equivalent prior studies including some research work.
Type: Scholarship
Value: Australian $22,500 per year tax free in fortnightly instalments. The scholarship also covers travel to Canberra from within Australia
Length of Study: 3 years
Frequency: Annual
Application Procedure: Apply online at http://applyonline.anu.edu.au/. Applications are only open to Indigenous Australians.
Additional Information: Please write to scholarships@anu.edu.au for more information about scholarship.

College of Business and Economics Graduate Scholarship

Eligibility: Open to an applicant who is an Australian citizen, a New Zealand Citizen or an Australian Permanent Resident; who has completed a Bachelors degree and achieved at least a 75 per cent average over the last two years of study; who has been offered admission to a Graduate Diploma or Master program offered by the College and who is applying to commence graduate study with the College for the first time.
Level of Study: Graduate
Type: Scholarship
Value: The Scholarship covers the full (100 per cent) domestic student tuition fee for one year (or part time equivalent) of a graduate program offered by the College
Frequency: Annual
No. of awards offered: Up to 1
Application Procedure: There will be a call for applications each year by the ANU College of Business and Economics.Applicants must provide a statement of aims of no more than 500 words to support their application.
Additional Information: The scholarship is available for two semesters of study (either full time or part time) only, subject to satisfactory academic progress and continued enrolment in a Graduate Diploma or Masters program offered by the ANU College of Business and Economics.

College of Business and Economics Graduate Scholarship for an Aboriginal or Torres Strait Islander Students

Purpose: The Scholarship aims to offset some of the financial burden of graduate coursework study. Theawards are competitive and based on demonstrated need.
Eligibility: To be eligible for consideration applicants must provide proof of Australian Aboriginal or Torres Strait Islander descent; identify as an Australian Aboriginal or Torres Strait Islander and be accepted as such by the communities in which she/he has been or with which she/he is associated; have previously completed an undergraduate bachelors degree, and be applying tocommence graduate study at the ANU College of Business and Economics for the first time;and be successful, or will be successful, in receiving on offer of admission into the first year of a graduate program offered by the ANU College of Business and Economics.
Type: Scholarship
Value: The Scholarship will cover 100 per cent of the scholar's domestic tuition fees (DTF). The Scholarship does not provide exemption from payment of reading and study materials,accommodation or any other cost associated with studying
Frequency: Annual
Application Procedure: There will be a call for applications each year by the ANU College of Business and Economics. Applications may consist of a prescribed form and a personal statement of need or motivation, which will be submitted to the College.
Additional Information: All enquiries regarding the scholarship and changes, deferment and further information afteracceptance should be made to Marketing and Development Office, ANU College of Business and Economics.

College of Business and Economics: International Graduate Scholarship

Eligibility: The Scholarship shall be available to an applicant who is an International Students as defined under the terms of the Education Services for Overseas Students Act 2000. Dependants of the staff of diplomatic or consular missions are also eligible; who has received an offer of admission into a Graduate Diploma or Masters by coursework program offered by the College to commence in teaching period for which they are applying for this award; who will commence a Graduate Diploma or Masters by coursework program within the College for the first time.
Type: Scholarship
Value: The Scholarship provides sponsorship for half (50 per cent) of the international student tuition fee for up to one year of a Graduate Diploma or Masters program offered by the College
Frequency: Annual
No. of awards offered: Up to 10
Application Procedure: There will be a call for applications each year by the ANU College of Business and Economics.Applicants must complete all components of the application and provide supporting documentswherever necessary.
Additional Information: One scholarship will be awarded to a student from Latin America, Europe, Asia, South East Asia and Eurasia while the remaining five scholarships will be open to applicants from all regions of the world. The scholarship is available for two semesters of full-time study, subject to satisfactory academic progress and continued enrolment in a Graduate Diploma or Masters program offered by the College.

College of Business and Economics: Neil Vousden Scholarship in the Master of Economics

Subjects: Any degree offered by the ANU College of Business and Economics.
Eligibility: The Scholarship shall be available to Domestic and International Students Students who meet the entry requirements for the Master of Economics program offered by the College. Students who are deemed by the selection committee to possess excellent qualities of academic ability and leadership, as demonstrated by his/her academic record, work experience and, at the option of the selection committee, a personal interview. A scholar's financial needs will also be taken into consideration by the selection committee.
Type: Scholarship
Value: The Scholarship is valued at 66 per cent of the Australian Postgraduate Award amount for the year in which the scholarship is awarded. The scholar is responsible for making payment of all his/her tuition fees by the prescribed date, as promulgated by the University from time to time. Scholars are responsible for the cost of books, study materials, accommodation and all other costs of study
Frequency: Annual
No. of awards offered: 1
Closing Date: December
Additional Information: The Scholarship amount is payable in two equal instalments, to be paid following the Census Date of each semester of study. Please write to mdo.cbe@anu.edu.au or check at http://cbe.anu.edu.au/scholarships for more information.

GSIA: Hedley Bull Scholarship in the Master of Arts (International Relations)

Subjects: Arts.
Level of Study: Postgraduate
Type: Scholarship
Value: Full-fee waiver for three full-time sessions (semesters). This amounts to 48 units of graduate coursework and a 24-unit thesis session
Frequency: Annual
No. of awards offered: Up to 4
Application Procedure: When applying for the MA (International Relations), eligible applicants for the Hedley Bull Scholarship should include a covering letter indicating their interest in being considered for the scholarship and making the case for being granted one. The covering letter together with the application for the MA (International Relations) (including transcripts and three academic references) will be considered as the full application for the scholarship. There is no need to provide a separate application.
Closing Date: October 12th

Indigenous Australian Graduate Scholarship

Eligibility: Applicants must identify as an Indigenous Australian and be enrolling in or enrolled in a graduate program of study at the ANU. Applicants must hold a Bachelors degree with at least upper second-class honours, or equivalent prior studies including some research work.
Type: Scholarship
Value: Australian $22,500 per year tax free in fortnightly instalments. The scholarship also covers travel to Canberra from within Australia
Length of Study: 3 years
Frequency: Annual
No. of awards offered: 1
Application Procedure: Applications are only open to Indigenous Australians. Complete the online applciation form, available at http://applyonline.anu.edu.au.
Closing Date: October 31st
Additional Information: Please contact to scholarships@anu.edu.au for more information.

Indigenous Reconciliation PhD Scholarship

Purpose: To support the development of future Australian Indigenous leaders to advance indigenous reconciliation.
Eligibility: Applicants must identify as an Indigenous Australian; and be enrolling in or enrolled in PhD program of study at the ANU. Applicants must hold a Bachelors degree with at least upper second-class honours, or equivalent prior studies including some research work.
Type: Scholarship
Value: Australian $22,500 per year tax free in fortnightly instalments. The scholarship also covers travel to Canberra from within Australia
Length of Study: 3 years full-time with a possible maximum extension of 6 months
Frequency: Annual
No. of awards offered: 1
Application Procedure: Complete the online application form, available at http://applyonline.anu.edu.au.
Closing Date: October 31st
Additional Information: Please contact to scholarships@anu.edu.au for more information.

Indigenous Students' Practical Legal Training Scholarship

Eligibility: Applicants must be Indigenous Australians; must meet the admission requirements of the Graduate Diploma of Legal Practice and; must have completed a Bachelor of Laws (LLB) or equivalent.
Type: Scholarship
Value: Tuition fee
Frequency: Annual
No. of awards offered: Up to 2
Application Procedure: Form available on the College website.
Closing Date: October 8th
Additional Information: Applicants must demonstrate how the scholarship will assist their personal and professional development. The Selection Committee may take into account the applicant's academic record, other achievements, and, if relevant, personal circumstances. Please contact to lwsa@law.anu.edu.au for more information.

James Rice Postgraduate Award: ANU College of Medicine, Biology and Environment

Subjects: Medical sciences.
Eligibility: Open to Australian and New Zealand citizens; permanent residents of Australia; international students.
Type: Award
Value: Australian $23,728 per year tax free, in fortnightly payments
Length of Study: 3 years full-time, with a possible extension of 6 months
Frequency: Annual
Application Procedure: Online applications for admission and scholarship can be made at http://students.anu.edu.au/apply_online.asp (recommended).
Closing Date: August 31st (international applicants) or October 31st (domestic applicants)
Additional Information: Please contact to research.scholarships@anu.edu.au.

Legal Workshop Indigenous Student Scholarship Scheme

Eligibility: Applicants must be Indigenous Australians; must meet the admission requirements of the Graduate Diploma of Legal Practice and; must have completed a Bachelor of Laws (LLB) or equivalent.
Type: Scholarship
Value: Tuition fee
No. of awards offered: Up to 2

National Security College Advancement Scholarship

Subjects: Available for any degrees offered by the ANU National Security College.
Eligibility: The Scholarship shall be available to an applicant who is an Australian citizen, a New Zealand citizen or an Australian Permanent Resident; has completed the Master of National Security Policy with a minimum average of 75 per cent in the core courses, and 70 per cent overall; has been offered admission to the Master of Advanced National Security Policy.
Type: Scholarship
Value: The Scholarship covers the full domestic tuition fee for the 24 unit sub-thesis component of the Master of Advanced National Security Policy. A travel grant of up to $3500, reviewed annually. The Scholarship does not cover any necessary admissions and deposit fees, the payment of reading and study materials, living expenses (other than approved travel per diems associated with the travel grant), accommodation or any other costs associated with studying
Frequency: Annual
No. of awards offered: 2
Application Procedure: Please print and complete application form, available at website.
Closing Date: July 20th
Additional Information: The Scholarship is available for one semester of full-time study or two semesters of part-time study only. Please write to national.security.college@anu.edu.au and check website at http://nsc.anu.edu.au/grad_studies.php for more information..

National Security College Entry Scholarship

Subjects: Degrees offered by the ANU National Security College.
Eligibility: The Scholarship shall be available to an applicant who is of Australian Aboriginal and/or Torres Strait Islander descent; has completed a Bachelors degree; has been offered admission to the Graduate Certificate in National Security Policy or Master of National Security Policy Is not the recipient of a College sponsored place.
Level of Study: Postgraduate
Type: Scholarship
Value: The Scholarship covers the full domestic tuition fee for up to 24 units of College core courses. The Scholarship does not cover any necessary admissions and deposit fees, the payment of reading and study materials, living expenses, accommodation or any other costs associated with studying
No. of awards offered: 1
Additional Information: The Scholarship is available in the first two semesters of study only.

National Security College Entry Scholarship for Aboriginal and Torres Strait Islander Students

Eligibility: The Scholarship shall be available to an applicant who is of Australian Aboriginal and/or Torres Strait Islander descent; has completed a Bachelors degree; has been offered admission to the Graduate Certificate in National Security Policy or Master of National Security Policy; and is not the recipient of a College sponsored place.
Type: Scholarship
Value: The Scholarship covers the full (100 per cent) domestic tuition fee for up to 24 units of College core courses. The Scholarship does not cover any necessary admissions and deposit fees, the payment of reading and study materials, living expenses, accommodation or any other costs associated with studying
Frequency: Annual
No. of awards offered: 1
Application Procedure: Please print and complete application form, available on college website.
Closing Date: July 20th
Additional Information: The Scholarship is available in the first two semesters of study only. Please contact to national.security.college@anu.edu.au or check at http://nsc.anu.edu.au/grad_studies.php for more information.

National Security College Entry Scholarship for Aboriginal and Torres Strait Islander Students

Subjects: Any degrees offered by the ANU National Security College.
Eligibility: The Scholarship shall be available to an applicant who is of Australian Aboriginal and/or Torres Strait Islander descent; has completed a Bachelors degree; has been offered admission to the Graduate Certificate in National Security Policy or Master of National Security Policy; is not the recipient of a College sponsored place.
Type: Scholarship
Value: The Scholarship covers the full domestic tuition fee for up to 24 units of College core courses. The Scholarship does not cover any necessary admissions and deposit fees, the payment of reading and study materials, living expenses, accommodation or any other costs associated with studying
No. of awards offered: 1
Application Procedure: Please print and complete application form, availabe at website.
Closing Date: July 20th
Additional Information: The Scholarship is available in the first two semesters of study only. Please write to national.security.college@anu.edu.au and check at http://nsc.anu.edu.au/grad_studies.php for more information.

NICTA Scholarships

Subjects: Engineering and information sciences and mathematical sciences.
Purpose: These scholarships are funded by the National ICT Australia (NICTA) centre of excellence for PhD study in a NICTA Research Program.
Eligibility: Applicants must have a Bachelor degree with first class honours, or equivalent and research proposal related to a NICTA research program; Australian and New Zealand citizens; Permanent residents of Australia; International Students.
Type: Scholarship
Value: Australian $23,728 per year tax free, in fortnightly payments. Tuition fee scholarships are available for International students
Length of Study: 3 years full-time, with a possible extension of 6 months
Frequency: Annual
Application Procedure: Application forms and general information are available online at www.nicta.com.au.
Additional Information: Students can apply any time, however they must also apply for any government or university scholarships for which they are eligible. The main university scholarship rounds close: October 31st (Domestic students) and August 31st (International students).

Prime Minister's Australia Asia Awards

Subjects: All subjects.
Purpose: The Awards aim to build deep and enduring education and professional linkages between Australia and Asia. They also provide an opportunity for Australian universities and the Australian Government to work in partnership to identify and reward high calibre scholars.
Eligibility: Australian applicants must be citizens and/or permanent residents of Australia. Australian Award Holders must retain Australian Citizenship and/or Australian Permanent Residency status for the duration of their Award program. In the event that their Australian citizenship/residency status lapses, the Award will be revoked or terminated from the date of effect. The Award Holder may continue their program but they will no longer be funded to do so.
Level of Study: Doctorate, Postgraduate
Type: Award
Value: The value of the awards is available in Section 5.1 of the 2013 Applicant Guidelines booklet, available at http://www.deewr.gov.au/International/EndeavourAwards/Documents/2013ApplicantGuidelines.pdf
Frequency: Annual
No. of awards offered: 40 (20 undergraduate and 20 postgraduate)
Application Procedure: All applications must be submitted through Endeavour Online at http://www.deewr.gov.au/International/Endea-

vourAwards/Pages/Apply.aspx. Please note that this online application form is only accessible when the round is open (April to June).
Closing Date: June 30th
Additional Information: Award Holders also have the exciting opportunity to undertake an internship to gain occupational and work related knowledge and skills; and build networks with industry and other organisations in the region. Twenty scholarships will also be awarded to the top international Endeavour Postgraduate Award recipients from the ten participating countries in Asia each year.

School of Art Graduate Materials Awards
Subjects: Visual Arts.
Purpose: Several awards are available annually to assist all new and continuing students enrolled in a graduate program at the School of Art. These awards provide funding for materials and processes to enable studetns to pursue a proposed program of study.
Eligibility: Open to Australian and New Zealand citizens; Permanent residents of Australia; International Students.
Type: Award
Value: A stipend of Australian $1,000 per year tax free
Frequency: Annual
Application Procedure: Application forms and further information may be obtained from Coordinator, Visual Arts Graduate Program, School of Art, The Australian National University.
Closing Date: Please contact college
Additional Information: Eligible programs are PhD, MPhil, Grad Dip Art, MVA, MDA, MNMA. Please contact to enquiries.arts@anu.edu.au for more information.

Supplementary Scholarship: ANU College of Physical & Mathematical Sciences
Subjects: Astronomy and astrophysics.
Eligibility: Applicants must hold a Bachelors degree with at least upper second-class honours, or a Masters degree with a research component or equivalent; Australian and New Zealand citizens; Permanent residents of Australia; International Students.
Type: Scholarship
Value: Australian $5,000 (current rate) per year tax free, in fortnightly payments
Length of Study: 3 years full-time, with a possible extension of 6 months
Frequency: Annual
Application Procedure: As this scholarship is awarded to all RSAA PhD students, it is not necessary to complete a separate application form.
Closing Date: October 31st each year
Additional Information: Please contact to graduate.enquiries@mso.anu.edu.au for more information.

T B Millar Scholarship in the Master of Arts (Strategic Studies) or the Master of Strategic Affairs
Eligibility: Open to new applicants to the Centre's Master coursework programs who intend to study full-time. Applicants must meet the admission requirements of the Master coursework programs offered by SDSC.
Type: Scholarship
Value: Tuition fees for the two core courses (or their equivalents) for a total of 24 units of coursework (out of a total of 48 units for the Master of Strategic Affairs and 72 units for the Master of Arts (Strategic Studies))
Frequency: Annual
No. of awards offered: 3
Application Procedure: When applying for a Master coursework program offered by SDSC, eligible applicants for theT.B. Millar Scholarship should submit the following: a cover letter indicating their interest in being considered for the scholarship and making the case for being granted one; a copy of your academic transcript/s; a current curriculum vitae; and two academic written references (or arrange to have these sent on your behalf to arrive by the closing date).
Closing Date: October 30th and March 30th each year
Additional Information: The scholarship only applies to the core courses or their equivalents, undertaken while enrolled in an SDSC Master coursework program. If a scholarship holder transfers to another ANU degree including another program in SDSC then the scholarship ceases to apply. Please contact to Administrator at gssd.administrator@anu.edu.au.

Yuill Scholarship
Subjects: Law.
Purpose: to Support the International Court of Justice Traineeship Program.
Eligibility: The scholarship applicant must be: a final year student in a Bachelor of Laws program in the ANU College of Law; or a final year student in a Juris Doctor program in the ANU College of Law; or enrolled in a Master of Laws program in the ANU College of Law; or a recent graduate of the Bachelor of Laws, Juris Doctor or Master of Laws program from the ANU College of Law. The Bachelor of Laws program may be undertaken as a single or a combined degree.
Type: Scholarship
Value: $25,000
Frequency: Annual
No. of awards offered: 1
Closing Date: As advised on ANU College of Law website
Additional Information: The scholarship will apply for the duration of the traineeship (9 months in total). Application process and criteria is set out in the Conditions of Award. Applications can be sent electronically or in hard copy. Electronic applications can be sent to karen.heuer@anu.edu.au. Hard copy applications should be mailed to Office of the Dean of Law, ANU College of Law.

AUSTRALIAN PAIN SOCIETY (APS)

Secretariat Office, Dc Conferences Pty Ltd, PO Box 637, North Sydney, NSW, 2059, Australia
Tel: (61) 2 9016 4343
Fax: (61) 2 9954 0666
Email: aps@apsoc.org.au
Website: http://www.apsoc.org.au/

The Australian Pain Society (APS) is a non-profit organisation and is directed by an elected honorary council. The APS is anxious to enlist as members, other professionals interested in supporting the society in its goal of alleviating the suffering of our fellow citizens. Active membership in the APS is open to all health care professionals engaged in pain research or in the diagnosis and management of pain syndromes.

PhD Scholarships
Subjects: Health and life sciences.
Purpose: To improve the education, research and development, diagnosis and treatment of all forms of pain and to support education and research in pain.
Eligibility: Open to permanent residents and citizens of Australia and New Zealand citizens and applicants with permanent humanitarian visa.
Level of Study: Doctorate
Type: Scholarship
Value: Australian $23,000 per year. In addition to this, the Australian Pain Society (APS) will provide financial support enabling the candidate to attend the Society's Annual Scientific Meeting.
Length of Study: 3 years
Frequency: Annual
Country of Study: Australia
No. of awards offered: 2
Application Procedure: Further information about the PhD scholarship, including the conditions of award, can be obtained from the APS secretariat.
Closing Date: September 21st
Funding: Commercial
Contributor: Mundipharma Australia and Janssen-Cilag Australia and New Zealand

AUSTRALIAN RESEARCH COUNCIL (ARC)

GPO Box 2702, Canberra, ACT, 2601, Australia
Tel: (61) 2 6287 6600
Fax: (61) 2 6287 6601
Email: ncgp@arc.gov.au
Website: www.arc.gov.au
Contact: Dr Laura Dan, Director, Program Operations

The Australian Research Council (ARC) is a statutory authority within the Australian Government's Innovation, Industry, Science and Research portfolio. The ARC advises the Government on research matters, manages the National Competitive Grants Program, a significant component of Australia's investment in research and development, and has responsibility for the Excellence in Research for Australia initiative.

ARC Discovery Projects: Australian Laureate Fellowships

Subjects: All areas of research except clinical medicine or dentistry.
Purpose: The Australian Laureate Fellowships provide opportunities for world-class researchers and research leaders to key positions in Australia and create new rewards and incentives for the application of their talents.
Eligibility: Open to researchers with excellent research records, strong leadership and mentoring skills.
Level of Study: Postdoctorate
Type: Fellowship
Value: Average funding Australian $2,618,641 over 5 years
Length of Study: 5 years
Frequency: Annual
Country of Study: Australia
No. of awards offered: 15
Application Procedure: Applicants must submit applications through an Australian eligible organization.
Closing Date: January
Funding: Government
No. of awards given last year: 17
No. of applicants last year: 141
Additional Information: The recipient must legally reside predominantly in Australia for the duration of the Fellowship, except where ARC approval has been granted.

ARC Discovery Projects: Discovery Early Career Researcher Award (DECRA)

Subjects: All areas except medical and dental research, defined as research and/or training primarily and substantially aimed at understanding or treating a human disease or human health condition.
Purpose: To provide more focused support for researchers and create more opportunities for early-career researchers in both teaching and research, and research-only positions.
Eligibility: Researchers who have been awarded a PhD within five years or, commensurate with significant career interruption for maternity or parental leave; carer's responsibility; illness; international post doctoral studies; or non-research employment have been awarded a PhD within eight years, of the closing date for the scheme are eligible to apply.
Level of Study: Doctorate
Type: Award
Value: Up to Australian $125,000 per year
Length of Study: 3 years (full-time)
Frequency: Annual
Country of Study: Australia
No. of awards offered: 200
Application Procedure: Applicants must submit applications through an Australian eligible organization.
Closing Date: March
Funding: Government
No. of awards given last year: 277
No. of applicants last year: 2,170
Additional Information: The recipient must hold an appointment at the Administering Organization for the duration of the Award. The recipient must legally reside predominantly in Australia for the duration of the Award, except where ARC approval has been granted.

ARC Discovery Projects: Discovery Indigenous Award (DIA)

Purpose: To support fundamental research and research training by Indigenous Australian researchers as individuals and as teams.
Eligibility: The applicant must be an Indigenous Australian researcher.
Level of Study: Doctorate
Type: Award
Value: Australian $85,000–180,000
Frequency: Annual
Country of Study: Australia
Application Procedure: Applicant must submit applications through an Australian eligible organization.
Closing Date: June
Funding: Government
No. of awards given last year: 5
No. of applicants last year: 20
Additional Information: The recipient must legally reside predominantly in Australia for the life of the project.

ARC Discovery Projects: Discovery Outstanding Researcher Award (DORA)

Purpose: The DORAs provide opportunities for mid to late career research-only and teaching and research adademics.
Eligibility: There are no restrictions in relation to time since award of PhD, and selection is based on the needs of the project in addition to the excellence of the researcher.
Level of Study: Postdoctorate
Type: Award
Value: Australian $128,632–182,792
Length of Study: Up to 3 years (full-time)
Frequency: Annual
Country of Study: Australia
No. of awards offered: Up to 70
Application Procedure: Applicants must submit applications through an Australian eligible organization.
Closing Date: March
Funding: Government
No. of awards given last year: 26
No. of applicants last year: 463
Additional Information: The recipient must legally reside predominantly in Australia for the duration of the Award, except where ARC approval has been granted.

ARC Discovery Projects: Future Fellowships

Subjects: Any subject.
Purpose: To promote research in areas of critical national importance by giving outstanding researchers incentives to conduct their research in Australia. The aim of *Fellowships* is to attract and retain the best and brightest mid-career researchers.
Eligibility: Applicants must have been awarded their PhD between 5 and 15 years before date of proposal submission.
Level of Study: Postdoctorate
Type: Fellowship
Value: Australian $133,582–189,826
Length of Study: 4 years, full-time
Frequency: Annual
Country of Study: Australia
No. of awards offered: 1,000
Application Procedure: Applicants must submit applications through an Australian eligible organization.
Closing Date: November
Funding: Government
No. of awards given last year: 203
No. of applicants last year: 670
Additional Information: The Future Fellowship must legally reside predominantly in Australia for the duration of the Future Fellowship, except where ARC approval has been granted.

ARC Linkage Projects: Australian Postdoctoral Fellow Industry (APD)

Subjects: All areas of research except clinical medicine or dentistry.
Purpose: To encourage and develop long-term strategic research alliances between higher education organizations and other organizations by fostering opportunities for postdoctoral researchers to pursue internationally competitive research in collaboration with organizations outside the higher education sector.
Eligibility: Applicants must have submitted their PhD thesis before the commencement of the fellowship.
Level of Study: Postdoctorate
Type: Fellowship
Value: Australian $83,136
Length of Study: 3 years
Frequency: Annual
Country of Study: Australia

No. of awards offered: 30
Application Procedure: Applicants must submit applications through an Australian eligible organization.
Closing Date: November
Funding: Government
No. of awards given last year: 8
No. of applicants last year: 31
Additional Information: An APDI Fellowship must be taken on a full-time basis to work on the approved Project at the Host organization..

Super Science Fellowships

Purpose: To attract and retain outstanding early-career researchers in three key areas: space and astronomy; marine and climate; and future industries.
Type: Fellowship
Value: Up to $72,500 a year plus 28 per cent on-costs
Length of Study: 3 years
Frequency: Annual
No. of awards offered: 100
Application Procedure: Please check at http://www.arc.gov.au/ncgp/ssf/ssf_default.htm for application procedure.
Additional Information: The Super Science Fellowships scheme does not support Medical and Dental research. Please check website for more information.

AUSTRALIAN SPORTS COMMISSION (ASC)

PO Box 176, Belconnen, ACT, 2616, Australia
Tel: (61) 02 6214 1111
Fax: (61) 02 6251 2680, 02 6214 1836
Email: recruitment@ausport.gov.au
Website: www.ausport.gov.au

The Australian Sports Commission (ASC) is responsible for implementing the Australian Government's national sports policy, which is based on a sports philosophy of excellence and participation. It promotes an effective national sports system that offers improved participation in quality sports activities by all Australians and helps the talented and motivated to reach their potential excellence in sports performance. Its work is guided by the Australian Government's national sports policy, Building Australian Communities through Sport (BACTS).

Biomechanics Postgraduate Scholarship (General Sports)

Subjects: Sports.
Purpose: To provide an opportunity for graduates with degrees with a major emphasis in biomechanics to have experience in the application of biomechanics to enhance elite sports performance.
Eligibility: Applicants must have tertiary qualification in science, human movement, mathematics or a related area, should have an interest and/or understanding of research principles and some knowledge of biomechanical systems and equipment.
Level of Study: Graduate
Type: Scholarship
Value: Australian $21,434 per year
Length of Study: 48 weeks
No. of awards offered: 3
Application Procedure: Applicants must submit a covering letter, a statement of experience and curriculum vitae along with the application form.
Closing Date: September 29th

For further information contact:

Tel: 02 6214 1659
Email: recruitment@ausport.gov.au
Contact: Dale Barnes

Indigenous Sporting Excellence Scholarships

Subjects: Sports.
Purpose: To give indigenous sportspeople the opportunity to improve their sporting performance at an elite level.
Eligibility: Applicants must be over 12 years of age, representing their state in national competition or Australia internationally within sport or the school sport system, a coach with level 1 or level 2 accreditation, a sports trainer with level 1 accreditation, a sports official with accreditation, competing in a sport that is recognized by the Australian Sports Commission.
Type: Scholarship
Value: Australian $500
No. of awards offered: 100
Application Procedure: Check website for further details.
Closing Date: May 31st
Additional Information: Athletes, coaches, sports trainers and officials who receive this scholarship are also eligible to apply for the Elite Indigenous Travel and Accommodation Assistance Program if they are selected for a state representative team attending national championships or an Australian team competing internationally.

National Coaching Scholarship Program

Subjects: Sports.
Purpose: To provide opportunities for potential and current elite coaches to develop skills and knowledge to coach effectively in high-performance programs.
Eligibility: Applicants should be a coach. They are expected to be integrally involved in the elite program in which they are placed.
Level of Study: Coach
Type: Scholarship
Value: Varies
Length of Study: 1–2 years
Application Procedure: Check website for further details.
Closing Date: Check website for further details.
Funding: Government

Performance Analysis Scholarship

Subjects: Sports.
Purpose: To provide performance analysis services to AIS sport programmes through coaches as directed and as required work on projects relating to enhancement of knowledge and coach education.
Eligibility: Applicants must have a tertiary qualification in computer science, software engineering, information technology, human movement or a related area and an interest and/or understanding of research principles.
Type: Scholarship
Value: Australian $21,434 per year
Length of Study: 48 weeks
Application Procedure: Applicants must write a covering letter referencing the position title, prepare a thorough (but concise) statement that focuses on the relevant experience, curriculum vitae that summarize the qualifications including contact details for two referees and can be submitted by email.
Closing Date: September 29th

For further information contact:

Tel: 02 6214 1659
Email: recruitment@ausport.gov.au
Website: www.ausport.gov.au
Contact: Dale Barnes

Postgraduate Scholarship Program–Physiology (Quality Control)

Subjects: Physiology.
Purpose: To provide opportunity for a graduate whose primary role will be to assist in the area of quality control under the direction of the laboratory manager.
Eligibility: Applicants must have a tertiary qualification in science or a related area, an interest and/or understanding of research principles and an experience in an administrative role and good computing skills.
Level of Study: Graduate
Type: Scholarship
Value: Australian $21,434 per year
Length of Study: 54 weeks
No. of awards offered: 1
Application Procedure: Applicants must submit a covering letter, a statement of experience and curriculum vitae along with the application form.
Closing Date: September 29th
Contributor: Australian Sports Commission

Additional Information: Terms and conditions are subject to change. Always confirm details with scholarship provider before applying.

For further information contact:

Tel: 02 6214 1564
Email: recruitment@ausport.gov.au
Contact: Marilyn Dickson, Manager

Postgraduate Scholarship–Biomechanics (Swimming)

Subjects: Sports.
Purpose: To provide the opportunity for graduates of degrees with major emphasis in biomechanics and to have experience in the application of biomechanics to enhance elite sports performance.
Eligibility: Applicants must have a degree in biomechanics or human movement sciences, an interest and/or understanding of research principles, experience in biomechanics services and knowledge of and experience in a competitive swimming environment.
Level of Study: Graduate
Type: Scholarship
Value: Australian $21,434 per year
Length of Study: 48 weeks
Application Procedure: Applicants must submit a covering letter, a statement of experience and curriculum vitae along with the application form.
Closing Date: September 29th

For further information contact:

Tel: 02 6214 1732
Email: recruitment@ausport.gov.au
Contact: Clare Jones

Postgraduate Scholarship–Physiology (Biochemistry/ Haematology)

Subjects: Physiology.
Purpose: To provide an opportunity for a graduate whose primary role will be to assist in the area of biochemistry and haematology under the direction of the biochemistry/haematology manager.
Eligibility: Applicants must have a degree in medical laboratory science or biological sciences and an interest and/or understanding of research principles, good computer skills.
Level of Study: Postgraduate
Type: Scholarship
Value: Australian $21,434 per year
Length of Study: 50 weeks
Application Procedure: Applicants must write a covering letter referencing the position title, prepare a thorough (but concise) statement that focuses on the relevant experience, curriculum vitae that summarize the qualifications including contact details for two referees and can be submitted by email.
Closing Date: September 29th

For further information contact:

Tel: (02) 6214 1700
Email: recruitment@ausport.gov.au
Website: www.ausport.gov.au
Contact: Graeme Allbon

Sport Leadership Grants for Women Program

Subjects: Sports.
Purpose: To provide women with an opportunity to undertake sport leadership training.
Eligibility: Applicants must be indigenous women, women in disability sport, women from culturally and linguistically diverse backgrounds and women in general sport leadership.
Type: Grant
Value: Up to Australian $5,000 for individuals and up to Australian $10,000 for incorporated organizations
Application Procedure: Check website for further details.
Closing Date: April 29th
Funding: Government

For further information contact:

Tel: (02) 6214 7994
Email: leadershipgrants@ausport.gov.au

Sports Physiology Postgraduate Scholarship–Fatigue and Recovery

Subjects: Physiology.
Purpose: To offer an Honours graduate in science or a related field the opportunity to complete a scholarship in physiology (fatigue and recovery).
Eligibility: Applicants must have an Honours Degree in science or a related area, basic skills in conducting routine physiological testing procedures, good computing skills, outstanding organizational skills and a high level of initiative.
Level of Study: Graduate
Type: Scholarship
Value: Australian $19,890 per year
Length of Study: 1 year
Application Procedure: Applicants must submit a covering letter, a statement of experience and curriculum vitae along with the application form.
Closing Date: April 6th

For further information contact:

Tel: 02 6214 1589
Email: Recruitment@ausport.gov.au
Website: www.ausport.gov.au
Contact: Shona Halson

AUSTRALIAN–AMERICAN FULBRIGHT COMMISSION

PO Box 9541, Deakin, ACT 2600, Australia
Tel: (61) 2 6260 4460
Fax: (61) 2 6260 4461
Email: lwilson@fulbright.com.au; fulbright@fulbright.com.au
Website: www.fulbright.com.au
Contact: Ms Lyndell Wilson, Programme Manager

The Australian–American Fulbright Commission is a non-profit organization in Australia, established through a binational treaty between the Australian and United States governments in 1949 under the auspices of the United States Educational Foundation (USEFA) in Australia. The mission of the Commission is to further mutual understanding between the people of Australia and the United States through educational and cultural exchange.

Fulbright Postdoctoral Fellowships

Subjects: All subjects.
Purpose: To enable those who have recently completed their PhD to conduct postdoctoral research, to further their professional training or lecture at a university in the United States of America.
Eligibility: Open to Australian citizens by birth or naturalization. Those holding dual citizenship, i.e. citizenship of both United States of America and Australian are not eligible. Applicants should have recently completed their PhD, normally less than 3 years prior to application, although those who have completed their PhD 4 or 5 years prior to application will be considered.
Level of Study: Postdoctorate
Type: Fellowship
Value: Up to US$40,000
Length of Study: 3–12 months
Frequency: Annual
Study Establishment: A university, college or research establishment or reputable private practice
Country of Study: United States of America
No. of awards offered: 1
Application Procedure: Applicants must complete and submit an application form along with three reference reports, already included in the application pack, documentation of citizenship, and qualifications. Further information and application packs are available from the website.
Closing Date: August 31st
Funding: Government
Additional Information: Fulbright Postdoctoral Scholars who also receive a salary related to their postdoctoral position in the U.S., at or above the value of the Fulbright stipend, will be granted a set amount of USD 20,000.

Fulbright Postgraduate Scholarship in Science and Engineering

Subjects: Science and engineering.
Purpose: To enable candidates to undertake an approved course of study for an American higher degree, or to engage in research relevant to an Australian higher degree.
Eligibility: Open to Australian citizens. Those with dual United States of America and Australia citizenship are not eligible.
Level of Study: Doctorate, Postgraduate
Value: US$40,000
Length of Study: 8–12 months funded or up to 4 years unfunded
Frequency: Annual
Study Establishment: An accredited institution
Country of Study: United States of America
No. of awards offered: 1
Application Procedure: Applicants must complete and submit an application form along with three reference reports, already included in the application pack, documentation of citizenship and qualifications. Further information and application packs are available from the website.
Closing Date: August 31st
Funding: Commercial, government
Contributor: Billiton Pvt. Ltd.

Fulbright Postgraduate Scholarships

Subjects: All subjects.
Purpose: To enable students to undertake an approved course of study for an American higher degree or equivalent, or to engage in research relevant to an Australian higher degree.
Eligibility: Open to Australian citizens by birth or naturalization. Those holding dual United States of America and Australian citizenship are not eligible.
Level of Study: Doctorate, Postgraduate, Research
Type: Scholarships
Value: US$40,000
Length of Study: 8–12 months funded, renewable for up to 5 years unfunded
Frequency: Annual
Study Establishment: An accredited institution
Country of Study: United States of America
No. of awards offered: Up to 15
Application Procedure: Applicants must complete an application form and submit this with three reference reports, already included in the application pack, and documentation of citizenship and qualifications. Naturalized citizens must provide a certificate of Australian citizenship with their application, and native-born Australians must provide a copy of their birth certificate. Further information and application packs are available from the website.
Closing Date: August 31st
Funding: Government
Additional Information: As the award does not include any provision for maintenance payments, applicants must be able to demonstrate that they have sufficient financial resources to support themselves and any dependants during their stay in the United States of America.

Fulbright Professional Scholarship

Subjects: All professional fields. Programmes should include an academic as well as a practical aspect.
Purpose: These awards are available to professionals from public and private sectors (junior to middle level staff poised for advancement to a senior level research or undertaking a programme of professional development in the United States of America.
Eligibility: Open to resident Australian citizens with a record of achievement poised for advancement to a senior management or policy role. Those holding dual United States of America and Australian citizenship are not eligible.
Level of Study: Professional development
Type: Scholarship
Value: Up to US$30,000
Length of Study: 3–4 months. Programmes of longer duration may be proposed, but without additional funding.
Frequency: Annual
Country of Study: United States of America
No. of awards offered: Up to 8
Application Procedure: Applicants must complete and submit an application form along with three reference reports, already included in the application pack, documentation of citizenship and qualifications. Further information and application packs are available from the website.
Closing Date: August 31st
Funding: Government

Fulbright Professional Scholarship in Vocational Education and Training

Subjects: Vocational education.
Purpose: To enable candidates to visit institutions or organizations and people in the United States of America from their own field.
Eligibility: Open to Australian citizens employed in the vocational education and training sector. Those holding dual United States of America and Australian citizenship are not eligible.
Level of Study: Professional development
Type: Scholarship
Value: Up to Australian $30,000
Length of Study: 3–4 months
Frequency: Annual
Country of Study: United States of America
No. of awards offered: 1
Application Procedure: Applicants must complete and submit an application form along with three reference reports, already included in the application pack, documentation of citizenship and qualifications. Further information and application packs are available from the website.
Closing Date: August 31st
Funding: Government
Contributor: The Australian National Training Authority
No. of awards given last year: 1

Fulbright Senior Scholarships

Subjects: All subjects.
Purpose: To allow candidates to teach, undertake research, be an invited speaker or visit institutions within their field.
Eligibility: Open to Australian citizens by birth or naturalization. Those holding dual United States of America and Australian citizenship are not eligible. Applicants should be either scholars of established reputation working in an academic institution, who intend to teach or research in the United States of America, leaders in the arts, e.g. music, drama, visual arts or senior members of the academically based professions who are currently engaged in the private practice of their profession.
Level of Study: Professional development
Type: Scholarships
Value: Up to US$30,000
Length of Study: 3–6 months
Frequency: Annual
Study Establishment: A university, college, research establishment or reputable private organization
Country of Study: United States of America
No. of awards offered: 2
Application Procedure: Applicants must complete an application form and submit this with three reference reports, already included in the application pack, documentation of citizenship and qualifications. Further information and application packs are available from the website. Naturalized citizens must provide a certificate of Australian citizenship with their application, and native-born Australians must provide a copy of their birth certificate.
Closing Date: August 31st
Funding: Government

AUSTRIAN ACADEMY OF SCIENCES

Institute of Limnology, Dr. Ignaz Seipel-Platz 2, 1010, Vienna, Austria
Tel: (43) 623 240 79
Fax: (43) 623 235 78
Email: gerold.winkler@oeaw.ac.at; webmaster@oeaw.ac.at
Website: www.oeaw.ac.at/limno
Contact: Mr Regina Brandstätter, IPGL Officer

With 1100 employee and 60 research institutions the Austrian Academy of Sciences is the leading organization promoting non-university academic research in Austria.The Institute of Limnology performs ecological research on inland waters to understand the structure, functions and dynamics of freshwater ecosystems.The IPGL office acts as a hub for postgraduate training, research and international networking.

Austrian Academy of Sciences, MSc Course in Limnology and Wetland Ecosystems

Subjects: Aquatic systems, Environmental Sciences.
Purpose: To understand the structure and functioning of aquatic and wetland ecosystems for the conservation of biodiversity and sustainable management of natural resources. To acquire skills for interacting with stakeholders, managers and policy makers in the development of best practices.
Eligibility: Open to candidates from developing countries who are maximum 35 years of age, have a good working knowledge of English and have an academic degree in science, agriculture or veterinary medicine from a university or any other recognized Institute of Higher Education. Applicants should have 3 years practical experience in at least one special subject in their field of professional training. All applications are considered on their individual merits.
Level of Study: Postgraduate
Type: Scholarship
Value: US$1,350 paid monthly to cover food, lodging and personal needs plus free tuition, health insurance, study material and equipment for laboratory work, field work and travelling expenses
Length of Study: 18 months
Frequency: Annual
Study Establishment: Institute for Limnology, Mondsee; Institute UNESCO-IHE, Delft, The Netherlands; Egerton University, Kenya; Czech Academy of Sciences, Trebon, Czech Republic; Austrian Universities and Federal Institutes in Austria
No. of awards offered: 4
Application Procedure: Applicants must obtain application forms from the website. Filled application forms can be sent to Institute for Limnology of the Austrian Academy of Sciences.
Closing Date: End of January
Funding: Government
Contributor: The Austrian Development Co-operation
No. of awards given last year: 4
No. of applicants last year: 90
Additional Information: No provisions are made for dependants. It is strongly advised that dependants do not accompany fellows due to frequent moves during the course. Fellows must also provide their own transportation to and from Austria.

For further information contact:

IPGL-Course, Institute for Limnology of the Austrian Academy of Sciences, Mondseestrasse 9, A-5310 Mondsee, Austria
Tel: (43) 6232/4079
Fax: (43) 6232/3578
Email: ipgl.mondsee@oeaw.ac.at

AUSTRIAN SCIENCE FUND (FWF)

Haus der Forschung, Sensengasse 1, Vienna, 1090, Austria
Tel: (43) 1 505 67 40
Fax: (43) 1 505 67 39
Email: office@fwf.ac.at
Website: www.fwf.ac.at
Contact: Scientific Administrator

The Austrian Science Fund (FWF) is Austria's central body for the promotion of basic research. It is equally committed to all branches of science and in all its activities it is guided solely by the standards of the international scientific community. Its mission is the promotion of high-quality basic research, education and training through research and scientific culture and knowledge transfer.

Elise Richter Program

Subjects: All subjects.
Purpose: To provide support for outstanding female scientists and researchers.
Eligibility: The candidate should possess appropriate postdoctoral experience, international scientific publications, and preparatory work related to the proposed research project.
Level of Study: Postdoctorate
Type: Research
Value: €15,000 (project specific costs), €66,680 (personal costs) and €1,950 (lump sum per child per year)
Length of Study: 12–48 months
Frequency: Twice a year
No. of awards offered: Varies
Closing Date: October 22nd and December 14th

For further information contact:

Tel: 43 1 505 67 40 ext. 8503
Email: susanne.menschik@fwf.ac.at
Contact: Susanne Menschik

Erwin Schrödinger Fellowship with Return Phase

Subjects: All subjects.
Purpose: To offer citizens the opportunity to work in leading research institutions and research programmes abroad, and to facilitate access to new areas of science and research for the fellows to later contribute to the scientific development in Austria.
Eligibility: Open to highly qualified Austrian citizens up to the age of 35 who have completed doctoral studies, have international scientific publications, and have invitation from the foreign research institution.
Level of Study: Postdoctorate
Type: Fellowship
Value: Fellowship abroad: depends on the destination, at an average €34,000 per year, tax free, for the return phase: contract of employment with Senior-Postdoc salary + €10,000 per year
Length of Study: 10–24 months without return phase respectively, 16–36 months including return phase (return phase = 6–12 months).
Study Establishment: Universities or research institutions
No. of awards offered: Varies
Application Procedure: Applicants must complete an application form, available from the Austrian Science Fund, from the website or by email. All necessary details can be found on the website www.fwf.ac.at/en/projects/erwin-schroedinger.html
Closing Date: Applications are accepted at any time
Funding: Government
No. of awards given last year: 55
No. of applicants last year: 117

Hertha Firnberg Research Positions for Women

Subjects: All subjects.
Purpose: To ensure a maximum support for female scientists and researchers starting academic careers at universities.
Eligibility: Open to female scientists up to the age of 40 who are residents of Austria. Applicant should have completed doctoral studies and have international scientific publications.
Level of Study: Postdoctorate
Type: Position (Employment)
Value: €60,610 personnel costs per year plus €10,000 for material, travel, assistance
Length of Study: 3 years
Frequency: Twice a year
Study Establishment: Any university
Country of Study: Austria
No. of awards offered: Varies
Application Procedure: Applicants must complete an application form, available from the Austrian Science Fund, from the website or by email. All necessary details can be found on the website www.fwf.ac.at/en/projects/firnberg.html.
Closing Date: October 22nd and December 14th
Funding: Government
No. of awards given last year: 11
No. of applicants last year: 43

Lise Meitner Program

Subjects: All subjects.
Purpose: To enhance the quality of scientific know-how in Austria's scientific community by supporting highly qualified researchers from abroad who could contribute to the scientific development of an Austrian research institution by working at it.

Eligibility: Open to highly qualified foreign scientists up to the age of 40.
Level of Study: Postdoctorate
Type: Position (Employment)
Value: €60,610 per year (Postdoctorate); €66,680 per year (Senior Postdoctorate) plus €10,000 for material, travel, and assistance.
Length of Study: 1–2 years
Study Establishment: Universities or research institutions
Country of Study: Austria
No. of awards offered: Varies
Application Procedure: Applicants must complete an application form, available from the Austrian Science Fund, from the website or by email. All necessary details can be found on the website: www.fwf.ac.at/en/projects/meitner.html.
Closing Date: Applications are accepted at any time.
Funding: Government
No. of awards given last year: 35
No. of applicants last year: 86

START Program

Subjects: All subjects.
Purpose: To provide highly promising young researchers of any discipline with the means to plan their research work on a long-term basis and with sufficient financial security.
Eligibility: Open to applicants who possess at least 2 and at most 10 years of postdoctoral experience at the time of application. In addition it is desirable that candidates have completed a research stay abroad of at least one year.
Level of Study: Postdoctorate
Value: Minimum €800 up to maximum €1.2 Mio
Length of Study: 6 years, with an interim review after 3 years.
No. of awards offered: Up to 6
Closing Date: September 20th
Funding: Government
Contributor: Federal Ministry for Science and Research (BMWF)

For further information contact:

Tel: 43 1 505 67 40 ext 8605
Email: mario.mandl@fwf.ac.at
Contact: Mario Mandl

Translational Brainpower

Subjects: All subjects.
Purpose: To support the integration of highly qualified scientists and researchers from abroad into research projects at the interface between basic and applied research in Austria.
Eligibility: Open to candidates who possess high scientific quality at international level.
Level of Study: Research
Value: According to the project; costs for the international partner (salary, max. €9,460 per month; travel costs, travel and subsistence costs).
Length of Study: Maximum 9 months
No. of awards offered: Varies
Application Procedure: Submit application to: FWF Der Wissenschaftsfonds, Haus der Forschung, Sensengasse 1, 1090 Wien.
Closing Date: March 22nd

For further information contact:

Tel: 43 1 505 67 40 ext. 8602
Email: milojka.gindl@fwf.ac.at
Contact: Milojka Gindl

Wittgenstein Award

Subjects: All subjects.
Purpose: To provide highly qualified researchers of any discipline with a maximum of freedom and flexibility in carrying out their research work.
Eligibility: The candidate should possess international recognition in the field and be employed in an Austrian research organization and should be aged 55 or under at the time of nomination.
Level of Study: Professional development, Research
Type: Research
Value: Up to €1.5 million per award
Length of Study: 5 years
Frequency: Annual
No. of awards offered: 1–2
Closing Date: September 20th
Funding: Government
Contributor: Federal Minister for Science

For further information contact:

Tel: 43 1 505 67 40 ext 8605
Email: mario.mandl@fwf.ac.at
Contact: Mario Mandl

AUSTRO-AMERICAN ASSOCIATION OF BOSTON

67 Bridle Path, Sudbury, MA, 01776, United States of America
Tel: (1) 781 283 2255
Email: thansen@wellesley.edu
Website: http://www.austria-boston.org/
Contact: Professor Thomas S. Hansen, Chairman of Scholarship Committee

Membership of the Austro-American Association of Boston is open to any individual interested in any aspect of Austrian history, economy, culture, politics and tourism. The association conducts meetings and get togethers focussing on events and experiences related to Austria.

A-AA Austrian Studies Scholarship Award

Purpose: Project areas may include history, literature, art, architecture, folk customs, music and contemporary life.
Eligibility: All undergraduate and graduate students at New England colleges and Universities are eligible to apply.
Level of Study: Graduate, Undergraduate
Type: Award
Value: $1500
No. of awards offered: 1
Application Procedure: There is no application form. Applicants should submit the following information: a detailed project proposal, including the reason for selecting it; curriculum vitae; and two confidential letters of support from faculty members who know the applicant well and can comment on the feasibility of the project.
Closing Date: March 1st
Additional Information: Please feel free to email any or submissions or inquiries to the co-chair of the scholarship committee, Dr. Johann Nittmann at Johann.Nittmann@cavium.com.

For further information contact:

Department of German, Wellesley College, Wellesley, Massachusetts, 02481, United States of America
Tel: (1) 781 283 2255
Email: thansen@wellesley.edu
Contact: Professor Thomas Hansen

Austro-American Association of Boston Stipend

Subjects: Austrian cultural studies, history, folklore, literature, music, fine and applied arts, and film.
Purpose: To promote the understanding and dissemination of Austrian culture.
Eligibility: Junior faculty members and students enrolled in a college in New England.
Level of Study: Unrestricted
Type: Stipendiary
Value: US$1,500
Study Establishment: Any in New England
Country of Study: United States of America
No. of awards offered: 1
Application Procedure: Applicants must submit a detailed description of the project including the reasons for selecting it, a curriculum vitae and two letters of recommendation from faculty members who know the applicant well and can comment on the feasibility of the project.
Closing Date: April 1st
Funding: Private
Contributor: Association members
No. of applicants last year: 1

Additional Information: The award is limited to individuals living or studying in New England. Projects funded in the past have included the preparation of musical or dramatic performances, the facilitation of appropriate publications and research trips to Austria. Culture is defined to include the humanities and the arts. The recipient may be asked to present the results of the project at an event of the Austro-American Association. The award may not be used to pay tuition fees at a college or university in New England.

For further information contact:

Wellesley College, Wellesley, Mass., 02481, United States of America
Email: thansen@wellesley.edu
Contact: Professor Thomas Hansen, Department of German

AWSCPA

Administrative Offices, 136 South Keowee Street, Dayton, OH, 45402, United States of America
Tel: (1) 937 222 1872
Fax: (1) 937 222 5794
Email: info@awscpa.org
Website: www.awscpa.org

The American Woman's Society of CPA provides annual scholarships to women working towards an accounting degree as well as to those working towards their Certified Public Accountant License.

AWSCPA National Scholarship

Subjects: Accounting.
Purpose: To fund women working towards an accounting degree.
Eligibility: Applicants must meet the minimum educational requirements to sit for the CPA exam within 1 year of the award of the scholarship. Applicant must have a 3.0 GPA in accounting and a 3.0 GPA overall.
Level of Study: Postgraduate
Type: Scholarship
Value: Complete Becker CPA Review Course worth $3,065.
Frequency: Annual
No. of awards offered: 1 per year
Application Procedure: Applicant must either be a U.S. citizen or a permanent resident of the United States. Applicants must provide a validated transcript from their college and submit an essay no longer than 1,000 words explaining why they are deserving of the scholarship. Apply through website.
Closing Date: May 11th
Funding: Foundation
Contributor: AWSCPA members and supporters and Becker Professional Education (Platinum Sponsor)
No. of awards given last year: 1
No. of applicants last year: 30

BACKCARE

16 Elmtree Road, Teddington, Middlesex, TW11 8ST, England
Tel: (44) 20 8977 5474
Fax: (44) 20 8943 5318
Email: info@backcare.org.uk
Website: www.backcare.org.uk
Contact: Mr Sash Newman, Chief Executive

BackCare is a national charity dedicated to educating people about how to avoid preventable back pain and support those living with back pain. BackCare provides education and information through its publications, telephone helpline, local branches and website. It also funds research and campaigns to raise the profile of issues surrounding back pain.

BackCare Research Grants

Subjects: Studies related to back pain.
Purpose: To reduce the incidences of disability from back pain and to improve its treatment by gaining, through research, a better understanding of its manifestation and causes.
Eligibility: Open to appropriately qualified and experienced persons.
Level of Study: Postgraduate
Type: Project grant
Value: Varies, dependent on funds available
Length of Study: Up to 2 years
Frequency: Dependent on funds available
Study Establishment: Suitable establishments
Country of Study: United Kingdom
No. of awards offered: Varies, depends on funds available
Application Procedure: Applicants must refer to the website for details of the application procedure.
Closing Date: Please check the website for details.
Funding: Trusts, individuals, private
No. of awards given last year: 4
No. of applicants last year: 20

THE BANFF CENTRE

107 Tunnel Mountain Road, Box 1020, Banff, AB, T1L 1H5, Canada
Tel: (1) 403 762 6100, 403 762 6180
Fax: (1) 403 762 6444
Email: arts_info@banffcentre.ca
Website: www.banffcentre.ca
Contact: Office of the Registrar

The Banff Centre is a catalyst for creativity, with a transformative impact on those who attend the programmes, conferences and events. Our alumni create, produce and perform works of art all over the world, lead our institutions, organizations and businesses, and play significant roles in our cultural, social, intellectual and economic well-being, and in the preservation of our environment.

Banff Centre Scholarship Fund

Subjects: Studio art, photography, ceramics, performance art, video art, theatre production and design, stage management, opera, singing, dance, drama, music, writing, creative non-fiction and cultural journalism, publishing, media arts, television and video, audio recording, computer applications, research, audio engineering work study, theatre production, design stage management work study, Aboriginal arts programmes in dance training, programme publicity and theatre production work study and screenwriting.
Purpose: To provide financial assistance to deserving artists for a residency at The Banff Centre.
Eligibility: Open to advanced students who have been accepted for a residency at The Banff Centre.
Level of Study: Professional development, Postgraduate
Type: Grant
Value: Residency fee and on-campus accommodation and meals.
Length of Study: Varies
Frequency: Annual
Study Establishment: The Banff Centre, Arts Programming
Country of Study: Canada
No. of awards offered: Varies
Application Procedure: Applicants must submit a completed application form, accompanied by requested documentation. See website for details.
Closing Date: Varies according to programme
Funding: Government, private
Contributor: Individual donations and The Banff Centre revenues
No. of awards given last year: 1,000

Paul D. Fleck Fellowships in the Arts

Subjects: Arts.
Purpose: To allow artists in all disciplines to participate in independent artists residencies at the Banff centre to create new work or collaborate with other arts programmes at the centre.
Eligibility: Nomination.
Level of Study: Research, Postgraduate, Professional development
Type: Fellowship/Scholarship
Value: Tuition, accommodation, meals and in approved circumstances, a travel award.
Frequency: Annual
Study Establishment: Arts programming, The Banff Centre
Country of Study: Canada
No. of awards offered: Varies
Application Procedure: See website for details.
Closing Date: Spring. See website for details.
Funding: Individuals, private, trusts

No. of awards given last year: Varies
No. of applicants last year: Varies

BANGOR UNIVERSITY

Research Office, Bangor University, 9th Floor, Alun Roberts Building, Deiniol Road, Bangor, Gwynedd, LL57 2UW, Wales
Tel: (44) 12 4835 1151
Fax: (44) 12 4837 0451
Email: aos033@bangor.ac.uk
Website: www.bangor.ac.uk

Bangor University is the principal seat of learning, scholarship and research in North Wales. Established in 1884, the University attaches considerable importance to research training in all disciplines and offers research studentships of a value similar to those of other United Kingdom public funding bodies.

Anniversary Research Scholarships

Subjects: All subjects.
Purpose: To celebrate our 125th anniversary in 2009, Bangor University launched a five year programme of postgraduate expansion. As a continuation of this strategic expansion, in 2012/2013 we will be offering further PhD research scholarships. This investment in research forms part of the University's strategy to grow research excellence, and to further enhance our dynamic research environment. We are an ambitious university, committed to expanding our research capacity and our international standing.
Eligibility: The Bangor University Anniversary Research Scholarships scheme offers opportunities to the very best students to work with our leading academic figures and rising stars.
Level of Study: Postgraduate
Type: Scholarship
Value: All fees for three years and provide an annual stipend and an annual research allowance
Length of Study: 3 years
Application Procedure: For details on how to apply, check at http://www.bangor.ac.uk/scholarships/apply.php.en.

ATM Postgraduate Scholarships

Subjects: Creative studies and media, chemistry, english, ocean studies, sport, health and exercise sciences, welsh and modern languages.
Purpose: Access To Masters is pan-Wales ESF Convergence project lead by Swansea University. ATM supports Taught Masters places on courses, which are linked to the Welsh Assembly Government's priority sectors.
Eligibility: For those pursuing a Masters in the following areas: ICT, energy and environment, advanced materials and manufacturing, creative industries, life sciences and financial and professional services.
Level of Study: Doctorate, Postgraduate
Type: Scholarship
Application Procedure: Please visit the website for more information: www.bangor.ac.uk/atm/.
Closing Date: Please contact the institution
Contributor: Convergence European Social Fund (ESF) through the Welsh Assembly Government

Bursaries

Subjects: Electronic engineering and computer science.
Purpose: To provide financial support, based on academic performance, for full-time masters students of all courses.
Eligibility: Open to applicants who have already been offered a place in one of the postgraduate degree courses.
Level of Study: Postgraduate
Type: Bursary
Value: £1,000
Frequency: Dependent on funds available
Study Establishment: Bangor University
Country of Study: United Kingdom
No. of awards offered: Unspecified
Application Procedure: Refer to the website for further details.
Contributor: Bangor University

Doctoral Scholarships

Purpose: To provide financial support to those undertaking full-time and research degrees.
Eligibility: Refer to the website for further details.
Type: Studentship and scholarship
Value: £12,000 per year
Frequency: Annual
Study Establishment: Bangor University
Country of Study: United Kingdom
No. of awards offered: Unspecified
Application Procedure: Refer to the website.
Contributor: Bangor University anfd UK Research Councils

Gold and Silver Scholarships

Subjects: Banking, Mangement, Business and Finance, and Law.
Purpose: To provide financial support to full-time students on all MSc, MBA and MA programmes.
Eligibility: Open to applicants who wish to apply for a postgraduate MBA or MA degree programme included in the scholarship scheme.
Level of Study: MBA, Postgraduate
Type: Scholarship
Value: SENRGY Gold Scholarship £5,000 per year. SENRGY Silver Scholarship £2,000 per year
Length of Study: 1 year
Frequency: Annual
Study Establishment: Bangor University
Country of Study: United Kingdom
Application Procedure: There is no application form for scholarships. Candidates who wish to be considered for the awards should include a letter listing their main academic and personal achievements together with a short essay on why they have chosen to study at Bangor and a curriculum vitae. For further information contact at law.pg@bangor.ac.uk.
Closing Date: March 1st (Interim deadline); July 1st (Final deadline)
Funding: Government

For further information contact:

Bangor Business School, Bangor University, Bangor, Gwynedd, LL57-2DG, Wales
Tel: (44) 1248 382644
Fax: (44) 1248 383228
Email: b.hamilton@bangor.ac.uk
Website: www.bbs.bangor.ac.uk
Contact: Bethan Hamilton-Hine Scholarships (Gold and Silver)

International Entrance Scholarship

Subjects: All subjects.
Purpose: Scholarships are available for outstanding International students applying for postgraduate study.
Eligibility: The scholarship is not open to UK/EU applicants.
Level of Study: Postgraduate
Type: Scholarship
Value: Successful applicants who have been offered a place to study at Bangor University will be awarded a £500 scholarship; £1,000 will be awarded to applicants who have achieved a 2.1 (or equivalent) in their bachelors programme; and £2,000 will be awarded to applicants who have achieved a 1st class honours (or equivalent) in their bachelors programme £5,000 will be awarded to a limited number of exceptional applicants.
Application Procedure: Please visit http://www.bangor.ac.uk/international/future/scholarship_pg.php for more details.
Closing Date: June 1st

Knowledge Economy Skills Scholarships (KESS)

Subjects: All subjects.
Purpose: Knowledge Economy Skills Scholarships (KESS) is a major European Convergence programme led by Bangor University on behalf of the HE sector in Wales. Benefiting from European Social Funds (ESF), KESS will support collaborative research projects (Research Masters and PhD) with external partners based in the Convergence area of Wales (West Wales and the Valleys). KESS will run from 2009 until 2014 and will provide 400+ PhD and Masters places.
Eligibility: Suitable for students living in the convergence area and who are eligible to work in the area after their study period.

Level of Study: Postgraduate, Doctorate
Type: Scholarship
Frequency: Annual
Application Procedure: Details of application can be found here: www.higherskillswales.co.uk.
Closing Date: Please check website for details

MA/MPhil/PhD Scholarships

Subjects: Art and design, history, humanities, languages, literature and creative writing, media studies and publishing, music and performing arts, theology and religious studies.
Eligibility: Open to students seeking places on MA, MPhil or PhD courses within the College of Arts and Humanities or in any of its research units or institutes.
Level of Study: Postdoctorate, Doctorate, Postgraduate
Type: Scholarship
Value: £2,000, subject to ratification
Frequency: Annual
Study Establishment: Bangor University
Country of Study: United Kingdom
No. of awards offered: Varies
Application Procedure: Refer to the website. For further details contact: s.lee@bangor.ac.uk.
Closing Date: June 29th

MANTAIS Masters Scholarship for Welsh Medium Study

Type: Scholarship
Value: £1,500
Frequency: Annual
Closing Date: July 15th
Additional Information: Please check website for more information.

Mr and Mrs David Edward Memorial Award

Subjects: All subjects offered by the University.
Purpose: To support doctoral studies in any subject area.
Eligibility: Open to holders of a relevant First Class (Honours) Degree or, exceptionally, Upper Second Class (Honours) Degree, who are classified as United Kingdom or European Union students for fee purposes.
Level of Study: Postgraduate, Doctorate
Type: Award
Value: No less than that of a Research Council or British Academy Research Studentship, including fees.
Length of Study: 1 year, renewable for a maximum of a further 2 years if satisfactory progress is maintained.
Frequency: Dependent on funds available
Study Establishment: Bangor University
Country of Study: United Kingdom
No. of awards offered: 1
Application Procedure: Nominations are made by the University and nominees are invited to submit formal applications.
Closing Date: See website for details.
Funding: Private
No. of awards given last year: 1
No. of applicants last year: 20

MSc Bursaries

Subjects: Sports and exercise psychology, sports and exercise physiology, sports science and exercise rehabilitation.
Eligibility: Open to applicants with good second class honours degree in sports science or health and to students with a 2:2 degree or a degree from a different academic area will also be considered.
Level of Study: Postgraduate
Type: Bursary
Value: £2,500 (UK/EU students); £3,500 (non-EU international students)
Length of Study: 1 year (full-time); 2 years (part-time); 30 weeks full-time (diploma)
Frequency: Dependent on funds available
Study Establishment: Bangor University
Country of Study: United Kingdom
No. of awards offered: 5–10
Application Procedure: Complete a Postgraduate Application Form with a four page (maximum) CV or resumé. Refer to the website for further details or mail to postgraduate@bangor.ac.uk.
Closing Date: June 30th
Contributor: Bangor University
Additional Information: £1,000 internal bursaries to former SHES (or related disciplines) students (who have 1st class undergraduate degree). £1,000 internal assistantships in addition to the other bursaries aimed at the very best students.

For further information contact:

Tel: 01248 383493
Email: mscsport@bangor.ac.uk/
Website: www.shes.bangor.ac.uk/
Contact: James Hardy

NERC Studentships

Subjects: Applied physical oceanography and marine environment protection.
Purpose: To support students who wish to pursue studies related to oceanography.
Eligibility: Open to applicants who already have, or expect to obtain, a first class or upper second class honours degree in an appropriate subject.
Level of Study: Doctorate, Postgraduate
Type: Studentships and bursaries
Value: Stipend plus fees for UK nationals; fees only for EU nationals; UK residency qualification may allow funding for non-EU nationals. Refer NERC website for further details.
Study Establishment: Bangor University
Country of Study: United Kingdom
No. of awards offered: 4 (for Applied physical oceanography), 5 (for Marine environment protection)
Application Procedure: Applicants must submit a completed application form, a curriculum vitae (with work and academic experience) and a covering letter with reasons for applying for the funding.
Closing Date: March 23rd
Contributor: NERC

Open PhD Studentships

Subjects: English, history, Welsh history and archaeology, linguistics and English language, modern languages, music, theology and religious studies, Welsh creative industries.
Eligibility: Open to candidates who have applied unsuccessfully to a UK funding council (e.g. the AHRC or the ESRC) to study at Bangor.
Level of Study: Doctorate
Type: Studentship
Value: Fees plus maintenance grant
Frequency: Dependent on funds available
Application Procedure: Applicants must submit a scholarship application form along with a summary of your proposed research project in up to 500 words.
Closing Date: June

For further information contact:

Email: s.lee@bangor.ac.uk

PhD Studentships

Subjects: Law.
Purpose: To fund research training at the PhD level.
Eligibility: Open to candidates classified as United Kingdom and European Union students for fee purposes who have attained a First Class (Honours) Degree or, exceptionally, an Upper Second Class (Honours) Degree or equivalent.
Level of Study: Doctorate, Postgraduate
Type: Studentship
Value: Equal to that of a British Research Council Studentship
Length of Study: 1 year, renewable for a maximum of 2 additional years
Frequency: Annual
Study Establishment: Bangor University
Country of Study: United Kingdom
No. of awards offered: Unspecified
Application Procedure: Applicants must contact the relevant department of proposed study. The department will nominate the most worthy and eligible students.
Closing Date: Refer to the website

Funding: Private
Contributor: Drapers Trust/Thomas Hovells Law PhD scholarship
No. of awards given last year: 11
No. of applicants last year: 90

Postgrad Solutions Bursary
Eligibility: Open to students from anywhere in the world.
Type: Bursary
Value: £500
Frequency: Annual

Postgraduate Widening Access Bursaries
Purpose: The awards are aimed specifically at widening access to full-time taught postgraduate Masters courses and are open to UK students who meet specific criteria.
Eligibility: Applicant must be a current Bangor University student / Bangor graduate, and have / expect to achieve a 2.1 or above in your undergraduate degree, and have applied / are considering applying for a full-time PGT Masters course (excluding MBAs and PGCEs) at Bangor University for entry in September, and not be in receipt of another bursary / scholarship for the PGT Masters course; must meet the specific postcode criteria for these awards.
Level of Study: Graduate
Type: Bursary
Value: £3,000 each
Frequency: Annual
No. of awards offered: 2
Additional Information: Please check website for more information.

Santander Taught Postgraduate Scholarships
Subjects: All subjects.
Purpose: The Santander Group will be awarding a number of one year undergraduate and taught postgraduate scholarships to current Bangor University students.
Eligibility: The scholarship fund aims to reward the most academically gifted students from countries that are supported by the Santander Universidades scheme. The award will be given to students from the following 11 countries: Argentina, Brazil, Chile, Colombia, Mexico, Portugal, Puerto Rico, Spain, Uruguay and Venezuela. To be eligible for the postgraduate scholarship you will have to have studied within a University which is part of the Santander Universidades Scheme.
Type: Scholarship
Value: All studentships are for one year only and vary from £3,000–4,166
Application Procedure: Application forms and guidance notes for the Santander Scholarship Scheme are available on the University website - http://www.bangor.ac.uk/scholarships/santander.php.en.
Closing Date: December 10th
Additional Information: For further information or if you have any questions about the scheme, please contact Academic Registry.

For further information contact:

Email: k.chidley@bangor.ac.uk
Contact: Mrs Karen Chidley, Academic Registry

Sir William Roberts Scholarship
Subjects: Animal care and veterinary science, food sciences.
Purpose: To fund research training at the PhD level.
Eligibility: Open to candidates classified as home/EU based who have attained a First Class (Honours) Degree or Upper Second Class (Honours) Degree. These scholarships are allocated to the School of The Environment and Natural Resources and the School of Biological Sciences.
Level of Study: Doctorate, Postgraduate
Type: Scholarships and fellowships
Value: Equal to that of a British Research Council Research Studentship
Length of Study: 1 year, renewable for a maximum of 2 additional years
Study Establishment: Bangor University
Country of Study: United Kingdom
No. of awards offered: 2
Application Procedure: Applicants must contact the Head of the relevant School.
Closing Date: June 1st
Funding: Private
No. of awards given last year: 2
No. of applicants last year: 20

Three-year PhD studentship
Subjects: Media and digital communication, journalism, writing, film studies, theatre and performance, creative practice.
Eligibility: Applicants with a strong background in broadcast journalism and/or media/broadcast communication are welcome to apply, and should have a first or upper second-class honours degree in journalism, media, communication, linguistics or a related subject. We would normally expect applicants to have successfully completed a Masters degree or have equivalent industry experience.
Level of Study: Doctorate
Type: Studentship
Value: £10,000 per year plus UK fees
Length of Study: 3 years
Frequency: Annual
Application Procedure: Applicants should complete the university's application form for postgraduate study, which can be found here: http://www.bangor.ac.uk/courses/postgrad/research/apply_research.php.en. Please make sure you write 'Drapers PhD studentship' on the form under 'area of research' in Section B, and include a detailed research proposal on a separate sheet, related to the broad topic described here.
Closing Date: February 28th
Contributor: Thomas Howell's Educational Fund for North Wales
Additional Information: Informal enquiries can be made to Dr Stephanie Marriott (stephanie.marriott@bangor.ac.uk) or to Dr Astrid Ensslin, Director of Graduate Studies for the School of Creative Studies and Media (stephanie.marriott@bangor.ac.uk).

BATTEN DISEASE SUPPORT AND RESEARCH ASSOCIATION

1175 Dublin Road, Columbus, Ohio, 43215, United States of America
Tel: (1) 800 448 4570
Fax: (1) 740 927 7683, 866 648 8718
Email: bdsra1@bdsra.org
Website: www.bdsra.org
Contact: Mr Lance W Johnston, Executive Director

The Batten Disease Support and Research Association provides information, education, medical referrals and support to families that have children with NCL or Batten Disease. The Association also provides funding for research into Batten Disease.

Batten Disease Support and Research Association Research Grant Awards
Subjects: NCL or Batten Disease in the areas of genetics, biochemistry, molecular biology and related areas with the eventual goal of developing a viable treatment.
Purpose: To support work that identifies genes, proteins, enzymes or additional NCLs and the development of novel therapeutic treatments.
Eligibility: There are no eligibility restrictions.
Level of Study: Doctorate, Postdoctorate, Research
Value: Varies
Length of Study: Research is for 1 year and doctorate and postdoctorate are up to 3 years
Frequency: Annual
Country of Study: Any country
Application Procedure: RFP (request for proposals) will be posted on website: www.bdsra.org, follow instructions contained in RFP.
Closing Date: May 15th. See website for complete information.
Funding: Private
No. of awards given last year: 8
No. of applicants last year: 25

BBC WRITERSROOM

1st Floor, Grafton House, 379-381 Euston Road, London, NW1 3AU, United Kingdom
Email: writersroom@bbc.co.uk
Website: www.bbc.co.uk/writersroom/opportunity

BBC Writersroom identifies and champions new writing talent and diversity across BBC Drama, Entertainment and Children's programmes. Writersroom is constantly on the lookout for writers of any age and experience who show real potential for the BBC. It invests in new writing projects nationwide and builds creative partnerships, including work with theatres, writer's organizations and film agencies across the country.

Alfred Bradley Bursary Award

Subjects: Drama.
Purpose: To encourage and develop new radio writing talent in the BBC North region.
Eligibility: Open to people who live in the North of England and who have not had a previous network radio drama commission.
Level of Study: Professional development
Type: Bursary
Value: Up to UK£5,000, a 12-month development mentorship with a Radio Drama Producer, and have your work produced on BBC Radio 4.
Frequency: Every 2 years
Country of Study: United Kingdom
No. of awards offered: 1
Application Procedure: Applicants must send completed afternoon play scripts for consideration. Details published on www.bbc.co.uk/writersroom.
Closing Date: September 15th
Funding: Corporation, private
Contributor: BBC
Additional Information: There is a change of focus for each award, e.g. previous years have targeted comedy, drama, verse drama, etc. In 2004, the brief was for a play suitable for the afternoon play slot.

Writers-in-Residence with NCH, the Children's Charity

Subjects: All subjects.
Purpose: To enable young writers to work in residence with selected NCH projects during their 'Growing Strong' campaign, which focuses on young people's emotional well-being, inner-strength and confidence.
Eligibility: Applicants must be talented writers with professional experience.
Type: Studentships and bursaries
Value: £3,000
Length of Study: 3 months
Country of Study: United Kingdom
No. of awards offered: 5
Application Procedure: Applicants must write a covering letter explaining why he/she would like to take part in the residency and a brief outline of a proposed project idea, a curriculum vitae detailing their relevant professional experience and a writing sample (full script) and send hard copies marked 'NCH Writer in Residence' to BBC Writersroom.
Closing Date: October 19th
Additional Information: Residencies will be based in London, Glasgow, Belfast, Cardiff, and Liverpool.

BEIT TRUST (ZIMBABWE, ZAMBIA AND MALAWI)

PO Box CH 76, Chisipite, Harare, Zimbabwe
Tel: (263) 4 496132
Fax: (263) 4 494046
Email: beitrust@africaonline.co.zw
Website: www.beittrust.org.uk
Contact: T M Johnson, Representative

Beit Trust Postgraduate Scholarships

Subjects: All subjects.
Purpose: To support postgraduate study or research.
Eligibility: Open to persons under 30 years of age or 35 years for medical doctors, who are university graduates domiciled in Zambia, Zimbabwe or Malawi. Applicants must be nationals of those countries.
Level of Study: Postgraduate
Type: Scholarship
Value: A variable personal allowance and fees plus books, clothing, thesis and departure allowances
Length of Study: 1–3 years depending on course sought
Frequency: Annual
Study Establishment: Approved universities and other institutions in South Africa, Britain and Ireland
No. of awards offered: 10
Application Procedure: Applicants must complete an application form.
Closing Date: August 31st
Funding: Private
No. of awards given last year: 8
No. of applicants last year: 400
Additional Information: Zambian applicants should contact the BEIT Trust UK office. Zimbabwe and Malawi applicants should contact the Zimbabwe office.

For further information contact:

The BEIT Trust, BEIT House, Grove Road, Woking, Surrey, GU21 5JB, England
Tel: (44) 01483 772 575
Fax: (44) 01483 725 833
Email: enquiries@beittrust.org.uk

BELGIAN AMERICAN EDUCATIONAL FOUNDATION (BAEF)

Marie-Claude Hayoit, Egmontstraat 11 Rue d'Egmont, Brussels, 1000, Belgium
Tel: (32) 2 513 59 55
Fax: (32) 2 672 53 81
Email: mail@baef.be
Website: www.baef.be

The BAEF is a nonprofit organization, funded by the general public under United States Law, and engaged in fostering the higher education of deserving Belgians and Americans.

BAEF Alumni Award

Subjects: Human sciences, social sciences, exact sciences, biomedical sciences, and applied sciences.
Purpose: To encourage young researchers in human sciences, social sciences, exact sciences, biomedical sciences, and applied sciences.
Eligibility: Open to Belgian researchers or to a researcher who is affiliated with a Belgian university since at least two years. The applicant should be below the age of 36 years.
Level of Study: Postgraduate
Type: Award
Value: €5,000
Frequency: Annual
No. of awards offered: Varies
Application Procedure: A completed application form along with 10 copies of curriculum vitae and summary of complete scientific work must be sent.
Closing Date: March 1st
Contributor: Belgian American Eductional Foundation
Additional Information: The Alumni Award is rotated among various scientific disciplines over a period of five years. Year 1: the human sciences; Year 2: the social sciences; Year 3: the exact sciences; Year 4: the biomedical sciences; Year 5: the applied sciences.

Fellowships for Biomedical Engineering Research

Subjects: Biomedical Engineering.
Purpose: To provide a stipend covering living and travel expenses and including health insurances.
Eligibility: Open to candidates who hold a university degree of a regular second cycle from a Belgian institution; should be nominated by a Rector or a Dean of Faculty from a recognized Belgian university; should have an outstanding academic record; should have good

command of the English language, and engage in doctoral or postdoctoral biomedical engineering research in the USA.
Level of Study: Doctorate, Postdoctorate
Type: Fellowship
Value: $40,000 (non-renewable) and health insurance.
Length of Study: 1 year
Frequency: Annual
Country of Study: United States of America
Application Procedure: Check website for details.
Closing Date: October 31st
Funding: Foundation

Fellowships for Research in the U.S.A

Subjects: All subjects.
Purpose: To support advanced study or research.
Eligibility: Open to Belgian citizen holding a university degree of a regular second cycle or doctorate from a Belgian institution with an outstanding academic record and with a good command of the English language.
Type: Fellowship
Value: US$21,000 (for pre-doctoral fellows) and US$35,000 (for postdoctoral fellows) (non renewable) and health insurance.
Length of Study: 1 year
Frequency: Annual
Country of Study: Belgium
No. of awards offered: 8
Application Procedure: A completed application form along with three letters of recommendation must be sent.
Closing Date: October 31st
Additional Information: Postdoctoral applicants should have no more than 3 years of postdoctoral research experience at the time of the selection interview.

Fellowships for Study or Research in Belgium

Subjects: All subjects.
Purpose: To encourage advanced study or research.
Eligibility: Open to applicants who are citizens of the United States, either with a Masters degree or equivalent degree, or working towards a PhD or equivalent degree. Preference is given to applicants under the age of 30 with a reading and speaking knowledge of Dutch, French, or German, and must reside in Belgium during the tenure of their fellowship.
Level of Study: Doctorate, Postgraduate
Value: $25,000 and health insurance
Length of Study: 6–12 months
Frequency: Annual
Country of Study: Belgium
No. of awards offered: Up to 8
Application Procedure: For additional information contact the Foundation.
Closing Date: October 31st
Funding: Foundation
Additional Information: Applicants should make their own arrangements to register or affiliate with a Belgian university or research institution.

Postdoctoral Fellowships for Biomedical or Biotechnology Research

Subjects: Biomedical or Biotechnology Research.
Purpose: To cover living and travel expenses and includes health insurance.
Eligibility: Open to candidates who are Belgian citizens, holding a university degree of a regular second cycle from a Belgian institution; have an outstanding academic record; have a good command of the English language, and engage in doctoral or postdoctoral biomedical research or biotechnology research in the USA.
Level of Study: Doctorate, Postdoctorate
Value: $35,000 and health insurance.
Length of Study: 1 year
Frequency: Annual
Country of Study: United States of America
Application Procedure: Check website for details.
Closing Date: October 31st
Funding: Foundation
Contributor: D. Collen Research Foundation vzw and BAEF
Additional Information: Fellows who have their own means to finance their research in the USA may apply for an Honorary Fellowship of the Foundation.

BERTHOLD LEIBINGER STIFTUNG

Johann-Maus-Strasse 2, D-71254 Ditzingen, Germany
Tel: (49) 7156 303 35201
Fax: (49) 7156 303 35205
Email: innovationspreis@leibinger-stiftung.de
Website: www.leibinger-stiftung.de
Contact: Mr Sven Ederer, Project Manager

Berthold Leibinger Innovation Prize

Subjects: Applied laser technology (application or generation of laser light).
Purpose: To promote the advancements in the application or generation of laser light.
Eligibility: Open to individuals and project groups who have completed an innovative scientific or technical development work on applying or generating laser light.
Level of Study: Unrestricted
Type: Prize
Value: €30,000 (First Prize), €20,000 (Second Prize), and €10,000 (Third Prize)
Frequency: Every 2 years
No. of awards offered: 3
Application Procedure: Applicants must submit a completed application form, short documentation of up to ten pages in accordance with stipulated structure, biography and explanation describing the context of work. Up to eight nominees are invited to present their work in the jury session. Candidates can also be suggested by a third party. Suggestions must include the reasons for the project's merit.
Closing Date: December 1st
Funding: Private
Contributor: Berthold Leibinger
No. of awards given last year: 3
No. of applicants last year: Approx. 30
Additional Information: The foundation bears travelling expenses for nominees to jury session and for prize winners to the prize ceremony.

For further information contact:

Berthold Leibinger Stiftung, Innovation Prize, 71252 Ditzingen, Germany

BETA PHI MU HEADQUARTERS

Florida State University, College of Communicationand Information, 101H, Louis Shores Building, 142 Collegiate Loop,, Tallahassee, FL, 32306-2100, United States of America
Tel: (1) 850 644 3907
Fax: (1) 850 644 9763
Email: betaphimuinfo@admin.fsu.edu
Website: www.beta-phi-mu.org
Contact: John Paul Walters, Program Director

Beta Phi Mu is a library and information studies honor society, founded in 1948, with over 35,000 graduates of the ALA-initiated accredited professional programmes. Beta Phi Mu was founded at the University of Illinois by a group of leading librarians and library educators. Aware of the notable achievements of honour societies in other professions, they believed that such a society would have much to offer librarianship and library education.

ALA Beta Phi Mu Award

Type: Award
Value: US$1,000 and a citation of achievement
Frequency: Annual
Closing Date: December 1st
Funding: Trusts
Contributor: Beta Phi Mu International Library Science Honorary Society
Additional Information: Dr. Mary Wagner, professor of library and information science at St. Catherine University in St. Paul, Minn., has

been selected the recipient of the American Library Association's 2012 Beta Phi Mu Award.

Blanche E Woolls Scholarship for School Library Media Service

Subjects: Library science.
Purpose: To assist a new student who plans to become a school media specialist.
Eligibility: Open to new students who have not completed more than 12 hours by Autumn. Applicants must be accepted into an ALA-accredited programme.
Level of Study: Graduate
Type: Scholarship
Value: US$2,250
Frequency: Annual
Study Establishment: An ALA-accredited school
Country of Study: United States of America
No. of awards offered: 1
Application Procedure: All applications with supporting material must be submitted online at www.beta-phi-mu.org.
Closing Date: March 15th
No. of awards given last year: 1

Eugene Garfield Doctoral Dissertation Fellowship

Subjects: Library and information science.
Purpose: To fund library and information science doctoral students who are working on their dissertations.
Eligibility: All requirements for the degree except the writing and defense of dissertation must have been completed.
Level of Study: Doctorate
Type: Fellowship
Value: US$3,000
Frequency: Annual
Study Establishment: Florida State University
Country of Study: United States of America
No. of awards offered: 6
Application Procedure: All applications must be submitted online at www.beta-phi-mu.org.
Closing Date: March 15th
Funding: Foundation

Eugene Garfield Doctoral Dissertation Scholarship

Subjects: Library science and information studies.
Purpose: To support library and information science doctoral students who are working on their dissertations.
Eligibility: Applicants must be doctoral students who have completed their coursework. Scholarships will be awarded based on the usefulness of the research topic to the profession.
Level of Study: Doctorate
Type: Scholarship
Value: US$3,000 each
Length of Study: 1 year
Frequency: Annual
Country of Study: Any country
No. of awards offered: 6
Application Procedure: Applicants must provide a 300-word abstract of dissertation, a curriculum vitae, a letter from their Dean or Director approving a topic and a work plan for the study.
Closing Date: March 15th
Funding: Private
No. of awards given last year: 1

Frank B Sessa Scholarship

Subjects: Library science or information studies.
Purpose: To enable the continuing professional education of a Beta Phi Mu member.
Eligibility: Open to Beta Phi Mu members only.
Level of Study: Professional development
Type: Scholarship
Value: US$1,500
Frequency: Annual
Country of Study: Any country
No. of awards offered: Varies
Application Procedure: All applications must be submitted online at www.beta-phi-mu.org.
Closing Date: March 15th
Funding: Private
No. of awards given last year: 1

Harold Lancour Scholarship For Foreign Study

Subjects: Library science.
Purpose: To assist a librarian or library school student to undertake short-term research in a foreign country.
Eligibility: Open to nationals of any country.
Level of Study: Unrestricted
Type: Scholarship
Value: US$1,750
Frequency: Annual
Country of Study: Any country
No. of awards offered: 1
Application Procedure: Applicants must write to Beta Phi Mu for further details, enclosing a stamped addressed envelope. All required documents must be submitted online. Further details are also available from the website.
Closing Date: March 15th
Funding: Private
No. of awards given last year: 1
No. of applicants last year: 10

Sarah Rebecca Reed Scholarship

Subjects: Library and information science.
Purpose: To assist a new student in library and information science at an ALA-accredited school.
Eligibility: Open to beginning students who have not completed more than 12 hours by Autumn. Applicants must also be accepted into an ALA-accredited programme and provide five references. Nationals of any country can apply.
Level of Study: Graduate
Type: Scholarship
Value: US$2,250
Frequency: Annual
Study Establishment: An ALA-accredited school
Country of Study: United States of America
No. of awards offered: 1
Application Procedure: All applications must be submitted online at www.beta-phi-mu.org.
Closing Date: March 15th
Funding: Private
No. of awards given last year: 1
No. of applicants last year: 30

BIAL FOUNDATION

Avenida da Siderurgia Nacional A, S. Mamede do Coronado, 4745-457, Portugal
Tel: (351) 22 986 6100
Fax: (351) 22 986 6199
Email: info@bial.com
Website: www.bial.com
Contact: Luis Portela, Chairman

The BIAL Foundation, a non-profit institution, was set up in 1994 with the aim of encouraging and supporting research focused on humans. It manages the BIAL award, one of the most distinguished awards for Health in Europe, and the BIAL Fellowship Programme, which focuses largely on psychophysiology and parapsychology.

BIAL Award

Subjects: Medical sciences and clinical medicine.
Purpose: To award intellectual written work in the subject area of health. To award work of a high quality and scientific relevance in clinical practice.
Eligibility: At least one of the authors must be a physician.
Level of Study: Graduate, MBA, Postdoctorate, Postgraduate, Predoctorate, Professional development, Research, Doctorate
Type: Prize
Value: €340
Frequency: Every 2 years
No. of awards offered: 6

Application Procedure: Applications, written in either portuguese or English should be sent as a PDF file, without restrictions, to fundacao@bial.com, by October 31st. In addition, two printed copies must be sent by mail to BIAL Foundation. Further requirements are listed on its regulation, which will be forwarded to prospective applicants on request or can be downloaded from www.bial.com.
Closing Date: October 31st
Funding: Private
Contributor: The BIAL Foundation
No. of awards given last year: 5
No. of applicants last year: 63

THE BIBLIOGRAPHICAL SOCIETY

c/o Institute of English Studies, University of London, Senate House, Malet Street, London, WC1E 7HU, United Kingdom
Email: Secretary@BibSoc.org.uk
Website: www.bibsoc.org.uk
Contact: Dr Margaret L Ford, Honorary Secretary

Founded in 1892, the Bibliographical Society is the senior learned society dealing with the study of the book and its history.

Antiquarian Booksellers Award

Subjects: Book trade and history of publishing.
Purpose: To support research into the history of the book trade and publishing industry.
Eligibility: Applicants may be of any age or nationality and need not be members of the Society.
Level of Study: Research
Type: Grant
Value: Up to US$1,500
Frequency: Annual
No. of awards offered: 1
Application Procedure: Application form (obtainable from the society's administrator), supported by letters from two referees familiar with the applicant's work.
Closing Date: January 11th
Funding: Private
Contributor: Antiquarian Booksellers Association
No. of awards given last year: 1
Additional Information: Successful applicants will be asked to report briefly on the progress of their project by December of the same year.

Barry Bloomfield Bursary

Subjects: Research.
Purpose: To honour Barry Bloomfield by supporting research in bibliography, particularly pertaining to book history in areas of the former British Empire.
Eligibility: Applicants may be of any age or nationality and need not be members of the society.
Level of Study: Research
Type: Bursary
Value: Up to UK£2,000
Frequency: Annual
No. of awards offered: 1
Application Procedure: Application form (obtainable from the Society's administrator) supported by letters from two referees familiar with the applicant's work.
Closing Date: January 10th
Funding: Private
Additional Information: Successful applicants will be asked to report briefly on the progress of their project by December of the same year.

Bibliographical Society Minor Grants

Subjects: Bibliographic research.
Purpose: To support bibliographic research projects by providing small grants for specific purposes.
Eligibility: Applicants may be of any age or nationality and need not be members of the society.
Level of Study: Research
Type: Grant
Value: UK£50–200
Frequency: Annual
Application Procedure: Details available on website.
Closing Date: January 10th

Falconer Madan Award

Subjects: Any subject connected with Oxford or available to research specifically in an Oxford library.
Purpose: To support a scholar's research in an Oxford library.
Eligibility: Applicants may be of any age or nationality.
Level of Study: Research
Value: Up to UK£500 plus eligibility for accommodation at Wolfson College, Oxford
Frequency: At the discretion of the Society
Study Establishment: An Oxford library
Country of Study: United Kingdom
No. of awards offered: 1
Application Procedure: Application form (obtainable from the Society's administrator) supported by letters from two referees familiar with the applicant's work.
Closing Date: January 10th
Funding: Private
Contributor: Oxford Bibliographical Society
Additional Information: Successful applicants will be asked to report briefly on the progress of their project by December of the same year.

The Fredson Bowers Award

Subjects: Bibliographic research.
Purpose: To support a bibliographic research project in the name of Fredson Bowers.
Eligibility: Applicants may be of any age or nationality and need not be members of the society.
Level of Study: Research
Type: Award
Value: US$1,500
Frequency: Annual
No. of awards offered: 1
Application Procedure: Application form (obtainable from the society's administrator) supported by letters from two referees familiar with the applicant's work.
Closing Date: January 10th
Funding: Private
Contributor: The Bibliographical Society of America
Additional Information: Successful applicants will be asked to report briefly on the progress of their project by December of the same year.

Katharine F. Pantzer Jr Research Fellowship

Subjects: To support bibliographical or back-historical study of the printed book in the hand-press period, i.e. up to c.1830.
Purpose: To support bibliographical research in honour of Katharine F Pantzer Jr.
Eligibility: Applicants may be of any age or nationality and need not be members of the society.
Level of Study: Graduate, Research
Type: Fellowship
Value: Up to £4,000
Frequency: Annual
No. of awards offered: 1
Application Procedure: Application form (obtained on the society's webiste) supported by letters from two references familiar with the applicant's work.
Closing Date: January 11th
Funding: Private
Additional Information: Successful applicants will be asked to report briefly on the progress of their project by the end of the same year.

BIBLIOGRAPHICAL SOCIETY OF AMERICA (BSA)

PO Box 1537, Lenox Hill Station, New York, NY, 10021, United States of America
Tel: (1) 212 734 2500
Fax: (1) 212 452 2710
Email: bsa@bibsocamer.org
Website: www.bibsocamer.org
Contact: Ms Michele Randall, Executive Secretary

The Bibliographical Society of America (BSA) invites applications for its annual short-term fellowships, which supports bibliographical inquiry as well as research in the history of the book trades and in publishing history.

BSA Fellowship Program

Subjects: Books or manuscripts as historical evidence. Topics may include establishing a text or studying the history of book production, publication, distribution, collecting or reading.
Purpose: To support bibliographical inquiry and research in the history of the book trades and publishing.
Eligibility: This programme is open to applicants of any nationality.
Level of Study: Postdoctorate, Doctorate, Postgraduate
Type: Fellowship
Value: US$2,000
Length of Study: 1 month
Frequency: Annual
Country of Study: Any country
No. of awards offered: 10
Application Procedure: Applicants must complete an application form. The original plus six photocopies must be posted to the Executive Secretary of the Fellowship Committee at the Bibliographical Society of America.
Closing Date: December 15th
Funding: Private
No. of awards given last year: 10
Additional Information: For applications now only available on website visit the BSA's website.

THE BIOCHEMICAL SOCIETY

Charles Darwin House, 12 Roger Street, London, WC1N 2JU, England
Tel: (44) 207 685 2400
Fax: (44) 207 685 2470
Email: alison.mcwhinnie@biochemistry.org
Website: www.biochemistry.org
Contact: Miss Alison McWhinnie, Head of Human Resources and Corporate Affairs

Serving biochemistry and biochemists since 1911, the aim of The Biochemical Society is to promote the advancement of the science of biochemistry. It does so in the context of cellular and molecular life sciences. Through its regular scientific meetings with special interest groups, its publishing company Portland Press Limited and its policy, professional and education contacts, the Society provides a forum for current research to be shared.

The Biochemical Society General Travel Fund

Subjects: The Society helps scientists become established and maintain their status to promote Biochemistry in the UK and international communities.
Purpose: To assist scientists of the British Biochemical Society who wish to attend scientific meetings, or make short visits to other laboratories.
Eligibility: New members may apply for their first travel grants after they have been a member of the British Biochemical Society for 1 year. Applicants will not be eligible if they have been awarded a Travel Grant from the Society during the previous 2 years. No age or career stage limitations. Post-graduate research students should provide a reference from their Head of Department.
Level of Study: Doctorate, Postdoctorate, Postgraduate, Predoctorate, Research
Type: Travel grant
Value: Up to £750. The committee may consider awarding more for exceptionally well-argued and well-supported cases describing high quality research from outstanding scientists or those with recognized potential.
Frequency: 7 times a year
Country of Study: Any country
No. of awards offered: Varies
Application Procedure: Please apply online through the website www.biochemistry.org. Applications should be for meetings that will be taking place at least 1 month after the closing date. The Travel Grants Committee meets seven times a year.
Closing Date: January 1st, March 1st, May 1st, June 1st, July 1st, September 1st, and November 1st
Funding: Private
Contributor: Biochemical society
No. of awards given last year: Varies
No. of applicants last year: Varies
Additional Information: Please refer to the website www.biochemistry.org for more details on "guidelines to application".

For further information contact:

Website: www.biochemistry.org

The Krebs Memorial Scholarship

Subjects: Biochemistry or allied biomedical science.
Purpose: To help candidates who wish to study for a PhD degree in biochemistry or in an allied biomedical science, but whose otherwise very promising research careers have been interrupted for quite extraordinary non-academic reasons totally beyond their control and who do not qualify for an award from public funds.
Eligibility: Open primarily to PhD students, but a post-doctoral fellowship might be considered for a candidate whose circumstances merit such consideration.
Level of Study: Doctorate, Postdoctorate, Postgraduate, Predoctorate
Type: Scholarship
Value: A maintenance grant and all necessary fees
Length of Study: 1 year, maximum of 3 years
Frequency: Every 2 years
Study Establishment: Any British university
Country of Study: United Kingdom
No. of awards offered: 1
Application Procedure: Application form is available in the autumn and can be downloaded from www.biochemistry.org or by contacting the Head of Human Resources and Corporate Affairs. Applications should be completed and forwarded through the Head of Department concerned, who should be able to place the applicant in the top 5 per cent of PhD candidates. Two references are required.
Closing Date: March 31st
Funding: Private
No. of awards given last year: 1
No. of applicants last year: Varies
Additional Information: Please refer to the website www.biochemistry.org for more details.

For further information contact:

Website: www.biochemistry.org

Visiting Biochemist Bursaries

Subjects: Biochemistry
Purpose: To support research capacity building in developing parts of the world e.g. Eastern Europe, Africa and the Middle East.
Eligibility: Open to an overseas researcher on a laboratory visit to the sponsor's research facility or to go out to visit a research facility in a developing region or to enable a member give seminars and undergraduate or postgraduate training.
Level of Study: Unrestricted
Type: Bursary
Value: Up to UK£2,000
Frequency: 7 times a year
No. of awards offered: Varies
Application Procedure: Please apply online at www.biochemistry.org.
Closing Date: January 1st, March 1st, May 1st, June 1st, July 1st, September 1st, November 1st
Funding: Private
Contributor: The Biochemical Society

BIOTECHNOLOGY AND BIOLOGICAL SCIENCES RESEARCH COUNCIL (BBSRC)

Polaris House, North Star Avenue, Swindon, Wiltshire, SN2 1UH, United Kingdom
Tel: (44) 1793 413200
Fax: (44) 1793 413201
Email: postdoc.fellowships@bbsrc.ac.uk
Website: www.bbsrc.ac.uk

BBSRC is the UK's principal funder of basic and strategic biological research. It is a non-departmental public body, one of the seven Research Councils supported through the Science and Innovation Group of the Department for Innovation, Universities and Skills (DIUS). It supports research and research training in universities and research centres throughout the UK, including BBSRC-sponsored institutes, and promotes knowledge transfer from research to applications in business, industry and policy, and public engagement in the biosciences. It funds research in some exciting areas including genomics, stem cell biology, and bionanotechnology.

David Phillips Fellowships

Subjects: All subjects.
Purpose: To support outstanding early-career scientists who wish to establish themselves as independent researchers.
Eligibility: Open to candidates who have a minimum of 3 years and no more than 10 years of active postdoctoral research experience.
Level of Study: Postdoctorate
Type: Fellowship
Value: Personal salary and a significant research support grant to support the costs of the research.
Length of Study: 5 years
No. of awards offered: 10
Application Procedure: All applications must be submitted online. For details mail to postdoc.fellowships@bbsrc.ac.uk or visit or check www.bbsrc.ac.uk/funding/fellowships.
Closing Date: November 14th
Additional Information: Fellowships are awarded under full economic costing (fEC). Further queries, please contact Postgraduate Training and Research Career Development Branch.
Important: applicants should ensure proposals are submitted to their host institution's Je-S submitter/approval pool well in advance (a minimum of 5 working days) of the published deadline.

For further information contact:

Email: postdoc.fellowships@bbsrc.ac.uk
Website: www.bbsrc.ac.uk/funding/fellowships

THE BIRLA INSTITUTE OF TECHNOLOGY & SCIENCE (BITS)

BITS Pilani, Pilani Campus, Vidhya Vihar Campus, Pilani, Rajasthan, 333031, India
Tel: (91) 91 01 596 245073
Fax: (91) 91 01596 244183
Email: lkm@bits-pilani.ac.in
Website: http://discovery.bits-pilani.ac.in

The Birla Institute of Technology & Science (BITS) is an all India institute for higher education. An important aspect of education at BITS is its institutionalized linkages with the industry. The Institute attaches great importance to university industry alliances. The primary motive of BITS is to train young men and women able and eager to create and put into action such ideas, methods, techniques and information.

BITS HP Labs India PhD Fellowship

Subjects: Information and communication technologies.
Purpose: To aid aspiring and deserving students for research in the area of information and communication technologies (ICT) relevant to fast-growing markets such as India.
Eligibility: Open to candidates with a higher degree in computer science, interested in doing Doctoral work in web based print distribution, data on paper, intelligent storage management, distributed operating systems and related areas.
Level of Study: Doctorate
Type: Fellowship
Value: A monthly stipend of Indian Rupees 25,000, full tuition fees, and an additional monthly allowance of Indian Rupees 25,000 to cover living expenses for the duration the candidate spends at HP Labs India, Bangalore.
Length of Study: 1–3 years
Frequency: Annual
Study Establishment: BITS Pilani
Country of Study: India
No. of awards offered: Varies
Application Procedure: Downloadable application forms are available at www.bits-pilani.ac.in.
Closing Date: May 15th
Funding: Corporation
Contributor: HP Labs India
No. of awards given last year: 2
No. of applicants last year: 180
Additional Information: Additional information can be got from www.hpl.hp.com/india or www.bits-pilani.ac.in.

BIRTH DEFECTS FOUNDATION

Newlife Foundation, Newlife Centre, Hemlock Way, Cannock, Staffordshire, WS11 7GF, England
Tel: (44) 01543 462777
Fax: (44) 01543 468999
Email: info@newlifecharity.co.uk
Website: www.newlifecharity.co.uk
Contact: Mrs Patrice McDonald

Newlife Foundation, more recently known as BDF Newlife, is a United Kingdom registered charity whose mission is to improve child health, aid families and raise awareness. Newlife is committed to funding basic, clinical and ethically approved research into the causes, prevention and treatment of birth defects.

Newlife Foundation-Full and Small Grants Schemes

Subjects: Aetiology, prevention, and treatment of birth defects.
Purpose: To fund basic and clinical research into the aetiology, prevention, and treatment of Birth Defects.
Level of Study: Graduate, Postgraduate, Research
Type: Fellowship
Value: Varies. Small grants up to UK£15,000, full grants up to UK£120,000
Length of Study: Varies
Frequency: Annual, Small Grants available throughout the year
No. of awards offered: Varies
Application Procedure: Applicants must fill in an application form for the full grant. Application for the small grant is by a brief proposal and then by application form if the proposal is of interest to the Foundation.
Closing Date: Full grants, October 7th; small grants throughout the year, February 24th.
Contributor: Charitable trading activity.
No. of awards given last year: 4 full grants, 6 small grants
No. of applicants last year: Over 30

THE BLAKEMORE FOUNDATION

1201 3rd Avenue, Suite 4800, Seattle, WA, 98101-3266, United States of America
Tel: (1) 206 359 8778
Fax: (1) 206 359 9778
Email: blakemore@perkinscoie.com
Website: www.blakemorefoundation.org
Contact: Mr Griffith Way, Trustee

The Blakemore Foundation was established in 1990 by Thomas and Frances Blakemore. The Blakemore Foundation makes grants for the advanced study of East and Southeast Asian languages and to improve the understanding of Asian fine art in the United States.

Blakemore Freeman Fellowships for Advanced Asian Language Study

Subjects: Asian language (Chinese, Vietnamese, Japanese, Indonesian, Thai, Korean, Khmer, Burmese).
Purpose: To provide encourage the advanced study of Asian languages.
Eligibility: Applicants must be citizens or permanent residents of the United States. The candidates must be pursuing an academic, professional or business career that involves the regular use of a modern East or Southeast Asian language.
Level of Study: Advanced study
Type: Fellowships

Value: Tuition or tutoring fees and stipend for travel, living and study expenses.
Length of Study: 1 year
Frequency: Annual
Study Establishment: Inter-University Center (IUC) for Japanese Language Studies in Yokohama, Japan or Inter-University Program (IUP) for Chinese Language Studies at Tsinghua University or International Chinese Language Program (ICLP) at National Taiwan University or similar program in other countries of East and SE Asia.
Country of Study: Asia
No. of awards offered: Varies
Application Procedure: Applicants must download the application form from the website.
Closing Date: December 30th
Funding: Foundation
No. of awards given last year: 15
No. of applicants last year: 165

Blakemore Refresher Grants: Short-Term Grants for Advanced Asian Language Study

Subjects: Asian language (Chinese, Vietnamese, Japanese, Indonesian, Thai, Korean, Khmer, Burmese).
Purpose: To encourage the advanced study of Asian languages.
Eligibility: Open to former Blakemore Fellows, professors teaching in an Asian field at a university or college in the United States or professionals working in an Asian field.
Level of Study: Advanced Study
Type: Grant
Value: Tuition and a maintenance stipend for travel, living and study expenses
Length of Study: Summer or semester
Frequency: Annual
Study Establishment: Inter-University Center (IUC) for Japanese Language Studies in Yokohama, Japan or Inter-University Program (IUP) for Chinese Language Studies at Tsinghua University or International Chinese Language Program (ICLP) at National Taiwan University or similar programs in other countries of East and SE Asia.
Country of Study: Asia
No. of awards offered: Varies
Application Procedure: Applications can be downloaded from the website.
Closing Date: December 30th
Funding: Foundation
No. of awards given last year: 1
No. of applicants last year: Varies

THE BLUE MOUNTAINS HOTEL SCHOOL (BMHS)

PO Box 905, Crows Nest, NSW, 2065, Australia
Tel: (61) 9437 0300
Fax: (61) 9437 0299
Email: communications@sha.cornell.edu
Website: www.hotelschool.com.au

The Blue Mountain Hotel School is a university level institution, recognized and accredited in Australia by the vocational education and training accreditation board.

BMHS Hospitality and Tourism Management Scholarship

Subjects: Hospitality and tourism management.
Purpose: To encourage students in the field of hospitality and tourism.
Eligibility: Open to Australian or New Zealand citizen who is eligible for entry into the firstyear in February or July under normal entry criteria.
Type: Scholarship
Value: Australian $15,000 towards study costs
Length of Study: 2–5
Frequency: Annual
No. of awards offered: 2
Application Procedure: Applications should arrive no later than September 30th for the February enrollment and no later than April 30th for the July enrollment.
Closing Date: September 30th and April 30th

For further information contact:

The Principal, Blue Mountains Hotel School, PO Box 905, Crows Nest, NSW, 2065

BOEHRINGER INGELHEIM FONDS

Foundation for Basic Research in Medicine, Schusterstr, 46–48, 55116 Mainz, Germany
Tel: (49) 6131 275 080
Fax: (49) 6131 275 0811
Email: secretariat@bifonds.de
Website: www.bifonds.de
Contact: Dr Kerstin Terrenoire, Communications Officer

The Boehringer Ingelheim Fonds is a public foundation – an independent, non-profit institution – for the promotion of basic research in biomedicine. It awards fellowships and excellent to up-and-coming junior scientists and makes every attempt to assist them and create an atmosphere conductive to creative research.

MD Fellowships (BIF)

Subjects: Basic research in biomedicine.
Purpose: To provide gifted young medical students with the opportunity of pursuing ambitious research projects in internationally renowned laboratories.
Eligibility: Candidates must be no older than 24 years and must have gained good marks in the intermediate exam (Physikum). All nationalities are eligible to apply, provided they are studying medicine in Germany and change their working place (city and institution) for the MD thesis project. The thesis must be an experimental project concerned with basic biomedical research.
Level of Study: Doctorate
Type: Fellowship
Value: €900–1200 per month. In most countries, fellows are paid an additional flat-rate of 150 euros per month for project-related costs (please refer to conditions for further details).
Length of Study: 10–12 months; an extension of up to 3 months is possible
Country of Study: Germany
No. of awards offered: Up to 15 per year
Closing Date: At least 3 months before beginning of MD thesis project
Funding: Foundation
Contributor: Boehringer Ingelheim Foundation
Additional Information: Fellowship holders may receive travel allowances for the participation in scientific meetings.

PhD Fellowships (BIF)

Subjects: Basic research in biomedicine.
Purpose: Promote basic research in biomedicine by funding and fostering the best young scientists who are given a comprehensive support in addition to a competitive monthly stipend. The award addresses young researchers who wish to pursue an ambitious PhD project in an internationally leading laboratory.
Eligibility: Candidates must not be older than 27 years for doctoral research. All nationalities are eligible to apply. European citizens are supported while in Europe and overseas; non-European citizens receive support while conducting research in Europe. The PhD project must be ambitious and pursued in internationally leading laboratories.
Level of Study: Doctorate, 3
Type: Fellowship
Value: €1,550 per month plus an additional flat rate sum of €150 per month to cover minor project-related costs, spouse allowance and/or child care allowance. A supplement for the respective countries may be added
Length of Study: 2 years for doctoral research (extension possible up to 1 additional year)
Frequency: 3 times per year
No. of awards offered: 50 per year
Application Procedure: Applications must be submitted online and a paper version in English. Applications must be written by the applicants personally; however, consultation with scientific supervisors is recommended.
Closing Date: February 1st, June 1st and October 1st
Funding: Foundation

Contributor: Boehringer Ingelheim Foundation
No. of awards given last year: 50
No. of applicants last year: 700
Additional Information: The foundation sponsors participation in scientific conferences and offers additional trainings, e.g. in communication. Also fellows can apply for travel allowances to go to research-oriented conferences and seminars.

BOLOGNA CENTER OF THE JOHNS HOPKINS UNIVERSITY

Via Belmeloro 11, 40126 Bologna, Italy
Tel: (39) 051 291 7811
Fax: (39) 051 228 505
Email: admission@jhubc.it
Website: www.jhubc.it
Contact: Ms Bernadette O'Toole, Assistant Registrar

The Bologna Center is an integral part of the Paul H. Nitze School of Advanced International Studies (SAIS), one of the leading United States of America graduate schools devoted to the study of international relations. The programme seeks to merge the wisdom of universities, business and labour with the knowledge and expertise of those presently engaged in government, foreign affairs and international economic practice.

Paul H. Nitze School of Advanced International Studies (SAIS) Financial Aid and Fellowships

Subjects: International economics, European and Middle East studies, international relations and international development. In addition to fundamental courses, international economics covers European economic integration, environmental and resource economics, commercial policies, corporate finance, economic development and public sector economics. European studies examines history, economics, contemporary politics and culture, as well as demographic and enlargement issues. International relations explores international law, international non-governmental organizations, human rights, conflict management, ethnic conflict and security issues.
Purpose: To facilitate graduate study in International relations.
Eligibility: Open to students who have completed their first university degree. Students who are in the process of completing their first degree may apply provided they obtain the degree prior to entry to the Bologna Center in the Autumn. All candidates must have an excellent command of written and spoken English and ideally have some background knowledge in economics, history, political or other social sciences. All fellowships and financial aid awards are based on need as well as academic merit.
Level of Study: Postgraduate
Type: Fellowships and financial aid
Value: Varies. Grants may cover partial or full tuition. Maintenance stipends are rarely provided
Frequency: Annual
Study Establishment: The Bologna Center of the Johns Hopkins University and the Paul H. Nitze School of Advanced International Studies
Country of Study: Italy
No. of awards offered: 2
Application Procedure: Applicants must submit an application form and financial aid application. Certain donor organizations require a separate application. Admission and financial aid for citizens of United States of America and permanent residents are administered by SAIS in Washington and all enquiries from United States of America students should be addressed to the Admissions Office in Washington. Financial aid and admission for non-United States of America students is administered in Bologna and all enquiries from non-United States of America students should be addressed to the Registrar's Office in Bologna.
Closing Date: The deadline for United States of America applicants is January 7th and February 1st for non-United States of America students
Funding: Commercial, corporation, individuals, private, trusts, foundation, government
No. of applicants last year: 600
Additional Information: Courses are also offered in the United States of America foreign policy, as well as Latin American, African and Middle East issues. Language instruction is offered in Arabic, English, French, German, Portuguese, Spanish and Russian. Special fellowships administered by the Bologna Center on behalf of other donor organizations have certain restrictions, which vary depending upon the donor. Many of the fellowships available to non-United States of America students are provided by government ministries and other European organizations and are reserved for citizens of the country providing the fellowship.

For further information contact:

United States of America Citizens: Admissions Office, 1740 Massachusetts Avenue North West, Washington, DC 20036, United States of America
Email: admission.sais@jhu.eduNon-United States of America Citizens: Registrar's Office, Bologna Center, Via Belmeloro 11, 40126 Bologna, Italy
Email: admission@jhubc.it

THE BOSTON SOCIETY OF ARCHITECTS (BSA)

BSA Space, 290 Congress Street, Second floor, Boston, MA 02210-1038, United States of America
Tel: (1) 617 361 4000
Fax: (1) 617-951-0845
Email: bsa@architects.org
Website: http://www.architects.org/

The Boston Society of Architects (BSA) is the regional and professional association of over 3,000 architects and 1,000 affiliate members. The BSA's affiliate members include engineers, contractors, clients or owners, public officials, other allied professionals, students and lay people. The BSA administers many programmes that enhance the public understanding of design as well as the practice of architecture.

Rotch Travelling Scholarship

Subjects: Architecture.
Purpose: To provide young architects with the opportunity to travel and study in foreign countries.
Eligibility: Open to United States of America architects who will be under 35 years of age on January 1 of the competition year and a degree from an accredited U.S. school of architecture and one year of full-time professional experience in a Massachusetts architecture firm as of January 1 of the competition year.
Level of Study: Professional development
Type: Scholarship
Value: $37,000
Length of Study: 8 months
Frequency: Annual
Country of Study: Other
No. of awards offered: 1–2
Application Procedure: Applicants must complete an application form, available on written request.
Closing Date: January
Funding: Private
No. of applicants last year: 34
Additional Information: The scholar is selected through a two-stage design competition. The first year of professional experience required should be completed prior to the beginning of the preliminary competition. Scholars are required to return to the United States of America after the duration of the scholarship and submit a report of their travels.

BRADFORD CHAMBER OF COMMERCE AND INDUSTRY

Devere House, Vicar Lane, Little Germany, Bradford, Yorkshire, BD1 5AH, England
Tel: (44) 1274 772777
Fax: (44) 1274 771081
Email: Julie.snook@bradfordchamber.co.uk
Website: www.bradfordchamber.co.uk
Contact: Julie Snook, Financial Controller

The Bradford Chamber of Commerce and Industry represents member companies in the Bradford and district area. It works with local partners to develop the economic health of the district and has a major voice within the British Chamber of Commerce movement in order to promote the needs of local business on a national basis.

John Speak Trust Scholarships

Subjects: Modern languages.
Purpose: To promote British trade abroad by assisting people in perfecting their basic knowledge of a foreign language.
Eligibility: Open to British nationals intending to follow a career connected with the export trade in the United Kingdom. Applicants must be over 18 years of age with a sound, basic knowledge of at least one language.
Level of Study: Professional development
Type: Scholarship
Value: Contribution towards living expenses and an amount towards the cost of travel
Length of Study: Between 3 months and 1 full academic year abroad depending on the circumstances and each candidate's level of knowledge of the language. It is non-renewable
Frequency: Annual
Study Establishment: A recognized college or university
No. of awards offered: 10
Application Procedure: Applicants must complete an application form and undertake an interview.
Closing Date: February 28th, May 31st or October 31st
Funding: Private
No. of awards given last year: 10
No. of applicants last year: 14

BRAIN RESEARCH INSTITUTE

Florey Institute of Neuroscience and Mental Health, Melbourne Brain Centre - Austin campus, 245 Burgundy Street, Heidelberg, VIC, 3084, Australia
Tel: (61) 3 9035 7000
Fax: (61) 3 9496 4071
Email: BRI@brain.org.au
Website: http://www.brain.org.au/

The Brain Research Institute (BRI) was established at Austin Health, Melbourne, Australia in 1996. It supports collaboration between specialities in order to develop a better understanding of how a healthy or diseased brain functions. It is an affiliated institution of The University of Melbourne, an administering institution of the National Health & Medical Research Council and a member of Research Australia.

BRI PhD Scholarships

Subjects: Engineering and technology information, computing and communication sciences, medical and health sciences or physical sciences.
Purpose: To encourage competitive research in understanding the structure and function of the human brain.
Eligibility: Open to candidates who have obtained Honours 1 or equivalent, or Honours 2a or equivalent.
Level of Study: Doctorate
Type: Scholarship
Value: Varies
Length of Study: 3 years
Frequency: Annual
Country of Study: Australia
Application Procedure: Applicants must send a curriculum vitae, academic transcripts, expression of interest for area of research and details of two academic referees.

For further information contact:

Email: scholarships@brain.org.au
Contact: Karen van Nugteren, Scholarships Officer

BRANDEIS UNIVERSITY

415 South Street, Waltham, MA, 02453, United States of America
Tel: (1) 781 736 2000
Email: goldfiel@brandeis.edu
Website: www.brandeis.edu

Brandeis University is a community of scholars and students united by their commitment to the pursuit of knowledge and its transmission from generation to generation. As a research university, Brandeis is dedicated to the advancement of the humanities, arts, and social, natural and physical sciences. As a liberal arts college, Brandeis affirms the importance of a broad and critical education in enriching the lives of students and preparing them for full participation in a changing society, capable of promoting their own welfare, yet remaining deeply concerned about the welfare of others.

International Visiting Scholar Awards

Subjects: Arts.
Purpose: For one academic year in any liberal arts subject.
Eligibility: Open to candidates from any country aged 18–23, with strong academic records in their home country. English proficiency and 600 TOEFL are required.
Type: Scholarship
Value: $19,380
Frequency: Annual
Study Establishment: Brandeis University
Country of Study: United States of America
No. of awards offered: 3
Application Procedure: Check website for further details.
Closing Date: February 1st
Funding: Government

For further information contact:

Waltham, Massachusetts, United States of America
Tel: (1) 02254 911
Website: www.brandeis.edu
Contact: Kutz Hall

BREAST CANCER CAMPAIGN

Clifton Centre, 110 Clifton Street, London, EC2A 4HT, England
Tel: (44) 020 7749 4114
Fax: (44) 20 7749 3701
Email: research@breastcancercampaign.org
Website: www.breastcancercampaign.org
Contact: Research Grants Administrator

Breast Cancer Campaign aims to beat breast cancer by funding innovative world-class research into breast cancer throughout the UK and Republic of Ireland. The charity currently funds 100 grants worth over £15 million.

Breast Cancer Campaign PhD Studentships

Subjects: Breast cancer research including prevention, causes, diagnosis, treatment and management.
Purpose: To attract new and highly qualified science graduates into a career of breast cancer research.
Eligibility: See website for details.
Level of Study: Graduate, Postdoctorate
Type: Studentship
Value: Around £90,000 for three years (inc PhD stipend and tuition fees)
Length of Study: 3 years
Frequency: Annual
Country of Study: United Kingdom, Republic of Ireland
Application Procedure: Application forms can be downloaded from the website www.breastcancercampaign.org
Closing Date: March – Refer to website for the exact date
Funding: Individuals, trusts

Breast Cancer Campaign Project Grants

Subjects: Breast cancer research including prevention, causes, diagnosis, treatment and management.
Purpose: To support innovative research into breast cancer.

Eligibility: Open to candidates working in universities, medical schools/teaching hospitals and research institutes within the UK and Republic of Ireland
Level of Study: Research, Postdoctorate
Type: Grant
Value: UK£85,000 per year for up to 3 years
Length of Study: 3 years
Frequency: Annual
Country of Study: United Kingdom, Republic of Ireland
No. of awards offered: Varies
Application Procedure: Application forms can be downloaded from the website: www.breastcancercampaign.org
Closing Date: January
Funding: Individuals, trusts

Breast Cancer Campaign Scientific Fellowships

Subjects: Breast cancer research including prevention, causes, diagnosis, treatment and management.
Purpose: To provide an opportunity for postdoctoral scientists to become independent researchers specializing in the breast cancer field and to undertake research of the highest quality.
Eligibility: Open to candidates who are working in universities, medical schools/teaching hospitals and research institutions within the UK and the Republic of Ireland
Level of Study: Postdoctorate, Professional development, Research
Type: Fellowship
Value: Up to £550,000 for one year
Length of Study: 5 years
Frequency: Annual
Country of Study: United Kingdom, Republic of Ireland
Application Procedure: Application forms can be downloaded from the website www.breastcancercampaign.org
Closing Date: January (refer to website for exact date)

Breast Cancer Campaign Small Pilot Grants

Subjects: Breast cancer research including prevention, causes, diagnosis, treatment and management.
Purpose: To support established scientists to investigate and develop new ideas in the field of breast cancer research.
Eligibility: Open to candidates working in universities, medical schools/teaching hospitals and research institutes within the UK and Republic of Ireland.
Level of Study: Postdoctorate, Research
Type: Grant
Value: Up to UK£20,000 for one year
Length of Study: Up to 1 year
Frequency: Annual
Country of Study: United Kingdom, Republic of Ireland
No. of awards offered: Varies
Application Procedure: Application forms can be downloaded from the website: www.breastcancercampaign.org
Closing Date: March 13th
Funding: Individuals, private, trusts

BRIAN MAY SCHOLARSHIP

ffrench.commercial lawyers, PO box 2656, Southport, QLD, 4215, Australia
Tel: (61) 07 5591 7555
Fax: (61) 07 55917 450
Email: bms@ffrenchlegal.com
Website: www.brianmayscholarship.org

The Trust is a charitable testamentary trust established under the will of the late Brian May, Australias leading film composer to finance promising Australian film composers.

The Brian May Scholarship

Subjects: Creative arts.
Purpose: To provide financial assistance to promising Australian film composers to study film scoring.
Eligibility: Candidates must be ordinarily resident in Australia. Candidates must hold a Bachelor's Degree from a University, preferably in music composition.
Level of Study: Postgraduate
Type: Scholarship
Value: Australian $80,000
Frequency: Every 2 years
Study Establishment: Thornton School of Music
Country of Study: United States of America
No. of awards offered: 1
Application Procedure: Contact the trust or apply online.
Closing Date: November 30th
Funding: Trusts
Contributor: The Brian May Trust
No. of awards given last year: 1

THE BRITISH ACADEMY

10 Carlton House Terrace, London, SW1Y 5AH, England
Tel: (44) 20 7969 5200
Fax: (44) 20 7969 5300
Email: chiefexec@britac.ac.uk
Website: www.britac.ac.uk
Contact: Ms Jane Lyddon, Assistant Secretary (International Relations)

The British Academy is the premier national learned society in the United Kingdom devoted to the promotion of advanced research and scholarship in the humanities and social sciences.

British Academy 44th International Congress of Americanists Fund

Subjects: Latin American studies.
Purpose: To enable British scholars to visit Latin America or Latin American scholars to visit Britain.
Eligibility: Open to support British or Latin American scholars. Applicants must be resident in the UK.
Level of Study: Postdoctorate
Type: Travel grant
Value: Awards do not generally exceed UK£1,000
Frequency: Annual
No. of awards offered: Varies
Closing Date: November
Funding: Private

British Academy Ancient Persia Fund

Subjects: Iranian, Central Asian studies in the pre-Islamic period.
Purpose: To encourage and support the study of Iranian or Central Asian studies in the pre-Islamic period. Grants are offered towards travel costs.
Eligibility: Preference will be given to scholars undertaking archaeological research or engaged in an archaeological project. Postdoctoral scholars of any nationality are eligible to apply.
Level of Study: Postdoctorate
Type: Travel grant
Value: Awards do not generally exceed UK£500
Length of Study: Tenable for 1 year
Frequency: Annual
Country of Study: Any country
No. of awards offered: 1–2
Closing Date: November
Funding: Private
Contributor: in memory of the distinguished Russian scholar Vladimir G Lukonin
Additional Information: For United Kingdom residents, an application to this fund for travel costs may be combined with an application for a small research grant for other costs, up to a total of UK£5,000.

British Academy Awards, Grants & Fellowships

Subjects: Various.
Purpose: Various awards, grants and fellowships - see website for further details.
Level of Study: Postdoctorate
Frequency: Varies
Application Procedure: Applications for grants are made through e-GAP system on the British Academy website.
Additional Information: Eligible nationals - Varies.

British Academy Elie Kedourie Memorial Fund

Subjects: Middle Eastern, modern European history and political thought.
Purpose: To promote the study of Middle Eastern, modern European history or history of political thought.
Eligibility: Awards are offered to support any aspect of research, including travel and publication.
Level of Study: Postdoctorate
Value: Up to GB £1,000
Frequency: Annual
Country of Study: Any country
No. of awards offered: 1–2
Closing Date: November
Funding: Private
Additional Information: Funds are not available to support travel to or attendance at conferences, workshops or seminars, either in the United Kingdom or elsewhere.

British Academy International Partnership and Mobility Scheme

Subjects: Humanities and social sciences.
Purpose: To support the development of research partnerships between the UK and other areas of the world.
Eligibility: Principal applicant must be resident in the Union Kingdom and be of postdoctoral or equivalent status. Postgraduate students are not eligible to apply.
Level of Study: Postdoctorate
Type: Grant
Value: £10,000
Length of Study: 1 or 3 years
Frequency: Annual
No. of awards offered: Up to 50
Application Procedure: Applications must be made via the British Academy online application system.
Closing Date: February
Funding: Government

British Academy Postdoctoral Fellowships

Subjects: Academic career development in humanities and social science.
Purpose: To enable outstanding young scholars to obtain experience of research and teaching in the university environment, which will strengthen their curriculum vitae and improve their prospects of obtaining permanent posts by the end of the fellowship.
Eligibility: Applicants must have obtained their doctorate within 3 years prior to the evaluating date and must not have held an established teaching post in an Institute of Higher Education. Applicants must be United Kingdom nationals, or nationals of any country who have a doctorate from United Kingdom university or European Union nationals resident in the United Kingdom.
Level of Study: Postdoctorate
Type: Fellowship
Value: Salary equivalent to relevant early career plus limited research expences, payment towards mentors time and indirect costs
Length of Study: Tenable for 3 years and not renewable
Frequency: Annual
Country of Study: United Kingdom
No. of awards offered: Up to 45
Application Procedure: Applications must be made via the British Academy online application system.
Closing Date: October
Funding: Government
No. of awards given last year: See the website
No. of applicants last year: 593

British Academy Sino-British Fellowship Trust

Subjects: Humanities and social sciences.
Purpose: To support individual or co-operative research projects.
Eligibility: Research may be conducted either in the United Kingdom or China, or in both countries and must involve person to person contact. Preference for applications that will help to achieve sustainable development regarding environmental issues and pollution.
Level of Study: Postdoctorate
Type: Fellowship
Value: Up to £10,000
Length of Study: 3 years
Frequency: Annual, As per availability of funds
No. of awards offered: 3–4
Closing Date: November
Funding: Private
Contributor: Sino-British Fellowship Trust (SBFT)
Additional Information: Successful applications will be forwarded to the SBFT for approval in the Autumn of each year. It should be noted that the Academy will be unable to offer support should the SBFT decline to confirm funding. Priority funding will be provided to projects which help to achieve sustainable development in the problems arising from environmental issues and pollution.

British Academy Small Research Grants

Subjects: Humanities and social sciences.
Purpose: For original research at postdoctoral level.
Eligibility: Applicants must be resident in the United Kingdom and be of postdoctoral or equivalent status. Postgraduate students are not eligible to apply.
Level of Study: Postdoctorate
Type: Grant
Value: Maximum UK£10,000
Length of Study: 2 years
Frequency: Annual, As per availability of funds
Country of Study: Any country
No. of awards offered: Approx. 500
Application Procedure: All applications should demonstrate that Academy funds are sought for a clearly defined, discrete piece of research, which will have an identifiable outcome on completion of the Academy-funded component of the research.
Closing Date: November
Funding: Private

British Academy Stein-Arnold Exploration Fund

Subjects: History, geography and arts.
Purpose: To encourage research into the antiquities or historical geography or early history or arts of those parts of Asia that come within the sphere of the ancient civilizations of India, China and Iran, including Central Asia.
Eligibility: Research should be as far as possible by means of exploratory work. Candidates must be British or Hungarian subjects.
Level of Study: Postdoctorate
Type: Funding support
Value: Not to exceed UK£2,500
Frequency: Annual
No. of awards offered: 3–4
Closing Date: February
Funding: Private

Elisabeth Barker Fund

Subjects: Recent European history, particularly of Eastern and Central Europe.
Purpose: To support studies in recent European history, particularly the history of Central and Eastern Europe.
Eligibility: Open to scholars of postdoctoral or equivalent status ordinarily resident in the United Kingdom. Candidates need not be British nationals. Applications must be made by a British resident and not a foreign scholar.
Level of Study: Postdoctorate
Value: Up to UK£1,000
Frequency: Annual
No. of awards offered: Up to 6
Closing Date: November
Funding: Private

Neil Ker Memorial Fund

Subjects: Western medieval manuscripts, particularly those of British interest.
Purpose: To promote the study of Western medieval manuscripts.
Eligibility: Candidates should be of postdoctoral status, or have comparable experience. Normally grants will only be given for monographs, secondary works, editions or studies of documents, texts or illustrations, that include analysis of the distinctive features of original manuscripts.

Level of Study: Postdoctorate
Type: Funding support
Value: Approx. UK£2,000
Frequency: Annual
Country of Study: Any country
No. of awards offered: 3–4
Closing Date: February
Funding: Private

Thank-Offering to Britain Fellowships

Subjects: Topics of an economic, industrial, social, political, literary or historical character relating to the British Isles. Preference will be given to projects in the modern period.
Purpose: To fund a research fellowship.
Eligibility: Open to persons ordinarily resident in the United Kingdom and of postdoctoral status. Candidates should be in mid-career and must be employed at a United Kingdom university in an established teaching post.
Level of Study: Postdoctorate
Type: Fellowship
Value: Equivlent to a junior lecturers' salary. The award pays for a replacement to undertake the teaching and administrative duties of the award holders for 1 year
Length of Study: Normally for 1 year
Frequency: Dependent on funds available
Country of Study: United Kingdom
No. of awards offered: 1
Closing Date: November
Funding: Private
Contributor: Association of Jewish Refugees
No. of awards given last year: 1
No. of applicants last year: 74

BRITISH ASSOCIATION FOR AMERICAN STUDIES (BAAS)

American Studies, School of Humanities, Keele University, Staffordshire, ST5 5BG, United Kingdom
Tel: (44) 44 0 1782 732000
Email: jo.gill@baas.ac.uk
Website: www.baas.ac.uk
Contact: Professor Ian Bell, Chair, BAAS Awards Sub-Committee

The British Association for American Studies (BAAS), established in 1955, promotes research and teaching in all aspects of American studies. The Association organizes annual conferences and specialist regional meetings for students, teachers and researchers. The publications produced are The Journal of American Studies with Cambridge University Press, BAAS Paperbacks with Edinburgh University Press and British Records Relating to America in Microform with Microform Publishing.

The Ambassador's Awards

Subjects: History, literature, film, politics or any other related or inter-related discipline.
Level of Study: Postgraduate
Type: Award
Value: £1,000 for postgraduate award; £500 for undergraduate award; £250 for school essay prize
Frequency: Annual
No. of awards offered: 3 (1 prize in each category)
Application Procedure: Must submit essays to judging panel.
Closing Date: January 23rd
Contributor: Embassy Sponsored

For further information contact:

American Studies, School of Humanities, Keele University, Keele, Staffs, ST5 5BG, United Kingdom
Contact: Professor Ian Bell, Chair, BAAS Awards Sub-Committee

The Arthur Miller Centre First Book Prize

Purpose: Recognise best first book.
Eligibility: Open to BAAS Members.
Level of Study: Research
Type: Prize
Value: £500
Frequency: Annual
No. of awards offered: 1
Application Procedure: Must submit a book.
Closing Date: March 1st
Funding: Foundation
Contributor: UEA
Additional Information: Those interested in entering a book for consideration should submit four copies, including publication details, to Those interested in entering a book for consideration should submit four copies, including publication details to The Arthur Miller Centre Prize Committee.

For further information contact:

Arthur Miller Centre for American Studies, University of East Anglia, Norwich, NR4 7TJ, United Kingdom
Contact: The Arthur Miller Centre Prize Committee

The Arthur Miller Centre Prize

Purpose: To recognise best American studies article published in a given year.
Eligibility: Open to BAAS members.
Level of Study: Research
Type: Prize
Value: £500
Frequency: Annual
No. of awards offered: 1
Application Procedure: Enter article.
Closing Date: March 1st
Funding: Foundation
Contributor: University of East Anglia
No. of awards given last year: 1
No. of applicants last year: Classified
Additional Information: Those interested in entering an article for consideration should submit three copies of the essay, including publication details to The Arthur Miller Centre Prize Committee.

For further information contact:

School of English and American Studies, University of East Anglia, Norwich, NR4 7TJ, United Kingdom
Contact: The Arthur Miller Centre Prize Committee

BAAS Book Prize

Subjects: American studies.
Eligibility: To be eligible for the BAAS Book Prize, books must have been published in English between January 1st, 2010 and December 31st, 2010 and authors must be members of BAAS.
Type: Prize
Value: £500
Closing Date: December 17th

For further information contact:

American Studies, School of Humanities, Keele University, Keele, Staffs, ST5 5BG, United Kingdom
Contact: Professor Chair, BAAS Awards Sub-Committee

BAAS Founders' Research Travel Awards

Subjects: American history, politics, society, literature, art, culture, etc.
Purpose: To offer assistance for short-term visits to the United States during the year 2011–2012 to scholars in the UK who need to travel to conduct research, or who have been invited to read papers at conferences on American Studies topics.
Eligibility: Open to BAAS Members.
Level of Study: Postdoctorate
Type: Award
Value: £750
Frequency: Annual
No. of awards offered: 5
Application Procedure: Applicant must submit an application form.
Closing Date: December
Funding: Individuals, private
Contributor: American Embassy

No. of awards given last year: 5
No. of applicants last year: Classified
Additional Information: For enquiries about the awards, contact at awards@baas.ac.uk.

For further information contact:

American Studies, School of Humanities, Keele University, Keele, Staffs, ST5 5BG, United Kingdom
Email: i.f.a.bell@ams.keele.ac.uk
Contact: Professor Ian Bell, Chair, BAAS Awards Sub-Committee

BAAS Honorary Fellowship

Subjects: American studies.
Purpose: To recognise American studies academics who have made an outstanding contribution to the association, to their institution(s), and to the American studies community in general over the course of a distinguished career. This is a Lifetime Achievement Award.
Eligibility: Open to BAAS members.
Level of Study: Postdoctorate
Type: Fellowship
Frequency: Annual
Country of Study: United Kingdom
No. of awards offered: 1
Application Procedure: Nomination and application form.
Closing Date: December 1st
Funding: Private, trusts
Contributor: American Embassy
No. of awards given last year: 1
No. of applicants last year: Classified

BAAS Monticello Teachers' Fellowships

Subjects: American studies.
Purpose: Workshops.
Eligibility: Applicants must have at least three-years' teaching experience, and teach A Level or advanced higher materials relevant to the fellowships.
Level of Study: Professional development
Type: Fellowships
Value: Approx. $2000
Frequency: Annual
Study Establishment: Monticello
Country of Study: United States of America
No. of awards offered: 1
Application Procedure: Applicant must submit application form.
Closing Date: February
Funding: Foundation
Contributor: Monticello, BAAS and American Embassy
No. of awards given last year: 1
No. of applicants last year: Classified

For further information contact:

School of Arts & Social Sciences, Northumbria University, Lipman Building, Newcastle upon Tyne, NE1 8ST, United Kingdom
Contact: Dr Sylvia Ellis, BAAS Awards Sub-Committee

BAAS Postgraduate Essay Prize

Purpose: It is awarded for the best essay-length piece of work on an American Studies topic written by a student currently registered for a postgraduate degree at a university or equivalent institution in Britain.
Level of Study: Postgraduate
Type: Prize
Value: £500
Frequency: Annual
No. of awards offered: 1
Application Procedure: Must submit essays to a judging panel.
Closing Date: February 11th
Contributor: Embassy Sponsored
Additional Information: For enquiries about the prize, contact awards@baas.ac.uk.

BAAS Postgraduate Short Term Travel Awards

Subjects: Culture and society of the United States of America.
Purpose: To fund travel to the United States of America for short-term research projects.
Eligibility: Open to residents in the United Kingdom. Preference is given to young postgraduates and to members of BAAS.
Level of Study: Postdoctorate, Postgraduate, Professional development, Doctorate
Type: Award
Value: UK£750
Frequency: Annual
Country of Study: United States of America
No. of awards offered: 5–10
Application Procedure: Applicants must complete an application form.
Closing Date: December 10th
Funding: Foundation
Contributor: American Embassy
No. of awards given last year: 6
No. of applicants last year: 35
Additional Information: Successful candidates must write a report and acknowledge BAAS assistance in any related publication.

For further information contact:

American Studies, School of Humanities, Keele University, Keele, Staffs, STS 5BG, United Kingdom
Contact: Professor Ian Bell, Chair, BAAS Awards Sub-Committee

Eccles Centre European Postgraduate Awards in North American Studies

Purpose: Research in American studies.
Level of Study: Doctorate
Type: Research
Value: £800 for travel and other expenses connected with the research visit to London
Length of Study: Varies
Frequency: Annual
Study Establishment: British Library
Country of Study: United Kingdom
No. of awards offered: 2
Application Procedure: Applicant must submit application form.
Closing Date: January
Funding: Private
Contributor: Eccles Centre
No. of awards given last year: 2
No. of applicants last year: Classified
Additional Information: For details of Eccles Centre activities, check at http://www.bl.uk/ecclescentre.

Eccles Centre Postgraduate Awards in North American Studies

Purpose: Research in American studies.
Level of Study: Doctorate
Type: Research
Value: £600 for travel and other expenses connected with the research visit to London
Length of Study: Varies
Frequency: Annual
Study Establishment: British Library
Country of Study: United Kingdom
No. of awards offered: 5
Application Procedure: Applicant must submit application form.
Closing Date: January
Funding: Private
Contributor: Eccles Centre
No. of awards given last year: 5
No. of applicants last year: Classified

Eccles Centre Visiting European Fellow in North American Studies

Purpose: Research in American studies.
Level of Study: Postdoctorate
Type: Research
Value: £2,500 for travel and other expenses connected with the research visit to London
Length of Study: Varies
Frequency: Annual
Study Establishment: British Library

Country of Study: United Kingdom
No. of awards offered: 1
Application Procedure: Applicant must submit applciation form.
Closing Date: January
Funding: Private
Contributor: Eccles Centre
No. of awards given last year: 1
No. of applicants last year: Classified

Eccles Centre Visiting Fellows in North American Studies

Subjects: Research in American studies.
Level of Study: Postdoctorate
Type: Research
Value: £2,500 for travel and other expenses connected with the research visit to London
Length of Study: Varies
Frequency: Annual
Study Establishment: British Library
Country of Study: United Kingdom
No. of awards offered: 3
Application Procedure: Applicant must submit an application form.
Closing Date: January
Funding: Private
Contributor: Eccles Centre
No. of awards given last year: 3
No. of applicants last year: Classified

Eccles Centre Visiting Professor in North American Studies

Subjects: American studies research.
Eligibility: Applicant must be a postdoctoral scholar resident in the USA or Canada whose research, in any field of North American Studies, entails the use of the British Library collection.
Level of Study: Postdoctorate
Type: Research
Value: £7,000 for travel and other expenses connected with the research visit to London
Length of Study: Varies
Frequency: Annual
Study Establishment: British Library
No. of awards offered: 1
Application Procedure: Applicant must submit an application form.
Closing Date: January
Funding: Private
Contributor: Eccles Centre
No. of awards given last year: 1
No. of applicants last year: classified

BRITISH ASSOCIATION FOR CANADIAN STUDIES (BACS)

2 Ancroft Southmoor, Berwick-upon-Tweed, TD15 2TD, England
Tel: (44) 020 7862 8687
Fax: (44) 20 7117 1875
Email: bacs@canadian-studies.org
Website: www.canadian-studies.net
Contact: Ms Jodie Robson

In response to the growing academic interest in Canada, the British Association for Canadian Studies (BACS) was established in 1975. Its aim is to foster teaching and research on Canada and Canadian issues by locating study resources in Britain, facilitating travel and exchange schemes for professorial staff and ensuring that the expertise of Canadian scholars who visit the United Kingdom is put to effective use. Principal activities include the publication of The British Journal of Canadian Studies and the BACS Newsletter, and the organization of the Association's annual multidisciplinary conference, which attracts scholars from Canada and Europe as well as from the United Kingdom.

Prix du Québec Award

Subjects: Humanities and social sciences.
Purpose: To assist British academics carrying out research related to Québec.
Eligibility: Open to citizens or long-term residents of the United Kingdom.
Level of Study: Doctorate, Postdoctorate, Professional development
Type: Award
Value: UK£2,000 each
Study Establishment: Universities, research institutions and schools
Country of Study: Canada
No. of awards offered: 2 (1, doctoral and postdoctoral students and 1, full-time teaching staff)
Application Procedure: Applicants must contact Jodie Robson, Administrator of BACS, for application guidelines.
Closing Date: February 15th
Funding: Government
Contributor: The Office of the Government of Québec in the United Kingdom
No. of awards given last year: 2
Additional Information: The award also seeks to encourage projects that incorporate Québec in a comparative approach. The Québec component must be more than 50 per cent.

BRITISH ASSOCIATION OF PLASTIC RECONSTRUCTIVE AND AESTHETIC SURGEONS (BAPRAS)

The Royal College of Surgeons, 35-43 Lincoln's Inn Fields, London, WC2A 3PE, England
Tel: (44) 20 7831 5161
Fax: (44) 20 7831 4041
Email: secretariat@bapras.org.uk
Website: www.bapras.org.uk
Contact: Ms Angela Rausch, Administrator

Founded in 1946 as British Association of Plastic Surgeons. The objective of the association is to relieve sickness and to protect and preserve public health by the promotion and development of Plastic Surgery. A name change to British Association of Plastic Reconstructive and Aesthetic Surgeons (BAPRAS) took effect in July 2006.

BAPRAS Barron Prize

Subjects: DVDs made on any subject that should run for a maximum of 20 minutes will be judged on their technical quality, production and editing, clinical content, educational value and the amount of information, visual and auditory, which is presented in the given time.
Purpose: To award the best DVD submitted by any member of the association or plastic surgery trainee.
Eligibility: Open to all BAPRAS members and UK based plastic surgery trainees.
Level of Study: Unrestricted
Type: Prize
Value: £500
Frequency: Annual
Application Procedure: Applicants must post DVDs and covering letters to the BAPRAS secretariat. No application form required.
Closing Date: August 10th
No. of awards given last year: 1
No. of applicants last year: 4
Additional Information: The selection will be made by the Education & Research Sub-Committee of the BAPRAS at their meeting in September. The name of the winner will be included in the BAPRAS handbook and a certificate will be presented.

BAPRAS European Travelling Scholarships

Purpose: To enable Specialist Registrars from UK to visit any plastic surgical centre in Europe.
Eligibility: Specialist Registrars (4–6) enrolled on a recognised training programme with the Specialist Advisory Committtee in plastic surgery are eligible to apply. Preference will be given to trainees travelling abroad without other financial awards and those applying for funding prior travel. Candidates seeking funds to travel abroad in paid jobs are less preferred.
Type: Scholarship
Value: £1,000
Frequency: Annual
Study Establishment: Plastic Surgery Units

Country of Study: European Union
No. of awards offered: 6
Application Procedure: Applicants should complete an application form, submit a proposed itinerary that should be detailed and give costs and the reasons for particular visits and a curriculum vitae (maximum length of two pages) to the Chairman of the Education and Research Sub-Committee, BAPRAS
Closing Date: December 10th
No. of applicants last year: 1

BAPRAS Student Bursaries

Subjects: Plastic surgery.
Purpose: To help medical students to cover expenses of travel and research related to plastic surgery.
Eligibility: Medical students in the United Kingdom.
Level of Study: Predoctorate
Type: Bursary
Value: UK£500
Length of Study: Varies
Frequency: Annual
Study Establishment: Hospital plastic surgery units or research laboratories
Country of Study: Any country
No. of awards offered: 20
Application Procedure: Application forms are available on request from the BAPRAS or from the BAPRAS website: www.bapras.org.uk
Closing Date: December 10th
Funding: Private
Contributor: BAPRAS
No. of awards given last year: 20
No. of applicants last year: 46

BAPRAS Travelling Bursary

Subjects: Plastic surgery.
Purpose: To enable a plastic surgeon in the United Kingdom to study new techniques abroad.
Eligibility: Open to members of the Association who are either specialist registrars in years 4–6, enrolled in a recognized training programme or who have not had more than 3 years as consultant plastic surgeons.
Level of Study: Professional development
Type: Bursary
Value: Up to UK£5,000
Length of Study: Varies
Frequency: Annual
Study Establishment: Any approved hospital plastic surgery units
Country of Study: Any country
No. of awards offered: 5
Application Procedure: Applicants must complete an application form and submit it with a proposed itinerary giving details of costs and reasons for wanting to attend a particular unit. A curriculum vitae of not more than two pages must also be submitted.
Closing Date: December 10th
Funding: Private
Contributor: BAPRAS
No. of awards given last year: 4
No. of applicants last year: 8

Paton/Masser Memorial Fund

Subjects: Plastic surgery research.
Purpose: To provide funds towards research projects.
Eligibility: Open to consultants and specialist registrars in plastic surgery working in the British Isles.
Level of Study: Research
Value: UK£5,000
Length of Study: Varies
Frequency: Annual
Study Establishment: A hospital or research laboratory
Country of Study: United Kingdom
No. of awards offered: Varies
Application Procedure: Application forms are available on the BAPRAS website: www.bapras.org.uk, or can be obtained by e-mailing a request to the secretariat@bapras.org.uk
Closing Date: December 31st
Funding: Private
Contributor: BAPRAS
No. of awards given last year: 1
No. of applicants last year: 3

Travelling Bursaries for Presentation at Overseas Meetings

Purpose: To cover the expenses for overseas travel by a consultant or trainee to present papers at international meetings.
Eligibility: Applicants must submit an application form to the Chairman of the Education and Research Sub-Committee, BAPRAS.
Type: Travel award
Value: £600
Frequency: Annual
Study Establishment: Various
Country of Study: Any country
No. of awards offered: 4
Application Procedure: Applicants must submit application form, abstract of paper to be presented and letter of acceptance to BAPRAS secretariat.
Closing Date: December 10th
Contributor: BAPRAS
No. of awards given last year: 6
No. of applicants last year: 9

THE BRITISH COUNCIL

10 Spring Gardens, London, SW1A 2BN, United Kingdom
Tel: (44) (0) 161 957 7755
Fax: (44) (0) 20 7389 6347
Email: general.enquiries@britishcouncil.org
Website: www.britishcouncil.org

The British Council is the United Kingdom's public diplomacy and cultural organization working in more than 100 countries, in arts, education, governance and science. The British Council promotes the diversity and creativity of British society and culture. The Foreign and Commonwealth Office provides The British Council with the core grant-in-aid.

Commonwealth Scholarships

Purpose: For students from the developed Commonwealth to study in the United Kingdom.
Eligibility: Please check at http://www.britishcouncil.org/india-scholarships-commonwealth-scholarships.htm for eligibility criteria.
Level of Study: Doctorate
Type: Scholarship
Value: Student visas for the UK will be arranged gratis by the British Council
No. of awards offered: 16
Application Procedure: See British Council website.
Closing Date: October 7th
Funding: Government

For further information contact:

Department of Higher Education, External Scholarship Division, ES.4 Section, West Block-1, Wing-6, 2nd Floor, R.K.Puram, New Delhi, 110066, India
Tel: (91) 11 26172491/26172492
Email: delhi.scholarship@in.britishcouncil.org

Entente Cordiale Scholarships for Postgraduate Study

Subjects: All subjects.
Purpose: To financially support those undertaking postgraduate study.
Eligibility: Applicants must be postgraduate students who can prove their academic success and their interest in consolidating Franco-British links. The candidate must fulfil all the criteria below to be eligible:
as the scheme is Franco-British you'll be expected to be of French nationality and/or to have completed your higher education in France to bac + 3 level;
you must have good english language skills (most UK Higher Education institutions require a minimum IELTS score of 6.5 for admission onto postgraduate courses);
you must be intending to undertake one year of study or research in

the UK that will not contribute to a French doctorate (co-tutelle); you will need an excellent track record illustrating your achievements, both academically and otherwise, and evidence that you will become a leader in your chosen field;
as the scheme is aimed at young professionals at the beginning of their career, you must not be over 35 years old when you apply.
Level of Study: Postgraduate
Type: Scholarship
Value: £10,000
Length of Study: 1 year
Frequency: Annual
Study Establishment: The University of Cambridge, St Edmund's College
Country of Study: United Kingdom
No. of awards offered: 20
Application Procedure: Applicants must obtain details of the application procedure from the British Council.
Closing Date: April 19th
Contributor: Offered in collaboration with the United Kingdom's Foreign and Commonwealth Office (FCO)

For further information contact:

British Council, 9–11 rue de Constantine, Paris, 75007, France
Tel: (33) 1 49 55 73 43180
Fax: (33) 1 47 05 77 02

The Goa Education Trust (GET) Scholarships

Subjects: Media management, architecture, computing and design, TV documentary, music therapy, architectural conservation and human rights, communication systems and signal processing and innovation technology and the law.
Purpose: To provide opportunities for dynamic young men and women of Goan origin who have demonstrated academic excellence and extra-curricular achievements to study or train in the United Kingdom.
Eligibility: Open to Indian nationals, domiciled and resident in Goa or born of Goan parents, who are not more than 30 years of age at the time of applying for the scholarship and have an excellent academic track record. Candidates must have confirmed admission for any technical/vocational/academic course of study in the United Kingdom for up to 1 year.
Level of Study: Postgraduate
Type: Scholarship
Value: Full or part tuition fees
Length of Study: 1 year
Frequency: Annual
Country of Study: United Kingdom
Application Procedure: A completed application form must be sent by post.
Closing Date: May 15th

For further information contact:

British Council Division, British Deputy High Commission, 901, 9th Floor, Tower1, One Indiabulls Centre, 841, Senapati Bapat Marg, Elphinstone Road (West), Mumbai, 400 013, India
Tel: (91) 22 67486748
Email: mumbai.enquiry@in.britishcouncil.org
Contact: GET Scholarships

Marshall Scholarships

Subjects: Any subject.
Purpose: To finance young Americans of high ability to study for a degree in the United Kingdom.
Eligibility: Open to US citizens who graduated with a first degree from an accredited four-year university or college in the United States by the start of the scholarship tenure; graduated with a cumulative GPA of at least 3.7 (A-); have formulated a feasible program of study to culminate in a second degree within two years.
Type: Scholarship
Value: Includes university fees, living allowance, round trip to the UK, support of a dependant spouse, annual book grant, annual thesis grant and research and daily travel grant.
Frequency: Annual
Country of Study: United Kingdom
No. of awards offered: Up to 40
Closing Date: Early October
Funding: Government
Contributor: British Council

For further information contact:

11766 Wilshire Boulevard, Suite 1200, Los Angeles, CA 90025 6538
Tel: 996 3028
Email: Losangeles@marshallscholarship.org
Website: www.marshallscholarship.org/
Contact: Alison Snyder, British Consulate-General

Saltire Scholarships

Subjects: Priority subject areas for Scotland's Saltire Scholarships are science, technology, the arts and creative industries, financial services, and clean and renewable energy.
Purpose: To provide the opportunity for bright, talented and hard working individuals to live, work and study in Scotland.
Eligibility: Applicants should: not have completed an undergraduate degree in Scotland (4 + years duration); have a conditional offer of a place at a Scottish university on an eligible course; be a citizen of Canada, the People's Republic of China, India or the USA; ensure you can meet the costs of living and remaining tuition fees; complete the online application process and send or post your application along with any supporting documents to your selected institution by the closing date. If posting your application please ensure you leave enough time so that it arrives by the closing date.
Level of Study: Masters
Type: Scholarship
Value: £2,000
Length of Study: 1 year
Frequency: Annual
Country of Study: Scotland
No. of awards offered: Up to 200
Closing Date: May 31st Midnight (GMT)
Funding: Government
Contributor: British Council
No. of awards given last year: 196
Additional Information: Please check at http://www.talentscotland.com/Students for more information.

BRITISH ECOLOGICAL SOCIETY (BES)

Charles Darwin House, 12 Roger Street, London, WC1N 2JU, England
Tel: (44) 20 7685 2500
Fax: (44) 20 7685 2501
Email: info@britishecologicalsociety.org; grants@britishecologicalsociety.org
Website: www.britishecologicalsociety.org

As a learned society and registered charity, the British Ecological Society (BES) is an independent organization receiving little outside funding. The aims of the Society are to promote the science of ecology through research, publications and conferences and to use the findings of such research to educate the public and to influence policy decisions that involve ecological matters. The BES is an active and thriving organization with something to offer anyone with an interest in ecology. Academic journals, teaching resources, meetings for scientists and policy makers, career advice and grants for ecologists are just a few of the societies areas of activity.

Ecologists in Africa

Subjects: Ecology. This grant provides support for ecologists in Africa to carry out innovative ecological research. The BES recognises that ecologists in Africa face unique challenges in carrying out ecological research and this grant is designed to provide them with support to develop their skills, experience and knowledge base as well as making connections with ecologists in the developed world.
Purpose: To support excellent ecological science in Africa by funding services and equipment.
Eligibility: Applicants must be a scientist and a citizen of a country in Africa or its associated islands; should have at least an MSc or equivalent degree; be working for a university or research institution in Africa (including field centres, NGOs, museums etc.) that provides

basic research facilities; must carry out the research in a country in Africa or its associated islands.
Level of Study: Postdoctorate, Research
Type: Project grant
Value: Up to UK£10,000
Length of Study: 1 year
Frequency: Annual
Country of Study: Any country
Application Procedure: Candidates must apply online through the BES office.
Closing Date: September 17th
No. of awards given last year: 6
No. of applicants last year: 68
Additional Information: Further information is available on request or from the website.

Outreach Grant

Subjects: Ecology. The Society defines ecology as the scientific study of the distribution, abundance and dynamics of organisms, their interactions with other organisms and their physical environment. It is therefore essential that applications promote and engage the public with the science of ecology. Grants will not be awarded for purely nature conservation purposes or any activity that does not promote the science of ecology.
Purpose: To promote ecological science to a wide audience.
Eligibility: Awards are open to individuals and organisations to organise public engagement events in ecology. This includes, but is not limited to, members of the BES, researchers, schools, museums, libraries and community groups. Applications will not be considered for: staff salaries; Purchase of apparatus, including computers, cameras etc., unless these are integral to the application. Applications from museums and schools are welcome but projects must involve significant outreach beyond schools. Projects aimed solely at delivering curriculum to school children will not be considered.
Level of Study: Professional development
Type: Project grant
Value: Up to £2,000
Frequency: Biannual
Country of Study: Any country
Application Procedure: Applicants must complete an online application form through the BES website.
Closing Date: March 4th
Contributor: British Ecological Society
No. of awards given last year: 4
No. of applicants last year: 34
Additional Information: Published papers and reports to other organizations should include an acknowledgement of the support from the BES. Other conditions may apply. The Coalbourn Trust is an independent trust that looks to the BES to nominate suitable projects for funding. Recommendations for funding will be made from among the applicants for Small Ecological Project Grants. All applicants for Small Ecological Project Grants will automatically be eligible for funding from the Coalbourn Trust. Further information is available on request or from the website. This is currently being reviewed, please check the webiste for up-to-date information.

Research Grant Large

Subjects: Ecology. We support projects where there is a clear ecological science focus to the work; any other aspects, i.e. sociology, economics, etc., must be clearly integrated into the ecology and scientific goals of the project. The grants are to support work of the highest international standard and applicants need to show how the work will advance ecological science. The objectives of the grant are to provide funding: - for new and innovative ecological research - for pump priming projects - to help early career ecologists to establish an independent research career in ecology.
Purpose: To support scientific ecological research where there are limited alternative sources of funding. Early career ecologists can apply for funding up to £20,000.
Eligibility: These grants are given to individuals, not organisations. Only current BES members may apply. They are limited to ecologists at an early stage in their career. In the absence of extenuating circumstances, 'early career' is defined as less than 5 years post- PhD or -D.Phil. experience. The investigator must have a PhD or D.Phil., or have submitted a thesis that will have been examined by the time the application is submitted. The investigator must demonstrate that the project is distinct from other current research activities, or if there is any overlap, that it would substantially extend their existing research into new and distinct areas. The investigator should undertake at least part of the research him/herself. Normally, funding is not given to individuals who already have full time permanent academic positions in universities or equivalent positions in research institutes. Those who have received an early career grant are ineligible from applying again. Applicants cannot use a Research Grant to fund a component of a larger study. Applicants are responsible for obtaining all relevant permits and permissions required to undertake the proposed work. Applications for projects based outside the UK should demonstrate liaison with collaborating organisations, local environmental agencies, NGOs and/or communities.
Type: Grant
Value: Up to £20,000
Frequency: Biannual
Application Procedure: Applicants must apply online through the BES website. Further information is available on request and from the website.
Closing Date: March 4th

Research Grant Small

Subjects: Ecology. We support projects where there is a clear ecological science focus to the work; any other aspects, i.e. sociology, economics, etc., must be clearly integrated into the ecology and scientific goals of the project. The grants are to support work of the highest international standard and applicants need to show how the work will advance ecological science. The objectives of the grant are to provide funding for new and innovative ecological research, pump priming projects and to help early career ecologists to establish an independent research career in ecology.
Purpose: To support scientific ecological research where there are limited alternative sources of funding. Small projects can be awarded up to £5,000.
Eligibility: These grants are given to individuals, not organisations. Only current BES members may apply. Applicants cannot use a Research Grant to fund a component of a larger, already funded study. Funding is not available for work that will form part of a degree or thesis. Applications to take part in an expedition will not be considered. However, we will consider applications for a stand alone research project which is taking place during an expedition. Applicants are responsible for obtaining all relevant permits and permissions required to undertake the proposed work. Applications for projects based outside the UK should demonstrate appropriate liaison with collaborating organisations, local environmental agencies, NGOs and/or communities. BES will not award more than one grant to any one applicant in any one year, and no more than three grants in any five year period. Failure to submit a satisfactory report at the end of a grant will mean the grantee is ineligible to apply for further grants.
Type: Grant
Value: Up to £5000
Frequency: Biannual
Application Procedure: A completed application form along with supporting statements from the UK host institution and the applicants institution should be submitted.
Closing Date: March 4th

Training & Travel Grant

Subjects: Ecology.
Purpose: To help students and postgraduate research assistants (RAs) or their equivalents to meet the costs of attending upcoming INTECOL Congress; network and publicise their research by presenting their work at upcoming INTECOL Congress. This grant is to support the training and development of students and postgraduate research assistants (RAs) or their equivalent.
Eligibility: All applicants are required to: have at least a BSc or equivalent degree; be working or studying at a university or research institution (including field centres, NGOs, museums, etc.) that provide research facilities; be working in scientific areas within the remit of the BES (the science of ecology) and of relevance to upcoming INTECOL; be a member of the BES; be giving a presentation at the conference; be a student, postgraduate research assistant (RA) or their equivalent.
Level of Study: Postdoctorate, Postgraduate
Type: Grant

Value: Grants of £750 are available for students from countries defined as low income or lower middle income according to the World Bank. Students from other countries may apply for a grant up to a maximum of £300.
Application Procedure: Applicants must complete an online application form, available from the BES website. Awarded on a first-come first-served basis.
Closing Date: By first-come, first-served
Contributor: British Ecological Society
Additional Information: Applications are accepted only online through the BES website.

BRITISH FEDERATION OF WOMEN GRADUATES (BFWG)

4 Mandeville Courtyard, 142 Battersea Park Road, London, SW11 4NB, England
Tel: (44) 20 7498 8037
Fax: (44) 20 7498 5213
Email: awards@bfwg.org.uk
Website: www.bfwg.org.uk
Contact: Secretary

The British Federation of Women Graduates (BFWG) provides opportunities to women in education and public life. BFWG works as part of an international organization to improve the lives of women and girls, fosters local, national and international friendship and offers scholarships for third year doctorat research.

British Federation of Women Graduates Scholarships

Subjects: All subjects.
Purpose: To recognise academic excellence in women doctoral (PhD, DPhil, etc) students during their third year of doctoral research at British (England, Wales, Scotland) higher education institutions.
Eligibility: Eligibility criteria (see www.bfwg.org.uk) include a range of dates for the start of the student's doctoral course.
Level of Study: Doctorate
Type: Prize
Value: £2,500–6,000
Length of Study: Student must be in her third year (or part time equivalent) at the time of giving out the awards (October each year)
Frequency: Annual
Study Establishment: University or other Higher Education Institution. Overseas distance learners are not eligible
Country of Study: England, Scotland and Wales
No. of awards offered: 5–10
Funding: Trusts
Contributor: British Federation of Women Graduates
Additional Information: Please check website for further details.

International Federation of University Women & National Member Organisations of IFUW

Subjects: All subjects.
Purpose: To assist in promoting postgraduate study or training relevant to career progress. Thus, variable awards for support of postgraduate study or training.
Eligibility: Some require membership of IFUW (see www.ifuw.org) to be eligible to apply.
Level of Study: Postgraduate
Type: Prize grant
Value: Variable, but none supports a complete period of study. Most £500–3,000 or equivalent in relevant local currency
Length of Study: Awards vary but most look for some life experience since the first degree
Study Establishment: University or other Higher Education Institution
No. of awards offered: Variable but usually under 20
Funding: Private, trusts
Contributor: Members of the international Federation of University Women
Additional Information: Details of the awards can be found on the IFUW webpages for IFUW awards and on the incorporated pages for awards given by IFUW national affiliates. Country of study varies depending on the award but range of awards covers the world.

THE BRITISH INSTITUTE AT ANKARA (BIAA)

10 Carlton House Terrace, London, SW1Y 5AH, United Kingdom
Tel: (44) 020 7969 5204
Fax: (44) 020 7969 5401
Email: biaa@britac.ac.uk
Website: www.biaa.ac.uk
Contact: Claire McCafferty, London Administrator

BIAA aims to support, promote, facilitate, and publish British research focused on Turkey and the Black Sea littoral in all academic disciplines within the arts, humanities, and social sciences and to maintain a centre of excellence in Ankara focused on the archaeology and related subjects of Turkey.

BIAA Research Scholarship

Subjects: Turkey and Black Sea littoral in any disciplines of the arts, humanities, and social sciences.
Purpose: To conduct their own research at doctoral level.
Eligibility: Open to candidates who hold a Master's degree and have a demonstrable connection to UK academia.
Level of Study: Postgraduate, Research
Type: Research scholarship
Value: £800 per month and the cost of one return flight between the United Kingdom and Turkey
Length of Study: 9 months
Frequency: Annual
Country of Study: Turkey
No. of awards offered: 1
Application Procedure: Deadline for applications is mid-September. Check website for further details.

For further information contact:

British Institute at Ankara, Ankara, Turkey
Tel: (90) 020 7969 5204
Fax: (90) 020 7969 5401
Email: biaa@britac.ac.uk
Website: www.biaa.ac.uk
Contact: Claire McCafferty

BIAA Study Grants

Subjects: Research in the fields of the arts, humanities, and the social sciences related to Turkey and the Black Sea littoral.
Purpose: To support doctoral or postdoctoral research in the fields of the arts, humanities, and the social sciences related to Turkey and the Black Sea littoral.
Eligibility: Open to postgraduate students or postdoctoral scholars based in a British university.
Level of Study: Graduate, Postgraduate, Postdoctorate
Type: Grant
Value: £500 per month for basic subsistence and accommodation and an airfare of £300
Length of Study: Upto 3 months
Frequency: Twice a year
Study Establishment: BIAA
Country of Study: Turkey
Application Procedure: Applicants must complete a copy of the form downloaded from the Institute's website, supported by a reference from the supervisor of postgraduate research or, for postdoctoral applicants, another suitable academic (who are not members of the BIAA Research Committee or staff of the Institute) and send the completed form to the London address.
Closing Date: April 1st
Additional Information: Do not enclose any attachments unless they have been specifically asked for by BIAA.

BIAA/SPHS Fieldwork Award

Subjects: Turkey and Black Sea relating to Hellenic studies.
Purpose: To support fieldwork by a postgraduate on Turkey or the Black Sea region relating to Hellenic studies.
Eligibility: Open to postgraduate students based at a British university. The study should be undertaken in Turkey and the Black Sea region and should relate to Hellenic studies.
Level of Study: Postgraduate

Type: Grant
Value: Up to UK£400
Frequency: Annual
Study Establishment: British Institute of Ankara/Society for the Promotion of Hellenic Studies
Country of Study: Turkey
No. of awards offered: 1–2
Application Procedure: Applicants should submit an application form to the London address.
Closing Date: April 1st
Additional Information: Shortlisted candidates will be interviewed in London in May.

Martin Harrison Memorial Fellowship

Subjects: Archaeology.
Purpose: To assist junior Turkish archaeologists, who are not able to take advantage of travelling, working in any area of the archaeology of Anatolia from Prehistory to the Ottoman period, to visit the United Kingdom, especially Oxford, in connection with their research work.
Eligibility: Open to Turkish citizens residing in Turkey who have completed at least 2 years of postgraduate research and at most held a doctorate for 5 years, working in any area of the archaeology of Anatolia (from the Prehistoric to the Ottoman period).
Level of Study: Doctorate, Postgraduate, Research
Type: Short-term fellowship
Value: £1,500 and travel expenses from and to Turkey
Length of Study: 6–13 weeks
Frequency: Annual
Study Establishment: University of Oxford
Country of Study: United Kingdom
No. of awards offered: 1
Application Procedure: Completed applications, including a curriculum vitae, should be sent to the Turkey address.
Closing Date: March 31st
Funding: Foundation
Contributor: Martin Harrison Fund for living expenses and British Institute at Ankara (BIAA) for travel
Additional Information: The selection will be made on the basis of the applicant's academic record, coherent research proposal, ability to benefit from libraries and scholars in Oxford, and a working knowledge of spoken and written English.

For further information contact:

The British Institute at Ankara, 24 Tahran Caddesi, Kavaklidere, Ankara, TR 06700, Turkey
Tel: (90) 90 312 427 54 87
Fax: (90) 90 312 428 01 59
Email: ggirdivan@biaatr.org

BRITISH INSTITUTE FOR THE STUDY OF IRAQ (GERTRUDE BELL MEMORIAL)

10 Carlton House Terrace, London, SW1Y 5AH, United Kingdom
Tel: (44) 207 969 5274
Fax: (44) 207 969 5401
Email: bisi@britac.ac.uk
Website: www.bisi.ac.uk
Contact: Mrs Joan Porter MacIver, Administrator

The British Institute for the Study of Iraq promotes, supports and undertakes research and public education relating to Iraq. Its coverage includes anthropology, archaeology, geography, history, languages and related disciplines within the arts, humanities and social sciences from the earliest times until the present. BISI has over 800 members and subscribers to its journal Iraq. Members may also subscribe to the International Journal of Contemporary Iraqi Studies.

The British Institute for the Study of Iraq Grants

Subjects: Humanities or social sciences.
Purpose: To support research and the organisation of academic conferences on Iraq.
Eligibility: Open to United Kingdom residents who are graduates, postgraduates or students doing postgraduate and graduate work at a UK institution or on an exception basis working on BISI research projects.
Level of Study: Doctorate, Graduate, Postgraduate, Postdoctorate
Type: Grant
Value: Usually up to UK£4,000, depending on the nature of the research
Length of Study: 1 academic year
Frequency: Annual, Twice a year
No. of awards offered: Varies
Application Procedure: The institute considers applications for individual research grants once a year in February. Information and application forms are available from either the Administrator or the website. Two academic references are required.
Closing Date: January 10th, March 1st and October 1st
Funding: Government, private, trusts
No. of awards given last year: 12
Additional Information: Details of the British Institute for the Study in Iraq are available on the website www.bisi.ac.uk. Grantees will be required to provide a written report of their work and abstracts from these reports will be published in future issues of the institute's newsletter. Individual research and travel grants are offered.The country of study is Iraq.

BRITISH INSTITUTE IN EASTERN AFRICA

PO Box 30710, Nairobi, GPO 00100, Kenya
Tel: (254) 20 434 7195/3190
Fax: (254) 20 434 3365
Email: office@biea.ac.uk
Website: www.biea.ac.uk
Contact: Dr David Anderson, Director

The British Institute in Eastern Africa (BIEA) exists to promote research in the humanities and social sciences. BIEA is based in Nairobi, but supports work across Eastern Africa, and is one of the schools and institutes supported by the British Academy.

British Institute in Eastern Africa Graduate Attachments

Subjects: Humanities and social sciences.
Purpose: To provide opportunities for research experience to recent university graduates.
Eligibility: Open to recent graduates from UK and East African universities.
Level of Study: Postgraduate
Type: Studentship
Value: Covers travel and subsistence
Length of Study: 3–6 months
Frequency: Annual
Study Establishment: The British Institute in Eastern Africa
No. of awards offered: Up to 12
Application Procedure: Candidates must submit a letter of application to the Director, BIEA with the names of two academic referees and a curriculum vitae.
Closing Date: March 31st
Funding: Government
Contributor: British Academy
No. of awards given last year: 13
No. of applicants last year: 72
Additional Information: Small grants and assistance may be offered on a discretionary basis to scholars of other nationalities. Archaeology students may be required to assist in excavation carried out by the Institute's staff. Details of activities are published in the *Archaeology Abroad* bulletin and in the Institute's annual report, copies of which are available on request.

British Institute in Eastern Africa Minor Grants

Subjects: Humanities and social sciences.
Purpose: To assist with the costs of research projects in Eastern Africa.
Eligibility: Open to applicants from United Kingdom and Eastern Africa.
Level of Study: Postgraduate
Type: Grant
Value: Up to UK£1,000
Frequency: Biannual
Study Establishment: The British Institute in Eastern Africa

No. of awards offered: Up to 10
Application Procedure: Application forms can be downloaded from the BIEA website.
Closing Date: January 31st
Funding: Government
Contributor: British Academy
No. of awards given last year: 22
No. of applicants last year: 45
Additional Information: Applicants must contact the Director for further information on relevant topics likely to receive support. Those awarded grants will be required to keep the Institute regularly informed of the progress of their research, to provide a preliminary statement of accounts within 18 months of the award dates and to provide the Institute with copies of all relevant publications. They are encouraged to discuss with the Director the possibility of publishing their results in the Institute's journal, *Azania*. Results for the Minor Grants Award may be expected within 2 months of either May 30th or November 30th. Those awarded grants are required to become members of the BIEA.

For further information contact:

Email: pjlane@insightkenya.com

BRITISH LIBRARY

96 Euston Road, London, NW1 2DB, England
Tel: (44) 20 7412 7702
Fax: (44) 20 7412 7780
Email: Customer-Services@bl.uk
Website: www.bl.uk
Contact: Mr Peter Barber, Map Librarian

Edison Fellowship

Purpose: The British Library is offering on a competitive basis an Edison Fellowship. This may be held as a full- or part-time appointment. Proposals will be considered which treat any aspect of the history of recording and the performance of western art music.
Level of Study: Research
Type: Fellowship
Value: £5,000
Length of Study: No longer than 4 months
Frequency: Annual
Application Procedure: The applicant is required to submit a curriculum vitae; a brief research proposal (not exceeding 750 words), together with a thesis or dissertation prospectus where available if the research is being conducted in the context of an advanced degree; and a proposed timescale for the project, whether a period of continuous residence, or regular visits to the British Library on a weekly or a monthly basis.
The proposal should include reference to any previous research on recordings carried out by the applicant, and make clear how the proposed research relates to the Library's collections of recordings. The applicant must also arrange to have two confidential letters of recommendation sent to the Curator.
Closing Date: January 28th
Funding: Commercial
Additional Information: The Library reserves the right to make no award in the event that no suitable applications are received.

For further information contact:

The British Library, 96 Euston Road, London, NW1 2DB, United Kingdom
Contact: Jonathan Summers, Classical Music Curator

Helen Wallis Fellowship

Subjects: The history of cartography, preferably with an international dimension, history.
Purpose: To promote the extended and complementary use of the British Library's book and cartographic collections in historical investigation.
Eligibility: Applicants should write for details.
Level of Study: Postdoctorate, Doctorate, Postgraduate
Type: Fellowship
Value: Up to UK£300
Length of Study: 6–12 months
Frequency: Annual
Study Establishment: British Library, London
Country of Study: United Kingdom
No. of awards offered: 1
Application Procedure: Applicants must submit a letter indicating the proposed period and outlining the research project together with a full curriculum vitae and give three references.
Closing Date: May 1st
Funding: Private
No. of awards given last year: 1
Additional Information: The award honours the memory of Dr Helen Wallis, OBE (1924–1995), Map Librarian at the British Museum and then the British Library between the years 1967–1986. Further information can be found on the website.

For further information contact:

British Library Map Library, 96 Euston Road, London, NW1 2DB
Tel: 20 7412 7525
Contact: Map Librarian

BRITISH LUNG FOUNDATION

73-75 Goswell Road, London, EC1V 7ER, England
Tel: (44) 20 7688 5555
Fax: (44) 20 7688 5556
Email: info@blfservices.co.uk
Website: www.lunguk.org
Contact: Julia Heidsta, Research Manager

The British Lung Foundation provides information to the public on lung conditions and all aspects of lung health. The Foundation provides support to those who live with a lung condition every day of their lives through the Breathe Easy Club, a nationwide network of local voluntary support groups, and finds solutions to lung disease by funding world-class medical research.

British Lung Foundation Project Grants

Subjects: Respiratory diseases.
Purpose: To promote medical research into the prevention, diagnosis and treatment of all types of lung diseases.
Eligibility: Open to graduates working within the United Kingdom who have relevant research experience. The principal applicant must be based in a research centre in the United Kingdom.
Level of Study: Postdoctorate, Doctorate, Postgraduate, Predoctorate, Professional development, Research
Type: Project grant
Value: UK£200,000
Length of Study: Up to 3 years
Frequency: Annual
Study Establishment: An approved research centre
Country of Study: United Kingdom
No. of awards offered: Approx. 10
Application Procedure: Applicants must complete an application form, available from the British Lung Foundation.
Closing Date: February 25th
Funding: Commercial, private
Contributor: Voluntary donations
No. of awards given last year: Approx. 10
No. of applicants last year: 132

BRITISH MEDICAL ASSOCIATION (BMA)

BMA Research Grants, BMA House, Tavistock Square, London, WC1H 9JP, England
Tel: (44) 207 383 6755
Fax: (44) 20 7383 6383
Email: info.sciencegrants@bma.org.uk
Website: www.bma.org.uk

The BMA is a voluntary professional association with over two-thirds of practising UK doctors in membership and an independent trade union dedicated to protecting individual members and the collective interests of doctors.

Doris Hillier Research Grant

Subjects: Research into rheumatism and arthritis (and every 3rd year into Parkinson's disease).
Purpose: To assist and support research.
Eligibility: Open to registered medical practitioners in the United Kingdom who are BMA members.
Level of Study: Research
Type: Research grant
Value: UK£55,000
Length of Study: 3 years
Frequency: Annual
Country of Study: United Kingdom
No. of awards offered: 1
Application Procedure: Applicants must complete an online application form.
Closing Date: March 15th at 17.00
Funding: Trusts
No. of awards given last year: 1
No. of applicants last year: 7
Additional Information: Grants are advertised from December, on the BMA website and in the *British Medical Journal*.

H C Roscoe Research Grant

Subjects: Research into the elimination of the common cold and/or other viral diseases of the human respiratory system.
Purpose: To assist and support research.
Eligibility: Open to members of the BMA and research scientists working in association with a BMA member.
Level of Study: Research
Type: Research grant
Value: UK£50,000
Length of Study: 3 years
Frequency: Annual
No. of awards offered: 1
Application Procedure: Applicants must complete an online application form.
Closing Date: March 15th at 17.00
Funding: Trusts
No. of awards given last year: 1
No. of applicants last year: 7
Additional Information: Grants are advertised from December, on the BMA website and in the *British Medical Journal*.

Helen H Lawson Research Grant

Subjects: Research into rehabilitation in stroke care.
Purpose: To assist and support research.
Eligibility: Open to registered medical practitioners in the United Kingdom who are BMA members.
Level of Study: Research
Type: Research grant
Value: UK£50,000
Length of Study: 3 years
Frequency: Annual
Country of Study: United Kingdom
No. of awards offered: 1
Application Procedure: Applicants must complete an online application form.
Closing Date: March 15th at 17.00
Funding: Trusts
No. of awards given last year: 1
No. of applicants last year: 8
Additional Information: Grants are advertised from December, on the BMA website and in the *British Medical Journal*.

The James Trust Research Grant

Subjects: Research into asthma.
Purpose: To assist and support research.
Eligibility: Open to registered medical practitioners in the United Kingdom who are BMA members.
Level of Study: Research
Type: Research grant
Value: UK£55,000
Length of Study: 3 years
Frequency: Annual
Country of Study: United Kingdom
No. of awards offered: 1
Application Procedure: Applicants must complete an online application form.
Closing Date: March 15th at 17.00
Funding: Trusts
No. of awards given last year: 1
No. of applicants last year: 6
Additional Information: Grants are advertised from December, on the BMA website and in the *British Medical Journal*.

Joan Dawkins Research Grant

Subjects: To assist research into repetitive head injury in sport.
Purpose: To assist and support research.
Eligibility: Open to registered medical practitioners in the United Kingdom. Research scientists may also apply. Projects must relate to the United Kingdom.
Level of Study: Research
Type: Research grant
Value: UK£55,000
Length of Study: 3 years
Frequency: Annual
Country of Study: United Kingdom
No. of awards offered: 1
Application Procedure: Applicants must complete an online application form.
Closing Date: March 15th at 17.00
Funding: Trusts
No. of awards given last year: 1
No. of applicants last year: 10
Additional Information: Grants are advertised from December, on the BMA website and in the *British Medical Journal*.

Josephine Lansdell Research Grant

Subjects: Research in the field of heart disease.
Purpose: To assist and support research.
Eligibility: Open to registered medical practitioners in the United Kingdom who are BMA members.
Level of Study: Research
Type: Research grant
Value: UK£50,000
Length of Study: 3 years
Frequency: Annual
Country of Study: United Kingdom
No. of awards offered: 1
Application Procedure: Applicants must complete an online application form.
Closing Date: March 15th at 17.00
Funding: Trusts
No. of awards given last year: 1
No. of applicants last year: 14
Additional Information: Grants are advertised from December, on the BMA website and in the *British Medical Journal*.

Kathleen Harper

Subjects: Varies each year.
Purpose: To assist and support research.
Eligibility: Registered medical practitioners in the United Kingdom who are also BMA members.
Value: Approx. UK£30,000
Length of Study: 3 years
Frequency: Annual
Application Procedure: Applicants should complete an online application form.
Closing Date: Mid-March
Funding: Private
No. of awards given last year: 1
Additional Information: Grants are advertised from December, on the BMA website and from January in the *British Medical Journal*.

Margaret Temple Research Grant

Subjects: Research into schizophrenia.
Purpose: To assist and support research.
Eligibility: Open to medical practitioners in United Kingdom. Research scientists may also apply. Projects must relate to the United Kingdom.

Level of Study: Research
Type: Research grant
Value: UK£55,000
Length of Study: 3 years
Frequency: Annual
Country of Study: United Kingdom
No. of awards offered: 1
Application Procedure: Applicants must complete an online application form.
Closing Date: March 15th at 17.00
Funding: Trusts
No. of awards given last year: 2
No. of applicants last year: 14
Additional Information: Grants are advertised from December, on the BMA website and in the *British Medical Journal*.

Strutt and Harper Grant

Subjects: Research into terminal care for non-cancer patients.
Purpose: To assist and support research.
Eligibility: Registered medical practitioners in the UK, who are BMA members.
Level of Study: Research
Type: Research grant
Value: UK£40,000
Length of Study: 3 years
Frequency: Annual
Country of Study: United Kingdom
No. of awards offered: 1
Application Procedure: Applicants must complete an online application form.
Closing Date: March 15th at 17.00
Funding: Trusts
No. of awards given last year: 1
No. of applicants last year: 5
Additional Information: Grants are advertised from December, on the BMA website and in the *British Medical Journal*.

TP Gunton

Subjects: Research into public health relating to cancer.
Purpose: To assist and support research.
Eligibility: Open to both medical practitioners and research scientists in the United Kingdom.
Level of Study: Research
Type: Research grant
Value: UK£40,000
Length of Study: 3 years
Frequency: Annual
Country of Study: United Kingdom
No. of awards offered: 1
Application Procedure: Applicants must complete an online application form.
Closing Date: March 15th at 17.00
Funding: Trusts
Additional Information: Grants are advertised from December, on the BMA website and in the *British Medical Journal*.

Vera Down Research Grant

Subjects: Research into neurological disorders.
Purpose: To assist and support research.
Eligibility: Open to registered medical practitioners in the United Kingdom who are BMA members.
Level of Study: Research
Type: Research grant
Value: UK£55,000
Length of Study: 3 years
Frequency: Annual
Country of Study: United Kingdom
No. of awards offered: 1
Application Procedure: Applicants must complete an online application form.
Closing Date: March 15th at 17.00
Funding: Trusts
No. of awards given last year: 1
No. of applicants last year: 19
Additional Information: Grants are advertised from December, on the BMA website and in the *British Medical Journal*.

BRITISH MOUNTAINEERING COUNCIL (BMC)

177-179 Burton Road, West Dibsbury, Manchester, M20 2BB, United Kingdom
Tel: (44) 0161 445 6111
Fax: (44) 0161 445 4500
Email: office@thebmc.co.uk
Website: www.thebmc.co.uk

The BMC is the representative body that exists to protect the freedoms and promote the interests of climbers, hill walkers and mountaineers.

Alpine Ski Club Kenneth Smith Scholarship

Subjects: Ski research.
Purpose: To assist skiers and mountaineers in improving their touring and ski mountaineering skills and qualifications.
Level of Study: Professional development
Type: Scholarship
Value: UK£600
Length of Study: 1 year
Frequency: Annual
No. of awards offered: 2
Application Procedure: Contact organization.
Closing Date: October 31st
Funding: Trusts
Contributor: Kenneth Smith Trust

For further information contact:

The ASC Awards Sub-Committee, 22 Hatton Court, Hatton of Fintray, Aberdeenshire, AB21 OYA
Contact: Mrs Jay Turner

BMC Grant

Purpose: To fund innovative-style ascents in the greater mountain ranges by professional mountaineers.
Level of Study: Professional development
Type: Grant
Value: UK£1,000
Length of Study: 1 year
Frequency: Annual
No. of awards offered: Varies
Application Procedure: Contact the British Mountaineering Council.
Closing Date: Check website or contact organisation

BRITISH ORTHODONTIC SOCIETY

British Orthodontic Society, 12 Bridewell Place, London, EC4V 6AP, United Kingdom
Tel: (44) 020 7353 8680
Fax: (44) 020 7353 8682
Email: d.bearn@dundee.ac.uk
Website: www.bos.org.uk
Contact: Mr David Bearn, Chairman

BOS Clinical Audit Prize

Subjects: Orthodontics.
Purpose: Awarded annually at BOC to the best reports published in the Clinical Effectiveness Bulletin (CEB) of the BOS in each year.
Eligibility: Any member of the BOS.
Level of Study: Professional development
Type: Prize
Value: First prize £500, second prize £350 and third prize £150
Frequency: Annual
Application Procedure: A published article in the Clinical Effectiveness Bulletin of the BOS. All articles published in CEB are automatically entered. Article must demonstrate:
1. Clearly articulated audit question and standard
2. Well-designed appropriate methodology
3. Sound data analysis

4. Well presented and coherent
5. Relevant conclusions.
Funding: Corporation

The Chapman Prize in Orthodontics

Subjects: Orthodontics.
Purpose: To award the best published article by a member of the BOS in a calendar year. Authors submit a paper for consideration for the prize by January 31st of the year after publication.
Eligibility: Any member of the BOS.
Level of Study: Professional development
Type: Prize
Value: £1,200
Frequency: Annual
Application Procedure: A published article on an orthodontic or allied subject. The article should be submitted electronically as a pdf file of the published article and the application form (downloadable word document). Articles for consideration must be submitted in the month of January of the year following publication.
Article demonstrates:
1. Clearly articulated research question
2. Well-designed appropriate study methodology
3. Appropriate outcomes reported
4. Sound data analysis
5. Well presented and coherent
6. Relevant conclusions
7. Likely impact on orthodontic knowledge/practice.
Closing Date: January 31st
Funding: Corporation
Additional Information: The winner should be prepared to present their article at the British Orthodontic Conference, at the discretion of the BOC Chairman..

Dental Directory Practitioner Group Prize

Subjects: Orthodontics.
Purpose: Awarded for clinical excellence to a member of the Practioner Group who presents the best treated case. Only one entry of one case per member per year is accepted. Cases must have been treated solely by the entrant either in hospital or practice.
Eligibility: Members of the PG of the BOS.
Level of Study: Professional development
Type: Prize
Value: £500 of Dental Directory vouchers
Frequency: Annual
Application Procedure: Clinical records for one treated case displayed at the British Orthodontic Conference.
1. Models and case records should show the names or initials of the patient but not the presenter's name.
Applicants are advised to ensure that they retain duplicate models and case records, as the security of submitted records cannot be guaranteed.
2. Presenters are advised to seek the consent of the patient and/or their guardian to the cases being shown at the BOC.
3. Presenters may not advertise any orthodontic appliances or treatment techniques.
4. One entry of one case per member per year is accepted. Cases must have been treated solely by the entrant either in hospital or practice.
Cases are judged on difficulty, clinical management and presentation.
Closing Date: August 13th
Funding: Corporation
Contributor: Dental Directory

Hawley Russell Research and Audit Poster Prizes

Subjects: Orthodontics.
Purpose: Awarded to the best research poster and the best audit poster displayed at the British Orthodontic Conference.
Eligibility: Any member of the BOS.
Level of Study: Professional development, Research
Type: Prize
Value: £400
Frequency: Annual
Application Procedure: A poster displayed at the British Orthodontic Conference. Abstract submitted to BOC Poster display organiser and, if accepted, the poster displayed at the BOC.
Closing Date: July 1st (each year)
Funding: Corporation
Contributor: Hawley Russell

Orthocare UTG Prize

Subjects: Orthodontics.
Purpose: The prize is open to any member of the British Orthodontic Society who has successfully completed a UK University Master's programme or equivalent within 13 months of the conference. A candidate is not allowed to enter the same project for both the UTG Research Prize and BOC Poster Prizes in the same year.
Eligibility: The prize is open to any member of the TGG of British Orthodontic Society.
Level of Study: Postgraduate
Type: Prize
Value: First prize £600; second prize £400 and third prize £200
Frequency: Annual
Application Procedure: Awarded to the best presentation at the University Teachers Group Research Session held annually at the British Orthodontic Conference of a project completed as part of a recent Master's course.
The presentation is judged using the following criteria:
1. Presentation: confidence of the presenter, and their familiarity with the topic and all aspects of the research.
2. Content: visual presentation of the presentation, its content and informativeness.
3. Question and Answer: ability of the presenter to think on their feet, and their breadth and depth of knowledge.
4. Impact Factor of Research: relevance of the research within its field, the importance of the question being asked, and the possibility of the research moving the field forward.
Closing Date: May 31st
Funding: Corporation
Contributor: OrthoCare

Research Protocol Award

Subjects: Orthodontics.
Purpose: The award is based on the submission of a Master's (or equivalent) study protocol including a review of the relevent literature by a TGG member in the first 18 months of an orthodontic training programme.
Eligibility: Any member of the TGG of the BOS in the first 18 months of an orthodontic training programme.
Level of Study: Postgraduate
Type: Award
Value: £700 to the winner and £500 to the winner's supervisor's academic department
Frequency: Annual
No. of awards offered: 1
Application Procedure: Awarded to the best presentation at the University Teachers Group Research Session held annually at the British Orthodontic Conference of a project completed as part of a recent Master's course. The protocol should be submitted electronically as a word document on the downloadable template (maximum file size 500kb) with the completed application form (downloadable word document).
Protocol demonstrates:
1. Clearly articulated research question
2. Well-designed appropriate study methodology
3. Appropriate outcomes
4. Sound data analysis plan
5. Well presented and coherent
6. Likely impact on orthodontic knowledge/practice.
Closing Date: Late February
Funding: Corporation

BRITISH RETINITIS PIGMENTOSA SOCIETY (BRPS)

RP Fighting Blindness, PO Box 350, Buckingham, Buckinghamshire, MK18 1GZ, England
Tel: (44) 12 8082 1334
Email: info@rpfightingblindness.org.uk
Website: www.rpfightingblindness.org.uk
Contact: Mrs Julie Child, Senior Fundraiser

RP Fighting Blindness is a membership organization with branches throughout the United Kingdom. The charity aims to raise funds for scientific research to provide treatments leading to a cure for retinitis pigmentosa. The charity provides a support and information service to anyone affected by RP.

RP Fighting Blindness Research Grants

Subjects: Retinitis pigmentosa.
Purpose: To financially support research into treatments leading to a cure for retinitis pigmentosa.
Eligibility: Please contact the charity.
Level of Study: Postgraduate
Type: Research grant
Value: Varies
Length of Study: Varies
Frequency: Twice a year
Country of Study: Any country
No. of awards offered: Varies
Application Procedure: Applicants must submit their application to the RP Fighting Blindness office.
Closing Date: March 10th, September 10th
Funding: Individuals, private, trusts

BRITISH SCHOOL AT ATHENS

52 Souedias Street, Athens, 106 76, Greece
Tel: (30) 211 102 2800
Fax: (30) 211 102 2803
Email: admin@bsa.ac.uk
Website: www.bsa.gla.ac.uk
Contact: Assistant Director

The British School at Athens promotes research into the archaeology, architecture, art, history, language, literature, religion and topography of Greece in ancient, medieval and modern times. It consists of the Library, Fitch Laboratory for Archaeological Science, Archive, Museum, hostel and a second base at Knossos for research and fieldwork.

The Elizabeth Catling Memorial Fund for Archaeological Draughtmanship

Purpose: To encourage excellence in archaeological drawing, including the preparation of finished drawings for publication. It is hoped that awards will help individuals to improve their standards of draughtsmanship and also enable the preparation of a larger number of drawings, of higher quality, than might otherwise have been possible.
Eligibility: Individual applicants must show that drawings are an essential part of their research. Furthermore, although not a precondition, it is hoped that they may be draughtsmen themselves. Applications from project directors, who may also apply during the course of a field campaign, are limited to unexpected expenses that are not provided for in the project's budget, such as extra maintenance costs to enable a draughtsman to draw unforeseen material and finds.
Level of Study: Predoctorate
Type: Funding support
Value: £200
Frequency: Annual
No. of awards offered: 3
Application Procedure: Candidates should submit letters of application to the School's London office by post in four copies or by e-mail. Letters should not be longer than two pages and should include a statement of the purposes of the application and a budget and timetable for the proposed work, together with the name and address of a referee whom the awarding panel(s) may consult. Applications may be made for but are not limited to, grants towards the maintenance costs of longer stays at museums and other study centres so as to achieve work that would not otherwise have been attempted. Recipients of awards must have been admitted as Students of the School for the appropriate Session before receiving their grants, and must submit a short report on the use of the grant to the London office.
Closing Date: April 1st
Additional Information: The Fund does not support printing expenses, or site drawings such as plans and sections, or computer graphics.

For further information contact:

British School at Athens, Senate House, Malet Street, London, WC1E 7HU
Email: bsa@sas.ac.uk

Hector and Elizabeth Catling Bursary

Subjects: Greek studies including the archaeology, art, history, language, literature, religion, ethnography, anthropology or geography of any period and all branches of archaeological science.
Purpose: To assist travel, maintenace costs and for the purchase of scientific equipments.
Eligibility: Open to researchers of British, Irish or Commonwealth nationality.
Level of Study: Doctorate, Postdoctorate, Postgraduate, Research
Type: Bursary
Value: £500 per bursary to assist with travel and maintenance costs incurred in fieldwork, to pay for the use of scientific or other specialized equipment in or outside the laboratory in Greece or elsewhere and to buy necessary supplies
Frequency: Annual
Study Establishment: The British School at Athens
No. of awards offered: 1–2
Application Procedure: Applicants must submit a curriculum vitae and state concisely the nature of the intended work, a breakdown of budget, the amount requested from the Fund and how this will be spent. Applications should include two sealed letters of reference. Bursary holders must submit a short report to the Committee upon completion of the project.
Closing Date: January 1st
Funding: Private
Additional Information: The bursary is not intended for publication costs, and cannot be awarded to an excavation or field survey team.

The John Morrison Memorial Fund for Hellenic Maritime Studies

Purpose: To further research into all branches of Hellenic maritime studies of any period.
Type: Funding support
Value: £500
Frequency: Annual
No. of awards offered: 1–2
Application Procedure: Candidates should submit letters of application to the School's London office by post in four copies or by e-mail. Letters should not be longer than two pages and should include a statement of the purposes of the application and a budget and timetable for the proposed work, together with the name and address of a referee whom the awarding panel(s) may consult. Applications may be made for but are not limited to, grants towards the maintenance costs of longer stays at museums and other study centres so as to achieve work that would not otherwise have been attempted. Recipients of awards must have been admitted as Students of the School for the appropriate Session before receiving their grants, and must submit a short report on the use of the grant to the London office.
Closing Date: April 1st
Additional Information: Grants may also be available from the Fund for buying maritime books and journals for the School's Library.

For further information contact:

British School at Athens, Senate House, Malet Street, London, WC1E 7HU
Email: bsa@sas.ac.uk

The Richard Bradford McConnell Fund for Landscape Studies

Subjects: All disciplines of the arts, humanities and sciences (or any combination of them).
Purpose: To assist research in the interaction of place and people in Greece and Cyprus at any period(s).
Type: Funding support

Value: £400
Frequency: Annual
Application Procedure: Candidates should submit letters of application to the School's London office by post in four copies or by e-mail. Letters should not be longer than two pages and should include a statement of the purposes of the application and a budget and timetable for the proposed work, together with the name and address of a referee whom the awarding panel(s) may consult. Applications may be made for but are not limited to, grants towards the maintenance costs of longer stays at museums and other study centres so as to achieve work that would not otherwise have been attempted. Recipients of awards must have been admitted as Students of the School for the appropriate Session before receiving their grants, and must submit a short report on the use of the grant to the London office.
Closing Date: April 1st
Contributor: Richard Bradford Trust

For further information contact:

British School at Athens, Senate House, Malet Street, London, WC1E 7HU
Email: bsa@sas.ac.uk

The Vronwy Hankey Memorial Fund for Aegean Studies

Purpose: To support research in the prehistory of the Aegean and its connections with the East Mediterranean.
Eligibility: Preference may be given to younger Students.
Type: Funding support
Value: £500 are available for the expenses (including, but not limited to, attending conferences to present papers, photography, and travel to museums and sites)
Application Procedure: Candidates should submit letters of application to the School's London office by post in four copies or by e-mail. Letters should not be longer than two pages and should include a statement of the purposes of the application and a budget and timetable for the proposed work, together with the name and address of a referee whom the awarding panel(s) may consult. Applications may be made for but are not limited to, grants towards the maintenance costs of longer stays at museums and other study centres so as to achieve work that would not otherwise have been attempted. Recipients of awards must have been admitted as Students of the School for the appropriate Session before receiving their grants, and must submit a short report on the use of the grant to the London office.
Closing Date: April 1st

For further information contact:

British School at Athens, Senate House, Malet Street, London, WC1E 7HU
Email: bsa@sas.ac.uk

BRITISH SCHOOL AT ROME (BSR)

The British Academy, 10 Carlton House Terrace, London, SW1Y 5AH, United Kingdom
Tel: (44) 20 7969 5202
Fax: (44) 20 7969 5401
Email: bsr@britac.ac.uk
Website: www.bsr.ac.uk
Contact: Dr Gill Clark, at (Registrar)

The British School at Rome (BSR) is an interdisciplinary research centre for the humanities, visual arts and architecture. Each year, the School offers a range of awards in its principal fields of interest. These interests are further promoted by lectures, conferences, publications, exhibitions, archaeological research and an excellent reference library.

Abbey Fellowships in Painting

Subjects: Painting.
Purpose: To give mid-career artists the opportunity of working in Rome.
Eligibility: Open to mid-career painters with an established record of achievement. Applicants must be citizens of the United Kingdom or United States of America or have been resident in either country for at least 5 years.
Level of Study: Doctorate, Postdoctorate, Postgraduate, Professional development, Research
Type: Fellowship
Value: UK£800 per month plus board and lodging
Length of Study: 3 months
Frequency: Annual
Study Establishment: The British School at Rome
Country of Study: Italy
No. of awards offered: 3
Application Procedure: Applicants must complete an application form and pay an application fee.
Closing Date: Mid-January
Funding: Private
Contributor: The Abbey Council
No. of awards given last year: 3

For further information contact:

Abbey Awards, 1st Lukes Court, 136 Falcon Road, London, SWII 2LP, United Kingdom
Contact: The Administrator

Abbey Scholarship in Painting

Subjects: Painting.
Purpose: To give exceptionally promising early career painters the opportunity to work in Rome.
Eligibility: Open to citizens of the United Kingdom and United States of America and to those of any other nationality provided that they have been resident in either country for at least 5 years.
Level of Study: Postdoctorate, Postgraduate, Doctorate, Graduate
Type: Scholarship
Value: UK£700 per month plus board and lodging
Length of Study: 9 months
Frequency: Annual
Study Establishment: The British School at Rome
Country of Study: Italy
No. of awards offered: 1
Application Procedure: Applicants must complete an application form and pay an application fee.
Closing Date: Mid-January
Funding: Private
Contributor: The Abbey Council
No. of awards given last year: 1

For further information contact:

Abbey Awards, 1st Lukes Court, 136 Falcon Road, London, SWII ZLP, United Kingdom
Contact: The Administrator

Arts Council Northern Ireland Fellowship

Subjects: Visual arts.
Eligibility: Visual artists resident in Northern Ireland.
Level of Study: Doctorate, Postgraduate, Professional development, Graduate, Postdoctorate
Value: UK£500 per month plus board and lodging
Length of Study: 6 months
Frequency: Every 2 years
Study Establishment: The British School at Rome
Country of Study: Italy
No. of awards offered: 1
Application Procedure: Application form must be completed.
Funding: Government
Contributor: Arts Council Northern Ireland
Additional Information: Please check website for more details.

For further information contact:

The Arts Council of Northern Ireland, MacNeice House, 77 Malone Road, Belfast, BT9 6AQ

Balsdon Fellowship

Subjects: Archaeology, art history, history, society and culture of Italy from prehistory to the modern period.

Purpose: To enable senior scholars engaged in research to spend time in Rome to further their studies.
Eligibility: Open to established scholars normally in a post in a university of the United Kingdom. Applicants must be British or Commonwealth citizens, or must be studying or have studied at postgraduate level in a higher education institution in the UK, having completed not less than 4 years of residence in the UK, or must hold a post in a higher education institution in the UK.
Level of Study: Postdoctorate, Professional development, Research
Type: Fellowship
Value: Board and lodging
Length of Study: 3 months
Frequency: Annual
Study Establishment: The British School at Rome
Country of Study: Italy
No. of awards offered: 1
Application Procedure: Applicants must complete an application form.
Closing Date: Mid-January
Funding: Private
Contributor: A bequest to the British School at Rome
No. of awards given last year: 1

Derek Hill Foundation Scholarship

Subjects: Painting and drawing.
Purpose: To encourage artists for whom the use of paint and/or drawing is important to the development of their work.
Eligibility: Open to those who are of British or Irish nationality who will be aged 24 years or over on September 1st of the academic year in which the award would be taken up.
Level of Study: Postdoctorate, Postgraduate, Professional development
Type: Scholarship
Value: Approx. UK£950 per month plus board and lodging at the British School in Rome
Length of Study: 3 months
Frequency: Annual
Study Establishment: The British School at Rome
Country of Study: Italy
No. of awards offered: 1
Application Procedure: Applicants must complete an application form and pay an entry fee.
Closing Date: Late January
Funding: Foundation
Contributor: Derek Hill Foundation
No. of awards given last year: 1

Giles Worsley Travel Fellowship

Subjects: Architecture and architectural history.
Purpose: To enable an architect or architectural historian to spend 3 months in Rome studying an architectural topic of his choice.
Eligibility: Open to those who are of British nationality or who have been living and studying in Britain for at least the last 3 years.
Level of Study: Postdoctorate, Postgraduate, Professional development
Type: Fellowship
Value: Approx. £700 per month plus full board and lodging at the British School at Rome
Length of Study: 3 months
Frequency: Annual
Study Establishment: The British School at Rome
Country of Study: Italy
No. of awards offered: 1
Application Procedure: Applicants must submit a curriculum vitae, a statement indicating the subject of their proposal and arrange for two references to be sent.
Closing Date: January/February
Funding: Private
No. of awards given last year: 1

Hugh Last Fellowship

Subjects: Classical antiquity.
Purpose: To enable established scholars to collect research material concerning classical antiquity.
Eligibility: Open to established scholars normally in a post at a United Kingdom university. Applicants must be British or Commonwealth citizens, or must be studying or have studied at postgaduate level in a higher education institution in the UK, having completed not less than four years of residence in the UK; or must hold a post in a higher education institution in the UK.
Level of Study: Postdoctorate, Professional development, Research
Type: Fellowship
Value: Board and lodging at the British School at Rome
Length of Study: 3 months
Frequency: Annual
Study Establishment: The British School at Rome
Country of Study: Italy
No. of awards offered: 1
Application Procedure: Applicants must complete an application form.
Closing Date: Mid-January
Funding: Private
Contributor: A bequest to the British School at Rome
No. of awards given last year: 2

Paul Mellon Centre Rome Fellowship

Subjects: The Grand Tour and Anglo-Italian cultural and artistic relations.
Purpose: To assist research on grand tour subjects or on Anglo-Italian cultural and artistic relations.
Eligibility: Open to established scholars in the United Kingdom, United States of America or elsewhere. Applicants should be fluent in Italian.
Level of Study: Doctorate, Graduate, Postdoctorate, Postgraduate, Professional development, Research
Type: Fellowship
Value: Full board at the British School at Rome. For independent scholars, the fellowship offers a stipend of UK£6,000 plus travel to and from Rome. For scholars in full-time university employment, the fellowship offers an honorarium of UK£2,000, travel to and from Rome and a sum of UK£6,000 towards replacement teaching costs for a term at the Fellow's home institution
Length of Study: 3 months
Frequency: Annual
Study Establishment: The British School at Rome
Country of Study: Italy
No. of awards offered: 1
Application Procedure: Applicants must contact the Paul Mellon Centre for Studies in British Art for details.
Closing Date: January 14th
Funding: Private
Contributor: The Paul Mellon Centre for Studies in British Art
No. of awards given last year: 2

For further information contact:

The Paul Mellon Centre for Studies in British Art, 16 Bedford Square, London, WC1B 3JA, England
Email: grants@paul-mellon-centre.ac.uk
Website: www.paul-mellon-centre.ac.uk
Contact: The Grants Administrator

Rome Awards

Subjects: Archaeology, art history, history, society and culture of Italy from prehistory to the modern period.
Purpose: To enable persons engaged in research at a pre- or early post-doctoral level to spend time in Rome to further their studies.
Eligibility: Applicants must be British or Commonwealth citizens, or must be studying or have studied at postgraduate level in a higher education institution in the UK, having completed not less than four years of residence in the UK; or must hold a post in a higher education institution in the UK. Applicants normally will have begun a programme of research in the general field for which the award is being sought, whether or not registered for a higher degree. Awards are not normally suitable for people in established posts. Preference may be given to applicants attached to, registered at or working at a university in the UK or Commonwealth.
Level of Study: Postdoctorate, Research, Doctorate, Graduate, Postgraduate, Predoctorate
Type: Scholarship

Value: Board and lodging at the British School at Rome, UK£150 per month, plus a one-off travel grant of £180
Length of Study: 3 months
Frequency: Annual
Study Establishment: The British School at Rome
Country of Study: Italy
No. of awards offered: Varies
Application Procedure: Applicants must complete an application form.
Closing Date: Mid-January
Funding: Private
No. of awards given last year: 4

Rome Fellowship

Subjects: Archaeology, art history, history, society and culture of Italy from prehistory to the modern period.
Purpose: To enable those who are at an early post-doctoral stage of their career to launch a major piece of postdoctoral research.
Eligibility: Applicants must be British or Commonwealth citizens, or must be studying or have studied at postgraduate level in a higher education institution in the UK, having completed not less than 4 years of residence in the UK. Successful applicants will need to have been awarded their doctorate prior to taking up the award. Preference may be given to applicants attached to, registered at or working at a university in the UK or Commonwealth. Applicants normally should have submitted their Doctorate not more than two years previous to the closing date for applications.
Level of Study: Postdoctorate
Type: Fellowship
Value: UK£475 per month plus board and lodging at the British School at Rome
Length of Study: 9 months
Frequency: Annual
Study Establishment: The British School at Rome
Country of Study: Italy
No. of awards offered: Varies
Application Procedure: Applicants must complete an application form.
Closing Date: Mid-January
Funding: Private
No. of awards given last year: 2

Rome Fellowship in Contemporary Art

Subjects: Artists working in any discipline.
Purpose: To give early-career artists with a significant track record or exhibiting the opportunity to work in Rome.
Eligibility: Open to British or Commonwealth citizens; and to those who have been working professionally or studying at postgraduate level for at least the last 3 years in the UK or Commonwealth.
Level of Study: Postdoctorate, Postgraduate, Professional development, Research
Type: Fellowship
Value: UK£1,500 per month plus board a lodging
Length of Study: 3 months
Frequency: Annual
Study Establishment: The British School at Rome
Country of Study: Italy
No. of awards offered: 1
Application Procedure: Applicants must complete an application form and pay an application fee.
Closing Date: Late January
Funding: Private

Rome Scholarships in Ancient, Medieval and Later Italian Studies

Subjects: Archaeology, art history, history, society and culture of Italy from prehistory to the modern period.
Purpose: To enable persons engaged in research, at a predoctoral level, to spend time in Rome to further their studies.
Eligibility: Applicants must be British or Commonwealth citizens, or must be studying or have studied at postgraduate level in a higher education institution in the UK, having completed not less than 4 years of residence in the UK, or must hold a post in a higher education institution in the UK. Applicants must have begun a programme of research in the general field for which the scholarship is being sought, whether or not registered for a higher degree. Preference may be given to applicants attached to, registered at or working at a university in the UK or Commonwealth.
Level of Study: Graduate, Doctorate, Postgraduate, Predoctorate, Research
Type: Scholarship
Value: UK£444 plus board and lodging at the British School at Rome
Length of Study: 9 months
Frequency: Annual
Study Establishment: The British School at Rome
Country of Study: Italy
No. of awards offered: Varies
Application Procedure: Applicants must complete an application form.
Closing Date: Mid-January
Funding: Private
No. of awards given last year: 3

Sainsbury Scholarship in Painting and Sculpture

Subjects: Painting and sculpture, with drawing.
Purpose: To give promising and ambitious painters and sculptors the opportunity to work in Rome.
Eligibility: Open to United Kingdom citizens and to those who have been working professionally or studying at postgraduate level for at least the last 5 years in the United Kingdom. Applicants must be under 30 on October 1st in the year in which they would begin to hold the scholarship.
Level of Study: Postgraduate, Research, Doctorate, Graduate, Postdoctorate
Type: Scholarship
Value: UK£500 per month plus board, lodging and a travel grant of UK£1,200
Length of Study: 1 year
Frequency: Annual
Study Establishment: The British School at Rome
Country of Study: Italy
No. of awards offered: 1
Application Procedure: Applicants must complete an application form and pay an application fee.
Closing Date: Late January
Funding: Trusts
Contributor: The Linbury Trust
No. of awards given last year: 1

BRITISH SKIN FOUNDATION

4 Fitzroy Square, London, W1T 5HQ, England
Tel: (44) 020 7391 6341
Fax: (44) 020 7391 6099
Email: admin@britishskinfoundation.org.uk
Website: www.britishskinfoundation.org.uk
Contact: Mrs Sarah Clinch, Office Manager

The British Skin Foundation exists to support research and education into skin diseases. Working closely with patient support groups as well as many of the country's leading dermatology departments, the foundation aims to help the 8 million people in the United Kingdom who suffer with a serious skin condition.

British Skin Foundation Large Grants

Subjects: Skin diseases in the UK.
Purpose: To support anybody wishing to carry out United Kingdom or Republic of Ireland-based research into skin disease.
Eligibility: Open to anyone wishing to carry out United Kingdom or Republic of Ireland-based research into skin disease.
Level of Study: Unrestricted
Value: UK£62,000–81,000
Length of Study: 1–3 years
Frequency: Annual
Country of Study: United Kingdom
Application Procedure: Application forms are available from the website www.britishskinfoundation.org.uk
Closing Date: August 30th
Funding: Commercial, foundation, individuals, private, trusts
No. of awards given last year: 7
No. of applicants last year: 51

British Skin Foundation Small Grants

Subjects: Skin diseases in the UK.
Purpose: To support anybody wishing to carry out United Kingdom or Republic of Ireland-based research into skin disease.
Eligibility: Open to anyone wishing to carry out United Kingdom or Republic of Ireland-based research into skin disease.
Level of Study: Unrestricted
Value: Up to UK£10,000
Length of Study: 1 year
Frequency: Annual
Country of Study: United Kingdom
Application Procedure: Application forms are available from the website www.britishskinfoundation.org.uk
Closing Date: April 27th
Funding: Individuals, commercial, foundation, private, trusts
No. of awards given last year: 8
No. of applicants last year: 38

BRITISH SOCIETY FOR ANTIMICROBIAL CHEMOTHERAPY

British Society for Antimicrobial Chemotherapy, Griffin House, 53 Regent Place, Birmingham, B1 3NJ, United Kingdom
Tel: (44) 0121 236 1988
Fax: (44) 0121 212 9822
Email: tguise @bsac.org.uk
Website: www.bsac.org.uk
Contact: Ms Tracey Guise, Executive Director

BSAC is an inter-professional organisation with 40 years of experience and achievement in antibiotic education, research and leadership. It is dedicated to saving lives through appropriate use and development of antibiotics now and in the future.

BSAC Education Grants

Subjects: Antimicrobial chemotherapy.
Purpose: The Education Fund is designated for research projects and initiatives of benefit to the field of antimicrobial chemotherapy.
Level of Study: Postgraduate, Research
Type: Grant
Value: £5,000–30,000
Length of Study: Up to 1 year
Frequency: Dependent on funds available, Ongoing
Application Procedure: Applications must include an assessment of the likely impact of the project that is proposed.
Closing Date: Check website for updated details
Funding: Foundation

BSAC Overseas Scholarship

Subjects: Antimicrobial chemotherapy.
Purpose: Overseas Scholarships are to enable workers from other countries the opportunity to work in UK Departments for up to six months.
Level of Study: Postgraduate, Professional development
Type: Scholarship
Value: £1,000 per calendar month for up to 6 months. The host institution will receive a consumables grant of £200 per calendar month for the duration of the scholarship
Length of Study: 6 months
Frequency: Annual
Country of Study: United Kingdom
No. of awards offered: Varies
Application Procedure: Successful applicants are required to submit a 500 word written report to the Secretary of the Grants Committee on completion of their project, and to forward details of any publications arising from the work undertaken. Applications for Overseas Scholarships should be made to the Society's HQ using the form provided on the BSAC website.
Closing Date: December 1st
Funding: Foundation
Additional Information: Excludes applicants from UK.

For further information contact:

Email: tguise@bsac.org.uk

BSAC PhD Studentship

Subjects: Antimicrobial chemotherapy.
Purpose: The PhD studentship scheme is designed to ensure a flow of first-class students into the field of antimicrobial chemotherapy, providing them with an excellent training in research.
Eligibility: Awarded or expected first class or high upper second class degree, or MSc with merit or distinction. The application must be made by an established investigator who will be the supervisor who has previously received three or less studentships and may be for named or unnamed students.
Level of Study: Postgraduate
Type: Scholarship
Value: Up to a maximum of £25,000 per year for the duration of the grant, may include: student stipend; tuition fees (set by the research institution); research consumables, directly attributable to the project
Length of Study: Up to 4 years
Frequency: Every 2 years
No. of awards offered: 1
Application Procedure: Completed application form comprising: student, names of two supervisors and institution details; detailed research proposal; statement detailing the scientific techniques for which the student will receive training; budget detailing annual costs (stipend, tuition fees, consumables); training record for each named supervisor. Accompanying documents (supervisor): letter of support from the Head of Department; brief CV of the supervisors (2 A4 sides max). Accompanying documents (student): record of candidate's academic performance; full CV of the candidate and statement of career intentions; two academic letters of reference.
Closing Date: Check website for updated details
Funding: Foundation
Additional Information: For non-UK university degrees: evidence from Graduate Office that degree held conforms to Fist class degree or higher.

BSAC Project Grants

Subjects: Antimicrobial chemotherapy.
Purpose: Grants are awarded to help new projects, support completion of an existing project, introduce a novel technique for existing work or funding trainees for projects/training.
Level of Study: Research
Type: Grant
Value: Up to £10,000 (maximum of £5,000 for funding trainees)
Length of Study: Up to 1 year in duration
Frequency: Annual
No. of awards offered: Varies
Application Procedure: Candidates are expected to provide full justification of project grant funds. Applications should be made using the official application form – details can be found on the website.
Closing Date: December 1st
Funding: Foundation
Additional Information: Please contact to chrisburley@bsac.org.uk for further information.

BSAC Research Grants

Subjects: Antimicrobial chemotherapy.
Purpose: To provide financial support in mechanisms of antibacterial action, mechanisms of antibacterial resistence, antiviral resistance, antiviral, antifungals, antibiotic methods, antibiotic prescribing, antibiotic therapy, antiparistics, evidence based medicine/systematic reviews.
Level of Study: Research
Type: Grant
Value: Maximum value of £50,000
Length of Study: Up to 1 year
Frequency: Annual
No. of awards offered: Varies
Application Procedure: Applications should be made using the official application form - details can be found on the website.
Closing Date: December 1st
Funding: Foundation
Additional Information: Please contact to chrisburley@bsac.org.uk for further information.

BSAC Travel Grants

Subjects: Antimicrobial chemotherapy.
Purpose: The society awards a number of travel grants to individuals to attend the annual meetings of ECCMID and ICAAC.
Eligibility: Travel grants are restricted to BSAC Members resident in the UK or overseas. Grants will be awarded to individuals attending the conference to give oral or poster presentations.
Level of Study: Professional development
Type: Travel grant
Value: ECCMID: maximum value of £1,000, number dependant on funds available; ICAAC: maximum value of £1,500, number dependant on funds available
Frequency: Annual
No. of awards offered: Varies
Application Procedure: Please submit one electronic copy of the application form (found on website) and the following attachments to tguise@bsac.org.uk.
1. Copy of abstract submitted to Scientific Committee
2. Copy of letter of acceptance of the abstract by the Scientific Committee
3. Brief curriculum vitae (maximum x2 A4 sides).
Closing Date: ECCMID: March 25th; ICAAC: July 29th
Funding: Foundation
Additional Information: Applicants who have received a travel grant are not eligible to apply the following year.

Terry Hennessey Microbiology Fellowship

Subjects: Antimicrobial chemotherapy.
Purpose: The Terry Hennessey Microbiology Fellowship offers a young investigator, working in the field of infectious diseases, a travel grant to present a paper/poster at the Annual Interscience Conference on Antimicrobial Agents and Chemotherapy (ICAAC) Meeting in the USA.
Eligibility: Applicant normally be under the age of 35.
Level of Study: Professional development, Research
Type: Fellowship
Value: £1,500
Frequency: Annual
No. of awards offered: 1
Application Procedure: By completion of the application form (found on website), which must be submitted electronically to tguise@bsac.org.uk. Postal applications will not be considered.
Closing Date: End July
Funding: Foundation

THE BRITISH SOCIETY FOR HAEMATOLOGY

100 White Lion Street, London, N1 9PF, United Kingdom
Tel: (44) 020 7713 0990
Fax: (44) 020 7837 1931
Email: info@b-s-h.org.uk
Website: www.b-s-h.org.uk
Contact: Daphne Harvey, Business Manager

The British Society of Haematology advances the practice and study of haematology and to facilitate contact between persons interested in haematology. It offers scientific scholarships, is an active participant of the International Society of Haematology (ISH) and the International Council for Standardization in Haematology (ICSH) and publishes regular bulletins. The society also provides financial support to regional and national scientific meetings.

Annual Scientific Meeting Scholarships for Haematology Professionals

Subjects: Haematology.
Purpose: To support attendance at the British Society for haematology annual scientific meeting.
Eligibility: Open to clinical scientists, biomedical scientists, academic scientists, PhD students and nurse practitioners working in the United Kingdom.
Level of Study: Doctorate
Type: Scholarship
Value: Up to UK£500 to support registration, travel and accommodation
Frequency: Annual
Country of Study: United Kingdom
No. of awards offered: 40
Application Procedure: Check website for further details.
Closing Date: March 1st
Funding: Private

BRITISH SOCIETY FOR MIDDLE EASTERN STUDIES

Institute for Middle Eastern & Islamic Studies, Durham University, Al-Qasimi Building, Elvet Hill Road, Durham, DH1 3TU, United Kingdom
Tel: (44) 0191 33 45179
Fax: (44) 0191 33 45661
Email: a.l.haysey@durham.ac.uk
Website: www.dur.ac.uk/brismes
Contact: BRISMES Administrative Office

The Abdullah Al-Mubarak Al-Sabah Foundation BRISMES Scholarships

Purpose: The purpose of the scholarships is to encourage more people to pursue postgraduate studies in disciplines related to the Middle East in British universities.
Eligibility: To qualify you must be a paid-up member of BRISMES (student membership suffices) but the time you apply..
Level of Study: Postgraduate
Type: Scholarship
Value: £2,000
Length of Study: 1 academic year
Frequency: Annual
Country of Study: United Kingdom
No. of awards offered: 2
Application Procedure: Submit an application of 600–1,000 words, by email to the BRISMES research committee This should include a sketch of the overall research topic, and a description of the purpose for which the grant would be used. Also you must obrain a brief supporting statement from a supervisor. Applications should be sent to Institute for Middle Eastern & Islamic Studies.
Closing Date: March 31st
Funding: Foundation

For further information contact:

Institute for Middle Eastern & Islamic Studies, Durham University, Al-Qasimi Building, Elvet Hill Road, Durham, DH1 3TU
Tel: 0191 33 45179
Fax: 0191 33 45661
Email: a.l.haysey@dur.ac.uk

MA Scholarship

Purpose: BRISMES offers an annual Master's scholarship for taught Master's study at a UK institution. The Master's programme can be in any discipline but should include a majority component specifically relating to the Middle East.
Eligibility: Preference will be given to candidates resident in the European Union, and to institutions who are members of BRISMES.
Level of Study: Doctorate
Type: Scholarship
Value: £1,200
Frequency: Annual
Country of Study: United Kingdom
Application Procedure: Applications should be forwarded by the Director of the Master's programme concerned, to the BRISMES Administrative Office, and should include: a supporting statement from the course Director not exceeding 500 words; the programme syllabus; a statement by the candidate not exceeding 500 words; the candidate's CV and transcript of previous academic results; two academic references. Applications should be sent to Institute for Middle Eastern & Islamic Studies.
Closing Date: March 31st
Funding: Foundation

For further information contact:

Institute for Middle Eastern & Islamic Studies, Durham University, Al-Qasimi Building, Elvet Hill Road, Durham, DH1 3TU

Tel: 0191 33 45179
Fax: 0191 33 45661
Email: a.l.haysey@dur.ac.uk

Research Student Awards

Purpose: BRISMES offers Research Awards to research students based in the UK working on a Middle Eastern studies topic.
Eligibility: To qualify you must have completed your first year of doctoral research and be a paid-up member of BRISMES (student membership suffices) by the time you apply.
Level of Study: Research, Doctorate
Type: Grant
Value: £1,000
Frequency: Annual
Country of Study: United Kingdom
Application Procedure: Submit an application of 600–1,000 words, by email to the Research Committee, Email: a.l.haysey@durham.ac.uk – this should include a sketch of your overall research topic, and a description of the purpose for which the grant would be used. You must also obtain a brief supporting statement from your supervisor.
Closing Date: March 31st
Funding: Foundation

BRITISH SOCIETY FOR PARASITOLOGY

87 Gladstone Street, Bedford, MK41 7RS, United Kingdom
Tel: (44) 01234 211015
Fax: (44) 01234 211015
Email: cathy@bsp.uk.net
Website: www.bsp.uk.net
Contact: Cathy Fuller, BSP Secretariat

Ann Bishop Travelling Award

Purpose: The purpose of the award will be to provide members with funds to allow travel in pursuit of their academic interests in parasitology, providing an opportunity to undertake field research, visit overseas institutions and/or visit endemic areas of disease.
Eligibility: Applicants should be PhD students in their final year of study or should have recently (within the last 2 years) completed their PhD. Applicants should be BSP members at the time of application.
Level of Study: Postdoctorate
Type: Travel award
Value: The society will provide support of up to £2,000 which should cover the costs of travel and subsistence for not less than two weeks
Frequency: Annual
Application Procedure: Applications should include a curriculum vitae of the applicant (on a single A4 page) which should include the applicant's BSP membership number and details of any other financial support received from the Society; a description (on a single A4 page) of the proposed work including a description of how the proposed work relates to the applicants current project and an outline of the anticipated outputs of the work; signed letters of support from the supervisor (who should also be a BSP member) and from the host institution; a budget for the proposed trip including details of the cost of travel, accommodation and subsistence. The Ann Bishop award is intended to cover only the costs incurred by the applicant and does not cover any research costs; a signed and dated declaration stating: 'I have read and understood the conditions of the award'.
Closing Date: January 13th
Funding: Private

C.A. Wright Memorial Medal

Subjects: The medal is awarded for contributions to the discipline of parasitology in the broadest sense.
Purpose: The recipient is a scientist in mid-career who, it is considered, will confirm their already outstanding achievements to become a truly distinguished future leader of their field.
Eligibility: Nominations are invited from members of BSP for this award for which the following conditions apply: each nomination is made by a proposer who must be a bona fide paid-up member of BSP; all currently serving officers and members of Council are excluded from acting as proposers for candidates; no currently serving officers or members of Council may be nominated as candidates; nominations must be made in writing to the Hon General Secretary, presenting a case for the awarding of the medal to the nominee; candidates must be under 50 years of age on the 1st April of the year the award is made; candidates must be fully paid-up members of BSP of not less than three year's standing.
Type: Prize
Frequency: Annual
Application Procedure: Nominations must be made in writing.
Closing Date: January 13th
Funding: Private

For further information contact:

Contact: Cathy Fuller, BSP Secretariat

Garnham Expeditionary Scholarship

Subjects: The intention of this scholarship is to promote field parasitology (collection of data or samples) under any difficult conditions.
Purpose: This award aims to give parasitologists at an early stage in their careers (undergraduates or PhD students) the opportunity to undertake field studies in parasitology. It is particularly aimed at candidates who wish to undertake studies under demanding conditions.
Eligibility: Applicants at an early stage of their interest in parasitology (undergraduates or PhD students in their first or second years of study) are especially encouraged. Applicants need not be members of the BSP but must have a written letter of support from a staff member of their institution who is a BSP member.
Level of Study: Doctorate, Graduate, Postdoctorate, Postgraduate, Predoctorate, Research
Type: Scholarship
Value: The Society will provide support of up to £1,000
Frequency: Annual
Application Procedure: Applications should include a curriculum vitae of the applicant (on a single A4 page); a description (on a single A4 page) of the proposed project; a letter of support from a colleague (who must be a BSP member) and from the host institution; a budget for the proposed trip including details of the cost of travel, accommodation and subsistence; a signed and dated declaration stating, 'I have read and understood the conditions of the award'.
Closing Date: January 13th, please check website for further details.
Funding: Private

Spring Meeting Travel Awards

Purpose: These awards give financial support to student members of the Society, to facilitate their participation at the Society's annual Spring meeting.
Eligibility: Applicants must be members of the BSP at the time of application. For first year students, an application to join the BSP must be submitted before or when applying for support. The applicant must be presenting an oral paper or poster at the conference. First year PhD students are exempt from this rule. The financial support of the BSP should be acknowledged in any talk or poster presentation made by the student at the meeting. Council considers this to be important because this allows BSP membership to see how Society funds are being used.
Level of Study: Unrestricted
Type: Award
Value: £200–600
Application Procedure: Applicants must submit an online application form.
Funding: Private
Additional Information: Please check website or contact society for updated information.

BRITISH SOCIOLOGICAL ASSOCIATION (BSA)

Bailey Suite, Palatine House, Belmont Business Park, Belmont, Durham, DH1 1TW, England
Tel: (44) (0) 191 383 0839
Fax: (44) (0) 191 383 0782
Email: enquiries@britsoc.org.uk
Website: www.britsoc.co.uk

The British Sociological Association (BSA) is the learned society and professional association for sociology in Britain. The Association was founded in 1951 and membership is drawn from a wide range of backgrounds, including research, teaching, students and practitioners in many fields. The BSA provides services to all concerned with the promotion and use of sociology and sociological research.

BSA Support Fund

Subjects: Sociology.
Purpose: To enable members of the association pursue their research interests by way of fieldwork/interview costs, conference attendance in the United Kingdom and overseas (including non-BSA events), thesis production costs.
Eligibility: Applicants must reside in the UK; be a BSA member in the UK Concessionary category and have limited income (income of less than £14,000) for the membership year; and be a BSA member for a minimum 12 months prior to application.
Level of Study: Postgraduate, Research, Unrestricted
Type: Grant
Value: Up to UK£250
Frequency: Annual
Country of Study: United Kingdom
No. of awards offered: Approx. 70
Application Procedure: The application form can be downloaded from the BSA website: www.britsoc.co.uk/ students/SupportFund.htm. Applications are processed upon receipt and applicants can expect a response within 14 days.
Closing Date: There is no closing date. Applications are considered as and when received by the Support Fund Committee.
Funding: Private
Contributor: British Sociological Association
No. of awards given last year: 76
No. of applicants last year: 91

THE BRITISH UNIVERSITIES NORTH AMERICA CLUB

BUNAC 16 Bowling Green Lane, London, EC1R 0QH, United Kingdom
Tel: (44) 20 7251 3472
Fax: (44) 20 7251 0215
Email: scholarships@bunac.org.uk
Website: www.bunac.org.uk
Contact: Jill Tabuteau, Senior Manager

BUNAC is a leader in the field of international work and travel exchange programmes. BUNAC offers an ever-increasing range of programmes worldwide, BUNAC administers the scholarship on behalf of the British Universities North America Club. BUNAC is dedicated to serving students and other young people everywhere by providing opportunities to live and work abroad legally.

BUNAC Educational Scholarship Trust (BEST)

Subjects: All subjects. Some awards are specifically for sports and geography-related courses.
Purpose: To help further transatlantic understanding.
Eligibility: Open to citizens of the United Kingdom who have graduated from a United Kingdom university within the last 5 years. Must be undertaking a postgraduate course in the US or Canada (cannot already have started the course or be part of an exchange programme).
Level of Study: Postgraduate
Type: Scholarship
Value: Up to approx. US$10,000 per award
Length of Study: 3 months to 3 years
Frequency: Annual
Country of Study: United States of America or Canada
No. of awards offered: Up to 10
Application Procedure: Applicants must complete an application form, available from the BUNAC website from January of each year. Shortlisted applicants will be called for a face to face interview in London in May or June.
Closing Date: Mid-March
Funding: Trusts
No. of awards given last year: 8
No. of applicants last year: 50

BRITISH VETERINARY ASSOCIATION

7 Mansfield Street, London, W1G 9NQ, England
Tel: (44) 20 7636 6541
Fax: (44) 20 7908 6349
Email: bvahq@bva.co.uk
Website: www.bva.co.uk
Contact: Mrs Helena Cotton, Media Officer

The British Veterinary Association's chief interests are the standards of animal health and veterinary surgeons' working practices. The organization's main functions are the development of policy in areas affecting the profession, protecting and promoting the profession in matters propounded by government and other external bodies and the provision of services to members.

Harry Steele-Bodger Memorial Travelling Scholarship

Subjects: Veterinary science and agriculture.
Purpose: To further the aims and aspirations of the late Harry Steele-Bodger.
Eligibility: Open to graduates of veterinary schools in the United Kingdom or the Republic of Ireland who have been qualified for not more than 3 years, and to penultimate or final-year students at those schools.
Level of Study: Graduate, Postgraduate
Type: Scholarship
Value: Approx. UK£1,100
Frequency: Annual
Study Establishment: A veterinary or agricultural research institute or some other course of study approved by the governing committee
Country of Study: Any country
No. of awards offered: 1–2
Application Procedure: Applicants must complete an application form, available on request.
Closing Date: April 10th
Funding: Private
No. of awards given last year: 2 (award divided)
No. of applicants last year: 5
Additional Information: Recipients must be prepared to submit a record of their study abroad.

For further information contact:

Contact: Helena Cotton, Media Officer

BROAD MEDICAL RESEARCH PROGRAM (BMRP)

The Eli and Edythe Broad Foundation, 10900 Wilshire Boulevard, 12th Floor, Los Angeles, CA, 90024-6532, United States of America
Tel: (1) 310 954 5091
Fax: (1) 310 954 5092
Email: info@broadmedical.org
Website: www.broadmedical.org
Contact: Dr Elizabeth Luna, Grants Administrator

The Eli and Edythe Broad Foundation established the Broad Medical Research Program (BMRP) for Inflammatory Bowel Disease (IBD) Grants in 2001. The BMRP funds innovative and early exploratory clinical and basic research projects that will improve diagnosis, therapy, or prevention of IBD and will lead to long-term funding by more traditional granting agencies.

Broad Medical Research Program for Inflammatory Bowel Disease Grants

Subjects: Understanding, treating and preventing IBD (Crohn's disease and ulcerative colitis).
Purpose: The BMRP is interested in providing funding for clinical or basic research in IBD that will improve the lives of patients with IBD by stimulating innovative early stage research that opens avenues for the diagnosis therapy and prevention of these diseases.
Eligibility: Open to non-profit organizations, such as universities, hospitals and research institutes. There are no other eligibility restrictions. In addition to experienced IBD researchers, the BMRP encourages applications from well-trained scientists who are not presently working in IBD to apply their knowledge, expertise and

techniques to IBD research. Interdisciplinary collaboration is strongly encouraged.
Level of Study: Research
Type: Research grant
Value: Budgets should be commensurate with the scope of the work. Those who will need significantly more than US$150,000 per year should contact the BMRP before preparing their letters of interest
Length of Study: 1–2 years, with possible renewal
Frequency: Continuous; letters of interest accepted year round (no deadlines)
Country of Study: Any country
No. of awards offered: Varies
Application Procedure: Applicants must submit a brief letter of interest of up to three pages and also visit the website for further information. Investigators whose letters of interest appear to fit the BMRP's aims will be invited to submit full proposals.
Closing Date: There are no deadlines for receipt of letters of interest
Funding: Foundation
Contributor: Eli and Edythe L Broad
No. of awards given last year: 18
No. of applicants last year: 100

BROADCAST EDUCATION ASSOCIATION (BEA)

Scholarship Committee, 344 Moore Hall, Central Michigan University, Mount Pleasant, MI 48859, United States of America
Tel: (1) 989 774 3851
Fax: (1) 989 774 2426
Email: orlik1pb@cmich.edu
Website: www.beaweb.org
Contact: Dr Peter B Orlik, Scholarship Chair

The Broadcast Education Association (BEA) is the professional association for professors, industry professionals and graduate students interested in teaching and research related to television, radio and the electronic media industry.

Broadcast Education Association Abe Voron Scholarship

Subjects: Radio.
Purpose: To assist study towards a career in radio.
Eligibility: Open to individuals who can show substantial evidence of superior academic performance and potential to be an outstanding radio professionals.
Level of Study: Unrestricted
Type: Scholarship
Value: US$5,000
Frequency: Annual
Study Establishment: BEA member institutions
No. of awards offered: 1
Application Procedure: Applicants must obtain an official application form from the BEA or from campus faculty. Applicants should refer to the website www.beaweb.org for more details.
Closing Date: October 11th
Funding: Private
Contributor: The Abe Voron Committee
No. of awards given last year: 1
No. of applicants last year: 55

Broadcast Education Association Alexander M Tanger Scholarship

Subjects: Broadcasting.
Purpose: To assist study for a career in any area of broadcasting.
Eligibility: The applicant must be able to show substantial evidence of superior academic performance and potential to be an outstanding electronic media professional.
Level of Study: Unrestricted
Type: Scholarship
Value: US$5,000
Frequency: Annual
Study Establishment: BEA member institutions
No. of awards offered: 1
Application Procedure: Applicants must obtain an official application form from the BEA or from campus faculty. Applicants should refer to the website www.beaweb.org for more details.
Closing Date: October 11th
Funding: Private
Contributor: Alexander M Tanger
No. of awards given last year: 1
No. of applicants last year: 60

Broadcast Education Association John Bayliss Scholarship

Subjects: Broadcasting.
Purpose: To assist study in radio.
Eligibility: The applicant must be able to show substantial evidence of superior academic performance and potential to be an outstanding radio professional.
Level of Study: Unrestricted
Type: Scholarship
Value: US$1,500 each
Frequency: Annual
Study Establishment: BEA member institutions
No. of awards offered: 1
Application Procedure: Applicants must obtain an official application form from the BEA or campus faculty. Applicants should refer to the website www.beaweb.org for more details.
Closing Date: October 11th
Funding: Private
Contributor: John Bayliss Foundation
No. of awards given last year: 1
No. of applicants last year: 40

Broadcast Education Association Richard Eaton Scholarship

Subjects: Broadcasting.
Purpose: To assist those who are studying towards a career in broadcasting.
Eligibility: The applicant must be able to show substantial evidence of superior academic performance and potential to be an outstanding electronic media professional.
Level of Study: Unrestricted
Type: Scholarship
Value: US$1,500
Frequency: Annual
Study Establishment: BEA member institutions
No. of awards offered: 1
Application Procedure: Applicants must obtain an official application form from the BEA or campus faculty. Applicants should refer to the website www.beaweb.org for more details.
Closing Date: October 11th
Funding: Private
Contributor: Richard Eaton Foundation
No. of awards given last year: 1
No. of applicants last year: 40

Broadcast Education Association Two Year College Scholarship

Subjects: Electronic media.
Purpose: To assist study towards an electronic media career.
Eligibility: The applicant must be able to show substantial evidence of superior academic performance and potential to be an outstanding electronic media professional. There should be compelling evidence that the applicant possesses high integrity and a well articulated sense of personal and professional responsibility. The applicant must be studying at, or have studied at, a BEA member two-year campus.
Level of Study: 2 or 4 year institutions
Type: Scholarship
Value: US$1,500 each
Frequency: Annual
Study Establishment: BEA member institutions
No. of awards offered: 2
Application Procedure: Applicants must obtain an official application form from the BEA or campus faculty. Applicants should refer to the website www.beaweb.org for more details.
Closing Date: October 11th
Funding: Private
Contributor: Sponsored by the Broadcast Education Association
No. of awards given last year: 2
No. of applicants last year: 22

Broadcast Education Association Vincent T Wasilewski Scholarship

Subjects: Broadcasting.
Purpose: To assist graduate study in any area of broadcasting.
Eligibility: The applicant must be able to show substantial evidence of superior academic performance and potential to be an outstanding electronic media professional. Available to graduate students only
Level of Study: Graduate
Type: Scholarship
Value: US$2,500
Frequency: Annual
Study Establishment: BEA member institutions
No. of awards offered: 1
Application Procedure: Applicants must obtain an official application form from the BEA or campus faculty. Applicants should refer to the website www.beaweb.org for more details.
Closing Date: October 11th
Funding: Private
Contributor: Patrick Communications Corporation
No. of awards given last year: 1
No. of applicants last year: 30

Broadcast Education Association Walter S Patterson Scholarships

Subjects: Broadcasting.
Purpose: To assist study towards any area of broadcasting.
Eligibility: The applicant must be able to show substantial evidence of superior academic performance and potential to be an outstanding radio professional.
Level of Study: Unrestricted
Type: Scholarship
Value: US$1,750 each
Frequency: Annual
Study Establishment: BEA member institutions
No. of awards offered: 2
Application Procedure: Applicants must obtain an official application form from the BEA or campus faculty. Applicants should refer to the website www.beaweb.org for more details.
Closing Date: October 12th
Funding: Private
Contributor: National Association of Broadcasters (NAB)
No. of awards given last year: 2
No. of applicants last year: 56

BROOKHAVEN NATIONAL LABORATORY

Brookhaven Women in Science, PO Box 5000, Upton, NY, 11973-5000, United States of America
Tel: (1) 631 344 8000
Email: greenb@bnl.gov
Website: www.bnl.gov
Contact: Ms Loralie Smart

Brookhaven National Laboratory is a multi-programme national laboratory operated by Brookhaven Science Associates for the United States Department of Energy. The Laboratory's broad mission is to produce excellent science in a safe, environmentally benign manner with the co-operation, support and appropriate involvement of many communities.

Renate W Chasman Scholarship

Subjects: Natural sciences, engineering and mathematics.
Purpose: To encourage women whose education was interrupted to pursue formal studies or a career in the natural sciences, engineering or mathematics.
Eligibility: Open to re-entry women residing in Nassau County, Suffolk County, Brooklyn or Queens, who must be citizens of the United States of America or permanent residents. They must be currently enrolled in or have applied for a degree-orientated programme at an accredited institution.
Level of Study: Postgraduate
Type: Scholarship
Value: US$2,000
Frequency: Annual
Country of Study: Any country
No. of awards offered: 1
Application Procedure: Applicants must submit a completed application, academic record, letters of reference and a short essay on career goals.
Closing Date: April 1st
Funding: Private
No. of awards given last year: 1
Additional Information: Please write to the given address for further information. Application forms are also available in PDF format on the website.

For further information contact:

PO Box 183, Upton, NY, 11973
Contact: Chasman Scholarships

BROWN UNIVERSITY

University Hall, 2nd Floor, Providence, Rhode Island, RI 02912, United States of America
Tel: (1) 401 863 9800
Fax: (1) 401 863 1961
Email: Linda_Dunleavy@Brown.edu
Website: www.brown.edu
Contact: Linda Dunleavy, Associate Dean of the College for Fellowships & Pre-Law

Long Term Fellowships

Purpose: To support scholars and writers whose work considers the early history of the Americas, including all aspects of European, African and Native American experience.
Eligibility: Applicants must be American citizens or have been resident in the United States for the three years immediately preceding the application deadline.
Level of Study: Postdoctorate, Predoctorate, Research
Type: Fellowship
Value: US$4,200 per month
Length of Study: 5–10 months
Frequency: Annual
Study Establishment: John Carter Brown Library
Country of Study: United States of America
Application Procedure: Applicants must see website for details.
Closing Date: December 15th
Contributor: National Endowment for the Humanities (NEH), Andrew W Mellon Foundation, Reed Foundation

Short Term Fellowships

Purpose: To support scholars and writers whose work considers the early history of the Americas, including all aspects of European, African and Native American experience.
Eligibility: Open to US and foreign scholars engaged in pre or post-doctoral research.
Level of Study: Postdoctorate, Predoctorate, Research
Type: Fellowship
Value: US$2,100 per month
Length of Study: 2–4 months
Frequency: Annual
Application Procedure: Applicants must see website for more details.
Closing Date: January 3rd

BUDAPEST INTERNATIONAL MUSIC COMPETITION

Philharmonia Budapest, Alkotmany u.31 1/2, Budapest, H-1054, Hungary
Tel: (36) 1 266 1459, 302 4961
Fax: (36) 1 302 4962
Email: liszkay.maria@hu.inter.net
Website: www.filharmoniabp.hu
Contact: Ms Maria Liszkay, Secretary

The Budapest Music Competition has been held since 1933. Competitions in different categories alternate annually.

Budapest International Music Competition

Subjects: Musical performance.
Eligibility: Open to young artists of all nationalities who are under 32 years of age.
Level of Study: Professional development
Type: Competition
Value: Up to €25,000
Frequency: Annual
Country of Study: Any country
No. of awards offered: 3
Application Procedure: Applicants must complete an application form to be submitted with other required documentation and should contact the office for further information.
Closing Date: May 1st
Funding: Government
No. of awards given last year: 3
No. of applicants last year: 70

Carl Flesch International Violin Competition

Subjects: Musical Performance.
Eligibility: Open to young artists of Europe, or from all over the world if they can certify at least two years of musical studies in Europe.
Level of Study: Professional development
Type: Competition
Value: Up to €7,000
Frequency: Every 3 years
Application Procedure: Applicants must complete an application form to be submitted with other required documentations and should contact the office for further information.
Closing Date: February 1st
Funding: Foundation
No. of awards given last year: 3
No. of applicants last year: Approx. 70
Additional Information: Country of study is Europe.

International Flute Competition

Subjects: Musical performance.
Purpose: To provide thorough and professional support to young musicians.
Eligibility: The competition is open to piano players of all nationalities who were born on or after 1 January 1979.
Level of Study: Professional development
Type: Competition
Value: €8,000 for 1st prize, €6,000 for 2nd prize and €4,000 for 3rd prize. Beyond these prizes concert engagements and special prizes will be offered to the winners by foreign and Hungarian institutions, festivals and concert organizers
Frequency: Annual
Application Procedure: The application form should be send by mail to Philharmonia Budapest Concert and Festival Agency (H-1054 Budapest, Alkotmány u. 31), or by e-mail (liszkay.maria@hu.inter.net) till May 1st.
Closing Date: May 1st
Funding: Government
No. of awards given last year: 3
Additional Information: The application is valid only after the entry fee has been paid in and the letter of confirmation has been received. The entry fee will not be refunded to contestants who withdraw from the competition. Candidates will be informed of the acceptance of their application by June 10th.

THE BUPA FOUNDATION

Bupa House, 15-19 Bloomsbury Way, London, WC1A 2BA, United Kingdom
Tel: (44) 20 7656 2591
Fax: (44) 20 7656 2708
Email: Bupafoundation@Bupa.com
Website: www.bupafoundation.co.uk
Contact: Lee Saunders, Registrar

The Bupa Foundation is an independent medical research charity that funds medical research to prevent, relieve and cure sickness and ill health.

Bupa Foundation Annual Specialist Grant

Subjects: The subject of study may change each year.
Purpose: To provide project funding for studies in a specified area of the Foundation's interests.
Eligibility: The competition is open to those based in the UK, Australia, Denmark, Hong Kong, New Zealand, Saudi Arabia, Spain and Thailand. Entries must be compliant with the local health and safety legislation if applicable. The Foundation will seek peer reviews from the home country of each shortlisted entry. Researchers and health professionals working for public or private organisations may apply for Bupa Foundation specialist grants for UK-based projects. Study for higher or further degrees, medical electives, educational courses, seminars, conferences, although valuable activities, are not eligible for themed grants. The Foundation will consider specialist grant applications for the creation of research reviews in the field specified in the current year's theme.
Level of Study: Project
Type: Grant
Value: Up to UK£750,000 for a project over 1–3 years
Frequency: Annual
Application Procedure: Apply online at bupafoundation.co.uk and one original signed copy to be posted to the Bupa Foundation.
Closing Date: Check website
Contributor: Bupa Foundation
No. of awards given last year: 4
No. of applicants last year: 31
Additional Information: Please see www.bupafoundation.co.uk for details.

For further information contact:

The Bupa Foundation
Tel: 020 7656 2591
Email: bupafoundation@bupa.com
Contact: Mrs Lee Saunders, Registrar

Bupa Foundation Medical Research Grant for Health at Work

Subjects: To encourage promotion of good health by not only making people aware of healthy behaviour but also by motivating them to practice it.
Purpose: To support research into the feasibility and potential value of workplace conditions for health promotion and active management of employee health.
Eligibility: Open to health professionals and health researchers.
Level of Study: Doctorate, Postdoctorate, Postgraduate, Research, Project funding
Type: Project grant
Value: Restricted by project need only.
Length of Study: A maximum of 3 years
Frequency: Biannual
Country of Study: United Kingdom
No. of awards offered: Varies
Application Procedure: For all queries contact Lee Saunders, the Foundation's Registrar. Apply online towww.bupafoundation.co.uk
Closing Date: October (Guestimate ONLY)
Funding: Foundation
Contributor: Bupa Foundation
No. of awards given last year: 1
No. of applicants last year: 5
Additional Information: For more specific information please see their website.

For further information contact:

Email: bupafoundation@bupa.com
Contact: Mrs Lee Saunders

Bupa Foundation Medical Research Grant for Information and Communication

Subjects: Health information and communication.

Purpose: To support research designed to enhance partnership between health professionals and public/patients.
Eligibility: Open to health professional and health researchers.
Level of Study: Doctorate, Postdoctorate, Postgraduate, Research
Type: Project grant
Value: Restricted by project needs only.
Length of Study: Maximum of 3 years
Frequency: Twice per year
Country of Study: United Kingdom
No. of awards offered: Varies
Application Procedure: For all queries contact Lee Saunders, the Foundation's Registrar. Apply online towww.bupafoundation.co.uk
Closing Date: October (Guestimate ONLY)
Funding: Foundation
Contributor: Bupa Foundation
No. of awards given last year: 6
No. of applicants last year: 13

For further information contact:

Email: bupafoundation@bupa.com
Contact: Mrs Lee Saunders

Bupa Foundation Medical Research Grant for Preventive Health

Subjects: Preventive health.
Purpose: To support research for preventive health projects in all health environments from epidemiology to health maintenance.
Eligibility: Open to health professionals and health researchers.
Level of Study: Doctorate, Postdoctorate, Postgraduate, Research
Type: Project grant
Value: Restricted by project needs only
Length of Study: A maximum of 3 years
Country of Study: United Kingdom
No. of awards offered: Varies
Application Procedure: For all queries contact Lee Saunders, the Foundation's Registrar. Apply online to www.bupafoundation.co.uk
Closing Date: October
Funding: Foundation
Contributor: Bupa Foundation
No. of awards given last year: 7
No. of applicants last year: 12

Bupa Foundation Medical Research Grant for Surgery

Subjects: Surgery.
Purpose: To support research into surgical practices, outcomes and new surgical techniques.
Eligibility: Open to health professionals and health researchers.
Level of Study: Doctorate, Postdoctorate, Postgraduate, Research
Type: Project grant
Value: Restricted by project needs only
Length of Study: Maximum of 3 years
Frequency: Biannual
Country of Study: United Kingdom
No. of awards offered: Varies
Application Procedure: For all queries contact Lee Saunders, the Foundation's Registrar. Apply online towww.bupafoundation.co.uk
Closing Date: July 31st for November intake and October for February intake
Funding: Foundation
Contributor: Bupa Foundation
No. of awards given last year: 6
No. of applicants last year: 19

Bupa Foundation Medical Research Grant for Work on Older People

Subjects: Prevention, treatment and palliative care of mental ill health in older people.
Purpose: To support research aimed at preventing, treating and caring for mental ill-health in older people.
Eligibility: Open to health professionals and health researchers.
Level of Study: Doctorate, Postdoctorate, Postgraduate, Research
Type: Project grant
Length of Study: Maximum of 3 years
Frequency: 6 months
Country of Study: United Kingdom
No. of awards offered: Varies
Application Procedure: For all queries contact Lee Saunders, the Foundation's Registrar. Apply online.
Closing Date: July 31st for November intake and October for February intake
Funding: Foundation
Contributor: Bupa Foundation
No. of awards given last year: 4
No. of applicants last year: 8

THE CAMARGO FOUNDATION

1, avenue Jermini, 13260 Cassis, France
Tel: (33) 4 42 01 11 57
Fax: (33) 4 42 01 36 57
Email: apply@camargofoundation.org
Website: www.camargofoundation.org
Contact: Emily Roberts, Applications Coordinator

The Camargo Foundation maintains a study centre to assist scholars who wish to pursue projects in the humanities and social sciences related to French and Francophone cultures, and to support projects by visual artists, photographers, filmmakers, video artists, media artists, composers and writers.

Camargo Fellowships

Subjects: Humanities and social sciences, visual arts, music composition and creative writing.
Purpose: To assist scholars who wish to pursue projects in the humanities and social sciences related to French and Francophone cultures, and to support projects by visual artists, photographers, filmmakers, video artists, media artists, composers and writers. This interdisciplinary residency program is intended to give fellows the time and space they need to realise their projects.
Eligibility: Open to members of university and college faculties who wish to pursue special studies while on leave from their institutions, independent scholars working on specific projects and graduate students whose academic residence and general examination requirements have been met and for whom a stay in France would be beneficial in completing the dissertation required for their degree. The award is also open to writers, visual artists, photographers, filmmakers, video artists, multimedia artists and composers with specific projects to complete.
Level of Study: Doctorate, Postgraduate, Professional development
Type: Fellowship
Value: US$1,500
Length of Study: Varies
Frequency: Annual
Study Establishment: The Camargo Foundation, study centre in Cassis
Country of Study: France
No. of awards offered: Varies, approx. 20–26
Application Procedure: Applicants must apply online, submitting a completed application form, a curriculum vitae, a detailed description of their project of up to 1,000 words in length and three letters of recommendation by individuals familiar with the applicant's professional work. At least two of the letters should come from persons outside the applicant's own institution; graduate students are exempt from this requirement. Artists should submit 10 JPEGs showing samples of their work, composers should submit a score, MP3 or compact disc and writers should send 10–20 pages of text. For further information applicants should log on to the Foundation's website at: www.camargofoundation-apply.org
Closing Date: January 12th of the following academic year
Funding: Private
Contributor: The Jerome Hill endowment
Additional Information: A written report will be required at the end of the stay. Each fellow must give a presentation of their project, followed by a discussion. Fellows are required to attend all project presentations and discussions.

THE CANADA COUNCIL FOR THE ARTS

350 Albert Street, PO Box 1047, Ottawa, ON, K1P 5V8, Canada
Tel: (1) 613 566 4414 ext 5060/800 263 5588
Fax: (1) 613 566 4390
Email: sarah.rushton@canadacouncil.ca
Website: www.canadacouncil.ca
Contact: Ms Sarah Rushton

The Canada Council for the Arts is a national agency that provides grants and services to professional Canadian artists and art organizations in dance, media arts, music, theatre, writing and publishing, inter-arts and the visual arts.

Canada Council Grants for Professional Artists

Subjects: Art: dance, music, theatre, media arts, visual arts, creative writing and inter-arts.
Purpose: To help professional Canadian artists pursue professional development and/or independent artistic creation or production.
Eligibility: Open to Canadian citizens or permanent residents of Canada who have finished their basic training in the arts and/or are recognized as professionals within their own disciplines.
Level of Study: Postgraduate, Professional development
Value: Canadian $3,000–60,000
Frequency: Varies
Country of Study: Any country
No. of awards offered: Varies
Closing Date: Varies
Funding: Government
Additional Information: Interested Canadian individuals should see the website for detailed information on the grants offered in each discipline.

Canada Council Michael Measures Prize

Subjects: Music.
Purpose: To recognize promising young performers of classical music.
Type: Prize
Value: $15,000
Frequency: Annual
Study Establishment: Canada Council for the Arts
Country of Study: Canada
No. of awards offered: 1
Funding: Private
Additional Information: Please refer to the National Youth Orchestra's website http://www.nyoc.org.

Canada Council Travel Grants

Subjects: Arts: dance, music, theatre, media art, visual art, creative writing and inter-arts.
Purpose: To enable Canadian artists to travel on occasions important to their professional careers.
Eligibility: Open to Canadian citizens or permanent residents of Canada who have finished their basic training in the arts and are recognized as professionals within their own disciplines.
Type: Travel grant
Value: A maximum of Canadian $2,500 to cover travel costs
Frequency: Varies
Country of Study: Any country
No. of awards offered: Varies
Funding: Government
Additional Information: Interested Canadian individuals should see the website for detailed information on the grants offered within each discipline.

J B C Watkins Award

Subjects: Architecture, music, and theatre.
Purpose: To allow Canadian artists to pursue graduate study outside Canada in theatre, architecture and music.
Eligibility: Open to Canadian artists who are graduates of a Canadian university or postsecondary art institution or training school in the above subjects.
Level of Study: Postgraduate
Type: Grant
Value: Canadian $5,000
Frequency: Annual
No. of awards offered: Varies
Application Procedure: Applicants must see the Canada Council's website for details. www.canadacouncil.ca/prizes/jbc_watkins
Closing Date: Varies
Funding: Private
Additional Information: All eligible candidates in the grants to individual program within the discipline of music, theatre or architecture will automatically be considered. Country of study - any country other than Canada. Preference is given to those wishing to carry out their studies in Denmark, Norway, Sweden or Iceland.

Killam Prizes

Subjects: Health sciences, natural sciences, engineering, social sciences, and humanities.
Purpose: Intended to honour eminent Canadian scholars actively engaged in research in Canada in universities, hospitals, research and scientific institutes or other equivalent or similar institutions.
Eligibility: Only Canadian citizens are eligible for this honour, and the prizes are awarded only to living candidates. To be nominated for the prizes, candidates must have made a substantial and distinguished contribution, over a significant period, to scholarly research in Canada. Their outstanding achievements must have been clearly demonstrated already, and they are expected to make further contributions to the scholarly and scientific heritage of Canada.
Level of Study: Postgraduate
Type: Prize
Value: Canadian $100,000
Length of Study: Varies
Frequency: Annual
Study Establishment: Universities, hospitals, research institutes or scientific institutes
Country of Study: Canada
No. of awards offered: 5
Application Procedure: Candidates may not apply on their own behalf; they must be nominated by an expert in their field. Information is available on the Canada Council website www.canadacouncil.ca/prizes/killam
Closing Date: June 15th
Funding: Private
Contributor: Killam Trust
No. of awards given last year: 5
Additional Information: Detailed guidelines for the application process are available in the website.

Killam Research Fellowships

Subjects: Humanities, social sciences, natural sciences, health sciences, engineering and studies linking any of the disciplines within these broad fields.
Purpose: To support Canadian scholars of exceptional ability engaged in advanced research projects.
Eligibility: Open to Canadian citizens or permanent residents of Canada. Killam Research Fellowships are aimed at established scholars who have demonstrated outstanding ability through substantial publications in their fields over a period of several years.
Level of Study: Postgraduate
Type: Fellowship
Value: Canadian $70,000 per year, paid to the university or research institution which employs the fellow
Length of Study: 2 years
Frequency: Annual
No. of awards offered: Varies
Application Procedure: There are no hard copy application forms: applicants must submit their requests through the Canada Council's online application system at killam.canadacouncil.ca
Closing Date: May 15th
Funding: Private
Contributor: Killam Trust

Robert Fleming Prize

Subjects: Composition in classical music.
Purpose: To encourage young Canadian composers.
Eligibility: It is intended to encourage the career development of young composers and is awarded to the most talented Canadian music composer in the competition for Canada Council Grants to Professional musicians in classical music.

Level of Study: Postgraduate
Type: Prize
Value: Canadian $2,000
Length of Study: Up to 1 year
Frequency: Annual
Country of Study: Any country
No. of awards offered: 1
Closing Date: March 1st
Funding: Government
Additional Information: Artists may not apply for this prize. All successful candidates in the Canada Council Grants to professional musicians in classical music are considered automatically.

THE CANADIAN ASSOCIATION FOR GRADUATE STUDIES (CAGS)

301-260, St-Patrick Street, Ottawa, ON, K1N 5K5, Canada
Tel: (1) 613 562 0949
Fax: (1) 613 562 9009
Email: info@cags.ca
Website: www.cags.ca

Canadian Association for Graduate Studies (CAGS) brings together 52 Canadian universities with graduate and post graduate programmes and 3 national graduate student associations. Its mandate is to promote graduate and post graduate education and research in Canada.

CAGS UMI Dissertation Awards

Subjects: Engineering, medical sciences and the natural sciences, fine arts, humanities and social sciences.
Purpose: To recognize Canadian Doctoral dissertations that make unusually significant and original contributions to the related academic field.
Eligibility: Open to students whose dissertation is completed and accepted by a Canadian graduate school.
Level of Study: Doctorate
Type: Award
Value: $1,500 prize, a Citation Certificate and travel expenses of up to $1,500 to attend the CAGS Annual Conference
Frequency: Annual
Country of Study: Canada
No. of awards offered: 2
Application Procedure: Applications must be submitted by a Canadian university.
Closing Date: Fabruary 15th
Funding: Private, corporation
Contributor: University Microfilms International
No. of awards given last year: 2
No. of applicants last year: 50

CANADIAN ASSOCIATION FOR THE PRACTICAL STUDY OF LAW IN EDUCATION (CAPSLE)

c/o Secretariat (Lori Pollock), 37 Moultrey Crescent, Georgetown, ON, L7G 4N4, Canada
Tel: (1) 905 702 1710
Fax: (1) 905 873 0662
Email: info@capsle.ca
Website: www.capsle.ca

CAPSLE is a national organization whose aim is to provide an open forum for the practical study of legal issues related to and affecting the education system and its stakeholders.

CAPSLE Fellowship

Subjects: Law.
Purpose: To provide an open forum for the practical study of legal issues affecting education.
Eligibility: Open to Canadian citizens or landed immigrants enrolled in a faculty of law or a Graduate School of Education at Canadian university.
Level of Study: Postgraduate
Type: Fellowship
Value: Canadian $5,000
Frequency: Annual
Study Establishment: Any accredited university or institution
Country of Study: Canada
Application Procedure: See the website.
Closing Date: April 15th
No. of awards given last year: 1
No. of applicants last year: 3
Additional Information: The successful candidate is invited to speak at the annual CAPSLE conference and to publish in our newsletter of CAPSLE or conference proceedings.

For further information contact:

CAPSLE, 37 Moultrey Crescent, Georgetown, Ontario, CANADA, L7G 4N4

CANADIAN ASSOCIATION OF BROADCASTERS (CAB)

PO Box 627, Station B, Ottawa, ON, K1P 5S2, Canada
Tel: (1) 613 233 4035
Fax: (1) 613 233 6961
Email: cab@cab-acr.ca
Website: www.cab-acr.ca
Contact: Vanessa Dawson, Special Events and Projects Co-ordinator

The Canadian Association of Broadcasters (CAB) is the collective voice of Canada's private radio and television stations and speciality services. The CAB develops industry-wide strategic plans, works to improve the financial health of the industry, and promotes private broadcasting's role as Canada's leading programmer and local service provider.

BBM Scholarship

Subjects: Statistical and quantitative research methodology.
Purpose: To ensure that there is an investment in the development of individuals, skilled and knowledgeable in research, who may be of future benefit to the Canadian broadcasting industry.
Eligibility: Open to students enrolled in a graduate studies programme, or in the final year of an Honours degree with the intention of entering a graduate programme, anywhere in Canada. Candidates must have demonstrated achievement in, and knowledge of, statistical and/or quantitative research methodology in a course of study at a Canadian university or postsecondary institution.
Level of Study: Graduate
Type: Scholarship
Value: Canadian $2,500–4,000 plus a commemorative certificate
Frequency: Annual
Country of Study: Canada
No. of awards offered: 1
Application Procedure: Applicants must complete an application form and submit a 250-word essay outlining their interest in audience research. Application forms are available from the website. Three references should be attached to the completed application form, including one from the course director.
Closing Date: June 30th
Funding: Private
Contributor: The BBM Bureau of Measurement and the Canadian Association of Broadcasters

CANADIAN BAR ASSOCIATION (CBA)

500-865 Carling Avenue, Ottawa, ON, K1S 5S8, Canada
Tel: (1) 613 237 2925
Fax: (1) 613 237 0185
Email: info@cba.org
Website: www.cba.org
Contact: Christine Sopora, Project Officer Commnucations

The Canadian Bar Association (CBA) represents more than 37,000 lawyers across Canada. Offers national perspective on legal issues, federal legislation and trends in law. Provides explanations, analysis and commentary on all areas of law from practising lawyers, academics and in-house counsel. CBA is dedicated to improvement in the law, the administration of justice, lawyer education and advocacy in the public interest.

Viscount Bennett Fellowship

Subjects: Law.
Purpose: To encourage a high standard of legal education, training and ethics.
Eligibility: Open to Canadian citizens only.
Level of Study: Postgraduate
Type: Fellowship
Value: Canadian $40,000
Length of Study: 1 year
Frequency: Annual
Study Establishment: An approved institution
Country of Study: Any country
No. of awards offered: 1
Application Procedure: Please refer to www.cba.org/cba/awards/viscount_bennett for full details and an application form.
Closing Date: November 15th
Funding: Trusts
No. of awards given last year: 1
No. of applicants last year: Average - 48

CANADIAN BLOOD SERVICES (CBS)

1800 Alta Vista Drive, Ottawa, ON, K1G 4J5, Canada
Tel: (1) 613 739 2300
Fax: (1) 613 731 1411
Email: elaine.konecny@bloodservices.ca
Website: www.bloodservices.ca
Contact: Ms Elaine Konecny, Program Assistant, Research & Development

Canadian Blood Services (CBS) is a non-profit, charitable organization whose sole mission is to manage the blood system for Canadians. CBS collects approx. 900,000 units of blood annually and processes it into components and products that are administered to thousands of patients each year.

CBS Graduate Fellowship Program

Subjects: Blood transfusion science focusing on aspects of the collection and preparation of blood from volunteer donors as well as on the biological materials derived from blood or their substitutes obtained through biotechnology. Research may encompass a broad variety of disciplines including, but not restricted to, epidemiology, surveillance, social sciences, blood banking, immunohaematology, haematology, infectious diseases, immunology, genetics, protein chemistry, molecular and cell biology, clinical medicine, laboratory sciences, virology, bioengineering, process engineering or biotechnology.
Purpose: To attract and support young investigators to initiate or continue training in the field of blood or blood products research.
Eligibility: Open to graduate students who are undertaking full-time research training leading to a PhD degree. Students registering solely for a Master's degree will not be considered and only those demonstrating acceptance into a PhD programme will receive continued support. Candidates must have completed sufficient academic work to be admitted in good standing to a graduate school by the time the award is to take effect, or be already engaged in a PhD programme. Applicants possessing a medical degree but not licensed to practice medicine in Canada are eligible to apply for this award providing they meet the above criteria.
Level of Study: Graduate
Type: Fellowship
Value: Canadian $21,000 per year plus a yearly research and travel allowance of Canadian $1,000 per year
Length of Study: Up to 4 years. The initial term is for 2 years, with the option for a 2 year renewal. Renewals must be requested in the form of a complete new application
Frequency: Biannual
Country of Study: Canada
Application Procedure: Candidates are required to submit a completed application form (GFP-01) that is available either from the website, from or the main address.
Closing Date: November 15th
Funding: Government
No. of awards given last year: 7
No. of applicants last year: 15

For further information contact:

Program Assistant, R&D, Canadian Blood Services, 1800 Alta Vista Drive, Ottawa, ON, K1G 4J5
Tel: 613 739 2230
Fax: 613 739 2201
Email: elaine.konecny@blood.ca
Contact: Elaine Konecny

CBS Postdoctoral Fellowship (PDF)

Subjects: Transfusion science. The CBS has active research programmes within transfusion science emphasizing platelets, stem cells, plasma proteins, infectious disease, epidemiology and transfusion practice.
Purpose: To support Fellows working with CBS-affiliated research and development groups across Canada and to foster careers related to transfusion science in Canada.
Eligibility: Candidates must hold a recent PhD or equivalent research degree or an MD, DDS, DVM, plus a recent research degree in an appropriate health field (minimum of a MSc) or equivalent research experience, neither must be registered for a higher degree at the time of acceptance of the award and nor undertake formal studies for such a degree during the period of appointment.
Level of Study: Postdoctorate, Professional development
Type: Fellowship
Value: The value of each fellowship is related to the major degree(s) and experience that the applicant holds. The fellowship offers a stipend based on current Medical Research Council rates for each of the 3 years as well as a 1st year research allowance of Canadian $10,000
Length of Study: 1–3 years
Frequency: Annual
Country of Study: Canada
No. of awards offered: 6
Application Procedure: Applicants must complete CBS Form RD40. Applications must be made through and with the support of a CBS-affiliated scientist. Application forms and guidelines are available from any of the CBS centres or from the main address.
Closing Date: July 2nd
Funding: Government
No. of awards given last year: 4
No. of applicants last year: 8

CANADIAN BUREAU FOR INTERNATIONAL EDUCATION (CBIE)

220 Laurier West, Suite 1550, Ottawa, ON, K1P 5Z9, Canada
Tel: (1) 613 237 4820
Fax: (1) 613 237 1073
Email: scholarships-bourses@cbie.ca
Website: www.cbie.ca

The Canadian Bureau for International Education (CBIE) is a national non-profit association comprising educational institutions, organizations and individuals dedicated to internal education and intercultural training. CBIE's mission is to promote the free movement of learners and trainees across national borders.

Canadian Commonwealth Scholarship Plan

Subjects: Science, law, environment studies, economics, sociology, geography, electronics and education.
Purpose: To offer scholarships to citizens of other Commonwealth countries to study in Canada.
Eligibility: Open to candidates who are not more than 40 years of age with completion of tertiary education from an English medium. Please check website for more details.
Level of Study: Postgraduate
Type: Scholarship

Value: Canadian $7,500 for graduate students for a period of 4 months and $10,000 for graduate students for a period of 5–6 months
Length of Study: 12–36 months
Frequency: Annual
Country of Study: Canada
Application Procedure: Further information available on the website www.scholarships.gc.ca.
Closing Date: Please see the website
Contributor: Commonwealth Scholarship Commission
Additional Information: Eligible coutries – Asia-Pacific: Bangladesh, India, Kiribati, Malaysia, Maldives, Nauru, Pakistan, Papua New Guinea, Samoa, Singapore, Solomon Islands, Sri Lanka, Tonga, Tuvalu, Vanuatu; Africa: Botswana, Cameroon, Gambia, Ghana, Kenya, Lesotho, Malawi, Mauritius, Mozambique, Namibia, Nigeria, Rwanda, Seychelles, Sierra Leone, South Africa, Swaziland, Tanzania, Uganda, Zambia.

Canadian Studies Postdoctoral Fellowship

Subjects: Area and cultural studies.
Purpose: To enable young academics who have completed a doctoral thesis to visit a Canadian or foreign university with a Canadian studies programme for a teaching or research fellowship.
Eligibility: Open to students with a doctoral degree.
Level of Study: Postdoctorate
Type: Fellowship
Value: Canadian $2,500 per month plus the cost of a return airline ticket for a maximum of $10,000
Length of Study: 1 year
Frequency: Annual
Study Establishment: Any accredited Canadian university
Country of Study: Canada
Application Procedure: Application form and a recommendation from the national Canadian studies Association must be submitted. For country specific information and addresses please see the website www.scholarships.gc.ca/pdrfcountries-en.html
Closing Date: November 23rd
Contributor: Foreign Affairs and International Trade Canada

For further information contact:

Website: www.scholarships.gc.ca/pdrfcountries-en.html

Organization of American States (OAS) Fellowships Programs

Subjects: Human development.
Purpose: To fund education of Canadian residents and nationals in other American nations.
Eligibility: Open to Canadian residents and nationals.
Type: Fellowship
Value: US$30,000.00 per academic year, which includes tuition, benefits, and administrative costs.
Length of Study: 1–2 years
Frequency: Annual
Country of Study: United States of America
Application Procedure: A completed application form must be submitted. Please see the website www.scholarships.gc.ca
Closing Date: For more information please see the website
Funding: Government
Additional Information: A new competition is expected to be announced in January.

CANADIAN CANCER SOCIETY RESEARCH INSTITUTE (CCSRI)

Suite 300, 55 St. Clair Avenue W., Toronto, Ontario, M4V 2Y7, Canada
Tel: (1) 416 961 7223
Fax: (1) 416 961 4189
Email: research@cancer.ca
Website: www.cancer.ca/research
Contact: Carol Bishop, Assistant Director, Research Operations

The Canadian Cancer Society (CCS) is the largest non-government funder of cancer research in Canada. The CCS provides support for research and related programmes undertaken at Canadian universities, hospitals and other research institutions.

CCS Impact Grants

Subjects: Cancer research.
Purpose: To stimulate Canadian investigators in a very broad spectrum of research.
Eligibility: A researcher who is designated as the Principal Investigator must be based in, or formally affiliated with, but not necessarily receive salary support form, an eligible Canadian Host Institution such as a university, research institute or health care agency. Graduate students, postdoctoral fellows, research associates, technical support staff, or investigators based outside of Canada are not eligible to be a Principal Investigator.
Type: Research grant
Value: Awards will be granted for the purchase and maintenance of animals, for expendable supplies, minor items of equipment, for payment of graduate students, postdoctoral fellows and technical and professional assitants, and for research travel and permanent equipment. These grants do not provide for personal salary support of the principal investigator and/or co-applicants or for institutional overhead costs.
Length of Study: 1–5 years
Frequency: Annual
Study Establishment: Universities or other institutions
Country of Study: Canada
No. of awards offered: Varies
Application Procedure: Applicants must complete an online application form. Further details are available on the website.
Closing Date: February 1st (Letter of Intent)
Funding: Private
Contributor: The Terry Fox Foundation and the Canadian Cancer Society
No. of awards given last year: 11
No. of applicants last year: 89
Additional Information: Further information is available on the website. Grants will be awarded to projects deemed worthy of support, provided that the basic equipment and research facilities are available in the institution concerned and that it will provide the necessary administrative services. Grants are made only with the consent and knowledge of the administrative head of the institution at which they are to be held and applications must be countersigned accordingly.

CCS Innovation Grants

Subjects: Cancer research.
Purpose: To stimulate Canadian investigators in a very broad spectrum of research.
Eligibility: A researcher who is designated as the Principal Investigator must be based in, or formally affiliated with, but not necessarily receive salary support from, an eligible Canadian Host Institution such as a university, research institute or health care agency. Graduate students, postdoctoral fellows, research associates, technical support staff, or investigators based outside Canada are not eligible to be a principal investigator.
Type: Research grant
Value: Awards will be granted for the purchase and maintenance of animals, for expendable supplies, minor items of equipment, for payment of graduate students, postdoctoral fellows and technical and professional assistants, and for research travel and permanent equipment. These grants do not provide for personal salary support of the principal investigator and/or co-applicants or for institutional overhead costs
Length of Study: 2–3 years
Frequency: Biannual
Study Establishment: Universities or other institutions
Country of Study: Canada
No. of awards offered: Varies
Application Procedure: Applicants must complete an online application form. Further details are available on the website.
Closing Date: October 1st and February 15th
Funding: Private
Contributor: The Canadian Cancer Society
No. of awards given last year: 37
No. of applicants last year: 204

Additional Information: Further information is available on the website. Grants will be awarded to projects deemed worthy of support, provided that the basic equipment and research facilities are available in the institution concerned and that it will provide the necessary administrative services. Grants are made only with the consent and knowledge of the administrative head of the institution at which they are to be held and applications must be countersigned accordingly.

CCS Travel Awards

Subjects: Cancer research.
Purpose: To provide financial assistance by helping to defray the travel costs associated with making a scientific presentation as a first author or presenter at a conference, symposium or other appropriate professional gathering.
Eligibility: Applicants must be students enrolled in a MD/PhD program at a Canadian institution and be in the final phase of their studies. Candidates must be attending a conference for the purpose of presenting data from a cancer-related project on a first author or presenter basis. Postdoctoral fellows must be within 5 years of attaining their PhD Medical resident/clinical fellow must be within 5 years of attaining their MD.
Level of Study: Graduate, Postgraduate
Type: Travel grant
Value: Up to Canadian $2,000
Frequency: 3 times per year (January/May/September)
Country of Study: International
No. of awards offered: Up to 60 per calendar year
Application Procedure: Applicants must refer to the website.
Closing Date: January 15th, May 15th, December 15th
Funding: Private
Contributor: The Canadian Cancer Society
Additional Information: Further information is available from the website.

Knowledge to Action Grant

Subjects: Cancer research.
Purpose: Intended to provide funding for research projects that build on existing cancer research finding and aim to improve outcomes and experiences through KT for people and populations at risk, patients, their families and communities across the cancer trajectory.
Eligibility: A researcher who is designated as the Principal Investigator (as per other CCS awards).
Level of Study: Research
Type: Grant
Value: $100,000
Length of Study: 1–2 years
Frequency: Annual
Study Establishment: Universities or Other Institutions
Country of Study: Canada
No. of awards offered: 3–4
Application Procedure: Applicants must complete an online application form. Further details are available on the website.
Closing Date: December 15 (letter of intent) and February 1st (full application)
Funding: Private
Contributor: The Canadian Cancer Society
Additional Information: Please check website for further information.

Quality of Life Research Grant

Subjects: Cancer research.
Purpose: To support quality of life research that has the potential to make a significant impact on the burden of disease in patients, survivor and caregiver.
Eligibility: A researcher who is designated as the Principal Investigator (as per other CLS awards).
Level of Study: Research
Type: Grant
Value: $100,000 per year for up to 3 years (max. $300,000)
Length of Study: 3 years
Frequency: Annual
Study Establishment: Universities of Other Institutions
Country of Study: Canada
No. of awards offered: 6–8 per year
Application Procedure: Applicants must complete an online application form. Further details are available on the website.
Closing Date: July 1st (letter of intent) and November 1st (full application)
Funding: Private
Contributor: The Canadian Cancer Society
No. of awards given last year: 6
No. of applicants last year: 63
Additional Information: Please check website for further information.

CANADIAN CENTENNIAL SCHOLARSHIP FUND

Canadian Women's Club, MacDonald House, 1 Grosvenor Square, London, London, W1K 4AB, England
Tel: (44) 20 7258 6344
Fax: (44) 20 7258 6637
Email: applications@canadianscholarshipfund.co.uk
Website: www.canadianscholarshipfund.co.uk
Contact: The Bursar

The Canadian Centennial Scholarship fund gives annual awards to Canadian men and women who are already studying in the United Kingdom. Scholarships are awarded on the basis of high academic standards, financial need and relevance of the proposed course of study to Canada.

Canadian Centennial Scholarship Fund

Subjects: All subjects.
Purpose: To assist Canadians studying in the United Kingdom and are exclusively for students who have already commenced their studies in the UK.
Eligibility: Canadian citizens currently enrolled in a United Kingdom programme of studies are eligible to apply.
Level of Study: Doctorate, Graduate, MBA, Postgraduate, Professional development
Type: Scholarship
Value: UK£500–3,000
Length of Study: 1 year. Recipients may re-apply
No. of awards offered: 12–15
Application Procedure: Candidates submit a written application and those short listed are interviewed. Applications can be downloaded from our website.
Closing Date: March 8th
Funding: Trusts
Contributor: Maple Leaf Trust
No. of awards given last year: 20
No. of applicants last year: 103

CANADIAN CYSTIC FIBROSIS FOUNDATION (CCFF)

2221 Yonge Street, Suite 601, Toronto, ON, M4S 2B4, Canada
Tel: (1) 416 485 9149
Fax: (1) 416 485 0960
Email: info@cysticfibrosis.ca
Website: www.cysticfibrosis.ca
Contact: Manager, Research Programs

Since 1960, the Canadian Cystic Fibrosis Foundation (CCFF) has worked to provide a brighter future for every child born with cystic fibrosis. Through its research and clinical programmes, the Foundation helps to provide outstanding care for affected individuals, while pursuing the quest for a cure or control.

CCFF Clinic Incentive Grants

Subjects: Cystic fibrosis.
Purpose: To enhance the standard of clinical care available to Canadians with cystic fibrosis, by providing funds to initiate a comprehensive programme for patient care, research and teaching or to strengthen an existing programme.
Eligibility: Canadian hospitals and medical schools are eligible to apply. Applicants must demonstrate the regional need for specialized clinical care for cystic fibrosis, the need of the institution for assistance and its plans to attract complementary funding from other sources to develop a complete cystic fibrosis programme, the potential for the

development of a comprehensive programme for care, clinical research and teaching and the desire to collaborate with the CCFF and other Canadian cystic fibrosis clinics.
Level of Study: Research
Type: Grant
Value: Honorarium and travel allowance
Length of Study: 1 year, renewable on an annual basis
Frequency: Annual
Country of Study: Canada
Application Procedure: Applicants must complete an application form. Late applications will be subject to a penalty equal to 10 per cent of the value of the award. This penalty will be deducted from the clinic director's honorarium.
Closing Date: October 1st

CCFF Fellowships

Subjects: Cystic fibrosis.
Purpose: To support basic or clinical research training in areas of the biomedical or behavioural sciences pertinent to cystic fibrosis.
Eligibility: Applicants must hold a PhD or MD. Medical graduates should have already completed basic residency training and must be eligible for Canadian licensure. Equitable consideration will be given to Fellowship applicants from outside of Canada, who intend to return to their own country on completion of a fellowship.
Level of Study: Postgraduate
Type: Fellowship
Value: Dependent upon academic qualifications and research experience
Length of Study: 2 years, renewable for up to 1 more year
Frequency: Annual
Study Establishment: An approved university department, hospital or research institute in Canada
Country of Study: Canada
Application Procedure: Applicants must arrange to send three letters of recommendation, one of which should be from the applicant's current or most recent supervisor.
Closing Date: October 1st
No. of awards given last year: 9
No. of applicants last year: 30

CCFF Research Grants

Subjects: Cystic fibrosis.
Purpose: To facilitate scientific investigation of all aspects of cystic fibrosis.
Eligibility: A principal investigator should hold a recognized, full-time faculty appointment in a relevant discipline at a Canadian university or hospital. Under exceptional circumstances and at the discretion of the Research Subcommittee, applications from other individuals may be evaluated on a case-by-case basis.
Level of Study: Doctorate
Type: Research grant
Value: Determined by the Medical or Scientific Advisory Committee following a detailed review of the applicant's proposed budget
Length of Study: Usually 1 or 2 years or, in a limited number of instances, 3 years
Frequency: Annual
Study Establishment: A Canadian institution
Country of Study: Canada
Application Procedure: Applicants must write for details. Incomplete or late applications will be returned to the applicant.
Closing Date: August 1st (Notice of intent to apply) and October 1st (application deadline)
Funding: Foundation
No. of awards given last year: 18
No. of applicants last year: 44
Additional Information: Investigators are eligible to hold more than one Research Grant. No more than one initial application may be submitted to a single competition, and it is a requirement that the focus of a second grant be clearly delineated from the first one. The specific aims of a second grant should represent new approaches to the cystic fibrosis problem and not an extension of an existing research programme.

CCFF Scholarships

Subjects: Cystic fibrosis.
Purpose: To provide salary support for a limited number of exceptional investigators, offering them an opportunity to develop outstanding cystic fibrosis research programmes, unhampered by heavy teaching or clinical loads. Intended to attract gifted investigators to cystic fibrosis research.
Eligibility: Open to holders of an MD or PhD degree who are sponsored by the Chairman of the appropriate department and by the Dean of Faculty. They may have recently completed training or be established investigators wishing to devote major research effort to cystic fibrosis. The beginning investigator should have demonstrated promise of ability to initiate and carry out independent research and the established investigator should have a published record of excellent scientific research.
Level of Study: Doctorate, Postgraduate
Type: Scholarship
Value: Salary support, which is dependent on the qualifications and experience of the successful candidate, will be determined by prevailing Canadian scholarship rates and the nominating university.
Length of Study: 3 years, renewable for an additional 2 years on receipt of a satisfactory progress report. In no case will an award be for more than 5 years
Frequency: Every 2 years
Study Establishment: Any approved university, hospital or research institute
Country of Study: Canada
No. of awards offered: Varies, subject to availability of funds
Application Procedure: Application form is available at http://www.cysticfibrosis.ca/en/research/ApplicationForms.php. Applicants must write for details.
Closing Date: October 1st
Funding: Foundation
Additional Information: Applications will be accepted only in odd-numbered years.

CCFF Senior Scientist Research Training Award

Subjects: Cystic fibrosis.
Purpose: To provide support to a limited number of cystic fibrosis investigators by offering them an opportunity to obtain additional training that will enhance their capacity to conduct research directly relevant to cystic fibrosis.
Eligibility: Applicants must have held a recognized, full-time faculty appointment in a relevant discipline at a Canadian university or hospital for at least six years.
Level of Study: Postgraduate
Type: Training award
Value: The amount of the award, a maximum of $15,000
Length of Study: 3 months–1 year
Frequency: Annual
Study Establishment: An approved university department or hospital in Canada
Country of Study: Canada
Application Procedure: Applications must be received by the Foundation no later than the deadline. Incomplete or late applications will be returned to the applicant.
Closing Date: April 1st and October 1st
Funding: Foundation
Additional Information: This award can be used for sabbatical support for qualified individuals.

CCFF Small Conference Grants

Subjects: Cystic fibrosis.
Purpose: To support small conferences that are focused on subjects of direct relevance to cystic fibrosis and to facilitate the exchange of special expertise between larger, university-based cystic fibrosis clinics and smaller, more remote clinics.
Eligibility: Open to clinic directors and CCFF-funded investigators.
Level of Study: Professional development
Type: Grant
Value: Grants to conferences will be up to a maximum of Canadian $2,500 and grants for the exchange of expertise will not normally exceed Canadian $1,000
No. of awards offered: Dependent on availability of funds
Application Procedure: Applicants must make applications in the form of a letter. For medical and/or scientific conferences, the application should indicate who is organizing and attending the

conference, and the specific topics and purpose of the conference. For inter-clinic exchanges, the application should specify the proposed arrangements for, and the specific purpose of the exchange.
Closing Date: Applications may be submitted at any time, but the Foundation should be consulted in advance with respect to the availability of funds
No. of awards given last year: 1
No. of applicants last year: 1
Additional Information: Grants are available on a first-come, first-served basis. Frequency of application from any particular individual or group should be reasonable.

CCFF Special Travel Allowances

Subjects: Cystic fibrosis.
Purpose: To enable CCFF-funded fellows and students to attend and participate in scientific meetings related to cystic fibrosis.
Eligibility: Open to Cystic Fibrosis Canada-supported fellows and students, for each year of their award, upon written application and pending the availability of funds.
Level of Study: Doctorate, Postgraduate
Type: Award
Value: Up to Canadian $1,200 annually
Length of Study: As determined by seminar length
Frequency: Annual
Study Establishment: Appropriate seminar
Country of Study: Any country
Application Procedure: Applications should be made in the form of a letter and must be submitted prior to travel.
Closing Date: Any time, but the Foundation must be consulted in advance
Funding: Foundation
No. of awards given last year: 21
No. of applicants last year: 21

CCFF Studentships

Subjects: Cystic fibrosis.
Purpose: To support highly qualified graduate students who are registered for a higher degree, and who are undertaking full-time research training in areas of the biomedical or behavioural sciences relevant to cystic fibrosis.
Eligibility: Applicants must be highly qualified graduate students who are registered for a higher degree and who are undertaking full-time research training in areas relevant to cystic fibrosis, or highly qualified students who are registered in a joint MD/MSc or MD/PhD programme. Equitable consideration will be given to studentship applicants from outside of Canada who intend to return to their own country on completion of a studentship.
Level of Study: Doctorate, Postgraduate
Type: Studentship
Value: Salary and cost-of-living award at the discretion of the Medical/Scientific Advisory Committee
Length of Study: 2–3 years (Master's level) and 2–5 years (Doctorate level)
Frequency: Annual
Study Establishment: Studentships are tenable only at Canadian universities
Country of Study: Canada
Application Procedure: The Foundation sponsors a studentship competition in October and April. Candidates for initial awards are eligible to apply to either competition. Similar to all CCFF grants, studentships are subject to the availability of funds, and the availability of funds is generally more certain with respect to the October competition.
Closing Date: October 1st
Funding: Foundation
No. of awards given last year: 13
No. of applicants last year: 34
Additional Information: Studentships are awarded for studies at the Master's or doctoral level. If a student receiving support for studies leading to a Master's degree elects to continue to a doctorate degree, he or she must reapply for an initial CCFF studentship at the doctoral level.

CCFF Transplant Centre Incentive Grants

Subjects: Cystic fibrosis.
Purpose: To enhance the quality of care available to cystic fibrosis transplant candidates by providing eligible centres with supplementary funding.
Eligibility: Open to any Canadian lung transplant centre that currently has one or more individuals with cystic fibrosis listed for transplant. Please note that under no circumstances will funding be provided to more than one transplant centre in the same city. Applicants must demonstrate how funds awarded would serve to enhance the quality of care available to patients in their centre.
Level of Study: Research
Type: Grant
Value: Determined in accordance with a formula that takes account of the number of patients assessed, accepted and followed preoperatively, transplanted and followed postoperatively in a given centre during the calendar year ending December 31st of the year preceding the application
Length of Study: 1 year
Frequency: Annual
Country of Study: Canada
Application Procedure: Applicants must contact the Foundation. Applicants must provide a rationale for the funding request and a detailed report on patient care and research within the lung transplant programme. All applications will be adjudicated by the Clinic Subcommittee of the Canadian Cystic Fibrosis Foundation.
Closing Date: October 1st, please contact Foundation for update
Additional Information: Late applications will be subject to a penalty equal to 10 per cent of the value of the award.

CCFF Visiting Scientist Awards

Subjects: Cystic fibrosis.
Purpose: To enable senior investigators from abroad who are invited to engage in cystic fibrosis research at a Canadian institution to travel to Canada or to assist junior or senior investigators who wish to work in another laboratory in Canada or abroad. This experience should, in some way, benefit the Canadian cystic fibrosis research effort.
Eligibility: A senior investigator can be considered such if he or she has attained at least the position of an associate professor, or has 6 years of equivalent experience.
Level of Study: Doctorate, Professional development
Type: Award
Value: Varies
Length of Study: Varies
Frequency: Dependent on funds available
Country of Study: Any country
Application Procedure: Applicants must send an application letter, accompanied by supporting letters from the head of the appropriate department of the host university. Supporting letters should also be provided by the Head of the Department and the Dean of the Faculty of the applicant's own university.
Closing Date: Applications are accepted at any time. Please consult with Foundation in advance

Clinical Fellowships

Subjects: Cystic fibrosis care.
Purpose: Intended for those individuals who have already obtained their clinical training and wish to pursue training in CF care, and also acquire the competencethat would allow them to participate in clinical trails.
Eligibility: Canadian citizens or permanent reidents who have an MD degree, have recently completed their clinical training and have obtained medical licensure in Canada are eligible to apply.
Level of Study: Research, MD degree
Type: Fellowship
Value: Depends on academic qualifications and/or research experience
Length of Study: 2 years, fellows can apply for 1 year renewal for further research training
Country of Study: Canada and abroad
No. of awards offered: 1
Application Procedure: Applicants must submit three letters of recommendation, one of which should be from the applicants' current or most recent supervisor. The application form should also include a description of the proposed research and clinical training program and official transcripts of the applicants' complete academic record. Applications that do not include these documents will be rejected.

Closing Date: October 1st
No. of awards given last year: 1
No. of applicants last year: 1

Clinical Projects Grants

Purpose: Intended to bolster the CCFF's commitment to clinical research and to promote the further development of the Clinical Studies Network. It provides a mechanism whereby the most important ideas for clinical studies can be forged into viable protocols and strategies of urgent clinical relevance can be pursued (seed money for small, start-up pilot projects).
Eligibility: Awarded at the discretion of the leaders of the Medical/ Scientific Advisory Committee in consultation with the Clinical Studies Network.
Level of Study: Research
Type: Grant
Value: Up to $15,000
Frequency: Dependent on funds available
No. of awards offered: Varies
Application Procedure: Applications should be made in the form of a letter in which a clear hypothesis is detailed, relevance to CF is demonstrated, and a brief budget is outlined.
Closing Date: February 1st (Notice of Intent to apply) and April 1st (Full Application deadline)
Additional Information: Applications may be submitted at any time but the Foundation should be consulted in advance with respect to the availability of funds.

Travel Supplement Grants for the European CF Society Conference

Purpose: To assist a limited number of individuals who plan to play an active role in ECFS conferences.
Eligibility: Applicants must be CCFF-funded investigators, Canadian CF clinic directors or clinicians, or Canadian CF clinic coordinators and allied health professionals. Applicants must show evidence of active participation at the conference.
Type: Grant
Value: Up to Canadian $2,000 per person
Frequency: Annual
No. of awards offered: Up to 5
Application Procedure: Applications should include a brief budget and frequency of application from any particular individual should be reasonable (once every 4 years). Completed application form and supporting documentation can be submitted by email or hard copy.
Closing Date: February 1st (for Basic Science Conference) and April 1st (for ECFS Conference)
No. of awards given last year: 5
No. of applicants last year: 5
Additional Information: Application form is available at http://www.cysticfibrosis.ca/en/research/TravelSupplementECFs.php. Incomplete or late applications will be returned to the applicant.

For further information contact:

Email: researchprograms@cysticfibrosis.ca

Visiting Allied Health Professional Awards

Subjects: Cystic fibrosis.
Purpose: To support allied health professionals from abroad who are invited to engage in CF clinical observation or activity at a Canadian institution or Canadian allied health professionals who wish to visit another clinic in Canada or abroad.
Eligibility: Applicants must be associated with a recognized CF clinic, and be an active member in CF clinical care. It is intended that this experience will in some way benefit Canadian CF clinical care.
Type: Award
Frequency: Dependent on funds available
Country of Study: Canada or abroad
No. of awards offered: Varies
Application Procedure: Applications should be made in the form of a letter, accompanied by a supporting letter from the head of the appropriate department of the applicants' institution. A supporting letter signed by the department head of the host institution should also be provided.
Closing Date: Application by letter at any time throughout the year
Additional Information: Applications may be submitted at any time, but the Foundation should be consulted in advance with respect to the availability of funds.

Visiting Clinician Awards

Subjects: Cystic fibrosis.
Purpose: To support clinicians from abroad who are invited to engage in CF clinical observation or activity at a Canadian institution or Canadian clinicians who wish to visit another clinic in Canada or abroad.
Eligibility: Applicants must be associated with a recognized CF clinic, and be an active member in CF clinical care. It is intended that this experience will in some way benefit Canadian CF clinical care.
Type: Award
Frequency: Dependent on funds available
No. of awards offered: Varies
Application Procedure: Applications should be made in the form of a letter, accompanied by a supporting letter from the head of the appropriate department of the applicants' institution. A supporting letter signed by the department head of the host institution should also be provided.
Closing Date: Application by letter at any time throughout the year
Additional Information: Applications may be submitted at any time, but the Foundation should be consulted in advance with respect to the availability of funds.

CANADIAN EMBASSY (USA)

Foreign Affairs and International Trade Canada, 501 Pennsylvania Ave NW, Washington, DC, 20001, United States of America
Tel: (1) 202 682 1740
Fax: (1) 202 682 7726
Email: enqserv@dfait-maeci.gc.ca
Website: www.canadianembassy.org
Contact: Daniel Abele, Academic Relations Officer

Canadian Embassy (USA) Research Grant Program

Subjects: Business and economic issues, Canadian values and culture, communications, environment, national and international security or natural resources e.g. energy, fisheries, forestry and trade.
Purpose: To assist individual scholars or a group of scholars in writing an article length manuscript of publishable quality and reporting their findings in scholarly publications.
Eligibility: Open to full-time faculty members at accredited 4 year United States colleges and universities, as well as scholars at American research and policy planning institutes who undertake significant research projects concerning Canada, Canada and the United States, or Canada and North America. Recent PhD recipients who are citizens or permanent residents of the United States are also eligible to apply.
Level of Study: Postgraduate
Type: Programme grant
Value: Up to US$15,000; applicants whose project focuses on the priority topics listed above and who can demonstrate matching funds from others sources may request funding up to US$20,000;
Frequency: Annual
Study Establishment: An accredited 4 year college or university
Country of Study: United States of America
Application Procedure: Applicants must provide 6 copies of the following in this order: the completed application form, a concise proposal of 4–8 pages which will identify all members of the research' team, if a team project, and specify each member's affiliation and role in the study, identify the key issues or the main theoretical problem, describe and justify the appropriate methodology, present a general schedule of research activities, indicate clearly both' the nature and scope of the projects contribution to the advancement of Canadian Studies, include a detailed budget including all other funding sources and a description of anticipated expenditures. A curriculum vitae, and the names and addresses of two scholars from whom the applicants will solicit recommendations should also be included. Application forms are available on request.

Closing Date: November 1st
Funding: Government
Additional Information: The Research Grant Program promotes research in the social sciences and humanities with a view to contributing to a better knowledge and understanding of Canada and its relationship with the United States or other countries of the world.

For further information contact:

Tel: 202 682 7717
Email: daniel.abele@dfait-maeci.gc.ca
Contact: Dan Abele, Academic Relations Officer

Canadian Embassy Faculty Enrichment Program

Subjects: Priority topics include bilateral trade and economics, Canada United States border issues, cultural policy and values, environment, natural resources, energy issues and security co-operation, projects that examine Canadian politics, economics, culture and society as well as Canada's role in international affairs.
Purpose: To provide faculty members with the opportunity to develop or redevelop courses with substantial Canadian content that will be offered as part of their regular teaching load, or as a special offering to select audiences in continuing or distance education.
Eligibility: Open to full-time, tenured or tenure track faculty members at accredited four year United States colleges and universities. Candidates should be able to demonstrate that they are already teaching, or will be authorized to teach, courses with substantial Canadian content (33 per cent or more). Team teaching applications are welcome. Applicants are ineligible to receive the same grant in two consecutive years or to receive two individual category Canadian Studies grants in the same grant period.
Type: Programme
Value: Up to US$6,000; Applicants may request an additional US $5,000 specifically to support student travel to Canada;
Frequency: Annual
Country of Study: United States of America
No. of awards offered: Varies
Application Procedure: Applicants must contact the organization for an application form.
Closing Date: December 1st
Additional Information: The Embassy especially encourages the use of new Internet technology to enhance existing courses, including the creation of instructional websites, interactive technologies and distance learning links to Canadian Universities.

For further information contact:

Tel: 202 682 7717
Email: daniel.abele@dfait-maeci.gc.ca
Contact: Dan Abele, Academic Relations Officer

Canadian Embassy Graduate Student Fellowship Program

Subjects: Business and economic issues, Canadian values and culture, communications, environment, national and international security or natural resources e.g. energy, fisheries, forestry and trade.
Purpose: To assist graduate students in conducting part of their doctoral research in Canada to acquire a better knowledge and understanding of Canada or its relationship with the US and other countries of the world.
Eligibility: Open to full-time doctoral students at accredited 4 year colleges and universities in the United States or Canada whose dissertations are related in substantial part to the study of Canada, Canada and the United States or Canada and North America. Candidates must be citizens or permanent residents of the United States and should have completed all doctoral requirements except the dissertation when they apply for a grant.
Level of Study: Graduate
Type: Fellowship
Value: Up to US$8,000
Frequency: Annual
Study Establishment: An accredited four year college or university
Country of Study: Other
Application Procedure: Applicants must provide six copies of the following in the order listed: the completed application form, a concise letter of three to four pages which will explain clearly the present status of the candidate's doctoral studies, describe the candidate's study plans in Canada, list Canadian contacts such as Scholars, research institutes, academic institutions or libraries, state clearly the exact number of months for which financial support is needed, provide a complete and detailed budget, indicate what other funding sources are available, give the names and addresses of two referees, one of which must be the dissertation advisor, contain the dissertation prospectus which must identify the key issues or the main theoretical problem, justify the methodology and indicate clearly the nature of the dissertation's contribution to the advancement of Canadian Studies. An unofficial transcript of grades, a curriculum vitae and proof of United States citizenship or permanent residency must also be included. Application forms are available on request.
Closing Date: October 31st
Funding: Government
Additional Information: The Graduate Student Fellowship Program promotes research in the social sciences and humanities with a view to contributing to a better knowledge and understanding of Canada and its relationship with the United States or other countries of the world.

For further information contact:

501 Pennsylvania Avenue, NW, Washington, DC, 20001
Tel: 202 682 7727
Email: daniel.abele@dfait-maeci.gc.ca
Contact: Dan Abele, Academic Relations Officer

Conference Grant Program

Subjects: Social science and humanities.
Purpose: To assist an institution in holding a conference and publishing the resulting papers and proceedings in a scholarly fashion.
Eligibility: Open to US institutions and universities who wish to undertake a conference on Canada–US issues.
Level of Study: Postgraduate
Type: Grant
Value: Applicants may request funding up to $20,000; Applicants whose project focuses on the priority topics listed above and who can demonstrate matching funds from others sources may request funding up to US$20,000;
Frequency: Annual
Country of Study: United States of America
Application Procedure: Applicants must complete the online application form at http://www.international.gc.ca/studies-etudes/grantconf-subconf.aspx?view=d.
Closing Date: June 30th
Funding: Government
Contributor: Foreign Affairs Canada
Additional Information: Proposals must be submitted in English or French and budget figures must be in Canadian dollars.

Program Enhancement Grant

Subjects: International relations.
Purpose: To encourage innovative projects that promote awareness among students and the public about Canada–US relations.
Eligibility: Open to US colleges, research institutions and universities who wish to undertake professional academic activities.
Level of Study: Postgraduate
Type: Grant
Value: Up to US$18,000 per year
Length of Study: 1 year
Frequency: Annual
Country of Study: United States of America
Application Procedure: A completed online application form must be submitted.
Closing Date: June 15th

CANADIAN FEDERATION OF UNIVERSITY WOMEN (CFUW)

331 Cooper Street, Suite 502, Ottawa, ON, K2P 0G5, Canada
Tel: (1) 613 234 8252 extn 104
Fax: (1) 613 234 8221
Email: cfuwfls@rogers.com; fellowships@cfuw.org
Website: www.cfuw.org
Contact: Betty A Dunlop, CFUW Fellowships Program Manager

Found in 1919, the Canadian Federation of University Women (CFUW) is a voluntary, non-partisan, non-profit, self-funded bilingual organization of 10,000 women university graduates. CFUW members are active in public affairs, working to raise the social, economic, and legal status of women as well as to improve education, the environment, peace, justice and human rights.

Canadian Home Economics Association (CHEA) Fellowship

Subjects: Home economics.
Purpose: To provide funding for studying one or more aspects in the field of home economics, at the masters or doctoral level in Canada.
Eligibility: Open to female Canadian citizens. Candidates should hold a Bachelor's degree or its equivalent from a recognized university, not necessarily in Canada.
Level of Study: Postgraduate
Type: Fellowship
Value: Canadian $6,000
Frequency: Annual
Study Establishment: A recognized university
Country of Study: Canada
No. of awards offered: 1
Application Procedure: Applicants must complete an application form, available from the Federation website.
Closing Date: November 1st
Funding: Private
No. of awards given last year: 1
No. of applicants last year: 12
Additional Information: The fellowship is not renewable.

CFUW 1989 École Poytechnique Commemorative Awards

Subjects: All subjects. (The applicant must justify the relevance of her work to women).
Purpose: To provide funding for graduate studies in any field.
Eligibility: Open to female Canadian citizens or women who have permanent residence prior to the submission of an application. Candidates should hold a Bachelor's degree or its equivalent from a recognized university, not necessarily in Canada.
Level of Study: Postgraduate
Type: Fellowship
Value: Canadian $7,000 and $5,000
Frequency: Annual
Study Establishment: A recognized university
Country of Study: Any country
No. of awards offered: 2
Application Procedure: Applicants must complete an application form, available from the Federation website.
Closing Date: November 1st
Funding: Private
No. of awards given last year: 1
No. of applicants last year: 64
Additional Information: The fellowship is not renewable.

CFUW Beverley Jackson Fellowship

Subjects: All subjects.
Purpose: To provide funding for graduate studies at an Ontario university.
Eligibility: Open to female Canadian citizens. Candidates should hold a Bachelor's degree or its equivalent from a recognized university, not necessarily in Canada. The applicant must be over the age of 35.
Level of Study: Postgraduate
Type: Fellowship
Value: Canadian $2,000
Frequency: Annual
Study Establishment: A recognized university
Country of Study: Canada
No. of awards offered: 1
Application Procedure: Applicants must complete an application form, available from the Federation website.
Closing Date: November 1st
Funding: Private
Contributor: UWC North York funds
No. of awards given last year: 1
No. of applicants last year: 10
Additional Information: The fellowship is not renewable.

CFUW Bourse Georgette Lemoyne

Subjects: All subjects. The applicant must be studying in French and writer the Statement of Intent (Section I) of the application in French.
Purpose: To provide funding for graduate study in any field at a Canadian university.
Eligibility: Open to female Canadian citizens. Candidates should hold a Bachelor's degree or its equivalent from a recognized university, not necessarily in Canada.
Level of Study: Postgraduate
Type: Fellowship
Value: Canadian $5,000
Frequency: Annual
Study Establishment: A recognized university
Country of Study: Canada
No. of awards offered: 1
Application Procedure: Applicants must complete an application form, available from the Federation website.
Closing Date: November 1st
Funding: Private
No. of awards given last year: 1
No. of applicants last year: 33
Additional Information: The fellowship is not renewable.

CFUW Dr Alice E. Wilson Awards

Subjects: All subjects.
Purpose: Awarded to mature students returning to graduate studies in any field, with special consideration given to those returning to study after at least three years.
Eligibility: Open to female Canadian citizens or women who have permanent residence prior to the submission of an application. Candidates should hold a Bachelor's degree or its equivalent from a recognized university, not necessarily in Canada.
Level of Study: Postgraduate
Type: Fellowship
Value: Canadian $5,000
Frequency: Annual
Study Establishment: A recognized university
Country of Study: Any country
No. of awards offered: 4
Application Procedure: Applicants must complete an application form, available from the Federation website.
Closing Date: November 1st
Funding: Private
No. of awards given last year: 4
No. of applicants last year: 90
Additional Information: The fellowship is not renewable.

CFUW Dr Marion Elder Grant Fellowship

Subjects: All subjects.
Purpose: To provide funding for full-time courses of studies at any level of a doctoral program.
Eligibility: Open to female Canadian citizens. Candidates should hold a Bachelor's degree or its equivalent from a recognized university, not necessarily in Canada, and be a full-time student in her doctoral programme.
Level of Study: Doctorate
Type: Fellowship
Value: Canadian $11,500
Frequency: Annual
Study Establishment: A recognized university
Country of Study: Any country
No. of awards offered: 1
Application Procedure: Applicants must complete an application form, available from the Federation website.
Closing Date: November 1st
Funding: Private
Contributor: CFUW Wolfville funds
No. of awards given last year: 1
No. of applicants last year: 54
Additional Information: The fellowship is not renewable.

CFUW Elizabeth Massey Award

Subjects: Visual Arts (for example, painting or sculpture; or Music.
Purpose: To provide funding for postgraduate studies in the visual arts, such as painting or sculpture; or in music. The award is tenable in Canada or abroad.

Eligibility: Open to female Canadian citizens. Candidates should hold a Bachelor's degree or its equivalent from a recognized university, not necessarily in Canada.
Level of Study: Postgraduate
Type: Award
Value: Canadian $4,000
Frequency: Annual
Study Establishment: A recognized university
Country of Study: Any country
No. of awards offered: 1
Application Procedure: Candidates must complete an application form, available from the Federation website.
Closing Date: November 1st
Funding: Private
Contributor: The Massey Family Funds
No. of awards given last year: 1
No. of applicants last year: 9
Additional Information: The fellowship is not renewable.

CFUW Margaret Dale Philp Award

Subjects: Humanities or social sciences and Canadian history.
Purpose: To provide funding for graduate studies in the humanities or social sciences. Special consideration given to study in Canadian history only as a deciding factor, all these being equal.
Eligibility: Open to female Canadian citizens. Candidates should hold a Bachelor's degree or its equivalent from a recognized university, not necessarily in Canada.
Level of Study: Postgraduate
Type: Award
Value: Canadian $3,500
Frequency: Annual
Study Establishment: A recognized university
Country of Study: Canada
No. of awards offered: 1
Application Procedure: Candidates must complete an application form, available from the Federation website.
Closing Date: November 1st
Funding: Private
Contributor: CFUW Kitchener-Waterloo funds
No. of awards given last year: 1
No. of applicants last year: 13
Additional Information: The fellowship is not renewable.

CFUW Margaret McWilliams Predoctoral Fellowship

Subjects: All subjects.
Purpose: To provide funding for full-time doctoral study.
Eligibility: Open to female Canadian citizens or women who have permanent residence prior to the submission of an application. Candidates should hold a Bachelor's degree or its equivalent from a recognized university, not necessarily in Canada, and be a full-time student at an advanced stage, i.e. at least 1 year into her doctoral programme.
Level of Study: Doctorate
Type: Fellowship
Value: Canadian $11,000
Frequency: Annual
Study Establishment: A recognized university
Country of Study: Any country
No. of awards offered: 1
Application Procedure: Applicants must complete an application form, available from the Federation website.
Closing Date: November 1st
Funding: Private
No. of awards given last year: 1
No. of applicants last year: 102
Additional Information: The fellowship is not renewable. There is a new Toll-Free number in the US & Canada: 1-888-220-9606.

CFUW Memorial Fellowship

Subjects: Science, mathematics, or engineering.
Purpose: To provide funding for postgraduate degree studies in science, mathematics, or engineering.
Eligibility: Open to female Canadian citizens. Candidates should hold a Bachelor's degree or its equivalent from a recognized university, not necessarily in Canada.
Level of Study: Postgraduate, Masters
Type: Fellowship
Value: Canadian $8,000
Frequency: Annual
Study Establishment: A recognized university
Country of Study: Any country
No. of awards offered: 1
Application Procedure: Applicants must complete an application form, available from the Federation website.
Closing Date: November 1st
Funding: Private
No. of awards given last year: 1
No. of applicants last year: 34
Additional Information: The fellowship is not renewable.

Ruth Binnie Fellowship

Subjects: Home economics.
Purpose: To provide funding for Master's studies with a focus on one or more aspects of home economics.
Eligibility: Open to female Canadian citizens. Candidates should hold a Bachelor's degree or its equivalent from a recognized university, not necessarily in Canada.
Level of Study: Postgraduate, Masters
Type: Fellowship
Value: Canadian $6,000
Frequency: Annual
Study Establishment: A recognized university
Country of Study: Any country
No. of awards offered: 1
Application Procedure: Applicants must complete an application form, available from the Federation website.
Closing Date: November 1st
Funding: Private
No. of awards given last year: 1
No. of applicants last year: 17
Additional Information: The fellowship is not renewable.

CANADIAN HOSPITALITY FOUNDATION

300 Adelaide Street East, Suite 339, Toronto, ON, M5A 1N1, Canada
Tel: (1) 416 363 3401
Fax: (1) 416 363 3403
Email: chf@theohi.ca
Website: www.thechf.ca/contact

The Canadian Hospitality Foundation is Canada's largest industry driven source of scholarships for students pursuing careers in the foodservice/hospitality industry.

Canadian Hospitality Foundation Scholarship

Subjects: Hotel and food administration and hospitality and tourism management.
Eligibility: Open only to Canadian citizens or Landed Immigrants. Eligible to students in two and one year college programs.
Level of Study: Postgraduate
Type: Scholarship
Value: Canadian $1,000–3,500
Frequency: Annual
Study Establishment: University of Calgary, University of Guelph, Ryerson Polytechnical Institute or Mount Saint Vincent University
Country of Study: Canada
No. of awards offered: 7
Application Procedure: More information and application forms available online.
Closing Date: March 20th

CANADIAN INSTITUTE FOR ADVANCED LEGAL STUDIES

P.O.Box 43538, Leaside Post Office, 1601 Bayview Avenue, Toronto, ON, M4G 4G8, Canada
Tel: (1) 416 429 3292
Fax: (1) 416 429 9805
Email: info@canadian-institute.com
Website: www.canadian-institute.com
Contact: Mrs Lynn Morrison, Executive Secretary

The Canadian Institute for Advanced Legal Studies conducts legal seminars for judges and lawyers in Cambridge, England and Strasbourg, France.

The Right Honorable Paul Martin Sr. Scholarship

Subjects: Law.
Purpose: To study for an LLM at the University of Cambridge.
Eligibility: Open to graduates who have been awarded a law degree from a three or four year program at a faculty of law in a Canadian university in the four years before the candidate will commence his or her studies at the University of Cambridge (supported by The Right Honourable Paul Martin Sr. Scholarship). An applicant must be accepted into the University of Cambridge and a college of the University of Cambridge for graduate studies in law in order to receive this scholarship, although such acceptance need not be confirmed at the tiem of the application for the scholarship nor at the time that the Institute provides the candidate with notice that he or she has been selected to receive the scholarship.
Level of Study: Postgraduate
Type: Scholarship
Value: Canadian $23,000
Length of Study: 1 year
Frequency: Annual
Study Establishment: The University of Cambridge
Country of Study: England
No. of awards offered: 2
Application Procedure: Applications must include curriculum vitae; a personal statement indicating why the applicant wishes to undertake graduate studies in law at the University of Cambridge and why the applicant is suited to undertake such studies; a copy of transcripts for undergraduate and graduate studies, for studies in law and for a Bar Admissions Course, as applicable; and a maximum of three letters of references.
Closing Date: December 31st
Funding: Private
No. of awards given last year: 2
No. of applicants last year: 25
Additional Information: The scholarship may be held with another small award as approved by the Institute.

For further information contact:

Canadian Institute for Advanced Legal Studies, 1601 Bayview Avenue, Ontario, Toronto, M4G 4G8
Tel: 416 429 3292
Fax: 416 429 9805
Email: info@canadian-institute.com
Contact: Mr Randall J Hofley, Vice President

CANADIAN INSTITUTE FOR NORDIC STUDIES

200 Arts Building, University of Alberta, AB, Edmonton, T6G 2E6, Canada
Email: cins@ualberta.ca
Website: www.ualberta.ca/
Contact: Linda Nøstbakken, Chair of the Board, CINS

The Canadian Institute for Nordic Studies (CINS) was established in 1987 to promote multidisciplinary academic and cultural interest in the Nordic countries of Denmark, Finland, Iceland, Norway and Sweden, including the Faeroe Islands and Greenland.

CINS Graduate Scholarship

Subjects: Fine arts, humanities, natural, physical, applied and social sciences and more.
Purpose: To provide financial assistance to students who wish to pursue higher studies.
Eligibility: Open to Canadian citizens or landed immigrants, who have completed a Bachelor's degree from a Canadian university or college with high scholastic achievement. Applicants must be in residency at the Nordic destination for a minimum of 6 months and provide a written report to CINS no later than 6 months after completing the proposed programme of study.
Level of Study: Postgraduate
Type: Scholarship
Value: Canadian $5,000
Frequency: Annual
No. of awards offered: Usually 1 award per year
Application Procedure: Applicants should include the following in their application: contact details, citizenship status, social insurance number and date of birth, current academic status with formal transcripts, written acceptance from the host Nordic institution and reference letters.
Country of study - applicant must be a student in one of the Nordic countries: Denmark, Finland, Iceland, Norway, Sweden, the Faeroe Islands, and Greenland.
Closing Date: October 15th
No. of awards given last year: 2
Additional Information: Applicants can study at any recognized institution granting earned degrees at the postbaccalaureate level in the applicant's field of study and located in one of the Nordic countries: Denmark, Finland, Iceland, Norway, Sweden, the Faroe Islands and Greenland.

CANADIAN INSTITUTE OF GEOMATICS (CIG)

900 Dynes Road, Suite 100 D, Ottawa, ON, K2C 3L6, Canada
Tel: (1) 613 224 9851
Fax: (1) 613 224 9577
Email: exdircig@magma.ca
Website: www.cig-acsg.ca
Contact: David R Stafford, Executive Director

The Canadian Institute of Geomatics (CIG) was founded in 1882. CIG has evolved to be a non-profit, scientific and technical association and represents the largest and most influential geospatial knowledge network in Canada. Over 50 per cent of its members are senior managers and researchers in government and private sectors, academic and NGO organizations.

The Hans Klinkenberg Memorial Scholarship

Subjects: Geomatics sciences.
Purpose: The Hans Klinkenberg Memorial Scholarship Fund provides scholarships to students in the Geomatics sciences at technical institutes and community colleges in Canada.
Eligibility: Applicants must be a Canadian citizen or a landed immigrant.
Type: Scholarship
Value: Awards range from Canadian $500–2,000
Frequency: Annual
Country of Study: Canada
No. of awards offered: 2
Application Procedure: Application can be downloaded from www.cig-acsg.ca.
Closing Date: February 15th
Funding: Trusts
Contributor: Hans Klinkenberg Memorial Fund
No. of awards given last year: 2
No. of applicants last year: 5

CANADIAN INSTITUTE OF UKRAINIAN STUDIES (CIUS)

University of Alberta, 4-30 Pembina Hall, Edmonton, AB, T6G 2H8, Canada
Tel: (1) 780 492 2972
Fax: (1) 780 492 4967
Email: cius@ualberta.ca
Website: www.cius.ca
Contact: Ms Bohdan Klid, Assistant Director/Media Relations

The Canadian Institute of Ukrainian Studies (CIUS) is part of the University of Alberta under the jurisdiction of the University's Vice President of Research. It was founded in 1976 in order to provide an institutional home to develop Ukrainian scholarship and Ukrainian language education in Canada. It also supports such studies internationally, through organizing research and scholarship in Ukrainian and Ukrainian and Canadian studies, by publishing books and a scholarly journal, developing materials for Ukrainian language education largely for the bilingual school programme, and organizing conferences, lectures and a seminar series. Policy is developed by the director in consultation with CIUS units, programme directors and an advisory council.

CIUS Research Grants

Subjects: Ukrainian or Ukrainian and Canadian studies in history, literature, language, education, social sciences and library sciences.
Purpose: To fund research by scholars on Ukrainian or Ukrainian-Canadian topic in the humanities and social sciences.
Eligibility: Please write for details.
Level of Study: Postdoctorate, Research
Type: Grant
Value: Up to Canadian $8,000
Length of Study: 1 year
Frequency: Annual
Country of Study: Any country
No. of awards offered: 1
Application Procedure: Candidates may request an application form and guide either from the main address or by email, or download from them the website.
Closing Date: March 1st
Funding: Private
No. of awards given last year: 38

For further information contact:

Canadian Institute of Ukrainian Studies, 430 Pembina Hall, University of Alberta, Edmonton, Alberta, Canada, T6G 2H8
Tel: 780 492 2973
Fax: 780 492 4967

Helen Darcovich Memorial Doctoral Fellowship

Subjects: Ukrainian or Ukrainian and Canadian topics in education, history, law, humanities, social sciences, women's studies and library sciences.
Purpose: To aid students to complete a thesis on a Ukrainian or Ukrainian and Canadian topic in education, history, law, humanities, social sciences, women's studies or library sciences.
Eligibility: Open to qualified applicants of any nationality. For non-Canadian applicants, preference will be given to students enrolled at the University of Alberta.
Level of Study: Doctorate
Type: Fellowship
Value: Up to Canadian $12,500
Length of Study: 1 academic year
Frequency: Annual
Study Establishment: Any approved Institute of Higher Education
Country of Study: Any country
No. of awards offered: 1
Application Procedure: Applicants may write to the main address for application form and guide. Application forms can also be downloaded from the website or received by email.
Closing Date: March 1st
Funding: Private
Contributor: The Helen Darcovich Memorial Endowment Fund
No. of awards given last year: 3
Additional Information: Only in exceptional circumstances may an award be held concurrently with other awards.

For further information contact:

Canadian Institute of Ukrainian Studies, 430 Pembina Hall, University of Alberta, Edmonton AB, Canada, T6G 2H8
Tel: 780 492 2973
Fax: 780 492 4967

John Kolasky Memorial Fellowship

Subjects: Research in social sciences or humanities specializing in Ukrainian studies.
Purpose: To allow scholars from Ukraine to undertake research in Canada.
Eligibility: Limited to scholar from Ukraine.
Level of Study: Postdoctorate
Type: Fellowship
Value: Canadian $7,500–30,000
Length of Study: 3–12 months
Frequency: Annual
No. of awards offered: 1 or more
Closing Date: March 1st
Funding: Private
No. of awards given last year: Up to 3
No. of applicants last year: 30

Marusia and Michael Dorosh Master's Fellowship

Subjects: Ukrainian or Ukrainian and Canadian topic in education, history, law, humanities, social sciences, women's studies and library sciences.
Purpose: To aid a student to complete a thesis on a Ukrainian or Ukrainian and Canadian topic in education, history, law, humanities, social sciences, women's studies or library sciences.
Eligibility: Open to qualified applicants of any nationality. For non-Canadian applicants, preference will be given to students enrolled at the University of Alberta.
Level of Study: Graduate
Type: Fellowship
Value: Up to Canadian $10,000
Length of Study: 1 academic year
Frequency: Annual
Study Establishment: Any approved Institute of Higher Education
Country of Study: Any country
No. of awards offered: 1
Application Procedure: Applicants may write to the main address for application form. Information and application forms can also be obtained by email or downloaded from the website.
Closing Date: March 1st
Funding: Private
Contributor: The Marusia and Michael Dorosh Endowment Fund
No. of awards given last year: 2
Additional Information: Only in exceptional circumstances may an award be held concurrently with other major awards.

For further information contact:

Canadian Institute of Ukrainian Studies, 430 Pembina Hall, University of Alberta, Edmonton AB, Canada, T6G 2H8

Neporany Doctoral Fellowship

Subjects: Awarded to one or more doctoral students specializing on Ukraine in political science, economics and related fields (social sciences and political, economic, and social history).
Purpose: To fund research of doctoral students writing dissertation in Ukrainian studies.
Eligibility: Applicants must be a PhD student writing a PhD thesis on Ukrainian studies.
Level of Study: Doctorate
Type: Fellowship
Value: Ranges from Canadian $5,000–20,000
Length of Study: 1 academic year
Frequency: Annual
Country of Study: Any country
No. of awards offered: 1 or more
Application Procedure: Applicants must write for further details to the main address. Information can be obtained by email or downloaded from the website.
Closing Date: March 1st
Funding: Foundation, private
Contributor: The Osyp and Josaphat Neporany Educational Fund
No. of awards given last year: 1

For further information contact:

Canadian Institute of Ukrainian Studies, 430 Pembina Hall, University of Alberta, Edmonton AB, Canada, T6G 2H8

CANADIAN INSTITUTES OF HEALTH RESEARCH (CIHR)

160 Elgin Street, 9th Floor, Address Locator 4809A, Ottawa, ON, K1A 0W9, Canada
Tel: (1) 613 954 1968
Fax: (1) 613 954 1800
Email: info@cihr-irsc.gc.ca
Website: www.cihr-irsc.gc.ca
Contact: Ms Karen Spierkel, Communications & Marketing Director

CIHR Canadian Graduate Scholarships Doctoral Awards

Purpose: To provide special recognition and support to students who are pursuing a doctoral degree in a health-related field in Canada.
Eligibility: Candidates are expected to have an exceptionally high potential for future research achievement and productivity. Complete eligibility criteria are available on the CIHR website.
Level of Study: Postgraduate, Doctorate
Type: Award
Value: Canadian $17,500–35,000
Length of Study: Maximum of 3 years
Frequency: Annual
Study Establishment: A Canadian Institution
Country of Study: Canada
Application Procedure: Applicants must complete an application form in accordance with programme guidelines, available on the website.
Closing Date: Vary mid-September to early-October
Funding: Government
No. of awards given last year: 875
Additional Information: This funding program will be administered through the CIHR Doctoral Award competition, with the top candidates meeting eligibility criteria below receiving a CGS award.

CIHR Doctoral Research Awards

Subjects: General medical sciences and health sciences.
Purpose: To provide recognition and funding to students early in their academic research career, providing them with an opportunity to gain research experience. To provide a reliable supply of highly skilled and qualified researchers.
Eligibility: Open to Canadian citizens and permanent residents of Canada at the time of application, who are engaged in full-time research training in a graduate school. Please check the website for further details.
Level of Study: Doctorate, Graduate
Type: Award
Value: An annual stipend of Canadian $21,000 for awards held inside Canada and Canadian $26,000 for awards held outside Canada, and Canadian $1,000 annual research allowance. Awards are valued in Canadian dollars and are taxable.
Length of Study: Maximum of 3 years
Frequency: Annual
Study Establishment: Universities or research institutions
Country of Study: Canada and abroad
No. of awards offered: Varies
Application Procedure: Applicants must complete an application form in accordance with programme guidelines, available on the CIHR website.
Closing Date: October 15th
Funding: Government
Contributor: CIHR
No. of awards given last year: 389
No. of applicants last year: 930
Additional Information: Please consult the CIHR website for the complete programme description.

CIHR Fellowships Program

Subjects: Applicants must hold, or be completing, a PhD, health professional degree or equivalent. The health professional degree must be in a regulated health profession such as medicine, dentistry, pharmacy, optometry, veterinary medicine, chiropractic, nursing or rehabilitative science which requires at least a Bachelor's degree to be eligible for licensure in Canada.
Purpose: To provide support for highly qualified candidates at the post PhD or post health professional degree stages to add to their experience by engaging in health research either in Canada or abroad.
Eligibility: Candidates must hold or be completing a PhD or health professional degree. Candidates with more than 3 years of post-PhD training by the competition deadline are not eligible to apply. Candidates may not hold more than 3 years of federal to undertake post PhD studies. Please consult CIHR's website for full eligibility requirements. www.cihr-irsc.ge.cale/22340.html
Level of Study: Graduate, Research, Postdoctorate, Postgraduate, Professional development, Doctorate
Type: Fellowship
Value: The annual stipend level for those with a PhD degree (or equivalent) is $40,000 per year. The following stipend levels apply to health professionals who hold licensure (full or educational) in Canada at the time of taking up the award. The stipend level is dependent upon the number of years of research or clinical training completed since obtaining the health professional degree. Upon completion of two years of postgraduate research training, the awardees may be eligible to receive a stipend increase to the higher level. (Updated: 2008-09-03) Less than 2 years of research or clinical training experience: $40,000 Two or more years of research or clinical training experience: $50,000 The stipend for health professionals who do not hold licensure in Canada is $21,000 per year (i.e., equivalent to a Doctoral Research Award). Upon completion of two years of postgraduate research training the stipend may increase to $40,000 per year. For awards held outside Canada $5,000 is added to the annual stipend. Stipends are valued in Canadian dollars and are taxable.
Length of Study: 5 years maximum for health professionals intending to proceed to a PhD degree, 4 years maximum for health professionals who do not intend to proceed to a PhD degree, 3 years maximum for those with a PhD degree or a PhD and health professional degree
Frequency: Biannual
Study Establishment: Universities or research institutions
Country of Study: Any country
No. of awards offered: Varies
Application Procedure: Applicants must submit a training module, a curriculum vitae module for both the candidate and the supervisor(s), official transcripts of the candidate's graduate and/or professional training including proof of any degrees completed, proof of Canadian licensure, three assessments from persons under whom the candidate has studied and a letter of support from the proposed supervisor of foreign candidates and proof of residency/citizenship for Canadians wishing to hold their award outside of Canada.
Closing Date: February 1st and October 1st
Funding: Government
No. of awards given last year: 139
No. of applicants last year: 1,074
Additional Information: Consult the CIHR website, www.cihr-irsc.gc.ca for the full programme description.

CIHR MD/PhD Studentships

Subjects: General medical sciences.
Purpose: To promote promising students embarking on a combined MD or PhD programme at approved Canadian Universities.
Eligibility: Candidates for this Studentship Award must be enrolled in a combined MD/PhD programme at one of the approved Canadian institutions. Research supervisors should normally be holders of operating grants or salary funding obtained through a CIHR peer review process.
Level of Study: Graduate, Doctorate, Research
Type: Grant
Value: A stipend of Canadian $21,000 per year plus a yearly research allowance of $1,000 is provided
Length of Study: Maximum of 6 years
Frequency: Annual
Study Establishment: The universities of British Columbia, Calgary, Manitoba, McGill, Memorial, Montreal, Toronto, Western Ontario, Alberta
Country of Study: Canada
Application Procedure: Applicants must be nominated by the director of the MD/PhD programme at each institution.
Closing Date: November 2nd
Funding: Government
Contributor: CIHR
Additional Information: For further information please contact the CIHR or refer to the website.

CANADIAN LIBRARY ASSOCIATION (CLA)

Scholarships Jury, 1150 Morrison Drive,Suite 400, Ottawa, ON, K2H 8S9, Canada
Tel: (1) 613 232 9625 x 322
Fax: (1) 613 563 9895
Email: info@cla.ca
Website: www.cla.ca
Contact: Judy Green

The Canadian Library Association works to maintain a tradition of commitment to excellence in library education and to advance continuing research in the field of library and information science.

CLA Dafoe Scholarship

Subjects: Library and information studies.
Eligibility: Open to Canadian citizens and landed immigrants, commencing studies for their first professional library degree at an ALA-accredited institution.
Level of Study: Postgraduate
Type: Scholarship
Value: Canadian $5,000
Length of Study: 1 year
Frequency: Annual
Study Establishment: An accredited library school
Country of Study: United States of America or Canada
No. of awards offered: 1
Application Procedure: Applicants must complete an application form. Applicants must submit transcripts, references and proof of admission to a library school.
Closing Date: May 1st
Funding: Commercial
No. of awards given last year: 1

CLA H.W. Wilson Scholarship

Subjects: Library and information studies.
Purpose: To support students who wish to pursue higher studies in library and information studies.
Eligibility: Open to candidates who are commencing studies for their first professional library degree at an ALA-accredited institution. Should have Canadian citizenship or hold a landed immigrant status.
Level of Study: Postgraduate
Type: Scholarship
Value: Canadian $2,000
Length of Study: 1 year
Frequency: Annual
Country of Study: Canada
Application Procedure: Applicants are required to complete CLA Scholarship application forms and supply transcripts, reference and proof of admission to a library school. Scholarship applications are reviewed by a committee of members of the CLA.
Closing Date: May 1st
Funding: Foundation
No. of awards given last year: 1

CLA Library Research and Development Grants

Subjects: Library and information sciences.
Purpose: To support members of the Canadian Library Association for theoretical and applied research in the related fields.To encourage and support research undertaken by practitionares in the field of library and information services.To promote research in the field of library and information services by and/or about Canadians.
Eligibility: Open to personal members of the Canadian Library Association.
Level of Study: Postgraduate
Type: Grant
Value: Canadian $1,000
Frequency: Annual
Country of Study: Canada
No. of awards offered: 1
Application Procedure: Applicants must submit grant applications via emails and MS word document in either French or English containing contact details, description of the research project, duration of the project, detailed assessment of costs and statement of other grants/awards received. Proposals should be submitted via email.
Closing Date: February 28th

For further information contact:

c/o Canadian Library Association, 1150 Morrison Drive, Suite 400, Ottawa, ON, K2H 8S9
Tel: 613 232 9625 ext 322
Fax: 613 563 9895
Email: info@cla.ca
Contact: CLA Research & Development Grant

CANADIAN LIVER FOUNDATION

3100 Steeles Avenue East, Suite 801, Markham, ON, L3R 8T3, Canada
Tel: (1) 416 491 3353
Fax: (1) 905 752 1540
Email: clf@liver.ca
Website: www.liver.ca
Contact: National Director of Health Promotion and Patient Services

The Canadian Liver Foundation provides support for research and education into the causes, diagnosis, prevention and treatment of diseases of the liver.

Canadian Liver Foundation Graduate Studentships

Subjects: Hepatology, chemistry and biochemistry.
Purpose: To enable academically superior students to undertake full-time studies in a Canadian university in a discipline relevant to the objectives of the Foundation.
Eligibility: Candidates must be accepted into a full-time university graduate science programme in a medically related discipline related to a Master's or doctoral degree, and hold a record of superior academic performance in studies relevant to the proposed training.
Level of Study: Doctorate, Graduate, Postgraduate
Type: Studentship
Value: Canadian $20,000 per year
Length of Study: 2 years
Country of Study: Canada
No. of awards offered: Dependent on availability of funds
Application Procedure: Applicants must submit application forms along with supporting documents. Application forms can be obtained from the applicant's institution or from the Canadian Liver Foundation website.
Closing Date: April 2nd
Funding: Private
Additional Information: A student supported by the Foundation must not hold a current stipend award from another granting agency.

Canadian Liver Foundation Operating Grant

Subjects: Hepatology.
Purpose: To support research projects directed towards a defined objective.
Eligibility: Open to hepatobiliary research investigators who hold an academic appointment in a Canadian university or affiliated institution.
Level of Study: Research
Type: Grant
Value: Up to Canadian $60,000 per year
Length of Study: 2 years
Country of Study: Canada
No. of awards offered: 6
Application Procedure: Applicants must submit application forms along with supporting documentation. Application forms can be obtained from the applicant's institution or from the Canadian Liver Foundation website.
Closing Date: April 2nd
Funding: Private
Additional Information: Three operating grants may be awarded to researchers in Canada whose projects are related to hepatobiliary research; two operating grants may be awarded to researchers in the province of Alberta whose projects are related to hepatobiliary research; and one operating grant may be awarded to a researcher in Canada whose research project is related to liver cancer.

THE CANADIAN NATIONAL INSTITUTE FOR THE BLIND (CNIB)

1929 Bayview Avenue, Toronto, ON, M4G 3E8, Canada
Tel: (1) 800 563 2642
Fax: (1) 416 480 7700
Email: info@cnib.ca
Website: www.cnib.ca
Contact: Ms Barbara J. Marjeram, Corporate Secretary

CNIB is a nationwide, community-based, registered charity committed to public education, research and the vision health of all Canadians.

CNIB provides the services and support necessary to enjoy a good quality of life while living with vision loss. Founded in 1918, CNIB reaches out to communities across the country, offering access to rehabilitation training, innovative consumer products and peer support programs as well as one of the world's largest libraries for people with a print disability. CNIB supports research to advance knowledge in the field of vision health. Our research program funds projects that focus on ways to cure, treat and prevent eye disease, and improve the quality of life for people with vision loss.

CNIB Baker Applied Research Fund

Subjects: Research focused on the social, educational, and cultural needs of Canadians who are blind or visually impaired.
Purpose: To promote non-medical applied research that will enhance the life of the blind or visually impaired.
Eligibility: Open to residents of Canada enrolled in graduate study in Canada, and includes a co-applicant who is either a supervisor or mentor with an academic appointment in Canada, or a supervisory position at a healthcare facility. Refer to the website for complete details.
Level of Study: Research
Type: Grant
Value: Up to Canadian $40,000 plus travel and publications costs up to $2,000
Length of Study: One year
Frequency: Annual
Country of Study: Canada
No. of awards offered: Varies
Application Procedure: Applicants must apply online no later than January 15th of each year. The application will be reviewed by a multidisciplinary review committee, and decisions will be finalized by April 30th.
Closing Date: January 15th
Funding: Private
Additional Information: Please check this website in May/June for a further update.

For further information contact:

Tel: 416 486 2500 ext. 7622
Fax: 416 480 7059
Email: shampa.bose@cnib.ca
Contact: Shampa Bose, Executive Assistant and Research Coordinator

CNIB Baker Fellowship Fund

Subjects: Ophthalmology and optometry.
Purpose: CNIB's Baker Fellowships are awarded annually for postgraduate training in ophthalmic subspecialties.
Eligibility: Open to Canadians for research or study in Canada, or abroad if returning to practice in Canada, with priority given to university teaching.
Level of Study: Postgraduate, Professional development, Research
Type: Fellowship or Grant
Value: Up to Canadian $40,000
Length of Study: 1–2 years
Frequency: Annual
Country of Study: Any country
No. of awards offered: Varies
Application Procedure: Applicants must apply online no later than January 15th to be eligible for consideration. Fellowship funds are provided quarterly commencing July 1st of the following year. Successful applicants are normally advised of their award in April.
Closing Date: January 15th
Funding: Private
No. of awards given last year: 13
Additional Information: Please check this website in May/June for a further update.

For further information contact:

Tel: 416 486 2500 ext. 7622
Fax: 416 480 7059
Email: shampa.bose@cnib.ca
Contact: Shampa Bose, Executive Assistant and Research Coordinator

CNIB Baker New Researcher Fund

Purpose: To provide one-year grants to encourage new investigations that may lead to the prevention of vision loss. It is intended to benefit new investigators (within 5 years after an academic faculty appointment) by giving them experience and results which can assist them in further grant applications and pilot investigations.
Eligibility: Applicants must be residents of Canada and research must be conducted primarily in Canada.
Level of Study: Postdoctorate, Professional development, Research
Type: Grant
Value: Up to Canadian $40,000
Length of Study: 1 year
Frequency: Annual
Country of Study: Canada
No. of awards offered: Varies
Application Procedure: Apply online at www.cnib.ca/en/research/funding/eabaker-researcher.
Closing Date: January 15th
Funding: Private
Additional Information: Please check this website in May/June for a further update.

For further information contact:

Tel: 416 486 2500 ext. 7622
Fax: 416 480 7059
Email: shampa.bose@cnib.ca
Contact: Shampa Bose, Executive Assistant and Research Coordinator

CNIB Winston Gordon Award

Subjects: Product Development assistive technology for the blind.
Purpose: The award is given for significant advances in the field of technology benefiting people with vision loss.
Eligibility: The significant advances in, or application of, technology must have occurred within 10 years of nomination. The device or application must have a documented benefit to people who are blind or visually impaired. The award may be presented to an individual, group, or organization, including corporations and academic institutions.
Type: Award
Value: 24-carat gold medal
Country of Study: Any country
No. of awards offered: 1
Application Procedure: To submit a nomination, please write a letter to the Winston Gordon Committee nominating the individual or group for its products or services, and explaining how the nominee meets or surpasses the eligibility criteria and matches the goals of the award.
Closing Date: June 30th
Funding: Private

For further information contact:

Winston Gordon Award Committee, CNIB, 1929 Bayview Avenue, Toronto, ON, M4G 3E8, Canada
Fax: (1) 416 480 7000
Email: shampa.bose@cnib.ca
Contact: Shampa Bose, Grants and Awards Coordinator

The E. (Ben) & Mary Hochhausen Access Technology Research Award

Subjects: Research awards may be applied to: research projects, study at centers of excellence in Archnology, fellowships, development of prototypes and development costs of bringing important new products to market.
Purpose: To encourage research in the field of access technology for people living with vision loss.
Eligibility: Applications are accepted from any country in the world.
Level of Study: Research
Type: Research award
Value: Up to Canadian $10,000
Country of Study: International
No. of awards offered: 1
Application Procedure: Please check at http://www.cnib.ca/en/research/funding/hochhausen/ for more information.
Closing Date: September 30th
Funding: Private
No. of awards given last year: 1

For further information contact:

CNIB, 1929 Bayview Avenue, Toronto, ON, M4G 3E8, Canada
Tel: (1) 416 486 2500 ext 7622
Fax: (1) 416 480 7000
Email: shampa.bose@cnib.ca
Contact: Trustees, The Hochhausen Fund

Gretzky Scholarship Foundation for the Blind Youth of Canada

Subjects: The Gretzky family continue a tradition of assisting the blind youth of Canada to pursue their academic and lifelong dreams.
Purpose: To provide scholarships to eligible blind and visually impaired students planning to study at the post-secondary level.
Eligibility: All applicants must be blind or visually impaired, a graduate from secondary school entering their first year of post-secondary education, and a Canadian citizen.
Level of Study: Post secondary for blind or visually impaired students.
Type: Scholarship
Value: Canadian $3,000–5,000 each
Frequency: Annual
Country of Study: Canada
No. of awards offered: 20
Application Procedure: All documents requested in the application form must be included with your application. Please send application and documents to Kim Kohler.
Closing Date: May 31st
Funding: Private
No. of awards given last year: 23

For further information contact:

955256 Canning Road, R.R. 2, Paris, Ontario, N3L 3E2, Canada
Tel: (1) 519 458 8665
Fax: (1) 519 458 8609
Email: Kim.Kohler@cnib.ca
Contact: Kim Kohler, Walter and Wayne Gretzky Scholarship Foundation

Ross Purse Doctoral Fellowship

Subjects: The fellowship will be awarded for research in social sciences, engineering and other fields of study that are immediately relevant to the field of vision loss.
Purpose: To encourage and support theoretical and practical research and studies at the postgraduate or doctoral level in the field of vision loss in Canada.
Eligibility: Applications will be considered from persons studying at a Canadian University or college, or at a foreign University, where a commitment to work in the field of vision loss in Canada for at least 2 years can be demonstrated.
Level of Study: Doctorate, Postgraduate
Type: Fellowship
Value: Up to Canadian $12,500
Length of Study: 2 years
Frequency: Annual
Country of Study: Any country
No. of awards offered: 1
Application Procedure: Please send completed applications to Research Coordinator.
Closing Date: April 2nd
Funding: Private

For further information contact:

CNIB, 1929 Bayview Avenue, Toronto, ON, M4G 3E8, Canada
Tel: (1) 416 486 2500 ext. 7622
Fax: (1) 416 480 7000
Email: shampa.bose@cnib.ca
Contact: Shampa Bose, Research Coordinator, Grants, Awards & Scholarship Program

Tuck MacPhee Award

Subjects: Ophthalmology and optometry.
Purpose: To provide a 1-year grant in support of research in macular degeneration.
Eligibility: Open to all researchers, however applicants must be residents of Canada and research must be conducted primarily in Canada. Trainees (residents, fellows, post graduate students) are not eligible to apply.
Level of Study: Research
Type: Grant
Value: Up to Canadian $35,000
Length of Study: One year
Frequency: Annual
Country of Study: Canada
No. of awards offered: Varies
Application Procedure: Applicants must apply online no later than January 15th of each year. The application will be reviewed by a multi-disciplinary review committee, and decisions will be finalized by April 1st.
Closing Date: January 15th
Funding: Private
Additional Information: For any inquiries please contact Shampa Bose, Executive Assistant and Research Coordinator.Please check this website in May/June for a further update.

CANADIAN NURSES FOUNDATION (CNF)

50 Driveway, Ottawa, ON, K2P 1E2, Canada
Tel: (1) 613 237 2159
Fax: (1) 613 237 3520
Email: info@cnf-fiic.ca
Website: www.cnf-fiic.ca
Contact: CNF Scholarship Co-ordinator

To advance nursing knowledge and improve healthcare by providing research grants, awards, and scholarships to Canadian nurses and nursing students. We raise funds for our activities through diverse partnerships with responsible organizations and individuals who share our goals.

CNF Scholarships and Fellowships

Subjects: All nursing specialities. Several awards are identified for neurosurgery, oncology, community health nursing, epidemiology, gerontology, child or family healthcare, nursing administration, occupational health, dialysis nursing, home care nursing and aplastic anaemia.
Purpose: To assist Canadian nurses pursuing further education and research.
Eligibility: Open to Canadian nurses, or nurses with permanent Canadian resident status.
Level of Study: Doctorate, Graduate, Postgraduate, Predoctorate, Professional development, Research, Baccalaureate
Type: Scholarships, fellowships, bursaries
Value: Canadian $1,000–6,000
Length of Study: 1 year
Frequency: Annual
Country of Study: Canada
No. of awards offered: Varies
Application Procedure: Applicants must visit the website for the application forms, criteria and requirements at www.cnf-fiic.ca.
Closing Date: March 31st
Funding: Private
Contributor: Corporations, other foundations, individuals
No. of awards given last year: 61
No. of applicants last year: 277
Additional Information: Recipients must provide a progress report, and a final report at the end of the their academic year.

CANADIAN SOCIETY FOR CHEMICAL TECHNOLOGY

The Chemical Institute of Canada (CIC), 130 Slater Street, Suite 550, Ottawa, ON, K1P 6E2, Canada
Tel: (1) 613 232 6252 ext 223
Fax: (1) 613 232 5862
Email: awards@cheminst.ca, gthirlwall@cheminst.ca
Website: www.chem-tech.ca
Contact: Gale Thirlwall, Awards Manager

The Canadian Society for Chemical Technology is the national technical association of chemical and biochemical technicians and technologists with members across Canada who work in industry, government or academia. The purpose of the Society is the advancement of chemical technology, the maintenance and improvement of practitioners and educators and the continual evaluation of chemical technology in Canada. The Society hopes to maintain a dialogue with educators, government and industry, to assist in the technology content of the education process of technologists, to attract qualified people into the professions and the Society, to develop and maintain high standards and enhance the usefulness of chemical technology to both the industry and the public.

CIC Fellowships

Subjects: Chemistry, chemical engineering, chemical technology.
Purpose: A senior class of membership that recognizes the merits of CIC members who have made outstanding contributions.
Eligibility: Member of CIC, in good standing for at least 10 years. Candidates should have made contribution to the discipline in 4 years or one outstanding contribution in one area in particular. Details at www.chemist.ca/fellowship.
Frequency: Annual
No. of awards offered: Multiple
Application Procedure: Nomination form, Letters of support must be from a member of CIC for a minimum of 10 years.
Closing Date: October 1st

CANADIAN SOCIETY FOR CHEMICAL TECHNOLOGY

Suite 550, 130 Slater Street, Ottawa, ON, K1P 6E2, Canada
Tel: (1) 613 232 6252 ext 223
Fax: (1) 613 232 5862
Email: gthirlwall@cheminst.ca
Website: www.cheminst.ca
Contact: Gale Thirlwall, Awards Manager

Norman and Marion Bright Memorial Award

Subjects: Chemical technology.
Purpose: To reward an individual who has made an outstanding contribution in Canada to the furtherance of chemical technology.
Eligibility: Open to chemical sciences technologists or persons from outside the field who have made significant or noteworthy contribution to its advancement.
Type: Award
Value: A framed certificate, together with an honorarium of $500
Frequency: Annual
Country of Study: Canada
No. of awards offered: 1
Application Procedure: Applicants must complete a nomination form and submit alongwith it (1) a curriculum vitae, (2) a bio and citation and (3) letters of support.
Closing Date: December 1st
Funding: Trusts
Contributor: CIC Chemical Education Fund
Additional Information: Award winners are welcome to submit papers at either the Canadian Society for Chemistry or Canadian Society for Chemical Engineering.

CANADIAN SOCIETY FOR CHEMISTRY (CSC)

130 Slater Street, Suite 550, Ottawa, ON K1P 6E2, Canada
Tel: (1) 613 232 6252 ext 223
Fax: (1) 613 232 5862
Email: awards@cheminst.ca, gthirlwall@cheminst.ca
Website: www.chemistry.ca
Contact: Gale Thirlwall, Awards Manager

The Canadian Society for Chemistry (CSC), one of three constituent societies of The Chemical Institute of Canada, is the national scientific and educational society of chemists. The purpose of the CSC is to promote the practice and application of chemistry in Canada.

Award for Research Excellence in Materials Chemistry

Subjects: Chemistry.
Purpose: To recognize outstanding contribution to materials chemistry while working in Canada.
Eligibility: Candidates must be with 15 years of their first independent appointment.
Level of Study: Research
Type: Fellowship
Value: Up to $1,000 travel costs for award tour, framed scroll
Frequency: Annual
Application Procedure: Please check website for details.
Closing Date: July 2nd
Funding: Private

Boehringer Ingelheim Doctoral Research Award

Subjects: Organic chemistry.
Purpose: For a Canadian citizen or landed immigrant whose PhD thesis in the field of organic or bioorganic chemistry who was formally accepted by a Canadian university in the 12 month period preceding the nomination deadline and whose doctoral research is judged to be of outstanding quality.
Eligibility: Applicant's PhD thesis in the field of organic or bioorganic chemistry was formally accepted by a Canadian university in the 12-month period preceding the nomination deadline of July 2nd and whose doctoral research is judged to be of outstanding quality.
Level of Study: Postdoctorate, Postgraduate
Value: A framed scroll, Canadian $2,500 cash
Frequency: Annual
No. of awards offered: 1
Application Procedure: Curriculum vitae, nomination form, brief synopsis of PhD thesis and 2 letters of support.
Closing Date: July 2nd
Funding: Corporation
Contributor: Boehringer Ingelheim (Canada) Ltd.
No. of awards given last year: 1

CCUCC Chemistry Doctoral Award

Subjects: Chemistry.
Purpose: To recognize outstanding achievement and potential in research by a graduate student.
Eligibility: Open to graduate student whose PhD thesis in chemistry was formally accepted by a Canadian university in the 12-month period preceding the nomination deadline.
Level of Study: Postgraduate
Type: Award
Value: A framed scroll, $2,000 and one-year membership to the society
Frequency: Annual
Application Procedure: Applicants must submit one original and four copies of the nomination package to the awards manager. Applicants must visit website for more details on this.
Closing Date: September 15th
Funding: Private
Contributor: Canadian Council of University Chemistry Chairs
No. of awards given last year: 1

Ichikizaki Fund for Young Chemists

Subjects: Synthetic organic chemistry.
Purpose: To provide financial assistance to young chemists who are showing unique achievements in basic research by facilitating their participation in international conferences or symposia.
Eligibility: Open to members of the Canadian Society for Chemistry or the Chemical Society of Japan who have not passed their 34th birthday as of December 31st of the year in which the application is submitted, who have a research speciality in synthetic organic chemistry and are scheduled to attend an international conference or symposium directly related to synthetic organic chemistry within 1 year of submission.
Level of Study: Doctorate, Postdoctorate, Postgraduate, Professional development
Value: The maximum value of any one award is Canadian $10,000. Successful applicants may re-apply in subsequent years, provided the cumulative total of all awards does not exceed Canadian $15,000
Frequency: Annual
Study Establishment: Ichikizaki Fund
Country of Study: Any country

Application Procedure: Applicants must submit an application, including a curriculum vitae, copies of recent research papers, the title and brief description of the conference that the applicant wishes to attend, the title and abstract, if available, of the research paper that the applicant intends to present and a proposed budget. Applications from graduate students must be accompanied by a letter of reference from the research supervisor. Resume, research papers, description of conference attending, title and abstract and budget and letter of reference.
Closing Date: December 31st for conferences scheduled between March 1st and February 28th of the following year
Funding: Foundation
Additional Information: The number of applicants to be recommended by the Society is limited to 10 per year. Although the awards are intended primarily for established researchers, applications from postgraduate students and postdoctoral Fellows will be considered. However, only one application per year from a graduate student can be recommended to the Fund.

CANADIAN SPACE AGENCY

6767 Route de l'Aéroport, Saint-Hubert, QC, J3Y 8Y9, Canada
Tel: (1) 450 926 4800
Fax: (1) 450 926 4352
Email: dave.kendall@space.gc.ca
Website: www.asc-csa.gc.ca
Contact: David Kendall, Director General, Space Science

The Canadian Space Agency (CSA) was established in 1989 by the Canadian Space Agency Act. The agency operates like a government department.

Canadian Space Agency Supplements Postgraduate Scholarships

Subjects: Space science.
Purpose: To foster advanced studies in space science by offering a supplement to the regular National Science and Engineering Research Council (NSERC) postgraduate scholarships.
Eligibility: Open to graduate and permanent resident and citizen of Canada engaged in Masters or Doctoral studies in the natural sciences or engineering, or intend to pursue such studies in the following year, is successful in obtaining a NSERC postgraduate scholarship (PGS) or a Canada graduate scholarships (CGS-Master's).
Level of Study: Postgraduate
Type: Scholarship
Value: $7,500 per year for one year for master's students and up to two years for doctoral students
Length of Study: 2 years
Frequency: Annual
Country of Study: Canada
No. of awards offered: 6
Application Procedure: Candidates should apply to the NSERC postgraduate scholarship or Canada graduate scholarship programs by completing Form 200. After reviewing the forms, notification of award will be sent to the selected applicants.
Closing Date: May 1st
Funding: Government
Contributor: National Science and Engineering Research Council
Additional Information: A candidate who is in receipt of a scholarship from federal sources other than NSERC will not be eligible for this supplement.

CANADIAN THORACIC SOCIETY (CTS)

The Lung Association, National Office, 1750 Courtwood Cres, Suite 300, Ottawa, ON, K2C 2B5, Canada
Tel: (1) 613 569 6411 ext. 270
Fax: (1) 613 569 8860
Email: ctsinfo@lung.ca
Website: www.lung.ca/cts
Contact: Grants Management Officer

The Canadian Thoracic Society (CTS) is the medical section of the Canadian Lung Association. It advises the Association on scientific matters and programmes including policies regarding support for research and professional education. The CTS provides a forum whereby medical practitioners and investigators may join in the study of thoracic diseases and other medical fields that may come within the scope of the Lung Association. The CTS's objectives are to maintain the highest professional and scientific standards in all aspects of respiratory diseases, to collect, interpret and distribute scientific information, to encourage epidemiological, clinical and other scientific studies in the prevention, diagnosis and treatment of respiratory diseases and to stimulate and support undergraduate, postgraduate and continuing medical education in respiratory diseases.

CTS Research Fellowship Program

Subjects: Pulmonary disease.
Purpose: To support research training in pulmonary disease.
Eligibility: Applicants must be Canadian citizens or permanent residents of Canada. Candidates for the award must have obtained an MD or PhD degree or the equivalent and must not hold a university-level academic position. Those expected to receive a PhD degree within the following year are eligible to apply but may not begin the fellowship until the PhD requirements have been completed. CLA Fellows may not work on projects that have not been approved by the appropriate institutional ethics committees.
Level of Study: Postdoctorate, Postgraduate
Type: Fellowship
Length of Study: 2 years, with a possibility of renewal for a further year
Frequency: Annual
Country of Study: Canada
Application Procedure: Applicants must submit applications on CIHR forms.
Closing Date: January, check website for updates
Funding: Commercial, government
Contributor: The Canadian Lung Association, the Canadian Institutes of Health Research, Industry Partners, e.g. Glaxo Smithkline Inc. Merck Frosst Can, Bayer Inc, Boehringer Ingelheim and Astrazeneca
Additional Information: Recipients are selected based on priority ratings provided by the CIHR and are subject to the approval of the Canadian Thoracic Society and the Canadian Lung Association (CLA) Board of Directors. Applicants are screened to ensure proposed research areas are appropriate to the goals of the CLA. Fellowships are awarded in each case for research training in a specific institution, and may not be transferred without the explicit approval of both institutions involved.

For further information contact:

Canadian Institutes of Health Research, 410 Laurier Avenue West, 9th Floor, Address Locator 4209A, Ottawa, ON, KIA 0W9, Canada
Website: www.cihr.ca

CANADIAN WATER RESOURCES ASSOCIATION (CWRA)

CWRA Membership Services, 9 Corvus Court, Ottawa, ON, K2E 7Z4, Canada
Tel: (1) 613 237 9363
Fax: (1) 613 594 5190
Email: services@aic.ca
Website: www.cwra.org

The Canadian Water Resources Association (CWRA) is a national organization of individuals and organizations interested in the management of Canada's water resources. The membership is composed of private and public sector water resource professionals including managers, administrators, scientists, academics, students and users. CWRA has branch organizations in 9 provinces and members throughout Canada and beyond.

CWRA Dillon Consulting Scholarship/Ken Thomson Scholarship

Subjects: Applied, natural or social science aspects of water resources.
Purpose: These scholarship are available to graduate students whose programs of study focus upon applied, natural, or social science aspects of water resources.
Eligibility: Open to Canadian citizens or landed immigrants attending a Canadian university or college who are enrolled in full-time graduate studies in any discipline.

Level of Study: Postgraduate
Type: Scholarship
Value: Canadian $2,000(1) $5,000(1) $1,500(3)
Frequency: Annual
Country of Study: Canada
No. of awards offered: 5
Application Procedure: Candidates must submit a 500 word statement that outlines the applicant's research project and its relevance to sustainable water resources, transcripts, reference letters, a statement from the programme chairman or director endorsing the application form that programme along with the completed application form. Applications are available from the award office of any university.
Closing Date: February 13th
Funding: Commercial, foundation
Contributor: Dillon Consulting Ltd
No. of awards given last year: 4
Additional Information: All candidates will receive a 1 year membership in the CWRA.

THE CANCER COUNCIL N.S.W.

Research Strategy Unit, New South Wales Cancer Council, 153 Dowling Street, Woolloomooloo NSW 2011, PO Box 572, Kings Cross, NSW 1340, Australia
Tel: (61) 02 9334 1900
Fax: (61) 02 9326 9328
Email: rong@nswcc.org.au
Website: www.cancercouncil.com.au
Contact: Mr Ron Gale, Administrative Assistant

The Cancer Council NSW is one of the leading cancer charity organizations in New South Wales. Its mission is to defeat cancer and is working to build a cancer-smart community. In building a cancer-smart community, the Council undertakes high-quality research and is an advocate on cancer issues, providing information and services to the public and raising funds for cancer programmes.

The Cancer Council NSW Research Project Grants

Subjects: All aspects of cancer that elucidate its origin, cause and control at a fundamental and applied level. Grants are open to all research disciplines relevant to cancer including behavioural, biomedical, clinical, epidemiological, psychosocial and health services.
Purpose: To provide flexible support for cancer researchers.
Eligibility: Open to researchers working in NSW institutions
Level of Study: Unrestricted
Type: Project grant
Value: Generally a maximum of Australian $120,000 per year
Length of Study: Up to 3 years
Frequency: Annual
Study Establishment: An approved institution in New South Wales
Country of Study: Australia
No. of awards offered: Varies
Application Procedure: Applicants must complete an application form, available on request or from the website. Applications are submitted through the researcher's institution to NHMRC. Applicants must also complete a supplementary question form and a consumer review form.
Closing Date: March 19th
Funding: Private
Contributor: Community fund-raising
No. of awards given last year: 19 grants awarded in 2007
No. of applicants last year: 106

For further information contact:

NHMRC, GPO Box 9848, Canberra, ACT, 2601, Australia

THE CANCER COUNCIL SOUTH AUSTRALIA

202 Greenhill Road, Eastwood, SA, 5063, Australia
Tel: (61) 8 8291 4111
Fax: (61) 8 8291 4122
Email: cc@cancersa.org.au
Website: www.cancersa.org.au

The Cancer Council South Australia is a community-based charity independent of government control that has developed since 1928 with the support of South Australians. The Foundation's mission is to pursue the eradication of cancer through research and education on the prevention and early detection of cancer, thus enhancing the quality of life for people living with cancer.

PhD Scholarships

Subjects: Cancer research.
Purpose: To support cancer researchers in South Australia through the provision of research and senior research fellowships.
Eligibility: Applicant must be a student judged to be the best applicant from University of Adelaide, Flinders University or University of South Australia, who is commencing PhD studies. The applicant must not be currently enrolled in a PhD, must be eligible for the Research Training Scheme and must not have been previously enrolled for a Research Degree. Students are eligible to apply for the scholarship if they are enrolled in the Faculty or Division of Health Sciences at their institution and if their PhD topic is in an area of cancer research.
Level of Study: Postgraduate
Type: Scholarship
Value: Equivalent to the value of the stipend for an APA award
Length of Study: 3 years
Frequency: Annual
Country of Study: Australia
No. of awards offered: 1
Application Procedure: Applicants must contact the relevant Scholarships Offices of The University of Adelaide, University of South Australia and Flinders University for further information and closing dates.
Closing Date: See the website

For further information contact:

Tel: 08 8291 4297
Email: npolglase@cancersa.org.au
Contact: Nicole Polglase, Executive Assistant Research and Development

Research Project Grants

Subjects: Any scientific or medical field directly concerned with the cause, diagnosis, prevention and treatment of cancer.
Purpose: To assist postgraduate research workers undertaking research into cancer.
Eligibility: Open to postgraduate research workers who show promise of establishing themselves or to those who have already established themselves in the field of cancer research.
Level of Study: Postdoctorate
Type: Research grant
Value: Up to $100,000 inclusive of $25,000 maintenance
Length of Study: 1 year
Frequency: Annual
Study Establishment: An appropriate research organization
Country of Study: Australia
No. of awards offered: Approx. 20
Application Procedure: Visit our website for details www.cancersa.org.au
Closing Date: April 5th
Funding: Private
Contributor: South Australian community
No. of awards given last year: 19
No. of applicants last year: 48

THE CANCER RESEARCH SOCIETY, INC.

625 President-Kennedy Avenue, Suite 402, Montréal, QC, H3A 3S5, Canada
Tel: (1) 514 861 9227
Fax: (1) 514 861 9220
Email: grants@src-crs.ca
Website: www.cancerresearchsociety.ca
Contact: Mr Andy Chabot, Executive Director

The Cancer Research Society, founded in 1945, is a national organization that devotes its funds exclusively to research on cancer.

The Society is committed to funding basic cancer research or seed money for original ideas. The funds are allocated in the form of grants to universities and hospitals across Canada.

Cancer Research Society (Canada) Operating Grants

Subjects: Fundamental research on cancer.
Purpose: To provide support for new or continuing research activities by independent scientists or groups of investigators in the field of cancer in Canada.
Eligibility: Candidates must hold an academic position on the staff of a Canadian university.
Level of Study: Research
Type: Grant
Value: Canadian 60,000 yearly, for 2 years to cover the cost of research. No equipment or travel is permitted
Frequency: Annual
Study Establishment: Universities and their affiliated institutions
Country of Study: Canada
No. of awards offered: 30–50
Application Procedure: Applicants must visit the website for details of application procedures.
Closing Date: February 15th
Funding: Foundation, individuals, private
No. of awards given last year: 47
No. of applicants last year: 205
Additional Information: Due to the contribution of donors and partners, competitions addressing specific areas of cancer research are available. Recent examples include 'Restricted fund' cancer type.

GRePEC (Research and Prevention Group on Environment-Cancer)

Subjects: Fundamental research on environment-cancer.
Purpose: To develop research on the links between environment and cancer in Quebec, Canada.
Eligibility: Open to team researchers with academic appointment from Quebec University. This program encourage multidisciplinary and multi-institutional projects.
Level of Study: Research
Type: Grant
Value: Between $350,000 and $1,000,000 yearly for 3 years
Frequency: Annual
Study Establishment: Universities and their affiliated institutions
Country of Study: Canada
No. of awards offered: 1
Application Procedure: Applicants must visit the CRS website for details of application procedure.
Closing Date: July 11th
Funding: Government, individuals, private
No. of awards given last year: 1
No. of applicants last year: 5

CANCER RESEARCH UK CAMBRIDGE RESEARCH INSTITUTE

Angel Building, 407 St John Street, London, EC1V 4AD, United Kingdom
Tel: (44) 01223 404 209
Fax: (44) 01223 404 208
Email: publicaffairs@cancer.org.uk
Website: www.cambridgecancer.org.uk
Contact: Fellowship and Studentship Enquiries

The Institute aims to link the laboratory to the clinic with a multi-disciplinary approach to cancer-focussed research.

PhD Studentships

Subjects: Cancer research/oncology.
Purpose: To provide postgraduate research opportunities with a comprehensive programme of training and support.
Eligibility: Applications are invited from recent graduates or final year undergraduates who hold or expect to gain a first/upper second class honours degree or equivalent from any recognised university world-wide.
Level of Study: Doctorate
Type: Studentship
Value: Annual stipend for the duration of their three/four year research project. Stipend amounts are reviewed periodically to take into account inflation, cost of living, etc.
Length of Study: 3-4 years
Frequency: As available
Study Establishment: Cambridge Research Institute
Country of Study: United Kingdom
Application Procedure: Advertised in popular journals and on the Institute's website.
Closing Date: See the website
Funding: Foundation
Contributor: Cancer Research UK and University of Cambridge

Post-doctoral Fellowships

Subjects: Cancer research/oncology.
Purpose: To developing research scientists interested in shaping their career in a prestigious institute.
Eligibility: Open to all postdoctoral oncologists.
Level of Study: Postdoctorate
Type: Fellowship
Value: £29,650–34,189 per year inclusive, depending on experience
Length of Study: 2–5 years
Frequency: As available
Study Establishment: Cambridge Research Institute
Country of Study: United Kingdom
Application Procedure: See website for vacancies.
Closing Date: See the website
Funding: Foundation
Contributor: Cancer Research UK and University of Cambridge

CANON COLLINS TRUST

22 The Ivories, 6 Northampton Street, London, N1 2HY, England
Tel: (44) 020 7354 1462
Fax: (44) 020 7359 4875
Email: info@canoncollins.org.uk, victoria@canoncollins.org.uk
Website: www.canoncollins.org.uk
Contact: Victoria Reed, Scholarships Officer

Canon Collins Trust South African Scholarships Programme

Subjects: Any subject.
Purpose: To help build the human resources necessary for economic, social and cultural development in the southern African region and to develop an educated and skilled workforce that can benefit the wider community.
Eligibility: Applicants must have completed either a university degree or a three year post-matric diploma at a recognised training college or a B.Tech at a former technikon. Academic achievement, work experience, motivation, relevance of the course to southern Africa's needs, and commitment to community and country are all factors in selection. Open to students from students from South Africa, Namibia, Botswana, Swaziland, Lesotho, Zimbabwe, Zambia, Malawi, Angola and Mozambique.
Level of Study: Research, MBA
Type: Scholarships
Value: Maintenance and tuition fees, or partial grants
Length of Study: 1 year
Frequency: Annual
Application Procedure: Application form can be downloaded from website between April and August. Two copies of the application form for study in South Africa must be completed and returned by post, together with the required enclosures, in good time for August.
Closing Date: August

Canon Collins Trust UK Scholarships Programme

Subjects: Any subject.
Purpose: To help build the human resources necessary for economic, social and cultural development in the southern African region and to develop an educated and skilled workforce that can benefit the wider community.
Eligibility: Applicant must be commited to returning to southern Africa upon completion of their studies, and using the knowledge, training and skills acquired in the UK for the general benefit of their home community and country. Academic achievement, work experience,

motivation, relevance of the course to southern Africa's needs, and commitment to community and country are all factors in selection. Open to students from students from South Africa, Namibia, Botswana, Swaziland, Lesotho, Zimbabwe, Zambia, Malawi, Angola and Mozambique.
Level of Study: MBA, Research
Type: Scholarship
Value: Maintenance and tuition fees, or partial grants
Length of Study: 1 year
Frequency: Annual
Country of Study: United Kingdom
No. of awards offered: 20–30
Application Procedure: Application form can be downloaded from website between August and December, then sent to relevant office.
Closing Date: December 10th
No. of awards given last year: 21
No. of applicants last year: 450
Additional Information: Applications are assessed by a scholarships selection committee comprised of the Chief Executive, the Scholarships Programme Manager, trustees and academics.

Distance Learning MBA Programme in Partnership with Edinburgh Business School

Subjects: Management - distance learning.
Purpose: To support life long learning amongst communities where it will make a positive contribution towards their development, and give individuals valuable post-graduate education which will enhance their day to day work.
Eligibility: Open to disadvantaged individuals who have the motivation and capacity to undertake the Distance Learning MBA. Successful candidates will be already in work, or self-employed, and at a point in their life and career where this MBA will make maximum impact for themselves and their community, although they may otherwise have been unable to undertake it due to the lack of funds. The impact of the MBA on their lives and that of their community will be crucial in determining the outcome of their application.
Level of Study: MBA
Type: Scholarship
Value: Contact relevant office for details
Frequency: Annual
Study Establishment: Edinburgh Business School - distance learning MBA
Application Procedure: Application form can be downloaded from website.
Closing Date: March 31st
Additional Information: This scholarship is currently closed for applications. Please check for updates in March 2013.

The Graça Machel Scholarship Programme

Subjects: Health, Education, Science & Technology, Economics & Finance, Development
Purpose: To provide female students with scholarships to equip them to take up leadership roles for the benefit of their community, nation and region.
Eligibility: Open to female students from Lesotho, Swaziland, Malawi, Mozambique and South Africa. Applicant must have at least 2 years of relevant work experience.
Type: Scholarships
Value: Maintenance allowance, travel, health insurance and tution fees. Contact relevant office for details
Length of Study: 1–2 years
Frequency: Annual
No. of awards offered: 60
Application Procedure: Send enquiries to jean@canoncollins.org.uk.
Closing Date: For UK: December 10th; For South Africa: August 11th

THE CANON FOUNDATION IN EUROPE

Bovenkerkerweg 59-61, 1185 XB Amstelveen, Netherlands
Tel: (31) 20 5458934
Fax: (31) 20 7128934
Email: foundation@canon-europe.com
Website: www.canonfoundation.org
Contact: Mrs Suzy Cohen, Secretary

The Canon Foundation is a non-profit, grant-making philanthropic organization founded to promote, develop and spread science, knowledge and understanding, in particular, between Europe and Japan.

Canon Foundation Research Fellowships

Subjects: All subjects.
Purpose: To contribute to scientific knowledge and international understanding, in particular between Europe and Japan.
Eligibility: Open to Japanese and European nationals only.
Level of Study: Doctorate, Postdoctorate, Postgraduate, Research
Type: Fellowship
Value: A maximum of award of €27,500
Length of Study: Up to 1 year
Frequency: Annual
No. of awards offered: 10–15
Application Procedure: Applicants must complete an application form, which is to be submitted with two reference letters, a curriculum vitae, a list of papers, two photographs and copies of certificates of higher education.
Closing Date: September 15th
Funding: Corporation, private
Contributor: Canon Europa NV
No. of awards given last year: 12
No. of applicants last year: 176

CANTERBURY HISTORICAL ASSOCIATION

c/o History Department, University of Canterbury, Private Bag 4800, Christchurch, New Zealand
Tel: (64) 3 364 2104
Fax: (64) 3 364 2003
Email: david.monger@canterbury.ac.nz
Website: www.hums.canterbury.ac.nz/hist/
Contact: Dr David Monger, Secretary

The Canterbury Historical Association (founded 1922, but in recess between 1940 and 1953) aims to foster public interest in all fields of history by holding meetings for the discussion of historical issues, and to promote historical research and writing through its administration of the J M Sherrard Award in New Zealand local and regional history.

J M Sherrard Award

Subjects: New Zealand regional and local history writing.
Purpose: To foster high standards of scholarship in New Zealand regional and local history.
Eligibility: Open to qualified applicants from New Zealand only. Major awards are normally restricted to substantial monograph length publications that meet scholarly standards. Small-scale works, biographies and family histories are not eligible.
Level of Study: Unrestricted
Type: Prize
Value: New Zealand $1,000
Country of Study: New Zealand
No. of awards offered: Varies
Application Procedure: No application is required, as judges assess all potential titles appearing in the New Zealand National Bibliography.
Funding: Private
No. of awards given last year: Not Awarded
Additional Information: The prize money is often divided among two or three finalists. A commendation list is also published.

CARDIFF UNIVERSITY

Student Recruitment and Web Division, Cardiff University, Deri House, 2-4 Park Grove, Wales, Cardiff, CF10 3PA, United Kingdom
Tel: (44) 29 2087 0084
Fax: (44) 29 2087 0085
Email: graduate@cardiff.ac.uk
Website: www.cardiff.ac.uk/postgraduate

Cardiff University is recognized in independent government assessments as one of the United Kingdom's leading teaching and research universities. Founded by Royal Charter in 1883, the University today combines impressive modern facilities and a dynamic approach to teaching and research with its proud heritage of service and achievement. Having gained national and international standing, Cardiff University's vision is to be a world-leading university and it's mission is to pursue research, learning and teaching of international distinction and impact.

Cardiff University Postgraduate Studentships

Subjects: All subjects.
Purpose: To support postgraduate study and training.
Eligibility: Usually, applicants require a First or Upper Second Class Honours Degree, although there are a few exceptions. Most studentships are available to United Kingdom and European Union students only, although in some subjects non-European Union students are also considered for school awards. Specific awards may have further eligibility criteria.
Level of Study: Postgraduate
Type: Studentships and bursaries
Value: Varies
Length of Study: Generally 1 year for Master's schemes and 3 years for PhD studentships
Frequency: Annual
Study Establishment: Cardiff University
Country of Study: United Kingdom
No. of awards offered: Approx. 150 per year
Application Procedure: Applicants will need to have received an offer of a place to study before they can apply for financial support. For further information please visit the website www.cardiff.ac.uk/postgraduate/pgfunding. Applicants can also contact the school in which they are interested in studying.
Closing Date: Variable, but usually around May or June each year. Please see the online listings for details
Additional Information: Cardiff University also has a strong track record of obtaining funding from the UK Research Councils to support postgraduate study. Subject-specific enquiries should be directed to the relevant schools. For more general enquiries, please contact the Postgraduate Recruitment Office by email.

CARNEGIE MELLON UNIVERSITY

Torrens Building, 220 Victoria Square,, Pittsburgh, PA, Adelaide, 5000, Australia
Tel: (61) 08 8110 9900
Fax: (61) 08 8211 9444
Email: admissions@cmu.edu.au
Website: www.heinz.cmu.edu.au

Carnegie Mellon is home to the world's leading experts in a range of fields. From computing to the arts to the environment to biotechnology, the students, faculty and staff of the University are shaping the future with a strong focus on finding practical answers to complex problems. These scholarships are available to full-time students.

Carnegie Mellon University–Aus AID Scholarships

Subjects: Public policy, management and information technology.
Purpose: To support a limited number of students from countries where Australia has a bilateral aid program, to undertake a Master's Degree at the H. John Heinz III College at Carnegie Mellon University's campus in Adelaide, Australia.
Eligibility: Open to students who have already been offered a place at Carnegie Mellon University in either of the two courses available under this program.
Type: Scholarship
Value: Full tuition fee, return economy airfares, a contribution to living expenses, and basic medical insurance
Length of Study: 1 year
Frequency: Annual
Study Establishment: Carnegie Mellon University
Country of Study: Australia
No. of awards offered: Varies
Application Procedure: Check website for further details.
Closing Date: See the website
Funding: Government
Contributor: Government of Australia

THE CARNEGIE TRUST FOR THE UNIVERSITIES OF SCOTLAND

Andrew Carnegie House, Pittencrieff Street, Dunfermline, Fife, KY12 8AW, Scotland
Tel: (44) 1383 724990
Fax: (44) 1383 749799
Email: jgray@carnegie-trust.org
Website: www.carnegie-trust.org
Contact: Ms Jackie Gray, Assistant Secretary

The Carnegie Trust for the Universities of Scotland, founded in 1901, is one of the many philanthropic agencies established by Andrew Carnegie. The trust aims to offer assistance to students, to aid the expansion of the Scottish universities and to stimulate research. See also the entry for the Caledonian Scholarships.

Caledonian Scholarship

Subjects: All subjects.
Purpose: To support postgraduate research in any subject.
Eligibility: Open to persons possessing a First Class (Honours) Degree from a Scottish university. Scholarship to be held at an institution in Scotland.
Level of Study: Doctorate, Postgraduate
Type: Scholarship
Value: UK£15,100 plus tuition fees and allowances
Length of Study: Up to 3 years subject to annual renewal
Frequency: Annual
Study Establishment: Any University
Country of Study: Scotland
No. of awards offered: 1 or 2
Application Procedure: Application forms available on website or from trust office.
Closing Date: March 15th
No. of awards given last year: 2
No. of applicants last year: 171
Additional Information: This award is considered along with Carnegie Scholarships.

Carnegie Research Grants

Subjects: All subjects in the universities' curriculum.
Purpose: To support personal research projects or aid in the publication of books likely to benefit the universities of Scotland.
Eligibility: Open to full-time members of staff of Scottish universities and in exceptional cases to graduates of Scottish universities.
Level of Study: Postdoctorate, Postgraduate, Professional development, Research, Only exceptionally
Type: Grant
Value: Varies according to requests but the maximum is UK£2,200
Length of Study: Up to 3 months
Frequency: 3 meetings per year
Country of Study: Any country
No. of awards offered: Varies
Application Procedure: Applicants must complete an application form, available from the Trust office or on the Trust's website.
Closing Date: January 15th, May 15th or October 15th prior to Executive Committee meetings in February, June and November
Funding: Private
No. of awards given last year: 248
No. of applicants last year: 278

Carnegie Scholarships

Subjects: All subjects in the university's curriculum.
Purpose: To provide 3 years financial support (fees plus stipend) for completion of PhD.
Eligibility: Open to candidates possessing a First Class (Honours) Degree from a Scottish university.
Level of Study: Doctorate
Type: Scholarship
Value: UK£15,200 per year (fees plus stipend)
Length of Study: Up to 3 years
Frequency: Annual
Study Establishment: Scottish University
Country of Study: Scotland

No. of awards offered: 13
Application Procedure: Applicants must be nominated by a senior member of staff at a Scottish university and an application form completed, available from the Trust office or on Trust's website.
Closing Date: March 15th
Funding: Private, trusts
No. of awards given last year: 16
No. of applicants last year: 119
Additional Information: Scholarship is to support postgraduate research leading to a PhD. Applications for 1 year postgraduate courses are not eligible.

Carnegie-Cameron Taught Post-Graduate Bursaries

Subjects: All subjects in the university's curriculum.
Purpose: To qualified and deserving, industrious and ambitious candidate, who would derive particular benefit from a one-year, taught, postgraduate degree which he or she would be unlikely to enjoy without the award.
Eligibility: Applicants must be Scottish by birth, descent (at least one parent born in Scotland) or have been continuously resident in Scotland for a period of at least 3 years for the purpose of secondary or tertiary education in Scotland. Vacations, periods of absence through illness and periods spent outside Scotland as part of a Scottish educational course shall not be taken into account in determining whether the residence has been continuous.
Level of Study: Postgraduate
Type: Bursary
Value: Payment of the tuition fees
Length of Study: 1-year full-time or 2-year part-time, taught, postgraduate degree course at the awarding university
Frequency: Annual
Study Establishment: Any Scottish university
Country of Study: Scotland
Application Procedure: A completed application form, together with a copy of your degree transcript, two academic references and proof of either Scottish birth, descent or residency should be provided. Applications are distributed by the Scottish Universities and should be returned to the university contact.
Closing Date: Please refer to the website
Funding: Private, trusts
No. of awards given last year: 65
No. of applicants last year: 425
Additional Information: The bursaries are awarded directly by the universities and candidates wishing to be considered for these bursaries should make application to the university where they wish to study and not to the Trust.

Henry Dryerre Scholarship

Subjects: Medical and veterinary physiology.
Purpose: To support postgraduate research.
Eligibility: Open to European citizens holding a First Class (Honours) Degree in Physiology.
Level of Study: Doctorate, Postgraduate, Predoctorate
Type: Scholarship
Value: Tuition fees, research costs and travel expenses and a maintenance grant of UK£15,100
Length of Study: 3 years full-time research
Frequency: Every 3 years
Study Establishment: A Scottish institution
Country of Study: Scotland
No. of awards offered: 1
Application Procedure: Applicants must be nominated by a professor, reader, or lecturer at a Scottish university.
Closing Date: March 15th
Additional Information: The scholarships are administered by the Carnegie Trust for the Universities of Scotland on behalf of The Royal Society of Edinburgh.

Largen Research Grants

Subjects: All subjects.
Purpose: Support for research collaboration between Scottish Universities.
Eligibility: Academic staff at Scottish Universities.
Level of Study: Research
Type: Grant
Value: £40,000
Length of Study: 6 months to 1 year
Frequency: Annual
Study Establishment: Any Scottish University
Country of Study: Any country
No. of awards offered: 7–10 years
Application Procedure: Applicants must fill application form.
Closing Date: February 1st
Funding: Trusts
No. of awards given last year: 7
No. of applicants last year: 25

St. Andrew's Society of New York Scholarship

Subjects: All subjects.
Purpose: Financial support towards postgraduate study in New York and surrounding area.
Eligibility: Students who are Scottish by birth and graduates of a Scottish University at Oxford or Cambridge.
Level of Study: Postgraduate
Type: Scholarship
Value: $20,000
Length of Study: 1 year
Frequency: Annual
Study Establishment: Any University in New York and surrounding area
Country of Study: United States of America
No. of awards offered: 2
Application Procedure: Applicants must fill application form.
Closing Date: January 25th
Funding: Trusts
No. of awards given last year: 2
No. of applicants last year: 10
Additional Information: Please check website for more details.

CATHOLIC ACADEMIC EXCHANGE SERVICE (KAAD)

Katholischer Akademischer, Ausländer-Dienst, Hausdorffstrasse 151, Bonn, 53129, Germany
Tel: (49) 228 91758 0
Fax: (49) 228 91758 58
Email: zentrale@kaad.de
Website: www.kaad.de

Catholic Academic Exchange Service (KAAD) provides financial and civic educational support as well as pastoral assistance for high-potential post-graduate scholars from Africa, Asia, Latin America, the Middle East and Eastern Europe. The KAAD has been registered as a charity of the German Catholic Church since 1958-today it is the largest Catholic Organization offering scholarships in the area of International educational collaboration in the world.

Research Scholarships: Programme I

Subjects: All subjects.
Purpose: To support candidates from developing nations, who are still in their home countries, for doctoral and postdoctoral research.
Eligibility: Applicants must be young academics from Asia, Africa, Latin America, Near and Middle East or Eastern Europe, with a commitment to return to their home country upon completion of their research stay.
Level of Study: Postgraduate, Doctorate, MBA, Postdoctorate, Research
Type: Scholarship
Value: In accordance with KAAD scholarship guidelines
Length of Study: 1 year (extendable up to 3 years)
Frequency: Annual
Study Establishment: A German university
Country of Study: Germany
Application Procedure: Application forms are available on request. Applications can be submitted to the KAAD partner organizations in the home country, which in turn will propose the applicants to the KAAD.
Closing Date: January 15th and June 15th
Contributor: Catholic Academic Exchange Service (KAAD)

Research Scholarships: Programme II

Subjects: All subjects.
Purpose: To support candidates from developing nations, who are already in Germany and are in an advanced stage of their research and whose research is not yet promoted by KAAD.
Eligibility: Applicants must be young academics from Asia, Africa, Latin America, Near and Middle East or Eastern Europe, with a commitment to return to their home country upon completion of their research stay.
Level of Study: Doctorate, MBA, Postdoctorate, Postgraduate
Type: Scholarship
Value: In accordance with KAAD scholarship guidelines
Length of Study: 1 year (extendable up to 3 years)
Frequency: Annual
Study Establishment: A German university
Country of Study: Germany
Application Procedure: Application forms are available on request. Applications can be submitted to the KAAD partner organizations in Germany, which in turn will propose the applicants to the KAAD.
Closing Date: January 15th and June 15th
Contributor: Catholic Academic Exchange Service (KAAD)

CATHOLIC LIBRARY ASSOCIATION (CLA)

205 West Monroe Street, Suite 314, Chicago, IL, 60606 5061, United States of America
Tel: (1) 855 739 1776, 312 739 1776
Fax: (1) 312 739 1778
Email: cla@cathla.org
Website: www.cathla.org

The Catholic Library Association (CLA) represents all segments of the library community. Members strive to initiate, foster and encourage any activity or library programme that will promote literature and libraries, not only of a Catholic nature, but also of an ecumenical spirit.

Rev Andrew L Bouwhuis Memorial Scholarship

Subjects: Library science.
Purpose: To encourage promising and talented individuals to enter librarianship and to foster advanced study in the library profession.
Eligibility: Open to individuals who have been accepted into a graduate school programme, show promise of success based on collegiate record and who demonstrate the need for financial aid.
Level of Study: Graduate, Postgraduate
Type: Scholarship
Value: US$1,500
Frequency: Annual
Country of Study: United States of America
No. of awards offered: 1
Application Procedure: Applicants must complete an application form, available at www.cathla.org or on request. Please send a stamped addressed envelope.
Closing Date: February 1st
Funding: Private
No. of awards given last year: 1

For further information contact:

Scholarship Committee, Catholic Library Association, 205 West Monroe Street, Suite 314, Chicago, IL, 60606 5061

Sister Sally Daly – Junior Library Guild Grant

Subjects: Library science.
Purpose: To enable a new CLA/CLSS member to attend the Association's annual convention.
Eligibility: Only new members of the Catholic Library Association (CLA)/Children's Library Services Section (CLSS) are eligible.
Type: Grant
Value: US$1,500
Frequency: Annual
No. of awards offered: Varies
Application Procedure: Applications may be obtained by writing to the Association or downloaded from the Association's website www.cathla.org.
Closing Date: February 1st
Contributor: Junior Library Guild
No. of awards given last year: 1
No. of applicants last year: 2

For further information contact:

Sister Sally Daly Memorial Grant, Catholic Library Association, 205 West Monroe Street, Suite 314, Chicago, IL, 60606 5061

CDS INTERNATIONAL, INC.

440 Park Avenue South, New York, NY, 10016, United States of America
Tel: (1) 212 497 3500
Fax: (1) 212 497 3535
Email: info@culturalvistas.org
Website: www.cdsintl.org

CDS International, Inc. is a non-profit organization that administers work exchange programmes. CDS International's goal is to further the international exchange of knowledge and technological skills, and to contribute to the development of a pool of highly trained and interculturally experienced business, academic and government leaders.

Alfa Fellowship Program

Subjects: Business, economics, journalism, law and public policy.
Purpose: To expand networks of American, British and Russian professionals and develop greater intercultural understanding and advancing US/Russian and British/Russian relations.
Eligibility: US or British citizen; 25–35 years old at the application deadline; Russian proficiency is preferred; qualified candidates with fluency in a second language may be considered; graduate-level degree or equivalent training in business, economics, journalism, law, public policy or government; a least 2 years of relevant work experience.
Level of Study: Graduate
Type: Fellowship
Value: Travel, free housing, monthly stipends, and insurance
Frequency: Annual
Application Procedure: Applicants must submit an application form.
Closing Date: December 1st
Funding: Private

Baden-Württemberg Stipendium "Work Immersion Study Program" (WISP)

Subjects: All subjects.
Purpose: To support students to gain practical work experience in their career field, improve their German language skills and experience German culture firsthand.
Eligibility: Candidates for WISP must meet the following eligibility requirements:
US citizen or permanent resident; 18–27 years of age; one semester of German instruction by program start; enrolled in an associate degree program at a community or technical college at the time of application; minimum of one year of study toward associate degree completed by program start; prior experience in target internship field through a summer or part-time job, volunteer position, or prior internship.
Value: Monthly stipend of €300. See http://www.cdsintl.org/fellowshipsabroad/wisp.php for more details
Length of Study: 3 months
Frequency: Annual
Application Procedure: Applicants must submit an online application form.
Closing Date: December 1st
Funding: Private

Congress Bundestag Youth Exchange for Young Professionals

Subjects: Business, technical, computer science, social and service fields.
Purpose: To foster the exchange of knowledge and culture between German and American youth, while providing career-enhancing theoretical and practical work experience.

Eligibility: Open to citizens of the United States of America and permanent residents aged 18–24 years who have well-defined career goals and related part or full-time work experience. Applicants must be able to communicate and work well with others, have maturity enabling them to adapt to new situations, an intellectual curiosity and a sense of diplomacy.
Level of Study: Professional development
Type: Scholarship
Value: Full-year scholarship including international airfare and partial domestic transportation, language training and study at a German professional school, seminars, including transportation and insurance, host family stay
Length of Study: 1 year: 2-month language; 4-month study; 5-month internship
Frequency: Annual
Study Establishment: A field-specific postsecondary professional school
Country of Study: Germany
No. of awards offered: 75
Application Procedure: Apply online at www.cdsintl.org/cbyx
Closing Date: December 1st
Funding: Government
Contributor: US Congress and German Bundestag
No. of awards given last year: 75
No. of applicants last year: 250–350
Additional Information: Participants must have US$300–350 pocket money per month. During the year of the award, American exchange students will have the opportunity to improve their skills through formal study and work experience. The programme also includes intensive language instruction and housing with a host family or in a dormitory.

Émigré Memorial German Internship Program

Subjects: EMGIP is ideal for students planning on pursuing careers at a regional level of government in the US or Canada, or who have an interest in a specific policy issue such as the environment, education and/or healthcare.
Purpose: To support students to gain government work experience, improve their advanced German language skills and learn about German culture firsthand.
Eligibility: Candidates for EMGIP must meet the following eligibility requirements:US or Canadian citizen; undergraduate and graduate students enrolled at accredited American or Canadian colleges and universities before, during and after the program may apply. US citizens who have graduated are also eligible, so long as their internships begin within three months of graduation; 18–30 years of age; high-intermediate German skills (oral and written). Candidates must be able and willing to communicate in German and possess a good command of professional vocabulary in their field; minimum of two years of university level studies in a field related to one of the following: international relations, public administration, political science, law, economics, european studies with an emphasis on Germany, German or German Studies, with a minor in one of the fields listed here; some relevant work experience (e.g. internship, volunteer work, summer job). It is important that a candidate knows how to adjust to a professional environment and how to use theoretical skills in the workplace.
Level of Study: Graduate, Unrestricted
Type: Internship
Value: Monthly stipends to ensure a total monthly salary of €670 for US citizens (Please note: monthly stipends for Canadian citizens cannot be guaranteed—this is dependent on the Landtag). This will cover basic expenses such as housing, local transportation and food
Length of Study: 1–3 months
Frequency: Thrice a year
Application Procedure: Applicants must submit an online application form.
Closing Date: October 1st for Spring; December 1st for early Summer; March 31st for Fall
Funding: Private

Robert Bosch Foundation Fellowship Program

Subjects: Business administration, journalism, law, public policy and closely related fields.
Purpose: To support young Americans the opportunity to complete a high-level professional development program in Germany.
Eligibility: Candidates for the Robert Bosch Foundation Fellowship Program must meet the following requirements:
US citizen; 23–34 years old at the application deadline; at least two years of relevant work experience; graduate degree or equivalent training in business administration, journalism, law, public policy, international relations or a closely related field; evidence of outstanding professional performance and community involvement; no German language skills are required at time of application; however, the willingness and commitment to participate in language training based on the results of an evaluation at the selection meeting is essential; most Bosch fellows are required to complete four months of private tutoring in the US (up to 8 hours per week) and 3 months of intensive language training in Berlin prior to the start of the program. All language training is funded by Robert Bosch Stiftung.
Level of Study: Unrestricted, Graduate
Type: Fellowship
Value: €2,000 per month stipend. See http://www.cdsintl.org/fellowshipsabroad/bosch.php for more details
Length of Study: 9 months
Frequency: Annual
No. of awards offered: 20
Application Procedure: Application form and supporting documents.
Closing Date: October 15th
Funding: Private

Transatlantic Renewable Energy Fellowship

Subjects: Environmental and energy fields.
Purpose: To build an international network of future leaders in renewable energy and environmental fields as well as to increase transatlantic cooperation on climate and energy issues.
Eligibility: Candidates must be enrolled in a US university or be a US citizen; must be 32 years old or younger at the application deadline; at least 3 years of study in one of the following fields: technical fields (electrical, industrial and mechanical engineering; information technology; production, manufacturing, logistics and supply chain management; geography; and meteorology); sciences (physics, material sciences and chemistry); design and development (regional and urban planning; architecture); liberal arts (international relations, environmental policy and environmental economics); business (general business administration, international business and public relations); two years of experience in renewable energy or a related field, preferably in a professional capacity. Should be able to demonstrate initiative and ambassadorial skills.
Level of Study: Graduate
Type: Scholarship
Value: Fellows will receive €1,100 monthly (for 3 months) as well as an international travel allowance up to €500, a travel allowance in Germany up to €200, insurance, and all seminar-related costs
Frequency: Annual
Application Procedure: Applicant must submit an online application form.
Closing Date: May
Funding: Private
Additional Information: Please check website for updated information.

CEC ARTSLINK

291 Broadway, 12th Floor, New York, NY, 10007, United States of America
Tel: (1) 212 643 1985
Fax: (1) 212 643 1996
Email: info@cecartslink.org
Website: www.cecartslink.org

CEC Artslink is an international arts service organization. Our programmes encourage and support exchange of artists and cultural managers between the United States and Central Europe, Russia and Eurasia. We believe that the arts are a society's most deliberate and complex means of communication.

ArtsLink Independent Projects

Subjects: Performing, design, media, literary and visual arts.
Purpose: To provide funding to artists and arts managers who propose to undertake projects in the United States in collaboration with a US non-profit arts organization.

Eligibility: Candidates must be citizens of, and reside in, an eligible countries Albania, Armenia, Azerbaijan, Belarus, Bosnia and Herzegovina, Bulgaria, Croatia, Czech Republic, Estonia, Georgia, Hungary, Kazakhstan, Kosovo, Kyrgyzstan, Latvia, Lithuania, Macedonia, Moldova, Mongolia, Montenegro, Poland, Romania, Russia, Serbia, Slovak Republic, Slovenia, Tajikistan, Turkmenistan, Ukraine and Uzbekistan. There are no age limitations. Arts managers must be affiliated with an organization in the non-commercial sector.
Type: Fellowship
Value: Up to US$5,000
Length of Study: 1 year
Frequency: Annual
Country of Study: United States of America
Application Procedure: Complete online application form.
Closing Date: December 3rd
Funding: Private, trusts
No. of awards given last year: 5

For further information contact:

CEC ArtsLink, 435 Hudson Street, 8th Floor, New York, NY 10014
Email: al@cecartslink.org

ArtsLink Projects

Subjects: Performing Arts, visual and media arts.
Purpose: To support US artists, curators, presenters and non-profit arts organizations undertaking projects in Eastern and Central Europe, Russia, Central Asia and the Caucasus.
Eligibility: Open to citizens of eligible countries: Albania, Armenia, Azerbaijan, Belarus, Bosnia and Herzegovina, Bulgaria, Croatia, Czech Republic, Estonia, Georgia, Hungary, Kazakhstan, Kosovo, Kyrgyzstan, Latvia, Lithuania, Macedonia, Moldova, Mongolia, Montenegro, Poland, Romania, Russia, Serbia, Slovak Republic, Slovenia, Tajikistan, Turkmenistan, Ukraine, and Uzbekistan.
Level of Study: Postgraduate
Type: Fellowship
Value: Up to US$10,000
Length of Study: 1 year
Frequency: Annual
Application Procedure: Complete online application form.
Closing Date: January 15th (Performing arts and literature application) and January 15th (Visual and media arts application)
Funding: Trusts, private
No. of awards given last year: 10

For further information contact:

Tel: 212-643-1985
Email: al@cecartslink.org

ArtsLink Residencies

Subjects: Literature, Performing arts, Visual and Media Arts.
Purpose: To create opportunities for artists and communities across the US to share artistic practices with artists and arts managers from abroad and engage in dialogue that advances understanding across cultures.
Eligibility: Applicants must be the citizens of, and reside in, eligible countries: Albania, Armenia, Azerbaijan, Belarus, Bosnia and Herzegovina, Bulgaria, Croatia, Czech Republic, Estonia, Georgia, Hungary, Kazakhstan, Kosovo, Kyrgyzstan, Latvia, Lithuania, Macedonia, Moldova, Mongolia, Montenegro, Poland, Romania, Russia, Serbia, Slovak Republic, Slovenia, Tajikistan, Turkmenistan, Ukraine and Uzbekistan.
Level of Study: Postgraduate
Type: Fellowship
Value: See http://www.cecartslink.org/grants/artslink_residencies/
Length of Study: 5 weeks
Frequency: Annual
Country of Study: United States of America
No. of awards offered: 14–16
Application Procedure: Complete online application form.
Closing Date: October 15th
Funding: Private, trusts
Contributor: ArtsLink Residencies are funded through public and private sources including CEC ArtsLink, the National Endowment for the Arts, the Trust for Mutual Understanding, the Ohio Arts Council, the Kettering Fund and the Milton and Sally Avery Arts Foundation with additional support from the Polish Cultural Institute and the Romanian Cultural Institute.
No. of awards given last year: 16

For further information contact:

Email: al@cecartslink.org

CENTER FOR CREATIVE PHOTOGRAPHY (CCP)

The University of Arizona, 1030 North Olive Road, Tucson, AZ, 210103, United States of America
Tel: (1) 520 621 7968
Fax: (1) 520 621 9444
Email: info@ccp.library.arizona.edu
Website: www.creativephotography.org/

The Center for Creative Photography (CCP) is an archive and research centre located on the University of Arizona campus.

CCP Ansel Adams Research Fellowship

Subjects: Curating/Research.
Purpose: To promote and support research on the Center's photograph, archive and library collections.
Eligibility: Open to researchers from any discipline who are engaged in studies that require an extended period of research in the collections of the Center.
Level of Study: Research
Type: Fellowship
Value: US$5,000
Length of Study: 2–4 weeks
Frequency: Annual
Country of Study: United States of America
Application Procedure: Applicants must send a cover letter along with 5 copies each of a curriculum vitae and a statement detailing the applicant's research interests.
Closing Date: See the organization website

For further information contact:

Center for Creative Photography, 1030 N. Olive Road, Tucson, AZ 85721-0103
Fax: 520 621 9444
Email: cass@ccp.library.arizona.edu.
Contact: Cass Fey, Curator of Education

THE CENTER FOR CROSS-CULTURAL STUDY (CC-CS)

Spanish Studies Abroad, The Center for Cross-Cultural Study, 446 Main Street, Amherst, MA, 01002 2314, United States of America
Tel: (1) 413 256 0011
Fax: (1) 413 256 1968
Email: info@spanishstudies.org
Website: www.spanishstudies.org

The CC-CS provides unique learning experiences for students in a true cross-cultural exchange by inviting them to expand their worldview through intense immersion in Seville, Havana and Cordoba. The CC-CS has developed it's reputation from an emphasis on the personal growth of students.

CC-CS Scholarship Program

Subjects: Cultural studies, Spanish studies.
Purpose: To fund continuing excellence in Spanish studies.
Eligibility: Open to all students enrolled on the cross-cultural scholarship programme in Spain, Argentina and Cuba.
Level of Study: Doctorate, Postdoctorate, Postgraduate
Type: Scholarship
Value: Up to US$2,500
Length of Study: 1 year
Frequency: Annual
Study Establishment: The center for cross-cultural study

Country of Study: Argentina
No. of awards offered: Varies
Application Procedure: Submit application accompanied by an original essay in Spanish, Portuguese and English, and a faculty recommendation.
Closing Date: 60 days prior to taking up a past
No. of awards given last year: 11

CENTER FOR DEFENSE INFORMATION (CDI)

1779 Massachusetts Avenue North West, Washington, DC, 20036 2109, United States of America
Tel: (1) 202 332 0600
Fax: (1) 202 462 4559
Email: info@cdi.org
Website: www.cdi.org
Contact: Development Director

The Center for Defense Information (CDI) provides responsible, non-partisan research and analysis on the social, economic, environmental, political and military components of national and global security, and aims to educate the public and inform policy makers about these issues. The organization is staffed by retired senior government officials and knowledgeable researchers and is directed by Dr Bruce G Blair.

CDI Internship

Subjects: Weapons proliferation, military spending, military policy, diplomacy and foreign affairs.
Purpose: To support the work of CDI's senior staff while gaining exposure to research, issues and communications related to national security and foreign policy.
Eligibility: There are no eligibility restrictions. Paid internships are available for nationals of the United States of America and legal immigrants.
Level of Study: Unrestricted
Type: Internship
Value: US$1,000 per month
Length of Study: 3–5 months
Study Establishment: CDI
Country of Study: Any country
No. of awards offered: Varies
Application Procedure: Applicants must submit a curriculum vitae, covering letter, brief writing sample, transcript and two letters of recommendation.
Closing Date: July 1st for the Autumn, October 15th for the Spring and March 1st for the Summer
Funding: Private
No. of awards given last year: 12
No. of applicants last year: 200

For further information contact:

Center for Defense Information, 1779 Massachusetts Avenue, N.W., Washington, DC, 20036-2109, United States of America
Fax: (1) 202 462 4559
Email: internships@cdi.org
Contact: Internship Coordinator

World Security Institute Internship

Subjects: Policy issues, including weapons proliferation, military spending and reform, diplomacy and foreign affairs, small aims trade, terrorism, missile defense and space weaponization.
Purpose: To support work of one of the World Security Institute's four divisions: the Center of Defense Information, Azimuth Media, International Media, or International Programs.
Eligibility: Open to recent graduates, graduate students, and highly qualified undergraduates with a strong interest in military policy, national security, foreign affairs, and related public policy issues who are willing to undertake some small administrative tasks. Although course work in these areas is not required, strong writing capabilities, prior experience in CDI's issue areas, and solid computer skills are appreciated. U.S. citizenship is not required.
Level of Study: Postgraduate
Type: Internship
Value: US$1,000 per month unless otherwise noted in the internship descriptions. In some cases, by pre-arrangement, interns may earn academic credit.
Length of Study: 1 year
Frequency: Annual
Study Establishment: World Security Institute
No. of awards offered: 18
No. of awards given last year: 16

CENTER FOR HELLENIC STUDIES

3100 Whitehaven Street NW, Washington, DC, 20008, United States of America
Tel: (1) 202 745 4400
Fax: (1) 202 332 8688
Email: fellowships@chs.harvard.edu
Website: http://chs.harvard.edu
Contact: Lanah Koelle, Programs Coordinator

The Center for Hellenic Studies (Trustees for Harvard University) offers residential and non-residential fellowships for professional scholars in ancient Greek studies.

Center for Hellenic Studies and Deutshes Archaologisches Institut Joint Fellowships

Subjects: Ancient Greek studies including archaeology, art history, epigraphy, history and interdisciplinary research.
Purpose: To encourage and support scholarship of the highest quality on ancient Greek civilization.
Eligibility: Open to scholars and teachers of ancient Greek studies with a PhD degree or equivalent qualification and some published work.
Level of Study: Postgraduate
Type: Fellowship
Value: Up to US$17,000 from the CHS and €10,353.66 from the DAI, plus housing and travel to and from the DAI and CHS
Length of Study: 9 months from September–May, non-renewable
Frequency: Annual
Study Establishment: The center for Hellenic studies, Washington DC and Deutsches Archaologisches Institut, Berlin
Country of Study: United States of America or Germany
No. of awards offered: 2
Application Procedure: Application must submit an application form, a curriculum vitae, a description of the research project, publication samples, and three letters of recommendation. Enquiries about eligibility and early applications are encouraged.
Closing Date: October
Funding: Private
No. of awards given last year: 2
No. of applicants last year: 13

Center for Hellenic Studies Fellowships

Subjects: Ancient Greek studies including archaeology, art history, epigraphy, history, literary criticism, philology, pedagogical applications and interdisciplinary research.
Purpose: To provide selected classics scholars an academic year or less free of other responsibilities to work on a publishable project. To support collaborative proposals and proposals that use advanced information technology in the study of the ancient Greek world.
Eligibility: Open to scholars and teachers of Ancient Greek studies with a PhD degree or equivalent qualification and some published work.

Level of Study: Postdoctorate
Type: Fellowship
Value: Up to US$17,000, plus private living quarters and a study at the Center building. Limited funds up to US$1,000 for research expenses and research related travel expenses are available
Length of Study: Up to 4 months from September–May, non-renewable and Up to 4.5 months from September–January or January–May
Frequency: Annual
Study Establishment: The Center for Hellenic Studies, Washington, DC
Country of Study: United States of America
Application Procedure: Applicants must submit an application form, a curriculum vitae, a description of the research project, publication

samples and three letters of recommendation. Enquiries about eligibility and early applications are encouraged. Applicants who are unable to stay for the full academic year may apply for a one-semester fellowship or a non-residential fellowship.
Closing Date: October
Funding: Private
No. of awards given last year: 14 semester residential fellowships, 10 non-residential fellowships
No. of applicants last year: 110

CENTER FOR PHILOSOPHY OF RELIGION

418 Malloy Hall, University of Notre Dame, Notre Dame, IN, 46556, United States of America
Tel: (1) 574 631 7339
Email: cpreligion@nd.edu/philreligion@nd.edu
Website: www.nd.edu/˜cprelig

The Center for Philosophy of Religion at the University of Notre Dame was established in 1976 in order to promote, support and disseminate scholarly work in the philosophy of religion and Christian philosophy. The center aims to promote work concerned with the traditional topics and questions that fall under the rubric of the philosophy of religion: the theistic proofs, the rationality of belief in God, the problem of evil, the nature of religious language and the like. At least as important, however, is the Center's effort to support and encourage the development and exploration of specifically Christian and theistic philosophy, the sort of philosophy which takes Christianity (or, more broadly, theism) for granted and then proceeds to work on philosophical questions and problems from that perspective. As one of the world's leading Catholic institution, the University of Notre Dame provides an ideal home for such work.

The Alvin Plantinga Fellowship

Subjects: Philosophy of religion.
Purpose: To provide time for reflection and writing to a distinguished senior scholar whose work is in the forefront of current research in the philosophy of religion and Christian philosophy.
Eligibility: Open to the senior scholar whose work is in the forefront of current research in the philosophy of religion and Christian philosophy.
Level of Study: Research
Type: Stipendiary
Value: $60,000
Frequency: Annual
No. of awards offered: 1
Application Procedure: Applications should include a complete curriculum vitae, three letters of recommendation, a statement of no more than three pages decribing the project and one published or unpublished paper.
Closing Date: February 1st
Contributor: Center's endowment and College of Arts and Letters at Notre Dame
No. of awards given last year: 1
No. of applicants last year: 15

Center for Philosophy of Religion's Postdoctoral Fellowships

Subjects: Philosophy of religion.
Purpose: To support those whose tenure at the Center would allow them to grow and make progress in the Center's areas of interest and subsequently disseminate and expand such work through their own teaching and writing.
Eligibility: Open to those whose tenure at the Center would allow them to grow and make progress in the Center's areas of interest.
Level of Study: Postdoctorate
Type: Fellowship
Value: The Alvin Plantinga Fellowship ($60,000), awarded to a distinguished senior scholar; up to two Research Fellowships ($40,000–50,000, depending on rank); the Frederick J. Crosson Fellowship ($45,000) reserved for foreign scholars and those outside the field of philosophy; and one Visiting Graduate Fellowship ($20,000) awarded to a graduate student in philosophy with research interests in the philosophy of religion. All fellows will receive up to $2,000 reimbursement for moving expenses, as well as up to $2,000 for research-related expenses.
Frequency: Annual
No. of awards offered: 5
Application Procedure: Applications should include a complete curriculum vitae, three letters of recommendation, a statement of no more than three pages decribing the project and one published or unpublished paper.
Closing Date: February 1st
Contributor: Center's endowment and College of Arts and Letters at Notre Dame
No. of awards given last year: 2
No. of applicants last year: 50

Center for Philosophy of Religion's Visiting Graduate Fellowship

Subjects: Philosophy of religion.
Purpose: To support research in philosophy of religion or Christian philosophy.
Eligibility: Applicant must be a graduate.
Level of Study: Graduate
Type: Fellowship
Value: $20,000
Frequency: Annual
No. of awards offered: 1
Application Procedure: Applications should include a complete curriculum vitae, three letters of recommendation, a statement of no more than three pages decribing the project and one published or unpublished paper.
Closing Date: February 1st
Contributor: Center's endowment and College of Arts and Letters at Notre Dame
No. of awards given last year: 2
No. of applicants last year: 10

The Frederick J. Crosson Fellowship

Subjects: Philosophy of religion.
Purpose: To support a foreign scholar (especially one outside the Anglo-American philosophical community) or to a scholar outside the field of philosophy (e.g. a theologian).
Eligibility: See the organization website
Level of Study: Research
Type: Fellowship
Value: $45,000
Length of Study: 1 year
Frequency: Annual
No. of awards offered: 1
Application Procedure: Applications should include a complete curriculum vitae, three letters of recommendation, a statement of no more than three pages decribing the project and one published or unpublished paper.
Closing Date: February 1st
Contributor: Center's endowment and College of Arts and Letters at Notre Dame
No. of awards given last year: 1
No. of applicants last year: 12

CENTRAL ASIA RESEARCH AND TRAINING INITIATIVE (CARTI)

Open Society Foundations, International Higher Education Support Program, Central Asia and Caucasus Research and Training Initiative, Október 6. u.12, Budapest, 1051, Hungary
Tel: (36) 1 882 3854
Fax: (36) 1 882 3112
Email: carti@osi.hu
Website: www.soros.org

Central Asia Research and Training Initiative (CARTI) is a regional higher education support programme of the Open Society Institute, a private operating and grantmaking foundation. CARTI promotes the development of indigenous capacities for original scholarly and academic work and internationalization of scholarship in the region of Central Asia including, but not limited to, the post Soviet states of Central Asia and Mongolia.

CARTI Junior Fellowships

Subjects: Humanities and social sciences.

Purpose: To support young individuals in early stages of their formal Doctoral studies (such as aspirantura) and focus on development of ideas and skills for high-quality research work.
Eligibility: Open to candidates holding a Master's degree or equivalent and formally registered at a Doctoral studies programme (PhD or equivalent programme) from Afghanistan, Kazakhstan, Mangolia, Tajikistan, Turkmenistan or Uzbekistan.
Level of Study: Doctorate
Type: Fellowships
Value: Varies
Length of Study: Up to 2 years
Frequency: Annual
Application Procedure: Applicants can download the application form from the website.
Closing Date: November 1st (online application)
Additional Information: The programme is open to citizens who are also residents of Afghanistan, Kazakhstan, Kyrgyzstan, Mongolia, Tajikistan, Turkmenistan or Uzbekistan.

CENTRAL QUEENSLAND UNIVERSITY

Building 5 Bruce Highway, Rockhampton, QLD 4702, Australia
Tel: (61) 07 4923 2607/ 7 4930 9000
Fax: (61) 07 4923 2600/ 7 4923 2100
Email: research-enquiries@cqu.edu.au
Website: www.research.cqu.edu.au

The Central Queensland University (CQU) is committed to excellence in research and innovation with a particular emphasis on issues that affect the region. CQU achieves relevance in its research goals through linkages with industry, business, government and the community and through collaboration with national and international researchers and research networks. CQU provides a range of exciting and relevant research opportunities for Masters and PhD candidates and is committed to excellence and quality in the research training experience of its candidates.

CQ University Australia Postgraduate Research Award

Subjects: Any subject.
Purpose: To support the research higher degree programs of Masters and PhD.
Eligibility: Open to international residents from any country apart from New Zealand who have first class honours or equivalent.
Level of Study: Postgraduate, Graduate, Research
Type: Research award
Value: Australian $23,728
Length of Study: 3 years (Doctorate) and 2 years (Masters)
Frequency: Annual
Study Establishment: Central Queensland University
Country of Study: Australia
No. of awards offered: 7
Application Procedure: Check website for further details.
Closing Date: October 31st
Funding: Government
Additional Information: This scholarship is paid fortnightly for the period of up to 2 years (Masters) or up to 3 years (doctorate). Open for applications from July 1st.

For further information contact:

Office of Research, Building 32, CQUniversity, Rockhampton, QLD 4702, Australia
Tel: (61) 07 4923 2602
Fax: (61) 07 4923 2600
Website: www.research.cqu.edu.au/FCWViewer/view.do?page=297
Contact: Kerne Thompson, Executive Office

CENTRE FOR ASIA-PACIFIC INITIATIVES (CAPI)

Sedgewick Building, Room C128, University of Victoria, PO Box 1700, STN CSC, Victoria, BC, V8W 2Y2, Canada
Tel: (1) 250 721 7020
Fax: (1) 250 721 3107
Email: capi@uvic.ca
Website: www.capi.uvic.ca
Contact: Lansdowne Associate Director, Helen

The Centre for Asia-Pacific Initiatives (CAPI) was established in 1987 as an important element of the University of Victoria's (Canada) plan to expand and strengthen its links with universities and other institutions in the Asia Pacific region, especially with China, Japan, Southeast Asia, Korea and the developing island states of the Southwest Pacific.

CAPI Student Fellowship for Thesis Research

Subjects: Languages and research on the Asia-Pacific region.
Purpose: To encourage excellence in research and in the study of languages and research of the Asia-Pacific region.
Eligibility: Open to any student enrolled at the University of Victoria who are working towards a Master's degree or PhD.
Level of Study: Graduate, Research, Biannual
Type: Fellowships
Value: Canadian $2,500
Length of Study: 1 year
Frequency: Annual
Study Establishment: Centre for Asia-Pacific Initiatives
Application Procedure: Applicants must submit their curriculum vitae, a 2-page description of the proposed research project and activities, plus a letter of support from the faculty supervisor.
Closing Date: March 15th
No. of awards given last year: 4

CENTRE FOR THE HISTORY OF SCIENCE, TECHNOLOGY AND MEDICINE (CHSTM)

The University of Manchester, Second Floor, Simon Building, Brunswick Street, Manchester, M13 9PL, England
Tel: (44) 0 161 275 5850
Fax: (44) 0 161 275 5699
Email: chstm@manchester.ac.uk
Website: www.manchester.ac.uk/chstm

The Centre for the History of Science, Technology and Medicine (CHSTM) maintains teaching and research programmes of the highest standards. It acts as a focus for the history of science, technology and medicine in the northwest of England. CHSTM houses a Welcome Unit for the History of Medicine and the National Archive for the History of Computing.

AHRC Studentships

Subjects: History.
Purpose: To support students working for their MSc and/or PhD in the history of science, technology and medicine.
Eligibility: Applicant must be a doctorate or post-graduate.
Level of Study: Doctorate, Postgraduate
Type: Studentships and bursaries
Value: See http://www.chstm.manchester.ac.uk/postgraduate/research/funding/index.aspx
Length of Study: 1 and/or 3 years
Frequency: Annual
Study Establishment: CHSTM
Country of Study: United Kingdom
Application Procedure: The AHRB deadline is May 1st. In order to ensure completion of paperwork and prompt submission of applications, the CHSTM deadline for AHRB forms is April 12th. We expect to work closely with applicants as they complete the forms, so early contact with CHSTM staff is advisable.
Closing Date: May 1st
Funding: Government
Contributor: Arts and Humanities Research Council

For further information contact:

Website: https://je-s.rcuk.ac.uk

Wellcome Trust Studentships

Subjects: History, and medicine and surgery
Purpose: To support applicants whose main interests are in the history of medicine.
Eligibility: Applicant should be a UK/European Economic Area (EEA) national with (or be in your final year and expected to obtain) a first- or

upper-second-class honours degree or an equivalent EEA graduate qualification.
Level of Study: Graduate, Research, Doctorate, Postgraduate
Type: Studentships and bursaries
Value: £18,053–19,903
Frequency: Annual
Study Establishment: CHSTM
Country of Study: United Kingdom
No. of awards offered: 5
Application Procedure: Applicants must complete the university application form (with two references), the Wellcome Trust Studentship application form and submit a curriculum vitae and samples of written work.
Closing Date: May 1st
Funding: Foundation
Contributor: Wellcome Trust
Additional Information: Applicants are encouraged to discuss their application informally with Professor Michael Worboys and to submit their applications as soon as possible.

CERIES (CENTRE DE RECHERCHES ET D'INVESTIGATIONS EPIDERMIQUES ET SENSORIELLES)

20 rue Victor Noir, Neuilly-sur-Seine, 92200, France
Tel: (33) 146 434 900
Fax: (33) 146 434 600
Email: contact@ceries.com
Website: www.ceries.com

CERIES (Centre de Recherches et d'Investigations Epidermiques et Sensorielles or Centre for Epidermal and Sensory Research and Investigation) is the healthy skin research centre of Chanel.

CERIES Research Award

Subjects: The biology and physiology of healthy skin and/or its reactions to environmental factors.
Purpose: To honour a scientific researcher for a fundamental or clinical research project in the field of healthy skin.
Eligibility: There are no eligibility restrictions.
Level of Study: Research
Value: €40,000
Length of Study: 1 year
Frequency: Annual
Country of Study: Any country
Application Procedure: Applicants must consult the website.
Closing Date: June 1st
Funding: Private
Contributor: Chanel
No. of awards given last year: 1
No. of applicants last year: 26

For further information contact:

CE.R.I.E.S. Research Award, 20 rue Victor Noir, 92521 NEUILLY sur Seine Cedex, FRANCE,
Email: chanelrt.award@ruderfinnasia.com
Contact: Claire BERNIN-JUNG / Marie-Hélène LAIR

CERN EUROPEAN ORGANIZATION FOR NUCLEAR RESEARCH

CH-1211, Geneva, 23, Switzerland
Tel: (41) 22 76 761 11
Fax: (41) 22 76 765 55
Email: recruitment.service@cern.ch
Website: www.cern.ch

CERN European Laboratory for Particle Physics is the world's leading laboratory in its field, that being the study of the smallest constituents of matter and of the forces that hold them together. The laboratory's tools are its particle accelerators and detectors, which are among the largest and most complex scientific instruments ever built.

CERN Summer Student Programme

Subjects: Physics, computing and engineering.
Purpose: To awaken the interest of undergraduates in CERN's activities by offering them hands-on experience during their long summer vacation.
Eligibility: Open to all interested students who have completed at least 3 years of full-time studies at university level.
Value: Travel allowance and a daily stipend
Length of Study: 8–13 weeks
Study Establishment: CERN
Country of Study: Switzerland
Application Procedure: A completed application and curriculum vitae along with 2 references must be submitted to CERN.
Closing Date: January 7th

For further information contact:

CERN Recruitment Services via the e-recruitment system
Email: jkrich@umich.edu

CERN Technical Student Programme

Subjects: Accelerator physics, computing, mathematics, engineering, geotechnics, instrumentation for accelerators and particle physics experiments, low temperature physics and superconductivity, materials science, radiation protection, environmental and safety engineering, solid state, surface physics and ultra-high vacuum.
Purpose: To provide placements for students who are specializing in different technical fields.
Eligibility: Open to applicants attending an educational establishment in a CERN member state and following a full-time course in one of the subjects listed, at university or advanced technical level. Students must be less than 30 years of age at the time of the Selection Committee meeting. Candidates must be nationals of the member states of CERN. Students specializing in theoretical or experimental particle physics are not eligible for the programme.
Value: A monthly living allowance to cover the expenses of a single person in the Geneva area. A health insurance for illnesses and accidents of professional or non-professional nature. Joining expenses (on a lump sum basis).
Length of Study: Appointments can last for 6 consecutive months, but mostly 1 year. Appointments can start throughout the year
Study Establishment: The European Laboratory for Particle Physics
Country of Study: Switzerland
No. of awards offered: Approx. 80–90
Application Procedure: Applications must be made electronically via the website.
Closing Date: March 5th
Funding: Government
No. of awards given last year: Approx. 80–90
No. of applicants last year: Approx. 240
Additional Information: The official languages of CERN are English and French. A good knowledge of at least one of these languages is essential. CERN member states include Austria, Belgium, Bulgaria, the Czech Republic, Denmark, Finland, France, Germany, Greece, Hungary, Italy, the Netherlands, Norway, Poland, Portugal, Slovakia Republic, Spain, Sweden, Switzerland and the United Kingdom.

CERN-Japan Fellowship Programme

Subjects: LHC data analysis and physics
Purpose: To support young researchers who are interested in LHC data analysis and physics studies.
Eligibility: Applicants should be nationals or permanent residents of Japan and have a doctorate for applicants in experimental or phenomenological physics and/or accelerator science. Candidates who are currently preparing a PhD are eligible to apply. However, they are expected to have obtained their PhD by the time they take up their appointment at CERN.
Level of Study: Doctorate
Type: Fellowship
Value: Covers travel expense and insurance coverage
Length of Study: Up to 3 years
Frequency: Annual
Application Procedure: A completed electronic application form along with a curriculum vitae should be submitted.
Closing Date: December 1st
Contributor: CERN

For further information contact:

Recruitment Service, Human Resource Department, CERN, Geneva 23, CH-1211, Switzerland
Email: recruitment.science@cern.ch

THE CHARLES AND ANNE MORROW LINDBERGH FOUNDATION

2150 Third Avenue North, Suite 310, Anoka, MN, 55303-2200, United States of America
Tel: (1) 763 576 1596
Fax: (1) 763 576 1664
Email: info@lindberghfoundation.org
Website: www.lindberghfoundation.org

Charles and Anne Morrow Lindbergh believed that balancing technology and the environment was vital to sustaining a healthy quality of life. The Lindbergh Foundation is committed to putting balance into action by giving research grants and awards to individuals whose scientific and educational innovations address important environmental issues around the world.

Lindbergh Grants

Subjects: Aviation, aerospace, conservation of natural resources including animal plant and water resources,general conservation including land, air, energy etc., education including humanities/ education, exploration, health and population sciences, adaptive technologies and waste minimizaton and management. Emphasis on Aviation projects which overlap on one of these other categories. No new application for 2013 funding are being accepted.
Purpose: To support innovative projects that foster the environment and keep the planet in balance.
Eligibility: Open to individuals for research or public education projects, not affiliated organizations for institutional programs. The Foundation does not provide support for overhead costs of organizations, tuition or scholarships. The Foundation welcomes candidates who may or may not be afiiliated with an academic, non-profit or for-profit organization. Candidates for grants are not required to hold any graduate or postgraduate academic degrees. The Lindbergh Grants Program is international in scope. All letters, applications, endorsers reports, and required progress and final reports must be submitted in English.
Level of Study: Unrestricted
Type: Research grant
Value: Up to $10,580
Length of Study: 1 year, but in exceptional cases up to 2 years
Frequency: Annual
Country of Study: Any country
No. of awards offered: Approx.8–10
Application Procedure: Applications must be submitted according to the relevant guidelines found at website. Six copies of the application by mail and one pdf by email must be sent to the Foundation's office.
Closing Date: The second Thursday in June
Funding: Foundation, individuals
No. of applicants last year: 98
Additional Information: Please check the Foundation's website www.lindberghfoundation.org for further information.

CHARLES BABBAGE INSTITUTE (CBI)

211 Andersen Library, University of Minnesota, 222 21st Avenue South, Minneapolis, MN, 55455, United States of America
Tel: (1) 612 624 5050
Fax: (1) 612 625 8054
Email: cbi@umn.edu
Website: www.cbi.umn.edu

The Charles Babbage Institute (CBI) is a research centre dedicated to promoting the study of the history of computing, its impact on society and preserving relevant documentation. CBI fosters research and writing in the history of computing by providing fellowship support, archival resources and information to scholars, computer scientists and the general public.

Adelle and Erwin Tomash Fellowship in the History of Information Processing

Subjects: The history of computing and information processing.
Purpose: To advance the professional development of historians in the field.
Eligibility: Open to graduate students whose dissertations deal with a historical aspect of information processing. Priority will be given to students who have completed all requirements for the doctoral degree except the research and writing of the dissertation.
Level of Study: Doctorate
Type: Fellowship
Value: US$14,000
Length of Study: 1 year
Frequency: Annual
Country of Study: Any country
No. of awards offered: 1
Application Procedure: Applicants must send their curriculum vitae, a five page statement and justification of the research problem, and a discussion of methods, research materials and evidence of faculty support for the project. Applicants should also arrange for three letters of reference and certified transcripts of graduate school credits to be sent directly to the Institute.
Closing Date: January 15th
Funding: Private

For further information contact:

Charles Babbage Institute University of Minnesota 103 Walter Library 117 Pleasant Street, SE, Minneapolis, MN 55455
Tel: 624-5050
Fax: 625-8054
Email: nels0307@umn.edu.
Contact: R. Arvid Nelsen, CBI Archivist

CHARLES DARWIN UNIVERSITY (CDU)

Orange 1, Casuarina campus, Charles Darwin University, PO Box 795, Darwin, NT, Alice Springs NT 0871, Australia
Tel: (61) 08 8946 6442
Fax: (61) 08 8959 5343
Email: scholarships@cdu.edu.au
Website: www.cdu.edu.au
Contact: Professor Robert Wasson, Scholarships Officer

The Charles Darwin University (CDU) offers programmes from certificate level to PhD, incorporating the full range of vocational education courses. CDU has a distinctive research profile, reflecting the priorities appropriate to its location. It is a participating member of several CRCs.

ARC Australian Postgraduate Award – Industry

Subjects: Agriculture, forestry and fishery; arts and humanities; education and teacher training; Engineering; fine and applied arts; medical sciences; natural sciences; recreation, welfare, protective services; social and behavioural sciences.
Purpose: To provide industry-oriented research training and enable postdoctoral researchers to pursue internationally competitive research opportunities in collaboration with industry.
Eligibility: See http://www.cdu.edu.au/research/students/scholarships/internal.html#APA
Level of Study: Doctorate, Foundation programme, Research, Master by Research
Type: Scholarship
Value: $24,653 per year. Additional benefits include paid sick leave, maternity/paternity leave, relocation and a thesis allowance.
Length of Study: 2–3 years
Frequency: Annual
Country of Study: Australia
No. of awards offered: Approx. 15–20 per year
Application Procedure: Completion and submission of an application: www.cdu.edu.au/research/office/applicationkit.html.
Closing Date: March 31st
Funding: Government
Contributor: Australian Government
No. of awards given last year: 20
No. of applicants last year: 50

Additional Information: Enquiries and requests for additional information may be directed to the CDU Research Degrees Administration Officer by email: research@cdu.edu.au. Intending international applicants should contact the CDU International Office by email: international@cdu.edu.au.
Eligibility of other countries conditional upon meeting eligibility criteria.

For further information contact:

Research Scholarships, Office of Research and Innovation, Charles Darwin University, Ellengowan Drive, Darwin, NT 0909, Australia

International Postgraduate Research Scholarships

Subjects: Agriculture, forestry and fishery; arts and humanities; education and teacher training; Engineering; fine and applied arts; medical sciences; natural sciences; recreation, welfare, protective services; social and behavioural sciences.
Purpose: To support annual course costs plus the cost of an Overseas Student Health Cover policy.
Eligibility: International applicants: An Australian bachelor degree with first class honours, or an Australian master degree with a substantial research component, or an equivalent level of academic attainment, and meet Australian international student visa requirements, and meet minimum English entry requirements. Full details are available at: www.cdu.edu.au/research/students/admissions.html.
Level of Study: Doctorate, Postgraduate, Research, Master by research
Type: Scholarship
Value: Annual course fees plus health cover
Length of Study: 2–3 years
Frequency: Annual
Country of Study: Australia
No. of awards offered: Approximately 2 per year
Application Procedure: Completion and submission of an application: www.cdu.edu.au/research/office/applicationkit.html.
Closing Date: March 31st
Funding: Government
Contributor: Australian Government
No. of awards given last year: 2
No. of applicants last year: 14
Additional Information: Enquiries and requests for additional information may be directed to the CDU Research Degrees Administration Officer by email: research@cdu.edu.au. Intending international applicants should contact the CDU International Office by email: international@cdu.edu.au.

For further information contact:

Research Scholarships, Office of Research and Innovation, Charles Darwin University, Ellengowan Drive, Darwin, NT 0909, Australia

CHARLES STURT UNIVERSITY (CSU)

Locked Bag 588, Wagga Wagga, NSW 2678, Australia
Tel: (61) 02 6933 2000
Fax: (61) 02 6933 2639
Email: inquiry@csu.edu.au
Website: www.csu.edu.au

CSU is one of the leading Australian universities for graduate employment and largest provider in distance education. Utilizing our expertise in distance education, CSU provides educational opportunities to students around the world. Around 36,000 students undertake their choice of study with CSU on one of our campuses, from home, their workplace or anywhere around the globe.

Academic Staff RHD Workload Support Scheme

Subjects: All subjects.
Purpose: To assist academic staff of the University to obtain a research higher degree qualification (Research Masters or PhD) or a research professional doctorate in areas of strategic importance to the institution.
Eligibility: This scheme is open to all academic staff of Charles Stuart Univeristy.
Level of Study: Postgraduate, Research
Value: A formal workload allocation during candidature plus, for staff enrolled in a CSU RHD program, tuition fees, student resource funds and supervision funds to the Faculty
Frequency: Annual
Application Procedure: Applicants must submit an application form.
Closing Date: No deadline
Additional Information: For further information please see the website: http://www.csu.edu.au/research/support/researchers/funding/internal/ashdwss

Australian Postgraduate Awards

Subjects: All subjects.
Purpose: To financially support postgraduate students of exceptional research promise in Master or Doctoral programs at Charles Sturt University.
Eligibility: Awards will only be available to those who are: Australian citizens and New Zealand citizens; have been granted permanent resident status by October 31st; have lived in Australia continuously for at least 12 months prior to October 31st; have completed at least four years of tertiary education studies at a high level of achievement; have obtained First Class Honours or equivalent results; will undertake a Master's (Honours) or Doctoral degree in 2011; are enrolling as full-time students or, in exceptional circumstances, be granted approval by CSU for a part-time award; have had their enrolment into the proposed higher degree programme accepted by CSU.
Level of Study: Postgraduate, Research
Type: Award
Value: Stipend and allowances (varies)
Length of Study: 2-3 years
Frequency: Annual
Study Establishment: Charles Sturt University
Country of Study: Australia
Application Procedure: Applicants must submit an application form.
Closing Date: October 31st
Funding: Government

Charles Sturt University Postgraduate Research Studentships (CSUPRS)

Subjects: All subjects.
Purpose: To support high quality research students in Masters or Doctoral programs at Charles Sturt University.
Eligibility: Open to the candidates who hold or expect to hold, at least a Bachelor degree with upper second class honours or a qualification deemed equivalent from CSU.
Level of Study: Graduate, Research
Type: Studentship
Value: $22,500 stipend plus allowances
Frequency: Annual
No. of awards offered: Up to 8
Application Procedure: Scholarship application form can be downloaded from the website. Send in the filled application to the center with original referee report and five copies of their report.
Closing Date: October 30th
Additional Information: Offers of scholarships cannot be made to candidates until their enrolment as Research Higher Degree students has been approved by the Board of Graduate Studies.

For further information contact:

Postgraduate Scholarships, Center for Research & Graduate Training, Charles Sturt University, Locked Bag 588, Wagga Wagga, NSW 2678, Australia
Tel: (61) 02 6933 4162
Email: pgscholars@csu.edu.au

Commercialisation Training Grant Scheme

Subjects: All subjects.
Purpose: Aims to provide Research Higher Degree students with a fully accredited course designed to enhance their professional capacity and skills, and provide commercial application of base concepts developed in research management.
Eligibility: To be eligible to receive a CTS place, a student must:
a. Be an Australian citizen, a New Zealand Citizen or an Australian permanent resident.
b. Have completed a minimum of one year full time equivalent of their

Research Higher Degree. (NOTE: Part-Time Higher Degree students are not eligible to receive the stipend).
c. Have the support of their Principal Supervisor.
d. Not have previously completed CTS training or training consistent with CTS requirements.
Level of Study: Postgraduate, Research
Type: Grant
Value: Maximum Australian $8,000
Length of Study: 15 months
Frequency: Annual
Application Procedure: Applicants must submit an application form.
Closing Date: March 23rd and July 23rd

International Postgraduate Research Studentships (IPRS)

Subjects: All subjects.
Purpose: To attract top quality international postgraduate students to areas of research strength in Australian higher education institutions.
Eligibility: Applicants should hold, or expect to hold, at least a Bachelor degree with upper second class honours or a qualification deemed equivalent.
Level of Study: Postgraduate, Research
Type: Scholarship
Value: The scholarship will cover tuition fees payable for each year of the course
Frequency: Annual
Country of Study: Australia
No. of awards offered: Up to 8
Application Procedure: Applicants must submit an application form.
Closing Date: October 29th
Funding: Government

Writing Up Awards – Postgraduate Students

Subjects: All subjects.
Purpose: To improve publication rates by providing a modest income to higher degree by research students following submission of thesis to enable the preparation of articles or books based on their postgraduate thesis.
Eligibility: For Charles Sturt University Masters by research or PhD candidates who either are about to submit a thesis or have just submitted a thesis for examination and have not yet qualified to graduate.
Level of Study: Postgraduate, Research
Type: Award
Value: Maximum Australian $5,000
Frequency: Twice a year
Application Procedure: Applicants must submit an application form.
Closing Date: May 23th and November 21st
Additional Information: This award does not provide financial support to postgraduate student to write their thesis.

THE CHARLES WALLACE TRUST

The Charles Wallace Trust, 4 Dorville Crescent, London, W6 0HJ, United Kingdom
Tel: (44) 020 8741 0836
Email: timbutchard@wallace-trusts.org.uk
Website: www.wallace-trusts.org.uk/
Contact: Mr Tim Butchard, Secretary

The Charles Wallace Bangladesh Trust - Doctoral Busaries

Purpose: Awards are granted to individual students already in the UK who are normally in the final year, or anticipating the final year, of their PhDs, and who need additional funding to help them complete their studies.
Eligibility: The Trust is not primarily a hardship fund and all applicants are required to demonstrate academic excellence as well as financial need.
Level of Study: Doctorate
Type: Bursary
Value: £1,500 at the maximum
Frequency: Biannual
Country of Study: United Kingdom
Application Procedure: Applicants must complete the Trust's Application Form and submit it, either in hard copy or as an email attachment, to the Secretary of the Trust. The application should be accompanied by a supporting letter on headed paper from the applicant's supervisor. Other documents testifying to the applicant's background and achievements to date should be kept to a minimum.
Closing Date: May 1st and November 1st
Funding: Trusts

The Charles Wallace Bangladesh Trust - Professional Training Bursaries

Subjects: Disciplines are unrestricted but the trustees reserve the right to assess the usefulness of the training both to Bangladesh and to the individual concerned..
Purpose: Limited financial support to enable mid-career professionals.
Eligibility: Normally aged between 35 and 45. Preference is given to candidates who have had little or no prior training or experience outside Bangladesh.
Level of Study: Professional development
Type: Bursary
Value: £750 (Maximum)
Length of Study: The eligible courses must last at least 2 weeks
Frequency: Biannual
Application Procedure: Applicants must complete the Trust's Application Form and submit it, either in hard copy or as an email attachment, to the Secretary of the Trust. The application should be accompanied by a supporting letter on headed paper from the applicant's supervisor. Other documents testifying to the applicant's background and achievements to date should be kept to a minimum.
Closing Date: May and November
Funding: Trusts

The Charles Wallace Burma Trust - Postgraduate Student Busaries

Purpose: Awards are granted to individual students undertaking, or about to undertake, postgraduate courses in the UK at Master's or Doctoral level and who need additional funding to help them cover the cost of their studies.
Level of Study: Postgraduate
Type: Bursary
Value: Our maximum grant is £1,500, so applicants must have funds from other sources to cover most of their expenses.
Frequency: Biannual
Country of Study: United Kingdom
Application Procedure: The trust is not primarily a hardship fund and all applicants are required to demonstrate academic excellence as well as financial need. Applicants must complete the trust's application form and submit it, either in hard copy or as an email attachment, to the Secretary of the Trust. The application should be accompanied by a supporting letter on headed paper from the applicant's course leader or supervisor in the United Kingdom. Applicants who have not yet commenced their UK courses should show written evidence of acceptance, and full details of their sources of finance, as well as a letter of reference from a senior academic source in Burma/Myanmar. Documents testifying to the applicant's background and achievements to date should be kept to a minimum.
Closing Date: May 1st and November 1st
Funding: Trusts

The Charles Wallace Burma Trust - Visiting Fellowships

Subjects: Intended for those holding management posts in the following subject areas: development management and disaster relief; environmental management; governance and human rights law; media production.
Purpose: To enable at least two Burmese professionals to undertake short visits to the UK each year in order to broaden their professional knowledge, skills and contacts. They are not intended to facilitate formal training.
Eligibility: Eligible candidates are Burmese nationals, residing in Burma. They are normally junior or mid-career professionals and practitioners, aged between 30 and 50. Candidates must have a working knowledge of the english language adequate for their requirements. Candidates who have never travelled abroad for study or professional purposes will have a modest advantage in the selection process.
Level of Study: Professional development

Type: Fellowship
Value: An all-inclusive monthly stipend of £1,250 is offered by the trust, also a return economy air fare. There is no provision for course or bench fees
Length of Study: The duration of a fellowship varies from 4 weeks to a maximum of 3 months
Frequency: Annual
Country of Study: United Kingdom
Application Procedure: The key preliminary step is that candidates identify a UK-based partner or host institution, and obtain an invitation letter. Secondly, an application form, acquired from the website of the British Council in Burma (www.britishcouncil.org/burma), must be completed and submitted as instructed. For further information please contact the British Council's information desk (enquiries@mm.britishcouncil.org) or email the Secretary of the Trust in London (timbutchard@wallace-trusts.org.uk).
Funding: Trusts

The Charles Wallace India Trust

Purpose: The Charles Wallace India Trust gives grants to Indians in the early or middle stages of their careers who are living in India and working or studying in the arts, heritage conservation or the humanities.
Level of Study: Doctorate, Postgraduate, Postdoctorate, Professional development
Type: Grants and fellowships
Value: Dependent upon award: funding towards arts and heritage conservation; funded fellowships; grants towards short research or professional visits; grants towards Doctoral study costs.
Frequency: Annual
Country of Study: United Kingdom
Application Procedure: See British Council website for more details: www.britishcouncil.org/india-scholarships-cwit.htm. For further information please email at cwit@in.britishcouncil.org.
Closing Date: November 25th and refer to website as deadlines differ
Funding: Trusts

The Charles Wallace Pakistan Trust - Open Visiting Fellowships

Purpose: To enable Pakistani men and women to undertake short visits to the UK in order to broaden their professional knowledge, skills, and contacts.
Eligibility: Eligible candidates are Pakistani nationals, residing in Pakistan. They are normally junior or mid-career professionals or academics aged between 30 and 50, working in the following disciplines: humanities, arts, and creative industries; social sciences and social development; the environmental and health sciences. Candidates who have never travelled abroad for study or professional purposes will have a modest advantage in the selection process.
Level of Study: Doctorate, Professional development
Type: Fellowship
Value: The fellowships will normally take the form of two principal activities: professional familiarization and interaction and study and research. An all-inclusive monthly stipend of £1,250 is offered by the trust, also a return economy air fare, but there is no provision for course or bench fees
Length of Study: The duration of a fellowship varies from 3 weeks to a maximum of 3 months
Frequency: Annual
No. of awards offered: 15
Application Procedure: The key preliminary step is that candidates identify a UK-based partner or host institutions, and obtain an invitation letter from them. Secondly, an application form, downloaded from the website of the British Council in Pakistan (www.britishcouncil.org.pk), must be completed and submitted as instructed in advance of the British Council's annual deadline. Interviews of short-listed candidates will take place in Islamabad, Lahore and Karachi in April of each year.
Closing Date: Mid-March
Funding: Trusts

The Charles Wallace Pakistan Trust - Reserved Visiting Fellowships

Purpose: The fellowships enable Pakistani academics and professionals to undertake short working visits to these institutions with the aim of broadening their professional knowledge, skills and contacts.
Level of Study: Postgraduate
Type: Fellowship
Value: An all-inclusive monthly stipend of £1,250 is offered by the Trust, together with return economy air fares, but there is no provision for course or bench fees.
Length of Study: The duration of a fellowship varies from 3 weeks to a maximum of 3 months
Frequency: Annual
Study Establishment: University of London, Oxford University, Edinburgh University
Country of Study: United Kingdom
No. of awards offered: 4
Application Procedure: The selection criteria adopted by these partner institutions vary and each has its own application requirements. Common to all is the need for a full curriculum vitae, and a clear statement of what the applicant proposes to achieve during the fellowship period. In every case, the selection is made by the host institution but must be endorsed by the British Council in Pakistan and by the Charles Wallace Pakistan Trust in the UK.
Funding: Trusts

The Charles Wallace Pakistan Trust - Visiting Artists

Subjects: Currently, these are mainly drawn from the visual arts, but other art forms also qualify for support.
Purpose: Enables arts practitioners from Pakistan to spend time in the UK on arts residencies or for training and familiarization purposes.
Level of Study: Professional development
Type: Grant
Value: An all-inclusive monthly stipend of £1,250 will be paid by the trust, also a return economy air fare.
Length of Study: The duration of stay in the UK can vary from 3 weeks to 3 months
Frequency: Annual
Study Establishment: The Prince's School of Traditional Arts, Gasworks
Application Procedure: Apply directly to the Trust's Secretary (timbutchard@wallace-trusts.org.uk).
Funding: Trusts

The Charles Wallace Pakistan Trust Doctoral Busaries

Subjects: The disciplines eligible for support are restricted to the following:
The humanities, arts, and creative industries
The social sciences, and social development
The environmental and health sciences.
Purpose: Twice a year, in June and December, awards are granted to individual students already in the UK who are normally in the final year, or anticipating the final year, of their PhDs, and who need additional funding to help them complete their studies.
Eligibility: All applicants are required to demonstrate academic excellence as well as financial need.
Level of Study: Doctorate
Type: Bursary
Value: Our maximum grant is £1,500, so applicants must have funds from other sources to cater for most of their needs.
Frequency: Biannual
Country of Study: United Kingdom
Application Procedure: Applicants must complete CWPT's application form and submit it, either in hard copy or as an email attachment, to the Secretary of the Trust. The application should be accompanied by a supporting letter on headed paper from the applicant's supervisor. Other documents testifying to the applicant's background and achievements to date should be kept to a minimum.
Closing Date: Mid-May or mid-November
Funding: Trusts

THE CHARLIE TROTTER CULINARY EDUCATION FOUNDATION

816 West Armitage, Chicago, IL 60614, United States of America
Tel: (1) 773 248 6228
Fax: (1) 773 248 6088
Email: info@charlietrotters.com
Website: www.charlietrotters.com/about/foundation.asp

Charlie Trotter's is regarded as one of the finest restaurants in the world, dedicated to excellence in the culinary arts. It has been instrumental in establishing new standards for fine dining. Its main goal is to educate and expose the youth to the great culinary arts in as many ways as possible. The Charlie Trotter Culinary Education Foundation, a non-profit organization, has been established to promote culinary arts among youth. The foundation is involved in awarding scholarships to students who are seeking careers in the culinary arts and working with Chicago-area youth to promote the enthusiastic quest for education as well as an interest in the cooking and food.

Charlie Trotter's Culinary Education Foundation Culinary Study Scholarship

Subjects: Cooking.
Eligibility: Open to an Illinois resident at the time of application.
Type: Scholarship
Value: US$5,000 cash scholarship for a pre-enrolled student
Length of Study: 1 year
Frequency: Annual
Country of Study: United States of America
Application Procedure: Check website for further details.
Closing Date: March 1st
Funding: Foundation, private
Contributor: Charlie Trotter's

For further information contact:

The Culinary Trust Scholarship Program P.O. Box 273, New York, NY 10013, United States of America
Tel: (1) 646 224 6989
Email: cholarships@theculinarytrust.com
Website: www.theculinarytrust.org
Contact: Amy Blackburn, Director of Administration

CIAT CHARTERED INSTITUTE OF ARCHITECTURAL TECHNOLOGISTS

397 City Road, London, EC1V 1NH, United Kingdom
Tel: (44) 020 7278 2206
Fax: (44) 020 7837 3194
Email: careers@ciat.org.uk
Website: http://www.ciat.org.uk/en/awards/John_Newey_Education_Foundation/
Contact: Holly Banks, Education and CPD Administrator

CIAT John Newey Education Foundation

Subjects: Architecture, building and planning and engineering.
Eligibility: Applicants must be members of CIAT, except chartered, and studying in full or part-time education on an approved course and facing hardship.
Level of Study: Foundation programme, Graduate, Postgraduate
Type: Bursary
Value: £500 each
Length of Study: 1 year
Frequency: Annual
No. of awards offered: 2
Application Procedure: Supporting testimony from the course tutor with an endorsing signature from the Head of Department.
Closing Date: March
Funding: Foundation

CHEMICAL HERITAGE FOUNDATION (CHF)

315 Chestnut Street, Philadelphia, PA, 19106, United States of America
Tel: (1) 215 925 2222
Fax: (1) 215 925 1954
Email: info@chemheritage.org
Website: www.chemheritage.org
Contact: Ashley Augustyniak, Fellowship Co-ordinator

The Beckman Center for the History of Chemistry is the historical unit of the Chemical Heritage Foundation (CHF), which is located in Philadelphia. The Center is devoted to preserving, making known and applying the history of the chemical and molecular science technologies and associated industries.

Dissertation Fellowships

Subjects: History of chemical sciences, technologies and industries.
Purpose: To fund graduate students at the PhD dissertation stage who are pursuing research in the chemical histories.
Eligibility: Open to scholars pursuing research on the history of the chemical sciences and must be a graduate student at the PhD dissertation stage.
Level of Study: Postdoctorate, Doctorate
Type: Fellowship
Value: US$26,000
Length of Study: 9 months
Frequency: Annual
Study Establishment: Chemical Heritage Foundation
Country of Study: United States of America
Application Procedure: Applicants must apply online at the website www.chemheritage.org.
Closing Date: February 15th
Funding: Private
No. of awards given last year: 4

Glenn E and Barbara Hodsdon Ullyot Scholarship

Subjects: The history of science.
Purpose: To advance understanding of the importance of the chemical sciences to the public's welfare.
Eligibility: Open to writers, journalists, educators and historians.
Level of Study: Doctorate, Postdoctorate, Postgraduate
Type: Scholarship
Value: US$6,000
Length of Study: A minimum of 2 months
Frequency: Annual
Study Establishment: Chemical Heritage Foundation
Country of Study: United States of America
No. of awards offered: 1
Application Procedure: Applicants must apply online at the website www.chemheritage.org.
Closing Date: February 15th
Funding: Private
No. of awards given last year: 1
Additional Information: Applications are invited from scholars, science writers and journalists.

Postdoctoral Fellowship

Subjects: History of chemical sciences, technologies and industries.
Purpose: To support historical research by PhD scholars focused on history of chemistry, technology and industry.
Eligibility: Open to a scholar with a PhD who will carry out historical research on the history of chemistry.
Level of Study: Postdoctorate
Type: Fellowship
Value: US$45,000
Length of Study: 9 months
Frequency: Annual
Study Establishment: Chemical Heritage Foundation
Country of Study: United States of America
Application Procedure: Applicants must apply online at the website www.chemheritage.org.
Closing Date: February 15th
Funding: Private

Research Travel Grants

Eligibility: Travel grant applicants must reside more than 75 miles from Philadelphia to be eligible. No more than one travel grant per person per fiscal year (July 1st to June 30th) can be awarded. Grants must be taken within one year of the award or the grantee must request an extension or reapply.
Type: Grant
Value: Travel grants are $750 per week and are intended to help defray the costs of travel and accommodation.
Application Procedure: There is no deadline for travel grant applications. Travel grant applications can be submitted at any time and are assessed by an internal CHF review committee. Please allow

for two weeks after submission for notification of the committee's decision.
Closing Date: No deadline
Additional Information: Travel grant applications must be submitted electronically, as Word or PDF files, to travelgrants@chemheritage.org.

Short Term Fellowship

Subjects: History of the chemical sciences and technologies.
Purpose: To fund scholars who are pursuing research on history of the chemical and molecular sciences, technologies, and industries.
Eligibility: Open to scholars pursuing research on the history of the chemical sciences.
Level of Study: Doctorate, Postdoctorate
Type: Fellowship
Value: US$3,000 per month
Length of Study: 1–4 months
Frequency: Annual
Study Establishment: Chemical Heritage Foundation
Application Procedure: Applicants must apply online at the website www.chemheritage.org.
Closing Date: February 15th
Funding: Private
Additional Information: Applicants must demonstrate a specific need to use the primary research collections in the CHF Library for their research.

Société de Chimie Industrielle (American Section) Fellowship

Subjects: The history of science.
Purpose: To stimulate public understanding of the chemical industries, using both terms in their widest sense.
Eligibility: Applications are encouraged from writers, journalists, educators and historians of science, technology and business.
Level of Study: Doctorate, Postdoctorate, Postgraduate
Type: Fellowship
Value: US$10,000
Length of Study: A minimum of 3 months
Frequency: Annual
Study Establishment: Chemical Heritage Foundation
Country of Study: United States of America
No. of awards offered: 1
Application Procedure: Applicants must apply online at the website www.chemheritage.org.
Closing Date: February 15th
Funding: Private
No. of awards given last year: 1
Additional Information: Multimedia, popular book projects and Web-based projects are encouraged.

THE CHEMICAL INSTITUTE OF CANADA

Suite 550, 130 Slater Street, Ottawa, ON, K1P 6E2, Canada
Tel: (1) 613 232 6252 ext 223
Fax: (1) 613 232 5862
Email: gthirlwall@cheminst.ca
Website: www.cheminst.ca
Contact: Gale Thirlwall, Awards Manager

The Chemical Institute of Canada (CIC) is the umbrella organization for three Constituent Societies - the Canadian Society for Chemistry (CSC), the Canadian Society for Chemical Engineering (CSChE) and the Canadian Society for Chemical Technology (CSCT). The CIC establishes strategic direction and identifies synergies in matters of common interest to the Constituent Societies, to enhance the image of the chemical sciences and engineering with all sectors of the public and to deliver common services to individual members.

CIC Award for Chemical Education

Subjects: Chemistry and chemical engineering.
Purpose: To recognize a person who has made outstanding contributions in Canada to education at the post-secondary level in the field of chemistry or chemical engineering.
Level of Study: Professional development
Type: Award
Value: A framed scroll, a cash prize of Canadian $1,000 and up to Canadian $400 for travel expenses
Frequency: Annual
Country of Study: Canada
No. of awards offered: 1
Application Procedure: Applicants must be nominated. The applicant should submit a curriculum vitae with letters of support and the CIC nomination.
Closing Date: July 2nd
Funding: Private
Contributor: CIC Chemical Education Fund
No. of awards given last year: 1

CIC Catalysis Award

Subjects: Chemistry/chemical engineering.
Purpose: To recognize an individual who has made a distinguished contribution to the field of catalysis while resident in Canada.
Level of Study: Research
Type: Award
Value: A rhodium-plated silver medal and travel expenses to present the Award Lecture
Frequency: Every 2 years
Country of Study: Canada
No. of awards offered: 1
Application Procedure: Applicants must be nominated. They should submit (1) nomination form, (2) curriculum vitae, (3) bio and citation, (4) letters of support.
Closing Date: October 1st (odd years only)
Funding: Foundation
Contributor: Catalysis Foundation

CIC Macromolecular Science and Engineering Lecture Award

Subjects: Macromolecular science and engineering
Purpose: To recognize an individual who has made a distinguished contribution to macromolecular science or engineering.
Level of Study: Research
Type: Award
Value: A framed scroll, a cash prize, and travel expenses
Frequency: Annual
Country of Study: Canada
No. of awards offered: 1
Application Procedure: Applicants must be nominated. The applicant must submit a nomination form, curriculum vitae, bio and citation and letters of support.
Closing Date: July 2nd
Funding: Private
Contributor: NOVA Chemicals Limited
No. of awards given last year: 1

CIC Medal

Subjects: Chemistry, chemical engineering and chemical technology.
Purpose: To recognize a person who has made an outstanding contribution to the science of chemistry or chemical engineering in Canada.
Level of Study: Research
Type: Award
Value: A medal and travel expenses
Frequency: Annual
Country of Study: Canada
No. of awards offered: 1
Application Procedure: Applicants must be nominated. The applicant must submit (1) nomination form, (2) curriculum vitae, (3) bio and citation, (4) letters of support.
Closing Date: July 2nd
Funding: Private
No. of awards given last year: 1

CIC Montreal Medal

Subjects: Chemistry, chemical engineering and chemical technology.
Purpose: To honour a person who has shown significant leadership in or outstanding contribution to the profession of chemistry or chemical engineering in Canada.
Eligibility: Open to administrative contributions within the Chemical Institute of Canada and other professional organizations that con-

tribute to the advancement of the professions of chemistry and chemical engineering. Contributions to the sciences of chemistry and chemical engineering are not considered. Administrative contributions to the CIC, contributions by chemical educators and by staff members of chemical industries and single individual exploits which contribute to the advancement of the chemical profession.
Level of Study: contribution within the chemical community
Type: Award
Value: A medal and travel expenses if required
Frequency: Annual
Country of Study: Canada
No. of awards offered: 1
Application Procedure: Applicants must be nominated. The applicant must submit an application form, curriculum vitae, bio and citation and letters of support.
Closing Date: July 2nd
Funding: Private
Contributor: Montréal CIC Local Section
No. of awards given last year: 1

Environmental Division Research and Development Award

Subjects: Environmental Chemistry or Environmental Chemical Engineering.
Purpose: To award distinguished contributions to the field of Environmental Chemistry or Environmental Chemical Engineering.
Eligibility: Open to any scientist or engineer residing in Canada who has made distinguished contributions to research and/or development in the fields of environmental chemistry or environmental chemical engineering.
Type: Award
Value: A framed scroll, $1,000 and travel expenses
Frequency: Annual
No. of awards offered: 1
Application Procedure: Applicants must submit electronically nomination package to the Awards manager.
Closing Date: July 2nd
Funding: Private
Contributor: P. Beaumier

CHIANG CHING KUO FOUNDATION FOR INTERNATIONAL SCHOLARLY EXCHANGE

13F, 65 Tun Hwa South Road Section 2, Taipei, 106-ROC, Taiwan
Tel: (886) 2 2704 5333
Fax: (886) 2 2701 6762
Email: cckf@ms1.hinet.net
Website: www.cckf.org

The Chiang Ching Kuo Foundation for International Scholarly Exchange is a non-profit organization headquartered in Taipei, the capital of the Republic of China. The Foundation was established in 1989 in honour of the late President Chiang Ching kuo. The main objective of the Foundation is to promote the study of Chinese culture and society, broadly defined.

Chiang Ching Kuo Foundation Doctoral Fellowships

Subjects: Chinese studies in the field of humanities and social sciences.
Purpose: To financially support Doctoral candidates while writing their dissertations.
Eligibility: Open to applicants who have completed all other requirements for their PhD degree except the dissertation. Candidates must not be employed or receive grants from other sources.
Level of Study: Doctorate
Type: Fellowships
Value: Up to US$15,000
Frequency: Annual
Application Procedure: Applicants need to submit a 1 page summary of the proposed project, budget, curriculum vitae and detailed description of the proposed project along with the application form. Application forms are available online.
Closing Date: October 15th
Funding: Private, commercial

For further information contact:

Email: cckf@ms1.hinet.net

Chiang Ching Kuo Foundation for International Scholarly Exchange Publication Subsidies

Subjects: Academic works, periodicals, and journals.
Purpose: To assist in the final stages of publishing academic works.
Eligibility: Open to scholars in the final stages of publishing academic works. Applications from scholars affiliated with institutions in Taiwan must involve cooperation with one or more scholars from other countries. Applicants for publication subsidies must be affiliated with a university or other academic institution.
Type: Grant
Value: New Taiwan $1,000,000
Application Procedure: Applicants must use the application forms provided directly from the Foundation Secretariat. Three copies of the application and supporting documents must be submitted by registered mail to the Secretariat. In addition, electronic version of all application materials must be enclosed on diskette or sent as e-mail attachment to: cckf@ms1.hinet.net with heading "Application Materials from (Name)" in the header of the message.
Closing Date: September 15th and January 15th
Funding: Trusts

For further information contact:

The Chiang Ching-kuo Foundation for International Scholarly Exchange
Email: cckfnao@aol.com

Chiang Ching–Kuo Foundation for Scholarly Exchange Eminent Scholar Lectureship

Subjects: Any subject.
Purpose: To sponsor eminent foreign scholars to come to Taiwan to take up lectureships or positions as visiting scholars.
Eligibility: Open to eminent scholars invited by universities or academic institutions of Taiwan.
Level of Study: Lectureship
Type: Lectureship/Prize
Value: New Taiwan $2,000,000
Length of Study: 1 year
Frequency: Annual
Study Establishment: Universities or academic institutions in Taiwan
Country of Study: Taiwan
Application Procedure: Applicants must use the application forms provided by the Foundation. The application must be sent by registered mail to the Secretariat. Electronic version of all application materials must be enclosed on diskette or sent as e-mail attachment to: cckf@ms1.hinet.net with heading "Application Materials from (Name)" in the header of the message. Applications are accepted from June 1st.
Closing Date: October 15th
Funding: Foundation
Additional Information: Project directors who are currently receiving Foundation aid are ineligible to apply. Project directors may not submit more than one application.

For further information contact:

Email: cckf@ms1.hinet.net

THE CHICAGO TRIBUNE

435 North Michigan Avenue, Chicago, IL, 60611, United States of America
Tel: (1) 312 222 1922
Fax: (1) 312 222 3751
Email: jwoelffer@tribune.com
Website: www.chicagotribune.com

The Chicago Tribune is the Midwest's leading newspaper. The Chicago Tribune Literary Awards are part of a continued dedication to readers, writers and ideas.

Nelson Algren Awards

Subjects: Short fiction.
Purpose: To award writers of short fiction.
Eligibility: This Contest is open to legal residents of the 50 United States or DC ages eighteen years and older at the time of entry. Employees (and the employees' immediate family members living in the same household) of the Sponsor and its advertising companies, parent companies, affiliates, subsidiaries, promotion and delivery contractors and/or public relations companies, are not eligible to participate. This Contest is Void where Prohibited By Law
Level of Study: Unrestricted
Type: Award
Value: One $5,000 prize and three runner-up prizes of $1,500
Frequency: Annual
Country of Study: Any country
No. of awards offered: 4
Application Procedure: Applicants must send a stamped addressed envelope with a request for written guidelines. The competition will begin accepting entries from November 1st.
Closing Date: February
Funding: Corporation
Additional Information: Please contact to Chicago Tribune for latest updates.

For further information contact:

Chicago Tribune, Nelson Algren Awards, 435 N. Michigan Avenue, TT200, Chicago, IL 60611
Email: printersrow@tribune.com

CHILDREN'S LITERATURE ASSOCIATION

1301 W. 22nd Street, Suite 202, Oak Brook, IL, 60523, United States of America
Tel: (1) 630 571 4520
Fax: (1) 708 876 5598
Email: info@childlitassn.org
Website: www.childlitassn.org
Contact: Administrator

The Children's Literature Association is an international organization whose mission is to encourage high standards of criticism, scholarship, research and teaching in children's literature.

ChLA Beiter Graduate Student Research Grant

Subjects: Children's literature.
Purpose: To fund proposals of original scholarship with the expectation that the undertaking will lead to a publication or a conference presentation and contribute to the field.
Eligibility: Winners must be, or become, members of the Children's Literature Association. Previous recipients are not eligible to reapply until the third year from the date of the first award.
Level of Study: Graduate
Type: Grant
Value: From US$500–1,500, which may be used to purchase supplies and materials e.g. books and videos, and as research support e.g. photocopying, or to underwrite travel to special collections or libraries
Frequency: Annual
Country of Study: Any country
No. of awards offered: 1–10
Application Procedure: Applicants must submit their application online including email address, academic institution and status, the expected date of their degree, a detailed description of the research proposal, a curriculum vitae and two letters of reference, one of which must be from the applicant's dissertation or thesis advisor. See www.childlitassn.org for full details.
Closing Date: February 1st
Funding: Private
No. of awards given last year: 7
No. of applicants last year: 14
Additional Information: Applicants should visit the website for further details. If applicants wish to receive guidelines by mail, a stamped addressed envelope must be provided.

ChLA Faculty Research Grant

Subjects: Children's literature.
Purpose: To award proposals dealing with criticism or original scholarship with the expectation that the undertaking will lead to publication and make a significant contribution to the field of children's literature in the area of scholarship or criticism.
Eligibility: Applicants must be, or become, members of the Children's Literature Association.
Level of Study: Postgraduate, Research, Predoctorate, Doctorate, Postdoctorate
Type: Grant
Value: Individual awards may range US$500–1,500 and may be used only for research-related expenses such as travel to special collections or materials and supplies. Funds are not intended for work leading to the completion of a professional degree
Frequency: Annual
Country of Study: Any country
No. of awards offered: 1–10
Application Procedure: Applicants must submit their application online and a curriculum vitae. Applications must include the applicant's name, address, telephone number and email address, details of the academic institution the applicant is affiliated with and a detailed description of the research proposal, not exceeding three single spaced pages, and indicating the nature and significance of the project, where it will be carried out and the expected date of completion. See www.childlitassn.org for full details.
Closing Date: February 1st
Funding: Private
No. of awards given last year: 5
No. of applicants last year: 18
Additional Information: In honour of the achievement and dedication of Dr Margaret P Esmonde, proposals that deal with critical or original work in the areas of science fantasy or science fiction for children or adolescents will be awarded the Margaret P Esmonde Memorial Grant. Applicants should visit the website for further details. If applicants wish to receive guidelines by mail, a stamped addressed envelope must be provided.

CHINESE AMERICAN MEDICAL SOCIETY (CAMS)

41 Elizabeth Street, Suite 600, New York, NY, 10013, United States of America
Tel: (1) 212 334 4760
Fax: (1) 646 304 6373
Email: jlove@camsociety.org
Website: www.camsociety.org
Contact: Dr H H Wang, Executive Director

The Chinese American Medical Society (CAMS) is a non-profit, charitable, educational and scientific society that aims to promote the scientific association of medical professionals of Chinese descent. It also aims to advance medical knowledge and scientific research with emphasis on aspects unique to the Chinese and to promote the health status of Chinese Americans. The Society makes scholarships available to medical dental students and provides summer fellowships for students conducting research in health problems related to the Chinese.

CAMS Scholarship

Subjects: Medical or dental studies.
Purpose: To help defray the cost of study.
Eligibility: Open to Chinese Americans, or Chinese students who are residing in the United States of America. Applicants must be full-time medical or dental students at approved schools within the United States of America and must be able to show academic proficiency and financial hardship.
Level of Study: Doctorate
Type: Scholarship
Value: US$1,500–2,500
Frequency: Annual
Country of Study: United States of America
No. of awards offered: 3–5
Application Procedure: Applicants must complete an application form and send it together with a letter for the Dean of Students verifying good standing, two to three letters of recommendation, a personal statement, a curriculum vitae and a financial statement. Application forms can also be downloaded from the website.

Closing Date: April 30th
Funding: Private
Contributor: Membership and fund-raising
No. of awards given last year: 7
No. of applicants last year: 13

For further information contact:

CAMS Scholarship Committee, 41 Elizabeth Street, Suite 403, NY 10013
Tel: (212) 334 4760
Contact: Jerry Huo, Chairman

CHINOOK REGIONAL CAREER TRANSITIONS FOR YOUTH

Room B310, 1701 - 5 Avenue South, Lethbridge, AB, T1J 0W4, Canada
Tel: (1) 403 328 3996
Fax: (1) 403 320 2365
Email: mvennard@pallisersd.ab.ca
Website: http://www.careersteps.ca

The Chinook regional career transitions for youth aims to improve the school-to-work transitions for students, promoting lifelong learning and coordinating and implementing career development activities and programming for youth.

Alberta Blue Cross 50th Anniversary Scholarships

Subjects: All subjects.
Purpose: To assist young Albertans pursuing post-secondary studies across the province.
Eligibility: Open to applicants who are registered Indian, Inuit, or Melis and are residents of Alberta and have financial need.
Level of Study: Postgraduate
Type: Scholarship
Value: Canadian $375–1,250
Frequency: Annual
Country of Study: Canada
No. of awards offered: 63
Application Procedure: A completed application form must be sent. For further information, please check website www.ab.bluecross.ca.
Closing Date: September 20th

For further information contact:

Alberta Blue Cross Corporate Offices 10009-108 Street NW, Edmonton, AB T5J 3C5
Fax: 780 498 8096

Robin Rousseau Memorial Mountain Achievement Scholarship

Subjects: Mountain leadership and safety.
Purpose: To bring about awareness of ways to improve safety in the mountains.
Eligibility: Applicants must be Alberta residents and active in the mountain community; and plan to study in any recognized Mountain Leadership and Safety program.
Level of Study: Professional development
Type: Scholarship
Value: Course fee
Frequency: Annual
No. of awards offered: 1
Application Procedure: A completed application form must be sent.
Closing Date: January 30th

For further information contact:

Alberta Scholarship Programs Box 28000 Stn Main, Edmonton, AB T5J 4R4
Tel: 780 427 8640
Fax: 780 427 1288
Email: scholarships@gov.ab.ca

Terry Fox Humanitarian Award

Subjects: Social services.
Purpose: To encourage voluntary humanitarian work.
Eligibility: Open to Canadian citizens who are not more than 25 years of age.
Level of Study: Professional development
Type: Scholarship
Value: Canadian $3,500–7,000
Frequency: Annual
No. of awards offered: 20
Application Procedure: A completed application form must be submitted.
Closing Date: February 1st

For further information contact:

The Terry Fox Humanitarian Award Program, AQ 5003, 8888 University Drive, Burnaby, BC V5A 1S6
Website: http://terryfoxawards.ca/english/

Toyota Earth Day Scholarship Program

Subjects: Environmental community service.
Purpose: To encourage community service.
Eligibility: Open to students who have achieved academic excellence and distinguished themselves in environmental community service and extracurricular and volunteer activities.
Level of Study: Professional development
Type: Scholarship
Value: Canadian $5,000
Frequency: Annual
No. of awards offered: 20
Application Procedure: Application form available online.
Closing Date: February 15th

For further information contact:

Toyota Earth Day Scholarship Program, III Peter Street, Suite 503, Toronto, ON M5V 2H1
Email: scholarship@earthday.ca

CHOIRS ONTARIO

Choirs Ontario A-1422 Bayview Avenue, Toronto, ON, M4G 3A7, Canada
Tel: (1) 416 923 1144
Fax: (1) 416 929 0415
Email: info@choirsontario.org
Website: www.choirsontario.org
Contact: Melva Graham

Choirs Ontario is an arts service organization dedicated to the promotion of choral activities and standards of excellence. Established in 1971 as the Ontario Choral Federation, Choirs Ontario provides services to choirs, conductors, choristers, composers, administrators and educators as well as anyone who enjoys listening to the sound of choral music. Choirs Ontario operates with the financial assistance of the Ministry of Culture, the Ontario Arts Council, the Trillium Foundation, the Toronto Arts Council and numerous foundations, corporations and individual donors.

Ruth Watson Henderson Choral Competition

Subjects: Singing, treble voice choirs.
Purpose: To reward new choral works.
Eligibility: Open to all Canadian Citizens and landed immigrants. Participating composers must reside in Canada. There is no age limit.
Level of Study: Unrestricted
Type: Award
Value: Cash prize of Canadian $1,000 and a concert performance with one of Toronto's leading choirs
Length of Study: 4–6 minutes
Frequency: Every 2 years
Country of Study: Canada
Application Procedure: Applicants must submit four legible photocopies of the score (not original manuscripts). More than one entry may be submitted, but each entry must be accompanied by a separate entry form and a fee of Canadian $20. The composer's name must not appear on any score. Scores will be returned if a stamped addressed envelope is included. Application forms can be downloaded from the website www.choirsontario.org/ruthwatsonhenderson.html
Closing Date: September 30th
Funding: Private
Contributor: Choirs Ontario

Additional Information: Ruth Watson Henderson, one of Canada's foremost musicians, is internationally known both as a composer and pianist. Her compositions have been commissioned, performed and recorded worldwide.

CHRONIC DISEASE RESEARCH FOUNDATION (CDRF)

St Thomas' Hospital, 4st Floor, South Wing Block D, Westminster Bridge Road, London, SE1 7EH, England
Tel: (44) 20 7633 9790
Fax: (44) 20 7922 8154
Email: christel.barnetson@cdrf.org.uk
Website: www.cdrf.org.uk
Contact: Mrs Christel Barnetson, Chief Administrator

The Chronic Disease Research Foundation (CDRF) was established to look at new ways of exploring the genetics of diseases associated with ageing. Its mission is to target those common diseases such as osteoporosis, arthritis, back pain, migraine, asthma and diabetes, inherited from our parents, and prevent and alleviate them now and for future generations. Its principal focus is on comparative studies of identical and non-identical twins, undertaken at the Twin Research Unit of St Thomas' Hospital.

CDRF Project Grants

Subjects: Medicine and surgery.
Purpose: To provide funds for researchers studying the genetic basis of common chronic diseases associated with ageing in developed countries.
Eligibility: There are no restrictions specified. For more details, see the organization website.
Level of Study: Research, Postgraduate
Type: Award/Grant
Value: UK£30,000–150,000
Length of Study: 2–3 years
Frequency: Dependent on funds available
Country of Study: United Kingdom
No. of awards offered: Dependent on availability of funds
Application Procedure: Applicants must submit a preliminary proposal of no more than one side of A4-size paper including an outline of the proposal, a list of principal aims and objectives and scale of funding. If the CDRF's panel of experts consider the project to be of relevance, applicants must then submit a full grant proposal.
Closing Date: No deadline
Funding: Private
No. of awards given last year: None

CDRF Research Fellowship

Subjects: Medicine and surgery.
Purpose: To promote postgraduate education and enable the charity to carry out further research projects.
Eligibility: See the organization website
Level of Study: Postgraduate, Research
Type: Scholarships and fellowships
Value: UK£30,000–150,000
Length of Study: 2–5 years
Frequency: Dependent on funds available
Country of Study: United Kingdom
No. of awards offered: Dependent on availability of funds
Application Procedure: Applicants must submit a preliminary proposal of no more than one side of an A4-size paper including an outline of the proposal, a list of principal aims and objectives and the scale of funding. If the CDRF's panel of experts consider the project to be of relevance, applicants must then submit a proposal for a full grant.
Closing Date: See the organization website
Funding: Private
No. of awards given last year: 2

THE CIRCULATION FOUNDATION

35-43 Lincoln's Inn Fields, London, WC2A 3PE, England
Tel: (44) 7304 4779
Fax: (44) 7430 9235
Email: info@circulationfoundation.org.uk
Website: www.circulationfoundation.org.uk

The Circulation Foundation aims to provide research funding to find cures, better treatments and improve diagnosis of vascular disease. The Foundation also hopes to raise awareness of the disease's prevalence and impact and to provide information and support to sufferers, their families and friends.

Owen Shaw Award

Subjects: Medicine and surgery.
Purpose: To devise better methods of helping patients to attain early mobilization.
Eligibility: Open to all those with an interest in amputee rehabilitation.
Level of Study: Unrestricted
Type: Award/Grant
Value: UK£3,000
Length of Study: 1 year
Frequency: Annual
Study Establishment: Any restricted research establishment
Country of Study: United Kingdom
No. of awards offered: 1
Application Procedure: Candidates must complete an outline proposal form, available from the Foundation.
Closing Date: June 5th
Funding: Private, trusts
Contributor: Owen Shaw
No. of awards given last year: 1
No. of applicants last year: 8

CITY UNIVERSITY, LONDON

Northampton Square, London, EC1V 0HB, United Kingdom
Tel: (44) 0 20 7040 5060
Fax: (44) 0 20 7040 5070
Email: enquiries@city.ac.uk
Website: www.city.ac.uk

The City University, London has developed into an innovative, forward-looking centre of excellence, with a well-deserved reputation in professional and business education and research. Today, it is ready and equipped to face the educational and professional challenges of the knowledge economy. It has close contacts with the leading professional institutions and with business and industry, both at home and abroad. Their professional links are reflected in their teaching and research staff.

Davis and Lyons Bursaries in Music

Subjects: Music and performing arts.
Purpose: To support students to acquire postgraduate degree in music at city.
Eligibility: Open to all students (UK/EU and overseas, full-time and part-time) beginning any postgraduate programme in music at city.
Type: Studentships and bursaries
Value: UK£1,000
Length of Study: 1 year
Frequency: Annual
Country of Study: United Kingdom
No. of awards offered: 2
Application Procedure: Check website, www.city.ac.uk/music/
Closing Date: March 1st
Additional Information: An announcement of the Davis and Lyons Bursaries in music outcome will be made in late July.

CLARA HASKIL COMPETITION

Case Postale 234, 31 rue du Conseil, CH-1800 Vevey, Switzerland
Tel: (41) 21 922 6704
Fax: (41) 21 922 6734
Email: info@clara-haskil.ch
Website: www.regart.ch/clara-haskil
Contact: Mr Patrick Peikert, Director

The Clara Haskil Competition exists to recognize and help a young pianist whose approach to piano interpretation is of the same spirit that constantly inspired Clara Haskil, and that she illustrated so perfectly.

Clara Haskil International Piano Competition

Subjects: Piano and music.

Purpose: To recognize and financially help a young pianist.
Eligibility: Open to pianists of any nationality and either sex who are no more than 27 years of age.
Level of Study: Postgraduate
Type: Prize
Value: Swiss Francs 25,000
Frequency: Every 2 years
Country of Study: Any country
No. of awards offered: 1
Application Procedure: Applicants must pay an entry fee of Swiss Franc 200.
Closing Date: May 15th
Funding: Corporation, international office, trusts
Contributor: Fondation Nestlé pour l'Art
No. of awards given last year: 2
No. of applicants last year: Approx. 150
Additional Information: The competition is usually held during the last weeks of August or the beginning of September.

For further information contact:

International Piano Competition, Concours Clara Haskil
Tel: 41 21 922 67 04
Fax: 41 21 922 67 34
Email: info@clara-haskil.ch

THE CLAUDE LEON FOUNDATION

17 Ivy Lane, Tokai, Cape Town, 7945, South Africa
Tel: (27) 21 712 7221
Fax: (27) 86 614 5915
Email: postdocadmin@leonfoundation.co.za
Website: www.leonfoundation.co.za
Contact: Mrs Gale Minnaar

The Claude Leon Foundation is a charitable trust, resulting from a bequest by Claude Leon (1887–1972). A founder and managing director of the Elephant Trading Company, a wholesale business based in Johannesburg, Claude Leon also helped develop several well-known South African companies, including Edgars, OK Bazaars and the mining house Anglo Transvaal (later Anglovaal). He served for many years on the Council of the University of the Witwatersrand, which in 1971 awarded him an honorary Doctorate of Law. The university postdoctoral fellowship award programme is now in its thirteenth year, and has as its goal the building of research capacity at South African universities.

Postdoctoral Fellowships (Claude Leon)

Subjects: Science, engineering, medical sciences.
Purpose: To fund postdoctoral research.
Eligibility: Open to South African and foreign nationals. Preference will be given to candidates who have received their doctoral degrees in the last 5 years and are currently underrepresented in South African science, engineering and medical science.
Level of Study: Postdoctorate
Type: Fellowship
Value: Rand 175,000
Length of Study: 2 years
Frequency: Annual
Study Establishment: South African universities
Country of Study: South Africa
Application Procedure: Application forms are available from the website or the Postdoctoral Fellowships Administrator.
Closing Date: May 31st
Funding: Foundation
Contributor: The Claude Leon Foundation
No. of awards given last year: 70
No. of applicants last year: 230

CLEMSON UNIVERSITY

Clemson University, Clemson, SC, 29634, United States of America
Tel: (1) 864 656 3311
Fax: (1) 864 656 5344
Email: finaid@clemson.edu
Website: www.clemson.edu
Contact: Associate Director

Uemson University is a selective, public, land-grant university, which is committed to world-class teaching, research and public service in the context of general education, student development and continuing education.

Clemson Graduate Assistantships

Subjects: All subjects.
Purpose: To assist graduates to be employed as a graduate assistant.
Eligibility: A student must possess at least a bachelor's degree and be enrolled in a graduate degree program. To be eligible for any graduate appointment, a graduate student must satisfy the appropriate minimum graduate level enrollment requirement of nine credit hours during each semester and three credit hours in each summer session.
Level of Study: MBA
Type: Graduate assistantship
Value: The assistantships pay stipends starting at US$6.18 per hour. The pay depends upon job duties and candidate qualifications. In addition, graduate assistants are granted partial remission of academic fees and enjoy some benefits provided to the University faculty and staff. Graduate assistantship presently pay US$1,044 per semester in tuition and fees
Length of Study: 2 years
Frequency: Dependent on funds available, on an annual or 9-month basis
Study Establishment: Clemson University
Country of Study: United States of America
No. of awards offered: Varies
Application Procedure: Applicants must contact the various university departments or offices for information or submit a general application with a curriculum vitae to the MBA office.
Closing Date: February 1st

CLEVELAND INSTITUTE OF MUSIC

11201 East Boulevard, Cleveland, OH 44106, United States of America
Tel: (1) 216 791 5000
Fax: (1) 216 795 3141
Email: info@cim.edu
Website: www.cim.edu
Contact: Kristie Gripp, Director of Financial Aid

The Cleveland Institute of Music's mission is to provide talented students with a professional, world-class education in the art of music. The Institute ranks among the top tier music schools across the nation, granting degrees up to the doctoral level. More than 80 per cent of the Institute's alumni perform in major national and international orchestras and opera companies, while others hold prominent teaching positions.

Cleveland Institute of Music Scholarships and Accompanying Fellowships

Subjects: Music.
Eligibility: Candidates for the accompanying fellowships should have a Bachelor of Music Degree or equivalent and must be proficient in English.
Level of Study: Graduate, Postgraduate
Type: Scholarships and fellowships
Value: US$1,000–38,000 for scholarships and US$1,000–3,000 for accompanying fellowships. No travel grants are provided
Length of Study: 1 academic year for scholarships or from August to the following May for accompanying fellowships. Scholarships are renewable
Frequency: Annual
Study Establishment: The Cleveland Institute of Music
Country of Study: United States of America
No. of awards offered: Approx. 425 scholarships and 15 accompanying fellowships
Application Procedure: Applicants must apply online.
Closing Date: December 1st
Funding: Private

THE COLLEGE OF OPTOMETRISTS

42 Craven Street, London, WC2N 5NG, United Kingdom
Tel: (44) 0 20 7839 6000
Fax: (44) 0 20 7839 6800
Email: optometry@college-optometrists.org
Website: www.college-optometrists.org

The College of Optometrists, founded in 1980, is a registered charity incorporated by Royal Charter in 1995. It is the single successor body to the British Optical Association, founded in 1895, and The Scottish Association of Opticians, formed in 1921.

College of Optometrists Postgraduate Scholarships

Subjects: Optometry.
Purpose: To support research work in optometry or a closely related subject.
Eligibility: Students must be Fellows or Members of the College, although this requirement does not apply to the project Supervisor.
Level of Study: Postgraduate
Type: Scholarships
Value: £19,400 (within London) or £17,180 (outside London)
Length of Study: Up to 3 years
Frequency: Annual
Country of Study: United Kingdom
No. of awards offered: 11
Closing Date: March 31st
Additional Information: For a copy of the regulations, the conditions of awards and application forms, contact murtagh@college-optometrists.org or through Tel: 020 7766 4364.

For further information contact:

The College of Optometrists 42 Craven Street, London, WC2N 5NG, United Kingdom
Tel: (44) 020 7766 4364
Contact: Martin Cordiner

COLLEGEVILLE INSTITUTE FOR ECUMENICAL AND CULTURAL RESEARCH

14027 Fruit Farm Road, Box 2000, Collegeville, MN, 56321, United States of America
Tel: (1) 320 363 3366
Fax: (1) 320 363 3313
Email: staff@collegevilleinstitute.org
Website: www.collegevilleinstitute.org
Contact: Donald Ottenhoff, Executive Director

The Institute for Ecumenical and Cultural Research seeks to discern the meaning of Christian identity and unity in a religiously and culturally diverse nation and world and to communicate that meaning for the mission of the church and the renewal of human community. The Institute is committed to research, study, prayer, reflection and dialogue, in a place shaped by the Benedictine tradition of worship and work.

Bishop Thomas Hoyt Jr Fellowship

Subjects: Ecumenical and cultural research.
Purpose: To provide the Institute's residency fee to a North American person of colour writing a doctoral dissertation, in order to help the churches to increase the number of persons of colour working in ecumenical and cultural research.
Eligibility: Open to a North American, Canadian or Mexican person of colour writing a doctoral dissertation within the general area of the Institute's concern.
Level of Study: Postgraduate
Type: Fellowship
Value: US$5,000 per year
Length of Study: 1 academic year
Frequency: Annual
Study Establishment: The Institute
Country of Study: United States of America
No. of awards offered: 1 each year (or 2 if for semesters)
Application Procedure: Applicants must apply in the usual way to the Resident Scholars Programme (see separate listing). If invited by the admissions committee to be a Resident Scholar, the person will then be eligible for consideration for the Hoyt Fellowship.
Closing Date: November 1st and February 1st
Funding: Private
No. of awards given last year: 1
No. of applicants last year: 1

For further information contact:

Tel: 320-363-3367
Email: dottenhoff @ collegevilleinstitute.org
Contact: Donald B. Ottenhoff, Director

COLLEGIO CARLO ALBERTO FOUNDATION (CCAF)

Via Real Collegio 30, Moncalieri, Turin, 10024, Italy
Tel: (39) 11 670 5000
Fax: (39) 11 670 5082
Email: segreteria@carloalberto.org
Website: www.collegiocarloalberto.it

The Collegio Carlo Alberto Foundation (CCAF) was legally constituted on April 27, 2004, at the joint initiative of the Compagnia di San Paolo Foundation and the University of Turin. The Foundation aims to promote, manage and develop, in conjunction with the University of Turin, research and postgraduate education in the fields of economics, finance and law, as well as in related disciplines.

CCAF Junior Research Fellowship

Subjects: Economics and finance.
Purpose: To help in the creation of a stimulating interactive environment with Doctoral students and junior researchers.
Eligibility: Open to candidates who have obtained a PhD.
Level of Study: Postdoctorate
Type: Fellowships
Value: €50,000 per year
Length of Study: 2 years
Frequency: Annual
Country of Study: Italy
No. of awards offered: 2
Application Procedure: Applicants must send a curriculum vitae, a completed research paper, a 1–2 page research proposal and 2 letters of reference.
Closing Date: December 10th

COLT FOUNDATION

New Lane Havant, Hampshire, PO9 2LY, England
Tel: (44) 23 9249 1400
Fax: (44) 23 9249 1363
Email: jackie.douglas@uk.coltgroup.com
Website: www.coltfoundation.org.uk

The primary interest of the Colt Foundation is the promotion of research into medical and environmental problems created by commerce and industry and is aimed particularly at discovering the cause of illnesses arising from conditions at the place of work. The Foundation also makes grants to students taking higher degrees in related subjects.

Colt Foundation PhD Fellowship

Subjects: Medical and natural sciences including public health and hygiene, sports medicine, biological and life sciences, physiology or toxicology.
Purpose: To encourage the young scientists of the future.
Eligibility: Open to any student proposing to take a PhD in the correct subject area in a United Kingdom university or college.
Level of Study: Doctorate
Type: Fellowship
Value: The stipend rate for the first year is £12,000 (£13,000 inside London), rising with inflation for the following two years. UK fees will be paid as incurred, together with a sum to cover research expenses.
Length of Study: 3 years

Frequency: Annual
No. of awards offered: 3
Application Procedure: Applicants must visit the website where details are posted in August each year.
Closing Date: October 17th
Funding: Foundation
No. of awards given last year: 3

COLUMBIA UNIVERSITY

405 Low Library, MC 4335, 535 West 116th Street, New York, NY, 10027, United States of America
Tel: (1) 212 854 3830
Fax: (1) 212 854 0274
Email: support@ei.columbia.edu
Website: www.earth.columbia.edu

The Earth Institute at Columbia University brings together talent from throughout the University to address complex issues facing the planet and its inhabitants, with particular focus on sustainable development and the needs of the world's poor.

Marie Tharp Visiting Fellowships

Subjects: Geosciences, social sciences, engineering and environmental health sciences.
Purpose: To provide an opportunity for women scientists to conduct research at one of the related departments within the Earth Institute.
Eligibility: Open to women candidates who have obtained their PhD and are citizens of the United States.
Level of Study: Doctorate, Research
Type: Fellowships
Value: US$25,000
Length of Study: 3 months
Frequency: Annual
Country of Study: United States of America
Application Procedure: Applicants must submit a 3 page proposal, a curriculum vitae, a proposed budget and complete contact information of 3 references.
Closing Date: March 31st
Additional Information: All application materials may be submitted by mail or by email.

For further information contact:

ADVANCE at The Earth Institute at Columbia University Lamont-Doherty Earth Observatory of Columbia University, United States of America
Email: novikova@ldeo.columbia.edu
Contact: Natasha Novikova , Program Coordinator

THE COMMONWEALTH FUND

1 East 75th Street, New York, NY, 10021, United States of America
Tel: (1) 212 606 3800
Fax: (1) 212 606 3500
Email: grants@cmwf.org.
Website: www.cmwf.org

The Commonwealth Fund of New York is a philanthropic foundation established in 1918. The Fund supports independent research on health and social issues and makes grants to improve healthcare practice and policy.

The Commonwealth Fund/Harvard University Fellowship in Minority Health Policy

Subjects: Health policy, public health and management, with special programme activities on minority health issues.
Purpose: To create physician-leaders who will pursue careers in minority health policy.
Eligibility: Open to physicians who are citizens of the United States of America and who have completed their residency. Additional experience beyond residency is preferred. Applicants must demonstrate an awareness of, or interest and experience in dealing, with the health needs of minority populations, strong evidence of past leadership experience, as related to community efforts and health policy and the intention to pursue a career in public health practice, policy, or academia.
Level of Study: Graduate, Research, Postgraduate, Professional development
Type: Fellowship
Value: US$50,000 stipend, full tuition, health insurance, books, travel, and related program expenses,including financial assistance for a practicum project.
Length of Study: 1 year
Frequency: Annual
Study Establishment: Harvard Medical School
Country of Study: United States of America
No. of awards offered: Up to 5
Application Procedure: Applications available online at the website: www.cmwf.org/fellowships
Closing Date: January 3rd
Funding: Foundation

For further information contact:

Minority Faculty Development Program, Harvard Medical School, 164 Longwood Avenue, 2nd Floor, Boston, MA, 02115-5818, United States of America
Tel: (1) 617 432 2922
Email: mfdp_cfhuf@hms.harvard.edu
Website: www.mfdp.med.harvard.edu/fellows_faculty/cfhuf/index.htm
Contact: JOAN Y. REEDE, Director, CFHUF

Harkness Fellowships in Healthcare Policy

Subjects: Healthcare policy and health services research.
Purpose: To encourage the professional development of promising healthcare policy researchers and practitioners who will contribute to innovation in healthcare policy and practice in the United States of America and their home countries.
Eligibility: Open to individuals who have completed a Master's degree or PhD in health services or health policy research. Applicants must also have shown significant promise as a policy-orientated researcher or practitioner, e.g. physicians or health service managers, journalists and government officials, with a strong interest in policy issues. Candidates should also be at the research Fellow to senior lecturer level, if academically based; be in their late 20s to early 40s, and have been nominated by their department chair or the director of their institution.
Level of Study: Postgraduate, Professional development, Research
Type: Fellowship
Value: Up to US$107,000
Length of Study: Up to 1 year. A minimum of 6 months must be spent in the United States of America
Frequency: Annual
Study Establishment: An academic or other research policy institution
Country of Study: United States of America
No. of awards offered: 3 for Australia and New Zealand 3 for Germany, 2 from the Netherlands, 1 from Switzerland, and 5 from the United Kingdom
Application Procedure: Applicants must complete a formal application available online at the website www.cmwf.org/fellowships Applicants must be submitted via email.
Closing Date: September 17th
Funding: Private
Contributor: The Commonwealth Fund
Additional Information: Successful candidates will have a policy orientated research project, on a topic relevant to the Fund's programme areas. Projects will be supervised by senior researchers and each Fellow will be expected to produce a publishable report contributing to a better understanding of health policy issues.

For further information contact:

Associate Professor & Director (Australia), Center for Health Economics Research & Evaluation, University of Sydney, Mallett Street Campus, 88 Mallett Street, Level 6, Building F, Camperdown, NSW, 2050, Australia
Tel: (61) 2 9351 0900
Fax: (61) 2 9351 0930
Email: sylviab@pub.health.usyd.edu.au
Contact: Dr Jane HallPolicy Representative, Executive Director (New Zealand), New Zealand-United States Educational Foundation, PO Box 3465, Wellington, New Zealand
Tel: (64) 4 722 065

Fax: (64) 4 995 364
Email: jennifer@fulbright.org.nz
Contact: Ms Jennifer M Gill

Packer Fellowships

Subjects: Health policy issues in Australia and the United States of America, and shared lessons of both policies.
Purpose: To enable Fellows to gain an in-depth understanding of the Australian health care system and policy process, recent reforms, and models for best practice, thus enhancing their ability to make innovative contributions to policymaking in the United States, to improve the theory and practice of health policy in Australia and the United States by stimulating the cross-fertilization of ideas and experience and to encourage ongoing health policy collaboration and exchange between Australia and the United States by creating a network of international health policy experts..
Eligibility: Open to Accomplished, mid-career health policy researchers and practitioners including academics, physicians, decision makers in managed care and other private organizations, federal and state health officials and journalists.
Level of Study: Research
Type: Fellowship
Value: Up to Australian $55,000 for terms of 6–0 months, with a minimum stay of 6 months in Australia required
Length of Study: Up to 10 months
Frequency: Annual
Study Establishment: Suitable establishment in Australia
Country of Study: Australia
No. of awards offered: 2
Application Procedure: Applicants must submit a formal application, including a project proposal that falls within an area of mutual policy interest, such as: healthcare quality and safety, the private/public mix of insurance and providers, the fiscal sustainability of health systems, the healthcare workforce and investment in preventive care strategies. Applications are available online (www.cmwf.org/fellowships) and must be submitted via email.
Closing Date: August 15th
Funding: Government
Additional Information: In Australia: Director; International Strategies Branch Portfolio Strategies Division Department of Health and Ageing MDP 85 GPO Box 9848 Canberra ACT 2601; Tel: 011 61 2 6289 4593; Fax: 011 61 2 6289 7087. Australian-American Health Policy Fellowships is the successor of the Packer Policy Fellowship Program, which ran from 2003 2009. Email at packerpolicyfellowship@health.gov.au.

For further information contact:

Email: ro@cmwf.org
Website: www.cmwf.org/fellowships
Contact: Robin Osborn, Vice President and Director

COMMONWEALTH SCHOLARSHIP COMMISSION IN THE UNITED KINGDOM

c/o The Association of Commonwealth Universities, Woburn House, 20–24 Tavistock Square, London, WC1H 9HF, United Kingdom
Tel: (44) 20 7380 6700
Fax: (44) 20 7387 2655
Email: info@cscuk.org.uk
Website: www.dfid.gov.uk/cscuk
Contact: Ms Natasha Lokhun, Communications Officer

The Commonwealth Scholarship Commission (CSC) in the United Kingdom is responsible for managing Britain's contribution to the Commonwealth Scholarship and Fellowship Plan (CSFP). The CSC makes available seven types of award and supports around 700 awards in total annually.

Commonwealth Academic Fellowship

Subjects: Fellowships are offered in all subjects. Applications are considered according to the following selection criteria: academic merit of the candidate; the quality of the proposal; the likely impact of the work on the development of the candidate's home country.
Purpose: Fellowships for mid-career academics from developing countries to spend three months at a UK university to network and update knowledge and skills related to their academic subject and responsibilities. A small number of awards may be granted for collaborative research which is already underway between a Fellow's employing institution and an institution in the UK
Eligibility: Applicants must be Commonwealth citizens, refugees, or British protected persons; be permanently resident in a developing Commonwealth country; have been employed for at least five years as an academic staff member of a university in a developing Commonwealth country; be in the employment of the nominating university; AND hold a doctorate; OR, in the fields of medicine and dentistry, have been qualified as a doctor or dentist for at least ten years.
Level of Study: Postdoctorate
Type: Fellowship
Value: Return airfare to the UK; research support grant fixed according to subject; personal maintenance allowance; grants towards the expenses of preparing reports and other written work, and study travel; initial arrival allowance.
Length of Study: 3 months
Frequency: Annual
Study Establishment: Any approved UK university or higher education institution
Country of Study: United Kingdom
Application Procedure: Applications must be submitted using the CSC's Electronic Application System (EAS), and submitted to and endorsed by an approved nominating body. For more information, visit http://bit.ly/cscuk-apply.
Closing Date: December 7th
Funding: Government
Contributor: Department for International Development
Additional Information: Applications must be made via a UK university or an approved nominating body. The CSC is unable to accept nominations from other organisations or applications directly from individuals, and these will not be acknowledged.

For further information contact:

Commonwealth Scholarship Commission in the United Kingdom c/o The Association of Commonwealth Universities, Woburn House, 20-24 Tavistock Square, London, WC1H 9FH, United Kingdom
Website: www.cscuk.org.uk/apply/academic_fellowships.asp

Commonwealth Distance Learning scholarship

Subjects: Scholarships are offered for study on specific Master's courses selected by the CSC for their demonstrable relevance to the development of candidates' home countries.
Purpose: To support candidates from developing Commonwealth countries to study UK Master's degree courses while living in their home countries.
Eligibility: Applicants must be Commonwealth citizens, refugees or British protected persons; be permanently resident in a developing Commonwealth country; normally hold a first degree of upper second class standard, or higher qualification – in certain cases we will consider a lower qualification and sufficient relevant experience. Commonwealth Distance Learning Scholarships may not be held concurrently for more than one course.
Level of Study: Postgraduate
Type: Scholarship
Value: Approved tuition and examination fees
Length of Study: 3–4 years
Frequency: Annual
Study Establishment: UK universities (by distance), some in partnership with institutions in developing commonwealth countries
Country of Study: Home Country
No. of awards offered: Up to 15 scholarships per course, dependant on the quality of the candidates
Application Procedure: Universities are invited to submit expressions of interest. Successful course providers are then invited to submit a formal bid for support; the CSC then decides on the number of scholarships to be allocated to a particular course. Providers will be requested to recruit candidates and forward their applications. Please check website http://bit.ly/cscuk-apply for details.
Closing Date: July
Funding: Government
Contributor: Department for International Development

No. of awards given last year: 165
No. of applicants last year: 250
Additional Information: Applications must be made via a UK university. The CSC is unable to accept nominations from other organisations or applications directly from individuals, and these will not be acknowledged.

Commonwealth Professional Fellowship

Subjects: Agriculture/fisheries/forestry, education, engineering/science/technology, environment, governance and public health.
Purpose: To support mid-career professionals from developing Commonwealth countries to spend period (typically three months) with a UK host organisation working in their field for a programme of professional development
Eligibility: Applicants must be Commonwealth citizens, refugees or British protected persons, and must be permanently resident in a developing Commonwealth country; have at least five years' relevant work experience in a profession related to the subject of the application, by the proposed start of the Fellowship; not have undertaken a Commonwealth Professional Fellowship in the last five years; not be seeking to undertake an academic programme of research or study in the UK. Academics are eligible to apply for the scheme, but only to undertake programmes of academic management, not research or courses relevant to their research subject.
Level of Study: Professional development
Type: Fellowship
Value: Return airfare to the UK; living allowance for the Fellow; initial arrival allowance; funding support (at a flat rate) to the host organisation; capped budget for attendance at conferences, on short courses; and other eligible costs.
Length of Study: Between one and six months, though typically three months. Justification is required for awards of more than three months
Frequency: Two application rounds annually
Study Establishment: Professional, charitable, public and private sector organisations based in the UK
Country of Study: United Kingdom
Application Procedure: Applications must come from an organisation in the UK willing to set up a programme and host the Fellow(s), or have the agreement of a separate organisation in the UK to act as the host (if the latter is the case, a letter will need to be supplied at the time of application to confirm the agreement). Organisations wishing to nominate a Fellow or Fellows will need to set up a suitable programme and identify the Fellow(s) themselves. For more information, visit http://bit.ly/cscuk-apply.
Closing Date: Please check at http://bit.ly/cscuk-apply
Funding: Government
Contributor: Department for International Development
Additional Information: Applications must be made via a host organisation. The CSC is unable to accept nominations from other organisations or applications directly from individuals, and these will not be acknowledged.

For further information contact:

Website: http://www.cscuk.org.uk/apply/eas.asp

Commonwealth Scholarship

Subjects: All subjects. Applications are considered according to the following selection criteria: academic merit of the candidate; the quality of the proposal; for those candidates from developing Commonwealth countries, the likely impact of the work on the development of the candidate's home country; for those candidates from developed Commonwealth countries, the potential of the candidate to lead in the pursuit of global excellence in research and knowledge.
Purpose: To support PhD and Master's study, for candidates from Commonwealth countries other than the UK.
Eligibility: Applicants must be Commonwealth citizens, refugees, or British protected persons; be permanently resident in a Commonwealth country other than the United Kingdom, the Channel Islands or the Isle of Man; hold a first degree of upper second class Honours standard (or above); or a second class degree and a relevant postgraduate qualification, which will normally be a Master's degree; AND, in the fields of medicine and dentistry, have been qualified as a doctor or dentist for between five and ten years.
Level of Study: Doctorate, Postgraduate
Type: Scholarship
Value: Return airfare to the UK; approved tuition and examination fees; personal maintenance allowance; grants towards the expenses of preparing a thesis or dissertation, study travel, and fieldwork costs (if applicable); initial arrival allowance; family allowance for Scholars on awards longer than 18 months.
Length of Study: 1–3 years maximum, dependent on type of degree
Frequency: Annual
Study Establishment: UK universities or higher education institutions which have agreed to provide tuition fee contributions
Country of Study: United Kingdom
Application Procedure: Applications must be submitted using the CSC's Electronic Application System (EAS), and submitted to and endorsed by an approved nominating body. For more information, visit http://bit.ly/cscuk-apply.
Closing Date: Applications must be completed using the EAS by December 7th
Funding: Government
Contributor: Department for International Development (for developing Commonwealth countries), and the Department for Business, Innovation and Skills and the Scottish Government (for developed Commonwealth countries), in conjunction with UK universities.
Additional Information: Applications must be made via an approved nominating body. The CSC is unable to accept nominations from other organisations or applications directly from individuals, and these will not be acknowledged.

Commonwealth Shared Scholarship

Subjects: Subjects must be demonstrably relevant to the economic, social or technological development of the candidate's home country.
Purpose: To support Master's study for candidates from developing Commonwealth countries who would not otherwise be able to study in the UK. Scholarships are jointly funded by UK universities.
Eligibility: Applicants must be nationals of (or permanently domiciled in) a developing Commonwealth country, and not currently be living or studying in a developed country; hold a first degree at either first or upper second class level; be sufficiently fluent in English to pursue the course; have not previously studied for one year or more in a developed country; be able to confirm in writing that neither they or their families would otherwise be able to pay for the proposed course of study; be willing to confirm that they will return to their home country as soon as their period of study is complete.
Level of Study: Postgraduate
Type: Scholarship
Value: Return airfare to the UK; approved tuition and examination fees; personal maintenance allowance; grant towards the expenses of preparing a thesis or dissertation.
Length of Study: 1 year
Frequency: Annual
Study Establishment: UK universities which have agreed to participate in the Shared Scholarship scheme
Country of Study: United Kingdom
Application Procedure: Applications must be submitted using the CSC's Electronic Application System (EAS), and submitted to the UK university at which the candidate wishes to study. For more information, visit http://bit.ly/cscuk-apply.
Closing Date: May 31st
Funding: Government
Contributor: Department for International Development and UK universities
Additional Information: Applications must be made via a UK university. The CSC is unable to accept nominations from other organisations or applications directly from individuals, and these will not be acknowledged.

For further information contact:

Website: http://www.cscuk.org.uk/docs/DistanceLearning1b.pdf

Commonwealth Split-Site Doctoral Scholarship

Subjects: Scholarships are offered in all subjects. Applications are considered according to the following selection criteria: academic merit of the candidate; the quality of the proposal; for those candidates from developing Commonwealth countries, the likely impact of the work on the development of the candidate's home country; for those candidates from developed Commonwealth countries, the potential of

the candidate to lead in the pursuit of global excellence in research and knowledge.
Purpose: To support one year's study at a UK university as part of a PhD being undertaken in your home country, under the joint supervision of a home country and UK supervisor.
Eligibility: Applicants must be Commonwealth citizens, refugees, or British protected persons; be permanently resident in a Commonwealth country other than the United Kingdom, the Channel Islands or the Isle of Man; be registered for a PhD at their home country university; hold a first degree of upper second class Honours standard (or above); or a second class degree and a relevant postgraduate qualification, which will normally be a Master's degree; AND, in the fields of medicine and dentistry, have been qualified as a doctor or dentist for between five and ten years.
Level of Study: Doctorate
Type: Scholarship
Value: Full tuition fees for one year at the UK host university; stipend for up to one year in the UK; return airfare; and other allowances. The scholarship does not support the period of study at the home country university.
Length of Study: 12 months, which can be taken at any stage during the doctoral study (providing this is justified in the study plan), and can be divided into two or more periods.
Frequency: Annual
Study Establishment: UK universities or higher education institutions which have agreed to provide tuition fee contributions. The final qualification obtained will be from the home country university, rather than the UK university.
Country of Study: United Kingdom
Application Procedure: Applications must be submitted using the CSC's Electronic Application System (EAS), and submitted to and endorsed by an approved nominating body. For more information, visit http://bit.ly/cscuk-apply.
Closing Date: December 7th
Funding: Government
Contributor: Department for International Development (for developing Commonwealth countries), and the Department for Business, Innovation and Skills and the Scottish Government (for developed Commonwealth countries), in conjunction with UK universities
Additional Information: Applications must be made via an approved nominating body. The CSC is unable to accept nominations from other organisations or applications directly from individuals, and these will not be acknowledged.

CONCORDIA UNIVERSITY

1455 De Maisonneuve Blvd. W., Montréal, QC, H3G 1M8, Canada
Tel: (1) 514 848 2424
Fax: (1) 514 848 2812
Email: verret@vax2.concordia.ca
Website: www.concordia.ca
Contact: Ms Patricia Verret, Graduate Awards Manager

Concordia University is the result of the 1974 merger between Sir George Williams University and Loyola College. The University incorporates superior teaching methods with an interdisciplinary approach to learning and is dedicated to offering the best possible scholarship to the student body and to promoting research beneficial to society.-

Bank of Montréal Pauline Varnier Fellowship

Subjects: Business and comerce.
Purpose: To support graduate students to acquire higher degree in the fields of business and commerce.
Eligibility: Open to Canadian citizens and permanent residents of Canada who must be registered full-time in a graduate programme.
Level of Study: MBA
Type: Fellowship
Value: Canadian $10,000 per year
Length of Study: Renewable for 1 year
Frequency: Annual
Study Establishment: Concordia University
Country of Study: Canada
No. of awards offered: 1
Application Procedure: Applicants must submit a completed application form, three letters of recommendation and official transcripts of all university studies by the closing date.
Closing Date: February 1st
Funding: Private
No. of awards given last year: 1
Additional Information: Academic merit is the prime consideration in the granting of the awards.

Concordia University Graduate Fellowships

Subjects: All subjects.
Purpose: To support research and encourage academic excellence.
Eligibility: Open to graduates of any nationality. Candidates must be planning to pursue a full-time Master's or doctoral study at the University and maintain a minimum CGPA of 3.7 or greater.
Level of Study: Postgraduate
Type: Fellowship
Value: Canadian $10,000 at the Master's level non-renewable and Canadian $10,800 per year at the doctoral level
Length of Study: Masters: 1 year (3 terms); non-renewable; PhD: 3 years (9 terms); renewable
Frequency: Annual
Study Establishment: Concordia University
Country of Study: Canada
No. of awards offered: Approx. 38
Application Procedure: Applicants must submit a completed application form, three letters of recommendation and official transcripts of all university studies by the closing date.
Closing Date: Consult the Faculty/Program to which you are seeking admission concerning their deadline date
Funding: Government, private
No. of awards given last year: 25
No. of applicants last year: 1,050
Additional Information: Academic merit is the prime consideration in the granting of the award.
All new admissions will be considered for these awards.

David J Azrieli Graduate Fellowship

Subjects: All subjects.
Purpose: To support gradute students to undrtake higher degree.
Eligibility: Open to full-time graduate students entering or currently registered in a Master's or Doctoral program, without citizenship restriction. Candidates must maintain a minimum CGPA of 3.7 or greater.
Level of Study: Postgraduate
Type: Fellowship
Value: Canadian $15,000 per year
Length of Study: 1 year, non-renewable
Frequency: Annual
Study Establishment: Concordia University
Country of Study: Canada
No. of awards offered: 1
Application Procedure: Applicants must submit a completed application form, three letters of recommendation and official transcripts of all university studies by the closing date.
Closing Date: February 1st
Funding: Private
No. of awards given last year: 1
No. of applicants last year: 1050
Additional Information: Academic merit is the prime consideration in the granting of the award. Tenure of the award starts in September

J W McConnell Memorial Fellowships

Subjects: All subjects.
Purpose: To support research and encourage academic excellence.
Eligibility: Open to Canadian citizens and permanent residents of Canada who are planning to pursue full-time Master's or doctoral study at the University. Fellowships are awarded on academic merit.
Level of Study: Postgraduate
Type: Fellowship
Value: Canadian $10,000 at the Master's level and Canadian $10,800 per year at the doctoral level
Length of Study: A maximum of 3 terms at the Master's level, non-renewable and 9 terms at the doctoral level, renewable, calculated from the date of entry into the programme
Frequency: Annual
Study Establishment: Concordia University
Country of Study: Canada

No. of awards offered: Approx. 9
Application Procedure: Applicants must submit a completed application form, three letters of recommendation and official transcripts of all university studies by the closing date.
Closing Date: Consult the Faculty/Program to which you are seeking admission concerning their deadline date.
Funding: Private
No. of awards given last year: 15
No. of applicants last year: 800
Additional Information: Academic merit is the prime consideration in granting the awards.
All new admissions will be considered for awards.

John W O'Brien Graduate Fellowship

Subjects: All subjects.
Eligibility: Open to full-time graduate students of any nationality. Candidates must be planning to pursue a full-time Master's or doctoral study at the University.
Type: Fellowship
Value: Canadian $4,000 per term non-renewable
Length of Study: A maximum of 3 terms
Frequency: Annual
Study Establishment: Concordia University
Country of Study: Canada
No. of awards offered: 1
Application Procedure: Applicants must submit a completed application form, three letters of recommendation and official transcripts of all university studies by the closing date.
Closing Date: February 1st
Funding: Private
No. of awards given last year: 1
No. of applicants last year: 1,050
Additional Information: Academic merit is the prime consideration in the granting of awards.

Stanley G French Graduate Fellowship

Subjects: All subjects.
Purpose: To support research and encourage excellence.
Eligibility: Open to graduates of any nationality. Candidates must be planning to pursue full-time Master's or doctoral study at the University.
Level of Study: Postgraduate
Type: Fellowship
Value: $3,033 per term at the Master's level and $3,533 at the Doctoral level
Length of Study: A maximum of 3 terms. Not renewable
Frequency: Annual
Study Establishment: Concordia University
Country of Study: Canada
No. of awards offered: 1
Application Procedure: Applicants must submit a completed application form, three letters of recommendation and official transcripts of all university studies by the closing date.
Closing Date: February 1st
Funding: Private
No. of awards given last year: 1
No. of applicants last year: 1,050
Additional Information: Academic merit is the prime consideration in the granting of awards.

CONSERVATION LEADERSHIP PROGRAMME

Conservation Leadership Programme, Birdlife International, Wellbrook Court, Girton Road, Cambridge, Cambridgeshire, CB3 0NA, England
Tel: (44) 12 2327 7318
Fax: (44) 12 2327 7200
Email: clp@birdlife.org
Website: www.conservationleadershipprogramme.org
Contact: The Programme Manager

Since 1985, the Conservation Leadership Programme has supported and encouraged international conservation projects that address global conservation priorities at a local level. This is achieved through a comprehensive system of advice, training and awards. The programme is managed through a partnership between BP, FFI, CI, WCS and Birdlife International.

Future Conservationist Awards

Subjects: Biodiversity conservation.
Purpose: To develop leadership capacity amongst emerging conservationists to address the most pressing conservation issues of our time.
Eligibility: The project must address a globally recognized conservation priority, involve people, have host government approval, be run by teams of at least three people, be student-led, have over 50 per cent students registered, last for less than 1 year and take place in Africa, Asia Pacific, Middle East, Eastern Europe, Latin America or the Caribbean.
Level of Study: Graduate, Postgraduate, Doctorate
Type: Award
Value: Up to $15,000
Length of Study: Projects should be less than 1 year in length
Frequency: Annual
Country of Study: This is a global programme
No. of awards offered: Up to 30
Application Procedure: Application forms are available from the website. Applications should be made electronically.
Closing Date: July
Funding: Private
Contributor: BP, BirdLife International, Conservation International, WildLife Conservation Society, and Fauna and Flora International
No. of awards given last year: 29
No. of applicants last year: 360
Additional Information: The next call for application will be in July. Keep up to date on funding deadlines via the CLP Facebook page.

CONSERVATION TRUST

National Geographic Society, 1145 17th Street NW, Washington, DC, 20090 8249, United States of America
Email: conservationtrust@ngs.org
Website: www.nationalgeographic.com/conservation

The objective of the Conservation Trust is to support conservation activities around the world as they fit within the mission of the National Geographic Society. The trust will fund projects that contribute significantly to the preservation and sustainable use of the Earth's biological, cultural, and historical resources.

National Geographic Conservation Trust Grant

Subjects: Conservation.
Purpose: To support cutting programmes that contribute to the preservation and sustainable use of the Earth's resources.
Eligibility: Applicants must provide a record of prior research or conservation action. Researchers planning work in foreign countries should include at least one local collaboration as part of their research teams. Grants recipients are excepted to provide the National Geographic Society with rights of first refusal for popular publication of their findings.
Level of Study: Research
Type: Research grant
Value: US$15,000–20,000
Frequency: Annual
No. of awards offered: Varies
Application Procedure: Apply online.
Closing Date: 8 months prior to anticipated field dates. See http://www.nationalgeographic.com/explorers/grants-programs/conservation-trust-application/ for more details
Funding: Trusts
Contributor: National Geographic Society

For further information contact:

Conservation Trust, National Geographic Society, 1145 17th Street NW, Washington, DC, 20090-8249
Email: conservationtrust@ngs.org

CONSULATE GENERAL OF SWEDEN

445 Park Avenue, 21st floor, New York, NY, 10022, United States of America
Tel: (1) 212 888 3000
Fax: (1) 212 888 3125
Email: newyork@consulateofsweden.org
Website: www.swedennewyork.com

The Consulate General of Sweden in New York represents Sweden in the United States, specifically in the New York area. Its broad mission is to provide assistance to Swedes and to promote Swedish interests in the United States.

Bicentennial Swedish-American Exchange Fund Travel Grants

Subjects: Area/ethnic studies, foreign language, social sciences, humanities, business/consumer services and education.
Purpose: To provide financial support for faculty, researchers and professionals to study in Sweden.
Eligibility: Open to citizens or permanent residents of the United States.
Level of Study: Postgraduate, Professional development, Research
Type: Grant
Value: Up to 30,000 SEK
Length of Study: 2–4 weeks
Frequency: Annual
Country of Study: Sweden
No. of awards offered: 2–4
Application Procedure: Applicants must submit 2 letters of recommendation and a detailed project plan. Application forms are available online.
Closing Date: November 15th
Contributor: Swedish Institute, Stockholm, Sweden

For further information contact:

Website: www.studyinsweden.se

COOLEY'S ANEMIA FOUNDATION

330 Seventh Avenue, #200, New York, NY, 10001, United States of America
Tel: (1) (800) 522 7222
Fax: (1) (212) 279 5999
Email: s.buczynski@cooleysanemia.org
Website: www.cooleysanemia.org

The Cooley's Anemia Foundation is dedicated to serving people afflicted with various forms of thalassemia, most notably the major form of this genetic blood disease, Cooley's anemia/thalassemia major.

Cooley's Anemia Foundation Research Fellowship

Subjects: Clinical or basic research related to thalassemia. Applications on topics such as cardiac and endocrine complications of iron overload, hepatitis C, osteoporosis, bone marrow transplantation, iron chelation and gene therapy are encouraged.
Purpose: To promote an increased understanding of Cooley's anemia, develop improved treatment and achieve a final cure for this life-threatening genetic blood disorder.
Eligibility: Fellows must have adequate preceptorship and guidance by an experienced investigator. The application is expected to be the original work of the candidate, but should reflect the close advice of the interested and involved sponsor. Applicants who are Fellows must have an MD, PhD or equivalent degree, and must not hold a faculty position. Applicants who are junior faculty must have an MD, PhD or equivalent degree, and must have completed less than 5 years at the assistant professor level at the time the applications are due.
Level of Study: Postgraduate
Type: Fellowship
Value: US$32,500
Length of Study: 1 year, renewable for 1 year
Frequency: Annual
Study Establishment: Any suitable establishment
Country of Study: Any country
No. of awards offered: 10–15
Application Procedure: Qualified applicants should download an application form from the website.
Closing Date: February 6th
Funding: Private
Contributor: Cooley's Anemia Foundation
No. of awards given last year: 11
Additional Information: The foundation seeks to make an extraordinary commitment towards recruiting doctors to pursue a career investigating thalassemia, especially due to the relatively small patient base in the United States of America.

CORE

3 St Andrew's Place, London, NW1 4LB, England
Tel: (44) 20 7486 0341
Fax: (44) 029 7487 3734
Email: info@corecharity.org.uk
Website: www.corecharity.org.uk
Contact: Alice Kington, Research Awards Coordinator

Core, the Digestive Disorders Foundation, supports research into the cause, prevention and treatment of digestive disorders, including digestive cancers, ulcers, irritable bowel syndrome, inflammatory bowel disease, diverticulitis, liver disease and pancreatitis. Core also provides information for the public that explains the symptoms and treatment of these and other common digestive conditions.

Core Fellowships and Grants

Subjects: Gastroenterology, such as basic or applied clinical research into normal and abnormal aspects of the gastrointestinal tract, liver and pancreas, and the prevention of and treatment for digestive disorders.
Purpose: To provide funding for gastroenterological research.
Eligibility: Open to applicants resident within the United Kingdom. Fellowship projects must contain an element of basic science training.
Level of Study: Doctorate, Postdoctorate, Postgraduate, Research
Type: Fellowship or Grant
Value: Research Fellowships: UK£50,000 per year salary and UK£10,000 per year consumablesDevelopment Grants: UK£50,000 total
Length of Study: 1–3 years
Frequency: Dependent on funds available
Study Establishment: Recognized and established research centres
Country of Study: United Kingdom
No. of awards offered: Varies
Application Procedure: Applicants must complete an application form for consideration in a research competition. Details are available from the website.
Closing Date: Varies
Funding: Commercial, individuals, private, trusts
Contributor: Charitable donations
No. of awards given last year: 4
No. of applicants last year: Varies
Additional Information: Conditions are advertised in the medical press. Research grants are awarded for specific projects in the same field of interest.

CORNELL UNIVERSITY

Campus Information and Visitor Relations, Day Hall Lobby, Cornell University, Ithaca, NY, 14853, United States of America
Tel: (1) 607 255 9274
Email: humctr-mailbox@cornell.edu
Website: www.arts.cornell.edu/sochum
Contact: Megan Dirks, Program Administrator

Cornell University is a learning community that seeks to serve society by educating the leaders of the future and extending the frontiers of knowledge. The university aims to pursue understanding beyond the limitations of existing knowledge, ideology and disciplinary structure, and to affirm the value of the cultivation and enrichment of the human mind to individuals and society.

Mellon Postdoctoral Fellowships

Subjects: Arts and humanities.
Purpose: To support the early development of scholars who show promise of distinguished research careers.
Eligibility: Open to citizens of the United States of America and Canada and permanent residents who have completed requirements for a PhD before the application deadline and within the last 5 years.
Level of Study: Postdoctorate
Type: Fellowship
Value: US$45,000
Length of Study: 3 years
Frequency: Annual
Study Establishment: Cornell University
Country of Study: United States of America
No. of awards offered: 3
Application Procedure: For application information: www.arts.cornell.edu/sochum/fellowships.html.
Closing Date: January (check website for postmarked deadlines)
No. of awards given last year: 2
No. of applicants last year: 150
Additional Information: While in residence at Cornell, postdoctoral Fellows have department affiliation, limited teaching duties and the opportunity for scholarly work. Areas of specialization change each year. Mellon Postdoctoral Fellowships are available in three areas of specialization: department of classics, department of comparative literature and department of music.

Society for the Humanities Postdoctoral Fellowships

Subjects: Humanities.
Eligibility: Open to holders of a PhD degree who have at least 1 or 2 years of teaching experience at the college level. Applicants should be scholars with interests that are not confined to a narrow humanistic specially and whose research coincides with the focal theme for the year. Fellows of the society devote most of their time to research writing, but they are encouraged to offer a weekly seminar related to their special projects.
Level of Study: Postdoctorate
Type: Fellowship
Value: US$45,000
Length of Study: 1 academic year
Frequency: Annual
Study Establishment: Cornell University
Country of Study: United States of America
No. of awards offered: 6–7
Application Procedure: Please check at www.arts.cornell.edu/sochum/fellowships.html for application information.
Closing Date: Postmarked on or before October 1st
No. of awards given last year: 7
No. of applicants last year: 180
Additional Information: Information about this year's theme is available upon request. The 2013–2014 Focal Theme is occupation. From space and time to practice and politics.

THE COSTUME SOCIETY

150 Aldersgate Street, London, EC1A 4AB, United Kingdom
Email: info@costumesociety.org.uk
Website: www.costumesociety.org.uk
Contact: Sylvia Ayton

The Costume Society, a registered charity was formed in 1965 to promote the study and preservation of significant examples of historic and contemporary costumes and with a constitutional aim of providing education in dress studies.

The Costume Society Museum Placement Award

Purpose: To support students seeking museum work experience with a clothing/fashion/dress/costume collection and to help UK museums accomplish projects essential to the care, knowledge and interpretation of these types of collections. The award has been introduced to fund a student volunteer working on a clothing/fashion/dress/costume-related project in a public museum collection in the United Kingdom. The museum project/work experience should involve at least one of the following activities: documentation, numbering objects, preparing mannequins, mounting garments for display or photography, improving storage. Other appropriate object-related museum activities will be considered. The placement must be for a minimum of two months, either full or part-time.
Eligibility: The volunteer should be a student (minimum 2nd year) or graduate of an appropriate UK university course, such as dress/fashion history, museum studies, fashion design, theatre costume design, history, social history, art history.
Level of Study: Postgraduate, Graduate
Value: Up to £1,000
Length of Study: 2 months
Frequency: Annual
Country of Study: United Kingdom
No. of awards offered: 1
Application Procedure: The curator/administrator shall submit the application form with the name of the proposed volunteer and his/her CV and a proposal of not more than 500 words detailing the work the volunteer will be engaged in, its benefits to the museum and to the volunteer.
Closing Date: April 1st
Contributor: The Costume Society
No. of awards given last year: 1
No. of applicants last year: 7
Additional Information: Full details of the award are published on the Society's website www.costumesociety.org.uk and in *Costume*, the annual journal of the society.

For further information contact:

Costume Society Museum Placement Award Co-ordinator, Textiles & Fashion, V&A Museum, Cromwell Road, London, SW7 2RL
Contact: Jenny Lister

The Costume Society Patterns of Fashion Award

Subjects: Historic and contemporary dress.
Purpose: To support a student on a theatre–wardrobe course at the graduate or post-graduate level who produces a reconstruction of a garment from a pattern in Janet Arnold's books *The Patterns of Fashion* to a standard which reflects that presented in the books.
Eligibility: The award is open to United Kingdom students studying costume-related higher education courses which involve the design and realization of costume.
Level of Study: Graduate, Postgraduate
Type: Grant
Value: UK£500 plus assisted travel for presentation for submitted work
Length of Study: As applicable to the course
Frequency: Annual
Study Establishment: Open to all courses applicable.
Country of Study: United Kingdom
No. of awards offered: 1
Application Procedure: Applicants must submit photographs of a finished garment reconstructed from a pattern selected from one of the *Patterns of Fashion* books by Janet Arnold. The application must be supported by the Head of Department and Academic Supervisor of the course for which the applicant is enrolled.
Closing Date: April 30th
Funding: Trusts
Contributor: The Costume Society
No. of awards given last year: 1
Additional Information: Full details of the Award are published on the Society's website www.costumesociety.org.uk and in *Costume*, the annual journal of the society.

For further information contact:

The Awards Co-ordinator, The Patterns of Fashion Award, The Costume Society, 150 Aldersgate Street, London, EC1A 4AB, England

The Costume Society Student Bursary

Subjects: Historic and contemporary dress.
Purpose: To offer a postgraduate student full attendance at the Costume Society's annual conference/symposium exclusive of transport.
Eligibility: Open to UK students at the graduate and postgraduate level engaged in research directed towards a dissertation or thesis on the history and theory of dress. The research should reflect the theme of the Costume Society's current symposium or be an object-based project on the history of dress.

Level of Study: Graduate, Postgraduate
Type: Bursary
Value: Full-time attendance at the 3-day symposium inclusive of accommodation, meals and lecturer visits upto a maximum of £400
Length of Study: The award offers an intensive three days of study
Frequency: Annual
Study Establishment: Not specified
Country of Study: United Kingdom
No. of awards offered: 1
Application Procedure: The applicant should submit a curriculum vitae and a proposal of no more than 200 words identifying the subject area of the proposed research. The applicant should specify the institution and course attended and the names of the Head of Department and the academic supervisor who will be required as referees for awarding the bursary.
Closing Date: April 30th
Funding: Trusts
Contributor: The Costume Society
No. of awards given last year: 1
Additional Information: Full details of the Award are published on the Society's website www.costumesociety.org.uk and with *Costume*, the annual journal of the society.

For further information contact:

The Student Bursary, The Costume Society, c/o Moore Stephens, 150 Aldersgate Street, London, EC1A 4AB
Contact: The Awards Co-ordinator

The Costume Society Yarwood Award

Subjects: The history of historic and contempory dress.
Purpose: The award is made to a student on a specified MA course in the history of dress or theatre wardrobe design.
Eligibility: MA courses in the history of dress and theatre wardrobe design.
Level of Study: Postgraduate
Type: Award
Value: Up to UK£500
Frequency: Annual
Study Establishment: An academic institution is selected for a 3-year period
Country of Study: United Kingdom
No. of awards offered: 1
Application Procedure: Applications should be made by course leaders in writing to the Yarwood Award Sub-Committee giving details of the course accompanied by a course prospectus. Members of the Sub-Committee will wish to visit the course and meet the staff.
Closing Date: March 31st
Funding: Trusts
Contributor: The Costume Society
No. of awards given last year: 1
Additional Information: The Award is offered at a specified academic institution for a period of 3 years and will subsequently be offered to another academic institution for a similar period so that all MA courses involving the history of dress or theatre wardrobe design may benefit. The Yarwood Award is made in consultation with the course leader following the presentation of final year students' research leading to a final major project. The Award may be used for travel, research materials and publishing costs.

For further information contact:

The Costume Society, 150 Aldersgate Street, London, EC1A 4AB
Contact: The Yarwood Award Co-ordinator

THE COSTUME SOCIETY OF AMERICA (CSA)

Costume Society of America, 390 Amwell Road, Suite 402, Hillsborough, NJ, 08844, United States of America
Tel: (1) 908 359 1471
Fax: (1) 908 450 1118
Email: national.office@costumesocietyamerica.com
Website: www.costumesocietyamerica.com
Contact: Administrative Assistant

The Costume Society of America (CSA) advances the global understanding of all aspects of dress and appearance. The Society seeks as members those who are involved in the study, education, collection, preservation, presentation and interpretation of dress and appearance in past, present and future societies.

CSA Adele Filene Travel Award

Subjects: Cultural heritage, museum studies and related areas.
Purpose: To assist student members in their travel to the CSA National Symposium to present an accepted paper or poster.
Eligibility: Open to current students with CSA membership who have been accepted to present a juried paper or poster at the CSA National Symposium.
Level of Study: Unrestricted
Type: Travel grant
Value: Up to US$500
Frequency: Annual
Country of Study: United States of America
No. of awards offered: 1–3
Application Procedure: Applicants must send three letters of support with a copy of the juried abstract and a one-page letter of application.
Closing Date: March 15th
No. of awards given last year: 4
No. of applicants last year: 3

CSA Stella Blum Student Research Grant

Subjects: North American costume.
Purpose: To support research projects on North American costume by CSA student members.
Eligibility: Open to student members of the Society, who are enrolled on a degree programme at an accredited institution.
Level of Study: Doctorate, Graduate, Postdoctorate, Postgraduate, Predoctorate
Type: Grant
Value: Up to US$2,000 plus a travel component of up to $500 to attend National Symposium to present the completed research
Frequency: Annual
Study Establishment: An accredited institution
Country of Study: United States of America
No. of awards offered: 1
Application Procedure: Candidates must complete an application form, available upon request.
Closing Date: May 1st
No. of awards given last year: 1
No. of applicants last year: 4
Additional Information: The award will be given based on merit rather than need. Judging criteria will include creativity and innovation, specific awareness of and attention to costume matters, impact on the broad field of costume, awareness of the interdisciplinary nature of the field, ability to successfully implement the proposed project in a timely manner and faculty adviser recommendation.

For further information contact:

National Office, Costume Society of America, 203 Towne Centre Drive, Hillsborough NJ 08844, 800-CSA-9447

CSA Travel Research Grant

Subjects: Textile and fashion design, museum studies and related areas.
Purpose: To allow an individual to travel to collections for research purposes.
Eligibility: Applicants must be current, non-student CSA members and must have held membership for 2 years or more. Applicants must give proof of work in progress and indicate why the particular collection is important to the project.
Level of Study: Professional development
Value: Up to US$1,500
Frequency: Annual
Country of Study: Any country
No. of awards offered: 1
Application Procedure: Applicants must send a letter of application of no more than two pages and include the name of the collection and projected date of visit, a description of the project underway, evidence of work accomplished to date, reasons for visiting the designated collection, projected completion date of project, what audience the project will be directed to, as well as a current curriculum vitae.

Closing Date: January 15th
No. of awards given last year: 1
No. of applicants last year: 3

THE COUNCIL FOR BRITISH RESEARCH IN THE LEVANT (CBRL)

The British Academy, 10 Carlton House Terrace, London, SW1Y 5AH, England
Tel: (44) 20 7969 5296
Fax: (44) 20 7969 5401
Email: cbrl@britac.ac.uk
Website: www.cbrl.org.uk
Contact: Penny McParlin, UK Administrative Secretary

In 1998, the British Institute at Amman for Archaeology and History and the British School of Archaeology in Jerusalem amalgamated to create the Council for British Research in the Levant (CBRL). The CBRL promotes the study of the humanities and social sciences as relevant to the countries of the Levant (Cyprus, Israel, Jordan, Lebanon, Palestinian Territories and Syria).

CBRL Pilot Study Award

Subjects: Humanities and social sciences subjects, e.g. archaeology, economics, geography, historical studies, legal studies, languages and literature, linguistics, music, philosophy, politics, social anthropology, sociology and theology or religious studies.
Purpose: To enable postdoctoral scholars to undertake initial exploratory work or feasibility study as a preliminary to making applications for major funding to other bodies.
Eligibility: Applicants must be of British nationality or ordinarily resident in the United Kingdom, Isle of Man or the Channel Islands.
Level of Study: Postdoctorate, Research
Type: Research award
Value: The value of individual awards does not normally exceed UK£7,500 and in most cases will be below that level
Frequency: Annual
Study Establishment: Council for British Research in the Levant
No. of awards offered: Varies
Application Procedure: Applicants must complete an application form, available from the United Kingdom Secretary at the main address or from the website.
Closing Date: December 1st
Funding: Government
Contributor: The British Academy
No. of awards given last year: 1
No. of applicants last year: 5

CBRL Travel Grant

Subjects: Humanities and social sciences subjects, e.g. archaeology, economics, geography, historical studies, legal studies, languages and literature, linguistics, music, philosophy, politics, social anthropology, sociology and theology or religious studies.
Purpose: To cover the travel and subsistence costs of students, academics and researchers undertaking reconnaissance tours or smaller research projects in the countries of the Levant.
Eligibility: Applicants must be of British nationality, a citizen of the European Union or ordinarily resident in the United Kingdom, Isle of Man or the Channel Islands, or registered for a full-time undergraduate or postgraduate degree in a United Kingdom university.
Level of Study: Unrestricted
Type: Travel grant
Value: A maximum of UK£800
Frequency: Annual
Study Establishment: Council for British Research in the Levant
No. of awards offered: Varies
Application Procedure: Applicants must complete an application form, available from the United Kingdom Secretary at the main address or from the CBRL website.
Closing Date: January 15th
Funding: Government
Contributor: The British Academy
No. of awards given last year: 14
No. of applicants last year: 35

CBRL Visiting Research Fellowships

Subjects: Humanities and social sciences subjects, e.g. archaeology, economics, geography, historical studies, legal studies, languages and literature, linguistics, music, philosophy, politics, social anthropology, sociology and theology or religious studies.
Purpose: To enable individuals to spend a period of time based at one or more of the CBRL's institutes to conduct research.
Eligibility: Applicants must be of British nationality or ordinarily resident in the UK, Isle of Man or the Channel Islands, or registered on a full-time doctoral degree in a UK university.
Level of Study: Doctorate, Postdoctorate, Postgraduate, Research
Type: Fellowship
Value: Fellowship will provide: a return airfare between the UK and Levant, a subsistence and free hostel accomodation whilst resident at CBRL;s overseas institutes
Frequency: Annual
Study Establishment: Council for British Research in the Levant
No. of awards offered: Varies
Application Procedure: Applicants must complete an application form, available from the UK Secretary at the main address or from the website.
Closing Date: January 15th
Funding: Government
Contributor: The British Academy
No. of awards given last year: 8
No. of applicants last year: 13

Team-Based Fieldwork Research Award

Subjects: Humanities and social science, e.g. archaeology, economics, geography, historical studies, languages, linguistics, music, philosophy, politics, social anthropology, sociology and theology, religious stuides.
Purpose: To support team-based fieldwork projects focused on either Wadi Faynan area or eastern Badia region of Jordan.
Eligibility: Applicants must be of British nationality or ordinarily resident in the United Kingdom, Isle of Man or Channel Islands.
Level of Study: Postdoctorate, Research
Type: Research award
Value: Up to UK£5,000 per year for two or three years of fieldwork
Study Establishment: Council for British Research in the Levant
No. of awards offered: Varies
Application Procedure: Applicants must complete an application form, available from website.
Closing Date: December 1st
Funding: Government
Contributor: British Academy

COUNCIL FOR INTERNATIONAL EXCHANGE OF SCHOLARS (CIES)

1400 K Street, NW, Suite 700, Washington, DC, 20005, United States of America
Tel: (1) 202 686 4000
Fax: (1) 202 362 3442/202 686 4029
Email: apprequest@cies.iie.org/scholars@iie.org
Website: www.cies.org

The Council for International Exchange of Scholars (CIES) is a private, non-profit organization that facilitates international exchanges in higher education. Under a co-operative agreement with the United States of America Department of State Bureau of Educational and Cultural Affairs, it assists in the administration of the Fulbright Scholar Program for faculty and professionals. CIES is affiliated with the Institute of International Education.

Fulbright Distinguished Chairs Program

Subjects: American studies (history, politics and literature), humanities, law, social sciences, computer science and e-commerce, business and management, fine arts, mass communications and journalism.
Purpose: To increase mutual understanding between the people of the United States of America and other countries and to promote international educational co-operation.
Eligibility: Open to citizens of the United States of America who hold a PhD or equivalent qualification. Candidates should have a prominent

record of scholarly achievement. See http://www.cies.org/Chairs/Eligibility.htm for more details.
Level of Study: Postdoctorate
Value: Varies by country. See http://catalog.cies.org/DistinguishedChairAwards.aspx for more details.
Length of Study: 3 months to 1 academic year
Frequency: Annual
No. of awards offered: Approx. 30
Application Procedure: Applicants must submit a project statement and an eight-page curriculum vitae, and visit the website www.cies.org for more information.
Closing Date: August 1st
Funding: Government, private
No. of awards given last year: 30

Fulbright Specialist Program

Subjects: Anthropology, archaeology, business administration, communications and journalism, economics, education, environmental science, information technology, law, library science, political science, public administration, sociology, social work, United States of America studies, urban planning, agriculture, applied linguistics/TEFL, peace and conflict resolution studies, biology education, chemistry education, engineering education, mathematics education, physics education.
Purpose: To offer short-term grants and encourage new types of activities in the Fulbright context. The program also aims to advance mutual understanding, establish long-term co-operation and create opportunities for institutional linkages.
Eligibility: Open to citizens of the United States of America with a PhD or comparable professional qualifications.
Level of Study: Postdoctorate, Professional development
Type: Grant
Value: International economy fare travel and approved related expenses plus a $200 per day honorarium. In-country lodging, meals and travel provided by requesting institution
Length of Study: 2–6 weeks
Frequency: Annual
Country of Study: Other
No. of awards offered: Varies
Application Procedure: Applicants must complete the online application, available on the CIES website.
Closing Date: Applications and grants are processed on a rolling basis, consult CIES website for peer review calendar
Funding: Government
Contributor: US Department of State
Additional Information: Successful candidates are expected to lecture, lead seminars, work with foreign counterparts on curriculum and program and institutional development. Applicants must contact fulspec@iie.org for more information.

COUNCIL FOR THE ADVANCEMENT OF SCIENCE WRITING, INC. (CASW)

PO Box 910, Hedgesville, WV 25427, United States of America
Tel: (1) 304 754 6786
Email: diane@casw.org
Website: www.casw.org
Contact: Ms Diane McGurgan, Administrator

The CASW is a group of distinguished journalists and scientists committed to improving the quality of science news reaching the general public.

Taylor/Blakeslee Fellowships for Graduate Study in Science Writing

Subjects: Journalism.
Purpose: To support graduate study in science writing.
Eligibility: Applicants must have a degree in science or journalism and must convince the CASW selection committee of their ability to pursue a career in science writing for the general public.
Level of Study: Postgraduate
Type: Fellowship
Value: A maximum of US$5,000
Length of Study: 1 year
Frequency: Annual
Country of Study: United States of America
No. of awards offered: 2–4
Application Procedure: Applicants must submit three collated sets of a completed application form, a curriculum vitae, a transcript of undergraduate studies if a student, three faculty recommendations or employer recommendations, three samples of writing on 8.5 by 11 inch sheets only and a short statement of career goals.
Closing Date: July 1st
Funding: Private
No. of awards given last year: 4
No. of applicants last year: 16–20
Additional Information: Science writing is defined as writing about science, medicine, health, technology and the environment for the general public via the mass media.

COUNCIL OF AMERICAN OVERSEAS RESEARCH CENTERS (CAORC)

PO Box 37012, MRC 178, Washington, DC, 20013 7012, United States of America
Tel: (1) 202 633 1599
Fax: (1) 202 786 2430
Email: fellowships@caorc.org
Website: www.caorc.org

Council of American Overseas Research Centers (CAORC) serve as a base for virtually every American scholar undertaking research in the host countries. The members have centres in many locations across the world.

CAORC Multi-Country Research Fellowship Program for Advanced Multi-Country Research

Subjects: Humanities, social sciences or allied natural sciences.
Purpose: To advance higher learning and scholarly research and to conduct research of regional or trans-regional significance.
Eligibility: Applicants must have obtained a PhD or be established postdoctoral scholars. The candidate should be a citizen of the United States. Preference will be given to Candidates examining comparative and/or cross-regional research.
Level of Study: Doctorate, Postdoctorate, Research
Type: Fellowships
Value: Up to $10,500
Frequency: Annual
No. of awards offered: Approx. 10
Application Procedure: The application can be downloaded from the website. To obtain hard copy of the application, please contact CAORC.
Closing Date: January 15th
Contributor: US State Department
No. of awards given last year: 9
No. of applicants last year: 120
Additional Information: Scholars must carry out research in at least one of the countries that host overseas research centres. Please check website for further information www.caorc.org

COUNCIL OF SCIENTIFIC & INDUSTRIAL RESEARCH (CSIR)

Anusandhan Bhawan, 2 Rafi Marg, New Delhi, 110001, India
Tel: (91) 011 2373 7889
Fax: (91) 011 2371 0618
Email: headhrdg@csirhrdg.res.in
Website: www.csir.res.in

Council of Scientific & Industrial Research (CSIR) is a premier national research and development organization in India. It is among the world's largest publicly funded research and development organization. Human Resource Development Group, a division of CSIR, realises this objective through various grants, fellowship, schemes, etc.

CSIR Senior Research Associateship (SRA) Scheme

Subjects: Agriculture sciences, chemical sciences, earth, atmosphere, ocean and planetary sciences, life sciences, material sciences, mathematical statistics, operation research and computer

sciences, physical sciences, engineering sciences, medical sciences and social sciences and humanities.
Purpose: To provide temporary placement to highly qualified, but unemployed Indian nationals including those returning from abroad, to carry out independent research.
Eligibility: Open to Indian citizens who are not more than 40 years of age and have obtained M.Tech/ME/MD/MS/MVSc/MPharma/PhD in any branch of science/PhD in social sciences and humanities. The candidate should have followed it by 2 years of research/teaching experience or PhD in engineering/technology. See http://csirhrdg.res.in/sraform.pdf for more details.
Level of Study: Predoctorate, Research, Doctorate, Postdoctorate
Type: Fellowship
Value: Indian Rupees 21,000–25,810 per month
Length of Study: 3 years. No extension.
Frequency: Annual
Country of Study: India
Application Procedure: Applicants must download the complete application form from the website.
Closing Date: See the organization website for details.

For further information contact:

The Head, Human Resource Development Group, CSIR Complex, Library Avenue, Pusa, New Delhi, 110012, India

COUNCIL OF SUPPLY CHAIN MANAGEMENT PROFESSIONALS (CSCMP)

333 East Butterfield Road, Suite 140, Lombard, IL, 60148, United States of America
Tel: (1) 630 574 0985
Fax: (1) 630 574 0989
Email: membership@cscmp.org
Website: www.cscmp.org
Contact: Kathleen Hedland, Director Education and Roundtable Services

The Council of Supply Chain Management Professionals (CSCMP) is a non-profit organization of business personnel who are interested in improving their logistics management skills. CSCMP works in co-operation with private industry and various organizations to further the understanding and development of the logistics concept. This is accomplished through a continuing programme of organized activities, research and meetings designed to develop the theory and understanding of the logistics process, promote the art and science of managing logistics systems, and foster professional dialogue and development within the profession.

CSCMP Distinguished Service Award

Subjects: Supply chain management and logistics.
Purpose: To provide honor to an individual for acievement in supply chain management.
Eligibility: All individuals who have made contributions to the field of supply chain management are eligible for the DSA. This includes practitioners with responsibilities in a functional area of supply chain management, consultants, and educators—anyone who has made a significant contribution to the advancement of supply chain management.
Type: Award
Frequency: Annual
Closing Date: April 30th

CSCMP Doctoral Dissertation Award

Subjects: Any supply chain function.
Purpose: To encourage research leading to advancement of the theory and practice to supply chain management.
Eligibility: Open to all candidates whose doctoral dissertation demonstrates signified originality and contributes to the logistics knowledge base. See http://cscmp.org/downloads/public/education/awards/dda-guidelines.pdf for details.
Level of Study: Postdoctorate
Value: $5,000
Frequency: Annual
Closing Date: May 1st

CSCMP George A Gecowets Graduate Scholarship Program

Subjects: Logistics management.
Purpose: To acknowledge the importance of logistics in a tangible way, while emphasising the Council's commitment to promote the art and science of managing logistics systems.
Eligibility: Applicants must be planning to pursue a career in logistics management, be a senior at an accredited 4 year college or university, and already be enrolled in the first year of a logistics or logistics related Master's degree programme.
Level of Study: Graduate
Type: Scholarship
Value: $2,000
Frequency: Annual
No. of awards offered: 15
Application Procedure: Applicants must submit a completed application, official college transcripts, Graduate Record Examination scores or Graduate Management Admission Test scores, and notification of any changes in address, school enrolment, or other pertinent information. The Citizens' Scholarship Foundation of America (CSFA) will then send a complete application package upon request.
Closing Date: April 1st
Funding: Private
Contributor: The Council of Logistics Management
Additional Information: The Council wishes to make high potential students aware of the tremendous opportunities and challenges that await them in a career in logistics management, as the last 10 years have seen an exponential increase in the importance of the logistics manager.

For further information contact:

Council of Supply Chain Management Professionals, George A Gecowets Graduate Scholarship Program, Scholarship Management Services CSFA, 1505 Riverview Road, PO Box 297, St Peter, MN, 56082, United States of America

COUNCIL ON FOREIGN RELATIONS (CFR)

The Harold Pratt House, 58 East 68th Street, New York, NY, 10065, United States of America
Tel: (1) 212 434 9400
Fax: (1) 212 434 9800
Email: fellowships@cfr.org
Website: www.cfr.org

The Council on Foreign Relations (CFR) is dedicated to increasing America's understanding of the world and contributing ideas to United States of America foreign policy. The Council accomplishes this mainly by promoting constructive debates and discussions, clarifying world issues and publishing *Foreign Affairs*, the leading journal on global issues.

CFR International Affairs Fellowship Program in Japan

Subjects: International relations.
Purpose: To cultivate the United States of America's understanding of Japan and to strengthen communication between emerging leaders of the two nations.
Eligibility: Open to citizens of the United States of America aged 27–45 who have not had prior substantial experience in Japan. Fellows will be drawn from academia, government institutions, the business community and the media. The programme does not fund pre- or postdoctoral scholarly research, work towards a degree or the completion of projects on which substantial progress has been made prior to the fellowship period. Knowledge of the Japanese language is not a requirement.
Level of Study: Professional development
Type: Fellowship
Value: Living expenses in Japan plus international transportation, health and travel insurance and necessary research expenses
Length of Study: 3 months to 1 year
Frequency: Annual
Country of Study: Japan

No. of awards offered: 2–5
Application Procedure: Application is primarily by invitation, on the recommendation of individuals in academic, government and other institutions who have occasion to know candidates particularly well suited for the experience offered by this fellowship. Others who inquire directly and who meet preliminary requirements may also be invited to apply without formal nomination. Those invited to apply will be forwarded application materials.
Closing Date: September 30th is the deadline for nominations and November 19th is the application deadline. Nominations and applications will also be accepted out of cycle
Funding: Private
Contributor: Hitachi Limited
No. of awards given last year: 3
No. of applicants last year: 6
Additional Information: While the Fellow is not required to produce a book, article or report, it is hoped that some written output will result.

For further information contact:

Fellowship Affairs, Council on Foreign Relations, 58 East 68th Street, New York, NY, 10065
Tel: 212 434 9489
Fax: 212 434 9870
Email: Fellowships@cfr.org

CFR International Affairs Fellowships

Subjects: Important problems in international affairs and their implications for the interests and policies of the United States of America, foreign states or international organizations.
Purpose: To bridge the gap between analysis and action in foreign policy by supporting a variety of policy studies and active experiences in policy making.
Eligibility: Open to United States citizens aged 27–35. While a PhD is not a requirement, successful candidates should generally hold advanced degrees and possess a solid record of work experience. The programme does not fund pre- or postdoctoral research, work towards a degree, or the completion of projects for which substantial progress has been made prior to the fellowship period.
Level of Study: Professional development, Research
Type: Fellowship
Value: Stipend of $85,000
Length of Study: 1 year
Frequency: Annual
Country of Study: Any country
No. of awards offered: 8–12
Application Procedure: Application is primarily by invitation, on the recommendation of individuals in academic, government and other institutions who have occasion to know candidates particularly well suited for the experience offered by this fellowship. Others who enquire directly and who meet preliminary requirements may also be invited to apply without formal nomination. Those invited to apply will be forwarded application materials.
Closing Date: September 30th is the deadline for nominations and November 19th is the application deadline.
Funding: Private
No. of awards given last year: 11
No. of applicants last year: 30
Additional Information: While the Fellow is not required to produce a book, article or report, it is hoped that some written output will result.

For further information contact:

Tel: 212 434 9489
Email: fellowships@cfr.org.

Edward R Murrow Fellowship for Foreign Correspondents

Subjects: Issues in international affairs and their implications for the interests and policies of the United States of America, foreign states or international organizations.
Purpose: To help the Fellow increase his or her competency in reporting and interpreting events abroad and to give him or her a period of nearly a year of sustained analysis and writing, free from the daily pressures that characterize journalistic life.
Eligibility: Open to any correspondent, editor or producer for radio, television, a newspaper or a magazine widely available in the United States of America who has covered international news. Eligibility is limited to those individuals who are authorized to work in the United States and who will continue to be authorized for the duration of the fellowship.
Level of Study: Professional development
Type: Fellowship
Value: A stipend equivalent to the salary relinquished, not to exceed US$65,000 for 9 months
Length of Study: 9 months
Frequency: Annual
Study Establishment: The Council headquarters in New York City
No. of awards offered: 1
Application Procedure: Application is primarily by nomination. A nomination letter must be submitted to the main address. The nomination letter may be submitted by a Council member, a former or current Murrow Fellow, the candidate's employer, or the candidates themselves. The nomination letter should confirm the candidate's eligibility as well as provide a brief description of their background and why the nominator believes the candidate to be an appropriate prospect for the Fellowship. For those candidates who choose to nominate themselves, their letter should address the same aforementioned issues in addition to providing a copy of their most recent curriculum vitae. Nominees who meet the criteria of the programme will then be forwarded an application form.
Closing Date: February 4th is the deadline for nominations and March 14th is the application deadline.
Funding: Private
No. of awards given last year: 1
No. of applicants last year: 10

COUNCIL ON LIBRARY AND INFORMATION RESOURCES (CLIR)

1707 L Street, NW Suite 650, Washington DC, 20036, United States of America
Tel: (1) 202 939 4750/4751
Fax: (1) 202 939 4765
Email: abishop@clir.org
Website: www.clir.org
Contact: Alice Bishop, Special Projects Associate

CLIR is an independent, nonprofit organization that forges strategies to enhance research, teaching, and learning environments in collaboration with libraries, cultural institutions, and communities of higher.

Postdoctoral Fellowship in Academic Libraries

Subjects: Humanities, social sciences, sciences.
Purpose: To provide a unique opportunity to develop expertise in the new forms of scholarly research and the information resources.
Eligibility: Open to candidates who have received a PhD in the field of humanities, social sciences, sciences.
Level of Study: Postdoctorate
Type: Fellowships
Value: Varies
Length of Study: 1–2 years
Frequency: Annual
Country of Study: United States of America
Application Procedure: Applicants can download the application form and the reference form from the website and 2 copies of completed application forms along with 2 copies of curriculum vitae, 3 letters of reference and graduate school transcripts are to be submitted online.
Closing Date: December 31st
Funding: Private

For further information contact:

Email: postdoc@clir.org

COUNCIL ON SOCIAL WORK EDUCATION (CSWE)

1701 Duke Street, Suite 200, Alexandria, VA, 22314 3457, United States of America
Tel: (1) 703 683 8080
Fax: (1) 703 683 8099
Email: info@cswe.org
Website: www.cswe.org

The Council on Social Work Education (CSWE) provides national leadership and a forum for collective action designed to ensure the preparation of competent and committed social work professionals. Founded in 1952, CSWE is a non-profit, tax exempt, national organization representing 2,700 individual members as well as 650 graduate and undergraduate programmes of professional social work education. CSWE's goals include improving the quality of social work education, preparing competent human service professionals and developing new programmes to meet the demands of the changing services delivery systems.

CSWE Doctoral Fellowships in Social Work for Ethnic Minority Students Preparing for Leadership Roles in Mental Health and/or Substance Abuse

Subjects: Mental health or substance abuse.
Purpose: To equip ethnic minority individuals for the provision of leadership, teaching, consultation, training, policy development and administration in mental health or substance abuse programmes and to enhance the development and dissemination of knowledge that is required for the provision of relevant clinical and social services to ethnic minority individuals and communities.
Eligibility: Applicants must be citizens or permanent residents of the United States of America, including, but not limited to, persons who are American Indian or Alaskan Native, Asian or Pacific Islander, Chinese, East Indian and other South Asians, Filipino, Hawaiian, Japanese, Korean or Samoan, black or Hispanic, e.g. Mexican or Chicano, Puerto Rican, Cuban, Central or South American. This programme is open to students who have a Master's degree in social work and who will begin full-time study leading to a doctoral degree in social work or who are currently enrolled as full-time students in a doctoral social work programme.
Level of Study: Graduate
Type: Fellowship
Value: Monthly stipend, tuition assistance, and when available funds to assist with travel to scientific meetings and other skill building training opportunities.
Length of Study: 1 year, award is renewable for up to 3 years upon reapplication if the fellow maintains satisfactory progress
Study Establishment: Schools of Social Work
Country of Study: United States of America
No. of awards offered: Varies
Application Procedure: Applicants must write to the CSWE for an application pack and further information or visit the website.
Closing Date: February 28th
Funding: Government
Contributor: The Substance Abuse and Mental Health Services Administration
Additional Information: Applicants should demonstrate potential for assuming leadership roles, as well as potential for success in doctoral studies and commitment to a career in providing mental health and/or substance abuse services to ethnic minority clients and communities.

CSWE Doctoral Fellowships in Social Work for Ethnic Minority Students Specializing in Mental Health

Subjects: Mental health research.
Purpose: To educate leaders of the nation's next generation of mental health researchers.
Eligibility: Applicants must be citizens of the United States of America or permanent residents, including, but not limited to, persons who are American Indian or Alaskan Native, Asian or Pacific Islander, Chinese, East Indian and other South Asians, Filipino, Hawaiian, Japanese, Korean or Samoan, black or Hispanic, e.g. Mexican or Chicano, Puerto Rican, Cuban, Central or South American. This programme is open to students who have a Master's degree in social work and who will begin full-time study leading to a doctoral degree in social work or are currently enrolled as full-time students in a doctoral social work programme.
Level of Study: Graduate
Type: Fellowship
Value: Monthly stipends to help defray living expenses. Tuition support provided according to the National Insititute of Health tuition formula.
Length of Study: 1 year, although the award is renewable upon reapplication if the Fellow maintains satisfactory progress towards degree objectives and funding is available
Study Establishment: Schools of Social Work
Country of Study: United States of America
No. of awards offered: Varies
Application Procedure: Applicants must write to the CSWE for an application pack and further information or visit the website.
Closing Date: March 15th
Funding: Government
Contributor: The Division of Epidemiology and Services Research, NIMH
Additional Information: Applicants should demonstrate potential for, and interest in, mental health research, as well as potential for success in doctoral studies and commitment to a career in mental health research.

For further information contact:

CSWE Minority Research Fellowship Program, 1600 Duke Street, Suite 300, Alexandria, VA 22314-3421

THE COUNTESS OF MUNSTER MUSICAL TRUST

Wormley Hill, Godalming, Surrey, GU8 5SG, England
Tel: (44) 14 2868 5427
Email: admin@munstertrust.org.uk
Website: www.munstertrust.org.uk
Contact: Mrs Kathy Butler Ure, Secretary

The Countess of Munster Musical Trust provides financial assistance towards the cost of studies and maintenance of outstanding postgraduate students who merit further training at home or abroad. Each year, the Trust is able to offer a small number of interest-free loans for instrument purchase to former beneficiaries.

Countess of Munster Musical Trust Awards

Subjects: Musical studies.
Purpose: To enable students, selected after interview and audition, to pursue a course of specialist or advanced performance studies.
Eligibility: Open to United Kingdom or British Commonwealth citizens who are aged 18–24 years (for instrumentalists, conductors and composers) or under 28 (for singers) who show outstanding musical ability and potential.
Level of Study: Postgraduate, Professional development, Private Lessons
Type: Grant
Value: By individual assessment to meet tuition fees and maintenance according to need, usually up to UK£6,000
Length of Study: 1 year, with the possibility of renewal
Frequency: Annual
Country of Study: Any country
No. of awards offered: Up to 60
Application Procedure: Applicants must complete an application form and will have to attend an audition and interview.
Closing Date: Application forms must be received between January and February 14th for awards to go through in September
Funding: Private
No. of awards given last year: 55
No. of applicants last year: 250

THE COURTAULD INSTITUTE OF ART

Somerset House, Strand, London, WC2R 0RN, United Kingdom
Tel: (44) 020 7872 0220
Email: galleryinfo@courtauld.ac.uk
Website: www.courtauld.ac.uk
Contact: Dr Gareth Morgan, Registrar Office

Andrew W Mellon Foundation/Research Forum Postdoctoral Fellowship

Subjects: A fellowship for an early career researcher in the field of modern or contemporary art.
Purpose: This fellowship will give the fellow the opportunity to pursue a research project while gaining teaching experience in a research environment and working in close collaboration with senior scholars to deliver interdisciplinary M.A. courses.
Eligibility: Applicants must be at an early stage of their career, not currently holding or having held a permanent university post and having received a doctorate within three years of taking up the award (and no later than December).
Level of Study: Postgraduate, Research
Type: Fellowship
Value: £28,138 per year
Frequency: Annual
Application Procedure: Applicants are asked to submit (1) a covering letter explaining the candidate's specific interest in the fellowship; (2) a completed application form; (3) two letters of recommendation, including one from the candidate's supervisor – these can be sent separately; and (4) an equal opportunities monitoring form..
Closing Date: October 5th

Association of Art Historians Fellows

Purpose: To subsidise participation in the annual conference of the Association of Art Historians.
Level of Study: Other
Type: Grant
Value: £100 to £200
Frequency: Annual
Application Procedure: Students who have papers accepted at a conference session should apply to the Research Forum with an abstract of the paper and a copy of the application sent to the Association of Art Historians.
Closing Date: January 31st

Caroline Villiers Research Fellowship

Subjects: Research proposals for the fellowship are welcomed from researchers and practitioners from diverse disciplines relating to the study and conservation of works of art..
Purpose: The purpose of the fellowship is to promote research in the interdisciplinary field of technical art history: the application of technical, scientific and/or historical methods, together with close observation, to the study of the physical nature of the work of art in relation to issues of making, change, conservation and/or display.
Level of Study: Research, Unrestricted
Type: Fellowship
Closing Date: Spring

The Marc de Montalembert Grant

Subjects: The grant is for a project which can be linked to candidate's professional training, but must be realised outside the usual professional or academic framework. A preference will be shown for projects having to do with Byzantium, the medieval period, or the contemporary Mediterranean world.
Purpose: To discover other Mediterranean cultures and to get a sense of their richness and diversity.
Eligibility: The scholar must be under the age of 30 and from a Mediterranean country.
Level of Study: Research, Unrestricted
Type: Grant
Value: €7,000
Frequency: Annual
Closing Date: December 31st

PhD Studentships in Tudor and Jacobean Artistic Practice

Subjects: The doctoral thesis would examine the materials and techniques used for portrait painting by Anglo-Netherlandish and Netherlandish émigré artists working in Britain.
Purpose: To support an ongoing research project on Tudor and Jacobean artistic practice called Making Art in Tudor Britain. This project is based on the collections of the National Portrait Gallery in collaboration with the University of Sussex and The Courtauld Institute of Art..
Level of Study: Doctorate, Research
Type: Studentship
Value: £13,590 per year
Frequency: Annual
Application Procedure: Potential candidates are required to register their interest by providing a CV and an outline of their of research interests relevant to the studentship..
Closing Date: October 31st

Romney Society Bursary

Subjects: The art of the eighteenth century with particular reference to the life and times of George Romney (1734–1802) and his contemporaries.
Purpose: To offer a platform to students and academics in their first posts who otherwise may not have the opportunity to be published and to encourage those with an interest in the art of the eighteenth century with particular reference to the life and times of George Romney (1734–1802) and his contemporaries.
Eligibility: The area of study should be an aspect of the life or work of George Romney or any contemporary linked to Romney (provided the link is part of the study).
Level of Study: Research, Unrestricted
Type: Bursary
Value: £500 or $900
Frequency: Other
Application Procedure: Applications may be made at any time and should be in the form of a proposal of not more than 150 words.
Closing Date: Applications may be made at any time

Terra Foundation for American Art International Essay Prize

Subjects: American art (circa 1500–1980).
Purpose: The aim of the award is to stimulate and actively support non-US scholars working on American art, foster international exchange of new ideas and create a broad, culturally comparative dialogue on American art.
Eligibility: To be eligible, essays should focus on historical American painting, sculpture, prints, drawings, decorative arts, photography or visual culture of the same period. Preference will be given to studies that address American art within a cross-cultural context as well as new ways of thinking about American art. Manuscripts previously published in a foreign language are eligible if released within the last two years. For scholars from English-language countries, only unpublished manuscripts will be considered.
Level of Study: Research
Type: Award
Value: US$500
Frequency: Annual
Application Procedure: The length of the essay (including endnotes) shall not exceed 8,000 words with approximately 12 illustrations. Manuscripts submitted in foreign languages should be accompanied by a detailed abstract in English. Six copies of the essay, clearly labelled "2010 Terra Foundation for American Art International Essay Prize," along with the scholar's name, mailing address, institutional affiliation, e-mail address and fax number must be received.
Closing Date: January 15th

Terra Foundation for American Art Postdoctoral Teaching Fellowship at The Courtauld Institute of Art

Subjects: The award will enable a recent postdoctoral scholar to teach at The Courtauld Institute of Art and to undertake a major research project intended for publication. The fellow will teach one course in the first year and two courses in the second year on selected American art topics.
Purpose: This fellowship is part of an initiative of the Terra Foundation that aims to develop international interest, knowledge and scholarship in the field of historical American art. The scheme offers an early career researcher in the field of historical American art the possibility of gaining experience of research and teaching in a university environment, which will enhance his or her curriculum vitae, improve his or her prospects of obtaining permanent teaching posts, and further the knowledge of American art.

Eligibility: Applicants are expected to be at an early stage of their career, not currently holding, or having held a permanent university post and having received a doctorate within the three-year period prior to taking up the award.
Level of Study: Postdoctorate, Research
Type: Fellowship
Value: £27,885 per year
Application Procedure: Applicants are asked to submit (1) a completed application form, with two letters of recommendation, including one from the candidate's supervisor (these can be sent separately); (2) an equal opportunities monitoring form.
Closing Date: January 15th

Terra Foundation for American Art research Travel Grants

Subjects: American art or transatlantic artistic relations prior to 1980.
Purpose: The Terra Foundation for American Art Research Travel Grants support travel to the United States for research projects that concern American art or transatlantic artistic relations prior to 1980.
Eligibility: Nationals of all European countries as well as non-EU nationals enrolled in European universities can apply for these grants according to their level of study.
Level of Study: Doctorate, Postdoctorate, Research
Type: Grant
Value: Three grants of US$6,000 each will be offered to researchers at doctoral level.
Three grants of US$9,000 each will be offered to postdoctoral researchers who have been awarded their doctorate within the past ten years
Frequency: Annual
Application Procedure: An official Terra Foundation Travel Grant application form. The applicant's curriculum vitae. A description of the applicant's research project demonstrating the need for a sojourn in the United States of America. Two letters of recommendation (for doctoral candidates, one of these letters should be by the dissertation advisor).
Closing Date: January 15th

Terra Summer Residency in Giverny 2010

Subjects: These fellowships are awarded to artists who have completed their studies at Master's level (or its equivalent) and doctoral students engaged in research on American art or transatlantic artistic exchange.
Purpose: The Terra Summer Residency in Giverny provides artists and scholars with an opportunity for the independent study of American art within a framework of interdisciplinary exchange and dialogue, and in a site rich in cultural significance.
Level of Study: Postgraduate, Research, Doctorate
Type: Residency
Value: Terra Summer Residency fellows are awarded a stipend of $5,000, and artists receive an additional $300 for the purchase of materials.
Frequency: Annual
No. of awards offered: 10
Application Procedure: Applicants must submit an application form.
Closing Date: January 15th

CRANFIELD UNIVERSITY

School of Applied Sciences, Bedfordshire, Cranfield, MK43 OAL, England
Tel: (44) 1234 754086
Fax: (44) 1234 754109
Email: appliedsciences@cranfield.ac.uk
Website: www.cranfield.ac.uk/sas
Contact: Vicky Mason, Online Marketing Manager

The School of Applied Sciences is recognised globally for its multidisciplinary approach to teaching and research in the key areas of manufacturing, materials, and environmental science and technology. Our focus is on fundamental research and its application, together with teaching, to meet the needs of industry and society.

Department of Agriculture and Rural Development (DARD) for Northern Ireland

Subjects: Agricultural and environmental engineering, environmental diagnostics and management, environmental management for business, geographical information management, waste and resource management, land management, water management and environmental engineering.
Purpose: To assist candidates to obtain the necessary qualifications and experience to fit them for advisory, teaching, research and other technical work in agricultural and food industries and development of the economy and social infrastructure of rural areas.
Eligibility: Open to full time students only.
Level of Study: Postgraduate
Type: Studentship
Length of Study: 1 year
Frequency: Annual
Study Establishment: Cranfield University, School of Applied Sciences
Country of Study: United Kingdom
Application Procedure: Application form can be downloaded from www.dardni.gov.uk
Funding: Government
Contributor: DARD

Douglas Bomford Trust

Subjects: Agricultural and environmental engineering, environmental diagnostics and management, environmental management for business, geographical information management, land management, water management, waste and resource management, water and wastewater engineering, water and wastewater technology and environmental engineering.
Purpose: To advance knowledge, understanding, practice, competence and capability in the application of engineering and physical science to agriculture, horticulture, forestry, amenity, and allied land basal and biological activities for sustainable benefit of the environment and mankind.
Eligibility: Individuals must demonstrate a long-term commitment to the areas of concern to the trust and have some connection with the United Kingdom through nationality, residency, or place of learning/registration.
Level of Study: Postgraduate
Value: £500–1,000 depending on application.
Length of Study: 1 year
Frequency: Annual
Study Establishment: Cranfield University, School of Applied Sciences
Country of Study: United Kingdom
Application Procedure: Please check the website www.dbt.org.uk for details.
Funding: Trusts
Contributor: Douglas Bomford Trust
Additional Information: Address for application: website: www.dbt.org.uk and headed "Student Award Application"

Environmental Issues Award

Subjects: Land management, geographical information management, water management, water science, environmental management for business, environmental diagnostics and management, environmental engineering and civil engineering.
Purpose: To support educational programmes dealing with environmental issues or support environmental project work.
Eligibility: Applicants must be members of the Institution of Mechanical Engineers.
Level of Study: Postgraduate
Value: Up to £1,000
Length of Study: 1 year
Frequency: Annual
Study Establishment: Cranfield University, School of Applied Sciences
Country of Study: United Kingdom
Application Procedure: Applicants must apply online at www.imech.org/awards.
Closing Date: 2012
Funding: Trusts
Contributor: IMechE

Additional Information: Please check at www.imech.org/about us/ Scholarships and Awards/professional for more information and industry grants/Environmental Issues Award

Geoplan Scholarship

Subjects: Geographical information management.
Purpose: To provide financial support to selected graduates wishing to undertake the MSc geographical information management.
Level of Study: Postgraduate
Type: Scholarship
Length of Study: 1 year
Frequency: Annual
Study Establishment: Granfield University, School of Applied Sciences
Country of Study: United Kingdom
No. of awards offered: 2
Closing Date: August 31st
Funding: Commercial
Contributor: Geoplan

The Lorch MSc Student Bursary

Subjects: Availabe to students wishing to study full-time MSc Water and Wastewater Engineering or MSc Water Management.
Purpose: To assist postgraduate study.
Eligibility: Applicants should be UK citizens and possess a minimum 2:1 UK Honours degree in Engineering or Physical Sciences or related discipline, and have been offered a place on the 1-year full-time MSc in Water and Wastewater Engineering or Water and Wastewater Technology.
Level of Study: Postgraduate
Type: Bursary
Value: UK£5,000 plus tuition fees
Length of Study: 1 year
Frequency: Annual
Study Establishment: Cranfield University, School of Applied Sciences
Country of Study: United Kingdom
No. of awards offered: 1
Application Procedure: Applicants must apply directly to the university.
Closing Date: July
Funding: Foundation
Contributor: The Lorch Foundation
No. of awards given last year: 1
Additional Information: The bursary is provided by the Lorch Foundation, a charitable institution founded to support and promote education and research in the field of water purification and related sciences for the benefit of mankind. The successful applicant will undertake thesis research on processes of water purification and industrial effluent recycling as part of the MSc programme.

Panasonic Trust Fellowships

Subjects: Water and wastewater engineering and waste and resource management.
Purpose: To provide financial support to selected graduate engineers wishing to undertake full time Masters courses in subjects related to the environment and sustainability.
Eligibility: Applicants should be UK or EU citizen, have a UK permanent residence status. Have several years experience working at a professional level in engineering. Have membership at any grade of a professional engineering institution accredited by the engineering council as a licenced number institution.
Level of Study: Postgraduate
Type: Fellowship
Value: £8,000
Length of Study: 1 year
Frequency: Annual
Study Establishment: Cranfield University, School of Applied Sciences
Country of Study: United Kingdom
No. of awards offered: 1
Application Procedure: Application form can be downloaded from www.panasonic.net/awards/appform.aspx or contact the Panasonic Trust.
Closing Date: August
Funding: Trusts
Contributor: The Royal Academy of Engineering

For further information contact:

Tel: 44 (0) 20 7222 2688/44 (0) 20 7766 0600

Royal Commission Industrial Design Studentships

Subjects: MDes Innovation and Creativity in Industry.
Purpose: To stimulate industrial design capability among the country's most able science and engineering graduates.
Eligibility: Open to applicants with good first degree in engineering or science and an offer of a place on the MDes Innovation and Creativity in Industry at Cranfield.
Level of Study: Postgraduate
Type: Studentship
Value: £10,000
Length of Study: 1 year
Frequency: Annual
Study Establishment: Cranfield University, School of Applied Sciences
Country of Study: United Kingdom
No. of awards offered: 6
Application Procedure: Applications can be downloaded from www.royalcommission1851.org.uk/ind_des.html.
Closing Date: April 26th
Funding: Trusts
Contributor: Royal Commision

Society for Underwater Technology (SUT)

Subjects: Offshore and ocean technology.
Purpose: To sponsor gifted students in Marine, Science and Engineering to meet industry's critical shortage of suitably qualified entrants.
Eligibility: First degree in an engineering or science subject.
Level of Study: Postgraduate
Type: Scholarship
Value: Up to £4,000
Length of Study: 1 year
Frequency: Annual
Study Establishment: Cranfield University, School of Applied Sciences
Country of Study: United Kingdom
No. of awards offered: 5–10
Application Procedure: Please check the website www.sut.org.uk
Closing Date: July 31st
Funding: Corporation
Contributor: SUT
Additional Information: Applications are to be sent by post not electronically.

For further information contact:

SUT, 80 Coleman Street, London,

Utilities and Service Industries Training (USIT)

Subjects: Economics for natural resources and environmental management, environmental diagnostics and management, environmental management for business, water and wastewater engineering, water management and environmental engineering.
Purpose: To support students attending an established academic course in the UK, which is relevant to one water utility industry. To assist individuals to benefit from opportunities that would not normally be available to them from employer's.
Level of Study: Postgraduate
Value: Up to £7,500
Length of Study: 1 year
Frequency: Annual
Study Establishment: Cranfield University, School of Applied Sciences
Country of Study: United Kingdom
Application Procedure: Application form can be downloaded from the website www.usit.org.uk. Applicants must submit a short paper to support the application.
Funding: Foundation
Contributor: USIT
No. of applicants last year: Many high competition

CRIMINOLOGY RESEARCH COUNCIL (CRC)

GPO Box 2944, Canberra, ACT 2601, Australia
Tel: (61) 2 6260 9216
Fax: (61) 2 6260 9218
Email: crc@aic.gov.au
Website: www.criminologyresearchcouncil.gov.au
Contact: Administrator

The Criminology Research Council (CRC) funds methodologically sound research in the areas of sociology, psychology, law, statistics, police, judiciary, corrections, mental health, social welfare, education and related fields. The research to be conducted is policy-orientated, and research outcomes should have the potential for application nationally or in other jurisdictions.

CRC Grants

Subjects: Criminological research in the areas of sociology, psychology, law, statistics, police, judiciary and corrections, etc. From time to time the Council will call for research in specific areas.
Purpose: To support research in Australia.
Eligibility: Open to Australian residents or visitors (actual or intending) who are pursuing or intend to pursue studies of consequence to the furtherance of criminological research in Australia. Grants are not likely to be given for assistance with research leading to the award of postgraduate degrees.
Level of Study: Doctorate, Postdoctorate
Type: Grant
Value: Variable
Length of Study: Usually 1 year, with a possibility of renewal for up to 3 years
Frequency: Annual
Country of Study: Australia
No. of awards offered: Approx. 6
Application Procedure: Applicants must complete an application form, available from the CRC.
Closing Date: August 15th
Funding: Government
No. of awards given last year: 3
No. of applicants last year: 45
Additional Information: The Council does not ordinarily consider applications involving travelling expenses outside Australia. Meetings are held in March, July and November. The November meeting is for general grants funding in any area the council deems relevant.

THE CROHN'S AND COLITIS FOUNDATION OF AMERICA

386 Park Avenue South, 17th Floor, New York, NY, 10016-8804, United States of America
Tel: (1) 800 932 2423
Fax: (1) 212 779 4098
Email: grants@ccfa.org
Website: www.ccfa.org
Contact: Ms Joseph O'Keefe, Director of Grants Administration

The Crohn's and Colitis Foundation of America is a non-profit, voluntary health organization dedicated to improving the quality of life for persons with Crohn's disease or ulcerative colitis. It supports basic and clinical scientific research to find the causes and cure for these diseases, provides educational programmes for patients, medical and healthcare professionals and the general public, alongside offering supportive services to patients, their families and friends.

Crohn's and Colitis Foundation Career Development Award

Subjects: Crohn's disease and ulcerative colitis.
Purpose: To stimulate and encourage innovative research that is likely to increase our understanding of the aetiology, pathogenesis, therapy and prevention of Crohn's disease and ulcerative colitis (IBD).
Eligibility: Candidates should hold an MD and must have 5 years of experience (with 2 years of research relevant to IBD).
Level of Study: Research, Postdoctorate
Type: Award
Value: Not to exceed US$90,000 per year
Length of Study: 1–3 years
Study Establishment: An approved research institute
Country of Study: United States of America
No. of awards offered: Varies
Application Procedure: The full application must be submitted electronically via proposal central and paper (master) of the full application must be sent to the CCFA National office.
Closing Date: January 14th or July 1st
Funding: Corporation, foundation, individuals, private, trusts
No. of awards given last year: 10
No. of applicants last year: 51

For further information contact:

Crohn's & Colitis Foundation of America, Research and Scientific Programs Department, 386 Park Avenue South – 17th Floor, New York, NY 10016-8804

Crohn's and Colitis Foundation Research Fellowship Awards

Subjects: Crohn's disease and ulcerative colitis (IBD).
Purpose: To stimulate and encourage innovative research that is likely to increase our understanding of the aetiology, pathogenesis, therapy and prevention of Crohn's disease and ulcerative colitis (IBD).
Eligibility: Applicants must hold an MD, PhD or equivalent with at least 2 years of research experience.
Level of Study: Postdoctorate, Research
Type: Award
Value: Not to exceed US$58,250 per year
Length of Study: 1–3 years
Frequency: Annual
Study Establishment: An approved research institute
Country of Study: United States of America
Application Procedure: The full application must be submitted electronically via the IGAM and paper (master) of the full application must be sent to the CCFA National office.
Closing Date: January 14th and July 1st
Funding: Corporation, foundation, individuals

Crohn's and Colitis Foundation Senior Research Award

Subjects: Crohn's disease and ulcerative colitis (IBD).
Purpose: To stimulate and encourage innovative research that is likely to increase our understanding of the aetiology, pathogenesis, therapy and prevention of Crohn's disease and ulcerative colitis (IBD).
Eligibility: Applicants should be researchers who hold an MD, PhD or equivalent.
Level of Study: Research, Postdoctorate
Type: Research award
Value: Up to US$117,000 direct cost per year plus indirect cost of 10 per cent of direct cost. Total: $128,700 per year
Length of Study: 1–3 years
Frequency: Annual
Study Establishment: An approved research institute
Country of Study: Any country
Application Procedure: The full application must be submitted electronically via the IGAM and paper (master) of the full application must be sent to the CCFA National office.
Closing Date: January 14th and July 1st
Funding: Corporation, foundation, individuals

Crohn's and Colitis Foundation Student Research Fellowship Awards

Subjects: Crohn's disease and ulcerative colitis (IBD).
Purpose: To stimulate and encourage innovative research that is likely to increase our understanding of the aetiology, pathogenesis, therapy and prevention of Crohn's disease and ulcerative colitis (IBD).
Eligibility: Applicants should be a medical student or graduate student studying at an accredited North American institution.
Level of Study: Graduate, Postgraduate
Value: US$2,500 per year
Length of Study: At least 10 weeks
Frequency: Annual
Study Establishment: An approved research institute
Country of Study: United States of America

Application Procedure: The full application must be submitted electronically via the IGAM and paper (master) of the full application must be sent to the CCFA National office.
Closing Date: March 15th
Funding: Corporation, foundation, individuals

THE CROSS TRUST

PO Box 17, 25 South Methven Street, Perth, Perthshire, PH1 5ES, Scotland
Tel: (44) 17 3862 0451
Fax: (44) 17 3863 1155
Email: crosstrust@mccash.co.uk
Website: www.thecrosstrust.org.uk
Contact: Ms Kathleen Carnegie, Assistant Secretary

The aim of the Cross Trust is to provide opportunities to young people of Scottish birth or parentage to extend the boundaries of their knowledge of human life. Proposals are to be of demonstrable merit from applicants with a record of academic distinction.

Cross Trust Grants

Subjects: Any approved subject.
Purpose: To enable young of Scottish birth or parentage people to extend the boundaries of their knowledge of human life.
Eligibility: Open to applicants of Scottish birth or parentage and must demonstrate thesis.
Level of Study: Unrestricted
Type: Grant
Value: Varies
Length of Study: Varies
Frequency: Annual
Study Establishment: An approved institute
Country of Study: Any country
No. of awards offered: Varies
Application Procedure: Applicants must complete an application form.
Closing Date: There may only be 3 meetings. Please check deadlines on website.
Funding: Private
No. of awards given last year: 121
No. of applicants last year: 347
Additional Information: Awards will only be considered from postgraduate students who have part funding in place from another organization. The Trust may support the pursuit of studies or research.

For further information contact:

McCash & Hunter, 25 South Methven Street, Perth, Perthshire, PH1 5ES, Scotland
Website: www.crosstrust.org.uk
Contact: The Secretaries

THE CROUCHER FOUNDATION

Suite 501, Nine Queen's Road Central, Hong Kong
Tel: (852) 2 736 6337
Fax: (852) 2 730 0742
Email: cfadmin@croucher.org.hk
Website: www.croucher.org.hk
Contact: Ms Elaine Sit, Administrative Officer

Founded to promote education, learning and research in the areas of natural science, technology and medicine, the Croucher Foundation operates a scholarship and fellowship scheme for individual applicants who are permanent residents of Hong Kong wishing to pursue doctoral or postdoctoral research overseas. The Foundation otherwise makes grants to institutions only.

Croucher Foundation Fellowships and Scholarships

Subjects: Natural science, medicine and technology.
Purpose: To enable selected students of outstanding promise to devote themselves to full-time postgraduate study or research in approved academic institutions outside Hong Kong.
Eligibility: Open to permanent residents of Hong Kong. Fellowships are intended for recent PhD graduates and not for the funding of career vacancies in universities. Scholarships are intended for those undertaking PhD studies.
Level of Study: Postdoctorate, Doctorate
Type: Scholarships, fellowships
Value: UK£22,100 per year for fellowships, and UK£12,300 per year and tution fees for scholarships, plus airfare assistance and other allowances
Length of Study: 1–2 years for fellowships, 1–3 years for scholarships
Frequency: Annual
Country of Study: Outside Hong Kong
No. of awards offered: Approx. 20–25
Application Procedure: Applicants can apply online at www.croucher.org.uk
Closing Date: November 15th
Funding: Private
No. of awards given last year: 17
No. of applicants last year: 92
Additional Information: Eligible to the nationals of Hong Kong.

THE CULINARY INSTITUTE OF AMERICA (CIA)

The Culinary Institute of America, 1946 Campus Drive, Hyde Park, NY 12538-1499, United States of America
Tel: (1) 845 452 9430
Email: admissions@culinary.edu
Website: www.ciachef.edu

The Culinary Institute of America (CIA) is a private, non-profit college dedicated to providing the world's best professional culinary education. CIA has been setting the standard for excellence in professional culinary education. The faculty, facilities, and academic programmes are offered at our campuses in Hyde Park, New York and St Helena, California.

The Culinary Institute of America Scholarship – Greystone

Subjects: Cooking.
Type: Scholarship
Value: US$5,000
Length of Study: 30 weeks
Frequency: Annual
Study Establishment: The Culinary Institute of America
Country of Study: United States of America
No. of awards offered: 1
Funding: Private

For further information contact:

Website: www.ciaprochef.edu

THE CULINARY TRUST

PO Box 273, New York, NY, 10013, United States of America
Tel: (1) 646 224 6989
Fax: (1) 888 345 4666
Email: scholarships@theculinarytrust.org
Website: www.theculinarytrust.com

The Culinary Trust has been the philanthropic partner to over 4,000 members of the International Association of Culinary Professionals (IACP) for over 20 years. The Trust solicits, manages and distributes funds for educational and charitable programmes related to the culinary industry in many areas.

Centro Culinario Ambrosia Mexican Cuisine Scholarship

Subjects: Mexican cuisine.
Purpose: To provide continuing education in Mexican cuisine.
Eligibility: Open to an experienced cook.
Level of Study: Unrestricted
Type: Scholarship
Value: Includes tuition, supplies, uniforms, and cutlery
Length of Study: 15 weeks
Frequency: Annual
No. of awards offered: 1

Application Procedure: Check website for further details.
Additional Information: Scholarship is valid from July 1st. Transportation and accommodations are not provided.

For further information contact:

Website: www.ambrosia.com.mx

Cuisinart Culinary Arts Scholarship

Subjects: Cuisine art.
Purpose: To provide financial assistance to individuals interested in preparing for a career in the culinary arts.
Eligibility: Check website for further details.
Level of Study: Unrestricted
Type: Scholarship
Value: Open to student enrolled in a Culinary Certificate or Degree programme at any nationally accredited culinary school
Length of Study: $1,500
Frequency: Annual
No. of awards offered: 1
Application Procedure: Scholarship valid from July 1st.
Closing Date: See the website

For further information contact:

Website: www.cuisinart.com

Culinary Arts Scholarship

Subjects: Culinary arts.
Purpose: To offer an assortment of funding opportunities available for the educational pursuits of beginning and/or currently enrolled culinary students as well as long-time culinary professionals seeking funding for continuing education short-courses and/or conducting independent study and research.
Eligibility: Open to a candidate who has pre-enrolled toward the 6–10 month Culinary Arts or Pastry & Baking Arts Diploma Program.
Type: Scholarship
Value: $5,000
Length of Study: 6–10 months
Frequency: Annual
Study Establishment: The Institute of Culinary Education New York
No. of awards offered: 1
Application Procedure: Check the website for further information.
Closing Date: March 1st
Additional Information: Scholarship may only be awarded to a student prior to their enrollment at The Institute of Culinary Education.

For further information contact:

Admissions Department, Institute of Culinary Education (ICE)
Tel: 212 847 0757
Website: www.iceculinary.com

Harry A. Bell Travel Grant

Subjects: Culinary research.
Purpose: For travel and research to food writers during the pre-contract phase of their book proposal.
Eligibility: Open to writers who are in the pre-contract phase of their book.
Level of Study: Unrestricted
Type: Grant
Value: $3,000–4,000
Frequency: Annual
No. of awards offered: Varies
Application Procedure: Check the website for further details.
Closing Date: May 31st

For further information contact:

Website: www.theculinarytrust.com

The Julia Child Endowment Fund Scholarship

Subjects: Culinary arts.
Purpose: To support a career professional to conduct independent study and research in France, as it relates to French food, wine, history, culture and traditions. This programme also encourages, enables and assists aspiring students and career professionals to advance their knowledge of the culinary arts.
Eligibility: Open to applicants who have 2 years of food service experience.
Level of Study: Professional development
Type: Scholarship
Value: US$5,000
Frequency: Annual
Country of Study: France
Application Procedure: Applicants are required to submit a 2 page essay illustrating their culinary goals along with 2 letters of professional reference.
Closing Date: December 15th
Funding: Trusts

For further information contact:

Website: www.theculinarytrust.org

The Julia Child Fund at the Boston Foundation Independent Study Scholarship

Subjects: French food, wine and culinary art.
Purpose: For independent study in France on French food, wine, and culinary disciplines.
Eligibility: Open to a career professional doing research in writing and teaching related to French food, wine and culinary disciplines.
Level of Study: Professional development
Type: Scholarship
Value: $5,000
Frequency: Annual
Country of Study: France
No. of awards offered: 1
Application Procedure: Check the website for further details.
Closing Date: See the website.

For further information contact:

Website: www.tbf.org

L'Academie de Cuisine Culinary Arts Scholarship

Subjects: Culinary arts.
Purpose: To financially prospective students prospective students for the Culinary Arts Program each year.
Eligibility: Open to a student pre-enrolled for the 12 months Culinary Arts or Pastry Arts Certificate Program.
Type: Scholarship
Value: $5,000
Length of Study: 1 year
No. of awards offered: 1
Application Procedure: Check the website for further details.
Closing Date: December 15th
Additional Information: Scholarship is valid for enrollment during July or October only.

For further information contact:

Tel: 646 224 6989
Email: scholarships@theculinarytrust.org
Website: www.lacademie.com/www.theculinarytrust.org
Contact: Amy Blackburn, Director of Administration for The Culcinary Trust

Zwilling, J.A. Henckels Culinary Arts Scholarship

Subjects: Culinary arts.
Purpose: To provide financial assistance to students from designated states who are interested in pursuing a degree in the culinary arts.
Eligibility: Open to any pre-enrolled student, currently enrolled student or career professional toward any culinary arts degree or certificate program at any nationally accredited culinary school.
Level of Study: Postgraduate
Type: Scholarship
Value: $5,000
No. of awards offered: 1
Application Procedure: Check the website for further details.
Closing Date: December 15th
Contributor: Zwilling, J.A. Henckels Trust

For further information contact:

Email: scholarships@theculinarytrust.org
Website: www.jahenckels.com

CURTIN UNIVERSITY OF TECHNOLOGY

Office of Research and Development, GPO Box U1987, Perth, Western Australia, 6845, Australia
Tel: (61) 8 9266 7331
Fax: (61) 8 9266 2605
Email: research_scholarships@curtin.edu.au
Website: http://www.curtin.edu.au/

Curtin University of Technology is a world class, internationally focused, culturally diverse institution. They foster tolerance and encourage the development of the individual. Their programmes centre around the provision of knowledge and skills to meet industry and workplace standards. A combination of first rate resources, staff and technology makes Curtin a forerunner in tertiary education both within Australia and internationally.

APA(I) – Innovation, Competition and Economic Performance

Subjects: Computing / Information Technology (IT) and Economics / Finance.
Purpose: To encourage students to undertake a Higher Degree by Research within the the Centre for Research in Applied Economics (CRAE).
Eligibility: Candidates must be Australian citizens or permanent residents or New Zealand citizens, should hold or are expected to hold a First Class Honours Degree or its equivalent and must meet Curtin University of Technology's requirements for admission to a PhD.
Level of Study: Graduate
Type: Competition
Value: Australian $25,118 per year
Length of Study: 3 years with the possibility of an extension of up to six months
No. of awards offered: 1
Application Procedure: Candidates must forward the completed application for admission to a higher degree by research to the Centre for Research into Applied Economics (CRAE).
Closing Date: March 31st

For further information contact:

The Centre for Research into Applied Economics (CRAE), Curtin Business School, Curtin University of Technology, GPO Box U1987, Perth, WA 6845
Tel: 61 8 9266 2035
Email: H.Bloch@exchange.curtin.edu.au
Contact: Professor Harry Bloch

Australian Biological Resources Study Postgraduate Scholarship

Subjects: Agricultural science. See http://scholarships.curtin.edu.au/scholarship.cfm?id=529 for more details
Purpose: To foster research training compatible with ABRS and national research priorities.
Eligibility: Applicants must be Australian citizens or permanent residents, must hold a First or Upper Second Class Honours or equivalent degree in an appropriate discipline and be enrolled as a full-time student in a PhD degree at an Australian institution.
Level of Study: Graduate
Type: Scholarship
Value: Australian $22,500 in 2010 (by $10,000 per year)
Country of Study: Australia
No. of awards offered: 1
Application Procedure: Applicants must submit the application to ABRS through the host institution. The application form will then be submitted to the ABRS Advisory Committee for consideration and assessment using the selection criteria. The individual selected as most worthy of funding will be awarded the scholarship.
Closing Date: October 26rd

For further information contact:

Australian Biological Resources Study, GPO Box 787
Tel: 02 6250 9554
Fax: 02 6250 9555
Email: abrs.grants@environment.gov.au
Website: www.environment.gov.au/biodiversity/abrs/admin/training/index.html
Contact: Business Manager

Curtin Business School Doctoral Scholarship

Subjects: Economics/finance, human resources, legal studies/politics, management/administration, marketing/public relations.
Purpose: To enable doctoral (PhD, DBA) students to study at the Curtin Business School.
Eligibility: Applicants must have completed at least four years of tertiary education studies at a high level of achievement and have First/Upper Second Class Honours or equivalent results. See http://scholarships.curtin.edu.au/scholarship.cfm?id=52 for more details
Level of Study: Postgraduate
Type: Scholarship
Value: $25,000
Length of Study: Up to 3 years
Frequency: Annual
No. of awards offered: 3
Application Procedure: Applicants can download the application form and obtain further information from the website.
Closing Date: December 31st

For further information contact:

CBS HDR Unit, Curtin University of Technology, GPO Box U 1987, PERTH, WA 6845, Australia
Tel: (61) 61 08 9266 4301
Email: j.boycott@curtin.edu.au
Contact: Ms Jo Boycott, Research Student Coordinator

Curtin University Postgraduate Scholarship (CUPS)

Subjects: All subjects.
Purpose: To assist with general living costs.
Eligibility: Applicants must be Australian or New Zealand citizens or Australian permanent residents and must have completed four years of higher education studies at a high level of achievement and must hold, or are expected to obtain, First Class Honours or equivalent results; be enrolled in or accepted to enrol in a Higher Degree by Research as a full-time student in the previous year in which thw award is to be given.
Level of Study: Graduate, Postgraduate
Type: Scholarship
Value: Varies, an annual living allowance of $23,728 was given last year. This stipend is indexed annually and is tax-free unless taken on a part-time basis.
No. of awards offered: 40
Application Procedure: Check website for further details.
Closing Date: October 31st

For further information contact:

Office of Research & Development, Curtin University of Technology, GPO Box U1987, PERTH, WA 6845, Australia
Tel: (61) 08 9266 4906
Fax: (61) 08 9266 3793
Email: research_scholarships@curtin.edu.au
Website: http://scholarships.curtin.edu.au/
Contact: Manager, Scholarships

Establishing the Source of Gas in Australia's Offshore Petroleum Basins–Scholarship

Subjects: Chemistry, geochemistry, geology.
Purpose: To develop an isotopic method to analyze gases in fluid inclusions and to establish the source of gas in Australia's offshore petroleum basins.
Eligibility: Applicants must have First Class Honours or equivalent science degree, preferably in chemistry/geology/geochemistry. Interests in analytical organic chemistry, laboratory skills in trace analysis, wet chemical methods, GC/GCMS or GC–IRMS instrumentation and awareness of stable isotopic concepts is desirable.
Level of Study: Postgraduate
Type: Scholarship
Value: At least $20,000 per year
No. of awards offered: 1
Application Procedure: Check website for further details.

Closing Date: December 31st
Contributor: The Stable Isotope and Molecular Biogeochemistry Research Group, Geoscience Australia, GFZ

For further information contact:

Stable Isotope and Molecular Biogeochemistry Group, Centre for Applied Organic Geochemistry, Department of Applied Chemistry, Curtin Universitiy of Technology, GPO Box U1987, Perth, WA 6845
Tel: 61 08 9266 2474
Fax: 61 08 9266 2300
Email: K.grice@curtin.edu.au
Website: www.caog.chemistry.curtin.edu.au
Contact: Professor Kliti Grice

French–Australian Cotutelle

Subjects: All subjects.
Purpose: To support the development of the double doctoral degree 'Cotutelle' between Australia and France.
Eligibility: Applicants must be PhD students (of any nationality) enrolled in a Cotutelle project between a French and an Australian university; should not have benefited from the French Embassy Cotutelle grant in previous years and should be registered with FEAST-France.
Level of Study: Postgraduate
Type: Grant
Value: $2,500
No. of awards offered: 10
Application Procedure: Applicants must provide the French Embassy with the completed application form and a copy of the Cotutelle convention.
Closing Date: December 8th

For further information contact:

Higher Education Attache, Embassy of France in Australia
Email: Stephane.GRIVELET@diplomatie.gouv.fr
Website: www.ambafrance-au.org
Contact: Mr Stephane GRIVELET

The General Sir John Monash Awards

Subjects: All subjects. See http://scholarships.curtin.edu.au/scholarship.cfm?id=153 for more details.
Purpose: To enable them to undertake postgraduate study abroad at the world's best Universities, appropriate to their field of study.
Eligibility: Applicants must be Australian citizens who have graduated from an Australian University with outstanding levels of academic achievement. See http://scholarships.curtin.edu.au/scholarship.cfm?id=153 for more details.
Level of Study: Graduate
Type: Award
Value: Australian $50,000 per year
Length of Study: 3 years
Frequency: Annual
No. of awards offered: 8
Application Procedure: Check website for further details.
Closing Date: August 31st

For further information contact:

The General Sir John Monash Foundation, Level 1, Bennelong House, 9 Queen Street, Victoria, Melbourne, 3000, Australia
Tel: (61) 613 9620 2428
Email: peter.binks@monashawards.org
Website: www.monashawards.org
Contact: Dr Peter Binks, Chief Executive Officer

Gowrie Research Scholarship

Subjects: All subjects. See http://scholarships.curtin.edu.au/scholarship.cfm?id=68 for details
Purpose: To assist members of the defence forces and direct descendants of members of the defence force to aid with their research study.
Eligibility: Candidates must be the Australian citizen or permanent resident and a graduate who have completed a course of tertiary education at other recognised Australian institutions.
Level of Study: Graduate
Type: Scholarship
Value: $4,000 per year
Length of Study: 2 years
No. of awards offered: 1
Application Procedure: Candidates must submit applications in duplicate on the prescribed form. If space is insufficient, a separate statement should be added.
Closing Date: October 31st
Contributor: Gowrie Scholarship Trust Fund

For further information contact:

The Gowrie Scholarship Fund Trust, 3/32 Beaconsfield Road, Mosman, NSW 2088, Australia
Contact: The Secretary

Hunter Postgraduate Scholarship

Subjects: Alzheimer's disease.
Purpose: To support a PhD student undertaking research in an area relevant to understanding the causes of Alzheimer's disease.
Eligibility: Candidates must be PhD students undertaking research in an area relevant to understanding the causes of Alzheimer's disease.
Level of Study: Postgraduate
Type: Scholarship
Value: $23,000 per year
Length of Study: 3 years
Application Procedure: Check website for further details.
Closing Date: October 31st

For further information contact:

Tel: 02 6254 7233
Email: aar@alzheimers.org.au
Website: www.alzheimers.org.au/content.cfm?infopageid=3102
Contact: Anna Conn

Scots Australian Council Scholarships

Subjects: Humanities. See http://scholarships.curtin.edu.au/scholarship.cfm?id=112 for details
Purpose: To develop lasting links between young Scots and Australians by offering outstanding graduates and young professionals the opportunity to study at a Scottish university.
Eligibility: Applicants must be Australian citizens or Australian permanent residents or New Zealand citizens or on permanent Humanitarian Visa. They must be indigenous or Torres Strait Islander students or students with a disability or students from rural or regional areas or mature students or sole parents or current students or prospective students.
Level of Study: Postgraduate
Type: Scholarship
Value: £12,000
No. of awards offered: 1
Application Procedure: Check website for further details.
Closing Date: January 14th
Contributor: Scottish universities, Scottish business and industry, British Foreign and Commonwealth Office

For further information contact:

The Scots Australian Council, 19 Dean Terrace, Edinburgh, EH4 1NL, Scotland
Email: scholarships@scotsoz.org
Contact: The Secretary

Sediment and Asphaltite Transport by Canyon Upwelling – Top Up Scholarship

Subjects: Chemistry, geochemistry, geology.
Purpose: To investigate the role of upwelling currents in transporting material across the continental slope of the Morum Sub-Basin, southern Australia using an integrated geological, oceanographic and organic geochemical approach.
Eligibility: Applicants must be Australian and New Zealand residents, First Class Honours or equivalent science degree holders, preferably in chemistry/geology/geochemistry. Interests in analytical organic chemistry, laboratory skills in trace analysis, wet chemical methods, GC/GCMS or GC-IRMS instrumentation; awareness of stable isotopic concepts is desirable.

Level of Study: Graduate
Type: Scholarship
Value: See the organization website
No. of awards offered: 1
Application Procedure: Applicants must forward their interests, curriculum vitae and names of two referees to Stable Isotope and Molecular Biogeochemistry Group, Centre for Applied Organic Geochemistry, Department of Applied Chemistry.
Closing Date: See the organization website
Contributor: The Stable Isotope and Molecular Biogeochemistry Research Group, Adelaide University, a petroleum industry partner

For further information contact:

Stable Isotope and Molecular Biogeochemistry Group, Centre for Applied Organic Geochemistry, Department of Applied Chemistry, Curtin Universitiy of Technology, GPO Box U1987, Perth, WA 6845
Tel: 61 08 9266 2474
Fax: 61 08 9266 2474
Email: K.grice@curtin.edu.au
Website: www.caog.chemistry.curtin.edu.au
Contact: Professor Kliti Grice

Water Corporation Scholarship in Biosolids Research

Subjects: Agriscience, environmental science.
Purpose: To investigate the potential impacts to soil and plants following the agricultural land application of alum-dosed wastewater sludge.
Eligibility: Applicants must be Australian Citizens, Australian permanent residents or must hold an Australian permanent Humanitarian Visa. They must hold a relevant degree from a recognized University in the preferred fields of Agriculture, Environmental Science or the equivalent and demonstrate a high level in their Honours project or equivalent.
Level of Study: Graduate, Postgraduate
Type: Scholarship
Value: $23,400 per year
Length of Study: 3 years for a doctoral program and 2 years for a masters program
No. of awards offered: 1
Application Procedure: Check website for further details.
Closing Date: October 31st
For further information contact:

Muresk Institute, Curtin University of Technology, GPO Box U1987, PERTH, WA 6845
Email: D.Pritchard@curtin.edu.au
Contact: Dr Deborah Pritchard

DAAD

German Academic Exchange Service, 1 Southampton Place, London, WCIA 2DA, England
Tel: (44) 20 7831 9511
Fax: (44) 20 7831 8575
Email: info@daad.org.uk
Website: www.london.daad.org.uk
Contact: Ms Judie Cole

Research Grants for Postgraduate and Doctoral Studies

Subjects: All subjects.
Purpose: To support research for Doctoral studies.
Eligibility: Open to the candidates with Masters Degree and PhD.
Level of Study: Doctorate, Postgraduate
Type: Grant
Value: €1,000 per month plus travel cost and a health insurance allowance.
Length of Study: 1 year
Frequency: Annual
Country of Study: Germany
Application Procedure: Please check website for details.
Closing Date: Check website
Funding: Government

Research Internships in Science and Engineering (RISE)

Subjects: Check website for details.
Purpose: To offer research internships to students in science and engineering.
Level of Study: Undergraduate
Type: Grant
Value: €650 per month
Length of Study: 1.5 to 3 years
Frequency: Annual
Country of Study: Germany
Funding: Government
Additional Information: Please visit www.daad.de/rise for the complete program description and application guidelines.

Study Scholarships for Graduates of All Disciplines

Subjects: All subjects.
Purpose: Study scholarships are awarded to provide foreign graduates of all disciplines with opportunities to complete a post-graduate or Master's course at a state (public) or state-recognised German higher education institution and to gain a degree in Germany (Master's/Diploma).
Eligibility: Open to excellently qualified junior teaching staff from state universities and from private universities and researchers from state-run research institutions in any field of study/research who hold a first degree (Bachelors Degree) with some teaching/research experience. Check website for further details.
Level of Study: Postgraduate
Type: Scholarship
Value: €750 plus travel and luggage cost and health insurance allowance along with a study and research allowance. Tuition fees cannot be paid by the DAAD. Furthermore, the DAAD will pay a study and research allowance and, where appropriate, a rent subsidy and family allowance.
Length of Study: 10–24 months (initially, scholarships are awarded for one academic year and can be extended for students with good study achievements to cover the full length of the chosen degree course).
Application Procedure: Detailed information on what application papers need to be submitted can be found on the application form for "Research Grants and Study Scholarships" on www.daad.de/en/form.
Additional Information: Any teaching staff predominantly or exclusively involved in non-degree programs is not eligible. Even if an applicant has qualified for a scholarship after the interview, the DAAD scholarship will only be granted after admission to the study program chosen.

DALHOUSIE MEDICAL RESEARCH FOUNDATION

The Dalhousie Medical Research Foundation, 1-A1 Sir Charles Tupper Medical Building, 5850 College Street, Halifax, NS, B3H 4R2, Canada
Tel: (1) 902 494 3502
Fax: (1) 902 494 1372
Email: dmrf@dal.ca
Website: www.dmrf.ca

Dalhousie University is one of Canada's leading universities, Dalhousie is widely recognized for outstanding academic quality and teaching, and a broad range of educational and research opportunities.

Dalhousie Medical Research Foundation Fellowships

Subjects: Basic or clinical science, medical research, health law, bioethics, medical informatics, population health, medical education, and medical humanities.
Purpose: To acknowledge the exemplary work in basic medical research and clinical medical research by the junior researchers.
Eligibility: Open to outstanding junior researchers.
Level of Study: Postgraduate, Medical research
Type: Award
Value: Canadian $7,000 salary support
Length of Study: 1–3 years
Frequency: Annual
Study Establishment: Dalhousie University
Country of Study: Canada
No. of awards offered: 3

Application Procedure: See website for admission process.
Closing Date: See website for exact information.
Funding: Foundation
Contributor: Donations
Additional Information: The DMRF funds three ongoing fellowships: William M. Sobey Fellowship in Cardiology, DMRF Clinical Research Fellowship, and A.K. Reynolds Memorial Postdoctoral Fellowship.

DAMON RUNYON CANCER RESEARCH FOUNDATION

One Exchange Plaza, 55 Broadway, Suite 302, New York, NY, 10006, United States of America
Tel: (1) 212 455 0520
Fax: (1) 212 455 0529
Email: awards@damonrunyon.org
Website: www.damonrunyon.org
Contact: Cait Ahearn, Programs Associate

The Damon Runyon Cancer Research Foundation selects the most brilliant early career scientists and provides them with funding to pursue innovative cancer research.

Damon Runyon Clinical Investigator Award

Subjects: Understanding the causes and mechanisms of cancer and developing more effective cancer therapies and preventions.
Purpose: To support young scientists conducting patient-orientated cancer research. To provide outstanding young physicians with the resources and training structure essential to becoming independent clinical investigators.
Eligibility: Open to a U.S. citizen or permanent legal resident and must have received an MD or MD/PhD degree(s) from an accredited institution and be board-eligible. Each applicant must be nominated by his/her institution. Applications will only be accepted from institutions that have been invited to submit them by the DRCRF. No more than two Damon Runyon Clinical Investigators will be funded to work with the same Mentor at any given time.
Level of Study: Research, Professional development
Type: Award
Value: Funding will amount to $450,000 over three years. Each year the researcher will receive $150,000. This amount can be used for a variety of scientific needs including the investigators stipend (up to $100,000), salaries for professional/technical personnel, special equipment, supplies. Clinical investigators chosen to receive a continuation grant will receive $150,000 per year for an additional 2 years.
Length of Study: Up to 5 years
Frequency: Annual
Study Establishment: Suitable facilities within the United States of America
Country of Study: United States of America
No. of awards offered: 5
Application Procedure: Application form can be downloaded from the website and must be completed in all respects.
Closing Date: February 15th
Funding: Private
Contributor: Eli Lilly, Pfizer, Genentech and Merck
No. of awards given last year: 5
No. of applicants last year: 50
Additional Information: In addition, the Foundation will retire up to $100,000 of any medical school debt owed by the awardee. New for 2011: partnership with the NIH/NCI. This partnership opens access to the NIH Clinincal Center to Damon Runyon Clinical Investigators.

Damon Runyon Fellowship Award

Subjects: Understanding the causes and mechanisms of cancer and developing more effective cancer therapies and preventions.
Purpose: To encourage all theoretical and experimental research relevant to the study of cancer and the search for cancer causes, mechanisms, therapies and prevention.
Eligibility: Open to US citizens or foreign candidates applying to do their research in the United States of America. Also, legal residents or citizens of the United States of America working abroad. Candidates must be at the beginning of their 1st full-time postdoctoral fellowship. Candidate cannot be in the sponsor's laboratory for more than 1 year.
Level of Study: Postdoctorate
Type: Fellowship
Value: Fellows are eligible to apply for additional support at the end of the 3 years – The Dale F. Frey Award for Breakthrough Scientists. US $50,000 per year for level I funding (for physician scientists who have completed residencies, clinical training and are board eligible, US $60,000 will be given per year). All fellows will receive a US$2,000 per year expense allowance as well.
Length of Study: 3 years
Frequency: Biannual
Study Establishment: Suitable accredited establishment within the United States of America
Country of Study: United States of America
No. of awards offered: 30–40 per year
Application Procedure: Applicants should download an application from the website. Fellowship awards are to be approved by the Board of Directors of the Damon Runyon Cancer Research Foundation acting upon the recommendation of the fellowship award Committee.
Closing Date: March 15th and August 15th. See the weblink.
Funding: Private
Contributor: HHMI, Robert Black Charitable Foundation, Merck
No. of awards given last year: 33
No. of applicants last year: 333
Additional Information: D R fellows are eligible to apply for the Dale F. Frey Award for Breakthrough Scientists in the third year of their award. The award is US$100,000 paid over one year (must be expanded within 2 years of initial award date).

THE DAVIES CHARITABLE FOUNDATION

245 Alwington Place, Kingston, ON, K7L 4P9, Canada
Tel: (1) 613 546 4000
Fax: (1) 613 546 9130
Email: daviesfoundation@cogeco.ca
Website: www.daviesfoundation.ca

The Davies Charitable Foundation is a registered, non-profit, charitable organization founded in 1990 by Michael R.L. Davies, former owner and publisher of the Kingston Whig Standard. The purpose of the Foundation is to support individuals and organizations within the local district in the areas of the arts, education, health and sports. Since its inception the Davies Charitable Foundation has donated over $7.5 million to over 400 individuals and institutions.

The Davies Charitable Foundation Fellowship

Subjects: All subjects.
Purpose: To support a native of the Kingston, Ontario, area at the peak of academic excellence.
Eligibility: Open to candidates born in the Kingston area or have resided in the area for at least 5 years prior to their 20th birthday and must have been accepted into a postdoctoral or fellowship position at the university of their choice.
Level of Study: Postdoctorate
Type: Fellowship
Value: Canadian $10,000
Length of Study: 1 year
Frequency: Annual
Country of Study: Canada
No. of awards offered: 1
Application Procedure: Application form can be downloaded from the website.
Closing Date: April 15th
Funding: Foundation
Contributor: Davies Charitable Foundation

For further information contact:

The Davies Charitable Foundation, 245, Alwington Place, ON, Kingston, K7L 4P9, Canada
Tel: (1) 613-546-4000
Email: daviesfoundation@cogeco.ca
Website: www.daviesfoundation.ca

THE DAYTON AREA GRADUATE STUDIES INSTITUTE (DAGSI)

3155 Research Boulevard Suite 205, Kettering, OH, 45420, United States of America
Tel: (1) 937 781 4000
Fax: (1) 937 781 4005
Email: edownie@dagsi.org
Website: www.dagsi.org
Contact: Dr Elizabeth Downie, Director

The Dayton Area Graduate Studies Institute (DAGSI) is a consortium of graduate engineering schools at the University of Dayton, a private institution, Wright State University, a state assisted institution and the Air Force Institute of Technology, a federal institution. It integrates and leverages the combined resources of the partnership, including faculty, facilities, equipment and other assets of the institutions.

DAGSI Research Fellowships

Subjects: Engineering and computer science.
Purpose: To financially support full-time students pursuing a research-based Doctoral study.
Eligibility: Open to candidates with a minimum of Bachelor's degree in engineering, computer science or a related field.
Level of Study: Doctorate
Type: Fellowships
Value: US$28,000 plus full tution
Length of Study: 3 years
Frequency: Annual
No. of awards offered: 15
Application Procedure: Applicants can download the application form from the website.
Closing Date: December 15th
Additional Information: DAGSI encourages women, minorities and persons with disabilities to participate in the fellowship.

DEAKIN UNIVERSITY

221 Burwood Highway, Burwood, Victoria, 3125, Australia
Tel: (61) 03 9244 6100
Fax: (61) 03 9244 8796
Email: dconnect@deakin.edu.au
Website: www.deakin.edu

Established in the 1970s, Deakin University is one of Australia's largest universities providing all the resources of a major university to more than 32,000 award students. The University's reputation for excellent teaching and innovative course delivery has been recognized through many awards over the past few years.

Coltman Prize

Subjects: Biomedical science.
Purpose: To recognize outstanding achievement and ability within the student's particular area of research.
Eligibility: Open to students undertaking research in biomedical sciences.
Level of Study: Postgraduate
Type: Prize
Value: Australian $3,000 and a framed certificate.
Frequency: Annual
Country of Study: Australia
Application Procedure: Check website for further details.
Contributor: Dr Kay and Mrs Barbara

Edward Wilson Scholarship for Graduate Diploma of Journalism

Subjects: Journalism.
Purpose: To assist students pursuing journalism.
Eligibility: Open to Australian citizens undertaking postgraduate studies in journalism.
Level of Study: Postgraduate
Type: Scholarship
Value: Tuition fees
Length of Study: 2 years
Frequency: Annual
Country of Study: Australia
No. of awards offered: 1
Application Procedure: A completed application form must be submitted.
Closing Date: October 7th

Helen Macpherson Smith Scholarship in Arts and Entertainment Management

Subjects: Arts and entertainment management.
Purpose: To financially support outstanding students.
Eligibility: A female undertaking postgraduate studies in either the Graduate Certificate or Masters of Arts and Entertainment Management course or the arts and entertainment specialisation in the Master of Business Administration.
Level of Study: Postgraduate
Type: Scholarship
Value: Australian $6,000
Length of Study: 1 year (for four credit points of study).
Frequency: Annual
Country of Study: Australia
No. of awards offered: 1
Application Procedure: Check website for further details.
Closing Date: March 2nd. See the weblink.
Additional Information: Must be a holder of an Australian permanent humanitarian visa.

For further information contact:

Website: http://www.deakin.edu.au/current-students/study-information/scholarships/helen-macpherson-smith-arts

The Isi Leibler Prize

Subjects: Area and cultural studies.
Purpose: To advance knowledge of multiculturism and community relations in Australia.
Eligibility: Open to students who have submitted a postgraduate or doctoral thesis at the university.
Level of Study: Doctorate, Postgraduate
Type: Award
Value: $1,000
Frequency: Annual
Study Establishment: Deakin University
Country of Study: Australia
Application Procedure: Completed application form plus additional information must be submitted. For further details see the Website.
Closing Date: January 14th
Additional Information: For students who completed their course in Trimesters 1 and 2, applications for all student prizes will close on December 16th.

For further information contact:

Vice-Chancellor's Prizes Committee, Academic Administrative Services Division, Geelong Waterfront Campus, Deakin University, Geelong, Vic 3217, Australia
Tel: (61) 03 5227 2333
Website: http://www.deakin.edu.au/current-students/study-information/awards/postgraduate/isi-leibler.php

Rex Williamson Prize

Subjects: Chemistry.
Purpose: To award students showing the best research potential and academic merit.
Eligibility: Open to students enrolled within the school of biological and chemical sciences.
Level of Study: Postgraduate
Type: Award
Value: Australian $5,000 and a framed certificate.
Frequency: Annual
Country of Study: Australia
Application Procedure: Check website for further details.
Closing Date: December 19th. Check website for further details.

Tennis Australia Prize

Subjects: Sport management.
Purpose: To recognize outstanding achievement.
Eligibility: Open to outstanding students of Master's/Graduate certificate of business.

Level of Study: Postgraduate
Type: Award
Value: Australian $750
Frequency: Annual
Country of Study: Australia
Application Procedure: A completed application form must be submitted.
Closing Date: October 31st

DEBRA INTERNATIONAL

DebRA House, 13 Wellington Business Park, Dukes Ride, Crowthorne, Berkshire, RG45 6LS, England
Tel: (44) 13 4477 1961
Fax: (44) 13 4476 2661
Email: debra@debra.org.uk, john.dart@debra.org.uk
Website: www.debra.org.uk
Contact: John Dart, COO

DebRA International is the national charity working on behalf of people with the genetic skin blistering condition, Epidermolysis bullosa (EB).

DebRA International Research Grant Scheme

Subjects: Epidermolysis bullosa.
Purpose: To fund research into epidermolysis bullosa (EB).
Eligibility: Applicants must be productive postdoctorates, usually with a track record as a principal investigator.
Level of Study: Postdoctorate, Research
Type: Project grant
Value: Maximum £80,000 per year.
Length of Study: Grants for projects are usually for 1–3 years.
Frequency: Twice a year
Country of Study: Any country
No. of awards offered: Varies
Application Procedure: Application form can be downloaded from the website www.debra-international.org or from the DebRA UK Office.
Closing Date: March 15th and September 15th
Funding: Private
Contributor: Charitable funding
No. of awards given last year: 8
No. of applicants last year: 22

DEMOCRATIC NURSING ORGANIZATION OF SOUTH AFRICA (DENOSA)

605 Church Street, Pretoria, 0001, PO Box 1280, Pretoria, 0001, South Africa
Tel: (27) 12 343 2315/6/7
Fax: (27) 12 344 0750
Email: info@denosa.org.za
Website: www.denosa.org.za
Contact: Executive Director

The Democratic Nursing Organization of South Africa (DENOSA) is a professional organization and labour union for nurses in South Africa.

DENOSA Bursaries, Scholarships and Grants

Subjects: Nursing.
Purpose: To encourage postbasic studies at a South African teaching institution.
Eligibility: Open to members of the organization in good standing who hold the required registered nursing qualifications.
Level of Study: Doctorate, Graduate, MBA, Professional development, Postgraduate, Research
Type: Bursary
Value: Varies
Length of Study: 3–4 years
Frequency: Annual
Study Establishment: A South African teaching institution
Country of Study: South Africa
No. of awards offered: Varies
Application Procedure: Applicants must complete an application form.
Closing Date: January 31st
Funding: Individuals
Contributor: Donor funding
No. of awards given last year: 153
No. of applicants last year: 173

THE DENMARK-AMERICA FOUNDATION

Nørregade 7A, 1165 København K, Denmark
Tel: (45) 3532 4545
Fax: (45) 3332 5323
Email: daf-fulb@daf-fulb.dk
Website: www.wemakeithappen.dk
Contact: Ms Marie Monsted, Executive Director

The Denmark-America Foundation was founded in 1914 as a private foundation, and today its work remains based on donations from Danish firms, foundations and individuals. The Foundation offers scholarships for studies in the United States of America at the graduate and postgraduate university level and also has a trainee programme.

Denmark-America Foundation Grants

Subjects: All subjects.
Purpose: To further understanding between Denmark and the United States of America.
Eligibility: Open to Danes and Danish-American citizens.
Level of Study: Graduate, MBA, Postdoctorate, Postgraduate, Professional development, Research
Type: Bursary
Value: Varies
Length of Study: 3–12 months
Frequency: Annual
Country of Study: United States of America
No. of awards offered: Varies
Application Procedure: Applicants must complete a special application form, available by contacting the secretariat.
Closing Date: Check website for further details.
Funding: Private
No. of awards given last year: 34–35
No. of applicants last year: 250

DEPARTMENT OF BIOTECHNOLOGY, MINISTRY OF SCIENCE AND TECHNOLOGY

Indian Institute of Science, Bangalore, 560 012, India
Email: kmbc@biochem.iisc.ernet.in
Website: www.dbtindia.nic.in
Contact: Professor K Muniyappa, Coordinator DBT-PDF Program

The Government of India, Ministry of Science and Technology established the Department of Biotechnology in 1986 to give a new impetus to the development of the field of modern biology and biotechnology in India. The Department has made significant achievements in the growth and application of biotechnology in the broad areas of agriculture, health care, animal sciences, environment and industry.

Postdoctoral Fellowship Programme

Subjects: Biotechnology.
Purpose: To train scientists in the frontier areas of research in biotechnology at institutions in India which are engaged in major biotechnological research activities.
Eligibility: Open to Indian citizens who have obtained a PhD in science, engineering or MD or MS in any area of medicine with research interests in biotechnology and life sciences. Those who have already submitted the PhD/MD/MS theses are also eligible to apply. The applicants should preferably be below the age of 40 years and 45 years in case of female candidates.
Level of Study: Research
Type: Fellowships

Value: Indian Rupees 16,000–18,000 per month and a research contingency grant of Indian Rupees 50,000 per year, payable to the host institution. Candidates who are yet to be awarded their PhD/MD/MS degree, if selected, will be paid Indian Rupees 15,000 per month until being awarded the degree. The Fellows will also be entitled to HRA and other benefits.
Length of Study: 2 years
Frequency: Annual, Biannual
Country of Study: India
No. of awards offered: 75
Application Procedure: Candidates must submit their application along with their curriculum vitae, list of publications (attach reprints of 2 important papers), copies of certificates, 1 page synopsis of PhD/MD thesis, 2 letters of recommendation (academic). The applicants are advised to propose their place of work, name of the supervisor, enclose 1 page synopsis of proposed research, which must be compatible with the ongoing research of the proposed supervisor, and his/her consent, for availing the fellowship. They should also enclose a declaration stating that if selected for the fellowship, they will complete the tenure of the fellowship.
Closing Date: October 15th
Funding: Government
Additional Information: In exceptional cases, based on the progress of research, and overall importance of the project of the Fellows, the fellowship may be extended upto 5 years.

DEPARTMENT OF EDUCATION SERVICES

22 Hasler Road, Osborne Park, Perth, WA, 6017, Australia
Tel: (61) 8 9441 1900
Fax: (61) 8 9441 1901
Email: des@des.wa.gov.au
Website: www.des.wa.gov.au

The Department of Education Services provides policy advice to the Minister for Education and Training and supporting universities, non-government schools and international education providers and in some cases individual students and teachers through scholarship programmes in Western Australia

Western Australian Government Japanese Studies Scholarship

Subjects: Japanese studies.
Purpose: To provide students with the opportunity to spend 1 year studying at a tertiary institution in Japan.
Eligibility: Candidates must have Australian citizenship, or evidence that Australian citizenship status will be approved prior to departure for Japan; be a student of a higher education institution in Western Australia, or an institution of equivalent standing, and have completed at least two years of full-time study (or equivalent of part-time study) in an appropriate Japanese language course; or be a graduate from a university, having a reasonable command of the Japanese language and developed an interest in Japan through employment or further studies.
Level of Study: Postgraduate
Type: Scholarship
Value: The scholarship includes a return airfare, an initial payment of $3,000 for fees and other expenses and a monthly maintenance allowance of ¥226,600.
Length of Study: 1 year
Frequency: Annual
Country of Study: Japan
Closing Date: July
Funding: Government

DEPARTMENT OF INFRASTRUCTURE (DOI)

GPO Box 2797, Melbourne, Vic, 3001, Australia
Tel: (61) 3 9655 6666
Fax: (61) 3 9095 4096
Email: lori-ann.dalton@transport.vic.gov.au
Website: www.transport.vic.gov.au
Contact: Ms Lori-Ann Dalton, Policy Officer

The Department of Infrastructure (DOI) is the lead provider of essential infrastructure in Victoria, with responsibility for transport, ports and marine, freight, information and communication technology, major development, energy and security.

DOI Women in Freight, Logistics and Marine Management Scholarship

Subjects: Freight, Logistics or Marine-related fields.
Purpose: To prepare women across Victoria for management positions within the freight, marine and logistics industries and help address the gender imbalance in an industry that has traditionally been male domonated.
Eligibility: Open to female candidates who are commencing or completing a PhD, Master's or postgraduate degree in the freight, logistics and marine fields. Candidates must be an Australian citizen and reside in Victoria, Australia to be eligible for the scholarship.
Level of Study: Doctorate, Postgraduate
Type: Scholarship
Value: Australian $10,000
Length of Study: 1 year (full-time), 2 years (part-time)
Frequency: Annual
Study Establishment: University of Victoria
Country of Study: Australia
No. of awards offered: 1
Application Procedure: Applicants must send a completed application form along with curriculum vitae, academic record and a statement about suitability for the scholarship. Refer the website for full details of the application process.
Closing Date: March 25th
Funding: Government
No. of awards given last year: 1

For further information contact:

Department of Infrastructure, GPO Box 2797, Melbourne, Vic 3001

DOI Women in Science, Engineering, Technology and Construction Scholarship

Subjects: Science, engineering, technology and construction.
Purpose: To provide encouragement and support to women with excellent technical and business skills and support women who wish to enter a non-traditional field of study.
Eligibility: Open to women who are undertaking postgraduate studies in science, engineering, technology and construction.
Level of Study: Postgraduate
Type: Scholarship
Value: $10,000
Length of Study: 1 year
Frequency: Annual
Country of Study: Australia
No. of awards offered: 3
Closing Date: December

DEPARTMENT OF INNOVATION, INDUSTRY AND REGIONAL DEVELOPMENT

GPO Box 4509, Melbourne, Vic, 3001, Australia
Tel: (61) 3 9651 9999
Fax: (61) 3 9651 9770
Email: innovation@diird.vic.gov.au
Website: www.diird.vic.gov.au

The Office of Science and Technology at the Department of Innovation, Industry and Regional Development, supports the ongoing development and advancement of a scientifically and technologically advanced Victoria.

Victoria Fellowships

Subjects: Engineering, science, innovation or technology.
Purpose: To offer support and encouragement to aspiring students to broaden their experience and develop networks. The fellowship also provides an opportunity for recipients to develop commercial ideas.
Eligibility: Open to candidates who are either currently employed or enrolled in post-graduate studies in Victoria in a field relating to

science, engineering or technology and Australian citizens or hold permanent residence in Australia and a current resident of Victoria.
Level of Study: Postgraduate, Professional development
Type: Fellowships
Value: Australian $18,000
Length of Study: 1 year
Frequency: Annual
Country of Study: Australia
No. of awards offered: Up to 6
Application Procedure: A completed application form should be submitted.
Closing Date: April (Check website for closing date)
Funding: Government
Contributor: Government of Victoria
No. of awards given last year: 6

For further information contact:

Australian Academy of Technological Sciences and Engineering (ATSE), Melbourne, Vic, Australia
Tel: (61) 3 9864 0905/9655 1040
Email: vicprize.fellows@atse.org.au
Contact: Helen Vella, Manager

DEPARTMENT OF SCIENCE AND TECHNOLOGY (DST)

Technology Bhavan, New Mehrauli Road, New Delhi, 110-016, India
Tel: (91) 11 2656 7373/26962819
Fax: (91) 11 2686 4570/26862418
Email: neerajs@nic.in
Website: www.dst.gov.in
Contact: Mr Neeraj Sharma, Scientist

Department of Science and Technology (DST) Government of India was established in May 1971 with the objective to promote new areas of science and technology and to play the role of a nodal department for organizing, co-ordinating and promoting science and technology activities in the country.

DST Swarnajayanti Fellowships

Subjects: Life sciences, physical sciences, chemical sciences, earth and atmospheric sciences, mathematical sciences and engineering sciences.
Purpose: To provide special assistance and support to a select number of young scientists, with a proven track record and enable them to pursue basic research in frontier areas of science and technology.
Eligibility: Open to Indian nationals who have obtained PhD in science, engineering, Masters in engineering or technology or MD in medicine. The fellowship is open to scientists between 30 and 40 years of age.
Level of Study: Research, Doctorate
Type: Fellowship
Value: A fellowship of Rs. 25,000/- per month for five years.The fellowship will be provided in addition to the salary they draw from their parent Institution. In addition to fellowship, grants for equipments, computational facilities, consumables, contingencies, national and international travel and other special requirements, if any, will be covered based on merit.
Length of Study: 5 years
Frequency: Annual
Study Establishment: Any science and technology institute.
Country of Study: India
Application Procedure: The complete application form is available online. Contact: 011 26590370; e-mail: milind@nic.in or check the website: http://serc-dst.org/ for further details.
Closing Date: April 15th
Funding: Government
Additional Information: The fellowships are not institution specific, are very selective and have close academic monitoring. In addition to fellowship, grants for equipment, computational and communication facilities, consumables, contingencies, administrative support, national and international travel and other special requirements will be covered.

DESCENDANTS OF THE SIGNERS OF THE DECLARATION OF INDEPENDENCE (DSDI)

7157 SE Reed College Place, Portland, OR, 97202-8354, United States of America
Email: registrar@dsdi1776.com
Website: www.dsdi1776.com
Contact: Mr J Alexander, Registrar General

Descendants of the Signers of the Declaration of Independence Scholarships

Subjects: All subjects.
Purpose: To financially assist descendants of the signers of the Declaration of Independence (those who prove eligibility and become members of this Society) to pursue their goals in higher education.
Eligibility: Open to proven direct lineal descendants of a signer of the Declaration of Independence as measured by active membership in DSDI and a society member number. Proof of lineage must be established before an application is sent. Applicants must give the name of their ancestor signer in their first communication or they will not receive a reply. Applicants must be attending an accredited post secondary course full-time.
Level of Study: Unrestricted
Type: Scholarship
Value: US$3,000–9,000, paid directly to the institution. Funds may be applied towards any costs chargeable to the students college account, ie. room, board, books, fees, tuition
Frequency: Annual
No. of awards offered: 50 (minimum)
Application Procedure: Applicants must visit the website www.dsdi1776.com for more details.
Closing Date: February 15th
Funding: Private
No. of awards given last year: 56
No. of applicants last year: 102
Additional Information: Only those with proven descent (a Society member number) will receive an application. Competition among eligible applicants is based on merit.

DEUTSCHE FORSCHUNGSGEMEINSCHAFT (DFG)

Kennedyallee 40, 53175 Bonn, Germany
Tel: (49) 0228 885 1
Fax: (49) 0228 885 2777
Website: www.dfg.de

The DFG is a central, self-governing research organization, which promotes research at universities and other publicity financial research institutions in Germany. The DFG serves all branches of science and the humanities by funding research projects and facilitating cooperation among researchers.

Albert Maucher Prize

Subjects: Geosciences.
Purpose: To promote outstanding young scientists and scholars in the field of geosciences.
Eligibility: Open to promising young scientists and scholars up to the age of 35 years who are German nationals or permanent residents of Germany.
Level of Study: Postdoctorate
Type: Award
Value: €10,000 each
Length of Study: Varies
Frequency: Every 2 years
Study Establishment: Approved universities or research institutions
Country of Study: Germany
No. of awards offered: 1–2
Application Procedure: Applicants must write for details or visit the website. Application is by nomination.
Contributor: Professor Albert Maucher

For further information contact:

Tel: 228 885 2012

Email: Kristian.Remes@dfg.de
Contact: Dr Kristian Remes

Bernd Rendel Prize in Geoscience

Subjects: Geoscience – geologists, mineralogists, geophysicists, oceanographers, geodesists.
Purpose: For the young geoscientists who have graduated, but do not yet hold a doctorate, and who have demonstrated great potential in their scientific career. The award must be used for scientific purposes, e.g. enabling prizewinners to attend international conferences and congresses.
Eligibility: Open to researchers from natural-science oriented fields in geoscience, researchers from humanities-oriented branches of geography are not eligible.
Level of Study: Predoctorate
Type: Prize
Value: €1,000
Frequency: Annual
No. of awards offered: 4
Application Procedure: Nominations may be submitted either by the researchers themselves, or by any researcher or academic working in a closely related field. Detailed information on the nominee's research to date (e.g. thesis, manuscripts, special publications) and future research plans, tabular curriculum vitae, list of publications, copies of certificates, statement on the proposed use of the prize money should be provided.
Contributor: The Bernd Rendel Foundation, which is administered by the Donors' Association for the Promotion of Sciences and Humanities in Germany

For further information contact:

Tel: 0228 885 2328
Email: birgit.scheibner-muenker@dfg.de
Contact: Dr Birgit Scheibner-Münker, Programme Officer

Communicator Award

Purpose: This personal award is presented to researchers who have communicated their scientific findings to the public with exceptional success.
Value: €50,000
Frequency: Annual
No. of awards offered: 1 (person or group)
Application Procedure: Proposals for candidates can be put forward by researchers who are capable of assessing both the communication effort and professional qualification of the nominee(s).
For further information contact:

Email: Jutta.hoehn@dfg.de
Contact: Jutta Höhn

Copernicus Award

Subjects: All subjects.
Purpose: To promote young researchers to further advance research and contribute to the German–Polish research cooperation.
Eligibility: Open to outstanding researchers in Germany and Poland who work at universities or research institutions.
Type: Award
Value: €100,000 (donated in equal shares)
Length of Study: 5 years
Frequency: Every 2 years
No. of awards offered: 2 (1 in Germany and 1 in Poland)
Application Procedure: A completed application form along with the required documents should be submitted.
Contributor: The Foundation for Polish Science and the DFG

For further information contact:

Tel: (228) 885 2663
Email: Philip.Thelen@ dfg.de
Contact: Dr Philip Thelen, Programme Officer International Affairs

DFG Collaborative Research Centres

Subjects: All subjects.
Purpose: To promote long-term co-operative research in universities and academic research.
Eligibility: Open to promising groups of German nationals and permanent residents of Germany.
Level of Study: Postdoctorate, Research
Type: Research grant
Value: Dependent on the requirements of the project
Length of Study: Up to 12 years
Study Establishment: Universities and academic institutions
Country of Study: Germany
No. of awards offered: Varies
Application Procedure: Applicants must write or visit the website for further information. Applications must be formally filed by the universities.
Closing Date: No submission deadline
Additional Information: A list of collaborative research centres is available in Germany only from the DFG.

For further information contact:

Tel: (228) 885 2356
Email: alexandra.kluetsch@dfg.de
Contact: Alexandra Klütsch
Tel: 228 885 2312
Email: petra.hammel@dfg.de
Contact: Petra Hammel

DFG Individual Research Grants

Subjects: All subjects.
Purpose: To foster the proposed research projects of promising academic scientists or scholars.
Eligibility: Open to promising researchers and scholars who are German nationals or permanent residents of Germany.
Level of Study: Doctorate
Type: Grant
Value: Dependent on the requirements of the project
Length of Study: Based on individual project needs
Study Establishment: Universities
Country of Study: Any country
No. of awards offered: Varies
Application Procedure: Applicants must submit a proposal for a research project. Applicants must write for more details or visit the website.
Closing Date: Applications are accepted at any time

DFG Mercator Programme

Subjects: All subjects.
Purpose: The DFG offers the Mercator Programme to enable Germany's research universities to invite highly qualified scientists and academics working abroad to complete a DFG-funded stay at their institutes.
Eligibility: Open to foreign scientists whose individual research is of special interest to research and teaching in Germany.
Level of Study: Postdoctorate
Type: Fellowship
Value: Dependent on the duration of the stay
Length of Study: 3–12 months
Frequency: Annual
Study Establishment: German universities
Country of Study: Germany
No. of awards offered: Varies
Application Procedure: A proposal must be submitted by the university intending to host the guest professor.

For further information contact:

Tel: 49 (228) 885 2232
Email: cora.laforet@dfg.de
Contact: Cora Laforet

DFG Priority Programme

Subjects: All subjects.
Purpose: To promote proposals made by interested groups of scientists in selected fields.
Eligibility: Open to interested groups of scientists from Germany or any country participating in the scheme.
Level of Study: Postdoctorate, Research
Type: Grant

Value: The Senate decides on the financial ceiling for each programme
Length of Study: Up to 6 years
Frequency: Annual
Study Establishment: Universities or academic establishments
Country of Study: Germany
No. of awards offered: 30
Application Procedure: Applicants must write or visit the website for further information. Priority programmes are operated through calls for proposals, with all applications subject to open panel review, usually after discussion with the applicants.
Closing Date: November 15th

DFG Research Training Groups

Subjects: All subjects.
Purpose: To promote high-quality graduate studies at the doctoral level through the participation of graduate students recruited through countrywide calls in research programmes.
Eligibility: Open to highly qualified graduate and doctoral students of any nationality.
Level of Study: Postgraduate, Predoctorate
Type: Grant
Length of Study: Upto 9 years
Frequency: Annual
Study Establishment: Any approved university
Country of Study: Germany
No. of awards offered: Varies
Application Procedure: Applications should be submitted in response to calls. For further information applicants must visit the website.
Closing Date: April 1st and October 1st, preliminary version to be submitted 3 months prior to these dates
Additional Information: A list of graduate colleges presently funded is available (in Germany only) from the DFG.

For further information contact:

Tel: 228 885 288
Email: sebastian.granderath@dfg.de
Contact: Dr Sebastian Granderath

DFG Research Units

Subjects: All subjects.
Purpose: To promote intensive co-operation between highly qualified researchers in one or several institutions in fields of high scientific promise.
Eligibility: Open to interested groups of German nationals and permanent residents of Germany.
Level of Study: Research, Postdoctorate
Type: Research grant
Value: Dependent on the requirements of the project
Length of Study: Up to 6 years
Frequency: Annual
Study Establishment: An approved university
Country of Study: Germany
No. of awards offered: Varies
Application Procedure: Applicants must submit proposals to the Senate of the DFG. They may write or visit the website for further information.
Closing Date: No submission deadline
Additional Information: A list of currently operating research groups is available in Germany only from the DFG.

Emmy Noether Programme

Subjects: All subjects.
Purpose: To give outstanding young scholars the opportunity to obtain the scientific qualifications needed to be appointed as a lecturer.
Eligibility: Open to promising young postdoctoral scientists within 5 years of receiving their PhD, who are up to 30 years of age and who are German nationals or permanent residents of Germany.
Level of Study: Postdoctorate
Type: Project grant
Value: For the 2 years of research spent abroad the candidate will receive a project grant in keeping with the requirements of the project including an allowance for subsistence and travel. For the 3 years of research spent at a German university or research institution the candidate will receive a project grant
Length of Study: 5 years
Frequency: Annual
Study Establishment: Universities or research institutions
Country of Study: Any country
No. of awards offered: 100
Application Procedure: Applicants must complete an application form. For further information applicants must write or visit the website.
Closing Date: Applications may be submitted at any time.

For further information contact:

Germany
Tel: (49) (0) 228 885 3008
Email: Verfahren-Nachwuchs@dfg.de

The Eugen and Ilse Seibold Prize

Subjects: Humanities, social science, law, economics, natural sciences, engineering and medicine.
Purpose: To promote outstanding young scientists and scholars who have made significant contributions to the scientific interchange between Japan and Germany.
Eligibility: Open to outstanding young German or Japanese scholars.
Level of Study: Postdoctorate
Type: Prize
Value: €10,000
Length of Study: Varies
Frequency: Every 2 years
Study Establishment: Universities or research institutions
No. of awards offered: 2 (1 in Germany and 1 in Japan)
Application Procedure: Applicants must write for details or visit the website. Application is by nomination.
Closing Date: August 31st

For further information contact:

Tel: (228) 885 2346
Fax: (228) 885 2550
Email: Joerg.Schneider@dfg.de
Contact: Dr Jörg Schneider, Head of Division International Affairs

European Young Investigator Award

Purpose: To enable and encourage outstanding young researchers from all over the world, to work in an European environment for the benefit of the development of European science and the building up of the next generation of leading European researchers.
Eligibility: The program is open to scientists of all disciplines and is open to candidates throughout the world.
Type: Prize
Value: Up to €1.25 million
Length of Study: 5 years
Application Procedure: Applicants should provide a completed application form, letters of recommendation and the letter of support from the host institution to the DFG.
Closing Date: November 30th
Contributor: The European Union Research Organisations Heads of Research Councils (EuroHORCS)
Additional Information: For further information log on to: www.dfg.de/en/news/scientific_prizes/euryi_award/index.html

For further information contact:

Tel: (228) 885 2845
Email: Anjana.Buckow@dfg.de
Contact: Dr Anjana Buckow

Excellence Initiative

Subjects: All subjects.
Purpose: To promote top-level research and improve the quality of German universities and research institutions in general, thus making Germany a more attractive research location, and more internationally competitive and focussing attention on the outstanding achievements of German universities and the German scientific community.
Eligibility: The precise conditions for receiving funding were defined in accordance with the criteria specified by the federal and state governments.
Type: Funding support

Length of Study: 5 years
Contributor: German federal and state governments
Additional Information: The three funding lines of the initiative: graduate schools to promote young scientists, clusters of excellence to promote top-level research, institutional strategies to promote top-level university research. For more details log on to www.dfg.de/en/research_funding/coordinated_programmes/excellence_initiative/

For further information contact:

Tel: 228 885 2254
Email: beate.konze-thomas@dfg.de
Contact: Dr Beate Konze-Thomas

Gottfried Wilhelm Leibniz Prize

Subjects: All subjects.
Purpose: To promote outstanding scientists and scholars in German universities and research institutions.
Eligibility: Open to outstanding scholars in German universities.
Level of Study: Predoctorate, Research
Type: Research grant
Value: €2.5 million per award
Length of Study: 5 years
Frequency: Annual
Study Establishment: Any approved university or research institution
Country of Study: Germany
No. of awards offered: Up to 10
Application Procedure: Applicants must write for details or visit the website. Application is by nomination. Nominations are restricted to selected institutions such as DFG member organizations or individuals e.g. former prize winners or chairpersons of DFG review committees.
Additional Information: A list of prize winners is available in Germany only from the DFG.

For further information contact:

Tel: (228) 885 2726
Email: Ursula.Rogmans-Beucher@dfg.de
Contact: Ursula Rogmans-Beucher

Heinz Maier–Leibnitz Prize

Subjects: All subjects.
Purpose: To promote outstanding young scientists at the doctorate level.
Eligibility: Open to promising young scholars up to 33 years of age, who are German nationals or permanent residents of Germany.
Level of Study: Doctorate, Postdoctorate
Type: Award
Value: €16,000 per award
Length of Study: Varies
Frequency: Annual
Study Establishment: Any approved university or research institution
Country of Study: Germany
No. of awards offered: 6
Application Procedure: Applicants must write for details or visit the website. Application is by nomination.
Closing Date: August 31st
Funding: Government
Contributor: The Federal Ministry of Education and Research

For further information contact:

Tel: (228) 885 2835
Email: Annette.Lessenich@dfg.de
Contact: Annette Lessenich, Legal Director Quality Assurance and Programme Development

Heisenberg Programme

Subjects: All subjects.
Purpose: To promote outstanding young and highly qualified reseachers.
Eligibility: Open to high-calibre young scientists up to the age of 35 years who are German nationals or permanent residents of Germany.
Level of Study: Postdoctorate
Type: Scholarship
Value: Varies
Length of Study: 5 years
Frequency: Annual
Study Establishment: Any approved university or research institution
Country of Study: Germany
No. of awards offered: Varies
Application Procedure: Applicants must submit a research proposal, a detailed curriculum vitae, copies of degree certificates, a copy of the thesis, a letter explaining the choice of host institution, a list of all previously published material and a letter outlining financial requirements in duplicate. For further information applicants must contact the DFG.
Closing Date: Applications are accepted at any time

For further information contact:

Tel: (0) 228 885 2398
Email: paul.heuermann@dfg.de
Contact: Paul Heuermann

Reinhart Koselleck Projects

Subjects: Research projects that are highly innovative and risky in a positive sense, which are not supported by other programmes of DFG.
Purpose: To enable outstanding researchers with a proven scientific track record to pursue exceptionally innovative, higher-risk projects.
Eligibility: Researchers who hold or are eligible to hold professorships, especially at universities, and who have an outstanding curriculum vitae and great scientific potential.
Type: Funding support
Value: €500,000–12,50,000
Length of Study: 5 years
Closing Date: June 1st
Contributor: Deutsche Forschungsgemeinschaft
Additional Information: For more details log on to www.dfg.de/en/research_funding/individual_grants_programme/reinhart_koselleck_-projects/index.html

Ursula M. Händel Animal Welfare Prize

Subjects: Animal welfare.
Purpose: To award scientists who make, through research, a significant contribution to the welfare of animals.
Eligibility: Open to scientists who aim at improving the welfare of animals through research.
Level of Study: Postdoctorate
Type: Award
Value: €25,000
No. of awards offered: Varies
Application Procedure: A completed application form and required documents must be submitted.
Funding: Trusts
Contributor: Mrs. Ursula M. Händel

For further information contact:

Tel: +49 (0) 228 885 2658
Email: Sonja.Ihle@dfg.de
Contact: Dr Sonja Ihle

The Von Kaven Awards

Subjects: Instrumental mathematics.
Purpose: The award is granted as a fellowship or as a support for research in the field of instrumental mathematics (including the von Kaven Prize and the von Kaven Research Award).
Eligibility: Persons who meet the general eligibility criteria stipulated by the DFG within the individual grants programme.
Type: Award
Value: €15,000 (€10,000 von Kaven Prize; €5000 von Kaven Research Award)
Frequency: Annual
Application Procedure: Nominations for the von Kaven (Prize and) award may be made by the members of the mathematics review board, its previous chairs and other DFG committee members in the field of mathematics (such as senators and members of the senate committee working in the field of mathematics). It is not possible to apply directly for the von Kaven (Prize and) award.
Closing Date: January 31st

For further information contact:

Tel: 49 (228) 885 2567
Email: frank.kiefer@dfg.de
Contact: Dr Frank Kiefer

DIABETES RESEARCH & WELLNESS FOUNDATION

101–102, Northney Marina, Hayling Island, Hampshire, PO11 0NH, England
Tel: (44) 23 9263 7808
Fax: (44) 23 9263 6137
Email: research@drwf.org.uk
Website: www.drwf.org.uk
Contact: Sarah Brown, Grants Administrator

The Diabetes Research & Wellness Foundation was established in 1998 to fund research into finding a cure for diabetes. Each year this goal becomes more important as the number of people diagnosed continues to rise. The organization hopes to make diabetes a thing of the past, and, until then, alleviate its awful complications.

DRWF Open Funding

Subjects: Endocrinology or diabetes.
Purpose: To encourage research into the complications and cure of diabetes.
Level of Study: Doctorate, Postgraduate, Research
Type: Grant
Value: Up to UK£20,000
Length of Study: 1 year
Frequency: Annual
Study Establishment: A recognized institution or research group in the United Kingdom
Country of Study: United Kingdom
No. of awards offered: Up to 6, depending on grant amount
Application Procedure: Applications should be no more than 4 sides of A4 paper, typed using single-line spacing and an 11 or 12 point clearly readable font. They should include (as appropriate): applicant's name, qualifications, present post and contact details; name and address of the institution(s) where the work will be carried out; head of department/institution and major participants in the project; signed verification of funding application by HOD, outline of the proposed research comprising title, research question, relevance to diabetes, expected outcome; lay summary of the research question; any additional information to support the application; amount of funding requested, with a general breakdown of costs; and a (brief) curriculum vitae of the main applicant on separate single sheet of A4.
Closing Date: August 27th
Funding: Commercial, individuals, private
No. of awards given last year: 4
No. of applicants last year: 33
Additional Information: Notification of awards in November. For further details contact 023 92 636135 and e-mail: sarah.brown@drwf.org.uk.

DRWF Research Fellowship

Subjects: Endocrinology or diabetes.
Purpose: To encourage research into the complications or cure of diabetes.
Eligibility: Open to suitable candidates who are working at an institution within the United Kingdom in an established position.
Level of Study: Doctorate, Postdoctorate, Research
Type: Fellowship
Value: Up to UK£55,000 each for both non-clinical and clinical fellowship
Length of Study: Up to 3 years
Frequency: Annual
Study Establishment: A recognized institution or research group in the United Kingdom
Country of Study: United Kingdom
No. of awards offered: 1
Application Procedure: Applicants must undergo a three-stage selection procedure starting with a pre-application. This is a single side of A4 paper, with single-line spacing and a clearly readable font in 11 or 12 points. The pre-application must include the applicant's name, qualifications, contact details and present post. The pre-application must also include the name of the head of the group, see open funding addition where the grant will be held, the post held or expected post to be held within the group and the relevant contact details of the group. There should also be a 300-word abstract of the proposed research work including the title, a research question of approximately 300 words, relevance to diabetes, expected outcome and any additional information to support the application, but no references. Lastly, a brief curriculum vitae of the applicant on a separate single sheet of A4 paper must be provided. Successful applicants at the pre-selection stage are required to submit a full application by August.
Closing Date: November 13th (for Non-Clinical Fellowship); April 27th (for Clinical Fellowship). See for web site.
Funding: Commercial, individuals, private
No. of awards given last year: 1
No. of applicants last year: 10
Additional Information: Fellowships are alternated between clinical and non-clinical, year by year. Final interviews of selected candidates are held in October. The recipient of the fellowship is expected to take it up early in the following year. For further details e-mail: sarah.brown@drwf.org.uk or check website.

DIABETES UK

Macleod House, 10 Parkway, London, NW1 7AA, England
Tel: (44) 020 7424 1000
Fax: (44) 020 7424 1001
Email: victoria.king@diabetes.org.uk
Website: www.diabetes.org.uk
Contact: Dr Victoria King, Research Manager

Diabetes UK's overall aim is to help and care for both people with diabetes and those closest to them, to represent and campaign for their interests and to fund research into diabetes. Diabetes UK continues to encourage research into all areas of diabetes.

Diabetes UK Equipment Grant

Subjects: Endocrinology, diabetes and subjects relevant to diabetes.
Purpose: To enable the purchase of a specific large item of multi user equipment necessary for diabetes related research projects.
Eligibility: Open to suitably qualified members of the medical or scientific professions who are resident in the United Kingdom.
Level of Study: Postdoctorate
Type: Grant
Value: Up to £100,000
Length of Study: 1 year–3 years
Frequency: Twice a year
Country of Study: United Kingdom
No. of awards offered: Varies
Application Procedure: Candidates must complete an application form that will be assessed by a peer review. Please write or telephone for details. Details can be found on the website.
Closing Date: February 1st and December 1st
Funding: Private
Contributor: Voluntary contributions
No. of awards given last year: 4
Additional Information: Applicants should be in receipt of substantial grant funding either from Diabetes UK or other funding bodies.

Diabetes UK Project Grants

Subjects: Endocrinology, diabetes and subjects relevant to diabetes.
Purpose: To provide funding for a well-defined research proposal of timeliness and promise that, in terms of the application, may be expected to lead to a significant advance in our knowledge of diabetes.
Eligibility: Open to suitably qualified members of the medical or scientific professions who are resident in the United Kingdom. Applicants must have a tenured post or be able to demonstrate that they will have a salary and position at the Institution they are applying from for the lifetime of the grant.
Level of Study: Postdoctorate
Type: Project grant
Value: There is no limit to research expenses that may be requested, however, all requests must be fully justified.
Length of Study: 1–3 years
Frequency: Twice a year
Country of Study: United Kingdom
No. of awards offered: Varies
Application Procedure: Candidates must complete an application form, which will be assessed by a peer review and should write or

telephone for details. Details can be found and the application forms downloaded from the website.
Closing Date: February 1st and December 1st
Funding: Private
Contributor: Voluntary contributions
No. of awards given last year: 25
No. of applicants last year: 120
Additional Information: Candidates applying for £500,000 or more must contact the office in the first instance and should normally hold substantial funding from Diabetes UK. Candidates applying for £15,000 or less should follow the Guidelines for Small Grant Applications.

Diabetes UK Small Grant Scheme

Subjects: Endocrinology, diabetes and subjects relevant to diabetes.
Purpose: To enable research workers to develop new ideas in the field of diabetes research.
Eligibility: Open to suitably qualified members of the medical or scientific professions who are resident in the United Kingdom.
Level of Study: Postdoctorate
Type: Grant
Value: A maximum of UK£15,000
Length of Study: 1–3 years
Frequency: Rolling
Country of Study: United Kingdom
No. of awards offered: Varies
Application Procedure: Applicants must complete an application form which will be assessed by a peer review within 6–8 weeks. Please write or telephone for details. Details can also be found on the website.
Closing Date: February 1st and December 1st.
Funding: Private
Contributor: Voluntary contributions
No. of awards given last year: 8
No. of applicants last year: 25

DIRKSEN CONGRESSIONAL CENTER

2815 Broadway, Pekin, IL, 61554, United States of America
Tel: (1) 309 347 7113
Fax: (1) 309 347 6412
Email: info@dirksencenter.org
Website: www.dirksencenter.org

The Dirksen Congressional Center sponsors educational and research programmes to help people understand better the United States of America Congress, its members and leaders and the public policies it produces.

Dirksen Congressional Research Award

Subjects: Political science and government.
Purpose: To fund the study of the United States of America Congress.
Eligibility: Open to citizens or residents of the United States of America. Awards are to individuals only. No institutional overhead charges are permitted. Political scientists, historians, biographers, scholars of public administration or American studies, and journalists are among those eligible.
Level of Study: Doctorate, Postdoctorate, Research
Type: Research grant
Value: Up to US$3,500.
Length of Study: Varies
Frequency: Annual
Study Establishment: Unrestricted
Country of Study: United States of America
No. of awards offered: 10–15 per year
Application Procedure: Candidates should visit the Center's website for application information. Candidates are responsible for showing the relationship between their work and the awards program guidelines.
Closing Date: March 1st
Funding: Private
No. of awards given last year: 12
No. of applicants last year: 70 +
Additional Information: You may send the application as a word or pdf attachment to an e-mail directed to Frank Mackaman at fmackaman@dirksencenter.org.

DONATELLA FLICK CONDUCTING COMPETITION

PO Box 34227, London, NW5 1XP, England
Fax: (44) 0207 584 6880
Website: www.conducting.org

The Donatella Flick Associazione organizes the Donatella Flick Conducting Competition, which, in association with the London Symphony Orchestra, aims to help advance career opportunities for young conductors. The award subsidizes study and concert engagements for the winner who will work as Assistant Conductor with the London Symphony Orchestra for 1 year.

Donatella Flick Conducting Competition

Subjects: Conducting.
Purpose: To assist a young conductor in establishing an international conducting career.
Eligibility: Open to conductors who are between the ages 18 and 35 and are citizens of member countries of the European Union.
Level of Study: Professional development
Type: Prize
Value: Award of £15,000 and opportunity to become Assistant Conductor of the LSO for up to one year.
Length of Study: 1 year
Frequency: Every 2 years
Study Establishment: London Symphony Orchestra
Country of Study: Any country
Application Procedure: Applicants must complete an application form and submit this with references specific to the competition, as well as videos and other supporting documentation such as other prizes, reviews, a curriculum vitae, etc.
Closing Date: April 12th
Funding: Private
Contributor: Mrs Donatella Flick
Additional Information: Entry is by recommendation, documentation and supporting video. Finalists are then selected for audition, and three finalists conduct a public concert. The course of study of entrants must be approved by the organizing committee.

DOSHISHA UNIVERSITY

International Center, Office of International Students, Karasuma-Higashi-iru Imadegawa-dori, Kamigyo-ku Kyoto, 602-8580, Japan
Tel: (81) 75 251 3257
Fax: (81) 75 251 3123
Email: ji-intad@mail.doshisha.ac.jp
Website: www.doshisha.ac.jp/english
Contact: Chieko Toboku

Located in the heart of Kyoto, Doshisha University occupies 5 separate campuses and is home to over 29,000 students engaged in both undergraduate and graduate studies. As one of Japan's most highly esteemed educational institutions, Doshisha offers students a wide ranging liberal arts education as well as studies in business and science.

Doshisha University Doctoral-Program Young Researcher Scholarship

Purpose: To support young researchers who hold future promise and display a strong passion toward academic research.
Eligibility: Students with a passion for academic research who have received a recommendation from one of the graduate schools and match one of the following profiles: (1) Students enrolled in doctoral programs at the Graduate Schools of Theology Letters; Social Studies, Law, Economics, Commerce, Policy and Management, Culture and Information Science, Science and Engineering, Life and Medical Sciences, Health and Sports Science, Psychology and Global Studies, who are aiming to acquire a doctoral degree and are under 34 years of age at the time of admission. (2) Students enrolled in an integrated program (master's and doctoral programs) (except the Graduate School of Brain Science) for a minimum of two years, who are aiming to acquire a doctoral degree and are under 32 years of age at the time of enrollment.
Level of Study: Doctorate, Integrated Doctoral
Type: Scholarship

Value: Amount equivalent to annual school fees (including admission fees at the time of enrollment, tuition fee for educational support and lab/practical fees)
Length of Study: 1 year (renewable for up to the standard number of years required for graduation)
Frequency: Annual
Study Establishment: Doshisha University
Country of Study: Japan
No. of awards offered: To all qualified students in each graduate school unless these exceed scholarship availability
Application Procedure: The scholarship is awarded on the basis of recommendations from the graduate schools and cannot be applied for individually.

For further information contact:

Section for Scholarships, Department of Student Support Services
Email: ji-kosei@mail.doshisha.ac.jp

Doshisha University Graduate School Reduced Tuition Special Scholarships for Self-Funded International Students

Purpose: To enable international students to concentrate on their studies free from financial concerns.
Eligibility: Those who satisfy one of the following qualifications are eligible: (1) who have passed the entrance examination for international students and who hold a 'college student' visa prescribed in the 'Emigration and Immigration Management and Refuge Recognition Law', at the time of enrollment. (2) Who have passed the entrance examination for international students and who hold a 'Permanent Resident' visa etc. (3) Who are enrolled in Doshisha regardless of the type of admission (type of entrance examination) and who hold a 'College Student' visa.
Level of Study: Doctorate, MBA, Postgraduate, Integrated Doctoral
Type: Scholarship
Value: Annual tuition fees
Length of Study: 2 years (renewable for up to the standard number of years required for graduation)
Frequency: Annual
Study Establishment: Doshisha University
Country of Study: Japan
No. of awards offered: Approx. 20 per cent of international students
Application Procedure: The Scholarship is awarded based on Doshisha's criteria without application.

For further information contact:

Office of International Students, International Center
Email: ji-intad@mail.doshisha.ac.jp

Doshisha University Graduate School Scholarship

Purpose: To provide for students enrolled in master's or doctoral programs experiencing difficulty meeting educational costs required for them to continue their academic research activities.
Eligibility: Graduate Students (Regular students). Please note Law School, Business School, and Graduate School of Brain Science students may not apply. Students who have been enrolled at school longer than the standard number of years for course completion (a leave of absence is not counted as) may not apply. Students selected to recieve the following scholarships may not apply: Doshisha University Doctoral-Program Young Researcher Scholarship, Japanese Government (MEXT) Scholarship, Doshisha University Graduate School Reduced Tuition Special Scholarship for Self-Funded International Students.
Level of Study: Doctorate, Postgraduate, Integrated Doctoral
Type: Scholarship
Value: Half of the total annual tuition fee
Length of Study: 1 year
Frequency: Annual
Study Establishment: Doshisha University
Country of Study: Japan
No. of awards offered: To be made based on recommendations from each graduate school
Application Procedure: Eligible applicants are required to submit an application to the Section for Scholatship by specified date.
No. of awards given last year: 16

For further information contact:

Section for Scholarships, Department of Student Support Services
Email: ji-kosei@mail.doshisha.ac.jp

Doshisha University Reduced Tuition Scholarships for Self-Funded International Students

Purpose: To enable international students to concentrate on their studies free from financial concerns.
Eligibility: Those who satisfy one of the following qualifications are eligible: (1)who have passed the entrance examination for international students and who hold a 'College Student' visa prescribed in the 'Emigration and Immigration Management and Refuge Recognition Law', at the time of enrollment; (2)who have passed the entrance examination for international students and who hold a 'Permanent Resident' visa etc.; (3)who are enrolled in Doshisha regardless of the type of admission (type of entrance examination), and who hold a 'College Student' visa.
Level of Study: Doctorate, MBA, Postgraduate, Integrated Doctoral
Type: Scholarship
Value: Equivalent to 50 per cent of tuition, equivalent to 30 per cent of tuition, to be made based on Doshisha's criteria
Length of Study: 2 years (renewable for up to the standard number of years required for graduation)
Frequency: Annual
Study Establishment: Doshisha University
Country of Study: Japan
No. of awards offered: Approx. 80 per cent of international students
Application Procedure: The scholarship is awarded based on Doshisha's criteria without application.
Additional Information: Nationals of Japan are not eligible.

For further information contact:

Office of International Students, International Center
Email: ji-intad@mail.doshisha.ac.jp

Graduate School of Brain Science Special Scholarship

Subjects: Brain science.
Eligibility: Open to doctorate students with academic excellence and depending on the candidate's contribution to the field. Those who have passed the entrance examination at the Graduate School of Brain Science and are under the age of 32 (for 3rd-year transfer students under the age of 34) at the time of enrollment.
Level of Study: Integrated Doctoral
Type: Scholarship
Value: Equivalent to the total amount of annual educational costs (including the admission fee, which is charged at the time of enrollment only, tuition, fee for educational support and lab/practical fees)
Length of Study: 1 year (Renewable for up to 5 years (for transfer students, for up to 3 years))
Frequency: Annual, Integrated Doctoral
Study Establishment: Doshisha University
Country of Study: Japan
No. of awards offered: To all qualified students
Application Procedure: As the initial registration procedure, eligible applicants are required to remit the registration fee and submit an application for the 'Graduate School of Brain Science Special Scholarship' by specified date.

For further information contact:

Office for Graduate School of Brain Science
Email: jt-nkgjm@mail.doshisha.ac.jp

DR HADWEN TRUST FOR HUMANE RESEARCH

Suit 8, Portmill House, Portmill Lane, 18 Market place, Hitchin, Herts, SG5 1DS, England
Tel: (44) 1462 436819
Fax: (44) 1462 436844
Email: info@drhadwentrust.org
Website: www.drhadwentrust.org
Contact: Grants Manager

The Dr Hadwen Trust for Humane Research is a registered charity, established in 1970, to promote research into techniques and procedures to replace the use of animals in biomedical research, teaching and testing.

Dr Hadwen Trust Research Assistant or Technician

Subjects: The development, validation or implementation of a technique or procedure that would replace one currently using animals in medical research.
Purpose: To provide additional scientific or technical support for a research project to replace animal experiments
Eligibility: Open to applications from UK-based researchers who are based at any UK research establishments, including higher education institutions, hospital/NHS trusts, research council establishments, charity laboratories and industry.
Level of Study: Doctorate, Postdoctorate
Type: Research
Value: Salary for research assistant or technician plus an allowance for consumables
Length of Study: 3 years
Frequency: Annual
Study Establishment: Varies
Country of Study: United Kingdom
No. of awards offered: Varies
Application Procedure: Applicants must make initial enquiries by contacting the Dr Hadwen Trust. Applications must be made by the senior researcher who will oversee the work. Preliminary applications are usually invited in Autumn. Details are posted on the website.
Closing Date: See website for details
Funding: Private
No. of awards given last year: 5
No. of applicants last year: 125
Additional Information: For further details and policy information see website www.drhadwentrust.org.

Dr Hadwen Trust Research Fellowship

Subjects: The development, validation or implementation of a technique or procedure that would replace one currently using animals.
Purpose: To attract and retain talented young scientists in non-animal research fields. The funds provide personal support and a contribution to direct research costs for research to replace animal experiments.
Eligibility: Open to applications from UK-based researchers who are based at any UK research establishment, including higher education institutions, hospital/NHS trusts, research council establishments, charity laboratories and industry.
Level of Study: Postdoctorate
Type: Fellowship
Value: Usually a maximum of £135,000 total for 3 years, to cover salary, consumables or small items of equipment.
Length of Study: 3 years maximum
Study Establishment: Varies
Country of Study: United Kingdom
No. of awards offered: Varies
Application Procedure: Applicants must make initial enquiries by contacting the Dr Hadwen Trust. Application forms must be submitted by a senior researcher who will oversee the work. Preliminary applications are usually invited in Autumn. Details and guidelines are posted on the website.
Closing Date: See for website.
Funding: Private
Contributor: Public donations
No. of awards given last year: 5
No. of applicants last year: 40
Additional Information: Further policy information and application details are available on the website www.drhadwentrust.org.

DUBLIN INSTITUTE FOR ADVANCED STUDIES

10 Burlington Road, Dublin, 4, Ireland
Tel: (353) 1 614 0100
Fax: (353) 1 668 0561
Email: registrar@admin.dias.ie
Website: www.dias.ie
Contact: Cecil Keaveney, Registrar

The Dublin Institute for Advanced Studies is a statutory corporation established in 1940, under the Institute for Advanced Studies Act of that year. It is a publicly funded independent centre for research in basic disciplines. Research is currently carried out in the fields of Celtic studies, theoretical physics and cosmic physics including astronomy, astrophysics and geophysics.

Dublin Institute for Advanced Studies Scholarship in Astronomy, Astrophysics and Geophysics

Subjects: Astronomy, astrophysics and geophysics.
Purpose: To enable training in advanced research methods in the fields of astronomy, astrophysics and geophysics.
Eligibility: Open to candidates from any country.
Level of Study: Graduate, Postgraduate, Predoctorate, Doctorate, Postdoctorate
Type: Scholarship
Value: Please contact the Institute for details.
Length of Study: 1 year
Frequency: Annual
Study Establishment: The Dublin Institute for Advanced Studies
Country of Study: Ireland
No. of awards offered: 3
Application Procedure: Applicants must complete an application form, available upon request.
Closing Date: Please write to the Institute for details.
Funding: Government
Contributor: State
No. of awards given last year: 2
No. of applicants last year: 12

Dublin Institute for Advanced Studies Scholarship in Celtic Studies

Subjects: Celtic studies.
Purpose: To enable training in advanced research methods in the field of Celtic studies.
Eligibility: Open to nationals of any country.
Level of Study: Predoctorate, Doctorate, Postdoctorate, Postgraduate
Type: Scholarship
Value: Please contact the Institute for details.
Length of Study: 1 year
Frequency: Annual
Study Establishment: The Dublin Institute for Advanced Studies
Country of Study: Ireland
No. of awards offered: 3
Application Procedure: Applicants must complete an application form, available upon request.
Closing Date: May 15th. See the website.
Funding: Government
No. of awards given last year: 3
No. of applicants last year: 10

Dublin Institute for Advanced Studies Scholarship in Theoretical Physics

Subjects: Physics.
Purpose: To enable training in advanced research methods in the field of theoretical physics.
Eligibility: Open to candidates of any country.
Level of Study: Postdoctorate
Type: Scholarship
Value: €16,000 (for predoctoral) and €20,000 (for postdoctoral).
Length of Study: 1 year
Frequency: Annual
Study Establishment: The Dublin Institute for Advanced Studies
Country of Study: Ireland
No. of awards offered: 5
Application Procedure: Applicants must complete an application form, available upon request.
Closing Date: Please write to the Institute for details.
Funding: Government
Contributor: State
No. of awards given last year: 3
No. of applicants last year: 30

DUKE UNIVERSITY

2127 Campus Drive, PO Box 90065, Durham, NC, 27708, United States of America
Tel: (1) 919 681 3257
Fax: (1) 919 668 0434
Email: grm@duke.edu
Website: www.gradschool.duke.edu
Contact: Co-ordinator

The Duke University ideally has a small number of superior students working closely with esteemed scholars. It has approximately 2,200 graduate students enrolled there, working with more than 1,000 graduate faculty members.

James B. Duke Fellowships

Subjects: All subjects.
Purpose: To pursue a programme leading to the PhD in the Graduate School at Duke University.
Level of Study: Doctorate
Type: Fellowship
Value: US$5,000 stipend.
Length of Study: 4 years
Frequency: Annual
Country of Study: United States of America
Closing Date: February 28th. See the website.
Additional Information: Its objective is to aid in attracting and developing outstanding scholars at Duke.

For further information contact:

Contact: Director of Graduate Studies

DUMBARTON OAKS: TRUSTEES FOR HARVARD UNIVERSITY

1703, 32nd Street North West, Washington, DC, 20007, United States of America
Tel: (1) 202 339 6400
Fax: (1) 202 339 6419
Email: DumbartonOaks@doaks.org
Website: www.doaks.org
Contact: Fellowship Programme Manager

Dumbarton Oaks houses important research and study collections in the areas of Byzantine and Medieval studies, landscape architecture studies and pre-Columbian studies. While the gallery holds exhibitions and the gardens are open to the public, the research facilities exist primarily to serve scholars who hold appointments at Dumbarton Oaks.

Dumbarton Oaks Fellowships and Junior Fellowships

Subjects: Byzantine civilization in all its aspects, including the late Roman and Early Christian period and the Middle Ages generally, studies of Byzantine cultural exchanges with the Latin West, Slavic and Near Eastern countries, Pre-Columbian studies of Mexico, Central America and Andean South America and Garden and Landscape studies, including garden industry, landscape architecture, and related disciplines.
Purpose: To promote study and research or to support writing of doctoral dissertations in the fields of Byzantine studies, Garden and Landscape studies, and Pre-Columbian studies.
Eligibility: Junior fellowships are open to persons of any nationality who have passed all preliminary examinations for a higher degree and are writing a dissertation. Candidates must have a working knowledge of any languages required for the research. Fellowships are open to scholars of any nationality holding a PhD or relevant advanced degree and wishing to pursue research on a project of their own at Dumbarton Oaks.
Level of Study: Postdoctorate, Doctorate
Type: Fellowships
Value: US$27,000 per year for unmarried junior fellowships, US $47,000 per year for a fellow from abroad accompanied by family members. Both junior and regular Fellows receive furnished accommodation or a housing allowance and US$2,100, if needed, to assist with the cost of bringing and maintaining dependants in Washington plus an expense account of US$1,000 for approved research expenditure during the academic year. Fellows are also provided with travel assistance. Travel expense reimbursement for the lowest available airfare, up to a maximum of $1,300, may be provided for Fellows and Junior Fellows if support cannot be obtained from other sources (such as a Fulbright travel grant).
Length of Study: Up to 1 academic year of full-time study, non-renewable
Frequency: Annual, also summer
Study Establishment: Dumbarton Oaks
Country of Study: United States of America
No. of awards offered: 10–11 fellowships in Byzantine studies and 4–6 in each of the other fields
Application Procedure: Applicants must apply online at www.doaks.org
Closing Date: November 1st of the academic year preceding that for which the fellowship is required
Funding: Private
No. of awards given last year: 35
No. of applicants last year: 200
Additional Information: Dumbarton Oaks also awards a limited number of Summer fellowships, projects grants, predoctoral residencies, and postdoctoral stipends. Please see www.doaks.org for further details.

DUTCH MINISTRY OF FOREIGN AFFAIRS

Bezuidenhoutseweg 67, The Hague, PO Box 20061, 2500 EB The Hague, Netherlands
Tel: (31) 31 70 3486486
Fax: (31) 31 70 3484848
Email: dsi-my@minbuza.nl
Website: www.minbuza.nl

Dutch foreign policy is driven by the conviction that international cooperation brings peace and promotes security, prosperity, and justice. It is bound by the obligation to promote Dutch interests abroad as effectively and efficiently as possible. To do so, the Netherlands needs a worldwide network of embassies, consulates, and permanent representations to international organisations. The activities, composition, and size of each mission depend on its host country and region. Embassies and consulates-bilateral missions-concern themselves with relations between the Netherlands and other countries.

Netherlands Fellowship Programme

Subjects: All subjects.
Purpose: To support mid-career professionals nominated by their employers.
Eligibility: Open to candidates who are employed by an organization other than a large industrial, commercial and/or multinational firm, must be nationals of one of the 57 selected countries (see 'Eligible countries'), must declare that they will return to their home country immediately after they complete the master programme, must have gained admission to a TU/e master course, which is on the NFP course list and have sufficient mastery of the English language. Priority is given to female candidates and to candidates coming from sub-Saharan Africa.
Level of Study: Postgraduate
Type: Fellowships
Value: Full-cost scholarship (including international travel, monthly subsistence allowance, tuition fee, books, and health insurance).
Length of Study: 2 years
Frequency: Annual
Country of Study: Netherlands
Application Procedure: Applicants must apply for an NFP fellowship through the Netherlands embassy or consulate in their own country by completing an NFP Application Form and submitting it together with all the required documents and information to the embassy or consulate. Then the Embassy checks and sends the forms to the Nufficchecks. Nuffic decides how many fellowships will be available for each program and sends TU/e the list of NFP candidates.
Closing Date: There are several application deadlines, check website for details.
Funding: Government

Contributor: Dutch Ministry of Foreign Affairs
Additional Information: Eligible countries: Afghanistan, Albania, Armenia, Autonomous Palestinian Territories, Bangladesh, Benin, Bhutan, Bolivia, Bosnia–Hercegovina, Brazil, Burkina Faso, Cambodia, Cape Verde, China*, Colombia, Costa Rica, Cuba, Ecuador, Egypt, El Salvador, Eritrea, Ethiopia, Georgia, Ghana, Guatemala, Guinea–Bissau, Honduras, India, Indonesia, Iran, Ivory Coast, Jordan, Kenya, Macedonia, Mali, Moldova, Mongolia, Mozambique, Namibia, Nepal, Nicaragua, Nigeria, Pakistan, Peru, Philippines, Rwanda, Senegal, South Africa, Sri Lanka, Suriname, Tanzania, Thailand, Uganda, Vietnam, Yemen, Zambia, Zimbabwe. For a more detailed list of criteria, please check website.

For further information contact:

International Relations Office, Education and Student Service Center
Tel: 31 (0) 40 247 4690
Fax: 31 (0) 40 244 1692
Email: io@remove-this.tue.nl
Website: www.nuffic.nl/nfp/

EARLY AMERICAN INDUSTRIES ASSOCIATION

P.O. Box 524, Hebron, MD, 21830-0524, United States of America
Email: execdirector@eaiainfo.org
Website: www.eaiainfo.org
Contact: Ms John H. Verrill, Executive Director

The Early American Industries Association seeks to encourage the study and better understanding of early American industries in the home, in the shop, on the farm and on the sea. It also wishes to discover, identify, classify and exhibit obsolete tools, implements and mechanical devices that were used in early America.

Early American Industries Association Research Grants Program

Subjects: Early American industrial development, including craft practices, industrial technology and identification and use of obsolete tools, implements and mechanical devices used prior to 1900.
Purpose: To encourage research leading to a publication, exhibition or audio-visual material for educational purposes.
Eligibility: Open to citizens or permanent residents of the United States of America. Individuals may be either sponsored by an institution or engaged in self-directed projects.
Level of Study: Doctorate, Graduate, Postdoctorate, Postgraduate, Predoctorate, Research
Type: Grant
Value: Up to US$2,000
Length of Study: 1 year, non-renewable
Frequency: Annual
Country of Study: United States of America
No. of awards offered: 3–5
Application Procedure: Applicants must submit a completed application form plus three letters of recommendation.
Closing Date: March 15th
Funding: Private
Contributor: Membership dues and donations
No. of awards given last year: 3
No. of applicants last year: 10
Additional Information: Awards may be used to supplement existing financial awards. Successful applicants are required to file a project report on forms supplied by the Association. These are not scholarship funds.

EARTHWATCH INSTITUTE

114 Western Avenue, Boston, MA, 02134, United States of America
Tel: (1) 978 461 0081
Fax: (1) 978 461 2332
Email: research@earthwatch.org
Website: www.earthwatch.org/research
Contact: Gitte Venicx, Research Program Manager

Earthwatch Institute supports diverse research projects of high scientific merit worldwide that address critical environmental and social issues at local, national and international levels. Researchers are given both funding and field assistance from layperson volunteers. Volunteers are recruited by Earthwatch, who pay for the opportunity to assist them in the field.

Earthwatch Field Research Grants

Subjects: Disciplines include, but are not limited to, anthropology, archaeology, biology, botany, cartography, conservation, ethnology, folklore, geography, geology, hydrology, marine sciences, meteorology, musicology, nutrition, ornithology, restoration, sociology and sustainable development.
Purpose: To provide grants for field research projects that can constructively utilize teams of non-specialist field assistants in accomplishing their research goals.
Eligibility: Earthwatch supports doctoral and post-doctoral researchers and in some instances researchers with equivalent experience supported by a scientific advisor are eligible to apply.
Level of Study: Postdoctorate, Postgraduate, Research, Doctorate
Type: Grant
Value: £16,800–42,000
Length of Study: 3 years study with 4–5 teams
Frequency: Annual
Study Establishment: Research sites
Country of Study: Any country
Application Procedure: Applicants must complete an application form, which can be obtained by visiting the website. Concept notes must be submitted 18 months prior to field dates.
Closing Date: Please check the website
Funding: Corporation, foundation, private
Contributor: Volunteers' contributions

For further information contact:

Email: research@earthwatch.org
Website: www.earthwatch.org/research

EAST ASIA INSTITUTE (EAI)

909 Sampoong B/D, Eulji-ro 158, Jung-gu, Seoul, 100-786, Korea
Tel: (82) 2 2277 1683 (ext. 112)
Fax: (82) 2 2277 1684
Email: fellowships@eai.or.kr
Website: www.eai.or.kr
Contact: Young-Hwan Shin, Executive Director

East Asia Institute (EAI), based in Seoul, Korea, was founded in May 2002, as an independent and non-partisan organization devoted to research, publication and education on public policy, institutions and East Asian affairs. The EAI strives to become a prominent think tank in Korea.

EAI Fellows Program on Peace, Governance and Development in East Asia

Subjects: Political science, international relations and sociology.
Purpose: To encourage interdisciplinary research with a comparative perspective in the study of East Asia.
Eligibility: Open to tenured, tenure-track as well as non-tenured East Asian professors, based in the United States.
Level of Study: Research
Type: Fellowships
Value: US$8,000–10,000
Length of Study: 3 weeks
Frequency: Annual
No. of awards offered: 5
Application Procedure: Applicants can download the application form from the website. A completed application cover sheet, applicant data form along with letters of recommendation and a curriculum vitae are to be submitted.
Closing Date: August. See the website.
Funding: Foundation
No. of awards given last year: 6
No. of applicants last year: 18

EAST LOTHIAN EDUCATIONAL TRUST

Finance Department, John Muir House, Haddington, East Lothian, EH41 3HA, Scotland
Tel: (44) 16 2082 7436
Fax: (44) 16 2082 7446
Email: eleducationaltrust@eastlothian.gov.uk
Website: www.eastlothian.gov.uk
Contact: Kim Brand, Clerk

The East Lothian Educational Trust provides grants to individuals who are undertaking studies, courses or projects of an educational nature, including scholarships abroad and educational travel. Applicants must be residents of East Lothian.

East Lothian Educational Trust General Grant

Subjects: All subjects, but must be of an educational nature.
Purpose: To provide supplementary support to individuals who undertake studies.
Eligibility: Open to residents of East Lothian, excluding Musselburgh, Wallyford and Whitecraig.
Level of Study: Unrestricted
Type: Grant
Value: Variable
Length of Study: Unrestricted
Frequency: Annual
No. of awards offered: Varies
Application Procedure: Applicants must complete an application form.
Closing Date: August 10th and November 10th
Funding: Private
No. of awards given last year: 125
No. of applicants last year: Approx. 155

For further information contact:

Department of Corporate Services, John Muir House, East Lothian, Haddington, EH41 3HA, United Kingdom

THE ECONOMIC AND SOCIAL RESEARCH CONSORTIUM (CIES)

Calle Antero Aspillaga, 584 San Isidro, Lima, 27, Peru
Tel: (511) 421 2278 ext 113
Email: postmaster@cies.org.pe
Website: www.cies.org.pe

The Economicc and Social Research Consortium (CIES) is a private umbrella organization of 34 institutions, private and public universities, research centres and the National Statistics Institute. It seeks to strengthen the economic and social research community by supporting it's research as well as promoting it's use as a tool for decision making in government, civil society and academic community. Its final goal is to contribute to the development of Peru by improving the level of the national debate concerning key policy options for economic and social development.

CIES Fellowship Program

Subjects: Economics and social sciences.
Purpose: To facilitate the development of long-term partnerships between Canadian institutions and CIES members.
Eligibility: Open to citizens or permanent residents of Canada who are proficient in Spanish up to a academic level and must be a regular full-time faculty member or researcher at a recognized Candian university or research institute or full-time graduate student at a Candian university or at a recognized Canadian research institute.
Level of Study: Postdoctorate, Postgraduate
Type: Fellowships
Value: Up to Canadian $8,000 for justifiable field research expenses and airfare to Peru
Length of Study: Minimum 30 days
Frequency: Annual
Study Establishment: Canadian university or research institute
Country of Study: Canada or abroad
No. of awards offered: 3
Application Procedure: Applicants must submit their application form, research proposal, curriculum vitae, letter of support, reference letter, authorized transcripts, proof of citizenship, proof of Spanish language competence and budget. Applications are available online.
Closing Date: June 30th
Additional Information: For further information see website: www.cies.org.pe.

ECONOMIC AND SOCIAL RESEARCH COUNCIL (ESRC)

Polaris House, North Star Avenue, Swindon, Wiltshire, SN2 1UJ, England
Tel: (44) 17 9341 3000
Fax: (44) 17 9341 3001
Email: ptd@esrc.ac.uk
Website: www.esrc.ac.uk
Contact: Ms Zoë Grimwood, Research Training & Development

The Economic and Social Research Council (ESRC) is an independent, government-funded body set up by royal charter. The mission of the ESRC is to promote and support, by any means, high-quality basic, strategic and applied research and related postgraduate training in the social sciences. It also aims to advance knowledge and provide trained social scientists who meet the needs of users and beneficiaries, thereby contributing to the economic competitiveness of the United Kingdom, the effectiveness of public services and policy and quality of life. ESRC also provides advice, disseminates knowledge and promotes public understanding of the social sciences.

ESRC 1+3 Awards and +3 Awards

Subjects: Social sciences.
Purpose: To promote social science research and postgraduate training. The ESRC aims to provide continuous support for high quality postgraduate training and research on issues of importance to business, the public sector and the government.
Eligibility: Open to United Kingdom or European Community nationals with a First or Upper Second Class (Honours) Degree in any subject, or a United Kingdom professional qualification acceptable to the ESRC as of degree standard plus 3 years of subsequent full-time, relevant professional work experience. Candidates must have ordinarily been resident in the United Kingdom throughout the 3-year period preceding the date of application.
Level of Study: Postgraduate
Type: Studentship
Value: ESRC 1+3 awards cover fees and/or maintenance, depending on the student's situation, circumstances and the type of award.
Length of Study: Up to 3 years
Frequency: Annual
Study Establishment: ESRC-recognized institutional outlets and courses
Country of Study: United Kingdom
No. of awards offered: Varies
Application Procedure: Applicants must complete an application form. Information sheets and application forms are available from February each year and must be collected from the social science department of any university or Institute of Higher Education or career guidance outlet. Forms are available from the website. Studentships are allocated under the quota system.
Closing Date: July 20th
Funding: Government
No. of awards given last year: 85
No. of applicants last year: 465
Additional Information: The 1 refers to the 1-year Master's and the 3 refers to the 3-year PhD.

EDMUND NILES HUYCK PRESERVE, INC.

PO Box 189, Rensselaerville, NY, 12147, United States of America
Tel: (1) 518 797 3440
Fax: (1) 518 797 3440
Email: info@huyckpreserve.org
Website: www.huyckpreserve.org/
Contact: Mr

The Edmund Niles Huyck Preserve is a 2,000-acre nature preserve and biological research station with a newly expanded laboratory and housing for 20. The habitat is the north-eastern hardwood Hemlock forest with lakes, streams, bogs and plantations.

Edmund Niles Huyck Preserve, Inc. Graduate and Postgraduate Grants

Subjects: Ecology, behaviour evolution and natural resources of the area and conservation biology.
Purpose: To promote scientific research on the flora and fauna of the Huyck Preserve and its vicinity.
Eligibility: Open to all nationalities. Awards are made without regard to sex, colour, religion, ethnic origin or academic affiliation of the applicant, and support is based solely on the quality of the proposed research and its appropriateness to the natural resources and facilities of the Preserve.
Level of Study: Graduate, Doctorate, Postdoctorate, Postgraduate
Type: Grant
Value: A maximum of US$250,000 plus laboratory space and lodging (renewable)
Length of Study: Varies
Frequency: Annual
Study Establishment: The Preserve
Country of Study: United States of America
No. of awards offered: 10
Application Procedure: Applicants must complete an application form, available on written request or online. Proposals must contain an abstract of not more than 200 words describing the background and significance of the proposal. A literature cited section should be included and an up-to-date curriculum vitae provided. The researcher should submit three references that deal specifically with their proposed work. Please see the website www.huyckpreserve.org for further details.
Closing Date: Second Friday in March
Funding: Private
No. of awards given last year: 6
No. of applicants last year: 12

EDUCATION AND RESEARCH FOUNDATION FOR THE SOCIETY OF NUCLEAR MEDICINE (SNM)

1850 Samuel Morse Drive, Reston, VA, 20190, United States of America
Tel: (1) 1 703 708 9000
Email: tpinkham@erfsnm.org
Website: erf.snm.org
Contact: Theresa Pinkham, Executive Director

The Society of Nuclear Medicine (SNM) is an international, scientific and professional organization founded in 1954 to promote the science, technology and practical application of nuclear medicine. Its 16,000 members are physicians, technologists and scientists specializing in the research and practice of nuclear medicine.

Cassen Post-Doctoral Fellowships

Subjects: Nuclear medicine.
Purpose: To provide financial support and attract scientists from other fields to study nuclear medicine.
Eligibility: Open to citizens of the United States.
Level of Study: Postgraduate
Type: Fellowships
Value: US$25,000
Frequency: Annual
Study Establishment: Society of Nuclear Medicine
Country of Study: United States of America
No. of awards offered: 2
Application Procedure: Applicants must send in reference letters and research proposal.
Closing Date: November 1st

For further information contact:

The Education and Research Foundation, c/o Sue Weiss, CNMT Executive Director, 6500, Appaloosa Ave., Forest Lake, MN, 55025, United States of America

SNM Pilot Research Grants in Nuclear Medicine/ Molecular Imaging

Subjects: Health and medical sciences and nuclear science.
Purpose: To support Master's or PhD students to start research in nuclear medicine.
Eligibility: Open to basic and clinical scientists in early stages of their career.
Level of Study: Doctorate, Postgraduate
Type: Grant
Value: US$25,000
Frequency: Annual
Study Establishment: Society of Nuclear Medicine
Country of Study: United States of America
No. of awards offered: 2
Application Procedure: Applicants must submit a completed application form along with abstract of project proposal and budget proposal.
Closing Date: February 20th

For further information contact:

SNM Development Office, 1850 Samuel Morse Drive,, Reston, VA, 20190, United States of America
Tel: (1) 703 652 6795
Email: nmitchell@snm.org

EDUCATION NEW ZEALAND

Level 9, 15 Murphy St Thorndon, PO Box 12041, Wellington, 6144, New Zealand
Tel: (64) 4 472 0788
Fax: (64) 4 471 2828
Email: info@educationnz.govt.nz
Website: www.newzealandeducated.com/scholarships
Contact: Miss Camilla Swan, Scholarships Manager

Education New Zealand (ENZ) is the Government Agency responsible for promoting New Zealand to the world.We create strategies and programmes alongside New Zealand's education sector, government agencies and governments overseas that increase and broaden our international education activities.

New Zealand International Doctoral Research Scholarships (NZIDRS)

Subjects: All subjects.
Purpose: To provide financial support to top achieving international students seeking doctoral degrees by research in New Zealand universities
Eligibility: The candidate must hold an 'A' average or equivalent (GPA 8.00/9.00) in their prior tertiary level studies and meet the requirements for entry into a research-based doctoral programme at a New Zealand University.
Level of Study: Doctorate
Type: Research scholarship
Value: Full tuition fees student services levies, an annual living allowance of New Zealand $25,000, travel allowance up to New Zealand $2000, annual health insurance allowance up to New Zealand $600, an establishment allowance of New Zealand $500 and a book and thesis allowance of New Zealand $800
Length of Study: 3 years
Frequency: Annual
Study Establishment: All New Zealand Universities
Country of Study: New Zealand
No. of awards offered: 10
Application Procedure: The candidate must complete the application form in English and attach supporting documents as stipulated within the NZIDRS application form etc. Application can be downloaded from www.newzealandeducated.com
Closing Date: July 15th
Funding: Government
Contributor: New Zealand Government
No. of awards given last year: 10
No. of applicants last year: 290

For further information contact:

(For courier services) Scholarships Manager Education New Zealand, Level 9, Classic House, 15 Murphy Street, Thorndon, Wellington, NZ 6144, New Zealand

Website: www.newzealandeducated.comEducation New Zealand Trust, PO Box 10-500, Wellington, 6143, New Zealand
Contact: (For Post Services) Scholarships Manager

EDUCATIONAL TESTING SERVICE (ETS)

660 Rosedale Road, Princeton, NJ, 08541-0001, United States of America
Tel: (1) 609 921 9000
Fax: (1) 609 734 5410
Email: ldelauro@ets.org
Website: www.ets.org
Contact: Ms Linda J DeLauro

The Educational Testing Service (ETS) is a non-profit organization whose goal is to help advance quality and equity in education by providing fair and valid assessments, research, and related services.

ETS Harold Gulliksen Psychometric Fellowship Program

Subjects: Educational measurement, psychometrics, and statistics.
Purpose: To increase the number of well-trained scientists in educational measurement, psychometrics, and statistics.
Eligibility: Open to candidates who are enrolled in a doctoral program at the time of application and have completed all the coursework toward the PhD, and be at the dissertation stage of their program.
Level of Study: Predoctorate
Type: Fellowship
Value: US$19,000 (stipend), US$8,000 (tuition fees, and work-study program commitments), and a small grant for the purchase of equipment or software.
Length of Study: 1 year
Frequency: Annual
Country of Study: United States of America or other countries if appropriate
No. of awards offered: Varies
Application Procedure: Refer website for further information about the application procedures.
Closing Date: February 1st. See the website.
Funding: Private
Contributor: ETS
No. of awards given last year: 1
No. of applicants last year: 10
Additional Information: During the academic year selected fellows at their universities participate in a research project under the supervision of an academic mentor and in consultation with an ETS research scientist. During the summer, fellows are invited to participate in the Summer Internship program for graduate students working under the guidance of an ETS researcher.

For further information contact:

Email: internfellowships@ets.org
Website: www.ets.org/research/fellowships.html

ETS Postdoctoral Fellowships

Subjects: Measurement theory, validity, natural language, processing and computational linguistics, cognitive psychology, learning theory, linguistics, speech recognition and processing, teaching and classroom research, and statistics.
Purpose: To provide research opportunities to individuals who hold a doctorate in education and related fields, and to increase the number of women and minority professionals conducting research in educational measurement and related fields.
Eligibility: Open to applicants who have received their doctoral degree within the past three years. Selections will be based on the candidate's scholarship, the technical strength of the proposed topic of research, and the explicit objective of the research and its relationship to ETS research goals and priorities.
Level of Study: Postdoctorate
Type: Fellowship
Value: US$5,000 one-time relocation incentive for round-trip relocation expenses, consistent with ETS guidelines, will be reimbursed upon presentation of receipts.
Length of Study: Up to 2 years, renewable after the first year by mutual agreement
Frequency: Annual, renewable
Country of Study: United States of America
No. of awards offered: Up to 3
Application Procedure: Refer the ETS website for further details. All application materials should be sent electronically as attachments.
Closing Date: February 1st. See the website.
Funding: Private
Contributor: ETS
No. of awards given last year: 1
No. of applicants last year: 15

For further information contact:

Email: internfellowships@ets.org

ETS Summer Internship Program for Graduate Students

Subjects: Measurement theory, validity, natural language, processing and computational linguistics, cognitive psychology, learning theory, linguistics, speech recognition and processing, teaching and classroom research, and statistics, and international large scale assessments.
Purpose: To provide research opportunities to individuals enrolled in a doctoral program and to increase the number of women and underrepresented minority professionals conducting research in educational and related fields.
Eligibility: Open to graduate students who are pursuing a doctorate in a relevant discipline and have completed 2 years of coursework towards a PhD or EdD by June 1st of the internship year. The main criteria for selection will be scholarship and the match of applicant interests with participating ETS staff. Affirmative action goals will also be considered.
Level of Study: Predoctorate
Type: Internship
Value: $6,000 salary and $3,000 for relocation and housing allowance
Length of Study: June–July (8 weeks)
Frequency: Annual
Country of Study: United States of America
Application Procedure: Refer the ETS website for further details. All application materials should be sent electronically as attachments.
Closing Date: February 1st
Funding: Private
Contributor: ETS
No. of awards given last year: 18
No. of applicants last year: 180

For further information contact:

Email: internfellowships@ets.org

ETS Sylvia Taylor Johnson Minority Fellowship in Educational Measurement

Subjects: Measurement theory, validity, natural language, processing and computational linguistics, cognitive psychology, learning theory, linguistics, speech recognition and processing, teaching and classroom research, statistics and minority issues in education.
Purpose: To promote excellence, to encourage original and significant research for early career scholars and to provide talented minority scholars an opportunity to carry out independent research under the mentorship of ETS senior researchers. Studies focused on issues concerning the education of minority students are especially encouraged.
Eligibility: Open to applicants who have received their doctoral degree within the past 10 years and who are citizens or permanent residents of the United States of America. Selections will be based on the applicant's record of accomplishment, and proposed topic of research. Applicants should have a commitment to education and an independent body of scholarship that signals the promise of continuing outstanding contributions to educational measurement.
Level of Study: Postdoctorate
Type: Fellowship
Value: Salary is competitive. $5,000 one-time relocation incentive for round-trip relocation expenses. In addition, limited relocation expenses, consistent with ETS guidelines, will be reimbursed.
Length of Study: Up to 2 years, renewable after the first year by mutual agreement
Frequency: Annual
Country of Study: United States of America
No. of awards offered: 1

Application Procedure: Refer the ETS website for further details. All application materials should be sent electronically as attachments.
Closing Date: January 1st
Funding: Private
Contributor: ETS
No. of awards given last year: 1
No. of applicants last year: 15
Additional Information: Through her research, extensive writings and service to the educational community as an educator, editor, counsellor, committee member and collaborator during her lifetime, Sylvia Taylor Johnson had a significant influence in educational measurement and assessment nationally. In honour of Dr Johnson's important contributions to the field of education, the ETS has established the Sylvia Taylor Johnson Minority Fellowship in educational measurement.

For further information contact:

Email: internfellowships@ets.org

THE EDWARD F ALBEE FOUNDATION, INC.

14 Harrison Street, New York, NY, 10013, United States of America
Tel: (1) 212 226 2020
Fax: (1) 212 226 5551
Email: info@albeefoundation.org
Website: www.albeefoundation.org
Contact: Mr Jakob Holder, Foundation Secretary

The Edward F Albee Foundation provides residence and working space to writers and visual artists at its facilities in Montauk, New York. The residency is offered at no charge to the participants and imposes no obligations, except diligent application to their work and respect for the privacy of others.

William Flanagan Memorial Creative Persons Center

Subjects: Writing, painting, sculpting and musical composition.
Purpose: To provide accommodation.
Eligibility: Open to artists and writers in need who have displayed evidence of their talent.
Level of Study: Unrestricted
Value: Accommodation only
Length of Study: 6 months (mid May–mid October); individual residencies are 4 weeks and 6 weeks.
Frequency: Annual
Study Establishment: The William Flanagan Memorial Creative Persons Center in Montauk, Long Island
Country of Study: United States of America
No. of awards offered: 20
Application Procedure: Applicants must complete an application form. Forms are available upon request and should be accompanied by a stamped addressed envelope. Other materials are also required, and applicants should write for further details.
Closing Date: January 1–March 1st
Funding: Private
No. of awards given last year: 20
No. of applicants last year: 300
Additional Information: The environment is communal and residents are expected to do their share in maintaining the conditions of the Center.

EDWIN O. REISCHAUER INSTITUTE OF JAPANESE STUDIES

1730 Cambridge Street, Cambridge, MA, 02138, United States of America
Tel: (1) 617 495 3220
Fax: (1) 617 496 8083
Email: tgilman@fas.harvard.edu
Website: rijs.fas.harvard.edu/fellowships/index.php
Contact: Dr Theodore J Gilman, Associate Director

The Edwin O. Reischauer Institute of Japanese Studies at Harvard University supports research on Japan and provides a forum for related academic activities and the exchange of ideas. It seeks to stimulate scholarly and public interest in Japan and Japanese studies at Harvard and around the world.

Harvard Postdoctoral Fellowships in Japanese Studies

Subjects: Japanese studies.
Purpose: To aid Japanese studies to recent PhDs of exceptional promise, to give them the opportunity to turn their dissertation into publishable manuscripts.
Eligibility: Open to candidates who have received their PhD degree in Japanese studies in any area of the humanities or social sciences.
Level of Study: Postdoctorate
Type: Fellowships
Value: US$50,000 and health insurance coverage; Postdoctoral fellows will be provided office space, and access to the libraries and resources of Harvard University.
Length of Study: 10 months
Frequency: Annual
No. of awards offered: 4
Closing Date: January 2nd
No. of awards given last year: 4
Additional Information: Residence in the Cambridge/Boston area and participation in Institute activities are required during the appointment. Please check further information at http://rijs.fas.harvard.edu/fellowships/postdoctoral.php.

For further information contact:

Reischauer Institute, Harvard University, 1730 Cambridge Street, Room S233, Cambridge, MA, 02138, United States of America
Tel: (1) 495 3220
Email: rijs.fas.harvard.edu/fellowships/index.php
Website: www.fas.harvard.edu/˜rijs
Contact: Dr Theodore J Gilman, Associate Director Postdoctoral Fellowships

THE EGULLET SOCIETY FOR CULINARY ARTS & LETTERS

20 East 93rd Street #1B, NY 10128, United States of America
Tel: (1) 212 828 0133
Email: sponsors@eGullet.org
Website: www.egsociety.org

The mission of the eGullet Society for Culinary Arts & Letters is to increase awareness and knowledge of the arts of cooking, eating and drinking, as well as the literature of food and drink. The Society sponsors the opportunity to present offerings to a rarefied audience by supporting the program services of the eGullet Society. The membership of the Society represents an audience of dedicated food enthusiasts who are knowledgeable, enthusiastic and care deeply about food.

The eGullet Society for Culinary Arts & Letters Culinary Journalist Independent Study Scholarship

Subjects: Culinary arts.
Purpose: To conduct independent study and research worldwide; designed to further writing on an original and innovative culinary topic.
Eligibility: Open to career journalist who demonstrates commitment to advancing his or her skills as a writer and whose work is primarily based on food, wine or some other aspects of gastronomy and the culinary arts.
Value: US$5,000.
Length of Study: 1 year
No. of awards offered: 1
Application Procedure: Applicants are required to include a project proposal that demonstrates true literary merit in both promise and achievement at writing; an itemized budget detailing the use of this award; a tentative travel schedule with dates and locations; and a current curriculum vitae.
Closing Date: June 30th
Funding: Private
Contributor: Jonathan Day and Melissa Taylor, as well as the eGullet Society general fund.
Additional Information: For further information check the website.

The eGullet Society for Culinary Arts & Letters Humanitarian Scholarship

Subjects: Culinary arts.
Purpose: To cover the expenses of a displaced victim of Hurricane Katrina, towards any culinary degree or certificate program at any accredited domestic or foreign culinary school.
Eligibility: Applicants must be pre-enrolled students, currently enrolled students, or career professionals.
Type: Scholarship
Value: US$5,000
Length of Study: 1 year
Application Procedure: Check website for further details.
Closing Date: June 30th
Funding: Private

The eGullet Society for Culinary Arts & Letters Matthew X. Hassett Memorial Culinary Arts Scholarship

Subjects: Culinary arts.
Purpose: To currently enrolled student or career professional, toward any culinary degree or certificate program at any accredited domestic or foreign culinary school.
Eligibility: Candidates must be pre-enrolled student, currently enrolled student or career professional, toward any culinary degree or certificate program at any accredited domestic orforeign culinary school.
Type: Scholarship
Value: US$5,000.
Length of Study: 1 year
No. of awards offered: 4
Application Procedure: Check website for further details.
Closing Date: June 30th
Funding: Private
Contributor: James and Dora Hassett in memory of their son, eGullet Society staff member Matthew X. Hassett

The eGullet Society for Culinary Arts & Letters Professional Chef Independent Study Scholarship

Subjects: Culinary arts.
Purpose: To conduct independent study on culinary arts worldwide.
Eligibility: Open to a professional chef with a demonstrated commitment to advancing his or her skills as a chef or pastry chef.
Type: Scholarship
Value: US$5,000.
Length of Study: 1 year
No. of awards offered: 1
Application Procedure: Check website for further details.
Closing Date: June 30th
Funding: Private

ELECTORAL COMMISSION NEW ZEALAND

Level 6, Greenock House 39, The Terrace, PO Box 3050, Wellington, 6140, New Zealand
Tel: (64) 4 474 0670
Fax: (64) 4 474 0674
Email: helena@elections.govt.nz
Website: www.elections.org.nz

The Electoral Commission New Zealand is an independent Crown entity, which registers political parties and party logos. It also receives registered parties annual returns of donations and returns of election expenses and allocates election broadcasting time and funds to eligible political parties. The Commission also encourages and conducts public education on electoral matters.

Wallace Scholarships

Subjects: Specific subjects are set each year, all are in the general areas of electoral participation.
Purpose: To encourage research work that will be useful in designing electoral education and information programmes and help raise public awareness of electoral issues.
Eligibility: Scholarships are for research as part of a New Zealand university degree.
Level of Study: Research
Type: Scholarships
Value: New Zealand $500–2,000.
Length of Study: usually 1 year
Frequency: Annual
Country of Study: New Zealand
Application Procedure: Applicants must send a 1 page research proposal, letter of endorsement from a academic supervisor and contact details and enrollment qualifications.
Closing Date: February 2nd
Funding: Government
No. of awards given last year: 3
No. of applicants last year: 6
Additional Information: All queries should be directed to Dr Helena Catt at catt@elections.govt.nz or phone 04 474 0676.

For further information contact:

Electoral Commission
Email: Wellingtonorcatt@elections.govt.nz

ELIMINATION OF LEUKAEMIA FUND (ELF)

The Director, ELF, Regent House, 291 Kirkdale, Sydenham, London, SE26 4QD, United Kingdom
Tel: (44) 20 8778 5353
Fax: (44) 20 8778 7117
Email: elffund@ukonline.co.uk
Website: www.leukaemia-elf.org.uk

The Elimination of Leukaemia Fund's mission is to advance the cure and treatment of leukaemia and related blood disorders. This mission is implemented in four ways explained below. ELF is a major funder of leukaemia research at King's College Hospital, London, and is also funding work at a number of other major centres including the Institute of Child Health, Great Ormond Street Hospital and Belfast City Hospital. ELF favours 'patient-centred' work so that there is an immediate or near future benefit to sufferers of leukaemia and the related blood disorders.

Elimination of Leukaemia Fund Travelling and Training Fellowships

Subjects: Oncology and haematology.
Purpose: To enable doctors, nurses, clinical scientists and related health professionals working in the UK to advance their knowledge and expertise in the treatment of, or research into, leukaemia and related blood diseases.
Eligibility: Suitably qualified persons working in the field of haematological malignancies who wish to visit other departments or attend meetings in the UK or overseas.
Level of Study: Research
Type: Fellowship
Value: Travelling fellowships to attend meetings will not normally exceed UK£1,500 and training fellowships to visit departments include a contribution to subsistence that will not normally exceed £4,000.
Frequency: Biannual
Application Procedure: Successful applicants will be notified within one month. Application forms are available on written application.
Closing Date: working day in August, November, February and May, please contact organization for updates

EMBASSY OF FRANCE IN AUSTRALIA

6 Perth Avenue, Yarralumla, Canberra, ACT, 2600, Australia
Tel: (61) 0 2 6216 0100
Fax: (61) 0 2 6216 0132
Email: education@ambafrance-au.org
Website: www.ambafrance-au.org
Contact: Higher Education Attaché

The Embassy of France in Australia supports the partnership between French and Australian Universities and offers grants and scholarships to help the students' mobility.

French Government Postgraduate Studies Scholarships

Subjects: Engineering (general) 6.1 political sciences and government 17.2.
Purpose: To offers scholarships for one year university studies of master degree in France.
Eligibility: For a first year of Master: A valid BA or BSc or equivalent at least.
For a second year of Master: A master already completed.
Level of Study: Postgraduate
Type: Scholarships
Value: Tuition fees, accommodation and travel to france
Length of Study: 2 years
Frequency: Annual
Study Establishment: Institut D'etudes Politiques De Paris or the "N +i" network of engineering schools
Country of Study: France
No. of awards offered: 8
Application Procedure: Candidates must submit an application after their admission to the postgraduate programme of a French university. Application forms are available from the French Embassy website www.ambafrance-au.org
Closing Date: May 31st
Funding: Government
No. of awards given last year: 2

For further information contact:

Ambassade De France, 6 Perth Avenue, Yerralumla, Service De Cooperation, et d'Action Culturelle, ACT 2600, Australia

EMBASSY OF JAPAN IN AUSTRALIA

Embassy of Japan, 112 Empire Circuit, Yarralumla, ACT, 2600, Australia
Tel: (61) 2 6273 3244
Fax: (61) 2 6273 1848
Email: cultural@japan.org.au
Website: www.au.emb-japan.go.jp/
Contact: Ms Eriko Prior, Monbukagakusho Scholarship Co-ordinator

Japanese Government (Monbukagakusho) Scholarships In-Service Training for Teachers Category

Subjects: Teacher training.
Eligibility: Open to Australians under 35 years of age who are university or teacher training graduates currently in active service in primary or secondary schools, or who are on the staff at teacher training institutions or educational administrative institutions. Applicants must have at least 5 years of experience in their terms of service. University academic staff members should not be selected as grantees.
Level of Study: Postgraduate
Type: Scholarship
Value: Return airfare plus ¥150,000 per month (Subject to change).
Length of Study: 18 months
Frequency: Annual
Study Establishment: A Japanese university
Country of Study: Japan
No. of awards offered: 1–2
Application Procedure: Applicants must complete an application form available from the Embassy of Japan in their own country. Applications are not available from the main organization.
Closing Date: March 13th
Funding: Government
No. of awards given last year: 2
No. of applicants last year: 6
Additional Information: Applicants must be willing to study the Japanese language.

Japanese Government (Monbukagakusho) Scholarships Research Category

Subjects: Humanities, social sciences, literature, history, aesthetics, law, politics, economics, commerce, pedagogy, psychology, sociology, music and fine arts, natural sciences, pure science, engineering, agriculture, fisheries, pharmacology, medicine, dentistry and home economics.
Eligibility: Open to Australian graduates under 35 years of age.
Level of Study: Doctorate, Postgraduate
Type: Scholarship
Value: Return airfare and allowance of 150,000 yen to 153,000 yen per month.
Length of Study: 18–24 months
Frequency: Annual
Study Establishment: A Japanese university
Country of Study: Japan
No. of awards offered: Approx. 17
Application Procedure: Applicants must complete an application form available from the Embassy of Japan in their own country. Applications are not available from the main organization.
Closing Date: June 14th
Funding: Government
No. of awards given last year: 15
No. of applicants last year: 50
Additional Information: Applicants must be willing to study the Japanese language.

EMBASSY OF JAPAN IN PAKISTAN

PO Box 1119, 53-70, Ramna 5/4 Diplomatic Enclave 1, Islamabad, 44000, Pakistan
Tel: (92) 51 907 2500
Fax: (92) 51 907 2352
Email: japanembculture@dslplus.net.pk
Website: www.pk.emb-japan.go.jp

The Japan Exchange and Teaching (JET) Programme

Subjects: International relations.
Purpose: To enhance the mutual understanding and relations that currently exist between Japan and Pakistan.
Eligibility: Open to applicants who are interested in Japan and who are below the age of 40 years.
Type: Scholarship
Value: ¥125,000 per year. In additon, (2,000 or 3,000 yen per month will be added in case that the recipient research in a designated area.)
Frequency: Annual
Country of Study: Japan
Application Procedure: A completed application form must be submitted.
Closing Date: March 10th

ENGINEERS CANADA

180 Elgin Street Suite 1100, Ottawa, ON, K2P 2K3, Canada
Tel: (1) 613 232 2474
Fax: (1) 613 230 5759
Email: awards@engineerscanada.ca
Website: www.engineerscanada.ca

Engineers Canada is the national organization of the provincial and territorial associations and ordre that regulate the practice of engineering in Canada.

Engineers Canada's National Scholarship Program

Subjects: Engineering and non engineering.
Purpose: To reward excellence in the Canadian engineering profession and support advanced studies and research.
Eligibility: Open to citizens or permanent residents of Canada who are registered as professional engineers in good standing with a provincial/territorial engineering association/order.
Level of Study: Graduate, Doctorate, MBA, Postgraduate, Research
Type: Scholarships
Value: Canadian $70,000 in total
Frequency: Annual
Country of Study: Canada and abroad
No. of awards offered: 7
Application Procedure: Applicants must contact Marc Bourgeois for further details.
Closing Date: March 1st
Contributor: TD Insurance Meloche-Monnex Insurance and Manulife Financial
No. of awards given last year: 7

No. of applicants last year: Approx. 50–55
Additional Information: Postdoctoral Fellows are not eligible to apply.

For further information contact:

Tel: 613 232 2474 ext 238
Email: marc.bourgeois@engineerscanada.ca
Contact: Marc Bourgeois, Director, Communications

ENGLISH-SPEAKING UNION (ESU)

Dartmouth House, 37 Charles Street, London, W1J 5ED, England
Tel: (44) 20 7529 1550
Fax: (44) 20 7495 6108
Email: esu@esu.org
Website: www.esu.org
Contact: Head of Cultural Programmes

The English-Speaking Union (ESU) is an independent, non-political educational charity with members throughout the world, promoting international and human achievement through the worldwide use of the English language.

ESU Chautauqua Institution Scholarships

Subjects: Art (painting, ceramics and sculpture), music education, literature and international relations and drama.
Purpose: To enable teachers from the United Kingdom to study at the Chautauqua Institution's Summer School.
Eligibility: Open to teachers from the United Kingdom with a particular interest in the arts.
Level of Study: Professional development
Type: Scholarship
Value: UK£850 plus board, room, tuition and lecture sessions at the Summer School
Length of Study: 2–6 weeks
Frequency: Annual
Study Establishment: Chautauqua Institution's Summer School
Country of Study: United States of America
No. of awards offered: 1
Application Procedure: Online application forms must be completed available from www.esu.org or by emailing education@esu.org.
Closing Date: November
Funding: Private
No. of awards given last year: 1
No. of applicants last year: 6

ESU Music Scholarships

Subjects: Music.
Purpose: To enable musicians of outstanding ability to study at summer schools in the United States of America, Canada, France and United Kingdom.
Eligibility: Candidates must be students or graduates from a recognized United Kingdom conservatory or university music department.
Level of Study: Professional development
Type: Scholarship
Value: Tuition, board and lodging and relevant flight costs
Length of Study: 2–9 weeks, depending on the particular scholarship
Frequency: Annual
Study Establishment: Summer school
No. of awards offered: 10
Application Procedure: Applications must be supported by a teacher's reference.
Closing Date: Late October
Funding: Private, commercial
Contributor: Private trust funds
No. of awards given last year: 6
No. of applicants last year: 50

ESU Travelling Librarian Award

Subjects: Library and Information science.
Purpose: To encourage United States of America and United Kingdom contacts in the library world and establish links between pairs of libraries.
Eligibility: Open to professionally qualified United Kingdom and information professionals.
Level of Study: Professional development
Type: Award
Value: Upto £3,000. Board and lodging and relevant flight costs
Length of Study: A minimum of 3 weeks
Frequency: Annual
Country of Study: United States of America
No. of awards offered: 1
Application Procedure: Candidates must submit a curriculum vitae and a covering letter explaining why they are the ideal candidates for the award.
Closing Date: April 27th
Funding: Commercial, private
Contributor: The English-Speaking Union and The Chartered Institute of Library and Information Professionals
No. of awards given last year: 1
No. of applicants last year: 16
Additional Information: Candidates should contact the Librarian by telephone or email: library@esu.org

Lindemann Trust Fellowships

Subjects: Astronomy, chemistry, engineering, geology, geophysics, mathematics, physics and biophysics.
Purpose: To allow postdoctoral research to be carried out at a university in the United States of America.
Eligibility: Open to United Kingdom and Commonwealth citizens who are graduates of a United Kingdom university and to United Kingdom and Commonwealth citizens who are pursuing postgraduate research at a United Kingdom university, although are not graduates of that institution. Preference is given to those who have demonstrated their capacity for original research.
Level of Study: Postdoctorate, Postgraduate
Type: Fellowship
Value: US$30,000 stipend per year.
Length of Study: 1 year
Frequency: Annual
Study Establishment: A university
Country of Study: United States of America
No. of awards offered: 2–3
Application Procedure: Online application forms must be completed available from www.esu.org or by emailing education@esu.org.
Closing Date: February 18th
Funding: Private
No. of awards given last year: 3
No. of applicants last year: 22
Additional Information: Fellows are not required to work for an American degree, but are expected to be attached to a university, college or seat of advanced learning and technical repute in the United States of America. The place of study and research programme must be approved by the Committee. A limited amount of teaching as an adjunct to research activities is not excluded.

ENTENTE CORDIALE SCHOLARSHIPS

French Cultural Department, 23 Cromwell Road, London, SW7 2EN, England
Tel: (44) 20 7073 1312
Fax: (44) 20 7073 1326
Email: entente.cordiale@ambafrance.org.uk
Website: www.ambascience.co.uk/entente-cordiale
Contact: Administrative Officer

Launched by an agreement between the United Kingdom and French governments in 1995, the Entente Cordiale Scholarships enable outstanding British postgraduates to study or carry out research on the other side of the Channel, with a view to dispel preconceived ideas and promote good relations between the two countries.

Bourses Scholarships

Subjects: All subjects.
Purpose: To allow individuals to study or carry out research in France.
Eligibility: Open to British citizens.
Level of Study: Postgraduate

Type: Scholarship
Value: UK£8,000 for students living in Paris and UK£7,500 for those studying outside Paris for the one-year award, UK£3,000 for 3 months, UK£6,000 for 6 months
Length of Study: 3 months, 6 months or 1 year
Frequency: Annual
Study Establishment: Approved universities or grande écoles
Country of Study: France
No. of awards offered: 10
Application Procedure: Applicants must complete an application form, available from the website.
Closing Date: March 15th
Funding: Private
Contributor: Blue Circle (Lafarge), BP, Kingfisher PLC, EDF Energy, UBS, Xerox, Paul Minet, Sir Patrick Sheehy Schlumberger, Vodafone, Rolls Royce, Parthenon Trust
No. of awards given last year: 7
No. of applicants last year: 60
Additional Information: Scholarships are also awarded to French postgraduates to study in the United Kingdom. Interested parties should contact the British Council in Paris.

ENTOMOLOGICAL SOCIETY OF CANADA (ESC)

393 Winston Ave, Ottawa, ON, K2A 1Y8, Canada
Tel: (1) 613 725 2619
Fax: (1) 613 725 9349
Email: entos.can@bellnet.ca
Website: www.esc-sec.ca
Contact: Office Manager

The Entomological Society of Canada (ESC) is one of the largest and oldest professional societies in Canada. Founded in Toronto on April 16, 1863, the Society was open to all students and lovers of entomology. ESC is a dynamic force in promoting research, disseminating knowledge of insects and encouraging the continued participation of all lovers of entomology in the most fascinating of all natural sciences. It is especially well known for its widely distributed and used publications.

Graduate Research Travel Scholarship

Subjects: Scientific studies on insects or other related terrestrial arthropods.
Purpose: To help students increase the scope of the graduate training.
Eligibility: Open to applicants who are full-time graduate student, studying at a Canadian university and are pursuing scientific studies on insects or other related terrestrial arthropods.
Level of Study: Graduate
Type: Scholarship
Value: Canadian $2,000
Frequency: Annual
Country of Study: Canada
No. of awards offered: 1
Closing Date: February 16th
No. of awards given last year: 1
Additional Information: The scientific merit of each application will be evaluated by a committee that has the option of sending specific projects out for external review by experts in the field.

For further information contact:

Department of Zoology, University of British Columbia, 6270 University Boulevard, Vancouver, BC V6T 1Z4, Canada
Contact: Dr Judith Myers, Chair, ESC Student Awards Committee

John H. Borden Scholarship

Subjects: Entomology and integrated pest management.
Purpose: To financially support students who are studying integrated pest management with an entomological emphasis.
Eligibility: Open to postgraduate students of Integrated Pest Management.
Level of Study: Postgraduate
Type: Scholarship
Value: Canadian $1,000.
Frequency: Every 2 years
Country of Study: Canada
No. of awards offered: 1
Application Procedure: Applicants must submit their application form, curriculum vitae, transcripts and reference letters. Application forms are available online.
Closing Date: February 16th

For further information contact:

Email: Floate@agr.gc.ac
Contact: Dr Judith Myers, Chair, ESC Students Award Committe

ENVIRONMENTAL LEADERSHIP PROGRAM

P.O. BOX 907, Greenbelt, MD, 20768-0907, United States of America
Email: info@elpnet.org
Website: www.elpnet.org

The Environmental Leadership Program (ELP) inspires visionary, action-oriented and diverse leadership to work for a just and sustainable future. ELP nurtures a new generation of environmental leaders characterized by diversity, innovation, collaboration and effective communications. ELP addresses the needs of relatively new environmental activists and professionals.

Environmental Leadership Fellowships

Subjects: Environmental leadership.
Purpose: To build the leadership capacity of the environmental field's most promising and emerging practitioners.
Eligibility: Open to citizens of the United States only.
Level of Study: Postgraduate
Type: Fellowship
Value: US$750 which includes room and board for the 3 overnight retreats, participation in 10 days of training and community building and access to our network of over 480 Senior Fellows
Length of Study: 2 years
Frequency: Annual
Country of Study: United States of America
No. of awards offered: 20–25
Closing Date: April 2nd

ENVIRONMENTAL RESEARCH AND EDUCATION FOUNDATION (EREF)

3301 Benson Drive, Suite 301 Raleigh, North Carolina, 27609, United States of America
Tel: (1) 919 861 6876
Fax: (1) 919 861 6878
Email: scholarships@erefdn.org
Website: www.erefdn.org
Contact: Dr Bryan Staley, President

The Environmental Research and Education Foundation funds and direct scientific research studies and educational intiatives, including internships and graduate level scholarships, to improve solid waste management practices and create a more sustainable world.

EREF Scholarships in Solid Waste Management Research and Education

Subjects: Waste management, environmental science, industrial ecology and waste services.
Purpose: To recognize excellence in masters, doctoral or postdoctoral students studying some aspect of solid waste management.
Eligibility: Open to full-time masters, doctoral or postdoctoral students and researchers irrespective of race, religion, national or ethnic origin, citizenship or disability. Applicant must have a clearly demonstrated interest in waste management research.
Level of Study: Doctorate, Graduate, Postdoctorate, Postgraduate
Type: Scholarship

Value: Up to US$12,000
Length of Study: Up to 3 years
Frequency: Annual
Country of Study: United States of America & Canada
Application Procedure: Application forms can be downloaded from the website www.erefdn.org
Closing Date: May 8th
Funding: Foundation
No. of awards given last year: 3
No. of applicants last year: 20
Additional Information: Country of study - United States of America, Canada, United Kingdom, and Europe.

EPILEPSY ACTION

New Anstey House, Gate Way Drive, Yeadon, Leeds, LS19 7XY, United Kingdom
Tel: (44) 11 3210 8800
Fax: (44) 11 3391 0300
Email: research@epilepsy.org.uk
Website: www.epilepsy.org.uk
Contact: Margaret Rawnsley, Research Administration Officer

Epilepsy Action is the largest member-led epilepsy organization in the United Kingdom. As well as campaigning to improve epilepsy services and raise awareness of the condition, we offer assistance to people in a number of ways including a national network of branches, volunteers, free telephone and an email helpline.

Postgraduate Research Bursaries (Epilepsy Action)

Subjects: Social, healthcare and psychological aspects of epilepsy.
Purpose: To support postgraduate research in the social and medical aspects of epilepsy and all non-laboratory research into epilepsy.
Eligibility: Students should be registered for a postgraduate degree or study at a United Kingdom University.
Level of Study: Predoctorate, Research, Doctorate, Postdoctorate, Postgraduate
Type: Bursary
Value: UK£1,500
Length of Study: Varies
Frequency: Annual
Study Establishment: A university in UK
Country of Study: United Kingdom
No. of awards offered: 3
Application Procedure: Applicants must contact the Research Administration Office.
Closing Date: October
Funding: Foundation
Contributor: Organization's own funds
No. of awards given last year: 3
No. of applicants last year: 9

Postgraduate Research PhD Studentship (Epilepsy Action)

Subjects: Social, healthcare and psychological aspects of epilepsy.
Purpose: To support postgraduate research in the social and medical aspect of epilepsy and all non-laboratory research into epilepsy.
Eligibility: Proposals must be submitted by a supervisor.
Level of Study: Doctorate
Type: Studentship
Value: Up to £75,000
Length of Study: 3 years
Frequency: Annual
Study Establishment: UK university
Country of Study: United Kingdom
No. of awards offered: 1
Application Procedure: Applicants must contact the Resrarch Administration Office.
Closing Date: October
Funding: Foundation
No. of awards given last year: 1
No. of applicants last year: 3

EPILEPSY FOUNDATION (EF)

8301 Professional Place, Landover, MD, 20785-7223, United States of America
Tel: (1) 800 332 1000
Fax: (1) 301 577 2684
Email: ResearchWebSupport@EFA.org
Website: www.epilepsyfoundation.org
Contact: Ms Cassandra Richard, Research Co-ordinator

The Epilepsy Foundation (EF) is a national, charitable organization, founded in 1968 as the Epilepsy Foundation of America. It is the only organization wholly dedicated to the welfare of people with epilepsy and to working on their behalf through research, education, advocacy and service.

Behavioral Sciences Postdoctoral Fellowships

Subjects: Epilepsy research relevant to the behavioural sciences. Appropriate fields of study include sociology, social work, psychology, anthropology, nursing, political science and others fields relevant to epilepsy research and practice.
Purpose: To offer qualified individuals the opportunity to develop expertise in epilepsy research through a training experience or involvement in an epilepsy research project.
Eligibility: Open to individuals who have received their doctoral degree in a field of the behavioural sciences by the time the fellowship commences and desire additional postdoctoral research experience in epilepsy. Applications from women and minorities are encouraged.
Level of Study: Postdoctorate
Type: Fellowship
Value: Up to US$40,000 maximum for one year.
Length of Study: 1 year
Frequency: Annual
Study Establishment: An approved facility
Country of Study: United States of America
No. of awards offered: 1
Application Procedure: Candidates must complete an application form, available from the Foundation. Candidates may also visit the website research page for details.
Closing Date: March 21st. See the website.
Funding: Private
No. of awards given last year: 1
Additional Information: The closing date for applications may vary from year to year. Candidates should email: cmorris@efa.org for details.

Postdoctoral Research Fellowships

Subjects: Basic or clinical epilepsy.
Purpose: To offer qualified individuals the opportunity to develop expertise in epilepsy research through involvement in an epilepsy research project.
Eligibility: Open to physicians and neuroscientist PhDs who desire postdoctoral research experience. Preference is given to applicants whose proposals have a paediatric or developmental emphasis. Research must address a question of fundamental importance. A clinical training component is not required. Applications from women and minorities are encouraged.
Level of Study: Postdoctorate
Type: Fellowships
Value: US$45,000 maximum for one year.
Length of Study: 1 year
Frequency: Annual
Study Establishment: A facility where there is an ongoing epilepsy research programme
Country of Study: United States of America or Canada
No. of awards offered: Varies
Application Procedure: Candidates must complete an application form, available from the Foundation. Candidates may also look at the website research page for details.
Closing Date: August 31st
Funding: Private

Research Grants (EF)

Subjects: Basic biomedical, behavioural and social science.
Purpose: To support basic and clinical research that will advance the understanding, treatment and prevention of epilepsy.

Eligibility: Open to United States of America researchers. Priority is given to investigators just entering the field of epilepsy, to new or innovative projects or to investigators whose research is relevant to developmental or paediatric aspects of epilepsy. Applications from women and minorities are encouraged, while applications from established investigators with other sources of support are discouraged. Research grants are not intended to provide support for postdoctoral Fellows.
Level of Study: Postgraduate, Postdoctorate
Type: Grant
Value: Up to US$50,000 one year (maximum $100,000 for two years)
Length of Study: 1 year
Frequency: Annual
Country of Study: United States of America
No. of awards offered: Varies
Application Procedure: Candidates must complete an application form, available from the Foundation. Candidates may also visit the website research page for details.
Closing Date: August 31st
Funding: Private
Additional Information: The closing date for applications may vary from year to year. Candidates should email: cmorris@efa.org for details.

EPILEPSY RESEARCH UK

PO Box 3004, London, W4 4XT, England
Tel: (44) 20 8995 4781
Fax: (44) 20 8995 4781
Email: info@eruk.org.uk
Website: www.epilepsyresearch.org.uk
Contact: Ms Delphine Van der Pauw, Research & Information Executive

Epilepsy Research UK promotes and supports basic and clinical and qualitative scientific research into the causes, treatment and prevention of epilepsy. Application are encouraged on all aspects of Epilepsy including basic and social science, clinical management and holistic management of patients.

Epilepsy Research UK Fellowship

Subjects: All aspects of Epilepsy including basic and social science, clinical management and holistic management of patients.
Purpose: To support a researcher in the field of epilepsy.
Eligibility: Open to researchers resident in the United Kingdom and affiliated to an academic institution in the United Kingdom. Applicants must be graduates in medicine or in one of the sciences allied to medicine.
Level of Study: Research, Unrestricted
Type: Fellowship
Value: UK£250,000
Length of Study: 1–3 years
Frequency: Annual
Country of Study: United Kingdom
No. of awards offered: 3
Application Procedure: Application is a two-stage process: A preliminary application form must be completed and submitted. If shortlisted, a full application form must be completed. Further details from our website www.epilepspyresearch.org.uk. An interview may be required.
Closing Date: Last Friday in September
Funding: Individuals, private, trusts
No. of awards given last year: 2
No. of applicants last year: 12
Additional Information: Financial correspondent must be in UK for any application. Please check at www.epilepsyresearch.org.uk/apply-for-funding/ for further information.

Epilepsy Research UK Pilot Grant in Epilepsy

Subjects: All aspects of epilepsy including basic and social science, clinical managment and holistic management of patients. Main outcome to generate pilot data for further study.
Purpose: To support a researcher in the field of epilepsy.
Eligibility: Open to researchers resident in the United Kingdom and affiliated to an academic institution in the United Kingdom. Applicants must be graduates in Medicine or in one of the sciences allied to Medicine.
Type: Pilot Grant
Value: £30,000 Maximum
Length of Study: Up to 2 years
Frequency: Annual
Country of Study: United Kingdom
Application Procedure: Application is a one-stage process. One application form must be completed and submitted. Further details from the website www.epilepsyresearch.org.uk. Apply at www.epilepsyresearch.org.uk/apply-for-funding.
Closing Date: Last Firday in September
Funding: Individuals
Additional Information: This is a new stream. Financial correspondent must be UK-based for any application.

Epilepsy Research UK Research Grant

Subjects: All aspects of Epilepsy including basic and social science, clinical management and holistic management of patients.
Purpose: To support a researcher in the field of epilepsy.
Eligibility: Open to researchers resident in the United Kingdom. And affiliated to an academic institution in the United Kingdom. Applicants must be graduates in medicine or in one of the sciences allied to medicine.
Level of Study: Research, Unrestricted
Type: Project
Value: A maximum of UK£150,000
Length of Study: Upto 3 years
Frequency: Annual
Country of Study: United Kingdom
No. of awards offered: 3
Application Procedure: Application is a two-stage process: A preliminary application form must be completed and submitted. If shortlisted, a full application form must be completed. Further details from the website www.epilepsyresearch.org.uk
Closing Date: Last Friday in September
Funding: Individuals, private, trusts
No. of awards given last year: 4 project awards
No. of applicants last year: 53
Additional Information: Financial correspondent must be UK-based for any application. Please see for further information atwww.epilepsyresearch.org.uk/apply-for-funding/.

THE ERIC THOMPSON TRUST

The Royal Philharmonic Society, 10 Stratford Place, London, W1C 1BA, England
Tel: (44) 20 7491 8110
Fax: (44) 20 7493 7463
Email: ett@royalphilharmonicsociety.org.uk
Website: www.etorgantrust.co.uk
Contact: Mr David Lowe, Clerk to The Trustees

The Eric Thompson Trust aims to provide modest grants to help aspiring professional organists. Preference will be given to students seeking assistance towards specific projects, rather than continuing academic tuition, e.g. summer schools or special lessons in addition to normal studies and opportunities to play on historical instruments in the context of further study.

Eric Thompson Charitable Trust for Organists

Subjects: To promote all aspects of organ music by awarding grants to students of the organ and by ecouraging the performance and publication of organ music.
Purpose: To provide aspiring professional organists with financial assistance for special studies such as summer schools, travel and subsistence for auditions or performance or other incidental costs incurred in their work.
Eligibility: Some professional training as an organist is required.
Level of Study: Professional development
Value: Determined by the Trustees, but normally limited to a contribution towards costs
Frequency: Twice a year
Country of Study: Any country
No. of awards offered: Varies

Application Procedure: An application form can be downloaded from the website. Applicants must send full details of their needs together with information on their training and career, two written references from organists of good standing in the profession and other relevant material to the Clerk to the Trustees.
Closing Date: November 30th or May 31st for consideration in December and June, respectively
Funding: Private
Contributor: Personal and corporate donors
No. of awards given last year: 4
No. of applicants last year: 8
Additional Information: An annual scholarship is also offered for a two-week period of study in the Netherlands with organ-builders Flentrop. The successful candidate will also have the opportunity of lessons with two eminent Dutch organists.

ESADE

MBA Office, Avenue d'Espluges 92–96, Barcelona, E-08034, Spain
Tel: (34) 93 280 6162
Fax: (34) 93 204 8105
Email: mba@esade.edu
Website: www.esade.edu
Contact: Ms Jordi Mora Pintado, Financial Aid & Operations Director

ESADE is an independent nonprofit university institution, founded in 1958 in Barcelona when a group of entrepreneurs and Jesuit Society members joined forces. Since 1995, it has formed part of the Ramon Llull University. ESADE's academic activity takes place on its Barcelona, Madrid and Buenos Aires campuses. The three main areas it focusses on are education, research and social dialogue.

ESADE MBA Scholarships

Subjects: 4.14 MBA (12, 15 or 18 months)
Purpose: To assist full-time MBA students with tuition fees award high potential candidates.
Eligibility: Depending on the scholarship: enrolled students: (Fellowships + Impact); Admitted students: (Direct, Merit and Commitment Scholarships + Excellence). Scholarships are awarded, restricted by merit achivement, geographical area, sector of activity and need base.
Level of Study: MBA
Type: Fellowship/Scholarship
Value: Direct scholarships: up to 40% of total tution fees; Scholarship for excellence: 50% of tution fees; Merit and Commitment Scholarships: up to 20% of tution fees; Impact Scholarships: up to 10% of toal tution fees; Fellowships: variable; depending on length.
Length of Study: 12, 15, 18 months
Frequency: Annual
Study Establishment: ESADE Business School
Country of Study: Spain
Application Procedure: Only application for: – Merit and Commitment Scholarship; – Fellowship. Other scholarships are decided.
Closing Date: June 30th
Funding: Foundation
Contributor: ESADE Foundation, ESCADE MBA scholarship fund
No. of awards given last year: 30
No. of applicants last year: Approx. 90
Additional Information: www.esade.edu/mba

For further information contact:

Email: financialaid@esade.edu

EUROPEAN CALCIFIED TISSUE SOCIETY

PO Box 337, Patchway, Bristol, BS32 4ZR, United Kingdom
Tel: (44) 1454 610255
Fax: (44) 1454 610255
Email: ects@ectsoc.org
Website: www.ectsoc.org

The European Calcified Tissue Society is the major organization in Europe for researchers and clinicians working on calcified tissues and related fields.

ECTS Career Establishment Award

Subjects: Calcified tissue and related fields.
Purpose: To assist newly appointed faculty members in launching a successful research career.
Eligibility: Applicants must be members of the ECTS, within 3 years of being appointed to their first faculty position as an independent researcher and must be working in an area of research relevant to the aims of ECTS.
Level of Study: Set-up grant not dependent on study
Type: Award
Value: Dependent on funds available
Frequency: Dependent on funds available
No. of awards offered: Dependent on funds available
Application Procedure: Applicants must complete an application form, available from the website.
Closing Date: November
Funding: Foundation
Contributor: ECTS
No. of awards given last year: 1
No. of applicants last year: 6

ECTS Exchange Scholarship Grants

Subjects: Calcified tissue and related fields.
Purpose: To enable researchers to spend time in another laboratory to learn new techniques.
Eligibility: Applicants must be members of the ECTS.
Level of Study: Professional development
Type: Scholarship
Value: Depends on anticipated expenses
Frequency: Dependent on funds available
No. of awards offered: Dependent on funds
Application Procedure: Applicants must complete an application form, available from the website.
Closing Date: There is no deadline
Funding: Foundation
Contributor: ECTS
No. of awards given last year: 4
No. of applicants last year: 5

ECTS PhD Studentship

Subjects: Calcified tissue and related fields.
Purpose: To assist European PhD students with expenses incurred for their PhD project.
Eligibility: Applicants must be members of the ECTS. PhD student commencing or clearing their first year of study.
Level of Study: Predoctorate, Professional development
Type: Scholarship
Value: Dependent on funds available
Length of Study: 3 years
No. of awards offered: Up to 4
Application Procedure: Applicants must complete an application form, available from the website.
Closing Date: November
Funding: Foundation
Contributor: ECTS
No. of awards given last year: 1
No. of applicants last year: 22

ECTS Postdoctoral Fellowship

Subjects: Calcified tissue and related fields.
Purpose: To assist European postdoctoral fellows with expenses relating to their own research project.
Eligibility: Applicants must be a member of the ECTS, within 10 years of gaining MD or PhD and must be based in Europe.
Level of Study: Professional development
Type: Fellowship
Value: Dependent on funds available
Length of Study: 2 years
No. of awards offered: Up to 3
Application Procedure: Applicants must complete an application form, available from the website.
Funding: Foundation
Contributor: ECTS
No. of awards given last year: 1
No. of applicants last year: 20

ECTS/AMGEN Bone Biology Fellowship

Subjects: Relevant areas of bone biology.
Purpose: To enable scientists who are at the beginning of their research career and who wish to conduct research into basic or clinical aspects of bone disease.
Eligibility: Open to any scientist or clinician member of ECTS, working in the field of bone biology.
Level of Study: Research
Type: Fellowship
Value: €100,000, of which 50 per cent shall be deployed to cover salary costs
Length of Study: 3 years
No. of awards offered: 1
Application Procedure: Applicants must complete an application form available from the website.
Closing Date: November
Funding: Commercial
Contributor: AMGEN
No. of awards given last year: 1
No. of applicants last year: 20
Additional Information: All applications must be supported by the Head of the host laboratory.

EUROPEAN LEAGUE AGAINST RHEUMATISM

Seestrasse 240, CH 8802 Kilchberg (Zürich), Switzerland
Tel: (41) 44 716 30 30
Fax: (41) 44 716 30 39
Email: eular@eular.org
Website: www.eular.org
Contact: Elly Wyss, EULAR Training Bursaries

EULAR Research Grants

Subjects: Research into diagnostic and therapeutic aspects of rheumatic diseases.
Purpose: To support work programmes that improve the quality of care in the field of rheumatology.
Level of Study: Research
Type: Project grant
Value: €30,000 per year
Length of Study: Up to 3 years
Frequency: Annual
Application Procedure: Applications for grants should include: an abstract, a rationale of the need and relevance of the project, an account of strategic objectives, resources, budget and organization, a description of the implementation and relevance of the project for EULAR, and a list of references of participating university/hospital centres relating to the topic of the proposal.
Closing Date: November 14th
Contributor: European League Against Rheumatism
Additional Information: For further information contact the EULAR Secretariat.

EUROPEAN MOLECULAR BIOLOGY ORGANIZATION (EMBO)

PO Box 1022.40, D-69012 Heidelberg, Germany
Tel: (49) 622 188 910
Fax: (49) 622 188 91200
Email: embo@embo.org
Website: www.embo.org
Contact: Mr Yvonne Kaul, Communications Officer

The European Molecular Biology Organization (EMBO) was established in 1964 to promote biosciences in Europe. Today EMBO supports transnational mobility, training and exchange through initiatives such as fellowships, courses, workshops and its young investigator activities.

EMBO Installation Grants

Subjects: Molecular biology and disciplines relying on molecular biology.
Purpose: To strengthen science in participating EMBC member states (Croatia, the Czech Republic, Estonia, Portugal Poland and Turkey).
Eligibility: Eligible scientists should have an excellent publication record and should have spent at least two consecutive years prior to the application deadline, outside the country in which they are applying to establish their lab. The applicant should be negotiating a full-time position at an institute/university in participating member states by the date of the application. Candidates who are planning to establish independent labs in the country that they are applying for can be in that country for 2 years at the time of the deadline.
Level of Study: Postdoctorate
Type: Grant
Value: €50,000
Length of Study: 3–5 years
Frequency: Annual
Study Establishment: For a new laboratory in a participating member state.
No. of awards offered: Varies
Application Procedure: Applicants must visit the organization's website www.embo.org for details of the application forms, letters of reference and online application.
Closing Date: April 15th
Funding: Government
Contributor: The 27 EMBC member states
No. of awards given last year: 7
No. of applicants last year: 27
Additional Information: Country of study: Participating member states

EMBO Long-Term Fellowships in Molecular Biology

Subjects: Molecular biology and disciplines relying on molecular biology.
Purpose: To support international research careers
Eligibility: Open to holders of a doctoral degree. EMBO fellowships are not awarded for exchanges between laboratories within one country. Applicants must be nationals from a European Molecular Biology Conference (EMBC) member state or be wishing to travel to a EMBC member state.
Level of Study: Postdoctorate, Biannual
Type: Fellowship
Value: A return travel allowance for the Fellow and any dependants plus a stipend and dependants' allowance
Length of Study: 1 year, renewable for a further year
Frequency: Twice a year
Study Establishment: A suitable laboratory
Country of Study: Any country
No. of awards offered: 250–300 per year
Application Procedure: Please see the EMBO website (www.embo.org/fellowships)
Closing Date: February 15th and August 15th
Funding: Government
Contributor: The 27 EMBC member states
No. of awards given last year: 263
No. of applicants last year: 1668
Additional Information: The following countries form the EMBC: Austria, Belgium, Croatia, the Czech Republic, Denmark, Estonia, Finland, France, Germany, Greece, Hungary, Iceland, Ireland, Israel, Italy, the Netherlands, Norway, Poland, Portugal, the Slovak Republic, Slovenia, Spain, Sweden, Switzerland, Turkey and the United Kingdom. Special provision is also made for applications involving Cyprus. For further information, email: fellowships@embo.org

EMBO Short-Term Fellowships in Molecular Biology

Subjects: Molecular biology and disciplines relying on molecular biology.
Purpose: To advance molecular biology research by helping scientists to visit another laboratory with a view to applying a technique not available in the home laboratory and to foster collaboration.
Eligibility: Please see the EMBO website (www.embo.org/fellowships)
Level of Study: Postdoctorate, Postgraduate, Predoctorate, Research, Doctorate
Type: Fellowship
Value: Return travel for the Fellow and a daily subsistance for the duration of the fellowship

Length of Study: 2 weeks–3 months
Frequency: Ongoing process
Study Establishment: A suitable laboratory
No. of awards offered: Varies
Application Procedure: Application form and guidelines are available on the EMBO website.
Closing Date: There is no deadline, but applications should be made at least 3 months before proposed start date
Funding: Government
Contributor: The 27 EMBC member states
No. of awards given last year: 219
No. of applicants last year: 421
Additional Information: Country of study: Any country, depending where the home laboratory is.

EMBO Young Investigator

Subjects: Molecular biology and disciplines relying on molecular biology.
Purpose: To promote the development of research in Europe and Israel.
Eligibility: Applicants should: be leading their first independent laboratory for at least 1 and not more then 4 years in an EMBC member state; have at least 2 years of post-PhD scientific experience, have an excellent track record; be working in the very broadly defined area of Molecular Biology; be supported by sufficient funds to run their laboratories; have published at least one last author paper after establishing an independent laboratory. Only in exceptional cases will applications from scientists over 40 years in age be considered. Extended eligibility for female candidates with children.
Level of Study: Postdoctorate, Professional development
Type: Grant
Value: €15,000
Length of Study: 3 years
Frequency: Annual
Study Establishment: The applicant's own independent laboratory
No. of awards offered: Varies
Application Procedure: Applicants must visit the website for details on Application forms, letters of reference and on-line application.
Closing Date: April 1st
Funding: Government
Contributor: The 27 EMBC member states
No. of awards given last year: 22
No. of applicants last year: 164

EUROPEAN RESEARCH CONSORTIUM FOR INFORMATICS AND MATHEMATICS (ERCIM)

2004, Route des Lucioles BP 93, F 06902, Sophia Antipolis, Cedex, France
Tel: (33) 4 92 38 50 10
Fax: (33) 4 92 38 50 11
Email: contact@ercim.org
Website: www.ercim.org

The European Research Consortium for Informatics and Mathematics (ERCIM) aims to foster collaborative work within the European research community and to increase co-operation with European industry.

Alain Bensoussan Fellowship Programme

Subjects: Applications of numerical mathematics in science, biomedical informatics, constraints technology and applications.
Purpose: To enable bright young scientists from all over the world to work on challenging problems as Fellows of leading European research centres.
Eligibility: Open to candidates who have obtained a PhD, and are fluent in English.
Level of Study: Research
Type: Fellowship
Value: Varies by country
Length of Study: 24 months
Frequency: Annual
Application Procedure: Applicants must submit the application form online.
Closing Date: April 30th
Additional Information: Not only are researchers from academic institutions encouraged to apply, but also scientists working in the industry. All queries should be directed to Emma Lière at emma.liereercim.org.

EUROPEAN SOCIETY OF SURGICAL ONCOLOGY (ESSO)

Avenue E. Mounier 83, B-1200, Brussels, Belgium
Tel: (32) (0)2 775 02 01
Fax: (32) (0)2 775 02 00
Email: info@essoweb.org
Website: www.esso-surgeonline.be
Contact: Secretariat

ESSO was founded to advance the art, science and practice of surgery for the treatment of cancer. ESSO endeavours to ensure that the highest possible standard of surgical treatment is available to cancer patients throughout Europe by organizing congresses, granting fellowships and publishing the *EJSO*.

ESSO Training Fellowships

Subjects: Surgical oncology.
Purpose: To provide young surgeons a chance to spend time in another specialist centre to either expand their experience or learn new techniques.
Eligibility: Open to applicants who are specialists/specializing in surgery (or in any other medical discipline where cancer surgery is performed). Applicants must be less than 40 years of age. European applicants may choose to visit European or non-European units, while non-European applicants must choose to visit a European center.
Level of Study: Postdoctorate
Type: Fellowship
Value: €2,000 for standard fellowships and €10,000 for the major international training fellowship
Length of Study: 1 month for standard fellowships and upto 1 year for the major training fellowship
Frequency: Annual
Country of Study: Any country
No. of awards offered: 10 standard and 1 major training fellowships
Application Procedure: Applicants must submit a full curriculum vitae with their application, together with a note of their career intentions. Applicants should also outline what they hope to gain from the training fellowship, including what specific experience is sought and how this will fit in with the applicant's career development. Applicants should provide details as to which institution they wish to visit, together with details of the clinical or research training opportunities that the department can offer. A letter of support from the applicant's head of department must be included and this can be in the form of a reference. A letter of support from the head of the department they wish to visit must also be supplied, indicating that the department to be visited will be in a position to provide the experience required by the applicant.
Closing Date: October 15th
No. of awards given last year: 11
No. of applicants last year: 18
Additional Information: Applicants must be or become ESSO members.

For further information contact:

Email: carine@esso-surgeonline.org
Website: www.esso-surgeonline.org
Contact: Ms Carine Lecoq, ESSO Administrator

EUROPEAN SOUTHERN OBSERVATORY (ESO)

Karl-Schwarzschild-Strasse 2, D-85748 Garching bei Muenchen, Germany
Tel: (49) 893 200 60
Fax: (49) 893 202 362
Email: vacancy@eso.org
Website: www.eso.org
Contact: Mr Roland Block, Head of Personnel Department

The European Southern Observatory (ESO) is an intergovernmental organization for research in astronomy. At present ESO is operating the Very Large Telescope (VLT) at Cerro Paranal in Chile, the world's most powerful facility for optical astronomy, and La Silla Observatory.

ESO Fellowship

Subjects: Astronomy and astrophysics.
Purpose: To provide a unique opportunity to learn and participate in the process of observational astronomy while pursuing a research programme.
Level of Study: Postdoctorate
Type: Fellowship
Value: A basic monthly salary of not less than €2,918, to which is added an expatriation allowance as well as some family allowances, if applicable. The Fellow will also have an annual travel budget for scientific meetings, collaborations and observing trips
Length of Study: 1 year, with a possible extension to 3 years in Garching. Fellowships in Chile are for 1 year with a possible extension to 4 years
Frequency: Annual
Study Establishment: The European Southern Observatory
No. of awards offered: 6–9
Application Procedure: Applicants must visit the ESO website for an application form and further information.
Closing Date: October 15th
Funding: Government
Additional Information: Fellowships begin between April and October of the year in which they are awarded. Selected Fellows can join ESO only after having completed their doctorate.

EUROPEAN SYNCHROTRON RADIATION FACILITY (ESRF)

6 rue Jules Horowitz, BP 220, Grenoble Cedex, 38043, France
Tel: (33) 4 76 88 20 00
Fax: (33) 4 76 88 20 20
Email: recruitment@esrf.fr
Website: www.esrf.fr
Contact: Ms Bénédicte Henry Canudas, Head of Recruitment

The European Synchrotron Radiation Facility (ESRF) supports scientists in the implementation of fundamental and applied research on the structure of matter in fields such as physics, chemistry, crystallography, Earth science, biology, medicine, surface science and materials science.

ESRF Postdoctoral Fellowships

Subjects: Physics, biology, chemistry, mineralogy and crystallography, computer engineering and accelerators science.
Purpose: To enable postdoctoral fellows develop their own research programme and motivate them to collaborate with external users.
Eligibility: Preference is given to PhD students who obtained their PhD less than 3 year ago.
Level of Study: Postdoctorate
Value: €3,409 each month, plus a possible expatriation allowance of up to €425 each month. These amounts correspond to a gross remuneration and are subject to social charges and income tax in France
Length of Study: 2–3 years
Frequency: Dependent on funds available
Country of Study: France
No. of awards offered: Up to 20
Application Procedure: Applicants must complete an application form, available on www.esrf.eu.
Closing Date: Individual deadlines exist for each position. Please contact the organization
Funding: International Office
Contributor: Public funds from 19 countries, mostly European
No. of awards given last year: 27
No. of applicants last year: 400
Additional Information: Member countries are Belgium, Denmark, Finland, France, Germany, Italy, the Netherlands, Norway, Spain, Sweden, Switzerland and the United Kingdom. New associated members are the Czech Republic, Israel, Portugal and the Republic of Hungary, Poland, Austria and Slovakia.

For further information contact:

Email: recruitment@esrf.fr
Website: www.esrf.fr/jobs

ESRF Thesis Studentships

Subjects: Physics, biology, chemistry, mineralogy and crystallography, computer engineering and accelerators science. The ESRF proposes subjects related to the use of synchrotron radiation or synchrotron or storage ring technology.
Purpose: To enable grant holders pursue a PhD at the ESRF and to enable young scientists acquire knowledge of the use of synchrotron radiation or its generation.
Eligibility: Preference is given to member-country nationals, but other nationals may be accepted for the PhD's positions.
Level of Study: Doctorate
Value: €2,280 per month. These amounts correspond to a gross remuneration and are subject to social charges and income tax in France
Length of Study: 2–3 years
Frequency: Dependent on funds available
Study Establishment: Universities
Country of Study: France
No. of awards offered: Up to 10
Application Procedure: Applicants must complete an application form, available on www.esrf.eu.
Closing Date: There is an individual deadline for each position
Funding: International Office
Contributor: Public funds from 19 countries, mainly European
No. of awards given last year: 20
No. of applicants last year: 300
Additional Information: Member countries are Belgium, Denmark, Finland, France, Germany, Italy, the Netherlands, Norway, Spain, Sweden, Switzerland and the United Kingdom. Newly associated members are the Czech Republic, Israel, Portugal and the Republic of Hungary, Poland, Austria and Slovakia.

For further information contact:

Email: recruitment@esrf.fr
Website: www.esrf.fr/jobs

EUROPEAN UNIVERSITY INSTITUTE (EUI)

Via dei Roccettini 9, I-50014 San Domenico di Fiesole, Italy
Tel: (39) 55 4685 322
Fax: (39) 55 468 5444
Email: ken.hulley@eui.eu
Website: www.eui.eu
Contact: Mr Kenneth Hulley, Administrator

The European University Institute's (EUI) main aim is to make a contribution to the intellectual life of Europe. Created by the European Union member states, it is a postgraduate research institution, pursuing interdisciplinary research programmes on the main issues confronting European society and the construction of Europe.

EUI Postgraduate Scholarships

Subjects: History and civilization, economics, law or political and social sciences.
Purpose: To provide the opportunity for study leading to the doctorate of the Institute.
Eligibility: Open to nationals of European Union member states. Candidates must possess a good Honours Degree or its equivalent and have a good working knowledge of English. Nationals of countries other than the European Union may be admitted to the Institute subject to scholarship agreements being in place.
Level of Study: Doctorate
Type: Scholarship
Value: Varies, €1,150–1,500
Length of Study: 1 year, renewable for up to an additional 3 years
Frequency: Annual
Study Establishment: EUI
Country of Study: Italy
No. of awards offered: Approx. 150

Application Procedure: Applications must be submitted online at www.eui.eu
Closing Date: January 31st
Funding: Government
Contributor: Member states of the European Union
No. of awards given last year: 150
No. of applicants last year: 1,600

Fernand Braudel Senior Fellowships

Subjects: Economics, history and civilization, law, political and social sciences.
Purpose: Research.
Eligibility: Established academics with international reputation.
Level of Study: Postdoctorate
Type: Fellowship
Value: €3,000 per month
Length of Study: 3–10 months
Frequency: Annual
Study Establishment: The EUI
Country of Study: Italy
No. of awards offered: 15
Application Procedure: Applicants must complete an online application form available via the internet at www.eui.eu
Closing Date: March 30th/September 30th
Contributor: The EU Commission/Members states of the European Union
No. of applicants last year: 100

Jean Monnet Fellowships

Subjects: Economics, history and civilization, law, political and social sciences.
Purpose: To encourage postdoctoral research.
Eligibility: Open to candidates with a doctoral degree at an early stage of their academic career.
Level of Study: Postdoctorate
Type: Fellowship
Value: €1,250–2,000 per month depending on whether the applicant is on a paid sabbatical or not
Length of Study: 1 or 2 years
Frequency: Annual
Study Establishment: The EUI
Country of Study: Italy
No. of awards offered: 25
Application Procedure: Applicants must complete an online application form available via the Internet at www.eui.eu
Closing Date: October 25th
Contributor: The EU commission Member states of the European Union
No. of awards given last year: 25
No. of applicants last year: 360

Max Weber Fellowships

Subjects: Economics, history and civilization, law, political and social sciences.
Purpose: To encourage postdoctoral research.
Eligibility: Open to candidates with a doctoral degree at an early stage of their academic career.
Level of Study: Postdoctorate
Type: Fellowship
Value: €1,250–2,000 per month depending on whether the applicant is on a paid sabbatical or not
Length of Study: 1 or 2 years
Frequency: Annual
Study Establishment: The EUI
Country of Study: Italy
No. of awards offered: 40
Application Procedure: Applicants must complete an online application form available via the internet at www.eui.eu
Closing Date: October 25th
Contributor: The EU Commission/Members State of European Union
No. of awards given last year: 40
No. of applicants last year: 1,040

EVANGELICAL LUTHERAN CHURCH IN AMERICA (ELCA)

Division for Ministry, 8765 West Higgins Road, Chicago, IL, 60631-4195, United States of America
Tel: (1) 773 380 2700
Fax: (1) 773 380 1465
Email: pwilder@elca.org
Website: www.elca.org
Contact: Mr Pat Wilder, Executive Secretary

ELCA Educational Grant Program

Subjects: Theological studies.
Eligibility: Open to members of the Evangelical Lutheran Church in America who are enrolled in an accredited graduate institution for study in a PhD, EdD or ThD programme in a theological area appropriate to seminary teaching. Priority is given to women and minority students.
Level of Study: Doctorate
Type: Grant
Value: Grants up to $4,000 per individual, per year are awarded
Length of Study: Grants are awarded for a maximum of 4 years with a 5th-year award for the dissertation
Frequency: Annual
Country of Study: United States of America
No. of awards offered: 40–65
Application Procedure: Applications are available online at www.elca.org/dm/te/grants.html in January. Two recommendations are required for each applicant.
Closing Date: April 15th
Funding: Private
No. of awards given last year: 65
No. of applicants last year: 72

THE EXPLORERS CLUB

Explorers Club, 46 East 70th Street, New York, NY 10021, United States of America
Tel: (1) 212 628 8383
Fax: (1) 212 288 4449
Email: asstmgr@explorers.org
Website: www.explorers.org

The Explores Club funds projects with scientific purpose to broaden our knowledge of the universe through remote travel and exploration.

Explorers Club Exploration Fund

Purpose: To support scientific expeditions.
Level of Study: Graduate, Postdoctorate
Type: Grant
Value: US$500–1,500
Frequency: Annual
No. of awards offered: Varies
Application Procedure: Download application form at www.explorers.org/index.php/expeditions/funding/expedition-grants.
Closing Date: November 15th
Funding: Trusts
No. of awards given last year: 56267
No. of applicants last year: 220

For further information contact:

Yugoslavia
Contact: Annie Lee, Member Services Director

Youth Activity Fund

Purpose: To support scientific expeditions.
Eligibility: Full time high school or undergraduate students.
Level of Study: High school and undergraduate
Type: Grant
Value: US$500–1,500
Frequency: Annual
No. of awards offered: Varies
Application Procedure: Download application form at: www.explorers.org/index.php/expeditions/funding/expedition_grants.
Closing Date: November 15th

Funding: Trusts
No. of awards given last year: 23
No. of applicants last year: 28

For further information contact:

Contact: Annie Lee, Member Services Director

F BUSONI FOUNDATION

Conservatorio Statale di Musica 'C Monteverdi'-Piazza Domenicani, 25-PO Box 368, I-39100 Bolzano, Italy
Tel: (39) 047 197 6568
Fax: (39) 047 132 6127
Email: info@concorsobusoni.it
Website: www.concorsobusoni.it
Contact: Ms Silvia Torresin, Secretary

The Busoni International Piano Competition was first held in 1949 to commemorate the 25th anniversary of the death of composer Ferruccio Busoni. The aim of the competition is to create a forum for Busoni's music as well as for promising young pianists.

Foundation Busoni International Piano Competition

Subjects: Piano performance.
Purpose: To award excellence in piano performance.
Eligibility: Open to pianists of any nationality between 16 and 30 years of age.
Level of Study: Unrestricted
Type: Prize
Value: The 1st prize is €22,000 plus 60 important concert contracts, the 2nd prize is €10,000, the 3rd prize is €5,000, the 4th prize is €4,000, the 5th prize is €3,000 and the 6th prize is €2,500. There are also other special prizes
Frequency: Every 2 years
Country of Study: Italy
No. of awards offered: 10
Application Procedure: Applicants must complete and submit an application form with a birth certificate, reports or certificates of study, a brief curriculum vitae and documentation of any artistic activity. Three recent photographs, the entrance fee and written evidence of any prizes and international competitions should also be included.
Closing Date: May 31st
Funding: Commercial, government, private
Contributor: The Municipality of Bolzano
No. of awards given last year: 10
No. of applicants last year: 150
Additional Information: The competition lasts for 2 years, with the pre-selection phase taking place in the 1st year.

FAIRBANK CENTER FOR CHINESE STUDIES

Harvard University, CGIS South Building, 1730 Cambridge Street Cambridge, MA, 02138, United States of America
Tel: (1) 617 495 4046/8120
Fax: (1) 617 496 2420
Email: lydiac@fas.harvard.edu
Website: www.fas.harvard.edu/˜fairbank/
Contact: Lydia Chen, Associate Director

The Fairbank Center for Chinese studies was founded in 1955 by Professor John King Fairbank. It supports research by offering public lectures and conferences, supporting visiting scholars, awarding a limited number of Postdoctoral fellowships, and maintaining a specialized library.

An Wang Postdoctoral Fellowship

Subjects: Chinese studies.
Purpose: To support outstanding scholarship in Chinese studies by recent PhDs.
Eligibility: Open to candidates who have obtained their PhD within the past five years.
Level of Study: Postdoctorate
Type: Fellowships
Value: US$50,000
Length of Study: 1 year
Frequency: Annual
Country of Study: China
No. of awards offered: 2
Application Procedure: Applicants may download the application form from the Fairbank Center website. The completed application form along with required supporting documents must be mailed.
Closing Date: January 6th
Additional Information: Those who have received their PhD from the Harvard University are normally not considered for this fellowship.

FANCONI ANEMIA RESEARCH FUND, INC.

1801 Willamette Street, Suite 200, Eugene, OR, 97401, United States of America
Tel: (1) 541 687 4658
Fax: (1) 541 687 0548
Email: info@fanconi.org
Website: www.fanconi.org
Contact: Ms Mary Ellen Eiler, Executive Director

To support research into effective treatments and a cure for Fanconi anaemia.

Fanconi Anemia Research Award

Subjects: Fanconi anaemia.
Purpose: To support research into effective treatments and a cure for Fanconi anaemia.
Eligibility: There are no restrictions on eligibility in terms of nationality, residency, age, gender, sexual orientation, race, religion or politics.
Level of Study: Doctorate, Postdoctorate
Type: Award
Value: Varies
Length of Study: 1–2 years
Country of Study: Any country
No. of awards offered: Unlimited
Application Procedure: Applicants must email to obtain information and application forms.
Closing Date: Ongoing
Funding: Foundation
No. of awards given last year: 9
No. of applicants last year: 17
Additional Information: The Internal Revenue Service has confirmed that the Fund is not a private foundation for the purposes of tax-exempt donations but a public charitable organization under 501(c) 3 of the Internal Revenue Code.

For further information contact:

Email: info@fanconi.org

FEDERATION OF EUROPEAN MICROBIOLOGICAL SOCIETIES (FEMS)

Keverling Buismanweg 4, 2628 CL Delft, Netherlands
Tel: (31) 15 269 3920
Fax: (31) 15 269 3921
Email: fems@fems-microbiology.org
Website: www.fems-microbiology.org
Contact: Dr D Van Rossum, Executive Officer

The Federation of European Microbiological Societies (FEMS) is devoted to the promotion of microbiology in Europe. FEMS advances research and education in the science of microbiology within Europe, by encouraging joint activities and facilitating communication among microbiologists, supporting meetings and laboratory courses and publishing books and journals.

FEMS Fellowship

Subjects: Microbiology.
Purpose: To foster transnational research in microbiology and to enable young scientists to pursue a short-term research project in another European country.

Eligibility: The award is restricted to members of FEMS member societies.
Level of Study: Postdoctorate, Doctorate, Graduate, Postgraduate, Predoctorate, Professional development, Research
Type: Fellowship
Value: A maximum of €4,000
Length of Study: A maximum of 3 months
No. of awards offered: Approx. 50
Application Procedure: Applicants must complete and submit an application form to a society that is a member of FEMS. The delegate of the member society will handle the application and submit it to the Federation for funding. FEMS will then make a decision on the application. Addresses of the Federation's delegates are published on the website.
Closing Date: December 1st and June 15th
Funding: Foundation
No. of awards given last year: 35
No. of applicants last year: 37

FFWG

43 Fern Road Storrington, Pulborough, West Sussex, RH20 4LW, United Kingdom
Tel: (44) 1903 7467 23
Email: valconsidine@btinternet.com
Website: http://ffwg.org.uk/
Contact: Mrs J V Considine, Co. Secretary

FfWG is the registered Trading Name for the BFWG Charitable Foundation. FfWG seeks to promote the advancement of education and the promotion of higher education of women graduates by offering grants to help women graduates with their living costs while registered for study or research at institutions in Great Britain.

FfWG Foundation Grants and Emergency Grants

Subjects: All subjects.
Purpose: To financially assist female graduates registered for study or research at an approved Institute of Higher Education within Great Britain.
Eligibility: Main Foundation Grants: Open to female graduates who are in their final year or writing-up year. Foundation Grants are awarded to female students in their final year of a PhD. There is no restriction on nationality or age.Emergency Grants: Open to female graduates engaged in study or research at Institutes of Higher Education, who face an unexpected financial crisis.
Level of Study: Doctorate, Postdoctorate, Postgraduate
Type: Grant
Value: Foundation Grants are up to £4,000 and Emergency Grants are up to £1,500. (these values are being reviewed)
Length of Study: Courses that exceed 1 full year in length
Frequency: Annual, Annually for main Foundation Grants and twice a year for Emergency Grants.
Study Establishment: Approved Institutes of Higher Education
Country of Study: Great Britain
No. of awards offered: Approx. 50 Foundation Grants and approx. 50 Emergency Grants
Application Procedure: For Foundation grants, applicants must complete an application form and submit it with two references and a brief summary of the thesis, if applicable. Requests for application forms must be made by e-mail. For Emergency grants requests for application forms must be made by e-mail. Applications for foundation grants must pay admin fee of £12 by internet banking.
Closing Date: Foundation Grants - April 4th; Emergency Grants - February 9th, May 31st
Funding: Private
Contributor: Investment income
No. of awards given last year: 44 Foundation Grants and 48 Emergency Grants
No. of applicants last year: Several hundreds

For further information contact:

Email: jean.c@blueyonder.co.uk

Theodora Bosanquet Bursary

Subjects: English Literature/History.
Purpose: This bursary is offered to women postgraduate students whose research in History or English Literature requires a short residence in London in the summer.
Eligibility: Open to female postgraduate students of any age from the United Kingdom and overseas.
Level of Study: Graduate
Type: Bursary
Value: Will be decided by the BFWG Charitable Foundation trading as FFWG
Length of Study: Up to 4 weeks
Frequency: Annual
Study Establishment: Any approved Institute of Higher Education in Great Britain
Country of Study: Great Britain
Application Procedure: Applicants must request an application form by e-mail or download it from the website. Forms should then be returned either by email or post enclosing a large stamped self-addressed envelope or international reply coupons. The envelope should be marked TBB.
Closing Date: October 31st
Funding: Private
Contributor: Investment income
No. of awards given last year: 2

For further information contact:

BFUG Charitable Foundation, Larkfield, Aylesford, Kent, 13 Brookfield Avenue, ME20-6RU
Tel: 017 3232 1139
Contact: The Grants Administrator

FIGHT FOR SIGHT

5th Floor, 9-13 Fenchurch Buildings, Fenchurch street, London, EC3M 5HR, United Kingdom
Tel: (44) 0207 624 3900
Fax: (44) 0207 488 3041
Email: info@fightforsight.org.uk
Website: http://fightforsight.org.uk
Contact: Dolores M Conroy, Director of Research

Fight for Sight is dedicated to funding pioneering research to prevent sight loss and eye disease at leading UK universities and hospitals. Current research projects include age-related macular degeneration, glaucoma, cataract, diabetic retinopathy, corneal disease, and childhood blindness.

Fight for Sight Awards

Subjects: Ophthalmology.
Purpose: For the prevention of sight loss and treatment of eye disease.
Eligibility: Open to suitably qualified individuals working in a higher education institute in the UK. But the charity may fund UK-based teams undertaking research overseas.
Level of Study: Doctorate, Postdoctorate, Postgraduate, Research
Type: Award
Value: Determined each year.
Length of Study: 2, 3 or 5 years; small grant awards for up to 12 months and all other funding opportunities up to 3 years
Frequency: Annual
Country of Study: United Kingdom
No. of awards offered: 26 awards made, 6 of these were small grant awards of UK£15,000, and 3 awards for new lectures for £20,000 each
Application Procedure: See website www.fightforsight.org.uk for details.
Closing Date: Various
Funding: Trusts, commercial, individuals, private
No. of awards given last year: 16
No. of applicants last year: 121 abstract applicants, 54 shortlisted to submit full applications
Additional Information: The British Eye Research Foundation merged with Fight for Sight in 2005. Studies may be undertaken abroad but the award holder must be attached to a United Kingdom

university or higher education institution. For up to date information on all funding opportunities can be found at http://www.fightforsight.org.uk/apply-for-funding.
Type of award - Project grants, PhD studentships, clinical fellowships and early career investigator awards.

FINE ARTS WORK CENTER IN PROVINCETOWN, INC.

24 Pearl Street, Provincetown, MA 02657, United States of America
Tel: (1) 508 487 9960
Fax: (1) 508 487 8873
Email: general@fawc.org
Website: www.fawc.org
Contact: Ms Margaret Murphy, Executive Director

Established in 1968 in historic Provincetown, Masachusetts, the Fine Arts Work Center offers seven-month fellowships to emerging visual artists and creative writers. Fellows are provided with apartments, studios, and a monthly stipend, and are asked only that they focus solely on their creative work while in residence.

Fine Arts Work Center in Provincetown Fellowships

Subjects: Visual arts and creative writing (fiction and poetry).
Purpose: To provide selected emerging visual artists and creative writers a significant amount of time to focus solely on their creative work in a supportive community of their peers.
Eligibility: The fellowship programme is open to all emerging visual artists and creative writers. Fellows are chosen based on the quality of creative work submitted.
Level of Study: Unrestricted
Type: Fellowship
Value: US$750 per month, plus housing and studio space
Length of Study: 7 months
Frequency: Annual
Study Establishment: Provincetown, MA
Country of Study: United States of America
No. of awards offered: 20 fellowships annually (10 for visual arts and 10 for writing)
Application Procedure: Application forms may be downloaded from the website, www.fawc.org, or may be obtained by written request (include SASE).
Closing Date: February 1st for visual artists and December 1st for writers
Funding: Corporation, foundation, government, individuals, private
No. of awards given last year: 20
No. of applicants last year: 1,100
Additional Information: The Center is a working community, not a school.

FLORIDA FEDERATION OF GARDEN CLUBS (FFGC)

1400 South Denning Drive, Winter Park, FL 32789-5662, United States of America
Tel: (1) 647 7016
Fax: (1) 647 5479
Email: office_manager@ffgc.org
Website: www.ffgc.org

The FFGC is the first state garden club on the Internet, which features FFGC activities, scholarships, youth programmes, floral design courses, arrangements, educational opportunities, tours, shows and special events of the various affiliated garden clubs in the state of Florida.

FFGC Scholarship in Ecology

Subjects: Ecology.
Purpose: To provide financial assistance to students in field of ecology.
Eligibility: Open to applicants who are residents of Florida and who demonstrate financial needs.
Level of Study: Postgraduate
Type: Scholarship
Value: US$1,500
Frequency: Annual
Country of Study: United States of America
No. of awards offered: 13
Application Procedure: Check website for details
Closing Date: May 1st
Contributor: Florida Federation of Garden Clubs
No. of awards given last year: 9

FFGC Scholarship in Environmental Issues

Subjects: Environmental studies.
Purpose: To support students in the field of environmental studies.
Eligibility: Open to students who are residents of Florida.
Level of Study: Postgraduate
Type: Scholarship
Value: Up to $3,500.
Frequency: Annual
Country of Study: United States of America
No. of awards offered: 13
Application Procedure: A completed application form must be submitted.
Closing Date: May 1st
Contributor: Florida Federation of Garden Clubs

FONDATION DES ETATS-UNIS

15 boulevard Jourdan, Paris, 75014, France
Tel: (33) 1 53 80 68 80
Fax: (33) 1 53 80 68 99
Email: administration@feusa.org
Website: www.feusa.org
Contact: Mr Terence Murphy, Director

For the past 75 years the Fondation des Etats-Unis has been welcoming American and International Students during their studies in Paris.

Harriet Hale Woolley Scholarships

Subjects: Visual fine arts and music.
Purpose: To support the study of visual fine arts and music in Paris.
Eligibility: Open to citizens of the United States of America, who are 21–35 years of age and have graduated with high academic standing from a US college, university or professional school of recognized standing. Preference is given to mature students who have already completed graduate study. Applicants should provide evidence of artistic or musical accomplishment. Applicants should have a good working knowledge of French, sufficient to enable the student to benefit from his or her study in France. Grants are for those doing painting, printmaking or sculpture and for instrumentalists, not for research in art history, musicology or composition, nor for students of dance or theatre. Successful candidates propose a unique and detailed project related to their study, which requires a 1-year residency in Paris.
Level of Study: Doctorate, Graduate, Postgraduate, Predoctorate
Type: Scholarship
Value: A stipend of €8,500.
Length of Study: 1 academic year
Frequency: Annual
Country of Study: France
No. of awards offered: Up to 4
Application Procedure: For a complete description of the scholarship including a list of general requirements, an application checklist and an application form, please visit: www.feusa.org, chapter 'Nos activités culturelles', sub-chapter 'The Harriet Hale Woolley Scholarship'.
Closing Date: October 1st to June 30th
Funding: Private
No. of awards given last year: 4
No. of applicants last year: 25

For further information contact:

Tel: 1 53 80 68 87
Email: culture@feusa.org
Contact: Miss Elizabeth Askren-Brie, Attachée culturelle

FONDATION FYSSEN

194 Rue de Rivoli, F-75001 Paris, France
Tel: (33) 1 42 97 53 16
Fax: (33) 1 42 60 17 95
Email: secretariat@fondation-fyssen.org
Website: www.fondation-fyssen.org
Contact: Mrs Nadia Ferchal, Director

The aim of the Fyssen Foundation is to encourage all forms of scientific enquiry into cognitive mechanisms, including thought and reasoning, that underlie animal and human behaviour, their biological and cultural bases and phylogenetic and ontogenetic development.

Fondation Fyssen Postdoctoral Study Grants

Subjects: Disciplines relevant to the aims of the Foundation such as ethology, palaeontology, archaeology, anthropology, psychology, ethnology, neurobiology.
Purpose: To assist French or foreign postdoctoral researchers, under 35 years of age, to work on topics in keeping with the Foundation's goals.
Eligibility: Open to a first post-doctorate with a PhD of less than two years on September 1st of the year of application. Open to French or foreign research scientists holder of a French doctorate (PhD) and attached to a laboratory in France who wish to work in laboratories abroad (except country of origin or joint supervision) and foreign or French research scientists holder of a foreign doctorate (PhD) and attached to a foreign laboratory who wish to work in French laboratories. Applicants should be under 35 years of age.
Level of Study: Postdoctorate
Type: Study grant
Value: €25,000
Length of Study: 1 or 2 years
Frequency: Annual
Application Procedure: Applicants must complete an application form, available from the Secretariat of the Foundation or from the website and send the same within the due date along with 15 copies.
Closing Date: End of March
Funding: Private
No. of awards given last year: Around 40

International Prize

Subjects: Neuropsychology
Purpose: To encourage a scientist who has conducted distinguished research in the areas supported by the Foundation.
Eligibility: Applicants are requested to visit the website www.fondation-fyssen.org/International Prize for eligibility information.
Value: €60,000
Frequency: Annual
Application Procedure: Candidates cannot apply directly but should be proposed by recognized scientists. Proposals for candidates should consist of (i) curriculum vitae, (ii) a list of publications, (iii) a summary (4 pages maximum) of the research. The proposal should be submitted in 14 copies to Secrétariat de la Fondation Fyssen.
Closing Date: October 31st
Funding: Private

FOOD AND DRUG LAW INSTITUTE (FDLI)

1155, 15th Street, NW, Suite 800, Washington, DC, 20005, United States of America
Tel: (1) 202 371 1420
Fax: (1) 202 371 0649
Email: comments@fdli.org
Website: www.fdli.org
Contact: Ms Rita M. Fullem, Vice President-Programs Publications

The Food and Drug Law Institute (FDLI) is a non-profit educational association dedicated to advancing public health by providing a neutral forum for a critical examination of the laws, regulations and policies related to drugs, medical devices, other healthcare technologies and food.

H Thomas Austern Memorial Writing Competition–Food and Drug Law Institute(FDLI)

Subjects: Current issues relevant to the food and drug field including relevant case law, legislative history and other authorities, particularly where the United States Food and Drug Administration is involved. Additional topic possibilities are available from the website.
Purpose: To encourage law students interested in the areas of law affecting foods, drugs, devices, cosmetics and biologics.
Eligibility: Entrants must currently be enrolled in a JD programme at any of the United States of America law schools. Anyone currently enrolled in a Juris Doctorate program in any US college or university.
Level of Study: Postgraduate
Value: Two first prizes of US$4,000; two second prizes of US$1,000
Frequency: Annual
Country of Study: United States of America
No. of awards offered: 4
Application Procedure: Applicants must submit a typewritten, double-spaced paper on 8.5 by 11 inch paper or submit a Word document elect ronically. The cover sheet must list the applicant's full name, address and telephone number, law school and year, and the date of submission of the paper. Papers must not exceed 40 pages in length, including footnotes for shorter paper competition. There is a 100 page limit for papers longer than 41 pages for long paper competition.
Closing Date: May 31st
Funding: Private
Contributor: Association funds and Association member dues
No. of awards given last year: 4
No. of applicants last year: Approx. 50
Additional Information: Winning papers will be considered for publication in the *Food and Drug Law Journal*.

FOOD SCIENCE AUSTRALIA

11 Julius Avenue, Riverside Corporate Park, North Ryde, NSW, 2113, Australia
Tel: (61) 2 9490 8333
Fax: (61) 2 9490 8499
Email: fsacontact@csiro.au
Website: www.foodscience.csiro.au

Food Science Australia is Australia's largest and most diversified food research organisation and a joint venture of CSIRO and the Victorian Government. They are committed to turning scientific research into innovative solutions for the food industry in Australia and overseas.

Food Science Australia CSIRO Postgraduate Scholarship Program

Subjects: Science and engineering.
Purpose: To provide opportunities in science and engineering for outstanding graduates who enroll each year at Australian tertiary institutions as full-time postgraduate students for research leading to the award of a PhD.
Eligibility: Open to doctoral students who have Australian Permanent Residency or citizenship and also who gain or expect to gain an Australian Postgraduate Award (APA) or equivalent university award.
Level of Study: Doctorate, Postgraduate
Type: Scholarship
Country of Study: Australia
Application Procedure: Check website for further details.
Closing Date: October/November each year
Funding: Government

FOREIGN AND COMMONWEALTH OFFICE

British Council Information Centre, Bridgewater House, 58 Whitworth Street, Manchester, M1 6BB, United Kingdom
Email: info@fco.gov.uk
Website: www.fco.gov.uk

Chevening Scholarships

Subjects: Any subject (though priority is given to science and innovation, new and renewable energy resources and energy security,

global environmental issues, science policy, sustainable development, human rights and political science).
Purpose: The Chevening Scholarships programme is funded by the Foreign and Commonwealth Office (FCO) in the United Kingdom and administered by the British Council. The programme offers outstanding graduates and young professionals the opportunity to study at UK universities.
Eligibility: Candidates should have a strong undergraduate degree (emphasis is placed on applications with First or Second Class Honours) intend to study a one-year taught Masters or between 3 and 12 months research towards a PhD in the UK.
Level of Study: Postgraduate
Type: Scholarship
Value: A return economy airfare to the UK, all compulsory academic fees, monthly stipend, book allowance and thesis allowance
Length of Study: 3–12 months
Country of Study: United Kingdom
Funding: Government
Additional Information: All applicants will be informed of the outcome of their application by mid-December. At this time short-listed candidates will be invited to be interviewed in January or February 2010. The interview dates and times are organised by the British High Commission.

For further information contact:

Website: www.chevening.com

FOULKES FOUNDATION

37 Ringwood Avenue, London, N2 9NT, England
Tel: (44) 20 8444 2526
Fax: (44) 20 8444 2526
Website: www.foulkes-foundation.org
Contact: M Foulkes, The Registrar

The aim of the Foulkes Foundation Fellowship is to promote medical research by providing financial support for postdoctoral science graduates who need a medical degree before they can undertake medical research, and similarly for medical graduates who need a science PhD degree.

Foulkes Foundation Fellowship

Subjects: All aspects of medical research, especially the areas of molecular biology and biological sciences.
Purpose: To promote research by providing financial support for postdoctoral study in a clinical medicine degree.
Eligibility: Open to recently qualified scientists who have a PhD or equivalent and intend to contribute to medical research.
Level of Study: Postdoctorate
Type: Fellowship
Value: Varies depending on individual need for personal maintenance only. Fellowships do not cover fees
Length of Study: Up to 3 years
Frequency: Annual
Country of Study: United Kingdom
No. of awards offered: Varies
Application Procedure: Application forms may be obtained by post, sending a stamped self-addressed envelope, or by email at registrar@foulkes-foundation.org
Closing Date: March 15th
Funding: Private
No. of awards given last year: 5
No. of applicants last year: 75
Additional Information: Applicants must be undertaking medical training in the United Kingdom.

FOUNDATION FOR ANESTHESIA EDUCATION AND RESEARCH (FAER)

200 First Street South West, WF6-674, Rochester, MN, 55905, United States of America
Tel: (1) 507 266 6866
Fax: (1) 507 284 0291
Email: schrandt.mary@mayo.edu
Website: www.faer.org
Contact: Ms Mary Schrandt, Associate Director

The Foundation for Anesthesia Education and Research (FAER) strives to foster progress in anaesthesiology, critical care, pain and all areas of perioperative medicine. The organization aims to generate new knowledge that advances health and patient care by facilitating the career development of anaesthesiologists dedicated to research and education.

FAER Mentored Research Training Grant (MRTG)

Subjects: Anaesthesiology.
Purpose: To allow the applicant to become an independent investigator.
Eligibility: Applicants must be instructors or assistant professors who are within 10 years of their initial appointment.
Level of Study: Postdoctorate
Type: Grant
Value: US$75,000 in the 1st year and US$100,000 in the 2nd.
Length of Study: 2 years
Frequency: Annual
Application Procedure: Applicants must visit the website www.faer.org.
Closing Date: February 15th

FAER Research Education Grant

Subjects: Anaesthesiology.
Purpose: To improve the quality and productivity of education and research.
Eligibility: Open to anaesthesiology residents or faculty.
Level of Study: Postgraduate
Type: Grant
Value: US$50,000 in the 1st year and US$50,000 in the 2nd year
Length of Study: 2 years
Frequency: Annual
Application Procedure: Applicants must visit the website.
Closing Date: February 15th

FAER Research Fellowship Grant

Subjects: Anaesthesiology.
Purpose: To provide significant training in research techniques and scientific methods.
Eligibility: Open to anaesthesiology residents after CA-1 training.
Level of Study: Postdoctorate
Type: Fellowship
Value: US$75,000
Length of Study: 1 year
Frequency: Annual
Application Procedure: Applicants must visit the website.
Closing Date: February 15th

FOUNDATION FOR DIGESTIVE HEALTH AND NUTRITION

4930 Del Ray Avenue, Bethesda, MD, 20814, United States of America
Tel: (1) 301 222 4002
Fax: (1) 301 222 4010
Email: awards@fdhn.org
Website: www.fdhn.org
Contact: Ms Wykenna S.C.Vailor, Research Awards Manager

The Foundation for Digestive Health and Nutrition is the foundation of the American Gastroenterological Association (AGA), the leading professional society representing gastroenterological and heptatologists worldwide. It is separately incorporated and governed by a distinguished board of AGA physicians and members of the lay public. The Foundation raises funds for research and public education in the prevention, diagnosis, treatment and cure of digestive diseases. Along with the AGA, it conducts public education initiatives related to digestive diseases. The Foundation also administers the disbursement of grants on the behalf of the AGA and other funders.

AGA Fellowship to Faculty Transition Awards

Subjects: Medical science, specifically gastroenterology and hepatology.
Purpose: To prepare physicians for independent research careers in digestive diseases.

Eligibility: Applicants must be MDs or MD/PhDs currently in a gastroenterology-related fellowship, at a North American institution and committed to academic careers. They should have completed at least two years of research training at the start of this award. Women and minority investigators are strongly encouraged to apply. Applicants must be AGA Trainee Members or be sponsored by an AGA Member at the time of application.
Level of Study: Postgraduate
Type: Award
Value: US$40,000 per year
Length of Study: 2 years
Frequency: Annual
Country of Study: The United States of America, Canada or Mexico
No. of awards offered: 2
Application Procedure: Applications can be downloaded from the AGA Foundation website. The completed application, letters of support or commitment and other documents must be submitted as one PDF document, titled by the applicant's last name and first initial only. Hard copies are not permitted. For further information visit the AGA Foundation website.
Closing Date: August 31st
Funding: Private
Contributor: The AGA
No. of awards given last year: 4
No. of applicants last year: 8
Additional Information: The award provides salary support for additional full-time research training in basic science to acquire modern laboratory skills. The additional two years of research training provided by the award would broaden the scope of investigative tools available to the recipient, generally in basic disciplines such as cell or molecular biology, or immunology. A complete financial statement and scientific progress report are required annually and upon completion of the programme. All publications arising from work funded by this programme must acknowledge support of the award.

For further information contact:

United States of America
Tel: (1) 301 222 4012
Email: awards@fdhn.org
Website: www.fdhn.org

AGA June and Donald O Castell, MD, Esophageal Clinical Research Award

Subjects: Oesophageal diseases.
Purpose: To support investigators who have demonstrated a high potential to develop an independent, productive research career.
Eligibility: Applicants must have an MD or PhD equivalent to hold a full-time faculty position at a United States of America or Canadian university or professional institute and be members of the AGA. The recipient must be at or below the level of assistant professor, and his/her initial appointment to the faculty position must have been within 7 years of the time of application. This award is not intended for Fellows, but for juniors who have demonstrated unusual promise, have some record of accomplishment in research and have established independent research programmes at the time of the award. Candidates must devote at least 50 per cent of their efforts to research related to oesophageal function or diseases. Applicants may not simultaneously apply for an AGA Research Scholar Award, AGA Fiterman Foundation Basic Research Award or AGA/Elsevier Research Initiative Award.
Level of Study: Graduate
Value: $25,000
Length of Study: 1 year
Frequency: Annual
Country of Study: United States of America
No. of awards offered: 1
Application Procedure: Electronic applications only.
Funding: Private

AGA R Robert and Sally D Funderburg Research Scholar Award in Gastric Biology Related to Cancer

Subjects: Gastric mucosal cell biology, regeneration and regulation of cell growth, inflammation, genetics of gastric carcinoma, epidemiology of gastric cancer, etiology of gastric epithelial malignancies, or clinical research in the diagnosis of gastric carcinoma.
Purpose: To support active, established investigators in the field of gastric biology who enhance the fundamental understanding of gastric cancer pathobiology in order to ultimately develop a cure for the disease.
Eligibility: Applicants must hold faculty positions at accredited North American institutions and must have established themselves as independent investigators in the field of gastric biology. Women and minority investigators are strongly encouraged to apply. Applicants must be members of the AGA at the time of application submission.
Level of Study: Postgraduate
Type: Award
Value: US$100,000
Length of Study: 2 years
Frequency: Annual
Country of Study: The United States of America, Canada or Mexico
No. of awards offered: 1
Application Procedure: Applications can be downloaded from the AGA Foundation website. The completed application, letters of support or commitment and other documents must be submitted as one PDF document, titled by the applicant's last name and first initial only. Hard copies are not permitted. For further information visit the AGA Foundation website.
Closing Date: August 31st
Contributor: The AGA, the late R Robert and the late Sally D Funderburg
No. of awards given last year: 1
No. of applicants last year: 4

For further information contact:

United States of America
Tel: (1) 301 222 4012
Email: awards@fdhn.org
Website: www.fdhn.org

AGA Research Scholar Awards

Subjects: Gastroenterology and hepatology.
Purpose: To enable young investigators to develop independent and productive research careers in digestive diseases by ensuring that a major proportion of their time is protected for research.
Eligibility: Candidates must hold an MD, PhD, or equivalent degree and a full-time faculty positions at North American universities or professional institutes at the time of commencement of the award. They must be members of the AGA at the time of application submission. The award is for young faculty, who have demonstrated unusual promise and have some record of accomplishment in research. Candidates must devote at least 70 per cent of their efforts to gastrointestinal tract or liver-related research. Women, minorities and physician/scientist investigators are strongly encouraged to apply.
Level of Study: Graduate
Type: Research grant
Value: US$90,000 per year.
Length of Study: 2 years
Frequency: Annual
Country of Study: United States of America
No. of awards offered: 5
Application Procedure: Applications can be downloaded from the AGA Foundation website. The completed application, letters of support or commitment and other documents must be submitted as one PDF document, titled by the applicant's last name and first initial only. Hard copies are not permitted. For further information visit the AGA Foundation website.
Closing Date: September 7th
Funding: Private
No. of awards given last year: 4
No. of applicants last year: 40
Additional Information: A complete financial statement and scientific progress report are required upon completion of the programme. All publications arising from work funded by this programme must acknowledge the support of the award. Awardees must submit their work for presentation at Digestive Disease Week during the last year of the award.

For further information contact:

United States of America
Tel: (1) 301 222 4012
Email: awards@fdhn.org
Website: www.fdhn.org

AGA Student Research Fellowship Awards

Subjects: Research related to the gastrointestinal tract, liver or pancreas.
Purpose: To stimulate interest in research careers in digestive diseases by providing salary support for research projects.
Eligibility: Applicants must be students at accredited North American institutions, may not hold similar salary support awards from other agencies. Women and minority students are strongly encouraged to apply.
Level of Study: Graduate, Postgraduate, Professional development
Type: Award
Value: US$2,500 per year
Length of Study: 10 weeks
Frequency: Annual
Country of Study: The United States of America, Canada or Mexico
No. of awards offered: 10
Application Procedure: Applications can be downloaded from the AGA Foundation website. The completed application, letters of support or commitment and other documents must be submitted as one PDF document, titled by the applicant's last name and first initial only. Hard copies are not permitted. For further information visit the AGA Foundation website.
Closing Date: March 23th
Funding: Private
Contributor: The AGA
No. of awards given last year: 12
No. of applicants last year: 48

For further information contact:

Tel: 301 222 4012
Email: awards@fdhn.org
Website: www.fdhn.org

Elsevier Pilot Grant

Subjects: Medical science, specifically gastroenterology and hepatology.
Purpose: To provide non-salary funds for new investigators to help them establish their research careers or to support pilot projects that represent new research directions for established investigators. The intent is to stimulate research in gastroenterology- or hepatology-related areas by permitting investigators to obtain new data that can ultimately provide the basis for subsequent grant applications of more substantial funding and duration.
Eligibility: Applicants must possess an MD or PhD degree or equivalent and must hold faculty positions at accredited North American institutions. In addition, they must be AGA members at the time of application submission. Women and minorities are strongly encouraged to apply.
Level of Study: Postdoctorate, Postgraduate, Predoctorate, Research
Type: Grant
Value: US$25,000
Length of Study: 1 year
Frequency: Annual
Country of Study: The United States of America, Canada or Mexico
No. of awards offered: 1
Application Procedure: Applications can be downloaded from the AGA Foundation website. The completed application, letters of support or commitment and other documents must be submitted as one PDF document, titled by the applicant's last name and first initial only. Hard copies are not permitted. For further information visit the AGA Foundation website.
Closing Date: January 13th
Funding: Private
Contributor: The AGA
No. of awards given last year: 1
No. of applicants last year: 22

For further information contact:

United States of America
Tel: (1) 301 222 4012
Email: awards@fdhn.org
Website: www.fdhn.org

FOUNDATION FOR JEWISH CULTURE

PO Box 489, New York, NY, 10011, United States of America
Tel: (1) 212 629 0500
Fax: (1) 212 629 0508
Email: grants@jewishculture.org
Website: www.jewishculture.org

The Foundation for Jewish Culture (formerly the National Foundation for Jewish Culture) is the leading advocate for Jewish cultural creativity and preservation in America. Since 1960, it has nurtured new generations of writers, filmmakers, artists, composers, choreographers and scholars. The Foundation invests in creative individuals in order to nurture a vibrant and enduring Jewish identity, culture and community. Its goals are achieved through the provision of grants, recognition awards, networking opportunities and professional development services. They collaborate with cultural institutions, Jewish organizations, consortia and funders to support the work of these artists and scholars. The Foundation also educates and builds audiences to provide meaningful Jewish cultural experiences to the American public, and advocates for the importance of Jewish culture as a core component of Jewish life.

Maurice and Marilyn Cohen Fund for Doctoral Dissertation Fellowships in Jewish Studies

Subjects: Jewish studies.
Purpose: To encourage study and research in the various disciplines related to Judaica and Jewish life.
Eligibility: Open to citizens or permanent residents of the United States who have completed all requirements for the PhD degree except the dissertation and are proficient in a Jewish language.
Level of Study: Doctorate
Type: Fellowships
Value: US$16,000 one academic year.
Length of Study: 1 year
Frequency: Annual
Country of Study: United States of America
Application Procedure: Please see the website www.jewishculture.org for current application and guidelines.
Closing Date: December 14th
Funding: Foundation, private

FOUNDATION FOR PHYSICAL THERAPY

1111 North Fairfax Street, Alexandria, VA, 22314, United States of America
Tel: (1) 800 875 1378 ext 8505
Fax: (1) 703 706 8587
Email: foundation@apta.org
Website: www.foundation4pt.org
Contact: Abegail Matienzo, Communications Assistant

The Foundation for Physical Therapy is an independent, non-profit organisation founded to support the physical therapy profession's research needs in the areas of scientific research, clinical research and health services research.

Florence P Kendall Doctoral Scholarships

Subjects: Physical therapy, rehabilitation medicine, neuroscience, sports medicine, paediatrics, medical sciences and social or preventative medicine.
Purpose: To assist physical therapists or physical therapist assistants with outstanding potential for doctoral studies in the 1st year of study towards a doctorate.
Eligibility: Open to candidates who possess a license to practice physical therapy, or as a physical therapist assistant in the United States of America and fulfill specific requirements with regard to research experience.
Level of Study: Doctorate
Type: Scholarship
Value: $5,000
Length of Study: 1 year
Frequency: Annual
Country of Study: United States of America
No. of awards offered: Varies

Application Procedure: Applicants must apply online. Guidelines, instructions and access to the online system are available in the website.
Closing Date: Not currently open to applications.
Funding: Private
No. of awards given last year: 4
No. of applicants last year: 15

New Investigator Fellowships Training Initiative (NIFTI)

Subjects: Physical therapy, rehabilitation medicine, neuroscience, sports medicine, paediatrics, medical sciences and social or preventative medicine and health services research.
Purpose: To fund doctorally-prepared physical therapists as developing researchers and improve their competitiveness in securing external funding for future research.
Eligibility: Open to candidates who possess a license to practice physical therapy, or as a physical therapist assistant in the United States of America, have received the required postprofessional doctoral degree no earlier than 5 years prior to the year of application or, for those already holding a postprofessional doctorates, a professional education degree in physical therapy no earlier than 5 years prior to the year of application. Candidates must also have completed a research experience as part of their postprofessional doctoral education.
Level of Study: Postdoctorate
Type: Fellowship
Value: US$78,000 (NIFTI); US$73,000 - Health Services Research (NIFTI-HSR) plus US$5,000 stipend
Length of Study: 2 years
Frequency: Annual
Country of Study: United States of America
No. of awards offered: Varies
Application Procedure: Applicants must apply online. Guidelines, instructions and access to online system are available in the website.
Closing Date: January 18th, Noon, ET
Funding: Private
No. of awards given last year: 2
No. of applicants last year: 7

Promotion of Doctoral Studies (PODS) Scholarships

Subjects: Physical therapy, rehabilitation medicine, neuroscience, sports medicine, paediatrics, medical sciences and social or preventative medicine.
Purpose: To fund doctoral students who, having completed 1 full year of coursework, wish to continue their coursework or enter the dissertation phase.
Eligibility: Open to candidates who possess a license to practice physical therapy, or as a physical therapist assistant in the United States of America and who are enrolled as students in a regionally accredited postprofessional, doctoral programme. The content of this programme should have a demonstrated relationship to physical therapy. Applicants must also be able to demonstrate continuous progress towards the completion of their postprofessional doctoral programme in a timely fashion and with a commitment to further the physical therapy profession through research and teaching within the United States of America and its territories.
Level of Study: Doctorate
Type: Scholarship
Value: Two levels, at USD 7,500 or USD 15,000
Length of Study: 1 year
Frequency: Annual
Country of Study: United States of America
No. of awards offered: Varies
Application Procedure: Applicants must apply online. Guidelines, instructions and access to the online system are available in the website.
Closing Date: January 18, Noon, ET
Funding: Private
No. of awards given last year: 21
No. of applicants last year: 46

Research Grants (FPT)

Subjects: Physical therapy, rehabilitation medicine, neuroscience, sports medicine, paediatrics, medical sciences and social or preventative medicine.
Purpose: The purpose of the Foundation's Research Grant programme is to fund research studies in specific areas initiated by emerging investigators.
Eligibility: Open to citizens or permanent residents of the United States of America or its territories who possess a license to practice physical therapy or as a physical therapist assistant. Projects to be completed in fulfilment of requirements for an academic degree are not eligible to be funded by a Foundation Research Grant. A doctoral student in the latter stage of the dissertation phase of his/her programme may submit an application, but must provide evidence of completion of the degree by October 15th. In addition, the proposed study must differ substantially from any thesis research being conducted by graduate assistant(s) to be supported by this Research Grant.
Level of Study: Research
Type: Grant
Value: US$40,000 (salary, fringe benefits, and direct expenses only).
Length of Study: The grant period of performance may be one or two years. No overhead is allowed.
Frequency: Annual
Country of Study: United States of America
No. of awards offered: Varies
Application Procedure: Applicants must apply online. Guidelines, instructions and access to the online system are available in the website.
Closing Date: Not currently open to applications
Funding: Private
No. of awards given last year: 3
No. of applicants last year: 10
Additional Information: Guidelines and application forms are available online in the spring at www.foundation4pt.org. A paper version of the RFP is available from the Foundation.

FOUNDATION FOR SCIENCE AND DISABILITY, INC.

1700 SW 23rd Dr, Gainesville, FL, 32608, United States of America
Tel: (1) 352 374 5774
Fax: (1) 352 374 5804
Email: richard.mankin@ars.usda.gov
Website: http://stemd.org
Contact: Dr Richard Mankin, Chair, Student Grants

The Foundation for Science and Disability aims to promote the integration of scientists with disabilities into all activities of the scientific community and of society as a whole, and to promote the removal of barriers in order to enable students with disabilities to choose careers in science.

Foundation for Science and Disability Student Grant Fund

Subjects: Engineering, mathematics, medicine, natural sciences and computer science.
Purpose: To increase opportunities in science for physically disabled students at the graduate or professional level.
Eligibility: Open to candidates from the United States of America.
Level of Study: Doctorate, Postgraduate
Type: Grant
Value: US$1,000
Length of Study: 1 year
Frequency: Annual
Country of Study: United States of America
No. of awards offered: 1–3
Application Procedure: Applicants must submit a completed application form, copies of official college transcripts, a letter from the research or academic supervisor in support of the request and a second letter from another faculty member.
Closing Date: December 1st
Funding: Private
No. of awards given last year: 1
No. of applicants last year: 7
Additional Information: The award may be used for an assistive device or instrument, or as financial support to work with a professor on an individual research project or for some other special need.

FRANK KNOX MEMORIAL FELLOWSHIPS

3 Birdcage Walk, Westminster, London, SW1H 9JJ, England
Tel: (44) 20 7222 1151
Fax: (44) 20 7222 7189
Email: annie@kennedytrust.org.uk
Website: www.frankknox.harvard.edu
Contact: Ms Annie Thomas, Secretary

The Frank Knox Memorial Fellowships were established at Harvard University in 1945 by a gift from Mrs Annie Reid Knox, widow of the late Colonel Frank Knox, to allow students from the United Kingdom to participate in an educational exchange programme.

Frank Knox Fellowships at Harvard University

Subjects: Arts, sciences including engineering and medical sciences, business administration and management, design, divinity, education, law, public administration and public health.
Eligibility: Open to citizens of the United Kingdom who, at the time of application, have spent at least 2 of the last 4 years at a university or university college in the United Kingdom and will have graduated by the start of tenure. Fellowships are not awarded for postdoctoral study and no application will be considered from persons already in the United States of America.
Level of Study: Graduate, Doctorate, MBA, Postgraduate, Professional development
Type: Fellowship
Value: US$24,000 plus tuition fees. Unmarried fellows may be accommodated in one of the university dormitories or halls
Length of Study: 1 academic year. Depending on the availability of sufficient funds fellowships may be renewed for those fellows registered for a degree programme of more than 1 year
Frequency: Annual
Study Establishment: Harvard University
Country of Study: United States of America
No. of awards offered: 5
Application Procedure: Applications are to be made online at www.kennedytrust.org.uk and comprise an online form, a personal statement and two references to be submitted online by the closing date. At the same time, applicants must file an admissions application directly with the graduate school of their choice by the relevant closing date.
Closing Date: Early November - see website
Funding: Private
Contributor: The estate of the late Frank Knox
No. of awards given last year: 5
No. of applicants last year: 197
Additional Information: Travel Grants are not awarded, although in cases of extreme hardship applications can be made to Harvard University for travel cost assistance.

FRANKLIN AND ELEANOR ROOSEVELT INSTITUTE

The Roosevelt Institute, 4079 Albany Post Road, Hyde Park, NY 12538, United States of America
Tel: (1) 845 486 1150
Fax: (1) 845 486 1151
Email: info@rooseveltinstitute.org
Website: www.rooseveltinstitute.org
Contact: The Chairman, Grants Committee

The Roosevelt Institute is a nonprofit organisation devoted to carrying forward the legacy and values of Frankin and Eleanor Roosevelt by developing progressive ideas and bold leadership in the service of restoring America's health and security. We seek to reanimate the progressive movement through a three-pronged approach: generating compelling new ideas and bold visions, developing the next generation of progressive leadership and bringing forth the Rossevelt legacy.

Roosevelt Institute Research Grant

Subjects: Research on the Roosevelt years and clearly related subjects.
Purpose: To encourage younger scholars to expand their knowledge and understanding of the Roosevelt period and to give continued support to more experienced researchers who have already made a mark in the field.
Eligibility: Open to qualified researchers of any nationality with a viable plan of work. Proposals are recommended for funding by an independent panel of Scholars which reports to the Institute Board.
Level of Study: Doctorate, Graduate, Postdoctorate
Type: Research grant
Value: Up to US$2,500
Frequency: Annual
Study Establishment: The Franklin D Roosevelt Library, Hyde Park in New York
Country of Study: United States of America
No. of awards offered: 10–15
Application Procedure: Applicants must submit two copies of each of the following: an application front sheet, research proposal, relevance of holdings, travel plans, time estimate, curriculum vitae, three letters of reference and budget. Application forms and guidelines are available from the website or by emailing, faxing or writing to the Roosevelt Institute.
Closing Date: November 15th
Funding: Private

FRAXA RESEARCH FOUNDATION

10 Prince Place, Newburyport, MA, 01950, United States of America
Tel: (1) 978 462 1866
Email: info@fraxa.org, mbudek@fraxa.org
Website: www.fraxa.org
Contact: Melissa Budek, Office Manager

The FRAXA Research Foundation funds postdoctoral fellowships and investigator-initiated grants to support medical research aimed at the treatment of Fragile X Syndrome. FRAXA is particularly interested in preclinical studies of potential pharmacological and genetic treatments and studies aimed at understanding the function of the FMRI gene.

FRAXA Grants and Fellowships

Subjects: The treatment of Fragile X Syndrome and potential pharmacological and genetic treatments and studies aimed at understanding the function of the FMRI gene.
Purpose: To promote research aimed at finding a specific treatment for Fragile X Syndrome.
Eligibility: There are no eligibility restrictions.
Level of Study: Postdoctorate, Research
Type: Fellowships
Value: $45,000 per year for postdoctoral fellowship for 2 years (for salary, fringe benefits and consumable costs)
Length of Study: 1 year, renewable
Country of Study: Any country
No. of awards offered: 25–35 each year
Application Procedure: Candidates must complete an application form, available from the FRAXA Research Foundation or from the website. Potential candidates are welcome to submit a one-page initial inquiry letter describing the proposed research before submitting a full application.
Closing Date: February 1st
Funding: Private
No. of awards given last year: 29
No. of applicants last year: 60

THE FREDERIC CHOPIN SOCIETY

Plac Pilsudskiego, Warszawa, PL, 00078, Poland
Tel: (48) 22 826 81 90
Fax: (48) 22 827 95 89
Email: konkurs@chopin.pl
Website: www.konkurs.chopin.pl
Contact: Administrative Assistant

The Frederic Chopin Society organizes the International Chopin Piano Competition, the Scholarly Piano Competition for Polish Pianists, the Grand Prix du Disque Frederic Chopin, courses in Chopin's music interpretation and Chopin music recitals as well as running a museum and collection.

International Fryderyk Chopin Piano Competition

Subjects: Piano performance of Chopin's music.
Purpose: To recognize the best artistic interpretation of Chopin's music and to encourage professional development.
Eligibility: Open to pianists of any nationality, born between 1977 and 1988.
Level of Study: Unrestricted
Type: Prize
Value: First Prize €30,000 and gold medal; Second Prize €25,000 and silver medal; Third Prize €20,000 and bronze medal; Fourth Prize €15,000; Fifth Prize €10,000; Sixth Prize €7,000 plus special prizes. Check website for complete details
Frequency: Every 5 years
Country of Study: Any country
No. of awards offered: 6
Application Procedure: Applicants must complete and submit an application form attached with the rules.
Closing Date: December 1st
Funding: Government, private
No. of awards given last year: 14

FREDERICK DOUGLASS INSTITUTE FOR AFRICAN AND AFRICAN-AMERICAN STUDIES

University of Rochester, 302 Morey Hall, Rochester, NY, 14627-0440, United States of America
Tel: (1) 585 275 7235
Fax: (1) 585 256 2594
Email: fdi@mail.rochester.edu
Website: www.rochester.edu/college/aas
Contact: Janise Carmichael, Student Assistant

The Frederick Douglass Institute for African and African-American Studies was established in 1986 to promote the development of African and African-American studies and graduate education through advanced research at the University of Rochester. It has served as an interdisciplinary centre, its focus being on the social sciences, though not excluding the humanities and the natural sciences.

Frederick Douglass Institute Postdoctoral Fellowship

Subjects: Historical and contemporary topics on the economy, society, politics and culture of Africa and its diaspora. Broadly conceived projects on human and technological aspects of energy development and agriculture in Africa are welcomed.
Purpose: To support the completion of a project.
Eligibility: Open to scholars who hold a PhD degree in a field related to the African and African-American experience.
Level of Study: Postdoctorate
Type: Fellowship
Value: A stipend of US$40,000 as well as full access to the university's facilities and office space in the Institute. It also supports the completion of a research project for 1 academic year. Additional US$3,000 research and travel fund
Frequency: Annual
Country of Study: Any country
No. of awards offered: 1
Application Procedure: Applicants must submit a completed application, a curriculum vitae, a three- to five-page description of the project plus a short bibliography and a sample of published or unpublished writing on a topic related to the proposal. Three letters of recommendation that comment upon the value and feasibility of the work proposed are to be sent by referees.
Closing Date: December 31st
No. of awards given last year: 1
Additional Information: All Fellows receive office space in the Institute and opportunities to interact and collaborate with scholars of their respective disciplines within the University. Fellows must be in full-time residence during the tenure of their awards and are expected to be engaged in scholarly activity on a full-time basis. They must be available for consultation with students and professional colleagues, make at least two formal presentations based on their research and contribute generally to the intellectual discourse on African and African-American Studies.

Frederick Douglass Institute Predoctoral Dissertation Fellowship

Subjects: Historical and contemporary topics on the economy, society, politics and culture of Africa and its diaspora. Broadly conceived projects on human and technological aspects of energy development and agriculture in Africa are welcomed.
Purpose: To support the completion of a dissertation.
Eligibility: Open to graduate students of any university who study aspects of the African and African- American experience. Applicants must have completed and passed all required courses, any qualifying oral and/or written exams and have written at least one chapter of the dissertation, which then becomes part of the application package, to qualify for this award.
Level of Study: Predoctorate
Type: Fellowship
Value: A stipend of US$26,000 as well as full access to the university's facilities and office space in the institute
Frequency: Annual
Country of Study: Any country
Application Procedure: Applicants must complete and send the FDI fellowship application form, a curriculum vitae, an official transcript showing completion of all preliminary coursework and qualifying examinations, the dissertation prospectus, a sample chapter from the dissertation and three letters of recommendation to be sent out by the referees, including one from the dissertation supervisor assessing the candidate's prospects for completing the project within a year.
Closing Date: December 31st
No. of awards given last year: 1
Additional Information: All Fellows receive office space in the Institute and opportunities to interact and collaborate with scholars of their respective disciplines within the University. Fellows must be in full-time residence during the tenure of their awards and are expected to be engaged in scholarly activity on a full-time basis. They must be available for consultation with students and professional colleagues, make at least two formal presentations based upon their research and contribute generally to the intellectual discourse on African and African-American Studies.

FREIE UNIVERSITÄT BERLIN

Lansstrasse 7-9, Berlin, 14195, Germany
Tel: (49) 030 838 52702
Fax: (49) 030 838 52882
Email: jfkistip@zedat.fu-berlin.de
Website: www.jfki.fu-berlin.de/en/library/researchgrant
Contact: JFK Institut für Nordamerikastudien

The John F. Kennedy Institut of the Freie Universität Berlin is renowned for its innovative interdisciplinary research and its rigorous study programs. The institute is dedicated to the study of the United States and Canada in six disciplines: culture, history, literature, political science, sociology and economics.

Freie Universität Berlin John-F.-Kennedy-Institut für Nordamerikastudien Research Grants

Subjects: Culture, economy, geography, history, language, literature, politics and society.
Purpose: To financially assist scholars who are interested in conducting research on topics concerning the United States or Canada.
Eligibility: Open to scholars with permanent residence in a European country working on research projects on topics concerning the United States and/or Canada.
Level of Study: Research
Type: Research grant
Value: The full scholarship awards €921 monthly to doctoral candidates and €1,330 monthly to postdoctoral candidates. The guest scholarship complements the salary of guest scholars by €665. The theses scholarship offers €562 to candidates working on their final theses
Frequency: Annual
Study Establishment: Freie Universität Berlin, John-F.-Kennedy-Institut für Nordamerikastudien
Country of Study: Germany

Application Procedure: Applicants must submit the applications including completed application form, a letter of reference, curriculum vitae, a project proposal and a bibliography of works and references needed at the Institute.
Closing Date: October 31st, May 31st
Funding: Government
Contributor: The research grants are supported by the Freie University of Berlin, in conjunction with the United States Information Agency and the Canadian Embassy in Berlin
No. of awards given last year: 25
No. of applicants last year: 86

THE FRENCH CULINARY INSTITUTE

Office of Financial Aid, 462 Broadway, 4th Floor, New York, NY 10013, United States of America
Tel: (1) 888 324 2433
Email: finaid@frenchculinary.com
Website: www.frenchculinary.com

The French Culinary Institute accredited by the Accrediting Commission of Career Schools and Colleges of Technology (ACCSCT) launches the bright careers of the next generation of culinary leaders. Chefs, pastry chefs, bread bakers, sommeliers, restaurant owners, managers, and food writers can hone their craft and shape their dreams with the incomparable training, experience, and career connections of the institute.

The French Culinary Institute Italian Culinary Experience Scholarship

Subjects: Italian culinary.
Purpose: For experience in Italian culinary.
Eligibility: Open to a student pre-enrolled for the Italian Culinary Experience Diploma Program.
Type: Scholarship
Value: $5,000
Study Establishment: ALMA La Scuola Internazionale di Cucina Italiana, Parma and The French Culinary Institute, New York
No. of awards offered: 1
Application Procedure: Check website for further details.

FRIENDS OF ISRAEL EDUCATIONAL FOUNDATION

Academic Study Group, POB 42763, London, N2 0YJ, England
Tel: (44) 020 8444 0777
Fax: (44) 020 8444 0681
Email: info@foi-asg.org
Website: www.foi-asg.org
Contact: Mr John D A Levy

The Friends of Israel Educational Foundation and its sister operation, the Academic Study Group, aim to encourage a critical understanding of the achievements, hopes and problems of modern Israel, and to forge new collaborative working links between the United Kingdom and Israel.

Friends of Israel Educational Foundation Academic Study Bursary

Subjects: All subjects.
Purpose: To provide funding for British academics planning to pay a first research or study visit to Israel.
Eligibility: Open to research or teaching postgraduates. The Academic Study Group will only consider proposals from British academics who have already linked up with professional counterparts in Israel and agreed terms of reference for an initial visit.
Level of Study: Postdoctorate
Type: Bursary
Value: UK£300 per person
Frequency: Annual
Country of Study: Israel
No. of awards offered: 30
Application Procedure: Applicants must contact the organization. There is no application form.
Closing Date: November 15th or March 15th
Funding: Private
Contributor: Trusts and individual donations
No. of awards given last year: 10
No. of applicants last year: Approx. 50

Friends of Israel Educational Foundation Young Artist Award

Subjects: Fine arts.
Purpose: To enable a promising British artist to pay a working visit to Israel and prepare work for an exhibition on Israeli themes in the United Kingdom.
Eligibility: Open to promising young British painters, print makers and illustrators.
Level of Study: Professional development, Postgraduate
Type: Award
Value: Return air passage to Israel. A minimum of six weeks work and general volunteering on a kibbutz, with time for painting by the award winner. possibility of a ten-day placement at the prestigious Bezalel School of Art in Jerusalem. Free time to travel round the country. An eventual Exhibition of the artist's Israel portfolio in London.
Length of Study: A minimum of 2 months
Frequency: Annual
Study Establishment: A kibbutz
Country of Study: Israel
No. of awards offered: 1–2
Application Procedure: Applicants must submit a personal curriculum vitae, an academic letter of reference, a statement of reasons for wishing to visit Israel and a representative selection of work.
Closing Date: May 1st
Funding: Private
Contributor: Individual donations
Additional Information: Shortlisted candidates will be interviewed and artwork examined by a distinguished panel of judges.

Jerusalem Botanical Gardens Scholarship

Subjects: Botany and horticulture.
Purpose: To provide opportunities for botanists and horticulturists to work at the Jerusalem Botanical Gardens.
Eligibility: Preference is given to permanent residents of the United Kingdom who hold a degree in a relevant subject. Landscape architects with practical plant skills are also eligible.
Level of Study: Postgraduate, Professional development
Type: Scholarship
Value: Return airfare to Israel, subsidized accommodation in the vicinity of the Hebrew University campus and a subsistence allowance that covers the full placement. Participants receive no formal salary
Length of Study: 6–12 months
Frequency: Annual
Study Establishment: The Jerusalem Botanical Gardens
Country of Study: Israel
No. of awards offered: Varies
Application Procedure: Applicants must submit a curriculum vitae, an academic letter of reference, a statement of reasons for wishing to work in Jerusalem, two passport-sized photographs and a handwritten covering letter.
Closing Date: March 31st
Funding: Private
Contributor: Trusts and individual donations
No. of awards given last year: 4

FRIENDS OF JOSÉ CARRERAS INTERNATIONAL LEUKEMIA FOUNDATION

1100 Fairview Avenue North, D5-100 PO Box 19024, Seattle, WA, 98109-1024, United States of America
Tel: (1) 206 667 7108
Fax: (1) 206 667 6124
Email: friendsjc@carrerasfoundation.org
Website: www.carrerasfoundation.org
Contact: Administrator

The Friends of José Carreras International Leukemia Foundation funds Medical Research Fellowships.

Friends of José Carreras International Leukemia Foundation E D Thomas Postdoctoral Fellowship

Subjects: Medical sciences or leukaemia.
Purpose: To support research in the field of leukaemia or related haematological disorders.
Eligibility: Candidates must hold an MD or PhD degree and have completed at least 3 years of postdoctoral training but must be less than 10 years past their first doctoral degree when the award begins.
Level of Study: Postdoctorate
Type: Fellowship
Value: US$50,000 per year
Length of Study: 1 year, renewable for an additional 2 years
Frequency: Annual
Study Establishment: A suitable institution with the academic environment to provide adequate support for the proposal project
Country of Study: Any country
No. of awards offered: 1
Application Procedure: Applicants must complete an application form, available from the website. All applications must be typed, single spaced, in English and must follow the format specified in the application packet. Award announcements will be made by letter in January. Please do not contact the Foundation for results. Reapplication by unsuccessful candidates will be necessary for the following year.
Closing Date: November 2nd
Funding: Foundation
Contributor: Individual donors
No. of awards given last year: 1
No. of applicants last year: 15

FROMM MUSIC FOUNDATION

c/o Department of Music, Harvard University, Cambridge, MA, 02138, United States of America
Tel: (1) 617 495 2791
Fax: (1) 617 496 8081
Email: moncrief@fas.harvard.edu
Website: www.music.fas.harvard.edu
Contact: Ms Jean Moncrieff

The Fromm Music Foundation at Harvard University, founded by the late Paul Fromm in the fifties, has been located at Harvard since 1972. Over, the course of its existence, the foundation has commissioned over 300 new compositions and their performances, and has sponsored hundreds of new music concerts and concert series.

Fromm Foundation Commission

Subjects: Music composition.
Purpose: To support compositions by young and lesser known as well as established composers who are citizens or residents of the United States of America. The award includes a stipend for premiere performance of commissioned work.
Eligibility: Applicants must be citizens or residents of the United States of America.
Level of Study: Unrestricted
Value: US$10,000
Frequency: Annual
Country of Study: United States of America
No. of awards offered: Up to 12
Application Procedure: Applicants must obtain guidelines from the Fromm Music Foundation.
Closing Date: June 1st
Funding: Foundation, private
No. of awards given last year: 12
No. of applicants last year: 200

FULBRIGHT COMMISSION (ARGENTINA)

Viamonte 1653, 2 Piso, Buenos Aires, C1055 ABE, Argentina
Tel: (54) 11 4814 3561
Fax: (54) 11 4814 1377
Email: info@fulbright.com.ar
Website: http://www.fulbright.edu.ar/esp/index.asp
Contact: Melina Ginszparg, Educational Advisor

The Fulbright Programme is an educational exchange programme that sponsors awards for individuals approved by the J William Fulbright Board. The programme's major aim is to promote international co-operation and contribute to the development of friendly, sympathetic and peaceful relations between the United States and other countries in the world.

Fulbright Commission (Argentina) Awards for US Lecturers and Researchers

Subjects: All subjects except medical science.
Purpose: To enable United States lecturers to teach at an Argentine university for one semester, and to enable United States researchers to conduct research at an Argentine institution for 3 months.
Eligibility: Open to United States researchers and lecturers. Applicants must be proficient in spoken Spanish.
Level of Study: Professional development
Value: Varies according to professional experience
Length of Study: 3 months
Frequency: Annual
Country of Study: Argentina
Closing Date: July 31st
Funding: Government
Contributor: The United States of America and the Argentine governments

For further information contact:

The Council for International Exchange of Scholars, 3001 Tilden Street, Washington, DC, 20008-3009
Tel: (202) 686-4000
Email: info@ciesnet.cies.org

Fulbright Commission (Argentina) Master's Program

Subjects: All subjects except medical science.
Purpose: To support Argentines pursuing a Master's degree in the United States of America.
Eligibility: Open to Argentines only.
Level of Study: Postgraduate
Value: Round-trip international travel, monthly maintenance, health insurance, and university fees and tuition (partial or total, depending on the university)
Length of Study: 1-2 years
Frequency: Annual
Country of Study: United States of America
Application Procedure: Applicants must contact the Fulbright Commission in Argentina between February 1st and April 30th.
Closing Date: April 16th
Funding: Government, private
Contributor: The Binational Commission and private sources

Fulbright Institutional Linkages Program

Subjects: Social sciences, public administration, business, economics, law, journalism and communications and educational administration.
Purpose: To support educational partnerships between US universities and foreign post-secondary institutions.
Level of Study: Postgraduate
Frequency: Annual
Contributor: The US department of state's institutional linkage programmes

For further information contact:

Humphrey Fellowships and Institutional linkage Branch, Office of Global Educational Programs, Bureau of Educational and Cultural Affairs, SA-44, 301 4th street, Washington, DC, 20547

Fulbright Scholar-in Residence

Subjects: Education administration.
Purpose: To enable visiting scholars to teach in the US about their home country or world region.
Eligibility: Open to candidates with strong international interest and some experience in study abroad and exchange programmes.
Level of Study: Professional development
Type: Grant
Value: Fulbright funding plus salary supplement and in-kind support from the host institution

Length of Study: 1 year
Frequency: Annual
Application Procedure: Candidates must submit a Fulbright visiting scholar application form and a brief project statement.
Closing Date: October 15th

For further information contact:

Tel: (011) 4814-3561/62
Email: info@fulbright.edu.ar

Fulbright Teacher and Administrator Exchange Awards – Elementary & High School Administrator

Subjects: Educational Administration.
Purpose: To support US Administrators who wish to shadow an Argentine counterpart and then host a reciprocal visit of the Argentine counterpart in the US.
Eligibility: Open to both Argentine and US candidates with considerable expenses as school principals, vice principal or other high level administrative position within the education sector.
Level of Study: Professional development
Type: Grant
Length of Study: 3 weeks
Frequency: Annual
Closing Date: May 14th

For further information contact:

Fulbright Teacher Exchange Program, 600 Maryland Avenue, SW, Room 235, Washington, DC, 20024-2520, United States of America

Fulbright Teacher and Administrator Exchange Awards – Teacher Exchange Awards

Subjects: Educational administration.
Purpose: To enable US teachers to exchange jobs for one semester with Argentine colleagues.
Eligibility: Applicants must have at least 3 years of teaching experience after graduation.
Level of Study: Professional development
Type: Grant
Value: Round-trip international travel, salary supplement and health services
Length of Study: One semester
Frequency: Annual
Closing Date: closed

For further information contact:

Fullbright Teacher Exchange Program, 600 Maryland Avenue, SW, Room 235, Washington, DC, 20024-2520

Fulbright Teaching Assistant Awards

Purpose: To strengthen English language instruction at Argentine educational institutions by establishing a native speaker presence.
Eligibility: Open to candidates with Argentine citizenship who show interest in interacting closely with the host community.
Level of Study: Professional development
Type: Award
Value: Roundtrip ticket, monthly stipend and health insurance
Length of Study: 8 months
Frequency: Annual
No. of awards offered: 15
Application Procedure: Download application form from the website.
Closing Date: October 31st

Fulbright Teaching Assistant Awards (FTA)

Purpose: To provide young Spanish teachers of the United states with an opportunity to work as language assistants at Teacher Training Colleges.
Eligibility: Open to all Humphrey fellows.
Level of Study: Postdoctorate
Type: Grant
Frequency: Annual
Country of Study: Argentina
No. of awards offered: 15–20
Application Procedure: Download application form from the website.
Closing Date: September 1st
Funding: Government
Contributor: IIE and US Department of State

Hubert H. Humphrey Fellowship Program

Purpose: To help mid-career professionals for one year of non-degree study and professional internships in the United States.
Eligibility: Open to all alumni who have been home for at least 3 years.
Level of Study: Postdoctorate, Professional development
Type: Grant
Value: Tuition and fees, monthly stipend, health insurance, book stipend, roundtrip ticket, and special allowance for grant-related activities.
Length of Study: 1 year
Frequency: Annual
Country of Study: United States of America
Application Procedure: Download the application form from the website.
Funding: Government

For further information contact:

Institute of International Education, 1400 K street, N.W., Suite 650, Washington, DC, 20005, United States of America
Tel: (1) (202) 326-7701
Fax: (1) (202) 326-7702

Researcher Awards

Purpose: To enable Argentine junior and senior scholars to conduct research in the United States.
Eligibility: Open to United States of America citizens who hold a Bachelor's degree, are writing a Master's thesis or PhD dissertation and are proficient in Spanish.
Level of Study: Doctorate, Graduate, Postdoctorate, Postgraduate, Research
Type: Research grant
Length of Study: 3 months
Frequency: Annual
Country of Study: United States of America
No. of awards offered: 15–20
Application Procedure: Applicants must complete an application form.
Funding: Government
Contributor: The United States of America and Argentine government

Traditional Fulbright Student Awards

Purpose: To enable students working for their PhD dissertation or towards a Master's degree to carry out independent research.
Eligibility: Applications are restricted to US citizens who are proficient in spoken and written Spanish.
Level of Study: Doctorate, Postgraduate
Type: Award
Value: Round-trip international travel, monthly stipend and health insurance
Length of Study: 9 months
Frequency: Annual
Closing Date: closed

For further information contact:

US Student Programs Institute of International Education, 809 United Nations Plaza, New York, 10017 3580
Tel: 212 984 5330
Fax: 212 984 5325

FULBRIGHT TEACHER EXCHANGE

Academy for Educational Development, 1825 Connecticut Avenue, NW, Washington, DC, 20009-5721, United States of America
Tel: (1) 202 884 8228
Fax: (1) 202 884 8407
Email: fulbrightcte@aed.org
Website: www.fulbrightteacherexchange.org
Contact: Administrative Officer

Sponsored by the United States Department of State, the Fulbright Teacher Exchange arranges direct one-to-one classroom exchanges to 7 countries for teachers at the elementary and secondary levels.

Fulbright Teacher Exchange

Subjects: Education and cultural exchange.
Purpose: To promote cultural understanding between peoples of other countries and the people of the United States of America through educational exchange.
Eligibility: Open to classroom and teachers of all subjects and levels from elementary through high school. Applicants must be citizens of the United States of America, be fluent in English, have a current full-time teaching position, be in at least their 3rd year of teaching and hold a Bachelor's degree.
Level of Study: Professional development
Type: Grant
Value: Varies by country
Length of Study: 6 weeks–1 year
Frequency: Annual
Study Establishment: K-12 schools
Country of Study: Any country
No. of awards offered: Varies
Application Procedure: Applicants must submit a basic application that includes a two-page essay, three letters of recommendation, administrative approval and a peer interview.
Closing Date: October 15th
Funding: Government
No. of awards given last year: 65
No. of applicants last year: 300

FUND FOR THEOLOGICAL EDUCATION, INC.

825 Houston Mill Road, Suite 100, Atlanta, GA, 30329, United States of America
Tel: (1) 404 727 1450
Fax: (1) 404 727 1490
Email: fte@fteleaders.org
Website: www.fteleaders.org
Contact: Ms Kim Hearn, Director

The Fund for Theological Education advocates excellence and diversity in pastoral ministry and theological scholarship. Through our initiatives, we enable gifted young people throughout the Christian community to explore and respond to God's calling in their lives. We seek to be a creative, informed catalyst for educational and faith communities in developing their own capacities to nurture men and women for vocations in ministry and teaching. We also aim to awaken the larger community to the contributions of pastoral leaders and educators who act with faith, imagination and courage to serve the common good.

Congregational Fellowship

Subjects: Religion, theology, divinity.
Purpose: To provide financial aid and mentoring support in a Master of Divinity degree programme.
Eligibility: Open to applicants aged 35 or younger and entering a Master of Divinity degree programme in the fall semester. The award is to be used for education and living expenses.
Level of Study: Graduate
Type: Fellowship
Value: USD 1,000-5,000 (one-to-one match with contribution from student's congregation) plus conference attendance
Frequency: Annual
Study Establishment: A school accredited by the ATS of North America
Country of Study: United States of America or Canada
No. of awards offered: 40
Application Procedure: Applicants must be nominated by their congregations, who have made a $1,000–5,000 financial commitment to educational costs and must complete and submit an application form together with supporting documentation. Forms are available from the website.
Closing Date: April 1st
Funding: Private
No. of awards given last year: 40

Dissertation Fellowship for African Americans

Subjects: Religion and theology.
Purpose: To support African American students in PhD and ThD programmes in the final writing stages of their dissertation.
Eligibility: Open to African-American students in the final writing stages of their dissertation. The dissertation proposal must have been approved prior to application.
Level of Study: Doctorate
Type: Fellowship
Value: Up to USD 20,000
Length of Study: 1 year, non-renewable
Frequency: Annual
Study Establishment: Graduate or theological schools
Country of Study: United States of America
No. of awards offered: Up to 9
Application Procedure: Applicants must complete an application form, available from the programme office or the website.
Closing Date: February 1st
Funding: Private
Contributor: Lilly Endowment, Inc.
No. of awards given last year: 9
No. of applicants last year: Varies

Doctoral Fellowship for African-Americans

Subjects: Religion and theology.
Eligibility: Open to African-American students entering the 1st year of a PhD or ThD programme and studying at an ATS (Association of Theological Schools) accredited school or in another accredited graduate programme in religion/theology.
Level of Study: Doctorate
Type: Fellowship
Value: Up to USD 20,000
Length of Study: 1 year, with a possibility of renewal for a 2nd year
Frequency: Annual
Study Establishment: Graduate or theological schools
Country of Study: United States of America
No. of awards offered: Up to 9
Application Procedure: Applicants must complete an application form, available from the programme office or website.
Closing Date: March 1st
Funding: Private
Contributor: Lilly Endowment, Inc.
No. of awards given last year: 10
No. of applicants last year: Varies

Ministry Fellowship

Subjects: Religion, theology and divinity.
Purpose: To provide financial aid and to enrich theological education in a Master of Divinity degree programme.
Eligibility: Open to applicants aged 35 or younger in the second year of a Master of Divinity degree programme. The award is to be used for education and living expenses, and for the design and implementation of creative projects during the Summer.
Level of Study: Graduate
Type: Fellowship
Value: US$10,000 plus conference attendance
Frequency: Annual
Study Establishment: A school accredited by the ATS of North America
Country of Study: United States of America or Canada
No. of awards offered: 20
Application Procedure: Applicants must be nominated by their seminary dean or president, and must complete and submit an application form together with supporting documentation. Forms are available from the website.
Closing Date: March 1st
Funding: Private
No. of awards given last year: 20
No. of applicants last year: 130

North American Doctoral Fellowship

Subjects: Religion and theology.
Purpose: To support students from African American, Asian American, Hispanic American and Native American populations already in doctoral programmes leading towards completion of a PhD or ThD.

Eligibility: Open to students from targeted racial or ethnic groups at any point in their graduate programme, although preference is given to students further along in their programmes. Students must be currently enrolled in PhD or ThD programmes of religion or theology.
Level of Study: Doctorate
Type: Fellowship
Value: US$5,000–10,000
Length of Study: 1 year
Frequency: Annual
Country of Study: United States of America or Canada
No. of awards offered: 10–12 per year
Application Procedure: Applicants must complete an application form, available from the programme office or website.
Closing Date: March 1st
Funding: Private
Contributor: Varies
No. of awards given last year: 12
No. of applicants last year: Varies

Volunteers Exploring Vocation Fellowship

Subjects: Religion, theology and divinity.
Purpose: To provide financial aid and mentoring support in a Master of Divinity degree programme.
Eligibility: Open to applicants aged 35 or younger and entering a Master of Divinity degree programme in the fall semester. The award is to be used for education and living expenses.
Level of Study: Graduate
Type: Fellowship
Value: US$10,000 over three years plus conference attendance
Frequency: Annual
Study Establishment: A school accredited by the ATS of North America
No. of awards offered: 10
Application Procedure: Applicants must have participated in a year-long, volunteer service program within the past three years, and must complete and submit an application form together with supporting documentation. Forms are available from the website.
Closing Date: April 1st
Funding: Private
No. of awards given last year: 12

FUNGAL INFECTION TRUST

PO Box 482, Macclesfield, Cheshire, SK10 9AR, England
Tel: (44) 16 2550 0228
Email: secretary@fungalresearchtrust.org
Website: www.fungalresearchtrust.org
Contact: J C Morgan, Trustee

The Fungal Research Trust is a small charity that funds small research and travel grants and the aspergillus website (www.aspergillus.man.ac.uk). The website is the most comprehensive source of data on the aspergillus fungus and the diseases it causes.

Fungal Research Trust Travel Grants

Subjects: Fungal diseases and fungi.
Purpose: To enable researchers to attend national and international fungal meetings.
Eligibility: No restrictions.
Level of Study: Doctorate, Postdoctorate, Postgraduate, Professional development, Research
Type: Travel grant
Value: UK£1,000
Length of Study: Up to 1 week
Study Establishment: Various
No. of awards offered: Up to 4
Application Procedure: Applicants must submit a letter of application.
Closing Date: Applications are considered at any time, but 3 months notice before travel is required
Funding: Commercial, private
Contributor: Various
No. of awards given last year: 4
No. of applicants last year: 4

Additional Information: Advertisements for other awards are placed in the *Lancet*.

THE GARDEN CLUB OF AMERICA

14 East 60th Street 3rd Floor, New York, NY, 10022, United States of America
Tel: (1) 212 753 8287
Fax: (1) 212 753 0134
Email: judygow@comcast.net
Website: www.gcamerica.org
Contact: Judy Gow, Vice Chairman

The Garden Club of America stimulates the knowledge and love of gardening, shares the advantages of association by means of educational meetings, conferences, correspondence and publications, and restores, improves and protects the quality of the environment through educational programmes and action in the fields of conservation and civic improvement.

The Anne S. Chatham Fellowship

Subjects: Medicinal botany.
Purpose: To protect and preserve knowledge about the medicinal use of plants and thus prevent the disappearance of plants with therapeutic potential.
Eligibility: Open to candidates who are currently enrolled in PhD programmes or have obtained a PhD or a graduate degree.
Level of Study: Doctorate, Postdoctorate
Type: Fellowship
Value: US$4,000
Frequency: Annual
No. of awards offered: 1
Application Procedure: Applicants must submit an application letter, an abstract, a research proposal and a curriculum vitae.
Closing Date: February 1st

For further information contact:

Missouri Botanical Garden, PO Box 299, St Louis, MO, 63166-0299, United States of America
Tel: (1) 314 577 9503
Email: james.miller@mobot.org
Contact: Dr James S Miller

GENERAL BOARD OF HIGHER EDUCATION AND MINISTRY

PO Box 340007, Nashville, TN, 37203-0007, United States of America
Tel: (1) 615 340 7388
Fax: (1) 615 340 7377
Email: bkohler@gbhem.org
Website: www.gbhem.org
Contact: Dr Dena Cheatham, Office Administrator

The General Board of Higher Education and Ministry of the United Methodist Church prepares and assists those whose ministry in Christ is exercised through ordination, the diaconate, licencing or certification. It also provides general oversight and care for United Methodist Institutions of Higher Education and campus ministries as well as financial resources for students to attend Institutions of Higher Education through church offerings and investments.

Dempster Fellowship

Subjects: Theology.
Purpose: To increase the effectiveness of teaching in United Methodist schools of theology by assisting worthy PhD candidates who are committed to serving the church through theological education.
Eligibility: The fellowships are open only to members of the United Methodist Church who plan to teach in seminaries, or to teach one of the technological disciplines (Bible, church history, theology, ethics, and the arts of ministry) in universities or colleges. The applicant must have received the MDiv degree or its equivalent from one of the United Methodist seminaries, or be in a PhD programme or its equivalent at a university affiliated with a United Methodist seminary at the time the award is granted.

Level of Study: Doctorate
Type: Fellowship
Value: Up to US$30,000 over a 5 year period
Length of Study: 1 year, with a possibility of renewal at the discretion of the Committee on Awards
Frequency: Annual
Country of Study: Any country
No. of awards offered: 5
Application Procedure: Applicants must submit a completed application form, transcripts of all previous academic work, letters of reference, a term or other paper of essay length, Graduate Record Examination scores, summary statement of academic plans and a curriculum vitae. Further information is available on request.
Closing Date: October 15th
Funding: Private
Contributor: The United Methodist Church
No. of awards given last year: 10
No. of applicants last year: 32

GEOLOGICAL SOCIETY OF AMERICA (GSA)

3300 Penrose Place, PO Box 9140, Boulder, CO 80301-9140, United States of America
Tel: (1) 303 357 1000
Fax: (1) 303 357 1070
Email: awards@geosociety.org
Website: www.geosociety.org
Contact: Ms Diane C Lorenz, Program Officer, Grants, Awards, and Recognition

Established in 1888, the GSA is a non-profit organization dedicated to the advancement of the science of geology. GSA membership is for the generalist and the specialist in the field of geology and offers something for everyone.

Alexander & Geraldine Wanek Fund

Subjects: Earth and geological sciences.
Purpose: To support research projects.
Eligibility: Open to GSA members.
Level of Study: Research
Type: Funding support
Value: $3,300
Frequency: Annual
Closing Date: February 1st
No. of awards given last year: 2

Alexander Sisson Award

Subjects: Geology.
Purpose: To support research.
Eligibility: Open to candidates who wish to pursue studies in Alaska and the Caribbean.
Type: Award
Value: $2400
Frequency: Annual
Application Procedure: Applications are available online.
Closing Date: February 1st
Contributor: Geological Society of America
No. of awards given last year: 1

Bruce L. "Biff" Reed Award

Subjects: Geology.
Purpose: To support students pursuing geologic research.
Eligibility: Open to candidates who are enrolled in a US, Canadian, Mexican or Central American university or college.
Type: Award
Value: Minimum award amount: $500. Maximum award amount: $2000.
Frequency: Annual
Application Procedure: For further details contact the Program Officer-Grants, Awards and Recognition.
Closing Date: February 1st
Funding: Private
No. of awards given last year: 1

Charles A. & June R.P. Ross Research Fund

Subjects: Biostratigraphy.
Purpose: To support research projects.
Eligibility: Open to GSA members.
Level of Study: Postgraduate
Type: Funding support
Value: $1,300
Frequency: Annual
Application Procedure: Applications available online.
Closing Date: February 1st

Claude C. Albritton, Jr. Scholarships

Subjects: Earth science and archaeology.
Purpose: To encourage students who want to pursue higher studies in the field of Earth science and archaeology.
Level of Study: Postgraduate
Type: Scholarship
Value: $650–1,000
Frequency: Annual
Closing Date: March 1st
Funding: Foundation
Contributor: GSA foundation
No. of awards given last year: 1

For further information contact:

Institute for Applied Sciences, PO Box 13078, University of North Texas, Denton, TX 76203
Contact: Reid Feming

Gladys W Cole Memorial Research Award

Subjects: Investigation of the geomorphology of semi-arid and arid terrain in the United States of America and Mexico.
Purpose: To provide financial support for research.
Eligibility: Open to GSA members or Fellows aged 30–65 who have published one or more significant papers on geomorphology. Funds cannot be used to pay for work already accomplished, but previous recipients may reapply if additional support is needed to complete their work. All qualified applicants are urged to apply.
Level of Study: Postdoctorate
Type: Research grant
Value: USD 7,000
Frequency: Annual
Country of Study: Other
No. of awards offered: 1
Application Procedure: Applicants must complete an application form available from the website.
Closing Date: February 1st
Funding: Private
Contributor: Dr W Storrs Cole
No. of awards given last year: 1

Gretchen L. Blechschmidt Award

Subjects: Geological sciences.
Purpose: To support research by women interested in achieving a PhD.
Eligibility: Open to all women candidates who are GSA members and wish to achieve a PhD.
Level of Study: Postdoctorate, Postgraduate, Research
Type: Award
Value: $1,300
Frequency: Annual
Application Procedure: For additional information contact: Program Officer Grants, Awards and Recognition.
Closing Date: February 1st

GSA Research Grants

Subjects: Earth science.
Purpose: To provide partial support for Master's and doctoral thesis research.
Eligibility: Open to students attending colleges and universities within the United States of America, Canada, Mexico and Central America. Applicants must be members of the GSA in order to apply.
Level of Study: Research, Postgraduate
Type: Research grant
Value: Up to $2,500

Length of Study: 1 year, renewable
Frequency: Annual
Country of Study: Other
No. of awards offered: Varies
Application Procedure: Applicants must complete current application forms.
Closing Date: February 1st
Funding: Government, private
Contributor: GSA's Penrose and Pardee endowments, the National Science Foundation, industry, individual GSA members through the GEOSTAR and Research Grants funds, and numerous dedicated research funds that have been endowed at the GSA Foundation by members
No. of awards given last year: 220
No. of applicants last year: 633
Additional Information: Grants are awarded on the basis of the scientific merits of the problem, the capability of the investigator and the feasibility of the budget, and as an aid to a research project, not to sustain the entire cost. Students may receive the award once at the Master's level and once at the PhD level.

Harold T. Stearns Fellowship Award
Subjects: Geology of Pacific Island and the circum-pacific region.
Purpose: To support research projects.
Eligibility: Open to GSA members.
Level of Study: Research
Type: Award
Value: $500–3,000
Frequency: Annual
No. of awards offered: 1–4
Application Procedure: For more information contact the Program Officer-Grants, Awards and Recognition.
Closing Date: February 1st
No. of awards given last year: 3 scholarship(s), totalling $3,000

History of Geology Student Award
Subjects: Geology.
Purpose: To award best proposals for a history of Geology paper.
Eligibility: Open to applicants who are enrolled in a US, Canadian, Mexican or Central American university or college.
Level of Study: Research
Type: Grant
Value: $1,000
Frequency: Annual
Application Procedure: Applications available online.
Closing Date: May 1st
No. of awards given last year: 1

J. Hoover Mackin and Arthur D. Howard Research Grants
Subjects: Quaternary geology/geomorphology.
Purpose: To support outstanding student research.
Eligibility: Applicants must be GSA members.
Level of Study: Research
Type: Grant
Value: Upto $2,500
Frequency: Annual
No. of awards offered: 1–2
Application Procedure: Application forms are available online.
Closing Date: February 1st
No. of awards given last year: 1

John Montagne Fund
Subjects: Geomorphology.
Purpose: To support research.
Eligibility: Open to GSA members.
Type: Award
Frequency: Annual
Application Procedure: Applications are available online.
Closing Date: February 1st
Funding: Private
No. of awards given last year: 1

John T. Dillon Alaska Research Award
Subjects: Earth science.
Purpose: To support research on earth science problems.
Eligibility: Open to applicants who are GSA members.
Level of Study: Research
Type: Award
Value: $2,900
Frequency: Annual
No. of awards offered: 1
Closing Date: February 1st

Lipman Research Award
Subjects: Volcanology and petrology.
Purpose: To promote and support graduate research.
Eligibility: Open to applicants who are GSA members.
Level of Study: Research
Type: Award
Value: $3,500
Frequency: Annual
No. of awards offered: 1
Application Procedure: Contact the Program officer for further information.
Closing Date: February 1st
Funding: Private

Marie Morisawa Award
Subjects: Quaternary geology/geomorphology.
Purpose: To support outstanding student research (for women only).
Level of Study: Graduate, Research
Type: Fellowship
Value: $1,000
Frequency: Annual
No. of awards offered: 1
No. of awards given last year: 1
Additional Information: Please check at http://rock.geosociety.org/qgg/ for more information.

Parke D. Snavely, Jr. Cascadia Research Award Fund
Subjects: Geology.
Purpose: To support field-oriented graduate student research.
Eligibility: Open to applicants enrolled in a US, Canadian, Mexican or Central American university or college.
Level of Study: Research
Type: Award
Frequency: Annual
Application Procedure: For more information contact: Program Officer-Grants, Awards and Recognition.
Closing Date: February 1st

Robert K. Fahnestock Memorial Award
Subjects: Earth and geological science.
Purpose: To award applicants with the best application in the field of sediment transport.
Eligibility: Applicants must be enrolled in a US, Canadian, Mexican or Central American university or college and must be GSA members.
Level of Study: Research
Type: Award
Value: Minimum award amount: $500. Maximum award amount: $2,000.
Frequency: Annual
No. of awards offered: 1
Application Procedure: For more details contact the Program Officer-Grants, Awards and Recognition.
Closing Date: February 1st

Roy J. Shlemon Scholarship Awards
Subjects: Engineering geology.
Purpose: To award best research proposals.
Eligibility: Open to student members of the Engineering Geology Division.
Level of Study: Doctorate, Research
Type: Scholarship
Value: At least two $1,000 scholarships will be awarded; one for Master's level and one for Doctoral level research. Additional awards may be made at the discretion of the Roy J. Shlemon Scholarship Awards Committee.
Frequency: Annual
No. of awards offered: 4

Application Procedure: Application forms are available online.
Closing Date: March 15th
No. of awards given last year: 1
Additional Information: The scholarship award committee strongly encouraged women, minorities and persons with disabilities to participate in this programme.

For further information contact:

13376 Azores Avenue, Sylmar, CA 91342
Contact: Robert A. Larson

S.E. Dwornik Student Paper Awards

Subjects: Planetary geology.
Purpose: To encourage students to become involved with NASA and planetary science.
Eligibility: Open to all American students interested in planetary science.
Level of Study: Postgraduate
Type: Award
Value: USD 500
Frequency: Annual
No. of awards offered: 2
Application Procedure: Students must submit abstract of the paper along with the application form.
Closing Date: February 1st
No. of awards given last year: 2 prize(s), totalling $1,000.

W Storrs Cole Memorial Research Award

Subjects: Invertebrate micropalaeontology.
Purpose: To support research into invertebrate micropalaeontology.
Eligibility: Open to GSA members or Fellows aged 30–65 who have published one or more significant papers on micropalaeontology. Funds cannot be used for work already accomplished but recipients of previous awards may reapply if additional support is needed to complete their work. All qualified applicants are urged to apply.
Level of Study: Postdoctorate
Type: Research grant
Value: US$6,500
Frequency: Annual
Application Procedure: Applicants must write for further details and an application form or visit the website.
Closing Date: February 1st
Funding: Private
Contributor: Dr W Storrs Cole
No. of awards given last year: 1

GEORGE WALFORD INTERNATIONAL ESSAY PRIZE (GWIEP)

Flat 143, 6 Slington House, Rankine Road, Basingstoke, London, RG24 8PH, United Kingdom
Email: richenda@gwiep.net
Website: www.gwiep.net
Contact: Ms Richenda Walford, Trustee

The George Walford International Essay Prize (GWIEP) is a registered charity that awards a cash prize each year to the winner of an essay on the subject of systematic ideology.

George Walford International Essay Prize (GWIEP)

Subjects: Any subject.
Purpose: To award a prize to the best essay on systematic ideology.
Eligibility: Everyone is eligible, with the exception of the trustees and judges themselves. There are no bars regarding age, race, nationality or gender.
Level of Study: Unrestricted
Type: Prize
Value: UK£3,500
Length of Study: Varies
Frequency: Annual
Country of Study: Any country
No. of awards offered: 1
Application Procedure: Applicants must contact GWIEP for details. Information can be requested by mail though communication via email and the website is greatly preferred.
Closing Date: May 31st
Funding: Private
Contributor: The family of the late George Walford
No. of awards given last year: 1
Additional Information: For more information about the prize and systematic ideology, please visit the website.

GEORGIA LIBRARY ASSOCIATION (GLA)

PO Box 793, Rex, Georgia, GA 30273, United States of America
Tel: (1) 770 801 5330
Fax: (1) 770 801 5319
Email: kendalls@cobbcat.org
Website: gla.georgialibraries.org
Contact: Scholarship Committee Chair

The Georgia Library Association's (GLA) mission is to provide an understanding of the place that libraries should take in advancing the educational, cultural and economic life of the state, to promote the expansion and improvement of library service and to stimulate activities toward these ends.

Beard Scholarship

Subjects: Library science.
Purpose: To recruit excellent librarians for Georgia and provide financial assistance toward completing a degree in library science, for candidates who show strong potential to inspire and motivate their peers in the profession and in professional associations.
Eligibility: Open to United States citizens accepted for admission to a Master's programme at an American Library Association (ALA) accredited library school, who intend to complete the course of study within 2 years.
Level of Study: Postgraduate
Type: Scholarship
Value: US$1,000, paid in equal instalments at the beginning of each term, semester or quarter
Length of Study: 2 years
Frequency: Annual
Study Establishment: An ALA accredited school
Country of Study: United States of America
No. of awards offered: 1
Application Procedure: Applicants must submit an official form of application, proof of acceptance in an accredited library school and official transcripts of all academic work sent directly from each institution of higher education. Three letters of reference must also be sent directly from the referee. More information and application forms are available at gla.georgialibraries.org/scholarships.htm.
Closing Date: May 21st
Funding: Individuals
Contributor: GLA members
No. of awards given last year: 1
No. of applicants last year: 15
Additional Information: The Scholar is required to work in a library or library-related capacity in Georgia for 1 year following completion of the programme, or agree to pay back a pro-rated amount of the scholarship plus interest within a 2 year period.

Hubbard Scholarship

Subjects: Library science.
Purpose: To recruit excellent librarians for Georgia and provide financial assistance toward completing a degree in library science.
Eligibility: Open to United States citizens accepted for admission to a Master's programme at an American Library Association (ALA) accredited library school, who intend to complete the course of study within 2 years.
Level of Study: Postgraduate
Type: Scholarship
Value: US$3,000, paid in equal instalments at the beginning of each term, semester or quarter
Length of Study: 2 years
Frequency: Annual
Study Establishment: An ALA accredited school

Country of Study: United States of America
No. of awards offered: 1
Application Procedure: Applicants must submit an official form of application, proof of acceptance in an accredited library school and official transcripts of all academic work sent directly from each institution of higher education. Three letters of reference must also be sent directly from the referee. More information and application forms are available at gla.georgialibraries.org/scholarships.htm.
Closing Date: May 21st
Funding: Individuals
Contributor: GLA members
No. of awards given last year: 1
No. of applicants last year: 15
Additional Information: The Scholar is required to work in a library or library related capacity in Georgia for 1 year following completion of the programme, or agree to pay back a pro-rated amount of the scholarship plus interest within a 2 year period.

THE GERALDINE R. DODGE FOUNDATION

14 Maple Avenue, Post Office Box 1239, Morris town, NJ 07962-12 39, United States of America
Tel: (1) 973 540 8442
Fax: (1) 973 540 1211
Email: info@grdodge.org
Website: www.grdodge.org

The Geraldine R. Dodge Foundation was established in 1974 with funds from the will of Geraldine Rockefeller Dodge. The mission of the foundation is to support and encourage those educational, cultural, social and environmental values that contribute to making our society more humane and our world more livable.

Dodge Foundation Frontiers for Veterinary Medicine Fellowships

Subjects: Veterinary medicine.
Purpose: To provide opportunities to veterinary students to explore and bring new problem-solving perspectives to animal-related issues.
Eligibility: Open to candidates who are enrolled as full-time veterinary students at a US or Canadian college of Veterinary medicine.
Level of Study: Postgraduate
Type: Fellowship
Value: US$7,000
Frequency: Annual
Study Establishment: Any American Veterinary Medical Association accredited college of veterinary medicine
Country of Study: United States of America or Canada
Application Procedure: Check website for details
Closing Date: December 16th

For further information contact:

Contact: Michelle Knapik, Director for Environmental and Welfare of Animals

Dodge Foundation Teacher Fellowships

Subjects: Teacher training.
Purpose: To enable teachers to grow as educational leaders to better impact their scholars and communities.
Eligibility: Open to K-12 full-time teachers who are employed in New Jersey public and public charter schools in Camden Country, New Jersey.
Level of Study: Postgraduate
Type: Fellowship
Value: US$2,000–7,500 for individuals, and between US$5,000 and 10,000 for teams
Frequency: Annual
Country of Study: United States of America
Application Procedure: Application form and details available on the website.
Closing Date: December 1st
Funding: Foundation

Dodge Foundation Visual Arts Initiative

Subjects: Visual art.
Purpose: To provide a unique professional infrastructure and sprinted network for the artist/educator.
Eligibility: Open to candidates who have taught the visual arts for at least 3 years and will be continuing as visual arts teachers.
Level of Study: Professional development
Type: Grant
Value: US$5,000 plus US$2,000 for visual arts project
Frequency: Annual
Country of Study: United States of America
No. of awards offered: 20
Application Procedure: Check website for details
Closing Date: January 31st
Funding: Foundation

For further information contact:

Tel: 973 540 8443, ext. 118
Email: erastocky@grdodge.org
Contact: Elaine Rastocky

GERMAN ACADEMIC EXCHANGE SERVICE (DAAD)

New York Office, 871 United Nations Plaza, New York, NY, 10017, United States of America
Tel: (1) 212 758 3223
Fax: (1) 212 755 5780
Email: daadny@daad.org/ kim@daad.org
Website: www.daad.org
Contact: Myoung-Shin Kim, Program Officer

The German Academic Exchange Service (DAAD) is the German national agency for the support of international academic cooperation. It offers programmes and funding for students, faculty, researchers and others in higher education, providing financial support to over 67,000 individuals per year.

DAAD Research Grant

Subjects: All subjects.
Purpose: To provide opportunities to students to study and research in Germany.
Eligibility: Open to citizens of Canada and the United States who have obtained a graduate, Master's, Doctoral or postdoctoral degree. And any foreign nationals who have been studying full time in North America more than two years by the time of application.
Level of Study: Postdoctorate, Research, Doctorate
Type: Grant
Value: Approx €1,000 per month, health, insurance, travel subsidy
Length of Study: 1–10 months
Frequency: Annual
Country of Study: Germany
Application Procedure: Applicants must apply online. The completed application form along with a curriculum vitae, research proposal, transcripts and two letters of recommendation must be submitted.
Closing Date: November 15th
Funding: Government
Additional Information: Please check weblink: https://www.daad.org/gradresearch

DAAD Study Scholarship

Subjects: All subjects including performing/visual arts.
Purpose: To provide opportunities to students to study and research in Germany.
Eligibility: Open to graduating seniors, graduates, recent graduates with academic excellence and depending on the candidate's contribution to the field. And any foreign nationals who have been study full time in North America more than two years by the time of application.
Level of Study: Graduate, MBA, Postgraduate, Research
Value: Approx. €750 per month, health, insurance, travel subsidy
Length of Study: 1–2 years
Frequency: Annual
Country of Study: Germany

Application Procedure: Online application - the complete application form needed along with a resume, study/research proposal, transcripts and two recommendation letters.
Closing Date: November 1st for performing/visual arts; November 15th for regular fields
Funding: Government

GERMAN HISTORICAL INSTITUTE

1607 New Hampshire Avenue North West, Washington, DC, 20009-2562, United States of America
Tel: (1) 202 387 3355
Fax: (1) 202 483 3430
Email: fellowships@ghi-dc.org
Website: www.ghi-dc.org
Contact: Bryan Hart

The German Historical Institute is an independent research institute dedicated to the promotion of historical research in the Federal Republic of Germany and the United States of America. The Institute supports and advises German and American historians and encourages co-operation between them. It is part of the foundation Deutsche Geisteswissenschaftliche Institute im Ausland (DGIA).

Fritz Stern Dissertation Prize

Subjects: German history, history of Germans in North America, German-American relations.
Purpose: To award the two best doctoral dissertations submitted on German history, German-American relations or the history of Germans in North America. The winners are invited to the GHI to present their research at the annual symposium of the Friends, each November. Candidates are nominated by their dissertation advisers at a North American university during the previous academic year.
Eligibility: Open to PhDs from the United States of America.
Level of Study: Doctorate
Type: Prize
Value: $2,000 and reimbursement for the travel to Washington DC
Frequency: Annual
Country of Study: United States of America
No. of awards offered: 2
Application Procedure: Applicants must refer to the website for further information.
Closing Date: May 31st
Funding: Private
Contributor: Friends of the German Historical Institute
No. of awards given last year: 2

German Historical Institute Doctoral and Postdoctoral Fellowships

Subjects: Humanities and social sciences, comparative studies in social, cultural and political history, studies of German-American relations and transatlantic studies.
Purpose: To give support to German and United States of America doctoral and postdoctoral students working on topics related to the Institute's general scope of interest.
Eligibility: Open to German, EU and United States of America doctoral students. Applications from women and minorities are especially encouraged.
Level of Study: Doctorate, Postdoctorate
Type: Fellowship
Value: €1,700 for doctoral students; €3,200 for postdoctoral students from European Institutions America
Length of Study: Up to 6 months
Frequency: Annual, Bi annual
Country of Study: United States of America
No. of awards offered: Open
Application Procedure: Applicants must refer to the website for details.
Closing Date: April 15th
Funding: Government
No. of awards given last year: Varies
No. of applicants last year: Varies
Additional Information: All candidates are expected to evaluate source material in the United States of America that is important for their research. At the end of the scholarship they are required to report on their findings or give a presentation at the GHI.

German Historical Institute Summer Seminar in Germany

Subjects: German handwriting, German archives, German history and transatlantic studies.
Purpose: To introduce students to German handwriting of previous centuries by exposing them to a variety of German archives, familiarizing them with major research topics in German culture and history and encouraging the exchange of ideas among the next generation of United States of America scholars.
Eligibility: Open to United States of America doctoral students. Applications from women and minorities are especially encouraged.
Level of Study: Doctorate
Type: Scholarship
Value: All transportation and accomodations
Length of Study: 2 weeks
Frequency: Annual
Country of Study: Germany
Application Procedure: Applicants must refer to the website for details.
Closing Date: December 31st
Funding: Government
No. of awards given last year: Varies

German Historical Institute Transatlantic Doctoral Seminar in German History

Subjects: The Transatlantic Doctoral Seminar in German History (TDS) is an annual seminar organized by the German Historical Institute in Washington DC and Georgetown University. The seminar brings together doctoral students in German history from Europe and North America who are nearing completion of their doctoral degrees. We usually invite eight doctoral students from each side of the Atlantic to discuss their research projects. The discussions at the seminar are based on papers (in German or English) submitted in advance of the conference. The seminar is be conducted bilingually, in German and English
Purpose: To bring together young scholars from Germany and the United States of America who are nearing completion of their doctoral degrees. It provides an opportunity to debate doctoral projects in a transatlantic setting.
Eligibility: Open to doctoral students in German history at North American and European Universities. Applications from women and minorities are especially encouraged.
Level of Study: Doctorate
Type: Scholarship
Value: Travel and accomodation
Length of Study: 4 days
Frequency: Annual
Country of Study: Germany
No. of awards offered: 8
Application Procedure: Applicants must refer to the website for details.
Closing Date: January 15th
Funding: Government
Contributor: German Historical Institute Washington Georgetown University
No. of awards given last year: 16
No. of applicants last year: Varies

Medieval History Seminar

Subjects: Medieval history.
Purpose: The Medieval History Seminar is devoted to the latest research in the field of European medieval studies. Similar to the Transatlantic Doctoral Seminar, this programme invites 16 doctoral students from Europe and North America to discuss their dissertation projects with peers and senior scholars from both sides of the Atlantic.
Eligibility: Open to citizens of North America and Europe.
Level of Study: Doctorate
Type: Grant
Value: Travel and accomodation
Length of Study: 4 days
Frequency: Every 2 years
Country of Study: Germany
No. of awards offered: 16

Application Procedure: Applicants must refer to the website for details.
Closing Date: February 1st
Funding: Government
No. of awards given last year: 16
No. of applicants last year: Varies

GERMAN ISRAELI FOUNDATION FOR SCIENTIFIC RESEARCH AND DEVELOPMENT

G.I.F.—Verbindungsbuero, c/o Forschungszentrum f. Umwelt & Gesundheit (GSF), Postfach 1129, Oberschleissheim, 85758, Germany
Tel: (49) 89 3187 3106
Fax: (49) 89 3187 3365
Email: gif.leie@helmholtz-muenchen.de
Website: www.gifres.org.il
Contact: GIF Verbindungsbuero

German Israeli Foundation Young Scientist's Programme

Subjects: Natural sciences, social sciences and humanities.
Purpose: This new initiative aims to encourage young German and Israeli scientists to establish initial contacts with potential counterparts in Israel or Germany. An integral part of the programme will be a visit of at least 2–3 weeks to Germany or Israel, to give a presentation of their research activities and results and to meet possible partners for future co-operation.
Eligibility: Scientists below 40 years, within the first 7 years after receiving their PhD, MD or equivalent degree, and recognized staff members of a GIF-eligible institution with legal status are eligible to apply.
Level of Study: Research
Type: Grant
Value: Up to €40,000 for the 1-year programme, for project-related equipment, disposable materials, computer services, auxiliary personnel, foreign travel, reports and publications
Length of Study: 1 year
Study Establishment: Any GIF-eligible institution
Country of Study: Germany and Israel
Application Procedure: Application forms are available on request.
Closing Date: Application deadlines vary. For details see www.gifres.org.il
Additional Information: Application deadlines vary. For details see www.gifres.org.il

Research Grant (GIF)

Subjects: Natural sciences, social sciences and humanities.
Purpose: To promote doctoral studies as well as basic and applied research projects within the framework of co-operative research programmes.
Eligibility: GIF projects must involve active collaboration between Israeli and German scientists. Special consideration will be given to young scientists, partners applying for the first time, scientists from the former East Germany and new immigrants to Israel from the former Soviet Union. Scientists applying for grants must be recognized staff members of a GIF-eligible institution with legal status. The principal investigators must hold a doctoral degree or equivalent at the time of application.
Level of Study: Doctorate, Research
Type: Grant
Value: €225,000–600,000 per project
Length of Study: 3 years
Study Establishment: Any GIF-eligible institution
Country of Study: Germany and Israel
Application Procedure: Doctoral candidates can apply to the GIF project co-ordinators, either in Germany or Israel, to pursue their doctoral research within the framework of the co-operative research project.
Closing Date: Application deadlines vary. For details see website.
Contributor: German Israeli Foundation for Scientific Research and Development
Additional Information: The joint research programme must be presented as a single, co-ordinated proposal in which the roles and tasks of both groups are clearly defined. If institutional academic regulations permit the granting of fellowships, the GIF may support a fellowship for tasks within the research plan.

GERMAN MARSHALL FUND OF THE UNITED STATES (GMF)

1744 R Street NW, Washington, DC 20009, United States of America
Tel: (1) 202 683 2650
Fax: (1) 202 265 1662
Email: info@gmfus.org
Website: www.gmfus.org
Contact: Lea Rosenbohm, Administrative Assistant

The German Marshall Fund of the United States (GMF) is an American institution that stimulates the exchange of ideas and promotes co-operation between the United States and Europe in the spirit of the post war Marshall Plan. GMF was created in 1972 by a gift from Germany as a permanent memorial to Marshall Plan Aid.

GMF Journalism Program

Subjects: Journalism.
Purpose: To contribute to better reporting on transatlantic issues by both American and European journalist.
Eligibility: Open to American and European journalists who have an outstanding record in reporting on foreign affairs.
Level of Study: Postdoctorate, Professional development
Type: Fellowship
Value: US$2,000–25,000 and funds for travel
Frequency: Annual
Application Procedure: Applicants including a description of the proposed project, current resume and samples of previous work must be sent.
Funding: Foundation
Contributor: The German Marshall Fund

For further information contact:

Email: usoyez@gmfus.org
Contact: Ursula Soyez

Manfred Wörner Seminar

Purpose: To provide an opportunity to broaden professional networks.
Level of Study: Professional development
Value: Travel, accommodation and meals
Length of Study: 10 days
Frequency: Annual
Country of Study: Germany
No. of awards offered: 30
Closing Date: February 13th
Funding: Government
Contributor: The German Government

For further information contact:

Email: nhagen@gmfus.org
Website: www.gmfus.org/fellowships/manfred.cfm
Contact: Nicola Hagen, Program assistant

Marshall Memorial Fellowship

Subjects: Politics, government business, media and non-profit sector committed to strengthening the transatlantic relationship.
Purpose: To provide opportunities for emerging leaders from the United States and Europe to explore societies, institutions and people from the other side of the Atlantic.
Eligibility: Open to candidates who are citizens or permanent residents of one of the 15MMF countries.
Level of Study: Postdoctorate
Type: Fellowships
Length of Study: 3–4 weeks
Frequency: Annual
Application Procedure: Candidates are required to submit a written application and undergo an interview in person.
Closing Date: Varies
Contributor: German Marshall Fund

For further information contact:

Website: www.gmfus.org/fellowships/mmf.cfm

Peter R. Weitz Journalism Prize

Subjects: Journalism.
Purpose: To acknowledge outstanding coverage of transatlantic and European issues by American media.
Eligibility: Open to all journalists covering European issues for American newspapers, magazines, and online media, whether they are correspondents based in Europe or cover Europe from the United States.
Type: Award
Value: $10,000
Frequency: Annual
Application Procedure: A completed application form must be submitted.
Closing Date: February 28th
Funding: Foundation
Contributor: The German Marshall Fund

Transatlantic Community Foundation Fellowship

Subjects: International relations.
Purpose: To create strengthen networks of people and share international expenses.
Eligibility: Open to staff of American and European community foundation.
Level of Study: Postdoctorate, Professional development
Type: Fellowship
Value: Roundtrip airfare, a daily stipend and reimbursement for car rental expenses (varies)
Frequency: Annual
Application Procedure: Contact the organisation for application information.
Closing Date: Varies, please contact at web: gmfus.org/grants-fellowships
Funding: Foundation
Contributor: Charles Stewart Mott Foundation

For further information contact:

Website: www.gmfus.org/fellowships/tcff.cfm

Transatlantic Fellows Program

Subjects: Foreign policy, international security, trade and economic development, immigration and other topics important to transatlantic cooperation.
Purpose: To build important networks of policymakers analysts in the Euro-Atlantic community.
Eligibility: Open to senior policy-practitioners, journalists business-people and academics.
Level of Study: Postdoctorate
Frequency: Annual
Additional Information: Contact weblink:www.gmfus.org

For further information contact:

Website: www.gmfus.org/fellowships/taf.cfm

GERMAN STUDIES ASSOCIATION

Kalamazoo College, 1200 Academy Street, Kalamazoo, MI, 49006-3295, United States of America
Tel: (1) 269 267 7585
Fax: (1) 269 337 7251
Email: director@thegsa.org
Website: www.thegsa.org
Contact: David E. Barclay, Exectuive Director

The German Studies Association (GSA) is a non-profit educational organization that promotes the research and study of Germany, Austria and Switzerland. The GSA Endowment Fund provides financial support to Association projects, the annual conference, and general operations.

Berlin Program Fellowship

Subjects: Modern and contemporary German and European affairs.
Purpose: To support doctoral dissertation research as well as postdoctoral research leading to the completion of a monograph.
Eligibility: Applicants for a dissertation fellowship must be full-time graduate students who have completed all coursework required for the PhD and must have achieved ABD status by the time the proposed research stay in Berlin begins. Also eligible are United States of America and Canadian PhDs who have received their doctorates within the past 2 calendar years.
Level of Study: Doctorate, Postdoctorate
Type: Fellowship
Value: €1,100 per month for dissertation fellows, €1,400 per month for postdoctoral fellows
Length of Study: 10–12 months
Frequency: Annual
Study Establishment: Freie Universität Berlin
Country of Study: Germany
No. of awards offered: 12
Application Procedure: Applicants must submit a single application packet consisting of completed application forms, a proposal, three letters of reference, language evaluation(s) and graduate school transcripts. Proposals should be no longer than 2,500 words or 10 pages, followed by a one- or two-page bibliography or bibliographic essay.
Closing Date: December 1st
Contributor: Halle Foundation and the National Endowment for the Humanities

For further information contact:

Berlin Program for Advanced German and European Studies, Freie Universität Berlin, Garystrasse 45, D-14195, Berlin, Germany
Tel: (49) 30 838 56671
Fax: (49) 30 838 56672
Email: bprogram@zedat.fu-berlin.de
Website: www.userpage.fu-berlin.de/˜bprogram

THE GETTY FOUNDATION

1200 Getty Center Drive, Suite 800, Los Angeles, CA, 90049–1679, United States of America
Tel: (1) 310 440 7300
Fax: (1) 310 440 7703
Email: researchgrants@getty.edu
Website: www.getty.edu/grants
Contact: Grants Administration

The J Paul Getty Trust is a privately operating foundation dedicated to the visual arts and the humanities. The Getty supports a wide range of projects that promote research in fields related to the history of art, the advancement of the understanding of art and the conservation of cultural heritage.

The Conservation Guest Scholar Program

Subjects: All subjects.
Purpose: To support new ideas and perspectives in the field of conservation, with an emphasis on the visual arts and to provide an opportunity for professionals to pursue scholarly research in an interdisciplinary manner.
Eligibility: Open to established conservators, scientists, and professionals who have attained distinction in conservation and allied fields.
Level of Study: Research
Type: Grant
Value: A monthly stipend of $3,500 is awarded
Application Procedure: Applicants are required to complete and submit the online Getty Conservation Guest Scholar Grant application form (which includes completing an online information sheet, uploading a Project Proposal; Curriculum Vitae; Selected Bibliography, and a single optional writing sample) by 5:00 p.m. PST, November 1st.
Closing Date: November 1st

Additional Information: Please check at http://www.getty.edu/foundation/funding/residential/conservation_guest_scholars.html for more information.

Graduate Internships

Subjects: Curatorial, education, conservation, research, information management, public programs, and grant making.
Purpose: To support full-time positions for students who intend to pursue careers in fields related to the visual arts.
Eligibility: Open to applicants of all nationalities who have currently enrolled in a graduate program leading to an advanced degree in a field relevant to the internship(s).
Type: Scholarship
Value: $17,400 for 8 months and $26,000 for 12 months
Length of Study: 8 or 12 months
Application Procedure: Applications are available for viewing and printing in Portable Document Format (PDF) or by contacting the Getty Grant Program office.
Closing Date: December 3rd
Additional Information: Please check at http://www.getty.edu/foundation/funding/leaders/current/grad_internships.html for more information.

Library Research Grants

Subjects: All subjects.
Purpose: To provide partial, short-term support for costs relating to travel and living expenses to scholars.
Eligibility: Open to scholars of all nationalities and at any level who demonstrate a compelling need to use materials housed in the research library, and whose place of residence is more than 80 miles from the Getty Center.
Type: Research grant
Value: $500 to $2,500
Study Establishment: Getty Research Institute
Application Procedure: Applicants will be required to complete and submit the online Getty Library Research Grant application form (which includes uploading a Project Proposal; Curriculum Vitae; Selected Bibliography of Getty Research Library Collections you wish to consult; and Proposed Estimated Travel Costs) by 5:00 p.m. PST, November 1ST.
Closing Date: October 15th
Additional Information: Projects must relate to specific items in the library collection.

Postdoctoral Fellowships in Conservation Science

Subjects: Chemistry or physical sciences.
Purpose: To provide recent PhDs in chemistry or the physical sciences with experience in the GCI's Museum Research Laboratory.
Eligibility: Open to scientists of all nationalities who are interested in pursuing a career in conservation science and have received a PhD in chemistry/physical science and have excellent written and oral communication skills.
Level of Study: Research
Type: Fellowship
Value: US$29,300 per year
Length of Study: 2 years
Application Procedure: Applicants must complete and submit an online application which includes completing an online information form, and uploading a Statement of Interest in Conservation Science, Doctoral Dissertation Abstract, Curriculum Vitae or Resume, Writing Sample, and Degree Confirmation Letter. Applicants are also required to submit two confidential letters of recommendation in support of the their application.
Closing Date: November 1st
Additional Information: The successful candidate will have a record of scientific accomplishment combined with a strong interest in the visual arts. Please check at http://www.getty.edu/foundation/funding/residential/postdoctoral_fellowship_conservation_science.html for more information.

For further information contact:

Attn: Postdoctoral Fellowship in Conservation Science, The Getty Foundation, 1200 Getty Center Drive, Suite 800, Los Angeles, CA, 90049 1685, United States of America
Tel: (1) 310 440 7374
Fax: (1) 310 440 7703 (inquiries only)
Email: researchgrants@getty.edu

Research Grants for Getty Scholars and Visiting Scholars

Subjects: Arts, humanities, or social sciences.
Purpose: To provide a unique research experience.
Eligibility: Open to established scholars, artists, or writers of all nationalities who have attained distinction in their fields and also for those who are working in the arts, humanities, or social sciences.
Level of Study: Research
Type: Scholarship
Value: A stipend of up to $65,000 per year will be awarded based on length of stay, need and salary. The grant also includes an office at the Getty Research Institute or the Getty Villa, research assistance, an apartment in the Getty scholar housing complex and airfare to and from Los Angeles.
Application Procedure: Applications are available for viewing and printing in Portable Document Format (PDF) or by contacting the Getty Grant Program office.
Closing Date: November 1st
Contributor: Getty Research Institute
Additional Information: Applicants will be notified of the Research Institute's decision by the Spring.

Residential Grants at the Getty Center and Getty Villa

Subjects: Arts and humanities.
Purpose: To provide support for established scholars to undertake research related to a specific theme while in residence at the Getty Center and Getty Villa in Los Angeles.
Eligibility: Open to established scholars who are working on projects that address the given scholarly theme.
Level of Study: Postdoctorate, Postgraduate, Professional development, Research
Type: Grant
Value: Please contact the organization
Frequency: Annual
Country of Study: Any country
No. of awards offered: Varies
Application Procedure: Applicants must complete an application form. Additional information, detailed guidelines and application forms are available from the website or by contacting the Getty Grant Program office.
Closing Date: October 15th
Funding: Private

USA –Getty Foundation Research Grants for Predoctoral and Postdoctoral Fellowships

Subjects: Arts, humanities, or social sciences.
Purpose: To support emerging scholars to complete work on projects related to the Getty Research Institute's annual theme.
Eligibility: Open to scholars of all nationalities who are working in the arts, humanities, or social sciences. Predoctoral fellowship applicants must have advanced to candidacy by the time of the fellowship start date and expect to complete their dissertations during the fellowship period.
Level of Study: Postdoctorate, Predoctorate
Type: Fellowship
Value: $25,000 predoctoral fellows (9-month residency) and $30,000 for the postdoctoral fellows (9-month residency)
Length of Study: 1 year
Application Procedure: Applicants must have access to a web browser that allows cookies. The following browser versions or later are strongly recommended: Internet Explorer 7.0 (for PC); Safari 3.2 (for Mac); Mozilla Firefox 3.0. Applicants are required to complete and submit the online Getty Pre- and Postdoctoral Fellowship application form, which includes completing an online information form and uploading a Project Proposal, Doctoral Dissertation Plan or Abstract, Curriculum Vitae, Writing Sample, Selected Bibliography, and Confirmation Letter of Academic Status (candidacy or degree conferred) by 5:00 p.m. PST, November 1st.
Closing Date: November 1st
Additional Information: Please check at http://www.getty.edu/foundation/funding/residential/getty_pre_postdoctoral_fellowships.html for more information.

Villa Predoctoral and Postdoctoral Fellowships

Subjects: Arts, humanities, or social sciences.
Purpose: To provide support for emerging scholars to complete work on projects related to the Getty Villa's annual theme.
Eligibility: Open to scholars of all nationalities who are working in the arts, humanities, or social sciences. Predoctoral fellowship applicants must have advanced to candidacy by the time of the fellowship start date and expect to complete their dissertations during the fellowship period. Postdoctoral fellowship applicants must have received their PhD within the last 5 years.
Level of Study: Postdoctorate, Predoctorate
Type: Fellowship
Value: Predoctoral stipends are $25,000 for nine months. Postdoctoral stipends are $30,000 for one year or $55,000 for two years.
Length of Study: 1 year
Application Procedure: Applications are available for viewing and printing in Portable Document Format (PDF) or by contacting the Getty Grant Program office.
Closing Date: November 1st

GÉZA ANDA FOUNDATION

Bleicherweg 18, CH-8002, Zurich, Switzerland
Tel: (41) 44 205 1423
Fax: (41) 44 205 1429
Email: info@gezaanda.org
Website: www.gezaanda.org
Contact: Ms Ruth Bossart, Secretary General

The Géza Anda Foundation was established in 1978 in memory of the pianist, Géza Anda. It holds the Géza Anda Concours, an international piano competition, every 3 years, and awards prize to winners and special prizes, and providing an opportunity for the laureates to appear as soloists in concerts and recitals.

International Géza Anda Piano Competition

Subjects: Piano playing.
Purpose: To sponsor young pianists in the musical spirit of Géza Anda.
Eligibility: Open to pianists born after June 6th, 1983.
Level of Study: Unrestricted
Type: Prize
Value: Cash prizes of Swiss francs 60,000 and other benefits such as free concert management services for 3 years
Frequency: Every 3 years
Country of Study: Switzerland
No. of awards offered: 3
Application Procedure: Applicants must complete four rounds in the competition: a preselection, a recital, Mozart and a final concert with orchestra.
Closing Date: January 15th
Funding: Private
No. of awards given last year: 3 official awards and 5 special awards
Additional Information: Next competition will be held from June 6–16.

GILBERT MURRAY TRUST

Department of International Relations, LSE, Houghton Street, London, WC2A 2AE, England
Tel: (44) 18 6555 6633
Email: p.c.wilson@lse.ac.uk
Website: http://icls.sas.ac.uk
Contact: Dr Peter Wilson, Honorary secretary, International Studies Committee

The International Studies Committee of the Gilbert Murray Trust is a small foundation dedicated to the promotion of international education, especially with regard to the theory and practice of the United Nations.

Gilbert Murray UN Study Awards

Subjects: International affairs or international law.
Purpose: To study the purposes and work of the United Nations.
Eligibility: Open to persons of any nationality who are, or who have been, students at a university or similar institution in the United Kingdom. Candidates should currently be taking or should have recently taken part in a course of international affairs or international law and must not be over 25 years of age, although consideration will be given to those over that age in special cases.
Level of Study: Postgraduate, Undergraduate
Type: Award
Value: UK£500
Frequency: Annual
Country of Study: Any country
No. of awards offered: 6–10
Application Procedure: Applicants must submit five copies, typed and on one side only, of a letter of application, a curriculum vitae, an outline of their intention with regard to a future career, full particulars of the purpose for which the award would be used and a supporting testimonial from a person capable of judging the candidate's ability to use the award profitably. Applicants must be currently registered on, or have recently completed, a degree programme at a UK University.
Closing Date: April 1st
Funding: Private
Contributor: Small individual contributions
No. of awards given last year: 10
No. of applicants last year: 17
Additional Information: Awards are only given to support a specific project, such as a research visit to the headquarters of an international organization, or to a particular country, or a short research course at an institution abroad that will assist the applicant in his or her study of international affairs in relation to the purpose and work of the United Nations. The UK Study Awards are not intended as general financial support for the study of international affairs.

GILCHRIST EDUCATIONAL TRUST (GET)

20 Fern Road, Storrington, Pulborough, West Sussex, RH20 4LW, England
Tel: (44) 01903 746723
Email: gilchrist.et@blueyonder.co.uk
Website: www.gilchristgrants.org.uk
Contact: Mrs J.V Considine, Secretary

Gilchrist Educational Trust awards grants to: individuals who face unexpected financial difficulties, which may prevent completion of a degree or higher education course; organizations if it seems likely that a project for which funds are sought will fill an educational gap or make more widely available for a particular aspect of education or learning; British expeditions proposing to carry out research of a scientific nature abroad.

GET Grants - Adult Study Grant

Subjects: All subjects.
Purpose: To promote the advancement of education and learning.
Eligibility: Open to students in the United Kingdom who have made proper provision to fund a degree or higher education course but find themselves facing unexpected financial difficulties which may prevent completion of it; students who are required to spend a short period abroad as part of their course. 1. Rejoining Expedition Grants 2. Organisational Grants.
Level of Study: Doctorate, Graduate, Postgraduate, Research
Type: Travel grant
Value: £500–1,000 (Individual Awards)
Frequency: Dependent on funds available, Monthly
Study Establishment: Any university
Country of Study: United Kingdom
No. of awards offered: Varies
Application Procedure: Application forms must be completed in all cases. For grants to individuals, please contact The Grants Officer on gilchrist.et@blueyonder.co.uk, or write to: 13 Brookfield Avenue, Larkfield, Aylesford ME20 6RU. For other grants contact the Secretary of the Trust on valconsidine@btinternet.com or write to her at 43 Fern Road, Storrington, Pulborough, West Sussex, RH20 4LW
Closing Date: Last day of February for organizations and expeditions. No deadline for applications from individuals
Funding: Private
No. of applicants last year: Approx. 85 (Organization Grants), approx. 21 (Expedition Grants), many for Adult study and Travel Grants

Additional Information: These are awarded in four different categories as 1. Organisational Grants - Academic Educational Projects (£1,000–4,000); 2. Expedition Grants - teams undertaking Scientific Research (£500–2,000); 3. Adult Study Grants - Unexpected financial hardship whilst studying (£500–1,000); 4. Travel Grants - Students who need to study in another country

Gilchrist Fieldwork Award

Subjects: All scientific subjects.
Purpose: To fund a period of fieldwork by established scientists or academics.
Eligibility: Open to teams wishing to undertake a field season of over 6 weeks in relation to one or more scientific objectives. Teams should consist of not more than 10 members, most of whom should be British and holding established positions in research departments at universities or similar establishments. The proposed research must be original and challenging, achievable within the timetable and preferably of benefit to the host country or region.
Level of Study: Research
Type: Grant
Value: UK£15,000
Frequency: Every 2 years
Country of Study: Any country
No. of awards offered: 1
Application Procedure: Send proposal to Secretary.
Closing Date: February 26th in even-numbered years
Funding: Private
No. of awards given last year: 1
No. of applicants last year: 12
Additional Information: The award is competitive.

For further information contact:

c/o RGS, 1 Kensington Gore, London, SW7 2AR, England
Email: grants@rgs.org

THE GILO CENTER FOR CITIZENSHIP, DEMOCRACY AND CIVIC EDUCATION

Faculty of Social Sciences, The Hebrew University of Jerusalem, Mount Scopus, Jerusalem, 91905, Israel
Tel: (972) 02 5883048
Fax: (972) 02 5324339
Email: gilocentre@savion.huji.ac.il
Website: http://gilocenter.mscc.huji.ac.il

The Gilo Center for Citizenship, Democracy and Civic Education at the Hebrew University of Jerusalem was established in 2001, with the generous support of the Gilo Family Foundation in order to promote research on citizenship and democracy in Israel and actively encourage its study and praxis within the education system. The establishment of the Center was made possible by the joint effort of the Hebrew University and Department of Political Science with the Ministry of Education.

The Yitzhak Rabin Fellowship Fund for the Advancement of Peace and Tolerance

Subjects: Advancement of peace and tolerance.
Purpose: To promote research on issues pertaining to citizenship, democracy and civic education by awarding excellent young researchers in their advanced studies.
Eligibility: Open to applicants enrolled in an accredited doctoral or postdoctoral programme focusing on areas relating to the pursuit of peace and/or to the enhancement of peaceful forms of social life. Open to one Canadian applicant, one Jewish and one Arab Israeli.
Level of Study: Doctorate, Postdoctorate, Postgraduate
Type: Fellowship
Value: $15,000
Length of Study: 1 year
Frequency: Annual
Country of Study: Israel
No. of awards offered: 3
Application Procedure: Applicants must provide a research essay or a completed project.
Closing Date: March 21st
Additional Information: The selected Fellow commits to a period of normally 1 month during which he or she spends time with the Rabin Scholars in seminars and/or field trips, as well as lectures, and undertakes pertinent research within the network of the Hebrew University.

For further information contact:

Email: inquiry@cfhu.org

GLADYS KRIEBLE DELMAS FOUNDATION

Gladys Krieble Delmas Foundation, 275 Madison Ave, 33rd floor, New York, NY, 10016-1101, United States of America
Tel: (1) 212 2687 0011
Fax: (1) 212 687 8877
Email: info@delmas.org
Website: www.delmas.org
Contact: Professor Julian Gardner, Honorary Secretary

The Gladys Krieble Delmas Foundation promotes the advancement and perpetuation of humanistic enquiry and artistic creativity by encouraging excellence in scholarship and in the performing arts, and by supporting research libraries and other institutions that preserve the resources that transmit this cultural heritage.

Gladys Krieble Delmas Foundation Grants

Subjects: The history of Venice, the former Venetian empire and contemporary Venetian society and culture. Disciplines of the humanities and social sciences are eligible areas of study, including but not limited to art, architecture, archaeology, theatre, music, literature, political science, economics and law.
Purpose: To promote research into Venice and the Veneto.
Eligibility: Open to citizens and permanent residents of the United States of America who have some experience in advanced research. Graduate students must have fulfilled all doctoral requirements except for completion of the dissertation. The dissertation proposal must, however, have been approved by the time of application for the grant. There is also a programme for scholars from Commonwealth countries.
Level of Study: Postdoctorate, Predoctorate
Type: Grant
Value: US$500–19,900 depending on the length of study. At the discretion of the trustees and advisory board of the Foundation, funds may be made available for aid on the publication of results
Length of Study: Up to 1 academic year
Frequency: Annual
Country of Study: Italy
No. of awards offered: Usually 15–25
Application Procedure: Applicants must complete an application form. Instruction sheets and forms are available from the website.
Closing Date: December 15th
Funding: Private
No. of awards given last year: 19
No. of applicants last year: 32

GLASGOW EDUCATIONAL AND MARSHALL TRUST

21 Beaton Road, Glasgow, G41 4NW, United Kingdom
Tel: (44) 0141 423 2169
Fax: (44) 0141 424 1731
Email: enquiries@gemt.org.uk
Website: www.gemt.org.uk

The Glasgow Educational and Marshall Trust is a charitable trust that meets quarterly and awards bursaries to, among others, mature students and postgraduate students. It also awards grants to aid travel.

Glasgow Educational and Marshall Trust Bursary

Subjects: All subjects.
Purpose: To offer financial support to those who have lived, or are currently living within the Glasgow Municipal Boundary.

Eligibility: Applicants must have a minimum of 5 years of residence within the Glasgow Municipal Boundary, as it was prior to 1975. Years spent within the Boundary purely for the purpose of study do not count.
Level of Study: Unrestricted
Type: Bursary
Value: UK£100–1,000
Frequency: Dependent on funds available
Country of Study: United Kingdom
Application Procedure: Applicants must complete and submit an application form together with two written references prior to the start of the course. No retrospective awards are available.
Closing Date: July 31st
Funding: Private
No. of awards given last year: 58
No. of applicants last year: 59

THE GLASGOW SCHOOL OF ART (GSA)

167 Renfrew Street, Glasgow, Scotland, G3 6RQ, United Kingdom
Tel: (44) 0 141 353 4500
Fax: (44) 0 141 353 4746
Email: welfare@gsa.ac.uk
Website: www.gsa.ac.uk

The Glasgow School of Art, internationally recognized as one of Britain's foremost higher education institutions for the study and advancement of fine art, design and architecture.GSA is a small, specialist and highly focused international community of artists, designers and architects and, as a prospective student, visitor, research partner or supporter of the School, and one can find the GSA an immensely stimulating and creative place.

DOG Digital Scholarship

Subjects: Related to digital design studio.
Purpose: To help students with fees or maintenance.
Eligibility: Open to individuals who meet the academic entry requirements for study at the GSA and hold an offer letter (either conditional or unconditional) to study.
Level of Study: Postgraduate
Type: Scholarships, fellowships
Value: £2,750 to cover fees or maintenance
Length of Study: 2 years
Frequency: Annual
Study Establishment: The Glasgow School of Art
Country of Study: United Kingdom
No. of awards offered: 1
Application Procedure: Applicants should complete the standard application form and financial need form which can be downloaded from the GSA website.
Additional Information: A candidate may apply only for two awards.

For further information contact:

Website: www.gsa.ac.uk/scholarships

Governors International Postgraduate Scholarships

Subjects: All subjects.
Eligibility: Open to applicants who are non EU (students paying full international fees). Applicants should be exceptional students undertaking one-year Master's programmes. A report is required at the end of each academic year.
Level of Study: Postgraduate
Type: Scholarships and fellowships
Value: £2,000 to cover fees
Length of Study: 1 year
Frequency: Annual
Country of Study: United Kingdom
No. of awards offered: 8
Application Procedure: Applicants should complete the standard application form and financial need form which can be downloaded from the GSA website.
Contributor: The Glasgow School of Art
Additional Information: A candidate may apply only for two awards.

For further information contact:

Website: www.gsa.ac.uk/scholarships

Grace and Clark Fyfe Architecture Masters Scholarships

Subjects: Architecture.
Eligibility: Open to individuals who meet the academic entry requirements for study at the GSA and hold an offer letter (either conditional or unconditional) to study MArch (Taught).
Level of Study: Postgraduate
Type: Scholarship
Value: £3,085 to cover fees or maintenance
Length of Study: 1 year
Frequency: Annual
Study Establishment: The Glasgow School of Art
Country of Study: United Kingdom
No. of awards offered: 1
Application Procedure: Applicants should complete the standard application form and financial need form which can be downloaded from the GSA website.
Additional Information: A candidate may apply only for two awards.

For further information contact:

Website: www.gsa.ac.uk/scholarships

Leverhulme Scholarships for Architecture

Subjects: Architecture.
Eligibility: Open to UK citizens who meet the academic entry requirements for study at the GSA and hold an offer letter (either conditional or unconditional) to study architecture specializing in urban building and design, creative urban practices, advanced computing and visualization or energy and environmental studies. Candidates will be expected to develop a PhD proposalduring their programme of study.
Level of Study: Postgraduate
Type: Scholarship
Value: £9,000 per year to cover fees and maintenance (students have home fees paid with balance given as a stipend)
Length of Study: 2 years
Frequency: Annual
Study Establishment: The Glasgow School of Art
Country of Study: United Kingdom
No. of awards offered: 2
Application Procedure: Applicants should complete the standard application form and financial need form which can be downloaded from the GSA website.
Additional Information: A candidate may apply only for two awards. Candidates are expected to develop a PhD proposal during their course of study.

For further information contact:

Website: www.gsa.ac.uk/scholarships

Leverhulme Scholarships for Masters of Fine Art

Subjects: Fine arts.
Eligibility: Open to UK citizens who meet the academic entry requirements for study at the GSA and hold an offer letter (either conditional or unconditional) to study Master of fine arts. For the scholarship to continue to year 2, students must have satisfactorily completed year 1.
Level of Study: Postgraduate
Type: Scholarship
Value: £9,000 per year to cover fees and maintenance (students have home fees paid with balance given as a stipend)
Length of Study: 2 years
Frequency: Annual
Study Establishment: The Glasgow School of Art
Country of Study: United Kingdom
No. of awards offered: 2
Application Procedure: Applicants should complete the standard application form and financial need form which can be downloaded from the GSA website.

Additional Information: A candidate may apply only for two awards. For the scholarship to continue to year 2, students must have satisfactory completed year 1.

For further information contact:

Website: www.gsa.ac.uk/scholarships

Mackendrick Scholarship

Subjects: Painting.
Eligibility: Open to individuals who meet the academic entry requirements for study at the GSA and hold an offer letter (either conditional or unconditional) to study postgraduate painting. Applicants must submit one painting based on the theme 'The City of Glasgow'. For the scholarship to continue to year 2, students must have satisfactorily completed year 1.
Level of Study: Postgraduate
Type: Scholarship
Value: £2,500 per year to cover fees or maintenance
Length of Study: 1 or 2 years, depending on the length of the course
Frequency: Annual
Country of Study: United Kingdom
No. of awards offered: 1
Application Procedure: Applicants should complete the standard application form and financial need form which can be downloaded from the GSA website.
Closing Date: Check with Glasgow School of Art (GSA)
Contributor: The Glasgow School of Art
Additional Information: A candidate may apply only for two awards. For the scholarship to continue to year 2, students must have satisfactorily completed year 1.

For further information contact:

Website: www.gsa.ac.uk/scholarships

Sir Harry Barnes Scholarship

Subjects: All subjects.
Eligibility: Open to individuals who meet the academic entry requirements for study at the GSA and hold an offer letter (either conditional or unconditional) to study.
Level of Study: Postgraduate
Type: Scholarship
Value: £1,500 to cover maintenance
Length of Study: 1 year
Frequency: Annual
Study Establishment: The Glasgow School of Art
Country of Study: United Kingdom
No. of awards offered: 1
Application Procedure: Applicants should complete the standard application form and financial need form which can be downloaded from the GSA website.
Closing Date: Check with Glasgow School of Art (GSA)
Additional Information: A candidate may apply only for two awards.

For further information contact:

Website: www.gsa.ac.uk/scholarships

Tetsuya Mukai Scholarship

Subjects: All subjects.
Eligibility: Open to Japanese (i.e., those paying full international fees) and UK students to undertake taught or research postgraduate study at the GSA.
Level of Study: Postgraduate
Type: Scholarship
Value: £3,000 to cover fees
Length of Study: 1 year
Frequency: Annual
Study Establishment: The Glasgow School of Art
Country of Study: United Kingdom
No. of awards offered: 1
Application Procedure: Applicants should complete the standard application form and financial need form which can be downloaded from the GSA website.
Closing Date: Check with Glasgow School of Art (GSA)
Additional Information: A candidate may apply only for two awards.

For further information contact:

Website: www.gsa.ac.uk/scholarships

Weavers Postgraduate Textiles Scholarships

Subjects: Textiles.
Eligibility: Open to individuals who meet the academic entry requirements for study at the GSA and hold an offer letter (either conditional or unconditional) to study MDes textiles as fashion.
Level of Study: Postgraduate
Type: Scholarship
Value: £1,500 to cover fees or maintenance
Length of Study: 1 year
Frequency: Annual
Study Establishment: The Glasgow School of Art
Country of Study: United Kingdom
No. of awards offered: 1
Application Procedure: Applicants should complete the standard application form and financial need form which can be downloaded from the GSA website.
Closing Date: Check with Glasgow School of Art (GSA)
Additional Information: A candidate may apply only for two awards.

For further information contact:

Website: www.gsa.ac.uk/scholarships

THE GOLDSMITHS' COMPANY

Goldsmiths' Hall, Foster Lane, London, EC2V 6BN, England
Tel: (44) 20 7606 7010
Fax: (44) 20 7606 1511
Email: education@thegoldsmiths.co.uk
Website: www.thegoldsmiths.co.uk
Contact: The Assistant Clerk

The Goldsmiths' Company is one of the Great Twelve Companies of the City of London. It has been responsible for hallmarking since 1300 and today operates the Assay Office London and supports the craft and industry of silversmithing and precious metal jewellery.

Goldsmiths' Company Science for Society Courses

Subjects: Genetics, particle physics, complementary medicine, astrophysics, sustainable development, materials science, and mathematics.
Purpose: To provide teachers of A levels with first-hand, practical experience of the theory that they teach.
Eligibility: Open to United Kingdom science teachers of secondary age children, but teachers from other disciplines are also accepted.
Level of Study: Professional development
Value: Free tuition, accommodation and travel after joining
Length of Study: 1 week in July
Frequency: Annual
Study Establishment: Various locations around the United Kingdom
Country of Study: United Kingdom
No. of awards offered: Approx. 120 vacancies each year
Application Procedure: Please write for details.
Closing Date: May 1st
Funding: Private

GRADUATE INSTITUTE OF INTERNATIONAL STUDIES, GENEVA

Rue de Lausanne 132, PO Box 36, Genève 21, CH-1211, Switzerland
Tel: (41) 22 908 5700
Fax: (41) 22 908 5710
Email: info@hei.unige.ch
Website: www.hei.unige.ch

The Graduate Institute of International Studies, best known as HEI, was founded in 1927 as one of the first instituions in the world dedicated to the study of international relations. A small and selective institute with about 1,100 undergraduate and graduate students from over 90 countries, HEI owes its reputation to the quality of its cosmopolitan faculty, the strength of its core disciplines (economics, history, law and political science), its policy-relevent approach to international affairs, and its bilingual English-French education programmes.

Graduate Institute of International Studies (HEI-Geneva) Scholarships

Subjects: History and international politics, international economics, international law and political science.
Eligibility: Open to any applicant who can prove sound knowledge of the French language and sufficient prior study in political science, economics, law or modern history through the presentation of a college or university degree.
Level of Study: Doctorate, Postgraduate
Type: Scholarship
Value: Swiss francs 18,000
Length of Study: 1 year, possibly renewable
Frequency: Annual
Study Establishment: Graduate Institute of International Studies, Geneva
Country of Study: Switzerland
No. of awards offered: 50
Application Procedure: Applicants must contact the Institute for details.
Closing Date: January 15th (for financial assistance scholarship request), May 30th (for currents students)
Funding: Government
Contributor: The Canton of Geneva and the Swiss Confederation
No. of awards given last year: 24
No. of applicants last year: 49
Additional Information: Scholars are exempt from Institute fees, but not from the obligatory fees of the University of Geneva, which confers the degree.

GRAINS RESEARCH AND DEVELOPMENT CORPORATION (GRDC)

PO Box 5367, Kingston, ACT 2604, Australia
Tel: (61) 2 6166 4500
Fax: (61) 2 6166 4599
Email: grdc@grdc.com.au
Website: www.grdc.com.au
Contact: Ms Contracts Coordinator

The Grains Research and Development Corporation is a statutory authority established to plan and invest in R&D for the Australian grains industry. Its primary objective is to support effective competition by Australian grain growers in global grain markets, through enhanced profitability and sustainability. Its primary business activity is the allocation and management of investment in grains R&D.

GRDC Grains Industry Research Scholarships

Subjects: Fields of high priority to the grains industry.
Purpose: To encourage postgraduate training in disciplines that contribute to the research, development and extension priorities of the GRDC and the Australian grains industry.
Eligibility: Candidates must be Australian citizens or permanent residents, must undertake full-time postgraduate study, must receive acceptance at a recognised research institution and must gain an Australian Postgraduate Scholarship or an equivalent base scholarship through a host university.
Level of Study: Doctorate, Postgraduate
Type: Scholarship
Value: A full annual operating budget of Australian $10,000 and an annual tax free top-up of Australian $16,875.
Length of Study: 3 years
Frequency: Annual
Study Establishment: Any university with a record of achievement for full-time research in the subject area leading to a DPhil.
Country of Study: Australia
No. of awards offered: Several
Application Procedure: Applicants are required to complete the GRDC's Grains Industry Research Scholarship (GRS) Application form available from the GRDC website. They must also arrange for three referee reports to be submitted separately.
Closing Date: Last Wednesday of October in each calendar year.
Funding: Government
Contributor: The government and Australian grain growers
No. of awards given last year: 13
No. of applicants last year: Approx. 30
Additional Information: The Corporation's 5-year research and development plan outlines the objectives and programmes to be covered. Copies of this may be obtained from the Secretariat.

GRDC Industry Development Awards

Subjects: Grains research and development.
Purpose: To fund study tours or training within Australia or overseas.
Eligibility: Open to Australian grain growers or groups working directly with growers for study tours, travel or other forms of training approved by the GRDC.
Level of Study: Unrestricted
Type: Award
Value: Up to Australian $15,000
Length of Study: Up to 6 months
Frequency: Annual
Country of Study: Any country
No. of awards offered: Several
Application Procedure: Applicants must submit five copies of the nominee's curriculum vitae, details of the proposed programme, the names, positions and locations of the proposed collaborators, approximate dates for the programme and details of any internal travel directly related to the proposed programme. A proposed budget, including the cost of international and internal travel, and expected accommodation and living expenses, an indication of other forms of support available to the nominee, evidence that the proposed collaborators are agreeable to the programme, supporting comments from two referees and a covering letter should also be included.
Closing Date: The first Wednesday of April and the last Wednesday of October in each calendar year.
Contributor: The government and Australian grain growers
Additional Information: On completion of the award, a report must be given to the Board. Preference may be given to applicants who have access to matching funds.

THE GREAT BRITAIN-CHINA EDUCATIONAL TRUST

15 Belgrave Square, London, SW1X 8PS, England
Tel: (44) 020 7235 6696
Fax: (44) 020 7245 6885
Email: contact@gbcc.org.uk
Website: www.gbcc.org.uk
Contact: L M Rivkin, Administrator

The Great Britain-China Educational Trust provides top-up grants to Chinese nationals in the final stages of their PhD courses. They should demonstrate their intention to return to China after completing their research.

Chinese Language Award

Subjects: Language study - Chinese, any dialect.
Purpose: Study of Chinese language for graduates who wish to make use of the language in their careers.
Eligibility: Graduates at undergraduate level minimum; UK citizens, normally resident in UK.
Level of Study: Foundation programme
Type: Grant
Value: Up to £2000
Length of Study: Up to 2 years
Frequency: Biannual
Application Procedure: Download application form from the website and the same return by post.
Closing Date: April 10th
Funding: Trusts
Contributor: Great Britain-China Educational Trust (GBCET)
No. of awards given last year: 10
No. of applicants last year: 33

Additional Information: Country of study - UK, Taiwan, China (in Hong Kong), Singapore. Chinese language as a major component of their undergraduate degree.

Chinese Student Awards

Subjects: All subjects.
Purpose: To provide a top-up grant to Chinese nationals in the final stages of their PhD courses in United Kingdom.
Eligibility: Open to Chinese candidates (from the PRC inc., Hong Kong) studying for a PhD in any subject and British postgraduate students (UK citizens only) giving conference papers in China, or traveling to China to pursue essential doctoral research.
Level of Study: Doctorate, Graduate, Postgraduate
Type: Award
Value: UK£1,500–2,000
Frequency: Twice a year
Country of Study: United Kingdom
No. of awards offered: Varies, usually 8 per meeting
Application Procedure: Check website for further details.
Closing Date: Check with website
Funding: Trusts
Contributor: Sino-British Fellowship Trust, the Universities' China Committee in London, and the Han Suyin Trust
No. of awards given last year: 18
No. of applicants last year: 100
Additional Information: Eligibility criteria vary very slightly for those people studying in the relevant subject area. British students on PhD courses that are related to China, may also appy for their grant, in order to subsidise field work in China or travel to China to present a conference paper.

GREAT MINDS IN STEM

3900 Whiteside Street, Los Angeles, CA 90063-1615, United States of America
Tel: (1) 323 292 0997
Fax: (1) 323 262 0946
Email: mvillafana@greatmindsinstem.org
Website: www.greatmindsinstem.org
Contact: Kathy Borunda Barrera, The Administrator

HENAAC Scholars Program

Purpose: To award scholarships to graduate and undergraduate science, technology, engineering and mathematics (STEM) students.
Eligibility: Full-time students majoring in STEM, with a minimum grade of 3.0 in their GPA, and active in student and community organizations and attending school in the United States.
Level of Study: Doctorate, Predoctorate
Type: Scholarship
Value: US$500–5,000 but varies
Frequency: Annual
Country of Study: United States of America
No. of awards offered: Approx. 108
Application Procedure: Application forms can be downloaded from the website www.greatmindsinstem.org
Closing Date: April 30th
Funding: Corporation, foundation, individuals
No. of awards given last year: 108

GREEK MINISTRY OF NATIONAL EDUCATION AND RELIGIOUS AFFAIRS

Greek Ministry of Education and Religious Affairs, Cultural and Sport, Directorate of International Relations in Education, Maroussi, 15180, Greece
Tel: (30) 00302103442469, 00302103443129
Fax: (30) 00302103442469
Email: des-a@minedu.gov.gr, des-art@minedu.gov.gr
Website: www.minedu.gov.gr
Contact: Directorate General of European and International Affairs

The Ministry's department of scholarships grants exclusively scholarships to students and Ph holders from developing countries through the OECD's D.A.C.

Scholarships for a Summer Seminar in Greek Language and Culture

Subjects: Greek language.
Purpose: To allow nationals from the Balkans, Eastern Europe, Asia and Africa to study Greek language.
Eligibility: Applicants must be nationals of Albania, Armenia, Azerbaijan, Bosnia & Herzegovina, China, Egypt, Ethiopia, FYROM, Georgia, India, Indonesia, Iran, Iraq, Jordan, Kazakhstan, Korea, Lebanon, Moldove, Montenegro, Mongolia, Pakistan, Palestine, Russia, Serbia, Sudan, Syria, Thailand, Tunisia, Turkey, Ukraine or Uzbekistan. Applicants should be foreign students/foreign professors of, Greek languages, or even foreign students/foreign professors of different fields who wish to omprove their level of Greek language.
Level of Study: One month Greek language seminar
Type: Scholarship
Value: The scholarships covers: accomodation, meals, tuition fees, small personal expenses, in case of an emergency medical care, visits to archaeological sites, museums, as well as instructive material
Frequency: Dependent on funds available
Country of Study: Greece
Closing Date: Check website
Funding: Government

Scholarships for Greek Language Studies in Greece

Subjects: Greek language.
Purpose: To allow nationals from the Balkans, Eastern Europe, Asia and Africa to study Greek Language.
Eligibility: Applicants must be nationals of Albania, Armenia, Azerbaijan, Bosnia and Herzegovina, China, Egypt, Ethiopia, FYROM, Georgia, India, Indonesia, Iran, Iraq, Jordan, Kazakhstan, Korea, Lebanon, Moldova, Montenegro, Mongolia, Pakistan, Palestine, Russia, Serbia, Sudan, Syria, Thailand, Tunisia, Turkey, Ukraine or Uzbekistan. Applicants should be students of Greek language abroad and have an excellent knowledge of modern Greek. Applicant must be foreign citizens, not of Greek orgin under 35 years of age.
Level of Study: Postgraduate, 1 month Greek language summer seminar
Type: Scholarship
Value: €650 for initial expenses, stay permit dues, tuition fees, free medical care and €550 for living expenses
Length of Study: Up to 1 year
Frequency: Dependent on funds available
Study Establishment: Aristotle University of Thessaloniki, School of Modern Greek Language
Country of Study: Greece
No. of awards offered: Up to 10
Application Procedure: Applicants should complete an application form, as well as provide letters of recommendation and relevant documentation.
Closing Date: April 30th
Funding: Government
No. of awards given last year: 11
Additional Information: School of Modern Greek Language: for the postgraduate scholarship for Greek language studies that last up to one academic year.

Scholarships for Postdoctoral Studies in Greece

Subjects: Any subject.
Purpose: To allow nationals from the Balkans, Eastern Europe, Asia and Africa to undergo postdoctoral research in Greece.
Eligibility: Applicants must be nationals of Albania, Armenia, Azerbaijan, Bosnia and Herzegovina, China, Egypt, Ethiopia, FYROM, Georgia, India, Indonesia, Iran, Iraq, Jordan, Kazakhstan, Korea, Lebanon, Moldova, Montenegro, Mongolia, Pakistan, Palestine, Russia, Serbia, Sudan, Syria, Thailand, Tunisia, Turkey, Ukraine or Uzbekistan. Applicants should have an excellent knowledge of Greek or French or English language. Applicant must be of foreign nationality and hold a doctorate degree and their first degree must be from a foreign university.
Level of Study: Postdoctorate
Type: Scholarship
Value: €800 for initial expenses, stay permit dues, tuition fees, free medical care and €700 for living expenses
Length of Study: 3 months
Country of Study: Greece

No. of awards offered: Up to 5
Application Procedure: Applicants should complete an application form as well as provide letters of recommendation and relevant documentation. Applications should be submitted to the Greek embassies.
Closing Date: April 30th
Funding: Government
No. of awards given last year: 4

Scholarships for Postgraduate Studies in Greece

Subjects: Any subject.
Purpose: To allow nationals from the Balkans, Eastern Europe, Asia and Africa to study in Greece.
Eligibility: Applicants must be nationals of Albania, Armenia, Azerbaijan, Bosnia and Herzegovina, China, Egypt, Ethiopia, FYROM, Georgia, India, Indonesia, Iran, Iraq, Jordan, Kazakhstan, Korea, Lebanon, Moldova, Montenegro, Mongolia, Pakistan, Palestine, Russia, Serbia, Sudan, Syria, Thailand, Tunisia, Turkey, Ukraine or Uzbekistan. Applicants must have a graduate degree from a foregin universit, an excellent knowledge of modern Greek, or English or French. Applicant must be of foreign citizens, not of Greek origin under 35 years of age.
Level of Study: Postgraduate, Graduate
Type: Scholarship
Value: €650 for initial expenses, stay permit dues, tuition fees, free medical care, and €550 for living expenses
Length of Study: 2 year MA plus 1 year Greek Language study
Frequency: Dependent on funds available
Country of Study: Greece
No. of awards offered: Up to 10
Application Procedure: Applicants should complete an application form, as well as provide letters of recommendation and relevant documentation. Applications should be submitted to Greek embassies.
Closing Date: April 30th
Funding: Government
No. of awards given last year: 11

Scholarships Granted by the GR Government to Foreign Citizens

Subjects: All subjects.
Purpose: To support candidates who wish to study or conduct research project in Greek Universities, or summer seminars of Greek language and culture, postgraduate studies.
Eligibility: Applicants must be nationals of China, Belgium, Bulgary France, Germany, Serbia, Syria, Turkey. Applicants should have an excellent knowledge of Greek or French or English language.
Applicant must be of foreign nationality of Estonia, Israel, Croatia, Cyprus, Luxembourg, Mexico, Norway, netherlands, hungary, Poland, Romania, Slovakia, Slovenia, Czech Republic, and Finland.
Level of Study: Postdoctorate, Research, Doctorate, Graduate, MBA, Postgraduate, Predoctorate, 1 month seminar of Greek language and culture
Type: Scholarship
Value: €550 per month, €500 lump sum for establishment expenses, €150 for transport expenses, exemption from tuition fees (only in selected master's degree) plus all expenses for summer seminars except travel expenses.
Length of Study: Varies
Frequency: Annual
Study Establishment: Greek public universities
Country of Study: Greece
No. of awards offered: Up to 350
Application Procedure: Check with Ministry of Education or Ministry of Foreign Affairs in Individual country. Applicants must apply through their home countries. Find out about deadlines of their own countries but also refer to the webiste.
Closing Date: March 31st
Funding: Government
Contributor: Greek Ministry of Education and Religious Affairs, Culture and Sport
No. of awards given last year: 40
No. of applicants last year: 60
Additional Information: Information about eligible countries, number of scholarships and the way to apply is renewed every year and candidates can find next year's decision by the end of December at www.minedu.gov.gr.

For further information contact:

Directorate of Studies and Student's Welfare, 00302103442365,
Tel: 00302103443469, 00302103443451
Email: foitmer.yp@ minedu.gov.gr
Website: www.minedu.gov.gr
Contact: Evi Zigra, Director

GRIFFITH UNIVERSITY

Nathan campus Griffith University, 170 Kessels Road, Nathan, QLD, 4111, Australia
Tel: (61) (07) 3735 3870
Fax: (61) (07) 3735 7957
Email: scholarships@griffith.edu.au
Website: www.gu.edu.au

In the pursuit of excellence in teaching, research and community service, Griffith University is committed to innovation, bringing disciplines together, internationalization, equity and social justice and lifelong learning, for the enrichment of Queensland, Australia and the international community.

Griffith University Postgraduate Research Scholarships

Subjects: All subjects.
Purpose: To provide financial support for candidates undertaking full-time research leading to the award of the degree of MPhil or PhD.
Eligibility: Open to any person, irrespective of nationality. Applicants must have completed, or expect to complete, a Bachelors degree with honours equivalent to first class and be intending to enrol full-time in an approved doctoral or research Master's programme. Applicants must have achieved First Class Honours or equivalent.
Level of Study: Postgraduate
Type: Scholarship
Value: Australian $24,653 per year
Length of Study: Up to 2 years for a research Master's programme and up to 3 years for a doctoral programme, with a possible extension of up to 6 months for the doctoral students, subject to satisfactory progress.
Frequency: Annual
Study Establishment: Griffith University
Country of Study: Australia
No. of awards offered: Varies
Application Procedure: Applicants must complete an application form.
Closing Date: Please check website
Additional Information: This scholarship does not cover tuition fees.

Jackson Memorial Fellowship

Subjects: The application of the social, political, economic, environmental or technological sciences to the analysis and resolution of substantial policy issues at the national or regional levels.
Purpose: To consolidate links with a variety of institutions in South East Asia and to provide funding to facilitate visits to Griffith University by faculty staff of the Association of South East Asian Institutions of Higher Learning (ASAIHL) member institutions.
Eligibility: Open to senior members of faculty staff of the ASAIHL member institutions.
Level of Study: Professional development
Type: Fellowship
Value: Australian $2,500 payable to the fellow to assist with travel to and from Australia; Australian $350 stipend per week paid to the fellow during the period of residence up to a total of $4,200;
remaining costs associated with the visit, up to a total of Australian $1,200
Frequency: Annual
Study Establishment: Griffith University
Country of Study: Australia
No. of awards offered: 1
Application Procedure: Applicants must apply through the heads of their employing institutions.
Closing Date: October 25th
Funding: Private

No. of awards given last year: 1
No. of applicants last year: Varies

Sir Allan Sewell Visiting Fellowship

Subjects: All subjects offered by Griffith University.
Purpose: To commemorate the distinguished service of Sir Allan Sewell to Griffith University by offering awards to enable visits by distinguished scholars engaged in academic work who can contribute to research and teaching in one or more areas of interest at a faculty or college of the university.
Eligibility: Open to researchers of any nationality.
Level of Study: Professional development
Type: Fellowship
Value: Australian $8,000 ($6,000 of which will be contributed by the Research Committee and $2,000 from the host research centre, school, or group).
Frequency: Annual
Study Establishment: Griffith University
Country of Study: Australia
No. of awards offered: Up to four
Application Procedure: Applicants must be invited to apply by faculties or colleges of the University.
Closing Date: October 25th
Funding: Private
No. of awards given last year: Varies
No. of applicants last year: Varies

GROUPE DE RECHERCHE SUR LE SYSTÈME NERVEUX CENTRAL

Faculty of Medicine, Universite de Montréal, PO Box 6128, Station Centre-Ville, Montréal, QC, H3C 3J7, Canada
Tel: (1) 514 343 6269
Fax: (1) 514 343 5850
Email: louis-eric.trudeau@umontreal.ca, sonia.gosselin@umontreal.ca
Website: www.grsnc.umontreal.ca
Contact: Dr Trevor Drew, Director

Founded in 1991, the Groupe de Recherche sur le Système Nerveux Central (GRSNC) is a multidisciplinary research group that includes researchers from several departments within the Faculties of Medicine and Dentistry at the Universite de Montréal. It receives funding from both the University and from the provincial government to support its research infrastructure. It organizes an annual international symposium and has weekly research seminars by invited speakers.

Herbert H Jasper Fellowship

Subjects: Neurology and neurosciences.
Purpose: To enable the use of the exceptional research facilities of the Groupe de Recherche sur le Système Nerveux Central of the University of Montréal.
Eligibility: Open to Canadian citizens or permanent residents.
Level of Study: Postdoctorate
Type: Fellowship
Value: Canadian $45,000 per year
Length of Study: 2 years
Frequency: Annual
Study Establishment: Group de Recherche sur le Système Nerveux Central, University of Montréal
Country of Study: Canada
No. of awards offered: 1
Application Procedure: Applicants must complete an application form, which can be obtained from the website or by writing to the Fellowship Committee.
Closing Date: January 31st
Funding: Government
No. of awards given last year: 1
Additional Information: The fellowship provides the opportunity for the recipient to work closely with the investigator of his or her choice within a large active group of neuroscientists who are members of the group.

THE GRUNDY EDUCATIONAL TRUST

Jefford Cottage, 3 Parkside Lane, Ropley, Hants, S024 0BB, England
Tel: (44) 1962 773118
Email: alicia.hardy@hotmail.co.uk
Website: www.grundyeducationaltrust.org.uk
Contact: Mrs A Hardy, Secretary to the Trustees

The Grundy Educational Trust was established in 1991 to advance education by providing or assisting in the provision of graduate and postgraduate awards to students for research and higher learning at selected institutions in the United Kingdom, namely, Surrey University, Loughborough University, Nottingham University, Imperial College, London, and University of Manchester (UMIST).

Grundy Educational Trust

Subjects: Technologically or scientifically based disciplines in industry and commerce.
Purpose: To assist in covering maintenance costs while obtaining postgraduate or second degrees.
Eligibility: Open to United Kingdom citizens under 30 years of age.
Level of Study: Postgraduate
Type: Award
Value: Up to UK£4,500
Length of Study: 1–5 years
Frequency: Annual
Study Establishment: Surrey University, Loughborough University, Imperial College, London, Nottingham University and University of Manchester (UMIST)
Country of Study: United Kingdom
No. of awards offered: Up to 16
Application Procedure: Applicants must apply through the five selected universities only and not directly.
Closing Date: March 30th
Funding: Private

GUILLAIN-BARRÉ SYNDROME SUPPORT GROUP

Woodholme House, Heckington Business Park, Station Road, Sleaford, Heckington, NG34 9JH, England
Tel: (44) 01529 469910
Fax: (44) 01529 469915
Email: admin@gbs.org
Website: www.gbs.org.uk
Contact: Administration Officer

The Guillain-Barré Syndrome Support Group provides emotional support, personal visits and comprehensive literature to patients and their relatives and friends. The Group also educates the public and the medical community about the Support Group and maintains their awareness of the illness. The Group fosters research into the causes, treatment and other aspects of the illness and encourages fund raising and support for its activities.

Guillain-Barré Syndrome Support Group Research Fellowship

Subjects: Any aspect of Guillain-Barré syndrome (GBS) or related diseases including chronic inflammatory demyelinating polyradiculoneuropathy (CIDP).
Purpose: To advance research into the prevention and cure of GBS and CIDP.
Level of Study: Doctorate, Postgraduate, Professional development, Research
Type: Fellowship
Value: Up to UK£65,000
Length of Study: Up to 3 years
Frequency: Dependent on funds available
Study Establishment: Any suitable hospital, university laboratory or department
No. of awards offered: 1
Application Procedure: Applicants must write for an application form.
Closing Date: Please contact the organization

Contributor: Members' donations, fund raising and trust funds
No. of awards given last year: 1
Additional Information: Further information is available on request.

THE GYPSY LORE SOCIETY

5607 Greenleaf Road, Cheverly, MD, 20785, United States of America
Tel: (1) 301 341 1261
Fax: (1) 301 341 1261
Email: headquarters@gypsyloresociety.org
Website: www.gypsyloresociety.org
Contact: Ms Sheila Salo, Treasurer

The Gypsy Lore Society, an international association of persons interested in Gypsy Studies, was formed in the United Kingdom in 1888. The Gypsy Lore Society, North American Chapter, was founded in 1977 in the United States of America and since 1989, has continued as the Gypsy Lore Society. The Society's goals include the promotion of the study of the Gypsy peoples and analogous itinerant or nomadic groups, dissemination of information aimed at increasing understanding of Gypsy culture in its diverse forms and establishment of closer contacts among Gypsy scholars.

Gypsy Lore Society Young Scholar's Prize in Romani Studies

Subjects: Any topic in Romani (Gypsy) studies.
Purpose: To recognize outstanding work by young scholars in Romani (Gypsy) studies.
Eligibility: Graduate students beyond the 1st year of study and PhD holders no more than 3 years beyond the degree. An unpublished paper not under consideration for publication is eligible for this award as well as self-contained scholarly articles of publishable quality that treat a relevant topic in an interesting and insightful way.
Level of Study: Postdoctorate, Doctorate, Graduate
Type: Cash prize
Value: US$500
Frequency: Varies
Study Establishment: Any
Country of Study: Any country
No. of awards offered: 1
Application Procedure: Submission file format is rich text file (RTF, PDF, MS word compatible). Files bigger than 5 MB should be presented on CD to the postal address below. A cover sheet should be included with the title of the paper, the author's name, affiliation, mailing, email address, telephone and fax number, date of entrance into an appropriate program or of awarding of the PhD, and US social security number, if the author has one. The applicant's name should appear on the cover sheet only.
Closing Date: October 30th
Funding: Corporation
Contributor: Gypsy Lore Society

For further information contact:

Gypsy Lore Society Prize Competition, Institute of Musicology, Hungarian Academy of Sciences, H-1250 Budapest, Pf 28, Hungary
Email: kovalcsik@zti.hu
Contact: Katalin Kovalcsik

H.E.A.R.T UK - THE CHOLESTEROL CHARITY

7 North Road, Maidenhead, Berkshire, SL6 1PE, United Kingdom
Tel: (44) 1628 777 046
Fax: (44) 1628 628 698
Email: cr@heartuk.org.uk
Website: www.heartuk.org.uk
Contact: Cathy Ratcliffe , Deputy Director

HEART UK sponsors research in the field of hyperlipidaemia, atherosclerosis (including coronary heart disease, stroke, peripheral arterial disease) with a special emphasis on genetic hyperlipidaemias including familial hypercholesterolaemia, familial combined hyperlipidaema and genetic hypertriglyceridaemia syndromes. Awards are open to basic scientists, professions allied to medicine and medical graduates to assist in travel for the purposes of either direct costs involved in research or to visit other laboratories to learn new techniques.

Conference Awards

Subjects: The charity's Annual Conference.
Eligibility: Open to students, junior clinical, scientific and paramedical staff. Applicants who have been successful in obtaining a conference grant in the last 2 years are not eligible to apply.
Value: Cover travel and registration to the charity's Annual Conference
Application Procedure: Submit your request by email to the Secretariat office (wheldonevents@btconnect.com) detailing why you should be given a grant to attend in no more than 200 words. Attach a supporting letter from your head of department explaining why the travel grant is needed and cannot be locally funded. All applicants will be notified no later than mid April and advised if their application has been successful.
Closing Date: March 24th

HAEMATOLOGY SOCIETY OF AUSTRALIA AND NEW ZEALAND (HSANZ)

145 Macquarie Street, Sydney, NSW 2000, Australia
Tel: (61) 2 9256 5456
Fax: (61) 2 9252 0294
Email: hsanz@hsanz.org.au
Website: www.hsanz.org.au
Contact: Lexy Harris, Administrative Officer

Haematology Society of Australia promote, foster, develop and assist the study and application of haematology. Its main aim is to promote improved standards, interest and research in all aspects of haematology and opportunities for meeting others in related fields of interest and discussing matters of common interest. It nourishes interest haematology amongst other interested persons including regional and international bodies.

Baikie Award

Subjects: Haematology.
Purpose: To recognize the best presentation (either oral or poster) at the annual scientific meeting by a new investigator who is a financial member of the society.
Eligibility: Open to new investigators who were awarded their postgraduate qualification (MSc, FRACP, FRCPA or PhD) within the past 5 years.
Level of Study: Doctorate, Postgraduate, Research
Type: Award
Value: Australian $3,000 plus Baikie Medal
No. of awards offered: 1–2
Application Procedure: Applications will be called for in the registration brochure of the annual meeting with the closing date the same as the abstract submission. The award is announced on the last day of the annual meeting.
Closing Date: Please check website

Hsanz Educational Grants

Subjects: All subjects.
Purpose: To support trainee members of the society to attend and present at an international meeting.
Eligibility: Applicants need to be trainee Members of the Society.
Level of Study: Postgraduate
Type: Grant
Value: Australian $3,000 each
No. of awards offered: 4
Application Procedure: Check website for further details.
Closing Date: October 11th
Additional Information: All applications will be considered by HSANZ Council and the successful applicants will be announced at the annual meeting.

Hsanz Travel Grant

Subjects: Science and medicine.

Purpose: To assist members (scientists and medical) with attendance at annual scientific meetings.
Eligibility: Applicants must submit an abstract and register for the meeting.
Level of Study: Research
Type: Grant
Value: Up to Australian $1,000
No. of awards offered: 20
Application Procedure: Tenable at the HSANZ Annual Scientific Meeting. Please check the website address.
Closing Date: Please check website

New Investigator Scholarships

Subjects: Haematology.
Purpose: To enhance the scientific stature of Australian and New Zealand haematology by providing the opportunity for medical or science graduates who are undertaking advanced training in haematology to gain experience in acknowledged centres of excellence.
Eligibility: Open for medical or science graduates who are undertaking advanced training in haematology.
Level of Study: Graduate, Postgraduate, Research
Value: AUD50,000 (tax free status if enrolled for a full-time higher degree)
Length of Study: 1 year
Frequency: Annual
No. of awards offered: Up to 5 Scholorships
Application Procedure: Check website for further details.
Closing Date: June 18th
Contributor: AMGEN, Bayer Schering Pharma, Novartis. HSANZ matches sponsors' funds dollar for dollar

Roche Haematology Fellowship

Subjects: All subjects.
Purpose: To support clinical or translational research, or other project initiatives of benefit to the haematology community within Australia.
Eligibility: Open to individuals undertaking advanced training in medical oncology or haematology, or to more senior oncologists or haematologists with limited research experience.
Level of Study: Postgraduate, Research
Type: Grant
Value: Australian $50,000
No. of awards offered: 1
Application Procedure: Check the website for further details.
Closing Date: June 18th

For further information contact:

Website: www.cosa.org.au

HAGLEY MUSEUM AND LIBRARY

PO Box 3630, Wilmington, Delaware, DE 19807-0630, United States of America
Tel: (1) 302 658 2400
Fax: (1) 302 655 3188
Email: clockman@hagley.org
Website: www.hagley.org
Contact: Ms Carol Ressler Lockman, Center Coordinator

Located along the Brandywine River on the site of the first du Pont black powder works, the Hagley Museum and Library provide a unique glimpse into American life at home and at work in the 19th century. Set among more than 230 acres of trees and flowering shrubs, Hagley offers a diversity of restorations, exhibits and live demonstrations for visitors of all ages.

Hagley Museum and Library Grants-in-Aid of Research

Subjects: History of business, technology and science and the library, archival and artifact collections of the Hagley Museum and Library research grants.
Purpose: To assist with travel and living expenses while using the research collections.
Eligibility: Open to degree candidates and advanced scholars of any nationality. Research must be relevant to Hagley's collections.
Level of Study: Doctorate, Graduate, Postdoctorate, Predoctorate
Type: Grant
Value: Up to US$1,600 per month
Length of Study: 2–8 weeks
Frequency: Quarterly
Study Establishment: The Library
Country of Study: United States of America
No. of awards offered: Varies
Application Procedure: Applicants must submit a completed application form with a five-page proposal.
Closing Date: March 31st, June 30th or October 29th
Funding: Private
Contributor: Foundation funds
Additional Information: Candidates may apply for research in the imprint, manuscript, pictorial and artefact collections of the Hagley Museum and Library. In addition the resources of the 125 libraries in the greater Philadelphia area will be at the disposal of the visiting scholar. The Research Fellowship is to be used only in the Hagley Library.

Henry Belin du Pont Dissertation Fellowship in Business, Technology and Society

Subjects: Business and technology.
Purpose: To aid students whose research on important historical questions would benefit from the use of Hagley's research collections.
Eligibility: Open to graduate students or PhD candidates.
Level of Study: Predoctorate, Graduate
Type: Fellowship
Value: US$6,000
Length of Study: 4 months
Frequency: Annual
Study Establishment: The Center for the History of Business, Technology and Society at Hagley
Country of Study: United States of America
No. of awards offered: 3
Application Procedure: Applicants must submit an application dossier including a dissertation prospectus, a statement concerning the relevance of Hagley's research collections to the project and at least two letters of recommendation. Writing samples are also welcome. Potential applicants are strongly encouraged to consult with Hagley staff prior to submitting their dossier.
Closing Date: November 15th
Funding: Private
No. of awards given last year: 2
No. of applicants last year: 5
Additional Information: Recipients are expected to have no other obligations during the term of the fellowship, to maintain continuous residence at Hagley for its duration and to participate in events organized by Hagley's Center for the History of Business, Technology and Society. Towards the end of the residency the recipient will make a presentation at Hagley based on research conducted during the Fellowship. Hagley should also receive a copy of the dissertation, as well as any publications aided by the Fellowship.

Henry Belin du Pont Fellowship

Subjects: Areas of study relevant to the library's archival and artefact collections.
Purpose: To enable scholars to pursue advanced research and study in the library, archival, pictorial, and artifact collections of the Hagley Museum and Library.
Eligibility: Open to applicants who have already completed their formal professional training. Consequently, degree candidates and persons seeking support for degree work are not eligible to apply. Applicants must not be residents of Delaware and preference will be given to those whose travel costs to Hagley will be higher. Research must be relevant to Hagley's collections.
Level of Study: Doctorate, Postdoctorate, Predoctorate
Type: Fellowship
Value: US$400/week for recipients who reside more than 50 miles from Hagley, and US$200/week for those within 50 miles.
Length of Study: Maximum of 8 weeks
Frequency: Quarterly
Study Establishment: The Library
Country of Study: United States of America
No. of awards offered: Varies

Application Procedure: Applicants must submit a completed application form with a five-page proposal.
Closing Date: March 31st, June 30th or October 31st
Funding: Private
Contributor: Foundation funds
Additional Information: Fellows must devote all their time to study and may not accept teaching assignments or undertake any other major activities during the tenure of their fellowships. At the end of their tenure, Fellows must submit a final report on their activities and accomplishments. As a centre for advanced study in the humanities, Hagley is a focal point of a community of scholars. Fellows are expected to participate in seminars, which are conducted periodically, as well as attend colloquia, lectures, concerts, exhibits and other public programmes offered during their tenure. Research fellowships are to be used in the Hagley Library only, not as scholarships for college.

THE HAGUE ACADEMY OF INTERNATIONAL LAW

Peace Palace, Carnegieplein 2, The Hague, NL-2517 KJ, Netherlands
Tel: (31) 70 302 4242
Fax: (31) 70 302 4153
Email: registration@hagueacademy.nl
Website: www.hagueacademy.nl
Contact: Sophie de Seze, Communication and publications Assistant

The Hague Academy organizes summer courses in public and private international law, with the aim of furthering scientific and advanced studies of the legal aspects of international relations. The summer courses take place over a period of six weeks in July and August. Please visit our website at www.hagueacademy.nl for all pertinent information.

Centre for Studies and Research

Subjects: Private or public international law.
Purpose: To bring together young international lawyers of a high standard from all over the world to undertake original research and work under the direction of professors on a common general theme determined each year by the Academy.
Eligibility: No restriction.
Level of Study: Postdoctorate
Type: Scholarship
Value: €35 (daily allowance) and reimbursement of half of the travel expenses upto a maximum of €910
Length of Study: 3 weeks
Frequency: Annual
Study Establishment: Hague Academy of International Law
Country of Study: Netherlands
No. of awards offered: 24 (12 English speaking, 12 French speaking)
Application Procedure: Online registration form available on the website of the Academy from November 1st to April 1st.
Closing Date: April 1st
Funding: Private
Contributor: Governments
No. of awards given last year: 24
No. of applicants last year: 40

Hague Academy of International Law Seminar for Advanced Studies

Subjects: International private or public law. Each year a subject is chosen.
Purpose: A limited number of scholarships can be awarded to candidates from developing countries and countries in transition. The programme is addressed to practitioners with a legal education including basic training in international law.
Eligibility: Practitioners with a legal education including basic training in international law.
Level of Study: Unrestricted
Type: Scholarship
Value: Applicants should contact Academy for details
Length of Study: 1 week
Frequency: Annual
Country of Study: Netherlands
Application Procedure: Online registration form available on the website of the Academy from May to August.
Closing Date: August 31st
Funding: Foundation, government
Contributor: The Hague Academy of International Law
No. of awards given last year: 10
No. of applicants last year: 75

Hague Academy of International Law/Doctoral Scholarships

Subjects: Private or Public International law.
Purpose: To aid individuals with the completion of their theses through research assistance at the Peace Palace Library.
Eligibility: Open to doctoral candidates from developing countries who reside in their home country and do not have access to scientific sources. No more age restriction.
Level of Study: Doctorate
Type: Scholarships
Value: €35 (daily allowance) and reimbursement of half of the travel expenses upto a maximum of €910
Length of Study: 2 months
Frequency: Annual
Country of Study: The Netherlands
No. of awards offered: 2 (for English) and 2 (for French speaking participants)
Application Procedure: Applicants must submit their applications with a letter of recommendation from the professor under whose direction the thesis is being written. The thesis may be concerned with either private or public international law and the title should be mentioned.
Closing Date: March 1st
Funding: Government
Contributor: The Hague Academy of International Law
No. of awards given last year: 4
No. of applicants last year: 30

Hague Academy of International Law/Scholarships for Summer Courses

Subjects: International private or public law.
Purpose: To assist students with living expenses, including the registration fee, during summer courses.
Eligibility: Open to candidates who have not yet received an Academy scholarship. Applicants must have sufficient knowledge of English or French. No more age restriction.
Level of Study: Doctorate, Graduate
Type: Scholarship
Value: Applicants should contact the Academy for details. Scholars are exempt from registration fees and examination fees.
Length of Study: 3 weeks
Frequency: Annual
Study Establishment: The Hague Academy of International Law
Country of Study: The Netherlands
No. of awards offered: Varies, 100
Application Procedure: Online registration form, available from 1st November to 1st March.
Closing Date: March 1st
Funding: Private
Contributor: Foundations, institutions and personalities
No. of awards given last year: 110
No. of applicants last year: 600

THE HAMBIDGE

PO Box 339, Rabun Gap, GA, 30568, United States of America
Tel: (1) 706 746 5718
Fax: (1) 706 746 9933
Email: center@hambidge.org
Website: www.hambidge.org
Contact: The Residency Director

The Hambidge Center's primary function is an artist residency programme with the following aims: to provide artists with time and space to pursue their work, to enhance their communities' art environment, provide public accessibility and to protect and sustain the natural environment, land and endangered species. The Center is set in 600 acres of mountain and valley terrain with waterfalls and nature trails.

Hambidge Residency Program Scholarships

Subjects: Any field or discipline of creative work.
Purpose: To provide applicants with an environment for creative work in the arts and sciences.
Eligibility: Open to qualified applicants in all disciplines who can demonstrate seriousness, dedication and professionalism. International residents are welcome. The Fulton County Arts Council Fellowship is open to residents of Fulton County, Georgia only.
Level of Study: Unrestricted
Type: Fellowship
Value: US$200 per week (of the US$1250 per week cost)
Length of Study: 2 weeks to 2 months
Frequency: Dependent on funds available
Study Establishment: Hambidge
Country of Study: United States of America
Application Procedure: Applicants must submit an application form and work samples to the centre marked for the attention of the Residency Program. The application form can be downloaded from the website.
Closing Date: January 15th, April 15th or September 15th
Funding: Foundation, government, private
No. of awards given last year: 102
Additional Information: The scholarships that are offered by the Center are the Nellie Mae Rowe Fellowship, the Fulton County Arts Council Fellowship and teaching fellowships at public or independent schools.

Nellie Mae Rowe Fellowship

Subjects: Any field or discipline of creative work.
Purpose: The scholarship was established to serve the memory of Nellie Mae Rowe, to recognize the creativity of those artists who come to Hambidge in her name and to encourage the artistic growth of African-American visual artists.
Eligibility: African-American
Level of Study: Unrestricted
Type: Fellowship
Value: All fees
Length of Study: 2 weeks
Frequency: Annual
Study Establishment: Hambidge
Country of Study: United States of America
No. of awards offered: 1
Application Procedure: Applicants must submit an application and work samples to the Center. The application form can be downloaded from the website.
Closing Date: January 15th, April 15th, September 15th
Contributor: Judith Alexander
No. of awards given last year: 1

Rabun Gap-Nacoochee School Teaching Fellowship

Subjects: Any field or discipline of creative work.
Purpose: To develop the calibre of creative thinkers.
Level of Study: Unrestricted
Type: Fellowship
Value: All fees
Frequency: Dependent on funds available
Study Establishment: Hambidge
Country of Study: United States of America
No. of awards offered: 1

HAND WEAVERS, SPINNERS & DYERS OF ALBERTA (HWSDA)

Valerie Forcese, 4951 Viceroy Drive NW, Calgary, AB, T3A 0V2, Canada
Tel: (1) 780 672 2551
Fax: (1) 780 672 5887
Email: studioword@studioword.com
Website: www.hwsda.org

The HWSDA is an exciting network of fibre artisans whose objectives are to foster and promote the development of fibre arts in the province of Alberta for both amateur and professional crafts people.

HWSDA Scholarship Program

Subjects: Textile design.
Purpose: To gain more knowledge in the field of weaving, spinning, dying, felting, basketry.
Eligibility: Open to HWSDA members who are involved in the art of textile design and who need financial help.
Level of Study: Professional development
Type: Scholarship
Value: Up to US$600 in total
Length of Study: 2 years
Frequency: Annual
Country of Study: Canada
Application Procedure: A complete application form and proposals should be submitted to the Vice President.
Closing Date: April 30th
Funding: Foundation
Contributor: The memorial scholarship fund
No. of awards given last year: 3
No. of applicants last year: 3

HARNESS TRACKS OF AMERICA

4640 East Sunrise Drive, Suite 200, Tucson, AZ 85718, United States of America
Tel: (1) 520 529 2525
Fax: (1) 520 529 3235
Email: info@harnesstracks.com
Website: www.harnesstracks.com
Contact: Jennifer Foley, Manager of Web Development

Harness Tracks of America, Inc. is an association of the finest harness racing establishments in the world, dedicated to the advancement and progress of the sport.

Harness Track of America Scholarship

Subjects: Sports.
Purpose: To provide financial assistance to young people actively engaged in the harness racing industry.
Eligibility: Open to applicants with active harness racing involvement.
Level of Study: Graduate, Postgraduate, Professional development
Type: Scholarship
Value: US$5,000
Length of Study: 1 year
Frequency: Annual
Country of Study: United States of America or Canada
No. of awards offered: 4 per cent of the applications
Application Procedure: A completed application form and official academic transcripts must be submitted.
Closing Date: May 15th
Funding: Private
No. of awards given last year: 5
No. of applicants last year: 31

THE HARRY FRANK GUGGENHEIM FOUNDATION (HFG)

25 West 53rd Street, 16th Floor, New York, NY, 10019, United States of America
Tel: (1) 646 428 0971
Fax: (1) 646 428 0981
Email: info@hfg.org
Website: www.hfg.org
Contact: Administrative Assistant

The Harry Frank Guggenheim Foundation (HFG) sponsors scholarly research on problems of violence, aggression and dominance. HFG provides both research grants to established scholars and dissertation fellowships to graduate students during the dissertation writing year. The *HFG Review of Research* is published and a report ID published occasionally every 3 years.

HFG Foundation Dissertation Fellowships

Subjects: Natural and social sciences, humanities.
Purpose: To financially support the research students and increase understanding of the causes, manifestations and control of violence, aggression and dominance. Highest priority is given to research that can increase understanding and amelioration of urgent problems of violence, aggression and dominance in the modern world.
Eligibility: Open to PhD candidates of any nationality who are in the writing stage of their dissertation.
Level of Study: Doctorate, Postdoctorate, Research
Type: Fellowships
Value: US$20,000 each
Frequency: Annual
Country of Study: United States of America
No. of awards offered: 10 or more
Application Procedure: Applicants must submit their application form, abstract, advisor's letter, curriculum vitae, a list of any relevant publications, transcripts and research plan. Application forms are available on the website or on request from the foundation.
Closing Date: February 1st
Funding: Foundation

HFG Research Program

Subjects: The social, behavioural and biological sciences. Research that is related to the Foundation's programme will be considered regardless of the disciplines involved.
Purpose: To promote understanding of the human social condition through the study of the causes and consequences of dominance, aggression and violence.
Eligibility: Open to individuals or institutions in any country.
Level of Study: Postdoctorate, Predoctorate
Value: $15,000 to $40,000 a year
Length of Study: 1 or 2 years
Frequency: Annual
Country of Study: Any country
No. of awards offered: 15–35 per year
Application Procedure: Applicants must submit an application form and research proposal along with a curriculum vitae and budget request. Application materials are available by contacting the Foundation.
Closing Date: August 1st
Funding: Foundation
No. of awards given last year: 15
No. of applicants last year: 200
Additional Information: The Foundation operates a programme of specific and innovative study and research. Proposals should be for a specific project and should describe well-defined aims and methods, not general institutional support.

THE HARRY S TRUMAN LIBRARY INSTITUTE

500 West US Highway 24, Independence, MO, 64050, United States of America
Tel: (1) 816 268 8200
Fax: (1) 816 268 8295
Email: lisa.sullivan@nara.gov
Website: www.trumanlibrary.org
Contact: Lisa Sullivan, Grants Administrator

The Harry S Truman Library Institute is a non-profit partner of the Harry S Truman Library. The institute's purpose is to foster the Truman Library as a centre for research and as a provider of educational and public programmes.

Harry S Truman Library Institute Dissertation Year Fellowships

Subjects: The public career of Harry S Truman and the history of the Truman administration.
Purpose: To encourage historical scholarship in the Truman era.
Eligibility: Open to graduates who have completed their dissertation research and are ready to begin writing. Dissertations must be on some aspect of the life and career of Harry S Truman or of the public and policy issues that were prominent during the Truman years.
Level of Study: Graduate, Postgraduate
Type: Fellowship
Value: US$16,000, payable in two instalments
Length of Study: 1 year
Frequency: Annual
Country of Study: United States of America
No. of awards offered: 1–2
Application Procedure: Application forms are available from the website.
Closing Date: February 1st
Funding: Private
Additional Information: Recipients will not be required to come to the Truman Library but will be expected to furnish the Library with a copy of their dissertation.

THE HASTINGS CENTER

21 Malcolm Gordon Road, Garrison, NY, 10524, United States of America
Tel: (1) 845 424 4040
Fax: (1) 845 424 4545
Email: mail@thehastingscenter.org
Website: www.thehastingscenter.org
Contact: Ms Lori P Knowles, Executive Vice President

The Hastings Center is an independent, non-profit research and educational institute that studies ethical, social and legal issues in medicine, the life sciences, health policy and environment policy.

Hastings Center International Visiting Scholars Program

Subjects: academia, medicine, law and the media, as well as the humanities, the sciences and the professions.
Purpose: To enable international visiting scholars to conduct independent research on issues in or related to bioethics.
Eligibility: Open to international scholars.
Level of Study: Doctorate, Graduate, Postdoctorate, Postgraduate, Predoctorate, Professional development, Research, Unrestricted
Type: Grant
Value: Assistance with accommodation costs is available based on need
Length of Study: Usually 2–8 weeks
Frequency: Annual
Study Establishment: The Hastings Center
Country of Study: United States of America
No. of awards offered: Varies
Application Procedure: Applicants must visit the website for applications. They must submit an approximately 750-word proposal describing research interests and work plan for proposed stay, a curriculum vitae and the names and addresses of two references.
Closing Date: Applications are accepted at any time, but should be submitted at least 2 months prior to the proposed stay
Funding: Private
No. of awards given last year: 13
No. of applicants last year: 15
Additional Information: Participation in the ongoing activities of the Center such as conferences, seminars and workshops is encouraged.

HATTORI FOUNDATION

7 Exton Street, London, SE1 8UE, England
Tel: (44) 20 7620 3053
Fax: (44) 20 7620 3054
Email: admin@hattorifoundation.org.uk
Website: www.hattorifoundation.org.uk
Contact: Mrs Sarah C Dickinson, Administrator

The chief aim of the Hattori Foundation is to encourage and assist exceptionally talented young instrumental soloists or chamber ensembles who are British nationals or resident in the United Kingdom, and whose talent and achievement give promise of an international career.

Hattori Foundation Awards

Subjects: Instrumental, solo performance and ensembles.
Purpose: To assist young instrumentalists of exceptional talent in establishing a solo or chamber music career at international level.

Eligibility: Open to British or foreign nationals aged 21–27 years studying full-time in the United Kingdom. Candidates should be of postgraduate performance level.
Level of Study: Postgraduate, Professional development
Type: Award
Value: No pre-determined amounts. The grant is based on the requirements of the approved project
Length of Study: Varies
Frequency: Annual
Country of Study: British Nationals can study in any country. Foreign nationals must be resident in the United Kingdom only
No. of awards offered: Up to 20
Application Procedure: Applicants must submit a completed application form with reference forms and a 30 minute performance (recital) on cassette tape or compact disc.
Closing Date: April 30th
Funding: Private
Contributor: Hattori family
No. of awards given last year: 13
No. of applicants last year: 60
Additional Information: Grants may be made for study, concert experience and international competitions, but course fees and the purchase of instruments are not funded. Projects must be submitted for approval and discussion with the Director of Music and the trustees. Auditions take place in June and are in two stages.

HAYSTACK MOUNTAIN SCHOOL OF CRAFTS

PO Box 518, Deer Isle, ME, 04627, United States of America
Tel: (1) 207 348 2306
Fax: (1) 207 348 2307
Email: haystack@haystack-mtn.org
Website: www.haystack-mtn.org
Contact: Virgnia H B Aldrich, Development Director

The Haystack Mountain School of Crafts studio program in the arts offers 1 and 2 week workshops in a variety of craft and visual media including blacksmithing, clay, wood, glass, metals, textiles, and graphics.

Haystack Scholarship

Subjects: Fine crafts.
Purpose: To allow craftspeople of all skill levels to study at Haystack sessions for 1 or 2 week periods. Technical assistant and work study positions as well as minority scholarships and fellowships are awarded.
Eligibility: Open to nationals of any country, who are 18 years or older. Technical Assistant Scholarship: 1 year of graduate specialization or the equivalent in the craft area for which is requested. Work study Scholarship: Intended for those who show high promise in their craft field. Criteria include stated financial need, commitment to and growing knowledge of the craft area for which application is made, and the ability to work in a supportive, close-knit community. Minority Scholarship: Haystack awards up to six full scholarships to students of colour. Same criteria as work study scholarships.
Level of Study: Postgraduate, Research, Unrestricted
Type: Scholarships and fellowships
Value: US$600–1,800
Length of Study: 1 and 2 week sessions
Frequency: Annual
Country of Study: Any country
No. of awards offered: 100
Application Procedure: Applications available on the website or by contacting the school. Applicants must include references and supporting materials in their application.
Closing Date: March 1st
Funding: Foundation, individuals, private
No. of awards given last year: 100
No. of applicants last year: 300
Additional Information: Technical assistants are responsible for assisting the instructor and for shop maintenance and organization. Expected to be familiar with general technical requirements of the particular studio/medium. Responsibilities take precedence, but there is ample time for personal work and study. Work study and minority scholarship students will be assigned periodic tasks in the kitchen, or around the school campus. Assigned tasks will not exceed three hours daily, and students have ample time for personal work and study in the studio.

HEALTH RESEARCH BOARD (HRB)

Research & Development for Health, 73 Lower Baggot Street, Dublin, 2, Ireland
Tel: (353) 1 234 5000
Fax: (353) 1 661 2335
Email: hrb@hrb.ie
Website: www.hrb.ie
Contact: The Research Grants Manager

The Health Research Board (HRB) comprises 16 members appointed by the Minister of Health, with eight of the members being nominated on the co-joint nomination of the universities and colleges. The main functions of the HRB are to promote or commission health research, to promote and conduct epidemiological research as may be appropriate at national level, to promote or commission health services research, to liase and co-operate with other research bodies in Ireland and overseas in the promotion of relevant research and to undertake such other cognate functions as the Minister may from time to time determine.

All-Ireland Institute for Hospice and Palliative Care Fellowships

Subjects: Clinical, epidemiological, public health, statistics, health economics, social science, operational and management disciplines.
Purpose: To enable graduates with some appropriate relevant experience to pursue a career in health devices and research in Ireland.
Eligibility: Candidates must normally hold a primary degree in a discipline relevant to health services research, have acquired appropriate postgraduate experience in the field of health services and research, have support from an approved academic department or centre, have obtained the prior approval of a head of department for the research study being proposed and be Irish citizens or graduates from overseas with a permanent Irish resident status.
Level of Study: Postgraduate
Type: Fellowship
Value: Please consult the organization salary on a postdoctorate scale up to €7,500 per year for consumables
Length of Study: The maximum period of the award will be 3 years,
Frequency: Annual
Study Establishment: Institutions approved by the Board, such as teaching hospitals, universities, research institutes and health boards in Ireland
Country of Study: Ireland
No. of awards offered: Varies
Application Procedure: Applicants must complete an online application form, available from the website.
Closing Date: September 30th
Funding: Government
No. of awards given last year: 4
No. of applicants last year: 20

Clinician Scientist Award for Clinical Health Professionals

Subjects: Word class research clinical and translational research with a strong relevance to human health.
Purpose: To release outstanding medically or professionally qualified researchers in the health professions from some or all of their service commitment to conduct.
Eligibility: Medical consultants in the Irish health system or senior clinicians in health related disciplines who are qualified to hold a post in the Irish health service.
Level of Study: Research
Type: Award
Value: €1.5 million
Length of Study: 5 years
Frequency: Annual
Study Establishment: Any Irish teaching hospital or academic institution

Country of Study: Ireland
No. of awards offered: 2–3
Application Procedure: Online application form available from website. Full applications from invited applicants only.
Closing Date: September 28th
Funding: Government
No. of awards given last year: 1
No. of applicants last year: 5

HRB Project Grants-General

Subjects: Biomedical sciences, public health and epidemiology, health services research or health research.
Purpose: To facilitate research in biomedical sciences, public health, epidemiology and health service research.
Eligibility: Postdoctoral researches can apply for their own salary support his or her speciality should be within the range of disciplines stated in the subject index. Applicants must reside in the Republic of Ireland and grants are tenable in this country.
Level of Study: Doctorate, Postdoctorate
Type: Project funding
Value: Please consult the organization. €100,000 per year if employing salaried researcher, €75,000 per year if training PhD student
Length of Study: Up to 3 years
Frequency: Annual
Study Establishment: An Irish academic institution of research
Country of Study: Ireland
No. of awards offered: Varies
Application Procedure: Applicants must complete an online application form, available from the website.
Closing Date: November 2nd
Funding: Government
No. of awards given last year: 74
No. of applicants last year: 339

HRB Summer Student Grants

Subjects: Medical, biomedical, dental science, health service and science.
Purpose: To develop interest in research and give the student the opportunity to become familiar with research techniques.
Eligibility: Open to students from medical, dental science, biomedical or health service-related disciplines.
Level of Study: Undergraduate
Type: Grant
Value: €250 per week for up to 8 weeks
Length of Study: 8 weeks
Frequency: Annual
Study Establishment: A university, research hospital or institution
Country of Study: Ireland
No. of awards offered: Varies
Application Procedure: Applicants must complete an online application form, available from the website.
Closing Date: January 26th
Funding: Government
No. of awards given last year: 50
No. of applicants last year: 100

HRB Translational Research Programmes

Purpose: To enable researchers to establish and support teams working full-time or extensive or long-term research programmes that have a clear link to patient care; to support the development of clinical research in Ireland; to improve patient outcomes and to contribute to the creation of IPR.
Eligibility: Open to candidates who hold a post in an established academic research centre, have an outstanding track record and have at least 5 years of research experience.
Value: €1.5 million
Length of Study: 5 years
Frequency: Annual
Study Establishment: Any Irish academic or research institution
Country of Study: Ireland
Application Procedure: Online application form available from the HRB website.
Closing Date: March 26th
Funding: Government
No. of awards given last year: 4
No. of applicants last year: 12

THE HEART AND STROKE FOUNDATION

Suite 1402, 222 Queen Street, Ottawa, ON, K1P 5V9, Canada
Tel: (1) 613 569 4361 ext 275
Fax: (1) 613 569 3278
Email: research@hsf.ca
Website: www.heartandstroke.ca

The Heart and Stroke Foundation is involved in eliminating heart disease and stroke and reducing their impact through the advancement of research and its application, and advocacy for the promotion of healthy living. It is a federation of 10 provincial foundations, led and supported by a force of more than 140,000 volunteers.

Career Investigator Award

Subjects: Cardiology.
Purpose: To support established independent researchers who wish to make research their full-time career.
Eligibility: Applicants must possess an MD, PhD, or equivalent degree and working in the field of cardiovascular and/or cerebrovascular disease. Applicants must have achieved national recognition at the time of the first application.
Level of Study: Postgraduate
Type: Scholarship
Value: Stipend $81,500 per year. $1,500 per year for travel and minimum $48,282 for scientific purpose
Application Procedure: Applicants must send the application form along with the transcript, essay references and a self-addressed stamped envelope and must provide proof of national recognition.
Closing Date: August 31st
Contributor: Heart and Stroke Foundations of Ontario and British Columbia and the Yukon
Additional Information: For more details see website

For further information contact:

Research Department
Tel: 613 569 4361 ext. 327
Fax: 613 569 3278
Email: lhodgson@hsf.ca
Contact: Lise Hodgson, Administrative Assistant

Dr Andres Petrasovits Fellowship in Cardiovascular Health Policy Research

Subjects: Cardiology.
Eligibility: Please visit the website www.hsf.ca/research/application/index.html
Level of Study: Postgraduate
Type: Fellowship
Value: $70,000
Length of Study: 3 years
No. of awards offered: 1
Application Procedure: Check website www.hsf.ca/research/application/index.html
Closing Date: Check with website

For further information contact:

Research Department
Tel: 613 569 4361 ext. 327
Fax: 613 569 3278
Email: lhodgson@hsf.ca
Contact: Lise Hodgson, Administrative Assistant

Grants-in-Aid of Research and Development

Subjects: Cardiology.
Purpose: To support researchers in projects of experimental nature in cardiovascular or cerebrovascular development.
Eligibility: Open for full-time medical student.
Level of Study: Postgraduate
Type: Grant
Value: Approx. $33,000,000
Frequency: Every 3 years

Application Procedure: Applicants must send the application form along with the transcript, essay references and a self-addressed stamped envelope.
Closing Date: August 31st

For further information contact:

Research Department
Tel: 613 569 4361 ext. 327
Fax: 613 569 3278
Email: lhodgson@hsf.ca
Contact: Lise Hodgson, Administrative Assistant

Heart and Stroke Foundation of Canada Doctoral Research Award

Subjects: Cardiology.
Purpose: To support individuals enrolled in a PhD program and undertaking full-time research training in the stroke field.
Eligibility: Applicants must be Canadians studying abroad or in Canada or for foreign visitors to Canada. The fellowship is open to citizens of United States of America.
Level of Study: Postgraduate
Type: Fellowship
Value: Varies
Country of Study: Canada
Application Procedure: Applicants must send the application form along with the transcript, essay references and a self-addressed stamped envelope.
Closing Date: November 1st

For further information contact:

Research Department
Tel: 613 569 4361 ext. 327
Fax: 613 569 3278
Email: lhodgson@hsf.ca
Website: http://www.hsf.ca/research/en/personnel-award-programs
Contact: Lise Hodgson, Administrative Assistant

Heart and Stroke Foundation of Canada New Investigator Research Scholarships

Subjects: Cardiology.
Eligibility: Open to candidates who possess a MD, PhD, or equivalent degree and working in the field of cardiovascular and/or cerebrovascular disease.
Level of Study: Postgraduate
Type: Scholarship
Value: Maximum $30,000
No. of awards offered: 8–10
Application Procedure: Applicants must send the application form along with the transcript, essay references and a self-addressed stamped envelope.
Closing Date: September 1st

For further information contact:

Research Department
Tel: 613 569 4361 ext. 327
Fax: 613 569 3278
Email: lhodgson@hsf.ca
Contact: Lise Hodgson, Administrative Assistant

Heart and Stroke Foundation of Canada Research Fellowships

Subjects: Cardiology.
Eligibility: Applicants must possess a full-time degree for study towards an MSc or PhD.
Level of Study: Postgraduate
Type: Fellowship
Value: $25,998 (minimum) and $33,426 (maximum)
Country of Study: Canada
No. of awards offered: 10–20
Application Procedure: Applicants must send the application form along with the transcript, essay references and a self-addressed stamped envelope.
Closing Date: November 1st

For further information contact:

Research Department
Tel: 613 569 4361 ext. 327
Fax: 613 569 3278
Email: lhodgson@hsf.ca
Contact: Lise Hodgson, Administrative Assistant

HEART RESEARCH UK

Suite 12D, Joseph's Well, Leeds, LS3 1AB, England
Tel: (44) 11 3234 7474
Fax: (44) 11 3297 6208
Email: mail@heartresearch.org.uk
Website: www.heartresearch.org.uk
Contact: Helen Wilson, Senior Research Officer

Heart Research UK funds pioneering medical research into the prevention, treatment and cure of heart disease. Heart Research UK is a visionary charity leading the way in funding ground-breaking, innovative medical research projects at the cutting edge of science into the prevention, treatment and cure of heart disease. There is a strong emphasis on clinical and surgical projects and young researchers. Heart Research UK encourages and supports original health lifestyle initiatives exploring novel ways of preventing heart disease in all sectors of the community.

Heart Research UK Novel and Emerging Technologies Grant

Subjects: Research focusing on the development of new and innovative technologies to diagnose, treat and prevent heart disease and related conditions.
Purpose: To support ground-breaking, innovative medical research into the prevention, treatment and cure of heart disease and related conditions.
Eligibility: Graduates or those holding a suitable professional qualification. Research must be carried out in the United Kingdom at a university, hospital or other recognized research institution.
Level of Study: Research, Unrestricted
Type: Project grant
Value: Up to UK£200,000
Length of Study: Up to 3 years
Frequency: Annual
Study Establishment: Centres of health and educational establishments
Country of Study: United Kingdom
No. of awards offered: 1
Application Procedure: Information and application forms available on Heart Research UK website www.heartresearch.org.uk
Closing Date: See website
Funding: Corporation, foundation, individuals, private, trusts
Contributor: Voluntary funds from supporters and grant-making trusts
No. of awards given last year: 2
No. of applicants last year: 19
Additional Information: Appropriate approaches would include tissue and bio-engineering, development and evaluation of new therapeutic devices, bioimaging, nanotechnology, biomaterials, genomic and proteomic approaches, computational biology and bioinformatics.

Heart Research UK Translational Research Project Grants

Subjects: Translational research projects that convert fundamental research into clinical benefits.
Purpose: To support ground-breaking, innovative medical research into prevention, treatment and cure of heart disease and related conditions.
Eligibility: Graduates or those holding a suitable professional qualification. Research must be carried out in the United Kingdom at a university, hospital or other recognized research institution.
Level of Study: Research, Unrestricted
Type: Project grant
Value: Up to £150,000
Length of Study: Up to 3 years
Frequency: Annual

Study Establishment: Centres of health and educational establishments
Country of Study: United Kingdom
No. of awards offered: Varies
Application Procedure: Information and application forms available on Heart Research UK website www.heartresearch.org.uk
Closing Date: February 1st
Funding: Corporation, foundation, individuals, private, trusts
Contributor: Voluntary funding from supporters and grant-making trusts
No. of awards given last year: 4
No. of applicants last year: 29
Additional Information: Grants are for research which efficiently transfers innovative discoveries into practical tools to prevent, diagnose and treat cardiovascular disease.

HEBREW IMMIGRANT AID SOCIETY (HIAS)

333 Seventh Avenue, 16th Floor, New York, NY, 10001, United States of America
Tel: (1) 212 613 1358
Fax: (1) 212 967 4356
Email: scholarship@hias.org
Website: www.hias.org
Contact: Antonio Mateiro , Program Assistant

The Hebrew Immigrant Aid Society (HIAS) is the oldest international and refugee resettlement agency in the United States of America, dedicated to assisting persecuted and opressed people worldwide and delivering them to safe havens. HIAS has helped more than 4.5 million people in its 126 years of existence.

HIAS Scholars Award

Subjects: All subjects.
Purpose: To help HIAS-assisted refugees and asylees pursue higher education and to help them develop the leadership skills that will enable them to impact their communities.
Eligibility: Open to HIAS-assisted refugees and asylees in the United States of America. U.S. applicants must have completed one year (i.e. two semesters) at a U.S. secondary or post-secondary institution immediately prior to applying. Applicants must be enrolled in a U.S. post-secondary (e.g. college, graduate school, technical school, professional degree program, etc.) the fall and spring semester and the academic year for which the award was given.
Level of Study: Doctorate, Graduate, Postdoctorate, Professional development, Unrestricted, MBA, Postgraduate, Predoctorate, Research
Type: Fellowship
Value: $4,000 for U.S. students and $51,000 for Israeli students
Length of Study: 1 year must be completed in a US school prior to the beginning of award and applicant must be poised to start another academic year
Frequency: Annual
Country of Study: United States of America
No. of awards offered: Varies (approx. 60)
Application Procedure: Applicants must complete an online application form. Application forms are available at bias.org/scholarship at the time of the scholarship competition, which usually begins sometime between mid-January and early February and end around mid-March of each year.
Closing Date: February 29th
Funding: Individuals
No. of awards given last year: 60
No. of applicants last year: 170
Additional Information: Applicants are judged on finanical need, academic record and community service and leardership experience and potential. The HIAS scholarship award in Israel has a different deadline, award amount, specifications, etc. For more information, visit our website www.bias.org/scholarship.

HIAS Scholarship Awards Competition

Subjects: All subjects.
Purpose: To help HIAS-assisted refugees and asylees in pursuing higher education.
Eligibility: Open to HIAS-assisted refugees and asylees in the United States of America. United States of America applicants must have completed 1 year, i.e. two semesters, at a United States of America high school, college, or graduate school. The student must have immigrated after January 1, 1992.
Level of Study: Doctorate, Graduate, MBA, Postdoctorate, Postgraduate, Predoctorate, Professional development, Research, Unrestricted, and trade programs
Type: Scholarship
Value: $1,000 for Israeli students and $4,000 (Increased U.S. scholarship awards)
Length of Study: 1 year must be completed in a US School prior to beginning the award and applicant must be poised to start another academic year
Frequency: Annual
Country of Study: United States of America
No. of awards offered: Varies (150 approx.)
Application Procedure: Applicants must complete an official application form, which must be filled online. Application forms are available from mid-November to mid-February of each year.
Closing Date: March 15th
Funding: Individuals, private
No. of awards given last year: 131
No. of applicants last year: 420
Additional Information: Applications are judged on financial need, academic scholarship and community service. The HIAS Scholarship Awards Competition in Israel has a different deadline, award amount, specifications, etc. For further information visit the website www.hias.org/scholarships/apply.html

HEINRICH-BÖLL FOUNDATION

Gabriele Tellenbach, Heinrich-Böll-Stiftung, Studienwerk, Schumannstr. 8, Berlin, 10117, Germany
Tel: (49) 30 285 34 0
Fax: (49) 30 285 34 109
Email: info@boell.de
Website: www.boell.de

Heinrich-Boll Foundation, affiliated with the Green Party, is a legally independent political foundation working in the spirit of intellectual openness. Its primary objective is to support political education both within Germany and abroad, thus promoting democratic involvement, socio-political activism and cross-cultural understanding.

Heinrich Böll Foundation Doctoral Scholarships

Subjects: All subjects.
Purpose: To provide support to students pursuing their Doctoral studies.
Eligibility: Open to those who have either fulfilled the Doctorate entry requirements of a state or state-recognized university or college in Germany or have a foreign university or college degree.
Level of Study: Doctorate, Postgraduate
Type: Scholarship
Value: Varies
Length of Study: 1–3 years
Frequency: Annual
Country of Study: Germany
Application Procedure: Please check the website for details.
Closing Date: March 1st
Funding: Foundation
Contributor: German Ministry of Foreign Affairs

Heinrich Böll Foundation Scholarships for Postgraduate Studies

Subjects: All subjects.
Purpose: To provide support to students pursuing their postgraduate studies (MA/MSc).
Eligibility: Open to applicants who have completed their undergraduate studies (bachelor's degree, diploma, magister, state examination).
Level of Study: Postgraduate
Type: Scholarship
Value: Varies
Length of Study: 2 years

Frequency: Twice a year
Country of Study: Germany
Application Procedure: Refer the foundation website for further details.
Closing Date: March 1st and September 1st
Funding: Foundation
Contributor: German Ministry of Foreign Affairs

THE HENRY MOORE INSTITUTE

74 The Headrow, Leeds, LS1 3AH, United Kingdom
Tel: (44) 113 246 7467
Fax: (44) 113 246 1481
Email: kirstie@henry-moore.ac.uk
Website: www.henry-moore-fdn.co.uk/hmi
Contact: Kirstie Gregory, Research Programme Assistant

The Henry Moore Institute is a world-recognised centre for the study of sculpture in the heart of needs an award-winning exhibitions venue, research centre, library and sculpture archive. The institute hosts exhibitions, conferences and lectures, as well as developing research to expand the understanding and scholarship of historical and contemporary sculpture.

Henry Moore Institute Research Fellowships

Subjects: Sculpture, both historical, and contemporary.
Purpose: To enable scholars to use the Institute's facilities, which include the sculpture collection, library, archive and slide library, to assist them in researching their particular field.
Eligibility: There are no restrictions.
Level of Study: Doctorate, Postdoctorate, Postgraduate, Research
Type: Fellowship
Value: Accommodation, travel and daily living expenses
Length of Study: 1 month
Frequency: Annual
Study Establishment: The Henry Moore Institute
Country of Study: United Kingdom
No. of awards offered: 4
Application Procedure: Applicants must send a letter of application, a proposal (maximum 1,000 words) and a curriculum vitae. Visit the website www.henry-moore.org.
Closing Date: January 9th
Funding: Foundation
Contributor: The Henry Moore Foundation
No. of awards given last year: 4
No. of applicants last year: 80

For further information contact:

Email: kirstie@henry-moore.org
Contact: Kirstie Gregory

Henry Moore Institute Senior Research Fellowships

Subjects: Any aspect of sculpture. Fellows are asked to make a small contribution to the research programme in Leeds in the form of a talk or a seminar.
Purpose: Senior fellowships are intended to give established scholars (working on any aspect of sculpture) time and space to develop a research project free from their usual work commitments.
Level of Study: Doctorate, Postdoctorate
Type: Fellowship
Value: Fellowships provide accommodation, travel expenses and a per diem
Length of Study: 3–6 weeks
Frequency: Annual
Study Establishment: Henry Moore Institute
Country of Study: United Kingdom
No. of awards offered: Up to 2 senior fellowships
Application Procedure: Full details are available from the website www.henry-moore.ac.uk. Applicants can also contact the institute at its address for details. An applicant must send a curriculum vitae and a proposal along with his or her letter of application.
Closing Date: January 9th
Funding: Foundation
Contributor: Henry Moore Foundation
No. of awards given last year: 1
No. of applicants last year: 15

Additional Information: Research fellowships are also available. The institute offers the possibilty of presenting finished research in published form as a seminar or as a small exhibition.

For further information contact:

Tel: 0113 246 7467
Email: kirstie@henry-moore.org
Contact: Kirstie Gregory, Research Programme Assistant

THE HERB SOCIETY OF AMERICA, INC.

9019 Kirtland Chardon Road, Kirtland, OH, 44094, United States of America
Tel: (1) 440 256 0514
Fax: (1) 440 256 0541
Email: herbs@herbsociety.org
Website: www.herbsociety.org
Contact: Ms Michelle Milks, Office Administrator

The aim of the Herb Society of America Inc. is to promote the knowledge, use and delight of herbs through educational programmes, research and sharing the experience of its members with the community.

Herb Society of America Research Grant

Subjects: Herbal projects.
Purpose: To further the knowledge and use of herbs and to contribute the results of study and research to the records of horticulture, science, literature, history, art or economics.
Eligibility: Open to persons with a proposed programme of scientific, academic or artistic investigation of herbal plants.
Level of Study: Unrestricted
Value: Up to US$5,000
Length of Study: Up to 1 year
Frequency: Annual
Country of Study: Any country
No. of awards offered: 1–2
Application Procedure: Applicants must submit an application clearly defining all their research in 500 words or less and a proposed budget with specific budget items listed. Requests for funds will not be considered unless accompanied by five copies of the application form and proposal.
Closing Date: January 31st
Contributor: Members
Additional Information: Finalists will be interviewed.

HSA Grant for Educators

Subjects: Herbal projects.
Purpose: To enhance herbal education in school systems, in communities, or in any public forum.
Level of Study: Unrestricted
Value: Up to US$5,000
Length of Study: Up to 1 year
Frequency: Annual
Country of Study: Any country
Application Procedure: Please submit application cover sheet, statement of qualifications, a comprehensive descriptions of the programme, and detailed budget (8 copies).
Closing Date: December 31st
Contributor: Members
No. of awards given last year: 1
Additional Information: Finalists will be interviewed.

HERBERT HOOVER PRESIDENTIAL LIBRARY ASSOCIATION

302 Parkside Drive, PO Box 696, West Branch, IA, 52358, United States of America
Tel: (1) 319 643 5327
Fax: (1) 319 643 2391
Email: info@hooverassociation.org
Website: www.hooverassociation.org
Contact: Ms Delene McConnaha, Promotions & Academic Programs Manager

The Herbert Hoover Presidential Library Association is a private, non-profit support group for the Herbert Hoover Presidential Library Museum and National Historic Site in West Branch, Iowa.

Herbert Hoover Presidential Library Association Travel Grants

Subjects: American history, journalism, political science and economic history.
Purpose: To encourage the scholarly use of the holdings, and to promote the study of subjects of interest and concern to Herbert Hoover, Lou Henry Hoover and other public figures.
Eligibility: Open to current graduate students, postdoctoral students and qualified scholars. Priority is given to well-developed proposals that utilize the resources of the Library, have the greatest likelihood of publication and subsequently, greatest likelihood of use by educators, students and policy makers.
Level of Study: Doctorate, Graduate, Postdoctorate, Postgraduate, Predoctorate, Professional development, Research
Type: Travel grant
Value: US$500–1,500 to cover the cost of a trip to the Library. There is no money available for any purpose other than to defray the expense of travel to West Branch, IA
Length of Study: Varies by individual
Frequency: Annual
Study Establishment: The Herbert Hoover Presidential Library-Museum in West Branch, IA
Country of Study: United States of America
No. of awards offered: Varies
Application Procedure: Applicants must submit a completed application form, a project proposal of up to 1,200 words and three letters of reference, mailed separately. The application form can be obtained from the website.
Closing Date: March 1st
Funding: Private
No. of awards given last year: 7
No. of applicants last year: 10
Additional Information: For archival holdings information please contact the Hoover Library on (1) 319 643 5301, email: hoover.library@nara.gov or visit the website: www.hoover.archives.gov

HERBERT SCOVILLE JR PEACE FELLOWSHIP

322 4th Street, NE, Washington, DC, 20002, United States of America
Tel: (1) 202 446 1565
Fax: (1) 202 543 6297
Email: scoville@clw.org
Website: www.scoville.org
Contact: Paul Revsine, Program Director

The Herbert Scoville Jr Peace Fellowship was established in 1987 to provide college graduates with the opportunity to gain a Washington perspective on key issues of peace and security.

Herbert Scoville Jr Peace Fellowship

Subjects: Arms control and disarmament.
Purpose: To provide a unique educational experience to outstanding graduates, which will allow them to develop leadership skills that can serve them throughout a career in arms control or a related area of public service, to contribute to the work of the participating arms control and disarmament organizations and to continue the work of Herbert Scoville Jr.
Eligibility: Open to United States of America college graduates with experience or interest in arms control, disarmament, international security and/or peace issues. A fellowship is awarded periodically to a foreign national from a country of arms proliferation concern to the United States of America.
Level of Study: Postgraduate
Type: Fellowship
Value: $2,600 per month and health insurance, plus travel expenses to Washington, DC. $1,000 per fellow to attend relevant conferences or meetings that could cover travel, accommodations, and registration fees
Length of Study: 6–9 months
Country of Study: United States of America
Application Procedure: Applicants must telephone, write or consult the website for information on application requirements.
Closing Date: January 20th and October 1st

THE HEREDITARY DISEASE FOUNDATION (HDF)

3960 Broadway 6th Floor, New York, NY, 10032, United States of America
Tel: (1) 212 928 2121
Fax: (1) 212 928 2172
Email: carljohnson@hdfoundation.org
Website: www.hdfoundation.org
Contact: Carl D Johnson

The Hereditary Disease Foundation (HDF) was formed in 1968. HDF spearheaded the Venezuela Collaborative Huntington's Disease Project, which led to the identification in 1983 of a genetic marker for Huntington's disease. The HDF offers support for research projects that will contribute to identifying and understanding the basic defect of Huntington's disease.

John J. Wasmuth Postdoctoral Fellowships

Subjects: Trinucleotide expansions, animal models, gene therapy, neurobiology and development of the basal ganglia, cell survival and death and intercellular signalling in striatal neurons.
Purpose: To support research on Huntington's disease.
Level of Study: Postdoctorate, Research
Type: Fellowship
Value: Up to US$56,000
Length of Study: 1 year
Frequency: Annual
Country of Study: United States of America
Application Procedure: A completed application form must be submitted online.
Closing Date: October 15th
Funding: Foundation

HERIOT-WATT UNIVERSITY

Postgraduate Admissions Office, Edinburgh, EH14 4AS, United Kingdom
Tel: (44) 131 449 5111
Email: edu.liaison@hw.ac.uk
Website: www.hw.ac.uk
Contact: Fiona Watt, Wider Access Assistant

Heriot-Watt University, one of the oldest higher education institutions in the UK, is Scotland's most international university. Our six academic schools and two postgraduate institutes offer research opportunities and postgraduate taught programmes in science and engineering, business, languages and design. We disburse over £6M in fee and stipend scholarships annually.

African Scholarship Programme

Subjects: MBA (distance learning).
Purpose: To give 250 people across Africa the opportunity to study the Edinburgh Business School Distance Learning MBA programme.
Eligibility: Applicants from Sub-Saharan Africa.
Level of Study: MBA
Type: Scholarship
Value: Full fees
Length of Study: Variable
Frequency: Annual
Study Establishment: Heriot-Watt University
No. of awards offered: 50 each year until 2015
Application Procedure: www.canoncollins.org.uk/scholarships.
Closing Date: Regular deadlines – see website
Funding: Trusts
Contributor: Heriot-Watt University and Canon Collins Trust

Additional Information: Country of study is Scotland, United Kingdom.

Alumni Scholarship Scheme

Subjects: All subjects.
Purpose: To assist all Heriot-Watt alumni with postgraduate tuition fees.
Eligibility: Open to all Heriot-Watt Alumni who have previously been registered for one year or more on a Heriot-Watt degree course.
Level of Study: Postgraduate
Type: Scholarship
Value: 20 per cent of the tuition-fee level. For part-time or other modes of study the value is pro-rata to the above level, up to a maximum of 20 per cent of a one-year fee.
Length of Study: 1 year
Frequency: Annual
Study Establishment: Heriot-Watt University
No. of awards offered: Varies
Application Procedure: Alumni applicants should contact the relevant School or Postgraduate Institute to confirm if the proposed course of study is eligible for Scholarship, or for further information.
Additional Information: Exclusions are: EBS courses, IPE courses (though they may offer a Scholarship for students studying in Orkney), MSc Actuarial Science students studying on a course offered jointly with another university (Erasmus Mundus courses and several courses joint with Scottish Universities), students studying with a learning partner students studying in Dubai, research degrees (e.g. PhD, MPhil).
Country of study is Scotland, United Kingdom.

Carnegie Cameron Taught Postgraduate Bursaries

Subjects: All subjects.
Purpose: Provide financial assistance to Scottish nationals.
Eligibility: Applicants must be Scottish by birth, have at least one parent born in Scotland or have been continuously resident in Scotland for a period of three years for the purpose of secondary or tertiary education. Candidates will normally have a first class honours degree.
Level of Study: Postgraduate
Type: Bursary
Value: The Bursary covers the full costs of tuition fees
Length of Study: 1 year
Frequency: Annual
Study Establishment: Heriot-Watt University
No. of awards offered: 3
Application Procedure: Further information available on www.carnegie-trust.org.
Funding: Trusts
Contributor: Carnegie Trust
Additional Information: The bursary covers the full cost of tuition fees.
Country of study is Scotland, United Kingdom.

Commonwealth Scholarship and Fellowship Plan

Subjects: All subjects.
Purpose: To allow commonwealth citizens to study in the UK or other commonwealth countries.
Eligibility: Commonwealth citizens.
Level of Study: Postgraduate, Research
Type: Scholarships and fellowships
Frequency: Annual
Country of Study: United Kingdom or Commonwealth
No. of awards offered: 1,000 worldwide
Application Procedure: Apply early through Commonwealth Scholarship Agency in country of residence.

For further information contact:

Website: www.cscuk.org.uk

DFID Shared Scholarship Scheme

Subjects: Available only for specified full-time masters courses in architectural engineering, urban and regional planning, water resources.
Purpose: To assist students from developing Commonwealth countries to come to the UK for a 1-year taught Master's degree when they would otherwise be financially unable to do so.
Eligibility: The award has an upper age limit of 35 years with preference given to applicants under 30 years of age. Government employees are ineligible. Awards made must be relevant to the economic, scientific and social development of the applicant's home country which must be a part of the Commonwealth.
Level of Study: Postgraduate
Value: Tuition fees plus a maintenance grant, arrival allowance and return air fare
Length of Study: 1 year
Frequency: Annual
Study Establishment: Heriot-Watt University
No. of awards offered: 1
Application Procedure: Applicants should apply through their school of study.
Closing Date: April 16th
Contributor: Association of Commonwealth Universities
Additional Information: Country of study is Scotland, United Kingdom.

International Scholarships Programme

Purpose: Financial assistance to high calibre overseas students on Masters courses or undertaking research.
Level of Study: Postgraduate, Research, Masters
Type: Scholarship
Frequency: Annual
Study Establishment: Heriot-Watt University
No. of awards offered: Varies
Application Procedure: Please contact the School directly for further information, an application form and deadlines.
Contributor: Heriot-Watt University
Additional Information: Scholarship applications can only be made once an offer to study on the course has been issued and awards are made on a competitive basis.
Country of study is Scotland, United Kingdom.

International Scholarships Scheme

Subjects: Selected taught Masters programmes.
Purpose: To support high-calibre students pursue postgraduate study at Heriot-Watt University.
Eligibility: Open to all suitably qualified overseas students on a competitive basis.
Level of Study: Postgraduate
Type: Scholarship
Value: Value varies by school but offers a partial fee remission
Length of Study: 1 year
Frequency: Annual
Study Establishment: Heriot-Watt University
No. of awards offered: Up to 50
Application Procedure: Please contact your school directly for advice on making an application.
Closing Date: Varies
Additional Information: Scholarship applications can only be made once an offer to study on the programme has been issued.
Country of study is Scotland, United Kingdom.

James Watt Fee Scholarships

Subjects: All subjects.
Purpose: Assist suitably qualified UK and EU students to undertake research.
Eligibility: UK and EU students on a competitive basis.
Level of Study: Research
Type: Scholarship
Value: In excess of £10,000 over course of study
Length of Study: 3 years
Frequency: Annual
Study Establishment: Heriot-Watt University
No. of awards offered: Varies
Application Procedure: Contact School of Study.
Contributor: Heriot-Watt University
Additional Information: Country of study is Scotland, United Kingdom.

James Watt Scholarships

Subjects: All research areas.
Purpose: To support suitably qualified students undertake research activities at Heriot-Watt University.
Eligibility: All students. Only the most able students are likey to be successful. If unsuccessful you may be considered for a partial scholarship or fees-only scholarship.
Level of Study: Research
Type: Scholarship
Value: Full university fees plus a contribution to maintenance costs of at least £10,000 per year for up to 3 years
Length of Study: Up to 3 years
Frequency: Annual
Study Establishment: Heriot-Watt University
No. of awards offered: Variable
Application Procedure: Please contact your school for further details.
Closing Date: April 30th, apply for a study place by March 1st.
Additional Information: James Watt Scholarship provides full University fees and a maintenance contribution of at least £10,000 per year for up to 3 years, partial scholarship provides full fees and maintenance contribution of around £2,000 per year for up to 3 years, fees-only scholarship provides full fees for up to 3 years. Country of study is Scotland, United Kingdom.

Mexican Scholarships

Subjects: Science, engineering and technology.
Purpose: Financial assistance for Mexican students in science, engineering and technology.
Eligibility: Mexican citizens.
Level of Study: Postgraduate
Type: Scholarship
Value: Tuition fees and living costs
Frequency: Annual
Study Establishment: Heriot-Watt University
No. of awards offered: Limited
Application Procedure: Contact Bob Tuttle.
Funding: Government
Contributor: Heriot-Watt and CONACYT (Mexican National Council for Science and Technology)
Additional Information: Country of study is Scotland, United Kingdom.

For further information contact:

Tel: 0131 451 3746
Email: b.tuttle@hw.ac.uk

Music Scholarships

Subjects: All subjects.
Purpose: To support musicians in obtaining a postgraduate qualification whilst developing their musical skills.
Eligibility: All instrumentalists and vocalists who have been accepted for a course.
Level of Study: Postgraduate
Value: Free music tuition up to value of £400 per year
Length of Study: 1 year
Frequency: Annual
Study Establishment: Heriot-Watt University
No. of awards offered: Many
Application Procedure: Please contact your school for further details.
Additional Information: Country of study is Scotland, United Kingdom.

School of Textiles and Design Awards

Subjects: Taught Master's programmes in fashion and textiles, research in art and design.
Purpose: Assist with payment of fees for the Master's programme.
Eligibility: All applicants who have been offered a place to study.
Level of Study: Postgraduate
Type: Scholarship
Value: Tiered system of fee reductions, 40–10 per cent; also various awards of between £500 and £2,000
Frequency: Annual
Study Establishment: Heriot-Watt University
No. of awards offered: Varies
Application Procedure: Contact School of Textiles and Design.
Contributor: Heriot-Watt University and various benefactors
Additional Information: Country of study is Scotland, United Kingdom.

For further information contact:

Website: www.tex.hw.ac.uk

Scotland's Saltire Scholarships

Subjects: All subjects.
Purpose: To assist students from China, India, Canada and USA to study Maters courses in Scotland.
Eligibility: Students from China, India, Canada and USA.
Level of Study: Postgraduate
Type: Scholarship
Value: Up to £2,000
Length of Study: 1 year
Frequency: Annual
Study Establishment: Heriot-Watt University
No. of awards offered: Up to 8
Application Procedure: Applications can be made online at www.scotlandscholarship.com.
Funding: Government
Additional Information: Eligibility applies to nationals of China also. Country of study is Scotland, United Kingdom.

Sports Scholarships

Subjects: All subjects.
Purpose: To support atheletes in obtaining a postgraduate qualification whilst continuing to develop their sporting prowess.
Eligibility: All students competing at a national level in any sport. Specific awards are also available in football.
Level of Study: Postgraduate
Type: Scholarship
Value: Up to £2,000
Length of Study: 1 year
Frequency: Annual
Study Establishment: Heriot-Watt University
No. of awards offered: 30
Application Procedure: Applications should be made to the Sports Scholarship Co-ordinator if an offer of a place to study at Heriot-Watt University is held.
Contributor: Heriot-Watt University Alumni Association and various partner sports organisations
Additional Information: Country of study is Scotland, United Kingdom.

THE HERTZ FOUNDATION

2300 First Street, Suite 250, Livermore, CA, 94550, United States of America
Tel: (1) 925 373 1642
Fax: (1) 925 373 6329
Email: askhertz@hertzfoundation.org
Website: www.hertzfoundation.org
Contact: Ms Linda Kubiak, Fellowship Administrator

The Fannie and John Hertz Foundation runs a national competition for graduate fellowships in the applied physical, biological and engineering sciences.

The Graduate Fellowship Award

Subjects: Applied physical, biophysical and engineering sciences.
Purpose: To support students of outstanding potential in the applied physical, biological and engineering sciences.
Eligibility: Open to citizens or permanent residents of the United States of America who have received a Bachelor's degree by the start of tenure and who propose to complete a programme of graduate study leading to a PhD. Students who have commenced graduate study are also eligible. The Foundation does not support candidates pursuing joint PhD and professional degree programmes.
Level of Study: Doctorate
Type: Fellowship

Value: US$31,000 per 9-month academic year, plus cost-of education allowance
Length of Study: 1 academic year and may be renewed annually for up to 5 years
Frequency: Annual
Study Establishment: Specific universities listed on the website
Country of Study: United States of America
No. of awards offered: 15
Application Procedure: Applicants must complete a Hertz application form, four reference reports on the supplied specific forms and official transcripts of all college work must be submitted. The application form is available from the Foundation's website.
Closing Date: October 28th
Funding: Private
No. of awards given last year: 15
No. of applicants last year: Approx. 550
Additional Information: $5,000 per year additional stipend for Fellows with dependent children.

HERZOG AUGUST LIBRARY

PO Box 1364, D-38299, Wolfenbuettel, Germany
Tel: (49) 5331 808 213
Fax: (49) 5331 808 266
Email: bepler@hab.de
Website: www.hab.de
Contact: Dr Gillian Bepler

The Herzog August Library is an independent research library devoted to the study of the cultural history of Europe from the middle ages to the early modern period. Its rich book manuscript holdings were founded in the 17th century and have survived intact until today.

Findel Scholarships and Schneider Scholarships

Subjects: History and related disciplines.
Purpose: The Herzog August Library administers the doctoral fellowships funded by the Dr Guenther Findel Foundation and Rolf and Ursula Schneider Foundation towards the advancement of history and other related disciplines.
Eligibility: Outstanding doctoral candidates from all over the world, whose research requires an intensive use of the rich collections at Herzog August Library, may apply.
Level of Study: Doctorate
Type: Scholarship
Value: €700 per month. Additionally, the Herzog August Library provides accommodation for Fellows in its guest house
Length of Study: 3–6 months
Frequency: Biannual
Study Establishment: Herzog August Library
Country of Study: Germany
Application Procedure: An application form is available on request.
Closing Date: April 1st and October 1st
Funding: Foundation
Contributor: Findel and Schneider Foundations
No. of awards given last year: 15
Additional Information: If researchers are already supported by another grant/fellowship, the foundation will only provide free accommodation, but no subsistence allowance.

Herzog August Library Fellowship

Subjects: History and all related disciplines.
Purpose: Awards to post-doctoral researchers whose projects are based on the historical book and manuscript holdings of the Wolfenbuettel Library.
Eligibility: Qualified researchers whose projects are partly or fully based on the library's holdings.
Level of Study: Research
Type: Fellowship
Value: €1,800 per month
Length of Study: 2–12 months
Frequency: Annual
Study Establishment: Herzog August Library
Country of Study: Germany
No. of awards offered: 20
Application Procedure: Application forms available on request.
Closing Date: January 31st
Funding: Government
Contributor: State of Lower Saxony
No. of awards given last year: 20
No. of applicants last year: 80
Additional Information: Fellowship holders have a residence requirement in Wolfenbuttel during their tenure. The library has its own guest accomodation.

HIGHER EDUCATION COMMISSION

Islamabad, H-9, Pakistan
Tel: (92) 51 9257651 60
Fax: (92) 51 9290128
Email: info@hec.gov.pk
Website: www.hec.gov.pk

The Higher Education Commission has been set up to facilitate the development of the universities of Pakistan to be world-class centers of education, research and development.

International Research Support Initiative Program

Subjects: All subjects.
Purpose: To provide a training programme.
Eligibility: Open to candidates who have completed their PhD studies and who are under the age of 45 years.
Level of Study: Doctorate
Type: Scholarship
Value: Travel costs (up to Rs 115,000), for the monthly stipend and the bench fee check the website http://www.hec.gov.pk/InsideHEC/Divisions/HRD/Scholarships/ForeignScholarships/ISSIP/Pages/Default.aspx
Length of Study: 6 months
Frequency: Annual
Application Procedure: A completed application form, which is available on the website, must be sent.
Closing Date: November 21st
Funding: Government
Additional Information: The candidates will have to enter into a bond with HEC to serve the country at least for 3 years.

Overseas Scholarship for MS (Engineering) in South Korean Universities

Subjects: Engineering.
Purpose: To create a critical mass of highly qualified engineering manpower in high-tech field.
Eligibility: Open to Pakistan and AJK nationals who are below the age of 35 years.
Level of Study: Doctorate
Type: Scholarship
Value: US$9,600 and round-trip expenses from South Korea
Length of Study: 24 months
Frequency: Annual
Country of Study: South Africa
No. of awards offered: 250
Application Procedure: A completed application form along with all other requirements must be sent.
Funding: Government

For further information contact:

HRD Divison Higher Education Commission
Contact: Reznana Siddiqui, Project Director (SK/FFSP)

Partial Support for PhD Studies Abroad

Subjects: All subjects.
Purpose: To financially support Pakistan students who are in the final stage of completion of their PhD studies aborad.
Eligibility: Open to candidate who are Pakistan nationals who are studying abroad.
Level of Study: Doctorate
Type: Scholarship
Value: Up to US$15,000
Frequency: Annual
No. of awards offered: 50

Application Procedure: A completed application form along with photocopies of all academic documents must be sent.
Funding: Government
Additional Information: An awardee is required to execute a bond with HEC to serve Pakistan for 2 years.

HILDA MARTINDALE EDUCATIONAL TRUST

Royal Holloway University of London, Egham, Surrey, TW20 0EX, England
Tel: (44) 17 8427 6158
Fax: (44) 17 8443 7520
Email: hildamartindaletrust@rhul.ac.uk
Contact: Miss Sarah Moffat, Administrator to the Trust

The Hilda Martindale Educational Trust was set up by Miss Hilda Martindale in order to help women of the British Isles with the costs of vocational training for any profession or career likely to be of use or value to the community. Applications are considered annually by six women trustees.

Hilda Martindale Trust Exhibitions

Subjects: All subjects where women are under represented.
Purpose: To assist with the costs of training or professional qualifications in areas in which women are under represented.
Eligibility: Open to women of the British Isles only. Assistance is not given to short courses, courses abroad, elective studies or access courses. Trust can only offer funding to women pursuing training/ qualifications in areas in which women are under represented.
Level of Study: Doctorate, Graduate, Postdoctorate, Postgraduate, Predoctorate, Professional development
Type: Grant
Value: Up to £3,000
Length of Study: 1 year
Frequency: Annual
Study Establishment: Any establishment approved by the trustees
Country of Study: United Kingdom
No. of awards offered: 5–6
Application Procedure: Applicants must complete two copies of an application form, which must be obtained from and returned to the Secretary to the Trustees.
Closing Date: March 1st for the following academic year
Funding: Private
Contributor: Private trust
No. of awards given last year: 8
No. of applicants last year: 27
Additional Information: Late or retrospective applications will not be considered.

For further information contact:

C/o College Secretary's Office, RHUL, TW20 OEX
Contact: Miss Clare Munton, Secretary to the Hilda Martindale Trust

THE HINRICHSEN FOUNDATION

2-6 Baches Street, London, N1 6DN, England
Email: hinrichsen.foundation@editionpeters.com
Website: www.hinrichsenfoundation.org.uk
Contact: L E Adamson, Administrator

It was founded in 1976 by Mrs Carla Eddy Hinrichsen to ensure the continuation of the tradition established by the Hinrichsen family as the proprietors of Edition Peters the music publishers, established more than 200 years ago in the German city of Leipzig.

Hinrichsen Foundation Awards

Subjects: Contemporary music composition, performance and research.
Purpose: To promote the written areas of music.
Eligibility: Preference will be given to United Kingdom applicants and projects taking place in the United Kingdom. Grants are not given for recordings, for the funding of commissions, for degree or other study courses or for the purchase of instruments or equipment.
Type: Award
Value: Varies
Frequency: Dependent on funds available
No. of awards offered: Varies
Application Procedure: Applicants must submit a completed application form along with two references.
Closing Date: Applications are accepted at any time
Additional Information: Grants programme was temporarily suspended.

For further information contact:

The Hinrichsen Foundation, 2–6 Baches Street, London, N1 6DN

HISTORY OF SCIENCE SOCIETY (HSS)

Executive Office, 440 Geddes Hall, University of Notre Dame, Notre Dame, IN, 46556, United States of America
Tel: (1) 574 631 1194
Fax: (1) 574 631 1533
Email: info@hssonline.org
Website: www.hssonline.org
Contact: Mr Robert J Malone, Executive Director

The History of Science Society (HSS) is the world's largest society dedicated to understanding science, technology, medicine and their interactions with society within their historical context.

The George Sarton Medal

Subjects: History of science.
Purpose: To honour an outstanding historian of science for lifetime scholarly achievement.
Eligibility: Open to candidates who have devoted their entire career to the field of the history of science.
Level of Study: Postgraduate
Type: Award
Value: The George Sarton Medal
Frequency: Annual
Application Procedure: A completed nomination form that is available online must be sent.
Closing Date: Please contact the HSS Executive Office at: info@hssonline.org
Funding: Foundation
Contributor: Dibner Fund
No. of awards given last year: 1

Joseph H. Hazen Education Prize

Subjects: Science education.
Purpose: To promote exemplary teaching and educational service in the history of science.
Eligibility: Open to applicants who have made outstanding contributions to the teaching of history sciences.
Level of Study: Postgraduate
Type: Award
Value: US$1,000
Frequency: Annual
Application Procedure: A completed application form along with the nominee's curriculum vitae must be sent.
Closing Date: April 1st
Funding: Foundation
No. of awards given last year: 1
No. of applicants last year: 10

Margaret W. Rossiter History of Women in Science Prize

Subjects: Medicine, technology, social and national sciences.
Purpose: To recognize an outstanding book (in odd–numbered years) or article (in even–numbered years) on the history of women in science.
Eligibility: Open to authors of books/articles that have been published no more than 4 years before the year of award.
Level of Study: Postgraduate
Type: Prize
Value: US$1,000
Frequency: Annual
No. of awards offered: 1
Application Procedure: A completed nomination form, available on the website, must be sent.

Closing Date: April 1st
Funding: Foundation
No. of awards given last year: 1
No. of applicants last year: 25

The Nathan Reingold Prize

Subjects: History of science.
Purpose: To recognize an outstanding student essay in the history of science and its cultural influences.
Eligibility: An original and unpublished article on the history of science and its cultural influences written by a graduate student enrolled at any college, university, or institute of technology.
Level of Study: Doctorate, Predoctorate
Type: Prize
Value: US$500 and up to US$500 towards travel reimbursement
Frequency: Annual
No. of awards offered: 1
Application Procedure: A complete application form along with an electronic copy of the essay and proof of student status must be submitted.
Closing Date: June 1st
Funding: Foundation
Contributor: Friends and family of Nathan Reingold
No. of awards given last year: 1
No. of applicants last year: 12

Pfizer Award

Subjects: History of science.
Purpose: To honour outstanding books related to the history of science.
Eligibility: Open to authors of books that are published in the three preceding calendar years.
Level of Study: Postgraduate
Type: Award
Value: US$2,500 and a medal
Frequency: Annual
Application Procedure: A completed nomination form that is available on the website must be sent.
Closing Date: April 1st
Funding: Corporation
No. of awards given last year: 1
No. of applicants last year: 65

Suzanne J. Levinson Prize

Subjects: History of the life sciences or natural history.
Purpose: To be awarded for an outstanding book on the history of the life sciences and natural history.
Eligibility: Books on the history of natural history or evolutionary theory published 4 years prior to the award year.
Type: Prize
Value: $1,000
Frequency: Every 2 years, Biannual
No. of awards offered: 1
Application Procedure: Nominations can be downloaded from the website.
Closing Date: April 1st
Funding: Foundation
No. of awards given last year: 1
No. of applicants last year: 40

Watson Davis and Helen Miles Davis Prize

Subjects: Writing.
Purpose: To promote a book that helps in public understanding of the history of science.
Eligibility: Open to authors of books published in the last 3 years.
Level of Study: Postgraduate
Type: Prize
Value: US$1,000
Frequency: Annual
Application Procedure: A completed nomination form that is available on the website must be submitted.
Closing Date: April 1st
Funding: Private
No. of awards given last year: 1
No. of applicants last year: 70

HONDA FOUNDATION

2nd Floor, Honda Yaesu Bldg, 6–20 Yaesu 2-chome, Chuo-ku, Tokyo, 104, Japan
Tel: (81) 3 3274 5125
Fax: (81) 3 3274 5103
Email: h_info@hondafoundation.jp
Website: http://www.hondafoundation.jp/en/
Contact: Yoichi Harada, Managing Director

The Honda Foundation was established in December 1977 to contribute to the creation of true human civilization on the basis of the philosophy of the late Mr Soichiro Honda, the founder of Honda Motor Co. Ltd.

Honda Prize

Subjects: Ecotechnology.
Purpose: To acknowledge the efforts of an individual or group who contribute new ideas which may lead the next generation in the field of ecotechnology.
Eligibility: Open to individuals or an organization, irrespective of nationality.
Type: Prize
Value: The prize includes a diploma and medal with a prize of ¥10,000,000.
Frequency: Annual
No. of awards offered: 1
Application Procedure: Recommenders approved by the Honda Foundation are able to recommend.
Closing Date: March 31st
Funding: Foundation
Contributor: The late Mr Soichiro Honda, the founder of Honda Motor Co. Ltd
No. of awards given last year: 1
No. of applicants last year: 1
Additional Information: Ecotechnology is a new concept that harmonizes the progress of technology and civilization, rather than pursuing technology designed solely for efficiency and profit.

HONG KONG JOCKEY CLUB MUSIC AND DANCE FUND

Home Affairs Bureau, 13/F, West Wing, Central Government Offices, 2 Tim Mei Avenue, Tamar, Hong Kong
Tel: (852) 3509 7083
Fax: (852) 2802 4893
Email: info@hkjcmdf.org.hk
Website: http://www.hkjcmdf.org.hk/eng/index.html

Hong Kong Jockey Club Music and Dance Fund as a contributer of the Hong Kong Arts Festival, the Club provides grants to the Festival each year with the aim of bringing new arts and cultural experiences to local audiences as well as building a growing international reputation for Hong Kong.

Hong Kong Jockey Club Music and Dance Fund Scholarship

Subjects: Music and dance.
Purpose: To enable exceptionally talented young musicians and dancers to pursue an integrated programme of post-diploma studies, post-graduate studies or professional training in music or dance studies outside Hong Kong at renowned institutions or to undertake less formal studies, projects or creative work outside Hong Kong.
Eligibility: Applicants for the scholarship must be permanent Hong Kong residents who have resided in Hong Kong for 7 continuous years immediately preceding the application period. Applicants must not be above 35 years of age and must hold a relevant degree or post-form 7 tertiary qualification in either music or dance specifically.
Level of Study: Postgraduate, Professional development
Type: Scholarship
Value: Tuition fee, return flight tickets to the intended country of study and subsistence allowance.
Length of Study: 2 years
Frequency: Annual

Application Procedure: Application forms can be downloaded from the website.
Closing Date: February 22nd
Additional Information: For further information, please call the Secretariat during office hours at 2594 5628 (Ms Wong) or 2594 5621 (Mr Leung).

THE HOROWITZ FOUNDATION FOR SOCIAL POLICY

PO Box 7, Rocky Hill, NJ, 08553 0007, United States of America
Tel: (1) 609 921 1479
Fax: (1) 732 445 3138
Email: mcurtis@transactionpub.com
Website: www.horowitz-foundation.org
Contact: Ms Mary E. Curtis, The Chairman

The Horowitz Foundation for Social Policy was established to support the advancement of research and understanding in the social sciences including: psychology, anthropology, sociology, economics, and political science. The Foundation assists individual scholars at the early stages of their career who require small grants to complete their dissertations.

Donald R. Cressey Award

Purpose: Awarded for work done at the empirical level that has direct implication for changes in criminal justice and penology practices.
Type: Award
Additional Information: Please contact the organisation for more information.

Eli Ginzberg Award

Subjects: Social sciences, including anthropology, area studies, economics, political science, psychology, sociology and urban studies as well as newer areas such as evaluation research.
Purpose: To support a project involving solutions to major urban health problems in urban settings.
Eligibility: Open to nationals of any country. Candidates may solicit support for final work on a dissertation including travel funds.
Level of Study: Doctorate
Type: Grant
Value: $7500 ($5000 initially and an additional $2500 upon receipt of a final report or a copy of the research).
Length of Study: 1 year
Frequency: Annual
Study Establishment: Rutgers University/Transaction Publishers
Country of Study: Any country
No. of awards offered: 1
Application Procedure: Applicants are not required to be U.S. citizens or U.S. residents. Candidates may propose new projects, and they may also solicit support for research in progress, including final work on a dissertation, supplementing research in progress, or travel funds. Awards are only open to aspiring PhDs at the dissertation level whose project has received approval from their appropriate department head/university. Grants are normally made for one year on a non-renewable basis. Awards will be made to individuals, not institutions, and if processed through an institution, a waiver for overhead is requested. A copy of the product of the research is expected no later than one year after completion. Upon receipt an additional $2500 will be paid. Recipients are expected to acknowledge assistance provided by the Foundation in any publication resulting from their research. Awards are publicized in appropriate professional media and on the Foundation website, www.horowitz-foundation.org.
Closing Date: January 31st
Funding: Corporation, private
No. of awards given last year: 15
No. of applicants last year: Approx. 300

Harold D. Lasswell Award

Subjects: Social sciences, including anthropology, area studies, economics, political science, psychology, sociology and urban studies as well as newer areas such as evaluation research.
Purpose: To support policy-related projects in international relations and foreign affairs.
Eligibility: Open to nationals of any country. Candidates may solicit support for final work on a dissertation including travel funds.
Level of Study: Doctorate
Type: Grant
Value: $7500 ($5000 initially and an additional $2500 upon receipt of a final report or a copy of the research).
Length of Study: 1 year
Frequency: Annual
Study Establishment: Rutgers University/Transaction Publishers
Country of Study: Any country
No. of awards offered: 1
Application Procedure: Applicants are not required to be U.S. citizens or U.S. residents. Candidates may propose new projects, and they may also solicit support for research in progress, including final work on a dissertation, supplementing research in progress, or travel funds. Awards are only open to aspiring PhDs at the dissertation level whose project has received approval from their appropriate department head/university. Grants are normally made for one year on a non-renewable basis. Awards will be made to individuals, not institutions, and if processed through an institution, a waiver for overhead is requested. A copy of the product of the research is expected no later than one year after completion. Upon receipt an additional $2500 will be paid. Recipients are expected to acknowledge assistance provided by the Foundation in any publication resulting from their research. Awards are publicized in appropriate professional media and on the Foundation website, www.horowitz-foundation.org.
Closing Date: January 31st
Funding: Corporation, private
No. of awards given last year: 15
No. of applicants last year: Approx. 300
Additional Information: The cover sheet in the application is most important, as it is the basis for the initial screening of prospects.

John L Stanley Award

Subjects: Social sciences, including anthropology, area studies, economics, political science, psychology, sociology and urban studies as well as newer areas such as evaluation research.
Purpose: To support a work that seeks to expand our understanding of the political and ethical foundations of policy research.
Eligibility: Open to nationals of any country. Candidates may solicit support for final work on a dissertation including travel funds.
Level of Study: Doctorate
Type: Grant
Value: $7,500 ($5,000 initially and an additional $2,500 upon receipt of a final report or a copy of the research).
Length of Study: 1 year
Frequency: Annual
Country of Study: Any country
No. of awards offered: 1
Application Procedure: Applicants are not required to be U.S. citizens or U.S. residents. Candidates may propose new projects, and they may also solicit support for research in progress, including final work on a dissertation, supplementing research in progress, or travel funds. Awards are only open to aspiring PhDs at the dissertation level whose project has received approval from their appropriate department head/university. Grants are normally made for one year on a non-renewable basis. Awards will be made to individuals, not institutions, and if processed through an institution, a waiver for overhead is requested. A copy of the product of the research is expected no later than one year after completion. Upon receipt an additional $2,500 will be paid. Recipients are expected to acknowledge assistance provided by the Foundation in any publication resulting from their research. Awards are publicized in appropriate professional media and on the Foundation website, www.horowitz-foundation.org.
Closing Date: January 31st
Funding: Private
No. of awards given last year: 15
No. of applicants last year: 300
Additional Information: The cover sheet in the application is most important, as it is the basis for the initial screening of prospects.

Joshua Feigenbaum Award

Subjects: Social sciences, including anthropology, area studies, economics, political science, psychology, sociology and urban studies as well as newer areas such as evaluation research.
Purpose: To support empirical research on policy aspects of the arts and popular culture, with special reference to mass communication.
Eligibility: Open to nationals of any country. Candidates may solicit support for final work on a dissertation including travel funds.
Level of Study: Doctorate
Type: Grant
Value: $7,500 ($5,000 initially and an additional $2,500 upon receipt of a final report or a copy of the research).
Length of Study: 1 year
Frequency: Annual
Country of Study: Any country
No. of awards offered: 1
Application Procedure: Applicants are not required to be U.S. citizens or U.S. residents. Candidates may propose new projects, and they may also solicit support for research in progress, including final work on a dissertation, supplementing research in progress, or travel funds. Awards are only open to aspiring PhDs at the dissertation level whose project has received approval from their appropriate department head/university. Grants are normally made for one year on a non-renewable basis. Awards will be made to individuals, not institutions, and if processed through an institution, a waiver for overhead is requested. A copy of the product of the research is expected no later than one year after completion. Upon receipt an additional $2,500 will be paid. Recipients are expected to acknowledge assistance provided by the Foundation in any publication resulting from their research. Awards are publicized in appropriate professional media and on the Foundation website, www.horowitz-foundation.org.
Closing Date: January 31st
Funding: Private
No. of awards given last year: 15
No. of applicants last year: 300
Additional Information: The cover sheet in the application is most important, as it is the basis for the initial screening of prospects.

Martinus Nijhoff Award

Subjects: Social sciences, including anthropology, area studies, economics, political science, psychology, sociology and urban studies as well as newer areas such as evaluation research.
Purpose: To support policy implications of scientific, technological and medical research.
Eligibility: Open to nationals of any country. Candidates may solicit support for final work on a dissertation including travel funds.
Level of Study: Doctorate, Unrestricted
Type: Grant
Value: $7,500 ($5,000 initially and an additional $2,500 upon receipt of a final report or a copy of the research).
Length of Study: 1 year
Frequency: Annual
Country of Study: Any country
No. of awards offered: 1
Application Procedure: Applicants are not required to be U.S. citizens or U.S. residents. Candidates may propose new projects, and they may also solicit support for research in progress, including final work on a dissertation, supplementing research in progress, or travel funds. Awards are only open to aspiring PhDs at the dissertation level whose project has received approval from their appropriate department head/university. Grants are normally made for one year on a non-renewable basis. Awards will be made to individuals, not institutions, and if processed through an institution, a waiver for overhead is requested. A copy of the product of the research is expected no later than one year after completion. Upon receipt an additional $2,500 will be paid. Recipients are expected to acknowledge assistance provided by the Foundation in any publication resulting from their research. Awards are publicized in appropriate professional media and on the Foundation website, www.horowitz-foundation.org.
Closing Date: January 31st
Funding: Private
No. of awards given last year: 15
No. of applicants last year: 300
Additional Information: The cover sheet in the application is most important, as it is the basis for the initial screening of prospects.

Robert K Merton Award

Subjects: Social sciences including anthropology, area studies, economics, political science, psychology, sociology and urban studies as well as newer areas such as evaluation research.
Purpose: To support studies in the relation between social theory and public policy.
Eligibility: Open to nationals of any country. Candidates may solicit support for final work on a dissertation including travel funds.
Level of Study: Doctorate
Type: Grant
Value: $7,500 ($5,000 initially and an additional $2,500 upon receipt of a final report or a copy of the research).
Length of Study: 1 year
Frequency: Annual
Country of Study: Any country
No. of awards offered: 1
Application Procedure: Applicants are not required to be U.S. citizens or U.S. residents. Candidates may propose new projects, and they may also solicit support for research in progress, including final work on a dissertation, supplementing research in progress, or travel funds. Awards are only open to aspiring PhDs at the dissertation level whose project has received approval from their appropriate department head/university. Grants are normally made for one year on a non-renewable basis. Awards will be made to individuals, not institutions, and if processed through an institution, a waiver for overhead is requested. A copy of the product of the research is expected no later than one year after completion. Upon receipt an additional $2,500 will be paid. Recipients are expected to acknowledge assistance provided by the Foundation in any publication resulting from their research. Awards are publicized in appropriate professional media and on the Foundation website, www.horowitz-foundation.org.
Closing Date: January 31st
Funding: Private
No. of awards given last year: 15
No. of applicants last year: 300
Additional Information: The cover sheet in the application is most important, as it is the basis for the initial screening of prospects.

HORSERACE BETTING LEVY BOARD (HBLB)

Parnell house, 25 Wilton road, London, SW1V 1LW, England
Tel: (44) 20 7333 0043
Fax: (44) 20 7333 0041
Email: equine.grants@hblb.org.uk
Website: www.hblb.org.uk
Contact: Equine Grants Team

The Horserace Betting Levy Board (HBLB) operates in accordance with the Betting, Gaming and Lotteries Act 1963. It assesses and collects contributions from bookmakers and the Horserace Totalisator Board and uses these for the advancement of equine veterinary science and education and other improvements within the horseracing industry.

Horserace Betting Levy Board Senior Equine Clinical Scholarships

Subjects: Equine veterinary studies with emphasis on the Thoroughbred horse.
Purpose: To support postgraduate equine veterinary clinical training.
Eligibility: Open to holders of veterinary degrees recognised by the RCVS, who have at least 2 years experience in veterinary practice.
Level of Study: Postgraduate
Type: Scholarship
Value: £20,060 in 1st year and a contribution to expenses related to the scholarship £9,660
Length of Study: Up to 4 years, subject to satisfactory progress
Frequency: Annual
Study Establishment: At any of the University veterinary schools, or in a joint venture with any appropriate university department, research institute or veterinary practice in the United Kingdom
Country of Study: United Kingdom

No. of awards offered: Usually 1, sometimes 2 per year
Application Procedure: The study establishment must submit applications in appropriate forms.
Closing Date: Please check with the HBLB for further details.
Funding: Government
No. of awards given last year: 1
No. of applicants last year: Not limited
Additional Information: Awards normally commence on October 1st. The study establishment is responsible for the appointment of clinical scholars.

Horserace Betting Levy Board Veterinary Research Training Scholarship

Subjects: Equine veterinary medicine or science, with emphasis on the Thoroughbred.
Purpose: To support postgraduate equine veterinary research training.
Eligibility: Open to holders of a veterinary degree who wish to undertake full-time training in research in the equine veterinary field leading to a PhD.
Level of Study: Postgraduate
Type: Scholarship
Value: UK£19,100 stipend in 1st year, with increments for 2nd and 3rd years. UK£4,000 (accountable) per year for fees and expenses, and UK£5,200 (unaccountable) per year to the department in which the holder works. Scales are reviewed annually.
Length of Study: Up to 3 years, subject to satisfactory progress
Frequency: Annual
Study Establishment: Any of the University veterinary schools, any appropriate university department or research institute in the United Kingdom
Country of Study: United Kingdom
No. of awards offered: Up to 2 per year
Application Procedure: Applicants must submit applications in appropriate forms.
Closing Date: Please check with the HBLB for further details
Funding: Government
No. of awards given last year: 2
No. of applicants last year: Maximum 3 per institution
Additional Information: The study establishment is responsible for the appointment of research scholars. Awards normally commence on October 1st.

HORTICULTURAL RESEARCH INSTITUTE

1200 G ST NW, Suite 800, Washington, DC, 20005, United States of America
Tel: (1) 202 695 2474
Fax: (1) 888 761 7883
Email: info@hriresearch.org
Website: www.hriresearch.org
Contact: Ms Teresa A Jodon, Executive Director

The aim of the Horticultural Research Institute is to direct, fund, promote and communicate research that increases the quality and value of plants, improves the productivity and profitability of the nursery and landscape industry and protects and enhances the environment.

Horticultural Research Institute Grants

Subjects: Nursery and landscape industry, especially woody and perennial landscape plants, their production, marketing, landscape, water management or the environment.
Purpose: To support necessary research for the advancement of the nursery, greenhouse and landscape industry.
Eligibility: Open to nationals and permanent residents of the United States of America and Canada. Candidates must submit an appropriate project that the Institute feels is deserving of support.
Level of Study: Unrestricted
Type: Grant
Value: Varies
Length of Study: Varies
Frequency: Annual
Study Establishment: State or federal research laboratories, land grant universities, forest research stations, botanical gardens and arboreta
Country of Study: United States of America
No. of awards offered: Varies
Application Procedure: Applicants must submit the application electronically.
Closing Date: May 31st
Funding: Private
Contributor: Nursery and landscape firms, as well as state and regional nursery and landscape associations
No. of awards given last year: 18
No. of applicants last year: 105
Additional Information: Please see the website for further details http://www.hriresearch.org/index.cfm?page=Content&categoryID=167.

HOSPITAL FOR SICK CHILDREN RESEARCH TRAINING CENTRE (RESTRACOMP)

555 University Avenue, Toronto, ON, M5G 1X8, Canada
Tel: (1) 416 813 1500
Fax: (1) 416 813 7311
Email: jennifer.ng@sickkids.ca
Website: www.sickkids.on.ca
Contact: N. Ramsundar

RESTRACOMP Research Fellowship

Subjects: Paediatric research, biomedical research.
Purpose: To provide funds to postgraduate students or fellows seeking research training.
Eligibility: Open to those nominated by the active senior staff of the Research Institute of the Hospital for Sick children. Postdoctoral trainees at the hospital for Sick Children, working under sick kids scientific staff.
Level of Study: Graduate, Postdoctorate, Postgraduate
Type: Fellowship
Value: Up to Canadian $33,750 per year
Length of Study: 2 years
Frequency: Biannual
Study Establishment: The Hospital for Sick Children, Toronto
Country of Study: Canada
No. of awards offered: 20–30
Application Procedure: Applicants must submit a completed application form.
Closing Date: April 3rd
Funding: Foundation
No. of applicants last year: 30 per cent success rate

HOUBLON-NORMAN FUND

Bank of England, Threadneedle Street, London, EC2R 8AH, England
Tel: (44) 20 7601 4878
Fax: (44) 20 7601 5460
Email: MA-HNGfund@bankofengland.co.uk
Website: www.bankofengland.co.uk/about/fellowships/index.htm
Contact: Miss Emma-Jayne Coker, Business Support Unit

Houblon-Norman Fellowships/George Fellowships

Subjects: Economics and finance.
Purpose: To promote research into and disseminate knowledge and understanding of the working, interaction and function of financial and business institutions in the United Kingdom and elsewhere and the economic conditions affecting them.
Eligibility: Open to distinguished research workers as well as younger postdoctoral or equivalent applicants of any nationality. Preference will be given to the United Kingdom and European Union nationals.
Level of Study: Postdoctorate
Type: Fellowship
Value: The value of a fellowship is dependent on the candidate's circumstances and will be of such amount as seems necessary for

undertaking the work. It might take the form of payment to the individual's employer
Length of Study: 1 month to 1 year
Frequency: Annual
Study Establishment: The Bank of England
Country of Study: United Kingdom
No. of awards offered: Varies
Application Procedure: Applicants must complete an application form advertised through one Economist, one Royal economic society and our website, where an application form can be found www.bankofengland.co.uk/education/fellowships/index.htm
Closing Date: As advertised in the press
No. of awards given last year: 3
No. of applicants last year: 16

HUDSON RIVER FOUNDATION (HRF)

17 Battery Place, Suite 915, New York, NY 10004, United States of America
Tel: (1) 212 924 7667
Fax: (1) 212 924 8325
Email: info@hudsonriver.org
Website: www.hudsonriver.org
Contact: Grants Management Officer

The Hudson River Foundation (HRF) supports scientific research, education and projects to enhance public access to the Hudson River. The purpose of the Foundation is to make science integral to the decision-making process with regard to the Hudson River and its watershed and to support competent stewardship of this extraordinary resource.

Hudson River Graduate Fellowships

Subjects: Research on the resources, key species, toxic substances, abundances of key organisms, dynamics of Hudson River trophic webs, hydrodynamics, sediment transport, public policy and social science of the Hudson Bay River.
Purpose: To fund research fellowships to advanced graduate students conducting research on the Hudson River system.
Eligibility: Applicants must be in an accredited graduate level, and must have a thesis advisor and a research plan approved by the applicant's institution.
Level of Study: Doctorate, Postgraduate
Type: Full-time research fellowship
Value: US$11,000 stipend plus US$1,000 expenses for Master's and US$15,000 stipend plus US$1,000 expenses for doctorate
Length of Study: 1 year
Frequency: Annual
Study Establishment: Any
Country of Study: Any country
No. of awards offered: Up to 4
Application Procedure: Applicants must supply a description, timetable, statement of significance and relevance, estimate of the cost, curriculum vitae and two letters of recommendation. The original and ten copies of the proposal must be forwarded to the Science Director at the main organization address.
Closing Date: March 21st
Funding: Private
No. of awards given last year: 1
No. of applicants last year: 5

Tibor T Polgar Fellowship

Subjects: All aspects of the environment relating to the Hudson River. Previous projects have studied hydrodynamics, larval fish, zooplankton, terrapins, landscape ecology, nutrients and public policy.
Purpose: To fund Summer research on the Hudson River.
Eligibility: There are no eligibility restrictions.
Level of Study: Graduate
Type: Fellowship
Value: US$3,800 and limited research funds
Length of Study: From May–June to August–September
Frequency: Annual
Country of Study: United States of America
No. of awards offered: 8 (every summer)
Application Procedure: Applicants must submit the original and five copies of their application, which must include letters of interest from the student and of support from the sponsor, a short description of the research project including its significance of between four and six pages, a detailed timetable for the completion of the project, a detailed budget with estimated cost of supplies, travel and other expenses, and the student's curriculum vitae. Because of the training and educational aspects of this programme, each potential fellow must be sponsored by a primary advisor. The advisor must be willing to commit sufficient time for supervision of the research and to attend at least one meeting to review the progress of the research. Advisors will receive a stipend of US$500.
Closing Date: February 11th
Funding: Private
No. of awards given last year: 8
No. of applicants last year: 17
Additional Information: The objectives of the programme are to gather important information on all aspects of the river and to train students in conducting estuarine studies and public policy research. Polgar Fellowships may be awarded for studies anywhere within the Hudson basin.

HUMANE RESEARCH TRUST

Brook House, 29 Bramhall Lane South, Bramhall, Stockport, Cheshire, SK7 2DN, England
Tel: (44) 161 439 8041
Fax: (44) 161 439 3713
Email: info@humaneresearch.org.uk
Website: www.humaneresearch.org.uk
Contact: Jane McAllister, Trust Administrator

The Humane Research Trust is a registered charity encouraging and supporting new medical research which does not include the use of animals, with the objectives of advancing the diagnosis and treatment of disease in humans. The Trust encourages scientists to develop innovative alternatives to the use of animals and eliminate the suffering of animals, which occurs in medical research and testing.

Humane Research Trust Grant

Subjects: Humane research.
Purpose: To encourage scientific programmes where the use of animals is replaced by other methods.
Eligibility: Open to established scientific workers engaged in productive research. Nationals of any country are considered but for the sake of overseeing, projects should be undertaken in a United Kingdom establishment.
Level of Study: Unrestricted
Type: Grant
Value: Varies
Length of Study: Varies
Frequency: Dependent on funds available
Study Establishment: Various
Country of Study: United Kingdom
No. of awards offered: Varies
Application Procedure: Applicants must complete an application form, available on the website www.humaneresearch.org.uk.
Closing Date: Varies
Funding: Private
Contributor: Supporters and legacies
No. of awards given last year: 3
No. of applicants last year: 13
Additional Information: The Trust is a registered charity and donations are encouraged. Please contact to Administrator, Mr. K Wright.

For further information contact:

Contact: K. Wright, Administrator

HUMANITIES RESEARCH CENTRE (HRC)

Australian National University (ANU), SRWB#120. MCoy Circuit, Canberra, ACT, 0200, Australia
Tel: (61) 2 6125 4357
Fax: (61) 2 6125 1380
Email: leena.messina@anu.edu.au
Website: www.anu.edu.au/hrc
Contact: Ms Leena Messina, Programs Manager

The Humanities Research Centre (HRC) was established in 1972, specifically to stimulate humanities research and debate at the Australian National University (ANU), within Australia and beyond.

HRC Visiting Fellowships

Subjects: Applications are particularly welcomed from scholars with interests in one or more of the HRC's Research Platforms: Humanities, biography and society, creativity and human rights, historical re-enactment and public memory.
Purpose: To provide scholars with time to pursue their own work in congenial and stimulating surroundings.
Eligibility: Open to candidates of any nationality who are at the postdoctoral level.
Level of Study: Postdoctorate
Type: Fellowship
Value: Return economy airfare up to Australian $3,000, plus accommodation
Length of Study: 12 weeks
Frequency: Annual
Study Establishment: The Humanities Research Centre at the Australian National University
Country of Study: Australia
No. of awards offered: 10–15
Application Procedure: Applicants must complete a formal application, available from the website.
Closing Date: March 31st
Funding: Government
No. of awards given last year: 16
No. of applicants last year: 60
Additional Information: Fellows are required to spend all of their time in residence at the Centre, but are encouraged to visit other institutions. Please refer to the website for further information.

For further information contact:

Website: http://hrc.anu.edu.au/news

THE HUNTINGTON

Committee on Fellowships, The Huntington, 1151 Oxford Road, San Marino, CA, 91108, United States of America
Tel: (1) 626 405 2194
Fax: (1) 626 449 5703
Email: cpowell@huntington.org
Website: www.huntington.org
Contact: Carolyn Powell, Assistant to Director of Research

The Huntington is an independent research center with holdings in British and American history, literature, art history, and the history of science.

Barbara Thom Postdoctoral Fellowships

Subjects: British and American history, literature, art history, and history of science.
Purpose: To support non-tenured faculty members who are revising a manuscript for publication.
Eligibility: Applicants must have received their PhD between 2008 and 2010.
Level of Study: Postdoctorate, Postgraduate
Type: Fellowship
Value: US$50,000
Length of Study: 9–12 months
Frequency: Annual
Study Establishment: The Huntington
Country of Study: United States of America
No. of awards offered: 2
Application Procedure: Applicants must contact the Committee on Fellowships.
Closing Date: Applications are accepted between October 1st and November 30th.
Funding: Private
No. of awards given last year: 3
No. of applicants last year: 15

Huntington Short-Term Fellowships

Subjects: British and American history, literature, art history, and history of science.
Purpose: To enable outstanding scholars to carry out significant research in the collections of the Library and Art Gallery, by assisting in balancing the budgets of such persons, on leave at reduced pay and living away from home.
Eligibility: Open to nationals of any country who have demonstrated, to a degree commensurate with their age and experience, unusual abilities as scholars through publications of a high order of merit. Attention is paid to the value of the project and the degree to which the special strengths of the Library and Art Gallery will be used.
Level of Study: Doctorate, Postdoctorate, Postgraduate
Type: Fellowship
Value: US$3,000 per month
Length of Study: 1–5 months
Frequency: Annual
Study Establishment: The Huntington
Country of Study: United States of America
No. of awards offered: Approx. 100, depending on funds available
Application Procedure: Applicants must contact the Committee on Fellowships.
Closing Date: Applications are accepted between October 1st and November 30th.
Funding: Private
No. of awards given last year: 105
No. of applicants last year: 320
Additional Information: Fellowships are available for work towards doctoral dissertations.

Mellon Fellowship

Subjects: British and American history, literature, art history, and history of science.
Purpose: To support scholarship in a field appropriate to the Huntington's collections.
Eligibility: Preference will be given to scholars who have not held a major award in the 3 years preceding the year of this award. Applicants must have received the PhD by June of 2012.
Level of Study: Postdoctorate, Postgraduate
Type: Fellowship
Value: US$50,000
Length of Study: 9–12 months
Frequency: Annual
Study Establishment: The Huntington
Country of Study: United States of America
No. of awards offered: 2
Application Procedure: Applicants must contact the Committee on Fellowships.
Closing Date: Applications are accepted between October 1st and November 30th.
Funding: Foundation
No. of awards given last year: 2
No. of applicants last year: 20

National Endowment for the Humanities Fellowships

Subjects: British and American history, literature, art history, and history of science.
Purpose: To support scholarship in a field appropriate to the Huntington's collections.
Eligibility: Preference will be given to scholars who have not held a major award in the 3 years preceding the year of this award. Applicants must have received the PhD by June of 2012.
Level of Study: Postdoctorate, Postgraduate
Type: Fellowship
Value: US$50,400
Length of Study: 9–12 months
Frequency: Annual
Study Establishment: The Huntington
Country of Study: United States of America
No. of awards offered: 3
Application Procedure: Applicants must contact the Committee on Fellowships.

Closing Date: Applications are accepted between October 1st and November 30th.
Funding: Government
No. of awards given last year: 3
No. of applicants last year: 92

HUNTINGTON'S DISEASE ASSOCIATION

Suite 24, Liverpool Science Park, Innovation Centre 1, 131 Mount Pleasant,, Liverpool, L3 5TF, England
Tel: (44) 0151 331 5444
Fax: (44) 0151 331 5444
Email: info@hda.org.uk
Website: www.hda.org.uk
Contact: Karen Crowder

The Association provides a team of regional care advisers, who offer care, advice, support and education to families and professionals who care for people with Huntington's Disease.

HDA Research Project Grants

Subjects: Furthering the understanding of Huntington's Disease, improving its treatment or otherwise improving the quality of life for patients and their careers.
Purpose: To support research projects on Huntington's Disease in a direct way. Preference is given to small 'pump priming grants' likely to lead to support from a major funding body.
Eligibility: Open to suitably qualified researchers of any nationality.
Level of Study: Postgraduate, Research
Type: Project grant
Value: Up to UK£100,000
Length of Study: Unspecified
Frequency: Annual
Study Establishment: A suitable institution in England or Wales
Application Procedure: Applicants should apply to the main organization. Please check the website for more information.
Closing Date: May 31st
Funding: Private
No. of awards given last year: 1
No. of applicants last year: 9

HDA Studentship

Subjects: To further understanding into Huntington's Disease, improving its treatment or otherwise improving the quality of life for patients and their carers.
Purpose: To support a postgraduate student undertaking research into Huntington's Disease.
Level of Study: Postgraduate
Type: Studentship
Value: Up to UK£90,000–100,000
Length of Study: Up to 3 years
Frequency: Annual
Study Establishment: An institution in England or Wales
Application Procedure: Applicants should apply to the main organization.
Closing Date: April
Funding: Private
No. of awards given last year: 1
No. of applicants last year: 9

THE HURSTON-WRIGHT FOUNDATION

12138 Central Avenue, Suite 209, Bowie, MD, 20721, United States of America
Tel: (1) 301 459 2108
Email: info@hurstonwright.org
Website: www.hurstonwright.org
Contact: Elizabeth Williams, Programs and Communications Director

The Foundation was established in September 1990 by novelist Marita Golden. Our mission is to develop, nurture and sustain the world community of writers of African descent.

The Hurston-Wright Award for College Writers

Subjects: Fiction or nonfiction writing in any genre.
Purpose: To support students of African descent enrolled full time as undergraduate or graduate students in any college or university in the United States of America.
Eligibility: Students of African descent from any area of the diaspora.
Level of Study: Graduate, Postgraduate
Type: Scholarship
Value: US$1,000 or US$500
Length of Study: Varies
Frequency: Annual
Study Establishment: Any suitable college or university
Country of Study: United States of America
No. of awards offered: 3
Application Procedure: Applicants must submit an application form to the main organization. Please check the website for details.
Closing Date: January 20th
Funding: Private
No. of awards given last year: 3

Hurston-Wright Legacy Award

Subjects: Works by published writers of African descent from any area of the diaspora.
Purpose: To support published writers of African descent in furthering their art.
Eligibility: Writers of African descent from any area of the diaspora.
Level of Study: Professional development, Unrestricted
Type: Scholarship
Value: US$10,000 or US$5,000
Length of Study: Varies
Frequency: Annual
No. of awards offered: 9
Application Procedure: Book must be submitted by the publisher with permission of the writer.
Closing Date: November 22nd
Funding: Private
No. of awards given last year: 9
Additional Information: Include with each application a $25 nonrefundable application fee (in the form of a money order or check only).

The Hurston-Wright Writer's Week Scholarships

Subjects: Fiction or nonfiction writing in any genre.
Purpose: To support promising writers attending Writer's Week.
Eligibility: Students of African descent from any area of the diaspora.
Level of Study: Professional development
Type: Scholarship
Value: Up to US$1,100
Length of Study: Duration of the workshop
Frequency: Annual
Study Establishment: Howard University, Washington, DC
Country of Study: United States of America
No. of awards offered: Variable
Application Procedure: Applicants must write a brief letter of no more than two paragraphs stating their financial situation and the amount of assistance they are requesting.
Closing Date: April 29th
Funding: Private

IBM CORPORATION

1 New Orchard Road, Armonk, New York, NY, 10504 1722, United States of America
Tel: (1) 877 426 6006
Fax: (1) 866 722 9226
Email: ews@us.ibm.com
Website: www.ibm.com

IBM stands today at the forefront of a worldwide industry that is revolutionizing the way in which enterprises, organizations and people operate and thrive. IBM strives to lead in the invention, development and manufacture of the industry's most advanced information technologies, including computer systems, software, storage systems and microelectronics.

IBM Herman Goldstine Postdoctoral Fellowship

Subjects: Mathematics and computer science.
Purpose: To provide scientists of outstanding ability an opportunity to advance their scholarship as resident department members at the Research Center.
Eligibility: Open to candidates who have obtained a PhD or expect to receive a PhD before the fellowship commences in the second half of 2012.
Level of Study: Research
Type: Fellowship
Value: US$95,000–115,000
Length of Study: 1 year
Frequency: Annual
Country of Study: United States of America
No. of awards offered: 2
Application Procedure: Applicants can download the application form from the website. The completed application form, curriculum vitae and abstract of PhD dissertation must be sent.
Closing Date: January 5th
Additional Information: Applications shall be accepted through email at goldpost@watson.ibm.com

IBM PhD Fellowship Program

Subjects: All subjects.
Purpose: To honour exceptional PhD students in an array of focus areas of interest to IBM and fundamental to innovation.
Eligibility: Open to students nominated by a faculty member. They must be enrolled full-time in a college or university PhD programme and they should have completed at least 1 year of study in their Doctoral programme at the time of their nomination.
Level of Study: Doctorate
Type: Fellowships
Value: US$17,500
Length of Study: 3 years
Frequency: Annual
Application Procedure: All nominations for the IBM PhD Fellowship must be submitted by faculty electronically over the web on a standardized form. The nomination form will be available on the IBM PhD Fellowship nomination website from September 19th to October 31st.
Closing Date: October 31st
Additional Information: Non-US citizens who wish to participate in an internship in the United Stated must obtain work authorization under the specifics of their particular visa.

ICMA CENTRE

Henley Business School, The University of Reading, Whiteknights Park, PO Box 242, Reading, Berkshire, RG6 6BA, England
Tel: (44) 11 8378 8239
Fax: (44) 11 8931 4741
Email: admin@icmacentre.rdg.ac.uk
Website: www.icmacentre.ac.uk
Contact: Mrs Lucy Hogg, Marketing Manager

Part of Henley Business School, the ICMA Centre offers a range of undergraduate, postgraduate and executive education, research and consultancy for the financial markets. The practical application of finance theory is one of the Centre's key advantages and is achieved through the use of its three state-of-the-art dealing rooms.

ICMA Centre Doctoral Scholarship

Subjects: Asset pricing, behavioural finance, commodity markets and derivaties, corporate finance and governance, credit risk and credit derivatives, equity and futures markets, financial engineering, financial regulationfootball finance, hedge funds, hedging strategies, high frequency econometrics, market microstructure, mergers and acquisitions, option pricing, pension schemes, portfolio management and performance assessment, quantitative finance, real estate finance and options, historical finance, liquidity risk, volatility and correlation modelling.
Purpose: To support research in the field.
Eligibility: Open to candidates with an excellent academic background who have completed, or who are in the process of completing, a Master's degree with grade averages at distinction level, in a course containing a significant proportion of finance.
Level of Study: Doctorate
Type: Scholarship
Value: Up to £15,000 paid quarterly plus PhD fee waiver
Length of Study: 3 years
Frequency: Annual
Study Establishment: ICMA Centre, University of Reading
Country of Study: United Kingdom
No. of awards offered: Up to 20
Application Procedure: Applicants must complete an application form, available from the Centre.
Closing Date: Please check website: http://www.icmacentre.ac.uk
Funding: Commercial, government, private
Contributor: ICMA Centre and the University of Reading
No. of awards given last year: 9
No. of applicants last year: 82

IEEE (INSTITUTE OF ELECTRICAL AND ELECTRONICS ENGINEERS, INC.) HISTORY CENTER

Rutgers University, 39 Union Street, New Brunswick, NJ, 08901, United States of America
Tel: (1) 732 562 5450
Fax: (1) 732 932 1193
Email: ieee-history@ieee.org
Website: www.ieee.org/about/history_center/fellowship.html
Contact: Mr Robert Colburn, Research Co-ordinator

The mission of the IEEE History Center is to preserve, research and promote the history of information and electrical technologies.

IEEE Fellowship in Electrical History

Subjects: The history of electrical engineering and computer technology.
Purpose: To support graduate work in the history of electrical engineering.
Eligibility: Open to suitably qualified graduate students, or post-doctoral candidate studying the history of electrical or computing technologies.
Level of Study: Doctorate, Postdoctorate, Postgraduate
Type: Fellowship
Value: US$17,000 plus US$3,000 research budget
Length of Study: 1 year
Frequency: Annual
Study Establishment: A college or university of recognized standing
Country of Study: Any country
No. of awards offered: 1
Application Procedure: Applicants must submit a completed application, transcripts, three letters of recommendation and a research proposal. Application materials can be downloaded from the website.
Closing Date: February 1st
Funding: Corporation
No. of awards given last year: 1
No. of applicants last year: 12
Additional Information: The fellowship is made possible by a grant from the IEEE Life Member Fund and is awarded by the IEEE History Committee. Application materials available on the website.

IESE BUSINESS SCHOOL

IESE Business School - Barcelona Campus, Avenida Pearson, 21, Barcelona, 08034, Spain
Tel: (34) 93 253 4200
Fax: (34) 93 253 4343
Email: info@iese.edu
Website: www.iese.edu

IESE Business School seeks to impact the management profession by offering high-quality learning to students and senior executives from around the world. Our programs are designed and delivered by faculty who are recognized for their dedication to teaching and research, with close ties with international business community.

The Cámara de Comercio Scholarship

Subjects: Business management.
Eligibility: Preference is given to students from developing countries.
Level of Study: MBA
Type: Scholarship
Value: All tuition fees
Length of Study: 1 year
Frequency: Annual
Study Establishment: IESE Business School
Country of Study: Spain
No. of awards offered: 2
Application Procedure: See website.
Closing Date: April 26th
Funding: Corporation
Contributor: Cámara de Comercio, Industria, Navegació de Barcelona

Fundación Ramón Areces Scholarship

Subjects: Business management.
Eligibility: Open to applicants of Spanish nationality only.
Level of Study: MBA
Type: Scholarship
Value: All tuition fees
Length of Study: 1 year
Frequency: Annual
Study Establishment: IESE Business School
Country of Study: Spain
Application Procedure: See website.
Closing Date: April 26th
Funding: Foundation
Contributor: Fudación Ramón Arces

IESE AECI/Becas MAE

Subjects: Business management.
Eligibility: Preference is given to candidates from nations designated as priority countries by the Director de la Cooperacion Espanola.
Level of Study: Professional development
Type: Scholarship
Value: All living expenses
Length of Study: 1 year
Frequency: Annual
Study Establishment: IESE Business School
Country of Study: Spain
No. of awards offered: Varies
Application Procedure: Check website for details.
Funding: Government
Contributor: The Spanish Ministry of Foreign Affairs

For further information contact:

Website: www.becasmal.es

IESE Alumni Association Scholarships

Subjects: Business management and MBA.
Purpose: To reward candidates who have demonstrated excellent work experience and personal merit.
Eligibility: For future MBA students who have demonstrated exceptional work experience and personal merit.
Level of Study: MBA
Type: Scholarship
Value: 50 per cent of tuition fees
Length of Study: 1 year
Frequency: Annual
Study Establishment: IESE Business School
Country of Study: Spain
No. of awards offered: 4
Application Procedure: Contact Admission's office.
Closing Date: June 28th

For further information contact:

MBA Admissions Department, Avda. Pearson 21, Barcelona, 08034, Spain

IESE Donovan Data Systems Anniversary

Subjects: Business management.
Purpose: To commemorate 30 years of DDS partnership with the advertising industry.
Eligibility: Preference is given to students interested in pursuing a career in advertising, marketing or communications.
Level of Study: Professional development
Type: Scholarship
Value: All agreed costs
Length of Study: 1 year
Frequency: Annual
Study Establishment: IESE Business School
No. of awards offered: 1
Application Procedure: See website.
Funding: Corporation
Contributor: Donovan Data Systems
Additional Information: The scholarship alternates each year between a European studying in North American and a North American Studying in Europe.

For further information contact:

Website: www.donovanadata.com

IESE Private Foundation Scholarships

Subjects: Business management and MBA.
Eligibility: Priority is given to students from developing countries.
Level of Study: MBA
Type: Scholarship
Value: All agreed costs
Length of Study: 1 year
Frequency: Annual
Study Establishment: IESE Business School
Country of Study: Spain
No. of awards offered: 4
Application Procedure: See website.
Closing Date: April 26th
Funding: Private

IESE Trust Scholarships

Subjects: Business management.
Eligibility: Outstanding academic records, excellent professional experience and personal merit.
Level of Study: Professional development
Type: Scholarship
Value: Cover 25 per cent or 50 per cent of tuition fees
Length of Study: 1 year
Frequency: Annual
Study Establishment: IESE Business School
Country of Study: Spain
No. of awards offered: 20–30
Application Procedure: Contact Admissions Office.
Funding: Trusts

For further information contact:

MBA Admissions Department, Avda. Pearson, 21, Barcelona, 08034, Spain

ONCE Foundation Scholarships

Purpose: IESE and Fundación ONCE offer an annual scholarship for physically disabled individuals who would like to carry out the Executive MBA programme. These scholarships underscore IESE's aim of supporting the integration of those with physical disabilities in the business world, as well as high level education for this group within society.
Eligibility: In order to qualify for these scholarships, applicants must be recognized legally as being 33 per cent disabled. A copy of the candidate's official Certificate of Disability is required.
Type: Scholarship
Value: Up to €37,290, which is deducted from the total cost of the Executive MBA tuition
Closing Date: Please refer to the webiste
Additional Information: A committee that includes members of Fundación ONCE and IESE's EMBA program will select the scholarship recipient. Scholarship holders will be announced a few weeks after submitting all required documentation.

Scholarships for Female Students
Type: Scholarship
Value: €22,600 which will be deducted from tuition fees (up to 40 per cent of tuition fees)
Application Procedure: Open to female students only.
Closing Date: Please refer to the website
Additional Information: It is advised to all applicants to initiate the admissions process sufficiently in advance, in order to secure a place in the program.

ILLINOIS TEACHERS ESOL & BILINGUAL EDUCATION (ITBE)

PMB 232 8926 N. Greenwood, Niles, IL, 60714-5163, United States of America
Tel: (1) 312 409 4770
Email: awards@itbe.org
Website: www.itbe.org

The ITBE is a non-profit organization of individuals involved in professional development legislation, government issues and specialist interest groups for the teaching of English to speakers of other languages and bilingual education.

ITBE Graduate Scholarship
Subjects: Bilingual education and teaching English to speakers of other languages.
Eligibility: a graduate student presently enrolled full or part-time in an accredited college or university program in TESOL or bilingual education or a practicing ESL/bilingual education professional with concrete plans to enroll in relevant graduate coursework
Level of Study: Postgraduate
Type: Scholarship
Value: US$1,000
Length of Study: 1 year
Frequency: Annual
Application Procedure: Submit application form and follow further instructions on the website.
Closing Date: January 15th

For further information contact:

c/o Albany Park Community Center, 5101 N Kimball Ave, 2nd Floor, Chicago, IL, 60625
Contact: Britt Johnson, Illinois TESOL-BE Awards Chair

INDIA HABITAT CENTRE

Visual Arts Galley, Lodhi Road, New Delhi, 110-003, India
Tel: (91) 11 24682001/09
Fax: (91) 11 24682010
Email: info@indiahabitat.org
Website: www.indiahabitat.org

The India Habitat Centre was conceived to provide a physical environment that would serve as a catalyst for a synergetic relationship between individuals and institutions working in diverse habitat related areas and, therefore, maximize their total effectiveness.

India Habitat Centre Fellowship for Photography
Subjects: Photography.
Purpose: To promote photography as an art form.
Eligibility: Open to Indian nationals who are between 21 and 40 years of age and who do not hold any other fellowship.
Level of Study: Professional development
Type: Fellowship
Value: INR 1,20,000
Frequency: Annual
Country of Study: India
Application Procedure: Applicants must send a project summary, curriculum vitae and reference letters.
Closing Date: August 31st

For further information contact:

Visual Arts Gallery
Email: alkapande@indiahabitat.org
Contact: Dr Alka Pande

INDIAN COUNCIL OF SOCIAL SCIENCE RESEARCH (ICSSR)

35 Firozshah Road, New Delhi, 110-001, India
Tel: (91) 11 2617 9849
Fax: (91) 11 2617 9836
Email: info@icssr.org
Website: www.icssr.org
Contact: Deputy Director

The Indian Council of Social Science Research (ICSSR) is an autonomous organization, funded by the Indian Government to promote research in the social sciences. It provides grants in aid for research projects, fellowships and study grants for young people and gives publication grants. It established the National Social Science Documentation Centre and Archives for providing information to social scientists.

Doctoral Fellowships
Subjects: Social science.
Purpose: To provide opportunities for social scientists to engage themselves in full-time research on important themes of their choice or to write books about their research.
Eligibility: Open to applicants who are not above 35 years of age, who hold a Master's degree in social sciences from a recognized university with a minimum overall aggregate of 55 per cent. Applicants must be registered for PhD in social sciences and cleared the MPhil, National Eligibility Test.
Level of Study: Doctorate
Type: Fellowships
Value: Indian Rupees 16,000 per month (unemployed scholars) and salary protection (employed scholars). Contingency grant of Indian Rupees 15,000 per year
Frequency: Annual
Country of Study: India
No. of awards offered: 55
Application Procedure: Application forms can be downloaded from www.icssr.org

General Fellowships
Subjects: Humanities and social science.
Purpose: To encourage promising and potential scholars in further research.
Eligibility: Open to candidates who are below the age of 50 years, who have shown significant promise and potential for research.
Level of Study: Postgraduate, Research
Type: Fellowship
Value: INR 6,000 per month (unemployed scholars) and salary protection (employed scholars). Contingency grant of INR 12,000 per year
Length of Study: 2 years
Frequency: Annual
Country of Study: India
Closing Date: June 15th

National Fellowships
Subjects: Social science.
Purpose: To enable eminent social scientists, who have made outstanding contributions to research in their respective fields, to further continue their academic work.
Eligibility: Open to social scientists of eminence, preferably below the age of 70 years.
Level of Study: Research
Type: Fellowships
Value: INR 55,000 per month plus a contingency grant of INR 60,000 per year
Length of Study: 2 years
Frequency: Annual
Country of Study: India
No. of awards offered: 6

Senior Fellowships
Subjects: Social science and humanities.
Purpose: To encourage professional social scientists who have their PhD and quality publications in the form of books and papers in professional journals to their credit.

Eligibility: Open to scientists who are not above 65 years of age and who hold a PhD. Social workers, journalists and civil servants known for their academic interests with record of publications may be considered.
Level of Study: Postgraduate, Research
Type: Fellowship
Value: INR 8,000 per month (unemployed scholars) and salary protection (employed scholars). Contingency grant of INR 36,000 per year
Length of Study: 2 years
Frequency: Annual
Country of Study: India
Closing Date: June 15th

INDIAN EDUCATION DEPARTMENT

Government of India, Ministry of Human Resource Department, Shastri Bhavan, New Delhi, 110001, India
Tel: (91) 11 23383936
Fax: (91) 11 23381355
Email: webmaster.edu@nic.in
Website: www.education.nic.in

The origin of the Indian Education Department, Government of India, dates back to pre-independence days when for the first time a separate Department was created in 1910 to look after education. However, soon after India achieved its independence, a full fledged ministry of Education was established.

Agatha Harrison Memorial Fellowship (St Antony's College, Oxford)
Subjects: Modern Indian studies in history, economics,and political science.
Eligibility: Open to the residents of India with 60 per cent marks at Master's Degree level, PhD Degree in the subject field or published works of equivalent merit and minimum 3 years teaching experience at graduate/postgraduate level. Age between 30 and 40 years.
Level of Study: Postgraduate
Type: Fellowship
Value: UK£18,580 + UK£5,956 allowance and economy class air passage (both ways) borne by the Government of India
Length of Study: 1 year
Application Procedure: Applicants must send typed application with attested copies of testimonials, programme of study, photograph and other documents on plain paper.
Funding: Government
Contributor: The Government of India

For further information contact:

Es. 4 Section (Scholarships), Ministry of Human Resource Development, Department of Education, (ES 4) External Scholarships Division, A1/w3 Curzon Road Barracks, K G Marg, New Delhi, 110 001, India
Contact: Section Officer

Belgium Scholarships
Subjects: Agronomy, environmental science, and photonics.
Eligibility: Open to the candidates up to the age of 35 who are graduates in agronomy/micro-electronics/chemical eng/technology/metallurgy/vet. science/environmental studies with 60 per cent or above marks with two years experience in the field for postgraduate studies.
Level of Study: Graduate
Type: Scholarship
Value: €84,000 per month plus reimbursement of tuition fee
Length of Study: 10 months
No. of awards offered: 3
Application Procedure: Applicants must send typed application with attested copies of testimonials, programme of study and other documents on plain paper by notified date.
Closing Date: March 19th
Additional Information: extensive medical and third party liability insurance and Traveling expenses from India to Belgium and back to be borne either by the candidate or his/her employer/sponsor.

For further information contact:

ES. 5, Ministry of Human Resource Development, Department of Higher Education, A2/w4, Curzon Road Barracks, KG Marg, New Delhi, 110 001, India
Contact: Section officer

China Scholarships
Subjects: Chinese language & literature, fine arts (painting & sculpture), botany, environmental science, plant breeding & genetics, political science, sericulture and agronomy.
Eligibility: Open to the Indian nationals below 40 years who have 2–3 year Cert/Dip in basic Chinese Language from a recognized Institution or University, have degree in fine arts with 60 per cent and 60 per cent for other subjects at postgraduate level with work research experience of 2 years.
Level of Study: Graduate
Type: Scholarship
Value: The Chinese Government pays a sum of Yuans 2000 per month to all senior advanced students, Yuans 1400 per month to ordinary graduate students and Yuans 1100 per month to undergraduate students. In addition, Government of India is paying supplementary grant of Yuans 1170 per month Government of China will also provide boarding, lodging, medical care and bear expenditure on tuition and other fees etc
Length of Study: 1–4 years
No. of awards offered: 25
Application Procedure: Applicants must send the application duly sponsored by the employers (if employed) furnishing particulars (as per notified format) by the prescribed date.
Closing Date: March 20th
Contributor: Government of China in association with the government of India

For further information contact:

ES. 3 Section, Ministry of Human Resource Development, Department of Higher Education, External Scholarship Division, A1/w3, Curzon Road Barracks, KG Marg, New Delhi, 110 001, India
Contact: Section Officer

Commonwealth Scholarship/Fellowships in UK
Subjects: All subject.
Eligibility: Open to the Indian national residing in India completing tertiary education in English medium and graduated first Master's Degree as per requirement within the last 10 years as on October of the in-take year. For more details, check website.
Level of Study: Doctorate, Postgraduate
Type: Fellowship/Scholarship
Value: Tourist-class air passage (both ways) fee, with adequate maintenance and other allowance
Length of Study: 1 year for postgraduation, 6 months for clinical training and 3 years for PhD
Application Procedure: Check the website for further details.
Closing Date: October 12th
Additional Information: Candidate should give one page academic justification for pursuing Master Degree course in United Kingdom.

For further information contact:

ES. 1 Section(Scholarships), Ministry of Human Resource Development, Department of Higher Education, (ES 1) External Scholarships Division, A1/W3 Curzon Road Barracks, K G Marg, New Delhi, 110 001, India
Contact: Section Officer

CZECH Scholarships
Subjects: Agriculture (for PhD), electrical/electronics engineering/mechanical engineering (for Master's and PhD).
Eligibility: Uniformly good academic record with at least 60 per cent marks in relevant subject at Bachelor's degree for Masters study and 60 per cent or more at Master's level in science/engineering/technology in the related subjects or equivalent qualification in the subject selected or in the allied field.
Level of Study: Doctorate, Postgraduate
Type: Scholarship
Value: 7,000 CZK for Master's and 7,500 CZK for Doctor's per month.

Length of Study: As per rules of Czech Universities
No. of awards offered: 4 (2 for PhD and 2 for Master's)
Application Procedure: Application, duly sponsored by the employers (if employed), furnishing particulars (as per notified format) may be submitted on plain paper by the prescribed date.
Additional Information: A Summary of study proposal needs to be furnished. Applicants who have graduated from Czech Universities will be given preference over other candidates.

For further information contact:

ES-3 Section, Ministry of Human Resource Development, Department of Higher Education, External Scholarship Division, A1/W3, Curzon Road Barracks, KG Marg, New Delhi, 110 001, India
Tel: (91) 11 23382458, 23386401
Contact: Section officer

Erasmus Mundus Scholarship Programme

Subjects: All subjects.
Purpose: For the benefit of Indian students.
Eligibility: Open to Indian nationals who are graduates from recognized institutions or universities.
Level of Study: Graduate
Type: Scholarships
Value: Covers airfare and living expenses
Length of Study: 1–2 years duration depending on the subject areas of study
Application Procedure: Applicants must apply directly to the universities/consortium of universities constituted under the Erasmus Mundus Programme.
Closing Date: February 18th
Contributor: European Union (EU)

For further information contact:

Email: eac-info@cec.eu.int
Website: www.europa.eu.int/comm/education/programmes/socrates/erasmus/students_en.html

Germany Scholarships

Subjects: Mechanical engineering, electronics and communication engineering, bio-pharmacology, metallurgy, environmental science, bio-technology, agriculture and forestry, veterinary sciences and animal husbandry, electrical engineering.
Eligibility: Open to the candidates with 60 per cent marks at Master's Degree in the subject/related field, sociology and economics candidates possessing adequate knowledge of German language with 2 years experience in teaching, research or practical training after obtaining the prescribed qualification.
Level of Study: Doctorate, Postgraduate
Type: Fellowship
Value: DM 1,800 per month and other allowances
Length of Study: 1 year
No. of awards offered: 7
Application Procedure: Applicants must send typed application, duly sponsored by employer, with research programme, attested copies of testimonials and other documents submitted on plain paper (as per notified format) by the prescribed date.
Additional Information: Period of fellowship is preceded by a compulsory 4 months German language course to be conducted at one of the branches of the Geothe Institute in Germany. Preference is given to candidates having contacts with German Professors or placement at the German Institute.

For further information contact:

ES.1 Section, Ministry of Human Resource Development, Department of Education (ES-I), External Scholarship Division, Curzon Road Barracks, K.G. Marg, New Delhi, 110 001, India
Contact: Section Officer

Greece Scholarships

Subjects: Mathematics, political philosophy/political thought.
Eligibility: Open to the candidates below 40 years who have Master's Degree with 60 per cent or more marks with 2 years teaching/research/practical experience. Candidates should be pursuing studies in universities or research centres.
Level of Study: Postgraduate
Type: Scholarship
Value: 150,000 drachmas. Check website for more details
Length of Study: 10 months
No. of awards offered: 1
Application Procedure: Applicants must send typed format along with photocopies of certificates through employer, if employed.
Closing Date: Please check website
Additional Information: Travelling expenses from India to Greece and back to be borne by the candidate or his/her sponsor/employee.

For further information contact:

ES. 5 Section, Ministry of Human Resource Development, Department of Education (ES-5), A1/W3, Curzon Road Barracks, K.G. Marg, New Delhi, 110 001, India

Ireland Scholarships

Subjects: Environmental resources management, community health, remedial and special education and developmental studies.
Eligibility: Open to the candidates up to the age 30 having First Class Bachelor's Degree in related subjects along with good knowledge about Ireland, and have stayed for 2 years in India after return from abroad for study/training.
Level of Study: Postgraduate
Type: Fellowship
Value: Fellowships cover return travel from India, university fee, living allowance in addition to supplement for books, preparation of thesis, warm clothing, etc.
Length of Study: 1 year
No. of awards offered: 6
Application Procedure: Applicants must send the typed application with attested copies of testimonials, photograph, plans for employment, and other documents on plain paper in prescribed format by notified date.

For further information contact:

ES.1 Section, Ministry of Human Resource Development, Department of Education, A.1/W.3, Curzon Road Barracks, K.G. Marg, New Delhi, 110 001, India
Contact: Section Officer

Israel Scholarships

Subjects: Economics, business management, mass communication, environment studies, Judaism, Hebrew language, history of the Jewish people, agriculture, chemistry, biology, nano-biology and Middle East Studies.
Eligibility: No age limit. Candidates who have 55 per cent or more marks for humanities, 60 per cent or more for agriculture.
Level of Study: Graduate
Type: Scholarship
Value: Paid by Israeli Government to cover stay, tuition fees, health insurance, etc
Length of Study: 8 months
No. of awards offered: 6 (4 for 8 months research and 2 for P.D.)
Application Procedure: Application, duly sponsored by the employers (if employed) furnishing particulars (as per notified format) may be submitted on plain paper, by the prescribed date.
Funding: Government
Contributor: Israeli Government
Additional Information: Passage cost from India to Israel and back is payable by the candidate/employer. Candidates who are staying abroad will not be considered. Candidate who are already abroad for more than 6 months for study/research/training are eligible to apply only if they stayed in India for 2 consecutive years after their return from abroad.

For further information contact:

ES. 3 Section, Ministry of Human Resource Development, Department of Higher Education, External Scholarship Division, A.1/W.3, Curzon Road Barracks, K.G.Marg, New Delhi, 110 001, India
Contact: Section Officer

Italy Scholarships

Subjects: Fashion technology, economics and management, cultural heritage and restoration, information and communication technology, environment, energy, biotechnology and microelectronics.

Eligibility: Open to the candidates of age not more than 35 years who have Graduate or equivalent degree with 60 per cent marks in the subject opted from a recognized university or a Master or equivalent degree with 60 per cent marks with specialization in the subject opted from a recognised University.
Level of Study: Postgraduate
Type: Scholarship
Value: €700 per month, which is reduced to 50% for scholarship holders of categories B, C, D and E.
Length of Study: 1 year
No. of awards offered: Check website for details
Application Procedure: Application in prescribed format with attested copies of testimonials, photograph, programme of study/research, fitness certificate, two letters of introduction from Indian academic authorities and other documents may be submitted or plain paper. Employed persons are required to send their applications through their employer.
Additional Information: Candidates who have received offer/invitation or secured admission in the above specified subjects in any University or Institution in Italy will be prefferred. Such candidates should produce before the Selection Committee a document proving the offer/invitation/admission.Some courses in the above subject are taught in English and Knowledge of Italian Language is not considered a condition for admission. However, the decision of the Universities/Institutions where such courses are held shall be final in this regard. No international/national passage is offered by the Government of Italy for any of these scholarships. No international travel cost to and fro Italy or local travel cost while in Italy for any Category of the above Scholarships will be paid either by the Government of Italy or by the Government of India. Therefore, an undertaking to the effect that the international travel cost to and fro Italy and local passage while in Italy (if any) will be borne by the candidates or his/her employers/sponsors should also be submitted along with application.

For further information contact:

ES. 5 Section, Ministry of Human Resource Development, Department of Higher Education, A.2/W.4, Curzon Road Barracks, K.G. Marg, New Delhi, 110 001, India
Contact: Section Officer

Japan Scholarships

Subjects: Information technology, optical fibre communication, quality and reliability engineering (as applied to various systems like electricity and power, railways, heavy industry, heavy electronics), electronics & communication, robotics, laser technology, biotechnology, Japanese language and literature, fisheries, Japanese studies, earthquake engineering, environmental science, architecture, aerospace engineering, material science/engineering.
Eligibility: Japanese language: Open to the Indian nationals residing in India of age below 35 years. Candidates who have enrolled in the postgraduate degree in Japanese language or completed undergraduate degree in Japanese language or completed Level 2 of JLPT conducted in Japan foundation and have completed at least 3 years of Japanese language study in a university. For more details, check website address.Japanese studies: Master's degree in the relevant field with minimum 60 per cent marks.research studies (PhD): Masters Degree in the relevant field with minimum 60 per cent marks. Candidate having practical research/ teaching / work experience after obtaining the prescribed qualification as on 30/05/2008 would be preferred. Students who have Bachelor's degree in science or arts or engineering and have completed 15 years or 16 years curriculum (i.e. 12+3 or 12+4) are also eligible for the Research Students Scholarship of the Japanese government provided that they have minimum 60 per cent marks.
Level of Study: Doctorate, Graduate
Type: Scholarship
Value: Covers study allowance, fee, part payment of medical expenses, accommodation, etc. Return air ticket is provided by the Japanese Government.
Length of Study: 18–24 months (may be extended)
No. of awards offered: 30
Application Procedure: Applicants must send typed application on plain paper as per format notified, along with attested photocopies of the testimonials, photograph, programme of study/research and other documents. Employed candidates are required to send applications through their employers with NOC.
Funding: Government
Contributor: Government of Japan
Additional Information: Return air ticket is provided by the Japanese Government.

For further information contact:

ES. 3 Section, Ministry of Human Resource Development, Department of Higher Education, External Scholarship Division, A.1/W.3, Curzon Road Barracks, K.G. Marg, New Delhi, 110 001, India
Tel: (91) 11 23382458, 23386401
Contact: Section Officer

Mexican Government Scholarship

Subjects: Spanish History, Literature, Linguistics, Economics, Demography, Asia & Africa.
Eligibility: Open to Indian nationals residing India. Age Below 45 years. Candidates should have a degree/PhD from University. Additional Requirements: Sufficient knowledge of geography, culture and heritage of Mexico; 2 years consecutive stay in India after return from abroad for study/training/specialisation.
Level of Study: Doctorate, Postgraduate
Type: Scholarship
Value: Monthly maintenance allowance, Registration and college fee, medical insurance to be paid by Government of Mexico. Travel expenses from India to Mexico payable by the Govt. of Mexico.
Length of Study: 1–2 years
No. of awards offered: 6
Application Procedure: Typed application (in prescribed format as notified) with attested copies of testimonials may be submitted on plain paper. Employed persons are required to send their applications through their employers within due date as notified.
Closing Date: September 17th
Funding: Government
Contributor: Government of Mexico

For further information contact:

ES.5 Section, Department of Higher Education, A.2/W.4, Curzon Road Barracks, K.G. Marg, New Delhi, 110 001, India
Contact: The Section Officer

Mongolia Scholarships

Subjects: Mongolian language and Mongolian study.
Eligibility: Open to the Indian nationals residing within the country are below 45 years. They should have Bachelor's degree along with good knowledge of English, and have stayed in India for 2 years consecutively after return from abroad for study/training.
Level of Study: Postgraduate
Type: Scholarship
Value: Course fees, living allowance, and free accommodation in hostel for the duration of stay. Travel expenses form India to Mongolia and back are payable by the candidate or his/her /employer/sponsor.
Length of Study: 1 semester
No. of awards offered: 2
Application Procedure: Applicants must send a typed application duly sponsored by employer, with attested copies of testimonials, photograph, programme of study/research and other documents on plain paper by the prescribed date.
Additional Information: Advance applications may be considered provisionally pending sponsorship by employer within 2 weeks.

For further information contact:

ES. 1 Section(Scholarships), Ministry of Human Resource Development, Department of Education (ES 1), External Scholarships Division, A1/W3, Curzon Road Barracks, K.G. Marg, New Delhi, 110 001, India
Contact: Section Officer

N.C.P.E.D.P. Rajiv Gandhi Postgraduate Scholarship Scheme 2005

Subjects: Medicine, surgery, engineering, architecture, management, business administration,social work, applied psychology, clinical psychology, etc.
Purpose: To enable disabled students with limited means to receive education or professional training at postgraduate and doctoral levels.

Eligibility: Open to Indian nationals between 18 and 35 years of age. The scholarship may be awarded to students with the disabilities as recognized by N.C.P.E.D.P. The candidate should be either pursuing or should have gained admission to a full-time course in an Indian university established by law or in a recognized equivalent institution.
Level of Study: Doctorate, Postgraduate
Type: Scholarship
Value: Rs 1,200 per month
Frequency: Annual
No. of awards offered: Varies
Application Procedure: Applicants must apply to the National Centre for Promotion of Employment for Disabled People.
Funding: Government
Additional Information: A scholarship will be provided for the entire duration of the approved course. Scholarship money will be released every 3 months. N.C.P.E.D.P. reserves the right to change the scheme and/or amend the rules without any notice. The income of the candidate or his parents/guardians should not exceed Rs 5,000 per month.At the discretion of the awarding authority, a scholarship may also be awarded for aprofessional course of Indira Gandhi National Open University (IGNOU) or any other recognised Open University.

For further information contact:

National Centre for Promotion of Employment for Disabled People, A-77, South Extension, Part II, New Delhi, 110 049
Tel: +91 11 26265647, 26265648
Email: education@ncpedp.org
Website: www.ncpedp.org

Narotam Sekhsaria Foundation Scholarship (Tribal)

Subjects: Engineering, technology, and sciences.
Purpose: For upliftment of SC candidates.
Eligibility: Open to candidates with a Bachelor's degree (minimum 50 per cent and 2 years of relevant work experience) and Master's degree (2 years teaching/professional experience), or MPhil. For PhD, minimum 5 years of teaching/professional experience is required. Age limit applicable is 35 years.
Level of Study: Doctorate, Graduate, Postgraduate
Type: Scholarship
Value: A fellowship of Rs 26,000 per month is offered and a contingency grant of Rs 10,00,000 per year is also offered
Length of Study: 2 years Masters degree, 3 years PhD, and 1–6 months postdoctoral research
Country of Study: India
No. of awards offered: 10
Application Procedure: Check the website for further details.
Funding: Government
Additional Information: Income limits is Rs 18,000 per month for all categories.

For further information contact:

Information Facilitation Centre, Ministry of Environment and Forests, Paryavaran Bhawan, CGO Complex, Lodi Road, New Delhi, 110003, India
Website: www.envfor.nic.in

Narotam Sekhsaria Foundation Scholarships

Subjects: Pure sciences, applied sciences, social sciences and humanities, law, architecture, and management.
Purpose: To offer various scholarships for pursuing higher education in different streams.
Eligibility: Open to Graduates who have an excellent background of academic and extra curricular activities for pursuing postgraduate studies in India or abroad. The upper age limit is 30 years.
Level of Study: Postdoctorate
Type: Postgraduate scholarships
Value: Up to Rs 10,00,000
Country of Study: India
Application Procedure: Application forms can be downloaded from the website.
Closing Date: March 20th
Funding: Foundation

For further information contact:

Narottam Sekhsaria Foundation, 46, Maker Chambers III, Nariman Point, Mumbai, Maharashtra, 400021, India
Website: http://pg.nsfoundation.co.in/Home/Scholarship

National Doctoral Fellowships

Subjects: Various fields of science and technology.
Eligibility: Open to candidates who are below 35 years as on August 31st of the year of application. A relaxation of 5 years for SC/ST, physically handicapped, and women. The applicant should possess a First Class Master's Degree in areas of technical education such as ME, M Tech, M Pharm, M Arch, MBA, MS by research or equivalent or should have scored a minimum of 60 per cent aggregate marks in master's degree examination or equivalent if class is not declared.
Level of Study: Doctorate, Postdoctorate
Type: Fellowship
Value: Rs 12,000 per month, a contingency grant of Rs 25,000 per year, and overhead charges of Rs 20,000 per year to be paid to the host institution
Length of Study: 3 years plus an extension up to 1 year
Frequency: Annual
No. of awards offered: 50
Application Procedure: Short-listed candidates are called for interview in September/October. Final selection is made from merit list prepared on the basis of candidate's performance in interview.
Closing Date: July
Funding: Government
Contributor: All India Council For Technical Education (AICTE)
Additional Information: Short-listed candidates will be called for interview sometime in September/October. The candidate should be in the first year of full-time doctoral programme in one of the host institutions listed in the Annexure.

For further information contact:

All India Council For Technical Education, Indira Gandhi Sports Complex, I.P. Estate, New Delhi, 110 002, India
Tel: (91) 011-23392506/63/64/65/68/71/73/74/78
Fax: (91) 011-2339255
Website: www.aicte.ernet.in

New Zealand Scholarship

Subjects: Education, primary health, sustainable rural livelihoods (including livestock, forestry, fisheries and horticulture).
Eligibility: Open to the Indian nationals residing in India who have completed their Master's degree with 60 per cent marks in opted subject field.
Level of Study: Doctorate
Type: Scholarship
Value: Travel expenses. Expenditure on board, lodging, books, study material, tuition fee, internal travel related to studies and medical care is met by donor country.
Application Procedure: Applicants must send the application typed in English on plain paper as per format notified, with attested photocopies of the certificates, through their employer, if employed, within the date as notified along with the name of the scholarship scheme and the country, namely, United Kingdom or New Zealand to the below address.
Additional Information: Applicants must have knowledge of India and of the donor country, and should have stayed for 2 consecutive years in India after return from abroad for study/training/specilization. Candidates qualifying from universities which do not award class/ division, requirements in lieu of First Class 60 per cent marks. For candidates who are doing PhD/MPhil after completion of Master's degree, actual period of research is taken into consideration as experience.

For further information contact:

ES.4 Section (Scholarships), Ministry of Human Resource Development, Department of Higher Education,(ES 4), External Scholarships Division, A2/W4, Curzon Road Barracks, K.G. Marg, New Delhi, 110 001, India
Contact: Section Officer

Norway Scholarships

Subjects: All subjects.
Eligibility: Open to the Indian national residents. Please check website
Level of Study: Postgraduate, Predoctorate
Type: Scholarship
Value: Please see website for details.
Length of Study: Varies
No. of awards offered: Varies
Application Procedure: Typed application with attested copies of testimonials, programme of study, photograph and other documents should be submitted on plain paper (as per format notified) by prescribed date. Employed persons should send their applications duly sponsored. Advanced application may not be considered. Academic year starts from August every year.
Additional Information: Advanced applications may not considered. Enclose a letter of invitation from a host institution in Norway where the candidate wishes to study or carry out research.Enclose a description of the project/purpose of the planned study/research stay in Norway.Enclose a recent letter of invitation from (or on behalf of) a member of the scientific staff of a Norwegian host institution – valid for the planned study/research stay in Norway(www.studyinnorway.no) Enclose a recent letter of recommendation from home institution with relevance to the planned study/research stay in Norway.Have good working knowledge of English or Norwegian (Danish/Swedish)Focus the professional programme around one host institution.

For further information contact:

ES. 5 Section (Scholarships), Ministry of Human Resource Development, Department of Higher Education, A2/W4, Curzon Road Barracks, K.G. Marg, New Delhi, 110 001, India
Contact: Section Officer

Portugal Scholarships

Subjects: Portuguese language and culture.
Eligibility: Open to the university teachers of age upto 45 who possess a Master's Degree in the concerned subject, know Portuguese language and be willing to undertake research in the Portuguese language and culture and have secured admissions or acceptance in a Portuguese Institute or university.
Level of Study: Postgraduate, Research
Type: Scholarship
Value: €500 (equivalent to INR 21,000) per month. €450 (equivalent INR 19,000. First month grant will be €825 equivalent to INR 34,500)
Length of Study: 8–12 months
No. of awards offered: 1 scholarship per month for postgraduate studies, 2 scholarships for research work for the period not exceeding 12 months and 6 scholarships of 8 months duration each for pursuing annual course of Portuguese language and culture
Application Procedure: Applicants must send typed application on plain paper as per format notified, along with attested photocopies of the testimonials, programme of study/research, medical fitness report, duly sponsored, if employed.
Funding: Government
Contributor: Government of Portugal
Additional Information: Advanced application may be considered if the 'NOC' from employer can be produced at the time of interview.

For further information contact:

ES.5 Section (Scholarships), Ministry of Human Resource Development, Department of Education, (ES 5) External Scholarships Division, A1/W3, Curzon Road Barracks, K.G. Marg, New Delhi, 110 001, India
Contact: Section Officer

Postgraduate Indira Gandhi Scholarship Scheme for Single Girl Child

Subjects: All subject.
Purpose: To support higher education of girls who happen to be the only child in their families and also to provide incentive for the parents to observe small family norm.
Eligibility: Open to any single girl child of her parents who has taken admission in Master's degree programmes in any recognized university or a postgraduate college.
Level of Study: Graduate, Postgraduate
Type: Scholarship
Value: Rs 2,000 per month. No tuition fees will be charged
Length of Study: 2 years
Frequency: Annual
No. of awards offered: 1,200
Application Procedure: Check the website address for further details.
Closing Date: October 3rd
Funding: Government
Contributor: The University Grants Commission
Additional Information: A student leaving the studies mid-way will have to take prior approval from the University Grants Commission by submitting an application along with justification through the concerned university and will have to refund the whole amount and the concerned institution will be responsible for this.

For further information contact:

University Grants Commission (UGC), Selections and Awards Bureau, Delhi University South Campus, Benito Juarez Marg, New Delhi, 110021, India

Shastri Indo-Canadian Fellowships

Subjects: Faculty research/enrichment, doctoral research, pilot project awards, visiting lectureships for study in the areas of women & development, Canadian studies, development & environment, social & economic reform, private sector development, social sciences/ humanities.
Eligibility: Open to Indian students/professors who are permanent citizens of India. Please check details on the website for different Indo-Canadian fellowships.
Level of Study: Postgraduate
Type: Fellowship
Value: Fellowships cover return travel from India, University fee, living allowance in addition to supplement for books, preparation of thesis, warm clothing
Length of Study: 4 years (PhD) and 2 years (Masters)
Frequency: Annual
Application Procedure: Applicants must send typed application with attested copies of testimonials, photograph, plans for employment, and other documents on plain paper in prescribed format by notified date (employed persons are expected to send their applications through their employer).For details contact The Vice-President, Shastri Indo-Canadian Institute, 5 Bhai Vir Singh Marg, New Delhi-110001.
Closing Date: Please refer to the website for details

INDIAN INSTITUTE OF SCIENCE BANGALORE (IISC)

Bangalore, 560012, India
Tel: (91) 91 - 80 - 2293 2004/2228/2001
Fax: (91) 1 80 2334 1683, 1 80 236 00683
Email: regr@admin.iisc.ernet.in
Website: www.iisc.ernet.in
Contact: The Registrar

The Indian Institute of Science (IISc) was started in 1909 through the pioneering vision of J N Tata. Since then, it has grown into a premier institution of research and advanced instruction, with more than 2,000 active researchers working in almost all frontier areas of science and technology.

IISc Kishore Vaigyanik Protsahan Yojana Fellowships

Subjects: Science, engineering and medicine.
Purpose: To assist students in realizing their potential and to ensure that the best scientific talent is developed for research and growth in the country.
Eligibility: Open to Indian citizens.
Level of Study: Graduate, Postdoctorate, Postgraduate, Predoctorate, Research
Type: Fellowship
Value: INR 4,000–7,000 per month and contingency grants of INR 16,000–28,000
Frequency: Annual
Study Establishment: Indian Institute of Science, Bangalore

Country of Study: India
Application Procedure: Applicants can download the application form from the website.
Closing Date: September

For further information contact:

Indian Institute of Science, Bangalore, Kishore Vaigyanik Protsahan Yojana, 560 012, India
Tel: (91) 80 2360 1008, 80 2293 2976
Email: kvpy@admin.iisc.ernet.in
Website: www.kvpy.org.in/main/fellowship.htm
Contact: The Convener

INDIAN INSTITUTE OF TECHNOLOGY (IIT)

Department of Computer Science and Engineering, Kanpur, UP, 208016, India
Tel: (91) 512 259 7338/7638
Fax: (91) 512 259 7586
Email: pgadm@cse.iitk.ac.in
Website: www.cse.iitk.ac.in
Contact: Harish Karnick, Professor and Head

Indian Institute of Technology (IIT) imparts training to students to make them competent, motivated engineers and scientists. The Institute not only celebrates freedom of thought, cultivates vision and encourages growth, but also inculcates human values and concern for the environment and the society.

Infosys Fellowship for PhD Students

Subjects: Computer science and engineering.
Purpose: To support those interested in pursuing the PhD programme in the Department of Computer Science and Engineering at IIT Kanpur.
Eligibility: Open to deserving students who have a MTech/ME in any branch of engineering and who have secured admission into the PhD programme.
Level of Study: Postgraduate
Type: Fellowship
Value: Rs 2.25 and Rs 2.50 lakhs per year. Out of this grant, Rs 1.8 lakhs (Rs 15,000 per month) will paid as stipend, remaining money can be utilized by the fellow for purchase of books, journals, payment of tuition fee, and travel for domestic and international conference attendance
Length of Study: 3.5–4 years
Frequency: Annual
Study Establishment: IIT Kanpur
Country of Study: India
No. of awards offered: 2
Application Procedure: The applicant must submit a separate application form to the Deptartment of Computer Science and Engineering.
Closing Date: Please check at: www.cse.iitk.ac.in

For further information contact:

Kanpur, Uttar Pradesh, 208016, India
Email: pgadm@cse.iitk.ac.in
Contact: Admissions In-Charge (PhD) Computer science and engineering department, Indian Institute of Technology

THE INDIANAPOLIS STAR

307N Pennsylvania Street, PO Box 145, Indianapolis, IN, 46206-0145, United States of America
Tel: (1) 317 444 4000
Email: rpulliam@indystar.com
Website: www.indystar.com

The Indianapolis Star celebrated its 100th anniversary on June 6, 2003. The brainchild of Muncie industrialist George F. McCulloch, The Star challenged the two existing morning newspapers, the Journal and the Sentinel.

Pulliam Journalism Fellowship

Subjects: Reporting, news design and graphics and photojournalism.
Purpose: To support newspaper journalism.
Eligibility: Open to candidates who have obtained graduate degrees.
Level of Study: Professional development
Type: Fellowships
Value: US$5,500
Length of Study: 10 weeks
Frequency: Annual
Country of Study: United States of America
No. of awards offered: 20
Application Procedure: Applicants can download the application form from the website. The completed application form along with samples of the best published writings, transcript of college credits, 3 letters of recommendation and a recent photograph must be sent.
Closing Date: November 1st
Additional Information: Please call Russ Pulliam in Indianapolis at 317 444 6001 or Bill Hill in Phoenix at 602 444 4368 for any further information.

For further information contact:

Website: www2.indystar.com/pjf
Contact: Russell B Pulliam, Pulliam Fellowship Director

INDICORPS

3418 Highway 6 South, Suite B309, Houston, TX, 77082, United States of America
Tel: (1) +91-93776-99950, +91-98337-62666
Email: info@indicorps.org
Website: www.indicorps.org

Indicorps is a non-partisan, non-religious, non-profit organization that encourages Indians around the world to actively participate in India's progress.

Indicorps Fellowship

Subjects: Social work with grassroots service organizations in India.
Purpose: To implement projects that are in the organizations' and India's best interest.
Eligibility: Open to Indian citizens only.
Level of Study: Professional development
Type: Fellowship
Value: Varies
Length of Study: 1–2 years
Frequency: Annual
Country of Study: India
No. of awards offered: 10–15
Application Procedure: Applicants must apply online.
Closing Date: March 1
Additional Information: For further information about the application procedure mail to apply@indicorps.org

For further information contact:

Indicorps, 3418 Highway 6 South Suite B #309 Houston, Texas, United States of America
Tel: (1) 281 617 1057

INSEAD

Boulevard de Constance, F-77305 Fontainebleau Cedex, France
Tel: (33) 1 60 72 40 00
Fax: (33) 1 60 74 55 00
Email: mba.europe@insead.edu
Website: www.insead.edu/mba
Contact: Ms Irina Schneider-Maunoury, Senior Manager, MBA Financing

INSEAD is widely recognized as one of the most influential business schools in the world. With its second campus in Asia to complement its established presence in Europe, INSEAD is setting the pace in globalizing the MBA. The 1-year intensive MBA programme is focused on international general management.

INSEAD Admiral Scholarship

Purpose: To supports MBA participants at INSEAD who are selected either on the basis of merit at the time of admission, or, who have demonstrated financial need and submitted an application accordingly.
Eligibility: All admitted candidates will be considered on the basis of merit as per the quality of their admission application. If a merit award is not allocated, then all candidates who qualify and have applied for need-based scholarships will be considered instead.
Type: Scholarship
Value: Up to €15,000
Frequency: Annual
No. of awards offered: 1 award per class
Additional Information: Please check website for more information.

INSEAD Alcatel-Lucent scholarship

Eligibility: Chinese candidates with the required profile (above)may apply. One or two candidates may also be invited for an interview (in person, by video or by telephone) with an Alcatel-Lucent representative in Asia in March (for the July Class) or September (for the December Class). Preference will be given to candidates who demonstrate financial need. If feasible, the INSEAD Alcatel-Lucent scholars will participate in a one-day event organised by Alcatel-Lucent for their scholars in France during the MBA Programme.
Type: Scholarship
Value: €20,000
No. of awards offered: 1 or 2 per class
Closing Date: February 11th for the July class and August 18th for the December class
Additional Information: Please check at http://mba.insead.edu for more information.

INSEAD Alexis and Anne-Marie Habib Foundation scholarship

Eligibility: Open to Lebanese nationals with strong academic credentials and demonstrated financial need.
Type: Scholarship
Value: Up to €15,000
No. of awards offered: 1 per class
Closing Date: February 11th the July class (starts September) and August 18th for the December class (starts January)
Additional Information: Please check at website for more details.

INSEAD Alumni Fund (IAF) Diversity Scholarship(s)

Subjects: MBA.
Purpose: To assist candidates admitted to the MBA programme.
Eligibility: Open to applicants from emerging or developing countries.
Level of Study: MBA
Type: Scholarship
Value: €5,000–15,000
Frequency: Annual
Study Establishment: INSEAD
No. of awards offered: Varies
Application Procedure: Applicants must complete a specific assignment, details of which are available from the website.
Closing Date: February 11th for July class and August 18th for December class
Contributor: Alumni

For further information contact:

Website: www.mba.insead.edu/scholarships

INSEAD Alumni Fund (IAF) Robin Hood Scholarship

Subjects: Candidates from emerging markets who have difficulty raising funds for the program.
Purpose: To assist MBA candidates from emerging and developing countries.
Eligibility: MBA.
Level of Study: MBA
Type: Scholarship
Value: €10,000
Frequency: Annual
Study Establishment: INSEAD
No. of awards offered: 2
Application Procedure: Apply on-line. Details are available on INSEAD'S website.
Closing Date: February 11th for the July class (starts in September) and August 18th for the December class (starts in January)
Funding: Individuals
Contributor: Alumni
No. of awards given last year: 2
Additional Information: Country of study: France, Singapore. Address for application: On-line.

INSEAD Alumni Fund (IAF) Special Profile Scholarship(s)

Purpose: For their class reunions, three classes - '68, '70 and '87J wish to empower exceptional individuals through the INSEAD experience. These awards of varying amounts will go to individuals who would not otherwise be able to attend INSEAD. By way of their backgrounds, work experience and personal profiles these candidates will bring an additional and unique diversity to INSEAD as well as to their own communities in the future as INSEAD alumni.
Eligibility: Only candidates who feel confident that they can fulfill the 'special' or 'different' category should apply. A strong preference will be given to those who demonstrate financial need.
Type: Scholarship
Value: €10,000 up to €25,000
Frequency: Annual
Application Procedure: Complete the application on line and answer the appropriate essays questions.
Closing Date: February 13th each year for the July class (Rounds 1 and 2); August 22nd each year for the December class (Rounds 1 and 2)
Additional Information: Please check website for further information.

INSEAD Alumni Fund (IAF) Women's Scholarship(s)

Subjects: MBA.
Purpose: The IAF Women's Scholarships support INSEAD's commitment to bring outstanding women professionals to the MBA Programme and to increase representation of women in leadership positions in the business community.
Eligibility: INSEAD seeks bright, dynamic and motivated women who are making significant achievements in their professional and/or personal lives. Merit scholarships will be awarded to recognize these outstanding women. Their financial situation may also be taken into consideration.
Type: Scholarship
Value: €5,000–15,000
Frequency: Annual
No. of awards offered: 10–15 awards per class
Application Procedure: To be considered for these scholarships please submit your application on line before the specified deadlines. Candidates will also be considered for the INSEAD Judith Connelly Delouvrier Scholarships.
Closing Date: February 13th each year for the July class (starts in September); August 22nd each year for the December class (starts in January)
Additional Information: Please check website for further information.

INSEAD Andrew Hordern Endowed Scholarship '91D

Subjects: To assist candidates who do not have traditional MBA backgrounds, e.g. artists, lawyers, scientists, not for profit etc.
Purpose: Non-traditional MBA backgrounds.
Eligibility: Open to MBA students.
Level of Study: MBA
Type: Scholarship
Value: €10,000
Frequency: Annual
Study Establishment: INSEAD
No. of awards offered: 1
Application Procedure: Applicants must apply online.
Closing Date: February 11th for the July class (starts in September) and August 18th for the December class (starts in January)
Funding: Individuals
Contributor: Alumni
Additional Information: Countries of study are France and Singapore.

INSEAD Andy Burgess Endowed Scholarship for Social Entrepreneurship

Purpose: To provide one scholarship per year for a participant in each January class, in perpetuity, to deserving candidates admitted to the INSEAD MBA programme, who through their experience prior to INSEAD can demonstrate a commitment to social entrepreneurship.
Eligibility: Candidates who demonstrate experience or commitment to social entrepreneurship. Entrepreneurial potential will be assessed, and preference will be given to candidates who demonstrate financial need.
Type: Scholarship
Value: Up to €10,000
Frequency: Annual
No. of awards offered: 1
Application Procedure: Complete the INSEAD online application. The successful candidate may expect to meet the donor during the program.
Closing Date: August 18th for December class (starts January)

INSEAD Belgian Alumni and Council Scholarship Fund

Subjects: MBA.
Purpose: To assist MBA participants.
Eligibility: Open to candidates of merit of Belgian nationality and those who have lived in Belgium for at least 5 years. Priority will be given to admitted applicants who intend to return to Belgium after their MBA.
Level of Study: MBA
Type: Scholarship
Value: €6,000
Frequency: Annual
Study Establishment: INSEAD
No. of awards offered: 2
Application Procedure: Applicants must complete a specific assignment, details of which are available from the organization or from the website.
Closing Date: February 11th for the September class of the same year and August 18th for the January class of the following year
Contributor: Alumni and the Belgian Council

INSEAD Bischoff Family Endowed Scholarship

Purpose: To support future participants from southern Africa and South-East Asia who demonstrate both merit and financial need.
Eligibility: Candidates admitted to the INSEAD MBA Programme from these two regions who stand out in terms of previous academic achievement and who require financial assistance. Candidates from countries bordering South Africa (as well as Mauritius) and from South-East Asian countries (except for Korea and Japan) are eligible.
Level of Study: Postgraduate
Type: Scholarship
Value: €10,000
Frequency: Annual
Application Procedure: To be considered for this scholarship, please complete the Need-based Scholarship application.
Closing Date: February 13th each year for the July class (starts August); August 22nd each year for the December class (starts January)
Additional Information: Please check website for further information.

INSEAD Børsen/Danish Council Scholarship

Subjects: MBA.
Purpose: To assist MBA participants.
Eligibility: Open to candidates of Danish nationality, admitted to the INSEAD MBA programme.
Level of Study: MBA
Type: Scholarship
Value: Up to €15,000
Frequency: Annual
Study Establishment: INSEAD
Country of Study: Other
No. of awards offered: 2
Application Procedure: Applicants must complete an application form, available from the website.
Closing Date: February 11th for the September class of the same year and August 18th for the January class of the following year
Contributor: The Danish Council/Børsen

INSEAD Canadian Foundation Scholarship

Subjects: MBA.
Purpose: To provide financial assistance and scholarships to deserving Canadians admitted to the INSEAD MBA programme.
Eligibility: Open to candidates of Canadian nationality, preferably resident in Canada, who have been admitted to the INSEAD MBA programme and who intend to return to Canada.
Level of Study: MBA
Type: Scholarship
Value: Up to Canadian $10,000
Frequency: Annual
Study Establishment: INSEAD
No. of awards offered: Varies
Application Procedure: Applicants must submit the following in support of their application: a covering letter requesting a scholarship specifying which campus the applicant is applying to, a budget detailing the need for financial assistance including current and expected sources of funding, a copy of a completed INSEAD admission form with essay and supporting documents, a copy of reference letters submitted in support of application to INSEAD, a copy of Graduate Management Admissions Test results, a copy of university transcripts and a copy of confirmation of admission to INSEAD.
Closing Date: May 31st for candidates admitted to the September intake of the same year and November 15th for candidates admitted to the January class of the following year
Funding: Foundation
Contributor: Alumni
Additional Information: The Canadian INSEAD Foundation is a non-profit corporation whose purpose is to encourage Canadian students to develop an international business understanding and perspective.

For further information contact:

The Board of Trustees, Canadian INSEAD Foundation, c/o Richard Tarte, Société générale de financement du Québec, 600 de La Gauchetière Street, West Suite 1700, Montréal, QC, H3B 4L8, Canada
Tel: (1) 514 876 9290 ext 2171

INSEAD Citi Foundation Scholarship(s)

Purpose: To foster an interest in the financial services industry.
Eligibility: Candidates from Eastern Europe, the Middle East and Africa may apply. In addition, they will need to demonstrate their interest in the financial services industry, particularly within corporate and investment banking. Preference will be given to candidates who are able to demonstrate their need for financial assistance.
Type: Scholarship
Value: Approx. US$5,000
Frequency: Annual
No. of awards offered: 2 or 3 per class
Closing Date: February 11th for the July class (starts in September) and August 18th for the class December (starts in January)

INSEAD Deepak and Sunita Gupta Endowed Scholarship

Subjects: For candidates from emerging markets who demonstrate proven financial need.
Purpose: For students from emerging markets.
Eligibility: Open to MBA students.
Level of Study: MBA
Type: Scholarship
Value: Up to €25,000
Length of Study: 1 year
Frequency: Annual
Study Establishment: INSEAD
No. of awards offered: 2
Application Procedure: Applicants must apply online through INSEAD'S website.
Closing Date: February 11th for July class (starts in September) and August 18th for December class (starts in January)
Funding: Individuals
Contributor: Alumina
No. of awards given last year: 1
Additional Information: Countries of study are France and Singapore.

INSEAD El Wakil-Lahham Scholarships

Subjects: MBA.
Purpose: To help build human capacity and to create economic and social value in countries that make up the Southern Mediterranean Rim. They will be awarded to meritorious MBA participants at INSEAD who demonstrate limited financial means and who are citizens of countries from the Southern Mediterranean Rim i.e. Morocco, Algeria, Tunisia, Libya, Egypt, Lebanon, Syria, (Jordan) and Turkey. Where applicable, first consideration will be given to Egyptian nationals and women participants from the countries within this geography.
Eligibility: Candidates from Morocco, Algeria, Tunisia, Libya, Egypt, Lebanon, Syria, (Jordan) and Turkey will be considered.
Type: Scholarship
Value: Up to €25,000
Frequency: Annual
Application Procedure: To be considered for this award, please apply under INSEAD Needbased Scholarships.
Closing Date: February 13th for July class (starts in September); August 22nd for December class (starts in January)
Additional Information: Please check website for further information.

INSEAD Eli Lilly and Company Innovation Scholarship

Subjects: MBA.
Eligibility: Open to students of merit who demonstrate the capacity for innovative thinking and actions. Nationals from Africa, Asia, Central and Eastern Europe, Middle East, Central and South America, Turkey and Canada may apply.
Level of Study: MBA
Type: Scholarship
Value: Partial tuition
Frequency: Annual
Study Establishment: INSEAD
No. of awards offered: 2 per class
Application Procedure: Applicants must complete a specific assignment, details of which are available from the organization or from the website.
Closing Date: May 5th for the September intake of the same year and September 30th for the January intake of the following year
Contributor: Eli Lilly Foundation
Additional Information: Eli Lilly creates and delivers innovative pharmaceutical-based healthcare solutions that enable people worldwide to live longer, healthier and more active lives.

INSEAD Elmar Schulte Diversity Scholarship

Subjects: MBA.
Purpose: To encourage diversity in the INSEAD MBA programme.
Eligibility: Open to candidates from non-traditional MBA backgrounds who have been admitted to the programme.
Level of Study: MBA
Type: Scholarship
Value: Varies
Frequency: Annual
Study Establishment: INSEAD
No. of awards offered: €10,000
Application Procedure: Applicants must complete an application form, available from the website.
Closing Date: May 5th for the September intake of the same year and September 30th for the January intake of the following year
Contributor: Alumni

INSEAD Elof Hansson Scholarship Endowed Fund

Subjects: MBA.
Purpose: To assist MBA participants.
Eligibility: Open to candidates of Swedish nationality who have been admitted to the INSEAD MBA programme.
Level of Study: MBA
Type: Scholarship
Value: €6,000
Frequency: Annual
Study Establishment: INSEAD
No. of awards offered: 2
Application Procedure: Applicants must complete an application form, available from the website.
Closing Date: February 11th for the September class of the same year and August 18th for the January class of the following year
Funding: Foundation
Contributor: The Elof Hansson Foundation

INSEAD Esa Hietala Endowed Scholarship

Purpose: To assist Finnish candidates to fund the program.
Eligibility: Open to MBA students.
Level of Study: MBA
Type: Scholarship
Value: €20,000
Length of Study: 1 year
Frequency: Annual
Study Establishment: INSEAD
No. of awards offered: 1
Application Procedure: Applicants must apply online.
Funding: Individuals
Contributor: Alumni
No. of awards given last year: 1
Additional Information: Countries of study are France and Singapore. Eligible to nationals of West European Countries and Finland.

INSEAD Forte Foundation Fellowship

Eligibility: Candidates for the Forte Fellowship should exhibit exemplary leadership in one or more ways: academic leadership; team leadership; community leadership; creative leadership. In addition INSEAD will look for measurable academic and personal achievement as well as involvement in issues related to the advancement of women. As Forte is currently a mainly US-based network, it would be an advantage for a candidate to show how she can benefit from it.
Type: Fellowship
Value: €15,000
Frequency: Annual
No. of awards offered: 2
Closing Date: February 11th for the July class (starts in September) and August 18th for the December class (starts in January)
Additional Information: Please check website for more information.

INSEAD Freddy Salem Scholarship(s)

Purpose: Candidates for the INSEAD scholarships will need to demonstrate: (a) academic achievement and promise (b) aptitude for business and financial management, and (c) a strong likelihood of spending the better part of their working careers in West Africa. In awarding these scholarships, INSEAD shall particularly seek candidates from Nigeria, Ghana, Togo and Benin.
Eligibility: Candidates must be nationals from a West Africa country (preferably Nigeria, Ghana, Togo and Benin) and will have spent most of their lives and received most or part of their education in Africa. Preference will be given to candidates who require proven financial assistance.
Type: Scholarship
Value: Up to €20,000
Frequency: Annual
Application Procedure: To be considered for this scholarship please refer to the INSEAD Needbased Scholarship pages and submit your application and background material accordingly.
Closing Date: February 13th for July class (starts in September); August 22nd for December class (starts in January)
Additional Information: Please check website for further information.

INSEAD Giovanni Agnelli Endowed Scholarship

Subjects: MBA.
Purpose: To support MBA participants.
Eligibility: Open to Italian candidates of high merit, admitted to the INSEAD MBA programme.
Level of Study: MBA
Type: Scholarship
Value: Up to €14,000
Frequency: Annual
Study Establishment: INSEAD
No. of awards offered: 1–2
Application Procedure: Applicants must complete an application form, available from the website.
Closing Date: February 11th for the September class of the same year and August 18th for the January class of the following year
Contributor: Fiat group

Additional Information: This endowed scholarship is offered by the Fiat Group.

INSEAD Goldman Sachs Scholarship for African Nationals

Purpose: The Goldman Sachs scholarships for African nationals at INSEAD is designed to give candidates from African countries access to a world-class MBA education. The scholarship winners will be allocated a Goldman Sachs mentor throughout the 10-month MBA programme.
Eligibility: The scholarship is open to all candidates from African countries studying at INSEAD on the Full-time MBA Programme. Successful candidates must demonstrate their desire to work in Africa and explain why building business in Africa is important to them.
Type: Scholarship
Value: Up to €15,000 per class
Frequency: Annual
Closing Date: February 13th for the July class (starts in September); August 22nd for the December class (starts in January)
Additional Information: Please check website for further information.

INSEAD Greek Friends Scholarship

Purpose: The Greek Friends of INSEAD Scholarship Fund was created in 2008 with donations from a number of INSEAD alumni and senior businessmen and women close to the school. They are interested in supporting and enhancing the Greek community at INSEAD and thereby the alumni network in Greece.
Eligibility: Greek nationals residing in Greece who have been admitted to the full-time MBA programme may apply. They must also demonstrate that they require financial assistance to complete year.
Type: Scholarship
Value: Up to €25,000
Frequency: Annual
No. of awards offered: 1
Application Procedure: Candidates must complete the INSEAD Needbased Scholarship application to be considered for this award.
Closing Date: February 13th for the July class (starts in September); August 22nd for the December class (starts in January)
Additional Information: Please check website for further information.

INSEAD Greendale Foundation Scholarship

Subjects: MBA.
Purpose: The Trustees of the Greendale Foundation wish to provide access to the INSEAD MBA programme to disadvantaged Southern and East Africans who are committed to developing international management expertise in Africa and who plan their careers in the Southern and East African regions.
Eligibility: Candidates must be nationals from Kenya, Malawi, Mozambique, South Africa (disadvantaged backgrounds), Tanzania, Uganda, Zambia or Zimbabwe (from disadvantaged backgrounds)who have undergone the major part of their education in Southern or East Africa. Preference will be given to those who have worked there prior to INSEAD. The candidate's financial situation will also be taken into consideration.
Type: Scholarship
Value: €35,000
Frequency: Annual
Application Procedure: Please check website for details.
Closing Date: February 13th each year for the July class (starts in September); August 22nd for the December class (starts in January)
Additional Information: Scholarship recipients must return to work in these African regions within 3 years of graduation; in the event this condition is not met, the recipients will be asked by INSEAD to refund their scholarship within the fourth year after completion of the MBA Programme.

INSEAD Groupe Galeries LaFayette Endowed Scholarship for Women

Purpose: This Scholarship targets women participants at INSEAD who embrace the group's values of commitment to serving customers, regardless of their social background and ethnic origin.
Eligibility: Women candidates who have been admitted to the INSEAD MBA programme who identify with the goals and ambitions of Groupe Galeries Lafayette. Preference will be given to women from regions where the Group is or will be present i.e. Western Europe, Middle East and Asia.
Type: Scholarship
Value: €10,000
Frequency: Annual
Closing Date: February 11th each year for the July class (starts in September) and August 18th each year for the December class (starts in January)
Additional Information: Please check website for more information.

INSEAD Henry Grunfeld Foundation Scholarship

Subjects: MBA.
Purpose: To aid MBA students who can demonstrate a commitment to a career in investment banking.
Eligibility: Open to participants from a United Kingdom background with an interest in pursuing a career in investment banking.
Level of Study: MBA
Type: Scholarship
Value: Up to €12,500
Frequency: Annual
Study Establishment: INSEAD
Country of Study: Other
No. of awards offered: 2 or 3
Application Procedure: Applicants must complete a specific assignment, details of which are available from the organization or from the website.
Closing Date: February 11th for the September class of the same year and August 18th for the January class of the following year
Contributor: The Henry Grunfeld Foundation
Additional Information: Henry Grunfeld was a co-founder of S G Warburg, the United Kingdom investment bank that became one of the largest securities firms in the world, combining merchant banking, securities broking and market-making.

INSEAD Ian Potter '93D & Family Endowed Scholarship

Purpose: For candidates from East and South Asia.
Eligibility: Open to MBA students.
Level of Study: MBA
Type: Fellowships
Value: €10,000
Length of Study: 1 year
Frequency: Annual
Study Establishment: INSEAD
Application Procedure: Applicants must apply through INSEAD's website.
Closing Date: February 11th each year for the July class (starts August) and August 18th each year for the December class (starts January)
Funding: Individuals
Contributor: Alumni
Additional Information: Countries of study are France and Singapore.

INSEAD Jacques Nasser Endowed Leadership Scholarship

Eligibility: Candidates who are citizens of an Arab country in the Middle East (including Egypt) or whose parents are/were citizens of one of these countries. Leadership potential will be assessed and a preference will be given to candidates who demonstrate financial need.
Level of Study: MBA
Type: Scholarship
Value: Up to €10,000
Frequency: Annual
No. of awards offered: Up to 2
Closing Date: February 11th for the July class (starts in September) and August 18th for the December class (starts in January).
Additional Information: Please check website for more information.

INSEAD Jewish Scholarship

Subjects: MBA.
Purpose: An INSEAD alumnus offers scholarships to Jewish students admitted to the INSEAD MBA Programme; a limited number of small awards are made each year. In keeping with the spirit and tradition behind this award, the winners are encouraged to give back by making

their own donation to scholarships at INSEAD through the Alumni Fund within a few years after graduation.
Eligibility: Jewish students who can justify difficulty in raising sufficient finances for their living expenses.
Type: Scholarship
Value: Up to €5,000
Frequency: Annual
Closing Date: February 13th each year for the July class (starts in September); August 22nd each year for the December class (starts in January)
Additional Information: Please check website for further information.

INSEAD Judith Connelly Delouvrier Endowed Scholarship

Subjects: MBA.
Purpose: To support women undertaking the MBA.
Eligibility: Open to deserving American women admitted to the September MBA programme.
Level of Study: MBA
Type: Scholarship
Value: Up to US$15,000
Frequency: Annual
Study Establishment: INSEAD
No. of awards offered: 1
Application Procedure: Applicants must complete a specific assignment, details of which are available from the website.
Closing Date: February 11th for the September class of the same year and August 18th for January class of the same year
Contributor: Alumni
Additional Information: This scholarship is offered in memory of Judith Connelly Delouvrier, wife of Phillippe Delouvrier, an INSEAD MBA of 1977, who was a victim of the TWA Flight 800 tragedy in 1996.

INSEAD L'Oréal Scholarship

Subjects: MBA.
Purpose: To foster creativity, diversity and entrepreneurial spirit within the MBA population.
Eligibility: Open to candidates of any nationality who demonstrate a capacity for creativity, innovation and entrepreneurial activity and who can demonstrate financial need.
Level of Study: MBA
Type: Scholarship
Value: Partial tuition fees
Length of Study: 1 year
Frequency: Annual
Study Establishment: INSEAD
No. of awards offered: 2 (per year) and 1 (per intake)
Closing Date: February 11th for the September class of the same year and August 18th for the January class of the following year
Contributor: L'Oréal

INSEAD Lister Vickery Memorial Scholarship

Subjects: MBA.
Purpose: This scholarship, which is meant to assist MBA participants from Eastern Europe and Central Asian countries, was created in memory of Lister Vickery (MBA '66) by two of his INSEAD classmates and close friends. Over the years other classmates, former recipients of the award and members of Vickery family have all contributed to this scholarship.
Eligibility: Candidates from Eastern Europe or Central Asia who show potential for pursuing a career in industry in the region. Preference will be given to a candidate who demonstrates financial need.
Type: Scholarship
Value: €5,000
Frequency: Annual
Application Procedure: To be considered for this scholarship please apply under INSEAD Needbased Scholarships.
Closing Date: February 13th for July class (starts in September); August 22nd for December class (starts in January)
Additional Information: Please check website for further information.

INSEAD Louis Franck Scholarship

Subjects: MBA.
Eligibility: Open to candidates of United Kingdom nationality admitted to INSEAD. Financial need is neither a necessary nor a sufficient condition for being granted an award. Nevertheless, the candidate's financial position will be taken into account, and awards will not necessarily be granted to the best candidates if there is a sound candidate who is in financial need. Selected scholars are required to write a thesis or report, the subject of which is to be agreed upon with the trustees, and is to be presented to the trustees within 3 months of graduation.
Type: Scholarship
Value: Up to €15,000
Frequency: Annual
Study Establishment: INSEAD
No. of awards offered: Up to 8
Application Procedure: Applicants must complete a specific assignment, details of which are available from the website.
Closing Date: February 11th for the September class of the same year and August 18th for the January class of the following year
Funding: Private
Contributor: The Louis Franck Trust
Additional Information: This scholarship was established in 1983 by Louis Franck, who served for many years on the Board of INSEAD.

INSEAD Louis Vuitton Scholarship

Purpose: The INSEAD Louis Vuitton scholarship is targeted at Chinese and Indian participants at INSEAD. By offering INSEAD participants a scholarship, Louis Vuitton Malletier aims to encourage the awareness of the challenges and opportunities for the luxury goods industry in the region.
Eligibility: To be eligible for the INSEAD Louis Vuitton Scholarship, candidates must first be admitted to the INSEAD MBA programme. They must be a national of China or India, have lived or studied in one of these countries and ideally completed their first professional experience there as well. Preference will be given to candidates who require proven financial assistance. Experience in retail is a plus.
Type: Scholarship
Value: Up to €25,000
Frequency: Annual
Closing Date: February 13th for the July class (starts in August); August 22nd for the December class (starts in January)
Additional Information: Please check website for further information.

INSEAD Marguerre Endowed Scholarship(s) for Entrepreneurial Talent

Purpose: The endowed fund will provide one or two scholarships per class to deserving candidates admitted to the INSEAD MBA programme who through their experience prior to INSEAD can demonstrate entrepreneurial talent.
Eligibility: Candidates with entrepreneurial talent and preference will be given to those who demonstrate financial need.
Type: Scholarship
Value: €11,000 if two awards or €22,000 if only one award per class
Frequency: Annual
No. of awards offered: 1 or 2 per class
Closing Date: February 11th for July class (starts in September) and August 18 for December class (starts in January)

INSEAD MBA '70 Special Profile Endowed Scholarship

Subjects: MBA.
Purpose: The '70 class wishes to support well-qualified individuals from poor backgrounds who would not otherwise be able to attend INSEAD and thereby empower them through the INSEAD experience. Such special profiles can be candidates from socially, economically and/or politically challenged countries.
Eligibility: Candidates admitted to the MBA Programme with the above mentioned profiles will be considered. Financial need will be taken into consideration.
Type: Scholarship
Value: Up to €15,000
Frequency: Annual
Application Procedure: To be considered for this award, candidates must complete the INSEAD Alumni Fund (IAF) Special Profiles Scholarship application.

Closing Date: February 13th for the July class (starts September); August 22nd for the December class (starts January)
Additional Information: Please check website for further information.

INSEAD MBA '87J Special Profile Endowed Scholarship

Subjects: MBA.
Purpose: The award will be made once a year to a candidate whose background, work experience and personal profile will bring an additional and unique diversity to INSEAD and in the future to their own communities as INSEAD alumni. Such special profiles can be candidates from socially, economically and/or politically challenged countries; a commitment to the not-for-profit sector; career in technological innovation that improves the environment. Candidates from emerging countries requiring financial assistance will also be considered.
Eligibility: Candidates admitted to the the MBA Programme with the above mentioned profiles will be considered. Financial need will be taken into consideration.
Type: Scholarship
Value: Up to €10,000
Frequency: Annual
Application Procedure: To be considered for this award candidates must complete the INSEAD Alumni Fund (IAF) Special Profile Scholarship application.
Closing Date: February 13th for the July class (starts September); August 22nd for the December class (starts January)
Additional Information: Please check website for further information.

INSEAD MBA '89D Endowed Scholarship

Subjects: MBA.
Purpose: The '89D class wishes to support well-qualified individuals who would not otherwise be able to attend INSEAD and thereby empower them through the INSEAD experience. Candidates requiring financial assistance will be considered.
Eligibility: Candidates admitted to the MBA Programme with the above mentioned profiles will be considered. Financial need will be taken into consideration.
Type: Scholarship
Value: Up to €10,000
Frequency: Annual
Application Procedure: To be considered for this scholarship, please complete the Needbased Scholarship application.
Closing Date: February 13th for the July class (starts September); August 22nd for the December class (starts January)
Additional Information: Please check website for further information.

INSEAD Mette Roed Heyerdahl Memorial Scholarship

Purpose: For women from Nepal, Uganda, Palestine, Turkey and Middle East.
Eligibility: Open to MBA students.
Level of Study: MBA
Type: Fellowships
Value: €12,500
Length of Study: 1 year
Frequency: Annual
Study Establishment: INSEAD
Application Procedure: Applicants must apply through INSEAD's website.
Closing Date: August 18th for December class (starts in January); February 11th for July class (starts in September)
Funding: Individuals
Contributor: Alumni
No. of awards given last year: 2
Additional Information: Countries of study are France and Singapore.

INSEAD Nelson Mandela Endowed Scholarship

Purpose: To assist African MBA candidates (sub-saharan Africa).
Eligibility: Open to MBA students.
Level of Study: MBA
Type: Scholarship
Value: €20,000
Length of Study: 1 year
Frequency: Annual
Study Establishment: INSEAD
No. of awards offered: 2
Application Procedure: Applicants must apply through INSEAD'S website.
Closing Date: February 11th for July class (starts in August) and August 18th for December class (starts in January)
Funding: Individuals
Contributor: Alumni
No. of awards given last year: 2
Additional Information: Countries of study are France and Singapore.

INSEAD Orange Endowed Scholarship(s) for Emerging Markets

Purpose: This endowed scholarship will support emerging market participants at INSEAD in perpetuity. Candidates need to be motivated to pursue their career in one of the regions where Orange operates. Experience or an interest in telecommunications is a plus.
Eligibility: Candidates admitted to the MBA programme from Armenia, Brazil, Central African Republic, Dominican Republic, Egypt, India, Kenya, Madagascar, Moldova, Niger, Romania, Russia, Slovakia, South Africa, United Arab Emirates (Dubai) and Uganda will be considered. Preference will also be given to candidates who demonstrate financial need. Knowledge of French would be a plus.
Type: Scholarship
Value: Up to €20,000
Frequency: Annual
Application Procedure: Applicants must submit their application online.
Closing Date: February 11th for July class (starts in September) and August 18th for December class (starts in January)
Additional Information: Please check website for more information.

INSEAD Pereira Endowed Scholarship

Purpose: For Portuguese applicants.
Eligibility: Open to MBA students.
Level of Study: MBA
Type: Scholarship
Value: €10,000
Length of Study: 1 year
Frequency: Annual
Study Establishment: INSEAD
No. of awards offered: 2
Application Procedure: Applicants must apply online through INSEAD's website.
Closing Date: February 11th for July class (starts in September) and August 18th for december class (starts in January)
Funding: Individuals
Contributor: Alumni
No. of awards given last year: 2
Additional Information: Countries of study are France and Singapore. Eligible to nationals of West European countries and Portugal.

INSEAD Russian Alumni Scholarship

Subjects: MBA.
Purpose: This scholarship has been created by a group of Russian alumni in support of INSEAD's efforts to attract top talent from Russia to the MBA programme. They will be awarded to candidates whose educational and professional experience is merit worthy. The alumni wish that the candidates who benefit from their support will return and be successful in Russia.
Eligibility: Russian nationals who have already been admitted to the MBA programme will be eligible. Candidates whose education and professional experience has a Russian focus and who demonstrate a strong potential for success in Russia will be considered. Preference will be given to candidates with proven financial need.
Type: Scholarship
Value: Up to €10,000
Frequency: Annual
Application Procedure: To be considered for this award please apply for the Needbased Scholarships.
Closing Date: February 13th for July class (starts in September); August 22nd for December class (starts in January)
Additional Information: Please check website for further information.

INSEAD Russian Alumni Scholarship

Subjects: MBA.
Purpose: This scholarship has been created by a group of Russian alumni in support of INSEAD's efforts to attract top talent from Russia to the MBA programme. They will be awarded to candidates whose educational and professional experience is merit worthy. The alumni wish that the candidates who benefit from their support will return and be successful in Russia.
Eligibility: Russian nationals who have already been admitted to the MBA programme will be eligible. Candidates whose education and professional experience has a Russian focus and who demonstrate a strong potential for success in Russia will be considered. Preference will be given to candidates with proven financial need.
Type: Scholarship
Value: Up to €10,000
Frequency: Annual
Application Procedure: To be considered for this award please apply for the Needbased Scholarships.
Closing Date: February 13th for July class (starts in September); August 22nd for December class (starts in January)
Additional Information: Please check website for further information.

INSEAD Ryoichi Sasakawa Young Leaders Fellowship Fund (Sylff)

Subjects: MBA.
Purpose: Sylff would like to encourage candidates of any nationality to broaden their knowledge and enhance their career opportunities through the INSEAD MBA programme. One award will be granted per class.
Eligibility: Candidates of any nationality who show strong leadership in a cross-cultural setting. Preference will be given to candidates who require financial assistance.
Type: Fellowship
Value: €10,000
Frequency: Annual
Closing Date: February 13th each year for the July class (starts in September); August 22nd each year for the December class (starts in January)
Additional Information: Please check website for further information.

INSEAD Ryoichi Sasakawa Young Leaders Fellowship Fund (Sylff)

Subjects: Social sciences and humanities.
Eligibility: Candidates of any nationality who show strong leadership in a cross-cultural setting. Preference will be given to candidates who require financial assistance.
Level of Study: Graduate
Type: Fellowship
Value: €10,000
Frequency: Annual
No. of awards offered: 1 award per class
Closing Date: February 11th each year for the July class (starts in September) and August 22nd each year for the December class (starts in January)
Contributor: Nippon Foundation
Additional Information: Please check website for more information.

INSEAD Sam Akiwumi Endowed Scholarship - '07D

Purpose: To assist candidates from Africa, preferably Ghanaians.
Eligibility: Open to MBA students.
Level of Study: MBA
Type: Scholarship
Value: €10,000
Length of Study: 1 year
Frequency: Annual
Study Establishment: INSEAD
Application Procedure: Applicants must apply online through INSEAD's website.
Closing Date: February 11th for July class (starts in August) and August 18th for December class (starts in January)
Funding: Individuals
Contributor: Alumni
No. of awards given last year: 2
Additional Information: Countries of study are France and Singapore.

INSEAD Sasakawa (SYLFF) Scholarships

Subjects: MBA.
Purpose: To encourage candidates to broaden their knowledge and enhance their career leadership through the INSEAD MBA programme.
Eligibility: Open to candidates of any nationality. The awards will be made on a competitive basis.
Level of Study: MBA
Type: Scholarship
Value: €10,000 depending on the number and quality of applications
Frequency: Annual
Study Establishment: INSEAD
No. of awards offered: 1 or more per intake
Application Procedure: Applicants must complete a specific assignment, details of which are available from the website.
Closing Date: February 11th for the September class of the same year and August 18th for the January class of the following year
Funding: Private
Contributor: The Sasakawa Young Leaders Fellowship Fund (SYLFF)

INSEAD Sisley-Marc d'Ornano Scholarship

Subjects: MBA.
Purpose: To support young graduates seeking further education in order to contribute to the economic development of Poland.
Eligibility: Open to Polish nationals admitted to the INSEAD MBA programme who demonstrate a commitment to work in Poland for 3 years after the INSEAD MBA programme. The winner of the scholarship will agree to take up a professional activity in Poland for at least 3 years, and if not, the candidate is obliged to reimburse the scholarship.
Level of Study: MBA
Type: Scholarship
Value: €25,000
Frequency: Annual
Study Establishment: INSEAD
No. of awards offered: 1
Application Procedure: Applicants must submit an essay addressing the following question: Give the main reason for your applying for the scholarship and describe your aspirations for your future career development. Scholarship applications may be submitted with the admissions application form. Application forms are available from the website.
Closing Date: February 11th for the September intake of the same year and August 18th for the January intake of the following year
Funding: Private
Contributor: Sisley
Additional Information: This scholarship is offered in memory of the late Marc d'Ornano, who lost his life in a car accident while at the start of an excellent career.

INSEAD Swedish Council Scholarship(s)

Subjects: MBA.
Purpose: To foster the pipeline of international business leaders in Sweden, the Council has created an MBA scholarship at INSEAD to reward such talent. The awards will be merit based, through preference will be given to candidates demonstrating financial need.
Eligibility: Candidates must be Swedish nationals and clearly demonstrate an interest for a career linked to business with Sweden. Candidates may not cumulate this award with any other INSEAD or external scholarships. If this is the case, the INSEAD scholarship will be reallocated.
Type: Scholarship
Value: Up to €15,000
Frequency: Annual
Application Procedure: All applications submitted to the INSEAD Elof Hansson Scholarship will be considered.
Closing Date: February 13th each year for the July class (starts in September); August 22nd each year for the December class (starts in January)
Additional Information: Please check website for further information.

INSEAD Syngenta Endowed Scholarship(s) for Emerging Country Leadership

Eligibility: Candidates must be a national of an emerging economy and have spent a substantial part of their lives in the developing world, in education and/or professional experience. Experience in industry will be considered a plus. Preference will be given to candidates who require proven financial assistance.
Type: Scholarship
Value: €22,500
No. of awards offered: 2 per class
Application Procedure: Applicants must complete online application and submit their essays.
Closing Date: February 11th for the July class and August 18th for the December class
Additional Information: Please check website for more information.

INSEAD Theo Vermaelen Scholarship

Subjects: MBA.
Purpose: INSEAD professor, Theo Vermaelen, the Schroders Chaired Professor of International Finance and Asset Management, has chosen to fund one scholarship per year with matched funds from INSEAD for Iraqi participants on the full-time MBA programme at INSEAD. He believes these scholarships will make a small, but hopefully significant contribution to improving the lives of at least a few people from Iraq and that the holder of these scholarships will be their country's leaders in the future.
Eligibility: Candidates must be Iraqi nationals, but not necessarily live in Iraq. Consideration will be given to both merit and financial need.
Type: Scholarship
Value: Up to full-tuition
Frequency: Annual
Closing Date: February 13th, May 10th, August 22nd, October 5th
Additional Information: Please check website for further information.

INSEAD Weston Scholarship(s)

Purpose: The Garfield Weston Foundation is a UK-based, general grant-giving charity endowed by the late W. Garfield Weston and members of his family. The Foundation has helped a wide range of organisations with projects in the Arts, Community, Education, Welfare, Medical, Social, Religion, Youth and Environment. The Garfield Weston Scholarships at INSEAD are based on merit i.e. the most deserving candidates whether from an academic, career potential or other relevant perspective.
Eligibility: Based on the quality of admission applications (no scholarship application is required) one or two awards will be made per intake. Only candidates who have been admitted in Round 1 and/ or Round 2 will be considered.
Type: Scholarship
Value: Up to €15,000 each
Frequency: Annual
No. of awards offered: Up to 2 per class
Application Procedure: No separate application is required. The holders of the awards will be required to submit an essay on their INSEAD experience at the end of the programme. Notification on the decision: successful candidates will receive the notification on the award approximately one week after admission. If you do not receive an e-mail on the award within that time frame please consider that you have not been selected for the Weston scholarship.
Closing Date: No deadline
Additional Information: Please check website for further information.

INSTITUT DE RECHERCHE ROBERT-SAUVÉN SANTÉT EN SÉCURITÉ DU TRAVAIL (IRSST)

505, De Maisonneuve Ouest, Montréal, QC, H3A 3C2, Canada
Tel: (1) 514 288 1551
Email: grants@irsst.qc.ca
Website: www.irsst.qc.ca

Institut de recherche Robert-Sauvén santét en sécurité du travail (IRSST), established in Quebec since 1980, is a scientific research organization known for the quality of its work and the expertise of its personnel. The Institute is a private, non-profit agency.

IRSST Graduate Studies Scholarships

Subjects: Occupational health and safety.
Purpose: To support Master's and doctoral students who wish to acquire research training in the occupational health and safety field.
Eligibility: Open to students who are registered full-time in a Master's or doctoral programme and have obtained a cumulative average of B + for all of their undergraduate studies.
Level of Study: Doctorate, Postgraduate
Type: Scholarship
Value: $14,100 per year. In addition, a scholarship recipient whose training and research program is outside Canada is reimbursed for the amount exceeding the first $750 in annual tuition fees; the cost of travelling to the training and research location, representing the cost of one round-trip economy airplane ticket or one round trip by car, for each year of the effective period of the scholarship (maximum of two years)
Length of Study: 2–3 years
Frequency: Annual
Closing Date: Third Wednesday of October
Contributor: The CSST provides most of the Institute's funding from the contributions it collects from the employers

INSTITUT FRANÇAIS D'AMÉRIQUE

Department of History, CB 31, UNC-CH, Chapel Hill, NC, 27599-3195, United States of America
Tel: (1) 919 962 2032
Fax: (1) 919 962 5457
Email: cmaley@email.unc.edu
Website: www.unc.edu/depts/institut
Contact: Professor Jay Smith

The mission of the Institut Français de Washington is to promote the American study of French culture, language, history and society, and to encourage the work of teachers, scholars and students in these fields. The Institute also sponsors events to foster public understanding of French-American relations. The IFW provides funds for fellowships, prizes, and conferences that serve this mission.

Edouard Morot-Sir Fellowship in French Studies

Subjects: French studies in the areas of art, economics, history, history of science, linguistics, literature or social sciences.
Eligibility: Open to those in the final stages of a PhD dissertation or who have held a PhD for no longer than 3 years before the application deadline.
Level of Study: Doctorate, Postdoctorate
Type: Fellowship
Value: US$1,500
Length of Study: At least 1 month
Frequency: Annual
Country of Study: France
No. of awards offered: 1
Application Procedure: Applicants must write a maximum of two pages describing the research project and planned trip and enclose a curriculum vitae. A letter of recommendation from the dissertation director is required and a letter from a specialist in the field for assistant professors.
Closing Date: January 15th
Funding: Foundation, private
No. of awards given last year: 3
No. of applicants last year: 90
Additional Information: Awards are for maintenance during research in France and should not be used for travel.Please check website www.unc.edu/depts/institut for further information.

Gilbert Chinard Fellowships

Subjects: French studies in the areas of art, economics, history, history of science, linguistics, literature or social sciences.
Eligibility: Open to those in the final stages of a PhD dissertation or who have held a PhD for no longer than 3 years before the application deadline.
Level of Study: Doctorate, Postdoctorate
Type: Fellowship
Value: US$1,500
Length of Study: At least 1 month

Frequency: Annual
Country of Study: France
No. of awards offered: 2
Application Procedure: Applicants must write a maximum of two pages describing the research project and planned trip and enclose a curriculum vitae. A letter of recommendation from the dissertation director is also required for PhD candidates and a letter from a specialist in the field for assistant professors.
Closing Date: January 15th
Funding: Private
No. of awards given last year: 1
No. of applicants last year: 28
Additional Information: Awards are for maintenance during research in France and should not be used for travel.

Harmon Chadbourn Rorison Fellowship

Subjects: French studies in the areas of art, economics, history, history of science, linguistics, literature or social sciences.
Eligibility: Open to those in the final stages of a PhD dissertation or who have held a PhD for no longer than 3 years before the application deadline.
Level of Study: Doctorate, Postdoctorate, Postgraduate
Type: Fellowship
Value: US$1,500
Length of Study: At least 1 month
Frequency: Every 2 years
Country of Study: France
No. of awards offered: 1
Application Procedure: Applicants must write a maximum of two pages describing the research project and planned trip and enclose a curriculum vitae. A letter of recommendation from the dissertation director is required for PhD candidates and a letter from a specialist in the field for assistant professors.
Closing Date: January 15th
Funding: Foundation, private
No. of awards given last year: 1
No. of applicants last year: 27
Additional Information: Awards are for maintenance during research in France and should not be used for travel.

THE INSTITUT MITTAG-LEFFLER

Auravägen 17, SE-18260 Djursholm, Sweden
Tel: (46) 8 6220560
Fax: (46) 8 6220589
Email: info@mittag-leffler.se
Website: www.mittag-leffler.se
Contact: Inger Halvarsson, Visitor Programme Administrator

Institut Mittag-Leffler is a Nordic research institute for mathematics, under the auspices of the Royal Swedish Academy of Sciences, created by Gösta and Signe Mittag-Leffler, who donated their house, library and fortune to the Academy in 1916.

Institut Mittag-Leffler Grants

Subjects: Mathematics. Programs are of semester length (September 1st to December 15th and January 15th to May 30th, respectively).
Eligibility: Open to recent PhDs and advanced graduate students. Preference will be given to applications for long stays.
Level of Study: Graduate, Postdoctorate
Type: Grant
Value: 13,000–16,000 Swedish Kronor per month; travel expenses to and from Stockholm, accommodation free of charge and office space
Length of Study: There are two different topics during one academic year
Frequency: Annual, semi-annual
Country of Study: Sweden
No. of awards offered: Varies
Application Procedure: In addition to the completed application form, applicants should send a short description of the candidate's research interests and plans, copies of the applicant's papers and preprints and two or (preferably) three letters of recommendation. See the website for further details: www.mittag-leffler.se/programs/1213/grants.php.
Closing Date: January 12th
Additional Information: For further information, please turn to Inger Halvarsson at Institut Mittag-Leffler or email to secretary@mittag-leffler.se

INSTITUTE FOR ADVANCED STUDIES IN THE HUMANITIES

University of Edinburgh, Hope Park Square, Edinburgh, EH8 9NW, Scotland
Tel: (44) 13 1650 4671
Fax: (44) 13 1668 2252
Email: iash@ed.ac.uk
Website: www.iash.ed.ac.uk
Contact: Ms Anthea Taylor, Institute Administrator

The Institute for Advanced Studies in the Humanities aims to promote scholarship in the humanities, and, wherever possible, to foster interdisciplinary enquiries. This is achieved by means of fellowships awarded for the pursuit of relevant research and by the public dissemination of findings in seminars, lectures, conferences, exhibitions, cultural events and publications.

David Hume Fellowship

Purpose: To enable research in Edinburgh in any aspect of Hume studies.
Eligibility: Open to Scholars of any nationality holding a doctorate or offering equivalent evidence of aptitude for Advanced studies.
Level of Study: Postdoctorate
Type: Fellowship
Value: £10,000
Length of Study: 6 months
Frequency: Annual
Study Establishment: Institute for Advanced Studies in the Humanities, University of Edinburgh
Country of Study: United Kingdom
No. of awards offered: 1
Application Procedure: Application form can be downloaded from Institute's website.
Closing Date: February 28th
Funding: Trusts

IASH Visiting Research Fellowships

Subjects: Any discipline within the humanities and social sciences, but priority will be given to those whose work falls within the scope of one of the institute's research themes.
Purpose: To promote advanced research within the field and also to sponsor interdisciplinary research.
Eligibility: Open to scholars of any nationality holding a doctorate or offering equivalent evidence of aptitude for advanced studies.
Level of Study: Postdoctorate
Type: Fellowship
Length of Study: 2–6 months
Frequency: Annual
Study Establishment: The Institute for Advanced Studies in the Humanities at the University of Edinburgh
Country of Study: United Kingdom
No. of awards offered: 15
Application Procedure: Applicants must complete an application form, available from the Institute. Candidates should advise their referees to write on their behalf directly to the Institute. A minimum of two and a maximum of three references are required. At least one referee should come from outside the institution of the applicant.
Closing Date: February 28th
Funding: Private
No. of awards given last year: 15
No. of applicants last year: 16
Additional Information: Fellows have a private office at the Institute, near the library and within easy reach of the National Library of Scotland, the Central City Library, the National Galleries and Museums, the Library of the Society of Antiquaries in Scotland and the National Archives of Scotland.

IASH-SSPS Visiting Research Fellowships

Subjects: Social and political science.
Purpose: To encourage outstanding research, international scholarly collaboration and networking activities.
Eligibility: Open to scholars of any nationality holding and doctorate or equivalent professional qualification.
Level of Study: Postdoctorate
Type: Fellowship
Value: Contributions to travel and accomodation costs
Length of Study: 2–4 months
Frequency: Annual
Study Establishment: Institute for Advanced Studies in the Humanities, University of Edinburgh
Country of Study: United Kingdom
No. of awards offered: 3
Application Procedure: Application form available on Institute's website.
Closing Date: February 28th
Funding: Private
No. of awards given last year: 3
No. of applicants last year: 7
Additional Information: Fellows will be expected to work in collaboration with one or more member of school of social and political science, University of Edinburgh.

Postdoctoral Bursaries

Subjects: Humanities and social sciences.
Purpose: To support candidates in any area of the Humanities and Social Sciences, whose work falls within the scope of one of the Institute for Advanced Studies current research themes or across disciplinary boundaries in the Humanities.
Eligibility: Applicants must have been awarded a doctorate, normally within the last three years, and should not have held a permanent position at a university, or a previous fellowship at the Institute for Advanced Studies. Those who have held temporary and/or short-term appointments are eligible to apply.
Level of Study: Postdoctorate
Type: Bursary
Value: Up to UK£10,000
Length of Study: 3–9 months
Frequency: Annual
Study Establishment: Institute for Advanced Studies in the Humanities, University of Edinburgh
Country of Study: Scotland
No. of awards offered: 10
Application Procedure: Application form can be downloaded from the Institute's website.
Closing Date: July - exact date to be confirmed
Funding: Trusts
No. of awards given last year: 14
No. of applicants last year: 40
Additional Information: Check website for further details.

INSTITUTE FOR ADVANCED STUDIES ON SCIENCE, TECHNOLOGY AND SOCIETY (IAS-STS)

Kopernikusgasse 9, Graz, 8010, Austria
Tel: (43) 316 813909-34
Fax: (43) 316 810274
Email: info@sts.tugraz.at
Website: www.sts.tugraz.at
Contact: Günter Getzinger, Acting Director

In 1999 the Inter University Research Centre for Technology, Work and Culture (IFZ) launched the IAS-STS in Graz, Austria. It promotes the interdisciplinary investigation of the links and interaction between science, technology and society as well as research on the development and implementation of socially and environmentally sound, sustainable technologies.

IAS-STS Fellowship Programme

Subjects: Gender (technology and environment), technology studies, information and communication technologies and society, technology assessment, participatory technology design, sustainable consumption and production, genetics and biotechnology, energy and climate.
Purpose: To give the students the opportunity to explore issues.
Eligibility: Applicants must hold an academic degree.
Level of Study: Doctorate, Postdoctorate, Postgraduate, Research
Type: Fellowships
Value: €940 per month
Length of Study: Up to 9 months
Frequency: Annual
Study Establishment: IAS-STS
Country of Study: Austria
No. of awards offered: 5
Application Procedure: Application forms can be download from the website.
Closing Date: June 30th
Funding: Government
Contributor: Styrian Government, City of Government
No. of awards given last year: 5
No. of applicants last year: 50

INSTITUTE FOR ADVANCED STUDY

Einstein Drive, Princeton, NJ, 08540, United States of America
Tel: (1) 609 734 8000
Fax: (1) 609 924 8399
Email: contactus@ias.edu
Website: www.ias.edu
Contact: Ms Christine Ferrara, Senior Public Affairs Officer

The Institute for Advanced Study is an independent, private institution whose mission is to support advanced scholarship and fundamental research in historical studies, mathematics, natural sciences and social science. It is a community of scholars where theoretical research and intellectual enquiry are carried out under the most favourable conditions.

Institute for Advanced Study Postdoctoral Residential Fellowships

Subjects: Social science, history, astronomy, astrophysics, theoretical physics, mathematics, theoretical computer science or theoretical biology.
Purpose: To support advanced study and scholarly exploration.
Eligibility: There are no restrictions on eligibility.
Level of Study: Postdoctorate
Type: Fellowship
Value: US$40,000–65,000
Length of Study: Generally 1 year
Frequency: Annual
Country of Study: United States of America
No. of awards offered: Approx. 190
Application Procedure: Applicants must complete an application. Materials are available from the school administrative officers.
Closing Date: Varies, but is between November 1st and December 15th
Funding: Individuals, corporation, foundation, government
No. of awards given last year: Approx. 190
No. of applicants last year: Approx. 2,400

THE INSTITUTE FOR CLINICAL SOCIAL WORK

401 South State Street, Suite 822, Chicago, IL, 60605, United States of America
Tel: (1) (312) 935 4232
Fax: (1) (312) 935 4255
Email: info@icsw.edu
Website: www.icsw.edu
Contact: Mr John Dowdy, Manager of Strategic Operations and Financial Aid

Elisabeth Jacobs Scholarship

Subjects: Clinical social work.
Purpose: To support promising students who are dedicated to working with families challenged by poverty, immigration or trauma.

Eligibility: Open to full time students who have successfully completed their first year of studies, who have financial need and have been nominated by faculty.
Level of Study: Doctorate
Type: Scholarship
Frequency: Annual
Study Establishment: The Institute for Clinical Social Work
Country of Study: United States of America
No. of awards offered: Varies
Application Procedure: Applicants must contact the financial aid administrator.

INSTITUTE FOR HUMANE STUDIES (IHS)

3351 Fairfax Dr, MSN 1C5, Arlington, VA, 22201, United States of America
Tel: (1) 703 993 4880
Fax: (1) 703 993 4890
Email: abrand@gmu.edu
Website: www.theihs.org
Contact: Ms Amanda Bland, Director of Academic Programs

The Institute for Humane Studies (IHS) is a unique organization that assists graduate students worldwide with a special interest in individual liberty. IHS awards over US$400,000 a year in scholarships to students from universities around the world. They also sponsor the attendance of hundreds of students at free summer seminars and provide various forms of career assistance. Through these and other programmes, IHS and a network of faculty associates promote the study of liberty across a broad range of disciplines, encouraging understanding, open enquiry, rigorous scholarship and creative problem-solving.

Hayek Fund for Scholars

Subjects: Social sciences, law, the humanities, journalism.
Purpose: To help offset expenses for participating in professional conferences and job interviews.
Eligibility: Open to graduate students and untenured faculty members and postdoctoral students.
Level of Study: Postgraduate
Value: US$750
Frequency: Annual
Country of Study: Any country
Application Procedure: For application requirements visit the website.

For further information contact:

George Mason University, 4400 University Drive, Fairfax, VA, 22030, United States of America
Tel: (1) 703 323 1055
Fax: (1) 703 425 1536
Contact: Keri Anderson, Programme Director

IHS Humane Studies Fellowships

Subjects: A variety of fields, including economics, philosophy, law, political science, history, and sociology.
Purpose: To support outstanding students with a demonstrated interest in the classical liberal tradition intent on pursuing an intellectual and scholarly career.
Eligibility: Open to graduate students who have enrolled for the next academic year at accredited colleges and universities.
Level of Study: Graduate, Postgraduate
Type: Fellowship
Value: $2,000–15,000
Frequency: Annual
Country of Study: Any country
Application Procedure: Applicants must complete and submit an application form with three completed evaluations, three essays, official test scores, official transcripts and a term paper or writing sample. Applications can be downloaded at www.theihs.org/hsf
Closing Date: December 31st
Funding: Private
No. of awards given last year: 195

IHS Summer Graduate Research Fellowship

Subjects: The humane sciences, e.g. history, political and moral philosophy, political economy, economic history, legal and social theory.
Purpose: To give students who share an interest in scholarly research in the classical liberal tradition the opportunity to work on a thesis chapter or a paper of publishable quality and to participate in interdisciplinary seminars under the guidance of a faculty supervisor.
Eligibility: Open to graduate students in the humanities, social sciences and law who intend to pursue academic careers and who are currently pursuing research in the classical liberal tradition.
Level of Study: Doctorate, Graduate, Postgraduate
Type: Fellowship
Value: $5,000 stipend plus a travel and housing allowance to attend the two conferences
Frequency: Annual
Country of Study: United States of America
No. of awards offered: 8–10
Application Procedure: Applicants must submit a proposal, curriculum vitae, a copy of Graduate Record Examination scores or Law School Admission Test scores and transcripts, a writing sample and reference details. Visit the website for further information.
Closing Date: February 15th
Funding: Private

THE INSTITUTE FOR SUPPLY MANAGEMENT (ISM)

PO Box 22160 Tempe, Tempe, AZ, 85285-2160, United States of America
Tel: (1) 480 752 6276
Fax: (1) 480 752 7890
Email: ssturzl@ism.ws
Website: www.ism.ws
Contact: Valerie Gryniewicz, Manager, Education

The Institute for Supply Management (ISM) is a non-profit association that provides national and international leadership in purchasing and supply management research and education. ISM provides more than 40,000 members with opportunities to expand their professional skills and knowledge.

ISM Dissertation Research Grant

Subjects: Purchasing materials and supply management.
Purpose: To financially assist individuals in preparation for a career in the field, for university teaching and to encourage research.
Eligibility: Open to doctoral candidates who are pursuing a PhD or DBA in purchasing, business, logistics, management, economics, industrial engineering or a related field and who are at the dissertation stage. Applicants must be enrolled in an accredited United States of America university are eligible for the award.
Level of Study: Doctorate
Type: Grant
Value: Up to US$12,000
Frequency: Annual
Country of Study: United States of America
No. of awards offered: Up to 4
Application Procedure: (1) Letter of application, signed by the candidate. (2) Official transcripts from the candidate's current university. (3) Proposal abstract, maximum 25 pages, including a literature search and a research design. The proposal must specifically discuss problem statement or hypothesis; research methodology, including data sources, collection, and analysis; significance/value of the research in purchasing/supply management. (4) A letter from the candidate's major advisor, stating that the dissertation topic is acceptable. (5) Three letters of recommendation (one letter may serve as the letter requested in #4 above) from professors or administrators familiar with the applicant's research capabilities. The letters should be sealed independently within the application package. (6) A curriculum vitae, which may include a list of research in progress, accomplishments in academe, honors, awards, and relevant work experience. The ISM Doctoral Grant Program seeks to notify all applicants by mid-June whether they received a grant. Please check at http://www.ism.ws/education/content.cfm?ItemNumber = 791 for detailed information.

Closing Date: January 31st
Funding: Private
Contributor: ISM
No. of awards given last year: 3
No. of applicants last year: 20
Additional Information: Upon successful completion of the research, the ISM will be interested in the publication of material from the study. Nominations are invited from departments of economics, management, marketing and business administration at United States of America universities offering a doctoral degree in appropriate fields.

ISM Professional Research Development Grant

Purpose: This program is designed to support assistants and young associate professor with terminal degree that are teaching in the field of a significant research track in supply management and associated areas. The goal is to help competitively selected faculty build a research and publication field and establish themselves in the profession (i.e. academic institutions, professional organisation, research fund granting bodies such as CAPS Research etc.). This would include CAPS research. Faculty selected will generally have two and eight years of teaching and research experience and have successful publication experiments beyond the doctoral degree.
Eligibility: Open to assistant professors, associate professors or equivalent who have demonstrated exceptional academic productivity in research and teaching. Candidates are chosen from those who can help produce useful research that can be applied to the advancement of purchasing and supply management. Candidates must be full-time faculty members within or outside the United States of America and be present or past members of ISM committees, groups, forums or affiliated organizations. An assistant professor should have 3 or more years of post-degree experience. Previous awardees are ineligible.
Level of Study: Postdoctorate
Type: Grant
Value: Up to US$10,000
Frequency: Annual
Country of Study: United States of America
No. of awards offered: 1
Application Procedure: (1) Letter of application explaining qualifications of the grant. (2) Research proposal of no more than five pages, including problem statement or hypothesis; research methodology, with data sources, collection and analysis; and value to the field of supply management. (3) Curriculum vitae, including works in progress.
Closing Date: January 31st
Funding: Private
Contributor: ISM
No. of awards given last year: 1
No. of applicants last year: 5–10
Additional Information: It is expected that the recipients will present the results of their research at an ISM forum, e.g. research symposium, ISM Annual International Purchasing Conference and/or an ISM publication such as *The Journal of Supply Chain Management*.

For further information contact:

c/o Robert A Kemp, PhD, CPM, Institute for Supply Management, PO Box 22160, Tempe, United States of America
Email: kempr@mchsi.com
Contact: ISM Doctoral Grant Committee

INSTITUTE FOR WORK AND HEALTH

481 University Avenue Suite 800, Toronto, ON, M5G 2E9, Canada
Tel: (1) 416 927 2027
Fax: (1) 416 927 4167
Email: info@iwh.on.ca
Website: www.iwh.on.ca

Institute for Work and Health is an independent, non-profit research organization whose mission is to conduct and share research with workers, labourers, employers, clinicians and policy makers to promote, protect and improve the health of working people.

S. Leonard Syme Training Fellowships in Work and Health

Subjects: Work and health.
Purpose: To financially support young researchers at the Master's or doctoral level who intend studying in the field of work and health.
Eligibility: Open to candidates who are enrolled at an Ontario university that has a formal affiliation with the Institute for Work & Health. Candidates who are part-way through their programme of study will also be considered. Preference will be given to candidates whose research interests include understanding the social determinants of health and illness in work environments, and/or evaluating workplace interventions to improve health and/or the associated measurement issues.
Level of Study: Doctorate, Postgraduate
Type: Fellowship
Value: Major award of up to $15,000 and a minor award of up to $5,000
Length of Study: 1 year
Frequency: Annual
Study Establishment: Several universities
Country of Study: Canada
Application Procedure: Applicants must provide a completed application form, a 300-word statement of their research interests and a 200-word statement of their career objectives, a reference letter and curriculum vitae. Application form is available online.
Closing Date: June 2nd
Additional Information: Please check website (http://www.iwh.on.ca/syme) for availability and latest updates.

For further information contact:

Website: www.iwh.on.ca
Contact: Ms Lyudmila Marsurova

INSTITUTE OF ADVANCED LEGAL STUDIES (IALS)

Institute of Advanced Legal Studies, Charles Clore House, 17 Russell Square, London, WC1B 5DR, England
Tel: (44) 20 7862 5800
Fax: (44) 20 7862 5850
Email: ials.administrator@sas.ac.uk
Website: www.ials.sas.ac.uk
Contact: Margaret Wilson, Institute Manager

The Institute of Advanced Legal Studies (IALS) plays a national and international role in the promotion and facilitation of legal research. It possesses one of the leading research libraries in Europe and organizes a regular programme of conferences, seminars and lectures. It also offers postgraduate taught and research programmes and specialized training courses.

IALS Visiting Fellowship in Law Librarianship

Subjects: Law and library science.
Purpose: To enable experienced law librarians, who are undertaking research in, appropriate fields, to relate their work to activities in which the Institutes own library is involved.
Eligibility: Open to experienced law librarians from any country.
Level of Study: Unrestricted
Type: Fellowship
Value: Fellowships can consist of or include a period working with Institute library staff or be a period of research based in a research carrel
Length of Study: A minimum of 2 months and a maximum of 1 year
Frequency: Annual
Study Establishment: The IALS
Country of Study: United Kingdom
No. of awards offered: 1
Application Procedure: Applicants must submit a full curriculum vitae, the names, addresses and telephone numbers of two referees and a brief statement of the research programme to be undertaken to the Administrative Secretary.
Closing Date: Applications may be considered at any time of the year
No. of awards given last year: 1
No. of applicants last year: 1

IALS Visiting Fellowship in Legislative Studies

Subjects: Law.
Purpose: To enable individuals in the field to undertake research.

Eligibility: Open to established academics and practitioners from any country. This award is not available for postgraduate research.
Level of Study: Unrestricted
Type: Fellowship
Value: Non-stipendary
Length of Study: A minimum of 3 months and a maximum of 1 year
Frequency: Annual
Study Establishment: The IALS
Country of Study: United Kingdom
No. of awards offered: 1
Application Procedure: Applicants must submit a full curriculum vitae, the names, addresses and telephone numbers of two referees and a brief statement of the research programme to be undertaken.
Closing Date: January 31st for the following academic year
No. of awards given last year: None
No. of applicants last year: 1

IALS Visiting Fellowships

Subjects: Law: legal skills, legal profession, legal education, legal implementation studies, company and commercial law, financial services law, access to legal information.
Purpose: Visiting fellowships are designed for persons already established in their own field of activity who are undertaking work within fields covered by or adjacent to the Institute's own research programmes or interests.
Eligibility: Open to nationals of any country who are established legal scholars and are undertaking research in appropriate fields.
Level of Study: Unrestricted
Type: Fellowship
Value: Non-stipendary
Length of Study: A minimum of 3 months and a maximum of 1 year
Frequency: Annual
Study Establishment: The IALS
Country of Study: United Kingdom
No. of awards offered: Up to 6
Application Procedure: Applicants must submit a full curriculum vitae, the names, addresses and telephone numbers of two referees and a brief statement of the research programme to be undertaken.
Closing Date: January 31st for the following academic year
No. of awards given last year: 7
No. of applicants last year: 16
Additional Information: This award is not available for postgraduate research.

THE INSTITUTE OF CANCER RESEARCH (ICR)

Genetic Epidemiology Building, 15 Cotswold Road, Belmont, Sutton, Surrey, SM2 5NG, England
Tel: (44) 20 8643 8901 ext 4253
Fax: (44) 20 8643 6940
Email: emma.pendleton@icr.ac.uk
Website: www.icr.ac.uk
Contact: Sarah Goodwin, Registry Project Manager

Over the past 100 years, the Institute of Cancer Research (ICR) has become one of the largest, most successful and innovative cancer research centres in the world. The Institute and the Royal Marsden NHS Trust exist side by side in Chelsea and on a joint site at Sutton, and this close association allows for maximum interaction between fundamental laboratory work and clinical environment.

ICR Studentships

Subjects: Cancer research.
Purpose: Research degree studentships.
Eligibility: First class or upper second class undergraduate degree required in a relevant subject.Overseas equivalent level: experience of lab or research work, a TOEFL score of 650 or IELTS score of 7.0 – or equivalent language assessment.
Level of Study: Doctorate, Postdoctorate
Type: Studentship
Value: £19,500 (inner London - Chester Beatty Laboratories, Fulham Road) or £18,180 (outer London - Sutton Campus), is increased annually in-line with the increase in cost-of-living.
Length of Study: Up to 4 years
Frequency: Annual
Study Establishment: The Institute of Cancer Research, University of London
Country of Study: United Kingdom
No. of awards offered: 20–30
Application Procedure: See website www.icr.ac.uk/phds
Funding: Government, trusts
Contributor: Cancer Research UK/Wellcome Trust, Medical Research Council
No. of awards given last year: 28
No. of applicants last year: 300
Additional Information: A limited number of Institute postdoctoral fellowships are offered from time to time as vacancies occur.

INSTITUTE OF CURRENT WORLD AFFAIRS

1779 Massachusetts Ave. NW, Suite 615, Washington, DC, 2003, United States of America
Tel: (1) 202 364 4068
Fax: (1) 202 364 0498
Email: icwa@icwa.org
Website: www.icwa.org
Contact: Steven Butler, Executive Director

Institute of Current World Affairs Fellowships

Subjects: International affairs.
Purpose: To enable young adults of outstanding promise and character to study and write about areas or issues of the world outside the United States of America.
Eligibility: Open to individuals up to the age of 36 who have finished their formal education. Applicants must have a good command of spoken and written English, and have "completed the current phase of their formal education" before the date of application.
Level of Study: Postgraduate, Professional development
Type: Fellowship
Value: The Institute provides fellows with sufficient funding to allow them and their families to live in good health and reasonable comfort.
Length of Study: Minimum 2 years
Frequency: Dependent on funds available, Biannual
Country of Study: Any country
No. of awards offered: 2
Application Procedure: Applicants must write to the Executive Director and briefly explain their personal background and the professional experience that would qualify them in the Institute's current areas of concern, details of which are available upon request. They should also describe the activities they would like to carry out during the 2 years overseas. This initial letter is followed by a more detailed written application process and must be completed prior to the deadline.
Closing Date: February 1st for summer applicants; September 1st for winter applicants
Funding: Private
Additional Information: Fellowships are not awarded to support work toward academic degrees nor to underwrite specific studies or research projects. The Institute is also known as the Crane-Rogers Foundation.

INSTITUTE OF EDUCATION

20 Bedford Way, London, WC1H 0AL, England
Tel: (44) 20 7612 6000
Email: info@ioe.ac.uk
Website: www.ioe.ac.uk
Contact: Josie Charlton, Head of Marketing and Development

Founded in 1902, the Institute of Education is a world-class centre of excellence for research, teacher training, higher degrees and consultancy in education and education-related areas of social science. Our pre-eminent scholars and talented students from all walks of life make up an intellectually rich and diverse learning community.

Nicholas Hans Comparative Education Scholarship

Subjects: Comparative education.
Purpose: To assist a well-qualified student to study for a PhD in comparative education at the Institute of Education.
Eligibility: Candidates must be registered Institute students not normally resident in the United Kingdom.
Level of Study: Doctorate
Type: Scholarship
Value: Full-time tuition fees
Length of Study: 3–7 years
Frequency: Annual
Study Establishment: Institute of Education
Country of Study: United Kingdom
No. of awards offered: 1
Application Procedure: Candidates are required to submit an extended essay of 25,000–30,000 words, based upon their research or proposed research, that exemplifies, extends or develops by critique the concerns of Nicholas Hans in comparative education.
Closing Date: June 1st
Funding: Trusts
Contributor: Trust fund based upon money left in the will of Nicholas Hans' widow

For further information contact:

Email: mailto:p.kelly@ioe.ac.uk
Contact: Patricia Kelly

INSTITUTE OF EUROPEAN HISTORY

Alte Universitätsstrasse 19, D-55116 Mainz, Germany
Tel: (49) 613 1393 9350
Fax: (49) 613 1393 5326
Email: ieg4@ieg-mainz.de
Website: www.ieg-mainz.de
Contact: Dr Denise Kratzmeier

The Institute of European History in Mainz, founded in 1950, is dedicated to the promotion of interdisciplinary historical research that focuses on European communication and transfer processes since 1450. Its research groups and international fellows focus on the interplay of religious, political and social phenomena relating to these processes.

Leibniz Institute of European History Fellowships

Subjects: Research on the historical foundations of Europe from early modern period to contemporary history, particularly their religion, political and social dimensions. Projects dealing with European communication and transfer processes or project focusing on theology, chruch history and intellectual history are particulary welcome.
Purpose: To support young scientists in the completion of their doctoral work or in the execution of shorter postdoctoral projects. Participation in the Institute's research groups is particularly welcome.
Eligibility: Doctoral Fellowships are open to holders of a Master's degree and to Fellows in the advanced stages of graduate work in history, theology or other historical subjects. Applicants must have successfully completed their comprehensive oral examinations. Postdoctoral Fellowships are open to applicants who have completed PhD.
Level of Study: Postdoctorate, Doctorate
Type: Fellowship
Value: A monthly stipend
Length of Study: 6–12 months
Frequency: Annual
Study Establishment: The Leibniz Institute of European History
Country of Study: Germany
No. of awards offered: 10–30 doctoral fellowships and 2 postdoctoral fellowships
Application Procedure: Applicants must contact the directors of the institute. For the application form, details and deadlines see the website: www.ieg-mainz.de/stipendienprogramm. Please use the application form including CV, list of publications and a description of PhD thesis. Copies of university transcripts and two letters of support are additionally required.
Closing Date: February and August
Funding: Government
No. of awards given last year: 20
No. of applicants last year: 100

INSTITUTE OF FOOD TECHNOLOGISTS (IFT)

525 W. Van Buren, Suite 1000, Chicago, IL, 60607, United States of America
Tel: (1) 312 782 8424
Fax: (1) 312.782.8348
Email: info@ift.org
Website: www.ift.org
Contact: Dr Robert A. Vitas

The Institute of Food Technologists (IFT), founded in 1939, is a non-profit scientific society with 29,000 members working in food science, technology and related professions in industry, academia and government. IFT's mission is to advance the science and technology of food through the exchange of knowledge. As a society for food science and technology, IFT brings a scientific perspective to the public discussion of food issues.

IFT Foundation Graduate Scholarships

Subjects: Food technology and food science.
Purpose: To encourage and support outstanding research.
Eligibility: Open to current graduates pursuing a course of study leading to an MS or PhD degree. Candidates must possess an above-average interest in research together with demonstrated scientific aptitude.
Level of Study: Doctorate, Graduate, Postgraduate
Type: Scholarship
Value: Varies
Frequency: Annual
Study Establishment: Any educational institution that is conducting fundamental investigations in the advancement of food science and technology
Country of Study: United States of America or other countries if appropriate
Application Procedure: Applicants must visit IFT foundation website for details.
Closing Date: April 2nd
Funding: Commercial, individuals, private
Contributor: Contributors include General Mills, Inc., Edlong Dairy Flavors, Nutraceuticals and Functional Foods Division, Education Division, Food Laws and Regulation Division, and Proctor & Gamble Company

Marcel Loncin Research Prize

Subjects: Chemistry/Physics/Engineering research applied to food processing and the improvement of food quality.
Purpose: To provide funds for research in food processing and the improvement of food quality.
Eligibility: Open to all individuals who are capable of conducting research.
Level of Study: Research
Type: Prize
Value: US$50,000 and a plaque
Application Procedure: A completed application form accompanied by a grant proposal and a biographical sketch must be submitted.
Closing Date: December 1st
Funding: Foundation
Contributor: Institute of Food Technologists

INSTITUTE OF HISTORICAL RESEARCH (IHR)

University of London, Senate House, Malet Street, London, WC1E 7HU, England
Tel: (44) 20 7862 8740
Fax: (44) 20 7862 8745
Email: ihr.reception@sas.ac.uk
Website: www.ihr.sas.ac.uk
Contact: Director

The Institute of Historical Research (IHR) is a centre for advanced study in history. It is the meeting place for scholars from around the world, housing the largest open access collection of primary sources for historians in the United Kingdom, administering research and providing courses, seminars and conferences.

The Annual Pollard Prize

Purpose: The Pollard Prize is awarded annually for the best paper presented at an Institute of Historical Research seminar by a postgraduate student or by a researcher within one year of completing the PhD..
Eligibility: Applicants are required to have delivered a paper at an IHR seminar during the academic year in which the award is made. Papers should be fully footnoted, although it is not necessary at this stage to follow Historical Research house style. All papers submitted must be eligible for publication..
Level of Study: Postdoctorate, Research
Type: Prize
Value: Fast track publication in the prestigious IHR journal, Historical Research, and £200 of Blackwell books. A variable number of runner up prizes will be awarded, depending on the quality of applications in any given year
Frequency: Annual
Application Procedure: Submissions should be supported by a reference from a convenor of the appropriate seminar.
Closing Date: May 31st
No. of awards given last year: 1

Conrad and Elizabeth Russell Postgraduate Emergency Hardship Fund

Subjects: History.
Purpose: The Conrad and Elizabeth Russell Postgraduate Emergency Hardship Fund exists to support PhD candidates (in History) who meet with sudden and unexpected hardship.
Eligibility: Applicants should meet all the following criteria: Have met with sudden and unexpected hardship. Be members of the IHR community (usually defined as being either a regular attender at IHR seminars or a regular user of the IHR as a reader). Be registered for a PhD in History at a British or North American university. Be resident in London, whether temporarily for the purposes of their research or as their normal place of residence. London is broadly defined as within the Greater London area. Have applied (wherever possible) to their own university or college for assistance before turning to the fund.
Level of Study: Doctorate
Type: Funding support
Value: £1,000 will normally be available for distribution during any one year and an individual student will not normally receive an award of more than £500.
Frequency: Dependent on funds available
Application Procedure: Applicants must submit an application form.
Closing Date: July 31st

The David Bates, Alwyn Ruddock, and IHR Friends' Bursaries

Purpose: Applications are invited from doctoral students registered at universities in the United Kingdom for bursaries to undertake research trips to London archives.
Eligibility: The bursaries are intended for students who are not registered at London-based institutions and who do not live within Greater London.
Level of Study: Doctorate, Research
Type: Bursary
Value: Up to £500
Frequency: Annual
Application Procedure: Applicants must submit an application form.
Closing Date: July 1st

The Huguenot Scholarship

Subjects: The study of any activity of the French, the Dutch, the Flemish or the Walloon Protestants from the 16th century to the present, in any geographical area. "Activity" will be interpreted in the widest sense.
Purpose: The award will be made to a student working for a higher degree on a Huguenot subject.
Level of Study: Research
Type: Scholarship
Value: £2,500
Frequency: Annual
Application Procedure: Candidates should ensure that they supply, in addition to their application form, two confidential references from academic referees in sealed envelopes.
Closing Date: July 1st
Funding: Trusts

IHR Past and Present Postdoctoral Fellowships in History

Subjects: History.
Purpose: To fund 1 year of postdoctoral research.
Eligibility: Applicants may be of any nationality and their PhD may have been awarded in any country. Those who have previously held another postdoctoral fellowship will normally not be eligible. The fellowship may not be held in conjunction with any other award. Fellowships will begin on October 1st each year and it is a strict condition of these awards that a PhD thesis should have been submitted by that date.
Level of Study: Postdoctorate
Type: Fellowship
Value: Approx. UK£20,000
Length of Study: 1 year
Frequency: Dependent on funds available
Study Establishment: IHR
Country of Study: United Kingdom
No. of awards offered: 2 (may vary according to funds)
Application Procedure: Applicants must complete an application form, available from the Fellowship Officer in early January.
Closing Date: April 5th
Funding: Private
Contributor: The Past and Present Society and IHR
No. of awards given last year: 2
No. of applicants last year: 200

Isobel Thornley Research Fellowship

Subjects: Medieval history, modern history or contemporary history.
Purpose: To help candidates at an advanced stage of a PhD to complete their doctorates.
Eligibility: Open to nationals of any country, but only to those who are registered for a PhD at the University of London.
Level of Study: Doctorate
Type: Fellowship
Value: UK£10,000
Length of Study: 1 year
Frequency: Dependent on funds available
Study Establishment: IHR
Country of Study: United Kingdom
No. of awards offered: 1
Application Procedure: Applicants must complete an application form, available from the Fellowship Assistant in early January.
Closing Date: March 4th
Funding: Private
Contributor: Isobel Thornley Bequest
No. of awards given last year: 1
No. of applicants last year: 80

Jacobite Studies Trust Fellowship

Subjects: Research into the Stuart Dynasty in Britain and Ireland, and in exile, from the departure of James II in 1688 to the death of Henry Benedict Stuart in 1807.
Purpose: To support six months of postdoctoral research into any aspect of Jacobite History.
Eligibility: Open to Doctorate students with academic excellence and depending on the candidate's contribution to the field. Open without regard to nationality or academic affiliation.
Level of Study: Postdoctorate
Type: Fellowship
Value: £7,500
Length of Study: 6 months
Frequency: Annual
No. of awards offered: 2
Application Procedure: Details at: www.history.ac.uk/awards.
Closing Date: March 4th
Funding: Trusts
Contributor: Jacobite Studies Trust

The Parliamentary History Prize

Subjects: The parliamentary history of Britain, England and Wales, Ireland, Scotland or British colonial assemblies..
Purpose: The award is offered for the best essay submitted on any aspect of the parliamentary history of Britain, England and Wales, Ireland, Scotland or British colonial assemblies.
Eligibility: Candidates must normally not at the date of submission be over the age of 35 (exception may be made for candidates with unusual academic CVs), and must submit a brief essay with their entry. The essay must be a genuine work of original research, not hitherto published or accepted for publication..
Level of Study: Postgraduate, Research
Type: Prize
Value: £400
Frequency: Annual
Application Procedure: Essay and CV..
Closing Date: June 1st

Royal History Society Fellowship

Subjects: Medieval history, modern history and contemporary history.
Purpose: To help candidates at an advanced stage of a PhD to complete their doctorates.
Eligibility: Open to nationals of any country.
Level of Study: Doctorate
Type: Fellowship
Value: Approx. UK£10,000
Length of Study: 1 year
Frequency: Dependent on funds available
Study Establishment: IHR
Country of Study: United Kingdom
No. of awards offered: 1
Application Procedure: Applicants must complete an application form, available from the Fellowship Officer in early January.
Closing Date: March
Funding: Private
Contributor: The Royal Historical Society
No. of awards given last year: 1

Scouloudi Fellowships

Subjects: Medieval history, modern history and contemporary history.
Purpose: To help candidates at an advanced stage of a PhD to complete their doctorates.
Eligibility: Only open to United Kingdom citizens or to candidates with a first degree from a United Kingdom university.
Level of Study: Doctorate
Type: Fellowship
Value: UK£10,000
Length of Study: 1 year
Frequency: Dependent on funds available
Study Establishment: IHR
Country of Study: United Kingdom
No. of awards offered: 5
Application Procedure: Applicants must complete an application form, available from the Fellowship Officer in early January.
Closing Date: March 2nd
Funding: Private
Contributor: The Scouloudi Foundation
No. of awards given last year: 4
No. of applicants last year: 80

Scouloudi Historical Awards

Purpose: The purpose for these awards are as a subsidy towards the cost of publishing a scholarly book or article, or an issue of a learned journal in the field of history. To pay for research, and other expenses, to be incurred in the completion of advanced historical work, which the applicant intends subsequently to publish. This does not include expenses incurred in the preparation of a thesis for a higher degree.
Eligibility: Awards are not available to those registered for undergraduate or postgraduate courses or degrees.
Level of Study: Doctorate, Postdoctorate, Research
Type: Award
Value: Up to £1,000
Frequency: Annual
Application Procedure: Applicants must submit an application form.
Closing Date: March 8th

The Sir John Neale Prize in Tudor History

Subjects: 16th Century in England.
Purpose: The Neale Prize is awarded annually to a historian in the early stages of his/her career.
Eligibility: Candidates must either be registered for a higher degree at a British institution or have been registered for such a degree at a British institution within the last three years.
Level of Study: Postgraduate, Research
Type: Prize
Value: £1,000
Frequency: Annual
Application Procedure: An essay and an application form. Essays should be no more than 8,000 words including footnotes, on a theme related to Tudor history. Three double-spaced copies should be submitted.
Closing Date: April 15th
No. of applicants last year: 2

INSTITUTE OF IRISH STUDIES

Queen's University Belfast, 53-67 University Road, Belfast, BT7 1NF, Northern Ireland
Tel: (44) 28 9097 3386
Fax: (44) 28 9097 3388
Email: irish.studies@qub.ac.uk
Website: www.qub.ac.uk/iis

The Institute of Irish Studies at Queen's University was established in 1965 and was one of the first of its kind. It is one of the leading centres for research-based teaching in Irish studies and is an internationally renowned centre of interdisciplinary Irish scholarship attracting academics from all over the world.

Institute of Irish Studies Research Fellowships

Subjects: Any field of Irish studies.
Purpose: To promote research.
Eligibility: Candidates must hold at least a Second Class (Honours) Degree, have research experience and a viable research proposal.
Level of Study: Postdoctorate
Type: Fellowship
Value: UK£28,000
Length of Study: 1 year
Frequency: Annual
Study Establishment: The Institute of Irish Studies, Queen's University Belfast
Country of Study: Northern Ireland
No. of awards offered: Up to 3
Application Procedure: Applicants must see the website for details. Awards are usually advertised in February to March.
Closing Date: Varies
Funding: Government
No. of awards given last year: 3
No. of applicants last year: 50

Mary McNeill Scholarship in Irish Studies

Subjects: Irish studies.
Eligibility: Open to well-qualified students enrolled in the 1-year MA course in Irish studies at Queen's University. Applicants must be citizens of the United States of America or Canada and be enrolled as overseas students in this course.
Level of Study: Postgraduate
Type: Scholarship
Value: UK£3,000
Length of Study: 1 year
Frequency: Dependent on funds available
Study Establishment: Queen's University Belfast
Country of Study: Northern Ireland
No. of awards offered: 1
Application Procedure: Application form can be downloaded from the website.
Closing Date: June 1st
No. of awards given last year: 1

THE INSTITUTE OF MATERIALS, MINERALS AND MINING

1 Carlton House Terrace, London, SW1Y 5DB, United Kingdom
Tel: (44) 01302 320486
Fax: (44) 01302 380900
Email: graham.woodrow@iom3.org
Website: www.iom3.org/index.htm
Contact: Dr M Urquhart, PA to Deputy Chief Executive

The Institute of Materials, Minerals and Mining (IOM3) was officially recognised by the UK's Privy Council on 26 June 2002, created from the merger of The Institute of Materials (IOM) and The Institution of Mining and Metallurgy (IMM). The Institute intends to be the leading international professional body for the advancement of materials, minerals and mining to governments, industry, academia, the public and the professionals.

Bosworth Smith Trust Fund

Subjects: Metal mining and non-ferrous extraction metallurgy or mineral dressing.
Purpose: To assist research.
Eligibility: Open to applicants who possess a degree in a relevant subject.
Level of Study: Postgraduate
Value: Approx. UK£5,500 to cover working expenses, visits to mines and plants in connection with research and the purchase of apparatus
Length of Study: 1 year
Frequency: Annual
Study Establishment: An approved university
Country of Study: United Kingdom
No. of awards offered: Varies
Application Procedure: Applicants must complete an application form, available on request.
Closing Date: March 26th

Edgar Pam Fellowship

Subjects: All subjects within field of interest ranging from explorative geology to extractive metallurgy.
Eligibility: Open to young graduates resident in Australia, Canada, New Zealand, South Africa or the United Kingdom who wish to undertake advanced study or research in the United Kingdom.
Level of Study: Postgraduate
Type: Fellowship
Value: UK£2,000
Length of Study: 1 year
Frequency: Annual
Study Establishment: Approved universities
Country of Study: United Kingdom
No. of awards offered: 1
Application Procedure: Applicants must complete an application form, available on request.
Closing Date: March 26th

G Vernon Hobson Bequest

Subjects: Mining geology.
Purpose: To advance the teaching and practice of geology as applied to mining.
Eligibility: Open to university staff throughout the United Kingdom.
Level of Study: Professional development
Value: Approx. UK£1,300 to cover travel, research or other objects in accordance with the terms of the bequest
Frequency: Annual
Country of Study: United Kingdom
No. of awards offered: More than 1
Application Procedure: Applicants must complete an application form, available on request.
Closing Date: March 26th

Mining Club Award

Subjects: Mineral industry operations.
Purpose: To enable candidates to study in the United Kingdom or overseas, to present a paper at an international minerals industry conference or to assist the candidate in attending a full-time course of study related to the minerals industry outside the United Kingdom.
Eligibility: Open to British citizens aged 21–35 years who are actively engaged in full or part-time postgraduate study or employment in the minerals industry.
Level of Study: Postgraduate, Professional development
Type: Award
Value: Approx. UK£1,500
Frequency: Annual
Country of Study: Any country
No. of awards offered: Varies
Application Procedure: Applicants must complete an application form, available on request.
Closing Date: March 26th

Stanley Elmore Fellowship Fund

Subjects: Extractive metallurgy and mineral processing.
Purpose: To provide funds for research that is related to metallurgy and mineral processing.
Level of Study: Doctorate, Postdoctorate
Type: Fellowship
Value: UK£14,000
Length of Study: 1 year
Frequency: Annual
No. of awards offered: 2
Closing Date: March 26th

The Tom Seaman Travelling Scholarship

Subjects: Mining and/or related technologies.
Purpose: To assist the study for an aspect of engineering in the minerals industry.
Eligibility: Open to candidates who are training or have been trained for a career in mining.
Level of Study: Postgraduate, Professional development
Type: Scholarship
Value: Up to UK£5,500
Frequency: Annual
Application Procedure: A completed application form must be submitted.
Closing Date: March 26th
Additional Information: Check the website for further details (www.iom3.org/content/scholarships-bursaries).

For further information contact:

The Institute of Materials, Minerals and Mining
Tel: (0) 1302 320486
Fax: (0) 1302 380900
Contact: Dr GJM Woodrow, Deputy Chief Executive

THE INSTITUTE OF SPORTS AND EXERCISE MEDICINE

30 Devonshire Street, London, W1G 6PU, England
Tel: (44) 20 7288 5292
Email: d.patterson@ucl.ac.uk
Website: www.fsem.ac.uk
Contact: Miss Diana Meynell, Secretary

The Institute of Sports and Exercise Medicine is a postgraduate medical institute, which was established to develop research, teaching and treatment in sports medicine. It offers annual awards to medical practitioners and runs courses on different aspects of this specialist subject. In 2007 it became the research arm of the Faculty of Sport and Exercise Medicine (UK), with a remit to promote sport and exercise medicine research throughout the UK.

Duke of Edinburgh Prize for Sports Medicine

Subjects: Sports medicine in the community.
Purpose: To promote postgraduate work and signify standards of excellence.
Eligibility: Open to medical practitioners in the United Kingdom.
Level of Study: Postgraduate
Type: Prize
Value: Varies, but usually a substantial cash prize
Frequency: Annual
Country of Study: United Kingdom

No. of awards offered: Varies
Application Procedure: Applicants must write for an entry or nomination form in the first instance.
Closing Date: Varies annually. The exact date is specified in the conditions of entry
Funding: Private
No. of awards given last year: 0
No. of applicants last year: 0

Sir Robert Atkins Award

Subjects: Sports medicine.
Purpose: To increase medical support and active involvement in the field and to recognize a doctor who has provided the most consistently valuable medical, clinical or preventive service to a national sporting organization or sport in general.
Eligibility: Open to medical practitioners in the United Kingdom.
Level of Study: Postgraduate
Type: Award
Value: Varies, but usually a substantial cash prize
Frequency: Annual
Country of Study: United Kingdom
No. of awards offered: 1
Application Procedure: Applicants must write for an entry or nomination form in the first instance.
Closing Date: Varies
Funding: Private
No. of awards given last year: 1
No. of applicants last year: 1

INSTITUTE OF TRANSPERSONAL PSYCHOLOGY

1069 E. Meadow Circle, Palo Alto, CA, 9430, United States of America
Tel: (1) 650 493 4430 ext. 271
Fax: (1) 650 493 6835
Email: askinnerjones@itp.edu
Website: www.itp.edu
Contact: Ms Ann Skinner-Jones

Center for Sacred Feminine (CSF) 2011–12 Scholarship

Subjects: Transpersonal Psychology.
Purpose: To fund research to further the education and awareness of the Sacred Feminine, gender and feminist studies, and transpersonal psychology.
Eligibility: Awarded on a study's disciplinary emphasis on the Divine Feminine, gender studies, and Feminist studies that includes some element of the sacred in them, and/or the project's contribution to the field of transpersonal psychology and the Divine Feminine.
Level of Study: Doctorate, MBA
Type: Scholarship
Value: All scholarship awards are in the form of tuition reduction applied directly to ITP student accounts in equal installments in the Fall, Winter, and Spring
Frequency: Annual
Study Establishment: Institute of Transpersonal Psychology
Country of Study: United States of America
No. of awards offered: 35
Application Procedure: Application form and other documents are available in the organization website.
Closing Date: April 25th
Contributor: The Center for the Divine Feminine (CDF)

ITP Institute Scholarship

Subjects: Transpersonal Psychology.
Purpose: The scholarship program is intended as a bridge to assist needy students in managing tuition increases.
Eligibility: Open to any student enrolled in an Institute program. The awards for the scholarship are need based, with secondary consideration given to diversifying the student body and to students demonstrating high potential to advance the field of transpersonal psychology.
Level of Study: MBA, Postdoctorate, Research
Type: Scholarship
Value: $500–1,500
Length of Study: 1 year
Frequency: Annual
Study Establishment: Institute of Transpersonal Psychology
Country of Study: United States of America
No. of awards offered: Varies
Application Procedure: Applicants must submit Institute Application Form, Cover Letter and Essay Student Aid Report, Global Scholarship Application Form and Residential Scholarship Application Form.

President's Diversity Scholarship

Subjects: Transpersonal Psychology.
Purpose: To support new incoming doctoral students of economic need and/or culturally diverse backgrounds.
Eligibility: New students who have been accepted into an Institute program and continuing students may qualify for the President's Scholarship. The awards are need-based with an emphasis on diversifying the student body. An award may be granted in additional years upon reapplication for the President's Scholarship by the March 1st priority deadline. Awards are limited to students enrolled full-time in an Institute Program.
Level of Study: Doctorate
Type: Scholarship
Value: 25 per cent tuition reduction
Length of Study: Up to 4 years
Frequency: Annual
Study Establishment: Institute of Transpersonal Psychology
Country of Study: United States of America
Application Procedure: Application form is available on website. Income verification statement (for non-U.S. citizens only), Student Aid Report (FAFSA), and a cover letter describing the student's qualifications for the scholarship (two pages maximum) (U.S. Citizens only) should be attached with the application form.
Closing Date: June 1st

Residential PhD African American Scholarship

Subjects: Transpersonal Psychology.
Purpose: To assist with PhD tuition costs.
Eligibility: Applicants must be African American in full-time third year of the Residential PhD Program.
Level of Study: Doctorate
Type: Scholarship
Value: $1,000 per quarter (maximum of $3,000 per year) as tuition reduction and applied directly to student accounts in the Fall, Winter and Spring quarters
Frequency: Annual
Study Establishment: Institute of Transpersonal Psychology
Country of Study: United States of America
Application Procedure: To be considered for this scholarship, all applicants must complete both the application for this scholarship and the Free Application for Federal Student Financial Aid (FAFSA). Please contact the Financial Aid office financial_aid@itp.edu or 650 493 4430, extn. 241) for more information on the FAFSA.
Closing Date: June 1st

INSTITUTE OF TURKISH STUDIES (ITS)

Georgetown University, Intercultural Center 305R, Washington, DC, 20057 1033, United States of America
Tel: (1) 202 687 0295
Fax: (1) 202 687 3780
Email: dcc@turkishstudies.org
Website: www.turkishstudies.org
Contact: David C Cuthell, Director

The Institute of Turkish Studies (ITS) was founded and incorporated in the District of Columbia in 1982. It is the only non-profit, private educational foundation in the United States that is exclusively dedicated to the support and development of Turkish Studies in United States higher education.

Dissertation Writing Grants for Graduate Students

Subjects: Social sciences and humanities.
Purpose: To fund advanced students who have finished the research stage of their dissertation.

Eligibility: Applicants must be graduate students in any field of the social sciences and/or humanities who are U.S. citizens or permanent residents at the time of the application, currently enrolled in a PhD degree program in the United States, and expecting to complete all PhD requirements except their dissertation by March.
Level of Study: Doctorate
Type: Grant
Value: US$5,000–15,000
Length of Study: 1 academic year
Frequency: Annual
Application Procedure: A complete application must include the two-page grant application cover sheet completed in full, a project proposal (maximum six double-spaced pages), a budget statement, three letters of recommendation sent directly to ITS, an updated curriculum vitae and academic transcripts of all graduate work.
Closing Date: March 8th
Funding: Private
Additional Information: Decisions on applications will be announced in May. For further details check the website http://turkishstudies.org/grants/grants_competition.shtml.

Post-doctoral Summer Travel Grants

Subjects: Ottoman and modern Turkish Studies.
Purpose: To provide partial support for travel and research to Turkey.
Eligibility: Open to citizens or permanent residents of the United States who currently live/work in the United States. The candidates must have obtained a PhD in humanities or social sciences.
Level of Study: Research
Type: Grant
Value: Maximum award is round-trip airfare to Turkey.
Length of Study: 4 weeks
Frequency: Annual
Country of Study: Turkey
Application Procedure: Applicants can download the application cover sheet from the website. The completed cover sheet along with a project proposal, budget, letters of recommendation and curriculum vitae must be submitted.
Closing Date: March 8th
Additional Information: Application forms and supporting materials submitted by fax will not be accepted. Please see the website for further details http://turkishstudies.org/grants/grants_competition.shtml.

Summer Language Study Grants in Turkey for Graduate Students

Subjects: Turkish language.
Purpose: To fund summer travel to Turkey for language study in preparation for graduate research.
Eligibility: Applicants must be graduate students in any field of the social sciences or humanities, currently enrolled in a university in the United States of America. Open to citizens of the United States of America or permanent residents.
Level of Study: Graduate
Type: Grant
Value: US$2,000–3,000
Length of Study: Minimum of 2 months
Frequency: Annual
Study Establishment: An established Ottoman or Turkish language training facility
Application Procedure: A complete application must include the two-page grant application cover sheet completed in full, a project proposal (maximum three double-spaced pages), a budget statement, three letters of recommendation sent directly to ITS, an updated curriculum vitae and academic transcripts of all graduate work.
Closing Date: March 8th
Funding: Private
Additional Information: Decisions on applications will be announced in May. For further details check the website.

Summer Research Grants in Turkey for Graduate Students

Subjects: Social sciences and humanities.
Purpose: To fund summer travel to carry out projects.
Eligibility: Applicants must be graduate students in any field of the social sciences or humanities in the United States of America, currently not engaged in dissertation writing. Applicants must be citizens of the United States of America or permanent residents.
Level of Study: Graduate, Postgraduate
Type: Grant
Value: US$1,000–3,000
Length of Study: Varies, minimum of 2 months
Frequency: Annual
Application Procedure: A complete application must include the grant application cover sheet completed in full, a project proposal (maximum five double-spaced pages), a detailed budget stating the amount requested from ITS, three letters of recommendation sent directly to ITS, an updated curriculum vitae and academic transcripts of all graduates work.
Closing Date: March 11th
Funding: Private
Additional Information: Decisions on applications will be announced in May. For further details check the website http://turkishstudies.org/grants/grants_competition.shtml.

Turkish Studies Academic Conference Grant

Subjects: Ottoman and modern Turkish studies.
Purpose: To support and encourage the development of research, scholarship and learning in the field of Turkish Studies in the United States.
Eligibility: Open to postdoctoral scholars in the United States who study aspects of the Republic of Turkey (post-1922). The applicants must be citizens or permanent residents of the United States and affiliated with a university in the United States.
Level of Study: Doctorate, Graduate, Postdoctorate
Type: Grant
Value: US$10,000
Length of Study: 1 year
Frequency: Annual
Application Procedure: Applicants can download the application cover sheet from the website. The completed cover sheet must be sent along with a project proposal, budget, letters of recommendation, curriculum vitae and academic transcripts.
Closing Date: Ongoing process (Check with website)
Additional Information: Application forms and supporting materials submitted by fax will not be accepted. Please see the website for further details http://turkishstudies.org/grants/grants_competition.shtml.

THE INSTITUTION OF CIVIL ENGINEERS

1 Great George Street, Westminster, London, SW1P 3AA, England
Tel: (44) 20 7665 2193
Email: quest.awards@ice.org.uk
Website: www.ice.org.uk/quest
Contact: QUEST Coordinator

QUEST Institution of Civil Engineers Continuing Education Award

Subjects: Civil engineering.
Purpose: This award provides funding of up to £1,500 for ICE members returning to full or part time education.
Eligibility: Open to ICE graduate and professionally qualified members of at least two years standing. Applicants must also have at least two years industrial experience.
Level of Study: Postgraduate, MBA
Type: Award
Value: Up to UK£1,500
Frequency: Annual
Country of Study: United Kingdom
No. of awards offered: Approx. 10–15
Application Procedure: Applicants must complete an application form, arrange for a reference form to be completed by someone suitable and provide a one page employment history. Forms are available to download at www.ice.org.uk/questcea.
Closing Date: April 26th

QUEST Institution of Civil Engineers Travel Awards

Subjects: Civil engineering, environmental engineering, transportation and agricultural engineering.

Purpose: This award provides funding of up to £1,500 for ICE members and up to £6,000 for a group of ICE members to undertake an activity that furthers their professional development and which involves overseas travel.
Eligibility: Open to ICE graduate and professionally qualified members of at least two years standing.
Level of Study: Graduate, Postgraduate
Type: Travel award
Value: Up to £6,000 for groups
Frequency: Annual
Country of Study: Any country
No. of awards offered: Approx. 25
Application Procedure: Complete an application form and arrange for a reference form to be completed by someone suitable. Form can be downloaded from www.ice.org.uk/questtravel.
Closing Date: April 26th, September 28th

THE INSTITUTION OF ENGINEERING AND TECHNOLOGY (IET)

Michael Faraday House, Six Hills Way, Stevenage, Hertfordshire, SG1 2AY, England
Tel: (44) 1438 313 311
Fax: (44) 1438 765 526
Email: awards@theiet.org
Website: www.theiet.org/awards
Contact: T Miller, Scholarships Coordinator

With more than 150,000 members spanning 127 countries, and offices in Asia, Europe and America, the IET is Europe's leading organization of engineering and tecnical professionals. Join us for the exchange of ideas, the sharing of knowledege and the positive promotion of science, engineering and technology around the world.

Hudswell International Research Scholarships

Subjects: Electrical, electronic, information technology, manufacturing engineering and related disciplines.
Purpose: To assist members of the IET with advanced research work, leading to the award of a doctorate, to be undertaken outside the applicant's home country.
Eligibility: Applicants should be members of the IET and must have commenced their studies prior to applying for this scholarship
Level of Study: Doctorate, Postgraduate
Type: Scholarship
Value: UK£5,000
Length of Study: 1 year
Frequency: Annual
Study Establishment: Internationally recognized universities or research establishments with a high reputation for research
Country of Study: Any country
No. of awards offered: 1
Application Procedure: Applicants should complete the online application form at www.theiet.org/postgradawards.
Closing Date: April 4th
Funding: Trusts

IET Postgraduate Scholarship for an Outstanding Researcher

Subjects: Engineering and Technology.
Purpose: To assist IET members with research studies.
Eligibility: Applicants should be members of the IET and must have commenced their studies prior to applying for this scholarship.
Level of Study: Doctorate, Postgraduate
Type: Scholarship
Value: UK£10,000
Length of Study: 1 year
Frequency: Annual
Country of Study: Any country
No. of awards offered: 1
Application Procedure: Applicants should complete the online application form at www.theiet.org/postgradawards.
Closing Date: April 4th
Funding: Trusts

IET Postgraduate Scholarships

Subjects: Engineering and Technology.
Purpose: To assist IET members with research studies.
Eligibility: Applicants should be members of the IET and must have commenced their studies prior to applying for this scholarship
Level of Study: Doctorate, Postgraduate, Research
Type: Scholarship
Value: UK£2,500
Length of Study: 1 year
Frequency: Annual
Study Establishment: University in the United Kingdom
Country of Study: United Kingdom
No. of awards offered: 2
Application Procedure: Applicants should complete the online application form at www.theiet.org/postgradawards.
Closing Date: April 4th
Funding: Trusts
No. of awards given last year: 2

IET Travel Awards

Subjects: Engineering and Technology.
Purpose: To support IET members wishing to travel abroad, attend conferences, work in industry or undertake study tours.
Eligibility: Open to IET members.
Level of Study: Unrestricted
Type: Travel grant
Value: UK£500
Length of Study: N/A
Frequency: Annual
Country of Study: Any country
No. of awards offered: 12 per year
Application Procedure: Applicants should complete the online application form at www.theiet.org/ambition.
Closing Date: Deadlines for applications can be found at www.theiet.org/ambition.
Funding: Trusts

Leslie H Paddle Scholarship

Subjects: Electronic and radio engineering.
Purpose: To assist IET members with research studies.
Eligibility: Applicants should be members of the IET and must have commenced their studies prior to applying for this scholarship
Level of Study: Doctorate, Postgraduate
Type: Scholarship
Value: UK£5,000
Length of Study: 1 year
Frequency: Annual
Country of Study: United Kingdom
No. of awards offered: 1
Application Procedure: Applicants should download an application form from www.theiet.org/postgradawards.
Closing Date: April 4th
Funding: Trusts
No. of awards given last year: 1
Additional Information: Applicants should complete the online application form at www.theiet.org/postgradawards.

INSTITUTION OF MECHANICAL ENGINEERS (IMECHE)

1 Birdcage Walk, Westminster, London, SW1H 9JJ, England
Tel: (44) (0) 20 7222 7899
Fax: (44) (0) 20 7222 4557
Email: enquiries@imeche.org
Website: www.imeche.org.uk
Contact: The Prizes and Awards Officer

The Institution of Mechanical Engineers (IMechE) was founded in 1847 by engineers. They formed an institution to promote the exchange of ideas and encourage individuals or groups in creating inventions that would be crucial to the development of the world as a whole. Now, over 150 years later, IMechE is one of the largest engineering institutions in the world, with over 88,000 members in 120 countries.

Donald Julius Groen Prize
Subjects: Engineering.
Purpose: To award the author of outstanding papers or for outstanding achievements in the group's sphere of activity.
Eligibility: Open to authors of papers or those who have achievements of a sufficiently high standard to warrant the award of an IMechE prize. As a general rule, but with certain exceptions, grants are normally awarded only to members of the Institution.
Type: Prize
Value: UK£250
Frequency: Annual
No. of awards offered: 1
Application Procedure: Applicants must contact the Institution of Mechanical Engineers for details.
Funding: Private
Additional Information: Please see the website for details http://www.imeche.org/knowledge/industries/tribology/prizes-and-awards/donald-julius-groen-prize and email awards@imeche.org for any query.

James Clayton Award
Subjects: Mechanical engineering.
Purpose: To enable the recipient to pursue advanced postgraduate studies or programmes of research.
Eligibility: Open to IMechE members who hold an accredited engineering degree or who have satisfied the academic requirements for IMechE membership by other means. Applicants must not have had less than 2 years of acceptable professional training in mechanical engineering.
Level of Study: Postgraduate
Type: Grant
Value: Up to UK£1,000 per year
Length of Study: Up to 3 years
Frequency: Annual
Study Establishment: An approved centre
Country of Study: United Kingdom
No. of awards offered: Approx. 10
Application Procedure: Applicants must complete and submit an application form with three references.
Closing Date: Applications can be made throughout the year using the online application form.
Funding: Private
Additional Information: A report is required within 3 months of the completion of the project. Please see the website for details.

James Clayton Overseas Conference Travel for Senior Engineers
Subjects: Mechanical engineering.
Purpose: To assist members of the Institution who have been invited to contribute in some way to a conference or who could be expected to make a significant contribution to the aims of a conference by their attendance.
Eligibility: Open to IMechE members over the age of 40 years.
Level of Study: Professional development
Type: Travel grant
Value: Up to UK£1,000
Country of Study: Any country
No. of awards offered: Varies
Application Procedure: Applicants must submit a completed application form with 3 references.
Funding: Private
Additional Information: A report is required three months after the conference. Please see the website for details.

James Clayton Postgraduate Hardship Award
Subjects: Mechanical engineering.
Purpose: To assist outstanding postgraduates who experience hardship while undertaking courses of advanced study, training or research work on a course approved by the Institution.
Eligibility: Open to candidates who have completed a degree course in mechanical engineering accredited by IMechE and who have gained graduate membership of IMechE.
Level of Study: Postgraduate
Type: Grant
Value: Up to UK£1,000
Length of Study: 1 year
Frequency: Annual
Country of Study: United Kingdom
No. of awards offered: Up to 3
Application Procedure: Applicants must complete and submit an application form with three references.
Closing Date: Applications can be made throughout the year using the online application form.
Funding: Private
Additional Information: A report is required 3 months after the activity has been completed. Please see the website for further details.

James Watt International Gold Medal
Subjects: Mechanical engineering.
Purpose: To award an eminent engineer who has attained worldwide recognition in mechanical engineering.
Eligibility: The award is open to IMechE members of all grades as well as non members.
Type: Prize
Value: Gold medal
Frequency: Every 2 years, (odd-numbered years)
No. of awards offered: 1
Application Procedure: Applicants must write for details.
Closing Date: March 31st
Funding: Private
Additional Information: This award is the premier international award of the Institution. Please email awards@imeche.org or call +44 (0) 12 8471 7887 if you have any queries or experience problems downloading the documents.

INTEL CORPORATION

2200 Mission College Blvd, Santa Clara, CA, 95054 1549, United States of America
Tel: (1) 408 765 8080
Fax: (1) 408 765 3804
Email: scholarships@intel.com
Website: www.intel.com

Intel Corporation is committed to maintaining and enhancing the quality of life in the communities where the company has a major presence.

Intel Public Affairs Russia Grant
Subjects: Science, mathematics, environmental students and technology education.
Purpose: To support further study programmes with educational and technological components in Russia.
Eligibility: Each request will be evaluated on the basis of the services offered and the programme's impact on the community and the potential for Intel employee involvement.
Type: Grant
Frequency: Annual
Country of Study: Russia
No. of awards offered: Varies
Application Procedure: Apply online or contact the office.
Funding: Corporation
Contributor: Intel Corporation
For further information contact:

30 Turgenev Street, Novgorod, Nizhny Novgorod, 603024, Russia
Tel: (7) (831) 296 94 44
Email: paris@intel.com
Contact: Mr Evgeny Zakablukovsky, Russia Community and Regional Government Relations Manager

INTENSIVE CARE SOCIETY (ICS)

Churchill House, 35 Red Lion Square, London, WC1R 4SG, England
Tel: (44) 020 7280 4350
Fax: (44) 020 7280 4369
Email: shaba@ics.ac.uk
Website: www.ics.ac.uk
Contact: Shaba Haque, Research Grant and Visiting Scholarship Enquiries & Educational Events Team Leader

The Intensive Care Society (ICS) is a charitable organization promoting advances in the care of the critically ill. This is largely accomplished through educational means and promoting research activity.

ICS Visiting Fellowship

Subjects: Medicine.
Purpose: To provide the cost of travel to ICS members who wish to travel to an institution other than their own, either within the UK, or overseas.
Eligibility: Open to all members of ICS.
Level of Study: Doctorate, Postdoctorate
Type: Fellowship
Value: Up to UK£5,000
Application Procedure: A written proposal of not more than 1,000 words should be submitted, outlining the proposed use of the grant, appropriate costs and the benefits accruing from the visit.
Closing Date: June 25th

ICS Young Investigator Award (Research Grants)

Subjects: Any aspect of intensive care medicine and care of the critically ill.
Purpose: To promote research and intended to be pump-priming grants that will lead to further and more substantial support from other grant giving bodies
Eligibility: Applicants must be ICS members. Please see the website for details.
Level of Study: Unrestricted
Type: Research grant
Value: £15,000
Frequency: Dependent on funds available
Country of Study: Any country
No. of awards offered: Varies
Application Procedure: Applications should be made to the Research Committee using the standard application form available from the ICS website. One Application to be sent electronically plus one original signed copy posted to the ICS offices. For any queries please contact jenny@ics.ac.uk
Closing Date: March 1st
Funding: Private
Additional Information: Further information is available on the Society's website http://www.ics.ac.uk/foundation_home/grants/new_investigator_award.

INTERNATIONAL AGENCY FOR RESEARCH ON CANCER (IARC)

150 Cours Albert Thomas, F-69372 Lyon Cedex 08, France
Tel: (33) 4 72 73 84 48
Fax: (33) 4 72 73 80 80
Email: fel@iarc.fr
Website: www.iarc.fr
Contact: Ms Eve Elakroud, Administrative Assistant IARC Fellowship Programme

The International Agency for Research on Cancer (IARC) is part of the World Health Organization. IARC's mission is to co-ordinate and conduct research into the causes of human cancer and the disease's mechanisms and to develop scientific strategies for cancer control. The Agency is involved in both epidemiological and laboratory research and disseminates scientific information through publications, meetings, courses and fellowships.

IARC Postdoctoral Fellowships for Training in Cancer Research

Subjects: Epidemiology (including genetic and molecular), biostatistics, bioinformatics, and areas related to mechanisms of carcinogenesis including molecular and cell biology, molecular genetics, epigenetics, and molecular pathology. There is an emphasis on interdisciplinary projects.
Purpose: To provide training in cancer research to junior scientists from any country. However applications from candidates from low- and medium-resource countries or from applicants from any parts of the world but with projects realted to low- and medium-resurce coutnries are encouraged.
Eligibility: Applicants are eligible from any country. Candidates should have finished their doctoral degree (PhD) within 5 years of the closing date for application. The working languages at IARC are English and French. Candidates must be proficient in English at a level sufficient for scientific communication. Candidates must contact the host group of their choice at IARC before application in order to establish a proposed programme of mutual interest. Candidates currently working as IARC-funded postdoctoral scientists, or candidates who have worked at the Agency for a period greater than six months cannot be considered.
Level of Study: Postdoctorate
Type: Fellowship
Value: Travel for the Fellow and for dependents if accompanying the Fellow for at least 8 months; an annual stipend of approx. €32,000, net of tax; an annual family allowance of US$400 for spouses and US $450 for each child; and health insurance covered
Length of Study: 2 years, the 2nd year being subject to satisfactory appraisal
Frequency: Annual
Study Establishment: IARC in Lyon, France
Country of Study: France
No. of awards offered: Approx. 8
Application Procedure: Applicants must complete and submit an application form.
Closing Date: November 30th
Contributor: IARC regular budget
No. of awards given last year: 11
No. of applicants last year: 38
Additional Information: In principle, applicants should provide reasonable assurance that they will return to a post in their home country at the end of the fellowship and to continue their work in cancer research. Please also see the IARC Education and Training Website for more details. http://www.iarc.fr/en/education-training/index.php

INTERNATIONAL ANESTHESIA RESEARCH SOCIETY

100 Pine Street, Suite 230, San Francisco, CA, 94111, United States of America
Tel: (1) 415 296 6900
Fax: (1) 415 296 6901
Email: info@iars.org
Website: www.iars.org

An International society committed to improving clinical care, education and research in anaesthesia, pain management and perioperative medicine.

Frontiers in Anesthesia Research Award

Subjects: Anesthesiology.
Purpose: To foster innovation and creativity by an individual researcher in the field of anesthesiology.
Eligibility: Please see the website for details regarding eligibility at http://iars.org/awards/frontiers_rules.asp.
Level of Study: Unrestricted
Type: Research grant
Value: Up to US$750,000
Length of Study: 3 years
Frequency: Every 2 years
No. of awards offered: 1
Application Procedure: Applicants must submit a formal application to the IARS by the published deadline. Applicants must be a member of the IARS. Please see the website www.iars.org for complete application procedures.
Closing Date: Check with website
Funding: Corporation
Contributor: International Anesthesia Research Society
No. of applicants last year: 13–15
Additional Information: If you have questions regarding the Frontiers Award, contact the IARS awards department at awards@iars.org.

Teaching Recognition Award

Subjects: Anaesthesiology.
Purpose: To recognize anesthesiology faculty who have demonstrated outstanding teaching skills and significant contributions to the academic community.
Level of Study: Unrestricted
Type: Grant
Value: $1,000 prize to the recipient and a $15,000 educational grant to the recipient's institution, round-trip coach airfare (up to $500) to/from the meeting, two nights (room and tax) hotel accommodations at the headquarters hotel, two days per diem ($100 per day).
Frequency: Annual
Study Establishment: International Anesthesia Research Society
No. of awards offered: 2
Application Procedure: Application must be submitted to the IARS by the published deadline. Applicant must be a member of the IARS. For more details please see the website www.iars.org
Closing Date: January 25th
No. of awards given last year: 2
Additional Information: Two separate awards, Innovation in Education and Achievement in Education, are granted annually. Please see the website for further details http://www.iars.org/awards/2013%20TRA%20Innovation_Full%20Guidelines%20and%20Instructions.pdf.

INTERNATIONAL ASSOCIATION FOR THE STUDY OF INSURANCE ECONOMICS

53 Route de Malagnou, CH-1208 Geneva, Switzerland
Tel: (41) 22 707 6600
Fax: (41) 22 736 7536
Email: secretariat@genevaassociation.org
Website: www.genevaassociation.org

The International Association for the Study of Insurance Economics was established in 1973 for the purpose of promoting economic research in the sector of risk and insurance.

Ernst Meyer Prize

Subjects: Risk and insurance economics.
Purpose: To recognize research work that makes a significant and original contribution.
Eligibility: Open to professors, researchers or students of economics.
Level of Study: Doctorate, Unrestricted
Type: Prize
Value: CHF5,000
Frequency: Annual
Country of Study: Any country
No. of awards offered: 1
Application Procedure: Applications should be addressed to The Geneva Association, 'Ernst Meyer Prize', General Secretariat, secretariat@genevaassociation.org.
Closing Date: January 31st
Funding: Private
No. of applicants last year: 7
Additional Information: Please see the website for details http://www.genevaassociation.org/Home/Prizes_and_Grants.aspx.

Reseach Grants

Subjects: Risk management and insurance economics.
Purpose: To promote economic research.
Eligibility: Open to graduates involved in research for a thesis leading to a doctoral degree in economics.
Level of Study: Postgraduate
Type: Research grant
Value: CHF10,000
Length of Study: 10 months
Frequency: Annual
Country of Study: Any country
No. of awards offered: 2
Application Procedure: Applications for research grants must be accompanied by a curriculum vitae, a research proposal and letters of recommendation from two professors of economics.
Closing Date: November 30th (Check with website)
Funding: Private
Additional Information: The Association reserves the right to support research on other subjects for which applications are submitted. The Association also grants authors of university theses already submitted, dealing in depth with a subject in the field of risk and insurance economics, a subsidy of up to Swiss francs 3,000 towards printing costs. Please see the website for details http://www.genevaassociation.org/Home/Prizes_and_Grants.aspx.

Subsidies for Thesis

Subjects: Topics of interest in risk management or insurance.
Purpose: To defray printing costs of university theses.
Eligibility: Open to authors of university theses already submitted.
Level of Study: Doctorate, Postdoctorate
Type: Grant
Value: CHF3,000
Frequency: Annual
Application Procedure: Applications for subsidies for theses must be accompanied by a curriculum vitae, a description of the research undertaken and letters of recommendation from two professors of economics.
Closing Date: November 30th
Funding: Private
Additional Information: Please see the website for details http://www.genevaassociation.org/Home/Prizes_and_Grants.aspx.

INTERNATIONAL ASSOCIATION FOR THE STUDY OF OBESITY

Charles Darwin House, 12 Roger Street, London, WCIN 2JU , United Kingdom
Tel: (44) 20 7685 2580
Fax: (44) 20 7685 2581
Email: kate.baillie@iaso.org/enquiries@iaso.org
Website: www.iaso.org

The International Association for the Study of Obesity (IASO) aims to improve global health by promoting the understanding of obesity and weight-related diseases through scientific research and dialogue whilst encouraging the development of effective policies for their prevention and management. IASO is the leading global professional organization concerned with obesity, operating in over 50 countries around the world.

IASO Per Björntorp Travelling Fellowship Award

Subjects: Medicine and surgery.
Purpose: To provide travel grants to enable young researchers to attend the International Congress of Obesity.
Eligibility: Applicants for this award must demonstrate their financial need for such support to attend the Congress. Applicants must be an IASO member. There is no age limit.
Level of Study: Postdoctorate, Predoctorate, Research, Doctorate, Postgraduate
Type: Studentships and bursaries
Value: Up to $2,000 of return economy flights, congress Registration and hotel accommodation at the International Congress of Obesity
Frequency: 4 years
Study Establishment: Any
Country of Study: Any country
No. of awards offered: Varies
Application Procedure: Download the application form from website.
Closing Date: See the web site for details
Additional Information: For further information about the IASO Travelling Fellowships Award please write to: awards@iaso.org.

For further information contact:

28 Portland Place, London, W1B 1LY, United Kingdom
Tel: (44) 20 7467 9610
Fax: (44) 20 7636 9258
Email: kate.baillie@iaso.org
Contact: Kate Baillie

INTERNATIONAL ASSOCIATION OF FIRE CHIEFS (IAFC) FOUNDATION

4025, Fair Ridge Drive, Suite 300, Fairfax, VA, 22033 2868, United States of America
Tel: (1) 703 273 0911, 703 896 4822
Fax: (1) 703 273 9363
Email: foundation@iafc.org/jcooke@iafc.org
Website: www.iafc.org
Contact: Ms Jennifer Cooke, Staff

Each year, the International Association of Fire Chiefs (IAFC) Foundation co-ordinates a scholarship programme made possible through the generosity of corporations throughout the United States of America as well as donations from individuals and persons sponsoring a scholarship as a memorial to a friend or colleague.

IAFC Foundation Scholarship

Subjects: All subjects.
Purpose: To assist fire service personnel towards college degrees.
Eligibility: Open to any person who is an active member of a state, county, provincial, municipal, community, industrial or federal fire department who has demonstrated proficiency as a member. Dependants of members are not eligible. See http://www.iafcf.org/documents/2012SCHOLARSHIPAPPLICATIONFORMFOR-MIAFCF2-2012.pdf for more details
Level of Study: Doctorate, Graduate, MBA, Postdoctorate, Postgraduate, Predoctorate, Professional development
Type: Scholarship
Value: US$500–5,000
Frequency: Annual
Country of Study: Any country
No. of awards offered: 20
Application Procedure: Applicants must complete an application form. This includes a 250-word statement outlining reasons for applying for assistance and an explanation as to why the candidate thinks that the course will be useful in their chosen field of course description.
Closing Date: June 1st (Check with website)
Funding: Commercial, individuals, private
No. of awards given last year: 13
No. of applicants last year: 65
Additional Information: In evaluating the applications, preference will be given to those demonstrating need, desire and initiative.

INTERNATIONAL ASTRONOMICAL UNION (IAU)

UAI Secretariat, 98 bis, boulevard Arago, Paris, F-75014, France
Tel: (33) 1 43 25 83 58
Fax: (33) 1 43 25 26 16
Email: iau@iap.fr
Website: www.iau.org
Contact: Administrative Assistant

The mission of the International Astronomical Union (IAU), founded in 1919, is to promote and safeguard the science of astronomy in all its aspects through international co-operation. The IAU, through its 12 scientific divisions and 40 commissions covering the full spectrum of astronomy, continues to play a key role in promoting and co-ordinating worldwide co-operation in astronomy.

IAU Grants

Subjects: Astronomy and astrophysics.
Purpose: To provide funds to qualified individuals to enable them to visit institutions abroad. It is intended that the visitors have ample time and opportunity to interact with the intellectual life of the host institution. It is a specific objective of the programme that astronomy in the home country is enriched after the applicant returns.
Eligibility: Open to faculty members, staff members, postdoctoral Fellows or graduate students at any recognized educational or research institution.
Level of Study: Postgraduate, Research, Graduate, Postdoctorate
Value: One return economy fare between home and host institutions
Length of Study: At least 3 months at a single host institution
Country of Study: Any country
No. of awards offered: 12–15 per year
Application Procedure: Applicants must submit an application including a curriculum vitae, a plan of scientific activity, letters of support from the home and host institutions, information on responsibility for subsistence at the host institution, and information on the lowest available fare. Applications should be submitted in time for the Officers of the Commission to consult by post.
Closing Date: There is no deadline
Contributor: Academy of Sciences
No. of awards given last year: 15
No. of applicants last year: 30

For further information contact:

University of Virginia-University of Station, Box 3818, Charlottesville, VA, 22 903 0818, United States of America
Tel: (1) 434 924 7494
Fax: (1) 434 924 3104
Email: crt@viginia.edu
Contact: Dr Charles R Tolbert, PresidentUniversity of Toronto, Erindale College, Mississauga, ON, L5l 1C6, Canada
Tel: (1) 905 828 5351
Fax: (1) 905 828 5328
Email: jpercy@credit.erin.utoronto.ca
Contact: John R Percy, Vice-President

INTERNATIONAL ATOMIC ENERGY AGENCY (IAEA)

Vienna International Centre, PO Box 100, Vienna, 1400, Austria
Tel: (43) 1 2600 0
Fax: (43) 1 2600 7
Email: crp.research@iaea.org
Website: http://cra.iaea.org
Contact: Teresa Benson, Section Head - NACA

The IAEA is the world's center of cooperation in the nuclear field. It was set up as the world's "Atoms for Peace" in 1957 within the United Nations family. The agency works with its Member States and multiple partners worldwide to promote safe, secure and peaceful nuclear technologies.

Research Contracts (IAEA)

Subjects: Any scientific or technical field related to the peaceful uses of atomic energy and the use of radio-isotopes in agriculture, industries, medicine, research, etc.
Purpose: To encourage and assist research on the development and practical application of atomic energy for peaceful purposes throughout the world.
Eligibility: Institutions with research projects developed in line with the overall goals of the agency. Priority is normally given to proposals received from institutions in developing countries. (headed by young and female researchers)
Level of Study: Research
Type: Research
Value: Approx. €5,000 per year per contract
Length of Study: 1 year (extension possible up to 3 years)
Frequency: Annual
Application Procedure: Application forms are available on request or can be downloaded from http://cra.iaea.org. Research proposals could be submitted either based on a proposal made by the agency or a proposal developed by the research institute itself.
Closing Date: Proposals accepted throughout the year
No. of awards given last year: 914
No. of applicants last year: 1077
Additional Information: Research Proposals will be considered which involve nuclear technologies or applications and relate to the Agency programme. Only available to IAEA member states.

INTERNATIONAL CENTER FOR JOURNALISTS (ICFJ)

1616 H Street NW, Third floor, Washington, DC 20006, United States of America
Tel: (1) 202 737 3700
Fax: (1) 202 737 0530
Email: editor@icfj.org
Website: www.icfj.org

The International Center for Journalists (ICFJ), a non-profit, professional organization, promotes quality journalism worldwide in the belief that independent, vigorous media are crucial in improving the human condition.

Arthur F. Burns Fellowship Program

Subjects: Media and journalism.
Purpose: The program offers young print and broadcast journalists from each country the opportunity to share professional expertise with their colleagues across the Atlantic while working as 'foreign correspondents' for their hometown news organizations.
Eligibility: Open to U.S. and German journalists between the age of 21–37, who are employed by a newspaper, news magazine, broadcast station, news agency or who work freelance and/or online. Applicants must have demonstrated journalistic talent and a strong interest in U.S.-European affairs. German language proficiency is not required, but it is encouraged.
Level of Study: Professional development, Unrestricted
Value: Travel expenses and a stipend are provided
Length of Study: 2 months (August–September)
Frequency: Annual
No. of awards offered: 10 from each country
Application Procedure: Please refer to the website at http://www.icfj.org/arthur-f-burns-fellowship/apply-now.
Closing Date: March 1st
Funding: Corporation, foundation, individuals, private
Additional Information: Countries of study are Germany and USA.

Knight International Journalism Fellowships

Subjects: Journalism and media.
Purpose: To improve the free flow of news and information in the public interest around the world.
Level of Study: Postdoctorate
Type: Fellowship
Value: Work with media partners in developing countries to launch projects to improve the free flow of news and information in the public interest.
Length of Study: 1 year
Frequency: Annual
Study Establishment: International Center for Journalists
No. of awards offered: Approx. 22
Application Procedure: Apply online at www.icfj.org/our-work/knight.
Closing Date: February 15th and August 15th
Funding: Foundation
Contributor: John S. and James L. Knight Foundation
Additional Information: Contact the Center by fax or mail for detailed application and program guidelines. These aren't really awards. They are essentially contracted assignments.

The McGee Journalism Fellowship in Southern Africa

Subjects: Journalism and technical, management and business aspects of the media.
Purpose: To help journalists improve the skills and standards they need to carry out their work.
Eligibility: Open candidates of outstanding personal and professional achievement in journalism, with experience of teaching overseas, a readiness to work under difficult conditions and an interest in Southern Africa.
Level of Study: Postdoctorate
Type: Fellowship
Value: The fellowship covers all travel, housing, health insurance, living expenses and an honorarium of US$100 per day
Frequency: Annual
Study Establishment: An approved South African University
Country of Study: South Africa
No. of awards offered: 1
Application Procedure: Submit a completed application form, an essay of 500 words or less and three letters of personal or professional recommendation.
Closing Date: April 16th
Funding: Foundation
Contributor: McGee Foundation

INTERNATIONAL CENTRE FOR GENETIC ENGINEERING AND BIOTECHNOLOGY (ICGEB)

AREA Science Park, Padriciano 99, Trieste, 34149, Italy
Tel: (39) 040 375 71
Fax: (39) 040 226 555
Email: fellowships@icgeb.org
Website: www.icgeb.org
Contact: Human Resources Unit

The International Centre for Genetic Engineering and Biotechnology (ICGEB) is an organization devoted to advanced research and training in molecular biology and biotechnology, with special regard to the needs of the developing world. The component host countries are Italy, India and South Africa. The full member states of ICGEB are Afghanistan, Algeria, Argentina, Bangladesh, Bhutan, Bosnia and Herzegovina, Brazil, Bulgaria, Burundi, Cameroon, Chile, China, Colombia, Costa Rica, Côte d'Ivoire, Croatia, Cuba, Ecuador, Egypt, Eritrea, FYR Macedonia, Hungary, Iran, Iraq, Jordan, Kenya, Kuwait, Kyrgyzstan, Liberia, Libya, Malaysia, Mauritius, Mexico, Montenegro, Morocco, Nigeria, Pakistan, Panama, Peru, Poland, Qatar, Romania, Russia, Saudi Arabia, Senegal, Serbia, Slovakia, Slovenia, Sri Lanka, Sudan, Syria, Tanzania, Trinidad and Tobago, Tunisia, Turkey, United Arab Emirates, Uruguay, Venezuela and Vietnam.

The Arturo Falaschi ICGEB Flexible Fellowships

Subjects: Molecular medicine, tumour virology, yeast molecular genetics, bacteriology, protein structure and bioinformatics, molecular pathology, molecular immunology, biosafety, biotechnology development, human molecular genetics, molecular virology, mouse molecular genetics, neurobiology, protein networks, mammalian biology, malaria, recombinant gene products, immunology, structural and computational biology, virology, cancer genomics, plant molecular biology, plant transformation, insect resistance, synthetic biology and biofuels, molecular haematology, Cancer genomics, cancer molecular and cell biology, cellular immunology, cytokines and disease.
Purpose: To provide short-term training in genetic engineering and biotechnology for scientists from the member states of ICGEB, and to promote academic and industrial research in an international context.
Eligibility: Open to promising pre- and postdoctoral students, who are nationals of one of the member states of ICGEB.
Level of Study: Postdoctorate, Predoctorate
Type: Fellowship
Value: An allowance to cover travel costs as well as boarding and lodging
Length of Study: 6–12 years
Frequency: Annual
Study Establishment: ICGEB laboratories in Trieste, Italy; New Delhi, India; Cape Town, South Africa; ICGEB Outstation at Monterotondo (Rome), Italy.
No. of awards offered: Varies
Application Procedure: Applicants must submit a complete application through the ICGEB Liaison Officer of the applicant's country of origin. Application forms can be found on the website.
Closing Date: Applications are accepted at any time
Additional Information: For further information please refer to the website.

The Arturo Falaschi ICGEB Postdoctoral Fellowships

Subjects: Mammalian biology: virology, immunology, malaria, recombinant gene products, structural and computational biology. Plant biology: plant molecular biology, plant transformation, insect resistance, synthetic biology and biofuels; Bacteriology, Biosafety, Biotechnology Development, Cellular Immunology, Human Molecular Genetics, Molecular Immunology, Molecular Medicine, Molecular

Pathology, Molecular Virology, Mouse Molecular Genetics, Neurobiology, Protein Networks, Protein Structure and Bioinformatics, Tumour Virology, Yeast Molecular Genetics, Cancer Genomics, Cancer Molecular and Cell Biology, Cellular Immunology, Cytokines and Disease; Molecular Hematology.
Purpose: To provide long-term training in genetic engineering and biotechnology for scientists from the member states and to promote state-of-the art academic and industrial research training in an international context and for the scientific development of the Fellow's home country.
Eligibility: Open to promising postdoctoral or established research students under the age of 35, who are nationals of one of the ICGEB member states to carry out their study in India, Italy or South Africa.
Level of Study: Postdoctorate
Type: Fellowship
Value: US$19,000–31,000 per Fellow per year, depending on the place of study, as well as travel costs and medical insurance
Length of Study: Up to 2 years
Frequency: Annual
Study Establishment: ICGEB laboratories in Trieste, Italy; New Delhi, India; Cape Town, South Africa; and Outstation at Monterotondo (Rome), Italy.
No. of awards offered: Varies
Application Procedure: Applicants must submit the complete application through the respective National Liaison Officer in their country of origin.
Closing Date: March 31st. Italian nationals should send the complete application directly to ICGEB Trieste, by April 30th
Additional Information: For further information please refer to the website.

The Arturo Falaschi ICGEB Predoctoral Fellowships ICGEB Cape Town International PhD Programme

Subjects: Cancer genomics, cancer molecular and cell biology, cellular immunology, cytokines and disease.
Purpose: To offer postgraduate training with the aim of obtaining a PhD degree in the field of life sciences at the University of Cape Town, South Africa, in collaboration with the ICGEB.
Eligibility: Applicant must be below the age of 32 years and nationals of one of the ICGEB member states and must have a BSc (honours) degree.
Level of Study: Predoctorate
Type: Fellowship
Value: Stipends to cover the cost of normal living expenses (annually) for one person at the location of the individual host institute; fellowships are renewable for the following years providing that the PhD Programme requirements are fulfilled, travel cost to and from the host country and medical health insurance cover the PhD Programme requirements are fulfilled.
Length of Study: Up to 3 years
Frequency: Annual
Study Establishment: ICGEB laboratories in Cape Town
Country of Study: South Africa
No. of awards offered: Varies
Application Procedure: Applicants must refer to the website.
Closing Date: March 31st
Additional Information: For more information on this programme please refer to the website.

The Arturo Falaschi ICGEB Predoctoral Fellowships ICGEB New Delhi International PhD Programme

Subjects: Mammalian biology: virology, immunology, malaria, recombinant gene products, structural and computational biology. Plant biology: plant molecular biology, plant transformation, insect resistance, synthetic biology and biofuels.
Purpose: To offer postgraduate training with the aim of obtaining a PhD degree in the field of life sciences at the Jawaharlal Nehru University in New Delhi, in collaboration with the ICGEB.
Eligibility: Open to promising young students in possession of an MSc degree from a recognized university, who are nationals of one of the ICGEB member states. Indian nationals are not eligible to apply.
Level of Study: Postgraduate, Predoctorate
Type: Fellowship
Value: Stipends to cover the cost of normal living expenses (annually) for one person at thelocation of the individual host institute; fellowships are renewable for the following years providing that the PhD Programme requirements are fulfilled. Costs to Travel to and from the host country and medical health insurance cover.
Length of Study: Up to 3 years
Frequency: Annual
Study Establishment: ICGEB laboratories in New Delhi
Country of Study: India
No. of awards offered: Varies
Application Procedure: Applicants must refer to the website.
Closing Date: March 31st
Additional Information: For more information on this programme please refer to the website.

The Arturo Falaschi ICGEB Predoctoral Fellowships ICGEB Trieste International PhD Programme

Subjects: Molecular medicine, tumour virology, bacteriology, protein structure and bioinformatics, molecular pathology, molecular immunology, human molecular genetics, molecular virology, mouse molecular genetics, neurobiology, protein networks, yeast molecular genetics, cellular immunology, and molecular hematology.
Purpose: To enable promising young students to attend and complete the PhD programme at ICGEB Trieste in Italy. The programme is validated by the Open University, UK, and the University of Nova Gorica, Slovenia.
Eligibility: Open to students having a BSc (Hons) university degree under the age of 32 from any member state of the ICGEB.
Level of Study: Predoctorate
Type: Fellowship
Value: Stipends to cover the cost of normal living expenses (annually) for one person at the location of the individual host institute; fellowships are renewable for the following years providing that the PhD Programme requirements are fulfilled. Cost to cover travel to and from the host country and medical health insurance cover.
Length of Study: Up to 3 years
Frequency: Annual
Study Establishment: ICGEB laboratories in Trieste and Monterotondo (Rome)
Country of Study: Italy
No. of awards offered: Varies
Application Procedure: Applicants must refer to the website.
Closing Date: March 31st
Additional Information: For more information on this programme please refer to the website.

INTERNATIONAL CENTRE FOR PHYSICAL LAND RESOURCES

University of Ghent, Krijgslaan 281/S8, B-9000 Ghent, Belgium
Tel: (32) 9 264 4638
Fax: (32) 9 264 4991
Email: plrprog.adm@ugent.be
Website: www.plr.ugent.be
Contact: Professor E Van Ranst

The International Centre for Physical Land Resources has a long-standing tradition in academic formation and training in physical land resources, including soil science, soil survey, land evaluation, agricultural applications and eremology, e.g. dryland and desertification. Since 1997, the scope of the courses has been widened with courses on the non-agricultural use and application of physical land resources. Students can major in either soil science or land resources engineering. Teaching is provided by lecturers of the University of Ghent and of the Free University of Brussels (VUB).

Master Studies in Physical Land Resources Scholarship

Subjects: Fundamental soil science, soil genesis, prospection and classification, non-agricultural use and applications of land and soils, geotechnical engineering, soil mechanics and hydrogeology, management of physical and land resources, agricultural applications, soil fertility, soil erosion and conservation or land evaluation.
Purpose: To provide MSc training opportunities to nationals from developing countries.
Eligibility: Open to nationals of the developing world or non-European Union members.
Level of Study: Master

Type: Scholarship
Value: €1,200 per month
Length of Study: 2 years
Frequency: Annual
Study Establishment: Ghent University
Country of Study: Belgium
No. of awards offered: Approx. 2
Application Procedure: Applicants must complete an application form and submit this with certified diplomas and transcripts to the Programme Secretariat to obtain academic admission.
Closing Date: March 1st (Check with website)
Funding: Government, private
No. of awards given last year: 1
No. of applicants last year: 45
Additional Information: For any enquiry email at PLRprog.adm@UGent.be and see the website for more details.

INTERNATIONAL COLLEGE OF SURGEONS

1516 North Lakeshore Drive, Chicago, IL, 60610, United States of America
Tel: (1) 312 642 3555
Fax: (1) 312 787 1624
Email: max@icsglobal.org
Website: www.icsglobal.org
Contact: International Executive Director

Postgraduate Scholarships

Subjects: Surgery.
Purpose: To bring surgeons and surgical specialists of all nations, races, and creeds together, to promote surgical excellence for the benefit of all of mankind and to foster fellowship worldwide.
Eligibility: Applicants must have graduated from an accredited medical school, completed their residency and be licensed to practice surgery in their home country (documentation of licensure must be provided). You do not have to be a Fellow of ICS to be eligible.
Level of Study: Postgraduate
Value: See the college web site for details
Length of Study: Committee will consider the amount of time allocated to the program.
Study Establishment: Established treatment, research facilities or educational institutions
Application Procedure: All grant requests must be accompanied by a current CV, which includes such information as education, post graduate training, current hospital affiliations, surgical specialty, other research activity, publications and presentations, etc. Completed applications should be sent to the Executive Director.
Closing Date: Check with website
Contributor: Voluntary contributions, which are made to the College by Fellows and other interested persons
Additional Information: Please see the website for further details https://www.icsglobal.org/about/scholarship_grants/au_sg_apply.asp.

THE INTERNATIONAL DAIRY-DELI-BAKERY ASSOCIATION

IDDBA, 636 Science Drive, PO Box 5528, Madison, WI, 53711 1073, United States of America
Tel: (1) 608 310 5000
Fax: (1) 608 238 6330
Email: iddba@iddba.org
Website: www.iddba.org

Our mission is to expand our leadership role in promoting the growth and development of daily, deli and bakery sales in the food industry. Our vision is to be the essential resource for relevant information and services that add value across all food channels for the dairy, deli and bakery categories.

IDDBA Graduate Scholarships

Subjects: Culinary arts, baking/party arts, food service, business and marketing.
Purpose: To support employees of IDDBA-member companies.
Eligibility: Applicants must be a current full- or part-time employee of an IDDBA-member company with an academic background in a food-related field and have a 2.5 grade-point average on a 4.0 scale, or equivalent.
Level of Study: Postgraduate
Type: Scholarship
Value: US$100–1,000
Length of Study: 1 year
Frequency: Annual
Country of Study: United States of America
No. of awards offered: Varies
Application Procedure: Contact the Education Information Specialist.
Closing Date: January 1st, April 1st, July 1st, October 1st
Funding: Foundation
Contributor: IDDBA
Additional Information: For additional information or enquiries contact organization and email at scholarships@iddba.org. Please see the website for further details http://www.iddba.org/scholarships.aspx.

For further information contact:

Email: kpeckham@iddba.org

INTERNATIONAL DEVELOPMENT RESEARCH CENTRE (IDRC)

PO Box 8500, 150 Kent Street, Ottawa, ON, K1P 0B2, Canada
Tel: (1) 613 236 6163
Fax: (1) 613 236 4026
Email: cta@idrc.ca
Website: www.idrc.ca

The International Development Research Centre (IDRC) is a Canadian crown corporation created by the Canadian government to help communities in the developing world find solutions to social, economic and environmental problems through research.

The Bentley Cropping Systems Fellowship

Subjects: Agriculture, forestry or biology.
Purpose: To provide assistance to Canadian and a developing country's graduate students with a university degree in agriculture, forestry or biology, who wish to undertake post-graduate, applied on-farm research with co-operating farmers in a developing country.
Eligibility: Applicants must be Canadian citizens, permanent residents of Canada or citizens of a developing country who are enrolled full-time in a graduate program (Master's, doctoral, postdoctoral) at a recognized university in Canada or in a developing country for the duration of the award period.
Level of Study: Doctorate, Graduate, Postdoctorate, Postgraduate, Unrestricted
Type: Fellowship
Value: Up to Canadian $30,000. If there is strong evidence of significant potential benefits, the award may be extended upon re-application.
Length of Study: 1.5–2 years
Frequency: Every 2 years
Study Establishment: Universities
No. of awards offered: 1–2
Application Procedure: Applicants must complete and submit an application form with various supporting documents. For further information see the IDRC website.
Closing Date: October 1st
Funding: Private
No. of applicants last year: Varies
Additional Information: Please see the website for details http://www.idrc.ca/EN/Funding/WhoCanApply/Pages/Bentley-Cropping-Systems-Fellowship.aspx.

For further information contact:

The Bentley Fellowship, Fellowships and Awards, International Development Research Centre (IDRC), PO Box 8500, Ottawa, ON, K1G 3H9, Canada

Canadian Window on International Development

Subjects: Agriculture and environment.
Purpose: To support field research in Canada and one or more developing countries.
Eligibility: Open to Canadian citizens, permanent residents of Canada and a developing country's nationals. Applicants must be registered at a Canadian university. See website for more details.
Level of Study: Doctorate, Graduate, Postgraduate, Master's
Value: Up to Canadian $20,000
Length of Study: 3 months to 1 year
Frequency: Annual
Study Establishment: Universities
Country of Study: Canada
No. of awards offered: 2–3
Application Procedure: Applicants must complete an application form. Please refer to the website www.idrc.ca/awards for details.
Closing Date: April 1st
Funding: Government
No. of awards given last year: 1
No. of applicants last year: Varies

Community Forestry: Trees and People-John G Bene Fellowship

Subjects: Forestry or agroforestry with socialsciences.
Purpose: To assist Canadian graduate students in undertaking research on the relationship between forest resources and the social, economic, cultural and environmental welfare of people in developing countries.
Eligibility: Open to Canadian citizens and permanent residents who are registered at a Canadian university at the Master's or doctoral level. Applicants must have an academic background that combines forestry or agroforestry with social sciences.
Level of Study: Doctorate, Graduate, Postgraduate
Type: Fellowship
Value: Canadian $15,000
Length of Study: 3 months to 1 year
Frequency: Annual
Study Establishment: Universities
Country of Study: Canada
No. of awards offered: 1
Application Procedure: Applicants must submit a research proposal and various supporting documents as part of their application. See the website for details.
Closing Date: April 1st
Funding: Private
Contributor: Endowment
No. of awards given last year: 1
No. of applicants last year: Varies
Additional Information: Please see the website for more details http://www.idrc.ca/EN/Funding/WhoCanApply/Pages/John-G-Bene-Fellowship.aspx.

ECOPOLIS Graduate Research and Design Awards

Subjects: Environmental issues borne by the poor.
Purpose: To promote research and design projects that help lighten the environmental problems borne by the urban poor.
Eligibility: Open to Canadian citizens or permanent residents of Canada as well as citizens of developing countries.
Level of Study: Graduate, Doctorate, Master's
Type: Award
Value: Research awards – maximum of Canadian $20,000 covers justifiable field work expenses. Design awards – maximum of Canadian $40,000
Frequency: Annual
No. of awards offered: Up to 10 in total. Up to 5 – Research awards, Up to 5 – Design awards
Application Procedure: Visit the IDRC website for further information.
Closing Date: May 15th
Funding: Government
No. of awards given last year: 9
No. of applicants last year: Varies

IDRC Doctoral Research Awards

Subjects: Agriculture and environment
Purpose: To promote the growth of Canadian capacity in research on sustainable and equitable development from an international perspective.
Eligibility: Open to Canadian citizens and permanent residents and a developing country's nationals. Applicants must be enrolled in a Canadian university, have a research proposal that has been approved by the thesis supervisor and be affiliated with an institution or organization in the region where the research will take place.
Level of Study: Doctorate
Type: Award
Value: Up to Canadian $20,000
Length of Study: 3 months to 1 year
Frequency: Twice a year
Study Establishment: Universities. Normally, such research is conducted in Latin America, Africa, the Middle East or Asia
Country of Study: Canada
No. of awards offered: Varies (20-25per year)
Application Procedure: Applicants must complete and submit an application form with a research proposal and various supporting documents. Information on required documents is available on the IDRC website. You may only apply for the competition that is posted on the Competitions page.
Closing Date: April 1st and November 1st
Funding: Government
Contributor: The Canadian government
No. of awards given last year: 18–20
No. of applicants last year: Varies
Additional Information: Please see the website for further details http://www.idrc.ca/EN/Funding/WhoCanApply/Pages/Doctoral-Research-Awards.aspx.

IDRC Research Awards

Subjects: Social and Economic Policy, Agriculture and Environment, Information and Communication Technologies for Development, Innovation, Policy and Science, Research for Health Equity.
Purpose: The program is aimed at candidates who, through demonstrated achievements in academic studies, work, or research, have shown interest in creating and using knowledge from an international perspective.
Eligibility: Applicants must be Canadian citizens, permanent residents of Canada or citizens of a developing country and be enrolled in the postgraduate studies or have obtained a postgraduate degree.
Level of Study: Doctorate, Graduate, Master's
Type: Award
Value: Salary in a range from CA$36 754 to CA$42 548 per year. Some travel and research expenses will also be supported, up to a maximum of CA$10 000. The salary range and benefits for interns located in the regional offices may vary according to regional conditions.
Length of Study: 1 year
Frequency: Annual
No. of awards offered: Varies
Application Procedure: Applicants must submit a research proposal and various supporting documents as part of their application. Note that all applications must be sent online.
Closing Date: September 12th
Funding: Government
No. of awards given last year: 14
No. of applicants last year: 120
Additional Information: Candidates must conduct their research in areas corresponding to IDRC's research priorities. Please see the website for further details http://www.idrc.ca/EN/Funding/WhoCanApply/Pages/Internships-at-IDRC.aspx.

For further information contact:

Tel: 613 696 2098
Fax: 613 236 4026

THE INTERNATIONAL FEDERATION OF UNIVERSITY WOMEN (IFUW)

IFUW Headquarters, 10 rue de Lac, Geneva, CH-1207, Switzerland
Tel: (41) 22 731 23 80
Fax: (41) 22 738 04 40
Email: info@ifuw.org
Website: www.ifuw.org

The International Federation of University Women (IFUW) is a non-profit, non-governmental organization comprising graduate women working locally, nationally and internationally to advocate the improvement of the status of women and girls at the international level, by promoting lifelong education and enabling graduate women to use their expertise to effect change.

British Federation Crosby Hall Fellowship

Subjects: All subjects.
Purpose: To encourage advanced scholarship and original research.
Eligibility: Open to female applicants who are either members of one of IFUW's national federations or associations or, in the case of female graduates living in countries where there is not yet a national affiliate, independent members of IFUW. Applicants should be well started on a research programme and should have completed at least 1 year of graduate work.
Level of Study: Research, Doctorate, Postdoctorate
Type: Fellowship
Value: UK£2,500
Frequency: Dependent on funds available
Study Establishment: An approved Institute of Higher Education
Country of Study: United Kingdom
No. of awards offered: 1
Application Procedure: Applicants must apply through their respective federation or association. A list of IFUW national federations can be obtained from the website. IFUW independent members and international individual members must apply directly to the IFUW headquarters in Geneva.
Closing Date: The deadline varies by country, but normally falls between August and mid-September. Please ask your national headquarters for the exact deadline.
Funding: Private
No. of awards given last year: 1
Additional Information: Fellowship is reserved for women whose study or research will take place in Great Britain. Please see the website for further details http://www.ifuw.org/what/fellowships/international/.

The CFUW/A Vibert Douglas International Fellowship

Subjects: Conservation biology, ecology and evolution.
Purpose: To financially assist women in pursuing advanced research, study and training.
Eligibility: Open to graduate women who are either members of one of IFUW's 62 national federations and associations or, if living in a country where there is not yet a national affiliate, an independent member of IFUW, or an international individual member of IFUW.
Level of Study: Postdoctorate, Doctorate, Postgraduate
Type: Fellowship or Grant
Value: Canadian $12,000
Frequency: Dependent on funds available
Country of Study: Any country
No. of awards offered: 1
Application Procedure: Applicants must apply through their respective federations or association. A list of IFUW federations can be obtained from the website. IFUW independent memebers must apply directly to IFUW Headquarters in Geneva.
Closing Date: The deadline varies by country, but normally falls between August and mid-September. Please ask your national headquarters for the exact deadline.
Funding: Private
No. of awards given last year: 1
Additional Information: Please see the website for further details http://www.ifuw.org/what/fellowships/international/.

Dorothy Leet Grants

Subjects: All subjects.
Purpose: To enable recipients to carry out research, obtain specialized training essential to research or training in new techniques.
Eligibility: Open to female applicants who are either a member of one of IFUW's national federations or associations or, in the case of female graduates living in countries where there is not yet a national affiliate, an independent member of IFUW. Applicants should be well started on a research programme and should have completed at least 1 year of graduate work.
Level of Study: Doctorate, Graduate, Postdoctorate, Postgraduate
Type: Grant
Value: Swiss franc 4,500
Length of Study: A minimum of 2 months
Frequency: Dependent on funds available
Country of Study: Any country
No. of awards offered: Varies
Application Procedure: Applicants should apply through their respective federation or association. A list of IFUW national federations can be found on the IFUW website. IFUW independent members must apply directly to IFUW Headquarters in Geneva.
Closing Date: The deadline varies by country, but normally falls between August and mid-September. Please ask your national headquarters for the exact deadline.
Funding: Individuals, private
No. of awards given last year: 1
Additional Information: Further information is available from the website http://www.ifuw.org/what/fellowships/international/.

Ida Smedley MacLean Fellowship

Subjects: All subjects.
Purpose: To encourage advanced scholarship and original research.
Eligibility: Open to female applicants who are either members of one of IFUW's national federations or associations or, in the case of female graduates living in countries where there is not yet a national affiliate, independent members of IFUW. Applicants should be well started on a research programme and should have completed at least 1 year of graduate work.
Level of Study: Doctorate, Postdoctorate, Postgraduate
Type: Fellowship
Value: Swiss Francs 8,000
Length of Study: More than 8 months
Frequency: Dependent on funds available
Study Establishment: An approved Institute of Higher Education other than that in which the applicant received her education
Country of Study: Any country
No. of awards offered: Varies
Application Procedure: Applicants should apply through their respective federation or association. A list of IFUW national federations can be found on the IFUW website. IFUW independent members must apply directly to the IFUW Headquarters in Geneva.
Closing Date: The deadline varies by country, but normally falls between August and mid-September. Please ask your national headquarters for the exact deadline.
Funding: Individuals, private
No. of awards given last year: 1 of each fellowship
Additional Information: Please see the website for further details http://www.ifuw.org/what/fellowships/international/.

INTERNATIONAL FOUNDATION FOR ETHICAL RESEARCH (IFER)

53 West Jackson Boulevard, Suite 1552, Chicago, IL, 60604, United States of America
Tel: (1) 312 427 6025
Fax: (1) 312 427 6524
Email: ifer@navs.org
Website: www.ifer.org

The International Foundation for Ethical Research (IFER) supports the development and implementation of viable, scientifically valid alternatives to the use of animals in research, product testing and classroom education. IFER is dedicated to the belief that through new technologies and diligent research, solutions can be found that will create a better world for all, without using animals.

IFER Graduate Fellowship Program

Subjects: Sciences, humanities, psychology and journalism.
Purpose: The purpose of these Graduate Fellowships in Alternatives in Scientific Research is to provide monetary assistance to graduate students whose programs of study seem likely to have an impact in one or more of these areas.
Eligibility: Application is open to students enrolled in Master's and PhD programs in the sciences, humanities, psychology and journalism.

Level of Study: Graduate, Postgraduate
Type: Fellowship
Value: Up to US$12,500 in stipendiary support and up to $2,500 for supplies per year.
Length of Study: 1 year, renewable for up to 3 years based on eligibility and funding
Frequency: Annual
Country of Study: Any country
No. of awards offered: Varies
Application Procedure: Applicants must write for details or refer to the website www.ifer@navs.org
Closing Date: March 15th
Funding: Private
Additional Information: Please see the website for further details http://www.ifer.org/fellowships.html.

INTERNATIONAL FOUNDATION FOR SCIENCE (IFS)

Karlavägen 108, 5th Floor, Stockholm, SE 115 26, Sweden
Tel: (46) 8 545 818 00
Fax: (46) 8 545 818 01
Email: info@ifs.se
Website: www.ifs.se
Contact: Director

Founded in 1972, the International Foundation for Science (IFS), a non-government organisation, has its largest presence in developing countries where it contributes to the strengthening of capacity to conduct relevant and high-quality research on the management, use and conservation of biological resources and the environment in which these resources occur and upon which they depend.

IFS Research Grants

Subjects: Aquatic resources, animal production, crop science, forestry/agroforestry, food science, water resources, social science and natural products.
Purpose: To provide opportunities for young researchers to contribute to the generation of scientific knowledge.
Eligibility: Applicants must be citizens of a developing country that is eligible for IFS support, and carry out the research in an eligible country (this does not have to be the country of citizenship).
Level of Study: Postdoctorate, Research
Type: Research grant
Value: Up to US$12,000
Length of Study: 1–3 years
Frequency: Bi-Annual
No. of awards offered: Varies
Application Procedure: Applications can be downloaded from the IFS website.
Closing Date: January 27th
Funding: Foundation
Contributor: International Foundation for Science
No. of awards given last year: 263
No. of applicants last year: 1,500
Additional Information: Please see the website for further details http://ifs.se/IFS/Documents/Calls/IFS_Call_for_Individual_Applications_Dec2012.pdf.

INTERNATIONAL HARP CONTEST IN ISRAEL

16 Hanaziv Street, Tel Aviv, 67018, Israel
Tel: (972) 3 604 1808
Fax: (972) 3 604 1688
Email: harzimco@netvision.net.il
Website: www.harpcontest-israel.org.il

The International Harp Contest takes place in Israel every 3 years and is judged by a jury of internationally known musicians. It was founded in 1959, and since then, harpists from all over the world gather in Jerusalem to participate in the contest, the only one of its kind.

International Harp Contest in Israel

Subjects: Harp playing.
Purpose: To encourage excellence in harp playing.
Eligibility: Open to harpists of any nationality who are aged 35 years or younger.
Level of Study: Professional development
Type: Prize
Value: $5000.00 cash prize (See website for details)
Frequency: Every 3 years
Country of Study: Any country
No. of awards offered: 1
Application Procedure: Applicants must complete an application form and submit this with recommendations, a record of concert experience, curriculum vitae and birth certificate. There is a registration fee of US$150.
Closing Date: May 1st (Check with website)
Funding: Government, private, trusts
Contributor: Culture Authority, the Government of Israel, the Ministry of Culture, foundations and donors
No. of awards given last year: 6
No. of applicants last year: 36
Additional Information: Please see the website for further details http://www.harpcontest-israel.org.il/.

THE INTERNATIONAL HUMAN FRONTIER SCIENCE PROGRAM ORGANIZATION (HFSPO)

12 quai Saint-Jean, BP 10034, F-67080 Strasbourg Cedex, France
Tel: (33) 3 8821 5126
Fax: (33) 3 8821 5289
Email: rhuie@hfsp.org
Website: www.hfsp.org
Contact: Rosalyn Huie, Communications Assistant

The Human Frontier Science Program (HFSP) promotes basic research in the life sciences that is original, interdisciplinary and requires international collaboration. The support and training of young investigators is given special emphasis.

HFSPO Cross-Disciplinary Fellowships

Subjects: Life sciences, biology. The aim is to support basic research focused on elucidating the complex mechanisms of living organisms. The fields supported range from biological functions at the molecular level to higher brain functions.
Purpose: Cross-Disciplinary Fellowships are intended for postdoctoral fellows with a PhD in the physical sciences, chemistry, mathematics, engineering or computer science who wish to receive training in biology. The fellowships provide young scientists with up to 3 years of postdoctoral research training in an outstanding laboratory in another country. The conditions are the same as for Long-Term Fellowships.
Eligibility: For details, see HFSP website.
Level of Study: Postdoctorate
Type: Fellowship
Value: Approx. US$50,000 per year, including allowances for travel and research expenses
Frequency: Annual
Country of Study: Nationals of one of the supporting countries can apply to receive training in any country. Nationals of any other country must apply to train in a supporting country.
No. of awards offered: Varies
Application Procedure: Applications must be submitted online.
Closing Date: September
Funding: Government
Contributor: Member countries: Australia, Canada, France, Germany, India, Italy, Japan, New Zealand, Norway, Republic of Korea, Switzerland, United Kingdom, United States of America, European Union
No. of awards given last year: 5
No. of applicants last year: 67
Additional Information: Fellowships for up to 3 years. The 3rd year can be used to support 1 year of postdoctoral training in the home country and can be deferred for up to 2 years. Former awardees are eligible to apply for a Career Development Award upon repatriation to

their home country or when moving to another HFSP member country to help establish themselves as individual investigators.

HFSPO Long-Term Fellowships

Subjects: Life sciences, biology. The aim is to support basic research focused on elucidating the complex mechanisms of living organisms. The fields supported range from biological functions at the molecular level to higher brain functions.
Purpose: Long-Term Fellowships provide young scientists with up to 3 years of postdoctoral research training in an outstanding laboratory in another country.
Eligibility: Applicants must see the website for details.
Level of Study: Postdoctorate
Type: Fellowship
Value: Approx. US$50,000 per year, including allowances for travel and research expenses
Frequency: Annual
Country of Study: Nationals of one of the supporting countries can apply to receive training in any country. Nationals of any other country must apply to train in a supporting country.
No. of awards offered: Varies
Application Procedure: Applications must be submitted online through the website.
Closing Date: September
Funding: Government
Contributor: Member countries: Australia, Canada, France, Germany, India, Italy, Japan, New Zealand, Norway, Republic of Korea, Switzerland, United Kingdom, United States of America, European Union
No. of awards given last year: 80
No. of applicants last year: 680
Additional Information: Fellowships for up to 3 years. The 3rd year can be used to support 1 year of postdoctoral training in the home country and can be deferred up to 2 years. Former awardees are eligible to apply for a Career Development Award upon repatriation to their home country or when moving to another HFSP member country to help establish themselves as independent investigators.Nationals of one of the supporting countries can apply to receive training in any country. nationals of any of the country must apply to train in a supporing country.

HFSPO Program Grant

Subjects: Life sciences, biology. The aim is to support basic research focused on elucidating the complex mechanisms of living organisms. The fields supported range from biological functions at the molecular level to higher brain functions.
Purpose: To enable teams of independent researchers at any stage of their careers to develop new lines of research.
Eligibility: Independent investigators early on in their careers are encouraged to apply.
Level of Study: Research
Type: Collaborative research grant
Value: Up to US$450,000 per grant per year
Length of Study: Max. 3 years
Frequency: Annual
Country of Study: The Principal applicant must have his/her laboratory in a member country. Atleast one other team member must be located in another country.
No. of awards offered: Varies
Application Procedure: Application online, by letter of intent; submission of full application by invitation.
Closing Date: Spring. See website for details
Funding: Government
Contributor: Member countries: Australia, Canada, France, Germany, India, Italy, Japan, New Zealand, Norway, Republic of Korea, Switzerland, United Kingdom, United States of America, European Union
No. of awards given last year: 25
No. of applicants last year: 612 letters of intent 68 full applications received

HFSPO Young Investigator Grant

Subjects: Life sciences, biology. The aim is to support basic research focused on elucidating the complex mechanisms of living organisms. The fields supported range from biological at the molecular level to higher brain functions.
Purpose: Young investigator grants are awarded to teams of researchers, all of whom are within the first 5 years of obtaining an independent position (e.g. Assistant Professor, Lecturer or equivalent). They must also be within 10 years of receiving their PhD before the deadline for submission of the letter of intent.
Level of Study: Research
Type: Collaborative research grant
Value: Up to US$450,000 per grant, per year
Length of Study: Maximum 3 years
Frequency: Annual
Country of Study: The Principal applicant must have his/her laboratory in a member country. Atleast one other team member must be located in another country.
No. of awards offered: Varies
Application Procedure: Application online by letter of intent; submission of full application by invitation.
Closing Date: Spring. See website for details
Funding: Government
Contributor: Member countries: Australia, Canada, France, Germany, India, Italy, Japan, New Zealand, Norway, Republic of Korea, Switzerland, United Kingdom, United States of America, European Union
No. of awards given last year: 9
No. of applicants last year: 182 letters of intent and 27 full applications

INTERNATIONAL INSTITUTE FOR APPLIED SYSTEMS ANALYSIS (IIASA)

Schlossplatz 1 A-2361, Laxenburg, Austria
Tel: (43) 2236 807 0
Fax: (43) 2236 71 313
Email: ysspinfo@iiasa.ac.at/huber@iiasa.ac.at
Website: www.iiasa.ac.at

The International Institute for Applied Systems Analysis (IIASA) is a non-governmental research organization. It conducts inter-disciplinary scientific studies on environmental, economic, technological and social issues in the context of human dimensions of global change. It is located in Austria near Vienna.

IIASA Postdoctoral Program

Subjects: Environment, economics, technology and social issues.
Purpose: To enrich IIASA's intellectual environment and help achieve research programme goals and to encourage and promote the development of young researchers and offer them the opportunity to further their careers by gaining hands-on professional research experience in a highly international scientific environment.
Eligibility: Open to candidates who have an advanced university degree equivalent to a PhD.
Level of Study: Postdoctorate
Type: Funding support
Value: Allowance for relocation expenses to and from Laxenburg, limited support for business travel and salary
Length of Study: 1–2 years
Frequency: Annual
No. of awards offered: 2
Application Procedure: Candidates must fill a personal information form online. In addition to this a research plan, a discussion of the relevance, a letter of support and names of 3 referees should be mailed.
Closing Date: Febuary 28th
Additional Information: Please see the website for further details http://webarchive.iiasa.ac.at/Admin/PDOC/apply_iiasa_application-closed.html.

For further information contact:

Website: www.iiasa.ac.at
Contact: Barbara Hauser, Postdoctoral Co-ordinator

INTERNATIONAL INSTITUTE FOR MANAGEMENT DEVELOPMENT (IMD)

Chemin de Bellerive 23, PO Box 915, Lausanne, CH-1001, Switzerland
Tel: (41) 21 618 0111
Fax: (41) 21 618 0707
Email: mbainfo@imd.ch
Website: www.imd.ch/mba
Contact: Barbara Partin

The International Institute for Management Development (IMD), created by industry to serve industry, develops cutting-edge research and programmes that meet real world needs. Their clients include dozens of leading international companies and their experienced faculty incorporate new management practices into the small and exclusive MBA programme. With no nationality dominating, IMD is truly global, practical and relevant.

IMD MBA Alumni Scholarships

Subjects: MBA.
Purpose: To financially support applicants from Africa, Middle East, Asia, Latin America, Eastern Europe, Western Europe, North America and Oceania undertaking an MBA at IMD.
Eligibility: Candidates who have already applied to the full-time IMD MBA program and who are citizens, but not necessarily current residents Africa / Middle East, Asia, Eastern Europe, Latin America, Western Europe / North America / Oceania.
Level of Study: Graduate, MBA
Type: Scholarship
Value: Swiss Francs 30,000 towards tuition fees and book expenses
Frequency: Annual
Study Establishment: The International Institute for Management Development (IMD)
Country of Study: Switzerland
No. of awards offered: 5
Application Procedure: Applicants must complete and submit the IMD MBA application form for financial assistance and the MBA application form. In addition, applicants must submit an essay of 750 words on the topic: As a business leader which issue would you set as your first priority to address in your region. Why would you choose this issue? How would you personally address it?
Closing Date: September 30th
Funding: Private
Contributor: IMD Alumni Loan Fund
No. of awards given last year: 5
No. of applicants last year: 47
Additional Information: Scholarship essays, or questions, should be sent to mbafinance@imd.ch. Please see the website for further details http://www.imd.org/programs/mba/fees/scholarships/Alumni.cfm.

The IMD MBA Future Leaders Scholarships

Subjects: MBA.
Purpose: To financially support candidates with exceptionally strong leadership potential undertaking an MBA at IMD.
Eligibility: Candidates who have already applied to the full-time IMD MBA program and who demonstrate exceptionally strong leadership potential.
Level of Study: MBA
Type: Scholarship
Value: Swiss Franc 30,000 towards tuition fees and book expenses
Length of Study: 1 year
Frequency: Annual
Study Establishment: IMD
Country of Study: Switzerland
No. of awards offered: 2
Application Procedure: Applicants must submit an essay of maximum of 750 words on the topic 'Leadership in an era of globalization' and should contact the organization for details.
Closing Date: September 30th
Funding: Private
No. of awards given last year: 2
No. of applicants last year: 50
Additional Information: Scholarship essays, or questions, should be sent to mbafinance@imd.ch. Please see the website for further details http://www.imd.org/programs/mba/fees/scholarships/Future-Leaders.cfm.

Jim Ellert Scholarship

Subjects: MBA.
Purpose: To financially support candidates from China, Hong Kong, Estonia, Czech Republic, Bulgaria, Rumania, Hungary, Slovenia or Switzerland undertaking an MBA at IMD.
Eligibility: Candidates who have already applied to the full-time IMD MBA program and who demonstrate strong financial need (with preference given to women).
Level of Study: MBA
Type: Scholarship
Value: Swiss Francs 20,000
Length of Study: 1 year
Frequency: Annual
Study Establishment: IMD
Country of Study: Switzerland
No. of awards offered: 1
Application Procedure: Applicants must complete and submit the IMD MBA application form for financial assistance and the MBA application form. In addition, candidates must submit a 500-word essay on the topic: Why I would like to do an MBA at IMD.
Closing Date: September 30th
Funding: Private
No. of awards given last year: 1
No. of applicants last year: 12
Additional Information: Scholarship essays, or questions, should be sent to mbafinance@imd.ch. Please see the website for further details http://www.imd.org/programs/mba/fees/scholarships/Jim-Ellert.cfm.

Nestlé Scholarship for Women

Subjects: MBA.
Purpose: To financially support women applicants with financial need undertaking an MBA at IMD.
Eligibility: Women candidates who have already applied to the full-time IMD MBA program and who demonstrate financial need. Employees of Nestlé and its subsiduaries are not elegible.
Level of Study: Graduate, MBA
Type: Scholarship
Value: Swiss Francs 25,000 towards tuition and living expenses
Length of Study: 1 year
Frequency: Annual
Study Establishment: IMD
Country of Study: Switzerland
No. of awards offered: 1
Application Procedure: Applicants must complete and submit the IMD MBA application form for financial assistance and the MBA application form. In addition, applicants must submit a 750 word on the topic: Does Diversity in Management impact the bottom line? If so, how?
Closing Date: September 30th
Funding: Corporation
No. of awards given last year: 1
No. of applicants last year: 10
Additional Information: Scholarship essays, or questions, should be sent to mbafinance@imd.ch. Please see the website for further details http://www.imd.org/programs/mba/fees/scholarships/Nestle.cfm.

Staton Scholarship

Subjects: MBA.
Purpose: To financially support applicants from South America (excluding Brazil) undertaking an MBA at IMD.
Eligibility: Candidates from the above countries who have already applied to the full-time IMD MBA program, on condition that the candidate must return to South America for at least three years after graduation.
Level of Study: Graduate, MBA
Type: Scholarship
Value: US$50,000 towards tuition fees and book expenses
Length of Study: 1 year
Frequency: Annual
Study Establishment: IMD
Country of Study: Switzerland
No. of awards offered: 1

Application Procedure: Applicants must complete and submit the IMD MBA application form. In addition, applicants must submit a 750-word essay on the topic: The role of entrepreneurship in moving my country forward and my contribution to that goal.
Closing Date: September 30th
Funding: Private
Contributor: Woods Staton
No. of awards given last year: 0
No. of applicants last year: 2
Additional Information: It is a condition of the scholarship that candidates return to South America for at least 3 years after graduation. Scholarship essays, or questions, should be sent to mbafinance@imd.ch. Please see the website for further details http://www.imd.org/programs/mba/fees/scholarships/Staton.cfm.

Van Oord Scholarship

Subjects: MBA.
Purpose: Financially support applicants from developing countries strongly influenced by its seaside location, struggle against the water, maritime tradition, etc.
Eligibility: Candidates with financial need who have already applied to the full-time IMD MBA program and who are citizens of the above geographical areas. Preference is given to those who wish to return to their countries after graduation and to those with a technical background and / or experience related to marine engineering.
Level of Study: MBA
Type: Scholarship
Value: CHF 30,000 towards tuition fees and book expenses
Frequency: Annual
Study Establishment: IMD
Country of Study: Switzerland
No. of awards offered: 1
Application Procedure: Applicants must complete and submit the IND MBA application form for financial assistance and the MBA application form. In addition, candidates must submit 750 word essay on the topic: How the seaside location of my country can be better used to contribute to economic growth and sustainability.
Closing Date: September 30th
Contributor: Van Oord family
No. of awards given last year: 1
No. of applicants last year: 12
Additional Information: Scholarship essays, or questions, should be sent to mbafinance@imd.ch. Please see the website for further details http://www.imd.org/programs/mba/fees/scholarships/van-Oord-MBA-Scholarship.cfm.

INTERNATIONAL NAVIGATION ASSOCIATION (PIANC)

Bâtiment Graaf de Ferraris - 11ième étage, Blvd. du Roi Albert II, 20 - Boîte 3, B-1000 Brussels, Belgium
Tel: (32) 32 2 553 71 61
Fax: (32) 32 2 553 71 55
Email: info@pianc-aipcn.org
Website: www.pianc-aipcn.org
Contact: General Secretariat

The International Navigation Association (PIANC) is a worldwide non-political and non-profit technical and scientific organization of private individuals, corporations and national governments. PIANC's objective is to promote the maintenance and operation of both inland and maritime navigation by fostering progress in the planning, design, construction, improvement, maintenance and operations of inland and maritime waterways, ports and coastal areas for general use in industrialized and industrializing countries. Facilities for fisheries, sport and recreational navigation are included in PIANC's activities.

De Paepe-Willems Award

Subjects: The design, construction, improvement, maintenance or operation of inland and maritime waterways such as rivers, estuaries, canals, port, inland and maritime ports and coastal areas and related fields.
Purpose: To encourage young professionals to submit for presentation outstanding technical articles in the fields of interest to PIANC.
Eligibility: Applicant must not have reached the age of forty on December 31 of the year of submission of the paper, submit his/her paper, attached to the application form, to the General Secretariat of PIANC before the August 31 of the year preceding the granting of the Award and be the author or the lead author of a team.
Level of Study: Unrestricted
Type: Award
Value: The Award consists of an amount of € 5,000 and free membership of PIANC for a five-year period.
Frequency: Annual
Country of Study: Any country
No. of awards offered: 1
Application Procedure: Applicants must complete an application form, available on request from the PIANC General Secretariat or on the PIANC website, and submit this together with the article. Articles must be written by a single author, not have been previously published elsewhere, not exceed 12,000 words, be in type script, and in English or French with a summary in the same language. Articles may be accompanied by illustrations or diagrams.
Closing Date: August 31st
Funding: Government, private
Additional Information: The prize will be awarded to the individual candidate who submits the most outstanding article in the calendar year preceding the Annual General Assembly at which the prize is awarded, provided the article is judged to be of sufficiently high standard. The prize winner will be invited to present a commentary on his or her article during the General Assembly of the PIANC or during the Congress. In judging the articles the jury shall take into account their technical level, originality and practical value and the quality of presentation. Candidates are advised that the Bulletin is designed for readers with a wide range of engineering interests and highly specialized articles should be written with this in mind.

INTERNATIONAL PEACE SCHOLARSHIP FUND

PEO, 3700 Grand Avenue, Des Moines, IA, 50312, United States of America
Tel: (1) 515 255 3153
Fax: (1) 515 255 3820
Email: www.peointernational.org.
Website: www.peointernational.org
Contact: Ms Carolyn J Larson, Project Supervisor

We are a philanthropic and educational organization who offers grants, loans and scholarships for women.

PEO International Peace Scholarship

Subjects: All subjects.
Purpose: To support international women studying for graduate degrees in USA or Canada.
Eligibility: Applicants of any nationality may apply, with the exception of residents of the United States of America or Canada. Eligibility is based on financial need, nationality, degree, full-time status and residence. Students who hold permanent residency in the United States of America or Canada are ineligible.
Level of Study: Doctorate, Graduate, MBA
Type: Scholarship (grant-in-aid)
Value: US$10,000 maximum per year
Length of Study: A maximum of 2 years
Frequency: Annual
Country of Study: United States of America or Canada
No. of awards offered: Approx. 200
Application Procedure: Eligibility must be established before application material is sent. Eligibility information is available online at www.peointernational.org. Click on 'P.E.O Project/Philanthropies' and scroll down to the International Peace Scholarship Fund. Please read the qualifications and restrictions. If you feel you qualify, you may access the Eligibility Form after August 1st at this same site. The completed Eligibility Form must be submitted between August 15th and December 15th. If the applicant is deemed eligible, the application material will be sent.
Funding: Private
Contributor: PEO members
No. of awards given last year: 200

No. of applicants last year: 350
Additional Information: Scholarships cannot be used for travel, research dissertations, internships or practical training. Applicants must also have round-trip return travel expense guaranteed at the time of the application and promise to return to their own country within 60 days of completion of their studies, depending on visa status, unless approved for optional practical training (OPT). Each applicant has 60 days from the date her eligibility form is approved to submit her completed application material. The material includes the Application Form, Personal Statement (Progress report for renewals), Educational/Employment Resume, Recommendations and Return Travel Guarantee Form.

INTERNATIONAL READING ASSOCIATION

800 Barksdale Road, PO Box 8139, Newark, DE, 19714 8139, United States of America
Tel: (1) 302 731 1600 ext 423
Fax: (1) 302 731 1057
Email: research@reading.org
Website: www.reading.org

The International Reading Association seeks to promote high levels of literacy for all by improving the quality of reading instruction through studying the reading processes and teaching techniques, serving as a clearing house for the dissemination of reading research through conferences, journals and other publications and actively encouraging the lifetime reading habit.

Albert J Harris Award

Subjects: Reading and literacy.
Purpose: To recognize outstanding published works focused on the identification, prevention, assessment, or instruction of learners experiencing difficulty learning to read or write.
Eligibility: Open to all literacy professionals.
Level of Study: Research, Postgraduate
Type: Award
Value: US$800
Frequency: Annual
Country of Study: Any country
No. of awards offered: 1
Application Procedure: Applicants must obtain guidelines with specific information from the main address or by visiting the website.
Closing Date: September 1st
Funding: Private
Additional Information: For additional information, contact research@reading.org. Please see the website for further details http://www.reading.org/resources/AwardsandGrants/albert_j_harris_award.aspx.

Dina Feitelson Research Award

Subjects: Literacy.
Purpose: To recognize an outstanding empirical study that was published in English in a refereed journal that specifically reports on an investigation of aspects of literary acquisition such as phonemic awareness, the alphabetic principle, bilingualism, home influences on literacy development or cross-cultural studies of beginning reading.
Eligibility: Articles must have been published in a refereed journal within the past 18 months and may be submitted by the author or anyone else. Empirical studies involve the collection of original data from direct experimentation or observation, and articles that develop theory without data, secondary reviews of the literature or descriptions of the theory are not eligible for this competition. Nominees for this award do not need to be members of the International Reading Association.
Level of Study: Research, Unrestricted
Type: Award
Value: US$500
Frequency: Dependent on funds available
Country of Study: Any country
No. of awards offered: 1
Application Procedure: Applicants must obtain guidelines with specific information from the main address or by visiting the website.
Closing Date: September 1st
Funding: Private
Additional Information: For additional information, contact research@reading.org. Please see the website for further details http://www.reading.org/resources/AwardsandGrants/research_feitelson.aspx.

Elva Knight Research Grant

Subjects: Literacy and reading.
Purpose: To assist a researcher in a reading and literacy project that addresses significant questions about literacy instruction and practice.
Eligibility: Applicants must be members of the International Reading Association.
Level of Study: Postgraduate, Research
Value: Upto US$8,000
Length of Study: 2 years
Frequency: Annual
Country of Study: Any country
No. of awards offered: Up to 2 awards
Application Procedure: The applicants can complete and submit their applications online using a web-based grant management system. Applicants may apply for more than one research grant; however, you are eligible to win one research grant per award year.
Closing Date: November 5th
Funding: Private
Additional Information: Please see the website for further details http://www.reading.org/resources/AwardsandGrants/research_knight.aspx.

Helen M Robinson Grant

Subjects: Literacy education.
Purpose: To support doctoral students at the early stages of their dissertation research in the area reading and literacy.
Eligibility: Open to all doctoral students at the early stages of their dissertation research worldwide who are members of the International Reading Association.
Level of Study: Research, Doctorate
Type: Research award
Value: US$1,200
Frequency: Annual
Country of Study: Any country
No. of awards offered: 1
Application Procedure: Submit your application through the web-based grant management system.
Closing Date: November 5th
Funding: Private
Additional Information: For additional information, contact research@reading.org. Please see the website for further details http://www.reading.org/resources/AwardsandGrants/research_robinson.aspx.

International Reading Association Outstanding Dissertation of the Year Award

Subjects: Reading and literacy.
Purpose: To recognize dissertations in the field of reading and literacy.
Eligibility: Open to all doctoral students worldwide who are members of the International Reading Association and who have completed dissertations in any aspect of the field of reading or literacy between May 15 and May 14 of the previous year in which the award is to be given.
Level of Study: Doctorate
Type: Research award
Value: US$1,000
Frequency: Annual
Country of Study: Any country
No. of awards offered: 1
Application Procedure: Applicants must obtain guidelines with specific information from the main address or by visiting the website.
Closing Date: October 1st
Funding: Private
Additional Information: For additional information, contact research@reading.org. Please see the website for further details http://www.reading.org/resources/AwardsandGrants/research_outstanding.aspx.

International Reading Association Teacher as Researcher Grant

Subjects: Literacy.
Purpose: To support classroom teachers in their enquiries about literacy learning and instruction.
Eligibility: All applicants must be members of the International Reading Association and practicing pre-K-12 teachers with full-time teaching responsibilities, including librarians, classroom teachers and resource teachers. Applicants are limited to one proposal per year. There must be a span of 3 years before past grant recipients can apply for another Teacher as Researcher Grant.
Level of Study: Research
Type: Research award
Value: Up to US$4,000 maximum, but priority is given to smaller grants of between US$1,000 and 2,000
Frequency: Annual
Country of Study: Any country
No. of awards offered: Several
Application Procedure: Applicants must obtain guidelines with specific information from the main address or by visiting the website. Submit your application through the web-based grant management system.
Closing Date: November 5th
Funding: Private
Additional Information: For additional information, contact research@reading.org. Please see the website for further details http://www.reading.org/resources/AwardsandGrants/research_teacher_as_researcher.aspx.

Jeanne S Chall Research Fellowship

Subjects: Reading and literacy.
Purpose: To encourage and support doctoral research investigating issues in beginning research, readability, reading difficulty and stages of reading development.
Eligibility: Open to doctoral students who are members of the International Reading Association and are planning or beginning dissertations.
Level of Study: Doctorate, Research
Type: Fellowship
Value: US$6,000
Frequency: Annual
Country of Study: Any country
No. of awards offered: 1
Application Procedure: Applicants must obtain guidelines with specific information from the main address or by visiting the website. Submit your application through the web-based grant management system.
Closing Date: November 5th
Funding: Private
No. of awards given last year: 1
Additional Information: For additional information, contact research@reading.org. Please see the website for further details http://www.reading.org/resources/AwardsandGrants/research_chall.aspx.

Steven A Stahl Research Grant

Subjects: Literacy education
Purpose: To encourage and support promising graduate students in their research.
Eligibility: At least three years of teaching experience who is conducting classroom research (including action research) focused on improving reading instruction and children's reading achievement.
Level of Study: Graduate, Research
Type: Grant
Value: US$1,000
Frequency: Annual
Country of Study: Any country
Application Procedure: Submit your application through the web-based grant management system.
Closing Date: November 5th
Funding: Private
Additional Information: For additional information, contact research@reading.org. Please see the website for further details http://www.reading.org/resources/AwardsandGrants/research_stahl.aspx.

INTERNATIONAL RESEARCH AND EXCHANGE BOARD (IREX)

2121 K Street North West, Suite 700, Washington, DC, 20037, United States of America
Tel: (1) 202 628 8188
Fax: (1) 202 628 8189
Email: irex@irex.org
Website: www.irex.org

The International Research and Exchange Board (IREX) is an international non-profit organization specializing in education, independent media, internet development and civil society programmes. Through training, partnerships, education, research and grant programmes, IREX develops the capacity of individuals to contribute to their societies. Since its founding in 1968, IREX has supported over 15,000 students, scholars, policymakers, business leaders, journalists and other professionals.

IREX Individual Advanced Research Opportunities

Subjects: Policy-relevant research in the social sciences and humanities.
Purpose: To provide opportunities for scholars from the United States of America wishing to pursue research in the humanities and social sciences in Europe and Eurasia.
Eligibility: Must be US citizen, enrolled in a graduate degree program or currently be holding a graduate degree at the time of application and must be from eligible Countries of Research.
Level of Study: Research, Doctorate, Graduate, MBA, Postdoctorate, Postgraduate, Predoctorate, Professional development
Type: Grant
Value: Up to a maximum of US$40,000. Covers travel and visa fees, dollar stipend and a housing allowance
Length of Study: Up to 9 months
Frequency: Annual
Study Establishment: Appropriate institutions
No. of awards offered: Varies
Application Procedure: Applicants must visit the website for application forms and further information.
Closing Date: November 14th
Funding: Government, private
Contributor: The United States of America Department of State (Title VIII) and the IREX Scholar Support Fund
No. of awards given last year: 25
Additional Information: Applicants must study in one of the following countries – Albania, Armenia, Azerbaijan, Belarus, Bosnia and Herzegovina, Bulgaria, Croatia, Czech Republic, Estonia, Georgia, Hungary, Kazakhstan, Kosovo, Kyrgyzstan, Latvia, Lithuania, Macedonia, Moldova, Montenegro, Poland, Romania, Russia, Serbia, Slovakia, Slovenia, Tajikistan, Turkmenistan, Ukraine and Uzbekistan. Please see the website for further details http://www.irex.org/application/individual-advanced-research-opportunities-iaro.

IREX Short-Term Travel Grants

Subjects: Policy-relevant research.
Purpose: To provide opportunities for scholars from the United States of America to pursue research in the social sciences in Europe and Eurasia.
Eligibility: STG applicants must have a graduate degree (PhD, MA, MD, MBA, MFA, MPA, MPH, MLIS, MS, JD) at the time of application and must be a U.S. citizens. STG applicants may not be pursuing a degree at the time of application.
Level of Study: Postdoctorate, Postgraduate
Type: Travel grant
Value: International coach class roundtrip transportation from the US to the host country(ies) for the period of grant awarded; a monthly allowance for housing and living expenses, based on IREX's pre-established country-specific rates; travel visas; emergency evacuation insurance
Length of Study: Up to 8 weeks
Frequency: Annual
Country of Study: Eastern Europe and Eurasia
No. of awards offered: Varies

Application Procedure: Candidates must contact Amy Schulz, Program Officer, at stg@irex.org for details or application guidelines. Application forms can be downloaded from the website.
Closing Date: February 6th
Funding: Government
Contributor: The United States of America Department of State's Title VIII Program
No. of awards given last year: Approx. 40
No. of applicants last year: Varies
Additional Information: Candidates will be notified of award decisions approx. 8 weeks after the application deadline. Candidates must study in one of the following countries – Albania, Armenia, Azerbaijan, Belarus, Bosnia and Herzegovina, Bulgaria, Croatia, Czech Republic, Estonia, Georgia, Hungary, Kazakhstan, Kosovo, Kyrgyzstan, Latvia, Lithuania, Macedonia, Moldova, Montenegro, Poland, Romania, Russia, Serbia, Slovakia, Slovenia, Tajikistan, Turkmenistan, Ukraine and Uzbekistan.

INTERNATIONAL TROPICAL TIMBER ORGANIZATION (ITTO)

International Organizations Center, 5th floor, Pacifico-Yokohama 1-1-1, Minato-Mirai, Nishi-ku, Yokohama, 220 0012, Japan
Tel: (81) 45 223 1110
Fax: (81) 45 223 1111
Email: itto@itto.or.jp
Website: www.itto.or.jp

ITTO is an intergovernmental organization promoting the conservation and sustainable management, use and the trade of tropical forest resources.

International Tropical Timber Organization (ITTO) Fellowship Programme

Subjects: Forestry management including forest industry development and trade in forest products and services
Purpose: To promote human resource development and to strengthen professional expertise in tropical forestry.
Eligibility: Only nationals of ITTO member countries are eligible to apply, and fellowships are awarded mainly to nationals of developing member countries.
Level of Study: Professional development, Doctorate, Postgraduate, Research
Type: Fellowship
Value: Up to US$10,000
Frequency: Twice a year
Study Establishment: Varies
No. of awards offered: 50–60 per year
Application Procedure: Application form must be completed and sent with required documebts to the ITTO Secretariat by post.
Closing Date: March 5th
Funding: Government
Contributor: International Tropical Timber Organization
No. of awards given last year: 45
No. of applicants last year: 245
Additional Information: Eligibal activities include participation in international/regional conferences, short term training courses, training internships at industries, research and educational institutions, study tours and lecture/demonstratin tours; Small grants for post graduate studies. Please see the website for more details http://www.itto.int/feature20/.

INTERNATIONAL UNION FOR VACUUM SCIENCE AND TECHNOLOGY (IUVSTA)

84 Oldfield Drive, Vicars Cross, Chester, CH3 5LW, United Kingdom
Tel: (44) 1244 34 2675, 771 34 03525
Fax: (44) 7005 860135
Email: iuvsta.secretary.general@ronreid.me.uk
Website: www.iuvsta.org
Contact: Dr R J Reid, Secretary General

The International Union for Vacuum Science and Technology (IUVSTA) is a non-government organization whose member societies represent all vacuum scientists, engineers and technologists in their country.

Welch Scholarship

Subjects: Vacuum science.
Purpose: To encourage promising scholars who wish to study vacuum science, techniques or their application in any field.
Eligibility: Open to applicants of any nationality who hold the minimum of a Bachelor's degree, although preference is given to those holding a doctoral degree.
Level of Study: Doctorate, Postdoctorate, Postgraduate
Type: Scholarship
Value: US$15,000. The scholarship money is paid in three installments – one of $7,500 at the beginning, another of $7,000 six months after he/she has started work, and a third of $500 upon delivery of a final report after completion of work.
Length of Study: 1 year
Frequency: Annual
Study Establishment: An appropriate laboratory
Country of Study: Any country
No. of awards offered: 1
Application Procedure: Applicants must complete and submit an application form with a research proposal, a curriculum vitae and two letters of reference. More information and application forms can be obtained from the website.
Closing Date: April 15th
Funding: Private
Contributor: IUVSTA
No. of awards given last year: 1
No. of applicants last year: 6
Additional Information: Researchers who applied unsuccessfully for previous Welch Scholarships may apply again. Applications for renewal of the Scholarship are not accepted.

For further information contact:

Canadian Photorics Fabrication Centre, Institute for Microstructural Sciences, National Research Council, Building M-50, Montréal Road, Ottawa, ON, K1A 0R6, Canada
Email: Frank.Shepherd@nrc-cnrc.gc.ca
Website: http://iuvsta-us.org/iuvsta2/index.php?id=654
Contact: Dr FR Shepherd, Administrator Technical Manager

INTERNATIONAL UNION OF BIOCHEMISTRY AND MOLECULAR BIOLOGY (IUBMB)

University of Calgary, Department of Biochemistry & Molecular Biology, 3330 Hospital Drive NW, HM G72B, Calgary, AB T2N 4N1, Canada
Tel: (1) 403 220 3021
Fax: (1) 403 270 2211
Email: walsh@ucalgary.ca
Website: www.iubmb.org
Contact: Professor Michael P Walsh, IUBMB General Secretary

IUBMB seeks to advance the international molecular life sciences community by: Promoting interactions across the diversity of endeavours in the molecular life sciences, creating networks that transcend barriers of ethnicity, culture, gender, and economic status, creating pathways for young scientists to fulfil their potential, providing evidence-based advice on public policy, promoting the values, standards, and ethics of science and the free and unhampered movement of scientists of all nations.

Wood-Whelan Research Fellowships

Subjects: Biochemistry and molecular biology.
Purpose: To provide financial assistance to young biochemists and molecular biologists to carry out research and training in a laboratory other than their own.
Eligibility: Open to applicants who are residents of countries that are members of IUBMB and students or young researchers less than 35 years old. Retroactive applications will not be considered.
Level of Study: Graduate, Postdoctorate, Postgraduate, Research

Type: Fellowship
Value: Up to US$4,000. It covers travel and incidental costs, as well as living expenses
Length of Study: 1–4 months
Frequency: Annual
No. of awards offered: 10
Application Procedure: Applicants must submit a completed application form along with details of the research proposal, budget, curriculum vitae with a list of publications and letters of recommendation following the guidelines which can be found at www.iubmb.org. The original application should be sent by the applicant by email as PDF files.
Closing Date: At least two months before the proposed visit.
Contributor: The main sources of income for IUBMB are dues from adhering bodies (member societies) and revenue from publications
No. of awards given last year: 11
No. of applicants last year: 18
Additional Information: Travel should commence within 4 months of the award being made.

For further information contact:

Email: iqbal.parker@uct.ac.za
Contact: Professor M. Iqbal Parker

INTERNATIONAL UNIVERSITY OF JAPAN

777, Kokusai-cho, Minami Uonuma-shi, Niigata, 949-7277, Japan
Tel: (81) 25 779 1104
Fax: (81) 25 779 1180
Email: info@iuj.ac.jp
Website: www.iuj.ac.jp
Contact: Gretchen W Shinoda, Manager, Office of Student Services

The mission of the International University of Japan (IUJ) is to train leaders who can make contributions to the practical resolution of global problems facing people living in various countries and regions in the world, and who work in governments, companies, and NGOs to extend public and social benefits globally.

Sohei Nakayama Memorial Scholarship (Type A, B, C, D and S)

Subjects: MBA (1-year and 2-year tracks), MA in international relations, MA in international development, MA in international peace studies, MA in public management, MA in economics, Master in E-business management and MA in economics.
Purpose: To support those who study in International University of Japan (IUJ) in one of our master's programs (MA and MBA).
Eligibility: Open to all applicants who want to study in IUJ.
Level of Study: Graduate, MBA, Postgraduate
Type: Scholarship
Value: Tuition exemption (all or partial), stipends (depending on scholarship rank)
Length of Study: Up to 2 years
Frequency: Annual
Study Establishment: 1982
Country of Study: Japan
No. of awards offered: Varies
Application Procedure: When applicants apply to one of our seven Master's level degree programmes. Applicants also submit an online scholarship application available from the website http://www.iuj.ac.jp/admis/.
Closing Date: Contact the admissions office: same as application deadlines to degree programs
Funding: Foundation, government, individuals, corporation
Contributor: The Ministry of Education, youth and sport, IUJ, ADB, JICA, IMF, KMMF, WB, etc.
No. of awards given last year: Varies
No. of applicants last year: Varies
Additional Information: Many other scholarships are facilitated by IUJ for studying at IUJ. Please see our website at http://www.iuj.ac.jp/admis/scholarship.

INTERNATIONAL VIOLIN COMPETITION–PREMIO PAGANINI

Secretariat, Fondazione Teatro Carlo Felice, Passo Eugenio Montale 4, Genova, 16121, Italy
Tel: (39) 0105381 1
Fax: (39) 0105381 363
Email: paganini@carlofelice.it
Website: www.carlofelice.it
Contact: Ms Bianca Fusco, Paganini Competition Secretariat

The international violin competition "Premio Paganini" is a competition for young violinists between 16 and 30 years of age. It offers prizes for a total amount of €40,000 plus special prizes and an opportunity for the winner to play "Cannone" - Paganini's violin - at the Carlo Felice Theatre during the artistic season.

International Violin Competition–Premio Paganini

Subjects: Violin.
Purpose: To discover new talented young violinists and encourage them to spread the values which Paganini himself and his music stands for.
Eligibility: Violinists between 16 and 30 years of age
Level of Study: Unrestricted
Type: Prize
Value: 1st prize "Premio Paganini" €25,000.00 (indivisible), 2nd prize € 10,000.00 (indivisible), 3rd prize € 5,000.00 and the other finalists will receive a sum of € 1,500.00 each.
Frequency: Every 2 years
Country of Study: Italy
No. of awards offered: 3 awards and special prizes
Application Procedure: Applicants must send the application form by mail together with a CD and the documents required to the address indicated below. The application form and the rules of the competition may be obtained from obtained by writing to the Competition Secretariat or they can be downloaded at: www.carlofelice.it. A pre-selection will be made to enter the competition.
Closing Date: January 1st
Funding: Government
Contributor: Comune di Genova
No. of awards given last year: 2 plus prizes
No. of applicants last year: 51
Additional Information: As a part of an important cultural Paganinian project, the competition has been deferred. Please, contact paganini@carlofelice.it to get informed. Please see the website for further details http://www.wfimc.org/Webnodes/en/Web/Public/Competitions/Competition+info?org=16612.

INTERNATIONALER ROBERT-SCHUMANN-WETTBEWERB ZWICKAU

Stadtverwaltung Zwickau, PF 200933, 08009 Zwickau, Kulturamt, Germany
Tel: (49) (0) 375 83 4130
Fax: (49) (0) 375 83 4141
Email: kulturbuero@zwickau.de
Website: www.schumann-zwickau.de

International Robert Schumann Competition

Subjects: Piano performance and individual singing.
Purpose: To support the interpretation of the works of Robert Schumann.
Eligibility: Open to pianists up to the age of 30 and to individual singers up to the age of 32.
Level of Study: Professional development, 4 years
Type: Competition
Value: Piano - 3 prizes with a total amount of €22,500, Singers (female) - 3 prizes with a total amount of €22,500, Singers (male) - 3 prizes with a total amount of €22,500, Special prize of €3,000 will be awarded to the best lied pianist
Frequency: Every 4 years
Country of Study: Any country
No. of awards offered: 10
Application Procedure: Applicants must write for further details at e-mail kulturbuero@zwickau.de.

Closing Date: February 15th
Funding: Commercial, government
No. of awards given last year: 10
No. of applicants last year: 108
Additional Information: Please see the website for further details http://www.schumannzwickau.de/PDF/RSW-2012-Ausschreibung_en.pdf.

IOTA SIGMA PI

Microelectronics Technology, Lord Corporation, 110 Lord Drive, Cary, NC, 27511, United States of America
Tel: (1) 919 468 5979
Email: sara.paisner@lord.com
Website: www.iotasigmapi.info
Contact: Sara Paisner, Senior Scientist

Iota Sigma Pi, founded in 1902, is a National Honor Society that serves to promote the advancement of women in chemistry by granting recognition to women who have demonstrated superior scholastic achievement and high professional competence by election into Iota Sigma Pi.

Agnes Fay Morgan Research Award

Subjects: Chemistry and biochemistry.
Purpose: To acknowledge research achievements in chemistry or biochemistry.
Eligibility: Open to female applicants who are not more than 40 years of age.
Level of Study: Postgraduate
Type: Award
Value: The Award will consist of $500, a certificate, and membership in Iota Sigma Pi with a waiver of dues for 1 year
Frequency: Annual
Study Establishment: Any accredited institution
Country of Study: Any country
No. of awards offered: 1
Application Procedure: The nomination dossier must be sent electronically (preferably as a pdf) to Dr. Nancy Eddy Hopkins.
Closing Date: February 15th
Contributor: Iota Sigma Pi
Additional Information: Please see the website for further details.

For further information contact:

Tel: 504 862 3162
Email: nhopkin@tulane.edu
Website: http://www.iotasigmapi.info/ISPprofawards/ISPprofawards.html.
Contact: Dr Nancy Eddy Hopkins, Director for Professional Awards

Anna Louise Hoffman Award for Outstanding Achievement in Graduate Research

Subjects: Chemistry.
Purpose: To recognize outstanding achievement in chemical research.
Eligibility: The candidate must be a full-time (as defined by the nominee's institution) woman graduate student who is a candidate for a graduate degree in an accredited institution. The research presented by the candidate must be original research which can be described by one of the main chemical divisions (e.g., analytical, biochemical, inorganic, organic, physical, and/or ancillary divisions of chemistry). The nominee may be, but need not be, a member of Iota Sigma Pi.
Level of Study: Postgraduate
Type: Award
Value: The award will be $500, a certificate and a waiver of dues for one year
Frequency: Annual
Study Establishment: Any accredited institution
Country of Study: Any country
Application Procedure: The complete dossier must be sent electronically as a single file (pdf format is recommended) to Professor Jill Nelson Granger.
Closing Date: February 15th
Contributor: Iota Sigma Pi
Additional Information: Please see the website for further details.

For further information contact:

Sweet Briar College, Department of Chemistry, Sweet Briar, VA, 24595, United States of America
Tel: (1) 434 381 6166
Email: granger@sbc.edu
Website: http://www.iotasigmapi.info/ISPstudentawards/ISPstudentawards.htm.
Contact: Professor Jill Nelson Granger, Director for Student Awards

Gladys Anderson Emerson Scholarship

Subjects: Chemistry and biochemistry.
Purpose: To award excellence in chemistry or biochemistry.
Eligibility: Open to applicants who are members of Iota Sigma Pi.
Level of Study: Postgraduate
Type: Scholarship
Value: US$2,000 and a certificate
Frequency: Annual
Study Establishment: Any accredited institution
Country of Study: Any country
Application Procedure: The complete dossier must be sent electronically as a single file (pdf format is recommended) to Professor Jill Nelson Granger.
Closing Date: February 15th
Contributor: Iota Sigma Pi
Additional Information: Please see the website for further details.

For further information contact:

Sweet Briar College, Department of Chemistry, Sweet Briar, VA, 24595
Tel: 434 381 6166
Email: granger@sbc.edu
Website: http://www.iotasigmapi.info/ISPstudentawards/ISPstudentawards.htm
Contact: Professor Jill Nelson Granger, National Director for Student Awards

Iota Sigma Pi Centennial Award

Subjects: Chemistry, biochemistry.
Purpose: To award excellence in teaching chemistry, biochemistry or chemistry-related subjects.
Eligibility: Holds a teaching position at an institution that does not have a graduate program in her department or holds a teaching position that is for teaching undergraduates > 75% of her time at an institution that does have a graduate program in her department. The nominee may be, but need not be, a member of Iota Sigma Pi.
Level of Study: Postgraduate
Type: Award
Value: US$500, a certificate and membership in Iota Sigma Pi with a waiver of dues for 1 year
Frequency: Annual
Application Procedure: One copy of the nomination dossier must be sent electronically (preferably as a pdf) to Dr. Nancy Eddy Hopkins.
Closing Date: Feburary 15th
Contributor: Iota Sigma Pi
Additional Information: Please see the website for details.

For further information contact:

Tel: 504 862 3162
Email: nhopkin@tulane.edu
Website: http://www.iotasigmapi.info/ISPprofawards/ISPprofawards.html
Contact: Dr Nancy Eddy Hopkins, Director for Professional Awards

Iota Sigma Pi National Honorary Member Award

Subjects: Chemistry.
Purpose: To honour outstanding women chemists.
Eligibility: Open to female candidate with exceptional achievements in chemistry. Applicants may or may not be members of Iota Sigma Pi.
Type: Award
Value: US$1,500 a certificate and membership in Iota Sigma Pi with a lifetime waiver of dues
Frequency: Every 3 years
No. of awards offered: Varies

Application Procedure: One copy of the nomination dossier must be sent electronically (preferably as a pdf) to Nancy Eddy Hopkins.
Closing Date: February 15th
Additional Information: Please see the website for details.

For further information contact:

Email: nhopkin@tulane.edu
Website: http://www.iotasigmapi.info/ISPprofawards/ISPprofawards.html
Contact: Dr Nancy Eddy Hopkins, Director for Professional Awards

Members-at-Large (MAL) Reentry Award

Subjects: Chemistry.
Purpose: To recognize potential excellence in chemistry and related fields.
Eligibility: Open to a woman candidate with a degree at any level in chemistry or a related field at an accredited four-year college or university.
Level of Study: Postgraduate
Type: Award
Value: US$1,500, a certificate and a year's complimentary membership in Iota Sigma Pi
Frequency: Annual
Study Establishment: Any accredited institution
Country of Study: Any country
No. of awards offered: Varies
Application Procedure: A completed application form must be sent.
Closing Date: February 15th
Contributor: Iota Sigma Pi
Additional Information: Please see the website for details.

For further information contact:

Science - Chemistry, Richland Northeast High School, 7500 Brookfield Road, Columbia, SC, 29223
Website: http://www.ispmembersatlarge.com/members-at-large-reentry-award.html
Contact: Joanne Bedlek-Anslow, MAL National Coordinator

Violet Diller Professional Excellence Award

Subjects: Chemistry.
Purpose: To recognize significant accomplishments in academic, governmental or industrial chemistry.
Eligibility: Open to female applicants who have contributed to the scientific community or society on a national level.
Level of Study: Postgraduate
Type: Award
Value: US$1,000 a certificate and membership in Iota sigma Pi with a lifetime of dues
Frequency: Every 3 years
Application Procedure: One copy of the nomination dossier must be sent electronically (preferably as a pdf) to to Nancy Eddy Hopkins.
Closing Date: February 15th
Contributor: Iota Sigma Pi
Additional Information: Please see the website for further details.

For further information contact:

Email: nhopkin@tulane.edu
Website: http://www.iotasigmapi.info/ISPprofawards/ISPprofawards.html

IRISH RESEARCH COUNCIL FOR THE HUMANITIES AND SOCIAL SCIENCES (IRCHSS)

First Floor, Brooklawn House, Crompton Avenue (off Shelbourne Road), Ballsbridge, Dublin, 4, Ireland
Tel: (353) 0 1 231 5000
Fax: (353) 0 1 231 5009
Email: info@irchss.ie
Website: www.irchss.ie

Irish Research Council for the Humanities and Social Sciences (IRCHSS) was established in 2000, by the Minister for Education and Science in response to the need to develop Ireland's research capacity and skills base in a rapidly changing global environment where knowledge is the key to economic and social growth.

IRCHSS Postdoctoral Fellowship

Subjects: Humanities and social sciences.
Purpose: To encourage excellence and the highest standards in the humanities and social sciences.
Eligibility: Open to candidates of any nationality who have been awarded their Doctoral degrees within the past 5 years.
Level of Study: Postdoctorate
Type: Fellowship
Value: €31,745 per year
Length of Study: 1–2 years
Frequency: Annual
Closing Date: February 17th
Additional Information: Applicants should be, or expect to be affiliated with a recognised higher education institution during the academic year. Please see the website for further details http://test.irchss.ie/event/2012-02-17/postdoctoral-fellowship-2012-13.

IRCHSS Postgraduate Scholarship

Subjects: Humanities and social sciences.
Purpose: To facilitate the integration of Irish researchers in the humanities and social sciences within the European Research Area.
Eligibility: Open to citizens of Ireland or citizens of a Member State of the European Union who have been residing within Ireland for less than 3 years. The candidates should be registered as full-time postgraduate research students.
Level of Study: Postgraduate
Type: Scholarship
Value: €16,000 per year
Length of Study: 1 year
Frequency: Annual
Application Procedure: Applicants can download the application form from the website.
Closing Date: March 20th
Additional Information: Please see the website for further details http://test.irchss.ie/funding/postgraduate-funding.

IRISH-AMERICAN CULTURAL INSTITUTE (IACI)

1 Lackawanna Place # 1, Morristown, NJ, 07960, United States of America
Tel: (1) 973 605 1991
Fax: (1) 973 605 8875
Email: info@iaci-usa.org
Website: www.iaci-usa.org

The Irish-American Cultural Institute (IACI), a non-profit educational institute, is dedicated to preserving and promoting the highest standards of artistic development, education, research and entertainment in fostering the cultural understanding of Irish heritage in America. With international headquarters in Morristown, NJ, the Institute has a long history of supporting the arts and humanities through grants and awards as well as through programming. The Institute is strictly non-political and non-sectarian. Founded in 1962, the IACI is the sole United States of America organization with the distinction of having the President of Ireland as patron.

IACI Visiting Fellowship in Irish Studies at the National University of Ireland, Galway

Subjects: Irish studies.
Purpose: To allow scholars whose work relates to any aspect of Irish studies who wish to spend a semester at the University of Ireland, Galway.
Eligibility: Open to scholars who normally reside in the United States of America, and whose work relates to any aspect of Irish studies.
Level of Study: Postdoctorate, Research
Type: Fellowship
Value: US$4,000
Length of Study: At least 4 months
Frequency: Annual
Country of Study: Ireland

No. of awards offered: 1
Application Procedure: Applicants must complete an application form and submit this with a current curriculum vitae and list of publications. Application forms are available on request. Applications can be downloaded from the IACI website.
Closing Date: December 31st
Funding: Foundation
Contributor: Jointly funded with the National University International-Galway
Additional Information: The holder of the fellowship will be provided with services appropriate to a visiting faculty member during his or her time at NUI-Galway. There are certain relatively minor departmental responsibilities expected of the holder during his or her time at UCG, and certain other expectations regarding publication, upon completion of the fellowship. Please see the website for details http://www.iaci-usa.org/pdf/NUI-G%20Application%20Letter.pdf.

Irish Research Funds

Subjects: All subjects; historical research has predominated, but other areas of research will be given equal consideration.
Purpose: To promote scholarly enquiry and publication regarding the Irish–American experience.
Eligibility: Open to individuals of any nationality. Media production costs and journal subventions will not be considered for funding.
Level of Study: Postgraduate
Type: Grant
Value: US$1,000–5,000
Country of Study: Any country
No. of awards offered: Varies
Application Procedure: Applicants must complete an application form.
Closing Date: October 1st
Funding: Foundation

THE ISLAMIC DEVELOPMENT BANK (IDB)

PO Box 5925, Kingdom of Saudi Arabia, Jeddah, 21432, Saudi Arabia
Tel: (966) 2 6361400
Fax: (966) 2 6366871
Email: idbarchives@isdb.org
Website: www.isdb.org

The Islamic Development Bank is an international financial institution, which aims to foster the economic development and social progress of member countries and Muslim countries.

IDB Merit Scholarship for High Technology

Subjects: Science and technology.
Purpose: To encourage advanced studies/research in science and high technology areas.
Eligibility: Applicant must not be over (a) 35 years for PhD study, and (b) 40 years for Post-Doctoral research
Level of Study: Postgraduate
Type: Scholarship
Value: Tuition/bench fee, monthly living allowance, monthly family allowance (for PhD study only), clothing/books allowance (for PhD study only), installation allowance (for PhD study only), computer allowance (for PhD study only), conference/thesis preparation allowance and medical coverage. Round trip air-tickets to and from place of study.
Frequency: Annual
No. of awards offered: Varies
Application Procedure: Download the application form for 3-Year PhD Study or application form for Post-Doctoral Research. The applications must be submitted through the nominating institutions to the Office of the IDB Governor for the country. It is the Office of the IDB Governor that will forward the applications to the IDB and not the individual applicants or the nominating institutions. As for Egypt, applications should be forwarded through the Ministry of Higher Education, Egypt instead of Office of the IDB Governor. Applications sent directly to the IDB will not be considered.
Closing Date: December 31st
Additional Information: Please see the website for further details.

IDB Scholarship Programme in Science and Technology

Subjects: Science, technology, engineering and medicine.
Purpose: To assist IDB least developed member countries in the development of science and technology.
Eligibility: Open to candidates from least developed member countries, who are below the age of 30 years, a graduate in science/technology with a grade above Good in his/her academic career, nominated by an academic research institution of his/her country, not in receipt of another scholarship and medically fit and be willing to undergo medical tests after selection.
Level of Study: Postgraduate
Type: Scholarship
Value: The scholarship covers the tuition fees, living allowance, clothing and books allowances, computer allowance, conference allowance, medical coverage and a return air ticket.
Length of Study: 2 years
Frequency: Annual
Application Procedure: Application form can be downloaded. The applications must be submitted to the Office of the IDB Governor for the country. It is the Office of the IDB Governor that will forward the applications to the IDB and not the individual applicants or the nominating institutions. Applications sent directly to the IDB will not be considered.
Closing Date: December 31st
Additional Information: Please see the website for details.

IWHM BERNARD BUTLER TRUST FUND

37 Oasthouse Drive, Fleet, Hants, GU15 2UL, United Kingdom
Tel: (44) 1252 627748
Fax: (44) 1252 627748
Email: info@bernardbutlertrust.org
Website: www.bernardbutlertrust.org
Contact: G Porter, Trust Secretary

The Trust was established in 1998 from the assets of the Institution of Works and Highways Management after the merger of its professional activities with the Institution of Civil Engineers in 1994.

The Bernard Butler Trust Fund

Subjects: Civil and municipal engineering.
Purpose: To encourage men and women engaged in the civil and municipal engineering field to improve their education, training and professional standing together with aiding and promoting individuals/organizations to advance engineering training, safety and methods of working.
Eligibility: Those who can show practical and personal qualities needed to promote engineering in relation to civil and municipal engineering. See the website for more details.
Level of Study: Graduate, Postgraduate, Professional development, Research
Type: Grant
Value: UK£1,000 upwards depending on submission
Study Establishment: Varies
Country of Study: Worldwide
Application Procedure: Applicants can download an application form from the website or request form from the Trust Secretary. In addition to this an online application can also be made.
Closing Date: Check with website
Funding: Private
Contributor: Institution of Works and Highways Management
No. of awards given last year: 19
No. of applicants last year: 28
Additional Information: Please see the website for further details http://www.bernardbutlertrust.org/grants.php.

J N TATA ENDOWMENT

Mulla House, 4th Floor, 51, M.G. Road, Fort, Mumbai, 400001, India
Tel: (91) 022 6665 7643
Fax: (91) 022 2204 5432
Email: nbmody@sdtatatrust.com
Website: http://www.dorabjitatatrust.org/about/endowment.aspx
Contact: The Director

The J N Tata Endowment awards loan scholarships to scholars of conspicuous distinction for postgraduate, PhD or postdoctoral studies abroad in all fields. Mid-career professionals with an outstanding academic background and experience in the field, who are going abroad for further specialisation, are also considered for the scholarship.

J N Tata Endowment Loan Scholarship

Subjects: All subjects.
Purpose: To provide an opportunity to the gifted to pursue higher studies abroad in all disciplines and subjects.
Eligibility: Open only to Indian nationals. Applicants must be graduates of a recognised Indian university with a sound academic and extracurricular record. Deserving mid-career professionals are also eligible.
Level of Study: Postdoctorate, Professional development, Postgraduate, Doctorate, MBA, Research
Type: Loan scholarship
Value: Ranging between 60,000 and 4,00,000
Length of Study: Minimum 1 year; minimum 6 months for mid-career professionals
Frequency: Annual
Country of Study: Any country except India
No. of awards offered: 100+
Application Procedure: Applicants must complete an application form. Forms are issued against an application fee of Indian Rupees 100 only.
Closing Date: See website
Funding: Private, trusts
Contributor: Tata Trusts
No. of awards given last year: 138
No. of applicants last year: 869
Additional Information: Interviews are conducted between March and June for the Autumn semester and between October and December for the Spring semester. Eligible to nationals of India only. Contact J.N. Tata Endowment, Administrative Office, Mumbai or email at ashlesha.lotankar@sdtatatrust.com; sbjadhav@sdtatatrust.com. Please see the website for further details.

THE J.H. STEWART REID MEMORIAL FELLOWSHIP TRUST

Canadian Association of University Teachers, 2705 Queensview Drive, Ottawa, ON, K2B 8K2, Canada
Email: stewartreid@caut.ca
Website: http://stewartreid.caut.ca
Contact: Awards Officer

The J.H. Stewart Reid Memorial Fellowship Trust was founded to honour the memory of the 1st Executive Secretary of the Canadian Association of University Teachers.

J.H. Stewart Reid Memorial Fellowship Trust

Subjects: All subjects.
Purpose: To financially support students who wish to study further.
Eligibility: Open to Canadian citizens or landed immigrants, or those who have convention refugee status from April 30th of the year prior to application. Must be registered in doctoral programme and have first class academic record in the graduate programme and not hold scholarships that exceed in total $25,000 inclusive of the J. H. Stewart Reid Memorial Fellowship Trust.
Level of Study: Doctorate
Type: Fellowship
Value: Minimum Canadian $5,000
Length of Study: 1 year
Frequency: Annual
Study Establishment: Any Canadian university
Country of Study: Canada
No. of awards offered: 1
Application Procedure: See the website.
Closing Date: April 30th
Funding: Trusts
No. of awards given last year: 1
Additional Information: Applicants must be registered in a doctoral programme at a Canadian university and have completed their comprehensive examinations or equivalent and have had their doctoral theses proposal accepted by April 30th. The Fellowship shall be paid to the student or students in two equal installments in the months of September and January.

JACKI TUCKFIELD MEMORIAL GRADUATE BUSINESS SCHOLARSHIP FUND (JTMGBSF)

1160 NW 87th Street, Miami, FL, 33150 2544, United States of America
Tel: (1) 305 371 2711
Fax: (1) 305 371 5342
Email: saadya.rivera@dadecommunityfoundation.org
Website: www.jackituckfield.org
Contact: Scholarship Committee

On October 21, 1997, Drs Jack and Gloria Tuckfield established Jacki Tuckfield Memorial Graduate Business Scholarship Fund (JTMGBSF) at the non-profit, tax-exempt Dade Community Foundation to commemorate the vibrant life of their extraordinary daughter. Jacki Tuckfield Memorial Graduate Business Scholarship Fund (JTMGBSF) provides financial support to African-American residents of South Florida who are enrolled in Master's and Doctoral degree business programmes in Florida universities.The fund's mission is to improve the diversity of career professionals in the executive, administrative and managerial levels of south Florida's workforce by funding tuition scholarships.

Jacki Tuckfield Memorial Graduate Business Scholarship Fund

Subjects: Business, consumer services.
Purpose: To improve the diversity of career professionals employed in the executive, administrative and managerial levels of the South Florida workforce.
Eligibility: Open to full-time students. They must be African-American United States citizens of South Florida, enrolled in a graduate business programme at a Florida University.
Level of Study: Doctorate, Graduate, MBA, Postgraduate
Type: Scholarships
Value: Varies. For more details, see the organization website.
Frequency: Annual
Country of Study: United States of America
No. of awards offered: 20–30
Application Procedure: Applicants must submit their application form, transcript, interview, essay, reference letters, photograph and curriculum vitae. Application forms can be downloaded from the website www.jackituckfield.org or www.dadecommunityfoundation.org
Closing Date: January 3rd
Funding: Foundation
Contributor: JTMGBSF
No. of awards given last year: 30
Additional Information: JTMGBSF awards 256 graduate business tuition scholarships, totalling $261,000 to South Florida residents. This fund is administered by The Miami Foundation. Please see the website for more details http://www.jackituckfield.org/.

THE JACOB RADER MARCUS CENTER OF THE AMERICAN JEWISH ARCHIVES

3101 Clifton Avenue, Cincinnati, OH, 45220, United States of America
Tel: (1) 513 221 1875
Fax: (1) 513 221 7812
Email: kproffitt@huc.edu
Website: www.americanjewisharchives.org
Contact: Mr Kevin Proffitt, Director, Fellowship Programmes

The Marcus Center of the American Jewish Archives was founded by Dr Jacob Rader Marcus in 1947 in the aftermath of World War II and the Holocaust. It is committed to preserving a documentary heritage of the religious, organizational, economic, cultural, personal, social and family life of American Jewry. It contains nearly 5,000 linear feet of archives, manuscripts, newsprint materials, photographs, audio and video tapes, microfilm and genealogical materials.

The American Council for Judaism Fellowship

Purpose: The American Council for Judaism Fellowship supports scholarly historical research in a wide variety of areas that relate to the organisation's core interests: historical perspectives on the American Council for Judaism; the history of American Reform Judaism; the interrelationships and integration of Jewish ideals and a democratic society; the historical development of the concept of Americans of the Jewish faith and its coequal relationship with all faiths in America; and the historical study of Classical Reform Judaism. Above all, th ACJ Fellowship supports the work of scholars who are pursuing critical and scientific research that promises to deepen our understanding of the history and development of American Judaism and the American Jewish experience.
Level of Study: Doctorate, Postdoctorate, Postgraduate, Predoctorate
Type: Fellowship
Value: Award determined at the discretion of the selection committee
Country of Study: Any country
Additional Information: Please check website for further information.

Bernard and Audre Rapoport Fellowships

Subjects: American Jewish studies, preserving a documentary heritage of the religious, organizational, economic, cultural, personal, social and family life of American Jewry and imparting it to the next generation.
Eligibility: Open to postdoctoral candidates of any nationality.
Level of Study: Doctorate, Postdoctorate, Predoctorate, Postgraduate
Type: Fellowship
Value: Award is determined at the discretion of the selection committee
Length of Study: 1 month
Frequency: Annual
Study Establishment: The Archives
Country of Study: Any country
Application Procedure: Applicants can refer to the website for related information.
Closing Date: March 18th
Funding: Private

Bertha V. Corets Memorial Fellowship

Purpose: To allow students of the anti-Nazi movement, women's studies and related subjects to examine the Corets papers together with other related holdings in the collections of the AJA to learn not only about Bertha V. Corets, but study this important era in American history.
Level of Study: Doctorate, Postdoctorate, Postgraduate, Predoctorate
Type: Fellowship
Funding: Private
Additional Information: Please check website or contact organisation for further information.

Jacob Rader Marcus Center of the American Jewish Archives Fellowship Program

Subjects: American Jewish studies, including – but not limited to – the religious, organizational, economic, cultural, personal, social and family life of American Jewry.
Purpose: To provide fellowships for research and writing in some area of the American Jewish experience using the vast collection of the American Jewish Archives.
Eligibility: Applicants must submit a fellowship application together with a five-page (maximum) research proposal that outlines the scope of their project and lists those collections at the American Jewish Archives that are crucial to their research. Applicants should also submit two letters of support, preferably from academic colleagues. For graduate and doctoral students, one of these two letters must be from their dissertation advisor.
Level of Study: Doctorate, Postdoctorate, Postgraduate, Predoctorate, Professional development
Type: Fellowship
Value: Covers transportation and living expenses while in residence in Cincinnati.
Frequency: Annual
No. of awards offered: Varies
Application Procedure: Fellowship application can be downloaded from the website or a request can be made to have one sent via postal mail.
Closing Date: March 18th

For further information contact:

Tel: 513 221 7444, ext. 304
Contact: Kevin Proffitt, The Director of the Fellowship Program

The Joseph and Eva R. Dave Fellowship

Subjects: American Jewish studies, preserving a documentary heritage of the religious, organizational, economic, cultural, personal, social and family life of American Jewry and imparting it to the next generation.
Purpose: To facilitate research and writing using the vast collection at the American Jewish Archives, and to preserve a documentary heritage of the religious, organizational, economic, cultural, personal, social and family life of American Jewry and impart it to the next generation.
Eligibility: Open to ABDs.
Level of Study: Postdoctorate, Doctorate, Postgraduate, Predoctorate, Senior or independent scholars
Type: Fellowship
Value: Award is determined at the discretion of the selection committee
Length of Study: 1 month
Frequency: Annual
Study Establishment: The Archives
Country of Study: Any country
No. of awards offered: 1
Application Procedure: Applicants can refer to the website for related information.
Closing Date: March 18th in the year of proposed study
Funding: Private

Loewenstein-Wiener Fellowship Awards

Subjects: American Jewish studies, preserving a documentary heritage of the religious, organizational, economic, cultural, personal, social and family life of American Jewry and imparting it to the next generation.
Eligibility: Open to ABDs who have completed all but the dissertation requirement, and to postdoctoral candidates.
Level of Study: Postdoctorate, Predoctorate, Postgraduate, Doctorate
Type: Fellowship
Value: Award is determined at the discretion of the selection committee
Length of Study: 1 month
Study Establishment: The Archives
Country of Study: Any country
Application Procedure: Applicants can refer to the website for related information.
Closing Date: March 18th
Funding: Private

The Rabbi Harold D. Hahn Memorial Fellowship

Subjects: American Jewish studies, preserving a documentary heritage of the religious, organizational, economic, cultural, personal, social and family life of American Jewry and imparting it to the next generation.
Purpose: A perpetual scholarship created to enable scholars to conduct independent research in subject areas relating to the history of North American Jewry.
Eligibility: Open to ABDs.
Level of Study: Doctorate, Postdoctorate, Postgraduate, Predoctorate, Senior or independent scholars
Type: Fellowship
Value: Award is determined at the discretion of the Selection Committee
Length of Study: 1 month
Frequency: Annual
Study Establishment: The Archives
Country of Study: Any country
No. of awards offered: 1

Application Procedure: Applicants can refer to the website for related information.
Closing Date: March 18th in year of proposed study
Funding: Private

The Rabbi Joachim Prinz Memorial Fellowship

Subjects: American Jewish studies, preserving a documentary heritage of the religious, organizational, economic, cultural, personal, social and family life of American Jewry and imparting it to the next generation.
Purpose: To enable the recipient to conduct an extensive study of the Rabbi Joachin Prinz collection in preparation for a doctoral dissertation or other scholarly publication.
Eligibility: Open to ABDs.
Level of Study: Doctorate, Postdoctorate, Postgraduate, Predoctorate, Senior or independent scholars
Type: Fellowship
Value: Award is determined at the discretion of the selection committee
Length of Study: 1 month
Frequency: Annual
Study Establishment: The Archives
Country of Study: Any country
No. of awards offered: 1
Application Procedure: Applicants can refer to the website for related information.
Closing Date: March 18th in the year of the proposed study
Funding: Private
Contributor: Deutsche Bank American Foundation

Rabbi Levi A. Olan Memorial Fellowship

Subjects: American Jewish studies, preserving a documentary heritage of the religious, organizational, economic, cultural, personal, social and family life of American Jewry and imparting it to the next generation.
Eligibility: Open to ABDs.
Level of Study: Doctorate, Postdoctorate, Postgraduate, Predoctorate
Type: Fellowship
Value: Award is determined at the discretion of the selection committee
Length of Study: 1 month
Study Establishment: The Archives
Country of Study: Any country
Application Procedure: Applicants can refer to the website for related information.
Closing Date: March 18th

Rabbi Theodore S Levy Tribute Fellowship

Subjects: American Jewish studies, preserving a documentary heritage of the religious, organizational, economic, cultural, personal, social and family life of American Jewry and imparting it to the next generation.
Eligibility: Open to ABDs.
Level of Study: Doctorate, Postdoctorate, Postgraduate, Predoctorate, Senior or independent scholars
Type: Fellowship
Value: Award is determined at the discretion of the selection committee
Length of Study: 1 month
Frequency: Annual
Study Establishment: The Archives
Country of Study: Any country
No. of awards offered: 1
Application Procedure: Applicants can refer to the website for related information.
Closing Date: March 18th
Funding: Private

Starkoff Fellowship

Subjects: American Jewish studies, preserving a documentary heritage of the religious, organizational, economic, cultural, personal, social and family life of American Jewry and imparting it to the next generation.
Eligibility: Open to ABDs.
Level of Study: Postgraduate, Doctorate, Postdoctorate, Predoctorate
Type: Fellowship
Value: Award is determined at the discretion of the committee
Length of Study: 1 month
Frequency: Annual
Study Establishment: The Archives
Country of Study: United States of America
Application Procedure: Applicants can refer to the website for related information.
Closing Date: March 18th
Funding: Private

JACOB'S PILLOW DANCE

358 George Carter Road, Becket, MA, 01223, United States of America
Tel: (1) 413 243 9919
Fax: (1) 413 243 4744
Email: info@jacobspillow.org
Website: www.jacobspillow.org

Jacob's Pillow is America's premier dance festival. Founded in 1930s, the Pillow today is renowned not only for producing a premier festival, but also for its professional school, intern programme, archives of rare holdings, artist residencies and year-round community programmes.

Jacob's Pillow Intern Program

Subjects: The following fields related to Dance studies: business, development, education, general management, graphic design, marketing, photojournalism, technical theatre production, ticket services, video documentation, archives/preservation, operations.
Purpose: To offer on-the-job training and experience working alongside professional staff.
Eligibility: Open to all candidates without regard to age, gender or national or ethnic origin.
Level of Study: Professional development
Type: Internship
Value: US$500 stipend, free room and board, access to festival activities, and expense allowance
Length of Study: 3 months
Frequency: Annual
Country of Study: United States of America
No. of awards offered: 31
Closing Date: January 26th for priority consideration and February 26th for consideration
Funding: Government, corporation, foundation, individuals
No. of awards given last year: 32
No. of applicants last year: 142
Additional Information: Please see the website for details http://www.jacobspillow.org/education/internships/summer-internships/.

Jacob's Pillow Summer Festival Internship Program

Subjects: Archives/preservation, business, development/individuals, development/institutional, editorial/press, education, graphic design, operations, photojournalism, presenting, production, ticket services, house management and video documentation.
Purpose: To support those aspiring to professional careers in arts administration and technical theatre production. Offers on-the-job trainingand experience working alongside professional staff.
Eligibility: Open to all candidates without regard to age, gender, or national or ethnic origin.
Level of Study: Professional development
Type: Internship
Value: $500 stipend and $150 travel/sundry expense allowance
Length of Study: May to August
Frequency: Annual
Country of Study: United States of America
No. of awards offered: 31
Application Procedure: Visit the website www.jacobspillow.org/home/summer-internships.asp for application procedures and more information.

Closing Date: Januray 22nd for priority consideration; February 19th for consideration
Funding: Individuals, corporation, foundation, government
No. of awards given last year: 31
No. of applicants last year: 180
Additional Information: Interns are selected through a competitive process of written and phone interviews. Please see the website for further details or email at internprogram@jacobspillow.org.

JAMES AND GRACE ANDERSON TRUST

32 Wardie Road, Edinburgh, EH5 3LG, Scotland
Tel: (44) 13 1552 4062
Fax: (44) 13 1467 1333
Email: tim.straton@virgin.net
Contact: Mr Timothy D Straton, Trustee

The James and Grace Anderson Trust funds research into the cure or alleviation of cerebral palsy.

James and Grace Anderson Trust Research Grant

Subjects: Research into cure or alleviation of cerebral palsy or awards to individuals who suffer from Cerebral Palsy or groups that provide assistance to such persons.
Purpose: To advance by investigation, research or any other way, knowledge with regard to the causes of cerebral palsy and related conditions and, if possible, curing or alleviating the same.
Level of Study: Research
Type: Grant
Value: Maximum UK£25,000
Length of Study: 1–3 years
Country of Study: United Kingdom
Application Procedure: Application must be made in writing, giving full details of the research being carried out and the anticipated value. A copy of the ethical approval, if granted, should also be included. If an individual, or a group for an individual, details of purpose to which award will be put must be detailed.
Closing Date: April 15th and October 15th
Funding: Private
No. of awards given last year: 10
No. of applicants last year: 30
Additional Information: Grants will not be given to organizations that have no involvement with cerebral palsy nor individuals who do not suffer from cerebral palsy.

THE JAMES BEARD FOUNDATION

167 West 12th Street, New York, NY, 10011, United States of America
Tel: (1) 212 675 4984
Fax: (1) 212 645 1438
Email: info@jamesbeard.org
Website: www.jamesbeard.org

It is our mission to celebrate, preserve, nurture American's culinary heritage and diversity in order to elevate the appreciation of our culinary excellence.

James Beard Foundation Scholarship

Subjects: Culinary studies, pastry and baking, hotel and restaurant management, wine studies, nutrition, and food writing.
Purpose: To fund aspiring culinary professionals.
Eligibility: Applicant must be aspiring career in America's diverse culinary heritage and future.
Level of Study: Professional development
Type: Scholarship
Value: Cash awards and tuition waivers. Amount varies from year to year.
Length of Study: 1–4 years
Frequency: Annual
No. of awards offered: 50
Application Procedure: Supply two letters of reference, mailed with a completed application.
Closing Date: May 15th
Funding: Foundation
Contributor: James Beard Foundation
No. of awards given last year: 50

James Beard Scholarship II

Subjects: Fine cuisine.
Purpose: To offer unstinting help and encouragement to people embracing on a culinary career.
Level of Study: Professional development
Type: Scholarship
Value: US$500–5,000
Length of Study: 1 year
Frequency: Annual
No. of awards offered: 50
Application Procedure: Supply two letters of reference, mailed with a completed application.
Closing Date: May
Funding: Foundation
Contributor: The James Beard Foundation
No. of awards given last year: 50

JAMES COOK UNIVERSITY

Graduate Research School, Townsville, QLD, 4811, Australia
Tel: (61) 7 4781 4575
Fax: (61) 7 4781 6204
Email: GRS@jcu.edu.au
Website: www.jcu.edu.au
Contact: Dr Julie Funnell, Manager

James Cook University prides itself on its international reputation for research and discovery and teaching that is enhanced and enlivened by that research activity.

James Cook University Postgraduate Research Scholarship

Subjects: All disciplines.
Purpose: To encourage full-time postgraduate research leading to a Master's or PhD degree.
Eligibility: Open to any student who has attained at least an Upper Second Class (Honours) Bachelor's Degree.
Level of Study: Postgraduate
Type: Scholarship
Value: Australian $23,728
Length of Study: 3 years with a possible additional 6 months in exceptional circumstances for the PhD, or 2 years for the Master's programme
Frequency: Annual
Study Establishment: James Cook University
Country of Study: Australia
No. of awards offered: Up to 20
Application Procedure: Application form must be completed in all respects.
Closing Date: October 31st for Domestic (Australian) students and August 31st for International students

Noel and Kate Monkman Postgraduate Award

Subjects: Marine biology.
Purpose: To encourage full-time study towards an MSc or PhD degree in marine biology.
Eligibility: Open to Australian citizens or those with permanent resident status in Australia who hold, or are expecting to hold, an Upper Second Class (Honours) Degree, or its equivalent, in marine biology or a related science.
Level of Study: Postgraduate
Type: Scholarship
Value: To be determined
Length of Study: 3 years for the PhD or 2 years for the Master's programme
Frequency: Dependent on funds available
Study Establishment: James Cook University
Country of Study: Australia
No. of awards offered: 1

JAMES MADISON MEMORIAL FELLOWSHIP FOUNDATION

1613 Duke Street, Alexandria, VA, 22314, United States of America
Tel: (1) 800 525 6928
Fax: (1) 202 653 6045
Email: madison@act.org
Website: www.jamesmadison.gov
Contact: Mr Anne Marie Kanakkanatt, Office Manager

The mission of the James Madison Memorial Fellowship Foundation is to strengthen the teaching of the principles, framing and development of the United States of America Constitution in secondary schools.

James Madison Fellowship Program

Subjects: History, political science and education.
Purpose: To provide support to individuals who desire to become outstanding teachers of the American Constitution at the secondary school level.
Eligibility: Applicants must be citizens of the United States of America. Applicants must be a teacher or plan to become a teacher of American history, American government or social studies at the secondary school level (grades 7–12).
Level of Study: Graduate
Type: Fellowship
Value: Up to US$24,000
Length of Study: Up to 5 years
Frequency: Annual
Country of Study: United States of America
No. of awards offered: 60 plus
Application Procedure: Applications may be downloaded from the Foundation's website.
Closing Date: March 1st
Funding: Foundation, government, private
No. of awards given last year: 65
Additional Information: Please see the website for further details http://www.jamesmadison.gov/.

JAMES PANTYFEDWEN FOUNDATION

9 Market Street, Aberystwyth, Ceredigion, SY23 1DL, Wales
Tel: (44) 01970 612806
Fax: (44) 01970 612806
Email: pantyfedwen@btinternet.com
Website: www.jamespantyfedwenfoundation.org.uk
Contact: Mr Richard H Morgan, Executive Secretary

James Pantyfedwen Foundation Grants

Subjects: All subjects.
Purpose: To promote mainly postgraduate research.
Eligibility: Open to Welsh nationals, especially those who wish to train as ministers of religion of any denomination. The qualifying criteria for this are defined by the benefactor.
Level of Study: Postgraduate
Type: Grant
Value: Varies, usually linked to the cost of postgraduate tution fees (up to a maximum of UK£7,000)
Length of Study: Up to 3 years
Frequency: Annual
Country of Study: United Kingdom
No. of awards offered: Varies
Application Procedure: Please submit all applications by post. We do not invite the submission of forms electronically because in most cases we require materials in addition to the application form.
Closing Date: Check with website
Funding: Private
Contributor: Exclusive to private investment portfolio
No. of awards given last year: 346
No. of applicants last year: 210
Additional Information: For all enquiries relating to the work of the Foundation, application forms, etc please contact Richard Morgan (Executive Secretary) or Nel Williams (Office Secretary). Please see the website for further details http://www.jamespantyfedwenfoundation.org.uk/main_en.html.

JANSON JOHAN HELMICH OG MARCIA JANSONS LEGAT

Blommeseter, Norderhov, 3512, Hönefoss, Norway
Tel: (47) 3 213 5465
Fax: (47) 3 213 5626
Email: post@jansonslegat.no
Website: www.jansonslegat.no
Contact: Mr Reidun Haugen, Manager

Janson Johan Helmich Scholarships and Travel Grants

Subjects: All subjects.
Purpose: To support practical or academic training.
Eligibility: Open to qualified Norwegian postgraduate students with practical experience for advanced study abroad.
Level of Study: MBA, Research, Postgraduate, Professional development, Doctorate
Type: Scholarship
Value: A maximum of Norwegian Krone 100,000
Length of Study: 1 year
Frequency: Annual
Country of Study: Any country
No. of awards offered: 50
Application Procedure: Applicants must complete an application form.
Closing Date: March 15th
No. of awards given last year: 58
No. of applicants last year: 274

JAPAN SOCIETY FOR THE PROMOTION OF SCIENCE (JSPS)

5-3-1 Kojimachi, Chiyoda-ku, Tokyo, 102 0083, Japan
Tel: (81) 3263 9094
Fax: (81) 3263 1854
Email: gaitoku@jsps.go.jp
Website: www.jsps.go.jp

The Japan Society for the Promotion of Science (JSPS) is an independent administrative institution, established for the purpose of contributing to the advancement of science in all fields of the natural and social sciences and the humanities. The JSPS plays a pivotal role in the administration of a wide specrum of Japan's scientific and academic programmes.

JSPS Award for Eminent Scientists

Subjects: Humanities, social sciences and natural sciences.
Purpose: To enable Nobel Laureates and other leading scientists to come to Japan for the purpose of associating directly with younger Japanese researchers so as to mentor, stimulate and inspire them to greater achievements.
Eligibility: Foreign researchers such as Nobel laureates, who possess a record of excellent research achievements and who are mentors and leaders in their respective fields are eligible to apply.
Level of Study: Research
Type: Award
Value: Business class round-trip air tickets, ¥42,000 as per day stipend and a family allowance of ¥10,000
Length of Study: 1–3 years
Frequency: Annual
Study Establishment: Any eligible host institution (see the website for details)
Country of Study: Japan
No. of awards offered: Approx. 4
Application Procedure: JSPS invities universities and institutions in Japan that wish to invite a Nobel laureate or other leading scientists to their campus to submit an invitation plan. JSPS reviews the plan and decides whether or not to fund the award. JSPS does not accept applications directly. It is the host institution that submits the application and supports the invitee's stay in Japan.
Closing Date: December
Funding: Government
No. of awards given last year: 4
No. of applicants last year: 9
Additional Information: Please check website http://www.jsps.go.jp/english/e-awards/index.html for updated information.

JSPS Invitation Fellowship Programme for Research in Japan

Subjects: Humanities, social sciences, natural sciences.
Purpose: To enable Japanese scientists to invite their foreign colleagues to Japan to participate in co-operative work and other academic activities. This programme also aims to promote international co-operation in mutual understanding through scientific research.
Eligibility: The candidate must be a researcher with an excellent record of research achievements who, in principle, is employed full time at an overseas research institution and is a citizen of a country that has diplomatic relations with Japan. Senior scientists, university professors and other persons with substantial professional experience are welcome to apply. Japanese applicants must have lived abroad and been engaged in research for over 10 years.
Level of Study: Professional development, Research
Type: Fellowship
Value: Long-term fellowships: Round-trip air ticket, Monthly stipend ¥369,000, Domestic travel allowance, ¥100,000, Research expense ¥40,000; short-term fellowships: Round-trip air ticket, Per diem ¥18,000, Domestic travel allowance ¥150,000
Length of Study: 14–60 days for short-term and 2–10 months for long-term fellowships
Frequency: Annual
Study Establishment: Any eligible host institution (see the website for details)
Country of Study: Japan
Application Procedure: Applications for these programmes must be submitted to JSPS by a host scientist in Japan through the head of his/her university or institution. Foreign scientists wishing to participate in this programme are advised to establish contact with a Japanese or foreign-resident researcher in their field and to ask him/her to submit an application. An application form and all supporting documents must be submitted by the deadline.
Closing Date: Check with website
Funding: Government
Additional Information: JSPS will begin issuing calls for applications under the "short-term S" category of the Invitation Fellowship Programs for Research in Japan. Under this category, Nobel laureates and other overseas researchers with similar records of eminent achievements are invited to Japan for short-term visits during which they offer advice and cooperate in the overall research activities of the host institution and give lectures at other Japanese research institutions. Please see the website for further details http://www.jsps.go.jp/english/e-inv/index.html.

JSPS Postdoctoral Fellowships for Foreign Researchers

Subjects: Humanities, social sciences, natural sciences, engineering and medicine.
Purpose: To assist promising and highly qualified young foreign researchers wishing to conduct research in Japan.
Eligibility: Open to citizens of countries that have diplomatic relations with Japan. Applicants must hold a doctoral degree when the fellowship goes into effect which must have been received within the past 6 years.
Level of Study: Postdoctorate
Type: Fellowship
Value: Round-trip air ticket, monthly maintenance allowance of ¥362,000, settling-in allowance of ¥200,000 and overseas travel accident and sickness insurance coverage
Length of Study: 2 years but a minimum of 1 year
Frequency: Annual
Study Establishment: Universities and research institutions
Country of Study: Japan
No. of awards offered: 3 postdoctoral fellowship programs, each with different eligibility requirements
Application Procedure: Applicants must write for details. Application must be submitted to JSPS by the host researcher in Japan.
Closing Date: Check with website
Funding: Government
Additional Information: Please see the website for further details http://www.jsps.go.jp/english/e-fellow/index.html.

JSPS Postdoctoral Fellowships for North American and European Researchers (Short-term)

Subjects: Humanities, social sciences, natural sciences, engineering and medicine.
Purpose: To assist promising and highly qualified young foreign researchers wishing to conduct research in Japan.
Eligibility: Be a citizen or permanent resident of an eligible country (the US, canada, EU countries, Switzerland, Norway and Russia). Candidates must have obtained their doctoral degree at a university outside Japan within six years of the date the fellowship goes into effect, or must be currently enrolled in a doctoral course at a university outside Japan and scheduled to receive their PhD within two years.
Level of Study: Predoctorate, Postdoctorate
Type: Fellowship
Value: Round-trip air ticket, monthly maintenance allowance of ¥362,000 for PhD holder and ¥200,000 for non-PhD holder, settling-in allowance of ¥200,000 and overseas travel accident and sickness insurance coverage
Length of Study: 1 year but a minimum of 1 month
Frequency: Annual
Study Establishment: Universities and research institutions
Country of Study: Japan
No. of awards offered: Varies
Application Procedure: Applicants must write for details. Application must be submitted to JSPS by the host researcher in Japan.
Closing Date: Check with website
Funding: Government
Additional Information: Please see the website for further details http://www.jsps.go.jp/english/e-fellow/postdoctoral.html.

JSPS Summer Programme

Subjects: Humanities, social sciences and natural sciences.
Purpose: To provide opportunities for young pre- and postdoctoral researchers from the US, UK, France, Germany and Canada to receive an orientation on Japanese culture and research systems and to pursue research under the guidance of host researchers at Japanese universities and research institutes.
Eligibility: Candidates must hold a doctorate, which must have been received within 6 years prior to April 1st of the year of the summer programme or must be enrolled in a university graduate programme. The candidate should also possess the nationality, citizenship, permanent residence or equivalent status in one of the countries represented by the five designated nominating agencies (i.e. the United States of America, United Kingdom, France, Germany or Canada).
Level of Study: Doctorate, Postdoctorate, Predoctorate
Type: Fellowship
Value: International travel, maintenance allowances (¥534,000), Domestic Research Trip Allowance (¥58,500), insurance, research-related expenses at the host Institution (up to ¥100,000). Hotel room charges at Narita and Tokyo, and meals and accommodation charges at SOKENDAI will be covered separately
Length of Study: 2 months (June–August)
Frequency: Annual
Study Establishment: Eligible host institutes (see the website for details)
Country of Study: Japan
Application Procedure: Candidates must contact the nominating authority in their respective countries for detailed application procedures and consult the website for further details.
Closing Date: Check with website
Funding: Government
No. of awards given last year: 94
Additional Information: Please see the website for further details http://www.jsps.go.jp/english/e-summer/index.html.

JAPANESE AMERICAN CITIZENS LEAGUE (JACL)

National Headquarters, 1765 Sutter Street, San Francisco, CA, 94115, United States of America
Tel: (1) 415 345 1075, 415 921 5225
Fax: (1) 415 931 4671
Email: ncwnp@jacl.org/youthdir@jacl.org
Website: www.jacl.org
Contact: Scholarships Officer

The Japanese American Citizens League (JACL) was founded in 1929 to fight discrimination against people of Japanese ancestry. It is the largest and one of the oldest Asian American organizations in the United States of America. The JACL has over 24,500 members in 112 chapters located in 25 states, Washington, DC, and Japan. The organization operates within a structure of eight district councils, with headquarters in San Francisco, CA.

Abe & Esther Hagiwara Student Aid Award

Subjects: All subjects.
Purpose: To provide financial assistance to students who are JACL members.
Eligibility: Open to members of JACL who demonstrate financial needs.
Level of Study: Postgraduate
Type: Scholarship
Value: Minimum $1,000 and Maximum $5,000
Frequency: Annual
Application Procedure: See the website.
Closing Date: April 1st
Contributor: Japanese American Citizens League

Mike M. Masaoka Congressional Fellowship

Subjects: Public service.
Purpose: To financially support and develop leaders for public service.
Eligibility: Candidates must be U.S. citizens who are graduating college seniors or students in graduate or professional programs and a member of the JACL. Preference will be given to those who have demonstrated a commitment to Asian American issues, particularly those affecting the Japanese American community. Communication skills, especially in writing, are important.
Level of Study: Postgraduate, Professional development
Type: Fellowship
Value: $2,200–2,500 a month
Length of Study: 6–8 months
Frequency: Annual
Country of Study: United States of America
No. of awards offered: Varies
Application Procedure: Applicants must send a completed application form and a letter of reference to the JACL national headquarters.
Closing Date: May 20th
Funding: Foundation
Additional Information: Preference will be given to those who have demonstrated a commitment to Asian American issues, particularly those affecting the Japanese American community. Please see the website for further details http://www.jacl.org/leadership/masaoka.htm.

For further information contact:

Japanese American Citizens League Headquarters, Mike M. Masaoka Fellowship, 1850 M Street NW, Suite 1100, Washington, DC, CA, 20036, United States of America
Email: policy@jacl.org

Norman Y. Mineta Fellowship

Subjects: All subjects.
Purpose: To focus on public policy advocacy as well as programs of safety awareness in the Asian Pacific American (APA) community.
Eligibility: Open to the members of the JACL with 4 year degree from an accredited college or university having excellent writing, analytical, and computer skills.
Level of Study: Postgraduate
Type: Fellowship
Value: A $2,200 monthly stipend will be provided along with roundtrip airfare, courtesy of Southwest Airlines.
Length of Study: 6–10 months
No. of awards offered: Varies
Application Procedure: Interested applicants should submit a resume, a sample of writing, and names and contactinformation for two references to the Washington, D.C. office of the JACL at policy@jacl.org with 'Mineta Fellowship' in the subject line.
Closing Date: Check with website
Contributor: State Farm Insurance
Additional Information: Candidates must have ability to take directions and follow through with assignments, must work well with others, and have good interpersonal skills. Please see the website for further details http://www.jacl.org/leadership/mineta.htm.

For further information contact:

JACL, 1828 L Street, NW Suite 802, Washington, DC, 20036, United States of America
Tel: (1) 202 223 1240
Fax: (1) 202 296 8082
Email: dc@jacl.org
Contact: Floyd Mori, National Director

THE JEANNETTE RANKIN FOUNDATION

1 Huntington Road, Suite 701, Athens, GA, 30606, United States of America
Tel: (1) 706 208 1211
Fax: (1) 706 548 0202
Email: info@rankinfoundation.org
Website: www.rankinfoundation.org/
Contact: Scholarship Enquiries

Jeannette Rankin Foundation Grant

Subjects: All subjects.
Purpose: To financially support low-income women in their education.
Eligibility: Open to female candidates who are US citizens and 35 years of age or older. See website for more details.
Level of Study: Professional development
Type: Grant
Value: Up to US$2,000
Frequency: Annual
Country of Study: United States of America
No. of awards offered: 50
Application Procedure: Application form can be downloaded from the website.
Closing Date: March 1st
Funding: Private
Additional Information: Please see the website for further details http://www.rankinfoundation.org/pdfs/2013_JRF_Scholarship_Application_final.pdf.

JEWISH COMMUNITY CENTERS ASSOCIATION (JCCA)

520 8th Avenue, New York, NY, 10018, United States of America
Tel: (1) 212 532 4949 ext 246
Fax: (1) 212 481 4174
Email: info@jcca.org
Website: www.jccworks.com

The Jewish Community Centers Association (JCCA) of North America is the leadership network of, and central agency for, over 275 Jewish Community Centers, YM-YWHAs and camps in the United States of America and Canada, which annually serve more than one million members. The Association offers a wide range of services and resources to enable its affiliates to provide educational, cultural and recreational programmes to enhance the lives of North American Jewry. The JCCA is also the United States of America government-accredited agency for serving the religious and social needs of Jewish military personnel, their families and patients in Virginia hospitals through the JWB Chaplains Council.

JCCA Graduate Education Scholarship

Subjects: Social work, Jewish education, health, physical education, recreation, education and non-profit business administration.
Purpose: To provide merit-based financial aid for students to use towards an advanced degree that will lead to or enhance professional careers in the Jewish Community Center movement.
Eligibility: Open to applicants who have obtained a BA (Honours) Degree with a grade point average of at least 3.0 and a strong commitment to the Jewish Community Center Movement. It is preferred that applicants have knowledge of Jewish community practices, customs, rituals and organization.
Level of Study: MBA, Graduate
Type: Scholarship

Value: Up to US$10,000 for tuition costs
Length of Study: 1–2 years
Frequency: Annual
Country of Study: United States of America or Canada
No. of awards offered: 6–8
Application Procedure: Applicants must submit an application, reference letters, personal statement and transcripts. Application forms and information are available on the website. Applications are available online on: www.jccworks.com
Closing Date: February 1st
Funding: Private
No. of awards given last year: 7
No. of applicants last year: 50
Additional Information: Candidates must make the commitment of working at a Jewish Community Center for a minimum of two years following the completion of graduate work.

JILA (FORMERLY JOINT INSTITUTE FOR LABORATORY ASTROPHYSICS)

440 B, University of Colorado, Boulder, CO, 80309, United States of America
Tel: (1) 303 492 7789
Fax: (1) 303 492 5235
Email: jilavf@jila.colorado.edu
Website: www.colorado.edu
Contact: Programme Assistant

JILA's interests are at present research and applications in the fields of laser technology, opto-electronics, precision measurement, surface science and semiconductors, information and image processing, and materials and process science, as well as basic research in atomic, molecular and optical physics, precision measurement, gravitational physics, chemical physics, astrophysics and geophysical measurements. To provide an opportunity for persons actively contributing to these fields, JILA operates the Visiting Fellowship Programme as well as the Postdoctoral Research Associate Programme.

JILA Postdoctoral Research Associateship and Visiting Fellowships

Subjects: Laser technology, optoelectronics, precision measurement, surface science and semiconductors, information and image processing, and nanoscience.
Purpose: To support additional training beyond the PhD and sabbatical research.
Eligibility: There are no restrictions other than those that might be required by the grant that supports the research.
Level of Study: Postdoctorate, Professional development
Type: Fellowship
Value: Varies
Length of Study: Visiting fellowships are for 4–12 months and Postdoctoral Research Associateships are for 1 year or more
Frequency: Annual
Country of Study: United States of America
No. of awards offered: Varies
Application Procedure: Applicants should download an application form, complete it, and send it (along with requested supporting materials) to the Visiting Scientists Program Assistant. All materials may be submitted via email to secretary via email.
Closing Date: November 1st
Funding: Government
Contributor: Varies
Additional Information: Please see the website for further details http://jila.colorado.edu/content/postdoctoral-research-associates.

For further information contact:

Visiting Scientists Program, 440 UCB, Boulder, CO, 80309
Tel: 303 492 5749
Email: secretary@jila.colorado.edu

JOHN CARTER BROWN LIBRARY AT BROWN UNIVERSITY

Box 1894, Brown University, Providence, RI, 02912, United States of America
Tel: (1) 401 863 2725
Fax: (1) 401 863 3477
Email: JCBL_Information@Brown.edu
Website: www.jcbl.org

The John Carter Brown Library, an independently funded and administered institution for advanced research in history and the humanities, is located on the campus of Brown University. The Library supports research focused on the colonial history of the Americas, including all aspects of the European, African and Native American involvement.

Alexander O Vietor Memorial Fellowship

Subjects: Maritime history.
Purpose: To assist students conducting research into early maritime history.
Eligibility: Open to scholars engaged in predoctoral, postdoctoral or independent research. Graduate students must have passed their preliminary or general examinations at the time of application. Independent scholars may also apply.
Level of Study: Doctorate, Postdoctorate, Predoctorate
Type: Fellowship
Value: US$2,100 per month
Length of Study: 2–4 months
Frequency: Annual
Country of Study: United States of America
No. of awards offered: 1
Application Procedure: Applicants must complete an application form. Candidates should write to or email the Director.
Closing Date: December 15th
Funding: Private
No. of awards given last year: 1

Andrew W Mellon Postdoctoral Research Fellowship at Brown University

Subjects: Humanities, humanistically-oriented social sciences, and new fields with close ties to the humanities
Purpose: To support two year postdoctoral fellowships in the humanities, humanistically-oriented social sciences, and new fields with close ties to the humanities.
Eligibility: Open to postdoctoral scholars who have received their degrees from institutions other than Brown within the last 5 years.
Level of Study: Postdoctorate, Research
Type: Fellowship
Value: US$4,200 per month
Length of Study: 2 years
Frequency: Annual
Study Establishment: The John Carter Brown Library
Country of Study: United States of America
No. of awards offered: 2
Application Procedure: Applicants must complete an application form. Candidates should write to or email the Director.
Closing Date: October 5th
Funding: Government, private
Contributor: The Andrew W Mellon Foundation and the National Endowment for the Humanities

Center for New World Comparative Studies Fellowship

Subjects: Comparative history of the colonial Americas.
Purpose: To enable research with a definite comparative dimension relating to the history of the colonial Americas.
Eligibility: Open to scholars engaged in predoctoral, postdoctoral or independent research. Graduate students must have passed their preliminary or general examinations at the time of application.
Level of Study: Predoctorate, Doctorate, Postdoctorate
Type: Fellowship
Value: US$2,100 per month
Length of Study: 2–4 months
Frequency: Annual

Country of Study: United States of America
No. of awards offered: 2
Application Procedure: Applicants must complete an application form. Candidates should write to or email the Director.
Closing Date: December 15th
Funding: Private
No. of awards given last year: 1

Donald L. Saunders Fellowship

Subjects: History.
Purpose: To support research in the history.
Eligibility: Applicants must be citizens of the United States or have been in residence in the U.S. for three years immediately preceding the application deadline. Graduate students and foreign nationals are not eligible, but fellowships are otherwise open to scholars at any academic level and to independent researchers.
Type: Fellowship
Value: $4,200 per month
Length of Study: 5–10 months
Country of Study: United States of America
No. of awards offered: Varies
Closing Date: December 15th

Helen Watson Buckner Memorial Fellowship

Subjects: Colonial history of the Americas, North and South, including all aspects of the European, African, and Native American involvement.
Purpose: To assist scholars in any area of research related to the Library's holdings.
Eligibility: Open to scholars engaged in predoctoral, postdoctoral or independent research. Graduate students must have passed their preliminary or general examinations at the time of application.
Level of Study: Postdoctorate, Predoctorate, Research
Type: Fellowship
Value: US$2,100 per month
Length of Study: 2–4 months
Country of Study: United States of America
No. of awards offered: Varies
Application Procedure: Applicants must complete an application form. Candidates should write to or email the Director.
Closing Date: December 15th
Funding: Private
No. of awards given last year: 1

The InterAmericas Fellowship

Subjects: History of the West Indies and the Carribean basin.
Purpose: To study the history of exploration and discovery.
Eligibility: Applicants must be American citizens or have been resident in the United States for 3 years immediately preceding the application deadline. Graduate students are not eligible.
Type: Funding support
Value: $4,200 per month
Length of Study: 5–10 months
Country of Study: United States of America
No. of awards offered: Varies
Closing Date: December 15th

J. M. Stuart Fellowship

Subjects: Colonial history of the Americas, North and South, including all aspects of the European, African, and Native American involvement.
Purpose: To support graduate students in the research of the history.
Eligibility: Applicant must be a graduate student at Brown University.
Level of Study: Graduate
Type: Fellowship
Value: $2,100 per month
Length of Study: 2–4 months
Country of Study: United States of America
Closing Date: December 15th
Additional Information: Nominations should be sent by email to Margot Nishimura (margot_nishimura@brown.edu), Deputy Director and Librarian of the John Carter Brown Library, by December 15th. Please include name of student(s), area of research, dissertation title, and email address. Application instructions will be forwarded to the candidate upon receipt of nomination and verification by the Graduate School of academic standing. Completed applications will be due to the JCB by February 1st.

Jeannette D Black Memorial Fellowship

Subjects: History of cartography or a closely related area.
Purpose: To enable research into the history of cartography or a closely related area.
Eligibility: Open to scholars engaged in predoctoral, postdoctoral or independent research. Graduate students must have passed their preliminary or general examinations at the time of application.
Level of Study: Graduate, Doctorate, Postdoctorate, Predoctorate
Type: Fellowship
Value: US$2,100 per month
Length of Study: 2–4 months
Frequency: Annual
Country of Study: United States of America
No. of awards offered: 1
Application Procedure: Applicants must write to, or email the Director.
Closing Date: December 15th
Funding: Private
No. of awards given last year: 1

Library Associates Fellowship

Subjects: Colonial history of the Americas, North and South, including all aspects of the European, African, and Native American involvement.
Purpose: To assist scholars in any area of research related to the Library's holdings.
Eligibility: Open to scholars engaged in predoctoral, postdoctoral or independent research. Graduate students must have passed their preliminary or general examinations at the time of application.
Level of Study: Postdoctorate, Predoctorate, Research
Type: Fellowship
Value: US$2,100 per month
Length of Study: 2–4 months
Frequency: Annual
Country of Study: United States of America
No. of awards offered: 1
Application Procedure: Applicants must complete an application form. Candidates should write to, or email the Director.
Closing Date: December 15th
Funding: Private
Contributor: Associates of the John Carter Brown Library

The Marie L and William R. Hartland Fellowship

Subjects: Early maritime history.
Purpose: To assist scholars whose work is centered on the colonial history of the Americas, North and South, including all aspects of the European, African, and Native American involvement.
Eligibility: open to Americans and foreign nationals who are engaged in pre- or post-doctoral, or independent, research. Graduate students must have passed their preliminary or general examinations at the time of application.
Type: Fellowship
Value: $2,100 per month
Length of Study: 2–4 months
Country of Study: United States of America
Closing Date: December 15th

Maury A. Bromsen Fellowship

Subjects: Colonial Spanish American history.
Purpose: To support scholars whose work is centered on the colonial history of the Americas, North and South, including all aspects of the European, African, and Native American involvement.
Eligibility: Open to Americans and foreign nationals who are engaged in pre- or post-doctoral, or independent, research. Graduate students must have passed their preliminary or general examinations at the time of application.
Level of Study: Doctorate, Postdoctorate, Predoctorate
Type: Fellowship
Value: $2,100 per month
Length of Study: 2–4 months
Country of Study: United States of America
No. of awards offered: Varies
Closing Date: December 15th

Norman Fiering Fund

Subjects: Colonial history of the Americas, North and South, including all aspects of the European, African, and Native American involvement.
Purpose: To support scholars in any area of research related to the Library's holdings.
Eligibility: Open to scholars in any area of research related to the Library's holdings.
Type: Funding support
Value: $2,100 per month
Length of Study: 2–4 months
Country of Study: United States of America
No. of awards offered: Varies
Closing Date: January 3rd

For further information contact:

John Carter Brown Library, Box 1894, Providence, RI, 02912, United States of America
Tel: (1) 401 863 2725
Email: JCBL_Fellowships@Brown.edu
Contact: Director

Paul W McQuillen Memorial Fellowship

Subjects: Colonial history of the Americas, North and South, including all aspects of the European, African, and Native American involvement.
Purpose: To assist scholars in any area of research related to the Library's holdings.
Eligibility: Open to scholars engaged in predoctoral, postdoctoral or independent research. Graduate students must have passed their preliminary or general examinations at the time of application.
Level of Study: Postdoctorate, Predoctorate, Research
Type: Fellowship
Value: US$2,100 per month
Length of Study: 2–4 months
Frequency: Annual
Country of Study: United States of America
No. of awards offered: Varies
Application Procedure: Applicants must complete an application form. Candidates should write to, or email the Director.
Closing Date: December 15th
Funding: Private
No. of awards given last year: 1

R. David Parsons Long-Term Fellowship

Subjects: History of exploration and discovery.
Purpose: To study the history of exploration and discovery.
Eligibility: Open to Americans and foreign nationals who are engaged in pre- or post-doctoral, or independent, research. Graduate students must have passed their preliminary or general examinations at the time of application.
Type: Fellowship
Value: $4,200 per month
Length of Study: 5–10 months
Country of Study: United States of America
No. of awards offered: Varies
Closing Date: December 15th

Ruth and Lincoln Ekstrom Fellowship

Subjects: The history of women and the family in the Americas prior to 1825, including the question of cultural influences on gender formation.
Purpose: To sponsor historical research.
Eligibility: Open to scholars engaged in predoctoral, postdoctoral or independent research. Graduate students must have passed their preliminary or general examinations at the time of application.
Level of Study: Postdoctorate, Predoctorate, Research
Type: Fellowship
Value: US$2,100 per month
Length of Study: 2–4 months
Frequency: Annual
Country of Study: United States of America
No. of awards offered: Varies
Application Procedure: Applicants must complete an application form. Candidates should write to, or email the Director.
Closing Date: December 15th
Funding: Private
No. of awards given last year: 1

Touro National Heritage Trust Fellowship

Subjects: Some aspect of the Jewish experience in the Western hemisphere prior to 1830.
Purpose: To sponsor historical research.
Eligibility: Open to graduates of any nationality engaged in predoctoral, postdoctoral or independent research. Applicants must have passed their preliminary or general examinations at the time of application.
Level of Study: Doctorate, Postdoctorate, Predoctorate
Type: Fellowship
Value: US$2,100 per month
Length of Study: 2–4 months
Frequency: Annual
Country of Study: United States of America
No. of awards offered: Varies
Application Procedure: Applicants must complete and submit an application form. Candidates should write to or email the Director.
Closing Date: December 15th
Funding: Private
No. of awards given last year: 1
Additional Information: The Touro Fellow will be selected by an academic committee consisting of representatives from Brown University, the American Jewish Historical Society, Brandeis University, the Newport Historical Society and the John Carter Brown Library, as well as a representative of the Executive Committee of the Touro National Heritage Trust. The Touro Fellow must be prepared to participate in symposia or other academic activities organized by these institutions and may be called upon to deliver one or two public lectures.

JOHN E FOGARTY INTERNATIONAL CENTER (FIC) FOR ADVANCED STUDY IN THE HEALTH SCIENCES

Building 31, 31 Center Drive, MSC 2220, Bethesda, MD, 20892 2220, United States of America
Tel: (1) 301 496 2075
Fax: (1) 301 594 1211
Email: FICinfo@mail.nih.gov
Website: www.fic.nih.gov
Contact: Program Officer

The John E Fogarty International Center (FIC) for Advanced Study in the Health Sciences, a component of the National Institutes of Health (NIH), promotes international co-operation in the biomedical and behavioural sciences. This is accomplished primarily through long- and short-term fellowships, small grants and training grants. This compendium of international opportunities is prepared by the FIC with the hope that it will stimulate scientists to seek research enhancing experiences abroad.

Fogarty HIV Research Training Program

Subjects: HIV.
Purpose: To strengthen the human capacity to contribute to the ability of institutions in low- and middle-income countries (LMIC) to conduct HIV-related research on the evolving HIV-related epidemics in their country and to compete independently for research funding.
Eligibility: See http://www.fic.nih.gov/Programs/Pages/hiv-aids-research-training.aspx for details
Level of Study: Research
Type: Research grant
Value: See http://www.fic.nih.gov/Programs/Pages/hiv-aids-research-training.aspx for details
Frequency: Annual
No. of awards offered: Varies
Application Procedure: Applications are accepted from United States institutions in response to a specific request for applications. Individuals who wish to become trainees must apply to the project director of an awarded grant. Application forms are available from the website.

Closing Date: June 24th (letter of intent) and July 24th (application)
Funding: Government
Additional Information: Please see the website for further details http://www.fic.nih.gov/Programs/Pages/hiv-aids-research-training.aspx.

Fogarty International Research Collaboration Award (FIRCA)

Subjects: Biomedical and behavioural sciences.
Purpose: To foster international research partnerships between NIH supported United States scientists and their collaborators in regions of the developing world.
Eligibility: Open to principal investigators of a United States based NIH sponsored research project grant that will be active for at least 1 year beyond the submission date of the FIRCA application. It is also open to scientists affiliated with public and private research institutions in Africa, Asia, (except Japan, Singapore, South Korea and Taiwan), Central and Eastern Europe, Russia and the Newly Independent States of the Former Soviet Union, Latin America and the non-United States Caribbean, the Middle East and the Pacific Islands except Australia and New Zealand. The United States scientist will apply as principal investigator with a colleague from a single laboratory or research site in an eligible country.
Level of Study: Unrestricted
Value: Up to $50,000 per year
Length of Study: Up to 3 years
Frequency: Annual
Study Establishment: The foreign collaborator's research site
Country of Study: Other
No. of awards offered: Approx. 35 depending on funds available
Application Procedure: Applicants must submit applications on the grant application form PHS 398. Special instructions and conditions apply. Please refer to the website for further information.
Closing Date: Check with website
Funding: Government
Additional Information: Please see the website for further details http://www.fic.nih.gov/programs/pages/research-collaboration.aspx.

Global Health Research Initiative Program for New Foreign Investigators (GRIP)

Subjects: Medicine.
Purpose: To assist well-trained young investigators to contribute to health care advances in their home countries.
Eligibility: Open to all well-trained young investigators. To verify eligibility, new foreign investigators should review the answers to Frequently Asked Questions. Please contact Dr. Xingzhu Liu by email at xingzhu.liu@nih.gov with questions.
Level of Study: Postgraduate
Type: Grant
Value: US$50,000 per year
Length of Study: 5 years
Frequency: Annual
No. of awards offered: Varies
Application Procedure: Application form on request.
Closing Date: Check with website
Funding: Foundation
Contributor: Fugarty International Center
Additional Information: Please see the website for further details http://www.fic.nih.gov/programs/Pages/new-foreign-investigators.aspx.

For further information contact:

Email: butrumb@mail.nih.gov
Contact: Bruce Butrum, Grants Management Officer

John E Fogarty International Research Scientist Development Award

Subjects: Medical research.
Purpose: To forge working relationships between future heads of health research programmes in the United States of America and established researchers in developing countries that will lead to ongoing collaborations in the study of health problems of mutual interest.
Eligibility: Applicants must be American citizens or permanent residents, have a doctoral or medical degree, or the equivalent, in a health science field earned within the last 7 years. Applicants must have a demonstrated commitment and competence in health research, and have an invitation from a sponsor affiliated with an internationally recognized research facility in Africa, Asia (except Japan, Singapore, South Korea and Taiwan), Central and Eastern Europe, Russia and the Newly Independent States of the Former Soviet Union, Latin America and the non United States Caribbean, the Middle East and the Pacific Islands except Australia and New Zealand. Applications to work in institutions in Sub-Saharan Africa are especially encouraged. Applicants must have a United States sponsor or mentor at a research institution with ongoing collaborative research funding in one of the eligible countries listed above.
Type: Award
Value: Salary support of up to $ 75,000 or 50 per cent of salary (whichever is less) per year, Research support of up to $30,000 per year
Application Procedure: Applicants must refer to the website for further information and application forms.
Closing Date: March 4th
Additional Information: Please see the website for further details http://www.fic.nih.gov/Programs/Pages/research-scientists.aspx.

JOHN F AND ANNA LEE STACEY SCHOLARSHIP FUND

National Cowboy and Western Heritage Museum, 1700 Northeast 63rd Street, Oklahoma City, OK, 73111, United States of America
Tel: (1) 405 478 2250
Email: dianef@nationalcowboymuseum.org
Website: www.nationalcowboymuseum.org

In accordance with the will of the late Anna Lee Stacey, a trust fund has been created for the education of young men and women who aim to make art their profession.

John F and Anna Lee Stacey Scholarships

Subjects: Painting and drawing in the classical tradition of western culture.
Purpose: To foster a high standard in the study of form, colour, drawing, painting, design and technique, as these are expressed in modes showing patent affinity with the classical tradition of western culture.
Eligibility: Open to citizens of the United States of America of 18–35 years of age who are skilled in printing and drawing and devoted to the classical or conservative tradition of western culture.
Level of Study: Graduate, Postgraduate, High-School Graduate
Type: Scholarship
Value: A total of approx. US$5,000
Length of Study: 1 year
Frequency: Annual
Country of Study: Any country
No. of awards offered: Varies
Application Procedure: Applicants must complete and submit an application form with up to 10 digital images of their work. Digital images or slides and completed application forms should be sent by United States of America mail. Applicants should also enclose a recent photograph, a letter outlining plans and objectives and at least four letters of reference.
Closing Date: February 1st
Funding: Private
Contributor: A bequest from the John F and Anna Lee Stacey Foundation
No. of awards given last year: 5
No. of applicants last year: 100
Additional Information: The Committee does not maintain storage facilities, so applicants must not send digital images or any materials before November 1st. Each successful competitor will be required to submit a brief quarterly report together with digital images of their work and a more complete report at the termination of the scholarship. Please see the website for further details http://www.nationalcowboymuseum.org/education/staceyfund/default.aspx.

JOHN F. KENNEDY LIBRARY FOUNDATION

Columbia Point, Boston, MA, 02125, United States of America
Tel: (1) 866 514 1960
Fax: (1) 617 514 1600
Email: Kennedy.library@nara.gov
Website: www.jfklibrary.org
Contact: Fellowships Administrator

The John F. Kennedy Library Foundation is a non-profit organization that provides financial support, staffing and creative resources for the John F. Kennedy Presidential Library and Museum whose purpose is to advance the study and understanding of President Kennedy's life and career, and the times in which he lived and to promote a greater appreciation of America's political and cultural heritage, the process of governing and the importance of public service.

Arthur M. Schlesinger, Jr. Fellowship

Subjects: Political science and history.
Purpose: To financially support scholars in the production of substantial work on the foreign policy of the Kennedy years.
Eligibility: Open to citizens of the United States only.
Level of Study: Postgraduate
Type: Fellowships
Value: Up to US$5,000
Frequency: Annual
Country of Study: United States of America
No. of awards offered: Up to 2
Application Procedure: Applicants must submit application form, financial need analysis, essay, reference letters and curriculum vitae. Applicants are strongly encouraged to contact the Kennedy Library for information about its collections and holdings before applying.
Closing Date: August 15th
Funding: Foundation
Contributor: Schlesinger Fund
Additional Information: Proposals are invited from all sources, but preference will be given to applicants specializing in the work on the foreign policy of the Kennedy years especially with regard to the western hemisphere, or on Kennedy domestic policy, especially with regard to racial justice and to the conservation of natural resources. Preference is also given to projects not supported by large grants from other institutions. Please see the website for further details.

For further information contact:

John F. Kennedy Library Foundation, Columbia Point, Boston, MA, 02125, United States of America
Tel: (1) 617 514 1629
Fax: (1) 617 514 1625
Email: Kennedy.library@nara.gov
Contact: Grant and Fellowship Coordinator

Ernest Hemingway Research Grants

Subjects: Literature/English/writing, social sciences and humanities.
Purpose: To provide funds for the related costs incurred during research in the Hemingway Collection.
Eligibility: Open to citizens of the United States only.
Level of Study: Postgraduate
Type: Grant
Value: US$200–1,000
Frequency: Annual
Country of Study: United States of America
No. of awards offered: Varies
Application Procedure: Applicants must submit application form, financial need analysis, essay, reference letters and curriculum vitae.
Closing Date: November 1st (Check with website)
Funding: Foundation
Contributor: John F. Kennedy Library Foundation
Additional Information: Preference is given to dissertation research by PhD candidates working in newly opened or relatively unused portions of the Collection, but all proposals are welcome and will receive careful consideration. Please see the website for further details http://www.jfklibrary.org/About-Us/Job-Volunteer-Internship-Opportunities/Internships.aspx.

Kennedy Research Grants

Subjects: Social sciences, criminal justice/criminology, economics, history, political science, library and information sciences, humanities, education, literature/english/writing and architecture.
Purpose: To help defray living, travel, and related costs incurred while doing research in the textual and non-textual holdings of the library.
Eligibility: Open to citizens of the United States only.
Level of Study: Postgraduate
Type: Grant
Value: US$500–2,500
Frequency: Annual
Country of Study: United States of America
No. of awards offered: 15–20
Application Procedure: Applicants must submit application form, financial need analysis, essay, reference letters and curriculum vitae.
Closing Date: March 15th and August 15th
Funding: Foundation
Contributor: John F. Kennedy Library Foundation
Additional Information: Please see the website for further details.

Marjorie Kovler Research Fellowship

Subjects: Political science.
Purpose: To financially support scholars in the production of a substantial work in the area of foreign intelligence and the presidency or a related topic.
Eligibility: Open to citizens of the United States only.
Level of Study: Postgraduate
Type: Fellowship
Value: Up to US$2,500
Frequency: Annual
Country of Study: United States of America
No. of awards offered: 1
Application Procedure: Applicants must submit application form, financial need analysis, essay, reference letters and curriculum vitae.
Closing Date: August 15th
Funding: Foundation
Contributor: John F. Kennedy Library Foundation
Additional Information: Please see the website for further details.

JOHN SIMON GUGGENHEIM MEMORIAL FOUNDATION

90 Park Avenue, New York, NY, 10016, United States of America
Tel: (1) 212 687 4470
Fax: (1) 212 697 3248
Email: fellowships@gf.org
Website: www.gf.org

The John Simon Guggenheim Memorial Foundation is concerned with encouraging and supporting scholars and artists to engage in research in any field of knowledge and creation within the arts. The Foundation was established by United States Senator Simon Guggenheim and his wife as a memorial to their son who died on April 26, 1922.

Guggenheim Fellowships to Assist Research and Artistic Creation (Latin America and the Caribbean)

Subjects: Sciences, humanities, social sciences and creative arts.
Purpose: To further the development of scholars and artists by assisting them to engage in research in any field of knowledge and creation in any of the arts, under the freest possible conditions irrespective of race, colour or creed.
Eligibility: Open to citizens and permanent residents of countries of Latin America and the Caribbean who have demonstrated an exceptional capacity for productive scholarship or exceptional creative ability in the arts.
Level of Study: Postdoctorate, Professional development
Type: Fellowship
Value: Grants will be adjusted to the needs of Fellows, taking into consideration their other resources and the purpose and scope of their plans. The average grant is US$33,000
Length of Study: Ordinarily for 1 year, but in no instance for a period shorter than 6 consecutive months

Frequency: Annual
Country of Study: Any country
No. of awards offered: 36
Application Procedure: Applicants must complete an application form. Further information is available on the Foundation's website.
Closing Date: Sepember 15th (Check with website)
Funding: Private
Contributor: The Foundation
No. of awards given last year: 35
Additional Information: Members of the teaching profession receiving sabbatical leave on full or part salary are eligible for appointment, as are holders of other fellowships and of appointments at research centres. Fellowships are awarded by the Trustees upon nominations made by a committee of selection. Please see the website for further details http://www.gf.org/about-the-foundation/the-fellowship/.

Guggenheim Fellowships to Assist Research and Artistic Creation (USA and Canada)

Subjects: Sciences, humanities, social sciences and creative arts.
Purpose: To further the development of scholars and artists by assisting them to engage in research in any field of knowledge and creation in any of the arts, under the freest possible conditions irrespective of race, colour or creed.
Eligibility: Open to citizens and permanent residents of the United States of America and Canada who have demonstrated an exceptional capacity for productive scholarship or exceptional creative ability in the arts.
Level of Study: Postdoctorate, Professional development
Type: Fellowship
Value: Grants will be adjusted to the needs of Fellows, taking into consideration their other resources and the purpose and scope of their plans. The average grant is US$37,360
Length of Study: Ordinarily for 1 year, but in no instance for a period shorter than 6 consecutive months
Frequency: Annual
Country of Study: Any country
Application Procedure: Applicants must complete an application form. Further information is available on the Foundation's website.
Closing Date: September 15th
Funding: Private
Contributor: The Foundation
No. of awards given last year: 186
Additional Information: Members of the teaching profession receiving sabbatical leave on full or part salary are eligible for appointment, as are holders of other fellowships and of appointments at research centres. Fellowships are awarded by the Trustees upon nominations made by a committee of selection. Please see the website for further details http://www.gf.org/about-the-foundation/the-fellowship/.

JOHNS HOPKINS UNIVERSITY

615 N. Wolfe Street, Baltimore, MD, 21205, United States of America
Tel: (1) 410 516 3400, 410 955 1680
Fax: (1) 410 223 1603
Email: admiss@jhsph.edu
Website: http://webapps.jhu.edu/jhuniverse/information_about_hopkins/
Contact: Grants Co-ordinator

The vision of the Johns Hopkins Center for Alternatives to Animal Testing is to be a leading force in the development and use of reduction, refinement and replacement alternatives in research, testing and education to protect and enhance the health of the public.

CAAT Grants Programme

Subjects: Toxicology and immunotoxicology.
Purpose: To promote and support research in the development of in vitro and other alternative techniques.
Eligibility: No eligibility restrictions. Applicants' proposals must meet the goals of the CAAT Grants Program.
Level of Study: Unrestricted
Value: For proposols relating to toxicology: up to US$25,000; Proposals Relating to Developmental Immunotoxicology: up to US $50,000
Length of Study: 1 year
Frequency: Annual
Country of Study: United States of America
No. of awards offered: Approx. 12
Application Procedure: Applicants must complete a preproposal. After review, selected applicants are invited to submit a full application.
Closing Date: March 21st
Funding: Private
No. of awards given last year: 12
No. of applicants last year: 30
Additional Information: Please see the website for further details http://caat.jhsph.edu/programs/grants/.

For further information contact:

615 N Wolfe St W7032, Baltimore, MD, 21205, United States of America
Fax: (1) 410 614 2871
Email: caat@jhsph.edu
Contact: CAAT Grants Coordinator

Greenwall Fellowship Program

Subjects: Biomedical science, ethics, public health, health policy and clinical care.
Purpose: To provide an unparalleled opportunity for fellowship and faculty development training in bioethics and health policy.
Eligibility: Open to applicants who have Doctoral degrees in medicine, nursing, philosophy, law, public health, biomedical sciences, social sciences or a related field.
Level of Study: Postdoctorate
Type: Fellowship
Value: US$$122,003
Length of Study: 2 years
Frequency: Annual
Study Establishment: Johns Hopkins University
Country of Study: United States of America
No. of awards offered: Varies
Application Procedure: Applicants must send a cover letter, a personal statement describing why they want to be a Greenwall Fellow, a copy of their curriculum vitae, 3 reference letters, official copies of undergraduate and graduate/professional school transcripts and copies of their written and/or published work.
Closing Date: December 1st

For further information contact:

The John Hopkins Bermer Bioethics Institute, 100 North Charles Street, Suite 740, Baltimore, MS, 21201, United States of America
Email: fellows@ihsph.edu
Contact: Kathy Chen

JOSEPHINE DE KÁRMÁN FELLOWSHIP TRUST

PO Box 3389, San Dimas, CA, 91773, United States of America
Tel: (1) 909 592 0607
Email: info@dekarman.org
Website: www.dekarman.org
Contact: Ms Judy McLain, Secretary

The Josephine De Kármán Fellowship Trust was established in 1954 by the late Dr Theodore von Kármán, world renowned aeronautics expert and teacher, in memory of his sister, Josephine, who passed away in 1951. The purpose of this Fellowship programme is to recognize and assist students whose scholastic achievements reflect Professor von Kármán's high standards.

Josephine De Kármán Fellowships

Subjects: All subjects.
Purpose: To provide financial support to students entering their senior undergraduate year and graduate students entering the final year of a PhD program.
Eligibility: Open to students entering their senior undergraduate year and graduate students entering the final year of a PhD program.
Level of Study: Doctorate, Graduate

Type: Fellowships
Value: US$22,000 graduate students' final year in school and $14,000 for undergraduates' final year
Length of Study: 1 year
Frequency: Annual, Fall and Spring semesters
Country of Study: United States of America
No. of awards offered: 10
Application Procedure: Applicants must submit completed application form including transcripts and 2 letters of recommendation before the deadline date.
Closing Date: January 31st
Funding: Foundation
Contributor: Josephine De Kármán Fellowship Trust
Additional Information: Special consideration will be given to applicants in the humanities field. Funds are disbursed in two equal amounts through the Penn financial system in September and in February. Please see the website for further details http://www.dekarman.org/.

JUVENILE DIABETES FOUNDATION INTERNATIONAL/THE DIABETES RESEARCH FOUNDATION

26 Broadway, 14th Floor, New York, NY, 10004, United States of America
Tel: (1) 1 800 533 CURE (2873)
Fax: (1) 212 785 9595
Email: info@jdrf.org
Website: www.jdrf.org
Contact: Grant Adminstrator

JDRF remains dedicated to finding a cure for type 1 diabetes as our highest priority. In addition to focusing on specific research challenges and gaps that will lead to curing, better treating, and preventing type 1 diabetes. JDRF works to decrease barriers to commercial development of products for type 1 diabetes.

JDRF Career Development Award

Subjects: Diabetes: the prevention of diabetes and its recurrence, restoration of normal metabolism and the avoidance of and reversal of complications.
Purpose: To provide salary and research support for exceptional scientists in research related to diabetes who are beginning their careers as junior staff members.
Eligibility: Open to highly qualified researchers of all nationalities, holding a PhD, MD, DMD, DVM or equivalent at a college, university, medical school or other research facility who have directed their expertise to target JDRF research priority areas. Both new and established researchers are supported. Proposals must be for a scientific research project involving the cause, treatment, prevention and/or cure of diabetes and its complications.
Level of Study: Professional development
Type: Award
Value: Maximum of US$150,000 total costs per year for up to 5 years. Indirect costs cannot exceed 10 per cent of total costs
Length of Study: 5 years
Closing Date: Varies. For details see the JDRF website

JDRF Innovative Grant

Subjects: Diabetes: the prevention of diabetes and its recurrence, restoration of normal metabolism and the avoidance of and reversal of complications.
Purpose: To support highly innovative basic and clinical research that is at the developmental stage.
Eligibility: Open to highly qualified researchers of all nationalities, holding a PhD, MD, DMD, DVM or equivalent at a college, university, medical school or other research facility, who have directed their expertise to target JDRF research priority areas. Both new and established researchers are supported. Proposals must be for a scientific research project involving the cause, treatment, prevention and/or cure of diabetes and its complications.
Level of Study: Research
Type: Grant
Value: Maximum of US$100,000 in direct cost costs and indirect cost of 10 per cent for a total of US$110,000.
Length of Study: 1 year
Application Procedure: Please check website for further information http://www.jdrf.org/index.cfm?page_id=103207
Closing Date: Varies. For details see the JDRF website

JDRF Postdoctoral Fellowships

Subjects: Diabetes - the prevention of diabetes and its recurrence, restoration of normal metabolism and the avoidance of and reversal of complications.
Purpose: To attract qualified, promising scientists entering their professional careers in diabetes.
Eligibility: Open to highly qualified researchers of all nationalities, holding a PhD, MD, DMD, DVM or equivalent at a college, university, medical school or other research facility, who have directed their expertise to target JDRF research priority areas. Both new and established researchers are supported. Proposals must be for a scientific research project involving the cause, treatment, prevention and/or cure of diabetes and its complications.
Level of Study: Postdoctorate
Type: Fellowship
Value: Research allowance of $5,500 includes allowance for travel to scientific meetings (up to $2000 per year), journal subscriptions, books, training courses, etc. Personal computer costs are allowed
Length of Study: 2-3 years
Application Procedure: Please check website for further information http://www.jdrf.org/index.cfm?page_id=103207
Closing Date: Varies. For details see the JDRF website
Additional Information: The award is renewable for a second year pending submission and approval of a renewal application and progress report.

Strategic Research Agreement

Subjects: Life sciences and medicine.
Purpose: To support and fund research to find a cure for Type I diabetes.
Level of Study: Research
Type: Grant
Value: Varies
Length of Study: 3 years
No. of awards offered: Varies
Closing Date: February 1st
Additional Information: Please check website for further details.

For further information contact:

Email: robertg@trdf.technion.ac.il; rivkag@trdf.technion.ac.il
Contact: Robi and Rivka

KAISER FAMILY FOUNDATION (KFF)

2400 Sand Hill Road, Menlo Park, CA, 94025, United States of America
Tel: (1) 650 854 9400, 234 9220
Fax: (1) 650 854 4800/7465
Email: pduckham@kff.org
Website: www.kff.org

Kaiser Family Foundation (KFF) is a non-profit, private operating foundation that focuses on the major health care issues facing the nation. KFF is an independent voice and source of facts and analysis for policymakers, the media, the health care community and the general public.

The Kaiser Media Fellowships in Health

Subjects: Journalism.
Purpose: To help journalists and commentators do the best possible job of keeping the public informed about health issues at this critical time in the evolution of our health care system.
Eligibility: Open to any journalist, editor or producer specializing in health reporting. The applicants must be citizens of the United States or must work for an accredited US media organization.
Level of Study: Professional development
Type: Fellowship

Value: Up to 10 stipends will be awarded based on the length of the fellowship - up to US$50,000 for a 9-month fellowship. The programme also covers travel costs and computer equipment based on the needs of the project
Length of Study: 9 months
Country of Study: United States of America
No. of awards offered: 10
Application Procedure: There is no application form. For details of the procedure see the website.
Closing Date: March 3rd
Funding: Foundation
Additional Information: There is no age restriction, but typically fellows must be in the early to mid-career range, with at least 5 years experience as a journalist. Please see the website for further details http://www.kff.org/mediafellows/fellowshipsinhealth/index.cfm.

KATHOLIEKE UNIVERSITEIT LEUVEN

Oude Markt 13, Bus 5005, Leuven, 3000, Belgium
Tel: (32) 16 32 40 10
Fax: (32) 16 32 40 14
Email: csb@dir.kuleuven.be/info@kuleuven.be
Website: www.kuleuven.be

Situated at the heart of Western Europe, Katholieke Universiteit Leuven has been a centre of learning for almost 6 centuries. Founded in 1425, by Pope Martin V, Katholieke Universiteit Leuven bears the double honour of being the oldest extant Catholic university in the world and the oldest university in the Low Countries.

Special Doctoral Students Grants for Advanced K.U. Leuven Doctoral Students

Subjects: All subjects.
Purpose: To encourage fundamental and applied research in all academic disciplines.
Eligibility: Open to doctoral students from non-EEA countries who wish to complete their project at Katholieke
Level of Study: Doctorate
Type: Scholarships
Value: A part-time salary will be paid for half-time contracts.
Length of Study: 12 months
Frequency: Annual
No. of awards offered: 3
Closing Date: January 31st
Contributor: Katholieke Universiteit Leuven
Additional Information: Please see the website for further details http://www.kuleuven.be/research/funding/bof/personal/.

KAY KENDALL LEUKAEMIA FUND

Allington House, 1st Floor, 150 Victoria Street, London, SW1E 5AE, England
Tel: (44) 20 7410 0330
Fax: (44) 20 7410 0332
Email: info@skklf.org.uk
Website: www.kklf.org.uk/
Contact: Ms Helen McLeod, Fund Executive

The Kay Kendall Leukaemia Fund awards grants for research on aspects of leukaemia and for relevant studies on related haematological malignancies. Applications are welcomed for first class research on innovative proposals via the fellowship or project grant routes. Please see website for full details.

Kay Kendall Leukaemia Fund Research Fellowship

Subjects: Aspects of leukaemia or relevant studies on related haematological malignancies.
Purpose: To encourage researchers in the field of leukaemia research to submit applications.
Eligibility: Open to applicants of any nationality intending to work mainly in the United Kingdom. Applicants must hold a recognized higher degree, but need not be medically qualified.
Level of Study: Postdoctorate
Type: Fellowship
Value: Salary and laboratory expenses
Length of Study: Junior (3 years), Intermediate (4 years)
Frequency: Annual
Country of Study: United Kingdom
No. of awards offered: 2–3
Application Procedure: Applicants must submit a research proposal form and details of support from the intended United Kingdom institution.
Closing Date: Please see website for details.
Funding: Private
No. of awards given last year: Junior (2), Intermediate (1)
Additional Information: £17,000 per year to contribute to the cost of laboratory consumables and a further amount up to £1,000 per year to cover travel costs to meetings (for junior fellowship).

KENNAN INSTITUTE

Woodrow Wilson International Center for Scholars, One Woodrow Wilson Plaza, 1300 Pennsylvania Avenue North West, Washington, DC, 20004 3027, United States of America
Tel: (1) 202 691 4000
Fax: (1) 202 691 4247
Email: kennan@wilsoncenter.org
Website: www.wilsoncenter.org
Contact: Scholar Programs

The Kennan Institute for Advanced Russian Studies sponsors advanced research on the successor states to the USSR and encourages Eurasian studies with its public lecture and publication programmes, maintaining contact with scholars and research centres abroad. The Institute seeks to function as a forum where the scholarly community can interact with public policymakers.

Fulbright-Kennan Institute Research Scholarships

Subjects: Social sciences and humanities. Research proposals examining topics in Eurasian studies are eligible, with those topics relating to regional Russia, the NIS and contemporary issues especially welcome.
Purpose: To offer support to junior scholars studying the former Soviet Union, allowing them time and resources in the Washington DC area to work on their first published work or to continue their research.
Eligibility: Open to scholars and researchers who have at least two years postdoctoral (post-Kandidat) academic and research experience. Applicants must be eligible to obtain a J-1 exchange visitor visa.
Level of Study: Postdoctorate
Type: Scholarship
Value: US$3,300 per month plus research facilities, computer support and some research assistance
Length of Study: 6 months
Frequency: Dependent on funds available
Study Establishment: The Kennan Institute
Country of Study: United States of America
No. of awards offered: 5
Application Procedure: Applicants must complete an application form. The application must include a project proposal, publications list, bibliography, biographical data and three letters of recommendation specifically in support of the research to be conducted at the Institute. Applications received by fax or email will not be accepted.
Closing Date: July 15th (for Russian applicants) and October (for Ukrainian applicants)
Funding: Government
No. of awards given last year: 5
No. of applicants last year: 14
Additional Information: Please see the website for further details http://www.wilsoncenter.org/opportunity/fulbright-kennan-institute-research-scholarships.

Title VIII-Supported Short-Term Grant

Subjects: Eurasian studies in the social sciences and humanities. Social sciences and humanities focusing on the former Soviet Union, excluding the Baltic States.
Purpose: To support U.S. citizens whose research in the social sciences or humanities focuses on the former Soviet Union (excluding the Baltic States), and who demonstrate a particular need to utilize the library, archival, and other specialized resources of the Washington, DC area.

Eligibility: Open to academic participants with a doctoral degree or those who have nearly completed their dissertations. For non-academic participants, an equivalent level of professional development is required. Applicants can be citizens of any country, but must note their citizenship while applying.
Level of Study: Postdoctorate, Postgraduate, Predoctorate, Professional development, Research, Doctorate
Type: Scholarship
Value: $3200 for 31 days ($103.22 per day)
Length of Study: Up to 31 days
Frequency: Dependent on funds available
Study Establishment: The Kennan Institute
Country of Study: United States of America
No. of awards offered: 4 are available to non-United States of America citizens and 4 to citizens of the United States of America
Application Procedure: Applicants must submit a concise description of their research project of 700–800 words, a curriculum vitae, a statement on preferred dates of residence in Washington DC and two letters of recommendation specifically in support of the research to be conducted at the institute. No application form is required for short-term grants.
Closing Date: March 1st, June 1st, September 1st and December 1st
Funding: Government, private
No. of awards given last year: 40
No. of applicants last year: 97
Additional Information: For more information, please email kennan@wilsoncenter.org. Please see the website for further details http://www.wilsoncenter.org/opportunity/kennan-institute-short-term-grant.

KENNEDY MEMORIAL TRUST

3 Birdcage Walk, Westminster, London, SW1H 9JJ, England
Tel: (44) 20 7222 1151
Fax: (44) 20 7222 7189
Email: annie@kennedytrust.org.uk
Website: www.kennedytrust.org.uk
Contact: Ms Annie Thomas, Secretary

As part of the British national memorial to President Kennedy, the Kennedy Memorial Trust awards scholarships to British postgraduate students for study at Harvard University or the Massachusetts Institute of Technology. The awards are offered annually following a national competition and cover tuition costs and a stipend to meet living expenses.

Kennedy Scholarships

Subjects: All fields of arts, science, social science and political studies.
Purpose: To enable students to undertake a course of study in the United States of America.
Eligibility: Open to resident citizens of the United Kingdom who have been wholly or mainly educated in the United Kingdom. At the time of application, candidates must have spent at least 2 of the last 5 years at a university in the United Kingdom and must either have graduated by the start of tenure in the following year or have graduated not more than 3 years before the commencement of studies. Applications will not be considered from persons already in the United States of America.
Level of Study: Postgraduate, Graduate
Type: Scholarship
Value: At least US$23,500 to cover support costs, special equipment and some travel in the United States of America, plus tuition fees and travelling expenses to and from the United States of America. Students applying for two year masters or PhD programmes will normally secure the necessary funding for second or subsequent years within their institutution
Length of Study: 1 year
Frequency: Annual
Study Establishment: Harvard University and the Massachusetts Institute of Technology (MIT), Cambridge, MA
Country of Study: United States of America
No. of awards offered: 6–8
Application Procedure: Applications are made online at www.kennedytrust.org.uk and comprise an online form, a statement of purpose and two references which are to be submitted online before the closing data.
Closing Date: October 28th
Funding: Private
Contributor: Public donation
No. of awards given last year: 7
No. of applicants last year: 297
Additional Information: Scholars are not required to study for a degree in the United States of America, but are encouraged to do so if they are eligible and able to complete the requirements for it. Please see the website for further details http://www.kennedytrust.org.uk/display.aspx?id = 1173&pid = 243.

KIDNEY HEALTH AUSTRALIA

Level 1, 25 North Terrace, GPO Box 9993, Adelaide, SA, 5001, Australia
Tel: (61) 08 8334 7555
Fax: (61) 08 8334 7545
Email: research@kidney.org.au
Website: www.kidney.org.au
Contact: Medical Director's Office

Founded in 1968, the Australian Kidney Foundation's mission is to be recognized as the leading non-profit national organization providing funding for, and taking the initiative in, the prevention of kidney and urinary tract diseases.

Australian Kidney Foundation Biomedical Scholarships

Subjects: Medical and scientific kidney and urology-related research.
Purpose: To provide scholarships for individuals wishing to study full-time for the research degrees.
Eligibility: Open to Australian applicants who are graduates, or proposing to graduate in the current academic year. Part-time students are not eligible.
Level of Study: Doctorate, Postgraduate
Type: Scholarship
Value: Australian $24,000 for science and $35,000 for medical
Length of Study: 2 years or 3 years
Frequency: Annual
No. of awards offered: Varies
Closing Date: August 31st
Contributor: Kidney Health Australia

For further information contact:

Kidney Health Australia, GPO Box 9993, Adelaide SA, 5001, Australia
Contact: The Medical Director

Australian Kidney Foundation Medical Research Grants and Scholarships

Subjects: The functions and disease of the kidney, urinary tract and related organs.
Purpose: To support medical research.
Eligibility: Open to Australian citizens who are graduates of Australian medical schools or overseas graduates who are eligible for Australian citizenship and for registration as medical practitioners in Australia.
Level of Study: Doctorate, Postgraduate
Type: Scholarship
Value: Please contact the organization
Length of Study: Up to 3 years
Frequency: Annual
Study Establishment: Any approved medical centre, university or research institute
Country of Study: Australia
No. of awards offered: Up to 6
Application Procedure: See guidelines in website.
Closing Date: August 31st
Additional Information: Please see the website for further details http://www.kidney.org.au/HealthProfessionals/MedicalResearchFunding/tabid/633/Default.aspx.

For further information contact:

Email: research@kidney.org.au
Contact: Joanna Stoic, Medical Director's Office

Investigator Driven Research Grants and Scholars

Subjects: Multiple sclerosis research.
Purpose: To award investigators who have applied to the NHMRC for funding but have just missed the cut-off mark.
Eligibility: Open to projects that are ranked as worthy of funding.
Type: Scholarship
Frequency: Annual
No. of awards offered: 1
Contributor: Kidney Health Australia

Kidney Health Australia Seeding and Equipment Grants

Subjects: The functions or diseases of the kidney, urinary tract and related organs, or relevant problems, dialysis, transplantation, organ donation and research.
Purpose: To provide financial support for research projects related to the kidney and urinary tract.
Eligibility: Open to Australian citizens connected with Australian universities or medical centres with requisite research facilities.
Level of Study: Unrestricted
Type: Grant
Value: Up to Australian $15,000 per year
Length of Study: Up to 2 years
Frequency: Annual
Study Establishment: Any medical centre, university or research institute
Country of Study: Australia
No. of awards offered: 30–35
Application Procedure: Applicants must contact the Foundation for details.
Closing Date: June 30th

KIDNEY RESEARCH UK

Nene Hall, Lynch Wood Park, Peterborough, Cambridgeshire, PE2 6FZ, England
Tel: (44) 0845 070 7601
Fax: (44) 1733 704685
Email: grants@kidneyresearchuk.org
Website: www.nkrf.org.uk
Contact: Mrs Elaine Davies, Grants Manager

The Kidney Research UK aims to advance and promote research into kidney and renal disease. These may include epidemiological, clinical or biological approaches to relevant problems. All research must be carried out in the United Kingdom.

Kidney Research UK Non-clinical Senior Fellowships

Subjects: Renal medicine and related scientific studies.
Eligibility: Open to postdoctoral researchers in the biomedical field with evidence of independent research. Applicants may be of any nationality but project work must be carried out in the United Kingdom.
Level of Study: Postdoctorate, Professional development, Research
Type: Fellowship
Value: Up to UK£375,000 over a maximum of 5 years. Salary will be at the established academic level at the appropriate university scale. Includes a UK£24,000 per year bench allowance for consumables.
Length of Study: 3–5 years, subject to review in the 3rd year
Frequency: Annual
Study Establishment: Any institution
Country of Study: United Kingdom
No. of awards offered: Varies
Application Procedure: Applicants must complete an application form available from the Grants Department or website.
Closing Date: November 25th (Check with website)
Funding: Private
Contributor: Public donations
Additional Information: Please see the website for further details http://www.kidneyresearchuk.org/research/grant-types.php.

Kidney Research UK Research Project and Innovation Grants

Subjects: Renal medicine.
Purpose: To support both basic scientific and clinical research towards improving the understanding of renal disease, its causes, treatment and management.
Eligibility: Open to suitably qualified researchers of any nationality. Work must be carried out in the United Kingdom.
Level of Study: Unrestricted
Type: Project grant
Value: Up to UK£180,000 over 1–3 years for a full research project grant, and up to UK£40,000 over 1–2 years for an innovation grant.
Length of Study: 1–3 years
Frequency: Annual
Study Establishment: Any institution
Country of Study: United Kingdom
No. of awards offered: Varies
Application Procedure: Applicants must complete an application form available from the Grants Department or website.
Closing Date: March 2nd (Check with website)
Funding: Private
Contributor: Public donations
No. of awards given last year: 11
No. of applicants last year: 65
Additional Information: Please see the website for more details http://www.kidneyresearchuk.org/research/grant-types.php.

Kidney Research UK Training Fellowships/Career Development Fellowships

Subjects: Renal medicine and related scientific studies.
Purpose: To enable medical or scientific graduates undertake specialized training in renal research.
Eligibility: Open to medical candidates of immediate postregistration to registrar level and to science candidates with a PhD or DPhil and at least 2 years of postdoctoral experience. Project work must be carried out in the United Kingdom.
Level of Study: Postdoctorate, Professional development, Research
Type: Fellowship
Value: Up to £250,000 over a maximum of 3 years. Salary will be at the SHO/Specialist Registrar or the established academic level of the appropriate university scale. The fellowship includes a UK£15,000 per year bench allowance for consumables.
Length of Study: 1–3 years, subject to annual review
Frequency: Annual
Study Establishment: Any institution
Country of Study: United Kingdom
No. of awards offered: Varies
Application Procedure: Applicants must complete an application form available from the Grants Department, the website or via email.
Closing Date: November 25th (Check with website)
Funding: Private
Contributor: Public donations
No. of awards given last year: 3
No. of applicants last year: 22
Additional Information: Please see the website for more details http://www.kidneyresearchuk.org/research/grant-types.php.

PhD Studentships

Subjects: Renal medicine.
Purpose: To enable postgraduates to start a career in renal medicine by completing a course of training including submitting a PhD thesis.
Eligibility: Open to applicants of any nationality. Work must take place in the United Kingdom.
Level of Study: Postgraduate
Type: Studentship
Value: Up to £48,243, consumables fund of £2,000 per year
Length of Study: 3 years, subject to a satisfactory annual report
Frequency: Annual
Study Establishment: Any institution
Country of Study: United Kingdom
No. of awards offered: Varies (only offered to current grant holders)
Application Procedure: Applicants must complete an application form available from the Grants Department or website.
Closing Date: Check with website
Funding: Private
Contributor: Public donations
No. of awards given last year: 4
No. of applicants last year: 15
Additional Information: Those eligible to apply will be contacted directly. Please see the website for more details http://www.kidneyresearchuk.org/research/grant-types.php.

KING'S COLLEGE LONDON

Strand, London, WC2R 2LS, United Kingdom
Tel: (44) 020 7836 5454
Fax: (44) 020 7848 3460
Email: ceu@kcl.ac.uk
Website: www.kcl.ac.uk

King's College is one of the oldest and largest colleges of the University of London with 13,800 undergraduate students and some 5,300 postgraduates in 9 schools of study. It was founded in 1,829 as a university college in the tradition of the Church of England.

Australian Bicentennial Scholarships and Fellowships

Subjects: Any discipline
Purpose: To promote scholarship, intellectual links, and mutual awareness and understanding between the United Kingdom and Australia.
Eligibility: An applicant for a scholarship must be registered as a postgraduate student at an Australian tertiary institution, and must have at least an upper second class Honours degree.
Level of Study: Postgraduate
Type: Fellowship/Scholarship
Value: Up to UK£4,000 per year
Length of Study: Miminum 3 months
Frequency: Annual
Country of Study: Australia
No. of awards offered: Varies
Application Procedure: The application should include a curriculum vitae, the names of two referees, a statement of the research project (approximately 2 sides of A4) and where it is to be carried out.
Closing Date: April 5th
Additional Information: Please see the website for further details http://www.kcl.ac.uk/artshums/ahri/centres/menzies/scholarships/absf/absfa.aspx.

THE KOSCIUSZKO FOUNDATION

The Kosciuszko Foundation, Inc.,15 East 65th Street, New York, NY, 10065, United States of America
Tel: (1) 212 734 2130
Fax: (1) 212 628 4552
Email: addy@thekf.org/info@thekf.org
Website: www.thekf.org

The Kosciuszko Foundation, founded in 1925, is dedicated to promoting educational and cultural relations between the United States of America and Poland and increasing American awareness of Polish culture and history. In addition to its grants and scholarships, which total US$1 million annually, the Foundation presents cultural programmes including lectures, concerts and exhibitions, promotes Polish culture in the United States of America and nurtures the spirit of multicultural co-operation.

Chopin Piano Competition

Subjects: Piano performance of Chopin or other composers.
Purpose: To encourage highly talented students of piano to study and play works of Chopin and other Polish composers.
Eligibility: Open to students between the age of 16 and 22 who are citizens of the United States of America or full-time international students in the United States of America with a valid visa who wish to pursue a career in piano performance.
Level of Study: Unrestricted
Type: Prize
Value: First prize of US$5,000, a second prize of US$2,500 and a third prize of US$1,500. Scholarships may be awarded in the form of shared prizes
Frequency: Annual
Country of Study: United States of America
No. of awards offered: 3
Application Procedure: Applicants must complete an application form, available in December via the Kosciuszko Foundation's Cultural Department. Applications should be marked Chopin Piano Competition.
Closing Date: March 12th
No. of awards given last year: 3
No. of applicants last year: 15
Additional Information: The required repertoire is as follows: Chopin: one Mazurka of the contestant's choice and two major works; Szymanowski: one Mazurka of the contestant's choice; a major work by Bach, excluding The Well-Tempered Clavier; a complete sonata by Beethoven, Hadyn, Mozart or Schubert; a major 19th-century work (including Debussy, Ravel, Prokofiev and Rachmaninoff but excluding those already mentioned); and a substantial work by an American, Polish or Polish-American composer written after 1950. The competition is held on 3 consecutive days in mid-April.

For further information contact:

Chopin Piano Competition, Kosciuszko Foundation, 15 East 65th Street, New York, NY, 10065, United States of America
Tel: (1) 212 734 2130
Fax: (1) 212 628 4552

Dr Marie E Zakrzewski Medical Scholarship

Subjects: Studies towards an M.D. degree.
Purpose: To fund a young American woman of Polish ancestry for the 1st, 2nd or 3rd year of medical studies at an accredited school of medicine in the United States of America.
Eligibility: The applicant must be a woman of Polish descent who is a citizen of the United States of America or a Polish citizen with permanent residency status in the United States of America, entering the 1st, 2nd or 3rd year of studies towards an MD degree with a minimum grade point average of 3.0.
Level of Study: M.D.
Type: Scholarship
Value: US$3,500
Length of Study: 1 academic year
Frequency: Annual
Study Establishment: An accredited school of medicine in the United States of America
Country of Study: United States of America
No. of awards offered: 1
Application Procedure: Applicants must submit a tuition scholarship application form, a US$35 non-refundable application fee and supporting materials to the Kosciuszko Foundation. E-mailed and faxed materials will not be considered.
Closing Date: January 4th
Funding: Foundation, private
Contributor: Kosciuszko Foundation.
No. of awards given last year: 1
Additional Information: First preference is given to residents of the state of Massachusetts. Qualified residents of New England are considered if no first-preference candidates apply. Scholarship decisions in late May.

For further information contact:

Grants Department, Kosciuszko Foundation, Inc., 15 East 65th Street, New York, NY, 10065
Tel: 212 734 2130 ext. 210

Graduate Study and Research in Poland Scholarship

Subjects: All subjects.
Purpose: To enable American students to pursue a course of graduate or postgraduate study and research in Poland.
Eligibility: Graduate level students and university faculty members who are US citizens. Polish citizens are not eligible.
Level of Study: Graduate, Doctorate, Postgraduate, Research
Type: Grant
Value: 1,350 zloty per month for housing and living expenses provided by the Polish Ministry. Additional funding of $300 per month of approved study/research is awarded by the Kosciuszko Foundation
Length of Study: Up to 9 months, October to June
Frequency: Annual
Study Establishment: Accredited institutions falling under the jurisdiction of the Polish Ministry of Education and Science
Country of Study: Poland
No. of awards offered: Varies
Application Procedure: Applicants must submit an application form and supporting materials with a $50 non-refundable application fee. Applications are available in the Autumn, by mail or from the website www.thekf.org

Closing Date: January 5th
Funding: Foundation, government, private
Contributor: The Kosciuszko Foundation and the Polish Ministry of National Education
No. of awards given last year: 1
No. of applicants last year: 5
Additional Information: Applicant must submit a research proposal and invitation from the host institution. Invitation must specify period of research/study and the conditions that will apply to the candidate's term in Poland (access to archives, libraries, housing etc.). Scholarship decisions in June or July.

For further information contact:

Exchange Programs to Poland, Kosciuszko Foundation, 15 East 65th Street, New York, NY, 10065, United States of America
Tel: (1) 212 734 2130 ext. 210

Kosciuszko Foundation Tuition Scholarship Program

Subjects: Polish-American related issues and activities. All subjects are supported. American citizens (non-heritage) are supported when majoring in Polish subject areas.
Purpose: To provide financial aid to US students of Polish descent and to permanent residents (Polish citizen) for graduate-level studies.
Eligibility: Open to citizens of the United States of America of Polish descent and Polish citizens who are legal permanent residents of the United States of America, who are pursuing graduate studies in any field at a US Institute of Higher Education, or citizens of the United States of America (non-heritage) who are pursuing a major in Polish studies at the graduate level. All entrants must have a grade point average of 3.0.
Level of Study: Doctorate, Graduate
Type: Scholarship
Value: US$1,000–7,000
Length of Study: 1 academic year, renewable for a further academic year
Frequency: Annual
Study Establishment: Accredited institutions in the United States of America and certain programmes in Poland.
Country of Study: United States of America
No. of awards offered: Varies
Application Procedure: Applications available on-line from October through December. Applicants must submit application-supporting materials and a US$35 non-refundable application fee by the deadline date.
Closing Date: January 5th
Funding: Foundation
Contributor: Kosciuszko Foundation.
No. of awards given last year: 30
No. of applicants last year: 84
Additional Information: Information and guidelines are available all year round. A student can receive funding through the Tuition Scholarship programme no more than twice. Only one member per immediate family may receive a Tuition Scholarship during a given academic year. Scholarship decisions in late May.

For further information contact:

Tuition Scholarships, Kosciuszko Foundation, Inc., 15 East 65th Street, New York, NY, 10065
Tel: 212 734 2130 ext. 210

The Kosciuszko Foundation Year Abroad Program

Subjects: Polish language, history, literature and culture.
Purpose: To support the study of Polish language, history and culture at the undergraduate and graduate levels by US citizens.
Eligibility: Open to citizens of the United States of America who are undergraduate- or graduate- level students and who have a minimum GPA of 3.0.
Level of Study: Graduate, Postgraduate
Type: Scholarship
Value: 1,350 zloty per month for housing and living expenses from the Polish Ministry of Education and Science. Additional funding of $900 per semester is awarded by the Kosciuszko Foundation
Length of Study: 1–2 Semesters
Frequency: Annual
Study Establishment: Jagiellonian University Center for Polish Language and Culture, Krakow Poland
Country of Study: Poland
No. of awards offered: Varies
Application Procedure: Applicants must complete an application form, submit supporting materials and a $50 non-refundable application fee by the deadline date. Application forms are available on-line from October through December.
Closing Date: December 15th
Funding: Government, foundation
Contributor: Polish Ministry of Education and Science and the Kosciuszko Foundation.
No. of awards given last year: 5
No. of applicants last year: 13
Additional Information: Students may apply for one semester or a full academic year. Notification in June or July.

For further information contact:

Exchange Program to Poland, Kosciuszko Foundation, Inc., 15 East 65th Street, New York, NY, 10065
Tel: 212 734 2130 ext. 210

Marcella Sembrich Memorial Voice Scholarship Competition

Subjects: Polish music.
Purpose: To encourage young singers to study the repertoire of Polish composers and to honour the great Polish soprano, Marcella Sembrich.
Eligibility: Open to all singers preparing for professional careers who are citizens of the United States of America or international full-time students with valid student visas, at least 18 years of age and born after May 13, 1969.
Level of Study: Unrestricted
Type: Scholarship
Value: 1st prize:$3,000; 2nd prize: US$1,500; 3rd prize: US$1,000
Frequency: Every 2 years
Country of Study: Any country
No. of awards offered: 3
Application Procedure: Applicants must submit a competition application form and a non-refundable fee of US$35 with supporting documents, suggested programme and two copies of an audio cassette recording of approx. 10 min.
Closing Date: February 6th
Funding: Foundation, individuals, private

The Metchie J E Budka Award of the Kosciuszko Foundation

Subjects: Polish literature from the 14th century to 1939, and Polish history, the state, the nation and the people from 962 to 1956.
Purpose: To reward outstanding scholarly work in Polish literature, Polish history and Polish-American relations.
Eligibility: Applicants must be graduate students in colleges and universities of United States of America and Doctoral degree recipients from these institutions who apply during, or at the close of, the first 3 years of their postdoctoral scholarly careers.
Level of Study: Doctorate, Graduate
Type: Award
Value: US$3,000
Study Establishment: Colleges and universities of United States of America
Country of Study: United States of America
No. of awards offered: Varies
Application Procedure: Applicants must send four copies of each complete submission together with a cover letter.
Closing Date: July 20th
Funding: Private
Additional Information: Materials to be submitted: two original articles, or comparable material written in English in a form appropriate for publication or published in a refereed scholarly journal, or annotated translations into English from the original Polish of one or more significant works, which fall within the designated guidelines.

For further information contact:

The Metchie J. E. Budka Award, The Kosciuszko Foundation, 15 East 65th Street, New York, NY, 10065-6595
Tel: 212 734 2130
Fax: 212 628 4552

The Polish-American Club of North Jersey Scholarships

Subjects: All subjects.
Purpose: To financially aid full-time undergraduate and graduate students in the United States of America. Applicant must be a member of the Polish American Club of North Jersey.
Eligibility: Applicants must be citizens of the United States of America or permanent residents of Polish descent, active members of the Polish-American Club of North Jersey, have a minimum grade point average of 3.0 and be children or grandchildren of Polish-American Club of North Jersey members.
Level of Study: Graduate, undergraduate
Type: Scholarship
Value: US$5,00 to US$2,000
Length of Study: 1 academic year (September–May)
Frequency: Annual
Study Establishment: An accredited 4-year institution in the US
Country of Study: United States of America
No. of awards offered: Varies
Application Procedure: Applicants must complete a KF tution scholarship application form, available on the website from October to the end of December. A US$35 non-refundable application fee is required. Supporting materials and proof of Polish descent are required. E-mailed and faxed materials will not be considered.
Closing Date: January 15th
Funding: Foundation
Contributor: Polish American Club of North Jersey
No. of awards given last year: 5
No. of applicants last year: 6
Additional Information: Only one member per immediate family may receive a Polish-American Club of North Jersey Scholarship during any given academic year. Scholarship decisions in late May.

For further information contact:

Grants Department, Kosciuszko Foundation, Inc., 15 East 65th Street, New York, NY, 10065
Tel: 212 734 2130 ext. 210
Email: Addy@thekf.org

THE KRELL INSTITUTE

 Krell Institute, 1609 Golden Aspen Drive, Suite 101, Ames, IA, 50010, United States of America
Tel: (1) 515 956 3696
Fax: (1) 515 956 3699
Email: csgf@krellinst.org
Website: www.krellinst.org/www.krellinst.org/csgf

Krell Institute was founded in 1997. The goal of the Krell Institute has been to provide superior technical resources, knowledge and experience in managing technology-based education and information programmes. They have done just that, developing outstanding fellowship programmes, educational outreach programmes and information management and exchange programmes.

Department of Energy Computational Science Graduate Fellowship Program

Subjects: Computational science.
Purpose: Provides outstanding benefits and opportunities to students pursuing doctoral degrees in fields of study that use high performance computing to solve complex science and engineering problems.
Eligibility: Open to citizens and permanent residents of the United States, undergraduate seniors, 1st and 2nd year graduate student in a PhD programme.
Level of Study: Graduate, Research
Type: Fellowship
Value: $36,000 yearly stipend. Payment of full tuition and required fees. Yearly conferences. $5,000 academic allowance in the first fellowship year. $1,000 academic allowance each renewed year. 12-week research practicum. Renewable up to four years.
Frequency: Annual
Country of Study: United States of America
No. of awards offered: Varies
Application Procedure: Apply online through the website www.krellinst.org/csgf/application
Closing Date: January 8th
Funding: Government
Contributor: Department of Energy (Office of Science and National Nuclear Security Administration)
No. of awards given last year: 16
No. of applicants last year: 396
Additional Information: Please see the website for further details http://www.krellinst.org/csgf/about-doe-csgf.

THE KURT WEILL FOUNDATION FOR MUSIC

7 East 20th Street, New York, NY, 10003, United States of America
Tel: (1) 212 505 5240
Fax: (1) 212 353 9663
Email: kwfinfo@kwf.org
Website: www.kwf.org
Contact: Mr Brady Sansone, Office Manager

The Kurt Weill Foundation for Music is a non-profit, private foundation chartered to preserve and perpetuate the legacies of the composer Kurt Weill (1900–1950) and his wife, singer and actress Lotte Lenya (1898–1981). The Foundation awards grants and prizes, sponsors print and online publications, maintains the Weill-Lenya Research Center and administers Weill's copyrights.

Kurt Weill Foundation for Music Grants Program

Subjects: Performances of Kurt Weill's musical works, scholarly research projects, and educational initiatives directly related to Weill and/or Lotte Lenya. Funding categories include: Research and Travel; Kurt Weill Disseratation Fellowship; Publication Assistance; Educational Outreach; College/University Performance; Professional Performance; Broadcasts.
Purpose: To fund projects that aim to perpetuate Kurt Weill's artistic legacy.
Eligibility: There are no eligibility restrictions.
Level of Study: Unrestricted
Type: Grant
Value: For college and university production performances, a maximum of US$5,000; otherwise no restrictions on requested amounts
Frequency: Annual
Country of Study: Any country
No. of awards offered: Varies
Application Procedure: Applicants must complete an application form. Guidelines and forms are available at www.kwf.org or from the Foundation directly.
Closing Date: November 1st. Applications for support of major professional productions/festivals/exhibitions, etc., will be evaluated on a case-to-case basis without application deadlines.
Funding: Private

Kurt Weill Prize

Subjects: Music theatre since 1900.
Purpose: To encourage distinguished scholarship in the disciplines of music, theater, dance, literary criticism and history addressing music theater since 1900 (including opera).
Eligibility: Open to nationals of any country.
Level of Study: Unrestricted
Type: Prize
Value: US$5,000 for books, US$2,000 for articles
Frequency: Every 2 years
Country of Study: Any country
No. of awards offered: 2
Application Procedure: Applicants must submit five copies of their published work. Works must have been published within the 2 years preceding the award year.
Closing Date: April 30th
Funding: Private

KUWAIT FOUNDATION FOR THE ADVANCEMENT OF SCIENCE (KFAS)

Ahmad Al Jaber St. - Sharq - State of Kuwait, P.O.Box 25263, Safat, 13113, Kuwait
Tel: (965) 224 25898
Fax: (965) 224 15365
Email: Publicr@kfas.org.kw
Website: www.kfas.org

The Kuwait Foundation for the Advancement of Science (KFAS) aims to support efforts for modernization and scientific development within Kuwait by sponsoring basic and applied research, awarding grants to support and encourage research and awarding grants, prizes and recognition to enhance intellectual development. The KFAS also grants scholarships and fellowships for academic or training purposes, holds symposia and scientific conferences and encourages, supports and develops research projects and scientific programmes.

Islamic Organization for Medical Sciences Prize

Subjects: Medical practice and Islamic Medical Heritage.
Purpose: To support and promote scientific research in the field of Islamic medical sciences.
Eligibility: Applicant must be a citizen of Arab country.
Level of Study: Unrestricted
Type: Prize
Value: Kuwaiti Dinars 6,000 a KFAS shield and certificate of recognition
Frequency: Every 2 years
No. of awards offered: 2
Application Procedure: Nominations must be proposed by universities, scientific institutes, international organizations, individuals, past recipients of the prize and academic bodies.
Closing Date: December 31st
Funding: Private
Contributor: KFAS

KFAS Kuwait Prize

Subjects: Basic sciences, applied sciences, literature, arts, social and economic studies and Arabic and Islamic Heritage.
Purpose: To support scientific research and encouraging scholars and researchers in Kuwait.
Eligibility: Applicants should have a distinguished and innovative production of great importance in the specified field, published along the past 10 years, for which the nominee was not granted any other prize from any other entity.
Level of Study: Unrestricted
Type: Prize
Value: Kuwaiti Dinar 30,000, which is approx. US$100,000, for each prize, in addition to a Gold Medal, a KFAS Shield and a certificate of Recognition that explains the qualifications and importance of the winning work in short. The combined value of the prizes is more than US$1 million
Length of Study: 10 years
Frequency: Annual
No. of awards offered: 2 (One for Kuwaiti citizens and the other for citizens of Arab countries)
Application Procedure: Applicants must apply themselves or through non-political organizations and may also be recommended by search committees.
Closing Date: October 31st (Check with website)
Funding: Private
Contributor: KFAS
No. of awards given last year: 5
No. of applicants last year: 91
Additional Information: The area of specific specialization varies every year. Please see the website for further details http://www.kfas.org/kuwait-prize.html.

LA TROBE UNIVERSITY

Research Services, Melbourne, VIC, 3086, Australia
Tel: (61) 3 9479 1976
Fax: (61) 3 9479 1464
Email: rgs@latrobe.edu.au
Website: www.latrobe.edu.au/rgso
Contact: Manager, Research Students

La Trobe University is one of the leading research universities in Australia. The University has internationally regarded strengths across a diverse range of disciplines. It offers a detailed and broad research training programme and provides unique access to technology transfer and collaboration with end users of its research and training via its Research and Development Park.

Australian Postgraduate Awards

Subjects: Health sciences, humanities, social sciences, science, technology, engineering, law, management and education.
Purpose: To support research leading to Master's or Doctoral degrees.
Eligibility: Open to applicants who have completed at least 4 years of tertiary education studies with a high level of achievement, e.g. a First Class (Honours) Degree or its equivalent at an Australian university. Applicants must be Australian citizens or have permanent resident status. Applicants who have previously held an Australian Government Award (APA, APRA or CPRA) for more than 3 months are not eligible for an APA and applicants who have previously held an Australian Postgraduate Course Award (APCA) may apply only for an APA to support PhD research. The awards may be held concurrently with other non-Australian government awards.
Level of Study: Doctorate, Postgraduate, Research
Type: Award
Value: Australian $24,653 per year full-time (tax exempt) plus allowances
Length of Study: 2 years for the Master's and 3 years for the PhD. Periods of study already undertaken towards the degree will be deducted from the tenure of the award
Frequency: Annual
Study Establishment: Melbourne Campus, La Trobe University
Country of Study: Australia
No. of awards offered: 65
Application Procedure: Applicants must write for application kits, available from the school in which the candidate wishes to study. Applications must be submitted in duplicate to the Research and Graduate Studies Office.
Closing Date: October 31st
Funding: Government
No. of awards given last year: 55
Additional Information: Paid work may be permitted for up to a maximum of 8 hours per week for the full-time award or up to 4 hours per week for the part-time award. Please see the website for further details.

David Myers Research Scholarship

Subjects: All subjects.
Eligibility: Open to outstanding individuals of any nationality.
Level of Study: Postgraduate, Research
Type: Scholarship
Value: Australian $30,000 per year, a research support grant of up to $3,000 per year, and a $1,000 thesis allowance
Length of Study: 3 years full-time study for a PhD or 2 years full-time study for a Master's degree
Frequency: Annual
No. of awards offered: 5
Application Procedure: Check website for further details.
Closing Date: October 31st
No. of awards given last year: 5

International Postgraduate Research Scholarships (IPRS)

Subjects: Health sciences, humanities, social sciences, science, technology, engineering, law, management and education.
Purpose: To attract top quality overseas postgraduate students to areas of research in institutes of higher education and to support Australia's research efforts.
Eligibility: Open to suitably qualified overseas graduates (excluding New Zealand) eligible to commence a doctoral or Master's degree by research. The IPRS may be held concurrently with a university research scholarship and applicants for an IPRS are advised to apply for a La Trobe University Postgraduate Research Scholarship (LTUPRS). Applicants who have already commenced a Master's or PhD candidature or applicants for Master's candidature by coursework and a minor thesis are not eligible to apply for an IPRS. IPRSs are awarded on the basis of academic merit and research capacity alone.

Level of Study: Doctorate, Postgraduate, Research
Type: Scholarship
Value: Tuition fees
Length of Study: 2 years for the Master's and 3 years for the PhD
Frequency: Annual
Study Establishment: Melbourne Campus, La Trobe University
Country of Study: Australia
No. of awards offered: 6
Application Procedure: Applications should be submitted on the application form for International candidates available from the La Trobe International Office: email at international@latrobe.edu.au or from the school in which you wish to study.
Closing Date: September 30th
Funding: Government
Contributor: The Australian government
No. of awards given last year: 6
Additional Information: Applicants in most instances will become members of a research team working under the direction of senior researchers. Please see the website for further details.

La Trobe University Fee Remission Research Scholarship

Purpose: For international students commencing a research doctorate or research Master's degree.
Eligibility: Students who are offered an LTUPS or some other competitive scholarship for living expenses.
Level of Study: Research
Type: Scholarship
Value: Tuition fees
Length of Study: 4 years for doctoral and 2 years for Masters
No. of awards offered: 40
Application Procedure: See the website for details.
Closing Date: September 30th
No. of awards given last year: 40
Additional Information: Candidate is liable for overseas student health cover. Please see the website for further details http://www.latrobe.edu.au/international/fees/scholarships/research.

La Trobe University Postgraduate Research Scholarship

Subjects: Education, engineering, health sciences, humanities and social sciences, law, management, science and technology.
Purpose: To provide a living allowance for postgraduate research candidates.
Eligibility: Open to applicants of any nationality having qualifications deemed to be equivalent to an Australian First Class (Honours) Degree.
Level of Study: Doctorate, Postgraduate, Research
Type: Scholarship
Value: Australian $24,653 per year
Length of Study: Up to 2 years for the Master's and up to 3 years for the PhD
Frequency: Annual
Country of Study: Australia
No. of awards offered: Approx. 100
Application Procedure: Applicants must complete an application form, available directly from the department offering the relevant course of study.
Closing Date: September 30th for overseas applicants and October 31st for Australian citizens and permanent residents
No. of awards given last year: 100
Additional Information: Please see the website for further details http://www.latrobe.edu.au/research/downloads/La_Trobe_University_Postgraduate_Research_Scholarship.pdf.

LADY TATA MEMORIAL TRUST

TATA Limited, 18, Grosvenor Peace, London, SW1X 7HS, United Kingdom
Tel: (44) 0 20 7235 8281
Fax: (44) 0 20 7235 8727
Email: daphne@tata.co.uk
Website: www.icr.ac.uk/research/research_sections/haemato_oncology/4480.shtml

Established in 1932, the Trust has an international advisory committee, based in London, which supports research on Leukaemia worldwide, and spends four-fifths of its income on this initiative.

Lady Tata Memorial Trust Scholarships

Subjects: Leukaemia.
Purpose: To encourage study and research in diseases of the blood, with special reference to leukaemias and furthering knowledge in connection with such diseases.
Eligibility: Open to qualified applicants of any nationality. No restrictions regarding age or country.
Level of Study: Doctorate, Postdoctorate
Type: Scholarship
Value: £25,000–35,000 per year for International awards and Rs 16,000–25,000 for Indian awards
Length of Study: Studentship for 2–3 years
Frequency: Annual
Country of Study: Any country
No. of awards offered: Normally 15
Application Procedure: Applicants must submit eight copies of the application, including the application form, summary of proposed research, and two letters of reference.
Closing Date: January 31st (Check with website)
Funding: Private
Contributor: Tata Memorial Trust (India)
No. of awards given last year: 12
No. of applicants last year: 28
Additional Information: Please see the website for further details http://www.ladytatatrust.org/.

LAHORE UNIVERSITY OF MANAGEMENT SCIENCES (LUMS)

Lahore University of Management Sciences, D.H.A, Lahore Cantt., Lahore, 54792, Pakistan
Tel: (92) 92 42 111 11 5867 ext 2177, 2178
Fax: (92) 92 42 572 2591, 92 42 3589 6559
Email: zahoor@lums.edu.pk
Website: www.lums.edu.pk

The Lahore university of Management Sciences is a national university, which aims to provide rigorous academic and intellectual training and a viable alternative to education comparable to leading universities around the world.

Khushhali Bank - USAID Scholarship

Subjects: Management.
Purpose: To support talented students who have a high financial need.
Eligibility: Open to candidates studying at Lahore university of Management Science.
Value: Tuition fees, books and materials, lodging and transportation.
Study Establishment: LUMS
Country of Study: Pakistan
No. of awards offered: 22
Application Procedure: A completed application form must be sent.
Closing Date: January 25th
Contributor: Higher Education Commission

Unilever-LUMS MBA Fund for Women

Subjects: Management.
Purpose: To provide financial support for female candidates who aspire for a career in management.
Eligibility: Open to females belonging to unprivileged background and have per month income of Rs. 25,000 or less.
Level of Study: Postgraduate
Type: Scholarships
Value: 7-year interest-free loan
Length of Study: 2 years
Frequency: Annual
Country of Study: Pakistan
No. of awards offered: 5
Application Procedure: A completed application form must be sent.
Closing Date: See http://sdsb.lums.edu.pk/pages/mba_financial_aid.php for details
Contributor: Unilever Pakistan

LANCASTER UNIVERSITY

Student Services, Lancaster University, Bailrigg, Lancaster, LA1 4YW, England
Tel: (44) 1524 65201
Fax: (44) 1524 594868
Email: studentfunding@lancaster.ac.uk
Website: www.lancs.ac.uk/funding
Contact: Craig Lowe

Lancaster University is a campus university dedicated to excellence in teaching and research, offering a wide range of nationally and internationally recognized postgraduate courses. For updated information on all the University's funding opportunities, please see our website www.lancs.ac.uk/funding

Alumni Postgraduate Scholarships

Subjects: All subjects.
Eligibility: Applicants must have submitted their application for study in order to be considered for a studentship. Applicants for this award must be able to demonstrate a commitment to the life of the University or of the wider community.
Level of Study: Postgraduate
Type: Studentship
Value: £1,000
Frequency: Annual
No. of awards offered: 5
Application Procedure: In order to apply, you must complete an application form and submit your curriculum vitae and two references.
Closing Date: July 31st (Check with website)
Contributor: Lancaster University Alumni
Additional Information: Please see the website for further details http://www.lums.lancs.ac.uk/masters/funding/.

For further information contact:

Alumni & Development, Lancaster University, Lancaster, LA1 4YW, England
Tel: (44) 0 1524 592556
Email: s.nelhams@lancaster.ac.uk
Contact: Sally Nelhams

Bowland College - Willcock Scholarships

Subjects: All subjects.
Purpose: To provide financial assistance to undergraduate or postgraduate members of Bowland College who are in very good academic standing but who are facing long-term financial or other difficulties which are beyond their control and which jeopardise their continuing or commencing study at Lancaster University.
Eligibility: All Bowland College students or alumni who would have difficulty achieving their full academic potential because of financial hardship may apply to the scholarship fund.
Level of Study: Postgraduate
Type: Scholarship
Value: £2500
Frequency: Annual
Application Procedure: Applications should be submitted to the Senior College Advisor.
Closing Date: April 27th
Additional Information: Please see the website for further details http://lugrad.files.wordpress.com/2012/03/willcock.pdf.

For further information contact:

Bowland College Office, Lancaster University, Lancaster, LA1 4YT, England
Tel: (44) 0 1524 594506
Email: p.m.brown@lancaster.ac.uk
Contact: Robert Blake, Senior College Advisor

Cartmel College Scholarships

Subjects: All subjects.
Purpose: Each year, Cartmel College offers a limited number of awards to its present or alumni members who are unable to obtain adequate grants from other bodies.
Eligibility: All applications will be considered on their merit.
Level of Study: Postgraduate
Type: Scholarship
Value: £500
Length of Study: Awards are tenable for 1 year
Frequency: Annual
Closing Date: June 1st

For further information contact:

Student Services, University House, Lancaster University, Lancaster, LA1 4YW, England
Email: student.services@lancaster.ac.uk
Website: www.lancaster.ac.uk/sbs/download/forms

County College Scholarships

Subjects: All subjects.
Eligibility: Members of the County College who have completed a Lancaster degree and are proposing to commence postgraduate study at the university are eligible to apply to this scholarship fund. Selections are on the basis of academic merit and financial need.
Level of Study: Postgraduate
Type: Scholarship
Value: £1,000
Length of Study: Awards are tenable for one year
Frequency: Annual
No. of awards offered: 3
Closing Date: June 1st
Additional Information: Please see the website for further details.

For further information contact:

Student Services, University House, Lancaster University, Lancaster, LA1 4YW, England
Email: student.services@lancaster.ac.uk
Website: www.lancaster.ac.uk/sbs/download/forms

County College Studentship

Subjects: All subjects.
Purpose: To assist prospective students who are unable to obtain adequate grants from other bodies.
Eligibility: Open to applicants who are past members of County College.
Level of Study: Postgraduate
Type: Studentship
Value: £1,000
Frequency: Annual
Study Establishment: Lancaster University
Country of Study: United Kingdom
No. of awards offered: 3
Application Procedure: Application available from Student Funding Service website.
Closing Date: June 1st
Funding: Government
Additional Information: Please see the website for further details http://www.lancs.ac.uk/colleges/county/tutorial.html.

The Division of Health Research (DHR) Doctoral Scholarships

Subjects: Mental health, palliative care, organisational health and wellbeing and public health.
Purpose: The Division of Health Research is pleased to offer Postgraduate Research Scholarships to outstanding applicants to its Distance Learning PhD programmes.
Eligibility: Outstanding applicants to the Distance Learning PhD programmes in mental health, palliative care, organisational health and wellbeing and public health.
Level of Study: Doctorate, Postgraduate
Type: Scholarship
Value: £1,000
Length of Study: 1 year only
Frequency: Annual
No. of awards offered: 5
Application Procedure: In order to be considered for scholarship funding applications must be made via the electronic MyLancaster applications portal. An applicant must have received an offer of a place on one of the above doctoral schemes by the deadline date.
Closing Date: March 22nd

Additional Information: Please see the website for further details http://www.lancs.ac.uk/shm/study/doctoral_study/studentships/dhr/.

ESRC-Funded Research Studentships

Purpose: Lancaster University forms part of the ESRC's North West Doctoral Training Centre (NWDTC). A large number of studentships, covering tuition and maintenance, are available for those wishing to study in areas covered by the ESRC at Lancaster, Liverpool or Manchester.
Eligibility: The Faculty of Health and Medicine has Economic and Social Research Council (ESRC) funding for research studentships within the broad areas of i) Health and Well-being; and ii) Quantitative Methods. Applicants for the Health and Well-being pathway must be residents of the UK or the EU. Studentships in Quantitative Methods are open to candidates of any nationality. Some studentships are reserved for students seeking joint supervision across two or more partner institutions.
Type: Studentship
Value: Full funding for PhD study
Length of Study: 3 or 4 years
Application Procedure: Candidates for these studentships should provide a full curriculum vitae together with a covering letter explaining their motivation for undertaking PhD studies and an indication of any research areas or topics that would be of particular interest to them. Please send these materials to Dawn McCracken.
Closing Date: February 2nd (Check with website)
Additional Information: Please check at d.mccracken@lancaster.ac.uk for more information or visit website for further details http://www.lancs.ac.uk/shm/study/doctoral_study/studentships/esrc/.

Geoffrey Leech Scholarships

Subjects: The scholarships are available to students applying for the following Masters programmes delivered at Lancaster: MA in english language and contemporary literary studies; MA in language studies; MA in teaching english as a foreign language (Lancaster-based); MA in teaching english to speakers of other languages (Lancaster-based).
Purpose: To mark the retirement of Professor Geoffrey Leech, the Department of Linguistics & English Language has established a scholarship fund in his honour.
Eligibility: The Geoffrey Leech Scholarships are open to applicants who qualify to pay fees at the UK/EU-fee rate. The scholarships will be awarded on a competitive basis; details of criteria applied may be found on the departmental website.
Level of Study: Postgraduate
Type: Scholarship
Value: They cover the whole of the tuition fee at the appropriate rate.
Frequency: Annual
No. of awards offered: 2
Application Procedure: In order to be considered for these awards, applicants must write a formal request to be considered for the scholarship, in no more than 500 words to Postgraduate Secretary.
Closing Date: March 22nd
Additional Information: Please see the website for further details http://www.ling.lancs.ac.uk/study/masters/funding.htm.

For further information contact:

Lancaster, LA1 4YT, England
Email: m.f.wood@lancaster.ac.uk
Website: www.ling.lancs.ac.uk/study/masters/leech
Contact: Mrs Marjorie Wood, Postgraduate Secretary, Linguistics and English Language

Grizedale College Awards Fund

Subjects: All subjects.
Purpose: Present and former undergraduate members of the college who wish to commence postgraduate study at Lancaster are eligible to apply for awards from this fund.
Eligibility: The awards, in the form of grants, are made according to financial circumstances. The intended purpose of any grant awarded and any past or present contribution to college life are also taken into consideration.
Level of Study: Postgraduate
Type: Award
Value: £1,500
Frequency: Annual
Closing Date: June 1st
Additional Information: Please see the website for further details http://www.lusi.lancs.ac.uk/funding/Detail.aspx?AwardID=29.

For further information contact:

Grizedale College, Lancaster University, Lancaster, LA1 4YU, England
Tel: (44) 0 1524 592190
Email: b.glass@lancaster.ac.uk
Contact: Barbara Glass, College Administrator

Heatherlea Bursary

Subjects: MSc in ecology and environment or the MSc in conservation science.
Purpose: Heatherlea, one of Britain's leading wildlife-holiday operators, offers one annual bursary of £1000 to a student applying for and studying on either the MSc in ecology and environment or the MSc in conservation science.
Eligibility: The Environment Centre will consider suitable candidates automatically so students do not need to apply separately. This award is a bursary and will be awarded on the basis of the financial background/need of the applicant.
Level of Study: Postgraduate
Type: Bursary
Value: £1,000
Frequency: Annual
Additional Information: Please see the website for further details.

For further information contact:

Postgraduate Studies Office, Lancaster Environment Centre, Lancaster University, Lancaster, LA1 4YQ, England
Tel: (44) 0 1524 593478
Email: lec.pg@lancaster.ac.uk

Iredell Trust Bursary

Subjects: MA in historical research.
Purpose: The Iredell Trust provides funding for postgraduate students within the Department of History.
Eligibility: Applicants should have, or expect to obtain, at least a 2:1 Honours or its equivalent in History or a cognate discipline. Awards will be made on the basis of academic record, references and broad outline of dissertation/research area but will also seek to take into account personal circumstances.
Level of Study: Postgraduate
Type: Bursary
Value: UK/EU fees
Frequency: Annual
No. of awards offered: 2
Application Procedure: The closing date for applications is expected to be in June and all applicants accepted for postgraduate study in the Department by this date will be automatically considered. For confirmation of the details of these awards, please contact the Department of History.
Closing Date: June 1st
Additional Information: Please contact to Department of History at g.o'neill@lancaster.ac.uk.

Iredell Trust Scholarships

Subjects: Law for research (PhD).
Purpose: The Iredell Trust provides funding for postgraduate students within the Law School.
Eligibility: Holders of these awards are normally required to provide tutorial assistance for no more than 4–6 hours per week.
Level of Study: Postgraduate, Research
Type: Scholarship
Value: UK/EU fees
Frequency: Annual
No. of awards offered: 2
Application Procedure: The closing date for applications is expected to be in June and all applicants accepted for postgraduate study in the Department by this date will be automatically considered. For confirmation of the details of these awards, please contact the Department of History.
Closing Date: June 1st
Additional Information: Please contact Law School at e.jones@lancaster.ac.uk for more information.

Lancaster Faculty Bursaries

Subjects: Offered by the faculty of arts and social sciences.
Eligibility: Applicants must hold an excellent Bachelors or Masters-level degree, and must have been made an offer onto a full-time study programme leading to a Masters.
Level of Study: Doctorate
Type: Bursary
Value: £13,490–15,000
No. of awards offered: 12
Application Procedure: Please visit www.lancaster.ac.uk/pgfunding for further information and closing date details.
Additional Information: Please see the website for further details.

Lancaster Faculty Research Studentships

Subjects: Arts and social sciences, management school and science and technology.
Purpose: Three of the university's faculties offer Research Studentships.
Eligibility: Applicants must hold an excellent Bachelors or Masters-level degree, and must have been made an offer onto a full-time study programme leading to a PhD.
Level of Study: Postgraduate
Type: Studentship
Value: Full studentships pay a living stipend expected to be between £13,490–15,000 and also exempt students from paying tuition fees (at UK/EU-rate except where indicated otherwise)
Application Procedure: Please visit www.lancaster.ac.uk/pgfunding for further information and closing date details.
Additional Information: Please see the website for further details.

Lancaster MBA Scholarships (E-Fellows)

Subjects: All subjects.
Purpose: For German-speaking students, and sponsored exclusively for members of the German website e-fellows.net.
Eligibility: For all of these Lancaster MBA scholarships, we can only consider you once you hold an offer of a place on the programme. The selection process focuses on the leadership skills or potential demonstrated by candidates in their academic and/or professional life.
Level of Study: Postgraduate
Type: Scholarship
Value: Tuition waiver covering up to the equivalent of £8,250
Frequency: Annual
Application Procedure: Further details of the e-fellows scheme and its corporate sponsors may be found online at www.e-fellows.net.
Closing Date: May 14th
Additional Information: Please see the website for further details.

Lancaster MBA Scholarships (Leadership)

Subjects: All subjects.
Purpose: The Lancaster MBA aims to attract you - the leaders of the future.
Eligibility: For all of these Lancaster MBA scholarships, we can only consider you once you hold an offer of a place on the programme. Available for self-funded candidates from any part of the world. In awarding these scholarships we will be looking for evidence of exceptional leadership qualities in your career to date.
Level of Study: Postgraduate
Type: Scholarship
Value: Up to £5,000
Frequency: Annual
Closing Date: May 14th
Additional Information: Please check at http://www.lums.lancs.ac.uk/masters/mba/admissions/scholarships/ for more information.

Lancaster MBA Scholarships (Open)

Subjects: All subjects.
Eligibility: For all of these Lancaster MBA scholarships, we can only consider you once you hold an offer of a place on the programme. Selection for these scholarships will focus on how your past performance in your career to date or current circumstances justify scholarship support.
Level of Study: Postgraduate
Type: Scholarship
Value: Up to £3,000
Frequency: Annual
Application Procedure: Please visit the website of University for more details.
Closing Date: May 14th
Additional Information: Please see the website for further details.

Lancaster Pamphlets PhD Studentship

Subjects: History.
Purpose: The Lancaster Pamphlets publishing house provides two PhD studentships each year to students in the Department of History.
Eligibility: Applicants for these awards must have applied for a degree programme leading to a PhD award within the Department.
Level of Study: Postgraduate
Type: Studentship
Value: Each award includes a tuition fee exemption at UK/EU rate and a living stipend at Research Council rates (expected to be £13,490)
Frequency: Annual
No. of awards offered: 2
Application Procedure: You are advised to contact the Department for further information and guidance before submitting your application.
Additional Information: Please contact at g.o'neill@lancaster.ac.uk and see the website for further details.

Lancaster University Peel Studentship Trust

Subjects: All subjects.
Purpose: Lancaster students studying at postgraduate level in any subject can apply to this trust, which is unique to the University.
Eligibility: A pre-condition of application to the Peel Studentship Trust is that the applicant must be aged 21 or over on the first day of the first term for which they are seeking assistance. Both UK and non-UK students can apply for a grant. Applications are assessed on the basis of both academic merit and financial need.
Level of Study: Postgraduate
Type: Studentship
Value: Up to £2,500. Students can use such grants to contribute towards paying either tuition fees or living costs.
Length of Study: 1 year
Frequency: Annual
No. of awards offered: 30
Closing Date: June 1st
Contributor: The Peel Studentship Trust
Additional Information: Please see the website for further details.

For further information contact:

Student Services, University House, Lancaster University, Lancaster, LA1 4YW, England
Email: student.services@lancaster.ac.uk
Website: www.lancaster.ac.uk/sbs/download/forms

Peel Studentship Trust Award

Subjects: All subjects.
Purpose: To enable students who are unable to secure finance from other sources to study at Lancaster University.
Eligibility: Open to candidates of any nationality who have a place at Lancaster University and who will be over 21 years of age at the commencement of the course.
Level of Study: Postgraduate
Value: Up to £2,500
Frequency: Annual
Study Establishment: Lancaster University
Country of Study: United Kingdom
No. of awards offered: 25–30
Application Procedure: Applicants must complete an application form, available on request from the Student Support Office.
Closing Date: May 1st
Funding: Trusts
Contributor: Dowager Countess Eleanor Peel Trust
No. of awards given last year: 25
Additional Information: Please see the website for further details http://www.lusi.lancs.ac.uk/funding/Detail.aspx?AwardID=42.

Postgraduate Access Awards (UK-Fee Only)

Subjects: These awards are intended for UK students commencing postgraduate Masters and Diploma courses in October.

Purpose: Lancaster University is intending to offer up to six Postgraduate Access Awards from the Access to Learning Fund.
Eligibility: Applicants should normally hold at least a 2:1 Honours degree.
Level of Study: Postgraduate
Type: Award
Value: £500. These awards are intended to help towards living costs associated with undertaking full-time or part-time postgraduate study (where part-time equals at least 50 per cent of a full-time course). Awards are intended for students not otherwise in receipt of Government funding, but who are able to find finance for the remaining costs of their course.
Frequency: Annual
Closing Date: Late July
Contributor: Access to learning fund
Additional Information: Please see the website for further details.

For further information contact:

Student Services, University House, Lancaster University, Lancaster, LA1 4YW, England
Email: student.services@lancaster.ac.uk
Website: www.lancaster.ac.uk/sbs/download/forms

Robinson Scholarship

Subjects: Operational research and management science.
Purpose: Eddie Robinson is a graduate of the Operational Research programme who, after leaving Lancaster, has had a successful career in the United States with Mars Inc. He has generously sponsored a scholarship for an MSc in Operational Research & Management Science student from a developing country (as defined in the World Development Report published by the World Bank).
Eligibility: As the scholarship will not cover all the costs associated with studying, applicants should explain carefully their financial circumstances. This explanation should cover two aspects. Firstly, why is it that you need funding? Secondly, how will you finance the rest of the cost of your stay in Lancaster?
Level of Study: Postgraduate
Type: Scholarship
Value: £5,000
Frequency: Annual
Application Procedure: Please see the application details at http://www.lancs.ac.uk/studentservices/download/forms/.
Closing Date: May 1st
Additional Information: Please see the website for further details http://www.lums.lancs.ac.uk/masters/management-science/robinson/.

For further information contact:

Department of Management Science, Lancaster University, Lancaster, LA1 4YX, England
Email: g.rand@lancaster.ac.uk
Contact: Graham Rand, MSc Admissions Tutor

LANCASTER UNIVERSITY MANAGEMENT SCHOOL

Bailrigg, Lancaster, LA1 4YX, United Kingdom
Tel: (44) 1524 510752
Fax: (44) 1524 592417
Email: mba@lancaster.ac.uk
Website: www.lums.lancs.ac.uk

Lancaster University Management School (LUMS) is one of the only two United Kingdom to business schools have the coveted 6-star (6*) rating for research quality, and has been rated 'excellent' for teaching quality by the Higher Education Funding Council for England and Wales. LUMS is EQUIS-accredited and its MBA programmes are AMBA-accredited and winners of the Association of MBAs Student of the Year Award twice. In the Financial Times MBA rankings 2006, LUMS was placed joint 30th in the world. The Lancaster MBA is ranked 1st in Europe (and 2nd in the world) for the value for money it delivers to its students and alumini.

Lancaster MBA High Potential Scholarship

Subjects: Full-time MBA.
Purpose: To assist exceptional highly motivated candidates to pursue the full-time MBA.
Eligibility: Open to candidates who hold an offer of a place on the Lancaster MBA.
Level of Study: Postgraduate, Full-Time MBA
Type: Scholarship
Value: £13,000–26,000
Length of Study: 1 year
Frequency: Annual
Study Establishment: Lancaster University Management School
Country of Study: United Kingdom
No. of awards offered: Up to 5
Application Procedure: To apply for this scholarship you have to prepare a three slide PowerPoint presentation as part of your application and upload this to the application site. This will then be discussed at your interview. Please refer to the website for more details on presentation.
Closing Date: May 31st (Check with website)
Funding: Private
Additional Information: You must upload an application for the Open and High Potential Scholarships with your MBA application documents prior to interview. Please see the website for fruther details http://www.lums.lancs.ac.uk/masters/mba/admissions/scholarships/.

Lancaster MBA Open Scholarships

Subjects: Management.
Purpose: To provide financial aid to help with the costs of the Lancaster MBA.
Eligibility: Open to candidates who already hold an offer of a place or a Lancaster MBA.
Level of Study: Postgraduate, full-time MBA.
Value: up to £3,000
Length of Study: 1 year
Frequency: Annual
Study Establishment: Lancaster University Management School
Country of Study: United Kingdom
No. of awards offered: Dependent on demand and funds available
Application Procedure: Applicants should contact the MBA Office for an application form. Candidates must write a 1,000 word statement.
Closing Date: March 13th and May 15th (two rounds of scholarship selections)
Funding: Private
No. of awards given last year: None
No. of applicants last year: None
Additional Information: Please see the website for further details http://www.lums.lancs.ac.uk/masters/mba/admissions/scholarships/.

LE CORDON BLEU AUSTRALIA

Days Road, Regency Park, South Australia, 5010, Australia
Tel: (61) 618 8346 3700/61 8 8348 3000
Fax: (61) 618 8346 3755/61 8 8348 3081
Email: australia@cordonbleu.edu
Website: www.lecordonbleu.com

Le Cordon bleu, a global leader hospitality education, provides professional development for existing executives.

The Culinary Trust Scholarship

Subjects: Hospitality management.
Purpose: To provide financial assistance towards the Masters of International Hospitality Management in the Adelaide school.
Eligibility: Open to applicants who meet entry requirements of the Masters of International Hospitality Management.
Level of Study: Postgraduate
Type: Partial scholarship
Value: Australian $8,000
Length of Study: 6 months
Frequency: Annual
Study Establishment: Le Cordon Bleu Australia, Adelaide
Country of Study: Australia

No. of awards offered: 1
Application Procedure: Application form and full guidelines available from The Culinary Trust website.
Closing Date: March 1st
Funding: Private
Additional Information: Scholarship is for partial tuition of the Graduate Certificate only and does not include other fees associated with commencing or continuing with the program. Please see the website for further details http://www.cordonbleu.edu/sydney/scholarships/en.

The James Beard Foundation Award

Subjects: International Hotel & Restaurant Management and International Hospitality Management.
Purpose: To provide financial assistance.
Eligibility: Applicant must be a graduate.
Level of Study: Postgraduate
Type: Scholarship
Value: The Foundation is awarding a scholarship to cover the tuition of the Graduate Certificate in International Hospitality Management of the Master of International Hospitality Management at a value of Australian $12,440 and one for partial tuition of the MBA of Australian $5000.
Frequency: Annual
Country of Study: Australia
No. of awards offered: 2
Application Procedure: A copy of educational history statement of culinary goals, financial statement and two letters of reference must be submitted.
Closing Date: Check with website
Additional Information: Please see the website for further details http://www.lcblondon.com/london/scholarships/en.

For further information contact:

The Beard House 167 West 12th Street, New York, NY 10011
Website: www.jamesbeard.org

LE CORDON BLEU OTTAWA

453 Laurier Avenue East, Ottawa, ON, K1N 6R4, Canada
Tel: (1) 613 236 2433
Fax: (1) 613 236 2460
Email: ottawa@cordonbleu.edu
Website: www.lcbottawa.com

Le Cordon Bleu Ottawa Culinary Arts Institute has evolved from a Parisian culinary school, which has international culinary network with more than 30 schools in 15 countries. It has a team of over 80 distinguished Master chefs whose learning is crafted into the course of study. In addition to teaching, Le Cordon Bleu has expanded its activities to include culinary publications, video cassettes, TV series, and cooking equipment.It has incorporated its culinary expertise into the restaurant Le Cordon Bleu Signatures Restaurant and has also developed accessories and table settings with its sister company Pierre Deux–French Country.

Le Cordon Bleu Ottawa Culinary Arts Institute Basic Cuisine Certificate Scholarship

Subjects: Culinary arts.
Eligibility: Open to the career professionals.
Level of Study: Graduate
Type: Scholarship
Value: Full-tuition fees
No. of awards offered: 1

Le Cordon Bleu Paris Certificat De Perfectionnement Professionel En Cuisine Scholarship

Subjects: Cuisine.
Eligibility: Open to the graduates from Le Cordon Bleu Diplôme de Cuisine who have a moderate command of the French language.
Level of Study: Graduate
Value: $2,500, tuition fees only
No. of awards offered: 1
Application Procedure: Please refer to the website for details.

LEAGUE OF UNITED LATIN AMERICAN CITIZENS (LULAC)

LULAC National Office 1133 19th Street, NW, Suite 1000, Washington, DC, 20036, United States of America
Tel: (1) 202 833 6130
Fax: (1) 202 833 6135
Email: scholarships@lnesc.org
Website: www.lulac.org

League of United Latin American Citizens (LULAC) consists of approximately 115,000 members throughout the United States of America and Puerto Rico. It is the largest and oldest Hispanic organization in the United States of America. LULAC advances the economic condition, educational attainment, political influence, health and civil rights of Hispanic Americans through community-based programmes operating at more than 700 LULAC councils nationwide. The organization involves and serves all Hispanic nationality groups.

LULAC National Scholarship Fund

Subjects: All subjects.
Purpose: To fund Hispanic students attending colleges and universities.
Eligibility: Open to citizens or legal residents of the United States who have applied to or enrolled in a college, university or graduate school.
Level of Study: Postgraduate
Type: Scholarships
Value: US$250–2,000
Frequency: Annual
Country of Study: United States of America
No. of awards offered: Varies
Application Procedure: Send applications to the nearest LULAC Council in your state as listed on the participating council list (see below). Applications will not be accepted at the LNESC National Office in Washington, DC.
Closing Date: March 31st
Additional Information: For additional information, please email: scholarships@lnesc.org. Please see the website for further details http://lnesc.org/site/338/Scholarships/LULAC-National-Scholarship-Fund.

LEEDS INTERNATIONAL PIANOFORTE COMPETITION

Leeds International Piano Competition, The University of Leeds, Leeds, LS2 9JT, United Kingdom
Tel: (44) 113 244 6586
Fax: (44) 113 234 6106
Email: pianocompetition@leeds.ac.uk/info@leedspiano.com
Website: www.leedspiano.com

The Leeds International Pianoforte Competition is a member of the World Federation of International Music Competitions and the Alink/Argerich Foundation. It was founded in 1961 and since then has developed to become the world's greatest piano competition producing prizewinners who have gone onto successful international careers.

Henry Rudolf Meisels Bursary Awards

Subjects: Music and performing arts.
Purpose: To award competitors accepted in the first stage of the competition.
Eligibility: Open to candidates who are accepted to perform in the first stage of the competition.
Level of Study: Professional development
Type: Scholarships and fellowships
Value: UK£100
Frequency: Every 3 years
Study Establishment: The University of Leeds
Country of Study: United Kingdom
No. of awards offered: Varies
Application Procedure: Entry by competitive audition. Application forms should be submitted by the closing date.
Closing Date: February 1st

Contributor: Henry Rudolf Meisels Bequest
No. of awards given last year: 71
No. of applicants last year: 196
Additional Information: A non-refundable entrance fee of UK£50 must be paid no later than February 1st.

For further information contact:

Email: pianocompetition@leeds.ac.uk
Website: www.leedspiano.com

Leeds International Pianoforte Competition Award

Subjects: Music and piano.
Purpose: To promote the careers of the winners.
Eligibility: Open to young, talented, professional pianists, who were born on or after September 1, 1982.
Level of Study: Professional development
Type: Award
Value: UK£65,750
Frequency: Every 3 years, Once every 3 years
Study Establishment: University of Leeds
Country of Study: United Kingdom
No. of awards offered: 33
Application Procedure: Application forms, completed in English, and repertoire to be submitted before the closing date.
Closing Date: February 1st
Funding: Foundation, individuals, private, trusts
Contributor: Leeds International Pianoforte Competition
No. of awards given last year: 33
No. of applicants last year: 235
Additional Information: A non-refundable entrance fee of UK£50 must be paid no later than February 1, 2012.

LENTZ PEACE RESEARCH ASSOCIATION (LPRA)

University of Missouri-St Louis, 366 Social Sciences and Business Building (MC 58), One University Boulevard, St Louis, MO, 63121 4400, United States of America
Tel: (1) 314 516 5753
Fax: (1) 314 516 6757
Email: bob.baumann@umsl.edu
Website: www.cfis-umsl.com
Contact: Mr Robert Baumann, Board Member

The Lentz Peace Research Association, which Theodore F. Lentz founded originally in 1930 as the Character Research Institute, is the oldest continuously operating peace research center in the world.

Theodore Lentz Fellowship in Peace and Conflict Resolution Research

Subjects: International relations.
Purpose: To support research projects in peace and conflict resolution and to enable the recipient to teach an introductory peace studies course in the Fall semester and one course in the Spring semester at a selected university in the US.
Eligibility: A completed PhD is required and preference is given to graduates of university programmes in peace studies and conflict resolution. Graduates of political science, international relations and other social science programmes that specialize in peace and conflict resolution are also invited to apply.
Level of Study: Postdoctorate
Type: Fellowship
Value: Approx. US$23,400 plus university benefits plus US$1,000 travel and expense allowance
Length of Study: 9 months
Frequency: Dependent on funds available
Country of Study: United States of America
No. of awards offered: 1
Application Procedure: Applicants must send a curriculum vitae, a letter of application, evidence of completion of the PhD, three letters of recommendation and a research proposal of approximately 750 words to the Lentz Peace Research Association.
Closing Date: April 15th
Funding: Private
Contributor: The Lentz Peace Research Association
No. of awards given last year: 1
No. of applicants last year: 10
Additional Information: The fellowship is supported in part by a selected US university. The fellow must serve in residence at that university.

LEO BAECK INSTITUTE (LBI)

15 West 16th Street (Between 5th & 6th Avenues), New York, NY, 10011, United States of America
Tel: (1) 212 744 6400/212 294 8340
Fax: (1) 212 988 1305
Email: lbaeck@lbi.cjh.org
Website: www.lbi.org
Contact: Secretary

The Leo Baeck Institute (LBI) is a research, study and lecture centre, a library and repository for archival and art materials. It is devoted to the preservation of original materials pertaining to the history and culture of German-speaking Jewry.

David Baumgardt Memorial Fellowship

Subjects: Modern intellectual history of German-speaking Jewry.
Purpose: To provide financial support to scholars whose research projects are connected with the writings of Professor David Baumgardt or his scholarly interests.
Eligibility: Applicants must be affiliated with an accredited institution of higher education.
Level of Study: Postgraduate, Doctorate, Postdoctorate
Type: Fellowship
Value: Up to US$3,000
Length of Study: 1 year
Frequency: Annual
Study Establishment: The Leo Baeck Institute
Country of Study: United States of America
No. of awards offered: 1
Application Procedure: Applicants must submit an application form, a curriculum vitae and a full description of the research project. Doctoral students must submit official transcripts of graduate and undergraduate work, written evidence that they are enrolled in a PhD programme and two letters of recommendation, one by their doctoral adviser and one by another scholar familiar with the work. Post-doctoral candidates must submit evidence of their degree, of which transcripts are not required, and two letters of recommendation from two colleagues familiar with their research. To apply please download and fill out the application form (DOC) and mail it to the attention of Dr. Frank Mecklenburg at the LBI.
Closing Date: November 1st (Check with website)
Funding: Private
Contributor: The Leo Baeck Institute
No. of awards given last year: 1
No. of applicants last year: 2
Additional Information: Please see the website for further details http://www.lbi.org/about/fellowships/david-baumgardt-memorial-fellowship/.

Fritz Halbers Fellowship

Subjects: Culture and history of German-speaking Jewry.
Purpose: To provide financial assistance to scholars whose projects are connected with the culture and history of German-speaking Jewry.
Eligibility: Applicants must be enrolled in a PhD program at an accredited institution of higher education
Level of Study: Doctorate, Graduate, Postdoctorate, Postgraduate, Predoctorate
Type: Fellowship
Value: Up to US$3,000
Length of Study: 1 year
Frequency: Annual
Country of Study: United States of America
No. of awards offered: Varies
Application Procedure: Applicants must submit an application form, curriculum vitae and a full description of the research project. Doctoral students must submit official transcripts of graduate and undergraduate work, written evidence that they are enrolled in a PhD

programme and two letters of recommendation, one by their doctoral adviser and one by another scholar familiar with the work. Post-doctoral candidates must submit evidence of their degree, of which transcripts are not required, and two letters of recommendation from two colleagues familiar with their research. To apply please fill download and fill out the application form (DOC) and mail it to the attention of Dr. Frank Mecklenburg at the LBI.
Closing Date: November 1st (Check with website)
Funding: Private
Contributor: The Leo Baeck Institute
No. of awards given last year: 1
No. of applicants last year: 7
Additional Information: Please see the website for further details http://www.lbi.org/about/fellowships/fritz-halbers-fellowship/.

LBI/DAAD Fellowship for Research at the Leo Baeck Institute, New York

Subjects: Social, communal and intellectual history of German-speaking Jewry.
Purpose: To provide assistance to students for dissertation research and to academics for writing a scholarly essay or book.
Eligibility: Applicants must be US citizens, and PhD candidates or recent PhD's who have received their degrees within the preceding two years.
Level of Study: Doctorate, Graduate, Postdoctorate
Type: Fellowship
Value: US$2,000
Length of Study: 1 year
Frequency: Annual
Study Establishment: The Leo Baeck Institute
Country of Study: United States of America
No. of awards offered: 2
Application Procedure: Applicants must submit an application form, a curriculum vitae and a full description of the research project. Doctoral students must submit official transcripts of graduate and undergraduate work, written evidence that they are enrolled in a PhD programme and two letters of recommendation, one by their doctoral adviser and one by another scholar familiar with the work. Post-doctoral candidates must submit evidence of their degree, of which transcripts are not required, and two letters of recommendation from two colleagues familiar with their research.
Closing Date: November 1st
Funding: Private
Contributor: The Leo Baeck Institute
No. of awards given last year: 2
Additional Information: The fellowship holders must agree to submit a brief report on their research activities after the period for which the fellowship was granted. Please see the website for further details http://www.lbi.org/about/fellowships/.

LBI/DAAD Fellowships for Research in the Federal Republic of Germany

Subjects: Social, communal and intellectual history of German-speaking Jewry.
Purpose: To provide financial assistance to doctoral students conducting research for their dissertation and to academics in the preparation of a scholarly essay or book.
Eligibility: Applicants must be citizens of the United States of America and PhD candidates or recent PhDs who have received their degrees within the preceding 2 years.
Level of Study: Doctorate, Postdoctorate
Type: Fellowship
Value: A monthly stipend of approximately €1,000, health insurance and a flat rate subsidy for travel costs (US East: €775/West: 1,000).
Length of Study: 1 year
Frequency: Annual
Country of Study: Germany
No. of awards offered: 1–2
Application Procedure: Applicants must submit an application form, a curriculum vitae and a full description of the research project. Doctoral students must submit official transcripts of graduate and undergraduate work, written evidence that they are enrolled in a PhD programme and two letters of recommendation, one by their doctoral adviser and one by another scholar familiar with the work. Post-doctoral candidates must submit evidence of their degree, of which transcripts are not required, and two letters of recommendation from two colleagues familiar with their research. For any questions please email Dr. Frank Mecklenburg at fmecklenburg@lbi.cjh.org.
Closing Date: November 1st
Funding: Private
Contributor: The Leo Baeck Institute
Additional Information: The fellowship holders must agree to submit a brief report on their research activities upon conclusion of their fellowship. These awards are in conjunction with awards offered by the German Academic Exchange Service (DAAD) in New York, United States of America. Please see the website for further details http://www.lbi.org/about/fellowships/lbidaad-fellowships/.

LEUKAEMIA & LYMPHOMA RESEARCH

Leukaemia & Lymphoma Research, 39-40 Eagle Street, London, WC1R 4TH, England
Tel: (44) 020 7504 2200
Fax: (44) 020 7405 3139
Email: info@llresearch.org.uk
Website: www.llresearch.org.uk

Leukaemia research is devoted exclusively to leukaemia, Hodgkin's disease and other lymphomas, myeloma, myelodysplastic syndromes, aplastic anaemia and the myeloproliferative disorders. We are committed to finding causes, improving and developing new treatments and diagnostic methods as well as supplying free information booklets and answering written and telephone enquiries.

The Clinical Research Training Fellowship

Subjects: All life science disciplines. The research topic must be applicable to blood cancers.
Purpose: To train registrar grade clinicians in research and allow them to obtain a higher degree.
Eligibility: Open to researchers of any nationality who work and reside in the United Kingdom.
Level of Study: Research
Type: Fellowship
Value: £50K – £100K
Length of Study: 1-3 years
Frequency: 3 times per year
Study Establishment: Universities, medical schools, research institutes and teaching hospitals
Country of Study: United Kingdom
No. of awards offered: Varies
Application Procedure: Applicants must complete an application form.
Closing Date: March 4th, July 15th
Additional Information: Please see the website for further details.

For further information contact:

Tel: 020 7405 0101
Email: sdarling@lrf.org.uk
Website: http://leukaemialymphomaresearch.org.uk/research/researchers/funding-opportunities
Contact: Sara Darling

Gordon Piller PhD Studentships

Subjects: All life science disciplines. The research topic must be applicable to blood cancers.
Purpose: To train graduates in life sciences in research and allow them to obtain a PhD degree.
Eligibility: Open to graduates of any nationality who work and reside in the United Kingdom.
Level of Study: Postgraduate
Type: Studentship
Value: Stipend of £18,870 (London) or £17,340 (outside London) per year plus fees together with an allowance of £12,000 towards research costs
Length of Study: 3 years
Frequency: Annual
Study Establishment: Universities, medical schools and research institutes
Country of Study: United Kingdom
No. of awards offered: 4

Application Procedure: Applicants must be invited to apply and must complete the appropriate form.
Closing Date: March 2nd
No. of awards given last year: 4
Additional Information: Please see the website for further details http://leukaemialymphomaresearch.org.uk/research/researchers/funding-opportunities.

LEUKAEMIA FOUNDATION

230 Lutwyche Road, Windsor, QLD, 4030, Australia
Tel: (61) 07 3866 4049
Fax: (61) 07 3866 4011
Email: jridge@leukaemia.org.au
Website: www.leukaemia.org.au
Contact: Jacinta Ridge, Mission and Vision Project Officer

The Leukaemia Foundation has offices across Australia and it is the only national non-profit organization dedicated to the care and cure of patients and families living with leukaemia, lymphoma, myeloma and related blood disorders.

Leukaemia Foundation PhD Scholarships

Subjects: Leukaemia.
Purpose: To enhance the knowledge or treatment of haematological malignancies to improve the care of the patients and their families.
Eligibility: Open to candidates who are citizens or permanent residents of Australia.
Level of Study: Doctorate
Type: Scholarships
Value: $40,000 ($30,000 stipend + $10,000 consumables)/annum for three years.
Frequency: Annual
Country of Study: Australia
No. of awards offered: 5
Application Procedure: Applicants can download the application forms from the website http://www.leukaemia.com/web/research/fellowships_phd.php.
Closing Date: September 14th
Funding: Foundation
Contributor: Leukaemia Foundation
Additional Information: All queries should be directed to Dr Anna Williamson, General Manager, Research, Advocacy and Patient Care, Leukaemia Foundation, National Research Program, PO Box 2126, Windsor, Queensland 4030, Australia. Email: awilliamson@leukaemia.org.au. Please see the website for further details http://www.leukaemia.com/web/research/researchgrants_applications.php.

Leukaemia Foundation Postdoctoral Fellowship

Subjects: Leukaemia.
Purpose: To encourage and support young researchers and to foster cutting-edge research to improve the understanding of leukaemia and related malignancies and to benefit patients and families in the short-term or long-term.
Eligibility: Open to candidates who have obtained a PhD no more than 3 years before the closing date for applications. Their PhD must be obtained by December of the year of application.
Level of Study: Postdoctorate
Type: Fellowship
Value: $100,000 (75 per cent salary + 25 per cent consumables) per year
Frequency: Annual
Country of Study: Australia
No. of awards offered: 1 or 2
Application Procedure: Applicants can download the application form from the website.
Closing Date: July 27th
Funding: Foundation
Contributor: Leukaemia Foundation
Additional Information: All queries should be directed to Dr. Anna Williamson, General Manager, Reseach, Advocacy and Patient Care, Leukaemia Foundation, National Research Program, PO Box 2126, Windsor, Queensland 4030, Australia. Email: awilliamson@leukaemia.org.au. Please see the website for further details http://www.leukaemia.com/web/research/researchgrants_applications.php.

Senior Research Fellowship

Subjects: Leukaemia.
Purpose: This is the most prestigious of the leukaemia foundation's personal awards and will provide sustained support for an outstanding young researcher preferably between 3 and 10 years postdoctorate who has established an international reputation for research in the field of leukaemia lymphona, myeloma related blood cancers and disorders.
Level of Study: Postdoctorate
Type: Fellowships
Value: $200,000 per year
Length of Study: 5 years
Frequency: Annual
Country of Study: Austria
No. of awards offered: 1
Application Procedure: Via website.
Closing Date: July 2nd
Funding: Foundation
Contributor: The Leukaemia Foundation Australia
Additional Information: Please see the website for further details http://www.leukaemia.com/web/research/researchgrants_applications.php.

THE LEUKEMIA & LYMPHOMA SOCIETY OF CANADA (LLSC)

2 Lansing Square, Suite 804, Toronto, ON, M2J 4P8, Canada
Tel: (1) 877 668 8326
Fax: (1) 416 661 7799
Email: laila.ali@lls.org
Website: www.llscanada.org
Contact: CEO

The Leukaemia & Lymphoma Society of Canada's mission is to cure leukaemia, lymphoma, Hodgkin's disease and myeloma and to improve the quality of life of patients and their families. Since its founding, the Society has invested millions of dollars in research specifically targeting leukaemia, lymphoma and myeloma.

LLSC Awards

Subjects: Leukaemia, lymphoma (Hodgkin's and non-Hodgkin's) and myeloma.
Purpose: To support peer-reviewed research in leukaemia, lymphoma (Hodgkin's and non-Hodgkin's) and myeloma.
Eligibility: Open to applicants with a faculty appointment at a Canadian university.
Level of Study: Research, Postdoctorate
Type: Research grant
Value: Varies
Length of Study: 1–2 years
Frequency: Annual
Country of Study: Canada
No. of awards offered: Varies
Application Procedure: Applicants must complete an official online application available on the website https://proposalcentral.altum.com/.
Closing Date: February 1st
Funding: Private
No. of awards given last year: 13 - Operating grants; 9 - Studentships
No. of applicants last year: 60
Additional Information: A special grant is available for the study of CLL. Please see the website for further details http://www.llscanada.org/#/researchershealthcareprofressionals/academicgrants/.

LEUKEMIA AND LYMPHOMA SOCIETY

1311 Mamoroneck Avenue, Suite 310, White Plains, NY, 10605, United States of America
Tel: (1) 914 949 5213
Fax: (1) 914 949 6691
Email: researchprograms@lls.org
Website: www.lls.org
Contact: Director of Research Administration

The Leukemia and Lymphoma Society is a national voluntary health agency dedicated to the conquest of leukaemia, lymphoma and myeloma through research. Through its research programme, the Society hopes to encourage and promote research activity of the highest quality. In addition, the Society also supports patient aid, public and professional education and community service programmes.

Career Development Program

Subjects: Leukemia, lymphoma, Hodgkin's disease and myeloma research.
Purpose: To provide support for individuals pursuing careers in basic, clinical or translational research.
Eligibility: Applications may be submitted by individuals working in domestic or non-profit organizations such as universities, hospitals or units of state and local governments. International applicants are welcome to apply.
Level of Study: Postdoctorate
Type: Fellowship
Value: US$55,000–110,000 per year
Length of Study: 3–5 years, based on experience and training
Frequency: Annual
Study Establishment: International and domestic non-profit organizations, including hospitals, universities and research institutes
Country of Study: Any country
No. of awards offered: Varies
Application Procedure: Applicants can visit the website for details of the Career Development Award application packet at www.lls.org under "Researchers & Healthcare Professionals".
Closing Date: Check with website
Funding: Private
No. of awards given last year: 46
No. of applicants last year: 379
Additional Information: Special Fellow awards are for up to $65,000 per year for three years. Scholar awards are for up to $110,000 per year for five years. Please see the website for further details http://www.lls.org/#/researchershealthcareprofessionals/academicgrants/careerdevelopment/.

Specialized Center of Research

Purpose: To encourage collaboration among leaders in the field to work synergistically toward a common goal.
Eligibility: Open to Senior Investigators with academic excellence and depending on the candidate's contribution to the field.
Level of Study: Postdoctorate, Research
Type: Fellowships
Value: Up to $1,250,000
Length of Study: 5 years
Frequency: Annual
Study Establishment: International and domestic, non-profit organizations, including hospitals, universities, and research institutes
Country of Study: Any country
Application Procedure: Please visit www.lls.org for additional information. Select "Researchers & Healthcare Professionals" or you can e-mail us at researchprogramms@lls.org.
Closing Date: November 1st (For letter of intent) and March 15th (For application submission)
Funding: Private
No. of awards given last year: 2
No. of applicants last year: 18
Additional Information: Please see the website for further details http://www.lls.org/#/researchershealthcareprofessionals/academicgrants/specializedcenterresearch/.

Translational Research Program

Subjects: Leukaemia, lymphoma, Hodgkin's disease and myeloma research.
Purpose: To encourage and provide early stage support for clinical research.
Eligibility: Open to individuals working in domestic or foreign nonprofit organizations such as universities, hospitals or units of state and local governments. International applicants are welcome to apply.
Level of Study: Postdoctorate, Research
Value: Up to US$200,000 per year
Length of Study: 3 years, with the possibility of 2 additional years
Frequency: Annual
Study Establishment: International and domestic non-profit organizations, including hospitals, universities and research institutes
Country of Study: Any country
No. of awards offered: Varies
Application Procedure: Applicants must contact the Society or visit the website for details. www.lls.org under "Researchers & Healthcare Professionals" or e-mail us at researchprograms@lls.org
Closing Date: March 1st (For letter of intent) and March 15th (For application submission)
Funding: Private
No. of awards given last year: 36
No. of applicants last year: 217
Additional Information: Please see the website for further details http://www.lls.org/#/researchershealthcareprofessionals/academicgrants/translationalresearch/.

LEUKEMIA RESEARCH FOUNDATION (LRF)

Research Grants Administrator, Leukemia Research Foundation, 3520 Lake Avenue, Suite 202, Wilmette, IL, 60091 1064, United States of America
Tel: (1) 847 424 0600, 888 558 5385
Fax: (1) 847 424 0606
Email: Linda@lrfmail.org/info@LRFmail.org
Website: www.leukemia-research.org

The Leukemia Research Foundation (LRF) was established in 1946. It aims to conquer leukemia, lymphoma and myelodysplastic syndromes by funding research into their causes and cures and to enrich the quality of life of those touched by these diseases.

LRF New Investigator Research Grant

Subjects: Basic science and clinical science.
Purpose: To enable an investigator to initiate and develop a project sufficiently to obtain continued funding from national agencies.
Eligibility: Preference given to proposals that focus on leukemia, lymphoma and MDS. New Investigators are considered to be within seven (7) years of their first independent position. Years as a resident physician, fellow physician, or post-doctoral fellow are considered to be training years. Questions regarding eligibility should be directed to our Research Grants Administrator. See the website for details regarding eligibility.
Level of Study: Postgraduate, New Investigators
Type: Grant
Value: US$100,000
Length of Study: 1 year
Frequency: Annual
Country of Study: United States of America
No. of awards offered: 5–10
Application Procedure: Applicants must submit application form, references, self-addressed stamped envelope and 1 paragraph abstract in lay terms.
Closing Date: February 18th
Funding: Private
Contributor: Leukaemia Research Foundation
No. of awards given last year: 10
No. of applicants last year: 73
Additional Information: Please see the website for further details http://www.leukemia-research.org/page.aspx?pid=216.

For further information contact:

Leukemia Research Foundation, 820 Davis Street, Suite #420, Evanston IL, 60201, United States of America
Tel: (1) 847 424 0600
Fax: (1) 847 424 0606
Email: Info@LRFMail.org
Contact: Kelli Fitzgerald, Medical Grants Administrator

THE LEVERHULME TRUST

Research Awards Advisory Committee, 1 Pemberton Row, London, EC4A 3BG, England
Tel: (44) 0207 042 9861
Fax: (44) 0207 042 9889
Email: agrundy@leverhulme.ac.uk
Website: www.leverhulme.ac.uk
Contact: Mrs Anna Grundy, Grants Manger

The Trust, established at the wish of William Hesketh Lever, makes awards for the support of research and education. The Trust emphasises individuals and encompasses all subject areas.

Early Career Fellowship

Subjects: Any subject.
Purpose: To provide career development opportunities for those who are at a relatively early stage of their academic careers.
Eligibility: Applicants must: not yet have held a full-time established academic post in a UK University or camparable UK Institution, nor may fellows hold such post concurrently with the ECF; must hold a doctorate or have equivalent research experience by the time they take up the fellowhip. Those who are or have been registered for a doctorate may only apply if they have submitted their doctoral thesis by the closing date of 4pm on March 7th; must have had their doctoral viva more than 5 years from the closing date.
Level of Study: Postdoctorate
Type: Fellowships
Value: The trust will contribute 50 per cent of each fellow's total salary costs up to a maximum of £23,000 per year, with the balance paid by the host institution. Each fellow may request annual research expenses of up to £6,000 to further his or her research activiteies
Length of Study: 3 years
Frequency: Annual
Country of Study: United Kingdom
No. of awards offered: Approx. 80
Application Procedure: Applicants must complete an online application form that is accessible from the website.
Closing Date: March 7th
Funding: Private
No. of awards given last year: 84
No. of applicants last year: 714
Additional Information: Applicants must hold a degree from a UK higher education institution at the time of taking up the fellowship or at the time of application hold an academic position in the UK. Applications from those having an association with the UK academic community of less than 2 years' duration will be strengthened by a move of employing institution.

International Academic Fellowships

Subjects: All subjects.
Purpose: To provide established researchers with a concentrated period based in one or more research centres outside the UK.
Eligibility: Applicants must be resident in the UK; should have held an established post in a UK institution of higher education, or in a museum, art gallery or comparable institution for at least 5 years.
Level of Study: Postdoctorate, Professional development
Type: Fellowship
Value: Maximum value of £22,000
Length of Study: 3–12 months
No. of awards offered: Approx. 15
Application Procedure: Applicants must complete an online application form that is accessible from the website.
Closing Date: November 8th
Funding: Private
Additional Information: Applicants may not be registered for a degree, doctoral studies, or for professional or vocational qualifications.

Leverhulme Trust Emeritus Fellowships

Subjects: All subjects.
Purpose: To assist senior established researchers to complete a research project and prepare the results for publication.
Eligibility: Open to individuals who have retired or who are about to retire. Applicants must hold, or have recently held, teaching and/or research posts in universities or institutions of similar status in the United Kingdom. Applicants must also have an established record of research.
Level of Study: Postdoctorate
Type: Fellowship
Value: Up to UK£22,000 by individual assessment. The awards are to meet incidental costs and do not provide a personal allowance or pension supplementation
Length of Study: 3 months–2 years
Frequency: Annual
No. of awards offered: Approx. 35
Application Procedure: Applicants must complete an online application form that is accessible from the website www.leverhulme.ac.uk.
Closing Date: January 31st
Funding: Private
No. of awards given last year: 32
No. of applicants last year: 91

Leverhulme Trust Research Fellowships

Subjects: All subjects.
Purpose: To assist experienced researchers pursuing investigations who are prevented, by routine duties or other causes, from undertaking or completing a research programme.
Eligibility: Open to experienced researchers, particularly those who are or have been prevented by routine duties from completing a programme of original research. There are no restrictions on academic discipline and awards are not limited to those holding appointments in higher education. Applicants must be resident in the UK at the time of application.
Type: Fellowship
Value: Up to UK£45,000 by individual assessment
Length of Study: 3 months–2 years
Frequency: Annual
Country of Study: Any country
No. of awards offered: Approx. 90
Application Procedure: Applicants must complete an online application form that is accessible from the website www.leverhulme.ac.uk.
Closing Date: November 8th
Funding: Private
No. of awards given last year: 94
No. of applicants last year: 545
Additional Information: Applicants should be able to demonstrate experience and academic background sufficient to confirm their ability to complete the proposed programme of research. Applicants may not be registered for a degree, for doctoral studies or for professional or vocational qualifications.

Leverhulme Trust Study Abroad Studentships

Subjects: All subjects.
Purpose: To fund advanced study or research at a centre of learning in any overseas country, with the exception of the USA.
Eligibility: Applicants must have been resident in the UK for at least 5 years at the time of application and most hold an undergraduate degree. Applicants must hold a degree from aUK institution. This may be either the undergraduate degree or a further degree.
Level of Study: Doctorate, Graduate, MBA, Postdoctorate, Postgraduate, Predoctorate, Research
Type: Studentship
Value: UK£17,000 per year plus return airfare and other allowances at the discretion of the Committee
Length of Study: 1 or 2 years
Frequency: Annual
Study Establishment: Any centre of learning outside of the UK and USA
No. of awards offered: Approx. 15
Application Procedure: Applicants must complete an online application form that is accessible from the website www.leverhulme.ac.uk
Closing Date: January 7th
Funding: Private
No. of awards given last year: 15
No. of applicants last year: 90
Additional Information: Applicants should either be a student at the time of application or have been registered as a student within the last 8 years and should be able to demonstrate how their work would benefit from being conducted overseas rather than in the UK. Those wishing only to improve their foreign language skills are not eligible for this scheme.

LEWIS & CLARK

0615 SW Palatine Hill Road, Portland, OR, 97219, United States of America
Tel: (1) 503 768 7000
Email: sfs@lclark.edu
Website: www.lclark.edu

The College was founded by Presbyterian pioneers as Albany Collegiate Institute in 1867. The school moved to the former Lloyd Frank estate in Portland's southwest hills in 1942 and took the name Lewis & Clark College.

Mary Stuart Rogers Scholarship

Subjects: Teacher education.
Purpose: To support postgraduate students by funding study at Master's degree or PhD level.
Eligibility: Applicants must be a full-time junior or senior in the College of Arts and Sciences and apply during the academic year prior to the award year, carry a grade point average of at least 3.2, possess personal qualities of dedication, integrity, compassion, sensitivity, self-disciplineand leadership and Demonstrate financial need according to federal and institutional eligibility criteria.
Level of Study: Postgraduate
Type: Scholarship
Value: Up to US$5,000 for a single academic year
Frequency: Annual
Study Establishment: Lewis & Clark College
Country of Study: United States of America
No. of awards offered: Varies
Application Procedure: Applicants must contact the Department of Teacher Education.
Closing Date: March 9th
Additional Information: All recipients also receive a ring to commemorate their designation as a Rogers scholar. Please see the website for further details http://www.lclark.edu/offices/financial_aid/special_scholarships/.

THE LEWIS WALPOLE LIBRARY

PO Box 1408, Farmington, CT, 06034, United States of America
Tel: (1) 860 677 2140
Fax: (1) 860 677 6369
Email: walpole@yale.edu
Website: www.library.yale.edu/Walpole
Contact: Dr Margaret K Powell, W S Lewis Librarian and Executive Director

The Lewis Walpole Library is a research centre for the study of all aspects of primarily British 18th-century studies and is a prime centre for the study of Horace Walpole and Strawberry Hill.

Lewis Walpole Library Fellowship

Subjects: 18th-century British studies including history, literature, theatre, drama, art, architecture, politics, philosophy or social history.
Purpose: To fund study into any aspect of the Library's collection of 18th-century British prints, paintings, books and manuscripts.
Eligibility: Applicants should normally be pursuing an advanced degree or must be engaged in postdoctoral or equivalent research.
Level of Study: Research, Doctorate, Postdoctorate, Postgraduate, Predoctorate
Type: Fellowship
Value: US$2,100 plus travel expenses and on-site accommodation
Length of Study: 1 month
Frequency: Annual
Study Establishment: Lewis Walpole Library, Yale University
Country of Study: United States of America
No. of awards offered: More than 2
Application Procedure: Applicants must submit a curriculum vitae, a brief research proposal of up to three pages and two confidential letters of recommendation. Application materials may also be submitted electronically to margaret.powell@yale.edu.
Closing Date: January 23rd
Funding: Private
No. of awards given last year: 16
No. of applicants last year: 45
Additional Information: Please see the website for further details http://www.library.yale.edu/librarynews/2011/12/lewis_walpole_library_fellowsh_3.html.

LIBRARY & INFORMATION TECHNOLOGY ASSOCIATION (LITA)

50 East Huron Street, Chicago, IL, 60611, United States of America
Tel: (1) 800 545 2433 x4270
Fax: (1) 312 280 3257
Email: lita@ala.org
Website: www.lita.org
Contact: Grants Management Officer

The Library and Information Technology Association (LITA) provides a forum for discussion, an environment for learning and a programme for action on the design, development and implementation of automated and technological systems in the library and information science field.

LITA/Christian (Chris) Larew Memorial Scholarship in Library and Information Technology

Subjects: Library and information science, with an emphasis on information technology.
Purpose: To encourage the entry of qualified people into the library and information technology field.
Eligibility: Open to students at the Master's degree level in an ALA-accredited programme.
Level of Study: Postgraduate
Type: Scholarship
Value: US$3,000
Frequency: Annual
Study Establishment: An ALA-accredited programme
Country of Study: United States of America
No. of awards offered: 1
Application Procedure: Candidates must send a statement indicating their previous experience in the field and what he or she can bring to the profession. Application forms and instructions are available from the website.
Closing Date: March 1st (Check with website)
Contributor: The Electronic Business and Information Services (EBIS), a unit of Baker & Taylor, Inc.
Additional Information: Factors considered in awarding the scholarship are academic excellence, leadership, evidence of commitment to a career in library automation and information technology and prior activity and experience in those fields. Please see the website for further details http://www.ala.org/lita/awards/larew.

LITA/LSSI Minority Scholarship in Library and Information Technology

Subjects: Library science and information technology.
Purpose: To encourage the entry of qualified persons into the library automation field.
Eligibility: Open to citizens of the United States of America or Canada who are qualified members of a principal minority group: American Indian or Alaskan Native, Asian or Pacific Islander, African American or Hispanic.
Level of Study: Postgraduate
Type: Scholarship
Value: US$2,500
Frequency: Annual
Study Establishment: An ALA-accredited programme
Country of Study: United States of America
No. of awards offered: 1
Application Procedure: Applicants must write, phone or visit the website for application forms.
Closing Date: March 1st (Check with website)
Funding: Commercial
Contributor: Library Systems and Service, Inc.
Additional Information: Please see the website for further details http://www.ala.org/lita/awards/lssi.

LITA/OCLC Minority Scholarship in Library and Information Technology

Subjects: Library information science, with an emphasis on library automation.
Purpose: To encourage the entry of qualified minority persons into the library automation field who plan to follow a career in that field and

who evidence potential leadership in, and a strong commitment to, the use of automated systems in libraries.
Eligibility: Open to citizens of the United States of America or Canada who are qualified members of a principal minority group: American Indian or Alaskan Native, Asian or Pacific Islander, African American or Hispanic.
Level of Study: Postgraduate
Type: Scholarship
Value: US$3,000
Frequency: Annual
Study Establishment: An ALA-accredited programme
Country of Study: United States of America
No. of awards offered: 1
Application Procedure: Applicants must write, phone or visit the website for application forms.
Closing Date: March 1st (Check with website)
Funding: Commercial
Contributor: Online Computer Library Center
Additional Information: Factors considered in awarding the scholarship are academic excellence, leadership, evidence of commitment to a career in library automation and information technology and prior activity and experience in those fields. Please see the website for further details http://www.ala.org/lita/awards/oclc.

THE LIBRARY COMPANY OF PHILADELPHIA

1314 Locust Street, Philadelphia, PA, 19107, United States of America
Tel: (1) 215 546 3181
Fax: (1) 215 546 5167
Email: jgreen@librarycompany.org
Website: www.librarycompany.org
Contact: Fellowship Office

Founded in 1731, the Library Company of Philadelphia was the largest public library in America until the 1850s and contains printed materials on aspects of American culture and society in that period. It is a research library with a collection of 500,000 books, pamphlets, newspapers and periodicals, 75,000 prints, maps and photographs and 150,000 manuscripts.

The Library Company of Philadelphia And The Historical Society of Pennsylvania Visiting Research Fellowships in Colonial and U.S. History and Culture

Subjects: 18th- and 19th-century American social and cultural history, African American history, literary history or the history of the book in America.
Purpose: To offer short-term fellowships for research in residence in their collections.
Eligibility: The fellowship supports both postdoctoral and dissertation research. The project proposal should demonstrate that the Library Company and/or the Historical Society of Pennsylvania has a primary source central to the research topic. Candidates are encouraged to enquire about the appropriateness of a proposed topic before applying.
Level of Study: Doctorate, Postdoctorate
Type: Fellowship
Value: US$2,000
Length of Study: 1 month
Frequency: Annual
Study Establishment: An independent research library
Country of Study: United States of America
No. of awards offered: 25
Application Procedure: See website under fellowships.
Closing Date: March 1st. Fellows may take up residence at any time from the following June to May of the next year
Funding: Private
Contributor: The Andrew W Mellon Foundation, the Barra Foundation, the Albert M Greenfield Foundation and the McLean Contributionship
No. of awards given last year: 34
No. of applicants last year: 150
Additional Information: Fellows will be assisted in finding reasonably priced accommodation. International applications are especially encouraged since two fellowships, jointly sponsored with the Historical Society of Pennsylvania, are reserved for scholars whose residence is outside the United States of America. A partial catalogue of the Library's holdings is available through the website. This programme includes fellowships offered by the Library Company's programme in Early American Economy and Society and by the Balch Institute for Ethnic studies.

Library Company of Philadelphia Dissertation Fellowships

Subjects: 18th- and 19th-century American social, cultural and literary history, pre-1860 American economic and business history.
Purpose: To promote scholarship by offering long-term dissertation fellowships.
Eligibility: The fellowship supports dissertation research in the collections of the Library Company and other Philadelphia repositories.
Level of Study: Doctorate
Type: Fellowship
Value: US$10,000
Length of Study: 4.5 months (one semester)
Frequency: Annual
Study Establishment: An independent research library
Country of Study: United States of America
No. of awards offered: 4
Application Procedure: Candidates are encouraged to enquire about the appropriateness of a proposed topic before applying. See website under fellowships.
Closing Date: March 1st. Fellows may take up residence for either the following Fall or the Spring semester
Funding: Private
Contributor: Albert M Greenfield Foundation
No. of awards given last year: 4
No. of applicants last year: 21
Additional Information: Further information can be found on the website, which also includes fellowships offered by the Library Company's programme in early American economy and society.

Library Company of Philadelphia Postdoctoral Research Fellowship

Subjects: 18th- and 19th-century American social, cultural and literary history.
Purpose: To promote scholarship by offering long-term postdoctoral and advanced research fellowships.
Eligibility: The fellowship supports both postdoctoral and advanced research in the collections of the Library Company and often Philadelphia repositories. Applicants must hold a doctoral degree.
Level of Study: Postdoctorate, Research
Type: Fellowship
Value: US$18,900–20,000
Length of Study: 4.5 months (1 semester)
Frequency: Annual
Study Establishment: An independent research library
Country of Study: United States of America
No. of awards offered: 4
Application Procedure: Candidates are encouraged to enquire about the appropriateness of a proposed topic before applying. See website under fellowships.
Closing Date: November 2nd. Fellows may take up residence for either the following Fall or the Spring semester
Funding: Government, private
Contributor: National Endowment for the Humanities
No. of awards given last year: 2
No. of applicants last year: 50
Additional Information: Two of the fellowships are offered by the Library Company's programme in early American economy and society. Two are supported by the National Endowment for the Humanities.

THE LIFE SCIENCES RESEARCH FOUNDATION (LSRF)

Lewis Thomas Laboratory, Princeton University, Washington Road, Princeton, NJ, 08544, United States of America
Tel: (1) 410 467 2597
Email: sdirenzo@princeton.edu
Website: www.lsrf.org
Contact: Susan DiRenzo, Assistant Director

The Life Sciences Research Foundation (LSRF) solicits monies from industry, foundations and individuals to support postdoctoral fellowships in the life sciences. The LSRF recognizes that discoveries and the application of innovations in biology for the public's good will depend upon the training and support of the highest quality young scientists in the very best research environments. The LSRF awards fellowships across the spectrum of life sciences: biochemistry, cell, developmental, molecular, plant, structural, organismic population and evolutionary biology, endocrinology, immunology, microbiology, neurobiology, physiology and virology.

Life Sciences Research Foundation Postdoctoral Fellowships

Subjects: Biological and life sciences.
Purpose: To support postdoctoral fellowships across the spectrum of the life sciences.
Eligibility: Open to researchers of any nationality, who are graduates of medical or graduate schools in the biological sciences and who hold an MD or PhD degree. Awards will be based solely on the quality of the individual applicant's previous accomplishments and on the merit of the proposal for postdoctoral research.
Level of Study: Postdoctorate
Type: Fellowship
Value: US$57,000 per year. the salary scale begins at $43,000 for a first-year postdoctoral, $45,000 for a second year, and $47,000 thereafter. The fellow, not the advisor, will control expenditure of the remainder. It can be used for fringe benefits (up to $2,000/year), travel to the host institution, travel to visit the sponsor and to the LSRF annual meeting. However, its main purpose is to support the fellow's research expenses.
Length of Study: 3 years
Frequency: Annual
Study Establishment: Appropriate research institutions
No. of awards offered: 1
Application Procedure: Electronic submission only. Please check website.
Closing Date: October 1st
Funding: Private
No. of awards given last year: 16
No. of applicants last year: 820
Additional Information: LSRF Fellows must carry out their research at non-profit institutions. The fellowship cannot be used to support research that has any patent commitment or other kind of agreement with a commercial profit-making company. Please see the website for further details http://www.lsrf.org/pages/geninfo.htm.

LIGHT WORK

316 Waverly Avenue, Syracuse, NY, 13244, United States of America
Tel: (1) 315 443 1300
Fax: (1) 315 443 9516
Email: jjhoone@syr.edu
Website: www.lightwork.org
Contact: Jessica Reed, Promotions Coordinator

Light Work is an artist-run space that focuses on providing direct support for artists working in photography through its Artist-in-Residence Program, exhibitions, publications, and online permanent collection.

Light Work Artist-in-Residence Program

Subjects: Photography and digital imaging.
Purpose: To support and encourage the production of new work by emerging and mid-career artists.
Eligibility: Open to artists of any nationality working in photography with experience and demonstrable, serious intent in the field. Students are not eligible.
Level of Study: Professional development
Type: Residency
Value: US$4,000 stipend plus a darkroom and apartment
Length of Study: 1 month, non-renewable
Frequency: Annual
Country of Study: United States of America
No. of awards offered: 12–15
Application Procedure: Applicants must send a letter of intent describing in general terms the project or type work they would like to accomplish while in residence. In addition, 20 proof prints or digital images, a curriculum vitae, and a short statement about the work must be included with a stamped addressed envelope for the return of materials. There are no application forms.
Closing Date: Applications are accepted throughout the year
Funding: Government, private
No. of awards given last year: 12
No. of applicants last year: 300
Additional Information: Participants in the residency program are expected to use their month to pursue their own projects: photographing in the area, printing for a specific project or book, etc. Artists are not obligated to teach at our facility, though we hope that the artists are friendly and accessible to local artists. Please see the website for further details http://www.lightwork.org/residency/index.html.

LINK FOUNDATION

C/O Binghamton University Foundation, PO Box 6005, Binghamton, NY, 13902 6005, United States of America
Email: gahring@binghamton.edu
Website: www.binghamton.edu/link-foundation/
Contact: Martha J Gahring, Office Administrator

The Link Foundation was established in 1953 to perpetuate and enhance the recognized Link legacy of technical leadership and excellence established by the founders in their fields of interest.

Advanced Simulation and Training Fellowships

Subjects: To foster advanced level study in simulation and training research; to enhance and expand the theoretical and practical knowledge of how to train the operators and users of complex systems and how to simulate the real-world environments in which they function; and to disseminate the results of that research through lectures, seminars, and publications.
Purpose: To support research in the field of flight training and to other qualifying doctoral students studying in the simulation and training field at US universities and Canadian universities.
Eligibility: The candidate should be working full-time towards a degree in an established doctoral program at a US or Canadian institution.
Level of Study: Doctorate, Postdoctorate, Research
Type: Fellowships
Value: US$25,000
Length of Study: 1 year
Frequency: Annual
Study Establishment: Any accredited academic institution in the United States of America and Canada
Country of Study: United States of America & Canada
No. of awards offered: 3
Application Procedure: Application forms can be downloaded from the website.
Closing Date: January 16th (Check with website)
Funding: Foundation
No. of awards given last year: 2
Additional Information: Please see the website for further details http://www.ist.ucf.edu/link_foundation.htm.

For further information contact:

College of Aeronautics, Florida Institute of Technology, George M. Skurla Hall, 150 W. University Blvd, Melbourne, FL, 32826-0544, United States of America
Fax: (1) 321 674 7519
Contact: Dr Dona F Wilt, Assoc. Professor

Energy Fellowships

Subjects: Energy.
Purpose: To foster education and innovation in the area of societal production and utilization of energy.
Eligibility: Open to candidates working towards a PhD in an academic institution.
Level of Study: Doctorate
Type: Fellowships
Value: US$52,000
Length of Study: 2 years
Frequency: Annual
Study Establishment: Any accredited academic institution in the United States of America and Canada
Country of Study: United States of America or Canada
No. of awards offered: 3
Application Procedure: Applicants can download the application cover sheet from the website. The completed cover sheet along with project description, endorsement letter, letters of recommendation, budget and a curriculum vitae is to be submitted.
Closing Date: December 1st
Funding: Foundation
No. of awards given last year: 3
Additional Information: Fellowships are only tenable at United States and Canadian Universities.

For further information contact:

Link Foundation Energy Programs, Thayer School of Engineering, Dartmouth College, Hanover, NH, 03755, United States of America
Contact: Dr Lee R Lynd, Manager

Ocean Engineering and Instrumentation Fellowships

Subjects: Ocean engineering and instrumentation.
Purpose: To foster ocean engineering and ocean instrumentation research, to enhance both the theoretical and practical knowledge and applications of ocean engineering and instrumentation research, and to disseminate the results of that research through lectures, seminars and publications.
Eligibility: Open to candidates working full-time towards a PhD in an academic institution in the United States of America or Canada.
Level of Study: Doctorate, Predoctorate
Type: Fellowships
Value: US$25,000
Length of Study: 1 year
Frequency: Annual
Study Establishment: Any accredited academic institution in the United States of America and Canada
Country of Study: United States of America or Canada
No. of awards offered: 3
Application Procedure: Applicants can download the application cover sheet from the website. The completed cover sheet along with project description, endorsement letter, letters of recommendation, budget and a curriculum vitae is to be submitted.
Closing Date: January 2nd
Funding: Foundation
No. of awards given last year: 3
Additional Information: Fellowships are only tenable at United States and Canadian universities.

For further information contact:

Link Foundation OE and I Fellowship Program, Ocean and Mechanical Engineering, Florida Atlantic, 101 N. beach rd., Dania Beach, Atlantic, FL, 33004 3023, United States of America
Contact: Dr Karl von Ellenrieder, Administrator

THE LIONEL MURPHY FOUNDATION

GPO Box 4545, Sydney, NSW 2001, Australia
Tel: (61) 9223 5151, 9223 4315
Fax: (61) 9223 5267
Email: lmf@wran.com.au
Website: http://lionelmurphy.anu.edu.au/

The Lionel Murphy Foundation was established in 1986 to provide postgraduates scholarship opportunities for the study of law and/or science, or other disciplines where there are opportunities for some common good in Australia or overseas.

The Lionel Murphy Australian Postgraduate Scholarships

Subjects: All subjects. Please check website for details.
Purpose: To support candidates who wish to study further and wish to use their knowledge and ability to further common goal.
Eligibility: Open to Australian Citizens only.
Level of Study: Postgraduate
Type: Scholarship
Value: Australian $40,000
Length of Study: 1 year
Frequency: Annual
Study Establishment: Australian or overseas tertiary institution
Country of Study: Australia
No. of awards offered: 1
Application Procedure: See the website.
Closing Date: November 30th
Funding: Foundation
No. of awards given last year: 1–2
No. of applicants last year: Approx. 80

THE LISTER INSTITUTE OF PREVENTIVE MEDICINE

PO Box 1083, Bushey, HERTS, WD23 9AG, England
Tel: (44) 01923 801886
Fax: (44) 01923 801886
Email: secretary@lister-institute.org.uk
Website: www.lister-institute.org.uk
Contact: The Administrator

The Lister Institute of Preventive Medicine, originally founded in 1891, operates as a medical research charity whose sole function is to award research prizes to young clinical and non-clinical scientists working in the biological and biomedical sciences. The prizes are awarded on the basis of the quality of the proposed research and its potential implications.

Lister Institute Research Prizes

Subjects: Biomedical and biological science.
Purpose: To promote biomedical excellence in the United Kingdom through the support of postdoctoral scientific research into the causes and prevention of disease in man, thereby enhancing the state of public health.
Eligibility: Open to residents of the United Kingdom who have obtained a PhD, DPhil, MD or MBBCh, with membership of the Royal College of Physicians. Candidates must have a minimum of 3 and a maximum of 10 years postgraduate research experience. Applicants must have guaranteed support for the period of the award. The bulk of the work should be United Kingdom-based.
Level of Study: Research, Postdoctorate
Type: Research prize
Value: UK£200,000 to be spent in support of the recipient's research for a period of 3 years. The only restriction is 'no personal salaries'
Length of Study: 5 years
Frequency: Annual
Study Establishment: An employing United Kingdom university, research institute unit, hospital or charity laboratory
Country of Study: United Kingdom
No. of awards offered: 3–4
Application Procedure: Applicants must complete and submit an application form, curriculum vitae and two letters of reference.
Closing Date: December 2nd (Check with website)
Funding: Private
Contributor: Investment income
No. of awards given last year: 4
No. of applicants last year: 48
Additional Information: Research topics are of the applicant's own choosing, but are primarily laboratory-based and targeted at generating understanding and underpinning knowledge through fundamental research. Epidemiological, bioinformatics and small clinical projects may also be considered but not social research. Projects are assessed on scientific merit and potential. The awards are personal and are transferable within the United Kingdom. Please see the

website for further details http://www.lister-institute.org.uk/research-prizes.html.

LONDON GOODENOUGH ASSOCIATION OF CANADA

P.O. BOX 5896, STN A, Toronto, ON, M5W 1P3, Canada
Email: lgac@lgac.ca
Website: www.lgac.ca

The London Goodenough Association of Canada (LGAC) is an association of Canadians who lived as graduate students at Goodebnough College in Mecklenburgh Square, London. The LGAC offers member events and provides a Scholarship Programme for Canadian graduate students studying in London and staying in London House or William Goodenough, the Goodenough College residence halls.

London Goodenough Association of Canada Scholarship Program

Subjects: All subjects.
Purpose: To support Canadian nationals who wish to pursue their higher studies in London.
Eligibility: Open to candidates who are full-time students enrolled in an accredited graduate programme in London or undertaking theses research in London while enrolled elsewhere.
Level of Study: Postgraduate, Research
Type: Scholarships
Value: £4,500
Frequency: Annual
Study Establishment: The London Goodenough Association of Canada
Country of Study: United Kingdom
No. of awards offered: 6+
Application Procedure: Application form can be downloaded from the website. Candidates must also arrange to have all post-secondary institution transcripts and 3 letters of reference sent to the address below.
Closing Date: January 6th (Check with website)
Funding: Foundation, individuals
Additional Information: For further information contact Dr Kathleen McCrone at the above address. Please see the website for further details http://www.lgac.ca/scholarships.

For further information contact:

PO Box 5896, Station A, Toronto, ON, M5W 1P3, Canada
Email: admin@lgac.ca
Contact: Andrew Gray, Chair

THE LONDON MATHEMATICAL SOCIETY

De Morgan House, 57-58 Russell Square, London, WC1B 4HS, England
Tel: (44) 020 7637 3686
Fax: (44) 020 7323 3655
Email: lms@lms.ac.uk
Website: www.lms.ac.uk

The UK national learned society for the promotion and extension of mathematical knowledge, by means of publishing, grants, meetings and contribution to national debate on mathematics, research and education.

Cecil King Travel Scholarship

Subjects: All areas of mathematical research. Proposals must describe the intended programme of work and the benefits to be gained from the visit.
Purpose: To enable a young mathematician of outstanding promise to spend a period of 3 months undertaking study or reserach overseas.
Eligibility: Nationals of the UK or Republic of Ireland, having recently completed a doctoral degree at a UK university.
Level of Study: Postdoctorate, Postgraduate
Type: Scholarship
Value: Up to £5,000
Length of Study: 3 months
Frequency: Annual
Study Establishment: University or research institute
Country of Study: Outside the United Kingdom
No. of awards offered: 1
Application Procedure: Application forms are available on request from the society or can be downloaded from the website. Applications should be returned by post or e-mail to Duncan Turton.
Closing Date: March 8th
Funding: Trusts
Contributor: Cecil King Memorial Fund
No. of awards given last year: 1
No. of applicants last year: 5
Additional Information: Please see the website for further details http://www.lms.ac.uk/prizes/cecil-king-travel-scholarship.

For further information contact:

Email: education@lms.ac.uk
Contact: Duncan Turton, Education and Research Officer

LONDON METROPOLITAN UNIVERSITY

London Metropolitan University, 166-220 Holloway Road, London, N7 8DB, United Kingdom
Tel: (44) 0 20 7423 0000
Email: info@canoncollins.org.uk
Website: www.londonmet.ac.uk

London Metropolitan University is one of Britain's largest universities, which offers a wide variety of courses in a huge range of subject areas. The University aims to provide education and training that will help students to achieve their potential and London to succeed us a world city.

Canon Collins Trust Scholarships

Subjects: Public administration.
Purpose: The key aim of the Scholarships Programme is to help build the human resources necessary for economic, social and cultural development in the southern African region and to develop an educated and skilled workforce that can benefit the wider community. Canon Collins Trust scholarship holders are thus expected to use the knowledge, training and skills acquired through their studies to contribute positively to the development of their home country.
Eligibility: Open to nationals of South Africa, Namibia, Botswana, Swaziland, Lesotho, Zimbabwe, Zambia, Malawi, Angola and Mozambique who have been offered admission to the University.
Level of Study: Postgraduate
Type: Scholarship
Value: Full-fee or half-fee waiver and support in the form of stipend, fares and books
Frequency: Annual
Study Establishment: London Metropolitan University
Country of Study: United Kingdom
Application Procedure: Application should be sent to the Canon Collins Trust and the Trust will forward it to the university. Please see the website www.londonmet.ac.uk/ how to apply. Applications that have been emailed or faxed or those that have been received after the deadline will not be considered.
Closing Date: March 3rd for UK scholarships (Check with website)
Funding: Trusts
Contributor: Cannon Collins Educational Trust and Department of Applied Social Sciences
Additional Information: Canon Collins Trust provides scholarships to students from South Africa, Namibia, Botswana, Swaziland, Lesotho, Zimbabwe, Zambia, Malawi, Angola and Mozambique who wish to pursue a postgraduate degree (mainly Master's degrees of one or two years) in either the United Kingdom or South Africa. Please see the website for details http://www.canoncollins.org.uk/scholarships.html.

For further information contact:

22 The Ivories, 6 Northampton Street, London, N12HY, United Kingdom
Tel: (44) 20 7354 1462
Fax: (44) 20 7359 4875
Email: info@canoncollins.org.uk
Website: http://www.canoncollins.org.uk
Contact: UK Scholarships Programme Manager

ISH/London Metropolitan Scholarship Scheme

Subjects: All subjects.
Purpose: To support international students from selected countries with tuition fees and accommodation.
Eligibility: Open to students from Afghanistan, Armenia, Bhutan, Cameroon, Cuba, East Timor, Gambia, Iran, Indonesia, Jordan, Kazakhstan, Lebanon, Namibia, Nepal, Sri Lanka, Tanzania, Tibet, Uganda, Uzbekistan, Vietnam and Zimbabwe who have been offered admission to the University.
Level of Study: MBA, Postgraduate
Type: Scholarship
Value: Free tuition and accommodation
Length of Study: 1–2 years
Frequency: Twice per year
Study Establishment: London Metropolitan University
Country of Study: United Kingdom
No. of awards offered: 1–5
Application Procedure: Please see the website www.londonmet.ac.uk/scholarships. Applications that have been emailed or faxed or those that have been received after the deadline will not be considered.
Closing Date: May 31st and October 31st (Check with website)
Funding: International Office
Contributor: International Students House (ISH) and London Metropolitan University
No. of awards given last year: 1–5
Additional Information: If applicants would like more information about International Students House, please access their website at http://www.ish.org.uk/.

For further information contact:

Scholarships International Office, London Metropolin University, 166-220 Holloway Road, London, N7 8DB, United Kingdom
Email: scholarships@londonmet.ac.uk
Website: http://www.ish.org.uk/

London Met Postgraduate Scholarships

Subjects: Any subject.
Purpose: To support international students at postgraduate level.
Eligibility: Open to international students (non-EU) with an unconditional offer on one of the taught masters programme.
Level of Study: MBA
Type: Scholarship
Value: Free tuition
Length of Study: 1–2 years
Frequency: Bi annual
Study Establishment: London Metropolitan University
Country of Study: United Kingdom
No. of awards offered: Varies
Application Procedure: There are now TWO ways to complete your scholarship application form: may apply electronically, online and sending an application form by post to Scholarships, International Office, London Metropolitan University. See the website for details.
Closing Date: May 31st and October 31st (Check with website)
Funding: International Office
Contributor: London Metropolitan University
No. of awards given last year: 5–15
Additional Information: If you have any further questions you can please see frequently asked questions or email scholarships@londonmet.ac.uk.

For further information contact:

Scholarships International Office, London Metropolin University, 166-220 Holloway Road, London, N7 8DB, United Kingdom

LONDON SCHOOL OF BUSINESS & FINANCE

8/9 Holborn, London, EC1N 2LL, United Kingdom
Tel: (44) 20 7823 2303
Fax: (44) 20 7823 2302
Email: admissions@lsbf.org.uk/ info@lsbf.org.uk
Website: www.lsbf.org.uk

The Corporate Scholarship

Subjects: Business.
Purpose: This scholarship represents a real opportunity for high potential students to reach their professional career goals and to one day become boardroom leaders, CEOs and Presidents of the world's most recognizable brands and multinational corporations.
Eligibility: Applicant must have an undergraduate degree or above and a proven history of academic and/or professional excellence. Applicant must meet the English requirements of the programme they are applying for.Applicant must be able to prove that you have sufficient funds to pay the remaining course fees. Applicant must have already applied for a programme at LSBF.
Level of Study: MBA, Doctorate, Postgraduate
Type: Scholarship
Value: £1,000–8,000 (towards reducing tuition fees, not include a contribution to living costs, travel or other expenses)
Frequency: Rolling basis
Study Establishment: London School of Business & Finance
Country of Study: United Kingdom
No. of awards offered: Varies
Closing Date: January 29th
Additional Information: Size of awards vary according to each scholar's circumstances.

Diversity Scholarship

Subjects: Business
Purpose: To ensure students originate from varied backgrounds creating an opportunity to form global corporate networks
Eligibility: Show a proven history of academic excellence. Meet the English requirements of the programme they are applying for. Provide proof of sufficient funds to pay the remaining course fees. Applicants must have already applied for a programme at LSBF. Applicants must be classified as an international student and not residing in the UK..
Level of Study: Postgraduate, Doctorate, MBA
Type: Scholarship
Value: £1,000–8,000 (towards reducing tuition fees, not include a contribution to living costs, travel or other expenses)
Frequency: Rolling basis
Study Establishment: London School of Business & Finance
Country of Study: United Kingdom
No. of awards offered: Varies
Closing Date: August 28th and January 29th
Additional Information: Size of awards vary according to each scholar's circumstances.

For further information contact:

London School of Business and Finance, Postgraduate Admissions Office, 8/9 Holborn, London, EC1N 2LL, United Kingdom
Tel: (44) 0 207 823 2303
Fax: (44) 0 207 823 2302
Email: admissions@lsbf.org.uk

The Emerging Markets Scholarship

Subjects: Business.
Purpose: To enable exceptional candidates from emerging markets to study for a Masters degree in London.
Eligibility: Applicant must meet the English requirements of the programme they are applying for. Applicants must have already applied for a programme at LSBF. Applicants must be classified as an international student and not residing in the UK.
Level of Study: MBA, Doctorate, Postgraduate
Type: Scholarship
Value: £1,000–8,000
Frequency: Rolling basis

Study Establishment: London School of Business & Finance
Country of Study: United Kingdom
No. of awards offered: Varies
Application Procedure: Applicants must demonstrate how they will contribute to the economic development of their emerging market should they be awarded this scholarship. To apply, choose your scholarship and then email the completed corresponding application form to your programme consultant or scholarship@lsbf.org.uk.
Closing Date: August 31st and January 31st (Check with website)
Funding: Individuals
Contributor: Prince Michael of Kent GCVO
Additional Information: Size of awards vary according to each scholar's circumstances. Please see the website for further details http://www.lsbf.org.uk/students/fees-funding/scholarship-programmes.html.

The Royal Bank of Scotland International Scholarship

Subjects: Business
Purpose: It aims at bridging international boundaries by providing Chinese business professionals with an opportunity to study a globally recognised degree in one of the world's financial centres.
Eligibility: A national of the People's Republic of China, Hong Kong (SAR), Macau (SAR); A graduate with proven academic skills; Committed to contribute to the socio-economic development of the People's Republic of China. Established in a career, with a track record of excellence and achievement, and the prospect of becoming a leader in his/her chosen field; Have good English Language skills, as most UK Higher Education Institutions require a minimum IELTS of 6.5 for admission onto Postgraduate courses. Have sufficient funds to meet your tuition fees and living expenses, after taking account of the possible award of the Bank of Scotland International Scholarship.
Level of Study: Doctorate, MBA, Postgraduate
Type: Scholarship
Value: Cover tuition fees
Frequency: Rolling basis
Country of Study: United Kingdom
No. of awards offered: 2
Application Procedure: Application form available at www.lsbf.org.uk.
Closing Date: August 28th
Contributor: Bank of Scotland

For further information contact:

United Kingdom
Tel: (44) (0) 207 823 2303
Email: info@lsbf.org.uk
Contact: Daniela Pantica, Co-ordinator

The Women in Business Scholarship

Subjects: Business.
Purpose: Intended as a conduit encouraging more female boardroom leaders who will one day become CEOs and Presidents of the world's most recognizable brands and multinational corporations.
Eligibility: Applicant must be female. Applicant must have an undergraduate degree or above and a proven history of academic and/or professional excellence. Applicant must meet the English requirements of the programme they are applying for.Applicant must be able to prove that you have sufficient funds to pay the remaining course fees. Applicant must have already applied for a programme at LSBF.
Level of Study: MBA, Postgraduate, Doctorate
Type: Scholarship
Value: £1,000–8,000 (towards reducing tuition fees, not include a contribution to living costs, travel or other expenses)
Frequency: Rolling basis
Study Establishment: London School of Business & Finance
Country of Study: United Kingdom
No. of awards offered: Varies
Closing Date: January 29th (Check with website)
Additional Information: Please see the website for further details http://www.lsbf.org.uk/students/fees-funding/scholarship-programmes.html.

LONDON SCHOOL OF ECONOMICS AND POLITICAL SCIENCE (LSE)

The London School of Economics and Political Science, Houghton Street,, London, WC2A 2AE, United Kingdom
Tel: (44) 20 7405 7686
Fax: (44) 20 7107 5285
Email: c.s.lee2@lse.ac.uk
Website: www.lse.ac.uk

London School of Economics (LSE) was founded in 1895 by Beatrice and Sidney Webb. LSE has an outstanding reputation for academic excellence. LSE is a world class centre for its concentration of teaching and research across the full range of the social, political and economic sciences.

Sir Ratan Tata Fellowship

Subjects: Poverty, inequality, human development and social exclusion, quality of public life, regional disparities, identities – gender, ethnicity, language, economy and environment.
Purpose: To encourage research on contemporary social and economic concerns of South Asia.
Eligibility: Applicants should be scholars in the social sciences with experience of research on South Asia. They should hold a PhD.
Level of Study: Research
Type: Fellowship
Value: UK£1,500 per month
Length of Study: Up to 8 months
Frequency: Annual
Country of Study: United Kingdom
No. of awards offered: Varies
Application Procedure: Applications should include a curriculum vitae and an outline of proposed research and the names and addresses of 2 referees who are familiar with their work, to be contacted by the Chairman. Applications should be addressed to The Fellowships Selection Committee.
Closing Date: May 13th
Additional Information: The fellowship is not intended for students registered for a degree or diploma, nor is it intended for senior academics. Applications will not be accepted via email or fax. Please see the website for further details.

For further information contact:

The Asia Research Centre, London School of Economics & Political Science, Houghton Street, London, WC2A 2AE, United Kingdom
Email: arc@lse.ac.uk|
Website: http://www.lsbf.org.uk/students/fees-funding/scholarship-programmes.html
Contact: The Chairman, The Fellowships Selection Committee

LONDON STRING QUARTET FOUNDATION

36 Wigmore Street, London, W1U 2BP, England
Tel: (44) 1708 761423
Fax: (44) 1708 761423
Email: info@playquartet.com
Website: www.lsqf.com

The London String Quartet Foundation is a registered charity whose purpose is to promote the discovery and development of talent and audiences for the string quartet. The main activity of the Foundation is organizing the triennial London International String Quartet Competition.

London International String Quartet Competition

Subjects: Musical performance.
Purpose: To encourage young string quartets to develop further on the world stage and take part in this prestigious competition.
Eligibility: There are no restrictions except that each musician in the quartet must be under 35 years of age at the date of the final.
Level of Study: Unrestricted
Type: Prize
Value: Prizes total UK£27,500, and are split into five different awards of differing amounts and a menu of development prizes for three prize

winners. Concerts are also organized for the first three prize winners. See the website for details.
Frequency: Every 3 years
Country of Study: Any country
No. of awards offered: 6
Application Procedure: Applicants must complete an application form, available on the website.
Closing Date: September 1st in the year preceding the award
Funding: Commercial, foundation, individuals, private, trusts

LOREN L ZACHARY SOCIETY FOR THE PERFORMING ARTS

2250 Gloaming Way, Beverly Hills, CA, 90210, United States of America
Tel: (1) 310 276 2731
Fax: (1) 310 275 8245
Email: info@zacharysociety.org
Website: www.zacharysociety.org
Contact: Mrs Nedra Zachary, President, Director of Competition

The Loren L Zachary Society for the Performing Arts was founded in 1972 by the late Dr Loren L Zachary and Nedra Zachary. The purpose of the organization is to help further the careers of young opera singers by providing financial assistance. The Loren L Zachary National Vocal Competition, now in its 39th year, has helped over 200 singers to embark on international careers.

Loren L Zachary National Vocal Competition for Young Opera Singers

Subjects: Operatic singing.
Purpose: To assist in furthering the careers of young opera singers through competitive auditions with monetary awards.
Eligibility: Open to female singers between 21 and 33 years of age and male singers between 21 and 35 years of age who have completed operatic training and are fully prepared to pursue professional operatic stage careers. Applicants must reside in the United States of America, or Canada. Applicants who have, or have had, a 2 year contract in a European opera house are not eligible.
Level of Study: Professional development
Type: Competition
Value: US$10,000–12,000 for the top winner. Approx. US$50,000 is distributed among the finalists and the minimum award is US$1,000–$2,000
Frequency: Annual
Study Establishment: Must be thoroughly trained and be ready to pursue a professional operatic career
Country of Study: Any country
No. of awards offered: 10
Application Procedure: Applicants must complete an application form accompanied by a proof of age and an application fee of US$50. For application forms and exact dates, singers should send a stamped addressed business-sized envelope to the Society in November. Faxed requests will not be accepted. Applications and rules may be obtained by visiting www.zacharysociety.org
Closing Date: The deadline for the New York preliminary auditions is in January and the Los Angeles deadline is in March
Funding: Individuals, private, trusts
No. of awards given last year: 10
No. of applicants last year: 240
Additional Information: All applicants are guaranteed an audition provided application is completed correctly with application fee. Applicants must be present at all phases of the auditions. Recordings are not acceptable. Preliminary and semifinal auditions take place in New York in February, and in Los Angeles in March. The grand finals and awards distribution occurs on June 2nd in Los Angeles, CA.

LOS ALAMOS NATIONAL LABORATORY (LANL)

PO Box 1663, MS P219, Los Alamos, NM, 87545, United States of America
Tel: (1) 505 667 4866
Fax: (1) 505 665 6932
Email: bmontoya@lanl.gov
Website: www.lanl.gov

Los Alamos National Laboratory (LANL) is the largest institution in Northern New Mexico with more than 9,000 employees plus approximately 650 contractor personnel. From its origins as a secret Manhattan Project Laboratory, Los Alamos has attracted world-class scientists and applied their energy and creativity to solving the nation's most challenging problems.

Los Alamos Graduate Research Assistant Program

Subjects: Technical and scientific disciplines.
Purpose: To provide students with relevant research experience while they are pursuing a graduate degree.
Eligibility: Applicant must be a graduate.
Level of Study: Doctorate, Research
Type: Research
Value: US$33,300–44,600, including benefits, travel and moving expenses
Length of Study: Year or less
Frequency: Annual
No. of awards offered: Varies
Closing Date: Continuous
Additional Information: For further inquiries contact Brenda Montoya, 505/667 4866, bmontoya@lanl.gov. Please see the website for further details http://www.lanl.gov/careers/career-options/student-internships/graduate/index.php.

LOUGHBOROUGH UNIVERSITY

Leicestershire, LE11 3TU, United Kingdom
Tel: (44) 1509 263171
Email: international-office@lboro.ac.uk
Website: www.lboro.ac.uk

With 3,000 staff and 12,000 students Loughbrough, with its impressive 410 acre campus, is one of the largest university's in the UK. Our mission is to increase knowledge through research, provide the highest quality of educational experience and the widest opportunities for students, advance industry and the profession, and benefit society.

Design School Scholarships

Subjects: Design and technology.
Purpose: To assist students financially who want to study in the department.
Eligibility: Applicant must be a UK/EU citizen and a full-time student. The awards will be merit-based as well as taking account of financial means.
Level of Study: Postgraduate
Type: Scholarship
Value: UK£500
Frequency: Annual
Study Establishment: Loughborough University
Country of Study: United Kingdom
No. of awards offered: 10
Application Procedure: See website.
Closing Date: June 30th
Additional Information: Please see the website for details http://www.lboro.ac.uk/admin/ar/funding/pg/international/lu/index.htm.

For further information contact:

Email: r.i.campbell@lboro.ac.uk

Eli Lilly Scholarship

Subjects: Chemistry.
Purpose: To increase knowledge through research, provide the highest quality of educational experience and the widest opportunities for students, advance industry and the profession, and benefit society.
Eligibility: Applicant must be a postgraduate in chemistry. See the website for details.
Level of Study: Postgraduate
Type: Scholarships and fellowships
Value: UK£1,000
Length of Study: 1 year
Frequency: Annual
Study Establishment: Loughborough University
Country of Study: United Kingdom
No. of awards offered: 1

Application Procedure: See website.
Closing Date: See the website for details.

For further information contact:

Leicestershire,
Tel: 1509 263171
Email: l.e.child@lboro.ac.uk/international-office@lboro.ac.uk
Website: www.lboro.ac.uk

ESPRC Studentships

Subjects: Chemistry.
Purpose: To financially assist students to cover their living costs while undertaking a PhD.
Eligibility: Applicant must have been in full-time education in the UK throughout the 3 years preceeding the start date of PhD course.
Level of Study: Postgraduate
Type: Studentship
Value: The amount of funding is agreed each year by all the research councils and increase in line with inflation. Tuition fees are also paid
Length of Study: 1 year
Frequency: Annual
Study Establishment: Loughborough University
Country of Study: United Kingdom
No. of awards offered: 9–10
Application Procedure: See website.
Closing Date: March 7th (Check with website)
Funding: Government
Contributor: EPSRC
Additional Information: Please see the website for further details http://www.lboro.ac.uk/departments/phir/pg-research/funding/.

For further information contact:

Email: l.e.child@lboro.ac.uk

Industrial Design Studentship

Subjects: Industrial design.
Eligibility: Open to British nationals, resident in UK, intending to make a career in British Industry normally aged 21–24, but older candidates may apply.
Level of Study: Doctorate
Type: Studentship
Value: All tuition fees, stipend of UK£9,000 per year, allowance of UK£850 per year for materials and agreed travel costs
Length of Study: 1–3 years
Frequency: Annual
Study Establishment: Loughborough University
Country of Study: United Kingdom
Application Procedure: See website.
Funding: Foundation
Contributor: Royal Commission for the Exhibition of 1851

For further information contact:

Email: royalcom1851@imperial.ac.uk

IPTME (Materials) Scholarships

Subjects: Materials science.
Purpose: To increase knowledge through research, provide the highest quality of educational experience and the widest opportunities for students, advance industry and the profession, and benefit society.
Eligibility: Open to all full-time, self-funded, international fee status students who are not in receipt of any other university funding.
Level of Study: Postgraduate
Type: Scholarships and fellowships
Value: 25% of the programme tuition fee which will be credited to the student's tuition fee account
Length of Study: 1 year
Frequency: Annual
Study Establishment: Loughborough University
Country of Study: United Kingdom
No. of awards offered: 6
Application Procedure: See website.
Closing Date: March 1st
Funding: Foundation
Contributor: Institute of Polymer Technology and Materials Engineering

For further information contact:

Email: iptme@lboro.ac.uk

Jean Scott Scholarships

Subjects: Business and management studies.
Purpose: To suppport best qualified Loughborough University students.
Eligibility: Open to the best qualified Loughborough University students entering MSc programmes in the Department of Economics.
Level of Study: Postgraduate
Type: Scholarship
Value: A maximum of US$5,000
Length of Study: 1 year
Frequency: Annual
Study Establishment: Loughborough University
Country of Study: United Kingdom
No. of awards offered: 2
Application Procedure: See website.
Closing Date: See the university web site for details

For further information contact:

Email: msc.economics@lboro.ac.uk

Loughborough Sports Scholarships

Subjects: Athletics, cricket, football, golf, hockey, rugby, swimming, tennis and triathlon.
Purpose: To support elite athletes.
Eligibility: Open to stidents who have excelled at least at junior international level (or equivalent) in their sport and have fulfilled the normal academic requirements for either undergraduate or postgraduate entry.
Type: Scholarship
Value: Up to £3,000 towards tuition fees, £1,000 towards living expenses, £250 towards facility membership (where applicable) and free parking on campus
Frequency: Annual
Study Establishment: Loughborough University
Country of Study: United Kingdom
No. of awards offered: 2
Application Procedure: See website.
Closing Date: September (Check with website)
Additional Information: Please see the website for further details http://www.lboro.ac.uk/study/undergraduate/fees-finance/other-scholarships/.

For further information contact:

Tel: 01509 226108
Email: sports-scholars@lboro.ac.uk

Mathematical Sciences Scholarship

Subjects: Industrial mathematical modelling.
Level of Study: Postgraduate
Type: Scholarship
Value: 25% of the programme tuition fee which will be credited to the student's tuition fee account
Length of Study: 1–3 years
Frequency: Annual
Study Establishment: Loughborough University
Country of Study: United Kingdom
No. of awards offered: 3
Application Procedure: See website.
Additional Information: Please see the website for further details http://www.lboro.ac.uk/study/undergraduate/courses/departments/mathematics/.

For further information contact:

Email: maths-admissions@lboro.ac.uk

MBA Scholarships

Subjects: Business management.
Purpose: To support international Students who want to do MBA from University of Loughborough
Eligibility: Applicant may be a citizen of any country.

Level of Study: MBA
Type: Scholarship
Value: Varies
Length of Study: Varies
Frequency: Annual
Study Establishment: Loughborough University
Country of Study: United Kingdom
No. of awards offered: 10
Application Procedure: See website.
Closing Date: Check with website

For further information contact:

Email: exec.mba@lboro.ac.uk
Website: http://www.lboro.ac.uk/departments/sbe/mba/fees-funding/

Politics Scholarships

Subjects: Politics, international relations and european studies.
Level of Study: Postgraduate
Type: Scholarship
Value: UK£1,000
Frequency: Annual
Study Establishment: Loughborough University
Country of Study: United Kingdom
Application Procedure: See website.

School of Art and Design Scholarships

Subjects: Art and design.
Level of Study: Postgraduate
Type: Scholarship
Value: UK£1,000 (tuition fee)
Length of Study: 1–3 years
Frequency: Annual
Study Establishment: Loughborough University
Country of Study: United Kingdom
No. of awards offered: 10
Application Procedure: See website.

For further information contact:

Email: R.Turner@lboro.ac.uk
Website: http://www.lboro.ac.uk/admin/ar/funding/ug/international/ac/index.htm

School of Business and Economis Scholarships

Subjects: Finance, international and marketing management.
Purpose: To assist financially the students who want to pursue a career in finance, international and marketing management
Eligibility: Applicant must have a Upper Second or First Class Honours degree.
Level of Study: Postgraduate
Type: Scholarship
Value: UK£3,167–4,750
Length of Study: 1 year
Frequency: Annual
Study Establishment: Loughborough University
Country of Study: United Kingdom
No. of awards offered: 28
Application Procedure: See website.
Closing Date: See website for details

For further information contact:

Tel: 01509 228278, 228844 and 223291
Email: msc.management@lboro.ac.uk/f.l.baddeley@lboro.ac.uk

THE LOUISVILLE INSTITUTE

1044 Alta Vista Road, Louisville, KY, 40205 1798, United States of America
Tel: (1) 502 992 5432
Fax: (1) 502 894 2286
Email: info@louisville-institute.org
Website: www.louisville-institute.org
Contact: Executive Director

The Louisville Institute seeks to nurture inquiry and conversation regarding the character, problems, contributions and prospects of the historic institutions and commitments of American Christianity. In all of its work, the Louisville Institute is guided by its fundamental mission to enrich the religious life of American Christians and to encourage the revitalization of their institutions, by bringing together those who lead religious institutions with those who study them, so that the work of each might inform and strengthen the other.

Louisville Institute Dissertation Fellowship Program

Subjects: Religion/theology. Christianity in North America.
Purpose: To support the final year of writing on promising PhD and ThD dissertation projects dealing with aspects of American religious life that are related to the concerns of the Louisville Institute.
Eligibility: Applicants must be candidates for the PhD or ThD degree who have fulfilled all predissertation requirements, including approval of the dissertation proposal before the deadline and expect to complete the dissertation by the end of the following academic year.
Level of Study: Doctorate
Type: Fellowships
Value: US$22,000
Length of Study: 1 year
Frequency: Annual
Country of Study: United States of America
No. of awards offered: 7
Application Procedure: Applications should include applicant information and project summary form, dissertation fellowship programme, additional information form, dissertation adviser's letter of recommendation form and faculty letter of recommendation form.
Closing Date: February 1st
Funding: Foundation
No. of awards given last year: 7
No. of applicants last year: 87
Additional Information: All tuition, medical insurance and required fees are the responsibility of the student. Travel and lodging expenses for the seminar will be covered by the Louisville Institute. Please see the website for further details http://www.louisville-institute.org/Grants/Programs/dfdetail.aspx.

LOWE SYNDROME ASSOCIATION (LSA)

PO Box 864346, Plano, Texas, TX, 75086-4346, United States of America
Tel: (1) 972 733 1338
Email: info@lowesyndrome.org
Website: www.lowesyndrome.org
Contact: Deborah Jacobs, President

The Lowe Syndrome Association (LSA) is an international non-profit organization made up of families, friends and professionals dedicated to helping children with Lowe Syndrome and their families. Its main purposes are to foster communication among families, provide information and support research.

LSA Medical Research Grant

Subjects: Understanding and treatment of Lowe Syndrome.
Purpose: To support research projects that will lead to a better understanding of the metabolic basis of Lowe Syndrome, better treatments of the major complications of the disease and the prevention of and/or a cure for it.
Eligibility: Open to researchers who are affiliated with a non-profit institution.
Level of Study: Unrestricted
Type: Grant
Value: Varies, approx. US$30,000
Length of Study: 1 year
Frequency: Dependent on funds available
No. of awards offered: Varies
Application Procedure: Applicants must submit their application in writing. Specific instructions are available in the grant proposal guidelines document.
Closing Date: Varies
Funding: Private
Contributor: Members of the LSA and fund-raising events
No. of awards given last year: 1
No. of applicants last year: 1

THE LUISS UNIVERSITY OF ROME (LUR)

Research Center on Human Rights, Via Tommasini 1, Rome, 00 162, Italy
Tel: (39) 06 8650 6568
Fax: (39) 06 8650 6503
Email: infophd@luiss.it
Website: www.luiss.it

In 1974 a group of businessmen led by Umberto Agnelli decided to invest people and money in an innovative project for the education and training of a managerial class. Luiss adopted a new educational model along the lines of those to be found in the leading international universities.

LUR PhD Studentships in Political Theory

Subjects: Political theory.
Purpose: To support students enrolled in the Doctoral programme in political theory.
Eligibility: Open to applicants who have majored in an area of social sciences, human sciences or philosophy; however, applicants who have degrees in other disciplines will also be considered on the basis of their curriculum vitae and research proposals.
Level of Study: Doctorate
Type: Scholarships
Value: €900 per month
Length of Study: 3 years
Frequency: Annual
Study Establishment: The Luiss University of Rome (LUR)
Country of Study: Italy
No. of awards offered: Varies
Application Procedure: Students interested in the PhD programme must apply by completing the application form. Applications must abide by the format attached to the call for application. Applications must be addressed to the Dean of Luiss Guido Carli and sent to the address below.
Closing Date: September 1st (Check with website)
Funding: Foundation
Additional Information: Applicants must submit a writing sample representative of the best work in an area consistent with the core-issues of the programme. Please see the website for further details.

For further information contact:

Direzione Didattica e Ricerca Luiss Guido Carli-Viale Pola 12, Rome, 00198, Italy

THE MACDOWELL COLONY

100 High Street, Peterborough, NH, 03458, United States of America
Tel: (1) 603 924 3886
Fax: (1) 603 924 9142
Email: info@macdowellcolony.org, admissions@macdowellcolony.org
Website: www.macdowellcolony.org
Contact: Ms Courtney Bethel, Admissions Director

The MacDowell Colony was founded in 1907 to provide creative artists with uninterrupted time and seclusion to work and enjoy the experience of living in a community with gifted artists. Residencies are up to eight weeks for writers, and playwrights, composers, film and video makers, visual artists, architects and interdisciplinary artists. Artists in residence receive room, board and the exclusive use of a studio. There are no residency fees. A travel grant and limited artist grants are available to artists in residence, based on need.

MacDowell Colony Residencies

Subjects: Creative writing, visual arts, musical composition, film and video making, architecture and interdisciplinary arts, playwriting.
Purpose: To provide a place where creative artists can take advantage of uninterrupted work time and seclusion in which to work and enjoy the experience of living in a community of gifted artists.
Eligibility: Open to established and emerging artists in the fields specified.
Level of Study: Unrestricted
Type: Residency
Value: Up to US$1,000 for artists in need of financial assistance (stipends), plus limited travel grants for any artist based on need
Length of Study: Usually 4 weeks, maximum of 8 weeks
Frequency: Annual
Study Establishment: The Colony
Country of Study: United States of America
No. of awards offered: A total of 32 studios are available during each application period for individual residencies
Application Procedure: Applicants must write, telephone or refer to the website www.macdowellcolony.org/apply.html for information and to apply online.
Closing Date: Summer (January 15th), Autumn (April 15th), Winter-Spring (September 15th)
Funding: Private
No. of awards given last year: 250 residencies
No. of applicants last year: 2,500
Additional Information: The studios are offered for the independent pursuit of the applicant's art. No workshops or courses are given. Under subject Fine and Applied Arts, choreography is also included with Dance.

MACQUARIE UNIVERSITY

Balaclava Road, North Ryde, New South Wales, NSW, 2109, Australia
Tel: (61) 2 9850 7111
Fax: (61) 2 9850 7433
Email: tgreen@ling.mq.edu.au
Website: www.mq.edu.au

Established in 1964, Macquarie attracts students from all walks of life, including large numbers from overseas. As the new millennium dawned, Macquarie had conferred some 58,000 degrees, diplomas and postgraduate certificates.

Centre for Lasers and Applications Scholarships

Subjects: Experimental and theoretical laser studies.
Purpose: To enable holders to pursue a research programme leading to the degree of MSc or PhD in experimental and theoretical laser studies.
Eligibility: Open to the Australian citizens, permanent residents, or citizens of overseas countries.
Level of Study: Postgraduate, Research
Type: Scholarship
Value: Australian $19,231 (maximum per year). This award is to be used for living expenses.
Length of Study: 2 years (Masters) or 3 years (PhD)
Frequency: Annual
Study Establishment: Macquarie University
Country of Study: Australia
No. of awards offered: Varies
Application Procedure: Check website for further details.
Closing Date: Not specified
Additional Information: Please see the website for details.

For further information contact:

Centre for Lasers and Applications
Tel: 02 9850 8911/61 2 9850 8645
Fax: 61 2 9850 8799
Email: jpiper@ics.mq.edu.au/jim.piper@mq.edu.au
Website: www.mq.edu.au
Contact: Professor Jim Piper, Professor & Deputy Vice Chancellor

Macquarie University PhD Scholarship in Knowledge Acquisition and Cognitive Modelling

Subjects: Virtual training environment for risk assessment and spans the fields of cognitive modelling, knowledge acquisition, agent based systems for human learning, virtual reality and game technology.
Purpose: To enable students to undertake research in related fields.
Eligibility: Open to applicants who have completed, an Australian 4 year undergraduate degree with at least Second Class (Honours) division 1 in computing or a related field, or equivalent qualifications.
Level of Study: Postdoctorate, Postgraduate
Type: Scholarships

Value: Australian $19,231 plus Australian $5,000 top-up per year tax exempt
Length of Study: 3 years
Frequency: Annual
Study Establishment: Macquarie University
Country of Study: Australia
No. of awards offered: Up to 2
Application Procedure: Applicants must download the form from the website.
Closing Date: May 31st (Check with website)
Additional Information: Additional Information can be obtained from the Higher Degree Research Unit by phoning 61 02 9850 7277, by emailing pgschol@mq.edu.au or by downloading the form from the website.

For further information contact:

Higher Degree Research Unit, Cottage C4C, Macquarie University, North Ryde, NSW, 2109, Australia
Email: richards@mq.edu.au

MAKING MUSIC, THE NATIONAL FEDERATION OF MUSIC SOCIETIES (NFMS)

2-4 Great Eastern Street, London, EC2A 3NW, England
Tel: (44) 20 7422 8280
Fax: (44) 20 7422 8299
Email: info@makingmusic.org.uk
Website: www.makingmusic.org.uk/awards
Contact: Awards Administrator

Making Music, The National Federation of Music Societies (NFMS), is the leading umbrella group for the voluntary and semi-professional music sector. Making Music represents and supports over 2,800 music groups throughout the United Kingdom. Making Music groups also promote over 10,000 performances, spending over UK£10 million on professional artists per year and perform to audiences of around 1.6 million.

Philip and Dorothy Green Award for Young Concert Artists in Association with Making Music

Subjects: Music.
Purpose: To support young musicians at the start of their professional careers by providing a number of professional engagements with amateur orchestras, choir and voluntary promoting societies.
Eligibility: To apply candidates must be 21 or over, is a European Community or Commonwealth citizen and expect to be available for engagements in the UK from mid-2013 to mid-2015. The upper age limit for instrumentalists is 27 and for singers 31, on or before September 1st.
Level of Study: Professional development
Type: Award
Value: Up to 50 performance engagements with affiliated societies throughout the United Kingdom and Northern Ireland
Frequency: Annual
No. of awards offered: 4–6 (up to 50 engagements in total)
Application Procedure: Applicants must submit a completed application form by email. Application forms available from www.makingmusic.org.uk from October each year, with deadline in early January and auditions in February.
Closing Date: March 1st
Funding: Trusts
No. of awards given last year: 6
No. of applicants last year: 60+
Additional Information: There is also a bursary available each year of around £400, for which candidates may be considered. The Music for Alice bursary is awarded to a candidate applying for the Philip & Dorothy Green Award, and intended to be applied to a specific project. Please see the website for further details http://www.makingmusic.org.uk/our-work/awards/award-for-young-concert-artists.

MANITOBA LIBRARY ASSOCIATION (MLA)

606-100 Arthur Street, Winnipeg, MB, R3B 1H3, Canada
Tel: (1) 204 943 4567
Fax: (1) 866 202 4567/204 942 1555
Email: manitobalibrary@gmail.com
Website: www.mla.mb.ca

The Manitoba Association (MLA) is a provincial, voluntary, non-incorporated association that provides leadership in the promotion, development and support of library and information services in Manitoba.

Jean Thorunn Law Scholarship

Subjects: Library science.
Purpose: To encourage careers in library and information science.
Eligibility: Open to applicants who have completed 12 months of library employment in Manitoba. The applicants must be a resident of Manitoba or returning to Manitoba after graduation.
Level of Study: Postgraduate
Type: Scholarship
Value: Up to Canadian $3,000
Frequency: Annual
Country of Study: Canada
No. of awards offered: Varies
Application Procedure: A completed application form, which is available online, must be sent. see website www.mla.mb.ca
Closing Date: June 1st
Funding: Foundation, trusts
No. of awards given last year: 3
Additional Information: Please see the website for further details http://mla.mb.ca/awards-bursaries.

John Edwin Bissett Scholarship

Subjects: Library science.
Purpose: To encourage careers in library and information science.
Eligibility: Open to candidates attending an accredited library school. An applicant must be resident of Manitoba or returning to Manitoba on completion of the degree course.
Level of Study: Postgraduate
Type: Scholarship
Value: Up to Canadian $6,500
Frequency: Annual
Country of Study: Canada
No. of awards offered: Varies
Application Procedure: Please see the website www.mla.mb.ca
Closing Date: June 1st
Funding: Foundation, trusts
No. of awards given last year: 3
Additional Information: Please see the website for further details http://mla.mb.ca/awards-bursaries.

MARCH OF DIMES

1275 Mamaroneck Avenue, White Plains, NY, 10605, United States of America
Tel: (1) 914 997 4488
Fax: (1) 914 997 4560
Email: researchgrantssupport@marchofdimes.com
Website: www.marchofdimes.com/professionals
Contact: Research and Grants Administration

Four major problems threaten the health of America's babies, birth defects, infant mortality, low birth weight and lack of prenatal care. The goal of the March of Dimes Birth Defects Foundation is to eliminate these problems so that all babies can be born healthy.

Basil O'Connor Starter Scholar Research Award

Purpose: This award is designed to support young scientists just embarking on their independent research careers and is limited, therefore, to those holding recent faculty appointments. The applicants' research interests should be consonant with those of the Foundation.
Type: Award

Value: Up to 75,000 per year and are awarded for 2 years
Frequency: Annual
Application Procedure: Deans, Chairs of Departments, or Directors of Institutes/Centers should submit nominations for this award addressed to the Senior Vice President for Research and Global Programs.
Closing Date: March 15th
Additional Information: These grants do not cover the recipient's salary, but do provide salary support for technical help. Please email any enquiries regarding award.

For further information contact:

Email: ResearchGrantsSupport@marchofdimes.com
Website: http://www.marchofdimes.com/downloads/BOC_RFP_2013.pdf

March of Dimes Graduate Nursing Scholarship

Subjects: Nursing.
Purpose: To recognize and promote excellence in nursing care of mothers and babies.
Eligibility: Open to applicants who are registered nurses currently enrolled in a graduate or postgraduate programme in maternal-child nursing at the Master's or Doctorate level. Applicant must be a member of at least 1 of the following professional organizations: the Association of Women's Health, Obstetric and Neonatal Nurses, the American College of Nurse-Midwives or the National Association of Neonatal Nurses.
Level of Study: Postgraduate
Type: Scholarships
Value: US$5,000
Frequency: Annual
Country of Study: United States of America
No. of awards offered: 4
Application Procedure: Applicants need to submit their application form, essay, reference letters and curriculum vitae.
Closing Date: January 15th
Funding: Foundation
Contributor: Scholarship supported in part by an educational grant from Pampers
Additional Information: Scholarship recipients are announced in May each year. Please see the website for further details.

March of Dimes Research Grants

Subjects: Health and medical sciences.
Purpose: To support research scientists with faculty appointments or equivalent at universities, hospitals or research institutions.
Eligibility: Open to citizens of the United States and to employees or relatives of employees in the teaching industry.
Level of Study: Postgraduate
Type: Grant
Value: $103,431 (Average); $101,432 (Median); $55,655–145,099 (Range) per year
Frequency: Annual
Country of Study: United States of America
Application Procedure: Potential applicants should submit electronically the required administrative information and a Letter of Intent addressed to the Senior Vice President for Research and Global Programs summarizing the proposed studies via our online system.
Closing Date: April 30th
Funding: Foundation
Contributor: March of Dimes Birth Defects Foundation
Additional Information: Please see the website for further details http://www.marchofdimes.com/downloads/RESEARCH_RFP_2013.pdf.

Prematurity Research Initiative Program

Purpose: The March of Dimes seeks applications requesting grant support for projects related to causes of prematurity.
Type: Programme
Value: $144,574 (Average); $135,287 (Median) and $110,070–168,554 (Range)
Frequency: Annual
Application Procedure: Potential applicants should submit electronically the required administrative information and a Letter of Intent addressed to the Senior Vice President for Research and Global Programs summarizing the proposed studies via our online system.
Closing Date: April 15th
Additional Information: Please see the website for further details http://www.marchofdimes.com/research/prematurityresearch.html.

MARINE CORPS HERITAGE FOUNDATION

3800 Fettler Park Drive, Suite 104, Dumfries, VA, 22025, United States of America
Tel: (1) 703 432 5058
Fax: (1) 703 432 5054
Email: history.division@usmc. mil
Website: http://www.marineheritage.org/Fellowships.asp
Contact: Paul J. Weber, Coordinator Grants and Fellowships

The Marine Corps Historical Center houses the History Division. The Division provides reference services to the public and government agencies, maintains the Marine Corps' oral history collection and produces official histories of Marine activities, units and bases.

General Lemuel C. Shepherd, Jr. Memorial Dissertation Fellowship

Subjects: US military and naval history, as well as history and history-based studies in the social and behavioural sciences, with a direct relationship to the history of the United States Marine Corps.
Purpose: To fund a doctoral dissertation pertinent to Marine Corps history.
Eligibility: Awards are based on merit, without regard to race, creed, colour, or gender.
Level of Study: Doctorate
Type: Fellowship
Value: US$10,000
Frequency: Annual
Country of Study: United States of America
No. of awards offered: 1
Application Procedure: See the website.
Closing Date: May 1st
Funding: Foundation, private
Contributor: The Marine Corps Heritage Foundation
No. of awards given last year: 1 + 3 partial awards
No. of applicants last year: 5
Additional Information: Please see the website for further details http://www.marineheritage.org/Fellowships.asp.

Lieutenant Colonel Lily H Gridley Memorial Master's Thesis Fellowships

Subjects: Topics in United States of America military and naval history, as well as history and history-based studies in the social and behavioural sciences, with a direct relationship to the United States Marine Corps. This programme gives preference to projects covering the pre-1975 period where records are declassified or can be most readily declassified and made available to scholars.
Purpose: To award a number of fellowships to qualified graduate students working on topics pertinent to Marine Corps history.
Eligibility: Applicants must be actively enrolled in an accredited Master's degree programme that requires a Master's thesis.
Level of Study: Postgraduate
Type: Fellowship
Value: US$3,500
Length of Study: 1–3 years
Frequency: Annual
Country of Study: United States of America
No. of awards offered: 2
Application Procedure: Applicants must complete an application form, available at www.history.usmc.mil. The applicant is responsible for ensuring that all required documentation is mailed before the closing date.
Closing Date: May 1st
Funding: Foundation, private
Contributor: The Marine Corps Heritage Foundation
No. of awards given last year: 1 + 3 partial awards
No. of applicants last year: 4

Additional Information: The Marine Corps Heritage Foundation will notify all applicants individually of the application committee's decision not later than mid-July. Please see the website for further details http://www.marineheritage.org/Fellowships.asp.

Marine Corps History College Internships

Subjects: History.
Purpose: To offer opportunities for college students to participate on a professional level in the Marine Corps Historical Center's many activities. The programme aims to give promising and talented student interns a chance to earn college credits while gaining meaningful experience in fields in which they might choose to seek employment after school or pursue a vocational interest.
Eligibility: Applicants must be registered students at a college or university. While there are no restrictions on individuals applying for intern positions, it has been found that mature and academically superior students are most successful.
Level of Study: Postgraduate, Under graduate.
Type: Internship
Value: A small grant of daily expense money is provided. Any other costs of the internship must be borne by the student
Frequency: Annual
Study Establishment: Internships are served at the Marine Corps Historical Center in Quantico, Virginia
Country of Study: United States of America
No. of awards offered: Varies
Application Procedure: Please see teh website for details http://www.marineheritage.org/Internships.asp.
Funding: Foundation, private
Contributor: The Marine Corps Heritage Foundation
No. of awards given last year: 8
No. of applicants last year: 12
Additional Information: To learn more about the internship opportunities available or to obtain an application, please email Mr. Paul Weber at the History Division – paul.j.weber1@usmc.mil or Deputy Director Charles Grow of the National Museum of the Marine Corps – charles.grow@usmc.mil.

Marine Corps History Research Grants

Subjects: Topics in United States of America military and naval history, as well as history and history-based studies in the social and behavioural sciences, with a direct relationship to the United States Marine Corps. This programme gives preference to projects covering the pre-1975 period where records are declassified or can be most readily declassified and made available to scholars.
Purpose: To encourage research.
Eligibility: While the programme concentrates on graduate students, grants are available to other qualified persons. Applicants for grants should have the ability to conduct advanced study in those aspects of American military history and museum activities related to the United States Marine Corps.
Level of Study: Graduate, Research
Type: Grant
Value: US$400–3,000
Frequency: Annual
Country of Study: United States of America
No. of awards offered: 5
Application Procedure: See the website.
Funding: Foundation, private
Contributor: The Marine Corps Heritage Foundation
No. of awards given last year: 4
No. of applicants last year: 7
Additional Information: Please see the website for further details http://www.marineheritage.org/Grants.asp.

MARINES' MEMORIAL ASSOCIATION

609 Sutter Street, San Francisco, California, CA, 94102, United States of America
Tel: (1) 415 673 6672
Fax: (1) 415 441 3649
Email: michaelallen@marineclub.com
Website: www.marineclub.com

The Marines' Memorial Association is a non-profit veterans organization chartered to honor the memory of and commemorate the valor of Marines who have sacrificed in the nation's wars.

The Marine Corps Scholarship

Subjects: All subjects.
Purpose: To provide financial assistance to children of active and former members of the US Marines Corps.
Eligibility: Open to children of the MMA members scholastic aptitude, community involvement and civic spirit. Applicant mus be Planning to attend an accredited undergraduate college or vocational/technical institution in the upcoming academic year. Applicant must have A maximum family adjusted gross income for the 2011 tax year that does not exceed $90,000. Non-taxable allowances are not included in determining adjusted gross income. Applicant must have A GPA of at least 2.0.
Level of Study: Postgraduate
Type: Scholarship
Value: Varies
Length of Study: Renewable
Frequency: Annual
Country of Study: Any country
No. of awards offered: 4
Application Procedure: See the website.
Closing Date: April 30th
Funding: Foundation
Contributor: The Marine Corps Foundation
Additional Information: Please see the website for further details and if you have any questions, contact us by email or phone.

For further information contact:

Tel: 415 673 6672 ext 293 or ext 215
Email: Member@MarineClub.com
Website: http://mmanetcom.marineclub.com/page.aspx?pid=372

MARQUETTE UNIVERSITY THE GRADUATE SCHOOL

Holthusen Hall, Third Floor (campus map), PO Box 1881, Milwaukee, WI, 53201 1881, United States of America
Tel: (1) 414 288 7137
Fax: (1) 414 288 1902
Email: mugs@marquette.edu
Website: www.grad.marquette.edu
Contact: Thomas Marek, Student Services Co-ordinator

Marquette University is a private, Jesuit university situated about 100 miles north of Chicago. It is a graduate research institution with 18 doctoral and over 40 Master's and certificate programmes.

Alpha Sigma Nu Graduate Scholarship

Subjects: All subjects.
Purpose: To offer tuition scholarships for members of Alpha Sigma Nu.
Eligibility: Open to 1st-year graduate students who are members of Alpha Sigma Nu.
Level of Study: Graduate
Type: Scholarship
Value: The amount being offered varies for each individual based on different criteria and qualifications
Length of Study: 2 years
Frequency: Annual
Country of Study: United States of America
No. of awards offered: 2
Application Procedure: Applicants must send a letter of application to the Graduate School.
Closing Date: February 15th
Funding: Private
No. of awards given last year: 2
Additional Information: Please see the website for further details.

Greater Milwaukee Foundation's Frank Rogers Bacon Research Assistantship

Subjects: Electrical and computer engineering.

Purpose: To provide financial assistance in the form of research assistantships.
Eligibility: Open to full-time Master's and doctoral students.
Level of Study: Graduate
Type: Fellowship
Value: Up to a full scholarship and stipend. May also offer reimbursement for books or equipment needed for course work.
Length of Study: Varies
Frequency: Annual
Country of Study: United States of America
No. of awards offered: Varies depending on funds
Application Procedure: Applicants must contact the Department of Electrical and Computer Engineering at the University.
Closing Date: Interested students should write to the chair of the Department of Electrical and Computer Engineering for consideration.
Funding: Private
No. of awards given last year: 7
Additional Information: Please see the website for further details http://www.marquette.edu/grad/finaid_bacon.shtml.

Johnson's Wax Research Fellowship

Subjects: Engineering, chemistry and biology.
Purpose: To provide financial assistance for doctoral students.
Eligibility: Applicants must be admitted to a doctoral programme.
Level of Study: Doctorate
Type: Fellowship
Value: Approx. US$5,500 stipend and a tuition scholarship
Length of Study: 1 year
Frequency: Annual
No. of awards offered: 1
Application Procedure: Qualified applicants will be contacted by their department.
Closing Date: February 15th
Funding: Private
No. of awards given last year: 1
Additional Information: The award is offered to a different programme each year. Please see the website for further details http://www.marquette.edu/grad/finaid_jwax.shtml.

Marquette University Graduate Assistantship

Subjects: All subjects.
Purpose: To financially support teaching and research.
Eligibility: Open to full-time Master's and doctoral students.
Level of Study: Graduate
Type: Assistantship
Value: Stipends start at $13,517. In addition, up to 18 credits of tuition scholarship and health insurance are offered
Length of Study: 1 year
Frequency: Annual
No. of awards offered: 289
Application Procedure: Applicants must submit an application to the Graduate School by the deadline. The graduate bulletin or the website should be consulted for further details.
Closing Date: February 15th for Fall and November 15th for Spring
Funding: Private
Contributor: Marquette University
No. of awards given last year: 289
Additional Information: Please see the website for details http://www.marquette.edu/grad/finaid_assistantship.shtml.

Marquette University Women's Club Fellowship

Subjects: All subjects. The Graduate School selects a different graduate program each year for the fellowship.
Purpose: To support students who received their baccalaureate degrees from Marquette University.
Eligibility: Applicants must be admitted to a degree programme and have received a Bachelor's degree from Marquette University.
Level of Study: Graduate
Type: Fellowship
Value: US$2,000 stipend
Length of Study: 1 year
Frequency: Annual
No. of awards offered: 1
Application Procedure: Qualified students will be contacted by their department.
Closing Date: February 15th
Funding: Private
No. of awards given last year: 1
Additional Information: The award is offered to a different graduate programme each year. Please see the website for further details http://www.marquette.edu/grad/finaid_muwomens.shtml.

R. A. Bournique Memorial Fellowship

Subjects: Chemistry.
Purpose: To support summer research financially.
Eligibility: Open to graduate students in chemistry.
Level of Study: Graduate
Type: Fellowship
Value: Varies depending on the fund availability
Length of Study: The Summer
Frequency: Annual
No. of awards offered: 2
Application Procedure: Applicants must contact the Department of Chemistry.
Closing Date: Contact the Department of Chemistry
Funding: Private
No. of awards given last year: 2
Additional Information: Please see the website for further details http://www.marquette.edu/grad/finaid_bournique.shtml.

MARSHALL AID COMMEMORATION COMMISSION

Woburn House, 20-24 Tavistock Square, London, WC1H 9HF, United Kingdom
Tel: (44) 20 7380 6704/3
Fax: (44) 20 7387 2655
Email: apps@marshallscholarship.org
Website: www.marshallscholarship.org

The Marshall Aid Commemoration Commission is responsible for the selection and placement of recipients of Marshall scholarships from the United States of America to the United Kingdom. The first awards were presented in 1954.

Marshall Scholarships

Subjects: All subjects.
Purpose: To provide intellectually distinguished young Americans with the opportunity to study in the United Kingdom, and thus to understand and appreciate the British way of life.
Eligibility: Open to citizens of the United States of America who have graduated with a minimum grade point average of 3.7 or A from an accredited US college not more than 2 years previously. Recipients are required to take a degree at their United Kingdom university. Preference is given to candidates who combine high academic ability with the capacity to play an active part in the United Kingdom university.
Level of Study: Postgraduate
Type: Scholarship
Value: University fees, cost of living expenses, annual book grant, thesis grant, research and daily travel grants, fares to and from the United States and, where applicable, a contribution towards the support of a dependent spouse
Length of Study: 2 academic years, with a possible extension for a third year
Frequency: Annual
Study Establishment: Any suitable institution
Country of Study: United Kingdom
No. of awards offered: 40
Application Procedure: Applicants must submit an application form, university or college endorsement and four references. Information and application forms can be obtained from the British consulate in selected cities. Full details on application procedures can be obtained from the website. Application online: http:/www.marshallscholarship.org/applications
Closing Date: Early-October of year preceding tenure
Funding: Government
No. of awards given last year: 37
No. of applicants last year: 900

Additional Information: Please see the website for further details http://www.marshallscholarship.org/about/generalinfo.

For further information contact:

Georgia Pacific Center, Suite 3400, 133 Peachtree Street NE, Atlanta, GA, 30303, United States of America
Tel: (1) (404) 954-7708
Contact: British Consulate-GeneralBritish Consulate-General, 33 North Dearborn Street, Chicago, IL, 60602, United States of AmericaBritish Consulate-General, 1 Sansome Street, San Francisco, CA, 94104, United States of AmericaBritish Consulate-General, 11766 Wiltshire Boulevard, Suite 400, Los Angeles, CA, 90025, United States of AmericaBritish Consulate-General, Wells Fayo Plaza, 1000 Louisiana, No 1900, Houston, TX, 77002, United States of America

MARYLAND INSTITUTE COLLEGE OF ART (MICA)

1300 West Mount Royal Avenue, Baltimore, Maryland, 21217, United States of America
Tel: (1) 410 225 2256
Fax: (1) 410 225 5275
Email: graduate@mica.edu
Website: www.mica.edu
Contact: Mr Scott G Kelly, Associate Dean for Graduate Admission

The Maryland Institute College of Art (MICA) offers a Master of Fine Arts (MFA) degree in painting, sculpture, photographic and electronic media, graphic design community arts, curatorial practice, illustration practice. It also offers excellent art education degrees such as an MAT and an MA in Art Education. The college has private studios, a strong visiting artists program and excellent exhibition opportunities.

MICA Fellowship

Subjects: Fine and applied arts: art, design, art education, photo, general, community arts and curatorial practice.
Purpose: To serve as a tuition scholarship.
Eligibility: Each applicant is considered.
Level of Study: Graduate
Type: Fellowship
Value: 50 per cent tuition
Length of Study: 1–2 years
Frequency: Annual
Study Establishment: MICA
Country of Study: United States of America
Application Procedure: All accepted applicants are automatically considered.
Closing Date: January 15th
Funding: Private
Contributor: MICA
No. of awards given last year: 19
No. of applicants last year: 1,300

For further information contact:

MICA Graduate Admission, 131 West North Avenue, MD, Baltimore, 21201, United States of America

MICA International Fellowship Award

Subjects: Fine and applied arts.
Purpose: To serve as a tuition scholarship.
Eligibility: Open to all international students accepted to the Institute.
Level of Study: Graduate
Type: Fellowship
Value: 50 per cent of tuition
Length of Study: 1–2 years
Frequency: Annual
Study Establishment: MICA
Country of Study: United States of America
No. of awards offered: 1
Application Procedure: All accepted applicants are automatically considered.
Closing Date: January 15th
Funding: Private
Contributor: MICA
No. of awards given last year: 1
No. of applicants last year: 260

For further information contact:

MICA Graduate Admission, 131 West North Avenue, MD, Baltimore, 21201

MASSEY UNIVERSITY

0800 MASSEY, TXT 5222, New Zealand
Tel: (64) 09 414 0800, 64 6 350 5701
Fax: (64) 09 443 9704
Email: contact@massey.ac.nz
Website: www.massey.ac.nz

Massey University is a state funded university with a proud 80-year tradition of academic excellence. Its distance-learning programme delivers university qualifications to all parts of the country. It has a strong national and international reputation for the quality of research, research-led teaching, and the contributions of staff and graduates. It is a leader in advancing the economic, social, and cultural well being of the people of New Zealand.

C. Alma Baker Postgraduate Scholarship

Subjects: Agriculture, agriculture-related technologies or the study of rural society.
Purpose: To encourage students enrolling in Master's or doctoral thesis programmes.
Eligibility: Open to candidates who are graduates and citizens of New Zealand. Awards are available for those intending to undertake postgraduate research either in New Zealand or overseas. Scholarships will be based on academic achievement.
Level of Study: Graduate, Postgraduate
Type: Scholarship
Value: Maximum $ 13,000 a year for a Master's student and $20,000 a year for a doctoral student
Length of Study: 1 year for Master's and up to 3 years for a Doctoral programme
Frequency: Dependent on funds available
Country of Study: New Zealand
No. of awards offered: Up to 6
Application Procedure: Applicants must apply on the prescribed form to the Secretary, C. Alma Baker Trust. A certified copy of Academic Record, birth certificate, passport, or other proof of citizenship, and an outline of proposed research (not more than one page) must be enclosed with the application.
Closing Date: February 1st
Funding: Trusts
Additional Information: Do not send original documents as application and attachments will not be returned. Information provided will be used by the trust or its representatives only for awarding scholarships and may be subject to verification procedures as appropriate.

For further information contact:

C. Alma Baker Trust, c/o School of People, Environment & Planning Social Sciences Tower, Level 3, Massey University, Palmerston North, PN 331, New Zealand
Email: Contact@massey.ac.nz
Contact: Professor Barrie Macdonald, Secretary

Hurley Fraser Postgraduate Scholarship

Subjects: Applied sciences.
Purpose: To support postgraduate research in agriculture and horticulture.
Eligibility: Applicants must have enrolled for a full-time postgraduate degree or a diploma in one of the Applied Sciences. The award is based on the candidate's academic attainment.
Level of Study: Graduate
Type: Scholarship
Value: Up to $2,000 per year
Study Establishment: Massey University
Country of Study: New Zealand

Application Procedure: Applicants must apply to the Scholarships Office, Graduate Research School on forms (ASSC.3) available from Massey Contact or can be downloaded from the website.
Closing Date: March 10th
Contributor: John Alexander Hurley Scholarship and the Edith Fraser Agricultural and Horticultural Research Fund
Additional Information: The award shall be paid in May. Applications will not be accepted more than 3 months in advance of the closing date.

For further information contact:

Tel: 0800 627 739
Email: Contact@massey.ac.nz

Sir Alan Stewart Postgraduate Scholarships

Subjects: All subjects.
Purpose: To encourage new postgraduate enrolments from other tertiary institutions and to assist Massey students to progress from undergraduate to postgraduate study.
Eligibility: Applicant must be enrolled or be intending to enrol full-time or part-time, in the year the award is to be made, in their initial year of a first postgraduate programme, undertaken either internally or extramurally and must be a New Zealand citizen or permanent resident who has completed an undergraduate degree at a New Zealand university.
Level of Study: Graduate
Type: Scholarship
Value: $4,000 per year for full-time students and pro-rated for part-time students
Frequency: Annual
Study Establishment: Any Massey University campus
Country of Study: New Zealand
No. of awards offered: Up to 10
Application Procedure: Check website for further details.
Closing Date: December 1st
Additional Information: Consideration will normally only be given to students with a B+ average or better. Full-time students may hold the scholarship only once. Part-time students may re-apply to the maximum of $4,000 in a period of 4 years from first enrolment in the postgraduate programme. Applications will NOT be accepted more than 3 months in advance of the closing date.

For further information contact:

Massey University, Private Bag 11-222, Palmerston North, New Zealand
Email: Contact@massey.ac.nz

THE MATSUMAE INTERNATIONAL FOUNDATION

4-14-46 Kamiogi, Suginami-ku, Tokyo, 167-0043, Japan
Tel: (81) 3 3301 7600
Fax: (81) 3 3301 7601
Email: contact2mif@mist.dti.ne.jp
Website: www.mars.dti.ne.jp/mif
Contact: Mr S Nakajima, Secretary General

Matsumae International Foundation Research Fellowship

Subjects: Any subject, however, priority is given to the fields of natural science, engineering and medicine.
Purpose: To provide an opportunity for foreign scientists to conduct research at Japanese institutions.
Eligibility: Open to applicants under 49 years of age, of non-Japanese nationality who hold a Doctorate, and who have not previously been to Japan.
Level of Study: Professional development, Postdoctorate, Research
Type: Fellowship
Value: ¥200,000 is provided monthly for the purpose of payment of tution, expenses for research material etc. ¥100,000 is provided to assist with local travel expenses on arrival, the initial cost of lodging etc.
Length of Study: 3–6 months
Frequency: Annual
Study Establishment: Unrestricted
Country of Study: Japan
No. of awards offered: 13
Application Procedure: Applicants must obtain the current issue of the fellowship announcement from the Foundation.
Closing Date: August 31st
Funding: Private
Contributor: Charitable donations from individuals
No. of awards given last year: 15
No. of applicants last year: 87
Additional Information: Priority will be given to the fields of natural science, engineering and medicine.

MCDONNELL CENTER FOR THE SPACE SCIENCES

Washington University in St. Louis, Campus Box 1105, One Brookings Drive, St Louis, MO, 63130 4899, United States of America
Tel: (1) 314 935 5332
Fax: (1) 314 935 4134
Email: janf@wustl.edu/webmaster@levee.wustl.edu
Website: www.mcss.wustl.edu

The McDonnell Center for the Space Sciences supports and stimulates scientists to work on fundamental problems in Space Sciences and Astroparticle Physics transcending borders between scientific disciplines.

McDonnell Center Astronaut Fellowships in the Space Sciences

Subjects: Space sciences, physics, earth sciences, astronomy, chemistry and planetary studies.
Purpose: To encourage scholarship and research in graduate students interest in the space sciences.
Eligibility: Open to citizens of the United States of America who have applied for and been accepted into a graduate programme in the departments of physics or earth and planetary sciences or chemistry at Washington University, St Louis. Candidates must have an excellent academic background and a particular interest in the space sciences.
Level of Study: Doctorate
Type: Fellowship
Value: Full tuition plus a stipend
Length of Study: 3 years
Frequency: Annual
Study Establishment: Washington University
Country of Study: United States of America
No. of awards offered: 2
Application Procedure: Applicants who have been accepted into a Washington University graduate programme in physics, earth and planetary sciences, chemistry, biology or electrical engineering and who have exhibited an interest in the space sciences are eligible to be considered for a McDonnell Center Astronaut Fellowship
Closing Date: January 15th (Check with website)
Funding: Private
No. of awards given last year: 1
Additional Information: Please see the website for further details http://mcss.wustl.edu/opportunities/graduate_fellowship.

McDonnell Graduate Fellowship in the Space Sciences

Subjects: Space sciences, physics, earth sciences, astronomy, chemistry and planetary studies.
Purpose: To encourage scholarship and research in graduate students in the space sciences.
Eligibility: Open to candidates who have applied for and been accepted into a graduate programme in the department of physics or earth and planetary sciences at Washington University, St Louis. Candidates must have an excellent academic background and a particular interest in the space sciences.
Level of Study: Doctorate
Type: Fellowship
Value: Full tuition plus a stipend
Length of Study: 3 years
Frequency: Annual
Study Establishment: Washington University
Country of Study: United States of America

No. of awards offered: 4
Application Procedure: Applicants who have been accepted into a Washington University graduate programme in physics, earth and planetary sciences, chemistry, biology or electrical engineering and who have exhibited an interest in the space sciences are automatically considered for a McDonnell Center Fellowship.
Closing Date: January 15th (Check with website)
Funding: Private
No. of awards given last year: 1
Additional Information: Please see the website for further details http://mcss.wustl.edu/opportunities/graduate_fellowship.

MCGILL UNIVERSITY

845 Sherbrooke Street West, Montréal, Quebec, Que., H3A 2T5, Canada
Tel: (1) 514 398 4066/4455
Fax: (1) 514 398 2499
Email: graduate.admissions@mcgill.ca/Maxwellboulton@hotmail.com
Website: www.mcgill.ca

McGill University is Canada's best known university, renowned internationally for the highest standards in teaching and research and the outstanding record of achievement of professors and students. In fields like neurosciences, pain, cancer research and public policy to name but a few McGill is at the forefront of achievement nationally and internationally.

Boulton Fellowship

Subjects: Law, especially with significance to the Canadian legal system and legal community.
Purpose: To provide young scholars with an opportunity to pursue a major research project or to complete the research requirements for a higher degree.
Eligibility: Open to candidates who have completed the residency requirements for a doctoral degree in law.
Level of Study: Doctorate, Postdoctorate
Type: Fellowship
Value: Canadian $30,000–35,000 per year
Length of Study: 1 year
Frequency: Annual
Country of Study: Canada
No. of awards offered: 2
Application Procedure: Applicants must write for details.
Closing Date: February 1st
No. of awards given last year: 2
No. of applicants last year: 20
For further information contact:

Boulton Fund Administrators, Faculty of Law, McGill University, 3644 Peel Street, Montreal, QC, H3A 1W9, Canada
Email: staffappointments.law@mcgill.ca

MEDICAL LIBRARY ASSOCIATION (MLA)

65 East Wacker Place, Suite 1900, Chicago, IL, 60601 7246, United States of America
Tel: (1) 312 419 9094
Fax: (1) 312 419 8950
Email: mlaedo@mlahq.org
Website: www.mlanet.org
Contact: Carla J. Funk, CAE, Executive Director

The Medical Library Association (MLA) is organized exclusively for scientific and educational purposes, and is dedicated to the support of health sciences research, education and patient care. MLA fosters excellence in the professional achievement and leadership of health sciences libraries and information professionals to enhance the quality of healthcare, education and research.

Cunningham Memorial International Fellowship

Subjects: Medical librarianship.
Purpose: To provide foreign medical librarians the opportunity to participate in and attend the Annual Meeting of the Library Association.
Eligibility: Open to those with a baccalaureate degree and a library degree who are working in a medical library. The applicant must submit a signed statement from a home country official stating that he or she is guaranteed a position in a medical library upon returning home. A satisfactory score must be achieved on the Test of English as a Foreign Language competency examination. Nationals of the United States of America or Canada are not eligible.
Level of Study: Professional development
Type: Fellowship
Value: US$3,500 to cover travel and expenses to the MLA's Annual Meeting
Length of Study: 2–3 weeks
Frequency: Annual
Study Establishment: Medical libraries
Country of Study: United States of America or Canada
No. of awards offered: 1–2
Application Procedure: Applicants must submit a completed application form with three letters of reference in English, a project overview, a certificate of health, Test of English as a Foreign Language examination result and an audio or video tape.
Closing Date: December 1st
Funding: Private
No. of awards given last year: 1
Additional Information: Please see the website for further details http://www.mlanet.org/awards/grants/cunningham.html.

MLA Continuing Education Grants

Subjects: The theoretical, administrative and technical aspects of library and information science.
Purpose: To provide professional health science librarians with the opportunity to continue their education.
Eligibility: Open to citizens of the United States of America or Canada or permanent residents who are medical librarians with a graduate degree in library science and at least 2 years of work experience at the professional level.
Level of Study: Professional development
Type: Fellowship
Value: US$100–500
Length of Study: 1 year
Frequency: Annual
Country of Study: United States of America
No. of awards offered: 1
Application Procedure: Applicants must complete and submit an application form along with three references. Candidates should also identify a continuing education programme.
Closing Date: December 1st
Funding: Private
No. of awards given last year: 1
Additional Information: The award does not support work towards a degree or certificate. For more information about any of MLA's grants or scholarships, contact Maria Lopez, 312.419.9094 x15.

MLA Research, Development and Demonstration Project Award

Subjects: Health science librarianship and the information sciences.
Purpose: To provide support for research, development and demonstration projects that will help to promote excellence in the field of health sciences librarianship.
Eligibility: Open to members of the Association who have a graduate degree in library science, are practicing medical librarians with at least 2 years of experience at the professional level and are citizens or permanent residents of the United States of America or Canada. Grants will not be given to support an activity that is only operational in nature or that has only local usefulness.
Level of Study: Postgraduate
Type: Award
Value: US$100–1,000
Length of Study: 1 year
Frequency: Annual

Country of Study: United States of America or Canada
No. of awards offered: 1
Application Procedure: Applicants must submit a completed application form with three references, a detailed description of the project design and budget.
Closing Date: December 1st
Funding: Private
No. of awards given last year: 1
Additional Information: For more information about any of MLA's grants or scholarships, contact Maria Lopez, 312.419.9094 x15.

MLA Scholarship

Subjects: Library science.
Purpose: To provide an opportunity to study at an ALA-accredited library school.
Eligibility: Open to citizens and permanent residents of the United States of America or Canada who are entering an ALA-accredited library school or who have at least one half of the academic requirements of the programme to finish in the year following the granting of the scholarship.
Level of Study: Graduate
Type: Scholarship
Value: US$5,000
Length of Study: 1 academic year
Frequency: Annual
Country of Study: United States of America or Canada
No. of awards offered: 1
Application Procedure: Applicants must submit a completed application form with two letters of reference, official transcripts and a statement of career objectives.
Closing Date: December 1st
Funding: Private
No. of awards given last year: 1
Additional Information: For more information about any of MLA's grants or scholarships, contact Maria Lopez, 312.419.9094 x15.

MLA Scholarship for Minority Students

Subjects: Medical librarianship.
Purpose: To provide a minority student with the opportunity to begin or continue graduate study in the field of library and information science.
Eligibility: Open to African, American, Hispanic, Asian, Pacific Island or Native American students who are entering an ALA-accredited library school and have at least one half of the academic requirements of the programme to finish in the year following the granting of the scholarship.
Level of Study: Graduate
Type: Scholarship
Value: US$5,000
Length of Study: 1 academic year
Frequency: Annual
Study Establishment: An ALA-accredited school
Country of Study: United States of America or Canada
No. of awards offered: 1
Application Procedure: Applicants must submit a completed application form with two letters of reference, official transcripts and a statement of career objectives.
Closing Date: December 1st
Funding: Private
No. of awards given last year: 1
Additional Information: For more information about any of MLA's grants or scholarships, contact Maria Lopez, 312.419.9094 x15.

Thomson Reuters/MLA Doctoral Fellowship

Subjects: Medical librarianship and information science.
Purpose: To encourage superior students to conduct doctoral work in the field of health sciences librarianship or information sciences.
Eligibility: Open to citizens or permanent residents of the United States of America or Canada who are graduates of an ALA-accredited library school and are enrolled in a PhD programme with an emphasis on biomedical and health-related information science.
Level of Study: Doctorate
Type: Fellowship
Value: US$2,000
Length of Study: 1 year, non-renewable
Frequency: Every 2 years
Country of Study: United States of America or Canada
No. of awards offered: 1
Application Procedure: Applicants must submit a completed application form with two letters of reference, transcripts of graduate work completed, a summary of the project and a detailed budget plus a signed statement of terms and conditions.
Closing Date: December 1st
Funding: Private, commercial
Contributor: Thomson Scientific
No. of awards given last year: 1
Additional Information: The award supports research or travel applicable to the candidate's study and may not be used for tuition. For more information about any of MLA's grants, scholarships, or fellowships, contact Maria Lopez, 312.419.9094 x15.

MEDICAL RESEARCH COUNCIL (MRC)

David Phillips Building, Polaris House, North Star Avenue, Swindon, Wiltshire, SN2 1FL, England
Tel: (44) 20 7636 5422
Fax: (44) 20 7436 6179
Email: joaune.mccallum@headoffice.mrc.ac.uk
Website: www.mrc.ac.uk

The Medical Research Council (MRC) offers support for talented individuals who want to pursue a career in the biomedical sciences, public health and health services research. It provides its support through a variety of personal award schemes that are aimed at each stage in a clinical or non-clinical research career.

Clinician Scientist Fellowship

Subjects: Biomedical sciences.
Purpose: Aim to develop outstanding medically and other clinically qualified professionals who have gained a PhD/DPhil to establish themselves as independent researchers.
Eligibility: Open to hospital doctors, dentists and general practitioners, nurses, midwives and members of the allied health professions. All applicants must have obtained a PhD, DPhil, or MD in a basic science or clinical research project, or expect to have done so by the time they take up the award. Applications from existing MRC fellows are particularly welcome. There are no residence eligibility restrictions for these fellowships.
Level of Study: Postdoctorate, Research
Type: Fellowship
Value: Competitive personal salary support plus research support staff at the technical level, research expenses, capital equipment and a travel allowance for attendance at scientific conferences
Length of Study: Up to 5 years
Frequency: Annual
Study Establishment: A suitable university department or similar institution
Country of Study: United Kingdom
No. of awards offered: Varies
Application Procedure: Applicants must submit a personal application. Forms and further details are available from the MRC.
Closing Date: April 9th by 4 pm
Funding: Government
No. of applicants last year: 40
Additional Information: Organization head office is split into two locations: London and Wintshire. Please see the website for further details http://www.mrc.ac.uk/Fundingopportunities/Fellowships/Clinicianscientist/index.htm#top.

Jointly Funded Clinical Research Training Fellowship

Subjects: Biomedical sciences.
Purpose: To provide an opportunity for specialized or further research training within the United Kingdom leading to the submission of a PhD, DPhil or MD.
Eligibility: Open to members of the Royal College of Surgeons of Edinburgh wishing to pursue research at the PhD or MD level. Residence requirements apply.
Level of Study: Postgraduate, Research
Type: Fellowship

Value: An appropriate clinical academic salary will be provided along with a fixed sum for research expenses and a travel allowance for attendance at scientific conferences
Length of Study: Up to 3 years
Frequency: Annual
Study Establishment: A suitable university department or similar institution
Country of Study: United Kingdom
No. of awards offered: Up to 2
Application Procedure: Applicants must submit a personal application. Forms and further details are available from the MRC.
Closing Date: Applicants are advised to check the MRC website for details
Funding: Government
No. of awards given last year: 2
No. of applicants last year: 17
Additional Information: The MRC provides opportunities for additional clinical research training fellowships through collaborations with Royal Colleges and Charity funders. Any jointly funded fellowships will be offered under standard MRC terms and conditions and at the same funding level as any other MRC clinical research training fellowship. Not all jointly funded opportunities are available for each competition round; please see the competition deadlines for information on which funders are offering award for that competition. If you have any queries on MRC fellowships, please contact email: fellows@headoffice.mrc.ac.uk.

MRC Career Development Award

Subjects: Biomedical sciences.
Purpose: To award outstanding researchers who wish to consolidate and develop their research skills and make the transition from postdoctoral research and training to becoming independent investigators, but who do not hold established positions.
Eligibility: It is expected that all applicants will hold a PhD or MPhil in a basic science and will have at least 3 years of postdoctoral research experience.
Level of Study: Postdoctorate, Research
Type: Fellowship
Value: Competitive personal salary support plus research support staff at the technical level, research expenses, capital equipment and a travel allowance for attendance at scientific conferences
Length of Study: Up to 5 years
Frequency: Annual
Study Establishment: A suitable university department or similar institution
Country of Study: United Kingdom
No. of awards offered: Up to 8
Application Procedure: Applicants must submit a personal application. Forms and further details are available from the MRC.
Closing Date: May 21st by 4 pm
Funding: Government
No. of awards given last year: 13
No. of applicants last year: 86
Additional Information: Awards may occasionally be jointly funded with other bodies. Please contact the Fellowships Section at the MRC for further details. Please see the website for further details http://www.mrc.ac.uk/Fundingopportunities/Fellowships/Careerdevelopmentaward/index.htm.

MRC Clinical Research Training Fellowships

Subjects: Biomedical sciences.
Purpose: To provide an opportunity for specialized or further research training leading to the submission of a PhD, DPhil or MD.
Eligibility: Open to hospital doctors, dentists, general practitioners, nurses, midwives and allied health professionals. Residence requirements apply.
Level of Study: Postgraduate, Research
Type: Fellowship
Value: An appropriate clinical academic salary will be provided along with a fixed sum for research expenses and a travel allowance for attendance at scientific conferences
Length of Study: Up to 3 years
Study Establishment: A suitable university department or similar institution
Country of Study: United Kingdom
No. of awards offered: Up to 32
Application Procedure: Applicants must submit a personal application. Forms and further details are available from the MRC.
Closing Date: September 11th by 4 pm (First round) and January 15th by 4 pm (Second round)
Funding: Government
No. of awards given last year: 39
No. of applicants last year: 169
Additional Information: In addition to this scheme, the MRC also offers Joint Training Fellowships with the Royal Colleges of Surgeons of England and Edinburgh and the Royal College of Obstetricians and Gynaecologists. These awards are aimed at individuals whose long-term career aspirations involve undertaking academic clinical research. Fellowships may also be jointly funded with the MS Society or the Prior Group. Please contact the Fellowships Section at the MRC for further details and see the website http://www.mrc.ac.uk/Fundingopportunities/Deadlines/index.htm.

MRC Clinician Scientist Fellowship

Subjects: Biomedical sciences.
Purpose: To provide an opportunity for outstanding clinical researchers who wish to consolidate their research skills and make the transition from postdoctoral research and training to becoming independent investigators.
Eligibility: The scheme is open to hospital doctors, dentists, general practitioners, nurses, midwives and allied health professionals. All applicants must have obtained their PhD or MD in a basic science or clinical project, or expect to have received their doctorate by the time they intend to take up an award, and must not hold tenured positions.
Level of Study: Postdoctorate, Research
Type: Fellowship
Value: Competitive personal salary support plus research support staff at the technical level, research expenses, capital equipment and a travel allowance for attendance at scientific conferences
Length of Study: Up to 5 years
Frequency: Annual
Study Establishment: A suitable university department or similar institution
Country of Study: United Kingdom
No. of awards offered: Up to 7
Application Procedure: Applicants must submit a personal application. Forms and further details are available from the MRC.
Closing Date: April 9th by 4 pm
Funding: Government
No. of awards given last year: 10
No. of applicants last year: 40
Additional Information: Please see the website for further details http://www.mrc.ac.uk/Fundingopportunities/Fellowships/Clinician-scientist/index.htm.

MRC Industrial CASE Studentships

Subjects: Any biomedical science.
Purpose: To enhance links between academia and industry in the provision of high-quality research training.
Eligibility: Candidates should have graduated with a good Honours Degree from a United Kingdom academic institution in a subject relevant to the MRC's scientific remit. This should be an Upper Second Class (Honours) Degree or higher. The MRC will, however, consider qualifications or a combination of qualifications and experience that demonstrates equivalent ability and attainment, e.g. a Lower Second Class (Honours) Degree can be enhanced by a Master's degree. A copy of the regulations governing residence eligibility may be obtained from the Council.
Level of Study: Postgraduate
Value: A tax-free maintenance stipend depending on United Kingdom location, university tuition fees up to the current DFEE recommended limit plus college fees, where applicable. Awards also include a fixed sum for conference travel expenses and a support grant to the university department to help cover incidental costs of students' training. As a measure of interest and involvement the industrial company is expected to make a financial contribution to the cost of the studentship
Length of Study: Up to 4 years
Frequency: Annual

Study Establishment: Universities, medical schools, industry and other academic institutions
Country of Study: United Kingdom
No. of awards offered: Approx. 30–35
Application Procedure: The MRC does not make awards directly to students. Awards are made to industrial partners who apply for studentships by application. The individual or academic partner will then advertise for students to apply for their awards. Students who wish to apply for a studentship are advised to contact the department where they wish to study to see if it has an allocation of awards.
Closing Date: Check with website
Funding: Commercial
No. of awards given last year: 10
No. of applicants last year: 23
Additional Information: Applicants should ensure that they have read the MRC Industrial CASE Scheme Guidance Notes before completing their application. Please see the webssite for further details http://www.mrc.ac.uk/Fundingopportunities/Studentships/IndustrialCASE/MRC004608.

For further information contact:

Email: students@headoffice.mrc.ac.uk

MRC Senior Clinical Fellowship

Subjects: Biomedical sciences.
Purpose: Aim to develop outstanding medically and other clinically qualified professionals such that they become research leaders.
Eligibility: Open to nationals of any country. Applicants are expected to have proven themselves to be independent researchers, be well qualified for an academic research career and demonstrate the promise of becoming future research leaders. The scheme is open to hospital doctors, dentists, general practitioners, nurses, midwives and allied health professionals. Applicants must hold a PhD or MD in a basic science or clinical project and have at least 3 years of postdoctoral research experience.
Level of Study: Postdoctorate, Research
Type: Fellowship
Value: Competitive personal salary support is provided plus research support staff at the technical and postdoctoral level, research expenses, capital equipment and a travel allowance for attendance at scientific conferences
Length of Study: 5 years
Frequency: Annual
Study Establishment: A suitable university department or similar institution
Country of Study: United Kingdom
No. of awards offered: Up to 4
Application Procedure: Applicants must submit a personal application. Forms and further details are available from the MRC.
Closing Date: April 9th by 4 pm
Funding: Government
No. of awards given last year: 1
No. of applicants last year: 8
Additional Information: Please see the website for further details http://www.mrc.ac.uk/Fundingopportunities/Fellowships/Seniorclinical/MRC001824.

MRC Senior Non-Clinical Fellowship

Subjects: Biomedical sciences.
Purpose: To provide support for non-clinical scientists of exceptional ability to concentrate on a period of research.
Eligibility: Open to nationals of any country. Applicants are expected to have proven themselves to be independent researchers, be well qualified for an academic research career and demonstrate the promise of becoming future research leaders. Applicants should normally hold a PhD or DPhil in a basic science project, have at least 6 years of relevant postdoctoral research experience and not hold a tenured position.
Level of Study: Postdoctorate, Research
Type: Fellowship
Value: Competitive personal salary support is provided plus research support staff at the technical and postdoctoral level, research expenses, capital equipment and a travel allowance for attendance at scientific conferences
Length of Study: 7 years
Frequency: Annual
Study Establishment: A suitable university department or similar institution
Country of Study: United Kingdom
No. of awards offered: Up to 6
Application Procedure: Applicants must submit a personal application. Forms and further details are available from the MRC.
Closing Date: May 21st
Funding: Government
No. of awards given last year: 3
No. of applicants last year: 24
Additional Information: Please see the website for further details http://www.mrc.ac.uk/Fundingopportunities/Fellowships/Seniornonclinical/index.htm.

MRC Special Training Fellowship in Biomedical Informatics (Bioinformatics, Neuroinformatics and Health Informatics)

Subjects: Biomedical sciences.
Purpose: Aim at developing outstanding individuals who are seeking to move into the application of mathematical, statistical and computational methods to biomedical and health research problems.
Eligibility: The scheme is aimed at individuals from a variety of backgrounds such as non-biological as well as biological, non-clinical as well as clinical, and individuals with PhDs or MDs or with informatics research experience at the predoctoral level.
Level of Study: Postgraduate, Research
Type: Fellowship
Value: An appropriate academic salary will be provided along with a fixed sum for research expenses and a travel allowance for attendance at scientific conferences
Length of Study: 3 years
Frequency: Annual
Study Establishment: A suitable university department or similar institution
Country of Study: United Kingdom
No. of awards offered: Up to 5
Application Procedure: Applicants must submit a personal application. Forms and further details are available from the MRC.
Closing Date: September 19th by 4 pm
Funding: Government
No. of awards given last year: 5
No. of applicants last year: 23
Additional Information: Award holders are encouraged to apply for a PhD or MD if they do not already have one. Please see the website for further details http://www.mrc.ac.uk/Fundingopportunities/Fellowships/Specialtraining/MRC001828.

MRC/RCOG Clinical Research Training Fellowship

Subjects: Biomedical sciences.
Purpose: To encourage clinicians to become involved in research and to promote research of relevance to the Royal College of Obstetricians and Gynaecologists.
Eligibility: Open to members of the RCOG wishing to pursue research at PhD or MD level. Applicants must have a minimum of 1 year of experience in clinical obstetrics and gynaecology and hold part one membership of the college. Residence requirements apply.
Level of Study: Postgraduate, Research
Type: Fellowship
Value: £50K - £100K (Estimated total funds: £80,000)Pre-doctoral level - The award provides a competitive personal salary, up to Specialist Registrar but not including NHS Consultant level, a Research Training Support Grant of up to £10,000 per year (items must be detailed and justified), and an annual travel allowance of £450. Post-Doctoral level - The fellowship provides a competitive personal salary, up to Specialist Registrar but not including NHS consultant level, research expenses, and travel costs at an appropriate level for the research, under Full Economic costs (fEC). For full details of what funding includes at each level please see their website
Length of Study: 1–3 years
Frequency: Annual
Study Establishment: A suitable university or similar institution
Country of Study: United Kingdom
No. of awards offered: 1

Application Procedure: Applicants must contact the Fellowships Section, Research Career Awards of the MRC for details.
Closing Date: Check with website
Funding: Government
No. of awards given last year: 2
No. of applicants last year: 2
Additional Information: Please see the website for further details.

For further information contact:

Tel: 020 7670 5485
Email: fellows@headoffice.mrc.ac.uk

MEDICAL RESEARCH SCOTLAND

Turcan Connell, Princes Exchange, 1 Earl Grey Street, Edinburgh, EH3 9EE, Scotland
Tel: (44) 131 659 8800
Fax: (44) 131 228 8118
Email: enquiries@medicalresearchscotland.org.uk
Website: www.medicalresearchscotland.org.uk
Contact: The Trust Administrator

Medical Research Scotland is comprehensive in its support and is not focused on research into any one disease or disorder. It supports the broad spectrum of clinical and laboratory-based medical research, all aimed at improving the understanding of basic disease mechanisms, diagnosis, treatment or prevention of disease or advances in medical technology.

PhD Studentship

Subjects: Any of the biomedical, clinical, physical or engineering sciences, provided that the research addresses a question relevant to the cause, diagnosis, prevention or treatment of any disease, or to the development of medical technology.
Purpose: To provide a fully-funded PhD studentship which provides preparation for a career in science in a challenging market place, by incorporating enhanced and tailored academic and commercial training and experience.
Eligibility: Open to Scottish universities/research institutions working in conjunction with a trading company operating in Scotland involved in medically-relevant life sciences research, to deliver 4 high-quality PhD studentship for suitably highly-qualified and motivated graduates of any country.
Level of Study: Postgraduate
Type: Studentship
Value: Approx. €100,000
Length of Study: 4 years
Frequency: Annual
Study Establishment: Universities or research institutions recognised as Scott
Country of Study: Scotland
No. of awards offered: Up tp 10
Application Procedure: Applications must be submitted electronically on forms available from the website, from where full procedural information is available (www.medicalresearchscotland.org.uk/apply.htm).
Closing Date: May/June – Check website for exact dates
Contributor: Income from the original endowment fund, established when the charity came into being in 1953, invested and augmented by voluntary donations and bequests
No. of awards given last year: TBC (this is a new award scheme)
No. of applicants last year: 24
Additional Information: Applications must be submitted by a university and an appropriate company and not by prospective students.

MEET THE COMPOSER, INC.

90 John Street, Suite 312, New York, NY, 10038, United States of America
Tel: (1) 212 645 6949
Fax: (1) 212 645 9669
Email: mtc@meetthecomposer.org
Website: https://www.newmusicusa.org/grants/commissioning-music-usa/

Meet The Composer's mission is to increase artistic and financial opportunities for American composers by fostering the creation, performance, dissemination and appreciation of their music.

Commissioning Music/USA

Subjects: Music commissioning.
Purpose: To support the commissioning of new works.
Eligibility: Open to citizens of the United States of America only. Organizations that have been producing or presenting for at least 3 years are eligible and may be dance, chorus, orchestra, opera, theatre and music-theatre companies, festivals, arts presenters, public radio and television stations, internet providers, soloists and small performing ensembles of all kinds, e.g. jazz, chamber, new music, etc.
Level of Study: Professional development
Type: Grant
Value: Up to US$10,000–20,000
Frequency: Annual
Country of Study: United States of America
Application Procedure: Individuals cannot apply on their own. Host organizations must submit completed application forms and accompanying materials.
Closing Date: March 19th (Check with website)
Contributor: Offered in partnership with the National Endowment for the Arts
No. of awards given last year: 20–30
No. of applicants last year: 150–200
Additional Information: Please see the website for further details https://www.newmusicusa.org/grants/commissioning-music-usa/.

For further information contact:

Tel: ext 102
Email: swinship@newmusicusa.org
Contact: Scott Winship, Director of Grantmaking Programs

JP Morgan Chase Regrant Program for Small Ensembles

Subjects: Musical performance.
Purpose: To support small New York City-based ensembles and music organizations committed to performing the work of living composers and contemporary music.
Eligibility: Open to organizations focused primarily or exclusively on new music and living composers, improvisers, sound artists or singer/songwriters.
Level of Study: Professional development
Type: Grant
Value: US$1,000–5,000
Frequency: Annual
Country of Study: United States of America
Application Procedure: Applicants must contact the organization.
Closing Date: Contact organization
Funding: Commercial
Contributor: In partnership with JP Morgan Chase
No. of awards given last year: 15–20
No. of applicants last year: 75–100

MetLife Creative Connections

Subjects: Musical performance.
Purpose: To support US composers for the participation in public activities related to specific performances of their original music. Creative Connections aims to increase awareness and enhance the creative artist's role in society by strengthening the connections between living composers, performing musicians, presenters, communities and audiences.
Eligibility: Open to citizens of the United States of America only. Organizations may be choruses, dance, opera, theatre and music-theatre companies, symphonies, arts presenters, musical organizations, festivals, television production companies, radio stations and performing ensembles of all kinds, for example, jazz, chamber, new music etc. Awards are based solely on the overall quality of the application, which includes the merit of the composer, participation, and level of audience or community involvement.
Level of Study: Professional development
Type: Award
Value: Up to US$250–3,500
Country of Study: United States of America
No. of awards offered: Over 7,500

Application Procedure: Individuals cannot apply themselves, as performing organizations apply on behalf of the composer. Interested parties should contact the organization for further details.
Closing Date: Check with website
Contributor: MetLife
No. of awards given last year: 85
No. of applicants last year: 800–1,000
Additional Information: Please see the website for further details https://www.newmusicusa.org/grants/metlife-creative-connections/.

For further information contact:

New Music USA, 90 John St, Suite 312, New York, NY, United States of America
Website: www.newmusicusa.org

Music Alive

Subjects: Music.
Purpose: To offer financial and administrative support for composer-in-residence positions with orchestral ensembles.
Eligibility: Open to all League American Orchestra and youth member orchestras.
Level of Study: Professional development
Type: Residency
Value: US$50,000–100,000
Length of Study: 1-2 years with a minimum on-site presence of three weeks per season and a minimum performance of one work per season.
Frequency: Annual
Country of Study: United States of America
No. of awards offered: 4–8
Application Procedure: Applications are submitted jointly by an organization and composer who wish to work together.
Closing Date: April 12th (Check with website)
Contributor: Offered in partnership with the League American Orchestras
No. of awards given last year: 6
No. of applicants last year: 37
Additional Information: Please see the website for further details https://www.newmusicusa.org/grants/music-alive/.

MELVILLE TRUST FOR CARE AND CURE OF CANCER

Tods Murray LLP, Edinburgh Quay, 133 Fountain Bridge, Edinburgh, EH3 9AG, Scotland
Tel: (44) 131 656 2000
Fax: (44) 131 656 2020
Email: melvilletrust@todsmurray.com
Website: www.rdfunding.org.uk/queries/ListCharityDetails.asp?CharityID = 1450
Contact: The Secretary

Melville Trust for Care and Cure of Cancer Research Fellowships

Subjects: The care and cure of cancer.
Purpose: To fund innovative research work in the care or cure of cancer.
Eligibility: Applicants, who need not necessarily hold a medical qualification or have experience of research, should have formulated proposals for a research project which have been discussed with an established research worker in the field. The applicant should normally be under 30 years of age.
Level of Study: Research
Type: Fellowship
Value: UK£2,000
Length of Study: 1–3 years
Frequency: Annual
Study Establishment: One of the clinical or scientific departments in Lothian, Borders, Fife or Dundee
Application Procedure: Applicants must complete an application form and then be interviewed.
Closing Date: February 28th
Funding: Private
No. of awards given last year: 1
No. of applicants last year: 4

For further information contact:

Melville Trust for Care and Cure of Cancer, c/o Tods Murray LLP
Email: melvilletrust@todsmurray.com

MEMORIAL FOUNDATION FOR JEWISH CULTURE

50 Broadway, 34th Floor, New York, NY, 10004, United States of America
Tel: (1) 212 425 6606
Fax: (1) 212 425 6602
Email: office@mfjc.org
Website: www.mfjc.org
Contact: Dr Marc Brandriss, Associate Director

The Memorial Foundation for Jewish Culture is today committed to the creation, intensification, and dissemination of Jewish culture worldwide, the development of creative programs to meet the emerging needs of the Jewish communities globally, and to serving as a central forum for identifying and supporting innovative programs to insure the continuation of creative Jewish life wherever Jewish communities exist.

Memorial Foundation for Jewish Culture International Doctoral Scholarships

Subjects: Jewish studies.
Purpose: To assist in the training of future Jewish scholars for careers in Jewish scholarship and research, and to enable religious, educational and other Jewish communal workers to obtain advanced training for leadership positions.
Eligibility: Open to graduate students of any nationality who are specializing in a field of Jewish studies. Applicants must be officially enrolled or registered in a doctoral programme at a recognized university.
Level of Study: Doctorate
Type: Scholarship
Value: Up to US$10,000 a year
Length of Study: 1 academic year, renewable for a maximum of 4 years
Frequency: Annual
Study Establishment: A recognized university
Country of Study: Any country
No. of awards offered: Varies
Application Procedure: Applicants must write requesting an application form.
Closing Date: October 31st

Memorial Foundation for Jewish Culture International Fellowships in Jewish Studies

Subjects: A specialized field of Jewish studies that will make a significant contribution to the understanding, preservation, enhancement or transmission of Jewish culture.
Purpose: To allow well-qualified individuals to carry out independent scholarly, literary or artistic projects.
Eligibility: Open to recognized or qualified scholars, researchers or artists of any nationality who possess the knowledge and experience to formulate and implement a project in a specialized field of Jewish studies.
Level of Study: Unrestricted
Type: Fellowship
Value: Up to US$10,000 a year
Length of Study: 1 academic year, in exceptional cases renewable for a further year
Frequency: Annual
Country of Study: Any country
No. of awards offered: Varies
Application Procedure: Applicants must write requesting an application form.
Closing Date: October 31st

Memorial Foundation for Jewish Culture Scholarships for Post-Rabbinical Students

Subjects: Jewish studies.
Purpose: To assist in the training of future Jewish religious scholars and leaders and to assist newly ordained rabbis to obtain advanced training for careers as head of Yeshivot, as Dayanim and other leadership positions.
Eligibility: Open to recently ordained rabbis engaged in full-time studies at a Yeshiva, Kollel or a rabbinical seminary.
Level of Study: Unrestricted
Type: Scholarship
Value: Minimum award US$1,000 and maximum US$3,000
Length of Study: 1 year
Frequency: Annual
Country of Study: Any country
No. of awards offered: Varies
Application Procedure: Applicants must write for details.
Closing Date: October 31st
Additional Information: The Foundation also provides grants to bolster Jewish educational programmes in areas of need. Grants are awarded on the understanding that the recipient institution will assume responsibility for the programme following the initial limited period of Foundation support. Grants are made only for team or collaborative projects.

MEMORIAL UNIVERSITY OF NEWFOUNDLAND (MUN)

PO Box 4200, St Johns, NL, A1C 5S7, Canada
Tel: (1) 709 737 8000
Fax: (1) 709 737 4569
Email: info@mun.ca
Website: www.mun.ca

Located in Canada's most easterly province, Newfoundland and Labrador, Memorial University of Newfoundland (MUN) offers a diverse selection of Graduate programmes leading to diplomas, Master's and Doctoral degrees in the arts, sciences, professional and interdisciplinary areas of study. Their goal is to promote excellence in all aspects of Graduate education in order to assist students to fulfil their personal goals and to prepare for a productive career.

A. G. Hatcher Memorial Scholarship

Purpose: To provide financial assistance to students who have demonstrated high academic merit.
Eligibility: Open to applicants who have a high academic merit. Typically a minimum of a first class degree is required
Level of Study: Graduate
Type: Scholarship
Value: Canadian $15,000
Length of Study: 1 year
Frequency: Annual
No. of awards offered: Up to 3
Closing Date: June 1st
Additional Information: Please see the website for further details https://www.mun.ca/sgs/current/scholarships/internal_apply.php#AG.

The Dr. Ethel M. Janes Memorial Scholarship in Education

Subjects: Reading or language arts.
Purpose: To aid students who want to specialize in reading and language arts and to those who want to make a career in research and teaching in primary and elementary education.
Eligibility: Must be a full-time graduate student (not working more than 24 hours per week) in the area of Language and Literacy Studies for the Fall and Winter semesters.
Level of Study: Graduate
Type: Scholarship
Value: Canadian $2,000
Frequency: Annual
No. of awards offered: 1
Application Procedure: If you wish to be considered for this award please contact Darlene Flight (dflight@mun.ca) or deliver to Room 2007, with your request to be considered for this award as well as a brief summary (one or two paragraphs) indicating your contributions to the area of language and literacy studies.
Closing Date: December 4th
Additional Information: This scholarship will be awarded on the basis of academic standing in a first Memorial University of Newfoundland Education degree to a graduate student with a specialization in reading or language arts. In the event that in any given year no graduate student qualifies for the award, this scholarship will be awarded to an undergraduate student. Instalments of $1,000.00 each will be awarded in two successive academic terms; and the scholarship is renewable for two years, provided first-class standing is maintained. Please see the website for details http://www.mun.ca/educ/grad/awards_scholarships.php.

The Echos Du Monde Classique/Classical Views Internship Fund

Subjects: Classics.
Purpose: To support a full time graduate student in Classics to train as an editorial intern with EMC/CV.
Eligibility: The student must be from a rural community in Newfoundland and Labrador, express an interest in returning to work in rural Newfoundland, and be registered for full-time studies at Memorial University of Newfoundland.
Level of Study: Graduate
Type: Internship
Value: Canadian $10,000
Frequency: Annual
No. of awards offered: Varies
Closing Date: Check with website
Contributor: Jointly sponsored by the journal Echos du Monde Classique/Classical Views and the school of Graduate studies
Additional Information: Please see the website for details http://www.mun.ca/sgs/current/scholarships/internal_nominated.php#echos.

The Imperial Tobacco Canada Limited Graduate Scholarship in Business Studies

Subjects: Business studies.
Purpose: To assist students whose area of specialization is business studies.
Eligibility: Applicant must be entering full-time graduation and whose area of specialization is business studies.
Level of Study: Graduate
Type: Scholarship
Value: Canadian $3,000
Frequency: Annual
No. of awards offered: 2
Closing Date: Check with website
Contributor: Imperial Tobacco Canada Limited
Additional Information: Please see the website for details http://www.mun.ca/sgs/current/scholarships/internal_nominated.php#ImperialTobacco.

The Maritime Awards Society of Canada (MASC) Maritime Studies Scholarship

Subjects: Maritime studies.
Purpose: To aid outstanding students who wish to study further.
Eligibility: Contact the School of Graduate Studies for details
Level of Study: Doctorate, Postdoctorate, Predoctorate
Type: Scholarship
Value: Canadian $5,000
Frequency: Annual
Study Establishment: Memorial University of Newfoundland
No. of awards offered: 1
Closing Date: February 7th (Check with website)
Additional Information: Please see the website for further details https://www.mun.ca/sgs/current/scholarships/internal_apply.php#AG.

Maritime History Internship

Subjects: History.
Purpose: To support graduate students in history to train as an editorial intern with the journal The Northerner Mariner/Le Marin du nord.
Eligibility: Applicant must be a graduate student in history.

Level of Study: Graduate, Postdoctorate, Predoctorate
Type: Internship
Value: Canadian $12,000. It will be paid over a period of three successive academic semesters and may be renewed
Frequency: Annual
No. of awards offered: Varies
Closing Date: Check with website
Funding: Foundation
Contributor: Sponsored by Canadian Nautical Research Society and the School of Graduate Studies
Additional Information: Please see the website for further details http://www.mun.ca/sgs/current/scholarships/internal_nominated.php#maritime.

The National Scholarship in Ocean Studies at Memorial University of Newfoundland

Subjects: Chemistry, biochemistry, biology, mathematics and statistics, physics and physical oceanography, earth sciences, geography, economics or engineering.
Purpose: To financially assist outstanding PhD candidates in an aspect of ocean studies.
Eligibility: Open to applicants who are outstanding PhD candidates in an aspect of ocean studies.
Level of Study: PhD
Value: $18,000 and a one time grant of up to $2,000 may be made in support of travel to appropriate conference where the student is presenting research findings
Frequency: Annual
No. of awards offered: 1
Application Procedure: Application forms are available at the School of Graduate studies.
Closing Date: July 1st
Funding: Foundation
Additional Information: Please see the website for further details https://www.mun.ca/sgs/current/scholarships/internal_apply.php#AG.

The P. J. Gardiner Award for Small Business and Entrepreneurship

Subjects: Business Administration.
Purpose: To recognize student creativity, innovation and entrepreneurship as evidenced by the establishment or plan to establish a new venture.
Eligibility: Students at all levels of their Business Program (graduate or undergraduate) are eligible for the award. The applicant must submit a Venture Plan in the case of a proposed venture or a detailed description of a venture that they have already established.
Level of Study: Graduate
Value: Canadian $5,000
Frequency: Annual
No. of awards offered: Varies
Closing Date: Check with website
Funding: Foundation
Additional Information: Please see the website for http://www.mun.ca/sgs/current/scholarships/internal_nominated.php#P.J.Gardiner.

The Peter Gardiner Award for International Study

Subjects: Business studies.
Purpose: To support Business students to study at Memorial University of Newfoundland Harlow Campus or at another university outside Canada.
Eligibility: Applicant must be an udergraduate or a graduate student of business studies.
Level of Study: Graduate
Type: Award
Value: Canadian $2,500
Frequency: Annual
Study Establishment: Memorial University of Newfoundland Harlow Campus or at another university outside Canada
No. of awards offered: 2
Closing Date: Check with website
Funding: Foundation
Additional Information: Please see the website for further details http://www.mun.ca/sgs/current/scholarships/internal_nominated.php#PeterGardiner.

Sally Davis Award – Graduate Scholarship

Subjects: Peace and international understanding, literacy, labour movement, gun control and environment.
Purpose: To provide financial support and celebrate the memory and life work of Sally Davis.
Eligibility: Open to candidates who are full-time students in the Master of Women's Studies programme.
Level of Study: Graduate
Type: Scholarship
Value: $750
Frequency: Annual
Study Establishment: Memorial University of Newfoundland
Country of Study: Canada
No. of awards offered: 1
Closing Date: Check with website
Contributor: Family and friends of Sally
Additional Information: Please see the website for further details http://www.mun.ca/sgs/current/scholarships/internal_nominated.php#sallydavis.

School of Graduate Studies F. A. Aldrich Award

Subjects: All subjects.
Purpose: To enable students to undertake full time graduate programme based on academic merit and need only if all other things are equal.
Eligibility: Applicant must be enrolled in a full time graduate programme.
Type: Award
Value: $2,000
Frequency: Annual
No. of awards offered: Up to 3
Closing Date: Check with website
Additional Information: Please see the website for details http://www.mun.ca/sgs/current/scholarships/internal_nominated.php#sgsaldrich.

MENINGITIS RESEARCH FOUNDATION

Midland Way, Thornbury, Bristol, BS35 2BS, England
Tel: (44) 014 5428 1811
Fax: (44) 014 5428 1094
Email: gillianc@meningitis.org
Website: www.meningitis.org
Contact: Mrs Gillian Currie, Research Officer

Meningitis Research Foundation is a registered charity that supports an international programme of independently peer-reviewed research into the prevention, detection and treatment of meningitis and septicaemia. The Foundation also provides information for public and health professionals, runs medical and scientific meetings and provides support to people affected by the disease.

Meningitis Research Foundation Project Grant

Subjects: Meningitis and associated infections.
Purpose: To fight death and disability from meningitis and septicaemia by supporting research with the potential to produce results in immediate problem areas. Priority is given to work that is likely to bring clinical or public health benefits.
Eligibility: Meningitis Research Foundation research grants may be held in any country; nationality is not a restriction. Employees of commercial companies are not eligible to apply.
Level of Study: Research
Type: Project grant
Value: Up to UK£150,000 per year
Length of Study: Up to 5 years
Frequency: Annual
Study Establishment: Universities, research medical institutions and hospitals
Country of Study: Any country
No. of awards offered: Varies (usually 3–6)
Application Procedure: Applicants must submit a two- or three-page outline proposal by email summarizing the work planned, the approximate cost and the duration of the project. This must include approximately ten lines explaining: (a) how the work described falls within the Foundation's research strategy, (b) the potential for

clinical/public health benefit arising from the study and (c) why the applicant has chosen to apply to the Foundation for funding for this particular project. Based on the preliminary proposal, applicants may be invited to submit a full application. Applicants who are invited to apply in full submit a detailed research plan with other support documents. These are assessed by external referees and the Foundation's Scientific Advisory Panel. Please see www.meningitis.org for more details.
Closing Date: Please see www.meningitis.org for more details including exact dates for this year's grant round.
Funding: Commercial, corporation, foundation, government, individuals, private, trusts
Contributor: Voluntary donations from the public
No. of awards given last year: 3
No. of applicants last year: 43 proposals, 18 full applications

MESA STATE COLLEGE

PO Box 2647, 1100 North Avenue, Grand Junction, CO, CO 81501, United States of America
Tel: (1) 970 248 1020
Fax: (1) 970 248 1973
Email: admissions@mesastate.edu
Website: www.mesastate.edu
Contact: MBA Admissions Officer

Mesa State College, founded in 1925, is a comprehensive, liberal arts college that offers programmes at the Master's, baccalaureate, associate degree and certificate levels.

Brach, Louis and Betty Scholarship

Subjects: Natural science and mathematics.
Purpose: To provide financial assistance.
Eligibility: Open to full-time students who are in need of financial help.
Level of Study: Postgraduate
Type: Scholarship
Value: US$1,000
Length of Study: 1 year
Frequency: Annual
Study Establishment: Mesa State College
Country of Study: United States of America
Application Procedure: See the website.
Closing Date: February 15th
Additional Information: Please see the website for further details.

Bray Leadership Scholarship

Subjects: Business administration.
Eligibility: Open to full-time students pursuing a Bachelor of Business Administration degree.
Level of Study: Postgraduate
Type: Scholarship
Value: US$1,000 or as determined by funds available
Frequency: Annual
Study Establishment: Mesa State College
Country of Study: United States of America
Application Procedure: A completed application form must be submitted.
Closing Date: February 15th
Additional Information: Please see the website for further details.

Burkey Family Memorial Scholarship

Subjects: All subjects.
Purpose: To support full-time students of the College.
Eligibility: Open to candidates who are full-time seniors at the College with a minimum GPA of 3.0.
Level of Study: Postgraduate
Type: Scholarship
Value: US$500 or as determined by funds available
Length of Study: 1 year
Frequency: Annual
Study Establishment: Mesa State College
Country of Study: United States of America
Application Procedure: See the website.
Closing Date: February 15th
Additional Information: Please see the website for further details.

Edwin and Harriet Hawkins History of Mathematics Scholarship

Subjects: Mathematics.
Eligibility: Open to full-time students enrolled in the spring mathematics course.
Level of Study: Postgraduate
Type: Scholarship
Value: US$1,000
Frequency: Annual
Study Establishment: Mesa State College
Country of Study: United States of America
Application Procedure: See the website.
Closing Date: February 15th
Additional Information: Please see the website for further details.

Mesa State College Academic (Colorado) Scholarship

Subjects: Natural sciences and mathematics.
Eligibility: Open to students of Meta State College who are citizens of United States of America.
Level of Study: Postgraduate
Type: Scholarship
Value: Up to US$1,000
Length of Study: 1 year
Frequency: Annual
Study Establishment: Mesa State College
Country of Study: United States of America
Application Procedure: A completed application form, available on the website, must be submitted.
Closing Date: February 15th
Additional Information: Please see the website for further details.

MÉTIS CENTRE AT NAHO

220 Laurier Avenue, W. Suite 1200, Ottawa, ON, K1P 5Z9, Canada
Tel: (1) 613 237 9462
Fax: (1) 613 237 1810
Email: info@naho.ca/metiscentre@naho.ca
Website: www.naho.ca
Tel: 877 602 4445

The Métis Centre is a national, non-profit, Métis-controlled centre of the National Aboriginal Health Organization. The Métis Centre is dedicated to improving the mental, physical, spiritual, emotional and social health of all Métis in Canada through public education and health promotion.

Métis Centre Fellowship Program

Subjects: Health research.
Purpose: To financially support graduate students of Métis ancestry currently engaging in, or with an interest in, developing research on Métis population health.
Eligibility: Open to all students of Métis ancestry registered in a full-time graduate level programme at any Canadian university.
Level of Study: Postgraduate
Type: Fellowships
Value: Canadian $5,000
Frequency: Annual
Country of Study: Canada
No. of awards offered: 3
Application Procedure: There is no application form, the application package must consist of a cover letter, concise summary of the proposed research paper, official transcripts and letter of support from a faculty member familiar with the applicant's research, as well as from a Métis community member.
Closing Date: August 11th
Additional Information: Fellowship recipients will be required to submit a detailed methodology, literature review and research paper to the Métis Centre at NAHO within a specified timeframe. Please see the website for further details.

THE METROPOLITAN MUSEUM OF ART

1000 Fifth Avenue, New York, NY, 10028 0198, United States of America
Tel: (1) 212 535 7710
Fax: (1) 212 650 2253
Email: mmainterns@metmuseum.org
Website: www.metmuseum.org
Contact: Christina Long, Education Programs Associate

The Metropolitan Museum of Art is one of the world's largest and finest art museums. Its collections include more than two million works of art spanning 5,000 years of world culture, from prehistory to the present, and from every part of the world. As one of the greatest research institutions in the world, the Metropolitan Museum welcomes the responsibility to train future scholars and museum professionals. The Museum offers opportunities for students at several stages in their academic careers, from high school to the postgraduate level. Internships and apprenticeships can be either paid or unpaid, full- or part-time and can last from 9 weeks to an academic year.

The Cloisters Summer Internship for College Students

Subjects: Art history and related fields.
Purpose: To support students showing an interest in museum careers.
Eligibility: Any matriculated college student who will not graduate before August is eligible to apply. First- and second-year undergraduates are given special consideration.
Level of Study: Undergraduate
Type: Internship
Value: Approximately $2,925 ($9.29/hour, less applicable taxes and deductions)
Length of Study: 9 weeks, full-time from June to August
Frequency: Annual
Study Establishment: The Metropolitan Museum of Art
Country of Study: United States of America
No. of awards offered: 8
Application Procedure: Please visit http://www.metmuseum.org/research/internships-and-fellowships/internships/internships-for-college-and-graduate-students/the-cloisters-summer-internship-for-college-students for application procedures.
Closing Date: January 15th
Funding: Private
Contributor: The Winston Foundation, Inc.
Additional Information: Applicants must submit a one-time, non-refundable $50 processing fee. If this fee is a financial burden, send a request for a fee waiver with a brief explanation of need to cloistersinterns@metmuseum.org before submitting your internship application. All fee waiver requests will be reviewed carefully and granted based on each individual's specific circumstances. Please see the website for further details.

Editorial Internship in Education

Subjects: Design, education, art history or related humanities fields.
Purpose: To support students who wish to participate in the production of print and online publications created for families, teachers, students and the general museum public.
Eligibility: This internship is open to current college seniors who will have graduated by the beginning of the program and individuals who completed their undergraduate degrees no more than one year prior to the application deadline. Current freshmen, sophomores and juniors are not eligible to apply.
Level of Study: Postgraduate
Type: Internship
Value: Approximately $25,000 ($13.74/hour, less applicable taxes and deductions) and benefits currently offered to non-exempt full-time employees of the museum
Length of Study: 1 year
Frequency: Annual
Study Establishment: The Metropolitan Museum of Art
Country of Study: United States of America
No. of awards offered: 1
Application Procedure: Please visit http://www.metmuseum.org/research/internships-and-fellowships/internships/internships-for-college-and-graduate-students/paid-internships-for-college-and-graduate-students-at-the-main-building for application procedures.
Closing Date: January 10th
Funding: Private
Additional Information: Please see the website for further details.

Lifchez/Stronach Curatorial Internship

Subjects: Art history.
Purpose: To support students interested in a curatorial career.
Eligibility: This internship is open to current college seniors who will have graduated by the beginning of the program, current graduate students, and individuals who completed their undergraduate or graduate degrees no more than one year prior to the application deadline. Current freshmen, sophomores, and juniors are not eligible to apply.
Level of Study: Postgraduate
Type: Internship
Value: Approximately $16,500 ($12.09/hour, less applicable taxes and deductions) and benefits currently offered to non-exempt full-time employees of the museum
Length of Study: 9 months
Frequency: Annual
Study Establishment: The Metropolitan Museum of Art
Country of Study: United States of America
No. of awards offered: 1
Application Procedure: Please visit www.metmuseum.org/research/internships-and-fellowships/internships.
Closing Date: January 10th
Funding: Private
Contributor: Judith Lee Stronach and Raymond Lifchez
No. of awards given last year: 1
Additional Information: Please see the website for further details.

Summer Internships for College and Graduate Students

Subjects: Art history and related fields.
Purpose: To support students showing an interest in museum careers.
Eligibility: Please refer to website for eligibility requirements.
Level of Study: Graduate, Postgraduate
Type: Internship
Value: US$3,500 for graduate students; $3,250 for college students
Length of Study: 10 weeks, full-time June through August
Frequency: Annual
Study Establishment: The Metropolitan Museum of Art
Country of Study: United States of America
No. of awards offered: Varies
Application Procedure: Please visit www.metmuseum.org/research/internships-and-fellowships/internships/internships-for-college-and-graduate-students/paid-internships-for-college-and-graduate-students-at-the-main-building for application form.
Closing Date: January 10th
Funding: Private
Contributor: The Lebensfeld Foundation, The Billy Rose Foundation, Francine LeFrak Friedburg, the Solow Art and Architecture Foundation and the Ittleson Foundation, Inc.
Additional Information: Graduate interns work on projects related to the Museum's collection or to a special exhibition as well as administrative areas. Please see the website for further details.

The Tiffany & Co. Foundation Curatorial Internship in American Decorative Arts

Subjects: Art history and related fields.
Purpose: To support students showing an interest in American Decorative Arts.
Eligibility: This internship is open to individuals who will have completed a master's degree in a field related to American decorative arts by the beginning of the program, or who completed a master's degree in a field related to American decorative arts no more than one year prior to the application deadline.
Level of Study: Graduate, Postgraduate
Type: Internship
Value: US$25,000; stipend of US$3,000 available for research and educational travel
Length of Study: 1 year
Frequency: Annual
Study Establishment: The Metropolitan Museum of Art
Country of Study: United States of America

Application Procedure: See the website.
Closing Date: January 10th
Funding: Private
Contributor: The Tiffany & Co. Foundation
No. of awards given last year: 1
Additional Information: Please see the website for further details.

MICHIGAN SOCIETY OF FELLOWS

0540 Rackham Building, 915 E. Washington Street, Ann Arbor, MI, 48109 1070, United States of America
Tel: (1) 734 763 1259
Fax: (1) 734 647 8168
Email: society.of.fellows@umich.edu
Website: http://societyoffellows.umich.edu/
Contact: Administrative Specialist

The Michigan Society of Fellows with support from the Andrew W. Mellon Foundation promotes academic and creative excellence in the humanities and the arts, the social, physical and life sciences and the professions. The objective of the Society is to provide financial and intellectual support for individuals selected for outstanding achievement, professional promise and interdisciplinary interests.

Michigan Society of Fellows Postdoctoral Fellowships

Subjects: All subjects.
Purpose: To provide financial and intellectual support for individuals selected for outstanding achievement, professional promise and interdisciplinary interests.
Eligibility: Applicants must be near the beginning of their professional careers and have completed a PhD or comparable professional or artistic degree 3 years prior to application.
Level of Study: Postdoctorate
Type: Fellowship
Value: A stipend of US$52,000 per year
Length of Study: 3 years
Frequency: Annual
Study Establishment: The University of Michigan
Country of Study: United States of America
No. of awards offered: 8
Application Procedure: Applicants must complete an application form available from the website.
Closing Date: October 1st (Check with website)
Funding: Private
Contributor: The University of Michigan, Andrew W.Mellon Foundation
No. of awards given last year: 8
No. of applicants last year: 932
Additional Information: Please see the website for further details http://societyoffellows.umich.edu/the-fellowship/.

MICROSOFT RESEARCH

One Microsoft Way, Redmond, WA, 98052, United States of America
Tel: (1) 800 642 7676
Fax: (1) 425 93 936 7329
Email: latamint@microsoft.com
Website: www.research.microsoft.com

In 1991, Microsoft Corporation became the first software company to create its own computer science research organization. It has developed into a unique entity among corporate research laboratories, balancing an open academic model with an effective process for transferring its research to product development teams.

Microsoft Fellowship

Subjects: Computer science.
Purpose: To empower and encourage PhD students in the Asia-Pacific region to realize their potential in computer science-related research and to recognize and award outstanding PhD students.
Eligibility: Open to candidates who specialize in computer science, electronic engineering, information technology or applied mathematics and are in their first or second year of PhD programme and is enrolled as a PhD student by the time of the nomination and has spent 6 to 18 months working towards a PhD.
Level of Study: Research
Type: Fellowships
Value: 100 percent of the tuition and fees, a stipend to cover living expenses while in school (US$28,000), travel allowance to attend professional conferences or seminars (US$4,000). See the website for details.
Length of Study: 2 years
Frequency: Annual
Application Procedure: Applicants must send the completed application form (downloaded from the website), 2 recommendation letters, curriculum vitae and a video or Power Point presentation with audio introduction (on compact disk), including statement of purpose, previous, on-going and future projects, research interests and accomplishments.
Closing Date: October 9th
Additional Information: Previous Microsoft Fellows are excluded. Please see the website for further details.

For further information contact:

MS Fellow 2006 Committee Microsoft Research Asia 3F Beijing Sigma Center No 49 Zhichun Road, Beijing PR, Haidian District, 100080, China
Email: fellowRA@microsoft.com
Website: http://research.microsoft.com/en-us/collaboration/awards/apply-us.aspx.

Microsoft Research European PhD Scholarship Programme

Subjects: Intersection of computing and the sciences including biology, chemistry and physics.
Purpose: To recognize and support exceptional students who show the potential to make an outstanding contribution to science.
Eligibility: Applicant must have been accepted by a university in Europe to start a PhD or will have completed no more than one year of their PhD by October
Level of Study: Doctorate
Type: Scholarships
Value: €30,000 per year and a laptop with a range of software applications
Length of Study: 3 years
Frequency: Annual
Closing Date: September 20th (Check with website)
Additional Information: All queries shoud be sent via email and please see the website for further details.

For further information contact:

Email: msrphd@microsoft.com
Website: http://research.microsoft.com/en-us/collaboration/global/apply-europe.aspx

Microsoft Research India PhD Fellowships

Subjects: Computer science.
Purpose: To support PhD students at Indian universities while they pursue doctoral education in computer science and related areas.
Eligibility: Students must be enrolled in a PhD program at an Indian university, students must be in the first, second or third year of their PhD studies, students who are studying computer science, electrical engineering, electrical/electronics, communication engineering, mathematics and other fields closely related to computer science are eligible to apply.
Level of Study: Doctorate
Type: Fellowships
Value: The award comprises a monthly stipend that will be disbursed to the awardee over a maximum period of four years. This monthly sum can be used for tuition, books, stipends, and/or any other expenses. Fellow receives a laptop computer and a separate sum of Rs 2,50,000 (rupees two lakh fifty thousand only) for travel to conferences and seminars during the term of the fellowship
Length of Study: 4 years
Frequency: Annual
No. of awards offered: 5
Application Procedure: The application process is completely online. Sign up to create your application by using the online application system.
Closing Date: April 27th

Additional Information: The Fellows will also have the option of a 3–6 months internship at Microsoft Research India. Please see the website for further details http://research.microsoft.com/en-us/collaboration/global/india/phdfellowships.aspx.

MINDA DE GUNZBURG CENTER FOR EUROPEAN STUDIES (CES) AT HARVARD UNIVERSITY

27 Kirkland Street, Cabot Way, Cambridge, MA, 02138, United States of America
Tel: (1) 617 495 4303
Fax: (1) 617 495 8509
Email: ces@fas.harvard.edu
Website: www.ces.fas.harvard.edu

The Minda de Gunzburg Center for European Studies at Harvard is dedicated to fostering the study of European history, politics and society. Through graduates, who go on to teach others about Europe and to many other roles in society, the Center sustains America's knowledge base about Europe, an important contribution to international understanding in difficult times.

Graduate Dissertation Research Fellowship

Subjects: Cultural, economic, historical, intellectual, political or social trends or on public policy in contemporary Europe (1750–present).
Purpose: To fund students from Harvard and the Massachusetts Institute of Technology who wish to conduct dissertation research.
Eligibility: Harvard doctoral students and MIT doctoral students in the social sciences. Applicants must have completed two years of graduate school and have passed their general examinations, but they can be at any stage of research.
Level of Study: Doctorate, Research
Type: Fellowship
Value: US$24,000
Length of Study: 1 year
Frequency: Annual
Study Establishment: Harvard University
Country of Study: United States of America
No. of awards offered: Varies
Application Procedure: The application in pdf format is available on the CES website.
Closing Date: February 13th
Funding: Foundation
Contributor: Krupp Foundation and Minda de Gunzburg Center for European Studies
No. of awards given last year: 10
No. of applicants last year: 18
Additional Information: The fellowships cannot be deferred and must be used within the 12 month period for which they are awarded.

Minda de Gunzberg Graduate Dissertation Completion Fellowship

Subjects: History, social sciences and cultural studies in modern or contemporary Europe (1750–present).
Purpose: To support students conducting dissertation work on contemporary Europe.
Eligibility: Open to advanced Harvard and Massachusetts Institute of Technology doctoral students. Students must have completed two draft dissertation chapters and submitted them to their advisor at the time of application.
Level of Study: Doctorate
Type: Fellowship
Value: US$24,000
Length of Study: 1 year
Frequency: Annual
Study Establishment: Harvard University
Country of Study: United States of America
No. of awards offered: Varies
Application Procedure: Application form is available on the CES website.
Closing Date: February 17th
Funding: Foundation
Contributor: Krupp Foundation and Minda de Gunzburg Center for European Studies
No. of awards given last year: 2
No. of applicants last year: 14
Additional Information: The fellowships cannot be deferred and must be used within the 12 month period for which they are awarded. No other employment is allowed under the grant except serving as an adviser for an undergraduate senior thesis. Students who have previously received writing/completion grants from other sources are not eligible.

MINERVA STIFTUNG

Gesellschaft für die Forschung mbH, Hofgartenstraβe 8, D-80539, Munich, Germany
Tel: (49) 89 2108 1420
Fax: (49) 89 2108 1451
Email: langegao@gv.mpg.de
Website: www.minerva.mpg.de

Minerva Fellowships

Subjects: All subjects.
Purpose: To fund scientific visits of Israeli scholars and scientists and to promote Israeli-German scientific co-operation.
Eligibility: Applicants from all subjects at German and Israeli universities and research institutions are elligible to apply.
Level of Study: Postdoctorate, Research, Doctorate
Type: Fellowship
Value: €1,128 per month for scientists working on their PhD thesis, €2,100 per month for scientists (PhD) with 2 years working experience and €2,300 per month for scientists (PhD) with 5 years working experience
Length of Study: Minerva Fellowships are initially granted for a minimum of six months, but they can be extended for up to a maximum of twenty-four months. There is the possibility of extending them for a third year (for graduates only)
Study Establishment: A German university or research institute
Country of Study: Germany
Closing Date: January 15th and June 15th
Additional Information: Dependency allowance: up to € 256 per month for an accompanying spouse (if spouse has income less than € 409), Children's allowance: up to € 51 per month per child (if children accompany family to Germany for minimum six months).

For further information contact:

Minerva Fellowship Office, Max Planck Gesellschaft, Frau Sieglinde Reichardt, Hofgartenstr. 8, München, 80539, Germany

Minerva Short-Term Research Grants

Subjects: All subjects.
Purpose: To fund scientific visits of Israeli scholars and scientists and to promote Israeli–German scientific co-operation.
Eligibility: Open to applicants from all research facilities (in Israel - public or governmental research institutes) and universities in Germany and Israel. Applicants should not be older than 38 years. (Maternal leave will be taken into account with regard to the age limit).
Level of Study: Research
Type: Grant
Value: €300 per week for doctoral candidates and €425 per week for postdoctoral candidates. (Additional payments like travel grants and family allowances are possible, depending on the type of the fellowship/grant)
Length of Study: 1–8 weeks
Study Establishment: A German university or research institute
Country of Study: Germany
Application Procedure: Applications including letters of invitation and letters of recommendation must be sent by electronic mail and must reach the Minerva office by May 2nd.
Closing Date: May 2nd

For further information contact:

Minerva Foundation, Head Office, Munich, Germany
Tel: (49) 49 89 2108 1258
Email: nagel@gv.mpg.de
Contact: Michael Nagel

MINISTRY OF EDUCATION, SCIENCE AND CULTURE (ICELAND)

Árni Magnússon Institute for Icelandic Studies, Sigurur Nordal Office, Pingholtsstraeti 29, 121, Reykjavik, P. O. Box1220, Iceland
Tel: (354) 562 6050
Fax: (354) 5626263
Email: nordals@hi.is
Website: www.arnastofnun.is
Contact: Ms Gurún Laufey Gumundsdóttir, Project Manager

The University of Iceland is a state university, founded in 1911. The university serves a nation of approximately 320.000 people and provides instruction for some 14,000 students studying in 5 schools and 26 faculties.

Ministry of Education, Science and Culture (Iceland) Scholarships in Icelandic Studies

Subjects: Icelandic as a second language provides students with a good practical and theoretical basis for the Icelandic language and Icelandic culture. It facilitates further study in linguistics, literature, history, and translation.
Eligibility: Citizens of the following countries may apply: Austria, Belgium, Bulgaria, Canada, China, Croatia, the Czech Republic, Denmark, Estonia, Faroe Islands, Finland, France, Germany, Greenland, Holland, Hungary, Ireland, Italy, Japan, Latvia, Lithuania, Norway, Poland, Russia, Slovakia, Spain, Sweden, Switzerland, the United Kingdom and United States of America. The scholarships are intended for students of the Icelandic language. Preference will, as a rule, be given to a candidate under 35 years of age.
Level of Study: Unrestricted
Type: Scholarship
Value: Icelandic Krona ca.9,50,000 plus tuition
Length of Study: 8 months
Frequency: Annual
Study Establishment: The University of Iceland, Reykjavik
Country of Study: Iceland
No. of awards offered: ca. 18
Application Procedure: Students from these countries, with exception of students from USA and Canada, can apply directly to the Ami Magnusson Institute. Application must be made on special application form that can be found on the website http://arnastofnun.is/page/studentastyrkir_menntamalaraduneytis_en. There are also application information for students from USA and Canada.
Closing Date: March 1st
Funding: Government
No. of awards given last year: 18
No. of applicants last year: 80
Additional Information: A special committee in Iceland will decide on the outcome of the applications and its evaluation will be announced to the candidates by April. No one can be granted a scholarship more than 3 times. For further information, please contact at email: nordalsehi.is,

For further information contact:

Email: nordals@hi.is
Website: www.arnastofnun.is

MINISTRY OF ENVIRONMENT & FORESTS (MOEF) GOVERNMENT OF INDIA

Wild Life Preservation, Paryavaran Bhavan CGO Complex, Lodhi Road, New Delhi, 110003, India
Tel: (91) 11 2436 0605
Email: envisect@nic.in
Website: www.envfor.nic.in
Contact: The Director

The Ministry of Environment & Forests (MoEF) is an agency in the administrative structure of the Central Government, for the planning, promotion, co-ordination and overseeing the implementation of environmental and forestry programmes. The principal activities undertaken by MoEF are conservation and survey of flora, fauna, forests and wildlife, prevention and control of pollution, afforestation and regeneration of degraded areas and protection of environment in the frame-work of legislations.

Rajiv Gandhi Wildlife Conservation Award

Subjects: Wildlife conservation.
Purpose: To encourage individuals who will make a significant contribution in the field of wildlife and also a major impact on the protection and conservation of wildlife in the country.
Eligibility: Open to Indian citizens, any recognised institution, individual or Government and forest staff who are engaged in scientific work for the cause of protection and conservation of wildlife. There is no age limit.
Level of Study: Postgraduate
Type: Award
Value: Indian Rupees 1,00,000
Frequency: Annual
Country of Study: India
No. of awards offered: 2
Application Procedure: Any citizen of India can recommend a name. The nominations must be accompanied by the following details alongwith detailed Biodata (10 copies): Name & address of the nominee, Area of work, Significant contributions made by the nominee, Measurable impact or likely impact of the nominee's work in the field of wildlife.
Funding: Government

THE MINISTRY OF FISHERIES

ASB Bank House, 101-103 The Terrace, Wellington, PO Box 2526, New Zealand
Tel: (64) 0800 00 8333
Fax: (64) 6448 94 0720
Email: info@fish.govt.nz
Website: www.fish.govt.nz/en-nz/default.htm

Ministry of Fisheries PG Scholarships in Quantitative Fisheries Science

Subjects: Fisheries science.
Purpose: To allow graduate students to develop expertise in quantitative fisheries science and encourage postgraduate students to contribute to priority research areas identified by the New Zealand government
Eligibility: Open to applicants with majors or minors in mathematics, statistics, biology, economics or computer science.
Level of Study: Postgraduate
Type: Scholarship
Value: $30,000 per year for PhD and up to $20,000 per year for Masters
Length of Study: 3 years (PhD) and up to 2 years (Masters)
Frequency: Annual
No. of awards offered: Varies
Application Procedure: Applicants should contact Rebecca Lawton for details.
Closing Date: September 20th
Contributor: In collaboration with NIWA
Additional Information: Research is most likely to be carried out a NIWA facility. Preference will be given to New Zealand citizens.

For further information contact:

The Ministry of Fisheries
Tel: 04 819 4251
Email: rebecca.lawton@fish.govt.nz
Contact: Rebecca Lawton

MINTEK

Private Bag X3015, Randburg, 2125, South Africa
Tel: (27) 11 709 4111
Fax: (27) 11 793 2413
Email: info@mintek.co.za
Website: www.mintek.co.za

Mintek is the partially state-funded South African metallurgical research organization. Its mission is to serve the national interest through high-calibre research and development and technology transfer that promotes mineral technology, and to foster the establishment of small, medium and large industries in the field of minerals and products derived from them.

Mintek Bursaries

Subjects: Chemical, metallurgical and electrical engineering, light current and electronics, chemistry, with an emphasis on inorganic, physical or analytical chemistry, metallurgy, extraction and mineralogy, geology and physics.
Purpose: To promote the training of research workers for the minerals industry in general and to meet its own needs for technically trained people.
Eligibility: Open to graduates of any nationality who possess an appropriate 4-year degree or higher qualification. Knowledge of English is essential. Preference is given to South African citizens. Merit and excellence are the criteria for selection. The project should be a promising development that will contribute to Mintek's activities and is in line with its strategic purpose. The student must have a strong academic background and show the potential to develop expertise that would benefit Mintek, and the institution where the project is carried out should be a centre of excellence in the subject.
Level of Study: Postgraduate
Type: Bursary
Value: Bursaries cover the full payment of registration, tuition and residence fees, plus an allowance, in return for a commitment to work at Mintek on a month-for-month basis.
Length of Study: Up to 2 years. Extension of this period can be granted by the President of Mintek
Frequency: Annual
Study Establishment: Any university or technikons in fields that complement its own activities
Country of Study: South Africa
No. of awards offered: 35
Application Procedure: Applicants must write for details.
Closing Date: June 30th
Funding: Government
Additional Information: Candidates are requested to sign a contract before starting their project. Usually, Mintek insists on a service commitment from the bursar on the completion of his or her studies on a year-to-year basis. Bursars are paid a monthly bursary-salary and Mintek pays the institution an amount to cover the project's running costs.

MINTRAC NATIONAL MEAT INDUSTRY TRAINING ADVISORY COUNCIL LIMITED

Suite 2, 150 Victoria Road, Drummoyne, New South Wales, 2047, Australia
Tel: (61) 02 9819 6699
Fax: (61) 02 9819 6099
Email: mintrac@mintrac.com.au
Website: www.mintrac.com.au/index.php

Aims to provide highly valued services to the Meat Industry in the areas of education and training development and advocacy along with providing products and services in accordance with world's best practice, satisfying and rewarding career paths for industry participants at every level, and securing and maintaining the ongoing commitment of the industry and its financial supporters.

MINTRAC Postgraduate Research Scholarship

Subjects: Research related to meat industry.
Purpose: To offer scholarships to students undertaking Honours, Masters or Doctorate programs with research in the meat industry.
Eligibility: Open to the candidates who want to undertake postgraduate research degrees (for Masters and Doctorate students) in disciplines relevant to the meat industry.
Level of Study: Doctorate, Postgraduate
Type: Scholarship
Value: A stipend of $25,000 per year will be paid to full-time students, or $12,500 per year forpart-time students, proceeding to Masters or Doctorate for the duration of theirstudies. A stipend of $19,000 per year will be paid to full-time or $9,500 per year for part-time students proceeding towards an Honours qualification.
Frequency: Ongoing
Study Establishment: Australian Universities
Country of Study: Australia
Application Procedure: Applicants must send the applications along with academic transcript(s), Overview of R&D project, Supervisor's letter confirming acceptance of proposed study, Meat industry conformation letter (unless forwarded separately), and Referee's reports (unless forwarded separately).
Closing Date: December 14th
Funding: Private
Contributor: Red meat industry
Additional Information: A thesis allowance of $800 will be paid to the student on receipt of their thesis and précis and on the condition that progress reports have been received in a timely manner.

MISSOURI STATE UNIVERSITY

College of Business, Glass Hall 223, 901 South National Avenue, Springfield, MO, 65897, United States of America
Tel: (1) 417 836 5616
Fax: (1) 417 836 6636
Email: MBAProgram@MissouriState.edu
Website: www.mba.missouristate.edu
Contact: Christopher D Lynn, Interim MBA Director

Missouri State University's College of Business has earned the highest level of accreditation for its programs, ensuring you will be prepared for the regional, national and international job market. MBA students experience a program that is flexible, accessible, affordable and with a solid reputation for international excellence.

BKD Graduate Scholarship

Subjects: Awarded annually to a full-time student enrolled in the Master of Accountancy or MBA with an Accounting concentration, a cumulative GPA of 3.0 must demonstrate leadership.
Purpose: To assist a worthy graduate student in pursuing an MBA or Master of Accountancy in the College of Business.
Eligibility: Enrolled, full-time student, in the Master of Accountancy or MBA, with an Accounting concentration. Cumulative 3.0 GPA required. One must demonstrate leadership.
Level of Study: MBA
Type: Scholarship
Value: $1,500
Frequency: Annual
Study Establishment: Missouri State University
Country of Study: United States of America
No. of awards offered: 3
Application Procedure: Apply online through website.
Closing Date: March 1st
Funding: Private
Contributor: BKD Educational Fund Endowment
No. of awards given last year: 2
No. of applicants last year: 92
Additional Information: Please check at http://business.missouristate.edu.

Carr Foundation/MBA Scholarship

Subjects: Awarded annually to a student in the MBA program. Awards made with regard for financial need.
Purpose: To assist a worthy graduate student in pursuing a master's of business administration degree in College of Business.
Eligibility: Students must be enrolled in the MBA program and display financial need. Awarded annually to a student in the MBA program. Awards made with regard to financial need.
Level of Study: Graduate, MBA
Type: Scholarship
Value: $600
Length of Study: Varies
Frequency: Annual
Study Establishment: Missouri State University
Country of Study: United States of America
No. of awards offered: 1

Application Procedure: Applicants must contact the organization for application details. Apply online (check in November).
Closing Date: March 1st
Funding: Foundation, private
Contributor: Carr Foundation/MBA Scholarship Endowment
No. of awards given last year: 3
No. of applicants last year: 92
Additional Information: The scholarship is automatically renewed as long as the student maintains satisfactory academic progress; however, the recipient must reapply to renew. Please check more information at http://business.missouristate.edu.

Dr James C and Mary Lee Snapp Graduate Scholarship

Purpose: To assist a worthy graduate student in pursuing a master's degree in College of Business.
Eligibility: Awarded annually to a full-time graduate student in the College of Business administration, with an undergraduate and graduate GPA of 3.0. Financial need will be considered.
Level of Study: Graduate
Type: Scholarship
Value: $1,000
Frequency: Annual
Study Establishment: Missouri State University
Country of Study: United States of America
No. of awards offered: 1
Application Procedure: Apply online.
Closing Date: March 1st
Funding: Individuals, private
Contributor: Dr. James C and Mary Lee Snapp Graduate Scholarship Endowment
No. of awards given last year: 1
No. of applicants last year: 92
Additional Information: Award is renewable; student must reapply for full consideration with all other applicants. Please check more information at http://business.missouristate.edu.

For further information contact:

Website: www.coba.missouristate.edu

Dr. R Stephen Parker College of Business Scholarship

Subjects: Awarded annually to a full-time student in College of Business.
Purpose: To assist a worthy undergraduate or graduate student in pursuing a Master's or Bachelor's degree within the College of Business.
Eligibility: Applicant must be full-time undergraduate or graduate student of College of Business, pursuing a Master's of Bachelor's degree. Applicant must be from the St. Louis MO Area. Given with regard to financial need. Participation in the Woman Team Leadership Program of St. Louis preferred.
Level of Study: Graduate, MBA, Undergraduate
Type: Scholarship
Value: $2,000
Frequency: Annual
Study Establishment: Missouri State University
Country of Study: United States of America
No. of awards offered: 1
Application Procedure: Apply online.
Closing Date: March 1st
Funding: Private
Contributor: Annual Gifts Private Donor
No. of applicants last year: 4365
Additional Information: Please check at http://business.missouristate.edu.

John L. and Rita M. Bangs Scholarship

Subjects: Awarded annually to a full-time student seeking a degree in the College of Business, preference will be given to students majoring in Computer Information System or Marketing. Preference for academically excellent students. Presidential scholars are eligible.
Purpose: To assist a worthy undergraduate or graduate student in pursuing a bachelor's or master's degree in College of Business.
Eligibility: Open to graduate College of Business applicants who are academically talented. Preference will be given to Computer Information Systems and/or Marketing major. Awarded annually to a full-time student seeking a degree in the college of business. Preference will be given to students majoring in computer information systems or marketing. preference for academically excellent students presidential scholars not eligible.
Level of Study: Graduate, Postgraduate, Undergraduate
Type: Scholarship
Value: $2,000
Frequency: Annual
Study Establishment: Missouri State University
Country of Study: United States of America
No. of awards offered: 2
Application Procedure: Apply online (check in November for upcoming academic year).
Closing Date: March 1st
Funding: Individuals, private
Contributor: John L. and Rita M. Bangs Scholarship endownment
No. of awards given last year: 1
No. of applicants last year: 92
Additional Information: Award is renewable; student must reply for full consideration with all other applicants. Please check more information at http://business.missouristate.edu.

Marketing Employers Scholarship

Purpose: To assist a worthy undergraduate or graduate student in pursuing an MBA or a Bachelor's degree with emphasis in Marketing within College of Business.
Eligibility: One must be in the College of Business as an undergraduate student, with emphasis in Marketing OR as a graduate student, pursuing MBA with emphasis in Marketing. Selection made with regards to leadership a strong work ethic and overall GPA of 3.0 required.
Level of Study: Graduate, MBA, Undergraduate
Type: Scholarship
Value: $400
Frequency: Annual
Study Establishment: Missouri State University
Country of Study: United States of America
No. of awards offered: 1
Application Procedure: Apply online.
Closing Date: March 1st
Funding: Private
Contributor: Private donor - individual
No. of awards given last year: 1
No. of applicants last year: 4365
Additional Information: Please check at http://business.missouristate.edu for further information.

Missouri State University Graduate Assistantships

Subjects: All subjects.
Purpose: To assist students with expenses and to enhance learning while studying for advanced degrees at Missouri State University.
Eligibility: Applicants must be admitted to a graduate program at MSU to be eligible. A minimum grade point average of 3.00 on the undergraduate cumulative last 60 hours of undergraduate coursework or a minimum grade point average of 3.00 on 9 hours or more of graduate coursework is required. Graduate students who did not receive both their primary and secondary education where English was the primary language must meet the following requirements to qualify for graduate assistantships: successful completion of one semester of graduate studies at MSU, during which they complete cultural orientation to prepare them for a teaching appointment and pass an MSU-juried examination, in which the candidate must demonstrate his or her ability to interpret written English passages and to communicate orally in English in a classroom setting.
Level of Study: Graduate, MBA
Value: A minimum stipend of US$8,000 for the academic year (9 months). In a few situations, a stipend of $9,730 may be awarded
Length of Study: A maximum of 2 years (including Fall, Spring, and Summer)
Frequency: Annual
Study Establishment: Missouri State University
Country of Study: United States of America
No. of awards offered: Varies
Application Procedure: Applicants must submit an application directly to the department in which the assistantship is sought. It is

wise to check with the department before applying. Information from an applicant must include employment history, academic history and the addresses of referees. Applications are available from the Graduate College and on the website: http://graduate.missouristate.edu/assets/graduate/GA_application.pdf.
Funding: Government
Contributor: College of Business Administration
Additional Information: For further information visit the website or contact the Graduate College.

Raikos Scholarship

Subjects: Scholarship is awarded annually to students enrolled in the College of Business, undergraduate or graduate, with a cumulative 3.5 GPA.
Purpose: To assist a worthy undergraduate or graduate student in pursuing a Master's or Bachelor's degree in the College of Business.
Eligibility: Student must have graduated from a high school within 100 miles of Springfield, MO. Applicant must be enrolled full-time as an undergraduate or graduate student, with a cumulative 3.5 GPA.
Type: Scholarship
Value: $2,500–10,000
Frequency: Annual
Study Establishment: Missouri State University
Country of Study: United States of America
No. of awards offered: 1–4
Application Procedure: Apply online.
Closing Date: March 1st
Funding: Private
Contributor: Private donor - individual
Additional Information: Please check at http://business.missouristate.edu. This is renewable; must reapply.

The Robert W. and Charlotte Bitter Graduate Scholarship Endowment

Purpose: To assist a worthy graduate student in pursuing a master's of business administration or master's of accounting degree in College of Business.
Eligibility: Awarded annually to a student seeking an MBA or MACC, be enrolled in 12 hours or enrolled in 6 hours or more each semester if the student is a graduate assistant, have a combined formula score (200 x grade point average plus Graduate Management Admission Test score) of 1,100 or higher and minimum 3.3 graduate grade point average, have completed a minimum of 24 hours or 15 hours, if the student is a graduate assistant, and have completed all prerequisite courses or be currently enrolled in final prerequisite courses.
Level of Study: MBA, Mater's in accounting
Type: Scholarship
Value: $1,000
Length of Study: Varies
Frequency: Annual
Study Establishment: Missouri State University
Country of Study: United States of America
No. of awards offered: 1
Application Procedure: Applicants must contact the organization for application details. Apply online (Check in November).
Closing Date: May 1st - absolute; March 1st - priority deadline
Funding: Foundation, private
Contributor: The Robert W. and charlotte Bitter Graduate Scholarship Endowment
No. of awards given last year: 1
No. of applicants last year: 92
Additional Information: Not renewable. Please check morre information at http://business.missouristate.edu.

MIZUTANI FOUNDATION FOR GLYCOSCIENCE

Shinkawa Chuo Building 1F 1-17-24 Shinkawa,Chuo-ku, Tokyo, 104-0033 Japan, Tokyo, 100-0005, Japan
Tel: (81) 3 3555 1861
Fax: (81) 3 3555 1862
Email: info@mizutanifdn.or.jp
Website: www.mizutanifdn.or.jp
Contact: Takashi Kato, Executive Secretary

The Mizutani Foundation for Glycoscience undertakes major programmes such as the worldwide distribution of research grants to qualified glycoscientists for their outstanding basic research, assistance of international exchanges between Japanese and foreign glycosciences, contributions to glycoscience-related meetings in Japan and practice of other activities that are necessary to achieve the aims of the Foundation.

Mizutani Foundation for Glycoscience Research Grants

Subjects: Glycoscience.
Purpose: To contribute to human welfare, through the enhancement of glycoscience, by awarding grants for creative research in glycoscience conducted by domestic and overseas researchers, by awarding grants for international exchanges and for convening conferences in the field of glycoscience.
Eligibility: An applicant must (a) have a doctor's degree or its equivalent, (b) have a documented capability of performing independent studies and (c) be a member of a research institute where he or she can carry out the proposed project.
Level of Study: Doctorate, Postdoctorate
Value: The total budget for grants is 70,000,000 yen per year.
Length of Study: 1 year
Frequency: Annual
Country of Study: Any country
No. of awards offered: 10–15
Application Procedure: Applicant must make online registration, and also submit a set of application documents by post.
Closing Date: The term for applications is from July 1st to September 1st
Funding: Foundation
Contributor: Seikagaku Corporation
No. of awards given last year: 16
No. of applicants last year: 139

MODERN LANGUAGE ASSOCIATION OF AMERICA (MLA)

26 Broadway, 3rd Floor, New York, NY, 10004 1789, United States of America
Tel: (1) 646 576 5000
Fax: (1) 646 458 0030
Email: awards@mla.org
Website: www.mla.org
Contact: Reiser

The Modern Language Association of America (MLA) is a non-profit membership organization that promotes the study and teaching of language and literature in English and foreign languages.

Aldo and Jeanne Scaglione Prize for a Translation of a Literary Work

Subjects: Translation.
Purpose: To award an outstanding translation of a book-length literary work into English.
Eligibility: Open to translations published in the year preceding the year in which the award is given.
Type: Prize
Value: Cash award, certificate, and and a one-year membership in the association
Frequency: Every 2 years
No. of awards offered: 1
Application Procedure: Applicants must send six copies of the work. For detailed information about specific prizes, applicants should contact the MLA.
Closing Date: April 1st
Funding: Private

Aldo and Jeanne Scaglione Prize for Comparative Literary Studies

Subjects: Comparative literary and cultural studies.
Purpose: To recognize outstanding scholarly work in comparative literary studies.
Eligibility: Open to books published in the year preceding the year in which the award is given. Authors must be members of the MLA.

Level of Study: Postdoctorate
Type: Prize
Value: Cash award and certificate
Frequency: Annual
No. of awards offered: 1
Application Procedure: Applicants must send four copies of the work. For detailed information about specific prizes, applicants should contact the MLA.
Closing Date: May 1st
Funding: Private

Aldo and Jeanne Scaglione Prize for French and Francophone Literary Studies

Subjects: French and Francophone linguistic or literary studies.
Purpose: To recognize outstanding scholarly work.
Eligibility: Open to books published in the year preceding the year in which the prize is given. Authors must be members of the MLA.
Level of Study: Postdoctorate
Type: Prize
Value: Cash award and certificate
Frequency: Annual
No. of awards offered: 1
Application Procedure: Applicants must send four copies of the work. For detailed information about specific prizes, applicants should contact the MLA.
Closing Date: May 1st
Funding: Private

Aldo and Jeanne Scaglione Prize for Italian Studies

Subjects: Italian literature, culture or comparative literature involving Italy.
Purpose: To award an outstanding scholarly work.
Eligibility: Open to books published in the year preceding the year in which the award is given. Authors must be members of the MLA.
Level of Study: Postdoctorate
Type: Prize
Value: Cash award and certificate
Frequency: Every 2 years
No. of awards offered: 1
Application Procedure: Applicants must send four copies of the work. For detailed information about specific prizes, applicants should contact the MLA.
Closing Date: May 1st
Funding: Private

Aldo and Jeanne Scaglione Prize for Studies in Germanic Languages and Literatures

Subjects: The linguistics or literatures of any of the Germanic languages including Danish, Dutch, German, Norwegian, Swedish or Yiddish.
Purpose: To recognize an outstanding scholarly work.
Eligibility: Authors must be members of the MLA. Books must have been published in the 2 years preceding the year of the award.
Level of Study: Postdoctorate
Type: Prize
Value: Cash award and certificate
Frequency: Every 2 years
No. of awards offered: 1
Application Procedure: Applicants must send four copies of the work. For detailed information about specific prizes, applicants should contact the MLA.
Closing Date: May 1st
Funding: Private

Aldo and Jeanne Scaglione Prize for Studies in Slavic Languages and Literatures

Subjects: Linguistic or literary study of a work in a Slavic language.
Purpose: To recognize an outstanding scholarly work.
Eligibility: Open to books published no more than two years preceding the year in which the prize is given. Authors need not be members of the MLA.
Level of Study: Postdoctorate
Type: Prize
Value: Cash award, certificate, and a one-year membership in the association
Frequency: Every 2 years
No. of awards offered: 1
Application Procedure: Applicants must send four copies of the work. For detailed information about specific prizes, applicants should contact the MLA.
Closing Date: May 1st
Funding: Private

Aldo and Jeanne Scaglione Prize for Translation of a Scholarly Study of Literature

Subjects: Translation.
Purpose: To recognize an outstanding translation of a book-length work of literary history, literary criticism, philology and literary theory into English.
Eligibility: Open to books published no more than 2 years preceding the year in which the prize is given. Authors need not be members of the MLA.
Level of Study: Postdoctorate
Type: Prize
Value: Cash award, certificate, and a one-year membership in the association
Frequency: Every 2 years
No. of awards offered: 1
Application Procedure: Applicants must send four copies of the work. For detailed information about specific prizes, applicants should contact the MLA.
Closing Date: May 1st
Funding: Private

Fenia and Yaakov Leviant Memorial Prize in Yiddish Studies

Subjects: English translation of a Yiddish literary work and/or Yiddish literature and culture.
Purpose: To recognize an outstanding translation into English or an outstanding scholarly work in the field of Yiddish.
Eligibility: The prize is awarded alternately to a translation or scholarly work in the field of Yiddish.
Level of Study: Postdoctorate
Type: Prize
Value: Cash award, certificate, and a one-year membership in the association
Frequency: Every 2 years
No. of awards offered: 1
Application Procedure: Applicants must send four copies of the work. For detailed information about specific prizes, applicants should contact the MLA.
Closing Date: May 1st
Funding: Private

Howard R Marraro Prize

Subjects: Italian literature or comparative literature involving Italian.
Purpose: To award an outstanding scholarly work.
Eligibility: Authors must be members of the MLA, and the book must be published in the previous year
Level of Study: Postdoctorate
Type: Prize
Value: Cash award and certificate
Frequency: Every 2 years
No. of awards offered: 1
Application Procedure: Applicants must send four copies of the work. For detailed information about specific prizes, applicants should contact the MLA.
Closing Date: May 1st
Funding: Private

James Russell Lowell Prize

Subjects: Literary theory, media, cultural history and interdisciplinary topics.
Purpose: To recognize an outstanding literary or linguistic study, a critical edition of an important work or a critical biography.
Eligibility: Open to books published the year preceding the year in which the award is due to be given. Authors must be current members of the MLA.

Level of Study: Postdoctorate
Type: Prize
Value: Cash award and certificate
Frequency: Annual
No. of awards offered: 1
Application Procedure: Applicants must send six copies of the work. For detailed information about specific prizes, applicants should contact the MLA.
Closing Date: March 1st
Funding: Private

Katherine Singer Kovacs Prize

Subjects: Latin American or Spanish literatures and cultures.
Purpose: To recognize an outstanding book published in English or Spanish.
Eligibility: Open to books published the year preceding the year in which the prize is given. Competing books should be broadly interpretative works that enhance the understanding of the interrelations among literature, the arts and society.
Level of Study: Postdoctorate
Type: Prize
Value: Cash award and certificate
Frequency: Annual
No. of awards offered: 1
Application Procedure: Applicants must send six copies of the work. For detailed information about specific prizes, applicants should contact the MLA.
Closing Date: May 1st
Funding: Private

Kenneth W Mildenberger Prize

Subjects: Language, culture, literacy, and literature that has a strong application to the teaching of languages other than English.
Purpose: To support an outstanding scholary work.
Eligibility: Authors need not be members of the MLA. Open to books published in the year preceding the year in which the prize is given.
Level of Study: Postdoctorate
Type: Prize
Value: Cash award, certificate, and a one-year membership in the MLA
Frequency: Annual
No. of awards offered: 1
Application Procedure: Applicants must send four copies of the work. For detailed information about specific prizes, applicants should contact the MLA.
Closing Date: May 1st
Funding: Private
Additional Information: The prize is given for a research article in odd-numbered years and a book in even-numbered years.

Lois Roth Award for a Translation of Literary Work

Subjects: Translation.
Purpose: To recognize an outstanding translation of a book-length literary work into English.
Eligibility: Open to translations published in the year preceding the year in which the prize is given. Translators need not be members of the MLA.
Level of Study: Postdoctorate
Type: Prize
Value: Cash award, certificate, and one-year membership in the organization
Frequency: Every 2 years
No. of awards offered: 1
Application Procedure: Applicants must send six copies of the work. For detailed information about specific prizes, applicants should contact the MLA.
Closing Date: April 1st
Funding: Private

Mina P Shaughnessy Prize

Subjects: Language, culture, literacy, and literature that has a strong application to the teaching of English.
Purpose: To recognize an outstanding scholary book
Eligibility: Open to books published in the year preceding the year in which the prize is given. Authors need not be members of the MLA.
Level of Study: Postdoctorate
Type: Prize
Value: Cash award, certificate, and one-year membership in the organization
Frequency: Annual
No. of awards offered: 1
Application Procedure: Applicants must send four copies of the work. For detailed information about specific prizes, applicants should contact the MLA.
Closing Date: May 1st

MLA Prize for a Distinguished Bibliography

Subjects: Bibliography.
Purpose: To award an outstanding enumerative or descriptive bibliography.
Eligibility: Editors need not be members of the MLA. Open to books published during the 2 years preceding the year in which the prize is given.
Level of Study: Postdoctorate
Type: Prize
Value: Cash award, certificate, and one-year membership in the organization
Frequency: Every 2 years
No. of awards offered: 1
Application Procedure: Applicants must send four copies of the work. For detailed information applicants should contact the MLA.
Closing Date: May 1st
Funding: Private

MLA Prize for a Distinguished Scholarly Edition

Subjects: Works in any of the modern languages.
Purpose: To recognize an outstanding scholarly edition.
Eligibility: At least 1 volume must have been published during the 2 years preceding the year in which the award is given. Editors need not be members of the MLA. Editions may be single or multiple volumes. an edition should be based on an examination of all available relevant textual sources; the source texts and the edited text's deviations from them should be fully described; the edition should employ editorial principles appropriate to the materials edited, and those principles should be clearly articulated in the volume; the text should be accompanied by appropriate textual and other historical contextual information; the edition should exhibit the highest standards of accuracy in the presentation of its text and apparatus; and the text and apparatus should be presented as accessibly and elegantly as possible.
Level of Study: Postdoctorate
Type: Prize
Value: Cash award, certificate, and one-year membership in the association
Frequency: Every 2 years
No. of awards offered: 1
Application Procedure: Applicants must send four copies of the work. For detailed information about specific prizes, applicants should contact the MLA.
Closing Date: May 1st
Funding: Private

MLA Prize for a First Book

Subjects: Literary theory, media, cultural history or interdisciplinary topics.
Purpose: To recognize an outstanding literary or linguistic study, or a critical biography.
Eligibility: Open to books published in the year preceding the year in which the prize is given as the first book-length publication of a current MLA member.
Level of Study: Postdoctorate
Type: Prize
Value: Cash award and certificate
Frequency: Annual
No. of awards offered: 1
Application Procedure: Applicants must send six copies of the work. For detailed information about specific prizes, applicants should contact the MLA.
Closing Date: April 1st
Funding: Private

MLA Prize for Independent Scholars

Subjects: English or other modern languages and literatures.
Purpose: To encourage the achievements and contributions of independent scholars.
Eligibility: Open to books published in the year preceding the year in which the prize is given. At the time of publication of the book, the author must not be enrolled in a programme leading to an academic degree or hold a tenured, tenure-accruing or tenure-track position in postsecondary education. Authors need not be members of the MLA.
Level of Study: Postdoctorate
Type: Prize
Value: Cash award, certificate, and one-year membership in the association
Frequency: Annual
No. of awards offered: 1
Application Procedure: Applicants must send six copies of the work. For detailed information about specific prizes, applicants should contact the MLA.
Closing Date: May 1st
Funding: Private

MLA Prize in United States Latina and Latino or Chicana and Chicano Literary and Cultural Studies

Subjects: Chicana and Chicano, Latina and Latino literary or cultural studies.
Purpose: To award an outstanding scholarly study.
Eligibility: Open to authors who are members of the MLA. Study must be published in the preceding two years.
Level of Study: Postdoctorate
Type: Prize
Value: Cash award, certificate, and one-year membership in the association
Frequency: Annual
No. of awards offered: 1
Application Procedure: Applicants must send four copies of their work. For detailed information about specific prizes, applicants should contact the MLA.
Closing Date: May 1st
Funding: Private

Morton N Cohen Award for A Distinguished Edition of Letters

Subjects: Edition of letters in any of the modern languages.
Purpose: To encourage an outstanding edition of letters.
Eligibility: At least one volume must have been published during the 2 years preceding the year in which the award is given. Editors need not be members of the MLA. Editions may be single or multiple volumes.
Level of Study: Postdoctorate
Type: Award
Value: Cash award, certificate, and one-year membership in the association
Frequency: Every 2 years
No. of awards offered: 1
Application Procedure: Applicants must send four copies of the work. For detailed information about specific prizes, applicants should contact the MLA.
Closing Date: May 1st
Funding: Private

William Sanders Scarborough Prize

Subjects: Black American literature and culture.
Purpose: To recognize an outstanding scholarly book.
Eligibility: Books that are primarily translations will not be considered.
Level of Study: Postdoctorate
Type: Prize
Value: Cash award, certificate, and one-year membership in the association
Frequency: Annual
No. of awards offered: 1
Application Procedure: Applicants must send four copies of the work. For detailed information about specific prizes, applicants should contact the MLA.
Closing Date: May 1st

MOLECULAR MEDICINE IRELAND

Newman House, 85a St. Stephens Green, Dublin 2, Ireland
Tel: (353) 353 (0)1 4779 820
Fax: (353) 353 (0)1 4779 823
Email: info@molecularmedicineireland.ie
Website: http://www.molecularmedicineireland.ie/home
Contact: Dr Mark Watson

MMI Clinical and Translational Research Scholars Programme

Subjects: The 4-year programme combines advanced training through laboratory rotations, taught modules and industry placements, with a 3-year PhD research project in clinical and translational research.
Purpose: A cross-institutional structured 4-year PhD programme, integrate with industry and clinical research centres, which will prepare graduates for careers in industry, academic medical centres and as biomedical entrepreneur.
Eligibility: Applications are welcomed from candidates holding a first or upper second class honours (> or equal to 3.8 GPA) degree in an apropriate discipline. Applications are also invited from those with a master's degree in an appropriate discipline.
Level of Study: Doctorate
Type: Fellowship
Value: Annual stipend of €16,000 plus research and travel allowance and PhD fees (for EU students).
Length of Study: 4 years
Frequency: Dependent on funds available
Study Establishment: NUI Galway, Trinity College Dublin, University College Cork, University College Dublin
Country of Study: Ireland
Application Procedure: Please see MMI website (www.molecularmedicineireland.ie) for application details. Scholarships will be advertised here when they become availabe.
Funding: Government
Contributor: The proramme for research in third level institutions (PRTLI) cycle 5, and co-funded under the European regional development fund (ERDF)
No. of awards given last year: 20
No. of applicants last year: 123
Additional Information: Please check at www.molecularmedicineireland.ie/ctrsp for more information. Placements in academic research groups, in industry and in clinical research facilities.

MMI Clinical Scientist Fellowship

Subjects: Clinical and translational research.
Purpose: The objective of the MMI Clinician Scientist Fellowship (CSFP) is to train the next generation of clinician scientists (academic medical leaders) with the unique and specialized knowledge essential to fulfill Ireland's research needs in translational medicine.
Eligibility: Applications are invited from suitable qualified high calibre medical graduates.
Level of Study: Doctorate
Type: Fellowship
Value: Please see www.molecularmedicineireland.ie/csfp for more information
Length of Study: 3 years
Frequency: Dependent on funds available
Study Establishment: NUI Galway, Trinity College Dublin, University College Cork, University College Dublin, Royal College of Surgeons in Ireland.
Country of Study: Ireland
Application Procedure: Please check website.
Funding: Government
Contributor: The MMI Clinician Scientist Fellowship Programme is funded under the Programme for Research in Third Level Institution (PRTLI) cycle 4, and co-funded under the European Regional Development fund (EROD)
Additional Information: Please see www.molecularmedicineireland.ie/csfp for more information.

MONASH UNIVERSITY

Institute of Graduate Research, Building 3D, Clayton Campus, Wellington Road, Clayton, Victoria, 3800, Australia
Tel: (61) 3 9905 3009
Fax: (61) 3 9905 5042
Email: migr@monash.edu
Website: www.monash.edu.au/migr

Monash University is one of Australia's largest universities, with ten faculties covering every major area of intellectual activity, six campuses in Australia and an increasing global presence. Research at Monash covers the full spectrum from fundamental to applied research and ranges across the arts and humanities, social, natural, health and medical sciences and the technological sciences. The University is determined to preserve its strength in fundamental research, which underpins its successes in applied research, and to continue to make a distinguished contribution to intellectual and cultural life.

International Postgraduate Research Scholarships

Subjects: All subjects.
Purpose: To provide support for supervised full-time research at the Master's and doctoral level.
Eligibility: Open to graduates of any Australian or overseas university who hold a First Class (Honours) Bachelor's Degree or qualifications and/or research experience deemed equivalent by the University.
Level of Study: Doctorate
Type: Scholarship
Value: The full cost of tuition fees plus and overseas student health cover
Length of Study: Up to 2 years for the Master's degree and up to 3 years with the possibility of an additional 6 months extension for the doctoral degree
Frequency: Twice a year
Study Establishment: Monash University
Country of Study: Australia
No. of awards offered: 25
Application Procedure: Please see the website www.mrgs.monash.edu.au/scholarships for further details.
Closing Date: October 31st, May 31st
Funding: Government
No. of awards given last year: 25
Additional Information: International students must meet English language proficiency requirements.

Monash Graduate Scholarship

Subjects: All subjects.
Purpose: To provide support for supervised full-time research at the masters and doctoral levels.
Eligibility: Open to graduates of any Australian or overseas university who holds a First Class (Honours) Bachelors degree or qualifications and/or research experience deemed equivalent by the university.
Level of Study: Research, Doctorate, Postgraduate
Type: Scholarship
Value: A stipend and living allowance of Australian $24,653
Length of Study: Upto 2 years for the Master's degree and upto 3 years with a possible extension for 6 months for the doctoral degree
Frequency: Twice a year
Study Establishment: Monash University
Country of Study: Australia
No. of awards offered: 120
Application Procedure: Please see the website www.mrgs.monash.edu.au/scholarships for further details.
Closing Date: October 31st and May 31st
Funding: Government
No. of awards given last year: 120
Additional Information: International students must meet English Language Proficiency requirement.

Monash International Postgraduate Research Scholarship (MIPRS)

Subjects: All subjects.
Purpose: To provide support for supervised full-time research at the Master's and doctoral level.
Eligibility: Open to graduates of any Australian or overseas university who hold a First Class (Honours) Bachelor's Degree or qualifications and/or research experience deemed equivalent by the University.
Level of Study: Doctorate, Master's by Research
Type: Scholarship
Value: The award meets the full cost of international tuition fees and overseas student health cover (OSHC)
Length of Study: Up to 2 years for the Master's degree and up to 3 years with the possibility of an additional 6-month extension for the doctoral degree
Frequency: Twice a year
Study Establishment: Monash University
Country of Study: Australia
No. of awards offered: 50
Application Procedure: Please see the website www.mrgs.monash.edu.au/scholarships for further details.
Closing Date: October 31st and May 31st
Funding: Government
No. of awards given last year: 50
Additional Information: International students must meet English language proficiency requirements.

Monash University Silver Jubilee Postgraduate Scholarship

Subjects: Different subjects areawarded by rotation to faculties.
Purpose: To provide supervised full-time research at the Master's and doctoral level.
Eligibility: Open to graduates of any Australian or overseas university who hold a First Class (Honours) Bachelor's Degree or qualifications and/or research experience deemed equivalent by the University.
Level of Study: Doctorate, Master's by Research candidates
Type: Scholarship
Value: A stipend and living allowance of Australian $27,651
Length of Study: Up to 2 years for the Master's degree and up to 3 years with the possibility of an additional 6-month extension for the doctoral degree
Frequency: Twice a year
Study Establishment: Monash University
Country of Study: Australia
No. of awards offered: 1
Application Procedure: Please visit our website ww.mrgs.monash.edu.au/scholarships for further details.
Closing Date: October 31st
No. of awards given last year: 1
Additional Information: International students must meet English language proficiency requirements.

Sir James McNeill Foundation Postgraduate Scholarship

Subjects: Engineering, medicine, music and science.
Purpose: To enable a PhD scholar to pursue a full-time programme of research that is both environmentally responsible and socially beneficial to the community.
Eligibility: Open to graduates of any Australian or overseas university who hold a First Class (Honours) Bachelor's Degree or qualifications and/or research experience deemed equivalent by the University.
Level of Study: Doctorate
Type: Scholarship
Value: A stipend and living allowance of Australian $28,151
Length of Study: Up to 3 years, with the possibility of an additional 6-month extension
Frequency: Annual
Study Establishment: Monash University
Country of Study: Australia
No. of awards offered: 1
Application Procedure: Please see the website www.mrgs.monash.edu.au/scholarships for further details
Closing Date: October 31st
Funding: Trusts
No. of awards given last year: 1
Additional Information: International students must meet English language proficiency requirements.

Vera Moore International Postgraduate Research Scholarships

Subjects: All subjects.

Purpose: To provide support for supervised full-time research at the Master's and doctoral level.
Eligibility: Open to graduates of any Australian or overseas university who hold a First Class (Honours) Bachelor's Degree or qualifications and/or research experience deemed equivalent by the University.
Level of Study: Doctorate
Type: Scholarship
Value: The full cost of tuition fees plus a research allowance of up to Australian $550 per year, and overseas student health cover
Length of Study: Up to 2 years for the Master's degree and up to 3 years with the possibility of an additional 6 months extension for the doctoral degree
Frequency: Annual
Study Establishment: Monash University
Country of Study: Australia
No. of awards offered: 1
Application Procedure: Please see the website www.mrgs.monash.edu.au/scholarships for further details.
Closing Date: October 31st
Funding: Trusts
No. of awards given last year: 1
Additional Information: International students must meet English language proficiency requirements.

MONTANA STATE UNIVERSITY-BILLINGS

1500 University Drive, Billings, MT, 59101, United States of America
Tel: (1) 657 2188, 800 565 6782, 406 657 2011
Fax: (1) 657 1789
Email: finaid@msubillings.edu/webmaster@msubillings.edu
Website: www.msubillings.edu

The Montana State University-Billings is dedicated to the development of workforce capacity by providing top quality learning opportunities and services to meet a variety of career choices and customer needs by being responsive, flexible and market driven.

Abrahamson Family Endowed Scholarship

Subjects: All subjects.
Purpose: To provide financial assistance.
Eligibility: Open to full-time enrolled students who are residents of Montana or Wyoming. Applicant must have and maintain at least a 2.75 cumulative GPA or above.
Level of Study: Postgraduate
Type: Scholarship
Value: $1,000
Frequency: Annual
Study Establishment: Montana State University-Billings
Country of Study: United States of America
No. of awards offered: 4
Application Procedure: See website.
Closing Date: February 1st
Funding: Private

The Berg Family Endowed Scholarship

Subjects: Humanities area including English, philosophy and history.
Purpose: To provide financial assistance.
Eligibility: Applicant must have a minimum 3.0 GPA, be enrolled full-time, demonstrate financial need, and reside in a Montana community east of Billings with a population of less than 50,000.
Level of Study: Postgraduate
Type: Scholarship
Value: The Scholarship has a maximum award of $1,100.
Frequency: Annual
Study Establishment: Montana State University-Billings
Country of Study: United States of America
No. of awards offered: 2
Application Procedure: Complete details available on the website.
Closing Date: February 1st
Funding: Private
Contributor: The Billings Foundation

Briggs Distributing Co., Inc./John & Claudia Decker Endowed Scholarship

Subjects: All subjects.
Purpose: To provide financial assistance.
Eligibility: Applicant must have a minimum 3.0 GPA, be enrolled full-time, and demonstrate financial need.
Level of Study: Postgraduate
Type: Scholarship
Value: The Scholarship has a maximum award of $3,000.
Frequency: Annual
Study Establishment: Montana State University-Billings
Country of Study: United States of America
Application Procedure: See the website.
Closing Date: February 1st
Funding: Private
Additional Information: Preference will be given to Children of employers of Briggs Distributing Co., Inc.

The Bruce H. Carpenter Non-Traditional Endowed Scholarship

Subjects: All subjects.
Purpose: To provide financial assistance.
Eligibility: Applicant must have a minimum 3.5 GPA or an endorsement from a counselor or advisor, have a minimum combined SAT Reasoning score of 1530 (composite ACT score of 22), demonstrate financial need, and be enrolled full-time. Three letters of recommendation required.
Level of Study: Postgraduate
Type: Scholarship
Value: The maximum award given by the Bruce H. Carpenter Non-Traditional Endowed Scholarship is $3,100.
Frequency: Annual
Study Establishment: Montana State University-Billings
Country of Study: United States of America
Application Procedure: A completed application form along with 3 letters of recommendation and a copy of college transcript must be submitted.
Closing Date: February 1st
Funding: Private
Additional Information: Students interested in applying for the scholarship must attach to their scholarship application an essay describing their educational goals and objectives which includes a description of why they believe campus/community involvement is an important part of their growth and development and a brief description of the contributions they would like to make to the campus and community during the time of their undergraduate education.

Ellen Shields Endowed Scholarship

Subjects: All subjects.
Purpose: To assist a student with economic need.
Eligibility: Applicant must have a minimum 2.5 GPA, be enrolled in a four-year degree program.
Level of Study: Postgraduate
Type: Scholarship
Value: The Scholarship has a maximum award of $5,500.
Length of Study: 4 years
Frequency: Annual
Study Establishment: Montana State University-Billings
Country of Study: United States of America
No. of awards offered: 1
Application Procedure: See the website for details.
Closing Date: February 1st
Funding: Private

Energy Laboratories Chemistry Endowed Scholarship

Subjects: Chemistry.
Purpose: To encourage students who wish to major in chemistry.
Eligibility: Applicant must have a minimum 3.0 GPA and be enrolled full-time. Preference is given to a chemistry majors.
Level of Study: Postgraduate
Type: Scholarship
Value: The Scholarship has a maximum award of $2,100.
Frequency: Annual
Study Establishment: Montana State University-Billings
Country of Study: United States of America

Application Procedure: See the website.
Closing Date: February 1st
Funding: Private

Eric Robert Anderson Memorial Scholarship

Subjects: Arts.
Purpose: To support art students.
Eligibility: Applicant must have a minimum 3.0 GPA, be a paid member of Art Students League, and be enrolled full-time.
Level of Study: Postgraduate
Type: Scholarship
Value: The Scholarship has a maximum award of $1,000.
Frequency: Annual
No. of awards offered: 2
Application Procedure: Applicant must submits resume and artist's statement, three letters of recommendation, and at least five slides of five separate works.
Closing Date: February 1st
Funding: Private

For further information contact:

MSU-Billings Art Office Secretary, First floor, Liberal Arts Building

Frances S. Barker Endowed Scholarship

Subjects: All subjects.
Purpose: To support students with financial needs.
Eligibility: Applicant must have a minimum 3.0 GPA, be enrolled full-time, and demonstrate financial need.
Level of Study: Postgraduate
Type: Scholarship
Value: The Scholarship has a maximum award of $1,000.
Frequency: Annual
Study Establishment: Montana State University-Billings
Country of Study: United States of America
No. of awards offered: 3
Application Procedure: See the website.
Closing Date: February 1st
Funding: Private
Contributor: The MSU-Billings Foundation Athletics Scholarship Committee
Additional Information: Preference will be given to student athletes.

The Haynes Foundation Scholarships

Subjects: All subjects.
Purpose: To provide financial assistance.
Eligibility: Applicant must have a minimum 3.5 GPA, be enrolled full-time, and demonstrate financial need. Freshman applicant must have a minimum combined SAT I score of 1590 (composite ACT score of 23).
Level of Study: Postgraduate
Type: Scholarship
Value: The Scholarship has a maximum award of $2,000.
Frequency: Annual
Study Establishment: Montana State University-Billings
Country of Study: United States of America
No. of awards offered: 30
Application Procedure: See the website.
Closing Date: February 1st
Funding: Private
Additional Information: Scholarships are not available to music or music education majors or student athletes participating in men's basketball, women's basketball, or women's volleyball, or to a student who receives a full tuition waiver as a result of their participation in an intercollegiate sport.

Jarussi Sisters Endowed Scholarship

Subjects: Education.
Purpose: To assist students financially
Eligibility: Applicant must have a minimum 2.8 GPA, be enrolled full-time, demonstrate financial need, and graduate from a Carbon County high school.
Level of Study: Postgraduate
Type: Scholarship
Value: $5,000
Frequency: Annual
Study Establishment: Montana State University-Billing
Country of Study: United States of America
No. of awards offered: 2
Application Procedure: See the website.
Closing Date: February 1st
Funding: Private

John F. and Winifred M. Griffith Endowed Scholarship

Subjects: All subjects.
Purpose: To encourage students who demonstrate academic and leadership potential.
Eligibility: Open to candidates who are full-time students at Montana State University-Billings and have 3.0 cumulative GPA with at least a 3.25 GPA in major field of study. Students enrolled in the College of Technology or who are majoring in theatre are not eligible for scholarship.
Level of Study: Postgraduate
Type: Scholarship
Value: The Scholarship has a maximum award of $6,500
Frequency: Annual
Study Establishment: Montana State University-Billing
Country of Study: United States of America
No. of awards offered: 3
Application Procedure: See website.
Closing Date: February 1st
Funding: Private
Additional Information: Students receiving the scholarship are eligible to continue to receive the scholarship by annually reapplying. Renewal recipients will receive preference as long as the student shows progress towards their degree and improves their GPA in their chosen discipline.

Kenneth W. Heikes Family Endowed Scholarship

Subjects: Accounting and information system.
Purpose: To assist students who wish to study further financially.
Eligibility: Applicant must have a minimum 3.0 overall GPA and a minimum 3.25 GPA in major, be enrolled in the teacher education program, and demonstrate financial need.
Level of Study: Postgraduate
Type: Scholarship
Value: The Scholarship has a maximum award of $2,000
Frequency: Annual
Study Establishment: Montana State University-Billings
Country of Study: United States of America
No. of awards offered: 1
Application Procedure: See the website.
Closing Date: February 1st
Funding: Private

MSU-Billings Chancellor's Excellence Awards

Subjects: All subjects.
Purpose: To recognize academic achievement and leadership qualities.
Eligibility: Open to applicants with evidence of leadership and a record of community service.
Level of Study: Postgraduate
Type: Scholarship
Value: US$3,000 per year
Length of Study: 4 years
Frequency: Annual
Study Establishment: Montana State University-Billings
Country of Study: United States of America
Application Procedure: A completed Chancellor's Excellence Awards application must be submitted.
Closing Date: January 15th
Funding: Private
Additional Information: Chancellor Scholars who meet or exceed all awards criteria may have their awards renewed for 4 years for a maximum award of US$10,000.

For further information contact:

Chancellor's Excellence Awards Committee, Montana State University-Billings Foundation

MONTREAL NEUROLOGICAL INSTITUTE

3801 University Street, Montréal, QC, H3A 2B4, Canada
Tel: (1) 514 398 6644
Fax: (1) 514 398 8248
Email: fil.lumia@mcgill.ca
Website: www.mni.mcgill.ca
Contact: Ms Filomena Lumia, Administration Assistant

The Montreal Neurological Institute is dedicated to the study of the nervous system and neurological disorders. Its 70 principal researchers hold teaching positions at McGill University. The Institute is characterized by close interaction between basic science researchers and the clinicians at its affiliated hospital.

Jeanne Timmins Costello Fellowships

Subjects: Neurology, neurosurgery and neuroscience research and study.
Purpose: To support research.
Eligibility: Applicant must hold an M.D and/or PhD degree.Prior to submitting an application, applicants must contact a faculty member at the Montreal Neurological Institute and obtain their agreement to act as a postdoctoral supervisor. The faculty member must be identified on the application and will be required to provide a letter of support.
Level of Study: Doctorate, Postgraduate
Type: Fellowship
Value: Appointments are for one year with a value of $40,000 per year
Length of Study: 1 year, with the possibility of renewal for a further year
Frequency: Annual
Study Establishment: The Montreal Neurological Institute
Country of Study: Canada
No. of awards offered: 4
Application Procedure: Applicants must write for details.
Closing Date: October 31st
Funding: Private
Contributor: Jean Timmins Costello
No. of awards given last year: 4
Additional Information: Supports research and study in clinical and basic neuroscience. Up to four fellowships are awarded each year.

Preston Robb Fellowship

Subjects: Neurology, neurosurgery and neuroscience.
Purpose: To support research.
Eligibility: Open to researchers of any nationality.
Level of Study: Doctorate, Postdoctorate, Postgraduate
Type: Fellowship
Value: Initial appointment is for one year with a value of $40,000 per year.
Length of Study: 1 year
Frequency: Annual
Country of Study: Canada
No. of awards offered: 1
Application Procedure: Applicants must write for details.
Closing Date: October 31st
Funding: Private
Contributor: Preston Robb
No. of awards given last year: 1
Additional Information: Supports the training of a clinical fellow to work jointly with our basic and clinician scientists.

MORRIS K UDALL AND STEWART L UDALL FOUNDATION

130 South Scott Avenue, Tucson, AZ, 85701, United States of America
Tel: (1) 520 901 8500
Fax: (1) 520 670 5530
Email: info@udall.gov
Website: www.udall.gov

The Udall Foundation is dedicated to educating a new generation of Americans to preserve and protect their national heritage through studies in the environment and Native American health and tribal public policy. The Foundation is also committed to the principles and practices of environmental conflict resolution.

Environmental Public Policy and Conflict Resolution PhD Fellowship

Subjects: Environment.
Purpose: To provide funds for doctoral candidates whose research concerns US environmental public policy and environmental conflict resolution and who are entering their final year of writing the dissertation.
Eligibility: Open to scholars in all fields of study whose dissertation topic has significant relevance to US environmental public policy and environmental conflict resolution. The applicant must be a US national or citizen or have permanent residence in the US, be entering their final year of writing the dissertation, have necessary approval for the dissertation research proposal and completed all PhD coursework and passed all preliminary examinations. Furthermore the applicant must be enrolled at a US institution of higher education; US citizens attending universities outside the US are not eligible to apply. It is the Foundation's intent that work conducted during the Fellowship be done in the US.
Level of Study: Doctorate
Type: Fellowship
Value: The Udall Foundation awards two one-year fellowships of up to $24,000 to doctoral candidates whose research concerns U.S. environmental public policy and/or environmental conflict resolution and who are entering their final year of writing the dissertation.
Length of Study: 1 year
Frequency: Annual
Study Establishment: Any US university
Country of Study: United States of America
No. of awards offered: 2
Application Procedure: Application forms are available from the Foundation's website www.udall.gov
Closing Date: February 24th
Funding: Government
No. of awards given last year: 2
No. of applicants last year: 40
Additional Information: Please check website for updated information.

MOSCOW UNIVERSITY TOURO

20/12 Podsosensky Pereulok, Moscow, 105062, Russia
Tel: (7) 495 917 4169
Fax: (7) 495 917 5348
Email: admin@touro.ru
Website: www.touro.ru
Contact: Admissions Office

Moscow University Touro is an independent, non-profit institution of higher education, established in 1991, which promotes international-style business education in Russia.

Moscow University Touro Corporate Scholarships

Subjects: All subjects.
Purpose: To fund exceptionally talented students or employees with the intellectual potential to become future business leaders in Russia.
Eligibility: All applicants must possess a baccalaureate degree from an accredited college or university with a minimum GPA of 3.0. Applicants must submit transcripts of all undergraduate and graduate work. Mastery of written and spoken English is measured by the TOEFL and a minimum TOEFL score required 550 on a written test or 215 on CAT.
Level of Study: Postgraduate
Type: Scholarship
Length of Study: 1 year
Frequency: Annual
Study Establishment: Moscow University Touro
Country of Study: Russia
Application Procedure: Apply online or contact the admissions office.
Closing Date: August 1st
Funding: Corporation
Contributor: Russian and International Corporations

Moscow University Touro Universal Scholarships

Subjects: Any subject.
Purpose: To provide financial assistance to ready student with strong scholarship potential.
Eligibility: Only available to student studying full time at the Moscow campus.
Level of Study: Postgraduate
Type: Scholarship
Value: Payment of an approved tuition sum
Length of Study: 1 year
Frequency: Annual
Study Establishment: Moscow University Touro
Country of Study: Russia
Application Procedure: Apply online or contact the admissions office.
Closing Date: August 1st

For further information contact:

Tel: +7 (0) 95 917 4052
Contact: Eugenij Pouchinkin

MOTOR NEURONE DISEASE ASSOCIATION

PO Box 246, Northampton, Northamptonshire, NN1 2PR, England
Tel: (44) 16 0425 0505
Fax: (44) 16 0463 8289
Email: research@mndassociation.org
Website: www.mndassociation.org
Contact: Mrs Marion Reichle, Research Grants Coordinator

The Motor Neurone Disease Association supports research on fundamental aspects of motor neurone disease (MND) and on its management and alleviation. It provides information and advice to patients and carers, and runs a nationwide care service. MND paralyzes selectively or generally and is fatal, irreversible and at present, incurable.

MND PhD Student Award

Eligibility: Overseas applicants must have a project that is unique in concept and design and involves a significant aspect of collaboration with a United Kingdom institute.
Type: Project grant
Value: The Association will provide a student stipend of £12,500 per year (£13,500 in London), £7,500 per year for laboratory expenses and a total budget of £1,000, over 3 years, for conference attendance
Application Procedure: Applicants must submit an online summary of their proposal. The online application form starts with an eligibility questionnaire. Once the application is submitted it is considered by three members of our Biomedical Research Advisory Panel (BRAP). Full applications are invited thereafter and application forms provided. Full applications are considered by a minimum of two independent external referees and then by the BRAP. Please see website at www.mndassociation.org/research/for_reseachers for full details of the application processes, together with grant guidelines and terms and conditions. The summary application form will be available on our webiste www.mndassociation.org from March 23rd.
Closing Date: May 7th
No. of awards given last year: 4
No. of applicants last year: 8

MND PhD Studentship Award

Subjects: Research on all aspects of MND in all relevant disciplines.
Purpose: To attract promising science graduates to develop a career in MND related research and to support research aimed at understanding the causes of MND, elucidating disease mechanisms and facilitating the translation of therapeutic strategies from the laboratory to the clinic.
Eligibility: Restricted to researchers from United Kingdom laboratories.
Level of Study: Postdoctorate
Type: Studentship
Value: The Association will provide a student stipend of £12,500 per year (£13,500 in London), £7,500 per year for laboratory expenses and a total budget of £1,000, over 3 years, for conference attendance. The Association will also cover relevant tuition/bench fees at the rate payable by UK students.
Length of Study: 3 years
Frequency: Annual
Country of Study: United Kingdom
No. of awards offered: 1 to 2
Application Procedure: Applicants must submit a summary online of their proposal, which is first checked for eligibility and considered by 3 members of the research advisory panel (RAP). Full applications are invited thereafter and application forms are provided. Full applications are submitted to 2 or more independent referees and then to the RAP for consideration. Please see our research governance overview at www.mndassociation.org/research/for_researchers – for information on our grant application processes. The summary application form will be available on our website from March 23rd.
Closing Date: May 7th
Funding: Individuals, trusts
No. of awards given last year: 4
No. of applicants last year: 8
Additional Information: The studentships are awarded on the basis of scientific merit and the value of research training offered.

MND Research Project Grants

Subjects: Research into all aspects of MND in all disciplines.
Purpose: To understand and research the cause and effective treatments of MND. To fund research to the highest scientific merit and greatest clinical or translational relevance to MND.
Eligibility: Overseas applicants must have a project that is unique in concept or design and that involves significant aspect of collaboration with a United Kingdom institute.
Level of Study: Postdoctorate, Professional development, Research
Type: Grant
Value: UK£85,000 per year, maximum
Length of Study: 1–3 years
Frequency: Annual
Study Establishment: A research institution
No. of awards offered: Varies
Application Procedure: Applicants must submit a summary online of their proposal, which is first checked for eligibility and considered by three members of the research advisory panel (RAP). Full applications are invited thereafter and application forms are provided. Full applications are submitted to two or more independent referees and then to the RAP for consideration. Please see our research governance overview at www.mndassociation.org/research/for_researchers – for information on our grant application processes. The summary applicatio form will be available on our website from March 23rd.
Closing Date: October 22nd
Funding: Individuals, trusts
No. of awards given last year: 7
No. of applicants last year: 12

THE MOUNTAINEERING COUNCIL OF SCOTLAND (MCOFS)

The Old Granary, West Mill Street, Perth, PH1 5QP, Scotland
Tel: (44) 01738 493942
Fax: (44) 01738 442095
Email: info@mcofs.org.uk
Website: www.mcofs.org.uk
Contact: The Development Officer

MCofS administers the SportScotland grants for Scotland's mountaineers and is also the national governing body for sport climbing.

Mountaineering Council of Scotland Climbing and Mountaineering Bursary

Subjects: Climbing expeditions.
Purpose: To give financial support, and to encourage, climbers and mountaineers in a wide range of mountaineering and climbing disciplines to undertake their activities.
Eligibility: Applicants must be Scottish or resident in Scotland and aged 14 years or above.
Level of Study: Professional development, Research
Type: Grant

Value: UK£1,000-3,500
Length of Study: Event specific
Frequency: Annual
Country of Study: Any country
No. of awards offered: Approx. 5
Application Procedure: Request a bursary information pack from the Development Officer.
Closing Date: End February
Funding: Government
Contributor: sportscotland
No. of awards given last year: 6
No. of applicants last year: 6

MULTIPLE SCLEROSIS RESEARCH AUSTRALIA (MSRA)

PO Box 625, Chatswood, New South Wales, 2057, Australia
Tel: (61) 1300 356 467
Fax: (61) 0893 462 455
Email: sumsr@iinet.net.au
Website: http://www.msra.org.au/

In August 2004, Multiple Sclerosis (MS) Australia, the national organization of MS Societies, launched a new MS-specific research organization called Multiple Sclerosis Research Australia (MSRA). Its aim is to substantially lift the Australian MS research effort and accelerate the progress towards knowledge of a cause and cure for MS.

MSRA NHMRC Betty Cuthbert Scholarship

Subjects: Multiple sclerosis.
Purpose: To accelerate progress in research on cure for multiple sclerosis.
Eligibility: Open to candidates who have obtained a graduate degree.
Level of Study: Postgraduate
Type: Scholarships
Value: $26,200
Length of Study: 3 years
Frequency: Annual
No. of awards offered: Varies
Application Procedure: Applicants must get and application ID number prior to submitting your application. To get an application number please email the MSRA at grants@msra.org.au. Applicants must submit an application to both NHMRC and MSRA.
Closing Date: August 6th
Additional Information: Please check at http://www.msra.org.au/scholarships for more information. This scholarship is awarded jointly with NHMRC.

MULTIPLE SCLEROSIS SOCIETY OF CANADA (MSSOC)

175 Bloor Street East, Suite 700, North Tower, Toronto, ON, M4W 3R8, Canada
Tel: (1) 416 967 3024/416 922 6065
Fax: (1) 416 922 7538
Email: karen.lee@mssociety.ca/info@mssociety.ca
Website: www.mssociety.ca

The mission of the MSSOC is to be a leader in finding a cure for multiple sclerosis and enabling people affected by the disease to enhance their quality of life.

MSSOC Donald Paty Career Development Award

Subjects: Multiple sclerosis.
Purpose: To support the salary of an independent researcher whose research is relevant to MS.
Eligibility: This award is open to those that hold a doctoral degree (PhD, M.D. or equivalent) and who have recently completed their training in research and are in the early stages of independent research relevant to MS.Applicants must hold a Canadian university faculty appointment and either holds an operating grant from the MSSOC or another funding agency.
Level of Study: Postdoctorate
Value: The amount provided per year for the award is $50,000
Length of Study: 3 years
Frequency: Annual
Study Establishment: A Canadian school of medicine/recognized institution
Country of Study: Canada
No. of awards offered: Limited
Application Procedure: All Applicants for regular research grants are required to use the website https://www.mscanadagrants.ca/ for the completion of their proposal. All components of the application must be submitted through the online system. No hard copies of any documentation will be accepted.
Closing Date: October 1st
Funding: Private
No. of awards given last year: 1
No. of applicants last year: 5

MSSOC Postdoctoral Fellowship Award

Subjects: Multiple sclerosis and allied diseases.
Purpose: To encourage research.
Eligibility: Open to qualified persons holding an MD or PhD degree and intending to pursue research work relevant to multiple sclerosis and allied diseases. The applicant must be associated to an appropriate authority in the field he or she wishes to study.
Level of Study: Postdoctorate
Type: Fellowship
Value: The amount provided per year for the award is $39,000 for a PhD and $48,500 for an M.D.
Length of Study: 1 year, with the opportunity for 2 renewals at 1 year each.
Frequency: Annual
Study Establishment: A recognized institution which deals that problems relevant to multiple sclerosis
Country of Study: Other
No. of awards offered: Varies
Application Procedure: All Applicants for regular research grants are required to use the website https://www.mscanadagrants.ca/ for the completion of their proposal. All components of the application must be submitted through the online system. No hard copies of any documentation will be accepted.
Closing Date: October 1st
Funding: Private
No. of awards given last year: 31
No. of applicants last year: 26

MSSOC Research Grant

Subjects: Biomedical and clinical and population health.
Purpose: To fund research projects.
Eligibility: Open to researchers working in Canada or intending to return to Canada and who is Applicants are those that hold a doctoral degree (PhD, M.D. or equivalent) and who have recently completed their training in research and are in the early stages of independent research relevant to MS.Applicants must hold a Canadian university faculty appointment and either holds an operating grant from the MSSOC or another funding agency.
Level of Study: Postgraduate
Type: Research grant
Value: Up to Canada $100,000 per year
Length of Study: 1–3 years
Frequency: Annual
Study Establishment: A Canadian school of medicine or recognized institution
Country of Study: Any country
No. of awards offered: Varies
Application Procedure: All Applicants for regular research grants are required to use the website https://www.mscanadagrants.ca/ for the completion of their proposal. All components of the application must be submitted through the online system. No hard copies of any documentation will be accepted.
Closing Date: October 1st
Funding: Private
No. of awards given last year: 17
No. of applicants last year: 36

MSSOC Research Studentships

Subjects: Multiple sclerosis and allied diseases.
Purpose: To support qualified persons to pursue specialized training in multiple sclerosis and allied diseases.
Eligibility: Eligible to qualified persons holding other than a MD or PhD degree wishing further training in a specialized area related to research in multiple sclerosis and allied diseases.
Level of Study: Doctorate, Research
Type: Studentship
Length of Study: The maximum period of a studentship will be four years but under exceptional circumstances may be extended by one additional year
Frequency: Annual
No. of awards offered: Limited
Application Procedure: Applicants must submit the completed application form, academic transcripts and letters of reference to the chairman. For further details please refer the website.
Closing Date: October 1st
No. of awards given last year: 30
No. of applicants last year: 85
Additional Information: Applications directed towards understanding the pathogenesis and potential treatment of multiple sclerosis will receive priority.

MUSCULAR DYSTROPHY ASSOCIATION (MDA)

National Headquarters, 3300 E. Sunrise Drive, Tucson, AZ, 85718, United States of America
Tel: (1) 520 529 2000/1 800 572 1717
Fax: (1) 520 529 5454
Email: grants@mdausa.org
Website: www.mdausa.org

The Muscular Dystrophy Association (MDA) supports research into over 40 diseases of the neuromuscular system to identify the causes of, and effective treatments for, the muscular dystrophies and related diseases including spinal muscular atrophies and motor neuron diseases, peripheral neuropathies, inflammatory myopathies, metabolic myopathies and diseases of the neuromuscular junction.

MDA Grant Programs

Subjects: The muscular dystrophies and related diseases.
Purpose: To support research into over 40 diseases of the neuromuscular system, and to identify the causes of, and effective treatments for, the muscular dystrophies and related diseases.
Eligibility: Hold a Doctor of Medicine (M.D.), Doctor of Philosophy (PhD), Doctor of Science (D.Sc.) or equivalent degree (i.e. D.O.); Open to persons who are professional or faculty members at appropriate educational, medical or research institutions; who are qualified to conduct and supervise a programme of original research; who have access to institutional resources necessary to conduct the proposed research project and who hold an MD or PhD.
Level of Study: Postdoctorate
Type: Grant
Value: Funding levels for primary Research Grants are unlimited. Development grants are a maximum of $60,000 per year.
Length of Study: Maximum of 3 years
Frequency: Annual
Country of Study: Any country
No. of awards offered: Varies
Application Procedure: Applicants must complete the Request for Research Grant Application.
Closing Date: Pre-applications are due no later than December 15th or June 15th
Funding: Private
Contributor: Voluntary contributions
No. of awards given last year: 140
Additional Information: Proposals from applicants outside the United States of America will only be considered for projects of highest priority to the MDA. Other conditions apply. Research is sponsored under the following grant programmes: Neuromuscular Disease Research and Neuromuscular Disease Research Development. Further information is available from the website.

MUSCULAR DYSTROPHY CAMPAIGN OF GREAT BRITAIN AND NORTHERN IRELAND

61 Southwark Street, London, SE1 0HL, England
Tel: (44) 020 7803 4800
Fax: (44) 020 7401 3495
Email: research@muscular-dystrophy.org
Website: www.muscular-dystrophy.org
Contact: Mrs Jenny Versnel, Head of Research

The Muscular Dystrophy Campaign is a United Kingdom-based charity funding medical research and support services for people with muscular dystrophy and related conditions.

Muscular Dystrophy Research Grants

Subjects: Muscular dystrophy and allied neuromuscular conditions.
Purpose: To provide continuity of funding for high-quality research of central importance to the Muscular Dystrophy Campaign's research objectives.
Eligibility: The MDC will only fund United Kingdom-based research. Principal Investigators must hold a contract that extends beyond the duration of the grant at an institution approved by the MDC.
Level of Study: Postdoctorate, Research
Type: Research grant
Value: The range of awards for fellowships is typically between $30,000 and $50,000.
Length of Study: 3–4 years (PhD Studentship), 1–3 years (Project Grants)
Frequency: Annual
Study Establishment: Universities, hospitals or research institutions
Country of Study: United Kingdom
No. of awards offered: Varies
Application Procedure: Applicants must contact the Head of Research with a summary of proposed research on A4-size paper.
Closing Date: October
Funding: Private
No. of awards given last year: 10
No. of applicants last year: 30
Additional Information: Please contact on research@muscular-dystrophy.org for updated information.

MUSEUM OF COMPARATIVE ZOOLOGY, HARVARD UNIVERSITY

26 Oxford Street, Cambridge, MA, 02138, United States of America
Tel: (1) 617 495 2460
Fax: (1) 617 495 5667
Email: grants@oeb.harvard.edu
Website: www.mcz.harvard.edu

The Museum of Comparative Zoology was founded in 1859 through the efforts of Louis Agassiz (1807–1873). Agassiz, a zoologist from Neuchatel, Switzerland, served as the Director of the Museum from 1859 until his death in 1873. A brilliant lecturer and scholar, he established the Museum and its collections as a centre for research and education.

Ernst Mayr Travel Grant

Subjects: Animal systematics.
Purpose: To stimulate taxonomic work on neglected taxa.
Eligibility: This grant is open to all
Level of Study: Research, Unrestricted
Type: Travel grant
Value: Awards average about $1,000 each and may not exceed $1,500.
Length of Study: 1 year
Frequency: Twice a year
Study Establishment: An accredited museum
Country of Study: Any country
No. of awards offered: Varies
Application Procedure: Applicants must submit a short project description, itinerary, budget, a curriculum vitae and three letters of support. No proposal forms are required or provided. Proposals may be submitted through standard mail or electronically. Submissions

through standard mail must include five copies of all materials and be received by the closing date. Proposals submitted electronically must be created in Microsoft Word or plain ASCII text format. For additional information visit the website, www.oeb.harvard.edu/mayr-grant.htm or emailing grants@oeb.harvard.edu
Closing Date: October 15th and April 1st
Funding: Trusts
Contributor: Ernst Mayr
No. of awards given last year: 13
No. of applicants last year: 25
Additional Information: Announcements of awards will be made within 2 months of the closing date.

For further information contact:

OEB Administration, Museum of Comparative Zoology, Harvard University, 26 Oxford Street, Cambridge, MA, 02138, United States of America
Tel: (1) 617 495 2460
Contact: Catherine Weisel

THE MUSICIANS BENEVOLENT FUND

7-11 Britannia Street, London, WC1X 9JS, England
Tel: (44) 020 7239 9100
Fax: (44) 020 7713 8942
Email: awards@helpmusicians.org.uk
Website: www.helpmusicians.org.uk
Contact: Ms Susan Dolton, Director of Giring

The Musicians Benevolent Fund is unique (UK based) charity which provides essential help to musicians of all ages and genres. We support music professionals throughout their working lives when a crisis such as an accident or illness can have a devastating impact, and in later life, with the challenges that growing older can bring. We support talented, financially needy young musicians to enable them to fully develop their creativity and musical potential, ensuring they enter the profession with the best prospects of success. We are also working in partnership with a number of organizations across the UK who provide outstanding opportunities for musicians. We rely on donations and the generosity of music lovers and musicians to enable our vital wok to continue.

Emerging Excellence Awards

Subjects: These awards are open to emerging artists aged 18–30 to fund an important development opportunity, project or programme of activity that will have a lasting impact on their professional career.
Purpose: To help individual musicians (including composers) and groups in the early stages of their professional careers.
Eligibility: Applicants must be British or Irish or have lived in the UK for at least 5 consecutive years.
Level of Study: Twice a year
Type: Award
Value: £500–4,000
Application Procedure: A written description of the project must be provided with a budget, a reference and a recording demonstrating composition and/or performance.
Closing Date: January 6th
No. of awards given last year: 18
No. of applicants last year: 108

Maggie Teyte Prize and Miriam Licette Scholarships

Subjects: The Maggie Teyte Prize and Miriam Licette Scholarships were founded to promote the interpretation of French Melodie in the UK. The awards help female singers in the final stages of study and about to enter the profession.
Purpose: To assist a female student of French song.
Eligibility: Open to female singers in full-time postgraduate vocal study at British Conservatires or the National Opera Studio. Applicants must be able to demonstrate considerable success at this level. Applicants should be British or Irish or be living or studying full-time in the UK for 5 years of December 10th.
Level of Study: Postgraduate, Professional development
Type: Scholarship
Value: Maggie Teyte Prize and Miriam Licette Scholarship £3,500 / Miriam Licette Scholarship £1,000–1,500
Frequency: Annual
Country of Study: Any country
No. of awards offered: Up to 4 awards (1 Maggie Teyte Prize and up to 3 Miriam Licette Scholarships)
Application Procedure: Selected students will be asked to audition. Application is by nomination by heads of voice at the UK conservatoires only.
Closing Date: December/January
Funding: Private
No. of awards given last year: 4
No. of applicants last year: 20
Additional Information: The award is run in conjunction with the Maggie Teyte Prize.

Musicians Benevolent Fund Postgraduate Performance Awards

Subjects: The aim is to help caver maintenance costs for the following academic year and provide further development opportunities.
Purpose: To support outstandingly talented instrumentalists and singers at postgraduate level.
Eligibility: Open to singers and instrumentalists possessing the potential to become first-class performers, who are British or Irish or have been resident in the United Kingdom for 5 years. Applicants must be studying at postgraduate level in the UK for the academic year.
Level of Study: Postgraduate
Type: Bursary
Value: £1,000–5,000
Frequency: Annual
Country of Study: United Kingdom
No. of awards offered: 60
Application Procedure: Application is by nomination only from UK music colleges and conservatoires.
Closing Date: February 15th
Funding: Private
No. of awards given last year: 60
No. of applicants last year: 120
Additional Information: The following major awards are also available through these auditions: Ian Fleming Charitable Trust Music Education Awards, Professor Charles Leggett Awards, Emily English Scholarship, Maidment Scholarships, Myra Hess Scholarships, Sir Henry Richardson Scholarships, Eleanor Warren Award, Awards for Accompanists and Repetiteurs and Manoub Parikian Award.

Peter Whittingham Jazz Award

Subjects: The award supports emerging Jazz musicians and ensembles.
Purpose: To promote both composition and performance through an innovative jazz project, providing crucial help at the start of an artists carrier.
Eligibility: Open to emerging individual Jazz musician or group showing talent and innovation. Candidates must be Britisher Insh or have been resident in the UK for at least 5 years at the closing date. Applicants must be in their first two years of the profession/last year of studying.
Level of Study: Graduate, Postgraduate, Professional development
Type: Award
Value: £4,000
Frequency: Annual
No. of awards offered: 1
Application Procedure: A written description of the project must be provided together with a budget, a reference and a recording demonstrating composition and/or performance.
Closing Date: November
Funding: Private
No. of awards given last year: 1 main award, 2 development awards
No. of applicants last year: 10
Additional Information: Selected applicants will be asked to attend an interview and audition

Sybil Tutton Awards

Subjects: The musicians Benevolent fund offers Sybil Tutton Awards to outstandingly talented singers on full-time postgraduate opera courses to help with study costs. A special Richard Van Allen Award may be offered to a suitably outstanding male singer.

Purpose: To assist students on advanced postgradaute opera courses.
Eligibility: Nominees must be British or Irish or have lived in the UK for at least fire consecutive years at the closing date.
Type: Award
Value: £1,000–5,000
Frequency: Annual
Application Procedure: Application is by nomination only from head of voice at the UK conservatoires and the National Opera Studio.
Closing Date: May

MYASTHENIA GRAVIS FOUNDATION OF AMERICA

355 Lexington Avenue, 15th Floor, New York, NY, 10017, United States of America
Tel: (1) 800 541 5454
Fax: (1) 212 370 9047
Email: mgfa@myasthenia.org
Website: www.myasthenia.org
Contact: Ms Janet Golden, Chief Executive

The Myasthenia Gravis Foundation of America's mission is to facilitate the timely diagnosis and optimal care of individuals affected by myasthenia gravis and to improve their lives through programmes of patient services, public information, medical research, professional education, advocacy and patient care.

Myasthenia Gravis Nursing Research Fellowship

Subjects: Nursing.
Purpose: To support nursing students or professionals interested in studying problems encountered by patients with myasthenia gravis or related neuromuscular conditions.
Eligibility: Open to nursing students and professionals.
Level of Study: Graduate, Postgraduate, Research
Type: Research grant
Value: Maximum: $5,000
Length of Study: Short-term
Frequency: Dependent on funds available
Country of Study: Any country
No. of awards offered: 2–3
Application Procedure: Four copies: cover letter, proposed budget, curriculum vitae of applicant and sponsoring preceptor, proposed work plan and budget
Closing Date: March 15th
Funding: Private
No. of awards given last year: None
No. of applicants last year: None

For further information contact:

Research Grant Committee, Myasthenia Gravis Foundation, 1821 University Ave W, Suite S256, St Paul, MN, 55104
Website: http://www.myasthenia.org

Post-Doctoral Fellowship

Subjects: Research pertinent to myasthenia gravis that is concerned with neuromuscular transmission, immunology, molecular cell biology of the neuromuscular synapse and the aetiology and pathogenesis, diagnosis or treatment of the disease.
Purpose: To attract physicians, PhD scientists and allied health professionals into conducting research into myasthenia gravis or related conditions.
Eligibility: There are no eligibility restrictions. The research preceptor must be an established investigator at an institution in the United States of America, Canada or abroad deemed appropriate by the MGFA Medical/Scientific Advisory Board. The applicant must either be (a) a permanent resident of the United States of America or Canada who has been accepted to work in the laboratory of an established investigator at an institution in the United States of America, Canada or abroad or (b) a foreign national who has been accepted to work in the laboratory of an established investigator at an institution in the United States of America or Canada deemed appropriate by the Medical/Scientific Advisory Board of MGFA. Please refer to the website for more information http://www.myasthenia.org/research/
Level of Study: Postdoctorate
Type: Fellowship
Value: US$50,000
Length of Study: 1 year
Frequency: Annual
Country of Study: Any country
No. of awards offered: 2–4
Application Procedure: Applicants must submit a nine copies of the proposal, a lay summary, budget, applicant's and preceptor's curriculum vitae and letters of recommendation to Post-Doctoral Committee.
Closing Date: October 1st
Funding: Foundation, private
No. of awards given last year: 3
No. of applicants last year: 11
Additional Information: Applicants will be evaluated by The Fellowship Committee based on some criteria.

Student Research Fellowship

Subjects: Medical and Health Sciences.
Purpose: To promote research into the cause and cure of myasthenia gravis.
Eligibility: Must be a medical students or medical graduate students planning to conduct research on the basis of Myasthenia Gravis (MG), or related neuromuscular conditions All candidates and their sponsoring institutions must comply with policies governing the protection of human subjects, the humane care of laboratory animals, and the inclusion of minorities in study populations
Level of Study: Doctorate, Graduate, Research
Type: Fellowship
Value: $5,000
Length of Study: Short-term
Frequency: Annual
Country of Study: Any country
No. of awards offered: Up to 3
Application Procedure: Must submit letter of interest, summary of research proposal, budget, curriculum vitae, and letters of recommendation For further information, please visit the website http://www.myasthenia.org/Research/ResearchFellowships.aspx
Closing Date: March 15th
Funding: Private

For further information contact:

Website: http://www.myasthenia.org/Research/ResearchFellowships.aspx

NANTUCKET HISTORICAL ASSOCIATION (NHA)

15 Broad Street, PO Box 1016, Nantucket, MA, 02554, United States of America
Tel: (1) 508 228 1894
Fax: (1) 508 228 5618
Email: bsimons@nha.org/ask@nha.org
Website: www.nha.org

The Nantucket Historical Association (NHA) is the principal repository of Nantucket history, with extensive archives, collections of historic properties and art and artefacts that broadly illustrate Nantucket's past. The Research Library at the NHA contains a rich collection of primary and secondary sources that document all facets of Nantucket's cultural, social, economic and spiritual history for more than three centuries. More than 400 manuscript collections relate to Nantucket individuals and families, ships, businesses and trades, churches, schools and organizations. Searchable inventories of the library's manuscript, map and book holdings can be accessed via the NHA website. The NHA art and artefacts collections include paintings, drawings, prints, baskets, silver, whaling tools, scrimshaw, furniture and textiles. The particular strengths of the collection lie in artefacts that document Nantucket's whaling industry.

E Geoffrey and Elizabeth Thayer Verney Research Fellowship

Subjects: History.
Purpose: To enhance the public's knowledge and understanding of the heritage of Nantucket, Massachusetts.
Eligibility: Open to academics, graduate students and independent scholars in any field to conduct research in the collections of the NHA.
Level of Study: Graduate
Type: Residency
Value: A stipend of $300 per week, and reimbursement of travel expenses up to $600
Length of Study: 3 weeks
Frequency: Annual
Study Establishment: The NHA
Country of Study: United States of America
No. of awards offered: 1
Application Procedure: Applicants must send a full description of the proposed project, a curriculum vitae, the name of three references and an estimate of anticipated time and duration of stay.
Closing Date: January 1st
No. of awards given last year: 1

For further information contact:

Nantucket Historical Association, PO Box 1016, Nantucket, MA, 02554-1016, United States of America
Contact: Mr Ben Simons, Chief Curator and Editor

NANYANG TECHNOLOGICAL UNIVERSITY (NTU)

50 Nanyang Avenue, Singapore, 639798, Singapore
Tel: (65) 65 67911744/ 65 67904777
Fax: (65) 65 6791 8522
Email: fellows@ntu.edu.sg
Website: www.ntu.edu.sg

Nanyang Technological University (NTU) is an established, research intensive tertiary institution with a vision to be a global university of excellence in science and technology. Asian Communication Resource Centre (ACRC) is a regional centre of NTU, dedicated to develop information resources on different aspects of communication, information and media.

Alumni Scholarship

Subjects: Management.
Purpose: To provide financial support to alumni of Nanyang Technological University admitted to the MBA programme on a full-time or part-time basis.
Eligibility: Open to alumni of Nanyang Technological University admitted to the MBA programme on a full-time or part-time basis. Recipients of other scholarships or bursaries are not eligible to apply.
Level of Study: MBA
Type: Scholarship
Value: $3,000 per year
Length of Study: Each scholarship shall be tenable only for the academic year in which it is awarded
Frequency: Annual
Study Establishment: NTU
Country of Study: Singapore
No. of awards offered: 1
Application Procedure: Applicants must submit a completed application form (available at http://admissions.ntu.edu.sg/graduate/documents/alumnischolarship.pdf).
Closing Date: From July 30th to August 13th
Funding: Private
Contributor: NTU Alumni Fund
No. of awards given last year: None

For further information contact:

Nanyang Technological University, Student Services Centre, Level 3, 42 Nanyang Avenue, Singapore, 639815, Singapore
Contact: Ms Jessica Wee, Assistant Director, Graduate Studies Office,

APEC Scholarship

Subjects: Mangement.
Purpose: To allow outstanding candidates from the APEC member economies except Singapore to pursue the MBA programme at NTU on a full-time basis.
Eligibility: This scholarship is open to nationals of the following APEC member economies to pursue a full-time Master of Business Administration (MBA) programme at NTU:Australia, Brunei Darussalam, Canada, Chile, People's Republic of China, Hong Kong, Indonesia, Japan, Republic of Korea, Malaysia, Mexico, New Zealand, Papua New Guinea, Peru, Philippines, Russia, Chinese Taipei, Thailand, United States, Vietnam.At least two years of management or professional experience;proficiency in the English Language; an acceptable score in the Graduate Management Admission Test [GMAT]; should not be on paid employment or accept paid employment or concurrently hold any other scholarship, fellowship, bursary or top-up allowance during the prescribed period of the award.
Level of Study: MBA
Type: Scholarship
Value: Monthly stipend of Singaporean $1,400; book allowance of Singaporean $500; tuition fee, health insurance, examination fee and other approved fees, allowances and expenses; cost of one overseas Business Study Mission (for MBA candidates only) undertaken within the tenable period of the scholarship; cost of travel from home country to Singapore on award of the scholarship; cost of travel from Singapore to home country on successful completion of the Masters degree within the tenable period of the scholarship.
Length of Study: Full-time for 16 months
Frequency: Annual
Study Establishment: NTU
Country of Study: Singapore
No. of awards offered: 1–2
Application Procedure: Application details are available at the University website.
Closing Date: December
Funding: Government
Contributor: NTU
No. of awards given last year: 2
No. of applicants last year: 50
Additional Information: Applications open in November and close in December each year. Invitations for applications will be placed in the newspapers in the capital cities of the APEC countries and on the University's website. International students have to apply for a student's pass from Singapore immigration in order to pursue a full-time course of study in Singapore. The University will assist successful international applicants in their applications for the student's pass.

ASEAN Graduate Scholarship

Subjects: Management.
Purpose: To allow outstanding candidates from the ASEAN (Association of South East Asian Nations) member economies, except Singapore, to pursue the MBA programme at NTU on a full-time basis.
Eligibility: Open to nationals of the member countries of ASEAN (Brunei Darussalem, Cambodia, Indonesia, Laos, Malaysia, Myanmar, Philippines, Thailand and Vietnam) who have excellent academic record, at least 2 years of working experience, a very good command of the English language; Singaporeans, Singapore permanent residents and recipients of other scholarships or bursaries are not eligible to apply.
Level of Study: MBA
Type: Scholarship
Value: Monthly stipend of Singaporean $1,350.00; book allowance of Singaporean $500; tuition fee, health insurance, examination fee and other approved fees, allowances and expenses; cost of one overseas Business Study Mission (for MBA candidates only) undertaken within the tenable period of the scholarship; cost of travel from home country to Singapore on award of the scholarship; cost of travel from Singapore to home country on successful completion of the Masters degree within the tenable period of the scholarship.
Length of Study: Full-time for 16 months
Frequency: Annual
Study Establishment: NTU
Country of Study: Singapore
No. of awards offered: 1–2

Application Procedure: Application details available at the University website.
Closing Date: December
Funding: Government
No. of awards given last year: 1
No. of applicants last year: 50
Additional Information: Applications open in November and close in December each year. Invitations for applications will be placed in the newspapers in the capital cities of the ASEAN countries and on the University's website. International students have to apply for a student's pass from Singapore immigration in order to pursue a full-time course of study in Singapore. The University will assist successful international applicants in their applications for the student's pass.

Asian Communication Resource Centre (ACRC) Fellowship Award

Subjects: Communication and information research from Asian perspective.
Purpose: To encourage in-depth research, promote cooperation and support scholars who wish to pursue research in communication, information and ICT-related disciplines in Asia.
Eligibility: All applicants should posses or be working towards a postgraduate degree from a reputable academic institution and Applicants should be working on a research project in communication, media, information or related areas that would be able to exploit the materials in the ACRC.
Level of Study: Research, Postgraduate
Type: Fellowships
Value: Up to $1,500 (economy class return air ticket), on-campus accommodation will be provided and weekly allowance of $210 will be provided
Length of Study: 1–3 months
Frequency: Annual
No. of awards offered: 2
Application Procedure: Applicants can download the application form from the website and send in their completed application form along with a copy of their latest curriculum vitae.
Closing Date: October 1st

For further information contact:

School of Communication and Information Nanyang Technological University, 31 Nanyang Link, 637718, Singapore
Tel: (65) 6790 4577
Fax: (65) 6791 5214
Email: acrc_fellowship@ntu.edu.sg

The Lien Foundation Scholarship for Social Service Leaders

Subjects: Full-time or part-time local post-graduate studies in management and selected specialist fields at the National University of Singapore (NUS) and the Nanyang Technological University (NTU).
Purpose: To provide scholarships to support education and professional development.
Eligibility: The scholarship is open to candidates with academic excellence, notable performance record and the potential to take up leadership positions in voluntary welfare organizations.
Value: The award includes tuition fees, maintenance allowance (for full-time studies), book allowance and any other compulsory fees.
Application Procedure: For more information on the scholarship do visit the websites www.ncss.org.sg/lien, http://lienfoundation.org/ScholarshipSSL.htm. For more information on the scholarship contact Ms Ng Hwee Choon: ng_hwee_choon@ncss.gov.sg and Ms Pamela Biswas: Pamela_biswas@ncss.gov.sg.
Closing Date: June 22nd

Nanyang Fellows Scholarship

Subjects: Management.
Purpose: To allow outstanding candidates from the world, except Singapore, to pursue the Nanyang fellows programme at NTU on a full-time basis.
Eligibility: Open to international applicants seeking to pursue a full-time Nanyang Fellows MBA programme at NTU who have at least 8 years of post-graduation experience including 5 years of managerial experience. Singaporeans and Singapore Permanent Residents are not eligible to apply
Level of Study: MBA
Type: Scholarship
Value: Full or partial tuition fees for NTU and MIT. A monthly stipend of Singaporean $1200 and lodging with return airfare to MIT may also be awarded to outstanding candidates from the civil service.
Length of Study: 12 months
Frequency: Annual
Study Establishment: NTU and MIT
No. of awards offered: 10
Application Procedure: Application details are available at the university website.
Closing Date: January 31st
Funding: Government
No. of awards given last year: 10
No. of applicants last year: 500
Additional Information: Applications open in September/October and close in December each year.

NTU-MBA Scholarship

Subjects: Mangement.
Purpose: To enable outstanding overseas candidates to pursue the MBA programme on a full-time basis.
Eligibility: Open to international applicants. Singaporeans, Singapore permanent residents and recipients of other scholarships or bursaries are not eligible to apply.
Level of Study: MBA
Type: Scholarship
Value: Full or partial tuition fees of a full-time programme
Length of Study: 4 trimesters
Frequency: Annual
Study Establishment: NTU
Country of Study: Singapore
No. of awards offered: 4
Application Procedure: Applicants must submit only one application form for application to the NTU MBA and scholarship. Online application is available at www.nanyangmba.ntu.edu.sg/applynow.asp
Closing Date: End of February
Funding: Private
No. of awards given last year: 4
No. of applicants last year: 81
Additional Information: Applications open in October each year and close at the end of February each year. International students have to apply for a student's pass from Singapore immigration in order to pursue a full-time course of study in Singapore. The University will assist successful international applicants in their applications for the student's pass.

Singapore Education – Sampoerna Foundation MBA in Singapore

Subjects: Management.
Purpose: To help Indonesian citizens below 35 years pursue their MBA studies.
Eligibility: Applicant must be an Indonesian citizen under 35 years, hold a local Bachelor's degree from any discipline with a minimum GPA of 3.00 (on a 4.00 scale), · have a minimum of 2 year full-time professional work experience after the completion of the undergraduate degree, · currently not enrolled in graduate or post-graduate program, or obtained a Master's degree or equivalent; · not be a graduate from overseas tertiary institutions, unless was on a full scholarship, · not receive other equivalent award or scholarship offering similar or other benefits at the time of the award.
Value: Approx. US$70,000–150,000. This will cover GMAT and TOEFL/IELTS reimbursement, university application fee, student visa application fee, return airfares from Jakarta to the place of study, tuition fees for the duration of study, living allowance to support living costs during period of study and literature allowance to purchase textbooks required for study.
Closing Date: February 1st
Additional Information: For further information see www.nanyangmba.edu.sg/Admissions/FinancialAid.asp.

SPRING Management Development Scholarship (MDS)

Subjects: Management.
Purpose: To nurture the next generation of leaders for the trailblazer companies of tomorrow.

Eligibility: Those who are currently working in an SME or are interested to join one, are citizens or permanent residents of Singapore, have less than 5 years of working experience and successfully apply for one of the approved MBA programmes.
Value: Full-time/Part-time MBA: SPRING will provide grant value of up to 70% of tuition fees, and other related expenses for full-time MBA scholars up to a maximum qualifying cost of $52,000.
Study Establishment: Nanyang Business School
Country of Study: Singapore
Application Procedure: Applicants will have to go through a joint selection process by SPRING and the participating SME, serve a 3 month internship in the SME prior to embarking on the approved MBA course (performance must be deemed satisfactory by the SME), and serve a bond of up to 2 years in the SME upon completion of studies. For more information on government assistance programmes, please contact the Enterprise One hotline at Tel: (65) 6898 1800 or email enterpriseone@spring.gov.sg or visit their website at www.spring.gov.sg/mds.
Closing Date: See web site for details
Contributor: SPRING Singapore and small medium enterprises (SME)
Additional Information: For more information on government assistance programmes, please contact the Enterprise One hotline at Tel: (65) 6898 1800 or email enterpriseone@spring.gov.sg or visit their website at www.spring.gov.sg/mds.

For further information contact:

Tel: 65 6898 1800
Email: enterpriseone@spring.gov.sg
Website: http://www.spring.gov.sg/mds

NARSAD

60 Cutter Mill Road, Suite 404, Great Neck, NY, 11021, United States of America
Tel: (1) 516 829 0091
Fax: (1) 516 487 6930
Email: grants@narsad.org
Website: www.narsad.org

NARSAD the world's leading charity dedicated to mental health research raises and distributes funds for scientific research into the causes, cures, treatments and prevention of psychiatric brain and behaviour disorders. NARSAD is the largest donor-supported organization devoted exclusively to supporting scientific research on psychiatric disorders in the world.

NARSAD Distinguished Investigator Awards

Subjects: Neurobiology.
Purpose: To stimulate the development of key personnel and resources, to facilitate the rapid initiation of research in innovative areas, and to enable investigators to create unique scientific opportunities.
Eligibility: Open to senior researchers at the rank of professor or equivalent who maintain their own laboratory, and applicants must be a full professor (or equivalent), and maintain peer reviewed competitively funded scientific programs.
Level of Study: Research, Postdoctorate
Type: Award
Value: A one-year grant of $100,000 is provided for established scientists pursuing innovative projects in diverse areas of neurobiological research.
Length of Study: 1 year
Frequency: Annual
Country of Study: Any country
No. of awards offered: Varies
Application Procedure: Electronic applications are available at www.narsad.org.
Closing Date: Oct 8th
Funding: Private
Contributor: Donors
No. of awards given last year: 16
No. of applicants last year: 143
Additional Information: Further information is available from the website.

NARSAD Independent Investigator Awards

Subjects: Neurobiology.
Purpose: To facilitate innovative research opportunities in diverse areas of neurobiological research.
Eligibility: Applicants must have a doctoral-level degree (MD, PhD, etc.), be an associate professor (or equivalent), and have received, as PI, competitive research support at a national level such as NIH, NSF, or foundation support.
Level of Study: Postdoctorate, Research
Type: Grant
Value: Funding is for two years and is up to $50,000 per year.
Length of Study: 2 years
Frequency: Annual
Country of Study: Any country
No. of awards offered: Varies
Application Procedure: Electronic applications are accepted through the NARSAD website.
Closing Date: March 15th (Do not submit before October 15th.)
Funding: Private
Contributor: Donors
No. of awards given last year: 42
No. of applicants last year: 246
Additional Information: Further information is available from the website.

NARSAD Young Investigator Awards

Subjects: Schizophrenia, major mood disorders, other serious mental illnesses such as anxiety disorders, borderline personality disorder, suicide and children's psychiatric disorders.
Purpose: To enable promising investigators to either extend their research fellowship training or to begin careers as independent research faculty.
Eligibility: Open to investigators who have attained a Doctorate or equivalent degree, are affiliated with a specific research institution and who have a mentor or senior collaborator who is performing significant research in an area relevant to schizophrenia, depression or other serious mental illnesses. Candidates should be at the postdoctoral to assistant professor level.
Level of Study: Postdoctorate, Research
Type: Award
Value: Two year awards up to $60,000, or $30,000 per year are provided to enable promising investigators to either extend research fellowship training or begin careers as independent research faculty.
Length of Study: Up to 2 years
Frequency: Annual
Country of Study: Any country
No. of awards offered: Varies
Application Procedure: Electronic applications are accepted through the NARSAD website.
Closing Date: February 28th
Funding: Private
Contributor: Donors
No. of awards given last year: 201
No. of applicants last year: 822
Additional Information: This award is intended to support only advanced Fellows through assistant professors or their equivalent. Further information is available from the website.

THE NATIONAL ACADEMIES

500 5th Street NW, Washington, DC, 20001, United States of America
Tel: (1) 202 334 2000
Fax: (1) 202 334 1667
Email: infofell@nas.edu
Website: www.nationalacademies.org

The National Academies perform an unparalleled public service by bringing together committees of experts in all areas of scientific and technological endeavor. These experts serve pro bono to address critical national issues and give advice to the federal government and the public.

Christine Mirzayan Science & Technology Policy Graduate Fellowship Program

Subjects: Science, engineering, medicine, veterinary medicine, business and law.
Purpose: To engage students in science and technology policy.
Eligibility: Graduate students and postdoctoral scholars and those who have completed graduate studies or postdoctoral research within the last 5 years are eligible to apply.
Level of Study: Postgraduate
Type: Fellowship
Value: The stipend for a 10-week program is $5,300
Length of Study: 12 weeks
Frequency: Annual
Application Procedure: A completed application form must be submitted. Application forms are available on the website.
Closing Date: November 1st

Ford Foundation Dissertation Fellowships

Subjects: Check website for details.
Purpose: To achieve excellence in college and university teaching.
Eligibility: Open to all citizens or nationals of the United States regardless of race, national origin, religion, gender, age, disability, or sexual orientation, individuals with evidence of superior academic achievement and committed to a career in teaching and research at the college or university level, PhD or ScD degree candidates studying in an eligible research-based discipline at a US educational institution. Also individuals who have not earned a doctoral degree at any time, in any field.
Level of Study: Doctorate
Type: Award
Value: Stipend:$21,000 for one year. Expenses paid to attend one Conference of Ford Fellows. Access to Ford Fellow Liaisons, a network of former Ford Fellows who have volunteered to provide mentoring and support to current fellows
Length of Study: 9–12 months
Frequency: Annual
Country of Study: United States of America
No. of awards offered: This year the program will award approximately 20 dissertation fellowships.
Application Procedure: Applicants must register and establish a personal user ID and password. Check website for further details.
Closing Date: November 17th
Contributor: The National Research Council

For further information contact:

Fellowships Office, Keck 576, National Research Council, 500 Fifth Street, NW, WA, 20001
Tel: 202 334 2872
Email: infofell@nas.edu
Website: www.national-academies.org/fellowships

Ford Foundation Postdoctoral Fellowships

Subjects: Check website for further details.
Purpose: For achieving excellence in college and university teaching and to increase the diversity of the nation's college and university faculties by increasing their ethnic and racial diversity, to maximize the educational benefits of diversity, and to increase the number of professors who can and will use diversity as a resource for enriching the education of all students.
Eligibility: Open to all citizens or nationals of the United States regardless of race, national origin, religion, gender, age, disability, or sexual orientation, individuals with evidence of superior academic achievement and committed to a career in teaching and research at the college or university level. Individuals should hold a PhD or ScD degree in an eligible research-based field from a US educational institution.
Level of Study: Postdoctorate
Type: Fellowships
Value: One-year Stipend: $40,000, Employing Institution Allowance: $1,500 and expenses paid to attend one Conference of Ford Fellows.
Length of Study: 9–12 months
No. of awards offered: 18
Application Procedure: Applicants must register and establish a personal user ID and password. Check website for further details.
Closing Date: November 24th
Funding: Foundation, government
Contributor: National Research Council (NRC) on behalf of the Ford Foundation
Additional Information: Candidates demonstrating superior academic achievement according to the judgement panels will be awarded.

For further information contact:

Fellowships Office, Keck 576, National Research Council, 500 Fifth Street, NW, Washington, DC, 20001, United States of America
Tel: (1) 202 334 2872
Email: infofell@nas.edu
Website: www.national-academies.org/fellowships

Ford Foundation Predoctoral Fellowships

Subjects: Check website for further details.
Purpose: For achieving excellence in college and university teaching and increasing the diversity of the nation's college and university faculties to maximize the educational benefits of diversity, and to increase the number of professors who can use diversity as a resource for enriching the education of all students.
Eligibility: Open to all citizens or nationals of the United States regardless of race, national origin, religion, gender, age, disability, or sexual orientation, individuals with evidence of superior academic achievement and committed to a career in teaching and research at the college or university level, should enroll in or planning to enroll in an eligible research-based program leading to a PhD or ScD degree at a US educational institution and who have not earned a doctoral degree at any time, in any field.
Level of Study: Predoctorate
Type: Bursary and scholarship
Value: Annual stipend: $20,000. Award to the institution in lieu of tuition and fees: $2,000. Expenses paid to attend at least one Conference of Ford Fellows
Length of Study: 3 years
No. of awards offered: 40
Application Procedure: Applicants must register and establish a personal user ID and password. Check website for further details.
Closing Date: November 14th
Funding: Foundation, government
Contributor: National Research Council on behalf of the Ford Foundation
Additional Information: Predoctoral fellows are required to enroll full-time in a program leading to a PhD or ScD degree in an eligible field of study.

For further information contact:

Fellowships Office, Keck 576, National Research Council, 500 Fifth Street, NW, Washington, DC, 20001
Tel: 202 334 2872
Email: infofell@nas.edu
Website: www.national-academies.org/fellowships

Jefferson Science Fellowship

Subjects: Science, technology, and engineering (STE).
Purpose: To offset the costs of temporary living quarters in the Washington, DC area.
Eligibility: Applicants must be US citizens and holding a tenured faculty position at a US degree granting academic institution of higher learning. For terms and conditions as well as further details log on to the website.
Level of Study: Postgraduate
Value: The Jefferson Science Fellow will be paid a per diem of up to $50,000 by the U.S. Department of State and $10,000 will be made available to the Fellow for travel associated with their assignment(s).
Length of Study: 1 year
Frequency: Annual
No. of awards offered: 2
Application Procedure: A complete nomination/application package consists of nomination/application form in PDF format and in word format; curriculum vitae (limit 10 pages); statements of qualifications (limit 2 pages each); and at least three (3), and no more than five (5), letters of recommendation from peers of the nominee/applicant.
Closing Date: January 14th

Contributor: National Academies supported through a partnership between American philanthropic foundations, the US STE academic community, professional scientific societies, and the US Department of State
Additional Information: Applicants should notify their institution while applying and encourage them to initiate a JSF/MOU as described on the website. Incomplete nomination/application packages, or those received after the deadline, will not be reviewed.

For further information contact:

The National Academies, Fellowships office, 500 fifth street, NW, Keck 568, WA, 20001, United States of America
Tel: (1) 202 334 2643
Fax: (1) 202 334 2759
Email: jsf@nas.edu
Website: www.nationalacademies.org

National Energy Technology Laboratory Methane Hydrates Fellowship Program (MHFP)

Subjects: Chemistry (Methane Hydrate).
Purpose: To provide postgraduate and postdoctoral candidates opportunities for career development, largely of their own choice in the Methane Hydrates field that are compatible with the interests of the sponsoring laboratories and universities, and to contribute thereby to the overall efforts of NETL in their support in the development of Methane Hydrate Science.
Eligibility: Open to candidates holding appropriate prior degree for the level of fellowship they intend to pursue. An applicant's training, professional experience, and research experience may be in any appropriate discipline or combination of disciplines required for the proposed project. Each Methane Hydrate Program fellow will be closely affiliated with a Research Adviser at the host venue.
Level of Study: Doctorate, Postdoctorate, Postgraduate
Type: Fellowship
Value: Stipend Rates: Master's Level (Fellow) begins at $30,000 with a maximum 2-year tenure, PhD Level (Fellow) begins at $35,000 with a maximum 3-year tenure and Postdoctoral Level (Research Associate) begins at $60,000 with a maximum 2-year tenure.
Length of Study: 3 years
Application Procedure: Application must be submitted only in hard copy and sent by Express Delivery to the Associateship Programs office. After completing the WebRAP application, you must also mail your supporting documents (transcripts and references) to the same address.
Closing Date: February 1st and August 1st
Additional Information: Please check the website for further details. Please direct all Application inquiries directly to the Research Associateship Programs at rap@nas.edu or by phone at 202 334 2760.

For further information contact:

Associateship Programs, Keck 555, National Research Council, 500 Fifth Street, NW, Washington, DC, 20001, United States of America
Tel: (1) 202 334 2707
Email: ebasquest@nas.edu@nas.edu
Website: www.national-academies.org/rap
Contact: Dr Eric O Basques, Research Adviser

NATIONAL ALLIANCE FOR MEDIA ARTS AND CULTURE (NAMAC)

145 9th Street, Suite 230, San Francisco, CA, 94103, United States of America
Tel: (1) 415 431 1391
Fax: (1) 415 431 1392
Email: jack@namac.org
Website: www.namac.org

NAMAC, founded in 1980, is a non-profit association whose membership comprises a diverse mix of organizations and individuals dedicated to a common goal: the support and advocacy of independent film, video, audio and online/multimedia arts.

McKnight Artist Fellowship for Filmmakers

Subjects: Cinema and television.
Purpose: To honour the professional and artistic accomplishments of Minnesota artists working in the medium of film/video.
Eligibility: Filmmakers who were full-time or part-time students in any undergraduate film, video, or screenwriting related degree programs at any time in the past three years are not eligible; applicants must demonstrate a sustained level of professional accomplishment over a period of at least five years; and applicants must provide documentation to show that their work has been shown in at least two film festivals or curated exhibitions, one of them outside of Minnesota, OR has been broadcast nationally or internationally;
Level of Study: Postgraduate
Type: Fellowship
Value: Two $25,000 Fellowships for Minnesota filmmakers.
Frequency: Annual
Country of Study: United States of America
Application Procedure: See the website.
Closing Date: March 29th

McKnight Artists Fellowship for Screenwriters

Subjects: Screenwriting.
Purpose: To recognize screenwriters who have excelled in the artistic discipline of writing for cinema.
Eligibility: Open to resident screenwriters of Minnesota.
Level of Study: Postgraduate
Type: Fellowship
Value: Five $25,000 awards are presented annually to accomplished Minnesota writers and spoken word artists.
Frequency: Annual
Country of Study: United States of America
No. of awards offered: 2
Application Procedure: See the website.
Closing Date: October 26th

NATIONAL ASSOCIATION OF BLACK JOURNALISTS (NABJ)

1100 Knight Hall, Suite 3100 College Park, Maryland 20742, Adelphi, MD, 20783 1716, United States of America
Tel: (1) 301 405 0248, 301 445 7100
Fax: (1) 301 314 1714, 301 445 7101
Email: nabj@nabj.org
Website: www.nabj.org

NABJ is an organization of journalists, students and media-related professionals that provides quality programmes and services to black journalists worldwide.

Ethel Payne Fellowships

Subjects: Journalism.
Purpose: To fund journalists wanting international reporting experience through self-conceived assignments in Africa.
Eligibility: Applicants should be an NABJ member with 5 years journalism experience, full-time freelance.
Level of Study: Postdoctorate
Type: Fellowship
Value: $5,000
Study Establishment: University of Maryland
Country of Study: United States of America
No. of awards offered: 2
Application Procedure: A completed application should be submitted accompanied by an 800-word project proposal, a 300 word essay on the applicant' journalism experience, 3 samples of work published or aired and 2 letters of recommendation.
Closing Date: December 15th
Contributor: The National Association of black Journalists (NABJ)
No. of awards given last year: 2

THE NATIONAL ASSOCIATION OF COMPOSERS/USA

PO Box 49256, Barrington Station, Los Angeles, CA, 90049, United States of America
Tel: (1) 310 838 4465
Fax: (1) 310 838 4465
Email: info@music-usa.org/nacusa
Website: www.music-usa.org/nacusa
Contact: President

The National Association of Composers presents concerts of music by American composers throughout the United States of America, and sponsors a Young composers competition.

National Association of Composers Young Composers Competition

Subjects: Music composition.
Purpose: To foster the creation of new American concert hall music.
Eligibility: Open to nationals of any country between the ages of 18 and 30.
Level of Study: Unrestricted, Postdoctorate, Postgraduate
Type: Competition
Value: First Prize-$400, Second Prize-$100 and possible performance on a NACUSA concert.
Frequency: Annual
Country of Study: Any country
No. of awards offered: 2
Application Procedure: Applicants must send in their music as there are no application forms.
Closing Date: October 31st
Funding: Private
No. of awards given last year: 2
No. of applicants last year: 45

NATIONAL ASSOCIATION OF TEACHERS OF SINGING (NATS)

9957 Moorings Drive, Suite 401, Jacksonville, FL, 32257, United States of America
Tel: (1) 904 992 9101; Toll Free: 888 262 2065
Fax: (1) 904 262 2587
Email: info@nats.org
Website: www.nats.org

The National Association of Teachers of Singing (NATS) is now the largest association of teachers of singing in the world.NATS offers a variety of lifelong learning experiences to its members, such as workshops, intern programmes, master classes, and conferences, all beginning at the chapter level and progressing to national events.

NATS Art Song Competition Award

Subjects: Singing.
Purpose: To stimulate the creation of quality vocal literature through the cooperation of singer and composer.
Eligibility: Open to any composer whose submitted work is a song cycle, group of songs, or extended single song of approximately fifteen minutes in length (13–17 minutes acceptable); for single voice and piano; to a text written in English, for which the composer has secured copyright clearance (only text setting permission necessary); composed within the last 2 years, and who pays the competition entry fee.
Type: Cash prize
Value: $2,000 and $1,000, 1st and 2nd place respectively.
Study Establishment: Valdosta State University
Country of Study: United States of America
No. of awards offered: 1
Application Procedure: Check the website for further details.
Closing Date: December 1st

For further information contact:

Department of Music, Valdosta State University, 1500 N. Patterson Street, Valdosta, GA, 31698, United States of America
Website: www.nats.org/competitions.php
Contact: Dr Carol Mikkelsen

THE NATIONAL ATAXIA FOUNDATION

2600 Fernbrook Lane Suite 119, Minneapolis, MN, 55447, United States of America
Tel: (1) (763) 553 0020
Fax: (1) (763) 553 0167
Email: naf@ataxia.org
Website: www.ataxia.org
Contact: Susan A Hagen, Patient Services Director

The National Ataxia Foundation is dedicated to improving the lives of persons affected by ataxia through support, education, and research. In 1978, the Foundation first began direct funding of ataxia research through the NAF Research "Seed-Money" Program. Since that time, the Foundation has established two additional research programs including the NAF Young Investigator Award and the NAF Fellowship Award.

Ataxia Research Grant

Subjects: Hereditary and sporadic ataxias.
Purpose: To assist investigators in the early or pilot phase of their studies relevant to the cause, pathogenesis or treatment of the hereditary or sporadic ataxias.
Eligibility: Open to all nationals.
Level of Study: Postgraduate, Research
Type: Grant
Value: $5,000–15,000 range. Funding may be considered for up to $35,000 for projects deserving special consideration
Frequency: Annual
No. of awards offered: 1
Application Procedure: Application forms can be downloaded from the website.
Closing Date: August 15th; letter of intent due July 15th
Funding: Foundation
No. of awards given last year: Varies
No. of applicants last year: Varies

Research Fellowship Award

Subjects: Hereditary and sporadic ataxias.
Purpose: To promote and support research to find the cause, treatment, and cure for the hereditary and sporadic ataxias.
Eligibility: Applicants should have completed at least one year of post-doctoral training, but not more than two at the time of application, and should have shown a commitment to research in the field of ataxia. A letter from the mentor should outline a program of studies for the applicant, and delineate the candidate's plans.
Level of Study: Postdoctorate
Type: Fellowship
Value: $35,000
Length of Study: 1 year
Frequency: Annual
Application Procedure: The signed original plus 2 copies of the completed application should be mailed to the Foundation. Applications can be downloaded from www.ataxia.org
Closing Date: September 15th; letter of intent due August 15th
Funding: Foundation
No. of awards given last year: Varies
No. of applicants last year: Varies

Young Investigator Award

Subjects: Hereditary and sporadic ataxia.
Purpose: To encourage young clinical and scientific investigators to pursue a career in the field of ataxia research.
Eligibility: Open to candidates who have attained a MD or PhD degree, and have an appointment as a junior faculty member. Individuals as the Associate Professor level are not eligible. Clinicians must have finished their residency no more than 5 years prior to applying.
Level of Study: Doctorate, Postgraduate
Type: Award
Value: $35,000–50,000
Length of Study: 1 year
Frequency: Annual
Application Procedure: Applicants should submit a completed research application form with a list of past, present and pending funding including project title, funding source, and amount of funding.

Applications must also contain a letter of nomination from a faculty sponsor to make appropriate resources available to support the project.
Closing Date: September 1st; letter of intent to apply due August 1st
Funding: Foundation

NATIONAL BRAIN TUMOR SOCIETY

East coast office, 124 Watertown Street, Suite 2D, Watertown, MA, 02472-2500, United States of America
Tel: (1) 617 924 9997
Fax: (1) 617 924 9998
Email: grants@tbts.org
Website: www.braintumor.org
Contact: David Hurwitz, Director of Research Programs

The National Brain Tumor Society exists to find a cure for brain tumours. It strives to improve the quality of life of brain tumour patients and their families. It disseminates educational information. It raises funds to advance carefully selected scientific research projects, improve clinical care and find a cure.

NBTS Research Grant

Subjects: Brain tumours.
Purpose: To fund scientific research aimed at finding a cure for brain tumours, and to support students working on a doctoral thesis in the subject.
Level of Study: Doctorate, Postgraduate
Type: Scholarship
Value: $100,000 per year
Length of Study: 2 years
Country of Study: United States of America
Application Procedure: Applicants must check the website.
Funding: Private

NATIONAL BREAST CANCER FOUNDATION (NBCF)

Level 9, 50 Pitt St, Sydney 2000, GPO Box 4129, Sydney 2001, Australia
Tel: (61) 02 8098 4825
Fax: (61) 02 8098 4801
Email: info@nbcf.org.au
Website: www.nbcf.org.au

The ultimate goal of the National Breast Cancer Foundation (NBCF) is to raise enough money to fund a cure for breast cancer. The NBCF supports and promotes research into breast cancer, facilitates consumer participation in all aspects of their work, acts as an advocate for breast cancer research, and provides opportunities for all Australians to contribute to breast cancer research.

Early Career Fellowship

Subjects: Broadly in Breast cancer research.
Purpose: To support the salary and research of outstanding new investigators with a vision to expand the scale and scope of breast cancer research in Australia.
Eligibility: Applicant must have visa status to stay and work in Australia for the duration of the award. Applicant must meet all eligibility criteria sets out in the guidelines and application form.
Level of Study: Professional development, Research
Type: Fellowship
Value: Up to &170,000 per year
Length of Study: 4 years
Country of Study: Australia
Application Procedure: Please check website for details.
Closing Date: Refer to online information, updated in March/April each year
Funding: Foundation
Contributor: NBCF. Australian Community and Corporate
No. of awards given last year: 4
No. of applicants last year: 12
Additional Information: NBCF Board will decide when to \call for application for what type of grant schemes each year and end of February.

Infrastructure Grant

Subjects: Activities may be supported by NBCF infrastructure grant may be operating in any field of cancer research, including behavioural, biomedical, clinical, epidemiological, health economics, health services and psychosocial.
Purpose: To support for resources, assets, facilities and research activities that are necessary to support breast cancer research across thr entire continuum in Australia.
Eligibility: Applicant must reside in Australia for the duration of grant award and must meet all eligibility criterials outlined in guidelines and application forms.
Level of Study: Foundation programme, Research, Unrestricted
Value: Up to $250,000 per year
Length of Study: 5 years
Country of Study: Australia
No. of awards offered: 3
Application Procedure: Please check website for details.
Closing Date: End of March
Contributor: NBCF, Australian Community and Corporate
No. of awards given last year: 3
No. of applicants last year: 24
Additional Information: NBCF will meet in late February to decide when to call for application. NBCF website will be update when details are finalised.

National Collaborative Breast Cancer Research Grant Program

Subjects: To pool knowledge, expertise, resources and encourage collaboration nationally and internationally to accelerate research programs into prevention, treatment, causes and management of breast cancer.
Purpose: To foster wide collaboration between groups of investigators, pooling expertise knowledge and resources to enable acceleration intor Breast cancer research.
Eligibility: Applicant must reside in Australia if receive funding. Applicant must meet all eligibility criterials in guidelines and application forms.
Level of Study: Research, Unrestricted
Type: Research grant
Value: Up to $5,000,000 over 5 years
Length of Study: 5 years
Country of Study: Australia
No. of awards offered: 5
Application Procedure: Please check website for details.
Funding: Foundation
Contributor: NBCF, Australian Community and Corporations
No. of awards given last year: 1
No. of applicants last year: 7
Additional Information: NBCF will decide whether this grant scheme will be offer in which year. On temporary hold at this state.

NBCF Postdoctoral Fellowship

Subjects: All aspects of breast cancer research.
Purpose: To provide outstanding researchers who have recently completed a PhD with support to pursue their breast cancer research interest and increase their research capability.
Eligibility: Applicant be a permanent resident or citizen of Australia, hold an MD or PhD in a health-related field of research, be actively engaged in breast cancer research, and have no more than five years postdoctoral experience. Have relevant visa status for the tenure of the fellowship to work and stay in Australia.
Level of Study: Research, Postdoctorate
Type: Fellowship
Value: Up to he Fellows awarded will receive CCAF's contribution of $100,000 for the first year followed by up to 3 years of NBCF funding of $100,000 from the second year onwards. Other successful Fellows will receive NBCF funding of $100,000 per year throughout the term of the Fellowship.
Length of Study: 4 years
Country of Study: Australia
No. of awards offered: Up to 4
Application Procedure: The applications are judged under peer review by experts in the field for their scientific merit and the contribution to either new knowledge or building on existing knowledge of breast cancer.

Closing Date: Approx. July each year will advise when Board decides
Funding: Foundation
Contributor: Australian community and corporate
No. of awards given last year: 4
No. of applicants last year: 13

Novel Concept Awards

Subjects: Oncology (Breast cancer research) and other novel ideas into Breast Cancer research.
Purpose: To provide investigators with the opportunity to pursue serendipitous observations and explore new, innovative, and untested ideas.
Eligibility: Open to applicants undertaking research in the entire continuum of breast cancer research. Residing in Australia throughout the funding period. Must meet all eligibility criteria outlined in guidelines and application form. Please check at www.nbcf.org.au.
Level of Study: Unrestricted
Type: Research grant
Value: Maximum value of $100,000 per grant per year
Length of Study: 1–2 years
Country of Study: Australia
No. of awards offered: Up to 12
Application Procedure: Please contact the Research Administrator or check the website for further details.
Closing Date: June 20th
Funding: Foundation, trusts
Contributor: NBCF, Australian Community and Corporate
No. of awards given last year: 12
No. of applicants last year: 35
Additional Information: Each year NBCF board will decide when to call for application and closing dates in late February.

Pilot Study Grants

Subjects: Oncology (breast cancer research).
Purpose: To financially assist investigators to obtain preliminary data regarding methodology, effect sizes and possible findings relating to new research ideas relevant to breast cancer.
Eligibility: Applicants must be Australian citizens, or be graduates from overseas with permanent Australian resident status, must reside in Australia throughout the funding period and not under bond to any foreign government.
Level of Study: Research, Unrestricted
Type: Grant
Value: A maximum of $100,000 for up to 2 years.
Length of Study: Up to 2 years
Country of Study: Australia
No. of awards offered: Up to 5
Application Procedure: Check website for further details.
Closing Date: May 10th
Funding: Foundation
Contributor: Australian Community and Corporate
No. of awards given last year: 3
No. of applicants last year: 76
Additional Information: NBCF Board will decide whether to offer this grant scheme again when they meet annual in late February.

Practitioner Fellowship

Subjects: To encourage health professinals, doctors, clinicians, allied health, nurses to engaging into research in breast cancer treatment, management so on.
Purpose: To encourage health professionals to engage into research and to bring in their expertise in exchange with some time by providing support for this exchange.
Eligibility: Applicant must meet the eligibility criterial outline in the guidelines and application form. Applicant must reside in Australia for the duration of the award.
Level of Study: Professional development, Research, Unrestricted
Type: Fellowship
Value: Up to $150,000 FTE per year
Length of Study: 4 years
Country of Study: Australia
No. of awards offered: 2
Application Procedure: Please check website for details.
Closing Date: July each year if being offer
Funding: Foundation
Contributor: NBCF, Australian Community and Corporate
No. of awards given last year: 2
No. of applicants last year: 5
Additional Information: NBCF Board will decide when to call for application in February. NBCF website will be update when finalise.

Translational Grant

Subjects: To assist in researchers focusing on translation of research findings into practice to improve cancer preventions, early detection, screening, diagnosis, prognosis, treatment and survivorship. Strengthen research collaboration, networks to provide greater depth in the area of translational research.
Purpose: To support proposals that aim to translate breast cancer research from lab bench to patients and public health care.
Eligibility: Applicant must reside in Australia for the duration of grant award and must meet all eligibility criterial outlined in guidelines and application form.
Level of Study: Research, Unrestricted
Type: Grant
Value: Up to $250,000 per year
Length of Study: 5 years
Country of Study: Australia
Application Procedure: Please check at website for details.
Closing Date: Closed in October
Funding: Foundation
Contributor: NBCF, Australia Community and Corporate
No. of applicants last year: 10
Additional Information: NBCF Board will decide when to call for application in late February. NBCF website will be update when details are finalised.

THE NATIONAL BUREAU OF ASIAN RESEARCH (NBR)

1414 NE 42nd Street, Suite 300, Seattle, WA, 98105, United States of America
Tel: (1) 206 632 7370
Fax: (1) 206 632 7487
Email: nbr@nbr.org
Website: www.nbr.org
Contact: George F Russell

NBR is a nonprofit, nonpartisan research institution dedicated to informing and strengthening policy. NBR conducts advanced research on politics and security, economics and trade, and health and societal issues, with emphasis on those of interest to the United States.

The Next Generation: Leadership in Asian Affairs Fellowship

Subjects: China's energy insecurity, military modernization in Asia, early health policy, central Asia's changing geopolitics, globalization or Chinese economic development, trends in Islamic education in South Asia, China–Southeast Asia relations.
Purpose: To further the professional development of Asian specialists in the year just after the completion of their Master's degree.
Eligibility: Open to citizens or permanent residents of the United States who have obtained a Master's degree.
Level of Study: Research
Type: Fellowships
Value: Each fellow will receive a $32,500 fellowship award (with benefits), as well as a reimbursement for some relocation expenses.
Length of Study: 1 year
Frequency: Annual
Country of Study: United States of America
No. of awards offered: 4
Application Procedure: Candidates must submit an online cover letter, curriculum vitae, 750 word essay stating the purpose of applying and 3 written references.
Closing Date: January 15th
Funding: Corporation, foundation, government
No. of awards given last year: 3

For further information contact:

Email: nextgen@nbr.org

NATIONAL CENTER FOR ATMOSPHERIC RESEARCH (NCAR)

PO Box 3000, Boulder, CO, 80307 3000, United States of America
Tel: (1) 303 497 1328/303 497 1000
Fax: (1) 303 497 1646
Email: paulad@ucar.edu
Website: www.asp.ucar.edu

The National Center for Atmospheric Research (NCAR) is a national research center focused on atmospheric science.

NCAR Faculty Fellowship Programme

Subjects: Atmospheric sciences.
Purpose: To facilitate residency study at NCAR.
Eligibility: Open to all faculty employed full-time at a college or university. Diversity concerns will also be addressed in the selection process.
Level of Study: Postdoctorate, Doctorate
Type: Fellowships
Value: Living expenses and related costs associated with undergraduate students conducting research with faculty applicants
Length of Study: 3–12 months
Frequency: Annual
Study Establishment: The National Center for Atmospheric Research
Country of Study: United States of America
No. of awards offered: 5–10
Application Procedure: Apply online.
Closing Date: Oct 31st
Funding: Government
Additional Information: If you have any questions please contact Paula Fisher; and email: paulad@ucar.edu or please see the website.

NCAR Graduate Student Visitor Programme

Subjects: Atmospheric sciences.
Purpose: To provide NCAR staff opportunities to bring graduate students to NCAR for 3 to 12-month collaborative visits with the endorsement of their thesis advisors and in pursuit of their thesis research.
Eligibility: Applicant must be a NCAR staff member
Level of Study: Graduate
Type: Scholarship
Value: Stipend of $1500 per month for living costs and $750 per month for living expenses plus travel expenses.
Length of Study: 3 to 12 months
Frequency: Annual
Study Establishment: The National Center for Atmospheric Research
Country of Study: United States of America
No. of awards offered: Varies
Application Procedure: Applicants must apply through an advisor and NCAR scientist.
Closing Date: October 31st
Funding: Government
Additional Information: If you have any questions about this new programme, please contact Paula Fisher; email: paulad@ucar.edu or please visit the website.

NCAR Postdoctoral Appointments in the Advanced Study Program

Subjects: Atmospheric sciences.
Purpose: To assist and support research.
Eligibility: Open to those who have recently received their PhD. Foreign nationals may also apply.
Level of Study: Postdoctorate
Type: Fellowship
Value: Successful applicants currently receive a stipend of $57,500 in the first year, and $60,000 in the second year.Travel expenses to NCAR will be reimbursed according to UCAR relocation guidelines for the fellow and his or her family. ASP also provides allowances of $750 for moving and storing personal belongings, and a minimum of $3,500 for annual scientific travel support.
Length of Study: Up to 1 year, with a possibility of renewal for a further year
Frequency: Annual
Study Establishment: The National Center for Atmospheric Research
Country of Study: United States of America
No. of awards offered: 10
Application Procedure: Applicants must apply online at www.asp.ucar.edu
Closing Date: January 14th
Funding: Government
No. of awards given last year: 8
No. of applicants last year: 120
Additional Information: NCAR is an equal opportunity employer with an affirmative-action programme.

For further information contact:

1850 Table Mesa Drive, Boulder, CO, 80305, United States of America

THE NATIONAL COLLEGIATE ATHLETIC ASSOCIATION

700 W. Washington Street, PO Box 6222, Indianapolis, IN, 46206-6222, United States of America
Tel: (1) 317 917 6222
Fax: (1) 317 917 6888
Email: pmr@ncaa.org
Website: www.ncaa.org

The National Collegiate Athletic Association is the organization through which the nation's colleges and universities speak and act on athletics matters at the national level. It is a voluntary association of more than 1,265 institutions, conferences, organizations and individuals devoted to the sound administration of intercollegiate athletics.

Freedom Forum-NCAA Foundation sports journalism scholarships

Subjects: Arts and media.
Purpose: To foster freedoms of the press and speech while promoting quality sports journalism education at the college level.
Eligibility: Open to applicants in their junior year in an NCAA member institution, have career goals in sports journalism, and major in journalism or experience in sports journalism on campus.
Level of Study: Postgraduate
Type: Scholarship
Value: $3,000
Frequency: Annual
Study Establishment: NCAA member institution
No. of awards offered: 8
Application Procedure: Check website for further details.
Closing Date: December 9th

For further information contact:

Website: www.ncaa.org/leadership_advisory_board/programs.html
Contact: The NCAA Leadership Advisory Board

NCAA Postgraduate Scholarship Program

Subjects: Sports.
Purpose: To honour outstanding student-athletes who are also outstanding scholars.
Eligibility: Open to student-athletes enrolled at an NCAA member institution, in the last year of intercollegiate competition and with a minimum grade point average of 3.2 on a 4.0 scale or its equivalent.
Level of Study: Postgraduate
Type: Scholarship
Value: One-time grant of $7,500. This is not earmarked for a specific area of postgraduate study but the awardee must use it as a part-time or full-time graduate student in a graduate or professional school of an academically accredited institution within 3 years of winning the award
Frequency: Each Academic year
Country of Study: Any country
No. of awards offered: Up to 174
Application Procedure: Applicants must be nominated by their faculty athletics representative or designee. Student may be any

nationality but must be attending an NCAA member institution and be in final year of eligibility in the sport they are nominated.
Closing Date: December 9th for Fall sports; February 24th for Winter sports; and May 5th for Spring sports
Funding: Foundation
No. of awards given last year: 174
No. of applicants last year: 374

Walter Byers Postgraduate Scholarship Program

Subjects: Sports.
Purpose: Encouraging excellence in academic performance by student-athletes and in recognition of outstanding academic achievement and potential for success in postgraduate study.
Eligibility: Applicant must have undergraduate cumulative grade-point average of 3.500, be a graduating senior who have competed in intercollegiate athletics as a member of a varsity team at an NCAA member institution; have demonstrated that participation in athletics has been a positive influence on their personal and intellectual development, and have competed in intercollegiate athletics as a member of a varsity team at an NCAA member institution.
Level of Study: Postgraduate
Type: Scholarship
Value: $24,000 per academic year, may be renewed for a second year.
Frequency: Annual
No. of awards offered: Varies
Application Procedure: Check website for further details.
Closing Date: January 27th
Additional Information: Generally, applications are available on-line in October of each year and the completed applications are due back to the national office in late January. Please refer to the current application packet for the current year's due date.

For further information contact:

The National Collegiate Athletic Association, PO Box 6222, Indianapolis, Indiana, 46206 6222
Contact: The Walter Byers Scholarship Committee Staff Liaison

NATIONAL DAIRY SHRINE

PO Box 725, Denmark, WI 54208, United States of America
Tel: (1) 920 863 6333
Fax: (1) 920 863 8328
Email: info@dairyshrine.org
Website: www.dairyshrine.org
Contact: David R. Selner, Executive Director

Founded in 1949 by a small group of visionary dairy leaders, National Dairy Shrine brings together dairy producers, scientists, students, educators, marketers and others who share a desire to preserve the dairy heritage and keep the dairy industry strong by encouraging students to get included in dairy related studies.

Dairy Student Recognition Program Award

Subjects: Related to dairy: production agriculture, manufacturing, marketing, agriculture law, business, veterinary medicine and environmental sciences.
Purpose: To recognize graduating seniors planning a career related to dairy who have demonstrated leadership skills, academic ability and interest in dairy cattle.
Eligibility: Only two applicants per college or university are accepted for this contest in any given year.
Level of Study: Postgraduate, Graduate
Type: Award
Value: 1st $2,000; 2nd $1,500; 3rd–9th $1,000
Frequency: Annual
Country of Study: United States of America
No. of awards offered: 2 per college
Application Procedure: Applicants must submit a letter of recommendation from the department head or a faculty member who is familiar with the applicant's activities and academic achievements and an official transcript showing all courses along with the application form. Applicants must complete a form for the scholarship. Forms available at www.dairyshrine.org.
Closing Date: April 15th
Funding: Private
No. of awards given last year: 9
No. of applicants last year: 22
Additional Information: The total number of awards will be determined by the number and quality of applicants. Winners will be recognized at the National Dairy Shrine awards banquet.

Kildee Scholarship (Advanced Study)

Subjects: Related to dairy industry: production agriculture, manufacturing, marketing, agricultural law, business, veterinary medicine and environmental sciences.
Purpose: The scholarship is offered in honour of late H.H. Kildee, Dean Emeritus at Iowa State University, whose counsel, encouragement and teaching inspired many of today's dairy leaders, for graduate study to a student who excelled in dairy cattle judging.
Eligibility: The awards are based on rank in the national contests, academic standing, leadership ability, student activities, interest and experience with dairy cattle and future plans.
Level of Study: Graduate, Postgraduate
Type: Scholarship
Value: US$3,000
Length of Study: 4 semesters
Frequency: Annual
Country of Study: United States of America
No. of awards offered: 2
Application Procedure: Applicants must submit the application form along with 2 letters of recommendation, one from department head and other from a faculty member who is familiar with the applicant's activities and academic achievements. An official transcript showing all courses should also be submittted. Applications available at www.dairyshrine.org.
Closing Date: April 15th
Funding: Private
No. of awards given last year: 2
No. of applicants last year: 6
Additional Information: The winners will be recognized at the National Dairy Shrine award banquet.

THE NATIONAL EDUCATION ASSOCIATION (NEA) FOUNDATION

1201 16th Street, North West, Washington, DC, 20036, United States of America
Tel: (1) 202 822 7840
Fax: (1) 202 822 7779
Email: foundation_info@nea.org
Website: www.neafoundation.org

The NEA Foundation offers programs and grants that support public school educators' efforts to close the achievement gaps, increase student achievement, salute excellence in education and provide professional development.

NEA Foundation Learning and Leadership Application Grants

Subjects: All subjects.
Purpose: To support individuals participating in high-quality professional development experiences.
Eligibility: Open to practising public school classroom teachers, public school education support personnel and faculty and staff of public higher education institutions. Two or more collaborating educators may also apply for a group grant.
Level of Study: Professional development
Type: Grant
Value: The grant amount is $2,000 for individuals and $5,000 for groups engaged in collegial study.
Length of Study: 1 year
Frequency: Annual
Country of Study: United States of America
No. of awards offered: Up to 75
Application Procedure: Applicants should consult the website for details.
Closing Date: October 15th, February 15th, June 1st
Funding: Foundation

No. of awards given last year: 72
No. of applicants last year: 842

NEA Foundation Student Achievement Grants

Subjects: Education.
Purpose: To support collaborative efforts by two or more colleagues to develop and implement creative project-based learning that results in high student achievement.
Eligibility: Open to teams of two or more practising United States public school teachers in grades K–12, public school education support personnel, public higher education faculty and staff.
Level of Study: Postgraduate
Type: Grant
Value: $5,000
Length of Study: 12 months
Frequency: Annual
Country of Study: United States of America
Application Procedure: Applicants must consult the website for details.
Closing Date: June 1st and October 1st
Funding: Foundation
No. of awards given last year: 74
No. of applicants last year: 771

Student Achievement Grants

Subjects: Any subject.
Purpose: To promote collaborative, innovative ideas that lead to student achievement of high standards.
Eligibility: Open to teams of two or more practising United States public school teachers in grades K–12, public school education support personnel, public higher education faculty and staff. Preference will be given to National Education Association members, and to educators who serve economically disadvantaged and/or underserved students.
Level of Study: Postgraduate
Type: Grant
Value: $2,000 and $5,000
Length of Study: 12 months
Frequency: Annual
Country of Study: United States of America
Application Procedure: Applicants must consult the organization for details.
Closing Date: February 1st, June 1st and October 15th

NATIONAL FOUNDATION FOR INFECTIOUS DISEASES (NFID)

4733 Bethesda Avenue, Suite 750, Bethesda, MD, 20814, United States of America
Tel: (1) 301 656 0003
Fax: (1) 301 907 0878
Email: info@nfid.org
Website: www.nfid.org

The National Foundation for Infectious Diseases (NFID) is a non-profit, non-governmental organization whose mission is public and professional education and promotion of research on the causes, treatment and prevention of infectious diseases.

NFID Advanced Vaccinology Course Travel Grant

Subjects: Immunology, vaccinology.
Purpose: To defray expenses related to attending the course, registration, airfare, ground transportation, lodging, meals and incidentals.
Eligibility: Applicants must be recent postdoctoral graduates (doctoral degree within past 3 years)/physicians who have completed speciality or subspeciality training within the past 2 years, with a demonstrated interest in a career in vaccinology. The applicant must be conducting research or working in a recognized and accredited United States of America Institution of Higher Learning or in a government agency.
Level of Study: Professional development
Type: Travel grant
Value: US$4,000 each
Length of Study: 10 days
Frequency: Annual
Study Establishment: Les Pensières, Veyrier-du-Lac
Country of Study: France
No. of awards offered: 2
Application Procedure: Applicant must submit an application and two copies, containing a letter from the applicant, letter of support from the department chairman, curriculum vitae and a copy of the application to the advanced vaccinology course, to be sent to the institution's main address.
Closing Date: November 15th
Funding: Foundation
Contributor: NFID Aventis Pasteur

NFID Travelling Professorship in Rural Areas

Subjects: Infectious diseases education, antimicrobial resistance and antimicrobial stewardship, new and future antimicrobials, tuberculosis, adult and adolescent immunizations, etc.
Purpose: To provide support for the applicant to provide face-to-face infectious diseases education to practicing physicians in rural areas in the applicant's state of residence or primary practice.
Eligibility: Applicant must be board certified in infectious diseases and must be a citizen of the United States of America.
Level of Study: Professional development
Type: Professorship
Value: The grant will be in the amount of a one time $10,000 cash payment that may be used for travel expenses, handout production, and supplies. Personal honorarium will not exceed $5,000 of the grant.
Length of Study: 5 working days
Frequency: Annual
Country of Study: United States of America
No. of awards offered: 1
Application Procedure: Applicant must submit an original application and four copies including cover letter, curriculum vitae and proposal (three pages) to be sent to the institution.
Closing Date: January 6th
Funding: Foundation
Contributor: NFID Steven R. Mostow Endowment for Outreach Programs

THE NATIONAL GALLERY

Information Department, The National Gallery, Trafalgar Square, London, WC2N 5DN, United Kingdom
Tel: (44) 20 7747 2885
Fax: (44) 20 7747 2423
Email: information@ng-london.org.uk
Website: www.nationalgallery.org.uk

The National Gallery, London houses one of the largest collections of European painting in the world.

The Pilgrim Trust Grants

Subjects: Art history.
Purpose: To fund research into regional collections, which will make a contribution to the National Inventory Research Project.
Level of Study: Research
Type: Grant
Value: A maximum of UK£5,000
Frequency: Twice yearly
Country of Study: United Kingdom
No. of awards offered: Varies
Application Procedure: There is no application form. Please submit a project plan including a list of paintings to be researched and a detailed budget, together with a letter of support from a senior curator or other responsible person and a curriculum vitae for the person who carry out the research.
Closing Date: April 30th and October 30th
Funding: Trusts
Contributor: The Pilgrim Trust
Additional Information: Researchers are most welcome to spend all or some part of the research period in the National Gallery Library, but this is not obligatory.

NATIONAL HEADACHE FOUNDATION

820 N Orleans, Suite 217, Chicago, IL, 60610 3132, United States of America
Tel: (1) 888 NHF 5552
Fax: (1) 312 274 2650
Email: info@headaches.org
Website: www.headaches.org

The National Headache Foundation disseminates information, funds research, sponsors public and professional education programmes and has a nationwide network of support groups, with 20,000 members. The Foundation is the recognized authority in headache and head pain, and offers the award winning newsletter NHF Head Lines, patient education brochures.

National HeadacFoundation Research Grant

Subjects: Treatment and causes of headache.
Purpose: To encourage better understanding and treatment of headache and head pain.
Eligibility: Open to researchers in neurology and pharmacology departments in medical schools throughout the United States of America. Submissions from other departments and individual investigators are also welcome.
Level of Study: Postgraduate, Doctorate, Postdoctorate
Type: Research grant
Value: US$100,000.
Length of Study: 1 year
Frequency: Annual
Country of Study: United States of America
No. of awards offered: Varies, depending on funds available and the number of worthy projects submitted
Application Procedure: Applicants must complete an application form.
Closing Date: February 1st
Funding: Private
Contributor: Dues and donations
No. of awards given last year: 7
No. of applicants last year: 12

NATIONAL HEALTH AND MEDICAL RESEARCH COUNCIL (NHMRC)

NHMRC, GPO Box 1421, Canberra, Australian Capital Territory, ACT, 2601, Australia
Tel: (61) 2 6217 9000
Fax: (61) 2 6217 9100
Email: grantnet.help@nhmrc.gov.au
Website: www.nhmrc.gov.au
Contact: Executive Director

The National Health and Medical Research Council (NHMRC) (Australia) consolidates within a single national organization the often independent functions of research funding and development of advice. One of its strengths is that it brings together and draws upon the resources of all components of the health system, including governments, medical practitioners, nurses and allied health professionals, researchers, teaching and research institutions, public and private programme managers, service administrators, community health organizations, social health researchers and consumers.

Biomedical (C J Martin) Fellowships

Subjects: Biomedical sciences.
Purpose: To enable fellows to develop their research skills and work overseas on specific research projects within the biomedical sciences under nominated advisers.
Eligibility: Open to Australian citizens or graduates from overseas with permanent Australian resident status who are not under bond to any foreign government. Candidates should hold a Doctorate in a medical, dental or related field of research, be actively engaged in such research in Australia and have no more than 2 years postdoctoral experience at the time of application.
Level of Study: Postdoctorate
Type: Fellowship
Value: An allowance of Australian $5,000 per year is payable for research support, including conference travel, for the 2 year Australian portion of the Fellowship.
Length of Study: 4 years, the first 2 of which are to be spent overseas and the final 2 in Australia
Frequency: Annual
Study Establishment: Institutions approved by the NHMRC, such as teaching hospitals, universities and research institutes
Country of Study: Any country
No. of awards offered: Varies
Application Procedure: Application forms available from the website.
Closing Date: July 7th
Funding: Government
No. of awards given last year: 31
No. of applicants last year: 79

Biomedical (Dora Lush) and Public Health Postgraduate Scholarships

Subjects: Biomedical sciences and public health.
Purpose: To encourage science Honours or equivalent graduates of outstanding ability to gain full-time health and medical research experience.
Eligibility: Open to Australian citizens who have already completed a science Honours degree (or equivalent) at the time of submission of the application, science Honours graduates and unregistered medical or dental graduates from overseas, who have permanent resident status and are currently residing in Australia. The scholarship shall be held within Australia.
Level of Study: Postgraduate
Type: Scholarship
Value: Varies
Length of Study: 1 year, renewable for up to 2 further years
Frequency: Annual
Study Establishment: Institutions approved by the NHMRC, such as teaching hospitals, universities and research institutes
Country of Study: Australia
No. of awards offered: Varies
Application Procedure: Applicants should visit the website at www.nhmrc.gov.au/funding/schlorships.htm for details.
Closing Date: Varies
Funding: Government
No. of awards given last year: 51
No. of applicants last year: 133

For further information contact:

Centre for Research Management & Policy, NHMRC, GPO Box 9848, Canberra, Australian Capital Territory,

Biomedical Australian (Peter Doherty) Fellowship

Subjects: Biomedical sciences.
Purpose: To provide full-time training in basic biomedical sciences research in Australia.
Eligibility: Open to Australian citizens or graduates from overseas with permanent Australian resident status who are not under bond to any foreign government. Candidates should hold a Doctorate in a medical, dental or related field of research or have submitted a thesis for such by December in the year of application, be actively engaged in such research in Australia or overseas and have no more than 2 years postdoctoral experience at the time of application.
Level of Study: Postdoctorate
Type: Fellowship
Value: Fellowship salary packages at the Training Support Package level 1 is $62,250. An allowance of $5,000 per year is payable for research support, including conference travel.
Length of Study: 4 years
Frequency: Annual
Study Establishment: Institutions approved by the NHMRC, such as teaching hospitals, universities and research institutes
Country of Study: Australia
No. of awards offered: Varies
Application Procedure: Application forms available from the website.
Closing Date: Junly 7th
Funding: Government

No. of awards given last year: 30
No. of applicants last year: 100

Career Development Fellowship

Subjects: Any human health-related research area.
Purpose: To help researchers to conduct research that is internationally competitive and to develop a capacity for independent research.
Eligibility: Open to Australian citizens or permanent residents, normally between 3 and 9 years postdoctoral experience.
Type: Fellowship
Value: Australian $96,040–106,230 per year
Length of Study: 5 years
Frequency: Annual
Study Establishment: Institutions approved by NHMRC, such as teaching hospitals, universities and research institutes
Country of Study: Australia
No. of awards offered: Varies
Application Procedure: Application forms available from the website www.nhmrc.gov.au
Closing Date: July 18th
Funding: Government
No. of awards given last year: 54
No. of applicants last year: 434

Clinical (Neil Hamilton Fairley) Fellowship

Subjects: All health-related fields.
Purpose: To provide training in scientific research methods.
Eligibility: Open to Australian citizens or graduates from overseas with permanent Australian resident status who are not under bond to any foreign government. Candidates should hold a Doctorate in a health-related field of research or have submitted a thesis for such by December of the year of application, be actively engaged in such research in Australia and have no more than 2 years postdoctoral experience at the time of application.
Level of Study: Postdoctorate
Type: Fellowship
Value: Fellowship salary packages at the Training Support Package level 1 is currently at $62,250 and if appropriate, clinical loadings will be paid.
Length of Study: 4 years, the first 2 of which are to be spent overseas and the final 2 in Australia
Frequency: Annual
Study Establishment: Institutions approved by the NHMRC, such as teaching hospitals, universities and research institutes
Country of Study: Any country
No. of awards offered: Varies
Application Procedure: Application form available from the website.
Closing Date: July 7th
Funding: Government
No. of awards given last year: 4
No. of applicants last year: 10

NHMRC Early Career Fellowship

Subjects: Scientific research, including the social and behavioural sciences, that can be applied to any area of clinical or community medicine.
Purpose: To undertake research that is both of major importance in its field and of benefit to Australian health.
Eligibility: Open to Australian citizens or graduates from overseas with permanent Australian resident status, who are not under bond to any foreign government. Candidates should hold a Doctorate in a health-related field of research or have submitted a thesis for such by December of the year of application, be actively engaged in such research in Australia or overseas and have no more than 2 years postdoctoral experience at the time of application.
Level of Study: Postdoctorate
Type: Fellowship
Value: Funding is for 4 years at TSP1 Level, which currently is $67, 508
Length of Study: 4 years
Frequency: Annual
Study Establishment: Institutions approved by the NHMRC, such as teaching hospitals, universities and research institutes
Country of Study: Australia
No. of awards offered: Varies
Application Procedure: Application forms are available from the website.
Closing Date: May 3rd
Funding: Government
No. of awards given last year: 9
No. of applicants last year: 21

NHMRC Medical and Dental and Public Health Postgraduate Research Scholarships

Subjects: Medical, dental and public health research.
Purpose: To encourage medical and dental and public health graduates to gain full-time research experience.
Eligibility: Open to Australian citizens who are medical or dental and public health research graduates registered to practice in Australia, with the proviso that medical graduates can also apply during their intern year and that dental postgraduate research scholarships may be awarded prior to graduation provided that the evidence of high quality work is shown. Also open to medical and dental graduates from overseas who hold a qualification that is registered for practice in Australia, who have permanent resident status and are currently residing in Australia.
Level of Study: Postgraduate
Type: Scholarship
Value: Varies
Length of Study: 1 year, renewable for up to 2 further years
Frequency: Annual
Study Establishment: Institutions approved by the NHMRC such as teaching hospitals, universities and research institutes
Country of Study: Australia
No. of awards offered: Varies
Application Procedure: Available from the website at www.nhmrc.gov.au/funding/scholarships.htm
Closing Date: Varies
Funding: Government
No. of awards given last year: 6
No. of applicants last year: 12
Additional Information: The award is divided into two categories: Medical and Dental Public HealthPostgraduate Research Scholarships and Public HealthPostgraduate Research Scholarships.

NHMRC/INSERM Exchange Fellowships

Subjects: Biomedical sciences.
Purpose: To enable Australian Fellows to work overseas on specific research projects in INSERM laboratories in France and vice versa.
Eligibility: Open to Australian citizens and permanent residents, who are not under bond to any foreign government, who hold a Doctorate in a medical, dental or related field of research or have submitted a thesis for such by December in the year of application, are actively engaged in such research in Australia and have no more than 2 years postdoctoral experience at the time of application.
Level of Study: Postdoctorate
Type: Fellowship
Value: Australian $62,250
Length of Study: 4 years, the first 2 of which are to be spent in France and the final 2 in Australia
Frequency: Annual
Study Establishment: Institutions approved by the NHMRC, such as teaching hospitals, universities and research institutes, and INSERM laboratories in France
Country of Study: France or Australia
No. of awards offered: 1
Application Procedure: Applications forms available from the website.
Closing Date: July 7th
Funding: Government
No. of awards given last year: 1
Additional Information: This fellowship is awarded in association with l'Institut National de la Santé et de la Recherche Médicale (INSERM), France.

Public Health Australian Fellowship

Subjects: Public health.
Purpose: To provide full-time training in public health research in Australia.

Eligibility: Applicants should hold a Doctorate in a health-related field of research or have submitted a PhD by December in the year of application and have no more than 2 years postdoctoral experience. Applicant must also, for years 1 and 2, nominate a department [preferably institution] and research group other than that where the applicant's doctoral qualifications were obtained. Open to Australian citizens or permanent residents.
Level of Study: Postdoctorate
Type: Fellowship
Value: Australian $67,508 and Australian $5,000
Length of Study: 4 years
Frequency: Annual
Study Establishment: Institutions approved by the NHMRC, such as teaching hospitals, universities and research institutes
Country of Study: Australia
No. of awards offered: Varies
Application Procedure: Application forms available from the website.
Closing Date: Varies
Funding: Government
No. of awards given last year: 17
No. of applicants last year: 61

Public Health Overseas (Sidney Sax) Fellowships

Subjects: Public health.
Purpose: To provide full-time training overseas and in Australia in public health research.
Eligibility: Applicants should hold a Doctorate in a health-related field of research or have submitted a PhD by December in the year of application and have no more than 2 years postdoctoral experience. Open to Australian citizens or permanent residents.
Type: Fellowship
Value: Fellowship salary support packages at the Training Support Package level 1, that is $59,750, will be paid. Minimum cost airfares for the Fellow and dependants will be provided for direct travel to, and return from, the overseas centre. Additional overseas and Australian allowances are also payable.
Length of Study: 4 years
Frequency: Annual
Study Establishment: Institutions approved by the NHMRC, such as teaching hospitals, universities and research institutes
Country of Study: Any country
No. of awards offered: Varies
Application Procedure: Application forms available from the website.
Closing Date: June 14th
Funding: Government
No. of awards given last year: 3
No. of applicants last year: 11

Training Scholarship for Indigenous Health Research

Subjects: Indigenous Australian Health Research.
Purpose: To provide support for research training or training leading to research areas of particular relevance to Indigenous Australians.
Eligibility: Applicant must be an Australian citizen or Australian permanent resident, have made prior arrangements with the Head of Department or Institution in which they propose to study, provide a specific study plan within a clearly defined area and conduct research of potential benefit to Australia.
Level of Study: Postdoctorate
Type: Scholarship
Value: Depends on the qualification and current registration.
Length of Study: 1 year, renewable up to further 2 years
Frequency: Annual
Country of Study: Australia
No. of awards offered: Varies
Application Procedure: Available from the website at www.nhmrc.gov/funding/scholarships.htm
Closing Date: August 4th
Funding: Government
No. of awards given last year: 3
No. of applicants last year: 6

NATIONAL HEART FOUNDATION OF AUSTRALIA

Level 12/500 Collins Street VIC 3000, Melbourne, VIC, 3003, Australia
Tel: (61) 3 9329 8511
Fax: (61) 3 9326 3190
Email: research@heartfoundation.com.au
Website: www.heartfoundation.com.au
Contact: Frank Anastasopoulos, Research Program Manager

The National Heart Foundation of Australia is a non-government, non-profit health organization funded mostly by public donation. The Foundation's mission is to reduce suffering and death from heart, stroke and blood vessel disease in Australia. The Foundation funds biomedical, clinical and public health research, provides clinical leadership and develops health promotion strategies and initiatives.

National Heart Foundation of Australia Career Development Fellowship

Subjects: Cardiovascular disease and related disorders.
Purpose: To enable a senior Australian researcher of exceptional merit and proven record in the cardiovascular field to undertake independent research.
Eligibility: Applicants are expected to have spent at least two years actively contributing to cardiovascular research in Australia and applicants must be an Australian citizen, permanent resident or a New Zealand citizen.
Level of Study: Research
Type: Fellowship
Value: Contact the organization
Length of Study: 4 years, non-renewable
Frequency: Annual
Study Establishment: Universities, hospitals or research institutions
Country of Study: Australia
Application Procedure: Applicants must be nominated by the head of the host department or institution.
Closing Date: May 4th
Funding: Private
Additional Information: Refer to the website www.heartfoundation.org.au for the closing date for applications.

National Heart Foundation of Australia Overseas Research Fellowships

Subjects: Clinical, public health and biomedical research related to cardiovascular disease and related disorders.
Purpose: To support the training of Australian researchers seeking to begin or maintain a postdoctoral position overseas.
Eligibility: Applicants are expected to have at least two years of cardiovascular research experience and must be Australian citizens or have permanent residency status.
Level of Study: Postdoctorate
Type: Fellowship
Value: Australian $37,345–45,362
Length of Study: 3 years
Frequency: Annual
Study Establishment: Approved institutions
Country of Study: Other
Application Procedure: Includes tuition, travel, living expenses, dependents expenses, research expenses, housing, health insurance
Closing Date: May 31st
Funding: Private
Additional Information: Fellowships are awarded on the understanding that the Fellow will return to Australia to continue his or her career upon completion of the fellowship. Refer to the website www.heartfoundation.org.au for the closing dates for applications.

National Heart Foundation of Australia Postdoctoral Research Fellowship

Subjects: Cardiovascular disease and related disorders.
Purpose: To award graduates who have demonstrated expertise and significant achievement in cardiovascular research.
Eligibility: Applicants must be in their first 3 years post-PhD in any area of cardiovascular health research, including biomedical, clinical, public health and health service delivery.
Level of Study: Postdoctorate

Type: Fellowship
Value: Stipend of $75,000 pro rata per year
Length of Study: Up to 2 years
Frequency: Annual
Study Establishment: Universities, hospitals or research institutions
Country of Study: Australia
Application Procedure: Applicants must submit an application outlining a research proposal accompanied by the supervisor's reference and backing.
Closing Date: May 4th
Funding: Private
Additional Information: Refer to the website www.heartfoundation.org.au for the closing date for applications.

National Heart Foundation of Australia Postgraduate Biomedical Research Scholarship

Subjects: Cardiovascular disease and related disorders.
Purpose: To allow graduates to undertake a period of training in research under the full-time supervision and tuition of a responsible investigator.
Eligibility: Open to citizens and permanent residents of Australia, and to citizen of New Zealand.
Level of Study: Postgraduate
Type: Scholarship
Value: Contact the organization
Length of Study: Up to 3 years
Frequency: Annual
Study Establishment: Universities, hospitals or research institutions
Country of Study: Australia
Application Procedure: Applicants must submit an application outlining a research proposal accompanied by the supervisor's reference and backing.
Funding: Private
Additional Information: Refer to the website www.heartfoundation.org.au for the closing date for applications.

National Heart Foundation of Australia Postgraduate Clinical Research Scholarship

Subjects: Cardiovascular disease and related disorders.
Purpose: To enable medical graduates to undertake a period of training in research under the full-time supervision and tuition of a responsible investigator.
Eligibility: Open to citizens and permanent residents of Australia, and to citizens of New Zealand.
Level of Study: Postgraduate
Type: Scholarship
Value: Please contact the organization
Length of Study: 3 years
Frequency: Annual
Study Establishment: Universities, hospitals and research institutions
Country of Study: Australia
No. of awards offered: Varies
Application Procedure: Applicants must submit an application outlining a research proposal accompanied by the supervisor's reference and backing.
Closing Date: July 11th
Funding: Private
Additional Information: Refer to the website www.heartfoundation.org.au for the closing date for applications.

National Heart Foundation of Australia Postgraduate Public Health Research Scholarship

Subjects: Cardiovascular research and related disorders.
Purpose: To allow graduates to undertake a period of training in research under the full-time supervision and tuition of a responsible investigator.
Eligibility: Open to citizens and permanent residents of Australia, and to citizens of New Zealand.
Level of Study: Postgraduate
Type: Scholarship
Value: Please contact the organization
Length of Study: Up to 3 years
Frequency: Annual
Study Establishment: Universities, hospitals or research institutions
Country of Study: Australia
Application Procedure: Applicants must submit an application outlining a research proposal accompanied by the supervisor's reference and backing.
Closing Date: July 11th
Funding: Private
Additional Information: Refer to the website www.heartfoundation.org.au for the closing date for applications.

National Heart Foundation of Australia Research Grants-in-Aid

Subjects: Biomedical, clinical and public health research.
Purpose: To support research in the cardiovascular field.
Eligibility: Open to citizens and permanent residents of Australia. In any one year, an applicant may apply as a Chief Investigator A (CIA) on up to three GIA applications. However, CIAs may only be awarded one GIA per year.
Level of Study: Research
Type: Grant
Value: Australian $65,000 per year
Length of Study: Up to 2 years
Frequency: Annual
Study Establishment: An approved institution
Country of Study: Australia
Application Procedure: Applicants must submit an application outlining research proposal. References are also required.
Closing Date: March 9th
Funding: Private
Additional Information: Refer to the website www.heartfoundation.org.au for the closing date for applications.

NATIONAL HEART FOUNDATION OF NEW ZEALAND

9 Kalmia Street, Ellerslie, PO Box 17-160, Greenlane, Newmarket, Auckland, 1546, New Zealand
Tel: (64) 9 571 9191
Fax: (64) 9 571 9190
Email: info@heartfoundation.org.nz
Website: www.heartfoundation.org.nz
Contact: Professor Norman Sharpe, Medical Director

The National Heart Foundation of New Zealand aims to promote good health and to reduce suffering and premature death from diseases of the heart and circulation. Grant advertisement for 2009 included. Please refer to website www.heartfoundation.org.nz for further information on research funding.

National Heart Foundation of New Zealand Fellowships

Subjects: Application's are particularly encouraged in areas that align with the National Heart Foundation's strategic priority objectives. For the list of objectives please visit the website.
Purpose: To promote the aims of the National Heart Foundation of New Zealand.
Eligibility: Normally open to New Zealand graduates only.
Level of Study: Postgraduate
Type: Fellowship/Scholarship
Value: Varies according to the determination of the Scientific Committee and within an annual budget
Frequency: Annual
No. of awards offered: Varies
Application Procedure: Details available on the website www.heartfoundation.or.nz/research. The 'Guidelines for Research Applicants' is available as a PDF file.
Closing Date: June 1st
Funding: Private
No. of awards given last year: 6

National Heart Foundation of New Zealand Limited Budget Grants

Subjects: Applications are particularly encouraged in areas that align with National Heart Foundation's strategic priority objectives. For the list of objectives please visit the website.

Purpose: To further the aims of the National Heart Foundation of New Zealand.
Eligibility: Normally open to New Zealand graduates only.
Level of Study: Postgraduate
Type: Small project grant
Value: Varies, according to the determination of the Scientific Committee and within an annual budget
Frequency: Biannual
Country of Study: New Zealand
No. of awards offered: Varies
Application Procedure: Details are available on the website www.heartfoundation.org.nz
Closing Date: February 1st, June 1st and October 1st
Funding: Private
No. of awards given last year: 21
Additional Information: These grants cover small projects, e.g. less than New Zealand $15,000 and grants-in-aid.

National Heart Foundation of New Zealand Project Grants

Subjects: Applications are particularly encouraged in areas that align with the National Heart Foundation's strategic priority objectives. For the list of objectives please visit the website.
Purpose: To provide short-term support for a single individual or a small group working on a clearly defined research project, which will promote the aims of the National Heart Foundation of New Zealand.
Eligibility: Normally open to New Zealand graduates only.
Level of Study: Postgraduate
Type: Grant
Value: Varies, according to the determination of the scientific committee and within an annual budget
Frequency: Annual
Country of Study: New Zealand
No. of awards offered: Varies
Application Procedure: Details are available on the website www.heartfoundation.org.nz/research. The 'Guidelines for Research Applicants' is available as a pdf file.
Closing Date: March 1st
Funding: Private
No. of awards given last year: 6
No. of applicants last year: 28

National Heart Foundation of New Zealand Senior Fellowship

Subjects: Cardiovascular disease.
Purpose: To support graduates from New Zealand who have trained as cardiologists or other scientists working in the field of cardiovascular research.
Eligibility: Candidates must possess an appropriate postgraduate degree or diploma and may undertake work for a higher degree such as a PhD or MD. The maximum age for appointment will normally be 40 years. The Fellowship must be taken up within 12 months of the award.
Level of Study: Postgraduate
Type: Fellowship
Value: The Foundation may provide an initial Grant-in-Aid of up to $1,500 to enable research work to commence. In subsequent years, Senior Fellows may apply for working expenses not exceeding $1,000 annually with no single item to exceed $500.
Length of Study: Up to 3 years
Frequency: Every 3 years
Country of Study: New Zealand
No. of awards offered: 1
Application Procedure: Applicants should follow the format outlined in the revised 'A Guide to Applicants for Research and Other Grants' which is available from the website. Applications should be sent to Professor Norman Sharpe, Medical Director.
Closing Date: April 1st
Funding: Private
No. of awards given last year: 1
No. of applicants last year: 4
Additional Information: The fellowship must be taken up within 1 year of the award. Further research funding will require an application for project grant funds from the Foundation. Funding for conference expenses must be applied for separately.

National Heart Foundation of New Zealand Travel Grants

Subjects: Applications are particularly encouraged in areas that align with Nationaal Heart Foundation Strategic priority objectives.
Purpose: To enable medical or non-medical workers to travel in New Zealand or overseas for short-term study or to attend conferences.
Eligibility: Only available to early career researchers, who are Australian or New Zealand citizens (or permanent resident) based in Australia and working in the area of cardiovascular disease or related areas.
Level of Study: Postgraduate
Type: Grant
Value: Travel Grants are worth up to $2,000 and are exclusive of GST.
Frequency: 3 times per year
Country of Study: Any country
No. of awards offered: Varies
Application Procedure: Details are available on the website www.heartfoundation.org.nz/research. The 'Guidelines for Research Applicants' is available as a PDF file.
Closing Date: June 1st and October 1st
Funding: Private
No. of awards given last year: 21

NATIONAL INSTITUTE FOR LABOR RELATIONS RESEARCH (NILRR)

5211 Port Royal Road, Suite 510 Springfield, VA, 22151, United States of America
Tel: (1) 703 321 9606
Fax: (1) 703 321 7342
Email: research@nilrr.org
Website: www.nilrr.org

The National Institute for Labor Relations Research's (NILRR) primary function is to act as a research facility for the general public, scholars and students. It provides the supplementary analysis and research necessary to expose the inequities of compulsory unionism.

The Applegate/Jackson/Parks Future Teacher Scholarship

Subjects: Education.
Purpose: To support students to pursue studies in journalism and related majors.
Eligibility: Applicants are limited to graduate or undergraduate students majoring in education institutions of higher learning throughout the United States.
Level of Study: Doctorate, Graduate, MBA
Type: Scholarship
Value: $1,000
Length of Study: 1 year
Frequency: Annual
Country of Study: United States of America
No. of awards offered: 1
Application Procedure: Applicants must submit a complete formal application, a copy of the up-to-date transcript of grades and a typewritten essay of 500 words demonstrating an interest in, and knowledge of, the right to work principle as it applies to educators.
Closing Date: December 31st
Funding: Private
No. of awards given last year: 1
No. of applicants last year: 250

For further information contact:

Website: www.nilrr.org/node/11

William B. Ruggles Right to Work Journalism Scholarship

Subjects: Journalism and related majors.
Purpose: To support students to pursue studies in journalism and related majors.
Eligibility: Applicants are limited to graduate or undergraduate studentsmajoring in journalism or related majors, in institutions of higherlearning throughout the United States.
Level of Study: Graduate, Postgraduate
Type: Scholarship

Value: $2,000
Length of Study: 1 year
Frequency: Annual
Country of Study: United States of America
No. of awards offered: 1
Application Procedure: A completed formal application, a copy of the most up-to-date transcript of grades and a typewritten essay of approximately 500 words clearly demonstrating an interest in, and knowledge of, the Right to Work principle.
Closing Date: December 31st
Funding: Private
No. of awards given last year: 1
No. of applicants last year: 250

For further information contact:

Website: www.nilrr.org/ruggles1.htm

NATIONAL INSTITUTE OF GENERAL MEDICAL SCIENCES (NIGMS)

45 Center Drive, MSC 6200, Bethesda, MD, 20892 6200, United States of America
Tel: (1) 301 496 7301
Fax: (1) 301 402 0224
Email: info@nigms.nih.gov
Website: www.nigms.nih.gov
Contact: Ms Jilliene Mitchell, Information Development Specialist

The National Institute of General Medical Sciences (NIGMS) is one of the National Institutes of Health (NIH), the principal biomedical research agency of the Federal Government. NIGMS primarily supports basic biomedical research that lays the foundation for advances in disease diagnosis, treatment, and prevention.

MARC Faculty Predoctoral Fellowships

Subjects: Biomedical or behavioural sciences.
Purpose: Awards provide an opportunity for eligible faculty who lack the PhD degree (or equivalent) to obtain the research doctorate.
Eligibility: Open to full-time, permanent faculty members in a biomedically related science or mathematics programme who have been at a minority or minority-serving institution for at least 3 years at the time of application. Candidates must be enrolled in, or have been accepted into, a PhD or combined MD-PhD training programme in the biomedical or behavioural sciences. Candidates must intend to return to the minority institution at the end of the training period.
Level of Study: Postgraduate, Predoctorate
Type: Fellowship
Value: An applicant may request a stipend equal to his or her annual salary, but not to exceed the stipend of a level one postdoctoral Fellow. The applicant may also request tuition and fees as determined by the training institution as well as an allowance per year for training-related costs.
Frequency: Annual
Study Establishment: An institution in the United States of America
Country of Study: United States of America
No. of awards offered: Varies
Application Procedure: Applicants must write to the main address for details or telephone Dr Adolphus Toliver, at (1) 301 594 3900. Further details are also available from the website, www.nigms.nih.gov
Closing Date: April 5th and December 5th
Funding: Government
Additional Information: Closing Date for Application: Standard application deadlines are available at http://grants.nih.gov/grants/dates.htm

NIGMS Fellowship Awards for Minority Students

Subjects: Biomedical or behavioural sciences.
Purpose: These awards provide up to 5 years of support for research training leading to a PhD or equivalent research degree, a combined MD-PhD degree or another combined professional Doctorate-research PhD.
Eligibility: Open to highly qualified students who are members of minority groups that are underrepresented in the biomedical or behavioural sciences in the United States of America. These groups include African Americans, Hispanic Americans, Native Americans, including Alaska Natives and natives of the United States of America Pacific Islands.
Level of Study: Postgraduate, Predoctorate
Type: Fellowship
Value: NIGMS provides tuition, fees and up to $4,200 per 12-month period to the predoctoral fellow's sponsoring institution to help defray such trainee expenses as research supplies and equivalent. The tuition, fees and institutinal allowance are detailed at http://grants1.nih.gov/grants/guide/notice-files/NOT-OD-07-052.html.
Length of Study: Up to 5 years
Frequency: Annual
Study Establishment: An institution in the United States of America
Country of Study: United States of America
No. of awards offered: Varies
Application Procedure: Applicants must write to the main address for details or telephone Dr Adolphus Toliver, at (1) 301 594 3900. Further details are also available from the website, www.nigms.nih.gov
Closing Date: Febuary 10th
Funding: Government

NIGMS Fellowship Awards for Students With Disabilities

Subjects: Biomedical and behavioural sciences.
Purpose: These awards provide up to 5 years of support for research training leading to a PhD or equivalent research degree, combined MD-PhD degree or another combined professional doctorate/research PhD degree in the biomedical or behavioural science.
Eligibility: Open to principal investigators at domestic institutions holding an active NIGMS research grant, programme project grant, centre grant or co-operative agreement research programme with a reasonable period of research support remaining.
Level of Study: Doctorate, Graduate, Postdoctorate, Postgraduate, Predoctorate
Type: Fellowship
Value: NIGMS provides tuition, fees and up to $4,200 per 12-month period to the predoctoral fellow's sponsoring institution to help defray such trainee expenses as research supplies and equipment. The tuition, fees and institutional allowance are detailed at http://grants1.nih.gov/grants/guide/notice-files/NOT-OD-07-052.html.
Length of Study: Up to 5 years
Frequency: Annual
Study Establishment: An institution in the United States of America
Country of Study: United States of America
No. of awards offered: Varies
Closing Date: Febuary 10th
Funding: Government

NIGMS Postdoctoral Awards

Subjects: Biomedical and behavioural sciences.
Purpose: NIGMS welcomes NRSA applications from eligible individuals who seek postdoctoral biomedical research training in areas related to the scientific programmes of the Institute.
Eligibility: Open to applicants who have received the doctoral degree (domestic or foreign) by the beginning date of the proposed award.
Level of Study: Postdoctorate
Type: Award
Value: Up to US$51,036 per year, based on the salary of the applicant at the time of the award
Frequency: Annual
Study Establishment: The institutional setting may be domestic or foreign, public or private
Country of Study: Any country
No. of awards offered: Varies
Application Procedure: Applicants must write to the main address for details or telephone Dr Alison Cole, at (1) 301 594 3349. Further details are also available from the website, www.nigms.nih.gov
Funding: Government

NIGMS Research Project Grants (R01)

Subjects: Biomedical and behavioural sciences.
Purpose: To support a discrete project related to the investigator's area of interest and competence.
Eligibility: Research project grants may be awarded to nonprofit organisations and institutions; governments and their agencies;

occasionally, though rarely, to individuals who have access to adequate facilities and resources for conducting the research; and to profit-making organisations. Foreign institutions and international organisations are also eligible to apply for these grants.
Level of Study: Postgraduate
Type: Grant
Value: These grants may provide funds for reasonable costs of the research activity, as well as for salaries, equipment, supplies, travel and other related expenses
Frequency: Annual
Country of Study: United States of America
No. of awards offered: Varies
Application Procedure: Applicants must contact the Office of Extramural Outreach and Information Resources for details.
Funding: Government
For further information contact:

Office of Extramural Outreach and Information Resources, NIH, 6701 Rockledge Drive, MSC 7910, Room 6207, Bethesda, MD, 20892-7910, United States of America
Tel: (1) 301 435 0714
Email: grantsinfo@nih.gov

Research Supplements to Promote Diversity in Health-Related Research

Subjects: Biomedical and behavioural sciences.
Purpose: Principal investigators holding NIGMS research grants may request supplemental funds to improve the diversity of the research workforce by supporting and recruiting students and postdoctoral fellows from underrepresented racial and ethnic groups; individuals with disabilities; and individuals from socially, culturally, economically or educationally disadvantaged backgrounds that have inhibited their ability to pursue a career in health-related research.
Eligibility: Open to principal investigators at domestic institutions holding an active NIGMS research grant, programme project grant, centre grant or co-operative agreement research programme with a reasonable period of research support remaining.
Level of Study: Unrestricted
Type: Grant
Value: Varies
Length of Study: 2 years or more
Frequency: Annual
Study Establishment: An institution in the United States of America
Country of Study: United States of America
No. of awards offered: Varies
Application Procedure: Applicants must write to the main address for details or telephone Dr Anthony René, at (1) 301 594 3833. Further details are also available from the website, www.nigms.nih.gov
Funding: Government

NATIONAL INSTITUTE ON AGING

Building 31, Room 5C27, 31 Center Drive, MSC 2292, Bethesda, MD, 20892, United States of America
Tel: (1) 301 496 1752
Fax: (1) 301 496 1072
Email: mk46u@nih.gov
Website: www.nia.nih.gov
Contact: Dr Miriam Kelty, Associate Director

The National Institute on Aging conducts and supports research and research training in all areas of biological ageing, the neuroscience and neuropsychology of ageing, geriatrics and the social and behavioural sciences of ageing.

NIH Research Grants

Subjects: The biology of ageing, the neuroscience and neuropsychology of ageing, geriatrics and clinical gerontology and the social and behavioural sciences of ageing.
Purpose: To support research and training in the biological, clinical, behavioural and social aspects of ageing mechanisms and processes.
Eligibility: Varies, depending on the mechanism, but generally open to United States of America citizens only.
Level of Study: Postdoctorate
Type: Grants, fellowships, career development awards and institutional training awards
Value: Varies
Length of Study: 1–5 years
Frequency: Annual, 3 cycles a year
Country of Study: The United States of America or others depending on mechanisms
No. of awards offered: Varies
Application Procedure: Applicants must download the application form, available on the website. NIH is transitioning to electronic transmission through grants.gov (www.nih.gov)
Closing Date: Deadlines are staggered. Please see website
Funding: Government
Contributor: The United States of America government
Additional Information: Some award mechanisms are limited to citizens and permanent residents of the United States of America. Others are open to applicants from any country. Further information is available on the website.

NATIONAL LEAGUE OF AMERICAN PEN WOMEN, INC. (NLAPW)

1300 17th Street NW, Washington, DC, 20036-1973, United States of America
Tel: (1) 202 785 1997
Fax: (1) 202 452 6868
Email: nlapw1@juno.com
Website: www.americanpenwomen.org
Contact: Ms Elaine Waidelich, National Scholarship Chairperson

The National League of American Pen Women (NLAPW) exists to promote women in the creative arts including art, writing and music.

Virginia Liebeler Biennial Grants for Mature Women (Art)

Subjects: Art.
Purpose: To advance creative purpose in art.
Eligibility: Open to women over 35 years of age who wish to pursue special work in their field of art, letters or music. Applicants must be citizens of the United States of America. Current and past recipients are not eligible for this award.
Level of Study: Unrestricted
Type: Grant
Value: $1,000. The award may be used for college, framing, research or any creative purpose that furthers a career in the creative arts.
Frequency: Every 2 years
Country of Study: United States of America
No. of awards offered: 1
Application Procedure: Applicants must; (a) send proof of age and United States of America citizenship (copy of birth certificate, passport ID page, voter's registration card and a driver's license or a state ID, if possible). Driver's license alone is not proof of citizenship; (b) enclose a 1 page cover letter describing how they intend to use the grant money, if awarded the grant, and any relevant information about themselves; (c) enclose a US$8 check made payable to NLAPW with submission; (d) enclose a stamped addressed envelope/mailer with sufficient postage for the return of the submission; (e) provide their name, address, telephone number and e-mail address on the cover page of submission; submit three 46 or larger colour prints (no slides): oil, watercolor, acrylic, mixed media, original works on paper etc. submit three 46 or larger photos of sculpture; and for photography, submit three 46 or larger prints in colour or black and white.
Closing Date: October 1st
Funding: Private
Contributor: NLAPW
No. of awards given last year: 6
No. of applicants last year: 150
Additional Information: Interested parties must send a stamped addressed envelope to receive current information.

For further information contact:

1300 17th Street, Washington, DC, 20036, United States of America
Contact: Dr N.Taylor Collins

Virginia Liebeler Biennial Grants for Mature Women (Music)

Subjects: Music.
Purpose: To advance creative purpose in music.
Eligibility: Open to women over 35 years of age who wish to pursue special work in their field of art, letters or music. Applicants must be citizens of the United States of America. Current and past recipients are not eligible for this award.
Level of Study: Unrestricted
Type: Grant
Value: $1,000 (minimum)
Frequency: Every 2 years
Country of Study: United States of America
No. of awards offered: 1
Application Procedure: Applicants must; (a) send proof of age and United States of America citizenship (copy of birth certificate, passport ID page, voter's registration card and a driver's license or a state ID, if possible). Driver's license alone is not proof of citizenship; (b) enclose a one-page cover letter describing how they intend to use the grant money, if awarded the grant, and any relevant information about themselves; (c) enclose a US$8 check made payable to NLAPW with the submission; (d) enclose a stamped addressed envelope/mailer with sufficient postage for the return of submission; and (e) provide their name, address, telephone number and e-mail address on the cover page of their submission. Applicants must also submit two compositions: 3-minute minimum performance time, 5-minute maximum performance time.
Closing Date: October 1st
Funding: Private
Contributor: NLAPW
No. of awards given last year: 6
No. of applicants last year: 100
Additional Information: Interested parties must send a stamped addressed envelope to receive current information.

For further information contact:

202 E Manford avenue Avenue Adq, Ohio, 45810, United States of America
Contact: Dr M.J.Sunny Zank

Virginia Liebeler Biennial Grants for Mature Women (Writing)

Subjects: Writing.
Purpose: To advance creative purpose in writing.
Eligibility: Open to women over 35 years of age who wish to pursue special work in their field of art, letters or music. Applicants must be citizens of the United States of America. Current and past recipients are not eligible for this award.
Level of Study: Unrestricted
Type: Grant
Value: $1,000
Frequency: Every 2 years
Country of Study: United States of America
No. of awards offered: 1
Application Procedure: Applicants must (a) send proof of age and United States of America citizenship (copy of birth certificate, passport ID page, voter's registration card and a driver's license or a state ID, if possible). Driver's license alone is not proof of citizenship; (b) enclose a one-page cover letter describing how they intend to use the grant money, if awarded the grant, and any relevant information about themselves; (c) enclose a US$8 check made payable to NLAPW with their submission; (d) enclose a stamped addressed envelope/mailer with sufficient postage for the return of their submission; and (e) provide their name, address, telephone number and email address on the cover page of the submission. Applicants must submit either a published or unpublished manuscript in any or all of the following categories: article, drama, essay, first chapter of a novel, narrative outline of a complete novel, three poems, short-story or TV script. The entry is not to exceed 4,000 words.
Closing Date: October 1st
Funding: Private
Contributor: NLAPW
No. of awards given last year: 6
No. of applicants last year: 60
Additional Information: Interested parties must send a stamped addressed envelope to receive current information.

For further information contact:

1300 17th Street, Washington, DC, 20036, United States of America
Contact: Dr N.Taylor Collin

NATIONAL LIBRARY OF AUSTRALIA

Parkes Place, Canberra, ACT, 2600, Australia
Tel: (61) 2 6262 1111
Fax: (61) 2 6257 1703
Email: fellowships@nla.gov.au
Website: www.nla.gov.au
Contact: Dr Marie-Louise Ayres, Curator of Manuscripts

The National Library of Australia is responsible for developing and maintaining a comprehensive collection of Australian library materials and a strong collection of non-Australian publications, and for administering and co-ordinating a range of national bibliographical activities.

Harold White Fellowships

Subjects: There are few subject limitations but most fellowships fall within the categories of arts and humanities, fine and applied arts or social sciences.
Purpose: To promote the Library as a centre of scholarly activity and research, to encourage the scholarly and literary use of the collection and the production of publications based on them and to publicize the Library's collections.
Eligibility: Open to established scholars, writers and librarians from any country. Fellowships are not normally offered to candidates working for a higher degree.
Level of Study: Unrestricted
Type: Fellowship
Value: Australian $850 per week
Length of Study: 3–6 months
Frequency: Annual
Study Establishment: The National Library of Australia
Country of Study: Australia
No. of awards offered: 3–6
Application Procedure: Applicants must complete an application form available in the website www.nla.gov.au/grants/haroldwhite.
Closing Date: April 30th
Funding: Government
No. of awards given last year: 12
No. of applicants last year: 70
Additional Information: Normally, Fellows will be expected to give a public lecture and at least one seminar on the subject of their research during their tenure. At least three quarters of the fellowship time should be spent in Canberra.

THE NATIONAL MULTIPLE SCLEROSIS SOCIETY (NMSS)

733 3rd Avenue, 3rd Floor, New York, NY, 10017, United States of America
Tel: (1) 212 463 7787
Fax: (1) 212 986 7981
Email: patricia.olooney@nmss.org
Website: www.nationalmssociety.org
Contact: Grants Management Officer

The National Multiple Sclerosis Society (NMSS) is dedicated to ending the devastating effects of multiple sclerosis. Founded in 1946, NMSS supports more on multiple sclerosis (MS) research and provides professional education programmes and furthers MS advocacy efforts than any other MS organization in the world.

Harry Weaver Junior Faculty Awards

Subjects: Neurosciences related to multiple sclerosis.
Purpose: To enable highly qualified persons who have concluded their research training and have begun academic careers as independent investigators to undertake independent research.
Eligibility: Open to citizens of the United States of America holding a doctoral degree, and who have had sufficient research training at the pre- or postdoctoral levels to be capable of independent research.

Individuals who have already carried out independent research for more than 5 years are not eligible. Candidate must hold or have been offered an academic appointment at the assistant professor level at an approved university.
Level of Study: Professional development
Value: Approx. $75,000 per year. For other rewards please refer to the website.
Length of Study: 5 years
Frequency: Annual
Study Establishment: An approved university, professional or research institute
Country of Study: United States of America
No. of awards offered: Varies
Application Procedure: Applicants must complete an application form.
Closing Date: August 12th
Funding: Private
No. of awards given last year: 1
No. of applicants last year: 4
Additional Information: The candidate will not be an employee of the Society but rather of the institution. It is expected that the institution will develop plans for continuing the candidate's appointment and for continued salary support beyond the 5-year period of the award. Fellows may not supplement their salary through private practice or consultation, nor accept another concurrent award. The grantee institution holds title to all equipment purchased with award funds.

National Multiple Sclerosis Society Pilot Research Grants

Subjects: Multiple sclerosis.
Purpose: To provide limited short-term support of novel high-risk research.
Eligibility: Open to suitably qualified investigators.
Level of Study: Research
Type: Research grant
Value: Up to $40,000 in direct costs may be requested
Length of Study: 1 year
Frequency: Dependent on funds available
Country of Study: Any country
No. of awards offered: Varies
Application Procedure: Applicants must complete an application form.
Closing Date: Applications will be reviewed on a quarterly basis; deadines: April 3rd, July 3rd and October 2nd
Funding: Private
No. of awards given last year: 58
No. of applicants last year: 121
Additional Information: Grants are awarded to an institution to support the research of the principal investigator. Progress reports are required.

National Multiple Sclerosis Society Postdoctoral Fellowships

Subjects: Multiple sclerosis.
Purpose: To provide postdoctoral training that will enhance the likelihood of performing meaningful and independent research relevant to multiple sclerosis.
Eligibility: Open to unusually promising recipients of MD or PhD degrees. Foreign nationals are welcome to apply for fellowships in the United States of America only. The Society will consider applications from established investigators who seek support to obtain specialized training in some field in which they are not expert, when such training will materially enhance their capacity to conduct more meaningful research. United States of America citizenship is not required for training in United States of America institutions but applicants who plan to train in other countries must be citizens of the United States of America.
Level of Study: Postdoctorate
Type: Fellowship
Value: Varies according to professional status, previous training, accomplishments in research and the pay scale of the institution in which the training is provided. Fellowships may be supplemented by other forms of support, with prior approval
Length of Study: 1–3 years
Frequency: Annual
Study Establishment: An institution of the candidate's choice
Country of Study: Any country
No. of awards offered: Varies
Application Procedure: Applicants must complete an application form.
Closing Date: August 15th
Funding: Private
No. of awards given last year: 11
No. of applicants last year: 52
Additional Information: Fellows are not considered employees of the Society but rather of the institution where the training is provided. The fellowship is to be administered in accordance with the prevailing policies of the sponsoring institution. It is the responsibility of the applicant to make all the necessary arrangements for their training with the mentor and institution of their choice.

National Multiple Sclerosis Society Research Grants

Subjects: Multiple sclerosis, the cause, prevention, alleviation and cure.
Purpose: To stimulate, co-ordinate and support fundamental or applied clinical or non-clinical research.
Eligibility: Open to suitably qualified investigators.
Level of Study: Professional development
Type: Research grant
Value: Funds may be used to pay the salaries of associated professional personnel, technical assistants and other non-professional personnel in proportion to the time spent directly on the project, in whole or in part. Salaries are made in accordance with the prevailing policies of the grantee institution. If requested, other expenses such as travel costs and fringe benefits may also be paid
Length of Study: 3 years
Country of Study: Any country
No. of awards offered: Varies
Application Procedure: Applicants must complete an application form.
Closing Date: Febrauary 6th
Funding: Private
No. of awards given last year: 62
No. of applicants last year: 245
Additional Information: Grants are awarded to an institution to support the research of the principal investigator. Scientific equipment and supplies bought with grant funds become the property of the grantee institution. Progress reports are required and appropriate publication is expected.

NMSS Patient Management Care and Rehabilitation Grants

Subjects: Health and medical sciences, therapy/rehabilitation.
Purpose: To support investigators with an MD, PhD or equivalent degree to research patient management care and rehabilitation.
Eligibility: Open to citizens of the United States.
Level of Study: Doctorate, Postgraduate
Type: Grant
Value: US$2,00,000–3,00,000
Frequency: Annual
Country of Study: United States of America
No. of awards offered: 2–3
Closing Date: Early February and August
No. of awards given last year: 5
No. of applicants last year: 21

For further information contact:

Health Care Delivery and Policy Research
Contact: Nicholas LaRocca, Associate Vice President

NATIONAL ORCHESTRAL INSTITUTE

2110 Clarice Smith Performing Arts Center, University of Maryland, College Park, MD, 20742 1620, United States of America
Tel: (1) 301 405 2317
Fax: (1) 301 314 9504
Email: noi@umd.edu
Website: www.noimusic.com

The National Orchestral Institute at the University of Maryland School of Music offers an intensive 4-week experience in orchestral musicianship and professional development for musicians on the threshold of their careers. Distinguished musicians and conductors work closely with participants to polish ensemble skills and orchestral excerpts.

National Orchestral Institute Scholarships

Subjects: Orchestral performance and chamber music.
Purpose: To provide an intensive 4-week orchestral training programme to enable musicians to rehearse and perform under internationally acclaimed conductors and study with principal musicians of the United States of America's foremost orchestras in preparation for careers as orchestral musicians.
Eligibility: Open to advanced musicians between 18 and 28 years of age, primarily students and postgraduates of United States of America universities, conservatories and colleges. Others, however, are welcome to apply, but must appear for an audition at the centre. String players, including harpists, who live more than 200 miles away from an audition centre, may audition by tape.
Level of Study: Unrestricted
Type: Scholarship
Value: Full tuition, room and boardscholarship worth over US$4,000
Length of Study: 4 weeks
Frequency: Annual
Study Establishment: The University of Maryland
Country of Study: United States of America
No. of awards offered: Approx. 90
Application Procedure: Applicants must submit an application, curriculum vitae and (optional) letter of recommendation.
Closing Date: March
Funding: Government
Contributor: The University of Maryland
No. of awards given last year: 92
No. of applicants last year: 720
Additional Information: Personal auditions are required at one of the audition centres throughout the country.

THE NATIONAL ORGANIZATION FOR RARE DISORDERS (NORD)

PO Box 1968, 55 Kenosia Avenue, Danbury, CT, 06813 1968, United States of America
Tel: (1) 203 744 0100
Fax: (1) 203 798 2291
Email: lcataldo@rarediseases.org
Website: www.rarediseases.org
Contact: Ms Linda M Cataldo, Field Services Co-ordinator

The National Organization for Rare Disorders (NORD) is a federation of voluntary health organizations dedicated to helping people with rare (orphan) diseases and assisting the organizations that serve them. NORD is committed to the identification, treatment and cure of rare disorders through programmes of advocacy, education, research and service.

NORD/Roscoe Brady Lysosomal Storage Diseases Fellowships

Subjects: Genetics, new treatments and diagnostics, and/or epidemiology of lysosomal storage diseases in general, or a specific lysosomal storage disease.
Purpose: To assist physicians who desire to establish careers in lysosomal storage diseases and clinical medicine.
Eligibility: Open to all countries that adhere to the most recent guidelines for human subject protection as set forth by the NIH. Applicants should have earned an MD degree within the past 10 years.
Level of Study: Doctorate, Postdoctorate, Research
Type: Fellowship
Value: $50,000 to $70,000 per year
Length of Study: 1 year, but renewable for a 2nd year
Frequency: Annual
Country of Study: Any country
No. of awards offered: 1
Application Procedure: Application forms and required attachments may be obtained directly from the website.
Closing Date: May 9th
Funding: Private
Contributor: Public donations
No. of awards given last year: 3
No. of applicants last year: 14

NATIONAL PHYSICAL SCIENCE CONSORTIUM (NPSC)

USC-RAN 3716 South Hope Suite 348, Los Angeles, CA, 90007-4344, United States of America
Tel: (1) 213 743 2409, 800 854 6772
Fax: (1) 213 743 2407
Email: jpowell@usc.edu
Website: www.npsc.org
Contact: Dr James L Powell, Executive Director

Established in 1987, the National Physical Science Consortium (NPSC) is headquartered in Los Angeles. It is a unique partnership between industry, government agencies and laboratories and higher education.

NPSC Fellowship in Physical Sciences

Subjects: Physical sciences.
Purpose: To increase the number of qualified citizens of the United States in the physical sciences and related engineering fields, emphasizing recruitment of a diverse applicant pool of women and historically underrepresented minorities.
Eligibility: Open to students who have at least 3.0 grade point average and are graduating seniors or graduate student (up to 2nd year) enrolled in a PhD programme.
Level of Study: Graduate, Doctorate
Type: Fellowship
Value: The charge to the employer for each student supported is $26,000 annually, of which $20,000 is the student stipend and $6,000 is NPSC's fee to support its operations.
Length of Study: 2–6 years
Frequency: Annual
Country of Study: United States of America
Application Procedure: Candidates must apply online.
Closing Date: November 30th
Additional Information: Members of underrepresented groups are encouraged to apply.

NATIONAL RADIO ASTRONOMY OBSERVATORY

520 Edgemont Road, Charlottesville, VA, 22903 2475, United States of America
Tel: (1) 434 296 0211
Fax: (1) 434 296 0278
Email: borahood@nrao.edu
Website: www.nrao.edu
Contact: Billie Orahood

The National Radio Astronomy Observatory designs, builds and operates the world's most sophisticated and advanced radio telescopes (the VLA, VLBA, GBT and ALMA), providing scientists from around the world the means to study all aspects of astronomy from planets in our Solar System to the most distant galaxies.

Jansky Fellowship

Subjects: Astronomy radio astronomy instrumentation, computation and theory.
Purpose: To provide an opportunity for young scientists to establish themselves as independent researchers so that they may more effectively compete for permanent positions. The placement of fellows at institutions other than the NRAO will help foster closer scientific ties between the NRAO and the US astronomical community. Annual Jansky Fellows symposia are planned to ensure close contact among all Fellows and the NRAO.
Eligibility: Open to astronomers, physicists, electrical engineers and computer specialists. Preference will be given to recent PhD recipients.
Level of Study: Postdoctorate
Type: Fellowship

Value: Starting salary of $63,000 per year with a research budget of $10,000 per year for travel and computing requirements. In addition, page charge support, as well as vacation accrual, health insurance, and a moving allowance are provided, as well as up to $3,000 per year to defray local institutional costs
Length of Study: 2 years, with a possibility of renewal for 1 further year
Frequency: Annual
Study Establishment: The Observatory's centres in Charlottesville, VA; Green Bank, WV; and Socorro, NM
Country of Study: Any country
No. of awards offered: Up to 3 appointments will be made annually for positions at any of the NRAO sites (Socorro, NM; Green Bank, WV; and Charlottesville, VA). Jansky Fellows are encouraged to spend time at universities working with collaborators during the course of their fellowship. In addition, up to 3 Jansky Fellowship appointments will be made annually for positions that may be located at a US university or research institute.
Application Procedure: Candidates normally commence in September or October. There is no application form. The initial letter should include a statement of the individual's research interests together with his or her own appraisal of his or her qualifications for carrying out research. Candidates should be single-sided with no staples. The application should have 3 letters of recommendation sent directly to the NRAO.
Closing Date: November 1st
Funding: Government
Contributor: The National Science Foundation
No. of awards given last year: 6
No. of applicants last year: 63
Additional Information: Research associates may formulate and carry out investigations either independently or in collaboration with others.

NATIONAL RESEARCH COUNCIL (NRC)

The National Academies, 500 Fifth Street NW, Washington, DC, 20001, United States of America
Tel: (1) 202 334 2000
Email: infofell@nas.edu
Website: www.nationalacademies.org

The National Research Council (NRC) is part of the National Academies, which also comprise the National Academy of Sciences, National Academy of Engineering and Institute of Medicine. They are private, non-profit institutions that provides science, technology and health policy advice under a congressional charter.

Ford foundation dissertation fellowships

Subjects: Literature/English/writing, history, foreign language, religion/theology, social sciences, political science, communications, physical sciences and mathematics and engineering-related technologies.
Purpose: To financially support underrepresented minorities in research-based fields of study.
Eligibility: Open to all citizens or nationals of the United States regardless of race, national origin, religion, gender, age, disability or sexual orientation (must have become a U.S. citizen by November 9th). Individuals with evidence of superior academic achievement (such as grade point average, class rank, honors or other designations). Individuals committed to a career in teaching and research at the college or university level, PhD or ScD degree candidates studying in an eligible research-based discipline at a U.S. educational institution andIndividuals who have not earned a doctoral degree at any time, in any field.
Level of Study: Postgraduate, Research
Type: Fellowships
Value: $21,000, and expenses paid to attend one Conference of Ford Fellows.
Frequency: Annual
Country of Study: United States of America
No. of awards offered: Approx. 20
Application Procedure: Applicants must submit their application form, transcript, essay and reference letters.
Closing Date: November 24th

NATIONAL RESEARCH COUNCIL OF CANADA (NRC)

1200 Montreal Road, Building M-58, Ottawa, ON, K1A 0R6, Canada
Tel: (1) 613 993 9101
Fax: (1) 613 952 9907
Email: info@nrc-cnrc.gc.ca
Website: www.nrc-cnrc.gc.ca
Contact: Research Associates Co-ordinator

The National Research Council of Canada (NRC) is a dynamic, nationwide research and development organization committed to helping Canada realize its potential as an innovative and competitive nation.

NRC Research Associateships

Subjects: Biological sciences, biotechnology, chemistry, molecular sciences, chemical engineering and process technologies, electrical engineering, astrophysics, industrial materials research, construction, mechanical engineering, aeronautics, physics, photonics, microstructural sciences, plant biotechnology, biochemistry, microbiology or advanced structural ceramics, bioinformatics, fuel cells, genomics, nanotechnology, nutraceuticals, proteomics, ocean engineering, aerospace, measurement standards, communication and technologies.
Purpose: To give promising scientists and engineers an opportunity to work in a challenging research environment during the early stages of their research careers.
Eligibility: Open to nationals of any country, although preference will be given to Canadians and permanent residents of Canada. Applicants should have acquired a PhD in natural science or a Master's degree in an engineering field within the last 5 years or should expect to obtain their degree before taking up the associateship. Selections will be made on a competitive basis with a demonstrated ability to perform original research of high quality in the chosen field as the main criterion.
Level of Study: Postgraduate
Value: The current annual PhD recruiting rate is $52,940 (plus allowance of $8,000).
Frequency: Annual
Study Establishment: Laboratories in the National Research Council of Canada
Country of Study: Any country
No. of awards offered: Approx. 50
Application Procedure: Applicants must fill out the online application form, available on the website www.careers-carrieres.nrc-cnrc.gc.ca
Closing Date: Applications are accepted at any time
Funding: Government
No. of awards given last year: 50
No. of applicants last year: 400
Additional Information: Salaries are revised annually. Further information is available from the website.

NATIONAL SCIENCE FOUNDATION (NSF)

Division of Earth Sciences, 4201 Wilson Boulevard, Arlington, VA, 22230, United States of America
Tel: (1) 703 292 5111
Fax: (1) 703 292 9025
Email: info@nsf.gov
Website: www.nsf.gov
Contact: Division Director

The National Science Foundation (NSF) supports research in the areas of geology, geophysics, geochemistry, paleobiology and hydrology, including interdisciplinary or multidisciplinary proposals that may involve one or more of these disciplines.

NSF Doctoral Dissertation Improvement Grants

Subjects: Biological sciences.
Purpose: To provide partial support to Doctoral dissertation research and allow Doctoral candidates to conduct research in specialized facilities or field settings away from the home campus.
Eligibility: Open to students who have advanced to candidacy for a PhD degree to be eligible to submit a proposal. Also US institutions that are eligible for awards from the NSF may submit proposals.
Level of Study: Doctorate
Type: Grant
Value: Approx. $2.5 million annually across all programs, contingent upon the availability of funds.
Length of Study: 2 years
Frequency: Annual
Country of Study: United States of America
Closing Date: November 8th (full proposal)
Contributor: National Science Foundation

NATIONAL SEA GRANT COLLEGE

NOAA/Sea Grant, R/SG 1315 East-West Highway SSMC-3, Eleventh Floor, Silver Spring, MD, 20910, United States of America
Tel: (1) 301 734 1066
Fax: (1) 301 713 0799
Email: oar.sg.fellows@noaa.gov
Website: www.seagrant.noaa.gov
Contact: Chelsea Lowes, Program Officer

Sea Grant is a nationwide network administered through the National Oceanic and Atmospheric Administration (NOAA) of 30 university-based programmes that work with coastal communities. The organizations research and programmes promote better understanding, conservation and use of America's coastal resources. In short, Sea Grant is science serving America's coasts.

Dean John A Knauss Fellowship Program

Subjects: Marine resource.
Purpose: To provide a unique educational experience to students who have an interest in ocean, coastal and Great Lakes resources and in the national policy decisions affecting those resources.
Eligibility: Open to all students regardless of citizenship. Applicants must be graduates or professionals in a marine or aquatic-related field at a United States accredited institution of higher education.
Level of Study: Postgraduate
Type: Fellowship
Value: US$52,500 per student
Length of Study: 1 year
Frequency: Annual
Country of Study: United States of America
Application Procedure: Please visit www.seagrant.noaa.gov/knauss
Closing Date: February 17th
Funding: Government

Sea Grant/NOAA Fisheries Fellowship

Subjects: Population dynamics and marine resource economics.
Purpose: To financially support and encourage qualified applicants to pursue careers in either population dynamics and stock assessment or in marine resource economics and also to increase available expertise related to these fields.
Eligibility: Applicants have to be PhD students in population dynamics or marine resource economics or related disciplines concentrating on the conservation and management of living marine resources.
Level of Study: Doctorate
Type: Fellowship
Value: $38,500 per year
Length of Study: 2–3years
Frequency: Annual
Closing Date: January 25th
Funding: Government

For further information contact:

Email: terry.smith@noaa.gov

THE NATIONAL SOCIETY OF HISPANIC MBAS (NSH MBA)

450 East John Carpenter Freeway, Suite 200, Irving, TX, 75062, United States of America
Tel: (1) 214 596 9338
Fax: (1) 214 596 9325
Email: jfarlinger@scholarshipamerica.org
Website: www.nshmba.org
Contact: Mr J Farlinger

The National Society of Hispanic MBAs, founded in 1988, is a non-profit organization. It exists to foster Hispanic leadership through management education and professional development.

NSHMBA Scholarship Program

Subjects: Business administration.
Purpose: To provide financial assistance to outstanding Hispanics pursuing a Master's degree.
Eligibility: Open to citizens of the United States or legal permanent residents of Hispanic heritage with a grade point average of 3.0 on a 4.0 scale and are enrolled or plan to enrol in a Master's degree programme.
Level of Study: MBA
Type: Scholarship
Value: Awards range from $2,500–10,000, and the top five full-time applicants who demonstrate financial need will receive an award for $10,000.
Frequency: Annual
Closing Date: April 30th
Additional Information: All queries should be directed to NSHMBA Scholarship Programme, Scholarship America – Julie Aretz, One Scholarship Way, PO Box 297, St Peter, MN 56082, United States of America, Tel: 507 931 1682.

NATIONAL TRAPPERS ASSOCIATION (NTA)

NTA Headquarters, 2815 Washington Avenue, Bedford, IN, 47421, United States of America
Tel: (1) 812 277 9670
Fax: (1) 812 277 9672
Email: ntaheadquarters@nationaltrappers.com
Website: www.nationaltrappers.com

The NTA, established in 1959, is an organization of dedicated individuals who have joined together to promote and protect the appropriate conservative use of our fur bearing species.

Charles L. Dobbins Memorial Scholarship

Subjects: Wildlife management.
Purpose: To encourage students majoring in a field of study pertaining to wildlife management or related topics.
Eligibility: Open to applicants who are members of a state or national trappers association.
Level of Study: Postgraduate
Type: Scholarship
Length of Study: US$500
Frequency: Annual
No. of awards offered: 12
Application Procedure: A completed application form must be submitted.
Closing Date: July 1st
Funding: Individuals

For further information contact:

National Trappers Association, 4170 St Clair, Fallon, NV 89406
Contact: Jim Curran

NATIONAL UNION OF TEACHERS (NUT)

Strategy & Communications, Hamilton House, Mabledon Place, London, WC1H 9BD, England
Tel: (44) 020 7388 6191
Fax: (44) 020 7387 8458
Email: a.bush@nut.org.uk
Website: www.teachers.org.uk
Contact: Ms Angela Bush

NUT Page Scholarship

Subjects: A specific aspect of American education relevant to the recipient's own professional responsibilities.
Purpose: To promote the exchange of educational ideas between Britain and America.
Eligibility: Open to teaching members of the NUT aged 25–60 years, although 25–55 is preferred.
Level of Study: Professional development
Type: Scholarship
Value: Each up to UK£1,700 pro rata daily rate with complete hospitality in the United States of America provided by the English-Speaking Union of the United States of America.
Length of Study: 2 weeks. The scholarship must be taken during the American academic year, which is September–May
Frequency: Annual
Country of Study: United States of America
No. of awards offered: 2
Application Procedure: Applicants must complete an application form. An outline and synopsis of the project must accompany the form along with a curriculum vitae and scholastic and personal testimonials.
Closing Date: December 22nd
Funding: Private
Contributor: NUT
No. of awards given last year: 2
No. of applicants last year: 100
Additional Information: The scholarship is limited to the individual teacher and neither the spouse nor partner can be included in the travel, accommodation or study arrangements. Recipients are required to report on their visit to teacher groups and educational meetings in the United States of America and on their return home.

NATIONAL UNIVERSITY OF IRELAND GALWAY

Postgraduate Admission Office, University Road, Galway, Ireland
Tel: (353) 91 524411
Fax: (353) 91 494501
Email: infoit.nuigalway.ie
Website: www.nuigalway.ie
Contact: Mairead Faherty

Charles Parsons Energy Research Award

Subjects: Microbial and biocatalytic fuel cell research.
Purpose: To focus on investigation and optimization of electron transfer reactions in biological fuel cells that can generate energy from diverse substrates. To focus the research on applications of pure- and mixed-culture microbial fuel cells, and biocatalytic enzyme-based fuel cells.
Eligibility: Open to engineering graduates.
Level of Study: Doctorate
Type: Research award
Value: salary scale €55,000 to €80,486 per year for researchers, stipend of €18,000 per year plus tuition fees for phd studentship and undergraduate engineering students €1,500 per month.
Application Procedure: Applicants should include a curriculum vitae and the names of two academic referees.
Additional Information: The research will involve liaison with international collaborators, bench research and reporting. To this end, good inter-personal, written communication and networking skills are advantageous.

For further information contact:

Email: donal.leech@nuigalway.ie
Contact: Dr Dónal Leech

Galway Scholarship

Subjects: Humanities, languages and literatures, and social and behavioral sciences.
Eligibility: Open to students who are in their first year of PhD programme. Students must register for one of the College's Structured PhD programmes. The award is not restricted by age, nationality or residence.
Level of Study: Doctorate
Type: Scholarship
Value: €15,000. If a candidate is in receipt of fees from another source, the award is €10,275
Length of Study: 4 years
Frequency: Annual
No. of awards offered: 20
Application Procedure: A completed application form and other required information must be sent. Please refer to the website for details.
Closing Date: See website

For further information contact:

Room 340, College of Arts, Social Sciences, and Celtic Studies, Arts Millennium Building, Galway, Ireland
Website: arts.scholarships@nuigalway.ie
Contact: Mairead Faherty

Helen M Moran Scholarship

Eligibility: The Scholarship will be awarded to one applicant from one of the following categories: a first-year student registered for any of the primary degree courses at National University of Ireland, Galway and who has demonstrated a high level of achievement in Gaelic Football with the potential to become a University representative player; a postgraduate student registered for any Master-level postgraduate course at National University of Ireland, Galway and who has demonstrated a high level of achievement in Gaelic Football at University representative level; an outstanding graduate registered for the MEd. degree course or MA degree, by research, in education at National University of Ireland, Galway.
Type: Scholarship
Frequency: Annual
No. of awards offered: 1
Application Procedure: Please contact university for further information.
Additional Information: Preference will be given to suitably qualified students who are also natives of Waterford.

NATIONAL UNIVERSITY OF IRELAND, MAYNOOTH

Research Support Office, Auxilia House, North Campus, NUI, Maynooth, Co. Kildare, Ireland
Tel: (353) 1 7086000
Fax: (353) 1 6289063
Email: research.support@nuim.ie
Website: www.nuim.ie

Following two centuries of internationally renowned scholarly activity on the Maynooth campus, the National University of Ireland, Maynooth was established under the 1997 Universities Act as an autonomous member of the federal structure known as the National University of Ireland. With approximately 8,400 registered students, NUI, Maynooth has 26 academic departments which are organized into three Faculties: Arts, Celtic Studies and Philosophy; Science and Engineering; and Social Sciences. Building on a tradition of scholarship and excellence in all aspects of its teaching, learning, and research activities, within the liberal arts and sciences tradition NUI, Maynooth is committed to being a first-class research-led centre of learning and academic discovery.

John and Pat Hume Postgraduate Scholarships

Subjects: Arts, humanities, social sciences, sciences and engineering.
Purpose: To build on excellence in areas across the arts, humanities, social sciences, sciences and engineering.
Eligibility: Applicants must have a First or Upper Second-Class Honours Primary Degree (or equivalent) from Ireland, the EU or from

any overseas university and intend to pursue a PhD degree at the University. Those who have commenced a research degree at NUI Maynooth prior to application will not be eligible.
Level of Study: Postgraduate, Research
Type: Scholarship
Value: €5,000 per year plus payment of fees at EU level. In some cases an additional fund of €3,000 is also provided to the student researcher for activities undertaken in support of the Department including tutorials and laboratory demonstration.
Length of Study: Up to 3 years
Frequency: Annual
Study Establishment: NUI Maynooth
Country of Study: Ireland
No. of awards offered: 38
Application Procedure: Applicants must first make contact with a NUI Maynooth department or centre to discuss their suitability for a PhD programme. A list of departmental contacts is available on the website. Application for the scholarship can then be filed.
Closing Date: May 6th
No. of awards given last year: 30
Additional Information: Supplement the scholarship with an additional €3000 for tutorial or demonstrating duties.

For further information contact:

Office of Research and Graduate Studies, NUI Maynooth, Ireland
Tel: (353) 1 708 6018
Fax: (353) 1 7083359
Email: pgdean@nuim.ie
Website: http://graduatestudies.nuim.ie

NATIONAL UNIVERSITY OF SINGAPORE (NUS)

21 Lower Kent Ridge Road, Singapore, 119077, Singapore
Tel: (65) 6516 6666
Fax: (65) 6775 9330
Email: gradenquiry@nus.edu.sg
Website: www.nus.edu.sg
Contact: Grants Management Officer

NUS aspires to be a dynamic connected knowledge community imbued with a no walls culture that promotes the free flow of talent and ideas. Individual members of our community enjoy access to diverse opportunities for intellectual and professional growth and in twin add value to NUS becoming a global knowledge enterprise.

Asian Development Bank-Japan Scholarship Program

Subjects: Public policy.
Purpose: To find further study in public policy implementation.
Eligibility: Open to residents of Asian Development Bank member countries currently enrolled at NUS. Upon completion of their study programmes, scholars are expected to contribute to the economic and social development of their home countries. Check website for further details.
Level of Study: Postgraduate
Type: Scholarship
Value: Singaporean $250 per semester (one-time book allowance), tuition, health insurance, examination and other approved fees. Cost of travel from home country to Singapore on award of the scholarship and from Singapore to home country on graduation.
Length of Study: 2 years for Master in public policy
Frequency: Annual
Study Establishment: Lee Kuan Yew School of Public Policy, National University of Singapore
Country of Study: Singapore
No. of awards offered: 5
Application Procedure: Application form with 3 pieces of your photograph attached, certificate of citizenship (or a copy of your valid passport), research and/or work experience, two confidential letters of recommendation
Closing Date: July 12th
Funding: Government
Contributor: Government
No. of awards given last year: 3
No. of applicants last year: 350

CapitaLand LKYSPP Scholarship

Subjects: Public administration.
Purpose: To find further study in public administration implementation.
Eligibility: Open to applicants of ASEAN countries (except Singapore), India, or the People's Republic of China.
Level of Study: Postgraduate
Type: Scholarship
Value: A monthly stipend, one-time book allowance, one-time settling-in allowance, shared housing, tuition, health insurance, examination and other approved fees. Cost of travel from home country to Singapore on award of the scholarship and from Singapore to home country on graduation
Length of Study: 1 year for Master in Public Administration
Frequency: Annual
Study Establishment: Lee Kuan Yew School of Public Policy, National University of Singapore
Country of Study: Singapore
No. of awards offered: 2
Application Procedure: Apply online.
Funding: Corporation
Contributor: CapitaLand
Additional Information: Please check website or contact university for updated information.

Hefner Scholarship

Subjects: Public administration.
Purpose: To find further study in public administration implementation.
Eligibility: Open to applicants of People's Republic of China or the United States of America.
Level of Study: Postgraduate
Type: Scholarship
Value: A monthly stipend, one-time book allowance, one-time settling-in allowance, shared housing, tuition, health insurance, examination and other approved fees. Cost of travel from home country to Singapore on award of the scholarship and from Singapore to home country on graduation
Length of Study: 1 year for Master in Public Administration
Frequency: Annual
Study Establishment: Lee Kuan Yew School of Public Policy, National University of Singapore
Country of Study: Singapore
No. of awards offered: 2
Application Procedure: Apply online.
Closing Date: Refer to website
Funding: Private
Contributor: Hefner

Law/Faculty Graduate Scholarship (FGS)

Subjects: Law.
Purpose: To reward an outstanding student of the faculty of Law.
Eligibility: Outstanding applicants of any nationality (including Singapore citizens and permanent residents) may be awarded the FGS to pursue the LLM coursework degrees: LLM, LLM (Asian Legal Studies), LLM (Corporate & Financial Services Law), LLM (Intellectual Property & Technology Law), LLM (International & Comparative Law), LLM (Maritime Law).
Level of Study: Postgraduate
Type: Scholarship
Value: The scholarship will cover tuition fees.
Frequency: Annual
Study Establishment: National University of Singapore
Country of Study: Singapore
Additional Information: Terms of award are subject to change without prior notice.

Lee Kong Chian Graduate Scholarships

Subjects: Any subject.
Purpose: To reward proven academic excellence, leadership and exceptional promise.
Eligibility: The Scholarships are open to students of all nationalities who gain admission to any PhD programme at the University.
Level of Study: Postdoctorate
Type: Scholarship

Value: A monthly stipend of Singaporean $3,300, tuition, examination fees and other approved fees at NUS, an annual book allowance of Singaporean $500, a one-off air travel allowance of 2 return tickets of up to Singaporean $4,000 (only for overseas students subject to a maximum of Singaporean $2,000 per ticket), and a one-off laptop allowance of Singaporean $1,500.
Length of Study: The award is tenable for 1 year in the first instance; but subject to the scholar's satisfactory progress, it may be renewed each semester. The maximum period of award is 4 years.
Frequency: Annual
Study Establishment: National University of Singapore
Country of Study: Singapore
No. of awards offered: Upto 5
Application Procedure: Apart from the other supporting documents required for PhD admission, candidates interested in the Scholarship must also submit a personal essay and a record of co-curricular activities or community service.
Closing Date: November 15th
Funding: Foundation
Contributor: Lee Foundation

Lee Kuan Yew School of Public Policy Graduate Scholarships (LKYSPPS)

Subjects: Public policy and public administration.
Purpose: To find further study in public policy and administration implementation.
Eligibility: Open to all nationalities (except Singapore).
Level of Study: Postgraduate
Type: Scholarship
Value: A monthly stipend, a one-time book allowance, a one-time settling-in allowance, shared housing, tuition, health insurance, examination and other approved fees, cost of travel from home country to Singapore on award of the scholarship and from Singapore to home country on graduation.
Length of Study: 1 year (for public administration) and 2 years (for public policy)
Frequency: Annual
Study Establishment: Lee Kuan Yew School of Public Policy, National University of Singapore
Country of Study: Singapore
Application Procedure: Apply online.
Closing Date: Refer to website
Funding: Government
Contributor: Government

For further information contact:

Email: LKYSPPmpp@nus.edu.sg

Lien Foundation Scholarship for Social Service Leaders

Subjects: Public administration.
Purpose: To find further study in public administration implementation.
Eligibility: Applicants who are staff of VWOs, applicants who are members of the public, should be a Singapore Citizen or Permanent Residen
Level of Study: Postgraduate
Type: Scholarship
Value: The award includes tuition fees, maintenance allowance, book allowance, and any other compulsory fees.
Length of Study: 1 year for Master in Public Administration
Frequency: Annual
Study Establishment: Lee Kuan Yew School of Public Policy, National University of Singapore
Country of Study: Singapore
Application Procedure: Apply online.
Closing Date: June 22nd
Funding: Private
Contributor: Hefner

National University of Singapore Research Scholarships

Subjects: Any subject.
Purpose: To reward outstanding graduates for research leading to a higher degree at the university.
Eligibility: Applicants must be university graduates with at least a Class II Honours degree or equivalent and, at the time of award of the Scholarship, must have been offered admission as a candidate for a full-time higher degree by research at NUS.
Level of Study: Postgraduate
Type: Scholarship
Value: Monthly stipends for Singapore Citizens, Singapore Permanent Residents and International Students are Singaporean $2,500, Singaporean $2,200 and Singaporean $2,000 respectively
Length of Study: 2–4 years
Frequency: Annual
Study Establishment: National University of Singapore
Country of Study: Singapore
Additional Information: Research scholars may be asked to assist in departmental work for which they can earn up to Singapore $16,000 (gross) per year at the current rate of remuneration. However, please note that the university does not guarantee employment to research scholars upon completion of their candidature.

NUS Graduate Scholarship for Asean Nationals

Subjects: Engineering, computing, science, law, dentistry, medicine, arts & social sciences, design and environmental studies.
Purpose: To enable a successful scholar to pursue full-time study.
Eligibility: The University offers these Scholarships each academic year to citizens/permanent residents of a member country of ASEAN (except Singapore) to finance the pursuit of the LLM coursework degrees: LLM, LLM (Corporate & Financial Services Law), LLM (Intellectual Property & Technology Law), LLM (International & Comparative Law). Applicants must have all of the following: an excellent academic record (at least be in the top 10 per cent of the class); a very good command of the English Language (a minimum iBT TOEFL score of 100 or IELTS of 7)); at least 2 years of relevant working experience; LLM (International & Comparative Law).
Level of Study: Postgraduate
Type: Scholarship
Value: The scholar will be provided with: a monthly stipend of Singaporean $1,350 throughout the period of the award (inclusive of the University's vacations); a one-time book allowance of $500; tuition, health insurance, examination fees, sports fee and other approved fees, allowances and expenses at NUS and throughout the period of the award; cost of economy travel directly from the home country to Singapore upon award of the scholarship; and cost of economy travel directly from Singapore to the home country upon successful completion of the course of study leading to the award of the degree.
Length of Study: 2–3 years
Frequency: Annual
Study Establishment: National University of Singapore
Country of Study: Singapore
No. of awards offered: 40
Application Procedure: The Registrar will invite applications at the appropriate time. Applications must be submitted on prescribed forms available from the respective Faculty/School.
Funding: Government

OCBC International Master in Public Policy Scholarship

Subjects: Public policy.
Purpose: To find further study in public policy implementation.
Eligibility: Students pursuing a Master's degree in the Public Policy Programme and Citizens of Malaysia, Indonesia and China
Level of Study: Postgraduate
Type: Scholarship
Value: A monthly living allowance of $1,350, one-time settling-in allowance of $500, one-time book allowance of $500
Length of Study: 2 years for Master in Public Policy
Frequency: Annual
Study Establishment: Lee Kuan Yew School of Public Policy, National University of Singapore
Country of Study: Singapore
No. of awards offered: 3
Application Procedure: Apply online.
Closing Date: Refer to website
Funding: Corporation
Contributor: OCBC

Rodamas – LKYSPP Scholarship

Subjects: Public administration.

Purpose: PT Rodamas Company and the Lee Kuan Yew School of Public Policy offer a joint scholarship to students who are citizens of Indonesia.
Eligibility: Open to Indonesian applicants.
Level of Study: Postgraduate
Type: Scholarship
Value: A monthly stipend, one-time book allowance, one-time settling-in allowance, shared housing, tuition, health insurance, examination and other approved fees. Cost of travel from home country to Singapore on award of the scholarship and from Singapore to home country on graduation
Length of Study: 1 year for Master in Public Administration
Frequency: Annual
Study Establishment: Lee Kuan Yew School of Public Policy, National University of Singapore
Country of Study: Singapore
No. of awards offered: 1
Application Procedure: Apply online.
Closing Date: Refer to website
Funding: Corporation
Contributor: PT Rodamas

Sequislife – LKYSPP Scholarship

Subjects: Public administration.
Purpose: To find further study in public administration implementation.
Eligibility: Open to applicants from Indonesia.
Level of Study: Postgraduate
Type: Scholarship
Value: A monthly stipend, one-time book allowance, one-time settling-in allowance, shared housing, tuition, health insurance, examination and other approved fees. Cost of travel from home country to Singapore on award of the scholarship and from Singapore to home country on graduation
Length of Study: 1 year for Master in Public Administration
Frequency: Annual
Study Establishment: Lee Kuan Yew School of Public Policy, National University of Singapore
Country of Study: Singapore
No. of awards offered: 1
Application Procedure: Apply online.
Closing Date: Refer to website
Funding: Corporation
Contributor: PT Sequislife

Singapore-MIT Alliance Graduate Fellowship

Subjects: Any subject.
Purpose: The SMA Graduate Fellowship is established by the Singapore Ministry of Education in January 2009 to attract the best and most talented PhD students from Singapore, the region and beyond, and educate them to be future leaders in the areas of science and technology. The selection of candidates will take place twice a year, in time for the start of the semesters in August and January.
Eligibility: The Scholarships are open to students of all nationalities who gain admission to any PhD programme at the University whose research interest fits within one or more of the projects currently being carried out in one of the SMART Interdisciplinary Research Groups (IRGs).
Level of Study: Graduate, Postgraduate
Type: Fellowship
Value: A monthly stipend of S$3,200; Tuition fees at NUS; and Scholarship allowance of up to $12,000 to help cover the expenses associated with a six-month research residency at MIT.
Length of Study: The award is tenable for 1 year in the first instance; but subject to the scholar's satisfactory progress, it may be renewed each semester. The maximum period of award is four years.
Frequency: Annual
Study Establishment: National University of Singapore and Nanyang Technological University
Country of Study: Singapore
Application Procedure: Applicants must apply separately to both MIT and NUS/NTU for the dual degrees and only to NUS or NTU for direct Phd degree; applicants must also apply directly to SMA for an SMA Graduate Fellowship.
Closing Date: Between January and March
Funding: Government
Contributor: A*Star, Economic and Development Board (EDB), Ministry of Education (MOE), National University of Singapore (NUS) and Nanyang Technological University (NTU)
No. of applicants last year: 120

For further information contact:

Tel: 6516 4787
Fax: 6775 2920
Email: smart@nus.edu.sg
Website: www.sma.nus.edu.sg

Standard Chartered LKYSPP Scholarship

Subjects: Public administration and public policy.
Purpose: To find further study in public administration and public policy implementation.
Eligibility: Open to ASEAN countries (except Singapore), India, the People's Republic of China or the United Arab Emirates.
Level of Study: Postgraduate
Type: Scholarship
Value: A monthly stipend, one-time book allowance, one-time settling-in allowance, shared housing, tuition, health insurance, examination and other approved fees. Cost of travel from home country to Singapore on award of the scholarship and from Singapore to home country on graduation
Length of Study: 1 year for Master in Public Administration and, 2 years for Master in Public Policy
Frequency: Annual
Study Establishment: Lee Kuan Yew School of Public Policy, National University of Singapore
Country of Study: Singapore
No. of awards offered: 2
Application Procedure: Apply online.
Closing Date: Refer to website
Funding: Corporation
Contributor: Standard Chartered

Temasek Scholarship

Subjects: Public policy.
Purpose: To find further study in public policy implementation.
Eligibility: Open to nationals of permanent residents of member countries of ASEAN (except Singapore) and APEC (except Singapore).
Level of Study: Postgraduate
Type: Scholarship
Value: Tuition fees (based on MOE tuition grants rates), a living allowance of S$500 per month, a housing allowance of up to $350 per month (where applicable) and one way economy air fare to Singapore of up to S$1800 (where applicable).
Length of Study: 2 years for Master in public policy
Frequency: Annual
Study Establishment: Lee Kuan Yew School of Public Policy, National University of Singapore
Country of Study: Singapore
Closing Date: December 1st to February 29th (Polytechnic Applications); December 1st to March 9th (University Applications)
Funding: Government
Contributor: Government

NATIONAL WILDLIFE FEDERATION (NWF)

National Wildlife Federation P.O. Box 1583, Reston, VA, 22116 1583, United States of America
Tel: (1) 1 800 822 9919
Email: campus@nwf.org
Website: www.nwf.org
Contact: Director of Research

For more than a decade, National Wildlife Federation (NWF) has been helping to transform the nation's college and university campuses into living models of an ecologically sustainable society and training a new generation of environmental leaders. Campus Ecology supports and promotes positive and practical conservation projects on campus and beyond to protect wildlife by restoring habitat and slowing global warming.

Campus Ecology Fellowship Program

Subjects: Building design, composting, dining services, energy, landscaping, management systems, purchasing, transportation and waste reduction or water.
Purpose: To provide the opportunity to students to pursue their vision of an ecologically sustainable future, identify and implement innovative greening initiatives as relevant to their campus and community.
Eligibility: Open to former NWF interns following 1 year from their final work date.
Level of Study: Postgraduate
Type: Fellowships
Value: Up to $1,000
Frequency: Annual
Country of Study: United States of America
Closing Date: January 15th
Contributor: National Wildlife Federation

NATURAL ENVIRONMENT RESEARCH COUNCIL (NERC)

Polaris House, North Star Avenue, Swindon, Wiltshire, SN2 1EU, England
Tel: (44) 017 9341 1500
Fax: (44) 017 9341 1501
Email: aval@nerc.ac.uk
Website: www.nerc.ac.uk/funding
Contact: Dr A E Allman, Process Manager

The Natural Environment Research Council (NERC) is one of the seven United Kingdom Research Councils that fund and manage research in the United Kingdom. NERC is the leading body in the United Kingdom for research, survey, monitoring and training in the environmental sciences. NERC supports research and training in universities and in its own centres, surveys and units.

NERC Independent Research Fellowships (IRF)

Subjects: Sciences of the natural science.
Purpose: To develop scientific leadership among the most promising early-career environmental scientists, by giving all Fellows five years' support, which will allow them sufficient time to develop their research programmes, and to establish international recognition.
Eligibility: Open to any nationality, and may be held in any area of the NERC remit, but the fellowship must be based at an eligible UK Research Organisation. Applicants may not have a permanent academic position in a university or equivalent organization. Applicants must expect to submit their PhD thesis before the fellowship interview would take place (April following the closing date) and, if successful, would not be able to take up the fellowship until the intent to award the PhD has been confirmed by the awarding university. Applicants may have up to a maximum of eight years of full-time postdoctoral research experience between the PhD certificate date and the closing date of the fellowship competition to which they are applying. The eight year window is based on full-time working. Where applicants have worked part-time or had research career breaks, the eight year window would be extended accordingly.
Type: Fellowship
Value: Includes 80 per cent of the full economic cost (FEC) of the proposal. NERC will provide funding for the fellow's salary costs in line with the agreed pay scales at the time of the award, with provision for future years
Length of Study: 5 years
Frequency: Annual
Study Establishment: Universities and other approved research institutes
No. of awards offered: 20
Application Procedure: Please refer to the Research Grants and Fellowships Handbook at http://www.nerc.ac.uk/funding/application/researchgrants/.
Closing Date: Early November (please check NERC website for exact date)
Funding: Government

NERC Research (PhD) Studentships

Subjects: Environmental sciences.
Purpose: To enable students to undertake research in particular scientific area under the guidance of named supervisors leading to the submission of a PhD thesis.
Eligibility: Open to British and EU citizens. EU citizens can only receive fees unless they have been resident in the UK for the preceding 3 years in which case they are eligible for a full award. Other individuals with settled status in the UK may also be eligible. Candidates must hold an honours degree in an appropriate branch of science or technology. For research studentships (PhD/MPhil) this should be a first or upper second class honours degree.
Level of Study: Postgraduate
Type: Studentship
Value: Stipend/maintenance grant now is £13,590 (£15,590 in London)
Length of Study: Up to 3.5 years, part of which may be spent at an overseas institution
Frequency: Annual
Study Establishment: Any approved Institute of Higher Education
Country of Study: Other
No. of awards offered: Approx. 300
Application Procedure: Individuals need to apply directly to departments that have NERC funding. Check website for further details.
Funding: Government
No. of awards given last year: 300
Additional Information: The list of departments with NERC PhD funding can be found at: www.nerc.ac.uk/funding/available/postgrad/awards/

NATURAL HISTORY MUSEUM

Cromwell Road, London, SW7 5BD, England
Tel: (44) 20 7942 5011
Email: l.wylde@nhm.ac.uk
Website: www.nhm.ac.uk/science
Contact: Mrs G Maldar, Liaison Officer

The Natural History Museum's mission is to maintain and develop its collections and use them to promote the discovery, understanding, responsible use and enjoyment of the natural world.

Synthesys Visiting Fellowship

Subjects: Biological and Earth science
Purpose: To provide access for researchers so that they can undertake short visits to utilize the facilities of 20 major museums and botanic gardens within Europe, including the Natural History Museum and its associates, the Royal Botanical Gardens, Kew and the Royal Botanic Garden Edinburgh.
Eligibility: Open to applicants from the European Union member states, and associated and accession states.
Level of Study: Research, Unrestricted, Doctorate, Graduate, Postdoctorate, Postgraduate, Predoctorate
Type: Fellowship
Value: International travel, accommodation, local travel and subsistence along with all access and facility costs
Length of Study: Up to 60 working days
Frequency: Annual
Study Establishment: The Natural History Museum, London
Country of Study: United Kingdom
No. of awards offered: 30 per call
Application Procedure: Online applications must be completed. Website: www.synthesys.info
Closing Date: Please refer to the website
Funding: Government
Contributor: The European Union Programme
No. of awards given last year: 526
No. of applicants last year: 220
Additional Information: During the visit the user is assigned a host, according to their speciality. The role of the host is to familiarize the user with the department, collections and facilities and to give training where necessary. The visits are often collaborative, in which case the host will work directly with the user.

For further information contact:

Website: www.synthesys.info

NATURAL SCIENCES AND ENGINEERING RESEARCH COUNCIL OF CANADA (NSERC)

350 Albert Street, Ottawa, ON, K1A 1H5, Canada
Tel: (1) 613 995 4273
Fax: (1) 613 992 5337
Email: claire.mcaneney@nserc-crsng.gc.ca
Website: www.nserc.ca
Contact: Corporate Account Executive

NSERC is Canada's instrument for promoting and supporting university research in the natural sciences and engineering, other than the health sciences. NSERC supports both basic university research through discovery grants and project research through partnerships among universities, governments and the private sector as well as the advanced training of highly qualified people.

Canada Postgraduate Scholarships (PGS)

Subjects: Natural sciences and engineering.
Eligibility: Open to a Canadian citizen or a permanent citizen of Canada, with a university degree in science or engineering, intending to pursue year full-time graduate study and research at the Master's or Doctorate level in one of the areas supported by NSERC with a first-class average in each of the last two completed years of study.
Level of Study: Postgraduate
Type: Fellowship
Value: Canadian $17,300 (Masters) per year for 1 year and Canadian $21,000 (Doctoral) per year for a period of 24–36 months.
Length of Study: 1 year awards.
Frequency: Annual
Country of Study: Canada
No. of awards offered: 2
Application Procedure: Check website for further details.
Closing Date: October 15th
Contributor: Natural Sciences and Engineering Research Council of Canada (NSERC)

For further information contact:

National Sciences and Engineering Research Council of Canada (NSERC), Scholarships and Fellowships Division, 350 Albert Street (for courier mailings, add 10th Floor), Ontario, Ottawa, K1A 1H5, Canada
Fax: (1) 613 996 2589
Email: schol@nserc.ca
Website: www.nserc.gc.ca

NSERC Postdoctoral Fellowships

Subjects: Engineering and natural sciences.
Purpose: To provide support to a core of the most promising researchers at a pivotal time in their careers. The fellowships are also intended to secure a supply of highly qualified Canadians with leading edge scientific and research skills for Canadian industry, government and universities.
Eligibility: Open to Canadian citizens or permanent residents residing in Canada, who have recently received, or will shortly receive a PhD from a Canadian university in one of the fields of research that NSERC supports.
Level of Study: Postdoctorate
Type: Fellowship
Value: Canadian $40,000 per year
Length of Study: 1 year, renewable for 1 additional year
Frequency: Annual
Study Establishment: A university or research institution of the Fellow's choice
Country of Study: Any country
No. of awards offered: 200–260
Application Procedure: Applicants must complete Form 200. Information is available on request.
Closing Date: October 15th
Funding: Government
No. of awards given last year: 254
No. of applicants last year: 1,097
Additional Information: The information provided here is subject to change. Please visit the NSERC's website for up-to-date information.

NETHERLANDS ORGANIZATION FOR INTERNATIONAL CO-OPERATION IN HIGHER EDUCATION (NUFFIC)

Nuffic, Po Box 29777, 2502 LT, The Hague, Netherlands
Tel: (31) 70 4260260
Fax: (31) 70 4260399
Email: nuffic@nuffic.nl
Website: www.nuffic.nl
Contact: Ms Rosalien van Santen, Information Officer

Since its founding in 1952, the Netherlands Organization for International Co-operation in Higher Education (NUFFIC) has been an independent, non-profit organization. Its mission is to foster international co-operation in higher education. Special attention is given to development co-operation.

NFP Netherlands Fellowships Programme for Development Co-operation

Subjects: All subjects offered by the Institutes for International Education in the Netherlands.
Purpose: To develop human potential through education and training mainly in the Netherlands with a view to diminishing qualitative and quantitative deficiencies in the availability of trained manpower in developing countries.
Eligibility: Open to candidates who have the education and work experience required for the course as well as an adequate command of the language in which it is conducted. This is usually English but sometimes French. The age limit is 40 for men and 45 for women. It is intended that candidates, upon completion of training, return to their home countries and resume their jobs. When several candidates with comparable qualifications apply, priority will be given to women. Candidates for a fellowship must be nominated by their employer and formal employment should be continued during the fellowship period.
Level of Study: Postgraduate
Type: Fellowship
Value: Normal living expenses, fees and health insurance. International travel expenses are provided only when the course lasts 3 months or longer
Length of Study: The duration of the course
Country of Study: The Netherlands
No. of awards offered: Varies
Application Procedure: Applicants must contact the Netherlands Embassy in their own country for information on nationality eligibility and on the application procedure. Information on the courses for which the fellowships are available can be obtained from the website.
Funding: Government
Additional Information: As a rule the candidate's government is required to state its formal support, except in the case of certain development-orientated non-government organizations. Further information is available from the website www.studyin.nl

NUFFIC-NFP Fellowships for Master's Degree Programmes

Subjects: Any subject on the list of eligible Master's degree programmes.
Purpose: To allow candidates to receive a postgraduate education and to earn either an MA, MSc or Professional Master's degree.
Eligibility: Applicant must be a national of one of 57 developing countries and have been admitted by a Dutch institution to one of the Master's degree programmes on the course list. Applicants who have received their education in any language other than English must provide International English Language Testing System scores (at least 5.5) or Test of English as a Foreign Language scores (at least 550).
Level of Study: Postgraduate
Type: Fellowship
Value: Monthly allowance of €485–970, cover cost of living, tuition fee
Length of Study: 9 months to 2 years
Frequency: Annual
Study Establishment: A Dutch institution
Country of Study: The Netherlands
Application Procedure: Applicants may apply for an NFP fellowship through the Netherlands embassy or consulate in their own country. To do this, applicants must complete an NFP Master's degree programme application form and submit it together with all the

required documents and information to the embassy or consulate well before the deadline. Forms can be obtained from the embassy or consulate or downloaded from the website, www.nuffic.nl/nfp
Closing Date: February1st and May 1st
Additional Information: Applicants may not be employed by a large industrial, commercial and/or multinational firms.

For further information contact:

Contact the Netherlands embassy or consulate in home country

NUFFIC-NFP Fellowships for PhD Studies

Purpose: To allow candidates to pursue a PhD at one of 18 Dutch universities and institutes for international education.
Eligibility: Candidate must be a national of one of 61 developing countries and have been admitted to a Dutch institution as a PhD fellow. Priority will be given to female candidates and candidates from sub-Saharan Africa.
Level of Study: Doctorate
Type: Fellowship
Value: Monthly allowance (€595–1190)
Length of Study: 4 years
Frequency: Annual
Study Establishment: Any one of 18 Dutch universities and institutes for international education. See list on website
Country of Study: The Netherlands
Application Procedure: After being accepted for admission to a Dutch institution, the candidate may submit a request for a PhD fellowship. Applicant must present a completed NFP PhD study application form to the Netherlands embassy or consulate in his/her own country. The application must be accompanied by the necessary documentation and by a research proposal that is supported by the supervisor(s). Form can be downloaded from website www.nuffic.nl/nfp
Closing Date: February 1st
Additional Information: A large portion of the PhD research must take place in the candidate's home country.

For further information contact:

Contact the Netherlands embassy or consulate in home country

THE NETHERLANDS ORGANIZATION FOR SCIENTIFIC RESEARCH (NWO)

Lann van Nieuw Oost Indie 300, PO Box 93138, The Hague, NL-2509 AC, Netherlands
Tel: (31) 70 344 0640
Fax: (31) 70 385 0971
Email: nwo@nwo.nl
Website: www.nwo.nl
Contact: F.A.O. Grants Department

The Netherlands Organization for Scientific Research (NWO) is the central Dutch organization in the field of fundamental and strategic scientific research. NWO encompasses all fields of scholarship and consequently plays a key role in the development of science, technology and culture in the Netherlands. NWO is an independent organization that acts as the national research council in the Netherlands. NWO is the largest national sponsor of fundamental scientific research undertaken in the 13 Dutch universities and provides many types of funding for research driven by intellectual curiosity.

Dutch Russian Scientific Cooperation Programme

Subjects: Scientific research.
Purpose: To give a strong impulse to the scientific collaboration between the two countries.
Eligibility: Open to all talented Dutch researchers together with a Russian counterpart.
Level of Study: Postgraduate
Type: Fellowship
Value: €500,000
Length of Study: 5 years
No. of awards offered: 2
Closing Date: September 24th
Contributor: The Netherlands Organization for Scientific Research

Rubicon Programme

Subjects: Scientific research.
Purpose: To encourage talented researchers at Dutch Universities to dedicate themselves to a career in postdoctoral research.
Eligibility: Open to researchers from all scientific disciplines engaged in PhD research or obtaining a PhD in the last 12 months
Level of Study: Postdoctorate
Type: Grant
Value: €5.3 million a year
Length of Study: Up to 2 years
Application Procedure: A completed application form to be submitted via NOW's electronic submission system Iris.
Closing Date: September 1st
Contributor: The Netherlands Ministry of Education, Culture and Science
Additional Information: Total amount for each 2011 Rubicon round is €2.9 million.

For further information contact:

Tel: (0)70-3440 565
Email: rubicon@nwo.nl
Website: www.nwo.nl/rubicon
Contact: Coordinator Rubicon

WOTRO Integrated Programmes

Subjects: Development relevant research beyond MDG or thematic in the fields of poverty and hunger; global health and health systems; sustainable environment; global relationships.
Purpose: To support excellent demand-driven interdisciplinary research programmes to the benefit of development and societal issues in the South.
Eligibility: Applications must be submitted by a senior researcher affiliated at a Dutch academic institute together with a senior researcher from a developing country; the research team must include at least one (postdoc or PhD) researcher from a developing country; researchers must have appropriate degrees.
Level of Study: Doctorate, Postdoctorate
Type: Grant
Value: Salary costs, living allowances, communication and research costs to a maximum of €700,000
Length of Study: Up to 5 years (combining at least 2 doctorate and/or postdoctorate projects)
Frequency: Biannual
Application Procedure: Twice a year preliminary applications are selected. Selected applicants receive a grant for elaborating the application into a full proposal, in joint collaboration with scientific and non-scientific stakeholders.
Closing Date: April and September
Contributor: WOTRO-Science for Global Development
Additional Information: Awarded programmes can usually start one year after submitting the preliminary application.

THE NEUROBLASTOMA SOCIETY

2 Caesar Court, Moss Street, York, YO23 1DD, England
Tel: (44) 01904 633744
Email: chairman@neuroblastoma.org.uk
Website: www.neuroblastoma.org.uk
Contact: Mr Stephen Smith, Chairman and Grant Administration

The Neuroblastoma Society was started in 1982 by a group of parents with children affected by neuroblastoma. The aim is to raise money to fund medical research towards better treatment and an eventual cure for this aggressive childhood tumour. The Society also aims to offer support to families affected by the disease.

Neuroblastoma Society Research Grants

Subjects: Paediatric oncology; diagnostic and therapeutic radiography; genetic research and gene therapy; immunotherapy; all specifically related to neuroblastoma.
Purpose: To fund clinical research towards improvements in treatment and a cure for neuroblastoma.
Eligibility: There are no age or nationality restrictions but candidates must be based in the United Kingdom or Republic of Ireland.
Type: Grant

Value: Up to UK£50,000 per year, depending on nature of research
Length of Study: 2–3 years
Frequency: Every 2 years
Study Establishment: A reputable research institution, usually a university or hospital
Country of Study: United Kingdom
No. of awards offered: 3–4
Application Procedure: Applicants must apply to the Society for the terms of grant application. Application forms available on website.
Closing Date: Mid January
Funding: Private
Contributor: Members and supporters of the Society
No. of awards given last year: 6
No. of applicants last year: 16
Additional Information: Please check further details at www.neuroblastoma.org.uk.

THE NEW JERSEY STATE FEDERATION OF WOMEN'S CLUBS OF GFWC

NJSFWC Headquarters, 55 Labor Center Way, New Brunswick, NJ, 08901, United States of America
Tel: (1) 732 249 5474
Email: njsfwc@njsfwc.org
Website: www.njsfwc.org

The New Jersey State Federation of Women's Clubs is the largest volunteer women's service organization in the state and a member of the General Federation of Women's Clubs, which provides opportunities for education, leadership training and community service through participation in local clubs and enabling members to make a difference in the lives of others.

Margaret Yardley Fellowship

Subjects: All subjects.
Purpose: To help female students who are in financial need.
Eligibility: Open to female graduate or doctoral students, whose residence is in New Jersey, USA.
Level of Study: Doctorate, Postgraduate
Type: Fellowship
Value: $5,000 towards tuition for study in a United States institution.
Frequency: Annual
No. of awards offered: 6–8
Application Procedure: Applicants must request applications prior to February 1st and requests must include a self-addressed, stamped envelope.
Closing Date: March 1st
Funding: Private
No. of awards given last year: 7
No. of applicants last year: 9

THE NEW JERSEY WATER ENVIRONMENT ASSOCIATION (NJWEA)

PO Box 1212, Fair Lawn, New Jersey, NJ, 07410, United States of America
Tel: (1) 201 296 0021
Fax: (1) 201 296 0031
Email: strom@aesop.rutgers.edu
Website: www.njwea.org

The New Jersey Water Environment Association (NJWEA) is a non-profit educational organization dedicated to preserving and enhancing the water environment. NJWEA was founded in 1915 and is one of the oldest organization in the United States of America. With a membership of 2,800 engineers, operators, scientists, students and other professionals, the NJWEA is an environmental leader in New Jersey.

NJWEA Scholarship Award Program

Subjects: Environmental science and engineering.
Purpose: To further encourage highly capable individuals to continue studies in environmental science or engineering with a strong component in one or more areas of water pollution control and environmental protection or hazardous waste management.
Eligibility: Open to students who are enrolled in a full-time programme leading to a degree in environmental science, environmental engineering or a closely related field with an emphasis on appropriate technical aspects of environmental protection or water pollution control.
Level of Study: Postgraduate
Type: Scholarship and award
Value: $2,000–2500 and totaling over $43,500 annually
Frequency: Annual
Country of Study: United States of America
No. of awards offered: 23
Application Procedure: Application forms can be downloaded from the website.
Closing Date: March 1st
Contributor: New Jersey Water Enviroment Association
Additional Information: The awards are based on academic performance, merit, scholastic ability and demonstrated interest in environmental science or engineering.

NEW SOUTH WALES ARCHITECTS REGISTRATION BOARD

Level 2, 156 Gloucester Street, Sydney, NSW, 2000, Australia
Tel: (61) 2 9241 4033
Fax: (61) 2 9241 6144
Email: mail@architects.nsw.gov.au
Website: www.architects.nsw.gov.au
Contact: Ms Mae Cruz, Deputy Registrar

Byera Hadley Travelling Scholarships

Subjects: Architecture.
Purpose: To allow candidates to undertake a course of study, research or other activity approved by the Board as contributing to the advancement of architecture.
Eligibility: Open to graduates or students of four accredited schools of architecture in New South Wales. Applicants must be Australian citizens.
Level of Study: Graduate, Postgraduate, Research
Type: Scholarship
Value: The Byera Hadley Travelling Scholarships enable winners to undertake a course of travel, study, research or other activity approved by the Board, which contributes to the advancement of architecture.The value of each Scholarship will range between $15,000 and $20,000
Frequency: Annual
Country of Study: Any country
No. of awards offered: 7
Application Procedure: Applicants must write for details.
Closing Date: July 30th
Funding: Private
Contributor: A bequest from the estate of the late Byera Hadley, an Australian architect
No. of awards given last year: 6
No. of applicants last year: 15
Additional Information: A report suitable for publication must be submitted within 12 months of the date of the award.

NSW Architects Registration Board Research Grant

Subjects: Any architectural topic approved by the board.
Purpose: To provide assistance to those wishing to undertake research on a topic approved by the Board to contribute to the advancement of architecture.
Eligibility: Open to candidates who are registered as architects in New South Wales.
Level of Study: Professional development
Value: Australian $20,000
Length of Study: 1 year
Frequency: Every 2 years
Country of Study: Australia
No. of awards offered: 1
Application Procedure: Applicants must write for details.
Closing Date: March 8th
Funding: Government
No. of awards given last year: None

Additional Information: A report is to be submitted upon completion of tenure.

NEW SOUTH WALES MINISTRY OF THE ARTS

Level 9 St James Centre, 111 Elizabeth Street PO Box A226, Sydney, NSW, 1235, Australia
Tel: (61) 1800 358 594, 02 8218 2222
Fax: (61) 2 92284722
Email: mail@arts.nsw.gov.au
Website: www.arts.nsw.gov.au

New South Wales Ministry of the Arts works closely with the State's 8 major cultural institutions, providing policy advice to Government on their operations.

Western Sydney Artists Fellowship

Subjects: Creative arts.
Purpose: To encourage artists and students in the field of creative arts.
Eligibility: Open to applicants who are residents of Western Sydney or whose practice is located primarily in Western Sydney.
Level of Study: Postgraduate
Type: Fellowship
Value: Australian $5,000–25,000
Length of Study: 1 year
Frequency: Annual
Study Establishment: New South Wales, Sydney Western Suburbs
Country of Study: Australia
Closing Date: September

NEW YORK FOUNDATION FOR THE ARTS (NYFA)

20 Jay Street, 7th floor, Brooklyn, NY, 11201, United States of America
Tel: (1) 212 366 6900
Fax: (1) 212 366 1778
Email: fellowships@nyfa.org
Website: www.nyfa.org

New York Foundation for the Arts (NYFA), founded in 1971, helps artists turn inspiration into art by giving more money and support to individual artists and arts organizations than any other comparable institution in the United States of America.

NYFA Artists' Fellowships

Subjects: Computer arts, crafts, cross-disciplinary/performative work, film, non-fiction literature, poetry,printmaking/drawing/artists' books and sculpture.
Purpose: To help the arts to flourish in New York State as well as nationally.
Eligibility: Should be a resident of New York State for at least two years prior to the application deadline and should not be enrolled in a degree program of any kind.
Level of Study: Professional development
Type: Fellowships
Value: US$7,000
Frequency: Annual
Application Procedure: Applicants must apply online.
Closing Date: December 16th

NEW YORK STATE HISTORICAL ASSOCIATION

PO Box 800, 5798 State Highway 80, Cooperstown, NY, 13326, United States of America
Tel: (1) 607 547 1400
Fax: (1) 607 547 1404
Email: mason@nysha.org
Website: www.nysha.org
Contact: Ms Catherine Mason, Assistant Editor

The mission of the New York State Historical Association is to instil and cultivate, in a broad public audience, an informed appreciation of the diversity of the American past, especially as represented and exemplified by the history of New York State, in order to better understand the present.

Dixon Ryan Fox Manuscript Prize

Subjects: The history of New York state.
Purpose: To honour the best unpublished book-length monograph.
Eligibility: Open to nationals of any country. Biographies may be included. Fiction and poetry are not eligible entries.
Level of Study: Unrestricted
Type: Prize
Value: US$3,000 plus assistance in publishing
Length of Study: Dependent on the book length
Frequency: Annual
No. of awards offered: 1
Application Procedure: Applicants must submit two copies of the manuscript, typed and double-spaced with at least 1 inch margins.
Closing Date: March 15th
Funding: Private
No. of awards given last year: 0
No. of applicants last year: 6

NEW ZEALAND'S INTERNATIONAL AID AND DEVELOPMENT AGENCY (NZAID)

Ministry of Foreign Affairs and Trade 195 Lambton Quay, Wellington 5045, New Zealand
Tel: (64) 4 439 8000
Fax: (64) 4 439 8515
Email: enquiries@nzaid.govt.nz
Website: www.nzaid.govt.nz

NZAID is a government agency responsible for delivering New Zealand's Official Development Assistance (ODA) and for advising Ministers on development assistance policy and operations. Its main purpose is to give a distinctive profile and new focus to ODA programme. Its team of development specialists aim in reducing poverty. NZAID helps to eliminate poverty and fulfilling basic needs through development partnerships, particularly in the Pacific region, and also supports projects in Asia, Africa and Latin America.

Commonwealth Scholarships

Subjects: All subjects.
Purpose: To make a significant contribution to the development of candidate's home country for elimination of poverty and to address the human resource development needs of developing countries.
Eligibility: Open to candidates who have strong academic merit and preference will be given to candidates who nominate fields of study relevant to the development of their home country.
Level of Study: Postgraduate
Type: Scholarships
Value: Varies
Length of Study: 2 years (Masters) and 4 years (Doctorate)
Frequency: Annual
Country of Study: New Zealand
No. of awards offered: 10
Application Procedure: For more details, see Commonwealth Scholarships and Fellowships Plan (CSFP) website.
Closing Date: March 19th
Funding: Government
Contributor: New Zealand's International Aid and Development Agency (NZAID) and administered by the New Zealand Vice-Chancellors' Committee

New Zealand Development Scholarships (NZDS)

Subjects: All subjects.
Purpose: To enable the students to study in areas of developmental relevance and make a significant impact on return to their home countries.
Eligibility: Open to candidates who are citizens of an eligible developing country and not have New Zealand or Australian residence status or citizenship (excluding the Cook Islands), must normally be living in home country, have a B grade average (or higher) on most

recently completed tertiary qualification, must not currently be studying, and not have been a recipient of any NZAID or other government scholarship within the last 12 months.
Level of Study: Postgraduate
Type: Scholarships
Value: Scholarships cover tuition fees, enrolment/orientation fees, return economy travel, medical insurance, and an allowance to meet course and basic living costs
Frequency: Annual
Country of Study: New Zealand
No. of awards offered: 48
Application Procedure: Check the website for specific details.
Closing Date: May to June
Funding: Government
Contributor: The New Zealand Agency for International Development on behalf of the New Zealand Ministry of Foreign Affairs and Trade (MFAT)
Additional Information: NZDS has two categories, public (40 awards) and open (8 awards). A candidate can only apply under one of the categories. Please check website for more information: www.aid.govt.nz/scholarships/

New Zealand Regional Development Scholarships

Subjects: All subjects.
Purpose: For Pacific Islanders to gain knowledge and skills in priority fields of study so that they can directly contribute to the sustainable development of key sectors in their home country.
Eligibility: Open to candidates who are citizens of a participating country (i.e. permanent residents are not eligible to apply), must be aged 17 or over before the scholarship start date, not hold another scholarship, or have held a New Zealand or Australian Government scholarship in the preceding 24 months at the time of application, satisfy the admission requirements of the education institution, and be able to take up the scholarship in the academic year for which it is offered.
Level of Study: Postgraduate
Type: Scholarships
Value: Tuition and enrolment fees, scholarship, related travel, an establishment allowance and a basic stipend.
Length of Study: 2 years
Frequency: Every 2 years
Application Procedure: Applicants must apply directly to a nominating authority in their home country, nominating authority undertakes an initial screening of applications and sends shortlisted candidates to the NZRDS.
Funding: Government
Contributor: New Zealand Government's Official Development Assistance programme and is administered by NZAID. AusAID and NZAID jointly fund Regional Development Scholarships for the Cook Islands

Short-Term Training Awards (NZ)

Subjects: All subjects.
Purpose: To provide short-term vocational and/or skills courses or work attachments.
Eligibility: Open for the candidates of Pacific Islands and Latin America (subject to ongoing review and amendment).
Type: Scholarships
Value: Payment of fees, return economy air fares, an establishment grant, a basic living allowance, and provision for health care
Length of Study: Up to 1 year
No. of awards offered: Varies
Application Procedure: Applicants must apply to the nominating authority in their home country responsible for selecting candidates for study or training abroad under development country programmes. Check website for further details.

NEWBERRY LIBRARY

60 West Walton Street, Chicago, IL, 60610-3380, United States of America
Tel: (1) 312 943 9090
Fax: (1) 312 255 3680
Email: research@newberry.org
Website: www.newberry.org
Contact: Research and Education

The Newberry Library, open to the public without charge, is an independent research library and educational institution dedicated to the expansion and dissemination of knowledge in the humanities. With a broad range of books and manuscripts relating to the civilizations of Western Europe and the Americas, the Library's mission is to acquire and preserve research collections of such material, and to provide for and promote their effective use by a diverse community of users.

American Society for Eighteenth-Century Studies (ASECS) Fellowship

Subjects: Arts and humanities from 1660 to 1815.
Purpose: To support scholars wanting to use the Newberry's collections.
Eligibility: A fellowship holder must be an ABD graduate student or post-doctoral, holding the PhD. or equivalent degree at the time of the application. Applicants must be members of the ASECS at the time of the award.
Level of Study: Doctorate, Postdoctorate
Type: Fellowship
Value: $2,000 per month
Length of Study: 1 month
Frequency: Annual
Study Establishment: The Newberry Library
Country of Study: United States of America
No. of awards offered: 1
Application Procedure: Applicants must submit a completed application form, a description of the project and three letters of reference.
Closing Date: February 6th
Funding: Private
Contributor: ASECS
No. of awards given last year: 1
No. of applicants last year: 34

Arthur Weinberg Fellowship for Independent Scholars and Researchers

Subjects: Humanities.
Purpose: To assist scholars working outside the academy who have demonstrated excellence through publishing and are working in a field appropriate to the Newberry's collections.
Eligibility: Open to all scholars but preference is given to those working on historical issues related to social justice or reform.
Level of Study: Unrestricted
Type: Fellowship
Value: US$2,000
Length of Study: 1 month
Frequency: Annual
Study Establishment: The Newberry Library
Country of Study: United States of America
No. of awards offered: Varies
Application Procedure: Applicants must write for details. Application forms can be downloaded from the website.
Closing Date: February 6th
Funding: Private
No. of awards given last year: 1
No. of applicants last year: 6

Audrey Lumsden-Kouvel Fellowship

Subjects: Portuguese, Spanish, and Latin American studies are especially welcome, as are translation projects
Purpose: To enable scholars to use the Newberry's extensive holdings. The fellowship is intended to encourage scholars to pursue research at the Newberry during sabbaticals.
Eligibility: Open to postdoctoral scholars wishing to carry out extended research. Applicants must plan to be in continuous residence for at least 3 months. Preference will be given to projects focusing on romance cultures.
Level of Study: Postdoctorate
Type: Fellowship
Value: US$4,200 per month
Length of Study: 6 months
Frequency: Annual
Study Establishment: The Newberry Library
Country of Study: United States of America
No. of awards offered: 1

Application Procedure: Applicants must write for details. Application forms can be downloaded from the website.
Closing Date: December 12th
Funding: Private
No. of awards given last year: 1
No. of applicants last year: 40

Frances C Allen Fellowships

Subjects: Humanities and social sciences.
Purpose: To encourage women of American Indian heritage in their studies through financial support.
Eligibility: Open to women of American Indian heritage who are pursuing an academic programme in any graduate or preprofessional field.
Level of Study: Graduate, Postgraduate
Type: Fellowship
Value: US$2,000 per month
Length of Study: 1–12 months
Frequency: Annual
Study Establishment: The Newberry Library
Country of Study: United States of America
No. of awards offered: Varies
Application Procedure: Applicants must write for details. Application forms can be downloaded from the website.
Closing Date: February 6th
Funding: Private
Contributor: The Frances C Allen Fund
No. of awards given last year: 2
No. of applicants last year: 4
Additional Information: Allen Fellows are expected to spend a significant part of their tenure in residence at the Newberry's D'Arcy McNickle Centre for American Indian History.

Frederick Burkhardt Residential Fellowships for Recently Tenured Scholars

Subjects: Humanities.
Purpose: To support long-term unusually ambitious projects in humanities and related social sciences.
Eligibility: Open to recently tenured humanists – scholars who will have begun their first tenured contracts by the aplication deadline but began their first tenured contracts no earlier than the fall semester or quarter. An applicant must be employed in a tenured position at a degree-granting academic institution in the US, remaining so for the duration of the fellowship.
Level of Study: Postdoctorate
Type: Fellowship
Value: $75,000
Length of Study: 1 academic year (normally 9 months)
Study Establishment: The Newberry Library
Country of Study: United States of America
No. of awards offered: 10
Application Procedure: Applicants must submit online application. It must include completed application form, proposal, bibliography, publications list, three reference letters and one institutional statement.
Closing Date: September 28th
Funding: Private
Contributor: The Andrew W. Mellon Foundation

For further information contact:

Website: http:/ofa.acls.org

Herzog August Bibliotek Wolfenbüttel Fellowship

Subjects: Applicants for long- and short-term fellowships at the Newberry may also ask to be considered for this joint fellowship providing an additional two-month fellowship in Wolfenbüttel, Germany. The proposed project should link the collections of both libraries; applicants should plan to hold both fellowships sequentially to ensure continuity of research.
Purpose: To sponsor additional study at the Herzog August Bibliotek in Wolfenbüttel, Germany.
Level of Study: Doctorate, Postgraduate
Type: Fellowship
Value: UK£1050, plus up to £600 for travel expenses
Length of Study: 2 months
Frequency: Annual
Study Establishment: Herzog August Bibliotek
Country of Study: Germany
No. of awards offered: 1
Funding: Private
Contributor: Herzog August Bibliotek
Additional Information: Application deadline: December 12, for linked long-term fellowship; February 6.

Institute for the International Education of Students Faculty Fellowships

Subjects: Humanities.
Eligibility: Open to faculty members from any IES centre.
Level of Study: Postdoctorate
Type: Fellowship
Value: $1,200 plus travel and lodging expenses
Length of Study: 2 month
Frequency: Annual
Study Establishment: The Newberry Library
Country of Study: United States of America
No. of awards offered: 2
Application Procedure: Applicants must submit an application, project description, curriculum vitae and letters of reference. Applicants should visit the wesite for further details and application forms.
Closing Date: February 6th
Funding: Private
Contributor: Institute for the International Education of Students
No. of awards given last year: 2
No. of applicants last year: 3

Lester J Cappon Fellowship in Documentary Editing

Subjects: Editing and archiving.
Purpose: To support historical editing projects based on Newberry materials.
Eligibility: Open only to postdoctoral scholars (not necessarily outside Chicago)
Level of Study: Postdoctorate
Type: Fellowship
Value: US$2,000 per month
Length of Study: Varies
Frequency: Annual
Study Establishment: The Newberry Library
Country of Study: United States of America
No. of awards offered: 1
Application Procedure: Applicants must write for details. Application forms can be downloaded from the website.
Closing Date: February 6th
Funding: Private
Contributor: Lester J Cappon
No. of awards given last year: 2
No. of applicants last year: 4
Additional Information: Applicants need not be from outside the Chicago area.

Lloyd Lewis Fellowship in American History

Subjects: Any field of American history appropriate to the collections of the Newberry Library.
Purpose: To pursue projects in any area of American history appropriate to the Newberry's collections.
Eligibility: Open to postdoctoral scholars (no citizenship restrictions)
Level of Study: Postdoctorate
Type: Fellowship
Value: $4,200 per month
Length of Study: 6–11 months
Frequency: Annual
Study Establishment: The Newberry Library
Country of Study: United States of America
No. of awards offered: Varies
Application Procedure: Candidates must write for details. Application forms can be downloaded from the website.
Closing Date: December 12th
Funding: Private
Contributor: The Lloyd Lewis Memorial Fund
No. of awards given last year: 2
No. of applicants last year: 56

Additional Information: Lewis Fellows participate in the Library's scholarly community through regular participation in seminars, colloquia and other events. Lewis Fellowships may be combined with sabbaticals or other stipendiary support. Candidates may ask to be considered for NEH and Mellon Fellowships at the time of their application. Applicants must request at least four and no more than 12 months of support.

Midwest Modern Language Association (MMLA) Fellowship

Subjects: Humanities.
Purpose: To support for work in residence at the Newberry.
Eligibility: Open to scholars with current MMLA membership at time of application and through the period of the fellowship.
Level of Study: Postdoctorate, Doctorate
Type: Fellowship
Value: $2,000 per month
Length of Study: 1 month
Frequency: Annual
Study Establishment: The Newberry Library
Country of Study: United States of America
No. of awards offered: 1
Application Procedure: Candidates must submit an application, project description, curriculum vitae and letters of reference. Candidates should visit the website for further details and application forms.
Closing Date: February 6th
Funding: Private
Contributor: Midwest Modern Language Association
No. of awards given last year: 1
No. of applicants last year: 9

National Endowment for the Humanities (NEH) Fellowships

Subjects: Any field appropriate to the Library's collections.
Purpose: To support projects in any field appropriate to the Library's collections
Eligibility: Open to citizens of the United States of America or foreign nationals who have been resident in the United States of America for 3 years, who are established scholars at the postdoctoral level or its equivalent. Preference is given to candidates who have not held major fellowships for 3 years preceding the proposed period of residency.
Level of Study: Postdoctorate
Type: Fellowships
Value: US$4,200 per month
Length of Study: 6–11 months
Frequency: Annual
Study Establishment: The Newberry Library
Country of Study: United States of America
No. of awards offered: Varies
Application Procedure: Candidates must write for an application form. Completed application forms must include all letters of reference. Application forms can be downloaded from the website.
Closing Date: December 12th
Funding: Government
Contributor: The NEH
No. of awards given last year: 3
No. of applicants last year: 119
Additional Information: Candidates may combine this award with sabbatical or other stipendiary support. Scholars conducting research in American history may also ask to be considered for the Lloyd Lewis Fellowship at the time of their application.

Newberry Library British Academy Fellowship for Study in Great Britain

Subjects: Humanities.
Purpose: To allow an individual to study in the United Kingdom in any field in which the Newberry's collections are strong.
Eligibility: Open to established scholars at the postdoctoral level or equivalent. Preference is given to readers and staff of the Newberry Library and to established Scholars who have previously used the Newberry Library.
Level of Study: Postdoctorate
Type: Fellowship
Value: A stipend of UK£1,350 per month while in the United Kingdom
Length of Study: Up to 3 months
Frequency: Annual
Country of Study: United Kingdom
No. of awards offered: Varies
Application Procedure: Applicants must write for details. Application forms can be downloaded from the website.
Closing Date: January 12th
Funding: Private
Contributor: The British Academy
No. of awards given last year: 1
No. of applicants last year: 9
Additional Information: The home institution is expected to continue to pay the Fellow's salary.

Newberry Library Ecole des Chartes Exchange Fellowship

Subjects: Renaissance studies.
Purpose: To enable a graduate student to study at the Ecole des Chartes in Paris.
Eligibility: Preference is given to graduate students at institutions in the Renaissance Centre Consortium.
Level of Study: Doctorate
Type: Fellowship
Value: Varies, but provides a monthly stipend and free tuition
Length of Study: 3 months
Frequency: Annual
Study Establishment: The Ecole des Chartes
Country of Study: France
No. of awards offered: Varies
Application Procedure: Applicants must write for details. Application forms can be downloaded from the website.
Closing Date: December 12th
Funding: Private
Contributor: The Ecole des Chartes
No. of awards given last year: 1
No. of applicants last year: 13
Additional Information: The Ecole des Chartes is the oldest institution in Europe specializing in the archival sciences, including palaeography, bibliography, textual editing and the history of the book.

Newberry Library Short-Term Fellowship in the History of Cartography

Subjects: The history of cartography.
Purpose: To financially support work in residence at the Newberry on projects related to the history of cartography.
Eligibility: Open to established scholars of any nationality, but limited to scholars who live outside the Chicago area.
Level of Study: Doctorate, Postdoctorate
Type: Fellowship
Value: US$2,000 per month
Length of Study: 1 week–2 months
Frequency: Annual
Study Establishment: The Newberry Library
Country of Study: United States of America
No. of awards offered: 1–2
Application Procedure: Applicants must submit an application, project description, curriculum vitae and letters of reference. Applicants must visit the website for further details and application forms.
Closing Date: February 6th
Funding: Private
Contributor: Arthur Holzheimer
No. of awards given last year: 2
No. of applicants last year: 14

Newberry Library Short-Term Resident Fellowships for Individual Research

Subjects: Any field appropriate to the Library's collections.
Purpose: To provide access to Newberry's collections for those who live beyond commuting distance from Chicago.
Eligibility: Open to nationals of any country who hold a PhD degree or have completed all requirements for the degree except the dissertation. Awards are limited to scholars from outside the Chicago area.
Level of Study: Doctorate, Postdoctorate
Type: Fellowship
Value: US$2,000 per month

Length of Study: 1 week–2 months
Frequency: Annual
Study Establishment: The Newberry Library
Country of Study: United States of America
No. of awards offered: Varies
Application Procedure: Applicants must write for details. Application forms can be downloaded from the website.
Closing Date: February 6th
Funding: Private
No. of awards given last year: 19
No. of applicants last year: 188

Northeast Modern Language Association (NEMLA) Fellowship

Subjects: Humanities.
Purpose: To support residential research at the library by members of NEMLA.
Eligibility: Open to scholars with current NEMLA membership at the time of application and through the period of the fellowship. Preference will be given to projects focusing on materials written in French, German, Italian or Spanish.
Level of Study: Doctorate, Postdoctorate
Value: $2,000 per month
Length of Study: 1 month
Frequency: Annual
Study Establishment: The Newberry Library
Country of Study: United States of America
No. of awards offered: 1
Application Procedure: Applicants must submit an application, project description, curriculum vitae and letters of reference. Applicants should visit the website for further details and application forms.
Closing Date: February 6th
Funding: Private
Contributor: Northeast Modern Language Association
No. of applicants last year: 8

Susan Kelly Power and Helen Hornbeck Tanner Fellowship

Subjects: American Indian heritage.
Purpose: To support residential research.
Eligibility: Open to applicants with American Indian heritage.
Level of Study: Doctorate, Postdoctorate
Type: Fellowship
Value: US$2,000 per month
Length of Study: Up to 2 months
Frequency: Annual
Study Establishment: The Newberry Library
Country of Study: United States of America
No. of awards offered: 1
Application Procedure: Applicants must visit the website for application details and forms.
Closing Date: February 6th
Funding: Private
No. of awards given last year: 1
No. of applicants last year: 9

NEWBY TRUST LIMITED

Hill Farm, Froxfield, Petersfield, Hampshire, GU32 1BQ, England
Tel: (44) 1730 827557
Email: info@newby-trust.org.uk
Website: www.newby-trust.org.uk
Contact: Miss W Gillam, Company Secretary

The Newby Trust Limited is a grant-giving charity working nationally, whose principal aims are to promote medical welfare, education and training and the relief of poverty.

Newby Trust Awards

Subjects: All subjects.
Purpose: Grants are provided directly to universities or institutions in the United Kingdom to support education at the postgraduate or postdoctoral level.
Eligibility: Application available to students showing exceptional promise, and who are accepted onto one of our MA pathways - dependant on the number and quality of applications, we may offer a combination of partial and full-fee waivers.
Level of Study: Predoctorate, Professional development, Doctorate, Postdoctorate, Postgraduate
Type: Grant
Frequency: Annual
Study Establishment: United Kingdom universities or institutions
Country of Study: United Kingdom
No. of awards offered: Varies
Application Procedure: Individual applications are not accepted.
Closing Date: Apr 11th
Funding: Private
Contributor: Funds from the Trust
Additional Information: Applications from individuals are no longer accepted.

NEWCASTLE UNIVERSITY

Manager, Student Financial Support, Newcastle University, 6 Kensington Terrace, Newcastle upon Tyne, NE1 7RU, United Kingdom
Tel: (44) 19 1222 6000
Fax: (44) 19 1222 5219
Email: international-scholarships@ncl.ac.uk
Website: www.ncl.ac.uk
Contact: Mrs Rencesova Financial Support Coordinator, Irena

The Newcastle University, established in Newcastle in 1834, is one of the UK's leading universities and is known for its quality of teaching, outstanding research, and works with the regional and local communities, business and industry.

Alumni Tuition Fee Discount

Subjects: All subjects offered by the University, but some conditions apply.
Purpose: To provide a discount for Newcastle University graduates.
Eligibility: Candidates for discounts must already have been offered a place to study at Newcastle University. Candidates must be graduates of the university i.e. Alumni - already have completed a degree here and be continuing with another one. Candidates must be self-financing.
Level of Study: Doctorate, Graduate, MBA, Postgraduate, Research
Type: Other
Value: 10 per cent of the tuition fees
Length of Study: Duration of programme
Frequency: Annual
Study Establishment: Newcastle University
Country of Study: United Kingdom
Application Procedure: All eligible applicants who are offered a place to study at Newcastle University are considered for this discount. Applicants must check the website for details or contact the Student Financial Support Team. See webpages www.ncl.ac.uk/postgraduate/funding/search/list/atfd for further information:
Closing Date: September
Contributor: Newcastle University

International Family Discount (IFD)

Subjects: All subjects offered by the University.
Purpose: To provide a discount for international students.
Eligibility: Candidates for discounts must already have been offered a place to study at Newcastle University. Also candidates must have a close family relative who is currently studying at Newcastle University or who has studied here in the past. See the webpages for details: www.ncl.ac.uk/students/wellbeing/finance/funding/nonukstudents/scholarships/undergraduate.
Level of Study: Doctorate, Foundation programme, Graduate, MBA, Postgraduate, Predoctorate, Research, Postdoctorate
Type: Other
Value: 10 per cent of the tuition fees
Length of Study: Duration of programme
Frequency: Annual
Study Establishment: Newcastle University
Country of Study: United Kingdom
Application Procedure: All eligible applicants who are offered a place to study at Newcastle University are invited to apply for one of

these discounts. Applicants must check the website for details or contact the Student Financial Support Team.
Closing Date: September
Contributor: Newcastle University
No. of awards given last year: 25

Newcastle University International Postgraduate Scholarship (NUIPS)

Subjects: All subjects offered by the University.
Purpose: To provide a partial scholarship for international students.
Eligibility: Candidates for scholarships must already have been offered a place to study at Newcastle University. See webpages: www.ncl.ac.uk/postgraduate/funding/search/list/nuips.
Level of Study: Postgraduate
Type: Partial scholarship
Value: UK£1,500 per year
Length of Study: Duration of programme
Frequency: Annual
Study Establishment: Newcastle University
Country of Study: United Kingdom
No. of awards offered: Approx. 67
Application Procedure: All eligible applicants who are offered a place to study at Newcastle University are invited to apply for one of these scholarships. Applicants must check the website for details or contact the Student Financial Support Team.
Closing Date: May 22nd
Contributor: Newcastle University
No. of awards given last year: 60
No. of applicants last year: 800

Overseas Research Studentship Award

Eligibility: Applicants must hold an offer (conditional or unconditional) to study in the Faculty and must be self-funding (at least partially) or hold a studentship from a UK charity.
Level of Study: Doctorate, Postgraduate
Type: Award
Value: Approximately £5,000–9,000 per year. The award covers the difference between fees for UK/EU students and international students.
Length of Study: 3 years (PhD) or 4 year (masters/PhD), with annual renewal being subject to satisfactory progress
Application Procedure: Applications should be made at Faculty ORS Scheme Application Form, available at http://www.ncl.ac.uk/fms/postgrad/funding/ors.htm.
Closing Date: April 26th
Additional Information: It is essential that applicants have an identifiable source of funding for the remainder of the fees and subsistence expenses.

For further information contact:

Email: s.j.yeaman@ncl.ac.uk
Contact: Professor S J Yeaman, Director of International Postgraduate Studies

NIJMEGEN CENTRE FOR MOLECULAR LIFE SCIENCES (NCMLS)

259 NCMLS, PO Box 9101, Nijmegen, 6500-HB, Netherlands
Tel: (31) 24 3610 707
Fax: (31) 24 3610 909
Email: info@ncmls.ru.nl
Website: www.ncmls.nl
Contact: A Cohen, Science Manager

Nijmegen Centre for Molecular Life Sciences (NCMLS) is a leading multidisciplinary research school within the domain of molecular mechanisms of disease and particularly in the fields of molecular medicine, cell biology and translational research.

NCMLS Tenure Track Fellowship

Subjects: Molecular life sciences.
Purpose: To promote innovation in academic research by giving creative and talented researchers the opportunity to conduct their own research within the context of Radboud University Nijmegen Medical Centre (RUNMC).
Eligibility: Preferred candidates should have obtained a PhD degree within the past 8 years and should have recent international experience. An excellent track record, sustained external funding and a robust research line are required.
Level of Study: Research
Type: Fellowship
Value: Maximum of €800,000 in 5 years.
Length of Study: 5 years
Frequency: Annual
No. of awards offered: 1
Application Procedure: Applicants can download the expression of interest form from the website www.umcn.nl/fellows or www.ncmls.eu.
Closing Date: Open all year
No. of awards given last year: 2
No. of applicants last year: 20
Additional Information: All queries should be directed to Dr A Cohen, Fellowship Admissions Officer.

NOAA COASTAL SERVICES CENTER

2234 South Hobson Avenue, Charleston, SC, 29405 2413, United States of America
Tel: (1) 843 740 1200
Fax: (1) 843 740 1224
Email: csc.fellowships@noaa.gov
Website: www.csc.noaa.gov
Contact: Margaret Vanderwilt, Fellowship Programme Manager

The NOAA Coastal Services Center is an office within the National Oceanic and Atmospheric Administration that is devoted to serving the nation's state and local coastal resource management programmes.

NOAA Coastal Management Fellowship

Subjects: Natural resource management and environmental-related studies.
Purpose: To provide financial support to students with state coastal zone programmes to work on projects proposed by the state.
Eligibility: Open to students who will shortly complete a Master's, Doctoral or professional degree programme in natural resource management or environmental related studies at an accredited university in the United States.
Level of Study: Postgraduate
Type: Fellowships
Value: US$34,000
Length of Study: 2 years
Frequency: Annual
Country of Study: United States of America
No. of awards offered: 6
Application Procedure: Applicants must send in their transcript, interview references and curriculum vitae.
Closing Date: January 25th
Contributor: NOAA Coastal Services Center
Additional Information: Students from a broad range of environmental programmes are encouraged to apply.

NOAA Coral Reef Management Fellowship

Subjects: Geography, natural resources, surveying, surveying technology, cartography and geographic information science.
Purpose: To provide assistantship to highly qualified recipients of Bachelor's and Master's degrees with hosts from coral management programmes within the United States Flag Pacific and Caribbean Islands.
Eligibility: Applicants must have a master's degree and two years of experience or a bachelor's degree and four years of experience. They must be U.S. citizen or U.S. permanent resident.
Level of Study: Postgraduate
Type: Fellowships
Value: Varies depending upon location, and will remain the same for the duration of the fellowship.
Length of Study: 2 years
Frequency: Annual
Country of Study: United States of America
No. of awards offered: 5

Contributor: NOAA Coastal Services Centre
Additional Information: Please check for more information: http://www.oakmgmt.com/careers.html.

NORTH DAKOTA UNIVERSITY SYSTEM

600 East Boulevard Avenue, Department 215 10th Floor, State Capitol, Bismarck, ND, 58505 0230, United States of America
Tel: (1) 701 328 2960
Fax: (1) 701 328 2961
Email: ndus.office@ndus.nodak.edu
Website: www.ndus.nodak.edu
Contact: Ms Rhonda Schauer, Director

The North Dakota University System, governed by the State Board of Higher Education, comprises 11 public campuses.

North Dakota Indian Scholarship Program

Subjects: All subjects.
Purpose: To assist American Indian students in obtaining a college education by providing scholarships
Eligibility: The applicant must be an enrolled member of a federally recognized Indian tribe and a resident of North Dakota and must have been accepted for admission at an institution of higher learning or state vocational education program within North Dakota. Recipients must be enrolled full-time and have a grade point average above 2.0.
Level of Study: Doctorate, Postgraduate
Type: Scholarship
Value: Grants may range from $800 to $2,000
Frequency: Annual
Country of Study: United States of America
No. of awards offered: 239
Application Procedure: Applicants must complete an application form.
Closing Date: July 15th
Funding: Government
Contributor: The State
No. of awards given last year: 239
No. of applicants last year: 500

For further information contact:

Contact: Ms

NORTH WEST CANCER RESEARCH FUND

22 Oxford Street, Liverpool, L7 7BL, England
Tel: (44) 15 1709 2919
Fax: (44) 15 1708 7997
Email: nwcrf@btclick.com
Website: www.cancerresearchnorthwest.co.uk
Contact: Mr A W Renison, General Secretary

North West Cancer Research Fund Research Project Grants

Subjects: All types of cancer, the mechanisms by which they arise and the way they exert their effects.
Purpose: To support fundamental research into the cause of cancers and the mechanisms by which cancers arise and exert their effects.
Eligibility: Open to candidates undertaking cancer research studies at one of the universities named below in the Northwest. Grants are only available for travel costs associated with currently funded 3-year cancer research projects. No grants are awarded for buildings or for the development of drugs.
Level of Study: Research
Type: Project
Value: Approx. UK£35,000 per year
Length of Study: Usually 3 years
Frequency: Dependent on funds available
Study Establishment: The University of Liverpool, Lancaster University and the University of Wales, Bangor
Country of Study: North West England, North and Mid-Wales
Application Procedure: The NWCRF Scientific Committee meets twice a year. All applications are subject to peer review.
Closing Date: April 1st and October 1st
Funding: Individuals, private
Contributor: Voluntary donations
No. of awards given last year: 10
No. of applicants last year: 50

For further information contact:

NWCRF Scientific Committee Department of Medicine, Duncan Building, Daulby Street, Liverpool, L69 3BX, England
Email: ricketts@liverpool.ac.uk
Contact: The Secretary

NORWAY - THE OFFICIAL SITE IN THE UNITED STATES

2720 34th Street NW, Washington, DC, 20008, United States of America
Tel: (1) 202 333 6000
Fax: (1) 212 754 0583
Email: cg.newyork@mfa.no
Website: www.norway.org
Contact: Grants and Scholarships Department

American-Scandinavian Foundation Scholarships (ASF)

Subjects: All subjects.
Purpose: To encourage Scandinavians to undertake advanced study and research programmes in the United States.
Eligibility: Applicants must be Scandinavians over the age of 21, applicants from Denmark must be between 21 and 30 years. Applicants can be young professionals who have completed their formal education with related work experience in Scandinavia or Europe, or students who are currently enrolled in colleges and universities who need to undertake an internship. They should be fluent in English.
Level of Study: Postgraduate
Value: The American-Scandinavian Foundation (ASF) offers over $500,000
Length of Study: 1 year
Frequency: Annual
Country of Study: United States of America
No. of awards offered: Varies
Application Procedure: Applicants must complete an application on ASF application forms, available on request from the Foundation.
Funding: Foundation

For further information contact:

58 Park Ave, New York, NY, 10016, United States of America
Tel: (1) 212 879 9779
Fax: (1) 212 249 3444
Email: grants@amscan.org

John Dana Archbold Fellowship Program

Subjects: All subjects.
Purpose: To support educational exchange between the United States of America and Norway.
Eligibility: Open to citizens of the United States of America citizens aged 20–35, in good health and of good character. Qualified applicants must show evidence of a high level of competence in their chosen field, indicate a seriousness of purpose, and have a record of social adaptability. There is ordinarily no language requirement.
Level of Study: Postdoctorate, Postgraduate, Professional development, Research
Type: Fellowship
Value: Up to US$5,000. Individual grants vary, depending on the projected costs. Note that there will be no tuition at the University of Oslo. The maintenance stipend is sufficient to meet expenses for a single person. The travel allowance covers round trip airfare to Oslo
Length of Study: 1 year
Frequency: Annual
Study Establishment: The University of Oslo
No. of awards offered: 2
Application Procedure: Applicants must write to the Nansen Fund, Inc. for an application form.
Closing Date: February 1st

Funding: Private
Additional Information: The University of Oslo International Summer School offers orientation and Norwegian languages courses 6 weeks before the start of the regular academic year. For Americans, tuition is paid. Attendance is required. Americans visit Norway in even-numbered years and Norwegians visit the United States of America in odd-numbered years. For further information please contact the Nansen Fund, Inc.

For further information contact:

The Nansen Fund, Inc., 77 Saddlebrook Lane, Houston, TX, 77024, United States of America
Tel: (1) 713 680 8255
Email: nacc@net1.net

The Norway-America Association Awards

Subjects: All subjects.
Eligibility: Open to Norwegian who wish to study in the United States on the graduate level, must have completed their Bachelor's Degree before applying for these scholarships. The applicants must also be members of the Norway-America Association, and the membership fee is 200 NOK/year.
Level of Study: Graduate, Research
Type: Award
Value: $2,000–20,000
Application Procedure: Check website for further details.
Closing Date: September 22nd
Additional Information: Norwegians who reside in Norway and plan to return to Norway after graduation are given preference in the selection process, and they cannot have studied for four or more years in the United States. Please contact the American-Scandinavian Foundation or the Norway-America Association directly in order to find the appropriate scholarship.

For further information contact:

The Norway–America Association, Rådhusgaten 23B, Oslo, 0158, Norway
Tel: (47) 47 23357160
Fax: (47) 47 23357175
Email: info@noram.no

The Norway-America Association Graduate & Research Stipend

Subjects: All subjects.
Purpose: To give the student substantial financial support for 1 year of studies in the United States.
Eligibility: Open to Norwegians and members of the Norway-America Association, and he/she must pay 250 NOK in administrative fees and are currently living in Norway, and intend to return to Norway after their graduation.
Level of Study: Graduate, Research
Value: The awards range $2,000–25,000
Application Procedure: Check website for further details.
Closing Date: September 22nd

For further information contact:

The Norway–America Association, Rådhusgaten 23B, Oslo, 0158, Norway
Tel: (47) 23357160
Fax: (47) 23357175
Email: info@noram.no
Website: www.noram.no/norsk/graduate_no.html

Norwegian Marshall Fund

Subjects: Science and humanities.
Purpose: To provide financial support for Americans to come to Norway to conduct postgraduate study or research in areas of mutual importance to Norway and the United States of America, thereby increasing knowledge, understanding and strengthening the ties of friendship between the two countries.
Eligibility: Open to citizens of the United States of America, who have arranged with a Norwegian sponsor or research institution to pursue a research project or programme in Norway. Under special circumstances, the awards can be extended to Norwegians for study or research in the United States of America.
Level of Study: Postgraduate
Type: Research grant
Value: $1,500–4,500 or NOK 10,000–30,000
Length of Study: Varies
Frequency: Annual
Study Establishment: Norwegian universities
Country of Study: Norway
No. of awards offered: 5–15
Application Procedure: Applicants must contact the Norway-America Association to receive an application. Application forms must be typewritten either in English or Norwegian and submitted in duplicate, including all supplementary materials. Each application must also be accompanied by a letter of support from the project sponsor or affiliated research institution in Norway. There is an application fee of Norwegian Krone 350.
Closing Date: March 15th
Funding: Government

For further information contact:

The Norway-America Association, Rådhusgaten 23 B, Oslo, N-0158, Norway
Tel: (47) 23357160
Fax: (47) 23357175
Email: info@noram.no
Website: www.noram.no

Norwegian Thanksgiving Fund Scholarship

Subjects: Fisheries, geology, glaciology, astronomy, social medicine and Norwegian culture.
Purpose: To provide eligible students to pursue their studies.
Eligibility: The applicant must be a US citizen doing graduate level work in Norway. The student must be working on social medicine, Norwegian culture, fisheries, geology, glaciology or astronomy at a Norwegian university.
Level of Study: Graduate
Type: Scholarship
Value: Up to US$3,000
Frequency: Annual
Country of Study: Norway
No. of awards offered: 1
Application Procedure: Candidates must contact the American Scandinavian Foundation.
Closing Date: March 15th
Funding: Individuals
Contributor: Former Norwegian students and friends

For further information contact:

The American Scandinavian Foundation, 58 Park Avenue, New York, NY, 10016, United States of America
Tel: (1) 212 879 9779
Fax: (1) 212 249 3444

The Professional Development Award

Subjects: All subjects.
Purpose: To help established professionals with a higher education who want to study within their own field of interest.
Eligibility: Open to Norwegian professionals who worked for at least 3 years after finishing his or her education as well as planning on doing special research or further study in their fields.
Level of Study: Postgraduate
Type: Award
Value: The award amount can vary but there is a recommended minimum of $250 and maximum of $1,000 with only one award per student within a given academic year.
Frequency: Annual
No. of awards offered: 2
Application Procedure: Check website for further details.
Additional Information: Candidates must be invited to apply for this award. There is also an administrative fee of 250 NOK, which must be deposited in bank account with number 7878.05.23025. Please contact the American-Scandinavian Foundation or the Norway-America Association directly in order to find the appropriate scholarship.

For further information contact:

The Norway–America Association, Cheryl Storø, Rådhusgaten 23B, Oslo, 0158, Norway
Tel: (47) 23357160
Fax: (47) 23357175
Email: info@noram.no

The Torskeklubben Stipend

Subjects: All subjects.
Purpose: To promote Norwegian-American relations through helping Norwegians come to the United States to study.
Eligibility: Open to Norwegians and must already be accepted at the Graduate School at the University of Minnesota before applying for the award.
Level of Study: Graduate
Value: A stipend of $15,000 for the academic year. For recipients without another source of tuition support, such as an assistantship, the Graduate School Fellowship Office will provide a Tuition Scholarship for full-time study for the academic year.
Frequency: Annual
Application Procedure: Application forms are available upon request from the Norway-America Association and the Graduate School at the University of Minnesota or it can be downloaded from the website.
Closing Date: March 1st

For further information contact:

The Norway–America Association, Rådhusgaten 23B, Oslo, 0158, Norway
Tel: (47) 23357160
Fax: (47) 23357175
Email: info@noram.no
Website: www.grad.umn.edu/fellowships/norwegian_citizens

NOTTINGHAM TRENT UNIVERSITY

Nottingham Business School, Newton Building (7th Floor), Burton Street, Nottingham, NG1 4BU, United Kingdom
Tel: (44) (0)115 941 8418
Email: charlotte.wix@ntu.ac.uk
Website: www.ntu.ac.uk
Contact: Charlotte Wix, MBA Administrator

Masters Scholarship

Subjects: Courses covered in the scheme are the MSc management programme and a selection of marketing, finance and business courses.
Purpose: Nottingham Business School offers the following competitive Scholarships for our full-time Masters courses.
Eligibility: Open to UK, EU and International students on a full-time Master's course.
Level of Study: Postgraduate
Type: Scholarship
Value: Up to £4,250 (UK/EU) and £6,300 (International), 50 per cent of fees
Frequency: Annual
No. of awards offered: 56 (UK/EU) and 14 (International)
Application Procedure: To apply please download the relevant documents from the website.
Closing Date: April 30th and July 30th
Additional Information: Please check at http://ntu.ac.uk/nbs/courses/postgraduate_professional/fees_funding/index.htmlor contact at nbs.enquiries@ntu.ac.uk for more information.

MBA Scholarships

Subjects: All subjects.
Eligibility: Open to students on a Nottingham Business School's full-time Masters of Business Administration.
Level of Study: Postgraduate
Type: Scholarship
Value: Nottingham Business School will be offering either one full scholarship of £14,800 or two part scholarships of £7,400, depending on applications and the judgment of the panel.
Frequency: Annual
Application Procedure: To apply please download the relevant documents from the website.
Closing Date: October 21st
Additional Information: Please check at http://ntu.ac.uk/nbs/courses/postgraduate_professional/fees_funding/index.html and contact at nbs.enquiries@ntu.ac.uk for more information.

PhD Studentships

Subjects: Art and design, architecture, design and the built environment, biomedical sciences / biosciences, business law and social sciences, communication, culture and media studies, english and physical sciences / technology.
Purpose: Nottingham Trent University has an outstanding reputation for our commitment to research that shapes lives and society. As part of our continued pledge to invest in research excellence we are delighted to announce the availability of 12 fully-funded PhD bursary places.
Eligibility: Open to students on various selected courses at the University.
Level of Study: Doctorate
Type: Scholarship
Value: The studentships will pay UK/EU fees and provide a maintenance stipend linked to the RCUK rate £13,590 per year for up to three years.
Frequency: Annual
No. of awards offered: 12
Closing Date: March 23rd
Additional Information: Please check at http://www.ntu.ac.uk/research/research_degrees/studentships/index.html for more information.

NURSES EDUCATIONAL FUNDS, INC.

304 Park Avenue South, 11th Floor, New York, NY, 10010, United States of America
Tel: (1) 212 590 2443
Fax: (1) 212 590 2446
Email: info@n-e-f.org
Website: www.n-e-f.org

Nurses Educational Funds, Inc. is an independent, non-profit organization that grants scholarships to registered nurses for graduate study. It is governed by a board of trustees of nursing and business leaders, and is supported by contributions from corporations, foundations, nurses and individuals interested in the advancement of nursing.

Nurses' Educational Funds Fellowships and Scholarships

Subjects: Administration, supervision, education, clinical specialization and research. Any nursing-related field.
Purpose: To provide the opportunity to registered nurses who seek to qualify through advanced study in a degree programme.
Eligibility: Baccalaureate-prepared registered nurses residing in any one of the 50 states and the District of Columbia who are pursuing either a Master's or doctoral degree in nursing are eligible to apply for NEF scholarship monies.
Level of Study: Doctorate, Graduate
Type: Fellowship
Value: Please contact the organization
Frequency: Annual
Country of Study: United States of America
No. of awards offered: Varies
Application Procedure: Application available on the website.
Closing Date: March 1st
Funding: Corporation, individuals, private, trusts
No. of awards given last year: 24
No. of applicants last year: 130

OHIO ARTS COUNCIL

30 E. Broad St., 33rd Floor, Columbus, OH, 43215-3414, United States of America
Tel: (1) 614 466 2613
Fax: (1) 614 466 4494
Email: webmaster@oac.state.oh.us
Website: www.oac.state.oh.us

The Ohio Arts Council is a state agency that funds and supports quality arts experiences to strengthen Ohio communities culturally, educationally and economically. It was created in 1965 to foster and encourage the development of the arts and assist the preservation of Ohio's cultural heritage.

Ohio Arts Council Individual Excellence Awards

Subjects: Interdisciplinary and performance art, criticism and music composition.
Purpose: To recognize and support the contributions of working artists to the cultural enrichment of the state.
Eligibility: Open to residents of Ohio who have lived in the state continuously for 1 year before the deadline and continue to remain an Ohio resident during the term of the award. Applicants cannot be students enrolled in any degree or certificate granting programme.
Level of Study: Postgraduate
Type: Fellowship
Value: $5,000
Frequency: Annual
Country of Study: United States of America
Application Procedure: Applicants must submit application form B and supporting documents. Guidelines and application forms are available on the OAC website or upon request by writing to the address. All forms must be submitted to the OAC in printed format, rather than electronically.
Closing Date: September 1st
Funding: Government

OMOHUNDRO INSTITUTE OF EARLY AMERICAN HISTORY AND CULTURE

PO Box 8781, Williamsburg, VA, 23187 8781, United States of America
Tel: (1) 757 221 1114
Fax: (1) 757 221 1047
Email: sdmaso@wm.edu
Website: http://oieahc.wm.edu
Contact: Ms Sally D Mason, Senior Assistant to the Director

The Omohundro Institute of Early American History and Culture publishes books in its field of interest, the *William and Mary Quarterly* and a biannual newsletter, *Uncommon Sense*. It also sponsors conferences and colloquia and annually awards a 2-year NEH postdoctoral fellowship and offers a 1-year Omohundro Institute postdoctoral research fellowship.

Omohundro Institute Postdoctoral Fellowship

Subjects: The history and culture of North America's indigenous and immigrant peoples during the colonial, revolutionary and early national periods of the United States of America, and the related histories of Canada, the Caribbean, Latin America, the British Isles, Europe and Africa from 1500 to 1815.
Purpose: To revise a dissertation into a first book that will make a major contribution to the field of early American history and culture, for publication by the Institute.
Eligibility: Foreign nationals must have been in continuous residence in the United States for 3 years immediately preceding the date of their application for the Fellowship in order to be eligible for NEH funding
Level of Study: Postdoctorate
Type: Fellowship
Value: US$50,400
Length of Study: 2 years
Frequency: Annual
Study Establishment: The Omohundro Institute of Early American History and Culture
Country of Study: United States of America
No. of awards offered: 1
Application Procedure: Applicants must complete an application form, available on written request or from the website.
Closing Date: November 5th
Funding: Government
Contributor: The NEH and the College of William and Mary
No. of awards given last year: 1
No. of applicants last year: 25

Omohundro Institute Postdoctoral Research Fellowship

Subjects: The history and culture of North America's indigenous and immigrant peoples during the colonial, revolutionary and early national periods of the United States of America, and the related histories of Canada, the Caribbean, Latin America, the British Isles, Europe and Africa from 1500 to 1815.
Purpose: To revise the applicant's book manuscript into a first book that will make a distinguished contribution to scholarship, with publication by the Institute intended.
Eligibility: Applicants must not have previously published a book or have a book under contract, and must also have received their PhD at least 1 year prior to the application deadline.
Level of Study: Postdoctorate
Type: Fellowship
Value: US$55,000
Length of Study: 1 year, residential recommended but not mandatory
Frequency: Annual
Study Establishment: The Omohundro Institute of Early American History and Culture
Country of Study: United States of America
No. of awards offered: 1
Application Procedure: Applicants must complete an application form and submit this with a completed manuscript. Application forms are available on request and from the website.
Closing Date: November 5th
Funding: Private
Contributor: The Omohundro Institute of Early Americal History and Culture
No. of awards given last year: 1
No. of applicants last year: 9
Additional Information: Please see the website for further details http://oieahc.wm.edu/fellowship/submission/index.cfm.

ONCOLOGY NURSING SOCIETY FOUNDATION (ONS)

125 Enterprise Drive, Pittsburgh, PA, 15275, United States of America
Tel: (1) 412 859 6100
Fax: (1) 412 859 6162
Email: customer.service@ons.org
Website: www.ons.org
Contact: Director of Research

The mission of the Oncology Nursing Society (ONS) is to promote excellence in oncology nursing and quality cancer care. ONS works to fulfil this mission by providing nurses and healthcare professionals with access to the highest quality educational programmes, cancer care resources, research opportunities and networks for peer support.

Aventis Research Fellowship

Subjects: Nursing, health and medical sciences.
Purpose: To financially support short-term postdoctorate oncology-specific research training.
Eligibility: Open to candidates who are registered nurses with interest in oncology and a completed Doctoral degree in nursing or a related discipline.
Level of Study: Postgraduate
Type: Fellowship
Value: Up to $10,000 awarded to cover transportation, lodging, tuition, and other expenses. Up to $1,700 to attend the ONS Congress in following year is also awarded
Frequency: Annual
Country of Study: United States of America
No. of awards offered: 1
Closing Date: June 1st
Funding: Foundation

Oncology Nursing Society Foundation Research Grant Awards

Subjects: Nursing.
Purpose: To financially support the principal investigator actively involved in some aspect of care, education or research for patients with cancer.
Eligibility: The principal investigator must be actively involved in some aspect of cancer patient care, education, or research, and be

PhD- or DNSc-prepared or a student working toward one of those research degrees. Funding preference is given to projects that involve nurses in the design and conduct of the research activity and that promote theoretically based oncology practice. Membership in ONS is not required for eligibility.
Level of Study: Postgraduate
Type: Grant
Value: Up to $25,000 each
Frequency: Annual
Country of Study: United States of America
No. of awards offered: 1
Closing Date: September 1st (letter of intent); October 1st (online application)
Funding: Foundation
Additional Information: Preference is given to projects that involve nurses in the design and conduct of the research activity and that promotes theoretically based oncology practice. For more information, contact the ONS Foundation Research Department.

ONS Breast Cancer Research Grant

Subjects: Oncology.
Purpose: To provide funding to support the ONS research agenda, which focuses on areas where gaps exist in the knowledge base for oncology nursing practice.
Eligibility: Open to the principal investigator who is actively involved in some aspect of cancer patient care, education or research.
Level of Study: Research
Type: Grant
Value: US$100,000 for 2 years for the research team
Length of Study: 2 years
Frequency: Annual
Country of Study: United States of America
Funding: Foundation
Contributor: ONS Foundation
Additional Information: Funding preference is given to projects that involve nurses in the design and conduct of the research activity.

ONS Novartis Pharmaceuticals Post-Master's Certificate Scholarships

Subjects: Oncology nursing.
Purpose: To financially assist nurses in furthering their education.
Eligibility: Open to applicants who are enrolled in (or applying to) a postmaster's nurse practitioner certificate academic credit-bearing programme in an NLN or CCNE accredited school of nursing. The candidates must also have a previous Master's degree in nursing and a current license in and commitment to oncology nursing.
Level of Study: Postgraduate
Type: Scholarship
Value: US$3,000
Frequency: Annual
Country of Study: United States of America
Application Procedure: Applicants must submit their application form and transcripts along with the application fee.
Closing Date: February 1st
Additional Information: Please see the website for further detailshttp://usascholarship-grant.blogspot.in/2009/01/novartis-pharmaceuticals-post-masters.html.

THE ONTARIO COUNCIL ON GRADUATE STUDIES (OCGS)

180 Dundas Street West Suite 1100, Toronto, ON, M5G 1Z8, Canada
Tel: (1) 416 979 2165 ext 212
Fax: (1) 416 979 8635
Email: kpanesar@cou.on.ca
Website: http://ocgs.cou.on.ca

The Ontario Council on Graduate Studies (OCGS) is an affiliate of the Council of Ontario Universities (COU). OCGS strives to ensure quality research and education across Ontario. In order to achieve this, OCGS conducts quality reviews of research programmes that have been proposed for implementation in Ontario's universities. It also performs quality reviews of existing programmes on a 7 year cycle.

Women's Health Scholars Awards 2013-2014

Subjects: Women's health.
Purpose: To financially support research in and study of women's health at Ontario universities.
Eligibility: Open to students registered full-time in a Master's or Doctoral graduate programme at an Ontario university and sponsored and endorsed by a Dean of graduate studies at his or her university. For a postdoctoral award, applicants must be engaged in full-time research at an Ontario university at the time of taking up the award.
Level of Study: Postgraduate, Doctorate, Postdoctorate
Type: Award
Value: Master's Awards - $18,000 plus $1,000 research allowance; Doctoral Awards - $20,000 plus $2,000 research allowance; Post-doctoral Awards - $40,000 plus $5,000 research allowance
Length of Study: 2 year term for Master's and up to 3 years for Doctoral and postdoctoral programmes
Frequency: Annual
Country of Study: Canada
No. of awards offered: up to two awards (each of the Master's, Doctoral and Postdoctoral)
Application Procedure: Applicants must submit their application form, curriculum vitae, a statement of research to be undertaken during the period of graduate or postdoctoral study, transcripts and reference letters. Applications are available from the Deans office of Ontario universities or from the website.
Closing Date: January 30th
Funding: Foundation
Additional Information: please see the website for further detailshttp://www.wlu.ca/page.php?grp_id=36&p=6301

ONTARIO FEDERATION OF ANGLERS & HUNTERS (OFAH)

4601 Guthrie Drive PO Box 2800, Peterborough, ON, K9J 8L5, Canada
Fax: (1) 705 748 9577
Email: ofah@ofah.org
Website: www.ofah.org
Tel: 705 748 6324

The Ontario Federation of Anglers & Hunters (OFAH), Canada's leading conservation organization, is a non-profit, registered charity, which is dedicated to protecting woodland and wetland habitat, conserving precious fish and widlife stocks and promoting outdoor education.

OFAF/OFAF Zone 6 Wildlife Research Grant

Subjects: Wildlife research work.
Purpose: To financially support students who wish to pursue their research work in wildlife research.
Level of Study: Research
Type: Fellowship
Value: Canadian $2,000
Frequency: Annual
Country of Study: Canada
Application Procedure: Applicants must submit a complete application form, research proposal consisting of abstract, introduction, methods, results anticipated, literature cited and budget, a curriculum vitae, transcripts and letter from supervising professor supporting the intended research projects.
Closing Date: First Friday in January

OFAH/Oakville and District Rod & Gun Club Conservation Research Grant

Subjects: Conservation of natural resources.
Purpose: To financially support students who wish to pursue their research work in conservation research.
Level of Study: Research
Type: Research grant
Value: Canadian $2,000
Frequency: Annual
Country of Study: Canada
Application Procedure: Applicants must submit a completed application form, research proposal consisting of abstract,

introduction, methods, research anticipated, literature cited and budget, curriculum vitae, transcripts and letter from supervising professor supporting the intended research project.
Closing Date: First Friday in January

OFAH/Toronto Sportsmen's Show Sport Fisheries Research Grant

Subjects: Fisheries research work.
Purpose: To financially support students who wish to pursue their research work in fisheries research.
Level of Study: Research
Type: Fellowships
Value: Canadian $2,000
Frequency: Annual
Country of Study: Canada
Application Procedure: Applicants must submit a complete application form, research proposal consisting of abstract, introduction, methods, results anticipated, literature cited and budget, curriculum vitae, transcripts and letter from supervising professor supporting the intended research projects.
Closing Date: First Friday in January

ONTARIO PROBLEM GAMBLING RESEARCH CENTRE (OPGR)

150 Research Park Lane, Suite 104, Guelph, ON, N1G-4T2, Canada
Tel: (1) 519 763 8049 ext 226
Fax: (1) 519 763 8521
Email: erika@opgrc.org
Website: www.gamblingresearch.org
Contact: Erika Veri Levett, Grants Officer

In April 2000, Ontario Problem Gambling Research Centre (OPGR) was created as an arms length funding agency. The Ontario Government dedicates 2 per cent of the gross revenue from slot operations at the province's charity casinos and racetracks to a Problem Gambling Strategy.

OPGR Studentship Awards

Subjects: Problem gambling research.
Purpose: To develop the capacity in Ontario to conduct problem gambling research.
Eligibility: Candidates must be Canadian citizens or permanent residents of Canada, registered in a full time Masters or Doctorate level program at an Ontario University. Consideration will be given to applicants awaiting acceptance. Applicants must be under the supervision of at least one researcher (preferably their supervisor) who has a full-time academic appointment at a University. The applicant's research may be from any discipline, provided the research proposed is substantially related and relevant to problem gambling.Applicants must be free from tri-council (SSHRC/NSERC/CIHR) or University sanction for financial or research misconduct.
Level of Study: Doctorate, Postgraduate
Type: Studentship
Value: Masters program - $17,500 (per year for 2 years) plus $10,000 = maximum $27,500 Doctoral program - $20,000 (per year for 2 years) plus $10,000 = maximum $30,000
Length of Study: 2 years for a Doctoral degree, 2 years for a Master's degree
Frequency: Annual
Country of Study: Canada
Application Procedure: Candidates must provide a completed application form, cover page, proposal, official transcripts, curriculum vitae, reference letter and confirmation letter from host university.
Closing Date: May 16th (Check with the website)
Contributor: Ontario Problem Gambling Research Centre
Additional Information: This award can be pro-rated for up to 5 years provided that all requirements for the Doctoral programme will be completed by that date. Preference will be given to permanent Ontario residents studying at Ontario universities. Please see the website for further detailshttp://www.gamblingresearch.org/content/default.php?id=4332.

OPEN SOCIETY

Cambridge House, 5th Floor, 100 Cambridge Grove, Hammersmith, London, W6 0LE, United Kingdom
Tel: (44) 020 8563 8566
Fax: (44) 220 7031 0247
Email: kedwards@sorosny.org.
Website: www.soros.org
Contact: Celine Keshishian, Network Scholarship Programs

Open Society Fellowship

Subjects: The Open Society Foundations work to build vibrant and tolerant democracies whose governments are accountable to their citizens. Among the Foundations' core areas of concern are human rights, government transparency, the promotion of civil society and social inclusion. Project themes should cut across these areas of interest. Applicants are encouraged to explore this website to acquaint themselves with the panoply of themes and geographic areas that fall within the Foundations' purview.
Purpose: The Open Society Fellowship supports individuals seeking innovative and unconventional approaches to fundamental open society challenges. The fellowship funds work that will enrich public understanding of those challenges and stimulate far-reaching and probing conversations within the Open Society Foundations and in the world.
Eligibility: The Open Society Fellowship chooses its fellows from a diverse pool of applicants that includes journalists, activists, academics, and practitioners in a variety of fields. Applicants should possess a deep understanding of their chosen subject area and a track record of professional accomplishment.
Level of Study: Research
Value: Full-time fellows based in the United States will receive a stipend of $80,000 or $100,000, depending on work experience, seniority, and current income. Stipends will be prorated for part-time fellows. For fellows based elsewhere, appropriate adjustments will be made to reflect the cost of living in those countries. The stipend does not necessarily equal the applicant's current salary. In certain cases, fellows will receive additional financial support to enable them to meet the residency expectation
Frequency: Biannual
Application Procedure: In evaluating each proposal, the selection committee weighs three factors: the applicant, the topic of the project, and the work product.
All interested applicants should complete the online application form at https://oas.soros.org/oas and submit supporting materials for consideration.
Closing Date: February 1st
Funding: Trusts
Additional Information: Please see the website for further detailshttp://www.mladiinfo.com/2012/11/23/open-society-fellowship-2013/

OPPENHEIM-JOHN DOWNES MEMORIAL TRUST

50 Broadway, London, Westminster, SW1H OBL, England
Tel: (44) 20 7227 7000
Fax: (44) 20 7222 3480
Email: lindasnape@bdb-law.co.uk
Website: www.oppenheimdownestrust.org/
Contact: Grants Management Officer

The Oppenheim-John Downes Memorial Trust makes annual awards in December to deserving artists of any kind unable to pursue their vocation by reason of poverty. Awards are restricted to persons over 30 years of age, who are natural born British subjects (S 34 of the Race Relations Act applies).

Oppenheim-John Downes Trust Grants

Subjects: Arts.
Purpose: To assist artists, musicians, writers, inventors, singers, actors and dancers of all descriptions who are unable to pursue their vocation by reason of their poverty.
Eligibility: Open to artists over 30 years of age, born in the British Isles of British parents and grandparents born after 1900. These

qualifications are mandatory and applicants who do not qualify in all respects should not apply.
Level of Study: Unrestricted
Type: Grant
Value: maximum award would be £1,000 but, in exceptional circumstances, the Trustees may consider making a larger award.
Frequency: Annual
Country of Study: Any country
No. of awards offered: Approx. 30–40
Application Procedure: Applicants must write for details.
Closing Date: October 15th
Funding: Private
No. of awards given last year: 59
No. of applicants last year: 222

OREGON COLLEGE OF ART & CRAFT (OCAC)

8245 SW Barnes Road, Portland, OR, 97225, United States of America
Tel: (1) 503 297 5544
Fax: (1) 503 297 9651
Email: admissions@ocac.edu
Website: www.ocac.edu
Contact: Sara Black, Extension Programme Director

Oregon College of Art & Craft (OCAC) is dedicated to excellence in teaching art through craft, contributing significantly to the continuity of contemporary craft as an artistic expression. The college traces its origins to 1907 when Julia Hoffman founded the Arts and Crafts Society to educate the public on the value of arts and crafts in daily life through art classes and exhibitions featuring the best examples of American crafts. Today OCAC is an accredited independent craft college offering studio classes in book arts, ceramics, drawing/painting, fibres, metals, photography and wood.

OCAC Junior Residency

Subjects: The college offers residencies that rotate through each of the studios: book arts, ceramics, drawing, fibres, metals, photography and wood.
Purpose: To encourage outstanding emerging artists to pursue a focused project in a stimulating art environment.
Eligibility: Open to citizens or permanent residents of the United States. Artists working in drawing/painting or fibres will be considered in the fall and those working in metals and photography for spring residency. The college defines emerging artists as postgraduates (post-MFA preferred) with less than 5 years experience as an exhibiting artist. Candidates from culturally diverse backgrounds are encouraged to apply.
Level of Study: Professional development
Type: Fellowship
Value: $1,200 fellowship, up to $500 reimbursement for travel to and from OCAC, up to $500 for materials and up to $100 for shipping the completed work
Length of Study: September–December or January–May
Frequency: Annual
Study Establishment: Oregon College of Arts and Craft
Country of Study: United States of America
No. of awards offered: 2 in fall and 2 in spring
Application Procedure: Applicants can download the application form from the website.
Closing Date: March 1st
Funding: Foundation
Contributor: The Collins Foundation
No. of awards given last year: 4
No. of applicants last year: 23
Additional Information: Housing on campus, $500 travel reimbursement, $500 materials reimbursement and a $1,200 fellowship stipend are provided to each junior resident. In addition, residents are invited to participate in a group show of resident work in the fall of the following year. Please see the website for further detailshttp://www.residencyunlimited.org/opportunites/2009/09/oregon-college-of-art-craft-junior-residency/.

OREGON STUDENT ASSISTANCE COMMISSION (OSAC)

1500 Valley River Drive, Suite 100, Eugene, OR, 97401, United States of America
Tel: (1) 541 687 7400
Email: awardinfo@mercury.osac.state.or.us
Website: www.osac.state.or.us

Oregon Student Assistance Commission (OSAC) administers a variety of State of Oregon, Federal and privately funded student financial aid programmes for the benefit of Oregonians attending institutions of postgraduate education. This agency was formerly known as the Oregon State Scholarship Commission.

Lawrence R. Foster Memorial Scholarship

Subjects: Health and medical sciences.
Purpose: To offer one time award to students enrolled or planning to enrol in a public health degree programme.
Eligibility: Open to applicants working in the public health field or pursuing a graduate degree in public health. Applicants must be residents of Oregon.
Level of Study: Postgraduate
Type: Scholarships
Value: US$1,000 to 10,000
Frequency: Annual
Country of Study: United States of America
Application Procedure: Applicants must submit an application form, transcript, financial need analysis, essay, 3 references and activity chart.
Closing Date: March 1st

For further information contact:

Tel: 800 452 8807
Contact: Director of Grant Programmes

ORENTREICH FOUNDATION FOR THE ADVANCEMENT OF SCIENCE, INC. (OFAS)

855 Route 301, Cold Spring, NY, 10516 9802, United States of America
Tel: (1) 845 265 4200
Fax: (1) 845 265 4210
Email: ofase@orentreich.org
Website: www.orentreich.org
Contact: N. F. Durr, Director for Scientific Affairs

The Orentreich Foundation for the Advancement of Science, Inc. (OFAS) is an operating private foundation that performs its own research and collaborates on projects of mutual interest.

OFAS Grants

Subjects: Areas of interest to the Foundation, including ageing, dermatology, endocrinology and serum markers for human diseases.
Purpose: To allow an individual to conduct collaborative biomedical research or research at the Foundation.
Eligibility: Open to applicants at or above the postgraduate level in science or medicine at an accredited research institution in the United States of America. There are, however, no citizenship restrictions.
Level of Study: Unrestricted
Type: Grant
Value: Varies, depending on the needs, nature and level of OFAS interest in the project
Frequency: Annual
Country of Study: United States of America
Application Procedure: Applicants must submit an outline of proposed joint or collaborative research, including, as a minimum, a brief overview of the current research in the field of interest, a statement of scientific objectives, a protocol summary, a curriculum vitae of the principal investigator, funding needed and total estimated project funding. Applications are reviewed quarterly. OFAS is usually the initiator of joint projects.
Closing Date: Applications may be submitted at any time

Funding: Foundation, private
Additional Information: Individuals who have a research question relating to a human disease or disease prevention factor for which there is adequate scientific basis for a serum marker to justify the use of the Serum Treasury, should submit a brief overview of a proposal. Researchers will be asked for additional information after the initial screening process. Proposals will be evaluated on an ongoing basis. In certain cases where the research applies directly to the primary interests of OFAS, limited grants to fund collaborative studies are available.

THE ORGANIZATION OF AMERICAN HISTORIANS (OAH)

112 N Bryan Ave, Bloomington, IN, 47408-4141, United States of America
Tel: (1) 812 855 7311
Fax: (1) 812 855 0696, 812 856 3340
Email: awards@oah.org
Website: www.oah.org
Contact: Award and Prize Committee Coordinator

The Organization of American Historians (OAH) was founded in 1907 as the Mississippi Valley Historical Association and originally focused on the history of the Mississippi Valley. Now, national in scope and with approx. 11,000 members and subscribers, it is a large professional organization created and sustained for the investigation, study and teaching of American history.

Avery O Craven Award

Subjects: The coming of the Civil War, the Civil War years or the Era of Reconstruction, with the exception of works of purely military history.
Purpose: For the most original book on the coming of the Civil war, civil war years, or the era of reconstruction, with the exception of works of purely military history.
Type: Award
Value: US$200
Frequency: Annual
No. of awards offered: 1
Application Procedure: Applicants must visit the website www.oah.org/activities for complete application requirements. There is no standard application form and no application fee. Publishers are encouraged to enter one or more books in the competition.
Closing Date: October 1st
Contributor: OAH
No. of awards given last year: 1
No. of applicants last year: 57
Additional Information: The exception of works of purely military history recognizes and reflects the Quaker convictions of Avery Craven, President of the Organization of American Historians (1963–1964).

Binkley-Stephenson Award

Purpose: For the best scholarly article published in *The Journal of American History* during the preceding calendar year.
Type: Award
Value: US$500
Frequency: Annual
Application Procedure: Applicants must visit the website www.oah.org/activities for complete application requirements. There is no standard application form and no application fee.
Closing Date: N/A
No. of awards given last year: 1
Additional Information: All articles published in the *The Journal of American History* during the preceding calendar year are considered.

David Thelen Award

Purpose: To expose Americanists to scholarship originally published in a language other than english, to overcome the language barrier that keeps scholars apart.
Type: Award
Value: US$500 and the winning article will be printed in The Journal of American History
Frequency: Every 2 years
Application Procedure: Please refer to the website http://oah.org/awards/awards.thelen.index.html.
No. of awards given last year: 1
No. of applicants last year: 5
Additional Information: Please check website for further details.

Ellis W Hawley Prize

Subjects: Political economy, politics, or institutions of the United States in its domestic or international affairs, from the civil war to the present.
Purpose: For the best book-length historical study of the political economy, politics, or institutions of the United States in its domestic or international affairs, from the civil war to the present.
Eligibility: Eligible works shall include book-length historical studies, written in English and published during a given calendar year.
Type: Prize
Value: US$500
Frequency: Annual
No. of awards offered: 1
Application Procedure: Applicants should visit the website for complete application requirements. There is no standard application form and no application fee.
Closing Date: October 1st
No. of awards given last year: 1
No. of applicants last year: 72

Erik Barnouw Award

Subjects: American history.
Purpose: To recognize outstanding reporting or programming on network television, cable television or in a documentary film, concerned with American history, the study of American history, and/or the promotion of history.
Level of Study: Professional development
Type: Award
Value: US$500 (if one film is selected); US$250 each (if two films are selected)
Frequency: Annual
No. of awards offered: 1–2
Application Procedure: Applicants should visit the website for complete application requirements. There is no standard application form and no application fee. Companies are encouraged to enter one or more films in the competition.
Closing Date: December 1st
No. of awards given last year: 1
No. of applicants last year: 21

Frederick Jackson Turner Award

Subjects: American history.
Purpose: For an author's first book dealing with significant phase of American history.
Eligibility: The work must be the first book-length study of history published by the author. If the author has a PhD, he or she must have received it no more than 7 years prior to the submission of the manuscript for publication. The work must be published in the calendar year before the award is given.
Type: Award
Value: US$1,000
Frequency: Annual
No. of awards offered: 1
Application Procedure: Applicants should visit the website www.oah.org/activities for complete application requirements. There is no standard application form and no application fee.
Closing Date: October 1st
No. of awards given last year: 1
No. of applicants last year: 101

Huggins-Quarles Dissertation Award

Purpose: For graduate students of color at the dissertation research stage of their PhD programme.
Eligibility: Open to minority graduate students at the dissertation research stage of their PhD.
Level of Study: Postgraduate
Type: Award
Value: US$1,200 if one recipient is selected and US$600 each if two recipients are selected

Frequency: Annual
No. of awards offered: 1–2
Application Procedure: Applicants should visit the website www.oah.org/activities for complete application requirements. There is no standard application form and no application fee.
Closing Date: December 1st
No. of awards given last year: 1
No. of applicants last year: 9

James A Rawley Prize

Subjects: The history of race relations in the United States of America.
Purpose: To reward a book dealing with the history of race relations in the United States of America.
Type: Prize
Value: US$1,000
Frequency: Annual
No. of awards offered: 1
Application Procedure: Applicants should visit the website www.oah.org/activities for complete application requirements. There is no standard application form and no application fee. Publishers are encouraged to enter one or more books in the competition.
Closing Date: October 1st
No. of awards given last year: 1
No. of applicants last year: 117

The Japan Residencies Program

Purpose: To facilitate scholarly dialogue and contribute to the expansion of scholarly networks among students and professors of American history in America and Japan.
Eligibility: Applicants must be members of the OAH, have a PhD and be scholars of American history.
Value: Round-trip airfare to Japan, housing (if the host university cannot offer housing, applicants are expected to pay hotel expenses from the daily stipend) and modest daily expenses
Frequency: Annual
Application Procedure: Please refer to the website http://oah.org/programs/residencies/index.html.
No. of awards given last year: 2
No. of applicants last year: 28

Lawrence W. Levine Award

Subjects: American cultural history.
Purpose: To recognize scholarly and professional achievement in the field of American cultural history.
Eligibility: Open to applicants of any nationality.
Type: Award
Value: $1,000
Frequency: Annual
No. of awards offered: Varies
Application Procedure: Applicants must visit the website www.oah.org/activities for complete application requirements. There is no standard application form and no application fee. Publishers are encouraged to enter one or more books in the competition.
Closing Date: October 1st
No. of awards given last year: 1
No. of applicants last year: 108

Lerner-Scott Dissertation Prize

Subjects: United States of America women's history.
Purpose: For the best doctoral dissertation in US women's history.
Level of Study: Postdoctorate
Type: Prize
Value: US$500
Frequency: Annual
No. of awards offered: 1
Application Procedure: Applicants must visit the website www.oah.org/activities for complete application requirements. There is no standard application form and no application fee.
Closing Date: October 1st
No. of awards given last year: 1
No. of applicants last year: 16

Liberty Legacy Foundation Award

Subjects: American history and the civil rights movement.
Purpose: For the best book on any historical aspect of the struggle for civil rights in the United States, from the nation's founding to the present.
Eligibility: Open to writers and publishers writing a book on any historical aspect of the struggle for human civil rights in the United States of America. Each entry must be published within a specified time period prior to the application deadline.
Level of Study: Professional development
Type: Competition
Value: US$800
Frequency: Annual
Country of Study: United States of America
No. of awards offered: 1
Application Procedure: Applicants must visit the website www.oah.org/activities for complete application requirements. There is no standard application form and no application fees. Publishers are encouraged to enter one or more books in the competition.
Closing Date: October 1st
No. of awards given last year: 1
No. of applicants last year: 49
Additional Information: This award was inspired by the OAH President Darlene Clark Hine's call in her 2002 OAH presidential address for more research on the origins of the civil rights movement in the period before 1954.

Louis Pelzer Memorial Award

Subjects: Any period or topic in the history of the United States of America.
Purpose: For the best essay in American history by a graduate student.
Type: Award
Value: The winning essay will be published in *The Journal of American History*. The organization offers a prize of US$500
Frequency: Annual
Application Procedure: Applicants must visit the website www.oah.org/activities for complete application requirements. There is no standard application form and no application fee.
Closing Date: December 1st
No. of awards given last year: 1
No. of applicants last year: 25

Merle Curti Award

Subjects: American social and intellectual history.
Purpose: To recognize books in the fields of American social and intellectual history.
Type: Award
Value: US$1,000 if one book is selected; US$500 each if two books are selected
Frequency: Annual
No. of awards offered: 1–2
Application Procedure: Applicants must visit the website www.oah.org/activities for complete application requirements. There is no standard application form and no application fee. Publishers are encouraged to enter one or more books in the competition.
Closing Date: October 1st
Contributor: OAH
No. of awards given last year: 2
No. of applicants last year: 153

OAH Darlene Clark Hine Award

Subjects: African-American women's and gender history.
Purpose: To honor the author and publisher of the best book on African-American women's and gender history.
Eligibility: Any author whose book on African-American women's and gender history is published between January 2008 and December 2009, and the publisher of the book.
Level of Study: Unrestricted
Type: Award
Value: US$1,000
No. of awards offered: 1
Application Procedure: The author/publisher must send one copy of the book to each committee member of the organization before October 2009 in order to be considered for the award.
Closing Date: October 1st
No. of awards given last year: 1
No. of applicants last year: 17

OAH-IEHS John Higham Travel Grants

Purpose: For graduate students to be used towards costs of attending the OAH/IEHS annual meeting.
Type: Travel grant
Value: US$500 for attending the OAH/IEHS annual meeting
Frequency: Annual
No. of awards offered: 3
Application Procedure: Applicants must visit the website www.oah.org/activities for complete application requirements. There is no standard application form and no application fee.
Closing Date: December 1st
No. of awards given last year: 3
No. of applicants last year: 6

Richard W Leopold Prize

Subjects: Foreign policy, military affairs and the historical activities of the federal government.
Purpose: To improve contacts and interrelationships within the historical profession where an increasing number of history-trained scholars hold distinguished positions in governmental agencies.
Eligibility: Applicant must have been employed in a government position for at least five years.
Level of Study: Professional development
Type: Prize
Value: US$1,500
Frequency: Every 2 years
Application Procedure: Please refer to the website http://oah.org/awards/awards.leopold.index.html.
No. of awards given last year: 1
No. of applicants last year: 12

Tachau Teacher of the Year Award

Subjects: History.
Purpose: To recognize the contributions made by precollegiate and classroom teachers to improve history education.
Eligibility: Applicants must be precollegiate teachers engaged at least half time in history teaching, with exceptional ability in initiating projects that involve students in historical research. They should also demonstrate ability in working with museums, historical preservation societies or other public history associations, and in publishing or presenting scholarship that advances education or knowledge.
Level of Study: Professional development
Type: Award
Value: US$500, a one-year OAH membership, a one-year subscription to the *OAH Magazine of History* and a complimentary registration for the annual meeting. If the winner is an OAH member, the award will include a one-year renewal of membership in the awardee's usual membership category. The winner's school will receive a certificate.
Frequency: Annual
No. of awards offered: 1
Application Procedure: Applicants must visit the website www.oah.org/activities for complete application requirements. There is no standard application form and no application fees.
Closing Date: December 1st
No. of awards given last year: 1
No. of applicants last year: 8

ORTHOPAEDIC RESEARCH AND EDUCATION FOUNDATION (OREF)

6300 N River Road, Suite 700, Rosemont, IL, 60018 4261, United States of America
Tel: (1) 847 698 9980
Fax: (1) 847 698 7806
Email: communications@oref.org
Website: www.oref.org
Contact: Mrs Jean McGuire, Vice President Grants

In 1955, leaders of the major professional organizations in the speciality, the American Orthopaedic Association, the American Academy of Orthopaedic Surgeons and the Orthopaedic Research Society, established the Orthopaedic Research and Education Foundation (OREF) as a means of supporting research and building the scientific base of clinical practice. Today, the Foundation raises over US$5 million per year and holds a unique place in medicine in the United States of America.

OREF Career Development Grant

Subjects: Orthopaedic surgery.
Purpose: To encourage a commitment to scientific research in orthopaedic surgery.
Eligibility: Applicants must be orthopaedic surgeons, and are not eligible if they are holders of the NIH ROI award. PhD scholars may apply if they are affiliated with an orthopaedic department.
Level of Study: Professional development
Type: Grant
Value: Up to US$75,000 per year
Length of Study: 3 years
Frequency: Dependent on funds available
Country of Study: United States of America
No. of awards offered: 2–3
Application Procedure: Applicants must make a formal application together with letters of recommendation.
Closing Date: September 14th
Funding: Private
Contributor: Orthopaedic surgeons
No. of awards given last year: 1
No. of applicants last year: 12

OREF Clinical Research Award

Subjects: Orthopaedics.
Purpose: To recognize outstanding clinical research related to musculoskeletal disease or injury.
Eligibility: Restricted to members of the American Academy of Orthopaedic Surgeons, the Orthopaedic Research Society (ORS), the Canadian Orthopaedic Association or the Canadian Orthopaedic Research Society. Alternatively, candidates may be sponsored by a member.
Level of Study: Professional development
Type: Award
Value: US$20,000
Length of Study: 1 year
Frequency: Annual
Country of Study: United States of America or Canada
No. of awards offered: 1
Application Procedure: Applicants must submit an original manuscript.
Closing Date: July 1st (Please check with the website)
Funding: Private
Contributor: Orthopaedic surgeons
No. of awards given last year: 1
No. of applicants last year: 9

OREF Prospective Clinical Research Grant

Subjects: Orthopaedics.
Purpose: To provide funding for promising prospective clinical proposals.
Eligibility: Applicants must be orthopaedic surgeons. PhD scholars may apply if they are affiliated with an orthopaedic department.
Level of Study: Professional development
Type: Research grant
Value: Up to US$150,000
Length of Study: 3 years
Frequency: Dependent on funds available
Study Establishment: A medical centre
Country of Study: United States of America
No. of awards offered: 2
Application Procedure: Applicants must make a formal application.
Closing Date: October 15th (Please check with the website)
Funding: Private
Contributor: Orthopaedic surgeons
No. of awards given last year: 2
No. of applicants last year: 24

OREF Resident Clinician Scientist Training Grants

Subjects: Orthopaedics.
Purpose: To encourage the development of research interests for residents and Fellows.
Eligibility: Applicants must be orthopaedic surgeon residents or orthopaedic Fellows in an approved residency programme in the United States of America.
Level of Study: Professional development
Type: Grant
Value: US$20,000
Length of Study: 1 year
Frequency: Annual
Study Establishment: A medical centre
Country of Study: United States of America
No. of awards offered: 9–12
Application Procedure: Applicants must make a formal application.
Closing Date: September 15th (Please check with the website)
Funding: Private
Contributor: Orthopaedic surgeons
No. of awards given last year: 11
No. of applicants last year: 54

OREF/Musculoskeletal Transplant Foundation Research Grant

Subjects: Sports medicine, surgery, rheumatology and treatment techniques.
Purpose: To encourage new investigators by providing seed money and start-up funding.
Eligibility: Open to orthopaedic surgeons who are principal investigators (PI) or co-PIs. PIs cannot have NIH ROI awards.
Level of Study: Postdoctorate
Type: Research grant
Value: Up to US$100,000
Length of Study: 2 years
Frequency: Annual
Study Establishment: A medical centre
Country of Study: United States of America
No. of awards offered: 9–12
Application Procedure: Applicants must make a formal application.
Closing Date: October 15th (Please check with the website)
Funding: Private
Contributor: Orthopaedic surgeons and corporations
No. of awards given last year: 7
No. of applicants last year: 58

OUR WORLD-UNDERWATER SCHOLARSHIP SOCIETY

PO Box 4428, Chicago, IL 60680, United States of America
Tel: (1) 630 969 6690
Fax: (1) 630 969 6690
Email: vicepresident-EU@owuscholarship.org
Website: www.owuscholarship.org

The Our World Underwater Scholarship Society in an organization dedicated to promoting education in the underwater world.

Our World-Underwater Scholarship Society Rolex Scholarships

Subjects: Underwater-related disciplines.
Purpose: To provide hands-on introduction to underwater and other aquatic-related endeavors.
Eligibility: Open to applicants who are certified scuba divers and above the age of 21 years. Refer to the website for further details.
Level of Study: Professional development
Type: Scholarship
Value: The maximum cash amount for the North American Rolex Scholarship is $25,000, for the European Rolex Scholarship the amount is £16,000 and for the Australasian Scholarship Australian $30,000.
Length of Study: 1 year
Frequency: Annual
No. of awards offered: 3
Application Procedure: Application form along with the required documents must be submitted before the deadline.
Closing Date: December 31st
Funding: Corporation, individuals
Contributor: Rolex Watch, USA and Rolex Watch, Geneva
No. of awards given last year: 3

PAINTING AND DECORATING CONTRACTORS OF AMERICA (PDCA)

1801 Park 270 Drive, Suite 220, St Louis, MO, 63146, United States of America
Tel: (1) 800 332 7322
Fax: (1) 314 514 9417
Email: ihoren@pdca.org
Website: www.pdca.org

The PDCA was established in 1884 by a group of contractors to devise a means for assuring the public of the skill, honorable reputation, and probity of master painters.

A E Robert Friedman PDCA Scholarship

Subjects: All subjects.
Purpose: To provide scholastic and vocational aid.
Eligibility: Open to applicants who are under the age of 26 and are nominated by a PDCA member
Level of Study: Postgraduate
Type: Scholarship
Frequency: Annual
No. of awards offered: 2
Application Procedure: See the website.
Closing Date: September 17th
Additional Information: Over $310,000 has been contributed to the Fund

PALOMA O'SHEA SANTANDER INTERNATIONAL PIANO COMPETITION

Calle Hernán Cortés 3, Santander, E-39003, Spain
Tel: (34) 94 231 1451
Fax: (34) 94 231 4816
Email: concurso@albeniz.com
Website: www.fundacionalbeniz.com
Contact: A Kaufmann, Secretariat General

The Paloma O'Shea Santander International Piano Competition is one of the best rated competitions in the world. It provides an opportunity for exceptionally talented pianists to enhance their careers. The jury is composed of renowned musicians in order to ensure that grants are made in a fair and unbiased manner.

Paloma O'Shea Santander International Piano Competition

Subjects: Piano performance.
Purpose: To give support to young pianists of exceptional talent.
Eligibility: Open to pianists of any nationality under 29 years of age.
Level of Study: Unrestricted
Type: Competition
Value: approximately US$135,000
Length of Study: July 25th–August 7th
Frequency: Every 4 years
Country of Study: Spain
No. of awards offered: 7
Application Procedure: Please refer to the website www.santanderpianocompetition.com.
Closing Date: November
Funding: Private, commercial, foundation, government
Contributor: Fundacion Albeniz
No. of awards given last year: 7
No. of applicants last year: 220
Additional Information: Please check website for further details at www.santanderpianocompetition.com

PAN AMERICAN HEALTH ORGANIZATION (PAHO) REGIONAL OFFICE OF THE WORLD HEALTH ORGANIZATION (WHO)

Regional Office for the Americas/Pan American Sanitary Bureau, 525 23rd Street NW, Washington, DC, 20037-2895, United States of America
Tel: (1) 202 974 3000
Fax: (1) 202 974 3663
Email: rgp@paho.org
Website: www.paho.org
Contact: Grants Administrator

The Pan American Health Organization (PAHO) is an international public health agency working to improve the health and living standards of the countries of the Americas. It serves as the specialized organization for health of the Inter-American System and as the regional office for the Americas of the World Health Organization.

PAHO Grants

Subjects: Public health studies.
Purpose: To contribute to the public health of the countries of the Americas (individual programme objectives vary). Programmes include graduate thesis grants, research training grants, regional research competitions (announced yearly) and special initiatives announced via the PAHO website.
Eligibility: Open to citizens and residents of Latin America and the Caribbean.
Level of Study: Postgraduate, Doctorate, Graduate, Research
Type: Research grant
Value: Please contact the organization
Study Establishment: Varies
Country of Study: Latin America or Caribbean countries only
No. of awards offered: Varies according to the programme
Application Procedure: Applicants must complete an application form. Guidelines and application forms are available from the website.
Closing Date: Varies according to the programme
Funding: Government
Contributor: Member states and their agencies

PARALYZED VETERANS OF AMERICA (PVA)

801,18th Street NW, Washington DC, 20006-3517, United States of America
Tel: (1) 800 555 9140
Fax: (1) 202 416 7652
Email: info@pva.org
Website: www.pva.org

The Paralyzed Veterans of America (PVA), a congressionally chartered veterans service organization founded in 1946, has developed a unique expertise on a wide variety of issues involving the special needs of the members-veterans of the armed forces who have experienced spinal cord injury or dysfunction.

PVA Educational Scholarship Program

Subjects: Spinal cord injury or dysfunction.
Purpose: To provide financial support for PVA members so that they can achieve their goals in the academic arena.
Eligibility: Open to all PVA members and their families.
Level of Study: Postgraduate
Type: Scholarship
Value: US$1,000
Frequency: Annual
No. of awards offered: 10
Application Procedure: Details are available on the website.
Closing Date: June 17th
Additional Information: We expect to offer new college scholarships next year. Check back on our website after April 1st, for complete information and a new application form.

PARAPSYCHOLOGY FOUNDATION, INC.

PO Box 1562, New York, NY, 10021 0043, United States of America
Tel: (1) 212 628 1550
Fax: (1) 212 628 1559
Email: office@parapsychology.org
Website: www.parapsychology.org
Contact: Vice President

Established in 1951, the Parapsychology Foundation acts as a clearinghouse for information about parapsychology. Essentially an administrative organization, it maintains one of the largest libraries to do with parapsychology, the Eileen J Garret Library, as well as supporting various programmes that include the library, a grant and scholarship programme, a conference and lecture programme, and a speaker's bureau and publishing programme.

Eileen J Garrett Scholarship

Subjects: Parapsychology.
Purpose: To assist students attending an accredited college or university in pursuing the academic study of the science of parapsychology.
Eligibility: Open to nationals of any country.
Level of Study: Unrestricted
Type: Scholarship
Value: US$3,000
Length of Study: 1 year
Frequency: Annual
Study Establishment: An accredited college or university
Country of Study: Any country
No. of awards offered: 1
Application Procedure: Applicants must submit samples of writings on the subject with an application form from the Foundation. Letters of reference are required from three individuals, familiar with the applicant's work and/or studies in parapsychology.
Closing Date: July 15th

For further information contact:

Email: office@parapsychology.org

Parapsychology Foundation Grant

Subjects: Parapsychology.
Purpose: To support original study, research and experiments in parapsychology.
Eligibility: Open to nationals of any country.
Level of Study: Unrestricted
Type: Grant
Value: Up to US$3,000
Length of Study: 1 year
Frequency: Annual
Country of Study: Any country
No. of awards offered: 10
Application Procedure: Applicants must contact the organization for details.
Closing Date: July 15th
Additional Information: Funding for the Foundation Grant is limited but applicants are still welcome to submit a proposal on the off chance that the programme will be reviewed.

PARKER B FRANCIS FELLOWSHIP PROGRAM

VA Puget Sound Health Care System, 1660S Columbian Way, 151L, Seattle, WA, 98108, United States of America
Tel: (1) 206 764 2219
Fax: (1) 617 277 2382
Email: deborah.snapp@gmail.com
Website: www.francisfellowships.org
Contact: Ms Deborah Snapp

A private foundation dedicated to creating current and future generations of well-rounded individuals who are creative, life-long learners, striving to achieve their fullest potential within their communities.

Parker B Francis Fellowship Program

Subjects: Pulmonary research.
Purpose: To support rising stars in the field of pulmonary research as they make the transition from postdoctoral trainees to independent researchers.
Eligibility: Ideally, open to applicants with between 2 and 7 years of postdoctoral research experience, published articles in leading journals and a clear trajectory in pulmonary research.
Level of Study: Postdoctorate
Type: Fellowship
Value: The total budget is limited to US$50,000 for the 1st year, US $52,000 for the 2nd and US$54,000 for the 3rd. These totals include a stipend plus fringe benefits and may include travel costs to a maximum of US$2,000. Direct research project costs and indirect costs are not covered.
Length of Study: 3 years
Frequency: Annual
Country of Study: United States of America, Canada or Mexico
No. of awards offered: 15
Application Procedure: Applicants must submit a completed application form, biographical sketch and a brief statement of their career goals. They must also provide a letter from their mentor evaluating the applicant's qualifications and indicating their career goals in the field of pulmonary research, three letters of recommendation, a summary of the past training record of the primary mentor (including names of former trainees and their current positions, sources and level of support with grants pending and the extent of equipment and space for research training available to the primary mentor and trainee) and signatures of the primary mentor, department or division head and the fiscal officer responsible for administering the grant on the face page.
Closing Date: October 15th (Please check with the website)
Funding: Private
Contributor: The Francis Family Foundation
No. of awards given last year: 15
No. of applicants last year: 69

THE PARKINSON'S UK

215 Vauxhall Bridge Road, London, SW1V 1EJ, England
Tel: (44) 20 7931 8080
Fax: (44) 20 7233 9908
Email: researchapplications@parkinsons.org.uk
Website: www.parkinsons.org.uk
Contact: Ms Bunia Gorelick, Research Grants Manager

Parkinson's UK is the largest charitable funder of Parkinson's research in the UK. So far, we've invested more than £55 million in groundbreaking research. By 2014, that figure will have reached £75million.

Parkinson's UK Career Development Awards

Subjects: Parkinson's disease.
Purpose: To support individuals in 2 ways. (1) Senior research fellowships - for individuals who wish to specialise in Parkinson's research by establishing their own research group. (2) Training fellowships - for individuals who wish to undertake research training relevant to Parkinson's, with the aim of achieving a higher research degree.
Eligibility: Grants are tenable only at a UK university, NHS trust, statutory social care organisation or other research institution.
Level of Study: Research
Type: Fellowship
Value: Up to £250,000
Length of Study: 3 years
Frequency: Annual
Study Establishment: UK university, NHS trust, statutory social care organisation or other research institution
Country of Study: United Kingdom
No. of awards offered: Varies
Application Procedure: Applications are made through our online system at https://research.parkinsons.org.uk.
Closing Date: December 2nd
Funding: Foundation, government
Contributor: Voluntary donations
No. of awards given last year: 3
No. of applicants last year: 13

Parkinson's UK Innovation Grants

Subjects: Parkinson's disease.
Purpose: To fund high-risk, high-reward research that tests new and unconventional hypotheses tackling major scientific or technical hurdles in the field of Parkinson's.
Eligibility: Principal applicants should hold employment contracts that extend beyond the period of the grant.
Level of Study: Research
Type: Grant
Value: £30,000
Length of Study: Up to 12 months
Frequency: 5 rounds per year
Study Establishment: UK university, NHS trust, statutory social care organisation or other research institution
Country of Study: United Kingdom
No. of awards offered: 10
Application Procedure: Applications are made through our online system at https://research.parkinsons.org.uk.
Closing Date: Please see website
Funding: Foundation
Contributor: Voluntary donations
No. of awards given last year: 10
No. of applicants last year: 62
Additional Information: There are 5 deadlines in a year.

Parkinson's UK PhD Studentship

Subjects: Parkinson's disease.
Purpose: To build research capacity related to Parkinson's by offering fully funded PhD studentships for outstanding students.
Eligibility: The PhD studentship scheme is not intended to cover funding gaps for PhD students whose studies have already started. Both home and overseas students with a first degree in a relevant discipline are eligible for this scheme. Supervisors should contact their university's research office regarding fees for overseas students. To allow recruitment of the most able students early in their final year, the final decision on the studentships will be made in December for appointment in summer/autumn of the following year.
Level of Study: Doctorate
Type: Studentship
Value: A stipend of £15,000 per year (or £16,000 per year if in London), full PhD tuition fees (at the UK/EU rate) and a contribution of up to £10,000 per year towards research costs. Students from outside of the EU are welcome to apply for this scheme on a full fee paying basis.
Length of Study: 3 years
Frequency: Annual
Study Establishment: Any UK university, NHS trust, statutory social care organisation or other research institution
Country of Study: United Kingdom
No. of awards offered: 3
Application Procedure: Applications are made by the potential student's supervisor (who must be based at a UK university, hospital or research institute) through our online system at https://research.parkinsons.org.uk.
Closing Date: June 17th
Funding: Foundation
Contributor: Voluntary donations
No. of awards given last year: 3
No. of applicants last year: 16

Parkinson's UK Project Grant

Subjects: Parkinson's disease.
Purpose: To support studies designed to answer a single question or a small group of related questions about an aspect of Parkinson's.
Eligibility: Grants are tenable only at a United Kingdom University or NHS Trust or statutory social care organization. Applicants should hold employment contracts that extend beyond the period of grant.
Level of Study: Research
Type: Project grant
Value: To include at least one salary plus consumables and equipment.
Length of Study: Up to 3 years

Frequency: Twice a year
Study Establishment: UK university, NHS trust, statutory social care organisation or other research institution
Country of Study: United Kingdom
No. of awards offered: Varies
Application Procedure: Applicants must complete an online application form available at https://research.parkinsons.org.uk.
Closing Date: October 7th
Funding: Foundation
Contributor: Voluntary donations
No. of awards given last year: 5
No. of applicants last year: 21
Additional Information: For further details visit the Charity's website.

PASTEUR FOUNDATION

420 Lexington Avenue Suite 1654, New York, NY, 10170, United States of America
Tel: (1) 212 599 2050
Fax: (1) 212 599 2047
Email: PasteurUS@aol.com
Website: www.pasteurfoundation.org
Contact: Caitlin Hawke, Executive Director

Pasteur Foundation is located in New York city and works to introduce the research conducted at the Institut Pasteur to the American public, to develop exchanges between Pasteurian and United States scientists and to raise funds for Pasteurian research.

Pasteur Foundation Postdoctoral Fellowship Program

Subjects: Biomedical science.
Purpose: To provide financial assistance and develop international scientific exchanges.
Level of Study: Postdoctorate
Type: Fellowship
Value: US$70,000 per year
Length of Study: 3 years
Frequency: Annual
Country of Study: France
Application Procedure: Applicants must submit a passport-size photo, reprints of publications, letters of recommendation, letter of support and curriculum vitae. See fellowship program on www.pasteurfoundation.org.
Closing Date: September 20th
For further information contact:

Institut Pasteur, 25 rue du Docteur Roux, 75724 Paris Cedex 15, France
Contact: Dr Claude Parsot, Secrétariat de la Direction de l'Evaluation Scientifique

PATERSON INSTITUTE FOR CANCER RESEARCH

The University of Manchester, Wilmslow Road, Withington, Withington, Manchester, M20 4BX, England
Tel: (44) 16 1446 3156
Fax: (44) 16 1446 3109
Email: enquiries@picr.man.ac.uk
Website: www.paterson.man.ac.uk

The Paterson Institute for Cancer Research is a Cancer Research UK-funded centre for cancer research and now forms a part of the University of Manchester. It carries out cancer research in a large number of areas and has 18 research groups and around 250 scientists.

Paterson 4-Year Studentship

Subjects: Molecular and cellular basis of cancer and translational cancer research.
Purpose: To support study towards a PhD.
Eligibility: Open to candidates who have obtained a First or Second Class (Honours) Bachelor of Science Degree. International First or 2.1 degree equivalent in biological science, medicine, or related subject.
Level of Study: Postgraduate, Doctorate
Type: Studentship
Value: UK£15,300 as stipend per year, university fees and bench fees
Length of Study: 4 years
Frequency: Annual, 3 times per year
Study Establishment: The University of Manchester
Country of Study: United Kingdom
No. of awards offered: Up to 10 per year
Application Procedure: www.paterson.man.ac.uk for details and download application form.
Closing Date: Refer the website
Funding: Private
Contributor: Cancer Research UK
No. of awards given last year: 8
No. of applicants last year: 320
Additional Information: All positions are advertised on the website. Self-funded students are accepted subject to qualifications and 3-year funding. Please use application form (website).
For further information contact:

Postgraduate Tutor, Paterson Institute
Email: gcowling@picr.man.ac.uk

THE PAUL & DAISY SOROS FELLOWSHIPS FOR NEW AMERICANS

400 West 59th Street 4th floor, New York, NY, 10019, United States of America
Tel: (1) 212 547 6926
Fax: (1) 212 548 4623
Email: pdsoros_fellows@sorosny.org
Website: www.pdsoros.org

The Soros Foundation was founded by Hungarian immigrants and American philanthropists in December 1997 with a charitable trust of 50 million dollars. Their reasons for doing so were several. They wished to give back to the country that had afforded them and their children such great opportunities and felt a fellowship programme was an appropriate vehicle.

Paul and Daisy Soros Fellowships for New Americans

Subjects: Engineering, medicine, law, social work, humanities, social sciences, sciences, fine arts and performing arts.
Purpose: To provide opportunities for continuing generations of able and accomplished New Americans to achieve leadership in their chosen fields.
Eligibility: Open to resident aliens, who have been naturalized as citizens of United States or whose parents are both naturalized citizens. Applicants must either have a Bachelor's degree or be in their final year of undergraduate study and must not be older than 30 years of age as of November 1st.
Level of Study: Graduate, Postgraduate, Doctorate
Type: Fellowships
Value: US$25,000 (paid in two installments) and a tuition grant of one-half the tuition cost of the United States graduate programme attended by the Fellow (up to US$20,000 per academic year)
Length of Study: 2 years
Frequency: Annual
Study Establishment: Any accredited graduate programme
Country of Study: United States of America
No. of awards offered: 30
Application Procedure: Applicants must submit a completed online application form, 2 essays on specified topics, 1–2 page curriculum vitae, 3 recommendation letters, transcripts and provide a photocopy of the test scores of Graduate Management Admission Test, MCAT, Graduate Record Examination and LSAT.
Closing Date: November 1st
Funding: Trusts
No. of awards given last year: 30
No. of applicants last year: 890
Additional Information: Individuals are expected to retain loyalty and a sense of commitment to their country of origin as well as to the United States, but is intended to support individuals who will continue to regard the United States as their principal residence and focus of national identity.

THE PAUL & DAISY SOROS FELLOWSHIPS FOR NEW AMERICANS

400 West 59th Street, 4th floor, New York, NY, 10019, United States of America
Tel: (1) 212 547 6926
Fax: (1) 212 548 4623
Email: pdsoros_fellows@sorosny.org
Website: www.pdsoros.org

The Paul & Daisy Soros Fellowship for New Americans

Subjects: Any subject.
Purpose: To provide opportunities for continuing generations of able and accomplished New Americans to achieve leadership in their chosen fields.
Eligibility: Open to New Americans: resident aliens (Green Card Holders) naturalized US citizens and/or children of 2 naturalized parents.
Level of Study: Postdoctorate
Type: Fellowship
Value: US$25,000& up to $20,000 in tuition support for each year
Length of Study: 2 years
Study Establishment: Any accredited graduate University in the United States
Country of Study: United States of America
No. of awards offered: 30
Application Procedure: Apply online.
Closing Date: November 12th
Funding: Private
Contributor: Paul and Daisy Soros
No. of awards given last year: 30
No. of applicants last year: 77

THE PAUL FOUNDATION

Apeejay House, 15 Park Street, Kolkata, West Bengal, 700 016, India
Tel: (91) 033 44035455
Fax: (91) 033 22299596
Email: thepaulfoundation@apeejaygroup.com
Website: www.thepaulfoundation.org

The Paul Foundation has been conceived to promote the individuals quest for intellectual excellence. To pursue higher studies in India and abroad, the Foundation supports those Indians who have shown ability in their specific areas of interest, to give them an opportunity to develop their potential and be recognized as leaders.

The Paul Foundation Postgraduate Scholarship

Subjects: Humanities, social sciences, basic sciences, applied sciences, law, management, medicine, engineering, fine arts and the performing arts.
Purpose: To encourage outstanding scholars who have the ability to produce thoughtful and thought-provoking work and make a difference to society.
Eligibility: Open to applicants who are Indian citizens and have completed their graduation from a UGC-recognized Indian university.
Level of Study: Doctorate, Postgraduate
Type: Scholarship
Value: To cover academic, living, and travel expenses for a maximum of 2 years.
Frequency: Annual
Country of Study: Any country
Application Procedure: A completed application form must be submitted.
Closing Date: February 28th
Funding: Foundation

THE PAULO CELLO COMPETITION

PL 1105, Helsinki, FIN-00101, Finland
Tel: (358) 40 5467079
Email: cello@paulo.fi
Website: www.cellocompetitionpaulo.org
Contact: Administrative Assistant

The Paulo Cello Competition organizes an international competition for cellists of all nations.

International Paulo Cello Competition

Subjects: Cello performance.
Eligibility: Open to cellists between 16 and 33 years of age.
Level of Study: Unrestricted
Type: Competition
Value: The 1st prize is €15,000, the 2nd prize is €12,000, the 3rd prize is €9,000 and the 4th, 5th and 6th prizes are €2,000
Frequency: Every 5 years
Country of Study: Finland
No. of awards offered: 6
Application Procedure: Applicants must write for a brochure, which contains details of the application and audition pieces.
Funding: Private
Contributor: The Paulo Foundation
No. of awards given last year: 6
No. of applicants last year: 77

PEMBROKE CENTER

172 Meeting Street Box l958, Brown University, Providence, RI, 02912, United States of America
Tel: (1) 401 863 2643/3466
Fax: (1) 401 863 1298
Email: pembroke_center@Brown.edu
Website: www.pembrokecenter.org
Contact: Kay Warren, Director

The Pembroke Center supports interdisciplinary research and teaching across the humanities and social sciences. With a focus on the human cost and potential of social change, the center's research agenda has a transnational perspective.

Pembroke Center Post-Doctoral Fellowships

Subjects: All subjects.
Purpose: To encourage scholars to pursue research in any discipline.
Eligibility: Open to scholars from all disciplines. Recipients may not hold a tenured position.
Level of Study: Postdoctorate
Type: Fellowships
Value: US$50,000 plus supplement for health insurance
Length of Study: 1 year
Frequency: Annual
Study Establishment: Brown University
Country of Study: United States of America
No. of awards offered: 3
Application Procedure: Apply online at https://secure.interfolio.com/apply/14921.
Closing Date: December
No. of awards given last year: 3
Additional Information: The Center particularly encourages under-represented and minority scholars to apply.

PEN AMERICAN CENTER

588 Broadway, Suite 303, New York, NY 10012, United States of America
Tel: (1) 212 334 1660
Fax: (1) 212 334 2181
Email: pen@pen.org, awards@pen.org
Website: www.pen.org
Contact: Paul W Morris, Awards Director

PEN American Center is a fellowship of writers dedicated to to advance literature, depend free expression and foster international fellowship. The American Center is the largest of 145 international PEN centers worldwide.

The PEN Translation Fund Grants

Subjects: Translations of works of fiction, creative nonfiction, poetry, and drama.
Purpose: The fund seeks to encourage translators to undertake projects they might not otherwise have had the means to attempt.
Eligibility: Book-length works that have not previously appeared in English in print or have appeared only in an outdated or flawed translation.

Level of Study: Unrestricted
Type: Grant
Value: US$2,000–4,000
Frequency: Annual
Country of Study: Any country
Application Procedure: All applications must include the cover sheet and items outlined at www.pen.org/awards, including the original and translated word, translator curriculum vitae, and artist's statement. Please send seven copies as instructed.
Closing Date: Between October 1st and February 1st (Early applications are strongly recommended)
Funding: Private
No. of awards given last year: 11
No. of applicants last year: 140
Additional Information: Anthologies with multiple translators, works of literary criticism and scholarly or technical texts do not qualify. Translators awarded grants by the fund are ineligible to reapply for 3 years after the year they receive a grant.

For further information contact:

PEN Literary Awards, PEN Translation Fund, PEN American Center
Email: awards@pen.org
Website: www.pen.org/awards

PEN Writer's Emergency Fund

Subjects: Professional writers in acute, emergency financial crisis.
Purpose: To assist professional published writers facing emergency situations.
Eligibility: Open to published professional writers and produced playwrights who have a traceable record of writing and publication.
Level of Study: Unrestricted
Type: Grant
Value: Up to US$2,000
Frequency: Dependent on funds available
Country of Study: United States of America & Canada
Application Procedure: Applicants must submit an application consisting of a two-page form, published writing samples, documentation of financial emergency, including bills, etc. and a professional curriculum vitae.
Closing Date: March 15th
Funding: Private
No. of awards given last year: 40
Additional Information: A separate fund exists for writers and editors with AIDS who are in need of emergency assistance. The funds are not for research purposes, to enable writers to complete unfinished projects, or to fund writing publications or organizations. Grants and loans are for unexpected emergencies only, for the support of working writers. PEN American Center also offers numerous annual awards to published writers to recognize distinguished writing, editing and translation.

PENINSULA SCHOOL OF MEDICINE AND DENTISTRY

The John Bull Building, Tamar Science Park, Research Way, Plymouth, PL6 8BU, England
Tel: (44) 01752 437474
Fax: (44) 01752 517842
Email: info@psmd.ac.uk
Website: www.pcmd.ac.uk

Peninsula Medical School and Peninsula Dental School have come together in The Peninsula College of Medicine and Dentistry, a partnership with the University of Exeter, University of Plymouth and the NHS in Devon and Cornwall. The Peninsula Medical School was established in 2000 and Peninsula Dental School was established in 2006. Postgraduate study, either at Masters level through taught programmes, or Doctorate level through research is available through the Peninsula College of Medicine & Dentistry Graduate School.

Peninsula College of Medicine and Dentistry PhD Studentships

Purpose: To attract PhD candidates of outstanding ability to join their exciting and rapidly expanding programme of internationally rated research.
Eligibility: Open to the suitably qualified graduates.
Level of Study: Doctorate
Type: Studentship
Value: £13,590 (Research Council Rate)
Frequency: Dependent on funds available
Study Establishment: Peninsula College of Medicine & Dentistry
Country of Study: England
Application Procedure: Check website for the details.
Closing Date: November 8th (See the details of specific studentship on website)
Contributor: Various sources

PENN ARTS & SCIENCES

University of Pennsylvania 3451 Walnut Street, Philadelphia, PA, 19104, United States of America
Tel: (1) 215 898 5000
Fax: (1) 215 746 5946
Email: humanities@sas.upenn.edu
Website: www.humanities.sas.upenn.edu

The Penn Humanities Forum promotes interdisciplinary collaboration across University of Pennsylvania departments and schools and between the University and the Philadelphia region. Each year, a broad topic sets the theme for a research seminar for resident and visiting scholars, courses and public events involving Philadelphia's cultural institutions.

Mellon Postdoctoral Fellowships in the Humanities

Subjects: Humanities.
Purpose: To support research by untenured junior scholars.
Eligibility: Open to junior scholars who, at the time of application, have received a PhD (degree must be in hand no later than December of the year preceding the fellowship) but have not held it for more than 8 years nor been granted tenure. Applicants may not be tenured during the year of the fellowship. Research proposals must relate to the Forum's annual topic of study and are invited in all areas of humanistic studies, except educational curriculum-building and the performing arts. Other requirements include residency at the University of Pennsylvania and the teaching of one freshman seminar in each of the two semesters.
Level of Study: Postdoctorate
Type: Fellowship
Value: $46,500 (stipend). The fellows will also have a one-time $2,500 budget for research support during the two years of their appointment, to be used for research travel, conference travel, publication expenses, or stipends to student research assistants. They will receive single health, dental, and life insurance.
Length of Study: 1 academic year, non-renewable
Frequency: Annual
Study Establishment: The University of Pennsylvania
Country of Study: United States of America
No. of awards offered: 5
Application Procedure: Applicants must complete an application form, which can be downloaded from the Forum's website.
Closing Date: November 30th
Funding: Foundation
No. of awards given last year: 5
No. of applicants last year: 200
Additional Information: Fellows may not normally hold other awards concurrently.

PERKINS SCHOOL OF THEOLOGY

Southern Methodist University, PO Box 750133, Dallas, TX, 75275 0133, United States of America
Tel: (1) 214 768 8436
Fax: (1) 214 768 2293
Email: theology@smu.edu
Website: www.smu.edu/perkins

Perkins School of Theology is one of the 13 seminaries of The United Methodist Church (and one of the only 5 university-related United Methodist theological schools), located in the heart of Dallas, Texas, with extension programmes in Houston/Galveston and San Antonio.

Diaconia Graduate Fellowships

Subjects: Theology.
Purpose: To supplement the financial resources of United Methodist students.
Eligibility: Open to consecrated diaconal ministers or ordained deacons and full-time Doctoral students.
Level of Study: Doctorate
Type: Fellowships
Value: US$10,000
Frequency: Annual
Application Procedure: Request for application forms can be sent to theology@smu.edu
Closing Date: February 1st
Contributor: Section of Deacons and Diaconal Ministries, General Board of Higher Education and Ministry, The United Methodist Church

For further information contact:

Diaconia Graduate Fellowships Section of Deacons and Diaconal Ministries PO Box 340007, Nashville, TN, 37203-0007
Tel: 615 340 7375
Email: www.sddm@gbhem.org

THE PERRY FOUNDATION

16 Sandgate Lane, Wandsworth Common, London, SW18 3JP, United Kingdom
Tel: (44) 020 8874 1460
Email: perry.gbennett@gmail.com
Website: www.perryfoundation.co.uk
Contact: Gordon Bennett, Comaony Secretary

The Perry Foundation offers research awards and postgraduate scholarships in agriculture and related disciplines. Research awards are offered to universities and institutes in the United Kingdom and are normally for a 3-year period. Postgraduate scholarships are offered to holders of First or Second Class Degrees and are for a 3-year period leading to a PhD. Both research awards and postgraduate scholarships must be undertaken at a university, college or research establishment in the United Kingdom.

Perry Postgraduate Scholarships

Subjects: The production and utilisation of crops for food and non-food uses, ecologically acceptable and sustainable farming systems, including, in particular, water and nutrient balances, integrated disease and pest control systems for both crops and livestock, socio-economic studies in the occupation and use of land, the rural economy and infrastructure and developments in marketing. Projects must be of definable benefit to United Kingdom agriculture.
Purpose: To enable postgraduates undertake research and investigative work in agriculture and related fields and to build up a pool of highly competent researchers in the United Kingdom.
Eligibility: Applicants must hold a First or Upper Second Class (Honours) Degrees in appropriate subjects, and must have been offered a place at a university, college or other establishment in the United Kingdom that will lead to the award of a PhD.
Level of Study: Doctorate
Type: Postgraduate scholarships
Value: UK£12,000 contribution per year
Length of Study: 3–4 years
Frequency: Annual
Study Establishment: Universities, colleges, research establishments and institutes in the United Kingdom
Country of Study: United Kingdom
No. of awards offered: 2–4 per year
Application Procedure: Applicants must write to or email the Foundation Secretary for an application form. Full details will be provided to suitable applicants.
Closing Date: October 31st
Funding: Foundation
No. of awards given last year: 3
No. of applicants last year: 100+
Additional Information: Projects must be of definable benefit to United Kingdom agriculture.

Perry Research Awards

Subjects: The production and utilization of crops for food and non-food uses, ecologically acceptable and sustainable farming systems, including, in particular, water and nutrient balances, integrated disease and pest control systems for both crops and livestock, socio-economic studies in the occupation and use of land, the rural economy and infrastructure and developments in marketing. Must be of definable benefit to United Kingdom agriculture.
Purpose: To support research projects at research establishments, universities and colleges in the United Kingdom and investigative work into agriculture and related fields.
Eligibility: Open to universities, colleges and research establishments in the United Kingdom
Level of Study: Research
Type: Research award
Value: Research awards are normally UK£15,000 per year maximum
Length of Study: Normally 3 years
Frequency: Annual
Study Establishment: Universities, colleges, institutes and research establishments
Country of Study: United Kingdom
No. of awards offered: 2–3, depending on the availability of funds
Application Procedure: Applicants must write to the Foundation Secretary for a brochure, which contains details of the application procedure and application forms. Applications submitted by individuals must be supported by their university, college, institute or other establishment.
Closing Date: October 31st
Funding: Private
No. of awards given last year: 1
No. of applicants last year: 20+
Additional Information: The research awards must be of definable benefit to United Kingdom agriculture.

PFIZER INC.

Pfizer MAP Program, MedPoint Communications, 1603 Orrington Ave, Suite 1900, Evanston, IL, 60201, United States of America
Tel: (1) 877 254 6953
Fax: (1) 847 425 7028
Email: MAPinfo@clinicalconnexion.com
Website: www.pfizermap.com
Contact: MAP Program Coordinator

As a reflection of our commitment to the advancement of healthcare, Pfizer Inc. is pleased to support medical innovation in a wide range of discipline through our Medical and Academic Partnership (MAP) grants and awards. Our Fellowships and Scholar Grants, which offer at career-building opportunities for academic researchers in basic, outcomes, and patient-oriented research, are key among these efforts. In addition, Pfizer Visiting Professorships continue to be a resource for in-depth, clinically focused exchange between medical scholars, host organizations and outside scholar-scientists. At Pfizer, we are proud to support the innovators and ideas that help make better treatments and cures possible.

ACCF/Pfizer Visiting Professorships in Cardiovascular Medicine

Subjects: Medicine.
Purpose: To provide opportunities for academic institutions to host a recognized expert for three days of educational exchange.
Eligibility: Open to U.S. medical schools and/or teaching hospitals.
Level of Study: Professional development
Type: Grant
Value: US$7,500
Length of Study: 3 days
Frequency: Annual
Country of Study: United States of America
No. of awards offered: 8
Application Procedure: See the website.
Closing Date: January 27th
Funding: Corporation

Contributor: The American College of Cardiology Foundation (ACCF) and Pfizer Inc.

AUAER/Pfizer Visiting Professorships in Urology

Subjects: Medicine.
Purpose: To provide opportunities for academic institutions to host a recognized expert for 3 days of educational exchange.
Eligibility: Open to US medical schools and/or teaching hospitals.
Level of Study: Postgraduate, Professional development
Type: Grant
Value: US$7,500
Frequency: Annual
Country of Study: United States of America
No. of awards offered: Up to 4
Application Procedure: See the website.
Closing Date: February 12th
Funding: Corporation
Contributor: The American Urological Association Education and Research Inc.

Pfizer Atorvastatin Research Awards Program

Subjects: Medicine.
Purpose: To support outstanding investigators at the early stages of their careers in academic research.
Eligibility: Open to US citizens who hold an MD or PhD.
Level of Study: Research
Type: Grant
Value: US$50,000 per year
Length of Study: 2 years
Frequency: Annual
No. of awards offered: 20
Application Procedure: A proposal must be submitted. See the website for further information.
Closing Date: March 17th
Funding: Corporation

For further information contact:

208 East 51st Street, PBM 173, New York, NY, 10022-6501, United States of America

Pfizer Fellowship in Biological Psychiatry

Subjects: Biological psychiatry.
Purpose: To provide training opportunities for promoting young physicians who wish to pursue research in an academic environment.
Eligibility: Open to applicants who are the citizens of the United States.
Level of Study: Postdoctorate
Type: Grant
Value: US$65,000 per year
Length of Study: 3 years
Frequency: Annual
Application Procedure: A complete form must be submitted.
Closing Date: January 10th

For further information contact:

UF college of medicine, 392-5398, United States of America
Contact: Dr Kristen Madsen, Director, Grants Programme Development

Pfizer Fellowships in Health Literary/ Clear Health Communication

Subjects: Medicine, health administration.
Purpose: To fund scientific research in health literary/clear health communication.
Eligibility: Open to US citizens who have undergone atleast 1 year of postdoctoral clinical training.
Level of Study: Doctorate, Professional development
Type: Fellowship
Value: Up to $100,000, paid over 2 years at $50,000 per year
Length of Study: 2 years
Frequency: Annual
No. of awards offered: 1
Application Procedure: See the website.
Closing Date: February 11th
Funding: Corporation

Pfizer Fellowships in Public Health Overview

Subjects: Health administration.
Purpose: To support the career development of faculty.
Eligibility: Open to US citizens who demonstrate that at least 75 per cent their professional time will be devoted to research.
Level of Study: Postgraduate, Professional development
Type: Grant
Value: Up to $100,000, paid over 2 years at $50,000 per year
Length of Study: 2 years
Frequency: Annual
No. of awards offered: Up to 2
Application Procedure: See the website.
Closing Date: February 10th

The Pfizer Fellowships in Rheumatology/Immunology

Subjects: Rheumatology and immunology.
Purpose: To provide training opportunities for promising young physicians who wish to pursue research in an academic environment.
Eligibility: Open to applicants who demonstrate a strong career interest in academic research in rheumatology and immunology.
Level of Study: Postgraduate, Postdoctorate
Type: Fellowship
Value: Up to $100,000 each, paid at $50,000 over 2 years
Length of Study: 3 years
Frequency: Annual
Application Procedure: See the website.
Closing Date: Please chekc website: http://www.pfizerfellowships.com/AwardDetails.aspx?AwardID=2268
Funding: Commercial

For further information contact:

VF College of Medicine, 392-5398
Contact: Dr Kristen Madsen, Director, Grants Programme Development

Pfizer International HDL Research Awards Program

Subjects: Biology.
Purpose: To support outstanding investigators in the field of HDL biology.
Eligibility: Applicant must hold an MD, a PhD, or the equivalent and hold the rank of instructor or assistant professor at the time of application.
Type: Grant
Value: US$100,000 per year
Length of Study: 2 years
Frequency: Annual
Application Procedure: See the website.
Closing Date: See the website
No. of awards given last year: 14

For further information contact:

International HDL Research Awards, 335 W. 16th Street., 4th Floor, New York, NY, 10011, United States of America
Contact: Grants Co-ordinator

Pfizer Visiting Professorship in Neurology

Subjects: Neurology.
Purpose: To bring new educational value to the institution.
Eligibility: Open to accredited US medical schools and/or affiliated teaching hospitals.
Level of Study: Professional development, Postgraduate
Type: Grant
Value: US$7,500 each
Length of Study: 3 days
Frequency: Annual
No. of awards offered: Up to 8
Application Procedure: See the website.
Closing Date: August 16th
Contributor: Pfizer Inc.

Pfizer Visiting Professorships in Diabetes

Subjects: Endocrinology.
Purpose: To bring new educational value to the institution.
Eligibility: Open to all accredited US medical schools (allopathic or osteopathic).
Level of Study: Postgraduate, Professional development
Type: Grant
Value: US$7,500 each
Length of Study: 3 days
Frequency: Annual
No. of awards offered: Up to 4
Application Procedure: See the website.
Closing Date: February 12th

Pfizer Visiting Professorships in Health Literacy

Subjects: Health administration.
Purpose: To facilitate in-depth, educationally focused visits by prominent experts to US healthcare organizations.
Eligibility: Open to US medical schools and/or teaching hospitals.
Level of Study: Professional development
Type: Grant
Value: US$7,500 each
Length of Study: 3 days
Frequency: Annual
No. of awards offered: 8
Application Procedure: See the website.
Closing Date: February 12th
Funding: Corporation

Pfizer Visiting Professorships in Oncology

Subjects: Oncology.
Purpose: To advance oncology.
Eligibility: Open to accredited US. Medical schools, teaching hospitals and/or academic or community cancer centers.
Level of Study: Postgraduate, Professional development
Type: Grant
Value: US$7,500
Length of Study: 3 days
Frequency: Annual
No. of awards offered: Up to 4
Application Procedure: See the website.
Closing Date: February 12th
Funding: Corporation

Pfizer Visiting Professorships in Pulmonology

Subjects: Pulmonology.
Purpose: To bring new educational value to the institution.
Eligibility: Open to accredited US. Medical schools and/or affiliated teaching hospitals.
Level of Study: Postgraduate, Professional development
Type: Grant
Value: US$7,500 each
Length of Study: 3 days
Frequency: Annual
No. of awards offered: Up to 4
Application Procedure: See the website.
Closing Date: Febraury 12th

Pfizer Visiting Professorships Program

Subjects: Medicine.
Purpose: To create opportunities for selected institutions to invite a distinguished expert for three days of teaching.
Eligibility: Open to accredited medical schools and/or affiliated teaching hospitals.
Level of Study: Postgraduate
Type: Grant
Value: US$7,500 each
Frequency: Annual
No. of awards offered: Up to 8
Application Procedure: Applications available online.
Closing Date: February 12th

PHARMACEUTICAL RESEARCH AND MANUFACTURERS OF AMERICA FOUNDATION (PHRMAF)

950 F Street, N.W., Suite 300, Washington, DC, 20004, United States of America
Tel: (1) 202 572 7756
Fax: (1) 202 572 7799
Email: foundation@phrma.org
Website: www.phrmafoundation.org
Contact: Ms Eileen M McCarron, Executive Director

The Pharmaceutical Research and Manufacturers of America Foundation (PhRMAF) is a non-profit organization, established in 1965 to promote public health through scientific and medical research. It provides funding for research and for the education and training of scientists and physicians who have selected pharmacology, pharmaceutics, toxicology, informatics or health outcomes as a career choice.

PhRMAF Paul Calabresi Medical Student Research Fellowship

Subjects: Pharmacology, toxicology and clinical pharmacology.
Purpose: To support medical/dental students who have substantial interests in research and teaching careers in pharmacology/clinical pharmacology and who are willing to work full time in a specific research effort within a pharmacology or clinical pharmacology unit.
Eligibility: A candidate must be enrolled in a United States medical/dental school and have finished at least 1 year of the school curriculum. Priority consideration will be given to those candidates who project strong commitments to careers in the field of clinical pharmacology. Applicants must be citizens or permanent residents of the US.
Level of Study: Graduate, Postgraduate, Predoctorate
Type: Fellowship
Value: Up to US$15,00 per month
Length of Study: 2 years
Frequency: Annual
Study Establishment: An accredited school of medicine or dentistry.
Country of Study: United States of America
Application Procedure: Requests for the Paul Calabresi Medical Student Research Fellowship are to be submitted online by the appropriate representative of the school or university.
Closing Date: February 1st
Funding: Private
No. of awards given last year: 2
Additional Information: Research projects involving animal subjects require a statement that the project will follow the guidelines set forth by the NIH Guide for the Care and Use of Laboratory Animals and that the project will be performed, reviewed and approved by a faculty committee of the university. The recipient school is expected to submit an annual report on the disposition of the funds awarded by PhRMAF. A final report is due within 60 days after the conclusion of the grant. These reports must be signed by the recipient's sponsor. Any publications, speeches, presentations and other materials that stem directly from the research supported by this grant must acknowledge the support of PhRMAF. 3 reprints of each publication should be forwarded to PhRMAF for further information visit the Foundation's website.

PhRMAF Postdoctoral Fellowships in Health Outcomes Research

Subjects: Health outcomes research, patient-reported outcomes or pharmacoeconomics.
Purpose: To support well-trained graduates from PharmD, MD and PhD programmes who seek to further develop and refine their research skills through formal postdoctoral training.
Eligibility: Open to full-time students who are citizens or permanent residents of the United States of America. Applicants must have a firm commitment from a university in the United States of America before applying for a PhRMAF award. The department's chair will be expected to verify the applicant's doctoral candidacy.
Level of Study: Graduate, Postdoctorate, Postgraduate
Type: Fellowship
Value: US$55,000 stipend per year
Length of Study: up to 2 years

Frequency: Annual
Study Establishment: An accredited school of medicine, pharmacy, dentistry, public health or nursing
Country of Study: United States of America
No. of awards offered: Varies
Application Procedure: Applications must include a research plan written by the applicant, the mentor's research record and a description of how the mentored experience will enhance the applicant's career development in health outcomes research. Applications are to be submitted online by the appropriate representative of the school or university to the Executive Director at PhRMAF. Detailed application requirements are stated on the Foundation's website where applicants can download an application form and read the specific requirements for each award.
Closing Date: February 1st
Funding: Private
No. of awards given last year: 2

PhRMAF Postdoctoral Fellowships in Informatics

Subjects: Informatics.
Purpose: To support well-trained graduates from PhD programmes who seek to further develop and refine their informatics research skills through formal postdoctoral training.
Eligibility: Open to full-time students who are citizens or permanent residents of the United States of America. Applicants must have a firm commitment from a university in the United States of America before applying for a PhRMAF award. The department's chair will be expected to verify the applicant's doctoral candidacy.
Level of Study: Postdoctorate
Type: Fellowship
Value: US$40,000 per year
Length of Study: 1–2 years
Frequency: Annual
Country of Study: United States of America
No. of awards offered: Varies
Application Procedure: Applicants must submit an application including a research plan written by the applicant, the mentor's research record and a description of how the mentored experience will enhance the applicant's career development in informatics. Applications are to be submitted online by the appropriate representative of the school or university to the Executive Director of at PhRMAF. Detailed application requirements are stated on the Foundation's website where the applicant can download an application form and read the specific requirements for each award.
Closing Date: September 1st
Funding: Private
Additional Information: Research projects involving animal subjects require a statement that the project will follow the guidelines set forth by the NIH Guide for the Care and Use of Laboratory Animals and that the project will be performed, reviewed and approved by a faculty committee of the university.

PhRMAF Postdoctoral Fellowships in Pharmaceutics

Subjects: Pharmaceutics.
Purpose: To encourage graduates to continue to develop and refine their pharmaceutics research skills through formal postdoctoral training.
Eligibility: Open to graduates from PhD programmes in pharmaceutics. Before an individual is eligible to apply for a PhRMAF award, the applicants must first have a firm commitment from a university in the United States of America. Applicants must be full-time students, and the department's chair is expected to verify the applicant's doctoral candidacy. All applicants must be citizens of the United States of America or permanent residents.
Level of Study: Postdoctorate, Postgraduate
Type: Fellowship
Value: US$40,000 stipend
Length of Study: 1–2 years
Frequency: Annual
Study Establishment: Schools of pharmacy in the United States of America
Country of Study: United States of America
No. of awards offered: Varies
Application Procedure: Applicants must visit the Foundation's website where detailed application requirements are stated and application forms can be downloaded. Applications must include a research plan written by the applicant, the mentor's research record and a description of how the mentored experience will enhance the applicant's career development in pharmaceutics. Applications are to be submitted online by the appropriate representative of the school or university to the Executive Director at PhRMAF.
Closing Date: September 1st
Funding: Private
Additional Information: Research projects involving animal subjects require a statement that the project will follow the guidelines set forth by the NIH Guide for the Care and Use of Laboratory Animals and that the project will be performed, reviewed and approved by a faculty committee of the university.

PhRMAF Postdoctoral Fellowships in Pharmacology/ Toxicology

Subjects: Pharmacology and toxicology.
Purpose: To facilitate career entry into pharmacology or toxicology at the level of postdoctoral training and to provide funding for recent graduates from PhD programmes who seek to develop research skills through formal postdoctoral training.
Eligibility: Open to applicants with a firm commitment from a university in the United States of America, prior to applying for a PhRMAF award. Applications must be submitted online by an accredited United States of America school and all applicants must be citizens of the United States of America or permanent residents.
Level of Study: Postdoctorate
Type: Fellowship
Value: US$40,000 stipend per year
Length of Study: 1–2 years
Frequency: Annual
Study Establishment: An accredited school of medicine, pharmacy, dentistry or veterinary medicine
Country of Study: United States of America
No. of awards offered: Varies
Application Procedure: Applicants must visit the Foundation's website where detailed application requirements are stated and application forms can be downloaded. Applications must be submitted online by the appropriate representative of the school or university to the Executive Director at PhRMAF. An application must include a research plan written by the applicant, a mentor's research record and a description of how the mentored experience will enhance the applicant's career development on pharmacology or toxicology.
Closing Date: September 1st
Funding: Private
No. of awards given last year: 2
Additional Information: Research projects involving animal subjects require a statement that the project will follow the guidelines set forth by the NIH Guide for the Care and Use of Laboratory Animals and that the project will be performed, reviewed and approved by a faculty committee of the university.

PhRMAF Predoctoral Fellowships in Health Outcomes Research

Subjects: Health outcomes research, patient-reported outcomes and pharmacoeconomics.
Purpose: To support a student's PhD doctoral programme after coursework has been completed and the remaining training activity is the student's research project.
Eligibility: Open to applicants who have a firm commitment from a university in the United States of America. Applicants must be full-time students and the department's chair is expected to verify the applicant's doctoral candidacy. All applicants must be citizens of the United States of America or permanent residents.
Level of Study: Postgraduate
Type: Fellowship
Value: A stipend of US$25,000 per year, up to US$1,000 per year may be used for expenses associated with thesis preparation.
Length of Study: 1–2 years
Frequency: Annual
Study Establishment: An accredited school of medicine, pharmacy, dentistry, public health or nursing
Country of Study: United States of America
No. of awards offered: Varies

Application Procedure: Applications are to be submitted online by the appropriate representative of the school or university to the Executive Director at PhRMAF. Detailed application requirements are stated on the Foundation's website where the applicant can download an application form and read the specific requirements for each award.
Closing Date: February 1st
Funding: Private
No. of awards given last year: 2
Additional Information: Research projects involving animal subjects require a statement that the project will follow the guidelines set forth by the NIH Guide for the Care and Use of Laboratory Animals and that the project will be performed, reviewed and approved by a faculty committee of the university.

PhRMAF Predoctoral Fellowships in Pharmaceutics

Subjects: Pharmaceutics.
Purpose: To support promising students during their thesis research.
Eligibility: The fellowship program of pre doctoral support is designed to assist full-time, in-residence PhD candidates in the fields of pharmacology or toxicology who are enrolled in USA schools of medicine, pharmacy, dentistry or veterinary medicine. The program supports full-time advanced students who will have completed the bulk of their pre-thesis requirements (at least two years of study) and are engaged in thesis research as PhD
Level of Study: Postgraduate
Type: Fellowship
Value: A stipend of US$20,000 per year, which includes up to US $1,000 for expenses associated with thesis research
Length of Study: 1–2 years
Frequency: Annual
Study Establishment: A school of pharmacy
Country of Study: United States of America
No. of awards offered: Varies
Application Procedure: Applicants must visit the Foundation's website where detailed application requirements are stated and application forms can be downloaded. Applications must be submitted online by the appropriate representative of the school or university to the Executive Director of at PhRMAF.
Closing Date: September 1st
Funding: Private
No. of awards given last year: 6
Additional Information: Research projects involving animal subjects require a statement that the project will follow the guidelines set forth by the NIH Guide for the Care and Use of Laboratory Animals and that the project will be performed, reviewed and approved by a faculty committee of the university.

PhRMAF Predoctoral Fellowships in Pharmacology/ Toxicology

Subjects: Pharmacology and toxicology.
Purpose: To support promising students during their thesis research.
Eligibility: Open to advanced students who have completed the bulk of their pre-thesis requirements and are starting their thesis research by the time the award is activated. Students just starting graduate school should not apply. Before an individual is eligible to apply for a PhRMAF award, the applicant must have a firm commitment from a university in the United States of America. The applicant must be a citizen or permanent resident of the United States of America.
Level of Study: Predoctorate
Type: Fellowship
Value: A stipend of US$20,000 per year, which includes up to $1,000 for expenses associated with thesis research.
Length of Study: 1–2 years
Frequency: Annual
Study Establishment: An accredited school of medicine, pharmacy, dentistry or veterinary medicine
Country of Study: United States of America
No. of awards offered: Varies
Application Procedure: Applicants must visit the Foundation's website where detailed application requirements are stated and application forms can be downloaded. Applications must be submitted online by the appropriate representative of the school or university to the Executive Director at PhRMAF.
Closing Date: September 1st
Funding: Private
No. of awards given last year: 7
Additional Information: Research projects involving animal subjects require a statement that the project will follow the guidelines set forth by the NIH Guide for the Care and Use of Laboratory Animals and that the project will be performed, reviewed and approved by a faculty committee of the university.

PhRMAF Research Starter Grants in Health Outcomes Research

Subjects: Health outcomes research, patient-reported outcomes and pharmacoeconomics.
Purpose: To support individuals beginning independent research careers in academia.
Eligibility: Open to applicants sponsored by the school or university at which the research is to be conducted. Applicants must be appointed to an entry level tenure track or equivalent permanent position in a department or unit responsible for pharmaceutical activities as part of its core mission. All applicants must be citizens or permanent residents of the United States of America.
Level of Study: Graduate, Postgraduate
Type: Grant
Value: US$60,000
Length of Study: 1 year
Frequency: Annual
Study Establishment: An accredited school of medicine, pharmacy, dentistry, public health or nursing
Country of Study: United States of America
No. of awards offered: Varies
Application Procedure: Applicants must visit the Foundations' website where detailed application requirements are stated and application forms can be downloaded. Applications must be submitted online by the appropriate representative of the school or university to the Executive Director of Development at PhRMAF. The description of an applicant's career goals and the departmental chair's description of institutional support for the applicant's salary are all important while evaluating an application.
Closing Date: February 1st
Funding: Private
No. of awards given last year: 3
Additional Information: Research projects involving animal subjects require a statement that the project will follow the guidelines set forth by the NIH Guide for the Care and Use of Laboratory Animals and that the project will be performed, reviewed and approved by a faculty committee of the university.

PhRMAF Research Starter Grants in Informatics

Subjects: Informatics.
Purpose: To offer support to new investigators beginning their independent research careers in academia at the faculty level.
Eligibility: Open to applicants sponsored by the school or university at which the research is to be conducted. Applicants must be appointed online to an entry level tenure track or equivalent permanent position in a department or unit responsible for informatics activities as part of its core mission. All applicants must be citizens or permanent residents of the United States of America.
Level of Study: Graduate, Postdoctorate, Postgraduate
Type: Grant
Value: US$60,000 per year
Length of Study: 1 year
Frequency: Annual
Country of Study: United States of America
No. of awards offered: Varies
Application Procedure: Applicants must visit the Foundation's website where detailed application requirements are stated and application forms can be downloaded. Applications must be submitted online by the appropriate representative of the school or university to the Executive Director at PhRMAF. The description of an applicant's career goals and the departmental chair's description of institutional support for the applicant's salary are all important while evaluating an application.
Closing Date: September 1st
Funding: Private
No. of awards given last year: 5
Additional Information: Research projects involving animal subjects require a statement that the project will follow the guidelines set forth by the NIH Guide for the Care and Use of Laboratory Animals and that

the project will be performed, reviewed and approved by a faculty committee of the university.

PhRMAF Research Starter Grants in Pharmaceutics

Subjects: Pharmaceutics.
Purpose: To offer support to new investigators beginning their independent research careers in academia.
Eligibility: Open to applicants sponsored by the school or university at which the research is to be conducted. Applicants must be appointed to an entry level tenure track or equivalent permanent position in a department or unit responsible for pharmaceutical activities as part of its core mission. All applicants must be citizens or permanent residents of the United States of America.
Level of Study: Graduate, Postdoctorate, Postgraduate
Type: Grant
Value: US$60,000 per year
Length of Study: 1 year
Frequency: Annual
Study Establishment: Schools of pharmacy
Country of Study: United States of America
No. of awards offered: Varies
Application Procedure: Applicants must visit the website where detailed application requirements are stated and application forms can be downloaded. Applications must be submitted online by the appropriate representative of the school or university to the Executive Director at PhRMAF. The description of an applicant's career goals and the departmental chair's description of institutional support for the applicant's salary are all important while evaluating an application.
Closing Date: September 1st
Funding: Private
Additional Information: Research projects involving animal subjects require a statement that the project will follow the guidelines set forth by the NIH Guide for the Care and Use of Laboratory Animals and that the project will be performed, reviewed and approved by a faculty committee of the university.

PhRMAF Research Starter Grants in Pharmacology/ Toxicology

Subjects: Pharmacology and toxicology.
Purpose: To support individuals beginning independent research careers in academia.
Eligibility: Open to applicants sponsored by the school or university at which the research is to be conducted. Applicants must be appointed to an entry level tenure track or equivalent permanent position in a department or unit responsible for pharmacology or toxicology activities as part of its core mission. All applicants must be citizens or permanent residents of the United States of America.
Level of Study: Graduate, Postgraduate, Postdoctorate
Type: Grant
Value: US$60,000 per year
Length of Study: 1 year
Frequency: Annual
Study Establishment: An accredited school of medicine, pharmacy, dentistry or veterinary medicine
Country of Study: United States of America
Application Procedure: Applicants must visit the website where detailed application requirements are stated and application forms can be downloaded. Applications must be submitted online by the appropriate representative of the school or university to the Executive Director of at PhRMAF. The description of an applicant's career goals and the departmental chair's description of institutional support for the applicant's salary are all important while evaluating an application.
Closing Date: September 1st
Funding: Private
No. of awards given last year: 3
Additional Information: Research projects involving animal subjects require a statement that the project will follow the guidelines set forth by the NIH Guide for the Care and Use of Laboratory Animals and that the project will be performed, reviewed and approved by a faculty committee of the university.

PhRMAF Sabbatical Fellowships in Health Outcomes Research

Subjects: Health outcomes research, patient-reported outcomes and pharmacoeconomics.
Purpose: To support faculty members at all levels with active research programmes and the opportunity to work at other institutions to learn new skills or develop new collaborations that will enhance their research and research training activities in health outcomes.
Eligibility: Open to citizens or permanent residents of the United States of America. Applicants are expected to have approval for a sabbatical leave from their home institution and to provide an endorsement from the mentor who will sponsor their visiting scientific activity. Matching funds must be provided through the university.
Level of Study: Professional development, Postdoctorate, Postgraduate
Type: Fellowship
Value: A stipend of up to US$40,000
Length of Study: 6 months–1 year
Frequency: Annual
Study Establishment: An accredited school of medicine, pharmacy, dentistry, public health or nursing
Country of Study: United States of America
No. of awards offered: Varies
Application Procedure: Applications are to be submitted online by the appropriate representative of the school or university to the Executive Director at PhRMAF. Detailed application requirements are stated on the Foundation's website where applicants can download an application form and read the specific requirements for each award.
Closing Date: February 1st
Funding: Private
Additional Information: Research projects involving animal subjects require a statement that the project will follow the guidelines set forth by the NIH Guide for the Care and Use of Laboratory Animals and that the project will be performed, reviewed and approved by a faculty committee of the university.

PhRMAF Sabbatical Fellowships in Informatics

Subjects: Informatics.
Purpose: To give faculty members at all levels with active research programmes an opportunity to work at other institutions and to develop new collaborations that will enhance their research and research training activities in informatics.
Eligibility: Applicants are expected to have approval for a sabbatical leave from their home institution and provide an endorsement from the mentor who will sponsor their visiting scientific activity. Matching funds must be provided through the university. All applicants must be citizens of the United States of America or permanent residents.
Level of Study: Postdoctorate, Postgraduate, Professional development
Type: Fellowship
Value: Up to US$40,000 stipend
Length of Study: 6 months–1 year
Frequency: Annual
Country of Study: United States of America
No. of awards offered: Varies
Application Procedure: Applications are to be submitted online by the appropriate representative of the school or university to the Executive Director of at PhRMAF. Detailed application requirements are stated on the Foundation's website where the applicant can download an application form and read the specific requirements for each award.
Closing Date: September 1st
Funding: Private
No. of awards given last year: 1
Additional Information: Research projects involving animal subjects require a statement that the project will follow the guidelines set forth by the NIH Guide for the Care and Use of Laboratory Animals and that the project will be performed, reviewed and approved by a faculty committee of the university.

PhRMAF Sabbatical Fellowships in Pharmaceutics

Subjects: Pharmaceutics.
Purpose: To enable pharmaceutics faculty members at all levels with active research programmes an opportunity to work at other institutions and to develop new collaborations that will enhance their research and research training activities in pharmaceutics.
Eligibility: Open to citizens of the United States of America or permanent residents. Applicants are expected to have approval for a sabbatical leave from their home institution and provide an endorse-

ment from the mentor who will sponsor their sabbatical activity. Matching funds must be provided through the university.
Level of Study: Postgraduate, Professional development, Postdoctorate
Type: Fellowship
Value: A stipend of up to US$40,000
Length of Study: 6 months–1 year
Frequency: Annual
Study Establishment: An approved institute
Country of Study: United States of America
Application Procedure: Applicants must visit the Foundation's website where detailed application requirements are stated and application forms can be downloaded. Applications must be submitted online by the appropriate representative of the school or university to the Executive Director of at PhRMAF.
Closing Date: September 1st
Funding: Private
Additional Information: Research projects involving animal subjects require a statement that the project will follow the guidelines set forth by the NIH Guide for the Care and Use of Laboratory Animals and that the project will be performed, reviewed and approved by a faculty committee of the university.

PhRMAF Sabbatical Fellowships in Pharmacology/ Toxicology

Subjects: Pharmacology and toxicology.
Purpose: To give faculty members at all levels with active research programmes an opportunity to work at other institutions to learn new skills or develop new collaborations that will enhance their research and research training activities in pharmacology or toxicology.
Eligibility: Eligible applicants must (1) hold a PhD degree or appropriate terminal doctorate and record of research accomplishment in a field of study logically or functionally related to the proposed post doctoral activities, (2) hold a faculty appointment that imparts eligibility for a sabbatical leave from their home institution, (3) have institutional approval of a sabbatical plan that includes partial salary that matches the PhRMA stipend, (4) hold an endorsement from a mentor who agrees to sponsor the applicant's visiting scientist activity, and (5) be a U.S. citizen or permanent resident.
Level of Study: Postdoctorate, Postgraduate, Professional development
Type: Fellowship
Value: Up to US$40,000 stipend
Length of Study: 6 months–1 year
Frequency: Annual
Study Establishment: An accredited school of medicine, pharmacy, dentistry or veterinary medicine
Country of Study: United States of America
No. of awards offered: Varies
Application Procedure: Applicants must visit the Foundation's website where detailed application requirements are stated and application forms can be downloaded. Applications must be submitted online by the appropriate representative of the school or university to the Executive Director of at PhRMAF.
Closing Date: September 1st
Funding: Private
Additional Information: Research projects involving animal subjects require a statement that the project will follow the guidelines set forth by the NIH Guide for the Care and Use of Laboratory Animals and that the project will be performed, reviewed and approved by a faculty committee of the university.

THE PHI BETA KAPPA SOCIETY

1606 New Hampshire Avenue NW, Washington, DC, 20009, United States of America
Tel: (1) 202 265 3808
Fax: (1) 202 986 1601
Email: info@pbk.org
Website: www.pbk.org

The Phi Beta Kappa Society has pursued its mission of fostering and recognizing excellence in the liberal arts and sciences since 1776.

Mary Isabel Sibley Fellowship

Subjects: French language or literature in even-numbered years and Greek language, literature, history or archaeology in odd-numbered years.
Purpose: To recognize female scholars who have demonstrated their ability to carry out original research.
Eligibility: Candidates must be unmarried women 25 to 35 years of age who have demonstrated their ability to carry on original research. They must hold a doctorate or have fulfilled all the requirements for a doctorate except the dissertation, and they must be planning to devote full-time work to research during the fellowship year. The award is not restricted to members of Phi Beta Kappa or to US citizens.
Level of Study: Doctorate, Postdoctorate, Postgraduate
Type: Fellowship
Value: US$20,000
Length of Study: 1 year, non-renewable
Frequency: Annual
Country of Study: Any country
No. of awards offered: 1
Application Procedure: Applicants must complete an application form, available from the website, and submit this with transcripts and references.
Closing Date: January 18th
Funding: Private
No. of awards given last year: 1
No. of applicants last year: 50

For further information contact:

Email: awards@pbk.org

The Walter J Jensen Fellowship for French Language, Literature and Culture

Subjects: French language, literature and culture.
Purpose: To help educators and researchers improve education in Standard French language, literature and culture and in the study of Standard French in the United States of America.
Eligibility: Candidates must be under 40 years of age and must be able to certify their career that will involve active use of the French language.
Level of Study: Postgraduate
Type: Fellowship
Value: US$10,000, with additional support available for airfare and, if applicable, support of dependants.
Length of Study: This 1-year-long fellowship includes 6 months of residence and study
Frequency: Annual
Country of Study: France
No. of awards offered: 1 per year
Application Procedure: Applicants must complete self-managed application form available in the website and submit it along with academic records, references, plans for study and proof of superior competence in French according to the standards established by the American Association for Teachers of French.
Closing Date: October 12th
Funding: Private
Contributor: Dr Walter J Jensen
No. of awards given last year: 1
No. of applicants last year: 15
Additional Information: Preference may be given to, though the eligibility is not restricted to, members of Phi Beta Kappa and teachers at the high school level or above. Standard French is defined to exclude a focus on Creole, Quebecois and other dialects.

PHILHARMONIA ORCHESTRA

Philharmonia Orchestra 6th Floor The Tower Building 11 York Road, London, SE1 7NX, United Kingdom
Tel: (44) 020 7921 3900
Fax: (44) 020 7921 3950
Email: orchestra@philharmonia.co.uk
Website: www.philharmonia.co.uk
Contact: Mr Martyn Jones, Administrator

Emanual Hurwitz Award for Violinists of British Nationality

Subjects: Musical performance on violin only.
Purpose: To reward exceptional musical talent.
Eligibility: Open to British nationals only.
Level of Study: Postgraduate
Type: Award
Value: £500
Frequency: Annual
Country of Study: Any country
No. of awards offered: 1
Application Procedure: Applicants must complete an application form and submit this with a stamped addressed envelope and a non-returnable registration fee of UK£10.
Closing Date: Check with the website.

For further information contact:

Tedward Cottage, Bradcutts Lane, Cookham, Berkshire, SL6 9EW
Contact: Jones Martyn, Administrator

John E Mortimer Foundation Awards

Subjects: Musical performance on all instruments.
Purpose: To reward exceptional musical talent.
Eligibility: Applicants must write for details.
Level of Study: Postgraduate
Type: Award
Value: Prizes of varying value.
Frequency: Annual
Country of Study: Any country
No. of awards offered: Varies
Application Procedure: Applicants must complete an application form and submit this with a stamped addressed envelope and a registration fee of UK£15.
Closing Date: December 1st

For further information contact:

Tedward Cottage, Bradcutts Lane, Cookham, Berkshire, SL6 9EW, United Kingdom
Contact: Martyn Jones, Administrator

June Allison Award

Subjects: Musical performance on woodwind only.
Purpose: To assist exceptional musical talent with specialist and advanced study, and to help bridge the gap between study and fully professional status.
Eligibility: Applicants must write for the details.
Level of Study: Postgraduate
Type: Award
Value: UK£500 plus recital
Frequency: Annual
Country of Study: Any country
No. of awards offered: 1
Application Procedure: Applicants must complete an application form and submit this with a stamped addressed envelope and a non-returnable registration fee of UK£10.
Closing Date: December 1st

For further information contact:

Tedward Cottage, Bradcutts Lane, Cookham, Berkshire, SL6 9EW
Contact: Martyn Jones, Administrator

Martin Musical Scholarships

Subjects: Musical performance.
Purpose: To assist exceptional musical talent with specialist and advanced study and to help in bridging the gap between study and fully professional status.
Eligibility: Open to practising musicians as well as students who are instrumental performers, including pianists, preparing for a career on the concert platform either as a soloist or orchestral player, and are of no more than 25 years of age. Preference is given to United Kingdom citizens.
Level of Study: Postgraduate
Type: Scholarship
Value: Check with the website.
Length of Study: 2 years, with a possibility of renewal
Frequency: Annual
No. of awards offered: Varies
Application Procedure: Applicants must complete an application form.
Closing Date: Check with the website.
Funding: Private
No. of awards given last year: 50
No. of applicants last year: 71
Additional Information: It is not the present policy of the Fund to support organists, singers, conductors, composers, academic students or piano accompanists. Please send your completed application, £20 application fee, stamped addressed envelope.

For further information contact:

Tedward Cottage, Bradcutts Lane, Cookham, Berkshire, SL6 9EW
Contact: Martyn Jones, Administrator

Reginald Conway Memorial Award for String Performers

Subjects: Musical performance on strings only.
Purpose: To reward exceptional musical talent.
Eligibility: Applicants must write for the details.
Level of Study: Postgraduate
Type: Award
Value: £500
Frequency: Annual
Country of Study: Any country
No. of awards offered: 1
Application Procedure: Applicants must complete an application form and submit this with a stamped addressed envelope and a non-returnable registration fee of UK£10.
Closing Date: February 1st

For further information contact:

Tedward Cottage, Bradcutts Lane, Cookham, Berkshire, SL6 9EW
Contact: Martyn Jones, Administrator

Sidney Perry Foundation Scholarship

Subjects: Musical performance.
Purpose: To support postgraduate study.
Eligibility: Open to nationals of any country.
Level of Study: Postgraduate
Type: Scholarship
Value: UK£18,000 (various awards)
Length of Study: Up to 2 years
Frequency: Annual
Country of Study: Any country
No. of awards offered: Varies
Application Procedure: Applicants must complete an application form and submit this with a stamped addressed envelope and a non-returnable registration fee of UK£10.
Closing Date: February 1st
No. of awards given last year: 3

For further information contact:

Tedward Cottage, Bradcutts Lane, Cookham, Berkshire, SL6 9EW
Contact: Martyn Jones, Administrator

PHILLIPS EXETER ACADEMY

20 Main Street, Exeter, NH, 03833-2460, United States of America
Tel: (1) 603 772 4311
Fax: (1) 603 777 4384
Email: kcurwen@exeter.edu
Website: www.exeter.edu
Contact: Dr Kathleen Curwen, Dean of Faculty

Phillips Exeter Academy is a private secondary school with over 1,000 students.

George Bennett Fellowship

Subjects: Creative writing.

Purpose: To allow a person commencing a career as a writer the time and freedom from material considerations to complete a manuscript in progress.
Eligibility: Preference is given to writers who have not published a book with a major commercial publisher. Works must be in English.
Level of Study: Unrestricted
Type: Fellowship
Value: US$13,650 per year (with housing and meals)
Frequency: Annual
Study Establishment: Phillips Exeter Academy, Exeter, NH
Country of Study: United States of America
No. of awards offered: 1
Application Procedure: Applicants must send a manuscript, together with an application form, personal statement and US$10.
Closing Date: December 1st for the following academic year
Funding: Private
No. of awards given last year: 1
No. of applicants last year: 150
Additional Information: Duties include being in residence for 1 academic year while working on the manuscript and informal availability to student writers. For further information, please visit the Academy's website.

THE PHILLIPS FOUNDATION

1 Massachusetts Avenue, NW Suite 620, Washington, DC, 20001, United States of America
Tel: (1) 202 250 3887 ext 609
Email: jfarley@thephillipsfoundation.org
Website: www.thephillipsfoundation.org
Contact: Vanessa Henderson, Executive Assistant

The Phillips Foundation was founded in 1990. Four years later, we inaugurated our Robert Novak Journalism Fellowship Program to award annual fellowships to young print and online journalists to undertake writing projects supportive of American culture, a free society, and free markets. Our new Fellows are introduced each May at an awards banquet at the National Press Club in Washington, DC before a distinguished audience of journalists, policy leaders and other influential figures.

The Robert Novak Journalism Fellowship Program

Subjects: Journalism (free market, history and law enforcement).
Purpose: To support working print and online journalist who share the same mission as the Foundation.
Eligibility: Open to US citizens who are working journalists with less than 10 years of professional experience in print and online journalism.
Level of Study: Professional development
Type: Fellowships
Value: US$50,000 for full-time grant, $25,000 for part-time grant
Length of Study: 1 year
Frequency: Annual
Country of Study: United States of America
Application Procedure: Applicants can download application forms from the website www.novakfellowships.org.
Closing Date: February 12th
Funding: Private
No. of awards given last year: 7
Additional Information: Apply online at www.novakfellowships.org.

THE PIERRE ELLIOTT TRUDEAU FOUNDATION

1514 Doctor Penfield Avenue 2nd Floor, Montreal, QC, H3G-1B9, Canada
Tel: (1) 514 938 0001
Fax: (1) 514 938 0046
Email: tfinfo@trudeaufoundation.ca
Website: www.trudeaufoundation.ca
Contact: Elise Comtois, Director of Corporate Services and Public Affairs

The Pierre Elliott Trudeau Foundation seeks to promote outstanding research in the social sciences and humanities and to foster a fruitful dialogue between scholars and policymakers in government, business, voluntary sector, professions and the arts community.

Trudeau Foundation Doctoral Scholarships

Subjects: Social sciences and humanities.
Purpose: To support doctoral candidates pursuing research of compelling present-day concern in areas touching upon 1 or more of the 4 themes of the Foundation: human rights and social justice, responsible citizenship, Canada and the world, humans and their natural environment.
Eligibility: Open to a Trudeau Scholar who is registered full-time in a doctoral level programme approved by the Foundation.
Level of Study: Doctorate
Type: Scholarships
Value: Canadian $40,000 per year
Length of Study: 4 years
Frequency: Annual
Country of Study: Canada
No. of awards offered: 15
Application Procedure: Applications to the Foundation must come from the applicant's academic institution as a result of an internal competition.
Closing Date: December 14th
Funding: Foundation
Contributor: Trudeau Foundation
No. of awards given last year: 15
Additional Information: An additional Canadian $20,000 per year will be available to support research-related travel approved by the Foundation, and to cover networking expenses associated with events and joint projects undertaken within the framework of the Foundation's programmes.

PLASTIC SURGERY EDUCATIONAL FOUNDATION (PSEF)

444 East Algonquin Road, Arlington Heights, IL, 60005-4664, United States of America
Tel: (1) 847 228 9900
Fax: (1) 800 766 4955
Email: cschmieden@plasticsurgery.org
Website: www.plasticsurgery.org
Contact: Ms Christine Schmieden

The Plastic Surgery Educational Foundation (PSEF) is the research arm of the American Society of Plastic Surgeons (ASPS) with the mission to develop and support domestic and international education, research and public service activities of plastic surgeons.

Pilot Research Grant

Subjects: Plastic surgery.
Purpose: To promote Plastic Surgery advancement and innovation. These grants provide 'seed' funding and are intended to allow researchers to conduct preliminary studies related to Plastic Surgery Science.
Eligibility: Open to plastic surgeons and holders of an MD or PhD working in plastic surgery. Residents, Fellows and non-members of ASPS require the sponsorship of a member or candidate of ASPS.
Level of Study: Postdoctorate, Postgraduate, Professional development
Type: Grant
Value: Research seed money up to US$10,000
Length of Study: 1 year
Frequency: Annual
Country of Study: United States of America or Canada
No. of awards offered: Varies
Application Procedure: Applicants must complete an application available on the website.
Closing Date: December 1st
Funding: Private
Contributor: The Plastic Surgery Educational Foundation

PSEF Scientific Essay Contest

Subjects: Plastic surgery.
Purpose: To recognise the authors of essays or articles that address timely and important plastic surgery topics. To encourage general

understanding of issues impacting the practice of plastic surgery, essays and articles submitted to the Scientific Essay Contest are written to communicate to a broad audience.
Eligibility: Open to persons involved in research in the field of plastic surgery.
Level of Study: Doctorate, Graduate, MBA, Postdoctorate, Postgraduate, Predoctorate, Professional development
Type: Prize
Value: US$500–3,000
Frequency: Annual
Country of Study: Any country
No. of awards offered: Varies
Application Procedure: Applicants must submit essays that contain the results of original clinical or basic science research in an area of importance to plastic and reconstructive surgery. Information on essay content and format are available on the website.
Closing Date: Check with the website.
Funding: Private
Contributor: Bernard G Sarnat MD, D Ralph Millard Plastic Surgery Society and the Plastic Surgery Educational Foundation

Research Fellowship Grant

Subjects: Plastic Surgery
Purpose: To encourage research and academic career development in plastic surgery. The research fellowship is intended to provide supplement salary support during a mentored research experience.
Eligibility: Open to residents planning to interrupt their training for a research experience or recent residency graduates wishing to supplement their clinical training with a research experience. Residents, Fellows and non-members of the ASPS require the sponsorship of a member or candidate for membership of the ASPS.
Level of Study: Postgraduate, Professional development
Type: Grant
Value: Upto US$50,000
Length of Study: 1 year
Frequency: Annual
Country of Study: United States of America or Canada
No. of awards offered: Varies
Application Procedure: Applicants must complete an application available on the website.
Closing Date: Check with the website.
Funding: Private
Contributor: PSEF

PLAYMARKET

PO Box 9767, Te Aro, Wellington, 6141, New Zealand
Tel: (64) 4 382 8462
Fax: (64) 4 382 8461
Email: info@playmarket.org.nz
Website: www.playmarket.org.nz
Contact: Director of Playmarket

Playmarket was founded in 1973 to empower New Zealand playwrights and promote their plays. Playmarket offers several playwriting competitions annually and offers agency services, script development and a bookshop.

Adam NZ Play Award

Subjects: Playwriting.
Eligibility: Open to New Zealand playwrights.
Type: Award
Value: US$5,000 for best play, US$1,000 for best play by a maori playwright, US$1,000 for best play by a Pasifika playwright; and US $1,000 for best play by a woman playwright
Frequency: Annual
No. of awards offered: 4
Application Procedure: Open application. Please visite the website for the details.
Closing Date: December 1st
Funding: Foundation
No. of awards given last year: 4
Additional Information: Please send all submissions and enquiries at scripts@playmarket.org.nz.

The Bruce Mason Playwriting Award

Subjects: Playwriting.
Purpose: To recognize achievement at the beginning of a career.
Level of Study: Unrestricted
Type: Award
Value: New Zealand $10,000
Length of Study: 1 year
Frequency: Annual
Country of Study: New Zealand
No. of awards offered: 1
Closing Date: October
Funding: Trusts
No. of awards given last year: 1
No. of applicants last year: Awarded: not by application

THE POINT FOUNDATION

5757 Wilshire Boulevard, Suite 370, Los Angeles, CA, 90036, United States of America
Tel: (1) 323 933 1234
Fax: (1) 866 397 6468
Email: info@pointfoundation.org
Website: www.pointfoundation.org

The Point Foundation provides financial support, mentoring and hope to meritorious students who are marginalized due to sexual orientation, gender expression or gender identity. The Foundation seeks the partnership of philanthropic individuals, corporations and foundations to supply financial support, professional guidance and a network of contacts for undergraduate, graduate and postgraduate students.

The Point Scholarship

Subjects: All subjects.
Purpose: To support outstanding LGBT students (lesbian, gay, bisexual and transgender) who are underprivileged, especially those who have been abandoned by family and other support systems because of their sexual orientation or gender identity.
Eligibility: Both LGBT and non-LGBT can apply provided they demonstrate leadership, scholastic achievement, participation in extracurricular activities and involvement in the LGBT community.
Level of Study: Postgraduate
Type: Scholarship
Value: Average scholarship award per scholar is $13,600
Frequency: Annual
Country of Study: United States of America
No. of awards offered: Varies
Application Procedure: Applicants must complete an application online before the stated deadline at www.pointfoundation.org
Closing Date: February 9th
Funding: Foundation
No. of awards given last year: 27
No. of applicants last year: 1,344

POLLOCK-KRASNER FOUNDATION, INC.

863 Park Avenue, New York, NY, 10075, United States of America
Tel: (1) 212 517 5400
Fax: (1) 212 288 2836
Email: grants@pkf.org
Website: www.pkf.org
Contact: Programme Officer

The Pollock-Krasner Foundation's mission is to aid, internationally, those individuals who have worked as professional artists over a significant period of time.

Pollock-Krasner Foundation Grant

Subjects: Painting, sculpting, print-making, mixed media and installation art.
Purpose: To aid, internationally, individual artists of artistic merit with financial need.
Eligibility: Applicants may be painters, sculptors, print-makers, mixed media or installation artists. The Foundation has no age or geographic limits. Commercial artists, photographers, film-makers, craft-makers and students are not eligible.

Level of Study: Professional development
Type: Grant
Frequency: Annual
Application Procedure: Applicants must write, fax or email to the Foundation for an application and guidelines.
Closing Date: There is no deadline as grants are awarded throughout the year
Funding: Private
Additional Information: The Foundation does not fund academic study.

THE POPULATION COUNCIL

Policy Research Division, One Dag Hammarskjold Plaza, 9th floor, New York, NY, 10017, United States of America
Tel: (1) 212 339 0500
Fax: (1) 212 755 6052
Email: pubinfo@popcouncil.org
Website: www.popcouncil.org

The Population Council is an international non-profit, non-governmental institution that seeks to improve the well being and reproductive health of current and future generations around the world and to help achieve a humane, equitable and sustainable balance between people and resources. The Council conducts biomedical, social science and public health research and helps build research capacities in developing countries.

Biomedical Fellowship Programs

Subjects: Basic and Translational Reproductive Sciences and HIV/AIDS Research.
Eligibility: Candidates for pre-doctoral biomedical fellowships must be engaged in graduate study leading to an advanced degree, MD, PhD, DVM or equivalent. Candidates for post-doctoral biomedical fellowships must have successfully completed an advanced degree, MD, PhD, DVM or equivalent.If the applicant is other than a US citizen or permanent resident, s/he must have a strong commitment to return to her/his own country.
Level of Study: Doctorate, Postdoctorate, Predoctorate, Professional development
Type: Fellowship
Length of Study: 1–2 years
Frequency: Dependent on funds available
Study Establishment: The Population Council's Centre for Biomedical Research
Country of Study: United States of America
No. of awards offered: Varies
Application Procedure: Contact the Center for Biomedical Research.
Closing Date: No deadline is set. Please check with the website.

For further information contact:

Center for Biomedical Research, 1230 York Avenue, New York, NY 10065

Fred H Bixby Fellowship Program

Subjects: Population studies in combination with a social science discipline (demography, economics, sociology, anthropology, geography) public health or biomedical sciences.
Purpose: To expand training opportunities by allowing population specialists and biomedical researchers from developing countries to work with experienced mentors in the population Council's international network of offices. Fellowships are awarded to professionals in the early stages of their careers who have demonstrated commitment to return to or stay in a developing country upon completion of their training programs to build capacity in local institutions.
Eligibility: Candidates must have recently completed (within the last 5 years) or anticipate completing a PhD or equivalent degree in the social sciences, public health or biomedical sciences. All applicants should have previous direct experience with either biomedical research, program research or policy-relevant social science research (preferably including one or more peer-reviewed publications). Applications must be legal citizens of a developing country and be proficient in english.
Level of Study: Postdoctorate, Professional development
Type: Fellowship
Value: A monthly stipend based on Fellowship location and years of experience and attendance at one professional meeting per year (including travel). An allowance for relocation and health insurance are also included
Length of Study: 1–2 years
Frequency: Annual
Study Establishment: One of the Population Council's international offices
No. of awards offered: Varies
Application Procedure: Applicants must submit an application form, research proposal and supporting documents in English. The application details and procedures can be found on the Bixby Fellowship website www.popcouncil.org/what/bixby.asp. Requests for information from the Fellowship Coordinator, which can be obtained by emailing to bixbyfellowship@popcouncil.org, should include a brief description of the candidate's academic and professional qualifications, a short statement about their research interests for the proposed Fellowship period and their curriculum vitae.
Closing Date: January 31st
Funding: Foundation
Contributor: Fred H Bixby Foundation
Additional Information: Selection will be based on the recommendation of the Fellowship Committee which consists of distinguished scholars in the field of population. Selection criterias will stress academic excellence, professional experience and prospective contribution to the population field. Prior to submitting a formal application to the Fellowship Office for consideration, Bixby applicants are required to seek sponsorship from at least one Population Council staff mentor, in consultation with the Fellowship coordinator.

PRADER-WILLI SYNDROME ASSOCIATION UK

125a London Road, Derby, DE1 2QQ, England
Tel: (44) 13 3236 5676
Fax: (44) 13 3236 0401
Email: admin@pwsa.co.uk
Website: http://pwsa.co.uk/
Contact: Administrative Assistant

Prader-Willi Syndrome Association UK provides support and information to people with Prader- Willi Syndrome.

PWSA UK Research Grants

Subjects: Prader-Willi syndrome.
Purpose: To improve understanding and treatment of Prader-Willi Syndrome.
Level of Study: Unrestricted
Frequency: Dependent on funds available
Country of Study: United Kingdom
Application Procedure: Applicants must write or call the PWSA (UK) for information.
Funding: Private

THE PREHISTORIC SOCIETY

Institute of Archaeology, University College London, 31-34 Gordon Square, London, WC1H 0PY, United Kingdom
Fax: (44) 20 7383 2572
Email: prehistoric@ucl.ac.uk
Website: www.prehistoricsociety.org
Contact: Ms Tessa Machling, Administrative Assistant

The Prehistoric Society is open to professionals and amateurs alike and has over 2,000 members worldwide. Its main activities are lectures, study tours and conferences and it publishes an annual journal (*PPS*) and a newsletter (*PAST*), which is published 3 times a year.

Prehistoric Society Conference Fund

Subjects: Archaeology, especially prehistoric.
Purpose: To finance attendance at international conferences.
Eligibility: Preference is given first to scholars from developing countries, whether they are members of the Society or not, then to

members of the Society not qualified to apply for conference funds available to university staff. Other members of the Society are also eligible.
Level of Study: Postgraduate
Type: Travel grant
Value: UK£200–300
Length of Study: 1 year, renewals are considered
Frequency: Annual
Country of Study: Any country
No. of awards offered: 2
Application Procedure: Applicants must contact the Honorary Secretary for an application form.
Closing Date: January 31st
Funding: Private
No. of awards given last year: 2
Additional Information: Recipients are required to submit a short report on the conference to *PAST*, the Society's newsletter and their papers for the Society's proceedings if these are not to be included in a conference volume.

Prehistoric Society Conference Fund

Subjects: Prehistoric archaeology.
Purpose: Its aim is to further the development of prehistory as an international discipline.
Eligibility: There are no eligibility restrictions.
Level of Study: Unrestricted
Type: Scholarship
Value: £200–300
Frequency: Annual
No. of awards offered: 2
Application Procedure: Applicants must complete an application form.
Closing Date: January 31st
Funding: Private

Prehistoric Society Research Fund

Subjects: Prehistoric archaeology.
Purpose: To further research in prehistory by excavation or other means.
Eligibility: Open to all members of the Society. The Society may make specific conditions relating to individual applications.
Level of Study: Unrestricted
Type: Grant
Value: £100–1000
Length of Study: 1 year, renewals are considered
Frequency: Annual
Country of Study: Any country
No. of awards offered: Varies
Application Procedure: Applicants must complete an application form and include the names of two referees in their application.
Closing Date: January 31st
Funding: Private
Additional Information: Awards are made on the understanding that a detailed report will be made to the Society as to how the grant was spent.

THE PRESIDENT'S COMMISSION ON WHITE HOUSE FELLOWSHIPS

c/o O.P.M-Sheila Coates, 1900 E.Street, NW, Room B431, Washington, DC, 20415, United States of America
Tel: (1) 202 395 4522
Fax: (1) 202 395 6179
Email: comments@whitehouse.gov
Website: www.whitehouse.gov/fellows
Contact: White House Fellowships

To maintain the healthy functioning of our system it is essential that we have a generous supply of leaders who have an understanding, gained first hand, of the challenges that our national government faces.

White House Fellowships

Subjects: Domestic and international policy studies.
Purpose: To offer exceptional young men and women first-handed experience working at the highest levels of the federal government.
Eligibility: Civilian employees of the Federal government are not eligible.
Level of Study: Postgraduate
Type: Fellowship
Value: A full-time, paid assistantship to the Vice President, Cabinet Securities, and other top-ranking government officials
Length of Study: 1 year
Frequency: Annual
Study Establishment: The White House
Country of Study: United States of America
Application Procedure: Application instructions are available on the website.
Closing Date: February 1st

PRESS GANEY ASSOCIATES INC.

404 Columbia Place, South Bend, IN, 46601, United States of America
Tel: (1) 800 232 8032
Fax: (1) 574 246 4913
Email: info@pressganey.com
Website: www.pressganey.com
Contact: Ms Kelly Leddy, Research Specialist

The Press Ganey Associates, Inc. is the healthcare industry's top satisfaction measurement and improvement firm, serving more than 5,900 healthcare facilities and processing nearly 7,000,000 surveys annually. As one of the industry's market leaders, the company offers one of the world's largest comparative databases and unparalleled benchmarking opportunities.

Press Ganey Best Place to Practice Award

Subjects: The subject of the grant is patient satisfaction in all aspects of the healthcare industry.
Purpose: To increase the systematic and rigorous study of patient satisfaction, through the funding of applied research that will identify best practices and that can be used throughout the healthcare industry.
Eligibility: There are no eligibility restrictions.
Level of Study: Research, Unrestricted
Value: US$10,000
Frequency: Annual
No. of awards offered: 5
Application Procedure: Applicants must visit the website or email Kelly Leddy at kleddy@pressganey.com for application procedures.
Closing Date: March 14th
Contributor: Press Ganey Associates
No. of awards given last year: 2
No. of applicants last year: 28

PRIMATE CONSERVATION INC.

1411 Shannock Road, Charlestown, RI, 02813-3726, United States of America
Tel: (1) 401 364 7140
Fax: (1) 401 364 6785
Email: nrowe@primate.org
Website: www.primate.org
Contact: Noel Rowe, Director

Primate Conservation Inc. gives small grants and matching funds for the conservation projects and studies of the least known and most endangered primates in their natural habitat.

Primate Conservation Inc. Grants

Subjects: Conservation and research projects on the least known and/or most endangered primates in habitat countries with wild populations.
Purpose: To protect and study the least known and/or most endangered primates in their natural habitats.
Eligibility: Applicants must be graduate students, qualified conservationists or primatologists.
Level of Study: Doctorate, Graduate, Postgraduate, Research

Type: Grant
Value: Average US$2,500 maximum US$5,000
Length of Study: 2 months to 2 years
Frequency: Biannual
No. of awards offered: Varies with each funding season
Application Procedure: Applicants must submit a grant proposal on forms that can be downloaded from the website, and submit three copies of the application to the institution's main address. Proposal must be typed, double spaced and in English. One copy of the proposal should be emailed to nrowe@primate.org
Closing Date: September 20th and Febuary 1st
Funding: Private
Contributor: Private donation
No. of awards given last year: 20
No. of applicants last year: 55

PRINCIPALITY OF LIECHTENSTEIN STATE EDUCATIONAL SUPPORT

PO Box 684, Principality of Liechtenstein, Vaduz, 9490, Switzerland
Tel: (41) 236 7681
Email: info@liechtenstein.li
Website: www.liechtenstein.li

The State provides educational assistance in the form of scholarships, loan, or contributions to expenses. Eligibility is determined on the basis of citizenship and the type of education.

Liechtenstein Scholarships and Loans

Subjects: All subjects.
Purpose: To awaken greater understanding of the Liechtenstein State by the domestic and foreign public.
Eligibility: Open to residents of Liechtenstein, expatriates or those whose mother, father, or spouse is a Liechtenstein citizen are eligible.
Level of Study: Postgraduate
Type: Scholarship
Frequency: Annual
Study Establishment: Liechtenstein Institute, Liechtenstein
Application Procedure: Applicants should first contact their chosen place of study.
Closing Date: No deadline
Funding: Government
Additional Information: All additional loans must be repaid within 6 years of completion of the course of study.

PROSTATE ACTION

6 Crescent Stables, 139 Upper Richmond Road, London, SW15 2TN, England
Tel: (44) 020 8788 7720
Fax: (44) 020 8789 1331
Email: ann.rolfe@prostateaction.org.uk
Website: www.prostateaction.org.uk
Contact: Rachel Culley, Research Officer

Vision - The eventual defeat of prostate disease.Mission - To beat prostate disease through research and education.Aims - Prostate disease affects one in two men during their lifetime. We are working to help men and their families beat prostate disease now and in the future.

Prostate Action Research Grants

Subjects: Research into all prostate diseases.
Purpose: To support research into all prostate diseases.
Eligibility: Available to English-speaking nationals of any country, of any mature age, resident in the United Kingdom and wishing to carry out research at a recognized United Kingdom hospital or institution into malignant or benign prostate disease. Bench space in a United Kingdom hospital/institution must be a prior condition for work to be carried out in the United Kingdom.
Level of Study: Professional development, Research
Value: 3 categories – £10,000, £25,000 and £50,000
Country of Study: United Kingdom
No. of awards offered: Dependent on funds available
Application Procedure: Advertisements appear once a year in *The British Journal Of Urology International* inviting applicants to describe research projects they wish to undertake and to indicate the level of funding sought. The advertisements explains the application procedure.
Closing Date: Specified in relevant advertisements
Funding: Individuals, trusts
Contributor: Charitable funds raised by Prostate Action
No. of awards given last year: Approx. 10
No. of applicants last year: Approx. 20

Prostate Action Training Grants

Subjects: Medical training relevant to the treatment of prostate cancer, benign prostatic hyperplasia and prostatitis.
Purpose: To support UK-based medical professionals for training relevant to the treatment of prostate diseases.
Eligibility: Available to English-speaking nationals of any country, of any mature age, resident in the United Kingdom and wishing to carry out training at a recognized United Kingdom or institution relevant to the treatment of malignant or benign prostate disease. A place at a United Kingdom hospital/institution is a prior condition.
Level of Study: Unrestricted
Type: Grant
Value: Up to £50,000 over a maximum of 2 years
Country of Study: United Kingdom
No. of awards offered: Dependent on the funds available
Application Procedure: Advertisements appear once a year in *The British Medical Journal* and *The British Journal Of Urology International* inviting applicants to describe training they wish to undertake and to indicate the level of funding sought. These advertisements explain the application procedure.
Closing Date: See website
Funding: Individuals, trusts
Contributor: Charitable funds raised by Prostate Action
No. of awards given last year: Approx. 5
No. of applicants last year: Approx. 5

PULITZER CENTER ON CRISIS REPORTING

1779 Massachusetts Avenue, Suite #615, Washington, DC, 20036, United States of America
Tel: (1) 202 332 0982
Email: info@pulitzercenter.org
Website: www.pulitzercenter.org
Contact: Nathalie Applewhite, Managing Director

The Pulitzer Center on Crisis Reporting, established in 2006, intends to be a leader in sponsoring the independent reporting that media organizations are increasingly less willing to undertake on their own. It works towards raising the standard of coverage of global affairs, and to do so in a way that engages both the broad public and government policy makers.

Pulitzer Center on Crisis Reporting Travel Grants

Subjects: Journalism
Purpose: To bring up issues that have gone unreported or under-reported in the mainstream American media.
Eligibility: Open to all journalists, writers or filmmakers, staff journalists as well as freelancers of any nationality.
Level of Study: Professional development
Type: Travel grant
Value: Most awards fall in the range of $2,000 to $10,000 but depending on project specifics may be as much as $20,000.
Frequency: Annual
Country of Study: United States of America
Application Procedure: Applicants must submit a proposed project, curriculum vitae, 3 recent writing samples and 3 references.
Closing Date: February 6th
Additional Information: Applications should be submitted by email to info@pulitzercenter.org

PYMATUNING LABORATORY OF ECOLOGY (PLE)

Department of Biological Sciences University of Pittsburgh, 4249 Fifth Avenue, Pittsburgh, PA, 15260, United States of America
Tel: (1) 412 624 4350
Fax: (1) 412 624 4759
Website: www.biology.pitt.edu/facilities/pymatuning
Contact: Dr Paula Grabowski, Department Chair

The Pymatuning Laboratory of Ecology (PLE) is a superb year-round facility that is part of Pitt's Department of Biology. Located in Northwestern Pennsylvania on the shores of beautiful Pymatuning Lake, PLE is dedicated to serving the academic community and visitors from around the world.

Leasure K. Darbaker Prize in Botany

Subjects: Botany
Purpose: To award funds for the pursuit of excellent graduate and recent postdoctoral research in botany.
Eligibility: Applicants must be graduate students, recent PhDs or senior researchers initiating new research at PLE.
Level of Study: Doctorate, Postdoctorate, Postgraduate
Type: Research grant
Value: US$500–1,500
Length of Study: 1 year
Frequency: Annual
Study Establishment: PLE
Country of Study: United States of America
No. of awards offered: 1–3
Application Procedure: Applicants must request guidelines from the PLE or refer to the website.
Closing Date: February 1st
Funding: Private
No. of awards given last year: 1
No. of applicants last year: 2

QUEEN ELISABETH INTERNATIONAL MUSIC COMPETITION OF BELGIUM

20 rue aux Laines, B-1000 Brussels, Belgium
Tel: (32) 2 213 4050
Fax: (32) 2 514 3297
Email: info@qeimc.be
Website: www.qeimc.be
Contact: Secretariat

The Queen Elisabeth International Music Competition of Belgium is a non-profit association, located in Brussels, whose principal aim is to organize major international competitions for music virtuosos. In this way, the competition participates in the Belgian and international music world, and gives its support to young musicians.

Queen Elisabeth International Music Competition of Belgium

Subjects: Music: piano, voice, violin and composition.
Purpose: To provide career support for young pianists, singers, violinists and composers.
Eligibility: Open to musicians of any nationality who are at least 17 years of age and not older than 30 years for violin, piano and singing and 40 years for composers. The competition is made up of a first round, a semi-final and a final round.
Level of Study: Unrestricted
Value: Prizes, awards and certificates along with cash prizes will be awarded
Frequency: Annual
Country of Study: Any country
No. of awards offered: 12
Application Procedure: Applicants must obtain an application form from the Secretariat of the Competition or via the website.
Closing Date: January 15th
Funding: Private
No. of applicants last year: Unrestricted
Additional Information: There are no master classes with jury members.

QUEEN MARGARET UNIVERSITY

Queen Margaret University Drive, Musselburgh, Edinburgh, EH21 6UU, United Kingdom
Tel: (44) 131 474 0000
Fax: (44) 131 474 0001
Email: rilo@qmu.ac. uk
Website: www.qmuc.ac.uk
Contact: Professor Anthony Cohen, Principal

Queen Margaret University provides vocationally relevant education in business and enterprise; drama and creative industries; health, including international health; and social sciences, media and communication. Its internationally recognized research activity informs our teaching. With around 4,500 students, our small size allows us to offer students a highly supportive environment.

SAAS Postgraduate Students' Allowances Scheme (PSAS)

Subjects: Cultural management programmes, art therapy, audiology and international health scheme.
Purpose: International Health Scheme.
Eligibility: United Kingdom and European Union nationals living in Scotland on the relevant date (conditions apply).
Level of Study: Postgraduate
Value: UK£3,400 towards tuition fees and living cost support, if applicable
Length of Study: 2 years, full-time
Frequency: Annual
Study Establishment: Queen Margaret University College
Country of Study: United Kingdom
No. of awards offered: 4
Application Procedure: Applicants must complete an application and send it to the SAAS, once nominated by the institution.
Closing Date: March 31st
Funding: Government
Contributor: Students Awards Agency for Scotland (SAAS)
No. of awards given last year: 2
Additional Information: Students cannot apply directly to the SAAS. They must have accepted an offer of a place and be nominated by the institution.

For further information contact:

Tel: 44 (0)131 474 0000
Email: rilo@qmu.ac.uk

QUEEN MARY, UNIVERSITY OF LONDON

Admissions and Research Student Office, Mile End Road, London, E1 4NS, England
Tel: (44) 20 7882 5555
Fax: (44) 20 7882 5588
Email: admissions@qmul.ac.uk
Website: www.qmul.ac.uk
Contact: Mr Peter Smith, Admissions Assistant

Queen Mary is the fourth largest college in the University of London. Located on an attractive campus, it has more than 8,000 students studying in four faculties plus St Bartholomew's and the Royal London School of Medicine and Dentistry. Of these, more than 1,600 are pursuing postgraduate courses or undertaking research.

Queen Mary, University of London Research Studentships

Subjects: Arts, sciences, engineering, social sciences, law, medicine and dentistry.
Purpose: To provide the opportunity for full-time research leading towards an MPhil or PhD.
Eligibility: Candidates will normally be expected to have a good first degree and a Masters degree in Politics, International Relations, or a related subject.
Level of Study: Research
Type: Studentship
Value: £15,590

Length of Study: 3 years full-time subject to a satisfactory academic report
Frequency: Annual
Study Establishment: Queen Mary University of London
Country of Study: United Kingdom
No. of awards offered: 20+
Application Procedure: Applicants must contact the Admission and Recruitment Office for further application details.
Closing Date: January 31st
Funding: Government
No. of awards given last year: 30
No. of applicants last year: 200+
Additional Information: These studentships are not available to existing Queen Mary research students.

THE QUEEN'S NURSING INSTITUTE

3 Albemarle Way, London, EC1V 4RQ, England
Tel: (44) 20 7549 1400
Fax: (44) 20 7490 1269
Email: rosemary.cook@qni.org.uk
Website: www.qni.org.uk
Contact: Anne Pearson, Practice Development Manager

The Queen's Nursing Institute works to support and develop new and best nursing practice and innovation in primary care. Through this support we want to ensure that patients receive the highest standard of nursing in the community. Primary care has always been the highest priority and the Queen's Nursing Institute firmly believes in working in partnership with nurses to achieve its overall objectives.

The Queen's Nursing Institute Fund for Innovation and Leadership

Subjects: Implementation of good practice, or a project or an idea, within the community.
Purpose: The QNI Fund for Innovation provides professional and financial support to community nurses wishing to undertake projects which improve services and/or develop practice in the care of patients at home and in the community.
Eligibility: Must be a qualified community nurse.
Level of Study: Professional development, Graduate, Postgraduate
Type: Grant
Value: Up to UK£5,000
Length of Study: 1 year
Frequency: Annual
Study Establishment: The Queen's Nursing Institute
Country of Study: United Kingdom
No. of awards offered: Varies, usually 6–10
Application Procedure: Applicants must submit an application proposal and a curriculum vitae.
Closing Date: October 17th (Please check the website for the details)
Funding: Private
Contributor: National Gardens Scheme
No. of awards given last year: 12

QUEENSLAND UNIVERSITY OF TECHNOLOGY (QUT)

Research Students Centre, GPO Box 2434, Brisbane, QLD, 4001, Australia
Tel: (61) 7 3138 2000
Fax: (61) 7 3138 1304
Email: research.enquiries@qut.edu.au
Website: www.rsc.qut.edu.au

QUT provides a career-oriented education which helps graduates find employment in their chosen career, in an environment which uses the latest technology to make learning stimulating and enjoyable. It provides information for students, staff and visitors about the resources of the University, its facilities and processes.

APAI Scholarships within Integrative Biology

Subjects: Plant science and biological sciences.
Eligibility: Open to citizens of Australia or permanent residents having Honours 1 Degree or equivalent.
Level of Study: Postgraduate
Type: Scholarship
Value: Australian $25,627
Length of Study: 3 years
Frequency: Annual
No. of awards offered: 2
Application Procedure: Check website for further details.
Closing Date: March 2nd

For further information contact:

School of Integrative Biology
Email: susanne.schmidt@uq.edu.au
Contact: Dr Susanne Schmidt, Senior Lecturer

ARC APAI – Alternative Engine Technologies

Subjects: Engineering and technology.
Purpose: The multidisciplinary nature of the project will provide thestudent with a significant intellectual challenge, to assimilate the requiredbackground research and to integrate this knowledge to achieve the aims of thecurrent project.
Eligibility: Open to citizens of Australia or New Zealand or permanent residents who have achieved Honours 1 or equivalent, or Honours 2a or equivalent.
Level of Study: Postdoctorate, Postgraduate
Type: Scholarship
Value: Australian $26,140 per year
Length of Study: 3 years
Frequency: Annual
Country of Study: Australia
No. of awards offered: 1
Application Procedure: Check website for further details.
Closing Date: September 28th

For further information contact:

Queensland University of Technology, School of Engineering Systems, GPO Box 2434, Brisbane, QLD, 4001, Australia
Tel: (61) 7 3138 5174
Email: rong.situ@qut.edu.au
Website: www.rsc.qut.edu.au/future/scholarships/APAI.jsp
Contact: Dr Rong Situ

Institute of Health and Biomedical Innovation Awards

Subjects: Biomedical engineering, engineering and technology, medical and health sciences or physical sciences.
Purpose: To support living expenses.
Eligibility: Open for citizens of Australia or permanent residents who have achieved Honours 1 or equivalent, or Honours 2a or equivalent.
Level of Study: Doctorate, Postgraduate
Type: Award
Value: Australian $36,140
Length of Study: 2 years (Masters) or 3 years (PhD)
Frequency: Annual
Country of Study: Australia
No. of awards offered: 4
Application Procedure: Check website for further details.
Closing Date: October 12th

For further information contact:

Queensland University of Technology, IHBI, QUT, GPO Box 2434, Brisbane, QLD, 4001, Australia
Tel: (61) 7 3138 6056
Fax: (61) 7 3138 6039
Email: s.winn@qut.edu.au
Website: www.rsc.qut.edu.au/studentsstaff/scholarships/arw_domestic.jsp
Contact: Stella Winn, Research Services Manager

THE RADCLIFFE INSTITUTE FOR ADVANCED STUDY

Byerly Hall, 10 Garden Street, Cambridge, MA, 02138, United States of America
Tel: (1) 617 495 8212
Fax: (1) 617 495 8136
Email: fellowships@radcliffe.edu
Website: www.radcliffe.edu
Contact: Administrator of Fellowships

The Radcliffe Institute for Advanced Study is a scholarly community where individuals pursue advanced work across a wide range of academic disciplines, professions and creative arts. Within this broad purpose, the Radcliffe Institute sustains a continuing commitment to the study of women, gender and society.

Radcliffe Institute for Advanced Study Fellowship Program

Subjects: All subjects.
Purpose: To support women and men of exceptional promise and demonstrated accomplishment, who wish to pursue independent work.
Eligibility: Open to female and male scholars in any field who gained a doctorate or appropriate terminal degree at least 2 years prior to appointment, or creative writers and visual or performing artists with a record of significant accomplishment and equivalent professional experience. Special eligibility requirements apply to creative artists.
Level of Study: Research, Postdoctorate
Type: Fellowship
Value: Up to US$70,000 for one year
Frequency: Annual
Study Establishment: Harvard University
Country of Study: United States of America
Application Procedure: Applicants must visit the website.
Closing Date: March
No. of awards given last year: 48
No. of applicants last year: 900 approx

RADIO TELEVISION DIGITAL NEWS FOUNDATION (RTDNF)

529 14th Street, NW, Suite 425, Washington, DC, 20045, United States of America
Tel: (1) 202 659 6510
Fax: (1) 202 223 4007
Email: jon@rtnda.org
Website: www.rtdna.org
Contact: Jon Wafalosky

The mission of the Radio Television Digital News Foundation (RTDNF) is to promote excellence in electronic journalism through research, education and professional training in four principal programme areas: journalistic ethics and practices, the impact of technological change on electronic journalism, the role of electronic news in politics and public policy and cultural diversity in the electronic journalism profession.

RTDNF Fellowships

Subjects: Electronic journalism.
Eligibility: To support young journalists in radio or television with up to 10 years of experience.
Level of Study: Postgraduate, Professional development
Type: Fellowship
Value: Up to US$2,500
Frequency: Annual
Country of Study: United States of America
No. of awards offered: 4
Application Procedure: Applicants must download the application form from www.RTDNA.org
Closing Date: May 13th
Funding: Private
Additional Information: The awards include: the Michele Clark Fellowship for Minority News Professionals, the Jacque I Minotte Health Reporting Fellowship, the Vada and Barney Oldfield National Security Fellowship and NS Bienstock Fellowship for Minority Journalists.

RADIOLOGICAL SOCIETY OF NORTH AMERICA, INC. (RSNA)

820 Jorie Boulevard, Oak Brook, IL, 60523 2251, United States of America
Tel: (1) 630 571 2670
Fax: (1) 630 571 7837
Email: swalter@rsna.org
Website: www.rsna.org/foundation
Contact: Mr Scott Walter, Assistant Director, Grant Administration

The Research and Education Foundation of the Radiological Society of North America (RSNA) provides grant support to medical students, residents, Fellows and full-time faculty members of departments of radiology, radiation oncology and nuclear medicine.

RSNA Education Scholar Grant Program

Subjects: Radiology or related disciplines.
Purpose: To provide funding opportunities for individuals with an active interest in radiologic education.
Eligibility: Applications are accepted from individuals throughout the world.
Level of Study: Unrestricted
Type: Grant
Value: Up to US$75,000 per year for up to 2 years (US$150,000 maximum) to be used as salary support for the Scholar.
Length of Study: Up to 2 years
Frequency: Annual
No. of awards offered: 1
Application Procedure: Candidates must complete an application form, available from the website http://rsna.org/foundation
Closing Date: January 15th
Funding: Foundation
Contributor: Individuals, private practice, corporate

RSNA Medical Student Grant Program

Subjects: Radiology and related disciplines.
Purpose: To make radiology research opportunities available for medical students early in their training and encourage them to consider academic radiology as a career option.
Eligibility: Open to full-time medical students at an accredited North American medical school.
Level of Study: Research
Type: Grant
Value: US$3,000 for an 10-week (minimum) research project, to be matched by the sponsoring department (US$6,000 total), as a stipend for the medical student.
Length of Study: 1 year
Frequency: Annual
Application Procedure: Applicants must complete an application form available from the website http://rsna.org/foundation
Closing Date: February 1st
Funding: Foundation
Contributor: Individuals, private practice, corporate

RSNA Research Resident/Fellow Program

Subjects: Radiology or related disciplines.
Purpose: To provide young investigators not yet professionally established in the radiological sciences an opportunity to gain further insight into scientific investigation and to develop competence in research techniques and methods.
Eligibility: Applicants must be a resident or fellow in a department of radiology, radiation oncology or nuclear medicine within a North American educational institutions at the time of application.
Level of Study: Postdoctorate, Doctorate
Type: Fellowship
Value: US$30,000 for a 1-year Research Resident project or US $50,000 for a 1-year Research Fellow project, to be used for salary and/or non-personnel research expenses.
Length of Study: 1 year
Frequency: Annual
No. of awards offered: 1

Application Procedure: Applicants must complete an application form, available from the website http://rsna.org/foundation
Closing Date: January 15th
Funding: Foundation
Contributor: Individuals, private practice, corporate

RSNA Research Scholar Grant Program
Subjects: Medical sciences.
Purpose: To support junior clinical faculty members and allow them to gain experience in research early in their academic careers.
Eligibility: Applicants must be within five years of initial faculty appointment in a department of radiology, radiation oncology or nuclear medicine within a North American institution.
Level of Study: Doctorate, Postdoctorate
Type: Award
Value: US$75,000 per year for two years (US$150,000 total), payable to the institution, to be used exclusively as a stipend for the scholar
Length of Study: 2 years
Frequency: Annual
No. of awards offered: 1
Application Procedure: Scholar applicants must be nominated by their host institution. Applicants must complete an application form available from the website http://rsna.org/foundation.
Closing Date: January 15th
Funding: Foundation
Contributor: Individuals, private practice, corporate

RSNA Research Seed Grant Program
Subjects: Diagnostic radiology, radiation oncology and nuclear medicine.
Purpose: To assist investigators in defining objectives and testing hypotheses before they apply for major grants from corporations, foundations or government agencies.
Eligibility: Applications are accepted from any country. Applicants must hold a full-time faculty position in an educational institution at the time the award commences and be in a department of diagnostic radiology, radiation oncology or nuclear medicine, having completed all advanced training.
Level of Study: Postdoctorate, Doctorate
Type: Research grant
Value: Up to US$40,000 for a 1-year project to support the preliminary or pilot phase of scientific projects.
Length of Study: 1 year
Frequency: Annual
Application Procedure: Applicants must complete an application form, available from the website http://rsna.org/foundation
Closing Date: January 15th
Funding: Foundation
Contributor: Individuals, private practice, corporate

REBECCA SKELTON FUND

PGR Support - REEO, University of Chichester, Bishop Otter Campus, College Lane,, Chichester, Sussex, PO19 6PE, England
Tel: (44) 01243 812137
Fax: (44) 01243 816080
Email: rebeccaskeltonfund@chi.ac.uk
Website: www.rebeccaskeltonfund.org
Contact: The Rebecca Skelton Fund Administrator

The fund provides financial assistance towards the cost of post-graduate dance study in experiential/creative work to include dance improvisation and those training methods such as Skinner Releasing Technique, Alignment Therapy, Feldenkrais Technique, Alexander Technique and other body-mind practices that focus on an inner awareness and use the proprioceptive communication system or an inner sensory mode.

The Rebecca Skelton Scholarship
Subjects: Music and performing arts
Purpose: To assist students to pursue a course of specific or advanced performance studies or an appropriate dance research and performance.
Eligibility: Open to anyone pursuing dance studies at postgraduate level.
Level of Study: Doctorate, Postdoctorate, Postgraduate, Professional development, Research
Type: Scholarships and fellowships
Value: UK£500
Frequency: Annual
Country of Study: United Kingdom
No. of awards offered: 1–4
Application Procedure: Application form on request.
Closing Date: January 12th
Funding: Foundation
Contributor: The Rebecca Skelton Fund
No. of awards given last year: 5
No. of applicants last year: 9
For further information contact:

Email: artsresearch@chi.ac.uk

REES JEFFREYS ROAD FUND

Merriewood, Horsell Park, Woking, Surrey, GU21 4LW, England
Tel: (44) 1483 750758
Fax: (44) 1483 750758
Email: briansmith@reesjeffreys.org
Website: www.reesjeffreys.org
Contact: Mr Brian Smith, Secretary

The Rees Jeffreys Road Fund makes grants for courses or research connected with roads and transportation. Within those subjects, it endows university teaching posts, pays bursaries for postgraduate students, sponsors research and contributes to research projects. It has a small budget for the provision of roadside rests and improving roadside environment.

Rees Jeffreys Road Fund Bursaries
Subjects: Transport scholarship and research that will lead to a better understanding of transport issues and offer the prospect of new thinking and ideas for dealing with contemporary transport issues.
Purpose: To facilitate postgraduate study or research into transport.
Eligibility: Open to candidates of any nationality who hold at least an Upper Second Class (Honours) Degree at a United Kingdom university.
Level of Study: Doctorate, Postgraduate
Value: Bursaries of £10,000
Length of Study: 1–2 years
Frequency: Annual
Study Establishment: Universities and research institutions
Country of Study: United Kingdom
No. of awards offered: Approx. 9
Application Procedure: Applicants must be recommended by the intended institution of study and submitted by that institution.
Closing Date: July 1st for MSc courses
Funding: Private
No. of awards given last year: 9
No. of applicants last year: 24
Additional Information: Bursaries for MSc (transport) students; annual contributions to PhD research students.

Rees Jeffreys Road Fund Research Grants
Subjects: Transport.
Purpose: To facilitate research projects into roads and transportation.
Level of Study: Research
Value: Up to £30,000
Frequency: Annual
Study Establishment: Universities and research institutions
Country of Study: United Kingdom
No. of awards offered: Approx. 10 per year
Application Procedure: No application form required.
Closing Date: There is no fixed deadline. Applications will be considered by the Trustees at 1 of the 5 meetings each year
Funding: Private
No. of awards given last year: 12
No. of applicants last year: 30

REGENT'S BUSINESS SCHOOL LONDON (RBS LONDON)

Inner Circle, Regent's Park, London, NW1 4NS, United Kingdom
Tel: (44) 20 7487 7505
Fax: (44) 20 7487 7425
Email: rbsl@regents.ac.uk
Website: www.rbslondon.ac.uk

Regent's Business School London (RBS London) is one of the fastest growing business schools in the UK and follows an innovative postgraduate curriculum. Central to RBS London's ethos is a practitioner's focus. Industry level analysis and face to face interaction with managers in the international business community bring a practical dimension to your learning.

RBS London Academic Excellence Scholarships

Subjects: Global management, global management (marketing) and global management (finance).
Eligibility: These merit awards are awarded to students with strong academic achievements and potential.
Level of Study: Foundation programme, Graduate, Postgraduate
Type: Scholarship
Value: Up to 50 per cent of tuition fees is paid in the form of scholarship.
Length of Study: 1 year
Frequency: Annual
Study Establishment: Regents Business School London
Country of Study: United Kingdom
No. of awards offered: 10
Application Procedure: Applicants must contact the school, or apply online and submit a 500 word statement for the attention of the scholarship committee stating why they should be considered for the award and what contribution they can make to the school.
Closing Date: There is no application deadline
Funding: International Office
Contributor: Individuals
No. of awards given last year: 7
No. of applicants last year: 30
Additional Information: Please contact the admission office for further information.

RBS London Work-Study Scholarships

Subjects: Business and management studies
Purpose: To provide financial support to the students in need.
Eligibility: These merit awards are awarded to students with strong academic achievements and potential, who can combine study and work without interfering with their academic progress.
Level of Study: Postgraduate, Professional development, Foundation programme
Type: Scholarships and fellowships
Value: Partial remission of tuition fees
Length of Study: 1 year
Frequency: Annual
Study Establishment: Regents Business School London
Country of Study: United Kingdom
No. of awards offered: 5–10 per year
Application Procedure: Apply online submit a 500-word statement for the attention of the scholarship committee.
Closing Date: There is no application deadline
Funding: International Office
No. of awards given last year: 5
No. of applicants last year: 10
Additional Information: Scholarship holders are required to work a specified number of hours per week (normally 10 hours, up to a maximum of 20 hours) during term time. Please contact the External Relations Office for further information.

For further information contact:

Regents Business School Candow (Admissions), External Relations, London, Regents College, Inner Circle, Regents Park, NW1 4NS, United Kingdom

THE REID TRUST FOR THE HIGHER EDUCATION OF WOMEN

1 Riverside Court, Collecton Crescent, Exeter, EX2 4BZ, England
Website: www.reidtrust.org.uk
Contact: Mrs Elena Isayev, Honorary Treasurer

The Reid Trust for the Higher Education of Women was founded in 1868 in connection with Bedford College for Women for the promotion and improvement of women's education. It is administered by a small committee of voluntary trustees.

Reid Trust for the Higher Education of Women

Subjects: All subjects.
Purpose: To promote the education of women.
Eligibility: Open to women educated in the United Kingdom who have appropriate academic qualifications and who wish to undertake further training or research in the United Kingdom.
Level of Study: Unrestricted
Type: Grant
Value: €250–1,000 each
Length of Study: Unrestricted
Frequency: Annual
Country of Study: United Kingdom
No. of awards offered: Usually 10–12
Application Procedure: Applicants must complete an application form, obtained by sending a stamped addressed envelope with a request or download a form from www.reidtrust.org.uk
Closing Date: May 31st
Funding: Private
No. of awards given last year: 12
No. of applicants last year: 60

For further information contact:

Amory Building, University of Exeter, EX4 4RJ
Contact: Elena Isayev

REMEDI

Winterflood Securities Ltd, The Atrium Building, Cannon Bridge, 25 Dowgate Hill, London, EC4R 2GA, United Kingdom
Tel: (44) 0207 384 2929
Fax: (44) 0207 731 8240
Email: rosie.wait@remedi.org.uk
Website: www.remedi.org.uk
Contact: R J Wait

REMEDI, founded in 1973, supports pioneering research into all aspects of disability and disease to improve the quality of life.

REMEDI Research Grants

Subjects: Diabetes, amputees, childhood eczema, osteoporosis, rehabilitation of the elderly, speech therapy, stroke, autism and head injury.
Purpose: To support pioneering research into all aspects of disability and disease. The Trustees are particularly interested in funding or part funding initial grants where applicants find it difficult to obtain funding through larger organizations.
Eligibility: Open to English-speaking applicants living and working in the United Kingdom.
Level of Study: Doctorate, Graduate, MBA, Postdoctorate, Postgraduate, Predoctorate, Professional development, Research, Unrestricted
Type: Research grant
Value: £1,000–60,000 as well as one major project award of up to £200,000
Length of Study: 1–4 years
Frequency: Twice a year
Study Establishment: A hospital or university
Country of Study: United Kingdom
No. of awards offered: Varies
Application Procedure: Applicants must email the Director with a summary, including costs and start date on one side of A4-size paper. If the Chairman considers the research project to be of interest the Director will email an application form to be completed. Completed

applications are sent to independent referees for peer review. The Trustees normally make awards twice a year in June and November.
Closing Date: Applications are accepted at any time
Funding: Private, trusts
Contributor: Trusts, companies and individuals
No. of awards given last year: 6
No. of applicants last year: 40
Additional Information: Grants will not be awarded for course fees, administration and university overheads. Trustees have decided to reduce the number of awards, but substantially increase the size of the grants.

For further information contact:

REMEDI, 14 Crondace road, London, SW6 4BB

REPORTERS COMMITTEE FOR FREEDOM OF THE PRESS

1101 Wilson Boulevard, Suite 1100, Arlington, VA 22209, United States of America
Tel: (1) 703 807 2100
Fax: (1) 703 807 2109
Email: info@rcfp.org
Website: www.rcfp.org
Contact: Ms Michele McMohan, Office Manager

The Reporters Committee for Freedom of the Press is a voluntary, unincorporated association of reporters and editors, dedicated to protecting the First Amendment interests of the news media. From its office in Arlington, Virginia, the Reporters Committee staff provide cost free legal defence and research services to journalists and their attorneys throughout the United States and also operate the FOI service centre to assist the news media with federal and state open records and open meetings issues.

Ethics & Excellence in Journalism Legal Fellowship

Subjects: Media law.
Purpose: To financially support a recent law school graduate to pursue a postgraduate study.
Eligibility: Candidates must have received a law degree no later than August and should plan to take a bar examination in any state by August. Strong legal research and writing skills are required and a background in news reporting is very strongly preferred.
Level of Study: Postgraduate, Professional development
Type: Fellowship
Value: US$40,000 plus fully paid health benefits
Length of Study: 1 year, September–August
Frequency: Annual
Country of Study: United States of America
No. of awards offered: 1
Application Procedure: There is no application form, but the applicants should submit a covering letter, contact details of three referees, a short legal writing sample and new clips or one non-legal writing sample may be submitted in lieu of news clips. Email applications will not be accepted. Applications must be sent to Lucy Daglish, Executive Director.
Closing Date: December 10th
Funding: Private
No. of awards given last year: 1
No. of applicants last year: 40
Additional Information: Legal fellows monitor significant developments in first amendment media law, assist with legal defense requests from reporters, prepare legal memoranda, amicus briefs and other special projects. They write for the Committee's publications, the quarterly magazine, *The News Media* and *The Law* and the bi-weekly newsletter, *News Media Update.*

Jack Nelson Legal Fellowship

Subjects: Media law.
Purpose: To financially support a recent law school graduate to pursue a postgraduate study.
Eligibility: To be eligible for the programme, candidates must have been granted a law degree no later than August. Significant experience in print or electronic news reporting and strong legal research and writing skills are required.
Level of Study: Postgraduate, Professional development
Type: Fellowship
Value: US$40,000 plus fully paid health benefits
Length of Study: 1 year (September to August)
Frequency: Annual
Country of Study: United States of America
No. of awards offered: 1
Application Procedure: There is no application form. Applicants must submit a covering letter, curriculum vitae, contact details of three referees, news clips and a short legal writing sample to the attention of Lucy Daglish, Executive Director.
Closing Date: December 10th
Funding: Private
No. of awards given last year: 1
No. of applicants last year: 40
Additional Information: The Fellow will monitor significant developments in first amendment media law, assist with legal defense requirements from reporters, prepare legal memoranda, amicus briefs and other special projects. He or she will also write for The Reporter's Committee's publications, the quarterly magazine, *The News Media* and *The Law* and the bi-weekly newsletter, *News Media Update.*

Robert R McCormick Tribune Foundation Legal Fellowship

Subjects: Media law.
Purpose: To financially support a law school graduate to pursue a postgraduate study.
Eligibility: Candidates must possess a law degree and be admitted to the Bar of any state. They must have a minimum of 2 or 3 years of postgraduate legal experience in a law firm, public interest group, government agency or judicial clerkship. Substantial experience in appellate brief writing is mandatory and strong legal research and writing skills is required. A background in news reporting is strongly preferred.
Level of Study: Postgraduate, Professional development
Type: Fellowship
Value: US$50,000 plus fully paid health benefits
Length of Study: 1 year (September to August)
Frequency: Annual
Country of Study: United States of America
No. of awards offered: 1
Application Procedure: There is no application form. Applicants must submit a covering letter, curriculum vitae and contact details of three referees, a sample appellate brief and news clips or another short non-legal writing sample to Lucy Daglish, Executive Director.
Closing Date: December 10th
Funding: Foundation, private
Contributor: Robert R McCormick Tribune Foundation
No. of awards given last year: 1
No. of applicants last year: 10
Additional Information: The Fellow will be expected to draft approx. six appellate amicus briefs in significant cases involving First Amendment media law issues during the fellowship. The legal Fellow will also monitor significant developments in media law, assist with responding to legal defence requests from reporters, prepare legal memoranda and other special projects. In addition, the legal Fellow will write for the Committee's publications, the quarterly magazine *The News Media* and *The Law* and the bi-weekly newsletter, *News Media Update.*

REPRESENTATION OF THE FLEMISH GOVERNMENT

Embassy of Belgium, 3330 Garfield Street NW, Washington, DC, 20008, United States of America
Tel: (1) 202 333 6900
Fax: (1) 202 338 4960
Email: washington@diplobel.fed.be
Website: www.diplobel.us
Contact: Bernard Geenen, Walloon Trade Commissioner

The Government of Flanders offers bursaries to permit students to continue their studies in Flanders.

Fellowship of the Flemish Community

Subjects: Study or research at universities conservatories of music, art, music, humanities, social and political sciences, law, economics, sciences and medicine.
Purpose: To enable students to continue their education in Flanders, Belgium.
Eligibility: Citizens of the United States of America who are already studying for a first or doctoral degree. Candidates cannot be older than 35 years at the time of the application's deadline. Fellowship students cannot have other Belgian sources of income.
Level of Study: Predoctorate, Doctorate, Graduate, Postdoctorate, Postgraduate
Type: Fellowship
Value: A monthly stipend of approximately €770 at a Flemish institution, reimbursement of tuition fees, health insurance and public liability insurance in accordance with Belgian law. There is no reimbursement of travel expenses
Length of Study: 10 months
Frequency: Annual
Study Establishment: Recognized institutes in Flanders
Country of Study: Belgium
No. of awards offered: 5
Application Procedure: Applicants must download an application form from the website or obtain one from the Cultural Officer at the Belgian Embassy in Washington DC.
Closing Date: January 31st
Funding: Government
Additional Information: The decision to grant a fellowship is made by the Ministry of the Flemish community after consultation with the prospective host institution.

RESEARCH CORPORATION FOR SCIENCE ADVANCEMENT

4703 E Camp Lowell Drive, Suite 201, Tucson, AZ, 85712, United States of America
Tel: (1) 520 571 1111
Fax: (1) 520 571 1119
Email: awards@rescorp.org
Website: www.rescorp.org
Contact: Editor, Science Advancement Programme

The Research Corporation (USA) was one of the first United States of America foundations, and is the only one wholly devoted to the advancement of academic science. An endowed organization, it makes grants totalling US$5–7 million annually for independently proposed research in chemistry, physics and astronomy at United States of America and Canadian colleges and universities.

Cottrell College Science Awards

Subjects: Physics, chemistry and astronomy.
Purpose: To support significant research that contributes to the advancement of science.
Eligibility: Open to faculty members at public and private Institutes of Higher Education in the United States of America or Canada. The principal investigator must have an appointment in a department of astronomy, chemistry or physics and the department must offer at least baccalaureate, but not doctoral degrees. The institution must demonstrate its commitment by providing facilities and opportunities for faculty and student research.
Level of Study: Doctorate
Type: Research award
Value: US$35,000 for two year
Length of Study: Up to 5 years, possibly renewable
Study Establishment: Public and private universities with non-PhD-granting departments of astronomy, chemistry or physics
Country of Study: United States of America or Canada
No. of awards offered: 100+
Application Procedure: Applicants must complete an application form and should visit the website for guidelines and application request forms.
Closing Date: January 15th or June 15th (brief pre-proposal)
Funding: Private
Contributor: Foundation endowment
No. of awards given last year: 93
No. of applicants last year: 262

Cottrell Scholar Awards

Subjects: Physics, chemistry and astronomy.
Purpose: To promote and support the university scholar model.
Eligibility: Open to faculty in the 3rd year of a first tenure-track position.
Level of Study: Doctorate
Type: Award
Value: US$75,000, which may be flexibly applied in keeping with the applicants' approved programme
Frequency: Annual
Study Establishment: Universities with PhD-granting departments of physics, chemistry and astronomy
Country of Study: United States of America or Canada
No. of awards offered: Varies
Application Procedure: Applicants must complete an application form and should visit the website for guidelines and application request forms.
Closing Date: August 1st
Funding: Private
Contributor: Foundation endowment
No. of awards given last year: 12
No. of applicants last year: 122
Additional Information: Candidates must provide both a research and teaching plan for peer review.

RESOURCES FOR THE FUTURE (RFF)

Resources for the Future, 1616 P Street NW, Suite 600, Washington, DC, 20036, United States of America
Tel: (1) 202 328 5000
Fax: (1) 202 939 3460
Email: williams@rff.org
Website: www.rff.org
Contact: Roberton Williams, Director of Academic Affairs

Resources for the Future (RFF) is a non-profit and non-partisan think tank located in Washington, DC that conducts independent research rooted primarily in economics and other social sciences on environmental and natural resource issues.

Gilbert F. White Postdoctoral Fellowship Program

Subjects: Social and policy sciences, environmental studies, energy and natural resources.
Purpose: To enable postdoctoral researchers to spend a year in residence conducting research in the social or policy sciences in areas related to the environment, energy or natural resources.
Eligibility: A strong economics background, or closely related discipline in the social or policy sciences, is required. Applicants must be postdoctorates.
Level of Study: Postdoctorate
Type: Fellowship
Value: Stipend based upon the current salary and an allowance of up to $1,000
Length of Study: 11 months
Frequency: Annual
Country of Study: United States of America
No. of awards offered: 1
Application Procedure: Applicants must submit a covering letter and curriculum vitae with a proposal relating to budget and three letters of recommendation. No application form is required. Application forms sent by fax or email will not be accepted.
Closing Date: February 22nd
No. of awards given last year: 2
Additional Information: Please refer to the website www.rff.org/about_rff/pages/fellowships_internships.aspx.

Joseph L. Fisher Doctoral Dissertation Fellowships

Subjects: Economics, policy sciences, environment and natural resources.
Purpose: To support PhD students in their last year of dissertation research in economics or other policy sciences on issues related to the environment, energy or natural resources.

Eligibility: Open to all nationalities. Students must be in the final year of dissertation research or writing.
Level of Study: Doctorate
Type: Award
Value: US$18,000
Length of Study: 1 year
Frequency: Annual
Country of Study: Any country
No. of awards offered: Varies
Application Procedure: Applicants must submit a letter of application, a curriculum vitae, graduate transcripts, a one-page abstract of the dissertation, a technical summary of the dissertation, a letter from the department chair and two letters of recommendation. Applications sent by fax or email will not be accepted.
Closing Date: February 22nd
Additional Information: Please refer to the website www.rff.org/about_rff/pages/fellowships_internships.aspx.

RFF Fellowships in Environmental Regulatory Implementation

Subjects: Environmental studies and natural resources.
Purpose: To support research that documents the implementation and outcomes of environmental regulations.
Eligibility: Open to scholars from universities and research establishments who have a doctorate or equivalent degree or professional research experience.
Level of Study: Postdoctorate
Type: Fellowship
Value: Minimum award amount: $50,000. Maximum award amount: $80,000
Length of Study: 1–2 years
Frequency: Annual
Country of Study: United States of America
No. of awards offered: 2
Application Procedure: Applicants must submit a preproposal no longer than two pages, single spaced, with 11-point or larger font and 1-inch margins, containing full details of the proposed project. Applicants whose preproposals are selected for further review will then be invited to submit final proposals. These are limited to ten pages and are to follow a similar format. They must include all of the information contained in the preproposal but should offer further detailed description of the proposed research, anticipated contribution of the project and the importance of the results. Applicants must also include a curriculum vitae with the applicant's educational background, professional experience, a list of most relevant publications and honours and awards received. In addition, final proposals must include three letters of recommendation from fellow faculty members or colleagues. Applications sent by fax or email will not be accepted.
Closing Date: January 4th
Funding: Private
No. of awards given last year: 3

RESUSCITATION COUNCIL (UK)

5th Floor, Tavistock House North, Tavistock Square, London, WC1H 9HR, England
Tel: (44) 20 7388 4678
Fax: (44) 20 7383 0773
Email: enquiries@resus.org.uk
Website: www.resus.org.uk
Contact: Dr Sara Harris, Assistant Director

The Resuscitation Council (UK) is the expert advisory body on the training and practice of resuscitation in the United Kingdom. It also actively pursues and promotes research in the field of resuscitation medicine.

Resuscitation Council Research Fellowships

Subjects: All aspects of the science, practice and teaching of resuscitation.
Purpose: To fund the salaries of research fellows based in the United Kingdom to conduct research into the science and practice of resuscitation medicine.
Eligibility: Open to research Fellows carrying out work in a suitable venue in the United Kingdom. But must be resident in United Kingdom.
Level of Study: Unrestricted
Type: Fellowship
Value: Basic salary level including National Insurance and Superannuation contributions. The salary would be at the appropriate point on the applicant's salary scale
Length of Study: Up to 2 years
Frequency: Dependent on funds available
Country of Study: United Kingdom
Application Procedure: Applicants should visit the website www.resus.org.uk for application procedures.
Closing Date: No closing dates. Applications are considered on a rolling basis.
Funding: Private
Contributor: The Resuscitation Council (UK)
No. of awards given last year: 1
No. of applicants last year: 2
Additional Information: Only electronic applications are accepted.

Resuscitation Council Research Grants

Subjects: All aspects of the science, practice and teaching of resuscitation techniques.
Purpose: To provide grants for capital cost, data analysis and administrative support for research into the science and practice of resuscitation medicine.
Eligibility: Applicants must be based in United Kingdom.
Level of Study: Unrestricted
Type: Grant
Value: Up to UK£20,000
Frequency: Dependent on funds available
Country of Study: United Kingdom
Application Procedure: Applicants should visit the website: www.resus.org.uk for application details.
Closing Date: No closing dates. Applications are considered on a rolling basis.
Funding: Private
Contributor: The Resuscitation Council, United Kingdom
No. of awards given last year: 1
No. of applicants last year: 5
Additional Information: Only electronic applications are accepted.

RHODES COLLEGE

2000 North Parkway, Memphis, TN, 38112-1690, United States of America
Tel: (1) 901 843 3000
Email: registrar@rhodes.edu
Website: www.rhodes.edu/

Rhodes, founded in 1848, seeks to graduate students with a lifelong passion for learning, a compassion for others, and the ability to translate academic study and personal concern into effective leadership and action in their communities and world.

Jacob K. Jarits Fellowships

Subjects: Humanities, arts and social sciences.
Purpose: Support for higher students in the areas of humanities, arts and social sciences.
Eligibility: Open to candidates who are US citizens, permanent residents of the US or citizens of any one of the freely associated states.
Type: Fellowship
Value: US$30,000 per year
Length of Study: Up to 4 years
Frequency: Annual
Study Establishment: Rhodes College
Country of Study: United States of America
No. of awards offered: 48
Application Procedure: Submit applications to the Rhodes Faculty post graduate scholarship committee 1 month prior to deadline.
Closing Date: September 15th (Internal Deadline)
Funding: Government
Contributor: The US Department of Education

For further information contact:

US Department of Education, DPE Teacher and student Development Programs Service Jacob K. Jarits Fellowships Program 1990 k street, N.W., 6th floor, Washington, DC 20006-8524
Tel: (202) 502-7542
Fax: (202) 502-7859
Website: www.ed.gov/programs/jacobjarits

Luce Scholars Program

Subjects: Any subject.
Purpose: To develop better cultural exchange between Asia and the US.
Eligibility: Open to US citizens with BA or equivalent degree, under 29 years of age with no significant exposure to Asia culture or study.
Level of Study: Postgraduate
Type: Scholarship
Length of Study: 1 year
Frequency: Annual
Country of Study: Asia
No. of awards offered: 15
Application Procedure: Requests for responsive or project grants should be submitted.
Closing Date: Early October (Internal Deadline)
Funding: Foundation
Contributor: Henry Luce Foundation

For further information contact:

The Henry Luce Foundation 111 West 50th street, New York, 10020
Tel: 212 489 7700
Website: www.hluce.org

Truman Scholarships

Subjects: All subject.
Purpose: To support college juniors preparing for leadership in public service.
Eligibility: Candidates should be US citizens and full-time students pursuing a Bachelor's degree.
Value: US$26,000
Frequency: Annual
Study Establishment: Rhodes College
Country of Study: United States of America
No. of awards offered: 75–80 from 600 applications
Application Procedure: Download application from the website.
Closing Date: December 3rd (Internal Deadline)
Funding: Foundation
Contributor: Truman Scholarship Foundation

For further information contact:

712 Jackson Place NW, Washington, DC 20006
Tel: (202) 395-4831
Fax: (202) 395-6995

Watson Fellowship

Subjects: Any subject
Purpose: To offer college graduates a year of independent study and travel outside the United States.
Eligibility: Candidates must be graduating seniors to apply; U.S. citizenship NOT required.
Type: Fellowship
Value: US$25,000
Frequency: Annual
No. of awards offered: Up to 40
Application Procedure: A fellowship form along with a project application proposal and details in not more than 5 pages should be submitted electronically.
Closing Date: Early October (Internal deadline)
Contributor: Thomas J. Watson Foundation

For further information contact:

Watson Fellowship Program, 293 South Main Street, Providence, RI 02903
Tel: 401 274 1952
Fax: 401 274 1954
Email: tjw@watsonfellowship.org
Contact: Thomas J

RHODES UNIVERSITY

PO Box 94, Grahamstown, 6139, South Africa
Tel: (27) 046 603 8055
Fax: (27) 046 622 8822
Email: research-admin@ru.ac.za
Website: www.ru.ac.za/research
Contact: Research Office

Rhodes University is a small university campus in Grahamstown with one of the highest research outputs per capita in South Africa. The University offers excellent undergraduate and postgraduate education, and fosters personal development and leadership as well as team, social and communication skills amongst its diverse student body.

Allan Gray Senior Scholarship

Subjects: Commerce (accounting, economics, management and information systems).
Purpose: To encourage and enable previously disadvantaged South Africans to pursue their studies at Honours and Master's levels.
Eligibility: Open to previously disadvantaged South African students with the appropriate qualifications.
Level of Study: Postgraduate
Type: Scholarship
Value: South African Rand 40,000 for Honours and Rand 50,000 for Master's
Length of Study: 1 year, renewable upon reapplication for Master's
Frequency: Annual
Study Establishment: Rhodes University, Grahamstown
Country of Study: South Africa
No. of awards offered: 10
Application Procedure: Applicants must submit a curriculum vitae and a full academic record and a covering letter of invitation indicating proposed degree and subject to be studied.
Closing Date: July 1st in the year preceding registration
Funding: Private
Contributor: Donor and investments
No. of awards given last year: 4
No. of applicants last year: Approx. 35 per year
Additional Information: Postgraduate scholars must assist with teaching and/or research duties up to a maximum of 6 hours a week in their department of study, without additional remuneration. Awards are for full-time study in attendance at Rhodes University only.

Andrew Mellon Foundation Scholarship

Subjects: Humanities, commerce, education, law, pharmacy and science.
Purpose: To encourage and enable previously disadvantaged people to pursue their studies at the Honours, Master's or Doctoral level at Rhodes University, as well as to enhance the recipients' ability to contribute to higher education in South Africa.
Eligibility: Open to previously disadvantaged individuals with the appropriate qualifications.
Level of Study: Postgraduate
Type: Scholarship
Value: South African Rand 35,000 for Honours, Rand 40,000 for Masters and Rand 60,000 for Doctoral
Length of Study: 1 year, renewable upon reapplication
Frequency: Annual
Study Establishment: Rhodes University, Grahamstown
Country of Study: South Africa
No. of awards offered: 25 at Honours level, 10 at Master's and Doctoral level
Application Procedure: Applicants must submit a curriculum vitae and a full academic record and a letter of invitation indicating proposed degree and subject to be studied.
Closing Date: July 1st in the year preceding registration
Funding: Private
Contributor: Donor and investments
No. of awards given last year: 25 at Honours level, 10 at Master's or Doctoral level
No. of applicants last year: Approx. 250 per year
Additional Information: Scholars must assist with teaching and/or research duties up to a maximum of 6 hours a week in their

department of study, without additional remuneration. Awards are for full-time study in attendance at Rhodes University only.

Henderson Postgraduate Scholarships

Subjects: Mathematical, physical, earth, life and pharmaceutical sciences, accountancy and information systems.
Purpose: To encourage and enable students to pursue their studies at the Master's and doctoral levels, and to enhance recipients' research abilities and produce internationally competitive graduates with an innovative, analytical, articulate and well-rounded desire to learn.
Eligibility: Open to South African citizens with appropriate qualifications.
Level of Study: Postgraduate
Type: Scholarship
Value: South African Rand 40,000 for Master's level and Rand 60,000 for Doctoral level
Length of Study: 1 year, renewable upon reapplication
Frequency: Annual
Study Establishment: Rhodes University, Grahamstown
Country of Study: South Africa
No. of awards offered: 10
Application Procedure: Applicants must submit a curriculum vitae and a full academic record and a covering letter indicating proposed degree and subject to be studied.
Closing Date: July 1st in the year preceding
Funding: Private
Contributor: Donor and investments
No. of awards given last year: 5
No. of applicants last year: Approx. 180
Additional Information: Scholars must assist with teaching and/or research duties up to a maximum of 6 hours a week in their department of study, without additional remuneration. Awards are for full-time study in attendance at Rhodes University only.

Hugh Kelly Fellowship

Subjects: All areas of science.
Purpose: To enable senior scientists to devote themselves to advanced work.
Eligibility: Open to suitable senior postdoctoral scientists with a very well-established research output record. Preference is given to candidates willing to accept appointments for at least 4 months. Applicants must be English speakers.
Level of Study: Postdoctorate, Research
Type: Fellowship
Value: Currently under review, but a package of approx. Rand 70,000. If the Fellow accepts an appointment for at least 4 months and is accompanied by a spouse, the spouse's air or rail fares will also be paid. University accommodation will be provided free of charge, but the provision of a telephone if available at the place of residence will be for the Fellow's personal account
Length of Study: Up to 1 year
Frequency: Annual
Study Establishment: Rhodes University, Grahamstown
Country of Study: South Africa
No. of awards offered: 1
Application Procedure: Applicants must complete an application form, available from the Dean of Research or from the website.
Closing Date: July 31st of the year preceding the award
Funding: Private
Contributor: Donor and investments
No. of awards given last year: 1
No. of applicants last year: 6
Additional Information: The Fellow will be required to present a concise report on the work completed at the conclusion of the term of the fellowship.

Hugh Le May Fellowship

Subjects: Humanities.
Purpose: To enable senior scholars of standing in the humanities to devote themselves to advanced work.
Eligibility: Open to scholars from the humanities with a well-established research output record. Applicants must be English speaking.
Level of Study: Research
Type: Fellowship
Value: A return economy class air ticket from the Fellow's place of residence, furnished University accommodation and a small monthly cash stipend, free use of the University Library and sports facilities
Length of Study: 3–4 months, extended by mutual agreement and subject to the availability of funds
Study Establishment: Rhodes University, Grahamstown
Country of Study: South Africa
Application Procedure: Application forms to be completed, available from the Research Office or the website.
Closing Date: July 31st
Funding: Private
Contributor: Donor and investments
No. of awards given last year: 1
No. of applicants last year: 5
Additional Information: The Fellow will not be expected to undertake any teaching studies but will be required to present a report upon the work undertaken at the conclusion of the term of the Fellowship.

Patrick and Margaret Flanagan Scholarship

Subjects: All subjects.
Purpose: To enable South African women graduates to attend a university in the United Kingdom in order to obtain a higher postgraduate qualification.
Eligibility: Open to women graduates, preferably those who are English-speaking applicants of South African descent. Selection is based initially on academic merit, as well as broader qualities of intellect and character.
Level of Study: Postgraduate
Type: Scholarship
Value: South African Rand 350,000 per year
Length of Study: 2 years
Frequency: Annual
Study Establishment: Rhodes University, Grahamstown
Country of Study: United Kingdom
No. of awards offered: 1
Application Procedure: Applicants must submit a curriculum vitae and full academic record. Candidates shortlisted will be sent an application form or additional information and referee reports.
Closing Date: August 1st of the year prior to registration abroad
Funding: Private
Contributor: Donor and investments
No. of awards given last year: 1
No. of applicants last year: Approx. 120

Rhodes University Postdoctoral Fellowship and The Andrew Mellon Postdoctoral Fellowship

Subjects: All subjects (Rhodes University Postdoctoral Fellowship); humanities (Andrew Mellon Postdoctoral Fellowship).
Purpose: To enable scholars to devote themselves to advanced work, that will closely complement existing programmes in the host department.
Eligibility: Open to any postdoctoral scholars of standing, with research publications to their credit and of exceptional merit. Applicants must be English speakers.
Level of Study: Postdoctorate
Type: Fellowship
Value: South African Rand 130,000 per year with an additional allocation of Rand 10,000 for travel or research
Length of Study: 1 year, but may be extended by mutual agreement, subject to the availability of funds
Frequency: Annual
Study Establishment: Rhodes University, Grahamstown
Country of Study: South Africa
No. of awards offered: 2–3
Application Procedure: Applicants must be nominated. Nominations should be made through heads of departments and directors of research institutes at Rhodes University. Nominations must include a full curriculum vitae, research proposal and the name of three referees who may be consulted. Further information may be obtained from the Research Office or from the website.
Closing Date: July 31st of the year preceding the award
Funding: Private

Contributor: Donor and investments
No. of awards given last year: 4 for Rhodes University Postdoctoral Fellowship, 4 for Andrew Mellon Postdoctoral Fellowship
No. of applicants last year: 32
Additional Information: Fellows are not expected to undertake teaching duties.

Rhodes University Postgraduate Scholarship

Subjects: Humanities, commerce, education, law, pharmacy, and science.
Purpose: To encourage and enable students to pursue their studies at the Master's and doctoral levels and enhance the recipients research abilities.
Eligibility: Open to students with the appropriate qualifications. Academic merit will override eligibility criteria.
Level of Study: Postgraduate
Type: Scholarship
Value: South African Rand 40,000 for Master's levels and Rand 60,000 for doctoral levels
Length of Study: 1 year, renewable upon reapplication
Frequency: Annual
Study Establishment: Rhodes University, Grahamstown
Country of Study: South Africa
No. of awards offered: 4
Application Procedure: Applicants must submit a curriculum vitae and a full academic record and a covering letter indicating degree and subject to be studied.
Closing Date: July 1st in the year preceding registration
Funding: Private
Contributor: Donor and investments
No. of awards given last year: 12
No. of applicants last year: Approx. 150
Additional Information: Scholars must assist with teaching and/or research duties up to a maximum of 6 hours per week in their department of study, without additional remuneration. Awards are for full-time study in attendance at Rhodes University only.

Ruth First Scholarship

Subjects: Humanitics, social science students whose fields of study are politics, sociology, philosophy, anthropology, economics, social policy, democracy studies, development studies, media studies, or studies in cog rate disciplines with a strong social and human rights orientation.
Purpose: The scholarship is intended to support candidates whose research is in the spirit of Ruth First's life and work, poses difficult social questions, and links knowledge and politics and scholarship and action.
Eligibility: For Masters or PhD study and black South African or Mozambican women candidates are particularly encouraged to apply.
Level of Study: Doctorate, Postgraduate
Type: Scholarship
Value: PhD R100,00 per year; Maters R80,000 per year
Frequency: Annual
Study Establishment: Rhodes University, Grahamsdown
Country of Study: South Africa
No. of awards offered: 1
Application Procedure: Applicants to submit a letter of motivation (2 pages or more) as to why they would be a potentially suitable candidate for this award and how their proposed research will be within the spirit of Ruth First's work. In addition, an academic transcript, academic curriculum vitae and a copy of an identity document is required.
Closing Date: August 31st
Funding: Private
Contributor: Donor and Investments
No. of awards given last year: 1
No. of applicants last year: 126

For further information contact:

Tel: 046 603 8755
Email: pgfinaid-admin@ru.ac.za
Website: www.ru.ac.za/research/postgraduates/funding

RICHARD III SOCIETY, AMERICAN BRANCH

2041 Christian Street, Philadelphia, PA, 19146, United States of America
Email: feedback@r3.org
Website: www.r3.org
Contact: Ms Laura Blanchard, Schallek Fellowships

The Richard III Society was founded in 1924 in England as The Fellowship of the White Boar and was renamed the Richard III Society in 1959. The American Branch was founded in 1961. Today, the Society has more than 4,000 members worldwide, and American Branch membership is more than 800.

William B Schallek Memorial Graduate Fellowship Award

Subjects: Medieval history.
Purpose: To support graduate study of 15th-century English history and culture.
Eligibility: Applicants must be citizens of the United States of America, have made an application for first citizenship papers or be permanent resident aliens enrolled at a recognized educational institution.
Level of Study: Doctorate
Type: Fellowship
Value: Dissertation awards of $2,000 each and dissertation fellowship of $30,000 annually
Length of Study: 1 year, renewals will be considered
Frequency: Annual
Study Establishment: Recognized and accredited degree-granting institutions
Country of Study: Any country
No. of awards offered: 5
Application Procedure: Applicants must visit the website for guidelines, lists of past awards, their topics and application forms.
Closing Date: Please check with the website.
No. of awards given last year: 3
No. of applicants last year: 7

RICS EDUCATION TRUST

Royal Institution of Chartered Surveyors, RICS, Parliament Square, London, SW1P 3AD, England
Tel: (44) 24 7686 8555
Fax: (44) 20 7334 3811
Email: contactrics@rics.org
Website: www.rics-educationtrust.org

The Royal Institution of Chartered Surveyors (RICS) is the professional institution for the surveying profession.

RICS Education Trust Award

Subjects: The theory and practice of surveying in any of its disciplines including general practice, quantity surveying, building surveying, rural practice, planning and development, land surveying or minerals surveying.
Eligibility: Open to chartered surveyors and others carrying out research studies in relevant subjects.
Level of Study: Unrestricted
Type: Research grant
Value: Up to UK£7,500
Frequency: Twice a year
Country of Study: Any country
Application Procedure: Applicants must complete an application form, available to download online at www.rics-educationtrust.org
Closing Date: September 30th or February 28th
Funding: Commercial
Contributor: RICS
No. of awards given last year: 20
No. of applicants last year: 40

RMIT UNIVERSITY

Info Corner - Office for Prospective Students, GPO Box 2476, Melbourne, VIC 3001, Australia
Tel: (61) 3 9925 2260
Fax: (61) 3 9925 3070
Email: study@rmit.edu.au
Website: www.rmit.edu.au

RMIT University is one of Australia's original and leading educational institutions producing some of Australia's most employable graduates. RMIT has an international reputation for excellence in work-relevant education professional and vocational education, high quality research, and engagement with the needs of industry and community.

Interior Design-Masters of Arts by Research

Subjects: Sculpture, film, theatre, journalism, visual arts, interior design and architecture.
Purpose: To offer a space within which candidates develop and contribute to the knowledge and possibilities of interior design.
Eligibility: Open to the candidates of any country who have a First Degree of RMIT with at least a credit average in the final undergraduate year or a deemed equivalent by RMIT to a First Degree of RMIT with at least a credit average in the final undergraduate year or evidence of experience.
Level of Study: Postgraduate
Length of Study: 2 years full-time (Masters) and 4 years part-time (PhD)
Application Procedure: Check website for further details.
Closing Date: October 31st
Funding: Government
Contributor: Commonwealth Government

For further information contact:

Tel: 03 9925 2819
Email: suzie.attiwill@rmit.edu.au
Contact: Ms Suzie Attiwill, Research Coordinator

ROB AND BESSIE WELDER WILDLIFE FOUNDATION

PO Box 1400, Sinton, TX, 78387, United States of America
Tel: (1) 361 364 2643
Fax: (1) 361 364 2650
Email: welderfoundation@welderwildlife.org
Website: www.welderwildlife.org
Contact: Director

The Rob and Bessie Welder Wildlife Foundation is a private non-profit foundation whose mission is to conduct research and education in wildlife management and closely related fields. The Welder Foundation operates a wildlife refuge on a commercial ranch in the midst of an active oil field.

Welder Wildlife Foundation Fellowship

Subjects: Wildlife ecology and management.
Purpose: To provide support to individual graduate student research.
Eligibility: Open to citizens of the United States of America or legal aliens registered at a United States of America university for a graduate degree. Priority is given to students who wish to work at the Welder Foundation Refuge or in the Coastal Bend Region of Texas.
Level of Study: Graduate, Doctorate
Type: Fellowship
Value: Stippened for full-time students are $1,400 per month for M.S. candidates and $1,600 per month for PhD candidates to cover living costs, tuition, fees and books
Length of Study: The duration of a graduate degree programme
Frequency: Annual
Study Establishment: Any legitimate wildlife management or wildlife ecological department at any accredited university
Country of Study: United States of America
No. of awards offered: 10 at any given time, approx. 3 each year
Application Procedure: Applicants must write for details.
Closing Date: October 1st for fellowships to begin in January
Funding: Private
Contributor: Private endowment
No. of awards given last year: 2–3
No. of applicants last year: 15–20

ROBERT BOSCH FOUNDATION

CDS International, Inc., 440 Park Avenue South, 2nd Floor, New York, NY, 10016, United States of America
Tel: (1) 212 497 3500
Fax: (1) 212 497 3535
Email: info@cdsintl.org
Website: www.cdsintl.org
Contact: Susana Lee, Assistant Program Officer

The Robert Bosch Foundation is one of the largest German industry foundations. Many of the foundation's international exchange programmes are aimed at providing young people with opportunities to improve their knowledge of other countries and cultures and to build up networks among future leaders in Europe and the United States of America.

Robert Bosch Foundation Fellowships

Subjects: Business administration, economics, public affairs, public policy, political science, law, journalism and mass communication.
Purpose: To promote the advancement of American and German-European relations, and to broaden the participants' professional competence and cultural horizons.
Eligibility: Open to citizens of the United States of America between the ages of 23 and 34 years with a graduate or professional degree or equivalent work experience in the above subject areas. Candidates must provide evidence of outstanding professional or academic achievement and a strong knowledge of the German language. For those candidates who are outstanding in other areas but lack sufficient knowledge of German, the Foundation will provide language training prior to programme participation.
Level of Study: Postgraduate, Professional development
Type: Fellowship
Value: €2,000 per month stipend. Extra funding is available for family, German language tutoring, health insurance and all program-related travel
Length of Study: 9 months (September–May)
Frequency: Annual
Country of Study: Germany
No. of awards offered: 20
Application Procedure: Applicants must complete an application form and attend a personal interview. Please visit the website www.cdsintl.org/bosch for more details.
Closing Date: Mid-October
Funding: Private
Contributor: Robert Bosch Foundation
No. of awards given last year: 20
No. of applicants last year: Varies each year
Additional Information: Programme participants receive internships in German institutions such as the Federal Parliament, private corporation headquarters, mass media and other elements within the framework of government or commerce. Internships will be at a high level, closely related to senior officials. The programme will follow the following schedule: an intensive course on German language, political, economic and cultural affairs, work experience, a visit to Berlin and the former East Germany, a visit to the European Economic Community (EEC) and NATO headquarters in Brussels, a group visit to France for an overview of the political, economic and cultural perspective of another European country and a final programme evaluation in Stuttgart. All activities are conducted in German.

THE ROBERT WOOD JOHNSON FOUNDATION

Route 1 & College Road East, PO Box 2316, Princeton, NJ, 08543 2316, United States of America
Tel: (1) 877 843 7953
Email: hss@nyam.org
Website: www.rwjf.org

Established in 1972 in memory of Robert Wood Johnson, who founded Johnson & Johnson. The mission of the Robert Wood Johnson Foundationis to improve the health and healthcare of all Americans. Our goal is clear: to help Americans lead healthier lives and get the care they need

The Robert Wood Johnson Health & Society Scholars Program

Subjects: A broad range of factors affecting the nation's health.
Purpose: To build the nation's capacity for research, leadership and action.
Eligibility: For eligibility information, please visit our website, www.healthandsocietyscholars.org
Level of Study: Postdoctorate
Type: Scholarship
Value: Stipend of US$80,000 annually
Length of Study: 2 years
Frequency: Annual
Study Establishment: One of six universities
Country of Study: United States of America
No. of awards offered: Up to 12
Application Procedure: The application for the Robert Wood Johnson Health and Society Scholars program is done through an online application system available on our website www.healthandsocietyscholars.org
Closing Date: Please check with the website.
Funding: Private
No. of awards given last year: 18

ROBERTO LONGHI FOUNDATION

Fondazione Roberto Longhi Via Benedetto Fortini 30, I-50125 Florence, Italy
Tel: (39) 55 658 0794
Fax: (39) 55 658 0794
Email: longhi@fondazionelonghi.it
Website: www.fondazionelonghi.it
Contact: Procuratore

The aim of the Roberto Longhi Foundation is to advance art historical studies. Applications are accepted annually from prospective Fellows who hold an advanced degree in art history. Fellows pursue individual and group research projects using the library and photo archive, participate in seminars and lectures by distinguished scholars at the Foundation and visit galleries, restorations and exhibitions.

Fondazione Roberto Longhi

Subjects: History of art.
Purpose: To aid those who want to seriously dedicate themselves to research in the history of art.
Eligibility: Open to citizens of Italy who possess a degree from an Italian university with a thesis in the history of art and to non-Italian citizens who have fulfilled the preliminary requirements for a doctoral degree in the history of art at an accredited university or an institution of equal standing. Students who have reached their 32nd birthday before the application deadline are not eligible.
Level of Study: Postgraduate, Doctorate
Type: Fellowship
Value: €600 per month
Length of Study: 9 months
Frequency: Annual
Study Establishment: The Foundation
Country of Study: Italy
No. of awards offered: Up to 10 or 12
Application Procedure: Applicants must submit an application containing their biographical data including place and date of birth, domicile, citizenship, transcript of undergraduate and graduate records, a copy of the degree thesis and of other original works, a curriculum vitae including knowledge of foreign languages both spoken and written, letters of reference from at least two persons of academic standing who are acquainted with the applicant's work, the subject of the research proposed and two passport size photographs.
Closing Date: May 15th
Funding: Private
Contributor: Private funds and capital endowment
No. of awards given last year: 9
No. of applicants last year: 16
Additional Information: Successful candidates must give an assurance that they will dedicate their full time to the research for which the fellowship is assigned. They may not enter into any connection with other institutions, they must live in Florence for the duration of the fellowship, excepting for travel required for their research. They may not exceed the periods of vacations fixed by the Institute and are required to attend seminars, lectures and other activities arranged by the Institute. The Fellows must, in addition, submit a written report at the end of their stay in Florence relating the findings of their individual research undertaken at the Longhi Foundation. Once approved by the scientific committee the Fellows research must be published only in the Foundations annual journal *Proporzioni*. Non-compliance with the above conditions will be considered sufficient grounds for the cancellation of a fellowship. Further information is available on request.

ROSL (ROYAL OVER-SEAS LEAGUE) ARTS

Over-Seas House, Park Place, St James's Street, London, SW1A 1LR, England
Tel: (44) 20 7408 0214 ext 219
Fax: (44) 20 7499 6738
Email: info@rosl.org.uk
Website: www.rosl.org.uk
Contact: Mandy Murphy, Administrative Assistant

The principal aim of the ROSL (Royal Over-Seas League) ARTS is to provide performance and exhibition opportunities for prize-winning artists and musicians early in their careers, bringing their work to the attention of the professional arts community, the media and the general public.

ROSL Annual Music Competition

Subjects: Musical performance, in four solo classes such as strings (including the harp and guitar), wind and percussion, keyboard, singers and to ensemble classes.
Purpose: To support and promote young Commonwealth musicians.
Eligibility: Open to the citizens of the United Kingdom and Commonwealth, including former Commonwealth countries, for instrumentalists and singers up to and including the age of 30 as at the date of the final concert.
Level of Study: Professional development
Type: Competition
Value: Over UK£60,000 in prizes, including a UK£10,000 gold medal and first prize and UK£10,000 for ensembles
Frequency: Annual
Country of Study: Any country
No. of awards offered: Varies
Application Procedure: Applicants must see the website: www.roslarts.org.uk
Closing Date: January 18th
Funding: Commercial, individuals, private, trusts
No. of awards given last year: 19
No. of applicants last year: 500

ROTARY INTERNATIONAL

One Rotary Center 1560 Sherman Avenue, Evanston, IL, 60201-3698, United States of America
Tel: (1) 847 866 3000
Fax: (1) 847 328 8554
Email: contact.center@rotary.org
Website: www.rotary.org
Contact: Administrative Assistant

Rotary International is a worldwide organization of business and professional leaders that provides humanitarian service, encourages high ethical standards in all vocations and helps build goodwill and peace in the world. Approximately 1.2 million Rotarians belong to more than 32,000 clubs in more than 200 countries and geographical areas.

Rotary Ambassadorial Scholar Program

Subjects: All subjects.
Purpose: To develop a sense of purpose as ambassadors of goodwill and to further international understanding and world peace.
Eligibility: Open to applicants who are citizens of a country in which there are Rotary clubs. Initial application must be made through a Rotary club in the applicant's legal or permanent residence or place of full-time study or employment. Applicant must be proficient in the language of the proposed host country.
Level of Study: Postgraduate
Type: Scholarships
Value: Up to US$25,000. Covers round-trip transportation, 1 month of intensive language training, room and boarding and some educational supplies
Length of Study: 1 academic year of full-time graduate study
Frequency: Annual
Country of Study: Any country in which there is a Rotary Club
Application Procedure: Applications for the scholarship programme must be made through a Rotary Club. Every Club is eligible to submit Ambassadorial Scholar applications to the District.
Closing Date: As early as 1 March or as late as 15 August
Additional Information: The scholarships are not appropriate for students seeking to continue studies already begun at a foreign institution.

Rotary Foundation Academic Year Ambassadorial Scholarships

Subjects: All subjects.
Purpose: To further international understanding and friendly relations among people of different countries.
Eligibility: Open to citizens of a country where there are Rotary clubs. The applicants should have completed more than 2 years of college-level coursework or equivalent professional experience before commencing their scholarship studies. Applicants must be proficient in the language of the proposed host country.
Level of Study: Unrestricted
Type: Scholarship
Value: Round-trip transportation, tuition, room and board expenses, some educational supplies and 1 month of language training if necessary, totalling up to US$25,000
Length of Study: 1 academic year
Frequency: Annual
Study Establishment: A study institution assigned by the Trustees of the Rotary Foundation
No. of awards offered: Varies
Application Procedure: Applicants should contact a local Rotary club for details.
Closing Date: As early as March 1st or as late as August 15th
Funding: Private
Additional Information: Scholars will not be assigned to study in areas of a country where they have previously lived or studied for more than 6 months. During the study year, scholars are expected to be outstanding ambassadors of goodwill through appearances before Rotary clubs, schools, civic organizations and other forums. Upon completion of the scholarship, scholars are expected to share the experiences of understanding acquired during the study year with the people of their home countries. Candidates should contact local Rotary clubs for information on the availability of particular scholarships. Not all Rotary districts are able to offer scholarships. Further details are available from the website.

THE ROYAL ACADEMY OF ENGINEERING

3 Carlton House Terrace, London, SW1Y 5DG, England
Tel: (44) 020 7766 0600
Fax: (44) 020 7930 1549
Email: ian.bowbrick@raeng.org.uk
Website: www.raeng.org.uk
Contact: Ian Bowbrick, Scheme Manager

The Royal Academy of Engineering's objectives may be summarized as the pursuit, encouragement and maintenance of excellence in the whole field of engineering in order to promote the advancement of the science, art and the practice of engineering for the benefit of the public.

ExxonMobil Excellence in Teaching Awards

Subjects: Chemical, petroleum and mechanical engineering, geology.
Purpose: To encourage able young engineering and Earth science lecturers to remain in the education sector in their early years.
Eligibility: Open to well-qualified graduates, preferably with industrial experience and full-time lecturing posts at Institutes of Higher Education in the United Kingdom. Applicants should have been in their current posts for at least 1 year. The post must include the teaching of chemical, petroleum or mechanical engineering to undergraduates through courses that are accredited for registration with professional bodies for qualifications such as chartered engineer. For applicants whose career path has been graduation at the age of 22, followed by academic or industrial posts, the age limit is generally 32 years (at the closing date). Older candidates who have taken time out, e.g. for industrial experience, parenthood or voluntary service, will also be considered. Applicants should preferably be chartered engineers, or of equivalent professional status, or should be making progress towards this qualification.
Level of Study: Postdoctorate
Type: Fellowship
Value: UK£5,000
Length of Study: 12 months
Frequency: Annual
Study Establishment: The applicant's current university in the United Kingdom
Country of Study: United Kingdom
No. of awards offered: Up to 6
Application Procedure: Applicants must complete an application form.
Closing Date: October 12th
Funding: Commercial
Contributor: Exxon Mobile
Additional Information: A brochure is available on request. Enquiries about Exxon mobile university contacts should be sent via email and please see the website for further details http://www.raeng.org.uk/research/univ/exxonmobil/default.htm.

For further information contact:

Email: bowbricki@raeng.co.uk

Panasonic Trust Awards

Subjects: Engineering, particularly new engineering developments and new technologies.
Purpose: To encourage the technical updating and continuous professional development of qualified engineers through courses provided by United Kingdom Institutes of Higher Education at the Master's level.
Eligibility: Open to United Kingdom citizens who are qualified at the degree level in engineering or a related discipline. HND, HNC, OND, ONC or City and Guilds Full Technological Certificate qualifications are acceptable as a minimum. Applicants must be members, at any grade, of an engineering institution, working at the professional level in engineering in the United Kingdom and have several years of experience working at this level. Preference is given to those undertaking part-time modular Master's courses. The intended course of study must be relevant to the applicant's current or future career plans.
Level of Study: Professional development
Type: Award
Value: £8,000
Length of Study: The duration of the course
Frequency: Dependent on funds available
Country of Study: United Kingdom
No. of awards offered: Varies
Application Procedure: Applicants must write to the main address for application instructions, the application form and guidelines for employers and course co-ordinators.
Closing Date: July 25th
Funding: Private
Additional Information: Employers are expected to support an application in writing. The Trustees hope that the employer, once approached, might pay the full fees for the course. However, if this is

not possible, a contribution from the employer is desirable. Applications must be supported by the course director or co-ordinator and confirm that the applicant is suitable for the course. Applications for grants of less than UK£1,000 should be submitted at least 4 weeks before the start of the course. Applications for larger grants should be submitted at least 6 weeks before the start of the course. Please see the website for further details http://www.panasonictrust.net/fellowships/default.aspx.

Royal Academy Engineering Professional Development

Subjects: Engineering.
Purpose: To ensure that the stills and knowledge of employees reflect the very latest in technological advances.
Eligibility: Open to UK citizens with a degree or HND/HNC in engineering or a closely allied subject. OND/ONC or City and Guilds Full Technological Certificate or NVQ level III qualifications are acceptable provided the individual has substantial industrial experience.
Level of Study: Professional development
Type: Grant
Value: £10,000 and £5,000 and prospective applicants should indicate for which level of award they are applying
Length of Study: 1 year
Frequency: Annual
Country of Study: United Kingdom
Application Procedure: For further information please contact the scheme manager Ian Bowbrick at the Academy.
Closing Date: October 24th
Additional Information: Please see the website for further details.

For further information contact:

Email: Ian.bowbrick@raeng.org.uk
Website: http://www.raeng.org.uk/education/professional/profdev/default.htm

Royal Academy Executive Engineering Programme

Subjects: Engineering.
Purpose: To ensure that top engineering graduates continue to develop as professional engineers in order to provide UK industry with the high caliber technical and commercial managers required by world class enterprises.
Eligibility: Open to permanent members of United Kingdom Institutes of Higher Education with an approved engineering qualification.
Level of Study: Postgraduate
Type: Scholarship
Value: All training programme fees
Length of Study: 1 year
Frequency: Annual
Study Establishment: An industrial or commercial company
Country of Study: United Kingdom
Application Procedure: Visit the programme website.
Closing Date: Offered all year round (Check with website)
Funding: Foundation
Contributor: Gatsby Charitable Foundation
Additional Information: Please see the website for further details http://www.raeng.org.uk/education/professional/eep/default.htm.

For further information contact:

Contact: Sandra Palmer

Royal Academy of Engineering Industrial Secondment Scheme

Subjects: All fields of engineering.
Purpose: To provide financial support for the secondment of academic engineering staff to industrial companies within the United Kingdom.
Eligibility: Open to permanent members of United Kingdom Institutes of Higher Education with an approved engineering qualification, teaching some aspects of engineering. Preference may be given to junior staff without previous industrial experience or to more senior members whose industrial experience may have taken place some years earlier.
Level of Study: Unrestricted
Type: Grant
Value: Varies
Length of Study: Usually 3–6 months
Frequency: Dependent on funds available
Study Establishment: An industrial or commercial company
Country of Study: United Kingdom
No. of awards offered: Varies
Application Procedure: Applicants must submit a completed application form with a curriculum vitae, personal statement outlining the nature and objectives of the proposed secondment, letter of support from the applicant's head of department, statement from the applicant's employer detailing the financial aspects of the application, statement from the host company confirming the agreed work programme and defining the benefits of the secondment to the company and detailing any contribution costs that the company may wish to make. Application forms are available from the main address.
Closing Date: February 15th (Check with website)
Funding: Government
No. of awards given last year: 18
No. of applicants last year: 26
Additional Information: The main objective is to obtain up-to-date industrial experience, to improve teaching capabilities and generally to foster academic industrial links. Contact Dr Imren Markes for further information. Please see the website for further details http://www.raeng.org.uk/research/univ/secondment/default.htm.

Royal Academy of Engineering MacRobert Award

Subjects: The successful development of innovation in engineering or the other physical sciences.
Purpose: To recognize and reward outstanding contributions relating to innovation in engineering.
Eligibility: Open to individuals, independent teams and teams working for a firm, organization or laboratory. There should be no more than five members in a team.
Level of Study: Unrestricted
Type: Award
Value: UK£50,000 and a gold medal
Frequency: Annual
Country of Study: United Kingdom
No. of awards offered: 1
Application Procedure: Applicants must submit a 200–500 word summary of the engineering achievement, supported by 15 copies of relevant technical documentation. Further details and rules and conditions are available from the Royal Academy of Engineering.
Closing Date: January 31st
Funding: Private
No. of awards given last year: 1
Additional Information: Please see the website for further details http://www.raeng.org.uk/prizes/macrobert/default.htm.

For further information contact:

Tel: 20 7766 0648
Contact: Sylvia Hampartumian

Royal Academy Sir Angus Paton Bursary

Subjects: Engineering for development and water and environmental management.
Purpose: To study water and environmental management.
Eligibility: The bursary supports a suitably qualified engineer study a full-time Masters' degree course specifically related to water resources engineering or some other environmental technology.
Level of Study: Postgraduate
Type: Bursary
Value: UK£8,000
Length of Study: 1 year
Frequency: Annual
Country of Study: United Kingdom
No. of awards offered: 1
Application Procedure: For further information please contact the scheme manager Ian Bowbrick at the Academy.
Closing Date: Offered all year round (Check with website)
Funding: Private
Contributor: Sir Angus Paton

For further information contact:

Email: ian.bowbrick@raeng.org.uk

Website: http://www.raeng.org.uk/education/professional/paton/default.htm
Contact: Ian Bowbrick

Royal Academy Visiting Professors Scheme

Subjects: Principles of engineering design, engineering design for sustainable development, integrated system design.
Purpose: To help universities to teach engineering design to undergraduates in a way that relates to real professional practice.
Level of Study: Doctorate, Postdoctorate, Research
Type: Fellowship
Length of Study: 1 year
Frequency: Annual
Country of Study: United Kingdom
No. of awards offered: 3
Application Procedure: Further information is available from the scheme manager Ian Bowbrick at the Academy.
Closing Date: Check with website
Additional Information: Please see the website for further details http://www.raeng.org.uk/education/vps/systemdesign/background.htm.

Sainsbury Management Fellowships

Subjects: Engineering.
Purpose: To enable young chartered engineers of the highest career potential to undertake MBA courses at European business schools.
Eligibility: Open to United Kingdom citizens, who hold a First or Upper Second Class (Honours) Degree in engineering or a closely allied subject, have chartered engineer status or are making substantial progress towards it, have the potential and ambition to achieve senior management responsibility at an early age and be aged 26–34 years at the commencement of the proposed MBA course.
Level of Study: Professional development, MBA, Postgraduate
Type: Fellowship
Value: Up to £30,000
Length of Study: 1 year
Frequency: Annual
Study Establishment: Awards are normally tenable at the following business schools: North America–Harvard, MIT, Stanford, Wharton, Columbia, Kellogg, University of Chicago; Europe–INSEAD (France), IMD (Switzerland), Erasmus (The Netherlands), IESE (Spain), SDA Bocconi (Italy)
Country of Study: Any country
No. of awards offered: 10
Application Procedure: Applicants must complete an application form, please contact the office for details.
Closing Date: Applications are accepted at any time
Funding: Private
Contributor: The Gatsby Charitable Foundation
No. of awards given last year: 10
No. of applicants last year: 46
Additional Information: Please see the website for further details.

For further information contact:

Tel: 20 7766 0600
Website: http://www.raeng.org.uk/education/professional/sainsbury/default.htm
Contact: Ian Bowbrick

Sainsbury Management Fellowships in the Life Sciences

Subjects: Life science.
Purpose: To support young scientists of high career potential to undertake activities related to their Personal Development Plans.
Eligibility: Open to UK citizens with a PhD in a bio-related subject, following a career compatible with the scheme's objectives in the UK. The applicants should be aged between 25 and 38 years.
Level of Study: Postgraduate
Type: Fellowship
Value: The award covers the cost of activities related to an individuals personal development plan, subject to approval
Frequency: Annual
Country of Study: Any country
Application Procedure: For further information please contact the scheme manager Ian Bowbrick at the Academy.
Closing Date: Offered all year round (Check with website)
Funding: Private
Contributor: Gatsby Charitable Foundation
Additional Information: Please contact ian.bowbrick@raeng.org.uk for further details.

ROYAL ACADEMY OF MUSIC

Marylebone Road, London, NW1 5HT, England
Tel: (44) 20 7873 7373
Fax: (44) 20 7873 7394
Email: go@ram.ac.uk
Website: www.ram.ac.uk
Contact: Sharon Moloney, Examinations and Data Records Officer

The Royal Academy of Music is a music college offering courses in music at postgraduate level. The Academy is part of the University of London.

Royal Academy of Music General Bursary Awards

Subjects: All relevant branches of music education and training.
Purpose: To help defray tuition fees and general living expenses for study at the Royal Academy of Music.
Eligibility: Open to any student offered a place at the Academy.
Level of Study: Postgraduate
Type: Scholarship/Bursary
Value: According to need and availability of funds
Length of Study: Normally for a complete academic year. Individual requirements may be imposed
Frequency: Annual
Study Establishment: The Royal Academy of Music
Country of Study: United Kingdom
No. of awards offered: Varies
Application Procedure: Applicants must complete an application form, which is sent automatically to all postgraduate students who are offered places.
Closing Date: January 31st
Funding: Commercial, individuals, private, trusts
No. of awards given last year: 74
No. of applicants last year: 493

ROYAL AERONAUTICAL SOCIETY

4 Hamilton Place, London, W1J 7BQ, England
Tel: (44) 20 7670 4300
Fax: (44) 20 7670 4309
Email: careers@aerosociety.com
Website: www.aerosociety.com
Contact: Careers Centre Manager

The Royal Aeronautical Society was founded in 1866 and is the only professional institution that covers all aspects of the aerospace industry including research, manufacture, operations and maintenance. Society membership unlocks a host of benefits for both the individual and organizations. Membership is open to anyone with an association or interest in aerospace.

Royal Aeronautical Society Centennial Scholarship Award

Subjects: Aeronautics and aerospace-related subjects such as aerospace engineering, safety, human factors, air transport management.
Purpose: To encourage the next generation of aerospace pioneers in the study of aeronautics and aerospace-related subjects and support of relevant projects to encourage young people into the field.
Eligibility: Open to students studying aerospace-related (e.g. aviation, aeronautical engineering, aeronautic-related) studies, and key organizations and research groups whose projects meet the aims and objectives of the fund.
Level of Study: Doctorate, Graduate, Postdoctorate, Postgraduate, Research, Relevant projects of national reach that meet the aims and objectives of the fund
Type: Scholarship
Value: As a guide, from UK£500–5,000

Length of Study: Most have already completed 1–2 years of relevant study
Frequency: Twice per year
Study Establishment: Not specified
Country of Study: Any country
No. of awards offered: Varies
Application Procedure: Applicants must complete an application form, available on request and provide all documentation specified in the form and accompanying guidelines.
Closing Date: May 31st
Funding: Individuals, commercial
Contributor: A number of corporate, private and individual donors
No. of awards given last year: 32
No. of applicants last year: Approx. 55

For further information contact:

Tel: 0 20 7670 4325
Email: rosalind.azouzi@aerosociety.com
Website: http://aerosociety.com/Careers-Education/centennial
Contact: Rosalind Azouzi, Careers and Education Manager

ROYAL ANTHROPOLOGICAL INSTITUTE

50 Fitzroy Street, London, W1T 5BT, England
Tel: (44) 20 7387 0455
Fax: (44) 20 7388 8817
Email: admin@therai.org.uk
Website: www.therai.org.uk
Contact: Amanda Vinson, Office Manager

The Royal Anthropological Institute is a non-profit registered charity. It is entirely independent, with a Director and a small staff accountable to the Council, elected annually from the Fellowship. Council and Committee members and the editorial team of the Institute's principal journal, the *Journal of the Royal Anthropological Institute* (incorporating MAN), give their services without remuneration.

Emslie Horniman Anthropological Scholarship Fund

Subjects: Anthropology.
Purpose: To provide predoctoral grants for fieldwork in anthropology with a preference for research outside the United Kingdom.
Eligibility: Open to citizens of the United Kingdom, Commonwealth or Irish Republic who are university graduates or who can satisfy the trustees of their suitability for the study proposed. Preference is given to applicants whose proposals include fieldwork outside the United Kingdom. Graduates who already hold a doctorate in anthropology are not eligible. Open to individuals only, as no grants are given to expeditions or teams.
Level of Study: Postgraduate, Predoctorate
Type: Scholarship
Value: UK£1,000–9,500
Frequency: Annual
Country of Study: Any country
No. of awards offered: Approx. 10
Application Procedure: Applicants must request an application form.
Closing Date: March 31st
Funding: Private
No. of awards given last year: 2
No. of applicants last year: 20
Additional Information: No grants are available for library research, university fees or subsistence in the United Kingdom.

ROYAL COLLEGE OF MIDWIVES

15 Mansfield Street, London, W1G 9NH, England
Tel: (44) 20 7312 3643
Fax: (44) 20 7312 3536
Email: info@rcm.org.uk
Website: www.rcm.org.uk
Contact: S E MacDonald, Education and Research Manager

The Royal College of Midwives is the major professional organization for midwives in the United Kingdom, and aims to contribute to the art and science of midwifery knowledge and practice. The RCM awards and scholarships provide opportunities for the development of good practice ultimately improving the care provided to women, their babies and families.

Mary Seacole Nursing Development and Leadership Award

Subjects: Midwifery, nursing, health visiting, health service needs of ethnic minority communities.
Purpose: To provide an opportunity for the development of professional practice, service development and leadership potential of black and ethnic minority midwives, nurses and health visitors.
Eligibility: Applicants must be a nurse and, midwife or health visitor. Application should reflect the example set by Mary Seacole.
Level of Study: Graduate, Research, Doctorate, Professional development
Type: Leadership awards
Value: UK£12,500, Development award UK£6,250
Length of Study: 1 year
Frequency: Annual
Country of Study: United Kingdom
No. of awards offered: 4
Application Procedure: Applicants must send a stamped addressed envelope to the Royal College of Nursing (RCN) to obtain details and an application form.
Closing Date: May 1st
Funding: Government
Contributor: Department of Health
No. of awards given last year: 1
No. of applicants last year: 3
Additional Information: Please see the website for further details.

For further information contact:

Bukola Samuel, Room 452C, Skipton House, Department of Health, 80 London Road, London, SE1 6LH, England
Contact: Award Officer, Mary Seacole Award,

RCM Annual Midwifery Awards

Subjects: There are nine categories of awards which celebrate the range of work and achievement of midwives midwifery education, midwifery management or leadership, innovations in midwifery, Members' Champion award, promotion of normal birth, outstanding contribution to care of newborns, award for turning vision into reality, midwifery impact on child development, promoting effective midwifery in community settings and Student Vision award.
Purpose: To recognize and celebrate innovation in midwifery practice, education and research.
Eligibility: Applicants may be individuals or small groups but should meet the criteria of 1 of the 10 categories. Check the website for complete details: www.rcm.org.uk/college/annual-midwifery-awards/
Level of Study: Professional development, Postgraduate, Research
Type: Award
Value: Varies (up to UK£20,000 in total)
Frequency: Annual
Country of Study: Projects Based in United Kingdom – though some requiring travel may involve external activity
No. of awards offered: 13
Application Procedure: Applicants must apply in writing or email to the address given below. The application must be accompanied by a 500-word description of the project. Short-listed candidates will be asked to attend an interview.
Closing Date: November 1st (check with website)
Funding: Commercial
Contributor: Several
No. of awards given last year: 12 awards in 12 categories plus a midwife award
No. of applicants last year: 82

For further information contact:

Gothic House, 3 The Green, Richmond, Surrey, TW9 1PL, England
Email: mail@chamberdunn.co.uk
Website: http://www.rcmawards.com/
Contact: Chamberlain Dunn Associates

Ruth Davies Research Bursary

Subjects: Midwifery.
Purpose: To promote and develop midwifery research and practice.

Eligibility: Open to practicing midwives who are RCM members, who have basic knowledge, skills and understanding of the research process, have access to research support in their trust or Institutes of Higher Education and who have been in practice for 2 years or more.
Level of Study: Postgraduate, Professional development, Doctorate, Graduate, Postdoctorate, Research, Predoctorate
Type: Bursary
Value: UK£5,000 per bursary
Length of Study: 1 year
Frequency: Annual
Country of Study: United Kingdom
No. of awards offered: 3
Application Procedure: Applicants must submit a succinct curriculum vitae covering the previous 5 years, a research proposal of no more than 2,500 words and letters of support from both employers and academics who are familiar with the applicant's work.
Closing Date: July 30th (check with website)
Funding: Commercial
Contributor: Bounty
No. of awards given last year: 3
No. of applicants last year: 4 shortlisted

For further information contact:

Tel: 0207 312 3463
Email: marlyn.gennace@rcm.org.uk
Contact: Mrs Marlyn Gennace, Ruth Davies Research Bursary Administrator

ROYAL COLLEGE OF MUSIC

Prince Consort Road, London, SW7 2BS, United Kingdom
Tel: (44) 20 7591 4300
Fax: (44) 20 7589 7740
Email: nicola.peacock@rcm.ac.uk
Website: www.rcm.ac.uk
Contact: Miss Nicola Peacock, International and Awards Officer

The Royal College of Music, London, provides specialized musical education and professional training at the highest international level for performers and composers. This enables talented students to develop the musical skills, knowledge, understanding and resourcefulness that will equip them to contribute significantly to musical life in the United Kingdom and internationally.

Royal College of Music Scholarships

Subjects: Music performance, composition and conducting.
Purpose: To recognize merit in music performance, composition or conducting.
Eligibility: All those auditioning in person for a place at the Royal College of Music are considered for a scholarship at the time of their audition. Scholarships are awarded on the basis of merit and potential.
Level of Study: Graduate, Postgraduate, Doctorate
Type: Scholarship
Value: Up to UK£24,750 (full fees-intensive masters)
Length of Study: 1–4 years
Frequency: Annual
Study Establishment: Royal College of Music
Country of Study: United Kingdom
No. of awards offered: Approx. 200
Application Procedure: Prospective students should apply to the RCM online at www.cukas.ac.uk. No separate application for scholarship is required.
Closing Date: October 1st
Funding: Private, trusts
No. of awards given last year: Approx. 200
No. of applicants last year: 1,800
Additional Information: Please see the website for further details http://www.ram.ac.uk/scholarships.

ROYAL COLLEGE OF NURSING (RCN)

20 Cavendish Square, London, W1G 0RN, England
Tel: (44) 20 7409 3333
Email: scholarships@rcn.org.uk
Website: www.rcn.org.uk
Contact: Ms Moira Lambert, Awards Officer

The RCN Foundation is an independent charity supporting nursing to improve the health and well-being of the public.

Ethicon Nurses and Practitioner Education Trust Fund

Subjects: Nursing, midwifery, and health visiting.
Purpose: To enable individual nurses and practitioners to further their professional development.
Eligibility: Nurses and midwives currently registered in the United Kingdom who will be implementing their learning and development in the United Kingdom.
Level of Study: Doctorate, Graduate, Postdoctorate, Postgraduate, Predoctorate, Professional development
Type: Grant
Value: Up to UK£1,000 each
Length of Study: Unrestricted
Frequency: Annual
Study Establishment: Unrestricted
Country of Study: Any country
No. of awards offered: Varies
Application Procedure: Applicants must contact Fitwise Drumcross Hall for an application form.
Closing Date: Marcg
Funding: Commercial
Contributor: Ethicon
No. of awards given last year: 20
No. of applicants last year: 44
Additional Information: These scholarships now come under the RCN Foundation. Please see attached organization profile.

For further information contact:

Fitwise Drumcross Hall, Bathgate, EH48 4JT, Scotland
Tel: (44) 01506 811077

Mary Seacole Leadership and Development Awards

Subjects: Nursing, midwifery and health visiting including public health, health policy, and health education.
Purpose: To provide funding for a project, or other educational/development activity that benefits the health needs of people from black and minority ethnic communities.
Eligibility: Open to nurses, midwives and health visitors in England.
Level of Study: Predoctorate, Professional development, Research, Doctorate, Graduate, Postdoctorate, Postgraduate
Type: Award
Value: Up to UK£6,250–12,500
Length of Study: Unrestricted
Frequency: Annual
Study Establishment: Unrestricted
Country of Study: England
No. of awards offered: 6
Application Procedure: Applications forms available between February and April. Email: bukola.samuel@dh.gsi.gov.uk for an application form, or visit www.rcn.org.uk/membership/scholarshipsawards to download an application forms.
Closing Date: May 11th
Funding: Government
Contributor: Department of Health
No. of awards given last year: 6
No. of applicants last year: 16

RCN Foundation Margaret Parkinson Scholarships

Subjects: Nursing and midwifery.
Purpose: To encourage graduates with a non-nursing degree to qualify as a nurse.
Eligibility: Open to graduates who have not yet started nurse training.
Level of Study: Postgraduate
Type: Scholarship
Value: Up to UK£1,000 each per year
Length of Study: Unrestricted
Frequency: Annual
Study Establishment: Universities
Country of Study: United Kingdom
No. of awards offered: Varies
Application Procedure: Applicants must send a stamped self-addressed envelope to the Awards Officer. Enquiries will be dealt with between November and January.

Closing Date: December 1st
Funding: Private
Contributor: Margaret Parkinson Scholarship Fund
No. of awards given last year: 4
No. of applicants last year: 46
Additional Information: These scholarships now come under the RCN Foundation. Please see attached organisation profile.

For further information contact:

Margaret Parkinson Scholarships, RCN Foundation, 20 Cavendish square, London, W1G 0RN
Contact: Awards Officer

RCN Foundation Professional Bursary Scheme
Subjects: The RCN Foundation is offering bursaries of up to £5,000 to support individuals in learning and development activities that contribute to their professional development. For example, post-graduate university study or professional short-courses. The learning from these activities should also enhance patient care and future practice.
Purpose: To support individuals in learning and development activities that contribute to their professional development.
Type: Bursary
Value: Up to £5,000
Frequency: Biannual
Country of Study: United Kingdom
Application Procedure: Check the RCN website for application form.
Closing Date: End of April
Funding: Foundation

ROYAL COLLEGE OF OBSTETRICIANS AND GYNAECOLOGISTS (RCOG)

27 Sussex Place, Regent's Park, London, NW1 4RG, United Kingdom
Tel: (44) 20 7772 6200
Fax: (44) 20 7723 0575
Email: mgoonewardene@rcog.org.uk
Website: www.rcog.org.uk
Contact: M Goonewardene, Awards Administrator

The Royal College of Obstetricians and Gynaecologists (RCOG) is dedicated to the encouragement of the study, and the advancement of science and practice of obstetrics and gynaecology.

American Gynecological Club/Gynaecological Visiting Society Fellowship
Subjects: Obstetrics and gynaecology.
Purpose: To enable the recipient to visit, make contact with and gain knowledge from a specific centre offering new techniques or methods of clinical management within the speciality.
Eligibility: Open to specialist registrars in the United Kingdom, and to junior Fellows or those in residency programmes in the United States of America.
Level of Study: Professional development
Type: Fellowship
Value: UK£500
Frequency: Annual
Country of Study: United Kingdom or United States of America
No. of awards offered: 1
Application Procedure: Applicants must contact the Awards Secretary for details or see website.
Closing Date: May 24th
No. of awards given last year: 1
No. of applicants last year: 3
Additional Information: Please see the website for further details http://www.rcog.org.uk/our-profession/careers/award/american-gynecological-clubgynaecological-visiting-society-fellowship-0.

The Calcutta Eden Hospital Annual Prize
Purpose: To provide knowledge from a specific centre offering new techniques of clinical management within O&G.
Eligibility: The prize will be awarded to first year medical students and junior doctors working in Foundation Year 1 in the UK and the Republic of Ireland.
Level of Study: Foundation programme, final year medical students
Type: Prize
Value: UK£300 book tokens
Frequency: Annual
Application Procedure: Applications can be obtained from awards administrator or download from website.
Closing Date: May 24th
Funding: Individuals
No. of awards given last year: 1
No. of applicants last year: 5
Additional Information: Please see the website for further details http://www.rcog.org.uk/our-profession/careers/award/calcutta-eden-hospital-annual-prize-0.

Eden Travelling Fellowship
Subjects: Obstetrics and gynaecology.
Purpose: To enable the recipient to gain additional knowledge and experience in the pursuit of a specific research project in which he or she is currently engaged.
Eligibility: Open to medical graduates of not less than 2 years standing from any approved university in the United Kingdom or Commonwealth.
Level of Study: Postgraduate
Type: Fellowship
Value: Up to UK£5,000, according to the project undertaken
Frequency: Annual
Study Establishment: Another department of obstetrics and gynaecology or of closely related disciplines
Country of Study: Any country
No. of awards offered: 1
Application Procedure: Applicants must include information on qualifications, areas of interest and/or publications in a specified area, centres to be visited with confirmation of arrangements from the head of that centre, estimated costs and the names of two referees.
Closing Date: May 24th
No. of awards given last year: 1
No. of applicants last year: 9
Additional Information: Please see the website for further details http://www.rcog.org.uk/our-profession/careers/award/eden-travelling-fellowship-obstetrics-and-gynaecology-0.

Endometriosis Millennium Fund Award
Subjects: Endometriosis.
Purpose: To stimulate and encourage research, clinical or laboratory based, or to encourage clinicians to acquire extra clinical skills to manage patients.
Eligibility: Open to members of the College or members of the RCOG trainees register who are residents and working within the United Kingdom.
Level of Study: Postgraduate
Value: Up to UK£5,000
Frequency: Annual
Country of Study: United Kingdom
No. of awards offered: More than 1
Application Procedure: Applicants must submit an application in the requisite form, which can be downloaded from the website or can be obtained from the Awards Secretary at the RCOG. Application forms must be accompanied by two references from supervisors or senior colleagues and written confirmation of availability of laboratory space or access to surgical training in the host institution must be provided with the application. An undertaking that a structured typewritten report will be provided at the end of the grant period and that the source of grant will be acknowledged in any related publications must also be submitted.
Closing Date: March 24th
Contributor: The Endometriosis Millennium Fund
No. of awards given last year: 1
No. of applicants last year: 6
Additional Information: The award may only be used for the purpose approved by the Assessment Committee. Please see the website for further details http://www.rcog.org.uk/our-profession/careers/award/endometriosis-millennium-fund-0.

Ethicon Travel Awards
Subjects: Obstetrics and gynaecology.
Purpose: To promote international goodwill in the speciality.

Eligibility: Open to RCOG members who have passed both Part I and II of the membership exams. Members from the United Kingdom must be trainees and overseas members must not be in independent practice.
Level of Study: Postgraduate
Type: Travel grant
Value: Up to UK£2,000
Frequency: Twice a year
No. of awards offered: Up to 15
Application Procedure: Applicants must complete an application form, available from the Awards Secretary or from the website.
Closing Date: May 24th
No. of awards given last year: 6
No. of applicants last year: 10
Additional Information: Travel must take place within 6 months of the award being made.

For further information contact:

Email: mgoonewardene@rcog.org.uk
Website: http://www.rcog.org.uk/our-profession/careers/award/ethicon-travel-award

Green-Armytage and Spackman Travelling Scholarship

Subjects: Obstetrics and gynaecology.
Purpose: To allow applicants to visit centres where work similar to their own is being carried out.
Eligibility: Open to Fellows and members of the College. Applicants should have shown a special interest in some particular aspect of obstetrical or gynaecological practice.
Level of Study: Postgraduate
Type: Scholarship
Value: £5,000–10,000
Frequency: Annual
Country of Study: Any country
No. of awards offered: 1
Application Procedure: Applicants must include information on qualifications, areas of interest and/or publications in a specified area, centres to be visited with confirmation from the head of that centre, estimated costs and the names of two referees.
Closing Date: May 24th (check with website)
No. of awards given last year: 1
No. of applicants last year: 6

Herbert Erik Reiss Memorial Case History Prize

Purpose: The prizes will be awarded to candidates who, in the opinion of the assessors, undertake the best presentation of a clinical case, including critical assessment and literature research, in a topic of obstetrics and gynaecology.
Eligibility: Open to medical students and junior doctors working in foundation years 1/2 in the UK and the Republic of Ireland.
Value: First prize is UK£400 and second prize is UK£200
Frequency: Annual
No. of awards offered: 2
Application Procedure: A maximum of 1,500 words with maximum of 10 references should be submitted. Please include a statement of contribution to the project and indicate the name and address of the supervisor. Only one submission per candidate is permitted.
Closing Date: May 24th
No. of awards given last year: 3
No. of applicants last year: 20
Additional Information: Please contact at mgoonewardene@rcog.org.uk for further details.

John Lawson Prize

Purpose: The prize will be awarded to a candidate who, in the opinion of the assessors, undertakes the best article on a topic of obstetrics or gynaecology derived from work carried out in Africa between the tropics of Capricorn and Cancer.
Eligibility: Candidature is not restricted to fellows and members of the college.
Value: UK£150
Frequency: Annual
Application Procedure: The record of the work can be submitted by way of an original manuscript, adequately referenced and written in a format comparable to that used for submission to a learned journal, or by means of a reprint of a published article.
Closing Date: May 24th
No. of awards given last year: 1
No. of applicants last year: 8
Additional Information: If joint authorship is involved then the candidate must identify his/her involvement in the publication.

For further information contact:

Email: mgoonewardene@rcog.org.uk
Website: http://www.rcog.org.uk/our-profession/careers/award/john-lawson-prize-0

Malcolm Black Travel Fellowship

Subjects: Obstetrics and gynaecology.
Purpose: To enable a College Member or Fellow of up to 5 years standing at the time of application, to travel either to the British Isles or from the British Isles abroad, for a period of time to extend postgraduate training courses or to visit centres of research or of particular expertise within the speciality of obstetrics and gynaecology.
Type: Travelling fellowship
Value: Travel and subsistence costs will be paid up to a maximum of UK£1,000
Frequency: Every 2 years
Closing Date: May 24th
No. of awards given last year: 1
No. of applicants last year: 5
Additional Information: Please see the website for further details.

Overseas Fund

Subjects: Obstetrics and gynaecology.
Purpose: To allow individuals to travel to the United Kingdom for further training.
Eligibility: Open to RCOG members or the equivalent, working overseas.
Level of Study: Professional development
Type: Travel grant
Value: Up to UK£2,500
Frequency: Annual
No. of awards offered: Up to 8
Application Procedure: Applicants must see the website.
Closing Date: May 24th
No. of awards given last year: 1
No. of applicants last year: 3

For further information contact:

Email: mgoonewardene@rcog.org.uk
Website: http://www.rcog.org.uk/our-profession/careers/award/overseas-fund-0

Peter Huntingford Memorial Prize

Subjects: Obstetrics and Gynaecology
Purpose: The prize is for presenting the best case history, clinical audit or a report of a research project in any aspect of fertility control in which the applicant is directly involved.
Eligibility: Open to doctors working in their foundation year or specialist training years 1 and/or 2 in the UK and the Republic of Ireland.
Value: First prize is UK£1,000 and second prize is UK£500
Frequency: Annual
No. of awards offered: 2
Application Procedure: A submission of no more than 1,500 words outlining the research with reference to publications (if any) is required. Please include a statement of the contribution to the project and indicate the name and address of the supervisor. Only one submission per candidate is permitted.
Closing Date: May 24th
Contributor: British Pregnancy Advisory Service (BPAS)
No. of awards given last year: 1
No. of applicants last year: 2

For further information contact:

Email: mgoonewardene@rcog.org.uk

Website: http://www.rcog.org.uk/our-profession/careers/award/peter-huntingford-memorial-prize-0

RCOG Bernhard Baron Travelling Scholarships

Subjects: Obstetrics and gynaecology.
Purpose: To expand the recipient's knowledge in areas in which he or she already has some experience.
Eligibility: Open to Fellows and members of the College.
Level of Study: Postgraduate
Type: Scholarship
Value: Up to UK£6,000
Frequency: Annual
Country of Study: Any country
No. of awards offered: 2
Application Procedure: Applicants must contact the Awards Secretary for details.
Closing Date: May 24th
No. of awards given last year: 2
No. of applicants last year: 8
Additional Information: Please see the website for further details mgoonewardene@rcog.org.uk.

For further information contact:

Contact: Marion Goonewardene, Awards Administrator

Richard Johanson Research Prize

Subjects: Obstetrics and gynaecology.
Purpose: To recognize the best research projects in any aspect of obstetrics and gynaecology.
Level of Study: Graduate
Value: First prize UK£250, second prize UK£100
Frequency: Annual
Country of Study: United Kingdom, Northern Ireland and Republic of Ireland
Application Procedure: Applicants should submit research outline no more than 1,500 words, with reference to publications (if any). The research must have been carried out during the applicant's undergraduate course and in a relevant department.
Closing Date: September 30th
Funding: Private
No. of awards given last year: 2
No. of applicants last year: 34
Additional Information: Submissions may by sent by email or post (four copies of the typewritten submission on A4-size paper). Applicants must state clearly which award they are applying for and include their supervisor's name and address.

Tim Chard Case History Prize

Subjects: Obstetrics and gynaecology.
Purpose: To reward students showing the greatest understanding of a clinical problem in obstetrics and gynaecology.
Level of Study: Graduate
Value: First prize UK£500, second prize UK£250, third prize UK£150
Frequency: Annual
Study Establishment: An accredited medical school
Country of Study: United Kingdom, Northern Ireland and Republic of Ireland
Application Procedure: Applicants should submit one case history with discussion no longer than 1,500 words with 10 biographical references that includes the name and address of the applicant's supervisor. Submissions may be sent by email or post (four copies of the typewritten submission on A4-size paper). Applicants must state clearly which award they are applying for, and make separate applications for each. Applications that do not comply with the stipulations will not be accepted.
Closing Date: May 24th
Funding: Private
Contributor: Bart's and the London School of Medicine
No. of awards given last year: 4
No. of applicants last year: 57
Additional Information: Please see the website for further details http://www.rcog.org.uk/our-profession/careers/award/tim-chard-case-history-prize.

Why Obs and Gynae? Prize

Subjects: Obstetrics and gynaecology.
Purpose: The college invites trainees currently in Foundation Years to tell your reasons as to why you have chosen a career in Obstetrics and Gynaecology.
Eligibility: For trainees currently in foundation years.
Level of Study: Foundation programme
Value: UK£200
Frequency: Annual
Closing Date: September 30th
No. of awards given last year: 1
No. of applicants last year: 23

William Blair-Bell Memorial Lectureships in Obstetrics and Gynaecology

Subjects: Obstetrics and gynaecology.
Purpose: To allow an individual to give a lecture.
Eligibility: Open to Fellows or members of not more than 2 years standing.
Value: Honorarium of UK£200
Frequency: Annual
Application Procedure: Applicants must contact the Awards Secretary.
Closing Date: May 25th
No. of awards given last year: 1
No. of applicants last year: 5

THE ROYAL COLLEGE OF OPHTHALMOLOGISTS

17 Cornwall Terrace, London, NW1 4QW, England
Tel: (44) 20 7935 0702
Fax: (44) 20 7935 9838
Email: training@rcophth.ac.uk
Website: www.rcophth.ac.uk
Contact: Susannah Grant, Deputy Head of Education and Training

The college is responsible for promoting high standards of professional practice, setting curricula and conducting examinations and providing professional support and advice for ophthalmologists. The college runs an annual scientific congress and seminars for opthalmologists and publishes a range of clinical guidelines.

Dorey Bequest & Sir William Lister Travel Awards

Subjects: Ophthalmology.
Purpose: To provide personal travel expenses to members and Fellows of the Royal College of Ophthalmologists who are travelling abroad for research or training.
Eligibility: Members and Fellows of the Royal College of Ophthalmologists.
Level of Study: Postgraduate
Type: Travel award
Value: Generally UK£300–600 per award
Frequency: Annual
Country of Study: United Kingdom
No. of awards offered: 2–4
Application Procedure: An application form must be completed and copied five times and submitted with five copies of the candidate's curriculum vitae.
Closing Date: September 25th
Funding: Private
No. of awards given last year: 2
No. of applicants last year: 7
Additional Information: Please see the website for further details http://www.rcophth.ac.uk/page.asp?section=320§ionTitle=Awards+and+Prizes.

Ethicon Foundation Fund Travel Award

Subjects: Ophthalmology.
Purpose: To prove financial assistance to members and Fellows of the Royal College of Ophthalmologists who are travelling abroad for research or training purposes.
Eligibility: Applicants must be members or Fellows of the Royal College of Ophthalmologists and affiliate members of Royal College of Ophthalmologists.

Level of Study: Postgraduate
Type: Travel award
Value: UK£300–1,000 per award
Frequency: Annual
Country of Study: United Kingdom
No. of awards offered: Varies
Application Procedure: An application form must be completed and copied five times and submitted with five copies of the candidates curriculum vitae.
Closing Date: October 30th
Funding: Commercial
Contributor: Ethicon Ltd.
No. of awards given last year: 7
No. of applicants last year: 11
Additional Information: The Committee favours applications which demonstrate that applicants are using their initiative to obtain clinical experience above and beyond that which they would derive from a routine exchange or secondment to an overseas centre. It is envisaged that the award will support more senior trainees towards the end of their training who have already gained some relevant experience in this country. This should not, however, discourage junior trainees with an outstanding project from applying. Please see the website for further details http://www.rcophth.ac.uk/page.asp?section=320§ionTitle=Awards+and+Prizes.

The Fight For Sight Award

Subjects: Ophthalmology.
Purpose: The award will be made in recognition of a significant piece of research (completed within 18 months prior to the closing date) in an area that falls within the remit of the charity. It is intended that the award will be used to further the education of the award winner by being sent on research-related activities including attendance at conferences and seminars on ophthalmology.
Eligibility: Applicants must be under 40 years of age.
Level of Study: Postgraduate
Value: UK£5,000
Frequency: Annual
Country of Study: United Kingdom
No. of awards offered: 1
Application Procedure: 5 copies each of the completed application form and the curriculum vitae, alongwith a supporting paper (published or otherwise) detailing a significant piece of research must be submitted.
Closing Date: February
Funding: Commercial
Contributor: Fight for Sight
No. of awards given last year: 1
No. of applicants last year: 212
Additional Information: Please see the website for further details http://www.rcophth.ac.uk/page.asp?section=320§ionTitle=Awards+and+Prizes.

The Keeler Scholarship

Subjects: Ophthalmology.
Purpose: To enable the scholar to study, research or acquire special skills, knowledge or experience at a suitable location in the United Kingdom or elsewhere for a minimum period of 6 months.
Eligibility: Fellows, members and affiliate members of the college are eligible to apply for the scholarship. The trustees will give special consideration to candidates intending to make a career in ophthalmology in the United Kingdom.
Level of Study: Postgraduate
Type: Scholarship
Value: Up to UK£30,000
Frequency: Every 2 years
Country of Study: United Kingdom or elsewhere
No. of awards offered: 1
Application Procedure: 5 copies each of the application form duly completed and the candidate's curriculum vitae should be submitted.
Closing Date: February
Funding: Commercial
Contributor: Keeler Ltd
No. of awards given last year: 1
No. of applicants last year: 16
Additional Information: Please see the website for further details http://www.rcophth.ac.uk/page.asp?section=320§ionTitle=Awards+and+Prizes.

The Pfizer Ophthalmic Fellowship

Subjects: Ophthalmology.
Purpose: To enable the scholar to study, research or acquire special skills, knowledge or experience at a suitable location in the United Kingdom or elsewhere for a minimum period of 6 months.
Eligibility: Applicants must be Fellows, members or affiliates of the Royal College of Ophthalmologists, in good standing.
Level of Study: Postgraduate
Type: Fellowship
Value: Up to UK£50,000 (to be divided between multiple recipients)
Frequency: Annual
Country of Study: United Kingdom or elsewhere
No. of awards offered: 1
Application Procedure: 5 copies each of the application form duly completed and the candidate's curriculum vitae should be submitted.
Closing Date: February
Funding: Commercial
Contributor: Pfizer Ltd.
No. of awards given last year: 1
No. of applicants last year: 9
Additional Information: Please see the website for further details http://www.rcophth.ac.uk/page.asp?section=320 §ionTitle=Awards+and+Prizes.

ROYAL COLLEGE OF ORGANISTS (RCO)

PO Box 56357, London, SE16 7XL, United Kingdom
Tel: (44) 5600 767208
Email: admin@rco.org.uk, andrew.mccrea@rco.org.uk
Website: www.rco.org.uk
Contact: Andrew McCrea, Director of Academic Development

The Royal College of Organists (RCO) is membership based. It promotes the art of organ playing as choral directing, and provides an organization with a library, events and examinations to further that object.

RCO Various Open Award Bequests

Subjects: Art and design
Purpose: To assist students who are training to become organists.
Eligibility: School and students in undergraduate and postgraduate education who are members of the college.
Level of Study: Postgraduate, Professional development, School and undergraduate
Type: Award
Value: Between UK£100 and UK£400 each
Frequency: Annual
Study Establishment: Various
No. of awards offered: Various
Application Procedure: Write for an application form.
Closing Date: April 18th (check with website for updated details)
Funding: Private

For further information contact:

Email: admin@rco.org.uk
Contact: The Registrar

ROYAL COLLEGE OF PAEDIATRICS AND CHILD HEALTH (RCPCH)

5-11 Theobalds Road, London, WC1X 8SH, England
Tel: (44) 020 7092 6000
Fax: (44) 020 7092 6001
Email: aaron.barham@rcpch.ac.uk
Website: www.rcpch.ac.uk
Contact: Mr Aaron Barham, Conference Organizer

The Royal College of Paediatrics and Child Health works to advance the art and science of paediatrics, to raise the standard of medical care provided to children, to educate and examine those concerned with health of children, and to advance the education of the public and, in particular, medical practitioners in child health.

Ashok Nathwani Visiting Fellowship

Subjects: Paediatrics.
Purpose: To enable young paediatricians from abroad to visit the United Kingdom.
Eligibility: Open to all Paediatricians.
Level of Study: Postdoctorate
Type: Fellowship
Value: A travel, accommodation and subsistence costs
Frequency: Annual
Country of Study: United Kingdom
Application Procedure: To apply for the Ashok Nathwani Fellowship, complete and return the application form by post (return address in application form).
Closing Date: October 14th (check with website for updated details)

Babes in Arms Fellowships

Subjects: Paediatrics.
Purpose: To support paediatricians researching into sudden infant death syndrome or associated problems.
Eligibility: Open to paediatricians in higher specialist praising.
Level of Study: Postdoctorate
Type: Fellowship
Value: Up to UK£3,000
Frequency: Annual
Closing Date: August 24th

Douglas Hubble Travel Bursary

Subjects: Paediatrics and child health.
Purpose: To provide financial support for young pediatricians, or paediatric research workers, who contribute to the British paediatric practice.
Eligibility: Open to all young paediatricians or paediatric research workers below consultant status.
Level of Study: Postgraduate
Type: Travel award
Frequency: Every 2 years
Closing Date: November 20th (check with website for updated details)
Funding: Foundation
Contributor: University of Birmingham and the Royal College of paediatrics and Child Health

For further information contact:

Royal College of Paediatrics and Child Health, 50 Hallam Street, London, W1W 6DE
Contact: Mrs Aaron Barham

Dr Michael Blacow Memorial Prize

Subjects: Paediatrics.
Purpose: To award the best plenary paper presented at the spring meeting.
Eligibility: Open to all trainee paediatricians but not given to same trainee paediatrician twice
Level of Study: Postgraduate
Type: Award
Value: UK£200
Frequency: Annual
Closing Date: Check with website for updated details
Funding: Trusts, foundation
Contributor: The British Paediatric Association

Heinz Visiting and Travelling Fellowships

Subjects: Paediatrics.
Purpose: To enable paediatricians from any part of the Commonwealth to visit the United Kingdom and to attend the Spring meeting of the College, or to enable British paediatricians to spend time in a developing country.
Eligibility: Open to young paediatricians from any commonwealth country.
Level of Study: Professional development
Type: Fellowship
Value: Air fares and living expenses
Length of Study: Up to 12 weeks for Visiting Fellowships and up to 3 months for Travelling Fellowships
Frequency: Annual
Country of Study: United Kingdom for the visiting fellowship and up to three months for travelling fellowship
No. of awards offered: Varies
Application Procedure: Application form can be downloaded from the website.
Closing Date: August 26th
Funding: Private, corporation
Contributor: H J Heinz Company
No. of awards given last year: 3
Additional Information: Please contact by email and see the website for further details.

For further information contact:

Email: aaron.barham@rcpch.ac.uk

RCPCH/VSO Fellowship

Subjects: Medicine and surgery
Purpose: To offer Specialist Trainees one-year placements working to improve child health in developing countries.
Eligibility: Open to specialist registrars undertaking type 1 training.
Level of Study: Postgraduate
Type: Fellowship
Value: Comprehensive training, living allowance, accommocation and health insurance. An in-country team of VSO staff and volunteers provide support while you are overseas.
Length of Study: 1 year
Frequency: Annual
Application Procedure: Contact VSO enquiries team for further information.
Closing Date: Check with website for updated details
Additional Information: Contact Jennie Snook for more information.

For further information contact:

Tel: 020 8780 7235
Email: jennie.snook@vso.org.uk
Contact: Jennie Snook

Young Investigator of the Year Medal

Subjects: Paediatrics.
Purpose: To award young medically qualified research worker for excellence in research.
Eligibility: Open to all young, medically qualified research workers.
Level of Study: Postgraduate
Type: Award
Value: UK£5,00 and medal plus UK£500 to the department in which they work for other expenses with their research
Frequency: Annual
No. of awards offered: 1
Application Procedure: All nominations are from the Head of Department. Direct applications are not accepted.
Closing Date: December 16th (check with website for updated etails)
Funding: Foundation
Contributor: Sport Aiding Medical Research for Kids (SPARKS)

ROYAL COMMISSION FOR THE EXHIBITION OF 1851

453, Sherfield Building, Imperial College, London, SW7 2AZ, England
Tel: (44) 20 7594 8790
Fax: (44) 20 7594 8794
Email: royalcom1851@imperial.ac.uk
Website: www.royalcommission1851.org.uk
Contact: Nigel Williams CEng, Secretary

The Royal Commission for the Exhibition of 1851 is an educational trust supporting innovation and creativity in science, technology and occasionally the arts, almost always at postgraduate level. Apart from the competitive schemes listed, assistance is sometimes given to suitable charities or individuals.

Royal Commission Industrial Design Studentship

Subjects: Industrial design.
Purpose: To fund engineering and science graduates for a postgraduate industrial design course.
Eligibility: Open to United Kingdom nationals only with a first degree in engineering, resident in UK; intending to make a career in British industry.
Level of Study: Postgraduate
Type: Studentship
Value: Stipend of £10,000 per year, allowance of £850 pa for materials and some travel expenses.
Length of Study: 1–2 years
Frequency: Annual
Study Establishment: Universities
Country of Study: Any country
No. of awards offered: 8
Application Procedure: Applicants must complete an application form and submit this with the required enclosures. Forms available from the commission or its website.
Closing Date: April 25th
Funding: Private
Contributor: The profits from the Great Exhibition
No. of awards given last year: 6
Additional Information: Please see the website for further details http://www.royalcommission1851.org.uk/ind_des.html.

Royal Commission Industrial Fellowships

Subjects: Industrial engineering and management.
Purpose: To allow able graduates working in industry to carry out research and development leading to a higher degree or other career milestone.
Eligibility: Open to resident in UK with a good degree in science or engineering who are working in or for a British-based company.
Level of Study: Doctorate, Postgraduate
Type: Fellowship
Value: £80,000 (University fees, 50% of Salary (or enhanced stipend for EngD candidates), £3,500 per year travelling expenses, £10,000 grant to University Department on completion).
Length of Study: 3 years
Frequency: Annual
Study Establishment: Any approved university/British company
Country of Study: United Kingdom
No. of awards offered: Approx. 8
Application Procedure: Applicants must complete an application form.
Closing Date: January 24th
Funding: Private
Contributor: The profits from the Great Exhibition
No. of awards given last year: 7
Additional Information: Please see the website for further details http://www.royalcommission1851.org.uk/ind_fellow.html.

Royal Commission Research Fellowship in Science and Engineering

Subjects: Pure and applied sciences including mathematics and any branch of engineering.
Purpose: To give postdoctorate scientists or engineers of exceptional promise the opportunity to conduct research of their own instigation.
Eligibility: Fellowship will normally be held in a UK institution except in exeptional circumstances. See 2012 General Regulation for full eligibility criteria available from www.royalcommission1851.org.uk.
Level of Study: Postdoctorate
Type: Fellowship
Value: £30,000 stipend payable for year 1, and £31,500 for years 2 and 3. In addition a London Weighting of £2,500 per year is payable in appropriate cases and final year fellows are also sponsored to attend a Royal Society communication skills course.
Length of Study: 3 years
Frequency: Annual
Study Establishment: Any approved university
Country of Study: Any country
No. of awards offered: Approx. 8
Application Procedure: Applicants must complete an application form online, available from the commission's website. Applications must be supported by professors of United Kingdom universities.
Closing Date: February 21st
Funding: Private
Contributor: The profits from the Great Exhibition
No. of awards given last year: 8
Additional Information: Please see the website for further details http://www.royalcommission1851.org.uk/res_fellow.html.

ROYAL GEOGRAPHICAL SOCIETY (WITH THE INSTITUTE OF BRITISH GEOGRAPHERS)

1 Kensington Gore, London, SW7 2AR, England
Tel: (44) 20 7591 3000
Fax: (44) 20 7591 3001
Email: grants@rgs.org
Website: www.rgs.org/grants
Contact: Joanne Shope, Grants Officer

The Royal Geographical Society (with the Institute of British Geographers) is the United Kingdom's learned society for geography and geographers and a professional body. It supports and promotes many aspects of geography including geographical research, education and teaching, field training and small expeditions, the public understanding and popularization of geography and the provision of geographical information.

30th International Geographical Congress Award

Subjects: Geography.
Purpose: To assist with the cost of attending an international geographical conference.
Eligibility: Applicants must be UK/EU nationals and must currently be employed by a UK Institute of Higher Education. Prefernce will be given to applicants within 6 years of completion of PhD. Attendance at AAG, CAG or the RGS-IBG Annual Conference is not eligible for support from this award.
Level of Study: Research
Type: Award
Value: Up to £750
Length of Study: Unspecified
Frequency: Annual
No. of awards offered: 5
Application Procedure: Applicants must download the guidelines and application form from the website www.rgs.org/grants or contact the grants officer.
Closing Date: February 22nd
No. of awards given last year: 3
No. of applicants last year: 11

Dudley Stamp Memorial Award

Purpose: To support PhD students or postdoctoral researchers in the early stages of their careers and to assist them in research or study travel leading to the advancement of geography and to international cooperation in the study of geography.
Eligibility: Applicants must be registered (full or part time) for a PhD at a UK Higher Education institution OR employed by a UK HEI within 6 years of completion of a doctoral degree. Applicants of any nationality are welcome to apply. Applicants registered on MPhil or other master courses are NOT eligible. These awards cannot provide funding solely for conference attendance.
Type: Award
Value: Up to £500
Frequency: Annual
No. of awards offered: Varies
Application Procedure: Completed application forms should be submitted by email to the Society's Grants Officer at grants@rgs.org. Application forms can be downloaded from www.rgs.org/dudleystamp.
Closing Date: February 22nd
Additional Information: Two referee statements are required. If you are currently a PhD student, one reference must be from your PhD supervisor. Please check website for detailed information.

Frederick Soddy Postgraduate Award

Eligibility: The awards support fieldwork/research in the UK or overseas. Applicants must be UK/EU nationals; must currently be

registered for a PhD. Applicants registered on MPhil or other Masters courses are not eligible; reapplication for the same project is not permitted; grants are availabe to individuals or small groups working together; re-application for the same project is not permitted; awards cannot be made retrospectively and costs that cannot be covered by the grant include travel, equipment, field assistance, expenses in the field.
Level of Study: Postgraduate
Type: Award
Value: Up to £6,000 each year
Length of Study: 3 years
Frequency: Annual
No. of awards offered: 1
Application Procedure: Applications should be submitted by email in word format to the Grants Officer at the society. Deadlines will be strictly enforced.
Closing Date: February 22nd
No. of awards given last year: 1
No. of applicants last year: 16
Additional Information: Applicants may only have one Society grant application pending at any one time. For more information, please read our frequently asked questions at www.rgs.org/grantfaqs.

From the Field' Goldsmith Awards

Purpose: To support secondary level geography teachers develop educational resources that will give pupils an insight into the research being carried out by geographers on some of the key global geographical issues - from climate change, to the loss of species, population growth or migration. The aim of the bursary is to develop educational resources that can be readily adopted by the school geography teaching community.
Eligibility: All applicants must be teaching secondary-age pupils in the UK.
Type: Award
Value: Up to £500
Frequency: Three times a year
Application Procedure: Applications should be submitted by email to the Grants Officer at the Society (grants@org.org). Applications must be submitted electronically in word format. Deadlines will be strictly enforced.
Closing Date: January 14th, April 23rd, and September 13th
Additional Information: Please check at www.rgs.org/grants for details.

Geographical Club Award

Subjects: Geography.
Purpose: To support a postgraduate student (Masters or PhD) undertaking geographical fieldwork.
Eligibility: Applicants must be UK/EU nationals and must currently be registered for a Masters or PhD at a UK Institute of Higher Education. Students who receive full funding from a Research Council or comparable levels of support from other sources with support for fieldwork/data collection are not eligible to apply.
Level of Study: Doctorate, Postgraduate
Type: Award
Value: £1000
Frequency: Annual
Country of Study: United Kingdom or elsewhere
No. of awards offered: 2
Application Procedure: Applicants must down load the application guidelines from the website www.rgs.org/grants or contact the grants officer. There is no application form.
Closing Date: November 23rd
No. of awards given last year: 2
No. of applicants last year: 20

Geographical Fieldwork Grants

Subjects: Geographical field work.
Purpose: To support UK-led research teams carrying out geographical field research and exploration overseas.
Eligibility: Open to multidisciplinary teams, rather than individuals, the majority of which must be British and over 19 years of age.
Level of Study: Postgraduate, Unrestricted, Undergraduate
Type: Grant
Value: Up to £3,000
Length of Study: Must be over 4 weeks
Frequency: Twice a year
Country of Study: Outside the United Kingdom
No. of awards offered: Varies (upto approx. 50)
Application Procedure: Applicants must complete an application form. These can be obtained from the Society's website or by writing to the Grants officer.
Closing Date: January 18th
No. of awards given last year: 30
No. of applicants last year: 50

Gilchrist Fieldwork Award

Subjects: Geography.
Purpose: To support original and challenging overseas fieldwork carried out by small teams of university academics and researchers.
Eligibility: The majority of team members must be British and hold established posts at University departments or equivalent research establishments. Team members may come from multiple establishments and teams must comprise up to 10 members. At least six weeks.
Level of Study: Research
Type: Grant
Value: £15,000
Length of Study: At least 6 weeks
Frequency: Every 2 years
No. of awards offered: 1
Application Procedure: Applicants must submit a written proposal. Guidelines are available on the website.
Closing Date: February 21st
No. of awards given last year: 1
No. of applicants last year: 12

Henrietta Hutton Research Grants

Subjects: Geographical field research.
Eligibility: Undergraduate/postgraduate geography students who are registered at a UK higher education institution.
Level of Study: Postgraduate, Undergraduate
Type: Grant
Value: UK£500
Length of Study: Above 4 weeks
Frequency: Annual
No. of awards offered: 2
Application Procedure: Applicants must submit a written proposal. Guidelines are available from the website www.rgs.org/grants
Closing Date: January 18th
No. of awards given last year: 2
No. of applicants last year: 21
Additional Information: Applicants who are members of other RGS-IBG supported teams may additionally apply for this award.

Hong Kong Research Grant

Subjects: Field research in The Greater China region, Taiwan, Macau, SAR and Hong Kong SAR, People's Republic of China.
Purpose: To support postgraduate research in greater China.
Eligibility: Open to candidates undertaking a PhD course. Field research in Greater China region.
Level of Study: Doctorate, Postgraduate
Type: Grant
Value: UK£2,500
Length of Study: Above 4 weeks
Frequency: Annual
Country of Study: Greater China
No. of awards offered: 1
Application Procedure: Applicants must submit a typed proposal. Guidelines are available from www.rgs.org/grants
Closing Date: November 23rd
Contributor: Royal Geographical Society Hong Kong
No. of awards given last year: 1
No. of applicants last year: 20

Journey of a Lifetime Award

Subjects: Travel and geography.
Purpose: To support an inspiring journey and encourage new radio broadcast talent.
Eligibility: Unrestricted. Candidates must be able to attend interviews in person.

Level of Study: Unrestricted
Type: Travelling bursary
Value: UK£4,000
Length of Study: January–July
Frequency: Annual
Country of Study: Any country
No. of awards offered: 1
Application Procedure: Applicants must submit a written proposal.
Closing Date: September 21st
Funding: Private
No. of awards given last year: 1
No. of applicants last year: 150
Additional Information: The winner is trained by the BBC and records their journey for a BBC Radio 4 documentary.

Monica Cole Research Grant

Subjects: Physical geography.
Level of Study: Postgraduate, Undergraduate
Type: Grant
Value: UK£1,000
Frequency: Annual
Country of Study: Outside the United Kingdom
No. of awards offered: 1
Application Procedure: Applicants must download guidelines from the website or contact the Grants officer. There is no application form.
Closing Date: January 18th
Funding: Private
No. of awards given last year: 1
No. of applicants last year: 5
Additional Information: Two refree statements are required.

Neville Shulman Challenge Award

Subjects: The geographical sciences and exploration.
Purpose: To further the understanding and exploration of the planet, while promoting personal development through the intellectual or physical challenges involved.
Level of Study: Unrestricted
Type: Grant
Value: UK£10,000
Length of Study: Unrestricted
Frequency: Annual
Country of Study: Any country
No. of awards offered: 1
Application Procedure: Applicants must submit a proposal, following the guidelines, to the Grants officer by the closing date. There is no application form. Details are available from the website.
Closing Date: November 30th
Funding: Private
No. of awards given last year: 1
No. of applicants last year: 7
Additional Information: Project can be desk or field based and can be carried out in the UK or overseas.

Peter Fleming Award

Subjects: Geographical research.
Purpose: For the advancement of geographical science.
Eligibility: Open to British individuals or teams. Applicants should either be working within a UK education institution or as an independent researcher within the UK. Applicants must hold a doctorate and have a proven track record.
Level of Study: Doctorate, Research
Type: Grant
Value: UK£9,000
Length of Study: Unspecified
Frequency: Annual
Country of Study: Any country
No. of awards offered: 1
Application Procedure: Applicants must submit a typed proposal.
Closing Date: November 23rd
Funding: Private
No. of awards given last year: 1
No. of applicants last year: 10
Additional Information: Equipment costs should not be more than 10 per cent of total. Conference attendance will not be funded. Projects can be desk, lab or field based and can be carried out in the UK or overseas.

Ralph Brown Expedition Award

Subjects: Marine or aquatic environments.
Purpose: To support and encourage objectives that involve the study of rivers, inland or coastal wetlands or shallow marine environments.
Eligibility: Applicant must be fellow or member of society.
Level of Study: Research
Type: Grant
Value: UK£18,500
Frequency: Annual
Country of Study: Any country
No. of awards offered: 1
Application Procedure: Applicants must contact the Grants Co-ordinator or refer to the website for detailed guidelines.
Closing Date: November 23rd
Funding: Private
No. of awards given last year: 1
No. of applicants last year: 10
Additional Information: The grant is awarded for the study of shallow marine environments, rivers, coral reefs and inland or coastal wetlands.

RGS-IBG Field Centre Grants

Purpose: To support field research on an important geographical topic at international field centres, preferably in some of the world's poorest countries.
Eligibility: Lead applicants must be UK nationals with a proven research record. Preference is given to research at field centres in some of the world's poorer countries.
Type: Grant
Value: £5,000
Frequency: Annual
Application Procedure: Applications should be submitted by email in word format to the Grants Officer at the Society. Deadlines will be strictly enforced. Please check further information at www.rgs.org/grants.
Closing Date: February 8th
Additional Information: Applicants may only have one Society grant application pending at any one time. For more information, please read our frequently asked questions at www.rgs.org/grantfaqs.

RGS-IBG Land Rover Bursary

Subjects: Geography.
Purpose: To promote a wider understanding and enjoyment of geography and to take the recipient beyond his or her normal limits and boundaries.
Eligibility: Must start and end in the UK. Teams of 2–4. Applicants must hold a clean UK driving licence and have 3 or more years of driving experience.
Level of Study: Unrestricted
Type: Grant
Value: Up to 30,000
Length of Study: Between May and October
Frequency: Annual
Country of Study: United Kingdom or elsewhere
No. of awards offered: 1
Application Procedure: Guidelines available on website. No application form.
Closing Date: November 30th
Funding: Corporation
No. of awards given last year: 1
No. of applicants last year: 30

RGS-IBG Postgraduate Research Awards

Subjects: Physical environment, conservation, sustainability, society, economy.
Purpose: For PhD students undertaking fieldwork/data collection. These awards are offered to individuals and aim to help students establish themselves in their particular field.
Eligibility: Currently registered for a PhD at a UK Higher Education Institution.
Level of Study: Postgraduate, PhD students
Type: Grant

Value: £2,000
Frequency: Annual
Country of Study: Any country
No. of awards offered: 6 (2 in each of the above subject area)
Application Procedure: Guidelines available on website. No application form.
Closing Date: November 23rd
No. of awards given last year: 6
No. of applicants last year: 30

Slawson Awards

Subjects: Geographical fieldwork involving key development issues with high social value, including projects in geography-related disciplines such as anthropology and economics.
Purpose: To assist students intending to carry out geographical field research in developing countries to increase knowledge, awareness and understanding of the world in which we live.
Eligibility: Applicants must normally be United Kingdom citizens, currently be registered for a PhD at a United Kingdom Institute of Higher Education and are encouraged to become Fellows of the Society.
Level of Study: Doctorate, Postgraduate
Type: Grant
Value: Normally up to UK£1,000–3,000
Frequency: Annual
Country of Study: Outside the United Kingdom, preferably in developing countries
No. of awards offered: 2–3
Application Procedure: Applicants must submit a proposal, following the guidelines, to the Grants officer by the closing date. There is no application form and more details are available from the website.
Closing Date: February 22nd
Funding: Private
No. of awards given last year: 2–3
No. of applicants last year: 18
Additional Information: Fieldwork must be outside UK and preferably in developing countries

Small Research Grants

Subjects: Geography.
Purpose: To provide grants for desk- or field-based research.
Eligibility: Applicants must be a fellow or member of the society. Preference to individual researchers within 10 years of the start of their PhD.
Level of Study: Postdoctorate, Research
Type: Grant
Value: Up to UK£3,000
Length of Study: Unspecified
Frequency: Annual
Country of Study: Unrestricted
No. of awards offered: Varies (4–10)
Application Procedure: An application form is available from the society's website or Grants officer.
Closing Date: January 18th
No. of awards given last year: 10
No. of applicants last year: 30

Thesiger-Oman Research Fellowship

Subjects: Research in arid lands: physical environment and human environment.
Purpose: To support research in arid environments.
Eligibility: Applicants must be affiliated to a higher education instituition anywhere in the world and have held a PhD for at least 3 years. Equipment should not exceed 10 per cent of budget.
Level of Study: Postdoctorate, Research
Type: Grant
Value: UK£8,000
Frequency: Annual
No. of awards offered: 1
Application Procedure: Guidelines can be downloaded from the website. No application form.
Closing Date: November 23rd
Contributor: His Majesty Qaboos Bin Said Alsaid, Sultan of Oman
No. of awards given last year: 1
No. of applicants last year: 10
Additional Information: One award is given to a project focused on the physical aspects of arid and semi-arid environments. Another award is given to a project focused on the human dimension of arid and semi-arid environments.

ROYAL HISTORICAL SOCIETY

University College London, Gower Street, London, WC1E 6BT, England
Tel: (44) 207 387 7532
Email: royalhistsoc@ucl.ac.uk
Website: www.royalhistoricalsociety.org
Contact: Susan Carr, The Executive Secretary

The Royal Historical Society has limited funds to assist postgraduate students (and some others) with expenses incurred in the pursuit of advanced historical research. These funds are intended principally for postgraduate students registered for a research degree at United Kingdom Institutions of Higher Education, both full-time and part-time.

Royal Historical Society Postgraduate Research Support Grant

Subjects: History.
Purpose: To assist postgraduate students with expenses incurred in the pursuit of advanced historical research.
Eligibility: Applicants must be postgraduate students registered for a research degree at a United Kingdom Institution of Higher Education, either full-time or part-time, or individuals who have completed doctoral dissertations within the last 2 years and who are not in full-time employment.
Level of Study: Doctorate, Postgraduate
Type: Grant
Value: £50–£500
Frequency: Annual
Study Establishment: United Kingdom Institutions of Higher Education
Country of Study: United Kingdom
Application Procedure: Applicants must download application form from website. Applications should include a well-costed budget, a proposal that indicates the scholarly value of the proposed project and letters of reference. Completed forms should be submitted by post.
Closing Date: Varies
Funding: Private
Additional Information: Preference will be given to members of the Royal Historical Society. Please see the website for further details http://www.royalhistoricalsociety.org/postgraduates.php.Please call if you experience any problems downloading any of the Society's application forms.

For further information contact:

Contact: The Administrative Secretary

ROYAL HORTICULTURAL SOCIETY (RHS)

80 Vincent Square, London, SW1P 2PE, England
Tel: (44) 0845 260 5000
Fax: (44) 1483 212382
Email: bursaries@rhs.org.uk
Website: www.rhs.org.uk/courses/bursaries
Contact: Secretary of RHS Bursaries Committee

The Royal Horticultural Society (RHS) is a membership charity holding a Royal Charter for horticulture. The Society promotes the science, art and practice of horticulture in all its branches through a wide range of educational, research and advisory activities. It also maintains some major gardens, shows and the internationally renowned Lindley Library.

Blaxall Valentine Trust Award

Subjects: Agricultural sciences, planning and surveying
Purpose: To help finance worldwide plant collecting in natural habitats and study expeditions that will provide real benefits to horticulture.

Eligibility: Open to applicants worldwide, but preference is given to United Kingdom and Commonwealth citizens. Applicants should satisfy the Society that their health enables them to undertake the project proposed. Financial sponsorship will be available to both professional and amateur horticulturists and consideration for an award is not restricted to RHS members. Proposals may be made by individuals or group of individuals.
Level of Study: Unrestricted
Type: Bursary
Value: Funds are limited. High-cost projects are expected to receive supplementary finance from other sources, including personal contributions
Frequency: Annual
Country of Study: Any country
Application Procedure: Applicants must complete an application form available on request. Application forms can be downloaded from the RHS website http://www.rhs.org.uk/Courses/Bursaries/The-Blaxall-Valentine-Award.
Closing Date: December 15th, March 31st, June 30th or September 30th (check with website for updated details)
Funding: Private
No. of awards given last year: 17

Coke Trust Awards

Subjects: Horticulture.
Purpose: To broaden professional gardeners' and student gardeners' knowledge, skills and experience, and to finance horticultural projects that demonstrate a distinct educational or historic value, as submitted by institutions, charities or gardens.
Eligibility: Submissions are welcomed from applicants worldwide, but preference is given to United Kingdom and Commonwealth citizens. Applicants should preferably be within the age bracket of 20–35 years and satisfy the Society that their health enables them to undertake the project proposed. Financial sponsorship will be available to both professional and amateur horticulturists and consideration for an award is not restricted to RHS members.
Level of Study: Unrestricted
Type: Bursary
Value: Funds are limited. High-cost projects are expected to receive supplementary finance from other sources, including personal contributions
Frequency: Annual
Country of Study: United Kingdom
No. of awards offered: Unlimited
Application Procedure: Applicants must complete an application form, available on request. Application forms can be downloaded from the RHS website www.rhs.org.uk/courses/bursaries.
Closing Date: December 15th, March 31st, June 30th or September 30th (check with website)
Funding: Private
No. of awards given last year: 53
No. of applicants last year: 69
Additional Information: Recipients must submit a brief factual report within 3 months of completion, along with an outline of achievements or difficulties, including any unusual problems, e.g. medical or political, and an account of expenses.

For further information contact:

RHS Garden Wisley, Woking, Surrey, GU23 6QB
Website: http://www.rhs.org.uk/Courses/Bursaries/Coke-Trust-Awards

The Dawn Jolliffe Botanical Art Bursary

Subjects: Horticulture.
Purpose: To assist a botanical artist with the cost of exhibiting at one of the RHS shows or to travel in order to paint plants in their habitats.
Eligibility: Preference is given to botanical artists who are not yet established. Please note that submission of an application is dependent on notification from the RHS Picture Commitee that an applicant's botanical artwork is of a sufficient standard. For full details please contact the secretary of the picture commitee Tel No. 0207 821 3051.
Level of Study: Unrestricted
Type: Bursary
Value: Restricted to £1,000 annually, to be divided between two or more botanical artists
Frequency: Annual
Country of Study: United Kingdom or elsewhere
No. of awards offered: 1–2
Application Procedure: Applicants must complete an application form available on request. Application forms can be downloaded from the RHS website www.rhs.org.uk/courses/bursaries.
Closing Date: June 30th (check with website)
Funding: Private
No. of awards given last year: 3
No. of applicants last year: 5
Additional Information: The award was set up by the RHS on behalf of Mr Brian Jolliffe, who established the award in memory of his wife Dawn Jolliffe, a keen botanical artist.

For further information contact:

RHS Garden Wisley, Woking, Surrey, GU23 6QB
Website: http://www.rhs.org.uk/Courses/Bursaries/The-Dawn-Jolliffe-Botanical-Art-Bursary

EA Bowles Memorial Bursary

Subjects: Horticulture.
Purpose: To aid a project regarding the horticultural work of EA Bowles, his garden at Myddelton House, Essex, or one of his particular horticultural interests, e.g., bulbs or alpines.
Eligibility: The award is open to students and trainees in horticulture and relevant related topics, e.g., garden history and design. There are no age restrictions.
Level of Study: Unrestricted
Type: Bursary
Value: UK£500
Frequency: Every 2 years
Country of Study: United Kingdom
No. of awards offered: Unlimited
Application Procedure: Applicants must complete an application form available on request. Application forms can be downloaded from the RHS website www.rhs.org.uk/courses/bursaries.
Closing Date: December 15th, March 31st, June 30th or September 30th (check with website)
Funding: Private
No. of awards given last year: None
Additional Information: The award was set up by the EA Bowles of Myddleton House Society in memory of EA Bowles VMH, one of the most eminent plantsmen of the first-half of the 20th century. Recipients must submit a brief factual report within 3 months of completion, along with an outline of achievements or difficulties, including any unusual problems e.g. medical or political and an account of expenses.

For further information contact:

RHS Garden Wisley, Woking, Surrey, GU23 6QB
Website: http://www.rhs.org.uk/Courses/Bursaries/The-E-A-Bowles-Memorial-Bursary

Gurney Wilson Award

Subjects: Horticulture.
Purpose: To help finance horticulture-related projects and to further the interests of horticultural education as the RHS council deems fit.
Eligibility: Open to the candidates of any age who are UK nationals and foreign nationals residing in the UK, subject to the general conditions of application and who should be able to satisfy the Society that their health enables them to take a project proposed. Financial sponsorship is available to both professional and amateur horticulturists and consideration for an award is not restricted to RHS members. Proposals may be made by individuals or groups.
Level of Study: Unrestricted
Type: Bursary
Value: Limited funds. High-cost projects are expected to receive supplementary finance from other sources, including personal contributions
Frequency: Annual
Country of Study: Any country
No. of awards offered: Unlimited

Application Procedure: Applicants must complete an application form, available on request. Application forms can be downloaded from the RHS website www.rhs.org.uk/courses/bursaries.
Closing Date: December 15th, March 31st, June 30th or September 30th (check with website)
Funding: Private
No. of awards given last year: 4
Additional Information: Recipients must submit a factual report within 3 months of completion, along with an outline of achievements or difficulties, including any unusual problems encountered, e.g., medical or political, and an account of expenses.

For further information contact:

RHS Garden Wisley, Woking, Surrey, GU23 6QB
Website: http://www.rhs.org.uk/Courses/Bursaries/Gurney-Wilson-Awards

Jimmy Smart Memorial Bursary

Subjects: Horticulture.
Purpose: To provide financial support to a working (employed) gardener for travel and/or accommodation during visits overseas with the prime intention of seeing or studying plants and/or trees growing in their natural habitats. Applications to visit Australia or New Zealand are particularly encouraged, but other gardening-related travel may also be funded.
Eligibility: The award is open to experienced working gardeners in the British Isles, particularly those whose work includes responsibilities in relation to National Collections. There is no age restriction.
Level of Study: Unrestricted
Type: Bursary
Value: UK£1,000 maximum
Frequency: Annual
Country of Study: Any country
No. of awards offered: Unlimited
Application Procedure: Applicants must complete an application form available on request. Application forms can be downloaded from the RHS website www.rhs.org.uk/courses/bursaries.
Closing Date: December 15th, March 31st, June 30th or September 30th (check with website)
Funding: Private
No. of awards given last year: 1
No. of applicants last year: N/A
Additional Information: Recipients must submit a brief factual report within 3 months of completion, along with an outline of achievements or difficulties, including any unusual problems, e.g. medical or political, and an account of expenses. Applicants to visit Australia or New Zealand are particularly encouraged. Additional support may be offered to gardeners spending time in either of these countries.

For further information contact:

RHS Garden Wisley, Woking, Surrey, GU23 6QB
Website: http://www.rhs.org.uk/Courses/Bursaries/The-Jimmy-Smart-Memorial-Bursary

Osaka Travel Award

Subjects: Horticultural-related study or work experience will be considered.
Purpose: To allow young people from the United Kingdom and Japan to study in each other's country and benefit from a cross-cultural exchange of ideas.
Eligibility: Open to British and Japanese citizens with a horticultural background. Applicants should preferably be within the age bracket of 20 and 35 years. Financial sponsorship will be available to both professional and amateur horticulturists and consideration for an award is not restricted to RHS members. The awards are made to individuals, not to groups or expeditions.
Level of Study: Unrestricted
Type: Bursary
Value: Funds are limited. High-cost projects are expected to receive supplementary finance from other sources, including personal contributions
Frequency: Annual
No. of awards offered: 1–2
Application Procedure: Applicants must complete an application form, available on request. Application forms can be downloaded from the RHS website www.rhs.org.uk/courses/bursaries.
Closing Date: December 15th, March 31st, June 30th or September 30th (check with website)
Funding: Private
No. of awards given last year: 2
Additional Information: Recipients must submit a brief factual report within 3 months of completion, along with an outline of achievements or difficulties and an account of expenses.

For further information contact:

RHS Garden Wisley, Woking, Surrey, GU23 6QB
Website: http://www.rhs.org.uk/Courses/Bursaries/The-Osaka-Travel-Award

Queen Elizabeth the Queen Mother Bursary

Subjects: Horticulture.
Purpose: To help finance related projects, e.g. study tours, horticulturally-related expeditions, minor research projects possibly working with an acknowledged expert on short-term research, taxonomic studies, specialized courses and programmes of study such as specific subject symposia. The award is reserved for proposals of particular excellence.
Eligibility: Submissions are welcomed from applicants worldwide but preference is given to United Kingdom and Commonwealth citizens. Applicants should preferably be within the age bracket of 20 and 35 years and satisfy the Society that their health enables them to undertake the project proposed. Financial sponsorship will be available to both professional and amateur horticulturists and consideration for an award is not restricted to RHS members. The awards are made to individuals, not to groups or expeditions.
Level of Study: Unrestricted
Type: Bursary
Value: Up to £3,000
Frequency: Variable
Country of Study: Any country
No. of awards offered: 1–3
Application Procedure: Applicants must complete an application form, available on request. Application forms can be downloaded from the RHS website www.rhs.org.uk/courses/bursaries.
Closing Date: December 15th, March 31st, June 30th or September 30th
Funding: Private
No. of awards given last year: None
No. of applicants last year: N/A
Additional Information: Recipients must submit a brief factual report within 3 months of completion, along with an outline of achievements or difficulties, including any unusual problems, e.g. medical or political, and an account of expenses. Recipients must also be prepared to give a lecture on their project at the Society's headquarters at Vincent Square in London, within 18 months of completing the project.

For further information contact:

RHS Garden Wisley, Woking, Surrey, GU23 6QB

The RHS Environmental Bursary

Subjects: Awarded for an application of outstanding horticultural merit that concentrates on environmental issues such as water-wise gardening, research into drought-resistant plants, sustainability and other climate change issues that have a direct relevance to horticulture.
Purpose: Established in 2008 with the principal aim of encouraging an environmentally focused horticultural project.
Eligibility: Application is open to UK nationals and to foreign nationals resident in the UK, subject to the general conditions of application. There are no age restrictions. Applicants should be able to satisfy the Society that their health enables them to undertake the proposed project. Financial scholarship is available to both professional and amateur horticulturists and consideration for an award is not restricted to RHS members. Proposals may be made by individual or groups.
Level of Study: Restricted to horticultural environmental issues
Type: Bursary

Value: Funds are limited. High-cost projects are expected to receive supplementary finance from other sources, including personal contributions
Frequency: Variable
Country of Study: Any country
No. of awards offered: Unlimited
Application Procedure: Applicants must complete an application form, available on request. Application forms can be downloaded from the RHS website www.rhs.org.uk/courses/bursaries.
Closing Date: December 15th, March 31st, June 30th or September 30th (check with website)
Funding: Private
No. of applicants last year: 1
Additional Information: Recipients must submit a factual report within 3 months of completion.

For further information contact:

RHS Garden Wisley, Woking, Surrey, GU23 6QB, United Kingdom

RHS General Bursary

Subjects: Horticulture.
Purpose: To provide financial support for travel and/or accommodation to aid a project related to the study of herbs.
Eligibility: Open to UK nationals and to foreign nationals resident in the UK of any age, subject to the general conditions of application. Financial sponsorship is open to students and trainees in horticulture. Proposals may be made by individuals or groups.
Level of Study: Unrestricted
Type: Bursary
Value: Limited funds. High-cost projects are expected to receive supplementary finance from other sources, including personal contributions
Frequency: Annual
Country of Study: Any country
No. of awards offered: Unlimited
Application Procedure: Applicants must complete an application form, available on request. Application forms can be downloaded from the RHS website www.rhs.org.uk/courses/bursaries.
Closing Date: December 15th, March 31st, June 30th or September 30th (check with website)
Funding: Private
Additional Information: Note - The Bullen, Mrs F Head and Lensbury (3 small bursary funds) were merged in 2010.

For further information contact:

RHS Garden Wisley, Woking, Surrey, GU23 6QB
Website: http://www.rhs.org.uk/Courses/Bursaries/The-RHS-General-Bursary

The RHS Interchange Fellowships

Subjects: The program enables a British graduate to attend the first year of a Master's degree course at an American university at the Garden Club of America's 'Interchange Fellow'. The Martin McLaren Horticultural Interchange Scholarship provides a non-credit work experience/study program in Great Britain for an American exchange student, selected by the GCA Scholarship Committee.
Purpose: To provide a reciprocal exchange of British and American graduate students, aged 27 or under, in horticulture, landscape architecture, botany or environmental studies. The intent of the program is to foster cultural understanding, promote horticultural studies and exchange information in this field.
Eligibility: US and UK graduates specializing in horticulture, botany, landscape architecture or environmental studies.
Level of Study: Postgraduate
Type: Fellowships
Value: US$30,000 for the UK student whilst undergoing study in the USA. For the US student undergoing horticultural practical work placements in the UK for the 10-month scholarship period, the award currently pays a monthly allowance of £900. In addition to this, other contingency expenses will be covered if appropriate
Length of Study: 10 months
Frequency: Annual
Country of Study: United Kingdom or United States of America
No. of awards offered: 1
Application Procedure: Applicants must complete an application form, available on request. Application forms can be downloaded from the RHS website www.rhs.org.uk/courses/bursaries.
Closing Date: October 31st
Funding: Private
No. of awards given last year: 1
No. of applicants last year: 5
Additional Information: Please see the website for further details http://www.rhs.org.uk/Courses/Bursaries/McClaren.

The Susan Pearson Bursary

Subjects: The practical application of theoretical horticultural principles to present day practices and the highest standards of professional practice.
Purpose: To sponsor horticultural trainees undertaking up to one year of practical work experience in a recognized garden open to public.
Eligibility: The proposed trainee must be either a British and/or Irish Citizen. See the website for details.
Level of Study: Unrestricted
Type: Bursary
Value: Up to £10,000
Frequency: Annual
Study Establishment: At host garden
Country of Study: United Kingdom
No. of awards offered: Unlimited
Application Procedure: Applications are to be submitted by a proposed host garden open to the public (e.g. such as those within the RHS Recommended Garden Scheme and be accompanied by an outline programme of the work the trainee will undertake and, where appropriate, details of any intended CPD/training courses. Application forms can be downloaded from the RHS website www.rhs.org.uk/courses/bursaries.
Closing Date: March 31st (check with website)
Funding: Private
Additional Information: Recipients must submit a factual report within 3 months of completion of the placement.

For further information contact:

RHS Garden Wisley, Surrey, Woking, GU23 6QB
Website: http://www.rhs.org.uk/Courses/Bursaries/The-Susan-Pearson-Bursary

The William Rayner Bursary

Subjects: Horticulture.
Purpose: To contribute significantly to horticultural career development.
Eligibility: Open to students and trainees in horticulture, along with working gardeners of any age who are UK nationals and foreign nationals residing in the UK, subject to the general conditions of application. Applicants should be able to satisfy the Society that their health enables them to take a project proposed.
Level of Study: Unrestricted
Type: Bursary
Value: Maximum £1,000 per year
Frequency: Annual
Country of Study: Any country
No. of awards offered: Unlimited
Application Procedure: Applicants must complete an application form, available on request. Application forms can be downloaded from the RHS website www.rhs.org.uk/courses/bursaries.
Closing Date: December 24th, March 31st, June 30th or September 30th (check with website)
Funding: Private
No. of awards given last year: 1
Additional Information: Recipients must submit a factual report within 3 months of completion, along with an outline of achievements or difficulties, including any unusual problems encountered e.g. medical or political, and an account of expenses.

For further information contact:

RHS Garden Wisley, Woking, Surrey, GU23 6QB
Website: http://www.rhs.org.uk/Courses/Bursaries/The-William-Rayner-Bursary

ROYAL IRISH ACADEMY

19 Dawson Street, Dublin, 2, Ireland
Tel: (353) 1 676 2570
Fax: (353) 1 676 2346
Email: admin@ria.ie
Website: www.ria.ie
Contact: Ms Laura Mahoney, Assistant Executive Secretary

The Royal Irish Academy is the senior institution in Ireland for both the sciences and humanities. It publishes a number of journals and monographs. It is Ireland's national representative in a large number of international unions, and through its national committees runs conferences, lectures and workshops. It also manages a number of long-term research projects. The Academy participates in the Royal Society European Science Exchange Programmes in pure and applied science, in the British Academy European Exchange Programmes in the humanities, and in the Austrian, Hungarian, or Polish academy exchange schemes in science and the humanities. Small grants for work in all disciplines are available annually from the Academy's own funds.

Royal Irish Academy Gold Medals

Subjects: Physical and mathematical sciences, engineering science, life sciences, environment and geosciences, humanities, and social sciences in a rotation basis for a 4-year cycle.
Purpose: To an individual who has made a distinguished scholarly contribution to their subject.
Eligibility: Open to people who are resident in either Northern Ireland or the Republic of Ireland and have made an internationally recognised, substantial and distinguished contribution to their subject whilst working in Ireland.
Level of Study: Postdoctorate
Value: A gold medal
Frequency: Annual
Country of Study: Ireland
No. of awards offered: 2
Application Procedure: Applicant must be nominated independently by two scientists who are familiar with his/her work. At least one of whom must be a member of the Academy, and/or a member of one of the Academy committees, and/or the Head of an Irish Higher Education Institution.
Closing Date: May each year
Funding: Government
No. of awards given last year: 2
Additional Information: Two awards are awarded each year. Awards for humanities and social sciences are awarded biennially. Please see the website for further details http://ria.ie/about/our-work/grants—awards.aspx.

Royal Irish Academy Postdoctoral Mobility Grants

Subjects: Humanities, social sciences and natural sciences.
Purpose: To provide small grants for short visits to any country supporting their primary research.
Level of Study: Postdoctorate, Professional development
Type: Fellowship
Value: Up to €2,500
Length of Study: 1–6 weeks
Frequency: Annual
Country of Study: Other
Application Procedure: Applicants must complete an application form, available from the Academy.
Closing Date: November 17th
Funding: Government
Additional Information: For further information, please contact grants@ria.ie.

ROYAL PHILHARMONIC SOCIETY (RPS)

10 Stratford Place, London, W1C 1BA, England
Tel: (44) 20 7491 8110
Fax: (44) 20 7493 7463
Email: admin@royalphilharmonicsociety.org.uk
Website: www.royalphilharmonicsociety.org.uk
Contact: General Administrator

The Royal Philharmonic Society (RPS) offers support for young musicians and composers, sponsorship of new music events, debate and discussion about the future of music and has an active commissioning policy. It recognizes achievement through the prestigious annual RPS Music Awards and with the Society's Gold Medal. Full information is available on the RPS website.

RPS Composition Prize

Subjects: Musical composition.
Purpose: To encourage young composers.
Eligibility: Open to past and present registered students of any conservatoire or university within the United Kingdom, of any nationality, under the age of 29. Former winners are not eligible.
Level of Study: Graduate, Postgraduate
Type: Prize
Value: Four £3,000 commissions and one £1,000 commission are awarded
Frequency: Annual
No. of awards offered: 4
Application Procedure: Applicants must complete an application form. There is a UK£20 entry fee but this is free to RPS members.
Closing Date: April 2nd
Funding: Private
No. of awards given last year: 3
No. of applicants last year: 70
Additional Information: Please see the website for further details http://royalphilharmonicsociety.org.uk/index.php/rps_today/news/2013_rps_composition_prize.

THE ROYAL SCOTTISH ACADEMY (RSA)

The Mound, Edinburgh, EH2 2EL, Scotland
Tel: (44) 131 225 6671
Fax: (44) 131 220 6016
Email: info@royalscottishacademy.org
Website: www.royalscottishacademy.org
Contact: Secretary

Founded in 1826, the RSA, Scotland's foremost body of artists, has promoted the works of leading contemporary painters, sculptors, printmakers and architects. It also gives practical and financial help to young artists through scholarships as well as the annual Student's Exhibition.

The Barns-Graham Travel Award

Subjects: Painting in any medium.
Purpose: To provide a travel and research opportunity for graduating and postgraduate students.
Eligibility: Entrants must be painters, printmakers or sculptors, entrants must either be graduating in 2013 or currently studying at postgraduate level at one of the following art schools in Scotland – Aberdeen, Dundee, Edinburgh, Glasgow, and Moray.
Level of Study: Postgraduate
Type: Scholarship
Value: £2,000
Length of Study: 3–6 months
Frequency: Annual
Study Establishment: Any of the main art colleges
Country of Study: Scotland
No. of awards offered: 1
Application Procedure: Applications forms and regulations will be available to download at the end of February.
Closing Date: June 12th
Funding: Private
Contributor: The Alastair Salvesen Trust
No. of awards given last year: 1
Additional Information: Please see the website for further details.

For further information contact:

Email: opportunities@royalscottishacademy.org
Website: http://www.royalscottishacademy.org/pages/scholarships_-detail.asp?id=27

The RSA John Kinross Scholarships to Florence

Subjects: Painting, sculpture, architecture and printmaking.

Purpose: To allow young artists from the established training centres in Scotland to spend time in Italy.
Eligibility: Open to all final year and postgraduate students from the four main colleges of art in Scotland.
Level of Study: Postgraduate
Type: Scholarship
Value: UK£2,000
Length of Study: 3 months
Frequency: Annual
Country of Study: Italy
No. of awards offered: 13
Application Procedure: Digital submissions accepted only. Please complete the accompanying application form. Your application form must be submitted electronically so a typed or electronic signature is acceptable.
Closing Date: March 29th
Funding: Private
Contributor: The Kinross Scholarship Fund, which is administered by the RSA
No. of awards given last year: 15
No. of applicants last year: 80

For further information contact:

Email: opportunities@royalscottishacademy.org
Website: http://www.royalscottishacademy.org/pages/scholarships_-detail.asp?id=37

The Rsa William Littlejohn Award for Excellence and Innovation in Water-Based Media

Subjects: Painting, sculpture, architecture and printmaking.
Purpose: To provide young professional artists who are Scottish or have studied in Scotland, with a period for personal development and the exploration of new directions.
Eligibility: Entrants must be working in water-based media (any pigment mixed with water), entrants must be born or have been resident in Scotland for at least 3 years, in the case of students applying, entrants must either be graduating in current year or studying at postgraduate level at one of the following art schools in Scotland (Aberdeen, Dundee, Edinburgh, Glasgow, and Moray).
Level of Study: Postgraduate
Type: Residency
Value: UK£2,000
Frequency: Annual
Study Establishment: Hospitalfield House, Arbroath
Country of Study: Scotland
Application Procedure: Applicants must contact the RSA.
Closing Date: June 12th
Funding: Private
Contributor: The Bequest Fund administered by the RSA

For further information contact:

Email: opportunities@royalscottishacademy.org
Website: http://www.royalscottishacademy.org/pages/scholarships_-detail.asp?id=31

ROYAL SOCIETY

6-9 Carlton House Terrace, London, SW1Y 5AG, United Kingdom
Tel: (44) 020 7451 2500
Website: http://royalsociety.org/JSPS-postdoctoral-fellowship-program/
Contact: The Grants Team

JSPS Postdoctoral Fellowship Program

Purpose: To provide the opportunity for highly qualified young researchers to conduct cooperative research with leading research groups in universities and other Japanese institutions. The Royal Society is provided with a quota for this scheme by the Japan Society for the Promotion of Science.
Eligibility: The research undertaken must be on a subject within the natural sciences, including: physics, chemistry, mathematics, computer science, engineering science, agricultural and medical research (non-clinical only). Applications for research in the social sciences should apply through The British Academy. The applicant must: have a PhD, or be likely to have a PhD by the time the funding starts; not have received their PhD more than six year prior to their application; be a citizen of the UK or European Union; and have a research proposal agreed with a Japanese host scientist.
Type: Fellowships
Value: Up to ¥1.5 million per year
Length of Study: 12–24 months
Frequency: Annual
Application Procedure: Applications are initially reviewed by members of the our panel with the most appropriate scientific expertise. Following this a shortlist is drawn up which is sent to Japan Society for the Promotion of Science (JSPS) for a final decision. Applicants will be notified of the result by JSPS.
Closing Date: February 28th
Additional Information: Please note that this scheme is offered by JSPS and is subject to their terms and conditions. Applicants should read through all the information offered on the JSPS website before applying to ensure they fully understand the terms of the award. Please check at http://royalsociety.org/grants/schemes/jsps-postdoctoral/ for more information.

THE ROYAL SOCIETY OF CHEMISTRY

Burlington House Piccadilly, London, W1J 0BA, England
Tel: (44) 0 20 7437 8656
Fax: (44) 0 20 7437 8883
Email: langers@rsc.org
Website: www.rsc.org
Contact: Mr S S Langer

The Royal Society of Chemistry is the learned society for chemistry and the professional body for chemists in the United Kingdom with 46,000 members worldwide. The Society is a major publisher of chemical information, supports the teaching of chemistry at all levels, organizes hundreds of chemical meetings a year and is a leader in communicating science to the public. It is now the United Kingdom National Adhering Organization (NAO) to the International Union of Pure and Applied Chemistry (IUPAC).

Corday-Morgan Memorial Fund

Subjects: Chemistry.
Purpose: To allow members of any established chemical society or institute in the Commonwealth to visit chemical establishments in another Commonwealth country.
Eligibility: Members of any established Chemical Society/Institute in the Commonwealth who wish to visit chemical establishments in the UK or another Conmmonwealth country.
Level of Study: Professional development
Type: Grant
Value: Up to a maximum of £ 500
Frequency: Four times a year
Country of Study: Any country
No. of awards offered: Up to 3
Application Procedure: Applicants must submit applications on the official form and will normally be considered within 1 month of receipt.
No. of awards given last year: 4
No. of applicants last year: 6
Additional Information: Applicants must be travelling to another country (not necessarily in the Commonwealth) and would normally stop en route to visit a third country that must be in the Commonwealth. Please see the website for details http://www.rsc.org/ScienceAndTechnology/Awards/CordayMorganPrizes/Index.asp.

Hickinbottom/Briggs Fellowship

Subjects: Organic chemistry.
Purpose: To assist research.
Eligibility: Open to applicants domiciled in the United Kingdom or Republic of Ireland. Candidates must already hold a PhD or equivalent in chemistry and be not more than 35 years of age on October 31st.
Level of Study: Postdoctorate
Type: Fellowship
Value: £2,000, a medal and a certificate
Length of Study: 3 years
Frequency: Dependent on funds available
Study Establishment: A British or Irish university or college

Country of Study: Other
No. of awards offered: 1
Application Procedure: Applicants must apply by application forms or be nominated.
Closing Date: January 15th
No. of awards given last year: 1
Additional Information: Please see the website for further details http://www.rsc.org/ScienceAndTechnology/Awards/HickinbottomAward/Index.asp.

J W T Jones Travelling Fellowship

Subjects: Chemistry.
Purpose: To promote international co-operation, to enable chemists to carry out short-term studies in well established scientific centres abroad and to learn and use techniques not accessible to them in their own country.
Eligibility: Open to members of the Royal Society of Chemistry who hold at least a Master's or PhD degree in chemistry or a related subject and are already actively engaged in research. Candidates must produce evidence that the theoretical and practical knowledge or training to be acquired in the foreign laboratory will be beneficial to their scientific development and must also return to their country of origin upon termination of the fellowship.
Level of Study: Professional development
Type: Fellowship
Value: Up to UK£5,000, designed to cover part or all of an economy class air or rail ticket and a subsistence allowance. It is expected that the institution of origin and/or the host institution will contribute to defray any remaining expenses incurred by the fellowship holder
Length of Study: Normally 1–3 months
Frequency: Four times a year
Country of Study: Any country
No. of awards offered: Approx. 6–8
Application Procedure: Applicants must apply for application forms, together with full details, from the International Affairs Officer.
Closing Date: January 1st, April 1st, July 1st, October 1st
No. of awards given last year: 7
No. of applicants last year: 12
Additional Information: Fellowships will not be awarded to attend scientific meetings. Applications will be considered by a Fellowship Committee and the holder will be required to submit a formal report on the work accomplished. Please see the website for further details http://www.rsc.org/ScienceAndTechnology/Funding/TravelGrants/JWTJonesFellowship.asp.

Royal Society of Chemistry Journals Grants for International Authors

Subjects: Chemistry.
Purpose: To allow international authors to visit other countries in order to collaborate in research, exchange research ideas and results, and to give or receive special expertise and training.
Eligibility: Open to anyone with a recent publication in any of the Society's journals. Those from the United Kingdom or Republic of Ireland are excluded.
Level of Study: Professional development
Type: Grant
Value: Up to £2500 cover travel and subsistence (but not research-related costs) and are available
Length of Study: Normally 1–3 months
Frequency: Four times a year
Country of Study: Any country
No. of awards offered: 100
Application Procedure: Candidates must apply for application forms, together with full details, from the International Affairs Officer.
Closing Date: January 1st, April 1st, July 1st or October 1st
No. of awards given last year: 83
No. of applicants last year: 107
Additional Information: Please see the website for further details http://www.rsc.org/ScienceAndTechnology/Funding/TravelGrants/InternationalAuthors.asp.

Royal Society of Chemistry Research Fund

Subjects: Chemistry and chemical education.
Purpose: To provide financial support to those working in less affluent institutions.
Eligibility: Open to members of the Society.
Level of Study: Professional development
Type: Grant
Value: Up to UK£2,000 for the purchase of chemicals, equipment or for running expenses of research
Length of Study: 1 year
Frequency: Annual
Country of Study: Any country
No. of awards offered: 20–30
Closing Date: October 31st
No. of awards given last year: 30
No. of applicants last year: 50
Additional Information: Funds are limited, in so preference will be given to those working less well-endowed institutions. The Selection Committee is especially anxious to see inventive applications of a 'pump priming' nature. Members in developing countries should note particularly that additional funds have been made available by the Society's International Committee to provide grants for successful applicants from such countries. Preference will be given to those able to cite collaborative research projects with United Kingdom institutions. Please see the website for further details http://www.rsc.org/ScienceAndTechnology/Funding/ResearchFund.asp.

Royal Society of Chemistry Visits to Developing Countries

Subjects: Chemistry.
Purpose: To promote international co-operation, specifically, to enable chemists to carry out short-term studies in well established scientific centres abroad and to learn and use techniques not accessible to them in their own country.
Eligibility: Open to members of the Society.
Level of Study: Professional development
Type: Grant
Value: Up to UK£500. The grants will complement, where appropriate, those for visits to Commonwealth countries, and funding would cover the additional travel costs involved, together with appropriate subsistence. Applicants should also see the Corday-Morgan Memorial Fund
Frequency: Four times a year
Country of Study: Any country
No. of awards offered: Approx. 4
Application Procedure: Applicants must submit applications on the official form and will normally be considered within 1 month of receipt.
Closing Date: January 1st, April 1st, July 1st and October 1st
No. of awards given last year: 2
No. of applicants last year: 2
Additional Information: Applicants must be travelling to another country and would normally stop en route to visit a third country. This must be a developing country. Please see the website for further details http://www.rsc.org/ScienceAndTechnology/Funding/TravelGrants/Stop-oversDeveloping.asp.

THE ROYAL SOCIETY OF EDINBURGH

22-26 George Street, Edinburgh, Scotland, EH2 2PQ, United Kingdom
Tel: (44) 131 240 5000
Fax: (44) 131 240 5024
Email: afraser@royalsoced.org.uk
Website: www.royalsoced.org.uk
Contact: Anne Fraser, Research Awards Manager

The Royal Society of Edinburgh awards research fellowships and scholarships to candidates based in Scotland.

Auber Bequest

Subjects: All subjects.
Purpose: To provide assistance for the furtherance of academic research.
Eligibility: Open to naturalized British citizens or individuals wishing to acquire British nationality who are over 60 years of age, resident in Scotland or England and are bona fide scholars engaged in academic, but not industrial, research. Applicants should not have been British nationals at birth, nor held dual British nationality, and must have since acquired British nationality.
Level of Study: Research

Value: Varies, not normally exceeding UK£3,000
Length of Study: Up to 2 years
Frequency: Every 2 years
Country of Study: Other
No. of awards offered: Varies
Application Procedure: Applicants must complete an application form, available from the Research Awards Manager or from the RSE website www.royalsoced.org.uk
Closing Date: Mid-January
Funding: Private
Contributor: The Auber Bequest
No. of awards given last year: 1
No. of applicants last year: 2
Additional Information: Please see the website for further details http://www.rse.org.uk/671_AuberBequest.html.

BP Trust Research Fellowships

Subjects: Mechanical engineering, chemical engineering, control engineering, solid state sciences, information technology, non-biological chemistry and geological sciences.
Purpose: To support independent research in specific fields.
Eligibility: Open to persons of all nationalities who have a PhD or equivalent qualification. Applicants should have 2–6 years postdoctoral research experience and must show they have a capacity for innovative research and a substantial volume of published work relevant to their proposed field of study.
Level of Study: Postdoctorate, Research
Type: Fellowship
Value: Salaries within the scale points 23–37 for research and analogous staff in Institutes of Higher Education with annual increments and superannuation benefits. Financial support towards expenses involved in carrying out the research is available, up to UK£6,000 for travel and subsistence each year.
Length of Study: Up to 5 years
Frequency: Every 2 years
Study Establishment: Any Institute of Higher Education in Scotland approved for the purpose by the Council of the Society
Country of Study: The Pacific Northwest or other regions of the North American Continent
No. of awards offered: 1
Application Procedure: Applicants must complete an application form, available from the Research Awards coordinator or the RSE website. Candidates must negotiate directly with the relevant head of department of the proposed host institution.
Closing Date: January 10th
Funding: Trusts
Contributor: BP Trust
Additional Information: Fellows will be expected to devote their full time to research and will not be allowed to hold other paid appointments without the express permission of the Council of the RSE. Please see the website for further details http://www.royalsoced.org.uk/648_BPResearchFellowships.html.

Cormack Postgraduate Prize

Subjects: Astronomy and astrophysics.
Purpose: To recognise and reward the most outstanding postgraduate student contribution to astronomical research in Scotland.
Level of Study: Postgraduate
Type: Prize
Value: £400
Frequency: Annual
Study Establishment: University of Scotland
Country of Study: Scotland
Closing Date: April 2nd
Funding: Trusts
No. of awards given last year: 1
No. of applicants last year: 5
Additional Information: Only one paper may be nominated for any one student, though there is no limit on the number of students nominated per institution. No student may be awarded the Prize on more than one occasion. Please see the website for further details http://www.rse.org.uk/585_PrizesMedals.html.

Cormack Small Astronomy Outreach Grant

Subjects: Astronomy and astrophysics.
Purpose: To develop public understanding of science activities.
Level of Study: Professional development
Type: Grant
Value: Approx. £250
Frequency: Annual
Study Establishment: Scottish University or Organisation
Country of Study: Scotland
Closing Date: October 1st
Funding: Trusts
No. of awards given last year: 2
No. of applicants last year: 3
Additional Information: Please see the website for further details http://www.royalsoced.org.uk/810_CormackSmallAstronomyOutreachGrants.html.

CRF European Visiting Research Fellowships

Subjects: Archaeology, art and architecture, economics and economic history, geography, history, jurisprudence, linguistics, literature and philology, philosophy and religious studies and social sciences.
Purpose: To create a two-way flow of visiting scholars in arts and letters and social sciences between Scotland and continental Europe.
Eligibility: Open to members of the academic staff of a Scottish Institute of Higher Education or equivalent continental European institution. Applicants from continental Europe must be nominated by members of staff from a Scottish Institute of Higher Education.
Level of Study: Professional development, Research
Type: Fellowship
Value: Up to UK£6,000 for visits of 6 months, which is reduced pro-rata for shorter visits, to cover actual costs of travel, subsistence and relevant study costs
Length of Study: 2–6 months
Frequency: Annual
Study Establishment: A Scottish Institute of Higher Education or a recognized Institute of Higher Education in a continental European country
Country of Study: Other
No. of awards offered: 6–8
Application Procedure: Applicants must complete an application form, available from the Research Awards Co-ordinator or the RSE website.
Closing Date: November 3rd
Funding: Private, trusts
Contributor: The Caledonian Research Fund
No. of awards given last year: 10
No. of applicants last year: 17
Additional Information: Successful applicants will be required to submit a report within 2 months of the end of the visit. Please see the website for further details http://www.royalsoced.org.uk/732_CRFEuropeanVisitingResearchFellowships.html.

Enterprise Fellowships funded by BBSRC

Subjects: Must be related to BBSRC funded project previously. For the commercialisation of BBSRC research.
Purpose: (As for Scottish Enterprise Fellowships) but throughout UK.
Eligibility: Academic and research staff and postgraduates with relevant experience are eligible to apply if employed by a – UK Higher Education Institution (HEI), or Institutes of BBSRC.
Level of Study: Postgraduate
Type: Fellowships
Value: A year's salary to provide time to develop a full business plan and seek investment. Access to mentors, business experts and professional advisors. Business training to help develop the required skills.
Length of Study: 1 year
Frequency: Annual, Biannual
Study Establishment: University of BBSRC funded Research Institute in UK
Country of Study: United Kingdom
No. of awards offered: Up to 4
Application Procedure: Docs are available on RSE website.
Closing Date: Spring or Autumn (April plus October)
Funding: Government
Contributor: BBSRC
No. of awards given last year: 3
No. of applicants last year: 8
Additional Information: Please see the website for further details.

Enterprise Fellowships funded by STFC

Subjects: Must be related to research projects previously funded by STFC. For the commercialisation of STFC research.
Purpose: As for RSE plus BBSRC fellowships.
Eligibility: Applicant must be connected to previously funded research project by STFC.
Type: Fellowship
Value: As for RSE plus BBSRC fellowships
Length of Study: 1 year
Frequency: Annual
Study Establishment: UK University or STFC funded Research Institute in UK
Country of Study: United Kingdom
No. of awards offered: 4 per year
Application Procedure: Docs are available on RSE website.
Closing Date: May 17th
Funding: Government
Contributor: STFC
No. of awards given last year: 2
No. of applicants last year: 5
Additional Information: Please see the website for further details http://www.stfc.ac.uk/Funding+and+Grants/1126.aspx.

International Exchange Programme – Bilateral

Subjects: All subjects.
Purpose: It facilitates the collaboration between researchers from Scotland and those based in institutions with which the RSE has a formal Memorandum of Understanding.
Eligibility: The Exchange programmes are open to Fellows and non-Fellows. Researchers travelling should have at least a PhD or equivalent at the time of travel. See the website for details.
Level of Study: Postdoctorate
Type: Small project grant
Value: Up to £3,000
Length of Study: 4 weeks
Frequency: Dependent on funds available
Study Establishment: Scottish university–to visit university or industry overseas. Scotland plus sister academy country overseas.
Application Procedure: Application docs are available on RSE website http://www.rse.org.uk/802_InternationalExchangeProgrammeBilateral.html.
Closing Date: February 15th (normally 3 times per year)
Funding: Government
Contributor: Scottish Government
No. of awards given last year: 60–70
No. of applicants last year: Varies at each round (11 at August round)

John Moyes Lessells Travel Scholarships

Subjects: All forms of engineering.
Purpose: To enable well-qualified engineering graduates of Scottish Institutes of Higher Education to study some aspect of their profession overseas.
Eligibility: Applicants must be graduates with an Honours or higher degree in engineering from a Scottish Institute of Higher Education or who are currently pursuing a postgraduate degree in engineering at a Scottish university.
Level of Study: Graduate, Postdoctorate, Postgraduate, Predoctorate, Professional development
Type: Scholarship
Value: £1,250 per month
Length of Study: Initially for 1 year but shorter periods or extension for a second year may be considered. Acceptance of a scholarship implies a willingness to spend at least 2 years in the UK following the period of tenure
Frequency: Annual
Country of Study: Other
No. of awards offered: Varies
Application Procedure: Applicants must complete an application form, available from the Research Awards Manager or the RSE website.
Closing Date: April (date to be confirmed)
Funding: Private, trusts
No. of awards given last year: 4
No. of applicants last year: 13
Additional Information: Please see the website for further details http://www.royalsoced.org.uk/973_JohnMoyesLessellsScholarships.html.

Research Workshops in Arts and Humanities

Subjects: Arts and humanities.
Purpose: To encourage collaborative investigation into a research proposition that is at an early stage of its development. A Workshop is regarded by the RSE as 'the coming together of scholars at the early stage of planning and developing a collaborative research initiative.'
Eligibility: Academic members or staff of Scottish universities or employees of Scottish cultural institutions working in research.
Level of Study: Professional development
Type: Research
Value: Up to £10,000
Length of Study: 1 year
Frequency: Dependent on funds available
Study Establishment: University or Cultural Institutional of Scotland
Country of Study: Scotland
Application Procedure: Application form and regulations on RSE website. Must be completed or submitted by the deadline.
Closing Date: October 1st
Funding: Government
Contributor: Scottish Government
No. of awards given last year: 6
No. of applicants last year: 14
Additional Information: Please see the website for further details http://www.royalsoced.org.uk/870_ResearchWorkshopsinArtsandHumanities.html.

Scottish Enterprise Fellowships

Subjects: Science and technology.
Purpose: To commercialise academic research in science plus technology areas.
Eligibility: Postgraduate level upwards
Level of Study: Postgraduate
Type: Fellowships
Value: Salary plus NI plus USS, business development fund of £10,000 and business training provided.
Length of Study: 1 year
Frequency: Biannual
Study Establishment: Scottish University or Research Institute
Country of Study: Scotland
Application Procedure: Application docs are available on RSE website. Closing dates in spring plus autumn each year.
Closing Date: November 5th
Funding: Government
Contributor: Scottish Enterprise
No. of awards given last year: 10
No. of applicants last year: 20
Additional Information: Please see the website for further details http://www.royalsoced.org.uk/636_ScottishEnterprise.html.

Scottish Government Personal Research Fellowships Co-funded by Marie Curie Actions

Subjects: All subjects.
Purpose: To encourage independent research in any discipline.
Eligibility: Open to persons of all nationalities who have a PhD or equivalent qualification. Applicants must have between 2–6 years of postdoctoral experience. They must also show that they have a capacity for innovative research and have a substantial volume of published work relevant to their proposed field of study.
Level of Study: Postdoctorate, Research
Type: Fellowship
Value: Annual stipends are within the scale points 23–37 for research and analogous staff in Institutes of Higher Education with annual increments and superannuation benefits. Expenses of up to UK£6,000 each year for travel and attendance at meetings or incidentals may be reimbursed. No support payments are available to the institution but Fellows may seek support for their research from other sources
Length of Study: Up to 5 years full-time research
Frequency: Dependent on funds available
Study Establishment: Any Institute of Higher Education, research institution or industrial laboratory approved for the purpose by the Council of the Society
Country of Study: Scotland

No. of awards offered: Up to 6.
Application Procedure: Applicants must complete an application form, available from the Research Awards Co-ordinator or the RSE website. Applicants should negotiate directly with the proposed host institution.
Closing Date: Febuary 14th (check with website)
Funding: Government
Contributor: The Scottish Government
Additional Information: Fellows may not hold other paid appointments without the express permission of the Council, but teaching or seminar work appropriate to their special knowledge may be acceptable. This fellowship is co-funded by Marie Curie actions.

Senior and Early Career Prizes

Subjects: 4 categories - Life sciences, physical sciences, art or humanities plus social sciences, business, education or public service.
Purpose: To recognise outstanding achievement at different career stages.
Level of Study: Postdoctorate
Type: Prize
Value: A medal is presented plus senior prize
Frequency: Annual
No. of awards offered: 8 (2 in each of 4 areas above)
Application Procedure: Only fellows of the RSE can make nominations for the RSE prizes.
Closing Date: November 23rd
Funding: Private
Contributor: RSE Prizes fund
Additional Information: Please see the website for further details http://www.royalsoced.org.uk/847_SeniorandEarlyCareerPrizes.html.

Small Research Grants in Arts and Humanities

Subjects: Arts and humanities.
Eligibility: As for workshops.
Level of Study: Professional development
Type: Research grant
Value: £500–7,500
Length of Study: 1 year
Frequency: Dependent on funds available
Country of Study: Scotland
Application Procedure: As for workshops.
Closing Date: October 1st
Funding: Government
Contributor: Scottish Government
No. of awards given last year: 15
No. of applicants last year: 26
Additional Information: Please check website for more information http://www.royalsoced.org.uk/871_SmallResearchGrantsinArtsandHumanities.html.

THE ROYAL SOCIETY OF MEDICINE (RSM)

1 Wimpole Street, London, W1G 0AE, England
Tel: (44) 20 7290 2900
Fax: (44) 20 7290 2989
Email: awards@rsm.ac.uk
Website: www.rsm.ac.uk
Contact: Awards Manager

The Royal Society of Medicine (RSM) provides academic services and club facilities for its members as well as publishing a monthly journal and an annual bulletin, and providing over 400 educational conferences and meetings per year.

Adrian Tanner Prize

Subjects: Clinical case reports should be submitted to reflect the multidisciplinary nature of the care for surgical patients.
Purpose: To encourage surgical trainees submit the best clinical case reports.
Eligibility: Open to all surgeons in training.
Value: UK£250
Frequency: Annual
Study Establishment: Royal Society of Medicine - Surgery section
Country of Study: United Kingdom
No. of awards offered: 1
Application Procedure: Applicants should download an application form from www.rsm.ac.uk/awards
Closing Date: March 28th
No. of awards given last year: 1
No. of applicants last year: 50

For further information contact:

Email: surgery@rsm.ac.uk
Website: http://www.rsm.ac.uk/academ/awards/

Alan Emery Prize

Subjects: Genetics.
Purpose: To reward the best published research article in Medical Genetics in the past 2 years.
Eligibility: Open to candidates in an accredited training or research post in the UK.
Type: Prize
Value: £500 or £300 or 1 year membership of Royal Society of Medicine
Frequency: Annual
Application Procedure: Candidates must submit full copy of the article, curriculum vitae or covering letter explaining the significance of the publication.
Closing Date: March 5th (check with website)
No. of awards given last year: 1
No. of applicants last year: 7

For further information contact:

Email: genetics@rsm.ac.uk
Website: http://www.rsm.ac.uk/academ/awards/

Cardiology Section Presidents Prize

Subjects: Cardiology.
Purpose: To reward original work for specialist registrar in cardiology.
Eligibility: Open to caridology trainees who have received all or part of their training at recognised centres of excellence in the UK. The subject of the presentation should represent original work.
Level of Study: Research
Type: Prize
Value: First prize – Commemoration medal and £1,000. Second prize – £500
Frequency: Annual
Study Establishment: Recognised centres of excellence in UK
Country of Study: United Kingdom
No. of awards offered: 2
Application Procedure: Candidates should submit an abstract of no more than 200 words.
Closing Date: May 8th
Contributor: Cardiology Section Funds
No. of awards given last year: 2

For further information contact:

Cardiology Section, Royal Society of Medicine, 1 Wimpole Street, London, WIG 0AE, United Kingdom
Email: cardiology@rsm.ac.uk
Website: http://www.rsm.ac.uk/academ/awards/

Catastrophes & Conflict Forum Medical Student Essay Prize

Subjects: Medicine and surgery
Eligibility: Open to candidates who are enrolled full-time at a UK medical school.
Level of Study: Postgraduate
Type: Prize
Value: £250 plus encouragement and advice on submitting the essay for publication in the JRSM.
Frequency: Annual
Study Establishment: Royal Society of Medicine
Country of Study: United Kingdom
No. of awards offered: 1
Application Procedure: Candidates should submit an essay no longer than 1,500 words, emailed in Word format.
Closing Date: March 1st

No. of awards given last year: 2
No. of applicants last year: 9

For further information contact:

Email: catastrophes@rsm.ac.uk
Website: http://www.rsm.ac.uk/academ/awards/

Clinical Forensic and Legal Medicine Section Poster Competition

Subjects: Clinical studies.
Purpose: To present a case or a poster in clinical studies.
Eligibility: Undergraduate students who have been working as part of courses leading to primary qualifications such as Legal Medicine Special Study Modules and electives for MBBS.
Level of Study: Trainee and students
Value: £200
Frequency: Annual
No. of awards offered: Varies
Application Procedure: Please visit the website www.rsm.ac.uk/academ/awards/index for application form details.
Closing Date: Check with website

For further information contact:

Email: forensic@rsm.ac.uk
Website: http://www.rsm.ac.uk/academ/awards/

Clinical Immunology & Allergy Section President's Prize

Subjects: Immunology or allergy.
Eligibility: Open to training grade doctors and young scientists (not above Specialist Registrar, Grade B Clinical Scientist or equivalent grade) with an immunological or allergy component of their clinical research.
Level of Study: Research, training graduate doctors and young scientists
Type: Prize
Value: 1st – £300 and 2nd – two prizes of £100
Frequency: Annual
Country of Study: United Kingdom
No. of awards offered: 3
Application Procedure: Check website for further details.
Closing Date: Please check the website.
No. of awards given last year: 3
No. of applicants last year: 10

For further information contact:

Email: immunology@rsm.ac.uk
Website: www.rsm.ac.uk/awards

Clinical Neurosciences Section Gordon Holmes Prize

Subjects: Clinical neurosciences.
Purpose: To award a research prize in clinical neurosciences.
Level of Study: Neuro trainee
Type: Prize
Value: £300
Frequency: Every 2 years
No. of awards offered: Varies
Application Procedure: Please check the website www.rsm.ac.uk/awards/index
Closing Date: Check with website

For further information contact:

Email: cns@rsm.ac.uk
Website: http://www.rsm.ac.uk/academ/awards/

Clinical Neurosciences Section Presidents Prize

Subjects: Neurology.
Purpose: To encourage clinical case presentation.
Eligibility: Open to trainees in neurosciences, including neurology, neurosurgery, neurophysiology, neuropathology or neuroradiology.
Level of Study: Trainee
Type: Prize
Length of Study: £300
Frequency: Every 2 years
No. of awards offered: Varies
Application Procedure: Applicants must submit one A4 page (up to 500 words) summary of research carried out. Those considered to be the best will be asked to give a 15 minute presentation.
Closing Date: Check with website

For further information contact:

Email: cns@rsm.ac.uk
Website: http://www.rsm.ac.uk/academ/awards/

Coloproctology Section John of Arderne Medal

Subjects: Coloproctology.
Purpose: To award the presenter of the best paper presented at the short papers meeting of the section of coloproctology. Applicants have to submit an abstract for presentation at the meeting.
Eligibility: Open to applicants of any nationality.
Level of Study: Professional development
Type: Award
Value: Approx. UK£600
Frequency: Annual
Study Establishment: Varies
Country of Study: Any country
No. of awards offered: 1
Application Procedure: Further details are available from the RSM administrator.
Closing Date: September and November (check with website)
Funding: Private
No. of awards given last year: 1
No. of applicants last year: 14–20

For further information contact:

Email: coloproctology@rsm.ac.uk
Website: http://www.rsm.ac.uk/academ/awards/

Dermatology Section Clinicopathological Meetings

Subjects: Clinicopathology.
Type: Prize
Value: £150
Frequency: Annual
Country of Study: United Kingdom
No. of awards offered: 1
Closing Date: August 10th (check with website)
Contributor: Royal Society of Medicine
No. of awards given last year: 1
No. of applicants last year: 32

For further information contact:

Email: dermatology@rsm.ac.uk
Website: http://www.rsm.ac.uk/academ/awards/

Epidemiology & Public Health Section Young Epidemiologists Prize

Subjects: Epidemiology and public health.
Purpose: To reward outstanding papers.
Eligibility: Any medical / non-medical epidemiologist or public health practitioner under the age of 40 years.
Level of Study: Unrestricted
Type: Award
Value: £250
Frequency: Annual
Country of Study: United Kingdom
No. of awards offered: 1
Application Procedure: Download the application form from website.
Closing Date: November 27th (check with website)
No. of awards given last year: 1
No. of applicants last year: 9

For further information contact:

Email: epidemiology@rsm.ac.uk
Website: http://www.rsm.ac.uk/academ/awards/

General Practice with Primary Healthcare Section John Fry Prize

Subjects: Primary healthcare.

Purpose: To award best examples of practice-based research involving members of the primary health and social community, demonstrating and promoting effective team work.
Eligibility: Open to candidates currently working in primary healthcare in the UK excluding members of the RSM section of GP Council.
Level of Study: Unrestricted
Type: Prize
Value: £300
Frequency: Annual
Country of Study: United Kingdom
No. of awards offered: 1
Application Procedure: Application form must be completed in all respects.
Closing Date: March 27th
Funding: Private
Contributor: John Fry
No. of awards given last year: 1
No. of applicants last year: 4

For further information contact:

RSM, 1 Wimpole Street, London, W1G 0AE
Email: gp@rsm.ac.uk
Website: http://www.rsm.ac.uk/academ/awards/
Contact: Gemma Lamb

Laryngology & Rhinology Section Travel and Equipment Grants

Subjects: Laryngology and rhinology.
Purpose: To assist with the cost of travel to overseas centres.
Eligibility: Open to senior registrars or consultants of not more than 2 years standing, who must be members of the section of laryngology and rhinology of the RSM.
Level of Study: Postdoctorate
Type: Scholarship
Value: UK£1,000
Frequency: Annual
Country of Study: Any country
No. of awards offered: 1
Application Procedure: Applicants must submit a paper to the RSM section of laryngology and rhinology. Further details are available from the RSM administrator.
Closing Date: April 19th
Funding: Commercial
Contributor: Karl Storz Endoscopy Limited
No. of awards given last year: 1
No. of applicants last year: 25
Additional Information: The recipient will be required to submit a brief report on the visit within 3 months of his or her return.

For further information contact:

Email: laryngology@rsm.ac.uk
Website: http://www.rsm.ac.uk/academ/awards/

Military Medicine Section Colt Foundation Research Prize

Purpose: To recognise the best abstract by a serving military medical officer.
Eligibility: Open to serving military officers in a training grade in general practice, a hospital or other speciality and are fellows of the RSM.
Level of Study: Postgraduate, serving military medical officers
Type: Prize
Value: £200 (first prize) and £100 each (second prize)
Frequency: Annual
Study Establishment: Royal Society of Medicine
Country of Study: United Kingdom
No. of awards offered: 1 (first prize) and 5 (second prizes)
Application Procedure: Candidates should email the abstracts. The abstracts will be shortlisted by the panel and judged.
Closing Date: November 1st
Funding: Trusts
Contributor: Colt Foundation
No. of awards given last year: 1
No. of applicants last year: 15

For further information contact:

Email: united.services@rsm.ac.uk
Website: http://www.rsm.ac.uk/academ/awards/

Nephrology Section Rosemarie Baillod Clinical Award

Eligibility: All doctors in training in nephrology and renal medicine at any grade.
Type: Research award
Value: £200
Frequency: Annual
Country of Study: United Kingdom
No. of awards offered: 1
Application Procedure: Please visit website for further details.
Closing Date: Check with website

For further information contact:

Email: nephrology@rsm.ac.uk
Website: http://www.rsm.ac.uk/academ/awards/

Occupational Medicine Section Malcolm Harrington Prize

Subjects: Occupational medicine.
Purpose: To award the work that is most likely to advance the study of occupational medicine in its broadest sense.
Eligibility: Open to occupational physician in training or within an year of achieving specialist accreditation.
Type: Prize
Value: £250
Frequency: Annual
Application Procedure: Candidates should submit an abstract of their own work (no longer than 200 words).
Closing Date: March 9th (check with website)
Funding: Private
Contributor: Professor Harrington
No. of awards given last year: 1
No. of applicants last year: 9

For further information contact:

Email: occupational@rsm.ac.uk
Website: http://www.rsm.ac.uk/academ/awards/

Oncology Section Sylvia Lawler Prize

Subjects: Oncology.
Purpose: To encourage scientists and clinicians in training to present the best scientific paper and best clinical paper on oncology.
Eligibility: All scientists and clinicians in training
Level of Study: Postgraduate
Type: Grant
Value: Two prizes of £500 to oral presenters and one prize of £50 voucher to poster presenter
Frequency: Annual
Study Establishment: Royal Society of Medicine
Country of Study: United Kingdom
No. of awards offered: 2
Application Procedure: Applicants should download an application form from the website, www.rsm.ac.uk/awards and submit the same via email to surgery@rsm.ac.uk
Closing Date: April 19th
No. of awards given last year: 1
No. of applicants last year: 50

For further information contact:

Email: oncology@rsm.ac.uk
Website: http://www.rsm.ac.uk/academ/awards/

Ophthalmology Section Travelling Fellowships

Subjects: Ophthalmology.
Purpose: To enable British ophthalmologists to travel abroad with the intention of furthering the study or advancement of ophthalmology, or to enable foreign ophthalmologists to visit the United Kingdom for the same purpose.
Eligibility: Open to ophthalmologists in the British Isles of any nationality who have not attained an official consultant appointment, nor undertaken professional clinical work or equivalent responsibility

for any substantial period before or during the execution of original work.
Level of Study: Professional development
Type: Fellowship
Value: UK£500–1,000
Frequency: Every 2 years
Study Establishment: Varies
Country of Study: Any country
No. of awards offered: Varies
Application Procedure: Applicants must apply to the academic administrator at RSM.
Closing Date: May 1st
No. of awards given last year: 4
No. of applicants last year: 6

For further information contact:

Email: ophthalmology@rsm.ac.uk
Website: http://www.rsm.ac.uk/academ/awards/

Orthopaedics Section President's Prize Papers

Purpose: To encourage research in the area of Orthopaedics.
Eligibility: Open to all orthopaedic trainees.
Level of Study: Postgraduate
Type: Prize
Value: Clinical paper prize – 1st prize of £300 and 2nd prize of £200
Frequency: Annual
Study Establishment: Royal Society of Medicine
Country of Study: United Kingdom
No. of awards offered: 1
Application Procedure: Candidates should submit abstracts no longer than 200 words. The abstracts will be judged by a panel of experts, shortlisted and asked to be presented at a meeting.
Closing Date: April 5th
No. of awards given last year: 1
No. of applicants last year: 10

For further information contact:

Email: orthopaedics@rsm.ac.uk
Website: http://www.rsm.ac.uk/academ/awards/

Otology Section Norman Gamble Grant

Subjects: Otology.
Purpose: To support specific research projects.
Eligibility: British citizens, both lay and medical
Level of Study: Unrestricted
Type: Prize
Value: UK£100
Frequency: Annual
Country of Study: Any country
No. of awards offered: 1
Application Procedure: Further details are available from the RSM administrator.
Closing Date: December 9th (check with website)
Funding: Private
No. of awards given last year: 1
No. of applicants last year: 4

For further information contact:

Email: otology@rsm.ac.uk
Website: http://www.rsm.ac.uk/academ/awards/

Paediatrics & Child Health Section Trainees Tim David Prize

Subjects: Paediatrics.
Purpose: To encourage research in the area of Paediatrics & Child Health.
Eligibility: Open to paediatric trainees.
Level of Study: Foundation programme, Postgraduate
Type: Prize
Value: First prize £150, one year's subscription to the RSM (worth up to £200) and membership of the Council of the Section of Paediatrics and Child Health for one year. There will also be a second prize of £100
Frequency: Annual
No. of awards offered: 2
Application Procedure: Candidates should submit abstracts.
Closing Date: April 22nd
Contributor: Paediatrics & Child Health Section, RSM
No. of awards given last year: 2
No. of applicants last year: 12

For further information contact:

Email: paediatrics@rsm.ac.uk
Website: http://www.rsm.ac.uk/academ/awards/

Palliative Care Section MSc/MA research prize

Eligibility: All students either currently studying, or within 18 months of completion, of MSc or MA in Palliative Medicine or an allied discipline, are eligible to apply.
Level of Study: Foundation programme, Postgraduate
Type: Prize
Value: First prize of £250, second prize of £100 and third prize of £50
Frequency: Annual
No. of awards offered: 2
Closing Date: September 17th
Contributor: Palliative Care Section

For further information contact:

Email: palliative@rsm.ac.uk
Website: http://www.rsm.ac.uk/academ/awards/

Psychiatry Section Mental Health Foundation Research Prize

Subjects: Psychiatry.
Purpose: To award an outstanding published paper.
Eligibility: Open to candidates practising medicine in the United Kingdom or the Republic of Ireland who are in training at any grade from senior house officer to senior registrar or equivalent. Applicants need not be members of RSM.
Level of Study: SHO or SpR
Type: Prize
Value: 1st prize UK£750, 2nd prize UK£100
Frequency: Annual
Country of Study: United Kingdom
No. of awards offered: 2
Application Procedure: Full copy of the published article, CV and covering letter explaining in their own words the significance of the publication to be submitted to the section of psychiatry.
Closing Date: Early January (check with website)
Funding: Private
No. of applicants last year: 1

For further information contact:

Email: psychiatry@rsm.ac.uk
Website: http://www.rsm.ac.uk/academ/awards/

Surgery Section Norman Tanner Prize and Glaxo Travelling Fellowship

Subjects: Oncology.
Purpose: To encourage clinical registrars submit the best clinical paper.
Eligibility: Open to all trainee oncologists.
Value: £250 plus the Norman Tanner Medal. Runner up – £250 Glaxo Travelling Fellowship
Frequency: Annual
Study Establishment: Royal Society of Medicine
Country of Study: United Kingdom
No. of awards offered: 1
Application Procedure: Applicants are requested to contact the Section Coordinator at oncology@rsm.ac.uk
Closing Date: April 27th (check with website)
No. of awards given last year: 12
No. of applicants last year: 30

For further information contact:

Email: surgery@rsm.ac.uk
Website: http://www.rsm.ac.uk/academ/awards/

Trainees' Committee The John Glyn Trainees' Prize

Purpose: The prize was established to promote best practise through high quality audit.
Eligibility: Open to trainees from any medical, dental, veterinary hospital or primary care speciality under the age of 40.
Level of Study: Foundation programme, Graduate, Postgraduate, Research
Type: Prize
Value: £300
Frequency: Annual
No. of awards offered: 1
Closing Date: April 12th (check with website)
Contributor: Royal Society of Medicine
No. of awards given last year: 1
No. of applicants last year: 40

For further information contact:

Email: trainees@rsm.ac.uk
Website: http://www.rsm.ac.uk/academ/awards/

Urology Section Professor Geoffrey D Chisholm CBE Communication Prize

Subjects: Urology.
Purpose: To reward the best abstract at the Short Papers Prize Meeting.
Eligibility: Open to members of the Urology section.
Level of Study: Trainee doctors
Type: Travel award
Value: Fully funded RSM travelling fellowship to the Urology Section's overseas winter scientific meeting in the next academic session
No. of awards offered: 3
Closing Date: March 8th
No. of awards given last year: 3
No. of applicants last year: 30

For further information contact:

Email: urology@rsm.ac.uk
Website: http://www.rsm.ac.uk/academ/awards/

Urology Section Professor John Blandy Essay Prize for Medical Students

Subjects: Urology.
Purpose: To enable the holder to enhance his or her knowledge and experience by visiting an overseas unit.
Eligibility: Medical students
Level of Study: Trainees
Type: Fellowship
Value: A bursary of £1,000 and an RSM award certificate
Frequency: Annual
No. of awards offered: 1
Closing Date: March 8th
Additional Information: You must be available on the May 16th for presentation your short paper to be eligible for this prize.

For further information contact:

Email: urology@rsm.ac.uk
Website: www.rsm.ac.uk

Urology Section Spring short papers prize

Subjects: Urology.
Purpose: To award the best short paper.
Type: Prize
Value: Fully funded RSM travelling fellowship to the Urology Section's overseas winter scientific meeting. Runner up prizes of a bursary towards the RSM overseas winter scientific meeting in the next academic session
Frequency: Annual
Closing Date: February 15th

For further information contact:

Email: urology@rsm.ac.uk
Website: www.rsm.ac.uk/awards

Urology Section Winter short papers prize (Clinical Uro-Radiological meeting)

Subjects: Urology.
Purpose: To reward the best clinicopathological short paper.
Eligibility: Urological and radiological trainees
Type: Prize
Value: Fully funded RSM travelling fellowship to the Urology Section's overseas winter scientific meeting in the next academic session. Runner up prizes of a bursary towards the RSM overseas winter scientific meeting in the next academic session
Frequency: Annual
No. of awards offered: 3
Closing Date: Check with website

For further information contact:

Email: urology@rsm.ac.uk
Website: http://www.rsm.ac.uk/academ/awards/

Venous Forum Spring Meeting Prizes

Purpose: To recognise the best original paper by a non-consultant.
Eligibility: Open to non-consultants.
Level of Study: Postgraduate
Type: Prize
Value: Prize – £250 (first), £200 (second), £150 (third) and £200 (poster)
Frequency: Annual
Study Establishment: Royal Society of Medicine
Country of Study: United Kingdom
No. of awards offered: 1
Application Procedure: Candidates should email abstracts. Short-listed candidates will be invited to present their papers.
Closing Date: March 13th
No. of awards given last year: 1
No. of applicants last year: 20

For further information contact:

Email: venous@rsm.ac.uk
Website: http://www.rsm.ac.uk/academ/awards/

ROYAL TOWN PLANNING INSTITUTE (RTPI)

41 Botolph Lane, London, EC3R 8DL, England
Tel: (44) 020 7929 9494
Fax: (44) 020 7929 9490
Email: judy.woollett@rtpi.org.uk
Website: www.rtpi.org.uk/

The Royal Town Planning Institute (RTPI) was founded in 1914 and is a registered charity. Its aim is to advance the science and art of town planning in all its aspects, including local, regional and national planning for the benefit of the public. The Institute is primarily concerned with maintaining high standards of competence and conduct within the profession, promoting the role of planning within the country's social, economic and political structures, and presenting the profession's views on current planning issues.

George Pepler International Award

Subjects: Town and country planning or some particular aspect of planning theory and practice.
Purpose: To enable young people of any nationality to visit another country for a short period to study.
Eligibility: Open to persons under 30 years of age of any nationality.
Level of Study: Professional development, Research
Type: Fees to performers
Value: Up to UK£1,500
Length of Study: Short-term travel outside the United Kingdom for United Kingdom residents or for visits to the United Kingdom for applicants from abroad
Frequency: Every 2 years
Country of Study: Any country
No. of awards offered: 1

Application Procedure: Applicants must submit a statement showing the nature of the study visit proposed, together with an itinerary. Application forms are available on request from the RTPI.
Closing Date: March 31st
Funding: Private
Contributor: Trust fund
No. of awards given last year: 1
No. of applicants last year: 20
Additional Information: At the conclusion of the visit the recipient must submit a report. Please see the website for further details http://www.rtpi.org.uk/events/awards/george-pepler-international-award/.

RSM ERASMUS UNIVERSITY

MBA Programmes, Burgemeester Oudlaan 50, 3062 PA Rotterdam, PO Box 1738, Rotterdam, 3000 DR, Netherlands
Tel: (31) 10 408 2222
Fax: (31) 10 452 9509
Email: info@rsm.nl
Website: www.rsm.nl
Contact: Denise Chasney van Dijk, Financial Aid Manager

As one of the world's top business schools, RSM is a centre of excellence and innovation for management education and research. Our goal is to empower individuals to succeed amid the complexities of modern international commerce and become the business leaders of tomorrow.

NFP Netherlands Fellowship Programme

Subjects: Business Management.
Purpose: To assist full-time MBA students financing MBA.
Eligibility: Open to residents of 57 countries as listed on the NUFFIC website with a preference for female candidates from sub-Saharan Africa.
Level of Study: MBA
Type: Fellowship
Value: €39,000 (full tuition), travel allowances and living expenses
Length of Study: 12 months
Frequency: Annual
Study Establishment: Erasmus University
Country of Study: Netherlands
No. of awards offered: 1
Application Procedure: Refer the website for further details.
Closing Date: May 1st
Funding: Government
Contributor: NUFFIC/Dutch Ministry of Foreign Affairs
No. of awards given last year: 1
No. of applicants last year: 4

RSM MBA Africa & Middle East Regional Scholarship

Subjects: Business Management.
Purpose: To assist candidates from the Africa & Middle East Region in financing their MBA study in the Netherlands.
Level of Study: MBA
Type: Scholarship
Value: 20 per cent tuition fee waiver
Length of Study: 12 months
Frequency: Annual
Study Establishment: RSM Erasmus University
Country of Study: Netherlands
No. of awards offered: 1
Funding: Private
Contributor: RSM Erasmus University

RSM MBA Asia & Australia Regional Scholarship

Subjects: Business Management.
Purpose: To assist candidates from the Asia & Australasia Region in financing their MBA study in the Netherlands.
Eligibility: The scholarship is open to high potential candidates who are a citizen or hold permanent residence status in one of the following listed countries: Australia, Bangladesh, Bhutan, Brunei, Burma, Cambodia, China, Fiji, Hong Kong, India, Indonesia, Japan, Kiribati, Laos, Macau, Malaysia, Micronesia, Mongolia, Nepal, New Zealand, Palau, Papua New Guinea, Philippines, Samoa, Singapore, Solomon Islands, Sri Lanka, Thailand, Timor-Leste, Tonga, Tuvalu, Vietnam, Yemen.
Level of Study: MBA
Type: Scholarship
Value: 20 per cent tuition fee waiver
Length of Study: 12 months
Frequency: Annual
Study Establishment: Erasmus University
Country of Study: Netherlands
No. of awards offered: 1
Application Procedure: Complete application form to be considered.
Closing Date: September 30th
Funding: Private
Contributor: RSM Erasmus University
No. of awards given last year: 1
No. of applicants last year: 36
Additional Information: Eligible to nationals of Asia and Australasia.

RSM MBA Central & Eastern Europe Regional Scholarship

Subjects: Business Managment.
Purpose: To assist candidates from the Central & Eastern Europe region in financing their MBA study in the Netherlands.
Eligibility: Open to high potential candidates who are a citizen or hold permanent residence status in one of the following listed countries: Albania, Armenia, Azerbaijan, Belarus, Bosnia and Herzegovia, Bulgaria, Croatic, Czez Republic, Estonia, Georgia, Hungary, Kazakhstan, Kosovo, Kyrgyzstan, Latvia, Lithuania, Macedonia, Moldova, Montenegro, Poland, Romania, Russia, Serbia, Slovakia, Slovenia, Turkey, Ukraine.
Level of Study: MBA
Type: Scholarship
Value: 20 per cent tuition fee waiver
Length of Study: 12 months
Frequency: Annual
Study Establishment: Erasmus University
Country of Study: Netherlands
No. of awards offered: 1
Application Procedure: Candidates complete application form.
Closing Date: October 30th
Funding: Private
Contributor: RSM Erasmus University
No. of awards given last year: 1
No. of applicants last year: 7
Additional Information: Eligible to nationals of Central and Eastern Europe.

RSM MBA Dean's Merit Award Scholarship

Subjects: Business Management.
Purpose: To assist high potential candidates finance their study.
Level of Study: MBA
Type: Scholarship
Value: Up to 40 per cent tuition fee waiver
Length of Study: 12 months
Frequency: Annual
Study Establishment: Erasmus University
Country of Study: Netherlands
No. of awards offered: Discretionary
Application Procedure: A separate application is not required as the RSM MBA Awards Committee will use the MBA Programme application materials to the Inernational full-time MBA Programme inculding application form, GMAT score, essays, interview assessment, references and CV to determine the recipient of the award.
Closing Date: Up to end November
Funding: Private
Contributor: RSM Erasmus University
No. of awards given last year: 4

RSM MBA Latin America & Caribbean Regional Scholarship

Subjects: Business Management.
Purpose: To assist candidates from the Latin America & Carribean region in financing their MBA study in the Netherlands.
Eligibility: Open to high potential candidates who are a citizen or hold permanent residence status in one of the following listed countries:

Antigua and Barbuda, Argentina, Aruba, Bahamas, Barbados, Belize, Bermudas, Bolivia, Bonaire, Brazil, Chile, Colombia, Costa Rica, Cuba, Curacao, Dominica, Dominican Republic, El Salvador, Ecuador, Grenada, Guatemala, Guyana, Haiti, Honduras, Jamaica, Mexico, Nicaragua, Panama, Paraguay, Peru, Saba, Saint Kitts and Nevis, Saint Lucia, Saint Vincent, Sint Eustatius, Sint Maarten, Suriname, Trinidad, Venezuela.
Level of Study: MBA
Type: Scholarship
Value: 20 per cent tuition fee waiver
Length of Study: 12 months
Frequency: Annual
Study Establishment: Erasmus University
Country of Study: Netherlands
No. of awards offered: 1
Application Procedure: Complete application form to be considered for this scholarship.
Closing Date: September 30th
Funding: Private
Contributor: RSm Erasmus University
No. of awards given last year: 1
No. of applicants last year: 12
Additional Information: Eligible to nationals of Latin America & Caribbean.

RSM MBA North America Regional Scholarship

Subjects: Business Management.
Purpose: To assist candidates from North America region in financing their MBA study in the Netherlands.
Eligibility: Open to high potential candidates who are citizen or hold permanent residence status in one of the following listed countries: Canada, United States of America.
Level of Study: MBA
Type: Scholarship
Value: 20 per cent tuition fee waiver
Length of Study: 12 months
Frequency: Annual
Study Establishment: Erasmus University
Country of Study: Netherlands
No. of awards offered: 1
Application Procedure: Complete application form is required.
Closing Date: October 30th
Funding: Private
Contributor: RSM Erasmus University
No. of awards given last year: 1
No. of applicants last year: 10

RSM MBA Sustainability Scholarship

Subjects: Business Management.
Purpose: To assist high potential candidates to finance their study and who are interested in sustainability.
Level of Study: MBA
Type: Scholarship
Value: €5,000
Length of Study: 12 months
Frequency: Annual
Study Establishment: Erasmus University
Country of Study: Netherlands
No. of awards offered: 1
Application Procedure: Please refer to the website for details.
Closing Date: November
Funding: Private
Contributor: RSM Erasmus University
No. of awards given last year: 1
No. of applicants last year: 23

RSM MBA Western Europe Regional Scholarship

Subjects: Business Management.
Purpose: To assist candidates from the Western Europe region in financing their MBA study in the Netherlands.
Eligibility: Open to high potential candidates who are a citizen or hold permanent residence status in one of the following listed countries: Andorra, Austria, Belgium, Cyprus, Denmark, Finland, France, Germany, Greece, Iceland, Ireland, Italy, Liechtenstein, Luxembourg, Malta, Monaco, Netherlands, Norway, Portugal, San Marino, Spain, Swaziland, Sweden, Switzerland, United Kingdom.
Level of Study: MBA
Type: Scholarship
Value: 20 per cent tuition fee waiver
Length of Study: 12 months
Frequency: Annual
Study Establishment: Erasmus University
Country of Study: Netherlands
No. of awards offered: 1
Application Procedure: Complete application form is required.
Funding: Private
Contributor: RSM Erasmus University
Additional Information: Eligible to nationals of Western Europe.

RSM MBA Women in Business Scholarship

Subjects: Business Management.
Purpose: To assist International full-time MBA students finance their study.
Level of Study: MBA
Type: Scholarship
Value: €5,000
Length of Study: 12 months
Frequency: Annual
Study Establishment: Erasmus University
Country of Study: Netherlands
No. of awards offered: 1
Application Procedure: Please refer to the website for further details.
Closing Date: November
Funding: Private
Contributor: RSM Erasmus University
No. of awards given last year: 2
No. of applicants last year: 13

RSM OneMBA Dean's Fund Assistance Award

Subjects: Business Management.
Purpose: To assist OneMBA candidates finance their study.
Level of Study: MBA
Type: Award
Value: €10,000 per award
Length of Study: 21 months
Frequency: Annual
Study Establishment: Erasmus University
Country of Study: Netherlands
No. of awards offered: Discretionary
Application Procedure: Please refer to the website for details.
Closing Date: September
Funding: Private
Contributor: RSM Erasmus University
No. of awards given last year: 1
No. of applicants last year: 2

RSM OneMBA Women in Business Scholarship

Subjects: Business Management.
Purpose: To assist OneMBA women students finance their study.
Level of Study: MBA
Type: Scholarship
Value: €10,000
Length of Study: 21 months
Frequency: Annual
Study Establishment: Erasmus University
Country of Study: Netherlands
No. of awards offered: 2
Application Procedure: Refer to the website for further details.
Closing Date: September
Funding: Private
Contributor: RSM Erasmus University
No. of awards given last year: 2
No. of applicants last year: 3
Additional Information: Committed to increasing the diversity of the OneMBA class, the Rotterdam School of Management, Erasmus University, offers the Women in Business scholarship to two women demonstrating exceptional business leadership potential.

RSM-NESO MBA in Taiwan Scholarship

Subjects: Business Management.
Purpose: To assist Taiwanese nationals in financing their MBA study in the Netherlands.
Eligibility: Open to candidates from Taiwan only.
Level of Study: MBA
Type: Scholarship
Value: €10,000 per award
Length of Study: 12 months
Frequency: Annual
Study Establishment: Erasmus University
Country of Study: The Netherlands
No. of awards offered: 2
Application Procedure: Please check website for further details.
Closing Date: October
Funding: Government
Contributor: Netherlands Education Support Office
No. of awards given last year: 2
No. of applicants last year: 5

RSPCA AUSTRALIA INC

PO Box 265, Deakin West, ACT 2600, Australia
Tel: (61) 02 6282 8300
Fax: (61) 02 6282 8311
Email: rspca@rspca.org.au
Website: www.rspca.org.au

The mission of RSPCA is to prevent cruelty to animals by actively promoting their care and protection and to be the leading authority in animal care and protection. The RSPCA shelters are one of the most visible parts of the Society's operations. RSPCA shelters receive more than 138,000 animals per year.

RSPCA Australia Alan White Scholarship for Animal Welfare Research

Subjects: General animal welfare issues.
Purpose: To encourage students to take an active interest in animal welfare issues, to support animal welfare research that might not otherwise attract funding, and to promote the objectives of the RSPCA within the research community.
Eligibility: Open to applicants who are enrolled in any accredited course at an Australian university. Applicants must also demonstrate a major commitment towards the involvement of animal welfare issues.
Level of Study: Postgraduate
Type: Scholarship
Value: Australian $8,200
Frequency: Annual
Country of Study: Australia
Application Procedure: Application forms can be downloaded from the website.
Closing Date: August 27th
Funding: Foundation
Contributor: RSPCA
Additional Information: For further information, please contact Chief Scientist.

RSPCA Australia Scholarship for Humane Animal Production Research

Subjects: Animal welfare issues in animal production.
Purpose: To encourage students to take an active interest in animal welfare issues, to support animal welfare research that might not otherwise attract funding, and to promote the objectives of the RSPCA within the research community.
Eligibility: Open to applicants who are enrolled in any accredited course at an Australian university. Applicants must also demonstrate a major commitment towards the involvement of animal welfare issues.
Type: Scholarship
Value: Australian $8,200
Frequency: Annual
Country of Study: Australia
Application Procedure: Application forms can be downloaded from the website.
Closing Date: August 27th
Funding: Foundation
Contributor: RSPCA
Additional Information: For further information, please contact Chief Scientist.

RUSSELL SAGE FOUNDATION

112 East 64th Street, New York, NY, 10065, United States of America
Tel: (1) 212 750 6000
Fax: (1) 212 371 4761
Email: info@rsage.org
Website: www.russellsage.org
Contact: Christopher Brogna, CFO

The Russell Sage Foundation is the principal American foundation devoted exclusively to research in the social sciences. Located in New York City, the Foundation is a research centre, a funding source for studies by scholars at other academic and research institutions and an active member of the nation's social science community.

Russell Sage Foundation Visiting Scholar Program

Subjects: Social sciences.
Purpose: To pursue writing and research.
Eligibility: Open to scholars in the social sciences. The Foundation particularly welcomes groups of visiting scholars who wish to collaborate on a specific project during their residence at the foundation. In order to develop these projects fully, support is sometimes provided for working groups prior to their arrival at the foundation. Awards are not made for the support of graduate degree work, nor for institutional support.
Level of Study: Doctorate, Postdoctorate
Type: Fellowship
Value: Varies
Length of Study: 1 academic year
Frequency: Annual
Study Establishment: The Foundation
Country of Study: Any country
No. of awards offered: Varies
Application Procedure: Applicants should consult the website or contact the organization for details.
Closing Date: September 30th
Funding: Foundation
No. of awards given last year: 21
No. of applicants last year: 110
Additional Information: Awardees are expected to offer the Foundation the right to publish any book-length manuscripts resulting from Foundation support. Please see the website for further details http://www.russellsage.org/about/contact-us#scholars.

RUTH ESTRIN GOLDBERG MEMORIAL FOR CANCER RESEARCH

PO Box 194, Springfield, Union, NJ, 07081, United States of America
Tel: (1) 908 686 5508
Email: goodfudgie@aol.com
Website: www.regm-cancer-research.us
Contact: Mrs Rhoda Goodman, Chairman

The Ruth Estrin Goldberg Memorial for Cancer Research is a foundation supporting cancer research. It gives grants to doctors undertaking research. Awarded to research being done in hospitals in NY, NJ, CT, PA.

Ruth Estrin Goldberg Memorial for Cancer Research

Subjects: Cancer research.
Purpose: To help fund cancer research in instituitions.
Eligibility: Open to candidates from the eastern United States of America, only New York, New Jersey, Pennsylvania, or Connecticut.
Level of Study: Unrestricted
Type: Quarterly payments
Value: US$10,000
Length of Study: 1 year
Country of Study: United States of America
No. of awards offered: 1
Application Procedure: Applicants must write for an application form and guidelines.

Closing Date: April 28th
Funding: Private
Contributor: Members and friends
No. of awards given last year: 1
No. of applicants last year: 10
Additional Information: The Ruth Estrin Goldberg Memorial for Cancer Research is a volunteer organization.

For further information contact:

653 Colonial Arms Road, Union, NJ, 07083, United States of America
Contact: Mrs Rhoda Goodman, Chairman

RYERSON UNIVERSITY

350 Victoria Street, Toronto, ON, M5B-2K3, Canada
Tel: (1) 416 979 5000
Fax: (1) 416 979 5292
Email: awards@ryerson.ca
Website: www.ryerson.ca
Contact: Dr Judith Sandys, International Relations

Ryerson University was founded in 1948 as Ryerson Institute of Technology, established primarily as a training ground for the growing workforce of a booming post war economy. The Institute was a novel alternative to the traditional apprenticeship system of technical learning.

Ryerson Graduate Scholarship (RGS)

Subjects: All subjects.
Purpose: To attract and retain excellent graduate students and support them financially.
Eligibility: Open to Master's candidates with a minimum grade point average of at least 3.67. Renewal will require a minimum 1st year grade point average of 3.67, with no grade below 2.67. For a Doctoral candidate a minimum grade point average of at least 3.67 in their Master's programme is required.
Level of Study: Doctorate, Postgraduate
Type: Scholarship
Value: Canadian $7,000
Length of Study: 2–3 years
Study Establishment: Ryerson University
Country of Study: Canada
Funding: Individuals, private
Additional Information: Students who did not receive an entry scholarship, but who perform at the level required for scholarship renewal will be eligible to be considered for an RGS after their 1st year in a Ryerson graduate programme. Please see the website for further details http://www.ryerson.ca/graduate/funding/.

S S HUEBNER FOUNDATION FOR INSURANCE EDUCATION

3000 Steinberg Hall - Dietrich Hall, 3260 Locust Walk, Philadelphia, PA, 19104-6302, United States of America
Tel: (1) 215 898 9631
Fax: (1) 215 573 2218
Email: hcalvert@wharton.upenn.edu
Website: www.huebnergeneva.org
Contact: Associate Director

The S S Huebner Foundation is an educational foundation with the objective of promoting education and research in risk management and insurance. It provides PhD fellowships for the study of risk management and insurance economics at the Wharton School of the University of Pennsylvania, and publishes books and working papers.

S S Huebner Foundation for Insurance Education Doctoral Fellowships

Subjects: Managerial science and applied economics, with a specialization in risk management and insurance economics.
Purpose: To increase the supply of college professors specializing in risk management and insurance economics.
Eligibility: To be eligible for a doctoral fellowship, an applicant must have obtained a baccalaureate degree from an accredited college or university.
Level of Study: Doctorate, Postdoctorate
Type: Fellowship
Value: Full tuition fees of the Wharton School of the University of Pennsylvania plus an annual living stipend of US$21,000
Length of Study: 4 years
Frequency: Annual
Study Establishment: The Wharton School of the University of Pennsylvania
Country of Study: United States of America
No. of awards offered: Varies
Application Procedure: Applicants must apply to the Wharton School doctoral programme for admission and to the Huebner Foundation for funding.
Closing Date: December 15th (Wharton admission), January 15th (Huebner funding)
Funding: Corporation, foundation, individuals
Contributor: Leading insurance companies in the United States and Canada
No. of awards given last year: 2
No. of applicants last year: 8
Additional Information: Candidates are required to certify that it is their intention to follow a teaching career in insurance and that they will major in insurance and risk management for a graduate degree. Applicants must take the admission test for graduate study in business. For information concerning these examinations, candidates should write directly to the Educational Testing Service (ETS). Applicants should apply separately and directly to the Wharton School Doctoral Programme Office for admission into the Insurance and Risk Management Doctoral Programme. Please see the website for further details http://www.huebnergeneva.org/huebner/fellowships.php.

SACRAMENTO STATE

CSUS 6000 J Street, Sacramento, CA, 95819, United States of America
Tel: (1) 916 278 6011
Fax: (1) 916 278 5199
Email: infodesk@csus.edu
Website: www.csus.edu
Contact: Timothy Hodson, Executive Director

Center for California Studies, CSU-Sacramento, California Legislature (CSUS) was founded in 1984. It is located on the capital campus of the California State University. Center for California Studies is a public service, educational support and applied research institute of CSUS. It is dedicated to promoting a better understanding of California's government, politics, people, cultures and history.

California Senate Fellows

Subjects: Public policy and politics.
Purpose: To expose people with diverse life experiences and backgrounds to the legislative process and provide research and other professional staff assistance to the Senate.
Eligibility: Open to candidates who have obtained a degree from a 4 year college or university.
Level of Study: Professional development
Type: Fellowships
Value: US$1,972 per month and full health, vision and dental benefits
Length of Study: 11 months
Frequency: Annual
Country of Study: United States of America
No. of awards offered: 18
Application Procedure: Applicants can download the application form from the website.
Closing Date: February 22nd
Funding: Government
Additional Information: For further information please contact David Pacheco, the program director, at 916 278 5408 (Sacramento State), 916 651 4160 (Senate) or email to david.pacheco@sen.ca.gov

For further information contact:

Tel: 916 278 6906
Email: calstudies@csus.edu
Website: http://www.csus.edu/calst/senate_fellows_program.html

Jesse M. Unruh Assembly Fellowship Program

Subjects: Public policy formation.
Purpose: To provide an opportunity for individuals of all ages, ethnic backgrounds and experiences to directly participate in the legislative process.
Eligibility: Applicants must have completed a Bachelor's degree by the end of Summer of the fellowship year. There are no preferred majors.
Level of Study: Professional development
Type: Fellowship
Value: US$1,972 per month and medical, dental and vision benefits
Length of Study: 11 months
Frequency: Annual
Study Establishment: Center for California Studies
Country of Study: United States of America
No. of awards offered: 18
Application Procedure: Applicants must download the complete application form from the website. Applicants must furnish academic, employment and activities data, unofficial transcripts from colleges attended, a personal statement, a policy statement on a specific topic contained in the application and 3 references.
Closing Date: February 22nd
Additional Information: Individuals with advanced degrees or those in mid-career are encouraged to apply.

For further information contact:

Tel: 916 278 6906
Email: calstudies@csus.edu.
Website: http://www.csus.edu/calst/assembly_fellowship_program.html

SAINT ANDREW'S SOCIETY OF THE STATE OF NEW YORK SCHOLARSHIPS, THE CARNEGIE TRUST FOR THE UNIVERSITIES OF SCOTLAND

Andrew Carnegie House, Pittencrieff Street, Dunfermline, Fife, KY12 8AW, Scotland
Tel: (44) 1383 724990
Fax: (44) 1383 749799
Email: jgray@carnegie-trust.org
Website: www.carnegie-trust.org
Contact: Jackie Gray, Assistant Secretary

The Carnegie Trust for the Universities of Scotland administrates on behalf of the Saint Andrew's Society of the State of New York Scholarships to students of Scottish descent or birth for study at a university in the United States of America, within a radius of 250 miles from New York City or the Washington DC area.

Saint Andrew's Society of the State of New York Scholarship Fund

Subjects: All subjects.
Purpose: To support advanced study exchanges between the United States of America and Scotland.
Eligibility: Open to newly qualified graduates of a Scottish university or of Oxford or Cambridge. Candidates are required to have a Scottish background. The possession of an Honours degree is not essential. Personality and other qualities will influence the selection.
Level of Study: Postgraduate
Type: Scholarship
Value: Up to US$20,000 each to cover university tuition fees, room and board and transportation expenses
Length of Study: 1 academic year
Frequency: Annual
Study Establishment: A university within 250 miles of New York City or the Washington DC area
Country of Study: United States of America
No. of awards offered: 2
Application Procedure: Applicants must write for details. Each Scottish university will screen its own applicants and nominate one candidate to go forward to the final selection committee to be held in Edinburgh in April. Oxford and Cambridge applicants should submit applications to the trust.
Closing Date: January 25th
Funding: Private
No. of awards given last year: 2
No. of applicants last year: 6
Additional Information: Only in unusual circumstances will the Society consider other locations. Thereafter, the scholar is expected to spend a little time travelling in United States of America before returning to Scotland. Applications should be made via the principal of the university attended(ing) in the case of the Scottish universities.

SAMUEL H KRESS FOUNDATION

174 East 80th Street, New York, NY, 10075, United States of America
Tel: (1) 212 861 4993
Fax: (1) 212 628 3146
Email: wyman@kressfoundation.org
Website: www.kressfoundation.org
Contact: Wyman Meers, Program Administrator

The Samuel H. Kress Foundation, since its creation in 1929, has devoted its resources almost exclusively to programmes related to European art. The Foundation devoted its resources to advancing the history, conservation, and enjoyment of the vast heritage of European art, architecture, and archaeology.

Samuel H Kress Foundation 2-Year Research Fellowships at Foreign Institutions

Subjects: Art history.
Purpose: To facilitate advanced dissertation research in association with a selected institute of art history in either Florence, Leiden, London, Munich, Paris, Rome.
Eligibility: Open to PhD candidates in the history of art for the completion of their dissertation research. Candidates must be citizens of the United States of America or matriculated at an institution in the United States of America.
Level of Study: Predoctorate
Type: Fellowship
Value: US$30,000 per year
Length of Study: 2 years
Frequency: Annual
Study Establishment: One of a number of art historical institutes in Florence, Leiden, London, Munich, Paris, Rome
No. of awards offered: 6
Application Procedure: Applicants must be nominated by their art history department. There is a limit of two applicant per department*. The Foundation does not accept grant materials by fax.
*Each art history department can nominate two individuals; however, only one nominee per art history department for each host institutions.
Closing Date: November 30th
Funding: Private
No. of awards given last year: 6
No. of applicants last year: Approx. 50
Additional Information: Applications submitted directly to host institution with a copy to Kress Foundation. Contact information available at www.kressfoundation.org.

Samuel H Kress Foundation Fellowships for Advanced Training in Fine Arts Conservation

Subjects: Specific areas of fine art conservation.
Purpose: To enable young American conservators to undertake post-MA advanced internships.
Eligibility: Open to those who have completed their academic training in conservation.
Level of Study: Postgraduate
Type: Fellowship
Value: US$32,000
Frequency: Annual
Study Establishment: Appropriate institutions
Country of Study: United States of America

No. of awards offered: 9
Application Procedure: Application procedures and contact information available at www.kressfoundation.org. Program administered on Kress Foundation's behalf by the American Institute for Conservation. Forms and guidelines for submission available on website.
Closing Date: March l0th
Funding: Private
No. of awards given last year: 9
No. of applicants last year: Approx. 25–30
Additional Information: Emphasis is on hands-on training. These grants are not for the completion of degree programmes. Enquiries should be directed to Wyman Meers.

For further information contact:

The Foundation of the American Institute for Conservation of Historic and Artistic Works, 1156 15th St, NW; Suite 320, Washington, DC 20005
Email: faicgrants@aic-faic.org

Samuel H Kress Foundation Interpretive Fellowships at Art Museums

Subjects: Kress interpretive fellowships provide competitive grants to American art museums which sponsor supervised internships in art museum education, with preference given to projects that promote collaboration between art museum educators and curators and advancement of the appreciation of pre-modern European art history.
Purpose: To encourage students to explore interpretive courses in art museums, whether as future museum educators or curators.
Eligibility: Open to individuals who have completed a degree (BA, MA or PhD) in art history, art education, studio art or museum studies and who are pursuing or contemplating graduate study or professional placement in these or related fields.
Level of Study: Graduate, Postgraduate
Type: Fellowships
Value: $30,000
Length of Study: 9–12 months
Frequency: Annual
Country of Study: United States of America
Application Procedure: Application guidelines available at www.kressfoundation.org.
Closing Date: April 1st
Funding: Private
No. of awards given last year: 6
No. of applicants last year: 20

THE SAN FRANCISCO FOUNDATION (SFF)

225 Bush Street, Suite 500, San Francisco, CA, 94104, United States of America
Tel: (1) 415 733 8500
Fax: (1) 415 477 2783
Email: info@sff.org
Website: www.sff.org

The San Francisco Foundation (SFF) is a leading agent of Bay Area philanthropy. They rank 7th in grant making and assets among the nation's community foundations. They cultivate a family of donors who share a commitment to the Bay Area. They give millions of dollars a year to build on community assets, respond to community needs and elevate public awareness.

Joseph Henry Jackson Literary Award

Subjects: Fiction (novel or short stories), nonfiction, prose and poetry.
Purpose: To support an author of an unpublished work in progress.
Eligibility: Open to residents of northern California or Nevada for 3 consecutive years who are between 20 and 35 years of age.
Level of Study: Postgraduate
Type: Award
Value: US$2,000
Frequency: Annual
Country of Study: United States of America
No. of awards offered: 2
Application Procedure: Applicants must submit manuscript of their unpublished work along with the application form, entry in a contest and stamped addressed envelope.
Closing Date: January 31st
Additional Information: In addition to the US$2,000 cash award, winners will be invited to participate in a public reading at Intersection for the Arts and the winning manuscripts will be permanently housed at UC Berkeley's Bancroft Library. Please see the website for further details http://www.sff.org/programs/collaborative-engagement/awards-programs/art-awards/literary-awards/.

Koshland Young Leader Awards

Purpose: It recognizes the next generation of leadership in community. Koshland Young Leaders are strongly motivated to achieve despite facing multiple challenges, such as economic and family responsibilities.
Eligibility: San Francisco public high school juniors. The most competitive candidates have at least a 3.25 cumulative or continually improving GPA, are college-bound, and embrace a commitment to strengthening their families and communities despite facing formidable life challenges.
Level of Study: Postgraduate
Type: Award
Value: $7,000
Length of Study: 2 years
Frequency: Every 2 years
Country of Study: United States of America
No. of awards offered: 8
Application Procedure: See the website.
Closing Date: February (four-part nomination process)
Contributor: San Francisco Foundation
Additional Information: Each winter, we invite teachers and counselors to nominate outstanding San Francisco public high school juniors for this award. If you have questions, please email or call Joshua Jones.

For further information contact:

Tel: 415 733 8587
Email: kyla@sff.org

SAN FRANCISCO STATE UNIVERSITY (SFSU)

1600 Holloway Avenue, San Francisco, CA, 94132, United States of America
Tel: (1) 415 338 2234/1111
Fax: (1) 415 338 0942
Email: mritter@sfsu.edu
Website: www.sfsu.edu

San Francisco State University (SFSU) is one of the nation's leading public urban universities. SFSU helps create and maintain an environment for learning that promotes respect for and appreciation of scholarship, freedom, human diversity and the cultural mosaic of the City of San Francisco. SFSU also provides a higher education for residents of the region and state, as well as the nation and world.

Robert Westwood Scholarship

Subjects: Arts, health, science and social services.
Purpose: To assist SFSU students who are living with HIV and plan to make a contribution in any field to communities affected by HIV.
Level of Study: Postgraduate
Type: Scholarship
Value: US$1,000
Frequency: Annual
Study Establishment: San Fransisco State University
Country of Study: United States of America
No. of awards offered: 2
Application Procedure: Applicants must submit a copy of the most recent SFSU academic transcript, along with a brief, typed essay discussing plans to incorporate academic work and degree at SFSU with service in the HIV community or in the area of HIV prevention.
Closing Date: May 7th
Additional Information: Applicants must submit a verification from the physician.

For further information contact:
Tel: 415 338 7339
Contact: Michael Ritter, Counseling and psychological services

SANSKRITI PRATISHTHAN

Head Office C-11 Qutab Institutional Area, New Delhi, 110-016, India
Tel: (91) 11 2696 3226, 2652 7077
Fax: (91) 11 2685 3383
Email: fellowships@sanskritifoundation.org
Website: www.sanskritifoundation.org

Sanskriti Pratishthan is a non-profit organization that was established in 1978. Sanskriti Pratishthan perceives its role as that of a catalyst, in revitalizing cultural sensitivity in contemporary times.

Kalakriti Fellowship in Indian Classical Dance

Subjects: Indian classical dance.
Purpose: To encourage young artists to develop their potential and enhance their skills through intensive practice and/or incorporating different facets of their art.
Eligibility: Open to Indian nationals in the age group of 25–40. The candidates should have at least 10 years of initial training in Indian classical dance. The Fellows would be required to have given at least 2–3 solo performances to his/her credit in recognized forums.
Level of Study: Professional development
Type: Fellowships
Value: Indian Rupees 50,000
Length of Study: 10 months
Frequency: Annual
Application Procedure: Applicants must send their 2 page curriculum vitae and a writeup of approx. 500 words, explaining their project. Full postal and telephone contact details together with any email id should be submitted to facilitate contact. Few samples of previous work, project or performances should be submitted. The names and contact addresses/telephones of 2 referees should also be sent.
Closing Date: June 30th
Funding: Foundation
Additional Information: The candidate should not be holding any other fellowship or working on any other project at the same time. Please see the website for further details http://www.sanskritifoundation.org/Kalakriti-Fellowship.htm.

Mani Mann Fellowship in Indian Classical Vocal Music

Subjects: Indian classical music.
Purpose: To encourage promising young artists to advance in their field. This fellowship will enable the recipient to have the resources and time to dedicate to the art.
Eligibility: Open to Indian nationals in the age group of 25–40. Applicants must hold a degree/diploma from a recognized university or institution in the field and/or the candidates should have at least 10 years of initial training in Indian classical music.
Level of Study: Professional development
Type: Fellowship
Value: Indian Rupees 1,00,000
Length of Study: 3–12 months
Frequency: Annual
Country of Study: India
Application Procedure: Candidates should send their 2 page curriculum vitae and a writeup of approximately 500 words explaining their project. Full postal and telephone contact details together with any email id should be submitted to facilitate contact. Few samples of previous work, project or performances should be submitted. The names and contact addresses/telephones of 2 referees should also be sent.
Closing Date: December 31st
Funding: Foundation
Additional Information: Please see the website for further details http://www.sanskritifoundation.org/Mani-Mann-Fellowship.htm.

Prabha Dutt Fellowship in Journalism

Subjects: Journalism.
Purpose: To encourage young mid-career women journalists to develop their potential by pursuing meaningful projects without having to work under the pressures of short deadlines.
Eligibility: Open to women candidates who are Indian nationals and between 25 and 40 years of age. It is exclusively for print journalists.
Level of Study: Research
Type: Fellowships
Value: Indian Rupees 1,00,000
Length of Study: 10 months
Frequency: Annual
Country of Study: India
Application Procedure: Applicants must send a two page curriculum vitae and a write-up of about 250–300 words explaining their project. Full postal and telephone contact details together with any email Id should be submitted to facilitate contact, 5 samples of work published should be submitted. The names and contact addresses/telephones of 2 referees should also be sent.
Closing Date: August 31st
Funding: Foundation
Additional Information: The candidate should not be holding another fellowship or working on any other project at the same time. Please see the website for further details http://www.sanskritifoundation.org/prabha-dutt-fellowship.htm.

SAVOY FOUNDATION

230 Foch Street, St Jean Sur Richelieu, Quebec, QC, J3B 2B2, Canada
Tel: (1) 450 358 9779
Fax: (1) 450 346 1045
Email: epilepsy@savoy-foundation.ca
Website: www.savoy-foundation.ca
Contact: Vivian Downing, Assistant to Vice President/Secretary

The Savoy Foundation's main activity is to support and encourage research into epilepsy.

Savoy Foundation Postdoctoral and Clinical Research Fellowships

Subjects: Medical and behavioural science, as they relate to epilepsy.
Purpose: To support a full-time research project in the field of epilepsy.
Eligibility: Candidates must be scientists or medical specialists with a PhD or MD.
Level of Study: Postdoctorate, Postgraduate, Research
Type: Research grant
Value: Canadian $30,000
Length of Study: 1 year (non-renewable)
Frequency: Annual
Country of Study: Canada
No. of awards offered: Varies
Application Procedure: Applicants must contact the Foundation or visit the website for application forms and further information.
Closing Date: January 15th
Funding: Foundation, private
Contributor: The Savoy Foundation endowments
No. of awards given last year: 2
No. of applicants last year: 9

Savoy Foundation Research Grants

Subjects: Medical and behavioural science, as they relate to epilepsy.
Purpose: To support further research into epilepsy.
Eligibility: Only available to clinicians and established scientists.
Level of Study: Postdoctorate, Postgraduate, Research
Type: Research grant
Value: Up to Canadian $25,000
Frequency: Annual
Country of Study: Canada
No. of awards offered: Varies
Application Procedure: Applicants must contact the Foundation or visit the website for application forms and further information.
Closing Date: January 15th
Funding: Foundation, private
Contributor: The Savoy Foundation endowments
No. of awards given last year: 5
No. of applicants last year: 19
Additional Information: The grant is only available to Canadian citizens or for projects conducted in Canada.

Savoy Foundation Studentships

Subjects: Biomedicine, neurology and epileptology.
Purpose: To support training and research in a biomedical discipline, the health sciences or social sciences related to epilepsy.
Eligibility: Candidates must have a good university record, e.g. a BSc, MD or equivalent diploma and have ensured that a qualified researcher affiliated to a university or hospital will supervise his or her work. Concomitant registration in a graduate programme is encouraged. The awards are available to Canadian citizens or for projects conducted in Canada.
Level of Study: Postgraduate, Predoctorate, Doctorate
Type: Studentship
Value: The stipend will be Canadian $15,000 per year. An annual sum of Canadian $1,000 will be allocated to the laboratory or institution as additional support for the research project
Length of Study: 1–4 years
Frequency: Annual
Country of Study: Canada
No. of awards offered: Varies
Application Procedure: Applicants must contact the Foundation or visit the website for application forms and further information.
Closing Date: January 15th
Funding: Foundation, private
Contributor: The Savoy Foundation endowments
No. of awards given last year: 6
No. of applicants last year: 24

SCHOOL OF ORIENTAL AND AFRICAN STUDIES (SOAS)

University of London, Thornhaugh Street, Russell Square, London, WC1H 0XG, England
Tel: (44) 20 7637 2388
Fax: (44) 20 7074 5089
Email: scholarships@soas.ac.uk
Website: www.soas.ac.uk
Contact: Miss Alicia Sales, Scholarships Officer, Registry

The School of Oriental and African Studies (SOAS) regards its role as advancing the knowledge and understanding of the cultures and societies of Asia and Africa and of the School's academic disciplines through high-quality teaching and research.

A K S Postgraduate Bursary in Korean Studies

Subjects: MA Korean studies, MA Korean literature, MPhil/PhD Korean Studies Research, MA History of Art.
Eligibility: Open to UK/EU and overseas applicants. See the website for details.
Level of Study: Postgraduate
Type: Bursary
Value: Up to £6,000 towards tuition fees
Length of Study: 1 year
Frequency: Annual
Study Establishment: SOAS
Country of Study: United Kingdom
No. of awards offered: 2
Application Procedure: See website www.soas.ac.uk/scholarships for details.
Closing Date: May 24th
Funding: Private
Contributor: Academy of Korean Studies
No. of awards given last year: 1
Additional Information: If you have any questions about the bursary application, please contact the Scholarships Officer.

For further information contact:

Email: ak49@soas.ac.uk
Website: http://www.soas.ac.uk/registry/scholarships/aks-postgraduate-bursary.html
Contact: Dr Anders Karlsson

Ahmad Mustafa Abu-Hakima Scholarship

Subjects: History of the modern Arab world.
Purpose: To offset tuition fees for a student taking a full-time master's programme which includes studying the history of the modern Arab world.
Eligibility: Open to UK/EU and overseas applicants undertaking full-time taught masters programme.
Level of Study: Postgraduate
Type: Scholarship
Value: £2,000
Length of Study: 1 year
Frequency: Annual
Study Establishment: SOAS
Country of Study: United Kingdom
No. of awards offered: 1
Application Procedure: See website www.soas.ac.uk/scholarships for details.
Closing Date: March 22nd
Funding: Private
No. of awards given last year: 1
Additional Information: Please see the website for further details http://www.soas.ac.uk/registry/scholarships/the-ahmad-mustafa-abu-hakima-bursary.html.

AHRC Studentships

Subjects: History, art, Asian languages and cultures, linguistics, middle east and African, music and religious studies.
Purpose: To support taught masters and research students.
Eligibility: Open to home and EU students.
Level of Study: Doctorate, Postgraduate
Type: Scholarship
Value: Maintenance plus approved tuition fees
Study Establishment: SOAS
Country of Study: England
Application Procedure: Scholarship application should be made on the appropriate scholarship application form which is available for download.
Closing Date: January 31st
Funding: Government
Additional Information: There are two types of studentship awards – a full studentship award for UK residents and a fees-only studentship award for EU residents. Please see the website for further details http://www.soas.ac.uk/registry/scholarships/ahrc-block-grant-partnership-studentships-bgp.html.

Bernard Buckman Scholarship

Subjects: MA in Chinese studies
Purpose: The scholarship is intended to off-set the tuition fees at the UK/EU rate.
Eligibility: Open to those candidates who qualify to pay for home or European Union tuition fees. Applicants must possess a good Honours Degree from a United Kingdom university or its equivalent.
Level of Study: Postgraduate
Type: Scholarship
Value: Home or European Union postgraduate fee
Length of Study: 1 year
Frequency: Annual
Study Establishment: SOAS
Country of Study: United Kingdom
No. of awards offered: 1
Application Procedure: You can apply for this scholarship via the online scholarship application form.
Closing Date: March 22nd
Funding: Private
No. of awards given last year: 1
No. of applicants last year: 7
Additional Information: For enquiries, please contact Scholarships Officer.

For further information contact:

Tel: 0 20 7074 5094/5091
Email: scholarships@soas.ac.uk
Website: http://www.soas.ac.uk/registry/scholarships/bernard-buckman-scholarship.html

FELIX Scholarship

Subjects: Oriental and African studies in archeology, area studies, economics, ethnomusicology, history, law, languages, linguistics, phonetics, politics, religious study, social anthropology and development studies.

Purpose: To support first class Indian students commencing a Master's programme or researching for a Doctoral degree at the School of Oriental and African Studies.
Eligibility: Open to applicants of any full-time taught Master's or MPhil/PhD programme, under 30 years of age, able to demonstrate financial need and would return to work in their home country after completion of studies.
Level of Study: Doctorate, Postgraduate
Type: Scholarship
Value: £12,316 per year plus tuition fees
Length of Study: 1–3 years
Frequency: Annual
Study Establishment: SOAS
Country of Study: United Kingdom
No. of awards offered: 7
Application Procedure: A Felix Scholarship application form is available for download from the download box at the top right or can be obtained from Scholarships Officer.
Closing Date: January 31st
Funding: Private
Contributor: Felix Scholarship Trust
No. of awards given last year: 6
Additional Information: One award is made each year to a non-Indian student from a developing country who demonstrates academic excellence and financial need. Please see the website for further details http://www.soas.ac.uk/registry/scholarships/felix-scholarships.html.

For further information contact:

Tel: 0 20 7074 5094/5091
Email: scholarships@soas.ac.uk

HSBC SOAS Scholarships

Subjects: Sinology or Chinese literature.
Purpose: To support UK or EU fee payers commencing a full-time master's course in Sinology or Chinese literature.
Eligibility: Applicants must possess or be about to complete a good honours degree, preferably first class, from a UK institution or overseas equivalent.
Level of Study: Postgraduate
Type: Scholarship
Value: £16,650 plus tuition fees at the home EU rate
Length of Study: 1 year
Frequency: Annual
Study Establishment: SOAS
Country of Study: United Kingdom
No. of awards offered: 2
Application Procedure: You can apply for this scholarship via the online scholarship application form. For enquiries, please contact Scholarships Officer.
Closing Date: March 22nd
Funding: Trusts
Contributor: HSBC educational trust
No. of awards given last year: 2

For further information contact:

Tel: 0 20 7074 5094/5091
Email: scholarships@soas.ac.uk
Website: http://www.soas.ac.uk/registry/scholarships/hsbc-soas-scholarships.html

Ouseley Memorial Scholarship

Subjects: Any programme which involves research requires the use of any Middle Eastern or Asian language.
Purpose: To encourage the study of Arabic, Persian, Hindustani and other Oriental languages
Eligibility: Open to UK/EU and overseas applicants. Applicants must refer to the organization website for detailed information.
Level of Study: Doctorate, Postgraduate
Type: Scholarship
Value: £6,000 for 1 year only
Frequency: Annual
Study Establishment: SOAS
Country of Study: United Kingdom
No. of awards offered: 1
Application Procedure: See the website.
Closing Date: January 31st
Funding: Private
No. of awards given last year: 1
No. of applicants last year: 20

For further information contact:

Tel: 0 20 7074 5094/5091
Email: scholarships@soas.ac.uk
Website: http://www.soas.ac.uk/registry/scholarships/ouseley-memorial-scholarship.html

SOAS Master's Scholarship

Subjects: A variety of taught Master's programmes.
Purpose: To provide financial assistance to study for a full-time taught Master's programme.
Eligibility: Open to UK/EU and overseas applicants who possess a First class Honours degree or equivalent.
Level of Study: Postgraduate
Type: Scholarship
Value: Approx. UK£15,000
Length of Study: 1 year, non-renewable
Frequency: Annual
Study Establishment: SOAS
Country of Study: United Kingdom
No. of awards offered: 11
Application Procedure: See the website.
Closing Date: March 22nd
Funding: Government
No. of awards given last year: 11
No. of applicants last year: 350

For further information contact:

Email: scholarships@soas.ac.uk
Website: http://www.soas.ac.uk/registry/scholarships/soas-masters-scholarships—faculty-of-arts-humanities.h

SOAS Research Scholarship

Subjects: The languages and cultures of Africa, East Asia, Near and Middle East, South Asia and South- East Asia, focusing on anthropology and sociology, art and archaeology, development studies, economics, ethnomusicology, financial and management studies, history, law, linguistics, political studies and the study of religions.
Purpose: To support full-time research study at SOAS.
Eligibility: Applicants must UK/EU and overseas and must possess or expect to be awarded a distinction in their Master's degree from a United Kingdom university or its equivalent.
Level of Study: Doctorate
Type: Scholarship
Value: £12,790
Length of Study: 3 years
Frequency: Annual
Study Establishment: SOAS
Country of Study: United Kingdom
No. of awards offered: 4
Application Procedure: Applicants must complete and submit an application form that can be downloaded from the website.
Closing Date: January 31st
Funding: Government
No. of awards given last year: 4
Additional Information: For enquiries, please contact Scholarships Officer.

For further information contact:

Email: scholarships@soas.ac.uk
Website: http://www.soas.ac.uk/registry/scholarships/soas-research-student-fellowship.html

Sochon Foundation Scholarship

Subjects: MA Korean studies, MA Korean literature, MPhil/PhD Korean studies research and MA History of Art
Purpose: For a student undertaking a full-time postgraduate programme in Korean studies.

Eligibility: Open to UK/EU and overseas applicants.
Level of Study: Postgraduate
Type: Scholarship
Value: UK£7,000
Length of Study: 1 year
Frequency: Annual
Study Establishment: SOAS
Country of Study: United Kingdom
Application Procedure: See the website.
Closing Date: May 24th
Funding: Foundation
No. of awards given last year: 1

For further information contact:

Email: scholarships@soas.ac.uk
Website: http://www.soas.ac.uk/registry/scholarships/sochon-foundation-scholarship.html

William Ross Murray Scholarship

Subjects: LLM.
Purpose: To support a student of high academic achievement from a developing country unable to pay overseas tuition fees and attending the full-time LLM degree at SOAS.
Eligibility: Applicants from a developing country who must have a high level of academic achievements preferably first class, from a UK institution or overseas equivalent.
Level of Study: Postgraduate
Type: Scholarship
Value: Overseas tuition fees. Free accommodation at International Student House and food vouchers
Length of Study: 1 year
Frequency: Annual
Study Establishment: SOAS
Country of Study: United Kingdom
No. of awards offered: 1
Application Procedure: See the website.
Closing Date: March 22nd
Funding: Foundation
No. of awards given last year: 1

For further information contact:

Email: scholarships@soas.ac.uk
Website: http://www.soas.ac.uk/registry/scholarships/william-ross-murray-scholarship.html

SCIENCE AND TECHNOLOGY FACILITIES COUNCIL (STFC)

Polaris House, North Star Avenue, Wiltshire, Swindon, SN2 1SZ, England
Tel: (44) 01793 442 000
Fax: (44) 01793 442 002
Website: www.stfc.ac.uk

STFC is keeping the UK at the forefront of international science and tackling some of the most significant challenges facing society such as meeting our future energy needs, monitoring and understanding climate change, and global security. The Council has a broad science portfolio and works with the academic and industrial communities to share its expertise in materials science, space and ground-based astronomy technologies, laser science, microelectronics, wafer scale manufacturing, particle and nuclear physics, alternative energy production, radio communications and radar.

PPARC Advanced Fellowships

Subjects: Particle physics, astronomy and solar system science.
Level of Study: Postdoctorate
Type: Fellowship
Value: £35,000 (approx. Canadian $72,150).
Frequency: Annual
Country of Study: United Kingdom
No. of awards offered: 12
Application Procedure: Applications for a PPARC Fellowship must be submitted using the Je-S system.
Closing Date: October 12th
Additional Information: You may undertake up to a maximum of six hours teaching, including preparation each working week.

For further information contact:

Email: Clare.Heseltine@pparc.ac.uk
Contact: Clare Heseltine

PPARC CASE Studentship

Subjects: Science, engineering.
Purpose: To give promising students experience outside a purely academic environment.
Eligibility: Advice on eligibility should be sought from the Registrar's Office.
Level of Study: Postgraduate
Type: Studentship
Value: Check with the website.
Length of Study: 3.5 years
Frequency: Annual
Country of Study: United Kingdom
Application Procedure: Contact the PPARC.
Closing Date: October 2nd
Funding: Commercial

PPARC CASE-Plus Studentship

Subjects: Science and engineering.
Purpose: To help students become more effective in promoting technology transfer.
Eligibility: Advice on eligibility should be sought from the Registrar's Office.
Level of Study: Postgraduate
Type: Studentship
Value: Up to £14,250
Length of Study: 3.5 years
Frequency: Annual
Country of Study: United Kingdom
Application Procedure: Contact the PPARC.
Closing Date: September 30th (Check with the website)
Funding: Commercial

PPARC Daphne Jackson Fellowships

Subjects: Particle physics, particle astrophysics, solar system science and astronomy.
Purpose: To enable high-level engineers and scientists to return to their professions after a career break for family or other reasons.
Eligibility: Open to promising engineers and scientists who have taken a career break.
Level of Study: Postdoctorate, Research
Type: Fellowship
Value: Dependent on age and experience
Length of Study: 2 years
Frequency: Annual
Study Establishment: Any academic institution acceptable to the PPARC
Country of Study: United Kingdom
No. of awards offered: Varies
Application Procedure: Applicants must contact Jennifer Woolley, Trust Director, or Sue Smith, Fellowship Administrator, for application forms and further information.
Closing Date: Please write for details.
Funding: Government
Additional Information: The Daphne Jackson Fellowship is also administered by the BBSRC and the EPSRC.

For further information contact:

The Daphne Jackson Trust, Department of Physics, University of Surrey, Guildford, Surrey, GU2 7XH, England
Tel: (44) 14 8368 9166

PPARC Gemini Studentship

Subjects: Astronomy.
Purpose: To enable promising South-American students to pursue a PhD in Great Britain.
Eligibility: Open to students from Argentina, Brazil or Chile only.

Level of Study: Doctorate
Type: Fellowship
Value: Agreed tuition costs and a maintenance allowance
Length of Study: 3 years
Frequency: Annual
Country of Study: United Kingdom
No. of awards offered: 3
Application Procedure: See website.
Closing Date: March 31st
Additional Information: Please check the website for further detailshttp://www.prospects.ac.uk/funding_award_details.htm?id=569.

PPARC Postdoctoral Fellowships

Subjects: Particle physics, astronomy and solar system science.
Level of Study: Postgraduate
Type: Fellowship
Value: All agreed salary, travel and subsistence, equipment, additional costs, and UK£1,500 stipend
Frequency: Annual
Country of Study: United Kingdom
No. of awards offered: 12
Application Procedure: Applications for a PPARC Fellowship must be submitted using the Je-S system.
Closing Date: October 15th
Additional Information: You may undertake up to a maximum of six hours teaching, including preparation each working week. In addition, please read the STFC fellowships general rules and regulations, as well as the rules and regulations specific to the STFC Postdoctoral Fellowships scheme.

PPARC Postgraduate Studentships

Subjects: Particle physics, particle astrophysics, solar system science and astronomy.
Eligibility: Open to postgraduates from the United Kingdom and European Union countries.
Level of Study: Postgraduate
Type: Studentship
Value: Stipend (excluding fees only students), approved fees, Research Training Support Grant, Conference and UK fieldwork element, Fieldwork expenses, Long Term Attachments, Other Allowances (where applicable).
Length of Study: Up to 3 years
Frequency: Annual
Study Establishment: Any academic institution that is acceptable to the PPARC
Country of Study: United Kingdom
No. of awards offered: Approx. 185
Application Procedure: Applicants must refer to the PPARC website for application information.
Closing Date: March 31st.
Funding: Government
Additional Information: Further information is available on request by emailing studentships@pparc.ac.uk

PPARC Spanish (IAC) Studentship

Subjects: Astronomy.
Purpose: To support students of the Instituto de Astrofiscia de Canarias (IAC), Tenerite, Spain pursuing a PhD in Great Britain.
Eligibility: Open to Spanish students of the Instituto de Astrofiscia de Canarias (IAC) only.
Level of Study: Doctorate
Type: Studentship
Value: Agreed tuition costs and a maintenance allowance
Length of Study: 3 years
Frequency: Annual
Country of Study: United Kingdom
No. of awards offered: 2
Application Procedure: See website.
Closing Date: March 31st
Additional Information: Please check with the website for further details.

SCIENCE FOUNDATION IRELAND

Wilton Park House, Wilton Place, Dublin 2, Ireland
Tel: (353) 1 6073200
Fax: (353) 1 607 3201
Email: info@sfi.ie
Website: www.sfi.ie
Contact: William C Harris, Director

SFI Investigator Programme Grants

Subjects: Biotechnology and information and communications technology (ICT).
Purpose: To support fields of science and engineering that underpin biotechnology and ICT.
Eligibility: Applicants must be distinguished researchers in biotechnology and ICT.
Level of Study: Research
Type: Grant
Value: €100,000–500,000 per year
Length of Study: 3–5 years
Frequency: Annual
Application Procedure: Application forms and support notes can be downloaded from the website.
Closing Date: Check with website
Funding: Foundation
Contributor: Science Foundation Ireland
Additional Information: Queries on the SESAME management system should first be directed to your Institute's Research Office. If still unresolved, queries should be emailed to sesame@sfi.ie. Please be aware that response time for queries may be up to 5 working days. For Programme related queries please contact via email.

For further information contact:

Email: investigators@sfi.ie
Website: http://www.sfi.ie/funding/funding-calls/closed-calls/sfi-investigators-programme-2012/

SCOTTISH RITE CHARITABLE FOUNDATION OF CANADA (SRFC)

4 Queen Street South, Hamilton, ON, L8P 3R3, Canada
Tel: (1) 905 522 0033
Fax: (1) 905 522 3716
Email: info@srcf.ca
Website: www.srcf.ca
Contact: Manager Information Services

The Scottish Rite Charitable Foundation of Canada (SRFC) is a private charitable foundation, funded by donations and bequests from the 26,000 members of the ancient and accepted SRFC. The SRFC labours for the benefit of all Canadians, regardless of race or creed. Over the years millions of dollars have been disbursed to assist dedicated researchers in a search for the causes and cure of intellectual impairment.

SRFC Graduate Student Research Awards

Subjects: Health and medical sciences.
Purpose: To support students registered in a Doctoral research programme focused on the physical-biological or social aspects of intellectual impairment.
Eligibility: Open to candidates who are enrolled in a Doctoral programme at a Canadian university or research hospital. Applicants must be citizens or a permanent resident of Canada.
Level of Study: Doctorate
Type: Research award
Value: Up to Canadian $10,000
Length of Study: 2 years
Frequency: Annual
Country of Study: Canada
Application Procedure: Application package consists of the application form and application guide. Application form is available online.
Closing Date: April 30th
Funding: Foundation

SRFC Major Research Grant for Biomedical Research into Intellectual Impairment

Subjects: Health and medical sciences.

Purpose: To support biomedical research into intellectual impairment.
Eligibility: Open to researchers who have or are offered at least a 3 year academic appointment at a Canadian university or research hospital. Applicants must be Canadian citizens or permanent resident.
Level of Study: Research
Type: Research grant
Value: Canadian $35,000
Length of Study: up to 3 years
Country of Study: Canada
No. of awards offered: 10
Application Procedure: Applicants must submit their application form and research proposal.
Closing Date: April 30th
Funding: Foundation
Additional Information: The focus of research should be on the causes and cure of the disease as opposed to the active treatment or palliative care.

SDA BOCCONI

School of Management - via Bocconi, 8, Milano, 20136, Italy
Tel: (39) 02 5836 6605/6606
Fax: (39) 02 5836 6638
Email: oriana.ghinato@sdabocconi.it
Website: www.sdabocconi.it/en
Contact: Oriana Ghinato, Secretary

SDA School of Management enjoys recognition as a leading management school at an international level. Its mission is to educate men and women to be ready to act anywhere in the world, using their knowledge and imagination.

MPM Partial Tuition Fee Waivers

Subjects: Master of Public Management (MPM) is described as an MBA but focused on the international public sector. MPM is designed to prepare individuals for careers in Governmaental organizations, international institutions, NGOs and private businesses working with the public sector. The MPM programme provides a learning environment for participants from all 5 continents and is taught entirely in English.
Purpose: To provide partial tuition fee waivers for deserving candidates.
Eligibility: The MPM is open to candidates from different backgrounds and age groups. However, candidates with a background/interest in political science or international affairs will get the most from the MPM. Work experience is preferred but not essential.
Level of Study: Postgraduate
Type: Scholarship
Length of Study: 1 year
Frequency: Annual
Study Establishment: SDA Bocconi, Bocconi University, Milan
Country of Study: Italy
Application Procedure: Applicants must send a completed application form which is available online at www.sdabocconi.it/mpm along with a curriculum vitae, transcripts of the GMAT/GRE/TOEFL (or other) scores, 2 reference letters and 4 passport size photographs, in a sealed envelope, to the course Secretary.
Closing Date: July (check with website)
Additional Information: MPM operates a rolling application process. Applications are assessed until the course is full. Candidates whose mother tongue is English or who have studied in institutions where the medium of instruction is English may request a waiver for the English language certificate such as TOEFL (or other).

For further information contact:

Room 115, SDA Bocconi, Via Bocconi, 8, Milan, 20136, Italy
Contact: Joanne Matthews, Secretary, MPM Master Division

SEMICONDUCTOR RESEARCH CORPORATION (SRC)

PO Box 12053, Research Triangle Park, NC, 27709 2053, United States of America
Tel: (1) 919 941 9400
Fax: (1) 919 941 9450
Email: students@src.org
Website: www.src.org
Contact: Sarah Jackson, Program Officer

The Semiconductor Research Corporation (SRC) is a consortium of about 60 semiconductor manufacturers and equipment makers. The SRC manages a research portfolio in major research universities throughout the world. At any given time it supports about 700 advanced degree students on contract research, 45 graduate Fellows and 15 Master's scholars. The SRC supports a pragmatic approach to developing student programmes and provides industry interactions and other opportunities for its students.

GRC Graduate Fellowship Program

Subjects: Electrical engineering, computer engineering, chemical engineering, mechanical engineering, materials science, physics and related areas.
Purpose: To support academically gifted doctoral students in research areas consistent with SRC goals.
Eligibility: Open to Doctorate students with academic excellence and depending on the candidate's contribution to the field. Recipients are required to be associated with an SRC-funded contract.
Level of Study: Doctorate
Type: Fellowship
Value: Please contact the organization
Length of Study: Up to 5 years or until completion of degree, whichever comes first
Frequency: Annual
Study Establishment: Universities having SRC-funded contracts. See src.org for details
Country of Study: United States of America
No. of awards offered: Varies
Application Procedure: Applicants must compute an application form. Applications are distribute through SRC-funded faculty in November and are due in early February. Application materials are also available at src.org.
Closing Date: February 15th
Funding: Commercial
Contributor: The semiconductor industry
No. of awards given last year: 9
No. of applicants last year: 100
Additional Information: Resources are available to assist qualified students in identifying suitable faculty with SRC-supported research. SRC contracts support precompetitive research in areas of interest to the semiconductor industry. Please see the website for further details http://www.src.org/student-center/fellowship/.

GRC Master's Scholarship Program

Subjects: Electrical engineering, computer engineering, chemical engineering, mechanical engineering, materials science, physics and related areas.
Purpose: To attract underrepresented minorities and women to disciplines of interest to the semiconductor industry.
Eligibility: Open to students having United States of America citizenship or permanent resident status. Students are also required to be from an underrepresented group, e.g. women, African American, Latino or Native American students must be pursuing or planning to pursue Master's research with an SRC-funded faculty.
Type: Scholarship
Value: Please contact the organization
Length of Study: 2 years or until completion of the Master's degree, whichever comes first
Frequency: Annual
Study Establishment: Universities having SRC-funded contracts. A list is available at the website
Country of Study: United States of America
No. of awards offered: Varies
Application Procedure: Applicants must complete an application form. Applications are distributed through SRC-funded faculty in November with a due date early in February. Applications are also distributed through other avenues outside the SRC community and applications are encouraged from non-SRC-funded colleges.
Closing Date: February 15th
Funding: Commercial
Contributor: The semiconductor industry
No. of awards given last year: 1
No. of applicants last year: 9
Additional Information: Recipients are required to be associated with an SRC-funded contract. Resources are available to assist

qualified students in identifying suitable faculty within SRC-funded universities. SRC contracts support precompetitive research in areas of interest to the semiconductor industry. Please see the website for further details http://www.src.org/student-center/fellowship/.

SEOUL NATIONAL UNIVERSITY

1 Gwanak-ro, Gwanak-gu, Seoul, 151 742, Korea
Tel: (82) 822 880 4447
Fax: (82) 822 880 4449
Email: snuadmit@snu.ac.kr
Website: www.useoul.edu
Contact: MBA Admissions Officer

Korea-Japan Cultural Association Scholarship

Subjects: Humanities and social-science.
Eligibility: Open to outstanding Japanese scholars receiving an education in Korea.
Level of Study: Postdoctorate, Postgraduate, Research
Type: Scholarship
Value: KRW 3,500,000
Length of Study: 1 academic year
Frequency: Annual
Study Establishment: Seoul National University
Country of Study: Korea
Application Procedure: Contact the Office of International Affairs.
Closing Date: March
Funding: Foundation
Contributor: Korea-Japan Cultural Association

For further information contact:

The Office of International Affairs
Tel: 2 880 8638
Fax: 2 880 8632
Email: sjlim@snu.ac.kr
Website: http://www.useoul.edu/apply/under/scholarships/external
Contact: Mr Sung Sub Yoon

Korean Government Scholarship

Subjects: Korean studies.
Purpose: To further international bilateral cultural agreement with Korea.
Eligibility: Applicants must be under 40 years of age as of September 1st.
Level of Study: Postdoctorate, Postgraduate, Research
Type: Scholarship
Value: All airfares, living expenses, tuition fees, research allowance, settling and repatriation allowance, language training expenses, dissertation publication costs and insurance
Frequency: Annual
Study Establishment: Seoul National University
Country of Study: Korea
Application Procedure: Contact the Korean Embassy in residing country or see website.
Closing Date: No deadline
Funding: Government

For further information contact:

Website: www.studyinkorea.go.kr

Overseas Korean Foundation Scholarship

Subjects: Korean studies.
Purpose: To support students with an outstanding academic record.
Eligibility: Preference is given to Korean students with majors related to Korean studies, in particular: language, literature, medicine, education or IT, which are beneficial to the development of Korea.
Level of Study: Postdoctorate, Postgraduate, Research
Type: Scholarship
Value: KRW 900,000
Length of Study: 1 year
Frequency: Annual
Application Procedure: Contact the Overseas Korean Foundation.
Closing Date: March–April

For further information contact:

Education Department, Overseas Korean Foundation, Seocho 2-dong, Seocho-gu, Seoul, (137-072)
Tel: 2 3415 0174
Fax: 2 3415 0118
Email: scholarship@okf.or.kr
Website: http://www.useoul.edu/apply/under/scholarships/external
Contact: Coordinator

SHELBY CULLOM DAVIS CENTER FOR HISTORICAL STUDIES

129 Dickinson Hall, Princeton University, Princeton, NJ, 08544-1017, United States of America
Tel: (1) 609 258 4997
Fax: (1) 609 258 5326
Email: jhoule@Princeton.EDU
Website: www.princeton.edu
Contact: Ms Jennifer Houle, The Manager

The Davis center for historical studies was founded in 1968 to assure the continuance of excellence in scholarship and the teaching of history at Princeton University.

Shelby Cullom Davis Center Research Projects, Research Fellowships

Subjects: History.
Purpose: To support research.
Eligibility: Applicants must have completed a PhD.
Level of Study: Postdoctorate
Type: Fellowship
Value: US$2,000 per semester
Length of Study: 1–2 semesters
Frequency: Annual
Study Establishment: Shelby Cullom Davis Center
Country of Study: United States of America
No. of awards offered: Varies
Application Procedure: Applications are available online.
Closing Date: December 1st
Funding: Private
No. of awards given last year: 7

SHORENSTEIN ASIA-PACIFIC RESEARCH CENTER (APARC)

Encina Hall, Room E301, 616 Serra Street, Stanford University, Stanford, CA, 94305-6055, United States of America
Tel: (1) 650 723 9741
Fax: (1) 650 725 2592
Email: sishi@stanford.edu
Website: www.aparc.stanford.edu

Shorenstein Asia-Pacific Research Center (APARC) is an important Stanford venue where faculty and students, visiting scholars and distinguished business and government leaders meet and exchange views on contemporary Asia and United States involvement in the region.

Shorenstein Fellowships in Contemporary Asia

Subjects: Contemporary political, economic and social change in the Asia-Pacific region.
Purpose: To financially support research and writing on Asia.
Eligibility: Open to candidates who have obtained a PhD.
Level of Study: Doctorate, Postdoctorate
Type: Fellowships
Value: A stipend rate of $50,000 plus $3,000 for research materials
Length of Study: 10 months
Frequency: Annual
Country of Study: Asia
No. of awards offered: 2
Application Procedure: Applicants should submit via email brief research statement (not to exceed five typed pages), which describes

the research and writing to be undertaken during the fellowship period as well as the proposed publishable product; Curriculum vitae and three letters of recommendation.
Closing Date: December 31st
Funding: Private
No. of awards given last year: 2
No. of applicants last year: 75+

For further information contact:

Tel: 650 723 2408
Email: shorensteinfellowships@stanford.edu
Website: http://aparc.stanford.edu/fellowships/shorenstein_fellowships_in_contemporary_asia
Contact: Victoria Kwong, Fellowship Coordinator

SIDNEY SUSSEX COLLEGE

Cambridge University, Sidney Street, Cambridge, CB2 3HU, England
Tel: (44) 1223 338800
Fax: (44) 1223 338884
Email: gradtutor@sid.cam.ac.uk
Website: www.sid.cam.ac.uk
Contact: Tutor for Graduate Students

Founded in 1596, Sidney Sussex College admits men and women as undergraduates and graduates. The college presently has 180 graduate students, including 100 working for the PhD degree. The college has excellent sporting, dramatic and musical facilities.

Evan Lewis Thomas Law Studentships

Subjects: Law and cognate subjects.
Purpose: To support students carrying out research or taking advanced courses.
Eligibility: There are no eligibility restrictions. Candidates must have shown proficiency in Law and Jurisprudence, normally by obtaining a university degree in Law by August 2010, and they must be or become candidates for the PhD Degree, the Diploma in Legal Studies, the Diploma in International Law, the MPhil Degree (1 year course) in Criminology, or the LL.M Degree. Students from other Cambridge Colleges may apply, but if successful they would be expected to transfer their membership to Sidney Sussex College. In the competition for the studentship, no preference will be given to candidates who nominate Sidney Sussex College as their college of first or second choice on their application form.
Level of Study: Doctorate, Postgraduate
Type: Bursary
Value: Between £1,000 and £3,000 per year, reduced to £1,000 if full funding is obtained from another source
Length of Study: 1–3 years
Study Establishment: The University of Cambridge
Country of Study: United Kingdom
No. of awards offered: Up to 5
Application Procedure: Sydney Sussex is implementing a new online application process. Please refer to the website www.sid.cam.ac.uk/postgrads/scholarships.
Closing Date: April 1st
Funding: Private
Contributor: Sidney Sussex College
Additional Information: For further information contact the Tutor for Graduate Students at gradtutor@sid.cam.ac.uk.

The Gledhill Research Studentship

Subjects: All subjects.
Purpose: To provide full support for research leading to a PhD degree.
Eligibility: Applicants must apply for a postgraduate place at the University of Cambridge. Students from other Cambridge colleges may apply, but if successful they would be expected to transfer their membership to Sidney Sussex College. In the competition for the studentship, no preference will be given to candidates who nominate Sidney Sussex as their college of first or second choice on their application form.
Level of Study: Doctorate
Type: Studentship
Value: £1,000 per year
Length of Study: 3 years
Frequency: Dependent on funds available
Study Establishment: The University of Cambridge
Country of Study: United Kingdom
No. of awards offered: 1
Application Procedure: Sydney Sussex is implementing a new online application process. Please see the college's website www.sid.cam.ac.uk/postgrads/scholarships for further details.
Closing Date: April 1st
Funding: Private
Contributor: Sidney Sussex College
Additional Information: For further information contact the Tutor for Graduate Students.

For further information contact:

Email: gradtutor@sid.cam.ac.uk.
Website: http://www.sid.cam.ac.uk/current/postgrads/scholarships/GledhillStudentship.html

SIGMA THETA TAU INTERNATIONAL

550 West North Street, Indianapolis, IN, 46202, United States of America
Tel: (1) 888 634 7575
Fax: (1) 317 634 8188
Email: research@stti.iupui.edu
Website: www.nursingsociety.org
Contact: Tonna M.Thomas, Grants Coordinator

Sigma Theta Tau International exists to promote the development, dissemination and utilization of nursing knowledge. It is committed to improving the health of people worldwide through increasing the scientific base of nursing practice. In support of this mission, the society advances nursing leadership and scholarship, and furthers the utilization of nursing research in healthcare delivery as well as in public policy.

Doris Bloch Research Award

Subjects: Nursing.
Purpose: To encourage qualified nurses to contribute to the advancement of nursing through research. Multidisciplinary and international research is encouraged.
Eligibility: Applicants must be a registered nurse with a current licence, must have received a Master's degree, must have submitted an application package, must be ready to start the research project and must have signed a Sigma Theta Tau International research agreement. Allocation of funds is based on the quality of the proposed research, the future promise of the applicant and the applicant's research budget. Applications from novice researchers who have received no other national research funds are encouraged and will receive preference for funding, other aspects being equal.
Level of Study: Postdoctorate, Predoctorate, Postgraduate, Doctorate, Master's prepared
Type: Research grant
Value: Up to US$5,000
Frequency: Annual
Country of Study: Any country
No. of awards offered: 1
Application Procedure: All applications must be submitted via the online submission system. A link to the submission system will be available mid-July.
Closing Date: December 1st
Funding: Private
Contributor: Sigma Theta Tau International
No. of awards given last year: 1
Additional Information: Please see the website for further details http://www.nursingsociety.org/Research/Grants/Pages/grant_bloch.aspx.

Rosemary Berkel Crisp Research Award

Subjects: Women's health, oncology and infant or child care.
Purpose: To support nursing research in the critical areas of women's health, oncology and pediatrics.
Eligibility: Open to registered nurses with a current licence who have a Master's or higher degree (those with baccalaureate degrees may

be co-investigators), have submitted a complete research application package, are ready to initiate the research project and are Sigma Theta Tau International members. Some preference is given to applicants residing in Illinois, Missouri, Arkansas, Kentucky and Tennessee. See website www.nursingsociety.org. Some preference is given to applicants residing in Illinois, Missouri, Arkansas, Kentucky or Tennessee.
Level of Study: Doctorate, Postdoctorate, Postgraduate, Predoctorate, Master's prepared
Type: Research award
Value: US$5,000
Frequency: Annual
Country of Study: Any country
No. of awards offered: 1
Application Procedure: All applications must be submitted via the online application system. See www.nursingsociety.org for information.
Closing Date: December 1st
Funding: Private
Contributor: The Harry L Crisp II and Rosemary Berkel Crisp Foundation to Sigma Theta Tau International's Research Endowment
No. of awards given last year: 1
Additional Information: The allocation of funds is based on a research project in the area of women's health, oncology or paediatrics that is ready for implementation, the quality of the proposed research, future potential of the application, appropriateness of the research budget and feasibility of the time frame. Please see the website for further details http://www.nursingsociety.org/Research/Grants/Pages/grant_crisp.aspx.

Sigma Theta Tau International Small Research Grants

Subjects: Nursing.
Purpose: To encourage nurses to contribute to the advancement of nursing through research.
Eligibility: Open to registered nurses with a current licence who have submitted a complete research application package, have a project ready for implementation, hold a Master's degree, are enrolled in a doctoral programme and have signed a Sigma Theta Tau International research agreement. See website www.nursingsociety.org
Level of Study: Doctorate, Postdoctorate, Postgraduate, Master's prepared
Type: Research grant
Value: Up to US$5,000
Frequency: Annual
Country of Study: Any country
No. of awards offered: 10–15
Application Procedure: All applications must be submitted via the online application system. See www.nursingsociety.org for information.
Closing Date: December 1st
Funding: Foundation, private
No. of awards given last year: 10–15
Additional Information: This grant has no specific focus; however, multidisciplinary, historical and international research is encouraged. The funding date is June 1st. Please see the website for further details http://www.nursingsociety.org/Research/Grants/Pages/small_grants.aspx.

Sigma Theta Tau International/Alpha Eta Collaborative Research Grant

Subjects: Nursing.
Purpose: To provide a research grant to Tau Lambda-at-Large chapter members for a new research project in Africa.
Eligibility: The applicant must be a member of the Tau Lambda-at-Large residing in Africa, must submit online a completed research application package and a signed research agreement, and should be ready to implement the research project when funding is received. The applicant should also submit a completed abstract to the Virginia Henderson International Nursing Research Library and credit grant research partners in all publications and presentations of the research, and a final report.
Level of Study: Doctorate, Postdoctorate, Postgraduate
Type: Research grant
Value: Up to US$2,000
Frequency: Annual
Country of Study: Africa
No. of awards offered: 1
Application Procedure: All applications must be submitted through the online submission system. See the website www.nursingsociety.org for information.
Closing Date: December 1st
Funding: Private
Contributor: Alpha Eta chapter of Sigma Theta Tau International
No. of awards given last year: 1
Additional Information: Please see the website for further details http://www.nursingsociety.org/Research/Grants/Pages/AlphaEta.aspx.

Sigma Theta Tau International/American Association of Critical Care Nurses

Subjects: Critical care nursing practice.
Purpose: To encourage qualified nurses to contribute to the advancement of nursing through critical care nursing practice research.
Eligibility: Open to registered nurses with a current licence who have received a Master's degree and submitted a grant proposal relevant to critical care nursing practice. Research must be related to critical care nursing practice. See website www.nursingsociety.org
Level of Study: Predoctorate, Doctorate, Postdoctorate, Postgraduate, Master's prepared
Type: Research grant
Value: Up to US$10,000
Frequency: Annual
Country of Study: Any country
No. of awards offered: 1
Application Procedure: For application please contact the American Association of Critical Care Nurses, Department of Research.
Closing Date: November 1st
Funding: Foundation, private
Contributor: The American Association of Critical Care Nurses and Sigma Theta Tau International
Additional Information: January 1st is the funding date.

For further information contact:

American Association of Critical Care Nurses, Department of Research, 101 Columbia, Aliso Viejo, CA, 92656-1491, United States of America
Tel: (1) 949 362 2000
Fax: (1) 949 362 2020
Email: www.aacn.org/grants
Website: http://www.nursingsociety.org/Research/Grants/Pages/grant_aacn.aspx

Sigma Theta Tau International/American Association of Diabetes Educators Grant

Subjects: Diabetes education and care.
Purpose: To encourage qualified nurses to contribute to the enhancement of and availability of quality through diabetes education and care through nursing research.
Eligibility: The applicant must be a registered nurse but team members may be from other disciplines. The principal investigator must also have received a Master's degree, and have the ability to complete the project in 1 year from the funding date. Preference will be given to Sigma Theta Tau International members, other qualifications being equal. The grant must be dedicated to diabetes education and care research.
Level of Study: Predoctorate, Doctorate, Postdoctorate, Postgraduate, Master's prepared
Type: Research grant
Value: Up to US$6,000
Frequency: Annual
Country of Study: Any country
No. of awards offered: 1
Application Procedure: For application please contact American Association of Diabetes Educators (AADE).
Closing Date: October 1st
Funding: Foundation, private
Contributor: The American Association of Diabetes Educators and Sigma Theta Tau International

Additional Information: January 1st is the funding date. Please check at website for more information.

For further information contact:

American Association of Diabetes Educators (AADE) Awards, 100 West Monroe Street, Suite 400, Chicago, IL, 60603, United States of America
Tel: (1) 312 424 2426
Fax: (1) 312 424 2427
Website: http://www.nursingsociety.org/Research/Grants/Pages/grant_aade.aspx

Sigma Theta Tau International/American Nurses Foundation Grant

Subjects: Any clinical topic.
Purpose: To encourage the research career development of nurses through the support of research conducted by beginning nurse researchers, or experienced nurse researchers who are entering a new field of study.
Eligibility: Open to registered nurses with a current licence and a Master's degree who have submitted a complete research application package, are ready to start the research project and have signed a Sigma Theta Tau International Research agreement. Preference will be given to Sigma Theta Tau International members, other qualifications being equal.
Level of Study: Postdoctorate, Predoctorate, Postgraduate, Doctorate, Master's prepared
Type: Research grant
Value: Up to US$7,500
Frequency: Annual
Country of Study: Any country
No. of awards offered: 1
Application Procedure: Applicants must write for an information booklet and application form. Proposals should be sent to the Sigma Theta Tau International headquarters in even-numbered years (for which applicants should use the Sigma Theta Tau International application), and to the American Nurses Foundation in odd-numbered years (for which applicants should use the American Nurses Foundation application).
Closing Date: May 1st
Funding: Foundation, private
Contributor: The American Nurses Foundation (ANF) and Sigma Theta Tau International
No. of awards given last year: 1
Additional Information: Allocation of funds is based on the quality of the proposed research, the future promise of the applicant and the applicant's research budget. October is the funding month. Please check at website for more information.

For further information contact:

8515 Georgia Avenue, Suite 400 West, Silver Spring, MD, United States of America
Tel: (1) 301 628 5227
Website: http://www.nursingsociety.org/Research/Grants/Pages/grant_anf.aspx

Sigma Theta Tau International/Association of Nurses in AIDS Care Grant

Subjects: HIV prevention, symptom management, promotion of self-care and adherence.
Purpose: To encourage research career development of nurses through support of clinically oriented HIV/AIDS research and increase the number of HIV studies being done by nurses.
Eligibility: Open to candidates who have obtained a master's degree and/or are enrolled in a doctoral programme, must be a registered nurse with current license. For further information please see the website.
Level of Study: Research
Type: Grant
Value: $2,500
Length of Study: 1 year
Frequency: Annual
No. of awards offered: 1
Application Procedure: All applications must be submitted via the online submission system. See website www.nursingsociety.org for information.
Closing Date: April 1st
Contributor: ANAC and Sigma Theta Tau International
No. of awards given last year: 1
Additional Information: Please see the website for further details http://www.nursingsociety.org/Research/Grants/Pages/anac_grant.aspx.

Sigma Theta Tau International/Association of Perioperative Registered Nurses Foundation Grant

Subjects: Perioperative nursing practice.
Purpose: To encourage nurses to conduct research related to perioperative nursing practice and contribute to the development of perioperative nursing science.
Eligibility: Applicants must be a registered nurse with a current license in the perioperative setting, or a registered nurse who demonstrates interest in or significant contributions to nursing practice. The principal investigator must have, as a minimum, a Master's degree in nursing. Applicants must submit a completed Association of Perioperative Registered Nurses (AORN) research application. Membership of either organization is acceptable, but not required.
Level of Study: Postdoctorate, Predoctorate, Doctorate, Postgraduate, Master's prepared
Type: Research grant
Value: US$5,000. Allocation of funds is based on the quality of the research, the future promise of the applicant and the applicant's research budget
Frequency: Annual
Country of Study: Any country
No. of awards offered: 1
Application Procedure: Applicants must write to the AORN for an application form and general instructions.
Closing Date: April 1st
Funding: Foundation, private
Contributor: The Association of Perioperative Registered Nurses and Sigma Theta Tau International
No. of awards given last year: 1
Additional Information: July is the funding month. Please check at website for more information.

For further information contact:

Association of Perioperative Registered Nurses (AORN), 2170 South Parker Road, Suite 300, Denver, CO, 80231-5711, United States of America
Tel: (1) 800 755 2676 ext. 277
Fax: (1) 303 750 2927
Email: sbeya@aorn.org
Website: http://www.nursingsociety.org/Research/Grants/Pages/grant_aorn.aspx

Sigma Theta Tau International/Canadian Nurses Foundation Grant

Subjects: Nursing care.
Purpose: To support research on nursing care issues and build nursing research capacity.
Eligibility: i) Must be a practicing Canadian registered nurse with a current license; ii) research must address at least one of the following nursing care practice priorities: supporting research that takes place in 'clinical' settings, where nurses provide care, including the community setting; supporting research that involves novice researchers; supporting research teams that are interdisciplinary; supporting research that is 'national', involving all provinces and research that involves under-resourced areas within all provinces; iii) be an active member of Sigma Theta Tau International; iv) submission of completed research application via STTI's online submission system must be completed by December 1st; v) highest reviewed submission will be forwarded to CNF for further funding consideration.
Level of Study: Research
Type: Grant
Value: Up to $5,000
No. of awards offered: 1

Application Procedure: All applications must be submitted via the online submission system. A link to the submission system will be available in mid-July.
Closing Date: December 1st
Contributor: Canadian Nurses Foundation and Sigma Theta Tau International
Additional Information: Applicants must be a practicing Canadian-registered nurse with a current license to apply. Please see the website for further details http://www.nursingsociety.org/Research/Grants/Pages/grant_cnf.aspx.

Sigma Theta Tau International/Emergency Nurses Association Foundation Grant

Subjects: Topics relating to the specialized practice of emergency nursing. All relevant subjects will be considered, although priority will be given to studies that relate to the Association's Research Initiatives, which include, but are not limited to, mechanisms to assure effective, efficient and quality emergency nursing care delivery systems, factors affecting healthcare cost, productivity and market forces to emergency services, ways to enhance health promotion and injury prevention, and mechanisms to assure quality and cost-effective educational programmes for emergency nursing.
Purpose: To encourage nursing research that will advance the specialized practice of emergency nursing.
Eligibility: Applicants must be a registered nurse, but team members may be from other disciplines. Applicants must have a Master's degree, submit a complete application with signed research agreement and be ready to or have already started the research project. See website www.nursingsociety.org
Level of Study: Doctorate, Postdoctorate, Postgraduate, Predoctorate, Master's prepared
Type: Research grant
Value: Up to US$6,000
Frequency: Annual
Country of Study: Any country
No. of awards offered: 1
Application Procedure: For application please contact Emergency Nurses Association (ENA) Foundation.
Closing Date: March 1st
Funding: Foundation, private
Contributor: The Emergency Nurses Association Foundation and Sigma Theta Tau International
No. of awards given last year: 1
Additional Information: July 1st is the funding date.

For further information contact:

Emergency Nurses Association (ENA) Foundation, 915 Lee Street, Des Plaines, IL, 60016-6569, United States of America
Tel: (1) 847 460 4100
Fax: (1) 847 460 4005
Website: http://www.nursingsociety.org/Research/Grants/Pages/grant_ena.aspx

Sigma Theta Tau International/Hospice and Palliative Nurses Foundation Grant

Subjects: Hospice and palliative care nursing.
Purpose: To encourage qualified nurses to contribute to the advancement of nursing care through research.
Eligibility: Open to candidates who have obtained a Master's or Doctoral degree or are enrolled in a doctoral programme. The candidate must be a registered nurse with current license and must sign a research grant agreement. For further information please see the website.
Level of Study: Doctorate, Postdoctorate, Postgraduate, Predoctorate, Master's prepared
Type: Research grant
Value: Up to US$10,000
Frequency: Annual
No. of awards offered: 1
Application Procedure: All applications must be submitted via the online submission system. See www.nursingsociety.org for further information.
Closing Date: April 1st
Funding: Foundation, private
Contributor: Sigma Theta Tau International and the Hospice and Palliative Nurses Foundation
No. of awards given last year: 1
Additional Information: The funding date is August 1st. Please see the website for further details http://www.nursingsociety.org/Research/Grants/Pages/grant_hpna.aspx.

Sigma Theta Tau International/Joan K. Stout, RN, Research Grant

Subjects: Nursing.
Purpose: To advance ongoing evidence-based study by nurse researchers on the impact of the practice of simulation education in schools of nursing and clinical care settings.
Eligibility: The applicant should be a registered nurse with a current license, have a Master's degree or be enrolled in a doctoral program. The applicant should be ready to implement the rearch project when funding is received, complete the project within 1 year of funding, submit a completed abstract to the Virginia Henderson International Nursing Research Library and a final report to Sigma Theta Tau International.
Level of Study: Postgraduate, Predoctorate
Type: Research grant
Value: Up to $5,000
Frequency: Annual
Country of Study: Any country
No. of awards offered: 1
Application Procedure: A completed research application must be submitted through the online submission system. See www.nursingsociety.org for information.
Closing Date: July 1st
Funding: Foundation
Contributor: Sigma Theta Tau International Foundation
No. of awards given last year: 1
Additional Information: Please see the website for further details http://www.nursingsociety.org/Research/Grants/Pages/JoanKStout.aspx.

Sigma Theta Tau International/Midwest Nursing Research Society Research Grant

Subjects: Multi-disciplinary, historical and international research.
Purpose: To encourage qualified nurses to contribute to the advancement of nursing through research.
Eligibility: Open to candidates who have obtained a master's degree and/or are enrolled in a doctoral programme. The candidate must also be a registered nurse with current license and be a Midwest Nursing Research Society and a honor society member in good standing. For further information please see the website.
Level of Study: Doctorate, Postdoctorate, Postgraduate, Predoctorate, Master's prepared
Type: Research grant
Value: Up to US$2,500
Frequency: Annual
No. of awards offered: 1
Application Procedure: All applications must be submitted via the online submission system. See www.nursingsociety.org for information.
Closing Date: April 1st
Funding: Foundation, private
Contributor: Sigma Theta Tau International and the Midwest Nursing Research Society
No. of awards given last year: 1
Additional Information: The funding date is August 1st. Please see the website for further details http://www.nursingsociety.org/Research/Grants/Pages/grant_mnrs.aspx.

Sigma Theta Tau International/National League for Nursing Grant (NLN)

Subjects: Nursing.
Purpose: To support research that advances the science of nursing education and learning through the use of technology in dissemination of knowledge.
Eligibility: The applicant must be a registered nurse with a current license, hold a master's or doctoral degree or be enrolled in a doctoral program. The applicant must complete a project within one year of funding and must sign a Research Grant Agreement. Preference will

be given to Sigma Theta Tau International and National League for Nursing members.
Level of Study: Doctorate, Postdoctorate, Postgraduate, Predoctorate
Type: Grant
Value: Up to US$5,000
Country of Study: Any country
No. of awards offered: 1
Application Procedure: All applications must be submitted via the online submission system. See www.nursingsociety.org for information.
Closing Date: June 1st
Funding: Individuals
No. of awards given last year: 1
Additional Information: Funding date is November 1st. Please see the website for further details http://www.nursingsociety.org/Research/Grants/Pages/grant_nln.aspx.

Sigma Theta Tau International/Oncology Nursing Society Grant

Subjects: Clinical oncology.
Purpose: To simulate clinically related oncology nursing research.
Eligibility: Open to registered nurses who are actively involved in some aspect of cancer patient care, education or research and hold a Master's degree. Preference will be given to Sigma Theta Tau members, other qualifications being equal. See website www.nursingsociety.org
Level of Study: Postdoctorate, Postgraduate, Predoctorate, Doctorate, Master's prepared
Type: Research grant
Value: Up to US$25,000
Frequency: Annual
Country of Study: Any country
No. of awards offered: 1
Application Procedure: Applicants must send proposals for this grant to the Oncology Nursing Society.
Closing Date: October 1st
Funding: Foundation, private
Contributor: The Oncology Nursing Society and Sigma Theta Tau International
Additional Information: The funding date is March 15th.

For further information contact:

Oncology Nursing Society, 125 Enterprise Drive, RIDC Park West, Pittsburgh, PA, 15275-1214, United States of America
Tel: (1) 412 859 6100
Fax: (1) 877 369 5497
Website: http://www.nursingsociety.org/Research/Grants/Pages/grant_ons.aspx

Sigma Theta Tau International/Rehabilitation Nursing Foundation Grant

Subjects: Rehabilitation nursing.
Purpose: To encourage research related to rehabilitation nursing.
Eligibility: The applicant must be a registered nurse in rehabilitation or a registered nurse who demonstrates interest in and significantly contributes to rehabilitation nursing. Proposals that address the clinical practice, educational or administrative dimensions of rehabilitation nursing are requested. Quantitative and qualitative research projects will be accepted for review. The principal investigator must have a Master's degree in nursing and an ability to complete the project within 2 years of initial funding. See website www.nursingsociety.org
Level of Study: Doctorate, Postdoctorate, Postgraduate, Predoctorate, Master's prepared
Type: Research grant
Value: Up to US$4,500
Frequency: Annual
Country of Study: Any country
No. of awards offered: 1
Application Procedure: Applicants must write to the Rehabilitation Nursing Foundation for details.
Closing Date: March 1st
Funding: Foundation, private
Contributor: The Rehabilitation Nursing Foundation and Sigma Theta Tau International
Additional Information: The funding month is the following January.

For further information contact:

Rehabilitation Nursing Foundation, 4700 West Lake Avenue, Glenview, IL, 60025, United States of America
Tel: (1) 847 375 4710
Fax: (1) 847 375 4710
Email: info@rehabnurse.org
Website: http://www.nursingsociety.org/Research/Grants/Pages/grant_rna.aspx

Sigma Theta Tau International/Southern Nursing Research Society Grant

Subjects: Nursing.
Purpose: To encourage qualified nurses to contribute to the advancement of nursing through research. Proposals for pilot and/or development research may be submitted for this grant.
Eligibility: The applicant must be a registered nurse with a current license, hold at the minimum a Master's degree or its equivalent and/or be enrolled for a doctoral program, and be a member in good standing of both the STTI and SNRS. He or she must submit online a completed research application package and a signed research agreement, and should be ready to implement the research project when funding is received; however, the research may not be conducted as part of an academic degree. The applicant should also submit a completed abstract to STTI's Virginia Henderson International Nursing Research Library, an abstract for publication to SNRS, and credit grant research partners in all publications and presentations of the research, and a final report to STTI.
Level of Study: Postgraduate, Predoctorate
Type: Research grant
Value: Up to US$5,000
Frequency: Annual
Country of Study: Any country
No. of awards offered: 1
Application Procedure: All applications must be submitted via the online submisson system. See the website www.nursingsociety.org for information.
Closing Date: April 1st
Funding: Foundation
Contributor: Sigma Theta Tau International Foundation and Southern Nursing Research Society
No. of awards given last year: 1
Additional Information: Please see the website for further details http://www.nursingsociety.org/Research/Grants/Pages/stti_snrs.aspx.

Sigma Theta Tau International/The Council for the Advancement of Nursing Science

Subjects: Nursing.
Purpose: To encourage qualified nurses to improve the health of the world's people through research. Proposals for clinical, educational or historical research, including plans for disseminating the research findings, may be submitted for the grant.
Eligibility: Preference will be given to a Sigma Theta Tau International member who holds at the minimum a Master's degree or its equivalent. He or she must submit online a completed research application package and should be ready to implement the research project when funding is received. The applicant should also submit a completed abstract to STTI's Virginia Henderson International Nursing Research Library, a completed project abstract for presentation to the Sigma Theta Tau International or the Council for the Advancement of Nursing Science, and credit grant research partners in all publications and presentations of the research.
Level of Study: Postgraduate, Predoctorate
Type: Research grant
Value: Up to $5,000
Frequency: Annual
Country of Study: Any country
No. of awards offered: 1
Application Procedure: Applicants must submit applications via the online submission system. See the website www.nursingsociety.org for information.
Closing Date: July 1st

Funding: Foundation
Contributor: Sigma Theta Tau International Foundation and The Council for the Advancement of Nursing Science
No. of awards given last year: 1
Additional Information: Applicants must be a member in good standing of both STTI and CANS.

Sigma Theta Tau International/Virginia Henderson Clinical Research Grant

Subjects: Clinical research.
Purpose: To encourage the research career development of clinically based nurses through support of clinically orientated research.
Eligibility: Open to registered nurses actively involved in some aspect of healthcare delivery, education or research in a clinical setting, who are Theta Tau International members, hold a Master's degree in nursing or are enrolled in a doctoral programme. The allocation of funds is based on a research project ready for implementation, the quality of the proposed research, the future potential of the applicant, appropriateness of the research budget and feasibility of the time frame.
Level of Study: Doctorate, Postdoctorate, Postgraduate, Predoctorate, Master's prepared
Type: Research grant
Value: US$5,000
Frequency: Every 2 years
Country of Study: Any country
No. of awards offered: 1
Application Procedure: All applications must be submitted via the online submission system. See www.nursingsociety.org for information.
Closing Date: December 1st in odd-numbered years
Funding: Private
Contributor: The Virginia Henderson Clinical Research Endowment Fund
Additional Information: June is the funding month. Please see the website for further details http://www.nursingsociety.org/Research/Grants/Pages/grant_VHL.aspx.

Sigma Theta Tau International/Western Institute of Nursing Research Grant

Subjects: Nursing.
Purpose: To encourage qualified nurses to contribute to the advancement of nursing through research.
Eligibility: Open to a registered nurse with a current license, who has a master's degree and/or is enrolled in a doctoral programme. The applicant must be a member of Western Institute of Nursing and Sigma Theta Tau International.
Level of Study: Doctorate, Postdoctorate, Postgraduate, Predoctorate, Master's prepared
Type: Grant
Value: Up to US$2,500
Frequency: Annual
Country of Study: Any country
No. of awards offered: 1
Application Procedure: All applications must be submitted via the online submission system. See www.nursingsociety.org for information.
Closing Date: December 1st
Funding: Foundation
Contributor: Sigma Theta Tau International and the Western Institute of Nursing
Additional Information: The funding date is April 1st. Please see the website for further details http://www.nursingsociety.org/Research/Grants/Pages/grant_win.aspx.

SILVERHILL INSTITUTE OF ENVIRONMENTAL RESEARCH AND CONSERVATION

60 Adelaide Street East, Suite 501, Toronto, ON, M5C 3E4, Canada
Email: peter@silverhillinstitute.com
Website: www.silverhillinstitute.com
Contact: Dr Peter Homenuck, RPP

The Silverhill Institute of Environmental Research and Conservation was established as a charitable foundation in 2004. This foundation was established according to the Canada Corporations Act. It supports demonstration projects to develop, maintain and preserve wood lots. It also supports wetland protection and restoration as means for conservation and contributing to diversity and the protection of groundwater and surface water resources, and carries out research on environmental issues of the day.

Silverhill Institute of Environmental Research and Conservation Award

Subjects: Geography, environmental studies, ecology, biology, natural resources, planning, forestry and water resources.
Purpose: To financially support students for thesis/research projects that have an applied environmental or conservation focus and that have the potential to contribute in a direct way to society and/or provide specific community benefit.
Eligibility: Open only to graduate students for research (thesis or major paper) that is applied in nature and can be demonstrated to have community benefit or applicability.
Level of Study: Postgraduate, Research
Type: Grant
Value: Canadian $2,500 for the Summer season. Smaller awards of $500 are also available.
Frequency: Annual
Study Establishment: University of Manitoba
Country of Study: Canada
No. of awards offered: Varies
Application Procedure: Download the grant application form in PDF or WORD format and return the completed form by the due date.
Closing Date: February 28th
Additional Information: The grant will be paid out as follows: Canadian $2,500 upon selection and Canadian $1,000 upon receipt of a copy of the finished theses/product. Please see the website for further details.

SIMON FRASER UNIVERSITY

8888 University Drive, Burnaby, BC, V5A 1S6, Canada
Tel: (1) 778 782 3708
Fax: (1) 778 782 4920
Email: pchhina@sfu.ca
Website: www.business.sfu.ca
Contact: Ms Preet Virk, Manager, Donor Relations

Named after explorer Simon Fraser, SFU opened on September 9, 1965. Taking only 30 months to grow from the idea stage into an almost-completed campus with 2,500 students it was dubbed the "Instant University". The original campus has grown into three vibrant campuses in Burnaby, Vancouver and Surrey and SFU's reputation has grown into one of the innovative teaching, research, and community outreach.

Joseph-Armand Bombardier Canada Graduate Scholarships (CGS) Master's Program

Subjects: All subjects.
Purpose: To seeks the develop research skills and assist in the training of highly qualified personnel by supporting students in the social sciences and humanities.
Eligibility: Open to those pursuing a first Graduate Degree. Be a citizen or permanent resident of Canada.
Level of Study: Graduate
Type: Scholarship
Value: $17,500 per year
Length of Study: 1 year
Frequency: Annual
Application Procedure: See the website.
Closing Date: Early November
Additional Information: Please see the website for further details.

Methanex Graduate Scholarship in International Marketing

Subjects: International business and marketing.
Purpose: To recognize and reward an outstanding student pursuing an MBA at the Segal Graduate School of Business.

Eligibility: Open to a student with a demonstrated focus on international business and marketing.
Level of Study: Graduate
Type: Scholarship
Value: $5,000
Frequency: Annual
Study Establishment: Segal Graduate School of Business
Country of Study: Canada
No. of awards offered: 1
Application Procedure: Check website for further information. Applications are accepted in January.
Closing Date: January 30th
Additional Information: Please see the website for further details http://beedie.sfu.ca/mba/apply/scholarships.php.

NSERC Industrial Post-Graduate Scholarships (IPS)

Subjects: Science and engineering.
Purpose: To encourage scholars to consider research careers in industry.
Eligibility: Open to highly qualified science and engineering graduates.
Level of Study: Graduate
Type: Scholarship
Value: $15,000 per year for up to 2 years plus company contribution of $6,000 minimum per year
Length of Study: Up to 2 years
Frequency: Annual
Country of Study: Canada
Application Procedure: Check website for further details.
Closing Date: Apply at any time (check details in website)

For further information contact:

Office of the Dean of Graduate Studies
Email: dcoburn@sfu.ca
Website: http://www.sfu.ca/dean-gradstudies/blog/scholarships/NSERC-IPS.html
Contact: Deena Coburn, Director, Administrative Services

NSERC/MITACS Joint Industrial Postgraduate Scholarship (IPS)

Subjects: Mathematics.
Eligibility: Open to any graduate student in Simon Fraser University (SFU) using mathematics in their research.
Level of Study: Doctorate, Graduate, Postgraduate
Type: Scholarship
Value: $22,500
Length of Study: Up to 3 years
Frequency: Annual
Study Establishment: Simon Fraser University (SFU)
Country of Study: Canada
Application Procedure: Check website address for further details.

Peter Legge Graduate Volunteer Leadership Award in Business

Subjects: Business administration.
Purpose: To provide financial support for graduate students pursuing or entering a degree at the Segal Graduate School of Business.
Eligibility: Open to graduate students pursuing a degree at the Segal Graduate School of Business.
Level of Study: Graduate
Type: Award
Value: $4,100
Frequency: Annual
Study Establishment: Segal Graduate School of Business
Country of Study: Bulgaria
Application Procedure: Check the website for further details http://www.sfu.ca/dean-gradstudies/awards/privateawards/all-awards/legge.html.
Closing Date: May 31st
Additional Information: Candidates must include the application for Private Graduate Scholarship along with the application, resume, cover letter, and letter of reference.

Phi Theta Kappa International Summit Scholarships

Subjects: All subjects.
Eligibility: Open to Phi Theta Kappa members with a minimum 3.75 Grade Point Average and have completed 30 credit hours.
Level of Study: Graduate
Type: Scholarship
Value: Canadian $3,500
No. of awards offered: Up to 3
Application Procedure: Candidates must apply through the website and also mail a package containing paper application form, original or photocopied, a one-page summary of achievements and activities, a copy of recent college transcript, and a copy of Phi Theta Kappa certificate of membership.
Closing Date: April 30th (for fall term), September 30th (for spring term), or January 31st (for summer term)
Additional Information: Students applying from outside Vancouver's Lower Mainland area may be eligible for a travel allowance of $500–1,000. Part-time students and students with a previous bachelor degree are not eligible.

For further information contact:

Entrance Scholarship Program, Student Recruitment, Student Services, MBC 3200, Burnaby, BC, V5A 1S6, Canada
Website: http://students.sfu.ca/financialaid/entrance/intles/es-apreq.html

Scotiabank Graduate Scholarship for Women Entrepreneurs

Subjects: Business administration.
Purpose: To provide financial support for a female student enrolling for MBA.
Eligibility: Open to academically excellent female students enrolled full or part-time in an MBA program.
Level of Study: Graduate
Type: Scholarship
Value: $5,000
Frequency: Annual
Application Procedure: Check website for further details http://www.sfu.ca/dean-gradstudies/awards/privateawards/all-awards/scotiabank.html.
Closing Date: May 30th
Additional Information: Candidates must submit a cover letter outlining entrepreneurial experience and a letter of reference from a business colleague or faculty member.Preference will be given to a candidate who is, or has been, an entrepreneur, or who plans to study entrepreneurship as part of her degree.

SSHRC Doctoral Fellowships

Subjects: All subjects.
Purpose: To doctoral support in the humanities and social sciences.
Eligibility: Open to those who are currently in attendance at Simon Fraser University.
Level of Study: Postgraduate
Type: Fellowship
Value: $20,000–35,000 per year
Length of Study: 1–4 years
Frequency: Annual
Application Procedure: Check website for further details.
Closing Date: November 2nd
Additional Information: Applications are ranked within subject areas by an SFU committee and recommended applications sent to SSHRC, which announces final results in April or May.

For further information contact:

Website: http://www.sfu.ca/dean-gradstudies/blog/scholarships/SSHRC-doctoral-2012.html

SSHRC Joseph-Armand Bombardier CGS Doctoral Scholarships

Subjects: Humanities and social sciences.
Purpose: To provide doctoral support in the humanities and social sciences.
Eligibility: Open to Canadian citizens or permanent residents, who have completed a Master's Degree or at least 1 year of doctoral study and pursuing full time studies leading to a first PhD or its equivalent.
Level of Study: Postgraduate, Predoctorate

Type: Scholarship
Value: $35,000 per year
Length of Study: Up to 3 years
Frequency: Annual
Country of Study: Canada
Application Procedure: Candidates can avail information from the Office of the Dean of Graduate Studies, or check the website.
Closing Date: November 5th
Additional Information: The deadline for applications to the appropriate SFU department is approximately October 15th. Please see the website for further details.

TCG International Graduate Scholarship in Business Administration

Subjects: Business administration.
Purpose: To support for a student in the MBA program.
Eligibility: Open to academically excellent (CGPA>3.5) graduates. Preference will be given to a BC resident.
Level of Study: Graduate
Type: Scholarship
Value: $10,000
Frequency: Annual
No. of awards offered: 1
Application Procedure: Check website for further details. Applications are accepted in September.
Closing Date: June 30th
Additional Information: Please see the website for further details http://www.sfu.ca/dean-gradstudies/awards/privateawards/all-awards/tcg-international.html.

Trudeau Foundation Doctoral Scholarship

Subjects: Social sciences and humanities.
Purpose: To support doctoral students pursuing research in one or more of the four themes: human rights and social justice, responsible citizenship, Canada and the world, and humans and their natural environment.
Eligibility: Open to outstanding students who are in their first or second year of a PhD programme.
Level of Study: Doctorate
Type: Scholarship
Value: $40,000 stipend plus $20,000 travel allowance
Length of Study: Up to 3 years
Frequency: Annual
No. of awards offered: 15
Application Procedure: Check website for further details.
Closing Date: November 7th
Additional Information: Research must be in one of the four themes of the foundation. Please see the website for further details http://www.sfu.ca/dean-gradstudies/blog/scholarships/Trudeau-2012.html.

Vanier Canada Graduate Scholarships (CGS)

Subjects: All subjects.
Purpose: The CGS Program provides financial support to outstanding eligible students pursuing master's or doctoral studies.
Eligibility: Open to candidates who are eligible to apply for the NSERC or SSHRC annual competition.
Value: $50,000
Length of Study: Up to 3 years
Application Procedure: See the website for details.
Closing Date: September 26th
Contributor: NSERC and SSHRC
Additional Information: Applicants who apply for the NSERC or SSHRC annual competitions are also considered.

SIR HALLEY STEWART TRUST

22 Earith Road, Willingham, Cambridge, Cambridgeshire, CB24 5LS, England
Tel: (44) 19 5426 0707
Email: email@sirhalleystewart.org.uk
Website: www.sirhalleystewart.org.uk
Contact: Mrs S M West, Trust Administrator

The Sir Halley Stewart Trust has a Christian basis and is concerned with the development of body, mind and spirit, a just environment and international goodwill. The Trust aims to promote innovative research activities or developments. See www.sirhalley stewart.org.uk

Sir Halley Stewart Trust Grants

Subjects: Medical, social or religious projects within certain priority areas.
Purpose: To assist pioneering research and development
Eligibility: Not open to general appeals, building, capital, running costs or personal education including educational and travel costs.
Level of Study: Doctorate, Postgraduate, Research
Type: Grant
Value: Salaries and relevant costs for innovative and imaginative researchers at the beginning of their careers in the region of UK£15,000–20,000
Length of Study: 2–3 years
Frequency: Dependent on funds available
Study Establishment: A United Kingdom charitable institution, e.g. a hospital, laboratory, university department or charitable organization
Application Procedure: Applicants must contact the Trust for further details. There is no application form. See website: www.sirhalleystewart.org.uk for the application process. Hard signed copies required by post. Preliminary enquiries available by email/phone call
Closing Date: Applications are accepted at any time
Funding: Private
No. of awards given last year: 41
No. of applicants last year: 503
Additional Information: Further information is available from the Trust's office. The website contains the most up-to-date details. Country of study: mainly United Kingdom but some overseas, Africa (west) given priority.

SIR JOHN SOANE'S MUSEUM FOUNDATION

1040 First Avenue, No. 311, New York, NY, 10022, United States of America
Tel: (1) 212 223 2012
Fax: (1) 860 435 8019
Email: info@soanefoundation.com
Website: www.soanefoundation.com
Contact: Charles A Miller-III, Executive Director

The Sir John Soane's Museum Foundation assists the Museum in London to further Soane's commitment to educate and inspire the general and professional public in architecture and the fine and decorative arts. Programmes have attracted students, collectors, architects, decorators and arts enthusiasts since 1991 to its events, lectures, tours and dinners.

Sir John Soane's Museum Foundation Travelling Fellowship

Subjects: Art, architecture and the decorative arts.
Purpose: To enable scholars to pursue research projects related to the work of Sir John Soane's Museum and its collections.
Eligibility: Open to candidates enrolled in a graduate degree programme in a field appropriate to the Foundation's purpose.
Level of Study: Graduate, Postgraduate
Type: Fellowship
Value: US$5,000
Frequency: Annual
Study Establishment: The choice of the fellowship recipient
Country of Study: Usually the United States of America or the United Kingdom
No. of awards offered: 1
Application Procedure: Applicants must submit a formal proposal of not more than five pages describing the goal, scope and purpose of the research project, in addition to three letters of recommendation. An interview may be required.
Closing Date: March 1st
Funding: Foundation, private
Contributor: The Board of Directors and the Advisory Board
No. of awards given last year: 1
No. of applicants last year: 6
Additional Information: At the end of each research project the award recipient must submit a written documentation or a sketch book

on the progress of the research as outlined in the original proposal with respect to goal, scope and allocation of funds, and give a lecture on the research, arranged by the Foundation. The scholar usually spends some time at Sir John Soane's Museum at 13 Lincoln's Inn Fields, London, studying architectural drawings, models and paintings. Please see the website for further details http://www.soanefoundation.com/fellowship.html.

SIR RICHARD STAPLEY EDUCATIONAL TRUST

PO Box 839, Richmond, Surrey, TW9 3AL, England
Email: admin@stapleytrust.org
Website: www.stapleytrust.org
Contact: Dr N Jachec

The Sir Richard Stapley Educational Trust awards grants to graduates studying for higher degrees without subject restriction. The grants are to cover shortfall incurred by the payment of tuition fees.

Sir Richard Stapley Educational Trust Grants

Subjects: Medical, dental and veterinary science and higher degrees in other subjects.
Purpose: To support postgraduate study.
Eligibility: Open to graduates holding a First Class (Honours) degree or an Upper Second Class (Honours) degree and who are more than 24 years of age on October 1st of the proposed academic year. Students in receipt of a substantial award from local authorities, the NHS Executive, Industry, Research Councils, the British Academy or other similar public bodies will not normally receive a grant from the Trust. Courses not eligible include electives, diplomas, placements, professional training and intercalated degrees. The Trust does not support students for full-time PhD studies beyond a 3rd year. Applicants must already be resident in the United Kingdom at the time of application.
Level of Study: Postgraduate
Type: Grant
Value: UK£300–1,000
Length of Study: Grants are awarded for 1 full academic year in the first instance
Frequency: Annual
Study Establishment: Any appropriate university
Country of Study: United Kingdom
No. of awards offered: Dependent on availability of funds
Application Procedure: Electronic applications are available in early January. The trust will consider either the first 300 complete applications or all applications received on or before March 31st. Applicants will be notified in June.
Closing Date: March 31st, or first 300 applicants
Funding: Commercial, private
No. of awards given last year: 239
No. of applicants last year: 300
Additional Information: Grants are only paid upon receipt of official confirmation of participation in the course as well as confirmation that a financial shortfall still exists. All matters concerning application are communicated by email and letter only.

For further information contact:

Email: admin@stapleytrust.org
Website: http://www.stapleytrust.org/wp/applications/
Email: admin@stapleytrust.org
Contact: Nancy Jachec, Administrator

SIR ROBERT MENZIES MEMORIAL FOUNDATION

210 Clarendon Street, East Melbourne, VIC, 3002, Australia
Tel: (61) 3 9419 5699
Fax: (61) 3 9417 7049
Email: menzies@vicnet.net.au
Website: www.menziesfoundation.org.au
Contact: Ms S K Mackenzie, General Manager

The Sir Robert Menzies Memorial Foundation is a non-profit, non-political organisation established in 1979 to perpetuate and honour the memory of Sir Robert Menzies by promoting excellence in medical and health research, education and post graduate scholarship by Australians.

The Robert Gordon Menzies Scholarship to Harvard

Subjects: All subjects.
Purpose: To encourage students who have gained admission to a Harvard graduate school.
Eligibility: Open to candidates who are Australian citizens who have graduated from an Australian university.
Level of Study: Postgraduate
Type: Scholarship
Value: US$60,000
Frequency: Annual
Study Establishment: Harvard University
Country of Study: United States of America
No. of awards offered: 1
Application Procedure: A completed application form with original academic transcripts must be submitted.
Closing Date: February
Funding: Foundation
Contributor: The Harvard Club of Australia, the Menzies Foundation and the Australian National University
No. of awards given last year: 2
No. of applicants last year: 25

For further information contact:

Council and Boards Secretariat, 1.09 Chancelry Bld 10, The Australian National University, Canberra, ACT 0200, Australia
Website: http://www.menziesfoundation.org.au/scholarships/scholarships.html#harvard
Contact: Scholarship Administrator and Selection Committee Secretary

Sir Robert Menzies Memorial Research Scholarships in the Allied Health Sciences

Subjects: Occupational therapy, speech pathology, physiotherapy, psychology, nursing, optometry and physical education.
Purpose: To allow an outstanding applicant to carry out doctoral research work that is likely to improve the health of Australians.
Eligibility: Open to Australian citizens of at least 5 years standing. Applicants will generally have completed the first year of their PhD project.
Level of Study: Doctorate, Postgraduate
Type: Scholarship
Value: Australian $27,500 free of income tax
Length of Study: 2 years
Frequency: Annual
Study Establishment: A tertiary institute with appropriate facilities
Country of Study: Australia
No. of awards offered: 2
Application Procedure: Applicants must complete and submit an application form, academic transcripts and other documents. Further information is available on request.
Closing Date: June 30th
Funding: Private
Contributor: The Sir Robert Menzies Memorial Foundation
No. of awards given last year: 2
No. of applicants last year: 35
Additional Information: Please see the website for further details http://www.menziesfoundation.org.au/scholarships/scholarships.html#harvard.

Sir Robert Menzies Memorial Scholarships in Engineering

Subjects: Engineering.
Purpose: To enable Australian citizens to pursue postgraduate studies in the United Kingdom, generally leading to a higher degree.
Eligibility: Open to Australian citizens of 5 years of standing with at least an Upper Second Class (Honours) Degree in engineering at the time of application.
Level of Study: Postgraduate
Type: Scholarship
Value: Tuition fees, return airfare, examination and other compulsory fees, an allowance of UK£12,250 per year (if enrolled at a university

outside London) and UK£14,250 per year (if enrolled at a London university)
Length of Study: 1–3 years
Frequency: Annual
Study Establishment: Universities in UK
Country of Study: United Kingdom
No. of awards offered: 1
Application Procedure: Applicants must complete and submit an application form along with academic transcripts and other documents.
Closing Date: July 31st
Funding: Private
Contributor: The Sir Robert Menzies Memorial Trust in London, the Australian Department of Education, Employment and Workplace Relations and the Menzies Foundation in Australia
No. of awards given last year: 1
No. of applicants last year: 12
Additional Information: Please see the website for further details http://www.menziesfoundation.org.au/scholarships/scholarships.html#harvard.

Sir Robert Menzies Memorial Scholarships in Law

Subjects: Law.
Purpose: To enable Australian citizens to pursue postgraduate studies in the United Kingdom, generally leading to a higher degree.
Eligibility: Open to Australian citizens of 5 years of standing with at least an Upper Second Class (Honours) Degree in law at the time of application.
Level of Study: Postgraduate
Type: Scholarship
Value: Tuition fees, return airfare, examination and other compulsory fees, an allowance of UK£12,250 per year (if enrolled at a university outside London) and UK£14,250 per year (if enrolled at a London university)
Length of Study: 1–2 years
Frequency: Annual
Study Establishment: Universities in UK
Country of Study: United Kingdom
No. of awards offered: 1
Application Procedure: Applicants must complete and submit an application form along with academic transcripts and other documents. Check website for details.
Closing Date: August 31st
Funding: Private
Contributor: The Sir Robert Menzies Memorial Trust in London, the Australian Department of Education, Employment and Workplace Relations, and the Menzies Foundation in Australia
No. of awards given last year: 1
No. of applicants last year: 35
Additional Information: Please see the website for further details http://www.menziesfoundation.org.au/scholarships/scholarships.html#harvard.

THE SKIDMORE, OWINGS AND MERRILL FOUNDATION

224 South Michigan Avenue, Suite 1000, Chicago, IL, 60604, United States of America
Tel: (1) 312 427 4202
Fax: (1) 312 360 4545
Email: somfoundation@som.com
Website: www.som.com
Contact: Ms Lisa Westerfield, Administrative Director

The mission of the Skidmore, Owings and Merrill Foundation is to identify and nuture emerging talent by sponsoring prestigious awards and traveling study grants to students of architecture, design, urban design and structural engineering. The SOM foundation identifies and supports individuals with the highest design aspirations, and enables them, through research and travel, to broaden their horizons and achieve excellence in their professional or academic careers. The foundation's goal is to instill in its fellows a heightened sense of reponsibility as future leaders in the design disciplines by offering them an opportunity to deepen their understanding of the complexities of the built environment.

SOM Prize and Travel Fellowship

Subjects: Architecture, design and urban design.
Purpose: To help young architects and designers to broaden their education and take an enlightened view of society's need to improve the built and natural environments.
Eligibility: Open to candidates of any citizenship who are graduating from a university in the United States of America. Fellowship winners must receive, prior to commencement of the fellowship, a Master's or Bachelor's degree in architecture. Students are eligible for nomination in the Spring of the academic year that they are due to graduate.
Level of Study: Graduate, Postgraduate
Type: Fellowship
Value: US$50,000 and US$20,000
Frequency: Annual
Study Establishment: Accredited schools, colleges or universities
Country of Study: United States of America
No. of awards offered: 1 grand prize and 1 travel fellowship
Application Procedure: Applicants must submit a portfolio with a proposed travel itinerary and a signed copyright release statement provided by the Foundation. Complete guidelines are available on request and on the website. Students must be nominated by the United States of America school of attendance.
Funding: Private
No. of awards given last year: 3
Additional Information: Please see the website for further details http://www.somfoundation.som.com/content.cfm/architecture_design_urban_design.

Structural Engineering Travelling Fellowship

Subjects: The role of aesthetics, innovation, efficiency and economy in the structural design of buildings, bridges and other structures.
Purpose: To foster an appreciation of the aesthetic potential inherent in the structural design of buildings, bridges and other major works of architecture and engineering.
Eligibility: Open to candidates of any citizenship who are graduating from a university in the United States of America with a Bachelor's, Master's or PhD degree in civil or architectural engineering and a specialization in structural engineering. Applicants must be nominated by the faculty and endorsed by the chair of the department from which they will receive the degree. Applicants must intend to enter the professional practice of structural engineering in the field of buildings or bridges.
Level of Study: Doctorate, Graduate, Postgraduate
Type: Fellowship
Value: US$10,000
Frequency: Annual
Study Establishment: Accredited schools
Country of Study: United States of America
No. of awards offered: 1
Application Procedure: Complete guidelines are available in the website.
Closing Date: January 28th
Funding: Private
No. of awards given last year: 1
Additional Information: The fellowship hopes to encourage an awareness of the visual impact of structural engineering among engineering students and their schools. It also helps to strengthen the connection between aesthetics and efficiency, economy and innovation in structural design. Please see the website for further details http://www.somfoundation.som.com/content.cfm/structural_engineering.

THE SMITH AND NEPHEW FOUNDATION

Heron House, 15 Adam Street, London, WC2N 6LA, England
Tel: (44) 20 7401 7646
Fax: (44) 20 7960 2358
Email: barbara.foster@smith-nephew.com
Website: www.snfoundation.org.uk
Contact: Ms Barbara Foster, Foundation Coordinator

To contribute to the development of a robust evidence base which will inform the identification of effective interventions for the management and prevention of skin breakdown and tissue integrity. To concentrate our effort on developing research capacity in this field.

Smith and Nephew Foundation Postdoctoral Nursing Research Fellowship

Subjects: Nursing care of patients in the field of skin breakdown and tissue integrity.
Purpose: To influence and develop evidence-based nursing interventions in the field of skin breakdown and tissue integrity.
Eligibility: Applications are invited from potential Fellows and the proposed host research team or unit. The host research team must be a university school, faculty or department of nursing and have a proven track record of research and development in the nursing care of patients with skin damage or tissue damage or vulnerability.
Level of Study: Postdoctorate
Type: Fellowship
Value: Up to UK£120,000
Length of Study: Up to 3 years
Frequency: Annual
Country of Study: United Kingdom
No. of awards offered: 1
Application Procedure: Application forms must be completed. Shortlisted candidates are interviewed in London.
Closing Date: Advertised usually at the beginning of March with a closing date approximately 6 weeks later
Funding: Foundation
Contributor: The Smith and Nephew Foundation
Additional Information: Research which will contribute to the evidence base for the nursing care of patients with skin or tissue damage and vulnerability, particularly in the following areas: Factors which influence the physiological response to tissue damage, the role of infection in delayed healing, factors which influence treatment decisions in skin/wound care, hidden cost of tissue damage. Applications which focus on research in other areas of skin, or tissue damage or vulnerability will also be considered. Please see the website for further details.

SMITHSONIAN ASTROPHYSICS OBSERVATORY (SAO)

60 Garden Street, Mail Stop 47, Cambridge, MA, 02138, United States of America
Email: predoc@cfa.harvard.edu
Website: www.cfa.harvard.edu
Contact: Christine A Crowley, Fellowship Program Coordinator

SAO Predoctoral Fellowships

Subjects: Astronomy, astrophysics, atomic and molecular physics, planetary science, radio and geoastronomy, solar and stellar physics and theoretical astrophysics.
Purpose: To allow students from other institutions throughout the world to undertake their thesis research at the SAO.
Eligibility: Applicants must have completed their preliminary course work and examinations and be ready to begin dissertation research at the time of the award.
Level of Study: Predoctorate
Type: Fellowship
Value: A stipend of US$30,228. Some funds may also be available for relocation, travel and other expenses
Length of Study: 6 months, with a possibility of renewal for up to a total of 3 years
Frequency: Annual
Study Establishment: The Smithsonian Astrophysics Observatory
Country of Study: United States of America
No. of awards offered: Varies
Application Procedure: Applicants must directly contact Smithsonian scientists in their area of interest to discuss possible research topics. Applicants must complete an application form available from the website.
Closing Date: Rolling admissions
No. of awards given last year: 7
No. of applicants last year: 15
Additional Information: Please see the website for further details http://www.cfa.harvard.edu/opportunities/fellowships/predoc/.

SMITHSONIAN ENVIRONMENTAL RESEARCH CENTER (SERC)

Smithsonian Institution, PO Box 28, 647 Contees Wharf Road, Edgewater, MD, 21037-0028, United States of America
Tel: (1) 443 482 2217
Email: gustafsond@si.edu
Website: www.serc.si.edu
Contact: Daniel E Gustafson, Jr, Professional Training & Volunteer Coordinator

Smithsonian Environmental Research Center (SERC) is the world's leading research center for environmental studies of the coastal zone. For over 40 years, SERC has been involved in critical research, professional training for young scientists and environmental education.

SERC Graduate Student Fellowship

Subjects: Global climate change, marine invasion biology, forest and wetland ecology, trace element and nutrient cycling, solar UV radiation, water quality, food web dynamics, coastal and upland ecosystems, and plant-herbivore interactions.
Purpose: To financially support student research at Smithsonian facilities or research stations.
Eligibility: Students must be formally enrolled in a graduate program of study at a degree granting institution must have completed at least one full-time semester. Intended for students who have not yet been advanced to candidacy if in a doctoral program.
Level of Study: Graduate
Type: Fellowship
Value: US$6,500
Length of Study: 10 weeks
Frequency: Annual
Application Procedure: Need to use Smithsonion online academic appointment system at http://solaa.si.edu/solaa/solaahome.html.
Closing Date: January 15th
Funding: Government
No. of awards given last year: 2
No. of applicants last year: 6

For further information contact:

Office of fellowships, Box 37012, 470 L'Enfant Plaza,, Washington, D. C, 20013-7012, United States of America

SERC Postdoctoral Fellowships

Subjects: Soil and water science, forestry, fishery, Marine biology, botany, zoology, and chemistry.
Purpose: To facilitate the Smithsonian's scholarly interactions with students and scholars at universities, museums and other research institutions around the world.
Eligibility: Open to scholars who held a PhD or equivalent degree for less than 7 years.
Level of Study: Postdoctorate, Predoctorate
Type: Fellowship
Value: US$45,000 annually plus research, healthcare and relocation allowances
Length of Study: 12 months
Frequency: Annual
Application Procedure: Need to use Smithsonian online academic appoint system at http://solaa.si.edu/solaa/solaahome.html.
Closing Date: January 15th
Funding: Government
No. of awards given last year: 6
No. of applicants last year: 46
Additional Information: The Smithsonian Institution does not discriminate on grounds of race, creed, sex, age, marital status, condition of disability or national origin of any applicant.Fellowships are renewed after first year if satisfactory progress is made.

For further information contact:

Office of Fellowships Smithsonian Institution PO Box 37012 470 L'Enfant Plaza, SW, Washington DC, 20013 7012, United States of America

SERC Predoctoral Fellowships

Subjects: Soil and water science, forestry, fishery, Marine biology, botany, zoology, and chemistry.

Purpose: To financially support research at Smithsonian facilities or field stations.
Eligibility: Open to doctoral candidates who have completed preliminary course work and examinations and have been advanced to candidancy.
Level of Study: Postdoctorate
Type: Fellowship
Value: US$30,000 per year
Length of Study: 3–12 months
Frequency: Annual
Application Procedure: Applicants can download the application form from the website www.si.edu/ofg/applications/sifell/sifellapp.htm. The completed application form, current curriculum vitae and a formal research proposal must be sent to the Smithsonian Office of Fellowships.
Closing Date: January 15th
Funding: Government
No. of awards given last year: 1
No. of applicants last year: 3
Additional Information: The Smithsonian Institution does not discriminate on grounds of race, creed, sex, age, marital status, condition of disability or national origin of any applicant.

For further information contact:

Office of Fellowships Smithsonian Institution PO Box 37012, 470 L'Enfant Plaza, SW, Washington DC, 20013 7012, United States of America

SERC Senior Fellowships

Subjects: Soil and water science, forestry, fishery, Marine biology, botany, zoology, and chemistry.
Purpose: To financially support research at Smithsonian facilities or field stations.
Eligibility: Open to applicants who have received their PhD or equivalent prior to January 15, 2005.
Level of Study: Research
Type: Fellowship
Value: US$45,000 annually plus research, healthcare and relocation allowances
Length of Study: 12 months
Frequency: Annual
Country of Study: United States of America
Application Procedure: Need to use Smithsonion online academic appointment system at http://solaa.si.edu/solaa/solaahome.html.
Closing Date: January 15th
Funding: Government
No. of awards given last year: 1
No. of applicants last year: 3
Additional Information: Senior fellowship applications may be submitted 2 years in advance.

For further information contact:

Office of Fellowships Smithsonian Institution PO Box 37012 470 L'Enfant Plaza, SW, Washington DC, 20013 7012, United States of America

SMITHSONIAN INSTITUTION-NATIONAL AIR AND SPACE MUSEUM

PO Box 37012, MRC 010, Washington, DC, 20013-7012, United States of America
Tel: (1) 202 633 2648
Fax: (1) 202 786 2447
Email: NASM-Fellowships@si.edu
Website: http://airandspace.si.edu/
Contact: Ms Collette Williams, Fellowships Programme Coordinator

A Verville Fellowship

Subjects: The history of aviation and space flights.
Purpose: To fund the analysis of major trends, developments and accomplishments in the history of aviation or space studies.
Eligibility: Open to all interested candidates who can demonstrate skills in research and writing. An advanced degree is not a requirement.
Level of Study: Postgraduate
Type: Fellowship
Value: An annual stipend of $55,000 will be awarded for a 12-month fellowship, with limited additional funds for travel and miscellaneous expenses
Length of Study: 9–12 months, normally starts between June 1st and October 1st
Country of Study: United States of America
No. of awards offered: 1
Application Procedure: Candidates can apply using the online form found at http://www.nasm.si.edu/forms/fellowapp.cfm.
Closing Date: January 15th
Additional Information: Residence in the Washington, DC metropolitan area during the fellowship term is a requirement of this fellowship. Please see the website for further details http://airandspace.si.edu/getinvolved/fellow/vfellow.cfm.

The Aviation Space Writers Foundation Award

Purpose: To support research on aerospace topics.
Type: Grant
Value: $5,000
Frequency: Every 2 years
Application Procedure: Candidates should submit the online Aviation Space Writers Foundation Award Application form, including 1) a maximum two-page, single-spaced proposal stating the subject of their research and their research goals; 2) a one to two-page curriculum vitae; and 3) a one-page detailed budget explaining how the grant will be spent.
Closing Date: January 15th
Additional Information: Award winners are required to provide a summary report in the form of a memorandum to Ms. Collette Williams that outlines how the grant was used to accomplish the goals of the project.

For further information contact:

Email: pisanod@si.edu
Website: http://airandspace.si.edu/getinvolved/fellow/writer_grant.cfm
Contact: Dr Dominick A. Pisano

Charles A Lindburgh Chair in Aerospace History

Eligibility: Open to senior scholars with distinguished records of publication who are at work on, or anticipate being at work on, books in aerospace history.
Type: Fellowship
Value: Replacement of salary and benefits up to a maximum of $100,000 a year. Research expenses and relocation are negotiable
Closing Date: January 15th
Additional Information: For more information, please contact David DeVorkin (DeVorkinD@si.edu) and see website.

For further information contact:

Email: PisanoD@si.edu
Website: http://airandspace.si.edu/getinvolved/fellow/lindfellow.cfm
Contact: Dominick A. Pisano

The Guggenheim Fellowships

Subjects: Historical research related to aviation and space.
Purpose: To promote research into, and writing about, the history of aviation and space flight.
Eligibility: Postdoctoral fellowships are open to applicants who have received a PhD degree or equivalent within 7 years of the beginning of the fellowship period. Predoctoral fellowships are open to applicants who have completed preliminary coursework and examinations and are engaged in dissertation research. All applicants must be able to speak and write fluently in English.
Level of Study: Postdoctorate, Predoctorate
Type: Fellowship
Value: An annual stipend of $30,000 for predoctoral candidates and $45,000 for postdoctoral candidates will be awarded, with limited additional funds for travel and miscellaneous expenses.

Length of Study: 3–12 months
Frequency: Annual
Study Establishment: A major portion of the research must be conducted at the Smithsonian Institution
Country of Study: United States of America
No. of awards offered: Varies
Application Procedure: Applicants must submit a summary description, research proposal, bibliography, estimated schedule, research budget, transcripts from all graduate institutions, English training with test scores and level of proficiency in reading, conversing and writing (if English is not the applicant's native language), a curriculum vitae and letters from 3 referees. 6 copies of the complete application must be submitted.
Closing Date: January 15th
Additional Information: Residence in the Washington, DC metropolitan area during the fellowship term is a requirement of the fellowship. Please see the website for further details http://airandspace.si.edu/getinvolved/fellow/gfellow.cfm.

Postdoctoral Earth and Planetary Sciences Fellowship

Subjects: Earth and planetary studies and global environment change.
Purpose: To support scientific research in earth and planetary studies. Scientists in the Center for Earth and Planetary Studies concentrate on geologic and geophysical research of the Earth and other terrestrial planets, using remote sensing data obtained from Earth-orbiting and interplanetary spacecraft.
Level of Study: Postdoctorate
Type: Fellowship
Value: Stipend
Application Procedure: Applicants must submit the application to NASM.
Additional Information: In years that the fellowship is offered, announcements will be made in the American Geophysical Union's professional publication EOS. Please see the website for further details http://airandspace.si.edu/getinvolved/fellow/cepsfell.cfm.

THE SNOWDON TRUST

Unit 18, Oakhurst Business Park, Southwater, Horsham, West Sussex, RH13 9RT, England
Tel: (44) 14 0373 2899
Email: info@snowdontrust.org.uk
Website: www.snowdontrust.org.uk
Contact: Paul Alexander, CEO

The Snowdon Trust provides grants of up to UK£2,000 to physically disabled students for disability-related costs of further or higher education or training in the United Kingdom.

Snowdon Trust Grants

Subjects: All subjects.
Purpose: To support physically impaired students for further or higher education or training in the United Kingdom.
Eligibility: Physically impaired students studying in the United Kingdom.
Level of Study: Postgraduate, Unrestricted
Type: Award
Value: UK£250–2,000 (UK£2,500 in exceptional circumstances)
Frequency: Annual
Study Establishment: All recognized further, higher education or training centres
Country of Study: United Kingdom
No. of awards offered: Up to 100 per year, depending on funds
Application Procedure: Applicants must submit an application form and at least four specialized references. The application form is available online from January to September.
Closing Date: May 31st
Funding: Private, trusts, commercial, foundation, individuals
Contributor: Trusts and commercial sponsors
No. of awards given last year: 105
No. of applicants last year: 134

SOCIAL SCIENCE RESEARCH COUNCIL (SSRC)

One Pierrepont Plaza, 15th Floor, 300 Cadman Plaza West, Brooklyn, NY, 11201, United States of America
Tel: (1) 212 377 2700
Fax: (1) 212 377 2727
Email: info@ssrc.org
Website: www.ssrc.org
Contact: Director

Founded in 1923, the Social Science Research Council (SSRC) is an independent, non-governmental, non-profit international association devoted to the advancement of interdisciplinary research in the social sciences. The aim of the organization is to improve the quality of publicly available knowledge around the world.

The Dissertation Proposal Development Fellowship (DPDF)

Subjects: Humanities and social sciences.
Purpose: To help graduate students in the humanities and social sciences formulate doctoral dissertation proposals.
Eligibility: Open to graduate students in the early phase of their research.
Level of Study: Predoctorate
Type: Fellowship
Value: Fellows up to US$5,000, research directors a stipend of US $10,000
Length of Study: 2 workshops and summer fieldwork funding
Frequency: Annual
No. of awards offered: 60
Application Procedure: Check website for further details - www.ssrc.org/programs/dpdf/.
Closing Date: February 18th
Funding: Foundation
Contributor: Andrew W. Mellon Foundation
No. of awards given last year: 60
No. of applicants last year: 450
Additional Information: Please see the website for further details http://www.ssrc.org/fellowships/dpdf-fellowship/.

Drugs, Security and Democracy Fellowship

Subjects: The fellowship seeks to develop a concentration of researchers who are interested in policy relevant outcomes and willing to become members of a global interdisciplinary network.
Purpose: Supports research on organised crime, drug policy and related topics across social sciences and related disciplines.
Eligibility: Open to graduate students at the dissertation phase who have an approved dissertation proposal, and recent PhD recipients who have completed the PhD within 5 years of January 2011.
Level of Study: Doctorate, Postdoctorate, Research
Value: Research and living expenses as necessary for the project, as well as travel expenses to research site and a pre-field and a post-field workshop
Length of Study: 3 months to 1 year
Frequency: Annual
No. of awards offered: Approx. 15
Application Procedure: Applicants are available on the SSRC website.
Closing Date: January 20th
Funding: Private
Contributor: Open Society Foundations
Additional Information: Please see the website for further details http://www.ssrc.org/fellowships/dsd-fellowship/.

ESRC/SSRC Collaborative Visiting Fellowships

Subjects: Social sciences (including history).
Purpose: To encourage communication and cooperation between social scientists in Great Britain and the Americas.
Eligibility: Open to PhD scholars in the Americas, ESRC-supported centres, and holders of large grants awards or professorial fellowships in Britain.
Level of Study: Doctorate, Research
Type: Fellowship
Value: Up to US$9,500

Length of Study: 1–3 months
Frequency: Annual
No. of awards offered: Approx. 18
Application Procedure: Check website for further details.
Closing Date: April 16th

For further information contact:

Email: international@esrc.ac.uk.
Website: http://www.ssrc.org/fellowships/esrc-ssrc-collaborative-visiting-fellowships/

Eurasia Program Fellowships

Subjects: Social sciences and humanities, with a specific focus on Eurasia.
Purpose: To allow advanced graduate students to devote to the intellectual development of their projects and to write-up the results of their research.
Eligibility: Applicants for Dissertation Write-up Fellowships must have attained ABD status (must have completed all requirements for the PhD degree except for the dissertation) by the fellowship start date. They must be citizens or permanent residents of the United States of America. The Fellow's home institution is expected to make a cost-sharing contribution of no less than 10 per cent of the fellowship award. Detailed information on eligibility criteria and conditions of awards will be available in the application materials.
Level of Study: Doctorate, Graduate
Type: Fellowship
Value: Up to US$25,000
Length of Study: Up to 1 year
Frequency: Annual
No. of awards offered: Varies
Application Procedure: Awards are made on the basis of evaluations and recommendations by the Title VIII Program Committee, an interdisciplinary committee composed of scholars of the region. The committee rewards proposals with clarity of argument, purpose, theory and method, written in a style accessible to readers outside the applicant's discipline. Applicants must submit a completed application, a narrative statement, transcripts, a course list and language evaluation form, and references. Full information is available online.
Funding: Government
Contributor: US Department of State under the Program for Research and Training on Eastern Europe and the Independent States of the Former Soviet Union (Title VIII)
No. of awards given last year: 4
No. of applicants last year: Approx. 40
Additional Information: No funding is available for research on the Baltic States.

For further information contact:

Email: eurasia@ssrc.org
Website: http://www.ssrc.org/fellowships/eurasia-fellowship/
Contact: SSRC, Eurasia Program

Japan Society for the Promotion of Science (JSPS) Fellowship

Subjects: Social sciences and humanities.
Purpose: To provide qualified researchers with the opportunity to conduct research at leading universities and other research institutions in Japan.
Eligibility: Applicants must be US citizens or permanent residents at the time of application and submit proof of affiliation with an eligible host research institution in Japan as part of the application packet. Permanent residents must provide a copy of a permanent resident card. Citizens of other countries are eligible for the short-term fellowship (1–12 months) if they have completed a Master's or PhD course at an institution of higher education in the US and, upon completing the course, have for at least three continuous years conducted high-level research at a university in the US. Applicants for long-term (12–24 months) fellowships must submit a copy of a PhD diploma dated no more than six years prior to applying. Check website for further details.
Level of Study: Doctorate, Postdoctorate
Type: Fellowship
Value: Round-trip airfare, insurance coverage for accidents and illness, a monthly stipend and settling-in allowance. Applicants will also be eligible for additional funds annually for research expenses for stays of 1–2 years and a domestic travel allowance for stays of 3–12 months
Length of Study: 1 month to 2 years
Frequency: Annual
Study Establishment: An approved institution
Country of Study: Japan
No. of awards offered: Up to 20
Application Procedure: Applications are available on www.ssrc.org/fellowships/jsps-fellowships
Closing Date: December
Funding: Government
Contributor: The Japan Society for the Promotion of Science
No. of awards given last year: 10–20
No. of applicants last year: 25–40
Additional Information: Fellows are selected by the Japan Society for the Promotion of Science based on nominations made by the SSRC Japan Advisory Board. Applicants will be notified of their nomination status by the following March. Successful applicants will be notified directly by JSPS in the summer.

For further information contact:

Tel: 212 37 2700
Email: japan@ssrc.org
Website: http://www.ssrc.org/fellowships/jsps-fellowship/

SSRC Abe Fellowship Program

Subjects: Social sciences and related fields relevant to any one or combination of (i) traditional and non-traditional approaches to security and diplomacy; (ii) global and regional economic issues; (iii) the role of civil society.
Purpose: To encourage international multidisciplinary research on topics of pressing global concern and to foster the development of a new generation of researchers who are interested in policy-relevant topics of long-range importance and who are willing to become key members of a bilateral and global research network built around such topics.
Eligibility: Open to citizens of Japan and the US and to other nationals who can demonstrate serious and long-term affiliations with research communities in Japan or the US. Applicants must hold a PhD or have attained an equivalent level of professional experience. Applications from researchers in non-academic professions are welcome.
Level of Study: Postdoctorate
Type: Fellowship
Value: Research and travel expenses as necessary for the completion of the research project in addition to limited salary replacement
Length of Study: Up to 1 year
Frequency: Annual
Study Establishment: An appropriate institution
Country of Study: United States of America
No. of awards offered: Approx. 14
Application Procedure: Applicants must submit an online application along with a writing sample, letter of reference and an optional language evaluation form.
Closing Date: September 1st
Funding: Foundation
Contributor: The Japan Foundation Center for Global Partnership
No. of awards given last year: 14
No. of applicants last year: 60–100
Additional Information: In addition to working on their research projects, Fellows will attend annual conferences and other events sponsored by the program, which will promote the development of an international network of scholars concerned with research on contemporary policy issues. Funds are provided by the Japan Foundation Center for Global Partnership. Further information is available by emailing abe@ssrc.org

For further information contact:

Email: abe@ssrc.org
Website: http://www.ssrc.org/fellowships/abe-fellowship/

SSRC International Dissertation Research Fellowship

Subjects: Non-US cultures and socities grounded in empirical and site-specific research (involving fieldwork, research in archival or manuscript collections, or quantitative data collection).

Purpose: To support distinguished graduate students in the humanities and social sciences conducting dissertation research outside the United States.
Eligibility: Open to full-time graduate students in the humanities and social sciences regardless of citizenship enrolled in doctoral programs in the United States. Applicants must complete all PhD requirements except on-site research by the time the fellowship begins.
Level of Study: Doctorate
Type: Fellowship
Value: Approx. US$20,000
Length of Study: 9–12 months
Frequency: Annual
No. of awards offered: Approx. 75
Application Procedure: Applications are available on SSRC website.
Closing Date: November
Funding: Private
Contributor: The Andrew W Mellon Foundation
Additional Information: Applicants must contact the programme for further information by emailing.

For further information contact:

Email: idrf@ssrc.org
Website: http://www.ssrc.org/fellowships/idrf-fellowship/

SOCIAL SCIENCES AND HUMANITIES RESEARCH COUNCIL OF CANADA (SSHRC)

350 Albert Street, PO Box 1610, Ottawa, ON, K1P 6G4, Canada
Tel: (1) 613 995 2694
Fax: (1) 613 992 2803
Email: awdad@sshrc-crsh.gc.ca
Website: www.sshrc.ca

The Social Sciences and Humanities Research Council of Canada (SSHRC) is the federal agency responsible for promoting and supporting research and research training in the social sciences and humanities in Canada. SSHRC supports research on the economic, political, social and cultural dimensions of the human experience.

Aid to Research and Transfer Journals

Subjects: Social sciences and humanities.
Purpose: To assist in the effective dissemination of original research findings and in the transfer of knowledge to practitioners.
Eligibility: Open to candidates who are either Canadian citizens or permanent residents of Canada at the time of application and are established researchers in the social sciences or in the humanities. They must also be a member of the sponsoring institution or organization and not be under SSHRC sanction for financial or research misconduct. Check SSHRC website for further details.
Level of Study: Research
Type: Grant
Value: Check SSHRC website for further details.
Frequency: Every 3 years
Country of Study: Canada
Application Procedure: Applicants must make the case for the funding requested. For full details on application visit the website.
Closing Date: June 30th
Funding: Government

Aid to Scholarly Journals

Subjects: Social sciences and humanities.
Purpose: To promote the sharing of research results by assisting the publication of individual works that make an important contribution to the advancement of knowledge.
Eligibility: Applicants must consult the Canadian Federation for the Humanities and Social Sciences website for eligibility requirements.
Level of Study: Postdoctorate, Research
Type: Grant
Value: Up to $30,000 per year
Length of Study: 3 years
Frequency: Annual
Application Procedure: Applicants must refer to the website or email the Humanities and Social Sciences Federation of Canada.
Closing Date: June 30th
Funding: Government
Additional Information: The program is administered on behalf of SSHRC by the Humanities and Social Sciences Federation of Canada.

For further information contact:

The Humanities and Social Sciences Federation of Canada, 151 Slater Street, Ottawa, ON, K1P 5H3, Canada
Tel: (1) 613 238 6112 ext 350
Email: secaspp@fedcan.ca

Aid to Small Universities

Subjects: Arts, humanities and social sciences.
Purpose: To enable the focused development of social sciences and humanities research capacity in small universities.
Eligibility: Open to institutions that have active degree-granting status for social sciences and humanities disciplines at the undergraduate level or beyond, and are institutional members of the Association of Universities and Colleges of Canada (AUCC), or institutional members of the AUCC and affiliated with an institution itself too large to be eligible for the ASU Program. Institutions applying must also have fewer than 250 full-time faculty in SSHRC fields and be independent of the federal government for the purpose of faculty employment status.
Level of Study: Research
Type: Grant
Value: Up to a maximum of Canadian $30,000
Length of Study: 3 years
Frequency: Every 3 years
Country of Study: Canada
Application Procedure: Applicants must contact the organization or visit the SSHRC website.
Closing Date: December 1st
Funding: Government
Additional Information: Please see the website for further details http://www.mta.ca/research_activities/development/small_com_index.html.

Aileen D Ross Fellowship

Subjects: Sociology.
Purpose: To support an outstanding SSHRC Doctoral Award or Postdoctoral Fellowship holder who is conducting research in sociology, especially on poverty.
Eligibility: Open to Canadian citizens or permanent residents living in Canada at the time of application who are not under SSHRC sanction resulting from financial or research misconduct, or already in receipt of SSHRC, NSERC or CIHR funding to undertake or complete a previous doctoral or combined MA or PhD degree. At the time of taking up the award, applicants must have completed either a Master's degree or at least 1 year of doctoral study, and be pursuing either full-time studies leading to a PhD or equivalent, with a research specialization in sociology and with the intention of pursuing an academic career. Candidates wishing to study at a foreign university may only do so if at least one of their previous degrees was earned in Canada.
Level of Study: Doctorate, Postdoctorate
Type: Fellowship
Value: Canadian $10,000 plus the value of the fellowship
Length of Study: 1 year
Country of Study: Canada
No. of awards offered: 1
Application Procedure: Applicants must indicate their interest on their doctoral or postdoctoral application form.
Closing Date: Postdoctoral applications are due October 6th. For applicants registered at a Canadian university: University sets the deadline. For all others: November 5th
Funding: Government, trusts
Additional Information: Preference will be given to postdoctoral applicants. Please see the website for further details http://www.sshrc-crsh.gc.ca/results-resultats/prizes-prix/fellowships-bourses-eng.aspx.

Bora Laskin National Fellowship in Human Rights Research

Subjects: Human rights, as relevant to Canada.

Purpose: To support research, preferably of a multidisciplinary or interdisciplinary nature, and to develop Canadian expertise in the field of human rights, with an emphasis on themes and issues relevant to the Canadian human rights scene.
Eligibility: Open to Canadian citizens or permanent residents of Canada. Preference will be given to applicants with at least 5 years of proven research experience in their field, as this fellowship is not intended for scholars just beginning their research careers. The successful candidate must not be under SSHRC sanction for financial or research misconduct.
Level of Study: Postdoctorate, Research
Type: Fellowship
Value: Canadian $45,000 plus a research and travel allowance of Canadian $10,000
Length of Study: 1 year, non-renewable
Frequency: Annual
No. of awards offered: 1
Application Procedure: Applicants must complete an application form, available from the SSHRC website.
Closing Date: October 1st
Funding: Government

Canadian Initiative on Social Statistics: Data Training Schools

Subjects: Social sciences.
Purpose: To promote research and training in the application of Canadian social statistics.
Eligibility: Eligible applicants include research groups, departments and centres that are operating in a Canadian university. Eligible participants include students and researchers at a variety of levels.
Level of Study: Doctorate, Graduate, Postdoctorate, Predoctorate, Research
Type: Grant
Value: Up to Canadian $50,000 per year
Length of Study: 3 years
Frequency: Annual
Application Procedure: Applicants must visit the SSHRC website for application forms and instructions.
Closing Date: December 7th

Joseph-Armand Bombardier Canada Graduate Scholarship Program (CGS): Master's Scholarship

Subjects: Social sciences and humanities.
Purpose: To support graduate students working towards a Master's degree at a Canadian university.
Eligibility: Open to citizens or permanent residents of Canada who are applying for, or registered in a Master's programme in the social sciences or humanities at a Canadian university.
Level of Study: Graduate, Predoctorate
Type: Scholarship
Value: Canadian $17,500
Length of Study: 1 year
Frequency: Annual
Country of Study: Canada
No. of awards offered: Several hundred
Application Procedure: Applicants must visit the SSHRC website for further information.
Closing Date: December 13th (direct application) and February 4th (university application)
Funding: Government
No. of awards given last year: Hundreds
No. of applicants last year: Thousands

Jules and Gabrielle Léger Fellowship

Subjects: The Crown and Governor General in a parliamentary democracy.
Purpose: To promote better understanding of the diverse contributions of the crown and its representatives to all aspects of Canadian life.
Eligibility: Open to Canadian citizens or permanent residents of Canada. Applicants must be affiliated with a post secondary institution not be under SSHRC sanction for financial or research misconduct. Preference will be given to applicants with at least five years of proven research experience in their field.
Level of Study: Postdoctorate, Research
Type: Fellowship
Value: Canadian $40,000, plus Canadian $10,000 for research and travel costs
Length of Study: 1 year, non-renewable
Frequency: Every 2 years
Study Establishment: A recognized Canadian post secondary institution
Country of Study: Any country
No. of awards offered: 1
Application Procedure: Applicants must visit the SSHRC website for full application details.
Closing Date: October 1st
Funding: Government

Knowledge Impact in Society

Subjects: Social sciences and humanities.
Purpose: To support university-based knowledge mobilization initiatives that will enable non-university stockholders to benefit from existing social sciences and humanities research.
Eligibility: The programme offers institutional grants, which are available only to universities eligible for SSHRC Institutional Grants. Applicants must check SSHRC website for full details.
Level of Study: Research
Type: Grant
Value: Up to Canadian $100,000 per year
Length of Study: Up to 3 years
Frequency: Annual
Application Procedure: Applicants must visit the website for application forms and instructions.
Closing Date: Visit website
Funding: Government

Northern Research Development Program

Subjects: Social sciences and humanities.
Purpose: To support research in and about the Canadian North, with emphasis on involving local stakeholders.
Eligibility: Open to individual researchers or groups of researchers affiliated with Canadian postsecondary institutions or from other organizations whose proposed activites include significant involvement of researchers affiliated with Canadian postsecondary institutions. For further details check the SSHRC website.
Level of Study: Research
Type: Grant
Value: Up to Canadian $40,000
Length of Study: Up to 2 years
Frequency: Annual
Country of Study: Canada
Application Procedure: Applicants must visit the SSHRC website for application information, forms and instructions.
Closing Date: November 8th
Funding: Government

Queen's Fellowships

Subjects: Canadian studies.
Purpose: To assist an outstanding candidate who intends to enter a doctoral program in the relevant field.
Eligibility: Open to Canadian citizens and permanent residents who, by the time of taking up the fellowship, will have completed 1 year of graduate study or all the requirements for the master's degree beyond the bachelor's (honours) degree or its equivalent, and will be registered in a program of studies leading to a PhD or its equivalent. This award is offered to one or two outstanding and successful doctoral fellowship candidates.
Level of Study: Doctorate
Type: Fellowship
Value: Tuition fees and travel to the main place of tenure and travel for research purposes
Length of Study: 1 year, non-renewable
Frequency: Annual
Study Establishment: A recognized university
Country of Study: Canada
No. of awards offered: 1–2
Application Procedure: Applicants cannot apply for this award, but are automatically eligible if they intend to study Canadian studies at a

Canadian university. The Queen's Fellowship is not a program, but a special award given to a doctoral fellow.
Funding: Government

Sport Participation Research Initiative: Research Grants

Subjects: Social sciences and humanities.
Purpose: To support policy-relevant research about participation in sport in Canada.
Eligibility: Open to individuals and research teams must be affiliated with Canadian postsecondary institutions who are working in one or more of the target areas of interest. For further information check the SSHRC website.
Level of Study: Research
Type: Grant
Value: Up to Canadian $100,000 per year
Length of Study: Up to 3 years
Frequency: Annual
Country of Study: Canada
Application Procedure: Applicants must visit the SSHRC website for application forms and instructions.
Closing Date: October 15th (Research Grants); October 6th (Postdoctoral Fellowship Supplements); check website for Doctoral Award Supplement
Funding: Government
Additional Information: Please see the website for further details http://www.sshrc-crsh.gc.ca/funding-financement/programs-programmes/sport_can-eng.aspx.

SSHRC Doctoral Awards

Subjects: Social sciences and humanities.
Purpose: To develop research skills and to assist in the training of highly qualified academic personnel by supporting students who demonstrate a high standard of scholarly achievement.
Eligibility: Open to Canadian citizens or permanent residents living in Canada at the time of application, who are not under SSHRC sanction resulting from financial or research misconduct, or already in receipt of SSHRC, NSERC or CIHR funding to undertake or complete a previous doctoral or combined MA/PhD degree. At the time of taking up the award applicants must have completed either a Master's degree or at least 1 year of doctoral study, and be pursuing either full-time studies leading to a PhD or equivalent. Candidates wishing to study at a foreign university may only do so if at least one of their previous degrees was earned in Canada.
Level of Study: Doctorate
Type: Fellowship/Scholarship
Value: SSHRC doctoral fellowships: Canadian $20,000 per year, up to and including the 4th year of doctoral study (tenable in Canada and abroad); CGS doctoral scholarship: Canadian $35,000 per year for 3 years (tenable only at Canadian universities)
Length of Study: 1–4 years
Frequency: Annual
Study Establishment: Any recognized universities
Country of Study: SSHRC doctoral fellowships: tenable in Canada or abroad; CGS doctoral scholarships: tenable in Canada only
No. of awards offered: Varies
Application Procedure: Applicants must complete an application form along with a detailed description. Forms are available from the SSHRC website.
Closing Date: For applicants registered at a Canadian university, the university sets the deadline; for all others, November 7th
Funding: Government
Additional Information: Please see the website for further details http://www.sshrc-crsh.gc.ca/funding-financement/index-eng.aspx#results-resultats.

SSHRC Institutional Grants

Subjects: Social sciences and humanities.
Purpose: To help universities develop and maintain a solid base of research and research-related activities.
Eligibility: Open to Canadian postsecondary institutions only.
Level of Study: Postgraduate
Type: Research grant
Value: Minimum Canadian $5,000 per year
Length of Study: 3 years
Frequency: Annual
Country of Study: Canada
No. of awards offered: Varies
Application Procedure: Applicants must complete an application form available from the SSHRC website.
Closing Date: December 1st
Funding: Government

SSHRC Postdoctoral Fellowships

Subjects: Social sciences and humanities.
Purpose: To support the most promising new scholars and to assist them in establishing a research base at an important time in their research career.
Eligibility: Open to candidates who are Canadian citizens or permanent residents living in Canada at the time of application. They must also be able to demonstrate skill in research, not be under SSHRC sanction resulting from financial or research misconduct and have earned their doctorate from a recognized university no more than 3 years prior to the competition deadline or have completed their degree within 6 years prior to the competition deadline, but have had their career interrupted or delayed for the purpose of child-rearing. Candidates must also have finalized arrangements for affiliation with a recognized university or research institution and have applied not more than twice before for a SSHRC postdoctoral fellowship. At the time of taking up the award, applicants must have completed all requirements for the doctoral degree, must intend to engage in full-time postdoctoral research for the period of the award and not hold or have held a tenure or tenure-track position.
Level of Study: Postdoctorate
Type: Fellowship
Value: Up to Canadian $38,000 per year, plus a research allowance of up to Canadian $5,000
Length of Study: 12-24 Months
Frequency: Annual
Study Establishment: For applicants who earned their doctorate at a Canadian University, there is no restriction on the location of the tenure
Country of Study: Any country, under certain conditions
No. of awards offered: Approx. 100
Application Procedure: Applicants must complete an application form available in the SSHRC website.
Closing Date: October 4th
Funding: Government
No. of awards given last year: 140
No. of applicants last year: 558
Additional Information: Applicants wishing to hold their award at a foreign university may do so only if their PhD was earned at a Canadian university. Please see the website for further details http://www.sshrc-crsh.gc.ca/funding-financement/index-eng.aspx#results-resultats.

SSHRC Strategic Research Grants

Subjects: Social sciences and humanities.
Purpose: To support targeted research and research-related activities in areas of national importance. SSHRC offers many separate programs to support strategic research. See the website for current offerings.
Eligibility: Varies as per requirements.
Level of Study: Doctorate, Postgraduate, Research
Type: Grant
Value: Up to $12,000 per year for master's student; up to $15,000 per year for doctoral students; up to $31,500 per year postdoctoral student
Length of Study: 1–5 Years
Frequency: Annual
Country of Study: Canada
Application Procedure: Applicants must visit the website for full application details.
Closing Date: Varies according to program
Funding: Foundation, government, trusts
Additional Information: Please see the website for further details http://www.sshrc-crsh.gc.ca/funding-financement/using-utiliser/grant_regulations-reglement_subventions/strat_grants-subventions_strat-eng.aspx.

Thérèse F Casgrain Fellowship

Subjects: Women and social change in Canada.
Purpose: To carry out research in the field of social justice, particularly in the defence of individual rights and the promotion of the economic and social interests of Canadian women.

Eligibility: Open to Canadian citizens or permanent residents. At the time of taking up the fellowship, the successful candidate must have obtained a doctorate or an equivalent advanced professional degree, as well as have proven research experience and not be under SSHRC sanction for financial or research misconduct.
Level of Study: Postdoctorate
Type: Fellowship
Value: Canadian $40,000, paid in three instalments, of which up to Canadian $10,000 may be used for travel and research expenses
Length of Study: 1 year, non-renewable
Frequency: Every 2 years
Country of Study: Canada
No. of awards offered: 1
Application Procedure: Applicants must complete an application form, available from the website. Application forms and instructions are available on the SSHRC website.
Closing Date: October 1st
Funding: Private
Contributor: The Thérèse F Casgrain Foundation
No. of awards given last year: 1
No. of applicants last year: 21
Additional Information: The fellowship was created by the Thérèse F Casgrain Foundation and is administered by the Social Sciences and Humanities Research Council. Affiliation with a university or an appropriate research institute or similar organization is desirable, but is not a condition of the award. Recipients must submit a final report to the Foundation outlining conclusions of the research, accompanied by a complete financial statement for all expenses.

SOCIAL WORKERS EDUCATIONAL TRUST

British Association of Social Workers, 16 Kent Street, Birmingham, West Midlands, B5 6RD, England
Tel: (44) 12 1622 3911
Fax: (44) 12 1622 4860
Email: swet@basw.co.uk
Website: www.socialworkerseducationaltrust.org
Contact: Ms Gill Aslett, Honorary Secretary

The Social Workers Educational Trust offers grants and scholarships to experienced social workers for postqualifying training or research, with the overall aim of improving social work practice in the United Kingdom.

Anne Cummins Scholarship

Subjects: Health-related social work.
Purpose: To support study or research on health-related social work practice.
Eligibility: Open to qualified social workers who have completed 2 years of work practice postqualification, resident in the United Kingdom.
Level of Study: Postgraduate, Professional development
Type: Scholarship
Value: UK£1,000–1,500
Frequency: Annual
Country of Study: United Kingdom
No. of awards offered: 1
Application Procedure: Applicants must complete an application form. Details and forms are available on written request from the Trust, at the Birmingham address.
Closing Date: May (check with website)
Funding: Private
No. of awards given last year: 1

SOCIETY FOR PROMOTION OF ROMAN STUDIES

Roman Society, Room 244, South Block, Senate House, Malet Street, London, WC1E 7HU, England
Tel: (44) 20 7862 8727
Fax: (44) 20 7862 8728
Email: office@romansociety.org
Website: www.romansociety.org
Contact: Dr Fiona Haarer, Secretary of Society

The Society for the Promotion of Roman Studies aims to promote the study of history, architecture, archaeology, language, literature and the art of Italy and the Roman Empire, including Roman Britain, from the earliest times to about 700 AD.

Hugh Last and Donald Atkinson Funds Committee Grants

Subjects: The history, archaeology, language, literature and art of Italy and the Roman Empire including Roman Britain.
Purpose: To assist in the undertaking, completion or publication of work that relates to any of the general scholarly purposes of the Roman Society.
Eligibility: Open to applicants of graduate, postgraduate or post-doctoral status or the equivalent, usually of United Kingdom nationality.
Level of Study: Graduate, Postdoctorate, Postgraduate, Research
Type: Grant
Value: Varies, but usually UK£200–1,500
Frequency: Annual
No. of awards offered: 20
Application Procedure: Applicants must ensure that two references are sent directly to the Society. Completion of an application form is essential.
Closing Date: January 15th
Funding: Private
No. of awards given last year: 16
No. of applicants last year: 28
Additional Information: Grants for the organization of conferences or colloquia and symposia will only be considered in exceptional circumstances.

SOCIETY FOR THE ARTS IN RELIGIOUS AND THEOLOGICAL STUDIES (SARTS)

United Theological Seminary of the Twin Cities, 3000 5th Street NW, New Brighton, MN, 55112, United States of America
Tel: (1) 651 255 6117, 651 255 6190
Fax: (1) 651 633 4315
Email: wyates@unitedseminary-mn.org
Website: www.artsmag.org
Contact: Wilson Yates

The Society for the Arts in Religious and Theological Studies (SARTS) was organized to provide a forum for scholars and artists interested in the intersections between theology, religion and the arts to share thoughts, challenge ideas, strategize approaches in the classroom and to advance the discipline in theological and religious studies.

Luce Fellowships

Subjects: Intersection of theology and art.
Purpose: To enhance and expand the conversation on theology and art.
Eligibility: Open to candidates teaching theology as a faculty member at an accredited postsecondary educational institution or graduate students
Level of Study: Graduate, Research
Type: Fellowships
Value: Awards are up to $3,000 each.
Length of Study: 1 year
Frequency: Annual
No. of awards offered: 2
Application Procedure: Applicants must submit an information sheet, curriculum vitae, a project abstract, a formal proposal, a budget and 2 letters of recommendation.
Closing Date: May 15th

For further information contact:

University of St Thomas, Mail JRC 153 2115 Summit Avenue, St Paul, MN, 55105, United States of America
Email: office@societyarts.org
Website: http://www.societyarts.org/fellowships/guidelines
Contact: Kayla Larson

THE SOCIETY FOR THE PROMOTION OF HELLENIC STUDIES

Senate House, South Block room 245, Malet Street, London, WC1E 7HU, England
Tel: (44) 20 7862 8730
Fax: (44) 20 7862 8731
Email: office@hellenicsociety.org.uk
Website: www-hellenicsociety.org.uk
Contact: Secretary

The Society for the Promotion of Hellenic Studies, generally known as the Hellenic Society, was founded in 1879 to advance the study of Greek language, literature, history, art and archaeology in the ancient, Byzantine and modern periods.

The Dover Fund

Subjects: Greek language and papyri.
Purpose: To further the study of the history of the Greek language in any period from the Bronze Age to the 15th century AD, and to further the edition and exegesis of Greek texts from any period within those same limits.
Eligibility: Open to currently registered research students and lecturers, teaching Fellows, research Fellows, postdoctoral Fellows and research assistants who are within the first 5 years of their appointment.
Level of Study: Postgraduate, Doctorate, Postdoctorate
Type: Grant
Value: Grants will be made for such purposes as: books, photography, visits to libraries, museums and sites.The sums awarded will vary according to the needs of the applicant, but most grants will be in the range £50–300
Frequency: Annual
Study Establishment: The Society for the Promotion of Hellenic Studies
Country of Study: United Kingdom
No. of awards offered: Varies
Application Procedure: Applicants must complete an application form, available from the Society. Applications should be marked for the attention of the Dover Fund and sent to the main address.
Closing Date: May 1st
Funding: Private
Contributor: Membership subscriptions
No. of awards given last year: 3
No. of applicants last year: 8
Additional Information: Please see the website for further details http://www.hellenicsociety.org.uk/grants/.

THE SOCIETY FOR THE PSYCHOLOGICAL STUDY OF SOCIAL ISSUES (SPSSI)

SPSSI Central Office, 208 I Street NE, Washington, DC, 20002-4340, United States of America
Tel: (1) 202 675 6956
Fax: (1) 202 675 6902
Email: spssi@spssi.org
Website: www.spssi.org
Contact: Alex Ingrams, Administrative Assistant

The Society for the Psychological Study of Social Issues (SPSSI) is an interdisciplinary, international organization of over 3,000 social scientists who share an interest in research on the psychological aspects of important social issues. The Society's goals are to increase the understanding of social issues through research and its dissemination and to support policy efforts consistent with such research.

Clara Mayo Grants

Subjects: Aspects of sexism, racism and prejudice.
Purpose: To support masters' theses or pre-dissertation research on aspects of sexism, racism, or prejudice.
Eligibility: Open to individuals who are SPSSI members and who have matriculated in graduate programs in psychology, applied social science, and related disciplines. A student who is applying for a Grants-In-Aid may not apply for the Clara Mayo award in the same award year. Applicants may submit only one Mayo application per calendar year. Proposals that include a college or university agreement to match the amount requested will be favored, but proposals without matching funds will also be considered.
Level of Study: Postgraduate, Graduate
Type: Research grant
Value: Up to US$1,000 (Proposals that include a college or university agreement to match the amount requested will be favored, but proposals without matching funds will also be considered.)
Frequency: Annual, Biannual
Country of Study: Any country
No. of awards offered: Up to 6
Application Procedure: The Application should include:1. A cover sheet stating title of thesis proposal, name of investigator, address, phone, and if possible, fax and e-mail;2. An abstract of no more than 100 words summarizing the proposed research;3. Project purposes, theoretical rationale, research methodology, and analytic procedures to be employed;4. Relevance of research to SPSSI goals and funding criteria;5. Status of human subjects review process (which must be satisfactorily completed before grant funds can be forwarded);6. Clear statement of type of degree program applicant is enrolled in (e.g., terminal master's program);7. Faculty advisor's recommendation, including certification that the proposal is for a master's thesis or for pre-dissertation research;8. Specific amount requested, including a budget;9. If available, an institutional letter of agreement to match the funds requested.Recommended length for points (1) through (4) of the application is 5–7 double-spaced, 12-point font, typed pages.
Closing Date: May 2nd and October 15th
Funding: Private
No. of awards given last year: 5
No. of applicants last year: 10
Additional Information: Incomplete applications will be returned. Late applications may be held until the next deadline. Please see the website for further details http://www.spssi.org/index.cfm?fuseaction=page.viewpage&pageid=727.

Louise Kidder Early Career Award

Subjects: Social and community psychology.
Purpose: To recognize social issues researchers who have made substantial contributions to the field early in their careers.
Eligibility: Nominees should be social issues investigators who have made substantial contributions to social issues research within 5 years of receiving a graduate degree and who have demonstrated the potential to continue such contributions. Nominees need not be current SPSSI members.
Level of Study: Postdoctorate, Professional development
Type: Award
Value: US$500 plus plaque
Frequency: Annual
Country of Study: Any country
No. of awards offered: 1
Application Procedure: The online application form should be completed and sent as an email attachment. For further details visit the website.Application should include a cover letter outlining the nominee's accomplishments to date and anticipated future contributions, the nominee's curriculumm vitae and three letters of support.
Closing Date: June 20th
Funding: Private
No. of awards given last year: 1
No. of applicants last year: 10
Additional Information: Late applications will be retained for the next year. The winner will be announced by August 1st. Please see the website for further details http://www.spssi.org/index.cfm?fuseaction=page.viewpage&pageid=725.

Otto Klineberg Intercultural and International Relations Award

Subjects: Intercultural and international relations.
Purpose: To recognize the best paper or article of the year on intercultural or international relations.
Eligibility: Entries can be either unpublished manuscripts, in press papers or books, or papers or books published no more than 18 months prior to the submission deadline. Entries cannot be returned. The competition is open to non-members, as well as members of

SPSSI, and graduate students are especially urged to submit papers. The originality of the contribution, whether theoretical or empirical, will be given special weight. Submissions from across the social sciences are encouraged, however the paper must clearly demonstrate its relevance for psychological theory and research in the domain of intercultural and international relations.
Level of Study: Doctorate, Graduate, Postdoctorate, Postgraduate
Type: Prize
Value: US$1,000
Frequency: Annual
Country of Study: Any country
No. of awards offered: 1
Application Procedure: The online application form should be completed and sent as an email attachment. For further details visit the website.
Closing Date: March 1st
Funding: Private
No. of awards given last year: 1
No. of applicants last year: 8
Additional Information: Late applications will be retained for the next year. Please see the website for further details http://www.spssi.org/index.cfm?fuseaction=page.viewpage&pageid=723.

SPSSI Applied Social Issues Internship Program

Subjects: The application of social principles to social issues in co-operation with a community or government organization, public interest group or other not-for-profit entity that will benefit directly from the project.
Purpose: To encourage research that is conducted in cooperation with a community or government organization, public interest group or other not-for-profit entity that will benefit directly from the project.
Eligibility: Open to college seniors, graduate students and 1st year postdoctorates in psychology, applied social science and related disciplines. Applicants must be SPSSI members.
Level of Study: Graduate, Postdoctorate, Postgraduate
Type: Grant
Value: US$300–2,500 to cover research costs, community organizing, Summer stipends, etc.
Frequency: Annual
Country of Study: Any country
No. of awards offered: Varies
Application Procedure: The Application should include: a) A 3–6 page proposal including the proposed budget and a cover sheet with your name, address, phone number, e-mail address and title of your proposal. If an intervention is planned, the proposal should carefully describe the theoretical rationale for the intervention, specifically how the effectiveness of the program will be assessed and the plan to disseminate the findings to relevant parties and policy makers. b) A short resume. c) A letter from a faculty sponsor/supervisor of the project, a statement concerning protection for participants if relevant and any funds that the sponsoring organization will use to support the intern's research. d) A letter from an organizational sponsor (waived if the applicant is proposing to organize a group) that endorses the intern's research activities, describes how the organization will potentially benefit from the work, and outlines any funds the organization will use to support the intern's research.
Closing Date: April 25th
Funding: Private
No. of awards given last year: 1
No. of applicants last year: 5
Additional Information: Late applications will be retained for the next year. Cost sharing by sponsoring department or organization is desirable. Please see the website for further details http://www.spssi.org/index.cfm?fuseaction=page.viewpage&pageid=637.

SPSSI Grants-in-Aid Program

Subjects: Scientific research in social problem areas related to the basic interests and goals of SPSSI and particularly those that are not likely to receive support from traditional sources.
Purpose: To support scientific research in social problem areas related to the basic interests and goals of SPSSI
Eligibility: Applicant must be a member of SPSSI. Applicants may submit only one application per deadline. If applied to the Clara Mayo Grant in the same award year he/she is not eligible for GIA. Individuals may submit a joint application.
Level of Study: Doctorate, Graduate, Postdoctorate, Postgraduate
Type: Grant
Value: Up to US$2,000 for postdoctoral work and up to US$1,000 for graduate student research that must be matched by the student's university.
Frequency: Biannual
Country of Study: Any country
No. of awards offered: Varies
Application Procedure: The Application should include:1. A cover sheet with your name, address, phone number, e-mail address and title of the proposal.2. An abstract of 100 words or less summarizing the proposed research.3. Project purposes, theoretical rationale, and research methodology and analytical procedures to be employed.4. Relevance of research to SPSSI goals and Grants-in-Aid criteria.5. Status of human subjects review process (which must be satisfactorily completed before grant funds can be forwarded).6. Resume of investigator (a faculty sponsor's recommendation must be provided if the investigator is a graduate student; support is seldom awarded to students who have not yet reached the dissertation stage).7. Specific amount requested, including a budget. For co-authored submissions, please indicate only one name and institution to which a check should be jointly issued if selected for funding.
Closing Date: May 15th and October 20th
Funding: Private
Contributor: The Sophie and Shirley Cohen Memorial Fund and membership contributions
No. of awards given last year: Approx. 10
No. of applicants last year: Approx.18
Additional Information: Late applications may be held until the next deadline. Proposals for highly timely and event-oriented research may be submitted at any time during the year to be reviewed within one month of receipt on an ad hoc basis. If yours is a time-sensitive application, please indicate that on the outside of the envelope.

For further information contact:

Email: awards@spssi.org
Website: http://www.spssi.org/index.cfm?fuseaction=page.viewpage&pageid=730

SPSSI Social Issues Dissertation Award

Subjects: Social issues in psychology or in a social science with psychological subject matter.
Purpose: To encourage excellence in socially relevant research.
Eligibility: Open to doctoral dissertations in psychology (or in a social science with psychology subject matter) accepted between March 1st of the year preceding that of application and March 1st of the year of application. In the award year (July 1 to June 30) an individual or group may only submit one paper to one SPSSI paper award (amongst Allport, Klineberg and Dissertation awards).
Level of Study: Postgraduate
Type: Prize
Value: First prize of US$1000 and a second prize of US$500
Frequency: Annual
Country of Study: Any country
No. of awards offered: 2
Application Procedure: A 500-word summary of the dissertation. The summary should include title, rationale, methods, and results of dissertation, as well as its implications for social problems. Please also include a cover sheet that states the title of your dissertation, your name, postal and e-mail addresses, phone number, and university granting the degree.
Closing Date: May 10th
Funding: Private
No. of awards given last year: 2
No. of applicants last year: 15
Additional Information: Online applications are the preferred method.Applicants will be notified of their status by July. Finalist will be asked to provide certification by the dissertation advisor of the acceptance date of the dissertation and four copies of dissertation. Final decision will be announced by September 1st.

For further information contact:

SPSSI, 208, Washington, DC, 20002-4340, United States of America
Website: http://www.spssi.org/index.cfm?fuseaction=page.viewpage&pageid=724

THE SOCIETY FOR THE SCIENTIFIC STUDY OF SEXUALITY (SSSS)

881 Third Street, Suite B-5, Whitehall, PA 18052, United States of America
Tel: (1) 610 443 3100
Fax: (1) 610 443 3105
Email: thesociety@sexscience.org
Website: www.sexscience.org
Contact: Mr David L Fleming, Executive Director

The Society for the Scientific Study of Sexuality (SSSS) is an international organization dedicated to the advancement of knowledge about sexuality. The Society brings together an interdisciplinary group of professionals who believe in the importance of both production of quality research and the clinical, educational and social applications of research related to all aspects of sexuality.

FSSS Grants-in-Aid Program

Subjects: Human sexuality.
Level of Study: Unrestricted
Type: Grant
Value: US$1,000 each
Length of Study: 1 year
Frequency: Annual
No. of awards offered: Varies
Application Procedure: For instructions and application guidelines see http://fsssonline.org/grant-instructions/.
Closing Date: February 1st and June 1st
Funding: Foundation
Contributor: The foundation for the scientific Study of sexuality
Additional Information: Preference will be given to research areasunlikely to receive support from other sources. Please see the website for further details http://www.sexscience.org/honors/fsss_grants_in_aid_program/.

SSSS Student Research Grants

Subjects: Human sexuality.
Eligibility: Applicants must be a member of the SSSS.
Level of Study: Doctorate, Postgraduate
Type: Scholarship
Value: US$1,000
Length of Study: 1 year
Frequency: Annual
No. of awards offered: 2
Application Procedure: Contact the society or check website for details.
Closing Date: February 1st and June 1st
Funding: Foundation
Contributor: The Foundation for the Scientific study of sexuality

For further information contact:

Email: mlpeters@sexscience.org
Website: http://www.sexscience.org/student_research_grants

SOCIETY FOR THE STUDY OF FRENCH HISTORY

School of History & Archives, Newman Building, University College Dublin, Belfield, Dublin, 4, Ireland
Tel: (353) 1 716 8151
Fax: (353) 1 334 462 927
Email: Sandy.Wilkinson@ucd.ie
Website: www.frenchhistorysociety.ac.uk
Contact: Dr Sandy Wilkinson, Secretary

The Society for the Study of French History was established to encourage research into French history. It offers a forum where scholars, teachers and students can meet and exchange ideas. It also offers bursaries for research and conferences to postgraduates in Britian and Ireland undertaking research into French history.

Society for the Study of French History Bursaries

Subjects: Any aspect of French history.
Purpose: To enable postgraduates undertake research in French history. To enable postgraduates attend conferences concerning French history.
Eligibility: Open to postgraduate students registered at a United Kingdom or Republic of Ireland university. Students at any level at postgraduate study may apply, provided that their dissertation is on some aspect of French history.
Level of Study: Postgraduate
Type: Bursary
Value: Up to UK£750
Length of Study: Dependent on the project for which the award is given
Frequency: Annual
Country of Study: France
No. of awards offered: Up to 20
Application Procedure: Applicants must give details of the research being pursued and the use to which the money would be put, along with the names of two referees. Applicants are responsible for writing to their referees. Successful applicants will be required to submit, in the first instance, a brief outline of their research proposal and then, after the research trip, a synopsis of their findings, both for publication on the Society's website. Application forms can be downloaded from the Society's website.
Closing Date: March 23rd
Funding: Private
Contributor: The Society for the Study of French History
No. of awards given last year: 11
No. of applicants last year: 13
Additional Information: Please see the website for further details http://www.frenchhistorysociety.ac.uk/grants.htm.

THE SOCIETY FOR THEATRE RESEARCH

c/o National Theatre Archive, 83-101 The Cut, London, SE1 8LL, England
Email: e.cottis@btinternet.com
Website: www.str.org.uk
Contact: Chairman, Research Awards Sub-Committee

The Society for Theatre Research was founded in 1948 for all those interested in the history and techniques of British theatre. It publishes annually one or more books, newsletters and three issues of the journal *Theatre Notebook*, holds lecture meetings and other events, and gives an annual book prize as well as grants for theatre research.

Society for Theatre Research Awards

Subjects: The history and practice of the British theatre, including music hall, opera, dance and other associated performing arts.
Purpose: To aid research into the history and practice of British theatre.
Eligibility: Applicants should normally be 18 years of age or over, but there is no other restriction on their status, nationality or the location of the research. Undergraduates should not apply.
Level of Study: Unrestricted
Value: £100–2,000
Frequency: Annual
Country of Study: Any country
No. of awards offered: 2 major awards and a number of lesser awards
Application Procedure: Applicants must write to the Chairman of the Research Awards Sub-Committee after October 1st of the preceding year for an application form and guidance notes. All applications and enquiries must be made by post to the Society's accommodation address.
Closing Date: February 7th
Funding: Private
Contributor: Members' subscriptions, donations and bequests
No. of awards given last year: 11
No. of applicants last year: 32
Additional Information: While applicants will need to show evidence of the value of the research and a scholarly approach, they are by no

means restricted to professional academics. Many awards, including major ones, have previously been made to theatre practitioners and amateur researchers who are encouraged to apply. The Society also welcomes proposals which in their execution extend the methods and techniques of historiography. In coming to its decisions, the Society will consider the progress already made by the applicants and the possible availability of other grants. Please see the website for further details http://www.str.org.uk/research/awards/index.html.

SOCIETY OF ACTUARIES

475 North Martingale Road, Suite 600, Schaumburg, IL, 60173-2226, United States of America
Tel: (1) 847 706 3500
Fax: (1) 847 706 3599
Email: eschulty@soa.org
Website: www.soa.org
Contact: Erika Schulty, Business Manager

The Committee on Knowledge Extension research of the Society of Actuaries, the Casuality Actuarial Society, The Actuarial Foundation of Canada, The Actuarial Foundation's Researchy Committee carry out research and education projects in actuarial science and study specific projects that could be advanced under this mechanism.

AFC/TAF/CKER/CAS Individual Grants Competition

Subjects: Actuarial science.
Purpose: To produce publications that will advance actuarial science, especially with regard to practical applications.
Eligibility: Individuals and groups may apply. Letters of Intent may come from practitioners, usually where the research project is not part of their employment; industry and university researchers; and academics. Graduate and undergraduate students are not eligible to apply individually, but may be part of a group in one of the above categories.
Level of Study: Unrestricted
Type: Grant
Value: Varies, approx. US$10,000–15,000
Length of Study: Projects should generally be of less than 1 year in duration
Frequency: Annual
Country of Study: Any country
No. of awards offered: Varies
Application Procedure: Applicants must submit a letter of intent and an application form.
Funding: Private
Contributor: The Society of Actuaries, The Actuarial Foundation, Actuarial Foundation of Canada and the Casualty Actuarial Society and individuals.
No. of awards given last year: 14
No. of applicants last year: 23
Additional Information: The project may be either theoretical or empirical in nature. A key criteria is that the project should have the potential to contribute significantly to the advancement of knowledge in actuarial science. The Actuarial Foundation, the CAS, the AFC and the SOA give preference to projects relating to current policy issues or having direct applications and those that further the basic or continuing education of actuaries. Proposals for innovative developments in actuarial education are also considered by The Actuarial Foundation, the CAS, the AFC and the SOA. More information is available from the website.

SOCIETY OF ARCHITECTURAL HISTORIANS (SAH)

1365 North Astor Street, Chicago, Illinois, 60610-2144, United States of America
Tel: (1) 312 573 1365
Fax: (1) 312 573 1141
Email: info@sah.org
Website: www.sah.org

SAH is an international not-for-profit membership organization that promotes the study and preservation of the built environment worldwide. The Society serves scholars, professionals in allied fields and the interested general public.

Beverly Willis Architecture Foundation Travel Fellowship

Subjects: Gender issues in the history of architecture, landscape architecture and associated fields.
Purpose: To advance the status of women in architecture.
Eligibility: Applicants must be members of the SAH.
Level of Study: Postdoctorate, Doctorate
Type: Fellowship
Value: US$1,500 travel stipend
Frequency: Annual
Country of Study: United States of America
No. of awards offered: 1
Application Procedure: Apply online.
Closing Date: September 5th (check with website)

Edilia and Francois-Auguste de Montequin Fellowship

Subjects: Spanish, Portugese and Ibero-American architecture.
Eligibility: Applicants must be members of the SAH.
Level of Study: Doctorate, Postdoctorate
Type: Fellowship
Value: US$2,000 for junior scholars awarded each year and US $6,000 for senior scholars awarded every 2 years
Length of Study: 1 year
Frequency: Every 2 years
No. of awards offered: 2
Application Procedure: Apply online.
Closing Date: September 5th
Funding: Foundation
Contributor: The Auguste de Montequin Foundation
Additional Information: Please see the website for further details.

George R. Collins Fellowship

Subjects: 19th or 20th century built environment.
Purpose: To enable a scholar to attend the annual meeting of the Society, held each April.
Level of Study: Doctorate, Postdoctorate, Postgraduate, Predoctorate
Type: Fellowship
Value: Up to $1000
Frequency: Annual
No. of awards offered: 1
Application Procedure: Applicants must complete an application form, available on request by writing to SAH for guidelines or visiting the SAH website.
Closing Date: September 5th
Funding: Private
No. of awards given last year: 1
Additional Information: Eligible to nationals of any country except United States of America.

Keepers Preservation Education Fund Fellowship

Subjects: Historic preservation.
Purpose: To enable a graduate student to attend the annual meeting of the Society, held each April.
Eligibility: Open to members of any nationality who are currently engaged in the study of historic preservation.
Level of Study: Postdoctorate, Doctorate, Postgraduate, Predoctorate
Type: Fellowship
Value: up to $1,000
Frequency: Annual
Country of Study: Any country
No. of awards offered: 1
Application Procedure: Applicants must write to SAH for application guidelines or download an application from the Society's website.
Closing Date: September 5th
Funding: Private
No. of awards given last year: 1
Additional Information: Please see the website for further details.

Rosann Berry Fellowship

Subjects: Architectural history or an allied field, e.g. city planning, landscape architecture, decorative arts or historic preservation.

Purpose: To enable a student engaged in advanced graduate study to attend the annual meeting of the Society.
Eligibility: Open to persons of any nationality who have been members of SAH for at least 1 year prior to the meeting, and who are currently engaged in advanced graduate study, normally beyond the Master's level, that involves some aspect of the history of architecture or of one of the fields closely allied to it.
Level of Study: Postdoctorate, Postgraduate
Type: Fellowship
Value: US$1,000 travel stipend
Frequency: Annual
Country of Study: Any country
No. of awards offered: 1
Application Procedure: Applicants must complete an application form, available on request by writing to SAH for guidelines or visiting the SAH website.
Closing Date: September 5th
Funding: Private
No. of awards given last year: 1
Additional Information: Please see the website for further details.

SAH Annual Meeting Fellowships for Scholars

Subjects: Europe in the 19th and 20th centuries as well as built environments worldwide (other than Europe) from ancient times to the present.
Purpose: To enable senior scholars and graduate students to attend the annual meeting of the society, held each April.
Eligibility: Applicants must be members of the SAH.
Level of Study: Doctorate, Postdoctorate, Postgraduate
Type: Fellowship
Value: Up to US$1,000 travel stipend
Frequency: Annual
Country of Study: United States of America
No. of awards offered: 10
Application Procedure: Apply online.
Closing Date: September 5th
Additional Information: Eligible to nationals of any country except United States of America.

SAH Fellowships for Independent Scholars

Purpose: To enable independent scholars to attend the annual meeting of the Society, held each April.
Level of Study: Doctorate, Graduate, Postdoctorate, Postgraduate
Type: Fellowship
Value: Up to $1000
Frequency: Annual
No. of awards offered: 2
Application Procedure: Applicants must complete an application form, available on request by writing to SAH for guidelines or visiting the SAH website.
Closing Date: September 5th
Funding: Private
No. of awards given last year: 2
Additional Information: Please see the website for further details.

SAH Study Tour Fellowship

Subjects: Internationally significant buildings and sites.
Eligibility: Applicants must be members of the SAH.
Level of Study: Doctorate
Type: Fellowship
Value: Varies
Length of Study: 5–10 days
Frequency: Annual
No. of awards offered: Varies
Application Procedure: Contact the Society.
Closing Date: Varies, check website

Sally Kress Tompkins Fellowship

Subjects: Architectural history and historic preservation.
Purpose: To enable an architectural history student to work as an intern on an Historic American Buildings Survey project, during the summer.
Eligibility: Open to architectural history and historic preservation students.
Level of Study: Doctorate, Postdoctorate, Postgraduate
Type: Fellowship
Value: US$10,000
Length of Study: 12 weeks
Frequency: Annual
Country of Study: United States of America
No. of awards offered: 1
Application Procedure: Applicants must submit an application including a sample of work, a letter of recommendation from a faculty member, and a United States Government Standard Form 171, available from HABS or most United States government personnel offices. Applications should be sent to the Sally Kress Tompkins Fellowship. Applicants not selected for the Tomkins Fellowship will be considered for other HABS Summer employment opportunities. For more information, please contact Lisa P. Davidson, HABS/HAER Co-ordinator.
Closing Date: December 31st
Funding: Government
No. of awards given last year: 1

For further information contact:

The Sally Kress Tompkins Fellowship, c/o HABS/HAER, National Park Service, 1849C Street NW, Washington, DC, 2270-20240, United States of America
Website: http://www.sah.org/jobs-and-careers/sah-fellowships-and-grants/research-fellowships

Samuel H. Kress Foundation Fellowships for Speakers

Subjects: Built environment of Europe from ancient times to the 19th century.
Eligibility: Applicants must be members of the SAH.
Level of Study: Doctorate, Postdoctorate
Type: Fellowship
Value: US$1,000 travel stipend
Frequency: Annual
Country of Study: United States of America
No. of awards offered: 2
Application Procedure: Apply online.
Closing Date: September 5th
Funding: Foundation
Contributor: Samuel H Kress Foundation
No. of awards given last year: 2
Additional Information: Eligible to nationals of any country except United States of America.

Scott Opler Membership Grant for Emerging Professionals

Subjects: Architectural history.
Eligibility: Applicants must be of the SAH.
Level of Study: Postgraduate, Postdoctorate
Type: Fellowship
Value: Up to $1,000 each
Length of Study: 1 year
Frequency: Annual
Country of Study: United States of America
No. of awards offered: 1
Closing Date: August 31st

For further information contact:

Website: http://www.sah.org/jobs-and-careers/sah-fellowships-and-grants/research-fellowships
Contact: Kathy Sturm, Meetings and Tours

Spiro Kostof Annual Meeting Fellowship

Subjects: Architectural history.
Purpose: To enable an advanced graduate student in architectural history to attend the annual meeting of the Society of Architectural Historians.
Eligibility: Open to Doctoral candidates only who have been members of the SAH for at least 1 year.
Level of Study: Doctorate, Predoctorate
Type: Fellowship
Value: US$1,000 travel stipend
Frequency: Annual
No. of awards offered: 1

Application Procedure: Applicants must write for an application form, available after June 1st by mail or by visiting the website.
Closing Date: September 5th
Funding: Commercial, private
Contributor: The Society of Architectural Historians
No. of awards given last year: 1

THE SOCIETY OF AUTHORS

84 Drayton Gardens, London, SW10 9SB, England
Tel: (44) 020 7373 6642
Fax: (44) 020 7373 5768
Email: info@societyofauthors.net
Website: www.societyofauthors.org
Contact: The Administration Office

The Society of Authors is a non-profit making organization, founded in 1884, to protect the rights and further the interests of authors, mainly by giving bursaries, advice and help to members. It administers prizes and grants for authors.

The Authors' Foundation Grants

Subjects: Natural history, landscape or the environment, travel writing.
Purpose: To provide grants to writers to assist them while writing books.
Eligibility: To apply the author must meet either of these conditions: the author has been commissioned by a commercial British publishers to write a work of fiction, poetry or non-fiction and needs funding. The author is without contractual commitment by a publisher but has least one book published already.
Level of Study: Professional development
Type: Grant
Value: UK£1,000–4,000
Length of Study: 1 year
Frequency: Biannual
Country of Study: United Kingdom
Application Procedure: Please check the website for more details http://www.societyofauthors.net/grants.
Closing Date: April 30th and September 30th
Funding: Foundation
Contributor: The Authors' Foundation
No. of awards given last year: 50
No. of applicants last year: 250
Additional Information: Editors, translators and screenwriters are not eligible.Specific grants also awarded every six months (for which all applicants will automatically be considered).

Elizabeth Longford Grants

Subjects: Historical biography.
Eligibility: Please check The Author's Foundation for specific eligibility details for this grant.
Level of Study: Professional development
Type: Grant
Value: UK£5,000
Frequency: Biannual
Country of Study: United Kingdom
Application Procedure: Please check the website for more details http://www.societyofauthors.net/grants.
Closing Date: April 30th and September 30th
Funding: Private
Contributor: Flora Fraser and Peter Soros
No. of awards given last year: 2

Great Britain Sasakawa Foundation Grants

Subjects: Authorship relating to Japanese culture or society.
Eligibility: Preference will be given to authors whose work helps to interpret modern Japan to the English-speaking world.
Level of Study: Professional development
Type: Grant
Value: UK£2,000
Length of Study: 1 year
Frequency: Annual
Country of Study: United Kingdom
Application Procedure: Contact Paula Johnson at the Authors' Foundation or check the website for more details http://www.societyofauthors.net/grants.
Closing Date: April 30th and September 30th
Funding: Foundation
Contributor: Sasakawa Foundation

The K. Blundell Trust Grant

Subjects: Fiction or non-fiction that has the aim of increasing social awareness.
Purpose: To support British authors under the age of 40 who need funding for important research, travel or other expenditure.
Eligibility: Applicants must be British by birth, resident in the UK, under the age of 40 and their work must contribute to the 'greater understanding of existing social and economic organization'. They must have had one book published.
Level of Study: Professional development
Type: Grant
Value: UK£1,000–4,000
Length of Study: 1 year
Frequency: Biannual
Country of Study: United Kingdom
Application Procedure: See the website for guidelines; submissions by fax or email are not acceptable.
Closing Date: April 30th and September 30th
Funding: Trusts
Contributor: K. Blundell Trust
No. of awards given last year: 8
No. of applicants last year: 25
Additional Information: No author may apply twice to the Trust within one 12-month period; grants will not be awarded to cover publication costs. Please see the website for further details http://www.societyofauthors.net/grants.

Michael Meyer Award

Subjects: Writing.
Eligibility: Please check The Author's Foundation for specific eligibility details for this grant.
Level of Study: Professional development
Type: Grant
Frequency: Annual, Biannual
Country of Study: United Kingdom
No. of awards offered: 1 in each award period
Application Procedure: Please check the website for more details http://www.societyofauthors.net/grants.
Closing Date: April 30th and September 30th
Funding: Trusts
Contributor: The late Michael Meyer
No. of awards given last year: 2

SOCIETY OF CHILDREN'S BOOK WRITERS AND ILLUSTRATORS (SCBWI)

8271 Beverly Boulevard, Los Angeles, CA, 90048, United States of America
Tel: (1) 323 782 1010
Fax: (1) 323 782 1892
Email: scbwi@scbwi.org
Website: www.scbwi.org
Contact: Mr Stephen Mooser, President

The Society of Children's Book Writers and Illustrators (SCBWI) is an organization of 22,000 writers, illustrators, editors, agents and publishers of children's books, television, film and multimedia.

Barbara Karlin Grant

Subjects: Children's picture books.
Purpose: To assist picture book writers in the completion of a specific project.
Eligibility: Open to both full and associate members of the Society who have never had a picture book published. The grant is not available for a project for which there is already a contract.
Level of Study: Unrestricted
Type: Grant
Value: The full grant is US$1,500 and the runner-up grant is US$500
Frequency: Annual
Country of Study: Any country
No. of awards offered: 1

Application Procedure: Applicants must write for details.
Closing Date: May 15th
Funding: Private
No. of awards given last year: 2
No. of applicants last year: 75

Don Freeman Memorial Grant-in-Aid

Subjects: Children's picture books.
Purpose: To enable picture-book artists to further their understanding, training and work in the picture-book genre.
Eligibility: Open to both full and associate members of the Society who, as artists, seriously intend to make picture books their chief contribution to the field of children's literature.
Level of Study: Unrestricted
Type: Grant
Value: The full grant is US$1,500 and the runner-up grant is US$500
Frequency: Annual
Country of Study: Any country
No. of awards offered: 2
Application Procedure: Applicants must submit an application to the Society. Receipt of the application will be acknowledged.
Closing Date: Application requests should be submitted by June 15th and completed applications should be submitted by February 10th
Funding: Private
No. of awards given last year: 2
No. of applicants last year: 48

SCBWI General Work-in-Progress Grant

Subjects: Children's literature.
Purpose: To assist children's book writers in the completion of a specific project.
Eligibility: Open to both full and associate members of the Society. The grant is not available for a project for which there is already a contract. Recipients of previous grants are not eligible to apply for any further SCBWI grants.
Level of Study: Unrestricted
Type: Grant
Value: The full grant is US$1,500 and the runner-up grant is US$500
Frequency: Annual
Country of Study: Any country
No. of awards offered: 1 full grant and 1 runner-up grant
Application Procedure: Applicants must write for details.
Closing Date: May 1st
Funding: Private
No. of awards given last year: 2
No. of applicants last year: 123

SCBWI Grant for a Contemporary Novel for Young People

Subjects: Children's literature.
Purpose: To assist children's book writers in the completion of a specific project.
Eligibility: Open to both full and associate members of the Society. The grant is not available for a project for which there is already a contract. Recipients of previous grants are not eligible to apply for any further SCBWI Grants.
Level of Study: Unrestricted
Type: Grant
Value: The full grant is US$1,500 and the runner-up grant is US$500
Frequency: Annual
Country of Study: Any country
No. of awards offered: 1 full grant and 1 runner-up grant
Application Procedure: Applicants must write for details.
Closing Date: May 1st
Funding: Private
No. of awards given last year: 2
No. of applicants last year: 110

SCBWI Grant for Unpublished Authors

Subjects: Children's literature.
Purpose: To assist children's book writers in the completion of a specific project.
Eligibility: Open to both full and associate members of the Society who have never had a book published. The grant is not available for a project for which there is already a contract. Recipients of previous grants are not eligible to apply for any further SCBWI Grants.
Level of Study: Unrestricted
Type: Grant
Value: The full grant is US$1,500 and the runner-up grant is US$500
Frequency: Annual
Country of Study: Any country
No. of awards offered: 1 full grant and 1 runner-up grant
Closing Date: May 1st
Funding: Private
No. of awards given last year: 2
No. of applicants last year: 45

SCBWI Nonfiction Research Grant

Subjects: Children's literature.
Purpose: To assist children's book writers in the completion of a specific project.
Eligibility: Open to both full and associate members of the Society. The grant is not available for a project for which there is already a contract. Recipients of previous grants are not eligible to apply for any further SCBWI Grants.
Level of Study: Unrestricted
Type: Grant
Value: The full grant is US$1,500 and the runner-up grant is US$500
Frequency: Annual
Country of Study: Any country
No. of awards offered: 1 full grant and 1 runner-up grant
Application Procedure: Applicants must write for details.
Closing Date: May 1st
Funding: Private
No. of awards given last year: 1
No. of applicants last year: 75

SOCIETY OF ENVIRONMENTAL TOXICOLOGY AND CHEMISTRY (SETAC)

229 South Baylen Street, 2nd Floor, Pensacola, FL, 32502, United States of America
Tel: (1) 850 469 1500
Fax: (1) 850 469 9778
Email: setac@setaceu.org
Website: www.setac.org

The Society of Environmental Toxicology and Chemistry (SETAC) is a non-profit, worldwide professional society comprised of individuals and institutions who support the development of principles and practices for protection, enhancement and management of sustainable environmental quality and ecosystem integrity.

Procter & Gamble Fellowship for Doctoral Research in Environmental Science

Subjects: Environmental sciences.
Purpose: To promote the advancement and application of scientific research, education in the environmental sciences and the use of science in environmental policy and decision making.
Eligibility: Open to Doctoral students whose research area and academic standing are consistent with the research topics.
Level of Study: Doctorate
Type: Fellowships
Value: US$15,000
Length of Study: 1 year
Frequency: Annual
Application Procedure: Applicants must submit electronically a description of the dissertation research, a curriculum vitae and a letter of support from the dissertation director.
Closing Date: September 1st
Additional Information: The award is paid to the recipient's institution, so it is necessary that an authorized representative approve the application on behalf of the institution and certify that none of the award will be spent on overhead.

For further information contact:

Europe Avenue de la Toison d'Or 67 B-1060, Brussels, FL, 32501-3367, Belgium
Tel: (32) 850 469 1500

Fax: (32) 850 469 9978
Contact: Bart Bosveld, Executive Director

SOCIETY OF EXPLORATION GEOPHYSICISTS FOUNDATION (SEG)

SEG Foundation PO 702740, Tulsa, OK, 74170-2740, United States of America
Tel: (1) 918 497 5500
Fax: (1) 918 497 5557
Email: scholarships@seg.org
Website: www.seg.org
Contact: SEG Scholarship Committee

The Society of Exploration Geophysicists Foundation (SEG) began a programme of encouraging the establishment of scholarship funds by companies and individuals in the field of geophysics in 1956. In 1963, the Foundation's activities were expanded to include grants-in-aid.

Geoscientists Without Borders

Subjects: Geophysics.
Purpose: To encourage and support scientific, educational and charitable activities of benefit to the general public, to geophysicists and to the geophysical community.
Value: Maximum US$50,000 per year
Length of Study: 1 or 2 year duration
Frequency: Annual
Application Procedure: Two phase application process.Phase I: (LOI) Project summary (twice a year) if selected it goes into phase II. Phase II: More detailed project description.
Closing Date: February 20th
Funding: Individuals, corporation
Contributor: SEG Foundation
Additional Information: Please see the website for further details http://www.seg.org/web/foundation/programs/geoscientists-without-borders.

SEG Scholarships

Subjects: Applied geophysics and related fields.
Purpose: To encourage careers in applied geophysics and related fields.
Eligibility: Open to citizens of any non-sanctioned country who are entering undergraduate or graduate level and have above average grades and an aptitude for geophysics.
Level of Study: Doctorate, Graduate, Undergraduate
Type: Scholarship
Value: US$500–14,000 per academic year. Average awards are approx. US$2,500
Length of Study: 1 academic year, may be eligible for renewal
Frequency: Annual
Country of Study: Any country
No. of awards offered: Varies
Application Procedure: Applicants must submit a completed application form accompanied by latest transcripts and 2 letters of recommendation from faculty members who are familiar with the applicant's academic work.
Closing Date: March 1st of the year in which the award is made
Funding: Private, commercial, corporation, individuals, trusts
No. of awards given last year: 119
No. of applicants last year: 600
Additional Information: More than US$400,000 was granted in scholarships during the last academic year. Please see the website for further details http://www.seg.org/web/foundation/programs/scholarship.

SOCIETY OF ORTHOPAEDIC MEDICINE

4th floor, 151 Dale Street, Liverpool, L2 2AH, England
Tel: (44) 0151 237 3970
Fax: (44) 0151 237 3971
Email: admin@somed.org
Website: www.somed.org

The Society of Orthopaedic Medicine is a non-profit making organization offering grants for work within musculoskeletal medicine.

SOM Research Grant

Subjects: Musculoskeletal and orthopaedic medicine.
Purpose: To assist those undertaking programmes of study that will increase knowledge in the field of orthopaedic medicine and enhance their professional development.
Eligibility: Applications are considered from those who may be undertaking a research degree leading to an MPhil, a research degree leading to a PhD/DProf, a pilot study or a presentation at a conference. Grants are also available for equipment. Must be member of society of Orthopaedic medicine.
Level of Study: Graduate, Postdoctorate, Predoctorate, Professional development
Value: Up to UK£5,000. UK£2,000 of this allocation will be available in the form of smaller grants of up to UK£500
Frequency: Annual
Country of Study: United Kingdom
No. of awards offered: Varies according to funds
Application Procedure: Applicants must complete an application form available from the organization.
Closing Date: July 23rd
Funding: Commercial
No. of awards given last year: 5
No. of applicants last year: 8

SOCIETY OF WOMEN ENGINEERS (SWE)

203 N. La Salle Street, Suite 1675, Chicago, IL, 60601, United States of America
Tel: (1) 1 877 793 4636
Fax: (1) 312 596 5252
Email: hq@swe.org
Website: www.swe.org
Contact: Ms Betty Shanahan, Executive Director

The Society of Women Engineers (SWE) was founded in 1950, and is a non-profit educational service organization. SWE is the driving force that establishes engineering as a highly desirable career aspiration for women. SWE empowers women to succeed and advance in those aspirations and be recognized for their life-changing contributions and achievements as engineers and leaders.

Society of Women Engineers' Scholarships

Subjects: The SWE Scholarship program provides financial assistance to women admitted to accredited baccalaureate or graduate programs, in preparation for careers in engineering, engineering technology and computer science.
Eligibility: Please check website for requirements and eligibility at www.SWE.org/scholarships.
Level of Study: Doctorate, Graduate, Postgraduate
Type: Scholarship
Value: $1,000–20,000
Country of Study: United States of America
No. of awards offered: Varies
Application Procedure: Apply online at website.
Closing Date: February 15th
Funding: Corporation, foundation
No. of awards given last year: 24 awards to masters and phD students
No. of applicants last year: 2,600 freshmen through PhD

SOCRATES SCULPTURE PARK

PO Box 6259, 32-01 Vernon Boulevard, Long Island City, NY, 11106, United States of America
Tel: (1) 718 956 1819
Fax: (1) 718 626 1533
Email: info@socratessculpturepark.org
Website: www.socratessculpturepark.org

Socrates Sculpture Park was an abandoned riverside landfill and illegal dumpsite until 1986 when a coalition of artists and community members, under the leadership of artist Mark di Suvero, transformed it into an open studio and exhibition space for artists and a neighbourhood park for local residents. Today it is an internationally renowned outdoor museum and artist residency programme that also serves as a vital New York City park offering a wide variety of public services.

Socrates Sculpture Park Emerging Artist Fellowship Program

Subjects: Sculpture.
Purpose: To provide artists with opportunities to create and exhibit large-scale work in a unique environment that encourages strong interaction between artists, artworks and the public.
Eligibility: Open to artists who are not yet well established, are New York State residents and are in need of financial assistance.
Level of Study: Professional development
Type: Fellowship
Value: US$5,000
Length of Study: 2–6 months
Frequency: Annual
Country of Study: United States of America
Application Procedure: Applicants can download the application form from the website. The completed application form along with a curriculum vitae, references, slide script and proposal must be submitted.
Closing Date: January 7th
Additional Information: Grants and fellowships are not available to artists who are enrolled in a school, college or university programme. Please see the website for further details http://www.socratessculpturepark.org/exhibitions/artist-opportunities/annual-submissions-guidelines.html.

SOIL AND WATER CONSERVATION SOCIETY (SWCS)

945 SW Ankeny Road, Ankeny, IA, 50023, United States of America
Tel: (1) 515 289 2331
Fax: (1) 515 289 1227
Email: sueann.lynes@swcs.org
Website: www.swcs.org

The Soil and Water Conservation Society (SWCS) fosters the science and the art of soil, water and related natural resource management to achieve sustainability. The SWCS promotes and practices an ethic recognizing the interdependence of people and the environment.

The Kenneth E. Grant Scholarship

Subjects: Soil science, water science and related natural resource management.
Purpose: To actively promote multi-disciplinary research.
Eligibility: An applicant for the research scholarship must be a member of SWCS; have demonstrated integrity, ability, and competence to complete the specified study topic; be eligible for graduate work at an accredited institution; and show reasonable need for financial assistance.
Level of Study: Postgraduate
Type: Scholarship
Value: US$1,300
Length of Study: 1 year
Frequency: Annual
No. of awards offered: 1
Application Procedure: Apply online.
Closing Date: February 12th
Additional Information: Please see the website for further details http://www.swcs.org/en/members_only/scholarships/.

SONS OF THE REPUBLIC OF TEXAS

1717 8th Street, Bay City, TX, 77414, United States of America
Tel: (1) 979 245 6644
Fax: (1) 979 244 3819
Email: srttexas@srttexas.org
Website: www.srttexas.org
Contact: Janet Knox, Administrative Assistant

Sons of the Republic of Texas seeks to perpetuate the memory and spirit of the men and women who settled Texas and won its independence through great personal sacrifice and dedication to the cause of freedom.

Presidio La Bahia Award

Subjects: History.
Purpose: To promote the suitable preservation of relics, appropriate dissemination of data and research into Texas heritage, with particular emphasis on the Spanish colonial period.
Eligibility: Open to all persons interested in the Spanish colonial influence on Texas culture.
Level of Study: Unrestricted
Type: Award
Value: The 1st prize is a minimum of US$1,200, and the 2nd and 3rd prizes are the divided balance from the total amount available of US $2,000 at the discretion of the judges
Frequency: Annual
Study Establishment: Any suitable institution
Country of Study: Any country
No. of awards offered: 3
Application Procedure: Applicants must submit four copies of published writings to the office. Galley proofs are not acceptable.
Closing Date: Entries are accepted from June 1st to September 30th
Funding: Private
No. of awards given last year: 1
Additional Information: Research writings have, in the past, proved to be the most successful type of entry. However, consideration will be given to other literary forms, art, architecture and archaeological discovery. For projects other than writing, contestants should furnish a description of the proposed entry, so that the Chairman may issue specific instructions.

Summerfield G Roberts Award

Subjects: History.
Purpose: To encourage literary effort and research about historical events and personalities during the days of the Republic of Texas 1836–1846, and to stimulate interest in the period.
Eligibility: Open to all writers.
Level of Study: Unrestricted
Type: Award
Value: US$2,500
Frequency: Annual
Study Establishment: Any suitable institution
Country of Study: Any country
No. of awards offered: 1
Application Procedure: Manuscripts must be written or published during the calendar year for which the award is given. There is no word limit. No entry may be submitted more than one time. The manuscripts must be mailed (5 copies, for the use of the judges) to the General Office of the Sons of the Republic of Texas.
Closing Date: January 15th
Funding: Private
No. of awards given last year: 1
Additional Information: The award was made possible through the generosity of Mr and Mrs Summerfield G Roberts.

THE SOROPTIMIST FOUNDATION OF CANADA

13311 Yonge Street, Suite 104, Richmond Hill, ON, L4E 3L6, Canada
Tel: (1) 905 773 9927
Email: chair@soroptimistfoundation.ca
Website: www.soroptimistfoundation.ca
Contact: Corinne Rivers, Treasurer

The Soroptimist Foundation was established in 1963 by incorporation under the laws of Canada, for charitable, scientific, literary and educational purposes.

Soroptimist Foundation of Canada Grants for Women

Subjects: All subjects.
Purpose: To financially assist women who wish to complete graduate studies for careers that will improve the quality of the lives of women and girls.
Eligibility: Open only to women candidates who are citizens or landed immigrants of Canada enrolled in an accredited Canadian

university and will work in Canada for a minimum of 2 years after completion of the degree course
Level of Study: Postgraduate
Type: Grant
Value: Canadian $7,500
Frequency: Annual
Country of Study: Canada
No. of awards offered: 4
Application Procedure: Applicants must send completed application forms and all required additional documents before the closing date. Please see the website www.soroptimistfoundation.ca for further details.
Closing Date: January 31st each year
Funding: Foundation
No. of awards given last year: 4
No. of applicants last year: 110

For further information contact:

For Western Canada: 3107 Highland Boulevard, North Vancouver, BC, V7R 0X5
Contact: Jean Violette

THE SOUTH AFRICAN INSTITUTE OF INTERNATIONAL AFFAIRS (SAIIA)

Jan Smuts House, PO Box 31596, Johannesburg, Braamfontein, 2017, South Africa
Tel: (27) 11 339 2021
Fax: (27) 11 339 2154
Email: info@saiia.org.za
Website: www.saiia.org.za
Contact: Mr Jonathan Stead, Director of Operations

The South African Institute of International Affairs (SAIIA) is an independent, non-governmental foreign policy think tank, whose purpose is to encourage wider and more informed interest in international affairs and public education and to focus on policy-relevant research.

SAIIA Bradlow Fellowship

Subjects: Social and behavioural sciences.
Purpose: To provide travel costs and a stipend to enable a senior scholar to reside at the Institute in order to research a subject of importance to South Africa's international relations.
Eligibility: Open to senior scholars.
Level of Study: Research
Type: Fellowship
Value: The stipend covers living expenses and a return economy-class airfare from the candidate's place of residence.
Length of Study: 3–6 months
Frequency: Annual
Study Establishment: SAIIA
Country of Study: South Africa
No. of awards offered: 1
Application Procedure: Applicants must send a curriculum vitae including a short research proposal of not more than 1,000 words to the Director of Studies.
Funding: Private
Contributor: The Bradlow Foundation
No. of awards given last year: 1

SAIIA Konrad Adenauer Foundation Research Internship

Subjects: Politics, international relations, journalism and economics.
Purpose: To enable research interns to enroll at the University of Witwatersrand for a Master's degree by coursework while simultaneously working at SAIIA.
Eligibility: Open to South African citizens under 30 years of age.
Level of Study: Postgraduate
Type: Internship
Value: Full bursary for tuition fees and a monthly stipend to cover living costs and accommodation
Length of Study: 10 months
Frequency: Annual
Study Establishment: SAIIA and the University of Witwatersrand
Country of Study: South Africa
No. of awards offered: 3
Application Procedure: Applicants must send a curriculum vitae, names and contact details of three referees, letter of motivation, outline of research interests in the field of international relations, three written references, June results and academic transcripts of previous degree and one example of written work not exceeding 3,000 words on a topic of choice.
Closing Date: October 8th
Funding: Foundation
Contributor: Konrad Adenauer Foundation
No. of awards given last year: 2
No. of applicants last year: 32

For further information contact:

Email: grobbelaarn@saiia.wits.ac.za

SOUTH AFRICAN MEDICAL RESEARCH COUNCIL (SAMRC)

Research Grants Administration, Francie van Zijl Drive, Parowvallei, Cape, PO Box 19070, Tygerberg, 7505, South Africa
Tel: (27) 21 938 0911
Fax: (27) 21 938 0200
Email: info@mrc.ac.za
Website: www.mrc.ac.za
Contact: Mrs Marina Jenkins, Research Grants Administration

The South African Medical Research Council's (SAMRC) brief is to undertake excellent research, which can be implemented to improve the health of all South Africans. Research priorities are regularly established through broad consultation. The SAMRC's mission is to improve the health, status and quality of life of the nation through excellence in scientific research.

SAMRC Local Postdoctoral Scholarships

Subjects: Health sciences.
Purpose: To create opportunities for recent postdoctoral scientists who are otherwise lost to the system and country.
Eligibility: Open to candidates who have completed a Doctorate research in any field in the health sciences.
Level of Study: Postdoctorate
Type: Scholarships
Value: Varies
Length of Study: 3 years
Frequency: Annual
Country of Study: South Africa
Application Procedure: Applicants must submit a research plan, institutional and ethics approval for the project, proof of source of funding for the research project and 3 referees reports.
Closing Date: August 31st
Additional Information: Please see the website for further details http://www.mrc.ac.za/researchdevelopment/localpostdocscholar.htm.

SAMRC Post MBChB and BChD Grants

Subjects: Health sciences.
Purpose: To provide an opportunity for full-time research training for medical doctors as part of their development towards a Master's degree.
Eligibility: Open to recent MBChB or BChD graduates who are registered with the health professions council and have completed their internship and community service and registered for a Master's degree in a specialist field or a research area.
Level of Study: Postgraduate
Type: Grant
Value: South African Rand 80,000 per year
Length of Study: 4 years
Frequency: Annual
Country of Study: South Africa
Application Procedure: Applicants must submit their applications to the MRC through the institution's postgraduate bursury office.
Closing Date: June 30th
Additional Information: Please see the website for further details http://www.mrc.ac.za/researchdevelopment/postscholar1.htm.

SOUTHEAST ASIA URBAN ENVIRONMENTAL MANAGEMENT APPLICATIONS (SEA-UEMA)

School of Environment, Resources and Development, Asian Institute of Technology PO Box 4, Klong Luang, Pathumthani, 12120, Thailand
Tel: (66) 02 5245777
Fax: (66) 02 5162126, 5248338
Email: uemapplications@ait.ac.th
Website: www.sea-uema.ait.ac.th

The Canadian International Development Agency and the Asian Institute of Technology have entered into a contribution agreement on urban environmental management. This is in the form of Southeast Asia Urban Environmental Management Applications (SEA-UEMA) Project, which aims to improve the urban environmental management policies and good practices in the region.

SEA-UEMA Post-doctoral Research Fellowships

Subjects: Urban environmental management practices and policies.
Purpose: To contribute to the improvements of urban environmental conditions in Southeast Asia.
Eligibility: Open to candidates from the Canadian ODA eligible Southeast Asian countries who hold a Doctoral degree.
Level of Study: Postdoctorate
Type: Fellowships
Value: Baht 50,000 per month plus Baht 1,50,000 as research fund and other benefits
Length of Study: 1 year
Frequency: Annual
Application Procedure: Applicants must submit an application form along with curriculum vitae, reprints of publications and a research proposal.
Closing Date: July 15th
Additional Information: For further inquiries please contact Mr Vikas Nitivattaranon, Assistant Professor. Email:bimal@ait.ac.th; Tel: 02 5246399.

SOUTHERN AFRICAN MUSIC RIGHTS ORGANIZATION (SAMRO) ENDOWMENT FOR THE NATIONAL ARTS

PO Box 31609, Johannesburg, Braamfontein, 2017, South Africa
Tel: (27) 11 712 8000
Fax: (27) 11 403 1934
Email: customerservices@samro.org.za
Website: www.samro.org.za
Contact: J C Otto, Liaison & Research Officer

Southern African Music Rights Organization (SAMRO) is Southern Africa's society of composers and lyricists, administering the performing, transmission and broadcasting rights in the musical works of its members and the members of its affiliated societies. Through the SAMRO Endowment for the National Arts (SENA), it encourages the developement of the arts by combining funding with support and advisory services to individuals and organisations to enable them to further their music education, composer careers and more.

SAMRO Intermediate Bursaries for Composition Study In Southern Africa

Subjects: Music.
Purpose: To support music composition study as a major subject in either the western art, choral or jazz popular music genres.
Eligibility: Open to citizens of South Africa, Botswana, Lesotho and Swaziland who have met the requirements for proceeding to the 3rd, 4th or honours year of a senior undergraduate degree or equivalent diploma course. Applicants must have been born after February 15th, 1974. For those entering any year of a Master's or doctoral degree, the age limit is 32. Older students are considered in special circumstances.
Level of Study: Postgraduate
Type: Bursary
Value: Varies
Length of Study: 1 year
Frequency: Annual
Study Establishment: A university, institute of technology or other recognized statutory institute of tertiary education approved by the trustees
Country of Study: South Africa, Botswana, Lesotho, or Swaziland
No. of awards offered: 7
Application Procedure: Applicants must complete an application form.
Closing Date: March 1st
Funding: Private
Contributor: SAMRO
No. of awards given last year: 4
No. of applicants last year: 5
Additional Information: Applicants must produce an official letter of acceptance for entering any year of a Master's or doctoral degree. Please see the website for further details http://www.samro.org.za/node/4570.

SAMRO Overseas Scholarship

Subjects: Music.
Purpose: To encourage music study at the postgraduate level in the western art/choral or jazz popular music genres.
Eligibility: Open to postgraduate students who are citizens of South Africa, Botswana, Lesotho or Swaziland. The age limit is 34 years.
Level of Study: Postgraduate
Type: Scholarship
Value: Rand 160,000 plus travel expenses of up to Rand 10,000
Length of Study: 2 years
Frequency: Annual
Study Establishment: An institute or educational entity approved by the SAMRO Endowment for the National Arts (SENA)
Country of Study: United Kingdom, Europe or North America
No. of awards offered: 2
Application Procedure: Applicants must complete an application form.
Closing Date: May 31st
Funding: Private
Contributor: SAMRO
No. of awards given last year: 2
No. of applicants last year: 13
Additional Information: Please see the website for further details.

For further information contact:

Tel: 011 712 8444/011 712 8417
Email: anriette.chorn@samro.org.za
Contact: Anriette Chorn

SAMRO Postgraduate Bursaries for Indigenous African Music Study

Subjects: Music.
Purpose: To encourage the study of indigenous African music at the postgraduate level in either the traditional, western art/choral or jazz popular music genres.
Eligibility: Open to postgraduate students who are citizens of South Africa, Botswana, Lesotho or Swaziland. The age limit is 40 years.
Level of Study: Postgraduate
Type: Bursary
Value: Rand 5,500
Length of Study: 5 years
Frequency: Annual
Study Establishment: A university or other recognized statutory institute of tertiary education approved by the trustees and situated in SAMRO's current territory of operation
Country of Study: South Africa, Botswana, Lesotho, or Swaziland
No. of awards offered: 6
Application Procedure: Applicants must complete an application form.
Closing Date: March 1st
Funding: Private
Contributor: SAMRO
No. of awards given last year: 4
No. of applicants last year: 5
Additional Information: Applicants must produce an official letter of acceptance from a recognized tertiary institute of learning. They must

have acceptance into the 1st year or any subsequent year of a postgraduate degree in indigenous African music at such an institute.

For further information contact:

Samro Endowment for the National Arts, P O Box 31609, BRAAMFONTEIN, 2017
Tel: 011 489 5000
Fax: 011 403 1934
Email: sena@samro.org.za
Website: http://www.samro.org.za/node/4570

THE SPENCER FOUNDATION

625 North Michigan Avenue Suite 1600, Chicago, IL, 60611, United States of America
Tel: (1) 312 337 7000
Fax: (1) 312 337 0282
Email: abrinkman@spencer.org
Website: www.spencer.org
Contact: Annie Brinkman, Grant Manager

The Spencer Foundation was established in 1962 by Lyle M. Spencer. The Foundation is committed to supporting high quality investigation of education through its research programmes and to strengthening and renewing the educational research community through its fellowship and training programmes and related activities.

Spencer Foundation Dissertation Fellowship Program

Subjects: Education.
Purpose: To encourage a new generation of scholars from a wide range of disciplines and professional fields to undertake research relevant to the improvement of education.
Eligibility: Open to candidates for the Doctoral degree at a graduate school within the United States of America.
Level of Study: Doctorate
Type: Fellowship
Value: US$25,000
Frequency: Annual
Country of Study: United States of America
No. of awards offered: Up to 25
Application Procedure: Applicants must apply online. In addition to the completed application form they must submit a list of publications/presentations, a dissertation abstract, a narrative discussion of the dissertation, a work plan, 2 letters of recommendation and a graduate transcript.
Closing Date: October 27th
Additional Information: Please see the website for further details http://www.spencer.org/content.cfm/fellowship-awards.

For further information contact:

Email: fellows@spencer.org

ST VINCENT'S INSTITUTE (SVI)

41 Victoria Parade, Fitzroy, VIC 3065, Australia
Tel: (61) 03 9288 2480
Fax: (61) 03 9416 2676
Email: grantsadmin@svi.edu.au
Website: www.svi.edu.au
Contact: Dr Rachel Mudge, Grants Officer

St Vincent's Institute (SVI) conducts programs of basic and clinical research into diseases that have a high impact on the community and is focused on exploring both disease causes and prevention, with a commitment to discovering practical and far-reaching solutions to diseases that impact on the everyday life of people around the world.

SVI Foundation Postgraduate Student Award

Subjects: Medical research.
Purpose: To provide valuable financial support for outstanding students commencing their PhD training at SVI.
Eligibility: Open to First Class (Honours) students commencing full time study towards a PhD at St Vincent's Institute. Students need to successfully apply for a full PhD stipend (APA, Dora Lush etc.) to be eligible for this award. This is intended to be a top up award for those students receiving full PhD scholarships.
Level of Study: Doctorate, Graduate
Type: Award
Value: $5,000 per year
Length of Study: 3 years
Frequency: Annual
Study Establishment: St Vincent's Institute
Country of Study: Australia
No. of awards offered: Up to 2
Application Procedure: Check website for further details.
Closing Date: October 31st
Funding: Foundation
No. of awards given last year: 2
No. of applicants last year: 7
Additional Information: Scholarships will be awarded on the basis of the applicant's academic excellence and research potential. An interview with the selection panel may be required.

THE STANLEY MEDICAL RESEARCH INSTITUTE

8401 Connecticut Avenue, Suite 200, Chevy Chase, MD, 20815, United States of America
Tel: (1) 301 571 0760
Fax: (1) 301 571 0769
Email: marter@stanleyresearch.org
Website: www.stanleyresearch.org
Contact: Ms Rhoda Marte

SMRI is a nonprofit organization that supports research designed to find better treatments for schizophrenia and bipolar disorder. Approximately 75 per cent of its annual expenditures are devoted to the direct clinical testing of new treatments. 25 per cent is earmarked for research on the causes of these illness.

Stanley Medical Research Institute Postdoctoral Research Fellowship Program

Subjects: Psychiatry and mental health-causes and treatment of schizophrenia and bipolar disorder.
Purpose: To attract top-quality scientists to specific areas of research in severe mental illness.
Eligibility: Open to researchers worldwide. Applicants must be a doctorate or equivalent professional and must not be more than 2 years past receipt of their doctoral degree.
Level of Study: Postdoctorate, Research
Type: Research fellowship
Value: Up to US$150,000 spread over 2 years including indirect costs of up to 15 per cent as part of the total grant budget
Length of Study: 2 years
Frequency: Annual
Country of Study: Any country
No. of awards offered: 1
Application Procedure: Interested applicants must apply online at the website. Detailed instructions for application are available in a printable format (PDF).
Closing Date: January 15th
Funding: Individuals, foundation
Additional Information: Top priority will be given to applications that address the following questions: (a) Synthesis or screening of compounds to be tested against molecular targets identified for servere mental illness (schizophrenia, bipolar disorder, servere depression). (b) Pharmacokinetic and toxicological characterization of potential drugs for servere mental illness. (c) Ethnopharmacology, i.e., discovery or characterization of potentially useful psychoactive compounds from non-Western medical traditions.

Stanley Medical Research Institute Research Grants Program

Subjects: Psychiatry and mental health: causes and treatment of schizophrenia and bipolar disorder.
Purpose: To support researchers at all levels of development in fields related to the cause and treatment of schizophrenia and bipolar

disorder, as well as from other areas of medicine and biology who wish to initiate new projects in this field.
Eligibility: Open to researchers worldwide. Applicants must be a doctorate or equivalent professional.
Level of Study: Research
Type: Research grant
Value: US$5,000 per year; indirect costs may be paid up to 15 per cent as part of the total grant budget
Length of Study: 2 years
Frequency: Annual
Country of Study: Any country
Application Procedure: Interested applicants must apply online at the SMRI website. Detailed instructions for application are available in a printable formate (PDF).
Closing Date: March 1st
Funding: Private, foundation, individuals

Stanley Medical Research Institute Treatment Trial Grants Program

Subjects: Psychology and mental health: treatments for schizophrenia and bipolar disorder.
Purpose: To support researchers and facilitate the direct testing of new treatments for schizophrenia and bipolar disorder. This programme supports projects involving human subjects and testing a therapeutic intervention (medication; device; putative medication including plant substances and nutritional; psychotherapy).
Eligibility: Open to researchers worldwide. Applicants must be a Doctorate or equivalent professional.
Level of Study: Research
Type: Research grant
Value: US$300,000 per year
Length of Study: 3 years
Frequency: Annual
Country of Study: Any country
Application Procedure: Interested applicants must apply online at the SMRI website. Detailed instructions for application are available in printable format (PDF).
Closing Date: October 1st
Funding: Foundation, individuals
Additional Information: Please check the master list of Stanley Medical Research Institute awarded trials on the website prior to completing the application. If we are supporting trials with the compound you propose, we are unlikely to support an additional trial until results are available.

STANLEY SMITH (UK) HORTICULTURAL TRUST

c/o Cambridge University Botanic Garden. Cory Lodge, 1 Brookside, Cambridge, CB2 1JE, England
Tel: (44) 12 2333 6299
Fax: (44) 12 2333 6278
Email: jc240@cam.ac.uk
Contact: Dr James Cullen, Director

The Stanley Smith (UK) Horticultural Trust supports projects that contribute to the development of the art and science of horticulture, i.e. garden conservation and restoration, education and training, research, publications and travel.

Stanley Smith (UK) Horticultural Trust Awards

Subjects: Horticulture. The Trust supports individual projects in all aspects (including training) of amenity horticulture and some aspects of commercial horticulture.
Eligibility: Open to institutions and individuals. All projects are judged entirely on merit and there are no eligibility requirements, but grants are not awarded for students to take academic or diploma courses of any kind.
Level of Study: Unrestricted
Type: Varies
Value: Varies
Length of Study: Dependent on the nature of the project
Frequency: Twice a year (spring, autumn)
Country of Study: Any country
No. of awards offered: Varies
Application Procedure: Applicants must apply to the Trust. Trustees allocate awards in Spring and Autumn.
Closing Date: February 15th and August 15th (check with website)
Funding: Private
Contributor: Donations
No. of awards given last year: 30
No. of applicants last year: 200

For further information contact:

Email: tdaniel@calacademy.org
Website: http://www.adminitrustllc.com/stanley-smith-horticultural-trust/

STATE LIBRARY OF NEW SOUTH WALES

Macquarie Street, Sydney, NSW, 2000, Australia
Tel: (61) 02 9273 1414
Fax: (61) 02 9273 1255
Email: library@sl.nsw.gov.au
Website: www.sl.nsw.gov.au
Contact: Mitchell Librarian

The State Library of New South Wales is the premier reference and research library in New South Wales. The Library consists of the State Reference Library and the Mitchell Library, which contains the renowned Australian research collections pertaining to the history of Australia and the Southwest Pacific region.

Blake Dawson Waldron Prize for Business Literature

Subjects: Australian corporate and commercial literature, histories, accounts and analyses of corporate affairs as well as biographies of business men and women.
Purpose: To encourage the highest standards of commentary in the fields of business and finance.
Eligibility: The author must be a living Australian citizen or hold permanent resident status. The work must have primary reference to business or financial affairs, business or financial institutions or people directly associated with business and financial affairs, be written in the English language, published in book form and consist of a minimum of 50,000 words.
Level of Study: Professional development
Type: Prize
Length of Study: Australian $30,000
Frequency: Annual
Study Establishment: State Library of New South Wales
Country of Study: Australia
Application Procedure: All nominations must be made on the appropriate form, be submitted with five copies of the nominated work and be accompanied by an entry fee of Australian $66 per title to be eligible for consideration. A separate form must be completed for each nomination. See State Library of NSW website www.sl.nsw.gov.au/awards/
Closing Date: September 25th (check with website)
Funding: Private

For further information contact:

Education and Client Liaison Branch, State Library of New South Wales, Australia
Email: smartin@sl.nsw.gov.au
Contact: Stephen Martin, Senior Project Officer

C H Currey Memorial Fellowship

Subjects: Australian history.
Purpose: To promote the writing of Australian history from original sources of information, preferably making use of the collection of the State Library of New South Wales.
Eligibility: Applicants may be residents or non-residents of Australia. Preference will normally be given to applications that support research on a topic or project that is not being pursued as part of a higher degree programme.
Level of Study: Professional development
Type: Fellowship
Value: Australian $20,000
Length of Study: 1 year
Frequency: Annual

Study Establishment: The State Library of New South Wales
Country of Study: Australia
Application Procedure: Applications should be made on forms available from the State Library or the website www.sl.nsw.gov.au/awards/
Closing Date: October 2nd
Additional Information: Please see the website for further details http://www.sl.nsw.gov.au/about/awards/currey.html.

C.H Currey Memorial Fellowships

Purpose: For the writing of Australian history from original sources, preferably making use of the State Library's resources.
Type: Fellowship
Value: $20,000
Length of Study: 12 months
Study Establishment: The State Library of New South Wales
Country of Study: Australia
Application Procedure: Applications for the C.H.Currey Memorial Fellowship should be made on forms available from the State Library or on the website www.sl.nsw.gov.au/about/awards/currey.html.
Closing Date: October 2nd
Funding: Private
Additional Information: Please see the website for further details http://www.sl.nsw.gov.au/about/awards/currey.html.

David Scott Mitchell Memorial Fellowship

Purpose: To encourage and support the use of the Mitchell Library's collections for the study and research of Australian history in writing and publication amongst scholars, researches and the wider community, including internationally.
Eligibility: Applicants may be residents or non-residents of Australia.
Level of Study: Professional development
Type: Research fellowship
Value: $12,000
Length of Study: 1 year
Frequency: Annual
Study Establishment: State Library of New South Wales
Country of Study: Australia
No. of awards offered: 2
Application Procedure: Applicants must complete an application form, available on request.
Funding: Trusts
Additional Information: Please see the website for further details http://www.sl.nsw.gov.au/about/awards/mitchell.html.

Dobbie Literary Award

Subjects: Writing.
Purpose: To recognise a first published work from an Australian female writer.
Eligibility: Open to a first published work of fiction or nonfiction classifiable as 'Life Writing' by a woman author.
Value: Australian $5,000
Frequency: Annual
Study Establishment: The State Library of New South Wales
Country of Study: Australia
Application Procedure: Applicants must complete an application form, available on request.
Funding: Private

For further information contact:

Tel: 1800 501 227
Email: philanthropy@perpetual.com.au
Website: http://www.sl.nsw.gov.au/about/awards/kibble.html

Jean Arnot Memorial Fellowship

Subjects: Librarianship.
Purpose: To reward an outstanding original paper of no more than 5,000 words on any aspect of librarianship by a female librarian or a female student of librarianship.
Eligibility: The author must be a professional librarian or a student at an Australian school of librarianship. The submitted paper must be an outstanding original paper on any aspect of librarianship.
Level of Study: Professional development
Type: Fellowship
Value: Australian $1,000
Length of Study: 1 year
Frequency: Annual
Study Establishment: State Library of New South Wales
Country of Study: Australia
No. of awards offered: 1
Application Procedure: Applicants must complete an application form, available on request.
Closing Date: March 15th (check with website)
Funding: Private
Contributor: National Council of Women of New South Wales Incorporated and the Australian Federation of Business and Professional Women's Association Inc.

For further information contact:

Collection Management Services and Mitchell Librarian, State Library of New South Wales, Australia
Email: eellis@sl.nsw.gov.au
Website: http://www.sl.nsw.gov.au/about/awards/arnot.html
Contact: Elizabeth Ellis, Assistant State Librarian

Kathleen Mitchell Award

Subjects: Writing, literature.
Purpose: To reward a novel by an Australian author.
Eligibility: Applicants must be an Australian author under 30 years of age.
Value: Australian $7,500
Frequency: Every 2 years
Study Establishment: The State Library of New South Wales
Country of Study: Australia
Application Procedure: Applicants must complete an application form, available on request.
Closing Date: Check with website
Funding: Private
Additional Information: Please see the website for further details http://www.sl.nsw.gov.au/about/awards/index.html.

Kibble Literary Awards

Subjects: Writing.
Purpose: To recognise the work of an established Australian female writer.
Eligibility: Open to women writers of a published book of fiction or nonfiction classifiable as 'Life Writing'.
Value: Australian $30,000
Frequency: Annual
Study Establishment: The State Library of New South Wales
Country of Study: Australia
Application Procedure: Applicants must complete an application form, available on request.
Closing Date: Check with website
Funding: Private

For further information contact:

Tel: 1800 501 227
Email: philanthropy@perpetual.com.au
Website: http://www.sl.nsw.gov.au/about/awards/kibble.html

Miles Franklin Literary Award

Subjects: Australian literature.
Purpose: To promote excellence in Australian literature.
Eligibility: Novel must be of the highest literary merit and must present Australian life in any of its phases. Novels submitted must have been published in the year of entry of the award.
Level of Study: Professional development
Type: Award
Value: Australian $42,000
Frequency: Annual
Application Procedure: Application forms can be downloaded from the website www.trustco.com.au/awards/miles_franklin.htm, see also State library of NSW website www.sl.nsw.gov.au/awards/
Closing Date: December 10th (check with website)
Funding: Individuals

For further information contact:

Cauz Group, Australia
Tel: (61) 02 9332 1559
Fax: (61) 02 9332 1298
Email: trustawards@cauzgroup.com.au
Website: http://www.sl.nsw.gov.au/about/awards/index.html
Contact: Petrea Salter

Milt Luger Fellowships

Subjects: Australian life, history and culture using the resources of the state library.
Purpose: For projects which investigate and document aspects of Australian life, history and culture.
Eligibility: Persons aged between 18 and 25 years.
Type: Fellowship
Value: US$5,000 and US$3,000
Study Establishment: The state library of NSW
Country of Study: Australia
No. of awards offered: 2 Awards
Application Procedure: Applicants must complete an application form, available on request. See State Library of NSW website www.sl.nsw.gov.au/awards/
Closing Date: To be confirmed
Funding: Private
For further information contact:

Mitchell Library Office, State Library of New South Wales, Macquarie Street, Sydney, Australia
Tel: (61) 02 9273 1467
Fax: (61) 02 9273 1245
Email: awards@sl.nsw.gov.au
Website: http://www.sl.nsw.gov.au/about/awards/index.html
Contact: Margaret Bjork

Nancy Keesing Fellowship

Subjects: Australian life and culture.
Purpose: To promote the State Library of New South Wales as a centre of research into Australian life and culture and to provide a readily accessible record of the fellowship project.
Eligibility: Applicants may be either residents or non-residents of Australia. Preference will normally be given to applications that support research on a topic or project that is not being pursued as part of higher degree programme.
Level of Study: Professional development
Type: Fellowship
Value: Australian $12,000
Length of Study: 1 year
Frequency: Annual
Study Establishment: State Library of New South Wales
Country of Study: Australia
Application Procedure: Applications must be made on forms available from the State Library or the website www.sl.nsw.gov.au/awards/
Closing Date: September 5th (check with website)
Funding: Private
Additional Information: Please see the website for further details http://www.sl.nsw.gov.au/about/awards/index.html.

National Biography Award

Subjects: Writing.
Purpose: To encourage the highest standards of writing in the fields of biography and autobiography and to promote public interest in biography and autobiography.
Eligibility: The author must be a living Australian citizen or hold permanent resident status. The work must be classified as either biography or autobiography, be written in the English language, published in book form and consist of a minimum of 50,000 words.
Level of Study: Professional development
Type: Award
Value: Australian $25,000
Frequency: Annual
Study Establishment: The State Library of New South Wales
Country of Study: Australia
Application Procedure: All nominations must be made on the appropriate form and be accompanied by an entry fee of Australian $55 to be eligible for consideration. Five copies of the nominated work must be submitted. A separate form must be completed for each nomination. See state library of NSW website www.sl.nsw.gov.au/awards/
Closing Date: November 5th (check with website)
Funding: Individuals, private
Contributor: Geoffrey Cains
Additional Information: Please see the website for further details http://www.sl.nsw.gov.au/about/awards/index.html.

STATISTICAL SOCIETY OF CANADA

209 - 1725 St. Laurent Blvd., Ottawa, Ontario, K1G 3V4, Canada
Tel: (1) 613 733 2662
Fax: (1) 613 733 1386
Email: info@ssc.ca
Website: www.ssc.ca
Contact: Sudhir Paul, Chair, Pierre Robillard Award

The Statistical Society of Canada provides a forum for discussion and interaction among individuals involved in all aspects of the statistical sciences. It publishes a newsletter, Liaison as well as a scientific journal, *The Canadian Journal of Statistics*. The Society also organises annual scientific meetings and short courses on professional development.

Pierre Robillard Award

Subjects: Statistics.
Purpose: To recognize the best PhD thesis defended at a Canadian university and written in a field covered by the Canadian Journal of Statistics.
Eligibility: Open to all postgraduates who have made a potential impact on the statistical sciences.
Level of Study: Doctorate
Type: Award
Value: A certificate, a monetary prize of Canadian $1,000 and 1 years membership of the Society
Frequency: Annual
Country of Study: Canada
No. of awards offered: 1
Application Procedure: Applicants must submit four copies of the thesis together with a covering letter from the thesis supervisor.
Closing Date: February 15th
No. of awards given last year: 1
No. of applicants last year: 8
Additional Information: The committee may decide that none of the submitted theses merits the award.

For further information contact:

Department of Mathematical and Statistical Sciences, University of Alberta, 632 Cab, Edmonton, AB, T6G 2G1, Canada
Tel: (1) 780 492 4230
Fax: (1) 780 492 6826
Email: kc.carriere@ualberta.ca
Contact: Dr Carrière Keumhee , Professor

STELLENBOSCH UNIVERSITY

Private Bay XI, Matieland, 7602, South Africa
Tel: (27) 21 808 9111
Fax: (27) 21 808 3822
Email: usbritz@sun.ac.za
Website: http://www.sun.ac.za/university/

The raison Dé of the chemistry of Stellenbosch is to create and sustain, in commitment to the academic ideal of excellent scholarly and scientific practice, an environment within which knowledge can be discovered, shared and applied for the benefit of the community.

Harry Crossley Master's and Doctoral Bursary

Subjects: Any subject, with the exception of theology and political science.
Purpose: To reward academically above-average studens.

Eligibility: To full-time students registered at Stellenbosch University in any postgraduate degree programme except theology and political science.
Level of Study: Postgraduate
Type: Bursary
Value: R75,000 (Honours), R80,000 (Master's), 90,000 (Doctoral)
Length of Study: 1 year
Frequency: Annual
Study Establishment: Stellenbosch University
Country of Study: South Africa
Application Procedure: The Application Form can be sourced from http://www.uct.ac.za/apply/funding/postgraduate/applications/.
Closing Date: October 15th
Funding: Foundation
Contributor: Harry Crossley Foundation
No. of awards given last year: 30
No. of applicants last year: 500
Additional Information: Please see the website further details.

For further information contact:

The Office for Postgraduate Student Funding, Postgraduate and International Office, Administrator Building Block A, Room A 2069
Tel: 021 808 4208/2957
Fax: 021 808 2739
Email: postgradfunding@sun.ac.za

Stellenbosch Merit Bursary Award

Subjects: Any subject.
Purpose: To reward academically above-average students.
Eligibility: Available to full-time students registered at Stellenbosch University in any postgraduate degree programme.
Level of Study: Postgraduate
Type: Bursary
Value: South African Rand 4,100–34,700
Length of Study: Up to 3 years
Frequency: Annual
Study Establishment: Stellenbosch University
Country of Study: South Africa
Application Procedure: Students must submit an application and a certified copy of a complete, official academic record.
Contributor: Stellenbosch University
No. of awards given last year: 340
No. of applicants last year: Approx. 700

For further information contact:

Office for postgraduate student funding, Postgraduate and International Office, Administration Building A, Room 2069
Tel: 021 808 4208
Fax: 021 808 2739
Email: postgradfunding@sun.ac.za

STOUT RESEARCH CENTRE, VICTORIA UNIVERSITY OF WELLINGTON

PO Box 600, Wellington, 6140, New Zealand
Tel: (64) 4 463 5305
Fax: (64) 4 463 5439
Email: stout-centre@vuw.ac.nz
Website: www.vuw.ac.nz/stout-centre
Contact: Dr Lydia Wevers, Director

The Stout Research Center was established in 1984 to encourage scholarly inquiry into New Zealand society, history and culture, and to provide a focus for the collegial atmosphere and exchange of ideas that enrich the quality of research.

J D Stout Fellowship

Subjects: New Zealand society, history and culture.
Purpose: To encourage research.
Eligibility: Open to distinguished scholars from New Zealand and abroad.
Level of Study: Postdoctorate
Type: Fellowship
Value: Within the Research Fellow scale depending upon the qualifications and experience of the applicant
Length of Study: 1 year
Frequency: Annual
Study Establishment: The Stout Research Centre
Country of Study: New Zealand
No. of awards offered: 1
Application Procedure: Applicants must write for details.
Closing Date: To be confirmed
Funding: Trusts
Contributor: Stout Trust
Additional Information: Please see the website for further details https://www.victoria.ac.nz/stout-centre/research-opportunities/jd-stout-info.

STROKE ASSOCIATION

Stroke House, 240 City Road, London, EC1V 2PR, England
Tel: (44) 20 7566 0300
Fax: (44) 20 7490 2686
Email: info@stroke.org.uk
Website: www.stroke.org.uk
Contact: Dr S Armstrong, Research Officer

The Stroke Association funds research into stroke prevention, treatment, rehabilitation, and long term care. It also helps stroke patients and their families directly through community services. It campaigns, educates and informs to increase knowledge of stroke at all levels of society and it acts as a voice for everyone affected by stroke.

The Stroke Association Junior Research Training Fellowships

Subjects: Stroke research.
Purpose: For outstanding nurse and allied health professional graduates intending to study for a PhD.
Eligibility: Open to nurses and allied health professionals, but consideration will be given to other health professionals. They will be awarded to departments that can demonstrate a track record and current participation in stroke research.
Level of Study: Postgraduate, Professional development
Type: Fellowship
Value: UK£35,000 per year for two years, with a discretionary extension of £35,000 for the third year
Length of Study: Up to 3 years
Frequency: Annual
Study Establishment: Suitable universities and hospitals
Country of Study: United Kingdom
No. of awards offered: 2
Application Procedure: Application forms are available from the website.
Closing Date: January 13th
Funding: Private, trusts
No. of awards given last year: 3
No. of applicants last year: 14
Additional Information: Applications are reviewed and shortlisted candidates are interviewed in early March.

The Stroke Association Research Project Grants

Subjects: Stroke research encompassing epidemiology, prevention, acute treatment, assessment and rehabilitation, psychology of stroke and stroke in ethnic minorities.
Purpose: To advance research into stroke.
Eligibility: Open to medically qualified and other clinically active researchers in the United Kingdom in the relevant fields. Applications are judged by peer review on their merit without limitations of age. Applicants can be from any country but must be based in the United Kingdom.
Level of Study: Postdoctorate, Research
Type: Project grant
Value: Up to £210,000 over three or five years
Length of Study: 1–3 years
Frequency: Twice a year
Study Establishment: A suitable university or hospital in the United Kingdom

Country of Study: United Kingdom
No. of awards offered: Approx. 50–70 ongoing at any point in time
Application Procedure: Application forms are available from the website.
Closing Date: February and July
Funding: Private, trusts
Contributor: Donations
No. of awards given last year: 8
No. of applicants last year: 54

The Stroke Association Senior Research Training Fellowships

Subjects: Stroke research.
Purpose: To support nurses or allied health professionals to embark on an independent career in academic stroke research.
Eligibility: Awarded to a department in the United Kingdom that can provide an educational programme and the expert supervision required to enable a specialist registrar to gain the appropriate clinical experience required for a career in stroke. Open to postdoctoral candidates from a nursing or allied health professional background. Medical professionals are not eligible to apply.
Level of Study: Postdoctorate, Professional development
Type: Fellowship
Value: UK£175,000 over three years
Length of Study: 3 years
Frequency: Annual
Study Establishment: Suitable universities and medical schools
Country of Study: United Kingdom
No. of awards offered: 2
Application Procedure: Application forms are available from the website.
Closing Date: January 13th
Funding: Private
No. of awards given last year: 1
No. of applicants last year: 3
Additional Information: Fellowships are assessed by peer review. Applications are reviewed and shortlisted candidates interviewed in early March.

SWANSEA UNIVERSITY

Singleton Park, Swansea, Wales, SA2 8PP, United Kingdom
Tel: (44) 0 1792 205678
Fax: (44) 0 1792 295157
Email: sro@swansea.ac.uk
Website: www.swan.ac.uk

Founded in 1920, Swansea University is one of the UK's leading universities for collaborative research with industry. We are a world-class, research led institution situated in the city of Swansea, in parkland, and next to a beach. We have taught and research postgraduate funding for UK, EU and International students: www.swansea.ac.uk/postgraduate/scholarships

College of Engineering: EngD Scholarship

Subjects: Materials engineering and mechanical engineering.
Purpose: To prepare students for research and industry leadership careers.
Eligibility: Open to Masters or Bachelor students with a First Class Honours or a good 2.1 degree in a suitable engineering, mathematical or scientific discipline. Students must have settled status in the UK and been ordinarily resident in the UK for 3 years prior to the start of the grant.
Level of Study: Doctorate, Graduate
Type: Studentship
Value: Full studentship of £16,500 per year plus tuition fees
Length of Study: 4 years
Frequency: Annual
Study Establishment: Swansea University
Country of Study: United Kingdom
No. of awards offered: 2
Application Procedure: Check the website www.swansea.ac.uk/engineering for further details.
Closing Date: Please contact organisation for details
Contributor: Swansea University and Industrial Partners
Additional Information: Please contact College of Engineering at engineering@swansea.ac.uk.

For further information contact:

Contact: Professor

The Eira Francis Davies Scholarship

Subjects: Health and medical related subjects.
Eligibility: Open to Non-EU, female student ordinarily resident in a developing country.
Level of Study: Postgraduate, Undergraduate
Type: Scholarship
Value: Full international tuition fee
Frequency: Annual
Study Establishment: Swansea University
Country of Study: United Kingdom
No. of awards offered: 1
Application Procedure: Complete application form available online at www.swansea.ac.uk.
Closing Date: June 1st
Contributor: Eira Francis Davies
No. of awards given last year: 1

International Excellence Scholarships

Subjects: All subjects.
Eligibility: Awards are available to postgraduate applicants from outside the EU. Other eligibility criteria may apply. Please contact us for details.
Level of Study: Postgraduate
Type: Scholarship
Value: £3,000
Length of Study: 1 year (Masters)
Frequency: Annual
Study Establishment: Swansea University
Country of Study: United Kingdom
No. of awards offered: 39
Application Procedure: Complete application form, available online at website.
Closing Date: June 4th
Additional Information: Please contact at international@swansea.ac.uk for further information. South America and India are also eligible countries.

PhD Fees-only Bursaries

Subjects: Normally available in all subject areas.
Eligibility: Open to good Master's graduates from the UK/EU who will be commencing PhD studies at Swansea University.
Type: Scholarship
Value: Covers UK/EU tuition fees
Length of Study: 3 years
Frequency: Annual
Study Establishment: Swansea University
Country of Study: United Kingdom
No. of awards offered: Approx. 10
Application Procedure: Please contact us for an application form.
No. of awards given last year: 10

For further information contact:

Postgraduate Admissions Office, Swansea University, Wales
Email: admissions-enquiries@swansea.ac.uk
Website: www.swansea.ac.uk/postgraduate

PhD Studentships

Subjects: Normally available in all subject areas. Please check our website www.swansea.ac.uk/postgraduate for details.
Eligibility: Open to good Masters graduates from the UK/EU who will be commencing PhD studies at Swansea University.
Level of Study: Doctorate, Postgraduate
Type: Scholarship
Value: Tuition fees plus maintenance grant of approx. £13,590 per year
Length of Study: 3 years
Frequency: Annual
Study Establishment: Swansea University

Country of Study: United Kingdom
No. of awards offered: Approx. 10
Application Procedure: Please contact us for an application form.
Closing Date: Throughout year
No. of awards given last year: 10

For further information contact:

Postgraduate Admissions Office, Swansea University, Wales
Email: admissions-enquiries@swansea.ac.uk

Swansea University Masters Scholarships

Subjects: Available in all academic schools.
Eligibility: Open to students from the UK/Eu who will be starting an eligible master's course at Swansea University for the first time in September.
Level of Study: Postgraduate
Type: Scholarship
Value: £2,900 towards tuition fees
Length of Study: 1 year full time or 2–3 years part-time
Frequency: Annual
Study Establishment: Swansea University
Country of Study: United Kingdom
No. of awards offered: 100
Application Procedure: You must complete an application form which is sent out with offer letter.
Closing Date: July
No. of awards given last year: 100

For further information contact:

Postgraduate Admissions Office, Swansea University, Wales
Email: admissions-enquiries@swansea.ac.uk

SWEDISH INFORMATION SERVICE

445 Park Avenue, 21st floor, New York, NY, 10022, United States of America
Tel: (1) 212 888 3000
Fax: (1) 212 888 3125
Email: generalkonsulat.new-york@foreign.ministry.se
Website: www.swedennewyork.com
Contact: Consulate General of Sweden

The Section for culture and Public Affairs of the Consulate General of Sweden in New York works to promote awareness in the United States of America of Swedish cultural achievement and advancement in scientific research and development, and contributes to the formation of public opinion and policy in an international context.

Bicentennial Swedish-American Exchange Fund

Subjects: Politics, public administration, working life, human environment, mass media, business and industry, education or culture.
Purpose: To provide an opportunity for those in a position to influence public opinion and contribute to the development of their society to make an intensive research trip to Sweden.
Eligibility: Applicants should be citizens or permanent residents of the United States of America. People who have made recurrent visits to or resided in Sweden will only be considered in exceptional circumstances. The grant may not be used to finance participation in conferences or regular ongoing vocational or academic courses. If co-applicants on the same project are selected, the grant will be divided between them. The grant may be used in conjunction with scholarships from other sources.
Level of Study: Research
Type: Travel grant
Value: Krona 30,000 or the equivalent in United States of America dollars to partially cover transportation and living expenses
Length of Study: 2–4 weeks intensive research
Frequency: Annual
Country of Study: Sweden
No. of awards offered: 2
Application Procedure: Application forms are available from the website or can be requested directly from the Swedish institute. Two letters of recommendation are also required.
Closing Date: November 15th
Funding: Government
Contributor: The Swedish Institute in Stockholm, Sweden
No. of awards given last year: 5
No. of applicants last year: 40
Additional Information: The project must be completed within 1 year of receipt of the grant. A report must be submitted to the Swedish institute 1 month after the research trip is completed. Award recipients are announced during the month of May.

For further information contact:

Website: www.studyinsweden.se

SWEDISH INSTITUTE (SI)

Slottsbacken 10 Box 7434, Stockholm, 103 91, Sweden
Tel: (46) 8 453 78 00
Fax: (46) 8 20 72 48
Email: si@si.se
Website: www.si.se
Contact: Rita Wikander, Program Officer

Swedish Institute SI is entrusted with the task to inform the world about Sweden and to organise exchanges with other countries in the spheres of culture, education, research and public life in general.

Swedish-Turkish Scholarship for Human Rights Law in Memory of Anna Lindh

Subjects: Human rights law.
Purpose: To advance the field of European studies.
Eligibility: Open to Turkish students only. Only for studies of Human Righs at Raoul Wallenberg Institute in Lund (Lund University).
Level of Study: Postdoctorate, Postgraduate
Type: Scholarship
Value: All tuition fees, plus travel allowance, plus livng expenses. Tution fees paid, SEK 5,000 per year as travel grant and SEK 8,000 monthly for living expenses (food, accomodation, books, etc.).
For PhD-students: Travel grant of SEK 5,000/year plus SEK 12,000 in monthly scholarship for living expenses.
For holders of PhD: Travel grant SEK 5,000 per year. Scholarship of SEK 15,000 per month.
Frequency: Annual
Country of Study: Sweden
No. of awards offered: 1–2 per year
Application Procedure: Apply online.
Closing Date: February 1st for master degree program; February 15th for PhD studies/research
Funding: Government
Contributor: Swedish government - Ministry of foreign affairs
No. of awards given last year: 1–2
No. of applicants last year: 10–20
Additional Information: This scholarship is part of the Swedish-Turkish Scholarship Program given by the Swedish Institute. Please visit at www.studyinsweden.se for more information about the criteria, how to apply, etc.

SWINBURNE UNIVERSITY OF TECHNOLOGY

PO Box 218, Hawthorn, VIC, 3122, Australia
Tel: (61) (03) 9214 8000
Fax: (61) (03) 9214 8637
Email: webmaster@swin.edu.au
Website: www.swinburne.edu.au/index.php
Contact: MBA Admissions Officer

It provides career-orientated education and as a university with a commitment to research. The University maintains a strong technology base and important links with industry, complemented by a number of innovative specialist research centres which attract a great deal of international interest. A feature of many Swinburne undergraduate courses is the applied vocational emphasis and direct industry application through Industry Based Learning (IBL) programs. Swinburne was a pioneer of IBL program which places students directly in industry for vocational employment as an integral part of the course structure. Swinburne is committed to the transfer of lifelong learning skills. It is heavily involved in international initiatives and plays a significant part in the internationalization of Australia's tertiary education system.

Australian Postgraduate Award (APA)

Subjects: All subjects.
Purpose: To support study towards a Higher Degree by Research.
Eligibility: Open to citizens of Australia, New Zealand citizens and Australian permanent residents.
Level of Study: Doctorate, Research
Type: Award
Value: A non-taxable indexed stipend of around $23,728 per year and a tuition fee scholarship (total value around $49k per year)
Length of Study: 3 years
Application Procedure: Check website for further details.
Closing Date: October 31st (to be confirmed)
Additional Information: Please see the website for further details.

Chancellor's Research Scholarship

Subjects: All subjects.
Purpose: To award students of exceptional research potential to undertake a higher degree by research (HDR).
Eligibility: Open to a local or an international student undertaking a higher degree by research (HDR) with Bachelor Degree with First Class Honours. For further details, please check the website.
Level of Study: Doctorate
Type: Research scholarship
Value: An annual stipend of $30,000, an Establishment Grant of up to $3,000 and up to $5000 for a 6 month overseas placement
Length of Study: 3 years
Frequency: Annual
Application Procedure: Check website for further details.

For further information contact:

Building 60Wm, Level 7, 60 William Street, Hawthorn campus
Tel: 9214 5547 or 9214 8744
Email: ehill@swin.edu.au, jamathews@swin.edu.au

Swinburne University Postgraduate Research Award (SUPRA)

Subjects: All subjects.
Eligibility: Open to citizens of all overseas countries except New Zealand who have completed at least 4 years (or equivalent) of tertiary education studies at a high level of achievement. Plus students must demonstrate English Language proficiency: IELTS.
Level of Study: Doctorate, Postgraduate
Type: Research scholarship
Value: A non-taxable indexed stipend of around $23,728 per year and a tuition fee scholarship (total value around $49k per year)
Length of Study: 3 years
Country of Study: Australia
No. of awards offered: 2
Application Procedure: Complete an application for admission to research higher degree candidature and scholarship and mail/courier or you can scan your application forms in and email them.
Closing Date: July and October (check with website)
Contributor: Australian Department of Education, Science and Training (DEST)
Additional Information: Please see the website for further details http://www.research.swinburne.edu.au/research-students/scholarships/supra.html.

Swinburne University Postgraduate Research Award (SUPRA)

Subjects: All subjects.
Purpose: To assist with general living costs.
Eligibility: Open to domestic or an international student undertaking a higher degree by research. Please check the website for further details.
Level of Study: Doctorate, Research
Type: Research award
Value: A non-taxable indexed stipend of around $23,728 per year and a tuition fee scholarship (total value around $49k per year)
Length of Study: 3 years (Research Doctorate) and 2 years (Research Masters)
Application Procedure: Please check website for further details.
Closing Date: May 30th (check with website)

For further information contact:

Building 60Wm, Level 7, 60 William Street, Hawthorn campus
Tel: 9214 5547 or 9214 8744
Email: ehill@swin.edu.au, jamathews@swin.edu.au
Website: http://www.research.swinburne.edu.au/research-students/scholarships/supra.html

Vice Chancellor's Centenary Research Scholarship (VCCRS)

Subjects: All subject.
Purpose: To assist with general living costs.
Eligibility: Open to domestic or an international student who have completed a Bachelor Degree with First Class Honours and are of exceptional research potential undertaking a higher degree by research (HDR). For further details, please check the website.
Level of Study: Research
Type: Research scholarship
Value: A non-taxable indexed stipend of around $23,728 per year and a tuition fee scholarship (total value around $49k per year)
Length of Study: 3 years
Application Procedure: Check website for further details.

For further information contact:

Building 60Wm, Level 7, 60 William Street, Hawthorn campus
Tel: 9214 5547 or 9214 8744
Email: ehill@swin.edu.au, jamathews@swin.edu.au
Website: http://www.research.swinburne.edu.au/research-students/scholarships/vcrs.html

SWISS FEDERAL INSTITUTE OF TECHNOLOGY ZÜRICH

International Instituitional Affairs, Rämistrasse 101, Zurich, CH-8092, Switzerland
Tel: (41) 44 632 1111
Fax: (41) 44 632 1010
Email: international@sl.ethz.ch, hagstroem@sl.ethz.ch
Website: www.master.ethz.ch
Contact: Anders Hagstrom

The Swiss Federal Institute of Technology Zurich is a science and technology university with an outstanding research record. Excellent research conditions, state-of-the-art infrastructure and an attractive urban environment add up to the ideal setting for creative personalities.

ETH Zurich Excellence Scholarship and Opportunity Award

Subjects: Architecture, engineering, (civil, mechanical, electrical, production, rural and surveying), computer science, materials science, chemistry, physics, mathematics, biology, environmental sciences, earth sciences, pharmacy, agriculture and forestry, international relations, political science.
Purpose: Full scholarships for tuition and cost of living for talented students of master's programs.
Eligibility: Open to graduates in one of the discipline represented at ETH Zurich with very good academic record.
Level of Study: Graduate, Master
Type: Scholarship
Value: Swiss Francs 1,750 per month plus tuition fees
Length of Study: 18–24 months
Frequency: Annual
Study Establishment: ETH Zurich
Country of Study: Switzerland
No. of awards offered: 50
Application Procedure: Applications must be made to the appropriate address. See www.master.ethz.ch for further information.
Closing Date: December 15th
Funding: Foundation, government

Contributor: ETH Zurich Foundation
No. of awards given last year: 36
No. of applicants last year: 430

For further information contact:

Contact: Anders Hagström

Sawiris Scholarship

Subjects: Topics limited to those covered by the chair of the ETH Zurich.
Purpose: Scholarships for doctoral students on research for the benefit of developing countries.
Eligibility: Open to Doctorate students with academic excellence and depending on the candidate's contribution to the field. Topics relevant to development; topic within scope of ETH Zurich; research aims to develop a product/method directly relevant for improving the lives of poor.people in developing countries.
Level of Study: Doctorate
Type: Fellowship
Value: CHF50,000 per year
Length of Study: 3 years
Frequency: Annual
Country of Study: Developing countries
No. of awards offered: 2
Application Procedure: Two-step procedure: (1) concept note, (2) full proposal. All details under the website given below.
Closing Date: October 31st
Funding: Foundation
Contributor: Sawiris Foundation for social development
No. of awards given last year: 2
No. of applicants last year: 22

For further information contact:

Email: elindberg@ethz.ch
Website: www.global.ethz.ch/r4d/
Contact: Emma Lindberg

SYNGENTA FOUNDATION

WRO-1002.11.52, Postfach CH-4002, Basel, Switzerland
Tel: (41) 61 323 5634
Fax: (41) 61 323 7200
Email: syngenta.foundation@syngenta.com
Website: www.syngentafoundation.com
Contact: Grants Enquiries

Syngenta Foundation Awards

Subjects: All subjects.
Eligibility: Open to candidates from African nations.
Type: Award
Value: Varies
Country of Study: Developing countries
No. of awards offered: Variable year to year
Application Procedure: Open
Additional Information: The Syngenta Foundation does not have a formal award mechanism. It awards people within the projects in developing countries on an ad hoc basis.

For further information contact:

Syngenta Foundation, Schwarzwalpallee 215, Bases, Switzerland, 4002

SYRACUSE UNIVERSITY

900 South Crouse Ave, Syracuse, NY, 13244, United States of America
Tel: (1) 315 443 1870
Fax: (1) 315 443 3423
Email: grad@syr.edu
Website: www.syr.edu

Syracuse University is a non-profit, private student research university. Its mission is to promote learning through teaching, research, scholarship, creative accomplishment and service.

African American Studies Graduate Fellowships

Subjects: Any subject.
Purpose: To support new continuing graduate students across disciplines whose work supports that of the African American/Pan African studies program and who will make a intellectual contribution to the life of the Department of African American studies.
Eligibility: Open to African American fellows enrolled in at least one three-credit graduate course each semester in the African American Studies program for the duration of their award.
Level of Study: Postgraduate, Doctorate
Type: Fellowship
Value: $13,040 for master's students, $21,805 for doctoral students
Frequency: Annual
No. of awards offered: 6
Application Procedure: Applicants must send their application along with a letter of intent and should indicate interest in this award.
Closing Date: January 1st

Hursky Fellowship

Subjects: Ukrainian language and literature, linguistics and culture.
Purpose: To a full-time graduate student of Ukranian background enrolled for the study of Ukrainian language and literature, Ukrainian linguistics and, culture.
Eligibility: Open to graduate students with a Ukrainian background enrolled in Maxwell School of Citizenship and Public Affairs or the College of Arts and Sciences, or any SU graduate whose area of study is the Ukraine or included Ukrainian topics.
Type: Fellowship
Value: stipend of $13,040 and a tuition scholarship for 24 credits for the academic year and the following summer
Frequency: Annual
Closing Date: January 1st

McNair Scholars Program

Purpose: To increase the number of low-income, first-generation and underrepresented minority college students who pursue and complete the doctoral degree.
Eligibility: Open to the candidates who were McNair Scholars at their undergraduate instituions.
Level of Study: Doctorate, Postgraduate
Type: Fellowship
Value: $13,040 for master's students, $21,805 for doctoral students
Frequency: Annual
No. of awards offered: 6
Closing Date: January 1st

STEM Doctoral Fellowship

Subjects: Science, technology, engineering and maths disciplines.
Purpose: To support doctoral students in the field of science, technology, engineering and maths from underrepresented group of US or its permanent residents.
Eligibility: Open to members of an underrepresented group who are US citizens or permanent residents.
Level of Study: Doctorate
Value: US$20,150 plus tuition scholarship
Frequency: Annual
No. of awards offered: 5
Closing Date: January 1st

Syracuse University Graduate Fellowship

Subjects: All subjects.
Purpose: To provide a full support package during a student's term of study.
Eligibility: Open to nationals of any country.
Level of Study: Unrestricted
Type: Fellowship
Value: US$13,040 stipend for Master, US$21,805 stipend for PhD plus tuition scholarship.
Length of Study: 1–6 years
Frequency: Annual
Study Establishment: Syracuse University
Country of Study: United States of America
No. of awards offered: Varies by school/college

Application Procedure: Applicants must apply through admission application.
Closing Date: January 1st
Funding: Private
Contributor: The Syracuse University Graduate School
No. of awards given last year: Varies
No. of applicants last year: 250

THE TAN KAH KEE FOUNDATION

Level 1, 43 Bukit Pasoh Road, 089856, Singapore
Tel: (65) 6222 6620
Fax: (65) 6462 1192
Email: tkkf@tkkfoundation.org.sg
Website: www.tkk.wspc.com.sg

The mission of The Tan Kah Kee Foundation is to carry on the charity works and to foster the Tan Kah Kee spirit in entrepreneurship and dedication to education. Over the years, the Foundation has been actively engaged in the promotion of education and culture. Today, its influence has been extended beyond the Chinese Community in Singapore and has begun to reach out to the region.

Tan Kah Kee Postgraduate Scholarship

Subjects: All subjects.
Purpose: To provide financial assistance to students pursuing their postgraduate studies.
Eligibility: Open to citizens and permanent residents of Singapore who are pursuing full-time Master's degree or PhD in any discipline, regardless of race or religion. Candidates are appraised on their academic achievements, outstanding personal character and bilingual capabilities.
Level of Study: Doctorate, Postgraduate
Type: Scholarship
Value: Singaporean $10,000 for overseas students, Singaporean $7,000 for local universities
Frequency: Annual
Application Procedure: Applicants can download the application form from the website http://www.tkkfoundation.org.sg/foundation/post_eng.shtml.
Closing Date: May 30th
Funding: Foundation
Contributor: Tan Kah Kee Foundation
Additional Information: Application is open in May every year. Shortlisted applicants will be informed to attend an interview in the month of June or July.

TANTE MARIE'S COOKING SCHOOL

271 Francisco Street, San Francisco, CA, 94133, United States of America
Tel: (1) 415 788 6699
Fax: (1) 415 788 8924
Email: peggy@tantemarie.com
Website: www.tantemarie.com

Tante Marie's Cooking School, located in San Francisco was founded as a full-time school in 1979. It is one of the first schools of fine cooking offering all-day, year-round classes for people who are serious about cooking well. Graduates from Tante Marie's have interesting and varied careers. In addition to offering professional courses for people wanting to begin a career in culinary or pastry, Tante Marie's welcomes interested avocational students in the Evening Series, Weekend Workshops, One-Day Workshops and Cooking Vacations. There are also cooking parties on weekend evenings where groups of up to 30 people cook together. The emphasis at Tante Marie's is in building confidence in the kitchen.

The Absolute taste Scholarship

Subjects: Cooking.
Purpose: To make the candidates learn cooking professionally.
Eligibility: All prospective students ordinarily resident in the British Isles who will have attained at least 16 years of age and be under 25 years of age on commencement of the course.
Level of Study: Unrestricted
Type: Scholarship
Value: Up to 100% of the course fee
Frequency: Annual
Country of Study: United States of America
Application Procedure: Check website for further details.
Closing Date: June 7th
Funding: Private

For further information contact:

Email: info@tantemarie.co.uk
Website: http://www.tantemarie.co.uk/news/scholarships_full.php?ID=5

THE TE PÔKAI TARA UNIVERSITIES NEW ZEALAND

PO Box 11915, Manners Street, Wellington, 6142, New Zealand
Tel: (64) 404 381 8500
Fax: (64) 404 381 8501
Email: kiri@nzvcc.ac.nz
Website: www.nzvcc.ac.nz
Contact: Kiri Manuera, Scholarships Manager

The New Zealand Vice Chancellors Committee (NZVCC) was established by the Universities Act 1961, which replaced the federal University of New Zealand with separate institutions. Today the Committee represents the interests of New Zealand's 8 universities. The NZVCC represents the interests of the New Zealand university system to government, its agencies and the public through a range of forums and communications from joint consultative groups to electronic and print publications.

The Association of University Staff Crozier Scholarship

Subjects: History, management, organization, economics, economic and social impact, sociology and pedagogy.
Purpose: To provide funds for individuals to undertake research towards an Honours, Master's or Doctoral degree and who are undertaking research or scholarly enquiry for a research project, theses or dissertation at a New Zealand or overseas university or research institution in the related fields.
Eligibility: Open to applicants who are citizens or permanent residents of New Zealand who have resided in New Zealand for at least 3 years. Applicants should also have completed the requirements for a Bachelors degree or equivalent in a field appropriate to their intended study at a New Zealand university.
Level of Study: Postgraduate
Type: Scholarship
Value: New Zealand $5,000
Length of Study: 3 Years
Frequency: Annual
Country of Study: New Zealand
No. of awards offered: 1
Application Procedure: Check the website.
Closing Date: October 1st
Contributor: New Zealand Vice Chancellors Committee Wellington

Cambridge Commonwealth Trust Prince of Wales Scholarship

Subjects: All subjects.
Purpose: To enable bright, young students of high academic ability to study at Cambridge University in Britain.
Eligibility: Open to citizens of New Zealand who wish to pursue a course of research leading to a PhD degree at Cambridge University. The candidate shoud have also applied for admission to Cambridge.
Level of Study: Doctorate, Research
Type: Scholarship
Value: The scholarship covers the University Composition Fee at the home rate along with a maintenance allowance
Length of Study: 3 years
Frequency: Annual
Study Establishment: Cambridge University
Country of Study: United Kingdom
Application Procedure: Check the website
Closing Date: October 1st

Claude McCarthy Fellowships

Subjects: All subjects.
Purpose: To enable graduates of a New Zealand university to undertake original work or research.
Eligibility: Open to any graduate of a New Zealand university.
Level of Study: Doctorate, Postgraduate
Type: Fellowship
Value: Varies. Funding is available for the year following application.
Length of Study: Usually no more than 1 year
Frequency: Annual
Country of Study: Any country
No. of awards offered: Varies, depending upon funds available, but usually 12–15
Application Procedure: Applicants must write for details.
Closing Date: August 1st
Funding: Private
Contributor: The Claude McCarthy Trust
Additional Information: Further information is available on request.

Dick and Mary Earle Scholarship in Technology

Subjects: Innovation and product development and bioprocess technology.
Purpose: To provide funds for individuals to undertake research towards a Masterate or Doctorate degree at a New Zealand university or research institution in related fields.
Eligibility: Open to candidates who are citizens or permanent residents of New Zealand, who have resided in New Zealand for at least 3 years and have completed the requirements for a BTech, BEng, BE degree or equivalent, with Honours at a New Zealand university.
Level of Study: Postgraduate
Type: Scholarships
Value: New Zealand $17,000 per year at Master's level and New Zealand $20,000 per year at PhD level
Frequency: Annual
Country of Study: New Zealand
No. of awards offered: 2
Closing Date: October 1st

Gordon Watson Scholarship

Subjects: International relationships and social and economic conditions.
Purpose: To enable the holder to study abroad questions of international relationships or social and economic conditions, at Masters or PhD level.
Eligibility: Candidates must be New Zealand citizens or permanent residents. Open to holders of an Honours degree, or a degree in theology from a university in New Zealand. Candidates must undertake to return to New Zealand after the scholarship period for not less than 2 years.
Level of Study: Postgraduate
Type: Scholarship
Value: New Zealand $12,000 per year
Length of Study: Up to 3 years
Frequency: Annual
Study Establishment: Any approved university
No. of awards offered: 1
Application Procedure: Candidates must write for details.
Closing Date: March 1st
Funding: Private
Contributor: The Gordon Watson Trust
Additional Information: Further information is available on request.

L. B. Wood Travelling Scholarship

Subjects: All subjects.
Purpose: To allow graduates to undertake doctoral studies in the United Kingdom.
Eligibility: Open to all holders of postgraduate scholarships from any faculty of any university in New Zealand, provided that application is made within 3 years of the date of graduation.
Level of Study: Doctorate
Type: Scholarship
Value: New Zealand $3,000 per year, as a supplement to another postgraduate scholarship
Length of Study: Up to 3 years
Frequency: Annual
Study Establishment: A university or institution of university rank
Country of Study: United Kingdom
No. of awards offered: 1
Application Procedure: Applicants must write for details.
Closing Date: March 1st
Funding: Private
Contributor: The L B Wood Trust
Additional Information: Further information is available on request.

Shirtcliffe Fellowship

Subjects: Arts, science, law, commerce and agriculture.
Purpose: To assist students of outstanding ability and character who are graduates of a university in New Zealand, in the continuation of their doctoral studies in New Zealand or the Commonwealth.
Eligibility: Candidates for a doctoral scholarship awarded by a New Zealand university; andwhose degree is awarded in any faculty or school which, if that degree had been available in 1935, would in the opinion of the NZVCC be expected to have been awarded following a course of study in one or other of the faculties of arts, science, law, commerce or agriculture.
Level of Study: Doctorate
Type: Fellowship
Value: New Zealand $5,000 as a supplement to the postgraduate scholarship emolument
Length of Study: Up to 3 years
Frequency: Annual
Study Establishment: A suitable Institute of Higher Education
No. of awards offered: 1
Application Procedure: Candidates must write for details.
Closing Date: March 1st
Funding: Private
Additional Information: Further information is available on request.

William Georgetti Scholarships

Subjects: All subjects.
Purpose: To encourage postgraduate study and research in a field that is important to the social, cultural or economic development of New Zealand.
Eligibility: Candidates must be New Zealand citizens or permanent residents. Open to graduates who have been resident in New Zealand for 5 years immediately before application and who are preferably aged between 21 and 28 years.
Level of Study: Postgraduate
Type: Scholarship
Value: Up to $20,000 per year for Masters study and $30,000 per year for doctoral study. For those students studying overseas the emolument shall be at a rate of up to NZ $45,000 per year.
Frequency: Annual
Study Establishment: Suitable universities
Country of Study: Any country
No. of awards offered: 1–4
Application Procedure: Applicants must write for details.
Closing Date: October 1st
Funding: Private
Contributor: The Georgetti Trust
Additional Information: Further information is available on request.

TEAGASC (IRISH AGRICULTURE AND FOOD DEVELOPMENT AUTHORITY)

Oak Park, Carlow, Ireland
Tel: (353) 59 917 0200
Fax: (353) 59 918 2097
Email: info@teagasc.ie
Website: www.teagasc.ie
Contact: Debbie Murphy

Teagasc (Irish Agriculture and Food Development Authority) is the parastatal body responsible for agricultural and food research, farm advisory services and farmer education in the Republic of Ireland. Its research programme includes foods, dairy cows, beef cattle, pigs, sheep, crops, horticulture, environment, rural economics and sociology at eight research centres.

Teagasc Walsh Fellowships

Subjects: Any subject relevant to food and agriculture in Ireland, e.g. animal sciences, plant sciences, physical or earth sciences, environment, economics and rural development.
Purpose: To support MSc and PhD projects on topics relevant to the overall Teagasc research programme on agriculture and food.
Eligibility: Applicants must be college faculty members who, in co-operation with Teagasc researchers, submit proposals relevant to the Teagasc programme on agriculture and food in Ireland. If successful, they then select postgraduate students for MSc or PhD programmes as Walsh Fellows. Applications are not accepted from individual students or for taught non-research postgraduate courses.
Level of Study: Doctorate, Postgraduate, Research
Type: Fellowship
Value: €21,000 per year to cover a postgraduate stipend and all fees. A limited provision for materials and travel is also available
Length of Study: Up to 2 years for an MSc, maximum of 4 years for a PhD
Frequency: Annual
Study Establishment: Any third-level college, in association with a Teagasc Research Centre
Country of Study: Ireland
No. of awards offered: Approx. 40
Application Procedure: Applicants must apply for an information brochure, which includes an application form, available on request. Students who wish to apply for a pre-awarded fellowship should check the Teagasc website (www.teagasc.ie) that contains a complete list of fellowships awarded with supervisor contact details.
Closing Date: September 24th, pre-proposals-September 10th. Please refer website
Funding: Government
Contributor: Teagasc's own resources, via the Irish Government and the European Union agri-food industry
No. of awards given last year: 40
No. of applicants last year: 120
Additional Information: The full list of awarded fellowships are posted on the Teagasc website (www.teagasc.ie) mid-April annually.

THE TECHNISCHE UNIVERSITEIT DELFT (TUD)

Post bus 5, 2600 AA Delft, Netherlands
Tel: (31) 15 2789111
Fax: (31) 15 2781855
Email: info@tudelft.nl
Website: www.tudelft.nl/msc

Founded in 1842, the Delft University of Technology is the oldest, largest, and most comprehensive technical university in the Netherlands. It is an establishment of both national importance and significant international standing. Renowned for its high standard of education and research, TU Delft collaborates with other educational establishments and research institutes, both within and outside of the Netherlands. TU Delft aims at being an 'interactive partner' to social issues, committed to answering its multifaceted demands and initiating changes to benefit people in the future.

The Shell Centenary Scholarship Fund, Netherlands

Subjects: All master programmes under the TSCSF scholarship scheme.
Purpose: To give students the opportunity to study at the TUD and gain skills that will make a long-term contribution to the further development of their countries.
Eligibility: Open to candidates who are nationals of and resident in any country other than the ones listed in 'Additional Information' and aged 35 or under, intending to study a subject that will be of significant value in aiding the sustainable development of their home country, fluent in spoken and written English, and neither a current nor former employee of the Royal Dutch/Shell Group of companies.
Level of Study: Postgraduate
Type: Scholarship
Value: Full-cost scholarship including tuition fees, international travel, living allowances and health insurance
Length of Study: 2 years
Frequency: Annual
Country of Study: Netherlands
No. of awards offered: 90
Application Procedure: Applicants must have been admitted to a MSc programme of TU Delft, the International Office will subsequently send you the application form by email, the International Office will check your application on the basis of the Royal Dutch/Shell criteria.
Closing Date: December 15th (check with website)
Contributor: TUD with support from The Shell Centenary Scholarship Fund (TSCSF)
Additional Information: Countries not eligible: Australia, Austria, Belgium, Canada, Cyprus, Czech Republic, Denmark, Estonia, Finland, France, Germany, Greece, Hungary, Iceland, Ireland, Italy, Japan, Latvia, Lithuania, Luxembourg, Malta, The Netherlands, New Zealand, Norway, Poland, Portugal, Slovenia, Slovakia, Spain, Sweden, Switzerland, United Kingdom and United States.

For further information contact:

International Office, Julianalaan 134, 2628 BL Delft, Netherlands
Tel: (31) 15 278 5690
Email: msc2@tudelft.nl
Website: http://id-scholarships.blogspot.in/2008/05/shell-centenary-scholarship.html

TEL AVIV UNIVERSITY (TAU)

PO Box 39040, Tel Aviv, 69978, Israel
Tel: (972) 0 3 640 8111
Email: tauinfo@post.tau.ac.il
Website: www.tau.ac.il

Tel Aviv University (TAU) was founded in 1956 and is located in Israel's cultural, financial and industrial heartland, TAU is the largest university in Israel and the biggest Jewish university in the world. TAU offers an extensive range of programmes in the arts and sciences.

TAU Scholarships

Subjects: History and contemporary music.
Purpose: To encourage innovative and interdisciplinary research that cuts across traditional boundaries and paradigms.
Eligibility: Open to candidates who have registered for their Doctoral or postdoctoral degree.
Level of Study: Doctorate, Postdoctorate
Type: Scholarships
Value: Varies
Frequency: Annual
No. of awards offered: Varies
Application Procedure: Applicants can download the application form from the website. The completed application form along with a curriculum vitae and one 2 page description of research project with a list of publications is to be sent.
Closing Date: Varies
Additional Information: Applications if sent by email, must be directed to ddprize@post.tau.ac.il

For further information contact:

The Lowy School for Overseas Students, Center Building, Tel-Aviv University, Ramat Aviv, Tel Aviv, 69978, Israel
Email: jkc@jackkentcookefoundation.org
Website: http://international.tau.ac.il/prospective-students/financial-terms/scholarships.html
Contact: Ms Smadar Fisher, Director, Dan David Prize

TENOVUS SCOTLAND

Small Research Grants, 232-242 St. Vincent Street, Glasgow, G2 5RJ, Scotland
Tel: (44) 14 1221 6268
Fax: (44) 12 9231 1433
Email: gen.sec@talk21.com
Website: www.tenovus-scotland.org.uk
Contact: I M'Fadzean, General Secretary

Tenovus Scotland supports innovative and pilot medical research projects carried out by young researchers who may not have a track record, across the full spectrum of Medicine and Dentistry.

Tenovus Scotland Small Research Grants

Subjects: Medicine, dentistry, medical sciences and allied areas.
Purpose: To foster high-quality research within the healthcare professions in Scotland.
Eligibility: Medical professionals of Scotland. (1) No restrictions on age although preference is for young researchers seeking to establish a track record. (2) Grants conditional on the work being carried out in a Scottish University/Teaching at an NHS Trust Hospital. (3) No restriction on nationality provided they meet the above criteria.
Level of Study: Research
Type: Grant
Value: Normally up to UK£10,000 or part there of
Frequency: Annual
Country of Study: Scotland
Application Procedure: Application forms must be filled, applications from investigators lacking support in the early stages of a new project are encouraged. Applications may be invited for salary support or for research studentships.
Closing Date: Edinburgh - September 15th, Grampian and Strathclyde - February 15th and September 15th, Tayside - May 1st and December 1st
Funding: Individuals, trusts
No. of awards given last year: 42
No. of applicants last year: 89
Additional Information: Please see teh website for further details http://www.tenovus-scotland.org.uk/ForResearchers.html.

THAILAND NATIONAL COMMISSION FOR UNESCO

Asia-Pacific Regional Bureau for Education, Mom Luang Pin Malakul Centenary Building, 920 Sukhumvit Road, Prakanong, Klongtoey, Bangkok, 10110, Thailand
Tel: (66) 2 628 56468
Fax: (66) 2 281 0953
Website: http://http://www.unescobkk.org/
Contact: Ms Chatuporn, Secretary General

Fulbright - Cambridge Scholarship

Subjects: All subjects.
Purpose: To pursue PhD study at the University of Cambridge.
Eligibility: US citizens are eligible to apply.
Level of Study: Doctorate
Type: Award
Value: Full fees and living stipend for duration of PhD programme
Frequency: Annual
Study Establishment: Cambridge
Country of Study: United Kingdom
No. of awards offered: 1
Application Procedure: Please check at http://us.fulbrightonline.org/applynow.html.
Closing Date: Mid-October
Funding: Government
Additional Information: Please check for more information at www.fulbright.co.uk.

THE FIELD PSYCH TRUST

301 Dixie Street, Carrollton, GA, 30117, United States of America
Tel: (1) 770 834 8143
Email: arichard@westga.edu
Website: www.fieldpsychtrust.org
Contact: Dr Anne C Richards, Trustee

The Field Psych Trust is a charitable trust honouring the professional life and contributions of psychologist/educator Dr Arthur W Combs. It provides grant funding to encourage graduate student research grounded in perceptual (field) psychology perspectives. It also supports the publication of manuscripts related to Dr Combs' professional life and work.

Field Psych Trust Grant

Subjects: As a psychological theory, perceptual (field) psychology is applicable to any subject area in which links between human experience, meaning and/or perception and human behavior can be explored.
Purpose: To encourage graduate student research exploring the history, contributions and further development of perceptual (field) psychology as related to the research and writings of Arthur W Combs
Eligibility: Open to graduate students in good standing through a competitive review process.
Level of Study: Postdoctorate, Predoctorate, Doctorate, Graduate
Type: Research grant
Value: Varies according to the itemized budget request of successful applicants and their projects. Awards range from US$500–1,500
Length of Study: Varies, although 1 year is preferable
Frequency: Biannually
Study Establishment: An accredited Institution of Higher Education
Country of Study: Any country
No. of awards offered: 5
Application Procedure: Applicants must complete an application form and submit references. Application forms can be found on the website. Applications are judged with respect to the relevance of the proposed project to the mission of the Field Psych Trust; substance, conceptual quality and clarity of the proposal; significance of the project in addressing matters of consequence to the human condition; and the degree of confidence that the prospective grant recipient has the ability to produce the proposed project.
Closing Date: January 31st and October 5th of each year
Funding: Private
Contributor: The estate of Arthur W Combs
No. of awards given last year: 1
No. of applicants last year: 1
Additional Information: Awards are subject to conditions, which are described, and include an obligation to submit a final report on conclusion of the project, which can take the form of a completed Master's thesis, research project report, doctoral dissertation or published manuscript. More information is available from the website, or by contacting Anne Richards at the main address.

THE ROYAL INSTITUTUE OF PHILOSOPHY

The Royal Institute of Philosophy, 14 Gordon Square, London, WC1H 0AR, United Kingdom
Tel: (44) 0207 387 4130
Email: j.garvey@royalinstitutephilosophy.org
Website: http://www.royalinstitutephilosophy.org
Contact: Dr James Garvey

Jacobsen Studentships

Subjects: Parts of philosophy: speculative metaphysics, critical metaphysics, the philosophy of mind, epistemology, philosophical logic, philosophy of life, determinism, the nature of life, evolution, moral philosophy, political and social philosophy.
Purpose: The Royal Institute of Philosophy, in accord with its own remit, seeks candidates whose work, while rigorous, avoids needless technicality and is in the tradition of philosophy as a humane discipline.
Eligibility: Applicants must have already completed at least one year of work for a higher degree in philosophy in a UK university by the start of the academic year. Preference may well be given to applicants nearing the end of a course of study leading to a doctorate.
Level of Study: Postgraduate
Type: Studentship
Value: £8,000 each
Length of Study: 1 year
Frequency: Annual
No. of awards offered: 8
Application Procedure: To apply, candidates must arrange for two academic referees, at least one from the institution at which they are currently studying, to write on their behalf direct to The Royal Institute of Philosophy. Candidates must also submit an application form. Application forms can be found at http://www.royalinstitutephilosophy.org/page/42.
Closing Date: June 15th
Additional Information: Please check the website for further details http://www.royalinstitutephilosophy.org/page/42.

Royal Institute of Philosophy Bursaries

Subjects: Philosophy.
Eligibility: Applicants must have already completed at least one year of work for a higher degree in philosophy in a UK university by the start of the academic year. Preference may well be given to applicants nearing the end of a course of study leading to a doctorate.
Type: Bursary
Value: £2,500 each
Length of Study: 1 year
Frequency: Annual
No. of awards offered: 8
Application Procedure: Candidates must arrange for two academic referees, at least one from the institution at which they are currently studying, to write on their behalf direct to The Royal Institute of Philosophy. Candidates must also submit an application form. Application forms can be found at http://www.royalinstitutephilosophy.org/page/42.
Closing Date: June 15th
Additional Information: Please check at http://www.royalinstitute-philosophy.org/page/42 for more information.

THIRD WORLD ACADEMY OF SCIENCES (TWAS)

TWAS Executive Director, ICTP Enrico Fermi Building, Room 108, Italy
Tel: (39) 040 2240 327
Fax: (39) 040 224559
Email: edoffice@twas.org
Website: www.twas.org
Contact: Professor Mohamed H A Hassan, Executive Director

The Third Word Academy of Sciences (TWAS) is an autonomous international organization that promotes and supports excellence in scientific research and helps build research capacity in the South.

CAS-TWAS Fellowship for Postdoctoral Research in China

Subjects: All areas of the natural sciences.
Purpose: To enable scholars who wish to pursue postdoctoral research to undertake research in laboratories or institutes of the Chinese Academy of Sciences.
Eligibility: Candidates must have a PhD and be nationals of a developing country other than China. They must also be regularly employed at a research or teaching institution in their home country. The maximum age limit is 40 years.
Level of Study: Postdoctorate
Type: Fellowship
Value: Covers food, accommodation and international travel, no provision for family members
Length of Study: Up to 1 year
Frequency: Annual
Study Establishment: CAS
Country of Study: China
No. of awards offered: Up to 15
Application Procedure: Applicants must send one copy of the application form to TWAS and three copies to CAS. Application forms can be obtained from the TWAS website.
Closing Date: August 31st (check with website)
Funding: Government
Contributor: Chinese Academy of Sciences (CAS) and TWAS
Additional Information: CAS has 5 academic divisions, 11 local branches, 84 research institutes and 3 universities or colleges, distributed throughout the country.

For further information contact:

Division of International Organization Programmes, Chinese Academy of Sciences, 52 Sanlihe Road, Beijing, 100864, China
Website: http://twas.ictp.it/prog/exchange/fells/fells-pdoc/cas-pdoc
Contact: Mr Wang Zhenyu, Deputy Director

CAS-TWAS Fellowship for Postgraduate Research in China

Subjects: All areas of the natural sciences.
Purpose: To carry out research towards the final year of a PhD programme in China.
Eligibility: Candidates must have a Master's degree in natural sciences, be nationals of a developing country other than China and be registered for a PhD in their home country. The maximum age limit is 35 years.
Level of Study: Postgraduate
Type: Fellowship
Value: Covers food, accommodation and international travel. There is no provision for family members
Length of Study: 1 year
Frequency: Annual
Study Establishment: CAS
Country of Study: China
No. of awards offered: Up to 20
Application Procedure: Applicants must send one copy of the application form to TWAS and three copies to CAS. Application forms can be obtained from the TWAS website.
Closing Date: August 31st
Funding: Government
Contributor: Chinese Academy of Sciences (CAS) and TWAS
No. of awards given last year: New
Additional Information: CAS has 5 academic divisions, 11 local branches, 84 research institutes and 3 universities or colleges distrubuted throughout the country.

For further information contact:

Division of International Organization Programmes, Chinese Academy of Sciences, 52 Sanlihe Road, Beijing, 100864, China
Website: http://twas.ictp.it/prog/exchange/fells/fells-pg/cas-pg
Contact: Mr Wang Zhengu, Deputy Director

CAS-TWAS Fellowship for Visiting Scholars in China

Subjects: All areas of the natural sciences.
Purpose: To pursue advanced research in the natural sciences.
Eligibility: Applicants must have a PhD, a regular research assignment and at least 5 years postdoctoral research experience. Chinese nationals are not eligible. The maximum age limit is 55 years.
Level of Study: Research
Type: Fellowship
Value: Varies
Length of Study: 1–3 months
Frequency: Annual
Study Establishment: CAS
Country of Study: China
No. of awards offered: Up to 15
Application Procedure: Applicants must send one copy of the application to TWAS and 3 copies to CAS. Application forms can be obtained from the TWAS website.
Closing Date: August 31st
Funding: Government
Contributor: Chinese Academy of Sciences (CAS) and TWAS
Additional Information: CAS has 5 academic divisions, 11 local branches, 84 research institutes and 3 universities or colleges distributed throughout the country.

For further information contact:

Division of International Organization Programmes, Chinese Academy of Sciences, 52 Sanlihe Road, Beijing, 100864, China
Website: http://twas.ictp.it/prog/exchange/fells/fellows-adv/cas-vis
Contact: Mr Wang Zhenyu, Deputy Director

CNPq-TWAS Doctoral Fellowships in Brazil

Subjects: All areas of the natural sciences.
Purpose: To enable scholars from developing countries (other than Brazil) to undertake research in Brazil.
Eligibility: Applicants must hold a Master's degree or equivalent, and be proficient in either English, French, Portuguese or Spanish. Open to nationals of a developing country other than Brazil. The maximum age limit is 30 years.
Level of Study: Postgraduate
Type: Fellowship

Value: Covers food, accommodation and international travel, no provision for family members
Length of Study: Up to 4 years
Frequency: Annual
Country of Study: Brazil
No. of awards offered: Up to 40
Application Procedure: Applicants must complete an application form, available on request or from the website www.twas.org
Closing Date: August 3rd
Funding: Government
Contributor: Brazilian ministry of science and technology, the Conselho Nacional de Desenvolvimento Cientifico e Tecnologico (CNPq) and TWAS
No. of awards given last year: New award
Additional Information: Please see the website for further details http://twas.ictp.it/prog/exchange/fells/fells-pg/bra-pg.

CNPq-TWAS Fellowships for Postdoctoral Research in Brazil

Subjects: All areas of the natural sciences.
Purpose: To enable scholars to pursue postdoctoral research in Brazil.
Eligibility: Applicants must have a PhD in the natural sciences, be proficient in English, French, Portuguese or Spanish and must be regularly employed at a research or teaching institution in their home country. Open to nationals of developing countries other than Brazil. The maximum age limit is 40 years.
Level of Study: Postdoctorate
Type: Fellowship
Value: Covers food, accommodation and international travel, no provision for family members
Length of Study: 6 months–1 year
Frequency: Annual
Country of Study: Brazil
No. of awards offered: Up to 10
Application Procedure: Applicants must complete an application form, available on request or from the website http://twas.ictp.it/prog/exchange/fells/fells-pdoc/bra-pdoc.
Closing Date: August 3rd
Funding: Government
Contributor: Brazilian Ministry of Science and Technology, the Conselho Nacional de Desenvolvimento Cientifico e Tecnologico (CNPq) and TWAS

CSIR (Council of Scientific and Industrial Research)/ TWAS Fellowship for Postgraduate Research

Subjects: Newly emerging areas of science and technology.
Purpose: To enable scholars from developing countries (other than India) who wish to pursue postgraduate research to undertake research in laboratories or institutes of the CSIR.
Eligibility: Candidates must have a Master's or equivalent degree in science or engineering and should be a regular employee in a developing country (other than India) and be holding a research assignment.
Level of Study: Postgraduate
Type: Fellowship
Value: Monthly stipend to cover for living costs, food and health insurance.
Length of Study: Up to 4 years
Frequency: Annual
Study Establishment: CSIR research laboratories or institutes
Country of Study: India
No. of awards offered: Varies
Application Procedure: One copy of the application should be sent to TWAS and three copies to CSIR. Application forms are available on request or from the website www.twas.org or www.ictp.trieste.it/˜twas/hg/csir_postgrad_form.html
Closing Date: June 1st
Funding: Government
Contributor: CSIR (India), the Italian Ministry of Foreign Affairs and the Directorate General for Development Co-operation
No. of awards given last year: 8
Additional Information: CSIR is the premier scientific organization of India, and has a network of research laboratories covering wide areas of scientific and industrial research.

For further information contact:

International S&T Affairs Directorate, Council for Scientific and Industrial Research (CSIR), Anusandhan Bhavan, 2 Rafi Marg, New Delhi, 110001, India
Fax: (91) 11 2371 0618
Email: rprasad@csir.res.in
Website: http://twas.ictp.it/prog/exchange/fells/fells-pg
Contact: Dr B K Ramprasad, Senior Deputy Advisor

CSIR (The Council of Scientific and Industrial Research)/ TWAS Fellowship for Postdoctoral Research

Subjects: Newly emerging areas of science and technology.
Purpose: To enable scholars from developing countries (other than India) who wish to pursue postdoctoral research to undertake research in laboratories or institutes of the CSIR.
Eligibility: The minimum qualification requirement is a PhD degree in science or technology. Applicants must be regular employees in a developing country (but not India) and should hold a research assignment.
Level of Study: Postdoctorate
Type: Fellowship
Value: Monthly stipend to cover for living costs, food and health insurance.
Length of Study: 6 to max. 12 months
Frequency: Annual
Study Establishment: CSIR research laboratories or institutes
Country of Study: India
No. of awards offered: Varies
Application Procedure: Applicants must complete an application form, available on request or from the website.
Closing Date: June 1st
Funding: Government
Contributor: CSIR (India), the Italian Ministry of Foreign Affairs and the Directorate General for Development Co-operation
No. of awards given last year: 4
Additional Information: CSIR is the premier civil scientific organization of India, which has a network of research laboratories covering wide areas of industrial research.

For further information contact:

Senior Deputy Advisor, CSIR (The Council of Scientific and Industrial Research), Anusandhan Bhavan, 2 Rafi Marg, New Delhi, 110001, India
Tel: (91) 11 331 6751
Fax: (91) 11 371 0618
Email: rprasad@csirhq.ren.nic.in
Website: http://rdpp.csir.res.in/csir_acsir/PDF/brochure_India.pdf
Contact: Dr B K Ramprasad

DBT-TWAS Biotechnology Fellowship for Postdoctoral Studies in India

Subjects: All areas of biotechnology.
Purpose: To support postdoctoral research in biotechnology in India.
Eligibility: Applicants must have a Master's degree in science or engineering or an equivalent degree and must be a national of a developing country. Indian nationals are not eligible. The maximum age limit is 40 years. Fellowships for PhD will be awarded only to candidates who are already registered for PhD in a university in their home country or willing to register in India
Level of Study: Postgraduate
Type: Fellowship
Value: Monthly stipend to cover for living costs, food and health insurance.
Length of Study: 1 year
Frequency: Annual
Study Establishment: More than 80 listed universities and research institutions
Country of Study: India
Application Procedure: Applicants must complete an application form, available on request or from the website http://dbtindia.nic.in/DBT-TWAS/DBT-Postdoc-new.pdf.
Closing Date: August 31st
Funding: Government
Contributor: Indian Department of Biotechnology (DBT) and TWAS

DBT-TWAS Biotechnology Fellowships for Postgraduate Studies in India

Subjects: All areas of biotechnology.
Purpose: To carry out research leading to a PhD in biotechnology.
Eligibility: Applicants must have a Master's degree in science, engineering or equivalent, must be a national of a developing country (except India) and must be registered for a PhD or be willing to register in India. The maximum age limit is 30 years.
Level of Study: Postdoctorate
Type: Fellowship
Value: Monthly stipend (INR 8,000 for the 1st and 2nd year, and INR 9,000 per month for the 3rd year) to cover for living costs, food and health insurance.
Length of Study: Up to 5 years
Frequency: Annual
Study Establishment: More than 80 listed universities and research institutions
Country of Study: India
No. of awards offered: Up to 40
Application Procedure: Applicants must complete an application form, available on request or from the website http://twas.ictp.it/prog/exchange/fells/fells-pg/dbt-pg.
Closing Date: August 31st
Funding: Government
Contributor: Indian Department of Biotechnology (DBT) and TWAS

ICSU-TWAS-UNESCO-UNU/IAS Visiting Scientist Programme

Subjects: All areas of science other than mathematics or physics.
Purpose: To provide institutions and research grants in the South, especially in least developed countries (LDCs), with the opportunity to establish long-term links with world leaders in science and help build scientific capacity in their country.
Eligibility: Candidate must be an internationally renowned expert.
Level of Study: Professional development, Research
Type: Consultancy
Value: Travel plus a US$500 honorarium. Local costs will be covered by the host institution
Length of Study: Minimum of 1 month
Frequency: Annual
Study Establishment: Teaching and research institutions
Country of Study: Any developing country, preference will be given to LDCs
Application Procedure: Applicants must complete an application form, available on request or from the website.
Closing Date: October 1st
Funding: Government, international office
Contributor: International Council for Science (ICSU), United Nations Educational, Scientific and Cultural Organization (UNESCO), United Nations University/Institute for Advanced Studies (UNU/IAS) and TWAS
No. of awards given last year: 16 visits
Additional Information: A similar programme for mathematics and physics is run by the Abdus Salam International Centre for Theoretical Physics (ICTP). See the website www.ictp.trieste.it/www-users/oea/vs for more information.

Support for International Scientific Meetings

Subjects: Agricultural, biological, chemical, engineering or geological and medical sciences.
Purpose: To encourage international scientific meetings in Third World countries.
Eligibility: Open to organizers of international scientific meetings in developing countries. Special consideration is given to those meetings that are likely to benefit the scientific community in the Third World and to promote regional and international co-operation in developing science and its applications to the problems of the Third World.
Level of Study: Postgraduate, Professional development
Type: Travel grant
Value: Up to US$3,000 for travel expenses of principal speakers from abroad and/or participants from the region
Frequency: Annual
Country of Study: Developing countries
No. of awards offered: Varies
Application Procedure: Applicants must complete an application form, available on request or from the website http://twas.ictp.it/prog/meetings/support-for-international-scientific-meetings.
Closing Date: June 1st for meetings held between January and June of the following year, and December 1st for meetings held between July and December of the following year
Funding: Government
Contributor: The Italian Ministry of Foreign Affairs and the Directorate General for Development Co-operation
No. of awards given last year: 30
Additional Information: Grants are not offered for meetings in Physics and Mathematics.

The Trieste Science Prize

Subjects: Biological sciences, chemical sciences, agricultural sciences, Earth, space, ocean and atmospheric sciences, engineering sciences, mathematics, medical sciences, physics and astronomy.
Purpose: To honour outstanding scientists living and working in developing countries.
Eligibility: Candidates must be nationals of developing countries living and working in the South. Individuals who have won the Nobel Prize, Tokyo/Kyoto Prize, Gafoord Prize or Abel Prize are not eligible.
Level of Study: Research
Type: Prize
Value: US$50,000 each
Frequency: Annual
Country of Study: Any developing country
No. of awards offered: 2
Application Procedure: Nomination forms must be downloaded from the website www.twas.org and accompanied by a 5–6 page biographical sketch outlining the nominee's major scientific achievements, pre-prints of up to 20 publications and a complete list of publications.
Closing Date: May 15th
Funding: Commercial, private
Contributor: Illycaffè, Trieste
Additional Information: Please see the website for further details http://twas.ictp.it/prog/prizes/trieste-science-prize.

TWAS Fellowships for Research and Advanced Training

Subjects: All fields of basic sciences.
Purpose: To enhance the research of young promising scientists, specifically those at the beginning of their research career, and to help them to foster links for future collaboration.
Eligibility: Open to nationals of developing countries with permanent positions in universities or research institutes in developing countries holding a PhD or equivalent. Candidates must not be older than 40 years and preference will be given to candidates from less developed countries.
Level of Study: Postdoctorate
Type: Fellowship
Value: Travel support and monthly subsistence of up to US$300. Living expenses are usually obtained from local sources
Length of Study: 3 months–1 year
Frequency: Annual
Country of Study: Developing countries
No. of awards offered: Varies
Application Procedure: Applicants must complete an application form, available on request or from the website.
Closing Date: October 1st
Funding: Government
Contributor: The Italian Ministry of Foreign Affairs and the Directorate General for Development Co-operation
Additional Information: Further information is available on the website http://twas.ictp.it/prog/exchange/fells/fellows-adv/twas-fells.

TWAS Prizes

Subjects: Medical sciences, biology, chemistry, mathematics and physics, agricultural sciences, engineering sciences and earth sciences.
Purpose: To recognize and support outstanding achievements made by scientists from developing countries. Prizes are awarded to those scientists whose research work has significantly contributed to the advancement of science.

Eligibility: Open to nationals of developing countries who are, as a rule, working and living in these countries. Consideration is given to proven achievements judged particularly by their national and international impact. Members of TWAS are not eligible for such awards.
Level of Study: Doctorate, Postdoctorate, Postgraduate, Professional development
Type: Prize
Value: US$15,000 each
Frequency: Annual
Country of Study: Developing countries
No. of awards offered: 8
Application Procedure: Applicants must be designated on the nomination form. The nomination must be accompanied by a one- to two-page biographical sketch of the nominee including their major scientific accomplishments, a list of twelve of the candidate's most significant publications as well as a complete list of publications and a curriculum vitae. Nominations for the awards are invited from all members of the TWAS as well as from academies, national research councils, universities and scientific institutions in developing countries and advanced countries. A nomination form is available on request or can be downloaded from the http://twas.ictp.it/prog/prizes/twas-prizes.
Closing Date: February 28th. Nominations received after the deadline will be considered in the next year
Funding: Government
Contributor: The Italian Ministry of Foreign Affairs and the Directorate General for Development Co-operation
No. of awards given last year: 8
Additional Information: The awards are usually presented on a special occasion, normally coinciding with the general meeting of the Academy and/or a general conference organized by the Academy. Recipients of awards are expected to give lectures about the work for which the awards have been made. Further information is available on the website.

TWAS Prizes to Young Scientists in Developing Countries

Subjects: Biology, chemistry, mathematics or physics, rotated annually.
Purpose: To enable science academies and research councils in developing countries to award prizes to scientists in their countries.
Eligibility: Open to academies and research councils in developing countries. The age limit for prize winners is 40 years.
Level of Study: Postgraduate
Type: Prize
Value: Usually US$2,000
Frequency: Annual
Country of Study: Developing countries
No. of awards offered: More than 30
Application Procedure: Applicants must write for details to info@twas.org or see the website http://twas.ictp.it/prog/prizes/twas-prizes-for-young-scientist-in-developing-countries.
Closing Date: Check with website
Funding: Government
Contributor: The Italian Ministry of Foreign Affairs and the Directorate General for Development Co-operation
No. of awards given last year: 23

TWAS Research Grants

Subjects: Biology, chemistry, mathematics and physics.
Purpose: To reinforce and promote scientific research in basic sciences in the Third World, to strengthen the endogenous capacity in science and to reduce the exodus of scientific talents from the South.
Eligibility: Applicants must be nationals of developing countries with an advanced academic degree, some research experience and must hold positions at universities or research institutions in developing countries.
Level of Study: Doctorate, Postdoctorate, Postgraduate, Professional development
Value: Up to US$10,000. Grants are to be used to purchase scientific equipment, consumable laboratory supplies and scientific literature (textbooks and proceedings only)
Length of Study: 1 year
Frequency: Annual
Country of Study: Developing countries
No. of awards offered: Varies
Application Procedure: Applicants must complete an application form available on request or from the website http://twas.ictp.it/prog/grants/research-grants. Applications must be submitted in English.
Closing Date: August 31st
Funding: Government
Contributor: The Italian Ministry of Foreign Affairs, the Directorate General for Development Co-operation and the Swedish Agency for Research Co-operation with Developing Countries
No. of awards given last year: 90
Additional Information: Further information is available on the request.

TWAS Spare Parts for Scientific Equipment

Subjects: Biology, chemistry and physics.
Purpose: The programme has been established in response to the current difficulty faced by several laboratories in the Third World to obtain badly needed spares and replacement parts for scientific equipment that is required for their experimental research.
Eligibility: Applicants must be research group leaders at universities or research institutes in developing countries.
Level of Study: Professional development
Type: Grant
Value: Up to US$1,000 including insurance and freight charges
Country of Study: Developing countries
No. of awards offered: Varies
Application Procedure: Applicants must first contact the suppliers and obtain a proforma invoice, valid for 3–6 months, including cost, insurance and freight charges for the items they require. Applicants must submit a completed application form with the proforma invoice from the supplier. Application forms are available on request or from the website.
Closing Date: Applications are accepted at any time
Funding: Government
Contributor: The Italian Ministry of Foreign Affairs and the Directorate General for Development Co-operation
No. of awards given last year: 30
Additional Information: Applications by email will not be accepted.

For further information contact:

c/o The Abdus Salam International Centre for Theoretical physics
Strada Costiera 11-34014, Trieste, Italy
Fax: (39) 39040224559
Website: http://www.ihep.ac.cn/library/lanmu/xiaoxi/spare.htm
Contact: Mrs M.T. Mahdavi

TWAS UNESCO Associateship Scheme

Subjects: Biology, chemistry, physics, mathematics, engineering, agricultural sciences, medical sciences and earth sciences.
Purpose: To alleviate the problem of isolated talented scientists in developing countries, and strengthen the research programmes of centres of excellence in the South.
Eligibility: Open to associates among the most eminent and promising researchers in developing countries. Special consideration is given to scientists from isolated institutions in developing countries.
Level of Study: Postdoctorate, Professional development
Value: Travel costs plus US$200 per month for incidental local expenses. The host centre provides local hospitality and research facilities
Length of Study: 3 years, plus the entitlement to visit the Centre twice for a period of 2–3 months each time. There is a possibility of renewal for a further 3 years depending on funds available
Frequency: Annual
Study Establishment: There are over 116 centres
Country of Study: Developing countries
No. of awards offered: Varies
Application Procedure: Applicants must complete an application form, available on request or from the website http://twas.ictp.it/prog/exchange/res-collab/assoc.
Closing Date: December 1st
Funding: Government
Contributor: UNESCO, the Italian Ministry for Foreign Affairs and the Directorate General for Development Co-operation

TWAS-S N Bose National Centre for Basic Sciences Postgraduate Fellowships in Physical Sciences

Subjects: Physical sciences.
Purpose: To carry out research leading to a PhD in the physical sciences.
Eligibility: Applicant must have a Master's degree in physics, mathematics or physical chemistry, must be a national of a developing country (other than India) and be employed at a research institution. The maximum age limit is 30 years.
Level of Study: Postgraduate
Type: Fellowship
Value: Monthly stipend (Rs 8,000 per month as well as free on-campus accommodation, Rs 2,400 will be paid towards house rent) to cover living costs, food and health insurance.
Length of Study: Up to 5 years
Frequency: Annual
Study Establishment: S N Bose National Centre for Basic Sciences
Country of Study: India
No. of awards offered: Up to 5
Application Procedure: Applicants must complete an application form, available on request or from the website.
Closing Date: August 31st
Funding: Government
Contributor: S N Bose National Centre for Basic Sciences, Kolkata, India and TWAS
Additional Information: For further information see the website http://twas.ictp.it/prog/exchange/fells/fells-pg/bose.

THURGOOD MARSHALL COLLEGE FUND (TMCF)

901 F Street NW, Suite 300, Washington, dc, 20004, United States of America
Tel: (1) 202 507 4851
Fax: (1) 202 652 2934
Email: emhall@tmcfund.org
Website: www.thurgoodmarshallfund.org

The Thurgood Marshall College Fund (TMCF) was established in 1987 to carry on Justice Marshall's legacy of equal access to higher education by supporting exceptional merit scholars attending America's public historically Black colleges and universities. More than 5000 Thurgood Marshall Scholars have graduated and are making valuable contributions to science, technology, government, human service, business, education and various communities.

Philip Morris USA Thurgood Marshall Scholarship

Subjects: Biology, business, chemistry, computer science, economics, engineering, finance and physics.
Purpose: To aid students who demonstrated financial need and the potential for success.
Eligibility: Open to citizens of the United States and current full-time students of Florida A&M, North Carolina A&T, Winston-Salem State, Howard University, Virginia State or Norfolk State Universities.
Level of Study: Postgraduate
Type: Scholarships
Value: Up to US$5,000
Length of Study: 1 year
Frequency: Annual
Country of Study: United States of America
No. of awards offered: 17
Application Procedure: A completed application form should be submitted to the TMSF College Coordinator.
Closing Date: Check with website
Funding: Foundation

For further information contact:

90 William Street, Suite 1203, New York, NY, 10038, United States of America
Website: http://www.thurgoodmarshallfund.net/scholarship/about-scholarships-program
Contact: Philip Morris

TMCF Scholarships

Subjects: Creative and performing arts.
Purpose: To financially support outstanding students.
Eligibility: Open to candidates who are academically exceptional in the creative and performing arts requiring financial help.
Type: Scholarship
Value: US$2,200 per semester and payment of tuition fees, accommodation and books
Frequency: Annual
Application Procedure: Completed applications must be submitted along with the required attachments.
Closing Date: Check with website
Funding: Foundation

For further information contact:

AIPLEF Scholarship, 80 Maiden Lane, Suite 2204, New York, NY 10038
Email: jessica.barnes@tmcfund.org
Website: http://www.thurgoodmarshallfund.net/scholarship/about-scholarships-program

TOKYU FOUNDATION FOR INBOUND STUDENTS

1-21-2 Dogenzaka, Shibuya Ku, Tokyo, 150-0043, Japan
Tel: (81) 3 3461 0844
Fax: (81) 3 5458 1696
Email: info@tokyu-f.jp
Website: www.tokyu-f.jp
Contact: Dr Takashi Izumi, Managing Director & Secretary General

The Tokyu Foundation for Inbound Students grant scholarships to postgraduate students studying in Japan from Asia-pacific areas.

Tokyu Scholarship

Subjects: All subjects.
Purpose: To promote international exchange by fostering the development of international goodwill between Japan and her neighbours in Asia and the Pacific and contributing to international co-operation and cultural exchange in the broadest possible sense.
Eligibility: Open to applicants from Asia or Pacific countries who will be able to explain about their research plan in Japanese. Please check website for complete qualifications.
Level of Study: Postgraduate
Type: Scholarship
Value: ¥160,000 per month per student; travel expenses for attending academic meetings in Japan will be provided
Length of Study: Up to 2 years
Frequency: Annual
Country of Study: Japan
No. of awards offered: 15–20
Application Procedure: Application forms and guidelines can be downloaded from the foundation's website: http://www.tokyu-f.jp/index1.htm (in Japanese only)
Closing Date: Between October 1st and November 1st. Applicants must abide by the deadline and submit application.
Funding: Commercial
Contributor: Tokyu Corporation
No. of awards given last year: 21
No. of applicants last year: 869
Additional Information: Applicants must travel to Japan at their own cost and be admitted to enter university postgraduate school. Please refer to the website for more information.

For further information contact:

Office for International Cooperation & Exchange (OICE), Graduate School of Agricultural and Life Sciences, The University of Tokyo
Tel: 81 3 5841 8122 Ext. 25485
Website: http://www.jpss.jp/en/scholarship/233/
Contact: Dr Neelam Ramaiah

TOLEDO COMMUNITY FOUNDATION

300 Madison Avenue, Suite 1300, Toledo, OH, 43604-1151, United States of America
Tel: (1) 419 241 5049
Fax: (1) 419 242 5549
Email: toledocf@toledocf.org
Website: www.toledocf.org

The Toledo Community Foundation, Inc. is a public, charitable foundation that exists to improve the quality of life in the Toledo region.

Charles Z. Moore Memorial Scholarship Fund

Subjects: Music.
Purpose: To encourage students pursuing a course of study in music with an emphasis on Jazz studies.
Eligibility: Open to residents of northwest Ohio or Southeast Michigan who demonstrate the talent, interest and ability needed to pursue the study of Jazz.
Level of Study: Professional development
Type: Scholarship
Frequency: Annual
Country of Study: United States of America
Application Procedure: A completed scholarship application form and support materials must be sent.
Closing Date: March 22nd
Additional Information: Please see the website for further details http://asoft4241.accrisoft.com/tcf/main/scholarships/.

Edith Franklin Pottery Scholarship

Subjects: Ceramic arts.
Purpose: To assist promising and accomplished potters in obtaining additional education or training in the ceramic arts.
Eligibility: Open to all applicants who are current residents of northwest Ohio Lenawee or Monroe Countries in Michigan with individual motivation, ability and potential.
Level of Study: Professional development
Type: Scholarship
Value: US$7,000
Frequency: Annual
Study Establishment: Any recognized college, university or nonprofit organization
Country of Study: United States of America
Application Procedure: A completed scholarship application form must be sent.
Closing Date: February 28th
Additional Information: Please see the website for further details http://asoft4241.accrisoft.com/tcf/main/scholarships/.

Harold W. Wott-IEEE Toledo Section Scholarship Fund

Subjects: Engineering.
Purpose: To support students studying in the engineering field.
Eligibility: Open to applicants attending institutions or universities in northwestern Ohio and southeastern Michigan.
Level of Study: Postgraduate
Type: Scholarship
Value: US$1,000
Frequency: Annual
Country of Study: United States of America
No. of awards offered: Varies
Application Procedure: A completed application form and required attachments and transcripts must be sent.
Closing Date: March 1st
Additional Information: Please see the website for further details http://asoft4241.accrisoft.com/tcf/main/scholarships/.

TOMSK POLYTECHNIC UNIVERSITY

Institute for International Education, 30 Lenin Avenue, Tomsk, 634050, Russia
Tel: (7) 3822 563304
Fax: (7) 3822 563299
Email: iie@tpu.ru
Website: www.iie.tpu.ru

Tomsk Polytechnic University was founded in 1896 and is the oldest technical educational institution in the Asian part of Russia. Since that time, the university scholars and graduates have greatly contributed to the Russian science, education, culture and industry development.

Tomsk Polytechnic University International Scholarship

Subjects: All subjects available at the university.
Purpose: To financially support outstanding students.
Eligibility: Open to candidates with an average of 80 per cent marks at graduation, who are not older than 40 years of age.
Level of Study: Doctorate, Postgraduate
Type: Scholarships
Value: Varies
Length of Study: 1–3 years
Frequency: Annual
Study Establishment: Tomsk Polytechnic University
Country of Study: Russia
No. of awards offered: 5
Application Procedure: Applicants can download the application form from the website. The completed application form and educational certificates must be submitted.
Closing Date: August 15th
Funding: Government, corporation
Contributor: Ministry of Russian Federation
No. of awards given last year: 3
No. of applicants last year: 50
Additional Information: Applications are to be submitted by fax at 3822 563304 or by email to iie@tpu.ru

TOURETTE SYNDROME ASSOCIATION, INC. (TSA)

42-40 Bell Boulevard, Suite 205, Bayside, NY, 11361-2820, United States of America
Tel: (1) 718 224 2999
Fax: (1) 718 279 9596
Email: ts@tsa-usa.org
Website: www.tsa-usa.org
Contact: Dr Kevin St P McNaught, Vice President, Medical & Scientific Programmes

The Tourette Syndrome Association, Inc. (TSA), founded in 1972, is the only national voluntary non-profit membership organization dedicated to identifying the cause, finding the cure and controlling the effects of Tourette Syndrome. Members include individuals with the disorder, their relatives and other interested, concerned people. The Association develops and disseminates educational material to individuals, professionals and agencies in the fields of healthcare, education and government, co-ordinates support services to help people and their families cope with the problems that occur with Tourette Syndrome and funds research that will ultimately find the cause of and cure for it and, at the same time, lead to improved medications and treatments.

TSA Research Grant and Fellowship Program

Subjects: Clinical and basic science relevant to Tourette Syndrome.
Purpose: To foster basic and clinical research related to the causes or treatment of Tourette Syndrome.
Eligibility: Open to candidates who have an MD, PhD or equivalent qualifications. Previous experience in the field of movement disorders is desirable, but not essential. Fellowships are intended for young postdoctoral investigators in the early stages of their careers.
Level of Study: Research, Postdoctorate
Type: Research grant
Value: Varies, depending upon the category and applicants's experience within that category, and is usually US$5,000–75,000. Postdoctoral grants (fellowships) are up to US$40,000
Length of Study: 1 year
Frequency: Annual
Study Establishment: Any institution with adequate facilities
Country of Study: Any country
No. of awards offered: Varies
Application Procedure: Applicants must submit a letter of intent briefly describing the scientific basis of the proposed project. For

further information and deadlines please see the http://www.tsa-usa.org/research.html.
Closing Date: October 0r November
Funding: Private
No. of awards given last year: 21
No. of applicants last year: 82
Additional Information: The Association provides up to 10 per cent of overhead or indirect costs within the total amount budgeted.

TOXICOLOGY EDUCATION FOUNDATION (TEF)

626 Admiral Drive, Ste. C, PMB 221, Annapolis, MD, 21401, United States of America
Tel: (1) 443 321 4654
Fax: (1) 443 321 8702
Email: tefhq@toxedfoundation.org
Website: www.toxedfoundation.org

The mission of TFE is to encourage, support and promote charitable and educational activities that increase the public understanding of toxicology.

Alleghery-ENCRC Student Research Award

Subjects: Toxicology.
Purpose: To support a student's thesis, dissertation and summer research project in toxicology and to encourage them to formulate and conduct meaningful research.
Eligibility: Open to students who are members in good standing of AE-SOT. The student's advisor must also be a member in good standing and submit a letter concerning availability.
Level of Study: Graduate
Type: Award
Value: Up to $1,000
Frequency: Annual
Country of Study: United States of America
No. of awards offered: 1
Application Procedure: Applicants must send four copies of completed application form alongwith a project description and budget.
Closing Date: May 30th

For further information contact:

CDC/NIOSH MS 2015, 1095 Willowdale Road, Morgatown, WV, 26505, United States of America
Email: LBattelli@cdc.gov
Contact: Lori Battelli

Colgate-Palmolive Grants for Alternative Research

Subjects: Reproductive and developmental toxicology, neurotoxicology, systemic toxicology, sensitization and acute toxicity.
Purpose: To identify and support efforts that promote, develop, refine or validate scientifically acceptable animal alternative methods to facilitate the safety assessment of new chemicals and formulations.
Level of Study: Research
Type: Research grant
Value: Plaque and maximum award of $40,000
Frequency: Annual
Country of Study: United States of America
Application Procedure: Application is available online. A research plan, budget, curriculum vitae and a letter from the institution must be sent.
Closing Date: October 9th
Funding: Private
Contributor: Colgate-Palmolive
No. of awards given last year: 5

Colgate-Palmolive Postdoctoral Fellowship Award in In Vitro Toxicology

Subjects: Toxicology.
Purpose: To advance the development of alternatives to animal testing in toxicological research.
Eligibility: Open to postdoctoral trainees employed by academic institutions. The applicants or postdoctoral advisors must be members or pending members of SOT.
Level of Study: Postdoctorate
Type: Fellowship
Value: Includes a stipend and research-related costs of up to US $38,500 per year
Length of Study: 1-3 year
Frequency: Every 2 years, alternate year
Country of Study: United States of America
Application Procedure: Applicants must submit their curriculum vitae, transcripts, research proposal, budget and recommendation letters. Applicants must also supply a description of the research to be performed.
Closing Date: October 9th
Contributor: Colgate-Palmolive
Additional Information: Preference is for applicants in their first year of study beyond the PhD, MD or DVM degree. Funding for second year is contingent upon satisfactory research progress.

For further information contact:

Website: www.toxicology.org/ai/af/awards.aspx

Food Safety SS Burdock Group Travel Award

Subjects: Food safety toxicology.
Purpose: To cover travel expenses for a student to attend the Annual Meeting.
Eligibility: Open to full-time graduate students with research interests in toxicology. Students in their early graduate training, who have not attended any SOT Annual Meeting are encouraged to apply.
Level of Study: Graduate
Value: Up to $500
Frequency: Annual
Country of Study: United States of America
No. of awards offered: 1
Application Procedure: Applicants must send a letter of request indicating that he/she is enrolled in good standing in a doctoral training programme. The applicant must also state how the research and training relate to food safety.
Closing Date: February 24th

For further information contact:

Email: rmatulka@burdockgroup.com
Website: http://www.toxicology.org/ISOT/SS/foodsafe/index.asp
Contact: Ray Matulka

Regulation and Safety SS Travel Award

Subjects: Toxicology.
Purpose: To help defray the costs of travel to the SOT meeting.
Eligibility: Open to students submitting a poster or making a presentation at the SOT meeting.
Type: Travel award
Value: $1500 each
Frequency: Annual
Country of Study: United States of America
No. of awards offered: 4–5
Application Procedure: Applicants must fill an application form and an abstract of work preserved.
Closing Date: December 15th

For further information contact:

Email: jtmacgror@earthlink.net
Website: http://www.toxicology.org/ISOT/SS/regulatorysafety/awards.asp
Contact: James TMacGregor

Robert L. Dixon International Travel Award

Subjects: Reproductive toxicology.
Purpose: To give financially assist students studying in the area of reproductive toxicology.
Eligibility: Open to applicants enrolled full-time in a PhD programme studying reproductive toxicology and are student members of SOT.
Level of Study: Doctorate, Graduate
Type: Award

Value: Includes a stipend of US$2,000 for travel costs to enable students to attend the International Congress of Toxicology meeting
Frequency: Every 3 years
Country of Study: United States of America and abroad
Application Procedure: Applicants must submit a completed application form, reference letter, graduate transcripts and lists of complete citations of the original work.
Closing Date: October 9th (check with website)
Contributor: Toxicology Education Foundation

For further information contact:

Email: tefhq@toxedfoundation.org
Website: http://www.toxedfoundation.org/dixon_award.html

THE TOYOTA FOUNDATION

37F, Shinjuku-Mitsui Building, 2-1-1, Nishi-Shinjuku, Shinjuku-Ku, Tokyo, 163-0437, Japan
Tel: (81) 3 3344 1701
Fax: (81) 3 3342 6911
Email: admin@toyotafound.or.jp
Website: www.toyotafound.or.jp

The Toyota Foundation was established in 1974 as a multi-purpose grant-making foundation. It provides financial assistance to carry out projects in Japan and other countries, mainly in the developing world, that address timely issues in a variety of fields.

The Toyota Foundation Research Programme

Subjects: All subjects.
Purpose: To support research.
Eligibility: Open to candidates of all nationalities studying a doctoral programme.
Type: Scholarship
Value: Varies
Length of Study: 1–2 years
Frequency: Annual
Application Procedure: Please see the website for details http://www.toyotafound.or.jp/english/02program/#header.
Closing Date: May 16th (via the Internet), May 12th (by postal mail)
Funding: Foundation
Contributor: The Toyota Foundation

TRANSPORTATION ASSOCIATION OF CANADA FOUNDATION

2323 St Laurent Boulevard, Ottawa, ON, K1G 4J8, Canada
Tel: (1) 613 736 1350, ext. 235
Fax: (1) 613 736 1395
Email: secretariat@tac-atc.ca
Website: www.tac-atc.ca
Contact: Ms Deb Cross, Secretary-Treasurer

The Transportation Association of Canada Foundation has a mandate to support the educational and research needs of the Canadian transportation industry.

TAC Foundation – Albert M. Stevens Scholarship

Subjects: Transportation-related disciplines.
Purpose: To contribute to maintain high quality transportation expertise in Canada.
Eligibility: Open only to the Canadian candidates or landed immigrants who are admissible to a postgraduate studies programme or already registered as full-time graduate students having a minimum grade point average of B.
Level of Study: Postgraduate
Type: Scholarship
Value: $5,000
Length of Study: 4 years
Application Procedure: Applicants must send application form (including a two page resume) to the TAC Foundation.
Closing Date: February 13th
Funding: Individuals
Contributor: Albert M. Stevens
Additional Information: Please see the website for further details http://www.tac-atc.ca/english/foundation/scholarships/index.cfm.

TAC Foundation – Cement Association of Canada Scholarship

Subjects: Transportation-related disciplines.
Purpose: To recognize the importance of education in the transportation field.
Eligibility: Open only to the Canadian candidates or landed immigrants who are admissible to a postgraduate studies program or already registered as full-time graduate students having a minimum grade point average of B.
Level of Study: Postgraduate
Type: Scholarship
Value: $5,000
Length of Study: 4 years
Application Procedure: Applicants must send application form (including a two page resume) to the TAC Foundation.
Closing Date: February 11th
Contributor: Cement Association of Canada

TAC Foundation – Delcan Corporation Scholarship

Subjects: Transportation-related disciplines.
Eligibility: Open only to the Canadian candidates or landed immigrants who are admissible to a postgraduate studies program or already registered as full-time graduate students having a minimum grade point average of B.
Level of Study: Postgraduate
Type: Scholarship
Value: $5,000
Length of Study: 4 years
Application Procedure: Applicants must send application form (including a two page resume) to the TAC Foundation.
Closing Date: February 11th (check with website)
Contributor: Delcan Corporation

TAC Foundation – EBA Engineering Consultants Ltd Scholarship

Subjects: Transportation engineering.
Eligibility: Open only to the Canadian candidates or landed immigrants who are admissible to a postgraduate studies program or already registered as full-time graduate students having a minimum grade point average of B.
Level of Study: Postgraduate
Type: Scholarship
Value: $5,000
Length of Study: 4 years
Application Procedure: Applicants must send application form (including a two page resume) to the TAC Foundation.
Closing Date: February 11th (check with website)
Contributor: EBA Engineering Consultants Limited
Additional Information: Preference is given in the areas of design, construction, maintenance and operation of roadway transportation systems in rural and urban environments.

TAC Foundation – IBI Group scholarship

Subjects: Transportation-related disciplines.
Purpose: To attract and prepare transportation planners, providers, funders, and users for a career in the transportation industry.
Eligibility: Open only to the Canadian candidates or landed immigrants who are admissible to a postgraduate studies program or already registered as full-time graduate students having a minimum grade point average of B.
Level of Study: Postgraduate
Type: Scholarship
Value: $4,500
Length of Study: 4 years
Application Procedure: Applicants must send application form (including a two page resume) to the TAC Foundation.
Closing Date: February 11th
Contributor: IBI Group

TAC Foundation – Municipalities Scholarship

Subjects: Transportation-related disciplines.

Purpose: To attract and prepare transportation planners, providers, funders, and users for a career in the transportation industry.
Eligibility: Open only to the Canadian candidates or landed immigrants who are admissible to a postgraduate studies program or already registered as full-time graduate students having a minimum grade point average of B.
Level of Study: Postgraduate
Type: Scholarship
Value: $3,000
Length of Study: 4 years
Application Procedure: Applicants must send application form (including a two page resume) to the TAC Foundation.
Closing Date: February 11th (check with website)
Contributor: Municipalities

TAC Foundation – Provinces and Territories Scholarship

Subjects: Transportation-related disciplines.
Purpose: To attract and prepare transportation planners, providers, funders, and users for a career in the transportation industry.
Eligibility: Open only to the Canadian candidates or landed immigrants who are admissible to a postgraduate studies program or already registered as full-time graduate students having a minimum grade point average of B.
Level of Study: Postgraduate
Type: Scholarship
Value: $5,000
Length of Study: 4 years
Application Procedure: Applicants must send application form (including a two page resume) to the TAC Foundation.
Closing Date: February 11th (check with website)
Contributor: Provinces and territories

TAC Foundation – Waterloo Alumni Scholarship

Subjects: Transportation-related disciplines.
Purpose: To continue the tradition of providing financial assistance to postgraduate students in the broad area of transportation.
Eligibility: Open only to the Canadian candidates or landed immigrants who are admissible to a postgraduate studies program or already registered as full-time graduate students having a minimum grade point average of B.
Level of Study: Postgraduate
Type: Scholarship
Value: $7,500
Length of Study: 4 years
Application Procedure: Applicants must send application form (including a two page resume) to the TAC Foundation.
Closing Date: February 11th (check with website)
Contributor: University of Waterloo Alumni who are past recipients of Transportation Association of Canada scholarships
Additional Information: Preference will be given to qualified candidates pursuing their work in the pavement field.

TAC Foundation – 3M Canada Bob Margison Memorial Scholarship

Subjects: Transportation-related disciplines.
Eligibility: Open only to the Canadian candidates or landed immigrants who are admissible to a postgraduate studies program or already registered as full-time graduate students having a minimum grade point average of B.
Level of Study: Postgraduate
Type: Scholarship
Value: $4,500
Length of Study: 4 years
Application Procedure: Check the website for further details.
Closing Date: February 11th (check with website)
Contributor: 3M Canada Bob Margison Memorial

TAC Foundation – HDR/iTRANS Scholarship

Subjects: Transportation-related disciplines.
Purpose: To foster education, innovation, and research in transportation planning and transportation engineering.
Eligibility: Open only to the Canadian candidates or landed immigrants who are admissible to a postgraduate studies program or already registered as full-time graduate students having a minimum grade point average of B.
Level of Study: Postgraduate
Type: Scholarship
Value: $5,000
Length of Study: 4 years
Application Procedure: Applicants must send application form (including a two page resume) to the TAC Foundation.
Closing Date: February 11th (check with website)
Contributor: HDR/iTRANS

TAC Foundation – Stantec Consulting Ltd Scholarship

Subjects: Transportation engineering.
Purpose: To encourage students to continue their postgraduate studies in the field of transportation engineering and to contribute to the cost-effective mobility upon which our society is based.
Eligibility: Open only to the Canadian candidates or landed immigrants who are admissible to a postgraduate studies program or already registered as full-time graduate students having a minimum grade point average of B.
Level of Study: Postgraduate
Type: Scholarship
Value: $5,000
Length of Study: 4 years
Application Procedure: Applicants must send application form (including a two page resume) to the TAC Foundation.
Closing Date: February 11th (check with website)
Contributor: Stantec Consulting Ltd

TAC Foundation Scholarships

Subjects: Road and transportation-related disciplines.
Purpose: To supports the educational needs of the Canadian transportation industry.
Eligibility: Open to Canadian citizens and landed immigrants who hold university degrees and who are acceptable to the university at which they plan to carry out their studies in the transportation field. See website for full details.
Level of Study: Graduate, Postgraduate
Type: Scholarship
Value: Canadian $3,000–5,000, with one Canadian $10,000 scholarship and one Canadian $7,500 scholarship
Length of Study: 1 year
Frequency: Annual
Study Establishment: Universities
Country of Study: Other
No. of awards offered: 15
Application Procedure: Online applications only.
Closing Date: February 13th
Funding: Commercial, government, private
No. of awards given last year: 39
Additional Information: Scholarships currently offered are from the DELCAN Corporation, Stantec Consulting Limited, provincial and territorial governments of Canada, EBA Engineering Consultants Limited, Dillon Consulting Limited, 3M Company, IBI Group, Armtec, HDR/iTRANS, Cement Association of Canada, Waterloo Alumni, Albert Stevens, McCormick Rankin Corporation. Please see the website for further details http://www.tac-atc.ca/english/foundation/scholarships/index.cfm.

TREE RESEARCH & EDUCATION ENDOWMENT FUND

Tree Fund, 552 So. Washington St., Suite 109, Naperville, IL, 60540, United States of America
Tel: (1) 630 369 8300
Fax: (1) 630 369 8382
Email: officemanager@treefund.org
Website: www.treefund.org
Contact: Ms M Janet Bornancin, Executive Director

To identify and fund projects and programmes that advance knowledge in the field of arboriculture and urban forestry that benefit people, trees and the environment. We support sustainable communities and environmental stewardship by funding research, scholarships and education programs essential to the discovery and dissemination of new knowledge in the fields of arboriculture and urban forestry.

Hyland R Johns Grant Program

Subjects: Arboricultural, urban and community forestry.
Purpose: To provide funding for research.
Eligibility: Open to qualified researchers of any nationality.
Level of Study: Research
Type: Research grant
Value: Up to US$25,000
Length of Study: 2–5 years
Frequency: Dependent on funds available
Country of Study: Any country
No. of awards offered: 1–7
Application Procedure: Applicants must complete an online application form.
Closing Date: April 1st
Funding: Foundation, private
No. of awards given last year: 2
No. of applicants last year: 30
Additional Information: Please see the website for further details http://fs2.formsite.com/TREEFund/HylandRJohnsGrant/index.html.

Jack Kimmel International Grant Program

Subjects: Arboricultural, urban and community forestry.
Purpose: To provide money to support projects.
Eligibility: Open to qualified researchers of any nationality.
Level of Study: Research
Type: Research grant
Value: Up to US$10,000. Funds cannot be used for expenses associated with attendance at colleges and universities, e.g. tuition, books or laboratory fees
Length of Study: 1–3 years
Frequency: Dependent on funds available
No. of awards offered: 1–2
Application Procedure: Applicants must complete a two-page application form, available online at www.treefund.org.
Closing Date: October 1st
Funding: Foundation, private
Contributor: Canadian Tree Fund
No. of awards given last year: 2
No. of applicants last year: 8
Additional Information: Please see the website for further details http://www.treefund.org/grants/research-grants.

John Z Duling Grant Program

Subjects: Arboricultural, urban and community forestry.
Purpose: To provide money to support projects.
Eligibility: Open to qualified researchers of any nationality.
Level of Study: Research
Type: Research grant
Value: A maximum of US$10,000. Funds cannot be used for expenses associated with attendance at colleges and universities, e.g. tuition, books or laboratory fees
Length of Study: 1–3 years
Frequency: Dependent on funds available
Country of Study: Any country
No. of awards offered: 1–10
Application Procedure: Applicants must complete a 2 page application form, available online at www.treefund.org.
Closing Date: October 1st
Funding: Foundation
No. of awards given last year: 2
No. of applicants last year: 28
Additional Information: Please see the website for further details http://www.treefund.org/grants/research-grants.

TRIANGLE COMMUNITY FOUNDATION

324 Blackwell Street, Suite 1220, Durham, NC, 27701-3690, United States of America
Tel: (1) 919 474 8370
Fax: (1) 919 941 9208
Email: info@trianglecf.org
Website: www.trianglecf.org

Triangle Community Foundation connects philanthropic resources with community needs, creates opportunity for enlightened change and encourages philanthropy as a way of life.

Shaver-Hitchings Scholarship

Purpose: To provide financial aid and honour individuals with a commitment to helping others in the area of drug and alcohol addiction.
Eligibility: Applicants must reside in Chatham, Durham, Orange or Wake countries, and be enrolled or planning to enroll in graduate school, a physician assistant programme of study or continue a programme in which the applicant is already enrolled. The student need not be pursuing a degree in addictive disorders, but must show demonstrated commitment to working with others in that field during or before graduate studies, preferably as a volunteer.
Level of Study: Graduate
Type: Scholarship
Value: US$1,500
Frequency: Annual
Country of Study: United States of America
Application Procedure: Application forms are available online.
Closing Date: March 15th
Funding: Foundation
Contributor: Triangle Community Foundation
No. of awards given last year: 1
No. of applicants last year: 6
Additional Information: The scholarship is available to any graduate student, physician's assistant or medical student in the Triangle area who has worked (preferably as a volunteer) to help others with alcoholism, drug abuse and addictive disorder treatment or with preventive education on the subject of addiction. Please see the website for further details http://www.trianglecf.org/grants_support/view_scholarships/shaver-hitchings_scholarship/.

TROPICAL AGRICULTURAL RESEARCH AND HIGHER EDUCATION CENTER (CATIE)

7170 Cartago, Turrialba, 30501, Costa Rica
Tel: (506) 2558 2000
Fax: (506) 2558 2060
Email: posgrado@catie.ac.cr
Website: www.catie.ac.cr
Contact: Dean of the Graduate School

The Tropical Agricultural Research and Higher Education Center (CATIE) is an international, non-profit, regional, scientific and educational institution. Its main purpose is research and education in agricultural sciences, natural resources and related subjects in the American tropics, with emphasis on Central America and the Caribbean.

Scholarship Opportunities Linked to CATIE's Postgraduate Program Including CATIE Scholarship Forming Part of the Scholarship-Loan Program

Subjects: Ecological agriculture, biotechnology and genetic resources, management and conservation of tropical forestry and biodiversity, tropical woodlands, tropical agroforestry, tropical crop protection and improvement, integrated watershed management and protected areas and environmental socioeconomics and rural enterprise development.
Purpose: To develop specialized intellectual capital in clean technology, tropical agriculture, natural resources management and human resources in the American tropics.
Eligibility: Priority is given to citizens of Belize, Guatemala, El Salvador, Honduras, Nicaragua, Panama, Costa Rica, Mexico, Venezuela, Colombia, the Dominican Republic, Bolivia and Paraguay.
Level of Study: Doctorate, Postgraduate
Type: Scholarship
Value: Tuition and fees
Length of Study: 2 years for a Master's degree and 3–4 years for a PhD
Frequency: Annual
Country of Study: Costa Rica
No. of awards offered: CATIE offers approximately 30 scholarships per year. Scholarships from other sources vary in origin and number. Information is available on the CATIE website.
Application Procedure: Applicants must undertake an admission process that constitutes 75 per cent for curricular evaluation and

25 per cent for a domiciliary examination. Please refer to the CATIE website for full instructions.
Closing Date: Applications are accepted at any time, but the evaluation deadline for Scholarship-Loan Program is late October. Other sources of financing have specific requirements available on CATIE website. These change considerably over time.
Funding: Foundation, government, international office, private
Contributor: ASDI, OAS, CATIE, DAAD, CONACYT (Mexico), Ford Foundation, Kellogg Foundation, Joint/Japan World Bank. USAID provided the original donation for the endowment financing the Scholarship-Loan Program, SENACYT and Belgium Cooperation
No. of awards given last year: 32 in the Scholarship-Loan Program. Over 25 students received funding from alternative sources.
No. of applicants last year: 350

THE TRUST COMPANY

Level 15, 20 Bond Street, GPO Box 4270, New South Wales, Sydney, NSW 2001, Australia
Tel: (61) 02 8295 8100
Fax: (61) 02 8295 8659
Email: scholarships@thetrustcompany.com.au
Website: thetrustcompany.com.au/philanthropy/awards

The Trust Company is manger at a number high profile awards and scholarships made possible through Charitable bequests/trusts. These include the Miles Franklin Literary Award, Kathleen Mitchell Award (literary), Portia Geach Memorial Award (for female artists), the Sir Robert William Askin Operatic Travelling Scholarship (for male singers), the Lady Mollie Isabelle Askin Ballet Travelling Scholarship, and the Marten Bequest Travelling Scholarship.

Lady Mollie Isabelle Askin Ballet Travelling Scholarship

Subjects: Contemporary or classical ballet.
Purpose: To support the advancement of culture and education in Australia and elsewhere. To reward Australian citizens of outstanding ability and promise in ballet.
Eligibility: Open to Australian citizens who are over the age of 17 and under the age of 30 at the closing date for entries for the award.
Level of Study: Unrestricted
Type: Scholarship
Value: Australian $20,000 in 2 years
Length of Study: More than 2 years
Frequency: Every 2 years
Country of Study: Any country
No. of awards offered: 3
Application Procedure: Applicants must complete an application form to be submitted with specified documents and enclosures.
Closing Date: September 24th
Funding: Private
Contributor: The Estate of Lady Mollie Isabelle Askin

Marten Bequest Travelling Scholarships

Subjects: Scholarships are available in each of the following categories, which rotate in two groups on an annual basis as follows: first category–architecture, ballet, instrumental music, painting, sculpture and singing. Second category–acting, instrumental music, painting, poetry, prose and singing.
Purpose: To augment a scholar's own resources towards affording him or her a cultural education by means of a travelling scholarship.
Eligibility: To be eligible for this award, you must be an Australian born and aged 21–35 (17–35 for ballet) at the closing date of entries.
Level of Study: Unrestricted
Type: Scholarship
Value: Australian $20,000
Length of Study: 2 years
Frequency: Annual
Country of Study: Any country
No. of awards offered: 6
Application Procedure: Applicants must complete an application form and submit this with a study outline and supporting material as required.
Closing Date: Please refer to the website
Funding: Private
Contributor: The estate of the late John Chisholm Marten
No. of awards given last year: 6

Miles Franklin Literary Award

Subjects: Writing.
Purpose: To reward the novel of the year that is of the highest literary merit and presents Australian life in any of its phases.
Eligibility: Refer to the application form. The novel must have been first published in any country in the year preceding the award. Biographies, collections of short stories, children's books and poetry are not eligible. All works must be in English.
Level of Study: Unrestricted
Type: Award
Value: Australian $60,000
Frequency: Annual
Country of Study: Any country
No. of awards offered: 1
Application Procedure: Applicants must complete an application form and send six copies of their novel.
Closing Date: December 13th
Funding: Private
Contributor: The estate of the late Miss S M S Miles Franklin
No. of awards given last year: 1
Additional Information: If there is no novel worthy of the prize, the award may be given to the author of a play.

For further information contact:

Email: trustawards@thetrustcompany.com.au stating
Website: http://www.thetrustcompany.com.au/philanthropy/award-s_and_scholarships/miles_franklin/

Portia Geach Memorial Award

Subjects: Fine and applied arts.
Purpose: To award the best portraits painted from life of a man or woman distinguished in art, letters or the sciences, by a female artist.
Eligibility: Entrants must be female Australian residents who are either Australian or British-born or naturalized. Works must be executed entirely in the previous year.
Level of Study: Unrestricted
Type: Award
Value: Australian $30,000
Frequency: Annual
Country of Study: Any country
No. of awards offered: 1
Application Procedure: Applicants must complete an application form and submit this with an entry fee and the work.
Closing Date: Please refer to the website
Funding: Private
Contributor: The estate of the late Miss Florence Kate Geach
No. of awards given last year: 1
Additional Information: The winning portrait and selected works are exhibited for 1 month at the S H Ervin Gallery in Sydney, Australia.

Sir Robert William Askin Operatic Travelling Scholarship

Subjects: Operatic singing.
Purpose: To support the advancement of culture and education in Australia and elsewhere. To reward male Australian citizens of outstanding ability and promise as an operatic singer.
Eligibility: Applicants must be male Australian citizens between the ages of 18 and 30 at the time of application.
Level of Study: Unrestricted
Type: Scholarship
Value: Australian $20,000 over 2 years
Length of Study: 2 years
Frequency: Every 2 years
Country of Study: Any country
No. of awards offered: 3
Application Procedure: Applicants must complete an application form to be submitted along with specified documents and enclosures.
Closing Date: Please refer to the website
Funding: Private
Contributor: The estate of Sir Robert William Askin

UCLA CENTER FOR 17TH AND 18TH CENTURY STUDIES AND THE WILLIAM ANDREWS CLARK MEMORIAL LIBRARY

10745 Dickson Plaza, 310 Royce Hall, Los Angeles, CA, 90095-1404, United States of America
Tel: (1) 310 206 8552
Fax: (1) 310 206 8577
Email: c1718cs@humnet.ucla.edu
Website: http://www.c1718cs.ucla.edu/
Contact: Fellowship Co-ordinator

The UCLA Center for 17th and 18th-century Studies provides a forum for the discussion of central issues in the field of early modern studies, facilitates research and publication, supports scholarship and encourages the creation of interdisciplinary, cross-cultural programmes that advance the understanding of this important period. The William Andrews Clark Memorial Library, administered by the Center, is known for its collections of rare books and manuscripts concerning 17th and 18th-century Britain and Europe, Oscar Wilde and the 1890s, the history of printing, and certain aspects of the American West.

Ahmanson and Getty Postdoctoral Fellowships

Subjects: Arts and humanities, religious studies and social sciences.
Purpose: To encourage participation by junior scholars in the Center's year-long interdisciplinary core programmes.
Eligibility: Open to postdoctoral scholars who have received their PhD in the last 6 years and whose research pertains to the theme as announced by the Library.
Level of Study: Postdoctorate
Type: Fellowship
Value: $39,264 for the three-quarter period together with paid medical benefits for scholar
Length of Study: 3 consecutive academic quarters
Frequency: Annual
Study Establishment: UCLA and the William Andrews Clark Memorial Library
Country of Study: United States of America
No. of awards offered: Up to 4
Application Procedure: Applicants must submit an application form, a curriculum vitae, a proposal statement, a bibliography and three letters of reference.
Closing Date: February 1st
Funding: Private
Contributor: The Ahmanson Foundation of Los Angeles and the Getty Trust
No. of awards given last year: 4
No. of applicants last year: 30
Additional Information: The award is theme-based and is announced each year. The series intends to be interdisciplinary, with emphasis on both literary and historical perspectives. Participating Fellows will be expected to make a substantive contribution to programme seminars. Please see the website for further details http://www.c1718cs.ucla.edu/.

ASECS (American Society for 18th-Century Studies)/ Clark Library Fellowships

Subjects: The Restoration and the 18th century.
Eligibility: Open to members of ASECS who are postdoctoral scholars and hold a PhD or equivalent at the time of application. The award is also open to advanced doctoral candidates who are members of ASECS.
Level of Study: Postdoctorate
Type: Fellowship
Value: US$2,500
Length of Study: 1 month
Frequency: Annual
Study Establishment: UCLA and the William Andrews Clark Memorial Library
Country of Study: United States of America
No. of awards offered: Varies
Application Procedure: Applicants must submit an application form, a curriculum vitae, a proposal statement, a bibliography and three letters of reference.
Closing Date: February 1st
Funding: Government
Contributor: ASECS and the Clark Library Endowment
No. of awards given last year: 1
No. of applicants last year: 40
Additional Information: Please see the website for further details http://www.c1718cs.ucla.edu/.

Clark Dissertation Fellowships

Subjects: Library.
Purpose: To support doctoral candidates whose dissertation involves extensive research in the library holding's.
Eligibility: Open to applicants whose dissertation involves extensive research in the library holding's.
Level of Study: Doctorate, Research
Type: Fellowship
Value: $18,000 plus fixed graduate fees
Length of Study: 1 year
Frequency: Annual
No. of awards offered: 1 or 2
Application Procedure: Applicants must visit the organization website for application procedure.
Closing Date: February 1st
Additional Information: Please see the website for further details http://www.c1718cs.ucla.edu/.

Clark Library Short-Term Resident Fellowships

Subjects: Research relevant to the Library's holdings.
Eligibility: Open to PhD scholars or equivalent who are involved in advanced research.
Level of Study: Postdoctorate
Type: Fellowship
Value: US$2,500 per month
Length of Study: 1–3 months
Frequency: Annual
Study Establishment: UCLA and the William Andrews Clark Memorial Library
Country of Study: United States of America
No. of awards offered: Varies
Application Procedure: Applicants must submit an application form, a curriculum vitae, a proposal statement, a bibliography and three letters of reference.
Closing Date: February 1st
Funding: Government, private
Contributor: The Ahmanson Foundation and the Clark Library Endowment
No. of awards given last year: 16
No. of applicants last year: 60
Additional Information: Please see the website for further details http://www.c1718cs.ucla.edu/.

Clark Predoctoral Fellowships

Subjects: Seventeenth- and eighteenth-century studies or one of the other areas represented in the Clark's collections.
Purpose: To support dissertation research.
Eligibility: Open to advanced doctoral students at the University of California whose dissertation concerns an area appropriate to the collections of the Clark Library.
Level of Study: Predoctorate
Type: Fellowship
Value: US$7,500
Length of Study: 3 months
Frequency: Annual
Study Establishment: UCLA and the William Andrews Clark Memorial Library
Country of Study: United States of America
No. of awards offered: Varies
Application Procedure: Applicants must submit an application form, a curriculum vitae, a proposal statement, a bibliography and three letters of reference.
Closing Date: February 1st
Funding: Private
Contributor: The Ahmanson Foundation
No. of awards given last year: 3
No. of applicants last year: 9

Additional Information: Please see the website for further details http://www.c1718cs.ucla.edu/.

Clark-Huntington Joint Bibliographical Fellowship

Subjects: Early modern literature and history and other areas where the sponsoring libraries have common strengths.
Purpose: To support bibliographical research.
Level of Study: Professional development, Postdoctorate
Type: Fellowship
Value: US$5,000
Length of Study: 2 months
Frequency: Annual
Study Establishment: The Clark Library and the Huntington Library
Country of Study: United States of America
No. of awards offered: 1
Application Procedure: Applicants must submit an application form, a curriculum vitae, a proposal statement, a bibliography and three letters of reference.
Closing Date: February 1st
Funding: Private
No. of awards given last year: 1
No. of applicants last year: 15

For further information contact:

William Andrews Clark Memorial Library, 2520 Cimarron Street, Los Angeles, CA, 90018-2098, United States of America
Website: http://www.c1718cs.ucla.edu/
Contact: Fellowship Co-ordinator

Graduate Student Research Assistantships

Purpose: To offer financial support to UCLA graduate students.
Eligibility: Open to research assistants who participate in research projects of core faculty members and take part in Center/Clark activities.
Level of Study: Research
Type: Assistantship
Value: Depends upon applicant's academic level and previous experience at UCLA
Frequency: Annual
Closing Date: February 1st
Additional Information: Please see the website for further details http://www.c1718cs.ucla.edu/.

Graduate Travel Grants

Subjects: Seventeenth and Eighteenth Century Studies and Oscar Wilde.
Purpose: To provide travel support for participation in professional conferences.
Eligibility: Open to graduate students at UCLA.
Type: Grant
Value: Up to $500 for domestic travel; up to $1,000 for foreign travel
Closing Date: February 1st
Additional Information: Please see the website for further details http://www.c1718cs.ucla.edu/.

Kanner Fellowship In British Studies

Subjects: British history and culture.
Eligibility: Open to both postdoctoral and predoctoral scholars.
Level of Study: Postdoctorate, Predoctorate
Type: Fellowship
Value: US$7,500
Length of Study: 3 months
Frequency: Annual
Application Procedure: Applicants must submit an application form, a curriculum vitae, proposal statement, a bibliography and three letters of reference.
Closing Date: February 1st
Funding: Private
Contributor: Penny Kanner
No. of awards given last year: 1
No. of applicants last year: 12
Additional Information: Please see the website for further details http://www.c1718cs.ucla.edu/.

UNION FOR INTERNATIONAL CANCER CONTROL (UICC)

62 route de Frontenex, 1207 Geneva, Switzerland
Tel: (41) 22 809 1842
Fax: (41) 22 809 1810
Email: vought@uicc.org, piguet@uicc.org
Website: www.uicc.org
Contact: Dr Tristan Piguet, Global Training Manager

UICC fellowships provide opportunities for professional development for cancer investigators, clinicians, nurses, and cancer society staff and volunteers.

American Cancer Society UICC International Fellowships for Beginning Investigators (ACSBI)

Subjects: Oncology.
Purpose: To facilitate cancer investigators and clinicians, who are in the early stages of their careers, to conduct cancer research projects into the pre-clinical, clinical, epidemiological, psychological, behavourial, health services, health policy, outcomes and cancer-control aspects of the disease.
Eligibility: Beginning investigators and clinicians in the early stages of their independent investigation have a minimum of two and a maximum of 10 years of postdoctoral experience after obtaining their MD or PhD degrees. Must possess a terminal, advanced degree with a desire to become an independent investigator. Applicants are to be in the early phases of their career and no longer under research mentoring. Please note that this is neither a first postdoctoral fellowship nor a clinical training fellowship. Candidates must hold an academic university or hospital position with an explicit commitment to return to home institute. Please note that fellowships are awarded only to individuals who will be conducting their research at not-for-profit institutions. Unsolicited applications will not be accepted from, nor will fellowship be awarded for, the support of research conducted at for-profit institutions.
Level of Study: Research
Type: Fellowship
Value: US$50,000 each for travel and stipend
Length of Study: 1 year
Frequency: Annual
Study Establishment: A suitable host institute abroad
Country of Study: Any country
No. of awards offered: 6–8 per year
Application Procedure: Online application process.
Closing Date: November 1st
Funding: Commercial
Contributor: American Cancer Society
No. of awards given last year: 4
No. of applicants last year: 16
Additional Information: Awards are conditional on the return of the fellow to the home institute at the end of fellowship and on the availability of appropriate facilitites and resources to apply the newly acquired skills.
*Terminal degree: is generally accepted as the highest degree in a field of study. An earned academic (or research) doctorate such as Doctor of Philosophy is considered the terminal degree in most academic fields of study in some countries. Many professional degrees are also considered terminal degrees because they are the highest professional degree in the field, even though 'higher' research degrees exist.

UICC Asia-Pacific Cancer Society Training Grants

Subjects: Oncology.
Purpose: To increase the capacity of volunteers and staff of voluntary cancer societies located in the Asia Pacific region by participating in, and learning from activities conducted by established collaborating cancer societies in the region.
Eligibility: Staff or accredited volunteers of voluntary cancer societies located in the Asia-Pacific region
Level of Study: Professional development, Unrestricted
Type: Grant
Value: US$1,800
Length of Study: 1 week

Frequency: Annual
Country of Study: India, Australia, Singapore
No. of awards offered: 5–10 per year
Application Procedure: Candidates must choose one of the projects offered and submit their curriculum vitae together with a "letter of justification" to the host society with a request for a formal invitation to participate in the specific project. The letter of justification must describe why a specific project was chosen, how candidates and their organizations would benefit from this experience and how they would disseminate the new skills upon return. On-line application on the www.apcasot.org under the "Apply now section".
Closing Date: Application open date: May 29th and Application closing date: September 21st
Funding: Trusts
Contributor: 1. William Rudder Memorial Fund (Australia) 2. The Cancer Council New South Wales, Sydney, Australia 3. The Cancer Council Queensland, Brisbane, Australia 4. Cancer Patients Aid Association, Mumbai, India 5. Singapore Cancer Society, Singapore
No. of awards given last year: 6
No. of applicants last year: 8
Additional Information: Selection results will be notified in mid-December. Awards are subject to the UICC general conditions for fellowships.

UICC International Cancer Research Technology Transfer Fellowships (ICRETT)

Subjects: Cancer control and prevention, epidemiology and cancer registration, public education and behavioural sciences.
Purpose: To facilitate the rapid international transfer of cancer research and clinical technology; to exchange knowledge and enhance skills in basic, clinical, behavourial and epidemiological areas of cancer research, and in cancer control and prevention; and to acquire up-to-date clinical management, diagnostic and therapeutic expertise.
Eligibility: Investigators and clinicians should be working in places where such teaching is not yet available and where the necessary facilities exist to apply and disseminate the new skills upon return. Qualified cancer investigators should be at the early stages in their careers, while clinicians should be well established in their oncology practice.
Level of Study: Professional development
Type: Fellowship
Value: US$3,400
Length of Study: 1–3 months with stipend support for 1 month
Study Establishment: A suitable host institute abroad
Country of Study: Any country
No. of awards offered: 120–150 per year
Application Procedure: Online application.
Closing Date: Any time
Funding: Commercial, government, private
Contributor: The fellowships are funded by a group of cancer institutes societies, leagues, associations and governmental agencies in North America, Europe and Australia
No. of awards given last year: 122
No. of applicants last year: 164
Additional Information: Candidates who are already physically present at the proposed host institute while their applications are under consideration are not eligible for a UICC fellowship.

UICC International Cancer Technology Transfer Training Workshops

Subjects: Basic, clinical, behavourial and epidemiological aspects of cancer research, cancer control and prevention, clinical management, and diagnostic and therapeutic skills.
Purpose: To facilitate teaching and training courses on cancer research.
Eligibility: An appropriately qualified and experienced cancer expert.
Level of Study: Teaching
Type: Fellowship
Value: A maximum of US$15,000 for travel and stipend for 3 international faculty members
Length of Study: 3–5 days
Frequency: Annual
Study Establishment: A suitable host institute in a resource-constrained abroad country
Country of Study: Any developing country
No. of awards offered: 10–20 per year
Application Procedure: Online application process – applicants should access the UICC fellowships website and check under the "Apply now" section under the left side selection menu.
Closing Date: Accepted at any time
Funding: Private
Contributor: The fellowships are funded by a group of cancer institutes, societies, leagues, associations and governmental agencies in North America, Europe and Australia.
No. of awards given last year: 12
No. of applicants last year: 16
Additional Information: Awards are subject to the UICC general conditions for fellowships.

UICC Yamagiwa-Yoshida Memorial International Cancer Study Grants

Subjects: Basic, translational or applied research
Purpose: To enable cancer investigators from any country to carry out bilateral research projects that exploit complementary materials or skills, including advanced training in experimental methods or special techniques.
Eligibility: Open to appropriately qualified investigators from any country who are actively engaged in cancer research. Candidates who are already physically present at the proposed host institute are not eligible.
Level of Study: Professional development, Research
Type: Grant
Value: The average stipend is US$10,000. If a Fellow's return home is delayed beyond the extra approved period, 50 per cent of the travel award, the return portion has to be reimbursed to the UICC. Calculation of travel and stipend awards are based on the candidate's estimates which are adjusted, if need be, to published fares and UICC scales. Travel awards contribute to the least expensive international return air fares or other appropriate form of transport. Travel estimates should not include costs for internal travel within the home or host countries. These and extra costs for visa, passports, airport taxes and insurance are the responsibility of the Fellow. No financial support is provided for dependants
Length of Study: 3 months. May be extended by their original duration, subject to written approval of the home and host supervisors. Funding for these additional periods may be secured from other funding agencies
Frequency: Annual
Study Establishment: A suitable host institute abroad
Country of Study: Any country
No. of awards offered: 15
Application Procedure: Online application.
Closing Date: January 15th for notification by mid April, July 1st for notification by mid October
Funding: Commercial, private
Contributor: Kyowa Hakko Kyoga Company Limited, Toray Industries, Inc. and the Japan National Committee for UICC
No. of awards given last year: 12
No. of applicants last year: 35
Additional Information: Awards are subject to the UICC general conditions for fellowships

UNITED CHURCH OF CHRIST (UCC)

Office of General Ministries, 700 Prospect Avenue, Cleveland, OH, 44115-1100, United States of America
Tel: (1) 216 736 3839
Fax: (1) 216 736 2237
Email: jeffersv@ucc.org
Website: www.ucc.org
Contact: Ministry Team, LCM

The United Church of Christ (UCC) was founded in 1957 as the union of several different Christian traditions. The UCC is one of the most diverse Christian churches in the United States of America.

William R Johnson Scholarship

Subjects: Theology.
Purpose: To affirm the long-standing conviction of the UCC that sexual orientation should not be a barrier to ordination.

Eligibility: Open to candidates enrolled in an American Theological School accredited seminary and have proof of both local membership in a UCC. Must be open about their sexual orientation.
Level of Study: Postgraduate
Type: Scholarship
Value: US$2,500
Frequency: Annual
Closing Date: April 1st
Funding: Individuals

UNITED NATIONS DEVELOPMENT PROGRAMME (UNDP)

One United Nations Plaza, New York, NY, 10017, United States of America
Tel: (1) 212 906 5000
Fax: (1) 212 906 5001
Email: ohr.recruitment.hq@undp.org
Website: www.undp.org

The UNDP launched the Local Initiative Facility for Urban Environment (LIFE) as a global pilot programme at the Earth summit in Rio de Janeiro in 1992. The programme uses environmental deprivation as the entry point for achieving sustainable human development.

UNDP's Local Initiative Facility for Urban Environment (LIFE) Global Programme

Subjects: Urban environmental development.
Purpose: To promote local-local dialogue and partnership between NGOs, CBOs, Local Governments and the private sector for improving conditions of the urban poor.
Eligibility: Application for need-based participatory, community-based projects in Urban poor communities.
Level of Study: Professional development
Type: Grant
Value: Up to US$50,000
Frequency: Annual
Funding: Government
Contributor: Life

UNITED STATES CENTER FOR ADVANCED HOLOCAUST STUDIES

United States Holocaust Memorial Museum, 100 Raoul Wallenberg Place South West, Washington, DC, 20024-2126, United States of America
Tel: (1) 202 488 0400 / 202 314 7802
Fax: (1) 202 479 9726
Email: vscholars@ushmm.org
Website: www.ushmm.org
Contact: Ms Jo-Ellyn Decker, Program Coordinator

The United States Holocaust Memorial Museum is the United States of America's national institution for the documentation, study and interpretation of Holocaust history, and serves as the country's memorial to the millions of people murdered during the Holocaust. The Center for Advanced Holocaust Studies fosters research in Holocaust and genocide studies.

United States Holocaust Memorial Museum Center for Advanced Holocaust Studies Visiting Scholar Programs

Subjects: History, political science, literature, philosophy, sociology, religion and other disciplines, as they relate to the study of the Holocaust.
Purpose: To support research and writing in the fields of Holocaust and genocide studies.
Eligibility: Fellowships are awarded to candidates working on their dissertations (ABD), postdoctoral researchers and senior scholars. Applicants must be affiliated with an academic and/or research institution when applying for a fellowship. Immediate postdocs and faculty between appointments will also be considered.
Level of Study: Doctorate, Postdoctorate
Type: Fellowship
Value: All awards include a monthly stipend (up to $3,500 per month), direct travel to and from Washington, DC and visa assistance, if necessary. The Museum provides office space, postage, and access to a computer, telephone, facsimile machine and photocopier
Length of Study: 3–9 months
Frequency: Annual
Country of Study: United States of America
No. of awards offered: Varies
Application Procedure: Applications must be submitted via an online application process. For more information, visit the Museum's website at www.ushmm.org/research/center/fellowship or email at www.visitingscholars@ushmm.org.
Closing Date: November. Please visit the website for exact details
Funding: Private
No. of awards given last year: 27
No. of applicants last year: Over 100

For further information contact:

Website: www.ushmm.org/research/center/fellowship

UNITED STATES INSTITUTE OF PEACE (USIP)

2301 Constitution Avenue, NW, Washington, DC, 20037, United States of America
Tel: (1) 202 457 1700
Fax: (1) 202 429 6063
Email: grant_program@usip.org
Website: www.usip.org
Contact: Ms Cornelia Hoggart, Senior Programme Assistant

The United States Institute of Peace (USIP) is mandated by the Congress to promote education and training, research and public information programmes on means to promote international peace and resolve international conflicts without violence. The Institute meets this mandate through an array of programmes, including grants, fellowships, conferences and workshops, library services, publications and other educational activities.

Jennings Randolph Program for International Peace Dissertation Fellowship

Subjects: A broad range of disciplines and interdisciplinary fields.
Purpose: To support dissertations that explore the sources and nature of international conflict, and strategies to prevent or end conflict and to sustain peace.
Eligibility: Open to applicants of all nationalities who are enrolled in an accredited college or university in the United States of America. Applicants must have completed all requirements for the degree except the dissertation by the commencement of the award.
Level of Study: Doctorate
Type: Fellowships
Value: US$20,000, which may be used to support dissertation writing or field research
Length of Study: 1 year
Frequency: Annual
Study Establishment: The student's home university or site of fieldwork
Country of Study: United States of America
No. of awards offered: 10
Application Procedure: Applicants must complete a web-based application form, available on the Institute's website.
Closing Date: January 5th
Funding: Government
No. of awards given last year: 10
Additional Information: The programme does not support work involving partisan political and policy advocacy or policy making for any government or private organization.

For further information contact:

Tel: 202 429 3853
Email: jrprogram@usip.org
Website: http://www.usip.org/fellows/scholars.html
Contact: Miss Shira Lowinger, Program Coordinator

Jennings Randolph Program for International Peace Senior Fellowships

Subjects: Preventive diplomacy, ethnic and regional conflicts, peacekeeping and peace operations, peace settlements, postconflict reconstruction and reconciliation, democratization and the rule of law, cross-cultural negotiations, United States of America policy in the 21st century and related subjects.
Purpose: To use the recipient's existing knowledge and skills towards a fruitful endeavour in the international peace and conflict management field, and to help bring the perspectives of this field into the Fellow's own career.
Eligibility: Open to outstanding practitioners and scholars from a broad range of backgrounds. The competition is open to citizens of any country who have specific interest and experience in international peace and conflict management. Candidates would typically be senior academics, but applicants who hold at least a Bachelor's degree from a recognized university will also be considered.
Type: Fellowship
Value: A stipend, an office with computer and voicemail and a part-time research assistant
Length of Study: Up to 10 months
Frequency: Annual
Study Establishment: USIP
Country of Study: United States of America
No. of awards offered: 8–12
Application Procedure: Applicants must complete a web-based application form, available on request from the Institute or from the website.
Closing Date: Early january (check with website)
Funding: Government
No. of awards given last year: 10
No. of applicants last year: 136

For further information contact:

Tel: 202 429 3886
Email: jrprogram@usip.org
Website: http://www.usip.org/grants-fellowships/jennings-randolph-senior-fellowship-program
Contact: Miss Shira Lowinger, Program Coordinator

USIP Annual Grant Competition

Subjects: Topic areas of interest to the Institute include, but are not restricted to, international conflict resolution, diplomacy, negotiation theory, functionalism and track-two diplomacy, methods of third party dispute settlement, international law, international organizations and collective security, deterrence and balance of power, arms control, psychological theories about international conflict, the role of non-violence and non-violent sanctions, moral and ethical thought about conflict and conflict resolution and theories about relationships among political institutions, human rights and conflict.
Purpose: To provide financial support for research, education and training, and the dissemination of information on international peace and conflict resolution.
Eligibility: The Institute may provide grant support to non-profit organizations and individuals from both the United States of America and other countries. These include institutions of post-secondary, community and secondary education, public and private education, training and research institutions and libraries. Although the Institute can provide grant support to individuals, it prefers that an institutional affiliation be established. The Institute will not accept applications that list members of the Institute's Board of Directors or staff as participants, consultants or project personnel. In addition, any application that lists the Institute as a collaborator in the project will not be accepted.
Level of Study: Postdoctorate, Research, Education, Training
Type: Grant
Value: Most awards fall in the range of US$50,000–120,000, although somewhat larger grants are also awarded. The amount of any grant is based on the proposed budget and on negotiations with successful applicants
Length of Study: 1–2 years
Frequency: Annual
No. of awards offered: Approx. 35
Application Procedure: Applicants must complete an online application form, http://www.usip.org/grants-fellowships/annual-grant-competition.
Closing Date: October 1st (check with website)
Funding: Government
No. of awards given last year: 23
No. of applicants last year: 460

USIP Priority Grant-making Competition

Subjects: Check website for specific details.
Purpose: To provide financial support for research, education and training, and the dissemination of information on international peace and conflict resolution.
Eligibility: Open to non-profit organizations and individuals from both the United States of America and other countries. These include institutions of post-secondary, community and secondary education, public and private education, training or research institutions and libraries. Although the Institute can provide grant support to individuals, it prefers that an institutional affiliation be established. The Institute will not accept applications that list as participants, consultants or project personnel members of the Institute's Board of Directors or staff. In addition, any application that lists the Institute as a collaborator in the project will not be accepted.
Level of Study: Postdoctorate, Research, Education, training
Type: Grant
Value: Most awards fall in the range of US$45,000–140,000. The amount of any grant is based on the proposed budget and on negotiations with successful applicants
Length of Study: 1–2 years
Frequency: Rolling deadline
No. of awards offered: Approx. 35
Application Procedure: Applicants must complete an application form, available on request from the Institute or from the website.
Closing Date: Check with website
Funding: Government
No. of awards given last year: 32
No. of applicants last year: 80
Additional Information: Please see the website for further details http://www.usip.org/grants-fellowships/priority-grant-competition.

UNITED STATES-INDIA EDUCATIONAL FOUNDATION (USIEF)

Fulbright House 12 Hailey Road, New Delhi, 110001, India
Tel: (91) 11 2332 8944/48
Fax: (91) 2332 9718
Email: info@usief.org.in
Website: www.usief.org.in
Contact: Programme Officer

The activities of the United States Educational Foundation in India (USEFI) may be broadly categorized as the administration of the Fulbright Exchange Fellowships for Indian and United States scholars and professionals, and the provision of educational advising services to help Indian students wishing to pursue higher education in the United States. USEFI also works for the promotion of dialogue among fulbrighters and their communities as an outgrowth of educational exchange.

East-West Center Asia Pacific Leadership Program

Subjects: Participants are very high-potential or current leaders from a broad array of countries and backgrounds, including science, business, development, politics, government, civil society, medicine, religious orders, art, finance, academia or research.
Purpose: To create a network of action, focused on building a peaceful, prosperous and just Asia Pacific community.
Eligibility: All participants have at least a Bachelor's degree and most have a Master's degree or at least five years of professional experience.
Level of Study: Postgraduate, Professional development
Type: Fellowship
Value: Approx. $15,000
Frequency: Annual
Application Procedure: Applicants must submit an application form.
Closing Date: March 1st

East-West Center Graduate Degree Fellowship Program

Subjects: Priority in the student selection process is given to applicants seeking degrees in fields of study related to research

themes at the East-West Center, focusing on topics in economics; environmental change, vulnerability and governance; politics and security; and population and health, at a local, national and/or regional level in the Asia Pacific region. The Center also welcomes applications in other fields of study on issues of common concern among the peoples and nations of Asia, the Pacific and the United States.
Purpose: The East-West Center (EWC) Graduate Degree Fellowship provides substantial funding towards Master's and Doctoral degrees for graduate students from Asia, the Pacific, and the US to participate in educational and research programs at EWC while pursuing graduate study at the University of Hawai'i.
Eligibility: Citizens or permanent residents of the United States
Level of Study: Postgraduate, Research
Type: Fellowship
Value: Fellowship provisions include tuition and fees, graduate residence hall room costs, health insurance, book allowance and partial living stipend
Length of Study: Master's degrees (up to 24-month fellowship) and Doctoral degrees (up to 48-month fellowship)
Frequency: Annual
Application Procedure: Applicants must submit an application form.
Closing Date: November 1st

For further information contact:

East-West Center, 1601 East-West Road, Honolulu, HI, 96848-1601, United States of America

East-West Center Jefferson Fellowship Program

Purpose: The Jefferson Fellowships Program of the East-West Center is a 21-day programme of professional dialogue and study tour travel for mid-career print, broadcast and online journalists from the US, Asia and the Pacific.
Eligibility: Working print, broadcast and online journalists in the United States, the Pacific Islands, and Asia with a minimum of five years of professional experience. Applicants must have the ability to communicate in English in a professional, multi-cultural environment.
Level of Study: Professional development
Type: Fellowship
Value: The grant covers economy class airfare (Honolulu, Hawaii), transportation, lodging, and meals.
Frequency: Annual
Application Procedure: Applicants must submit an application form.
Closing Date: February 15th
Funding: Foundation

Fulbright Classroom Teacher Exchange Program

Subjects: Designed for Indian secondary school teachers (9th to 12th grades) of English, mathematics, and science to participate in direct exchanges of positions with US teachers for semester.
Purpose: To provide Indian teachers the opportunity to work within a US school system and experience US society and culture similarly US teachers will work in an Indian school set up and share their expertise.
Eligibility: Open to Indian citizens resident in India at the time of application who have not been to the United States of America in the preceding 3 years. Applicants must have a high level of academic and professional achievement, proficiency in English and be in good health. Applicants must have a Master's degree in English, mathematics or science subjects with a teacher training degree and be a full-time teacher (9th to 12th grades) in a recognized secondary school for at least 5 years.
Level of Study: Postgraduate, Professional development, Secondary school
Type: Fellowship
Value: Round trip travel, moderate monthly stipend, health insurance
Length of Study: One semester
Frequency: Annual
Country of Study: United States of America
No. of awards offered: 7
Application Procedure: Applicants must complete an application form. Requests for application materials must state the applicant's academic and professional qualifications, date of birth, current position and grant category and be accompanied by a 7 by 10 inch stamped addressed envelope. Requests for application materials must be sent to the USEFI offices in the applicant's region. Alternatively, applicants can download information from the USEFI website.
Closing Date: September 16th
Funding: Government
Contributor: USEFI
No. of applicants last year: 109
Additional Information: Please see the website for further details http://www.usief.org.in/Fellowships/Fulbright-Teacher-Exchange-Program.aspx.

Fulbright-Hays Doctoral Dissertation Research Abroad (DDRA)

Subjects: Projects deepen research knowledge on and help the nation develop capability in areas of the world not generally included in US curricula. Projects focusing on Western Europe are not supported.
Purpose: This program provides grants to colleges and universities to fund individual doctoral students who conduct research in other countries, in modern foreign languages and area studies.
Eligibility: Citizen or national of the United States or is a permanent resident of the United States; Is a graduate student in good standing at an institution of higher education in the United States who, when the fellowship begins, is admitted to candidacy in a doctoral program in modern foreign languages and area studies at that institution; Is planning a teaching career in the United States upon graduation; and possesses adequate skills in the language(s) necessary to carry out the dissertation.
Level of Study: Predoctorate, Research
Value: Travel expenses, including excess baggage to and from the residence of the fellow to the host country of research. Maintenance and dependents allowances based on the cost of living in country(ies) of research for the fellow and his or her dependent(s). Project allowance for research-related expenses such as books, copying, tuition and affiliation fees, local travel and other incidental expenses; Health and accident insurance premiums; and administrative fee of $100 to applicant institution
Length of Study: 6–12 months
Frequency: Annual
Application Procedure: Applicants must submit an application form.
Closing Date: Check with website
Additional Information: Please visit http://www.ed.gov/programs/iegpsddrap/index.html for grant benefit details and application procedures.

Fulbright-Nehru Doctoral and Professional Research Fellowships

Subjects: The United States Special education, conflict resolution, museum studies, arts/culture management and heritage conservation.
Eligibility: Open to Indian citizens resident in India at the time of application who have not been to the United States of America in the previous 3 years. Applicants must have a high level of academic and professional achievement, proficiency in English and be in good health. Applicants must be registered for a PhD at an Indian institution at least 1 year prior to application. The fellowship is also open to professionals with postgraduate degrees who are working in areas where study of IPR or special education would be helpful to their institutions.
Level of Study: Predoctorate, Professional development
Type: Fellowship
Value: Maintenance in the United States of America, affiliation fees, health insurance and round trip economy class airfare
Length of Study: Up to 9 months
Frequency: Annual
Country of Study: United States of America
No. of awards offered: 6–7
Application Procedure: Application material can be downloaded from the website.
Closing Date: July 16th
Funding: Foundation
Contributor: USEFI
No. of awards given last year: 6
No. of applicants last year: 71

Fulbright-Nehru Environmental Leadership Program

Subjects: Environmental Information/Systems Reporting; Environmental Education; Environmental Policy, Regulations and Law;

Environmental Sciences and Toxicology; and Environmental Management.
Purpose: To provide mid-level Indian environment professionals with an opportunity for short-term practical training/internship and network with U.S. environmental organizations.
Eligibility: Open to Indian citizens resident in India at the time of application who have not been to the United States of America in the previous 3years. Applicants should possess a master's or a professional degree of at least four years duration and have at least five years of professional experience.
Level of Study: Professional development
Type: Fellowship
Value: Settling-in allowance, health insurance, round trip economy class airfare and professional allowance
Length of Study: Up to 4 months
Frequency: Annual
Country of Study: United States of America
No. of awards offered: 18–19
Application Procedure: Applications can be downloaded from the website www.usief.org.in.
Closing Date: August 15th
Funding: Foundation, government
Contributor: USIEF and GOI
No. of awards given last year: 15
No. of applicants last year: 80

Fulbright-Nehru Master's Fellowship for Leadership Development

Subjects: Public administration, economics, communication studies and environment.
Purpose: The programme is specially targeted at highly motivated individuals who demonstrate leadership qualities.
Eligibility: Open to Indian citizens resident in India at the time of application who have not been to the United States of America in the preceding 3 years. Applicants must have completed an equivalent of a US Bachelor's degree from a recognized Indian university with at least 55 per cent marks. Applicants should either possess a 4 year Bachelors degree or a completed master's degree, if the Bachelor's degree is of less than 4 years duration.
Level of Study: Postgraduate
Type: Fellowship
Value: Maintenance in the United States of America, affiliation fees, health insurance and round trip economy air fare, limited allowance for books, J-1 visa support
Length of Study: 1–2 years
Frequency: Annual
Country of Study: United States of America
No. of awards offered: 4–6
Application Procedure: Applicants must complete an application form. Requests for application materials must state the applicant's academic and professional qualifications, date of birth, current position and grant category and be accompanied by a 7 by 10 inch stamped addressed envelope. Requests for application materials must be sent to the USEFI offices in the applicant's region or alternatively can be downloaded from the website.
Closing Date: July 16th
Funding: Government
Contributor: USEFI
No. of applicants last year: 45
Additional Information: Further information is available on request from USEFI.

Fulbright-Nehru Postdoctoral Research Fellowship

Subjects: Priority fields are: agricultural sciences; economics; education; energy, sustainable development and climate change; environment; international relations; management and leadership development; media and communications with focus on public service broadcasting; public administration; public health; science and technology; study of India with focus on contemporary issues; study of the United States. For study of India or the study of the United States (American studies) the areas could include: language and literature, history, government, economics, society and culture, religion and film studies.
Purpose: These fellowships are designed for Indian scholars and professionals who are in the early stages of their research careers. The Postdoctoral Research Fellowships will provide opportunities to talented scholars and professionals to strengthen their research capacities, have access to some of the finest resources in their areas of interest, and help build long-term collaborative relationships with US faculty and institutions.
Eligibility: In addition to the general prerequisites: scholars and professionals should have a PhD degree from an Indian institution within the past four years. Applicants must have obtained their PhD degrees between July 2007 and July 2011; candidates must be published in reputed journals and demonstrate evidence of superior academic and professional achievement; should enclose a recent significant publication (copy of paper/article); if employed, applications should be routed through proper channel; preferably be 45 years of age or under.
Level of Study: Professional development, Research, Postdoctorate
Value: Round trip economy class airfare and professional allowance.
Length of Study: 12 months maximum
Frequency: Annual
Application Procedure: Applicants must submit an application form.
Closing Date: July 15th
Additional Information: Please see the website for further details http://www.usief.org.in/Fellowships/Fulbright-Nehru-Postdoctoral-Research-Fellowship.aspx.

Fulbright-Nehru Senior Research Fellowships

Subjects: The United States of America education, society and development, and management.
Purpose: To provide scholars with the opportunity to undertake research on contemporary issues and concerns.
Eligibility: Open to Indian citizens resident in India at the time of application who have not been to the United States of America in the preceding 3 years. Applicants must have a high level of academic and professional achievement, proficiency in English and be in good health. Applicants must be employed as full-time faculty members in an Indian college, university or research institution, or as a full-time professional at an Indian non-profit organization, hold a PhD degree or have equivalent published work and be under 50 years of age.
Level of Study: Postdoctorate
Type: Grant
Value: Round trip travel, monthly stipend, university affiliation fees, health insurance and a modest settling in allowance, dependent allowance and round trip travel for one dependent
Length of Study: Up to 8 months
Frequency: Annual
Study Establishment: A university or research institution
Country of Study: United States of America
No. of awards offered: 4–5
Application Procedure: Application information can be downloaded from the website.
Closing Date: July 16th
Funding: Foundation
Contributor: USEFI
No. of awards given last year: 4
No. of applicants last year: 162
Additional Information: Applicants must demonstrate the relevance of the proposed research to India and/or the United States, its applicability in India, its benefit to the applicant's institution and the need to carry it out in the United States.

Fulbright-Nehru Student Research

Subjects: Preferred subjects are: agricultural sciences; business studies, management and leadership development; economics; education; energy, especially alternative and renewable energy; environment; governance and democracy; law and civic engagement; media and communications; public administration; public health; science and technology; society and development; study of India; and study of the United States.
Purpose: This program encourages several categories of students to visit India – recent BS/BA graduates, Master's and Doctoral candidates and young professionals and artists – for personal development and international experience.
Eligibility: Candidates at all degree levels, developing professionals and artists.
Level of Study: Postgraduate, Professional development, Research
Value: Maintenance allowance, dependant maintenance allowance, incidental allowance,relocation allowance, material allowance, excess

baggage allowance, research allowance, round trip travel according to USIEF guidelines
Length of Study: 9 months
Frequency: Annual
Application Procedure: Applicants must submit an application form.

Fulbright-Nehru Visiting Lecturer Fellowships

Subjects: Humanities and social sciences.
Purpose: To allow scholars to share their expertise on contemporary issues significant to India and the United States of America.
Eligibility: Open to Indian citizens resident in India at the time of application who have not been to the United States of America in the preceding 3 years. Applicants must have a high level of academic and professional achievement, proficiency in English and be in good health. Applicants must be permanent full-time faculty members at an Indian college, university or research institute, hold a PhD degree or have equivalent published work, have at least 10 years of college or university level teaching experience and be no more than 50 years of age.
Level of Study: Postdoctorate
Type: Grant
Value: Round trip travel, monthly stipend, university fees, health insurance and a modest settling in allowance
Length of Study: 4–6 months
Frequency: Annual
Country of Study: United States of America
No. of awards offered: 2–3
Application Procedure: Applicants must complete an application form. Requests for application materials must state the applicant's academic and professional qualifications, date of birth, current position and grant category and be accompanied by a 7 by 10 inch stamped addressed envelope. Requests for application materials must be sent to the USEFI offices in the applicant's region or can be downloaded from the USEFI website.
Closing Date: July 15th
Funding: Foundation
Contributor: USEFI
No. of awards given last year: 8
No. of applicants last year: 44

International Fulbright Science and Technology Award

Subjects: Aeronautics and astronomics, aeronautical engineering, astronomy, planetary sciences, biology, chemistry, computer sciences, engineering-electrical, chemical, civil, mechanical, ocean and petroleum, environmental science, geology, earth and atmospheric sciences, information sciences engineering, materials science engineering, mathematics, neuroscience, brain and cognitive sciences, oceanography and physics.
Purpose: To enable deserving students to study in accredited institutions in the United States.
Eligibility: Open to applicants who have high level of academic or professional achievements and have completed equivalent of a United States Bachelor's degree from a recognized Indian university with at least 60 per cent marks. Applicants must possess a 4 year Bachelor's degree or a completed Master's degree. The candidates must also possess a valid Indian Passport.
Level of Study: Doctorate
Type: Scholarships
Value: A monthly stipend for up to 3 years, health insurance, round-trip air travel, J-1 visa for up to 5 years, books and equipment, research and professional conference allowances and tuition fees
Length of Study: 3 years
Frequency: Annual
Country of Study: United States of America
No. of awards offered: 40
Application Procedure: A completed application form must be sent before the deadline.
Closing Date: May 1st
Funding: Government
Contributor: The Bureau of Educational and Cultural Affairs of the United States Department

USEFI Fulbright-CII Fellowships for Leadership in Management

Subjects: Leadership in management.
Purpose: To enable business managers to attend a management programme in the United States of America.
Eligibility: Open to Indian business managers. Age limit, preferably not above 40, Nationality: Indian, Designed for: Mid-level Managers in Indian Industries.
Level of Study: Professional development
Type: Fellowship
Value: See website for details
Length of Study: 10 weeks
Frequency: Annual
Study Establishment: Carnegie Mellon University
Country of Study: United States of America
No. of awards offered: 5–6
Application Procedure: Applicants must complete an application form, available from the CII.
Closing Date: February 15th
Funding: Corporation, foundation
Contributor: USEFI and employers of selected Managers
No. of awards given last year: 8
No. of applicants last year: 49
Additional Information: For further information contact the Confederation of Indian Industry (CII).

For further information contact:

Confederation of Indian Industry, Mantosh Sondhi Centre, 23, Institutional Area, Lodi Road, New Delhi, 110 003, India
Tel: (91) 24629994-7 ext. 367
Fax: (91) 24601298, 24626149
Email: sudarsan@usief.org.in.
Contact: Ms S Rajeshwari, Executive Officer

USEFI Hubert H Humphrey Fellowships

Subjects: Agricultural development/agricultural economics; communications/journalism, drug abuse education, treatment and prevention, economic development educational planning and administration, finance and banking HIV/AIDS policy and prevention, human resource management, law and human rights, natural resources and environmental management, nonproliferation studies, public health policy and management, public policy analysis and public administration, technology policy and management, prevention of trafficking in persons and policy development and management for anti-trafficking efforts, teacher training and curriculum development, and urban regional planning.
Purpose: To bring accomplished mid-level professionals from developing countries to the United States for 10 months of non-degree graduate study and related professional experiences.
Eligibility: Open to Indian citizens resident in India at the time of application who have not been to the United States of America in the previous 3 years. Applicants must have a high level of academic and professional achievement, proficiency in English and be in good health. Preferably applicants will have a first class master's or a professional degree of at least 4 year's duration. Applicants must have at least 5 years of substantial professional experience in the respective field and not be more than 40 years of age. Applicants applying for the fellowships in drug abuse must hold a PhD or equivalent degree in the health, behavioural or social sciences or an MD.
Level of Study: Professional development
Type: Fellowship
Value: Tuition and fees, a monthly maintenance allowance, modest allowance for books and supplies, round trip international travel to the host institution and domestic travel to Washington, DC and/or Minnesota workshops
Length of Study: 10 months
Frequency: Annual
Country of Study: United States of America
No. of awards offered: Varies
Application Procedure: Applicants must complete an application form. Requests for application materials must state the applicant's academic and professional qualifications, date of birth, current position and grant category and be accompanied by a 7 by 10 inch stamped addressed envelope. Requests for application materials must be sent to the USEFI offices in the applicant's region. Alternatively, applications can be downloaded from the USEFI website.

Closing Date: July 1st
Funding: Government
Contributor: Bureau of Educational and Cultural Affairs, US Department of State, Washington, DC
No. of awards given last year: 9
No. of applicants last year: 57
Additional Information: Please see the website for further details http://www.usief.org.in/Fellowships/Hubert-H-Humphrey-Fellowship-Program.aspx.

UNIVERSITIES FEDERATION FOR ANIMAL WELFARE (UFAW)

The Old School, Brewhouse Hill, Wheathampstead, Hertfordshire, AL4 8AN, England
Tel: (44) 15 8283 1818
Fax: (44) 15 8283 1414
Email: ufaw@ufaw.org.uk
Website: www.ufaw.org.uk
Contact: Scientific Officer

The Universities Federation for Animal Welfare (UFAW) is an independent, scientific and educational animal welfare charity concerned with improving knowledge and understanding of animals' needs in order to promote high standards of welfare for farm, companion, laboratory and captive wild animals and those with which we interact in the wild.

UFAW 3Rs Liaison Group Research Studentship

Subjects: Laboratory animal studies.
Purpose: To support a promising graduate based at a research institute in the British Isles to undertake a three year programme of research leading to a degree at the doctorate level in any aspect of the replacement, reduction or refinement of laboratory animal use.
Eligibility: Open to nationals of any country studying in the United Kingdom.
Level of Study: Doctorate
Value: £75,000 over a three year period
Length of Study: 3 years
Frequency: Dependent on funds available
Country of Study: United Kingdom
Application Procedure: Application for this award is by a two-stage process. Initially, supervisors are required to submit a brief Concept Note, available from the UFAW website or by emailing the UFAW Scientific Officer at scioff@ufaw.org.uk. Following assessment of these concept notes, selected applicants are invited to submit more detailed proposals prepared jointly by the supervisor and the PhD candidate.
Closing Date: November 14th
Funding: Private
Additional Information: Please see the website for further details http://www.ufaw.org.uk/phhscResearchStudentship.php.

UFAW Animal Welfare Research Training Scholarships

Subjects: UFAW would particularly like to encourage applications in the following fields:development of methodologies aimed at elucidation of the neurological basis of sentience in animals; developments in approaches to alleviating welfare problems in farmed, companion and/or laboratory animals through breeding; developments in detection and alleviation of pain;developments of methods of welfare/quality of life assessment. However, UFAW does not wish to exclude potentially valuable projects in other aspects of animal welfare science and applications for work in other areas will also be considered.
Purpose: To encourage high-quality science likely to lead to substantial advances in animal welfare and to enable promising graduates to start a career in animal welfare science.
Eligibility: Open to nationals of any country studying in the United Kingdom.
Level of Study: Doctorate
Type: Scholarship
Value: Science graduates: £19,339 rising to £22,619 for the third year (London weighting £21,629 rising to £25,298 for the third year), veterinary graduates, £20,306 rising to £23,750 for the third year (London weighting £22,710 rising to £26,563 for the third year). Research costs up to £10,000 per year and approved tuition fees up to £3466 pa will be met
Length of Study: 3 years
Frequency: Dependent on funds available
Country of Study: United Kingdom
Application Procedure: Application for this award is by a two-stage process. Initially, supervisors are required to submit a brief Concept Note available from the UFAW website www.ufaw.org.uk or by emailing the UFAW Scientific Officer at scioff@ufaw,org.uk. Following assessment of these concept notes selected applicants will then be invited to submit more detailed proposals prepared jointly by the supervisor and PhD candidate.
Closing Date: Check with website
Funding: Private
Additional Information: Please see the website for further details http://www.ufaw.org.uk/AWResearchtrainingscholarship.php.

UFAW Animal Welfare Student Scholarships

Subjects: Research that is likely to provide new insight into the subjective mental experiences of animals relevant to their welfare and at understanding their needs and preferences, and also applied research aimed at developing practical solutions to animal welfare problems.
Purpose: To encourage students to develop their interests in animal welfare and their abilities for animal welfare research.
Eligibility: Applications are welcome from individuals studying at universities or colleges in the United Kingdom or an overseas institution at which there is a UFAW University Links representative. Students will usually be undertaking courses in the agricultural, biological, medical, psychological, veterinary or zoological sciences. Applicants need a nominated supervisor to oversee the project.
Level of Study: Graduate
Type: Scholarship
Value: UK£170 subsistence and UK£30 departmental expenses for each week of study
Length of Study: Maximum of 8 weeks
Frequency: Annual
No. of awards offered: Up to 15
Application Procedure: Applicants must complete a UFAW Animal Welfare Student Scholarship application form available for download from the UFAW website www.ufaw.org.uk or by emailing the UFAW Scientific Officer at scioff@ufaw.org.uk
Closing Date: February 28th
Funding: Private
Additional Information: Successful applicants must submit a written report to UFAW by November of the year in which the project was undertaken. Please see teh website for further details http://www.ufaw.org.uk/vacationScholarshipawards.php.

UFAW Companion Animal Welfare Award

Subjects: The UFAW Companion Animal Welfare Award scheme is an annual, international prize competition open to all individuals and organizations working in the field of companion animal welfare science and technology. Interested applicants are invited to apply to UFAW detailing how their innovation is likely to, or already has, led to welfare improvements in companion animals. UFAW would particularly like to encourage applications relating to the following fields: Genetic welfare problems; detection and alleviation of pain; diagnoses and treatment of painful disease. However, UFAW does not wish to exclude potentially valuable projects dealing with other aspects of companion animal welfare and applications in other relevant fields will also be welcomed.
Purpose: To recognize significant innovations or advances for the welfare of companion animals.
Eligibility: Applications are invited from individuals or organizations in the field of companion animal welfare science and technology. Applicants may be based in the UK or internationally.
Level of Study: Unrestricted
Value: UK£1,000
Frequency: Annual
Country of Study: Any country
Application Procedure: Applicants must complete a Companion Animal Welfare Award application form available for download from the UFAW website or by emailing the UFAW Scientific Officer at scioff@ufaw.org.uk. Applicants should present an evidence-based,

scientifically-informed case and the relevance to animal welfare should be made clear. Key points to be considered include the number of companion animals likely to benefit, and the severity and duration of the welfare problem addressed.
Closing Date: Check with website
Funding: Private

UFAW Research and Project Awards

Subjects: Full or part funding for self-contained projects or initial funding for studies that may lead to further investigation or welfare benefits. Awards can be used to supplement current work, to extend previous projects that promote animal welfare or to support non-research projects that promote animal welfare, such as preparation and publication of books and teaching materials.
Purpose: To encourage fundamental and high-quality research that is likely to lead to substantial improvements in animal welfare.
Eligibility: Open to nationals of any country.
Level of Study: Unrestricted
Type: Grant
Value: UK£3,500+ (rarely in excess of UK£10,000)
Length of Study: For the duration of the approved works
Frequency: Dependent on funds available
Country of Study: Any country
No. of awards offered: Varies
Application Procedure: Applicants must submit an application form available for download from the UFAW website www.ufaw.org.uk or by emailing the UFAW Scientific Officer at scioff@ufaw.org.uk
Closing Date: Any time
Funding: Private
Additional Information: Annual progress reports are required within 1 month of the anniversary of the start date and a final report must be submitted within 3 months of the completion date. Please see the website for further details http://www.ufaw.org.uk/ResearchandProjectawards.php.

UFAW SAWI Fund

Subjects: UFAW SAWI is keen to identify and support projects through which high quality science can lead to major animal welfare benefits in Israel. Funds are available via a number of routes including UFAW SAWI Small Project and Travel Awards, UFAW SAWI Research and Project Awards, UFAW SAWI Vacation Scholarship Awards, and a postgraduate scholarship for an Israeli student to undertake study in animal welfare science in the UK.
Purpose: To promote animal welfare in Israel.
Eligibility: Please visit the UFAW website for further information on individual SAWI grants.
Length of Study: Varies
Frequency: Dependent on funds available
Application Procedure: Application forms may be downloaded from the UFAW website (www.ufaw.org.uk) or by contacting the UFAW scientific officer at scioff@ufaw.org.uk
Funding: Private

UFAW Small Project and Travel Awards

Subjects: Through its Small Project and Travel Awards (up to £3,500) UFAW supports a variety of activities for the benefit of animal welfare. Applications may be made for: the purchase of equipment; organization of educational meetings, lectures and courses; publication, translation or transmission of information on animal welfare; and for other small projects in support of UFAW's objectives. Particularly welcomed are applications for pilot studies where there is a likelihood of successful completion leading to further, more substantial work. Please note, all applications are judged on their merits for animal welfare and (in the case of research applications) their scientific quality.
Purpose: To support research projects and other activities for the benefit of animal welfare.
Eligibility: Open to nationals of any country.
Level of Study: Unrestricted
Type: Grant
Value: Up to UK£3,500
Length of Study: For the duration of the approved works
Frequency: Dependent on funds available
Country of Study: Any country
Application Procedure: Applicants must complete a Small Project and Travel Award application form available for download from the UFAW website www.ufaw.org.uk or by emailing the UFAW Scientific Officer at scioff@ufaw.org.uk
Closing Date: Any time
Funding: Private
Additional Information: A report is required on completion of the project. Please see the website for further details http://www.ufaw.org.uk/UFAWSmallProjectsandtravelAwards.php.

UFAW Wild Animal Welfare Award

Subjects: Alleviating or preventing anthropogenic harm to the welfare of free-living wild animals including work to improve the humaneness of rodent 'pest' control, work toward the control of wildlife disease, of anthropogenic origin (infectitious or non-infectitious), improved methods of capture, handling or marking for field studies. and other aspects of wild animal welfare.
Purpose: To recognize significant innovations or advances for the welfare of wild animals.
Eligibility: Applications are invited from individuals or organizations working in relevant areas including research, management, conservation, control and reintroduction of wild animals. Applicants may be based in the UK or internationally.
Value: UK£1,000
Frequency: Annual
Application Procedure: Applicants must complete a Wild Animal Welfare Award application form, available for download from the UFAW website (http://www.ufaw.org.uk/wawa.php) or by emailing the UFAW Scientific Officer at scioff@ufaw.org.uk
Closing Date: December 7th (check with website)

UNIVERSITY COLLEGE BIRMINGHAM

Summer Row, Birmingham, B3 1JB, England
Tel: (44) 12 1604 1000
Fax: (44) 12 1608 7100
Email: Registry@bcftcs.ac.uk
Website: www.ucb.ac.uk
Contact: Student Scholarships

Academic Excellence Scholarships

Subjects: Hospitality, tourism and child care.
Purpose: To reduce tuition fee for outstanding students.
Eligibility: Applicants must demonstrate excellence in academic achievement.
Level of Study: Postgraduate
Type: Scholarship
Value: Up to UK£1,000
Length of Study: 1 year
Frequency: Annual
Country of Study: United Kingdom
No. of awards offered: 6
Closing Date: August 31st for Semester one or December 31st for Semester two
Contributor: University College Birmingham
Additional Information: Please see the website for further details http://www.ucb.ac.uk/services-and-support/scholarships/academic-excellence-scholarship.aspx.

Sporting Excellence Scholarships

Subjects: Hospitality, tourism and child care/development.
Purpose: To reduce tuition fee for outstanding students.
Eligibility: Applicants must demonstrate excellence in academic achievement.
Level of Study: Postgraduate
Type: Scholarship
Value: Up to UK£1,000 and up to £500 over the duration of the programme towards the cost of travel for training/events
Length of Study: 1 year
Frequency: Annual
Study Establishment: University College Birmingham
Country of Study: United Kingdom
No. of awards offered: 6
Additional Information: Please see the website for further details http://www.ucb.ac.uk/services-and-support/scholarships/sporting-excellence-scholarship.aspx.

Target Recruitment Partial Fee Waiver

Subjects: Tourism, hospitality and child care.
Purpose: To reduce tuition fee for new international students.
Eligibility: Applicants must refer to the website for details.
Level of Study: Postgraduate
Type: Scholarships
Value: Up to UK£1000
Length of Study: 1 year
Frequency: Annual
Study Establishment: University College Birmingham
Country of Study: United Kingdom
No. of awards offered: 7
Closing Date: Check the website
Additional Information: Applications can be considered for entry in Semester 1 (Late September/Early October) or Semester 2 (Late January/Early February). Applications should be made at least 2 months prior to the entry date. No applications are necessary. Partial Fee Waivers will be noted in all offer letters sent to applicants from relevant countries. For more information contact the International Student Office on international.

For further information contact:

Email: international@ucb.ac.uk
Website: http://www.ucb.ac.uk/services-and-support/scholarships/target-recruitment-partial-fee-waiver.aspx

UNIVERSITY COLLEGE DUBLIN (UCD)

UCD Research, UCD, Belfield, Dublin 4, Ireland
Tel: (353) 1 716 7777
Fax: (353) 1 269 7262
Email: derek.mchugh@ucd.ie
Website: www.ucd.ie
Contact: Mr Derek McHugh, Research Information Officer

University College Dublin (UCD) was founded in the mid 19th century and is a dynamic, modern university. It is committed to becoming one of the top 30 research universities in the European Union, where cutting edge research and scholarship will create a stimulating intellectual environment, the ideal surroundings for learning and discovery.

SRG Postdoctoral Fellowships

Subjects: Semantics, verification and software engineering.
Purpose: To develop into a world-class research group in systems and software technology.
Eligibility: Open to candidates who have obtained a PhD in computer science or mathematics.
Level of Study: Postdoctorate
Type: Fellowships
Value: €35,866–39,337
Length of Study: 18 months–3 years
Frequency: Annual
Study Establishment: University College Dublin (UCD)
Country of Study: Ireland
No. of awards offered: 2
Additional Information: For general information or applications, mail to aquigley@ucd.ie

For further information contact:

Website: http://srg.cs.ucd.ie

UNIVERSITY COLLEGE LONDON

Gower Street, London, WC1E 6BT, United Kingdom
Tel: (44) 20 7679 2000
Fax: (44) 20 7691 3112
Email: postmaster@ucl.ac.uk
Website: www.ucl.ac.uk

Just 175 years ago, the benefits of a university education in England were restricted to men who were members of the Church of England; University College London (UCL) was founded to challenge that discrimination. UCL was the first university to be established in England after Oxford and Cambridge, providing a progressive alternative to those institutions social exclusivity, religious restrictions and academic constraints. UCL is the largest of over 50 colleges and institutes that make up the federal University of London.

A C Gimson Scholarships in Phonetics and Linguistics

Subjects: Phonetics and linguistics.
Purpose: To financially support MPhil/PhD research.
Eligibility: United Kingdom, European Union and overseas students are eligible to apply.
Level of Study: Doctorate, Postgraduate, Research
Type: Scholarship
Value: UK£1,000
Frequency: Annual
Study Establishment: University College London
Country of Study: United Kingdom
No. of awards offered: Up to 2
Application Procedure: Applicants should write indicating their intention to compete for the bursaries to the department.
Closing Date: June 1st

For further information contact:

Department of Phonetics and Linguistics
Tel: 44 0 20 7679 4245
Email: n.wilkins@ucl.ac.uk
Contact: Natalie Wilkins

A J Ayer-Sumitomo Corporation Scholarship in Philosophy

Subjects: Philosophy.
Purpose: To financially support MPhil/PhD research.
Eligibility: All applicants who are admitted to research programmes in philosophy will automatically be considered for this award.
Level of Study: Doctorate, Postgraduate, Research
Type: Scholarship
Value: Up to UK£800
Frequency: Annual
Study Establishment: University College London
Country of Study: United Kingdom
No. of awards offered: Varies
Application Procedure: No separate application is required. All who are admitted to research programmes in philosophy will automatically be considered for the scholarship. Any queries should be directed to the department.
Additional Information: Decisions regarding this award will be made in September.

For further information contact:

Department of Philosophy
Tel: 44 0 20 7679 4451
Email: r.madden@ucl.ac.uk
Contact: Dr Rory Madden

Alfred Bader Prize in Organic Chemistry

Subjects: Organic chemistry.
Purpose: To financially support MPhil/PhD research.
Eligibility: Applicants must contact the department.
Level of Study: Doctorate, Postgraduate, Research
Type: Scholarship
Value: UK£1,000
Frequency: Annual
Study Establishment: University College London
Country of Study: United Kingdom
No. of awards offered: 1
Application Procedure: Applicants must contact the department.
Closing Date: Check the website for closing date
Additional Information: The award will be announced in October.

For further information contact:

Department of Chemistry
Tel: (44) 20 7679 4650
Fax: (44) 20 7679 7463
Email: m.l.jabore@ucl.ac.uk
Contact: Ms Mary Lau Jabore

Amelia Zollner IPPR/UCL Internship Award

Purpose: The Amelia Zollner IPPR/UCL Internship Award was founded in 2007 in memory of UCL student Ameial Zollner. IPPR and UCl co-fund an annual, approximately 3-month London-based IPPR internship reserved for a recent UCL graduate – as a stepping stone to working in policy or politics. The internship will normally run from October through December, but can take place at any time of the year.
Eligibility: 1. Applicants must be final year UCL students, about to graduate (Bachelors, Masters or Research degree) in any area, interested in and passionate about policy and politics, current affairs and social justice, and the work of IPPR, as well as enthusiastic about political research.2. Applicants can come from any country, but must be able to work in the UK (UK/EU nationals or holders of valid work permit).3. Applicants should have some research skills, and need to be committed to the aims of IPPR of wanting to build a fair, more democratic and environmentally sustainable world.
Level of Study: Graduate, Postgraduate, Research
Type: Award
Value: UK£5,000 or the equivalent of the current IPPR salary for the duration of the internship
Length of Study: 3 months
Frequency: Annual
Country of Study: United Kingdom
Application Procedure: Applications should be made direct to IPPR and submitted – preferably by – email – to intern@ippr.org
Closing Date: Refer website
No. of applicants last year: 4–8

For further information contact:

Internships – IPPR, 30-32 Southampton Street, London, WC2 7RA, United Kingdom
Tel: (44) 020 7470 6133
Email: k.holdway@ippr.org
Website: www.ippr.org.uk

Andrew Szmiola Postgraduate Scholarship

Purpose: To enable and encourage academically able students from Poland to pursue study at UCL.
Eligibility: Polish nationals currently resident in Poland.
Level of Study: Masters
Type: Scholarship
Value: £10,000
Length of Study: 1 year
Frequency: Annual
Study Establishment: University of London
Country of Study: United Kingdom
No. of awards offered: 3
Application Procedure: As per website.
Closing Date: March 18th
Funding: Private
Additional Information: This is first time the scholarship is offering.

Archibald Richardson Scholarship for Mathematics

Subjects: Pure mathematics.
Purpose: To financially support students to pursue MPhil/PhD research.
Eligibility: United Kingdom, European Union and overseas applicants are eligible to apply. All applicants who firmly accept a place for MPhil/PhD research in pure Mathematics department will be considered.
Level of Study: Research, Doctorate, Postgraduate
Type: Scholarship
Value: UK£3,000
Frequency: Annual
Study Establishment: University College London
Country of Study: United Kingdom
No. of awards offered: 1
Application Procedure: All applicants who firmly accept a place for MPhil/PhD research in pure mathematics will be considered automatically.
Closing Date: May 15th

For further information contact:

Department of Mathematics, United Kingdom
Tel: (44) (0) 20 7679 2839
Fax: (44) (0) 20 7383 5519
Email: h.higgins@ucl.ac.uk
Contact: Ms Helen Higgins

Ardalan Scholarship

Subjects: Preference will be given to students intending to study within the Department of economics.
Purpose: To assist Iranian nationals to undertake Master's study at UCL.
Eligibility: 1. Open to applicants who have successfully completed undergraduate level studies in Iran and are Iranian nationals. 2. Pursuing full-time gaught master's at UCL, within any department except religious studies.
Level of Study: Postgraduate
Type: Scholarship
Value: £20,000
Frequency: Annual
Study Establishment: University College London
Country of Study: United Kingdom
No. of awards offered: 1
Application Procedure: See website for details.
Closing Date: See website for details

Arts and Humanities Faculty Postgraduate Research Studentship

Subjects: The award is given for academic excellence. Applicants should be intending to pursue a course of study leading to a PhD degreee in the faculty of Arts and Humanities.
Purpose: To provide financial asssitance to students admitted for MPhil/PhD degree studies in the Faculty of Arts and Humanities at University College London.
Eligibility: Applicants should contact the Faculty of Arts and Humanities for full details on eligibility.
Level of Study: Postgraduate, Doctorate, Research
Type: Studentship
Value: For 1 year, each £5,000 (for 3); for 2 years, £6,500 for 2 overseas scholarships, subject to revision at the end of year 1. Value of Studentship is usually deducted from tuition fees at the start of the academic session.
Frequency: Dependent on funds available
No. of awards offered: Up to 3
Application Procedure: Applicants should refer to the website for further information.
Closing Date: March 16th
Contributor: Faculty of Arts and Humanities
No. of awards given last year: 5
No. of applicants last year: Approx. 10

For further information contact:

Website: www.ucl.ac.uk/ah/pages.studentships
Contact: Dr Stephanie, Bird

Bentham Scholarships

Subjects: Law.
Purpose: To financially support prospective LLM students.
Eligibility: Applicants must be overseas students from outside the European Union. An applicant must have accepted an offer (either conditional or unconditional) to read for the LLM at UCL to be eligible.
Level of Study: Postgraduate
Type: Scholarship
Value: UK£2,000
Frequency: Annual
Study Establishment: University College London
Country of Study: United Kingdom
No. of awards offered: 5
Application Procedure: There is no application procedure. All eligible students will automatically be considered if they have firmly accepted their offer of admission by May 31st.
Closing Date: Refer website
Additional Information: The scholarships will be based on academic merit. The faculty will only consider these applicants who have firmly accepted their offer of admission to the LLM by May 31st.

For further information contact:

Faculty of Laws, University College London, Bentham House, Endsleigh Gardens, London, WC1H 0EG, United Kingdom

Tel: (44) 20 7679 1441
Fax: (44) 20 7209 3470
Email: graduatelaw@ucl.ac.uk
Contact: The Graduate Officer

Bioprocessing Graduate Scholarship

Subjects: Biochemical engineering.
Purpose: To financially support study leading to an MPhil/PhD.
Eligibility: Open to students resident outside the United Kingdom and pursuing the MSc or MPhil or PhD research in the department of biomedical engineering.
Level of Study: Doctorate, Postgraduate
Type: Scholarship
Value: Up to a maximum of UK£11,000 per year. This sum can be set against tuition fees and/or be received as maintenance allowance payable in quarterly installments
Length of Study: Maximum of 4 calendar years
Frequency: Annual
Study Establishment: University College London
Country of Study: United Kingdom
No. of awards offered: 1
Application Procedure: Applicants must contact the department at the address given below.
Closing Date: February 15th
Funding: Trusts
No. of applicants last year: 8

For further information contact:

PhD Admissions Tutor, Department of Biochemical Engineering, University College London, Torrington Place, London, United Kingdom
Tel: (44) 20 7679 3796
Email: nigelth@ucl.ac.uk
Contact: Dr Paul Dalby

Brain Research Trust Prize

Subjects: Neurology or clinical neurosciences.
Purpose: To financially support MPhil/PhD research students.
Eligibility: Open to United Kingdom, European Union and overseas students. Overseas fee-paying students should be aware that only home tuition fee (EU rates) is included in the award.
Level of Study: Doctorate, Postgraduate, Research
Type: Studentship
Value: Stipend, tuition fees at UK/EU rate + travel budget
Length of Study: Up to 3 years
Frequency: Annual
Study Establishment: UCL Institute of Neurology
Country of Study: United Kingdom
Application Procedure: Applicants should contact the UCL Institute of Neurology. Please submit full curriculum vitae + references + statement of research interests indicating how these would complement projects on offer.
Closing Date: Refer website
Funding: Government, trusts
Contributor: The Brian Research Trust

For further information contact:

Cell Signalling Laboratory, Institute of Neurology, UCL, 1 Wakefield Street, London, WC1N 1PJ
Tel: 44 (0) 207 679 4031
Email: phdstudentship@ion.cl.ac.uk
Contact: Dr Jennifer Pocock

British Chevening/UCL Israel Alumni/Chaim Herzog Award

Subjects: All subjects. Preference given to public policy, economics and history.
Purpose: To financially support masters study.
Eligibility: Candidates will need to return to Israel for at least 3 years after completion. Preference will be given to: – candidates between 25–40; – Israeli nations currently living in Israel; – applicants who have never studied in the UK; – applicants who are graduates of an Israeli university/students who obtained 1st degree by summer 2004.
Level of Study: Postgraduate
Type: Scholarship
Value: Tuition fees, return airfare and living expenses
Frequency: Annual
Study Establishment: University College London
Country of Study: United Kingdom
Closing Date: Refer to website

For further information contact:

The British Council, PO Box 10304, 3 Shimshom Street, Jerusalem, 91102
Email: scholarships@britishcouncil.org.il
Website: www.britishcouncil.org.il

Chief Justice Scholarships for Students from India

Subjects: Law.
Purpose: To financially support prospective LLM students.
Eligibility: Applicants must be overseas students from India. An applicant must have accepted an offer (either conditional or unconditional) to read for the LLM at UCL to be eligible.
Level of Study: Postgraduate
Type: Scholarship
Value: UK£2,000
Frequency: Annual
Study Establishment: University College London
Country of Study: United Kingdom
No. of awards offered: 10
Application Procedure: There is no application procedure. All eligible students who have firmly accepted their offer of admission by May 31st will automatically be considered.
Additional Information: The scholarships will be based on academic merit. The faculty will only consider these applicants who have firmly accepted their offer of admission to the LLM by May 31st.

Clinical Psychology Awards

Subjects: Clinical psychology.
Purpose: To financially support doctorate study.
Eligibility: United Kingdom and European Union students are eligible to apply.
Level of Study: Doctorate
Type: Scholarship
Value: Salary, tuition fees and travelling expenses
Frequency: Annual
Study Establishment: University College London
Country of Study: United Kingdom
No. of awards offered: 40
Application Procedure: Applicants should contact the department via Clearing House, Leeds.
Closing Date: December 1st
Funding: Government
No. of awards given last year: 40
Additional Information: Clinical Psychology Awards are salaried positions with the Camden & Islington Community Mental Health Trust.

For further information contact:

The Clearing House for Postgraduate Courses in Clinical Psychology, Fairbairn House, 71–75 Clarendon Road, Leeds, LSZ 9PL
Tel: (44) 20 7436 4276
Website: www.leeds.ac.uk/chpccp

David Pearce Research Scholarship

Subjects: Economics.
Purpose: To provide financial assistance to students admitted for MPhil/PhD degree studies in the faculty of Arts and Humanities at UCL.
Eligibility: Applicants should contact the Department of Economics for full details on eligibility.
Level of Study: Postgraduate, Research
Type: Scholarship
Value: £15,590 plus full tuition fee waiver
Frequency: Annual
No. of awards offered: Up to 3
Application Procedure: Applicants should contact the Department for further information.
Closing Date: See website

No. of awards given last year: 3

For further information contact:

Website: www.econ.ucl.ac.uk

Dawes Hicks Postgraduate Scholarships in Philosophy

Subjects: Philosophy.
Purpose: To financially support postgraduate research.
Eligibility: United Kingdom, European Union and overseas students are eligible to apply.
Level of Study: Postgraduate, Research
Type: Scholarship
Value: Up to UK£5,000
Frequency: Annual
Study Establishment: University College London
Country of Study: United Kingdom
No. of awards offered: Varies
Application Procedure: No separate application is required. All who are admitted to research programmes in philosophy will automatically be considered for the scholarship. Any queries should be directed to the department.
Additional Information: Decisions regarding this award will be made in September.

For further information contact:

Department of Philosophy
Tel: 44 0 20 7679 4451
Email: r.madden@ucl.ac.uk
Contact: Dr Rory Madden

Department of Geography Teaching/Computing Assistantships

Subjects: Geography.
Purpose: To financially support teaching and computer assistants.
Eligibility: Applicants should be United Kingdom or European Union students. Overseas students who can pay the balance of fees to the higher rate, including those receiving an Overseas Research Students award for this purpose.
Level of Study: Postgraduate, Doctorate
Type: Assistantship
Value: Stipend based on that paid by the United Kingdom Research Councils and fees at the Home/European Union rate.
Frequency: Every 3 years
Study Establishment: University College London
Country of Study: United Kingdom
No. of awards offered: 2
Application Procedure: Applicants must contact the Department of Geography.
Additional Information: Please check website for updated information or details.

For further information contact:

Graduate Admissions, Deparment of Geography, University College of London, Pearson Building, Gower Street,, WCIE 6BT
Tel: 44 0 20 7679 0500
Website: www.geog.ucl.ac.uk/admission

Edwin Power Scholarship

Subjects: Applied mathematics.
Purpose: To support a graduate student to the Department of Mathematics for Master's study or PhD research in Applied Mathematics.
Eligibility: Must be a graduate student admitted to the Department of Mathematics for Master's study or MPhil/PhD research in applied mathematics.
Level of Study: Research, Postgraduate
Type: Scholarship
Value: UK£400, subject to annual renewal based on satisfactory progress.
Length of Study: 3 years
Frequency: Annual
Study Establishment: University College London
Country of Study: United Kingdom
No. of awards offered: 1
Application Procedure: Candidates wishing to apply for the Scholarship must send written notice of their intention to apply for the Scholarship to the Head of the Department of Mathematics at UCL.
Closing Date: August 15th

For further information contact:

Department of Mathematics, University College London, Gower Street, London, WC1E 6BT
Tel: 44 (0) 20 7679 2839
Fax: 44 (0) 20 7383 2839
Email: h.higgins@ucl.ac.uk
Contact: Ms Helen Higgins

Eleanor Grove Scholarships for Women Students

Subjects: German language and literature.
Purpose: To promote and encourage study and proficiency in German.
Eligibility: Candidate must be a female student of the college. The scholarship will be awarded to assist a student who is reading for a higher degree, or, in special cases, to enable a BA student to study German abroad. Candidates must have completed at least five terms of study with UCL's Arts and Humanities faculty.
Type: Scholarship
Value: £950
Frequency: Annual
Study Establishment: University college or elsewhere approved by the Faculty of Arts and Humanities
Country of Study: United Kingdom
No. of awards offered: 2
Application Procedure: Contact the Graduate Tutor.
Closing Date: May 25th
No. of applicants last year: 2

For further information contact:

The Graduate Tutor, Department of German, UCL
Email: german@ucl.ac.uk

Fielden Research Scholarship

Subjects: German language and literature.
Purpose: To financially support MPhil/PhD study.
Eligibility: Applicants must have applied for MPhil/PhD research in German language and literature.
Level of Study: Doctorate, Graduate, Postgraduate
Type: Scholarship
Value: UK£500
Frequency: Annual
Study Establishment: University College London
Country of Study: United Kingdom
No. of awards offered: 1
Application Procedure: Applicants must contact the department.
Closing Date: May 25th
No. of awards given last year: 1
No. of applicants last year: 1

For further information contact:

Graduate Tutor, Department of German
Email: german@ucl.ac.uk

Follett Scholarship

Subjects: Philosophy.
Purpose: To financially support MPhil/PhD research.
Eligibility: United Kingdom, European Union and overseas students are eligible to apply.
Level of Study: Doctorate, Postgraduate, Research
Type: Scholarship
Value: Up to UK£13,000 towards fees and/or maintenance
Frequency: Annual
Study Establishment: University College London
Country of Study: United Kingdom
Application Procedure: No separate application is required. Applicants who are admitted to research programmes in philosophy will automatically be considered for the scholarship. Any queries should be directed to the department.
Additional Information: Decisions regarding this award will be made in September.

For further information contact:

Department of Philosophy
Tel: (44) 20 7679 4451
Fax: (44) 20 7679 3336
Email: r.madden@ucl.ac.uk
Contact: Dr Rory Madden

Franz Sondheimer Bursary Fund

Subjects: Chemistry.
Purpose: To financially support MPhil/PhD research.
Eligibility: Preference is given to overseas applicants.
Level of Study: Doctorate, Postgraduate, Research
Type: Scholarship
Value: Approx. UK£2,000
Study Establishment: University College London
Country of Study: United Kingdom
No. of awards offered: 1
Application Procedure: Applicants must contact the department.
Closing Date: May 1st
No. of awards given last year: 1
No. of applicants last year: 2

For further information contact:

Department of Chemistry
Tel: (44) 20 7679 4650
Fax: (44) 20 7679 7463
Email: m.l.jabore@ucl.ac.uk
Contact: Ms Mary Lou Jabore

Frederick Bonnart-Braunthal Scholarship

Subjects: Tackling the causes and consequences of intolerance (religious, racial and cultural prejudices).
Purpose: To financially support students who plan to explore the nature of religious, racial and cultural prejudices, and to find ways of combating them.
Eligibility: Open to prospective MPhil/PhD candidates with a First Class (Honours) Degree or equivalent are eligible to apply.
Level of Study: Doctorate, Research
Type: Scholarship
Value: UK£15,000 per year
Length of Study: 3 years
Frequency: Annual
Study Establishment: University College London
Country of Study: United Kingdom
No. of awards offered: Up to 2
Application Procedure: Please check the website for further details about application procedures and deadlines.
Closing Date: See website
No. of awards given last year: 1

For further information contact:

Student Funding Office, Registrar's Division
Tel: (44) 20 7679 2005/4167
Website: www.ucl.ac.uk/scholarships

Gaitskell MSc Scholarship

Subjects: Economics.
Purpose: To financially support full-time Master's study in the Department of Economics.
Eligibility: Open to candidates who have applied for a place for graduate study at UCL and are not already receiving full financial support from other sources for fees and living costs.
Level of Study: Postgraduate
Type: Scholarship
Value: UK£5,000
Frequency: Annual
Study Establishment: University College London
Country of Study: United Kingdom
No. of awards offered: Up to 4
Application Procedure: Applications not needed. All applicants to the department are automatically considered.

For further information contact:

Department of Economics
Tel: (44) 20 7679 5861
Fax: (44) 20 7616 2775
Email: d.fauvrelle@ucl.ac.uk
Contact: Ms Daniella Fauvrelle

Gay Clifford Fees Award for Outstanding Women Students

Subjects: Any Master's programme in either the Faculty of Arts and Humanities or the Faculty of Social and Historical Sciences.
Purpose: To financially support postgraduate study.
Eligibility: Open to prospective female Master's degree students in the faculties of Arts and Humanities and Social and Historical Sciences with a First Class (Honours) undergraduate Degree or equivalent.
Level of Study: Postgraduate
Type: Scholarship
Value: UK£2,500 (Deducted from tuition fees)
Length of Study: 1 year
Frequency: Annual
Study Establishment: University College London
Country of Study: United Kingdom
No. of awards offered: 4
Application Procedure: Applicants should refer to the website for further information about the application procedures and deadlines.
Closing Date: See website
No. of awards given last year: 4
Additional Information: Awarded in memory of Gay Clifford (1943–1998), academic and poet.

For further information contact:

Student Funding Office, Registrar's Division
Tel: (44) 20 7679 2005/4167
Email: studentfunding@ucl.ac.uk
Website: www.ucl.ac.uk/scholarships

The George Melhuish Postgraduate Scholarship

Subjects: Philosophy.
Purpose: To financially support postgraduate research.
Eligibility: Open to United Kingdom, European Union and overseas applicants.
Level of Study: Postgraduate, Research
Type: Scholarship
Value: Up to UK£3,700
Frequency: Annual
Study Establishment: University College London
Country of Study: United Kingdom
No. of awards offered: Varies
Application Procedure: No separate application is required. Applicants who are admitted to research programmes in philosophy will automatically be considered for the scholarship. Any queries should be directed to the department.
No. of awards given last year: 2
Additional Information: Decision regarding this award will be made in September.

For further information contact:

Department of Philosophy
Tel: 44 0 20 7679 4451
Email: r.madden@ucl.ac.uk
Contact: Dr Rory Madden

Ian Karten Charitable Trust Scholarship (Hebrew and Jewish Studies)

Subjects: Hebrew and Jewish studies.
Purpose: To financially support postgraduate study.
Eligibility: Applicants must have applied for a place for graduate study at UCL in the Department of Hebrew and Jewish Studies.
Level of Study: Graduate, Postgraduate
Type: Scholarship
Value: UK£1,000 each

Frequency: Annual
Study Establishment: University College London
Country of Study: United Kingdom
No. of awards offered: 4
Application Procedure: Applicants should contact the department. If the applicants have not applied to UCL they must complete a graduate application form and enclose it with the scholarship application.
Closing Date: June 1st
Funding: Trusts

For further information contact:

Department of Hebrew and Jewish Studies
Tel: (44) 20 7679 3028
Fax: (44) 20 7209 1026
Email: n.f.lochery@ucl.ac.uk
Contact: Dr Neil Lochery

J J Sylvester Scholarship

Subjects: Mathematics.
Purpose: To financially support MPhil/PhD research.
Eligibility: United Kingdom, European Union and overseas students are eligible to apply.
Level of Study: Doctorate, Postgraduate, Research
Type: Scholarship
Value: Up to £3,000
Frequency: Annual
Study Establishment: University College London
Country of Study: United Kingdom
No. of awards offered: 1
Application Procedure: Applicants must contact the department.

For further information contact:

Department of Mathematics, United Kingdom
Tel: (44) 44 (0) 20 7679 2839
Fax: (44) 44 (0) 20 7383 5519
Email: h.higgins@ucl.ac.uk
Contact: Ms Helen Higgins

Jacobsen Scholarship in Philosophy

Subjects: Philosophy.
Purpose: To financially support MPhil/PhD research.
Eligibility: Open to United Kingdom, European Union and overseas students.
Level of Study: Doctorate, Postgraduate, Research
Type: Scholarship
Value: Up to UK£9,500
Frequency: Annual
Study Establishment: University College London
Country of Study: United Kingdom
Application Procedure: No separate application is required. Applicants who are admitted to research programmes in philosophy will automatically be considered for the scholarship. Any queries should be directed to the department.
Additional Information: Decisions regarding this award will be made in September.

For further information contact:

Department of Philosophy
Tel: 44 0 20 7679 7115
Email: r.madden@ucl.ac.uk
Contact: Dr Rory Madden

The James Lanner Memorial Scholarship

Subjects: Economics, environmental resource economics.
Purpose: To financially support students to pursue Master's in Economics.
Eligibility: Candidates should have applied for a place for graduate study at UCL. Open to prospective students of an MSc in the Department of Economics.
Level of Study: Postgraduate
Type: Scholarship
Value: Up to UK£5,000
Frequency: Annual
Study Establishment: University College London
Country of Study: United Kingdom
Application Procedure: Applicants to the department are automatically considered.
Additional Information: Preference is given to candidates who intend to specialize in international economics or an issues of economic policy in Europe.

For further information contact:

The Graduate Administrator, Department of Economics
Tel: (44) 20 7679 5861
Fax: (44) 20 7616 2775

Jean Orr Scholarship

Subjects: Masters student studying in one or more of the following departments - History, French, Italian, and Dutch.
Level of Study: Graduate
Type: Scholarship
Value: £7,000
Length of Study: 1 year
Frequency: Annual
Study Establishment: University of London
Country of Study: United Kingdom
No. of awards offered: 1
Application Procedure: No application is required. Candidates are selected by the departments based on the academic merit of their admission application.
No. of awards given last year: 1

John Carr Scholarship for Students from Africa and the Caribbean

Subjects: Law.
Purpose: To financially support prospective LLM students.
Eligibility: Applicants must be overseas students from Africa and the Caribbean. An applicant must have accepted an offer (either conditional or unconditional) to read for the LLM at UCL to be eligible.
Level of Study: Postgraduate
Type: Scholarship
Value: UK£2,000
Frequency: Annual
Study Establishment: University College London
Country of Study: United Kingdom
No. of awards offered: 10
Application Procedure: There is no application procedure. All eligible students who have accepted the offer of admission by March 31st will automatically be considered.
Additional Information: The scholarships will be based on academic merit. The faculty will only consider those applicants who have firmly accepted their offer of admission to the LLM by May 31st.

For further information contact:

The Graduate Officer, Faculty of Laws
Tel: 44 0 20 7679 1441
Fax: 44 0 20 7209 3470
Email: graduatelaw@ucl.ac.uk

John Stuart Mill Scholarship in Philosophy of Mind and Logic

Subjects: Philosophy.
Purpose: To financially support MPhil/PhD research.
Eligibility: Open to United Kingdom, European Union and overseas applicants.
Level of Study: Doctorate, Postgraduate, Research
Type: Scholarship
Value: Up to UK£1,400
Frequency: Annual
Study Establishment: University College London
Country of Study: United Kingdom
Application Procedure: No separate application is required. All who are admitted to research programmes in philosophy will automatically be considered for the scholarship. Any queries should be directed to the department.
Additional Information: Decision regarding this award will be made in September.

For further information contact:

Department of Philosophy
Tel: 44 0 20 7679 4451
Email: r.madden@ucl.ac.uk
Contact: Dr Rory Madden

Joseph Hume Scholarship

Subjects: Law.
Purpose: To financially support LLM students or MPhil/PhD research.
Eligibility: For all LLM or MPhil/PhD research students in the Department of Laws.
Level of Study: Doctorate, Postgraduate, Research
Type: Scholarship
Value: UK£1,600
Frequency: Annual
Study Establishment: University College London
Country of Study: United Kingdom
No. of awards offered: 1
Application Procedure: There is no application procedure. All eligible students will automatically be considered.
Closing Date: March 23rd

For further information contact:

Faculty of Laws
Tel: 44 0 20 7679 1441
Fax: 44 0 20 7209 3470
Email: graduatelaw@ucl.ac.uk
Contact: The Graduate Officer

Keeling Scholarship

Subjects: Philosophy.
Purpose: To financially support MPhil/PhD research.
Eligibility: United Kingdom, European Union and overseas students are eligible to apply.
Level of Study: Doctorate, Postgraduate, Research
Type: Scholarship
Value: United Kingdom/European Union tuition fees plus a bursary
Frequency: Annual
Study Establishment: University College London
Country of Study: United Kingdom
Application Procedure: No separate application is required. Applicants who are admitted to research programmes in philosophy will automatically be considered for the scholarship. Any queries should be directed to the department.
Additional Information: Decisions regarding this award will be made in September.

For further information contact:

Department of Philosophy
Tel: 44 0 20 7679 7115
Email: r.madden@ucl.ac.uk
Contact: Dr Rory Madden

Liver Group PhD Studentship

Subjects: Hepatology.
Purpose: To financially support students to pursue MPhil/PhD research.
Eligibility: Open to United Kingdom and European Union applicants holding a relevant first or upper second class.
Level of Study: Doctorate, Postgraduate, Research
Type: Scholarship
Value: Home student fees plus a maintenance allowance (1st year approx. UK£14,500)
Length of Study: 3 years
Frequency: Dependent on funds available
Study Establishment: Royal Free and University College Medical School, UCL-Hampstead campus
Country of Study: United Kingdom
No. of awards offered: 1
Application Procedure: Applicants should contact the department.
Closing Date: Refer website
Funding: Foundation
Contributor: The Liver Group Charity
No. of awards given last year: 1

For further information contact:

Centre for Hepatology, Department of Medicine (Royal Free Campus), Royal Free and University College Medicine School, Rowland Hill Street, Hampstead, London, NW3 2PF, United Kingdom
Tel: (44) 20 7433 2854
Fax: (44) 20 7433 2852
Email: c.selden@rfc.ucl.ac.uk
Contact: Dr Clare Selden

Margaret Richardson Scholarship

Subjects: German.
Purpose: To financially support MPhil/PhD study.
Eligibility: Open to applicants should have applied for a place for graduate study at UCL and have upper-second class degree or equivalent in German or a related field of study to enable them to do research work in this subject. Previous tenure of the scholarship does not debar a candidate from competing on a second occasion.
Level of Study: Graduate, Doctorate, Postgraduate
Type: Scholarship
Value: UK£2,800
Frequency: Annual
Study Establishment: University College London
Country of Study: United Kingdom
Application Procedure: Applicants should contact the department. If the applicant has not applied to UCL, they must complete a graduate application form and enclose it with the scholarship application. The application must give particulars of the research work which they intend to pursue in the event of the scholarship being awarded to them.
Closing Date: May 25th

For further information contact:

The Graduate Tutor, Department of German
Email: german@ucl.ac.uk

Master of the Rolls Scholarship For Commonwealth Students

Subjects: Law.
Purpose: To financially support prospective LLM students.
Eligibility: Applicants must be overseas students from the Commonwealth countries. An applicant must have accepted an offer (either conditional or unconditional) to read for the LLM at UCL.
Level of Study: Postgraduate
Type: Scholarship
Value: UK£2,000
Frequency: Annual
Study Establishment: University College London
Country of Study: United Kingdom
No. of awards offered: 10
Application Procedure: There is no application procedure. All eligible students who firmly accept the offer of admission by May 31st will be automatically considered.
No. of awards given last year: 1
Additional Information: The scholarships will be based on academic merit. The faculty will only consider those applicants who have firmly accepted their offer of admission to the LLM by May 31st.

For further information contact:

Faculty of Laws
Tel: 44 0 20 7679 1441
Fax: 44 0 20 7209 3470
Email: graduatelaw@ucl.ac.uk
Contact: The Graduate Officer

Master's Degree Awards in Archaeology

Subjects: Archaeology.
Purpose: To financially support MA and MSc programmes in the Institute of Archaeology.
Eligibility: Open to students on MA and MSc programmes in UCL's institute of archaeology.
Level of Study: Postgraduate
Type: Scholarship
Value: Approx. UK£1,000
Frequency: Annual

Study Establishment: University College London
Country of Study: United Kingdom
Application Procedure: Applicants must contact the department.
Closing Date: Mid-March, check the department website for details

For further information contact:

Institute of Archaeology
Tel: (44) 20 7679 7499
Fax: (44) 20 7383 2572
Email: l.daniel@ucl.ac.uk
Contact: Lisa Daniel

Mayer de Rothschild Scholarship in Pure Mathematics

Subjects: Pure mathematics.
Eligibility: Open to students of the college who have attended for not less than 5 terms a course of study in pure mathematics.
Type: Scholarship
Value: Approx. £275 per year
Frequency: Annual
No. of awards offered: 2
Application Procedure: For further information please visit the website.
Closing Date: May 15th
No. of awards given last year: 1

Monica Hulse Scholarship

Subjects: Mathematics.
Eligibility: Open to graduate students in the department of mathematics in the first year of their Master's course or PhD. The applicant should not be in receipt of any other special funding.
Level of Study: Postgraduate, Research, Doctorate
Type: Scholarship
Value: £1,000
Frequency: Annual
No. of awards offered: 1
Application Procedure: Applicants must send written notice of their intention prior to the start of the academic year alongwith an academic curriculum vitae.
Closing Date: May 31st
No. of awards given last year: 1

For further information contact:

Department of Mathematics, The Graduate Tutor
Email: h.higgins@ucl.ac.uk

NHS Bursaries

Subjects: Speech and language sciences.
Purpose: To financially support postgraduate study.
Eligibility: Open to United Kingdom and European Union applicants only who have applied for a place for graduate study at UCL.
Level of Study: Postgraduate
Type: Bursary
Value: United Kingdom/European Union tuition fees. United Kingdom residents will also normally be eligible for a means tested bursary
Frequency: Annual
Study Establishment: University College London
Country of Study: United Kingdom
Application Procedure: There is no separate bursary application form. Application procedure is an automatic process once an offer of a place has been made.
Additional Information: When an offer of a place has been made, the NHS Student Grants unit will contact the applicant directly.

For further information contact:

Tel: 0 20 7679 4202
Email: n.wilkins@ucl.ac.uk
Contact: Natalie Wilkins

Perren Studentship

Subjects: Astronomy.
Purpose: To financially support graduate study and research.
Eligibility: United Kingdom, European Union and overseas applicants are eligible to apply.
Level of Study: Postgraduate, Research
Type: Studentship
Frequency: Every 2 years
Study Establishment: University College London
Country of Study: United Kingdom
No. of awards offered: 1
Application Procedure: Applicants should provide particulars of their academic record and of the work that they intend to pursue to the department.
Closing Date: May 15th
Funding: Trusts
Contributor: Perren Fund

For further information contact:

Department of Physics and Astronomy
Tel: (44) 20 7679 473
Email: lahauestar@ucl.ac.uk
Contact: Professor Ofer Lahau

Physics and Astronomy Departmental Studentship Grant

Subjects: Physics or astronomy related study.
Purpose: To support MPhil/PhD physics and astronomy related study.
Eligibility: Students currently pursuing an MPhil/PhD in the Department of Physics and Astronomy.
Level of Study: Research
Type: Studentship
Length of Study: 3 years
Frequency: Every 3 years
Study Establishment: University College London
Country of Study: United Kingdom
Application Procedure: For information please contact the Departments Graduate Admission Tutor - ucapphd@ucl.ac.uk
Closing Date: Refer website
Funding: Government
Contributor: University College London

PTDF-UCL Nigeria Scholarships

Subjects: Related fields of engineering, geological sciences, environmental and energy studies.
Eligibility: Applicants must have a 2.1 in engineering, geosciences, science and environmental studies; have an NYSC discharge certificate; not more than 30 years for MSc and 40 years for PhD; be Nigerians.
Level of Study: Doctorate, Postgraduate
Type: Scholarship
Value: Fees and maintenance allowance
Frequency: Annual
Study Establishment: University College London
Country of Study: United Kingdom
Application Procedure: Applications for these awards should be made directly to the PTDF, following their established procedures. Applicants selected by PTDF, who are also offered a place at UCL, will then automatically qualify for the scholarship.
Closing Date: See website www.ptdf.gov.ng

For further information contact:

Website: www.ptdf.gov.ng

R B Hounsfield Scholarship in Traffic Engineering

Subjects: Traffic engineering.
Purpose: To financially support MPhil/PhD research in Traffic Engineering in the Department of Civil and Environmental Engineering.
Eligibility: Candidates should hold or expect to obtain a First Class (Honours) Degree or equivalent and should have been offered a place at UCL and have firmly accepted that offer or be intending to do so.
Level of Study: Postgraduate, Doctorate, Research
Type: Scholarship
Value: Not less than UK£120
Frequency: Dependent on funds available
Study Establishment: University College London
Country of Study: United Kingdom
Application Procedure: Enquiries should be directed to the department.
Closing Date: May 15th

Additional Information: If the applicant has not already applied to UCL, please complete a graduate application form and enclose it with the scholarship application.

For further information contact:

Centre for Transport Studies
Tel: (44) 20 7679 2710
Fax: (44) 20 7380 0986
Email: civeng.admissions@ucl.ac.uk
Contact: Professor R L Mackett

Richard Chattaway Scholarship

Subjects: History of modern warfare from 1870.
Purpose: To financially support MPhil/PhD study.
Eligibility: Open to applicants who have applied for a graduate study at UCL. The scholar selected must pursue research for a higher degree at UCL. Receipt of the scholarship must be acknowledged in any publication or research paper which has benefitted from it.
Level of Study: Doctorate, Postgraduate
Type: Scholarship
Value: UK£2,000
Frequency: Annual
Study Establishment: University College London
Country of Study: United Kingdom
No. of awards offered: 1
Application Procedure: Applicants must contact the department. If the applicants have not already applied to UCL they must complete a graduate application form and enclose it with the scholarship application. Applicants must include particulars of the research work they intend to pursue in the event of the scholarship being awarded to them.
Closing Date: May 15th
Funding: Individuals
Contributor: Private donation in honour of Richard Chattaway
No. of awards given last year: 1

For further information contact:

Department of History
Email: n.miller@ucl.ac.uk
Contact: Professor Nicola Miller, Head of the Department

Said Foundation/UCL Joint Scholarship

Subjects: All subjects.
Purpose: To financially support taught Master's study students.
Eligibility: Applicants must be Master's students from Iraq, Jordan, Lebanon, Palestine or Syria.
Level of Study: Postgraduate, Doctorate, Research
Type: Scholarship
Value: Full or partial support for tuition fees and maintenance, depending on financial need and availability of other grants to applicants.
Frequency: Annual
Study Establishment: University College London
Country of Study: United Kingdom
No. of awards offered: Up to 6
Application Procedure: Applicants should refer to the website for further information.
Closing Date: January 31st
No. of awards given last year: 1

For further information contact:

British Council Offices in East Jerusalem, Amnan, Berut, Gaza or Damascus
Tel: 0 20 7679 2005/4167
Website: www.ucl.ac.uk/scholarships

Santander Master's Scholarship

Purpose: To assist the most academically able students from leading universities in each of the Santander network countries to pursue a Master's programme at UCL.
Eligibility: Applicants should refer to the website for full details on eligibility.
Level of Study: Postgraduate
Type: Scholarship
Value: UK£2,000 or 5,000, depending on other funding held by the scholar
Frequency: Annual
No. of awards offered: Up to 10
Application Procedure: Applicants should refer to the website for further information on application procedures, forms and deadlines.
Closing Date: March 1st
No. of awards given last year: 9
No. of applicants last year: 57

Sir Frederick Pollock Scholarship for Students from North America

Subjects: LLM.
Purpose: To financially support prospective LLM students.
Eligibility: Applicants must be overseas students from North America. An applicant must have accepted an offer (either conditional or unconditional) to read for the LLM at UCL to be eligible.
Level of Study: Postgraduate
Type: Scholarship
Value: £2,000
Frequency: Annual
Country of Study: United Kingdom
No. of awards offered: Refer website
Application Procedure: There is no application procedure. All eligible students will be automatically considered.
Additional Information: The scholarships will be based on academic merit. The faculty will only consider those applicants who have firmly accepted their offer of admission to the LLM by May 31st.

For further information contact:

Faculty of Laws
Tel: 44 0 20 7679 1441
Fax: 44 0 20 7209 3470
Email: jane.ha@ucl.ac.uk
Contact: The Graduate Officer

Sir George Jessel Studentship in Mathematics

Subjects: Mathematics.
Purpose: To financially support MPhil/PhD research.
Eligibility: Applicants must be graduates of University College London. All applicants who firmly accept a place for MPhil/PhD research in the Mathematics Department will be considered.
Level of Study: Postgraduate, Research, Doctorate
Type: Studentship
Value: UK£1,800
Frequency: Annual
Study Establishment: University College London
Country of Study: United Kingdom
No. of awards offered: 1
Application Procedure: Applicants must contact the department.
Closing Date: Refer website

For further information contact:

Department of Mathematics
Tel: (44) 20 7679 2839
Fax: (44) 20 7383 5519
Email: h.higgins@ucl.ac.uk
Contact: Ms Helen Higgins

Sir James Lighthill Scholarship

Subjects: Applied mathematics.
Purpose: To financially support MPhil/PhD research.
Eligibility: United Kingdom, European Union and overseas students are eligible to apply. All applicants who firmly accept a place for MPhil/PhD research in the Mathematics Department will be considered.
Level of Study: Doctorate, Postgraduate, Research
Type: Scholarship
Value: £500 per year
Frequency: Annual
Study Establishment: University College London
Country of Study: United Kingdom
No. of awards offered: 1
Application Procedure: Applicants must contact the department.
Closing Date: Refer website

For further information contact:

Department of Mathematics
Tel: 20 7679 2839
Email: h.higgins@ucl.ac.uk
Contact: Ms Helen Higgins

Sir John Salmond Scholarship for Students from Australia and New Zealand

Subjects: Law.
Purpose: To financially support prospective LLM students.
Eligibility: Applicants must be overseas students from Australia and New Zealand. An applicant must have accepted an offer (either conditional or unconditional) to read for the LLM at UCL to be eligible.
Level of Study: Postgraduate
Type: Scholarship
Value: UK£2,000
Frequency: Annual
Study Establishment: University College London
Country of Study: United Kingdom
No. of awards offered: Refer website
Application Procedure: There is no application.
Closing Date: Refer website
Additional Information: The scholarships will be based on academic merit. The faculty will only consider these applicants who have firmly accepted their offer of admission to the LLM by May 31st.

For further information contact:

Faculty of Laws
Tel: 44 0 20 7679 1441
Fax: 44 0 20 7209 3470
Email: jane.ha@ucl.ac.uk
Contact: The Graduate Officer

SPDC Niger Delta Postgraduate Scholarship

Subjects: MSc chemical process engineering or MSc mechanical engineering or MSc civil engineering.
Purpose: To provide opportunities for postgraduate study in the UK for young students and professionals, who demonstrate both academic excellence and the potential to become leading professionals in the oil and associated industries.
Eligibility: The applicant must: (1) have obtained a degree of at least an equivalent standard to a UK Upper Second Class (Honours Degree); (2) be neither a current nor former employee (who have left employment less than 5 years before) of SPDC, the Royal Dutch Shell Group of Companies or Wider Perspectives Limited, or current employee's relatives; (3) not already have had the chance of studying in the UK or another developed country; (4) be aged between 21–30 years; (5) originate from one of the Niger Delta States in Nigeria, namely Rivers, Delta or Bayelsa and currently reside in Nigeria.
Level of Study: Postgraduate
Type: Scholarship
Value: Full tuition fee funding, maintenance allowance, return airfares and arrival allowance
Frequency: Annual
Study Establishment: University College London
Country of Study: United Kingdom
No. of awards offered: Up to 3
Closing Date: See website for details

For further information contact:

Student Funding Office, UCL, Gower Street, London, WC1E6BT, United Kingdom
Email: studentfunding@ucl.ac.uk

Sully Scholarship

Subjects: Psychology.
Purpose: To financially support MPhil/PhD research.
Eligibility: Open to the most outstanding candidate in the second year of their PhD research programme.
Level of Study: Doctorate, Postgraduate, Research
Type: Scholarship
Value: UK£2,200
Frequency: Annual
Study Establishment: University College London
Country of Study: United Kingdom
No. of awards offered: 1
Application Procedure: There is no separate application and the award is given to an outstanding student who is registered in the department's PhD programme and is in the second year of study.
Closing Date: Refer website
No. of awards given last year: 1

For further information contact:

Department of Psychology
Tel: 44 0 20 7679 5332
Fax: 44 0 20 7430 4276
Email: psychology-pg-enquiries@ucl.ac.uk
Contact: Head of the Department

Teaching Assistantships (Economics)

Subjects: Economics.
Purpose: To financially support MPhil/PhD students.
Eligibility: MPhil/PhD students who have successfully completed their 1st year in the department.
Level of Study: Doctorate
Type: Assistantship
Value: UK£11,000 as salary
Frequency: Annual
Study Establishment: University College London
Country of Study: United Kingdom
Application Procedure: Contact the Department of Economics for details.
Closing Date: Refer website

For further information contact:

Graduate Admissions Tutor
Email: d.fauvrelle@ucl.ac.uk

Ted Hollis Scholarship

Subjects: Wetland hydrology and conservation.
Purpose: To financially support a student pursuing research leading to a PhD.
Eligibility: Open to MPhil/PhD student pursuing research leading to a PhD in the field of Wetland Hydrology and Conservation at UCL Department of Geography.
Level of Study: Doctorate
Type: Scholarship
Value: Tuition fees at UK/EU rate and a maintenance allowance equivalent to the NERC
Length of Study: 3 years
Frequency: Every 3 years
Study Establishment: University College London
Country of Study: United Kingdom
No. of awards offered: 1
Application Procedure: Applicants must contact the department at the address given below. If the applicants have not applied to UCL they must complete a graduate application form and enclose it with the scholarship application. Candidates should send particulars of the research work they intend to pursue in the event of the scholarship being awarded to them.
Closing Date: Refer website

For further information contact:

Graduate Admissions, Department of Geography, University College London, Gower Street, London, WC1E 6BT, United Kingdom
Tel: (44) 20 7679 5500
Website: www.geog.ucl.ac.uk/admission

Thomas Witherden Batt Scholarship

Subjects: Life sciences and mathematical and physical sciences.
Purpose: To financially support students to pursue MPhil/PhD research.
Eligibility: Open to United Kingdom nationals only who are pursuing MPhil/PhD research in any department within the Faculty of Life Sciences and Faculty of Mathematical and Physical Sciences.
Level of Study: Doctorate, Postgraduate, Research
Type: Scholarship
Frequency: Annual

Study Establishment: University College London
Country of Study: United Kingdom
Application Procedure: Applicants should contact the department.
Closing Date: See website

For further information contact:

Faculty of Mathematical and Physical Sciences
Tel: (44) 20 7679 3359

UCL Alumni Scholarship

Subjects: Any Master's programme.
Purpose: To enable and encourage those who have gained a UCL undergraduate degree to pursue full-time Master's degree studies at UCL.
Eligibility: Have successfully completed or be currently completing undergraduate degree studies at UCL.
Level of Study: Postgraduate
Type: Scholarship
Value: £10,000
Frequency: Annual
Study Establishment: University College London
Country of Study: United Kingdom
No. of awards offered: 1
Application Procedure: See website for details.
Closing Date: See website for details

UCL CSC Research Student Visits Awards

Purpose: To support visotors Chinese research students admitted to UCL.
Eligibility: Applicants must be a Chinese national pursuing MPhil/PhD research at a recognised top Chinese University and have applied to UCL for admission as a visiting research student.
Level of Study: Postgraduate, Research
Type: Scholarship
Value: Full fees, maintenance allowance, flights and visa costs
Length of Study: 1 year
Frequency: Annual
No. of awards offered: Variable
Application Procedure: Applicants should refer to the website for further information on application procedures, forms and deadlines. The student must contact the UCL Department they are interested in visiting and ask them to apply on their behalf.
Closing Date: March 1st
No. of awards given last year: 3

UCL Global Excellence Scholarship

Purpose: To reward academic excellence among new UCL students.
Eligibility: Applicants should refer to the website for full details on eligibility.
Level of Study: Postgraduate
Type: Scholarship
Value: £5,000
Frequency: Annual
No. of awards offered: 8
Application Procedure: There is no application procedure. Students are nominated by their departments.
Closing Date: Refer website
No. of awards given last year: 7

UCL Graduate Research Scholarship

Subjects: Any subject.
Purpose: To attract high-quality students to undertake research at UCL.
Eligibility: Open to prospective MPhil/PhD candidates with a United Kingdom First Class (Honours) Degree or equivalent and first year MPhil/PhD candidates who have registered on or after the scholarship deadline. Applicants must be admitted to or currently registered at UCL for full-time or part-time MPhil/PhD or engineering research, and should hold or expect to achieve at least a UK 2:1 honours undergraduate degree or equivalent qualification.
Level of Study: Doctorate, Research
Type: Scholarship
Value: UK/EU fees plus annual maintenance allowance
Length of Study: 3 years
Frequency: Annual
Study Establishment: University College London
Country of Study: United Kingdom
No. of awards offered: 15
Application Procedure: Applicants should refer to the website for further information about the application procedures and deadlines.
Closing Date: Refer to website
No. of awards given last year: 15
Additional Information: Overseas fee-payers are strongly advised to apply for an Overseas Research Student (ORS) award, which can be held together with a Graduate School Research Scholarship. Please check website for updated information or details.

For further information contact:

Tel: (44) 20 7579 2005/4167
Website: www.ucl.ac.uk/scholarships

UCL Graduate Research Scholarships for Cross-Disciplinary Training

Subjects: All subjects.
Purpose: To support UCL students in acquiring research skills and knowledge from a different discipline that can be applied in their normal area of research.
Eligibility: Applicants must be admitted to or currently registered at UCL for full-time MPhil/PhD or Engineering research and in receipt of three years funding for their normal MPhil/PhD or Engineering programme.
Level of Study: Doctorate, Postgraduate
Type: Scholarship
Value: Tuition fees at the UK/EU rate plus maintanance allowance of £15,363
Length of Study: 1 year
Frequency: Annual
Study Establishment: University College London
Country of Study: United Kingdom
No. of awards offered: Up to 4
Application Procedure: Applicants should refer to the website for further details about the application procedures and deadlines.
Closing Date: Refer to the website
No. of awards given last year: 4

For further information contact:

Entrance Scholarships Office, Registrar's Division
Tel: (44) 20 7679 2005/4167
Website: www.ucl.ac.uk/scholarships

UCL Hong Kong Alumni Scholarships

Purpose: To provide financial aid for studies at UCL.
Eligibility: 1. Applicants must be holders of Hong Kong Permanent Identity Cards or Chinese citizens who have received full-time education in Hong Kong and/or China for no less than 5 of the 8 years immediately prior to the start of study at UCL.2. Applicants must hold an offer of admission to full-time undergraduate study at UCL – in any area except Medicine – which they have firmly accepted or intend to do so.3. Applicants must be self-financing and liable to pay tuition fees at the rate for overseas students.4. Applicants should be in financial need and unable to fund their planned studies at UCL without financial help.
Level of Study: Undergraduate
Type: Scholarship
Value: Up to UK£10,000 toward tuition fees for a maximum of four years
Frequency: Annual
Study Establishment: United College London
Country of Study: United Kingdom
No. of awards offered: 1
Application Procedure: Completed UG HK application form should be submitted along with required documentation as detailed in the form to UCL Scholarships.
Closing Date: April 1st
No. of awards given last year: 1
No. of applicants last year: 13
Additional Information: Interviews will be held in Hong Kong and applicants invited for interview will be required to present themselves for interview in Hong Kong at applicants' own expense.

For further information contact:

Student Funding Office, The Registry, University College London, Gower Street, London, WC1E 6BT, United Kingdom
Tel: (44) 0 20 7679 0004
Email: studentfunding@ucl.ac.uk

UCL Marshall Scholarships

Purpose: To strengthen the enduring relationship between the British and American people, their governments and their institutions.
Eligibility: 1. Applicants must fulfill the eligibility criteria required for a Marshall Scholarship and have been selected for a Marshall Scholarship. 2. Applicants must have been selected by the MACC Committees to study at UCL. 3. Applicants must have followed the usual admission procedures at UCL and be holding an offer of admission to pursue graduate studies in any subject available at UCL.
Level of Study: Graduate
Type: Scholarship
Value: Full tuition fees and maintenance costs
Length of Study: 2 years
Frequency: Annual
Study Establishment: University College London
Country of Study: United Kingdom
No. of awards offered: 1
Application Procedure: Application is available at the website www.marshallscholarship.org/
Closing Date: Refer website

UCL Overseas Research Scholarships

Purpose: To attract high-quality international students to the UK to undertake research.
Eligibility: Applicants should refer to the website for full details on eligibility.
Level of Study: Postgraduate, Research
Type: Scholarship
Value: Overseas fees minus UK/EU fees; UCL covers UK/EU fees where no further fee funding is held
Frequency: Annual
No. of awards offered: Up to 40
Application Procedure: Applicants should refer to the website for further information on application procedures, forms and deadlines.
Closing Date: Refer website
No. of awards given last year: 30

UCL UPC Progression Scholarships

Purpose: To encourage University Preparatory Certificate for Science and Engineering (UPCSE) and University Preparatory Certificate for Humanities and Social Sciences (UPCH) students to continue to undergraduate study at UCL and recruit new students into UCL's UPCSE and UPCH programmes.
Eligibility: Candidates will be enrolled on UCL's UPCSE or UPCH programme and admitted for undergraduate studies at UCL.
Level of Study: Undergraduate
Type: Scholarship
Value: UK£5,000 per year for the duration of the undergraduate programme. Scholarships will be applied to fees with any remainder paid as a maintenance allowance
Study Establishment: University College London
Country of Study: United Kingdom
No. of awards offered: 2
Application Procedure: All UCL UPCSE/UPCH students holding an offer of admission for undergraduate studies at UCL will be considered.
Closing Date: Refer website
No. of awards given last year: 2

UCL Wellcome Trust Vacation Scholarships

Purpose: To provide promising undergraduates 'hands-on' research experience during the summer vacation and to encourage them to consider a career in research.
Eligibility: Applicants must be undergraduate students registered at a university within the UK or Ireland (including UCL) who are currently enrolled in the middle year(s) of their undergraduate degree in a basic science, veterinary science or dentistry; or are currently enrolled between the end of their second year and the end of their penultimate year of a medical degree; and have secured a vacation research project placement at UCL in an area of biomedicine.
Level of Study: Undergraduate
Type: Scholarship
Value: Each award consists of a weekly stipend of a maximum of for a maximum of 8 weeks over the summer vacations. The weekly stipend rate is for scholarships for placements at UCL is UK£190.
Study Establishment: University College London
Country of Study: United Kingdom
Application Procedure: Completed application forms must be submitted as a hard copy with original signatures on behalf of the student by the project supervisor.
Closing Date: See website
Additional Information: Please check website for updated information or details.

UCL-UWC Undergraduate International Outreach Bursaries

Purpose: To enable graduates of a UWC, who are financially unable to study in the UK, to pursue full-time undergraduate studies at UCL.
Eligibility: 1. Applicants must be a final-year IB diploma student attending a United World College, or in case of candidates coming from Waterford KaMhlaba UWC, have completed their studies in the preceding calendar year. 2. Applicants must have – before or by January – applied for admission through UCAS to a full-time undergraduate degree programme of study at UCL. 3. Applicants must be liable to pay tuition fees at the rate applicable to Overseas students, as assessed by UCL. 4. Applicants must lack the financial means necessary to pursue undergraduate degree studies at UCL.
Level of Study: Undergraduate
Type: Bursary
Value: Each bursary will consist of full tuition fees and a maintenance allowance for the duration of the student's programme of study, as well as international economy air travel to/from the UK at the beginning and end of the bursary-holder's degree programme. The maintenance of allowance will normally rise each year in line with inflation but at the absolute discretion of UCL
Frequency: Annual
Study Establishment: University College London
Country of Study: United Kingdom
No. of awards offered: 2
Application Procedure: Application is by mutation. UWC Guidance Counsellors will invite up to three eligible students from each UWC to apply for the bursary.
Closing Date: March 15th
No. of awards given last year: 2
No. of applicants last year: 18

W M Gorman Graduate Research Scholarship

Subjects: Economics.
Purpose: To financially support students entering the first year of the MPhil/PhD degree in department of Economics.
Eligibility: Open to candidates who have applied for a place for graduate study at UCL.
Level of Study: Doctorate, Postgraduate
Type: Scholarship
Value: UK£15,000
Frequency: Annual
Study Establishment: University College London
Country of Study: United Kingdom
No. of awards offered: Up to 7
Application Procedure: Students must indicate why they wish to be considered for this scholarship on their admission application form (section 26).
No. of awards given last year: 9
Additional Information: The scholarship will be awarded to students who are not already receiving full financial support from other sources for fees and living costs, and will be, awarded on the basis of academic merit.

For further information contact:

Department of Economics
Tel: 20 7679 5861
Fax: 20 7916 2775
Email: d.fauvrelle@ucl.ac.uk
Contact: Ms Daniella Fauvrelle

UNIVERSITY INSTITUTE OF EUROPEAN STUDIES

Via Maria Vittoria 26, I-10123 Turin, Italy
Tel: (39) 011 839 4660
Fax: (39) 011 839 4664
Email: info@iuse.it
Website: www.iuse.it
Contact: Ms Maria Grazia Goiettina, Course Secretariat

The University Institute of European Studies promotes international relations and European integration by organizing academic activities. The Institute has a comprehensive library in international law and economics. Since 1952 the Institute has been a European Documentation Centre (EDC), thus receiving all official publications of European institutions.

Law & Business in Europe Fellowships

Subjects: The increasing impact of European law on business activities, together with the multi-faceted dimension of the single market after the enlargement, requires us to combine business and legal expertise. The programme combines European law, international business and advanced management, to confront with global issues that involve both legal and economic aspects.The Autumn School programme provides some insights about the role law has to play in resolving business issues. The course focuses, inter alia, on: business challenges for Europe after Lisbon, corporate finance and European tax regimes, investment banking, European competition policy, European innovation and firm performance.
Purpose: To allow students to attend the Law & Business in Europe post-graduate programme (Autumn School), jointly organised by the University Institute of European Studies and the Centre for Studies on Federalism.
Eligibility: Open to Italian and foreign graduates in law, business, economics, political science or with equivalent qualifications. Undergraduates in their last year of attendance may also be considered for admission. Applicants must be fluent in English (working language of the Course).
Level of Study: Postgraduate
Type: Fellowship
Value: Part of accommodation expenses and/or registration fees
Length of Study: Autumn School – 3 weeks
Frequency: Annual
Study Establishment: Collegio Carlo Alberto, Moncalieri (Turin)
Country of Study: Italy
No. of awards offered: Varies
Application Procedure: Candidates should fill in the application form, downloadable from the website http://lbeurope.iuse.it. Copies of the official degree certificates with the official lists of academic results, a complete curriculam vitae (including postal and e-mail adresses) and where the applicant's mother tongue is not English a certification of proficience in English must be enclosed within the application form.
Closing Date: September
Funding: Foundation
Contributor: Compagnia di San Paolo, Turin (Italy)
No. of awards given last year: 10
No. of applicants last year: 60
Additional Information: The programme (held in English) is designed for young graduates and professionals all over the world, and it combines academic teaching with a problem solving approach including practical cases presented by experts and a final workshop. The course faculty is composed by prominent experts, professors, senior officials from national and European institutions, lawyers, project consultants and practitioners. Full time attendance is required for the entire duration of the course.

THE UNIVERSITY OF ADELAIDE

National Wine Centre of Australia, Corner of Botanic and Hackney Road, Adelaide, SA, 5005, Australia
Tel: (61) 8 8303 4455
Fax: (61) 8 8303 7444
Email: student.centre@adelaide.edu.au
Website: www.adelaide.edu.au
Contact: Adelaide Graduate Centre

University of Adelaide was established in 1874 and has been amongst Australia's leading universities. Its contribution to the wealth and wellbeing of South Australia and Australia as a whole in all fields of endeavour has been enormous. The university is committed to producing graduates recognised worldwide for their creativity, knowledge and skills, as well as their culture and tolerance.

Adelaide Postgraduate Coursework Scholarships

Subjects: All subjects.
Eligibility: Open only to the citizens of Australia or permanent residents who have achieved Honours 1 or equivalent.
Level of Study: Postgraduate
Type: Scholarship
Value: Covers 50 per cent of the tuition fee costs
Length of Study: 2 years
Frequency: Annual
Study Establishment: University of Adelaide
Country of Study: Australia
No. of awards offered: 6
Application Procedure: Applicants must apply directly to the scholarship provider. Check website for further details.
Closing Date: Varies

For further information contact:

Adelaide Graduate Centre, Adelaide University, Adelaide, South Australia, 5005, Australia
Tel: (61) 08 8303 3044
Fax: (61) 08 8223 3394
Email: adrienne.gorringe@adelaide.edu.au
Website: http://www.adelaide.edu.au/scholarships/pgcoursework/

Adelaide Scholarships International

Subjects: All subjects.
Purpose: To attract high quality overseas postgraduate students to areas of research strength in the University of Adelaide to support its research effort.
Eligibility: Open only to the citizens of Australia or permanent residents who have achieved Honours 1 or equivalent.
Level of Study: Postgraduate
Type: Scholarships
Value: An annual living allowance of approximately $24,653, course tuition fees for two years for a Masters degree by Research and three years for a Doctoral research degree (an extension is possible for doctoral programs only).
Length of Study: 2 years (Masters) or 3 years (PhD)
Frequency: Annual
Study Establishment: University of Adelaide
Country of Study: Australia
No. of awards offered: 10
Application Procedure: Candidates must apply directly to the university. Check the website for further details.
Closing Date: April 12th
Contributor: Adelaide University

For further information contact:

Email: student.centre@adelaide.edu.au
Website: http://www.adelaide.edu.au/graduatecentre/scholarships/postgrad/international/asi.html

Chilean Bicentennial Fund Scholarships

Subjects: Agriculture food and wine, mining, biotechnology and biosciences, information and communication technologies, energy, environmental issues, health, education, aquaculture and veterinary science.
Purpose: Working in partnership with the Government of Chile to provide greater opportunities for Chilean students to pursue post-graduate studies in Australia.
Level of Study: Postgraduate
Type: Scholarship
Value: The scholarship provides living expenses while studying in Australia, full tuition fees, return economy air travel, health insurance for the duration of the program and english language proficiency training before the commencement of the program (when required)
Length of Study: Up to 4 years depending on level of study
Frequency: Annual

Study Establishment: University of Adelaide
Country of Study: Australia
Application Procedure: Information about the application process is available on the university website.
Closing Date: March late
Additional Information: Please see the website for further details http://www.adelaide.edu.au/graduatecentre/scholarships/postgrad/international/chilean.html.

Ferry Scholarship - UniSA

Subjects: Chemistry and physics.
Purpose: To promote study and research into the scientific fields of physics and chemistry.
Eligibility: Open only to the citizens of Australia below the age of 25 who have achieved Honours 1 or equivalent.
Level of Study: Postgraduate
Type: Scholarship
Value: Australian $7,500 per year
Length of Study: 1 year
Frequency: Annual
Study Establishment: Flinders University, The University of Adelaide, University of South Australia
Country of Study: Australia
No. of awards offered: 1
Application Procedure: Applicants must apply directly to the university. Check the website for further details.
Closing Date: March 31st
Funding: Individuals
Contributor: Late Cedric Arnold Seth Ferry

For further information contact:

University of South Australia, Australian Admissions and Scholarships
Tel: 8302 3967
Email: jenni.critcher@unisa.edu.au
Contact: Jenni Critcher

The G O Lawrence Scholarship

Subjects: Operative dentistry, crown and bridgework, endodontics, related dental materials, implantology and those parts of paedodontics which involved the above treatments.
Purpose: To enable students to undertake postgraduate study or research in the fields of operative dentistry, crown and bridgework, endodontics, related dental materials, implantology and those parts of paedodontics which involved the above treatments.
Eligibility: Open to candidates who are postgraduate and undertake study or research in fields of operative dentistry, crown and bridgework, endodontics, related dental materials, implantology and those parts of paedodontics which involved the above treatments. However, the relevance of the applicant's research topic is considered when deeming eligibility.
Level of Study: Postgraduate
Type: Scholarship
Value: $22,500 (tax free) per year
Application Procedure: Check website for further details.
Closing Date: October 31st

Grape and Wine Research and Development Corporation Honours Scholarships

Subjects: Viticulture.
Purpose: To help attract an increasing number of postgraduate students to the fields of wine and viticultural research.
Eligibility: Scholarships will be awarded on the basis of academic excellence, the quality of an applicant's curriculum vitae and the likelihood of the candidate's future involvement in the wine industry.
Level of Study: Postgraduate
Type: Scholarships
Value: Australian $6,000
Frequency: Annual
Country of Study: Australia
No. of awards offered: 7
Application Procedure: Application forms can be downloaded from the website.
Closing Date: December
Contributor: Australian Wine Industry and the Australian Government
Additional Information: Please see the website for further details http://www.adelaide.edu.au/scholarships/undergrad/gwrdc.html.

The Herbert Gill-Williams Scholarship

Subjects: Dentistry.
Purpose: To support a full-time candidate of outstanding merit to undertake further research studies in dentistry at Adelaide University.
Eligibility: Open to the full-time candidates who have Honours Degree of Bachelor of Science in dentistry and have been accepted into a programme of study leading to a research degree in the Dental School. Check website for further details.
Level of Study: Postgraduate
Type: Scholarship
Value: $23,728 per year
Length of Study: Up to 3 years
Frequency: Annual
Application Procedure: Check website for further details.
Closing Date: October 31st

The Oliver Rutherford-Turner Scholarship

Subjects: Dentistry.
Purpose: Assist students to undertake a higher degree in a field of dentistry.
Eligibility: Open only to the permanent residents of Australia and New Zealand with a 4-year undergraduate degree or equivalent, (First Class Honours degree or equivalent for Australians) with minimum English language proficiency. Check website for further details.
Level of Study: Postgraduate
Type: Scholarship
Value: Australian $22,860
Length of Study: 3 years
Application Procedure: Check website for further details.
Closing Date: October 31st
Funding: Private
Contributor: Bequest from the late Oliver Rutherford Turner

Scholarships in Plant Cell Physiology

Subjects: Plant cell physiology.
Purpose: To improve the nutritional qualities of crop plants allowing the fortification of animal and human diets without adversely affecting crop plant.
Eligibility: Open only to the citizens of Australia who have achieved Honours 1 or equivalent, or Honours 2a or equivalent.
Level of Study: Postgraduate
Type: Scholarship
Value: Australian $27,500 per year
Length of Study: 3 years
Frequency: Annual
Study Establishment: The University of Adelaide
Country of Study: Australia
No. of awards offered: 2
Application Procedure: Applicants must apply directly to the university. Check website for further details.
Closing Date: August 31st
Contributor: Adelaide University

For further information contact:

Plant Research Centre, University of Adelaide, PMB 1, Glen Osmond, South Australia, 5064, Australia
Tel: (61) 08 8303 8145
Email: matthew.gillam@adelaide.edu.au
Website: www.adelaide.edu.au/GSSO
Contact: Dr Matthew Gillam

UNIVERSITY OF ALBERTA

116 St. and 85 Ave., Edmonton, AB, T6G 2R3., Canada
Tel: (1) 780 492 3499
Fax: (1) 780 492 0692
Email: grad.awards@ualberta.ca
Website: www.gradstudies.ualberta.ca
Contact: Dana Dragon-Smith, Graduate Student Services Advisor

Opened in 1908, the University of Alberta has a long tradition of scholarly achievements and commitment to excellence in teaching, research and service to the community. It is one of Canada's five largest research-intensive universities, with an annual research income from external sources of more than Canadian $300 million. It participates in 18 of 21 of the Federal Networks of Centres of Excellence, which link industries, universities and governments in applied research and development.

Grant Notley Memorial Postdoctoral Fellowship

Subjects: The politics, history, economy or society of Western Canada or related fields.
Purpose: To encourage scholars of superior research ability who graduated within the last 3 years.
Eligibility: Open to citizens of any country, must be within 3 years post-PhD from the time of application submission, must have completed a doctoral degree or will do so in the immediate future, must have obtained doctorate degree from a university other than University of Alberta, must not currently hold a postdoc or be employed at the University of Alberta, must not hold or have held any other fellowships, must not hold faculty position, fellowship holders should be likely to contribute to the advancement of learning. Fellowships are only tenable at the University of Alberta. Applicants must visit the website for additional information.
Level of Study: Postdoctorate
Type: Fellowship
Value: Canadian $46,000 per year and a non-renewable research grant of Canadian $4,000
Length of Study: 2 years
Frequency: Annual
Study Establishment: The University of Alberta
Country of Study: Canada
No. of awards offered: 1
Application Procedure: Applicants must visit the website for application information.
Closing Date: December 15th (check with website)
Funding: Private
No. of awards given last year: 1
No. of applicants last year: 7

Izaak Walton Killam Memorial Scholarships

Subjects: All subjects.
Eligibility: Open to candidates of any nationality who are registered in, or are admissible to, a doctoral programme at the University. Scholars must have completed at least 1 year of graduate work prior to beginning the scholarship. Applicants must be nominated by the department in which they plan to pursue their doctoral studies.
Level of Study: Doctorate
Type: Scholarship
Value: Canadian $35,000; international students' differential fee is also paid
Length of Study: 2 years from May 1st or September 1st, subject to review after the 1st year
Frequency: Annual
Study Establishment: The University of Alberta
Country of Study: Canada
No. of awards offered: Approx. 12
Application Procedure: Applicants must visit the website for application information.
Closing Date: March 2nd for the submission of nominations from departments. Please check with the department for their internal deadline
Funding: Private
No. of awards given last year: 12
No. of applicants last year: 115

Queen Elizabeth II Scholarships-Doctoral

Subjects: All subjects.
Eligibility: Open to candidates registered full-time in a doctoral degree program during the tenure of the award. Applicants must have a GPA of 3.3 or greater and must be a Canadian citizens or permanent residents at the time of nomination.
Level of Study: Doctorate
Type: Scholarship
Value: Canadian $15,000 for commencement in September and Canadian $7,500 for commencement in January
Length of Study: 8 months from September 1st or January 1st (4 months)
Frequency: Annual
Study Establishment: The University of Alberta
Country of Study: Canada
No. of awards offered: Approx. 227
Application Procedure: Visit the website for application information.
Closing Date: July 15th
Funding: Government
No. of awards given last year: 217
No. of applicants last year: 732
Additional Information: Recipients must carry out a full-time research programme during the Summer months. Please see the website for further details http://www.grad.uwo.ca/current_students/student_finances/QEIIGSST_terms.html.

Queen Elizabeth II Scholarships-Master's

Subjects: All subjects.
Eligibility: Open to candidates registered full-time in a master's degree program during the tenure of the award. Applicants must have a GPA of 3.3 or greater and must be a Canadian citizens or permanent residents at the time of nomination.
Level of Study: Graduate
Type: Scholarship
Value: Canadian $10,800 for commencement in September, Canadian $5,400 for commencement in January
Length of Study: 8 months from September 1st or January 1st (4 months)
Frequency: Annual
Study Establishment: The University of Alberta
Country of Study: Canada
No. of awards offered: Approx. 223
Application Procedure: Applicants must visit the website for application information.
Closing Date: June 1st
Funding: Government
No. of awards given last year: 217
Additional Information: Recipients must carry out a full-time research programmes during the summer months. Please see the website for further details http://www.queensu.ca/sgs/forstudents/awardholders/2012-13_QEII-GSST_Guidelines.pdf.

Walter H Johns Graduate Fellowship

Subjects: All subjects.
Eligibility: Open to students registered full-time in a Graduate Degree programme who are receiving an eligible scholarship of less than $30,000 from NSERC, SSHRC, or CIHR.
Level of Study: Postgraduate, Graduate
Type: Fellowship
Value: Canadian $5,100
Application Procedure: Check website for further details.
Closing Date: September 5th and January 5th

For further information contact:

Faculty of Graduate Studies and Research, Killam Centre for Advanced Studies, 2-29 Triffo Hall
Website: http://www.gradstudies.ualberta.ca/awardsfunding/scholarships/walterhjohns/

THE UNIVERSITY OF AUCKLAND

Private Bag 92019, Auckland Mail Centre, Auckland, 1142, New Zealand
Tel: (64) 9 373 7999
Email: rfatialofa.patolo@auckland.ac.nz
Website: www.auckland.ac.nz

The University of Auckland is New Zealand's pre-eminent research-led University. Established in 1883, it has grown into an international centre of learning and academic excellence and now is the largest university in New Zealand. Its mission is to be an internationally recognised, research-led university, known for the excellence in teaching, research, and service to its local, national and international communities. It aims to be a vibrant and intellectually challenging place of learning and nurturing a community of scholars who share a passion for discovery, the advance of knowledge and human progress.

Anne Bellam Scholarship

Subjects: Music.
Purpose: To assist students to further their musical education overseas.
Eligibility: The candidate must be under 30 years of age and a citizen of New Zealand, must have compeleted or will complete in the year of application any degree or diploma in performance or any postgraduate music degree at the University of Auckland.
Level of Study: Postgraduate
Type: Scholarship
Value: Up to $30,000 for study overseas and up to $10,000 for study at the University of Auckland
Length of Study: 1 year
Frequency: Annual
Study Establishment: University of Auckland
Country of Study: New Zealand
Application Procedure: The candidate must supply, one week in advance of examination, an outline of his proposed study plans and itinerary.
Closing Date: September 30th
Funding: Government

For further information contact:

Faculty of Creative Arts and Industries: School of Music, Auckland, New Zealand
Email: scholarships@auckland.ac.nz
Website: www.auckland.ac.nz/scholarships

Arthington Davy Scholarship

Subjects: Any subject.
Purpose: To study and research in areas which will significantly contribute to the development of Tonga.
Eligibility: The candidate must be a Tongan citizen, born to Tongan parents and possess a first University degree.
Level of Study: Research
Type: Scholarship
Value: The Arthington Davy Scholarship may cover the cost of part of the cost of a postgraduate study or research programme
Study Establishment: Trinity College
Country of Study: United Kingdom
Application Procedure: The candidate must submit a complete curriculum vitae and academic record, proof of Tongan origin, details of the intended postgraduate study preferably with a letter of conditional acceptance from the University concerned, full details of tuition fees and living expenses and of finances available from the student's own resources or elsewhere and the names of two academic referees.
Closing Date: November 30th (for course commencing in March or April), May 31st (for course commencing in September or October)
Funding: Commercial, foundation, individuals, private

For further information contact:

Trinity College, Cambridge, CB2 1TQ, United Kingdom
Email: hf202@hermes.cam.ac.uk
Website: http://www.auckland.ac.nz/uoa/is-arthington-davy-scholarship#s2c5
Contact: Tutor for Advanced Studies

Asian Development Bank Japan Scholarship

Subjects: Envrionmental science, development studies, international business and engineering.
Eligibility: The candidate must possess a minimum English language requirement for entry into postgraduate study at the University of Auckland. IELTS with an overall score of 6.5 and no band less than 6.0 or a TOEFL paper-based 575 with a TWE of 4.5 or computer-based 233 with a TWE of 4.5.
Level of Study: Postgraduate
Type: Scholarship
Value: Tuition fee at the University of Auckland, Airfare from his or her home country to Auckland, New Zealand, Basic cost of living in Auckland, Health and medical insurance in New Zealand, and Airfare from Auckland, New Zealand, to the scholar's home country at the conclusion of his or her course of study.
Study Establishment: University of Auckland
Country of Study: New Zealand
Application Procedure: Applications can be filled online.
Closing Date: July 20th (check with website)
Funding: Government, private
Contributor: Asian Development Bank and Government of Japan

For further information contact:

The University of Auckland, Auckland International, Private Bag 92018, Auckland, New Zealand
Email: information@adbj.org
Website: http://www.adb.org/site/careers/japan-scholarship-program/main

International College of Auckland PhD Scholarship in Plant Sciences

Subjects: Plant Science.
Purpose: To assist eminent Chinese scholars from nominated areas of China to study plant sciences at the University of Auckland and to promote links between China and New Zealand in the field of plant sciences.
Eligibility: The candidate must possess a PhD in the field of plant science and who has paid the fees, or arranged to pay the fees, for full–time enrolment in the School of Biological Sciences.
Level of Study: Postdoctorate
Type: Scholarship
Value: $20,000 per year. The scholarship's emolument will be paid as a tuition/compulsory fees credit and the balance as a fortnightly stipend
Length of Study: 3 years
Frequency: Annual
Study Establishment: University of Auckland
Country of Study: New Zealand
No. of awards offered: 1
Application Procedure: The candidate must submit the completed application form along with the curriculum vitae and atleast two academic reference letters.
Closing Date: October 1st
Funding: Government
Contributor: The International College of Auckland

For further information contact:

Scholarships Office, Room 012, Clock tower, Ground floor, 22 Princes Street, Auckland Central, Auckland, New Zealand
Email: scholarships@auckland.ac.nz

NZ Development Scholarship – Public Category

Subjects: Any subject.
Eligibility: The candidate must be a possess a degree in any discipline.
Level of Study: Postgraduate
Type: Scholarship
Value: Covers tuition fee, airfare, cost of living, health and medical insurance, education of dependant children
Study Establishment: University of Auckland
Country of Study: New Zealand
Application Procedure: The candidate must submit his application form through his home nominating authority. The New Zealand Ministry of Foreign Affairs and Trade seeks placements and decides which institution, polytechnic or university the candidate may study at in New Zealand.
Closing Date: May or June each year
Funding: Government
Contributor: Ministry of Foreign Affairs and Trade, Government of New Zealand

NZ International Doctoral Research (NZIDRS) Scholarship

Subjects: Any subject.
Purpose: To provide financial support for postgraduate students from designated countries seeking doctoral degrees by research in New Zealand universities.

Eligibility: The candidate must hold an 'A' average or equivalent in their studies, meet the requirements for entry into a research-based doctoral degree programme at a New Zealand university.
Level of Study: Research, Doctorate
Type: Research scholarship
Value: Living allowance (NZ$25,000 per year), a travel allowance (NZ $2,000), a health insurance allowance (NZ$500), and a book and thesis allowance (NZ$800)
Length of Study: 3 years
Study Establishment: University of Auckland
Country of Study: New Zealand
No. of awards offered: 38
Application Procedure: The candidate must complete the application form in English and attach supporting documents as stipulated in the application form.
Closing Date: July 16th
Funding: Government
Contributor: Government of New Zealand

For further information contact:

The Education New Zealand Trust, PO Box 10-500, Wellington, New Zealand
Tel: (64) 4 4720788
Fax: (64) 4 4712828
Email: scholarships@educationnz.org.nz
Website: www.nzeducated.com/scholarships/documents/NZIDR-SApplicationform.doc
Contact: Scholarships Manager

NZAID Development Scholarship (NZDS) – Open Category

Subjects: Any subject.
Eligibility: The candidate must possess minimum English language requirements for entry into the University of Auckland postgraduate study. IELTS (International English Language Testing System Certificate) with an overall score of 6.5 and no band less than 6.0 or a TOEFL (Test of English as a Foreign Language) paper based 575 with a TWE of 4.5 or computer based 233.
Level of Study: Postgraduate
Type: Scholarship
Value: Tuition, enrollment/orientation fees, return economy fare travel, medical insurance and provision for students to meet course and basic living costs
Study Establishment: University of Auckland
Country of Study: New Zealand
Application Procedure: Application form can be downloaded from the website.
Closing Date: June 1st
Funding: Government
Contributor: New Zealand Agency for International Development and the Ministry of Foreign Affairs

For further information contact:

Auckland International Student Information Centre, Auckland, New Zealand
Tel: (64) 9 3737599 ext 87556
Fax: (64) 9 373 7405
Email: rfatialofa.patolo@auckland.ac.nz

Property Institute of New Zealand Postgraduate Scholarship

Subjects: Real property.
Purpose: To promote postgraduate study in the field of real property.
Eligibility: The candidate must possess a Master's degree or full-time PhD candidate and has the paid the fees, or arranged to pay the fees, for study in the Department of Property at Lincoln University, Massey University or The University of Auckland.
Level of Study: Doctorate
Type: Scholarship
Value: $1,500
Length of Study: 1 year
Frequency: Annual
Study Establishment: University of Auckland
Country of Study: New Zealand
No. of awards offered: 2
Closing Date: March 31st
Funding: Government
Contributor: Property Institute of New Zealand

For further information contact:

The Scholarships Office, The University of Auckland, Private Bag 92019, Auckland, New Zealand
Tel: (64) (09) 373 7599 ext 87494
Fax: (64) (09) 308 2309
Email: scholarships@auckland.ac.nz

Reardon Postgraduate Scholarship in Music

Subjects: Music.
Purpose: To honour and in memory of Daniel Patrick Reardon and Kathleen Mary Reardon.
Eligibility: The candidate must possess a degree or diploma with a specialisation in Performance in the year of the award.
Level of Study: Postgraduate
Type: Scholarship
Value: $4,500
Length of Study: 1 year
Frequency: Annual
Study Establishment: University of Auckland
Country of Study: New Zealand
No. of awards offered: 1
Closing Date: September 30th
Funding: Government
Contributor: Reardon Memorial Music Trust

For further information contact:

Faculty of Creative Arts and Industries: School of Music, University of Auckland, Auckland, New Zealand
Email: scholarships@auckland.ac.nz
Website: www.auckland.ac.nz/scholarships

University of Auckland Commonwealth Scholarship

Subjects: Any subject.
Purpose: The scholarships are available to students to be enrolled at the University of Auckland for the Degree of Doctor of Philosophy; another approved Doctorate or a Master's degree.
Eligibility: The candidate must be a citizen of Commonwealth of Nations, including Australian, British and Canadian citizens. The candidate must be tenable for a maximum of 36 months for a PhD candidate or 21 months for a Master's candidate.
Level of Study: Doctorate, Postgraduate
Type: Scholarship
Value: New Zealand $15,000 per year plus fees and other allowances
Study Establishment: University of Auckland
Country of Study: New Zealand
Application Procedure: The candidate must send the application for UA Commonwealth Scholarships must be made through the appropriate organization in the scholar's home country on the Commonwealth Scholarship application form.
Closing Date: Varies
Funding: Government

For further information contact:

Scholarships Office, Student Administration, The University of Auckland, Private Bag 92019, Auckland Mail Centre, Auckland, New Zealand
Tel: (64) 9 373 7599 ext 87494
Fax: (64) 9 308 2309
Email: scholarships @auckland.ac.nz
Website: http://www.auckland.ac.nz/uoa/university-of-auckland-commonwealth-scholarship

The University of Auckland Doctoral Scholarship

Subjects: Any subject.
Purpose: To assist and encourage students to pursue doctoral studies at The University of Auckland
Eligibility: The candidate must be a citizen or a permanent resident of New Zealand.
Level of Study: Doctorate

Type: Scholarship
Value: Up to $25,000 plus compulsory fees
Length of Study: 3 years
Frequency: Annual
Study Establishment: University of Auckland
Country of Study: New Zealand
No. of awards offered: Varies
Application Procedure: The candidate must fill the application form and send it to the Scholarships office. The selection will be made on the basis of merit.
Closing Date: Variable (refer to relevant faculty)
Funding: Government

For further information contact:

Scholarships Office, University of Auckland, Auckland, New Zealand
Email: scholarships@auckland.ac.nz
Website: www.auckland.ac.nz/scholarships

University of Auckland Fulbright Scholarship

Subjects: Any subject.
Purpose: To encourage and facilitate study for approved postgraduate degrees at the University of Auckland by candidates already selected to hold Fulbright Awards.
Eligibility: The candidate must be a citizen of the United States of America and intending to take up Fulbright Awards to study in New Zealand and should enrol for a full-time at the University of Auckland for an approved Master's or Doctoral degree.
Level of Study: Doctorate, Postgraduate
Type: Scholarship
Value: $15,000 per year plus research/tuition fees
Frequency: Annual
Study Establishment: University of Auckland
Country of Study: New Zealand
No. of awards offered: 3
Application Procedure: The candidate must send the completed application form to the Scholarships office.
Closing Date: No closing date
Funding: Government

For further information contact:

Scholarships Office, Student Administration, The University of Auckland, Private Bag 92019, Auckland Mail Centre, Auckland, 1142, New Zealand
Tel: (64) 9 373 7599 ext 87494
Fax: (64) 9 308 2309
Email: internationalscholarships@auckland.ac.nz
Website: http://www.auckland.ac.nz/uoa/university-of-auckland-fulbright-scholarship

The University of Auckland International Doctoral Fees Bursary

Subjects: All subject.
Purpose: To assist international students from all countries who wish to pursue doctoral studies.
Eligibility: Permanent citizens and residents of Australia and New Zealand are not eligible for the scholarship.
Level of Study: Doctorate
Type: Bursary
Value: New Zealand $ 25,000
Frequency: Annual
Study Establishment: University of Auckland
Country of Study: New Zealand
No. of awards offered: Up to 10
Application Procedure: The application form can be obtained from the Scholarships Office, University of Auckland.
Closing Date: August 1st
Funding: Government

For further information contact:

Scholarhips Office, University of Auckland, Private Bag 92019, Auckland Mail Centre, Auckland, 1142, New Zealand
Email: c.tuu@auckland.ac.nz
Website: www.auckland.ac.nz/uoa/for/currentstudents/money/

University of Auckland International Doctoral Scholarship

Subjects: Any subject.
Purpose: To assist international students from all countries who wish to pursue doctoral studies.
Eligibility: The scholarship is available to international students from all countries who wish to pursue Doctoral studies on a full-time basis. Permanent citizens and residents of Australia and New Zealand are not eligible for the scholarship.
Level of Study: Doctorate
Type: Scholarship
Value: New Zealand $25,000, in the form of a fortnightly stipend
Frequency: Annual
Study Establishment: University of Auckland
Country of Study: New Zealand
No. of awards offered: 4
Application Procedure: The application form can be obtained from the Scholarships Office, University of Auckland.
Closing Date: Variable
Funding: Government

For further information contact:

Scholarships Office, University of Auckland, Private Bag 92019, Auckland Mail Centre, Auckland, 1142, New Zealand
Email: c.tuu@auckland.ac.nz
Website: www.auckland.ac.nz/uoa/for/currentstudents/money/

University of Auckland Maori and Pacific Graduate Scholarships (Doctoral Study)

Subjects: Any subject.
Purpose: To assist Maori and Pacific students enrolled full-time at The University of Auckland for the Degree of Doctor of Philosophy or the research component of another approved Doctorate.
Eligibility: The candidate must be a Maori or Pacific student who are citizens or permanent residents of New Zealand.
Level of Study: Doctorate
Type: Scholarships
Value: Up to $25,000 plus compulsory fees
Length of Study: Up to 3 years
Frequency: Annual
Study Establishment: University of Auckland
Country of Study: New Zealand
No. of awards offered: Varies
Application Procedure: The candidate must fill up the application form and send it to the Scholarships office. Selection will be made on the basis of merit.
Closing Date: November 1st
Funding: Government

For further information contact:

Scholarships Office, University of Auckland, Auckland, New Zealand
Email: scholarships@auckland.ac.nz
Website: www.auckland.ac.nz/scholarships

University of Auckland Maori and Pacific Graduate Scholarships (Masters/Honours/PGDIP)

Subjects: Any subject.
Purpose: To assist and encourage Maori and Pacific students to pursue Masters, Honours and PGDip courses at The University of Auckland.
Eligibility: The candidate must be a Maori or Pacific student who are citizens or permanent residents of New Zealand.
Level of Study: Graduate, Postgraduate
Type: Scholarship
Value: Up to $10,000 plus compulsory fees
Length of Study: 1 year
Frequency: Annual
Study Establishment: University of Auckland
Country of Study: New Zealand
No. of awards offered: Varies
Application Procedure: The candidate must fill up the application form and send it to the Scholarships office. The Selection Committee will assess the application form.
Closing Date: November 1st

Funding: Government

For further information contact:

Scholarships Office, University of Auckland, Auckland, New Zealand
Email: scholarships@auckland.ac.nz
Website: www.auckland.ac.nz/scholarships

University of Auckland Masters/Honours/PGDIP Scholarships

Subjects: Any subject.
Eligibility: The candidate must be a citizen or a permanent resident of New Zealand. In case of a Master's degree, the candidate must be tenable until the date for completion of the requirements for the degree as specified in the General Regulations – Masters degrees.
Level of Study: Graduate, Postgraduate
Type: Scholarship
Value: Up to $10,000 per year plus compulsory fees
Length of Study: 1 year
Frequency: Annual
Study Establishment: University of Auckland
Country of Study: New Zealand
No. of awards offered: Varies
Application Procedure: The candidate must fill the application form and send it to the Scholarships office. Selection will be made on the basis of merit.
Closing Date: November 1st
Funding: Government

For further information contact:

Scholarships Office, University of Auckland, Auckland, New Zealand
Email: scholarships@auckland.ac.nz
Website: www.auckland.ac.nz/scholarships

University of Auckland Senior Health Research Scholarships

Subjects: Health.
Purpose: To attract health professionals to return to the University to study full-time for a PhD in a health-related field.
Eligibility: The candidate must be a citizen or a permanent resident of New Zealand and who have worked for 3 years as a health professional.
Level of Study: Doctorate
Value: $40,000 plus compulsory fees
Length of Study: 3 years
Frequency: Annual
Study Establishment: University of Auckland
Country of Study: New Zealand
No. of awards offered: 3
Application Procedure: The candidate must submit an application form and send it to the Scholarships office. Selection shall be made on the basis of merit.
Closing Date: November 1st
Funding: Government

For further information contact:

Scholarships Office, University of Auckland, Auckland, New Zealand
Email: scholarships@auckland.ac.nz
Website: www.auckland.ac.nz/scholarships

UNIVERSITY OF BALLARAT

Research Services, Mt Helen, VIC, 3350, Australia
Tel: (61) 3 5327 9508
Fax: (61) 3 5327 9602
Email: hdresearch@ballarat.edu.au
Website: www.ballarat.edu.au/ard/ubresearch
Contact: Sue Read

The University of Ballarat is Australia's only regional, multi-sector University acknowledged for its excellence in education, training, and research, committed to providing high quality services to students, the community, and the industry. It provides educational and training programs from apprenticeships, certificates and diplomas to post-graduate qualifications, masters, and doctorates by research. International students at the University come from over 25 different countries to participate in a diverse range of TAFE and higher education programmes. The University is proud of its track record in business innovation and entrepreneurship, research, consulting, and educational programs, and promoting new technology in products and services through scientific and industrial research.

Australian Postgraduate Award

Subjects: Behavioural and cognitive sciences, business and management, education, engineering and technology, human movement and sports science, information technology, computing and communication sciences, mathematical sciences, nursing, science, social sciences, humanities and arts.
Purpose: To support postgraduate students undertaking research in either a Doctorate or Masters by Research program.
Eligibility: Open to candidates who have a First Class (Honours) Degree or equivalent. The APA is open to candidates who have received a Masters by Research (for Doctorate applicants) and/or a Honours degree (First Class / H1A) or equivalent.
Level of Study: Doctorate, Postgraduate, Research, Masters by research
Type: Scholarship
Value: The annual value of the award is set by the Australian Government each year. $24,653 per year tax free in 2012.
Frequency: Annual
Study Establishment: The University of Ballarat
Country of Study: Australia
No. of awards offered: 8
Application Procedure: Application forms and further information about the Scholarships process can be found at:www.ballarat.edu.au/ard/ubresearch/hdrs/scholarships/index.shtml.
Closing Date: October 31st
Funding: Government
No. of awards given last year: 8
No. of applicants last year: 130
Additional Information: It is required that the successful applicant commence studies in the year the scholarship was awarded for. Studies should commence no earlier than February 1st, and no later than August 31st.

For further information contact:

PO Box 663, Ballarat, Victoria, 3350, Australia
Tel: (61) 03 5327 9508
Fax: (61) 03 5327 9602
Email: HDResearch@ballarat.edu.au
Website: www.ballarat.edu.au/ard/ubresearch/hdrs/scholarships/index.shtml
Contact: Sue Read

International Postgraduate Research Scholarship

Subjects: All subjects.
Purpose: To support overseas postgraduate students undertaking a Higher Degree by Research by providing course fees, overseas student health care (OSHC) and a stipend.
Eligibility: Open to citizens of all countries except Australia and New Zealand, to candidates with First Class (Honours) Degree or equivalent.
Level of Study: Research, Doctorate
Type: Scholarship
Value: The annual value of the award is set by the Australian Government. IPRS recipients are also provided with an APA stipend. In 2012, the APA stipend was $24,653 and tax free.
Length of Study: 2 years (Masters) and 3 years (PhD)
Frequency: Annual
Study Establishment: The University of Ballarat
Country of Study: Australia
No. of awards offered: 1
Application Procedure: Check website for further details. Application are open from January 1st.
Closing Date: October 31st
Funding: Government
No. of awards given last year: 1
No. of applicants last year: 50
Additional Information: The study should start no earlier than February 1st.

For further information contact:

PO Box 663, Ballarat, Victoria, 3350, Australia
Tel: (61) 03 5327 9508
Fax: (61) 03 5327 9602
Email: HDResearch@ballarat.edu.au
Website: www.ballarat.edu.au/ard/ubresearch/graduate_studies/index.shtml
Contact: Sarah McArthur

Special Overseas Student Scholarship (SOSS)

Subjects: All subjects.
Eligibility: Open to citizens of all countries except Australia, who have achieved First Class (Honours) or equivalent.
Level of Study: Doctorate, Research
Type: Scholarship
Value: The value of this award equates to the published course fee rates for the year in which study commences
Length of Study: 3 years
Frequency: Annual
Study Establishment: The University of Ballarat
Country of Study: Australia
No. of awards offered: 1
Application Procedure: Check website for further details. Applications are open from January 1st.
Closing Date: October 31st
Funding: Commercial, government
Contributor: Universy of Ballarat
Additional Information: The study should start no earlier than February 1st.

For further information contact:

PO Box 663, Ballarat, Victoria, 3350
Tel: 03 5327 9508
Fax: 03 5327 9602
Email: s.mcarthur@ballarat.edu.au
Website: www.ballarat.edu.au/ard/research/graduate_studies/index.shtml
Contact: Dr Sue Read

University of Ballarat Postgraduate Research Scholarship

Subjects: Behavioural and cognitive sciences, business and management, education, engineering and technology, human movement and sports science, information, computing and communication sciences, mathematical sciences, nursing, science, social sciences, humanities and arts.
Purpose: To assist students in postgraduate research.
Eligibility: Open to those who have achieved First Class (Honours) or equivalent. Awards are restricted to Australian citizens or those with residence status and current University of Ballarat IPRS Awardees.
Level of Study: Postgraduate, Doctorate, Research
Type: Scholarship
Value: The UBPRS is awarded at the same rate, and under the same conditions as an Australian Postgraduate Award (APA). In 2012, the APA rate is $24,653 and tax free per year
Frequency: Annual
Study Establishment: University of Ballarat
Country of Study: Australia
No. of awards offered: 6
Application Procedure: Check website for further details. Applications are from January 3rd.
Closing Date: October 31st
Funding: Government
No. of awards given last year: 6
No. of applicants last year: 97
Additional Information: This scholarship is paid fortnightly. The study should start no earlier than February 1st.

For further information contact:

PO Box 663, Ballarat, Victoria, 3350, Australia
Tel: (61) 03 5327 9508
Fax: (61) 03 5327 9602
Email: s.mcarthur@ballarat.edu.au
Website: www.ballarat.edu.au/ard/research/graduate_studies/index.shtml
Contact: Sue Read

UNIVERSITY OF BERGEN

Post Box 7800, N-5020 Bergen, Norway
Tel: (47) 55 580000
Fax: (47) 55 589643
Email: post@fa.uib.no
Website: www.uib.no

The university of Bergen is a young, modern university, which is Norway's International university. The academic profile of the university has two major focuses marine research and co-operation with developing countries.

The Holberg International Memorial Prize/The Holberg Prize

Subjects: Arts and humanities, social science, law, and theology.
Purpose: To honor scholars who have made outstanding, internationally recognized contributions to research interdisciplinary work.
Eligibility: Open to outstanding researchers from the relevant academic fields.
Level of Study: Postgraduate
Type: Award
Value: Norwegian Kroner 4.5 million (Approx. US$750,000)
Frequency: Annual
Application Procedure: See the website: www.holbergprisen.no/HP_prisen/en_hp_utlysning.html for details.
Closing Date: September 15th
Funding: Foundation

UNIVERSITY OF BIRMINGHAM

Student funding office, Edgbaston, Birmingham, West Midlands, B15 2TT, England
Tel: (44) 121 414 3142
Fax: (44) 121 414 6637
Email: j.e.bryan@bham.ac.uk
Website: www.as.bham.ac.uk/funding
Contact: Joanne Bryan, Assistant Director of Student Financial Support

The University of Birmingham is a leading research institution, offering a wide range of programmes, high teaching and research standards, and excellent facilities for academic work.

A E Hills Scholarship

Subjects: Any subject.
Purpose: To financially support students pursuing a higher degree at the University of Birmingham.
Eligibility: Open to all postgraduates studying at the University of Birmingham.
Level of Study: Postgraduate
Type: Scholarship
Value: The home rate of tuition fee and a maintenance stipend set at research council rates.
Length of Study: 1 year
Study Establishment: The University of Birmingham
Country of Study: United Kingdom
No. of awards offered: 1
Application Procedure: Schools are asked to nominate their top 2 candidates. Awards are based on academic performance.
Closing Date: End of May
Funding: Private
No. of awards given last year: 1
No. of applicants last year: 18

Adrian Brown Scholarship

Subjects: This prize is awarded to high-achieving students in the School of Biosciences and School of Chemical Engineering.
Purpose: To support students reading for a research degree degree at the University of Birmingham.
Eligibility: Open to all students reading a research degree in Chemical Engineering and Biosciences related to brewing sciences and technology.
Level of Study: Doctorate
Type: Scholarship

Value: The home rate of tuition fee and a maintanance stipend set at research council rates
Length of Study: Tenable for period of study subject to satisfactory progress
Study Establishment: The University of Birmingham
Country of Study: United Kingdom
No. of awards offered: 1
Application Procedure: Apply to the Chair of Biotechnology Management Committee. Contact the Student Funding Office at the university for details. Recommendation from the Head of School is required.
Funding: Private
No. of awards given last year: 1
No. of applicants last year: 1
Additional Information: Award is rotated between School of Biosciences and School of Chemical Engineering every 3 years.

AHRC Studentships

Subjects: Arts and humanities, selected areas within life and environmental sciences and medical and dental sciences.
Purpose: To support research among the postgraduate community.
Eligibility: UK/EU students may apply. EU students must meet residency criteria (living in the UK three years prior to start of the course) in order to be eligible for full support.
Level of Study: Postgraduate, Research
Type: Studentship
Value: Home rate of tuition fee and maintenance stipend set at research council rates.
Length of Study: Normally 3 years
Frequency: Annual
Study Establishment: University of Birmingham
Country of Study: United Kingdom
No. of awards offered: 13 Doctorate and 5 Masters
Application Procedure: Application forms can be downloaded from www.graduateschool.bham.ac.uk.
Closing Date: January
Funding: Government
Contributor: AHRC
No. of awards given last year: 15 Doctoral and 3 Masters
No. of applicants last year: 162
Additional Information: Application can be sent via e-mail to: researchcouncilfunding@contacts.bham.ac.ukClassics and Ancient History, English Literature, Culture & History, Architecture and Design, Italian Culture, Drama and Performing Arts, Film, Digital and Media Production

BBSRC Studentships

Subjects: Biological sciences.
Purpose: To support research among the postgraduate community.
Eligibility: UK/EU students.
Level of Study: Postgraduate, Research
Type: Studentship
Value: The home rate of tuition fee and maintenance stipend set at research council rates.
Length of Study: Normally 3–4 years
Frequency: Annual
Study Establishment: University of Birmingham
Country of Study: United Kingdom
No. of awards offered: Varies
Application Procedure: Nominations made by schools/colleges. Please visit our website for the contact details of each school: www.birmingham.ac.uk. Candidates should express their wish to be nominated.
Closing Date: Nomination between January and June
Funding: Government
Contributor: BBSRC
Additional Information: www.bbsrc.ac.uk

Computer Science PhD Studentship

Subjects: Computer Science.
Purpose: To support outstanding students reading for a research degree.
Eligibility: Every PhD application, if accepted, will automatically be considered for this scholarship.
Level of Study: Postgraduate, Research
Type: Scholarship
Value: The home rate of tuition fee and a maintenance stipend set at research council rates.
Length of Study: Varies (normally 3–4 years)
Frequency: Dependent on funds available
Study Establishment: University of Birmingham
Country of Study: United Kingdom
No. of awards offered: Varies
Application Procedure: All nominations are made by the School of Computer Science. Only students who have appeared for a phD place and have received a conditional or unconditional offer will be considered.
Closing Date: June 30th
Funding: Private
Additional Information: A strong PhD application which is relevant to the research close in the department and received within good time will assist in the nomination process. For more information visit: www.cs.bham.ac.uk

Dinshaw Bursary

Subjects: Theology.
Purpose: To support students reading for a research degree.
Eligibility: Bursaries are awarded on the basis of academic merit and financial need. They are mainly aimed at existing students. The bursary is open to international students, however, they are expected to have made full provision for their tuition and living costs before starting their course. International applicants can only be considered if they demonstrate a change in financial circumstances since registration.
Level of Study: Postgraduate, Research
Type: Bursary
Value: £1,000
Length of Study: 1 year
Frequency: Annual
Study Establishment: University of Birmingham
Country of Study: United Kingdom
No. of awards offered: 1
Application Procedure: Details can be obtained from the Student Funding Office or the Department of Theology + Religion. Nominations are made by the School.
Closing Date: December
Funding: Private
No. of awards given last year: 2
No. of applicants last year: 4
Additional Information: email: theology@contacts.bham.ac.uk, website: www.birmingham.ac.uk/calgs

Edna Pearson Studentship

Subjects: History.
Purpose: To support students reading for a research degree.
Eligibility: Open to candidates pursuing a full-time research Master's degree. Applicants must also apply to AHRC or equivalent.
Level of Study: Postgraduate
Type: Scholarship
Value: The home rate of tuition fee.
Length of Study: 1 year
Study Establishment: University of Birmingham
Country of Study: United Kingdom
No. of awards offered: 1
Application Procedure: Candidates must be nominated by the School of History & Cultures. Check website for further details or contact the college of Arts and Law Graduate School by email: calpg-research@contacts.bham.ac.uk
Closing Date: June
Funding: Private
Additional Information: www.birmingham.ac.uk/calgs

EPSRC Studentships

Subjects: Engineering and physical sciences.
Purpose: To support research among the postgraduate community.
Eligibility: Open to all UK/EU students.
Level of Study: Postgraduate, Research
Type: Studentship
Value: Majority are fees and maintenance. Home rate of tuition fee and maintenance stipend set at research council rates.
Length of Study: Varies (normally 3–4 years)
Frequency: Annual

Study Establishment: University of Birmingham
Country of Study: United Kingdom
No. of awards offered: Varies
Application Procedure: Nominations made by schools/colleges. Please visit our website for the contact details of each school: www.birmingham.ac.uk. Candidates should express their wish to be nominated.
Closing Date: Nomination between January and June
Funding: Government
Contributor: EPSRC
Additional Information: www.epsrc.ac.au

ESRC Studentships

Subjects: Economics and social sciences.
Purpose: To support research among the postgraduate community.
Eligibility: Open to home students established in UK residency. Home/EU students are eligible for full support. EU students who do not fulfil the residency criteria are eligible for tuition fee support only. Research in Economics and Advanced Quantitative Methods is open to International applicants as well as Home/EU.
Level of Study: Predoctorate, Doctorate
Type: Studentship
Value: Home rate of tuition fee and maintenance stipend set at research council rates.
Length of Study: PhD (3 years), Masters + PfD (1 + 3 years) and integrated PhD (4 years)
Frequency: Annual
Study Establishment: The University of Birmingham
Country of Study: United Kingdom
No. of awards offered: 14
Application Procedure: Nominations made by schools/colleges. Please visit our website for the contact details of each school: www.birmingham.ac.uk. Candidates should express their wish to be nominated.
Closing Date: January
Funding: Government
Contributor: ESRC
No. of awards given last year: 16
Additional Information: Please visit www.graduateschool.bham.ac.uk or www.esrc.ac.uk for more information.

Francis Corder Clayton Scholarship

Subjects: Arts, education, government and society, social policy.
Purpose: To financially support students pursuing a higher degree at the University of Birmingham.
Eligibility: Open to current full-time postgraduates undertaking a programme in an Arts subject.
Level of Study: Postgraduate
Type: Scholarship
Value: The home rate of tuition fee and a maintenance stipend set at research council rates.
Length of Study: 1 year
Study Establishment: The University of Birmingham
Country of Study: United Kingdom
No. of awards offered: 1
Application Procedure: Schools are asked to nominate their top 2 candidates. Awards are based on academic performance.
Closing Date: End of May
Funding: Private
No. of awards given last year: 1
No. of applicants last year: 3

George Henry Marshall Scholarship

Subjects: Preference given to those undertaking a history-based programme.
Purpose: To financially support students pursuing a higher degree at the University of Birmingham.
Eligibility: Open to students undertaking a history-based programme in the College of Arts and Law. Preference given to candidates from the West Midlands.
Level of Study: Postgraduate
Type: Scholarship
Value: The home rate at tuition fee and a maintenance stipend set at research council rates
Length of Study: 1 year
Study Establishment: The University of Birmingham
Country of Study: United Kingdom
No. of awards offered: 1
Application Procedure: Schools are asked to nominate their top two candidates. Awards are based on academic performance.
Closing Date: End of May
Funding: Private
No. of awards given last year: 1
No. of applicants last year: 1

Guest, Keen and Nettlefolds Scholarship

Subjects: Engineering.
Purpose: To support students reading for a research degree.
Eligibility: This scholarship is open to engineering graduates who wish to undertake postgraduate work in the University of Birmingham.
Level of Study: Postgraduate
Type: Scholarship
Value: Dependent on funds, usually £1,000
Length of Study: 1 year
Frequency: Dependent on funds available
Study Establishment: The University of Birmingham
Country of Study: United Kingdom
No. of awards offered: Variable according to available funds
Application Procedure: Nominations by each school within the College of Engineering and Physical Sciences. Please contact the relevant school for more information.
Closing Date: Nominations made in spring term
Funding: Private
No. of awards given last year: 4
No. of applicants last year: 4
Additional Information: www.birmingham.ac.uk/colleges/eps

Haywood Scholarship

Subjects: History of Art.
Purpose: To support students reading for a research degree.
Eligibility: Open to candidates pursuing research degree. Applicants must also apply to AHRC or equivalent.
Level of Study: Postgraduate, Research
Type: Scholarship
Value: The home rate of tuition fee
Length of Study: 1 year
Study Establishment: University of Birmingham
Country of Study: United Kingdom
No. of awards offered: 1
Application Procedure: (1) Candidates must be nominated by the School of Department of Art History. (2) Check website for further details or contact the College of Arts and Law Graduate School by email: calpg-research@contacts.bham.ac.au
Closing Date: July
Funding: Private
Additional Information: www.birmingham.ac.uk/calgs

Joseph Chamberlain Scholarship

Subjects: Open to full-time postgraduate students in the Business School; the School of Public Policy; the School of Social Sciences or the School of Engineering.
Purpose: To financially support students pursuing a higher degree at the university of Birmingham.
Eligibility: Open to candidates whose parents/guardians have been bona fide residents in the former west midlands country for 4 years prior to the start of the course, or those who fulfil the residency criteria themselves.
Level of Study: Postgraduate
Type: Scholarship
Value: The home rate of tuition fee and a maintenance stipend set at research council rates.
Length of Study: 1 year
Study Establishment: The University of Birmingham
Country of Study: United Kingdom
No. of awards offered: 1
Application Procedure: Schools are asked to nominate their top 2 candidates. Awards are based on academic performance
Closing Date: End of May
Funding: Private
No. of awards given last year: 1
No. of applicants last year: 4

Kirkcaldy Scholarship
Subjects: Business, government and society, philosphy, theology and religion, social policy.
Purpose: To financially support exisiting students pursuing a higher degree at the University of Birmingham.
Eligibility: Open to current full-time postgraduates undertaking a programme in the college of social sciences (except Education). Preference given to those who have been declared ineligible for public funds.
Level of Study: Postgraduate
Type: Scholarship
Value: The home rate of tuition fee and a maintenance stipend set at research council rates.
Length of Study: 1 year
Study Establishment: The University of Birmingham
Country of Study: United Kingdom
No. of awards offered: 1
Application Procedure: Schools are asked to nominate their top 2 candidates. Awards are based on academic performance.
Closing Date: End of May
Funding: Private
No. of awards given last year: 1
No. of applicants last year: 4

Leventis Scholarship
Subjects: Modern Greek studies.
Purpose: To support students pursuing full-time research degree.
Eligibility: Candidates should have a First Class or Upper Second Class Degree (or equivalent) in a relevant subject.
Level of Study: Postgraduate
Type: Scholarship
Value: Tuition fees Home/EU. Tuition fees (up to the value of €12,000 and at the rate applicable to UK and EU students)
Length of Study: 1 year
Frequency: Annual
Study Establishment: University of Birmingham
Country of Study: United Kingdom
No. of awards offered: 3
Application Procedure: Application forms can be downloaded from the College of Arts and Law Graduate School website: www.birmingham.ac.uk/schools/colgs
Closing Date: June
Funding: Private
Contributor: A.G. Leventis Foundation
Additional Information: For more information email: calpg-research@contacts.bham.ac.uk

MRC Studentships
Subjects: Medical education, chemistry, psychology, biosciences.
Purpose: To support research among the postgraduate community.
Eligibility: UK/EU students, though there are various criteria.
Level of Study: Postgraduate, Research
Type: Studentship
Value: Home rate of tuition and maintenance stipend set at research council rates
Length of Study: Varies (normally 3–4 years)
Frequency: Annual
Study Establishment: University of Birmingham
Country of Study: United Kingdom
No. of awards offered: Varies
Application Procedure: Nominations made by schools/colleges. Please visit our website for the contact details of each school: www.birmingham.ac.uk. Candidates should express their wish to be nominated.
Closing Date: Nomination between January and June
Funding: Government
Contributor: MRC
Additional Information: www.mrc.ac.uk

NERC Studentships
Subjects: Biosciences, computer science, earth science, environmental health and risk management, geography and environmental sciences, medical education.
Purpose: To support research among the postgraduate community.
Eligibility: UK/EU students, though there are various criteria.
Level of Study: Postgraduate, Research
Type: Studentship
Value: Home rate of tuition fee and maintenance stipend set at research council rates
Length of Study: Varies (normally 3–4 years)
Frequency: Annual
Study Establishment: University of Birmingham
Country of Study: United Kingdom
No. of awards offered: Varies
Application Procedure: Nominations made by schools/colleges. Please visit our website for the contact details of each school: www.birmingham.ac.uk. Candidates should express their wish to be nominated.
Closing Date: Nomination between January and June
Funding: Government
Contributor: NERC
Additional Information: www.nerc.ac.uk

Neville Chamberlain Scholarship
Subjects: Humanities subjects. Preference will be given to research in the political and social history of Great Britain, the British Empire and the Commenwealth during the 19th century.
Purpose: To financially support students pursuing a higher degree at the University of Birmingham.
Eligibility: Open to candidates undertaking a programme in the College of Arts and Law (except Law). Preference is given to international students
Level of Study: Postgraduate, Research
Type: Scholarship
Value: The home rate of tuition fee and a maintenance stipend set at research council rates
Length of Study: 1 year
Study Establishment: The University of Birmingham
Country of Study: United Kingdom
No. of awards offered: 1
Application Procedure: Schools are asked to nominate their top 2 candidates. Awards are based on academic performance.
Closing Date: End of May
Funding: Private
Contributor: The family of Neville Chamberlain
No. of awards given last year: 1
No. of applicants last year: 1

Paul Ramsay MSc Computer Science Bursary
Subjects: Computer Science.
Purpose: To support outstanding students with priority given to those from a low income background.
Eligibility: The bursary will be awarded to Home/EU Masters students in need of financial assistance and then on academic merit. International students may apply, but must prove that they have made adequate provision for studying in UK in order to be considered.
Level of Study: Postgraduate
Type: Bursary
Value: The home rate of tuition fee
Length of Study: 1 year
Frequency: Annual
Study Establishment: University of Birmingham
Country of Study: United Kingdom
No. of awards offered: 2
Application Procedure: Application forms can be obtained from the School of Computer Science or downloaded from the website: www.cs.bham.ac.uk.
Closing Date: Mid-July
Funding: Private
Contributor: Paul and Yuanbi Ramsay
No. of awards given last year: 2
No. of applicants last year: 19

Richard Fenwick Scholarship
Subjects: All subjects (except those under College of Medical and Dental Sciences).
Purpose: To financially support students pursuing a higher degree at the University of Birmingham.
Eligibility: Open to full-time postgraduates undertaking any programme (except those studying in the College of Medical + Dental Sciences).
Level of Study: Postgraduate

Type: Scholarship
Value: The home rate of tuition fee and a maintenance stipend set at research council rates
Length of Study: 1 year
Study Establishment: The University of Birmingham
Country of Study: United Kingdom
No. of awards offered: 1
Application Procedure: Schools are asked to nominate their top 2 candidates. Awards are based on academic performance.
Closing Date: End of May
Funding: Private
No. of awards given last year: 1
No. of applicants last year: 12

STFC Studentships

Subjects: Physics and astronomy – particle physics, astronomy, nuclear physics and facility development.
Purpose: To support research among the postgraduate community.
Eligibility: UK/EU students, though there are various criteria.
Level of Study: Postgraduate
Type: Studentship
Value: Home rate of tuition fee and maintenance stipend set at research council rates
Length of Study: Normally 3–4 years
Frequency: Annual
Study Establishment: University of Birmingham
Country of Study: United Kingdom
No. of awards offered: Varies
Application Procedure: Nominations made by schools/collegs. Please visit our website for the contact details of each school: www.birmingham.ac.uk. Candidates should express their wish to be nominated.
Closing Date: Nomination between January and June
Funding: Government
Contributor: STFC
No. of awards given last year: 22
Additional Information: www.stfc.ac.uk

TI Group Scholarship

Subjects: Studying on a programme within the School of Engineering and intend to pursue a scientific or technological career in the manufacturing industry.
Purpose: To support students wishing to pursue a career in manufacturing.
Eligibility: Priority will be given to students who supply evidence of financial need.
Level of Study: Postgraduate
Type: Scholarship
Value: £1,000 each
Length of Study: 1 year (renewable)
Frequency: Annual
Study Establishment: University of Birmingham
Country of Study: United Kingdom
No. of awards offered: 5
Application Procedure: Application form can be obtained from the Student Funding Office or the School of Engineering.
Closing Date: December
Funding: Private
No. of awards given last year: 2
No. of applicants last year: 56

University of Birmingham Alumni Scholarship

Subjects: Any subject.
Purpose: To financially support students pursuing a higher degree at the University of Birmingham.
Eligibility: Candidates must either be an alumnus or the child of an alumnus, of the university of Birmingham. Preference will be given to students reading a research degree.
Level of Study: Postgraduate, Research
Type: Scholarship
Value: The home rate of tuition fee and a maintenance stipend set at research council rates.
Length of Study: Up to 3 years
Study Establishment: The University of Birmingham
Country of Study: United Kingdom
No. of awards offered: 1
Application Procedure: Schools are asked to nominate their top 2 candidates. Awards are based on academic performance.
Closing Date: End of May
Funding: Private
No. of awards given last year: 1
No. of applicants last year: 13

UNIVERSITY OF BREMEN

Building SFG, area 3260, Enrique Schmidt road 7, Bibliothekstraβe, Bremen, 1-D 28359, Germany
Tel: (49) 421 218 60320
Fax: (49) 421 218 9860320
Email: arici@uni-bremen.de
Website: www.uni-bremen.de

University of Bremen was set up as a Science complex in the early 1970s but could gain importance only after 1980s when the mathematics professor Jurgen Timm was elected in 1982. The result is increasingly higher levels in the research rankings, national recognition, a number of endowment professorships, the profiling of interdisciplinary scientific focuses nine DFG sponsored collaborative research centres and, sensationally, the Research Center of Ocean Margins embedded in the earth sciences with the emphasis on Global Change in the Marine Realm; one of only three – initially – national research centers of the DFG (German Research Foundation). From 1996 until 2001 the University of Bremen (along with six other universities in Germany) has been participating in a pilot scheme for structural reform of university administration, funded by the Volkswagen Foundation.

Studentships in the Life Sciences Graduate Program

Subjects: Life sciences.
Purpose: To attract the very best students worldwide to our American-type BSc and MSc degree programs.
Eligibility: Applicants should have outstanding undergraduate records in the biological, chemical, or physical sciences and show great promise of successful careers in research.
Level of Study: Graduate
Value: Waiver of the study fees at IUB, and a monthly allowance to cover food, housing, and personal expenses, and are guaranteed for two years
No. of awards offered: 10
Application Procedure: Check website for details.
Closing Date: February 1st and May 1st
Funding: Private
Additional Information: Students who do well in the MSc stage can, after 18 months, directly transfer to their PhD thesis work without MSc thesis and examination. This is dependent on an offer of financial suppport from a faculty member.

University of Bremen PhD Studentship

Subjects: Social Sciences.
Purpose: To support its PhD fellows in achieving early scientific independence, it also provides funds for conducting, presenting and publishing research.
Eligibility: Open for the candidates who have obtained post graduation in MA or equivalent and have a good command on English.
Level of Study: Postgraduate
Value: €1,250 per month
Length of Study: 3 years
No. of awards offered: 15
Application Procedure: Applications can be downloaded from the website.
Closing Date: March 1st
Funding: Private
Contributor: Volkswagen Foundation
Additional Information: Funding is also available for empirical research and travel. As the University of Bremen intends to increase the proportion of female employees in science, women are particularly encouraged to apply. In case of equal personal aptitudes and qualification disabled persons will be given priority.

For further information contact:

University of Bremen, Postfach 330440, Bremen, 28334, Germany
Website: www.gsss.uni-bremen.de
Contact: Dr Steffen Mau

UNIVERSITY OF BRISTOL

Student Funding Office, Senate House (Ground Floor), Tyndall Avenue, Bristol, BS8 1TH, England
Tel: (44) 11 7928 9000
Fax: (44) 11 7331 7873
Email: student-funding@bris.ac.uk
Website: www.bristol.ac.uk
Contact: Ms Penny Rowe, Student Funding Advisor

The University of Bristol is committed to providing high-quality teaching and research in all its designated fields.

International Postgraduate Scholarships – Taught Master's Programmes

Subjects: All subjects.
Purpose: This programme provides plenty of offers for future research postgraduate researchers with access to studentships, supervision from world-class experts and excellent facilities on our PhD and doctoral programmes.
Eligibility: Open to candidates holding an offer for a one year taught postgraduate programme at the University of Bristol.
Level of Study: Postgraduate
Type: Scholarship (MSc)
Value: £2,000
Length of Study: One year
Frequency: Annual
Study Establishment: University of Bristol
Country of Study: United Kingdom
No. of awards offered: 10
Application Procedure: Check website for further details.
Closing Date: June 30th
Funding: Private

For further information contact:

International Recruitment Office, University of Bristol Union, Queens Road, Clifton, Bristol, Clifton, BS8 1LN, United Kingdom
Email: iro@bristol.ac.uk
Website: www.bristol.ac.uk/international/fees-finances/io-pg-scholarships.html
Contact: Penny Rowe, Student Funding Advisor

University of Bristol Postgraduate Scholarships

Subjects: Any research topic that is covered in the work of the department within the University.
Purpose: To recruit high-quality research students.
Eligibility: Open to new PhD research students with at least an Upper Second Class (Honours) Degree or equivalent. Home/European Union Overseas students must be registered as full-time. Overseas students are only awarded a scholarship if they are successful in obtaining an Overseas Research Student (ORS) award. All scholars must be registered and in attendance for a research degree at the University.
Level of Study: Doctorate
Type: Scholarship
Value: Basic reasearch council stipend rate plus tution fees - subject to revision
Length of Study: 3 years
Frequency: Annual
Study Establishment: University of Bristol
Country of Study: United Kingdom
No. of awards offered: 15 Home/European Union plus 12 Overseas (subject to confirmation)
Application Procedure: Applicants should contact the department where they intend to carry out their research in the first instance, and should look at the Student Funding Office website for up-to-date information on the application process either – www.bristol.ac.uk/studentfunding/overseas_pg/overseas_schols.html for overseas students or www.bristol.ac.uk/student funding/home_pg/schols.html for UK and EU students.
Closing Date: March 1st
Funding: Private
Contributor: University of Bristol
No. of awards given last year: 27 in total

THE UNIVERSITY OF CALGARY

Faculty of Graduate Studies, Earth Sciences Building, Room 720, 844 Campu place 2500 University Drive North West, Calgary, AB, T2N 1N4, Canada
Tel: (1) 403 220 4938
Fax: (1) 403 289 7635
Email: gsaward@ucalgary.ca
Website: www.grad.ucalgary.ca
Contact: Ms Connie Baines, Graduate Scholarship Officer

The University of Calgary is a place of education and scholarly inquiry. Its mission is to seek truth and disseminate knowledge and it aims to pursue this mission with integrity for the benefit of the people of Alberta, Canada and other parts of the world.

Alberta Law Foundation Graduate Scholarship

Subjects: Natural resources, energy and environmental law.
Eligibility: Open to full-time graduates who are registered in or admissible to a programme of studies leading to a Master's degree in the Faculty of Law at the University of Calgary.
Level of Study: Postgraduate
Type: Scholarship
Value: Canadian $7,400 (may be split into 2 awards of $3,700 each)
Length of Study: 1 year, non-renewable
Frequency: Annual
Study Establishment: The University of Calgary
Country of Study: Canada
No. of awards offered: 1 or 2
Application Procedure: Applicants must complete an application form, available from the Director of the graduate programme at the Faculty of Law.
Closing Date: December 15th
Additional Information: In cases where no suitable application is received no awards will be made.

Honourable N D McDermid Graduate Scholarship in Law

Subjects: Law.
Eligibility: Offered annually to students entering or enrolled in a full-time graduate program in the Faculty of Law at the University of Calgary. If, in the opinion of the University of Calgary Graduate Scholarship Committee, no suitable applications are received, no awards will be made.
Level of Study: Postgraduate
Type: Scholarship
Value: Canadian $9,125 each
Length of Study: 1 year
Frequency: Annual
Study Establishment: The University of Calgary
Country of Study: Canada
No. of awards offered: 2
Application Procedure: Applicants must complete an application form, available from the Dean's office. Awards will be recommended by a Committee of the Faculty of Law based upon academic excellence and proven ability in the area of study. The awards will be subject to the approval of the University of Calgary Graduate Scholarship Committee.
Closing Date: December 15th
Contributor: McDermid Law Fund
Additional Information: The scholarship is not renewable. In cases where no suitable applications are received the award will not be made.

Izaak Walton Killam Memorial Scholarships

Subjects: All subjects.
Purpose: This scholarship is awarded to doctoral students of outstanding caliber. They are highly ranked in the University of Calgary Faculty of Graduate Studies Killam Competition by the Killam Memorial Scholarship Committee.
Eligibility: Open to qualified graduates of any university who are admissible to a doctoral programme at the University of Calgary. Applicants must have completed at least 1 year of graduate study prior to taking up the award. The field of study is unrestricted except for the Arts as defined in the Canada Council Act. There is no citizenship restriction.
Level of Study: Postgraduate, Doctorate

Type: Scholarship
Value: Canadian $36,000 and includes a $3,000 research allowance
Length of Study: 2 years
Frequency: Annual
Study Establishment: The University of Calgary
Country of Study: Canada
No. of awards offered: 12 approx
Application Procedure: Applicants must complete an online application form, available from the University Of Calgary Faculty of Graduate Studies website at: www.grad.ucalgary.ca/funding/htm/scholarship_app_guidelines.htm
Closing Date: February 1st
No. of awards given last year: 4–5

Peter C Craigie Memorial Scholarship

Subjects: any subject.
Eligibility: Continuing undergraduate students in any faculty who have completed at least one year.Applicant should have a minimum grade point average of 2.60. Strong citizenship qualities and integrity as exemplified by the late Dr. Craigie.
Level of Study: Postgraduate
Type: Scholarship
Value: Canadian $1,600
Length of Study: full time
Frequency: Every 2 years
Study Establishment: The University of Calgary
Country of Study: Canada
No. of awards offered: 2
Application Procedure: Applicants must apply to the Faculty of Humanities in the first instance. Recommendations from the Faculty will be submitted for consideration and approval by the University Graduate Scholarship Committee.
Closing Date: May 1st
Contributor: Peter C. Craigie Memorial Scholarship Fund

Queen Elizabeth II Graduate Scholarships

Subjects: All subjects.
Eligibility: Candidates must be registered in, or admissible to, a programme leading to a Master's or doctoral degree. The award is restricted to Canadian citizens and landed immigrants.
Level of Study: Doctorate, Postgraduate
Type: Fellowship/Scholarship
Value: The Master's level scholarship consists of Canadian $10,800 per year and the doctoral level scholarship of Canadian $15,000 per year
Length of Study: 1 year, renewable in open competition
Frequency: Annual
Study Establishment: The University of Calgary
Country of Study: Canada
No. of awards offered: 60–70
Application Procedure: Applicants must complete an application form, available from the directors of graduate studies of the departments concerned.
Closing Date: March 5th
Additional Information: Students whose awards begin in May are expected to carry out a full-time research programme during the Summer months.

Sheriff Willoughby King Memorial Scholarship

Subjects: Prevention of family violence and the treatment of the victims of family violence.
Eligibility: Candidates must be registered in the Faculty of Graduate Studies at the University of Calgary, be in the second year of their graduate program (thesis MSW or PhD), and be a Canadian citizen. One award will be made to a candidate studying in the area of prevention and one award will be made to a candidate studying in the area of treatment. Awards will be made on the basis of academic excellence. A 10–15 page research proposal is required.
Level of Study: Postgraduate
Type: Scholarship
Value: Canadian $5,000
Length of Study: 1 year
Frequency: Annual
Study Establishment: The University of Calgary
Country of Study: Canada
No. of awards offered: 1 (area of treatment) and 1 (area of prevention)
Application Procedure: Applicants must apply to the Faculty of Social Work. Recommendations from the Faculty will be considered by the University Graduate Scholarship Committee at its annual meeting. Awards are made on the basis of academic excellence.
Closing Date: February 1st
Contributor: The estate of the late Sheriff Willoughby King
No. of awards given last year: 1 (area of treatment) and 1 (area of prevention)

University of Calgary Faculty of Law Graduate Scholarship

Subjects: Natural resources, energy and environmental law.
Eligibility: Open to full-time graduate students who are registered in or admissible to a programme of studies leading to a Master's degree in the Faculty of Law.
Level of Study: Postgraduate
Type: Scholarship
Value: Up to $10,000
Length of Study: 1 year, non-renewable
Frequency: Annual
Study Establishment: The University of Calgary
Country of Study: Canada
No. of awards offered: 1
Application Procedure: Applicants must complete an application form, available from the graduate programme director at the Faculty of Law. Awards will be recommended by a committee of the Faculty of Law based upon academic excellence.
Closing Date: December 15th
Additional Information: In cases where no suitable applications are received no awards will be made.

William H Davies Medical Research Scholarship

Subjects: Medicine.
Eligibility: Open to qualified graduates of any recognized university who will be registered in the Faculty of Graduate Studies at the University of Calgary. Successful candidates must conduct their research programme within the Faculty of Medicine.
Level of Study: Postgraduate
Type: Scholarship
Value: Canadian $3,000–11,000 depending on qualifications, experience and graduate programme
Length of Study: 4 months–1 year, renewable in open competition
Study Establishment: The University of Calgary
Country of Study: Canada
No. of awards offered: More than 1
Application Procedure: Applicants must apply to the Assistant Dean of Medical Science in the first instance. The Graduate Scholarship Committee will make the final decision based on departmental recommendations. Awards are made on the basis of academic excellence.
Closing Date: June 15th
Contributor: William H. Davies

UNIVERSITY OF CALIFORNIA, BERKELEY

Graduate Services, Graduate Fellowships Office, 318 Sproul Hall #5900, Berkeley, CA, 94720 5900, United States of America
Tel: (1) 510 642 6000
Fax: (1) 510 642 6000
Email: gradappt@berkeley.edu
Website: www.berkeley.edu

Founded in the wake of the gold rush by the leaders of the newly established 31st state, the university of California's flagship campus at Berkeley has become one of the preeminent universities in the world. Its early guiding lights, charged with providing education (both "practical" and "classical") for the state's people, gradually established a distinguished faculty (with 20 Nobel laureates to date), a stellar research library, and more than 350 academic programs.

Albert Newman Fellowship for Visually Impaired Students

Subjects: All subjects.
Purpose: To encourage substantially visually impaired graduate students who demonstrate scholastic achievement.
Eligibility: Applicant must be substantially visually impaired (documented by the Disabled Student's Program). Award will be based on scholastic Achievement. Applicants must be registered for Spring and Fall semesters (each year) when they receive the award.
Level of Study: Graduate
Type: Fellowship
Frequency: Annual
Application Procedure: Applicants must submit completed application and supporting documents.
Closing Date: April 13th

Bay Area Water Quality Fellowship

Purpose: This fellowship is intended to support scientific research into the following topics: the exposure or effect, if any, of organisms within the San Francisco Bay estuary to selenium, metals, and/or organic chemicals through food chain transfer the degree, if any to which sediments are a source of exposure of organisms within the San Francisco Bay estuary to selenium, metals, and/or organic chemicals other research proposed by the University that involves the effect of pollution on the San Francisco Bay estuary and/or its ecosystem.
Eligibility: Open to graduate students whose studies are related specifically to water quality issues that affect the San Francisco Bay.
Level of Study: Graduate
Type: Fellowship
Value: Recipients will receive spring in state tuition and fees (formerly called registration andeducational fees) as well as a stipend of up to $10,000.
Frequency: Dependent on funds available
Closing Date: November 1st

Chancellor's Dissertation-Year Fellowship

Subjects: Humanities and social sciences.
Purpose: To support outstanding students in the humanities and social sciences.
Eligibility: Applicants must be advanced to candidacy at the time of the award and expect to finish their dissertations during the fellowship year.
Level of Study: Doctorate
Type: Fellowship
Value: Awards will include a living stipend of $20,500, payment of fees, and a travel allowance; an additional $3,000 stipend will be paid to fellows who file their dissertations by May 16th.
Length of Study: 1 year
Frequency: Annual
Study Establishment: University of California–Berkeley
Country of Study: United States of America
No. of awards offered: 13
Application Procedure: The applicant must submit a letter from the department or group chair supporting the nomination, a letter from the dissertation adviser describing the importance of the dissertation project, the Report on Progress in Candidacy in the Doctoral Programme form signed by the dissertation adviser, a copy of the Chancellor's Dissertation Year Fellowship application, a detailed outline of the dissertation, a one page summary of the dissertation in non-technical language, a copy of the latest chapter completed or other substantive work as evidence of progress on dissertation and an unofficial copy of the University of California–Berkeley graduate transcript showing date of advancement to candidacy.
Closing Date: March 6th

Conference Travel Grants

Subjects: Masters and Ph.D students in all academic degree programs
Purpose: To allow students to attend professional conferences.
Eligibility: Applicant must be registered graduate students in good academic standing. They must be in the final stages of their graduate work and planning to present a paper on their dissertation research at the conference they are attending.
Level of Study: Doctorate, Graduate
Type: Grant
Value: Amount of the grant depends upon the location of conference (i.e. upto $400 within California; $600 elsewhere in North America, including Mexico or Canada; and $1,000 outside North America)
Frequency: Dependent on funds available
Study Establishment: University of California–Berkeley
Country of Study: United States of America
No. of awards offered: 1
Application Procedure: Applicants must submit an application form and one letter of support from their graduate advisor attesting to the academic merit of the trip. Applications can be obtained from the website.
Closing Date: 3 weeks before the date of travel

Dr and Mrs James C Y Soong Fellowship

Subjects: All subjects.
Purpose: To financially support graduate students from Taiwan.
Eligibility: Applicants must have graduated from a fully accredited, 4-year college or university in Taiwan, with a grade point average of 3.7 (A-) or higher, must be a citizen of the Republic of China and have lived in Taiwan consecutively for at least 10 years and must have demonstrated financial need in pursuit of advanced degrees.
Level of Study: Graduate
Type: Fellowship
Value: The fellowship may be renewed one time.
Length of Study: 1 year
Frequency: Annual
Study Establishment: University of California–Berkeley
Country of Study: United States of America
Application Procedure: Applicant must submit a completed application form, available from the website, a one-page typewritten statement of purpose, a copy of the University of California–Berkeley transcript, the international student financial information form and a letter of support from the department chair.
Closing Date: April 15th

Elizabeth Roboz Einstein Fellowship

Subjects: Neurosciences.
Purpose: To fund doctoral candidates in the neurosciences relating to human development.
Eligibility: Applicants must have demonstrated distinguished scholarship as well as the ability to conduct research at an advanced level.
Level of Study: Doctorate
Type: Fellowship
Value: Approx $3,000
Length of Study: 1 semester
Frequency: Annual
Study Establishment: University of California–Berkeley
Country of Study: United States of America
No. of awards offered: 2
Application Procedure: Applicants must submit a completed application form, two letters of recommendation in sealed envelopes, a copy of the latest University of California–Berkeley transcript and a one-page description of the research project.
Closing Date: November 11th

Foreign Language and Area Studies (FLAS) Fellowships

Subjects: Foreign languages, humanities, social sciences, area and international studies and professional fields.
Purpose: To ensure continued national competence in modern foreign languages and area and international studies.
Eligibility: Candidates must be graduate students and citizens, nationals or permanent residents of the United States of America.
Level of Study: Graduate, Predoctorate
Type: Award
Value: For graduate students, the Academic Year FLAS Fellowship will cover registration fees and provide a stipend of $15,500. And for undergraduate students, the Academic Year FLAS Fellowship will cover registration fees up to $10,000 and provide a stipend of $5,000.
Frequency: Annual
Study Establishment: University of California–Berkeley
Country of Study: United States of America
Application Procedure: Incoming students should apply using the Graduate Application for Admission and Fellowships, which should be submitted directly to their departments by the departmental deadlines.
Closing Date: January 23rd

Mentored Research Award

Subjects: All subjects.
Purpose: The purpose of this program is to assist doctoral students in acquiring sophisticated research skills by working under faculty mentorship on their own pre-dissertation research.
Eligibility: The applicant must be a citizen or permanent resident of the United States of America who demonstrates high academic potential and promise and whose background and life experiences enhance the diversity within the department or discipline. Generally for 3rd and 4th year students.
Level of Study: Doctorate
Type: Award
Value: $16,000 stipend plus payment of in-state fees
Length of Study: 1 year
Frequency: Annual
Study Establishment: University of California–Berkeley
Country of Study: United States of America
No. of awards offered: 13
Application Procedure: Each nominating department must submit a letter from the department or group chair supporting the nomination and confirming eligibility; a letter from the proposed mentor describing plans for mentoring the student and past mentoring experience; a letter from the student describing his or her academic progress to date, the nature of the research project in which she or he will be involved, its importance to the student's academic career, and how this award would help the student to achieve his or her goals; a copy of the Mentored Research Award application; and unofficial copies of all graduate transcripts.
Closing Date: March 6th

Paul J Alexander Memorial Fellowship

Subjects: Ancient history.
Purpose: To encourage the study of Byzantine, ancient and medieval history.
Eligibility: Advanced University of California, Berkeley graduate students studying in the general area of ancient history are invited to apply.
Level of Study: Graduate
Type: Fellowship
Value: Approx. $3,500
Application Procedure: Applicants must submit an application form, available from the website. The application must also include a letter of endorsement from the applicant's sponsor, a copy of the latest University of California–Berkeley transcript and a one-page description of the dissertation research.
Closing Date: November 11th
Additional Information: A student can receive this award only once during his/her academic career.

University of California Dissertation-Year Fellowship

Subjects: All subjects.
Purpose: To support the writing of a doctoral dissertation.
Eligibility: Applicants must be citizens or permanent residents of the United States of America who demonstrate high academic potential and promise and whose backgrounds and life experiences enhance the level of diversity within the department or discipline. Applicants must be advanced to candidacy for the PhD at the time of nomination.
Level of Study: Doctorate
Type: Fellowship
Value: Stipend of $20,500, payment of fees, and travel/research allowances; an additional $2,000 stipend will be paid to fellows who file their dissertations by May 16th, 2014
Length of Study: 1 year
Frequency: Annual
Study Establishment: University of California–Berkeley
Country of Study: United States of America
No. of awards offered: 12
Application Procedure: Each nominating department must submit a letter from the department or group chair supporting the nomination, a letter from the dissertation adviser describing the importance of the dissertation project, an outline of dissertation progress signed by the dissertation adviser, a copy of the UC Dissertation-Year Fellowship application, a detailed outline of the dissertation, a 1-page summary of the dissertation in non-technical language, a copy of the latest chapter completed or other substantive work as evidence of progress on the dissertation and unofficial copies of all graduate transcripts.
Closing Date: February 27th

UNIVERSITY OF CALIFORNIA, LOS ANGELES (UCLA)

Department of Sociology, 2201 Hershey, Los Angeles, CA, 90095-1551, United States of America
Tel: (1) 310 825 3232
Fax: (1) 310 825 4321
Email: grusky@ucla.edu
Website: www.ucla.edu
Contact: Mr Oscar Grusky

UCLA is a leader in many fields, pursuing its mission through excellence in education, research and service. Its faculty, students, and staff work together to advance knowledge in the sciences, humanities and professional fields, address contemporary issues and improve the quality of life.

Charles F Scott Fellowship

Eligibility: Applicants should be graduate students with baccalaureates from UCLA, consideration may be given to students with baccalaureates from other UC campuses.
Type: Fellowship
Value: Up to $15,000 each, from which fees are paid, for graduate students with baccalaureates from UCLA; consideration may be given to students with baccalaureates from other UC campuses.
No. of awards offered: Varies
Application Procedure: Applicants must provide evidence that they are enrolled in a course of study that prepares them for leadership in national, state or local governmental administration.

Dr Ursula Mandel Scholarship

Subjects: Scientific fields related, allied or of value to the medical field.
Eligibility: Applicants must have a doctorate as their degree objective. MD and DDS students are not eligible.
Level of Study: Doctorate
Value: Up to $15,000 each as fees
No. of awards offered: Varies

Eugene V Cota-Robles Fellowship

Subjects: College or university teaching and research.
Purpose: To provide access to higher education for students who might otherwise find it difficult or impossible to successfully pursue graduate study.
Eligibility: Candidates must be either US citizens or permanent residents and should demonstrate high potential and promise. Individuals from cultural, racial, linguistic, geographic and socio-economic backgrounds that are currently underrepresented in graduate education are especially encouraged to participate in the programme. Candidates must be nominated by their department/school. Students pursuing MD or DDS degrees are not eligible for this programme.
Level of Study: Postdoctorate
Type: Fellowship
Value: Stipend of & $20,000 plus registration fees and nonresident tuition (for the first year only) if necessary. During the second, third and fourth years it provides support in the form of a Graduate Research Mentorship Award on activation of the award by the student and the department
Length of Study: 4 years
No. of awards offered: Varies
Application Procedure: Candidates must complete both the Fellowship Application for Entering Graduate Student and the Diversity Fellowships–Supplemental Application and submitted along with Admission Application's Statement of Purpose essay on contributions to the University's diversity mission.
Contributor: University of California Office of the President and the UCLA Graduate Division

Gordon Hein Memorial Scholarship

Subjects: Any subject.
Purpose: To support blind graduate students.
Eligibility: Awards are made on the basis of the student's financial need during the fellowship year and academic record.
Level of Study: Graduate
Type: Scholarship
Value: $5,000 each
Frequency: Dependent on funds available
No. of awards offered: Varies
Application Procedure: Applicants must submit verification of their blindness (e.g. letter from a physician or from the Office for students with Disabilities) and a completed Free Application for Federal Student Aid (FAFSA) or UCLA Financial Statement with the application.

Graduate Opportunity Fellowship Program (GOFP)

Purpose: To provide access to higher education for students who might otherwise find it difficult or impossible to successfully pursue graduate study. Individuals from cultural, racial, linguistic, geographic and socioeconomic backgrounds that are currently underrepresented in graduate education are especially encouraged to apply.
Eligibility: Applicants must be either US citizens or permanent residents and should demonstrate high potential and promise. Individuals from cultural, racial, linguistic, geographic and socio-economic backgrounds that are currently underrepresented in graduate education are especially encouraged to participate in the programme. Students pursuing doctoral degrees (e.g. MD, PhD, DDS, etc.) are not eligible for this programme.
Level of Study: Postgraduate
Type: Fellowship
Value: $18,000 stipend plus registration fees
Length of Study: 1 year
No. of awards offered: Varies
Application Procedure: Applicants must complete both the Fellowship Application for Entering Graduate Student and the Diversity Fellowships-Supplemental Application and submitted along with Admission Application's Statement of Purpose essay on contributions to the University's diversity mission.

Karekin Der Avedisian Memorial Endowment Fund

Subjects: Armenian Studies
Eligibility: Applicants may show financial need (via copy of FAFSA web submission confirmation page or UCLA Financial Statement submitted at time of application) or outstanding academic ability.
Level of Study: Graduate
Type: Funding support
Value: $1,000
No. of awards offered: 1
Application Procedure: Applicants may show financial need (via a Free Application for Federal Students Aid (FAFSA) or a UCLA Financial Statement submitted at time of application) or outstanding academic ability.

Kasper and Siroon Hovannisian Fellowship

Subjects: Armenian studies with preference given to American history.
Eligibility: One has to be a graduate student with focus in Armenian Studies.
Type: Fellowship
Value: Up to $10,000 as fees
No. of awards offered: 1
Application Procedure: Applicants should provide a statement of their projected plan of study along with the application.

Malcolm R Stacey Memorial Scholarship

Subjects: Engineering.
Eligibility: Candidates should be Jewish graduate student in any area of engineering.
Level of Study: Graduate
Type: Scholarship
Value: Up to $5,000. Amount of award is based on financial need during the fellowship year, as determined by the Financial Aid Office
No. of awards offered: Varies
Application Procedure: Candidates should submit a completed FAFSA or UCLA Financial Statement with the application.
Closing Date: November 1st
Contributor: University of California Office of the President

Mangasar M Mangasarian Scholarship

Eligibility: Applicants must be of Armenian descent and provide evidence that at least one parent is Armenian. Awards are made on the basis of candidates' academic records.
Level of Study: Graduate
Type: Scholarship
Value: Up to $10,000 each as fees
Frequency: Dependent on funds available
No. of awards offered: Varies
Closing Date: March 2nd Midnight

Non resident Tuition Fellowships/Registration Fee Grants

Eligibility: Applicants must be enrolled in a full-time programme of study and may not be recipients of awards from federal, state or private foundations that provide tuition coverage. Non resident tuition fellowships are not available for students financially sponsored by foreign governments.
Type: Fellowship or Grant
Value: To cover either the non resident tuition or the registration fees and living expenses
No. of awards offered: Varies
Closing Date: December 15th
Additional Information: Contact the department for further details.

Paulson Scholarship Fund

Eligibility: Applicants must be nationals of Sweden.
Level of Study: Graduate
Type: Scholarship
Value: Up to $6,000
No. of awards offered: Varies

Rose and Sam Gilbert Fellowship

Eligibility: Applicants should be graduate students who attended UCLA as undergraduates for at least 2 years and participated on men's or women's athletic teams (intramural teams are not eligible).
Type: Fellowship
Value: Up to $10,000 each
No. of awards offered: Approx. 2
Closing Date: December 15th

Steven J Sackler Scholarship

Eligibility: Offered to graduate or undergraduate student who had or is experiencing cancer and who can demonstrate financial need for the fellowship year. The student must (1) provide a letter from a physician verifying the condition and (2) submit a copy of FAFSA web submission confirmation page or UCLA Financial Statement. Final selection is made by the Sackler family.
Level of Study: Graduate
Type: Scholarship
Value: Rs5,000
No. of awards offered: 1
Application Procedure: Applicants must provide a letter from a physician verifying the condition and submit a completed Free Application for Federal Student Aid (FAFSA) or a UCLA Financial Statement. Final selection will be made by the Sackler family.
Closing Date: December 15th
Contributor: Sackler family

UCLA Competitive Edge

Subjects: Science, technology, engineering, and mathematics.
Purpose: To provide awardees with research and professional development experiences to enhance their success in UCLA Doctoral in the fields of science, technology, engineering and mathematics (STEM) programmes.
Eligibility: Applicants must be admitted or entering Doctoral students in the fields of STEM with strong interest in pursuing a faculty or research position. Applicants must also be US citizens or permanent residents and, in accordance with the NSF, have backgrounds underrepresented in STEM doctoral programmes.
Level of Study: Doctorate
Type: Award

Value: Provides faculty-guided research and mentoring, as well as academic and professional workshops
Length of Study: 6 weeks
Contributor: National Science Foundation (NSF) Alliance for Graduate Education and the Professoriate (AGEP)
Additional Information: For further information contact at: (310) 825-3953 or via mail: aguzman@grad.ucla.edu

Werner R Scott Fund

Eligibility: Applicants should be Caucasian graduate students who are residents of Hawaii.
Type: Funding support
Value: Up to $8,000 as fees
No. of awards offered: Varies
Closing Date: March 2nd midnight

Will Rogers Memorial Fellowship

Purpose: To support graduate students with physical disabilities in any field of study.
Eligibility: Applicants should be graduate students with physical disabilities in any field of study.
Level of Study: Graduate
Type: Fellowship
Value: Up to $10,000 each as fees
No. of awards offered: Varies
Application Procedure: Applicants must submit verification of their physical disability (e.g. letter from a physician or from the Office for Students with Disabilities) with the application.

UNIVERSITY OF CAMBRIDGE

4, Mill Lane, Cambridge, CB2 1RZ, England
Tel: (44) 12 2333 2317
Fax: (44) 12 2333 2332
Email: dmh14@cam.ac.uk
Website: www.admin.cam.ac.uk
Contact: Hugo Hocknell, Student Registry

The University of Cambridge is a loose confederation of faculties, colleges and other bodies. The colleges are mainly concerned with the teaching of their undergraduate students through tutorials and supervisions and the academic support of both graduate and undergraduate students, while the University employs professors, readers, lecturers and other teaching and administrative staff who provide the formal teaching in lectures, seminars and practical classes. The University also administers the University Library.

Board of Graduate Studies: Allen, Meek and Reed Scholarship (Bursary)

Eligibility: Applicants must be registered as MPhil and must demonstrate their intention to continue to a PhD. Applications are restricted to full-time students. Scholars must already possess a degree from the University of Cambridge (including the Certificate of Advanced Study in Mathematics) or be about to graduate during the competition period. Open to students who are liable to pay the University Composition Fee at the 'Home' or 'EU' rate (i.e. eligible for the CHESS competition).
Type: Bursary
Value: Maximum £6,000
Length of Study: 1 year
Frequency: Annual
No. of awards offered: 10
Application Procedure: Should apply through the department.
Closing Date: July 1st
Contributor: The Allen, Meek and Read (AMR) Fund

Board of Graduate Studies: Dorothy Hodgkin Postgraduate Award

Eligibility: Appliants must have already applied for and been accepted for a PhD place at the University of Cambridge, must be a national of one of the eligible countries (Russia plus all countries on the DAC List of ODA Recipients), must be intending to start their PhD that coming October and must hold a high-grade qualification, at least the equivalent of a UK First Class (Honours) Degree, from a prestigious academic institution.
Type: Award
Value: Approx. £12,300 per year (university composition plus college fees and maintenance stipend)
Length of Study: 3 years
Frequency: Annual
No. of awards offered: Approx. 7
Application Procedure: There is no separate application form for this competition.
Closing Date: December 15th
Contributor: The UK Research Councils and Industrial Partners

For further information contact:

Email: kfw20@admin.cam.ac.uk

Cambridge International Scholarship Scheme (CISS)

Eligibility: Students must be liable to pay the overseas university composition fee. They must be engaged in a three-year research programme leading to the PhD (i.e. will be registered as PhD, "Probationary PhD" or "CPGS") starting in the academic year 2011–12. They must be engaged in full-time study and must have a high upper-second-class undergraduate honours degree from a UK Higher Education Institution, or an equivalent from an Overseas Institution.
Type: Scholarship
Value: Full cost of fees and maintenance for the duration of the course
Length of Study: 3 years
Frequency: Annual
No. of awards offered: 80
Closing Date: December 2nd

CHESS: MPhil Awards

Eligibility: Applicants must have applied for admission by the relevant deadline. They must be registered as liable to pay the University Composition Fee at the "Home" or "EU" rate. Students eligible for UK Research Council funding must have applied for such funding or be able to explain why an application to the Research Councils has not been made (e.g. no quota studentships available to the student's department). It is a condition that a successful CHESS student must accept Research Council funding if it is subsequently offered. Full-time and part-time students are eligible. Any subject area is eligible.
Level of Study: Predoctorate
Type: Award
Value: Approximately £8,000, depending on financial liability, for UK (and UK resident) students and approximately £5,500 for EU resident students
Length of Study: 1 year
Frequency: Annual
Application Procedure: There is no application form for this funding; as long as you have applied for admission/continuation by the deadline you will automatically be considered for this competition.
Additional Information: Please check website for more details.

CHESS: Phd Awards

Eligibility: You must have applied for admission through the Graduate Application Form (GRADSAF) or have applied to continue to the PhD. Students who started their PhD in the Lent or Easter term preceeding and were not considered in the previous year's competition will not automatically be entered but may request to be considered by their department. Please contact the graduate secretary by the end of January Applicants must be registered as liable to pay the University Composition Fee at the Home or EU rate. Students eligible for UK Research Council funding must have applied for such funding or be able to explain why an application to the Research Council has not been made (e.g. no quota studentships available to the student's department). Please check website for more details.
Level of Study: Doctorate
Type: Award
Length of Study: 3 years
Frequency: Annual
Application Procedure: There is no application form for this funding; as long as you have applied for admission/continuation by the deadline you will automatically be considered for this competition.

Christ's College: Emily & Gordon Bottomley Travel Fund

Purpose: To provide support for graduate students towards the cost of travel to conferences or travel which is in some way directly connected with their academic work.
Value: Travel allowance
Application Procedure: Application for an award should be submitted in writing to Mrs M. Stringer, Secretary to the Tutors for Graduate Students, in the College Office (Y4).
Closing Date: December 3rd, February 25th, May 27th
Additional Information: For further information, please refer to the website http://www.christs.cam.ac.uk/current-students/awards-grants/pg_awards_links/.

Christ's College: Travel & Research Fund

Eligibility: Applicant must be a graduate member of the college.
Type: Fees to performers
Value: Travel and other expenses

Christ's College: Whittaker Scholarship

Eligibility: Strong first class or high 2:1 degree, GPA above 3.7 (out of 4) or equivalent from a recognised university; Minimum GMAT score of 700; Preference may be given to candidates who can demonstrate a strong academic background.
Type: Scholarship
Value: £10,000
No. of awards offered: 1
Application Procedure: Candidates should apply for admission to the Cambridge MBA in the normal way.
Additional Information: For further details, please contact our Admissions Coordinator by phone (+44 [0] 1223 339561) or email at mba-admissions@jbs.cam.ac.uk.

Darwin College: Bursaries

Subjects: All subjects.
Eligibility: Only students coming into their first year of graduate study are eligible. For both the Philosophy Studentships and College Bursaries preference will be given to those who nominate Darwin as their college of first choice.
Type: Bursary
Value: £1,000
Application Procedure: Application forms can be obtained by writing to the Dean.
Additional Information: Please check website for updated information.

Darwin College: Philosophy Studentship

Subjects: Philosophy.
Eligibility: Preference will be given to UK students.
Level of Study: Graduate
Type: Studentship
Value: £2,000

Emmanuel College: Research Fellowships

Type: Fellowships
Value: Fees, living expenses and grant towards attending one or more conference during the course of study
Closing Date: June 30th
Additional Information: Please check website for latest updates.

Environmental Services Association Education Trust Studentships in Law and the Sciences

Subjects: Law and environmental sciences.
Purpose: For graduates to conduct or continue research leading to the PhD at Cambridge in Law whose area of research reflects in some way the mission and vision of the Environmental Services Association (www.esauk.org).
Eligibility: The award is open to candidates who are either already members of Trinity Hall or who have made Trinity Hall their college of first preference on the GAF form; they are tenable uniquely at Trinity Hall, and for as long as a student remains a member of Trinity Hall; are (usually first-class) Honours graduates of a respected university or other degree-awarding institution (including Cambridge); if not already graduates, they should have graduated by August; confirmation of awards may rely on satisfactory results in final degree examinations; have been provisionally accepted by their Faculty and by the Board of Graduate Studies to start their study the following academic year. Awards are only tenable by students who begin their course in the Michaelmas Term of the relevant academic year. If we have not received your application by the closing date and you have applied to us for funding, we will defer your application to our reserve list. This means that students should make their applications to the university as early as possible.
Bursaries and Studentships can be renewed, depending on the length of the initial award; for MPhil students supported with a Bursary or Studentship, reapplication is required; for PhD students, renewal is subject to review of diligence and progress, in the form of an annual progress report and accompanying letter of support from the supervisor.
Level of Study: Doctorate, Research
Type: Studentship
Value: Covers university composition fees at the home/EU rate and maintenance for a three-year period
Frequency: Annual
Study Establishment: Cambridge
No. of awards offered: 1
Application Procedure: Application forms for all awards are available from the Graduate Officer in the college.
Closing Date: March 31st
Funding: Trusts

For further information contact:

Trinity Hall, Cambridge, CB2 1TJ, United Kingdom
Contact: Graduate Officer

Faculty of Law: Arnold McNair Scholarship Scholarship

Subjects: International law.
Eligibility: The scholarship is open to any member of the university who has kept at least eight terms and who is a candidate for or has been classed in either Part IB or Part II of the Law Tripos in the year of application.
Type: Scholarship
Value: At least £5,000

Fitzwilliam College Graduate Scholarship

Subjects: All subjects.
Eligibility: Arts students may be preferred. Candidates should be already registered for, or applying for, a research degree course, at Fitzwilliam College.
Level of Study: Graduate
Type: Scholarship
Value: £1,250 per year
Length of Study: 1–3 years, yearly re-application required
Frequency: Annual
Country of Study: United Kingdom
Application Procedure: Application form available from college graduate admissions office at grad.admissions@fitz.can.ac.uk.
Closing Date: September 25th
No. of awards given last year: 1

For further information contact:

c/o graduate officer, Fitzwilliam College, Cambridge, CB3 0DG, United Kingdom

Fitzwilliam College: E D Davies Scholarship

Subjects: All subjects.
Eligibility: Candidates should be already registered for, or applying for, a research degree course, at Fitzwilliam College.
Level of Study: Graduate
Type: Scholarship
Value: £1,250 per year
Length of Study: 1–3 years, yearly re-application required
Frequency: Annual
Country of Study: United Kingdom
No. of awards offered: 2
Application Procedure: Application form available from College Graduate Admissions Office gradadmissions@fitz.cam.ac.uk
Closing Date: September 25th
No. of awards given last year: 2
No. of applicants last year: 25

For further information contact:

C/o Graduate officer, Fitzwilliam College, Cambridge, CB3 0DG, United Kingdom

Fitzwilliam College: Gibson Scholarship

Subjects: Theology.
Eligibility: Intention to work towards a doctorate in New Testament Studies required.
Level of Study: Graduate, Postgraduate
Type: Scholarship
Value: £1,000 per year
Length of Study: 1–3 years
Frequency: Annual
Country of Study: United Kingdom
No. of awards offered: 1
Application Procedure: Application form available from college graduate admission office gradadmissions@fitz.can.ac.uk
Closing Date: September 25th
No. of awards given last year: 1

For further information contact:

c/o Graduate officer, Fitzwilliam College, Cambridge, CB3 0DG, United Kingdom

Fitzwilliam College: Hirst-Player Scholarship

Subjects: Theology.
Eligibility: Students needing assistance with fees, who would otherwise be unable to study at Cambridge, eligible only. Students having intention to take Holy Orders in a Christian Church preferred. Students must be studying as a member of Fitzwilliam College during the period of the award.
Level of Study: Postgraduate
Type: Scholarship
Value: £2,000 maximum
Length of Study: 1–2 years
Frequency: Annual
Country of Study: United Kingdom
No. of awards offered: 1 or 2
Application Procedure: Application forms available from college graduate admissions office gradadmissions@fitz.can.ac.uk
Closing Date: September 25th
No. of awards given last year: 2

For further information contact:

c/o graduate officer, Fitzwilliam College, Cambridge, CB3 0DG, United Kingdom

Fitzwilliam College: Leathersellers' Graduate Scholarship

Subjects: Engineering.
Eligibility: Only home graduates from a British University are eligible. Physical or biological sciences or mathematics students also eligible. Studying as a member of Fitzwilliam College during the period of the award.
Level of Study: Doctorate
Type: Scholarship
Value: £3,000 per year
Length of Study: 1–3 years, subject to the approval of the Tutorial Committee
Frequency: Annual
Country of Study: United Kingdom
No. of awards offered: 4 (usually 1 or 2 per year)
Application Procedure: Application form available from college graduate admissions office gradadmissions@fitz.cam.ac.uk
Closing Date: June 13th
No. of awards given last year: 1
No. of applicants last year: 8

For further information contact:

c/o Graduate officer, Fitzwilliam college, Cambridge, CB3 0DG, United Kingdom

Fitzwilliam College: Shipley Scholarship

Subjects: Theology.
Eligibility: Applicant must be studying as a member of Fitzwilliam during the period of the award.
Level of Study: Graduate, Postgraduate
Type: Scholarship
Value: £1,250
Length of Study: 1 year
Frequency: Annual
Country of Study: United Kingdom
No. of awards offered: 1
Application Procedure: Application form available from college graduate admissions office grad.admissions@fitz.can.ac.uk
Closing Date: September 25th
No. of awards given last year: 1

For further information contact:

c/o Graduate officer, Fitzwilliam College, Cambridge, CB3 0DG, United Kingdom

Fitzwilliam PhD Studentship

Subjects: All subjects.
Eligibility: Fitzwilliam must be the applicant's first choice on university application form, and must have received a conditional offer from the university by the closing date for receipt of applications.
Type: Studentship
Value: Full stipend for maintenance for up to three and half years. It does not cover fees
Length of Study: 3–4 years
Frequency: Annual
No. of awards offered: 1
Application Procedure: Application form available from College Graduate Administration Officer at gradadmissions@fitz.cam.ac.uk.
Closing Date: April 30th
Additional Information: PhD study only.

For further information contact:

Fitzwilliam College, Cambridge, CB3 0DG, United Kingdom

Fitzwilliam Society JRW Alexander Law Book Grants

Subjects: Law.
Eligibility: All Fitzwilliam students on the LLM degree course are eligible.
Level of Study: Postgraduate
Type: Grant
Value: £100
Length of Study: 1 year
Frequency: Annual
Country of Study: United Kingdom
Application Procedure: No application is necessary.
Closing Date: 5
No. of awards given last year: 3
No. of applicants last year: 3

For further information contact:

Fitzwilliam College, Cambridge, CB3 0DG, United Kingdom

Fitzwilliam Studentship (1 year courses)

Subjects: All subjects.
Eligibility: Fitzwilliam must be the applicant's first choice on university application form, and must have received a conditional offer from the university by the closing date for receipt of applications.
Level of Study: Graduate
Type: Scholarship
Value: University and college fees, plus stipend for maintenance
Length of Study: 1 year
Frequency: Annual
Country of Study: United Kingdom
No. of awards offered: 1
Application Procedure: Application form available from College Graduate Administration Office gradadmissions@fitz.cam.ac.uk
Closing Date: March 31st
No. of awards given last year: 1
No. of applicants last year: 45
Additional Information: Does not include PhD, MBA, MFin or PGCE.

For further information contact:

Fitzwilliam College, Cambridge, CB3 0DG, United Kingdom

Girton College Graduate Research Scholarship

Subjects: All subjects covered by Cambridge University.
Eligibility: A first-class degree is almost always required and election will be conditional on the candidate being granted Graduate Student status by the University of Cambridge. The holder must become a member of the college and either be a candidate for a masters or a PhD degree.
Level of Study: Doctorate, Graduate, MBA, Postgraduate
Type: Scholarship
Value: University and college fees and some proportion of maintenance costs
Frequency: Annual
Study Establishment: University of Cambridge
Country of Study: England
No. of awards offered: 2 or 3
Application Procedure: Application forms can be downloaded from www.girton.cam.ac.uk/graduates/research-awards or may be obtained from the Graduate Secretary, Girton College, Cambridge, CB3 0JG, UK (email: graduate.office@girton.cam.ac.uk).
Closing Date: March 28th
No. of awards given last year: 3
No. of applicants last year: 100

Girton College Overseas Bursaries

Subjects: All subjects.
Eligibility: A first-class degree is almost always required and election will be conditional on the candidate being granted Graduate Student status by the University of Cambridge. The holder must become a member of the college.
Level of Study: Doctorate, Graduate, MBA, Postgraduate
Type: Bursary
Value: £200-1,000 per year
Frequency: Dependent on funds available
Study Establishment: University of Cambridge
Country of Study: England
Application Procedure: Application forms can be downloaded from http://www.girton.cam.ac.uk/graduates/research-awards or may be obtained from the Graduate Secretary, Girton College, Cambridge, CB3 0JG, UK (email: graduate.office@girton.cam.ac.uk).
Closing Date: March 28th

Girton College: Diane Worzala Memorial Fund

Subjects: Historical themes, with a preference for those reading British women's history.
Eligibility: Open to students researching historical themes with a preference for those researching British women's history.
Level of Study: Doctorate, Graduate, Postgraduate
Type: Award
Value: £480
Frequency: Annual
Study Establishment: University of Cambridge
Country of Study: England
Application Procedure: Application forms can be downloaded from http://www.girton.cam.ac.uk/graduate/research-awards or may be obtained from the Graduate Secretary, Girton College, Cambridge, CB3 0JG, UK (email: graduate.office@girton.cam.ac.uk).
Closing Date: March 28th

Girton College: Doris Russell Scholarship

Subjects: English.
Eligibility: Available to students of any nationality applying to undertake graduate studies in any area covered by the University's faculty of English.
Level of Study: Doctorate, Graduate, Postgraduate
Type: Scholarship
Value: Contribution towards maintenance costs
Frequency: Annual
Study Establishment: University of Cambridge
Country of Study: England
Application Procedure: Application forms can be downloaded from http://www.girton.cam.ac.uk/graduates/research-awards or may be obtained from the Graduate Secretary.
Closing Date: March 28th
Additional Information: For further information, please refer to the website www.darwin.cam.ac.uk/deanery/awards/shtml.

For further information contact:

Girton College, Cambridge, CB3 0JG, United Kingdom
Email: graduate.office@girton.cam.ac.uk
Contact: Graduate Secretary

Girton College: Doris Woodall Studentship

Subjects: Research in economics or an allied subject.
Eligibility: A first-class degree is almost always required and election will be conditional on the candidate being granted Graduate Student status by the University of Cambridge. The holder must become a member of the college.
Level of Study: Doctorate, Graduate, MBA, Postgraduate
Type: Studentship
Value: Between £750 and £5,000
Frequency: Annual
Study Establishment: University of Cambridge
Country of Study: England
Application Procedure: Application forms can be downloaded from http://www.girton.cam.ac.uk/graduates/research-awards or may be obtained from the Graduate Secretary, Girton College, Cambridge, CB3 0JG, UK (email: graduate.office@girton.cam.ac.uk).
Closing Date: March 28th

Girton College: Ida and Isidore Cohen Research Scholarship

Subjects: Modern Hebrew studies.
Eligibility: Open to students working in modern Hebrew studies. A first-class degree is almost always required and election will be conditional on the candidate being granted Graduate Student status by the University of Cambridge. The holder must become a member of the college.
Level of Study: Doctorate, Graduate, Postgraduate
Type: Scholarship
Value: Between £3,000 and £5,000
Frequency: Annual
Study Establishment: University of Cambridge
Country of Study: England
Application Procedure: Application forms can be downloaded from www.girton.cam.ac.uk/graduate/research-awards or may be obtained from the Graduate Secretary, Girton College, Cambridge, CB3 0JG, UK (email: graduate.office@girton.cam.ac.uk).
Closing Date: March 28th

Girton College: Irene Hallinan Scholarship

Subjects: All subjects.
Eligibility: A first-class degree is almost always required and election will be conditional on the candidate being granted Graduate Student status by the University of Cambridge. The holder must become a member of the college and either be a candidate for a masters or a PhD degree.
Level of Study: Doctorate, Graduate, MBA, Postgraduate
Type: Scholarship
Value: Between £3,000 and £6,000
Frequency: Annual
Study Establishment: University of Cambridge
Country of Study: England
No. of awards offered: 2 or 3
Application Procedure: Application forms can be downloaded from www.girton.cam.ac.uk/graduates/research-awads or may be obtained from the Graduate Secretary, Girton College, Cambridge, CB3 0JG, UK (email: graduate.office@girton.cam.ac.uk).
Closing Date: March 28th
No. of awards given last year: 3
No. of applicants last year: 100

Girton College: Maria Luisa de Sanchez Scholarship

Subjects: All subjects covered by University of Cambridge.
Eligibility: Applicants must be of Venezuelan nationality. A first-class degree is almost always required and election will be conditional on the candidate being granted Graduate Student status by the University of Cambridge. The holder must become a member of the college.
Level of Study: Doctorate, Graduate, MBA, Postgraduate
Type: Scholarship

Value: University and college fees, and some proportion of maintenance costs
Frequency: Annual
Study Establishment: University of Cambridge
Country of Study: England
Application Procedure: Application forms can be downloaded from www.girton.cam.ac.uk/graduate/research-awards or may be obtained from the Graduate Secretary, Girton College, Cambridge, CB3 0JG, UK (email: graduate.office@girton.cam.ac.uk).
Closing Date: March 28th
No. of awards given last year: 1
Additional Information: Eligible to nationals of Venezuela.

Girton College: Ruth Whaley Scholarship

Subjects: Anglo-saxon, norse and celtic, archaeology, history of art, architecture, landscape architecture, classics, liguistics, english, modern and medival languages, history, music philosophy.
Eligibility: Open to outstanding students of non-EU citizenship seeking admission to Girton. It is open to students following arts subjects. A first-class degree is almost always required and election will be conditional on the candidate being granted Graduate Student status by the University of Cambridge. The holder must become a member of the college.
Level of Study: Graduate, Doctorate, Postgraduate
Type: Scholarship
Value: Contribution towards living costs
Frequency: Annual
Study Establishment: University of Cambridge
Country of Study: England
Application Procedure: Application forms can be downloaded from www.girton.cam.ac.uk/graduates/research-awards or may be obtained from the Graduate Secretary.
Closing Date: March 28th
No. of awards given last year: 1
Additional Information: For further details, please refer to the website www.girton.cam.ac.uk/students/graduate-scholarships/.

For further information contact:

Girton College, Cambridge, CB3 0JG, United Kingdom
Email: graduate.office@girton.cam.ac.uk
Contact: Graduate Secretary

Girton College: Sidney and Marguerite Cody Studentship

Purpose: Period of travel and study in continental Europe of up to 12 months and normally of not less than 6 months.
Eligibility: Open to graduate members of any faculty except english who have completed less than nine terms in residence. A first-class degree is almost always required and election will be conditional on the candidate being granted Graduate Student status by the University of Cambridge. The holder must become a member of the college.
Level of Study: Doctorate, Graduate, Postgraduate
Type: Studentship
Value: Up to £3,000
Frequency: Annual
Study Establishment: University of Cambridge
Country of Study: England
No. of awards offered: 1
Application Procedure: Application forms can be downloaded from http://www.girton.cam.ac.uk/graduates/research-awards or may be obtained from the Graduate Secretary, Girton College, Cambridge, CB3 0JG, UK (email: graduate.office@girton.cam.ac.uk).
Closing Date: March 28th
No. of awards given last year: 1

Girton College: Stribling Award

Eligibility: Open to Girton students who are already members of the College, namely undergraduates coming into graduate status or current MPhil students who are going on to a PhD. A first-class degree is almost always required and election will be conditional on the candidate being granted Graduate Student status by the University of Cambridge.
Level of Study: Graduate, Doctorate, Postgraduate
Type: Award
Value: £1,000, normally in addition to a studentship or any other funding for fees and maintenance
Frequency: Annual
Study Establishment: University of Cambridge
Country of Study: England
No. of awards offered: 2
Application Procedure: Application forms can be downloaded from www.girton.cam.ac.uk/graduates/research-awards or may be obtained from the Graduate Secretary, Girton College, Cambridge, CB3 0JG, UK (email: graduate.office@girton.cam.ac.uk).
Closing Date: March 28th
No. of awards given last year: 2

Gonville & Caius College: Bauer Studentships

Purpose: To help gifted graduate students to whom funding would not otherwise be available, to undertake study at Cambridge.
Eligibility: Open both to existing graduate members pursuing an approved postgraduate course, or to candidates who are not already members of the college but who propose to register as graduate students in the University of Cambridge and follow an approved postgraduate course. Candidates will be expected to be of outstanding academic ability.
Level of Study: Postgraduate
Type: Studentship
Frequency: Annual
Closing Date: March 1st

Gonville & Caius College: Darlington Studentships

Eligibility: Open to candidates who are not already members of the college but who propose to register as graduate students in the University of Cambridge and follow an approved Master's postgraduate course.
Type: Studentship
Value: The value of the Studentship will be determined after considering successful candidates' income from any other sources.
No. of awards offered: Not being offered for 2012–2013
Closing Date: March 31st

Gonville and Caius College: WM Tapp Studentships

Subjects: Law.
Eligibility: Candidates must be graduates in law of any university in the United Kingdom or elsewhere, or be about to graduate. They will be expected to be of outstanding academic ability.
Type: Studentship
Value: Approved university and college fees will be paid, together with a maintenance award (currently £8,845 for candidates pursuing the LLM and £13,590 for candidates pursuing the PhD, respectively). An additional college studentship of £500 per year plus a contribution towards travel expenses
Frequency: Annual
Closing Date: December 31st

Graduate Hardship Awards

Eligibility: Eligibility is restricted to Graduate students who are not eligible to seek support from Lundgren Research Awards are formally registered for a research degree; that is, the PhD, MSc or MLitt degree have relied, during their course of research, on a substantial part of the required fees and living expenses being met from their own funds or other private resources.
Level of Study: Graduate
Type: Award
Frequency: Twice a year
Closing Date: March 31st or September 30th

Gurnee Hart Scholarship

Eligibility: Open to a graduate studying an arts subject with a preference for history and historical studies.Applicants should apply directly through the College Graduate Office and to have indicated Jesus College as their first choice in their application to the Board of Graduate Studies. Current Jesus students are also eligible.
Level of Study: Doctorate, Postgraduate
Type: Scholarship
Value: £600 per year
Closing Date: September 30th
Additional Information: Please check website for latest updates.

Henry Fawcett Memorial Scholarship in Mathematics

Subjects: Mathematics.
Purpose: For candidates intending to read for one of the one-year courses available in mathematical subjects.
Eligibility: The award is open to candidates who: are either already members of Trinity Hall or who have made Trinity Hall their college of first preference on the GAF form; they are tenable uniquely at Trinity Hall, and for as long as a student remains a member of Trinity Hall; are (usually first-class) Honours graduates of a respected university or other degree-awarding institution (including Cambridge); if not already graduates, they should have graduated by August; confirmation of awards may rely on satisfactory results in final degree examinations; have been provisionally accepted by their Faculty and by the Board of Graduate Studies to start their study the following academic year. Awards are only tenable by students who begin their course in the Michaelmas Term of the relevant academic year. If we have not received your application by the closing date and you have applied to us for funding, we will defer your application to our reserve list. This means that students should make their applications to the university as early as possible.
Bursaries and Studentships can be renewed, depending on the length of the initial award; for MPhil students supported with a Bursary or Studentship, reapplication is required; for PhD students, renewal is subject to review of diligence and progress, in the form of an annual progress report and accompanying letter of support from the supervisor.
Level of Study: Unrestricted
Type: Award
Value: £500
Frequency: Annual
Study Establishment: Cambridge
No. of awards offered: Varies
Application Procedure: Application forms for all awards are available from the Graduate Officer in the college.
Closing Date: March 31st
Funding: Private

For further information contact:

Trinity Hall, Cambridge, CB2 1TJ
Contact: Graduate Officer

Hughes Hall Awards: Scholarships and Bursaries

Eligibility: Open to current Hughes Hall undergraduate or postgraduate students applying for a higher course.
Type: Scholarship/Bursary
Value: College fee plus university composition fee at the PhD home/EU rate (as scholarship); college fees (as bursary)
No. of awards offered: 1 award as a scholarship and 4 awards as bursary
Closing Date: April 30th

Hughes Hall: Doris Zimmern HKU - Hughes Hall Scholarships

Eligibility: Those who have graduated from the University of Hong Kong, or have been admitted as postgraduate students there, applying to study at Hughes Hall for an MPhil or PhD in any subject.
Type: Scholarships
Value: The university composition fee at the overseas rate, up to a maximum of the fee for science courses; the college fee; a maintenance allowance at the standard rate set by the Cambridge Overseas Trust; and a settling-in fee
Application Procedure: Please contact the Development Office at Hughes Hall (development@hughes.cam.ac.uk) for further details.
Closing Date: December 1st
Additional Information: Please check website for latest updates.

Hughes Hall: Edwin S H Leong Hughes Hall Scholarships

Eligibility: Those who have graduated from the University of Hong Kong, or have been admitted as postgraduate students there, applying to study at Hughes Hall for a Master's degree or PhD in any subject.
Type: Scholarships
Value: The university composition fee at the overseas rate, up to a maximum of the fee for science courses; the college fee; a maintenance allowance at the standard rate set by the Cambridge Overseas Trust; and a settling-in fee. For courses where the university fee is above the science fee (e.g. MBA, MFin, Clinical Medicine) then the scholarship will not cover the full fee
Application Procedure: Please contact the Development Office at Hughes Hall (development@hughes.cam.ac.uk) for further details.
Closing Date: December 1st
Additional Information: Please check website for latest updates.

Hughes Hall: Elizabeth Cherry Bursary

Purpose: To augment the student's major source of funding where this falls short of the total sum required.
Eligibility: Students who wish to study for the PhD. Special consideration will be given to students in arts subjects.
Level of Study: Doctorate
Type: Bursary
Value: College fee
Frequency: Annual
No. of awards offered: 1
Closing Date: April 30th
Additional Information: Please check website for latest updates.

Hughes Hall: William Charnley Law Scholarships

Subjects: Law.
Eligibility: Open to students studying for the LLM or PhD in law. All students who have named Hughes Hall as their college of first choice are eligible to apply. Preference will be given to students resident in the UK or intending to practise in the legal profession in the UK.
Type: Scholarships
Value: £1,000
No. of awards offered: 2
Application Procedure: Applicants must complete the 'Ogden Trust Application Form' or 'Charnley Law Application Form' (availabe at http://www.hughes.cam.ac.uk/ScholarshipsBursariesHardshipAwards) and submit it to the Academic Office by the deadline. They must also ask their referee to complete the 'HH Combined Reference Form' and submit it by the same date.
Closing Date: June 15th
Additional Information: Please check at http://www.hughes.cam.ac.uk/ScholarshipsBursariesHardshipAwards for more information.

Humanitarian Trust Fund Studentship

Subjects: Public international law.
Eligibility: Open to candidates who have obtained, or are likely to obtain before the end of the academical year of their candidature a degree or a diploma at a university in the Commonwealth of Nations, the United States of America, the Continent of Europe, the former Union of Soviet Socialist Republics, at the Hebrew University of Jerusalem or at any other university or college approved by the electors for the purpose of this regulation. They must also produce evidence of their fitness to engage in advanced study.
Type: Studentship
Value: £1,000 although the electors may, in exceptional circumstances and where funds are available, award a more generous sum
Closing Date: January 1st

The Isaac Newton Trust Small Research Grants Scheme

Purpose: To allow teaching staff to employ junior-level research assistance on their research projects.
Value: The maximum individual grant should be £1,000, though it is envisaged that most grants will be awarded in the £500–£600 region. A total of £20,000 is available in each school each year
Closing Date: School of Arts and Humanities: February 13th; School of Humanities and Social Sciences: January 12th and April 13th
Additional Information: Please check at www.newtontrust.cam.ac.uk/research/smallgrant.html for more details.

Jesus College: Charles Rawlinson Graduate Choral Scholarships

Eligibility: Open to graduate students who have the experience and wish to participate fully in the college choirs either as organ scholar or choral scholar. Applicants should have indicated Jesus College as their first choice of college in their application to the Board of Graduate Studies. Current Jesus students are also eligible.
Type: Scholarships
Value: £500

No. of awards offered: 1 or 2
Application Procedure: Applicants should apply through the College Graduate Office.
Additional Information: Please check website for latest updates.

Judge Business School: Africa Regional Bursary

Subjects: MBA.
Eligibility: The applicant must be an African national and must meet the usual requirements for the Cambridge MBA.
Type: Bursary
Value: Up to £18,000
Additional Information: For further details, please contact the MBA Admissions staff via the Cambridge MBA Portal.

Judge Business School: Director's Scholarships

Subjects: MBA.
Purpose: To help outstanding candidates with the cost of the Cambridge MBA.
Eligibility: The scholarships are merit based; Recipients usually have an exceptional academic background, strong GMAT score, and have enjoyed considerable success in their careers to date.
Level of Study: MBA
Type: Scholarship
Value: UK£500
Length of Study: 1 year
Frequency: Annual
Study Establishment: Judge Business School, University of Cambridge
Country of Study: United Kingdom
No. of awards offered: Approx. 10
Application Procedure: All applicants who are offered a place on the MBA programme are automatically considered.
Closing Date: June 6th
Funding: Private
Contributor: Judge Business School
No. of awards given last year: 10
No. of applicants last year: 150 (all admitted students are considered)
Additional Information: As the number of scholarships is limited, candidates who apply early in the admissions round are more likely to receive this award.

Junior Research Fellowship Competition

Eligibility: Open to graduates, women and men, of any university who have recently or are about to complete their doctorates.
Level of Study: Postdoctorate
Type: Fellowship
No. of awards offered: 9
Closing Date: November 21st
Additional Information: Please check website for latest updates.

King's College: Graduate Studentships

Subjects: All subjects.
Purpose: To assist with fees.
Eligibility: The competition is open to any individual beginning a new graduate degree as amember of King's College.
Level of Study: Doctorate, Postgraduate
Type: Studentship
Value: £12,000 towards university and college fees plus a maintenance
Frequency: Annual
Study Establishment: University of Cambridge - King's College
Country of Study: United Kingdom
Application Procedure: All applicants must apply in the usual way to the University of Cambridge through its Board of Graduate Studies Board of Graduate Studies. It is important to note that only applicants conditionally offered a place by the Board can be offered a College place.
Closing Date: May
Additional Information: Please check website for latest updates.

King's College: Non-Stipendiary Junior Research Fellowships

Subjects: Sciences, mathematics, or engineering.
Purpose: These enable young researchers with external support (e.g. from the Royal Society, the British Academy or the Research Councils) to participate in College life as Fellows.
Level of Study: Research
Type: Fellowships
Frequency: Annual
Study Establishment: University of Cambridge - King's College
Country of Study: United Kingdom
Application Procedure: Details of the method of application are given at http://www.kings.cam.ac.uk/research/jrfs/non-stipend.html.
Closing Date: June 1st
Additional Information: Frequently Asked Questions available at http://www.kings.cam.ac.uk/research/jrfs/method-application.html.

Kings College: Stipendiary Junior Research Fellowships

Purpose: To support gifted young researchers.
Level of Study: Postdoctorate, Research
Type: Fellowship
Value: Permit complete freedom to carry out research within the academic environment of the college
Length of Study: 4 years
Frequency: Annual
Study Establishment: University of Cambridge - King's College
Country of Study: United Kingdom
Closing Date: October

For further information contact:

King's College, King's Parade, Cambridge, CB2 1ST, United Kingdom
Tel: (44) 01223 331100
Email: info@kings.cam.ac.uk
Website: http://www.kings.cam.ac.uk
Contact: Dr Keith Carne, Bursary

Le Bas Scholarsips (Bursary)

Subjects: Literature.
Purpose: To offer a limited number of bursaries to the best students in the study of literature and who intend to continue to a PhD.
Eligibility: Candidates must have registered for MPhil and must demonstrate their intention to continue to a PhD. Applications are restricted to full-time students, undertaking study in the field of literature. Open to students who are liable to pay the University Composition Fee at the 'Home' or 'EU' rate (i.e. eligible for the CHESS competition).
Type: Bursary
Value: Approx. £6,000
Length of Study: 1 year
Frequency: Annual
No. of awards offered: 5
Application Procedure: Candidates must apply through the department. Check website for further details.
Closing Date: March 31st
Contributor: Le Bas Trust Fund
Additional Information: Please check at http://www.mml.cam.ac.uk/gradstudies/funding-for-print.html for more information.

Leslie Wilson Research Studentship

Subjects: All subjects.
Eligibility: Applicants who have obtained, or who have a strong prospect of obtaining, a first-class Honours degree (or its equivalent), evidence of subsequent intellectual development will be taken into account, will be considered.
Level of Study: Doctorate
Type: Studentship
Value: A maximum award of £19,340 per year plus the increase in university and college fees will be made to a scholar who has no other sources of finance as follows: maintenance grant - £13,650 (estimate), university fees - £3,440 and college fees - £2,250 (estimate). The successful candidate will receive a minimum award of £500 per year regardless of their resources.
Frequency: Annual
Closing Date: May 2nd

Lord Morris of Borth-y-Gest Scholarship

Subjects: MPhil (one-year) course in criminology, international relations or linguistics.

Eligibility: The award is open to candidates who: are either already members of Trinity Hall or who have made Trinity Hall their college of first preference on the GAF form; they are tenable uniquely at Trinity Hall, and for as long as a student remains a member of Trinity Hall; are (usually first-class) Honours graduates of a respected university or other degree-awarding institution (including Cambridge); if not already graduates, they should have graduated by August; confirmation of awards may rely on satisfactory results in final degree examinations; have been provisionally accepted by their Faculty and by the Board of Graduate Studies to start their study the following academic year. Awards are only tenable by students who begin their course in the Michaelmas Term of the relevant academic year. If we have not received your application by the closing date and you have applied to us for funding, we will defer your application to our reserve list. This means that students should make their applications to the university as early as possible.
Bursaries and Studentships can be renewed, depending on the length of the initial award; for MPhil students supported with a Bursary or Studentship, reapplication is required; for PhD students, renewal is subject to review of diligence and progress, in the form of an annual progress report and accompanying letter of support from the supervisor.
Level of Study: Postgraduate, Research
Value: Up to £2,500
Frequency: Annual
Study Establishment: Cambridge
Application Procedure: Application forms for all awards are available from the Graduate Officer in the College.
Closing Date: March 31st
Funding: Private
Additional Information: Please check website for latest updates.

For further information contact:

Trinity Hall, Cambridge, CB2 1TJ, United Kingdom
Contact: Graduate Officer

Lucy Cavendish College: Becker Law Scholarships
Subjects: Law
Purpose: LLM
Eligibility: Open to women accepted to read for the LLM by the law Faculty at the University of Cambridge.
Level of Study: Postgraduate
Type: Scholarship
Value: £1,000 per year
Length of Study: Up to 3 years, conditional on satisfactory academic progress.
Frequency: Annual
Study Establishment: Lucy Cavendish College, University of Cambridge.
Country of Study: England
No. of awards offered: 3
Application Procedure: Please contact to the Secretary of the Studentship and Bursary Committee.
Closing Date: June 30th

For further information contact:

Lucy Cavendish College, Cambridge, CB3 0BU, United Kingdom
Email: st420@cam.ac.uk
Contact: Secretary, Studentship and Bursary Committee

Lucy Cavendish College: Dorothy and Joseph Needham Studentship
Purpose: For postgraduate research in Natural Sciences.
Eligibility: Open to a woman accepted to undertake postgraduate research in Natural Sciences at the University of Cambridge. Preference will be given to those undertaking research in Biochemistry or related fields.
Level of Study: Postgraduate
Type: Studentship
Value: £1,000 per year
Length of Study: Up to 3 years, conditional on satisfactory academic progress.
Frequency: Annual
Study Establishment: Lucy Cavendish College, University of Cambridge
Country of Study: England
No. of awards offered: 1
Application Procedure: Please contact to the Secretary of the Studentship and Bursary Committee.
Closing Date: June 30th

For further information contact:

Lucy Cavendish College, Cambridge, CB3 0BU, United Kingdom
Email: st420@cam.ac.uk
Contact: Secretary, Studentship and Bursary Committee

Lucy Cavendish College: Enterprise Studentship
Subjects: For MBA or Mphil in Bioscience Enterprise.
Eligibility: Open to women accepted on the MBA course and on the MPhil in Bioscience Enterprise. One is open to any student the other is only available to nationals of USA and Canada.
Level of Study: MBA, Postgraduate
Type: Studentship
Value: £1,000 per year
Frequency: Annual
Study Establishment: Lucy Cavendish College, University of Cambridge
Country of Study: England
Application Procedure: Please contact to the Secretary of the Studentship and Bursary Committee.
Closing Date: June 30th

For further information contact:

Lucy Cavendish College, Cambridge, CB3 0BU, United Kingdom
Email: st420@cam.ac.uk
Contact: Secretary, Studentship and Bursary Committee

Lucy Cavendish College: Evelyn Povey Studentship
Subjects: For postgraduate research in French Studies.
Eligibility: A woman accepted to undertake doctoral research in French Studies at the University of Cambridge.
Level of Study: Postgraduate, Research
Type: Studentship
Value: College fees
Length of Study: Up to 3 years, conditional on satisfactory academic progress.
Frequency: Annual
Study Establishment: Lucy Cavendish College, University of Cambridge
Country of Study: England
Application Procedure: Please contact to the Secretary of the Studentship and Bursary Committee.
Closing Date: June 30th

For further information contact:

Lucy Cavendish College, Cambridge, CB3 0BU, United Kingdom
Email: lcc-admin@lists.cam.ac.uk
Contact: Secretary, Studentship and Bursary Committee

Lucy Cavendish College: Lord Frederick Cavendish Studentship
Purpose: For doctoral research in any subject.
Eligibility: Open to women accepted to undertake doctoral research in any subject at the University of Cambridge.
Level of Study: Postgraduate
Type: Studentship
Value: college fees
Length of Study: Up to 3 years, conditional on satisfactory academic progress
Frequency: Annual
Study Establishment: Lucy Cavendish College, University of Cambridge
Country of Study: England
No. of awards offered: 1
Application Procedure: Please contact to the Secretary of the Studentship and Bursary Committee.
Closing Date: June 30th

For further information contact:

Lucy Cavendish College, Cambridge, CB3 0BU

Email: st420@cam.ac.uk
Contact: Secretary, Studentship and Bursary Committee

Lucy Cavendish College: Mastermann-Braithwaite Studentship

Subjects: Linguistics.
Eligibility: A woman accepted to undertake postgraduate research in Linguistics at the University of Cambridge.
Level of Study: Research, Postgraduate
Type: Studentship
Value: College fees
Length of Study: Up to 3 years, conditional on satisfactory academic progress.
Frequency: Annual
Study Establishment: Lucy Cavendish College, University of Cambridge
Country of Study: England
No. of awards offered: 1
Application Procedure: Please contact to the Secretary of the Studentship and Bursary Committee.
Closing Date: June 30th

For further information contact:

Lucy Cavendish College, Cambridge, CB3 0BU, United Kingdom
Email: st420@cam.ac.uk
Contact: Secretary, Studentship and Bursary Committee

Lundgren Research Awards

Eligibility: Open to Graduate Students who: are formally registered for the PhD degree; are ordinarily resident overseas (including EU countries); are pursuing research in a scientific subject (this includes architecture, biological anthropology, geography, management studies and maths).
Type: Award
Value: Usually to a maximum of £1,500
Frequency: Twice a year
Closing Date: March 31st or September 30th

Magdalene College: Clutton-Brock Scholarship

Subjects: All subjects.
Purpose: To financially support study towards a PhD and post-graduation.
Eligibility: The scholarships are open to citizens of Zimbabwe, normally under the age of 35 and normally resident in Zimbabwe, who already have, or expect to obtain before October 1st, a First Class or High Second Class Degree or its equivalent from a recognised university.
Preference will be given to students wishing to study subjects relevant to the development of Zimbabwe.
Level of Study: Doctorate, Postgraduate
Type: Scholarship
Value: The University Composition Fee (home rate: PhD; overseas rate: MPhil), approved college fees, a maintenance allowance sufficient for a single student and a contribution towards return economy airfare
Length of Study: Up to 3 years
Frequency: Annual
Study Establishment: Magdalene College, the University of Cambridge
Country of Study: United Kingdom
No. of awards offered: 2
Application Procedure: Further information, and application materials, are available in the University's Graduate Studies Prospectus available at: www.admin.cam.ac.uk/offices/gradstud/prospec/apply/index.html.
Applicants for admission to the University must complete and return a GRADSAF Application Form to the Board of Graduate Studies.
Closing Date: December 2nd
Contributor: Offered by the Government of Zimbabwe in collaboration with Magdalene College, Cambridge in honour of Guy Clutton-Brock, hero of Zimbabwe

Magdalene College: Donner Scholarship

Subjects: Matters relevant to Anglo-American relations.
Level of Study: Predoctorate
Type: Scholarship
Value: £10,000 per year
Frequency: Annual
Country of Study: United Kingdom
Closing Date: June 1st

For further information contact:

Atlantic Studies Programme, Centre of International Studies, First Floor, 17 Mill Lane, Cambridge, CB2 1RX, United Kingdom

Medical Elective Grants

Subjects: Clinical study.
Purpose: To contribute towards the cost of travel for clinical study on an approved medical elective course overseas.
Type: Travel grant
Value: £400 at the maximum
Application Procedure: Application for an award should be made in writing to Professor J H Gillard, Director of Studies in Clinical Medicine.

Monica Kornberg Memorial Fund

Subjects: Any research.
Purpose: To facilitate the research of graduate students.
Application Procedure: Application for an award should be submitted in writing to Mrs M. Stringer, Secretary to the Tutors for Graduate Students, in the College Office (Y4).
Closing Date: One of the following deadline: December 3rd; February 25th; May 27th
Additional Information: For further information, please refer to the website http://www.christs.cam.ac.uk/current-students/awards-grants/pg_awards_links/.

Mr and Mrs Johnson Ng Wai Yee Award

Subjects: All subjects.
Purpose: The award will provide a modest needs-based scholarship or bursary for promising undergraduate and graduate students in any subject, with preference being given to students from developing nations.
Eligibility: The award is open to candidates who: are either already members of Trinity Hall or who have made Trinity Hall their college of first preference on the GAF form; they are tenable uniquely at Trinity Hall, and for as long as a student remains a member of Trinity Hall; are (usually first-class) Honours graduates of a respected university or other degree-awarding institution (including Cambridge); if not already graduates, they should have graduated by August; confirmation of awards may rely on satisfactory results in final degree examinations; have been provisionally accepted by their Faculty and by the Board of Graduate Studies to start their study the following academic year.
Awards are only tenable by students who begin their course in the Michaelmas Term of the relevant academic year. If we have not received your application by the closing date and you have applied to us for funding, we will defer your application to our reserve list. This means that students should make their applications to the university as early as possible.
Bursaries and Studentships can be renewed, depending on the length of the initial award; for MPhil students supported with a Bursary or Studentship, reapplication is required; for PhD students, renewal is subject to review of diligence and progress, in the form of an annual progress report and accompanying letter of support from the supervisor.
Level of Study: Graduate, Undergraduate
Type: Scholarship/Bursary
Frequency: Annual
Study Establishment: Cambridge
Application Procedure: Application forms for all awards are available from the Graduate Officer in the college.
Closing Date: March 31st
Funding: Private

For further information contact:

Trinity Hall, Cambridge, CB2 1TJ, United Kingdom
Contact: Graduate Officer

Nabil Boustany Scholarships

Subjects: MBA.
Purpose: To enable a Lebanese national to attend the Cambridge MBA on a full scholarship.
Eligibility: Open to candidates of all nations although priority given to Lebanese nationals, who have obtained a good Honours Degree from a recognized university and have at least 3 years of full-time, real-world experience. Candidates will need to demonstrate a high intellectual potential, practical common sense and the ability to put ideas into action. They also need to be highly motivated with a strong desire to learn. Applicants will be asked to take the Test of English as a Foreign Language (TOEFL) where applicable and the Graduate Management Admission Test (GMAT).
Level of Study: MBA
Type: Scholarship
Value: UK£30,000
Length of Study: 2 years
Frequency: Every 2 years
Study Establishment: Judge Business School, University of Cambridge
Country of Study: United Kingdom
No. of awards offered: 1
Application Procedure: Applicants must email their curriculum vitae to admissions@boustany-foundation.org. Applicants must also be applying for the Cambridge MBA programme.
Closing Date: May 15th
Funding: Foundation
Contributor: The Nabil Boustany Foundation
No. of awards given last year: 1
No. of applicants last year: 4
Additional Information: The scholarship recipient is normally required to spend a summer internship, carrying no salary, within a Lebanese organization. Applicants must be accepted into the Cambridge MBA programme.

For further information contact:

1 avenue des Citronniers, Monte Carlo, 98000, Monaco
Fax: (377) 77 93 15 05 56
Email: info@boustany-foundation.org
Website: www.jbs.cam.ac.uk
Contact: Mr M Tamar

Newnham College: Ann Duncan Memorial Fund

Subjects: Modern and Medieval Languages Tripos or a research degree in modern languages, particularly those with special interest in contemporary French or Latin-American studies.
Purpose: To fund projects that the awarders think would have interested Ann Duncan.
Eligibility: In recent years awards have gone mainly to those whose travel plans have a distinctive social or intellectual purpose.
Level of Study: Research
Type: Award
Value: Varies
Study Establishment: University of Cambridge
Country of Study: England
No. of awards offered: Varies
Application Procedure: Once the application has been completed two emails will be sent to the named Director of Studies or Supervisor requesting a letter of support for the applicant. It is recommended that applicants pre-warn the person providing their letter of support that a request will be forthcoming.
Closing Date: May 13th
Additional Information: Please contact to Principal's Secretary for queries.

For further information contact:

Newnham College, Cambridge, CB3 9DF
Email: cb560@cam.ac.uk
Website: https://app.casc.cam.ac.uk/fas_live/duncan.aspx
Contact: Principal's Secretary

Newnham College: Junior Research Fellowships

Subjects: Humanities and social sciences, science, mathematics or engineering.
Purpose: To allow highly-talented women to pursue their research single-mindedly in a supportive academic environment.
Eligibility: Fellowships are normally awarded to women just finishing their PhDs, or those in their first post-doctoral appointment. All new Junior Research Fellows are encouraged to organise a Newnham-based research event - such as a conference or workshop - during their tenure.
Level of Study: Postdoctorate
Type: Fellowship
Value: Up to £10,000
Frequency: Annual
Study Establishment: University of Cambridge
Country of Study: England
No. of awards offered: 2
Closing Date: October 28th (humanities) and November 4th (science)
Additional Information: One fellowship is for humanities and social sciences, the other is for science, mathematics or engineering.

For further information contact:

Website: www.newn.cam.ac.uk/joining-newnham/research-fellowships

Newnham College: Phyllis and Eileen Gibbs Travelling Research Fellowships

Subjects: Biology, archaeology, social anthropology, or sociology, with preference for archaeology or biology.
Purpose: To provide assistance to women graduates who have considerable experience of research (normally beyond the doctoral level), who intend to undertake fieldwork projects outside the British isles (and normally not within their own countries of residence).
Eligibility: Fellows are required to produce a report to the College on their research, and are invited to present their work at a public lecture.
Level of Study: Doctorate
Type: Fellowship
Frequency: Annual
Study Establishment: University of Cambridge
Country of Study: United Kingdom
Closing Date: January 11th

For further information contact:

Website: www.newn.cam.ac.uk/at-newnham/research/travelling-fellowships

Nightingale Research Studentships

Subjects: Doctoral research at Cambridge in the arts, humanities, social sciences, or mental health.
Eligibility: The award is open to candidates who: are either already members of Trinity Hall or who have made Trinity Hall their college of first preference on the GAF Form; they are tenable uniquely at Trinity Hall, and for as long as a student remains a member of Trinity Hall; are (usually first-class) Honours graduates of a respected university or other degree-awarding institution (including Cambridge); if not already graduates, they should have graduated by August; confirmation of awards may rely on satisfactory results in final degree examinations; have been provisionally accepted by their Faculty and by the Board of Graduate Studies to start their study the following academic year. Awards are only tenable by students who begin their course in the Michaelmas Term of the relevant academic year. If we have not received your application by the closing date and you have applied to us for funding, we will defer your application to our reserve list. This means that students should make their applications to the university as early as possible.
Bursaries and Studentships can be renewed, depending on the length of the initial award; for MPhil students supported with a Bursary or Studentship, reapplication is required; for PhD students, renewal is subject to review of diligence and progress, in the form of an annual progress report and accompanying letter of support from the supervisor.
Level of Study: Doctorate, Postgraduate, Research
Type: Studentship
Frequency: Annual
Study Establishment: Cambridge
Application Procedure: Application forms for all awards are available from the Graduate Officer in the College.
Closing Date: March 31st
Funding: Private

Additional Information: Please check website for latest updates.

For further information contact:

Trinity Hall, Cambridge, CB2 1TJ, United Kingdom
Contact: Graduate Officer

Ogden Trust Science Education Awards

Eligibility: Students studying for the PGCE or MEd in science education and citizens of the UK, who nominate Hughes Hall as their college of first choice.
Type: Award
Value: £2,500
No. of awards offered: 3
Closing Date: April 30th
Additional Information: Please check website for latest updates.

Pembroke College: College Research Studentships

Subjects: All subjects.
Eligibility: Preference in awarding these studentships will be given to candidates who intend to register for a PhD degree at Pembroke. However, candidates registering to study for an MPhil will also be considered for an award if they are intending to carry on to a PhD after they have finished their MPhil.
Level of Study: Postgraduate
Type: Studentship
Value: College/University fees plus a maintenance allowance of £9,880 per year
Frequency: Annual
Study Establishment: Pembroke College
Country of Study: England
Application Procedure: All applicants for any of these awards must apply in the first instance to the Board of Graduate Studies for their University place. Candidates should indicate that they are applying for a Pembroke College award. In making awards preference will be given to those who nominate Pembroke as their College of first choice. All candidates are expected to apply for Research Council funding where appropriate, and for University CHESS funding, if they are eligible. The College will take into account candidates' income from other sources when making awards.
Closing Date: January 20th
Additional Information: Applicants should also complete a Pembroke Studentship form. This can be completed online, downloaded or obtained from the Graduate Secretary. The awards are conditional on the selected students being admitted as a registered Graduate Student by the Board of Graduate Studies with effect from 1 October each academic year. Early application is recommended.

For further information contact:

Pembroke College, Cambridge, CB2 1RF, United Kingdom
Tel: (44) 01223 338100
Fax: (44) 01223 338163
Email: tut@pem.cam.ac.uk
Website: www.pem.cam.ac.uk
Contact: Graduate Secretary

Pembroke College: Graduate Studentships in Arabic and Islamic Studies (including Persian)

Subjects: Arabic and Islamic studies.
Eligibility: Preference in awarding these studentships will be given to candidates who intend to register for a PhD degree at Pembroke. However, candidates registering to study for an MPhil will also be considered for an award if they are intending to carry on to a PhD after they have finished their MPhil.
Level of Study: Postgraduate
Type: Studentship
Value: College/university fees plus a maintenance allowance of £9,880 per year
Frequency: Annual
Study Establishment: Pembroke College
Country of Study: England
Application Procedure: All applicants for any of these awards must apply in the first instance to the Board of Graduate Studies for their University place. Candidates should indicate that they are applying for a Pembroke College award. In making awards preference will be given to those who nominate Pembroke as their College of first choice. All candidates are expected to apply for Research Council funding where appropriate, and for University CHESS funding, if they are eligible. The college will take into account candidates' income from other sources when making awards.
Additional Information: Applicants should also complete a Pembroke Studentship form. This can be completed online, downloaded or obtained from the Graduate Secretary. The awards are conditional on the selected students being admitted as a registered Graduate Student by the Board of Graduate Studies with effect from October 1st each academic year. Early application is recommended.

For further information contact:

Pembroke College, Cambridge, CB2 1RF, United Kingdom
Email: tut@pem.cam.ac.uk
Contact: Graduate Secretary

Pembroke College: M.Phil Studentship for Applicants from the Least Developed Countries

Subjects: All subjects.
Purpose: The College is offering this one-year studentship to enable the winner to study for an MPhil degree, or equivalent, at the University of Cambridge.
Eligibility: Eligibility is confined to nationals of the fifty 'Least Developed Countries' as defined by the United Nations. The other is for nationals of the fifty 'Least Developed Countries' as defined by the United Nations (check at www.un.org/special-rep/ohrlls/ldc/list), most of which are in Africa or South-East Asia.
Level of Study: Postgraduate
Type: Studentship
Value: University fees at least at the standard rate for Home/EU students, plus college fees and a maintenance figure of £9,880
Frequency: Annual
Study Establishment: Pembroke College, University of Cambridge
No. of awards offered: 1
Application Procedure: All applicants for any of these awards must apply in the first instance to the Board of Graduate Studies for their University place. Candidates should indicate that they are applying for a Pembroke College award. In making awards preference will be given to those who nominate Pembroke as their College of first choice. All candidates are expected to apply for Research Council funding where appropriate, and for University CHESS funding, if they are eligible. The College will take into account candidates' income from other sources when making awards.
Additional Information: Applicants should also complete a Pembroke Studentship form. This can be completed online, downloaded or obtained from the Graduate Secretary. The awards are conditional on the selected students being admitted as a registered Graduate Student by the Board of Graduate Studies with effect from October 1st each academic year. Early application is recommended.

For further information contact:

Pembroke College, Cambridge, CB2 1RF, United Kingdom
Email: tut@pem.cam.ac.uk
Contact: Graduate Secretary

Pembroke College: The Bethune-Baker Graduate Studentship in Theology

Subjects: Arabic and Islamic studies, law, and theology.
Eligibility: Open to candidates who intend to register for the PhD degree at the University of Cambridge.
Level of Study: Postgraduate
Type: Studentship
Value: College and university fees for three years
Frequency: Annual
Study Establishment: University of Cambridge
Country of Study: England
Application Procedure: All applicants for any of these awards must apply in the first instance to the Board of Graduate Studies for their University place. Candidates should indicate that they are applying for a Pembroke College award. In making awards preference will be given to those who nominate Pembroke as their College of first choice. All candidates are expected to apply for Research Council funding where appropriate, and for University CHESS funding, if they are eligible.

The College will take into account candidates' income from other sources when making awards.
Contributor: HM the Sultan of Oman and of Professor E.G. Browne
Additional Information: Applicants should also complete a Pembroke Studentship form. This can be completed online, downloaded or obtained from the Graduate Secretary. The awards are conditional on the selected students being admitted as a registered Graduate Student by the Board of Graduate Studies with effect from October 1st each academic year. Early application is recommended.

For further information contact:

Pembroke College, Cambridge, CB2 1RF
Email: tut@pem.cam.ac.uk
Contact: Graduate Secretary

Pembroke College: The Bristol-Myers Squibb Graduate Studentship in the Biomedical Sciences

Subjects: History, physics and the bio-medical sciences.
Eligibility: Open to candidates who intend to register for the PhD degree at the University of Cambridge.
Level of Study: Postgraduate
Type: Studentship
Value: The studentship will have a value sufficient to pay college fees for three years. Moreover, additional awards, of up to the equivalent of University fees for a Home student (£3,770 in 2011/12), may be made to individual applicants, depending on need and the availability of funds.
Frequency: Annual
Country of Study: England
No. of awards offered: 3
Application Procedure: All applicants for any of these awards must apply in the first instance to the Board of Graduate Studies for their University place. Candidates should indicate that they are applying for a Pembroke College award. In making awards preference will be given to those who nominate Pembroke as their College of first choice. All candidates are expected to apply for Research Council funding where appropriate, and for University CHESS funding, if they are eligible. The College will take into account candidates' income from other sources when making awards.
Additional Information: Applicants should also complete a Pembroke Studentship form. This can be completed online, downloaded or obtained from the Graduate Secretary. The awards are conditional on the selected students being admitted as a registered Graduate Student by the Board of Graduate Studies with effect from October 1st each academic year. Early application is recommended.

For further information contact:

Pembroke College, Cambridge, CB2 1RF, United Kingdom
Email: tut@pem.cam.ac.uk
Contact: Graduate Secretary

Pembroke College: The Grosvenor-Shilling Bursary in Land Economy

Subjects: All subjects.
Eligibility: Applicants must normally reside in Australia and hold a qualification from an Australian tertiary institution. There is no restriction as to the academic field.
Level of Study: Postgraduate
Type: Scholarship
Value: £500
Frequency: Annual
No. of awards offered: 2
Application Procedure: All applicants for any of these awards must apply in the first instance to the Board of Graduate Studies for their University place. Candidates should indicate that they are applying for a Pembroke College award. In making awards preference will be given to those who nominate Pembroke as their College of first choice. All candidates are expected to apply for Research Council funding where appropriate, and for University CHESS funding, if they are eligible. The College will take into account candidates' income from other sources when making awards.
Additional Information: Applicants should also complete a Pembroke Studentship form. This can be completed online, downloaded or obtained from the Graduate Secretary. The awards are conditional on the selected students being admitted as a registered Graduate Student by the Board of Graduate Studies with effect from October 1st each academic year. Early application is recommended.

For further information contact:

Pembroke College, Cambridge, CB2 1RF
Email: tut@pem.cam.ac.uk
Contact: Graduate Secretary

Pembroke College: The Lander Studentship in the History of Art

Purpose: The College is very pleased to be able to offer one studentship for an outstanding art historian, supported by the estate of Professor J.R. Lander.
Eligibility: Candidates must be applying to study for a PhD degree in the History of Art at the University of Cambridge, with Pembroke as first-choice college.
Level of Study: Postgraduate
Type: Studentship
Value: The studentship will, if necessary, pay university and college fees, at the home rate, plus a maintenance allowance of £9,880 per year, for a maximum of three years. Overseas candidates must find the difference between home and overseas fee rates by other means.
Frequency: Annual
Study Establishment: Pembroke College, University of Cambridge
Country of Study: England
No. of awards offered: 1
Application Procedure: All applicants for any of these awards must apply in the first instance to the Board of Graduate Studies for their University place. Candidates should indicate that they are applying for a Pembroke College award. In making awards preference will be given to those who nominate Pembroke as their College of first choice. All candidates are expected to apply for Research Council funding where appropriate, and for University CHESS funding, if they are eligible. The College will take into account candidates' income from other sources when making awards.
Additional Information: Applicants should also complete a Pembroke Studentship form. This can be completed online, downloaded or obtained from the Graduate Secretary. The awards are conditional on the selected students being admitted as a registered Graduate Student by the Board of Graduate Studies with effect from October 1st each academic year. Early application is recommended.

For further information contact:

Pembroke College, Cambridge, CB2 1RF, United Kingdom
Email: tut@pem.cam.ac.uk
Contact: Graduate Secretary

Pembroke College: The Monica Partridge Studentship

Subjects: All subjects.
Purpose: To offer a graduate studentship for a student from South-East Europe to study at Pembroke.
Eligibility: Open to the students of the nationals of Albania, Bosnia and Herzegovina, Bulgaria, Croatia, Greece, Kosovo, Macedonia, Montenegro and Serbia. Applications from students from Romania, Slovenia and Turkey will be considered if there is no suitable candidate from the countries listed above. Preference will be given to fund students studying for a PhD, but MPhil applicants intending to continue to a PhD will also be considered.
Level of Study: Postgraduate
Type: Studentship
Value: The studentship will have a value sufficient to cover college fees (£2,184) and maintenance (£9,880) for three years for a PhD student or, in the case of an MPhil student, one year.
Frequency: Annual
Study Establishment: Pembroke College, University of Cambridge
Country of Study: England
No. of awards offered: 1
Application Procedure: All applicants for any of these awards must apply in the first instance to the Board of Graduate Studies for their University place. Candidates should indicate that they are applying for a Pembroke College award. In making awards preference will be given to those who nominate Pembroke as their College of first choice. All candidates are expected to apply for Research Council funding where appropriate, and for University CHESS funding, if they are eligible.

The College will take into account candidates' income from other sources when making awards.
Additional Information: Applicants should also complete a Pembroke Studentship form. This can be completed online, downloaded or obtained from the Graduate Secretary. The awards are conditional on the selected students being admitted as a registered Graduate Student by the Board of Graduate Studies with effect from October 1st each academic year. Early application is recommended.

For further information contact:

Pembroke College, Cambridge, CB2 1RF, United Kingdom
Email: tut@pem.cam.ac.uk
Contact: Graduate Secretary

Pembroke College: The Nahum Graduate Studentship in Physics

Subjects: Physics.
Eligibility: Preference in awarding these studentships will be given to candidates who intend to register for a PhD degree at Pembroke. However, candidates registering to study for an MPhil will also be considered for an award if they are intending to carry on to a PhD after they have finished their MPhil.
Level of Study: Postgraduate
Type: Studentship
Value: College/University fees plus a maintenance allowance of £9,880 per year
Frequency: Annual
Study Establishment: Pembroke College
Country of Study: England
Application Procedure: All applicants for any of these awards must apply in the first instance to the Board of Graduate Studies for their University place. Candidates should indicate that they are applying for a Pembroke College award. In making awards preference will be given to those who nominate Pembroke as their College of first choice. All candidates are expected to apply for Research Council funding where appropriate, and for University CHESS funding, if they are eligible. The College will take into account candidates' income from other sources when making awards.
Additional Information: Applicants should also complete a Pembroke Studentship form. This can be completed online, downloaded or obtained from the Graduate Secretary. The awards are conditional on the selected students being admitted as a registered Graduate Student by the Board of Graduate Studies with effect from October 1st each academic year. Early application is recommended.

For further information contact:

Pembroke College, Cambridge, CB2 1RF, United Kingdom
Email: tut@pem.cam.ac.uk
Contact: Graduate Secretary

Pembroke College: The Pembroke Austrailian Scholarship

Subjects: All subjects.
Eligibility: Applicants must normally reside in Australia and hold a qualification from an Australian tertiary institution. There is no restriction as to the academic field.
Level of Study: Postgraduate
Type: Scholarship
Value: £500
Frequency: Annual
No. of awards offered: 2
Application Procedure: All applicants for any of these awards must apply in the first instance to the Board of Graduate Studies for their University place. Candidates should indicate that they are applying for a Pembroke College award. In making awards preference will be given to those who nominate Pembroke as their College of first choice. All candidates are expected to apply for Research Council funding where appropriate, and for University CHESS funding, if they are eligible. The College will take into account candidates' income from other sources when making awards.
Additional Information: Applicants should also complete a Pembroke Studentship form. This can be completed online, downloaded or obtained from the Graduate Secretary. The awards are conditional on the selected students being admitted as a registered Graduate Student by the Board of Graduate Studies with effect from October 1st each academic year. Early application is recommended.

For further information contact:

Pembroke College, Cambridge, CB2 1RF, United Kingdom
Email: tut@pem.cam.ac.uk
Contact: Graduate Secretary

Pembroke College: The Thornton Graduate Studentship in History

Subjects: History.
Eligibility: Preference in awarding these studentships will be given to candidates who intend to register for a PhD degree at Pembroke. However, candidates registering to study for an MPhil will also be considered for an award if they are intending to carry on to a PhD after they have finished their MPhil.
Level of Study: Postgraduate
Type: Studentship
Value: College/university fees plus a maintenance allowance of £9,880 per year
Frequency: Annual
Study Establishment: Pembroke College
Country of Study: England
Application Procedure: All applicants for any of these awards must apply in the first instance to the Board of Graduate Studies for their University place. Candidates should indicate that they are applying for a Pembroke College award. In making awards preference will be given to those who nominate Pembroke as their College of first choice. All candidates are expected to apply for Research Council funding where appropriate, and for University CHESS funding, if they are eligible. The College will take into account candidates' income from other sources when making awards.
Additional Information: Applicants should also complete a Pembroke Studentship form. This can be completed online, downloaded or obtained from the Graduate Secretary. The awards are conditional on the selected students being admitted as a registered Graduate Student by the Board of Graduate Studies with effect from October 1st each academic year. Early application is recommended.

For further information contact:

Pembroke College, Cambridge, CB2 1RF, United Kingdom
Email: tut@pem.cam.ac.uk
Contact: Graduate Secretary

Pembroke College: The Ziegler Graduate Studentship in Law

Subjects: Law.
Eligibility: Preference in awarding studentships is given to candidates who intend to register for a PhD degree at Pembroke.
Level of Study: Postgraduate
Type: Studentship
Value: College/University fees plus a maintenance allowance of £9,880 per year
Frequency: Annual
Study Establishment: Pembroke College
Country of Study: England
Application Procedure: All applicants for any of these awards must apply in the first instance to the Board of Graduate Studies for their University place. Candidates should indicate that they are applying for a Pembroke College award. In making awards preference will be given to those who nominate Pembroke as their College of first choice. All candidates are expected to apply for Research Council funding where appropriate, and for University CHESS funding, if they are eligible. The College will take into account candidates' income from other sources when making awards.
Additional Information: Applicants should also complete a Pembroke Studentship form. This can be completed online, downloaded or obtained from the Graduate Secretary. The awards are conditional on the selected students being admitted as a registered Graduate Student by the Board of Graduate Studies with effect from October 1st each academic year. Early application is recommended.

For further information contact:

Pembroke College, Cambridge, CB2 1RF, United Kingdom

Email: tut@pem.cam.ac.uk
Contact: Graduate Secretary

Peterhouse: Research Studentships

Subjects: All subjects.
Eligibility: Open to prospective PhD candidates.
Level of Study: Postgraduate
Type: Studentship
Frequency: Annual
Study Establishment: Peterhouse College
Country of Study: England
Application Procedure: Please contact at graduates@pet.cam.ac.uk for more information.
Closing Date: January 20th

For further information contact:

Peterhouse, Trumpington Street, Cambridge, CB2 1RD, United Kingdom
Email: graduates@pet.cam.ac.uk
Website: www.pet.cam.ac.uk
Contact: Graduate Admissions

Professor Peter Brown Memorial Bursary

Subjects: Computer science or mathematics.
Eligibility: Applicant must be an outstanding student in computer science or mathematics.
Type: Bursary
Value: £500
Closing Date: June 30th
Additional Information: Please check website for latest updates.

Research Grants

Eligibility: Candidates for the Research Grants must be applying to read for the degree of PhD in the University of Cambridge, or, if applying for a one-year course, intending that this should be followed immediately by registration for a PhD in Cambridge.
The award is open to candidates who: are either already members of Trinity Hall or who have made Trinity Hall their college of first preference on the GAF Form; they are tenable uniquely at Trinity Hall, and for as long as a student remains a member of Trinity Hall; are (usually first-class) Honours graduates of a respected university or other degree-awarding institution (including Cambridge); if not already graduates, they should have graduated by August; confirmation of awards may rely on satisfactory results in final degree examinations; have been provisionally accepted by their Faculty and by the Board of Graduate Studies to start their study the following academic year. Awards are only tenable by students who begin their course in the Michaelmas Term of the relevant academic year. If we have not received your application by the closing date and you have applied to us for funding, we will defer your application to our reserve list. This means that students should make their applications to the university as early as possible.
Bursaries and Studentships can be renewed, depending on the length of the initial award; for MPhil students supported with a Bursary or Studentship, reapplication is required; for PhD students, renewal is subject to review of diligence and progress, in the form of an annual progress report and accompanying letter of support from the supervisor.
Level of Study: Doctorate, Postgraduate, Research
Type: Research grant
Frequency: Annual
Study Establishment: Cambridge
Application Procedure: Application forms for all awards are available from the Graduate Officer in the College.
Closing Date: March 31st
Additional Information: Please check website for latest updates.

For further information contact:

Trinity Hall, Cambridge, CB2 1TJ, United Kingdom
Contact: Graduate Officer

Research Studentships

Eligibility: Candidates for the Research Studentships must be applying to read for the degree of PhD in the University of Cambridge, or, if applying for a one-year course, intending that this should be followed immediately by registration for a PhD in Cambridge. The award is open to candidates who: are either already members of Trinity Hall or who have made Trinity Hall their college of first preference on the GAF Form; they are tenable uniquely at Trinity Hall, and for as long as a student remains a member of Trinity Hall; are (usually first-class) Honours graduates of a respected University or other degree-awarding Institution (including Cambridge); if not already graduates, they should have graduated by August; confirmation of awards may rely on satisfactory results in final degree examinations; have been provisionally accepted by their Faculty and by the Board of Graduate Studies to start their study the following academic year. Awards are only tenable by students who begin their course in the Michaelmas Term of the relevant academic year. If we have not received your application by the closing date and you have applied to us for funding, we will defer your application to our reserve list. This means that students should make their applications to the University as early as possible. Bursaries and Studentships can be renewed, depending on the length of the initial award; for MPhil students supported with a Bursary or Studentship, reapplication is required; for PhD students, renewal is subject to review of diligence and progress, in the form of an annual progress report and accompanying letter of support from the supervisor.
Level of Study: Postgraduate, Research
Type: Award
Value: Up to six full-cost (at Home and EU rates) or more part-cost, studentships are awarded annually, covering some or all of the maintenance costs, University and College fees. Trinity Hall does not usually pay the full cost of overseas students who have not been awarded a Cambridge International Scholarship, with the exception of the Trinity Hall Overseas Studentship
Frequency: Annual
Study Establishment: Cambridge
No. of awards offered: 6
Application Procedure: Application forms for all awards are available from the Graduate Officer in the College.
Closing Date: March 31st
Additional Information: Please check website for latest update.

For further information contact:

Trinity Hall, Cambridge, CB2 1TJ, United Kingdom
Contact: Graduate Officer

Reuben Levy Travel Fund

Purpose: To assist graduate members of the college with the costs of travel which need not be connected with their academic work.
Type: Travel grant
Value: Cost of travel
Application Procedure: Application for an award should be submitted in writing to Mrs M. Stringer, Secretary to the Tutors for Graduate Students, in the College Office (Y4).
Closing Date: One of the following deadlines: December 3rd; February 25th; May 27th

Robinson College: Lewis Graduate Scholarship

Purpose: The College expects to award one Lewis Scholarship to a graduate student applying to read for a PhD degree in the humanities.
Eligibility: Open to all applicants who name Robinson College as their College of first choice on the Board of Graduate Studies Application Form for Admissions as a Graduate student, or are repared to change College if offered the scholarship. The scholarship is conditional on the candidate being offered a place at the University.
Level of Study: Postgraduate
Type: Scholarship
Value: The scholarship covers university and college fees at the Home/EU level and provides the student with a maintenance allowance up to the University recommended level (currently £11, 250)
Length of Study: The scholarship is tenable for up to 3 years, subject to satisfactory academic progress
Study Establishment: University of Cambridge
Country of Study: England
No. of awards offered: 1
Application Procedure: Applicants should send a curriculum vitae and details of their intended programme of research including no more

than one A4 page describing their proposed research project), together with details of other grant applications. Applications must be sumbitted by post to the Graduate Admissions Tutor.
Closing Date: March 31st
Additional Information: Please check website for latest updates.

For further information contact:

Robinson College, Cambridge, CB3 9AN, United Kingdom
Email: graduate-admissions@robinson.cam.ac.uk
Contact: Graduate Admissions Tutor

Robinson College: Yates-Unilever Scholarships

Subjects: Arts and humanities.
Purpose: Yates-Unilever Scholarships are offered annually to Graduate Students from overseas countries who register for a PhD degree in a science subject at the College.
Eligibility: Applications are usually made in the first year of study.
Level of Study: Postgraduate
Type: Scholarship
Value: In any one year the award is unlikely to exceed £300 to any student
Length of Study: Tenable for one year
Study Establishment: University of Cambridge
Country of Study: England
No. of awards offered: 3
Additional Information: For more information visit: http://www.robinson.cam.ac.uk/admissions/graduates.php, email graduate-admissions@robinson.cam.ac.uk.

Roosevelt Scholarship

Subjects: Matters relevant to Anglo-American relations.
Eligibility: Applicants who have a record of academic excellence consistent with the proposed field of study and who can demonstrate the potential to make a significant contribution to the broader life of Magdalene College, as well as to the wider interests of the Atlantic Studies Programme at the University of Cambridge, will be considered.
Level of Study: Predoctorate
Type: Scholarship
Value: A maximum individual maintenance award of no more than £10,140 (UK) or £10,465 (Canadian) will be made to a scholar. Rented accommodation in or near college will be made available to an unmarried scholar. A married scholar will be offered rented accommodation near to the college
Application Procedure: All candidates should complete a Roosevelt Scholarship application form which can be found at: http://www.magd.cam.ac.uk/admissions/postgraduate/scholarships-roosevelt.html.
Closing Date: January 31st (January 15th for electronic applications)

Sainsbury Bursary Scheme

Subjects: MBA.
Purpose: To support students engaged in charitable, voluntary or public sector work in areas such as housing, health and education, local economic development and social services, as it is difficult for such candidates to secure sponsorship from their employers.
Eligibility: Preference will be given to applicants from the United Kingdom but exceptional candidates working for international aid agencies based outside the United Kingdom will be given consideration. It is expected that applicants will contribute to the sectors in the United Kingdom after completion of the MBA. Candidates must have completed 3 years of work within the charitable, voluntary or public sector prior to submitting their application. They must have strong support from their employer, or evidence of an on-going commitment to the charitable or voluntary sectors. They must also show evidence of a career plan showing how they would use the skills and knowledge gained during the MBA course to develop their career within the charitable, voluntary or public sector.
Level of Study: MBA
Type: Bursary
Value: UK£14,000– 28,000 at the discretion of The Sainsbury Bursary Scheme Committee
Length of Study: 1 year
Frequency: Annual
Study Establishment: Judge Business School, University of Cambridge
Country of Study: United Kingdom
No. of awards offered: 5
Application Procedure: Applicants must submit a completed Cambridge MBA application (including references and supporting documents), with a separate email letter to "mba-admissions@jbs.cam.ac.uk" indicating that you would also like to apply for a Sainsbury Bursary and explaining how you meet the eligibility criteria above.
Closing Date: April 15th
Funding: Private
Contributor: The Monument Trust
No. of awards given last year: 3
No. of applicants last year: 3
Additional Information: The Monument Trust is one of the Sainsbury Family Charitable Trusts.

Seung Jun Lee Bursary

Subjects: Social sciences.
Eligibility: Applicant should be an outstanding student working towards a PhD in the social sciences.
Type: Bursary
Value: £3,000
Closing Date: June 30th
Additional Information: Please check website for latest updates.

St Catharine's College: MBA Benavitch Scholarships

Subjects: MBA.
Purpose: To encourage exceptional candidates to pursue the Cambridge MBA.
Eligibility: Candidates should apply for admission to the Cambridge MBA in the normal way, but must also send a covering letter with their application form explaining in not more than 300 words the contribution they expect to make to business and society in the future.
Level of Study: MBA
Type: Scholarship
Value: UK£10,000
Length of Study: 1 year
Frequency: Annual
Study Establishment: St Catharine's College, University of Cambridge
Country of Study: United Kingdom
No. of awards offered: 5
Application Procedure: First class degree, GPA above 3.7 (out of 4) or equivalent from a recognised university; minimum GMAT score of 700 and 3 years' full-time post-graduation work experience.
Closing Date: February 25th
Funding: Private
Contributor: The late Maurice Benavitch and his late wife Natalie
No. of awards given last year: 5
No. of applicants last year: 20
Additional Information: Recipients of Benavitch Scholarships will become members of St Catharine's College, Cambridge and be given the option to live in single, college-owned accommodation (some married accommodation may be offered, subject to availability).

For further information contact:

MBA Admissions, Judge Business School, Trumangton Street, Cambridge, CB2 1AG, United Kingdom

St Catherine's College: PhD Benavitch Scholarship

Subjects: PhD in management studies.
Eligibility: Open to doctorate students with academic excellence and depending on the candidate's contribution to the field. Candidates must hold an offer of a place to study a PhD in management studies at the University of Cambridge.
Level of Study: Doctorate
Type: Scholarship
Value: £9,000
Length of Study: 3 years
Frequency: Annual
Study Establishment: t Catherine's College, University of Cambridge
Country of Study: United Kingdom
No. of awards offered: 1
Application Procedure: Candidates should apply for the PhD to the University of Cambridge in the usual way, but in addition, submit a copy of their PhD application form (GRADSAF), along with their CV

and PhD reseach proposal to St Catherines College. They must also include two references to be enclosed in sealed envelopes.
Closing Date: February 28th for courses commencing the following October
Funding: Private
Contributor: The Late Maurice Benavitch and his late wife Natalie
No. of awards given last year: 1
Additional Information: Recipients of Benavitch Scholarships will become members of St Catharine's College, Cambridge.

For further information contact:

Benavitch PhD scholarship, St Catherine's College, Trumpington Street, Cambridge, CB2 1RL
Contact: The Graduate Administrator

St Catherine's College: Tunku Abdul Rahman PhD Scholarship

Subjects: Arts, humanities, and social sciences.
Purpose: To encourage the development of humanities and social sciences in Malaysia.
Eligibility: Applicants must be of Malaysian nationality. Applicants must already hold an offer of study from the University of Cambridge.
Level of Study: Doctorate, Research
Type: Scholarship
Value: Full fees and maintenance for 3 years. The value of any studentship awarded will be reduced appropriately to take account of any payments from other sources
Length of Study: 3 years
Frequency: Annual
Study Establishment: University of Cambridge
Country of Study: United Kingdom
No. of awards offered: 1
Application Procedure: Further details and application form can be downloaded from www.caths.cam.ac.uk/tunkuscholarship.
Closing Date: March 1st for courses commencing the following October
Funding: Government
Contributor: Government of Malyasia
No. of awards given last year: 1
Additional Information: Recipients of Tunku scholarships will become members of St Catharine's College, Cambridge.Eligible to nationals of Malaysia.

Thaddeus Mann Studentship – Eastern and Central Europe

Subjects: Preference will be given to candidates studying a scientific or engineering subject.
Eligibility: The award is open to candidates who: are either already members of Trinity Hall or who have made Trinity Hall their college of first preference on the GAF Form; they are tenable uniquely at Trinity Hall, and for as long as a student remains a member of Trinity Hall; are (usually first-class) Honours graduates of a respected university or other degree-awarding institution (including Cambridge); if not already graduates, they should have graduated by August; confirmation of awards may rely on satisfactory results in final degree examinations; have been provisionally accepted by their Faculty and by the Board of Graduate Studies to start their study the following academic year. Awards are only tenable by students who begin their course in the Michaelmas Term of the relevant academic year. If we have not received your application by the closing date and you have applied to us for funding, we will defer your application to our reserve list. This means that students should make their applications to the university as early as possible.
Bursaries and Studentships can be renewed, depending on the length of the initial award; for MPhil students supported with a Bursary or Studentship, reapplication is required; for PhD students, renewal is subject to review of diligence and progress, in the form of an annual progress report and accompanying letter of support from the supervisor.
Level of Study: Graduate, Postgraduate, Research
Type: Studentship
Frequency: Annual
Study Establishment: Cambridge
Application Procedure: Application forms for all awards are available from the Graduate Officer in the College.
Closing Date: March 31st
Funding: Private
Additional Information: Please check website for latest updates.

For further information contact:

Trinity Hall, Cambridge, CB2 1TJ, United Kingdom
Contact: Graduate Officer

Thomas Waraker Postgraduate Bursary in Law

Subjects: Law.
Eligibility: Candidates must propose either as an affiliated student to read for Part II of the Law Tripos, or as a graduate student to read for the LLM, or as a graduate student to read for a PhD in law. The award is open to candidates who: are either already members of Trinity Hall or who have made Trinity Hall their college of first preference on the GAF form; they are tenable uniquely at Trinity Hall, and for as long as a student remains a member of Trinity Hall; are (usually first-class) Honours graduates of a respected university or other degree-awarding institution (including Cambridge); if not already graduates, they should have graduated by August; confirmation of awards may rely on satisfactory results in final degree examinations; have been provisionally accepted by their Faculty and by the Board of Graduate Studies to start their study the following academic year. Awards are only tenable by students who begin their course in the Michaelmas Term of the relevant academic year. If we have not received your application by the closing date and you have applied to us for funding, we will defer your application to our reserve list. This means that students should make their applications to the university as early as possible. Bursaries and Studentships can be renewed, depending on the length of the initial award; for MPhil students supported with a Bursary or Studentship, reapplication is required; for PhD students, renewal is subject to review of diligence and progress, in the form of an annual progress report and accompanying letter of support from the supervisor.
Level of Study: Postgraduate
Value: £150–300
Frequency: Annual
Study Establishment: Cambridge
Application Procedure: Application forms for all awards are available from the Graduate Officer in the College.
Closing Date: March 31st
Funding: Trusts
Additional Information: Please check website for latest updates.

For further information contact:

Trinity Hall, Cambridge, CB2 1TJ, United Kingdom
Contact: Graduate Officer

Tidmarsh Studentship – Canadian citizens

Subjects: All subjects.
Eligibility: The award is open to candidates who: are either already members of Trinity Hall or who have made Trinity Hall their college of first preference on the GAF Form; they are tenable uniquely at Trinity Hall, and for as long as a student remains a member of Trinity Hall; are (usually first-class) Honours graduates of a respected university or other degree-awarding institution (including Cambridge); if not already graduates, they should have graduated by August; confirmation of awards may rely on satisfactory results in final degree examinations; have been accepted for a Cambridge International Scholarship; have been provisionally accepted by their Faculty and by the Board of Graduate Studies to start their study the following academic year. Awards are only tenable by students who begin their course in the Michaelmas Term of the relevant academic year. If we have not received your application by the closing date and you have applied to us for funding, we will defer your application to our reserve list. This means that students should make their applications to the university as early as possible.
Bursaries and Studentships can be renewed, depending on the length of the initial award; for MPhil students supported with a Bursary or Studentship, reapplication is required; for PhD students, renewal is subject to review of diligence and progress, in the form of an annual progress report and accompanying letter of support from the supervisor.
Level of Study: Graduate, Postgraduate, Research
Type: Studentship

Frequency: Every 3 years
Study Establishment: Cambridge
Application Procedure: Application forms for all awards are available from the Graduate Officer in the College.
Closing Date: March 31st
Funding: Trusts
Additional Information: Please check website for latest updates.

For further information contact:

Trinity Hall, Cambridge, CB2 1TJ, United Kingdom
Contact: Graduate Officer

Trinity Hall Overseas Studentship

Subjects: Awarded to an overseas student, applying to read for the degree of PhD in the University of Cambridge, or, if applying for a one-year course, intending that this should be followed immediately by registration for a PhD in Cambridge.
Eligibility: The award is open to candidates who: are an overseas student, applying to read for the degree of PhD in the University of Cambridge; are either already members of Trinity Hall or who have made Trinity Hall their college of first preference on the GAF form; they are tenable uniquely at Trinity Hall, and for as long as a student remains a member of Trinity Hall; are (usually first-class) Honours graduates of a respected university or other degree-awarding institution (including Cambridge); if not already graduates, they should have graduated by August; confirmation of awards may rely on satisfactory results in final degree examinations; have been provisionally accepted by their Faculty and by the Board of Graduate Studies to start their study the following academic year. Awards are only tenable by students who begin their course in the Michaelmas Term of the relevant academic year. If we have not received your application by the closing date and you have applied to us for funding, we will defer your application to our reserve list. This means that students should make their applications to the university as early as possible.
Bursaries and Studentships can be renewed, depending on the length of the initial award; for MPhil students supported with a Bursary or Studentship, reapplication is required; for PhD students, renewal is subject to review of diligence and progress, in the form of an annual progress report and accompanying letter of support from the supervisor.
Level of Study: Doctorate, Research
Type: Studentship
Value: A full-cost studentship, covering maintenance costs, university and college fees
Frequency: Annual
Study Establishment: Cambridge
Application Procedure: Application forms for all awards are available from the Graduate Officer in the college.
Closing Date: March 31st
Funding: Trusts
Additional Information: Please check website for latest updates.

For further information contact:

Trinity Hall, Cambridge, CB2 1TJ, United Kingdom
Contact: Graduate Officer

Trinity Hall: Brockhouse Studentship

Subjects: Candidates must be proposing to study an engineering-based subject.
Eligibility: The award is open to candidates who are either already members of Trinity Hall or who have made Trinity Hall their college of first preference on the GAF Form; they are tenable uniquely at Trinity Hall, and for as long as a student remains a member of Trinity Hall; are (usually first-class) Honours graduates of a respected university or other degree-awarding institution (including Cambridge); if not already graduates, they should have graduated by August; confirmation of awards may rely on satisfactory results in final degree examinations; have been provisionally accepted by their Faculty and by the Board of Graduate Studies to start their study the following academic year. Awards are only tenable by students who begin their course in the Michaelmas Term of the relevant academic year. If we have not received your application by the closing date and you have applied to us for funding, we will defer your application to our reserve list. This means that students should make their applications to the University as early as possible. Bursaries and Studentships can be renewed, depending on the length of the initial award; for MPhil students supported with a Bursary or Studentship, reapplication is required; for PhD students, renewal is subject to review of diligence and progress, in the form of an annual progress report and accompanying letter of support from the supervisor.
Level of Study: Postgraduate, Research
Type: Studentship
Frequency: Annual
Study Establishment: Cambridge
No. of awards offered: 1
Application Procedure: Application forms for all awards are available from the Graduate Officer in the College.
Closing Date: March 31st
Funding: Private
Additional Information: Please check website for latest updates.

For further information contact:

Trinity Hall, Cambridge, CB2 1TJ, United Kingdom
Contact: Graduate Officer

Trinity Hall: Chris McMenemy Scholarship in Development and Environmental Studies

Subjects: For students whose postgraduate studies contribute to the sustainable development of any underdeveloped region of the world.
Eligibility: The award is open to candidates who are either already members of Trinity Hall or who have made Trinity Hall their college of first preference on the GAF form; they are tenable uniquely at Trinity Hall, and for as long as a student remains a member of Trinity Hall; are (usually first-class) Honours graduates of a respected university or other degree-awarding institution (including Cambridge); if not already graduates, they should have graduated by August; confirmation of awards may rely on satisfactory results in final degree examinations; have been provisionally accepted by their Faculty and by the Board of Graduate Studies to start their study the following academic year. Awards are only tenable by students who begin their course in the Michaelmas Term of the relevant academic year. If we have not received your application by the closing date and you have applied to us for funding, we will defer your application to our reserve list. This means that students should make their applications to the university as early as possible. Bursaries and Studentships can be renewed, depending on the length of the initial award; for MPhil students supported with a Bursary or Studentship, reapplication is required; for PhD students, renewal is subject to review of diligence and progress, in the form of an annual progress report and accompanying letter of support from the supervisor.
Level of Study: Postdoctorate
Type: Scholarship
Value: £1,500
Frequency: Annual
Study Establishment: Cambridge
No. of awards offered: 1
Application Procedure: Application forms for all awards are available from the Graduate Officer in the college.
Closing Date: March 31st
Funding: Private

For further information contact:

Trinity Hall, Cambridge, CB2 1TJ, United Kingdom
Contact: Graduate Officer

Trinity Hall: Dr Clark's Theological Scholarship

Subjects: For students reading a theological subject.
Eligibility: The award is open to candidates who are either already members of Trinity Hall or who have made Trinity Hall their college of first preference on the GAF form; they are tenable uniquely at Trinity Hall, and for as long as a student remains a member of Trinity Hall; are (usually first-class) Honours graduates of a respected university or other degree-awarding institution (including Cambridge); if not already graduates, they should have graduated by August; confirmation of awards may rely on satisfactory results in final degree examinations; have been provisionally accepted by their Faculty and by the Board of Graduate Studies to start their study the following academic year. Awards are only tenable by students who begin their course in the Michaelmas Term of the relevant academic year. If we have not

received your application by the closing date and you have applied to us for funding, we will defer your application to our reserve list. This means that students should make their applications to the university as early as possible.

Bursaries and Studentships can be renewed, depending on the length of the initial award; for MPhil students supported with a Bursary or Studentship, reapplication is required; for PhD students, renewal is subject to review of diligence and progress, in the form of an annual progress report and accompanying letter of support from the supervisor.

Level of Study: Unrestricted
Type: Award
Value: £150–400
Frequency: Annual
Study Establishment: Cambridge
No. of awards offered: Varies
Application Procedure: Application forms for all awards are available from the Graduate Officer in the college.
Closing Date: March 31st
Funding: Private

For further information contact:

Trinity Hall, Cambridge, CB2 1TJ, United Kingdom
Contact: Graduate Officer

Vargas Scholarship

Eligibility: Only Venezuelan citizens studying for the PhD degree in medicine (or possibly related subjects) are eligible.
Type: Scholarship
Value: All approved fees and maintenance for three years
Length of Study: 3 years
Frequency: Annual
Additional Information: Please check at http://www.darwin.cam.ac.uk/deanery/awards.shtml for more information.

Wolfson College: Donald & Beryl O'May Studentship

Subjects: Arts and social sciences (including law).
Eligibility: A successful applicant may be a student at undergraduate (either mature or affiliated) or postgraduate level and must be a citizen of the United Kingdom or the Republic of Ireland. Preference will be given to candidates who have had a significant break in their studies at some stage since leaving school.
Level of Study: Postgraduate, Undergraduate
Type: Studentship
Value: £11,500
Study Establishment: Wolfson College
Country of Study: United Kingdom
Funding: Private

UNIVERSITY OF CAMBRIDGE (CAMBRIDGE COMMONWEALTH TRUST, CAMBRIDGE OVERSEAS TRUST, GATES CAMBRIDGE TRUST, CAMBRIDGE EUROPEAN TRUST AND ASSOCIATED TRUSTS)

Cambridge Trusts, Trinity College, Trinity Street, Cambridge, CB2 1TQ, United Kingdom
Tel: (44) 1223 351 449
Fax: (44) 1223 323 322
Email: info@overseastrusts.cam.ac.uk
Website: www.admin.cam.ac.uk

The Cambridge Commonwealth Trust and the Cambridge Overseas Trust (formerly the Chancellor's Fund) were established in 1982 by the University of Cambridge under the Chairmanship of his Royal Highness the Prime of Wales to provide financial assistance for students from overseas who, without help, would be unable to take up their places at Cambridge. Since 1982, the Cambridge Commonwealth Trust has brought 6,600 students from 51 countries to Cambridge, the Cambridge Overseas trust 4,252 students from 76 countries.

Arab-British Chamber Cambridge Scholarship

Subjects: All subjects relevants to their country's needs.
Purpose: To financially support those undertaking postgraduate study.
Eligibility: Applicants must be citizens of Algeria, the Comoro Islands, Djibouti, Egypt, Jordan, Lebanon, Mauritania, Morocco, Palestine, Somalia, Sudan, Syria, Tunisia or the Yemen. They must apply to the University of Cambridge and be offered a place at Cambridge in the normal way. All applicants must have a First Class or High Second Class (Honours) Degree or equivalent and normally be under 26. After completion of their studies, scholars must undertake to return to their country or to another member state of the Arab League.
Level of Study: Postgraduate
Type: Scholarship
Value: The University Composition Fee at the overseas rate, approved college fees, a maintenance allowance sufficient for a single student and a contribution towards return economy airfare
Length of Study: 1 year
Frequency: Annual
Study Establishment: The University of Cambridge
Country of Study: United Kingdom
No. of awards offered: 5
Application Procedure: Applicants must complete a preliminary application form, which can be obtained from local universities, offices of the British Council or the Trust. Completed forms must be returned to the main address. Shortlisted candidates will be sent forms for admission to the University of Cambridge.
Closing Date: January 30th
Contributor: Offered in collaboration with the Arab-British Chamber Charitable Foundation and the Foreign & Commonwealth Office

BAT Cambridge Scholarships and Bursaries

Subjects: All subjects.
Purpose: To financially support study towards a PhD and postgradution.
Eligibility: 1) Applicants to do a PhD or a one-year postgraduate course should already hold (or expect to hold by the time of taking up the award) a degree equivalent to a first-class or a high upper second from a UK university 2) Applicants for an undergraduate place to do a Bachelor degree must apply directly to a Cambridge College, submit completed application forms to the College by 15 October of the year prior to their intended year of entry and be offered a place by a College
Level of Study: Postgraduate, Doctorate
Type: Scholarships and fellowships
Value: The scholarships will normally be a substantial, means-tested contribution towards the University Composition Fee at the appropriate rate, approved College fees, a maintenance allowance sufficient for a single student, a contribution towards a return economy airfare by the cheapest available route
Frequency: Annual
Study Establishment: The University of Cambridge
Country of Study: United Kingdom
No. of awards offered: Varies
Application Procedure: The British American Tobacco Cambridge scholars will be selected by British American Tobacco, in consultation with the Cambridge Overseas Trust, from among those candidates who have been conditionally offered a place at Cambridge and nominated to British American Tobacco by the Cambridge Overseas Trust.
Closing Date: See the website
Contributor: Offered in collaboration with the British-American Tobacco (BAT) Company

BG Cambridge Scholarships

Subjects: All subjects.
Purpose: To support one-year postgraduate courses of study.
Eligibility: Applicants should already hold (or expect to hold by the time of taking up the award) a degree equivalent to a first-class from a UK university. Applicants should be the citizens of India, Trinidad & Tobago, Bolivia, Brazil, China, Kazakhstan, Tunisia, Madagascar, Nigeria, Egypt, Oman, Algeria, Chile, areas of Palestinian Authority, Libya, Malaysia, Philippines and Thailand. Four of the fourteen scholarships are a seperate partnership between BG Group and Cambridge Trusts, and are restricted as follows: two are for Nigerian nationals only, one is for Egyptian nationals only, one is for Omani

nationals only. Applicants must not be, or be related to, a BG Group employee.
Type: Scholarships
Value: Covers the University Composition Fee at the overseas rate, approved college fees, a maintenance allowance sufficient for a single student, a return economy airfare by the cheapest available route
Length of Study: 1 year
No. of awards offered: 6
Application Procedure: Suitably qualified candidates must apply for admission to the University of Cambridge by submitting a Graduate Application Form to the Board of Graduate Studies.
Contributor: The Cambridge Trusts

For further information contact:

Board of Graduate Studies, 4 Mill Lane, Cambridge, England, CB2 1RZ

Blyth Cambridge Commonwealth Trust Scholarship

Subjects: All except medicine and veterinary medicine.
Eligibility: Applicants should require financial assistance to study at the University of Cambridge. The scholarship is only available for citizens of Canada.
Level of Study: Graduate
Type: Scholarship
Value: University composition fee, college fee, annual stipend sufficient for a single student and one return economy airfare

Boustany Scholarship in Astronomy

Subjects: Astronomy.
Eligibility: The scholarship is available for citizens of any country outside the EU.
Level of Study: Doctorate
Type: Scholarship
Value: University composition fee, college fee, annual stipend and all expenses relating to internship
Frequency: Every 3 years

BP Cambridge Scholarships for Egypt

Subjects: Preference will be given to candidates who intend to pursue courses in petroleum-related or business-related subjects, but candidates proposing to study other subjects particularly relevant to Egypt's needs will also be eligible to apply.
Purpose: To financially support study towards a post graduation.
Eligibility: Preference will be given to candidates who intend to pursue courses in petroleum-related or business-related subjects, but candidates proposing to study other subjects particularly relevant to Egypt needs will also be eligible to apply. Applicants should normally have a first-class degree or its equivalent from a recognised university. Any person employed by BP Egypt at the time of application is not eligible to apply for the BP Cambridge Scholarships for Egypt. However, dependants of those employed by BP Egypt are welcome to apply through the normal competitive process
Level of Study: Postgraduate
Type: Scholarship
Value: The scholarships will normally cover up to the University Composition Fee at the appropriate rate, approved College fees, a maintenance allowance sufficient for a single student and a contribution towards a return economy airfare by the cheapest available route
Length of Study: 1 year
Frequency: Annual
Study Establishment: The University of Cambridge
Country of Study: United Kingdom
No. of awards offered: Up to 10
Application Procedure: Applicants must complete a preliminary application form, which can be obtained from local universities, offices of the British Council or the Trust. Completed forms must be returned to the main address. Shortlisted candidates will be sent forms for admission to the University of Cambridge. The preliminary application form can also be downloaded from www.admin.cam.ac.uk/offices/gradstud/admissions/forms/
Contributor: Offered in collaboration with BP

BP Chevening Cambridge Scholarships for Egypt

Subjects: Petroleum-related business and management studies; geology and geographical science relevant to Egypt's need.
Purpose: To provide financial assistance for students from overseas who, without help, would be unable to take up their places at Cambridge.
Eligibility: Candidate should be from Egypt. Preference will be given to candidates who intend to pursue one-year postgraduate courses of study in petroleum-related or business related subjects, but candidates proposing to study other subjects particularly relevant to Egypt's needs will also be eligible to apply.
Type: Scholarships
Value: University Composition Fee at the overseas rate, approved college fees, a maintenance allowance sufficient for a single student, a contribution towards a return economy airfare by the cheapest available route.
Length of Study: 1 year
No. of awards offered: Varies
Closing Date: See website for details.
Contributor: In collaboration with BP and the Foreign & Commonwealth Office

Britain-Australia Bicentennial Scholarships for Postgraduate Study

Subjects: All subjects.
Purpose: To financially support those undertaking postgraduate study.
Eligibility: Candidates must be citizens of Australia, must apply to the University of Cambridge and be offered a place at Cambridge in the normal way. All applicants must have a First Class or High Second Class (Honours) Degree or equivalent and normally be under 26.
Level of Study: Postgraduate
Type: Scholarships and fellowships
Value: Contribution to the overall costs
Length of Study: 1 year
Frequency: Annual
Study Establishment: Jesus College, the University of Cambridge
Country of Study: United Kingdom
No. of awards offered: 2
Application Procedure: Applicants must contact the Board of Graduate Studies.
Closing Date: January 30th
Contributor: Offered in collaboration with the Foreign and Commonwealth Office (FCO) and Jesus College, Cambridge

For further information contact:

The Board of Graduate Studies, 4 Mill Lane, Cambridge, Cambridgeshire, CB2 1RZ, England
Contact: The Secretary

British Chevening Cambridge Foundation Chile Scholarships.

Subjects: All subjects.
Purpose: To financially support study towards a PhD.
Eligibility: Applicants must be from Chile, must apply to the University of Cambridge and be offered a place at Cambridge in the normal way. All applicants must have a First Class or High Second Class (Honours) Degree or equivalent and normally be under 26. They must be successfully nominated for an Overseas Research Student (ORS) award.
Level of Study: Postgraduate
Type: Scholarships and fellowships
Value: The University Composition Fee at the overseas rate, approved college fees and a maintenance allowance sufficient for a single student and contribution towards return airfare
Length of Study: 1 year
Frequency: Annual
Study Establishment: The University of Cambridge
Country of Study: United Kingdom
No. of awards offered: 2
Application Procedure: candidates for the above awards must apply directly to the British Council in their home country. Candidates are reminded that they will also need to apply to the Cambridge Trusts on the Scholarship Application Form (SAF) in the usual way.
Closing Date: See the website
Contributor: Offered in collaboration with the Foreign and Commonwealth Office (FCO)

For further information contact:

The British Council, Eliodoro Yanez 832, Santiago de Chile, Chile

British Chevening Cambridge Scholarships for Postgraduate Study (Australia)

Subjects: All subjects.
Purpose: To financially support those undertaking postgraduate study.
Eligibility: Open for students from Australia. They should have a first-class or high second-class honours degree from a recognised university.
Level of Study: Postgraduate
Type: Scholarship
Value: The University Composition Fee at the overseas rate, approved college fees, a maintenance allowance sufficient for a single student and a contribution towards return economy airfare
Length of Study: 1 year
Frequency: Annual
Study Establishment: The University of Cambridge
Country of Study: United Kingdom
No. of awards offered: 2
Application Procedure: Applicants must write for information.
Closing Date: January 30th
Contributor: Offered in collaboration with the Foreign and Commonwealth Office (FCO)

For further information contact:

The Board of Graduate Studies, 4 Mill Lane, Cambridge, Cambridgeshire, CN2 1RZ, England
Contact: The Secretary

British Chevening Cambridge Scholarships for Postgraduate Study (Hong Kong)

Subjects: All subjects.
Purpose: To financially support those undertaking postgraduate study.
Eligibility: Applicants must be from Hong Kong, must apply to the University of Cambridge and be offered a place at Cambridge in the normal way. All applicants must have a First Class or High Second Class (Honours) Degree or equivalent and normally be under 26.
Level of Study: Postgraduate
Type: Scholarships and fellowships
Value: The University Composition Fee at the overseas rate, approved college fees, a maintenance allowance sufficient for a single student and a contribution towards return economy airfare
Length of Study: 1 year
Frequency: Annual
Study Establishment: The University of Cambridge
Country of Study: United Kingdom
No. of awards offered: Up to 8
Application Procedure: Applicants must complete a preliminary application form, which can be obtained from local universities, offices of the British Council or the main address. Completed forms must be returned to the main address. Shortlisted candidates will be sent forms for admission to the University of Cambridge and a scholarship application form. These forms must be returned to the Board of Graduate Studies at the address below.
Closing Date: February 28th
Contributor: Offered in collaboration with the Foreign and Commonwealth Office (FCO)

For further information contact:

The Board of Graduate Studies, 4 Mill Lane, Cambridge, Cambridgeshire, CB2 1RZ, England
Contact: The Secretary

British Chevening Cambridge Scholarships for Postgraduate Study (Mexico)

Subjects: All subjects.
Purpose: To financially support those undertaking postgraduate study.
Eligibility: Applicants must be from Mexico, must apply to the University of Cambridge and be offered a place at Cambridge in the normal way. All applicants must have a First Class or High Second Class (Honours) Degree or equivalent and normally be under 26.
Level of Study: Postgraduate
Type: Scholarships and fellowships
Value: The University Composition Fee at the overseas rate, approved college fees and a maintenance allowance sufficient for a single student
Length of Study: 1 year
Frequency: Annual
Study Establishment: The University of Cambridge
Country of Study: United Kingdom
No. of awards offered: 2
Application Procedure: Applicants for this scholarship must complete a preliminary application form, which can only be obtained from the British Council, Mexico City.
Closing Date: February 28th
Contributor: Offered in collaboration with the Foreign and Commonwealth Office (FCO)

For further information contact:

The British Council, Maestro Antonio Caso 127, Col San Rafael, Delegacion Cuauhtemoc, Apartado Postal 30-588, Mexico City, DF, 06470, Mexico

British Chevening Cambridge Scholarships for Postgraduate Study (Vietnam)

Subjects: All subjects.
Purpose: To financially support those undertaking postgraduate study.
Eligibility: Applicants must be from Vietnam, must apply to the University of Cambridge and be offered a place at Cambridge in the normal way. All applicants must have a First Class or High Second Class (Honours) Degree or equivalent and normally be under 26. They must be successfully nominated for an Overseas Research Student (ORS) award.
Level of Study: Postgraduate
Type: Scholarships and fellowships
Value: The University Composition Fee at the overseas rate, approved college fees, a maintenance allowance sufficient for a single student and a contribution towards return economy airfare
Length of Study: 1 year
Frequency: Annual
Study Establishment: The University of Cambridge
Country of Study: United Kingdom
No. of awards offered: Up to 2
Application Procedure: Applicants for these scholarships should apply directly to the British Embassy in Vietnam.
Closing Date: February 28th
Contributor: Offered in collaboration with the Foreign and Commonwealth Office (FCO)

For further information contact:

The British Embassy, 16 Ly Thuong Kiet, Hanoi, Vietnam

British Prize Cambridge Scholarships (Barbados and The Eastern Caribbean)

Subjects: All subjects relevant to their country's development.
Purpose: To financially support those undertaking postgraduate study.
Eligibility: Applicants must be from Barbados, Antigua & Barbuda, Dominica, Grenada Grenadines, St Kitts-Nevis, St Vincent, and have a First Class or High Second Class (Honours) Degree or equivalent. Normally they should be under 26.
Level of Study: Postgraduate
Type: Scholarships and fellowships
Value: The University Composition Fee at the overseas rate, approved college fees, maintenance allowance for a single student and a contribution towards return economy airfare
Length of Study: 1 year
Frequency: Dependent on funds available
Study Establishment: The University of Cambridge
Country of Study: United Kingdom
No. of awards offered: Up to 2
Application Procedure: Applicants for a place to do a PhD should apply for an ORS award and should normally be successfully

nominated for an ORS award or an ORS equivalent award, which meets the difference between the higher overseas rate and the lower domestic rate of the University composition fee.
Closing Date: See the website
Funding: Trusts
Contributor: Offered in collaboration with the Foreign and Commonwealth Office (FCO)

Brockmann Cambridge Scholarship

Subjects: All subjects.
Eligibility: The scholarship is only available for citizens of Mexico.
Level of Study: Postgraduate
Type: Scholarship
Value: University composition fee (in the case of applicants for the MBA and MFin, the scholarship will pay the University Composition Fee at the science rate, i.e. not the full Fee), college fee and annual stipend sufficient for a single student
Length of Study: 1 year
Additional Information: Scholars are required to return to work in Mexico within 2 years of completing their course in Cambridge.

Calbee Cambridge Scholarship

Subjects: All subjects.
Purpose: To financially support postgraduate study.
Eligibility: Offer for students from Japan. Preference being given to candidates who have been offered a place at Pembroke College.
Level of Study: Postgraduate
Type: Scholarships and fellowships
Value: Contribution towards fees
Length of Study: 1 year
Frequency: Annual
Study Establishment: The University of Cambridge
Country of Study: United Kingdom
No. of awards offered: 1
Application Procedure: Applicants must apply to the university of Cambridge and be offered a place at Cambridge in the normal way. Applicants should complete only one scholarship form, which will enable them to be considered for all awards for which they are eligible. Application forms can be downloaded from the website.
Closing Date: February 28th
Funding: Trusts
Contributor: Cambridge Trusts

Cambridge Assessment Scholarships

Subjects: English language, applied linguistics and education.
Purpose: To support research leading to the degree or Phd and one-year postgraduate courses.
Eligibility: Open to candidates who hold a first class honours degree or its equivalent from a recognised university.
Level of Study: Postgraduate
Type: Scholarships and fellowships
Value: University Composition Fee at the appropriate rate and college fees
Length of Study: 1 year
No. of awards offered: 2
Closing Date: See the website
Contributor: In collaboration with Cambridge Assessment

Cambridge Australia Scholarship

Subjects: All subjects.
Purpose: To financially support those undertaking postgraduate study through a number of award schemes.
Eligibility: Open to citizens of Australia. Applicants must apply to the University of Cambridge and be offered a place at Cambridge in the normal way. All applicants must have a First Class or High Second Class (Honours) Degree or equivalent and normally be under 26.
Level of Study: Doctorate, Postgraduate
Type: Scholarships and fellowships
Value: The University Composition Fee at the overseas rate, approved college fees, a maintenance allowance sufficient for a single student and a contribution towards return economy airfare
Frequency: Varies
Study Establishment: The University of Cambridge
Country of Study: United Kingdom
No. of awards offered: Varies
Application Procedure: Applicants must contact the Board of Graduate Studies.
Closing Date: January 30th
Contributor: Offered in collaboration with the Foreign and Commonwealth Office (FCO) and the Cambridge Australia Trust
Additional Information: Further information is available on request. For details of scholarships offered in collaboration with the Cambridge Australia Trust please see the website www.anu.edu.au/graduate/scholarships

For further information contact:

The Board of Graduate Studies, 4 Mill Lane, Cambridge, Cambridgeshire, CB2 1RZ, England
Contact: The Secretary

Cambridge Canada Scholarships for PhD Study

Subjects: All subjects.
Purpose: To financially support study towards a PhD.
Eligibility: Open to students from Canada. Applicants must apply to the University of Cambridge and be offered a place at Cambridge in the normal way. All applicants must have a First Class or High Second Class (Honours) Degree or equivalent and normally be under 26. They must be successfully nominated for an Overseas Research Student (ORS) award.
Level of Study: Doctorate, Predoctorate
Type: Scholarship
Value: The University Composition Fee at the home rate and approved college fees
Length of Study: Up to 3 years
Frequency: Annual
Study Establishment: The University of Cambridge
Country of Study: United Kingdom
No. of awards offered: Up to 5
Application Procedure: Application forms for the scholarship will be sent out to eligible candidates once the completed form for admission to the University of Cambridge has reached the Board of Graduate Studies.
Closing Date: See the website
Contributor: Cambridge Commonwealth Trust, Cambridge Overseas Trust, Gates Cambridge Trust, Cambridge European Trust and Associated Trusts

For further information contact:

The Board of Graduate Studies, 4 Mill Lane, Cambridge, Cambridgeshire, CB2 1RZ, England
Contact: The Secretary

Cambridge International Scholarship

Subjects: All subjects.
Purpose: To financially support students towards their Phd.
Eligibility: The scholarship is only available for citizens of countries outside the EU. Scholarships are offered to candidates who are highly ranked by their prospective Departments within the University, and are awarded on the basis of academic ability and research potential, examination results and references. The financial situation of candidates does not affect selection.
Level of Study: Doctorate
Type: Scholarship
Value: University composition fee, college fee and annual stipend sufficient for a single student
No. of awards offered: Varies
Closing Date: December 2nd
Additional Information: Please check website for latest updates.

Cambridge Mandela Scholarships for South Africa

Subjects: All subjects relevant to South Africa's needs.
Purpose: To financially support study towards a PhD and post-graduation.
Eligibility: Applicants must be from South Africa. Applicants must apply to the University of Cambridge and be offered a place at Cambridge in the normal way. All applicants must have a First Class or High Second Class (Honours) Degree or equivalent and normally be under 26. They must be successfully nominated for an Overseas Research Student (ORS) award.
Level of Study: Doctorate, Predoctorate

Type: Scholarship
Value: The University Composition Fee at the appropriate rate, approved college fees, a maintenance allowance sufficient for a single student and a contribution to return economy airfare
Length of Study: Up to 3 years
Frequency: Annual
Study Establishment: The University of Cambridge
Country of Study: United Kingdom
No. of awards offered: Up to 10
Application Procedure: Applicants must complete a preliminary application form, which can be obtained from local universities, offices of the British Council or the Trust. The preliminary application form can also be downloaded from www.admin.cam.ac.uk/offices/gradstud/admissions/forms/completed forms must be returned to the main address. Shortlisted candidates will be sent forms for admission to the University of Cambridge.
Contributor: Offered by the Malaysian Commonwealth Studies Centre, Trinity College, Cambridge, and the Cambridge University Press, Cambridge Assessment (formerly LES)
Additional Information: These scholarships are offered in honour of former South African President Nelson Mandela.

Cambridge Overseas Trust Scholarship

Subjects: All subjects.
Purpose: To support university graduates to pursue a one-year postgraduate course of study.
Eligibility: The scholarship is only available for citizens of countries outside the EU and the Commonwealth.
Type: Scholarship
Value: £6,000 or £12,000 depending on level of degree and financial requirement
Length of Study: 1 year
Closing Date: December 1st

Cambridge Rutherford Scholarship

Subjects: Science and technology.
Purpose: To support students to undertake Phd at the University of Cambridge.
Eligibility: Applicant must be the permanent resident of New Zealand.
Level of Study: Doctorate
Type: Scholarship
Value: £11,000 per year
Length of Study: 3 years
No. of awards offered: 2
Closing Date: August 1st

CAPES Cambridge Scholarship

Subjects: Science and technology.
Purpose: To support students to undertake Phd.
Eligibility: Preference will be given to applicants who already hold a Master's degree from an institution recognised by CAPES. The scholarship is only available for citizens of Brazil.
Level of Study: Doctorate
Type: Scholarship
Value: University composition fee, approved college fee, annual stipend sufficient for a single student and return travel
Length of Study: Up to 4 years
No. of awards offered: Up to 10
Closing Date: December 2nd
Contributor: CAPES Brazil

Charles Wallace India Trust

Subjects: Arts, Humanities, and Heritage Conservation.
Purpose: To financially assist postgraduate study for applicants who are not successful in winning a scholarship.
Eligibility: This scholarship is only available for citizens of India. Applicants for a place to do a PhD should apply for an ORS award and should be successfully nominated for an ORS award or an ORS equivalent award, which meets the difference between the higher overseas rate and the lower domestic rate of the University Composition Fee.
Level of Study: Any level
Type: Bursary
Value: Variable. Part Cost.
Frequency: Annual
Study Establishment: The University of Cambridge
Closing Date: See website for details.
Contributor: Charles Wallace India Trust

Charlie Perkins Scholarship at the University of Cambridge

Subjects: All subjects.
Purpose: To support students to undertake higher studies ar the University of Cambridge.
Eligibility: Applicants must be of Australian Aboriginal or Torres Strait Islander descent.
Level of Study: Doctorate, Predoctorate
Type: Scholarship
Value: University composition fee, approved college fee, annual maintenance allowance (suffcient for a single student), annual travel allowance of £1,000
Length of Study: Up to 4 years
No. of awards offered: 2
Closing Date: December 1st

Chevening Cambridge Scholarship for Koreans

Subjects: Human rights, conflict prevention/development studies and governance/public administration
Purpose: To support successful applicants to Masters courses at the University of Cambridge.
Eligibility: The scholarship is only available for citizens of South Korea. Applicants should demonstrate leadership potential, and require financial assistance to study at the University of Cambridge.
Level of Study: Postgraduate
Type: Scholarship
Value: University composition fee, college fee, annual stipend sufficient for a single student, one return economy airfare, IELTS testing (where necessary)
Length of Study: 1 year
Closing Date: December 2nd
Contributor: Foreign and Commonwealth Office and British Embassy Seoul

Churchill Research Studentship

Subjects: Engineering and chemical engineering.
Purpose: To support students to undertake Phd at the University of Cambridge.
Eligibility: The scholarship is available for citizens of any country outside the EU. Preference will be given to candidates who nominate Churchill as their first choice college.
Type: Studentship
Value: University composition fee, college fee and annual stipend sufficient for a single person
Closing Date: February 15th

CIALS Cambridge Scholarships

Subjects: Law.
Purpose: To support study towards the Master of Law (LLM) degree.
Eligibility: Open to graduates of Canadian law schools who have completed a Bachelor of Law degree on or before June 1st. Applicants must be Canadian citizens. Applicants must apply to the University of Cambridge and be offered a place at Cambridge in the normal way. All applicants must have a First Class or High Second Class (Honours) Degree or equivalent and normally be under 26.
Level of Study: Postgraduate
Type: Scholarship
Value: The scholarships will normally cover the University Composition Fee at the overseas rate, approved College fees. A maintenance allowance sufficient for a single student and a contribution towards a return economy airfare by the cheapest available route.
Also payable by the Commonwealth Scholarship Commission as applicable, thesis expenses of up to £225, study travel grant up to £100 and possible grant, upon application, towards excess baggage allowance for accompanied books up to 10k on return home
Length of Study: 1 year
Frequency: Annual
Study Establishment: The University of Cambridge
Country of Study: United Kingdom
No. of awards offered: 2

Application Procedure: Applicants must apply in writing to the Executive Director of the Canadian Institute for Advanced Legal Studies (CIALS).
Closing Date: December 31st
Contributor: Offered in collaboration with the CIALS

For further information contact:

The Canadian Institute of Advanced Legal Studies, 4 Beechwood Avenue, Ottawa, ON, K1L 8L9, Canada
Contact: Mr Frank E McArdle, Executive Director

Commonwealth Cambridge Scholarship

Subjects: All subjects.
Purpose: To offer the opportunity for individuals with proven academic merit to study towards a PhD and postgraduation at Cambridge.
Eligibility: This scholarship is available for citizens of any country that is a member of the Commonwealth. Applicants from developed countries of the Commonwealth (Australia, Bahamas, Brunei, Canada, Cyprus, Malta, New Zealand, Singapore) are only eligible for PhD scholarships. Scholars should be nominated for an Overseas Research Student (ORS) award.
Level of Study: Doctorate, Predoctorate
Type: Scholarship
Value: University Composition Fee, college fee, annual stipend and return airfare to/from the UK
Length of Study: Up to 3 years
Frequency: Annual
Study Establishment: The University of Cambridge
Country of Study: United Kingdom
No. of awards offered: Up to 15
Application Procedure: Candidates must apply to the local Commonwealth scholarship agency in their home country.
Closing Date: See the organization website
Contributor: Offered in collaboration with the Commonwealth Scholarship Commission in the United Kingdom

Commonwealth Shared Cambridge Scholarship

Subjects: Subjects relevant to the economic, social and technological development of the candidate's home country.
Purpose: To partly support those undertaking postgraduate study.
Eligibility: Open to citizens from developing countries of the Commonwealth. All applicants must have a First Class or High Second Class (Honours) Degree or equivalent and be under the age of 35 on October 1st of the year they are applying for, with priority given to those candidates under the age of 30. Applicants should not be presently living or studying in a developing country.
Level of Study: Postgraduate
Type: Scholarships and fellowships
Value: The University Composition Fee, college fee, annual stipend sufficient for a single student and contribution towards one return economy airfare
Length of Study: 1 year
Frequency: Annual
Study Establishment: The University of Cambridge
Country of Study: United Kingdom
No. of awards offered: 40
Application Procedure: Applicants must complete a preliminary application form, which can be obtained from local universities, offices of the British Council or the Trust. Completed forms must be returned to the main address. Shortlisted candidates will be sent forms for admission to the University of Cambridge. The preliminary application form can also be downloaded from www.admin.cam.ac.uk/offices/gradstud/admissions/forms/
Closing Date: February 28th
Contributor: Offered in collaboration with the Department for International Development (DFID)

CONACyT Cambridge Scholarship

Subjects: Biotechnology, energy and environment, engineering and applied sciences, computer science and information technology, applied mathematics.
Purpose: To support students to undertake Phd.
Eligibility: Applicants should already hold a Master's degree. The scholarship is only available for citizens of Mexico.
Level of Study: Doctorate
Type: Scholarship
Value: University composition fee, college fee and annual stipend sufficient for a single student
Closing Date: See the website for details.
Contributor: CONACyT

Corpus Christi ACE Scholarship

Subjects: Fields of conservation, development and environment.
Purpose: To support graduate students to undertake postgraduation at the University of Cambridge
Eligibility: The scholarship is available for citizens of any country outside the EU. Scholars must be offered a place at Corpus Christi College. The scholarships are mainly for candidates from the developing world, and preference may be given to candidates from Eastern Europe.
Level of Study: Graduate, Postgraduate
Type: Scholarship
Value: University composition fee, college fee and annual stipend sufficient for a single student
Length of Study: 1 year

Corpus Christi Research Scholarship

Subjects: Any subjects.
Purpose: To financially assist study towards a PhD.
Eligibility: Applicants must be from India, and already hold a degree equivalent to a first-class or a high upper second from a UK university.
Level of Study: Doctorate
Type: Scholarship
Value: The university composition fee at the appropriate rate, approved college fees, a maintenance allowance sufficient for a single student.
Study Establishment: Corpus Christi College, The University of Cambridge
Country of Study: United Kingdom
Application Procedure: Applicants for a place to do a PhD should apply for an ORS award and should normally be successfully nominated for an ORS award or an ORS equivalent award, which meets the difference between the higher overseas rate and the lower domestic rate of the university composition fee.
Closing Date: October 1st. See website for details.
Contributor: In collaboration with the Corpus Christi College

Croucher Cambridge International Scholarship

Subjects: Natural sciences, technology and medicine (excluding clinical medicine).
Purpose: To support students to undertake Phd and pursue their research abroad.
Eligibility: The scholarship is only available for citizens of Hong Kong.
Level of Study: Doctorate
Type: Scholarship
Value: University composition fee, college fee and annual stipend sufficient for a single student
Closing Date: See the website for details

CSC Cambridge Scholarship

Subjects: Priority subjects set each year by the China Scholarship Council.
Purpose: To support research leading to the degree of PhD.
Eligibility: Applicants should have a degree from a recognised university in the People's Republic of China.
Type: Scholarship
Value: University Composition Fee, college fee, annual stipend sufficient for a single student and contribution towards travel costs
No. of awards offered: 30
Application Procedure: Candidates must submit a copy of their GRADSAF form and supporting documents to the China Scholarship Council at the same time as they send their application for admission at the University of Cambridge to the Board of Graduate Studies, Cambridge.
Closing Date: February 20th
Contributor: In collaboration with the China Scholarship Council

David M. Livingstone (Australia) Scholarship

Subjects: All subjects.
Purpose: To support students to undertake 1-year postgraduate degree course at the University of Cambridge.
Eligibility: The scholarship is only available for citizens of Australia. Scholars must specify Jesus College as their first choice college.
Level of Study: Postgraduate
Type: Scholarship
Value: University composition fee and college fee
Length of Study: 1 year
Frequency: Annual
No. of awards offered: 2
Closing Date: March 31st
Contributor: Jesus College

For further information contact:

Board of Graduate Studies (The Cambridge Trust)
Tel: 44 01223760606
Fax: 44 01223338723
Email: admissions@gradstudies.cam.ac.uk

Developing World Education Fund Scholarships for PhD Study

Subjects: Any subjects.
Purpose: To financially support study towards a PhD.
Eligibility: For citizens from Bangladesh, China, India, Pakistan, Sri Lanka or Zambia. Applicants must apply to the University of Cambridge and be offered a place at Cambridge in the normal way. All applicants must have a degree equivalent to a First Class from a UK university, and normally be under 26.
Level of Study: Doctorate
Type: Scholarships and fellowships
Value: The University Composition Fee at the appropriate rate, approve College fees, a maintenance allowance sufficient for a single student, contribution towards an economy return airfare
Frequency: Varies
Study Establishment: The University of Cambridge
Country of Study: United Kingdom
No. of awards offered: Varies
Application Procedure: Applicants for a place to do a PhD should apply for an ORS award and should normally be successfully nominated for an ORS award or an ORS equivalent award, which meets the difference between the higher overseas rate and the lower domestic rate of the University Composition Fee.
Closing Date: See website for details.

Developing World Education Fund Scholarships for Postgraduate Study

Subjects: All subjects.
Purpose: To financially support those undertaking postgraduate study.
Eligibility: For citizens from Bangladesh, China, India, Pakistan, Sri Lanka, or Zambia. Applicants must apply to the University of Cambridge and be offered a place at Cambridge in the normal way. All applicants must have a degree equivalent to a First Class from a UK university and normally be under 26.
Level of Study: Postgraduate
Type: Scholarships and fellowships
Value: The University Composition Fee at the overseas rate, approved college fees, a maintenance allowance sufficient for a single student and a contribution towards return economy airfare
Length of Study: 1 year
Frequency: Varies
Study Establishment: The University of Cambridge
Country of Study: United Kingdom
No. of awards offered: Varies
Application Procedure: Applicants must complete a preliminary application form, which can be obtained from local universities, offices of the British Council or the Trust. Completed forms must be returned to the main address. Shortlisted candidates will be sent forms for admission to the University of Cambridge. The preliminary application form can also be downloaded from www.admin.cam.ac.uk/offices/gradstud/admissions/forms/
Closing Date: February 28th
Contributor: Offered in collaboration with the Developing World Education Fund

Dharam Hinduja Cambridge Scholarships and Bursaries

Subjects: Any subject.
Purpose: To financially support students from India.
Eligibility: Applicants from India should already hold a degree equivalent to a first class or a high upper second from a UK university.
Level of Study: Postgraduate, Any level
Type: Scholarship/Bursary
Value: he scholarships will normally cover up to the University Composition Fee at the appropriate rate, approved College fees and a maintenance allowance sufficient for a single student
Frequency: Varies
Study Establishment: The University of Cambridge
Country of Study: United Kingdom
No. of awards offered: 8
Application Procedure: Applicants for a place to do a PhD should apply for an ORS award and should be successfully nominated for an ORS award or an ORS equivalent award, which meets the difference between the higher overseas rate and the lower domestic rate of the university composition fee.
Closing Date: See website for details.
Contributor: In collaboration with Hinduja Cambridge Trust
Additional Information: For further information visit www.cam.ac.uk.

Dharam Hinduja Commonwealth Shared Scholarships

Subjects: All subjects relevant to the needs of India.
Purpose: To offer financial support.
Eligibility: Applicants must be from India, and must be under the age of 35 on October 1st, with priority given to those candidates under the age of 30. Applicants must not be employed by a national or local government department or by a parastatal organization, or at present be living or studying in a developed country. Priority will be given to candidates wishing to pursue a course of study related to the economic and social development of their country. Applicants should have a first class degree.
Level of Study: Postgraduate
Type: Scholarships and fellowships
Value: The University Composition Fee at the appropriate rate, approved college fees, a maintenance allowance sufficient for a single student and a contribution towards return economy airfare
Length of Study: 1 year
Frequency: Annual
Study Establishment: The University of Cambridge
Country of Study: United Kingdom
No. of awards offered: 4
Application Procedure: Applicants may obtain further details and a preliminary application form by writing before August 16th of the year before entry to the address given below with details of academic qualifications.
Closing Date: February 28th
Contributor: Offered in collaboration with the Hinduja Cambridge Trust and the Department for International Development

For further information contact:

The Nehru Trust for Cambridge University, Teen Murti House, Teen Murti Marg, New Delhi, 110011, India
Contact: The Joint Secretary

Doris Zimmern Scholarship

Subjects: All subjects.
Purpose: To support students to undertake 1-year postgraduate or PhD at the University of Cambridge.
Eligibility: Applicants should have a degree from the University of Hong Kong, or have been accepted for PhD studies at the University of Hong Kong. Scholars must be offered a place at Hughes Hall. The scholarship is only available for citizens of Hong Kong.
Level of Study: Doctorate, Postgraduate
Type: Scholarship
Value: University composition fee, college fee and annual stipend sufficient for a single student
Length of Study: 1 year
Closing Date: See the website for details

Edwin S H Leong Hughes Hall Scholarship

Subjects: All subjects.
Purpose: To support students to undertake 1-year postgraduate or PhD.
Eligibility: Applicants should normally have a first-class degree or equivalent from the University of Hong Kong, or been admitted as a postgraduate student to the University of Hong Kong. Scholarships are only tenable at Hughes Hall, and preference will be given to applicants naming Hughes Hall as first-choice college. The scholarship is only available for citizens of Hong Kong.
Level of Study: Doctorate, Postgraduate
Type: Scholarship
Value: University composition fee, college fee, annual stipend sufficient for a single student and contribution towards travel costs
Length of Study: 1 year
Closing Date: See the website for details

FCO-China Chevening Fellowships for Postgraduate Study (China)

Subjects: All subjects.
Purpose: To financially support those undertaking postgraduate study.
Eligibility: Open to citizens of China. Applicants must apply to the University of Cambridge and be offered a place at Cambridge in the normal way. All applicants must have a First Class or High Second Class (Honours) Degree or equivalent and normally be under 26.
Level of Study: Postgraduate
Type: Scholarships and fellowships
Value: The University Composition Fee at the overseas rate, approved college fees, a maintenance allowance sufficient for a single student and a contribution towards return economy airfare
Length of Study: 1 year
Frequency: Annual
Study Establishment: The University of Cambridge
Country of Study: United Kingdom
No. of awards offered: 3
Application Procedure: Applicants must complete a preliminary application form, which can be obtained from local universities, offices of the British Council or the Trust. Completed forms must be returned to the main address. Shortlisted candidates will be sent forms for admission to the University of Cambridge. The preliminary application form can also be downloaded from www.admin.cam.ac.uk/offices/gradstud/admissions/forms/
Closing Date: February 28th
Contributor: Offered in collaboration with the Foreign and Commonwealth Office (FCO)

First Canadian Donner Foundation Research Cambridge Scholarship

Subjects: All subjects.
Purpose: To financially support study towards a PhD.
Eligibility: Open to citizens of Canada or the USA, who excel in sport. Candidates must gain admission to Magdalene College, Cambridge in the normal way. All applicants must be successfully nominated for an Overseas Research Student (ORS) award.
Level of Study: Doctorate, Postgraduate, Predoctorate
Type: Scholarships and fellowships
Value: Up to £15,000 per year
Length of Study: 3 years or 1 year postgraduate
Frequency: Annual
Study Establishment: Magdalene College, the University of Cambridge
Country of Study: United Kingdom
No. of awards offered: 1
Application Procedure: Applicants for a place to do a PhD should apply for an ORS award and should normally be successfully nominated for an ORS award or an ORS equivalent award, which meets the difference between the higher overseas rate and the lower domestic rate of the university composition fee.
Closing Date: See the website
Contributor: Offered in collaboration with the Canadian Donner Foundation and Magdalene College and the Centre for International Studies

Gates Cambridge Scholarship

Subjects: All subjects.
Purpose: To support students to undertake postgraduate degee at the University of Cambridge.
Eligibility: See http://www.gatesscholar.org/apply/who-is-eligible.asp for details
Level of Study: Graduate, Postgraduate
Type: Scholarship
Value: Full cost of studying at Cambridge. The total amount of the award will vary depending on the applicant's home country, number of dependants and other factors. See http://www.gatesscholar.org/apply/value.asp for details
Frequency: Annual
Country of Study: United Kingdom
Application Procedure: All applicants apply simultaneously for admission as a graduate student to the University of Cambridge and a Gates Cambridge Scholarship using the one application pack.
Closing Date: October 15th (for US citizens currently undertaking a degree programme in the USA or currently residing in the USA), December 1st (for citizens of all countries other than the USA (except the United Kingdom) and US citizens currently residing outside of the USA). See http://www.gatesscholar.org/apply/application-timetable.asp for details
Funding: Foundation

George and Mary Vergottis Cambridge Bursaries

Subjects: All subjects.
Purpose: To financially support those undertaking postgraduate study.
Eligibility: Open to citizens of Greece.
Level of Study: Postgraduate, Doctorate
Type: Studentships and bursaries
Value: Varies
Length of Study: Varies
Frequency: Annual
Study Establishment: The University of Cambridge
Country of Study: United Kingdom
No. of awards offered: Varies
Closing Date: See the website
Contributor: In collaboration with The George and Mary Vergottis Cambridge Fund

Gita Wirjawan Graduate Fellowship

Subjects: All subjects.
Purpose: To support students to undertake a 1-year master's program at the University of Cambridge.
Eligibility: The scholarship is only available for citizens of Indonesia. Applicants should have (a) a confirmed acceptance at any faculty at the University of Cambridge; (b) an excellent academic record with a first degree equivalent to a good Second Class (Upper) Honors or a GPA of at least 3.5; (c) a very good command of the English language; and (d) assessed to have outstanding potential for leadership in government, business, or civil society after graduation.
Level of Study: Postgraduate
Type: Fellowship
Value: University composition fee, college fee, annual stipend sufficient for a single studentand return travel
Length of Study: 1 year
Frequency: Annual
No. of awards offered: 2
Closing Date: December 2nd
Additional Information: Please check website for latest updates.

Grace and Thomas C.H. Chan Scholarships

Subjects: Arts, humanities and social sciences.
Purpose: To support students to undertake Phd at the University of Cambridge.
Eligibility: Scholars must have been nominated for a University of Cambridge CISS award. The scholarship is only available for citizens of China.
Level of Study: Doctorate
Type: Scholarships
Value: University composition fee, college fee and annual stipend sufficient for a single student
Closing Date: See the website for details

Grosvenor Cambridge Scholarship

Subjects: Real Estate Finance, planning, growth and regeneration and environmental policy.
Purpose: To financially support those undertaking postgraduate study.
Eligibility: This scholarship is only available for citizens of China, and holders of Hong Kong and Macau SAR passports.
Level of Study: Postgraduate
Type: Scholarship
Value: £20,000
Length of Study: 1 year
Frequency: Annual
Study Establishment: The University of Cambridge
Country of Study: United Kingdom
No. of awards offered: 3
Closing Date: See the website for details.
Additional Information: Preference will normally be given to scholars with places at Pembroke College or Gonville & Caius College.

Guan Ruijun Memorial Bursary

Subjects: All subjects.
Purpose: To support postgraduate study in memory of Miss Guan Ruijun.
Eligibility: Preference will be given to students from Peking University. Applicants must have a First Class or High Second Class (Honours) Degree or equivalent and normally be under 26, from China.
Level of Study: Postgraduate
Type: Studentships and bursaries
Value: Part-cost contribution towards the fees of the bursary
Length of Study: 1 year
Frequency: Dependent on funds available
Study Establishment: Wolfson College, Cambridge
Country of Study: United Kingdom
No. of awards offered: 1
Application Procedure: Applicants must apply to the university of Cambridge and be offered a place at Cambridge in the normal way. Applicants should complete only one scholarship form, which will enable them to be considered for all awards for which they are eligible. Application forms can be downloaded from the website.
Closing Date: February 28th
Contributor: Wolfson College and Cambridge Trusts

Hamilton Cambridge International Scholarship

Subjects: All subjects.
Purpose: To support students to undertake Phd at the University of Cambridge.
Eligibility: The scholarship is available for citizens of any country outside the EU. And applicants should select Selwyn College as their first-choice College.
Type: Scholarship
Value: University composition fee, college fee and annual stipend sufficient for a single student
Closing Date: See the website for details.
Contributor: Selwyn College

Hughes Hall Cambridge International Scholarship

Subjects: All subjects.
Purpose: To support students to undertake Phd.
Eligibility: Applicants should be currently studying for a one-year postgraduate degree at the University of Cambridge, and should be members of Hughes Hall. The scholarship is available for citizens of any country outside the EU.
Type: Scholarship
Value: University composition fee, college fee and annual stipend sufficient for a single student
Closing Date: See the website for details
Additional Information: Please check website for latest updates.

Hutchison Whampoa Chevening Scholarship

Subjects: All subjects.
Purpose: To financially support those undertaking postgraduate study.
Eligibility: Open to students from China and Hong Kong. Applicants must apply to the University of Cambridge and be offered a place at Cambridge in the normal way. All applicants must have a First Class or High Second Class (Honours) Degree or equivalent.
Level of Study: Postgraduate
Type: Scholarship
Value: The University Composition Fee, approved college fees, a maintenance allowance sufficient for a single student and a contribution towards travel costs
Length of Study: 1 year
Study Establishment: The University of Cambridge
Country of Study: United Kingdom
No. of awards offered: Up to 21
Application Procedure: Applicants must complete a preliminary application form, which can be obtained from local universities, offices of the British Council or the Trust. Completed forms must be returned to the main address. Shortlisted candidates will be sent forms for admission to the University of Cambridge. The preliminary application form can also be downloaded from www.admin.cam.ac.uk/offices/gradstud/admissions/forms/
Closing Date: February 28th
Contributor: In collaboration with Hutchison Whampoa and the Foreign and Commonwealth Office (FCO)

ICBC Cambridge Scholarship

Subjects: Education.
Purpose: To financially support those undertaking postgraduate study.
Eligibility: Open to students from Chile.
Level of Study: Postgraduate
Type: Scholarships and fellowships
Value: The University composition fee at the overseas rate, approved college fees, a maintenance allowance sufficient for a single student, contribution towards an economy return airfare.
Length of Study: 1 year
Frequency: Annual
Study Establishment: The University of Cambridge
Country of Study: United Kingdom
No. of awards offered: 1
Application Procedure: Apply directly to the British Council in Chile. Applicants are reminded that they will also need to apply to the Cambridge Trusts on the Scholarship Application Form (SAF) in the usual way.
Closing Date: See website for details.
Contributor: Offered in collaboration with the Instituto Chileno Britanico de Cultura, the British Council, Chile and Cambridge Assessment (formerly the Local Examinations Syndicate), University of Cambridge.

IDB Cambridge International Scholarship

Subjects: Science and technology.
Purpose: To support students to undertake Phd.
Eligibility: The scholarship is available for citizens of members countries of the Islamic Development Bank: Afghanistan, Albania, Algeria, Azerbaijan, Bahrain, Bangladesh, Benin, Brunei, Burkina Faso, Cameroon, Chad, Comoros, Cote d'Ivoire, Djibouti, Egypt, Gabon, Gambia, Guinea, Guinea-Bissau, Indonesia, Iran, Iraq, Jordan, Kazakstan, Kuwait, Kyrgyzstan, Lebanon, Libya, Malaysia, Maldives, Mali, Mauritania, Morocco, Mozambique, Niger, Nigeria, Oman, Pakistan, Palestine, Qatar, Saudi Arabia, Senegal, Sierra Leone, Somalia, Sudan, Suriname, Syria, Tajikistan, Togo, Tunisia, Turkey, Turkmenistan, Uganda, United Arab Emirates, Uzbekistan and Yemen.
Level of Study: Doctorate
Type: Scholarship
Value: University composition fee, college fee, annual stipend sufficient for a single student and one return economy airfare
Closing Date: See the website for details.

Isaac Newton Trust European Research Studentships

Subjects: All subjects.
Purpose: To support research leading to a PhD.
Eligibility: Open to candidates from the European Union.
Level of Study: Doctorate
Type: Studentships and bursaries
Value: UK£2,000 per year
Length of Study: 3 years

Frequency: Annual
Study Establishment: The University of Cambridge
Country of Study: United Kingdom
No. of awards offered: 33
Application Procedure: Applicants must contact the Trust.
Closing Date: See the website
Contributor: Offered in collaboration with the Isaac Newton Trust and the Cambridge European Trust

Jawaharlal Nehru Memorial Trust Cambridge Scholarships

Subjects: All subjects.
Purpose: To financially support study towards a PhD.
Eligibility: Open to candidates from India. All applicants must be successful in winning an Overseas Research Student (ORS) award. Those who have, in addition to a First Class (Honours) Degree, a First Class Master's Degree or equivalent, may be given preference.
Level of Study: Doctorate
Type: Scholarship
Value: The University Composition Fee, approved college fees, annual stipend sufficient for a single student and contribution towards travel costs
Length of Study: 2 years
Frequency: Annual
Study Establishment: Trinity College, the University of Cambridge
Country of Study: United Kingdom
No. of awards offered: 1
Application Procedure: Applicants may obtain further details and a preliminary application form by writing before August 16th of the year before entry to the Joint Secretary of the Nehru Trust for Cambridge University, giving details of academic qualifications.
Closing Date: See the website for details
Funding: Trusts
Contributor: Offered in collaboration with the Jawaharlal Nehru Memorial Trust and Trinity College, Cambridge

For further information contact:

The Nehru Trust for Cambridge University, Teen Murti House, Teen Murti Marg, New Delhi, 110011, India
Contact: The Joint Secretary

Jawaharlal Nehru Memorial Trust Commonwealth Shared Scholarships

Subjects: All subjects.
Purpose: To offer financial support.
Eligibility: Open to citizens from India. All applicants must be under the age of 35 on October 1st with priority given to those candidates under the age of 30. They must not be employed by a national or local government department or by a parastatal organization, nor at present be living or studying in a developed country and not have undertaken studies lasting a year or more in a developed country. Priority will be given to candidates wishing to pursue a study related to the economic and social development of their country.
Level of Study: Postgraduate
Type: Scholarship
Value: The University Composition Fee, approved college fees, annual stipend sufficient for a single student and contribution towards travel costs
Length of Study: 1 year
Frequency: Annual
Study Establishment: The University of Cambridge, Trinity College
Country of Study: United Kingdom
No. of awards offered: 2
Application Procedure: Applicants may obtain further details and a preliminary application form by writing before August 16th of the year before entry to the Joint Secretary of the Nehru Trust for Cambridge University, giving details of their academic qualifications.
Closing Date: February 28th
Contributor: Offered in collaboration with the Jawaharlal Nehru Memorial Trust and the Commonwealth Scholarship Commission

For further information contact:

The Nehru Trust for Cambridge University, Teen Murti House, Teen Murti Marg, New Delhi, 110011, India
Contact: The Joint Secretary

Kapitza Cambridge Scholarships

Subjects: All subjects.
Purpose: To financially support study towards a PhD.
Eligibility: Open to students from countries of the former Soviet Union. Applicants must apply to the University of Cambridge and be offered a place at Cambridge in the normal way. All applicants must have a First Class degree or a master's degree or its equivalent and normally be under 26.
Level of Study: Postdoctorate
Type: Scholarships and fellowships
Value: The University Composition Fees at the overseas rate, approved college fees, a maintenance allowance sufficient for a single student and a contribution towards return economy airfare
Length of Study: 1 year
Frequency: Annual
Study Establishment: The University of Cambridge
No. of awards offered: Varies
Application Procedure: Applicants must complete a preliminary application form, which can be obtained from local universities, offices of the British Council or the Trust. Completed forms must be returned to the main address. Shortlisted candidates will be sent forms for admission to the University of Cambridge. The preliminary application form can also be downloaded from www.admin.cam.ac.uk/offices/gradstud/admissions/forms/
Closing Date: February 28th
Contributor: Offered in collaboration with Trinity College, Cambridge and Mr Stephen Anderman

Karim Rida Said Cambridge Scholarship for Postgraduate Study

Subjects: All subjects.
Purpose: To financially support those undertaking postgraduate study.
Eligibility: Applicants must be from Iraq, Jordan, Lebanon, Israel, Palestine or Syria. Applicants must apply to the University of Cambridge and be offered a place at Cambridge in the normal way. All applicants must have a First Class or High Second Class (Honours) Degree or equivalent and may be up to the age of 40. They must be successfully nominated for an Overseas Research Student (ORS) award. Priority given to those with at least 2 years of work experience.
Level of Study: Postgraduate
Type: Scholarships and fellowships
Value: The University Composition Fee at the overseas rate, approved college fees, a maintenance allowance sufficient for a single student and a contribution towards a return economy airfare
Length of Study: 1 year
Frequency: Annual
Study Establishment: The University of Cambridge
Country of Study: United Kingdom
No. of awards offered: 7
Application Procedure: Applicants must complete a preliminary application form, which can be obtained from local universities, offices of the British Council or the Trust. Completed forms must be returned to the main address. Shortlisted candidates will be sent forms for admission to the University of Cambridge. The preliminary application form can also be downloaded from www.admin.cam.ac.uk/offices/gradstud/admissions/forms/
Closing Date: February 28th
Contributor: Offered in collaboration with the Karim Rida Said Foundation
Additional Information: This scholarship is offered in memory of Karim Rida Said.

Kenneth Sutherland Memorial Scholarship

Subjects: All subjects.
Purpose: To support study towards a PhD.
Eligibility: Applicants must be from Canada and must be successfully nominated for an ORS award. All applicants must have a First Class or High Second Class (Honours) Degree or equivalent and normally be under 30. Preference will be given to applications in engineering.
Level of Study: Doctorate
Type: Scholarship
Value: The university composition fee at home rate and approved college fees
Length of Study: 3 years

Frequency: Dependent on funds available
Study Establishment: Jesus College, University of Cambridge
Country of Study: United Kingdom
No. of awards offered: 1
Application Procedure: Applicants must apply to the university of Cambridge and be offered a place at Cambridge in the normal way. Applicants should complete only one scholarship form, which will enable them to be considered for all awards for which they are eligible. Application forms can be downloaded from the website. Candidates must also apply for an ORS award.
Closing Date: February 28th
Funding: Trusts
Contributor: Cambridge Trust and Jesus college
Additional Information: Applicants should select Jesus College as their first choice.

Khazanah Cambridge Scholarship

Subjects: All subjects.
Eligibility: This scholarship is only available for citizens of Malaysia. Preference will be given to applications in business, bioscience and engineering.
Type: Scholarship
Value: University composition fee, college fee, annual stipend sufficient for a single student and one return economy airfare
Additional Information: Please check website for latest updates.

Lady Noon Cambridge Shared Scholarships

Subjects: All subjects.
Purpose: To financially support those undertaking postgraduate study.
Eligibility: Open to students from Pakistan. All applicants must be under the age of 35 on October 1st with priority given to those candidates under the age of 30. They must not be employed by a national or local government department or by a parastatal organization, nor at present be living or studying in a developed country. Priority will be given to candidates wishing to pursue a course of study related to the economic and social development of their own country.
Level of Study: Postgraduate
Type: Scholarship
Value: The University Composition Fee at overseas rate, approved college fees, a maintenance allowance sufficient for a single student and a contribution towards return economy airfare
Length of Study: 1 year
Frequency: Annual
Study Establishment: The University of Cambridge
Country of Study: United Kingdom
No. of awards offered: Varies
Application Procedure: Applicants must complete a preliminary application form, which can be obtained from local universities, offices of the British Council or the Trust. Completed forms must be returned to the main address. Shortlisted candidates will be sent forms for admission to the University of Cambridge. The preliminary application form can also be downloaded from www.admin.cam.ac.uk/offices/gradstud/admissions/forms/
Closing Date: February 28th
Contributor: Offered in collaboration with the Noon Educational Foundation and the Commonwealth Scholarship Commission

Link Foundation Cambridge Bursaries

Subjects: All subjects.
Purpose: To financially support those undertaking postgraduate study.
Eligibility: Open to citizens of New Zealand. Applicants must apply to the University of Cambridge and be offered a place at Cambridge in the normal way. All applicants must have a First Class or High Second Class (Honours) Degree or equivalent and normally be under 26.
Level of Study: Postgraduate
Type: Studentships and bursaries
Value: Part-cost bursaries
Length of Study: 1 year
Frequency: Annual
Study Establishment: The University of Cambridge
Country of Study: United Kingdom
No. of awards offered: 3
Application Procedure: Applicants must contact the Board of Graduate Studies.
Closing Date: February 28th
Contributor: Offered in collaboration with the Link Foundation for UK-New Zealand Relations (formerly known as the Waitangi Foundation) and the Foreign and Commonwealth Office (FCO)

For further information contact:

The Board of Graduate Studies, 4 Mill Hill, Cambridge, Cambridgeshire, CB2 1RZ, England
Contact: The Secretary

LMB Cambridge Scholarship

Subjects: Molecular biology.
Purpose: To financially support study towards a PhD.
Eligibility: Applicants must apply to the University of Cambridge and be offered a place at Cambridge in the normal way. All applicants must have a First Class or High Second Class (Honours) Degree or equivalent and normally be under 26.
Level of Study: Doctorate, Predoctorate
Type: Scholarship
Value: University Composition Fee, college fee, annual stipend sufficient for a single student, contribution towards travel costs
Length of Study: Up to 3 years
Frequency: Annual
Study Establishment: The Laboratory of Molecular Biology (LMB), the University of Cambridge
Country of Study: United Kingdom
No. of awards offered: Up to 5
Application Procedure: Candidates for LMB Cambridge Scholarships should apply directly to the LMB by: nominating up to four possible PhD supervisors and projects from the list of available projects as published by the LMB; providing a personal statement of not more than one page (for each proposed supervisor) explaining why they wish to work in that chosen area of research and their reasons for nominating that supervisor; providing a copy of an up-to-date curriculum vitae; and providing a copy of their transcripts.
Closing Date: December 13th
Contributor: Offered in collaboration with the Laboratory of Molecular Biology, Cambridge
Additional Information: This scholarship is available for citizens of any country outside the EU.

For further information contact:

The MRC Laboratory of Molecular Biology, Hills Road, Cambridge, Cambridgeshire, CB2 2QH, England
Website: www.mrc-lmb.cam.ac.uk
Contact: Director of Studies

Malaysian Commonwealth Scholarship

Subjects: Any subject.
Purpose: To financially support study towards graduation, postgraduation and PhD.
Eligibility: Open to outstanding calibre students from Malaysia and the countries of the Commonwealth.
Level of Study: Doctorate
Type: Scholarship/Bursary
Value: A means tested contribution towards fees only or towards the overall costs of studying at Cambridge
Frequency: Annual
Study Establishment: The Cambridge Commonwealth Trust
Country of Study: United Kingdom
No. of awards offered: Varies
Application Procedure: Applicants for a place to do a PhD should apply for an ORS award and should normally be successfully nominated for an ORS award or an ORS equivalent award, which meets the difference between the higher overseas rate and the lower domestic rate of the University Composition Fee.
Closing Date: See the website
Contributor: In collaboration with the Trustees of the Malaysian Commonwealth Studies Centre and of the Cambridge Commonwealth Trust

Mandela Magdalene College Scholarships for South Africa

Subjects: All subjects relevant to South Africa's needs.

Purpose: To financially support those undertaking postgraduate study and research.
Eligibility: Students must have been offered a place at Magdalene College, Cambridge and be citizens of South Africa. All applicants must have a First Class or High Second Class (Honours) Degree or equivalent and normally be under 35.
Level of Study: Postgraduate
Type: Scholarship
Value: The University Composition Fee, approved college fee, a maintenance allowance sufficient for a single student and a contribution to return airfare
Length of Study: 1 year
Frequency: Annual
Study Establishment: Magdalene College, the University of Cambridge
Country of Study: United Kingdom
No. of awards offered: Up to 3
Application Procedure: Applicants must complete a preliminary application form, which can be obtained from local universities, offices of the British Council or the Trust. Completed forms must be returned to the main address. Shortlisted candidates will be sent forms for admission to the University of Cambridge. The preliminary application form can also be downloaded from www.admin.cam.ac.uk/offices/gradstud/admissions/forms/
Closing Date: December 2nd
Contributor: Offered in collaboration with Magdalene College, Cambridge and Mr Chris von Christierson

MOHE Egypt Cambridge Scholarship

Subjects: All subjects.
Eligibility: Applicants should be working in a public university or research institute in Egypt at the time of application. The scholarship is only available for citizens of Egypt.
Level of Study: Doctorate
Type: Scholarship
Value: University composition fee, approved college fee, annual stipend sufficient for a single student and one return airfare
Contributor: Ministry of Higher Education & Scientific Research, Arab Republic of Egypt

Nanoscience/Engineering Cambridge Scholarship

Subjects: Nanoscience and engineering.
Eligibility: The scholarship is available for citizens of any country outside the EU.
Level of Study: Doctorate
Type: Scholarship
Value: University composition fee, college fee, annual stipend sufficient for a single student and contribution towards travel costs
Closing Date: December 2nd

National Research Foundation Cambridge Scholarships

Subjects: Any subject relevant to South Africa's identified research priority needs.
Purpose: To financially support research study courses leading towards PhD, in order to contribute towards the building of a highly skilled doctoral community in South Africa.
Eligibility: Open to students form South Africa. After completion of their degree, candidates must return to South Agrica for a period equivalent to the duration of support.
Level of Study: Doctorate
Type: Scholarships and fellowships
Value: A means tested scholarship, usually a contribution of UK£14,000 per year towards the overall annual costs of doing a PhD at Cambridge
Frequency: Annual
Study Establishment: The University of Cambridge
Country of Study: United Kingdom
No. of awards offered: Up to 5
Application Procedure: Applicants for a place to do a PhD should apply for an ORS award and should normally be successfully nominated for an ORS award or an ORS equivalent award, which meets the difference between the higher overseas rate and the lower domestic rate of the University Composition Fee.
Closing Date: See the website
Contributor: In collaboration with the National Research Foundation, South Africa

Nedcor Cambridge Awards

Subjects: Economics or Finance.
Purpose: To financially support study towards MPhil in Economics or Finance.
Eligibility: Open to students of South Africa doing a MPhil in Economics or Finance.
Level of Study: Postgraduate
Type: Award
Value: The award is a substantial cash prize from Nedcor which is awarded to the winner of the annual Nedbank Old Mutual Budget Speech competition. If the winner of the prize has been offered a place at Cambridge to read Economics or Finance, the cash prize will be tenable in conjunction with a substantial grant from the Cambridge Commonwealth Trust
Frequency: Dependent on funds available
Study Establishment: The University of Cambridge
Country of Study: United Kingdom
No. of awards offered: 1
Application Procedure: Participants in the Nedbank Old Mutual Budget Speech competition who wish to apply to Cambridge by the deadlines set by the Departments of Economics and the Judge Business School.
Closing Date: See website for details.
Contributor: In collaboration with Nedcor Bank, South Africa

Nehru Cambridge Scholarship

Subjects: Any subject.
Purpose: To financially support study towards a PhD.
Eligibility: This scholarship is only available for citizens of India.
Type: Scholarship
Value: The University Composition Fee at the appropriate rate, approved college fees, a maintenance allowance sufficient for a single student and a contribution towards a return economy airfare
Frequency: Annual
Study Establishment: Robinson College, The University of Cambridge
Country of Study: United Kingdom
No. of awards offered: Up to 8
Application Procedure: Applicants for a place to do a PhD should apply for an ORS award and should normally be successfully nominated for an ORS award or an ORS equivalent award, which meets the difference between the higher overseas rate and the lower domestic rate of the university composition fee.
Closing Date: See website for details.
Contributor: In collaboration with Nehru Trust for Cambridge University

Nestle Cambridge Scholarships

Subjects: Any subject.
Purpose: To financially support study towards a PhD.
Eligibility: Open to students from a disadvantaged background from the Commonwealth countries of Africa.
Level of Study: Doctorate
Type: Scholarship
Value: The University Composition Fee at the appropriate rate, approved College fees, a maintenance allowance sufficient for a single student and a contribution towards a return economy airfair
Frequency: Dependent on funds available
Study Establishment: Robinson College, The University of Cambridge
Country of Study: United Kingdom
No. of awards offered: 1
Application Procedure: Applicants for a place to do a PhD should apply for an ORS award and should normally be successfully nominated for an ORS award or an ORS equivalent award, which meets the difference between the higher overseas rate and the lower domestic rate of the University Composition Fee.
Closing Date: See website for details.
Contributor: In collaboration with Nestle Charitable Trust and Robinson College

Noon Educational Foundation Cambridge Scholarship

Subjects: Natural, applied and social sciences.
Purpose: To financially support graduate and postgraduate study.

Eligibility: Candidates should be nationals of and normally resident in Pakistan during the academic year in which they apply for an award. They must preferably have not already spent a full academic year or more studying in an Institution of Higher Education in the West, and must return to their home country at the end of their scholarship period to continue their studies/work there.
Level of Study: Graduate, Postgraduate
Type: Scholarship
Value: Varies
Study Establishment: University of Cambridge
Country of Study: United Kingdom
No. of awards offered: Up to 5
Application Procedure: Applicants must complete a preliminary application form, which can be obtained from local universities, offices of the British Council or the Trust. Completed forms must be returned to the main address. Shortlisted candidates will be sent forms for admission to the University of Cambridge. The preliminary application form can also be downloaded from www.admin.cam.ac.uk/offices/gradstud/admissions/forms/.
Closing Date: February 28th
Funding: Foundation, trusts
Contributor: Cambridge Trusts, Noon Educational Foundation, Foreign and Commonwealth Office and the Open Society Institute (OSI)

Open Society Foundations/University of Cambridge Scholarships

Subjects: Conservation Leadership, Education, Environmental Science, Environment, Society and Development, Environmental Policy, Multi-disciplinary Gender Studies, Public Health, Public Law
Eligibility: Applicants must be citizens of and resident in one of the following countries during the academic year in which they apply for the award: Belarus, Kazakhstan, Kyrgyzstan, Macedonia, Moldova, Palestine, Serbia, Turkmenistan and Ukraine.
Applicants who are temporarily out of the country (for a total period of less than three consecutive months) may be treated as being resident.
Demonstrate exceptional academic potential, familiarity with the chosen field of study and clear commitment to open society goals. This requirement will be treated as of considerbale importance in the selection process, and applicants should be sure to complete Section B (6) of the Graduate Admission and Scholarship Application Form (GRADSAF) fully and appropriately.
Applicants should be planning to return to their home country at the end of the period of study.
Already hold (or expect to hold by the time of taking up the award) a first degree of an equivalent standard to a good UK Second Class Honours Degree.
Applicants will be required to have passed an English proficiency test (TOEFL or IELTS) before being offered a scholarship. The minimum requirement for some of the courses available is in TOEFL 600 (250 in the computer-based test) or IELTS 7.0; however most courses require the higher minimum requirement of IELTS 7.5, and some IELTS 8.0. Please refer to the University of Cambridge Graduate Studies Prospectus for further details.
Level of Study: Postgraduate, Predoctorate
Type: Scholarships
Value: University Composition Fee at the overseas rate, approved college fees, a maintenance allowance sufficient for a single student, and one return economy airfare to the UK.
Application Procedure: Application forms (GRADSAF) are available from the Cambridge University website www.admin.cam.ac.uk/offices/gradstud/prospec/apply/applynow/index.html#paper. They are also available in printed form from the local representatives of Open Society Foundations.
Closing Date: November 25th

OSF Cambridge Scholarship

Subjects: Development studies, education, law and social anthropology.
Eligibility: The scholarship is available for citizens of Afghanistan, Bosnia & Herzegovina, Croatia, Kazakstan, Kosovo, Kyrgyzstan, Macedonia, Montenegro, Russian Federation, Serbia and Ukraine.
Level of Study: Postgraduate
Type: Scholarship
Value: University composition fee, college fee, annual stipend sufficient for a single student, one return economy airfare, and OSF pre-academic summer school.
Length of Study: 1 year
Closing Date: November 25th
Additional Information: Scholars are expected to return to their home country to work or study after completion of their course.

OSF Central Asia Cambridge Scholarship

Subjects: Education, environmental studies, law, and public health.
Eligibility: Scholars should intend to pursue academic work in their home country or region. The scholarship is available for citizens of Kazakstan, Kyrgyzstan, Turkmenistan, and Uzbekistan.
Level of Study: Doctorate
Type: Scholarship
Value: University composition fee, college fee, annual stipend sufficient for a single student, and contribution towards travel costs.
No. of awards offered: 10
Application Procedure: See website.
Closing Date: November 19th

OSI Chevening Cambridge Scholarships for Postgraduate Study

Subjects: Art and design and social sciences.
Purpose: To financially support those undertaking postgraduate study.
Eligibility: Open to nationals of, and normally residents in Albania, Azerbaijan, Bangladesh, Bosnia, Croatia, Kazakhstan, Kosovo, Macedonia, Montenegro, Pakistan, Tajikistan, Turkmenistan, Turkey, Ukraine, Yugoslavia. Applicants must apply to the University of Cambridge and be offered a place at Cambridge in the normal way. All applicants must have a First Class or High Second Class (Honours) Degree or equivalent and normally be under 26. Preference will be given to candidates who have not previously studied outside of their home country.
Level of Study: Postgraduate
Type: Scholarships and fellowships
Value: The University Composition Fee at the overseas rate, approved college fees, a maintenance allowance sufficient for a single student and a contribution towards return economy airfare
Length of Study: 1 year
Frequency: Annual
Study Establishment: The University of Cambridge
Country of Study: United Kingdom
No. of awards offered: Up to 34
Application Procedure: Applicants must complete a preliminary application form, which can be obtained from local universities, offices of the British Council or the Trust. Completed forms must be returned to the main address. Shortlisted candidates will be sent forms for admission to the University of Cambridge.
Closing Date: February 28th
Contributor: Offered in collaboration with the Open Society Institute (OSI) and the Foreign and Commonwealth Office (FCO)

OSI Middle East Cambridge Scholarship

Subjects: Environmental studies, public health, education and law.
Eligibility: Scholars should intend to pursue academic work in their home country or region. The scholarship is available for citizens of Egypt, Iraq, Jordan and Syria.
Level of Study: Doctorate
Type: Scholarship
Value: University composition fee, college fee, annual stipend sufficient for a single student, contribution towards travel costs and OSF Pre-Academic Summer School

Pakistan HEC Cambridge Scholarship

Subjects: Any subject, except those for which a clinical-rate tuition fee is charged.
Eligibility: The scholarship is only available for citizens of Pakistan.
Level of Study: Doctorate
Type: Scholarship
Value: University tuition fee, college fee, annual stipend sufficient for a single person and return travel between Pakistan and the UK
Closing Date: March 30th
Funding: Foundation

Contributor: Higher Education Commission, Pakistan
Additional Information: Successful candidates will be required to sign an undertaking to return to work in Pakistan for a specified length of time after completion of their degree in Cambridge.

Pegasus Cambridge Scholarships for Postgraduate Study

Subjects: Law.
Purpose: To financially assist students who have gained an offer of a place to read for the Master of Law degree (LLM).
Eligibility: Applicants must be from one of the following countries: Australia, Antigua & Barbuda, Barbados, Bermuda, Canada, Dominica, Grenada, Hong Kong, India, Jamaica, Kenya, New Zealand, Nigeria, St Kitts & Nevis, St Lucia, St Vincent & the Grenadines, Singapore, Trinidad & Tobago, or Zambia.
Level of Study: Postgraduate
Type: Scholarship
Value: The scholarships are held in conjunction with other awards from the Cambridge Commonwealth Trust and enable the successful applicants to spend three months in London on work placements at Clifford Chance and the Inner Temple after completing the LLM at Cambridgeand other sources
Frequency: Annual
Study Establishment: The University of Cambridge
Country of Study: United Kingdom
No. of awards offered: Up to 6
Application Procedure: Applicants must complete a preliminary application form, which can be obtained from local universities, offices of the British Council or the Trust. Completed forms must be returned to the main address. Shortlisted candidates will be sent forms for admission to the University of Cambridge.
Closing Date: February 28th
Contributor: Offered in collaboration with the Pegasus Scholarships Trust

Pemanda Monappa Scholarship

Subjects: Biological sciences (excluding medicine and veterinary medicine), computer science, economics, english literature, law, physical sciences and technology.
Eligibility: The scholarship is only available for citizens of India. Scholars should have a first-class first degree from a recognised university in Andhra Pradesh, Karnataka, Kerala or Tamil Nadu. Scholars should require financial assistance in order to study at the University of Cambridge.Scholars should be under the age of 25 at the time of applying.
Level of Study: Postgraduate
Type: Scholarship
Value: £12,000
Length of Study: 1 year
Contributor: Pemanda Monappa Scholarship Fund

Pexim Cambridge Scholarship

Subjects: Any subject which could enhance Serbia's and Macedonia's EU accession capacity and economic prosperity.
Eligibility: The scholarship is only available for citizens of Macedonia and Serbia. Successful applicants will be required to seek employment in their home country following completion of the scholarship, and will be assisted in this by the PEXIM Foundation and their respective governments. A PEXIM Cambridge scholarship recipient is required to work in his/her home country for at least two years for each year for which he/she receives support.
Level of Study: Postgraduate, Graduate, Predoctorate
Type: Scholarship
Value: A sum equal to the difference between the University of Cambridge's Home/EU fee and overseas fee. Each scholarship may also, based on an assessment of financial need, pay up to the remainder of the fees (university composition fee and approved college fee) and an annual stipend for living expenses
Contributor: PEXIM Foundation

PHFI Cambridge Scholarship

Subjects: Public health.
Eligibility: This scholarship is only available for citizens of India. Scholars should be selected by PHFI under its Future Faculty Programme. Scholars should spend their research phase in India, spending a percentage of their time on PHFI-related academic/teaching work. Scholars should commit to 5 years of teaching and research activities at PHFI or its institutes following completion of their studies.
Level of Study: Doctorate
Type: Scholarship
Value: University composition fee, college fee, annual stipend sufficient for a single student, visa fee and return airfare for two journeys between India and the UK and other expenses related to research period in India and thesis preparation
No. of awards offered: 1
Contributor: Public Health Foundation of India

Prince of Wales Scholarships for PhD Study

Subjects: All subjects.
Purpose: To enable graduates of high academic ability to study at Cambridge University in Britain. Also to financially support study towards a PhD.
Eligibility: Candidates must be citizens of New Zealand, must apply to the University of Cambridge and be offered a place at Cambridge in the normal way. All applicants must have a First Class or High Second Class (Honours) Degree or equivalent and normally be under 26. They must be successfully nominated for an Overseas Research Student (ORS) award.
Level of Study: Doctorate, Predoctorate
Type: Scholarship
Value: Determined by the Cambridge Commonwealth Trust in light of the financial circumstances of the applicant.
Length of Study: Up to 3 years
Frequency: Annual
Study Establishment: The University of Cambridge
Country of Study: United Kingdom
No. of awards offered: Varies
Application Procedure: Applicants must apply directly to the Scholarships Officer at their own university. Otherwise, they should apply directly to the New Zealand Vice Chancellor's Committee.
Closing Date: October 1st
Contributor: Offered in collaboration with the New Zealand Vice Chancellor's Committee

For further information contact:

The New Zealand Vice Chancellor's Committee, Level 11, 94 Dixon Street, Wellington 6034, PO Box 11-915, New Zealand
Contact: Scholarships Officer

Prince Philip Graduate Exhibitions

Purpose: To financially support study towards a PhD.
Eligibility: Students who have graduated from the Chinese University of Hong Kong and the University of Hong Kong. Applicants must apply to the University of Cambridge and be offered a place at Cambridge in the normal way. All applicants must have a First Class or High Second Class (Honours) Degree or equivalent and normally be under 26. They must be successfully nominated for an Overseas Research Student (ORS) award.
Type: Scholarships and fellowships
Value: The Prince Philip Graduate Exhibitions are normally means-tested, substantial part-cost awards towards the overall costs of the student.
Length of Study: Up to 3 years
Frequency: Annual
Study Establishment: The University of Cambridge
No. of awards offered: 2
Application Procedure: Applicants must complete a preliminary application form, which can be obtained from local universities, offices of the British Council or the Trust. Completed forms must be returnedto the main address. Shortlisted candidates will be sent forms for admission to the University of Cambridge. The preliminary application form can also be downloaded from www.admin.cam.ac.uk/offices/gradstud/admissions/forms/.
Contributor: Offered in collaboration with the Friends of Cambridge

Prince Philip Scholarship

Subjects: All subjects.
Eligibility: Successful applicants who complete a BA at the University of Cambridge may apply for a further scholarship to continue on to a

one-year taught postgraduate degree. The scholarship is only available for citizens of Hong Kong.
Level of Study: Graduate
Type: Scholarship
Value: University composition fee (subject to means-testing), college fee (subject to means-testing), £4,500 per year and return economy airfare
Contributor: Friends of Cambridge University in Hong Kong

PTDF Cambridge Scholarships

Subjects: Science-related subjects contributing to an improvement in technical manpower for the oil and gas industry..
Purpose: To financially support postgraduate study.
Eligibility: Candidates must be from Nigeria, must apply to the University of Cambridge and be offered a place at Cambridge in the normal way. All applicants must have a First Class or High Second Class (Honours) Degree or equivalent and normally be under 26. They must be successfully nominated for an Overseas Research Student (ORS) award. The students may apply to MPhil in Environment and Development, Environmental Engineering and Sustainable Development, Environmental Policy, Fluid Flow in Industry and the Environment, or Geographical Information Systems and Remote Sensing.
Level of Study: Postgraduate
Type: Scholarships and fellowships
Value: The university composition fee at the appropriate rate, approved college fees, a maintenance allowance sufficient for a single student and a contribution towards a return economy airfare
Length of Study: 1 year
Frequency: Annual
Study Establishment: University of Cambridge
Country of Study: United Kingdom
No. of awards offered: Varies
Application Procedure: Applicants must complete a preliminary application form, which can be obtained from local universities, offices of the British Council or the Trust. Completed forms must be returned to the main address. Shortlisted candidates will be sent forms for admission to the University of Cambridge.
Closing Date: February 28th
Funding: Trusts
Contributor: Offered in collaboration with Univation at the Robert Gordon University, Aberdeen

Queens' College Stephen Thomas Studentship

Subjects: Computer science and engineering.
Eligibility: The scholarship is available for citizens of any country outside the EU. Applicants should select Queens' College as their first choice college. Applicants should require financial assistance in order to study at the University of Cambridge.
Level of Study: Doctorate
Type: Studentship
Value: Up to maximum of £25,000 per year
No. of awards offered: 1 each year
Application Procedure: This scholarship programme allows for one student in residence each year, and an award will next be made to a student starting in next year.
Contributor: Queen's College

Queens' College Walker Studentship

Subjects: Arts, humanities and social sciences.
Eligibility: Applicants should select Queens' College as their first choice college. The scholarship is available for citizens of any country outside the EU..
Level of Study: Doctorate
Type: Studentship
Value: University composition fee, college fee and annual stipend sufficient for a single student
No. of awards offered: 1 each year
Application Procedure: The programme allows for one student in residence each year, and an award will next be made to a student starting in next year.

Rajiv Gandhi (UK) Foundation Cambridge Scholarship

Subjects: All subjects.
Purpose: To financially support study towards a PhD.
Eligibility: Applicants must be from India, must apply to the University of Cambridge and be offered a place at Cambridge in the normal way. All applicants must have a First Class (Honours) Degree. Those with a First Class Master's Degree may be given preference.
Level of Study: Doctorate, Postgraduate
Type: Scholarship
Value: University composition fee, college fee, annual maintenance sufficient for a single student
Length of Study: 2 years
Frequency: Annual
Study Establishment: The University of Cambridge
Country of Study: United Kingdom
No. of awards offered: 1
Application Procedure: Applicants must contact the organization.
Closing Date: February 28th
Contributor: Offered in collaboration with the Rajiv Gandhi Foundation

For further information contact:

The Nehru Trust for Cambridge University, Teen Murti House, Teen Murti Marg, New Delhi, 110011, India
Contact: The Joint Secretary

Raymond and Helen Kwok Research Scholarship

Subjects: Any subject relevant to People's Republic of China needs.
Purpose: To financially support study leading to a PhD.
Eligibility: Applicants must be from China, and must have been awarded a First Class (Honours) Degree or equivalent.
Level of Study: Doctorate
Type: Scholarship
Value: University composition fee, college fee, annual stipend sufficient for a single student
Length of Study: 3 years
Frequency: Annual
Study Establishment: Jesus College, The University of Cambridge
Country of Study: United Kingdom
No. of awards offered: 2
Application Procedure: Applicants must complete a preliminary application form, which can be obtained from local universities, offices of the British Council or the Trust. Completed forms must be returned to the main address. Shortlisted candidates will be sent forms for admission to the University of Cambridge. Applicants must also apply for an ORS award. The preliminary application form can also be downloaded from www.admin.cam.ac.uk/offices/gradstud/admissions/forms/
Closing Date: March 31st
Funding: Individuals, trusts
Contributor: Mr Raymond Kowk and Jesus College

The Right Honourable Paul Martin Sr Scholarship at Cambridge University

Subjects: Law.
Purpose: To support students studying Master of Law (LLM).
Eligibility: The scholarship is only available for citizens of Canada.
Level of Study: Postgraduate
Type: Scholarship
Value: University composition fee, college fee and annual stipend sufficient for a single student
No. of awards offered: 1
Closing Date: December 31st
Contributor: Canadian Institute for Advanced Legal Studies

Royal Government of Thailand Cambridge Scholarship

Subjects: All subjects.
Purpose: To financially support study towards a PhD and postgraduation.
Eligibility: Applicants must be from Thailand, must apply to the University of Cambridge and be offered a place at Cambridge in the normal way. All applicants must have a First Class or High Second Class (Honours) Degree or equivalent and normally be under 26. For PhD study they be successfully nominated for an Overseas Research Student (ORS) award.
Level of Study: Doctorate
Type: Scholarships and fellowships

Value: The University Composition Fee at the appropriate rate, approved college fees, a maintenance allowance sufficient for a single student and a contribution towards return economy airfare
Length of Study: Up to 3 years
Frequency: Annual
Study Establishment: The University of Cambridge
Country of Study: United Kingdom
No. of awards offered: 1
Application Procedure: Applicants must complete a preliminary application form, which can be obtained from local universities, offices of the British Council or the main address. Completed forms must be returned to the main address. Shortlisted candidates will be sent forms for admission to the University of Cambridge. These forms must be returned to the Board of Graduate Studies.
Closing Date: See the website
Contributor: Offered in collaboration with the Cambridge Thai Foundation and Civil Service Commission, Royal Government of Thailand

Saïd Foundation Cambridge Scholarship

Subjects: Any subject that is of use to the applicant's home country or the Middle East region.
Eligibility: Applicants should be Arab students of the eligible countries, should normally have at least two-years' work experience and should require financial assistance to study at the University of Cambridge. Successful candidates are requested to work for the development of the Middle East region after completion of their studies. The scholarship is available for citizens of Israel, Jordan, Lebanon, Palestine and Syria.
Level of Study: Doctorate, Postgraduate
Type: Scholarship
Value: University composition fee, college fee, annual stipend sufficient for a single student and contribution towards travel costs
Length of Study: 1 year
Contributor: Said Foundation

Santander Cambridge Scholarship

Subjects: All subjects.
Eligibility: This scholarship is available for citizens of Argentina, Brazil, Chile, Colombia, Mexico, Peru, Puerto Rico, Uruguay and Venezuela.
Level of Study: Postgraduate
Value: £10,000
Length of Study: 1 year
Contributor: Santander

Schlumberger Cambridge International Scholarship

Eligibility: The scholarship is available for citizens of any country outside the EU..
Level of Study: Doctorate
Type: Scholarship
Value: University composition fee, college fee and annual stipend sufficient for a single student
Closing Date: September 1st
Contributor: Schlumberger Gould Research Ltd
Additional Information: Applications are invited in subjects relevant to the work of the Schlumberger Cambridge Research Center.

Scott Polar Centenary Scholarship

Subjects: Antarctic studies at the Scott Polar Research Institute.
Eligibility: This scholarship is only available for citizens of New Zealand.
Level of Study: Doctorate
Type: Scholarship
Value: University composition fee, college fee, annual stipend sufficient for a single person and one return airfare between New Zealand and UK
Closing Date: January 20th

SGPC Cambridge Scholarship

Subjects: All subjects crucial to the development of higher education.
Eligibility: Scholars should be members of the Sikh community. The scholarship is only available for citizens of India..
Level of Study: Doctorate, Postgraduate
Type: Scholarship
Value: University composition fee, college fee and annual stipend sufficient for a single student
Length of Study: 1 year

Shell Centenary Cambridge Scholarships (Countries Outside of the Commonwealth)

Subjects: Economics, Environmental science and ecology and Social science.
Purpose: To financially support those undertaking postgraduate study.
Eligibility: Open to citizens from all countries except: Australia, Austria, Belgium, Canada, Denmark, Finland, France, Germany, Greece, Iceland, Ireland, Italy, Japan, Luxembourg, The Netherlands, New Zealand, Norway, Portugal, Spain, Sweden, Switzerland, UK or USA. Applicants must apply to the University of Cambridge and be offered a place at Cambridge in the normal way. All applicants must have a First Class or High Second Class (Honours) Degree or equivalent and normally be under 26.
Level of Study: Postgraduate
Type: Scholarships and fellowships
Value: The University Composition Fee at the overseas rate, approved college fees, a maintenance allowance sufficient for a single student and a contribution towards return economy airfare
Length of Study: 1 year
Frequency: Annual
Study Establishment: The University of Cambridge
Country of Study: United Kingdom
No. of awards offered: Up to 20
Application Procedure: Applicants must complete a preliminary application form, which can be obtained from local universities, offices of the British Council or the Trust. Completed forms must be returned to the main address. Shortlisted candidates will be sent forms for admission to the University of Cambridge. The preliminary application form can also be downloaded from www.admin.cam.ac.uk/offices/gradstud/admissions/forms/
Closing Date: February 28th
Contributor: Offered in collaboration with Shell Centenary Scholarship Fund

Smuts Cambridge International Scholarship

Subjects: All subjects.
Eligibility: The scholarship is available for citizens of any member country of the Commonwealth.
Level of Study: Doctorate
Type: Scholarship
Value: University composition fee, college fee and annual stipend (sufficient for a single person)

South African College Bursaries

Subjects: All subjects.
Purpose: To enable citizens of South and Southern Africa to study at the University of Cambridge.
Eligibility: Applicants must be from South or Southern Africa, must apply to the University of Cambridge and be offered a place at Cambridge in the normal way. All applicants must have a First Class or High Second Class (Honours) Degree or equivalent and normally be under 26. They must be successfully nominated for an Overseas Research Student (ORS) award.
Level of Study: Postgraduate
Type: Studentships and bursaries
Value: Part-cost bursaries
Length of Study: 1 year
Frequency: Annual
Study Establishment: The University of Cambridge
Country of Study: United Kingdom
No. of awards offered: Varies
Application Procedure: Applicants must complete a preliminary application form, which can be obtained from local universities, offices of the British Council or the Trust. Completed forms must be returned to the main address. Shortlisted candidates will be sent forms for admission to the University of Cambridge. The preliminary application form can also be downloaded from www.admin.cam.ac.uk/offices/gradstud/admissions/forms/
Closing Date: February 28th

Contributor: Offered in collaboration with Churchill College, Newnham College, Selwyn College, St Catherine's College and Sidney Sussex College, Cambridge
Additional Information: The bursaries are normally held in conjunction with other awards from the Cambridge Commonwealth Trust and other sources.

St Edmund's Duke of Edinburgh Scholarship

Subjects: All subjects.
Eligibility: The scholarship is available for citizens of any country outside the EU.
Level of Study: Doctorate
Type: Scholarship
Value: £2,000
Additional Information: Scholars should become members of St Edmund's College. They should already be in receipt of a scholarship from the Cambridge Trusts or some other public source.

Sun Hung Kai Properties - Kwoks' Scholarship

Subjects: All subjects.
Purpose: To financially support postgraduate study.
Eligibility: For students from China.
Level of Study: Postgraduate
Type: Scholarship/Bursary
Value: £24,000 per year
Length of Study: 1 year
Frequency: Annual
Study Establishment: The University of Cambridge
Country of Study: United Kingdom
No. of awards offered: Up to 15 scholarships, and a number of bursaries
Application Procedure: Applicants must complete a preliminary application form, which can be obtained from local universities, offices of the British Council or the Trust. Completed forms must be returned to the main address. Shortlisted candidates will be sent forms for admission to the University of Cambridge.
Closing Date: March 31st
Contributor: In collaboration wih the Sun Hung Kai Properties Limited
Additional Information: Scholars will be asked to sign an agreement to use and practise what they have learnt towards the future development of the relevant field in China, and to return to China to work in relevant fields for at least 5 years (PhD) or 3 years (MPhil).

Taiwan Cambridge Scholarship

Subjects: All subject except clinical medicine.
Eligibility: The scholarship is only available for citizens of Taiwan.
Level of Study: Doctorate
Type: Scholarship
Value: University composition fee, approved college fee and annual stipend sufficient for a single student

Tidmarsh Cambridge Scholarship for PhD Study

Subjects: All subjects.
Purpose: To financially support study towards a PhD.
Eligibility: Open to citizens of Canada. Applicants must apply to the University of Cambridge and be offered a place at Cambridge in the normal way. All applicants must have a First Class or High Second Class (Honours) Degree or equivalent and normally be under 26. They must have been successfully nominated for an Overseas Research Student (ORS) award.
Level of Study: Doctorate
Type: Scholarships and fellowships
Value: The University Composition Fee at the home rate, approved college fees and a maintenance allowance sufficient for a single student
Frequency: Dependent on funds available
Study Establishment: Trinity Hall, the University of Cambridge
Country of Study: United Kingdom
No. of awards offered: 1
Application Procedure: Application forms for the scholarship will be sent out to eligible candidates once the completed form for admission to the University of Cambridge has reached the Board of Graduate Studies.
Closing Date: February 28th
Contributor: A benefaction from Dr Evan Schulman

For further information contact:

The Board of Graduate Studies, 4 Mill Lane, Cambridge, Cambridgeshire, CB2 1RZ, England
Contact: The Secretary

TNK/BP Kapitza Cambridge Scholarships.

Subjects: All subjects.
Purpose: To financially support study towards a PhD.
Eligibility: Open to citizens of Russia and Ukraine. Preference may be given to candidates intending to pursue research in the broad fields of engineering, economics, management studies, law, mathematics and natural sciences. Applicants should normally have a first class honours degree and, preferably, a Masters degree or its equivalent from a recognised university. Applicants should apply for an Overseas Research Student (ORS) award.
Level of Study: Doctorate
Type: Scholarships and fellowships
Value: The University composition fee at the appropriate rate, approved college fees, a maintenance allowance sufficient for a single student, a contribution towards a return economy airfare, and an annual dicretionary allowance for study-related expenses
Frequency: Annual
Study Establishment: Trinity College, the University of Cambridge
Country of Study: United Kingdom
No. of awards offered: Up to 8
Application Procedure: Application forms for the scholarship will be sent out to eligible candidates once the completed form for admission to the University of Cambridge has reached the Board of Graduate Studies.
Closing Date: February 28th
Contributor: Offered in collaboration with the TNK/BP and Trinity College

UK-Germany Millenium Studentships

Subjects: All subjects.
Purpose: To financially support postgraduate study.
Eligibility: Applicants must be German nationals. All applicants must have a First Class or High Second Class (Honours) Degree or equivalent and normally be under 26.
Level of Study: Postgraduate
Type: Studentships and bursaries
Value: UK£2,000 per year
Length of Study: 1 year
Frequency: Annual
Study Establishment: St Edmund's College, Cambridge
No. of awards offered: 2
Application Procedure: Applicants must apply to the University of Cambridge and be offered a place at Cambridge in the normal way. Applicants should complete only one scholarship form, which will enable them to be considered for all awards for which they are eligible. Application form can be downloaded from the website.
Closing Date: See the website
Contributor: Cambridge Trusts and St Edmund's College
Additional Information: Please check website for latest updates.

UNDP Cambridge Scholarship

Subjects: All subjects.
Eligibility: Candidates must be female professionals employed full-time in the public sector in Vietnam and must be citizens of Vietnam.
Level of Study: Postgraduate
Type: Scholarship
Value: University composition fee, college fee, annual stipend sufficient for a single studentand one return economy airfare
Length of Study: 1 year
Additional Information: Please check website for latest updates.

University of Central Asia Cambridge Scholarship

Subjects: Physics, development studies, economics, biochemistry, chemistry, mathematics, sociology, Asian and Middle Eastern studies.
Eligibility: The scholarship is available for citizens of Afghanistan, China, India, Iran, Kazakstan, Kyrgyzstan, Mongolia, Pakistan, Tajikistan, Turkmenistan and Uzbekistan.
Level of Study: Doctorate, Postgraduate
Type: Scholarship

Value: University composition fee, college fee, annual stipend sufficient for a single student, contribution towards travel costs
Length of Study: 1 year
Contributor: University of Central Asia
Additional Information: Scholars should intend to work at the University of Central Asia, under its Central Asian Faculty Development Programme, on completion of their degree at Cambridge.

William and Margaret Brown Scholarship

Subjects: Engineering, natural sciences, physical sciences, and social sciences.
Purpose: To financially support study towards a PhD.
Eligibility: Open to students from Canada. Applicants must apply to the University of Cambridge and be offered a place at Cambridge in the normal way. All applicants must have a First Class or High Second Class (Honours) Degree or equivalent and normally be under 26. They must be successfully nominated for an Overseas Research Student (ORS) award.
Level of Study: Doctorate
Type: Scholarship
Value: University Composition Fee and college fee
Length of Study: Up to 3 years
Frequency: Dependent on funds available
Study Establishment: The University of Cambridge
Country of Study: United Kingdom
No. of awards offered: 1
Application Procedure: Application forms for the scholarship will be sent out to eligible candidates once the completed form for admission to the University of Cambridge has reached the Board of Graduate Studies.
Contributor: A benefaction from Dr Donald Pinchen
Additional Information: Please check website for latest updates.

For further information contact:

The Board of Graduate Studies, 4 Mill Hill, Cambridge, Cambridgeshire, CB2 1RZ, England
Contact: The Secretary

Wing Yip Cambridge Scholarships

Subjects: All subjects.
Purpose: For postgraduate courses of study.
Eligibility: Open to students from China. Applicants must be studying at Peking University or Tsinghua University.
Level of Study: Postgraduate
Type: Scholarship
Value: University Composition Fee and college fee
Length of Study: 1 year
Frequency: Annual
Study Establishment: Churchill College, The University of Cambridge
No. of awards offered: 2
Application Procedure: Applicants must apply to the University of Cambridge and be offered a place at Cambridge in the normal way. Applicants should complete only one scholarship form, which will enable them to be considered for all awards for which they are eligible. Application forms can be downloaded from the website.
Closing Date: February 28th
Contributor: Cambridge Trusts, Mr Wing Yip and the Education Section of the Chinese Embassy, London

Woolf Fisher Scholarship at Cambridge

Subjects: All subjects.
Purpose: To financially support study towards a PhD.
Eligibility: Open to students from New Zealand, and have attended a secondary school in New Zealand. Students should have graduated or expect to graduate from a university in New Zealand. Applicants for a place to do a PhD should apply for an ORS award and should normally be successfully nominated for an ORS award or an ORS equivalent award.
Level of Study: Doctorate
Type: Scholarship
Value: The University Composition Fee, college fee, annual stipend sufficient for a single student and one return economy airfare
Frequency: Annual
Study Establishment: Trinity College, The University of Cambridge
Country of Study: United Kingdom
No. of awards offered: Up to 3
Application Procedure: Candidates must apply on the prescribed forms to the Scholarships Manager, New Zealand Vice Chancellors Committee, PO Box 11-915, Wellington by September 1st in the year prior to that in which the scholarship will be taken up.
Closing Date: August 1st
Contributor: In collaboration with the Woolf Fisher Trust

World Bank Cambridge Scholarships for Postgraduate Study

Subjects: Social sciences.
Purpose: To financially support those undertaking postgraduate study.
Eligibility: Candidates must be nationals of a World Bank member country, be under the age of 45, with priority given to those candidates under 35, have or be about to obtain a First Class or High Second Class Degree from a recognized university in a development-related field. Other prerequisites include 2, but preferably 4–5 years of recent full-time professional experience in their home country or other developing country, usually in public service. Candidates must not hold resident status in the United States of America or another industrialized country. They should not hold a Master's degree or diploma or at present be studying towards a Master's degree or diploma from an industrialized country.
Level of Study: Postgraduate
Type: Scholarships and fellowships
Value: The University Composition Fee at the overseas rate, approved college fees, a maintenance allowance sufficient for a single student and a contribution towards return economy airfare
Length of Study: 1 year
Frequency: Annual
Study Establishment: The University of Cambridge
Country of Study: United Kingdom
No. of awards offered: Up to 20
Application Procedure: Applicants must contact the organization.
Closing Date: February 28th
Contributor: Offered in collaboration with the World Bank

UNIVERSITY OF CANTERBURY

College of Arts, University of Canterbury, Private Bag 4800, Christchurch, 8140, New Zealand
Tel: (64) 3 364 2426 ext 6426
Fax: (64) 3 364 2683
Email: michelle.payton@canterbury.ac.nz
Website: www.canterbury.ac.nz
Contact: Ms Michelle Payton, Human Resource Administrator

The University of Canterbury offers a variety of subjects in a few flexible degree structures, namely, first and postgraduate degrees in arts, commerce, education, engineering, fine arts, forestry, law, music and science. At Canterbury, research and teaching are closely related, and while this feature shapes all courses, it is very marked at the postgraduate level.

University of Canterbury Alumni Association Scholarships

Subjects: All subject.
Purpose: To provide opportunity to students who intend to enrol for full-time study at the University of Canterbury.
Eligibility: Awarded annually to the top students from nominated faculties.
Level of Study: Postgraduate
Type: Scholarship
Value: $5,000
Length of Study: 1 year
Frequency: Annual
No. of awards offered: 4
Application Procedure: No application necessary.
Closing Date: November 1st
Additional Information: The scholarships are awarded in two categories: (a) for students who intend to begin their first year of an undergraduate degree programme, who have completed or are completing a university entrance qualification, and who attended

school in the year of application; and (b) for students who intend to undertake the final year of an honours degree or any year of a postgraduate degree or diploma.

University of Canterbury and Creative New Zealand Ursula Bethell Residency in Creative Writing

Subjects: Creative writing, fiction, poetry, scriptwriting and literary nonfiction.
Purpose: To foster New Zealand writing by providing a full-time opportunity for a writer to work in an academic environment.
Eligibility: Open to authors of proven merit who are normally resident in New Zealand and to New Zealand nationals temporarily resident overseas.
Level of Study: Unrestricted
Type: Fellowship
Value: Emolument at the rate of New Zealand $52,600
Length of Study: Up to 1 year
Frequency: Dependent on funds available
Study Establishment: University of Canterbury
Country of Study: New Zealand
No. of awards offered: 1
Application Procedure: Applicants must submit details of published writing and work in progress, and include a proposal of work to be undertaken during the appointment.
Closing Date: Each August and September
Funding: Government
No. of awards given last year: 30
Additional Information: The appointment will be made on the basis of published or performed writing of high quality. Conditions of appointment should be obtained from the Human Resources Department before applying, available in August from: hr@arts.canterbury.ac.nz.

University of Canterbury Doctoral Scholarship – Students with Disabilities

Subjects: Any subject
Purpose: To provide an incentive for students with high academic achievement who have a significant disability.
Eligibility: Open to candidate with a disability that significantly impairs the ability to study and must be eligible to enrol for the degree of PhD on 10 December in the year of application.
Level of Study: Doctorate
Type: Scholarship
Value: $20,000 per year plus tuition fees (NZ domestic rate)
Length of Study: Up to 4 years
Application Procedure: Applicants must contact the scholarship office.
Closing Date: October 15th
Additional Information: The scholarship may not be held with a University of Canterbury Doctoral Scholarship. The general conditions for the scholarship are as for the University of Canterbury Doctoral Scholarship regulations.

University of Canterbury Doctoral Scholarships

Subjects: All subjects. A small number of Doctoral scholarships are reserved for certain colleges/faculties: Business and Economics; Creative Arts; Education; and Law.
Purpose: To support full-time or part-time study towards a PhD degree at the University of Canterbury.
Eligibility: Open to international students who meet the academic requirements for enrolment in a PhD.
Level of Study: Doctorate
Type: Scholarship
Value: $20,000 plus tuition fees at NZ domestic rate
Length of Study: 3 years
Frequency: Annual
Application Procedure: Applicants must contact the scholarship office.
Closing Date: May 15th and October 15th
Additional Information: Applicants are also considered for the following prestigious scholarships: Brownlie Scholarship; Roper Scholarship; Canterbury Scholarship (domestic students); and/or UC International Doctoral Scholarships (international students). No separate application is required.

University of Canterbury International Doctoral Scholarship

Subjects: Any subject.
Purpose: To provide support for international students with high academic achievement.
Eligibility: Applicants must be international students (New Zealand citizens and permanent residents are not eligible to apply). They must meet the academic requirements for enrolling in a PhD undertaking or planning to undertake full-time study.
Level of Study: Doctorate
Type: Scholarship
Value: $25,000 per year for thesis tuition fees at NZ domestic rate and economy return airfare.
Length of Study: Up to 3 years
No. of awards offered: 5
Application Procedure: The applicants must contact the scholarship office.
Closing Date: October 15th; May 15th

University of Canterbury Masters Scholarship – Students with Disabilities

Subjects: Any subject.
Purpose: To provide an incentive for students who have a significant disability with high academic achievement.
Eligibility: Open to candidates with disabilities. Candidate must be eligible to enrol for a Masters degree at the University of Canterbury.
Level of Study: Doctorate, Postdoctorate
Type: Scholarship
Value: $12,000 per year plus tuition fees (NZ domestic rate)
Length of Study: Up to 2 years
Application Procedure: Applicants must contact the scholarship office.
Closing Date: October 15th
Additional Information: The scholarship may not be held with a University of Canterbury Masters Scholarship. The general conditions for the scholarship are as for the University of Canterbury Masters Scholarship regulations.

University of Canterbury Masters Scholarships

Subjects: Any subject.
Purpose: To support a full-time or part-time study towards the research year of a Masters degree at the University of Canterbury.
Level of Study: Postgraduate
Type: Scholarship
Value: $12,000 per year plus tuition fees (NZ domestic rate)
Length of Study: 1 year
Frequency: Annual
No. of awards offered: 3
Application Procedure: Applicants must contact the scholarship office.
Closing Date: May 15th and October 15th

Wood Technology Research Centre – Postgraduate Scholarships

Subjects: Chemical engineering.
Purpose: To develop a computer model to simulate energy flow and energy efficiency in wood and wood product processing industry.
Level of Study: Postgraduate
Type: Scholarship
Value: $24,000 per year for PhD and $18,000 per year for ME
Length of Study: 3 years for PhD and one and half year for ME
Application Procedure: To apply or for further information on the above scholarships, please contact Dr. Shusheng Pang.
Closing Date: See website for details.
Contributor: University of Canterbury
Additional Information: Case studies will be conducted for manufacturing of Laminated Veneer Lumber (LVL) and Medium Density Fibreboard (MDF). The project will be conduced in collaboration with the University of Otago and a wood processing company.

For further information contact:

Wood Technology Centre, Department of Chemical and Process Engineering, University of Canterbury, Christchurch, New Zealand
Tel: (64) 3 364 2538

Fax: (64) 3 364 2063
Email: shusheng.pang@canterbury.ac.nz
Contact: Dr Shusheng Pang, Associate Professor and Director

UNIVERSITY OF CENTRAL LANCASHIRE (UCLAN)

University of Central Lancashire, Fylde Rd, Preston, PR1 2HE, United Kingdom
Tel: (44) 1772 201201
Fax: (44) 1772 201201
Email: uadmissions@uclan.ac.uk
Website: www.uclan.ac.uk

UCLan is the sixth largest university in the UK ranked in the leading third of the UK's modern universities with around 30,000 students studying at Preston and is the leading modern university in the North West of England. During the last 10 years the University has spent over £77 million in providing a modern, state-of-the-art learning environment, including campus-wide free Internet access via an extensive computer network, and a new library open seven days a week.

Sports and Arts Related Scholarships

Subjects: Art and design and sports science.
Purpose: To support artistic and sporting talent of the future in China.
Eligibility: Open to Chinese students applying to relevant UCLan undergraduate or postgraduate programme. Applicants must not be in receipt of a full fee scholarship from any other source.
Level of Study: Graduate
Type: Scholarship
Value: £2,000
No. of awards offered: 10
Application Procedure: Check website for further details or mail to cbmagee@uclan.ac.uk.
Closing Date: August 24th
Contributor: University of Central Lancashire (UCLan).
Additional Information: Scholarships will be awarded for prior academic and personal achievement in the chosen area of study based on information supplied in the scholarship application form. Students will also get the opportunity to discuss their academic background as well as apply for these scholarships on the same day therefore students are advised to bring their transcript, English language score and references with them.

THE UNIVERSITY OF CINCINNATI

Department of Electrical and Computer Engineering and Computer Science, 2600 Clifton Avenue, Cincinnati, OH, 45221-0030, United States of America
Tel: (1) 513 556 4756
Fax: (1) 513 556 6245
Email: dpa@ececs.uc.edu
Website: www.uc.edu
Contact: Dr Dharma P Agrawal, Admissions Committee

The University of Cincinnati offers students a balance of educational excellence and real world experience. Since its founding in 1819, University of Cincinnati has been the source of many discoveries creating positive change for society.

Ohio Board of Regents (OBR) Distinguished Doctoral Research Fellowship in Computer Science and Engineering

Subjects: Computer science.
Purpose: To provide opportunity to highly qualified researchers.
Eligibility: The candidate must be highly qualified and be a U.S. citizen.
Level of Study: Doctorate
Type: Fellowship
Value: US$24,000 per year plus tuition
Length of Study: 2 years
Frequency: Annual
Study Establishment: University of Cincinnati
Country of Study: United States of America
Closing Date: March 15th
Contributor: Ohio Board of Regents (OBR)
Additional Information: All queries should be directed to dpa@ececs.uc.edu

UNIVERSITY OF COLOMBO

College House, University of Colombo, 94 Cumaratunga Munidasa Mawatha, Colombo, 3, Sri Lanka
Tel: (94) 9411 2581835/2584695/2585509/2583818
Fax: (94) 11 2583810
Email: postmast@admin.cmb.ac.lk; registrar@admin.cmb.ac.lk
Website: www.cmb.ac.lk
Contact: Acting Registrar

University Research Grants

Subjects: Any subject.
Purpose: To inculcate research culture amongst staff/students and to upscale the high quality/impactresearch.
Eligibility: Collaborative research grant proposals: Senior or Mid-career academics. Scholarships or Fellowships: Academics, Research officers and Scientific Assistants in the permanent cadre of the university.
Level of Study: Research
Type: Grant
Value: SL Rs 1,000,000 (1.0Mn) as annual cost (collaborative research grant) and SL Rs 500,000 as annual cost (scholarships or fellowships).
Length of Study: 3 years
Frequency: Ongoing
Study Establishment: University of Colombo
Application Procedure: Application form can be found on the website.
Closing Date: September 1st
Additional Information: The country of study is Sri Lanka.

For further information contact:

Website: www.cmb.ac.lk/wp-content/uploads/2008/12/Application-for-Research-Grants.pdf

UNIVERSITY OF DELAWARE

Department of History, Newark, DE, 19716, United States of America
Tel: (1) 302 831 8226
Fax: (1) 302 831 1538
Email: pato@udel.edu
Website: www.udel.edu
Contact: Ms Patricia H Orendorf, Administrative Assistant

The Department of History offers MA and PhD programmes in American and European history and more limited graduate study Ancient, African, Asian, Latin American, and Middle Eastern history. In conjunction with these, it offers special programmes in the history of industrialization, material culture studies, American Civilization, and museum studies.

E Lyman Stewart Fellowship

Subjects: History.
Purpose: To provide a programme of graduate study leading to an MA or PhD degree for students who plan careers as museum professionals, historical agency administrators or seek careers in college teaching and public history.
Eligibility: Open to nationals of any country.
Level of Study: Doctorate, Predoctorate, Graduate
Type: Fellowship
Value: US$16,500 plus tuition
Study Establishment: University of Delaware
Country of Study: United States of America
No. of awards offered: 6–8
Application Procedure: Applicants must submit an application form, transcripts, Graduate Record Examination (GRE) scores, Test of English as a Foreign Language (TOEFL) scores where applicable, plus three letters of recommendation and a writing sample.
Closing Date: January 15th
Funding: Private

No. of awards given last year: 8
No. of applicants last year: 40
Additional Information: This is a residential programme.

Fellowships in the University of Delaware Hagley Program

Subjects: The history of industrialization (broadly defined to include business, economics, labour and social history) and the history of science and technology.
Purpose: To provide a programme of graduate study leading to an MA or PhD degree for students who seek careers in college teaching and public history.
Eligibility: Open to graduates of any nationality seeking degrees in American or European history or the history of science and technology.
Level of Study: Doctorate, Graduate, Predoctorate
Type: Fellowship
Value: US$16,500 for Master's and doctoral candidates. All tuition fees for university courses are paid
Length of Study: 1 year, renewable once for those seeking a terminal MA and up to three times for those seeking the doctorate
Study Establishment: University of Delaware
Country of Study: United States of America
No. of awards offered: Approx. 2–3
Application Procedure: Fellows are selected upon Graduate Record Examination scores, recommendations, undergraduate grade index, work experience and personal interviews.
Closing Date: January 15th
Funding: Private
No. of awards given last year: 2
No. of applicants last year: 10
Additional Information: This is a residential programme.

UNIVERSITY OF DUBLIN, TRINITY COLLEGE

Graduate Studies Office, Arts Building, Trinity College, College Green, Dublin, 2, Ireland
Tel: (353) 1 896 1166/353 1 896 1000/353 1 896 1999
Fax: (353) 353 1 896 1000
Email: gradinfo@tcd.ie
Website: www.tcd.ie/graduate_studies
Contact: Bernadette Curtis

The University of Dublin, Trinity College was founded in 1592 and is the oldest university in Ireland. Trinity College is the sole constituent college of the University. Trinity is now ranked in the top 20 European universities (13th) and 53rd in the world by international employers.

Claude and Vincenette Pichois Research Award

Subjects: French literature.
Purpose: To support research in 19th and/or 20th century french literature.
Eligibility: Open to a candidate holding a first-class or II.1 Honors Degree in French (or equivalent) and having a research project within the area of nineteenth and/or twentieth-century French Literature.
Level of Study: Research
Type: Scholarship
Value: €16,000 per year and EU fees.
Length of Study: 3 years (renewed annually)
Frequency: Every 3 years
Country of Study: Ireland
Application Procedure: Applications for this Award must be made on the College award form, together with a complete application for admission to the research register, inclusive of sealed letters of reference from two academic referees, each contained in a special envelope provided by the Graduate Studies Office. Applications should be mailed to Graduate Studies Office, Trinity College.
Closing Date: May 1st
Contributor: Claude and Vincenette Pichois Memorial Fund

Cluff Memorial Studentship

Subjects: History.
Purpose: To support postgraduate study in history.
Eligibility: Open to all candidates in history.
Level of Study: Research
Type: Studentship
Value: €2,285 per year
Length of Study: 1–3 years
Frequency: Dependent on funds available
Country of Study: Ireland
Application Procedure: Applications should be mailed to the Professor of Modern history, Trinity College Dublin.
Closing Date: July 31st
Funding: Individuals
Contributor: Mr. W.V. Cluff

E.C. Smith Scholarship in Pathology

Subjects: Pathology.
Purpose: To support research in pathology (including immunology, virology, and suchaspects of microbiology, haematology, and clinical biochemistry as arerelevant to disease in human beings).
Eligibility: Open to all candidates.
Level of Study: Research
Type: Scholarship
Value: €9,523 per year
Length of Study: 2 years
Frequency: Every 3 years
Country of Study: Ireland
Application Procedure: Applications should be mailed to the Graduate Studies Office, Trinity College.
Closing Date: See website for details.

Elrington Scholarship

Subjects: Theology.
Purpose: To support theological research.
Eligibility: Open to all candidates in theology and/or divinity and/or related academic disciplines.
Level of Study: Research
Type: Scholarship
Value: €3,174
Length of Study: 2 years
Frequency: Every 2 years
Country of Study: Ireland
Application Procedure: Applications should be mailed to the Professor of Theology, University of Dublin, Trinity College.
Closing Date: No later than the end of Trinity term in the year and every second year. See website for details.

Frances E. Moran Research Studentship

Subjects: Law.
Purpose: To support research in Irish law.
Eligibility: Open to all candidates.
Level of Study: Research
Type: Studentship
Value: €1,841–2,222 (tax-free) plus full fees cover
Length of Study: 1 year
Frequency: Annual
Country of Study: Ireland
No. of awards offered: 1
Application Procedure: Applications should be mailed to the Regius Professor of Laws, Trinity College.
Closing Date: See website for details.
Contributor: Trinity Trust
Additional Information: The successful candidate will be required to register for the degree of M.Litt. The successfulcandidate may be required to assist with tutorial work in the Law School for four hours per week.

Henry Flood Research Scholarship

Subjects: Irish folk and language studies.
Purpose: To support research in the area of Irish folk and language studies.
Eligibility: Open to all candidates.
Level of Study: Research
Type: Scholarship
Value: €8,000 plus EU fee cover per year
Length of Study: 1–2 years
Frequency: Annual

Country of Study: Ireland
Application Procedure: Applications should be mailed to the Professor of Irish, University of Dublin, Trinity College not later than the end of Trinity term.
Closing Date: The end of Trinity term. See website for details.

Home Hewson Scholarship

Subjects: Music, theatre, literature, and visual arts.
Purpose: To support studies in music, literature, theatre, and visual arts.
Eligibility: Open to all candidates.
Level of Study: Research
Type: Scholarship
Value: €2,539 (approx.) per year (value will depend at any given time on the interest available from the capital sum).
Length of Study: 1–3 years
Frequency: Dependent on funds available
Country of Study: Ireland
Application Procedure: Applications should be mailed to the Professor of Music, University of Dublin, Trinity College.
Closing Date: See website for details.
Additional Information: Please check website for latest updates.

Postgraduate Research Studentships

Subjects: Any subject.
Purpose: To support all research.
Eligibility: Open to new entrants and continuing students on the full-time PhD register.
Level of Study: Research
Type: Scholarship
Value: €8,000 plus full fees cover per year
Length of Study: 3 years
Frequency: Annual
Country of Study: Ireland
No. of awards offered: Varies
Application Procedure: Completed application form must be submitted to the organization online. Continuing students should contact their School directly.
Closing Date: Please check website.

Postgraduate Travelling Scholarship in Medicine and Surgery

Subjects: Medicine and Surgery in alternate years.
Purpose: To encourage younger graduates to undertake further work in specialised aspects of medicine and surgery, including the acquisition of modern techniques and the carrying out of research.
Eligibility: Open to all candidates.
Level of Study: Research
Type: Scholarship
Value: €22,220 plus the Sheppard Memorial Prize value €5,078, together with the SirJohn Banks medal in medicine or the Edward Hallaran Bennett medal in surgery.
Length of Study: 1 year
Frequency: Annual
Country of Study: Ireland
Application Procedure: Applications should be mailed to the Graduate Studies Office, Trinity College.
Closing Date: See website for details.
Contributor: John Banks Fund, E. Hallaran Bennett Fund, Bicentenary Fund, and Dr H. Hutchinson Stewart Fund
Additional Information: The successful candidate is expected to work for at least 9 months under the direction of a senior member of staff in a hospital or university department outside Dublin approved by the Board of the College.

Professor D.A. Webb Scholarship

Subjects: Botany.
Purpose: To support research in botany to understand fully, Irish plant variation, distribution and occurrence, in its wider European context.
Eligibility: Open to all candidates.
Level of Study: Research
Type: Scholarship
Value: €6,348
Length of Study: 1 year
Frequency: Dependent on funds available
Country of Study: Ireland
Application Procedure: Applications should be mailed to the Curator of the Herbarium, Trinity College.
Closing Date: See website for details.

R. B. McDowell Ussher Postgraduate Fellowships

Subjects: Modern history.
Purpose: To support research students.
Eligibility: Fellowship is available only to students registered for a Higher Degree by Research leading to a PhD in Modern History. Preference will be given to candidates intending to work on the history of Britain or of relations between Britain and Ireland.
Level of Study: Research
Type: Fellowship
Value: €16,680 per year
Length of Study: 3 years
Frequency: Annual
Study Establishment: School of Modern History, Trinity College Dublin, Ireland
Country of Study: Ireland
Application Procedure: Completed application must be submitted to Graduate Studies Office Trinity College Dublin.
Closing Date: See website for details.
Contributor: The London Trust for Trinity College Dublin

R.A.Q. O'Meara Research Fund

Subjects: Medicine.
Purpose: To support research in the field of cancer and allied disorders.
Eligibility: Open to all candidates.
Level of Study: Research
Type: Scholarship
Value: €10,158
Frequency: Every 2 years
Country of Study: Ireland
Application Procedure: Applications should be mailed to the Dean of the Health Sciences faculty.
Closing Date: See website for details.
Contributor: Marie Curie Memorial Foundation

White Postgraduate Fellowship

Subjects: Irish art history.
Purpose: To support research in Irish Art History.
Eligibility: Open to all candidates. Award is subject to annual review of academic progress.
Level of Study: Research
Type: Fellowship
Value: €15,000 maintenance per year, plus waiver of College annual fees (subject to specified annual limits).
Length of Study: 2 years for MLitt students, and 3 years for PhD.
Frequency: Annual
Country of Study: Ireland
Application Procedure: Applications should be mailed to Irish Art Research Centre, Trinity College. For more information mail to Irish Art Research Centre in the History of Art Department (triarc@tcd.ie).
Closing Date: See website for details.

UNIVERSITY OF EAST ANGLIA (UEA)

Faculty of Arts and Humanities, School of Literature and Creative Writing, Norwich, Norfolk, NR4 7TJ, England
Tel: (44) 16 03456161
Fax: (44) 16 03507728
Website: www.uea.ac.uk/lit/fellowships
Contact: Fellowship Administrator

The University of East Anglia (UEA) is organized into 23 schools of study encompassing arts and humanities, health, sciences and social sciences. These are supported by central service and administration departments.

Charles Pick Fellowship for South Asian Writers

Subjects: Fiction and nonfiction.

Purpose: To assist and support the work of a new and as yet unpublished writer or fiction or nonfictional prose and to give promising writers time to devote to the development of their talent. The fellowship would be for the purposes of completing a major work.
Eligibility: Open to applicants of all ages who are writers of fiction or nonfictional prose in English. Applicants must not yet have had a book published, but all applicants must provide a reference from either an editor, agent or accredited teacher of creative writing to be sent directly to the Charles Pick Fellowship. The writer should be from South Asia. The writer should be from South Asia (Afghanistan, Bangladesh, Bhutan, India, Kazakhstan, Kyrgyzstan, Maldives, Burma/Myanmar, Nepal, Pakistan, Sri Lanka, Turkmenistan, Tajikistan, Uzbekistan), but does not need to be domiciled there. The first Charles Pick South Asia Fellow was Shubhangi Swarup.
Type: Fellowship
Value: UK£10,000 plus free accommodation on the university campus
Length of Study: 6 months starting October 1st
Frequency: Annual
No. of awards offered: 1
Closing Date: January 31st
Funding: Private
No. of awards given last year: 1
No. of applicants last year: 200
Additional Information: Eligible to the nationals of South Asia. Please check website for further information.

For further information contact:

Schoold of Literature, Drama & Creative Writing, Norwich, UEA, NR4 7TJ, United Kingdom
Email: charlespickfellowship@uea.ac.uk
Contact: Natalie Mitchell, Senior Administrator

David T K Wong Fellowship

Subjects: Writing.
Purpose: To support promising writers in producing a work of fiction set in the Far East.
Eligibility: Open to all writers whose projects deal with some aspect of life in the Far East.
Level of Study: Professional development
Type: Fellowship
Value: UK£26,000
Length of Study: 9 months starting October 1st
Frequency: Annual
Study Establishment: UEA
Country of Study: United Kingdom
No. of awards offered: 1
Application Procedure: Applicants must send their completed applications to the Fellowship Administrator.
Closing Date: January 13th
Funding: Private
Contributor: David T K Wong
No. of awards given last year: 1
No. of applicants last year: 75

For further information contact:

Registry 3.15, University of East Anglia, Norwich, NR4 7TJ
Website: www.uea.ac.uk/lit/fellowships
Contact: The David Wong Fellowship Fellowship Administrator

THE UNIVERSITY OF EDINBURGH

Scholarships and Student Funding Services, University of Edinburgh, Old College, South Bridge, Edinburgh, EH8 9YL, Scotland
Tel: (44) 131 651 4070
Fax: (44) 131 651 4066
Email: scholarships@ed.ac.uk
Website: www.ed.ac.uk/student-funding
Contact: Robert Lawrie

The University of Edinburgh is an international centre of excellence in research and teaching, with outstanding resources and facilities for postgraduate students across the whole range of academic disciplines. More than 160 taught postgraduate programmes are available, with around 130 academic units offering subjects for degrees by research.

College of Medicine and Veterinary Medicine PhD Studentship

Subjects: A number of studentships for prospective PhD candidates who are required to select their research project from 30 available projects.
Eligibility: Applicants should hold, or expect to obtain a first or upper second class (Honours) degree or equivalent qualification in a relevant subject.
Level of Study: Doctorate
Type: Scholarship
Value: An annual stipend, payment of tuition fees at the United Kingdom/European Union rate, payment of bench fees, and an annual conference allowance
Study Establishment: The University of Edinburgh
Country of Study: Scotland
Application Procedure: Applicants are asked to select from the 30 projects on offer, and to submit a completed application form. Applications not using the application form may not be accepted. Applicants are also required to arrange for two confidential referee reports to be submitted by the closing date.
Closing Date: February 1st
Additional Information: For further information please check the website www.mvm.ed.ac.uk/gradschool/apply/funding.htm

For further information contact:

College of Medicine and Veterinary Medicine, University of Edinburgh, Queens Medical Research Institute, 47 Little France Crescent, Edinburgh, EH16 4TJ

Edinburgh Global Master's Scholarships

Subjects: All subjects.
Purpose: To support full-time study leading to a Master's degree in any discipline.
Eligibility: Open to one year Master's students for postgraduate study in any subject offered by the university.
Level of Study: Postgraduate
Type: Scholarship
Value: 15 awards of UK£5,000 and 25 awards of UK£3,000
Length of Study: 1 year
Frequency: Annual
Study Establishment: The University of Edinburgh
Country of Study: Scotland
No. of awards offered: 40
Application Procedure: Applications can be made online at www.ed.ac.uk/student-funding/masters.
Closing Date: April 1st
Contributor: University of Edinburgh

Edinburgh UK/EU Master's Scholarships

Subjects: Any subject offered by the University.
Purpose: To assist UK/EU students who have been accepted by the University to study on a one year full-time Master's programme in any subject area by covering a part of their tuition fees.
Eligibility: The scholarships will be awarded to UK/EU citizens who have been accepted for admission on a full-time basis for a postgraduate Master's programme in any subject offered by the University. Candidates must have or expect to obtain a UK first class or upper second class Honours degree or the overseas equivalent before applying for admission.
Level of Study: Graduate, Postgraduate
Type: Scholarship
Value: Up to £6,050
Length of Study: 1 year
Frequency: Annual
Study Establishment: University of Edinburgh
Country of Study: Scotland
No. of awards offered: 15
Application Procedure: Apply online at www.ed.ac.uk/student-funding/uk-masters.
Closing Date: May 1st

Malaysia Chevening Scholarship

Subjects: International law, economics, international and European politics, sustainable energy systems, environment and development,

environmental protection and management, environmental sustainability, e-science, high performance computing and informatics.
Purpose: To enable a Malaysian student to undertake a one-year taught Master's degree at the University of Edinburgh.
Eligibility: Please contact the British Council.
Level of Study: Graduate
Type: Scholarship
Length of Study: 1 year
Study Establishment: The University of Edinburgh
Country of Study: Scotland
Application Procedure: Please contact the British Council.

Principal's Career Development And Scholarship

Subjects: All subjects.
Level of Study: Doctorate
Type: Scholarship
Value: UK tuition fees and maintenance allowance broadly equivalent to those of the UK Research councils
Frequency: Annual
Study Establishment: The University of Edinburgh
Country of Study: Scotland
Additional Information: Please check website for further details.

University of Edinburgh College of Humanities and Social Science Research Studentships and Scholarships

Subjects: All subjects.
Eligibility: Students should hold, or expect to obtain, a First Class or Upper Second Class (Honours) Degree from the United Kingdom or the overseas equivalent. Candidates who have progressed to obtain a Master's degree with distinction will also be considered.
Level of Study: Doctorate, MPhil, MLitt
Type: Studentship and scholarship
Value: United Kingdom tuition fees and a maintenance allowance broadly equivalent to those of the United Kingdom Research Councils and the British Academy. Research costs may also be paid.Studentship value – Tuition fees of UK/Eu rate, annual stipend and research expenses. Scholarship value – Tuition fees (UK/EU rate) and research expenses.
Frequency: Annual
Study Establishment: The University of Edinburgh
Country of Study: Scotland
Application Procedure: Full details on how to apply can be found online at www.ed.ac.uk/student-funding/postgraduate
Closing Date: February 1st
Additional Information: For further information please see the website www.ed.ac.uk/student-funding/postgraduate

Wellcome Trust 4-Year PhD Programme Studentships

Subjects: Cell Biology.
Eligibility: Students should be from a life sciences background and should hold, or expect to obtain, at least an Upper Second Class (Honours) Degree.
Level of Study: Doctorate
Type: Studentship
Value: Tuition fees, research costs and maintenance allowance
Frequency: Annual
Study Establishment: The University of Edinburgh
Country of Study: Scotland
No. of awards offered: 5
Application Procedure: Please see the website www.wcb.ed.ac.uk/phd.
Closing Date: December 12th
No. of awards given last year: 5

For further information contact:

Email: karen.traill@ed.ac.uk
Contact: Karen Traill

UNIVERSITY OF ESSEX

Wivenhoe Park, Colchester, C04 3SQ, England
Tel: (44) 1206 873687
Fax: (44) 1206 872808
Email: pg scholarships@essex.ac.uk
Website: www.essex.ac.uk
Contact: Kaherine Free, Bursaries and Scholarships Officer

The University of Essex is founded nearly 50 years ago, the University of Essex is ranked ninth in the UK for research excellence, following the most recent Research Assessment Exercise. This allows us to offer world-class supervision and training opportunities, with our research students owrking at the heart of our internationally-acknowledged and well-connected research community. We are also one of the world's most internationally diverse universities, with more than one third of our 11,000 students coming from outside the UK, thus creating a diverse social and academic learning experience for all.

Access to Learning Fund

Subjects: All subjects.
Purpose: To support home students with study.
Eligibility: Applicants must be home students.
Level of Study: Unrestricted
Type: Grant
Value: According to individual needs
Frequency: Annual
Study Establishment: University of Essex
Country of Study: United Kingdom
Application Procedure: Applicants must contact Student Support.
Funding: Government
Contributor: ALF

AHRC Award for Dance, Drama and Performing Arts

Subjects: East is Acting School.
Purpose: To support postgraduate study.
Eligibility: Open to postgraduate students in East is Acting School.
Level of Study: Postgraduate
Type: Scholarship
Value: Fees plus maintenance for UK students. Fees only for EU students
Length of Study: 1 year
Frequency: Annual
Study Establishment: University of Essex
Country of Study: United Kingdom
Application Procedure: Please check website.
Closing Date: Mid-February
Contributor: University of Essex
Additional Information: Please check website for further information.

AHRC Award for English Language and Literature

Subjects: English language and linguistics. Literature, film and theatre studies.
Purpose: To support postgraduate study.
Eligibility: Open to postgraduate students in the Department of Language and Linguistics and the Department of Literature, Film and theatre studies.
Level of Study: Postgraduate
Type: Scholarship
Value: Covers fees and maintenance for UK students and fees only for EU students.
Length of Study: 1 year
Frequency: Annual
Study Establishment: University of Essex
Country of Study: United Kingdom
Application Procedure: Please check website.
Closing Date: Mid-February
Contributor: AHRC
Additional Information: Please check website for further information.

AHRC Award for Law

Subjects: Law.
Purpose: Support postgraduate study.
Eligibility: Postgraduate students in the School of Law.
Level of Study: Postgraduate

Type: Award
Value: Fees and maintenance for UK students. Fees only for EU students.
Length of Study: 1 year
Study Establishment: University of Essex
Country of Study: United Kingdom
Application Procedure: Please check website.
Closing Date: Mid-February
Contributor: AHRC
Additional Information: Please check website for further information.

AHRC Award for Philosophy

Subjects: Philosophy.
Purpose: To support postgraduate study.
Eligibility: Applicants must be postgraduate students.
Level of Study: Postgraduate
Type: Scholarship
Value: Covers fees and maintenance for UK students and fees only for EU students.
Length of Study: 1 year
Frequency: Annual
Study Establishment: University of Essex
Country of Study: United Kingdom
Application Procedure: Please check website.
Closing Date: Mid-February
Contributor: AHRC
Additional Information: Please check website for further information.

AHRC Doctoral Award for Art History

Subjects: Art history.
Purpose: Support postgraduate study.
Eligibility: Students applying for postgraduate study in the department.
Level of Study: Postgraduate
Type: Award
Value: Fees and maintenance for UK students, fees only for EU students.
Study Establishment: University of Essex
Country of Study: United Kingdom
Application Procedure: Please check website.
Closing Date: Mid-February
Contributor: AHRC
Additional Information: Please check website for further information.

AHRC Doctoral Award for History

Subjects: History from 1500 AD to present day.
Purpose: Support research students.
Eligibility: Applicants must be postgraduate students. Research must be humanities based.
Level of Study: Postgraduate
Type: Scholarship
Value: UK students fees and maintenance, EU students fees only.
Frequency: Annual
Study Establishment: University of Essex
Country of Study: United Kingdom
Application Procedure: Please check website.
Closing Date: Mid-February
Contributor: AHRC
Additional Information: Please check website for further information.

The Artellus Scholarships

Subjects: Refugee care.
Purpose: Support study in refugee care.
Eligibility: Students applying for MA/PhD in refugee care.
Level of Study: Postgraduate, Research
Type: Scholarship
Value: Up to £2,000 off set against tuition fees
Length of Study: One year Masters or three year PhD
Frequency: Dependent on funds available
Study Establishment: University of Essex
Country of Study: United Kingdom
Application Procedure: Applicants must visit the website.
Funding: Corporation
Contributor: Artellus Limited

BBSRC Studentship

Subjects: Biological sciences.
Purpose: To support postgraduate biological science students.
Eligibility: Open to applicants who are biological science postgraduates, but mathematics or computing graduates may be eligible depending on the project.
Level of Study: Postgraduate, Research
Type: Studentship
Length of Study: 3 years
Frequency: Annual
Study Establishment: University of Essex
Country of Study: United Kingdom
Application Procedure: Applicants must contact the department concerned.
Closing Date: Information on website when projects are advertised
Funding: Government
Contributor: BBSRC

Don Pike Award

Subjects: Philosophy of social science.
Purpose: Support students over the age of 21 working in one field of philosophy of social sciences at postgraduate or undergraduate level.
Eligibility: Applicants must be students working on philosophy of social science.
Level of Study: Unrestricted
Type: Scholarship
Value: £250 towards purchase of books and expenses in preparing a thesis or dissertation
Frequency: Annual
Study Establishment: University of Essex
Country of Study: United Kingdom
Application Procedure: Applicants must contact the department concerned.
Funding: Trusts
Additional Information: Please check website.

Drake Lewis Graduate Scholarship for Art History

Subjects: Art history.
Purpose: To support new MA students in art history.
Eligibility: Open to postgraduate applicants.
Level of Study: Postgraduate
Type: Scholarship
Value: £5,000
Frequency: Annual
Study Establishment: University of Essex
Country of Study: United Kingdom
No. of awards offered: Up to 3
Funding: Private
Contributor: Drake Lewis
Additional Information: Please check website for further information.

Drake Lewis Graduate Scholarship for Health and Human Sciences

Subjects: Public health and health studies.
Purpose: To support students on full-time masters in public health or health studies.
Eligibility: Open to postgraduate applicants.
Level of Study: Postgraduate
Type: Scholarship
Value: £5,000
Frequency: Annual
Study Establishment: University of Essex
Country of Study: United Kingdom
Application Procedure: Applicants must contact the department concerned.
Funding: Private
Contributor: Drake-Lewis
Additional Information: Please check website for further information.

EPSRC School of Computer Science and Electronic Engineering Research Studentship

Subjects: Electronic systems engineering, computer science, applied physics.
Purpose: To support postgraduate study.

Eligibility: Applicants must be graduates holding a good UK honours degree or overseas equivalent and preferably a Masters degree.
Level of Study: Doctorate, Postgraduate, Research
Type: Scholarship
Value: Fees and maintenance for UK and EU students who have been resident in the UK for 3 years prior to commencing the course. Fees only for other EU students.
Frequency: Annual
Study Establishment: University of Essex
Country of Study: United Kingdom
Application Procedure: Applicants must submit a formal application for PhD by research and state on the form that they wish to be considered for an EPSRC studentship.
Contributor: EPSRC
Additional Information: Please check website.

For further information contact:

School of Computer Science and Engineering, University of Essex, Wivenhoe Park, Colchester, CO4 3SQ
Contact: Postgraduate Research Administrator

ESRC Studentships

Subjects: Our Doctoral Training Centre offers 16 fully-funded ESRC studentships across 22 doctoral pathways.
Purpose: Support postgraduate study.
Eligibility: Students applying for PhD study.
Level of Study: Postgraduate, Research
Type: Studentship
Value: Fees and maintenance for Uu students, fees only for EU students.
Length of Study: 3 years PhD
Frequency: Annual
Study Establishment: University of Essex
Country of Study: United Kingdom
Application Procedure: Applicants must visit the website.
Closing Date: Mid-February
Funding: Government
Contributor: ESRC
Additional Information: Please check website.

Essex Rotary University Travel Grants

Subjects: All subjects.
Purpose: To support students to spend time studying abroad.
Eligibility: Applicant must be a student at the University of Essex.
Level of Study: Unrestricted
Type: Grant
Frequency: Annual
Study Establishment: University of Essex
Country of Study: United Kingdom
No. of awards offered: 2
Application Procedure: Applicants must send a statement (not exceeding 500 words) indicating what they would do with a Rotary Travel Grant. See website.
Funding: Trusts
Contributor: Rotary clubs in Essex

Essex Society for Family History Award

Subjects: Essex related local or family history.
Purpose: Support a postgraduate working on an Essex-related subject in local or family history.
Eligibility: A postgraduate working in relevant field.
Level of Study: Postgraduate
Type: Award
Value: £500
Frequency: Dependent on funds available
Study Establishment: University of Essex
Country of Study: United Kingdom
No. of awards offered: 1
Application Procedure: Applicants must visit the website.
Contributor: Essex Society for Family History

Essex/Fulbright Commission Postgraduate Scholarships

Subjects: All subject.
Purpose: For US citizens coming to the UK to do a one-year Masters.
Eligibility: A US graduate wishing to pursue a one-year Masters.
Type: Scholarship
Value: Tuition fee plus stipend
Length of Study: 1 year
Frequency: Annual
Study Establishment: University of Essex
Country of Study: United Kingdom
Application Procedure: Applicants must visit the website.
Contributor: Fulbright Commission

Friends of Historic Essex Fellowship Award

Subjects: History.
Purpose: Support postgraduates working on Essex services.
Eligibility: Postgraduate student.
Level of Study: Postgraduate
Type: Award
Value: Approx. £750
Frequency: Dependent on funds available
Study Establishment: University of Essex
Country of Study: United Kingdom
No. of awards offered: 1–2
Application Procedure: Applicants must visit the website.
Contributor: Friends of Historic Essex

Giulia Mereu Scholarships

Subjects: LLM International Human Rights Law and LLM International Human Rights and Humanitarian Law.
Purpose: Support a student on the University of Essex LLM in International Human Rights Law.
Eligibility: Applicants must be postgraduate students on the LLM in International Human Rights Law or LLM International Human Rights and Humanitarian Law at Essex and must hold a firm offer from the school.
Level of Study: Postgraduate
Type: Scholarship
Value: Equivalent to home/EU fees plus an allowance towards an internship
Length of Study: One year
Frequency: Annual
Study Establishment: University of Essex
Country of Study: United Kingdom
No. of awards offered: 1
Application Procedure: Applicants must contact the School of Law.
Funding: Trusts
Contributor: Family and friends of Giulia Mereu
No. of awards given last year: 1

Marshall Scholarships

Subjects: All research degrees.
Purpose: Support American students doing a research degree.
Eligibility: Applicants must be postgraduate. American students applying for full-time degree.
Level of Study: Postgraduate, Research
Type: Scholarship
Value: Tuition fees, living expenses, annual book grant, thesis grant, research and daily travel grant, fares to and from the US
Frequency: Annual
Study Establishment: University of Essex
Country of Study: United Kingdom
Application Procedure: See www.marshallscholarship.org
Closing Date: Mid-October of year before tenure
Funding: Foundation

Modern Law Review

Subjects: Topics in law broadly in the publishing interests of 'The Modern Law Review'.
Purpose: Supports law research students.
Eligibility: Law research students.
Value: £5,000–10,000
Frequency: Annual
Study Establishment: University of Essex
Country of Study: United Kingdom
Application Procedure: Applicants must visit the website.

National Federation of Business and Professional Women's Clubs Travel Grants

Subjects: All subjects.

Purpose: To enable full-time female postgraduates to travel for research/studies.
Eligibility: Female full-time postgraduates at the University of Essex.
Level of Study: Postgraduate
Type: Grant
Value: Varies
Frequency: Annual
Study Establishment: University of Essex
Country of Study: United Kingdom
No. of awards offered: 1
Application Procedure: See the University website.
Contributor: National Federation of Business and Professional Women's clubs

NERC Studentships

Subjects: Biological sciences.
Purpose: To support postgraduate biological science students.
Eligibility: Applicants must be biological science postgraduates.
Level of Study: Postgraduate, Research
Type: Studentship
Value: Depends on funds available
Length of Study: 3 years
Frequency: Annual
Study Establishment: University of Essex
Country of Study: United Kingdom
Application Procedure: Please contact the department concerned.
Closing Date: Information on website when projects are advertised
Funding: Government
Contributor: NERC
No. of awards given last year: 2
Additional Information: Details for this year have not yet been confirmed, so this information is a guide based on previous years.

OSF/University of Essex Scholarship in Human Rights

Subjects: Human rights.
Purpose: Support postgraduate study in human rights.
Eligibility: Students applying for MA theory and practice of human rights, MA human rights and cultural diversity, LLM international human rights law and LLM international human rights and humanitarian law who are from Afghanistan, Armenia, Azerbaijan, Georgia, Lebanon, Nepal, Palestine, Tajikistan, Turkmenistan or Uzbekistan.
Level of Study: Postgraduate
Type: Scholarship
Value: Tuition fees, examination fees, a monthly stipend for living expenses sufficient for a single student and other agreed allowances including one return economy airfare
Length of Study: 1 year
Frequency: Annual
Study Establishment: University of Essex
Country of Study: United Kingdom
Application Procedure: Applicants must visit the website.
Contributor: FCO

Peter and Michael Hiller Scholarships

Subjects: Refugee care.
Purpose: Support students studying refugee care at postgraduate level.
Eligibility: Students applying for MA/PhD in refugee care.
Level of Study: Postgraduate, Research
Type: Scholarship
Value: £6,500
Length of Study: One year for MA or three year for PhD
Study Establishment: University of Essex
Country of Study: United Kingdom
Application Procedure: Applicants must visit the website.
Contributor: The Peter and Michael Hiller Charitable Trust

Santander Masters Scholarships

Subjects: All subjects.
Purpose: To support students from Santander network countries to undertake further study.
Eligibility: Open to graduates residing in one of the Santander network countries who have an offer to study at Masters level.
Level of Study: Postgraduate
Type: Scholarships
Value: £5,000
Frequency: Annual
Study Establishment: University of Essex
Country of Study: United Kingdom
No. of awards offered: 10
Application Procedure: Please check website.
Funding: Corporation
Contributor: Santander

Santander Travel Bursaries

Subjects: All subject.
Purpose: Support study abroad opportunities.
Eligibility: Students undertaking study abroad at selected universities.
Level of Study: Postgraduate
Type: Bursary
Value: Varies
Length of Study: Full year or one semester
Frequency: Annual
Study Establishment: University of Essex
Country of Study: United Kingdom
Application Procedure: Please check website.
Funding: Corporation
Contributor: Santander

Sir Eric Berthoud Travel Grant

Subjects: All subjects.
Purpose: To support full-time students wishing to travel as part of their studies/research.
Eligibility: Applicants must be students at the University of Essex.
Level of Study: Unrestricted
Type: Grant
Frequency: Annual
Study Establishment: University of Essex
Country of Study: United Kingdom
Application Procedure: Applicants must visit the website for details.
Funding: Trusts
Contributor: Bequest from Sir Eric Berthoud

Sports Scholarship

Subjects: All subject.
Purpose: Support talented teams or individuals.
Eligibility: Athletes performing at national or international level at junior/senior level or who show evidence of having the potential to achieve these levels.
Level of Study: Unrestricted
Type: Scholarship
Value: Depends on funds available
Study Establishment: University of Essex
Country of Study: United Kingdom
Application Procedure: Please check website.

Tinson Fund Scholarships for Law

Subjects: Law.
Purpose: To support students from the former Soviet Bloc interested in studying postgraduate law.
Eligibility: Open to students from former Soviet Bloc countries, who have an offer on an LLM programme.
Level of Study: Postgraduate
Type: Scholarship
Value: Tuition fees
Frequency: Annual
Study Establishment: University of Essex
Country of Study: United Kingdom
Application Procedure: See website for details.
Closing Date: Mid-May
No. of awards given last year: 1

Travel Grants

Subjects: All subjects.
Purpose: To support students who need additional financial assistance.
Eligibility: Graduate students.
Level of Study: Postgraduate
Type: Grant

Frequency: Annual
Study Establishment: University of Essex
Country of Study: United Kingdom
No. of awards offered: 1–2
Application Procedure: Applicants can obtain application forms from Registry.
Contributor: University of Essex
No. of awards given last year: 1–2

University of Essex Cardiac Rehabilitation Bursary

Subjects: MSc cardiac rehabilitation.
Purpose: Pays for training as a qualified phase four cardiac rehabilitation instructor.
Eligibility: Students must be registered on course to apply.
Level of Study: Postgraduate
Type: Bursary
Value: Fees to train as a qualified phase four cardiac rehabilitation instructor, including travel and subsistence. Also partly covers tuition fees, depending on funds available
Length of Study: 1 year
Frequency: Annual
Study Establishment: University of Essex
Country of Study: United Kingdom
Application Procedure: Applicants must visit the website.
Funding: Government
Contributor: University of Essex
No. of awards given last year: 1
Additional Information: Please check website for further information.

University of Essex Department of Biological Sciences Studentships

Subjects: Biological sciences.
Purpose: To support postgraduate biological science students.
Eligibility: Open to applicants who are biological science postgraduates, Mathematics or computing graduates are also eligible depending on the project.
Level of Study: Postgraduate, Research
Type: Studentship
Value: Depends on funds available
Length of Study: 3 years
Frequency: Annual
Study Establishment: University of Essex
Country of Study: United Kingdom
Application Procedure: Applicants must contact the department concerned.
Closing Date: Information on website when projects are advertised in January
Funding: Government
Contributor: University of Essex
No. of awards given last year: 3
No. of applicants last year: 10

University of Essex Doctoral Scholarship

Subjects: Our University of Essex Doctoral Scholarships support talented PhD students to study at the University. These awards are available to students who can demonstrate an excellent academic background and the potential for future achievement, as well as finanical need. They are awarded for each year of study during the standard minimum period of enrolment for your degree, subject to satisfactory progress.
Purpose: To support PhD study.
Eligibility: Applicants must be prospective PhD students with an offer of a place.
Level of Study: Postgraduate
Type: Scholarship
Value: Fee waiver (partial fee waiver for overseas students) plus stipend, GTRA and training elements depending on the package offered.
Frequency: Annual
Study Establishment: University of Essex
Country of Study: United Kingdom
Application Procedure: Please check website.
Closing Date: March 1st
Contributor: University of Essex
Additional Information: Please check website for futher information.

University of Essex Silberrad Scholarships

Subjects: All subjects.
Purpose: To support graduates to pursue PhD study.
Eligibility: Open to applicants holding a degree from or about graduate from University of Essex who have an offer place at the University of Essex and are eligible to pay home/EU tuition fees.
Level of Study: Postgraduate, Research
Type: Scholarship
Value: Home/EU tuition fee and bursary element towards living costs
Length of Study: Up to 1 year
Frequency: Annual
Study Establishment: University of Essex
Country of Study: United Kingdom
No. of awards offered: 2
Application Procedure: Please check website.
Funding: Trusts
Contributor: Silberrad Estate

Winsten Scholarship

Subjects: Mathematical sciences.
Purpose: Support postgraduate study in maths.
Eligibility: Postgraduates applying for mathematical sciences.
Level of Study: Postgraduate
Type: Scholarship
Value: Depends on funds available but up to £4,000
Frequency: Annual
Study Establishment: University of Essex
Country of Study: United Kingdom
Application Procedure: Applicants must visit the website.

UNIVERSITY OF EXETER

Postgraduate Administration Office, Northcote House, The Queen's Drive, Exeter, Devon, EX4 4QJ, England
Tel: (44) 01392 722 207
Fax: (44) 01392 262 458
Email: gradfunding@exeter.ac.uk
Website: www.exeter.ac.uk/gradschool
Contact: Mrs Julie Gay, Scholarships Secretary

The University of Exeter combines a reputation of national and international excellence in research with a record of established excellence in teaching. Providing a superb environment in which to live and work, the University of Exeter is representative of the best in university education in the United Kingdom.

Andrew Stratton Scholarship

Subjects: Engineering and science.
Purpose: To help to support postgraduate study
Eligibility: Studying for taught Masters in Engineering or science International only
Level of Study: Postgraduate
Type: Scholarship
Value: Up to UK£1,000
Length of Study: 1 year
Frequency: Dependent on funds available
Study Establishment: The University of Exeter
Country of Study: United Kingdom
No. of awards offered: 1
Application Procedure: Applicants must contact Postgraduate Administration office for further details.
Closing Date: June 30th
No. of awards given last year: 1
No. of applicants last year: 8
Additional Information: Further information is available on request.

Anning Morgan Bursary

Subjects: All subjects.
Purpose: To financially support postgraduate study.
Eligibility: Open to students residing in Duchy of Cornwall prior to entry or during their postgraduate studies.
Level of Study: Postgraduate
Type: Bursary
Value: Up to £1,000
Length of Study: Maximum 2 years

Frequency: Dependent on funds available
Study Establishment: University of Exeter
Country of Study: United Kingdom
No. of awards offered: 1
Application Procedure: Application forms available from postgraduate administration office.
Closing Date: June 30th
No. of awards given last year: 2
No. of applicants last year: 11
Additional Information: Further information available on request.

For further information contact:

Contact: Julie Gay, Scholarships Secretary

British Council Awards and Scholarships for International Students

Subjects: All subjects offered by the University.
Purpose: To allow international students to pursue postgraduate study.
Eligibility: Open to candidates for research degrees who have obtained at least an Upper Second Class (Honours) Degree or its equivalent.
Level of Study: Postgraduate
Type: Scholarship
Value: Varies
Length of Study: Varies
Frequency: Annual
Study Establishment: The University of Exeter
Country of Study: United Kingdom
No. of awards offered: Varies
Application Procedure: Applicants must obtain details from the British Council representative in the applicant's own country.
Closing Date: Please contact the organization
Funding: Government
Additional Information: Further information is available on request.

Commonwealth Scholarship Plan

Subjects: All subjects.
Purpose: To allow students to pursue postgraduate study.
Eligibility: Open to students from Commonwealth countries who do not already hold scholarships from their own country.
Level of Study: Postgraduate
Type: Scholarship
Value: Varies, but includes payment of tuition fees
Frequency: Annual
Study Establishment: The University of Exeter
Country of Study: United Kingdom
No. of awards offered: Varies
Application Procedure: Applicants must apply well in advance in their country of permanent residence through the Commonwealth Scholarship Agency.
Closing Date: Please contact the organization
Funding: Government
Additional Information: Further information is available on request.

Cornwall Heritage Trust Bursary

Subjects: All subjects centering on Cornwall's heritage.
Purpose: To support the tuition fees of the applicants.
Eligibility: Open to one or more postgraduate students producing dissertations/theses centred on any aspect of Cornwall's heritage.
Level of Study: Postgraduate
Type: Bursary
Value: Up to UK£1,000
Length of Study: 1 year
Frequency: Dependent on funds available
Study Establishment: The University of Exeter
Country of Study: United Kingdom
No. of awards offered: 1 or more
Application Procedure: Applicants should contact Postgraduate administration office.
Closing Date: June 30th
Funding: Trusts
No. of applicants last year: 4
Additional Information: Further information is available on request.

Jonathan Young Scholarship

Subjects: Subjects covered by Business School.
Purpose: To finance travel in support of furtherance of studies of research.
Eligibility: Postgraduate/undergraduate in Business School or Department of History, Politics or Sociology.
Level of Study: Postgraduate, Undergraduate
Type: Scholarship
Value: Up to £1,000
Length of Study: 1 year
Frequency: Dependent on funds available
Study Establishment: University of Exeter
Country of Study: United Kingdom
No. of awards offered: 1 or more
Application Procedure: Application forms available from postgraduate administration office.
Closing Date: November 11th
Funding: Private
No. of awards given last year: 3
No. of applicants last year: 19

Tom Davis Scholarship

Subjects: All subject.
Purpose: Payment of part-time fees for Taught Masters Programme
Eligibility: Students ordinarily resident in Devon with preference given to those ordinarily resident in Exeter studying Masters degree.
Level of Study: Postgraduate
Type: Scholarship
Value: Part-time fees
Length of Study: 2 years maximum
Frequency: Dependent on funds available
Study Establishment: University of Exeter
Country of Study: United Kingdom
No. of awards offered: 1
Application Procedure: Application forms available from postgraduate administration office.
Closing Date: June 30th

University of Exeter Chapel Choir Choral and Organ Scholarship

Purpose: Annual scholarships offered to choral and organ practitioners to aid in recitals on behalf of the chapel choir.
Eligibility: Based on audition. The Director of Chapel Music will invite for competitive audition on the basis of applications demonstrating a high level of competence and experience plus details of two referees familiar with the applicant's ability.
Level of Study: Doctorate, Graduate, MBA, Postgraduate, Research
Type: Scholarship
Value: £400 per year for choral scholars, £700 per year for senior organ scholars, £300 per year for junior organ scholars
Length of Study: 1 year initially but may be renewed for the duration of study, where appropriate
Frequency: Annual
Study Establishment: The University of Exeter
Country of Study: United Kingdom
No. of awards offered: 10 – choral and 2 – organ
Application Procedure: Applicants can contact the Director for more information and request an application form.
Closing Date: February 13th
Funding: Corporation, individuals, trusts
No. of awards given last year: 10
No. of applicants last year: 60
Additional Information: From the 10 choral scholarships available, 4 are offered to sopranos, 2 each for other voice parts of alto (male and female), tenor and bass. The award for senior organ scholarship status will require recipients to direct the choir when required.

For further information contact:

Email: a.j.musson@exeter.ac.uk
Contact: Professor Anthony Musson, Director of Chapel Music

University of Exeter Departmental Research Scholarships

Subjects: All subjects.
Purpose: To fund MPhil and PhD study.

Eligibility: Open to all international applicants for PhD and MPhil/PhD programmes.
Level of Study: Doctorate, Postgraduate, Research
Type: Scholarship
Value: Full fees
Length of Study: 3 years
Frequency: Annual
Study Establishment: The University of Exeter
Country of Study: United Kingdom
No. of awards offered: Varies
Application Procedure: Applicants should make enquiries to the college in which they wish to study.
Contributor: University of Exeter

University of Exeter Graduate Research Assistantships

Subjects: All subjects.
Purpose: To assist students by offering them a top-quality scheme that offers excellent career development opportunities.
Eligibility: Open to MPhil and PhD students.
Level of Study: Doctorate, Postgraduate, Research
Type: Other
Value: Fees and maintenance at research council rates
Length of Study: 4 years
Frequency: Annual
Study Establishment: The University of Exeter
Country of Study: United Kingdom
No. of awards offered: Varies
Application Procedure: Applicants must visit the website of the college in which they wish to study.
Additional Information: Further details of this scheme are available from the website. Graduate research assistants are required to undertake 2 days of research a week for a research team in addition to their own research.

University of Exeter Graduate Teaching Assistantships (GTA)

Subjects: All subjects.
Purpose: To assist students by offering them a top-quality scheme that offers excellent career development opportunities.
Eligibility: Open to MPhil and PhD students.
Level of Study: Doctorate, Research, Postgraduate
Type: Other
Value: Fees and maintenance at research council rates
Length of Study: Up to 4 years
Frequency: Annual
Study Establishment: The University of Exeter
Country of Study: United Kingdom
No. of awards offered: Varies
Application Procedure: Applicants must visit the website of the school in which they wish to study.
Closing Date: Please contact the organization
Additional Information: Further details of this scheme are available from the university website.

University of Exeter Sports Scholarships

Subjects: Sports.
Purpose: To assist students of outstanding sporting ability who show evidence of achievement or potential at international or national level.
Eligibility: Performers from any sport considered, but emphasis is placed on cricket, golf, hockey, rugby, sailing and tennis. Both male and female atheletes in these sports. Emphasis is also placed on badminton, lacrosse, net ball and rowing.
Level of Study: Graduate, Postgraduate
Type: Scholarship
Value: Up to £2,000 (For all sporting expenses/discounted accommodation costs) as well as additional sports specific support and services totalling upto £3,500
Length of Study: 1 year initially, but may be renewed for an additional 2 years
Frequency: Annual
Study Establishment: The University of Exeter
Country of Study: United Kingdom
No. of awards offered: Varies
Application Procedure: Applicants must complete an online application form on scholarships/bursaries website.
Closing Date: March 31st
Funding: Commercial, individuals
Contributor: University of Exeter
No. of awards given last year: 29
No. of applicants last year: 260

For further information contact:

Website: www.ex.ac.uk/sport
Tel: n.e.beasant@ex.ac.uk
Contact: Dr N Beasant, Assistant Director of Sport

UNIVERSITY OF GLAMORGAN

Treforest Pontypridd, Wales, CF37 1DL, United Kingdom
Tel: (44) (0) 1443 654 450
Fax: (44) (0) 1443 654 050
Email: enquiries@glam.ac.uk
Website: www.glam.ac.uk
Contact: Enquiries and Admissions Unit

The University of Glamorgan is a dynamic institution with an exceptional record for academic excellence, teaching and research. Glamorgan offers first class teaching, excellent facilities and outstanding academic support.

Crawshays Rugby Scholarship

Subjects: All subjects.
Purpose: To support talented young rugby players through higher education at the University.
Eligibility: Open to applicants who are rugby players studying at the University of Glamorgan.
Level of Study: Postgraduate, postgraduate degree
Type: Scholarship
Value: UK£1,000
Length of Study: Up to 3 years
Frequency: Annual
Study Establishment: University of Glamorgan
Country of Study: United Kingdom
No. of awards offered: 1
Application Procedure: For further details contact the university.
Closing Date: September 1st
Contributor: Crawshay's Welsh RFC

For further information contact:

Sardis Road, Pontypridd RFC
Contact: Clive Jones, Director of Rugby

The University of Glamorgan Sports Scholarships

Subjects: Sports.
Purpose: To support potential elite athletes and sports people who are competing at national and international level in their chosen field.
Eligibility: Open to students studying on the Treforest or Glyntaff Campus.
Level of Study: Professional development
Type: Scholarship
Value: A no-strings cash award each academic year of £500, annual grant of £150 to use in the University's excellent sports centre
Frequency: Annual
Country of Study: United Kingdom
Application Procedure: Contact the university for further details.
Closing Date: December 1st

UNIVERSITY OF GLASGOW

University Avenue, 6 University Gardens, Glasgow, Lanarkshire, G12 8QQ, Scotland
Tel: (44) 0141 330 6828; (0) 141 330 2000
Fax: (44) 0141 330 2000
Email: e.queune@admin.gla.ac.uk
Website: www.gla.ac.uk
Contact: Emily Queune, Faculty of Arts Office

The University of Glasgow is a major research led university operating in an international context, which aims to provide education through the development of learning in a research environment, to undertake fundamental, strategic and applied research and to sustain and add value to Scottish culture, to the natural environment and to the national economy.

Adam Smith Research Foundation PhD Scholarships

Subjects: Public policy, governance and social justice, work, ethics and technology, people places and change, macroeconomics, business and finance, legal and political thought.
Purpose: To promote and sustain research within the UK, European and international arenas.
Eligibility: Open to candidates in any of the faculty's constituent departments: accounting and finance; central and East European studies; centre for drugs misuse research; economic and social history; economics; School of Law; management; politics, sociology, anthropology and applied social sciences; urban studies.
Level of Study: Doctorate
Type: Scholarship
Value: maintenance stipend equivalent to the Economic and Social Research Council (ESRC) award for a single person £13,290) and payment of tuition fees at the UK/EU rate of £3,390 or the overseas rate of £8,800
Length of Study: 3 years
Frequency: Annual
No. of awards offered: 1
Application Procedure: Candidates should apply for entry to PhD in a department in the faculty in the usual way www.gla.ac.uk/lbss/graduateschool/applications.html
Closing Date: May 10th
Funding: Foundation
Contributor: Adam Smith Research Foundation

Alexander and Dixon Scholarship (Bryce Bequest)

Subjects: English literature.
Eligibility: Open to the citizens of United Kingdom or a European Union national.
Level of Study: Doctorate
Type: Scholarship
Value: £3,500 fees only
Length of Study: 3 years
No. of awards offered: 2
Application Procedure: The application should consist of a 500–word case for support and a brief covering letter, including the proposed title of the thesis, the name(s) of the proposed supervisor(s), and give the applicant's e-mail and other contact details.
Closing Date: June 21st

For further information contact:

Department Office, Department of English Literature, University of Glasgow, Glasgow, G12 8QQ, United Kingdom
Email: a.macmillan@englit.arts.gla.ac.uk
Website: www.arts.gla.ac.uk/SESLL/EngLit/grad.htm
Contact: Anna Macmillan

Alexander and Margaret Johnstone Postgraduate Research Scholarships

Subjects: Arts.
Eligibility: Open to students intending a research degree in the faculty of arts in a department rated 5 or 5* in the research assessment exercise.
Level of Study: Doctorate
Type: Research scholarship
Value: Tuition fees at the Home/EU student rate, plus stipend of between £6,000 and £7,000
Length of Study: 3 years
Application Procedure: Check website for further details.
Funding: Government

For further information contact:

Faculty of Arts Office, University of Glasgow, 6 University Gardens, Glasgow, G12 8QQ, United Kingdom
Tel: (44) 0141 330 6828
Email: e.queune@admin.gla.ac.uk
Contact: Emily Queune

Bellahouston Bequest Fund

Subjects: Arts and science.
Eligibility: Open to postgraduate students undertaking a Masters Degree course in the faculty of Arts.
Level of Study: Postgraduate
Type: Scholarship
Value: £1,000
Length of Study: 1 year
Frequency: Annual
No. of awards offered: 3
Application Procedure: The candidate must contact the clerk of the faculty of arts. Check website for further information.
Closing Date: July 31st
Additional Information: Preference will be given to the Glaswegians.

For further information contact:

University of Glasgow, Glasgow, G12 8QQ, United Kingdom
Tel: (44) 0141 330 2000
Email: ugs@archives.gla.ac.uk
Contact: Clerk of the Faculty of Arts

British Federation of Women Graduates (BFWG)

Subjects: All subjects.
Purpose: To encourage applicants to become members of the Federation to help promote better links between female graduates throughout the world.
Eligibility: Open to female graduate with academic excellence. Research students of all nationalities who will be studying in the United Kingdom are eligible for the scholarship.
Level of Study: Research, Graduate
Type: Scholarship and award
Value: £1,000–3,000
Frequency: Annual
Country of Study: United Kingdom
No. of awards offered: 5–6
Application Procedure: Check website for further details.
Closing Date: March 30th
Funding: Private
Additional Information: Male graduates and female undergraduates are not eligible.

For further information contact:

4 Mandeville Courtyard, 142 Battersea Park Road, London, SW11 4NB, United Kingdom
Tel: (44) 020 7498 8037
Fax: (44) 020 7498 5213
Email: info@bfwg.org.uk
Website: www.bfwg.org.uk

The Catherine Mackichan Trust

Subjects: Scottish history.
Eligibility: Open to applications from academic centres worldwide, schools, colleges and individuals or groups.
Level of Study: Research
Type: Award
Value: £350
Length of Study: 1 year
Frequency: Annual
No. of awards offered: 4
Application Procedure: Check website for further details.
Funding: Trusts
Contributor: The Catherine Mackichan Trust

For further information contact:

Catherine Mackichan Trust, School of Scottish Studies, 27–29 George Square, Edinburgh, EH8, United Kingdom
Contact: I Fraser, Vice Chairman

Dorothy Hodgkin Postgraduate Awards

Subjects: Engineering and physical sciences, physics and astronomy and biological sciences.
Purpose: To support the highest calibre doctoral students from developing countries to undertake their PhD in the UK.

Eligibility: This award is only open to student nationals from India, China, Hong Kong, Brazil, South Africa, Russia and the developing world, as defined by the OECD.
Level of Study: Research
Type: Award
Value: £75,000
Length of Study: 3 years
Frequency: Annual
Study Establishment: Glasgow University
Country of Study: Scotland
No. of awards offered: 1–3
Application Procedure: Please check the website for details.
Funding: Commercial, government
Contributor: UK Research Councils and Industrial Partners
No. of awards given last year: 1
Additional Information: The future of this scheme is currently under review.

Eglington Fellowship for Postgraduate Study

Subjects: Art.
Eligibility: Open to candidates possessing a graduate degree.
Level of Study: Graduate
Type: Fellowship
Value: £500 per year
Length of Study: 3 years
Application Procedure: Check website for further details.
Closing Date: May 1st

For further information contact:

Faculty Office, 2nd floor, 6 University Gardens, Glasgow, G12 9QQ, United Kingdom
Website: www.gsah.arts.gla.ac.uk/html/eglington.html
Contact: Clerk of the Arts Faculty

Faculty of Medicine Studentships

Subjects: Medicine.
Eligibility: Open to candidates having an upper second or first class degree in a relevant subject. Overseas applicants will be required to obtain an ORS in addition to a scholarship or studentship awarded by the faculty or the division/section.
Level of Study: Research, Doctorate, Postgraduate
Type: Scholarships and fellowships
Study Establishment: University of Glasgow
Country of Study: United Kingdom
Application Procedure: Check website for details regarding postgraduate research and postgraduate taught programmes.
Closing Date: December 14th
Additional Information: Applicants should enclose a referee's report, degree transcripts, proof of English language proficiency, a curriculum vitae and an approval of title in principle.

For further information contact:

Wolfson Medical School Building, University Avenue, University of Glasgow, Glasgow, G12 8QQ, United Kingdom
Website: www.gla.ac.uk/faculties/medicine/gradschool/applications.html
Contact: Faculty of Medicine Graduate School

Faculty Studentships

Subjects: Adult education, higher education including teacher education and teachers work, innovation in curriculum policy and practice, critical and cultural perspectives on education.
Purpose: To provide funding for PhD students.
Eligibility: As per university eligibility for a PhD.
Level of Study: Doctorate
Type: Studentship
Value: Home fees plus stipend of £12,400
Length of Study: 3 years
Frequency: Annual
Country of Study: Scotland
No. of awards offered: 1
Application Procedure: Normal university procedures. Please check the website for applications.
Closing Date: February
Contributor: Faculty of Education
No. of awards given last year: 1
No. of applicants last year: 10
Additional Information: Applications are invited based on the faculty research strengths.

Glasgow Educational & Marshall Trust Award

Subjects: All subjects.
Purpose: To offer financial support to those who have lived, or are currently living within the Glasgow Municipal Boundary.
Eligibility: Open to the candidates who are above 18 years of age and a resident of Glasgow within one of the following post code areas: G1–5, G11/12, G14/15, G20, G22/23, G31, G34, G40–42, G45, 51.
Level of Study: Graduate, Research, MBA, Postdoctorate, Postgraduate, Predoctorate, Doctorate
Type: Studentships and bursaries
Value: £50–1,000
Length of Study: 1 year
Study Establishment: Glasgow Educational and Marshall Trust
Country of Study: United Kingdom
Application Procedure: Check website for further details.
Closing Date: April 30th
Contributor: Glasgow Educational and Marshall Trust
Additional Information: Mrs Avril Sloane (email: sloanea@hutchesons.org), 21 Beaton RoadGlasgow, G41 4NW, United Kingdom, Tel: UK (+44) 141 4334449, Fax: UK (+44) 141 424 1731

For further information contact:

Glasgow Educational and Marshall Trust, 21 Beaton Road, Glasgow, G41 4NW, United Kingdom
Tel: (44) 0141 423 2169
Fax: (44) 141 424 1731
Email: enquiries@gemt.org.uk
Website: http://www.gemt.org.uk/index.htm
Contact: Secretary and Treasurers

Henry Dryerre Scholarship in Medical and Veterinary Physiology

Subjects: animal care and veterinary science; biology and life sciences; health sciences; medicine and surgery
Eligibility: Open to candidates holding a degree of a Scottish University with first class honours or, if in their final year, to be expected to achieve first class honours.
Level of Study: Postgraduate
Type: Scholarships and fellowships
Value: Varies
Frequency: Every 3 years
Application Procedure: The candidate must submit the application form through a member of staff on the appropriate Henry Dryerre Nomination Form.

For further information contact:

The Carnegie Trust for the Universities of Scotland, Cameron House, Abbey Park Place, Dunfermline, KY12 7P
Website: www.carnegie-trust.org/our_schemes.htm
Contact: Assistant Secretary

James Watt Research Scholarships

Subjects: Engineering, support for PhD studies in engineering at the University of Glasgow.
Level of Study: Postgraduate
Type: Scholarships and fellowships
Value: £10,000 per year
Country of Study: United Kingdom
Application Procedure: Please see the website www.gla.ac.uk/faculties/engineering/gradschool
Closing Date: February 8th

John & James Houston Crawford Scholarship

Subjects: Veterinary medicine.
Purpose: To assist the advanced study or research into the bovine and equine animals.
Eligibility: Open to university graduates or qualified veterinary surgeons.
Level of Study: Postgraduate

Type: Scholarship
Length of Study: 1–2 years
Frequency: Every 2 years
Study Establishment: Institution approved by the faculty of veterinary medicine
Country of Study: United Kingdom
Application Procedure: Applications to be submitted to the clerk of the Faculty of Veterinary Medicine.
Closing Date: June 1st
Additional Information: Preference is given to graduates in veterinary medicine of the University of Glasgow.

For further information contact:

Faculty of Veterinary Medicine, University of Glasgow Veterinary School, Bearsden, Glasgow, G61 1QH, Scotland

Lord Kelvin/Adam Smith Postgraduate Scholarships

Subjects: All subjects.
Purpose: To enable the University to recruit outstanding postgraduate research students to a range of innovative, boundary-crossing research developments.
Eligibility: Open to postgraduate students.
Level of Study: Postgraduate
Type: Scholarship
Value: Stipend of £13,590. The project will benefit from £5,300 per year research costs.
Length of Study: 4 years
Frequency: Annual
No. of awards offered: 10
Application Procedure: Check website for further details.
Funding: Trusts

For further information contact:

United Kingdom
Email: lauren-currie@enterprise.gla.ac.uk
Contact: Lauren Currie, Postgraduate Research Secretary,

Mac Robertson Travel Scholarship

Subjects: All subjects.
Purpose: To provide funding which will enrich and further the awardholder's academic experience and research achievements.
Eligibility: Open to postgraduate research students.
Level of Study: Research, Postgraduate
Type: Scholarship
Value: £2,000–3,000
Length of Study: 1 year
Frequency: Annual
Study Establishment: Glasgow University
No. of awards offered: Varies each year
Application Procedure: Check website for further details.
Closing Date: May 3rd

For further information contact:

Postgraduate Research Office, 10 The Square, Glasgow, G12 8QQ, United Kingdom
Tel: (44) 0141 330 1989
Website: www.gla.ac.uk/scholarships/travelscholarships
Contact: Lauren Currie

Overseas Research Student Awards Scheme (ORS)

Subjects: All subjects.
Purpose: To support overseas research students of outstanding merit and research potential.
Eligibility: To students who are classed as international student for fee purposes.
Level of Study: Research
Type: Award
Value: The difference between home/EU fee element and international fees
Frequency: Annual
Study Establishment: Glasgow University
Country of Study: Scotland
No. of awards offered: 20
Application Procedure: Please check the website http://glas.ac.uk for applications and further details. Applications should be completed and returned to the relevant faculty office.
Closing Date: February 1st
Funding: Government

R. Harper Brown Memorial Scholarship

Subjects: All subjects.
Purpose: To honour the late R. Harper Brown and to assist in defraying the cost of an American (US) college student's study at a university in Scotland.
Eligibility: Open to the candidates who are graduating seniors in high school with an acceptance and intention to attend university in Scotland or a student in an accredited American (US) college or university looking for a study-abroad experience.
Level of Study: Graduate
Type: Scholarship
Application Procedure: Check website for further details.
Closing Date: Between January 1st and March 31st
Funding: Private
Contributor: The Illinois Saint Andrew Society

For further information contact:

Tel: 847 967 2725
Email: dforlow@yahoo.com
Website: http://www.chicago-scots.org/scholarship.html
Contact: David Forlow

Royal Historical Society: Postgraduate Research Support Grants

Subjects: History.
Purpose: To assist postgraduate students in the pursuit of advanced historical research.
Eligibility: The candidate must be a postgraduate student registered for a research degree at United Kingdom Institute of Higher Education.
Level of Study: Research, Graduate
Type: Award/Grant
Country of Study: United States of America
Application Procedure: Check website for further details.
Closing Date: January 14th, February 18th, May 6th, June 18th, September 9th and November 12th
Contributor: Royal Historical Society

For further information contact:

Website: www.rhs.ac.uk/postgrad.htm#grant

Saint Andrew's Society of the State of New York Scholarship Fund

Subjects: All subjects.
Eligibility: Open to candidates who are either graduates of a Scottish university or of Oxford or Cambridge and have completed their first Degree course.
Level of Study: Graduate
Type: Scholarship
Value: US$40,000
Frequency: Annual
Country of Study: United States of America
No. of awards offered: 2
Application Procedure: The candidate must arrange for references from two academic referees to be submitted in the appropriate referee forms.
Closing Date: January 6th
Funding: Trusts
Contributor: Saint Andrew's Society of the State of New York

For further information contact:

Senate Office, Gilbert Scott Building, University of Glasgow, University Avenue, Glasgow, G12 8QQ, United Kingdom
Tel: (44) (0)141 330 6063
Email: c.omand@admin.gla.ac.uk
Contact: Catherine Omand

Stevenson Exchange Scholarships

Subjects: All subjects.

Purpose: To promote friendly relations between the students of Scotland, Germany, France and Spain.
Eligibility: Open to current or recent students of French, German or Spanish universities who intend to study at any university in Scotland.
Level of Study: Postdoctorate
Type: Scholarship
Value: £200–4,000
Application Procedure: Check website for further details.
Closing Date: February 28th

For further information contact:

1 The Square, University of Glasgow, Glasgow, G12 8QQ, United Kingdom
Tel: (44) 0141 330 4241
Fax: (44) 0141 330 4045
Email: l.buchan@admin.gla.ac.uk
Contact: Linda Buchan, Exchange Co-ordinator

The Sue Green Bursary

Subjects: Archaeology.
Purpose: To support a postgraduate student in the Department of Archaeology at the University of Glasgow.
Eligibility: Open to postgraduate students.
Level of Study: Postgraduate, Research
Type: Bursary
Value: £500
Application Procedure: The candidate must send an application in the form of a short research proposal (c. 500 words) and an accompanying brief curriculum vitae. Check website for further details.
Closing Date: August 31st
Funding: Private
Contributor: The Sue Green Bursary

For further information contact:

Department of Archaeology, University of Glasgow, Glasgow, G12 8QQ, United Kingdom
Contact: Head of the Department

Synergy Scholarship

Purpose: To enhance existing synergy collaborations or assist in creating new areas of synergy collaborations between Glasgow and Strathclyde universities.
Eligibility: The project must be jointly supervised in one university and a second supervisor in the other university.
Level of Study: Research
Type: Scholarship
Value: £12,923 for maintenance
Length of Study: 3 years
Frequency: Annual
Study Establishment: University of Scotland
Country of Study: Scotland
No. of awards offered: 1–2
Application Procedure: Application forms are available in the website. Forms should be completed and returned to the relevant faculty office.
Closing Date: January 31st
Additional Information: The project must be jointly supervised in one university and a second supervisor in the other university. Please check the wesite for more details.

University of Glasgow Postgraduate Research Scholarships

Subjects: All subjects.
Purpose: To assist with research towards a PhD degree.
Eligibility: Open to candidates of any nationality who are proficient in English and who have obtained a first or an upper second class (Honours) Degree or equivalent.
Level of Study: Postgraduate
Type: Scholarship
Value: Please consult the organization
Length of Study: 3 year's maximum award
Frequency: Annual
Study Establishment: The University of Glasgow
Country of Study: Scotland
Application Procedure: Applicants must refer to the University's application for graduate studies form that covers application for admission and scholarship. Please refer to the notes for applicants issued with the application form for address details.
Closing Date: Please refer to institution website
Funding: Private
Contributor: Endowments
Additional Information: Scholars from outside the European Union will be expected to make up the difference between the home fee and the overseas fee.

For further information contact:

Postgraduate Research Office, Research & Enterprise, 10 The Square, University of Glasgow, Glasgow, G12 8QQ
Tel: 0141 330 1989
Fax: 0141 330 5856

William and Margaret Kesson Award for Postgraduate Study

Subjects: Arts.
Purpose: To enable a student to undertake study leading to a postgraduate degree in the faculty of arts.
Eligibility: Open to candidates of Scottish or English nationality possessing a graduate degree.
Level of Study: Graduate
Type: Scholarship
Length of Study: 3 years
Application Procedure: Check website for further details.
Closing Date: May 1st

For further information contact:

Faculty of Arts, United Kingdom
Contact: Clerk of the Faculty

William Barclay Memorial Scholarship

Subjects: Biblical studies, theology and church history or any subject falling within the faculty of the divinity.
Purpose: To provide an opportunity for a scholar to pursue full-time study or to undertake research.
Eligibility: Open to any suitably qualified graduate of theology from a university outside the United Kingdom.
Level of Study: Postgraduate
Type: Scholarship
Value: £3500
Frequency: Annual
Study Establishment: The Faculty of Divinity at the University of Glasgow
Country of Study: Scotland
Application Procedure: Applicants must request postgraduate study application material. The PG form is used for the Barclay application.
Funding: Private

For further information contact:

Faculty of Divinity, Glasgow, G12 8QQ, Scotland
Tel: (44) 141 330 6525
Fax: (44) 141 330 4943

William Ross Scholarship

Subjects: History.
Purpose: To encourage the extraction of Scottish material from archives outside Scotland relating to all aspects of the history of Scotland, the Scottish people and Scottish influence abroad.
Eligibility: Open to candidates possessing a degree in MLitt.
Level of Study: Research
Type: Scholarships and fellowships
Value: £1,000
Length of Study: 1 year
Frequency: Annual
No. of awards offered: 1
Application Procedure: The candidate must submit a letter outlining a dissertation research proposal including the planned topic and archival research plans.
Closing Date: July 1st
Contributor: Trustees of the Ross Fund

For further information contact:

9 University Gardens, University of Glasgow, Glasgow, G12 8HQ, United Kingdom
Email: c.leriguer@arts.gla.ac.uk
Contact: Christelle LeRiguer

Wingate Scholarships

Subjects: All subjects.
Purpose: To support creative or original work of intellectual, scientific, artistic, social or enviromental value.
Eligibility: Open for mature candidates and those from non-traditional academic backgrounds without any upper age limit.
Level of Study: Unrestricted
Type: Scholarship
Value: £6,500–10,000 in any one year
Length of Study: 1 year
Frequency: Annual
Application Procedure: Check website for further details.
Closing Date: February 1st

For further information contact:

Queen Anne Business Center, 28 Broadway, London, SW1H 9JX, United Kingdom
Website: www.wingate.org.uk/scholarships/overview.php

UNIVERSITY OF GUELPH

University Centre, Room 437, 50 Stone Road East, Guelph, ON, N1G 2W1, Canada
Tel: (1) 519 824 4120
Fax: (1) 519 767 1693
Email: immccorki@uoguelph.ca
Website: www.uoguelph.ca
Contact: Linda McCorkindale, Associate Registrar

The University of Guelph is renowned in Canada and around the world as a research-intensive and learner-centred institution and for its commitment to open learning, internationalism and collaboration. Their vision is to be Canada's leader in creating, transmitting and applying knowledge to improve the social, cultural and economic quality of life of people in Canada and around the world.

The Brock Doctoral Scholarship

Subjects: All subjects.
Purpose: To financially support Doctoral students to attain a high level of academic achievement and to make significant teaching and research contributions.
Eligibility: Open to students with sustained outstanding academic performance, evidence of strong teaching and research skills, demonstrated outstanding communication skills and excellent potential for research and teaching as assessed by the College Dean.
Level of Study: Doctorate
Type: Scholarship
Value: Up to $120,000 ($10,000 per semester for up to twelve semesters)
Length of Study: 6 years
Frequency: Annual
Study Establishment: University of Guelph
Country of Study: Canada
Application Procedure: Students entering a Doctoral programme should apply to their College Dean by February 1st with a curriculum vitae, which must then be forwarded to Graduate Program Services by February 15th, with the Dean's written assessment of the candidate's research and teaching potential attached.
Closing Date: February 15th
Additional Information: The Brock Doctoral Scholarship is one of the most prestigious Doctoral awards available at the University. It is hoped that award holders will be mentors for future Brock Doctoral Scholarship winners.

For further information contact:

Office of Registrarial Services, University of Guelph, Guelph, ON N1G 2W1, Canada
Email: sinclair@registrar.uoguelph.ca
Website: www.uoguelph.ca/graduatestudies/calendar/gradawards/gradawards-uwia.shtml

Dairy Farmer's of Ontario Doctoral Research Assistantships

Subjects: Animal and herd health, management systems for dairy cattle, nutrition, economic aspects of milk production and marketing or the processing, quality and use of dairy products.
Purpose: To provide research assistantship to outstanding students entering a Doctoral programme.
Eligibility: Open to Doctoral applicants, with at least a First Class (A) average in the most recently completed 2 years of academic study.
Level of Study: Research, Doctorate
Type: Assistantship
Value: Canadian $20,000
Length of Study: 3 years
Frequency: Annual
Study Establishment: University of Guelph
Country of Study: Canada
Application Procedure: The assistantship application letter from the student should include a 1 page research proposal and name the proposed graduate faculty adviser at the University of Guelph. The assistantship application letter and graduate application file will be circulated to, and selection made by, an intercollege committee that includes the Dean of Graduate Studies.
Closing Date: January 11th

THE UNIVERSITY OF HONG KONG

Pok Fu Lam Road, Hong Kong, China
Tel: (86) 852 2859 2111
Fax: (86) 852 2858 2549
Email: gradsch@hkucc.hku.hk
Website: www.hku.hk

The University of Hong Kong, as a pre-eminent international university in Asia, seeks to sustain and enhance its excellence as an institution of higher learning through outstanding teaching and world-class research, so as to produce well-rounded graduates with the abilities to provide leadership within the societies they serve.

Hui Pun Hing Scholarship for Postgraduate Research Overseas

Subjects: Sciences including medicine.
Purpose: To provide scholarship to a student who intends to pursue postgraduate research studies in the field of science, including medicine and who is expected to make a significant contribution to the development of the chosen specialization on his/her return from study abroad.
Eligibility: Open to students who have obtained a Bachelor's degree or a Master's degree or both at the University of Hong Kong including medicine and to those who intend to pursue postgraduate research studies in the field of Science.
Level of Study: Postgraduate, Research
Type: Scholarship
Value: Up to a maximum of HK $100,000
Length of Study: 3 years at most
Frequency: Annual
Study Establishment: The University of Hong Kong
Country of Study: China
No. of awards offered: 2
Application Procedure: Open application
Funding: Trusts
Contributor: Hui Pun Hing Endowment Fund
No. of awards given last year: 1
No. of applicants last year: 1

Peter Vine Postgraduate Law Scholarship

Subjects: Law
Purpose: To provide scholarships, (1) outstanding LLB graduates of the Faculty of Law of University of Hong Kong to undertake one year of full-time study overseas for the degree of Master of Laws or its equivalent and (2) for outstanding Law graduates from universities in Mainland China undertaking one year full-time study at the University

of Hong Kong in the Postgraduate Diploma in Common Law/Master of Common Law programme.
Eligibility: Open to LLB graduates of the University of Hong Kong and Law graduates from universities in Mainland China
Level of Study: Graduate, Postgraduate
Type: Scholarship
Value: HK $60,000 each for category (1), a total of not more than HK $144,000 for category (2)
Length of Study: 1 year
Frequency: Annual
Study Establishment: The University of Hong Kong
Country of Study: China
No. of awards offered: 5
Application Procedure: Open application but nomination from the Faculty of Law is required
Funding: Foundation
Contributor: The Peter Vine Charitable Foundation
No. of awards given last year: 2
No. of applicants last year: 2

THE UNIVERSITY OF KENT

Admissions and Partnership Services, The Registry, Canterbury, Kent, CT2 7NZ, United Kingdom
Tel: (44) 1227 764 000
Fax: (44) 1227 827 077
Email: scholarships@kent.ac.uk
Website: www.kent.ac.uk

The University of Kent is a UK higher education institution funded by the Higher Education Funding Council for England (HEFCE). The university provides education of excellent quality characterized by flexibility and inter disciplinarily and informed by research and scholarship, meeting the lifelong needs of diversity students.

Computing Laboratory Scholarship

Subjects: Theoretical computer science, systems, architecture, applied and interdisciplinary informatics cognitive systems, pervasive computing, information systems security, computing education.
Purpose: To support postgraduate studies towards a PhD.
Eligibility: Candidates are expected to hold an Upper Second Class (Honours) Degree.
Level of Study: Postgraduate, Doctorate
Type: Scholarship
Value: £1,000
Length of Study: 3 years
Frequency: Annual
Study Establishment: The University of Kent
Country of Study: United Kingdom
No. of awards offered: 5
Application Procedure: Applicants must enclose a covering letter and indicate their area of interest. See website www.cs.kent.ac.uk/research/pg/funding/bursary_app.html
Closing Date: June 15th
Funding: Government
No. of awards given last year: 4
No. of applicants last year: Approx. 150
Additional Information: Please contact the Computing Laboratory at computer-science@kent.ac.uk

EPSRC Doctoral Training Awards for Kent Business School

Subjects: All subjects within Kent Business School.
Purpose: To support research.
Eligibility: Home students eligible for maintenance grant and home fees award. EU students eligible for home fees only award.
Level of Study: Doctorate, Postgraduate
Type: Research scholarship
Value: Maintenance stipend plus fees at the home rate
Length of Study: Up to 3 years
Frequency: Annual, Dependent on funds available
Study Establishment: University of Kent
Country of Study: United Kingdom
No. of awards offered: 5
Application Procedure: Please refer to the website http://www.kent.ac.uk/scholarships/postgraduate/research_council/epsrc.html for details.
Closing Date: April 9th
Funding: Government
Contributor: The Engineering and Physical Sciences Research Council (EPSRC) is funded by the UK government through the Department for Universities, Innovation and Skills

EPSRC Doctoral Training Awards for School of Computing

Subjects: All subjects.
Purpose: All subjects within the School of Computing.
Eligibility: Open to the home students eligible for maintenance grant and home fees award. EU students are eligible for home fees only award.
Level of Study: Doctorate, Postgraduate
Type: Research scholarship
Value: Home fees and maintenance stipend
Length of Study: Up to 3 years
Frequency: Annual, Dependent on funds available
Study Establishment: The University of Kent
Country of Study: United Kingdom
No. of awards offered: 5
Application Procedure: Please refer to the website http://www.kent.ac.uk/scholarships/postgraduate/research_council/epsrc.html for details.
Funding: Government
Contributor: EPSRC is funded by the UK government through the Department for Universities, Innovation and Skills.
No. of awards given last year: 4

EPSRC Doctoral Training Awards for School of Engineering and Digital Arts

Subjects: All subjects within the School of Engineering and Digital Arts.
Purpose: To support research.
Eligibility: Home students eligible for maintenance grant and home fees award. EU students eligible for home fees only award.
Level of Study: Doctorate
Type: Research scholarship
Value: Home fees and maintenance stipend
Length of Study: Up to 3 years
Frequency: Dependent on funds available
Study Establishment: University of Kent
Country of Study: United Kingdom
Application Procedure: Please refer to the website http://www.kent.ac.uk/scholarships/postgraduate/research_council/epsrc.html for details.
Funding: Government
Contributor: The Engineering and Physical Sciences Research Council (EPSRC) is funded by the UK government through the Department for Universities, Innovation and Skills
No. of awards given last year: 1

EPSRC Doctoral Training Awards for School of Mathematics, Statistics and Actuarial Science

Subjects: All subjects within the School of Mathematics, Statistics and Actuarial Science.
Purpose: To support research.
Eligibility: Home students eligible for maintenance grant and home fees award. EU students eligible for home fees only award.
Level of Study: Doctorate
Type: Research scholarship
Value: Home fees and maintenance stipend
Length of Study: Up to 3 years
Frequency: Dependent on funds available
Study Establishment: University of Kent
Country of Study: United Kingdom
Application Procedure: Please refer to the website http://www.kent.ac.uk/scholarships/postgraduate/research_council/epsrc.html for details.
Funding: Government

Contributor: The Engineering and Physical Sciences Research Council (EPSRC) is funded by the UK government through the Department for Universities, Innovation and Skills
No. of awards given last year: 5

EPSRC Doctoral Training Awards for School of Physical Sciences

Subjects: All subjects within the School of Physical Sciences.
Purpose: To support research.
Eligibility: Open to the home students eligible for maintenance grant and home fees award. EU students eligible for home fees only award.
Level of Study: Doctorate
Type: Research scholarship
Value: Home fees and maintenance stipend
Length of Study: Up to 3 years
Frequency: Dependent on funds available
Study Establishment: The University of Kent
Country of Study: United Kingdom
Application Procedure: Applicants must complete an application form. Please refer to the website http://www.kent.ac.uk/scholarships/postgraduate/research_council/epsrc.html for details.
Funding: Government
Contributor: The Engineering and Physical Sciences Research Council (EPSRC) is funded by the UK government through the Department for Universities, Innovation and Skills.
No. of awards given last year: 5
No. of applicants last year: 1

Ian Gregor Scholarship

Subjects: Postcolonial studies, Dickens and Victorian culture, eighteenth century studies, english and american literature, creative writing, critical theory, medieval and early modern studies.
Purpose: To support a candidate registered for a taught MA programme in english.
Eligibility: Candidates are expected to hold at least an upper second class (Honours) degree or equivalent. Candidates should also have applied for an external scholarship, such as AHRC.
Level of Study: Graduate, Postgraduate
Type: Scholarship
Value: Home fees and maintenance stipend
Length of Study: 1 year
Frequency: Annual
Study Establishment: The University of Kent
Country of Study: United Kingdom
No. of awards offered: 1
Application Procedure: See webpages at http://www.kent.ac.uk/english/postgraduate/fund.htm
Closing Date: June 29th
Funding: Trusts
No. of awards given last year: 1

For further information contact:

School of English, Rutherford College Extension, University of Kent, Canterbury, Kent, CT2 7NX, England
Email: englishpg@kent.ac.uk
Contact: Claire Lyons, Administrative Assistant

Kent Law School Studentships and Bursaries

Subjects: Banking law, carriage of goods, company law, comparative law, computers and the law, criminal law and penology, critical legal studies, environmental law, emergency powers, European comparative and human rights law, family law, gender, sexuality and law, immigration law, intellectual property law, international law and human rights, international economic and trade law, labor law, law and multiculturalism, legal services, legal theory, modern legal history, multinationals and the law, and private law.
Purpose: To provide funding for 1 year in the first instance, extended to a maximum of 3 years (for registered students only) based on satisfactory progress (including upgrading to a PhD). The retention of the posts will be subject to a review of progress and performance in both research and teaching after the 1st year.
Eligibility: Candidates should hold an Upper Second Class (Honours) Degree or a good postgraduate taught degree in law.
Level of Study: Doctorate, Postgraduate, Research
Type: Studentship
Value: £13,590 and tuition fees paid at the home/EU rate (Up to £3,732 last year)
Length of Study: 1–3 years
Frequency: Annual
Study Establishment: The University of Kent
Country of Study: United Kingdom
No. of awards offered: 3
Application Procedure: Applicants must submit to the University's recruitment and admissions office a research proposal, curriculum vitae and covering letter with an application for their chosen research degree. They should also ensure that the recruitment and admissions office receives two referees' reports by the closing date for applications. Applications are available at http://records.kent.ac.uk/external/admissions/pg-application.php
Closing Date: April 13th
Funding: Private
Contributor: Kent Law School
No. of awards given last year: 2
No. of applicants last year: 20–30
Additional Information: Holders of the studentships will be expected to teach for a maximum of 4 hours per week in term time on an undergraduate law module, at the direction of the head of department. For further information contact the Kent Law school at kls-pgoffice@kent.ac.uk

For further information contact:

Kent Law School, University of Kent, Canterbury, Kent, CT2 7NS, England
Tel: (44) 01227 827949
Email: m.drakopoulou@kent.ac.uk
Contact: Ms Maria Drakopoulou, Director of Postgraduate Research

Maurice Crosland History of Science Studentship

Subjects: Grant is designed to support postgraduate research in the history of science and technology (excluding history of medicine). The emphasis is on the history of science and technology in cultural contexts (e.g. religion).
Purpose: To fund research to PhD level.
Eligibility: Open to qualified applicants of any nationality.
Level of Study: Doctorate, Postgraduate, Research
Type: Studentship
Value: Tuition fees at the home/EU rate and plus the potential for a subsistence allowance
Length of Study: 3 years
Frequency: Dependent on funds available
Study Establishment: The University of Kent
Country of Study: United Kingdom
No. of awards offered: 1
Application Procedure: Applicants must apply to Professor Crosbie Smith. Candidates should submit a detailed curriculum vitae together with a 500-word description of their preferred research topic. They may wish to discuss this informally with Professor Crosbie Smith first.
Closing Date: June 1st
Funding: Private
No. of awards given last year: 2
No. of applicants last year: 5

For further information contact:

Centre for History of Science, Technology & Medicine, Rutherford College, University of Kent at Canterbury, Canterbury, Kent, CT2 7NX, England
Tel: (44) 12 2776 4000
Fax: (44) 12 2782 7258
Contact: Professor Crosbie Smith

Tizard Centre Scholarship 2011

Subjects: Community care. learning disability, applied psychology, clinical psychology of learning disability, mental health.
Purpose: To support research.
Eligibility: Applicants should have, or expect to receive a very good honours or Master's degree in a relevant social science subject such as psychology, social policy, social work or sociology.
Level of Study: Doctorate, Postgraduate
Type: Scholarship
Value: £10,500

Length of Study: Up to 3 years
Frequency: Annual, dependent on funds
Study Establishment: University of Kent
Country of Study: United Kingdom
No. of awards offered: 1
Application Procedure: See webpages at: www.kent.ac.uk/scholarships/postgraduate/departmental/tizard.html.
Closing Date: January 1st
Funding: Commercial, government
No. of awards given last year: 2

University of Kent Business School Scholarships

Subjects: Accounting and finance, industrial relations, management, management science, marketing, operational research.
Purpose: To support research.
Eligibility: Candidates must hold a good Honours degree (first or 2i) or a Master's degree at merit or distinction in a relevant subject or equivalent. The quality of the candidate's proposal and the support of their proposed supervisor will also be taken into account.
Level of Study: Doctorate, Postgraduate
Type: Scholarship
Value: Home fees and maintenance stipend
Length of Study: Up to 3 years
Frequency: Annual, dependent on funds available
Study Establishment: University of Kent
Country of Study: United Kingdom
No. of awards offered: To be confirmed
Application Procedure: See wepbages at: www.kent.ac.uk/scholarships/postgraduate/departmental/kbs.html.
Closing Date: April
Funding: Government, commercial
No. of awards given last year: 5

University of Kent Centre for Journalism Scholarships

Subjects: Journalism.
Purpose: To support research.
Eligibility: Candidates must hold a good Honours degree (first or upper second class) or a Master's degree at merit or distinction in a relevant subject or equivalent. The scholarship competition is open to all postgraduate research applicants. UK, EU and overseas fee paying students as well as full-time and part-time postgraduate research students are invited to apply.
Level of Study: Doctorate, Postgraduate
Type: Scholarship
Value: Home fees and maintenance stipend
Length of Study: Up to 3 years
Study Establishment: University of Kent
Country of Study: United Kingdom
No. of awards offered: To be confirmed
Application Procedure: See website www.kent.ac.uk/scholarships/postgraduate/departmental/journalism.html.
Closing Date: April
Funding: Commercial, government
No. of awards given last year: 5

University of Kent Department of Biosciences

Subjects: Cancer research, cell and developmental biology, infectious diseases and protein science.
Purpose: To support research.
Eligibility: Primarily open to United Kingdom nationals although European Union citizens may qualify in special circumstances. Candidates should hold an Upper Second Class (Honours) Degree.
Level of Study: Postgraduate, Doctorate, Research
Value: Home tuition fees and a maintenance bursary at the same rate as that provided by the Research Councils
Length of Study: 3 years
Frequency: Annual
Study Establishment: The University of Kent
Country of Study: United Kingdom
No. of awards offered: 7
Application Procedure: Applicants should contact the Department of Biosciences for further information via email to bio-admin@kent.ac.uk
Closing Date: July 31st
Funding: Government
Contributor: BBSRC
No. of awards given last year: 7

University of Kent Department of Economics Bursaries

Subjects: Labour economics, money and development, international finance and trade, migration, defence and energy economics, macro economics, ganne theory and econometrics.
Purpose: To support research.
Eligibility: Candidates must have at least a taught Master's degree in economics.
Level of Study: Doctorate, Postgraduate, Research
Type: Bursary
Value: Up to the equivalent of home fees and/or a maintenance grant of UK£6,000
Length of Study: 3 years
Frequency: Dependent on funds available
Study Establishment: The University of Kent
Country of Study: United Kingdom
No. of awards offered: 2
Application Procedure: Applicants must enclose a separate letter proposing their wish to apply for the bursary along with their university PhD application form. A link to the online application form can be found on the department's website www.ukc.ac.uk/economics/students/postgrad/MPhil/PhD.html
Closing Date: May
Funding: Government
No. of awards given last year: 2
Additional Information: Candidates should note that 4–6 hours of teaching per week and acceptable progress in the programme of study will be expected of the student. The bursary will be subject to review each year.

University of Kent Department of Electronics Studentships

Subjects: Electronics, including image processing and vision, embedded systems, broadband and wireless communications and electronic instrumentation.
Purpose: To enable well-qualified students to undertake research programmes within the department.
Eligibility: Candidates are expected to hold an Upper Second Class (Honours) Degree or equivalent in an appropriate subject, and be nationals of one of the European Union countries.
Level of Study: Doctorate, Postgraduate, Research
Type: Studentship
Value: Research Council studentships are at a fixed rate determined annually by EPSRC. Departmental bursaries depend on individual circumstances
Length of Study: 3 years
Frequency: Annual
Study Establishment: The University of Kent
Country of Study: United Kingdom
No. of awards offered: Varies
Application Procedure: Applicants should contact the Department of Electronics.
Closing Date: June
Funding: Government
Contributor: EPSRC
No. of awards given last year: 5
No. of applicants last year: 10

For further information contact:

Department of Electronics, University of Kent, Canterbury, Kent, CT2 7NT, England
Email: ee-admissions-pg@kent.ac.uk
Contact: Professor J Z Wang

University of Kent Department of Politics and International Relations Bursary

Subjects: European studies, international relations and international political economy, intergovernmental co-operation in the European Union, political theory and law, politics and conflict analysis.
Purpose: To support research.
Eligibility: Open to candidates accepted for research study.
Level of Study: Doctorate, Postgraduate
Type: Bursary
Value: Home fees
Length of Study: 3 years
Frequency: Annual

Study Establishment: The University of Kent
Country of Study: United Kingdom
No. of awards offered: 5
Application Procedure: Applicants must apply by letter to the head of the department after being accepted for research study.
Closing Date: Applications may be submitted at any time. Deadline for following academic year is the end of March
No. of awards given last year: 5
No. of applicants last year: 7
Additional Information: A maximum of 6 hours of teaching per week will be required.

University of Kent Department of Psychology Studentship

Subjects: Social psychology, cognitive psychology, neuropsychology, health psychology, developmental psychology, forensic psychology.
Purpose: To support research studies.
Eligibility: Candidates should hold or expect to obtain at least an Upper Second Class (Honours) Degree.
Level of Study: Postgraduate, Research
Type: Studentship
Value: Tuition fees at the UK rate plus a maintenance bursary (at the same rate as provided by the ESRC)
Length of Study: 3 years
Frequency: Annual
Study Establishment: The University of Kent
Country of Study: United Kingdom
No. of awards offered: 1
Application Procedure: Applicants must complete an application form.
Closing Date: End of July or mid-August
Funding: Private
Contributor: The University of Kent
No. of awards given last year: 2
No. of applicants last year: 30–40
Additional Information: A maximum of 6 hours of teaching per week will be required.

For further information contact:

Department of Psychology, University of Kent, Canterbury, Kent, England
Tel: (44) 1227 823085
Email: rsg@kent.ac.uk
Contact: Dr Roger Giner-Sorolla, CSGP

University of Kent English Scholarship

Subjects: Accounting and finance, industrial relations, management, management science, marketing, operational research.
Purpose: To support research.
Eligibility: Candidates must hold a good Honours degree (First or 2:1) or a Master's degree at merit or distinction in a relevant subject or equivalent. The quality of the candidate's proposal and the support of their proposed supervisor will also be taken into account.
Level of Study: Doctorate, Postgraduate
Type: Scholarship
Value: Home fees and maintenance stipend
Length of Study: Up to 3 years
Study Establishment: The University of Kent
Country of Study: United Kingdom
Application Procedure: See webpages at: www.kent.ac.uk/scholarships/postgraduate/departmental/kbs.html.
Closing Date: April
Funding: Commercial, government
No. of awards given last year: 5

University of Kent Law School Studentship

Subjects: Critical commercial law, business law and regulation, criminal justice, environmental law, European and comparative law, gender and sexuality, health care law and ethics, law politics and culture, law and political economy, legal theories and philosophy, property law.
Purpose: To support research.
Eligibility: Applicants should normally have obtained or be about to obtain an undergraduate degree in Law of at least Upper Second Class Honours level (2:1 or equivalent from other countries), or a postgraduate degree in Law.
Level of Study: Doctorate, Postgraduate
Type: Scholarships
Value: Home fees (£3,732 last year) and maintenance stipend (£13,590 last year)
Length of Study: Up to 3 years
Frequency: Annual, dependent on funds available
Study Establishment: University of Kent
Country of Study: United Kingdom
No. of awards offered: To be confirmed
Application Procedure: See webpages at: www.kent.ac.uk/scholarships/postgraduate/departmental/law.html.
Closing Date: April 13th
Funding: Commercial, government
No. of awards given last year: 8

University of Kent Postgraduate Funding Awards

Subjects: Social policy, sociology, criminology, environmental, women studies or urban studies.
Purpose: To support research.
Eligibility: Candidates should hold a First Class or an Upper Second Class (Honours) Degree.
Level of Study: Doctorate, Postgraduate, Predoctorate, (not available to MA/MSc students)
Type: Award
Value: Up to 4 awards at UK£9,000 for full-time students, up to 10 awards at UK£4,000 for full-time students or UK£2,000 for a part-time student
Length of Study: 3 years
Frequency: Annual
Study Establishment: The University of Kent, School of Social Policy, Sociology and Social Research
Country of Study: United Kingdom
No. of awards offered: Up to 14
Application Procedure: Applicants must contact The Admissions Secretary, School of Social Policy, Canterbury, Kent, Sociology or Social Research, CT2 7NF, or email: socio-office@kent.ac.uk
Closing Date: June
Funding: Government
No. of awards given last year: 8
Additional Information: A maximum of 2 hours of teaching per week will be required.

University of Kent School of Anthropology and Conservation Scholarships

Subjects: Anthropology, Ethnobiology.
Purpose: To support research.
Eligibility: Candidates should hold a good Honours degree (first or upper second class).
Level of Study: Doctorate, Postgraduate
Type: Scholarship
Value: Home fees and maintenance stipend
Length of Study: Up to 3 years
Frequency: Annual, dependent on funds available
Study Establishment: University of Kent
Country of Study: United Kingdom
No. of awards offered: To be confirmed
Application Procedure: See webpages at: www.kent.ac.uk/scholarships/postgraduate/departmental/anthropology.html.
Closing Date: April
Funding: Commercial, government
No. of awards given last year: 6

University of Kent School of Architecture Scholarships

Subjects: Architecture.
Purpose: To support research.
Eligibility: Candidates must hold a good Honours degree (first or upper second class) or a Master's degree at merit or distinction in a relevant subject or equivalent.The scholarship competition is open to all postgraduate research applicants. UK, EU and overseas fee paying students as well as full-time and part-time postgraduate research students are invited to apply.
Level of Study: Postgraduate, Doctorate

Type: Scholarship
Value: Home fees and maintenance stipend
Length of Study: Up to 3 years
Frequency: Annual, dependent on funds available
Study Establishment: University of Kent
Country of Study: United Kingdom
No. of awards offered: To be confirmed
Application Procedure: See wepbages at: www.kent.ac.uk/scholarships/postgraduate/departmental/architecture.html.
Closing Date: April
Funding: Commercial, government
No. of awards given last year: 1

University of Kent School of Arts Scholarships

Subjects: Drama, Film Studies and History and Philosophy of Art.
Purpose: To support research.
Eligibility: Candidates are exptected to have completed or be in the process of completing an MA or equivalent in the area of research being applied for the scholarship competition is open to all new postgraduate research applicants. UK, EU and international fee-paying applicants are invited to apply. Where eligible, applicants for part-time post-graduate study may also apply.
Level of Study: Doctorate, Postgraduate
Type: Scholarship
Value: Home fees and maintenance stipend
Length of Study: Up to 3 years
Frequency: Annual, dependent on funds available
Study Establishment: University of Kent
Country of Study: United Kingdom
No. of awards offered: To be confirmed
Application Procedure: See wepbages at: www.kent.ac.uk/scholarships/postgraduate/departmental/arts.html.
Closing Date: April
Funding: Commercial, government
No. of awards given last year: 9

University of Kent School of Biosciences Scholarships

Subjects: Protein science, biomedicine research, cell and development biology, neuroscience and medical image computing.
Purpose: To support research.
Eligibility: Open to Home/EU students. Candidates should hold an Upper second class (Honours) degree.
Level of Study: Doctorate, Postgraduate
Type: Scholarship
Value: Home fees and maintenance stipend
Length of Study: Up to 3 years
Frequency: Annual, dependent on funds available
Study Establishment: University of Kent
Country of Study: United Kingdom
No. of awards offered: To be confirmed
Application Procedure: See wepbages at: www.kent.ac.uk/scholarships/postgraduate/departmental/biosciences.html.
Funding: Commercial, government

University of Kent School of Computing Scholarships

Subjects: Computational intelligence, computing education, future computing, information systems security, programming languages and systems.
Purpose: To support research.
Eligibility: Candidates are expected to hold an Upper Second Class (Honours) Degree.
Level of Study: Doctorate, Postgraduate
Type: Scholarship
Value: Home fees and maintenance stipend
Length of Study: Up to 3 years
Frequency: Annual, dependent on funds available
Study Establishment: University of Kent
Country of Study: United Kingdom
Application Procedure: See wepbages at: www.kent.ac.uk/scholarships/postgraduate/departmental/computing.html.
Closing Date: July 31st
Funding: Commercial, government
No. of awards given last year: 4

University of Kent School of Drama, Film and Visual Arts Scholarships

Subjects: Drama, including performance-making process and theory, theatre history and practice as research film studies, including varied aspects of film aesthetics, film theory and film history as well as research by practice.History and philosophy of art, including contemporary aesthetics and the history of art thoery, the photograph, and the historical interplay of image, theory and institutions from the renaissance to the present.
Purpose: To support research.
Eligibility: Candidates are expected to hold an Upper Second Class (Honours) Degree and be a citizen of one of the European Union countries.
Level of Study: Postgraduate, Research
Type: Bursary - fees only
Value: To cover home fees only.
Length of Study: The bursary is 1 year
Frequency: Annual, 1 initial bursary for 1 year only. Opportunity to reapply on a competitive basis
Study Establishment: The University of Kent
Country of Study: United Kingdom
No. of awards offered: 7
Application Procedure: Applicants must indicate their interest on the postgraduate application form.
Closing Date: July 31st
Additional Information: Some teaching or research assistant work may be required.

For further information contact:

Email: k.j.goddard@kent.ac.uk

University of Kent School of Economics Scholarships

Subjects: Economics, agri-environmental ecomonics.
Purpose: To support research.
Eligibility: Candidates must hold a good Master's degree at merit or distinction in economics and/or equivalent subjects. Candidates will be ranked on the basis of academic merit and their research project.
Level of Study: Doctorate, Postgraduate
Type: Scholarship
Value: Home fees and maintenance stipend
Length of Study: Up to 3 years
Frequency: Annual, dependent on funds available
Study Establishment: University of Kent
Country of Study: United Kingdom
No. of awards offered: To be confirmed
Application Procedure: See wepbages at: www.kent.ac.uk/scholarships/postgraduate/departmental/economics.html.
Closing Date: April
Funding: Commercial, government
No. of awards given last year: 2

University of Kent School of Engineering and Digital Arts Scholarships

Subjects: Broadband and wireless communications, digital media, image and information engineering, instrumentation control and embedded systems.
Purpose: To support research.
Eligibility: This Scholarship is awarded on the basis of academic excellence as assessed by the entry qualifications for each postgraduate programme.
Level of Study: Doctorate, Postgraduate
Type: Scholarship
Value: Home fees and maintenance stipend
Length of Study: Up to 3 years
Frequency: Annual, dependent on funds available
Study Establishment: University of Kent
Country of Study: United Kingdom
No. of awards offered: To be confirmed
Application Procedure: See wepbages at: www.kent.ac.uk/scholarships/postgraduate/departmental/eda.html.
Closing Date: April
Funding: Government, commercial
No. of awards given last year: 5

University of Kent School of European Culture and Languages Scholarships

Subjects: Classics and archaeology, comparative literature, french, german, philosophy and religious studies.
Purpose: To support research.
Eligibility: The candidates must hold a bachelor's (first or high 2:1) or master's degree (merit or distinction) and hold an offer of a place, or be a currently registered student for a research degree at the University of Kent.
Level of Study: Doctorate, Postgraduate
Type: Scholarship
Value: Home fees and maintenance stipend
Length of Study: Up to 3 years
Frequency: Annual, dependent on funds available
Study Establishment: University of Kent
Country of Study: United Kingdom
No. of awards offered: To be confirmed
Application Procedure: See webpages at: www.kent.ac.uk/scholarships/postgraduate/departmental/secl.html.
Closing Date: April
Funding: Commercial, government
No. of awards given last year: 4

University of Kent School of European Culture and Languages Scholarships

Subjects: European and Latin American literary, linguistic and cultural studies (French, German, Hispanic studies, Italian and comparative literary studies), philosophy (with special expertise in moral and political philosophy, aesthetics, philosophy of mind, philosophical logic and paradoxes, Bayesian epistemology and artificial intelligence), religious studies (modern theology, Christian ethics, mysticism and religious experience, religion and film, psychology of religion and cultural study of cosmology and divination) and the literature, history and archaeology of classical World (including Britain and Gaul).
Purpose: To support research.
Eligibility: Candidates are expected to hold a First Class (Honours) Degree or a minimum Upper Second Class (Honours) Degree and be a citizen of one of the European Union countries.
Level of Study: Postgraduate, Research
Type: Scholarships
Value: Offers three full fee bursaries for full-time and part-time research degrees. These bursaries classed as UK/EU for the purpose of fees, cover the payment of tuition fees (currently £3,350 per year full time and £1,710 for part time). Funding in the second and third years is subject to satisfactory progress
Length of Study: 3 years
Frequency: Annual
Study Establishment: The University of Kent
Country of Study: United Kingdom
No. of awards offered: Varies (up to 4)
Application Procedure: Application is via School of European Culture and Language website at www.kent.ac.uk/secl/researchcentres/graduate/funding.html
Closing Date: May 1st
Funding: Government
No. of awards given last year: 3
No. of applicants last year: 7
Additional Information: Some part-time teaching may be required.

University of Kent School of European Culture and Languages Studentships

Subjects: European and Latin American literary, linguistic and cultural studies (French, German, Hispanic studies, Italian and comparative literary studies), philosophy (with special expertise in moral and political philosophy, aesthetics, philosophy of mind, philosophical logic and paradoxes, Bayesian epistemology and artificial intelligence), religious studies (modern theology, Christian ethics, mysticism and religious experience, religion and film, psychology of religion and cultural study of cosmology and divination) and the literature, history and archaeology of classical World (including Britain and Gaul).
Purpose: To support research.
Eligibility: Candidates are expected to hold a First Class (Honours) Degree or a minimum Upper Second Class (Honours) Degree and be a citizen of one of the European Union countries.
Level of Study: Postgraduate, Research
Type: Studentship
Value: Offers three full fee bursaries for full-time and part-time research degrees. These bursaries classed as UK or EU for the purpose of fees, cover the payment of tuition fees (currently £3,350 per year full time and £1,710 for part time). Funding in the second and third years is subject to satisfactory progress
Length of Study: 3 years
Frequency: Annual
Study Establishment: The University of Kent
Country of Study: United Kingdom
No. of awards offered: Varies (up to 4)
Application Procedure: Application is via School of European Culture and Language website at www.kent.ac.uk/secl/researchcentres/graduate/funding.html
Closing Date: June 27th
Funding: Government
No. of awards given last year: 5
No. of applicants last year: 13

University of Kent School of History Scholarships

Subjects: Medieval and early modern religious history, history of medicine, modern British military history, environmental history, maritime history, political propaganda in the 20th century.
Purpose: To support research.
Eligibility: Candidates must hold a good bachelor's degree (first or high 2:1) or a master's degree (merit or distinction) in a relevant subject.The scholarship competition is open to all postgraduate research applicants. UK, EU and overseas fee paying students as well as full-time and part-time postgraduate research students are invited to apply.
Level of Study: Doctorate, Postgraduate
Type: Scholarship
Value: Home fees and maintenance stipend
Length of Study: Up to 3 years
Frequency: Annual, dependent on funds available
Study Establishment: University of Kent
Country of Study: United Kingdom
No. of awards offered: To be confirmed
Application Procedure: See webpages at: www.kent.ac.uk/scholarships/postgraduate/departmental/history.html.
Closing Date: April
Funding: Commercial, government
No. of awards given last year: 3

University of Kent School of Mathematics, Statistics and Actuarial Science (SMSAS)

Subjects: Mathematics or statistics.
Purpose: To support studies at the SMSAS.
Eligibility: Candidates for scholarships should hold a First Class (Honours) Degree in mathematics or a related subject.
Level of Study: Doctorate
Type: Studentship
Value: Up to £13,290 of tuition fees plus the maintenance bursary at the same rate as provided by the EPSRC
Length of Study: 3 years
Frequency: Dependent on funds available
Study Establishment: The University of Kent
Country of Study: United Kingdom
No. of awards offered: 3
Application Procedure: Candidates should complete an application form for postgraduate study and indicate that they wish to be considered for this award.
Closing Date: April 30th
Funding: Private
No. of awards given last year: 5
No. of applicants last year: 36
Additional Information: Holders of the studentships are required to undertake a small amount of teaching assistance. For more information, please email imspg-admiss@ukc.ac.uk or see the website www.kent.ac.uk/IMS

University of Kent School of Mathematics, Statistics and Actuarial Science Scholarships

Subjects: Mathematics, statistics or actuarial science.
Purpose: To support research.

Eligibility: Candidates should hold a good (first or upper second) Honours degree, or a Master's degree in a relevant subject.
Level of Study: Postgraduate, Doctorate
Type: Scholarship
Value: Home fees and maintenance stipend
Length of Study: Up to 3 years
Frequency: Annual, dependent on funds available
Study Establishment: University of Kent
Country of Study: United Kingdom
No. of awards offered: To be confirmed
Application Procedure: See wepbages at: www.kent.ac.uk/scholarships/postgraduate/departmental/maths.html.
Closing Date: April 20th
Funding: Government, commercial
No. of awards given last year: 5

For further information contact:

Cornwallis Building, Canterbury, Kent, CT2 7NF, England
Email: smsaspgadmin@kent.ac.uk

University of Kent School of Physical Sciences Scholarships

Subjects: Physical sciences, functional materials, applied optics or astrophysics and planetary science.
Purpose: To support research.
Eligibility: Candidates should hold a good (first or upper second class) degree, or a master's degree (merit or distinction) in a relevant subject or equivalent.
Level of Study: Doctorate, Postgraduate
Type: Scholarship
Value: £13,590 last year
Length of Study: Up to 3 years
Study Establishment: University of Kent
Country of Study: United Kingdom
No. of awards offered: To be confirmed
Application Procedure: See wepbages at: www.kent.ac.uk/scholarships/postgraduate/departmental/sps.html.
Closing Date: April
Funding: Commercial, government
No. of awards given last year: 1

University of Kent School of Physical Sciences Studentships

Subjects: Physical sciences, study in materials science, applied optics or astronomy and space science.
Eligibility: Candidates should hold an Upper Second Class (Honours) Degree or equivalent EU degree and be a citizen of one of the European Union countries or equivalent EU degree.
Level of Study: Doctorate, Postgraduate
Type: Studentship
Value: Home/EU fees plus maintenance at Research Council Rate
Length of Study: 3 years
Country of Study: United Kingdom
No. of awards offered: 4
Application Procedure: Applicants must contact Dr Chris Solomon, School of Physical Sciences or by email c.j.solomon@kent.ac.uk
Closing Date: June
Funding: Government, commercial
Contributor: The University of Kent
No. of awards given last year: 4
No. of applicants last year: 30

University of Kent School of Politics and International Relations Scholarships

Subjects: Politics and government, international relations, international conflict analysis.
Purpose: To support research.
Eligibility: These scholarships are available to Home/EU/Overseas students who have been made an offer by Kent for Mphil/PhD study.
Level of Study: Doctorate, Postgraduate
Type: Scholarship
Value: Home fees and maintenance stipend
Length of Study: Up to 3 years
Frequency: Annual, dependent on funds available
Study Establishment: University of Kent
Country of Study: United Kingdom
No. of awards offered: To be confirmed
Application Procedure: See wepbages at: www.kent.ac.uk/scholarships/postgraduate/departmental/politicsandir.html.
Closing Date: April
Funding: Commercial, government
No. of awards given last year: 2

University of Kent School of Psychology Scholarships

Subjects: Cognitive psychology, developmental psychology, forensic psychology, group processes and intergroup relations, and social psychology.
Purpose: To support research.
Eligibility: Candidates must hold a good Honours degree (first class or 2i) or a Master's degree at merit or distinction in Psychology. Non-British qualifications will be judged individually; we will generally require an overall result in the top two grading categories.
Level of Study: Doctorate, Postgraduate
Type: Scholarship
Value: Home fees and maintenance stipend
Length of Study: Up to 3 years
Study Establishment: University of Kent
Country of Study: United Kingdom
Application Procedure: Please check at www.kent.ac.uk/scholarships/postgraduate/departmental/psychology.html.
Closing Date: April
Funding: Commercial, government
No. of awards given last year: 5

University of Kent School of Social Policy, Sociology and Social Research Scholarships

Subjects: Social Policy, sociology, criminology.
Purpose: To support reserch.
Eligibility: Candidates should hold a good (first or upper second class) Honours degree or equivalent, in a relevant discipline.
Level of Study: Postgraduate, Doctorate
Type: Scholarship
Value: Home fees and maintenance stipend
Length of Study: Up to 3 years
Frequency: Annual, dependent on funds available
Study Establishment: University of Kent
Country of Study: United Kingdom
No. of awards offered: To be confirmed
Application Procedure: See wepbages at: www.kent.ac.uk/scholarships/postgraduate/departmental/sspssr.html.
Closing Date: February 10th
Funding: Commercial, government
No. of awards given last year: 3

University of Kent Sociology Studentship

Subjects: Sociology, social policy, criminology, environmental social science, social work and urban studies.
Purpose: To support research.
Eligibility: Candidates should hold a First Class or Upper Second Class (Honours) Degree.
Level of Study: Postgraduate, (Not available to MA/MSc students)
Type: Studentship
Value: Up to 2 awards at UK£9,250 for full-time students, up to 2 awards at UK£3,250 for full-time students or UK£2,000 for a part-time student, up to 2 awards at UK£12,600 plus £3,270 fees, 2 ESRC 1+3 awards
Length of Study: 3 years
Frequency: Annual
Study Establishment: The University of Kent
Country of Study: United Kingdom
No. of awards offered: Up to 9
Application Procedure: Candidates must contact: The Graduate Admissions Secretary, School of Social Research, Kent, CTZ 7NP England or email: socio-office@kent.ac.uk
Closing Date: March 20th
Funding: Government
No. of awards given last year: 9

THE UNIVERSITY OF LEEDS

Postgraduate Scholarships, Research Student Administration, Leeds, West Yorkshire, LS2 9JT, England
Tel: (44) 113 343 4077 ext 34077
Fax: (44) 113 343 3941
Email: pg-scholarships@leeds.ac.uk
Website: www.scholarships.leeds.ac.uk
Contact: Erika Smith, Senior Clerk

The University of Leeds aims to promote excellence and to achieve and sustain international standing in higher education teaching, learning and research, and to serve a wide range of student constituencies, social and professional communities and industrial, commercial and government agencies, locally, nationally and internationally.

Chevening-Beit Trust-University of Leeds Scholarships

Subjects: All subject.
Purpose: To provide postgraduate scholarships to international students of high calibre, who demonstrates both academic excellence and the potential to become leaders, decision makers and opinion formers in their own country.
Eligibility: Open to candidates from Malawi, Zambia and Zimbabwe who have obtained at least a UK upper second class honours degree or equivalent.
Level of Study: Postgraduate
Type: Scholarship
Value: Academic fee, maintenance allowance, return airfare and other allowances
Length of Study: 12 months
Frequency: Annual
Study Establishment: University of Leeds
Country of Study: United Kingdom
Application Procedure: Candidates should visit the Beit Trust website for details of how to apply at http://www.beittrust.org.uk/Scholarships.htm.
Closing Date: Please see the website.

China Scholarship Council - University of Leeds Scholarships

Subjects: All subject.
Purpose: To provide postgraduate scholarships for international research students of high calibre.
Eligibility: Open to citizens and permanent residents of China at the time of application. Applicants must hold at least a UK first class honours degree or equivalent undergradute degree and must be currently studying or working in specified Chinese universities and be commencing PhD study for the first time*. Applicants whose first language is not English must have already met the University's English language admissions requirements. There are limited number of scholarship places open to Chinese students currently in their last year of undergraduate or postgraduate study overseas.
Level of Study: Doctorate
Type: Scholarships
Value: Fees at international rate plus a maintenance allowance
Length of Study: Up to 3 years
Frequency: Annual
Study Establishment: University of Leeds
Country of Study: United Kingdom
Application Procedure: Applicants must satisfy the selection criteria set out by the China Scholarship Council, by completing the CSC application form and CSC employer reference form which can be found at www.csc.edu.cn. University application form must be completed and returned to the Postgraduate Scholarships office.
Closing Date: January 4th
Contributor: University of Leeds and China Scholarships Council

Collection of Small Endowed Awards

Subjects: Arts, business, education, social sciences, law, performance, visual arts, communications.
Purpose: To provide postgraduate scholarships for UK and EU Master's students of high calibre.
Eligibility: Candidates from the UK or other EU countries wishing to undertake study for a Master's degree. Candidates must hold at least a UK upper second class honours degree or equivalent.
Level of Study: Postgraduate
Type: Scholarship
Value: Fees at the UK/EU rate plus a maintenance allowance
Length of Study: 1 year
Study Establishment: University of Leeds
Country of Study: United Kingdom
Application Procedure: An application form must be completed and returned to the Postgraduate Scholarships Office by the relevant date.
Closing Date: See website
Funding: Private
Additional Information: The scholarship is subject to funding and will only be advertised on http://scholarships.leeds.ac.uk if there are available funds.

Commonwealth Shared Scholarship Scheme

Subjects: Subjects related to the economic, social and technological development of the student's home country.
Purpose: To provide scholarships to students of high academic calibre from developing Commonwealth countries.
Eligibility: Applicants must already have obtained a United Kingdom Upper Second Class (Honours) Degree or equivalent. Candidates must be nationals of, or permanently resident in, a developing Commonwealth country. Candidates whose first language is not English must have already met the University's English language requirement at the time of application.
Level of Study: Postgraduate
Type: Scholarship
Value: Academic fees, living expenses, other allowances, economy return airfares
Length of Study: 1 year
Frequency: Annual
Study Establishment: The University of Leeds
Country of Study: United Kingdom
Application Procedure: Applicants must complete an application form.
Closing Date: February 4th
Funding: Government, private

Emma and Leslie Reid Scholarship

Subjects: Diseases of the heart or brain.
Purpose: To provide postgraduate scholarships for UK and EU research students of high calibre.
Eligibility: Candidates must be from the UK or an EU country (or eligible to pay fees at the UK rate) and be commencing PhD study for the first time. Candidates must hold at least a UK upper second class honours degree or equivalent.
Level of Study: Doctorate
Type: Scholarship
Value: Fees at the UK/EU rate plus a maintenance allowance
Length of Study: Up to 3 years, subject to satisfactory progress
Frequency: Dependent on funds available
Study Establishment: University of Leeds
Country of Study: United Kingdom
Application Procedure: An application form must be completed and returned to the Postgraduate Scholarships office by the relevant date.
Closing Date: June 1
Funding: Private
Additional Information: The scholarship is subject to funding and will only be advertised on www.scholarships.leeds.ac.uk if there are available funds.

Fee Scholarship

Subjects: All subject.
Purpose: To provide postgraduate scholarships for UK and EU mastership students of high calibre.
Eligibility: Candidates from the UK or other EU countries wishing to undertake study for a master's degree.
Level of Study: Postgraduate
Type: Scholarship
Value: Academic fees at the University's standard UK/EU rate.
Length of Study: 12 months
Frequency: Annual
Study Establishment: University of Leeds
Country of Study: United Kingdom

Application Procedure: An Application form must be completed and returned to the Postgraduate Scholarships Office by the relevant date.
Closing Date: See website
Contributor: University of Leeds
Additional Information: Awards available for full- and part-time.

Frank Stell Scholarship

Subjects: Biology, agricultural science, social and political science.
Purpose: To provide postgraduate scholarships for UK research students of high calibre.
Eligibility: Candidates must be commencing PhD study for the first time and hold at least a UK upper second class honours degree or equivalent. Candidates should be resident, or have parents resident within the former administrative area of the County Council of the West Riding of Yorkshire.
Level of Study: Doctorate
Type: Scholarship
Value: Fees at the UK/EU rate plus a maintenance allowance
Length of Study: Up to 3 years, subject to satisfactory progress
Frequency: Dependent on funds available
Study Establishment: University of Leeds
Country of Study: United Kingdom
Application Procedure: An application form must be completed and returned to the Postgraduate Scholarships Office by the relevant date.
Closing Date: June 1st
Funding: Private
Additional Information: The scholarship is subject to funding and will only be advertised on www.scholarships.leeds.ac.uk if there are available funds.

Henry Ellison Scholarship

Subjects: Physics.
Purpose: To provide postgraduate scholarships for UK and EU research students of high calibre.
Eligibility: Candidates must be from the UK or an EU country (or eligible to pay fees at the UK rate) and be commencing PhD study for the first time. Candidates must hold at least a UK upper second class honours degree or equivalent and be a University of Leeds graduate.
Level of Study: Doctorate
Type: Scholarship
Value: Fees at the UK/EU rate plus a maintenance allowance
Length of Study: Up to 3 years, subject to satisfactory progress
Frequency: Dependent on funds available
Study Establishment: University of Leeds
Country of Study: United Kingdom
Application Procedure: An application form must be completed and returned to the Postgraduate Scholarships Office by the relevant date.
Closing Date: See website
Funding: Private
Additional Information: The award is available within the School of Physics and Astronomy. The scholarship is subject to funding and will only be advertised on http://scholarships.leeds.ac.uk if there are available funds.

John Henry Garner Scholarship

Subjects: Research on matters relating to chemical and biological surveys of rivers and streams and the pollution, prevention and purification of sewage and trade effluents in the area previously known as the West Riding of Yorkshire.
Purpose: To provide postgraduate scholarships for UK and EU research students of high calibre.
Eligibility: Candidates must be from the UK or an EU country (or eligible to pay fees at the UK rate) and be commencing PhD study for the first time. Candidates must hold at least a UK upper second class honours degree or equivalent.
Level of Study: Doctorate
Type: Scholarship
Value: Fees at the UK/EU rate plus a maintenance allowance
Length of Study: Up to 3 years, subject to satisfactory progress
Frequency: Dependent on funds available
Study Establishment: University of Leeds
Country of Study: United Kingdom
Application Procedure: An application form must be completed and returned to the Postgraduate Scholarships Office by the relevant date.
Closing Date: June 1st
Funding: Private
Additional Information: The scholarship is subject to funding and will only be advertised on www.scholarships.leeds.ac.uk if there are available funds.

Leeds International Research Scholarships (LIRS)

Subjects: All subjects.
Purpose: To provide postgraduate scholarships for international research students of high calibre.
Eligibility: Candidates must be liable to pay fees at the international rate and be commencing PhD study for the first time. Candidates will be required to hold a UK first class honours degree or equivalent undergraduate degree. Applicants whose first language is not English must have already met the University's English language admissions requirements at the time of application.
Level of Study: Doctorate
Type: Scholarship
Value: Fees at the international rate plus a maintenance allowance
Length of Study: Up to 3 years, subject to satisfactory progress
Frequency: Annual
Study Establishment: University of Leeds
Country of Study: United Kingdom
Application Procedure: An application form must be completed and returned to the Postgraduate Scholarships Office by the relevant date.
Closing Date: See website
Contributor: University of Leeds

Lund, Stephenson Clarke Scholarship

Subjects: Colour science.
Purpose: To provide postgraduate scholarships for UK and EU research students of high calibre.
Eligibility: Candidates must be from the UK or an EU country (or eligible to pay fees at the UK rate) and be commencing PhD study for the first time. Candidates must hold at least a UK upper second class honours degree or equivalent.
Level of Study: Doctorate
Type: Scholarship
Value: Fees at the UK rate plus a maintenance allowance
Length of Study: up to 3 years, subject to satisfactory progress
Frequency: Dependent on funds available
Study Establishment: University of Leeds
Country of Study: United Kingdom
Application Procedure: An application form must be completed and returned to the Postgraduate Scholarships Office by the relevant date.
Closing Date: June 1st
Funding: Private
Additional Information: This scholarship is subject to funding and will only be advertised on http://scholarships.leeds.ac.uk if there are available funds.

Lund, Stephenson Clarke Scholarship

Subjects: Colour science.
Purpose: To provide postgraduate scholarships for UK and EU research students of high calibre.
Eligibility: Applicants must be from UK or EU and be commencing PhD study for the first time. Candidates must hold at least a UK upper second class honours degree or equivalent.
Level of Study: Doctorate
Type: Scholarship
Value: Fees at the UK rate plus a maintenance allowance
Length of Study: Up to 3 years
Frequency: Annual
Study Establishment: University of Leeds
Country of Study: United Kingdom
No. of awards offered: 1
Application Procedure: Application form must be completed and returned to the Postgraduate Scholarships Office.
Closing Date: June 1st
Funding: Private
Additional Information: Applications can be send by email at pg_scholarships@leeds.ac.uk.

Mary and Alice Smith Memorial Scholarship

Subjects: Research into the prevention or cure of cancer or heart disease.

Purpose: To provide postgraduate scholarships for UK research students of high calibre.
Eligibility: Candidates must be British and be commencing PhD study for the first time. Candidates must hold at least a UK upper second class honours degree or equivalent.
Level of Study: Doctorate
Type: Scholarship
Value: Fees at the UK/EU rate plus a maintenance allowance
Length of Study: Up to 3 years, subject to satisfactory progress
Frequency: Dependent on funds available
Study Establishment: University of Leeds
Country of Study: United Kingdom
Application Procedure: An application form must be completed and returned to the Postgraduate Scholarships Office by the relevant date.
Closing Date: See website
Funding: Private
Additional Information: The scholarship is subject to funding and will only be advertised on www.scholarships.leeds.ac.uk if there are available funds.

Stanley Burton Research Scholarship

Subjects: Music or fine art.
Purpose: To provide postgraduate scholarships for UK/EU research students of high calibre.
Eligibility: Candidates must be from UK or an EU country and be commencing PhD study for the first time. Candidates must hold at least a UK upper second class honours degree or equivalent.
Level of Study: Doctorate
Type: Scholarship
Value: Fees at the UK/EU rate plus a maintenance allowance
Length of Study: Up to 3 years, subject to satisfactory progress
Frequency: Dependent on funds available
Study Establishment: University of Leeds
Country of Study: United Kingdom
Application Procedure: An application form must be completed and returned to the Postgraduate Scholarships Office by the relevant date.
Closing Date: June 1st
Funding: Private
Additional Information: The award is available within the School of Fine Art, History of Art and Cultural Studies and the School of Music. This scholarship is subject to funding and will only be advertised on http://scholarships.leeds.ac.uk if there are available funds.

Tetley and Lupton Scholarship

Subjects: All subject.
Purpose: To provide postgraduate scholarships for international mastership students of high calibre.
Eligibility: Open to international students commencing taught postgraduate study in any faculty. Applicants will be required to hold a UK first class honours degree or equivalent.
Level of Study: Postgraduate
Type: Scholarship
Value: Partial contribution towards academic fees
Length of Study: 12 months
Frequency: Annual
Study Establishment: University of Leeds
Country of Study: United Kingdom
Application Procedure: Application form must be completed and returned to the Postgraduate scholarships office.
Closing Date: March 1st

University of Leeds Arts and Humanities Research Scholarship

Subjects: Arts and humanities.
Purpose: To provide postgraduate scholarships for UK and EU research students of high calibre.
Eligibility: Applicants must be from UK or EU and be commencing PhD study for the first time. Candidates must hold at least a UK upper second class honours degree or equivalent.
Level of Study: Doctorate
Type: Scholarship
Value: Fees at the UK/EU rate plus a maintenance allowance
Length of Study: Up to 3 years
Frequency: Annual
Study Establishment: University of Leeds
Country of Study: United Kingdom
Application Procedure: Application form must be completed and returned to the Postgraduate Scholarships Office. Applicants not applying for an AHRC studentship must complete the University's AHRC studentship application form for internal selection purposes.
Closing Date: See website
Additional Information: Applicants applying for the AHRC Block Grant Partnership doctoral studentships will be automatically be considered for these awards.

University Research Scholarships

Subjects: All subjects.
Purpose: To provide postgraduate scholarships for UK and EU research students of high calibre.
Eligibility: Candidates must be from the UK or an EU country (or eligible to pay fees at the UK rate) and be commencing PhD study for the first time. Candidates must hold at least a UK upper second class honours degree or equivalent undergraduate degree.
Level of Study: Doctorate
Type: Scholarship
Value: Fees at the UK/EU rate plus a maintenance allowance
Length of Study: Up to 3 years, subject to satisfactory progress
Frequency: Annual
Study Establishment: University of Leeds
Country of Study: United Kingdom
Application Procedure: An application form must be completed and returned to the Postgraduate Scholarships Office by the relevant date.
Closing Date: Varies, according to faculty.
Contributor: University of Leeds
Additional Information: Please check the website for details.

White Rose Studentships

Subjects: Project based, subjects to be confirmed.
Purpose: To provide postgraduate scholarships for UK and EU research students of high calibre.
Eligibility: Candidates must be from the UK or an EU country (or eligible to pay fees at the UK rate) and be commencing PhD study for the first time. Candidates must hold at least a UK upper second class honours degree or equivalent.
Level of Study: Doctorate
Type: Scholarship
Value: Fees at the UK rate plus a maintenance allowance
Length of Study: Up to 3 years, subject to satisfactory progress
Frequency: Annual
Study Establishment: University of Leeds
Country of Study: United Kingdom
Application Procedure: An application form must be completed and returned to the Postgraduate Scholarships Office by the relevant date.
Closing Date: February 15th (check website also)
Funding: Private
Additional Information: The University of Leeds is offering a number of White Rose Research Studentships, in collaboration with the Universities of Sheffield and York, for students from the United Kingdom and from European Union Countries commencing full-time PhD study in October.

UNIVERSITY OF LEICESTER

University Road, Leicester, LE1 7RH, United Kingdom
Tel: (44) 0116 252 2522
Fax: (44) 0116 252 2200
Website: www.le.ac.uk

AHRC PhD Studentship in Archaeology or Ancient History

Purpose: To provide PhD studentship.
Eligibility: Open for the students of UK/EU and International Applicants.
Level of Study: Doctorate
Type: Scholarship
Value: Tuition fees with the annual conference/training allowance of £200.
Length of Study: 3 years
Frequency: Dependent on funds available

Study Establishment: University of Leicester, School of Archaeology plus Ancient History
Country of Study: United Kingdom
No. of awards offered: 1
Application Procedure: PhD application form.
Closing Date: March 31st
Contributor: AHRC
No. of awards given last year: 1
No. of applicants last year: 32

Graduate Teaching Assistant in Computer Science

Subjects: Computer science.
Purpose: Opportunity to combine postgraduate research with teaching duties (part-time).
Eligibility: Open to students registered with the University of Leicester for the full term of their contract. International students will be required to fund fees above the level of the fee waiver.
Level of Study: Doctorate
Type: Graduate assistantship
Value: Fee waiver of £3,732 per year; maintenance stipend of £13,590 per year (based on last year's levels)
Length of Study: 4 years
Frequency: Annual
Study Establishment: University of Leicester, Computer Science Department
Country of Study: United Kingdom
No. of awards offered: 4
Application Procedure: Opportunities are advertised on the University of Leicester recruitment website at http://www2.le.ac.uk/offices/jobs and computer science homepage at http://www2.le.ac.uk/departments/computer-science. Application forms are online.
Closing Date: January 10th (Please check with the website)
Contributor: University of Leicester
No. of awards given last year: 4

MA in Archaeology and Heritage

Purpose: To provide fee reduction bursary/studentship.
Level of Study: Postgraduate
Type: Scholarship
Value: £1,000 fee reduction in year one.
Length of Study: 2 years part-time (Distance Learning)
Frequency: Dependent on funds available
Study Establishment: University of Leicester
No. of awards offered: 2
Application Procedure: MA application form.
Closing Date: July 31st
No. of awards given last year: 2
No. of applicants last year: 3

MA in Archaeology of the Roman World Studentship

Subjects: Archeology.
Purpose: To provide fee reduction scholarship.
Level of Study: Postgraduate
Type: Scholarship
Value: £1,000 fee reduction - Home/EU; £2,000 fee reduction - International students
Length of Study: 1 year
Frequency: Dependent on funds available
Study Establishment: University of Leicester, School of Archaeology and Ancient History
Country of Study: United Kingdom
Application Procedure: MA application form.
Closing Date: July 31st
No. of applicants last year: 4

MA in Archaeology Studentship

Subjects: Archaeology with pathways in pre-history, bioarchaeology and medieval archaeology.
Purpose: Fee reduction studentship.
Level of Study: Postgraduate
Type: Scholarship
Value: £2,000 International fee reduction; £1,000 Home/EU fee reduction
Length of Study: 1 year
Frequency: Dependent on funds available
Country of Study: United Kingdom
Application Procedure: MA application form.
Closing Date: July 31st
No. of awards given last year: 2 (1 for Home/EU and 1 for International)
No. of applicants last year: 5

MA in Historical Archaeology Studentship

Subjects: Historical archaeology.
Purpose: To provide fee reduction studentship.
Level of Study: Postgraduate
Type: Studentship
Value: £1,000 fee reduction - Home/EU students/Distance Learning; £2,000 fee reduction - international students
Length of Study: 1–2 years Distance Learning
Frequency: Dependent on funds available
Study Establishment: University of Leicester, School of Archaeology plus ancient history
Country of Study: United Kingdom
Application Procedure: MA application form.
Closing Date: March 31st
No. of awards given last year: 1
No. of applicants last year: 8

MA in The Classical Mediterranean

Subjects: Archaeology of classical mediterranean.
Purpose: To provide fee reduction studentship.
Eligibility: Good first degree, second class or higher in ancient history, archaeology or cognate disciplines or the equivalent for overseas or European applicants.
Level of Study: Postgraduate
Type: Studentship
Value: £1,000 fee reduction - Home/EU/Distance Learning; £2,000 fee reduction - international students
Length of Study: 1–2 years Distance Learning
Frequency: Dependent on funds available
Study Establishment: University of Leicester, School of Archeology plus ancient history
Country of Study: United Kingdom
Application Procedure: MA application form.
Closing Date: July 31st
No. of awards given last year: 1
No. of applicants last year: 10

PhD Bursaries for International Students in Archaeology or Ancient History

Subjects: Archaeology/ancient history.
Purpose: To provide bursaries for PhD study.
Eligibility: Open to international students. Home/EU students are not eligible to apply.
Level of Study: Doctorate
Type: Scholarship/Bursary
Value: £3,000 fee reduction for 3 years
Length of Study: 3 years
Frequency: Dependent on funds available
Study Establishment: University of Leicester
Country of Study: United Kingdom
No. of awards offered: 1
Application Procedure: PhD application.
Closing Date: March 31st
No. of awards given last year: 1
No. of applicants last year: 12

PhD Fee Discount for International Students in Archaeology or Ancient History

Subjects: Archaeology/ancient history.
Purpose: To provide PhD study.
Eligibility: Open to international students. Home/EU students are not eligible to apply.
Level of Study: Doctorate
Type: Scholarship
Value: £6,518 fee reduction per year for 3 years - student pays Home/EU fee rate
Length of Study: 3 years
Frequency: Dependent on funds available

No. of awards offered: 1
Application Procedure: PhD application.
Closing Date: March 31st
No. of awards given last year: 1
No. of applicants last year: 8

PhD Studentship in Trans-Saharan Archaeology

Subjects: Archaeology.
Purpose: To help with stipend and provide scholarship.
Level of Study: Doctorate
Type: Scholarship
Value: Covers tuition fees (home/EU rate) plus annual stipend of up to £14,000 for 3 years
Length of Study: 3 years
Frequency: Dependent on funds available
Study Establishment: University of Leicester
No. of awards offered: 1
Closing Date: Check with the website.

PhD Studentships in Archaeology or Ancient History

Purpose: To provide studentship with stipends - PhD study.
Eligibility: International students would get reduction in fees of Home/EU rate.
Level of Study: Doctorate
Type: Scholarship
Value: £13,590 stipends for 3 years; Home/EU fees paid £3,732 for 3 years
Length of Study: 3 years
Frequency: Dependent on funds available
Study Establishment: University of Leicester
Country of Study: United Kingdom
No. of awards offered: 2
Application Procedure: PhD application.
Closing Date: May 31st
No. of awards given last year: 3
No. of applicants last year: 40

PhD Studentships in Computer Science

Subjects: Computer science.
Purpose: To support research organisations attract the best people into postgraduate research and training and to provide doctoral training grant support.
Eligibility: Please check at EPSRC website at http://www.epsrc.ac.uk/funding/students/pages/default.aspx.
Level of Study: Doctorate
Type: Studentship
Value: Fee waiver (£3,732 per year), Stipend (maintenance payment) £13,590 per year
Length of Study: 3 years
Frequency: Dependent on funds available, 4 intakes per year; January 1st, April 1st, July 1st and October 1st
Study Establishment: University of Leicester, Computer Science Department
Country of Study: United Kingdom
No. of awards offered: Up to 3
Application Procedure: Please refer to the University of Leicester website at http://www.le.ac.uk/departments/computer-science/postgraduate/research. Our PhD superiors have a range of interests and the aim is to align you with someone with expertise in your field. Details of our research themes are on the website.
Closing Date: Check with the website.
Funding: Government
Contributor: EPSRC (DTG)
Additional Information: Applications are considered throughout the year. Refer to advertisement on website.

For further information contact:

Email: fdv1@mcs.le.ac.uk
Contact: Dr Fer-Jan de Vries, PhD Admissions Tutor

PhD Studentships in Engineering

Subjects: PhD in engineering across various disciplines, including bioengineering, control and instrumention, electrical and electronic, embedded sytems, mechanics of materials and thermofluids.
Purpose: To support PhD student.
Level of Study: Research
Type: Studentship
Value: Stipend of £13,590 per year
Study Establishment: University of Leicester, Department of Engineering
Country of Study: United Kingdom
Closing Date: Check with the website.
No. of applicants last year: 3
Additional Information: Please check at le.ac.uk/engineering for more information.

PhD Studentships in Geology

Purpose: Geochemistry, geology, geophysics and seismology, mineralogy and crystallography, palaeontology and petrology.
Eligibility: Applicants must have a first-class or high upper second-class honours degree (or equivalent qualification) in geology, geophysics, earth sciences, physics, or a relevant discipline and meet the University's standard English language entry requirements.
Level of Study: Doctorate
Type: Studentship
Value: An annual tax-free stipend of at least £13,590
Length of Study: 3 years
Frequency: Annual
Study Establishment: Department of Geology, University of Leicester, Leicester LE1 7RH
Country of Study: United Kingdom
Application Procedure: Application through University website.
Closing Date: Check with the website.
Funding: Commercial, government

For further information contact:

Contact: Mark Williams, Postgraduate Tutor (Geology)

PhD Studentships in Mathematics

Purpose: Pure and applied mathematics, financial mathematics and actuarial science are considered.
Level of Study: Doctorate, Research
Type: Studentship
Value: Approx. £14,000 per year
Country of Study: United Kingdom
Application Procedure: A completed PhD application form and supporting documents as per the university requirements.
Closing Date: Check with the website.

UNIVERSITY OF LONDON INSTITUTE IN PARIS

9-11 rue de Constantine, F-75340 Paris Cedex 07, France
Tel: (33) 1 44 11 73 76
Fax: (33) 1 44 11 73 82
Email: c.brown@ulip.lon.ac.uk
Website: www.ulip.lon.ac.uk
Contact: Collette Brown, PA to the Dean

The University of London Institute in Paris is a United Kingdom Institute of Higher Education devoted to the study of French language, literature and culture.

Quinn, Nathan and Edmond Scholarships

Subjects: An area in which the institute can provide supervision in the field of humanities relating to France.
Purpose: To assist postgraduate research in France.
Eligibility: Open to graduates who demonstrate an outstanding academic record in the field of French and Comparative Studies.
Level of Study: Doctorate, Postdoctorate, Postgraduate
Type: Scholarship
Value: Depending on the funds available and the needs of the candidates, scholarships are of the order of UK£500 per month.
Length of Study: Nine months
Frequency: Dependent on funds available
Study Establishment: University of London Institute in Paris
Country of Study: France
No. of awards offered: Dependent on availability of funds

Application Procedure: Application forms can be downloaded from the website.
Closing Date: June 29th
Funding: Private
Contributor: Trust funds
No. of awards given last year: 2
Additional Information: Scholarships cannot be held concurrently with other major awards. These scholarships are intended for research and not for those following taught courses.both research and taught postgraduate courses.

UNIVERSITY OF LONDON, SCHOOL OF ADVANCED STUDY

Senate House, Malet Street, London, WC1E 7HU, United Kingdom
Tel: (44) 207 862 8844
Fax: (44) 207 862 8657
Email: sas.info@sas.ac.uk
Website: www.sas.ac.uk/fellowshipprogrammes.html

Henry Charles Chapman Visiting Fellowship

Subjects: Social sciences, 20th century history, Commonwealth relations, preferably aligned with a research interest of a current member of staff of the Institute of Commonwealth Studies.
Purpose: To offer an opportunity to academic staff of universities in the Commonwealth to undertake research in the social sciences relevant to the work of the institute.
Eligibility: Restricted to academic staff of universities in Commonwealth countries.
Level of Study: Research
Type: Fellowship
Value: Up to £4,000
Length of Study: 3 months–12 months
Frequency: Every 2 years
Study Establishment: Institute of Commonwealth Studies, University of London
Country of Study: United Kingdom
No. of awards offered: 1
Application Procedure: Download the application form from the website www.commonwealth.sas.ac.uk.
Closing Date: May 31st (in alternate years)
Funding: Trusts
Contributor: Henry Charles Chapman Trust Fund

Robin Humpreys Fellowship

Subjects: Latin America diplomacy.
Purpose: To further the study of diplomacy with and within Latin America.
Eligibility: Restricted to those with a background in international diplomacy.
Level of Study: Research
Length of Study: 1 year or 2 years
Frequency: Every 2 years
Study Establishment: Institute for the Study of the Americas, University of London
Country of Study: United Kingdom
Application Procedure: By invitation only.
Closing Date: No deadline – appointment by invitation only
Funding: Government
Contributor: No funds are offered for the fellowship
No. of awards given last year: 1
No. of applicants last year: 1
Additional Information: Fellowship awarded by invitation. No application procedure.

THE UNIVERSITY OF MANCHESTER

Oxford Road, Manchester, M13 9PL, United Kingdom
Tel: (44) 161 306 6000
Email: hr@manchester.ac.uk
Website: www.manchester.ac.uk

The University of Manchester is Britain's largest single site university with a proud history of achievement and an ambitious agenda for the future. The University has an exceptional record of generating and sharing new ideas, and the quality, breadth and volume of its research activity is unparalleled in Britain.

IDPM Taught Postgraduate Scholarship Scheme

Subjects: Development studies.
Eligibility: Open to citizens of any country (excluding Andorra, Australia, Austria, Belgium, Bermuda, Canada, Cyprus, Denmark, Faroe Islands, Finland, France, Germany, Greece, Hong Kong, Iceland, Ireland, Israel, Italy, Japan, Liechtenstein, Luxembourg, Macau, Monaco, Netherlands, New Zealand, Norway, Portugal, San Marino, Singapore, Slovenia, South Korea, Spain, Sweden, Switzerland, Taiwan, United Kingdom, United States, Vatican City) who are holding a first degree, have not previously studied for 1 year or more in the UK or any other developed country, and have received an offer to study from IDPM.
Level of Study: Postgraduate
Type: Scholarship
Value: UK£2,000
Frequency: Annual
No. of awards offered: 3
Application Procedure: Check website for further details.
Closing Date: June 30th

For further information contact:

University of Manchester, Harold Hankins Building, Room 5.31, Precinct Centre, Oxford Road, Manchester, M13 9PL, United Kingdom
Email: paul.arrowsmith@manchester.ac.uk
Website: www.sed.manchester.ac.uk/
Contact: Paul Arrowsmith, School of Environment & Development, Admissions Office

Manchester-China Scholarship Council Joint Postgraduate Scholarship Programme

Subjects: Telecommunication and information technology, life science and public health, material science and new material, energy sources and environment, engineering science and applied social sciences and WTO-related areas.
Purpose: To provide scholarships to the nationals of PR China who wish to pursue their PhD at the University of Manchester.
Eligibility: Open to students who are citizens and permanent residents of PR China and who hold a Master's degree from one of the 38 Chinese universities under 985 Programme.
Level of Study: Postgraduate, Doctorate
Type: Scholarships
Value: UK£4,800
Length of Study: 3 years
Frequency: Annual
Study Establishment: University of Manchester
Country of Study: United Kingdom
No. of awards offered: 10
Application Procedure: Applicants must complete the standard postgraduate application form and return it together with: Academic transcripts, English language qualification, 2 reference letters and a research proposal.
Closing Date: April 20th

UNIVERSITY OF MANITOBA

Department of English, 623 Fletcher Argue Building, Winnipeg, MB, R3T 2N2, Canada
Tel: (1) 204 474 6209
Fax: (1) 204 474 7659
Email: university_1@umanitoba.ca
Website: www.umanitoba.ca

The University of Manitoba is the province's largest, most comprehensive and only research intensive post secondary educational institution. It was founded in 1877. In a typical year, the university has an enrolment of 24,542 undergraduate students and 3,021 graduate students. The University offers 82 degrees, 51 at the undergraduate level. Most academic units offer Graduate studies programmes leading to Master's or Doctoral degrees.

University of Manitoba Graduate Fellowship

Subjects: English literature.
Purpose: To provide fellowships to students who wish to pursue higher studies in english literature.
Eligibility: Open to students who have maintained high grades and made satisfactory progress in full-time graduate programme.
Level of Study: Doctorate, Postgraduate
Type: Fellowships
Value: Canadian $16,000 for PhD, Canadian $12,000 for MA
Length of Study: 2–4 years
Frequency: Annual
Study Establishment: University of Manitoba
Country of Study: Canada
No. of awards offered: 2
Application Procedure: Application for these fellowships will be invited in September each year. Application forms are available on the Faculty of Graduate Studies website at www.umanitoba.ca/faculties/graduate_studies/funding/index.html
Closing Date: March 31st

University of Manitoba Teaching Assistantships

Subjects: English literature.
Purpose: To financially support students enrolled in full-time PhD programme.
Eligibility: Open to full-time students registered in a PhD programme who may apply for positions as lecturers or seminar leaders in the first year literature courses.
Level of Study: Doctorate
Type: Assistantship
Value: Varies
Length of Study: 3–4 years
Frequency: Dependent on funds available
Study Establishment: University of Manitoba
Country of Study: Canada
Application Procedure: To view current postings and obtain an application form students can log on to www.umanitoba.ca/admin/human_resources/employment
Closing Date: Check with the website.
Funding: Trusts

THE UNIVERSITY OF MELBOURNE

Scholarships Office, Melbourne, Melbourne, VIC, 3010, Australia
Tel: (61) (3) 8344 4000
Fax: (61) (3) 8344 5104
Email: pg-schools@unimelb.edu.au
Website: www.unimelb.edu.au

The University of Melbourne has a long and distinguished tradition of excellence in teaching and research. It is the leading research institution in Australia and enjoys a reputation for the high quality of its research programmes, consistently winning the largest share of national competitive research funding.

A.O. Capell Scholarship

Subjects: All subjects.
Eligibility: Open to APA and MRS candidates having completed tertiary studies that are equivalent to a 4 years Australian degree with a minimum result of First Class Honours and currently enrolled in, a research higher degree.
Level of Study: Postgraduate
Type: Scholarship
Value: $30,000 per year
Length of Study: 1 year
Frequency: Every 3 years
Study Establishment: University of Melbourne
Country of Study: Australia
No. of awards offered: 3
Application Procedure: Check website for further details.
Closing Date: October 31st
Funding: Private
Contributor: A.O. Capell Scholarship Fund, Stella Mary Langford Scholarship Fund, and the Henry & Louisa Williams Bequest

Baillieu Research Scholarship

Subjects: Medicine, law, commerce, economics and architecture.
Eligibility: Open to lineal descendants of Australian soldiers and sailors killed, blinded or permanently incapacitated while engaged in war service, with priority being given to World War I service.
Level of Study: Postgraduate
Type: Scholarship
Value: $23,728 per year (last year's rate) and other benefits as per the MRS
Frequency: Annual
Application Procedure: Applicants should lodge to the Melbourne Scholarships office the SCHOLS online application indicating that they wish to be considered for the Baillieu Research Scholarship and documents to verify that they are a lineal descendent of an Australian solider or sailor killed, blinded or permanently incapacitated while engaged in war service.
Closing Date: October 31st
Funding: Private
Contributor: Baillieu Research Scholarship Fund

Dairy Postgraduate Scholarships and Awards

Subjects: Dairy farming and dairy manufacturing.
Purpose: To enable students for research contributing to the technical areas relevant to dairy farming and dairy manufacturing operations.
Eligibility: Open to citizens or permanent residents of Australia.
Level of Study: Postgraduate
Type: Scholarship
Value: Australian $25,000 stipend plus Australian $3,000 per year
Frequency: Annual
Country of Study: Australia
Application Procedure: To apply for this scholarship one must apply direct to the faculty.
Closing Date: October 20th

For further information contact:

Dairy Australia Limited, Australia
Tel: (61) 03 9694 3810
Fax: (61) 03 9694 3701
Email: research@dairyaustralia.com.au
Website: www.dairyaustralia.com.au

Ernst and Grace Matthaei Research Scholarship

Subjects: Optics.
Eligibility: This scholarship is for study in Australia. Only citizens of Australia or New Zealand or AUS permanent resident can apply.
Level of Study: Postgraduate
Type: Scholarship
Value: $28,000 per year and other benefits as per the APA
Frequency: Annual
Country of Study: Australia
No. of awards offered: 1
Application Procedure: Applicants must indicate on SCHOLS, the online application service, their interest in being considered for the scholarship.
Closing Date: October 31st
Contributor: Ernst and Grace Matthaei Bequest

Fay Marles Scholarships

Subjects: Human rights.
Purpose: To provide financial support to students of indigenous Australian descent, with disabilities and whose academic career has been adversely affected.
Eligibility: Open to applicants who are citizens of Australia or New Zealand or Australian permanent residents. Applicants must provide evidence to show they meet at least 1 of the following criteria: student of indigenous Australian descent, student whose academic career has been adversely affected or student with a disability.
Level of Study: Postgraduate
Type: Scholarships
Value: Australian $23728 (minimum per award).
Length of Study: 2–3 years
Frequency: Annual
Study Establishment: University of Melbourne
Country of Study: Australia

No. of awards offered: 8
Closing Date: October 31st
Additional Information: For further information please visit www.services.unimelb.edu.au/scholarship

Fred Knight Research Scholarship

Subjects: All subjects.
Eligibility: Open to APA and MRS candidates having completed tertiary studies that are equivalent to a 4 years Australian degree with a minimum result of First Class Honours and currently enrolled in a research higher degree.
Level of Study: Postgraduate
Type: Scholarship
Value: Australian $8,000 for 1 year in addition to an APA or MRS
Frequency: Annual
Country of Study: Australia
No. of awards offered: 1 per year
Application Procedure: Check website for further details.
Closing Date: October 31st
Funding: Private
Contributor: Fred Knight Research Scholarship Fund

Grimwade Scholarship

Subjects: All subjects.
Eligibility: This scholarship is for study in Australia. Only citizens of Australia or New Zealand or AUS permanent resident can apply.
Level of Study: Doctorate
Type: Scholarship
Value: Australian $21,000 (maximum per year).
Frequency: Annual
No. of awards offered: 1
Application Procedure: Check website for further details.
Closing Date: October 31st
Funding: Private
Contributor: Russell and Mab Grimwade Miegunyah Fund

The Helen Macpherson Smith Scholarships

Subjects: Science, humanities and social sciences.
Eligibility: Open to outstanding women who are entering postgraduate study. Only citizens of Australia may apply.
Level of Study: Postgraduate
Type: Scholarship
Value: Australian $8,000 top-up payment (payable over one year) in addition to an APA or MRS
Frequency: Annual
Country of Study: Australia
No. of awards offered: 2
Application Procedure: Check website for further details.
Closing Date: October 31st
Funding: Trusts
Contributor: Helen Macpherson Smith Trust

Henry James Williams Scholarship

Subjects: All subjects.
Eligibility: Open to APA and MRS candidates having completed tertiary studies that are quivalent to a 4 year Australian degree with a minimum result of First Class Honours and currently enrolled in a research higher degree.
Level of Study: Postgraduate
Type: Scholarship
Value: $28,000 per year and provides other benefits as per the APA
Length of Study: 3 years
Frequency: Annual
No. of awards offered: 3
Application Procedure: Check website for further details.
Closing Date: October 31st
Funding: Private
Contributor: A.O. Capell Scholarship Fund, the Stella Mary Langford Scholarship Fundand Henry & Louisa Williams Bequest

Human Rights Scholarship

Subjects: Human rights.
Purpose: To support international and local students who will be undertaking postgraduate studies in the human rights field and who are able to demonstrate their commitment to the peaceful advancement of respect for human rights.
Eligibility: Open to applicants who are able to demonstrate their commitment to the peaceful advancement of respect for human rights extends beyond their academic studies, a high H2A (i.e. 78–79 per cent and above) minimum grade average and must be planning to commence or be currently enrolled in a postgraduate diploma, Masters by coursework, Doctorate by coursework or research higher degree in the human rights field at the University of Melbourne.
Level of Study: Postgraduate
Type: Scholarship
Value: Living allowance of $28,000 per year (2012 rate), relocation grant of $3,000, relocation allowance and thesis allowance of up to $420 (2012 rate) for Masters and $840 (2011 rate) for PhD and other Doctorate by research candidates
Frequency: Annual
Country of Study: Australia
No. of awards offered: 2
Application Procedure: Local applicants must submit to three complete sets (one original or certified and two copies) of the documents listed on the document checklist. International applicants must submit three sets (one original and two copies) of the International HRS Application Form.
Closing Date: October 31st

Melbourne International Fee Remission Scholarships

Subjects: All subjects offered by the University.
Purpose: To enable graduates to undertake a research higher degree in any discipline.
Eligibility: Please see www.services.unimelb.edu/scholarships/pgrad for full details.
Level of Study: Postgraduate, Research
Type: Scholarship
Value: Full fee remission as per the IPRS (but not OSHC)
Length of Study: Up to 2 years at the Master's level and up to 3 years at the PhD and research doctorate level. A 6-month extension is possible at the research doctorate level
Frequency: Annual
Study Establishment: The University of Melbourne
Country of Study: Australia
No. of awards offered: 150. Available only to international students
Application Procedure: International applicants who have not already commenced the course for which they seek scholarship will automatically be considered for scholarship if they receive an unconditional offer of a place in a research higher degree course. Information about applying for courses is available at www.services.unimelb.edu.au/admissions/apply/. A separate application form is not required.
Closing Date: October 31st
Funding: Government
Contributor: Scholarship fund
No. of awards given last year: 60
No. of applicants last year: 500

Melbourne Research Scholarships

Subjects: All subjects offered by the University.
Purpose: To enable graduates to undertake a research higher degree in any discipline.
Eligibility: Please see www.services.unimelb.edu.au/scholarships/pgrad for full details.
Level of Study: Doctorate, Postgraduate, Research
Type: Scholarship
Value: Living allowance of $23,728 per year (2012 rate), relocation grant of of $2,000, thesis allowance of $420 (2012 rate) for masters by research and up to $840 (2012 rate) for PhD candidates
Length of Study: Up to 2 years at the Master's level and up to 3 years at the PhD and research doctorate level. A 6-month extension is possible at the research doctorate level
Frequency: Annual
Study Establishment: The University of Melbourne
Country of Study: Australia
No. of awards offered: 220, of which approx. 95 may be awarded to international students
Application Procedure: International applicants who have not already commenced the course for which they seek scholarship will

automatically be considered for scholarship if they receive an unconditional offer of a place in a research higher degree course. Information about applying for courses is available at www.services.unimelb.edu.au/admissions/apply/. A separate application form is not required.
Closing Date: October 31st
Funding: Government
Contributor: Scholarship fund
No. of awards given last year: 210
No. of applicants last year: 1,000

PhD Scholarship: Research in Health Law and Policy
Subjects: Health law and policy.
Purpose: To research into policy issues at the intersection of the legal and health care systems in Australia.
Eligibility: Open to applicants holding an Honours Degree in law. Prior training in statistics or quantitative methods will be highly valued.
Level of Study: Doctorate
Type: Scholarship
Value: Australian $25,600 per year
Length of Study: 3 years
Frequency: Annual
Study Establishment: Melbourne Law School or School of Population Health
Country of Study: Australia
Application Procedure: Check website for further details.
Contributor: ARC

For further information contact:

Centre for Health Programs, Policy and Economics, School of Population Health
Tel: 8344 0710
Contact: Joy Yeadon

PhDs in Bio Nanotechnology
Subjects: Bio nanotechnology.
Eligibility: Open to those who have achieved Honours 2a or equivalent.
Level of Study: Postgraduate, Research
Type: Scholarship
Value: Australian $19,616 per year
Length of Study: 2 years (Masters) and 3 years (PhD)
Frequency: Annual
Study Establishment: University of Melbourne
Country of Study: Australia
No. of awards offered: 3
Application Procedure: Applicants must apply directly to the scholarship provider. Check website for further details.
Closing Date: July 21st
Contributor: University of Melbourne

For further information contact:

Chemical and Biomolecular Engineering, Australia
Email: fcaruso@unimelb.edu.au
Website: www.cnst.unimelb.edu.au
Contact: Professor Frank Caruso, Federation Fellow

Pratt Foundation Scholarship
Subjects: All subjects.
Eligibility: This scholarship is for study in Australia. Only citizens of Australia or AUS permanent resident can apply.
Level of Study: Postgraduate
Type: Scholarship
Value: Australian $28,000 per year; a relocation allowance of up to $1,520; and a thesis allowance of $420 for master students and $840 for PhD students.
Frequency: Annual
Country of Study: Australia
No. of awards offered: 1
Application Procedure: Check website for further details.
Closing Date: October 31st
Funding: Foundation
Contributor: Pratt Foundation

Rae and Edith Bennett Travelling Scholarship
Subjects: All subjects.
Purpose: To enable students and graduates of the University of Melbourne to undertake postgraduate study or research in the United Kingdom.
Eligibility: Open to students and graduates of the University of Melbourne who can demonstrate outstanding academic merit and promise.
Level of Study: Postgraduate
Type: Scholarship
Value: $40,000–60,000 per year
Frequency: Annual
Country of Study: Australia
No. of awards offered: 2
Application Procedure: Applicants must provide official transcript/s of all tertiary study, a curriculum vitae, an outline of the proposed study or research, evidence of acceptance into a postgraduate course, or research program at a tertiary institution in the United Kingdom, a completed Estimated Budget form and two academic references and one personal reference.
Closing Date: June 30th
Funding: Private
Contributor: Rae and Edith Bennett Travelling Scholarship Fund

Sir Arthur Sims Travelling Scholarship
Subjects: All subjects.
Purpose: To enable graduates of Australian universities to undertake postgraduate study or research in United Kingdom.
Eligibility: Open to graduates of Australian universities who can demonstrate outstanding academic merit and promise. Applicants must be born in Australia or have parents who have been Australian residents for 7 years or more.
Level of Study: Postgraduate, Research
Type: Scholarship
Value: Living allowance, tuition fees (approx. $20,000 per year)
Length of Study: 1–3
Country of Study: United Kingdom
Application Procedure: Check website for further the details.
Closing Date: Check with the website.
Funding: Private
Contributor: Sir Arthur Sims Traveling Scholarship Fund

For further information contact:

Website: www.jason.edu

Sir John and Lady Higgins Research Scholarship
Subjects: Industrial chemistry and biochemistry.
Purpose: For the development of the pastoral and agricultural industries.
Eligibility: Open to students undertaking a research higher degree in the fields of industrial chemistry and biochemistry.
Level of Study: Research
Type: Scholarship
Value: Living allowance (Australian $28,000 per year full time) and other benefits as per the APA
Length of Study: 2 years if masters or 3 years if PhD.
Frequency: Every 2 years
Country of Study: Australia
No. of awards offered: 1
Application Procedure: Applicants must indicate on SCHOLS, the online application service, their interest for this scholarship and explain how their research in industrial chemistry and biochemistry relates directly to the study and development of the pastoral and agricultural industries. Check website for more details.
Closing Date: October 31st

Sir Thomas Naghten Fitzgerald Scholarship
Subjects: Medical and health sciences.
Purpose: This scholarship is to support further surgical training in Australia or Overseas.
Eligibility: Open to candidates studied or currently studying at The University of Melbourne.
Level of Study: Postgraduate
Type: Scholarship
Value: Australian $12,000
Length of Study: 1 year

Frequency: Annual
Study Establishment: University of Melbourne
Country of Study: Australia
No. of awards offered: 1
Application Procedure: Check website for further details.
Closing Date: October 31st
Funding: Government

For further information contact:

The University of Melbourne, Parkville, Victoria, 3010, Australia
Email: jyv@unimelb.edu.au
Website: www.mdhs.unimelb.edu.au
Contact: Joan Vosen, Medicine Faculty

Stella Mary Langford Scholarship

Subjects: All subjects.
Eligibility: This scholarship is for study in Australia. Only citizens of Australia or New Zealand or AUS permanent resident can apply.
Level of Study: Unrestricted
Type: Scholarship
Value: Living allowance ($28,000 per year) and other benefits as per the APA
Length of Study: 2 years if masters or 3 years if PhD.
Frequency: Annual
Country of Study: Australia
No. of awards offered: 1
Application Procedure: Check website for further details.
Closing Date: October 31st
Funding: Private
Contributor: A.O. Capell Scholarship Fund, Henry & Louisa Williams Bequest, and Stella Mary Langford Scholarship Fund

Viola Edith Reid Bequest Scholarship

Subjects: Medicine.
Purpose: This scholarship is to support postgraduate study in any discipline of Medicine.
Eligibility: Open for study in Australia. There are no restrictions on citizenship.
Level of Study: Postgraduate, Research
Type: Scholarship
Value: Australia $21,174 maximum per award
Length of Study: 1 year
Frequency: Annual
Study Establishment: University of Melbourne
Country of Study: Australia
No. of awards offered: 1
Application Procedure: Check website for further details.
Closing Date: October 31st
Funding: Government

For further information contact:

The University of Melbourne, Parkville, Victoria, 3010, Australia
Tel: (61) 8344 4019
Fax: (61) 9347 7854
Email: jyv@unimelb.edu.au
Website: www.mdhs.unimelb.edu.au
Contact: Joan Vosen, Medicine Faculty

UNIVERSITY OF MICHIGAN

University of Michigan-Dearborn, Ann Arbor, MI, 48109, United States of America
Tel: (1) 734 764 1817
Fax: (1) 734 647 3081
Email: financial.aid@umich.edu
Website: www.umich.edu

University of Michigan is Internationally renowned for research and education. The University offers a wide variety of degree programmes for undergraduate and graduate students.

Knight-Wallace Fellowship

Subjects: Journalism.
Purpose: To provide financial aid, broaden perspectives, nurture intellectual growth and inspire personal transformation.
Eligibility: Open to full-time journalists with 5 years experience and whose work appears regularly as an employee or freelance in the United States.
Level of Study: Professional development
Type: Fellowship
Value: US$70,000– distributed as $8,750 monthly – from September through April. Stipend details vary for international fellows and are worked out on an individual basis.
Length of Study: 8 months
Frequency: Annual
Country of Study: United States of America
No. of awards offered: 12
Application Procedure: Applicants can download the application form from the website. The completed application form along with a study plan, autobiographical statement and work samples must be submitted.
Closing Date: February 1st
Additional Information: There are no academic prerequisites.

For further information contact:

Knight-Wallace Fellows Wallace House 620 Oxford Road University of Michigan, Ann Arbor, MI, 48104-2635, United States of America
Website: www.kwfellows.org
Contact: Charles R Eisendrath, Director

UNIVERSITY OF MONTANA (UM)

Financial Aid Office, 2nd Floor, Emma B. Lommasson Center, Griz Central, 32 Campus Drive, Missoula, MT 59812, United States of America
Tel: (1) 406 243 0211
Email: dss@umontana.edu
Website: www.umt.edu

University of Montana is a magnet not only for top-notch teachers and researchers, but also for students from across the country and around the globe. Students receive a high-quality, well-rounded education and training for professional careers in the University's three colleges – arts and sciences, forestry and conservation, and technology – and six schools – journalism, law, business, education, pharmacy and the fine arts. A city within a city – with its own eateries, stores, medical facilities, banking and postal services, and zip code – UM has an increasingly diverse population and rich culture. The University has nurtured a tradition of cultural and scientific exploration.

Erasmus Scholarships

Subjects: English, environmental studies, history, law, philosophy, political science, foreign languages, interdisciplinary studies.
Eligibility: Open to the full-time UM students of all nationalities including freshmen, transfer students and graduate students.
Level of Study: Graduate
Type: Scholarships
Value: $500–4,000
Length of Study: 1 year (renewable)
Frequency: Annual
No. of awards offered: 35
Closing Date: February 1st

UNIVERSITY OF NEBRASKA AT OMAHA (UNO)

College of Business Adminstration, University of Nebraska at Omaha, Omaha, 6001 Dodge Street, NE, 68182-0048, United States of America
Tel: (1) 402 554 2800
Fax: (1) 402 554 4036
Email: cba@unomaha.edu/mba
Website: www.mba.unomaha.edu
Contact: Lex Kaczmarek, MBA Program Director

The University of Nebraska at Omaha's (UNO) College of Business Administration offers a dynamic, challenging Master's programme designed to help students acquire the knowledge, perspective and skills necessary for success in the marketplace of today and tomorrow. The goal of the programme is to develop leaders who have the ability to incorporate change, use information technology to resolve problems and make sound business decisions. The curriculum focuses on results with an emphasis on how to excel in a rapidly changing world.

UNO Graduate Assistantships

Subjects: MBA, MACC, MA/ECM and MS/ECM.
Eligibility: Open to qualified students who are enrolled in a graduate degree programme.
Level of Study: Graduate, MBA
Value: A waiver of tuition costs up to 12 hours of graduate credit per semester
Frequency: Every 2 years
Study Establishment: UNO
Country of Study: United States of America
No. of awards offered: 1–3 each semester
Application Procedure: Applicants must make enquiries in their department about the availability of assistantships, the procedures for applying, and the details of when the application and supporting credentials should be on file in the department or school for consideration. Applicants must complete application for graduate assistantships and submit a curriculum vitae and supporting materials.
Closing Date: June 1st
No. of awards given last year: 1–3
No. of applicants last year: 40

UNIVERSITY OF NEVADA, LAS VEGAS (UNLV)

Graduate College, 4505 Maryland Parkway, Box 451010, Las Vegas, NV, 89154-1017, United States of America
Tel: (1) 702 895 3011
Fax: (1) 702 895 4180
Email: gradcollege@unlv.edu
Website: www.unlv.edu
Contact: Administrative Officer

UNLV Alumni Association Graduate Scholarships

Subjects: All subjects.
Purpose: To reward outstanding graduate students.
Eligibility: Applicants must have completed at least 12 credits of graduate study at UNLV, have a minimum undergraduate and graduate grade point average of 3.5 and enrol for six or more graduate credits in each semester of the scholarship year.
Level of Study: MBA, Graduate
Type: Scholarship
Value: US$1,500
Length of Study: 1 year
Frequency: Annual
Study Establishment: UNLV
Country of Study: United States of America
No. of awards offered: 9
Application Procedure: Applicants must telephone (1) 702 895 3320 or write for application forms or further information.
Closing Date: March 3rd

UNLV Graduate Assistantships

Subjects: All subjects.
Purpose: To offer financial assistance and support to students admitted to any graduate degree programme.
Eligibility: Open to students who have already been admitted to any graduate degree programme.
Level of Study: MBA, Graduate
Type: Assistantship
Value: A monthly stipend of $5,000 plus a waiver of all out-of-state tuition and a reduction in tuition fees
Length of Study: 1 year
Frequency: Annual
Study Establishment: UNLV
Country of Study: United States of America
No. of awards offered: Varies
Application Procedure: Applicants must send applications and all supporting materials to the Dean of the Graduate College.
Closing Date: March 1st or November 1st for the Spring assistantship. Applications may be accepted after this date in the event of an unexpected opening for the Autumn semester. On some rare occasions an assistantship is available for the Spring semester
Additional Information: Graduate assistants must carry a minimum of 6 semester hours of credit and are expected to spend 20 hours per week on departmental duties such as instruction or research.

UNLV James F Adams/GPSA Scholarship

Subjects: All subjects.
Purpose: To recognize the academic achievements of graduate students.
Eligibility: Applicants must have completed at least 12 credits of graduate study at UNLV, have a minimum undergraduate and graduate grade point average of 3.5 and enrol for six or more graduate credits in each semester of the scholarship year.
Level of Study: Graduate, MBA
Type: Scholarship
Value: US$1,000
Length of Study: Varies
Frequency: Annual
Study Establishment: UNLV
Country of Study: United States of America
No. of awards offered: 6
Application Procedure: Applicants must telephone (1) 702 895 3320, or write for application forms or further information.
Closing Date: March 3rd

UNIVERSITY OF NEW BRUNSWICK (UNB)

PO Box 4400 Station A, Fredericton, New Brunswick, E3B 5A3, Canada
Tel: (1) 506 453 4666
Fax: (1) 506 453 4599
Email: chantelo@unbsj.ca
Website: www.unb.ca

The University of New Brunswick (UNB) was founded in 1785. It is the oldest, public English language comprehensive university in Canada. The University offers over 60 graduate diploma and degree programmes in the faculties of arts, science, engineering, forestry and environmental management, computer science, kinesiology, nursing, business administration and education.

Dr William S. Lewis Doctoral Fellowships

Subjects: Science and humanities.
Purpose: To support incoming University of New Brunswick Doctoral students who have the potential to be regional, national and international leaders in research and the dissemination of knowledge.
Eligibility: Open to Doctorate students with academic excellence and depending on the candidate's contribution to the field.
Level of Study: Doctorate
Type: Fellowship
Value: Canadian $25,000
Length of Study: Up to 4 years
Frequency: Annual
Country of Study: Canada
No. of awards offered: Normally, one new award per year.
Closing Date: February 1st
Funding: Individuals

UNIVERSITY OF NEW ENGLAND (UNE)

Research Grants Office, Armidale, NSW, 2351, Australia
Tel: (61) 2 6773 3333
Fax: (61) 2 6773 3100
Email: research@une.edu.au
Website: www.une.edu.au

UNE is internationally recognized as one of the best teaching and research universities. Yearly, the university offers students more than $2.5 million in scholarships, prizes, and bursaries and more than $18 million for staff and students involved in research. It provides distance education for the students. Its scholars and scientists have established international reputations through their contributions in areas such as rural science, agricultural economics, educational administration, linguistics and archaeology.

A S Nivison Memorial Scholarship

Subjects: Pasture improvement, animal husbandry, farm management, wool research or promotion, water conservation and environmental protection.
Eligibility: Applicants must be a citizen or a permanent resident of Australia undertaking PhD or Research Masters at the University of England in one of the listed areas.
Level of Study: Doctorate, Postgraduate
Type: Scholarship
Value: $5,000
Length of Study: 1 year
Frequency: Annual
Country of Study: Australia
Application Procedure: Check website for further details.
Closing Date: April 30th
Funding: Private
Contributor: Nivision family

For further information contact:

Australia
Tel: (61) 61 2 6773 3745
Email: pgscholarships@une.edu.au
Contact: Belinda Keogh

CRC Spatial Information PhD Scholarship

Subjects: Agriculture.
Purpose: To produce long-lasting outcomes relating to understanding how complex decision-making processes can be improved using spatial and other data.
Eligibility: Applicants must hold a Class 1 or 2A Honours (or equivalent) Degree in a suitable discipline, and be a citizen or permanent resident of Australia. A valid driver's licence is also a necessary requirement.
Level of Study: Doctorate
Type: Scholarship
Value: Please check website
Frequency: Annual
Study Establishment: University of New England
Country of Study: Australia
Application Procedure: Applicants should send a letter outlining suitability for the position, accompanied by a brief curriculum vitae (including contact details of two referees) and a copy of academic transcripts.
Closing Date: Please check with the website.
Funding: Government

For further information contact:

Centre for Sustainable Farming Systems
Tel: 02 6773 2436
Email: jim.scott@une.edu.au
Contact: Professor Jim Scott

CRDC Postgraduate Scholarship

Subjects: Cotton research.
Purpose: To enhance the environmental, economic and social performance of the Australian cotton industry.
Eligibility: Applicants must be Australian citizens, studying at an Australian university and interested in working in the Australian cotton industry to pursue postgraduate studies relating to the cotton industry or its related community activities.
Level of Study: Postgraduate
Type: Scholarship
Value: $30,000 per year
Length of Study: 3 years
Frequency: Annual
Study Establishment: University of New England
Country of Study: Australia, New Zealand or South Africa
Application Procedure: Check with the website for further details.
Closing Date: End of January
Funding: Government
Contributor: Cotton Research and Development Corporation (CRDC)
Additional Information: Projects may relate to any field of cotton-related research.

For further information contact:

Cotton Research and Development Corporation, 2 Lloyd Street, Narrabri, New South Wales, Australia
Tel: (61) 02 6792 4088
Fax: (61) 02 6792 4400
Email: research@crdc.com.au
Website: www.crdc.com.au

PhD Scholarship in Animal Breeding

Subjects: Animal breeding.
Purpose: To investigate aspects of sow feed intake and its impact on reproductive performance and longevity.
Eligibility: Applicants should be well versed in statistics and/or animal breeding units at a tertiary level and computing and data analysis skills is highly desirable.
Level of Study: Doctorate
Type: Scholarship
Value: Australian $28,000 per year
Length of Study: 3 years
Frequency: Annual
Study Establishment: University of New England
Country of Study: Australia
Application Procedure: Check website une.edu.au/imp/courses/sciences/postgrad.php
Closing Date: June 30th
Contributor: Australian Pork CRC

For further information contact:

Tel: 02 6773 3788
Email: kbunter2@une.edu.au
Contact: Dr Kim Bunter

PhD Scholarship: Molecular Factors in Plant–Microbe Associations

Subjects: Molecular biology.
Purpose: To study the molecular aspect of the interaction between the fungal pathogen and the plant.
Eligibility: Applicants must hold a Class 1 or 2A Honours (or equivalent) Degree in a suitable discipline, and be an citizen or permanent resident of Australia.
Level of Study: Doctorate
Type: Scholarship
Value: $26,000 per year (tax free)
Length of Study: 3 years
Frequency: Annual
Study Establishment: University of New England
Country of Study: Australia
Application Procedure: Applicants should send a letter outlining their suitability for the position, accompanied by a brief curriculum vitae (including contact details of two referees) and a copy of their academic transcripts.
Funding: Government

For further information contact:

Molecular and Cellular Biology, School of Science and Technology
Tel: 02 6773 2708
Fax: 02 6773 3267
Email: lperegge@une.edu.au
Contact: Dr Lily Pereg-Grek

PhD Scholarship: Weed Ecology

Subjects: Ecology.
Purpose: To manage the species through a series of field and controlled environment experiments on emergence, growth, reproduction and spread of environment experiments on emergence, growth, reproduction and spread of fleabane species.
Eligibility: Applicants must hold a Class 1 or 2A Honours (or equivalent) Degree in a suitable discipline, and be an citizen or permanent resident of Australia.

Level of Study: Doctorate
Type: Scholarship
Value: $26,000 per year (tax free)
Length of Study: 3 years
Study Establishment: University of New England
Country of Study: Australia
Application Procedure: Applicants should send a letter outlining their suitability for the position accompanied by a brief curriculum vitae (including contact details of two referees) and a copy of their academic transcripts.
Closing Date: April 27th
Funding: Government

For further information contact:

School of Rural Science and Agriculture, University of New England, Armidale, New South Wales, Armidale, 2351, Australia
Tel: (61) 02 6773 3238
Email: bsindel@une.edu.au
Contact: Brian Sindel, Associate Professor

UNE Mary Dolan Memorial Travelling Scholarship

Subjects: Archaeology.
Purpose: To encourage students to travel to undertake work at an archaeological site as members of a team.
Eligibility: Applicants must currently be enrolled at the University of New England and must have completed at least 2 years of study at either undergraduate or postgraduate level, in disciplines of study ranging from the ancient world through the medieval world and on into the modern historical world.
Level of Study: Postgraduate
Type: Scholarship
Value: Up to $3,000
Length of Study: 1 year
Frequency: Annual
Study Establishment: University of New England
Country of Study: Australia
Application Procedure: Check website, www.prod.une.edu.au/research-services/forms/refereesreport.pdf
Closing Date: October 31st
Funding: Government

University of New England Postgraduate Equity Scholarship

Subjects: Science
Eligibility: Applicants must be citizens of Australia or New Zealand and be a Aboriginal or Torres Strait Islander, non-English speaking background person, student with a disability or woman from non-traditional area.
Level of Study: Postgraduate, Research
Type: Scholarship
Value: Australian $19,231 per year
Length of Study: 2 years (Masters) or 3 years (PhD)
Frequency: Annual
Study Establishment: University of New England
Country of Study: Australia, New Zealand or South Africa
No. of awards offered: This scholarship is offered to an unspecified number of people.
Application Procedure: Check website for further details.
Closing Date: March 22nd
Funding: Government

For further information contact:

University of New England, Armidale, NSW 2351, Australia
Email: aharris@une.edu.au
Website: www.une.edu.au/reseach-services/grants/
Contact: Thea Harris, Scholarships Administrative Assistant

University of New England Research Scholarship

Subjects: Any subject.
Eligibility: Applicants must have achieved Honours 1 or equivalent, or Masters or equivalent.
Level of Study: Doctorate, Postgraduate
Type: Scholarship
Value: Australian $19,231
Length of Study: 3 years (PhD) and 2 years (Masters)
Frequency: Annual
Study Establishment: University of New England
Country of Study: Australia
No. of awards offered: Varies
Application Procedure: Check the website, www.une.edu.au/research-services/grants/ for further details.
Closing Date: End of September or December
Funding: Government

For further information contact:

University of New England, Armidale, NSW 2351, Australia
Tel: (61) 6773 3571
Fax: (61) 6773 3543
Email: aharris@une.edu.au
Website: www.une.edu.au/research-services/grants/
Contact: Thea Harris, Scholarships Administrative Assistant

UNIVERSITY OF NEW MEXICO (UNM)

Department of Computer Science, Mail Stop MS Col 1130, 1 University of New Mexico, Albuquerque, NM, 87131, United States of America
Tel: (1) 505 277 0111
Fax: (1) 505 277 6927
Email: ucam@unm.edu1
Website: www.unm.edu
Contact: The Postgraduate Admissions Office

University of New Mexico (UNM), a Hispanic-Serving Institution, represents a wide cross-section of cultures and backgrounds. It was founded in 1889. UNM boasts outstanding faculty members and includes a Nobel Laureate, MacArthur Fellows and several members of the national academies.

New Mexico Information Technology Fellowships

Subjects: Computer science.
Purpose: To financially support students and strengthen New Mexico's technology base.
Eligibility: Open to candidates who have obtained an undergraduate degree with a grade point average of 3.5.
Level of Study: Doctorate, Postgraduate
Type: Fellowships
Value: US$25,000 per year
Length of Study: 2–4 years
Frequency: Annual
Country of Study: United States of America
Closing Date: February 15th

THE UNIVERSITY OF NEW SOUTH WALES (UNSW)

Scholarships and Financial Support, Sydney, NSW, 2052, Australia
Tel: (61) 02 9385 1000
Fax: (61) 02 938 500706
Email: scholarships@unsw.edu.au
Website: www.unsw.edu.au

University of New South Wales (UNSW) is one of Australia's leading research and teaching universities. UNSW takes great pride in the broad range and high quality of teaching programmes. UNSW's teaching gains strength, vitality and currency both from their research activities and from their international nature.

APAI Scholarship in Metallurgy/Materials

Subjects: Metallurgy, materials engineering.
Purpose: To undertake blast furnace research in collaboration with industry.
Eligibility: This scholarship requires candidates to have achieved Honours 1 or equivalent, or Honours 2a or equivalent.
Level of Study: Doctorate, Postgraduate
Type: Scholarship
Value: Stipend of Australian $25,000 per year
Length of Study: 3–3.5 years
Frequency: Annual

Study Establishment: School of Materials Science and Engineering
No. of awards offered: 7
Application Procedure: Check website for further details.
Closing Date: October 31st

For further information contact:

Email: a.yu@unsw.edu.au
Website: www.materials.unsw.edu.au/
Contact: Professor Aibing Yu, Scientia Professor & Federation Fellow

College of Fine Arts Postgraduate Research Scholarship

Subjects: Fine arts.
Purpose: To provide financial assistance to students pursuing a PhD.
Eligibility: Open to full-time PhD students from any country
Level of Study: Doctorate
Type: Scholarship
Value: Australian $22,500 per year
Length of Study: 3 years
Frequency: Annual
Study Establishment: University of New South Wales
Country of Study: Australia
Application Procedure: Application forms are available at the COFA Student Centre on the UNSW Scholarships website.
Closing Date: January 31st
Additional Information: To assist in the selection process, UNSW may also request applicants to provide, orally or in writing, further information. Applicants can also submit the application to Melanie Cheung at the COFA Student Centre.

For further information contact:

Tel: +61 2 9385 0614
Email: l.mitchell@unsw.edu.au
Contact: Leah Mitchell, Scholarships Coordinator

PhD Scholarship in Materials Science and Engineering

Subjects: Materials science and engineering.
Purpose: To support research on metal dusting.
Eligibility: Open to citizens of Australia or permanent residents holding high Honours Degree in science.
Level of Study: Postgraduate, Research
Type: Scholarship
Value: Australian $25,000–$30,000 per year (tax free).
Length of Study: 3–3.5 years
Frequency: Annual
Study Establishment: University of New South Wales
Country of Study: Australia
No. of awards offered: 3
Application Procedure: Check website for further details.
Closing Date: February 28th

For further information contact:

UNSW, Kensington, NSW, 2052, Australia
Tel: (61) 93854322
Fax: (61) 93855956
Email: d.young@unsw.edu.au
Website: www.materials.unsw.edu.au
Contact: David Young, (Professor) Science/Materials

PhD Scholarships in Environmental Microbiology

Subjects: Environmental microbiology, microbial genomics.
Purpose: To attract the nations strongest candidates capable of pursuing PhD studies in the genomics of environmental microorganisms.
Eligibility: This scholarship is for study in Australia for those who have achieved Honours 1 or equivalent, or Masters or equivalent. There are no restrictions on citizenship.
Level of Study: Doctorate, Research, Postgraduate
Type: Scholarship
Value: Australian $35,000
No. of awards offered: 3
Application Procedure: Check website for further details.
Closing Date: March 31st

For further information contact:

Email: r.cavicchioli@unsw.edu.au
Contact: Rick Cavicchioli

The Senior Artists from Asia College of Fine Arts Research Scholarship

Subjects: Fine arts.
Purpose: To provide financial assistance to full-time Master's (by research) students.
Eligibility: Open only to full-time Master's (by research) students who are citizens of an Asian country and normally resident in Asia.
Level of Study: Postgraduate
Type: Scholarship
Value: Payment of programme fees
Frequency: Annual
Study Establishment: University of New South Wales
Country of Study: Australia
Application Procedure: Application forms are available at the COFA Student Centre or on the UNSW Scholarships website.
Closing Date: October 31st

Vida Rees Scholarship in Pediatrics

Subjects: Medicine.
Purpose: To support Australian students to undertake research in paediatrics.
Eligibility: Open to applicants who are undertaking an Honours project or postgraduate research in paediatrics and selection will be based on academic merit, demonstrated ability and leadership qualities, potential to contribute to the wider life of the university and consideration of financial need.
Level of Study: Research, Postgraduate
Type: Scholarship
Value: Australian $3,000
Length of Study: 1 year
Frequency: Annual
Study Establishment: University of New South Wales
Country of Study: Australia
Closing Date: December 1st–February 18th

Viktoria Marinov Award in Art

Subjects: Creative arts.
Purpose: To financially assist female artists who are proposing to undertake the Master of Art or Master of Fine Arts course.
Eligibility: Open to female artists under the age of 35 years who are proposing to undertake the Master of Art or Master of Fine Arts course.
Level of Study: Postgraduate
Type: Award
Value: Australian $7,500
Length of Study: 1 year
Frequency: Annual
Study Establishment: New South Wales, Sydney City Central and Eastern Suburbs
Country of Study: Australia
No. of awards offered: 2
Closing Date: July 30th

For further information contact:

Email: j.elliot@unsw.edu.au
Contact: Joanna Elliot

THE UNIVERSITY OF NEWCASTLE

Research Division University of Newcastle, Callaghan, NSW 2308, Australia
Tel: (61) 02 4921 5000
Fax: (61) 02 4985 4200
Email: research@newcastle.edu.au
Website: www.newcastle.edu.au/research/rhd/
Contact: Office of Graduate Studies

The University of Newcastle is one of Australia's top ten research universities. The University has over 1,200 research degree candidates enrolled in five faculties, incorporating a wide range of disciplines including architecture, building, humanities, social sciences, education, economics, management, engineering, computer science, law, medicine, nursing, health sciences, music, drama and creative arts, physcial and natural sciences, mathematics and information technology. Scholarships are available to support research degree candidates in most disciplines. About 90 new scholarships are awarded each year.

Australian Government Postgraduate Awards

Subjects: All subjects.
Purpose: To support students undertaking full-time higher research degree programmes.
Eligibility: Applicants must have completed 4 years of full-time undergraduate study and gained a First Class (Honours) Degree or equivalent award.
Level of Study: Doctorate, Postgraduate, Research
Type: Scholarship
Value: Living allowance ($ 28,715 per year full time stipend), a relocation allowance and a thesis allowance.
Length of Study: 2 years full-time study for research Master's candidates or 3 years full-time for PhD candidates
Frequency: Annual
Study Establishment: Any university
Country of Study: Australia
No. of awards offered: Approx. 40
Application Procedure: Applicants must complete an application form from the Office of Graduate Studies or from the website.
Closing Date: December 1st
Funding: Government

Chemical Engineering Scholarship

Subjects: Chemical Engineering.
Purpose: To develop models capable of simulating temporal and spatial characteristics of rainfall fields over large river basins using novel approaches to hierarchical modelling, storm clustering, advection and calibration. The models will provide continuous simulation support for the design and assessment of water-related infrastructure.
Eligibility: Open only to the postgraduates who are the citizens of Australia or the permanent residents of Australia.
Level of Study: Postgraduate
Type: Scholarship
Value: Australian $25,118 (per year)
Length of Study: 2 years (Masters) and 3 years (PhD)
Frequency: Annual
Country of Study: Australia
No. of awards offered: 4
Application Procedure: Application form and the Research Higher Degree prospectus from can be downloaded from the website.
Closing Date: July 1st

For further information contact:

University of Newcastle, Australia
Tel: (61) 02 4921 6038
Email: George.Kuczera@newcastle.edu.au
Website: www.newcastle.edu.au/research/rhd/prospective.html
Contact: Professor George Kuczera

Commercialisation Training Scheme (CTS)

Subjects: All subjects.
Eligibility: Open to the citizens of Australia and New Zealand or its permanent residents who are enrolled in a research higher degree and will also be enrolled in a commercialization training programme or have a research higher degree thesis under examination with only 6 months full-time or part-time equivalent study remaining in the commercialization training programme.
Level of Study: Research
Type: Scholarship
Value: $5,281
Length of Study: 28 weeks
Application Procedure: Check website for further details.
Additional Information: Please check website for latest updates.

Endeavour International Postgraduate Research Scholarship

Subjects: All subjects.
Eligibility: Open to the residents of any country except Australia and New Zealand who have achieved Honours 1 or equivalent and meet the minimum English Language proficiency level.
Level of Study: Postgraduate, Research
Type: Scholarship
Value: Living allowance ($20,427 per year, last year), a relocation allowance and a thesis allowance
Length of Study: 2 years (Masters) and 3 years (PhD)
Frequency: Annual
Country of Study: Australia
No. of awards offered: 8
Application Procedure: Apply directly to the university. Applications can be downloaded from the website. For further details please visit www.jason.edu.au/
Closing Date: August 31st
Additional Information: Please check at www.newcastle.edu.au/research/rhd/scholarships.html for more details.

The Gowrie Trust Fund Research Scholarships

Subjects: All subjects.
Eligibility: Open to the members and children of the members of the Forces including, grandchildren or other lineal descendants of such members and to the residents of Australia and Australian soldiers who are the graduates of Australian universities or to others who have completed a course of tertiary education at other recognized institutions in Australia.
Level of Study: Research
Type: Research scholarship
Value: $4,000 per year
Length of Study: 2 years
Frequency: Annual
No. of awards offered: 1–2
Application Procedure: Check website for further details.
Closing Date: October 31st
Funding: Trusts
Contributor: The Gowrie Trust Fund

For further information contact:

The Gowrie Scholarship Trust Fund, 3/32 Beaconsfield Rd, Mosman, NSW, 2088, Australia
Tel: (61) 02 99603458
Contact: The Secretary

Palliative Care in Cancer Scholarship

Subjects: Health Research and Psycho-oncology.
Purpose: To undertake a PhD in behavioural cancer research.
Eligibility: Open to the candidates who have achieved Honours 1 or equivalent.
Level of Study: Postgraduate
Type: Scholarship
Value: Australian $19,616 per year
Length of Study: 3 years
Frequency: Annual
Country of Study: Australia
No. of awards offered: 1
Application Procedure: Apply direct to faculty. Check website for more details.
Closing Date: December 31st

For further information contact:

Website: www.newcastle.edu.au

PhD Scholarship in Coal Utilization in Thermal and Coking Applications

Subjects: Chemical engineering, mechanical engineering, and chemistry.
Purpose: To undertake research on particular coal properties which determine its utilization potential.

Eligibility: Open to the engineering and the science graduates who have an Honours Degree in chemical or mechanical engineering or chemistry.
Level of Study: Doctorate, Postgraduate
Type: Scholarship
Value: Australian $19,616 per year
Length of Study: 3 years
No. of awards offered: 1
Application Procedure: Check website for further details.
Closing Date: March 1st

For further information contact:

University of Newcastle
Tel: 61 2 49 21 6179
Email: Terry.Wall@newcastle.edu.au
Contact: Professor Terry Wall

Postgraduate Research Scholarship in Physics

Subjects: Physics.
Eligibility: Open to the residents of Australia and New Zealand or the permanent residents of Australia who have an Honours 1 or 2A or a Masters Degree.
Level of Study: Research, Postgraduate
Type: Scholarship
Value: $17,071 per year
Length of Study: 3 years
Frequency: Annual
No. of awards offered: 3
Application Procedure: Check website for more details.
Closing Date: July 1st
Contributor: ARC Discovery-projects

For further information contact:

School of Mathematical and Physical Sciences, Callaghan, NSW 2308, Australia
Tel: (61) 2 4921 6653
Fax: (61) 2 4921 6907
Email: vicki.keast@newcastle.edu.au
Website: www.newcastle.edu.au
Contact: Dr Vicki Keast

Scholarship in Cancer Prevention

Subjects: Cancer prevention.
Eligibility: Open to the candidates who have achieved Honours 1 or equivalent.
Level of Study: Research, Postgraduate
Type: Scholarship
Value: Australian $19,616 per year
Length of Study: 3 years
Frequency: Annual
Country of Study: Australia
No. of awards offered: 1
Application Procedure: Apply direct to faculty. Check website for further details.
Closing Date: December 31st

For further information contact:

Website: www.newcastle.edu.au

Space-Time Model-Rainfall Fields Scholarship

Subjects: Environmental engineering.
Purpose: Development and implementation of computer models for hydrologic applications.
Eligibility: Open to the candidates who have achieved Honours 1 or equivalent.
Level of Study: Research, Postgraduate
Type: Scholarship
Value: Australian $25,118 per year
Length of Study: 2 years if masters or 3 years if PhD
Frequency: Annual
Country of Study: Australia
No. of awards offered: 4
Application Procedure: Apply directly to the scholarship provider. Check website for further details.
Closing Date: March 1st

For further information contact:

Website: www.newcastle.edu.au

University of Newcastle International Postrgraduate Research Scholarship

Subjects: All subjects.
Eligibility: Open to the candidates from any country except Australia and New Zealand who have achieved Honours 1 or equivalent.
Level of Study: Postgraduate
Type: Scholarship
Value: Exemption for tuition fees, living allowance ($22,860 per year), payment of Overseas Students Health Cover
Length of Study: 2 years (Masters) or 3 years (PhD)
Frequency: Annual
Country of Study: Australia
No. of awards offered: 25
Application Procedure: Check website for further details.
Closing Date: August 31st

For further information contact:

Website: www.Jason.edu.au

University of Newcastle Postgraduate Research Scholarship (UNRS Central)

Subjects: All subjects.
Eligibility: Open to the residents of Australia and New Zealand or permanent residents who have achieved Honours 1 or equivalent and have completed at least 4 years of undergraduate study.
Level of Study: Postgraduate, Research
Type: Scholarship
Value: $23,728 per year full time stipend, $12,898 part time stipend
Length of Study: 2 years (Masters) and 3 years (PhD)
Frequency: Annual
Country of Study: Australia
No. of awards offered: 30
Application Procedure: Check website for further details.
Closing Date: October 31st

For further information contact:

Research Higher Degrees, The Chancellery Eastern Wing, University Drive, Callaghan, NSW 2308, Australia
Tel: (61) 02 4921 6537
Fax: (61) 02 4921 6908
Email: research@newcastle.edu.au
Website: www.newcastle.edu.au

UNIVERSITY OF NOTRE DAME: COLLEGE OF ARTS AND LETTERS

Office of the Dean, 100 O'Shaughnessy Hall, Notre Dame, IN, 46556, United States of America
Tel: (1) 574 631 7085
Fax: (1) 574 631 7743
Email: aldean@nd.edu
Website: http://al.nd.edu

University of Notre Dame: College of Arts and Letters offers one of the finest liberal arts educations in the nation. Its Division of the Humanities was recently ranked 12th among private universities, while the social sciences continue their ascent in the national rankings. College of Arts and Letters is the largest and oldest of the University's 4 colleges.

The Erskine A. Peters Dissertation Year Fellowship at Notre Dame

Subjects: Arts, humanities, social sciences and theological disciplines.
Purpose: To provide an opportunity for African American scholars at the beginning of their academic careers to experience life at a major Catholic research university.

Eligibility: Open to African-American Doctoral candidates who have completed all degree requirements with the exception of the dissertation.
Level of Study: Postgraduate, Research
Type: Fellowship
Value: US$30,000 stipend and US$2,000 research budget
Length of Study: 10 months
Frequency: Annual
Country of Study: United States of America
Application Procedure: Applicants may apply online.
Closing Date: November each year
No. of awards given last year: 3

For further information contact:

Department of Africana Studies
Tel: 574 631 5628
Fax: 574 631 3587
Email: astudies@nd.edu
Website: http://africana.nd.edu

THE UNIVERSITY OF NOTTINGHAM

Graduate School, University Park, Nottingham, Nottinghamshire, NG7 2RD, England
Tel: (44) 115 846 8400
Fax: (44) 115 846 7799
Email: graduate-school@nottingham.ac.uk
Website: www.nottingham.ac.uk/gradschool
Contact: Ms Claire Palmer, PG Funding Manager

The University of Nottingham is a community of students and staff dedicated to bringing out the best in all of its members. It aims to provide the finest possible environment for teaching, learning and research and has a well-known record of success.

University of Nottingham Doctoral Training Awards

Subjects: All subjects offered by the University.
Purpose: To promote research.
Eligibility: Open to graduates of all nationalities.
Level of Study: Postgraduate, Predoctorate
Type: PhD scholarship
Value: Minimum of UK£13,590 maintenance per year where appropriate, plus payment of fees at the home or European Union rate
Length of Study: 3–4 years leading to PhD submission given adequate academic progress
Frequency: Annual
Study Establishment: The University of Nottingham
Country of Study: United Kingdom
No. of awards offered: 75+
Application Procedure: Applicants must contact the individual schools for information. Applicants must apply to the school where they intend to study.
Closing Date: Please contact the individual schools for information
Contributor: University of Nottingham
No. of awards given last year: 75+
Additional Information: The scholarships are awarded internally to the schools and/or faculties. It is then up to those schools receiving awards to advertise the scholarship and set an application deadline.

University of Nottingham Weston Scholarships

Subjects: All subjects from a prescribed list.
Purpose: To provide promising students with full-time home or European Union fees.
Eligibility: There are no eligibility restrictions.
Level of Study: Postgraduate, Predoctorate
Type: Studentship
Value: Part Home or European Union fees only
Length of Study: Usually 1 year full-time
Frequency: Annual
Study Establishment: The University of Nottingham
Country of Study: United Kingdom
No. of awards offered: 4
Application Procedure: Applicants must submit applications to the individual schools and contact them for details. Applicants must apply to the school where they intend to study. Information about participating schools is at: www.nottingham.ac.uk/graduateschool/westonscholarships.
Closing Date: Please contact the individual schools for information
Funding: Trusts
No. of awards given last year: 4
Additional Information: The scholarships are awarded internally to four schools on a competitive basis. It is then up to those schools receiving awards to advertise the scholarship and set an application deadline.

THE UNIVERSITY OF OKLAHOMA SCHOOL OF ART AND ART HISTORY

520 Parrington Oval, FJC Room 202, Norman, Oklahoma, 73019, United States of America
Tel: (1) 405 325 2691
Fax: (1) 405 325 1668
Email: info@art.ou.edu/ art@ou.edu
Website: http://art.ou.edu

The School of Art is the largest, most comprehensive art school in Oklahoma, which provides excellent professional education and a focus for the study of visual arts. Additionally, the school is dedicated to promoting, pursuing and supporting creative activity and scholarly research in the visual arts.

Ben Barnett Scholarship

Subjects: Arts.
Eligibility: Any full-time Art majors admitted to either the MA or MFA degree program.
Level of Study: Postgraduate
Type: Scholarship
Value: $200–1000 per semester
Frequency: Annual
Study Establishment: School of Art, University of Oklahoma
Country of Study: United States of America
No. of awards offered: 4–10
Application Procedure: See the website.
Closing Date: March 1st

Francis Weitzenhoffer Memorial Fellowship

Subjects: History of art.
Eligibility: Full-time Art History students enrolled in the MA or PhD program.
Level of Study: Postgraduate
Type: Fellowship
Value: US$8,000
Length of Study: 1 year
Frequency: Annual
Study Establishment: School of Arts, University of Oklahoma
Country of Study: United States of America
No. of awards offered: 1
Application Procedure: See the website.
Closing Date: March 1st

Glenis Horn Scholarship

Subjects: Figurative Sculpture.
Purpose: To support art students who are majoring in Sculpture.
Eligibility: School of Art students majoring in Sculpture with an emphasis in figurative sculpture. Must be making satisfactory progress and be of high moral character.
Level of Study: Professional development
Type: Scholarship
Value: $2,500
Frequency: Annual
Study Establishment: School of Art, University of Oklahoma
Country of Study: United States of America
No. of awards offered: 1
Application Procedure: The School of Art Scholarship application may be downloaded from the website.
Closing Date: March 1st

Kim and Paul Moore Scholarship

Subjects: Sculpture.

Eligibility: Full-time Art majors admitted to the studio program with an emphasis in figurative sculpture; based on GPA and portfolio.
Level of Study: Graduate, Undergraduate
Type: Scholarship
Value: Varies
Frequency: Annual
Study Establishment: School of Art, University of Oklahoma
Country of Study: United States of America
No. of awards offered: Varies
Application Procedure: Contact the scholarship office.
Closing Date: March 1st

Selma Naifeh Memorial Scholarship

Subjects: Painting
Eligibility: Student must have State of Oklahoma residency majoring in Painting, must be in good academic standing, maintaining a GPA of at least 3.0 and enrolled full-time.
Level of Study: Graduate, Undergraduate
Type: Scholarship
Value: $2,500
Frequency: Annual
Study Establishment: School of Art, University of Oklahoma
Country of Study: United States of America
No. of awards offered: 1
Application Procedure: A completed application form must be sent to the scholarship office.
Closing Date: March 1st

UNIVERSITY OF OTAGO

Doctoral and Scholarships Office, PO Box 56, Dunedin, New Zealand
Tel: (64) 3 479 1100 ext 5291
Fax: (64) 3 479 5650
Email: university@otago.ac.
Website: www.otago.ac.nz
Contact: Mr Mel Adams, Scholarship Administrator

The University of Otago has over 17,000 students, most of whom are based at the Dunedin campus, which is the oldest campus in New Zealand. The University has four divisions: the Division of Commerce (School of Business), the Division of Health Sciences, the Division of Humanities and the Division of Science. The University has a School of Medicine in Christchurch and Wellington, and a campus in Auckland.

University of Otago Coursework Master's Scholarship

Subjects: All subjects.
Purpose: To fund course work-based Master's students studying at the University of Otago.
Eligibility: Open to applicants of any country but must be primarily resident in New Zealand during study.
Level of Study: MBA, Postgraduate, Coursework Master's degrees.
Type: Scholarship
Value: New Zealand $10,000 towards fees in the first instance
Length of Study: 1 year
Frequency: Annual
Study Establishment: University of Otago
Country of Study: New Zealand
No. of awards offered: 20
Application Procedure: Please visit www.otago.ac.nz/applynow for further details.
Closing Date: Applicants can apply anytime.
Additional Information: Source of funding: University.

University of Otago Doctoral Scholarships

Subjects: All subjects.
Purpose: To fund research towards a PhD degree at the University of Otago.
Eligibility: Open to applicants of any country but must be primarily resident in New Zealand during study.
Level of Study: Doctorate, Research
Type: Scholarship
Value: New Zealand $25,000 plus fees (excluding insurance and student services fees)
Length of Study: 3 years
Frequency: Annual
Study Establishment: The University of Otago
Country of Study: New Zealand
No. of awards offered: 180
Application Procedure: Please visit www.otago.ac.nz/applynow for further details.
Closing Date: Applicant can apply anytime.
No. of awards given last year: 180

University of Otago International Master's Scholarship

Subjects: All subjects.
Purpose: To assist international students in their master's thesis year of studies at the University of Otago.
Eligibility: Open to all international applicants intending to study at the University of Otago who would normally be charged international fees.
Level of Study: Postgraduate, Research
Type: Scholarship
Value: New Zealand $13,000, International tuition fees (excluding insurance and student services fees)
Length of Study: 1 year for their Master's thesis study.
Frequency: Annual
Study Establishment: The University of Otago
Country of Study: New Zealand
No. of awards offered: 4
Application Procedure: Please visit www.otago.ac.nz/applynow for further details.
Closing Date: Applicants can apply anytime
Contributor: The University of Otago
No. of awards given last year: 4 Master's
Additional Information: Source of funding is University.

University of Otago Master's Scholarships

Subjects: All subjects.
Purpose: To fund the thesis year of a Master's degree at the University of Otago.
Eligibility: Open to permanent residents or citizens of New Zealand or Australia, who are entering the thesis year of a Master's degree.
Level of Study: Postgraduate, Research
Type: Scholarship
Value: New Zealand $13,000 per year plus domestic fees (excluding student services fees)
Length of Study: 1 year
Frequency: Annual
Study Establishment: The University of Otago
Country of Study: New Zealand
No. of awards offered: 60
Application Procedure: Please visit www.otago.ac.nz/applynow for further details.
Closing Date: Applicants can apply anytime
No. of awards given last year: 60
Additional Information: Source of funding: University.

UNIVERSITY OF OXFORD

University Offices, Wellington Square, Oxford, Oxfordshire, OX1 2JD, England
Tel: (44) 18 6527 0000
Fax: (44) 18 6527 0708
Email: ben.nicholson@admin.ox.ac.uk
Website: www.ox.ac.uk
Contact: Mrs Ben Nicholas, Graduate Funding Administrator

African Studies: AHRC Block Grant Partnership Studentships: Research Master's Scheme

Subjects: African studies.
Eligibility: Open to UK applicants for African Studies. Other EU nationals are eligible for a fees-only award.
Level of Study: Graduate
Type: Studentship
Value: University fee, college fee and full living expenses
Length of Study: 1 year
Application Procedure: Please visit website for more details and how to apply.
Closing Date: January 18th

Additional Information: Please check at http://www.humanities.ox.ac.uk/prospective_students/graduates/funding/ahrc for more information.

African Studies: Ioma Evans Pritchard

Subjects: African studies.
Eligibility: Open to all applicants for MSc African Studies.
Level of Study: Graduate
Value: University fee, college fee and full living expenses
Length of Study: 1 year
Application Procedure: Please visit website for more details and how to apply.
Closing Date: January 18th
Additional Information: Please check at http://www.africanstudies.ox.ac.uk/ for more information.

African Studies: ORISHA

Subjects: MSc African studies.
Purpose: To assist graduate students with fees and travel expenses.
Eligibility: Open to all applicants for MSc African Studies.
Level of Study: Postgraduate
Type: Grant
Value: University fee, college fee and full living expenses
Length of Study: 1 year
Application Procedure: Please visit website for more details and how to apply.
Closing Date: January 20th
Additional Information: Please check website http://www.africanstudies.ox.ac.uk/ for further information.

Archaeology School: AHRC Block Grant Partnership Studentships: Doctoral Scheme

Subjects: DPhil politics.
Eligibility: Open to UK applicants for DPhil in Archaeology. Other EU nationals are eligible for a fees-only award.
Level of Study: Postgraduate, Research
Type: Studentship
Value: University fee, college fee and full living expenses
Length of Study: Up to 3 years
Application Procedure: Please visit website for more details and how to apply.
Closing Date: January 18th
Additional Information: Please check at http://www.humanities.ox.ac.uk/prospective_students/graduates/funding/ahrc for more details.

Archaeology School: AHRC Block Grant Partnership Studentships: Research Master's Scheme

Subjects: Various arts and humanities degrees.
Eligibility: Open to UK applicants to MSt and Mphil courses offered by the Departement of Archaeology. Other EU nationals are eligible for a fees-only.
Level of Study: Postgraduate, Research
Type: Studentship
Value: University fee, college fee and full living expenses
Length of Study: 1 years
Application Procedure: Applicants must visit the university website for application procedure.
Closing Date: January 18th
Additional Information: Please check at http://www.humanities.ox.ac.uk/prospective_students/graduates/funding/ahrc for more details.

Area Studies: China Centre Postgraduate Scholarship

Subjects: MSc Modern Chinese studies.
Purpose: To assist graduate students with fees and a living allowance.
Eligibility: Open to all graduate applicants to the MSc in Modern Chinses Studies.
Level of Study: Postgraduate
Type: Scholarship
Value: £3,300
Length of Study: 1 year
Frequency: Every 2 years
Application Procedure: Please visit website for more details and how to apply.
Closing Date: March 13th
Additional Information: Please check at http://www.ccsp.ox.ac.uk/prospective_students/msc_in_modern_chinese_studies/funding for more details.

Area Studies: Contemporary China Studies Departmental Scholarship

Subjects: MSc Modern Chinese studies.
Eligibility: Open to all graduate applicants to the MSc in Modern Chinses Studies.
Level of Study: Postgraduate
Type: Scholarship
Value: £3,300
Length of Study: 1 year
Application Procedure: Please visit website for more details and how to apply.
Additional Information: Please check at http://www.ccsp.ox.ac.uk/prospective_students/msc_in_modern_chinese_studies/funding for more details.

Area Studies: Contemporary India Departmental Award

Subjects: MSc Contemporary India Studies.
Eligibility: Open to all graduate applicants to the MSc in Contemporary India Studies.
Level of Study: Postgraduate
Type: Award
Value: £5,000 towards university fee
Length of Study: 1 year
Application Procedure: Please visit website for more details and how to apply.
Additional Information: Please check at http://www.southasia.ox.ac.uk for more details.

Area Studies: the Pran Nath Bahl memorial Scholarship

Subjects: MSc contemporay India.
Eligibility: Open to all graduate applicants to the MSc in Contemporary India Studies.
Level of Study: Postgraduate
Type: Scholarship
Value: £25,000 towards university fee and college fee
Length of Study: 1 year
Frequency: Annual
Application Procedure: Please visit website for more details and how to apply.
Additional Information: Please check at http://www.southasia.ox.ac.uk/prospective_students/msc_in_contemporary_india/funding more information.

Atmospheric, Oceanic & Planetary Physics: NERC & STFC-funded Doctoral Training Awards

Subjects: Atmospheric, oceanic & planetary physics.
Eligibility: Open to UK applicants for Atmospheric, Oceanic & Planetary Physics. Other EU nationals are eligible for a fees-only award.
Type: Training award
Value: University fee, college fee and full living expenses. Fees-only awards for non-UK, EU students.
Length of Study: Period of fee liability
Application Procedure: Please visit website for more information and how to apply.
Closing Date: January 20th
Additional Information: Please check at http://www2.physics.ox.ac.uk/study-here/postgraduates/atmospheric-oceanic-and-planetary-physics for more information.

Balliol College: A. G. Leventis Scholarship

Subjects: Classical Greek Studies.
Eligibility: Open to all graduate applicants in Classical Greek Studies.
Type: Scholarship
Value: Up to £6,000 per year
Length of Study: Up to 3 years
Application Procedure: Please see website for more details and how to apply.
Closing Date: January 18th

Additional Information: Please check at http://www.balliol.ox.ac.uk/graduate-admissions/scholarships for more information.

Balliol College: Balliol College Scholarships

Eligibility: Open to all graduate applicants from the EU with Research Council funding.
Level of Study: Graduate
Type: Scholarship
Value: University fee, college fee and full living expenses. Fees-only awards for non-UK, EU students
Length of Study: Up to 3 years
Application Procedure: Please see website for more details and how to apply.
Closing Date: January 18th
Additional Information: Please check at http://www.balliol.ox.ac.uk/graduate-admissions/scholarships for more information.

Balliol College: Eddie Dinshaw Scholarship

Subjects: Engineering, mathematics, economics, history, law, and the physical and biological sciences. (This includes biochemistry but excludes integrated immunology and the DPhil in psychiatry).
Eligibility: Open to all graduate applicants from India for relevant areas of study.
Level of Study: Postgraduate, Research
Type: Scholarship
Value: Full living expenses
Length of Study: Up to 3 years
Frequency: Annual
Application Procedure: Please see webiste for more details and how to apply.
Closing Date: January 18th
Additional Information: Eligible to nationals of India. Please check at http://www.balliol.ox.ac.uk/graduate-admissions/scholarships for more information.

For further information contact:

Email: graduate.admissions@balliol.ox.ac.uk
Contact: Tutor for Graduate Admissions

Balliol College: Foley-Bejar Scholarships

Eligibility: Open to all graduate applicants who were born in or who have one parent born in Mexico, Spain, or the Republic of Ireland, or who have a strong connection with Northern Ireland.
Type: Scholarship
Value: Full living expenses
Length of Study: Up to 3 years
Application Procedure: Please see website for more details and how to apply.
Closing Date: January 18th
Additional Information: Please check at http://www.balliol.ox.ac.uk/graduate-admissions/scholarships for more information.

Balliol College: Gregory Kulkes Scholarships in Law

Subjects: Law.
Purpose: To assist new students pursue studies in law at Balliol colllege.
Eligibility: Open to all graduate applicants in Law.
Level of Study: Postgraduate, Research
Type: Scholarship
Value: Up to £2,500 per year
Length of Study: Up to 3 years
Frequency: Annual
Study Establishment: Balliol College, University of Oxford
Country of Study: United Kingdom
No. of awards offered: Up to 2
Application Procedure: Please see website for more details and how to apply.
Closing Date: January 18th
Funding: Private
Contributor: Balliol College and the Clarendon Fund Award
Additional Information: Can be held in conjunction with the University Clarendon Awards.

For further information contact:

Tutor for Graduate Admissions, Balliol College, University of Oxford, Oxford, OX1 3BJ, England
Website: http://www.balliol.ox.ac.uk/graduate-admissions/scholarships

Balliol College: Hakeem Belo-Osagie Scholarship

Subjects: All subjects.
Purpose: To support graduate studies.
Eligibility: Open to students from any African country, with no subject restrictions.
Type: Scholarship
Value: Up to £9,000 per year
Length of Study: Up to 3 years
Frequency: Annual
Closing Date: January 20th
Additional Information: May be awarded in conjunction with Clarendon Fund or other award or scholarship. Please check at http://www.balliol.ox.ac.uk/graduate-admissions/scholarships for more information.

Balliol College: Jason Hu Scholarship

Eligibility: Open to all graduate applicants from Asia with preference for candidates from Taiwan and China.
Level of Study: Postgraduate, Research
Type: Scholarship
Value: £10,000 per year
Length of Study: Up to 3 years
Frequency: Annual
No. of awards offered: 5
Application Procedure: Please see website for more details and how to apply.
Closing Date: January 18th
Additional Information: It is awarded in conjunction with Clarendon Fund or other award or scholarship. Four scholarships can only be offered to students from Taiwan and China as a whole, one can be offered to candidates from any Asian country.

For further information contact:

Email: graduate.admissions@balliol.ox.ac.uk
Website: http://www.balliol.ox.ac.uk/graduate-admissions/scholarships
Contact: Tutor for Graduate Admissions

Balliol College: Marvin Bower Scholarship

Subjects: Management and business studies (except MBA), economics, international relations and mathematics.
Purpose: To assist new students pursue studies in law at Balliol College.
Eligibility: Open to all graduate applicants in relevant subject areas.
Level of Study: Postgraduate, Research
Type: Scholarship
Value: Up to £7,000 per year
Length of Study: Up to 3 years
Frequency: Dependent on funds available
Study Establishment: Balliol College, University of Oxford
Country of Study: United Kingdom
No. of awards offered: 1
Application Procedure: Please see website for more details and how to apply.
Closing Date: January 18th
Contributor: Balliol College and the Clarendon Fund Award
Additional Information: Can be held in conjunction with the University Clarendon Awards.

For further information contact:

Tutor for Graduate Admissions, Balliol College, University of Oxford, Oxford, OX1 3BJ, England
Website: http://www.balliol.ox.ac.uk/graduate-admissions/scholarships

Balliol College: Peter Storey Scholarship

Subjects: History.
Eligibility: Open to all graduate applicants to History from the UK with Research Council funding.

Level of Study: Postgraduate, Research
Type: Scholarship
Value: University fee, college fee and full living expenses. Fees-only awards for non-UK, EU students
Length of Study: Up to 3 years
Application Procedure: Please see website for more details and how to apply.
Closing Date: January 18th
Additional Information: Scholarship will be awarded to a candidate with AHRC Funding.

For further information contact:

Email: graduate.admissions@balliol.ox.ac.uk
Website: http://www.balliol.ox.ac.uk/graduate-admissions/scholarships
Contact: Tutor for Graduate Admissions

Balliol College: Phizackerley Senior Scholarships

Subjects: Phizackerley - medical sciences or closely allied sciences.
Eligibility: Open to UK graduates currently working in Oxford who are reading, or intend to read, for a DPhil in the Medical Sciences.
Type: Scholarship
Value: £1,750 per year plus additional benefits including dining rights
Length of Study: 2 years
Application Procedure: Please see website for more details and how to apply.
Closing Date: January 31st
Additional Information: Please check at http://www.balliol.ox.ac.uk/current-members for more information.

Balliol College: Singapore Law Scholarship

Subjects: Law.
Eligibility: Open to all graduate applicants in Law (preferably BCL) from Singapore.
Type: Scholarship
Value: Up to £5,000 per year
Length of Study: Up to 3 years
Application Procedure: Please see website for more details and how to apply.
Closing Date: January 18th
Additional Information: Please check at http://www.balliol.ox.ac.uk/graduate-admissions/scholarships for more information.

Blackfriars Scholarship

Subjects: Theology, philosophy and social sciences.
Eligibility: Open to all graduate applicants for theology, philosophy and social sciences.
Level of Study: Graduate, Undergraduate
Type: Scholarship
Value: £2,500
Length of Study: 1 year
Closing Date: June 30th
Additional Information: Please check at http://www.bfriars.ox.ac.uk/hall_intro.php for more information.

Business School (Saïd): Dean's Latin America and Africa Scholarship

Subjects: MBA.
Purpose: To support an exceptional MBA student from Latin America or Africa.
Eligibility: Open to all applicants for the MBA that are permanent residents of Latin America and Africa.
Level of Study: MBA, Postgraduate
Type: Scholarship
Value: £10,000
Length of Study: 1 year
Application Procedure: Please visit website for more details and how to apply.
Additional Information: Please check at http://www.sbs.ox.ac.uk/degrees/mba/scholarships/Pages/Deansscholarship.aspx for more information.

Business School (Saïd): EU Scholarship

Subjects: Master of Financial Engineering.
Eligibility: Open to all EU applicants for the MFE.
Level of Study: Postgraduate
Type: Scholarship
Value: £14,500
Length of Study: 1 year
Application Procedure: Please visit website for more details and how to apply.
Additional Information: Please check at http://www.sbs.ox.ac.uk/degrees/mba/scholarships/Pages/SBS.aspx for more details.

Business School (Saïd): OBA Australia Boston Consulting Group Scholarship

Subjects: Master of Business Administration and Master of Financial Engineering.
Purpose: To undertake a postgraduate business degree at the Saïd Business School.
Eligibility: Open to all applicants for the MBA and MFE that are permanent residents of Australia and New Zealand.
Level of Study: MBA, Postgraduate
Type: Scholarship
Value: Australian $40,000
Length of Study: 1 year
Application Procedure: Please visit website for more details and how to apply.
Additional Information: Please check at http://www.sbs.ox.ac.uk/degrees/mba/scholarships/Pages/OBAAustralia.aspx for more information.

Business School (Saïd): Skoll Scholarship

Subjects: Master of Business Administration.
Purpose: To give social entrepreneurs the knowledge, skills and networks they need to turn ideas into reality; and deepen their conviction for doing so.
Eligibility: Open to all applicants for the MBA.
Level of Study: MBA, Postgraduate
Type: Scholarship
Value: University fee, college fee and full living expenses
Length of Study: 1 year
Application Procedure: Please visit website for more details and how to apply.
Additional Information: Please contact department or check at http://www.sbs.ox.ac.uk/degrees/mba/scholarships/Pages/SkollScholarships.aspx for more information.

For further information contact:

Website: www.sbs.ox.ac.uk/MBA/Fees/Scholarships.htm

Chevening Scholarships

Subjects: All subjects.
Eligibility: Eligibility varies by country. Check website.
Level of Study: Postgraduate
Type: Scholarship
Value: University fee, college fee, and full living expenses
Length of Study: 1 year
Application Procedure: Please see website for more details and how to apply.
Additional Information: Please check at http://www.fco.gov.uk/en/about-us/what-we-do/scholarships/chevening/how-to-apply/ for more details.

China Oxford Scholarship Fund

Subjects: All subjects.
Eligibility: Open to graduate applicants from People's Republic of China (including Hong Kong).
Type: Funding support
Value: Various awards of differing value
Length of Study: 1 year
Application Procedure: Please see website for more details and how to apply.
Closing Date: April 15th
Additional Information: Please check at http://www.ox.ac.uk/feesandfunding/prospectivegrad/scholarships/university/chinaoxford/ for more details.

Chinese Scholarship Council-University of Oxford Scholarships

Eligibility: Open to graduate applicants that have graduated from one of the 211 Project universities and demonstrate both academic excellence and leadership potential.
Level of Study: Graduate
Type: Scholarship
Value: University fee, college fee, and full living expenses. Additional expenses for flights may be included
Length of Study: Up to 4 years coterminous with fee liability
Application Procedure: Please see website for more details and how to apply.
Closing Date: March 20th
Additional Information: Please check at http://www.ox.ac.uk/feesandfunding/prospectivegrad/scholarships/university/cme/ for more information.

Christ Church: Myers Scholarship for Australian Graduate Students

Subjects: All subjects.
Purpose: Established by Christ Church alumnus Allan Myers to enable Australian graduate students to pursue postgraduate studies at Christ Church.
Eligibility: Open to all graduate applicants in Law.
Level of Study: Postgraduate, Research
Type: Scholarship
Value: Up to £13,750 towards tuition and college fees
Length of Study: 1 year
Frequency: Annual
Study Establishment: Christ Church, University of Oxford
Country of Study: United Kingdom
No. of awards offered: 1
Application Procedure: Please see website for details of how to apply.
Closing Date: May 1st
Funding: Private
Additional Information: Preference given to students studying Law. Please check at http://www.chch.ox.ac.uk/admissions/graduate for more details.

For further information contact:

Tutor for Graduates' Secretary, Christ Church, Oxford, Oxfordshire, OX1 1DP, England
Website: www.chch.ox.ac.uk

Clarendon Fund Scholarships

Subjects: All subjects.
Purpose: To enable academically outstanding students to take up their places at Oxford.
Eligibility: Open to all graduate applicants. Consideration will be automatc; no application is required.
Level of Study: Research, Postgraduate
Type: Scholarships
Value: University fee, college fee, and full living expenses
Length of Study: Period of fee liability
No. of awards offered: 120
Application Procedure: Please see website for more details and how to apply.
Closing Date: January 18th
Additional Information: Please check at www.clarendon.ox.ac.uk/about for more information.

Classics Faculty: AHRC Block Grant Partnership Studentships: Doctoral Scheme

Eligibility: Open to UK and EU applicants for the DPhil degrees offered by the Faculty of Classics.
Level of Study: Graduate
Type: Studentship
Value: University fee, college fee, and full living expenses. Fees-only awards for non-UK, EU students
Length of Study: Period of fee liability
Application Procedure: Apply on the University application form.
Closing Date: January 18th
Additional Information: Please check at http://www.humanities.ox.ac.uk/prospective_students/graduates/funding/ahrc for more information.

Classics Faculty: AHRC Block Grant Partnership Studentships: Research Master's Scheme

Eligibility: Open to UK and EU applicants for the MSt and MPhil degrees offered by the Faculty of Classics.
Level of Study: Graduate
Type: Studentship
Value: University fee, college fee, and full living expenses. Fees-only awards for non-UK, EU students
Length of Study: Period of fee liability
Application Procedure: Apply on the University application form.
Closing Date: January 18th
Additional Information: Please check at http://www.humanities.ox.ac.uk/prospective_students/graduates/funding/ahrc for more information.

CNPq Scholarships

Eligibility: Funded through the National Council for Scientific and Technological Development of Brazil and open to all undergraduate and graduate applicants from Brazil for degrees in the Medical, Mathematical, Physical and Life Sciences.
Level of Study: Graduate
Type: Scholarship
Value: University fee, college fee, and full living expenses
Length of Study: Period of fee liability
Additional Information: This is a brand new scholarship and more information will available soon.

Commonwealth Scholarship and Fellowship Plan

Subjects: All subjects.
Purpose: To assist graduate students with fees and living allowance.
Eligibility: Open to all graduate applicants who are Commonwealth citizens and British protected persons permanently resident in a Commonwealth country (not UK).
Level of Study: Postgraduate, Research
Type: Scholarship
Value: University fee, college fee, and full living expenses
Length of Study: Period of fee liability
Application Procedure: Please see website for more details and how to apply.
Closing Date: January 18th
Additional Information: Preference is given to DPhil students. Please check at http://cscuk.dfid.gov.uk/ for more information.

Commonwealth Shared Scholarship Scheme (CSSS)

Subjects: All subjects.
Type: Scholarship
Value: University and college fees; full grant for living costs; return air travel to UK
Length of Study: 1 year
Closing Date: January 18th
Additional Information: Applicants should normally be under 35 at the time the award begins studying for a taught Master's degrees in subjects relating to the economic and social development of your home country. Not available to those living or studying in a developed country,.

For further information contact:

Website: http://www.admin.ox.ac.uk/studentfunding/scholarship_profiles/CSSS.shtml

Commonwealth/Chevening Scholarships for African Human Rights Advocates

Eligibility: Open to all graduate applicants for the MSc in International Human Rights that are from an African Commonwealth Country and are resident in one (though not necessarily their own) when applying.
Level of Study: Graduate
Type: Scholarship
Value: University fee, college fee, and living expenses during residential sessions. In addition, return air travel from the scholar's home country for each residential session

Length of Study: Period of fee liability
Application Procedure: Please see website for details of how to apply.
Additional Information: Please check at http://humanrightslaw.conted.ox.ac.uk/MStIHRL/fees/index.php#africanscho for more information.

Computer Science Departmental Funding

Subjects: Computer science.
Eligibility: Open to all graduate applicants for Computer Science.
Level of Study: Graduate
Type: Funding support
Value: University fee, college fee and full living expenses
Length of Study: 3 years
Application Procedure: Please visit website for more information and how to apply.
Additional Information: Please check at http://www.cs.ox.ac.uk/news/studentships.html for more information.

Computer Science EPSRC Doctoral Training Grant Studentships

Eligibility: Open to all EU applicants for DPhil Computer Science.
Level of Study: Graduate
Type: Studentship
Value: University fee, college fee and full living expenses
Length of Study: Up to 3 and a half years
Application Procedure: Please visit website for more information and how to apply.
Additional Information: Please check at http://www.cs.ox.ac.uk/news/studentships.html for more information.

Corpus Christi College: A E Haigh English Studentship

Eligibility: Open to all graduate applicants in English.
Level of Study: Postgraduate, Research
Type: Studentship
Value: £7,500 per year towards college fee and contribution towards living expenses
Length of Study: Period of fee liability
No. of awards offered: 2
Application Procedure: Please see website for details of how to apply.
Additional Information: Please check at http://www.ccc.ox.ac.uk/ for more information.

Corpus Christi College: A E Haigh Physics Studentship

Subjects: Physics.
Eligibility: Open to all graduate applicants in Physics.
Level of Study: Postgraduate, Research
Type: Studentship
Value: £7,500 per year towards college fee and contribution towards living expenses
Length of Study: Period of fee liability
Frequency: Annual
Application Procedure: Please see website for details of how to apply.
Additional Information: Please check at http://www.ccc.ox.ac.uk/ for more information.

Criminology: ESRC 1+3/ +3 Nomination (Quota) Studentship

Subjects: MSc criminology and criminal justice (research methods) or DPhil criminology
Purpose: To assist graduate students with fees and maintenance.
Eligibility: Open to all EU (including UK) graduate applicants in Criminology.
Level of Study: Postgraduate, Research
Type: Studentship
Value: University fee, college fee and full living expenses
Length of Study: Up to 4 years
Application Procedure: Please visit website for more details and how to apply.
Closing Date: March 8th
Additional Information: Please check at http://www.crim.ox.ac.uk/graduate/funding.htm for more information.

Dulverton Scholarships

Subjects: All subjects.
Purpose: To assist graduate students with fees and maintenance.
Eligibility: Open to applicants from selected countries to Masters courses.
Level of Study: Postgraduate, Research
Type: Scholarships
Value: University fee, college fee, and full living expenses
Length of Study: Period of fee liability
Application Procedure: Please see website for more details and how to apply.
Closing Date: January 18th
Additional Information: Candidates should be applying to start a one year Master's course within the University's medical sciences, mathematical, physical and life sciences, humanties and social sciences. Scholarships will be awarded on the basis of outstanding academic merit. Financial need may also be taken into consideration. Please check at http://www.ox.ac.uk/feesandfunding/prospectivegrad/scholarships/university/dulverton/for more information.

Economics - ESRC Quota Award

Subjects: MPhil and DPhil economics.
Purpose: To assist graduate students with fees and maintenance.
Eligibility: Open to all graduate applicants in Economics.
Level of Study: Doctorate, Research
Type: Award
Value: University fee, college fee, and full living expenses
Length of Study: Up to 4 years
Application Procedure: Applicants are strongly advised to submit their application for consideration at the first application deadline of January 18th. Places on the second application deadline will be limited.
Closing Date: January 18th
Additional Information: Please visit website for more details and how to apply http://www.economics.ox.ac.uk/index.php/graduate.

For further information contact:

Tel: (0) 1865 281162
Email: econgrad@economics.ox.ac.uk
Website: www.economics.ox.ac.uk/index.php/graduate
Contact: Julie Minns

Economics: Fee Waiver Award

Subjects: Economics.
Purpose: To support the finest aspiring economists from around the world.
Eligibility: Open to all applicants for Mphil Economics.
Level of Study: Postgraduate
Type: Award
Value: University feee and college fee
Length of Study: 2 years
Application Procedure: Applicants are strongly advised to submit their application for consideration at the first application deadline of January 18th. Places on the second application deadline will be limited.
Closing Date: January 18th
Additional Information: Please visit website for more details and how to apply http://www.economics.ox.ac.uk/index.php/graduate.

Education: Departmental Studentships

Eligibility: Open to all applicants for DPhil in Education.
Level of Study: Graduate
Type: Studentship
Value: Value of award varies. Please see website for more details
Length of Study: 3 years
Application Procedure: Please visit website for more details and how to apply.
Closing Date: March 8th
Additional Information: Please chech at http://www.education.ox.ac.uk/courses/d-phil/funding-opportunities/ for more information.

Education: ESRC

Subjects: DPhil education.
Purpose: To assist graduate students with fees and maintenance.
Eligibility: Open to all EU applicants for DPhil in Education.

Level of Study: Doctorate, Research
Type: Award
Value: University fee, college fee, and full living expenses.
Length of Study: 4 years
Application Procedure: Please visit website for more details and how to apply.
Closing Date: Januray 18th
Additional Information: Please check at http://www.education.ox.ac.uk/courses/d-phil/funding-opportunities/for more information.

Education: Routledge Scholarship

Subjects: MSc education (comparative and international education).
Eligibility: Open to all applicants for MSc in Education (Comparative and International Education).
Level of Study: Postgraduate
Type: Scholarship
Value: £6,000
Length of Study: 1 year
Application Procedure: Please visit website for more details and how to apply.
Closing Date: March 8th
Additional Information: Please check at http://www.education.ox.ac.uk/courses/d-phil/funding-opportunities/ for more information.

Edward Hall awards

Subjects: Archaeological science.
Eligibility: Open to all applicants to MSt in Archaeological Science. All applicants to course will be automatically considered.
Level of Study: Postgraduate
Type: Award
Value: University fee (at Home/EU rate)
Length of Study: 1 year
Application Procedure: No separate application process, automatic consideration of all applicants.
Closing Date: January 18th
Additional Information: Please visit website for more details.

Engineering Science: EPSRC Doctoral Training Grant Studentships

Subjects: Engineering science.
Purpose: To assist graduate students with fees and maintenance.
Eligibility: Open to all EU doctoral applicants in Engineering Science.
Level of Study: Research
Type: Studentship
Value: University fee, college fee and full living expenses
Length of Study: Up to 3.5 years
No. of awards offered: 5–8
Application Procedure: Please visit website for more information and how to apply at http://www.eng.ox.ac.uk/admissions/postgraduate/graduate-studentships-scholarships.
Additional Information: Between 5 and 8 studentships are available each year, dependent on the annual funding announcement by EPSRC. An EU national who has studied for an undergraduate degree at a UK university during the 3 years leading up to the application may be classed as a UK resident.

Engineering Science: Research Project Studentships (Various Sponsors) and EPSRC Industrial CASE

Subjects: Engineering science.
Purpose: To assist graduate students with fees and maintenance.
Eligibility: Open to all EU doctoral applicants in Engineering Science.
Level of Study: Research
Type: Studentship
Value: University fee, college fee, and full living expenses
Length of Study: Up to 3.5 years
No. of awards offered: 7–12
Application Procedure: Please visit website for more information and how to apply at http://www.eng.ox.ac.uk/admissions/postgraduate/graduate-studentships-scholarships.
Additional Information: Between 7 and 12 studentships are available each year, dependent on the annual funding announcement by EPSRC.

Engineering Science: Sloane-Robinson Scholarships

Subjects: Biomedical engineering.
Purpose: To assist graduate students with fees.
Eligibility: Open to applicants from developing countries (including Eastern Europe) for the MSc in Biomedical Engineering.
Level of Study: Research
Type: Scholarships
Value: $10,000 or its GBP equivalent
Length of Study: 1 year
Additional Information: Applicants must have demonstrated exceptional talent and promise in the field of biomedical engineering. Please check website for further information http://www.eng.ox.ac.uk/admissions/postgraduate/graduate-studentships-scholarships.

English Faculty: AHRC Block Grant Partnership studentships: Doctoral scheme

Eligibility: Open to UK and EU applicants for the DPhil degrees offered by the Faculty of English.
Level of Study: Graduate
Type: Studentship
Value: University fee, college fee, and full living expenses. Fees-only awards for non-UK, EU students
Length of Study: Period of fee liability
Application Procedure: Apply on the University application form.
Closing Date: January 18th
Additional Information: Please check at http://www.humanities.ox.ac.uk/prospective_students/graduates/funding/ahrc for more information.

English Faculty: AHRC Block Grant Partnership studentships: Research Master's scheme

Subjects: MSt in Film aesthetics and women's studies.
Eligibility: Open to UK and EU applicants for the MSt and MPhil degrees offered by the Faculty of English; MSt in Film Aesthetics, and MSt in Women's Studies.
Level of Study: Graduate
Type: Studentship
Value: University fee, college fee, and full living expenses. Fees-only awards for non-UK, EU students
Length of Study: Period of fee liability
Application Procedure: Apply on the University application form.
Closing Date: January 18th
Additional Information: Please check at http://www.humanities.ox.ac.uk/prospective_students/graduates/funding/ahrc for more information.

English Faculty: Cecily Clarke Studentship

Subjects: English literature.
Eligibility: Open to all applicants to English Medieval Studies, with preference to Middle English Philology. All students applying for English graduate courses will automatically be considered.
Level of Study: Postgraduate
Type: Studentship
Value: £12,000 per year
Length of Study: Up to 2 years
Closing Date: Januray 18th
Additional Information: All students applying for English graduate courses will automatically be considered. Candidates apply at the same time as they apply for admission to their postgraduate programme at the University of Oxford, using the same application form. To be considered candidates must apply by the January deadline. Please check at http://www.english.ox.ac.uk for more information.

English Faculty: E.K. Chambers Studentship

Subjects: English literature.
Purpose: To assist graduate students with fees and maintenance.
Eligibility: Open to graduate applicants in English Literature with good classical attainments (Classics 'A' Levels) who has graduated from a subject other than single honours English. Apply on the University application form.
Level of Study: Postgraduate
Type: Studentship
Value: University fee and living expenses
Length of Study: Period of fee liability

Closing Date: January18th
Additional Information: Please check at http://www.english.ox.ac.uk for more information.

English Faculty: Faculty Studentships

Subjects: English literature and language.
Purpose: To assist graduate students with fees and maintenance.
Eligibility: Open to all graduate applicants in English Literature & Language. Apply on the University application form.
Level of Study: Postgraduate
Type: Studentship
Value: University fee, and living expenses
Length of Study: Period of fee liability
Closing Date: January 18th
Additional Information: All students applying for English graduate courses will automatically be considered. Candidates apply at the same time as they apply for admission to their postgraduate programme at the University of Oxford, using the same application form. Please check at http://www.english.ox.ac.uk for more information.

English Faculty: J K Griffiths Studentship

Subjects: English Literature and Language.
Eligibility: Open to all graduate applicants with interests relating to the history of the book admitted to the MSt 650-1550. All students applying for the relevant MSt course by the January deadline will automatically be considered.
Level of Study: Postgraduate
Value: University fee, maintenance grant of around £3,000, and free accommodation. Scholarship held at St Hilda's College.
Length of Study: Up to 2 years
Closing Date: January 18th
Additional Information: Available to graduate students with interests relating to the history of the book admitted to the Mst 650–1550. All students applying for the relevant MSt course will automatically be considered if they have applied by the Janurary deadline. Please check at http://www.english.ox.ac.uk for more information.

English Faculty: News International Studentship

Subjects: MSt in English language.
Eligibility: Open to all applicants for the MSt in English Language.
Level of Study: Graduate
Type: Studentship
Value: University fee, and living expenses
Length of Study: Period of fee liability
Application Procedure: All students applying for the course by the January deadline will automatically be considered.
Closing Date: January 18th
Additional Information: Please check at http://www.english.ox.ac.uk for more information.

English Faculty: Violet Vaughan Morgan Studentship

Subjects: English literature and language.
Purpose: To assist graduate students with fees and maintenance.
Eligibility: Open to all graduate applicants in English Literature & Language. Apply on the University application form.
Level of Study: Postgraduate
Type: Studentship
Value: University fee and living expenses
Length of Study: Period of fee liability
Closing Date: January 18th
Additional Information: All students applying for English graduate courses will automatically be considered, if they have applied by the January deadline. Please check at http://www.english.ox.ac.ukfor more information.

EPSRC Doctoral Training Grant Studentships in Condensed Matter Physics, Atomic & Laser Physics and Theoretical Physics

Subjects: Condensed matter physics, atomic and laser physics, and theoretical physics.
Eligibility: Open to UK applicants for Condensed Matter Physics; Atomic & Laser Physics; Theoretical Physics. Other EU nationals are eligible for a fees-only award.
Level of Study: Graduate
Type: Studentship
Value: University fee, college fee and full living expenses
Length of Study: 3 years (with possible extension)
Application Procedure: Please visit website for more information and how to apply.
Additional Information: Please check at http://www.physics.ox.ac.uk/admissions/postgrad.htm for more information.

EPSRC Doctoral Training Grant Studentships in Condensed Matter Physics, Atomic & Laser Physics etc

Subjects: Condensed matter physics, atomic and laser physics, and theoretical physics.
Purpose: To assist graduate students with fees and maintenance.
Eligibility: Open to UK applicants for Condensed Matter Physics; Atomic & Laser Physics; Theoretical Physics. Other EU nationals are eligible for a fees-only award.
Level of Study: Research
Type: Studentship
Value: University fee, college fee and full living expenses
Length of Study: Period of fee liability
Application Procedure: Please visit website for more information and how to apply.
Closing Date: January 18th
Additional Information: Typically, we award about 10 DTA studentships each year, but this number depends on the annual funding announcement made by the sponsors. Candidates are advised to apply directly to the sub-departments where advertised on the website, or before the University Application Deadline 2. Please check at http://www.physics.ox.ac.uk/admissions/postgrad.htmfor more information.

ESRC Geography Pathway

Subjects: Geography.
Eligibility: Open to all EU (including UK) applicants for the MPhil in Geography intending to go on to DPhil study.
Level of Study: Graduate
Type: Grant
Value: University fee, college fee and full living expenses
Length of Study: Period of fee liability
Application Procedure: Please visit website for more details and how to apply.
Additional Information: Please check at http://www.geog.ox.ac.uk/news/studentships/ for more information.

Exeter College: Amelia Jackson Studentship

Subjects: All subjects.
Eligibility: Open to all graduate applicants. Please see website for details of how to apply.
Type: Studentship
Value: University fee, college fee, and full living expenses
Length of Study: Up to 3 years (renewed anually)
Closing Date: June 1st
Additional Information: Any graduate of Exeter College who is going on to read for a research degree. Please check at http://www.exeter.ox.ac.uk/currentstudents/finance/scholarships/for more information.

Exeter College: Arthur Peacocke Studentship

Subjects: Theology.
Eligibility: Open to all graduate applicants in Theology.
Level of Study: Research, Postgraduate
Type: Studentship
Value: University fee (at the Home/EU level), college fee, and full living expenses
Length of Study: Up to 3 years
Application Procedure: Please see website for details of how to apply.
Closing Date: April 1st
Additional Information: Please check at http://www.exeter.ox.ac.uk/currentstudents/finance/scholarships/ for more information.

Exeter College: Jonathan Wordsworth Scholarship

Subjects: English.
Eligibility: Open to all graduate applicants in English.

Level of Study: Postgraduate, Research
Type: Scholarship
Value: £5,000 per year
Length of Study: Up to 3 years
Application Procedure: Please see website for details of how to apply.
Closing Date: April 1st
Additional Information: Please check at http://www.exeter.ox.ac.uk/currentstudents/finance/scholarships/ for more information.

For further information contact:

Website: www.exeter.ox.ac.uk

Felix Scholarships

Subjects: All subjects.
Purpose: To enable first class Indian students to pursue graduate studies at the University of Oxford, University of Reading and the School of Oriental and African Studies, University of London (SOAS).
Eligibility: Open to all Indian graduate applicants under the age of 30 that have completed a first degree at an Indian university.
Level of Study: Postgraduate, Research
Type: Scholarships
Value: University fee, college fee, and full living expenses. Additional expenses for flights may be included.
Length of Study: Period of fee liability
No. of awards offered: 7
Application Procedure: Please see website for more details and how to apply.
Closing Date: Januray 18th
Additional Information: Eligible to the nationals of India. Please check at http://www.ox.ac.uk/feesandfunding/prospectivegrad/scholarships/university/felix/ for more information.

Fulbright Scholarships

Subjects: All subjects.
Purpose: To assist graduate students with fees.
Eligibility: Open to all US graduate applicants.
Level of Study: Postgraduate, Research
Type: Scholarships
Value: University fee, college fee, and full living expenses
Length of Study: 1 year
Application Procedure: Please see website for more details and how to apply.
Closing Date: January 18th
Additional Information: If you are currently enrolled in an undergraduate or graduate program at a US college or university, you must apply through the Fulbright Program Adviser (FPA) on your campus. Each institution sets its own campus deadline for Fulbright applications, which will be earlier than the IIE deadline. If you are not currently enrolled in a U.S. institution of higher learning or unable to apply through your home campus or alma mater, you may apply At-Large. This includes U.S. students studying at institutions outside of the U.S. or students attending institutions where there is not a Fulbright Program Adviser. Please check at http://www.cies.org/About.htm for more information.

Geography: Ancora Foundation Graduate Scholarship

Subjects: MSc in environmental change and management.
Eligibility: Open to all applicants for MSc Environmental Change and Management.
Level of Study: Graduate
Type: Scholarship
Value: Univeristy fee, college fee, and full living expenses (up to £28,500)
Length of Study: 1 year
Application Procedure: Please visit website for more details and how to apply.
Additional Information: Please check at http://www.eci.ox.ac.uk/teaching/msc/scholarships.php for more information.

Geography: Andrew Goudie Bursary

Subjects: Environmental change.
Eligibility: Open to all applicants for the MSc Environmental Change and Management.
Level of Study: Postgraduate
Type: Bursary
Value: £3,000
Length of Study: 1 year
Application Procedure: Please visit website for more details and how to apply.
Additional Information: Tenable at any college. Please check at http://www.eci.ox.ac.uk/teaching/msc/scholarships.phpfor more information.

Geography: Birkett Scholarship at Trinity College

Subjects: MSc environmental change and management.
Eligibility: Open to all graduate applicants.
Level of Study: Postgraduate
Type: Scholarship
Value: £3,500
Length of Study: 1 year
Frequency: Annual
Application Procedure: Please visit website for more details and how to apply at http://www.eci.ox.ac.uk/teaching/msc/scholarships.php.
Closing Date: Please contact the college/department for more information

Geography: Hitachi Chemical Europe Scholarship with Linacre College

Subjects: Environmental change and management.
Purpose: To assist graduate students with fees and maintenance.
Eligibility: Open to all applicants for MSc Environmental Change and Management from China, Central or South America. Please visit website for more details and how to apply.
Level of Study: Postgraduate
Type: Scholarship
Value: College fee and £1,000 towards living expenses
Length of Study: 1 year
Application Procedure: Please visit website for more details and how to apply.
Closing Date: Please contact the department
Additional Information: Please check at http://www.eci.ox.ac.uk/teaching/msc/scholarships.php for more information.

Geography: Joan Doll Scholarship

Subjects: MSc environmental change and management only.
Eligibility: Open to all applicants for MSc Environmental Change and Management. The award is tenable only at Green Templeton College.
Level of Study: Postgraduate
Type: Scholarship
Value: £1,500
Length of Study: 1 year
Application Procedure: Please visit website for more details and how to apply.
Additional Information: Tenable at Green Templeton College. Please check at http://www.eci.ox.ac.uk/teaching/msc/scholarships.php for more information.

Geography: NERC DPhil Quota

Subjects: Geography.
Purpose: To assist graduate students with fees and maintenance.
Eligibility: Open to all UK applicants for the DPhil in Geographer.
Level of Study: Doctorate, Research
Value: University fee, college fee and full living expenses.
Length of Study: 3 years
Application Procedure: Please visit website for more details and how to apply.
Closing Date: Please contact the department
Additional Information: Please check at http://www.ouce.ox.ac.uk/graduate/ for more information.

Geography: Norman and Ivy Lloyd Scholarship/ Commonwealth Shared Scholarship

Subjects: Environmental change.
Eligibility: Open to all graduate applicants from Commonwealth countries for Masters courses in Geography. Tenable at Linacre College.
Level of Study: Postgraduate

Type: Scholarship
Value: University fee, college fee, and full living expenses. Additional return air travel from the Scholar's home country.
Length of Study: 1 year
Application Procedure: Candidates wishing to be considered for this scholarship MUST first apply for the University's Commonwealth Shared Scholarship Scheme. They will then, automatically be considered for this scholarship. No separate application to the college is needed. Please visit website for more details and how to apply at http://www.eci.ox.ac.uk/teaching/msc/scholarships.php.
Additional Information: Tenable at Linacre College. This scholarship, when paired with a Commonwealth Shared Scholarship (CSS) provides full support for University and College fees, maintenance and return air travel from the scholar's home country.

Geography: Sir Walter Raleigh Postgraduate Scholarship

Subjects: Environmental change.
Eligibility: Open to all applicants for MSc Environmental Change and Management. The award is tenable only at Oriel College.
Level of Study: Postgraduate
Type: Scholarship
Value: £3,500
Length of Study: 1 year
Frequency: Annual
No. of awards offered: 1
Application Procedure: Any person wishing to be considered for this award should follow the application procedure for admission to the degree of Master of Science in Environmental Change and Management at Oxford University, nominating Colleges of preference as detailed in the application procedure. They must, additionally, download (from website) and complete the application form and return it to the address given below.
Additional Information: Please note that Oriel College only offers a small number of places per year to applicants for the MSc in Environmental Change and Management, and one of these places will be reserved for the applicant who has been awarded the Sir Walter Raleigh Scholarship.

For further information contact:

Academic Office, Oriel College,, Oxford, OX1 4EW
Tel: 01865 276520
Fax: 01865 286548
Email: admissions@oriel.ox.ac.uk
Website: http://www.eci.ox.ac.uk/teaching/msc/scholarships.php
Contact: The Admissions Officer

Geography: The Boardman Scholarship

Subjects: Environmental change.
Eligibility: Open to all applicants for MSc Environmental Change and Management.
Level of Study: Postgraduate
Type: Scholarship
Value: £5,000
Length of Study: 1 year
Frequency: Annual
No. of awards offered: 1
Application Procedure: Please visit website for more details and how to apply at http://www.eci.ox.ac.uk/teaching/msc/scholarships.php.
Additional Information: Tenable at any college.

Geography: Yungtai Hsu Scholarship

Purpose: To support an ECI MSc student committed to the environmental protection and development of China or Taiwan
Eligibility: Open to all applicants for MSc Environmental Change and Management that are committed to the environmental protection and development of China or Taiwan, who will work in this field upon completion of their course. Tenable only at St John's College.
Level of Study: Postgraduate
Type: Scholarship
Value: £15,000 towards university fee and college fee
Length of Study: 1 year
Frequency: Annual
Country of Study: China
Application Procedure: Please visit website for more details and how to apply.
Additional Information: Tenable at St John's College. Student must be committed to the environmental protection and development of China or Taiwan, who will work in this field upon completion of their course. Please check at http://www.eci.ox.ac.uk/teaching/msc/scholarships.phpfor more information.

Green Moral Philosophy Scholarship

Eligibility: Open to all graduate applicants for DPhil courses in the Faculty of Philosophy. All applicants who receive a place on any graduate course automatically considered.
Type: Scholarship
Value: The value of the scholarship is to be determined by the Board of Graduate Admissions. The scholarship may not be offered on a yearly basis
Length of Study: 1 year
Closing Date: January 4th
Additional Information: Please check website for more information.

Hanfling Scholarship

Eligibility: The scholarship is open to those who, at the time of application, (i) are in the second-year of the BPhil course and applying to transfer to the DPhil, (ii) have been admitted to the DPhil programme to begin as Probationary Research Students; or (iii) are first-year Probationary Research Students (or first-year DPhil students, if transferred from the BPhil).
Level of Study: Graduate
Type: Scholarship
Value: £3,400 per year
Length of Study: 1 year (renewable)
Closing Date: January 18th
Additional Information: Please check website for more information.

Hertford College: Carreras Senior Scholarship

Subjects: Natural sciences.
Eligibility: Open to DPhil applicants in the Natural Sciences.
Level of Study: Research
Type: Scholarship
Value: £4,000 per year with certain associated dining rights.
Length of Study: 2 years
Application Procedure: Please see website for details of how to apply.
Additional Information: Preference may be given to candidates who are reading for, or who intend to read for a doctoral degree. Please check at http://www.hertford.ox.ac.uk/advertised-postsfor more information.

Hertford College: Drapers' Senior Scholarship

Purpose: Arts and humanities.
Eligibility: Open to DPhil applicants in the Humanities.
Level of Study: Research
Type: Scholarship
Value: £4,000 per year with certain associated dining rights.
Length of Study: 2 years
Application Procedure: Please see website for details of how to apply.
Additional Information: Preference may be given to candidates who are reading for, or who intend to read for a doctoral degree. Please check at http://www.hertford.ox.ac.uk/advertised-postsfor more information.

Hertford College: Mann Senior Scholarship

Subjects: All subjects.
Eligibility: Open to DPhil applicants in all subjects.
Level of Study: Research
Type: Scholarship
Value: £4,000 per year with certain associated dining rights.
Length of Study: 2 years
Frequency: Annual
Application Procedure: Please see website for details of how to apply.
Additional Information: Please check at http://www.hertford.ox.ac.uk/advertised-posts for more information.

For further information contact:

Hertford College, Catte Street, Oxford, OX1 3BW, United Kingdom
Tel: (44) 0 1865 279 400
Fax: (44) 0 1865 279 466

Hertford College: Worshipful Company of Scientific Instrument Makers Senior Scholarship

Subjects: All subjects (Mathematical, physical, and life sciences).
Eligibility: Open to DPhil applicants in all subjects.
Level of Study: Research
Type: Scholarship
Value: £4,000 per year with certain associated dining rights.
Length of Study: 2 years
Application Procedure: Please see website for details of how to apply at http://www.hertford.ox.ac.uk/advertised-posts.
Additional Information: Funded by the Worshipful Company of Scientific Instrument Makers: Applicants are expected to be involved in the design of instrumentation.

Hill Foundation Scholarships

Subjects: All subjects.
Purpose: To enable Russians of very high academic ability to undertake a period of study at Oxford University before returning to develop their careers in the Russian Federation.
Eligibility: Open to all graduate applicants from selected countries.
Level of Study: Postgraduate, Research
Value: University fee, college fee, and full living expenses
Length of Study: Period of fee liability
No. of awards offered: Varies
Application Procedure: Apply at the same time as you apply to Oxford by selecting Hill Scholarship in the Funding Section of the University's Graduate Application Form. Please see website for more details and how to apply.
Closing Date: January 18th
Additional Information: Maximum age normally 35. If Israeli, the candidate must be first generation Israeli of Russian descent. Please contact at http://www.ox.ac.uk/feesandfunding/prospectivegrad/scholarships/university/hill/ for more information.

History Faculty: AHRC Block Grant Partnership Studentships: Doctoral scheme

Subjects: History and history of arts.
Eligibility: Open to UK and EU applicants for the DPhil degrees offered by the Faculty of History, including History of Art.
Level of Study: Graduate
Type: Studentship
Value: University fee, college fee, and full living expenses. Fees-only awards for non-UK, EU students
Length of Study: Period of fee liability
Application Procedure: Apply on the University application form.
Closing Date: January 18th
Additional Information: Please check at http://www.humanities.ox.ac.uk/prospective_students/graduates/funding/ahrc for more information.

History Faculty: AHRC Block Grant Partnership Studentships: Research Master's scheme

Subjects: History of art, MSt in film aesthetics and MSt in women's studies.
Eligibility: Open to UK and EU applicants for the MSt and MPhil degrees offered by the Faculty of History, including History of Art; MSt in Film Aesthetics; MSt in Women's Studies.
Level of Study: Graduate
Type: Studentship
Value: University fee, college fee, and full living expenses. Fees-only awards for non-UK, EU students
Length of Study: Period of fee liability
Application Procedure: Apply on the University application form.
Closing Date: January 18th
Additional Information: Please check at http://www.humanities.ox.ac.uk/prospective_students/graduates/funding/ahrc for more information.

International Development and Anthropology: Departmental Scholarship

Subjects: MSc in migration studies.
Eligibility: Open to all graduate applicants for MSc Migration Studies with a preference for those from Sub-Saharan Africa.
Level of Study: Graduate
Type: Scholarship
Value: University fee, college fee, and £8,000 towards living expenses
Length of Study: 2 years
Application Procedure: Please visit website for more details and how to apply.
Closing Date: March 8th
Additional Information: Please check at http://www.qeh.ox.ac.uk/ for more information.

International Development: Corpus Christi Graduate Studentship in Development

Subjects: Economics for development, forced migration, and global governance and diplomacy.
Purpose: To assist graduate students with fees and maintenance.
Eligibility: Open to all applicants for the MSc in Economics for Development, MSc in Refugee and Forced Migration Studies, MSc in Global Governance and Diplomacy, MSc in Migration Studies. Preference will be show to those from Sub-Saharan Africa.
Level of Study: Postgraduate
Type: Studentship
Value: University fee, college fee and £8,000 towards living expenses
Length of Study: 1 year
Application Procedure: Please visit website for more details and how to apply.
Closing Date: March 8th
Additional Information: Please contact at http://www.qeh.ox.ac.uk/ for more information.

International Development: Department of Economics Scholarship

Subjects: Economics for development.
Purpose: To assist graduate students with fees.
Eligibility: Open to all applicants for MSc in Economics for Development.
Level of Study: Postgraduate
Type: Scholarship
Value: Up to £20,000
Length of Study: 1 year
Application Procedure: Please visit website for more details and how to apply.
Closing Date: March 8th
Additional Information: MSc in Economics for Development only. Apply by the final course application deadline. Please check at http://www.qeh.ox.ac.uk/ for more information.

International Development: MSc in Forced Migration Scholarships

Subjects: Forced migration.
Purpose: To explore forced migration through a thesis, a group research essay, and a range of required courses.m
Eligibility: Open to all applicants for the MSc in Refugee and Forced Migration Studies.
Level of Study: Postgraduate
Type: Scholarships
Value: Value of award varies. Please see website for more details.
Length of Study: 1 year
Frequency: Annual
No. of awards offered: 2
Application Procedure: Please visit website for more details and how to apply.
Closing Date: March 8th
Additional Information: Two full or part bursaries are normally available for outstanding applicants. Apply by the final course application deadline. Please check at http://www.rsc.ox.ac.uk/ for more information.

International Development: QEH Scholarship

Subjects: Development studies.
Purpose: To assist graduate students with fees and maintenance.

Eligibility: Open to all graduate applicants for DPhil in International Development with a preference for those from Sub-Saharan Africa.
Level of Study: Research
Type: Scholarship
Value: University fee, college fee, and possibility of additional living expenses
Length of Study: Up to 3 years
Application Procedure: Apply by the final course application deadline. Please visit website for more details and how to apply.
Closing Date: March 8th
Additional Information: Please check at http://www.qeh.ox.ac.uk/ for more information.

International Development: QEH Scholarships

Subjects: Economics for development, forced migration, and global governance and diplomacy.
Purpose: To assist graduate students with fees and maintenance.
Eligibility: Open to all graduate applicants for MSc Economics for Development; MSc in Refugee and Forced Migration Studies; MSc in Global Governance and Diplomacy with a preference for those from Sub-Saharan Africa.
Level of Study: Research
Type: Scholarships
Value: University fee, college fee, and £8,000 towards living expenses
Length of Study: 1 year
Application Procedure: Apply by the final course application deadline. Please visit website for more details and how to apply.
Closing Date: March 8th
Additional Information: Please check at http://www.qeh.ox.ac.uk/ for more information.

Internet Institute: ESRC Studentship

Subjects: DPhil information, communication, and the social sciences.
Eligibility: Open to all UK applicants to the DPhil in Information, Communication, and the Social Sciences. Other EU nationals are eligible for a fees-only award.
Level of Study: Research
Type: Studentship
Value: University fee, college fee and full living expenses
Length of Study: Up to 3 years
Application Procedure: All prospective candidates must make their application through the University's Graduate Studies Office, and are advised to consult the University's Graduate Studies Prospectus beforehand. All applications are processed initially by the Graduate Studies Office, and must not be made directly to the OII. Please check at http://www.oii.ox.ac.uk/teaching/dphil/apply.cfm for more information.
Closing Date: January 18th
Additional Information: Please contact the department for more information.

Jenkins Memorial Fund Scholarships

Subjects: Humanities and social sciences.
Eligibility: Open to all Masters applicants that have studied at selected institutions.
Level of Study: Postgraduate, Research
Type: Scholarship
Value: £10,000 per year
Length of Study: 1 year
Frequency: Annual
No. of awards offered: Up to 4
Application Procedure: Apply at the same time as you apply to Oxford by selecting Jenkins Memorial in the Funding Section of the University's Graduate Application Form. Please see website for more details and how to apply.
Closing Date: January 18th
Additional Information: Priority will be given to students following a one-year Masters degree but candidates entering a two-year Masters degree or a Second BA degree may also be considered. Please check at http://www.ox.ac.uk/feesandfunding/prospectivegrad/scholarships/university/jenkins/ for more information.

Jesus College: Peter Thomason Graduate Bursary

Subjects: Mathematics or physical sciences.
Eligibility: Open to all applicants from a selected list of countries applying for a Masters by Research degree in mathematics or the physical sciences.
Level of Study: Graduate
Type: Bursary
Value: University fee, college fee and living expenses
Length of Study: 1 or 2 years
Application Procedure: Please see website for more details and how to apply.
Closing Date: March 8th
Additional Information: Please check at http://www.jesus.ox.ac.uk/current-students/peter-thomason-graduate-bursary-fund for more information.

Keble College: Gosden-Water Newton Scholarship

Subjects: All subjects.
Eligibility: Open to graduate applicants in all subjects.
Level of Study: Postgraduate, Research
Type: Award
Value: Up to £25,000 per year
Length of Study: Up to 3 years
Application Procedure: Applicants must be willing to be or be ordained. Please contact at College.Office@Keble.ox.ac.uk, or www.keble.ox.ac.uk for more information. Please see website for details of how to apply.
Closing Date: March 1st
Additional Information: Please check at http://www.keble.ox.ac.uk/admissions/graduate/graduate-scholarships-1for more information.

Keble College: Gwynne Jones Scholarship

Subjects: All subjects.
Purpose: To enable students from Sierra Leonne or those Yoruba-speaking countries to pursue a postgraduate research degree.
Eligibility: Open to all graduate applicants from Sierra Leone.
Level of Study: Postgraduate, Research
Type: Award
Value: Up to £6,000 per year towards living expenses
Length of Study: Up to 3 years
Application Procedure: Please see website for details of how to apply.
Closing Date: March 1st
Additional Information: Please check website for further information http://www.keble.ox.ac.uk/admissions/graduate/graduate-scholarships-1.

Keble College: Ian Palmer Scholarship

Subjects: Computer science.
Eligibility: Open to all graduate applicants in Information Technology.
Level of Study: Research
Value: £2,000 per year towards living expenses
Length of Study: Up to 3 years
Application Procedure: Please see website for details of how to apply.
Closing Date: March 1st
Additional Information: Please check at http://www.keble.ox.ac.uk/admissions/graduate/graduate-scholarships-1 for more information.

Keble College: Ian Tucker Award

Subjects: All subjects.
Eligibility: Open to graduate applicants in all subjects. PLEASE NOTE: preference will be given to those with prowess in Rugby Football.
Level of Study: Postgraduate, Research
Type: Award
Value: Up to £6,000 per year towards living expenses
Length of Study: 1 year
Frequency: Annual
Application Procedure: Please check details at http://www.keble.ox.ac.uk/admissions/graduate/graduate-scholarships-1.
Closing Date: March 1st
Additional Information: Please contact: College.Office@Keble.ox.ac.uk, or www.keble.ox.ac.uk for more information.

Kellogg College Scholarships

Eligibility: Open to all graduate applicants in subject area of either current Kellogg Fellows, Students or Research Centres.
Level of Study: Research, Postgraduate
Type: Scholarship

Value: College fee
Length of Study: Period of fee liability
Frequency: Annual
Application Procedure: Please see website for details and how to apply.
Closing Date: May 10th
Additional Information: Please check at http://www.kellogg.ox.ac.uk/kellogg-college-scholarships-2013 for more information.

Kellogg College: Progress Scholarship

Eligibility: Open to all applicants to a full-time DPhil that are currently full-time Masters students at Kellogg.
Level of Study: Graduate
Type: Scholarship
Value: College fee
Length of Study: 3 years
Closing Date: July 19th
Additional Information: Please check at http://www.kellogg.ox.ac.uk/kellogg-college-progress-scholarship-2013-14 for more information.

Kwok Scholarships

Subjects: Public policy.
Eligibility: Open to all graduate applicants that are reident in Hong Kong for the Master in Public Policy.
Level of Study: Graduate
Type: Scholarship
Value: University fee, college fee and full living expenses
Length of Study: 1 year
Application Procedure: Please visit website for more details and how to apply.
Closing Date: January 18th
Additional Information: Please check at http://www.bsg.ox.ac.uk/kwok-scholarships for more information.

Lady Margaret Hall: One-year Awards within each Division

Subjects: All subjects taken by the college.
Eligibility: Open to all graduate applicants.
Level of Study: Postgraduate, Research
Type: Award
Value: £2,000 per year plus additional benefits including accomodation and dining rights
Length of Study: Period of fee liability
Frequency: Annual
No. of awards offered: 4
Application Procedure: Please see website for details of how to apply.
Closing Date: Januray 18th
Additional Information: Please check at http://www.lmh.ox.ac.uk for more information.

Lady Margaret Hall: Faculty Studentships

Subjects: English, economics, physics, molecular chemistry and chemical biology, classics, modern languages, history, and medicine (wellcome).
Eligibility: Open to all graduate applicants to particpating faculties.
Level of Study: Postgraduate, Research
Type: Studentship
Value: £2,000 per year plus additional benefits including accomodation and dining rights
Length of Study: Up to 3 years
Frequency: Annual
No. of awards offered: Up to 9
Application Procedure: Please see website for more details and how to apply.
Closing Date: Januray 1st
Additional Information: This scholarship held in conjunction with a Faculty Studentship. Please check at http://www.lmh.ox.ac.uk for more information.

Lady Margaret Hall: Jex-Blake Scholarships

Subjects: All subjects taken by the college.
Eligibility: Open to all graduate applicants who are already student at Lady Margaret Hall.
Level of Study: Postgraduate, Research
Type: Scholarship
Value: £2,000 per year plus additional benefits including accomodation and dining rights
Length of Study: 1 year
No. of awards offered: 2
Application Procedure: Please see website for details of how to apply.
Closing Date: January 18th
Additional Information: Candidates must be members of the College who will continuing at LMH for a further degree. Please check at http://www.lmh.ox.ac.uk for more information.

Lady Margaret Hall: The LMH Studentship for Canadian Studies

Subjects: The subject of applicants' research must have a substantial Canadian content.
Eligibility: Open to all graduate applicants from the UK whose research will centre on Canada in some way.
Level of Study: Research
Type: Studentship
Value: £4,000 per year plus additional benefits including accomodation and dining rights
Length of Study: Period of fee liability
Frequency: Annual
Application Procedure: Submit scholarship application by the final University deadline. Please see website for details of how to apply.
Closing Date: Januray 18th
Additional Information: Please check at http://www.lmh.ox.ac.uk for more information.

Lady Margaret Hall: The Thomas Wong Esq. Residential Scholarship

Eligibility: Open to all graduate applicants to the Department of Education from China and developing countries.
Level of Study: Graduate
Type: Scholarship
Value: College Accommodation worth up to £4,000 and limited dining rights
Length of Study: 1 year
Application Procedure: Please see website for details of how to apply.
Closing Date: January 18th
Additional Information: Please check at http://www.lmh.ox.ac.uk for more information.

Lady Margaret Hall: Warr-Goodman Scholarship

Subjects: All subjects taken by the college.
Eligibility: Open to continuing research students at Lady margaret Hall.
Level of Study: Research
Type: Scholarship
Value: £2,000 per year plus additional benefits including accomodation and dining rights
Length of Study: One year with the possibility of renewal
Application Procedure: Please see website for details of how to apply.
Additional Information: Please check at http://www.lmh.ox.ac.uk for more information.

Latin American Centre: Ronaldo Falconer, LAC Scholarship, St Antony's College

Subjects: Latin American studies, public policy in Latin America, and Latin American studies.
Purpose: To assist graduate students with fees.
Eligibility: Open to all Cost Rican applicants for the MSc in Latin American Studies.
Level of Study: Postgraduate, Research
Type: Scholarship
Value: £9,000
Length of Study: 1 year
Application Procedure: Please visit website for more details and how to apply at http://www.lac.ox.ac.uk/funding.
Closing Date: March 8th
Additional Information: Candidate must be residents of Costa Rica (of any nationality) applying for graduate studies at the University of

Oxford. Preference is given to those candidates applying for the two-year MPhil in Latin American Studies.

Law Faculty: Fountain Court Chambers

Eligibility: Open to all EU graduate applicants in Law. Preference will be shown for candidates with an interest in pursuing a career in commercial law at the English bar.
Level of Study: Graduate
Type: Grant
Value: £10,000
Length of Study: 1 year
Application Procedure: Please visit website for more details and how to apply.
Closing Date: January 18th
Additional Information: Please check at http://www.law.ox.ac.uk/postgraduate/scholarships.php for more information.

Law Faculty: Four New Square Chambers

Eligibility: Open to all EU applicants for the BCL with a UK Qualifying Law Degree (QLD).
Type: Grant
Value: £10,000
Length of Study: 1 year
Application Procedure: Please visit website for more details and how to apply.
Closing Date: January 18th
Additional Information: Please check at http://www.law.ox.ac.uk/postgraduate/scholarships.php for more information.

Law Faculty: Graduate Assistance Fund Awards

Subjects: Legal studies and law.
Purpose: To assist graduate students with fees.
Eligibility: Open to all graduate applicants in Law courses.
Level of Study: Postgraduate, Research
Type: Funding support
Value: £10,000
Length of Study: 1 year
Application Procedure: Please visit website for more details and how to apply t http://www.law.ox.ac.uk/postgraduate/scholarships.php.
Closing Date: January 18th

Law Faculty: Paul Hastings Scholarship

Subjects: Law.
Eligibility: Open to all graduate applicants in Law courses. The selection process will take financial need into account.
Level of Study: Graduate
Type: Scholarship
Value: £10,000
Length of Study: 1 year
Application Procedure: Please visit website for more details and how to apply.
Closing Date: January 18th
Additional Information: Please check at http://www.law.ox.ac.uk/postgraduate/scholarships.php for more information.

Law Faculty: Pump Court Tax Chambers

Eligibility: Open to all EU graduate applicants for the BCL and MJur. Preference will be shown for candidates with an interest in pursuing a career in commercial law at the English bar.
Level of Study: Graduate
Type: Grant
Value: £10,000
Length of Study: 1 year
Application Procedure: Please visit website for more details and how to apply.
Closing Date: January 18th
Additional Information: Please check at http://www.law.ox.ac.uk/postgraduate/scholarships.php for more information.

Law Faculty: South Square Chambers

Eligibility: Open to all EU graduate applicants in Law. Preference will be shown for candidates with an interest in pursuing a career in commercial law at the English bar.
Level of Study: Graduate
Type: Grant
Value: £10,000
Length of Study: 1 year
Application Procedure: Please visit website for more details and how to apply.
Closing Date: January 18th
Additional Information: Please check at http://www.law.ox.ac.uk/postgraduate/scholarships.php for more information.

Law Faculty: The 3 Verulam Buildings of Gray's Inn

Subjects: Commercial law.
Eligibility: Open to all EU graduate applicants in Law. Preference will be shown for candidates with an interest in pursuing a career in commercial law at the English bar.
Level of Study: Graduate
Type: Grant
Value: £10,000
Length of Study: 1 year
Application Procedure: Please visit website for more details and how to apply.
Closing Date: January 18th
Additional Information: Please check at http://www.law.ox.ac.uk/postgraduate/scholarships.php for more information.

Law Faculty: The Essex Court Chambers Scholarship

Subjects: Commercial law.
Eligibility: Open to all graduate applicants for the BCL. Preference will be shown for candidates with an interest in pursuing a career in commercial law at the English bar.
Level of Study: Graduate
Type: Scholarship
Value: £10,000
Length of Study: 1 year
Application Procedure: Please visit website for more details and how to apply.
Closing Date: January 18th
Additional Information: Please check at http://www.law.ox.ac.uk/postgraduate/scholarships.php for more information.

Law Faculty: The Peter Birks memorial scholarship:

Subjects: Legal studies and law.
Purpose: To assist graduate students with fees.
Eligibility: Open to all graduate applicants.
Level of Study: Postgraduate, Research
Type: Scholarship
Value: £5,000
Length of Study: 1 year
Application Procedure: Please visit website for more details and how to apply at http://www.law.ox.ac.uk/postgraduate/scholarships.php.
Closing Date: January 18th

Law Faculty: Winter Williams Studentships

Subjects: BCL/MJur, MLF, M.St Legal Studies, M.Phil. Or D.Phil Law.
Purpose: To assist graduate students with fees.
Eligibility: Open to all graduate applicants in Law courses.
Level of Study: Postgraduate, Research
Type: Studentship
Value: £7,500
Length of Study: 1 year
Frequency: Annual
Application Procedure: Please visit website for more details and how to apply.
Closing Date: January 4th
Additional Information: Please check at http://www.law.ox.ac.uk/postgraduate/scholarships.php for more information.

Life Sciences Interface Doctoral Training Centre EPSRC Studentships

Subjects: MPLS.
Eligibility: Open to UK applicants for all MPLS subjects. Other EU nationals are eligible for a fees-only award.
Level of Study: Graduate

Type: Studentship
Value: University fee, college fee and full living expenses
Length of Study: 4 years
Application Procedure: Please visit website for more information and how to apply.
Additional Information: Please check at http://www.dtc.ox.ac.uk/ for more information.

Linacre College: A.J Hosier Studentship

Subjects: Husbandry; Agricultural Economics or Statistics; Applied Agricultural Science.
Purpose: To fund graduate study.
Eligibility: Open to UK/EU Honours graduates of a UK/EU University that are applying for Husbandry, Agricultural Economics or Statistics, or Applied Agricultural Science.
Level of Study: Postgraduate, Research
Type: Studentship
Value: Up to £4,000 per year
Length of Study: Period of fee liability
Study Establishment: Linacre College, University of Oxford
Country of Study: United Kingdom
No. of awards offered: Varies
Application Procedure: Applicants must contact the College Secretary for further details. Please see website for details of how to apply.
Additional Information: UK citizens who are Honours graduates of a UK University.

For further information contact:

Linacre College, Oxford, Oxfordshire, OX1 3JA, England
Website: http://www.linacre.ox.ac.uk/Admissions/Scholarships
Contact: College Secretary

Linacre College: Canadian National Scholarship

Subjects: All subjects.
Purpose: To enable suitably qualified Canadian nationality students intending to read for a post graduate degree.
Eligibility: Open to all graduate applicants from Canada. Please see website for details of how to apply.
Level of Study: Graduate, Postgraduate, Research
Type: Scholarship
Value: £5,000 per year
Length of Study: 1 year
Frequency: May not be available every year
Study Establishment: Linacre College, University of Oxford
Country of Study: United Kingdom
Application Procedure: Applicants must write for details or visit the website www.linacre.ox.ac.uk
Contributor: Canadian National
Additional Information: Please check at http://www.linacre.ox.ac.uk/Admissions/Scholarships for more information.

For further information contact:

The College Secretary, Linacre College, Oxford, Oxfordshire, OX1 3JA, England

Linacre College: Carolyn and Franco Gianturco Scholarship

Subjects: Musicology or theoretical chemistry.
Eligibility: Open to all graduate applicants in Musicology or Theoretical Chemistry.
Level of Study: Graduate
Type: Scholarship
Value: College fee
Length of Study: Up to 4 years (depending on period of fee liability)
Application Procedure: Please see website for details of how to apply.
Additional Information: Please check at http://www.linacre.ox.ac.uk/Admissions/Scholarships for more information.

Linacre College: David Daube Scholarship

Eligibility: Open to all graduate applicants for the BCL and MJur.
Level of Study: Postgraduate
Type: Scholarships
Value: £4,500 per year
Length of Study: Period of fee liability
Application Procedure: Please see website for details of how to apply.
Additional Information: Applicants should first secure a place on the Oxford BCL or MJur course and mark Linacre College as their chosen College. Please check at http://www.linacre.ox.ac.uk/Admissions/Scholarships for more information.

Linacre College: EPA Cephalosporin Scholarship

Subjects: Pathology research degree.
Eligibility: Open to all applicants in biological, medical and chemical sciences.
Level of Study: Postgraduate, Research
Type: Scholarship
Value: College fee
Length of Study: Period of fee liability
Application Procedure: Please see website for details of how to apply at http://www.linacre.ox.ac.uk/Admissions/Scholarships.
Additional Information: Please contact: Director of Graduate Studies, Sir William Dunn School of Pathology, Oxford OX1 3RE, UK. Website: administration@path.ox.ac.uk.

Linacre College: Heselton Legal Research Scholarship

Subjects: English law or European community law.
Eligibility: Open to all graduate applicants for a research degree in English Law or European Community Law.
Level of Study: Graduate
Type: Scholarship
Value: College fee
Length of Study: Period of fee liability
Application Procedure: Please see website for details of how to apply.
Additional Information: Please check at http://www.linacre.ox.ac.uk/Admissions/Scholarships for more information.

Linacre College: Hitachi Chemical Europe Scholarship

Subjects: Environmental change and management.
Purpose: To assist graduate students with fees and a living allowance.
Eligibility: Open to all graduate applicants from China and Central/South America for MSc Environmental Change & Management.
Level of Study: Postgraduate
Type: Scholarship
Value: College fee plus £1,000 towards living expenses
Length of Study: 1 year
Application Procedure: Please see website for details of how to apply.
Additional Information: Preference given to applicant from China or Central/South America.

For further information contact:

School of Geography, OUCE, Oxford,
Email: enquiries@ouce.ox.ac.uk
Website: http://www.linacre.ox.ac.uk/Admissions/Scholarships
Contact: Director of Graduate Studies

Linacre College: Mary Blaschko Graduate Scholarship

Subjects: Arts & humanities research.
Purpose: To enable European students to carry out research for 1 year in the department of pharmacology or the MRC anatomical neuropharmacology unit.
Eligibility: Open to all graduate applicants in Humanities.
Level of Study: Research
Type: Scholarship
Value: College fee
Length of Study: Period of fee liability
Frequency: Annual
Study Establishment: Linacre College, University of Oxford
Country of Study: United Kingdom
Application Procedure: Please see website for details of how to apply.
Funding: Private
Additional Information: Please check at http://www.linacre.ox.ac.uk/Admissions/Scholarships for more information.

Linacre College: Norman and Ivy Lloyd Scholarship
Subjects: MSc course.
Purpose: To enable a student from developing Commonwealth countries to undertake an MSc course in Environmental change and management.
Eligibility: Open to all graduate applicants from sub-Saharan Africa for Msc in Biodiversity Conservation and Management, MSc in Environmental Change and Management, MSc in Integrated Immunology, MSc in Global Health Science.
Level of Study: Postgraduate
Type: Scholarship
Value: University fee, college fee, and full living expenses
Length of Study: 1 year
Frequency: Annual
Study Establishment: Linacre College, University of Oxford
Country of Study: United Kingdom
No. of awards offered: Varies
Application Procedure: Candidates wishing to be considered for this scholarship must first apply for the University's CSS (http://www.admin.ox.ac.uk/studentfunding/scholarship_profiles/csss.shtml). Please see website for details of how to apply.
Funding: Private
Additional Information: This scholarship will be linked to a Commonwealth Shared Scholarship or University Clarendon award. Eligible to nationals of Africa. Please check at http://www.linacre.ox.ac.uk/Admissions/Scholarships for more information.

For further information contact:

Environmental Change Institute, 5 Mansfield Road, Oxford, Oxfordshire, OX1 3TB, England
Contact: Dr J Boardman

Linacre College: Rausing Scholarship in Anthropology
Subjects: Anthropology.
Eligibility: Open to all graduate applicants in Anthropology.
Level of Study: Research
Type: Scholarship
Value: College fee plus £2,000 towards living expenses
Length of Study: Period of fee liability
Application Procedure: There is no application form. Applications should consist of a detailed doctoral proposal of 3–4 pages, a curriculum vitae and two letters of reference, one of which should be provided by the student's current or prospective supervisor for the doctorate. Applications should be sent to the Director of Graduate Studies. Please see website for details of how to apply.
Additional Information: Please check at http://www.linacre.ox.ac.uk/Admissions/Scholarships for more information.

For further information contact:

c/o Ms Vicky Dean, Institute of Social and Cultural Anthropology, 51 Banbury Road, Oxford, OX2 6PE, United Kingdom
Contact: Director of Graduate Studies

Linacre College: Raymond and Vera Asquith Scholarship
Subjects: Humanities.
Eligibility: Open to all graduate applicants in Humanities.
Level of Study: Research
Type: Scholarship
Value: College fee
Length of Study: Period of fee liability
Application Procedure: Please see website for details of how to apply.
Additional Information: This scholarship is linked to Arts and Humanities Research Council Award. Please check at http://www.linacre.ox.ac.uk/Admissions/Scholarships for more information.

Linacre College: Ron and Jane Olson Scholarship in Refugee Studies
Subjects: MSc in refugee and forced migration studies.
Eligibility: Open to all graduate applicants for MSc degree in Forced Migration. Students are automatically considered.
Level of Study: Postgraduate
Type: Scholarship
Value: College fee plus £7,000 towards living expenses
Length of Study: 1 year
Frequency: Annual
Application Procedure: Candidates wishing to be considered for this scholarship must first apply for the University's Commonwealth Shared Scholarship Scheme.
Additional Information: This scholarship will be linked to University Clarendon award. Please check at http://www.linacre.ox.ac.uk/Admissions/Scholarships for more information.

Linacre College: Ruth and Neville Mott Scholarship
Eligibility: Open to all graduate applicants to research degrees in Musicology or Theoretical Chemistry.
Level of Study: Research, Postgraduate
Type: Scholarship
Value: College fee
Length of Study: Period of fee liability
Application Procedure: At the time of taking up the Scholarship, the Scholar must have been admitted as a graduate student by the Law Faculty of the University of Oxford and must be, or become, a member of Linacre College. Please see website for details of how to apply.
Additional Information: Please check at http://www.linacre.ox.ac.uk/Admissions/Scholarships for more information.

Lincoln College: Berrow Foundation Lord Florey Scholarships
Subjects: Medical, chemical or biomedical sciences.
Eligibility: Open to all graduate applicants of Swiss or Lichtenstein nationality that are students at, or have graduated within the last 5 years from, any Swiss universities including ETHZ and EPFL.
Level of Study: Postgraduate, Research
Type: Scholarship
Value: University fee (at the Home/EU level), college fee, and full living expenses
Length of Study: Period of fee liability
Application Procedure: Please see website for details of how to apply.
Closing Date: January 18th
Additional Information: Please check at www.lincoln.ox.ac.uk/ for more information. If you have any queries please contact the Rector's Office.

For further information contact:

Rector's Office, Lincoln College, Oxford, OX1 3DR
Email: rectors.office@lincoln.ox.ac.uk

Lincoln College: Berrow Foundation scholarships
Eligibility: Open to all graduate applicants of Swiss or Lichtenstein nationality that are students, or have recently graduated from, the relevant Swiss universities.
Level of Study: Postgraduate, Research
Type: Scholarship
Value: University fee (at the Home/EU level), college fee, and full living expenses
Length of Study: Period of fee liability
Application Procedure: Candidates must nominate Lincoln as first choice college on the graduate application form. Please see website for more details and how to apply.
Closing Date: January 18th
Additional Information: Please check at www.lincoln.ox.ac.uk/ for more information.

Lincoln College: Crewe Graduate Scholarships
Subjects: All subjects.
Eligibility: Open to all graduate applicants that show evidence of financial need, academic merit and potential for good college citizenship.
Level of Study: Postgraduate, Research
Type: Scholarship
Value: £5,000 per year
Length of Study: Period of fee liability
Application Procedure: Please see website for details of how to apply at www.lincoln.ox.ac.uk/.
Closing Date: June 3rd
Additional Information: Successful candidates will show evidence of both academic merit and financial need. Applicants must hold a place, or an offer of a place, at Lincoln College before applying.

Lincoln College: Lord Crewe Graduate Scholarships in the Humanities

Subjects: All graduate courses within the Humanites Division.
Eligibility: Open to graduates of any UK university for all courses within the Humanities.
Level of Study: Postgraduate, Research
Type: Scholarship
Value: £18,000 per year
Length of Study: Period of fee liability
Application Procedure: Please see website for details of how to apply.
Closing Date: January 18th
Additional Information: Please check at www.lincoln.ox.ac.uk/ for more information.

Lincoln College: Lord Crewe Graduate Scholarships in the Social Sciences

Subjects: All graduate courses within the Social Sciences Division.
Eligibility: Open to graduates of any UK university for all courses within the Social Sciences.
Level of Study: Postgraduate, Research
Type: Scholarship
Value: £18,000 per year
Length of Study: Period of fee liability
Application Procedure: Please see website for details of how to apply.
Closing Date: January 18th
Additional Information: Please check at www.lincoln.ox.ac.uk/ for more information.

Lincoln College: Menasseh Ben Israel Room

Subjects: All subjects.
Eligibility: Open to all graduate applicants that have graduated from an Israeli university. Preference will be show to graduants of Hebrew University, Jerusalem.
Level of Study: Research, Postgraduate
Value: Free accommodation for one academic year as occcupant of the Menasseh Ben Israel Room in college.
Length of Study: 1 year
Frequency: Annual
Application Procedure: Please see website for details of how to apply.
Closing Date: June 3rd
Additional Information: All applicants must nominate Lincoln as first choice college on the Graduate Application Form. Please check at www.lincoln.ox.ac.uk/ for more information.

Lincoln College: Polonsky Foundation Grants

Subjects: All subjects.
Purpose: To financially assist meritorious student.
Eligibility: Open to all overseas graduate applicants that show evidence of financial need, academic merit and potential for good college citizenship.
Level of Study: Postgraduate, Research
Type: Grant
Value: £5,300 per year
Length of Study: Period of fee liability
Frequency: Annual
No. of awards offered: 3
Application Procedure: Please see website for details of how to apply.
Closing Date: June 3rd
Additional Information: Successful candidates will show evidence of financial need, academic merit and potential for good College citizenship. All applicants must nominate Lincoln as first choice College on the Graduate Application Form. Please check at www.lincoln.ox.ac.uk/ for more information.

Linguistics Faculty: AHRC Block Grant Partnership Studentships: Doctoral scheme

Subjects: Linguistics, philology and phonetics.
Eligibility: Open to UK and EU applicants for the DPhil degrees offered by the Faculty of Linguistics, Philology and Phonetics.
Level of Study: Graduate
Type: Studentship
Value: University fee, college fee, and full living expenses. Fees-only awards for non-UK, EU students
Length of Study: Period of fee liability
Closing Date: January 18th
Additional Information: Please check at http://www.humanities.ox.ac.uk/prospective_students/graduates/funding/ahrc for more information.

Linguistics Faculty: AHRC Block Grant Partnership Studentships: Research Master's scheme

Subjects: Linguistics, phonetics and philology.
Eligibility: Open to UK and EU applicants for the MSt and MPhil degrees offered by the Faculty of Linguistics, Phonetics and Philology.
Level of Study: Graduate
Type: Studentship
Value: University fee, college fee, and full living expenses. Fees-only awards for non-UK, EU students
Length of Study: Period of fee liability
Application Procedure: Apply on the University application form.
Closing Date: January 18th
Additional Information: Please check at http://www.humanities.ox.ac.uk/prospective_students/graduates/funding/ahrc for more information.

Louis Dreyfus-Weidenfeld Scholarship and Leadership Programme

Subjects: Social sciences.
Eligibility: Open to all applicants for selected courses in the Social Sciences from selected countries.
Type: Scholarship
Value: University fee, college fee, and full living expenses
Length of Study: Period of fee liability
Application Procedure: Please see the website for more details and how to apply.
Closing Date: January 18th
Additional Information: Please check at http://www.ox.ac.uk/feesandfunding/prospectivegrad/scholarships/university/louisdreyfusweidenfeld/ for more information.

Magdalen College: Hichens MBA/MFE Awards

Subjects: MBA and MFE.
Purpose: To assist graduate students with accomodation.
Eligibility: Open to all graduate applicants in the MBA and MFE.
Level of Study: MBA, Postgraduate
Type: Award
Value: Free accommodation in college
Length of Study: 1 year
Frequency: Annual
No. of awards offered: 3
Application Procedure: Please see website for details of how to apply at http://www.magd.ox.ac.uk/admissions_graduate/scholarships.shtml.
Additional Information: Recipient(s) chosen by the economics tutors from applicants who have been accepted to read for the MBA/MFE at Magdalen.

Magdalen College: Hong Kong Scholarships

Type: Scholarship

Magdalen College: Perkin Research Studentship

Subjects: All subjects.
Eligibility: Open to all graduate applicants from Commonwealth countries for Chemistry.
Level of Study: Research
Type: Studentship
Value: £7,000 per year
Length of Study: Period of fee liability
Application Procedure: Please see website for details of how to apply.
Additional Information: Please check at http://www.magd.ox.ac.uk/admissions_graduate/scholarships.shtml for more information.

Magdalen College: The Mackinnon and the Tavella Stewart Scholarships

Subjects: All subjects.
Eligibility: Open to all graduate applicants that have completed their first degree at Magdalen College and can demonstrate financial need.
Level of Study: Postgraduate, Research
Type: Scholarship
Value: According to individual circumstances
Length of Study: 1 year with the possibility of renewal
Application Procedure: Please see website for details of how to apply.
Additional Information: Awarded to candidates with research potential who have been unable, after application, to obtain funding from other sources. Please check at http://www.magd.ox.ac.uk/admissions_graduate/scholarships.shtml for more information.

Magdalen College: Zvi and Ofra Meitar Magdalen Graduate Scholarships

Subjects: Sciences.
Eligibility: Open to all graduate applicants in the sciences.
Level of Study: Graduate
Type: Scholarship
Value: Up to £80,000 over four years towards fees and living expenses
Length of Study: Up to 4 years
Application Procedure: Please see website for details of how to apply.
Closing Date: January 18th
Additional Information: Please check at http://www.magd.ox.ac.uk/admissions_graduate/scholarships.shtml for more information.

Mansfield College: John Hodgson Theatre Research Fellowship

Subjects: Graduate research english.
Eligibility: Open to all graduate applicants in the English Faculty working in Theatre Studies.
Level of Study: Research
Type: Fellowship
Value: £4,000 per year
Length of Study: Up to 4 years (depending on period of fee liability)
Application Procedure: Please see website for details of how to apply.
Additional Information: Please check at http://www.mansfield.ox.ac.uk/current/prizes-scholarships.html for more information.

Mathematical Institute: EPSRC Doctoral Training Grant Studentships

Subjects: Mathematics.
Eligibility: Open to UK applicants for Mathematics. Other EU nationals are eligible for a fees-only award.
Level of Study: Research
Type: Studentship
Value: University fee, college fee and full living expenses
Length of Study: Up to 3.5 years
No. of awards offered: Approx. 12
Application Procedure: Applications are made online, different deadlines are used for EPSRC applications. Please visit website for more information and how to apply.
Closing Date: Closing date as advertised on departmental website
Additional Information: Please check at http://www.maths.ox.ac.uk/prospective-students/graduate/procedures/epsrc for more information.

Mathematical Institute: Research Project Studentships

Subjects: Mathematics.
Eligibility: Open to EU applicants for Mathematics.
Level of Study: Research
Type: Studentship
Value: University fee, college fee and full living expenses
Length of Study: Up to 3.5 years
Application Procedure: Please visit website for more information and how to apply.
Closing Date: Janurary 18th
Additional Information: These awards are individually advertised on the departmental website. Please check at http://www.maths.ox.ac.uk for more information.

Medical Sciences Graduate School Studentship

Subjects: Medical sciences.
Eligibility: Open to all applicants to graduate courses in the Medical Sciences will be considered for funding from a number of sources, including the Wellcome Trust, MRC, BBSRC, CR-UK, British Heart Foundation, the Clarendon Fund, Oxford Colleges and numerous charitable sources.
Level of Study: Graduate
Type: Studentship
Value: University fee, college fee, and full living expenses
Length of Study: Period of fee liability
Closing Date: January 4th
Additional Information: Please check at http://www.medsci.ox.ac.uk/graduateschool/finance-funding/oxfunding for more information.

Medieval & Mod Languages: Zaharoff Graduate Scholarship

Subjects: French.
Purpose: To assist graduate students with fees and maintenance.
Eligibility: All eligible doctoral applicants studying French will automatically be considered, irrespective of College choice, provided that they have applied for other funded awards (if eligible). In the event that the successful applicant is subsequently awarded grant funding of a higher value (e.g. AHRC) the scholarship will be withdrawn and awarded to the next best applicant).
Level of Study: Doctorate, Research
Type: Scholarship
Value: University fee (Home/EU rate), and living expenses
Length of Study: Period of fee liability
Application Procedure: Please check the website for further information http://www.mod-langs.ox.ac.uk/prizes/index.php#Zahschol.
Closing Date: January 18th
Additional Information: The award will go to the best applicant. In the event that the successful applicant is subsequently awarded grant funding of a higher value (e.g. AHRC) the scholarship will be withdrawn and awarded to the next best applicant. The successful candidate will be notified in April.

Medieval and Modern Langauges Faculty: AHRC Block Grant Partnership Studentships: Doctoral scheme

Subjects: Medieval and modern languages.
Eligibility: Open to UK and EU applicants for the DPhil degrees offered by the Faculty of Medieval and Modern Languages. Apply on the University application form.
Level of Study: Graduate
Type: Studentship
Value: University fee, college fee, and full living expenses. Fees-only awards for non-UK, EU students
Length of Study: Period of fee liability
Closing Date: January 18th
Additional Information: Please check at http://www.humanities.ox.ac.uk/prospective_students/graduates/funding/ahrc for more information.

Medieval and Modern Languages Faculty: AHRC Block Grant Partnership Studentships: Research Master's scheme

Subjects: Medieval and modern languages; MSt in film aesthetics; MSt in women's studies.
Eligibility: Open to UK and EU applicants for the MSt and MPhil degrees offered by the Faculty of Medieval and Modern Languages; MSt in Film Aesthetics; MSt in Women's Studies.
Level of Study: Graduate
Type: Studentship
Value: University fee, college fee, and full living expenses. Fees-only awards for non-UK, EU students
Length of Study: Period of fee liability
Application Procedure: Apply on the University application form.
Closing Date: January 18th

Additional Information: Please check at http://www.humanities.ox.ac.uk/prospective_students/graduates/funding/ahrc for more information.

Merton College: Barnett Award

Subjects: Law.
Eligibility: Open to all graduate applicants in Law.
Level of Study: Postgraduate, Research
Type: Scholarship
Value: £10,000 per year
Length of Study: Period of fee liability
Frequency: Annual
Application Procedure: Please see website for details of how to apply.
Closing Date: January 18th
Additional Information: Eligible candidates will be identified to the college by the department and the College will make its decision in late May based on the University application form. Please check at http://www.merton.ox.ac.uk/graduateadmissions/application.shtml#gs for more information.

Merton College: Barton Scholarship

Eligibility: Open to British graduate applicants for the BCL.
Level of Study: Postgraduate
Type: Scholarship
Value: £5,000 per year
Length of Study: 1 year
Application Procedure: Please see website for details of how to apply.
Closing Date: January 18th
Additional Information: Please check at http://www.merton.ox.ac.uk/graduateadmissions/application.shtml#gs for more information.

Merton College: Joint RCUK Award (50:50)

Subjects: Ancient history, classics, english language and literature, modern history, archaeology, medical sciences.
Eligibility: Open to all graduate applicants in Economics and rea Studies.
Level of Study: Research
Type: Award
Value: University fee, college fee, and full living expenses
Length of Study: Period of fee liability
No. of awards offered: 6
Application Procedure: Please see website for details of how to apply.
Closing Date: January 18th
Additional Information: One award in each category. Please check at http://www.merton.ox.ac.uk/graduateadmissions/application.shtml#gs for more information.

Merton College: Joint RCUK Award (50:50)

Subjects: Philosophy.
Eligibility: Open to all graduate applicants in DPhil Philosophy.
Level of Study: Graduate
Type: Award
Value: University fee, college fee, and full living expenses
Length of Study: Up to 4 years (depending on period of fee liability)
Application Procedure: Please see website for details of how to apply.
Closing Date: January 18th
Additional Information: Please check at http://www.merton.ox.ac.uk/graduateadmissions/application.shtml#gs for more information.

Merton College: Joint RCUK Award (50:50)

Subjects: Medical sciences.
Eligibility: Open to all graduate applicants for a DPhil in the Medical Sciences.
Level of Study: Graduate
Type: Award
Value: University fee, college fee, and full living expenses
Length of Study: Period of fee liability
Application Procedure: Please see website for details of how to apply.
Closing Date: January 18th
Additional Information: Please check at http://www.merton.ox.ac.uk/graduateadmissions/application.shtml#gs for more information.

Merton College: Joint RCUK Award (50:50)

Subjects: Economics and area studies.
Eligibility: Open to all graduate applicants in Economics and area Studies.
Level of Study: Graduate
Type: Award
Value: University fee, college fee and full living expenses
Length of Study: Period of fee liability
Application Procedure: Please see website for details of how to apply.
Closing Date: January 18th
Additional Information: Please check at http://www.merton.ox.ac.uk/graduateadmissions/application.shtml#gs for more information.

Merton College: Joint RCUK Award (50:50)

Subjects: Oriental studies.
Eligibility: Open to all graduate applicants for DPhil Oriental Studies.
Level of Study: Graduate
Type: Award
Value: University fee, college fee and full living expenses
Length of Study: Up to 4 years (depending on period of fee liability)
Application Procedure: Please see website for details of how to apply.
Closing Date: January 18th
Additional Information: Please check at http://www.merton.ox.ac.uk/graduateadmissions/application.shtml#gs for more information.

Merton College: Joint RCUK Award (50:50)

Subjects: Enlgish.
Eligibility: Open to all graduate applicants in DPhil English.
Level of Study: Graduate
Type: Award
Value: University fee, college fee and full living expenses
Length of Study: Up to 4 years (depending on period of fee liability)
Application Procedure: Please see website for details of how to apply.
Closing Date: January 18th
Additional Information: Please check at http://www.merton.ox.ac.uk/graduateadmissions/application.shtml#gs for more information.

Merton College: Maths & Stats Scholarship

Subjects: Mathematics and statistics.
Eligibility: Open to all DPhil applicants in Mathematics and Statistics.
Level of Study: Graduate
Type: Scholarship
Value: University fee, college fee and full living expenses
Length of Study: Period of fee liability
Application Procedure: Please see website for details of how to apply.
Closing Date: January 18th
Additional Information: Please check at http://www.merton.ox.ac.uk/graduateadmissions/application.shtml#gs for more information.

Merton College: Merton Lawyers' BCL Scholarship

Subjects: Law (BCL).
Eligibility: Open to British graduate applicants for the BCL.
Level of Study: Postgraduate
Type: Scholarship
Value: £5,000 per year
Length of Study: 1 year
Application Procedure: Please see website for details of how to apply.
Closing Date: January 18th
Additional Information: Please check at http://www.merton.ox.ac.uk/graduateadmissions/application.shtml#gs for more information.

Mica and Ahmet Ertegun Graduate Scholarship Programme in the Humanities

Subjects: Humanities.
Eligibility: Open to graduate applicants in the Humanities.
Level of Study: Graduate

Type: Scholarship
Value: University fee, college fee and full living expenses
Length of Study: Period of fee liabilty
Application Procedure: Candidates apply on the University application form.
Closing Date: January 4th
Additional Information: Please check at www.ox.ac.uk/ertegun/ for more information.

Music: AHRC Block Grant Partnership Studentships: Doctoral scheme

Subjects: Music.
Eligibility: Open to UK and EU applicants for the DPhil degrees offered by the Faculty of Music.
Level of Study: Graduate
Type: Studentship
Value: University fee, college fee, and full living expenses. Fees-only awards for non-UK, EU students
Length of Study: Period of fee liability
Application Procedure: Apply on the University application form.
Closing Date: January 18th
Additional Information: Please check at http://www.humanities.ox.ac.uk/prospective_students/graduates/funding/ahrc for more information.

Music: AHRC Block Grant Partnership Studentships: Research Master's scheme

Subjects: Music.
Eligibility: Open to UK and EU applicants for the MSt and MPhil degrees offered by the Faculty of Music.
Level of Study: Graduate
Type: Studentship
Value: University fee, college fee, and full living expenses. Fees-only awards for non-UK, EU students
Length of Study: Period of fee liability
Application Procedure: Apply on the University application form.
Closing Date: January 18th
Additional Information: Please check at http://www.humanities.ox.ac.uk/prospective_students/graduates/funding/ahrc for more information.

New College: 1379 Society Old Members Scholarship

Eligibility: Open to all graduate applicants for the BCL or MJur.
Level of Study: Graduate
Type: Scholarship
Value: £17,000 per year
Length of Study: Up to 3 years
Application Procedure: Please see website for details of how to apply.
Closing Date: January 18th
Additional Information: Please check at http://www.new.ox.ac.uk/graduate/scholarships for more information.

New College: 1379 Society Old Members Scholarship

Subjects: All subjects.
Purpose: To assist graduate students with fees.
Eligibility: Open to all Home/EU applicants for a research degree in any subject.
Level of Study: Research
Type: Award
Value: £10,000 per year
Length of Study: Period of fee liability
No. of awards offered: 6
Application Procedure: Please see website for details of how to apply at http://www.new.ox.ac.uk/graduate/scholarships.
Closing Date: Januray 18th

New College: 1379 Society Old Members Scholarship: The David Gieve Award

Subjects: Politics or economics.
Eligibility: Open to all graduate applicants for a research degree in politics or economics.
Type: Scholarship
Value: £17,000 per year
Length of Study: Up to 3 years
Application Procedure: Please see website for details of how to apply.
Closing Date: January 18th
Additional Information: Please check at http://www.new.ox.ac.uk/graduate/scholarships for more information.

New College: 1379 Society Old Members Scholarship: The Graeme Gates Award

Subjects: All subjects.
Eligibility: Open to all Home/EU applicants for a research degree in any subject.
Level of Study: Graduate
Type: Scholarship
Value: £5,000 per year
Length of Study: Period of fee liability
Application Procedure: Please see website for details of how to apply.
Closing Date: January 18th
Additional Information: Please check at http://www.new.ox.ac.uk/graduate/scholarships for more information.

New College: 1379 Society Old Members Scholarship: The Patrick Stables Award

Subjects: All subjects.
Eligibility: Open to all Home/EU applicants for a research degree in any subject.
Level of Study: Graduate
Type: Scholarship
Value: £5,000 per year
Length of Study: Period of fee liability
Application Procedure: Please see website for details of how to apply.
Closing Date: January 18th
Additional Information: Please check at http://www.new.ox.ac.uk/graduate/scholarships for more information.

New College: 1379 Society Old Members Scholarship: The Yeotown Scholarship in Science

Subjects: Science.
Eligibility: Open to all Home/EU applicants for a research degree in a science subject that candidates have not studied at Oxford previously.
Level of Study: Graduate
Type: Scholarship
Value: £10,000 per year
Length of Study: Up to 4 years (depending on period of fee liability)
Application Procedure: Please see website for details of how to apply.
Closing Date: January 18th
Additional Information: Please check at http://www.new.ox.ac.uk/graduate/scholarships for more information.

New College: Aso Group Scholarship

Subjects: All subjects.
Eligibility: Open to all Japanese graduate applicants applying for a research degree or intending to go on to one. Preference for applicants from Fukuoka Prefecture.
Type: Scholarship
Value: Up to £25,000
Length of Study: 1 year
Application Procedure: Please see website for details of how to apply at http://www.new.ox.ac.uk/graduate/scholarships.
Closing Date: January 18th
Additional Information: Students who have received their education in Japan, and are native speakers of Japanese, with a preference for applicants from Fukuoka Prefecture. Excluding MBA or Medicine.

Nissan Institute Bursary

Subjects: Modern Japanese studies.
Eligibility: Open to all graduate applicants for the MSc or Mphil in Modern Japanese Studies.
Level of Study: Graduate
Type: Bursary
Value: £3,000

Length of Study: 1 year
Application Procedure: Please visit website for more details and how to apply.
Closing Date: March 8th
Additional Information: Please check at http://www.nissan.ox.ac.uk/prospective_students/msc_modern_japanese_studies/funding for more information.

Noon Foundation / OSI / Chevening / Oxford Scholarships

Subjects: Medical science.
Eligibility: Open to Pakistani graduate applicants for all subjects except in the Medical Sciences.
Level of Study: Graduate
Type: Scholarship
Value: Full or partial awards to cover University and college fees and a full grant for living costs
Length of Study: Period of fee liability
Application Procedure: Please see website for more details and how to apply.
Closing Date: January 18th
Additional Information: Please check at http://www.ox.ac.uk/feesandfunding/prospectivegrad/scholarships/university/noon/ for more information.

Nuffield College: JK Swire Memorial Scholarship

Subjects: Social sciences.
Eligibility: Open to all graduate applicants in Social Sciences from China.
Level of Study: Postgraduate, Research
Type: Scholarship
Value: University fee, college fee, and full living expenses
Length of Study: Period of fee liability
Application Procedure: Please see website for details of how to apply.
Additional Information: Please check at http://www.nuffield.ox.ac.uk/general/prospectus/funding.aspx for more information.

Nuffield College: Nuffield Sociology Doctoral Studentships

Subjects: Sociology.
Purpose: To assist doctoral students with their university and their college fees.
Eligibility: Open to all graduate applicants for a DPhil in Sociology. Please note that this award is not available to Oxford students transferring from a Masters course.
Level of Study: Research, Doctorate
Type: Studentship
Value: University fee (at the Home/EU level), college fee, and full living expenses
Length of Study: Period of fee liability
Frequency: Annual
No. of awards offered: Up to 2
Application Procedure: Nuffield also has links with the Marshall Scholarship scheme, in the form of a Nuffield/Marshall Scholarship. The Scholarship provides full funding for students for a maximum of four years. Application should be made via the usual Marshall Scholarship application process.
Additional Information: Not available to Oxford students transferring from a Masters course. Please check at www.nuffield.ox.ac.uk/general/prospectus/funding.aspx.

Nuffield College: Nuffield Studentships

Subjects: Social sciences.
Purpose: To assist students in the payment of fees and provide support for maintenance.
Eligibility: Open to all graduate applicants in Social Sciences.
Level of Study: Postgraduate, Research
Type: Studentship
Value: Up to £18,000 per year (for home/EU students; £25,000 for overseas students)
Length of Study: Period of fee liability
No. of awards offered: Varies
Application Procedure: There is no separate application procedure.
Additional Information: Please check at http://www.nuffield.ox.ac.uk/general/prospectus/funding.aspx for more information.

Nuffield College: Nuffield/Marshall Scholarship

Subjects: Social science.
Eligibility: Open to all graduate applicants to the Social Sciences that are US citizens.
Level of Study: Postgraduate, Research
Type: Scholarship
Value: University fee, college fee, and full living expenses
Length of Study: Period of fee liability
Application Procedure: Candidates fill out the form and submit it through the system to their institution, the institution ensures that the application is complete and decides whether or not to endorse it. If the Institution decides to endorse the application they will add the letter of endorsement and will then submit it through the system to the appropriate regional committee. Candidates should contact the appropriate contact at their institution if they are considering applying for a Marshall Scholarship. Please see website for more details and how to apply.
Additional Information: Candidates may apply for either the one year or two year Scholarship not both. Any candidate found applying for both will automatically be disqualified. Please see the website http://www.marshallscholarship.org/applications/eligible for more information.

Oppenheimer Fund Scholarships

Eligibility: Open to all graduate applicants to degree-bearing courses from South Africa.
Level of Study: Graduate
Type: Scholarship
Value: Up to £6,500 per year towards fees or living expenses
Length of Study: Period of fee liability
Application Procedure: For more details and how to apply, please visit website.
Closing Date: January 18th
Additional Information: Please check at http://www.ox.ac.uk/feesandfunding/prospectivegrad/scholarships/university/oppenheimer/ for more information.

Oriel College: Frankel Memorial Scholarship

Subjects: Economics or political economy.
Eligibility: Open to all DPhil applicants, or those already undertaking research in, economics or political economics.
Level of Study: Research
Type: Scholarship
Value: £5,000 per year per year plus additional benefits including accomodation rights.
Length of Study: Period of fee liability
Application Procedure: Please see website for details of how to apply.
Additional Information: The successful applicant will be expected to transfer their College membership to Oriel College.

Oriel College: Oriel Graduate Scholarships

Subjects: All subjects.
Eligibility: Open to all current graduate students at the college.
Level of Study: Research, Postgraduate
Type: Scholarship
Value: Approximately £2,500 per year and acommodation rights
Length of Study: Period of fee liability
Application Procedure: Please see website for details of how to apply.
Additional Information: Scholars are entitled to dine free of charge at High Table once per week during term time. Please contact to Academic Assistant at academic.office@oriel.ox.ac.uk.

Oriel College: Sir Walter Raleigh Scholarship

Subjects: Environmental change and management.
Purpose: To assist a student studying at the Environmental Change Institute for an MSc degree in Environmental Change and Management.
Eligibility: Open to all applicants to the MSc in Environmental Change and Management.
Level of Study: Postgraduate
Type: Scholarship
Value: £4,000
Length of Study: 1 year

Frequency: Annual
No. of awards offered: 1
Application Procedure: Applicants can download an application form from the website www.ox.ac.uk/admissions/index.html or obtain one from the Academic Office or the Environmental Change Institute.
Additional Information: The successful applicant will be expected to transfer their application to Oriel College if they have already been accepted by another college.

For further information contact:

Academic Office or the Environmental Change Institute
Email: admissions@oriel.ox.ac.uk
Website: www.ox.ac.uk/admissions/index.html
Contact: Admissions Officer

Oriental Studies: AHRC Block Grant Partnership Studentships: Doctoral scheme

Subjects: Oriental studies.
Eligibility: Open to UK and EU applicants for the DPhil degrees offered by the Faculty of Oriental Studies.
Level of Study: Graduate
Type: Studentship
Value: University fee, college fee, and full living expenses. Fees-only awards for non-UK, EU students
Length of Study: Period of fee liability
Application Procedure: Apply on the University application form.
Closing Date: January 18th
Additional Information: Please check at http://www.humanities.ox.ac.uk/prospective_students/graduates/funding/ahrc for more information.

Oriental Studies: AHRC Block Grant Partnership Studentships: Research Master's scheme

Subjects: Oriental studies.
Eligibility: Open to UK and EU applicants for the MSt and MPhil degrees offered by the Faculty of Oriental Studies.
Level of Study: Graduate
Type: Studentship
Value: University fee, college fee, and full living expenses. Fees-only awards for non-UK, EU students
Length of Study: Period of fee liability
Application Procedure: Apply on the University application form.
Closing Date: January 18th
Additional Information: Please check at http://www.humanities.ox.ac.uk/prospective_students/graduates/funding/ahrc for more information.

Oriental Studies: Sasakawa Fund

Subjects: Japanese.
Eligibility: Open to all graduate applicants in Oriental Studies, or UK applicants whose work will require some time spent in Japan.
Level of Study: Postgraduate, Research
Type: Award
Value: £5,000 per year
Length of Study: Period of fee liability
No. of awards offered: Up to 2
Application Procedure: An application form is available at http://www.orinst.ox.ac.uk/general/grants/sasakawa_fund.html.
Closing Date: January 18th

For further information contact:

The Secretary of the Sasakawa Fund, The Oriental Institute, Pusey Lane, Oxford, OX1 2LE
Tel: 01865 278225
Fax: 01865 278190
Email: chris.williams@orinst.ox.ac.uk

ORISHA DPhil Scholarship (African Humanities Research Fund) at St Antony's College

Subjects: Humanities.
Eligibility: Open to all graduate applicants to the humanities with research proposals with a major focus on Africa. African students are strongly encoraged to apply.
Level of Study: Graduate
Type: Scholarship
Value: University fee, college fee and living expenses
Length of Study: Period of fee liability
Application Procedure: Please see the website for more details and how to apply.
Closing Date: January 18th
Additional Information: Please check at http://www.africanstudies.ox.ac.uk/dphil-scholarships#ORISHA%20DPhil for more information.

OSF Scholarships

Subjects: Humanities, social sciences and environmental sciences.
Purpose: To support students who are citizens of and ordinarily resident in Albania, Armenia, Azerbaijan, Bosnia and Herzegovina, Georgia, Kosovo, and Montenegro.
Eligibility: Open to visiting Masters students.
Level of Study: Postgraduate
Type: Scholarship
Value: University fee, college fee, and full living expenses. Additional expenses for flights may be included.
Length of Study: 1 year
Application Procedure: Contact your local coordinator who will assess your eligibility. If you are eligible, they will provide you with a reference number which will enable your application fee to be waived. Please send this number to www.graduate.ox.ac.uk/ask along with your full name, email address and course you are applying for to request the waiver.
Closing Date: January 18th
Additional Information: Please check at http://www.ox.ac.uk/feesandfunding/prospectivegrad/scholarships/university/osfoxford/ for more information.

OUCEA Studentships

Eligibility: Open to all applicants for DPhil in Education.
Level of Study: Graduate
Type: Studentship
Value: University fee, college fee and full living expenses
Length of Study: Up to 3 years
Application Procedure: Please visit website for more details and how to apply.
Closing Date: March 8th
Additional Information: Please check at http://www.education.ox.ac.uk/courses/d-phil/funding-opportunities/ for more information.

Oxford Centre for Asian Archaeology, Art and Culture Doctoral Studentships

Subjects: Project title is China and Inner Asia (c.1000–200BC): interactions that changed China.
Eligibility: Open to all applicants for DPhil Archaeology whose research will centre on China and Inner Asia (c.1000-200BC). Please visit website for more details and how to apply.
Level of Study: Postgraduate, Research
Type: Studentship
Value: University fee, college fee and full living expenses.
Length of Study: Up to 3 years
Frequency: Annual
Application Procedure: Applicants intending to start their course in October should notify Professor Jessica Rawson (jessica.rawson@merton.ox.ac.uk) as soon as possible of their intention to apply, providing email and postal addresses.
Additional Information: Candidates should have a first degree in either an archaeological subject or in Chinese language if Chinese is not the applicant's first language. Please check at http://www.arch.ox.ac.uk/studentships.html for more information.

Oxford Centre for Islamic Studies (OCIS) Graduate Scholarship

Eligibility: Open to UK graduate applicants from Muslim communities and applicants from developing countries in Asia and Africa for taught and research degrees in fields derived from or of relevance to the Islamic tradition.
Level of Study: Graduate
Type: Scholarship
Value: University fee, college fee and full living expenses
Length of Study: Period of fee liability

Application Procedure: Please see website for more details and how to apply.
Closing Date: January 18th
Additional Information: Please check at http://www.ox.ac.uk/feesandfunding/prospectivegrad/scholarships/university/ocis/ for more information.

Oxford Internet Institute Scholarship

Subjects: Information, communication and social sciences, social science of the internet.
Purpose: To assist graduate students with fees.
Eligibility: Open to all graduate applicants and current students on the OII DPhil and MSc programmes.
Level of Study: Postgraduate, Research
Type: Scholarship
Value: University fee
Length of Study: 1 year
Frequency: Annual
Application Procedure: All prospective candidates must make their application through the University's Graduate Studies Office, and are advised to consult the University's Graduate Studies Prospectus beforehand. All applications are processed initially by the Graduate Studies Office, and must not be made directly to the OII. Please visit website for more details and how to apply at http://www.oii.ox.ac.uk/teaching/dphil/apply.cfm
Closing Date: January 18th
No. of awards given last year: 7
No. of applicants last year: 43
Additional Information: Prospective candidates are strongly advised to apply in the first two deadlines if they are planning to apply for funding.

Oxford Kobe Scholarships

Subjects: All subjects.
Purpose: To enable Japanese graduate students to study at Oxford.
Eligibility: Open to all Japanese graduate applicants.
Level of Study: Postgraduate, Research
Type: Scholarships
Value: University fee, college fee, and full living expenses. Additional expenses for flights may be included.
Length of Study: Period of fee liability
Application Procedure: Apply at the same time as you apply to Oxford by selecting Oxford Kobe Scholarships in the Funding Section of the University's Graduate Application Form. Please see website for more details and how to apply.
Closing Date: January 18th
Additional Information: Eligible to nationals of Japan. Applicant must have secured a place on your chosen programme of study by the expected final decision date (March 17th). Please check at http://www.ox.ac.uk/feesandfunding/prospectivegrad/scholarships/university/kobe/ for more information.

Oxford Marshall Scholarships

Subjects: All subjects.
Purpose: To assist graduate students with fees and living allowance.
Eligibility: Open to all US graduate applicants.
Level of Study: Postgraduate, Research
Type: Scholarships
Value: University fee, college fee, and full living expenses
Length of Study: 2 years (possibly 3 years)
Frequency: Annual
Application Procedure: Please see website for more details and how to apply.
Additional Information: Please check at http://www.marshallscholarship.org for more information.

Oxford University Alumni Scholarship

Eligibility: Open to all applicants for the EMBA.
Level of Study: Graduate
Type: Scholarship
Value: 50 per cent of university fee
Length of Study: 2 years
Application Procedure: Please visit website for more details and how to apply.
Additional Information: Please check at http://www.sbs.ox.ac.uk/degrees/emba/Pages/alumnischolarship.aspx for more information.

Pembroke College: Atkinson Scholarship

Subjects: Any, but preferred subjects are classics, law, medicine, philosophy or theology.
Eligibility: Open to all graduate applicants from Australia. Preference for Classics, Law, Medicine, Philosophy or Theology.
Level of Study: Postgraduate, Research
Type: Scholarship
Value: £5,000 per year towards college fee and contribution towards living expenses. Plus additional benefits including dining rights.
Length of Study: Up to 4 years (depending on period of fee liability)
Application Procedure: Please see website for details of how to apply.
Closing Date: May 3rd
Additional Information: Please check at http://www.pmb.ox.ac.uk/Students/Graduate_Students/Scholarships_Awards/Graduate_Scholarships.php for more information.

Pembroke College: Gordon Aldrick Scholarship

Subjects: Chinese studies.
Eligibility: Open to all Chinese graduate applicants for research degrees in the area of cultural studies, literature, art or history.
Level of Study: Research
Value: £5,000 per year towards college fee and contribution towards living expenses.
Length of Study: Period of fee liability
Application Procedure: Please see website for details of how to apply at http://www.pmb.ox.ac.uk/Students/Graduate_Students/Scholarships_Awards/Graduate_Scholarships.php.
Closing Date: May 3rd

Pembroke College: Stanley Ho Scholarship

Subjects: Chinese studies.
Purpose: To assist applicants who are eligible to pay university and college fees and who have been accepted to read a research degree in Oxford University in Pembroke college.
Eligibility: Open to all Chinese graduate applicants for research dregrees.
Level of Study: Research
Type: Scholarship
Value: £5,000 per year towards college fee and contribution towards living expenses.
Length of Study: Period of fee liability
No. of awards offered: 1
Application Procedure: Intending candidates should apply in Section L of the Oxford University graduate application form or notify the Admissions & Acess Officer at Pembroke. Please see website for details of how to apply at http://www.pmb.ox.ac.uk/Students/Graduate_Students/Scholarships_Awards/Graduate_Scholarships.php.
Closing Date: May 3rd

Pembroke College: TEPCO Scholarship

Subjects: Japanese studies.
Eligibility: Open to all graduate applicants specialising in studies of Japanese literature, art or history. Please see website for details of how to apply at http://www.pmb.ox.ac.uk/Students/Graduate_Students/Scholarships_Awards/Graduate_Scholarships.php.
Level of Study: Research
Value: £5,000 per year towards college fee and contribution towards living expenses.
Length of Study: Period of fee liability
Closing Date: May 3rd

Philosophy Faculty: AHRC Block Grant Partnership Studentships: Doctoral scheme

Subjects: Philosophy.
Eligibility: Open to UK and EU applicants for the DPhil degrees offered by the Faculty of Philosophy.
Level of Study: Graduate
Type: Studentship
Value: University fee, college fee and full living expenses. Fees-only awards for non-UK, EU students

Length of Study: Period of fee liability
Application Procedure: Apply on the University application form.
Closing Date: January 4th
Additional Information: Please check at http://www.humanities.ox.ac.uk/prospective_students/graduates/funding/ahrc for more information.

Philosophy Faculty: AHRC Block Grant Partnership Studentships: Research Master's scheme

Subjects: Philosophy.
Eligibility: Open to UK and EU applicants for the MSt and BPhil degrees offered by the Faculty of Philosophy.
Level of Study: Graduate
Type: Studentship
Value: University fee, college fee and full living expenses. Fees-only awards for non-UK, EU students
Length of Study: Period of fee liability
Application Procedure: Apply on the University application form.
Closing Date: January 4th
Additional Information: Please check at http://www.humanities.ox.ac.uk/prospective_students/graduates/funding/ahrc for more information.

Physics: Joint Department-College scholarships

Subjects: Physics.
Purpose: To assist graduate students with fees and maintenance.
Eligibility: Open to all EU applicants in Physics.
Level of Study: Research
Type: Scholarships
Value: University fee, college fee and full living expenses
Length of Study: 3 years (with possible extension)
No. of awards offered: 1 or 2
Application Procedure: All applications should be made using the central University form available online on the University Admissions site. Please visit website for more information and how to apply.
Additional Information: Please check at http://www.physics.ox.ac.uk/admissions/postgrad.htm for more information.

Pirie-Reid Scholarship

Subjects: All subjects.
Purpose: To benefit Scottish students coming to Oxford.
Eligibility: Open to all Scottish graduate applicants.
Level of Study: Postgraduate, Research
Type: Scholarship
Value: University fee, college fee, and full living expenses
Length of Study: Period of fee liability
Application Procedure: Apply at the same time as you apply to Oxford by selecting Pirie Reid in the Funding Section of the University's Graduate Application Form. Please see website for more details and how to apply at http://www.ox.ac.uk/feesandfunding/prospectivegrad/scholarships/university/piriereid/.
Closing Date: January 18th
Additional Information: Eligible to nationals of Scotland. If you do not apply in full by the deadline, you will not be considered for the scholarship, even if you have selected Pirie Reid on the Graduate Application Form.

Plant Sciences: Christopher Welch Scholarship

Subjects: Biological sciences.
Purpose: To assist graduate students with fees and maintenance.
Eligibility: Open to UK applicants in the Biological Sciences.
Level of Study: Research
Type: Scholarship
Value: University fee (Home/EU rate), college fee, £13,500 towards living expenses per year
Length of Study: 2 years (with possible extension)
Application Procedure: Please visit website for more details and how to apply at http://dps.plants.ox.ac.uk/plants/students/postgraduates/default.aspx.
Closing Date: January 18th

Plant Sciences: DTP Studentships

Subjects: Plant sciences.
Eligibility: Ope to all graduate applicants to Plant Sciences.
Level of Study: Graduate
Type: Studentship
Value: University fee, college fee and full living expenses
Length of Study: 4 years
Application Procedure: Please visit website for more details and how to apply.
Additional Information: Please check at http://www.biodtp.ox.ac.uk/ for more information.

Plant Sciences: Newton Abraham Scholarship

Subjects: Biological sciences.
Purpose: To assist graduate students with fees and maintenance.
Eligibility: Open to UK applicants in the Biological Sciences.
Level of Study: Research
Type: Scholarship
Value: University fee, college fee and full living expenses
Length of Study: Up to 3.5 years
Application Procedure: Please visit website for more details and how to apply at http://dps.plants.ox.ac.uk/plants/students/postgraduates/default.aspx.

Politics and International Relations: AHRC BGP Studentships - Doctoral Awards

Subjects: Politics specialising in political theory or philosophy.
Eligibility: Open to all UK applicants to the DPhil in Politics specialising in political theory or philosophy. Other EU nationals are eligible for a fees-only award.
Level of Study: Graduate
Type: Studentship
Value: University fee, college fee and full living expenses
Length of Study: Up to 3 years
Application Procedure: Please visit website for more details and how to apply.
Closing Date: January 4th
Additional Information: Please check at http://www.politics.ox.ac.uk/index.php/student-funding/student-funding.html for more information.

Politics and International Relations: Departmental Bursaries

Eligibility: Open to all graduate applicants in Politics and International Relations and those on course and still liable for fees.
Level of Study: Research, Postgraduate
Type: Bursary
Value: £2,500
Length of Study: 1 year
Frequency: Annual
No. of awards offered: 12
Application Procedure: Please visit website for more details and how to apply.
Additional Information: Please check at http://www.politics.ox.ac.uk/index.php/student-funding/student-funding.html for more information.

Politics and International Relations: Departmental Studentships

Subjects: Politics and international relations.
Eligibility: Open to all graduate applicants in Politics and International Relations and those on course.
Level of Study: Postgraduate, Research
Type: Studentship
Value: Value of award varies. Please see website for more details.
Length of Study: Up to 3 years
Frequency: Annual
Application Procedure: Please visit website for more details and how to apply.
Closing Date: January 18th
Additional Information: Please check at http://www.politics.ox.ac.uk/index.php/student-funding/student-funding.html for more information.

Politics and International Relations: ESRC DTC +3 Studentships (Doctoral Awards)

Subjects: Dphil in Politics and International Relations.
Eligibility: Open to all UK applicants for DPhil Politics and Dphil International Relations. Other EU nationals are eligible for a fees-only award.

Level of Study: Graduate, Postdoctorate
Type: Studentship
Value: University fee, college fee and full living expenses
Length of Study: Up to 3 years
Application Procedure: Please visit website for more details and how to apply.
Closing Date: January 18th
Additional Information: Please check at http://www.politics.ox.ac.uk/index.php/student-funding/student-funding.html for more information.

Politics and International Relations: ESRC DTC 1+3 / 2 +2 Studentships (Masters leading to Doctorate)

Subjects: Masters course in Politics and International relations with intention of going on to doctoral study.
Eligibility: Open to all UK applicants for Masters course in Politics and International relations with intention of going on to doctoral study. Other EU nationals are eligible for a fees-only award.
Level of Study: Graduate, Predoctorate
Type: Studentship
Value: University fee, college fee and full living expenses
Length of Study: Up to 4 years
Application Procedure: Please visit website for more details and how to apply.
Closing Date: January 18th

Politics and IR: AHRC BGP Studentships - Research Master's Awards

Subjects: MSc in Political Theory Research and MPhil in Politics (Political Theory).
Eligibility: Open to all applicants for MSc in Political Theory Research and MPhil in Politics (Political Theory). Other EU nationals are eligible for a fees-only award.
Level of Study: Graduate, Postgraduate
Type: Studentship
Value: University fee, college fee and full living expenses
Length of Study: Period of fee liability
Application Procedure: Please visit website for more details and how to apply.
Closing Date: January 4th

Project Studentships in Computer Science, jointly with industry

Subjects: DPhil computer science.
Eligibility: Open to all EU applicants for DPhil Computer Science.
Level of Study: Graduate
Type: Studentship
Value: University fee, college fee and full living expenses
Length of Study: Up to 3.5 years
Application Procedure: Please visit website for more information and how to apply.
Additional Information: Please check at http://www.cs.ox.ac.uk/news/studentships.html for more information.

Project Studentships in Condensed Matter Physics, Atomic & Laser Physics and Theoretical Physics

Subjects: Condensed matter physics (experiment & theory); atomic and laser physics (experiment & theory).
Purpose: To assist graduate students with fees and maintenance.
Eligibility: Open to EU applicants for Condensed Matter Physics; Atomic & Laser Physics; Theoretical Physics.
Level of Study: Research
Type: Studentship
Value: University fee, college fee and full living expenses
Length of Study: 3 years (with possible extension)
Application Procedure: Candidates are advised to apply directly to the sub-departments where advertised on the website. Please visit website for more information and how to apply.
Closing Date: January 18th
Additional Information: These are for specific EPSRC funded projects, the number varies from year to year but is typically between 10–15 each year. Please check at http://www.physics.ox.ac.uk/admissions/postgrad.htmfor more information.

Queen's College: Florey EPA Studentship

Subjects: Medical, biological and chemical sciences.
Eligibility: Open to all graduate applicants from selected countires for study of Medical, Chemical or Biological Science.
Level of Study: Postgraduate, Research
Type: Studentship
Value: Half of university fee and full college fee
Length of Study: 1 year (with possibility of renewal)
Frequency: Annual
Application Procedure: Please visit website for more details and how to apply.
Additional Information: Eligible to nationals of Florey. Please check at http://www.queens.ox.ac.uk/admissions/postgraduates for more information.

Queen's College: George Oakes Senior Scholarship

Subjects: Any degree involving a thesis on some aspect of the culture of the USA.
Eligibility: Open to all graduate applicants whose study will concentrate on an aspect of the culture of the USA.
Level of Study: Postgraduate, Research
Type: Scholarship
Value: College fee
Length of Study: 1 year (with possibility of renewal)
Additional Information: Please contact the Tutor for Graduates for further information. Please check at http://www.queens.ox.ac.uk/admissions/postgraduate-admissions/ for more information.

Queen's College: Holwell Studentship

Subjects: Theology.
Eligibility: Open to all graduate applicants in Theology.
Level of Study: Postgraduate, Research
Type: Studentship
Value: College fee
Length of Study: 2 years (with possibility of renewal)
Frequency: Annual
Application Procedure: Please visit website for more details and how to apply.
Additional Information: The award is in conjunction with graduate studentship offered by Theology Faculty. Please check at http://www.queens.ox.ac.uk/admissions/postgraduate-admissions/ for more information.

REES Studentship

Eligibility: Open to all applicants for DPhil in Education.
Level of Study: Graduate
Type: Studentship
Value: University fee, college fee and full living expenses
Length of Study: Up to 3 years
Application Procedure: Please visit website for more details and how to apply.
Closing Date: March 8th
Additional Information: Please check at http://www.education.ox.ac.uk/courses/d-phil/funding-opportunities/ for more information.

Regent's Park College: Eastern European Scholarship

Subjects: Theology.
Eligibility: Open to all graduate applicants in Theology from Central and Eastern Europe. Preference is shown for members of the Baptist denomination.
Level of Study: Research, Postgraduate
Type: Scholarship
Value: College fee
Length of Study: Up to 3 years
Frequency: Annual
Application Procedure: Please visit website for more details and how to apply.
Closing Date: April 27th

For further information contact:

Regent's Park College, Pusey Street, Oxford, OX1 2LB, United Kingdom
Tel: (44) 01865 288120
Fax: (44) 01865 288121
Email: enquiries@regents.ox.ac.uk
Website: http://www.rpc.ox.ac.uk

Regent's Park College: Ernest Payne Scholarship
Subjects: Theology.
Eligibility: Open to all UK graduate applicants in Theology. Preference is shown for those preparing for the Baptist ministry.
Level of Study: Postgraduate, Research
Type: Scholarship
Value: College fee
Length of Study: 2 years
Application Procedure: Please visit website for more details and how to apply.
Closing Date: April 27th
Additional Information: Scholarship is awarded over 2 years, extendable in proportion for a 3rd year.

For further information contact:

Regent's Park College, Pusey Street, Oxford, OX1 2LB, United Kingdom
Tel: (44) 01865 288120
Fax: (44) 01865 288121
Email: enquiries@regents.ox.ac.uk
Website: http://www.rpc.ox.ac.uk

Regent's Park College: Henman Scholarship
Subjects: Theology.
Eligibility: Open to all graduate applicants in Theology.
Level of Study: Postgraduate, Research
Type: Scholarship
Value: Up to the amount of the college fee each year
Length of Study: Period of fee liability
Application Procedure: Please see website for details of how to apply.
Closing Date: April 26th
Additional Information: Eligible to overseas nationals.

For further information contact:

Email: larry.kreitzer@regents.ox.ac.uk)
Website: http://www.rpc.ox.ac.uk
Contact: Dr Larry Kreitzer, Graduate Studies Tutor

Regent's Park College: J W Lord Scholarship
Eligibility: Open to all overseas graduate applicants in Theology that are preparing to serve Christian churches in India, Hong Kong or China, or otherwise in Asia, Africa, Central and South America and the Caribbean.
Level of Study: Graduate
Type: Scholarship
Value: College fee
Length of Study: 2 years
Application Procedure: Please visit website for more details and how to apply.
Closing Date: April 27th
Additional Information: Please check at http://www.rpc.ox.ac.uk for more information.

Regent's Park College: J W Lord Scholarship
Subjects: Theology.
Eligibility: Open to all graduate applicants in Theology preparing to serve Christian churches in India, Hong Kong or China, or otherwise in Asia, Africa, Central and South America and the Caribbean.
Type: Scholarship
Value: Up to the amount of the college fee each year
Length of Study: Period of fee liability
Frequency: Annual
Application Procedure: Please see website for details of how to apply.
Closing Date: April 26th
Additional Information: Eligible to overseas countries.

For further information contact:

Regent's Park College, Pusey Street, Oxford, OX1 2LB, United Kingdom
Tel: (44) 01865 288120
Fax: (44) 01865 288121
Email: enquiries@regents.ox.ac.uk
Website: http://www.rpc.ox.ac.uk

Regent's Park College: Studentships of the Centre for Christianity and Culture
Subjects: Anthropology or Literature, etc.
Eligibility: Open to all graduate applicants in Theology who want to make connections between the field and other subjects e.g. Anthropology or Literature etc.
Type: Studentship
Value: College fee
Length of Study: Up to 3 years
Application Procedure: Please visit website for more details and how to apply.
Closing Date: April 27th
Additional Information: Please check at http://www.rpc.ox.ac.uk for more information.

Regent's Park College: Studentships of the Centre for Christianity and Culture
Subjects: Theology.
Eligibility: Open to all graduate applicants in Theology who want to make connections between it and other subjects e.g. Anthropology or Literature.
Level of Study: Postgraduate, Research
Type: Studentship
Value: Up to the amount of the college fee each year
Length of Study: Period of fee liability
Application Procedure: Please see website for details of how to apply.
Closing Date: April 26th
Additional Information: Please check at http://www.rpc.ox.ac.uk for more information.

Rhodes Scholarships
Subjects: All graduate subjects except MBA.
Purpose: To assist graduate students with fees and living allowance.
Eligibility: Eligibility varies by country.
Level of Study: Research, Postgraduate
Type: Scholarships
Value: University fee, college fee, and full living expenses
Length of Study: Period of fee liability
Application Procedure: Varies by country; please see website for more details and how to apply.
Closing Date: January 18th
Additional Information: MFE only tenable in 2nd year of Scholarship. Please check at http://www.rhodeshouse.ox.ac.uk/ for more information.

Ruskin School of Drawing: AHRC Block Grant Partnership Studentships: Doctoral Scheme
Eligibility: Open to UK and EU applicants for the DPhil degree offered by the Ruskin School.
Level of Study: Graduate
Type: Studentship
Value: University fee, college fee and full living expenses. Fees-only awards for non-UK, EU students
Length of Study: Period of fee liability
Application Procedure: Apply on the University application form.
Closing Date: January 18th
Additional Information: Please check at http://www.humanities.ox.ac.uk/prospective_students/graduates/funding/ahrc for more information.

Russian & East European Studies: AHRC Block Grant Partnership Studentships: Research Master's Scheme
Subjects: MSc and Mphil in Russian and East European studies.
Eligibility: Open to all UK applicants for the MSc and Mphil in Russian and East European Studies. Other EU nationals are eligible for a fees-only award.
Level of Study: Graduate
Type: Studentship
Value: University fee, college fee and full living expenses
Length of Study: Period of fee liability
Application Procedure: Please visit website for more details and how to apply.
Closing Date: January 18th

Additional Information: Please check at http://www.humanities.ox.ac.uk/prospective_students/graduates/funding/ahrc for more information.

Said Business School Scholarship
Subjects: Management research.
Purpose: To assist graduate students with fees.
Eligibility: Open to all applicants for the MBA.
Level of Study: Postgraduate
Type: Scholarship
Value: Up to £20,000
Length of Study: 1 year
Frequency: Annual
Application Procedure: The application process involves a number of steps. Candidates may apply online via Apply Yourself at http://www.sbs.ox.ac.uk/degrees/mba/Pages/process.aspx.
Additional Information: There are 3 application stages and applicants can apply in any of these. However, applicants are advised to apply as early as possible. Please check at http://www.sbs.ox.ac.uk/degrees/mba/Pages/FAQ.aspx#ans4 for FAQs.

For further information contact:

Admission Office, Said Business School, Park End Street, Oxford, OX1 1HP, United Kingdom
Tel: (44) (0)1865 278804
Fax: (44) (0)1865 288831
Website: http://www.sbs.ox.ac.uk/degrees/mba/scholarships/Pages/SBS.aspx

Santander DPhil Scholarship - Spanish Studies
Subjects: DPhil in Spanish or Spanish American Literary or Linguistic studies.
Eligibility: Open to applicants from selected countries for the DPhil in Spanish or Spanish American Literary or Linguistic Studies. Open to all graduate applicants from selected countries.
Level of Study: Graduate
Type: Scholarship
Value: £18,300 per year
Length of Study: Period of fee liability
Application Procedure: Please see website for more details and how to apply.
Additional Information: Please check at http://www.ox.ac.uk/feesandfunding/prospectivegrad/scholarships/university/santanderspanish/ for more information.

Santander Graduate Awards
Subjects: All subjects.
Eligibility: Open to Masters applicants from selected countries.
Level of Study: Postgraduate, Research
Type: Award
Value: £5,000
Length of Study: 1 year
Application Procedure: Apply at the same time as you apply to Oxford by selecting Santander Graduate Award in the Funding Section of the University's Graduate Application Form. Please see website for more details and how to apply.
Closing Date: January 18th
Additional Information: Please check at http://www.ox.ac.uk/feesandfunding/prospectivegrad/scholarships/university/santander/#Am%20I%20eligible for more information.

Scatcherd European Scholarships
Subjects: All subjects.
Purpose: To assist graduate students with fees and living allowance.
Eligibility: Open to all graduate applicants from selected countries.
Level of Study: Postgraduate, Research
Type: Scholarships
Value: University fee, college fee, and full living expenses
Length of Study: Period of fee liability
No. of awards offered: Varies
Application Procedure: There is no need to actively apply, as we will automatically consider those people who are awarded a fees-only Research Council studentship (EU students) or a Clarendon Scholarship. Please see website for more details and how to apply.
Closing Date: January 18th
Additional Information: Please email us through the Ask a Question tool at http://www.ox.ac.uk/feesandfunding/prospectivegrad/scholarships/university/scatcherd/.

For further information contact:

Website: http://www.ox.ac.uk/feesandfunding/prospectivegrad/scholarships/university/scatcherd/

School of Anthropology and Museum Ethnography Bursary
Subjects: Social anthropology.
Eligibility: Open to all graduate applicants for Social Anthropology.
Level of Study: Graduate
Type: Bursary
Value: Partial funding for Dphil
Length of Study: Up to 3 years
Application Procedure: Please visit website for more details and how to apply.
Additional Information: Please check at http://www.isca.ox.ac.uk/prospective-students/funding/school-grants/ for more information.

School of Archaeology Student Bursaries
Subjects: Archaeology and archaeological science.
Eligibility: Open to all applicants for DPhils in Archaeology and Archaeological Science.
Level of Study: Research
Type: Bursary
Value: £3,000 per year
Length of Study: 3 years
Application Procedure: No separate application process, automatic consideration of all applicants. Please visit website for more details and how to apply.
Closing Date: January 18th
Additional Information: Please check at http://www.arch.ox.ac.uk/graduate-prospective-funding.html for more information.

Social & Cultural Anthropology: ESRC
Subjects: Social anthropology.
Eligibility: Open to UK applicants for Social Anthropology. Other EU nationals are eligible for a fees-only award.
Level of Study: Graduate
Type: Grant
Value: University fee, college fee and full living expenses
Length of Study: Period of fee liability
Application Procedure: Please visit website for more details and how to apply.
Closing Date: January 18th
Additional Information: Please check at http://www.isca.ox.ac.uk/prospective-students/funding/ for more information.

Social Policy and Intervention: The Barnett Scholarship Fund
Subjects: DPhil in either social policy or social intervention.
Purpose: To assist graduate students with fees.
Eligibility: Open to all applicants for the DPhils in Social Policy and Social Intervention.
Level of Study: Research
Type: Scholarship
Value: £20,000 per year
Length of Study: 3 years
Application Procedure: Applications can be made by post or e-mail on the appropriate application form together with an up to date CV and a piece of recent written work (e.g. course essay, draft chapter, paper, journal article). Please visit website for more details and how to apply.

For further information contact:

Department of Social Policy and Intervention
Website: http://www.spi.ox.ac.uk/students/prospective/admissions/the-barnett-scholarship-fund.html
Contact: Bryony Groves

Sociology: Anthony Heath Scholarship
Subjects: DPhil in Sociology.
Eligibility: Open to all applicants for the DPhil in Sociology.

Level of Study: Graduate
Type: Scholarship
Value: University fee (at Home/EU rate), college fee and full living expenses
Length of Study: 3 years
Application Procedure: Please visit website for more details and how to apply.
Closing Date: March 1st
Additional Information: Please check at http://www.stx.ox.ac.uk/admissions/funding/graduate_studentship_in_sociol for more information.

Sociology: Departmental Bursaries

Subjects: MSC, Mphil, DPhil Sociology.
Purpose: To assist graduate students with fees.
Eligibility: Open to all graduate applicants in Sociology.
Level of Study: Postgraduate, Research
Type: Bursary
Value: University fee (at Home/EU rate)
Length of Study: 1 year
Frequency: Annual
No. of awards offered: Varies
Application Procedure: Please visit website for more details and how to apply.
Closing Date: June 1st
Additional Information: The Bursary is usually for the equivalent of home/EU fees. In return, Bursary students work for 3 hours per week for the Department, assisting with research or teaching. Please contact at enquiries@sociology.ox.ac.uk for further information.

For further information contact:

Website: http://www.sociology.ox.ac.uk/index.php/graduate/graduate-study.html

Sociology: ESRC Quota Studentships

Subjects: MSC, Mphil, DPhil Sociology.
Purpose: To support postgraduate training in the social sciences.
Eligibility: Open to all UK graduate applicants in Sociology. Other EU nationals are eligible for a fees-only award.
Level of Study: Postgraduate, Research
Type: Studentship
Value: University fee, college fee, and full living expenses
Length of Study: Up to 4 years
Frequency: Annual
Application Procedure: Please visit website for more details and how to apply at http://www.esrc.ac.uk/ESRCInfoCentre/opportunities/postgraduate/.
Closing Date: April 1st
Additional Information: Two nominations available from the department.

Sociology: ESRC Studentship Competition

Subjects: MSC, Mphil, DPhil Sociology.
Purpose: To support postgraduate training in the social sciences.
Eligibility: Open to all UK graduate applicants in Sociology. Other EU nationals are eligible for a fees-only award.
Level of Study: Postgraduate, Research
Type: Studentship
Value: University fee, college fee, and full living expenses
Length of Study: Up to 4 years
Frequency: Annual
Application Procedure: Please visit website for more details and how to apply at http://www.esrc.ac.uk/ESRCInfoCentre/opportunities/postgraduate/.
Closing Date: March 1st
Additional Information: One nomination available from the department.

Sociology: Nuffield Full Funded Studentships

Subjects: DPhil Sociology.
Purpose: To assist graduate students with fees and maintenance.
Eligibility: Open to all applicants for the DPhil in Sociology willing to apply to Nuffield College.
Level of Study: Doctorate, Research
Type: Studentship
Value: University fee (at Home/EU rate), college fee, and full living expenses
Length of Study: 3 years
Frequency: Annual
Application Procedure: Applicant must apply to Nuffield College. Applicants are expected to apply to appropriate external funding bodies for grants to cover their fees and maintenance. Please visit website for more details and how to apply at http://www.nuffield.ox.ac.uk/general/prospectus/funding.aspx.
Closing Date: May 1st
Additional Information: The University fees for doctoral students are payable for a total of nine terms. Application should be made via the usual Marshall Scholarship application process.

Sociology: Nuffield Partial Funded Studentships (variable number)

Subjects: DPhil Sociology.
Eligibility: Open to all applicants for the DPhil in Sociology willing to apply to Nuffield College.
Level of Study: Research
Type: Studentship
Value: Value of award varies. Please see website for more details.
Length of Study: 3 years
Frequency: Annual
Application Procedure: Applicant must apply to Nuffield College. Please visit at www.admin.ox.ac.uk/postgraduate/apply/ for online application form. Please visit website for more details and how to apply.
Closing Date: May 1st
Additional Information: Please check at http://www.nuffield.ox.ac.uk/general/prospectus/funding.aspx for more information.

St Anne's College: Graduate Development Studentship

Subjects: All subjects.
Eligibility: Open to all current DPhil students in the last two years of course.
Level of Study: Research
Type: Studentship
Value: College fee plus additional housing allowance
Length of Study: 1 year with the possibility of renewal
Frequency: Annual
Application Procedure: Please see website for details of how to apply at http://www.st-annes.ox.ac.uk/study/graduate/graduate_development_scholarships.html.
Closing Date: June 30th
Additional Information: Scholarships offer academic career development to doctoral students.

St Antony's College: Economic and Social Research Council Studentship

Eligibility: Open to all graduate applicants in the Social Sciences from the UK/EU.
Level of Study: Graduate
Type: Studentship
Value: University fee, college fee and living expenses
Length of Study: Period of fee liability
Application Procedure: Please see website for more details and how to apply.
Closing Date: January 18th
Additional Information: Please check at http://www.sant.ox.ac.uk/study/scholarships/esrc.html for more information.

St Antony's College: ENI Scholarship

Subjects: Development studies, economic and social history, economics, international relations, African studies, economics for development, global governance and diplomacy.
Purpose: The central objective of the Eni Scholars Programme is to provide future African leaders with the financial means to pursue world-class postgraduate training at the University of Oxford.
Eligibility: Open to graduate applicants from Angola, Ghana and Nigeria for selected courses.
Level of Study: Postgraduate
Type: Scholarship
Value: University fee, college fee, and full living expenses
Length of Study: 1 year

Application Procedure: Please see website for more details and how to apply.
Closing Date: March 31st
Additional Information: The programme aims to appoint one scholar in each of the following countries: Angola, Ghana and Nigeria. In the event that we cannot find a suitably qualified Scholar for one or more of these countries, the pool may be widened to the rest of sub-Saharan Africa.

St Antony's College: The Ronaldo Falconer Scholarship

Eligibility: Open to all graduate applicants who have already undertaken higher education in Costa Rica.
Level of Study: Graduate
Type: Scholarship
Value: University fee, college fee and full living expenses. Plus additional travel costs
Length of Study: Period of fee liability
Country of Study: Costa Rica
Application Procedure: Please see website for details of how to apply.
Additional Information: Please check at http://www.sant.ox.ac.uk/study/scholarships.html for more information.

St Antony's College: The Sir John Swire Scholarship

Eligibility: Open to all graduate applicants from North-east and South-east Asia.
Type: Scholarship
Value: University fee, college fee and full living expenses. Plus additional travel costs
Length of Study: Period of fee liability
Application Procedure: Please see website for details of how to apply.
Closing Date: March 31st
Additional Information: Please check at http://www.sant.ox.ac.uk/study/scholarships.html for more information.

St Antony's College: The Swire Cathay Pacific Scholarship (Hong Kong)

Subjects: Any subject in which the College specialises.
Purpose: To assist fresh graduates with their university and college fees.
Eligibility: Open to all graduate applicants from Hong Kong.
Level of Study: Postgraduate, Research
Type: Scholarship
Value: University fee, college fee, and full living expenses. Plus additional travel costs
Length of Study: Period of fee liability
No. of awards offered: 1
Application Procedure: Applications for the Scholarship should be made directly to the College Registrar by deadline. Please see website for details of how to apply.
Closing Date: March 31st
Additional Information: Eligible to nationals of Hong Kong. St Antony's College must be the applicant's first choice college.

For further information contact:

Website: http://www.sant.ox.ac.uk/study/scholarships.html
Contact: College Secretary, St Antony's

St Antony's College: The Swire Centenary and Cathay Pacific Scholarship (Japan)

Subjects: Any subject in which the College specialises.
Purpose: To assist fresh graduates with their university and college fees.
Eligibility: Open to all Japanese graduate applicants that have completed the majority of their education there.
Level of Study: Postgraduate, Research
Type: Scholarship
Value: University fee, college fee, and full living expenses. Plus additional travel costs.
Length of Study: Period of fee liability
Frequency: Annual
Application Procedure: Applications should consist of a letter of application addressed to the College Registrar, a CV, research/study proposal and an academic reference. There is no application form. Please see website for details of how to apply.
Closing Date: March 31st
Additional Information: Applicants for the Scholarship must have applied to the University for their course and College place by the University's deadline of January 20th.

For further information contact:

Email: registrar@sant.ox.ac.uk
Website: http://www.sant.ox.ac.uk/study/scholarships.html
Contact: College Registrar, St Antony's

St Catherine's College: College Scholarship (Arts)

Subjects: Humanities and social sciences.
Purpose: To assist graduates who are, or will be reading for an Oxford University DPhil, MLitt, or MSc by research degree.
Eligibility: Open to all graduate applicants in the Humanities and Social Sciences.
Level of Study: Graduate, Research
Type: Scholarship
Value: £2,000 per year plus additional benefits including accomodation and dining rights
Length of Study: Period of fee liability
Frequency: Annual
Study Establishment: St Catherine's College, University of Oxford
Country of Study: United Kingdom
No. of awards offered: Up to 3
Application Procedure: Please see website for details of how to apply at www.stcatz.ox.ac.uk for details.
Closing Date: March 11th

For further information contact:

St Catherine's College, Oxford, Oxfordshire, OX1 3UJ, England
Fax: (44) 186 527 1768
Email: academic.registrar@stcatz.ox.ac.uk
Website: http://www.stcatz.ox.ac.uk/content/graduate-scholarships
Contact: Academic Registrar

St Catherine's College: College Scholarship (Sciences)

Subjects: Sciences (mathematical, physical and life sciences, medical sciences).
Purpose: To assist students who are, or will be reading for an Oxford University DPhil, MLitt, or MSc by research degree.
Eligibility: Open to all graduate applicants in the MPLS and Medical Sciences divisions.
Level of Study: Research, Graduate
Type: Scholarship
Value: £2,000 per year plus additional benefits including accomodation and dining rights.
Length of Study: Period of fee liability
Frequency: Annual
Study Establishment: St Catherine's College, University of Oxford
Country of Study: United Kingdom
No. of awards offered: Up to 3
Application Procedure: Please see website for details of how to apply at www.stcatz.ox.ac.uk for details.

For further information contact:

St Catherine's College, Oxford, Oxfordshire, OX1 3UJ, England
Fax: (44) 186 527 1768
Email: academic.registrar@stcatz.ox.ac.uk
Website: http://www.stcatz.ox.ac.uk/content/graduate-scholarships
Contact: Academic Registrar

St Catherine's College: Great Eastern Scholarship

Subjects: All subjects.
Purpose: To assist students who are, or will be reading for an Oxford University DPhil, MLitt, or MSc by research degree.
Eligibility: Open to all graduate applicants from India.
Level of Study: Graduate, Research
Type: Scholarship
Value: £2,000 per year plus additional benefits including accomodation and dining rights
Length of Study: Period of fee liability

Frequency: Dependent on funds available
Study Establishment: St Catherine's College, University of Oxford
Country of Study: United Kingdom
No. of awards offered: 1
Application Procedure: Please see website for details of how to apply at website www.stcatz.ox.ac.uk for details.

For further information contact:

St Catherine's College, Oxford, Oxfordshire, OX1 3UJ, England
Email: academic.registrar@stcatz.ox.ac.uk
Website: http://www.stcatz.ox.ac.uk/content/graduate-scholarships
Contact: Academic Registrar

St Catherine's College: Leathersellers' Company Scholarship

Subjects: Biochemistry, chemistry, computing, earth sciences, engineering science, materials, mathematics, physics, plant sciences, statistics or zoology.
Purpose: To assist students who are, or will be reading for an Oxford University DPhil, MLitt, or MSc by research degree.
Eligibility: Open to all graduate applicants in selected science subjects that have studied at a European (including UK) university.
Level of Study: Research, Graduate
Type: Scholarship
Value: £3,000 per year plus additional benefits including accomodation and dining rights
Length of Study: Period of fee liability
Frequency: Annual
Study Establishment: St Catherine's College, University of Oxford
Country of Study: United Kingdom
No. of awards offered: Up to 3
Application Procedure: Please see website for more details and how to apply at www.stcatz.ox.ac.uk for details.

For further information contact:

St Catherine's College, Manor Road, Oxford, Oxfordshire, OX1 3UJ, England
Fax: (44) 186 527 1768
Email: academic.registrar@stcatz.ox.ac.uk
Website: http://www.stcatz.ox.ac.uk/content/graduate-scholarships
Contact: Academic Registrar

St Catherine's College: Magellan Prize

Subjects: Language, literature, culture, or history of the Portuguese-speaking world.
Purpose: To support best student beginning graduate studies in the language, literature, culture or history of the portuguese speaking world.
Eligibility: Open to all graduate applicants specialising in the study of the language, literature, culture, or history of the Portuguese speaking world.
Level of Study: Postgraduate, Research
Type: Prize
Value: £3,000 per year plus additional benefits including dining rights
Length of Study: 1 year
Frequency: Annual
Application Procedure: For application details, visit the website www.sant.ox.ac.uk/study/scholarships.shtml.
Additional Information: The Magellan Prize is awarded in each year to the best student beginning graduate studies in the University of Oxford in the language, literature, culture, or history of the Portuguese speaking world. The prize is associated with a non-stipendiary graduate scholarship at St Catherine's College.The prize is associated with a non-stipendiary graduat.

For further information contact:

Website: http://www.stcatz.ox.ac.uk/content/graduate-scholarships

St Catherine's College: Random House Scholarship

Subjects: Jewish studies.
Eligibility: Open to all graduate applicants specialising in Jewish Studies.
Level of Study: Postgraduate, Research
Type: Scholarship
Value: £2,000 per year plus additional benefits including accomodation and dining rights
Length of Study: Period of fee liability
Application Procedure: Please see website for details of how to apply.
Closing Date: March 11th
Additional Information: Please check at http://www.stcatz.ox.ac.uk/content/graduate-scholarships for more information.

St Catherine's College: Wilfrid Knapp Scholarship (Sciences)

Subjects: Medical science.
Eligibility: Open to all graduate applicants in the MPLS and Medical Sciences divisions.
Level of Study: Graduate
Type: Scholarship
Value: £5,000 per year towards college fee and contribution towards living expenses. Plus additional benefits including accomodation and dining rights
Length of Study: Period of fee liability
Application Procedure: Please see website for details of how to apply.
Additional Information: Please check at http://www.stcatz.ox.ac.uk/content/graduate-scholarships for more information.

St Cross College: E.P. Abraham Scholarships

Subjects: Chemical, biological/life and medical sciences.
Eligibility: Open to all graduate appliants for research degrees in the chemical, biological/life and medical sciences.
Level of Study: Research
Type: Scholarship
Value: £5,000 per year towards college fee, and the remainder towards living expenses
Length of Study: Up to 3 years
Closing Date: May 24th
Additional Information: Please check at http://www.stx.ox.ac.uk/prospective-students/funding-support/ep-abraham-scholarships-chemical-biologicallife-and-medical for more information.

St Cross College: Graduate Scholarship in Anthropology

Subjects: Anthropology.
Eligibility: Open to UK applicants to the DPhil in Anthropology. Other EU nationals are eligible for a fees-only award.
Level of Study: Postgraduate
Type: Scholarship
Value: University fee, college fee, and full living expenses
Length of Study: Up to 4 years
Application Procedure: Please visit website for more details and how to apply.
Closing Date: January 1st
Additional Information: Please check at http://www.stx.ox.ac.uk/prospective-students/funding-support/graduate-scholarship-anthropology for more information.

St Cross College: Graduate Scholarship in Archaeology

Subjects: Archaeology.
Purpose: To assist graduate students with fees.
Eligibility: Open to all applicants to the DPhil in Archaeology.
Level of Study: Doctorate, Postgraduate, Research
Type: Studentship
Value: College fee, and £7,000 towards living expenses per year
Length of Study: 3 years
Application Procedure: Applicants entering into DPhil should apply for a place in the School of Archaeology using the University's standard application form and must list St. Cross College as their first choice. Please see website for more details and how to apply at http://www.stx.ox.ac.uk/prospective-students/funding-support/graduate-scholarship-archaeology.
Closing Date: March 8th
Contributor: Oxford Journal of Archaeology (partly)
Additional Information: Applicants must list St Cross as their first choice college. The scholarship is awarded purely on the basis of academic merit. The successful scholar will be guaranteed to have a room in College accommodation (at the standard rent) for the first year of their course.

St Cross College: Graduate Scholarship in History

Subjects: History.
Eligibility: Open to all graudate applicants from the UK and EU applying to a DPhil in the History faculty.
Level of Study: Postgraduate, Research
Type: Scholarship
Value: University fee, college fee, and possible living expenses
Length of Study: Period of fee liability
Application Procedure: Please visit the website for more details and how to apply.
Closing Date: January 18th
Additional Information: Please check at http://www.stx.ox.ac.uk/prospective-students/funding-support/graduate-scholarship-history for more information.

St Cross College: Graduate Scholarship in Paediatrics

Subjects: DPhil in Paediatrics.
Eligibility: Open to all applicants for the DPhil in Paediatrics.
Type: Scholarship
Value: University fee and college fee
Length of Study: Period of fee liability
Application Procedure: Please visit website for more details and how to apply.
Closing Date: January 4th
Additional Information: Please check at http://www.stx.ox.ac.uk/prospective-students/funding-support/graduate-scholarship-paediatrics for more information.

St Cross College: Graduate Scholarship in Politics and International Relations

Subjects: DPhil Politics or DPhil International Relations.
Eligibility: Open to all applicants to DPhil Politics or DPhil International Relations.
Level of Study: Doctorate, Graduate
Type: Scholarship
Value: University fee, college fee and £12,000 living expenses
Length of Study: Period of fee liability
Closing Date: January 4th
Additional Information: Please check at http://www.stx.ox.ac.uk/graduate-scholarship-politics-and-international-relations for more information.

St Cross College: Hélène La Rue Scholarship in Musical Collections

Subjects: Humanities or social sciences.
Eligibility: The successful applicant may be based in the Faculty of Music or if working on other musical collections based in any relevant Faculty or Department including the Faculty of History and the School of Anthropology. Applicants must list St Cross College as their first choice college on their Graduate Admissions application in order to be eligible to apply for this scholarship. Preference may be given to a research topic related to the musical collections at the University, including those at the Ashmolean Museum, those at the Pitt Rivers Museum, the Bate Collection in the Faculty of Music and those held in any of the colleges.
Level of Study: Research
Type: Scholarship
Value: College fee for the first 3 years of study
Length of Study: 3 years
Frequency: Annual
Application Procedure: The Scholarship is tenable at St Cross College only. Application forms can be downloaded from http://www.stx.ox.ac.uk/hlr or requested by email using the Contact Form (choose the category Academic and Admissions) from the same weblink.
Closing Date: July 1st

For further information contact:

St Cross College, St Giles, Oxford, OX1 3LZ, United Kingdom
Website: http://www.stx.ox.ac.uk/hlr
Contact: Admissions & Academic Secretary

St Cross College: MPhil Scholarship in the Humanities and Social Sciences

Subjects: Humanities and social sciences.
Eligibility: Open to all applicants for MPhil degrees in the Humanities and the Social Sciences.
Level of Study: Graduate
Type: Scholarship
Value: College fee
Length of Study: Period of fee liability
Application Procedure: Please visit website for more details and how to apply.
Closing Date: March 8th
Additional Information: Please check at http://www.stx.ox.ac.uk/prospective-students/funding-support/mphil-scholarship-humanities-and-social-sciences for more information.

St Cross College: Paula Soans O'Brian Scholarship in the Arts and Humanities

Subjects: Humanities.
Eligibility: Open to all applicants for a research degree in the Humanities.
Level of Study: Research
Type: Scholarship
Value: £4,500 towards the college fee, with the remainder towards living expenses
Length of Study: Period of fee liability
Application Procedure: Please visit website for more details and how to apply.
Closing Date: March 8th
Additional Information: Please check at http://www.stx.ox.ac.uk/prospective-students/funding-support/paula-soans-obrian-scholarship-arts-and-humanities for more information.

For further information contact:

St Cross College, St Giles, Oxford, OX1 3LZ, United Kingdom
Contact: Admissions & Academic Secretary

St Cross College: Scholarship in Global Health Science

Subjects: MSc in Global Health Science.
Eligibility: Open to all applicants to the MSc in Global Health Science.
Level of Study: Graduate
Type: Scholarship
Value: College fee
Length of Study: 1 year
Application Procedure: Please see website for more details and how to apply.
Closing Date: June 7th
Additional Information: Please check at http://www.stx.ox.ac.uk/prospective-students/funding-support/st-cross-scholarship-global-health-science for more information.

St Cross College: The Robin & Nadine Wells Scholarship

Subjects: All subjects.
Purpose: To provide financial assistance to an academically meritorious graduate student who has been accepted into both an accredited one year's Masters programme at the University of Oxford and St Cross College and are unable to secure funding elsewhere.
Eligibility: Open to all applicants for one year Masters courses who is unable to secure funding elsewhere.
Level of Study: Research, Postgraduate
Type: Scholarship
Value: £5,000
Length of Study: 1 year
Frequency: Annual
Application Procedure: Please visit website for more details and how to apply.
Closing Date: June 21st
Additional Information: The successful scholar will be guaranteed to have a room in College accommodation (at the standard rent) for the first year of their course.

For further information contact:

St Cross College, St Giles, Oxford, OX1 3LZ, United Kingdom
Tel: (44) (0)1865 278458
Fax: (44) (0)1865 278484

Website: http://www.stx.ox.ac.uk/prospective-students/funding-support/robin-nadine-wells-scholarship
Contact: Admissions & Academic Secretary

St Cross College: Unilever Graduate Scholarship in the Sciences

Eligibility: Open to all graduate applicants ifor a DPhil n the sciences.
Level of Study: Research
Type: Scholarship
Value: College fee
Length of Study: Period of fee liability
Application Procedure: Please visit website for more details and how to apply.
Closing Date: May 24th
Additional Information: Please check at http://www.stx.ox.ac.uk/prospective-students/funding-support/unilever-graduate-scholarship-sciences for more information.

St Edmund Hall: Brockhues Graduate Award

Subjects: All subjects.
Eligibility: Open to all graduate research applicants.
Level of Study: Research
Type: Award
Value: £4,800 per year
Length of Study: Period of fee liability
Application Procedure: Please see website for details of how to apply.
Closing Date: January 20th
Additional Information: Please check at http://www.seh.ox.ac.uk/admissions/scholarships for more information.

St Edmund Hall: Emden Doctorow Graduate Scholarship

Subjects: All subjects.
Eligibility: Open to all graduate applicants. Preference for those applying for a DPhil in the Social Sciences.
Level of Study: Research
Type: Scholarship
Value: £4,800 per year
Length of Study: Period of fee liability
No. of awards offered: 2
Application Procedure: Please see website for details of how to apply.
Closing Date: January 20th
Additional Information: Please check at http://www.seh.ox.ac.uk/admissions/scholarships for more information.

St Edmund Hall: Justin Gosling Graduate Scholarship

Subjects: All subjects.
Eligibility: Open to all graduate research applicants.
Level of Study: Research
Type: Scholarship
Value: College fees
Length of Study: Period of fee liability
Application Procedure: Please see website for details of how to apply.
Closing Date: January 20th
Additional Information: Please check at http://www.seh.ox.ac.uk/admissions/scholarships for more information.

St Edmund Hall: Routledge St Edmund Hall Studentship

Subjects: Comparative and international education.
Eligibility: Open to all graduate applicants for an MSc in Education (Comparative & International Education).
Level of Study: Postgraduate
Type: Studentship
Value: £6,000 per year
Length of Study: Period of fee liability
Frequency: Annual
No. of awards offered: 1
Application Procedure: All those who apply by the March deadline are automatically considered – no separate application is necessary. The Routledge Scholarship cannot be used as evidence of funding to secure a place on the course. Please see website for details of how to apply.
Closing Date: January 20th
Contributor: St Edmund Hall
Additional Information: The scholarship is awarded on the basis of: (1) strength of academic qualifications; (2) fit between the proposed research project and the research interests of the Department and (3) financial need. Please check at http://www.education.ox.ac.uk/courses/studentships/ for more information.

St Edmund Hall: William R. Miller Postgraduate Award

Subjects: All subjects.
Purpose: To assist graduate students with fees and accomodation.
Eligibility: Open to all graduate research applicants.
Level of Study: Postgraduate, Research
Type: Award
Value: Free accommodation in college
Length of Study: Period of fee liability
Frequency: Annual
No. of awards offered: 3
Application Procedure: Please see website for details of how to apply.
Closing Date: January 20th
Additional Information: Please check at http://www.seh.ox.ac.uk/admissions/scholarships for more information.

St Hilda's College: Jeremy Griffiths Memorial Scholarship in the History of the Book in the British Isles before 1625

Subjects: MPhil in Medieval English.
Eligibility: Open to all graduate applicants for the MSt 650-1550 or MPhil in Medieval English.
Level of Study: Graduate, Postgraduate
Type: Scholarship
Value: University fee, college fee and full living expenses as well as free accommodation in college
Length of Study: 1 year
Application Procedure: Please see website for details of how to apply.
Additional Information: Please check at http://www.st-hildas.ox.ac.uk/index.php/graduate/scholarshipsandbursaries.html for more information.

St Hilda's College: New Zealand Bursaries

Subjects: All subject.
Eligibility: Open graduate students from New Zealand.
Level of Study: Postgraduate, Research
Type: Bursary
Value: Up to £2,000
Length of Study: 1 year, possible extension
Frequency: Annual
Additional Information: Graduate students from New Zealand. Please contact to Admissions Secretary at college.office@st-hildas.ox.ac.uk.

For further information contact:

St Hilda's College, Cowley Place, Oxford, OX4 1DY, United Kingdom
Tel: (44) 1865 276884
Fax: (44) 1865 276816

St Hilda's College: New Zealand Bursaries

Subjects: MSt in english (650–1550).
Eligibility: Open to all graduate applicants from New Zealand.
Level of Study: Postgraduate
Type: Scholarship
Value: Up to £2,000 per year
Length of Study: 1 year with the possibility of renewal
Frequency: Annual
Application Procedure: Please see website for details of how to apply.
Closing Date: Contact department

For further information contact:

St Hilda's College, Cowley Place, Oxford, OX4 1DY, United Kingdom
Tel: (44) 1865 276884
Fax: (44) 1865 276816
Email: college.office@st-hildas.ox.ac.uk.

Website: http://www.st-hildas.ox.ac.uk/index.php/graduate/scholarshipsandbursaries.html
Contact: Admissions Secretary

St John's College/RCUK Graduate Partnership Award (Humanities)

Subjects: Humanities.
Eligibility: Open to all graduate applicants in the Humanities.
Level of Study: Graduate
Type: Award
Value: University fee, college fee and full living expenses
Length of Study: Up to 4 years
Application Procedure: Please visit website for more details and how to apply.
Closing Date: January 18th
Additional Information: Please check at http://www.sjc.ox.ac.uk/ & http://www.ox.ac.uk/ for more information.

St John's College/RCUK Graduate Partnership Award (Medical Sciences)

Subjects: Medical sciences.
Eligibility: Open to all graduate applicants in the Medical Sciences.
Level of Study: Graduate
Type: Award
Value: University fee, college fee and full living expenses
Length of Study: Up to 4 years
Application Procedure: Please visit website for more details and how to apply.
Closing Date: January 18th
Additional Information: Please check at http://www.sjc.ox.ac.uk/ & http://www.ox.ac.uk/ for more information.

St John's College/RCUK Graduate Partnership Award (MPLS)

Subjects: Mathematical, physical and life sciences.
Eligibility: Open to all graduate applicants in the Mathematical, Physical and Life Sciences.
Level of Study: Graduate
Type: Award
Value: University fee, college fee and full living expenses
Length of Study: Up to 4 years
Application Procedure: Please visit website for more details and how to apply.
Closing Date: January 18th
Additional Information: Please check at http://www.sjc.ox.ac.uk/ & http://www.ox.ac.uk/ for more information.

St John's College/RCUK Graduate Partnership Award (Social Sciences)

Subjects: Social sciences.
Eligibility: Open to all graduate applicants in the Social Sciences.
Level of Study: Graduate
Type: Award
Value: University fee, college fee and full living expenses
Length of Study: Up to 4 years
Application Procedure: Please visit website for more details and how to apply.
Closing Date: January 18th
Additional Information: Please check at http://www.sjc.ox.ac.uk/ & http://www.ox.ac.uk/ for more information.

St John's College: Daniel Slifkin Studentship

Eligibility: Open to all BCL or MJur applicants.
Level of Study: Postgraduate
Type: Studentship
Value: University fee, college fee, and full living expenses
Length of Study: 1 year
Application Procedure: Please visit website for more details and how to apply.
Closing Date: January 18th
Additional Information: Please check at http://www.sjc.ox.ac.uk for more information.

St John's College: Dr Yungtai Hsu Studentship

Subjects: Environmental change and management only.
Eligibility: Open to all applicants for the MSc in Environmental Change & Management who are committed to the environmental protection and development of China or Taiwan and who will work in this field upon completion of their course.
Level of Study: Postgraduate
Type: Studentship
Value: £15,000
Length of Study: 1 year
Application Procedure: Please visit website for more details and how to apply.
Closing Date: January 18th
Additional Information: Please check at http://www.sjc.ox.ac.uk for more information.

Stanley Lewis Scholarships

Eligibility: Open to all DPhil applicants studying the history, politics, economics or society of Modern Israel.
Level of Study: Graduate
Type: Scholarship
Value: University fee and/or college fee and/or living expenses. Up to £16,200 per year subject to circumstances
No. of awards offered: Up to 3 years
Application Procedure: Please visit website for more details and how to apply.
Closing Date: March 8th
Additional Information: Please check at http://www.area-studies.ox.ac.uk/ for more information.

Statistics: EPSRC studentships

Subjects: Statistics.
Purpose: To assist graduate students with fees and maintenance.
Eligibility: Open to UK applicants for Statistics. Other EU nationals are eligible for a fees-only award.
Level of Study: Research
Type: Studentship
Value: University fee, college fee and full living expenses
Length of Study: Upto 3.5 years
Frequency: Annual
No. of awards offered: Up to 3
Application Procedure: Please visit website for more information and how to apply.
Contributor: EPSRC
Additional Information: Please check at http://www.stats.ox.ac.uk/prospective_students/research_degrees/DPhil_funding for more information.

Statistics: Teaching Assistant Bursaries/ Departmental studentships

Subjects: Statistics.
Purpose: To assist graduate students with fees and maintenance.
Eligibility: Open to all graduate applicants in Statistics.
Level of Study: Research
Type: Bursary
Value: University fee, college fee and full living expenses
Length of Study: Up to 3.5 years
Frequency: Annual
No. of awards offered: 3
Application Procedure: Please visit website for more information and how to apply at http://www.stats.ox.ac.uk/prospective_students/research_degrees/DPhil_funding.
Closing Date: Deadlines are advertised on departmental website
Additional Information: Several Departmental Studentships/Teaching Assistantships are available each year. These two types of award are separate but are normally held at the same time.

STFC Studentships in Particle Physics, Astrophysics, Theoretical Physics

Subjects: Particle physics (experiment and theory), astrophysics (experiment and theory) and atmospheric physics.
Purpose: To assist graduate students with fees and maintenance.
Eligibility: Open to all EU applicants in Particle Physics, Astrophysics and Atmospheric Physics.
Level of Study: Research

Type: Studentship
Value: University fee, college fee and full living expenses
Length of Study: 3 years (with possible extension)
Frequency: Annual
No. of awards offered: Up to 19
Application Procedure: Please visit website for more information and how to apply at http://www.physics.ox.ac.uk/admissions/postgrad.htm.
Closing Date: January 18th
Additional Information: Typically, we award about 19 STFC studentships a year: 9 to Particle Physics (experiment), 3 Particle Physics (theory) and 7 to Astrophysics (experiment and theory). The exact number depends on the funding announcement made by the sponsors.

Systems Approaches to Biomedical Sciences Industrial Doctorate Centre EPSRC Studentships

Eligibility: Open to UK applicants for all MPLS subjects. Other EU nationals are eligible for a fees-only award.
Level of Study: Graduate
Type: Studentship
Value: University fee, college fee and full living expenses
Length of Study: 4 years
Application Procedure: Please visit website for more information and how to apply.
Additional Information: Please check at http://www.dtc.ox.ac.uk/ for more information.

Systems Biology Doctoral Training Centre EPSRC Studentships

Eligibility: Open to UK applicants for all MPLS subjects. Other EU nationals are eligible for a fees-only award.
Level of Study: Graduate
Type: Studentship
Value: University fee, college fee and full living expenses
Length of Study: 4 years
Application Procedure: Please visit website for more information and how to apply.
Additional Information: Please check at http://www.dtc.ox.ac.uk for more information.

Talbot Scholarships

Eligibility: Open to all EU applicants for DPhil in Education.
Level of Study: Graduate
Type: Scholarship
Value: £5,000 per year
Length of Study: Up to 3 years
Application Procedure: Please visit website for more details and how to apply.
Closing Date: March 8th
Additional Information: Please check at http://www.education.ox.ac.uk/courses/d-phil/funding-opportunities/ for more information.

Theology Faculty Centre: AHRC Block Grant Partnership Studentships: Doctoral Scheme

Subjects: Theology.
Eligibility: Open to UK and EU applicants for the DPhil degrees offered by the Faculty of Theology.
Type: Studentship
Value: University fee, college fee and full living expenses. Fees-only awards for non-UK, EU students
Length of Study: Period of fee liability
Application Procedure: Apply on the University application form.
Closing Date: January 18th
Additional Information: Please check at http://www.humanities.ox.ac.uk/prospective_students/graduates/funding/ahrc for more information.

Theology Faculty Centre: AHRC Block Grant Partnership Studentships: Research Master's Scheme

Subjects: Theology.
Eligibility: Open to UK and EU applicants for the MSt and MPhil degrees offered by the Faculty of Theology.
Level of Study: Graduate
Type: Studentship
Value: University fee, college fee and full living expenses. Fees-only awards for non-UK, EU students
Length of Study: Period of fee liability
Application Procedure: Apply on the University application form.
Closing Date: January 18th
Additional Information: Please check at http://www.humanities.ox.ac.uk/prospective_students/graduates/funding/ahrc for more information.

Theology Faculty Centre: Theology Faculty Studentships

Subjects: Theology.
Eligibility: Open to all gradaute applicants in TheologyThe Theology.
Level of Study: Postgraduate, Research
Type: Studentship
Value: A variety of awards which may cover UK/EU course or college fees and a number of smaller bursaries.
Length of Study: 1 year
Frequency: Annual
Application Procedure: Please visit the website for more details and how to apply.
Closing Date: January 18th
Additional Information: The Theology Faculty has a number of studentships open to all Theology students (eligibility varies between the awards). Please check at http://www.theology.ox.ac.uk/student-funding/graduate2.html for more information.

Trinity College: Austin Farrer Memorial Scholarship

Subjects: Theology.
Eligibility: Open to all graduate applicants who have completed a Theology or Philosophy and Theorology degree at Oxford.
Level of Study: Postgraduate, Research
Type: Scholarship
Value: £2,000
Length of Study: 1 year
Application Procedure: Please see website for details of how to apply.
Closing Date: August 31st
Additional Information: Please check at http://www.trinity.ox.ac.uk/pages/admissions/loans-grants-and-bursaries.php for more information.

Trinity College: Birkett Scholarships in Environmental Studies

Subjects: Environmental change and management.
Purpose: To promote graduate education in the environment.
Eligibility: Open to all graduate applicants for the MSc in Environmental Change and Management.
Level of Study: Postgraduate, Research
Type: Scholarships and fellowships
Value: £7,000
Length of Study: 1 year
Frequency: Annual
Study Establishment: Trinity College, University of Oxford
Country of Study: United Kingdom
No. of awards offered: 2
Application Procedure: Candidates who wish to be considered for the scholarship should apply on the university application form. Please check at www.admin.ox.ac.uk/postgraduate for details. Please note interest on application form.
Closing Date: August 31st
Additional Information: Please check at http://www.trinity.ox.ac.uk/pages/admissions/loans-grants-and-bursaries.php for more information.

Trinity College: Cecil Lubbock Memorial Scholarship

Subjects: Humanities and social sciences.
Purpose: To promote research in philosophy.
Eligibility: Open to all UK graduate applicants for taught courses in the Humanities.
Level of Study: Doctorate, Postgraduate, Research
Type: Scholarship
Value: £10,000 per year
Length of Study: Period of fee liability
Study Establishment: Trinity college, University of Oxford

Country of Study: United Kingdom
No. of awards offered: 1
Application Procedure: Candidates who wish to be considered for the scholarship should apply on the university application form. Please see www.trinity.ox.ac.uk for details. Please note interest on application form, having checked college website to see whether the award is available for entry.
Closing Date: August 31st
Additional Information: Please contact to Academic Administrator or check at http://www.trinity.ox.ac.uk/pages/admissions/loans-grants-and-bursaries.php for more information.

Trinity College: M. B. Grabowski Fund Postgraduate Scholarship in Polish Studies

Subjects: Modern history or Modern languages for carrying out research in Polish Studies.
Purpose: To promote graduate studies in Polish.
Eligibility: Open to all graduate applicants in the Humanities whose study incorporates aspects of Polish or Slavonic studies.
Level of Study: Doctorate, Postgraduate, Research
Type: Scholarship
Value: £6,000 per year
Length of Study: Period of fee liability
Frequency: Dependent on funds available
Study Establishment: Trinity college, University of Oxford
Country of Study: United Kingdom
No. of awards offered: 1
Application Procedure: Candidates who wish to be considered for the scholarship should apply on the university application form. Please see www.admin.ox.ac.uk/postgraduate for further details.
Closing Date: August 31st
Additional Information: Please note interest on application form. Please check website for further details http://www.trinity.ox.ac.uk/pages/admissions/loans-grants-and-bursaries.php.

Trinity College: Michael and Judith Beloff Scholarship

Subjects: Civil law.
Purpose: To assist graduate students with fees.
Eligibility: Open to all graduate applicants for the BCL. Preference given those intending to practise at the Bar of England and Wales.
Level of Study: Postgraduate
Type: Scholarship
Value: £6,500
Length of Study: 1 year
Frequency: Annual
Application Procedure: Please note interest on application form. Please see website for details of how to apply.
Closing Date: August 31st
Additional Information: Please check at http://www.trinity.ox.ac.uk/pages/admissions/loans-grants-and-bursaries.php for more information.

Trinity College: Mitchell Scholarship

Eligibility: Open to all graduate applicants in Humanities and Social Sciences who are already members of Trinity College.
Level of Study: Postgraduate, Research
Type: Scholarship
Value: £10,000 per year
Length of Study: 1 year
Application Procedure: Please see website for details of how to apply.
Closing Date: August 31st
Additional Information: Please check at http://www.trinity.ox.ac.uk/pages/admissions/loans-grants-and-bursaries.php for more information.

Trinity College: Said MBA and EMBA Scholarships

Subjects: MBA/EMBA.
Eligibility: Open to all graduate applicants to the MBA and EMBA.
Level of Study: Postgraduate, Research
Type: Scholarships
Value: College fee
Length of Study: 1 year
Frequency: Annual
No. of awards offered: 2 for MBA and 1 for EMBA
Application Procedure: Please note interest on application form. Please see website for details of how to apply at http://www.trinity.ox.ac.uk/pages/admissions/loans-grants-and-bursaries.php.
Closing Date: August 31st

Trudeau Foundation Scholarship

Subjects: Humanities and social sciences.
Eligibility: Open to on-course Canadian DPhil students in Humanities and Social Sciences at Oxford. Open to all graduate applicants from selected countries.
Level of Study: Graduate
Type: Scholarship
Value: $40,000 (Canadian Dollars) per year
Length of Study: Period of fee liability
Application Procedure: Please see website for more details and how to apply.
Additional Information: Please check at http://www.ox.ac.uk/feesandfunding/prospectivegrad/scholarships/university/trudeau/#d.en.14782 for more information.

University College: Bartlett Scholarship

Subjects: All subjects.
Purpose: To assist graduate students with fees.
Eligibility: Open to all North American graduate applicants for a DPhil course.
Level of Study: Research
Type: Scholarships
Value: Up to £3,500 per year
Length of Study: Period of fee liability
Frequency: Annual
Application Procedure: Please see website for details of how to apply.
Additional Information: Please check at http://www.univ.ox.ac.uk/postgraduate/financial_1/scholarships_and_studentships/ for more information.

University College: Chellgren

Subjects: All subjects, but with a preference for economics.
Purpose: To assist graduate students with fees.
Eligibility: Open to all graduate applicants in Economics.
Level of Study: Postgraduate, Research
Type: Scholarship
Value: £4,000 per year
Length of Study: Period of fee liability
Frequency: Annual
Application Procedure: Please see website for details of how to apply.
Additional Information: Please check at http://www.univ.ox.ac.uk/postgraduate/financial_1/scholarships_and_studentships/ for more information.

University College: Loughman

Subjects: All subjects.
Purpose: To assist graduate students with fees.
Eligibility: Open to all graduate applicants that can demonstrate potential to make significant contributions to College life through extra-academic pursuits (arts, sports, community service, etc.).
Level of Study: Postgraduate, Research
Type: Scholarship
Value: £4,000 per year
Length of Study: Period of fee liability
Frequency: Annual
Application Procedure: Please see website for details of how to apply.
Additional Information: Please check at http://www.univ.ox.ac.uk/postgraduate/financial_1/scholarships_and_studentships/ for more information.

University College: Swire Scholarship (Taiwan)

Subjects: Non-clinical Science subjects.
Purpose: To assist graduate students with fees, maintenance allowance and travel costs.
Eligibility: Open to all Taiwanese graduate applicants for the selected subjects.
Level of Study: Research

Type: Scholarship
Value: University fee, college fee, full living expenses and economy return fare between Taiwan and London.
Length of Study: Period of fee liability
Application Procedure: Eligible candidates wishing to be considered for a Swire Scholarship must follow the application procedure detailed in the application information pack, available at http://www.univ.ox.ac.uk/postgraduate/financial_1/scholarships_and_studentships/.
Additional Information: Please check at http://www.univ.ox.ac.uk/postgraduate/financial_1/scholarships_and_studentships/ for more information.

University of Oxford Croucher Scholarship
Subjects: DPhil only. Natural sciences, technology or medicine.
Purpose: To provide students who are able to demonstrate academic excellence with the opportunity to pursue postgraduate research study at the University of Oxford.
Eligibility: Open to graduate applicants from Hong Kong for subjects in MPLS or Medical Sciences. Open to all graduate applicants from selected countries.
Level of Study: Research
Type: Scholarship
Value: University fee, college fee, and full living expenses
Length of Study: Period of fee liability
Frequency: Annual
No. of awards offered: 4
Application Procedure: Apply at the same time as you apply to Oxford by selecting University of Oxford Croucher Scholarships in the Funding Section of the University's Graduate Application Form. You must also apply separately to the Croucher Foundation for this scholarship. Full details on how to apply are available from the scholarships section of the Croucher Foundation website.
Closing Date: January 18th
Additional Information: Eligible to the national of HongKong. If you have any questions about this scholarship which are not answered above, please email us through the Ask a Question facility at http://www.ox.ac.uk/feesandfunding/prospectivegrad/scholarships/university/croucher/.

Wadham College: Donner-Gotlieb Scholarship
Eligibility: Open to all Canadian graduate applicants for the BCL or MJur.
Level of Study: Graduate
Type: Scholarship
Value: £19,000
Length of Study: 1 year
Application Procedure: Please see website for details of how to apply.
Additional Information: Please check at http://www.wadham.ox.ac.uk/student-life/scholarships/graduate-scholarship-in-law-2010-11.html for more information.

Wadham College: Norwegian Scholarship
Subjects: All subjects.
Eligibility: Open to all graduate applicants that are students or graduates of Oslo University.
Level of Study: Postgraduate, Research
Type: Scholarship
Value: University fee, college fee, and full living expenses
Length of Study: 1 year
Frequency: Annual
Application Procedure: Please see website for details of how to apply at http://www.wadham.ox.ac.uk/student-life/scholarships/scholarships-and-prizes.html.
Additional Information: Please contact at Iver.Neumann@nupi.no for more information.

Wadham College: Peter Carter Scholarship
Subjects: Law and finance.
Eligibility: Open to all graduate applicants for the BCL, Mjur or MSc in Law and Finance.
Level of Study: Graduate
Type: Scholarship
Value: £10,000 per year
Length of Study: 1 year
Application Procedure: Please see website for details of how to apply.
Additional Information: Please check at http://www.wadham.ox.ac.uk/student-life/scholarships/peter-carter-scholarship-in-law.html for more information.

Wadham College: Philip Wright Scholarship
Subjects: All subjects.
Eligibility: Open to former pupils of Manchester Grammar School.
Level of Study: Postgraduate, Research
Type: Scholarship
Value: Home fees, college fees and a maintenance grant of £8,000
Length of Study: 1 year, possible renewal
Application Procedure: Application forms are available at http://www.wadham.ox.ac.uk/student-life/scholarships/the-philip-wright-scholarship.html either in Word or PDF format. Please send the completed form, a full curriculum vitae and (for graduates applying for research work) a one page summary of your research proposal, either by post to the Tutor for Graduates.
Closing Date: Please contact college
Funding: Trusts
Contributor: Philip Wright Fund

For further information contact:

c/o Tutorial Office, Wadham College, Parks Road, Oxford, OX1 3PN
Email: admissions@wadh.ox.ac.uk
Contact: Tutor for Graduates

Waverley - African Studies Centre Joint Scholarship
Subjects: African studies.
Eligibility: Open to graduate applicants from Sub-Saharan Africa for the MSc in African Studies.
Level of Study: Postgraduate
Type: Scholarship
Value: University fee, college fee, and full living expenses. Additional expenses for flights may be included.
Length of Study: 1 year
Closing Date: January 18th
Additional Information: Please check at http://www.ox.ac.uk/feesandfunding/prospectivegrad/scholarships/university/waverley/ for more information.

Waverley DPhil Scholarship
Eligibility: Open to all applicants from sub-Saharan Africa to the relevant DPhil subjects.
Level of Study: Graduate
Type: Scholarship
Value: University fee, college fee and living expenses
Length of Study: Period of fee liability
Application Procedure: Please see the website for more details and how to apply.
Additional Information: Please check at http://www.africanstudies.ox.ac.uk/dphil-scholarships#Waverley%20DPhil for more information.

Weidenfeld - Hoffman Scholarship Scheme
Subjects: Social sciences.
Eligibility: Open to all applicants to selected courses in the Social Sciences from selected countries.
Level of Study: Graduate, Predoctorate
Type: Scholarship
Value: University fee, college fee and full living expenses
Length of Study: Period of fee liability
Application Procedure: Please see the website for more details and how to apply.
Closing Date: January 18th
Additional Information: Please check at http://www.ox.ac.uk/feesandfunding/prospectivegrad/scholarships/university/weidenfeld-hoffmann/ for more information.

Weidenfeld - Roland Berger Scholarship Scheme
Eligibility: Open to all graduate applicants from selected countries.
Level of Study: Graduate
Type: Scholarship
Value: University fee, college fee and full living expenses

Length of Study: Period of fee liability
Application Procedure: Please see website for more details and how to apply.
Closing Date: January 18th
Additional Information: Please check at http://www.ox.ac.uk/feesandfunding/prospectivegrad/scholarships/university/weidenfeld-rolandberger/ for more information.

Wolfson College D.Phil. Studentships in Classical Archaeology - Lorne Thyssen

Subjects: Archaeology (classical archaeology).
Level of Study: Research
Type: Studentship
Value: £20,000 per year
Length of Study: 1–3 years
Closing Date: January 18th
Additional Information: Please check at www.wolfson.ox.ac.uk/financial/ for more information.

Wolfson College D.Phil. Studentships in Classical Archaeology - Mougins Museum

Subjects: Archaeology (classical archaeology).
Eligibility: Open to all applicants for DPhil Archaeology (Classical Archaeology).
Level of Study: Research
Type: Studentship
Value: £20,000 per year
Length of Study: Up to 3 years
Application Procedure: Please visit website for more details and how to apply.
Closing Date: January 18th
Additional Information: Please check at http://www.wolfson.ox.ac.uk/scholarships for more information.

Worcester College: Martin Senior Scholarships

Subjects: All subjects.
Eligibility: Open to all graduate applicants from the EU that are current or previous members of the college.
Level of Study: Postgraduate, Research
Type: Scholarships
Value: Up to £4,000 per year
Length of Study: Period of fee liability
Frequency: Annual
Application Procedure: Please see website for details of how to apply.
Closing Date: March 1st
Additional Information: Please contact at graduate.enquiries@worc.ox.ac.uk for more information.

Worcester College: Ogilvie Thompson Scholarship

Subjects: All subjects.
Eligibility: Open to all graduate applicants that have taken final UG examinations at the college within the last two years and have not undertaken any graduate work at Oxford or elsewhere.
Level of Study: Postgraduate, Research
Type: Scholarship
Value: Up to £4,000 per year
Length of Study: 1 year
Application Procedure: Please see website for details of how to apply.
Closing Date: March 1st
Additional Information: Please contact at graduate.enquiries@worc.ox.ac.uk for more information.

Yousef Jameel Scholarship

Subjects: Islamic art and archaeology.
Eligibility: Open to all applicants for the MPhil or MSt in Islamic Art and Archaeology.
Type: Scholarship
Value: University fee, college fee and full living expenses
Length of Study: Period of fee liability
Closing Date: January 18th
Additional Information: Please check at http://www.ox.ac.uk/feesandfunding/prospectivegrad/scholarships/university/yousef%20jameel/ for more information.

UNIVERSITY OF PUNE

Institute of Bioinformatics & Biotechnology (IBB), Ganeshkhind, Pune, Maharashtra, 411007, India
Tel: (91) 20 25692039
Fax: (91) 20 25690087
Email: director@bioinfo.ernet.in
Website: www.unipune.ernet.in
Contact: Director

The University stands for humanism and tolerance, for reason for adventure of ideas and for the search of truth. It stands for the forward march of the human race towards even higher objectives. If the universities discharge their duties adequately then it is well with the nation and the people – Jawaharlal Nehru.

Department of Biotechnology (DBT) - Junior Research Fellowship

Subjects: Biotechnology and applied biology.
Purpose: To support candidates pursuing research in areas of biotechnology and applied biology.
Eligibility: Open to candidates from the centres supported by the DBT, New Delhi.
Level of Study: Research
Type: Fellowship
Value: 30,000 per fellow per year
Length of Study: 3–5 years
Frequency: Annual
Study Establishment: University of Pune
Country of Study: India
Application Procedure: A written application along with application fee of Rs 500 in the form of a DD in favour of – Registrar, University of Pune.
Contributor: Government of India

For further information contact:

Department of Biotechnology, University of Pune, Pune, Maharashtra, 411 007, India
Tel: (91) 020-25694952 (lab), 25692248
Email: jkpal@unipune.ernet.in, jkpal@hotmail.com
Contact: Professor Jayanta Kumar Pal, Co-ordinator, DBJ-JRF Programme

THE UNIVERSITY OF QUEENSLAND

Research and Postgraduate Studies, Cumbrae-Stewart Building, Brisbane, St Lucia, QLD, 4072, Australia
Tel: (61) 7 3365 1111
Fax: (61) 7 3365 4455/6941
Email: scholarships@research.uq.edu.au
Website: www.uq.edu.au

The University of Queensland has an outstanding profile in the Australian and international research community. It maintains a world-class, comprehensive programme of research and research training, underpinned by state-of-the-art infrastructure and a commitment to rewarding excellence. As one of Australia's premier universities, UQ attracts researchers and students of outstanding calibre.

The Accenture Scholarship in ITEE

Subjects: Information technology and electrical engineering.
Purpose: To financially assist students to study in the field of information technology and electrical engineering.
Eligibility: Applicants must be enrolled in an Honours programme in the Bachelor of Information Technology, Bachelor of Science or Bachelor of Engineering.
Level of Study: Postgraduate
Type: Scholarship
Value: Australian $1,500
Length of Study: 1 year
Frequency: Annual

Study Establishment: University of Queensland
Country of Study: Australia
No. of awards offered: 1
Closing Date: May
Funding: Foundation, government

Alumini Association – Elizabeth Usher Memorial Travelling Scholarship

Subjects: Any subject.
Purpose: To assist a research higher degree student to travel overseas to present a research paper or poster at an international conference.
Eligibility: Open to candidates who have graduated from the University of Queensland not more than 5 years ago. The applicants must be enrolled (full-time or part-time) for a PhD and must be in their second or full-time equivalent yera of study for the PhD.
Level of Study: Postgraduate
Type: Scholarship
Value: $2,500
Length of Study: 1 year
Frequency: Annual
Country of Study: Australia
No. of awards offered: 1
Application Procedure: Applicants must send a completed application form.
Closing Date: June 20th
Funding: Individuals
Contributor: Bequest from the estate of the late Elizabeth Catherine Usher AO
No. of awards given last year: 1
No. of applicants last year: 14
Additional Information: Consideration will be given to applicant's work during the entire postgraduate career, their aptitude for original research and the extent to which participation in this conference will benefit Australia.

For further information contact:

Research Scholarships, Research and Research Training Division, The University of Queensland, Brisbane, Queensland, 4072, Australia

APAI Scholarships within Integrative Biology

Subjects: Integrative biology.
Purpose: To characterize the newly generated polyploid trees using molecular, physiological and/or field investigations.
Level of Study: Doctorate, Postgraduate
Type: Scholarship
Value: $25,118 (per year)
Length of Study: 3 years
Frequency: Annual
No. of awards offered: 4
Application Procedure: Candidates must contact Dr Susanne Schmidt (susanne.schmidt@uq.edu.au), Dr Peer Schenk (p.schenk@uq.edu.au) or Professor Christa Critchley (c.critchley@uq.edu.au) for more information.
Closing Date: February 28th

Australian Postgraduate Awards

Subjects: Any subject.
Purpose: To provide a living allowance for research higher degree candidates to undertake a PhD or MPhil.
Eligibility: Open to candidates who hold or expect to hold a Bachelor's degree with Honours Class I or equivalent results. They must be an Australian citizen, permanent Australian resident or a citizen of New Zealand.
Level of Study: Postgraduate
Type: Scholarship
Value: $24,653 per year (2013 rate). This rate is indexed annually.
Length of Study: 2–3 years for Research Masters degree, with a possible extension of 6 months, for a Research Doctorate degree
Frequency: Annual
Country of Study: Australia
No. of awards offered: 140
Application Procedure: Applicants must send in completed application forms.
Closing Date: October 31st
Funding: Government
Contributor: Australian Government
No. of awards given last year: 186
No. of applicants last year: 430

For further information contact:

Research Scholarships, Research and Research Training Division, The University of Queensland QLD 4072, Brisbane, Australia

The Baillieu Research Scholarship

Subjects: Medicine, law, commerce, economics, architecture, planning.
Purpose: To assist a research higher degree student.
Eligibility: Applicants must be enrolled, or eligible to enrol, in a research higher degree in the disciplines of medicine, law, commerce, economics, architecture or planning; preferably be a lineal descendant of an Australian soldier or sailor who served in World War I and suffered death, blindness of total incapacity (applicants who do not meet this criteria are still eligible to apply); and not be more than 35 years of age at the time of application.
Level of Study: Postgraduate
Type: Scholarship
Value: $5,000
Length of Study: 1 year
Frequency: Annual
Country of Study: Australia
No. of awards offered: 1
Application Procedure: Applicants must send a completed application form.
Closing Date: Early August
Contributor: Established in 1954 by a gift of $9,677.42 under the provisions of the Repatriation Fund (Baillieu Gift) Act of 1937
No. of awards given last year: 1
No. of applicants last year: 8
Additional Information: Preference will be given to graduates of not more than five years standing.

For further information contact:

Research Scholarships, Office of Research and Postgraduate Studies, The University of Queensland QLD 4072, Brisbane, Australia

The Constantine Aspromourgos Memorial Scholarship for Greek Studies

Subjects: Greek studies.
Purpose: To assist a research higher degree student studying at least 1 area of Greek studies.
Eligibility: Open to candidates who have obtained their Bachelors or Masters degrees and are undertaking a postgraduate programme involving studies which pertain to at least one area of Greek studies.
Level of Study: Postgraduate
Type: Scholarship
Value: Approx. $4,000
Length of Study: 1 year
Frequency: Annual
Country of Study: Australia
No. of awards offered: 1
Application Procedure: Applicants must send a completed application form.
Closing Date: March 22nd
Funding: Individuals
Additional Information: The Scholarship is also open to candidates who are undertaking the programme as a student of another university acceptable to the committee, or this university, provided that some part of the programme involves studies at another university.

For further information contact:

Faculty of Arts, Forgan Smith Building, The University of Queensland QLD 4072
Tel: (07) 3365 1333
Email: arts@uq.edu.au
Contact: Executive Dean

Dr Rosamond Siemon Postgraduate Renal Research Scholarship

Subjects: Medical sciences.

Purpose: To support a research higher degree candidate to undertake multidisciplinary, collaborative research into renal disease, repair and regeneration.
Eligibility: Open to candidates who are enrolled or intend to enrol in a research higher degree at the University of Queensland and who demonstrate a high level of academic acheivement and ability.
Level of Study: Postgraduate
Type: Scholarship
Value: $30,000 per year (a stipend of $25,000 and a direct research cost allowance of $5,000)
Length of Study: 3 years and 6 months
Frequency: Annual
Country of Study: Australia
No. of awards offered: 1
Application Procedure: Applicants must send a proposed research project description, certified copies of academic transcripts, academic Curriculum vitae, including publications and 3 letters of recommendation.
Closing Date: August 31st
Funding: Individuals
Contributor: Dr Rosamond Siemon
Additional Information: Research Scholarships Refree Report Form can be used. This can be accessed from www.uq.edu.au/grad-school/scholarship-forms

For further information contact:

Research Scholarships, Office of Research and Postgraduate Studies, The University of Queensland QLD 4072, Brisbane, Australia
Contact: Professor Melissa Little

E.M.A and M.C Henker Postgraduate Medical Research Scholarship

Subjects: Medical Sciences.
Purpose: To assist a research higher degree student to undertake medical research.
Eligibility: Open to MBBS graduates of not more than 8 years standing seeking to undertake medical research. Preference shall be given to applicants who: enrol full-time PhD or MPhil degree; demonstrate that they are involved or have arranged involvement in an existing research project; and demonstrate an intention to transfer to an NHMRC Medical Postgraduate Research Scholarship at the termination of this scholarship.
Level of Study: Postgraduate
Type: Scholarship
Length of Study: Scholarship will be held for 1 year, but extension for a further year may be granted if progress has been excellent and there are good reasons for failure to find alternative funding
Frequency: Dependent on funds available
Country of Study: Australia
No. of awards offered: 1
Application Procedure: Application form must be completed.
Closing Date: July
Funding: Individuals
Contributor: Miss Edith M.A. Henker and Miss Minnie C. Henker
No. of awards given last year: 1
No. of applicants last year: 1
Additional Information: Scholarship shall consist of: a stipend commensurate with the stipend associated with an NH&MRC Medical Postgraduate Research Scholarship; and a school maintenance grant to support the scholar's research of an amount determined by the Head, School of Medicine for each year.

The E.S Cornwall Memorial Scholarship

Subjects: Engineering.
Purpose: The object of the scholarship is to enable UQ Engineering graduates to obtain special experience abroad in aspects of the electricity supply industry so that the industry in Australia may benefit by the knowledge and experience thus gained by them.
Eligibility: Preference will be given to applicants who have between 3 and 5 years industry experience.
Level of Study: Postgraduate
Type: Scholarship
Value: $3,000 per month, for not less than 9 months and not more than 18 months.
Length of Study: Period between 9 and 18 months, in accordance with a programme approved by the Advisory Commitee
Frequency: Dependent on funds available
Country of Study: Australia
No. of awards offered: 1
Application Procedure: Application form must be completed.
Closing Date: End September
Contributor: University of Queensland

Edwin Tooth Scholarship

Subjects: Medicine, public health.
Purpose: To enable a scholar to undertake full-time reserach leading to the degree of PhD.
Eligibility: Applicants must be eligible to enrol in a PhD. Research must be conducted within the biological and chemical sciences faculty or within the schools for medicine or population health.
Level of Study: Postgraduate
Type: Scholarship
Value: $34,036
Length of Study: 2 years initially with the possibility of a further year
Frequency: Dependent on funds available
Country of Study: Australia
No. of awards offered: 1
Application Procedure: Application form must be completed.
Closing Date: July 5th
Funding: Individuals
Contributor: Bequest in the will of Sir Edwin Tooth
No. of awards given last year: 1
No. of applicants last year: 8

Endeavour International Postgraduate Research Scholarship (IPRS)

Subjects: All subject.
Purpose: This scholarship covers tuition fees and health cover for International students undertaking a PhD or MPhil.
Eligibility: The IPRS programmes enables outstanding international students to undertake an MPhil or PhD in areas of research strength at Australian universities to gain experience with leading Australian researchers.
Level of Study: Postgraduate
Type: Scholarship
Value: $22,500 per year
Length of Study: For an initial period of 3 years for PhD and 2 years for MPhil study
Frequency: Annual
Country of Study: Australia
No. of awards offered: 30
Application Procedure: To apply just complete the International Student Application for Research Studies – MPhil or PhD from the university website paying attention to the section on scholarships.
Closing Date: October 31st
Funding: Government
Contributor: Australian Government Department of Education, Science and Training (DEST)
No. of awards given last year: 30
No. of applicants last year: 435
Additional Information: For more information please visit the Commonwealth Department of Education, Science and Training website. All enquiries from prospective international students regarding this award should be directed to the UQ International Education Directorate.

For further information contact:

International Admissions Section, The University of Queensland, Level 2, JD Story Building, Brisbane, Queensland, 4072, Australia
Contact: The Manager

Graduate School Research Travel Grant (GSRTG)

Subjects: Any subject.
Purpose: To fund students to travel to access resources in Australia or overseas that have enabled them to speed up progress on and enhance the quality of their thesis.
Eligibility: Applicants must be enrolled for a PhD or MPhil degree, must be confirmed PhD or MPhil candidates by the closing date – no exceptions, who have previously held a GSRTG are not eligible,

grants will not be made retrospectively. Applicants are ineligible if they are already undertaking the research travel at the time of application closing date and if the need for this research travel was identified in the initial research proposal.
Level of Study: Postgraduate
Type: Grant
Value: Australian $3000 for travel in the Pacific region (eg. Indonesia, New Zealand, Papua New Guinea) and $5000 for travel elsewhere in the world
Frequency: Bi annual
Country of Study: Australia
No. of awards offered: Up to 80
Application Procedure: Application form must be completed.
Closing Date: February 22nd (Round 1)
Contributor: University of Queensland
No. of awards given last year: Round 1: 42, Round 2: 37
No. of applicants last year: Round 1: 56, Round 2: 43
Additional Information: These grants do not fund essential research travel signalled in the initial research proposal since that should be met from other grants or school funds; GSRTGs are not intended for consultation and Conference travel will not be funded. Preference will be given to travel that was unforeseen at the time of confirmation of candidature. Travel to workshops where new specialist techniques are taught may be funded in special cases; these require a strong and well supported argument, clearly indicating how this training will benefit the timely submission of the thesis. Successful applicants must be enrolled full-time for the duration of their travel and travel must commence within 6 months of the grant being offered. Students enrolled in a PhD by Cotutelle, in International Collaborative Mode or in a similar international collaborative arrangement must satisfy the same eligibility criteria as all other GSRTG applicants.

Herdsman Fellowship in Medical Science

Subjects: Medicine, related health sciences.
Purpose: The fellowship is open to graduates in medicine or related health sciences enrolled full-time for a PhD on a topic related to the medical problems of the aged.
Eligibility: Applicants must be graduates in medicine or related health sciences, enrol full-time for a PhD, and be undertaking a research topic related to the medical problems of the aged.
Level of Study: Postgraduate
Type: Fellowship
Value: $22,860 per year
Length of Study: Fellowship shall initially be for 1 year but may be extended by the commitee for further terms of 1 year up to a total of 3 years
Frequency: As tenure falls vacant
Country of Study: Australia
No. of awards offered: 1
Application Procedure: Applications must consist of: covering letter addressing the Herdsman Fellowship Rules, in particular point 2, academic Curriculum vitae, 2 referee reports. No strict format is required; however the Research Scholarships generic Referee Report may be used.
Closing Date: August 31st
Contributor: Maintained by the income from a bequest of $260,000 from Mrs Rose Herdsman
No. of awards given last year: 1
No. of applicants last year: 1

For further information contact:

Faculty of Health Sciences, University of Queensland, Brisbane, 4072, Australia
Email: s.tett@pharmacy.uq.edu.au
Contact: Professor Susan Tett, Deputy Executive Dean and Director of Research

PhD Scholarship in Immunology and Immunogenetics

Subjects: Immunology.
Purpose: To provide the foundations for the development of treatments based on the genetic findings.
Eligibility: Open to a dynamic, intelligent and diligent PhD candidate (Australian or international) with either a clinical or a relevant basic science background to take forward the project.
Level of Study: Doctorate
Type: Scholarship
Value: Australian $25,000 per year
Length of Study: 3 years
Frequency: Annual
Study Establishment: The University of Queensland
Country of Study: Australia
No. of awards offered: 1
Application Procedure: Candidates must contact Prof. Brown for more information.
Closing Date: September 3rd
Additional Information: International applicants must cover tuition fees ($27,000 per year).

For further information contact:

Diamantina Institute for Cancer, Immunology & Metabolic Medicine, Level 4 Research Wing, Princess Alexandra Hospital, Ipswich Road
Tel: 07 3240 2870
Email: matt.brown@uq.edu.au
Contact: Professor Matt Brown

Queensland Cancer Fund PhD Scholarships

Subjects: Any involving cancer research.
Purpose: For any field related to cancer research.
Eligibility: Candidates must be ordinarily resident in Queensland, be an Australian citizen or hold a visa or passport allowing them to work or study in Queensland.
Level of Study: Postgraduate
Type: Scholarship
Value: $30,000(stipend amount to be indexed annually to CPI – no extension allowed)
Length of Study: 3 years
Frequency: Annual
Country of Study: Australia
No. of awards offered: 3
Application Procedure: Application forms and further information is available from the Queensland Cancer website: www.qldcancer.com.au/research/qcf_grantd/PhDScholarships.htm
Closing Date: Check with the website.
Funding: Corporation
Contributor: Queensland Cancer fund
No. of awards given last year: 2
No. of applicants last year: 2

R.N. Hammon Scholarship

Subjects: Science, engineering, medicine, dentistry, architecture, agriculture and veterinary science, and other fields of study.
Purpose: To assist Australian Aboriginal and/or Torres Strait Island students for further studies.
Eligibility: Open to Australian Aboriginal and/or Torres Strait Island students who have successfully completed at least one year of an undergraduate or postgraduate program and are enrolling on a full-time basis for a subsequent year of that program, or for a further program.
Level of Study: Postgraduate
Type: Scholarship
Value: $3,500
Frequency: Annual
Study Establishment: The University of Queensland, Queensland University of Technology, University of Southern Queensland, Central Queensland University, or Queensland Colleges of TAFE
Country of Study: Australia
Application Procedure: Candidates can download the application form and referee report form from the website.
Closing Date: March 15th
Additional Information: The Selection Committee shall take into account the academic merit or technical excellence, any other scholarship, bursary, award or benefit, whether governmental or otherwise, to which the applicant is entitled; and social and economic need.

For further information contact:

Tel: 07 33651984
Email: ugscholarships@uq.edu.au

Sister Janet Mylonas Memorial Scholarship

Subjects: Any involving cancer research.
Purpose: To assist a student undertaking a research higher degree involving cancer research.
Eligibility: Open to students who are commencing a programme at the University which either directly involves cancer research or will prepare the students to undertake cancer research in the future, and satisfy the Selection Committee that they are most likely to engage in cancer research during the tenure of the scholarship or in the future.
Level of Study: Postgraduate
Type: Scholarship
Value: Up to $33,500 per year.
Length of Study: 3 years for PhD and 2 years for master
Frequency: Annual
Country of Study: Australia
No. of awards offered: 1
Application Procedure: Applicants must submit an application form.
Closing Date: Check with the website.
Contributor: Income bequeathed to the university by late Stelios Demetrion Mylonas
No. of awards given last year: 1
No. of applicants last year: 4

Sustainable Tourism CRC – Climate Change PhD

Subjects: Commerce, management, tourism and services.
Purpose: To develop a tourism consumer decision-making model that focuses on climate change as a driver of consumer choice and apply it to Australian tourism market.
Eligibility: Open to candidates who have achieved First Class (Honours) Degree or equivalent.
Level of Study: Postgraduate, Research
Type: Scholarship
Value: Australian $19,930 per year
Length of Study: 3 years
Frequency: Annual
Study Establishment: The University of Queensland
Country of Study: Australia
No. of awards offered: 1
Application Procedure: Candidates must contact Jane Malady for application forms.
Contributor: Sustainable Tourism CRC and University of Queensland
Additional Information: For further information on topic and research proposal contact Prof Ballantyne at r.ballantyne@uq.edu.au.

For further information contact:

STCRC Education Program, Sustainable Tourism CRC
Tel: 61 7 5552 9063
Email: Jane@crctourism.com.au
Website: www.crctourism.com.au
Contact: Jane Malady

The UQ AHURI PhD Scholarship

Subjects: Any subject.
Purpose: The scholarship holder is required to undertake research on a topic related to the research and policy areas included in the AHURI research agenda.
Eligibility: Applicants must be an Australian citizen, permanent resident or New Zealand citizen at the application closing date; may be any age; must meet the University's English proficiency requirements; must enrol for a full-time research higher degree (part-time awards are available in exceptional circumstances); must hold or expect to hold a Bachelor's degree with Honours Class I, or a qualification deemed equivalent. This qualification must be in a relevant field; must have completed at least 4 years of full-time equivalent tertiary education.
Level of Study: Postgraduate
Type: Scholarship
Value: $22,500 per year
Length of Study: 3 years
Frequency: Annual
Country of Study: Australia
No. of awards offered: 1
Application Procedure: Application form.
Closing Date: August 20th
Contributor: University of Queensland
Additional Information: The student awarded the UQ AHURI PhD scholarship may also be recommended for an AHURI 'top-up' scholarship of $7,000 per year to supplement the AHURI PhD scholarship.

UQ International Research Award (UQIRA)

Subjects: Any subject.
Purpose: The UQIRA is funded by the University of Queensland and covers tuition fees and health cover for international students for an initial period of 3 years for PhD and 2 years for MPhil study. They are awarded during the annual Endeavour IPRS round.
Eligibility: As for the Endeavour International Postgraduate Research Scholarship (IPRS). Please see UQ website.
Level of Study: Postgraduate
Type: Award
Value: $22,500 per year
Length of Study: 3 years for PhD and 2 years for MPhil
Frequency: Annual
Country of Study: Australia
No. of awards offered: Approx. 20
Application Procedure: These scholarships are awarded during the Endeavour International Postgraduate Research Scholarship (IPRS) round. As with the IPRS, applications should be completed on the International Student Application for Research Studies – MPhil or PhD. All applicants considered for the IPRS will be automatically considered for a UQ International Research Award.
Closing Date: August 31st
Contributor: University of Queensland
No. of awards given last year: 19
No. of applicants last year: 21

For further information contact:

International Admissions Section, University of Queensland, Level 2, JD Story Building, Brisbane, Queensland, 4072, Australia
Contact: The Manager

UQ Research Scholarship (UQRS)

Subjects: All subject.
Purpose: It enables recently confirmed students who have demonstrated the strength of their research potential during their first year of candidature to work on their research higher degree full-time with scholarship support.
Eligibility: Check website for detailed eligibility criteria.
Level of Study: Postgraduate
Type: Scholarship
Value: Australian $23,728 per year
Length of Study: Up to 3 years
Frequency: Bi-annual
Country of Study: Australia
No. of awards offered: Approx. 20
Application Procedure: Students do not submit an application form, but must register an expression of interest with their school. Heads of School/Postgraduate Coordinators will complete a UQCS Nomination form to provide evidence in support of each nomination.
Closing Date: May and September
Contributor: University of Queensland
No. of awards given last year: 40 (Round 1) and 25 (Round 2)
No. of applicants last year: 44 (Round 1) and 31 (Round 2)

UQ Research Scholarships (UQRS)

Subjects: Any subject.
Purpose: To provide a living allowance for research higher degree candidates (PhD or MPhil).
Eligibility: Applicants may be of any age, must hold or expect to hold a Bachelor's degree with Honours Class I or equivalent results by December 31st and must be an Australian citizen, Australian permanent resident or New Zealand citizen as at the application closing date.
Level of Study: Postgraduate
Type: Scholarship
Value: Australian $24, 653 per year
Length of Study: 3 years
Frequency: Annual
Country of Study: Australia
No. of awards offered: 35

Application Procedure: Application form.
Closing Date: Check with the website.
Contributor: University of Queensland
No. of awards given last year: 38
No. of applicants last year: 433
Additional Information: The University of Queensland offers successful applicants who relocate more than 250 km to take up an APA or UQPRS a $1,100 Establishment Allowance. This is in addition to the Travel/Removal Allowance.

Venerable Archdeacon E L Hayes Postgraduate Scholarship

Subjects: Australian studies, literature and history.
Purpose: To provide support for research in the fields of Australian literature and/or historical sources.
Eligibility: Open to full-time internal students of the University of Queensland who are enrolled in a course leading to a Master's or Doctorate of Philosophy.
Level of Study: Postgraduate
Type: Scholarship
Value: $23,728 per year (relocation allowance up to $1,500; travel allowance; and thesis allowance up to $840).
Length of Study: 2 years for masters or 3 years for PhD, with a possible 6-month extension.
Frequency: Annual
Study Establishment: Brisbane Metropolitan and Ipswich West Moreton, Queensland
Country of Study: Australia
Closing Date: March 25th

THE UNIVERSITY OF READING

Whiteknights, PO Box 217, Reading, Berkshire, RG6 6AH, England
Tel: (44) 11 8987 5123
Fax: (44) 11 8931 4404
Email: student.recruitment@reading.ac.uk
Website: www.rdg.ac.uk
Contact: Student Financial Support Office

The University of Reading offers postgraduate taught and research degree courses in all the traditional subject areas except medical sciences. Vocational courses are also offered. Research work in many areas is of international renown.

University of Reading Arts and Humanities Studentship

Subjects: Research areas within those covered by the Faculty of Arts and Humanities.
Purpose: To enable students to obtain a doctoral degree.
Level of Study: Doctorate, Postgraduate
Type: Studentship
Value: Composition fee at the home standard rate plus a maintenance award related to that paid by the relevant United Kingdom Research Council
Length of Study: 3 years
Frequency: Annual
Study Establishment: The University of Reading
Country of Study: United Kingdom
No. of awards offered: Varies
Application Procedure: Applicants must complete an application form, available from the Faculty Office via email, faspg@reading.ac.uk. Eligible applicants are expected to apply for, and accept if offered, a scholarship from the relevant United Kingdom Research Council or an ORS award.
Closing Date: February 19th
Funding: Private

University of Reading Felix Scholarship

Subjects: All subjects.
Eligibility: Open to international postgraduate taught and research students who have Indian citizenship, under 30 years of age holding at least a first class honours degree and who have not studied outside India.
Level of Study: Postgraduate
Type: Scholarship
Value: £12,685 (tuition fees at the international rate and provide a stipend (maintenance grant) to cover living expenses)
Length of Study: 1–3 years
Frequency: Annual
Study Establishment: University of Reading
Country of Study: United Kingdom
No. of awards offered: 1
Application Procedure: Contact The Student Financial Support Office at the Carrington Building.
Closing Date: February 1st
Funding: Trusts
Contributor: The Felix Trust
No. of awards given last year: 6
No. of applicants last year: 120
Additional Information: For Additional information, please contact: (0) 118 378 4245

University of Reading General Overseas Scholarships

Subjects: All subjects, subject to the availability of appropriate supervision at the University.
Purpose: To enable students to obtain a doctoral degree.
Eligibility: Open to applicants who hold a first degree qualification and pay international fees.
Level of Study: Doctorate, Postgraduate
Type: Studentship
Value: Full tuition fees and maintenance at a rate related to that offered by the relevant United Kingdom Funding Council
Length of Study: Up to 3 years
Frequency: Annual
Study Establishment: The University of Reading
Country of Study: United Kingdom
No. of awards offered: 5
Application Procedure: Applicants must fulfil the eligibility criteria and express an interest in the scholarships to be considered. They are also requested to contact The Student Financial Support Office at the Carrington Building.
Closing Date: Please contact the University for ascertaining this information
Funding: Private
Contributor: University of Reading
No. of awards given last year: 3
No. of applicants last year: 135

University of Reading Graduate School for the Social Sciences Studentships

Subjects: Social sciences.
Purpose: To enable students to obtain a doctoral degree.
Level of Study: Doctorate, Postgraduate
Type: Scholarship
Value: Composition fee at the home standard rate plus a maintenance award related to the rate paid by the relevant Research Council; also some smaller awards.
Length of Study: Up to 3 years
Frequency: Annual
Study Establishment: The University of Reading
Country of Study: United Kingdom
No. of awards offered: Up to 10
Application Procedure: Applicants must complete an application form available from the Faculty Office via email, faspg@reading.ac.uk
Closing Date: February 19th
Funding: Private
Contributor: The University of Reading
No. of awards given last year: 4

University of Reading MSc Intelligent Buildings Scholarship

Subjects: Construction management and engineering.
Eligibility: In order to be considered for this Scholarship you must hold the offer of a place on the MSc Intellegent Buildings course.
Level of Study: Postgraduate
Type: Scholarship
Value: UK£3,000
Length of Study: 1 year
Frequency: Annual
Study Establishment: University of Reading

Country of Study: United Kingdom
No. of awards offered: 1
Application Procedure: Contact Gulay Ozkan, Programme Coordinator at the School of Construction Management and Engineering.
Closing Date: August 30th
Contributor: The Happold Trust
No. of awards given last year: 1
No. of applicants last year: 1

For further information contact:

University of Reading, Whiteknights, Reading, Berkshire, RG6 6AW, England
Tel: (44) (0) 118 378 6254
Email: g.ozkan@rdg.ac.uk
Website: www.reading.ac.uk/ib
Contact: Gulay Ozkan, Programme Coordinator

University of Reading Postgraduate Studentship

Subjects: All subjects, subject to the availability of appropriate supervision at the University.
Purpose: To enable students to obtain a doctoral degree.
Eligibility: Open to candidates holding a first degree qualification.
Level of Study: Doctorate, Postgraduate
Type: Studentship
Value: The composition fee at the home standard rate plus a maintenance award related to the relevant Research Council rate
Length of Study: Up to 3 years
Frequency: Annual
Study Establishment: The University of Reading
Country of Study: United Kingdom and Australia
No. of awards offered: 4
Application Procedure: Applicants must be nominated to the Financial Support Office. Full details are available on the Funding search page of The University of Reading website (www.reading.ac.uk).
Closing Date: March 15th
Funding: Private
Contributor: The University of Reading
No. of awards given last year: 4
No. of applicants last year: 90
Additional Information: Applications are assessed on the quality of their research proposal and shortlisted applicants will be asked to attend an interview.

UNIVERSITY OF REGINA

3737 Wascana Parkway, Regina, SK, S4S 0A2, Canada
Tel: (1) 306 585 4411
Fax: (1) 306 585 4893
Email: grad.studies@uregina.ca
Website: www.uregina.ca/gradstudies
Contact: Ms Ann Bishop, Faculty of Research Studies

Founded as Regina College in 1911, the University of Regina became an affiliated junior college of the University of Saskatchewan in 1925 and acquired degree granting status in 1959. The University achieved academic autonomy in 1974. The Faculty of Graduate Studies and Research offers a wide range of programmes such as social justice, culture and heritage, energy and the environment, health and informatics.

China-Canada Scholars Exchange Program

Subjects: All subjects.
Purpose: To support Canadian scholars and students who wish to study or do research in subject areas related to China in the Chinese universities that are open to Chinese Government Scholarship recipients.
Eligibility: Open to applicants who are faculty members at Canadian Universities or students who have a Bachelor's degree or are enrolled in a graduate programme.
Level of Study: Graduate, Postgraduate, Research
Type: Scholarship
Value: The scholarship covers the cost of basic living allowance, payment of tuition fees, on-campus accommodation, medical insurance and teaching and research materials
Length of Study: 4–12 months
Frequency: Annual
Application Procedure: Applicants must submit 5 copies of each of the following: a completed application form, a detailed study or research proposal indicating objectives, duration of the proposed stay, methodologies, 2 letters of reference, academic transcripts and a certificate of degree letter verifying the graduation time by the dean.
Closing Date: February 3rd
Contributor: China Scholarship Council
Additional Information: Preference will be given to those candidates whose research is related to the study of China.

For further information contact:

Education Office the Chinese Embassy in Canada 80 Cobourg Street, Ottawa, ON, K1N-8H1, Canada
Tel: (1) 613 789 6312
Fax: (1) 613 789 0262
Website: www.chinaembassycanada.org

UNIVERSITY OF SHEFFIELD

Western Bank, Sheffield, S10 2TN, United Kingdom
Tel: (44) 114 222 2000
Fax: (44) 114 222 3739
Email: grad.school@sheffield.ac.uk
Website: www.shef.ac.uk

The University's history dates back to 1828, when the Sheffield School of Medicine was founded, and the University Charter was granted in 1905. With 25,000 students, from 116 countries, and almost 6000 staff, it is one of the leading universities of United Kingdom.

Dorothy Hodgkin Postgraduate Awards

Subjects: All subjects.
Purpose: To enable top-quality applicants to undertake doctoral research at the University.
Eligibility: Open to prospective international students only.
Level of Study: Postgraduate, Research
Type: Studentship
Value: Up to a maximum of £13,650 per award to cover tuition fees and living expenses
Length of Study: 3 years
Frequency: Annual
Study Establishment: University of Sheffield
No. of awards offered: 3
Application Procedure: Candidates can check the website for further details.
Closing Date: June 25th

University of Sheffield Postgraduate International Scholarships

Subjects: Arts and sciences.
Purpose: To provide opportunities to Russian students of high academic standing.
Eligibility: Open to candidates who are Russian by birth or permanently domiciled in Russia and must hold an offer of a study place for entry.
Level of Study: Postgraduate
Type: Scholarships
Value: UK£2,000
Length of Study: 1 year
Frequency: Annual
Country of Study: Russia
Application Procedure: Candidates will be sent a postgraduate International Scholarships application form automatically if they are eligible.
Closing Date: February 1st
Additional Information: Applicants cannot apply for the scholarship before being offered a study place at Sheffield.

White Rose Studentships

Subjects: Plant biology.
Purpose: Collaborative research networks within the three White Rose Universities.

Eligibility: Open to United Kingdom, European Union, and international applicants who will register with the University for a PhD degree.
Level of Study: Postgraduate, Research
Type: Studentship
Value: Home/EU tuition fees, an annual maintenance grant of £13,590
Length of Study: 3 years
Frequency: Annual
Study Establishment: One of the three White Rose Universities
No. of awards offered: 1
Application Procedure: Candidates can check the website for further details.
Closing Date: Check with the website.
Additional Information: International applicants are only eligible if they can show sufficient funds to cover the difference between the United Kingdom and international students tuition fee.

For further information contact:

Email: s.beecroft@shef.ac.uk
Contact: Simon Beecroft

UNIVERSITY OF SOUTH AUSTRALIA

GPO Box 2471, Adelaide, SA, 5000, Australia
Tel: (61) 8 8302 6611/3615
Fax: (61) 8 8302 2466/3997
Email: research.international@unisa.edu.au
Website: www.unisa.edu.au

The University of South Australia is an innovative and successful institution with a distinctive profile. It is committed to educating professionals, creating and applying knowledge and serving the community.

Aboriginal Advancement League Study Grants

Subjects: All subjects.
Purpose: To help Australian Indigenous and Torres Strait Islander students of the University of South Australia and Flinders University who are enrolled in a postgraduate or Medical Degree.
Eligibility: Open to Australian Indigenous and Torres Strait Islander (permanent residents of South Australia) and students of the University of South Australia and Flinders University, who have enrolled in a postgraduate or Medical Degree.
Level of Study: Postgraduate
Type: Study grant
Value: $5,000 per year to a maximum of $10,000 (Full-Time)
Frequency: Annual
Study Establishment: University of South Australia and Flinders University
Country of Study: Australia
Application Procedure: Check website for further details.
Closing Date: March 31st
Additional Information: Part-time grants are also awarded on a pro rata basis. For further information please check www.unisa.edu.au/scholarship/Ab_Advancement_League_appform

The AFUW–SA Inc Trust Fund Bursary

Subjects: All subjects.
Purpose: To assist women undertaking postgraduate studies.
Eligibility: Open to women undertaking postgraduate degrees by coursework at Australian universities.
Level of Study: Postgraduate
Type: Bursary
Value: $4,000
Frequency: Annual
Study Establishment: Australian universities
Country of Study: Australia
Application Procedure: Candidates must apply on forms available from the AFUW website from October to February.
Closing Date: March 1st
Funding: Trusts
Contributor: The Australian Federation of University Women – South Australia Inc (AFUW–SA Inc) Trust Fund
Additional Information: All bursaries are envisaged primarily as short-term aids and must be used within 1 year of the date of award.

For further information contact:

GPO Box 634, Adelaide, 5001, Australia
Website: www.afuwsa-bursaries.com.au

The Diamond Jubilee Bursary

Subjects: All subjects.
Purpose: To assist men and women undertaking postgraduate studies.
Eligibility: Open to men and women undertaking a postgraduate degree by coursework at a South Australian university.
Level of Study: Postgraduate
Type: Bursary
Value: $4,000
Frequency: Annual
Study Establishment: Australian universities
Country of Study: Australia
Application Procedure: Candidates must apply on forms available from the AFUW website from October to February.
Closing Date: March 1st
Contributor: The Australian Federation of University Women – South Australia Inc (AFUW–SA Inc) Trust Fund
Additional Information: All bursaries are envisaged primarily as short-term aids and must be used within 1 year of the date of award.

For further information contact:

GPO Box 634, Adelaide, 5001, Australia
Website: www.afuwsa-bursaries.com.au

Division of Business Student Mobility Scholarships

Subjects: Business programme.
Purpose: To assist business students undertaking an international exchange (via the UniSA International Student Exchange Program) at a partner university.
Eligibility: Open to both undergraduate and postgraduate coursework students enrolled in a Division of Business program and are participating in exchange for the first time.
Level of Study: Postgraduate
Type: Scholarship
Value: $5,000 for Institutional Partner Scholarships, $2500 for Student Mobility Scholarships
Frequency: Annual
Application Procedure: For extended criteria and application details, please contact: Ms Sarah Oolyer-Braham.
Closing Date: January 30th

For further information contact:

Student Mobility and Academic Administration, Business Division Office
Tel: 08 8302 0880
Email: sarah.collyer-braham@unisa.edu.au
Contact: Ms Sarah Collyer-Braham, Administrative Officer

Donald Dyer Scholarship – Public Relations & Communication Management

Subjects: Public relations, communication management.
Purpose: To encourage research of an original nature leading to the advancement of knowledge in public relations and communication.
Eligibility: Open to candidates who have achieved First Class (Honours) or equivalent. Candidates from discipline areas such as public relations, communication, marketing or advertising are encouraged to apply.
Level of Study: Postgraduate, Research
Type: Scholarship
Value: Australian $22,000 per year(tax-free) Plus one return travel airfare between the candidate's home location and Adelaide, and organised by the University
Frequency: Annual
Study Establishment: University of South Australia
Country of Study: Australia
No. of awards offered: 1
Application Procedure: Check website for further details.
Closing Date: October 29th
Contributor: Bequest from the estate of the Late Sylvia Dyer

For further information contact:

School of Communication
Tel: 08 8302 4493
Fax: 08 8302 4745
Email: david.brittan@unisa.edu.au
Website: www.unisa.edu.au/cppc/DyerScholarship.asp
Contact: David Brittan, (Postgraduate Research) Education Administrator

Ferry Scholarship

Subjects: Physics and chemistry.
Purpose: Promoting study and research in physics and chemistry.
Eligibility: Open to Australian citizens under the age of 25 years on January 1st of the year of the award, who have completed at least 4 years of tertiary education studies, have a First Class (Honours) (or equivalent undergraduate degree), and have enrolled as full–time students for a Master's Degree or Doctorate by research in chemistry or physics.
Level of Study: Postgraduate
Type: Scholarship
Value: $7,500 per year
Length of Study: Up to 3 years (Doctorate) and 2 years (Masters Degree)
Frequency: Annual
Application Procedure: Applications can be filled online.
Closing Date: January 31st
Contributor: Bequest from the late Cedric Arnold Seth Ferry

For further information contact:

Australian Admissions and Scholarships, University of South Australia
Tel: 8302 3967
Email: jenni.critcher@unisa.edu.au
Website: www.unisa.edu.au/resdegrees/howtoapply/default.asp
Contact: Jenni Critcher

International Postgraduate Research Scholarships (IPRS)

Subjects: All subjects.
Purpose: To attract international postgraduate students to study for a higher degree by research in Australia and support Australia's research effort.
Eligibility: Available to international research candidates generally with first-class honours degree or equivalent.
Level of Study: Doctorate, Postgraduate
Type: Research scholarship
Value: $29,843 per year for 3 years, with the possibility of one six-month extension (2013 stipend rate)
Length of Study: 3 years
Frequency: Annual
Study Establishment: The University of South Australia
Country of Study: Any country
Application Procedure: Applicants must submit an application form, which can be obtained from the website.
Closing Date: August 31st (to commence in following year)
Funding: Government
Contributor: Department of Education, Science and Training (DEST)
No. of awards given last year: 6
No. of applicants last year: 160

The Jean Gilmore, Thenie Baddams and Daphne Elliot Bursaries

Subjects: All subjects.
Purpose: To assist women undertaking postgraduate studies.
Eligibility: Open to women enrolled for research towards a PhD or Masters at Australian universities.
Level of Study: Postgraduate
Type: Bursary
Value: $6,000
Frequency: Annual
Study Establishment: Australian universities
Country of Study: Australia
Application Procedure: Candidates must apply on forms available from the AFUW website from October to February.
Closing Date: March 3rd
Funding: Trusts
Contributor: The Australian Federation of University Women – South Australia Inc (AFUW–SA Inc) Trust Fund
Additional Information: All bursaries are envisaged primarily as short-term aids and must be used within 1 year of the date of award.

For further information contact:

GPO Box 634, Adelaide, 5001, Australia
Website: www.afuwsa-bursaries.com.au

Lewis O'Brien Scholarship

Subjects: Education, arts, and social sciences.
Purpose: To assist and encourage Aboriginal and Torres Strait Islander people in postgraduate study in a field of particular relevance and potential benefit to the Indigenous Australian community.
Eligibility: Open to Aboriginal and Torres Strait Islander people eligible to undertake a postgraduate program in the division of education, arts and social sciences.
Level of Study: Postgraduate
Type: Scholarship
Value: Maximum $10,000 per year ($2,500 will be paid on commencement and the remainder will be paid in instalments during the year subject to certification by the supervisor of satisfactory progress).
Frequency: Annual
Country of Study: Australia
Application Procedure: Candidates must contact Ms Jillian Mille for further information.
Closing Date: February 11th
Contributor: Division of Education, Arts and Social Sciences

For further information contact:

Tel: 08 8302 9151
Fax: 08 8302 7034
Email: jillian.miller@unisa.edu.au
Contact: Jillian Miller, Coordinator Indigenous Support Services

Margaret George Award

Subjects: Archival research, history.
Purpose: To encourage and facilitate use of National Archives collection by promoting archival research in Australia and encouraging scholarly use of its holdings.
Eligibility: Open to postgraduate degree holders, historians, academics, independent researchers, or journalists with a talent for research.
Level of Study: Postgraduate
Type: Award
Value: $10,000
Frequency: Annual
Country of Study: Australia
Application Procedure: Check website for further details.
Closing Date: June 30th each year
Additional Information: Successful applicants may undertake their award at any time from the date of the announcement of the award until June 30th the following year.

For further information contact:

National Archives of Australia in Canberra, Australia
Tel: (61) 02 6212 3986
Fax: (61) 02 6212 3699
Email: derina.mclaughlin@naa.gov.au
Website: www.naa.gov.au/about_us/margaret_george.html
Contact: Derina McLaughlin

The Padnendadlu Postgraduate Bursary

Subjects: All subjects.
Purpose: To assist women undertaking postgraduate studies.
Eligibility: Open to Indigenous Australian women undertaking postgraduate degrees at South Australian universities.
Level of Study: Postgraduate
Type: Bursary
Value: AUD3500 (maximum per award). This award is to be used for research, living expenses, travel and fees.
Frequency: Annual
Study Establishment: South Australian universities

Country of Study: Australia
Application Procedure: Candidates must apply on forms available from the AFUW website from October to February.
Closing Date: March 1st
Contributor: The Australian Federation of University Women – South Australia Inc (AFUW–SA Inc) Trust Fund
Additional Information: All bursaries are envisaged primarily as short–term aids and must be used within 1 year of the date of award.

For further information contact:

GPO Box 634, Adelaide, 5001, Australia
Website: www.afuwsa-bursaries.com.au

Trevor Prescott Memorial Scholarship

Subjects: All subjects.
Purpose: To help youth in the South Australian community to advance their careers through further postgraduate studies.
Eligibility: Open to students between 20 and 30 years of age who desire to do further postgraduate studies or equivalent.
Level of Study: Unrestricted
Type: Scholarship
Value: Up to $25,000 and may be divided between more than 1 recipient
Frequency: Annual
Application Procedure: Check website for further details.
Closing Date: June 30th
Funding: Foundation
Contributor: The Masonic Foundation Inc
Additional Information: Preference is not given to a Freemason or to a member of the family for the scholarship.

For further information contact:

The Masonic Foundation Inc, Australia
Tel: (61) 08 8443 9909
Fax: (61) 08 8443 9928
Email: masfound@senet.com.au
Website: www.freemasonrysaust.org.au/foundation.html

UNIVERSITY OF SOUTHAMPTON

University of Southampton University Road Southampton SO17 1BJ, Southampton, Hampshire, SO17 1BJ, England
Tel: (44) 0 23 8059 5000
Fax: (44) 0 23 8059 3131
Email: admissns@soton.ac.uk
Website: www.soton.ac.uk
Contact: Student Marketing Office

The University of Southampton was granted its Royal Charter in 1952. Today, the University is one of the United Kingdom's top ten research universities, offering a wide range of postgraduate taught and research courses in engineering, science, mathematics, law, arts, social sciences, medicine and health and life sciences.

University of Southampton Engineering Doctorate

Subjects: Engineering, physical sciences, and mathematics.
Purpose: To provide financial assistance to graduate students with their studies and research.
Eligibility: Open to candidates who hold a good Honours degree and are eligible for admission to the academic school in which they intend to study.
Level of Study: Doctorate
Type: Studentship
Value: Full-time £3,466, part-time £1,164
Length of Study: 4 years
Frequency: Annual
Study Establishment: The University of Southampton
Country of Study: United Kingdom
No. of awards offered: 10
Application Procedure: Applicants should make initial enquiries to the academic school in which they intend to study.
Funding: Commercial, government
Additional Information: Please check website for latest updates.

University of Southampton Postgraduate Studentships

Subjects: All subjects.
Purpose: To provide financial assistance to graduate students with their studies and research.
Eligibility: Open to candidates who hold a good Honours Degree and are eligible for admission to the academic school in which they intend to study.
Level of Study: Doctorate, MBA, Postgraduate, Research
Type: Studentship
Value: Varies (awards may cover tuition fees and/or maintenance)
Length of Study: The duration of the course of study and research
Frequency: Annual
Study Establishment: The University of Southampton
Country of Study: United Kingdom
No. of awards offered: Varies
Application Procedure: Applicants should make initial enquiries to the academic school in which study is to be undertaken.
Closing Date: Varies
Funding: Commercial, government
Additional Information: Applicants should contact each academic school for further details.

World Universities Network (WUN) International Research Mobility Scheme

Subjects: A wide range of subject areas reflecting the research interest of the WUN network.
Purpose: To support research visits to WUN partner universities in the United States of America, China and Europe.
Eligibility: Open to candidates currently registered for an MPhil or PhD degree in one of the University of Southampton's schools.
Level of Study: Doctorate, Postdoctorate, Postgraduate, Research
Type: Scholarship (travel and subsistence grant)
Value: Varies, usually up to around UK£5,000
Length of Study: The duration of the research visit, which is normally between 1 and 6 months
Frequency: Annual, annual budget but awards made twice a year in November and March
Study Establishment: Partner Universities
Country of Study: United States of America, China, Norway, Netherlands
No. of awards offered: Varies
Application Procedure: Applicants must make initial enquiries to the head of the academic school in which they are registered for their research degree. Information can also be obtained from the WUN coordinator via Email at elisa@soton.ac.uk
Closing Date: November 30th, March 15th
Contributor: The University of Southampton
No. of awards given last year: 8
No. of applicants last year: 10
Additional Information: For subject details see the website at www.wun.ac.uk

UNIVERSITY OF STRATHCLYDE

16 Richmond Street, Glasgow, G1 1XQ, Scotland
Tel: (44) 0 141 552 4400
Fax: (44) 0 141 552 0775
Email: cathy.bonner@strath.ac.uk
Website: www.strath.ac.uk
Contact: Postgraduate Research Office

Department of Chemical and Process Engineering PhD Studentship

Subjects: Chemical and process engineering.
Eligibility: Candidates should be highly motivated and have a First Class Honours degree in chemical engineering, physics or chemistry. An MSc/MEng in science or engineering and previous experience in the field of chemistry of materials would be an advantage. Students will engage in the Department's research seminar programme, and will have opportunities to attend national and international conferences. Other generic skills and courses are open to students, including scientific writing, presentation and careers workshops. International students must be proficient in english language (the University's entry requirements are IELTS 6.5, TOEFL 600 including

the test of written english, TOEFL 250 computer based test or TOEFL 90–95 internet based test).
Level of Study: Doctorate
Type: Studentship
Value: The award will cover UK/EU tuition fees and will pay a stipend of £13,590 per year (for 3 years). International students would have to pay the difference between the Home/EU and internaional fee.
Length of Study: 3 years
Frequency: Annual
Study Establishment: University of Strathclyde
Country of Study: Scotland
No. of awards offered: 1
Application Procedure: Please send your CV and a covering letter, indicating your previous experience and fields of interest and include the details of at least two academic referees to Dr S V Patwardhan.
Closing Date: May 30th

For further information contact:

Tel: 141 548 5786
Email: Siddharth.Patwardhan@strath.ac.uk

Department of Civil Engineering/David Livingstone Centre for Sustainability Excellence Awards

Subjects: Civil engineering.
Level of Study: Doctorate
Type: Award
Value: £1,000 award
Frequency: Annual
Study Establishment: University of Strathclyde
Country of Study: Scotland
No. of awards offered: 20 awards for MSc degrees (International students only) 6 awards for MRes degrees (home/EU students)
Application Procedure: Candidates should apply for their chosen course of study at www.strath.ac.uk/civeng/pg/. Please ensure that your application is complete and includes proof of previous degree, two references and proof of english language, if applicable. In addition, please add a one-page statement with these points: Why do you want to take this degree and how will the degree support your career goals? What makes you 'excellent' that justifies you getting one of these Excellence Awards?.
Closing Date: October
Additional Information: Please check at http://www.strath.ac.uk/civeng/pg/excellenceawards/ for further information.

Department of English Studies: Studentships

Subjects: MLitt in Literature, Culture and Place.
Purpose: The Department of English Studies is offering a limited number of awards to students planning to pursue its one-year taught Master's (Litt) programme in Literature, Culture and Place.
Eligibility: Open to home and overseas students, preference will be given to students intending to undertake PhD research in the Department on successful completion of their master's degree.
Level of Study: Doctorate
Type: Studentship
Value: The awards cover the cost of full-time fees or part-time fees (currently £3,235 / £1,620 for home students, and £8,695 / £4,350 for overseas students).
Frequency: Annual
Study Establishment: University of Strathclyde
Application Procedure: For information about the MLitt programme, visit the website. For further details, please email. There is no formal application procedure for these studentships.
Applicants are selected on merit by departmental selectors in mid-July before the academic year. It is important that all MLitt applicants should complete the MLitt application form as fully as possible. In particular, you should supply details in the blank space allowed for further information in support of your application. This space should be used to explain how your previous study has prepared you for this course, what interests you about the curriculum, and how the course relates to your eventual career aims or research aspirations. Make sure you specify your particular literary interests – e.g. in individual authors, period or topics.
Closing Date: July 1st

For further information contact:

Livingstone Tower, 26 Richmond Street,, Glasgow, G1 1XH, Scotland

Faculty of Humanities & Social Sciences One-Year Master's Scholarships

Subjects: English, French, Geography, History, Italian, Law, Politics, Psychology, Sociology and Spanish.
Purpose: Applications are invited for one-year Faculty Master's Scholarships in the Faculty of Humanities & Social Sciences. A list of eligible courses and research degrees in which these scholarships may be held is available from the Faculty Office website.
Eligibility: Scholarships are available on a competitive basis to well-qualified applicants.
Level of Study: Doctorate
Type: Scholarship
Value: Awards comprise full tuition fees for home/EU and overseas students
Frequency: Annual
Study Establishment: University of Strathclyde
Country of Study: Scotland
Application Procedure: There is no separate application form for the scholarship. To express an interest in being considered for a scholarship, please contact the relevant person – these are available from the faculty's website.
Closing Date: See website for details

Glasgow Cathedral Choral Scholarships

Subjects: Music (choral).
Eligibility: The Choral Scholarship, tenable at Glasgow Cathedral, is normally offered each year. The scholarship is open to men and women who are either registered or potential students of the University and whose registration at the University will last for 9 months or longer starting from the date at which the scholarship begins.
Level of Study: Postgraduate
Value: Stipend of £500 plus a sum of £125 (travelling expenses)
Frequency: Annual
Study Establishment: University of Strathclyde
Country of Study: Scotland
No. of awards offered: 1
Application Procedure: Candidates are required to attend for an interview at which they will be examined in sight-reading and aural perception. In addition candidates are required to prepare two solos of contrasting styles, suitable for singing in the Cathedral, for performance at the interview.
Closing Date: November 30th
Contributor: University of Strathclyde/Glasgow Cathedral
Additional Information: Please contact the department for address for applications.

International Scholarships

Subjects: All subjects.
Purpose: For new students who begin a one year full-time taught Master's course.
Eligibility: Students of outstanding academic calibre from overseas and EU students on premium fee courses.
Level of Study: Postgraduate
Type: Scholarship
Value: Awards of £4,000 and £6,000 are available
Frequency: Annual
Study Establishment: University of Strathclyde
Country of Study: Scotland
No. of awards offered: A number of awards are available of varying amounts
Application Procedure: Applications should be made online.
Closing Date: May 31st
Additional Information: Please check at http://ewds.strath.ac.uk/igo/Apply.aspx for further information.

Mac Robertson Travelling Scholarship

Subjects: All subjects.
Purpose: To provide funding that will enrich and further the award holder's academic experience and research achievements.

Eligibility: Applicants should be postgraduate research students currently registered at Strathclyde or Glasgow Universities.
Level of Study: Postgraduate, Research
Type: Scholarship
Value: Varies
Length of Study: Varies
Frequency: Annual
Study Establishment: University of Strathclyde
Country of Study: United Kingdom
No. of awards offered: Normally 2 or 3 per University
Application Procedure: Application forms can be downloaded from the website www.strath.ac.uk
Closing Date: May 4th, 4 pm
Funding: Individuals
Contributor: Mac Robertson
No. of awards given last year: 11
No. of applicants last year: 20
Additional Information: The aim of the scheme is to provide funding which will enrich and further the award holder's experience and research achievements.

For further information contact:

Email: cathy.bonner@mis.strath.ac.uk

Malawi Masters Law Scholarship

Subjects: LLM in international economic law; international law and sustainable development.
Eligibility: Normally an Honours degree in law (or, for the international law and sustainable development LLM, an Honours degree in any discipline, although some law content would be useful). Other qualifications (for both programmes) are recognised, especially where the applicant's work experience is relevant to the course.
Level of Study: Doctorate, Postgraduate
Value: Fee waiver and stipend of £10,000
Frequency: Annual
Study Establishment: University of Strathclyde
Country of Study: Scotland
No. of awards offered: 1
Application Procedure: Further details are available from Ms Linda Ion, Course Administrator: linda.ion@strath.ac.uk.
Closing Date: September

MSc in Marketing and International Marketing

Subjects: Business (international marketing and marketing).
Purpose: The MSc in International Marketing programme provides students with a unique opportunity to develop the new skills and perspectives which will be required by international business managers of the future.
Eligibility: A limited number of departmental bursaries are available to well-qualified students who have been made an offer for the MSc international marketing and MSc marketing programmes.
Level of Study: Postgraduate
Type: Bursary
Value: £3,000
Frequency: Annual
Study Establishment: University of Strathclyde
Country of Study: Scotland
Application Procedure: International marketing applciations should be sent to mscim.helpdesk@strath.ac.uk. Marketing applications should be sent to mscm.helpdesk@strath.ac.uk.
Closing Date: July 31st

Pakistan 50th Anniversary Fund

Subjects: Engineering, computing and business management.
Purpose: To provide an opportunity for well-qualified but needy Pakistan nationals to pursue 1-year MSc study in the United Kingdom.
Eligibility: Applicants should be Pakistan nationals with a First Class (Honours) Degree from a recognized Pakistan university. Candidates should have relevant work experience and have not previously studied in the UK.
Level of Study: Postgraduate
Type: Scholarship
Value: Tuition fees, living costs, return travel and other expenses
Length of Study: 1 year
Frequency: Annual
Study Establishment: University of Strathclyde
Country of Study: United Kingdom
No. of awards offered: Two awards will be offered for entry in September
Application Procedure: Application forms will be available from British Council offices in Pakistan, or can be downloaded from the website, normally in February.
Closing Date: February 29th
Funding: Foundation
Contributor: Scottish Pakistani Association
No. of awards given last year: 2

For further information contact:

McCance Building, 16 Richmond Street, Glasgow, G1 1XQ, Scotland
Contact: Ms Shirley Kirk , Advice Centre - Student Financial Support

PhD Studentships in Civil Engineering Structures

Subjects: Civil Engineering.
Level of Study: Doctorate, Research
Type: Studentship
Value: University fees plus a student stipendÂ (stipend level of Â£13,490)
Frequency: Annual
Study Establishment: University of Strathclyde
Country of Study: Scotland
No. of awards offered: 1
Application Procedure: For further information, please contact: Dr Mohamed Saafi (email: m.bensalem.saafi@strath.ac.uk) or if you have queries specifically regarding the application procedure, please contact Lisa Lyons (email lisa.lyons@strath.ac.uk).
Applications should be submitted through our online system, which is accessed via: www.strath.ac.uk/civeng/pg/mphilphd/. Please indicate your interest in the scholarship in the 'source of funding' box on the form.
Closing Date: October 29th

For further information contact:

Department of Civil Engineering, University of Strathclyde, John Anderson Building, Glasgow, G4 0NG

Scottish Overseas Research Students Award Scheme (SORSAS)

Subjects: All subjects.
Purpose: To assist overseas applicants and students who are seeking sources of financial assistance to pursue full-time postgraduate research study at the University of Strathclyde.
Eligibility: Candidates must have a First Class (Honours) Degree or equivalent. Candidates entering the 3rd year of a PhD are not eligible. Candidates from European Union countries are not eligible.
Level of Study: Research
Type: Scholarship
Value: The balance between the overseas and United Kingdom/ European Union rate for tuition fees
Length of Study: Up to 3 years
Frequency: Annual
Study Establishment: University of Strathclyde
Country of Study: United Kingdom
Application Procedure: Application forms can be downloaded from the website.
Closing Date: March 12th
Funding: Government
Contributor: HE Funding Councils
No. of awards given last year: 7
No. of applicants last year: 100

For further information contact:

Room 738, Livingstone Tower, 76 Richmond Street, Glasgow, G1 1XH
Contact: Graduate Office

Sports Bursaries

Eligibility: The Centre offers support to athletes in two main areas. There is a University Golf Programme supported by the R&A Foundation as well the University Sports Bursary Programme which is

run in conjunction with Glasgow City Council, Glasgow University and Glasgow Caledonian University for a wide range of sports. Applicants must have reached or demonstrated a particular level of standard.
Level of Study: Postgraduate
Type: Bursary
Value: Up to £1,000
Study Establishment: University of Strathclyde
Country of Study: Scotland
Application Procedure: Application forms may be downloaded from the Centre for Sport and Recreation's website around the first week in September.
Closing Date: September

For further information contact:

Website: www.strath.ac.uk/sport/sportsbursaries

University of Strathclyde Research Scholarships

Subjects: All subjects.
Purpose: To assist applicants and students who are seeking sources of financial assistance to pursue full-time research study at the University of Stratclyde.
Eligibility: Applicants should be research students of outstanding academic merit.
Level of Study: Research, Doctorate
Type: Scholarship
Value: Maintenance plus Fee
Length of Study: Up to 3 years
Frequency: Annual
Study Establishment: University of Strathclyde
Country of Study: United Kingdom
Application Procedure: Applicants seeking nomination for the awards should contact the department they wish to join. Existing research students should contact their head of department or supervisor.
Closing Date: There is no central deadline. Each faculty sets its own deadline dates
Funding: International Office
Contributor: University
No. of awards given last year: 25
Additional Information: For more information, visit the website www.strath.ac.uk.

UNIVERSITY OF SUSSEX

Postgraduate Office, Sussex House, Falmer, Brighton, East Sussex, BN1 9RH, England
Tel: (44) 12 7360 6755
Fax: (44) 12 7367 8335
Email: information@sussex.ac.uk
Website: www.sussex.ac.uk
Contact: Mr Terry O'Donnell

The University of Sussex is one of the United Kingdom's foremost research institutions. The University boasts a distinguished faculty that includes 17 Fellows of the Royal Society and four Fellows of the British Academy. The University has around 12,000 students, 25 per cent of whom are postgraduates.

Chancellor's International Scholarships

Subjects: Scholarships are available in all subjects at all schools (except the Institute of Development Studies and Brighton and Sussex medical school).
Purpose: To provide scholarships to new overseas fee-paying students at the postgraduate taught degree level.
Eligibility: Candidates must have applied for and accepted a place to start a postgraduate degree (e.g. MA, MSc, LLM, postgraduate studies and graduate diploma) at the University of Sussex.
Level of Study: Postgraduate, Graduate
Type: Scholarship
Value: 50 per cent off International Student tuition fee
Length of Study: Duration of the degree programme
Frequency: Annual
Study Establishment: University of Sussex
Country of Study: United Kingdom
No. of awards offered: 40
Application Procedure: Completed application forms should be returned by email, post or fax.
Closing Date: May 1st
Funding: Private
Additional Information: Scholarships are awarded on the basis of academic merit and potential.

For further information contact:

International and Study Abroad Office, Mantell Building, University of Sussex, Falmer, Brighton, BN1 9RF, United Kingdom
Fax: (44) 12 7367 8640
Email: cisscholarships@sussex.ac.uk

Economic and Social Development Scholarship

Subjects: Anthropolgy, gender studies, development studies, industrial relations, economics, human rights, education and migration studies.
Purpose: Intended for students intending to contribute to the development of their home countries.
Eligibility: Nationals from Brazil, India, Pakistan or Sri Lanka who have not previously studied in the UK are eligible. The applicants are liable to pay the overseas fees.
Level of Study: Postgraduate
Type: Scholarship
Value: Fee waiver of UK£5,000
Length of Study: 1 year Master's
Frequency: Annual
Study Establishment: University of Sussex
Country of Study: United Kingdom
Application Procedure: Application forms can be downloaded from the website www.sussex.ac.uk and sent to the university after filling in the same.
Closing Date: March 29th
No. of awards given last year: 2

Geoff Lockwood Scholarship

Subjects: All subjects.
Purpose: To encourage high-calibre graduate applications for MSc programmes.
Eligibility: The scholarship is only available to a fully self-financing UK MSc candidate who holds an offer of a place for admission.
Level of Study: Postgraduate
Type: Scholarship
Value: UK£1,000 per year for tuition fees
Length of Study: 1 year
Frequency: Annual
Study Establishment: University of Sussex
Country of Study: United Kingdom
No. of awards offered: 1
Application Procedure: Applicants should first apply for an MSc place in the school of Life Sciences and SPRU for admission. Eligible candidates are automatically considered. Applicants are strongly encouraged to provide a supporting statement.
Closing Date: July 20th
Contributor: George Lockwood
Additional Information: Applications received the closing date will be considered for an MSc place but not for the Geoff Lockwood scholarship.

SYLFF Fellowship

Subjects: Development studies, international relations, anthropology.
Purpose: Educating graduate students with high potential for future leadership in international affairs.
Eligibility: UK citizens or citizens of East European states listed on the website.
Level of Study: Postgraduate
Type: Fellowship
Value: Fee remission
Length of Study: 1 year
Frequency: Annual
Study Establishment: Sussex University
Country of Study: United Kingdom
No. of awards offered: 2
Closing Date: March 31st
Funding: Foundation

Contributor: Tokyo Foundation
No. of awards given last year: 2

USA Friends Scholarship

Subjects: Open to all subjects excludin IDS programmes.
Purpose: Rewarding academic merit.
Eligibility: Applicants must be US citizens.
Type: Scholarship
Value: US$5,000
Length of Study: 1 year
Frequency: Annual
Study Establishment: Sussex University
Country of Study: United Kingdom
No. of awards offered: Approx.4
Application Procedure: Application forms can be downloaded from the website www.sussex.ac.uk and sent to the university after filling in the same.
Closing Date: April 3rd
Funding: Individuals
No. of awards given last year: 4

THE UNIVERSITY OF SYDNEY

Scholarships Office, Jane Foss Russell Building, G02, Sydney, NSW, 2006, Australia
Tel: (61) 28627 8112
Fax: (61) 2 8627 8485
Email: research.training@sydney.edu.au
Website: www.sydney.edu.au
Contact: Mrs Carmen NG, Manager, Research Scholarships

The role of the University of Sydney is to create, preserve, transmit, extend and apply knowledge through teaching, research, creative works and other forms of scholarship. In carrying out its role, the University affirms its commitment to the values and goals of institutional autonomy, recognizes the importance of ideas and intellectual freedom to pursue critical and open enquiry, as well as social responsibility, tolerance, honesty and respect, as the hallmarks of relationships throughout the University community. It also understands the needs and expectations of those whom it serves and constantly improves the quality and delivery of its services.

Alexander Hugh Thurland Scholarship

Subjects: Agriculture.
Eligibility: Open to the graduates from other universities with relevant degree.
Level of Study: Doctorate, Postgraduate, Research
Type: Scholarship
Value: Australian $24,653 per year
Length of Study: 2 years for Masters by research candidates and 3 years with a possible 6-month extension for research doctoral candidates
Frequency: Dependent on funds available
Study Establishment: The University of Sydney
Country of Study: Australia
Application Procedure: Check website www.sydney.edu.au/agriculture for further details.
Funding: Trusts

For further information contact:

Faculty of Agriculture and Environment, Australian Technology Park, C81, The University of Sydney, NSW, 2006, Australia
Tel: (61) 2 8627 1002
Fax: (61) 2 8627 1099
Email: pg@agric.usyd.edu.au

Australian Postgraduate Award (APA)

Subjects: All subjects.
Purpose: To enable candidates with exceptional research potential to undertake a higher degree by research.
Eligibility: Open to Australian citizens and permanent residents, and citizens of New Zealand.
Level of Study: Doctorate, Postgraduate, Research
Type: Scholarship
Value: Australian $24,653 per year
Length of Study: 2 years for Research Master's by research candidates, and 3 years with a possible 6-month extension for Research Doctorate candidates
Frequency: Annual
Study Establishment: The University of Sydney
Country of Study: Australia
No. of awards offered: 340
Application Procedure: Applicants must complete a form, available from the Scholarships Office between late August and October for semester one, and between mid May and mid June for semester two. Forms can also be downloaded from the website or emailed on request.
Closing Date: October 31st for semester one and June 15th for semester two
Funding: Government
Contributor: Australian Government
No. of awards given last year: 340

International Postgraduate Research Scholarships (IPRS) Australian Postgraduate Awards (APA)

Subjects: All subjects.
Purpose: To support candidates with exceptional research potential.
Eligibility: Open to suitably qualified graduates eligible to commence a higher degree by research. Australia and New Zealand citizens and Australian permanent residents are not eligible to apply.
Level of Study: Doctorate, Postgraduate, Research
Type: Scholarship
Value: Tuition fees for IPRS, and an Australian Postgraduate Award for Australian $24,653 per year
Length of Study: 2 years for the Master's by research candidates, and 3 years with a possible 6-month extension for PhD candidates
Frequency: Annual
Study Establishment: The University of Sydney
Country of Study: Australia
No. of awards offered: 34
Application Procedure: Applicants must complete an application form for admission available from the International Office.
Closing Date: July 31st for Semester 1 and December 15th for Semester 2
Funding: Government
Contributor: Australian Government and University of Sydney
No. of awards given last year: 32

For further information contact:

International Office, Jane Foss Russell Building G02, Sydney, NSW, 2006, Australia
Tel: (61) 2 8627 8358
Fax: (61) 2 8627 8387
Email: infoschol@io.usyd.edu.au

University of Sydney - China Scholarship Council Early Career Research Scholars Scheme

Subjects: All subjects.
Eligibility: Open to the applicants from leading Chinese research universities who hold a doctorate and be citizens of the People's Republic of China and not be currently working or studying in Australia.
Level of Study: Postdoctorate, Research
Type: Fellowship
Value: Living expenses, international airfare, office accommodation
Length of Study: 1 year
Frequency: Dependent on funds available
Country of Study: Australia
No. of awards offered: 2
Application Procedure: Check website for further details.
Funding: Government
Contributor: University of Sydney
Additional Information: Preference will be given to applicants working in areas of study that meet Chinese and Australian. The website address is http://sydney.edu.au/scholarships/prospective/uos-csc-scholarships.shtml.

For further information contact:

John Woolley Building A20, The University of Sydney, NSW, 2006, Australia
Contact: International Development Officer

University of Sydney – China Scholarship Council Research Schemes

Subjects: Areas of study that meets Chinese and Australian national research priorities.
Eligibility: Open to the citizens of the People's Republic of China, who are currently not working or studying in Australia and will normally be from leading Chinese research universities, including 985 Program Universities.
Level of Study: Doctorate, Postgraduate, Research
Type: Studentship
Value: Living allowance (Australian $20,400 per year), tuition scholarship, and airfare
Length of Study: Up to 3 years
Frequency: Annual
Study Establishment: University of Sydney
Country of Study: Australia
No. of awards offered: Up to 11
Application Procedure: Check website for further details.
Closing Date: December 15th
Funding: Government
Contributor: Chinese government and the University of Sydney
Additional Information: Preference will be given to applicants working in areas of study that meet Chinese and Australian National research priorities. The website address is http://sydney.edu.au/scholarships/prospective/uos-csc-scholarships.shtml.

For further information contact:

John Woolley Building, A20, The University of Sydney, NSW, 2006, Australia
Contact: International Development Officer

University of Sydney International Scholarship (USydIS)

Subjects: All subjects.
Purpose: To attract top quality international postgraduate students to undertake research projects to enhance the University's research activities.
Eligibility: Open to eligible international postgraduate students eligible to commencea higher degree by research.
Level of Study: Doctorate, Postgraduate, Research
Type: Scholarship
Value: Tuition fees and living allowance of Australian $24,653 per year
Length of Study: 2 years for masters by research candidates, and 3 years with a possible 6-month extension for research doctorate extension.
Frequency: Annual
Study Establishment: University of Sydney
Country of Study: Australia
No. of awards offered: 40
Application Procedure: Applicants must complete an application for admission form available from the International Office.
Closing Date: July 31st for Semester 1 and December 15th for Semester 2
Contributor: University of Sydney
No. of awards given last year: 45

For further information contact:

International Office, Jane Foss Russell Building GO2, Sydney, NSW, 2006, Australia
Tel: (61) 2 8627 8358
Fax: (61) 2 8627 8387
Website: infoschol@io.usyd.edu.au

University of Sydney Postgraduate Award (UPA)

Subjects: All subjects.
Purpose: To enable candidates with exceptional research potential to undertake a higher degree by research.
Eligibility: Open to Australian citizens and permanent residents and New Zealand citizens.
Level of Study: Postgraduate, Research, Doctorate
Type: Scholarship
Value: Australian $24,653 per year
Length of Study: 2 years for the Research Master's by research candidates, and 3 years with a possible 6-month extension for Research Doctorate candidates
Frequency: Annual
Study Establishment: University of Sydney
Country of Study: Australia
No. of awards offered: 28
Application Procedure: Applicants must complete a form, available from the Scholarships Office between late August and October for semester one, and between mid May and mid June for semester two. Forms can also be downloaded from the website or emailed on request.
Closing Date: October 31st for semester one and June 15th for semester two
Funding: Government
Contributor: University of Sydney
No. of awards given last year: 28

UNIVERSITY OF TASMANIA

Private Bag 45, Hobart, TAS, 7001, Australia
Tel: (61) 3 6226 2766
Fax: (61) 3 6226 7497
Email: scholarships@research.utas.edu.au
Website: www.utas.edu.au
Contact: Graduate Research Unit

The University of Tasmania was officially founded on January 1st 1890, by an Act of the Colony's Parliament and was only the fourth university to be established in 19th century Australia. The university represents areas of significant research strengths and substantial teaching endeavours.

Victoria League for Commonwealth Friendship Medical Research Trust

Subjects: Research into dementia, heart complaints or cancer.
Purpose: To provide support and guidance to deserving candidates for medical research and related fields.
Eligibility: Open to students who have completed at least 4 years of tertiary education studies and have achieved at least an upper Second Class Honours Degree or equivalent and who are citizens or permanent residents of Australia.
Level of Study: Postgraduate
Type: Fellowship
Value: Australian $ 20,000
Length of Study: 3 years
Frequency: Every 3 years
Study Establishment: University of Tasmania
Country of Study: Australia
Closing Date: October 31st
Funding: Trusts
No. of awards given last year: 1

UNIVERSITY OF ULSTER

Research Office, Cromore Road, Coleraine, Co. Londonderry, BT52 1SA, Northern Ireland
Tel: (44) (28) 7032 4729
Fax: (44) (28) 7032 4905
Email: hj.campbell@ulster.ac.uk
Website: www.ulster.ac.uk
Contact: Mrs H Campbell, Administrative Officer

The University of Ulster is a dynamic and innovative institution, which is very proud of its excellent track record in the education and training of researchers. Our doctoral graduates can demonstrate outstanding achievements in advancing knowledge and making breakthroughs of relevance to the economic, social and cultural development of society.

Vice Chancellor's Research Scholarships (VCRS)

Subjects: All subjects.
Purpose: To assist candidates of a high academic standard to complete research degrees (PhD).
Eligibility: Applicants must have or expect to obtain the minimum of an upper second class Honours degree in a specific research area (as advertised). Applications are invited from UK, European Union and overseas students. Only candidates who are new applicants to PhD will be eligible.
Level of Study: Doctorate, Postgraduate
Type: Scholarship
Value: Fees and maintenance grant
Length of Study: Up to 3 years
Frequency: Annual
Study Establishment: University of Ulster
Country of Study: United Kingdom
No. of awards offered: Varies
Application Procedure: Applicants must complete an application form.
Closing Date: Check the website
Funding: Private
No. of awards given last year: 30
No. of applicants last year: 800
Additional Information: VCRS awards are not available for applicants for any other research degree or for any candidate who has previously registered for a PhD, or anyone who has already obtained a PhD.Further information is available on the website www.ulster.ac.uk/research study

UNIVERSITY OF UTAH

Admissions Office, 201 Presidents Circle, Room 201, Salt Lake City, Utah, SLC UT 84112, United States of America
Tel: (1) 801 581 7200
Fax: (1) 801 585 7864
Email: admissions@sa.utah.edu
Website: www.utah.edu/home/index.uofu

The mission of University College is to assist new, transfer and transitioning students, through academic advising, to develop and implement individual plans for achieving educational and life goals.

Liberal Education Scholarships

Subjects: All subjects.
Purpose: To support students chosen an approved program abroad to complete their studies in their field.
Eligibility: Open to candidates of all nationalities and ages who are in need of liberal education credit and have matriculated status, 3.0 cumulative GPA, completed 36 hours of residence at the University of Utah prior to departure, full-time status (at least 2 out of 3 quarters during the year prior to departure) 6 weeks study program, priority will be given to students who will be sophomores or juniors during their study abroad program.
Type: Scholarship
Value: $1,000–1,500
Frequency: Biannual
No. of awards offered: 12 (every 6 months)
Application Procedure: Completed applications should be sent to the center's address.
Closing Date: November (for program beginning in winter and spring), March (for summer and fall)

For further information contact:

159 University Union, Salt Lake City, Utah, 841 12, United States of America

THE UNIVERSITY OF WAIKATO

Gate 1 Knighton Road Private Bag 3105, Hamilton, 3240, New Zealand
Tel: (64) 7 856 2889
Fax: (64) 7 838 4300
Email: info@waikato.ac.nz
Website: www.waikato.ac.nz

The University of Waikato through high quality teaching and research aims to deliver a world class education and research portfolio and a full dynamic university experience which is distinctive in character and pursue strong international linkages to advance knowledge.

Acorn Foundation Eva Trowbridge Scholarship

Subjects: All subjects.
Purpose: To support the people of Tauranga and the Western Bay of Plenty community.
Eligibility: Adult students (25 years and over) studying at the University of Waikato's Tauranga campus and residing in the areas administered by Tauranga City Council or Western Bay of Plenty District Council are eligible for the award.
Type: Scholarship
Value: $3,000 per year
Length of Study: 1 year
Frequency: Annual
Country of Study: New Zealand
Closing Date: August 31st
Contributor: Acorn Foundation
Additional Information: The Scholarship will be paid in one lump sum to the successful applicant.

For further information contact:

The Scholarships Office, ITS Building, The University of Waikato, Private Bag 3105, Hamilton, 3240, New Zealand
Tel: (64) 07 838 4964 or 07 858 5195
Email: scholarships@waikato.ac.nz

Alan Turing Prize

Subjects: Computer science and mathematics.
Purpose: To encourage students to develop strong joint interests in Computer Science and Mathematics.
Eligibility: Open to students who have strong interest in computer science and mathematics.
Level of Study: Research
Type: Prize
Length of Study: 3 years
Frequency: Annual
Application Procedure: Check website for further details.
Contributor: Council of the University of Waikato

Alumini Master's Scholarship

Subjects: All subjects.
Purpose: To support a student who has graduated with a degree of the University of Waikato and is enrolled for a Masters Degree at this University in the year of tenure.
Eligibility: Open to New Zealand citizens or permanent residents who have qualified for a First Degree from the University of Waikato and be enrolled full-time for a Masters Degree at the University of Waikato in the year of tenure. The candidate must be in their final year of study for the degree.
Level of Study: Postgraduate
Type: Scholarship
Value: $5,500 plus actual tuition fees up to a maximum of $4,000
Length of Study: 1 year
Frequency: Annual
Application Procedure: Check website for further details.
Closing Date: October 31st
Additional Information: The Scholarship will be awarded to a student who demonstrates academic merit; who is active in University affairs and who contributes to the activities of the School or Faculty in which they are enrolled; who demonstrates willingness to maintain an active relationship with the Alumni programme; who demonstrates willingness to attend Alumni functions and promotional activities; who is considered to be a good ambassador for the University of Waikato and the Alumni Association.

For further information contact:

The Scholarships Office, ITS Building, The University of Waikato, Private Bag 3105, Hamilton, 3105, New Zealand
Tel: (64) 07 838 4964 or 07 858 5195
Email: scholarships@waikato.ac.nz

Chamber of Commerce Tauranga Business Scholarship

Subjects: Management Studies.
Purpose: For the benefit of members of the Tauranga Chamber of Commerce to assist the recipient to undertake study at postgraduate level.
Eligibility: Open to citizens or permanent residents of New Zealand having a tertiary or relevant professional qualification; must have a minimum of 5 years' relevant work experience; must own or be employed by a business or organization which is a member of the Tauranga, Chamber of Commerce; must have the support of his/her employer and currently not enrolled in a Postgraduate Diploma in Management Studies with the Waikato, Management School, University of Waikato. Check website for further details.
Level of Study: Postgraduate
Type: Scholarship
Value: The value of the scholarship is usually equivalent to 1 year's fees (a total of four papers)
Length of Study: Above 2 years
Frequency: Annual
Application Procedure: Check website for further details.
Closing Date: November 30th

For further information contact:

The Scholarships Office, ITS Building, The University of Waikato, Private Bag 3105, Hamilton, New Zealand
Tel: (64) 07 838 4964 or 07 858 5195
Email: scholarships@waikato.ac.nz

Evelyn Stokes Memorial Doctoral Scholarship

Subjects: Geography, Tourism and Environmental Planning.
Eligibility: Open to candidates enrolled or intending to enroll in an approved full-time programme of study at the University of Waikato in the years of tenure of the scholarship. Applicants who have undertaken study outside the University of Waikato must supply a full verified copy of their academic transcript(s) from their previous institution(s).
Level of Study: Doctorate
Type: Scholarship
Value: $5,000 per year
Length of Study: 3 years
Frequency: Annual
Study Establishment: University of Waikato
Country of Study: New Zealand
Application Procedure: Check website for further details.
Closing Date: October 31st in the year prior to that in which the award will be taken up
Funding: Private
Additional Information: The Scholarship will end on the completion of doctoral study, or after 3 years, whichever is the earlier date, provided that the candidate is enrolled during this time in an appropriate programme of studies. Completion takes place when the postgraduate studies committee has accepted the report of the examiners and recommends the awarding of the degree.

For further information contact:

The Scholarships Office, ITS Building, The University of Waikato, Private Bag 3105, Hamilton, New Zealand
Tel: (64) 07 838 4964 or 07 858 5195
Email: scholarships@waikato.ac.nz

The Faculty of Arts and Social Sciences Honours / Postgraduate Diploma Awards

Subjects: Arts, humanities and social sciences.
Purpose: To support students enrolling in a master's degree based on academic merit.
Eligibility: Open to applicants who are New Zealand citizens or permanent residents with high academic caliber and enrolling in a master's degree in the faculty of arts and social sciences.
Level of Study: Postgraduate
Type: Award
Value: $1,500
Frequency: Annual
Study Establishment: University of Waikato
Country of Study: New Zealand
No. of awards offered: 40
Application Procedure: For further details please check website www.waikato.ac.nz/research/scholarships
Closing Date: March 25th for A Semester Awards and July 22nd for B Semester Awards
Contributor: School of Science and Engineering

Faculty of Arts and Social Sciences Masters Thesis Awards

Subjects: Humanities, arts and social sciences.
Purpose: To assist students enrolling in a Master's degree in the faculty of Arts and Social sciences.
Eligibility: Open to all Master's enrolling in the faculty of arts and social sciences.
Level of Study: Postgraduate
Type: Award
Value: Up to $3,000
Length of Study: 1 year
Frequency: Annual
Study Establishment: University of Waikato
Country of Study: New Zealand
No. of awards offered: 20
Application Procedure: Application forms and information available on www.waikato.ac.nz/research/scholarships
Closing Date: March 25th for A Semester Awards and July 22nd for B Semester Awards
Contributor: The International Council for Canadian studies (ICCS)

Hilary Jolly Memorial Scholarship

Subjects: Fresh water ecology.
Purpose: To encourage research in the field of fresh water ecology at the University of Waikato at Master's or Doctoral level.
Eligibility: Open to candidates who are enrolled for a Master's degree at the university in the area of fresh water ecology.
Level of Study: Doctorate, Postgraduate
Type: Scholarship
Value: Master's–$6,000 plus tuition fees. Doctoral–$15,000 plus tuition fees
Length of Study: 2 years for Master's and 3 years for doctorate
Frequency: Dependent on funds available
Study Establishment: University of Waikato
Country of Study: New Zealand
No. of awards offered: 1
Application Procedure: A completed application form must be submitted to the scholarship office at the university. Application forms are available on www.waikato.ac.nz/research/scholarships
Closing Date: November 15th

The Michael Baldwin Memorial Scholarship

Subjects: All subjects.
Purpose: To commemorate Michael's commitment to teaching and to assist people in their educational endeavours.
Eligibility: Open to students who are citizens of Papua New Guinea. Preference will be given to applicants who can demonstrate a need for financial support in order to undertake tertiary study. Candidates must be current or intending students, who will be enrolled in an approved full–time programme of study at The University of Waikato in the year of tenure of the scholarship.
Type: Scholarship
Value: Up to $10,000
Length of Study: 1 year
Frequency: Annual
No. of awards offered: 1
Application Procedure: Applications should include a personal statement indicating how the scholarship will benefit the candidate's particular circumstances and aspirations and must give certified details of the candidate's academic history, accompanied by copies of two recent testimonials.
Closing Date: August 30th

New Zealand Federation of University Women Emmy Noether Prize in Mathematics

Subjects: Mathematics.
Eligibility: Open for the outstanding woman student studying first year mathematics in that year at the University.
Type: Prize

Value: $1,500
Frequency: Annual
Application Procedure: Check website for further details.
Contributor: Federation of University Women to the University of Waikato
Additional Information: All women students enrolled in first year mathematics courses in any 1 year will be considered without application as candidates for the award of the prize in that year.

Priority One Management Scholarship

Subjects: Management studies.
Eligibility: Open to citizens or permanent residents of New Zealand who have a tertiary or relevant professional qualification with a minimum of 5 years relevant work experience, own or be employed by a business or organization which is a member of Priority One, have the support of his/her employer, must not already be enrolled in a Postgraduate Diploma in Management Studies with the Waikato, Management School, University of Waikato, not have been a previous recipient of any Waikato Management School, University of Waikato.
Level of Study: Postgraduate
Type: Scholarship
Value: Equivalent to 1 year's fees
Length of Study: 2 years
Frequency: Annual
Application Procedure: Check website for further details.
Closing Date: November 30th

For further information contact:

The Scholarships Office, ITS Building, The University of Waikato, Private Bag 3105, Hamilton, New Zealand
Tel: (64) 07 838 4964 or 07 858 5195
Email: scholarships@waikato.ac.nz

Science Admission Fees Scholarships

Subjects: Science and engineering
Purpose: To provide financial assistance to science students.
Eligibility: Open to New Zealand citizens or permanent residents enrolling in science faculty.
Level of Study: Postgraduate
Type: Scholarship
Value: NZ $4,000
Frequency: Annual
Study Establishment: University of Waikato
Country of Study: New Zealand
No. of awards offered: 10
Application Procedure: Further information can be found on www.waikato.ac.nz/research/scholarships
Closing Date: January 14th

Science and Engineering Masters Fees Award

Subjects: Science and engineering.
Purpose: To support students pursuing Master's degree in school of science and engineering.
Eligibility: Open to New Zealand citizens or permanent residents enrolling for science master's study.
Level of Study: Postgraduate
Type: Scholarship
Value: $2,000 credited to fees
Frequency: Annual
Study Establishment: University of Waikato
Country of Study: New Zealand
No. of awards offered: 15
Application Procedure: Further information and application forms are available on www.waikato.ac.nz/research/scholarships
Closing Date: February 21st

Sir Edmund Hillary Scholarship Programme

Subjects: Any subject.
Purpose: To support students to achieve a high academic standard and excellence in sports or creative and performing arts.
Eligibility: Open to candidates who can demonstrate academic achievement and excellence in either a sporting code or a performing or creative arts.
Level of Study: Graduate, Postgraduate
Type: Scholarship
Value: Tuition fees and associated charges
Length of Study: 1 year, renewable for period of degree, up to maximum 4 years
Frequency: Annual
Study Establishment: University of Waikato
Country of Study: New Zealand
No. of awards offered: Varies
Application Procedure: Further information available on the website www.waikato.ac.nz/research/scholarships
Closing Date: October 31st
No. of awards given last year: 40

Ted Zorn Waikato Alumni Award For Management Communication

Subjects: Management studies.
Purpose: To provide an opportunity for peer recognition of graduates of the department who have, since their graduation, distinguished themselves in a field of management communication.
Eligibility: Open to candidates holding a responsible position in an organization or in a project, sustainability and/or workplace well being; must know the use of creativity and initiative in performing the responsibilities of the position. Check website for further details.
Level of Study: Postgraduate
Type: Award
Value: $1,000
Frequency: Annual
Application Procedure: Check website for further details.
Closing Date: December 31st

For further information contact:

Department of Management Communication, The University of Waikato, Private Bag 3105, Hamilton, New Zealand
Email: jbeaton@waikato.ac.nz
Contact: Jean Beaton

The University of Waikato Doctoral Scholarships

Subjects: All subjects.
Purpose: To support candidates applying for doctoral studies.
Eligibility: Open to New Zealand citizens or permanent residents enrolling in a doctoral degree.
Level of Study: Doctorate
Type: Scholarship
Value: $22,000 plus tuition fees per year.
Length of Study: 3 years
Frequency: Annual, Twice a year
Study Establishment: University of Waikato
Country of Study: New Zealand
Application Procedure: Contact the scholarships office at the university. Applications are available on www.waikato.ac.nz/research/scholarships
Closing Date: October 31st and April 30th
No. of awards given last year: 41

University of Waikato Masters Research Scholarships

Subjects: All subjects.
Purpose: To encourage research at the University, principally by assisting with course-related costs.
Eligibility: Open to citizens and permanent residents of New Zealand who have qualified for a first degree and be enrolled full-time for a first Masters or Master of Philosophy degree at the University of Waikato in the year of tenure.
Level of Study: Postgraduate
Type: Research grant
Value: $12,000, of which up to $3,500 is to be applied to tuition fees for the masters degree. The remainder ($8,500 in the case of a full Scholarship) will normally be paid out in two instalments
Length of Study: 1 year
Frequency: Annual
No. of awards offered: Up to 60 awards
Application Procedure: Check website for further details.
Closing Date: October 31st and April 30th annually
No. of awards given last year: 65
Additional Information: Should a student also hold another fees scholarship, the University of Waikato Masters Research Scholarship will pay the balance of any fees (up to $3,500).

For further information contact:

The Scholarships Office, ITS Building, The University of Waikato, Private Bag 3105, Hamilton, New Zealand
Tel: (64) 07 858 5136 or 07 858 5195
Email: scholarships@waikato.ac.nz

WMS International Exchange Scholarships

Eligibility: Applicants must have completed at least 1 year of study at the Waikato Management School and be eligible to apply for exchange programmes, must have been accepted into a University of Waikato exchange programme, should apply to institutions listed as recommended institutions, should be New Zealand citizens or permanent residents of New Zealand and should be full-time students at the Waikato Management School.
Level of Study: Postgraduate
Type: Scholarship
Value: $2,500 for students participating in exchange programmes in Europe; $2,000 for students participating in exchange programmes in USA, Canada and Mexico; and $1,500 for students participating in exchange programmes in Asia
Frequency: Annual, Twice a year
No. of awards offered: 10
Application Procedure: Applicants should submit an International Exchange Scholarship Application Form, an application letter and curriculum vitae.
Closing Date: August 15th for A Semester exchanges and March 15th for B Semester exchanges

For further information contact:

Scholarships Office, University of Waikato, ITS Building, Private Bag 3105, Gate 1, Knighton Road, Hamilton, New Zealand

UNIVERSITY OF WALES, LAMPETER

Lampeter Campus, Lampeter, Ceredigion, SA48 7ED, Wales
Tel: (44) 15 7042 2351
Fax: (44) 15 7042 3423
Email: t.rodervick@lamp.ac.uk
Website: http://www.trinitysaintdavid.ac.uk/en/
Contact: Ms Gwawr Davies, Registry Administrator

At Lampeter, research is greatly prized by our academic staff and we can offer a wide range of supervision in the humanities and social sciences. There is simply no other long-established university with our pedigree that operates on such an intimate scale.

Delahaye Memorial Benefaction

Subjects: Theology.
Purpose: To financially support a graduate of UWL who is accepted to read Honours Theology at Cambridge.
Eligibility: Graduates.
Level of Study: Graduate
Value: UK£100
Frequency: Annual
Study Establishment: Magdalene College, University of Cambridge
Country of Study: United Kingdom
No. of awards offered: 1
Application Procedure: Applications for postgraduate awards should be made to the Postgraduate Office, which writes, during the Michaelmas term, to all registered postgraduate students reminding them of the awards and enclosing the official application forms.
Closing Date: March 7th
No. of awards given last year: None
No. of applicants last year: None

Helen McCormack Turner Memorial Scholarship

Subjects: All subjects.
Purpose: The applicant must be a postgraduate student who graduated recently from the University of Wales, Lampeter.
Eligibility: The recipient must be less than 25 years of age on August 1st and must be registered to support graduates pursuing research for the degrees of MPhil or PhD.
Type: Scholarship
Value: UK£250
Frequency: Annual
No. of awards offered: 1
Application Procedure: Applications for postgraduate awards should be made to the Postgraduate Office, which writes, during the Michaelmas term, to all registered postgraduate students reminding them of the awards and enclosing the official application forms.
Closing Date: March 7th
No. of awards given last year: 1
No. of applicants last year: 1

Herbert Hughes Scholarship

Subjects: Theology.
Purpose: To support postgraduate or undergraduate students of any discipline who intend to serve in the Ministry of the Church in Wales.
Eligibility: Postgraduate or undergraduate. (Preference may be given to candidates from the parish of Silian, Ceredigion).
Type: Scholarship
Value: UK£250
No. of awards offered: 1
Application Procedure: Applications for postgraduate awards should be made to the Postgraduate Office, which writes, during the Michaelmas term, to all registered postgraduate students reminding them of the awards and enclosing the official application forms.
Closing Date: March 7th
No. of awards given last year: 1
No. of applicants last year: 1

Mary Radcliffe Scholarship

Subjects: Theology.
Purpose: To support graduates of UWL to study theology.
Eligibility: Applicants must be graduates of UWL.
Level of Study: Graduate
Type: Scholarship
Value: UK£175
Frequency: Annual
No. of awards offered: 1
Application Procedure: Applications for postgraduate awards should be made to the Postgraduate Office, which writes, during the Michaelmas term, to all registered postgraduate students reminding them of the awards and enclosing the official application forms.
Closing Date: March 7th
No. of awards given last year: 1
No. of applicants last year: 1

RHYS Curzon-Jones Scholarship

Subjects: Theology.
Purpose: To support a postgraduate student of any discipline who is a candidate for Holy Orders in the Church in Wales.
Type: Scholarship
Value: UK£50
Frequency: Annual
No. of awards offered: 1
Application Procedure: Applications for postgraduate awards should be made to the Postgraduate Office, which writes, during the Michaelmas term, to all registered postgraduate students reminding them of the awards and enclosing the official application forms.
Closing Date: March 7th
No. of awards given last year: None
No. of applicants last year: None

Ridley Lewis Bursary

Subjects: All subjects.
Purpose: To support a postgraduate student funding his/her studies in whole or in part from his/her own resources.
Eligibility: Applicants must be postgraduate students.
Type: Bursary
Value: UK£75
Frequency: Annual
No. of awards offered: 1
Application Procedure: Applications for postgraduate awards should be made to the Postgraduate Office, which writes, during the Michaelmas term, to all registered postgraduate students reminding them of the awards and enclosing the official application forms.
Closing Date: March 7th
No. of awards given last year: 1
No. of applicants last year: 29

W D Llewelyn Memorial Benefaction

Subjects: All subjects.
Purpose: To support graduates of UWL for further degrees or research at Lampeter or at other universities.
Eligibility: Applicants must be graduates of UWL.
Value: Six awards of UK£100 each
Length of Study: 1 year
Frequency: Annual
No. of awards offered: 6
Application Procedure: Applications for postgraduate awards should be made to the Postgraduate Office, which writes, during the Michaelmas term, to all registered postgraduate students reminding them of the awards and enclosing the official application forms.
Closing Date: March 7th
No. of awards given last year: 6
No. of applicants last year: 16

UNIVERSITY OF WALES, NEWPORT

Caerleon Campus, PO Box 179, Newport, NP18 3YG, Wales
Tel: (44) (01633) 432432
Fax: (44) (01633) 432046
Email: uic@newport.ac.uk
Website: www3.newport.ac.uk/

The University of Wales, Newport, has been involved in higher education for more than 80 years, and its roots go back even further to the first Mechanics Institute in the town, which opened in 1841.

Pilcher Senior Research Fellowship

Subjects: Welsh and Celtic studies.
Purpose: To support a student in advanced research in the fields of Welsh and Celtic studies.
Level of Study: Postdoctorate
Type: Fellowship
Frequency: As needed
Study Establishment: University of Wales
Country of Study: Wales
Application Procedure: Applicants must check with the website or the University.
Funding: Private
Contributor: The late Sophia Margaretta Pilcher

The Stott Fellowship

Subjects: Welsh and Celtic studies.
Purpose: To support a student in advanced research in the fields of Welsh and Celtic studies.
Level of Study: Postgraduate
Type: Fellowship
Frequency: As needed
Study Establishment: University of Wales
Country of Study: Wales
Application Procedure: Applicants must check with the website or the University.
Funding: Private
Contributor: Miss Muriel Stott of Colwyn Bay

UNIVERSITY OF WARWICK

Coventry, CV4 7AL, England
Tel: (44) 024 7652 3523
Fax: (44) 024 7646 1606
Website: www.warwick.ac.uk
Contact: Project Officer,, Postgraduate Scholarships

The University of Warwick offers an exciting range of doctoral, research-based and taught Master's programmes in the humanities, sciences, social sciences and medicine. In the 2001 Research Assessment Exercise, Warwick was ranked 5th in the United Kingdom for research quality. Postgraduate students make up around 35 per cent of Warwick's 18,000 students. The University is located in the heart of England, adjacent to the city of Coventry and on the border with Warwickshire.

Bangladesh and Pakistan Postgraduate Award (Warwick Manufacturing Group)

Subjects: WMG (Warwick Manufacturing Group) taught masters courses.
Purpose: To support Master's Bangladeshi and Pakistani students at WMG.
Eligibility: Applicants must be nationals of Bangladesh or Pakistan, classed as international fee-payer, not currently registered for a taught masters course at the University of Warwick and should have received an offer of a place from WMG at Warwick.
Level of Study: Postgraduate
Type: Scholarship
Value: UK£5,000
Length of Study: 1 year
Frequency: Dependent on funds available
Study Establishment: University of Warwick
Country of Study: United Kingdom
No. of awards offered: 3
Application Procedure: Applicants must complete an online application form.
Closing Date: June 30th
Contributor: University of Warwick
Additional Information: Non-renewable, deducted from tuition fees, for taught masters only.

For further information contact:

Website: www.warwick.ac.uk/go/scholarships

Chancellor's International Scholarships

Purpose: PhD candidates in any discipline.
Eligibility: Applicants for the scholarship must also be applying for a PhD at the University of Warwick. Applicants must expect to be 'overseas' students for fees purposes, but there is no other nationality criteria. Applicants may be from any discipline at Warwick.
Level of Study: Doctorate
Type: Scholarship
Value: £17,130 for overseas tuition fees and £13,726 as a maintenance stipend in line with RCUK rates
Length of Study: Maximum up to 3 years
Frequency: Annual
Country of Study: United Kingdom
No. of awards offered: 17
Closing Date: January 11th
No. of awards given last year: 17

Chancellor's Scholarship

Purpose: PhD candidates in any discipline.
Eligibility: Open to home, EU and overseas students from all disciplines at Warwick. For more details, please refer to our website: www2.warwick.ac.uk/services/academic office/gsp/scholarship/wprs.
Level of Study: Doctorate
Type: Scholarship
Value: The payment of academic fees at the Home/EU rate £3,900 and a maintenance grant in line with the UK Research Council stipend of £13,726
Frequency: Annual
No. of awards offered: 45
Application Procedure: Please refer to our website: www2.warwick.ac.uk/services/academic office/gsp/scholarship/wprs.
Closing Date: January 11th
No. of awards given last year: 45
Additional Information: Students and applicants who wish to apply for an AHRC Doctoral award should apply to the Chancellor's Scholarship competition and will automatically be considered for both competitions. Address for application: electronic applications only.

Colombia Postgraduate Awards (Warwick Manufacturing Group/Colfuturo)

Subjects: WMG (Warwick Manufacturing Group) taught masters courses.
Purpose: To support Master's Columbian students at WMG.
Eligibility: Applicants should be nationals of Columbia, not currently registered on a postgraduate course at the University, and should have received an offer of a place from WMG. Only students who have been awarded a COLFUTURO scholarship-loan are eligible.

Level of Study: Postgraduate
Type: Scholarship
Value: full tuition fees
Length of Study: 1 year
Frequency: Dependent on funds available
Study Establishment: University of Warwick UK
Country of Study: United Kingdom
No. of awards offered: 5
Application Procedure: Applicants must apply via COLFUTURO.
Contributor: WMG/COLFUTURO
Additional Information: Non-renewable, for taught masters only.

For further information contact:

Email: yosoyfuturo@colfuturo.com.co

COLOMBIA Postgraduate Awards (Warwick/Colfuturo)

Subjects: Any except MBA or WMG courses.
Purpose: To support Colubmian students on postgradaute courses at Warwick.
Eligibility: Applicants should be nationals of Columbia and classed as an international fee-payer. Applicants can be registered on an Undergraduate course at the University of Warwick, but should have received a place on a Postgraduate Taught Master's course at Warwick. Only students who have been awarded a COLFUTURO scholarship-loan are eligible.
Level of Study: Postgraduate
Type: Scholarship
Value: Tuition fees
Length of Study: 1 year
Frequency: Dependent on funds available
Study Establishment: University of Warwick
Country of Study: United Kingdom
No. of awards offered: Variable
Application Procedure: Application should be submitted via COLFUTURO.
Funding: International Office
Contributor: COLFUTURO
Additional Information: Non-renewable, for taught masters only.

For further information contact:

Email: yosoyfuturo@colfuturo.com.co

India Postgraduate Awards (Warwick Manufacturing Group)

Subjects: WMG (Warwick Manufacturing Group) taught masters courses.
Purpose: To support Master's Indian students at WMG.
Eligibility: Applicants must be nationals of India, classed as international fee-payer, not currently registered for a taught masters course at the University of Warwick and should have received an offer of a place from WMG at Warwick.
Level of Study: Postgraduate
Type: Scholarship
Value: Two awards of £5,000 each and five awards of £3,000
Length of Study: 1 year
Frequency: Dependent on funds available
Study Establishment: University of Warwick
Country of Study: United Kingdom
No. of awards offered: 7
Application Procedure: Applicants must complete an online application form.
Closing Date: June 30th
Contributor: University of Warwick
No. of awards given last year: 2
Additional Information: Non-renewable, deducted from tuition fees, for taught masters only.

For further information contact:

Website: www.warwick.ac.uk/go/scholarships

International Office Taught Masters Scholarship

Eligibility: Open to all Taught Masters students who have received an offer to study a full time one year Masters Degree at the the University of Warwick. Applicants must be considered as international fee payers.
Level of Study: Graduate, Postdoctorate
Type: Fellowships
Value: £17,000
Length of Study: 1 year
Frequency: Annual
Study Establishment: University of Warwick
Country of Study: United Kingdom
Closing Date: March 4th
Funding: International Office

Karim Rida Said Foundation Postgraduate Award (KRSF/Warwick)

Subjects: All subjects.
Purpose: To support students from the Middle Eastern region on Postgraduate courses at Warwick.
Eligibility: Applicants should be Jordanian, Iraqi, Lebanese, Palestinian or Syrian nationals and be resident in the Middle East. Applicants should meet all other eligibility criteria as set by KRSF and awards will only be offered to applicants who already hold an offer of a place at Warwick.
Level of Study: Postgraduate
Type: Scholarship
Length of Study: 1 year
Study Establishment: University of Warwick
Country of Study: United Kingdom
No. of awards offered: 1
Application Procedure: Applications are submitted via KRSF website www.krsf.org/whatwedo/masters.
Funding: International Office
Contributor: KRSF
No. of awards given last year: 2
Additional Information: Non-renewable, for taught masters only.

For further information contact:

International Office, University of Warwick, Coventry, CV4 8UW, United Kingdom
Tel: (44) 24 7652 2469
Email: j.c.inegbedion@warwick.ac.uk
Contact: Jon Inegbedion

Latin American Postgraduate Award (Warwick Manufacturing Group)

Subjects: WMG (Warwick Manufacturing Group) taught masters courses.
Purpose: To support Master's students from Latin America at WMG.
Level of Study: Postgraduate
Type: Scholarship
Value: £6,000 and £3,000 for all Latin American students
Length of Study: 1 year
Frequency: Dependent on funds available
Study Establishment: University of Warwick
Country of Study: United Kingdom
No. of awards offered: 2
Application Procedure: No separate application required.
Contributor: WMG
Additional Information: Non-renewable, deducted from tuition fees, for taught masters only.

Lord Rootes Memorial Fund

Subjects: Economic, environmental, social and technological problems.
Purpose: To support a student's or society's project.
Eligibility: Applicants must be either full-time or part-time postgraduate or undergraduate students belonging to a student society at the University of Warwick. Applicants must be on courses of at least one year in duration.
Level of Study: Unrestricted
Type: Grant
Value: £100–3,000
Length of Study: At least 1 year
Frequency: Annual
Study Establishment: University of Warwick
Country of Study: United Kingdom

No. of awards offered: Varies
Application Procedure: Applicants must submit an application form plus a detailed project proposal of up to six A4 pages including financial plan. Short-listed applicants are interviewed.
Closing Date: January 12th
Funding: Private
Contributor: Lord Rootes Memorial Fund
No. of awards given last year: 13
No. of applicants last year: 38

Mexico Postgraduate Award (Chevening/Brockmann/ Warwick)

Subjects: All subject (except MBA).
Purpose: To support Mexican students on a postgraduate course at Warwick.
Eligibility: Applicants should be nationals of Mexico, not currently registered at the University of Warwick and should have received an offer of a place at Warwick.
Level of Study: Postgraduate
Type: Scholarship
Value: Full tuition fees plus maintanence
Length of Study: 1 year
Study Establishment: University of Warwick
Country of Study: United Kingdom
No. of awards offered: 1
Application Procedure: Applicants must complete an application.
Closing Date: May 9th
Funding: Government
Contributor: The Foreign and Commonwealth Office and the Brockmann Foundation
No. of awards given last year: 1
Additional Information: Non-renewable.

For further information contact:

The British Council, Lope de Vega, 316, Col Chapultepec Morales, Delègacion Miguel Hidalgo, CP 11570, DF, Mexico
Tel: (52) 24 7657 2686
Email: ana_delcarmen@hotmail.com
Contact: Ana Galllegos, Adviser for Latin America - International Office

Middle East And North Africa Postgraduate Award (Warwick Manufacturing Group)

Subjects: WMG (Warwick Manufacturing Group) taught masters courses.
Purpose: To support Master's students from the Middle East and northern Africa region at WMG.
Eligibility: Applicants must be nationals of one of the country within the region of Middle East and Northern Africa, classed as international fee-payer, not currently registered for a taught masters course at the University of Warwick and should have received an offer of a place from WMG at Warwick.
Level of Study: Postgraduate
Type: Scholarship
Value: UK£5,000
Length of Study: 1 year
Frequency: Dependent on funds available
Study Establishment: University of Warwick
Country of Study: United Kingdom
No. of awards offered: 5
Application Procedure: Applicants must complete an online application form.
Closing Date: June 30th
Contributor: University of Warwick
No. of awards given last year: 1
Additional Information: Non-renewable, deducted from tuition fees, for taught masters only. Eligible to nationals of African Nations (North).

For further information contact:

Website: www.warwick.ac.uk/go/scholarships

Pakistan and Bangladesh Postgraduate Awards (Warwick Manufacturing Group)

Subjects: WMG (Warwick Manufacturing Group) taught masters courses.
Purpose: To support Master's Pakistani and Bangladeshi students at WMG.
Eligibility: Applicants should be nationals of Pakistan or Bangladesh, not currently registered on a postgraduate course at the University, and should have received an offer of a place from WMG.
Level of Study: Postgraduate
Type: Scholarship
Value: £3,000
Length of Study: 1 year
Frequency: Dependent on funds available
Study Establishment: University of Warwick
Country of Study: United Kingdom
No. of awards offered: 2
Application Procedure: Online application form on WMG website.
Closing Date: June 30th
Contributor: University of Warwick
Additional Information: Non-renewable, for taught masters only.

Taiwan Postgraduate Awards (Warwick Manufacturing Group)

Subjects: WMG (Warwick Manufacturing Group) taught masters courses.
Purpose: To support Master's Taiwanese students at WMG.
Eligibility: Applicants must be nationals of Taiwan, classed as international fee-payer, not currently registered for a taught masters course at the University of Warwick and should have received an offer of a place from WMG at Warwick.
Level of Study: Postgraduate
Type: Scholarship
Value: UK£5,000
Length of Study: 1 year
Frequency: Dependent on funds available
Study Establishment: University of Warwick
Country of Study: United Kingdom
No. of awards offered: 2
Application Procedure: Applicants must complete an online application form.
Closing Date: May 11th
Contributor: University of Warwick
No. of awards given last year: 2
Additional Information: Non-renewable, deducted from tuition fees, for taught masters only.

For further information contact:

Website: www.warwick.ac.uk/go/scholarships

Turkey Postgraduate Awards (Warwick Manufacturing Group)

Subjects: WMG (Warwick Manufacturing Group) taught masters courses.
Purpose: To support Master's Turkish students at WMG.
Eligibility: Applicants should be nationals of Turkey, not currently registered on a postgraduate course at the University of Warwick and should have received an offer of a place from WMG.
Level of Study: Postgraduate
Type: Scholarship
Value: UK£3,000
Length of Study: 1 year
Frequency: Dependent on funds available
Study Establishment: University of Warwick
Country of Study: United Kingdom
No. of awards offered: 15
Application Procedure: All applicants who have received an offer of a place from WMG will automatically be considered for the scholarship, no additional application is required.
Contributor: University of Warwick
No. of awards given last year: 15
Additional Information: Non-renewable, deducted from tuition fees, for taught masters only. Eligible to the nationals of Turkey.

For further information contact:

Website: www.warwick.ac.uk/go/scholarships

University of Warwick Music Scholarships

Subjects: Open to all disciplines/subjects to support extra-curricular music.

Purpose: To support students with outstanding musical ability.
Eligibility: Applicants should be undergraduate or postgraduate students, should have applied for admission to the University for a full-time scheme and have satisfied the entry requirements.
Level of Study: Unrestricted
Type: Scholarship
Value: UK£450 per year (with an additional subsidy on music tuition fees)
Length of Study: Up to 3 years
Study Establishment: University of Warwick
Country of Study: United Kingdom
No. of awards offered: 10
Application Procedure: Application forms are available online or from the Music Centre Secretary.
Closing Date: February 10th in proposed calendar year of entry
Funding: Private
No. of awards given last year: 4
No. of applicants last year: 40

Warwick French Department Bursaries

Subjects: French culture and thought, French studies.
Purpose: To support students registered for an MA in French studies.
Eligibility: Open to candidates who apply for admission to the MA programme through the normal application procedures via Warwick Graduate School. Candidates may be asked to attend an interview and/or to submit a sample of written work, a brief description of circumstances and/or motivations to departmental bursary, to the Department's Director of Graduate Studies at the time of application to the MA.
Level of Study: Postgraduate
Type: Scholarship
Value: Equivalent of standard fees at UK/EU rates
Length of Study: 1 year full-time/2 years part-time
Frequency: Dependent on funds available
Study Establishment: University of Warwick
Country of Study: United Kingdom
No. of awards offered: 3
Application Procedure: There is no separate application form. Applicants must apply through normal application procedures via Warwick Graduate School available on www2.warwick.ac.uk/study/postgraduate/.
Closing Date: March 12th
Contributor: University of Warwick
No. of awards given last year: 4
No. of applicants last year: 4

Warwick Postgraduate Research Scholarships

Eligibility: Open to Home, EU and Overseas students from all disciplines at Warwick. For more details, please refer to the website..
Level of Study: Doctorate
Type: Scholarship
Value: £3,390 for full-time students for the payment of academic fees at the Home/EU rate. A maintenance grant, in line with the UK Research Council stipend, of £13,290 for full-time award holders
Country of Study: United Kingdom
No. of awards offered: 45
Application Procedure: Please refer to the website www2.warwick.ac.uk/services/academicoffice/gsp/scholarship/wprs.
Closing Date: February 28th
Additional Information: Students and applicants who wish to apply for an AHRC doctoral award should apply to the WPRS competition and will automatically be considered for both competitions.

UNIVERSITY OF WATERLOO

Graduate Studies Office, 200 University Avenue West, Waterloo, ON, N2L 3G1, Canada
Tel: (1) 519 888 4567
Fax: (1) 519 884 8009
Email: hmussar@uwaterloo.ca
Website: www.uwaterloo.ca
Contact: Heidi Mussar, Manager, Graduate Studies Financial Aid Programs

The University of Waterloo has long been recognized as the most innovative university in Canada. The University is committed to advancing learning and knowledge through teaching, research, and scholarship in our faculties, colleges, and schools. Our six faculties include applied health sciences, arts, engineering, environmental studies, mathematics, and science.

Industrial Postgraduate Scholarship 1 (NSERC IPS)

Subjects: Science and engineering.
Purpose: To encourage scholars to consider research careers in industry where they will be able to contribute to strengthening Canadian innovation.
Eligibility: Open to citizen or permanent resident of Canada, with a degree in science or engineering from a university whose standing is acceptable to NSERC, have certification from the Dean of Graduate Studies that the applicant has a first-class academic standing (a grade of 'A'); and be pursuing full- or part-time graduate studies in the natural sciences or engineering at an eligible Canadian university.
Level of Study: Postgraduate
Type: Scholarship
Value: $15,000 per year for up to three years plus a minimum contribution from the sponsoring organization of $6,000 per year
Length of Study: 2 years
Application Procedure: The applicants should contact the Senior Manager, Graduate Studies Financial Aid Programs.
Closing Date: Please check website
Funding: Government
Contributor: NSERC

UNIVERSITY OF WESTERN AUSTRALIA

35 Stirling Highway, Crawley, WA,, 6009, Australia
Tel: (61) 8 9380 2490, 8 6488 6000
Fax: (61) 8 9380 1919, 8 6488 1380
Email: general.enquiries@uwa.edu.au Staff
Website: www.uwa.edu.au

Since its establishment in 1911, the University of Western Australia has helped to shape the careers of more than 75,000 graduates. Their success reflects the UWA's balanced coverage of disciplines in the arts, sciences and professions.

Postgraduate Top Up Scholarships in Offshore Engineering and Naval Architecture

Subjects: Offshore engineering and naval architecture.
Purpose: To provide students with quality professional training, education, research and development within the petroleum industry.
Eligibility: Open to International self-funding students or candidates who qualify for an Australian postgraduate award or university postgraduate award.
Level of Study: Postgraduate
Type: Scholarships
Value: The tax-free top-up scholarships are in the range of $5,000–20,000
Frequency: Annual
Study Establishment: University of Western Australia
Country of Study: Australia
No. of awards offered: 5
Closing Date: October 31st for Australian and New Zealand applications and August 31st for International applications

Sir Charles and Lady Court Music Fund

Subjects: Music.
Purpose: To assist music graduates to obtain more advanced musical experience outside Western Australia.
Eligibility: Open to graduates holding the degree of Bachelor of music or Bachelor of music education, (pass or honours) or Bachelor of arts with a major in music (pass or honours) of the University of Western Australia.
Level of Study: Postgraduate
Type: Award
Application Procedure: Check website for further details.
Funding: Trusts

Society of Petroleum Engineers Western Australia Scholarships

Subjects: Petroleum engineering.

Purpose: To encourage students to specialize in the fields of study that may lead to a professional career in petroleum engineering.
Eligibility: Open to citizens or permanent residents of Australia who are enrolled in full-time study in oil and gas engineering at the University of Western Australia.
Level of Study: Postgraduate
Type: Scholarships
Value: $3,750 per year
Length of Study: 1 year
Frequency: Annual
Study Establishment: University of Western Australia
Country of Study: Australia
No. of awards offered: 5
Closing Date: March 1st
Contributor: Western Australian Section of the Society of Petroleum Engineers (SPE-WA)
Additional Information: The award of scholarships will be based on the written applications only.

For further information contact:

Dept of Petroleum Engineering ARRC 26 Dick Perry Avenue, Kensington, Western Australia, 6151, Australia
Contact: Geoffrey Weir, SPE WA Scholarship Chairman

UNIVERSITY OF WESTMINSTER

309 Regent Street, London, W1B 2UW, England
Tel: (44) 0 20 7915 5511
Fax: (44) 0 20 7911 5858
Email: studentfinance@wmin.ac.uk, scholarships@wmin.ac.uk
Website: www.westminster.ac.uk

The University of Westminster is giving generous financial support to many of its students so that they can get on with their studies rather than worrying about finances. The scholarships listed are examples of what is available. Westminster now has the most valuable scholarship scheme of any UK university (£5 million per year). Full details are available from our website www.westminster.ac.uk/scholarships.

Alumni Scholarships

Subjects: All subjects.
Purpose: To encourage our graduates to continue to a Master's programme.
Eligibility: Open to graduates progressing to a full-time Masters Degree at the University of Westminster, with a minimum of a 2.1 in their Bachelor's degree.
Level of Study: Postgraduate
Type: Scholarship
Value: £2,000 tuition fee waivers
Frequency: One off fee waiver
Study Establishment: University of Westminster
Country of Study: United Kingdom
Application Procedure: Check website for further details - application necessary. Full details are available from the website www.westminster.ac.uk/scholarships.
Closing Date: November 1st
Contributor: University of Westminster
No. of awards given last year: 25
No. of applicants last year: 25 (eligible applicants)

For further information contact:

101, New Cavendish Street, London, W1W 6XH, United Kingdom

David Faddy Scholarships

Subjects: Photojournalism.
Eligibility: Open to full-time students holding an offer for the full-time MA photojournalism or MA photographic studies.
Level of Study: Postgraduate
Type: Scholarship
Value: £2,500 tuition fee waivers
Frequency: Annual
Study Establishment: University of Westminster
Country of Study: United Kingdom
Application Procedure: Check website for further details. Full details are available from the website www.westminster.ac.uk/scholarships.
Closing Date: May 31st
Funding: Private
Contributor: Richard Jenkins and Maureen Amar
No. of awards given last year: 2
No. of applicants last year: 6

For further information contact:

101, New Cavendish Street, London, W1W 6XH, United Kingdom

EU Acceding and Candidate Countries Scholarship

Subjects: All subjects.
Purpose: To support students from acceding and candidate countries, to study subjects helpful to their country's integration to the EU or to help their bid to join the EU.
Eligibility: Open to students from Bulgaria, Croatia, Romania and Turkey holding an offer for any full-time Master's degree.
Level of Study: Postgraduate
Type: Scholarship
Value: Full tuition fee waiver
Length of Study: 1 year
Frequency: Annual
Study Establishment: University of Westminster
Country of Study: United Kingdom
No. of awards offered: 1
Application Procedure: Check website for further details.
Closing Date: May 31st
Contributor: International Students House
No. of awards given last year: 1
No. of applicants last year: 20+

MA English Literature Scholarships

Subjects: English literature.
Eligibility: Open to citizens of North America and EU member states holding an offer for the full-time MA English literature.
Level of Study: Postgraduate
Type: Scholarship
Value: 50 per cent tuition fee waivers
Frequency: Annual
Study Establishment: University of Westminster
Country of Study: United Kingdom
No. of awards offered: 2
Application Procedure: Check website for further details.
Closing Date: May 31st
Funding: Government
Contributor: University of Westminster
No. of awards given last year: 2
No. of applicants last year: 10+

MA Fashion Design and Enterprise Scholarship

Subjects: Fashion design.
Purpose: To reduce the financial burden to well-qualified students.
Eligibility: Open to students of United Kingdom holding an offer for the full-time MA fashion design and enterprise.
Level of Study: Postgraduate
Type: Scholarship
Value: £5,000 covering fees and living expenses
Frequency: Annual
Application Procedure: Check website for further details.
Closing Date: May 31st
Funding: Government
No. of awards given last year: 2
No. of applicants last year: 10+

For further information contact:

35 Marylebone Road, London, NW1 5LS, England

MA Public Communication and Public Relations Scholarships

Subjects: Public relations.

Purpose: The University of Westminster is giving generous financial support to many of its students so that they can get on with their studies rather than worrying about finances.
Eligibility: Open to students holding an offer for the full-time or part-time course.
Level of Study: Postgraduate
Type: Scholarship
Value: Full tuition fee waiver for full-time student and part-time student
Frequency: Annual
Study Establishment: University of Westminster
Country of Study: United Kingdom
No. of awards offered: 2
Application Procedure: Check website for further details.
Closing Date: May 31st (full time) and August 1st (part time)
Funding: Government
Contributor: University of Westminster
No. of awards given last year: 2
No. of applicants last year: 20+

MBA Scholarships

Subjects: Business administration.
Eligibility: Candidates must hold an offer for the part-time MBA degree at the University.
Level of Study: Postgraduate
Type: Scholarship
Value: 50 per cent tuition fee waivers
Frequency: Annual
Study Establishment: University of Westminster
Country of Study: United Kingdom
Application Procedure: Check website for further details.
Closing Date: August 31st
Contributor: University of Westminster
No. of awards given last year: 20
No. of applicants last year: 50+

Media, Arts and Design Scholarship

Subjects: Media, arts and design.
Eligibility: Candidates must hold an offer for any part-time Masters degree in the School of Media, Arts and Design at the University.
Level of Study: Postgraduate
Type: Scholarship
Value: £2,000 tuition fee waiver.
Study Establishment: University of Westminster
Country of Study: United Kingdom
Application Procedure: Check website for further details.
Closing Date: May 31st (full time) and August 1st (part time)
Contributor: University of Westminster
No. of awards given last year: 2
No. of applicants last year: 50+

Part Fee Waiver Scholarships

Subjects: All subjects.
Purpose: To reward outstanding students.
Eligibility: Open to students from any country holding an offer for a full-time Masters Degree.
Level of Study: Postgraduate
Type: Scholarship
Value: £2,000 tuition fee waivers
Frequency: Annual
Study Establishment: University of Westminster
Country of Study: United Kingdom
Application Procedure: Check website for further details.
Closing Date: May 31st
Funding: Government
Contributor: University of Westminster
No. of awards given last year: 20
No. of applicants last year: 200+

Westminster Business School (WBS) Scholarships

Subjects: All subjects excluding MBA.
Purpose: To reward academic achievement.
Eligibility: Open to students holding an offer for any full-time Masters Degree, excluding MBA, within Westminster Business School.
Level of Study: Postgraduate
Type: Scholarship
Value: Full tuition fee waivers
Frequency: Annual
Study Establishment: University of Westminster
Country of Study: United Kingdom
No. of awards offered: 6
Application Procedure: Check website for further details.
Closing Date: May 31st
Contributor: University of Westminster
No. of awards given last year: 6
No. of applicants last year: 60+

UNIVERSITY OF YORK

Sally Baldwin Buildings, Block B, Heslington, York, YO10 5DD, United Kingdom
Tel: (44) 01904 324043
Fax: (44) 01904 324142
Email: student-financial-support@york.ac.uk
Website: www.york.ac.uk
Contact: Student Financial Support Unit

Annual Fund Scholarships

Purpose: Annual Fund scholarships are made possible by the generous donations of University of York Alumni.
Eligibility: In order to be eligible you must: have applied for and been offered a place on an MSc, MA, or MRes Masters programme. Have or expect to obtain a first or upper second class honours degree or equivalent prior to commencing the Masters degree. Have completed, or expect to have completed before commencing the Masters programme, an undergraduate programme at the University of York.
Level of Study: Postgraduate
Type: Scholarship
Value: Successful applicants will receive a £5000 stipend and a £5000 partial fee waiver
No. of awards offered: 2
Application Procedure: Please visit www.york.ac.uk/study/postgraduate/fees-funding/postgraduate/annual_fund/ for further details. Applications are submitted online.
Closing Date: April 30th
Contributor: University of York Alumni
Additional Information: Four Annual Fund Scholarships are available for this year for one year of full time registration on a Masters programme.

Overseas Continuation Scholarship (OCS)

Subjects: All subjects.
Purpose: For current University of York Masters students who are progressing to PhD studies at the University of York.
Eligibility: Students must be outstanding academically and have the support of their chosen department at York. You must hold an offer for PhD study and be a current University of York Masters student to be eligible to apply.
Level of Study: Doctorate
Type: Scholarship
Value: The scholarship is worth £5,000 in the first year of study as a deduction from tuition fees.
Application Procedure: Pleas visit www.york.ac.uk/study/international/fees-funding/scholarships/ to apply.
Closing Date: April 30th

Overseas Research Scholarship (ORS)

Subjects: All subjects.
Purpose: For applicants commencing PhD study at the University of York. Applicants must be liable to pay the overseas rate of tuition fee.
Eligibility: Students must be outstanding academically and have the support of their chosen department at York. You must hold an offer for PhD study to be eligible to apply.
Level of Study: Doctorate
Type: Scholarship
Value: The scholarship will pay the full overseas tuition fee and a stipend of £5,000 per year for each year of successful study.
Application Procedure: Please visit www.york.ac.uk/study/international/fees-funding/scholarships/ to apply.
Closing Date: April 30th

Scholarship for Overseas Students

Subjects: All subjects except MBBS (Medicine).
Purpose: For applicants commencing study of any subject (excluding students applying to the Hull York Medical School) at any level as a full-time student at the University of York. Applicants must be liable to pay the overseas rate of tuition fee.
Eligibility: This is a competitive scholarship based on academic merit and financial need. You must hold an offer for academic study to be eligible to apply.
Level of Study: Postgraduate
Type: Scholarship
Value: The scholarship is worth one-third or one-sixth of the overseas tuition fee for each year of successful study.
Frequency: Annual
Application Procedure: Please visit www.york.ac.uk/study/international/fees-funding/scholarships/ to apply.
Closing Date: April 30th

White Rose University Consortium Studentships

Subjects: Available across various schools and subjects, see website for further details.
Purpose: Each year the University of York collaborates with the Universities of Leeds and Sheffield to be able to offer a total of 12 studentships (3–4 studentships in each of the three Universities).
Eligibility: In order to be eligible you must: have applied for a place on a full time PhD programme in the relevant Department. Have or expect to obtain a first or upper second class honours degree or equivalent prior to commencing the PhD degree.
Level of Study: Doctorate
Type: Studentship
Value: A full Research Council equivalent stipend: £13,590 in 2012/13. Rates for 2013/4 were not set at time of publication. A fee waiver at the home/EU rate (Overseas candidates are welcome to apply but would need to fund the difference between home/EU fee rate and international fee rate.). A Research Support Grant: £900.
Frequency: Annual
No. of awards offered: 1
Application Procedure: Please visit www.york.ac.uk/study/postgraduate/fees-funding/postgraduate/white_rose/ for further details. Applications are submitted online.
Closing Date: April 30th

Wolfson Foundation Scholarships

Subjects: The Wolfson Postgraduate Scholarships will fund doctoral research in three disciplines that align closely with the Foundation's interests: history, literature and languages.
Purpose: The University is delighted to be offering Wolfson Scholarships in the Humanities for the second year as part of a national Arts funding scheme.
Eligibility: The Wolfson Postgraduate Scholarships in the humanities will be awarded to outstanding students who demonstrate the potential to make an impact on their chosen field. Wolfson Scholarships will be awarded solely on academic merit. In order to be eligible you must: have applied for and be in receipt of an offer of a place on a full time PhD programme in the relevant department (some departments may be able to accept applications on the basis of a programme application without an offer - please speak to your prospective department to confirm). Expect to begin your PhD studies in October. Have or expect to obtain a first or upper second class honours degree or equivalent prior to commencing the PhD. Have completed a masters level qualification before commencing the PhD.
Level of Study: Doctorate
Type: Scholarship
Value: The scholarships are worth £26,000 per year for up to three years of PhD study. Each scholarship will have a minimum stipend of £15,000 per year. Each scholarship will contribute to fees (up to £10,000). Any funds remaining after payment of the stipend and fees will be kept as a Research Training and Support Grant to be accessed by the award holder on request to support the work of the PhD
No. of awards offered: 3
Application Procedure: Please visit www.york.ac.uk/study/postgraduate/fees-funding/postgraduate/wolfson/ for further details. Applications are submitted online.
Closing Date: February 1st

US DEPARTMENT OF EDUCATION

International Education Programs Service, Language Resource Centers Program, 1990 K Street, N.W.,Rm. 6077, Washington, DC, 20006-8521, United States of America
Tel: (1) 202 502 7589
Fax: (1) 202 502 7860
Email: cynthia.dudzinski@ed.gov
Website: www.ed.gov
Contact: Cynthia Dudzinski

The US Department of Education is the agency of the federal government that establishes policy, administers and co-ordinates most federal assistance to education. The Department's mission is to serve American students and to ensure that all have equal access to education and to promote excellence in various schools.

Fulbright Hays Doctoral Dissertation Research Abroad Program

Subjects: All subjects.
Purpose: To assist PhD students in carrying out dissertation research in other countries.
Eligibility: Open to US citizens or nationals who are Doctoral candidates enrolled in a PhD programme in the United States.
Level of Study: Doctorate
Type: Scholarship
Value: The program expects to award 100 fellowships ranging from $15,000–60,000.
Length of Study: six to twelve months
Frequency: Annual
Country of Study: United States of America
Application Procedure: Applicants must submit a completed application form and proof of enrollment in a Doctoral programme.
Closing Date: November 2nd

For further information contact:

Website: www.ed.gov/offices/OPE/HEP/iegps

US-IRELAND ALLIANCE

2800 Clarendon Boulevard, Suite 502 West, Arlington, VA, 22201, United States of America
Tel: (1) 703 841 5492
Email: vargo@us-irelandalliance.org
Website: www.us-irelandalliance.org
Contact: Trina Vargo, President

The US-Ireland Alliance is a proactive, non-partisan, non-profit organization dedicated to consolidating existing relations between the United States and Ireland and building that relationship for the future.

George J. Mitchell Scholarships

Subjects: All subjects.
Purpose: To recognize outstanding young Americans who exhibit the highest standards of academic excellence, leadership and community service.
Eligibility: Open to American citizens between the ages of 18 and 30 who have academic standing sufficient to assure completion of a Bachelor's degree before they begin study under the Mitchell Scholarship..
Type: Scholarship
Value: The scholarship is non-renewable and covers tuition, room and board and provides a $12,000 stipend.
Length of Study: 1 year
Frequency: Annual
Study Establishment: Any accredited institution of higher learning
Country of Study: United Kingdom
No. of awards offered: 12
Application Procedure: Applications forms are available online.
Closing Date: February 10th
Funding: Foundation
Contributor: US-Ireland Alliance
No. of applicants last year: 20

US-IRELAND ALLIANCE

2800 Clarendon Blvd, Suite 502 West, Arlington, Virginia, VA 22201, United States of America
Tel: (1) 703 841 5843
Email: lamonte@mitchellscholars.org
Website: http://us-irelandalliance.org
Contact: Jennie LaMonte, Managing Director George Mitchell Scholars program

George J Mitchell Scholarship

Subjects: All subject.
Purpose: To introduce and connect generations of future American leaders to the island of Ireland, while recognizing and fostering intellectual achievement, leadership, and a commitment to public service and community.
Eligibility: Open to Mitchell Scholars between the ages of 18 and 30.
Level of Study: Postgraduate
Type: Scholarship
Value: Tuition, housing, a living expenses stipend, and an international travel stipend
Length of Study: 1 year
Frequency: Annual
Country of Study: Ireland
Application Procedure: Application form available from website. Applicants are required to provide no more and no less than five letters of recommendation, proof of US citizenship (passport/birth certificate), academic transcripts, completed online application, and a well-thought out 1000-word personal statement.
Closing Date: October 5th
Additional Information: Questions about the Mitchell Scholars Program can be addressed to Jennie LaMonte, Managing Director, at lamonte@mitchellscholars.org.

THE US-UK FULBRIGHT COMMISSION

188 Kirtling Street, London, SW8 5BN, England
Tel: (44) 20 7404 6880
Fax: (44) 20 7498 4023
Email: programmes@fulbright.co.uk
Website: www.fulbright.co.uk
Contact: Mr Michael Scott-Kline, Director

The US-UK Fulbright Commission has a programme of awards offered annually to citizens of the United Kingdom and United States of America.

Fulbright - Aberdeen University Award

Subjects: All subjects.
Purpose: US citizen to pursue postgraduate study in any field at Aberdeen.
Eligibility: US citizens are eligible to apply. Applicants should hold or expected to receive a Bachelor's degree in a relevant area.
Level of Study: Doctorate, Graduate, Postgraduate
Type: Award
Value: £12,000 living stipend plus full tuition fee waiver
Frequency: Annual
Study Establishment: Aberdeen University
Country of Study: United Kingdom
No. of awards offered: 1
Application Procedure: Please check at http://us.fulbrightonline.org/applynow.html.
Closing Date: October 17th
Funding: Government
Additional Information: Please check for more information at www.fulbright.co.uk.

Fulbright - Aberystwyth University International Relations Award

Subjects: International relations, international politics, international affairs.
Purpose: To pursue postgraduate study in International Relations at Aberystwyth.
Eligibility: US citizens are eligible to apply.Applicants should hold or expected to receive a Bachelor's degreein a relevant area.
Level of Study: Doctorate, Graduate, Postgraduate
Type: Award
Value: £12,000 plus full tuition fee waiver
Frequency: Annual
Study Establishment: Aberystwyth University
Country of Study: United Kingdom
No. of awards offered: 1
Application Procedure: Please check at http://us.fulbrightonline.org/applynow.html.
Closing Date: October 17th
Funding: Government
Additional Information: Please check for more information at www.fulbright.co.uk.

Fulbright - Birmingham Scholar Award

Subjects: All subjects.
Purpose: To pursue research at Birmingham.
Eligibility: Applicants must be a US citizen and should hold or expect to receive a PhD in a relevant area.
Level of Study: Postdoctorate, Research
Type: Award
Value: The award provides a monthly stipend of £2,375 paid in instalments directly to the grantee.Limited sickness and accident benefit coverage is also included.
Length of Study: 3
Frequency: Annual
Study Establishment: University of Birmingham
Country of Study: United Kingdom
No. of awards offered: 1
Application Procedure: Please check at www.cies.org.
Closing Date: August 1st
Funding: Government
Additional Information: Please check for more information at www.fulbright.co.uk.

Fulbright - Birmingham University Award

Subjects: All subjects.
Purpose: To pursue postgraduate study of the University of Birmingham, UK.
Eligibility: US citizen resident anywhere except the United Kingdom, and should hold or expect to receive a Bachelor's degree (or equivalent professional training or experience) in a relevant area.
Level of Study: Doctorate, Graduate, Postgraduate
Type: Award
Value: The grantee will receive a full tuition fee waiver from the University. In addition, the grantee will receive £12,000 which is intended as a contribution towards general maintenance costs (accommodation, travel, subsistence, etc) while in the UK.
Frequency: Annual
Study Establishment: University of Birmingham (UK)
Country of Study: United Kingdom
No. of awards offered: 1
Application Procedure: Please check at http://us.fulbrightonline.org/applynow.html.
Closing Date: Mid-October
Funding: Government
Additional Information: Please check for more information at www.fulbright.co.uk.

Fulbright - Cardiff Scholar Awards

Subjects: All subjects.
Purpose: Pursue research at Cardiff.
Eligibility: Applicant should be a US citizen (resident anywhere except the United Kingdom), andhold or expect to receive a PhD (or equivalent professional training or experience) in a relevant area.
Level of Study: Postdoctorate
Type: Award
Value: The award provides a monthly stipend of £2,500 paid in instalments directly to the grantee.
Length of Study: 4–6 months
Frequency: Annual
Study Establishment: Cardiff University
Country of Study: Wales
No. of awards offered: 1
Application Procedure: Please check at www.cies.org.

Closing Date: August 1st
Funding: Government
No. of awards given last year: 1
Additional Information: Please check more information at www.fulbright.co.uk.

Fulbright - Cardiff University Award

Subjects: All subjects.
Purpose: To pursue postgraduate study at Cardiff.
Eligibility: US Citizens are eligible to apply.
Level of Study: Doctorate, Graduate, Postgraduate, Predoctorate
Type: Award
Value: £12,000 plus tuition fees
Frequency: Annual
Study Establishment: Cardiff University
Country of Study: Wales
No. of awards offered: 1
Application Procedure: Please check at http://us.fulbrightonline.org/applynow.html.
Closing Date: October 17th
Funding: Government
No. of awards given last year: 1
Additional Information: Please check for more information at www.fulbright.co.uk.

Fulbright - Deafness Research UK Scholar

Subjects: All subjects.
Purpose: To enable a UK citizen to pursue research into prevention, diagnosis or treatment of all forms of hearing impairment for a period of 3–12 months.
Eligibility: Applicant must be a UK citizen (resident anywhere but the US), holding a doctoral degree or equivalent professional training or experience at the time of application.
Level of Study: Postdoctorate, Professional development, Research
Type: Award
Value: Up to £75,000 (converted at prevailing rate) paid in installments, limited sickness and accident benefit coverage, visa sponsership and processing.
Length of Study: 3–12 months
Frequency: Annual
Country of Study: United States of America
No. of awards offered: 2–3
Application Procedure: Formal applications are made online via www.fulbright.co.uk.
Closing Date: May 31st
Contributor: Deafness Research UK
Additional Information: Please check more information at www.fulbright.co.uk.

Fulbright - Durham Scholar Award

Subjects: All subjects.
Purpose: Pursue research at Durham's Institute of Advanced Studies (IAS).
Eligibility: The applicant must be a US citizen* (resident anywhere except the United Kingdom), and should hold or expect to receive a PhD (or equivalent professional training or experience) in a relevant area.
Level of Study: Postdoctorate
Type: Award
Value: One-off grant of £2,500 plus full tuition/fees and accomodation in Durham
Length of Study: 3 months
Frequency: Annual
Study Establishment: Durham University
Country of Study: United Kingdom
No. of awards offered: 1
Application Procedure: Please check website www.cies.org.
Closing Date: August 1st
Funding: Government
Additional Information: Please check at www.fulbright.co.uk.

Fulbright - Durham University / Powers Award

Subjects: All subjects.
Purpose: To pursue postgraduate study at Durham.
Eligibility: The candidate must be a US citizens and must hold or expect to receive a Bachelor's degree (or equivalent professional training or experience) in a relevant area.
Level of Study: Doctorate, Graduate, Postgraduate, Predoctorate
Type: Award
Value: £12,000 plus tuition fees
Frequency: Annual
Study Establishment: Durham University
Country of Study: United Kingdom
No. of awards offered: 1
Application Procedure: Please check at http://us.fulbrightonline.org/applynow.html.
Closing Date: October 17th
Funding: Government
No. of awards given last year: 1
Additional Information: Please check for more information at www.fulbright.co.uk.

Fulbright - East Anglia University Award

Subjects: All subjects.
Purpose: To pursue postgraduate study at USA.
Eligibility: US citizens are eligible to apply.
Level of Study: Doctorate, Graduate, Postgraduate, Predoctorate
Type: Award
Value: £12,000 plus tuition fees
Frequency: Annual
Study Establishment: UEA
Country of Study: United Kingdom
No. of awards offered: 1
Application Procedure: Please check at http://us.fulbrightonline.org/applynow.html.
Closing Date: October 17th
Funding: Government
No. of awards given last year: 1
Additional Information: Please check for more information at www.fulbright.co.uk.

Fulbright - Elsevier Award

Subjects: All disciplines plus subjects eligible but preference will be given to sciences.
Purpose: To pursue postgraduate study in the US.
Eligibility: Must be a UK citizen* (resident anywhere except the United States), and should hold or expect to obtain a minimum 2:1 honours undergraduate degree or the equivalent prior to your anticipated enrolment with a US university.
Level of Study: Doctorate, Graduate, Postgraduate, Predoctorate
Type: Award
Value: Scholarships cover tuition fees and provide a living stipend for one academic year.
Frequency: Annual
Country of Study: United States of America
No. of awards offered: 1
Application Procedure: Please check at www.fulbright.co.uk.
Closing Date: November 15th
Funding: Government

Fulbright - Essex University Award

Subjects: All subjects.
Purpose: To pursue postgraduate study at Essex.
Eligibility: US citizens are eligible to apply.
Level of Study: Doctorate, Graduate, Postgraduate, Predoctorate
Type: Award
Value: £12,000 plus tuition fees
Frequency: Annual
Study Establishment: Essex University
Country of Study: United Kingdom
No. of awards offered: 1
Application Procedure: Please check at http://us.fulbrightonline.org/applynow.html.
Closing Date: October 17th
Funding: Government
No. of awards given last year: 1
Additional Information: Please check for more information at www.fulbright.co.uk.

Fulbright - Exeter University Award

Subjects: All subjects.
Purpose: To pursue postgraduate study at Exeter.
Eligibility: US citizens are eligible to apply.
Level of Study: Doctorate, Graduate, Postgraduate, Predoctorate
Type: Award
Value: £12,000 plus tuition fees
Frequency: Annual
Study Establishment: Exeter
Country of Study: United Kingdom
No. of awards offered: 1
Application Procedure: Please check at http://us.fulbrightonline.org/applynow.html.
Closing Date: October 17th
Funding: Government
No. of awards given last year: 1
Additional Information: Please check for more information at www.fulbright.co.uk.

Fulbright - Exeter University Scholar Award

Subjects: All subjects.
Purpose: To pursue research at Exeter.
Level of Study: Postdoctorate
Type: Award
Value: £2,375 per month
Length of Study: 3–9 months
Frequency: Annual
Study Establishment: University of Exeter
Country of Study: United Kingdom
No. of awards offered: 1
Application Procedure: Please check at www.cies.org.
Closing Date: August 1st
Funding: Government
Additional Information: Please check for more information at www.fulbright.co.uk.

Fulbright - Fight for Sight

Subjects: All subjects.
Purpose: To enable a UK citizen to pursue research into prevention and treatment of blindness or eye disease for a period of 12 months.
Eligibility: Applicant must be a UK citizen (resident anywhere but the US) holding a doctoral degree or equivalent professional training or experience at the time of application.
Type: Award
Value: US dollar equivalent of £75,000 (converted at the prevailing rate), paid in installments
Length of Study: 1 year
Frequency: Annual
Country of Study: United States of America
Application Procedure: Formal applications are made online via www.fulbright.co.uk.
Closing Date: May 31st
Contributor: Fight for sight
No. of awards given last year: 1
Additional Information: Please check for more information at www.fulbright.co.uk.

Fulbright - Imperial College London Award

Subjects: All subjects.
Purpose: To pursue postgraduate study at Imperial College.
Eligibility: US citizen resident citizens are eligible to apply.
Level of Study: Doctorate, Graduate, Postgraduate, Predoctorate
Type: Award
Value: £12,500 plus tuition fees
Frequency: Annual
Study Establishment: Imperial College London
Country of Study: United Kingdom
No. of awards offered: 1
Application Procedure: Please check at http://us.fulbrightonline.org/applynow.html.
Closing Date: October 17th
Funding: Government
No. of awards given last year: 1
Additional Information: Please check for more information at www.fulbright.co.uk.

Fulbright - Institute of Education Award

Subjects: All subjects.
Purpose: To pursue postgraduate study at IOE.
Eligibility: All candidates must
Level of Study: Doctorate, Graduate, Postgraduate
Type: Award
Value: £13,000 plus tuition fees be US citizen resident and should hold or expect to receive a Bachelor's degree in a relevant area.
Frequency: Annual
Study Establishment: IOE
Country of Study: United Kingdom
No. of awards offered: 1
Application Procedure: Please check at http://us.fulbrightonline.org/applynow.html.
Closing Date: October 17th
Funding: Government
No. of awards given last year: 1
Additional Information: Please check for more information at www.fulbright.co.uk.

Fulbright - Kent University Award

Subjects: All subjects.
Purpose: To pursue postgraduate study at Kent.
Eligibility: US citizens are eligible to apply.
Level of Study: Doctorate, Graduate, Postgraduate, Predoctorate
Type: Award
Value: £12,000 plus tuition fees
Frequency: Annual
Study Establishment: University of Kent
Country of Study: United Kingdom
No. of awards offered: 1
Application Procedure: Please check at http://us.fulbrightonline.org/applynow.html.
Closing Date: October 17th
Funding: Government
Additional Information: Please check for more information at www.fulbright.co.uk.

Fulbright - Lancaster University Award in Science and Technology

Subjects: Science and technology fields offered by Lancaster University's faculty of science and technology.
Purpose: To pursue postgraduate study in a science/technology field at Lancaster.
Eligibility: US citizens are eligible to apply.
Level of Study: Doctorate, Graduate, Postgraduate, Predoctorate
Type: Award
Value: £12,000 plus tuition fees
Frequency: Annual
Study Establishment: Lancaster University
Country of Study: United Kingdom
No. of awards offered: 1
Application Procedure: Please check at http://us.fulbrightonline.org/applynow.html.
Closing Date: October 17th
Funding: Government
Additional Information: Please check for more information at www.fulbright.co.uk.

Fulbright - Lancaster University Scholar Award (stem)

Subjects: All subjects.
Purpose: To pursue research in science / technology at Lancaster.
Level of Study: Postdoctorate
Type: Award
Value: £2,375 per month
Length of Study: 3–6 months
Frequency: Annual
Study Establishment: Lancaster University
Country of Study: United Kingdom
No. of awards offered: 1
Application Procedure: Please check at www.cies.org.
Closing Date: August 1st
Funding: Government
Additional Information: Please check for more information at www.fulbright.co.uk.

Fulbright - Leicester University Award
Subjects: All subjects.
Purpose: To pursue postgraduate study of Leicester.
Eligibility: US citizens are eligible to apply.
Level of Study: Doctorate, Graduate, Postgraduate, Predoctorate
Type: Award
Value: £12,000 plus tuition fees
Frequency: Annual
Study Establishment: Leicester University
Country of Study: United Kingdom
No. of awards offered: 1
Application Procedure: Please check at http://us.fulbrightonline.org/applynow.html.
Closing Date: October 17th
Funding: Government
No. of awards given last year: 1
Additional Information: Please check for more information at www.fulbright.co.uk.

Fulbright - Liverpool University Award
Subjects: All subjects.
Purpose: To pursue postgraduate study at the University of Liverpool.
Eligibility: Applicant must be a US citizen (resident anywhere but the UK), and a graduating senior, holding a B.S./B.A. degree, master's or doctoral degree candidate, young professional or artist.
Level of Study: Doctorate, Graduate, Postgraduate
Type: Award
Value: Up to UK£20,000, limited sickness and accident benefit coverage as well as participation in a number of Fulbright Scholar events
Frequency: Annual
Study Establishment: University of Liverpool
Country of Study: United Kingdom
No. of awards offered: 1
Application Procedure: Please check at http://us.fulbrightonline.org/applynow.html.
Closing Date: Mid-October
Funding: Government
No. of awards given last year: 1
Additional Information: Please check for more information at www.fulbright.co.uk.

Fulbright - LSE Award
Subjects: All subjects.
Purpose: To pursue postgraduate study at LSE.
Eligibility: All candidates must be US citizen resident anywhere except the United Kingdom and should hold or expect to receive a Bachelor's degree in a relevant area.
Level of Study: Doctorate, Graduate, Postgraduate, Predoctorate
Type: Award
Value: £13,500 plus tuition fees
Frequency: Annual
Country of Study: United Kingdom
No. of awards offered: 1
Application Procedure: Please check at http://us.fulbrightonline.org/applynow.html.
Closing Date: October 17th
Funding: Government
Contributor: LSE
No. of awards given last year: 1
Additional Information: Please check for more information at www.fulbright.co.uk.

Fulbright - Northern Ireland Policy and Goverance Scholar Award
Purpose: To pursue research and / or lecture at Queen's University Belfast or Ulster University.
Eligibility: Candidates must be US citizen and should hold or expect to receive a PhD (or equivalent professional training or experience) in a relevant area
Level of Study: Postdoctorate
Type: Award
Value: Approx. £12,500
Length of Study: 6 months
Frequency: Annual
Study Establishment: QUB or Ulster
Country of Study: United Kingdom
No. of awards offered: 1–2
Application Procedure: Please check at www.cies.org.
Closing Date: August 1st
Funding: Government
No. of awards given last year: 1
Additional Information: Please check for more information at www.fulbright.co.uk.

Fulbright - Nottingham University Award
Subjects: All subjects.
Purpose: To pursue postgraduate study at Nottingham.
Eligibility: Candidate must be a US citizen and should hold a or expect to receive a Bachelor's degree (or equivalent professional training or experience) in a relevant area.
Level of Study: Doctorate, Postdoctorate, Postgraduate, Predoctorate
Type: Award
Value: £12,000 plus tuition fees
Frequency: Annual
Study Establishment: Nottingham University
Country of Study: United Kingdom
No. of awards offered: 1
Application Procedure: Please check at http://us.fulbrightonline.org/applynow.html.
Closing Date: October 17th
Funding: Government
Additional Information: Please check for more information at www.fulbright.co.uk.

Fulbright - Nottingham University Scholar Award
Subjects: All subject.
Purpose: To pursue postdoctoral research at Nottingham.
Eligibility: Candidate must be a US citizen (resident anywhere except the United Kingdom) andshould hold or expect to receive a PhD (or equivalent professional training or experience) in a relevant area before departure to the UK.
Level of Study: Postdoctorate
Type: Award
Value: £2,375 per month
Length of Study: 3–9 months
Frequency: Annual
Study Establishment: Nottingham University
Country of Study: United Kingdom
No. of awards offered: 1
Application Procedure: Please check at www.cies.org.
Closing Date: August 1st
Funding: Government
Additional Information: Please check more information at www.fulbright.co.uk.

Fulbright - Oxford Claredon Scholarship
Subjects: All subjects.
Purpose: To pursue DPhil at Oxford.
Eligibility: Candidate must be a US citizen resident anywhere except the United Kingdom.
Level of Study: Doctorate
Type: Award
Value: Offers generous financial support for one year of study in the UK. Full tuition fees plus a living stipend
Frequency: Annual
Study Establishment: Oxford
Country of Study: United Kingdom
No. of awards offered: 1–2
Application Procedure: Please check at http://us.fulbrightonline.org/applynow.html.
Closing Date: October 17th
Funding: Government
No. of awards given last year: 2
Additional Information: Please check for more information at www.fulbright.co.uk.

Fulbright - Roehampton Scholar Award
Subjects: All subjects.
Purpose: To pursue research at Roehampton University London.

Eligibility: Candidate must be a US citizen and should hold or expect to receive a PhD (or equivalent professional training or experience) in a relevant area.
Level of Study: Postdoctorate
Type: Award
Value: £2,375 per month
Length of Study: 3–9 months
Frequency: Annual
Study Establishment: Roehampton University London
Country of Study: United Kingdom
No. of awards offered: 1
Application Procedure: Please check at www.cies.org.
Closing Date: August 1st
Funding: Government
Additional Information: Please check for more information at www.fulbright.co.uk.

Fulbright - Roehampton University Award in Dance

Subjects: Dance.
Purpose: To pursue postgraduate study in dance at Roehampton University London.
Eligibility: A US graduate student pursuing a one-year (or the first year of a) master's degree programme in Dance at the University of Roehampton.
Level of Study: Graduate, Postgraduate
Type: Award
Value: £13,500 plus tuition fees
Frequency: Annual
Study Establishment: Roehampton University London
Country of Study: United Kingdom
No. of awards offered: 1
Application Procedure: Please check at http:/us.fulbrightonline.org/applynow.html.
Closing Date: October 17th
Additional Information: Please check for more information at www.fulbright.co.uk.

Fulbright - Royal College of Surgeons of England Research Award

Subjects: All subjects.
Purpose: To enable a UK citizen to pursue research (that does not include clinical work, laboratory-based or otherwise) into the development of new operative techniques, improvements in patient care and recovery and/or the causes of surgical conditions and how to treat them at any accredited US institution.
Eligibility: Applicant must be a UK citizen (resident anywhere but the US), be a member of The Royal College of Surgeons of England and hold or expect to a Bachelor of Medicine, Bachelor of Surgery, Master's, PhD or equivalent professional training or experience in a relevant area before departure to the US.
Level of Study: Postdoctorate, Professional development, Research
Type: Award
Value: US dollar equivalent of £2,350 (converted at prevailing rate) per month paid in installments
Length of Study: 3–12 months
Frequency: Annual
Country of Study: United States of America
No. of awards offered: 1
Application Procedure: Formal applications are made online via www.fulbright.co.uk.
Closing Date: November 15th
Contributor: Royal College of Surgeons of England
No. of awards given last year: 1
Additional Information: Please check for more information at www.fulbright.co.uk.

Fulbright - Royal Institution of Chartered Surveyors Scholar Award

Subjects: All subjects.
Purpose: To enable a UK citizen to pursue research in property, land or the built environment, at any accredited US institution.
Eligibility: Applicant must be a UK citizen (resident anywhere but the US), holding or expect to receive a Master's, PhD or equivalent professional training or experience in a relevant area before departure to the US.
Level of Study: Postdoctorate, Postgraduate, Predoctorate, Professional development, Research
Type: Award
Value: US dollar equivalent of £2,750 per month (up to a maximum of £22,000, converted at prevailing rate) paid in installment
Length of Study: 3–8 months
Frequency: Annual
Country of Study: United States of America
No. of awards offered: 1
Application Procedure: Formal applications are made online via www.fulbright.co.uk.
Closing Date: May 31st
Contributor: RICS
No. of awards given last year: 1
Additional Information: Please check for more information at www.fulbright.co.uk.

Fulbright - Scotland Visiting Professor at Aberdeen

Subjects: All subjects.
Purpose: To pursue research at Aberdeen.
Eligibility: Candidate must be a US citizen and should hold or expect to receive a PhD (or equivalent professional training or experience) in a relevant area.
Level of Study: Postdoctorate
Type: Award
Value: The award is for £27,500 plus accommodation (including for his/her spouse) paid in instalments directly to the grantee
Length of Study: 6 months
Frequency: Annual
Study Establishment: Aberdeen University
Country of Study: United Kingdom
No. of awards offered: 1
Application Procedure: Please check at www.cies.org.
Closing Date: August 1st
Funding: Government
Additional Information: Please check for more information at www.fulbright.co.uk.

Fulbright - Scotland Visiting Professorship at Edinburgh

Subjects: All subjects.
Purpose: To pursue research at University of Edinburgh's College of Humanities and Social Sciences.
Eligibility: Applicant must be a US citizen* (resident anywhere except the United Kingdom), and must hold or expect to receive a PhD (or equivalent professional training or experience) in a relevant area before departure to the UK.
Level of Study: Postdoctorate
Type: Award
Value: £27,500 plus accommodation
Length of Study: 6 months
Frequency: Annual
Study Establishment: Edinburgh University
Country of Study: United Kingdom
No. of awards offered: 1
Application Procedure: Please check at www.cies.org.
Closing Date: August 1st
Funding: Government
Additional Information: Please check for more information at www.fulbright.co.uk.

Fulbright - Scotland Visiting Professorship at Glasgow Digital Design Studio

Purpose: Offered to a US citizen to contribute to the developments and dissemination of 3D medical simulation and training that will seek to support the medical visualisation research being conducted at the Digital Design Studio.
Level of Study: Postdoctorate, Professional development, Research
Type: Award
Value: The award is for £25,000 plus accommodation.
Length of Study: 6 months
Frequency: Annual
Country of Study: United Kingdom
No. of awards offered: 1
Application Procedure: Please check at www.cies.org.
Closing Date: August 1st

Funding: Government
Additional Information: Please check for more information at www.fulbright.co.uk.

Fulbright - Scotland Visiting Professorship at the Glasgow Urban Lab

Purpose: To pursue research in affiliation with Glasgow School of Art's Urban Lab.
Eligibility: Candidate must be a US citizen and must hold a or expect to receive a PhD (or equivalent professional training or experience) in a relevant area.
Level of Study: Postdoctorate
Type: Award
Value: £27,500 plus accommodation
Length of Study: 6 months
Frequency: Annual
Study Establishment: Glasgow School of Art
Country of Study: Scotland
No. of awards offered: 1
Application Procedure: Please check at www.cies.org.
Closing Date: August 1st
No. of awards given last year: 1
Additional Information: Please check for more information at www.fulbright.co.uk.

For further information contact:

Contact: Lauren Jacobs, US Programme Coordinator

Fulbright - Sheffield Scholar Award

Subjects: All subject.
Purpose: To pursue research at Sheffield.
Eligibility: Applicant must be a US citizen* (resident anywhere except the United Kingdom), and should hold or expect to receive a PhD (or equivalent professional training or experience) in a relevant area before departure to the UK.
Level of Study: Postdoctorate
Type: Award
Value: £2,375 per month
Length of Study: 3–9 months
Frequency: Annual
Study Establishment: Sheffield University
Country of Study: United Kingdom
No. of awards offered: 1
Application Procedure: Please check at www.cies.org.
Closing Date: August 1st
Funding: Government
Additional Information: Please check more information at www.fulbright.co.uk.

Fulbright - Sheffield University Award

Subjects: All subjects.
Purpose: To pursue postgraduate study at Shelffield.
Eligibility: Applicant must be a US citizen and should hold or expect to receive a Bachelor's degree (or equivalent professional training or experience) in a relevant area.
Level of Study: Doctorate, Graduate, Predoctorate, Postgraduate
Type: Award
Value: £12,000 plus tuition fees
Frequency: Annual
Study Establishment: Sheffield University
Country of Study: United Kingdom
No. of awards offered: 1
Application Procedure: Please check at http://us.fulbrightonline.org/applynow.html.
Closing Date: October 17th
Additional Information: Please check for more information at www.fulbright.co.uk.

Fulbright - Southampton University Award

Subjects: All subjects.
Purpose: To pursue postgraduate study at Southampton.
Eligibility: Candidates must be a US citizen resident anywhere except the United Kingdom, and must hold or expect to receive a Bachelor's degree (or equivalent professional training or experience) in a relevant area before departure to the UK.
Level of Study: Doctorate, Graduate, Postgraduate, Predoctorate
Type: Award
Value: £12,000 plus tuition fees
Frequency: Annual
Study Establishment: University of Southampton
Country of Study: United Kingdom
No. of awards offered: 1
Application Procedure: Please check at http://us.fulbrightonline.org/applynow.html.
Closing Date: October 17th
Funding: Government
Additional Information: Please check for more information at www.fulbright.co.uk.

Fulbright - Strathclyde University Research Awards

Subjects: All subjects.
Purpose: To pursue postgraduate study by research at Strathclyde University.
Eligibility: Candidate must be a US citizen resident anywhere except the United Kingdom, and must hold or expect to receive a Bachelor's degree (or equivalent professional training or experience) in a relevant area before departure to the UK.
Level of Study: Research, Predoctorate, Doctorate, Graduate, Postgraduate
Type: Award
Value: £12,000 plus tuition fees
Frequency: Annual
Study Establishment: Strathclyde University
Country of Study: Scotland
No. of awards offered: Up to 5
Application Procedure: Please check at http://us.fulbright.org/applynow.html.
Closing Date: October 17th
Funding: Government
No. of awards given last year: 2
Additional Information: Please check for more information at www.fulbright.co.uk.

Fulbright - UCL Technology Entrepreneurship Award

Subjects: Technology entrepreneurship.
Purpose: To pursue a master's (MSe) in technology entrepreneurship at UCL.
Eligibility: Must be US citizen resident anywhere except the United Kingdom, and should hold or expect to receive a Bachelor's degree (or equivalent professional training or experience) in a relevant area before departure to the UK.
Level of Study: Postgraduate
Type: Award
Value: £13,500 plus tuition fees
Frequency: Annual
Study Establishment: University College London (UCL)
Country of Study: United Kingdom
No. of awards offered: 1
Application Procedure: Please check at http://us.fulbrightonline.org/applynow/html.
Closing Date: October 17th
Funding: Government
Additional Information: Please check for more information at www.fulbright.co.uk.

Fulbright - University of Newcastle upon Tyne Award

Subjects: All subjects.
Purpose: To pursue postgraduate study at Newcastle.
Eligibility: Must be US citizen resident anywhere except the United Kingdom, and should hold or expect to receive a Bachelor's degree (or equivalent professional training or experience) in a relevant area before departure to the UK.
Level of Study: Doctorate, Graduate, Postgraduate, Predoctorate
Type: Award
Value: £12,000 plus tuition fees
Frequency: Annual
Study Establishment: Newcastle University
Country of Study: United Kingdom
No. of awards offered: 1
Application Procedure: Please check at http://us.fulbrightonline.org/applynow.html.

Closing Date: October 17th
Funding: Government
Additional Information: Please check for more information at www.fulbright.co.uk.

Fulbright - University of the Arts London Award

Subjects: All subjects.
Purpose: To pursue postgraduate study at University of the Arts London.
Eligibility: Must be US citizen resident anywhere except the United Kingdom, and must hold or expect to receive a Bachelor's degree (or equivalent professional training or experience) in a relevant area before departure to the UK.
Level of Study: Doctorate, Graduate, Postgraduate, Predoctorate
Type: Award
Value: £13,500 plus tuition fees
Frequency: Annual
Study Establishment: University of the Arts, London
Country of Study: United Kingdom
No. of awards offered: 1
Application Procedure: Please check at http://us.fulbrightonline.org/applynow.html.
Closing Date: Mid-October
Funding: Government
No. of awards given last year: 1
Additional Information: Please check more information at www.fulbright.co.uk.

Fulbright - University of the Arts London Scholar

Subjects: All subjects.
Purpose: To pursue research at University Arts London.
Eligibility: Must be US citizen* (resident anywhere except the United Kingdom), and must hold or expect to receive a PhD (or equivalent professional training or experience) in a relevant area before departure to the UK.
Level of Study: Postdoctorate
Type: Award
Value: £2,375 per month
Length of Study: 3–9
Frequency: Annual
Study Establishment: University of the Arts London
Country of Study: United Kingdom
No. of awards offered: 1
Application Procedure: Please check at www.cies.org.
Closing Date: August 1st
Funding: Government
Additional Information: Please check more information at www.fulbright.co.uk.

Fulbright - University of York Scholar Award in History of Art

Subjects: Any subject.
Purpose: To support a US citizen in research and / or teaching / lecturing in any subject.
Eligibility: Must be US citizen* (resident anywhere except the United Kingdom), and should hold or expect to receive a PhD (or equivalent professional training or experience) in a relevant area before departure to the UK.
Level of Study: Postdoctorate, Professional development, Research
Type: Award
Value: £2,500 per month paid in installments
Length of Study: 3–8 months
Frequency: Annual
Country of Study: United Kingdom
Application Procedure: Please check at www.cies.org.
Closing Date: August 1st
Funding: Government
Additional Information: Please check for more information at www.fulbright.co.uk.

Fulbright - York University Award

Subjects: All subjects.
Purpose: To pursue postgraduate study at the University of York.
Eligibility: Must be US citizen resident anywhere except the United Kingdom, and should hold or expect to receive a Bachelor's degree (or equivalent professional training or experience) in a relevant area before departure to the UK.
Level of Study: Doctorate, Graduate, Postgraduate, Predoctorate
Type: Award
Value: £12,000 plus tuition fees
Frequency: Annual
Study Establishment: University of York
Country of Study: United Kingdom
No. of awards offered: 1
Application Procedure: Please check at http://us.fulbrightonline.org/applynow.html.
Closing Date: October 17th
Funding: Government
No. of awards given last year: 1
Additional Information: Please check more information at www.fulbright.co.uk.

Fulbright - York University Scholar Award

Subjects: All subjects.
Purpose: To pursue research at York.
Eligibility: Must be US citizen* (resident anywhere except the United Kingdom), and should hold or expect to receive a PhD (or equivalent professional training or experience) in a relevant area before departure to the UK.
Level of Study: Postdoctorate
Type: Award
Value: £2,500 per month
Length of Study: 3–8 months
Frequency: Annual
Study Establishment: University of York
Country of Study: United Kingdom
No. of awards offered: 1
Application Procedure: Please check at www.cies.org.
Closing Date: August 1st
Funding: Government
No. of awards given last year: 1
Additional Information: Please check for more information at www.fulbright.co.uk.

Fulbright All-disciplines Scholar Awards

Subjects: Any subject.
Purpose: To support outstanding UK professionals of academics to undertake lecturing, research or a combination of the two in any field, at any accredited US institution.
Eligibility: Open to UK citizens that have/expect to have a PhD or equivalent professional training or experience in a relevant area before departure to the US.
Level of Study: Postdoctorate, Professional development, Research
Type: Award
Value: $5,000 per month
Length of Study: 3–10 months
Frequency: Annual
Country of Study: United States of America
No. of awards offered: Approx. 5
Application Procedure: Applicants must visit the website www.fulbright.co.uk for applications.
Closing Date: November 15th
Funding: Government
No. of awards given last year: 7
Additional Information: Please check at www.fulbright.co.uk.

Fulbright Graduate Student Awards

Subjects: All subjects.
Purpose: To enable students to pursue postgraduate study or research in the United Kingdom.
Eligibility: Open to citizens of the United States of America, normally resident in the United States of America. Applicants must have a minimum grade point average of 3.5 and be able to demonstrate evidence of leadership qualities. Applicants should be adventurous and be able to demonstrate that they will maximize academic, social and cultural opportunities available in the United Kingdom. A full description of the selection criteria is available from the website.
Level of Study: Graduate, Doctorate, Postgraduate, Research
Type: Award
Value: Full tuition plus a living stipend of £12,000–12,500

Length of Study: 9 months (1 academic year)
Frequency: Annual
Study Establishment: Any approved Institute of Higher Education
Country of Study: United Kingdom
No. of awards offered: 4–7
Application Procedure: Applicants must submit a formal application via www.iie.org.
Closing Date: Mid-October
Funding: Government
Contributor: The United States and United Kingdom governments
No. of awards given last year: 10
No. of applicants last year: approx. 450
Additional Information: A telephone interview is required of short-listed candidates. Please check at www.fulbright.co.uk.

For further information contact:

Institute of International Education, 809 United Nations Plaza, New York, NY, 10017, United States of America
Tel: (1) 212 984 5466
Fax: (1) 212 984 5465
Website: www.iie.org
Contact: Student Program Division

Fulbright Northern Ireland Public Sector Scholar Awards

Purpose: To support those working in the public sector in Northern Ireland at senior management level (equivalent to grade 7 and above in the NICS) to pursue research and/or assess best practice affiliated with any US institution.
Eligibility: Open to UK citizen* (resident anywhere except the United States). Candidate must have worked in Northern Ireland Public Sector administrator at senior management level (equivalent to Grade 7 and above in the NICS).
Level of Study: Professional development, Research
Type: Award
Value: Up to $5,000 per month paid in installements
Length of Study: 3–5 months
Frequency: Annual
Country of Study: United States of America
No. of awards offered: 2
Application Procedure: Applicants must visit the website www.fulbright.co.uk for applications.
Closing Date: November 20th
Funding: Government
No. of awards given last year: 1
Additional Information: Please check at www.fulbright.co.uk.

Fulbright Police Research Fellowships

Subjects: Policing.
Purpose: To support active UK police officers and staff from all ranks to conduct research, pursue professional development and/or assess best practice affiliated with any US institution.
Eligibility: Minimum eligibility for this Award category: UK citizen* (resident anywhere except the United States), and active police officer or police staff.
Level of Study: Professional development, Research
Type: Award
Value: £3,600 per month paid in installments (at the prevailing rate)
Length of Study: 3–5 months
Frequency: Annual
Study Establishment: An approved United States of America Institute of Higher Education or police force
Country of Study: United States of America
No. of awards offered: 3
Application Procedure: Applicants must visit the website www.fulbright.co.uk
Closing Date: November 15th
Funding: Government
No. of awards given last year: 2
No. of applicants last year: 6
Additional Information: Please check at www.fulbright.co.uk.

For further information contact:

Fulbright House, 62 Doughty Street, London, WC1N 2JZ, England
Website: www.fulbright.co.uk

Fulbright Postgraduate Student Awards

Subjects: All subjects.
Purpose: To enable students to pursue postgraduate study or research in the United States of America.
Eligibility: Candidates must be UK citizens, should hold a minimum of an Upper Second Class (Honours) Degree (or equivalent) and demonstrate outstanding leadership qualities.
Level of Study: Postgraduate
Type: Award
Value: Tuition fees plus living stipend for one academic year
Length of Study: 9–12 months
Frequency: Annual
Study Establishment: An approved Institute of Higher Education
Country of Study: United States of America
No. of awards offered: Up to 30
Application Procedure: Applicants must submit a formal application. Applications can be obtained www.fulbright.co.uk website.
Closing Date: November 15th
Funding: Government
No. of awards given last year: 22
Additional Information: Please check at fulbright.co.uk.

Fulbright Scholar Award in Scottish Studies

Purpose: To support scholars academics, artists or professionals to undertake lecturing, carry out research relating to Scottish studies and develop institutional links with any accredited US institution.
Eligibility: Open to UK citizens that have/expect to have a PhD or equivalent professional training or experience in a relevant area before departure to the US.
Level of Study: Professional development
Type: Award
Value: $5,000 per month paid in installments
Length of Study: 3–8 months
Frequency: Annual
Country of Study: United States of America
No. of awards offered: 1
Application Procedure: Applicants must visit the website www.fulbright.co.uk for applications.
Closing Date: November 15th
Funding: Government
Contributor: Scottish Government
No. of awards given last year: 1
Additional Information: Please check at www.fulbright.co.uk.

Fulbright Scholar Awards

Subjects: All subjects.
Purpose: To enable scholars to carry out lecturing and research in the United Kingdom.
Eligibility: PhD requried.
Level of Study: Postdoctorate, Professional development, Research, Lecture
Type: Award
Value: $25,000 USD per year, or $12,500 for one semester
Length of Study: 3 months–10 months
Frequency: Annual
Study Establishment: An approved Institute of Higher Education
Country of Study: United Kingdom
No. of awards offered: Up to 5 awards
Application Procedure: Please check at www.cies.org.
Closing Date: November 15th
Funding: Government
No. of awards given last year: 5
Additional Information: Please check at www.fulbright.co.uk.

For further information contact:

Council for International Exchange of Scholars (CIES), 3007 Tilden Street North West, Suite 5M, Washington DC, 20008-3009, United States of America
Tel: (1) 202 686 6245
Website: www.cies.org

The Fulbright-Bristol University Award

Purpose: To enable a US citizen to pursue postgraduate study in the United Kingdom at the University of Bristol.

Eligibility: Applicant must be a US citizen (resident anywhere but the UK), and a graduating seniro, hold a BS/BA degree, Master's or doctoral degree candidate, young professional or artist.
Level of Study: Doctorate, Graduate, Postgraduate
Type: Award
Value: £10,000 towards living costs plus tuition fees
Frequency: Annual
Study Establishment: University of Bristol
Country of Study: United Kingdom
No. of awards offered: 1
Application Procedure: Please check at http://US.fulbrightonline.org/applynow.html.
Closing Date: Mid-October
No. of awards given last year: 1
Additional Information: For more information, please visit www.bristol.ac.uk/prospectus/postgraduate and http://us.fulbrightonline.org/program_country.html?id = 112 and www.fulbright.co.uk.

For further information contact:

University of Bristol, Senate House, Tyndall Avenue, Bristol, BS8 1TH, United Kingdom
Tel: (44) (0) 117 928 9000
Website: www.bristol.ac.uk

Fulbright-British Friends of Harvard Business School Awards

Subjects: MBA.
Purpose: To enable British candidates to participate in Harvard Business School MBA programmes.
Eligibility: Open to citizens of the United Kingdom. Applicants must hold the minimum of an Upper Second Class (Honours) Degree and be able to demonstrate leadership qualities.
Level of Study: Postgraduate, MBA
Type: Award
Value: Up to US$5,000
Length of Study: 9–12 months
Frequency: Annual
Study Establishment: Harvard Business School
Country of Study: United States of America
No. of awards offered: Up to 5 awards
Application Procedure: Applicants must complete an application form available in www.fulbright.co.uk
Closing Date: April 19th
Funding: Government, individuals
Contributor: The British Friends of Harvard Business School
No. of awards given last year: 2
Additional Information: Please check at www.fulbright.co.uk.

For further information contact:

Website: www.fulbright.co.uk

The Fulbright-Coventry Award in Design

Subjects: Design.
Purpose: To enable a US citizen to pursue a Master's degree in Design.
Eligibility: Applicant should be a US citizen (resident anywhere but the UK), and a graduating senior, holding a BS/BA degree, master's or doctoral degree candidate, young professional or artist.
Level of Study: Doctorate, Graduate, Postgraduate
Type: Award
Value: £12,000 intended as a contribution towards general maintenance costs in addition to a full tuition fee waiver
Frequency: Annual
Study Establishment: Coventry University
Country of Study: United Kingdom
No. of awards offered: 1
Application Procedure: Please visit the website http://us.fulbrightonline.org/applynow.html for applications.
Closing Date: October 17th
No. of awards given last year: 1
Additional Information: Please check at www.fulbright.co.uk.

For further information contact:

Website: www.coventry.ac.uk/courses/postgraduate-full-time-a-z/a/381

Fulbright-Diabetes UK Research Award

Subjects: Any diabetes related discipline.
Purpose: To support research into the clinical or biomedical aspects of diabetes or the social or economic conditions of sufferers at any accredited US institution.
Eligibility: Open to UK citizens that have a doctoral degree or equivalent professional training or experience at the time of application.
Level of Study: Postdoctorate, Professional development, Research
Type: Award
Value: £75,000 (converted at the prevailing rate) paid in installments
Length of Study: 1 year
Frequency: Annual
Country of Study: United States of America
No. of awards offered: 1
Application Procedure: Applicants must visit the website www.fulbright.co.uk for applications.
Closing Date: November 15th
Contributor: Diabetes UK
No. of awards given last year: 1
Additional Information: Please check at www.fulbright.co.uk.

The Fulbright-Glasgow University Award

Subjects: All subjects.
Purpose: To enable a US citizen to pursue postgraduate study in the United Kingdom at Glasgow University.
Eligibility: Applicant should be a US citizen (resident anywhere but the UK), and a graduating senior, hold a BS/BA degree, master's or doctoral degree candidate, young professional or artist.
Level of Study: Graduate, Postgraduate
Type: Award
Value: UK£20,000.From the £20,000 tuition fees will be deducted and the remainder will be paid in installments directly to the grantee.
Frequency: Annual
Study Establishment: Glasgow University
Country of Study: United Kingdom
No. of awards offered: 1
Application Procedure: Please check at http://us.fulbrightonline.org/applynow.html for application details.
Closing Date: Mid-October
No. of awards given last year: 1
Additional Information: For more information, visit www.gla.ac.uk/postgraduate/, http://us.fulbrightonline.org/program_country.html?id = 112 and www.fulbright.co.uk.

Fulbright-Hubert Humphrey Public Affairs Fellowship

Subjects: Any subject.
Purpose: To assist a UK civil servant to develop skills in operational and policy delivery through research and/or study.
Eligibility: Open to UK civil servants at Grade 7/A or above who can demonstrate experience appropriate to study at postgraduate level.
Level of Study: Professional development, Research
Type: Award
Value: Tuition fees, return airfare allowance ($2,000), accomodation and maintenance stipend ($2,000), computer and books subsidies ($1,000), professional development and affiliation travel allowances ($3,700)
Length of Study: 4 months
Frequency: Annual
Country of Study: United States of America
No. of awards offered: 1
Application Procedure: Applicants must visit the website www.fulbright.co.uk for applications.
Closing Date: May 31st
Contributor: Hubert Humphrey Institute of Public Affairs
No. of awards given last year: 1
Additional Information: Please check at www.fulbright.co.uk.

Fulbright-King's College London Scholar Award

Subjects: All subjects.
Purpose: To support research at King's College, London.
Eligibility: Applicant must be a PhD or equivalent professional/terminal degree (including a master's depending on the field) as appropriate at the time of application.
Level of Study: Postdoctorate, Professional development, Research

Type: Award
Value: The award provides a monthly stipend of £2,375 paid in instalments directly to the grantee.
Length of Study: 3–6 months
Frequency: Annual
Study Establishment: King's College London
Country of Study: United Kingdom
No. of awards offered: 1
Application Procedure: Please visit www.cies.org/us_scholars/us_awards for application details.
Closing Date: August 1st
No. of awards given last year: 1
Additional Information: For more information, visit www.cies.org/us_scholars/us_awards/index.html, www.kcl.ac.uk/ and www.fulbright.co.uk.

The Fulbright-Leeds University Award

Subjects: All subjects.
Purpose: To enable a US citizen to pursue postgraduate study in the United Kingdom at the University of Leeds.
Eligibility: Applicant must be a US citizen (resident anywhere but the UK), and a graduating senior, hold a BS/BA degree, master's or doctoral degree candidate, young professional or artist.
Level of Study: Doctorate, Graduate, Postgraduate
Type: Award
Value: £12,000 intended as a general maintenance stipend plus a full tuition fee waiver
Length of Study: 6 months
Frequency: Annual
Study Establishment: University of Leeds
Country of Study: United Kingdom
No. of awards offered: 1
Application Procedure: Please visit the website http://us.fulbrightonline.org/applynow.html
Closing Date: October 17th
Additional Information: For more information, please visit www.leeds.ac.uk, http://us.fulbrightonline.org/program_country.html?id=112 and www.fulbright.co.uk.

Fulbright-Leeds University Distinguished Chair Award

Subjects: All subjects.
Purpose: To contribute to the intellectual life of Leeds University through seminars, public lectures and curriculum development in any discipline.
Eligibility: Applicant must be US citizen and should have a PhD or equivalent professional/terminal degree (including a master's depending on the field) as appropriate at the time of application.
Level of Study: Postdoctorate, Professional development, Research
Type: Award
Value: £27,500 plus accommodation. Limited sickness and accident benefit coverage as well as participation in a number of Fulbright Scholar events including the Fulbright Forum in January are also included
Length of Study: 6 months
Frequency: Annual
Study Establishment: Leeds University
Country of Study: United Kingdom
No. of awards offered: 1
Application Procedure: Please visit the website www.cies.org/Chairs/Apply.htm
Closing Date: August 1st
No. of awards given last year: 1
Additional Information: For more information, please visit www.cies/org/Chairs, www.cies.org/Chairs/2009/area2.htm, www.leeds.ac.uk and www.fulbright.co.uk.

The Fulbright-Liverpool University Award

Subjects: All subjects.
Purpose: To enable a US citizen to pursue postgraduate study in the United Kingdom at the University of Liverpool.
Eligibility: Applicant must be a US citizen (resident anywhere but the UK), and a graduating senior, holding a BS/BA degree, master's or doctoral degree candidate, young professional or artist.
Level of Study: Doctorate, Graduate, Postgraduate
Type: Award
Value: Up to UK£20,000, limited sickness and accident benefit coverage as well as participation in a number of Fulbright Scholar events
Frequency: Annual
Study Establishment: University of Liverpool
Country of Study: United Kingdom
No. of awards offered: 1
Application Procedure: Please visit the website http://us.fulbrightonline.org/applynow.html
Closing Date: Mid-October
No. of awards given last year: 1
Additional Information: For more information, visit www.liverpool.ac.uk, http://us.fulbrightonline.org/program_country.html?id=112 and www.fulbright.co.uk.

Fulbright-Multiple Sclerosis Society Research Award

Subjects: All subjects.
Purpose: To enable a UK citizen to pursue research in clinical or biomedical aspects of MS or the social or economic conditions of sufferers, at any accredited US institution, for a minimum of 12 months.
Eligibility: Applicant must be a UK citizen (resident anywhere but the US), holding a doctoral degree or equivalent professional training or experience at the time of application.
Level of Study: Postdoctorate, Professional development, Research
Type: Award
Value: Approximately UK£75,000, plus limted sickness and accident benefit coverage, visa sponsorship and processing and participation in a number of Fulbright Scholar events
Length of Study: 1 year
Frequency: Annual
Country of Study: United States of America
No. of awards offered: 1
Application Procedure: Formal applications are made online via the US–UK Fulbright Commission: www.fulbright.co.uk
Closing Date: May 31st
Contributor: MS Society UK
No. of awards given last year: 1
Additional Information: Please check at www.fulbright.co.uk.

Fulbright-Queen's University Belfast Anglophone Irish Writing Scholar Award

Subjects: Creative Writing.
Purpose: To enable teaching/lecturing, writing and public readings of their own work at Queen's University Belfast.
Eligibility: Applicant must be a PhD or equivalent professional/terminal degree (including a master's depending on the field) as appropriate at the time of application.
Level of Study: Postdoctorate
Type: Award
Value: UK£15,000. Dedicated housing will be offered. Limited sickness and accident benefit coverag as well as participation in a number of Fulbright Scholar events including the Fulbright Forum in January are also included
No. of awards offered: 1
Application Procedure: Please visit the website www.cies.org/us_scholars/us_awards/
Closing Date: August 1st
Additional Information: For more information, visit www.cies.org/us_scholars/us_awards/index.html, www.qub.ac.uk/schools/SchoolofEnglish/ and www.qub.ac.uk/schools/SeamusHeaneyCentreforPoetry/

Fulbright-Queen's University Belfast Creative Writing Scholar Award

Subjects: Language and literature (Non-US), creative writing, poetry and drama.
Purpose: To support the research and teaching/lecturing of Anglophone Irish Literature from the 18th century through to contemporary writing at Queen's University Belfast.
Eligibility: Applicant must be a PhD or equivalent professional/terminal degree (including a master's depending on the field) as appropriate at the time of application.
Level of Study: Postdoctorate
Type: Award

Value: UK£15,000. Dedicated housing will be offered. Limited sickness and accident benefit coverage as well as participation in a number of Fulbright Scholar events including the Fulbright Forum in January are also included
Length of Study: 6 months
Frequency: Annual
Study Establishment: Queen's University Belfast
Country of Study: United Kingdom
No. of awards offered: 1
Application Procedure: Please visit the website www.cies.org/us_scholars/us_awards/Eligibility.htm for application details.
Closing Date: August 1st
No. of awards given last year: 1
Additional Information: For more information, visit www.cies.org/us_scholars/us_awards/index.html, www.qub.ac.uk/schools/SchoolofEnglish/, www.qub.ac.uk/schools/SeamusHeaneyCentreforPoetry/ and www.fulbright.co.uk.

Fulbright-Robertson Visiting Professorship in British History

Subjects: British history or Western civilization (any specialisation).
Purpose: To enable a British scholar to join the Department of history at Westminster College, Fulton, Missouri.
Eligibility: Must be UK citizen* (resident anywhere except the United States), must hold or expect to receive a PhD in a relevant area before departure to the US, and have at least one year's experience of lecturing to undergraduate students
Level of Study: Professional development
Type: Award
Value: $52,500 plus travel budget up to $10,000
Length of Study: 1 year
Frequency: Annual
Study Establishment: Westminster College, Fulton, Missouri
Country of Study: United States of America
No. of awards offered: 1
Application Procedure: Application forms can be obtained from www.fulbright.co.uk
Closing Date: November 15th
Funding: Private
Contributor: Westminster College, Fulton, Missouri
No. of awards given last year: 1

The Fulbright-Sussex University Award

Subjects: All subjects.
Purpose: To enable a US citizen to pursue postgraduate study in the United Kingdom at the University of Sussex.
Eligibility: Applicant must be a US citizen (resident anywhere but th UK), and a graduating senior, holding a BS/BA degree, master's or doctoral degree candidate, young professional or artist.
Level of Study: Doctorate, Graduate, Postgraduate
Type: Award
Value: Up to UK£20,000. Limited sickness and accident benefit coverage, and participation in a number of Fulbright Scholar events
Frequency: Annual
Study Establishment: University of Sussex
Country of Study: United Kingdom
No. of awards offered: 1
Application Procedure: Please visit the website http://us.fulbright-online.org/applynow.html for application details.
Closing Date: Mid-October
No. of awards given last year: 1
Additional Information: For more information, visit www.sussex.ac.uk/pgstudy, http://us.fulbrightonline.org/program_country.html?id = 112 and www.fulbright.co.uk.

The Fulbright-University College Falmouth Media Award

Subjects: Media.
Purpose: To enable a US citizen to pursue a postgraduate degree in media studies at University College, Falmouth.
Eligibility: Applicant must be a US citizen (resident anywhere but the UK), and a graduating senior, holding a BS/BA degree, master's or doctoral degree candidate, young professional or artist.
Level of Study: Doctorate, Graduate, Postgraduate
Type: Award
Value: UK£15,000. An additional allowance of up to £1,000 is available to support engagement with (travel) UK publishing and/or media companies. Limited sickness and accident benefit coverage as well as participation in a number of Fulbright Scholar events including the Fulbright Forum in January are also included
Frequency: Annual
Study Establishment: University College Falmouth
Country of Study: United Kingdom
No. of awards offered: 1
Application Procedure: Please visit the website http://us.fulbright-online.org/applynow.html for application details.
Closing Date: October 17th
No. of awards given last year: 1
Additional Information: For more information, please visit www.fulbright.co.uk.

The Fulbright-University College London Award

Subjects: All subjects.
Purpose: To enable a US citizen to pursue postgraduate study in the United Kingdom at the University College London.
Eligibility: Applicant must be a US citizen (resident anywhere but the UK), and a graduating senior, holding a BS/BA degree, master's or doctoral degree candidate, young professional or artist.
Level of Study: Doctorate, Graduate, Postgraduate
Type: Award
Value: £13,500 intended as a contribution towards general maintenance costs in addition to a full tuition fee cost
Frequency: Annual
Study Establishment: University College London
Country of Study: United Kingdom
No. of awards offered: 1
Application Procedure: Please visit the website http://us.fulbright-online.org/applynow.html
Closing Date: October 17th
No. of awards given last year: 1
Additional Information: For more information, please visit the website.

Fulbright-University of the Arts London Distinguished Chair Award

Subjects: Art, communications, history, philosophy.
Purpose: To pursue research, lecture, teach and/or contribute to the curriculum at TrAIN, a department of the University of the Arts London.
Eligibility: Applicant must be a PhD or equivalent professional/terminal degree (including a master's depending on the field) as appropriate at the time of application.
Level of Study: Postdoctorate
Type: Award
Value: UK£27,500 plus accommodation. Limited sickness and accident benefit coverage as well as participation in a number of Fulbright Scholar events including the Fulbright Forum in January are also included
Length of Study: 6 months
Frequency: Annual
Study Establishment: University of Arts London
Country of Study: United Kingdom
No. of awards offered: 1
Application Procedure: Please visit the website www.cies.org/Chairs/Apply.htm for application details.
Closing Date: August 1st
No. of awards given last year: 1
Additional Information: Please check at www.fulbright.co.uk.

The Fulbright-Warwick University Award

Subjects: All subjects.
Purpose: To enable a US citizen to pursue a postgraduate study in the United Kingdom in any field at the University of Warwick.
Eligibility: Applicant must be a US citizen (resident anywhere but the UK), and a graduating senior, holding a BS/BA degree, master's or doctoral degree candidate, young professional or artist.
Level of Study: Doctorate, Graduate, Postgraduate
Type: Award

Value: Up to UK£20,000, limited sickness and accident benefit coverage, as well as participation in a number of Fulbright Scholar events
Frequency: Annual
Study Establishment: University of Warwick
Country of Study: United Kingdom
No. of awards offered: 1
Application Procedure: Please check at http://us.fulbrightonline.org/applynow.html.
Closing Date: Mid-October
No. of awards given last year: 1
Additional Information: For more information, visit www.warwick.ac.uk/study/postgraduate and http://us.fulbrightonline.org/program_-country.html?id=112

International Fulbright Science and Technology PhD Awards

Subjects: Science or technology.
Purpose: To pursue a PhD in a science or technology field, at any accredited US university.
Eligibility: Open to UK citizens that have/expect to have a 2:1 honours degree or equivalent before departure to the US.
Level of Study: Doctorate, Graduate
Type: Award
Value: Full tuition and associated fees for up to 3 years, monthly stipend for up to 36 months; book, equipment, research and professional conference allowances, J-1 visa sponsorship for up to five years, round-trip airfare from home city to the host institution in the US, health and accident coverage, and specially tailored enrichment activities
Length of Study: 3–4 years
Frequency: Annual
Country of Study: United States of America
Application Procedure: Applicants must visit the website www.fulbright.co.uk for applications.
Closing Date: May 31st
Funding: Government
No. of awards given last year: 1
Additional Information: Please check at fulbright.co.uk.

UK Fulbright Alistair Cooke Award in Journalism

Subjects: Journalism.
Purpose: To finance the first year of postgraduate study for an aspiring journalist.
Eligibility: Must be US nationals
Level of Study: Postgraduate
Type: Award
Value: Up to UK£20,000
Length of Study: 9–12 months
Frequency: Annual
Country of Study: United States of America
No. of awards offered: 1–2
Application Procedure: Applicants must visit the website www.fulbright.co.uk for applications.
Closing Date: Late October
Funding: Government
No. of awards given last year: 2
Additional Information: Please check at www.fulbright.co.uk.

US Fulbright Alistair Cooke Award in Journalism

Subjects: Journalism.
Purpose: To finance an aspiring journalist, UK citizens.
Eligibility: Open to UK citizens that have/expect to have a 2:1 honours degree or equivalent before departure to the US
Level of Study: Postgraduate
Type: Scholarship
Value: Up to $25,000
Length of Study: 9 months
Frequency: Annual
Country of Study: United Kingdom
No. of awards offered: 1
Application Procedure: See website www.iie.org
Closing Date: May 31st
Funding: Government
No. of awards given last year: 1
Additional Information: Please check at www.fulbright.co.uk.

US-UK Fulbright Commission, Criminal Justice and Police Research Fellowships

Subjects: Policing and criminal justice.
Purpose: To enable active domestic police officers and staff to extend their professional expertise and experience in conducting research into any aspect of policing, in the UK.
Eligibility: Applicants should ideally hold a Bachelor's degree in criminal justice, police studies or a related discipline within the social sciences.
Level of Study: Professional development
Value: UK£10,000
Length of Study: Between 3 and 6 months
Frequency: Annual
Study Establishment: Institutes of Higher Education or a police force
Country of Study: United Kingdom
No. of awards offered: 1–2
Application Procedure: Applicants must complete an application form available from the Council for the International Exchange of Scholars (CIES) – www.cies.org
Closing Date: August 1st
Funding: Government
No. of awards given last year: 2
No. of applicants last year: 3
Additional Information: Please check at www.fulbright.co.uk.

For further information contact:

3007 Tilden Street, Suite 5M, Washington DC, 2008-3009, United States of America

VALPARAISO UNIVERSITY

1320 South Chapel Drive, Valparaiso, IN, 46383, United States of America
Tel: (1) 219 464 5317
Fax: (1) 219 464 5496
Email: lillyfellows.program@valpo.edu
Website: www.lillyfellows.org

Valparaiso University was founded in 1859 by Methodists, with approximately 4,000 students from most states and more than 40 countries. The university, which is nearly 150 years old offers more than 70 fields of study.

Lilly Postdoctoral Teaching Fellowship Program

Subjects: Arts and humanities.
Purpose: To strengthen the quality of church-related institutions of higher learning for the 21st century.
Eligibility: Open to new scholar-teachers who have obtained a PhD and are interested in the relationship between Christianity and the academic vocation and are seriously considering a career at a church-related college or university.
Level of Study: Postgraduate, Teaching
Type: Fellowships
Value: $46,800 plus standard benefits, a moving allowance, and an annual professional and travel allowance.
Length of Study: 2 years
Frequency: Annual
Country of Study: United States of America
No. of awards offered: 3
Application Procedure: Applicants must submit a curriculum vitae, a graduate transcript, 2 letters of recommendation and an essay or personal statement.
Closing Date: November 1st
Additional Information: All queries should be directed to lillyfellows.program@valpo.edu. For further information contact The Selection Committee Lilly Fellows Program at the above address.

VATICAN FILM LIBRARY

Mellon Fellowship Program, Pius XII Memorial Library, Saint Louis University, 3650 Lindell Boulevard, St Louis, MO, 63108 3302, United States of America
Tel: (1) 314 977 3090
Fax: (1) 314 977 3108
Email: vfl@slu.edu
Website: http://libraries.slu.edu/special/vfl/
Contact: Mrs Barbara J Channell, Secretary

The Vatican Film Library at Saint Louis University in St Louis, Missouri, is a microfilm repository for the Vatican Library manuscripts housed in Rome and a research centre for medieval and Renaissance manuscripts studies in general. The research collections contain a wide range of primary source manuscript materials on microfilm, microfiche and in digital reproduction, in addition to slides and printed facsimile editions ranging in date from the 5th to the 19th centuries, and a large number of incunabula and early-printed books also on microfilm and microfiche. At the core of these collections are the microfilmed copies of approximately three-quarters of the Vatican Library's Greek, Latin and Western European vernacular manuscripts, as well as Hebrew, Ethiopic and Arabic manuscripts.

Vatican Film Library Mellon Fellowship

Subjects: Classical languages and literature, palaeography, scriptural and patristic studies, history, philosophy and sciences in the Middle ages and the Renaissance and early Romance literature. There are also opportunities for supported research in the history of music, manuscript illumination, mathematics and technology, theology, liturgy, Roman and canon law or political theory.
Purpose: To assist scholars wishing to conduct research in the manuscript collections in the Vatican Film Library at Saint Louis University.
Eligibility: Open to candidates who are at the postdoctoral level, or graduate students formally admitted to PhD candidacy and working on their dissertation.
Level of Study: Graduate, Postdoctorate, Postgraduate, Predoctorate, Research, Doctorate
Type: Fellowship
Value:
The scholarship provides travel expenses and a living allowance of $2250 USD per month for two to eight weeks.
Length of Study: 2–8 weeks
Study Establishment: The Vatican Film Library
Country of Study: United States of America
Application Procedure: Applicants must write, in the first instance, to describe the topic of the planned research and to indicate the exact dates during which support is desired. A formal application should then be submitted, which should include a cover letter stating the title and proposed dates of research, a detailed statement of the project proposal of not more than two or three pages in length, a list of the manuscripts or other archival materials to be consulted, a selective bibliography of primary and secondary sources relating to the research topic and a curriculum vitae. Applications submitted by PhD candidates should also include a letter of recommendation from their advisor with reference to the applicant's palaeographical and language.
Closing Date: March 1st for research in June to August, June 1st for research in September to December and October 1st for research in January to May

VERNE CATT MCDOWELL CORPORATION

PO Box 1336, Albany, OR, 97321-0440, United States of America
Tel: (1) 541 924 0976
Contact: Ms NADINE WOOD, Business Manager

The Verne Catt McDowell Scholarship educates pastoral ministers of the Christian Church (Disciples of Christ) by providing supplementary financial grants to graduate theology students.

Verne Catt McDowell Corporation Scholarship

Subjects: Religion, theology and church administration.
Purpose: To provide supplemental financial grants to men and women for graduate theological education for ministry in the Christian Church (Disciples of Christ) denomination.
Eligibility: All scholarship candidates must be ministers ordained or studying to meet the requirements to be ordained as a minister in the Christian Church (Disciples of Christ). Candidates must be members of the Christian Church (Disciples of Christ) denomination. Preference is given to Oregon graduates and citizens of the United States of America.
Level of Study: Postgraduate
Type: Scholarship
Value: Average: $3,000
Length of Study: 1–3 years
Study Establishment: A graduate institution of theological education, accredited by the general assembly of the Christian Church (Disciples of Christ)
Country of Study: United States of America
No. of awards offered: 5
Application Procedure: Applicants must complete an application form and provide details of qualifications, transcripts, 3 references and state where they obtained information about the scholarship. An interview may be requested.
Closing Date: May 1st
Funding: Private
Contributor: I A McDowell
No. of awards given last year: 3
No. of applicants last year: 4

VERNON WILLEY TRUST-GUARDIAN TRUST COMPANY

PO Box 9, Christchurch, 8001, New Zealand
Tel: (64) 3 379 0645
Fax: (64) 3 366 7616
Email: gary_anderson@nzgt.co.nz
Website: www.nzgt.co.nz
Contact: Mr G Anderson, Client Manager

Vernon Willey Trust Awards

Subjects: The sheep and wool industry of New Zealand.
Purpose: To assist with research and education into the production, processing and marketing of wool and the general development of the industry for the national benefit of New Zealand.
Eligibility: Open to New Zealand citizens, permanent New Zealand residents or overseas researchers working in New Zealand.
Level of Study: Doctorate, Postdoctorate
Type: Fellowship
Value: Varies, usually between New Zealand $30,000–35,000
Length of Study: Up to 3 years
Frequency: Dependent on funds available
Country of Study: New Zealand
No. of awards offered: 1
Application Procedure: Applicants must complete an application form.
Closing Date: March 15th
Funding: Private
No. of awards given last year: 2
No. of applicants last year: 2
Additional Information: Applicants for financial grants must satisfy the Committee that their activities are of general or public benefit. The results of the research or studies are expected to be covered by material suitable for publication in recognized scientific or technical journals.

VICTORIA UNIVERSITY OF WELLINGTON

PO Box 600, Wellington, 6140, New Zealand
Tel: (64) 4 463 5113, 4 472 1000
Fax: (64) 4 496 5454, 4 499 4601
Email: scholarships-office@vuw.ac.nz
Website: www.vuw.ac.nz
Contact: Scholarship Office

Victoria University of Wellington is a thriving community of over 20,000 students of all nationalities. Victoria has a reputation for academic excellence and the calibre of its research. Scholarships are available in all disciplines. Victoria University is located in New Zealand's compact vibrant capital city. As one of the oldest universities in New Zealand, it has a tradition of teaching excellence and is one of New Zealand's leading research institutions and has produced Nobel Prize winning scholars. Scholarships are available in all disciplines.

Victoria Doctoral Scholarships

Subjects: All subjects.
Purpose: To provide financial assistance to Doctoral students.
Eligibility: Open to students who are about to commence their Doctoral studies. The selection is based on academic merit.
Level of Study: Doctorate, Postgraduate
Type: Scholarship
Value: New Zealand $23,500 stipend annually, plus tuition fees.
Length of Study: 3 years
Frequency: Annual, Every 4 months
Study Establishment: Victoria University of Wellington
Country of Study: New Zealand
Application Procedure: Application forms available from www.victoria.ac.nz/fgr
Closing Date: Victoria PhD Scholarships-March 1st, July 1st and November 1st.
Contributor: Victoria University of Wellington
No. of awards given last year: 115
No. of applicants last year: 350
Additional Information: Application for PhD scholarships and admission to the university is one process.

For further information contact:

Email: pg-research@vuw.ac.nz

VILLA I TATTI: THE HARVARD UNIVERSITY CENTER FOR ITALIAN RENAISSANCE STUDIES

Via di Vincigliata 26, 50135 Florence, Florence, Italy
Tel: (39) 055 603 251
Fax: (39) 055 603 383
Email: info@itatti.harvard.edu
Website: www.itatti.harvard.edu
Contact: Angela Lees, Administrative Assistant

Villa I Tatti is devoted to advanced study of the Italian Renaissance in all its aspects, the history of art, political, economic and social history, the history of science, philosophy and religion and the history of literature and music.

Craig Hugh Smyth Visiting Fellowship

Subjects: Fellowship to work at I Tatti for 3 months. The project must represent advanced research in the Italian Renaissance, broadly defined as the period ranging from the 13th to the 17th centuries. Subjects covered include the architecture, history, literature, material culture, music, philosophy, religion, science, or visual arts of Italy. Applications would also be welcomed from candidates working on the transmission and circulation of ideas, objects, and people during the Renaissance, into and beyond the Italian peninsula, or the historiography of the Italian Renaissance, including the rebirth of interest in the Renaissance in later periods. Museum professionals may apply to carry out research for a forthcoming exhibition or for the catalogue of permanent collections, but they can also apply for projects relating to their personal research interests.
Purpose: Advanced research in the Italian Renaissance.
Eligibility: Italian Renaissance scholars with limited research time, see website.
Level of Study: Doctorate, Postdoctorate, Predoctorate
Type: Residential fellowships
Value: Up to $5,000 per month plus a one-time supplement (Max. $1,500) towards relocation expenses
Length of Study: 3 months
Frequency: Annual
Country of Study: Italy
No. of awards offered: Varies
Application Procedure: Please check at http://itatti.harvard.edu/research/fellowships/craig-hugh-smyth-visiting-fellowship.
Closing Date: February 1st
Funding: Foundation
Contributor: The Andrew W. Mellon Foundation
No. of awards given last year: 3

I Tatti Fellowships

Subjects: Fellowship to work at I Tatti for one year. The project must represent advanced research in the Italian Renaissance, broadly defined as the period ranging from the 13th to the 17th centuries. Subjects covered include the architecture, history, literature, material culture, music, philosophy, religion, science or visual arts of Italy. Scholars can also apply to work on the transmission and circulation of ideas, objects, and people during the Renaissance, into and beyond the Italian peninsula, or the historiography of the Italian Renaissance, including the rebirth of interest in the Renaissance in later periods.
Purpose: Advanced research in the Italian Renaissance.
Eligibility: At the time of application, scholars must hold a PhD, dottorato di ricerca, or an equivalent doctorate. They must be conversant in either English or Italian and able to understand both languages. They should be in the early stages of their career, having received a PhD between 2002–2011 and have a solid background in Italian Renaissance studies.
Level of Study: Postdoctorate
Type: Residential fellowships
Value: US$50,000
Length of Study: 1 year
Frequency: Annual
Country of Study: Italy
No. of awards offered: 15
Application Procedure: Please check at http://itatti.harvard.edu/research/fellowships/i-tatti-fellowship.
Closing Date: October 15th
Funding: Private, trusts, foundation, individuals
No. of awards given last year: 15

Mellon Visiting Fellowship

Subjects: Fellowship to work at I Tatti for 3–6 months. The project must represent advanced research in the Italian Renaissance, broadly defined as the period ranging from the 13th to the 17th centuries. Subjects covered include the architecture, history, literature, material culture, music, philosophy, religion, science or visual arts of Italy. Applications would also be welcomed from candidates working on the transmission and circulation of ideas, objects, and people during the Renaissance, into and beyond the Italian peninsula or the historiography of the Italian Renaissance, including the rebirth of interest in the Renaissance in later periods.
Purpose: Advanced research in the Italian Renaissance.
Eligibility: Applicants must be Italian Renaissance scholars. The fellowship is designed for scholars from areas that have been under-represented at I Tatti. Preference will be given to junior scholars who teach, or plan to teach, in Asia, Islamic countries, Latin America, and the Mediterranean basin. Scholars working in North America, Northern Europe, Italy, France, or Australia are not eligible.
Level of Study: Doctorate, Postdoctorate
Type: Residential fellowships
Value: Up to $5,000 per month plus a one-time supplement (Max. $1,500) towards relocation expenses
Length of Study: 3–6 months
Frequency: Annual
Country of Study: Italy
No. of awards offered: Varies
Application Procedure: Please check at http://itatti.harvard.edu/research/fellowships/mellon-visiting-fellowship.
Closing Date: February 1st
Funding: Trusts
Contributor: The Andrew W. Mellon Foundation
No. of awards given last year: 5

THE VINAVER TRUST

45 Albert Street, Western Hill, Durham, DH1 4RJ, England
Tel: (44) 019 1380 8898
Email: geoffreybromiley@btinternet.com
Contact: Dr G.N. Bromiley, Secretary-Treasurer

The Eugène Vinaver Memorial Trust exists to promote research into Arthurian studies, as defined by the International Arthurian Society. It offers subventions to publishers to facilitate the publication of scholarly works; it also offers grants to postgraduate students pursuing research in the Arthurian field.

Barron Bequest

Subjects: Any field of Arthurian studies.
Purpose: To support postgraduate research in Arthurian studies.
Eligibility: Open to graduates of any university of the United Kingdom or the Republic of Ireland.
Level of Study: Postgraduate Research
Type: Grant
Value: Up to UK£1,250 towards academic fees
Length of Study: 1 year. Candidates may apply for further years on a basis of parity with those applying for the first time
Frequency: Annual
Country of Study: United Kingdom, Republic of Ireland
No. of awards offered: Varies
Application Procedure: For application details applicants must contact Professor J.H.M Taylor at the address given below.
Closing Date: April 30th
Funding: Private
Contributor: The Eugène Vinaver Memorial Trust
No. of awards given last year: 2
No. of applicants last year: 8

For further information contact:

Penruddock, Penrith, Cumbria, CA11 0QU, United Kingdom
Email: jane.taylor@durham.ac.uk
Contact: Professor Jane H M Taylor, Garth Head

VISUAL COMMUNICATIONS (VC)

120 Judge John Aiso Street Basement Level, Los Angeles, CA, 90012-3805, United States of America
Tel: (1) 213 680 4462
Fax: (1) 213 687 4848
Email: kennedy@vconline.org
Website: www.vconline.org
Contact: Kennedy Kabasares

Visual Communications (VC) is the premier Asian Pacific media arts center in the United States of America with a history of more than 30 years. It promotes intercultural understanding through the production, presentation and preservation of honest and sensitive stories about Asian Pacific people.

Armed with a Camera Fellowship

Subjects: Photography.
Purpose: To cultivate a new generation of Asian Pacific American media artists committed to preserving the legacy and vision of VC.
Eligibility: Open to candidates of Asian Pacific descent, who are 30 years of age or below, and are residents of California and have had previous work experience in the VC Film fest, Chili visions, any other VC exhibition and/or any film festival.
Level of Study: Professional development
Type: Fellowships
Value: US$500
Length of Study: 7 months
Frequency: Annual
No. of awards offered: 10
Closing Date: August 3rd

THE W.L. MACKENZIE KING MEMORIAL SCHOLARSHIPS

Curtis Building, University of British Columbia, 1822 East Mall, Vancouver, BC, V6T 1Z1, Canada
Tel: (1) 604 822 4564
Fax: (1) 604 822 8108
Email: mkingscholarships@law.ubc.ca
Website: www.mkingscholarships.ca
Contact: Professor Joost Blom

The Mackenzie King Scholarship Trust consists of two funds established under the will of the Right Honourable William Lyon Mackenzie King (1874–1950). Both scholarships are to support postgraduate study for graduates of Canadian universities.

Mackenzie King Open Scholarship

Subjects: All subjects.
Eligibility: Open to graduates of any Canadian university. Applicants should be persons of unusual worth or promise as awards are determined on the basis of academic achievement, personal qualities and demonstrated aptitudes. Consideration is also given to the applicant's proposed programme of postgraduate study.
Level of Study: Postgraduate
Type: Scholarship
Value: Canadian $10,000 but is subject to change
Length of Study: 1 year, non-renewable
Frequency: Annual
Country of Study: Any country
No. of awards offered: 1
Application Procedure: Applicants must complete an application form available from the website. Applications must be submitted to the Dean of Graduate Studies at the Canadian university from which the candidate most recently graduated.
Closing Date: February 1st
Funding: Private
No. of awards given last year: 1

Mackenzie King Travelling Scholarship

Subjects: International or industrial relations, including international aspects of law, history, politics and economics.
Purpose: To give Canadian students the opportunity to broaden their outlook and sympathies and to contribute in some measure to the understanding of the problems and policies of other countries.
Eligibility: Open to graduates of any Canadian university who propose to engage in postgraduate studies in the given fields in the United States of America or the United Kingdom.
Level of Study: Postgraduate
Type: Scholarship
Value: Canadian $11,000 but is subject to change
Length of Study: 1 year, non-renewable
Frequency: Annual
Study Establishment: Suitable institutions
Country of Study: United Kingdom or United States of America
No. of awards offered: 4
Application Procedure: Applicants must complete an application form available from the website. Applications must be submitted to the Dean of Graduate Studies at the Canadian university from which the candidate most recently graduated.
Closing Date: February 1st
Funding: Private
No. of awards given last year: 4

WALT DISNEY STUDIOS

Writing Fellowship Programme, 500 South Buena Vista Street, Burbank, CA, 91521-4016, United States of America
Tel: (1) 818 560 6894
Email: ABCWritingFellowship@disney.com
Website: www.disneyabctalentdevelopment.com

The Walt Disney Company is one of the largest media and entertainment corporations in the world. Founded on October 16, 1923, by brothers Walt and Roy Disney as a small animation studio, today it is one of the largest Hollywood studios and also owns 9 theme parks and several television networks, including the American Broadcasting Company (ABC).

The Walt Disney Studios and ABC Entertainment Writing Fellowship Program

Subjects: Feature film and television.
Purpose: To seek out and employ culturally and ethnically diverse new writers.
Eligibility: Open to all writers.
Level of Study: Professional development
Type: Fellowships

Value: Fellows will each be provided a salary of $50,000 for a one-year period tentatively
Length of Study: 1 year
Frequency: Annual
No. of awards offered: Up to 15
Application Procedure: Applicants can download the application form from the website.
Closing Date: June 23rd
Additional Information: Members with Writers Guild of America (WGA) credits are also eligible for this programme and can apply via Employment Access Department at 323 782 4648.

THE WARBURG INSTITUTE

University of London, Woburn Square, London, WC1H 0AB, England
Tel: (44) 20 7862 8949
Fax: (44) 20 7862 8955
Email: warburg@sas.ac.uk
Website: www.warburg.sas.ac.uk
Contact: Administrative Assistant

The Warburg Institute is concerned mainly with cultural history, art history and history of ideas, especially in the Renaissance. It aims to promote and conduct research on the interaction of cultures, using verbal and visual materials. It specializes in the influences of ancient Mediterranean traditions on European culture from the Middle Ages to the modern period.

Albin Salton Fellowship

Subjects: Cultural and intellectual history that led to the formation of a new worldview, understood in the broadest cultural, political and socioeconomic terms as Europe developed contacts with the world in the late medieval, renaissance and early modern periods and that world came into contact with Europe.
Purpose: To enable a young scholar spend 2 months at the Warburg Institute, pursuing research.
Eligibility: Fellowships are generally for young scholars in the early stages of their career; Candidates may be pre- or postdoctorates, but must have completed at least 1 year of research on their doctoral dissertation by the time they apply. Postdoctoral candidates, must normally have been awarded their doctorate within the preceding five academic years. If it was awarded before, then they should explain the reasons for any interruption in their academic career in a covering letter.
Level of Study: Doctorate, Postdoctorate
Type: Research fellowship
Value: £2,500
Length of Study: 2 months
Frequency: Annual
Study Establishment: The Warburg Institute, London
Country of Study: United Kingdom
No. of awards offered: 1
Application Procedure: Applications should be made by letter to the Director, enclosing a curriculum vitae, an outline of proposed research, particulars of grants received, if any, for the same project. The names and addresses of 2 or 3 academic referees who have agreed to support the application should also be submitted. Please visit www.warburg.sas.ac.uk/fellowships for full details.
Closing Date: Late November
Funding: Private
No. of awards given last year: 1

Brian Hewson Crawford Fellowship

Subjects: The classical tradition.
Purpose: To support research into any aspect of the classical tradition.
Eligibility: Fellowships are generally intended for scholars in the early stages of their career. Candidates may be pre or postdoctoral, but must have completed at least 1 year of research on their doctoral dissertation by the time they submit their application. Postdoctoral candidates must normally have been awarded their doctorate within the preceding 5 years. If their doctorate was awarded before this they should explain the reasons for any interruption in their academic career in a covering letter.
Level of Study: Doctorate, Postdoctorate
Type: Fellowship
Value: £2,500
Length of Study: 2 months
Frequency: Annual
Study Establishment: The Warburg Institute, London
Country of Study: United Kingdom
No. of awards offered: 1
Application Procedure: Applications should be made by letter to the Director enclosing a full curriculum vitae comprising name, date of birth, address (including email address) and present occupation, school and university education, degrees, teaching and research experience, list of publications, an outline of proposed research (of not more than two pages), particulars of grants received, if any, on the same subject and the names and addresses of 2 or 3 persons who have agreed to write, without further invitation, in support of the application. Please visit www.warburg.sas.ac.uk/fellowships for full details.
Closing Date: Late November
Funding: Private
No. of awards given last year: 1
Additional Information: Those employed as professor, lecturer or equivalent in a university or learned institution may normally hold an award only if they are taking unpaid leave for the whole of the period. The fellowship may not be held concurrently with another fellowship or award. Applications may be sent by post or fax, but not by e-mail.

Brill Fellowships at CHASE

Subjects: Any aspect of the relations between Europe and the Arab world from the Middle Ages to the 19th century.
Purpose: To carry out research on any aspect of the relations between Europe and the Arab world from the Middle Ages to the 19th century.
Eligibility: The fellowships are generally intended for scholars in the early stages of their careers. Candidates must have completed at least one year research on their doctoral dissertation by the time they submit their application for a fellowship and, if postdoctoral, must normally have been awarded their doctorate within the preceding 5 years. If their doctorate was awarded before this date, they should explain the reasons for any interruptions in their academic career in a covering letter.
Level of Study: Doctorate, Postdoctorate
Type: Fellowship
Value: £2,500 for 2 months; £3,500 for 3 months; £4,800 for 4 months
Length of Study: 2, 3, or 4 months
Frequency: Annual
Study Establishment: The Warburg Institute
Country of Study: United Kingdom
No. of awards offered: 1 or 2
Application Procedure: Applications should be made by letter to the Director enclosing a curriculum vitae along with an outline of proposed research project and details of grants received, if any for the same project. The names and addresses of 2 or 3 academic referees, who will write in support of the application should also be submitted. Please visit www.warburg.sas.ac.uk/fellowships for full details.
Closing Date: Late November
Contributor: Brill Publishing (Lelden) is sponsoring up to two annual research fellowships which have been made possible by the Sheikh Zayed Book Award for publishing, which Brill won in 2012

Frances A Yates Fellowships

Subjects: Any aspect of cultural and intellectual history with emphasis on the medieval and renaissance periods. Preference will be given to those areas of knowledge to which Dame Frances made a contribution.
Purpose: To promote research in any aspect of cultural and intellectual history.
Eligibility: Fellowships are generally intended for young scholars in the early stages of their career. Candidates may be pre- or postdoctorates, but must have completed at least one year of research on their doctoral dissertation before they apply. Postdoctoral candidates, must normally have been awarded their doctorate within the preceding five academic years. If their doctorate was awarded earlier, they should explain the reason for any interruption in their academic career in a covering letter.
Level of Study: Postdoctorate, Doctorate

Type: Fellowship
Value: For short-term fellowships: £2,500 for 2 months; 3,600 for 3 months; and £4,800 for 4 months
Length of Study: The long-term fellowship is 1–3 years, not normally renewable, and short-term fellowships are 2–4 months, non-renewable
Frequency: Annual
Study Establishment: The Warburg Institute, London
Country of Study: United Kingdom
No. of awards offered: Several short-term awards. Occasionally a long-term fellowship is offered.
Application Procedure: Applications should be made by letter to the Director, (enclosing) a curriculum vitae, outline of proposed research and particulars of grants received, if any, for the same subject. The names and addresses of three persons who have agreed to write, without further invitation, in support of the application should also be submitted. Please visit www.warburg.sas.ac.uk/fellowships for full details.
Closing Date: Late November
Funding: Private
No. of awards given last year: 7 (short-term)

Grete Sondheimer Fellowship

Subjects: Any aspect of cultural and intellectual history with the emphasis on the medieval and renaissance period.
Purpose: To promote research in any aspect of cultural and intellectual history.
Eligibility: The fellowships are generally intended for scholars in the early stages of their careers. Candidates must have completed at least 1 year research on their doctoral dissertation by the time they submit their application for a fellowship and, if postdoctoral, must normally have been awarded their doctorate within the preceeding 5 years. If their doctorate was awarded before this date, they should explain the reasons for any interruptions in their academic career in a covering letter.
Level of Study: Doctorate, Postdoctorate
Type: Fellowship
Value: UK£2,500
Length of Study: 2 months
Frequency: Annual
Study Establishment: The Warburg Institute, London
Country of Study: United Kingdom
No. of awards offered: 1
Application Procedure: Applications should be made by letter to the Director enclosing a curriculum vitae along with an outline of proposed research project and details of grants received, if any, for the same project. The names and addresses of 2 or 3 academic referees, who will write in support of the application should also be submitted. Please visit www.warburg.sas.ac.uk/fellowships for full details.
Closing Date: Late November
Funding: Private
No. of awards given last year: 1

Henri Frankfort Fellowship

Subjects: The intellectual and cultural history of the ancient Near East, with reference to society, art, architecture, religion, philosophy and science; the relations between the cultures of Mesopotamia, Egypt and the Aegean, and their influence on later civilizations.
Purpose: To promote research into the history of the ancient near east.
Eligibility: Fellowships are generally intended for young scholars in the early stages of their career. Candidates may be pre- or postdoctorates, but must have completed at least 1 year of research on their doctoral dissertation before they apply. Postdoctoral candidates must normally have been awarded their doctorate within the preceding five academic years. If their doctorate was awarded before this, candidates should explain the reason for any interruption in their academic career in a covering letter. Please visit www.warburg.sas.ac.uk/fellowships for full details.
Level of Study: Doctorate, Postdoctorate
Type: Fellowship
Value: £2,500 for 2 months and £3,600 for 3 months
Length of Study: 2–3 months
Frequency: Annual
Study Establishment: The Warburg Institute, London
Country of Study: United Kingdom
No. of awards offered: 1
Application Procedure: Applications should be made by letter to the Director (enclosing a curriculum vitae, outline of proposed research and particulars of grants received, if any, for the same subject). The names and addresses of two or three persons who have agreed to write, without further invitation, in support of the application should also be submitted.
Closing Date: Late November
Funding: Private
No. of awards given last year: 1

WARWICK BUSINESS SCHOOL

University of Warwick, Coventry, CV4 7AL, England
Tel: (44) 24 7652 4306
Fax: (44) 24 7652 3719
Email: enquiries@wbs.ac.uk
Website: www.wbs.ac.uk
Contact: Mr Alex Keeline, Marketing Manager

Warwick Business School is an international school with 330 staff and 4,500 students from 130 countries worldwide, and is accredited with management associations in North America, Europe and the United Kingdom. Its high-calibre research feeds into top-quality teaching on undergraduate, specialist Master's, doctoral, and MBA degrees.

ESRC Awards (For Doctoral Study)

Subjects: Finance, industrial, employment relations, knowledge management and networks, information systems and management, marketing, strategy and operational research. See the ESRC website for up-to-date details.
Purpose: To allow the candidates to pursue the WBS Doctoral Programme.
Eligibility: Open to students applying for entry into the full-time WBS Doctoral Programme subject to the research being relevant. Home students (tuition fees + stipend) and EU students (tuition fees only).
Level of Study: Doctorate
Value: All tuition fees and stipend set at UK Research Council. Refer to the Warwick Business School website for further details
Length of Study: 4 years (full-time)
Frequency: Annual
Study Establishment: Warwick Business School
Country of Study: United Kingdom
No. of awards offered: 3
Application Procedure: Via online application system.
Closing Date: July 31st
Funding: Government
No. of awards given last year: 3

ESRC PhD in Finance

Subjects: Finance.
Purpose: To allow candidates to pursue the WBS Doctoral Programme.
Eligibility: Open to students applying for entry into the full-time Warwick Business School PhD in Finance programme.
Level of Study: Doctorate
Value: Studentships will cover fees and a maintenance grant
Frequency: Annual
Study Establishment: Warwick Business School
Country of Study: United Kingdom
No. of awards offered: ESRC Quota awards
Application Procedure: Procedures vary accoding to the scholarship relevant to the area of interest.
Closing Date: April 30th
Funding: Government
Additional Information: Please check ESRC website for latest updates.

Warwick Busines School Scholarships (For Masters Study)

Subjects: Finance and economics, management, financial maths, marketing and strategy, international employment relations, information systems management and innovation management science and operational research, business analytics and consulting, management

and organizational analysis, industrial relations and managing human resources, accounting and finance, and finance.
Purpose: To allow candidates to pursue full-time and part-time specialist Masters courses.
Eligibility: Open to homeley students applying for full-time and part-time masters courses. Please refer to the website for eligibility criteria.
Level of Study: Postgraduate
Type: Scholarship
Value: UK£2,500–10,000
Length of Study: 1 year (full-time), 2 years (part-time)
Frequency: Annual
Study Establishment: Warwick Business School
Country of Study: United Kingdom
No. of awards offered: Varies
Application Procedure: All applicants to the courses will automatically be considered for the scholarships.
Closing Date: Rolling admissions
No. of awards given last year: 35
Additional Information: Please refer to the website for details.

Warwick Business School PhD in Finance Scholarships (for Doctoral Study)

Subjects: Finance.
Purpose: To allow candidates to pursue the WBS Doctoral Programme.
Eligibility: Open to students applying for entry into the full-time Warwick Business School PhD in Finance programme.
Level of Study: Doctorate
Value: Approx. £17,500 per year. Refer the Warwick Business School website for further details
Length of Study: 4 years (full time)
Study Establishment: Warwick Business School
Country of Study: United Kingdom
No. of awards offered: 3
Application Procedure: All applicants will be considered subject to receiving no other scholarship.
Closing Date: July 31st may consider strong applications after the deadline
No. of awards given last year: 3

Warwick Business School Scholarships (For Doctoral Study)

Subjects: Finance, industrial, employment relations, knowledge mangement, networks, operational research, information systems and management, marketing, and strategy.
Purpose: To allow candidates to pursue the WBS Doctoral Programme.
Eligibility: Open to students applying for entry into the full-time Warwick Doctoral Programme, subject to research area being relevant.
Level of Study: Doctorate
Type: Bursary and scholarship
Value: Approx. £17,500 per year. Refer to the Warwick Business School website for further details
Length of Study: 4 years (full-time)
Study Establishment: Warwick Business School
Country of Study: United Kingdom
No. of awards offered: 5
Application Procedure: All applicants will be considered subject to receiving no other scholarship.
Closing Date: July 31st may consider strong applications after the deadline.
No. of awards given last year: 5

Warwick Business School Scholarships (For Executive MBA)

Subjects: Warwick Executive MBA.
Purpose: To allow candidates to pursue the Warwick Executive MBA course.
Eligibility: Open to all applicants applying for the Warwick Executive MBA.
Level of Study: MBA
Type: Scholarship
Value: The scholarships provide either 25 or 50 per cent of the annual tuition fee (for the first year only)
Length of Study: 2.5–3 years
Study Establishment: Warwick Business School
Country of Study: United Kingdom
No. of awards offered: 14
Application Procedure: All applicants to the programme will automatically be considered for the scholarship.
Closing Date: Please see the website
No. of awards given last year: 6

Warwick Business School Scholarships (For Full-Time MBA Study)

Subjects: MBA by full-time study
Purpose: To allow candidates to pursue the Warwick MBA course.
Eligibility: Open to students applying for full-time.
Level of Study: MBA, Professional development
Type: Scholarship
Value: The scholarships provide either 25 or 50 per cent of the tuition fees, for full-time MBA applicants
Length of Study: 1 year
Study Establishment: Warwick Business School
Country of Study: United Kingdom
No. of awards offered: 25
Application Procedure: All applicants to the programme will automatically be considered for the scholarship.
Closing Date: Full-time July 31st
No. of awards given last year: 22

Warwick Business School Scholarships (for Global Energy MBA)

Subjects: Warwick Global Energy MBA.
Purpose: To allow candidates to pursue the Warwick Global Energy MBA.
Eligibility: Open to all students applying for Global Energy MBA.
Level of Study: MBA
Type: Scholarship
Value: The scholarships provide up to 50 per cent of the tuition fees, (for the first year only) up to a maximum of UK£10,400.
Length of Study: 3 years
Study Establishment: Warwick Business School
Country of Study: United Kingdom
No. of awards offered: 4
Application Procedure: All applicants to the programme will automatically be considered for the scholarship.
Closing Date: March 5th
No. of awards given last year: 2

Warwick Business School Scholarships (For MBA Study via Distance Learning)

Subjects: MBA by distance learning
Purpose: To allow candidates to pursue the distance-learning MBA course.
Eligibility: All self funding candidates for distance learning applying for the programme are eligible.
Level of Study: MBA
Type: Scholarship
Value: Up to 50 per cent of the annual tuition fee (for the first year only)
Length of Study: 2.5–3 years
Study Establishment: Warwick Business School
Country of Study: United Kingdom
No. of awards offered: 16
Application Procedure: All applicants to the Distance Learning MBA will automatically be considered.
Closing Date: Please see the website

WASHINGTON UNIVERSITY

Graduate School of Arts and Sciences, Box 1187, 1 Brookings Drive, St Louis, MO, 63130-4899, United States of America
Tel: (1) 314 935 5000
Fax: (1) 314 935 4887
Email: scholarships@wsu.edu
Website: www.artsci.wustl.edu
Contact: Ms Nancy P Pope, Associate Dean

Washington University has a diverse offering of events, disciplines, people, and resources that create unlimited possibilities for discovery and growth. Arts & Sciences signals a curriculum and place, a core of teaching, learning, and discovery at Washington University.

Mr and Mrs Spencer T Olin Fellowships for Women

Subjects: Architecture, arts and humanities, business, engineering, mathematics and science.
Purpose: To encourage women of exceptional promise to prepare for professional careers.
Eligibility: Open to female graduates of a baccalaureate institution in the United States of America who plan to prepare for a career in higher education or the professions. Applicants must meet the admission requirements of their graduate or professional school at Washington University. Preference will be given to those who wish to study for the highest earned degree in their chosen field, do not already hold an advanced degree, and are not currently enrolled in a graduate or professional degree programme.
Level of Study: Doctorate, Graduate
Type: Fellowship
Value: Provides $13,000–20,000 plus tuition per year
Length of Study: 1 year, renewable for up to 4 years, or until the completion of the degree programme, whichever comes first
Frequency: Annual
Study Establishment: Washington University
Country of Study: United States of America
No. of awards offered: Approx. 10
Application Procedure: Applicants must complete an application form. Finalists must be interviewed on campus at the expense of the University.
Closing Date: February 1st
Funding: Private
Contributor: The Monticello College Foundation
No. of awards given last year: 11
No. of applicants last year: 280
Additional Information: Candidates must also make concurrent application to the department or school of Washington University in which they plan to study.

Washington University Chancellor's Graduate Fellowship Program

Subjects: Arts and sciences, business, engineering and social work.
Purpose: To encourage students who are interested in becoming college or university professors, and who bring diversity to the campus environment.
Eligibility: Open to doctoral candidates. Applicants must meet the admission requirements of their graduate or professional school at Washington University, and provide evidence that they will contribute to diversity on our campus.
Level of Study: Doctorate, Graduate
Type: Fellowship
Value: Doctoral candidates will receive full tuition plus US$25,250 stipend and allowances
Length of Study: 5 years, subject to satisfactory academic progress
Frequency: Annual
Study Establishment: Washington University
Country of Study: United States of America
No. of awards offered: 5–6
Application Procedure: Applicants must complete an on-line application form. Finalists will be interviewed on the campus at the expense of the University.
Closing Date: January 25th
No. of awards given last year: 5
No. of applicants last year: 185
Additional Information: The fellowship includes other Washington University programmes providing final disciplinary training for prospective college professors.

THE WELDER WILDLIFE FOUNDATION

PO Box 1400, Sinton, TX, 78387, United States of America
Tel: (1) 361 364 2643
Fax: (1) 361 364 2650
Email: welderfoundation@welderwildlife.org
Website: www.welderwildlife.org
Contact: Terry Blankenship, Director

The Welder Wildlife Foundation is a private, non-profit foundation established in 1954. The Foundation has gained international recognition through its research programme. The primary purpose of the Foundation is to conduct research and education in the field of wildlife management and conservation and other closely related fields. As a private foundation their purpose and operation remain unhindered by outside political or institutional pressures.

Rob and Bessie Welder Wildlife Foundation's Graduate Research Scholarship Program

Subjects: Animal behaviour, biology, botany, conservation education, ecology, mammalogy, ornithology, parasitology, range science, veterinary pathology and wildlife sciences.
Purpose: To conduct research and education in the field of wildlife management and conservation and other closely related fields.
Eligibility: Applicant must have a minimum GRE score of 1100 and 3.0 GPA in the last two years of undergraduate or graduate work
Level of Study: Doctorate, Master of Science
Type: Scholarship
Value: $1,300 per month for full-time M.S. candidates and $1,400 per month for full-time PhD candidates to cover living costs, tuition, fees, and books.
Frequency: Annual
Country of Study: United States of America
Application Procedure: Applications and abbreviated proposals may be submitted in letter form and must be signed by a qualified member of the faculty at the parent university.
Closing Date: October 1st
Funding: Foundation, private
No. of awards given last year: 2
No. of applicants last year: 10

WELLBEING OF WOMEN

27 Sussex Place, London, NW1 4SP, England
Tel: (44) 20 7772 6400
Fax: (44) 20 7724 7725
Email: wellbeingofwomen@rcog.org.uk
Website: www.wellbeingofwomen.org.uk
Contact: Mrs Philip Matusavage, Research Manager

Wellbeing of Women is the only UK charity funding vital research into all aspects of reproductive health in three key areas; gynaecological cancers, pregnancy and birth, and quality of life problems.

Wellbeing of Women Entry-Level Scholarship

Subjects: All subjects of relevance to obstetrics, gynaecology.
Purpose: To provide Pump-Priming funds to enable trainees to be exposed to a research environment to obtain pilot data for bids for definitive funding.
Eligibility: Work must be undertaken in the UK. Applications must be from individuals who have not previously been involved in substantial research projects.
Level of Study: Graduate, Postgraduate
Type: Scholarship
Value: A maximum of UK£20,000
Length of Study: 1–3 years
Frequency: Annual
Study Establishment: A hospital or university
Country of Study: United Kingdom
No. of awards offered: Varies depending on the amount of disposable income
Application Procedure: Applicants must write for details or access the website.
Closing Date: September
Funding: Commercial, individuals, private, trusts
No. of awards given last year: 3
No. of applicants last year: 13

Wellbeing of Women Project Grants

Subjects: All subjects of relevance to obstetrics and gynaecology.
Purpose: To fund research projects.
Eligibility: Open to specialists in any obstetrics/gynaecology interrelated field.
Level of Study: Postdoctorate, Research

Type: Grant
Value: A maximum of UK£200,000 over 3 years
Length of Study: 1–3 years
Frequency: Annual
Study Establishment: A hospital or university
Country of Study: United Kingdom
No. of awards offered: Varies depending on the amount of disposable income
Application Procedure: Applicants must write for details or access the website.
Closing Date: February
Funding: Commercial, individuals, private, trusts
No. of awards given last year: 5
No. of applicants last year: 104

Wellbeing of Women Research Training Fellowship

Subjects: All subjects of relevance to obstetrics and gynaecology.
Purpose: To fund training in basic science or clinical research techniques and methodology for a medical graduate embarking upon a career in obstetrics and gynaecology.
Eligibility: Research training fellowships are to be undertaken in the UK.
Level of Study: Postdoctorate, Postgraduate, Research
Type: Fellowship
Value: The upper limit for this award is UK£200,000
Length of Study: 1–3 years
Frequency: Annual
Study Establishment: A hospital or university
Country of Study: United Kingdom
No. of awards offered: Varies depending on the amount of disposable income
Application Procedure: Applicants must write for details or access the website.
Closing Date: September
Funding: Individuals, private, trusts
No. of awards given last year: 1
No. of applicants last year: 18

WELLCHILD INTERNATIONAL

16 Royal Crescent, Cheltenham, Gloucester, GL50 3DA, England
Tel: (44) 0845 458 8171
Fax: (44) 12 4253 0008
Email: info@wellchild.org.uk
Website: www.wellchild.org.uk
Contact: The Administrator

WellChild is committed to getting sick children better, whatever their illness, through care, support and research.

WellChild Pump-Priming Grants

Subjects: Diseases in children.
Purpose: To enable a student to run a pilot study.
Level of Study: Doctorate, Postdoctorate, Postgraduate
Type: Grant
Value: Up to UK£30,000
Frequency: Annual
Study Establishment: Accredited institute in the United Kingdom
Country of Study: United Kingdom
Application Procedure: Applicants must download application and guidelines from the website.
Closing Date: September 26th
Funding: Private
Additional Information: Assessment by peer review and the Scientific Medical Advisory Committee.

WellChild Research Fellowships

Subjects: Diseases in children.
Purpose: To support clinicians working towards the development of their own research projects.
Level of Study: Doctorate, Postdoctorate, Postgraduate
Type: Fellowship
Value: £183,500
Length of Study: Up to 3 years
Frequency: Annual
Study Establishment: Accredited institute in the United Kingdom
Country of Study: United Kingdom
Application Procedure: Applicants must download application and guidelines from the website.
Closing Date: September 26th
Funding: Private
Additional Information: Assessment by peer review and the Scientific Medical Advisory Committee.

WELLCOME TRUST

Gibbs Building, 215 Euston Road, London, NW1 2BE, England
Tel: (44) 20 7611 8888
Fax: (44) 20 7611 8545
Email: contact@wellcome.ac.uk
Website: www.wellcome.ac.uk

The Wellcome Trust's mission is to foster and promote research with the aim of improving human and animal health. The Trust funds most areas of biomedical research and funds research in the history of medicine, biomedical ethics and public engagement of science.

Arts Awards

Subjects: The scheme aims to: stimulate interest, excitement and debate about biomedical science through the arts; examine the social, cultural, and ethical impact of biomedical science; support formal and informal learning; encourage new ways of thinking; encourage high quality interdisciplinary practice and collaborative partnerships in arts, science and/or education practice. All art forms are covered by the programme: dance, drama, performance arts, visual arts, music, film, craft, photography, creative writing or digital media. The Trust invites applications for projects which engage adult audiences and/or young people.
Purpose: Arts Awards support imaginative and experimental arts projects that investigate biomedical science.
Eligibility: The scheme is open to a wide range of people including, among others, artists, scientists, curators, filmmakers, writers, producers, directors, academics, science communicators, teachers, arts workers and education officers. Applicants are usually affiliated to organisations, but can apply as individuals. Organisations might include: museums and other cultural attractions; arts agencies; production companies; arts venues; broadcast media; schools; local education authorities; universities and colleges; youth clubs; community groups; research institutes; the NHS; and science centres. Partnership projects (between different people and organisations, e.g. scientists and ethicists, educators and artists) are welcomed. If this is the first time an organisation is applying to the Wellcome Trust an eligibility assessment will be carried out. For this assessment, the following documentation from the applying organisation should be submitted: articles of association; audited accounts from the previous two years; details of similar projects/grant funding received; confirmation that no funding has been received or is scheduled to be received from any tobacco company.
Level of Study: Research
Type: Award
Value: Funding can be applied for at two levels: (1)Small to medium-sized projects (up to and including £30,000). This funding can either be used to support the development of new project ideas, deliver small-scale productions or workshops, investigate and experiment with new methods of engagement through the arts, or develop new collaborative relationships between artists and scientists and (2) Large projects (above £30,000). This funding can be used to fund full or part production costs for large-scale arts projects that aim to have significant impact on the public's engagement with biomedical science. We are also interested in supporting high-quality, multi-audience, multi-outcome projects.Applicants can apply for any amount within the above boundaries, for projects lasting a maximum of three years
Frequency: Annual
Application Procedure: Application form for awards up to and including £30,000, preliminary application form for awards over £30,000.
Closing Date: For small to medium-sized projects (up to and including £30,000) deadlines are January 28th and April 28th. For large projects (above £30,000) the deadline is March 25th

Funding: Trusts
Additional Information: Applicants must be based in the UK or the Republic of Ireland and the activity must take place in the UK or the Republic of Ireland.

Broadcast Development Awards

Subjects: We are interested in funding individuals and organisations with brilliant early-stage ideas for TV, radio or new media projects. Our funding will enable these ideas to be developed into high impact, well-researched proposals to be utilised in securing a broadcast platform and/or further funding. A successful project would primarily be aimed at a mainstream UK and/or Republic of Ireland audience in the first instance, although the subject matter can be international.
Purpose: To support the development of broadcast proposals in any genre that engages the audience with issues around biomedical science in an innovative, entertaining and accessible way.
Eligibility: The proposal must primarily be aimed at a mainstream UK and/or Republic of Ireland audience in the first instance but the subject matter can be international. Applicants are usually affiliated to organisations, but can apply as individuals. The scheme is open to broadcast professionals and other organisations or individuals working on broadcast projects. Partnership between broadcasters and other professionals such as scientists, ethicists, educators etc are especially welcomed.
Level of Study: Research
Type: Award
Value: Up to £10,000, for a maximum of 1 year
Length of Study: 1 year
Frequency: Biannual
Application Procedure: Candidates should complete and submit an application form by the published deadline.
Closing Date: April 26th, July 19th and October 18th
Funding: Trusts
Additional Information: Applicants must be based in the UK or the Republic of Ireland, although other members of the project team can be based overseas.

Career Re-entry Fellowships

Subjects: Biomedical Science.
Purpose: This scheme is for postdoctoral scientists who have recently decided to recommence a scientific research career after a continuous break of at least two years.
Eligibility: The awards are open to individuals with a relevant connection to the European Economic Area (EEA). You should be a research scientist with at least two years' postdoctoral experience and intend to be based in a UK or Republic of Ireland organisation. You must have had a continuous career break of at least two years and should have either a strong research track record (if applying for up to four years' support) or demonstrated the potential for a strong research career prior to your break (if applying for two years' support). A two-year fellowship should provide sufficient training support to consolidate your potential. The proposed research should fall within the Wellcome Trust's normal funding remit. Resubmissions are not normally encouraged. If your application has been unsuccessful, please contact the Office for advice. You must have an eligible sponsoring laboratory in the UK or Republic of Ireland that will administer the fellowship for the duration of the award.
Level of Study: Professional development
Type: Fellowship
Value: It provides support that includes the fellow's salary, as determined by the host institution with an additional Trust enhancement, and Research expenses (consumables, animals, travel support to attend scientific meetings).
Frequency: Biannual
Application Procedure: A preliminary application form [Word 92kB] should be completed and submitted by the published deadline. It should be sent electronically (as a Word document), with the requested accompanying information, to the appropriate funding stream at the Trust (see website).
If successful, you will be shortlisted for interview.
Closing Date: May 20th, 5pm (preliminary applications); July 26th, 5pm (full applications); and November13th-15th (shortlisted candidate interviews)
Funding: Trusts

Clinical PhD Programmes

Subjects: Biomedical science.
Purpose: This is a flagship scheme aimed at supporting the most promising medically qualified clinicians who wish to undertake rigorous research training.
Eligibility: You should have demonstrated the potential to pursue a career as an academic clinician. It is anticipated that many applicants will have already commenced their specialist training, but this is not essential.
Level of Study: Postgraduate, Research
Type: Grant
Value: The duration may vary from Programme to Programme, but each provides a clinical salary, PhD registration fees at UK/EU student rate, research expenses, contribution towards travel and contribution towards general training costs.
Frequency: Annual
No. of awards offered: 7
Application Procedure: Students are recruited annually by the individual Programmes. Recruitment begins in the preceding January. If you are interested in applying you should contact the relevant Programme directly. Please see website for more details.
Closing Date: Varies
Funding: Trusts

Doctoral Studentships

Subjects: Medical history and humanities.
Purpose: This scheme enables scholars to undertake up to three years of full-time research on a history of medicine topic leading to a doctoral degree at a university in the UK or Republic of Ireland.
Eligibility: You should hold a Master's in the history of medicine or a Master's with strong emphasis on the history of medicine. The proposed project must be on a history of medicine topic. If specialist language skills are essential to undertake the research, a Master's in the language required may be acceptable (classical languages, Arabic, Chinese, etc.). Your application must be sponsored by a senior member of the department, unit or institute, or History of Medicine grantholder (current or former), who would supervise you if an award were made. Applications must be submitted through the host institution.
Level of Study: Postgraduate, Research
Value: Support is provided for up to three years, and includes: the student's stipend; a set amount to cover conference travel, research expenses and, where justified, the cost of overseas fieldwork; all compulsory university and college fees at the UK/Irish/Dutch home postgraduate student level; fees at the overseas rate will not be provided; institutions sponsoring candidates are expected to provide laptops and PCs as part of their postgraduate research-training infrastructure
Application Procedure: Preliminary applications should be made by e-mail or post by the published deadline, and should include: a brief CV with details of the Master's degree held; details of the research proposed (maximum of one page); a letter of support from the head of the department in which you will be working (this can be sent under separate cover); a letter of support from the supervisor.
Closing Date: April 2nd
Funding: Trusts

Four-year PhD Studentship Programmes

Subjects: Biomedical science.
Purpose: This is a flagship scheme aimed at supporting the most promising students to undertake in depth postgraduate training. Supporting specialised training provided in a range of important biomedical research areas:

1. developmental biology and cell biology
2. genetics, statistics and epidemiology
3. immunology and infectious disease
4. molecular and cellular biology
5. neuroscience
6. physiological sciences
7. structural biology and bioinformatics.

Eligibility: You should be a student who has, or expects to obtain, a first- or upper-second-class honours degree or equivalent.
Level of Study: Postgraduate, Research
Value: A stipend, PhD registration fees at UK/EU student rate, contribution towards laboratory rotation expenses in the first year,

research expenses for years two to four, contribution towards travel and contribution towards transferable-skills training.
Length of Study: Support provided for 4 years
Frequency: Various
No. of awards offered: Support 27 Programmes based in centres of excellence throughout the UK
Application Procedure: Students are recruited annually by the individual Programmes for uptake in October each year. Recruitment begins in the preceding December. If you are interested in applying you should contact the relevant Programme directly. Please see website for details.
Closing Date: October
Funding: Trusts

Health Innovation Challenge Fund

Purpose: This is a five-year parallel funding partnership between the Wellcome Trust and the Department of Health to stimulate the creation of innovative healthcare products, technologies and interventions, and facilitate their development for the benefit of patients in the NHS and beyond.
Eligibility: Please note that the 'Lead applicant' for all HICF awards must be a UK organisation or company.The following types of organisation (singly or in collaboration) will be eligible for funding: NHS organisations (including NHS Trusts and NHS Foundation Trusts), and equivalent UK authorities; universities, and research institutes and not-for profit organisations; start-up companies founded to capture and develop intellectual property of relevance to healthcare; biotechnology, pharmaceutical, bioinformatics, engineering or other companies. A collaboration between two or more of the entities detailed above is also eligible and encouraged where it strengthens the overall proposal.
Type: Award
Application Procedure: Applicants should submit a preliminary application (see the forms and guidance tab on this page) including the following information: an outline of the work packages that are to be undertaken using Wellcome Trust/Department of Health funding including details of specific milestones, objectives and deliverables; current validation of the concept, how it addresses a medical need, position on patient management pathway or disease algorithm; downstream route to launch, market introduction and adoption; sensitivity or risk analysis for the major hurdles; an overview of intellectual property and regulatory approval issues; eventual financial sustainability of the product line; an approximate breakdown of costs; justification for requesting Wellcome Trust/Department of Health funds - if the applicant is a company; details of all information which an applicant considers commercially sensitive or confidential.
Closing Date: September 2nd (Preliminary application); October mid (Shortlisting of proposals); January 9th (Full applications submitted); April 12th (Funding decisions).
Funding: Trusts
Additional Information: The Health Innovation Challenge Fund (HICF) is a £100 million, five-year parallel funding partnership between the Wellcome Trust and the Department of Health. The funders are collaborating to stimulate the creation of innovative healthcare products, technologies and interventions, and facilitate their development for the benefit of patients in the NHS and beyond. The HICF will have a succession of thematic calls for proposals, each selected to focus on unmet needs in healthcare relevant to the NHS, and will support innovative developments that are within three to five years of launch or adoption.

Integrated Training Fellowships for Veterinarians

Subjects: Biomedical science.
Purpose: This scheme provides support for veterinary graduates and undergraduates to develop a career in veterinary research, by providing funding to obtain a PhD and continue clinical training towards a relevant clinical certificate, diploma or postgraduate pathology qualification. The PhD should be laboratory based and may be undertaken within the veterinary school, but applications are particularly encouraged where the PhD is based in, or involves collaboration with, a basic biomedical science laboratory.
Eligibility: You must have completed, or be about to complete, a first degree in veterinary medicine or veterinary science (e.g. BMedVet, BMedSci, VetMB). An intercalated BSc is desirable and some experience in clinical practice, together with evidence of an ongoing interest in a research career, such as project work and/or summer school attendance, would also be advantageous. You are expected to undertake a research project that balances the provision of training with the opportunity to advance knowledge in a given area. Under this scheme, there is also the possibility of support for PhDs to be held during an undergraduate veterinary degree. Support for post graduation clinical training can also be provided. Please talk to your course supervisor in the first instance and then contact the Trust to discuss further. The awards are open to individuals with a relevant connection to the European Economic Area.
Level of Study: Postgraduate, Research
Type: Fellowship
Value: (1) A basic salary as determined by the host institution (2) Research expenses for three years (consumables, small pieces of equipment, animals where necessary) (3) Training (requests must be justified) (4)A travel allowance to attend scientific meetings (this is automatically provided as part of an award).
Length of Study: Up to 6 years
Frequency: Annual
Application Procedure: A preliminary application form should be completed and submitted by the published deadline. An electronic copy of the form (as a Word document) and each accompanying document should be emailed to vets@wellcome.ac.uk. One hard copy of the form (including signatures), should be sent to our postal address, marked for the attention of Veterinary Fellowships.
Please note that all Word documents will need to be sent in Word 2003 (or an earlier version). Successful candidates will be invited to submit a full application, to be returned to the Trust no later than the published deadline.
Closing Date: September 1st
Funding: Trusts

Intermediate Clinical Fellowships

Subjects: Biomedical science.
Purpose: This scheme is for medical, dental, veterinary or clinical psychology graduates who have had an outstanding start to their research career. It will enable successful candidates to continue their research interests at a postdoctoral level in an appropriate unit or clinical research facility.
Eligibility: The award is open to individuals with a relevant connection to the EEA. You should have previously undergone a period of research training and will have completed, or be about to complete, a higher degree.
You should have completed general professional training as defined by the relevant college.
1. Medical and dental candidates should either have a National Training Number (NTN) or Certificate of Completion of Specialist Training (CCST) or equivalent.
2. Veterinary candidates should have a degree in veterinary medicine (e.g. BVSc, BVM&S, BVMS, BVetMed, VetMB) and some experience in clinical practice and will have completed, or be about to complete, a higher research degree (preferably a PhD).
3. GPs are advised to contact the office to clarify their eligibility.
4. Clinical Psychologists must have obtained a professional Doctorate-level qualification in Clinical Psychology accredited by the British Psychological Society.
Level of Study: Postgraduate, Research
Type: Fellowship
Value: Fellowships are for up to four or five years, depending on situation. They provide research expenses (consumables, travel, support to attend scientific meetings) and the fellow's salary, set by the host institution according to age and experience. Requests for specific items of equipment, where relevant, may be considered, and research or technical assistance may be requested. However, a laboratory appropriate to the research proposed should be selected, and the necessary facilities required for the proposed research must be available to the candidate. Funding for a period of research abroad may be requested if scientifically justified, and we provide appropriate allowances for fellows based overseas
Frequency: Annual
Application Procedure: A preliminary application form [Word 161kB] should be completed and submitted at any time before the appropriate deadline. It should be sent electronically (as a Word document) to the appropriate funding stream at the Trust (see website). If your preliminary application is successful, you will be invited to submit a full

application by the published deadline. This will be peer reviewed and considered by the relevant Funding Committee. Shortlisted candidates will subsequently be invited to attend for interview at the Trust.
Closing Date: March 8th
Funding: Trusts

Intermediate Fellowships in Public Health and Tropical Medicine

Subjects: Biomedical science.
Purpose: This scheme enables high-calibre, mid-career researchers from low- and middle-income countries to establish an independent research programme. Fellows must be based primarily in a low- and middle-income country. Research projects should be aimed at understanding and controlling diseases (either human or animal) of relevance to local, national or global health. This can include laboratory based molecular analysis of field or clinical samples, but projects focused solely on studies in vitro or using animal models will not normally be considered under this scheme.
Eligibility: Applications are only accepted in the Public Health and Tropical Medicine Interview Committee remit. This covers research on infectious and non-communicable diseases within the fields of public health and tropical medicine that is aimed at understanding and controlling diseases (either human or animal) of relevance to local, national or global health. You must be a national or legal resident of a low- and middle-income country and should be either:
1. A graduate in a subject relevant to public health or tropical medicine (e.g. biomedical or social science, veterinary medicine, physics, chemistry or mathematics) with a PhD and three to six years' postdoctoral experience, or
2. A medical graduate with a higher qualification equivalent to membership of the UK Royal Colleges of Physicians (i.e. qualified to enter higher specialist training) or recognised as a specialist within a relevant research area, with three to six years' research experience.
You must have a relevant high-quality publication record and show potential to become a future scientific leader.
Applicants who do not have a PhD but who are educated to first degree or Master's level and have extensive research experience, as evidenced by their publication record, may be considered.
Level of Study: Postgraduate, Research
Type: Fellowship
Value: Fellowships are for up to five years (non-renewable) and provide support that includes: a basic salary for the fellow; research expenses (e.g. consumables, equipment, collaborative travel, research assistance, technical support); training costs where appropriate and justified; an inflation/flexible funding allowance and support to attend scientific meetings. Contributions to costs of the project which are directly incurred by the overseas institution may be provided.
Frequency: Three times per year
Application Procedure: You must complete and submit a preliminary application form [Word 236kB] by the published deadline. The form should be emailed to phatic@wellcome.ac.uk. Completed forms will normally be assessed within one month of the preliminary deadline. If the preliminary application meets the scheme's requirements, a full application will be invited.
Closing Date: March 1st (preliminary application), May 21st (Full application), December 25th-27th (Shortlisted candidate interviews)
Funding: Trusts

International Engagement Awards

Subjects: International Engagement Awards support projects that aim to achieve some or all of the following: to strengthen the capacity of people in low- and middle-income countries to facilitate public engagement with health research; to stimulate dialogue about health research and its impact on the public in a range of community and public contexts in low- and middle-income countries; to investigate and test new methods of engagement, participation, communication or education around health research; to promote collaboration on engagement projects between researchers and community or public organisations; to support Wellcome Trust funded researchers in low- and middle-income countries in engaging with the public and policy makers.
Projects could involve: communities and members of the public (particularly those affected by or involved in health research); science communicators, health and science journalists; healthcare professionals, educators, field workers, community workerspolicy and decision makers.
Purpose: Engaging with global health research.
Eligibility: The scheme is open to a wide range of people, including media professionals, educators, science communicators, health professionals and researchers in bioscience, health, bioethics and history. Partnership projects (between different people and organisations, e.g. scientists and media professionals, ethicists and community workers) are welcomed. Applicants must be based in listed low- and middle-income countries or in the UK working with partners in the low- and middle-income countries. The activity must primarily take place in one or more low- and middle-income countries and the primary goal must be to involve participants or engage audiences located in low- and middle-income countries. Applicants from listed restructuring countries in Europe and Asia are not eligible. We can only accept applications in the English language but we welcome projects that bring together people from different backgrounds who speak diverse languages. All projects must involve engagement with health research. Projects dealing purely with development research not related to health are not eligible. Please note also, that the scheme is not intended to support standard delivery of health education and promotion which does not focus on health research or involve health researchers. Applicants must be affiliated to organisations or institutions. Organisations might include: media organisations, research centres or research groups, community-based development organisations, education organisations. The International Engagement Awards will not fund traditional scientist-led health research. We may consider an application for participatory health research. This is research in which participants are supported to own and shape a research process, setting their own research questions and directing the research process. This type of research should not look like a consultatory exercise or health education but should aim to be collaborative process of enquiry in which the analysis is conducted and findings can be used by all participating parties. This could lead into circular processes of research and action.
Level of Study: Research
Type: Award
Value: Up to £30,000 for projects lasting a maximum of 3 years
Frequency: Biannual
Application Procedure: Please contact the International Engagement Awards office well in advance of the deadline to request an application form and to confirm the eligibility of your project.
Closing Date: February 8th
Funding: Trusts

International Senior Research Fellowships

Subjects: Biomedical science.
Purpose: This scheme supports outstanding researchers, either medically qualified or science graduates, who wish to establish a research career in an academic institution in selected European countries - Croatia, Czech Republic, Estonia, Hungary, Poland, Slovakian Republic and Slovenia.
Eligibility: You should have between five and ten years' research experience at a postdoctoral level or clinical equivalent and have a substantial record of publications in your chosen area of research in leading international journals. Your proposed research must be conducted at an academic institution in Croatia, the Czech Republic, Estonia, Hungary, Poland, Slovakian Republic or Slovenia. You need not be a national of the country in which you wish to hold the fellowship. We usually expect candidates to have spent a significant period of their postdoctoral (or equivalent) research career working outside their chosen country. However proposals can also be considered from those who have pursued successful careers entirely in-country. Applications are particularly encouraged from outstanding scientists working outside their own countries who wish to return home.
Level of Study: Postgraduate, Research
Type: Fellowship
Value: The fellowship is for five years, and provides (1) A salary, set according to age, experience and the appropriate academic scales (2) Essential costs of the research programme, including consumables, equipment, collaborative travel, research assistance and technical support if appropriate (3) A flexible funding allowance and support to attend scientific meetings, in addition to requested essential costs
Frequency: Annual

Application Procedure: You must complete and submit a preliminary application form [Word 218kB] by the published deadline. An electronic copy of the completed form (as a Word document) should be emailed to the appropriate funding stream (see website). You will be notified in writing of your success, or otherwise, in reaching the next round of the competition. Decisions will not be available by telephone. If you are successful you will be invited to submit a full application.
Closing Date: October 3rd
Funding: Trusts

Joint Basic and Clinical PhD Studentship Programmes

Subjects: Biomedical science.
Purpose: This is a flagship scheme aimed at supporting the most promising basic or medically qualified clinicians who wish to undertake both rigorous basic and clinical science research training. Successful candidates will develop their potential to become leading academics of the future within a structured and mentored training environment. Programmes will provide the individual trainee with opportunities to sample high-quality research environments before they develop a research proposal that is tailored to their individual interests.
Eligibility: Basic science applicants should have, or expect to obtain, a first- or upper-second-class honours degree or equivalent.Clinically qualified candidates should have demonstrated the potential to pursue a career as an academic clinician. It is anticipated that many applicants will have already commenced their specialist training but this is not essential.
Level of Study: Postgraduate, Research
Type: Studentship
Value: A stipend for basic candidates, a clinical salary for medically qualified candidates, PhD registration fees at UK/EU student rate, rsearch expenses, contribution towards travel and contribution towards transferable-skills training.
Length of Study: Support is provided for 4 years
Frequency: Annual
No. of awards offered: Two programmes have been established based in centres of excellence at Birmingham and the Institute of Cancer Research
Application Procedure: Students are recruited annually by the individual programmes. Recruitment begins in the preceding January. If you are interested in applying you should contact the relevant programme directly. Please see their website for more details.
Closing Date: January
Funding: Trusts

Master's Fellowships in Public Health and Tropical Medicine

Subjects: Biomedical science.
Purpose: This scheme strengthens scientific research capacity in low- and middle-income countries, by providing support for junior researchers to gain research experience and high-quality research training at Master's degree level. Research projects should be aimed at understanding and controlling diseases (either human or animal) of relevance to local, national or global health. This can include laboratory based molecular analysis of field or clinical samples, but projects focused solely on studies in vitro or using animal models will not normally be considered under this scheme.
Eligibility: You should be:
1. A national or legal resident of a low- and middle-income country, and hold a first degree in subject relevant to tropical medicine or public health (clinical or non-clinical)
2. At an early stage in your career, with limited research experience, but have a demonstrated interest in or aptitude for research.
Level of Study: Research, Postdoctorate, Postgraduate
Type: Fellowship
Value: This fellowship normally provides up to 30 months' support. A period of 12 months should normally be dedicated to undertaking a taught Master's course at a recognised centre of excellence, combined with up to 18 months to undertake a research project. While undertaking a Master's course, fellows will receive a stipend in accordance with the cost of living in the country in which he/she will be studying; travel costs and support for approved tuition fees. Master's training by distance learning is acceptable. Master's course fees will be paid according to the rate charged by the training institution.
Frequency: Biannual
Application Procedure: A completed application form [Word 1.58MB] should be submitted by the sponsor by the published deadline. The form should be emailed to phatic@wellcome.ac.uk. The application should include details of your sponsor's track record in training and a list of their other students at the institution. It must be supported by the head of the institution where the research will be based, and a career plan for the proposed candidate must be included.
Closing Date: November 23rd (for applications submissions)
Funding: Trusts

Master's Awards

Subjects: Medical history and humanities.
Purpose: This scheme enables scholars to undertake basic training in research and methods through a one-year Master's course in medical history and humanities.
Eligibility: You should have a minimum of an excellent upper-second-class honours degree (or equivalent) in a relevant subject. Applications will not be considered from those who have already received support for their postgraduate studies from another funding body.
Level of Study: Postgraduate, Research
Type: Award
Value: The award is for one year, and includes: the student's stipend; all compulsory university and college fees at the UK home post-graduate student level. Fees at the overseas rate will not be provided.
Frequency: Annual
Application Procedure: All enquiries about Master's Awards should be made directly to the relevant institution.
Closing Date: May 1st
Funding: Trusts

Medical History and Humanities Travel Grants

Subjects: Medical history and humanities.
Purpose: Travel grants fund short-term visits by scholars based outside the UK or the Republic of Ireland to one of these countries.
Eligibility: You must be based outside the UK or the Republic of Ireland and be applying to visit one of these countries. Visits under this scheme may be to consult libraries or archives and to exchange views or work with colleagues who have similar research interests. Study or lecture tours, meetings of a professional or vocational nature, workshops, symposia and international congresses are normally excluded. Experienced researchers need not be in academic life but will normally be expected to hold a doctorate or clinical qualification and have established a research interest in the medical humanities.
Level of Study: Research
Type: Grant
Value: The maximum award under this scheme is £1,500 (although a slightly higher limit may apply in the case of researchers from certain developing countries)
Frequency: Available throughout the year
Application Procedure: Hard copy of the application form including signatures should be sent to the Trust's postal address, marked for the attention of Grants Management - History of Medicine.
Closing Date: Applications may be submitted at any time during the year
Funding: Trusts

New Investigator Awards

Subjects: Biomedical science.
Purpose: To support world-class researchers who are no more than five years from appointment to their first academic position, but who can already show that they have the ability to innovate and drive advances in their field of study..
Eligibility: If you are based in the UK, Republic of Ireland or a low- or middle-income country: you should have an established academic post at an eligible higher education or research institution. By this we mean you are employed on a permanent, open-ended or long-term rolling contract, salaried by your host institution. You should be no more than five years from appointment to your first established academic post on the date you submit your main application. You are also eligible to apply if you have a written guarantee of an established academic post at your host institution, which you will take up by the start of the award. If you are based in a low- or middle-income country in sub-Saharan Africa, South East Asia or South Asia (with the

exception of India – see below). Please note also that you are eligible to apply if you fulfil the above eligibility criteria and are working within the Trust's broad science funding remit. If you are based in a low- or middle-income country other than in the territories mentioned above, please note that you are eligible to apply only if you are a researcher carrying out research in the fields of public health and tropical medicine aimed at understanding and controlling human and animal diseases of local, national and global health importance. The New Investigator Award scheme is not available to researchers in India (please see guidance on schemes offered by the Wellcome Trust/ Department of Biotechnology India Alliance), or in countries where we currently offer International Senior Research Fellowships. See website for other eligibility requirements.
Level of Study: Research
Type: Award
Value: An award can be worth anything up to £425k per year and for any duration up to a maximum of seven years. Please note that awards may not necessarily be made at the upper end of this range, and we expect costs to be suited to and justified by the proposed research. Covering cost such as: research expenses, including research assistance, animals, equipment and funding for collaborative activity, travel and subsistence for scientifically justified visits, overseas allowances where appropriate. The award does not include your salary costs.
Frequency: Annual
Application Procedure: Stage 1 - CV details check
Stage 2 - Main application
Stage 3 - Scientific review and shortlisting
Stage 4 - External peer review
Stage 5 - Interview.
Closing Date: varies. refer the website link: http://www.wellcome.ac.uk/Funding/Biomedical-science/Funding-schemes/investigator-awards/wtx059284.htm
Funding: Trusts
Additional Information: Please apply via the eGrants facility on the institution website.

New Wellcome Trust Four-Year PhD programme

Subjects: Biomedical science.
Purpose: To support new, innovative PhD studentship programmes training biomedical scientists.
Eligibility: The programme should be based in an eligible institution in the UK or Republic of Ireland.
The principal applicant should be the proposed director of the Programme who should be a recognised international leader in their field with a strong track record in postgraduate research training. These new programmes are intended for the support of basic scientists rather than clinicians.
Level of Study: Postgraduate, Research
Type: Grant
Value: 1. Four-years' stipend
2. University fees at home student rates
3. Contribution towards laboratory expenses in the first year
4. Research expenses for years two to four
5. A contribution towards travel
6. A contribution towards transferable skills training
Length of Study: 5 years
Frequency: Annual
No. of awards offered: 5
Application Procedure: An electronic copy of the preliminary application [Word 96kB] should be sent to 4yrphd@wellcome.ac.uk.
Closing Date: May 11th
Funding: Trusts

People Awards

Subjects: People Awards support projects that aim to achieve at least one of the following: stimulate interest, excitement and debate about biomedical science through various methods; support formal and informal learning about biomedical science; reach new audiences not normally engaged with biomedical science, as well as continuing to target existing audiences; examine the social, cultural, historical and ethical impact of biomedical science; encourage new ways of thinking about biomedical science; encourage high quality interdisciplinary practice and collaborative partnerships; investigate and test new methods of engagement, participation and education.
Purpose: To explore the impact of biomedical science on society, its historical roots, effects on different cultures, or the ethical questions that it brings, by supporting activities such as events, debates, exhibitions, art projects and drama productions related to biomedical science.
Eligibility: The scheme is open to a wide range of people, including mediators and practitioners of science communication; science centre/museum staff; artists; educators; health professionals; and academics in bioscience, social science, bioethics and history.
Applicants are encouraged to apply through an organisation rather than as individuals. If this is not possible, individuals can apply but they must demonstrate a strong track record in the area of their application.
Applications will also be accepted from commercial companies who would not otherwise be able to undertake the proposed work and where outputs would not be considered for commercial funding.
Organisations might include: museums and other cultural attractions; arts agencies; production companies; broadcast media; schools; local education authorities; universities and colleges; youth clubs; community groups; research institutes; the NHS; and science centres.
Partnership projects (between different people and organisations, e.g. scientists and ethicists, educators and artists) are welcomed.
If this is the first time an organisation is applying to the Wellcome Trust an eligibility assessment will be carried out. For this assessment, the following documentation from the applying organisation should be submitted: articles of association; audited accounts from the previous year; details of similar projects/grant funding received; confirmation that no funding has been received or is scheduled to be received from any tobacco company; standard health education and promotion projects, or projects dealing purely with non-biomedical sciences, are not eligible.
Level of Study: Research
Type: Award
Value: Applicants can apply for up to £30,000, for projects lasting a maximum of 3 years
Frequency: Biannual
Application Procedure: You should complete and submit an application form by the published deadline.
Closing Date: January 27th, April 26th, July 26th and October 25th.
Funding: Trusts
Additional Information: Applicants must be based in the UK or the Republic of Ireland and the activity must take place in the UK or the Republic of Ireland.

Pilot Grants

Subjects: Medical history and humanities.
Purpose: Pilot grants provide a mechanism for researchers to test research questions, develop methodologies, or explore collaborations with a view to constructing a competitive programme grant application.
Eligibility: Applicants should normally hold an established post in a university or institution in the UK or Republic of Ireland, and should have a good track record of research.
Level of Study: Research
Type: Grant
Value: A pilot grant can last up to 2 years, and provides: the salaries and associated costs for research assistants (if named, please include a full CV); funds to cover travel, equipment and other items essential for research; a set amount for the applicant(s) and any research assistants to attend conferences, seminars and other meetings of a scholarly nature; the salary of a temporary lecturer (in case an application for Research Leave is also made as part of the application)
Frequency: Thrice a year
Application Procedure: Preliminary applications should be made in writing, and include: a brief CV and full publication list; details of the proposed research (maximum of one page); the approximate cost of the proposal, broken down into equipment and project running expenses.
Closing Date: Preliminary applications should be submitted at least 6 weeks before the full application deadlines, which are March 1st, August 1st, December 1st
Funding: Trusts

Postdoctoral Training Fellowship for MB/PhD Graduates

Subjects: Biomedical science.
Purpose: This scheme provides a unique opportunity for the most promising newly qualified MB/PhD graduates, or those who have

achieved a high-quality PhD during or before starting their medical degree. It will enable successful candidates to make an early start in developing their independent research careers, by undertaking a period of postdoctoral training in the best laboratories in the UK and overseas, and can be tailored to allow them to continue their postgraduate clinical training.
Eligibility: The award is open to individuals with a relevant connection to the European Economic Area (EEA) who have either graduated with an MB/PhD or who have achieved a high-quality PhD in a relevant subject, either during or prior to commencing their initial medical, veterinary or dental degree. You should have completed your foundation training and have demonstrated significant progress towards gaining the core clinical competences that would be expected of a ST level Trainee/IATP Academic Clinical Fellow. 1. Medical and dental candidates should hold a National Training Number (NTN) or equivalent. 2. Veterinary candidates must have obtained their RCVS certificate or equivalent, e.g. CertSAM, CertVA, CertEP. 3. GPs and clinical psychologists are advised to contact the Trust to clarify their eligibility.
Level of Study: Postgraduate, Research
Type: Fellowship
Value: Fellowships are for up to four years (see 'Eligibility') and, depending on the duration of the fellowship, would not be expected to exceed £350,000. Fellowships provide: A salary set by the host institution, according to age and experience, research expenses (e.g. materials and consumables, animals, small items of equipment),travel and overseas subsistence and support to attend scientific meetings
Frequency: Annual
Application Procedure: A preliminary application form [Word 119kB] should be completed and submitted at any time before the published deadline. It should be sent electronically (as a Word document) to Dr Lucy Bradshaw (see website for contact details). If your preliminary application is successful, you will be invited to submit a full application. This will be reviewed and if successful you will be shortlisted for interview.
Closing Date: April 5th, Feb 11th, September 21st
Funding: Trusts

Principal Research Fellowships

Subjects: Biomedical science.
Purpose: This is the most prestigious of our personal awards and provides long-term support for researchers of international standing. Successful candidates will have an established track record in research at the highest level.
Eligibility: You should have an established track record in research at the highest level. This award is particularly suitable for exceptional senior research scientists currently based overseas who wish to work in the UK or Republic of Ireland.
Level of Study: Postgraduate, Research
Type: Fellowship
Value: Awards are for seven years in the first instance, and provide both a personal salary and research programme funding in full.After the first period of award, the fellowship will be subject to a competitive scientific review, which will subsequently occur on a rolling basis every five years
Frequency: Ongoing
Application Procedure: If you intend to apply you should contact us with a full CV, preferably 18 months in advance of the desired award date. You may not apply for more than one Wellcome Trust fellowship scheme at any one time.
Closing Date: Applicants can express interest at any time. Interview usually held in June and December.
Funding: Trusts

Programme Grants

Subjects: Medical history and humanities.
Purpose: Programme grants provide support for extensive or long-term research.
Eligibility: Applicants should normally hold an established post in a university or institution in the UK or Republic of Ireland, and should have a good track record of research.
Level of Study: Research
Type: Grant
Value: A programme grant normally lasts for 5 years, and provides the salaries and associated costs for research assistants (if named, please include a full CV); funds to cover travel, equipment and other items essential for research; a set amount for the applicant and any research assistants to attend conferences, seminars and other meetings of a scholarly nature. The level of support available depends on the needs of the programme, and the amount requested does not need to be above any particular threshold
Frequency: Thrice a year
Application Procedure: A preliminary application must be submitted, and should include: brief CVs of the applicant(s), including full publication lists and the source of their salary/salaries (e.g. HEFC/NHS); an outline of the work (up to 5 pages) explaining the background and aims of the project, and the reason for requesting longer-term support; brief CVs of any named research assistants; the approximate cost of the programme, broken down into salaries, equipment and project running costs; details of all current funding from the Trust and other bodies.
Closing Date: Preliminary applications should be sent in no later than the beginning of January, May or October
Funding: Trusts

R&D for Affordable Healthcare in India

Subjects: Projects covering any aspect of technology development for healthcare will be considered, including diagnostics, therapeutics, vaccines, medical devices and regenerative medicine. Proposals drawing on the disciplines of the physical sciences, maths and engineering, as well as biomedicine, are equally encouraged.
Purpose: For translational research projects that will deliver safe and effective healthcare products for India – and potentially other markets – at affordable costs.The objective of this initiative is to fund translational research projects that will deliver safe and effective healthcare products for India – and potentially other markets – at affordable costs. A key feature of the scheme is that it encourages innovations that bring together researchers from both the public and private sectors to extend access to care to the greatest numbers of beneficiaries, without compromising on quality.
Level of Study: Research
Type: Award
Value: Awards will be made by way of funding agreements that will be negotiated on a case-by-case basis. The principles of the Wellcome Trust Grant Conditions will apply. The terms and conditions of funding will be discussed with applicants individually. Typically, the agreements will contain a provision for the appropriate sharing of benefits. The funds available will be ring-fenced for the specified programme of work. Neither working capital nor building or refurbishment expenditure will be provided. Funding will be released in tranches against the attainment of pre-agreed project milestones.
Frequency: Twice a year
Application Procedure: In the first instance interested applicants should contact Dr Shirshendu Mukherjee to discuss their interest in funding via the Affordable Healthcare Initiative. Alternatively, applicants may complete a concept note and mail this directly to Dr Shirshendu Mukherjee.
Closing Date: January 31st
Funding: Trusts

Research Career Development Fellowships in Basic Biomedical Science

Subjects: Biomedical Science.
Purpose: To provide support for outstanding postdoctoral scientists based in academic institutions in the UK and Republic of Ireland (RoI).
Eligibility: You should have a relevant connection to the European Economic Area. You are expected to have science or veterinary qualifications and, at the preliminary application stage, should normally have between three and six years' research experience from the date of your doctoral degree (PhD viva). Due allowance will be given to those whose career has been affected for personal reasons. You must have made intellectual contributions to research that have been published in leading journals, and be able to demonstrate your potential to carry out independent research. The proposed research should fall within our normal funding remit. Resubmissions are not normally encouraged. If your application has been unsuccessful, please contact the Office for advice. You must have an eligible sponsoring host institution in the UK or Republic of Ireland (RoI) and an eligible sponsor who can guarantee space and resources for the tenure of any award.

Level of Study: Postdoctorate, Research
Type: Fellowship
Value: (1)A basic salary, as determined by the host institution, with an additional Wellcome Trust enhancement. (2) Research expenses, including research assistance if required (normally a graduate research assistant or technician; requests for additional research staff may be considered where fieldwork or clinical studies in a low- or middle-income country are proposed). (3) Overseas allowances where appropriate. (4)Travel and subsistence for scientifically justified visits of up to one year.
Length of Study: 5 years
Frequency: Annual
Application Procedure: A preliminary application form [Word 89kB] should be completed and submitted by the published deadline. It should be sent electronically (as a Word document), with the requested accompanying information, to the appropriate funding stream at the Trust (see website). If successful, you will be invited to submit a full application.
Closing Date: March 15th (Deadline for preliminary applications); June 10th, 5pm (Deadline for invited full applications)
Funding: Trusts

Research Expenses

Subjects: Medical history and humanities.
Purpose: This scheme supports experienced researchers who wish to carry out a modest programme of study on a specific topic in the medical humanities within the UK and Republic of Ireland. This research does not necessarily have to be 'historically grounded'. It also provides modest assistance with research expenses for self-funded, part-time and full-time postgraduate students working for a doctorate on a history of medicine topic.
Eligibility: Applicants must be based in the UK or Republic of Ireland. Experienced researchers in an established academic post will normally be expected to have written some publications in an appropriate field. Experienced researchers not in an established academic post are expected to possess a doctorate or clinical qualification, and have established a research interest in the medical humanities. Self-funded, part-time and full-time students must be registered for a doctoral degree at a university or other institution of higher education in the UK or Republic of Ireland.
Level of Study: Research
Type: Grant
Value: Funding may be provided for a maximum of two years. The normal maximum award payable to postdoctoral scholars is £5,000. For self-funded doctoral students, the normal maximum award payable is £3,000
Frequency: Available throughout the year
Application Procedure: Hard copy of the application form including signatures should be sent to the Trust's postal address, marked for the attention of Grants Management - History of Medicine.
Closing Date: Applications may be submitted at any time during the year
Funding: Trusts

Research Fellowships

Subjects: Medical history and humanities.
Purpose: This scheme supports individuals at all stages of their career not in established academic posts, wishing to undertake a period of research.
Eligibility: You are eligible to apply if you are a postdoctoral scholar who is not in a tenured or otherwise long-term established post. Fellowships must be held at a UK, Irish or low- or middle-income country institution.You will also be expected to have been awarded your PhD before you are eligible to apply. Applications from candidates who are still awaiting their viva by the time of the full application will not normally be accepted.
Level of Study: Research
Type: Fellowship
Value: An award will not normally exceed £250 000, exclusive of any standard Wellcome Trust allowances.Fellowships provide a salary, plus appropriate employer's contributions.Essential research expenses, including travel and fieldwork, are available, as is a set amount for travel to conferences, seminars and other meetings of a scholarly nature.
Frequency: Biannual
Application Procedure: Preliminary applications should be made in writing, and include: a brief CV and full publication list; details of research proposed (maximum of 1 page); a letter of support from the head of department in which you will be working; the approximate cost of the proposal, broken down into equipment and project running expenses.
Closing Date: Preliminary application deadlines are: June 20th and December 3rd; the full application deadline will be between six weeks and two months after the preliminary application deadline.
Funding: Trusts
Additional Information: The maximum duration of the awards is 3 years. The awards are full-time but can be tenable on a part-time basis if a case can be made that personal circumstances require this.

Research Leave Awards

Subjects: Medical history and humanities.
Purpose: These awards allow university staff in the UK or Republic of Ireland to be released from their teaching and administrative duties so they can undertake an uninterrupted period of full-time research.
Eligibility: You must be able to demonstrate that you are active in research, but have a high teaching/administrative load that is hampering research progress. You should be based in the UK or Republic of Ireland and must be sponsored by your head of department.
Level of Study: Research
Type: Award
Value: Awards are normally tenable for one to three years, and will provide the salary of a temporary lecturer (usually at a lesser level of seniority), research expenses and a travel allowance.
Frequency: Thrice times a year
Application Procedure: You should submit a preliminary application in writing, including: a brief CV, a full publication list and confirmation that your personal support is from the Higher Education Funding Council; details of the research proposed (maximum one page); details of hours spent on teaching and administration; the approximate cost of the proposal, broken down into staff salaries, equipment and running expenses.
Closing Date: Preliminary applications should be submitted at least six weeks before the full application deadline, which are March 1st, August 1st, December 1st
Funding: Trusts

Research Resources in Medical History

Subjects: Medical history and humanities - the next theme is Understanding the Brain.
Purpose: To provide funding for projects to catalogue and preserve significant collections of printed books and archives in the UK and Ireland. Applications must demonstrate the significance to the MHH research community and how collections fit within the themes identified in the Trust's Strategic Plan.
Eligibility: The scheme is open to any type of institution in the UK or Republic of Ireland, but not to individuals. Libraries, archives and repositories in all sectors are eligible. In exceptional circumstances, strategically important collections held in other countries might be eligible. Collaborative projects, which may be part-funded by other agencies or sources, will also be considered.
Level of Study: Research
Type: Grant
Value: Grants are normally between £10,000 and £100,000
Frequency: Thrice a year
Application Procedure: Preliminary applications should include the following: the completed application form; an explanation of how the collection contents fit within the current theme. Collections that address more than one theme can be considered (applicants should discuss their collections with Trust staff first for advice on the timing of their applications); a description of the size of the collection, type of material and physical condition; an estimate of the costs required and how the funds will be used, i.e. for preservation, cataloguing, digitisation; brief CVs of the principal applicants; any reports or information provided by initial scoping phases such as preservation needs assessments or preliminary sorting.
Closing Date: June 7th (Preliminary application deadline); October 1st (Full application deadline)
Funding: Trusts

Research Training Fellowships

Subjects: Biomedical Science.

Purpose: This scheme is for medical, dental, veterinary or clinical psychology graduates who have little or no research training, but who wish to develop a long-term career in academic medicine. Applications are encouraged from individuals who wish to undertake substantial training through high-quality research in an appropriate unit or clinical research facility, towards a PhD or MD qualification.

Eligibility: The fellowship is open to individuals with a relevant connection to the European Economic Area (EEA) for fellowships to be held in a UK or Republic of Ireland institution. Non-UK candidates should contact the office for advice before submitting an application.

1. Medical graduates must have passed the relevant exam for their specialty, e.g. MRCP, MRCS, MRCOphth/FRCOphth Part 1, MRCPsych, MRCOG Part 1, MRCPCH, FRCA Part 1. GPs are advised to contact the office to clarify their eligibility.
2. Dental candidates must have obtained MFD, MFDS, MGDS, MFGDP or equivalent.
3. Veterinary candidates should have a degree in veterinary medicine (e.g. BVSc, BVM&S, BVMS, BVetMed, VetMB) and some experience in clinical practice. An intercalated degree is desirable, but not essential.
4. Clinical psychology candidates must have obtained a professional Doctorate-level qualification in Clinical Psychology accredited by the British Psychological Society before taking up the award. Candidates are advised to contact the office to clarify their eligibility.

You are expected to undertake a high-quality research project that balances the provision of training with the opportunity to advance knowledge in a given area. A project based solely on a systematic review of a particular area is not suitable, unless it includes a significant element of methodological innovation.

Level of Study: Postgraduate, Research

Type: Fellowship

Value: Fellowships are normally for two to three years. In exceptional cases a fellowship may be for up to four years for those who wish to undertake a relevant Master's training or diploma course. All training requests must be fully justified in the application. Fellowships provide research expenses (consumables, travel, and support to attend scientific meetings) and a fellow's salary, set according to age, experience and our policy on enhancement

Frequency: Three times per year

Application Procedure: Application form is available from the website.

Closing Date: Feb 8th (full application deadline)

Funding: Trusts

Science Media Studentships

Subjects: These studentships offer financial support for two practising biomedical scientists to undertake a postgraduate qualification in Science Media Production at Imperial College London and to follow this with a six-month placement working in the broadcast industry.

Purpose: We aim to increase the crossover between science and the media and to enable bright, articulate and motivated scientists to explore a career in the broadcast industry.

Eligibility: Applicants should be practising biomedical scientists wishing to explore a career in the broadcast media. Applicants must have a PhD or equivalent, some experience of science communication, and a demonstrable aptitude for working with TV, radio or film.

Level of Study: Postdoctorate, Research

Type: Studentship

Value: The award will pay for tuition costs of one year plus a grant of £18,000 over the 18-month placement to cover basic living expenses

Frequency: Annual

Application Procedure: Application is through the Imperial College website.

Closing Date: February 25th

Funding: Trusts

Seeding Drug Discovery

Subjects: The aim is to develop drug-like, small molecules that will be the springboard for further research and development by the biotechnology and pharmaceutical industry in areas of unmet medical need.

Purpose: To facilitate early-stage small-molecule drug discovery. The awards help applicants with a potential drug target or new chemistry embark on a programme of compound discovery and/or lead optimisation.

Level of Study: Research

Type: Award

Value: Early-stage drug discovery projects are able to apply for funding for up to two years to facilitate screening of chemical compounds to identify one or more lead series of molecules. Late-stage projects, where a lead compound has already been identified, are able to apply for funding for up to four years, to support lead optimisation and preclinical development through to clinical trials.

Frequency: Twice a year

Application Procedure: A preliminary application form should be completed and returned to Technology Transfer by the published deadline. Applications will be considered at one of the two Seeding Drug Discovery Committee meetings in each 12-month period. Successful applicants will be shortlisted and invited to complete a full application.

Closing Date: November 8th and June 7th (Preliminary deadline); May 29–30th and October 24–25th (Funding decision for invited full proposal).

Funding: Trusts

Senior Fellowships in Public Health and Tropical Medicine

Subjects: Biomedical science.

Purpose: This scheme supports outstanding researchers from low- and middle-income countries to establish themselves as leading investigators at an academic institution in a low- and middle-income country location. This fellowship is the most senior of a series of career awards aimed at building sustainable capacity in areas of research that have the potential for increasing health benefits for people and their livestock in low- and middle-income countries. Research projects should be aimed at understanding and controlling diseases (either human or animal) of relevance to local, national or global health.

Eligibility: Applications are only accepted in the Public Health and Tropical Medicine Interview Committee remit. This covers research on infectious and non-communicable diseases within the fields of public health and tropical medicine that is aimed at understanding and controlling diseases (either human or animal) of relevance to local, national or global health. This can include laboratory based molecular analysis of field or clinical samples, but projects focused solely on studies in vitro or using animal models will not normally be considered under this scheme.

You must be a national or legal resident of a low- and middle-income country, and be either a:

1. Graduate in a subject relevant to public health or tropical medicine (for example; biomedical or social science, veterinary medicine, physics, chemistry or mathematics) with a PhD and at least five years' postdoctoral experience, or
2. Medical graduate with a higher qualification equivalent to membership of the UK Royal College of Physicians (i.e. qualified to enter higher specialist training), or be recognised as a specialist within a relevant research area, and have at least five years' research experience.

Applicants who do not have a PhD but who are educated to first degree or Master's level and have substantial research experience, as evidenced by their publication record, may be considered.

Level of Study: Postgraduate, Research

Type: Fellowship

Value: A basic salary; research expenses (e.g. consumables, equipment, collaborative travel, research assistance, technical support), training costs where appropriate and justified; an inflation/flexible funding allowance and support to attend scientific meetings; and contributions to costs of the project that are directly incurred by the overseas institution may also be provided.

Length of Study: Up to 5 years

Frequency: Thrice a year

Application Procedure: You are required to complete and submit a preliminary application form [Word 236kB] by the published deadline. The form should be emailed to phatic@wellcome.ac.uk.
Completed forms will normally be assessed within one month of the preliminary deadline. If your preliminary application meets the scheme's requirements, a full application will be invited.

Closing Date: March 4th, September 24th, Novemebr 26th

Funding: Trusts

Senior Investigator Awards

Subjects: Biomedical Science.

Purpose: To support exceptional, world-class researchers, who hold an established academic position and have a compelling long-term vision for their research. We will support researchers who have an international track-record of significant achievement, who have demonstrated the originality and impact of their research, and who are leading their field.

Eligibility: If you are based in the UK, Republic of Ireland or a low- or middle-income country: you should have an established academic post at an eligible higher education or research institution. By this we mean you are employed on a permanent, open-ended or long-term rolling contract, salaried by your host institution. You are also eligible to apply if you have a written guarantee of an established academic post at your host institution, which you will take up by the start of the award. If you are uncertain as to whether your employment status meets the above eligibility criteria, please contact the Trust for advice (see 'Contacts'). If you are based in a low- or middle-income country in sub-Saharan Africa, South East Asia or South Asia (with the exception of India – see below) please note also that: you are eligible to apply if you fulfil the above eligibility criteria and are working within the Trust's broad science funding remit. If you are based in a low- or middle-income country other than in the territories mentioned above, please note that: you are eligible to apply only if you are a researcher carrying out research in the fields of public health and tropical medicine aimed at understanding and controlling human and animal diseases of local, national and global health importance. The Senior Investigator Award scheme is not available to researchers in India (please see guidance on schemes offered by the Wellcome Trust/Department of Biotechnology India Alliance), or in countries where we currently offer International Senior Research Fellowships.

Level of Study: Research

Type: Award

Value: Range of £100,000–425,000 per year. Please note that awards may not necessarily be made at the upper end of this range, and we expect costs to be suited to and justified by the proposed research. Covers cost of research expenses, including research assistance, animals, equipment and funding for collaborative activity, travel and subsistence for scientifically justified visits, overseas allowances where appropriate. The award does not include your salary costs.

Frequency: Annual

Application Procedure: Stage 1 is CV details check, stage 2 is main application, stage 3 is scientific review and shortlisting, stage 4 is external peer review, and stage 5 is interview.

Closing Date: March 4th (full application); July 9-11th (shortlisted candidate interviews)

Funding: Trusts

Additional Information: Please apply via the eGrants facility on the institution website.

Senior Research Fellowships in Basic Biomedical Science

Subjects: Biomedical science.

Purpose: To provide support for outstanding postdoctoral scientists based in academic institutions in the UK and Republic of Ireland (RoI).

Eligibility: The fellowship is open to individuals with a relevant connection to the EEA.

You should have between five and normally ten years' research experience (from the date of your viva to the date of your preliminary application) at postdoctoral level, or veterinary equivalent, and have a substantial record of publications in your chosen area of research in leading international journals.

Candidates that do not hold an established post may apply to remain in their current laboratory, to return to one where they have worked before or to move to a new laboratory in the UK or RoI.

Candidates that hold an established post are not eligible to apply for a fellowship to be held at their current employing institution. However, we are willing to consider a preliminary application where a candidate wishes to move institution and is able to make an appropriate justification for the move.

The Trust does not normally accept resubmissions of full applications for its fellowships. Please contact the Office for further advice.

You must have an eligible sponsor and host institution in the UK or RoI who can guarantee space and resources for the tenure of the award.

Level of Study: Postdoctorate, Research

Type: Fellowship

Value: The fellowship is for five years in the first instance, and provides a basic salary, as determined by the host institution (normally up to £55,000 per year) with an additional Trust supplement of £12,500 per year; the essential costs of the research programme (e.g. consumables, equipment, research assistance, overseas allowances, collaborative travel and subsistence); an inflation and Flexible Funding Allowance; and support to attend scientific meetings.

Frequency: Annual

Application Procedure: A preliminary application form [Word 223kB] should be completed and submitted electronically (as a Word document) to the relevant funding stream (see website) no later than the published deadline. Full application forms will usually be sent to shortlisted candidates within one month of the preliminary deadline. In the full application, if invited, the host institution will be required to confirm that it will support a successful renewal of the fellowship under the shared funding arrangement for the full period of any renewal.

Closing Date: July 12th, March 4th and Novemeber 26th

Funding: Trusts

Senior Research Fellowships in Clinical Science

Subjects: Biomedical science.

Purpose: This scheme provides support for clinical investigators to further develop their research potential and to establish themselves as leading investigators in clinical academic medicine.

Eligibility: You must have a relevant connection to the EEA. If you are a non-UK candidate, please contact the Office for advice before submitting a preliminary application. You should be a clinical scientist with a medical, dental, veterinary or clinical psychology qualification and will normally have no more than 15 years' clinical and research experience from the date of your first medical, dental, veterinary or British Psychological Society-accredited psychology qualification. (Due allowance will be given to those whose career has been affected by a late start or interruption for personal/family reasons.) Successful candidates will have made significant progress towards establishing themselves as independent clinical investigators. A research degree (PhD/MD), together with evidence of advanced (postdoctoral) research training (typically at least three to five years), is expected. They will have published consistently in their chosen area of research, placing substantive papers in leading journals. Candidates will not normally hold a tenured academic post in a university in the UK or Republic of Ireland, or a consultant post in the NHS.

Level of Study: Postgraduate, Research

Type: Fellowship

Value: The fellowship is for five years in the first instance, and provides: Employment costs (including basic salary, employer's contributions, incremental progression, London weighting as applicable, and an allowance for inflation over future years); the essential costs of the research programme (e.g. consumables, equipment, collaborative travel, research assistance and technical support); an inflation and Flexible Funding Allowance and support to attend scientific meetings, in addition to the requested essential costs; and appropriate allowances to fellows based overseas.

Frequency: Annual

Application Procedure: A preliminary application form [Word 232kB] should be completed and submitted by the published deadline. It should be sent electronically (as a Word document), with the requested accompanying information, to the appropriate funding stream at the Trust (see website). Incomplete or incorrectly completed forms will not be accepted. Faxed applications will not be accepted. Please do not send any additional material. You will be notified in writing of your success, or otherwise, in reaching the next round of the competition. In some instances, we may recommend that candidates apply for an Intermediate Clinical Fellowship.

Closing Date: July 12th, March 4th, Novemeber 26th

Funding: Trusts

Short-term Research Leave Awards for Clinicians and Scientists

Subjects: Medical history and humanities.

Purpose: This scheme encourages research in the history of 20th-century medicine and medical science.It enables clinicians or scientists to undertake a short-term period of full-time research at a centre or department with academic expertise in medical history, to

learn the methods of historical scholarship and to explore the wider determinants and contexts of their own medical and scientific work.
Eligibility: You should be a clinician or scientist in mid-career, holding an established post to which you would return on completion of the award. You must be resident in the UK or Republic of Ireland. You should have a record of publication in medical or scientific journals.
Level of Study: Research
Type: Award
Value: Awards can last for up to six months. We will provide the salary of a locum or replacement lecturer for the duration of the award, and a set amount for travel to conferences
Frequency: Thrice a year
Application Procedure: You should submit a preliminary application in writing, including: a brief CV, including details of your salary support and a full publication list; details of the proposed research (one page maximum); a letter of support from the head of the department in which you would work; an approximate cost of the proposal.
Closing Date: Preliminary application deadlines are: June 20th (with a full application deadline of August 1st), December 1st (with a full application deadline of February 1st).
Funding: Trusts

Sir Henry Wellcome Postdoctoral Fellowships

Subjects: Biomedical science.
Purpose: To provide a unique opportunity for the most promising newly qualified postdoctoral researchers to make an early start in developing their independent research careers, working in the best laboratories in the UK and overseas.
Eligibility: These awards are open to individuals with a relevant connection to the European Economic Area. You must be in the final year of your PhD studies or have no more than one year of postdoctoral research experience from the date of your PhD viva to the full application submission deadline (e.g. if the full deadline is in February 2011, your viva should not have occurred prior to February 2010). Time spent outside the research environment will be taken into consideration. You must have an eligible sponsoring institution in the UK or Republic of Ireland that will administer the fellowship for the full duration of the award.
Level of Study: Postdoctorate, Research
Type: Fellowships
Value: Four year full-time fellowship. Provides an award of £250,000
Frequency: Annual
Application Procedure: You should complete and submit a preliminary application form by the published deadline. It should be sent electronically (as a Word document), with the requested accompanying information, to the relevant funding stream at the Trust. Your preliminary application will be assessed within four weeks of the submission deadline. If successful, you will be invited to submit a full application. Your full application will be peer reviewed by the relevant Funding Committee and, if successful, you will be shortlisted for interview.
Closing Date: May 20th, 5pm (Deadline for preliminary applications); July 26th, 5pm (Deadline for full applications) and Feb 1st (for preliminary applications submitted to November 9th)
Funding: Trusts

Society Awards

Subjects: Our aim is to encourage people of all ages and walks of life to learn about these developments and have an opportunity to consider, question and debate the implications and issues arising from such work. By inspiring, informing and involving whole communities, Society Awards enable people to consider and discuss issues that affect them, those close to them and the world in which they live. Projects should aim to achieve at least one of the following: stimulate interest, excitement and debate about biomedical science through various methods; examine the social, cultural, historical and ethical impact of biomedical science; encourage new ways of thinking about biomedical science.
Purpose: Society Awards are for ambitious and creative projects that engage people with developments in biomedical science on a regional or national scale.
Eligibility: The scheme is open to anyone with a good idea for engaging people with developments in biomedical science. This might include: mediators and practitioners of science communication; science centre/museum staff; artists; educators; health professionals; and academics in bioscience, social science, bioethics and history. Grants will normally be awarded through organisations, but individuals can apply. Organisations might include: venues attracting large audiences (e.g. museums, cultural attractions or nature attractions); arts agencies; production companies; schools; local education authorities; universities; youth clubs; community groups; research institutes; the NHS; and science centres. Partnership projects (between different people and organisations, e.g. scientists and ethicists, educators and artists) are welcomed. Please note that standard health education and promotion projects, or projects dealing purely with non-biomedical sciences, are not eligible. Large broadcast media projects are not eligible for consideration through the Society Awards. These projects can be considered through our Broadcast Strategy and the Large Broadcast Awards. Smaller broadcast media projects are eligible for funding either through the People Awards (for production costs) or the Broadcast Development Awards (for development costs).
Level of Study: Research
Type: Award
Value: Society Awards are for amounts over £30,000, for a maximum of three years
Frequency: Annual
Application Procedure: Please contact the Society Awards office well in advance of the preliminary deadline to discuss a potential application. You must complete and submit a preliminary application form by the published deadline.
Closing Date: March 28th
Funding: Trusts
Additional Information: Applicants must be based in the UK or the Republic of Ireland and the activity must take place in the UK or the Republic of Ireland.

Strategic Awards in Biomedical Science

Subjects: Biomedical science.
Purpose: Strategic Awards provide flexible forms of support to excellent research groups with outstanding track records in their field.
Eligibility: Applications will be considered from principal applicants who meet our eligibility criteria and are recognised international leaders in their field.
Level of Study: Research
Type: Award
Value: It provides equipment, support staff, consumables, training programmes, networking, biological, clinical or epidemiological research resources.Limited capital building or refurbishment essential to the programme can also be requested
Length of Study: Awards are normally for five years
Frequency: Ongoing
Application Procedure: You (prospective applicant) are required to submit a preliminary application, which should include the following information:
1. Your track record - you must complete the CV pages [Word 112kB] (these are questions 14 and 15 from the standard project grant application form)
2. High-level aims and objectives, and how the proposal addresses the strategic challenges in the Wellcome Trust's Strategic Plan for 2010–2020 (maximum of two pages)
3. Key targets, milestones and management structures, if appropriate (maximum of two pages)
4. Duration of support requested and outline costings broken down into main headings (e.g. staff, equipment)
5. A statement from the head of the institution, indicating how the proposal fits within the context of the institution's strategic vision and what financial commitment the institution will make to the group if the application is successful.
If your preliminary application is successful, you will be invited to submit a full application. The relevant form will be provided at this time.
Closing Date: Preliminary applications may be submitted at any time and are assessed on a rolling basis.
Funding: Trusts

Strategic Translation Awards

Subjects: For Strategic Translation Awards the Trust will normally actively participate in the stewardship of the project and lead on intellectual property management and exploitation. A wide range of

biomedical developments can be considered, including therapeutics, vaccines, diagnostics, enabling technologies (including research tools), medical devices and regenerative medicine.
Purpose: Strategic Translation Awards support research projects that are viewed as strategically important to the Wellcome Trust's mission. Technology Transfer at the Wellcome Trust proactively seeks applications from scientists who wish to work in partnership with the Trust to achieve commercialisation of their inventions. Compared with the Translation Awards, the Trust is more proactively engaged in project management, working alongside the institution or company involved. Strategic projects are exceptional projects that - due to the combination of potential high impact, risk, scale or complexity - warrant strategic status to provide a high level of momentum for the project.
Eligibility: Applications are welcome from research centres (non profit making). Businesses who undertake medical research can also apply. Projects that will be viewed as eligible will be addressing unmet needs in healthcare or in applied medical research.
Level of Study: Postgraduate, Research
Type: Award
Value: The important criterion is to develop the innovation to the point at which it can be adopted by another party. Awards will normally be for periods of two to three years, but can be longer in exceptional cases. Providing it is adequately justified, modest equipment purchase and maintenance costs may be included in an application. Building or refurbishment expenditure will not normally be considered. Applications may not include requests for academic institutional overheads.
Length of Study: two to three years, but can be longer in exceptional cases.
Application Procedure: Prospective applicants should first contact Technology Transfer staff at the Wellcome Trust to discuss their proposal. You will then be asked to submit a preliminary application, which will be considered for its strategic potential and the likely impact of the project downstream. If successful, you will be invited to submit a full application. The further progression of any strategic proposal will be dependent upon successful due diligence by the Wellcome Trust, and only those applications that are competitive will be taken forward to a Technology Transfer Strategy Panel decision. All funding decisions are made by the Strategy Panel.
Closing Date: There is an open call for applications. Prospective applicants should contact Technology Transfer to discuss their proposal.
Funding: Trusts

Support for Conferences, Symposia and Seminar Series

Subjects: Medical history and humanities.
Purpose: This scheme provides institutions with financial support for conferences (or a session within a conference), symposia, seminar series, etc.
Eligibility: You should be based at an eligible institution in the UK or Republic of Ireland. Awards are not normally made to individuals, so please name the institution to which the award should be made. Grants are not available for symposia held in association with established organisations with permanent staff, or to support large international meetings or learned societies overseas.
Type: Grant
Value: The Trust will consider making small contributions in the region of £1000–10 000
Frequency: Available throughout the year
Application Procedure: An application form needs to be completed.
Closing Date: Applications may be submitted at any time throughout the year.
Funding: Trusts

Training Fellowships in Public Health and Tropical Medicine

Subjects: Biomedical science.
Purpose: This scheme provides researchers from low- and middle-income countries - who are at an early stage in the establishment of their research careers - with opportunities for research experience and high-quality research training in public health and tropical medicine. Research projects should be aimed at understanding and controlling diseases (either human or animal) of relevance to local, national or global health. This can include laboratory-based molecular analysis of field or clinical samples, but projects focused solely on studies in vitro or using animal models will not normally be considered under this scheme.
Eligibility: Applications are only accepted in the Public Health and Tropical Medicine Interview Committee remit. This covers research on infectious and non-communicable diseases within the fields of public health and tropical medicine that is aimed at understanding and controlling diseases (either human or animal) of relevance to local, national or global health. You must be a national or legal resident of a low- and middle-income country and should be either:
1. A graduate in a subject relevant to public health or tropical medicine (e.g. biomedical or social science, veterinary medicine, physics, chemistry or mathematics) with a PhD and no more than three years' postdoctoral experience, or
2. A medical graduate with a higher qualification equivalent to membership of the UK Royal Colleges of Physicians (i.e. qualified to enter higher specialist training) and some initial research experience.
Applicants may also apply if they do not have a PhD, but have a clinical, basic or Master's degree and some initial research experience, with the expectation that they will register for a PhD.
Level of Study: Postgraduate, Research
Type: Fellowship
Value: It provides support that includes a basic salary for the fellow, research expenses (e.g. consumables, equipment, collaborative travel, research assistance, technical support) training costs where appropriate and justified, an inflation/flexible funding allowance and support to attend scientific meetings, and contributions to costs of the project that are directly incurred by the overseas institution may also be provided
Length of Study: 3 years
Frequency: Thrice a year
Application Procedure: You are required to complete and submit a preliminary application form [Word 236kB] by the published deadline. The form should be emailed to phatic@wellcome.ac.uk. Completed forms will normally be assessed within one month of the preliminary deadline. If the preliminary application meets the scheme's requirements, you will be invited to submit a full application.
Closing Date: March 1st, May 21st, November 25–27th
Funding: Trusts

Translation Awards

Subjects: Projects covering any aspect of technology development from a range of disciplines - including physical, computational and life sciences - will be considered. Projects must address an unmet need in healthcare or in applied medical research, offer a potential new solution, and have a realistic expectation that the innovation will be developed further by the market.
Purpose: Translation Awards are response-mode funding designed to bridge the funding gap in the commercialisation of new technologies in the biomedical area.
Eligibility: Projects must address an unmet need in healthcare or in applied medical research, offer a potential new solution, and have a realistic expectation that the innovation will be developed further by the market.
Institutions: eligible institutions are not-for-profit research institutions, including those funded by the Medical Research Council, Cancer Research UK, and Biotechnology and Biological Sciences Research Council, in the UK. Institutions are normally required to sign up to a short funding agreement and the Grant Conditions.
Companies: we are able to use our charitable monies to fund commercial companies to meet our charitable objectives through programme-related investment (PRI). For further details please refer to our policy on PRI. Companies will normally be expected to sign up to specific terms relating to the scheme.
Overseas organisations: UK organisations may contract or collaborate with overseas organisations. Although overseas organisations are not eligible for Translation Awards, some proposals may be invited for consideration as a Strategic Translation Award (including Seeding Drug Discovery). Overseas organisations should contact Technology Transfer staff about their proposed project in the first instance.
Principal applicants and coapplicants: applicants should normally hold a position of responsibility within the eligible organisation and be able to sign up to or comply with the conditions or terms of an award.
In addition, postdoctoral research assistants - whether seeking their own salary as part of the grant proposal, funded by the Wellcome Trust on another grant, or funded by another agency - are eligible for coapplicant status if they make a significant contribution to a research proposal and have agreement from their funding agency.

Other eligibility information:
Disciplines outside biomedicine – researchers from disciplines outside biomedicine can apply providing the application of research is designed to facilitate or meet a need in healthcare. For example, the application of physics, chemistry, computing, engineering and materials science to the development of medical products is entirely appropriate.
Healthcare need in an area that is not commercially attractive.
We are committed to the translation of research into practical healthcare benefits across the full spectrum of disease. Disease areas neglected by industry because of the lack of a return on investment pose a particular problem, but imaginative ways forward can sometimes be developed (e.g. public-private partnerships such as the Medicines for Malaria Venture).
Intellectual property rights (IPR)/publications – if there are any restrictions on IPR or publications arising from your research, you must provide a written statement that details them. Restrictions on intellectual property may affect your eligibility to apply to the Trust. Please refer to our Grant Conditions.
Level of Study: Research
Type: Award
Value: The important criterion is to develop the innovation to the point at which it can be adopted by another party. Providing it is adequately justified, modest equipment purchase and maintenance costs may be included in a Translation Award application. Building or refurbishment expenditure will not normally be considered. Applications may not include requests for academic institutional overheads. If you hold a tenured university post, you may not re-charge your salary (in full or part) to a Translation Award.
Frequency: Four times a year
Application Procedure: A preliminary application form must be completed and sent to Technology Transfer by the published deadline. Preliminary applications are subject to a triage for shortlisting for the full application stage. Applications will be considered by the Technology Transfer Challenge Committee (TTCC), which meets twice a year. Full applications will be invited following the triage meeting. Shortlisted applicants will be invited to submit a full application and will be subject to international peer review and due diligence. Applicants will be expected to make a presentation on their proposal to the TTCC. Unless otherwise advised, this will be at the next scheduled meeting of the TTCC.
Closing Date: Preliminary deadline: July 22nd, January 6th, July 20th; TTCC meeting (presentations by shortlisted applicants): March, July. All applications received by 17.00 GMT on the deadline date will be considered
Funding: Trusts

Translational Medicine and Therapeutics Programmes

Subjects: Biomedical science.
Purpose: This flagship scheme established four high-quality integrated research training programmes for clinicians in translational medicine and therapeutics. The programmes have been developed around a unique partnership between academic and industrial partners. Support for the programmes has been provided to the host institutions by GlaxoSmithKline, Wyeth Research, Roche, AstraZeneca, Sanofi-Aventis, Sirtris Pharmaceuticals and PTC Therapeutics.
Eligibility: You should have demonstrated the potential to pursue a career as an academic clinician. It is anticipated that many applicants will have already commenced their specialist training, but this is not essential.
Level of Study: Postgraduate, Research
Value: Includes a clinical salary, PhD registration fees at UK/EU rate, research expenses, contribution towards travel, and a contribution towards training costs.
Length of Study: Support varies
Frequency: Annual
No. of awards offered: Four programmes have been established, based in centres of excellence throughout the UK
Application Procedure: If you are interested in applying you should contact the relevant programme. Please see website for details.
Closing Date: October
Funding: Trusts

University Awards

Subjects: Medical history and humanities.
Purpose: This scheme allows universities to attract outstanding research staff by providing support for up to five years, after which time the award holder takes up a guaranteed permanent post in the university. A monograph and other substantial publications are expected to result from an award, so teaching and other non-research commitments are expected to be minimal during the period of full Wellcome Trust support.
Eligibility: You must be nominated by your prospective head of department and have an undertaking from the head of the institution, vice-chancellor, principal or dean that your personal support will be taken over by the institution at the end of the award.
Support is normally available only at lecturer level, although in exceptional cases awards to senior-lecturer level may be possible.
Level of Study: Research
Type: Award
Value: Up to five years' support is available, providing your full salary for three years, 50 per cent in the fourth year and 25 per cent in the fifth year. Travel expenses to attend meetings are provided for five years, but research expenses are provided for the first three years of the award only
Frequency: Thrice a year
Application Procedure: Initial enquiries about the scheme may be made by you (the potential candidate) or a department in an institution. These enquiries should be followed by a preliminary application from you by e-mail or post including an explicit statement from the head of the institution, vice-chancellor or dean demonstrating the institution's commitment to the history of medicine field, and a statement confirming that the institution will provide 50 per cent salary costs in year four, 75 per cent in year five and full salary thereafter; CV and full publication list; an outline of no more than two pages of the proposed project; a letter of support from the head of department, including a statement on your expected teaching/administrative load for the five-year period (this can be sent by separate cover); the approximate cost of the proposal, broken down into your salary, equipment and project running costs.
Closing Date: Preliminary application deadlines are: June 20th (with a full application deadline of August 1st), December 1st (with a full application deadline of February 1st)
Funding: Trusts

Value in People Awards

Subjects: Biomedical science.
Purpose: These awards help universities with the recruitment, career progression and retention of key academic and research staff. Awards are provided to the top 30 Trust-funded universities.
Eligibility: See a list of universities currently in receipt of funding on website.
Level of Study: Professional development
Type: Award
Value: We do not wish to be prescriptive about how funds are used, but as an example they could provide: salary funds for new recruits until a university post or fellowship becomes available; bridging funding for researchers on fixed-term contracts; short-term funding for non-biologists to work in biomedicine; funds for staff to attend short training or updating courses
Frequency: Various
Application Procedure: Awards are administered by the recipient universities.
Closing Date: Various
Funding: Commercial, corporation, foundation, government, individuals, international office, private, trusts

Veterinary Postdoctoral Fellowships

Subjects: Biomedical science.
Purpose: This scheme provides an opportunity for veterinary postdoctoral researchers to undertake high-quality research and develop their independence - in an appropriate research facility in a UK veterinary school - with a guarantee of a permanent post at the end of the fellowship.
Eligibility: You should have a bachelor's degree in veterinary medicine/science (e.g. BMedVet, BMedSci, VetMB) and will have completed, or be about to complete, a higher research degree (preferably a PhD). You should normally also have obtained a Royal College of Veterinary Surgeons (RCVS) certificate or equivalent (e.g. CertSAM, CertVA, CertEP). Previous research experience gained in a

laboratory environment beyond a veterinary school would also be advantageous. A strong collaborative link to a basic biomedical laboratory would be encouraged.
The awards are open to individuals with a relevant connection to the European Economic Area.
Level of Study: Postgraduate, Research
Type: Fellowship
Value: 1. A basic salary as determined by the veterinary school, with an additional Trust enhancement
2. Research expenses
3. Small items of equipment
4. Collaborative travel for scientifically justified visits
5. A travel allowance to attend scientific meetings (this is automatically provided as part of an award)
Length of Study: Up to 3 years
Frequency: Annual
Application Procedure: A preliminary application form should be completed and submitted by the published deadline. An electronic copy of the form (as a Word document) and each accompanying document should be emailed to vets@wellcome.ac.uk. One hard copy of the form (including signatures), should be sent to our postal address, marked for the attention of Veterinary Fellowships. Please note that all Word documents will need to be sent in Word 2003 (or an earlier version). If your preliminary application is successful, you will be invited to submit a full application. This will be reviewed and, if successful, you will be short-listed for interview.
Closing Date: November 12th
Funding: Trusts

Veterinary Research Entry Fellowships

Subjects: Biomedical science.
Purpose: This scheme provides one year's support for research-minded veterinarians to undertake a Master's degree by research.
Eligibility: You must have completed a first degree in veterinary medicine or veterinary science (e.g. BMedVet, BMedSci, VetMB), with a good track record of academic achievement during the course of your studies. Experience in clinical practice and a demonstrated interest in a research career, such as an intercalated degree, involvement in project work and/or summer school attendance would be advantageous. Applications are also welcomed from candidates who have had a career break and now wish to acquire research experience, with a view to considering a career in veterinary research. The awards are open to individuals with a relevant connection to the European Economic Area.
Level of Study: Doctorate, Postgraduate, Research
Type: Fellowship
Value: A basic salary as determined by the veterinary school; research expenses; Master's course fees; and a travel allowance
Length of Study: 1 year
Frequency: Annual
Application Procedure: An application form [Word 1.64MB] should be completed and submitted by the published deadline. Guidance notes [Word 203KB] accompany the form.
1. An electronic copy of the form (as a Word document) and each accompanying document should be emailed to vets@wellcome.ac.uk.
2. One hard copy of the form (including signatures), should be sent to our postal address, marked for the attention of Veterinary Fellowships.
Closing Date: February 7th
Funding: Trusts

Wellcome Trust and Howard Hughes Medical Institute Exchange Programme

Subjects: Biomedical science.
Purpose: The Wellcome Trust and Howard Hughes Medical Institute (HHMI) Exchange Programme promotes international collaborations among scientists funded by the Trust and HHMI. The programme provides training and career opportunities for members of Trust-funded teams to work with eligible HHMI laboratories.
Eligibility: In partnership with a HHMI investigator - or a group leader or fellow at the Janelia Farm Research Campus - the Exchange Programme is open to the following Trust grantholders:
1. Senior Research Fellows in Basic Biomedical Science
2. Senior Research Fellows in Clinical Science
3. Principal Research Fellows
4. Programme grantholders in a UK Wellcome Trust Centre
5. Investigators at the Wellcome Trust Sanger Institute.HMMI investigators and group leaders/fellows at the Janelia Farm Research Campus wishing to apply for an award should contact HHMI directly.
Level of Study: Research
Value: Funding is provided for between 3 and 12 months for a Trust-funded postdoctoral researcher to visit the laboratory of an HHMI investigator or group leader/fellow in the Janelia Farm Research Campus. Costs will be provided for the postdoctoral researcher's return flight to the USA and subsistence (an allowance of £1500 for each month of the proposed visit). Funding will be provided as a supplement to the award on which the postdoctoral researcher is supported and he/she should have salary support available on that award for the duration of the proposed exchange.
Frequency: Annual
No. of awards offered: 1
Application Procedure: Trust-funded applicants should complete an application form [Word 90kB] and return a hard copy to the relevant Scientific Programme Officer or funding stream (see website).
Closing Date: Applications can be made at any time and decisions will normally be made within six weeks of receipt.
Funding: Trusts

Wellcome Trust and NIH Four-year PhD Studentships

Subjects: Biomedical science.
Purpose: This scheme provides opportunities for the most promising postgraduate students to undertake international, collaborative four-year PhD training based in both a UK/Republic of Ireland (RoI) academic institution and the intramural campus of the National Institutes of Health at Bethesda (Maryland, USA).
Eligibility: You should be a UK/European Economic Area (EEA) national with (or be in your final year and expected to obtain) a first- or upper-second-class honours degree or an equivalent EEA graduate qualification. You must have:
1. A suitable doctoral supervisor at an eligible academic host institution in the UK or Republic of Ireland. The host institution must be able to confer doctoral degrees
2. A suitable supervisor at a NIH institute. The NIH supervisor should hold a tenured or tenured-track position for the proposed period of the award and should be willing to provide funding for the student whilst at the NIH.
Level of Study: Doctorate, Research
Type: Studentship
Value: The studentship is awarded for four years with support provided by the Wellcome Trust (in the UK/Republic of Ireland) and the NIH (in the USA).Our funding will provide support for the student's stipend, PhD fees, college fees (if required) and a contribution towards research costs.
Frequency: Annual
Application Procedure: The application form [1.1MB] should be completed and submitted by the closing date. An electronic copy (as a Word document) should be emailed to wtnih@wellcome.ac.uk. One signed hard copy should be addressed to the 'Wellcome Trust-NIH PhD studentships' at the Trust's postal address – see website.
Closing Date: February 18-19th (Shortlisted candidate interviews)
Funding: Trusts

Wellcome Trust Grants

Subjects: Biomedical sciences, from the basic sciences related to medicine to the clinical aspects of medicine and veterinary medicine. The Trust also operates a portfolio of schemes to support research in the history of medicine, biomedical ethics and the public engagement of science.
Purpose: To foster and promote research with the aim of improving human and animal health. The Trust aims to: (a) Support research to increase understanding of health and disease, and its societal context; (b) support the development and use of knowledge to create health benefit; (c) engage with society to foster an informed climate within which biomedical research can flourish; (d) foster a research community and individual researchers who can contribute to the advancement and use of knowledge; (e) promote the best conditions for research and the use of knowledge.
Eligibility: Eligible institutions in the UK and Republic of Ireland are normally universities, medical schools, or NHS Trusts. Institutions outside the UK and Republic of Ireland must confirm their status with the Wellcome Trust for research grants. Principal applicants are

established researchers who are applying from an eligible institution and are able to sign up to the Trust's grant conditions, and normally hold an academic or research post (or equivalent) and have at least 5 years' postdoctoral or equivalent research experience. Check website for further details.
Level of Study: Professional development, Doctorate, Postdoctorate, Postgraduate, Research
Type: Research grant or fellowship
Value: Varies
Length of Study: Varies
Frequency: Applications are considered throughout the year
No. of awards offered: Varies
Application Procedure: Applications can be submitted via online application system.
Closing Date: Please visit the website for scheme deadlines
Contributor: Endowment
No. of awards given last year: 951
No. of applicants last year: 2,736
Additional Information: The Wellcome Trust is one of the most richly endowed of all charitable institutions that fund general medical research in the United Kingdom. The Governors review their policy annually in response to proposals from their advisory committees and professional staff.

Wellcome Trust-Massachusetts Institute of Technology (MIT) Postdoctoral Fellowships

Subjects: Biomedical science.
Purpose: This scheme offers opportunities for postdoctoral scientists to undertake research at the interfaces between biology/medicine and mathematics, engineering, computer, physical or chemical sciences, firstly at MIT and then at a UK institution.
Eligibility: These awards are open to individuals with a relevant connection to the European Economic Area. If you have ever trained or worked in an academic or research institution in the United States, you are not eligible to apply to this scheme. You should be about to submit your doctoral thesis or have up to, but no more than, three years' postdoctoral experience from date of your PhD viva to the deadline for applications to the scheme.
Level of Study: Postgraduate, Research
Type: Fellowship
Value: The fellowship is for four years full-time, with the fellow based at MIT for two to three years (Phase 1) before returning to a host institution in the UK for the remainder of the award (Phase 2). Applicants are not expected to have made definitive arrangements for their return to the UK at the time of their initial application; the Trust acknowledges that any current plan may well be subject to change
Length of Study: 4 years
Frequency: Annual
Application Procedure: You must complete and submit an application using our web-based application system, eGrants, by the closing date.
Closing Date: July 12th, 5pm (Deadline for application); November 13-15th (Shortlisted candidate interviews)
Funding: Trusts

Wellcome Trust-POST Fellowships in Medical History and Humanities

Subjects: Medical history and humanities.
Purpose: This scheme enables a PhD student or junior fellow funded through the Wellcome Trust Medical History and Humanities (MHH) programme to undertake a three-month fellowship at the Parliamentary Office of Science and Technology (POST).
Eligibility: Applicants should be in the second or third year of their PhD or in the first year of a fellowship funded by the MHH Programme. POST is a strictly non-partisan organisation. Wellcome Trust-POST Fellows will be required to abstain from any lobbying or party political activity, and generally uphold the principles of parliamentary service, including a commitment to confidentiality, during their time with the Office. All provisionally selected candidates must sign a declaration to this effect. They must also receive security clearance from the parliamentary security authorities as a condition of finally taking up the fellowship.
Level of Study: Postdoctorate, Research
Type: Fellowship
Value: The successful applicant will receive a fully funded three-month extension to their PhD or fellowship award. While placements typically last three-months, they may be extended under exceptional circumstances. If the successful applicant is not within reasonable daily travelling distance to POST in London, the Wellcome Trust will consider paying travel and accommodation costs up to a maximum of £2000
Frequency: Annual
Application Procedure: An application should include the application form, your CV, a letter of support from your sponsor/supervisor and a summary of a proposed topic for a POST publication. The summary should be no longer than 1,000 words and should demonstrate: why you think this subject would be of particular parliamentary interest; how the training you have received and your research to date will enable you to carry out this work; your ability to write in a style suitable for a parliamentary (rather than an academic) audience.
Closing Date: November 23rd
Funding: Trusts

Wellcome-Beit Prize Fellowships

Subjects: Biomedical science.
Purpose: The Wellcome-Beit Prize Fellowships are intended to provide additional recognition for four outstanding biomedical researchers who have been awarded other Wellcome Trust fellowship funding. The awards were inaugurated in 2009 and replaced the Beit Memorial Fellowships for Medical Research.
Eligibility: Wellcome-Beit Prize Fellowships are considered during the interview process for Wellcome Trust Research Career Development Fellowships and Intermediate Clinical Fellowships. No separate application is required..
Level of Study: Research
Type: Fellowship
Value: £25,000 is awarded to each of four selected Research Career Development Fellows or Intermediate Clinical Fellows in addition to the salary and research expenses already to be funded by the Wellcome Trust. The £25,000 prize money can be used flexibly in support of the fellows' ongoing research
Frequency: Annual
No. of awards offered: 4
Application Procedure: None required.
Funding: Trusts

Wellcome-Wolfson Capital Awards in Biomedical Science

Subjects: Biomedical science.
Purpose: The Wellcome Trust and the Wolfson Foundation are pleased to announce a call for proposals for science-based capital projects that fall within the Trust's biomedical science remit. This scheme provides capital funding for large-scale projects (above £1 million), in partnership with the host institution.
Eligibility: Applicants should be researchers, normally based in the UK or Republic of Ireland, who fulfil our normal eligibility criteria for biomedical research grants.
Level of Study: Research
Type: Grant
Value: Awards over £1 million will be made
Frequency: Annual
Application Procedure: Prospective applicants are encouraged to contact us in the first instance to explore whether their proposal meets our criteria. This should be done well in advance of the deadline for preliminary applications.
Preliminary applications should include:
1. An outline project plan
2. Its key aims and objectives
3. How it fits with our strategy and the institutional strategy
4. The research groups that would benefit (including CVs of research team leaders and the source(s) of their research funding)
5. Details of funding requested from the funding organisations
6. A summary of other funding committed or sought
7. A supporting statement from the vice-chancellor (or equivalent) on behalf of the host institution
8. Evidence that the host institution has had preliminary discussions with local planning authorities where appropriate.

If successful, a full application will be invited. These should include a full business plan.
Closing Date: See website
Funding: Trusts

WENNER-GREN FOUNDATION FOR ANTHROPOLOGICAL RESEARCH

The Fellowships Office, 470 Park Avenue South, 8th Floor, New York, NY, 10016, United States of America
Tel: (1) 212 683 5000
Fax: (1) 212 683 9151
Email: inquiries@wennergren.org
Website: www.wennergren.org
Contact: Victoria Malkin, Anthropologist

The Wenner-Gren Foundation for Anthropological Research supports research, conferences, training, archiving and collaboration in all branches of anthropology, including cultural and social anthropology, ethnology, biological and physical anthropology, archaeology and anthropological linguistics, and in closely related disciplines concerned with human origins, development and variation.

Hunt Postdoctoral Fellowships

Subjects: Anthropology.
Purpose: To support the writing-up of already completed research.
Eligibility: Applicants must have a PhD or equivalent at the time of application and must have received a PhD or equivalent within ten years of the application deadline.Qualified scholars are eligible without regard to nationality, institutional, or departmental affiliation although preference is given to applicants who are untenured or do not yet have a permanent academic position.
Level of Study: Postdoctorate
Type: Fellowship
Value: Up to US$40,000
Frequency: Annual, Twice a year
No. of awards offered: Up to 8
Application Procedure: Applications can be downloaded from the website and must be submitted online.
Closing Date: November 1st and May 1st
No. of awards given last year: 9
No. of applicants last year: 89
Additional Information: Qualified scholars are eligible without regard to nationality or institutional affiliation.

Wenner-Gren Foundation Dissertation Fieldwork Grants

Subjects: Anthropology.
Purpose: To support basic research in anthropology and to ensure that the discipline continues to be a source of vibrant and significant work that furthers our understanding of humanity's cultural and biological origins, development, and variation.
Eligibility: Students must be enrolled in a doctoral program (or equivalent, if applying from outside the United States) at the time of application. Students of all nationalities are eligible to apply. Funding is to support research experiences only not tuition or writing of dissertation. Open to all individuals in a doctoral programme regardless of nationality or country of institution.
Level of Study: Doctorate
Type: Grant
Value: Dissertation Fieldwork Grants provide a maximum of US $20,000 and the Osmundsen Initiative supplement provides up to an additional $5,000 for a maximum grant of US$25,000
Frequency: Annual, Twice a year
Application Procedure: Applicants must complete a formal application on an up-to-date form that should be downloaded from the website and should be submitted online at www.wennergren.org.
Closing Date: May 1st and November 1st
No. of awards given last year: 109
No. of applicants last year: 662

Wenner-Gren Foundation Post-PhD Grants

Subjects: Cultural anthropology/physical anthropology/biological anthropology/linguistic anthropology plus archaeology.
Purpose: To support basic research in anthropology and to ensure that the discipline continues to be a source of vibrant and significant work that furthers our understanding of humanity's cultural and biological origins, development, and variation.
Eligibility: Open to individuals holding a PhD or equivalent degree to support individual research projects. Applicants can apply regardless of institutional affiliation, country of residence, or nationality.
Level of Study: Postdoctorate
Type: Grant
Value: Up to US$25,000
Frequency: Annual, Biannual
Application Procedure: Applicants must download an up-to-date form from the website.
Closing Date: May 1st
No. of awards given last year: 42
No. of applicants last year: 220

WESLEYAN UNIVERSITY

45 Wyllys Avenue, Middletown, Middlesex County, CT 06459, United States of America
Tel: (1) 860 685 2000
Fax: (1) 860 685 2171
Email: bkeating@wesleyan.edu
Website: www.wesleyan.edu
Contact: Ms Brenda Keating, Administrative Assistant

Wesleyan University offers instruction in 41 departments and programmes and 50 major fields of study and awards the Bachelor of Arts and graduate degrees. Master's degrees are awarded in 11 fields of study and doctoral degrees in 6. Students may choose from about 960 courses each year and may be asked to devise, with the faculty, some 1,500 individual tutorials and lessons.

Andrew W Mellon Postdoctoral Fellowship

Subjects: Humanities and humanistic social sciences.
Purpose: To provide scholars with free time to further their own work in a cross-disciplinary setting, and to associate them with a distinguished faculty.
Eligibility: Open to persons who have received their PhD within the last 4 years. Scholars who have received their PhD degree after June 2004 in any field of inquiry in the humanities or humanistic social sciences, broadly conceived, are invited to apply.
Level of Study: Postdoctorate
Type: Fellowship
Value: US$40,000
Frequency: Annual
Country of Study: Any country
No. of awards offered: 1
Application Procedure: Applicants must request a brochure detailing the application process. There is no formal application form. Applicants should refer to the Center for the Humanities website for instructions on how to apply.
Closing Date: January 10th
Funding: Private
No. of awards given last year: 2
No. of applicants last year: 285
Additional Information: The Fellow must reside in Middletown during the tenure of the fellowship, give one public lecture and teach one course for 20 students.

WILFRID LAURIER UNIVERSITY

75 University Avenue West, Waterloo, ON, N2L 3C5, Canada
Tel: (1) 519 884 1970
Fax: (1) 519 886 9351
Email: webmaster@wlu.ca
Website: www.wlu.ca
Contact: Mr Al Hecht, International Relations

Wilfrid Laurier University is well known for offering an extremely high quality academic experience as well as for cultivating a closely-knit undergraduate and graduate student population. Wilfrid Laurier University is committed to continuing to provide the educational experiences and environment that foster such development and nurture what can best be described as the "Laurier spirit".

Hans Viessmann International Scholarship

Subjects: All subjects.
Purpose: To assist students wanting to study in Germany.
Level of Study: Postgraduate
Type: Scholarship
Value: $900

Length of Study: 1 year
Frequency: Annual
Country of Study: Germany
Application Procedure: Contact University.
Closing Date: July 5th

President Marsden Scholarship

Subjects: All subjects.
Purpose: To reward a strong student who best exemplifies the mission of Laurier.
Eligibility: Applicant should be a full-time undergraduate or graduate students. Applicants must be Canadian citizens, Permanent Residents (landed immigrants) or Protected Persons
Level of Study: Postgraduate
Type: Scholarship
Value: Up to Canadian $1,400
Length of Study: 1 year
Frequency: Annual
Study Establishment: Laurier University
Country of Study: Canada
No. of awards offered: 1
Application Procedure: Apply online.
Closing Date: October 15th

Ross and Doris Dixon Special Needs Awards

Subjects: All subjects.
Purpose: To create a positive environment for students with special needs.
Eligibility: No restrictions.
Level of Study: Postgraduate
Type: Scholarship
Value: Varies
Length of Study: 1 year
Frequency: Annual
Study Establishment: Laurier University
Country of Study: Canada
No. of awards offered: 20
Application Procedure: Apply online.
Closing Date: September 30th
Funding: Trusts
Contributor: Ross and Doris Dixon

Viessmann/Marburg Travel Scholarship

Subjects: All subjects.
Purpose: To assist students wanting to study in Germany.
Level of Study: Postgraduate
Type: Scholarship
Value: €767
Length of Study: 1 year
Frequency: Annual
Study Establishment: An approved place of study in Marburg
Country of Study: Germany
Application Procedure: Contact University.
Closing Date: July 2nd

Wilfrid Laurier University Postdoctoral Fellowship

Subjects: All subjects.
Purpose: To provide full-time research opportunities for recent graduates who wish to pursue independent and collaborative research under the supervision of Laurier faculty.
Eligibility: Applicants must be within 5 years of the completion of all PhD requirements.
Level of Study: Postdoctorate
Type: Fellowship
Length of Study: 1–2 years
Study Establishment: Wilfrid Laurier University
Country of Study: Canada
Application Procedure: Self-funded applicants must submit a copy of their SSHRC or NSERC PDF application, curriculum vitae and a detailed research plan for each year. Applicants who wish to teach will be interviewed by the chair.

William and Marion Marr Graduate Award

Subjects: All subjects.
Purpose: To aid graduate students with special needs.
Level of Study: Postgraduate
Type: Scholarship
Value: Up to Canadian $2,000
Length of Study: 1 year
Frequency: Annual
Study Establishment: Laurier University
Country of Study: Canada
Application Procedure: Apply online.
Closing Date: October 17th
Funding: Private
Contributor: Dr and Mrs Marr

WLU Graduate Incentive Scholarships

Subjects: All subjects.
Purpose: To reward graduate students who are successful in major external scholarship competitions.
Level of Study: Postgraduate
Type: Scholarship
Value: $1,000 to $9,000 over a period of one to three years.
Length of Study: 1–3 years
Frequency: Annual
Study Establishment: Laurier University
Country of Study: Canada
No. of awards offered: Varies
Application Procedure: Contact University.
Closing Date: No deadline

WLU Graduate Scholarships

Subjects: All subjects.
Level of Study: Postgraduate
Type: Scholarship
Value: Canadian $1,000
Length of Study: 1 year
Frequency: Annual
Study Establishment: Laurier university
Country of Study: Canada
Application Procedure: Apply online.
Closing Date: October 11th
Additional Information: A scholarship is awarded on academic merit, not financial need.

WLU President's Centennial Scholarship

Subjects: All subjects.
Purpose: To reward significant contribution to the community as a volunteer or to the discipline as a scholar.
Eligibility: Full-time undergraduate students entering year 1. Minimum overall average of 95 per cent in best six Grade 12 U and/or Grade 12 M courses or Ontario Academic Credits (OACs) (or equivalent).
Level of Study: Postgraduate
Type: Scholarship
Value: $3,000 for first year and $5,000 after first year
Length of Study: 1 year
Frequency: Annual
Study Establishment: Laurier University
Country of Study: Canada
Application Procedure: Apply online.
Closing Date: February 28th
Funding: Trusts
Contributor: Dr Neale H Taylor

WLU Student International Travel Scholarship

Subjects: All subjects.
Purpose: To assist Laurier students with travel costs associated with an academic exchange programme.
Level of Study: Postgraduate
Type: Scholarship
Value: Varies
Length of Study: 1 year
Frequency: Annual
Study Establishment: An approved University or Institute
Application Procedure: Contact University.
Closing Date: July 2nd

WILLIAM HONYMAN GILLESPIE SCHOLARSHIP TRUST

Messrs Tods Murray LLP, Edinburgh Quay, 133 Fountain Bridge, Edinburgh, EH3 9AG, Scotland
Tel: (44) 131 656 2000
Fax: (44) 131 656 2020
Email: maildesk@todsmurray.com
Website: http://www.todsmurray.com/
Contact: Trustees

William Honyman Gillespie Scholarships

Subjects: Theology.
Purpose: To allow the recipient to engage in a full-time approved scheme of theological studies or research.
Eligibility: Open to graduates of a theological college of one of the Scottish universities.
Level of Study: Postgraduate
Type: Scholarship
Value: UK£1,000 per year
Length of Study: 2 years
Frequency: Annual
Study Establishment: An approved university or similar institution
Country of Study: Any country
No. of awards offered: Varies, usually 1 to 2
Application Procedure: Applicants must submit applications through the principal of the theological college of which the applicant is a graduate. Application guidelines are available from the Trust or the candidate's university department.
Closing Date: May 15th
Funding: Private
No. of awards given last year: 2
No. of applicants last year: 2

WILLIAM J. CUNLIFFE SCIENTIFIC AWARDS

Department of Dermatology, Venereology, Allergology and Immunology, Dessau Medical center, Auenweg 38, Dessau, 06847, Germany
Tel: (49) 340 501 4000
Fax: (49) 340 501 4025
Email: info@cunliffe-awards.de
Website: www.cunliffe-awards.de
Contact: Professor Dr Christos C. Zouboulis, Chairman of the Executive Committee

The William J Cunliffe Scientific Awards aim to recognize and encourage innovative and outstanding research in the areas of endocrine dermatology and skin pharmacology, conferring great benefit upon understanding the function of the pilosebaceous unit as well as the pathophysiology and treatment of its disease.

William J Cunliffe Scientific Awards

Subjects: The significance of skin, and especially of the pilosebaceous unit, as hormone target and endocrine organ, the development of new molecules to target skin diseases and the understanding of the molecular action of therapeutic compounds.
Purpose: The William J Cunliffe Scientific Awards aim to recognize and encourage innovative and outstanding research in the areas of endocrine dermatology and skin pharmacology.
Eligibility: Open to persons for any nationality. Living individuals or public or private institutions of any nation, are eligible for nomination.
Level of Study: Professional development
Type: Lectureship/Prize
Length of Study: Nominees will be considered for the year of their nomination
Frequency: Dependent on funds available
Country of Study: Worldwide
No. of awards offered: 1 per year (major)
Application Procedure: Applicants must complete an application form. Guidelines can be found on the website www.cunliffe-awards.org. Applicants should be nominated.
Closing Date: May 31st
Funding: Corporation

WILSON ORNITHOLOGICAL SOCIETY

Biology Department, Albion College, Albion, MI, 49224, United States of America
Tel: (1) 254 399 9636
Fax: (1) 254 776 3767
Email: DKennedy@albion.edu
Website: www.wilsonsociety.org
Contact: Dr Dale Kennedy, The Administrative Assistant

Founded in 1888 and named after Alexander Wilson, the father of American ornithology, the Wilson Ornithological Society publishes a scientific journal, the *Wilson Journal of Ornithology*, holds annual meetings, provides research awards and maintains an outstanding research library.

George A Hall/Harold F Mayfield Award

Subjects: Any aspect of ornithology.
Purpose: To encourage and stimulate research projects on birds, by amateurs and students.
Eligibility: Open to independent researchers without access to funds and facilities available at colleges, universities or governmental agencies. The award is restricted to non-professionals.
Level of Study: Unrestricted
Type: Award
Value: US$1,000
Frequency: Annual
Country of Study: Any country
No. of awards offered: 1
Application Procedure: An application form must be completed and submitted with three letters of recommendation and a research proposal. Forms are available from the website.
Closing Date: February 1st
Funding: Private

Louis Agassiz Fuertes Award

Subjects: Any aspect of ornithology.
Eligibility: Open to all ornithologists, although graduate students and young professionals are preferred. Any avian researcher is eligible.
Level of Study: Unrestricted
Type: Award
Value: US$2,500
Frequency: Annual
No. of awards offered: 2
Application Procedure: Application forms are available from the website.
Closing Date: February 1st
No. of awards given last year: 37

WINSTON CHURCHILL FOUNDATION OF THE USA

600 Madison Avenue, Suite 1601, New York, NY, NY 10022-1615, United States of America
Tel: (1) 212 752 3200
Fax: (1) 212 246 8330
Email: info@winstonchurchillfoundation.org
Website: www.winstonchurchillfoundation.org
Contact: Mr Peter C. Patrikis, Executive Director

The Winston Churchill Foundation of the United States was established in 1959 as an expression of American admiration for one of the great leaders of the free world. The foundations enables outstanding American students to attend graduate school at the University of Cambridge.

Winston Churchill Foundation Scholarship

Subjects: Engineering, mathematics, computer science and natural and physical sciences.
Purpose: To encourage the development of American scientific and technological talent and foster Anglo-American ties.
Eligibility: Open to citizens of the United States of America only. Applicants must be enrolled in one of 102 institutions participating in the programme, may not have enrolled a PhD and must be no older than 26 years at the time of taking up the scholarships.

Level of Study: Postgraduate, Predoctorate
Type: Scholarship
Value: Approx. US$45,000–50,000
Length of Study: The Scholarship is tenable from nine to twelve months, in accordance with the academic program.
Frequency: Annual
Study Establishment: Churchill College, the University of Cambridge
Country of Study: United Kingdom
No. of awards offered: 14
Application Procedure: Applicants must apply through their sponsoring institution.
Closing Date: November 13th
Funding: Foundation
No. of awards given last year: 14
No. of applicants last year: 100

WINSTON CHURCHILL MEMORIAL TRUST (AUS)

GPO Box 1536, Canberra City, ACT, 2601, Australia
Tel: (61) 26 2478 333
Fax: (61) 26 2498 944
Email: info@churchilltrust.com.au
Website: www.churchilltrust.com.au
Contact: Ms Louise Stenhouse, Senior Executive Officer, Finance and Administration

The Winston Churchill Memorial Trust (Aus) perpetuates and honours the memory of Sir Winston Churchill by awarding memorial Fellowships known as Churchill Fellowships.

Churchill Fellowships

Subjects: All subjects.
Purpose: To enable Australian citizens over the age of 18 to undertake an overseas research project of a kind that is not available in Australia.
Eligibility: All Australian citizens over the age of 18 years are eligible to apply. Applications are assessed on merit of the proposal and benefit to the Australian community at either a local, State or National level. Applications to further tertiary qualifications will not be eligible.
Level of Study: Unrestricted
Type: Fellowship
Value: Average of Australian $20,000, return economy airfare to the country or countries to be visited and a living allowance plus fees if approved
Length of Study: 4–6 weeks, depending upon the project
Frequency: Annual
Study Establishment: 1965
Country of Study: Any country
No. of awards offered: Over 100 per year
Application Procedure: Applicants must complete and submit an application form supported by 2 references. For an application form, contact the National Office or visit the website.
Closing Date: Late February
Funding: Private
No. of awards given last year: 115
No. of applicants last year: 1054
Additional Information: Applicants need to apply in their state or territory of residence. Details and addresses are contained in the application package.

For further information contact:

National Office, Australia
Tel: (61) 1800 777 231

WINSTON CHURCHILL MEMORIAL TRUST (UK)

South Door, Church House, Great Smith Street, London, SW1P 3BL, England
Tel: (44) 20 7799 1660
Fax: (44) 20 7799 1667
Email: office@wcmt.org.uk
Website: www.wcmt.org.uk
Contact: Ms S Matthews, Trust Office Manager

Each year the trust awards fellowships to 100 British citizens to carry out Travelling Fellowships overseas for the benefit of country, community and international goodwill. Different categories of fellowships are awarded each year, e.g. conservation of the environment, agriculture and horticulture, science and technology, adventure, exploration and leaders of expeditions.

Winston Churchill Memorial Trust (UK) Travelling Fellowships

Subjects: Approx. 10 categories that vary annually and are representative of culture, social and public service, technology, commerce and industry, agriculture and nature, recreation and adventure.
Purpose: To enable British citizens resident in the UK from all walks of life and all ages to travel abroad in pursuit of a worthwhile purpose and so to contribute more to their trade or profession, their community back in the UK.
Eligibility: Open to British citizens resident in the UK whose purposes must be covered by one of the categories chosen for the year. Not for 'Gap year' activities, courses, academic studies or student grants; this includes electives, degree placements, internships and post graduate studies, unless real and wiser benefits to others in the UK centre demonstrated.
Level of Study: Unrestricted
Type: Fellowship
Value: UK£5,600 approx. Dependent on scope of award
Length of Study: 4–8 weeks
Frequency: Annual
Country of Study: Any country
No. of awards offered: Approx. 100
Application Procedure: Applicants must complete an application form.
Closing Date: Mid-October
Funding: Private
Contributor: The public
No. of awards given last year: 100
No. of applicants last year: 1,059
Additional Information: Award winners are announced at the beginning of February, and Fellows may travel after April 1st.

WINTERTHUR

Winterthur Museum, Garden and Library, Route 52, Winterthur, DE, 19735, United States of America
Tel: (1) 800 448 3883
Email: tourinfo@winterthur.org
Website: www.winterthur.org
Contact: Rosemary T Krill, Senior Lecturer, Academic Programmes

Founded by Henry Francis du Pont, Winterthur is the premier museum of American Decorative Arts, reflecting both early America and the du Pont family's life here. Its 60-acre naturalistic garden is among the country's best, and its research library serves scholars from around the world.

Winterthur Dissertation Research Fellowships

Subjects: American history and art history.
Purpose: To encourage the use of Winterthur's collections for critical inquiry that will further the understanding of American history and visual and material culture.
Level of Study: Research
Type: Fellowship
Value: US$7,000–14,000
Length of Study: 1–2 semesters
Frequency: Annual
Country of Study: United States of America
Application Procedure: Applicants can download application form from the website.
Funding: Foundation
Contributor: Winterthur
No. of awards given last year: 4

Winterthur National Endowment for the Humanties Fellowships

Subjects: American history and art history.

Purpose: To encourage the use of Winterthur's collections for critical inquiry that will further the understanding of American history and visual and material culture.
Eligibility: Open to scholars who hold the PhD degree, pursuing advanced research and who are citizens of the United States or residents for 3 years.
Level of Study: Research, Postdoctorate
Type: Fellowships
Value: US$4200 per month
Length of Study: 4–9 months
Frequency: Annual
Country of Study: United States of America
Application Procedure: Applicants can download the application form from the website.
Closing Date: January 15th
Funding: Government
Contributor: National Endowment for the Humanities
No. of awards given last year: 2

For further information contact:

Office of Academic Programmes Winterthur Museum, Garden and Library, Country Estate, Winterthur, DE, 19735, United States of America
Email: rkrill@winterthur.org
Website: www.winterthur.org
Contact: Rosemary T Krill, Senior Lecturer

Winterthur Research Fellowships

Subjects: American history and art history.
Purpose: To encourage the use of Winterthur's collections for critical inquiry that will further the understanding of American history and visual and material culture.
Eligibility: Open to scholars pursuing advanced research.
Level of Study: Research
Type: Fellowship
Value: US$1,500
Length of Study: 1 month
Frequency: Annual
Country of Study: United States of America
Application Procedure: Applicants can download the application form from the website.
Closing Date: January 15th
Funding: Foundation
Contributor: Winterthur
No. of awards given last year: 21

For further information contact:

Office of Academic Programmes, Winterthur Museum, Garden and Library, Winterthur, DE 19735, United States of America
Email: rkrill@winterthur.org
Website: www.winterthur.org
Contact: Rosemary T Krill, Senior Lecturer Officer

THE WOLFSON FOUNDATION

8 Queen Anne Street, London, W1G 9LD, England
Tel: (44) 20 7323 5730
Fax: (44) 20 7323 3241
Website: www.wolfson.org.uk
Contact: The Chief Executive

The aim of the Wolfson Foundation is the advancement of the arts and humanities, science, health and education. Grants are given to back excellence and talent, generally through the funding of capital infrastructure.

Wolfson Foundation Grants

Subjects: Medicine and healthcare, including the prevention of disease, and the care and treatment of the sick and disabled, research, science, technology and education, and the arts and humanities, including libraries, museums, galleries, theatres, academies and historic buildings.
Eligibility: Open to registered charities and to exempt charities such as universities. Eligible applications from registered charities for contributions to appeals will normally be considered only when at least 50 per cent of that appeal has already been raised. Grants to universities for research and scholarship are normally made under the umbrella of designated competitive programmes in which vice chancellors and principals are invited to participate from time to time. Applications from university researchers are not considered outside these programmes. Grants are not made to private individuals.
Level of Study: Research, Postgraduate
Type: Grant
Value: The Trustees make several types of grants, which are not necessarily independent of each other. Capital Project Grants may contribute towards the cost of erecting a new building or extension, or of renovating and refurbishing existing buildings. Equipment Grants supply equipment for specific purposes and/or furnishing and fittings. Recurrent costs are not normally provided
Frequency: Twice a year
Country of Study: United Kingdom or Israel or The Commonwealth countries
No. of awards offered: Varies
Application Procedure: Applicants must submit in writing a brief outline of the project with one copy of the organization's most recent audited accounts for the past two years before embarking on a detailed proposal. Please see website at www.wolfson.org.uk for information before writing.
Closing Date: March 1st or September 1st
Funding: Private
No. of awards given last year: 330
No. of applicants last year: 1000

THE WOLFSONIAN-FLORIDA INTERNATIONAL UNIVERSITY

1001 Washington Avenue, Miami Beach, FL, 33139, United States of America
Tel: (1) 305 531 1001
Fax: (1) 305 531 2133
Email: research@thewolf.fiu.edu
Website: http://www.wolfsonian.org/
Contact: Mr Jonathan Mogul, Fellowship Co-ordinator

The Wolfsonian-Florida International University is a museum and research centre that promotes the examination of modern material culture. Through exhibitions, publications, scholarships, educational programmes and public presentations, the Wolfsonian strives to enhance the understanding of objects as agents and reflections of social, cultural, political and technological change. The collection includes works on paper, furniture, paintings, sculpture, glass, textiles, ceramics, books and many other kinds of objects.

Wolfsonian-FIU Fellowship

Subjects: Modern material and visual culture.
Purpose: To conduct research on the Wolfsonian's collection of objects and library materials from the period 1885 to 1945, including decorative arts, works on paper, books and ephemera.
Eligibility: The programme is open to holder of Master's or doctoral degrees, PhD candidates, and to other who have a record of significant professional achievement in relevant fields.
Level of Study: Doctorate, Postdoctorate, Professional development
Type: Fellowship
Value: Fellowships include a stipend, accommodations, and round-trip travel.
Length of Study: 3–5 weeks
Frequency: Annual
Study Establishment: The Wolfsonian-Florida International University
Country of Study: United States of America
No. of awards offered: Varies, approx. 5
Application Procedure: Applicants must complete an application form and submit this with three letters of recommendation. Contact the Fellowship Co-ordinator for details and application materials. Applicants may also download programme information and an application form from the website www.wolfsonian.fiu.edu/education/research/
Closing Date: December 31st
No. of awards given last year: 5
No. of applicants last year: 26

WOMEN BAND DIRECTORS INTERNATIONAL (WBDI)

7424 Whistlestop Drive, Austin, TX, 78749, United States of America
Tel: (1) 512 496 3591
Fax: (1) 512 841 3811
Email: dgorzycki@austin.rr.com
Website: www.womenbanddirectors.org
Contact: Diane Gorzycki, Scholarships Chair

Women Band Directors International (WBDI) is an organization in which every woman band director is represented at the international level, regardless of the length of her experience or the level at which she works. It is the only international organization for women band directors.

WBDI Scholarship Awards

Subjects: Music education.
Purpose: To support young college women presently preparing to be band directors.
Eligibility: Any female student enrolled in college and majoring in music education with the purpose of becoming a band director.
Level of Study: Unrestricted
Type: Award
Value: US$1,500
Frequency: Annual
Country of Study: United States of America
No. of awards offered: 5
Application Procedure: Download the application from our website- essay, 2 letters of recommendations transcript, statement of philosophy, completed application.
Closing Date: December 1st
Funding: Private
No. of awards given last year: 4
No. of applicants last year: 50 per year

WOMEN OF THE EVANGELICAL LUTHERAN CHURCH IN AMERICA (ELCA)

8765 W. Higgins Road, Chicago, IL, 60631-4101, United States of America
Tel: (1) 800 638 3522
Fax: (1) 773 380 1465
Email: info@elca.org
Website: www.womenoftheelca.org

The mission of ELCA is to mobilize women to act boldly on their faiths. As an organization with an anti-racist identity, it offers opportunities for growth by cross-cultural ministry and leadership development.

Amelia Kemp Scholarship

Subjects: All subjects.
Purpose: To provide assistance to women studying for a career other than the ordained ministry.
Eligibility: Open to female citizens of the United States who are members of the ELCA.
Level of Study: Professional development
Type: Scholarship
Value: US$600–1,000
Length of Study: 1 year
Frequency: Annual
Country of Study: United States of America
No. of awards offered: 1
Application Procedure: A completed application form must be submitted.
Closing Date: February 15th
Contributor: Lutheran Church in America

Arne Administrative Leadership Scholarship

Subjects: Church administration.
Purpose: To provide assistance to ELCA women interested in reaching the top of their field as an administrator.
Eligibility: Applicant must be a female U.S. citizen who holds a membership in the Evangelical Lutheran Church of America, and has completed a B.A. or B.S. degree or its equivalent.
Level of Study: Professional development
Type: Scholarship
Value: Upto US$2,000
Frequency: Annual
Application Procedure: See the website.
Closing Date: February 15th

Belmer/Flora Prince Scholarship

Subjects: Theology.
Purpose: To assist ELCA women as they prepare for ELCA service abroad.
Eligibility: Applicant must be a woman who is a member of the Evangelical Lutheran Church in America and is studying for service abroad.
Level of Study: Professional development
Type: Scholarship
Value: US$800–$1,000
Frequency: Annual
Country of Study: United States of America
No. of awards offered: 2
Application Procedure: Application forms are available on the website.
Closing Date: February 15th
Contributor: Lutheran Church in America

Herbert W. and Corinne Chilstrom Scholarship for Women Preparing for Ordained Ministry

Subjects: Theology.
Purpose: To provide assistance to ELCA women during their final year at an ELCA seminary.
Eligibility: Open to women candidates of the United States who are members of the ELCA.
Level of Study: Professional development
Type: Scholarship
Value: US$600–1,000
Frequency: Annual
Application Procedure: Application form available on the website.
Closing Date: February 15th

WOMEN'S RESEARCH AND EDUCATION INSTITUTE (WREI)

714 G Street S.E., Suite 200, Washington, DC, WA, 20003, United States of America
Tel: (1) 202 280 2720
Fax: (1) 202 293 4507
Email: WREI@WREI.org
Website: www.wrei.org
Contact: Fellowship Programme

Founded in 1977, Women's Research and Education Institute (WREI) is an independent, non-profit, non-partisan organization governed by a board of directors that includes leading Americans from many fields who are committed to equality for women. Its mission is to inform and help shape the public policy debate on issues affecting women and their roles in the family, workplace and public arena.

WREI Congressional Fellowships on Women and Public Policy

Subjects: Public policy.
Purpose: To encourage more effective participation by women in the formulation of public policy. To train feminist leaders in legislative procedure at the national level.
Eligibility: Open to students who are currently in, or have recently completed a graduate or professional degree programme at an accredited institution in the United States.
Level of Study: Graduate, MBA, Postgraduate
Type: Fellowship
Value: Stipend of approximately $1,450 per month, up to $500 for purchase of health insurance, reimbursement up to a maximum of $1,500 for the cost of tuition at their homeinstitutions

Frequency: Annual
Country of Study: United States of America
No. of awards offered: At least 5
Application Procedure: Application forms can be downloaded from the website www.wrei.org every year in January. Applications by email are encouraged. Please note that applications sent by fax are not accepted.
Closing Date: May 20th
Funding: Corporation, foundation, individuals
No. of awards given last year: 5 or more
No. of applicants last year: 35–40
Additional Information: Fellows are selected on the basis of academic competence as well as their demonstrated interest in the public policy process. They are expected to be articulate and adaptable and to have strong writing skills.

WOMEN'S STUDIO WORKSHOP (WSW)

722 Binnewater Lane PO Box 489, Rosendale, NY, 12472, United States of America
Tel: (1) 845 658 9133
Fax: (1) 845 658 9031
Email: info@wsworkshop.org
Website: www.wsworkshop.org
Contact: Ms Ann Kalmbach, Executive Director

The Women's Studio Workshop (WSW) is an artist-run workshop with facilities for printmaking, papermaking, photography, book arts and ceramics. WSW supports the creation of new work through studio residency and annual book arts grant programmes and an ongoing subsidized fellowship programme. WSW offers studio-based educational programming in the above disciplines through its annual Summer Arts Institute.

Artists Fellowships at WSW

Subjects: Intaglio, water-based silkscreen, photography, papermaking or ceramics, letterpress, book arts, ceramics.
Purpose: To provide a time for artists to explore new ideas in a dynamic and co-operative community of women artists in a rural environment.
Eligibility: Open to women artists only.
Level of Study: Unrestricted
Type: Fellowship
Value: The award includes on-site housing and unlimited access to the studios. Cost to artists will be US$200 per week, plus their own material
Length of Study: Each fellowship is 3–6 weeks long. Fellowship opportunities are from September to June
Frequency: Annual
Study Establishment: WSW
Country of Study: United States of America
No. of awards offered: 10–20
Application Procedure: Applicants must complete an application form, available on request or online at the website. One-sentence summary plus half-page description of proposed project, resume, 10 slides plus slide script, self addressed stampe envelope for return of materials.
Closing Date: March 15th or October 15th
Funding: Government, private
Contributor: Private foundations
No. of awards given last year: 25
No. of applicants last year: 100

Artists' Book Residencies at WSW

Subjects: Artists books.
Purpose: To enable artists to produce a limited edition book work at the Women's Studio Workshop.
Level of Study: Unrestricted
Type: Residency grant
Value: The grant includes a stipend of $350/week, materials up to $750, access to all studios, travel costs up to $250 within the Continental US, and housing.
Length of Study: 6–8 weeks
Frequency: Annual
Study Establishment: WSW
Country of Study: United States of America
No. of awards offered: Varies, usually 3–5
Application Procedure: Applicants must submit an application including a one-sentence summary followed by a half-page description of the proposed project, the medium or media used to print the book, the number of pages, page size, edition number, a structural dummy, materials budget, a curriculum vitae, 10 slides and a stamped addressed envelope for return of materials. Applications are reviewed by past grant recipients and a WSW staff artist. Applicants should write for an application form or download one from the website plus slide script including title, media, dimension and date.
Closing Date: November 15th
Funding: Government, private
Contributor: Private foundations
No. of awards given last year: 2
No. of applicants last year: 150

Studio Residency Grant at WSW

Subjects: Books, printmaking papermaking, photography and clay.
Purpose: To provide artists with time and resources to create a new body of work or to edition a new bookwork.
Eligibility: Open to all national and international applicants. Emerging artists are encouraged to apply.
Level of Study: Unrestricted
Type: Grant
Value: The grant includes a stipend of $350 per week, up to $500 toward materials used during the residency, travel costs up to $250 within the Continental US, housing, and unlimited studio use.
Length of Study: 6–8 weeks
Frequency: Annual
Study Establishment: WSW
Country of Study: United States of America
No. of awards offered: 2
Application Procedure: Applicants must submit an application form, a one-semester summary followed by a half-page project description on a separate sheet of paper, a curriculum vitae, ten slides of recent work, a slide script including title, medium, size and date and a stamped addressed envelope, for return of materials. Forms are available from the website.
Closing Date: April 1st
Funding: Foundation, government
No. of awards given last year: 2
No. of applicants last year: 80

WSW Hands-On-Art Visiting Artist Project

Subjects: Artist's books.
Purpose: To assist an emerging artist in the creation of a new artist's book, while also working with school children in WSW's studio-based Art-In-Education programme.
Eligibility: Open to all artists. Emerging artists are encouraged to apply.
Level of Study: Unrestricted
Type: Grant
Value: A stipend of up to US$400 per week for 10 weeks, plus materials of up to US$750 and housing plus travel costs
Length of Study: 8 weeks
Frequency: Annual
Study Establishment: WSW
Country of Study: United States of America
No. of awards offered: 2
Application Procedure: Applicants must apply through a two-part application. Artists apply to WSW, WSW juries and then applies to other funding sources. Artists must submit a one-sentence description plus half-page description of their intended artist book project, including details of the medium to be used for printing the book, number of pages, page size, edition size (100 preferred), a structural dummy, materials budget, a curriculum vitae, a one-page description of relevant work experience with young people, ten slides of recent work and a stamped addressed envelope.
Closing Date: November 15th
Funding: Foundation, government
No. of awards given last year: 2
No. of applicants last year: 30

WSW Internships

Subjects: Book arts, papermaking, printmaking, ceramics and photography plus arts administration.

Purpose: To provide opportunities for young artists to continue development of their work in a supportive environment, while learning studio skills and responsibilities.
Eligibility: Open to emerging and established female artists.
Level of Study: Unrestricted
Type: Internship
Value: A private room in our onsite housing and a stipend of $250/month.
Length of Study: 2–6 months
Frequency: Annual
Study Establishment: WSW
Country of Study: United States of America
No. of awards offered: 5
Application Procedure: Applicants must submit a curriculum vitae, 10 slides with slide list, 3 current letters of reference, a letter of interest that addresses the question of why an internship at WSW would be important and a stamped addressed envelope. Arts administration: 3 work samples, i.e. press releases, design samples, etc.
Closing Date: See website
Funding: Government, private
Contributor: Private foundations
No. of awards given last year: 6
No. of applicants last year: 150

THE WOODROW WILSON NATIONAL FELLOWSHIP FOUNDATION

P O Box 5281, Princeton, NJ, 08543-5281, United States of America
Tel: (1) 609 452 7007
Fax: (1) 609 452 0066
Email: marrero@woodrow.org
Website: www.woodrow.org
Contact: Ms Antoinette Marrero, Communications Associate

The Woodrow Wilson National Fellowship Foundation identifies and develops the best minds for the nation's most important challenges. The fellowships are awarded to enrich human resources, work to improve public policy, and assist organizations and institutions in enhancing practice in the US and abroad.

Charlotte W Newcombe Doctoral Dissertation Fellowships

Subjects: Religion, ethics, values, humanities and social sciences.
Purpose: To encourage new and significant study of ethical or religious values in all fields of humanities and social sciences.
Eligibility: Open to students enrolled in doctoral programmes in the humanities and social sciences at a university in the United States of America. Students must have completed all predissertation requirements before the application deadline.
Level of Study: Doctorate
Type: Fellowship
Value: US$25,000
Length of Study: 1 year, full-time
Frequency: Annual
Study Establishment: Any appropriate graduate school in the US
Country of Study: Any country
No. of awards offered: Minimum 21
Application Procedure: Applicants must visit the organization website at www.woodrow.org/newcombe
Closing Date: November
Funding: Private
No. of awards given last year: 21
No. of applicants last year: 550

Thomas R. Pickering Graduate Foreign Affairs Fellowship

Purpose: Prepare participants for a career as a Foreign Service Officer in the US Department of State.
Eligibility: U.S. citizen, minimum 3.2 GPA on 4.0 scale, entering a 2-year terminal master's degree program related to international affairs.
Level of Study: Graduate
Type: Fellowship
Value: $40,000
Length of Study: 2 years
Frequency: Annual
Country of Study: United States of America
No. of awards offered: 20
Application Procedure: Visit www.woodrow.org.
Closing Date: January 21st
Funding: Government
Contributor: U.S. Department of State
No. of awards given last year: 20

Woodrow Wilson Teaching Fellowship

Subjects: Teacher education, STEM (Science, Technology, Engineering, Mathematics) teaching, secondary school teaching
Eligibility: The Woodrow Wilson Teaching Fellowship seeks to attract talented, committed individuals with science, technology, engineering, and mathematics (STEM) backgrounds – including current undergraduates, recent college graduates, midcareer professionals, and retirees – into teaching in high-need secondary schools. A qualified applicant should demonstrate a commitment to the program and its goals; have US citizenship or permanent residency; have attained, or expect to attain by June 30th, a bachelor's degree from an accredited US college or university; have majored in and/or have a strong professional background in an STEM field; have achieved a cumulative undergraduate grade point average (GPA) of 3.0 or better on a 4.0 scale (negotiable for applicants from institutions that do not employ a 4.0 GPA scale).Note - Prior teaching experience does not exclude a candidate from eligibility. All applications are considered in their entirety and selection is based on merit.
Level of Study: Graduate
Type: Fellowship
Value: $30,000 stipend
Length of Study: 12–18 months plus 3 year teaching commitment
Frequency: Annual
Study Establishment: Fellowship is only available for use at specific schools in Indiana (Ball State University, Indiana University-Purdue University Indianapolis, Purdue University, and the University of Indianapolis); Michigan (Eastern Michigan University, Grand Valley State University, Michigan State University, University of Michigan, Wayne State University, Western Michigan University); and Ohio (John Carroll University, Ohio State University University of Akron and University of Cincinnati)
No. of awards offered: 280
Application Procedure: Online application procedure and supporting documents. See http://www.wwteachingfellowship.org.
Closing Date: February 15th
Funding: Foundation, government, private
Contributor: Ohio STEM, Lilly Endowment Inc., W K Kellogg Foundation, Choose Ohio First
No. of awards given last year: 221
No. of applicants last year: 482
Additional Information: University of Dayton and University of Toledo are also included in study establishment.

For further information contact:

Tel: 609 452 7007 ext. 141
Email: wwteachingfellowship@woodrow.org

WOODS HOLE OCEANOGRAPHIC INSTITUTION (WHOI)

266 Woods Hole Road, Woods Hole, MA, 02543-1050, United States of America
Tel: (1) 508 548 1400
Fax: (1) 508 457 2188
Email: information@whoi.edu
Website: www.whoi.edu/education
Contact: Janet Fields, Coordinator

The Woods Hole Oceanographic Institution is a private, independent, non-profit corporation dedicated to research and higher education at the frontiers of ocean science. Its primary mission is to develop and effectively communicate a fundamental understanding of the processes and characteristics governing how the oceans function and how they interact with the Earth as a whole.

CICOR Postdoctoral Scholar Fellowship in Coastal Oceanography, Climate or Marine Ecosystems

Subjects: Coastal ocean and near shore processes, the ocean's participation in climate and climate variability and marine ecosystem processes analysis.

Purpose: To build ties between WHOI investigators and colleagues at NOAA laboratories and to develop co-operative NOAA funded research at academic institutions in the Northeastern United States of America. The fellowship also aims to further the education and training of recent recipients of doctoral degrees in the marine sciences.
Eligibility: Applicants must have received their doctoral degree within the past 4 years, completed if full in biology, physics, microbiology, molecular biology, chemistry, geology, geophysics, oceanography, meteorology, engineering or mathematics. Candidates should have a command of the English language. Candidates holding a WHOI appointment at the post-PhD level the 12 months prior to the scholar application deadline are not eligible.
Level of Study: Postdoctorate
Type: Fellowship
Value: Minimum: $45,250; Average: $45,250; Maximum: $45,250
Length of Study: 18 months
Frequency: Annual
Study Establishment: Woods Hole Oceanographic Institution
Country of Study: United States of America
Closing Date: January 15th
Funding: Government
Additional Information: Award holders work in the laboratory under the general supervision of an appropriate member of the staff, but are expected to work independently on research problems of their own choice.

For further information contact:

Clark laboratory, MS #31, Woods Hole Oceanographic Institution, 266 Woods Hole Road, Woods Hole, Massachusetts, 02543-1541, United States of America
Contact: Academic Programs Office

NOSAMS Postdoctoral Fellowships in Marine Radiocarbon Studies and in Accelerator Mass Spectrometry

Subjects: Radiocarbon research.
Purpose: To attract outstanding individuals interested in either or both, studies of radiocarbon in oceanic systems and developments in accelerator mass spectrometry and related techniques. Typical projects include diverse studies of the biogeochemical cycling of carbon, the detection and tracing of pollutants in natural systems and paleoceanographic and paleoclimatic investigations of all kinds.
Eligibility: Applicants must have received their doctoral degree before taking up their appointment at Woods Hole. They should have related research interests. Applications from physicists, chemists and biologists as well as from oceanographers are specifically invited.
Level of Study: Postdoctorate
Type: Fellowship
Value: Stipend of $56,000 per year for an 18 month appointment along with health and dental insurance. Also, some support for travel expenses, equipment, supplies and special services available.
Length of Study: 18 months
Frequency: Annual
Study Establishment: Woods Hole Oceanographic Institution
Country of Study: United States of America
Closing Date: January 15th

For further information contact:

Clark Laboratory, MS #31, Woods Hole Oceanographic Institution, 266 Woods Hole Road, Woods Hole, Massachusetts, 02543-1541, United States of America
Contact: Academic Programs Office

WHOI Geophysical Fluid Dynamics (GFD) Fellowships

Subjects: Classical fluid dynamics, physical oceanography, engineering, geophysics, meteorology, astrophysics, planetary atmospheres, geological fluid dynamics, hydromagnetics, physics and applied mathematics.
Purpose: To bring together graduate students and researchers from a variety of fields who share a common interest in the non linear dynamics of rotating, stratified fields.
Eligibility: There are no eligibility restrictions.
Level of Study: Graduate
Type: Fellowships
Value: A stipend of US$5,600 and travel allowance
Length of Study: 10 weeks
Frequency: Annual
Study Establishment: Woods Hole Oceanographic Institution
Country of Study: United States of America
No. of awards offered: Up to 10
Application Procedure: Application forms may be obtained from the GFD section of the education website or by writing directly to the Fellowship Committee.
Closing Date: February 15th
Funding: Government
Contributor: The United States office of Naval Research and the United States National Science Foundation

WHOI Marine Policy Fellowship Program

Subjects: Economic, legal and policy issues that arise from use of the world's oceans.
Purpose: To provide support and experience to research fellows interested in marine policy issues, to provide opportunities for interdisciplinary application of social sciences and natural sciences to marine policy problems, and to conduct research and convey information necessary for the development of effective local, national and international ocean policy.
Eligibility: Applicants must have completed a doctoral level degree or possess equivalent professional qualifications through career experience. The center also welcomes experienced professionals who can arrange a leave or sabbatical.
Level of Study: Postdoctorate
Type: Fellowship
Value: Recipients will receive a stipend of $57,000 for a period of one year and are eligible for group health and dental insurance. In addition, modest research and travel funds will be made available.
Length of Study: 1 year
Frequency: Annual
Study Establishment: Woods Hole Oceanographic Institution
Country of Study: United States of America
No. of awards offered: Varies
Application Procedure: A completed application form, a statement of proposed research, a current curriculum vitae including educational background and work experience, transcripts of college and university records and at least 3 personal references.
Closing Date: January 5th

For further information contact:

Clarks laboratory, MS #31, Woods Hole Oceanographic Institution, Woods Hole, Massachusetts, 02543-1541, United States of America
Contact: Academic Programs Office

WHOI Postdoctoral Scholarship Program

Subjects: Applied ocean physics and engineering, biology, marine chemistry and geochemistry, geology and geophysics, physical oceanography.
Purpose: To further education and training of the applicant with primary emphasis placed on the individual's research promise.
Eligibility: Open to applicants who have received their doctoral degree within the past 2–3 years, completed in full. Candidate should have a command of the English language. Candidates holding a WHOI appointment at the post-PhD level during the 12 months prior to the Scholar application deadline are not eligible.
Level of Study: Postdoctorate
Type: Scholarship
Value: $56,500 per year
Length of Study: 18 months
Study Establishment: Woods Hole Oceanographic Institution
Country of Study: United States of America
No. of awards offered: 5–6 departmental awards, 4 institute awards
Closing Date: Jan 5th

For further information contact:

Clark Laboratory, MS #31, Woods Hole Oceanographic Institution, Woods Hole, Massachusetts, 02543-1541, United States of America
Contact: Academic Programs Office

WORCESTER POLYTECHNIC INSTITUTE (WPI)

100 Institute Road, Worcester, MA, MA 01609-2280, United States of America
Tel: (1) 508 831 5000
Fax: (1) 508 831 5776
Email: gaann@cs.wpi.edu
Website: www.wpi.edu
Contact: Dr Matthew Ward, GAANN Director

Worcester Polytechnic Institute (WPI) was founded in Worcester, MA, in 1865 and was one of the nation's earliest technological universities. From the very beginning, it has taken a unique approach to science and technology education. There are over 20 WPI's project centres throughout North America and Central America, Africa, Australia, Asia and Europe.

Graduate Assistance in Areas of National Need (GAANN) Award

Subjects: Computer science.
Purpose: To make important contributions to the computer science research community.
Eligibility: Open to citizens or permanent residents of the United States pursuing a PhD in computer science.
Level of Study: Doctorate
Type: Fellowships
Value: US$30,000 per year
Length of Study: 3 years
Frequency: Annual
Country of Study: United States of America
No. of awards offered: 5
Application Procedure: Applicants must send a letter of intent, curriculum vitae and a description of any coursework or research completed.
Closing Date: April 1st for consideration for the Fall semester or November 1st for the Spring semester

WORLD BANK INSTITUTE

1818 H Street, Washington, NW, DC 20433, United States of America
Tel: (1) 202 473 1000
Fax: (1) 202 477 6391
Email: pic@worldbank.org
Website: www.worldbank.org
Contact: Communications Officer

One of the largest sources of funding and knowledge for transition and development councils; The World Bank uses its financial resources, staff and extensive experience to help developing countries reduce poverty, increase economic growth and improve their quality of life.

Social Development Civil Society Fund

Subjects: Civil engagement.
Purpose: To strengthen the voice and influence of poor and marginalized groups.
Eligibility: Applicable for Civil society organizations based in a developing country and working on issues of development can apply for a grant. Civil society organizations must be in good standing and have a record of achievement.
Level of Study: Postgraduate
Type: Grant
Value: Each grant will range around $10,000
Frequency: Annual
Application Procedure: Guidelines and application forms available from the participating World Bank country office.
Closing Date: April 2nd
Contributor: The World Bank

World Bank Grants Facility for Indigenous Peoples

Subjects: Indigenous culture, intellectual property and human rights.
Purpose: To support sustainable and culturally appropriate development projector planed and implemented by and for Indigenous People.
Eligibility: Applicant must be an Indigenous Peoples' community or not-for-profit/non-governmental Indigenous Peoples' organization, must be legally registered in the country of grant implementation, the country must be eligible to borrow from the World Bank (IBRD and/or IDA). Applicant should have an established bank account in the name of the applicant organization and should demonstrate internal controls to govern the use of funds. Applicant should not have received a grant from the Grants Facility for Indigenous Peoples in the previous two years.
Level of Study: Professional development, Research
Type: Grant
Value: Proposed project budget requests should range between US $10,000 and US$30,000 and include a minimum contribution of 20% of the total project cost.
Frequency: Annual
Application Procedure: A complete application, not more than 10 pages, should be submitted.
Closing Date: November 15th
Contributor: The World Bank

For further information contact:

World Bank Grants Facility for Indigenous Peoples, Social Development Department, Mailstop MC 5-52b, World Bank, 1818 H Street, Washington, NW, DC 20433, United States of America
Fax: (1) 1 202 522 1669
Email: indigenouspeoples@worldbank.org

WORLD FEDERATION OF INTERNATIONAL MUSIC COMPETITIONS

104, rue de Carouge, Genève, CH - 1205, Switzerland
Tel: (41) 22 321 3620
Fax: (41) 22 781 1418
Email: info@beethoven-comp.at/fmcim(at)fmcim.org
Website: www.beethoven-comp.at
Contact: Ms

The artistic reputation of musicians is highly dependent upon the quality of their Beethoven interpretations. The International Beethoven Piano Competition in Vienna gives young pianists the possibility to demonstrate their musicianship and artistic maturity.

International Beethoven Piano Competition Vienna

Subjects: Piano.
Purpose: To encourage the artistic development of young pianists.
Eligibility: Open for pianists born between January 1, 1981 and December 31, 1996.
Level of Study: Unrestricted
Type: Competition
Value: The first prize is €8,000, a Boesendorfer Model 200 piano and engagements, the second prize is €6,000, the third prize is €4,500, and there are three further prizes of €2,000. All information subject to change.
Frequency: Every 4 years
Country of Study: Austria
No. of awards offered: 6
Application Procedure: Apply online through the website www.beethoven-comp.at
Closing Date: October 15th
Funding: Government, private
No. of awards given last year: 6 plus special prizes
No. of applicants last year: 207
Additional Information: Please refer to the website for more details.

For further information contact:

Competition office, University of Music and Performing Arts, Anton-von-Webern Platz 1, Vienna, A-1030, Austria
Tel: (43) 43 1 71155/5113
Fax: (43) 43 1 711 55/5199
Email: info@beethoven-comp.at
Website: www.beethoven-comp.at

THE WORLD PIANO COMPETITION

441 Vine Street, Suite 1030, Cincinnati, OH, 45202, United States of America
Tel: (1) 513 421 5342
Fax: (1) 513 421 2672
Email: wpc@cincinnatiwpc.org
Website: www.cincinnatiwpc.org
Contact: Founder/Artistic Director

The American Music Scholarship Association (AMSA) produces the annual world piano competition in Cincinnati. AMSA also provides outreach programmes to Cincinnati children (e.g. The Bach, Beethoven, and Brahms Club) and worldwide performances. The young artist division winners perform at Carnegie Hall and the gold medallist of the artist division performs at Lincoln Center's Alice Tully Hall.

AMSA World Piano Competition

Subjects: Musical performance on the piano.
Purpose: To encourage the careers of aspiring young pianists and expose them to the performances of great musicians.
Eligibility: Open to piano students of any nationality who are between the ages of 5 and 30.
Level of Study: Unrestricted
Type: Scholarship
Value: The Artist Division's first prize is US$10,000 plus a fully managed debut recital at the Lincoln Center in New York. The second prize is US$3,000, the third US$2,000, the fourth US$1,000, the fifth US$500 and the sixth US$300. The Young Artists Division grand prize at levels 9–12 is US$1,500
Country of Study: Any country
Application Procedure: Applicants must apply in compliance with the full competition rules and regulations, which are available on request and on the website.
Closing Date: January 1st
Funding: Private

WORLD WILDLIFE FUND (WWF)

US Headquarters 1250, NW 24th Street PO Box 97180, Washington, DC, 20090-7180, United States of America
Tel: (1) 202 293 4800
Email: membership@wwfus.org
Website: www.worldwildlife.org

World Wildlife Fund (WWF) is known to everyone by its panda logo, WWF leads international efforts to protect endangered species and their habitats. Now in its 5th decade, WWF works in more than 100 countries around the globe to conserve the diversity of life on earth.

WWF Fuller Fellowships

Subjects: Natural and social sciences.
Purpose: To help scientists to address research questions that will powerfully inform and improve the practice of biodiversity conservation.
Eligibility: Open to candidates who have obtained a Doctoral degree.
Level of Study: Postdoctorate
Type: Fellowships
Value: Fellows may receive up to $15,000 per year for expenses and the host institution may receive an allowance of up to $2,500 per year
Length of Study: 2 years
Frequency: Annual
No. of awards offered: 2
Application Procedure: Applicants can apply online.
Closing Date: Check website for the next deadline.
Additional Information: The fellowship is not applicable to employees of WWF-US or immediate family members of WWF-US employees.

WWF Prince Bernhard Scholarship for Nature Conservation

Subjects: Any field that is directly relevant to the delivery and promotion of conservation.
Purpose: To provide financial support to individuals who wish to pursue professional training or formal studies that will help them to contribute more effectively to conservation efforts in their country.
Eligibility: For individuals working in the field of conservation or associated disciplines directly relevant to the delivery and promotion of conservation. Open to candidates from developing countries, central and eastern Europe and the Middle East.
Level of Study: Postgraduate
Type: Scholarship
Value: Swiss Franc 10,000
Frequency: Annual
Closing Date: January 11th
Contributor: World Wide Fund for Natures international programme based in Gland, Switzerland

THE WORSHIPFUL COMPANY OF FOUNDERS

Founders' Hall, Number One, Cloth Fair, London EC1A 7JQ, London, EC1A 7JQ, England
Email: FoundersCompany@aol.com
Website: www.foundersco.org.uk
Contact: A.J. Gillett

Malcolm Ray Travelling Scholarship

Subjects: Materials engineering.
Purpose: To help candidates advance both personal skills and professional knowledge.
Eligibility: Open to applicants studying materials engineering and who are under the age of 28, and can put forward an imaginative and testing project outside the UK.
Level of Study: Postgraduate, Professional development
Type: Scholarship
Value: UK£3,000
Frequency: Every 2 years
No. of awards offered: 1
Closing Date: April 20th
Funding: Trusts
Contributor: Malcolm Ray Travelling Scholarship
No. of awards given last year: 2

THE WORSHIPFUL COMPANY OF MUSICIANS

6th Floor, 2 London Wall Buildings, London, EC2M 5PP, England
Tel: (44) 20 7496 8980
Fax: (44) 20 7588 3633
Email: clerk@wcom.org.uk
Website: www.wcom.org.uk
Contact: Ms Margaret Alford, Clerk

The Worshipful Company of Musicians supports young musicians particularly in the wilderness years between graduating and setting out on their musical careers.

Allcard Grants + Busenhart-Morgan-Evans Award

Subjects: Music.
Purpose: To support the advanced training of performers at home or abroad for string, voice or piano accompanists, and wind instruments in exceptional circumstances.
Eligibility: Open to individuals wishing to undertake a relevant training or research programme. The grants are not available for courses leading either to a first degree at a university or to a diploma at a college of music, and only in exceptional circumstances will assistance towards the cost of a 4th or 5th year at a college of music be considered. Grants are not available towards the purchase of instruments. Applicants must have studied at a British institution for at least 3 years.
Level of Study: Postgraduate
Type: Grant
Value: Up to UK£6,000
Frequency: Annual
Country of Study: Any country
No. of awards offered: 4
Application Procedure: Applicants must be nominated by principals or heads of music departments at the Royal Academy of Music, the Royal College of Music, the Guildhall School of Music, the Royal

Northern College of Music, the Royal Scottish Academy of Music, the Welsh College of Music, the Birmingham Conservatoire, the Trinity College of Music, City University, Huddersfield University, Goldsmiths or other university departments.
Closing Date: Applications are to be made after January 1st and before April 30th
Funding: Trusts
No. of awards given last year: 4
No. of applicants last year: 30

Carnwath Scholarship

Subjects: Music.
Purpose: To support young pianists.
Eligibility: Open to any person permanently resident in the United Kingdom and 21–25 years of age. The scholarship is intended only for the advanced student who has successfully completed a solo performance course at a college of music.
Level of Study: Postgraduate
Type: Scholarship
Value: UK£4,150 per year
Length of Study: Up to 2 years
Frequency: Every 2 years
Country of Study: United Kingdom
No. of awards offered: 1
Application Procedure: Applicants must be nominated by principals of the Royal Academy of Music, the Guildhall School of Music, the Royal Northern College of Music, the Royal Scottish Academy of Music, Trinity College of Music, London College of Music, the Welsh College of Music, the Birmingham School of Music or the Royal College of Music. No application should be made directly to the Worshipful Company of Musicians.
Closing Date: April 30th
Funding: Trusts

Goldman Award

Purpose: To support the advanced training of performers at home or abroad for string, voice, piano accompanists and wind instruments in exceptional circumstances.
Eligibility: Open to individuals wishing to undertake a relevant training or research programme. Applications must have studied at a British institution for at least 3 years.
Type: Grant
Value: Up to £2,000
Frequency: Annual
Application Procedure: By nomination only by principals or heads of music departments at the major music colleges/conservatories or university departments.
Closing Date: April 30th
No. of awards given last year: 1

John Clementi Collard Fellowship

Subjects: Music.
Eligibility: Open to professional musicians of standing and experience who show excellence in one or more of the higher branches of musical activity, such as composition, research and performance including conducting.
Level of Study: Postgraduate
Type: Fellowship
Value: UK£5,000 per year
Length of Study: 1 year
Frequency: Dependent on funds available, Approx. every 3 years
Country of Study: United Kingdom
No. of awards offered: 1
Application Procedure: Applicants can be nominated by professors of music at Oxford, Cambridge or London Universities, directors of the Royal College of Music, principals of the Royal Academy of Music, the Guildhall School of Music or the Royal Northern College of Music. Applications can be made directly to the Worshipful Company of Musicians.

Maisie Lewis Young Artists Fund & Concordia Foundation Artists Fund

Subjects: Musical performance.
Purpose: To assist young artists of outstanding ability who wish to acquire experience on the professional soloist concert platform.
Eligibility: Open to instrumentalists up to 27 years of age and to singers of up to 32 years of age.
Level of Study: Postgraduate
Value: Reimbursement of recitalists' expenses
Frequency: Annual
Country of Study: United Kingdom
No. of awards offered: 6 half recitals per year
Application Procedure: Applicants must complete an application form, available from October 15th.
Funding: Trusts
No. of awards given last year: 6
No. of applicants last year: 80
Additional Information: Auditions are held in March.

THE XEROX FOUNDATION

6th Floor/PO Box 4505, 45 Glover Avenue, Norwalk, CT 06856-4505, United States of America
Tel: (1) 800 275 9376
Email: D.Garvin.Byrd@xerox.com
Website: www.xerox.com
Contact: Dr Joseph M. Cahalan

Xerox Foundation is a US$15.7 billion technology and services enterprise that helps businesses deploy Smarter Document Management strategies and find better ways to work. Its intent is to constantly lead with innovative technologies, products and services that customers can depend upon to improve business results.

Xerox Technical Minority Scholarship

Subjects: Chemistry, engineering, optics, software, information management systems and physics and material science.
Purpose: To provide funding to minority students enrolled in one of the technical sciences or engineering disciplines.
Eligibility: Open to citizens of the United States or visa-holding permanent residents of African American, Asian, Pacific Island, Native American, Native Alaskan or Hispanic descent. Applicants must have grade point average of 3.0 or better.
Level of Study: Postgraduate
Type: Scholarship
Value: Scholarships amount vary from $1,000–10,000.
Frequency: Annual
Country of Study: United States of America
Application Procedure: Applicants must submit the completed application form along with a curriculum vitae.
Closing Date: September 30th

For further information contact:

Xerox Technical Minority Scholarship Programme office

YIVO INSTITUTE FOR JEWISH RESEARCH

15 West 16th Street, New York, NY, 10011-6301, United States of America
Tel: (1) 212 246 6080
Email: pglasser@yivo.cjh.org
Website: www.yivoinstitute.org
Contact: Dr Paul Glasser, Associate Dean, Senior Research Associate

YIVO Institute for Jewish Research was founded in 1925, in Vilna, Poland as the Yiddish Scientific Institute, the YIVO Institute for Jewish Research is dedicated to the history and culture of Ashkenazi Jewry and to its influence in the Americas.

Abraham and Rachela Melezin Fellowship

Subjects: Jewish educational networks in Lithuania.
Purpose: To support doctoral and postdoctoral research on Jewish educational networks in Lithuania, with emphasis on pre-war Vilna and the Vilna region.
Level of Study: Postdoctorate, Doctorate
Type: Fellowship
Value: US$1,500
Length of Study: 1–3 months

Frequency: Annual
Study Establishment: YIVO Library and Archives
Country of Study: United States of America
Application Procedure: Applicants must send a covering letter, curriculum vitae, research proposal and 2 letters of support through regular mail, fax or email.
Closing Date: December 31st
Additional Information: A written summary of one's research is required; a public lecture is optional.

For further information contact:

Email: pglasser@yivo.cjh.org
Contact: Dr Paul Glasser

Abram and Fannie Gottlieb Immerman and Abraham Nathan and Bertha Daskal Weinstein Memorial Fellowship

Subjects: Jews of Courland and Latvia.
Purpose: To support travel for PhD dissertation research in archives and libraries of the Baltic states with preference given to research on the Jews of Courland and Latvia.
Eligibility: For those engaged in PhD dissertation research in archives and libraries of the Baltic states with preference given to research on the Jews of Courland and Latvia.
Level of Study: Doctorate
Type: Fellowship
Value: US$2,000
Frequency: Every 2 years
Application Procedure: Applicants must send a cover letter, curriculum vitae, research proposal and 2 letters of support through regular mail, fax or email. A written summary of one's research is required.
Closing Date: December 31st

For further information contact:

Email: pglasser@yivo.cjh.org
Contact: Dr Paul Glasser

Aleksander and Alicja Hertz Memorial Fellowship

Subjects: Polish-Jewish history.
Purpose: To encourage research on Jewish-Polish relations and Jewish contributions to Polish literature and culture in the modern period.
Level of Study: Doctorate, Postdoctorate
Type: Fellowship
Value: US$1,500
Length of Study: 1–3 months
Frequency: Annual
Country of Study: United States of America
Application Procedure: Applicants must send their curriculum vitae, research proposal and 2 letters of support through regular mail, fax or email. A written summary of one's research is required.
Closing Date: December 31st

For further information contact:

Email: pglasser@yivo.cjh.org
Contact: Dr Paul Glasser, Chair Fellowship Committee

Dina Abramowicz Emerging Scholar Fellowship

Subjects: Eastern European Jewish studies.
Purpose: To support a significant scholarly publication that may encompass the revision of a doctoral dissertation.
Eligibility: Applicants are required to give a public lecture.
Level of Study: Postdoctorate
Type: Fellowship
Value: US$3,000
Length of Study: 1–3 months
Frequency: Annual
Application Procedure: Applicants must send their curriculum vitae, a research proposal and 2 letters of support through regular mail, fax or email.
Closing Date: December 31st

For further information contact:

Email: pglasser@yivo.cjh.org
Contact: Dr Paul Glasser, Chair, Fellowship Committee

Dora and Mayer Tendler Fellowship

Subjects: Jewish studies.
Purpose: To support graduate research in Jewish studies with preference given to research in YIVO collections.
Eligibility: Graduate applicants must carry out original research in the field of Jewish studies and give a written summary of the research carried out.
Level of Study: Doctorate, Graduate
Type: Fellowship
Value: US$3,000
Frequency: Annual
Study Establishment: YIVO collections
Country of Study: United States of America
Application Procedure: Applicants must send a cover letter, curriculum vitae, research proposal and 2 letters of support through regular mail, fax or email.
Closing Date: December 31st
Additional Information: A public lecture at the end of the tenure of the Fellowship is optional.

For further information contact:

Email: pglasser@yivo.cjh.org
Contact: Dr Paul Glasser, Chairman - Fellowship Committee

Joseph Kremen Memorial Fellowship

Subjects: Eastern European Jewish music, art and theater
Purpose: To financially assist researchers at the YIVO Archives and Library.
Eligibility: A written summary of one's research is required.
Level of Study: Postgraduate, Research
Type: Fellowship
Value: US$2,000
Frequency: Annual
Application Procedure: Applicants must send their curriculum vitae, research proposal and 2 letters of support by regular mail, fax or email.
Closing Date: December 31st

For further information contact:

Email: pglasser@yivo.cjh.org
Contact: Dr Paul Glasser, Chair Fellowship Committee

Maria Salit-Gitelson Tell Memorial Fellowship

Subjects: Lithuanian Jewish history.
Purpose: To support original doctoral or postdoctoral research in the field of Lithuanian Jewish history, the city of Vilnus in particular, at the YIVO Library and Archives.
Eligibility: Applicants must carry out original doctoral or postdoctoral research in the field of Lithuanian Jewish history and give a public lecture at the end of the tenure of the Fellowship.
Level of Study: Doctorate, Postdoctorate
Type: Fellowship
Value: US$1,500
Length of Study: 1–3 months
Frequency: Annual
Study Establishment: YIVO Library and Archives
Country of Study: United States of America
Application Procedure: Applicants must send a cover letter, curriculum vitae, research proposal and 2 letters of support through regular mail, fax or email.
Closing Date: December 31st

For further information contact:

Email: pglasser@yivo.cjh.org
Contact: Dr Paul Glasser, Chairman - Fellowship Committee

Natalie and Mendel Racolin Memorial Fellowship

Subjects: East European Jewish history.

Purpose: To support original doctoral or postdoctoral research in the field of East European Jewish history at the YIVO Library and Archives.
Eligibility: Applicants must carry out original doctoral or postdoctoral research in the field of East European Jewish history and give a public lecture at the end of the tenure of the Fellowship.
Level of Study: Doctorate, Postdoctorate
Type: Fellowship
Value: US$1,500
Length of Study: 1–3 months
Frequency: Annual
Study Establishment: YIVO Library and Archives
Country of Study: United States of America
Application Procedure: Applicants must send a cover letter, curriculum vitae, research proposal and 2 letters of support through regular mail, fax or email.
Closing Date: December 31st

For further information contact:

Email: pglasser@yivo.cjh.org
Contact: Dr Paul Glasser, Chairman - Fellowship Committee

Professor Bernard Choseed Memorial Fellowship

Subjects: East European Jewish studies.
Purpose: To financially support doctoral and postdoctoral students who conduct research.
Eligibility: Applicants are required to give a public lecture.
Level of Study: Doctorate, Postdoctorate
Type: Fellowship
Value: US$7,500
Length of Study: 1–3 months
Frequency: Annual
Country of Study: United States of America
Application Procedure: Applicants must submit a curriculum vitae, a research proposal and 2 letters of support through regular mail, fax or email.
Closing Date: Decemeber 31st

For further information contact:

Email: pglasser@yivo.cjh.org
Contact: Dr Paul Glasser, Chair, Fellowship Committee

Rose and Isidore Drench Memorial Fellowship

Subjects: American Jewish history with a focus on Jewish labor movement.
Purpose: To encourage research in American Jewish history.
Eligibility: Applicants are required to give a public lecture.
Level of Study: Postdoctorate, Doctorate
Type: Fellowship
Value: US$2,500
Length of Study: 1–3 months
Frequency: Annual
Application Procedure: Applicants must submit their curriculum vitae, a research proposal and 2 letters of support through regular mail, fax or email.
Closing Date: December 31st

For further information contact:

Email: pglasser@yivo.cjh.org
Contact: Dr Paul Glasser, Chair Fellowship Committee

Samuel and Flora Weiss Research Fellowship

Subjects: Polish Jewry or Polish-Jewish relations during the Holocaust period.
Purpose: To support research on the destruction of Polish Jewry or on Polish-Jewish relations during the Holocaust period.
Eligibility: Applicants must carry out original research on the destruction of Polish Jewry or on Polish-Jewish relations during the Holocaust period and give a written summary of the research carried out.The research should result in a scholarly publication.
Level of Study: Doctorate
Type: Fellowship
Value: US$2,500
Frequency: Annual
Application Procedure: Applicants must send a cover letter, curriculum vitae, research proposal and 2 letters of support through regular mail, fax or email.
Closing Date: December 31st
Additional Information: A public lecture at the end of the tenure of the Fellowship is optional.

For further information contact:

Email: pglasse@yivo.cjh.org
Contact: Dr Paul Glasser, Chairman - Fellowship Committee

Vivian Lefsky Hort Memorial Fellowship

Subjects: Yiddish literature.
Purpose: To support original doctoral or postdoctoral research in the field of Yiddish literature.
Eligibility: Applicants must carry out original doctoral or postdoctoral research in Yiddish literature and give a public lecture at the end of the tenure of the Fellowship.
Level of Study: Doctorate, Postdoctorate
Type: Fellowship
Value: US$2,000
Length of Study: 1–3 months
Frequency: Annual
Study Establishment: YIVO Library and Archives
Country of Study: United States of America
Application Procedure: Applicants must send a cover letter, curriculum vitae, research proposal and 2 letters of support through regular mail, fax or email.
Closing Date: December 31st

For further information contact:

Email: pglasser@yivo.cjh.org
Contact: Dr Paul Glasser

Vladimir and Pearl Heifetz Memorial Fellowship in Eastern European Jewish Music

Subjects: Eastern European Jewish Music.
Purpose: To assist undergraduate, graduate and postgraduate researchers defray expenses connected with research in YIVO's music collection at the YIVO Archives and Library.
Eligibility: Undergraduate, graduate and postgraduate researchers who will carry on research in YIVO's music collection at the YIVO Archives and Library.
Level of Study: Graduate, Postgraduate, Undergraduate
Type: Fellowship
Value: US$1,500
Frequency: Annual
Study Establishment: YIVO's music collection
Country of Study: United States of America
Application Procedure: Applicants must send a cover letter, curriculum vitae, research proposal and 2 letters of support through regular mail, fax or email.
Closing Date: December 31st
Funding: Foundation
Additional Information: A written summary of one's research is required; a public lecture is optional.

For further information contact:

Email: pglasser@yivo.cjh.org
Contact: Dr Paul Glasser, Chaiman-Fellowship Committee

Workmen's Circle/Dr Emanuel Patt Visiting Professorship

Subjects: Eastern European Jewish Studies.
Purpose: To support postdoctoral research at the YIVO Library and Archives.
Level of Study: Postdoctorate
Type: Fellowship
Value: US$5,000
Length of Study: 3 months
Frequency: Annual
Study Establishment: YIVO Library and Archives
Country of Study: United States of America

Application Procedure: Applicants must send a covering letter, curriculum vitae, research proposal and 2 letters of support through regular mail, fax or email.
Closing Date: December 31st
Additional Information: The visiting faculty member should give a public lecture at the end of the award's tenure.

For further information contact:

Email: pglasser@yivo.cjh.org
Contact: Dr Paul Glasser

YORKSHIRE SCULPTURE PARK

Bretton Hall, West Bretton, Wakefield, Yorkshire, WF4 4LG, England
Tel: (44) 01924 832631/01924 832528
Email: info@ysp.co.uk
Website: www.ysp.co.uk
Contact: Mr Peter Murray, Administrator

Yorkshire Sculpture Park is one of Europe's leading open-air art organizations, showing modern and contemporary work by leading United Kingdom and international artists.

Feiweles Trust Bursary

Subjects: Dance, music, art and sculpture.
Purpose: To support an artist at the beginning of his or her career, after training.
Level of Study: Postgraduate
Type: Bursary
Value: UK£10,000
Length of Study: 3 months
Frequency: Annual
Study Establishment: Workshops within the community
Country of Study: United Kingdom
Application Procedure: Applicants must send a letter, explaining exactly how the bursary would benefit the development of their career and a full curriculum vitae.
Closing Date: February 20th
Funding: Private
Contributor: Feiweles Trust
Additional Information: It is expected that the appointed artists will use this bursary experience to develop their own artistic practice. A different art form is supported each year.

YORKVILLE UNIVERSITY

1149 Smythe Street St, Fredericton, NB, E3B 3H4, Canada
Tel: (1) 506 454 1220
Fax: (1) 506 454 1221
Email: info@yorkvilleu.ca
Website: www.yorkvilleu.ca

Yorkville University, established in 2003 in Fredericton, New Brunswick, is private and non-denominational, specializing in practice-oriented graduate level academic programmes.

Jacob Markovitz Memorial Scholarship

Subjects: Counselling psychology.
Purpose: To support graduate students in the helping professions at a number of academic institutions.
Eligibility: Open to all students with Canadian citizenship or with landed immigrant status in Canada who are registered in a programme at the Yorkville University. Only first time applicants are eligible to apply.
Level of Study: Postgraduate
Type: Scholarship
Value: 50 per cent of first year tuition
Frequency: Annual
Study Establishment: Yorkville University
Country of Study: Canada
No. of awards offered: 2–4
Closing Date: November 26th
Funding: Trusts
Contributor: Jacob Markovitz Memorial Scholarship Fund
Additional Information: If granted a scholarship one must maintain a 3.0 average in each trimester for the duration of the scholarship grant.

ZONTA INTERNATIONAL FOUNDATION

1211 West 22nd Street, Suite 900, Oak Brook, IL, 60523, United States of America
Tel: (1) 630 928 1400
Fax: (1) 630 928 1559
Email: programs@zonta.org
Website: www.zonta.org
Tel: 190200 UT
Contact: Ms Programs Coordinator

The Foundation is a worldwide service organization of executives in business and the professions working together to advance the status of women. The Amelia Earhart Fellowship was established in 1938 in honour of Amelia Earhart, famed pilot and member of Zonta International.

Jane M Klausman Women in Business Scholarship

Eligibility: Women pursuing a business or business-related program who demonstrate outstanding potential in the field and are living or studying in a Zonta district/region.
Level of Study: MBA
Type: Scholarship
Value: US$1,000 (District scholarship) and US$7,000 (International scholarship)
Frequency: Annual
No. of awards offered: 32
Application Procedure: Applicants must contact a Club directly in the country were they are studying or living. Must contact an existing club in their country for an address and deadline details.
Closing Date: Generally in March to May
No. of applicants last year: 26

Zonta Amelia Earhart Fellowships

Subjects: Aerospace-related sciences and aerospace-related engineering.
Purpose: The Fellowships are granted to women pursuing PhD/Doctoral degrees in aerospace related sciences and engineering.
Eligibility: Open to women of any nationality who demonstrate a superior academic record in the field of aerospace-related sciences and engineering.
Level of Study: Doctorate
Type: Fellowship
Value: US$10,000
Frequency: Annual
Study Establishment: Any University
No. of awards offered: Approx. 35
Application Procedure: The Zonta International Amelia Earhart Fellowship committee reviews and recommends recipients to the Zonta International Board of Directors. All applicants will be notified of their status by the end of April.
Closing Date: November 15th
Funding: Foundation
No. of awards given last year: 35
No. of applicants last year: 120
Additional Information: Recipients are not permitted to defer the Fellowship although Zonta will consider a new application the following year.

SUBJECT AND ELIGIBILITY GUIDE TO AWARDS

AGRICULTURE, FORESTRY AND FISHERY

- General
- Agricultural business
- Agricultural economics
- Agriculture and farm management
- Agronomy
- Animal husbandry
  - Sericulture
- Crop production
- Fishery
  - Aquaculture
- Food science
  - Brewing
  - Dairy
  - Fish
  - Harvest technology
  - Meat and poultry
  - Oenology
- Forestry
  - Forest biology
  - Forest economics
  - Forest management
  - Forest pathology
  - Forest products
  - Forest soils
- Horticulture and viticulture
- Soil and water science
  - Irrigation
  - Soil conservation
  - Water management
- Tropical agriculture
- Veterinary science

ARCHITECTURE AND TOWN PLANNING

- General
- Architectural and environmental design
- Architectural restoration
- Landscape architecture
- Regional planning
- Rural planning
- Structural architecture
- Town planning

ARTS AND HUMANITIES

- General
- Archaeology
- Classical languages and literatures
  - Classical Greek
  - Latin
  - Sanskrit
- Comparative literature
- History
  - Ancient civilisations
  - Contemporary history
  - Medieval studies
  - Modern history
  - Prehistory
- Linguistics
  - Applied linguistics
  - Grammar
  - Logopedics
  - Phonetics
  - Psycholinguistics
  - Semantics and terminology
  - Speech studies
- Modern languages
  - African Languages
  - Afrikaans
  - Altaic languages
  - Amerindian languages
  - Arabic
  - Austronesian and oceanic languages
  - Baltic languages
  - Celtic languages
  - Chinese
  - Danish
  - Dutch
  - English
  - Eurasian and North Asian languages
  - European languages (others)
  - Finnish
  - Fino Ugrian languages
  - French
  - German
  - Germanic languages
  - Hebrew
  - Hungarian
  - Indian languages
  - Indic languages
  - Iranic languages
  - Italian
  - Japanese
  - Korean
  - Modern Greek
  - Norwegian
  - Portuguese
  - Romance languages
  - Russian
  - Scandinavian languages
  - Slavic languages (others)
  - Spanish
  - Swedish
- Native language and literature
- Philosophy
  - Ethics
  - Logic
  - Metaphysics
  - Philosophical schools
- Translation and interpretation
- Writing (authorship)

BUSINESS ADMINISTRATION AND MANAGEMENT

- General
- Accountancy
- Business and commerce
- Business computing
- Business machine operation
- Finance, banking and investment
- Human resources
- Institutional administration
- Insurance
- International business
- Labour/industrial relations
- Management systems
- Marketing
  - Public relations
- MBA
- Personnel management
- Private administration
- Public administration
- Real estate
- Secretarial studies

EDUCATION AND TEACHER TRAINING

- General
- Adult education
- Continuing education
- Educational science
  - Curriculum
  - Distance education
  - Educational administration
  - Educational and student counselling
  - Educational research
  - Educational technology
  - Educational testing and evaluation
  - International and comparative education
  - Philosophy of education
  - Teaching and learning
- Higher education teacher training
- Nonvocational subjects education
  - Education in native language
  - Foreign languages education
  - Humanities and social science education
  - Literacy education
  - Mathematics education
  - Physical education
  - Science education
- Pre-school education
- Primary education
- Secondary education
- Special education
  - Bilingual/bicultural education
  - Education of foreigners
  - Education of natives
  - Education of specific learning disabilities
  - Education of the gifted
  - Education of the handicapped
  - Education of the socially disadvantaged
- Staff development
- Teacher trainers education
- Vocational subjects education
  - Agricultural education
  - Art education
  - Commerce/business education
  - Computer education
  - Health education
  - Home economics education
  - Industrial arts education
  - Music education
  - Technology education

ENGINEERING

- General
- Aeronautical and aerospace engineering
- Agricultural engineering
- Automotive engineering
- Bioengineering and biomedical engineering
- Chemical engineering
- Civil engineering
- Computer engineering
- Control engineering (robotics)
- Electrical and electronic engineering
- Energy engineering
- Engineering drawing/design
- Environmental and Sanitary Engineering
- Forestry engineering
- Hydraulic engineering
- Industrial engineering
- Marine engineering and naval architecture
- Materials engineering
- Measurement/precision engineering
- Mechanical engineering
- Metallurgical engineering
- Mining engineering
- Nanotechnology
- Nuclear engineering
- Petroleum and gas engineering
- Physical engineering
- Production engineering
- Safety engineering
- Sound engineering
- Surveying and mapping science

FINE AND APPLIED ARTS

- General
- Art criticism
- Art history
  - Aesthetics
- Art management
- Cinema and television
- Dance
- Design
  - Display and stage design
  - Fashion design
  - Furniture design
  - Graphic design
  - Industrial design
  - Interior design
  - Textile design
- Drawing and painting
- Handicrafts
- Music
  - Conducting
  - Jazz and popular music
  - Music theory and composition
  - Musical instruments
  - Musicology
  - Opera
  - Religious music
  - Singing
- Photography
- Religious art
- Sculpture
- Theatre

HOME ECONOMICS

- General
- Child care/child development
- Clothing and sewing
- Consumer studies
- House arts and environment
- Household management
- Nutrition

LAW

- General
- Air and space law
- Canon law
- Civil law
- Commercial law
- Comparative law
- Criminal law
- European community law
- History of law
- Human rights
- International law
- Islamic law
- Justice administration
- Labour law
- Maritime law
- Notary studies
- Private law
- Public law

Administrative law
Constitutional law
Fiscal law

MASS COMMUNICATION AND INFORMATION SCIENCE

General
Communication arts
Documentation techniques and archiving
Journalism
Library science
Mass communication
Media studies
Museum management
Museum studies
Public relations and publicity
Radio/television broadcasting
Restoration of works of art

MATHEMATICS AND COMPUTER SCIENCE

General
Actuarial science
Applied mathematics
Artificial intelligence
Computer science
Statistics
Systems analysis

MEDICAL SCIENCES

General
Acupuncture
Biomedicine
Chiropractic
Dental technology
Prosthetic dentistry
Dentistry and stomatology
Community dentistry
Oral pathology
Orthodontics
Periodontics
Forensic medicine and dentistry
Health administration
Homeopathy
Medical auxiliaries
Medical technology
Medicine
Anaesthesiology
Cardiology
Dermatology
Endocrinology
Epidemiology
Gastroenterology
Geriatrics
Gynaecology and obstetrics
Haematology
Hepathology
Nephrology
Neurology
Oncology
Ophthalmology
Otorhinolaryngology
Paediatrics
Parasitology
Pathology
Plastic surgery
Pneumology
Psychiatry and mental health
Rheumatology
Tropical medicine
Urology
Venereology
Virology
Midwifery
Nursing
Optometry
Osteopathy
Pharmacy
Podiatry
Public health and hygiene
Dietetics
Social/preventive medicine
Sports medicine
Radiology
Rehabilitation and therapy
Traditional eastern medicine
Treatment techniques

NATURAL SCIENCES

General
Astronomy and astrophysics
Biological and life sciences
Anatomy
Biochemistry
Biology
Biophysics and molecular biology
Biotechnology
Botany
Embryology and reproduction biology
Genetics
Histology
Immunology
Limnology
Marine biology
Microbiology
Neurosciences
Parasitology
Pharmacology
Physiology
Plant pathology
Toxicology
Zoology
Chemistry
Analytical chemistry
Inorganic chemistry
Organic chemistry
Physical chemistry
Earth sciences
Geochemistry
Geography (Scientific)
Geology
Geophysics and seismology
Mineralogy and crystallography
Palaeontology
Petrology
Marine science and oceanography
Meteorology
Arctic studies
Arid land studies
Physics
Atomic and molecular physics
Nuclear physics
Optics
Solid state physics
Thermal physics

RECREATION, WELFARE, PROTECTIVE SERVICES

General
Civil security
Criminology
Environmental studies
Ecology
Environmental management
Natural resources
Waste management
Wildlife and pest management
Fire protection science
Leisure studies
Military science
Parks and recreation
Peace and disarmament
Police studies
Social welfare and social work
Social and community services
Sports
Sociology of sports
Sports management
Vocational counselling

RELIGION AND THEOLOGY

General
Church administration (pastoral work)
Comparative religion
Esoteric practices
History of religion
Holy writings
Religious education
Religious practice
Religious studies
Agnosticism and Atheism
Ancient religions
Asian religious studies
Christian religious studies
Islam
Judaic religious studies
Sociology of religion
Theology

SERVICE TRADES

General
Cooking and catering
Cosmetology
Hotel and restaurant
Hotel management
Retailing and wholesaling
Tourism

SOCIAL AND BEHAVIOURAL SCIENCES

General
Econometrics
Economic and finance policy
Economic history
Economics
Industrial and production economics
International economics
Taxation
Ancient civilisations (Egyptology, Assyriology)
Anthropology
Ethnology
Folklore
Cognitive sciences
Cultural studies
African American
African studies
American
Asian
Canadian
Caribbean
East Asian
Eastern European
European
Hispanic American
Indigenous studies
Islamic
Jewish
Latin American
Middle Eastern
Native American
Nordic
North African
Pacific area
South Asian
Southeast Asian
Subsahara African
Western European
Demography and population
Development studies
Geography
Heritage Preservation
International relations
Political science and government
Comparative politics
Psychology
Clinical psychology
Development psychology
Educational psychology
Experimental psychology
Industrial and organisational psychology
Personality psychology
Psychometrics
Social psychology
Rural studies
Sociology
Comparative sociology
Futurology
History of societies
Social institutions
Social policy
Social problems
Urban Studies
Women's studies

TRADE, CRAFT AND INDUSTRIAL TECHNIQUES

General
Building technologies
Electrical and electronic equipment and maintenance
Food technology
Graphic arts
Printing and printmaking
Publishing and book trade
Heating, and refrigeration
Laboratory techniques
Leather techniques
Mechanical equipment and maintenance
Metal techniques
Optical technology
Paper and packaging technology
Textile technology
Wood technology

TRANSPORT AND COMMUNICATIONS

General
Air transport
Marine transport and nautical science
Postal services
Railway transport
Road transport
Telecommunications services
Transport economics
Transport management

ANY SUBJECT

ANY SUBJECT

Any Country

African Nations

Australia

Canada

Caribbean Countries

East European Countries

European Union

Middle East

New Zealand

South Africa

United Kingdom

United States of America

West European Countries

AGRICULTURE, FORESTRY AND FISHERY

GENERAL

Any Country

African Nations

Australia

Canada

Caribbean Countries

East European Countries

European Union

Middle East

New Zealand

South Africa

United Kingdom

United States of America

West European Countries

AGRICULTURAL BUSINESS

AGRICULTURAL ECONOMICS

Any Country

African Nations

Australia

Canada

Caribbean Countries

East European Countries

European Union

Middle East

New Zealand

South Africa

United Kingdom

United States of America

African Nations

Australia

Canada

Caribbean Countries

East European Countries

European Union

Middle East

New Zealand

South Africa

United Kingdom

United States of America

West European Countries

SERICULTURE

Any Country

African Nations

Canada

Caribbean Countries

East European Countries

European Union

Middle East

South Africa

United Kingdom

CROP PRODUCTION

Any Country

African Nations

Australia

Canada

Caribbean Countries

East European Countries

FOOD SCIENCE

BREWING

DAIRY

FISH

HARVEST TECHNOLOGY

Any Country

Perry Postgraduate Scholarships 527
UFAW Animal Welfare Student Scholarships 654

Australia

Australian Postgraduate Award 674

Canada

OFAF/OFAF Zone 6 Wildlife Research Grant 515

European Union

Perry Postgraduate Scholarships 527

New Zealand

Australian Postgraduate Award 674

United Kingdom

Perry Research Awards 527

MEAT AND POULTRY

Any Country

Perry Postgraduate Scholarships 527
Teagasc Walsh Fellowships 633
UFAW Animal Welfare Student Scholarships 654

Australia

MINTRAC Postgraduate Research Scholarship 459

European Union

Perry Postgraduate Scholarships 527

New Zealand

Vernon Willey Trust Awards 835

United Kingdom

Perry Research Awards 527

OENOLOGY

Any Country

UFAW Animal Welfare Student Scholarships 654

FORESTRY

Any Country

Earthwatch Field Research Grants 279
Edmund Niles Huyck Preserve, Inc. Graduate and Postgraduate Grants 281
ETH Zurich Excellence Scholarship and Opportunity Award 629
Franklin Research Grant Program 80
Gilbert F. White Postdoctoral Fellowship Program 546
Hyland R Johns Grant Program 644
International Tropical Timber Organization (ITTO) Fellowship Programme 392
Jack Kimmel International Grant Program 644
John Z Duling Grant Program 644
Joseph L. Fisher Doctoral Dissertation Fellowships 546
National Geographic Conservation Trust Grant 248
Scholarship Opportunities Linked to CATIE's Postgraduate Program Including CATIE Scholarship Forming Part of the Scholarship-Loan Program 644
SERC Graduate Student Fellowship 603
SERC Postdoctoral Fellowships 603
SERC Predoctoral Fellowships 603
SERC Senior Fellowships 604
Silverhill Institute of Environmental Research and Conservation Award 598
Special Overseas Student Scholarship (SOSS) 675
Teagasc Walsh Fellowships 633

African Nations

The Bentley Cropping Systems Fellowship 383
IDRC Doctoral Research Awards 384
IFS Research Grants 386

Australia

Australian Postgraduate Award 674

Canada

The Bentley Cropping Systems Fellowship 383
Community Forestry: Trees and People-John G Bene Fellowship 384
IDRC Doctoral Research Awards 384
OFAH/Oakville and District Rod & Gun Club Conservation Research Grant 515

Caribbean Countries

The Bentley Cropping Systems Fellowship 383
IDRC Doctoral Research Awards 384
IFS Research Grants 386

East European Countries

The Bentley Cropping Systems Fellowship 383
IDRC Doctoral Research Awards 384

Middle East

The Bentley Cropping Systems Fellowship 383
IDRC Doctoral Research Awards 384
IFS Research Grants 386

New Zealand

Australian Postgraduate Award 674

South Africa

The Bentley Cropping Systems Fellowship 383
IDRC Doctoral Research Awards 384
IFS Research Grants 386

United Kingdom

Mr and Mrs David Edward Memorial Award 164
Perry Research Awards 527

United States of America

Environmental Public Policy and Conflict Resolution PhD Fellowship 468

West European Countries

Mr and Mrs David Edward Memorial Award 164

FOREST BIOLOGY

Any Country

Edmund Niles Huyck Preserve, Inc. Graduate and Postgraduate Grants 281
Future Conservationist Awards 248
Hyland R Johns Grant Program 644
International Tropical Timber Organization (ITTO) Fellowship Programme 392
Jack Kimmel International Grant Program 644
John Z Duling Grant Program 644
Oscar Broneer Traveling Fellowship 91
Silverhill Institute of Environmental Research and Conservation Award 598
Special Overseas Student Scholarship (SOSS) 675

African Nations

The Bentley Cropping Systems Fellowship 383

Caribbean Countries

East European Countries

European Union

Middle East

New Zealand

South Africa

United Kingdom

United States of America

West European Countries

WATER MANAGEMENT

Any Country

African Nations

Australia

Canada

Caribbean Countries

East European Countries

European Union

Middle East

New Zealand

South Africa

United Kingdom

United States of America

West European Countries

TROPICAL AGRICULTURE

Any Country

African Nations

Canada

Caribbean Countries

East European Countries

European Union

Middle East

South Africa

United Kingdom

West European Countries

VETERINARY SCIENCE

Any Country

Australia

Canada

East European Countries

European Union

United Kingdom

United States of America

West European Countries

ARCHITECTURE AND TOWN PLANNING

GENERAL

Any Country

African Nations

Australia

Canada

Caribbean Countries

East European Countries

European Union

Middle East

New Zealand

South Africa

United Kingdom

United States of America

West European Countries

ARCHITECTURAL AND ENVIRONMENTAL DESIGN

Any Country

ETH Zurich Excellence Scholarship and Opportunity Award 629
Franklin Research Grant Program 80
Hyland R Johns Grant Program 644
Jack Kimmel International Grant Program 644
John Z Duling Grant Program 644
MacDowell Colony Residencies 439
Ohio Arts Council Individual Excellence Awards 514
Stanley Smith (UK) Horticultural Trust Awards 623

African Nations

ECOPOLIS Graduate Research and Design Awards 384
International Postgraduate Research Scholarships (IPRS) 417

Canada

ECOPOLIS Graduate Research and Design Awards 384
Horticultural Research Institute Grants 344
International Postgraduate Research Scholarships (IPRS) 417
Jim Bourque Scholarship 114

Caribbean Countries

ECOPOLIS Graduate Research and Design Awards 384
International Postgraduate Research Scholarships (IPRS) 417

East European Countries

ECOPOLIS Graduate Research and Design Awards 384
International Postgraduate Research Scholarships (IPRS) 417

European Union

Collaborative Doctoral Awards 121
Geoplan Scholarship 259
Professional Preparation Master's Scheme 121
Research Preparation Master's Scheme 121

Middle East

ECOPOLIS Graduate Research and Design Awards 384
International Postgraduate Research Scholarships (IPRS) 417

South Africa

ECOPOLIS Graduate Research and Design Awards 384
International Postgraduate Research Scholarships (IPRS) 417

United Kingdom

Collaborative Doctoral Awards 121
Geoplan Scholarship 259
Hilda Martindale Trust Exhibitions 340
International Postgraduate Research Scholarships (IPRS) 417
Professional Preparation Master's Scheme 121
Research Preparation Master's Scheme 121
The RHS Environmental Bursary 568

United States of America

Environmental Public Policy and Conflict Resolution PhD Fellowship 468
Horticultural Research Institute Grants 344
International Postgraduate Research Scholarships (IPRS) 417

West European Countries

International Postgraduate Research Scholarships (IPRS) 417

ARCHITECTURAL RESTORATION

Any Country

ASCSA Advanced Fellowships 89
Earthwatch Field Research Grants 279
Keepers Preservation Education Fund Fellowship 614
Rosann Berry Fellowship 614
SAH Fellowships for Independent Scholars 615
Sir John Soane's Museum Foundation Travelling Fellowship 600
Spiro Kostof Annual Meeting Fellowship 615

African Nations

ECOPOLIS Graduate Research and Design Awards 384

Canada

ECOPOLIS Graduate Research and Design Awards 384

Caribbean Countries

ECOPOLIS Graduate Research and Design Awards 384

East European Countries

ECOPOLIS Graduate Research and Design Awards 384

Middle East

ECOPOLIS Graduate Research and Design Awards 384

South Africa

ECOPOLIS Graduate Research and Design Awards 384

United Kingdom

Hilda Martindale Trust Exhibitions 340

United States of America

ACC Fellowship Grants Program 125
American Academy in Rome Fellowships in Design Art 26
NEH Fellowships 91
Sally Kress Tompkins Fellowship 615

LANDSCAPE ARCHITECTURE

Any Country

Dumbarton Oaks Fellowships and Junior Fellowships 278
Franklin Research Grant Program 80
MacDowell Colony Residencies 439
Ohio Arts Council Individual Excellence Awards 514
Stanley Smith (UK) Horticultural Trust Awards 623

African Nations

Girton College: Ruth Whaley Scholarship 689
International Postgraduate Research Scholarships (IPRS) 417

Australia

Girton College: Ruth Whaley Scholarship 689

Canada

Girton College: Ruth Whaley Scholarship 689
Horticultural Research Institute Grants 344
International Postgraduate Research Scholarships (IPRS) 417

Caribbean Countries

Girton College: Ruth Whaley Scholarship 689
International Postgraduate Research Scholarships (IPRS) 417

East European Countries

International Postgraduate Research Scholarships (IPRS) 417

European Union

Collaborative Doctoral Awards 121
Geoplan Scholarship 259
Professional Preparation Master's Scheme 121
Research Preparation Master's Scheme 121

Middle East

Girton College: Ruth Whaley Scholarship 689
International Postgraduate Research Scholarships (IPRS) 417

New Zealand
Girton College: Ruth Whaley Scholarship 689

South Africa
Girton College: Ruth Whaley Scholarship 689
International Postgraduate Research Scholarships (IPRS) 417

United Kingdom
Collaborative Doctoral Awards 121
Geoplan Scholarship 259
Hilda Martindale Trust Exhibitions 340
International Postgraduate Research Scholarships (IPRS) 417
Professional Preparation Master's Scheme 121
Research Preparation Master's Scheme 121

United States of America
American Academy in Rome Fellowships in Design Art 26
Girton College: Ruth Whaley Scholarship 689
Horticultural Research Institute Grants 344
International Postgraduate Research Scholarships (IPRS) 417

West European Countries
International Postgraduate Research Scholarships (IPRS) 417

REGIONAL PLANNING

Any Country
ETH Zurich Excellence Scholarship and Opportunity Award 629
Franklin Research Grant Program 80

African Nations
George Pepler International Award 579
International Postgraduate Research Scholarships (IPRS) 417

Australia
George Pepler International Award 579

Canada
George Pepler International Award 579
International Postgraduate Research Scholarships (IPRS) 417
Public Safety Canada Research Fellowship Program in Honour of Stuart Nesbitt White 134
TAC Foundation Scholarships 643

Caribbean Countries
George Pepler International Award 579
International Postgraduate Research Scholarships (IPRS) 417

East European Countries
George Pepler International Award 579
International Postgraduate Research Scholarships (IPRS) 417

Middle East
George Pepler International Award 579
International Postgraduate Research Scholarships (IPRS) 417

New Zealand
George Pepler International Award 579

South Africa
George Pepler International Award 579
International Postgraduate Research Scholarships (IPRS) 417

United Kingdom
George Pepler International Award 579
International Postgraduate Research Scholarships (IPRS) 417

United States of America
Environmental Public Policy and Conflict Resolution PhD Fellowship 468
George Pepler International Award 579
International Postgraduate Research Scholarships (IPRS) 417

West European Countries
George Pepler International Award 579
International Postgraduate Research Scholarships (IPRS) 417

RURAL PLANNING

STRUCTURAL ARCHITECTURE

Any Country
ETH Zurich Excellence Scholarship and Opportunity Award 629
Franklin Research Grant Program 80
MacDowell Colony Residencies 439
Structural Engineering Travelling Fellowship 602

African Nations
ECOPOLIS Graduate Research and Design Awards 384
International Postgraduate Research Scholarships (IPRS) 417

Canada
ECOPOLIS Graduate Research and Design Awards 384
International Postgraduate Research Scholarships (IPRS) 417

Caribbean Countries
ECOPOLIS Graduate Research and Design Awards 384
International Postgraduate Research Scholarships (IPRS) 417

East European Countries
ECOPOLIS Graduate Research and Design Awards 384
International Postgraduate Research Scholarships (IPRS) 417

Middle East
ECOPOLIS Graduate Research and Design Awards 384
International Postgraduate Research Scholarships (IPRS) 417

South Africa
ECOPOLIS Graduate Research and Design Awards 384
International Postgraduate Research Scholarships (IPRS) 417

United Kingdom
Hilda Martindale Trust Exhibitions 340
International Postgraduate Research Scholarships (IPRS) 417

United States of America
International Postgraduate Research Scholarships (IPRS) 417

West European Countries
International Postgraduate Research Scholarships (IPRS) 417

TOWN PLANNING

Any Country
ETH Zurich Excellence Scholarship and Opportunity Award 629
Franklin Research Grant Program 80
RICS Education Trust Award 550

African Nations
ECOPOLIS Graduate Research and Design Awards 384
George Pepler International Award 579
International Postgraduate Research Scholarships (IPRS) 417

Australia
George Pepler International Award 579

Canada
ECOPOLIS Graduate Research and Design Awards 384

Caribbean Countries

East European Countries

Middle East

New Zealand

South Africa

United Kingdom

United States of America

West European Countries

ARTS AND HUMANITIES

GENERAL

Any Country

African Nations

Australia

Canada

Caribbean Countries

East European Countries

European Union

Middle East

New Zealand

South Africa

United Kingdom

United States of America

West European Countries

ARCHAEOLOGY

Any Country

African Nations

Australia

Canada

Caribbean Countries

European Union

Middle East

New Zealand

South Africa

United Kingdom

United States of America

West European Countries

CLASSICAL LANGUAGES AND LITERATURES

Any Country

Australia

SANSKRIT

Any Country

European Union

United Kingdom

COMPARATIVE LITERATURE

Any Country

African Nations

Australia

Canada

Caribbean Countries

East European Countries

European Union

Middle East

New Zealand

South Africa

United Kingdom

United States of America

West European Countries

HISTORY

Any Country

African Nations

Australia

Canada

Caribbean Countries

East European Countries

European Union

Middle East

New Zealand

South Africa

United Kingdom

United States of America

West European Countries

ANCIENT CIVILISATIONS

Any Country

African Nations

Australia

Canada

Caribbean Countries

East European Countries

European Union

Middle East

New Zealand

South Africa

United Kingdom

United States of America

West European Countries

CONTEMPORARY HISTORY

Any Country

African Nations

Australia

Canada

MEDIEVAL STUDIES

European Union

Middle East

New Zealand

South Africa

United Kingdom

United States of America

West European Countries

PREHISTORY

Any Country

Australia

Canada

European Union

New Zealand

United Kingdom

United States of America

West European Countries

LINGUISTICS

Any Country

African Nations

Australia

Canada

Caribbean Countries

European Union

Middle East

New Zealand

South Africa

United Kingdom

United States of America

West European Countries

APPLIED LINGUISTICS

Any Country

African Nations

Australia

Canada

Australia

European Union

New Zealand

United Kingdom

United States of America

West European Countries

SEMANTICS AND TERMINOLOGY

Any Country

Australia

European Union

New Zealand

United Kingdom

United States of America

West European Countries

SPEECH STUDIES

MODERN LANGUAGES

Any Country

African Nations

Australia

Canada

Caribbean Countries

East European Countries

European Union

Middle East

New Zealand

South Africa

United Kingdom

United States of America

West European Countries

AFRICAN LANGUAGES

Any Country

Research Preparation Master's Scheme 121

Middle East
Aberystwyth International Excellence Scholarships 4

New Zealand
Aberystwyth International Excellence Scholarships 4

South Africa
Aberystwyth International Excellence Scholarships 4

United Kingdom
Collaborative Doctoral Awards 121
Professional Preparation Master's Scheme 121
Research Preparation Master's Scheme 121

United States of America
Aberystwyth International Excellence Scholarships 4
Summer Language Study Grants in Turkey for Graduate Students 375
Title VIII Southeast European Language Training Program 55

FINNISH

Any Country
Lucy Cavendish College: Evelyn Povey Studentship 692
United States Holocaust Memorial Museum Center for Advanced Holocaust Studies Visiting Scholar Programs 649

European Union
Collaborative Doctoral Awards 121
Professional Preparation Master's Scheme 121
Research Preparation Master's Scheme 121

United Kingdom
Collaborative Doctoral Awards 121
Professional Preparation Master's Scheme 121
Research Preparation Master's Scheme 121

FINO UGRIAN LANGUAGES

Any Country
United States Holocaust Memorial Museum Center for Advanced Holocaust Studies Visiting Scholar Programs 649

European Union
Collaborative Doctoral Awards 121
Professional Preparation Master's Scheme 121
Research Preparation Master's Scheme 121

United Kingdom
Collaborative Doctoral Awards 121
Professional Preparation Master's Scheme 121
Research Preparation Master's Scheme 121

FRENCH

Any Country
Ahmanson and Getty Postdoctoral Fellowships 646
Aldo and Jeanne Scaglione Prize for French and Francophone Literary Studies 462
ASECS (American Society for 18th-Century Studies)/Clark Library Fellowships 646
Barron Bequest 837
Camargo Fellowships 200
Clark Library Short-Term Resident Fellowships 646
Clark Predoctoral Fellowships 646
Clark-Huntington Joint Bibliographical Fellowship 647
Claude and Vincenette Pichois Research Award 722
Doctoral Career Developement Scholarship (DCDS) Scheme 4
Jean Orr Scholarship 661
La Trobe University Postgraduate Research Scholarship 418
Lewis Walpole Library Fellowship 429
Mary Isabel Sibley Fellowship 533
Medieval & Mod Languages: Zaharoff Graduate Scholarship 778
Queen Mary, University of London Research Studentships 540
Society for the Study of French History Bursaries 613
United States Holocaust Memorial Museum Center for Advanced Holocaust Studies Visiting Scholar Programs 649
University of Bristol Postgraduate Scholarships 680
University of Kent School of European Culture and Languages Scholarships 742
University of Otago Coursework Master's Scholarship 761
University of Otago Doctoral Scholarships 761
University of Otago International Master's Scholarship 761
University of Southampton Postgraduate Studentships 806
Warwick French Department Bursaries 819
World Universities Network (WUN) International Research Mobility Scheme 806

African Nations
Aberystwyth International Excellence Scholarships 4
International Postgraduate Research Scholarships (IPRS) 417

Australia
Aberystwyth International Excellence Scholarships 4
French Government Postgraduate Studies Scholarships 285
University of Otago Master's Scholarships 761

Canada
Aberystwyth International Excellence Scholarships 4
Edouard Morot-Sir Fellowship in French Studies 364
Gilbert Chinard Fellowships 364
Harmon Chadbourn Rorison Fellowship 365
International Postgraduate Research Scholarships (IPRS) 417

Caribbean Countries
Aberystwyth International Excellence Scholarships 4
International Postgraduate Research Scholarships (IPRS) 417

East European Countries
International Postgraduate Research Scholarships (IPRS) 417

European Union
Collaborative Doctoral Awards 121
Professional Preparation Master's Scheme 121
Quinn, Nathan and Edmond Scholarships 748
Research Preparation Master's Scheme 121

Middle East
Aberystwyth International Excellence Scholarships 4
International Postgraduate Research Scholarships (IPRS) 417

New Zealand
Aberystwyth International Excellence Scholarships 4
University of Otago Master's Scholarships 761

South Africa
Aberystwyth International Excellence Scholarships 4
International Postgraduate Research Scholarships (IPRS) 417

United Kingdom
Collaborative Doctoral Awards 121
International Postgraduate Research Scholarships (IPRS) 417
Mr and Mrs David Edward Memorial Award 164
Professional Preparation Master's Scheme 121
Quinn, Nathan and Edmond Scholarships 748
Research Preparation Master's Scheme 121
University of Kent School of European Culture and Languages Scholarships 742
University of Kent School of European Culture and Languages Studentships 742

United States of America

Aberystwyth International Excellence Scholarships 4
Edouard Morot-Sir Fellowship in French Studies 364
Fulbright Teacher Exchange 308
Gilbert Chinard Fellowships 364
Harmon Chadbourn Rorison Fellowship 365
International Postgraduate Research Scholarships (IPRS) 417
Marshall Scholarships 181
The Walter J Jensen Fellowship for French Language, Literature and Culture 533

West European Countries

Galway Scholarship 497
International Postgraduate Research Scholarships (IPRS) 417
Mr and Mrs David Edward Memorial Award 164
University of Kent School of European Culture and Languages Scholarships 742
University of Kent School of European Culture and Languages Studentships 742

GERMAN

Any Country

Ahmanson and Getty Postdoctoral Fellowships 646
Aldo and Jeanne Scaglione Prize for Studies in Germanic Languages and Literatures 462
Arthur F. Burns Fellowship Program 381
ASECS (American Society for 18th-Century Studies)/Clark Library Fellowships 646
Austro-American Association of Boston Stipend 161
Barron Bequest 837
Clark Library Short-Term Resident Fellowships 646
Clark Predoctoral Fellowships 646
Clark-Huntington Joint Bibliographical Fellowship 647
Doctoral Career Developement Scholarship (DCDS) Scheme 4
Fielden Research Scholarship 659
Queen Mary, University of London Research Studentships 540
United States Holocaust Memorial Museum Center for Advanced Holocaust Studies Visiting Scholar Programs 649
University of Bristol Postgraduate Scholarships 680
University of Kent School of European Culture and Languages Scholarships 742
University of Otago Coursework Master's Scholarship 761
University of Otago Doctoral Scholarships 761
University of Otago International Master's Scholarship 761
University of Southampton Postgraduate Studentships 806
World Universities Network (WUN) International Research Mobility Scheme 806

African Nations

Aberystwyth International Excellence Scholarships 4

Australia

Aberystwyth International Excellence Scholarships 4
University of Otago Master's Scholarships 761

Canada

Aberystwyth International Excellence Scholarships 4

Caribbean Countries

Aberystwyth International Excellence Scholarships 4

European Union

AHRC Studentships 230
Collaborative Doctoral Awards 121
Professional Preparation Master's Scheme 121
Research Preparation Master's Scheme 121

Middle East

Aberystwyth International Excellence Scholarships 4

New Zealand

Aberystwyth International Excellence Scholarships 4
University of Otago Master's Scholarships 761

South Africa

Aberystwyth International Excellence Scholarships 4

United Kingdom

AHRC Studentships 230
Collaborative Doctoral Awards 121
Mr and Mrs David Edward Memorial Award 164
Professional Preparation Master's Scheme 121
Research Preparation Master's Scheme 121
University of Kent School of European Culture and Languages Scholarships 742
University of Kent School of European Culture and Languages Studentships 742

United States of America

Aberystwyth International Excellence Scholarships 4
Congress Bundestag Youth Exchange for Young Professionals 225
Marshall Scholarships 181

West European Countries

The Eugen and Ilse Seibold Prize 272
Galway Scholarship 497
Mr and Mrs David Edward Memorial Award 164
University of Kent School of European Culture and Languages Scholarships 742
University of Kent School of European Culture and Languages Studentships 742

GERMANIC LANGUAGES

Any Country

Ahmanson and Getty Postdoctoral Fellowships 646
Aldo and Jeanne Scaglione Prize for Studies in Germanic Languages and Literatures 462
Austro-American Association of Boston Stipend 161
Barron Bequest 837
Clark Library Short-Term Resident Fellowships 646
Clark Predoctoral Fellowships 646
Clark-Huntington Joint Bibliographical Fellowship 647
Queen Mary, University of London Research Studentships 540
United States Holocaust Memorial Museum Center for Advanced Holocaust Studies Visiting Scholar Programs 649
Workmen's Circle/Dr Emanuel Patt Visiting Professorship 870

European Union

Collaborative Doctoral Awards 121
Professional Preparation Master's Scheme 121
Research Preparation Master's Scheme 121

United Kingdom

Collaborative Doctoral Awards 121
Professional Preparation Master's Scheme 121
Research Preparation Master's Scheme 121

United States of America

Congress Bundestag Youth Exchange for Young Professionals 225

HEBREW

Any Country

Barron Bequest 837
Girton College: Ida and Isidore Cohen Research Scholarship 688
Jacob Hirsch Fellowship 91
SOAS Research Scholarship 588
United States Holocaust Memorial Museum Center for Advanced Holocaust Studies Visiting Scholar Programs 649

Canada
JCCA Graduate Education Scholarship 403

European Union
CBRL Travel Grant 252
Collaborative Doctoral Awards 121
Professional Preparation Master's Scheme 121
Research Preparation Master's Scheme 121

United Kingdom
CBRL Pilot Study Award 252
CBRL Travel Grant 252
Collaborative Doctoral Awards 121
Professional Preparation Master's Scheme 121
Research Preparation Master's Scheme 121

United States of America
JCCA Graduate Education Scholarship 403

HUNGARIAN

Any Country
United States Holocaust Memorial Museum Center for Advanced Holocaust Studies Visiting Scholar Programs 649

European Union
Collaborative Doctoral Awards 121
Professional Preparation Master's Scheme 121
Research Preparation Master's Scheme 121

United Kingdom
Collaborative Doctoral Awards 121
Professional Preparation Master's Scheme 121
Research Preparation Master's Scheme 121

United States of America
Fulbright Teacher Exchange 308
IREX Individual Advanced Research Opportunities 391

INDIAN LANGUAGES

Any Country
SOAS Research Scholarship 588

European Union
AHRC Studentships 230
Collaborative Doctoral Awards 121
Professional Preparation Master's Scheme 121
Research Preparation Master's Scheme 121

United Kingdom
AHRC Studentships 230
Collaborative Doctoral Awards 121
Professional Preparation Master's Scheme 121
Research Preparation Master's Scheme 121

INDIC LANGUAGES

IRANIC LANGUAGES

Any Country
British Academy Ancient Persia Fund 175
SOAS Research Scholarship 588

European Union
Collaborative Doctoral Awards 121
Professional Preparation Master's Scheme 121
Research Preparation Master's Scheme 121

United Kingdom
The British Institute for the Study of Iraq Grants 184
Collaborative Doctoral Awards 121
Professional Preparation Master's Scheme 121
Research Preparation Master's Scheme 121

ITALIAN

Any Country
Ahmanson and Getty Postdoctoral Fellowships 646
Aldo and Jeanne Scaglione Prize for Italian Studies 462
ASECS (American Society for 18th-Century Studies)/Clark Library Fellowships 646
Barron Bequest 837
Clark Library Short-Term Resident Fellowships 646
Clark Predoctoral Fellowships 646
Clark-Huntington Joint Bibliographical Fellowship 647
Craig Hugh Smyth Visiting Fellowship 836
Howard R Marraro Prize 462
I Tatti Fellowships 836
Jean Orr Scholarship 661
United States Holocaust Memorial Museum Center for Advanced Holocaust Studies Visiting Scholar Programs 649
University of Bristol Postgraduate Scholarships 680

European Union
AHRC Studentships 230
Collaborative Doctoral Awards 121
Professional Preparation Master's Scheme 121
Research Preparation Master's Scheme 121

United Kingdom
AHRC Studentships 230
Balsdon Fellowship 190
Collaborative Doctoral Awards 121
Professional Preparation Master's Scheme 121
Research Preparation Master's Scheme 121
Rome Awards 191
Rome Fellowship 192
Rome Scholarships in Ancient, Medieval and Later Italian Studies 192
University of Kent School of European Culture and Languages Scholarships 742
University of Kent School of European Culture and Languages Studentships 742

United States of America
Gladys Krieble Delmas Foundation Grants 319
Marshall Scholarships 181

West European Countries
University of Kent School of European Culture and Languages Scholarships 742
University of Kent School of European Culture and Languages Studentships 742

JAPANESE

Any Country
Harvard Postdoctoral Fellowships in Japanese Studies 283
Oriental Studies: Sasakawa Fund 782
Pembroke College: TEPCO Scholarship 783
SOAS Research Scholarship 588
United States Holocaust Memorial Museum Center for Advanced Holocaust Studies Visiting Scholar Programs 649
Western Australian Government Japanese Studies Scholarship 269

European Union
AHRC Studentships 230
Collaborative Doctoral Awards 121

SWEDISH

NATIVE LANGUAGE AND LITERATURE

PHILOSOPHY

Any Country

Australia

Canada

European Union

New Zealand

United Kingdom

United States of America

West European Countries

Galway Scholarship 497
University of Kent School of European Culture and Languages Scholarships 742
University of Kent School of European Culture and Languages Studentships 742

ETHICS

Any Country

ASCSA Advanced Fellowships 89
Camargo Fellowships 200
Charlotte W Newcombe Doctoral Dissertation Fellowships 864
CIHR Fellowships Program 214
Delahaye Memorial Benefaction 815
Helen McCormack Turner Memorial Scholarship 815
Herbert Hughes Scholarship 815
IAS-STS Fellowship Programme 366
Kanner Fellowship In British Studies 647
Mary Radcliffe Scholarship 815
Oscar Broneer Traveling Fellowship 91
RHYS Curzon-Jones Scholarship 815
Ridley Lewis Bursary 815
United States Holocaust Memorial Museum Center for Advanced Holocaust Studies Visiting Scholar Programs 649
University of Kent School of European Culture and Languages Scholarships 742
W D Llewelyn Memorial Benefaction 816

Canada

Harry Bikakis Fellowship 90
Oscar Broneer Traveling Fellowship 91

European Union

Collaborative Doctoral Awards 121
Professional Preparation Master's Scheme 121
Research Preparation Master's Scheme 121

United Kingdom

Collaborative Doctoral Awards 121
ESU Chautauqua Institution Scholarships 286
Professional Preparation Master's Scheme 121
Research Preparation Master's Scheme 121
University of Kent School of European Culture and Languages Scholarships 742
University of Kent School of European Culture and Languages Studentships 742

United States of America

Harry Bikakis Fellowship 90
Marshall Scholarships 181
NEH Fellowships 91
Oscar Broneer Traveling Fellowship 91

West European Countries

University of Kent School of European Culture and Languages Scholarships 742
University of Kent School of European Culture and Languages Studentships 742

LOGIC

Any Country

ASCSA Advanced Fellowships 89
Camargo Fellowships 200
John Stuart Mill Scholarship in Philosophy of Mind and Logic 661
Kanner Fellowship In British Studies 647
Oscar Broneer Traveling Fellowship 91
University of Kent School of European Culture and Languages Scholarships 742

Canada

Oscar Broneer Traveling Fellowship 91

European Union

Collaborative Doctoral Awards 121
Professional Preparation Master's Scheme 121
Research Preparation Master's Scheme 121

United Kingdom

Collaborative Doctoral Awards 121
Professional Preparation Master's Scheme 121
Research Preparation Master's Scheme 121
University of Kent School of European Culture and Languages Scholarships 742
University of Kent School of European Culture and Languages Studentships 742

United States of America

Marshall Scholarships 181
NEH Fellowships 91
Oscar Broneer Traveling Fellowship 91

West European Countries

University of Kent School of European Culture and Languages Scholarships 742
University of Kent School of European Culture and Languages Studentships 742

METAPHYSICS

Any Country

ASCSA Advanced Fellowships 89
Camargo Fellowships 200
Kanner Fellowship In British Studies 647
Oscar Broneer Traveling Fellowship 91
University of Kent School of European Culture and Languages Scholarships 742

Canada

Oscar Broneer Traveling Fellowship 91

European Union

Collaborative Doctoral Awards 121
Professional Preparation Master's Scheme 121
Research Preparation Master's Scheme 121

United Kingdom

Collaborative Doctoral Awards 121
Professional Preparation Master's Scheme 121
Research Preparation Master's Scheme 121
University of Kent School of European Culture and Languages Scholarships 742
University of Kent School of European Culture and Languages Studentships 742

United States of America

Marshall Scholarships 181
NEH Fellowships 91
Oscar Broneer Traveling Fellowship 91

West European Countries

University of Kent School of European Culture and Languages Scholarships 742
University of Kent School of European Culture and Languages Studentships 742

PHILOSOPHICAL SCHOOLS

Any Country

Ahmanson and Getty Postdoctoral Fellowships 646

Canada

European Union

United Kingdom

United States of America

West European Countries

TRANSLATION AND INTERPRETATION

Any Country

European Union

United Kingdom

United States of America

West European Countries

WRITING (AUTHORSHIP)

Any Country

African Nations

Australia

Canada

Caribbean Countries

East European Countries

European Union

Middle East

New Zealand

South Africa

United Kingdom

United States of America

West European Countries

BUSINESS ADMINISTRATION AND MANAGEMENT

GENERAL

Any Country

African Nations

Australia

Canada

Caribbean Countries

East European Countries

European Union

Middle East

New Zealand

South Africa

United Kingdom

United States of America

West European Countries

ACCOUNTANCY

Any Country

BKD Graduate Scholarship 459
Doctoral Career Developement Scholarship (DCDS) Scheme 4
Jacob's Pillow Intern Program 399
La Trobe University Postgraduate Research Scholarship 418
Rhodes University Postdoctoral Fellowship and The Andrew Mellon Postdoctoral Fellowship 549
School of Management and Business Masters Scholarships 5
UNO Graduate Assistantships 754
World Universities Network (WUN) International Research Mobility Scheme 806

African Nations

Aberystwyth International Excellence Scholarships 4

Australia

Aberystwyth International Excellence Scholarships 4
Australian Postgraduate Award 674

Canada

Aberystwyth International Excellence Scholarships 4

Caribbean Countries

Aberystwyth International Excellence Scholarships 4

European Union

University of Essex Silberrad Scholarships 729

Middle East

Aberystwyth International Excellence Scholarships 4

New Zealand

Aberystwyth International Excellence Scholarships 4
Australian Postgraduate Award 674

South Africa

Aberystwyth International Excellence Scholarships 4
Allan Gray Senior Scholarship 548
Henderson Postgraduate Scholarships 549

United Kingdom

Mr and Mrs David Edward Memorial Award 164
University of Essex Silberrad Scholarships 729

United States of America

Aberystwyth International Excellence Scholarships 4
AICPA Fellowship for Minority Doctoral Students 66
AICPA John L. Carey Scholarship 66
AICPA/Accountemps Student Scholarship 67
Congress Bundestag Youth Exchange for Young Professionals 225
Jacki Tuckfield Memorial Graduate Business Scholarship Fund 397
Kenneth W. Heikes Family Endowed Scholarship 467
Marshall Scholarships 181
Woodrow Wilson Teaching Fellowship 864

West European Countries

Mr and Mrs David Edward Memorial Award 164

BUSINESS AND COMMERCE

Any Country

The Cámara de Comercio Scholarship 349
Doctoral Career Developement Scholarship (DCDS) Scheme 4
ETS Summer Internship Program for Graduate Students 282
IESE AECI/Becas MAE 349
IESE Alumni Association Scholarships 349
IESE Donovan Data Systems Anniversary 349
IESE Private Foundation Scholarships 349
IESE Trust Scholarships 349
ISM Professional Research Development Grant 368
La Trobe University Postgraduate Research Scholarship 418
Marshall Memorial Fellowship 315
School of Management and Business Masters Scholarships 5
Title VIII-Supported Short-Term Grant 411

African Nations

Aberystwyth International Excellence Scholarships 4

Australia

Aberystwyth International Excellence Scholarships 4
Australian Postgraduate Award 674

Canada

Aberystwyth International Excellence Scholarships 4

Caribbean Countries

Aberystwyth International Excellence Scholarships 4

European Union

Fundación Ramón Areces Scholarship 349
University of Essex Silberrad Scholarships 729

Middle East

Aberystwyth International Excellence Scholarships 4

New Zealand

Aberystwyth International Excellence Scholarships 4
Australian Postgraduate Award 674

South Africa

Aberystwyth International Excellence Scholarships 4
Allan Gray Senior Scholarship 548

United Kingdom

Mr and Mrs David Edward Memorial Award 164
University of Essex Silberrad Scholarships 729

United States of America

Aberystwyth International Excellence Scholarships 4
Congress Bundestag Youth Exchange for Young Professionals 225
ETS Summer Internship Program for Graduate Students 282
Fulbright Specialist Program 253
ISM Dissertation Research Grant 367
Jacki Tuckfield Memorial Graduate Business Scholarship Fund 397
Marshall Scholarships 181
Philip Morris USA Thurgood Marshall Scholarship 639

West European Countries

Mr and Mrs David Edward Memorial Award 164

BUSINESS COMPUTING

Any Country

Jacob's Pillow Intern Program 399
John L. and Rita M. Bangs Scholarship 460
Rhodes University Postdoctoral Fellowship and The Andrew Mellon Postdoctoral Fellowship 549
University of Southampton Postgraduate Studentships 806
World Universities Network (WUN) International Research Mobility Scheme 806

Australia

Australian Postgraduate Award 674

New Zealand

Australian Postgraduate Award 674

United States of America

Congress Bundestag Youth Exchange for Young Professionals 225

BUSINESS MACHINE OPERATION

FINANCE, BANKING AND INVESTMENT

Any Country

African Nations

Australia

Canada

Caribbean Countries

East European Countries

European Union

Middle East

New Zealand

South Africa

United Kingdom

United States of America

West European Countries

HUMAN RESOURCES

INSTITUTIONAL ADMINISTRATION

Any Country

INSURANCE

Any Country

Canada

United Kingdom

United States of America

West European Countries

INTERNATIONAL BUSINESS

Any Country

African Nations

Australia

Canada

Caribbean Countries

Aberystwyth International Excellence Scholarships 4

European Union

Transatlantic Fellows Program 316
University of Essex Silberrad Scholarships 729

Middle East

Aberystwyth International Excellence Scholarships 4
Middle East And North Africa Postgraduate Award (Warwick Manufacturing Group) 818

New Zealand

Aberystwyth International Excellence Scholarships 4
Australian Postgraduate Award 674

South Africa

Aberystwyth International Excellence Scholarships 4

United Kingdom

John Speak Trust Scholarships 174
University of Essex Silberrad Scholarships 729

United States of America

Aberystwyth International Excellence Scholarships 4
Congress Bundestag Youth Exchange for Young Professionals 225
Fulbright Distinguished Chairs Program 252
Fulbright Specialist Program 253
Fulbright-Kennan Institute Research Scholarships 411
ISM Dissertation Research Grant 367
Marshall Scholarships 181
Transatlantic Fellows Program 316

LABOUR/INDUSTRIAL RELATIONS

Any Country

Mackenzie King Travelling Scholarship 837

East European Countries

Warwick Busines School Scholarships (For Masters Study) 839

European Union

Warwick Busines School Scholarships (For Masters Study) 839

United Kingdom

Mr and Mrs David Edward Memorial Award 164
Warwick Busines School Scholarships (For Masters Study) 839

United States of America

Fulbright Distinguished Chairs Program 252
Fulbright Specialist Program 253

West European Countries

Mr and Mrs David Edward Memorial Award 164
Warwick Busines School Scholarships (For Masters Study) 839

MANAGEMENT SYSTEMS

Any Country

Doctoral Career Developement Scholarship (DCDS) Scheme 4
ISM Professional Research Development Grant 368
Queen Mary, University of London Research Studentships 540
Rhodes University Postdoctoral Fellowship and The Andrew Mellon Postdoctoral Fellowship 549
School of Management and Business Masters Scholarships 5

African Nations

Aberystwyth International Excellence Scholarships 4

Australia

Aberystwyth International Excellence Scholarships 4
Australian Postgraduate Award 674

Canada

Aberystwyth International Excellence Scholarships 4
Sloan Industry Studies Fellowships 20

Caribbean Countries

Aberystwyth International Excellence Scholarships 4

East European Countries

Warwick Busines School Scholarships (For Masters Study) 839

European Union

Warwick Busines School Scholarships (For Masters Study) 839

Middle East

Aberystwyth International Excellence Scholarships 4

New Zealand

Aberystwyth International Excellence Scholarships 4
The Association of University Staff Crozier Scholarship 631
Australian Postgraduate Award 674

South Africa

Aberystwyth International Excellence Scholarships 4

United Kingdom

Warwick Busines School Scholarships (For Masters Study) 839

United States of America

Aberystwyth International Excellence Scholarships 4
AICPA/Accountemps Student Scholarship 67
Campus Ecology Fellowship Program 501
Congress Bundestag Youth Exchange for Young Professionals 225
ISM Dissertation Research Grant 367
Jacki Tuckfield Memorial Graduate Business Scholarship Fund 397
Sloan Industry Studies Fellowships 20

West European Countries

Warwick Busines School Scholarships (For Masters Study) 839

MARKETING

Any Country

Doctoral Career Developement Scholarship (DCDS) Scheme 4
Jacob's Pillow Intern Program 399
John L. and Rita M. Bangs Scholarship 460
MSc in Marketing and International Marketing 808
School of Management and Business Masters Scholarships 5
Sohei Nakayama Memorial Scholarship (Type A, B, C, D and S) 393

African Nations

Aberystwyth International Excellence Scholarships 4

Australia

Aberystwyth International Excellence Scholarships 4
Australian Postgraduate Award 674

Canada

Aberystwyth International Excellence Scholarships 4
Horticultural Research Institute Grants 344

Caribbean Countries

Aberystwyth International Excellence Scholarships 4

East European Countries

Warwick Busines School Scholarships (For Masters Study) 839

EDUCATION AND TEACHER TRAINING

GENERAL

Any Country

African Nations

Australia

Canada

Caribbean Countries

East European Countries

European Union

Middle East

New Zealand

South Africa

United Kingdom

United States of America

West European Countries

ADULT EDUCATION

Any Country

African Nations

Australia
Australian Postgraduate Award 674
Early Career Fellowship 428
National Collaborative Breast Cancer Research Grant Program 480
Pilot Study Grants 481
Practitioner Fellowship 481

Canada
International Postgraduate Research Scholarships (IPRS) 417

Caribbean Countries
International Postgraduate Research Scholarships (IPRS) 417

East European Countries
International Postgraduate Research Scholarships (IPRS) 417

Middle East
International Postgraduate Research Scholarships (IPRS) 417

New Zealand
Australian Postgraduate Award 674

South Africa
International Postgraduate Research Scholarships (IPRS) 417

United Kingdom
International Postgraduate Research Scholarships (IPRS) 417

United States of America
ETS Postdoctoral Fellowships 282
ETS Summer Internship Program for Graduate Students 282
ETS Sylvia Taylor Johnson Minority Fellowship in Educational Measurement 282
International Postgraduate Research Scholarships (IPRS) 417

West European Countries
International Postgraduate Research Scholarships (IPRS) 417

CONTINUING EDUCATION

EDUCATIONAL SCIENCE

Any Country
Camargo Fellowships 200
ETS Postdoctoral Fellowships 282
Faculty Studentships 733
IAU Grants 380
Lindbergh Grants 232
SERC Postdoctoral Fellowships 603
SERC Predoctoral Fellowships 603
SERC Senior Fellowships 604
University of Bristol Postgraduate Scholarships 680
University of Southampton Postgraduate Studentships 806
World Universities Network (WUN) International Research Mobility Scheme 806

European Union
All Saints Educational Trust Personal Scholarships 21
Collection of Small Endowed Awards 744

United Kingdom
All Saints Educational Trust Corporate Awards 21
All Saints Educational Trust Personal Scholarships 21
Collection of Small Endowed Awards 744
Kennedy Scholarships 412

United States of America
ETS Postdoctoral Fellowships 282
Japan Society for the Promotion of Science (JSPS) Fellowship 606
Kennedy Research Grants 408
SSRC Abe Fellowship Program 606

CURRICULUM

Any Country
Faculty Studentships 733
International Postgraduate Research Scholarship 674
Jacob's Pillow Intern Program 399
Lindbergh Grants 232

African Nations
International Postgraduate Research Scholarships (IPRS) 417

Australia
Australian Postgraduate Award 674

Canada
International Postgraduate Research Scholarships (IPRS) 417

Caribbean Countries
International Postgraduate Research Scholarships (IPRS) 417

East European Countries
International Postgraduate Research Scholarships (IPRS) 417

Middle East
International Postgraduate Research Scholarships (IPRS) 417

New Zealand
Australian Postgraduate Award 674

South Africa
International Postgraduate Research Scholarships (IPRS) 417

United Kingdom
International Postgraduate Research Scholarships (IPRS) 417
Mr and Mrs David Edward Memorial Award 164

United States of America
International Postgraduate Research Scholarships (IPRS) 417

West European Countries
International Postgraduate Research Scholarships (IPRS) 417
Mr and Mrs David Edward Memorial Award 164

DISTANCE EDUCATION

Any Country
Faculty Studentships 733
International Postgraduate Research Scholarship 674

Australia
Australian Postgraduate Award 674

New Zealand
Australian Postgraduate Award 674

EDUCATIONAL ADMINISTRATION

Any Country
ETS Summer Internship Program for Graduate Students 282
Fulbright Institutional Linkages Program 306
International Postgraduate Research Scholarship 674
Jacob's Pillow Intern Program 399

International Postgraduate Research Scholarships (IPRS) 417
Mr and Mrs David Edward Memorial Award 164

United States of America

Aberystwyth International Excellence Scholarships 4
ETS Postdoctoral Fellowships 282
ETS Summer Internship Program for Graduate Students 282
ETS Sylvia Taylor Johnson Minority Fellowship in Educational Measurement 282
International Postgraduate Research Scholarships (IPRS) 417
Nurses' Educational Funds Fellowships and Scholarships 513

West European Countries

International Postgraduate Research Scholarships (IPRS) 417
Mr and Mrs David Edward Memorial Award 164

EDUCATIONAL TECHNOLOGY

Any Country

ETS Postdoctoral Fellowships 282
ETS Summer Internship Program for Graduate Students 282
Faculty Studentships 733
International Postgraduate Research Scholarship 674
Lindbergh Grants 232

African Nations

International Postgraduate Research Scholarships (IPRS) 417

Australia

Australian Postgraduate Award 674

Canada

International Postgraduate Research Scholarships (IPRS) 417

Caribbean Countries

International Postgraduate Research Scholarships (IPRS) 417

East European Countries

International Postgraduate Research Scholarships (IPRS) 417

Middle East

International Postgraduate Research Scholarships (IPRS) 417

New Zealand

Australian Postgraduate Award 674

South Africa

International Postgraduate Research Scholarships (IPRS) 417

United Kingdom

International Postgraduate Research Scholarships (IPRS) 417

United States of America

ETS Postdoctoral Fellowships 282
ETS Summer Internship Program for Graduate Students 282
International Postgraduate Research Scholarships (IPRS) 417
Nurses' Educational Funds Fellowships and Scholarships 513

West European Countries

International Postgraduate Research Scholarships (IPRS) 417

EDUCATIONAL TESTING AND EVALUATION

Any Country

ASBAH Research Grant 128
ETS Postdoctoral Fellowships 282
ETS Summer Internship Program for Graduate Students 282
International Postgraduate Research Scholarship 674

African Nations

International Postgraduate Research Scholarships (IPRS) 417

Australia

Australian Postgraduate Award 674

Canada

International Postgraduate Research Scholarships (IPRS) 417

Caribbean Countries

International Postgraduate Research Scholarships (IPRS) 417

East European Countries

International Postgraduate Research Scholarships (IPRS) 417

Middle East

International Postgraduate Research Scholarships (IPRS) 417

New Zealand

Australian Postgraduate Award 674

South Africa

International Postgraduate Research Scholarships (IPRS) 417

United Kingdom

Educational Project Grant 116
International Postgraduate Research Scholarships (IPRS) 417

United States of America

ETS Postdoctoral Fellowships 282
ETS Summer Internship Program for Graduate Students 282
ETS Sylvia Taylor Johnson Minority Fellowship in Educational Measurement 282
International Postgraduate Research Scholarships (IPRS) 417

West European Countries

International Postgraduate Research Scholarships (IPRS) 417

INTERNATIONAL AND COMPARATIVE EDUCATION

Any Country

Camargo Fellowships 200
Education: Routledge Scholarship 767
Frederick Douglass Institute Predoctoral Dissertation Fellowship 304
Jennings Randolph Program for International Peace Senior Fellowships 650
St Edmund Hall: Routledge St Edmund Hall Studentship 792

African Nations

Nicholas Hans Comparative Education Scholarship 370

Australia

Nicholas Hans Comparative Education Scholarship 370

Canada

Nicholas Hans Comparative Education Scholarship 370

Caribbean Countries

Nicholas Hans Comparative Education Scholarship 370

East European Countries

Nicholas Hans Comparative Education Scholarship 370

European Union

Transatlantic Fellows Program 316

African Nations

Australia

Canada

Caribbean Countries

East European Countries

European Union

Middle East

New Zealand

South Africa

United Kingdom

United States of America

West European Countries

LITERACY EDUCATION

Any Country

African Nations

Australia

Canada

Caribbean Countries

East European Countries

Middle East

New Zealand

South Africa

East European Countries

FEMS Fellowship 295
Hastings Center International Visiting Scholars Program 330

Middle East

Aberystwyth International Excellence Scholarships 4
Hastings Center International Visiting Scholars Program 330

New Zealand

Aberystwyth International Excellence Scholarships 4
Australian Postgraduate Award 674
Hastings Center International Visiting Scholars Program 330
Hsanz Travel Grant 326

South Africa

Aberystwyth International Excellence Scholarships 4
Hastings Center International Visiting Scholars Program 330

United Kingdom

FEMS Fellowship 295
Goldsmiths' Company Science for Society Courses 321
Hastings Center International Visiting Scholars Program 330
Hilda Martindale Trust Exhibitions 340

United States of America

Aberystwyth International Excellence Scholarships 4
ETS Postdoctoral Fellowships 282
ETS Summer Internship Program for Graduate Students 282
NIH Research Grants 491
Paul and Daisy Soros Fellowships for New Americans 524
Woodrow Wilson Teaching Fellowship 864

West European Countries

FEMS Fellowship 295
Hastings Center International Visiting Scholars Program 330

PRE-SCHOOL EDUCATION

Any Country

ETS Postdoctoral Fellowships 282
ETS Summer Internship Program for Graduate Students 282
Faculty Studentships 733
International Postgraduate Research Scholarship 674
Special Overseas Student Scholarship (SOSS) 675

African Nations

International Postgraduate Research Scholarships (IPRS) 417

Australia

Australian Postgraduate Award 674

Canada

International Postgraduate Research Scholarships (IPRS) 417
JCCA Graduate Education Scholarship 403

Caribbean Countries

International Postgraduate Research Scholarships (IPRS) 417

East European Countries

International Postgraduate Research Scholarships (IPRS) 417

Middle East

International Postgraduate Research Scholarships (IPRS) 417

New Zealand

Australian Postgraduate Award 674

South Africa

International Postgraduate Research Scholarships (IPRS) 417

United Kingdom

International Postgraduate Research Scholarships (IPRS) 417

United States of America

ETS Postdoctoral Fellowships 282
ETS Summer Internship Program for Graduate Students 282
ETS Sylvia Taylor Johnson Minority Fellowship in Educational Measurement 282
International Postgraduate Research Scholarships (IPRS) 417
JCCA Graduate Education Scholarship 403

West European Countries

International Postgraduate Research Scholarships (IPRS) 417

PRIMARY EDUCATION

Any Country

ETS Postdoctoral Fellowships 282
ETS Summer Internship Program for Graduate Students 282
Faculty Studentships 733
International Postgraduate Research Scholarship 674
Special Overseas Student Scholarship (SOSS) 675
University of Southampton Postgraduate Studentships 806
World Universities Network (WUN) International Research Mobility Scheme 806

African Nations

International Postgraduate Research Scholarships (IPRS) 417

Australia

Australian Postgraduate Award 674

Canada

International Postgraduate Research Scholarships (IPRS) 417

Caribbean Countries

International Postgraduate Research Scholarships (IPRS) 417

East European Countries

International Postgraduate Research Scholarships (IPRS) 417

European Union

All Saints Educational Trust Personal Scholarships 21

Middle East

International Postgraduate Research Scholarships (IPRS) 417

New Zealand

Australian Postgraduate Award 674

South Africa

International Postgraduate Research Scholarships (IPRS) 417

United Kingdom

All Saints Educational Trust Personal Scholarships 21
International Postgraduate Research Scholarships (IPRS) 417
Mr and Mrs David Edward Memorial Award 164

United States of America

ETS Postdoctoral Fellowships 282
ETS Summer Internship Program for Graduate Students 282
ETS Sylvia Taylor Johnson Minority Fellowship in Educational Measurement 282
Fulbright Teacher Exchange 308
International Postgraduate Research Scholarships (IPRS) 417

West European Countries

International Postgraduate Research Scholarships (IPRS) 417
Mr and Mrs David Edward Memorial Award 164

SECONDARY EDUCATION

Any Country
ETS Postdoctoral Fellowships 282
ETS Summer Internship Program for Graduate Students 282
Faculty Studentships 733
International Postgraduate Research Scholarship 674
Special Overseas Student Scholarship (SOSS) 675
University of Southampton Postgraduate Studentships 806
World Universities Network (WUN) International Research Mobility Scheme 806

African Nations
International Postgraduate Research Scholarships (IPRS) 417

Australia
Australian Postgraduate Award 674

Canada
International Postgraduate Research Scholarships (IPRS) 417

Caribbean Countries
International Postgraduate Research Scholarships (IPRS) 417

East European Countries
FEMS Fellowship 295
International Postgraduate Research Scholarships (IPRS) 417

European Union
All Saints Educational Trust Personal Scholarships 21

Middle East
International Postgraduate Research Scholarships (IPRS) 417

New Zealand
Australian Postgraduate Award 674

South Africa
International Postgraduate Research Scholarships (IPRS) 417

United Kingdom
All Saints Educational Trust Personal Scholarships 21
FEMS Fellowship 295
International Postgraduate Research Scholarships (IPRS) 417
Mr and Mrs David Edward Memorial Award 164

United States of America
ETS Postdoctoral Fellowships 282
ETS Summer Internship Program for Graduate Students 282
ETS Sylvia Taylor Johnson Minority Fellowship in Educational Measurement 282
Fulbright Teacher Exchange 308
International Postgraduate Research Scholarships (IPRS) 417
James Madison Fellowship Program 401
The Walter J Jensen Fellowship for French Language, Literature and Culture 533

West European Countries
FEMS Fellowship 295
International Postgraduate Research Scholarships (IPRS) 417
Mr and Mrs David Edward Memorial Award 164

SPECIAL EDUCATION

Any Country
ETS Postdoctoral Fellowships 282
ETS Summer Internship Program for Graduate Students 282
Faculty Studentships 733
Snowdon Trust Grants 605
Special Overseas Student Scholarship (SOSS) 675

Australia
Australian Postgraduate Award 674

Canada
JCCA Graduate Education Scholarship 403

New Zealand
Australian Postgraduate Award 674

United States of America
ACRES Scholarship 54
ETS Postdoctoral Fellowships 282
ETS Summer Internship Program for Graduate Students 282
ETS Sylvia Taylor Johnson Minority Fellowship in Educational Measurement 282
JCCA Graduate Education Scholarship 403

BILINGUAL/BICULTURAL EDUCATION

Any Country
Doctoral Career Developement Scholarship (DCDS) Scheme 4
Neporany Doctoral Fellowship 213

African Nations
Aberystwyth International Excellence Scholarships 4

Australia
Aberystwyth International Excellence Scholarships 4

Canada
Aberystwyth International Excellence Scholarships 4

Caribbean Countries
Aberystwyth International Excellence Scholarships 4

Middle East
Aberystwyth International Excellence Scholarships 4

New Zealand
Aberystwyth International Excellence Scholarships 4

South Africa
Aberystwyth International Excellence Scholarships 4

United Kingdom
Mr and Mrs David Edward Memorial Award 164

United States of America
Aberystwyth International Excellence Scholarships 4
Fulbright Teacher Exchange 308

West European Countries
Mr and Mrs David Edward Memorial Award 164

EDUCATION OF FOREIGNERS

Any Country
AUC Arabic Language Fellowships 107

United Kingdom
Mr and Mrs David Edward Memorial Award 164

United States of America
Fulbright Teacher Exchange 308

West European Countries
Mr and Mrs David Edward Memorial Award 164

TECHNOLOGY EDUCATION

Any Country

Australia

New Zealand

United States of America

ENGINEERING

GENERAL

Any Country

African Nations

Australia

Canada

Caribbean Countries

East European Countries

European Union

Middle East

New Zealand

South Africa

United Kingdom

United States of America

West European Countries

AERONAUTICAL AND AEROSPACE ENGINEERING

Any Country

African Nations

Australia

Canada

Caribbean Countries

East European Countries

European Union

Middle East

South Africa

United Kingdom

United States of America

West European Countries

AGRICULTURAL ENGINEERING

Any Country

Teagasc Walsh Fellowships 633

African Nations

Canadian Window on International Development 384
IDRC Doctoral Research Awards 384
International Postgraduate Research Scholarships (IPRS) 417

Australia

Australian Postgraduate Award 674
Fulbright Postgraduate Scholarship in Science and Engineering 159

Canada

Canadian Window on International Development 384
CWRA Dillon Consulting Scholarship/Ken Thomson Scholarship 219
IDRC Doctoral Research Awards 384
International Postgraduate Research Scholarships (IPRS) 417

Caribbean Countries

Canadian Window on International Development 384
IDRC Doctoral Research Awards 384
International Postgraduate Research Scholarships (IPRS) 417

East European Countries

IDRC Doctoral Research Awards 384
International Postgraduate Research Scholarships (IPRS) 417

European Union

Department of Agriculture and Rural Development (DARD) for Northern Ireland 258

Middle East

IDRC Doctoral Research Awards 384
International Postgraduate Research Scholarships (IPRS) 417

New Zealand

Australian Postgraduate Award 674

South Africa

Canadian Window on International Development 384
IDRC Doctoral Research Awards 384
International Postgraduate Research Scholarships (IPRS) 417

United Kingdom

Department of Agriculture and Rural Development (DARD) for Northern Ireland 258
Douglas Bomford Trust 258
Environmental Issues Award 258
International Postgraduate Research Scholarships (IPRS) 417

United States of America

Air Force Summer Faculty Fellowship Program 92
DEED (Demonstration of Energy-Efficient Developments) Student Research Grant and Internship 87
Environmental Public Policy and Conflict Resolution PhD Fellowship 468
International Postgraduate Research Scholarships (IPRS) 417

West European Countries

International Postgraduate Research Scholarships (IPRS) 417

AUTOMOTIVE ENGINEERING

Any Country

John Moyes Lessells Travel Scholarships 574
University of Bristol Postgraduate Scholarships 680
University of Southampton Postgraduate Studentships 806
World Universities Network (WUN) International Research Mobility Scheme 806

African Nations

International Postgraduate Research Scholarships (IPRS) 417

Australia

Australian Postgraduate Award 674
Fulbright Postgraduate Scholarship in Science and Engineering 159

Canada

International Postgraduate Research Scholarships (IPRS) 417

Caribbean Countries

International Postgraduate Research Scholarships (IPRS) 417

East European Countries

International Postgraduate Research Scholarships (IPRS) 417

European Union

University of Southampton Engineering Doctorate 806

Middle East

International Postgraduate Research Scholarships (IPRS) 417

New Zealand

Australian Postgraduate Award 674

South Africa

International Postgraduate Research Scholarships (IPRS) 417

United Kingdom

International Postgraduate Research Scholarships (IPRS) 417
University of Southampton Engineering Doctorate 806

United States of America

International Postgraduate Research Scholarships (IPRS) 417
Winston Churchill Foundation Scholarship 859

West European Countries

International Postgraduate Research Scholarships (IPRS) 417

BIOENGINEERING AND BIOMEDICAL ENGINEERING

Any Country

A*STAR Graduate Scholarship (Overseas) 11
A*STAR International Fellowship 12
Action Medical Research Project Grants 8
Action Medical Research Training Fellowship 8
AIHS Graduate Studentship 14
Charles Parsons Energy Research Award 497
ESRF Postdoctoral Fellowships 293
ETH Zurich Excellence Scholarship and Opportunity Award 629
Fellowships for Biomedical Engineering Research 166
Hastings Center International Visiting Scholars Program 330
Hudswell International Research Scholarships 376
IEEE Fellowship in Electrical History 348
International Postgraduate Research Scholarship 674
John Moyes Lessells Travel Scholarships 574
Lindbergh Grants 232
Queen Mary, University of London Research Studentships 540
Singapore-MIT Alliance Graduate Fellowship 500
Welch Scholarship 392

African Nations

ESRF Thesis Studentships 293
Hastings Center International Visiting Scholars Program 330
International Postgraduate Research Scholarships (IPRS) 417

Australia

Australian Postgraduate Award 674
Biomedical (C J Martin) Fellowships 485

United States of America

West European Countries

CIVIL ENGINEERING

Any Country

African Nations

Australia

Canada

Caribbean Countries

East European Countries

European Union

Middle East

New Zealand

South Africa

United Kingdom

United States of America

West European Countries

COMPUTER ENGINEERING

Any Country

La Trobe University Postgraduate Research Scholarship 418
Queen Mary, University of London Research Studentships 540
SRG Postdoctoral Fellowships 656
Tomsk Polytechnic University International Scholarship 640
University of Bristol Postgraduate Scholarships 680
University of Kent School of Engineering and Digital Arts Scholarships 741
University of Southampton Postgraduate Studentships 806
Victoria Doctoral Scholarships 836

African Nations

Aberystwyth International Excellence Scholarships 4
ESRF Thesis Studentships 293
International Postgraduate Research Scholarships (IPRS) 417

Australia

Aberystwyth International Excellence Scholarships 4
Australian Postgraduate Award 674
ESRF Thesis Studentships 293
Fulbright Postgraduate Scholarship in Science and Engineering 159

Canada

Aberystwyth International Excellence Scholarships 4
ESRF Thesis Studentships 293
International Postgraduate Research Scholarships (IPRS) 417

Caribbean Countries

Aberystwyth International Excellence Scholarships 4
ESRF Thesis Studentships 293
International Postgraduate Research Scholarships (IPRS) 417

East European Countries

CERN Technical Student Programme 231
ESRF Thesis Studentships 293
International Postgraduate Research Scholarships (IPRS) 417

European Union

EPSRC School of Computer Science and Electronic Engineering Research Studentship 726
Microsoft Research European PhD Scholarship Programme 456
University of Kent Department of Electronics Studentships 739
University of Southampton Engineering Doctorate 806

Middle East

Aberystwyth International Excellence Scholarships 4
ESRF Thesis Studentships 293
International Postgraduate Research Scholarships (IPRS) 417

New Zealand

Aberystwyth International Excellence Scholarships 4
Australian Postgraduate Award 674
ESRF Thesis Studentships 293

South Africa

Aberystwyth International Excellence Scholarships 4
ESRF Thesis Studentships 293
International Postgraduate Research Scholarships (IPRS) 417

United Kingdom

CERN Technical Student Programme 231
EPSRC School of Computer Science and Electronic Engineering Research Studentship 726
ESRF Thesis Studentships 293
Frank Knox Fellowships at Harvard University 303
International Postgraduate Research Scholarships (IPRS) 417
Kennedy Scholarships 412
Mr and Mrs David Edward Memorial Award 164
University of Kent Department of Electronics Studentships 739
University of Southampton Engineering Doctorate 806

United States of America

Aberystwyth International Excellence Scholarships 4
Air Force Summer Faculty Fellowship Program 92
Congress Bundestag Youth Exchange for Young Professionals 225
DEED (Demonstration of Energy-Efficient Developments) Student Research Grant and Internship 87
ESRF Thesis Studentships 293
GRC Graduate Fellowship Program 591
GRC Master's Scholarship Program 591
International Postgraduate Research Scholarships (IPRS) 417
Marshall Scholarships 181
ONR Summer Faculty Research Program 93
Winston Churchill Foundation Scholarship 859

West European Countries

CERN Technical Student Programme 231
ESRF Thesis Studentships 293
International Postgraduate Research Scholarships (IPRS) 417
Microsoft Research European PhD Scholarship Programme 456
Mr and Mrs David Edward Memorial Award 164
University of Kent Department of Electronics Studentships 739

CONTROL ENGINEERING (ROBOTICS)

Any Country

Advanced Simulation and Training Fellowships 431
ASPRS Ta Liang Memorial Award 99
BP Trust Research Fellowships 573
Doctoral Career Developement Scholarship (DCDS) Scheme 4
ESRF Postdoctoral Fellowships 293
ETH Zurich Excellence Scholarship and Opportunity Award 629
Hudswell International Research Scholarships 376
John Moyes Lessells Travel Scholarships 574

African Nations

Aberystwyth International Excellence Scholarships 4
International Postgraduate Research Scholarships (IPRS) 417

Australia

Aberystwyth International Excellence Scholarships 4
Fulbright Postgraduate Scholarship in Science and Engineering 159

Canada

Aberystwyth International Excellence Scholarships 4
ASPRS Robert N. Colwell Memorial Fellowship 99
Horticultural Research Institute Grants 344
International Postgraduate Research Scholarships (IPRS) 417

Caribbean Countries

Aberystwyth International Excellence Scholarships 4
International Postgraduate Research Scholarships (IPRS) 417

East European Countries

International Postgraduate Research Scholarships (IPRS) 417

Middle East

Aberystwyth International Excellence Scholarships 4
International Postgraduate Research Scholarships (IPRS) 417

New Zealand

Aberystwyth International Excellence Scholarships 4

South Africa

Aberystwyth International Excellence Scholarships 4
International Postgraduate Research Scholarships (IPRS) 417

United Kingdom

International Postgraduate Research Scholarships (IPRS) 417

United States of America

Aberystwyth International Excellence Scholarships 4
Air Force Summer Faculty Fellowship Program 92
ASPRS Robert N. Colwell Memorial Fellowship 99

Horticultural Research Institute Grants 344
International Postgraduate Research Scholarships (IPRS) 417
Winston Churchill Foundation Scholarship 859

West European Countries

International Postgraduate Research Scholarships (IPRS) 417

ELECTRICAL AND ELECTRONIC ENGINEERING

Any Country

A*STAR Graduate Scholarship (Overseas) 11
A*STAR International Fellowship 12
The Accenture Scholarship in ITEE 797
Bursaries 163
ESRF Postdoctoral Fellowships 293
Essex Rotary University Travel Grants 727
ETH Zurich Excellence Scholarship and Opportunity Award 629
Greater Milwaukee Foundation's Frank Rogers Bacon Research Assistantship 442
Hudswell International Research Scholarships 376
IEEE Fellowship in Electrical History 348
IET Travel Awards 376
International Postgraduate Research Scholarship 674
John Moyes Lessells Travel Scholarships 574
Johnson's Wax Research Fellowship 443
Leslie H Paddle Scholarship 376
National Federation of Business and Professional Women's Clubs Travel Grants 727
NRC Research Associateships 495
Queen Mary, University of London Research Studentships 540
Singapore-MIT Alliance Graduate Fellowship 500
Sir Eric Berthoud Travel Grant 728
University of Bristol Postgraduate Scholarships 680
University of Essex Doctoral Scholarship 729
University of Kent School of Engineering and Digital Arts Scholarships 741
University of Southampton Postgraduate Studentships 806
Victoria Doctoral Scholarships 836
World Universities Network (WUN) International Research Mobility Scheme 806

African Nations

International Postgraduate Research Scholarships (IPRS) 417

Australia

Fulbright Postgraduate Scholarship in Science and Engineering 159

Canada

International Postgraduate Research Scholarships (IPRS) 417

Caribbean Countries

International Postgraduate Research Scholarships (IPRS) 417

East European Countries

CERN Technical Student Programme 231
International Postgraduate Research Scholarships (IPRS) 417

European Union

EPSRC School of Computer Science and Electronic Engineering Research Studentship 726
EPSRC Studentships 676
ESRC Studentships 677
University of Essex Silberrad Scholarships 729
University of Kent Department of Electronics Studentships 739
University of Southampton Engineering Doctorate 806

Middle East

International Postgraduate Research Scholarships (IPRS) 417

South Africa

International Postgraduate Research Scholarships (IPRS) 417

United Kingdom

Access to Learning Fund 725
CERN Technical Student Programme 231
EPSRC School of Computer Science and Electronic Engineering Research Studentship 726
EPSRC Studentships 676
ESRC Studentships 677
Frank Knox Fellowships at Harvard University 303
International Postgraduate Research Scholarships (IPRS) 417
Kennedy Scholarships 412
Mr and Mrs David Edward Memorial Award 164
University of Essex Silberrad Scholarships 729
University of Kent Department of Electronics Studentships 739
University of Southampton Engineering Doctorate 806

United States of America

Air Force Summer Faculty Fellowship Program 92
Congress Bundestag Youth Exchange for Young Professionals 225
DEED (Demonstration of Energy-Efficient Developments) Student Research Grant and Internship 87
Department of Energy Computational Science Graduate Fellowship Program 416
Essex/Fulbright Commission Postgraduate Scholarships 727
GRC Graduate Fellowship Program 591
GRC Master's Scholarship Program 591
International Postgraduate Research Scholarships (IPRS) 417
Marshall Scholarships 181
NPSC Fellowship in Physical Sciences 494
ONR Summer Faculty Research Program 93
Winston Churchill Foundation Scholarship 859

West European Countries

CERN Technical Student Programme 231
International Postgraduate Research Scholarships (IPRS) 417
Mr and Mrs David Edward Memorial Award 164
University of Kent Department of Electronics Studentships 739

ENERGY ENGINEERING

Any Country

ASHRAE Grants-in-Aid for Graduate Students 101
Earthwatch Field Research Grants 279
Energy Fellowships 432
ETH Zurich Excellence Scholarship and Opportunity Award 629
Hudswell International Research Scholarships 376
International Postgraduate Research Scholarship 674
John Moyes Lessells Travel Scholarships 574
Lindbergh Grants 232
Tomsk Polytechnic University International Scholarship 640

Australia

Australian Postgraduate Award 674
Fulbright Postgraduate Scholarship in Science and Engineering 159

European Union

NERC Research (PhD) Studentships 501

New Zealand

Australian Postgraduate Award 674

United Kingdom

NERC Research (PhD) Studentships 501
Society for Underwater Technology (SUT) 259

United States of America

Air Force Summer Faculty Fellowship Program 92
DEED (Demonstration of Energy-Efficient Developments) Student Research Grant and Internship 87

Environmental Public Policy and Conflict Resolution PhD Fellowship 468
ONR Summer Faculty Research Program 93
Winston Churchill Foundation Scholarship 859

ENGINEERING DRAWING/DESIGN

Any Country

International Postgraduate Research Scholarship 674
John Moyes Lessells Travel Scholarships 574

African Nations

International Postgraduate Research Scholarships (IPRS) 417

Australia

Australian Postgraduate Award 674
Fulbright Postgraduate Scholarship in Science and Engineering 159

Canada

International Postgraduate Research Scholarships (IPRS) 417
TAC Foundation Scholarships 643

Caribbean Countries

International Postgraduate Research Scholarships (IPRS) 417

East European Countries

International Postgraduate Research Scholarships (IPRS) 417

Middle East

International Postgraduate Research Scholarships (IPRS) 417

New Zealand

Australian Postgraduate Award 674

South Africa

International Postgraduate Research Scholarships (IPRS) 417

United Kingdom

International Postgraduate Research Scholarships (IPRS) 417
Royal Commission Industrial Design Studentships 259

United States of America

DEED (Demonstration of Energy-Efficient Developments) Student Research Grant and Internship 87
International Postgraduate Research Scholarships (IPRS) 417
Winston Churchill Foundation Scholarship 859

West European Countries

International Postgraduate Research Scholarships (IPRS) 417

ENVIRONMENTAL AND SANITARY ENGINEERING

Any Country

EREF Scholarships in Solid Waste Management Research and Education 287
International Postgraduate Research Scholarship 674
John Moyes Lessells Travel Scholarships 574
Lindbergh Grants 232
QUEST Institution of Civil Engineers Travel Awards 375
Small Research Grants 566
Space-Time Model-Rainfall Fields Scholarship 759
Tomsk Polytechnic University International Scholarship 640
Utilities and Service Industries Training (USIT) 259
World Universities Network (WUN) International Research Mobility Scheme 806

African Nations

The Bernard Butler Trust Fund 396
Canadian Window on International Development 384
International Postgraduate Research Scholarships (IPRS) 417

Australia

The Bernard Butler Trust Fund 396
Fulbright Postgraduate Scholarship in Science and Engineering 159

Canada

The Bernard Butler Trust Fund 396
Canadian Window on International Development 384
CWRA Dillon Consulting Scholarship/Ken Thomson Scholarship 219
Horticultural Research Institute Grants 344
International Postgraduate Research Scholarships (IPRS) 417
TAC Foundation Scholarships 643

Caribbean Countries

Canadian Window on International Development 384
International Postgraduate Research Scholarships (IPRS) 417

East European Countries

CERN Technical Student Programme 231
International Postgraduate Research Scholarships (IPRS) 417

European Union

The Bernard Butler Trust Fund 396
Department of Agriculture and Rural Development (DARD) for Northern Ireland 258
Panasonic Trust Fellowships 259

Middle East

The Bernard Butler Trust Fund 396
International Postgraduate Research Scholarships (IPRS) 417

New Zealand

The Bernard Butler Trust Fund 396

South Africa

The Bernard Butler Trust Fund 396
Canadian Window on International Development 384
International Postgraduate Research Scholarships (IPRS) 417

United Kingdom

The Bernard Butler Trust Fund 396
CERN Technical Student Programme 231
Department of Agriculture and Rural Development (DARD) for Northern Ireland 258
Douglas Bomford Trust 258
Environmental Issues Award 258
International Postgraduate Research Scholarships (IPRS) 417
Panasonic Trust Fellowships 259
Utilities and Service Industries Training (USIT) 259

United States of America

The Bernard Butler Trust Fund 396
DEED (Demonstration of Energy-Efficient Developments) Student Research Grant and Internship 87
Department of Energy Computational Science Graduate Fellowship Program 416
Environmental Public Policy and Conflict Resolution PhD Fellowship 468
Horticultural Research Institute Grants 344
International Postgraduate Research Scholarships (IPRS) 417
Winston Churchill Foundation Scholarship 859

West European Countries

The Bernard Butler Trust Fund 396
CERN Technical Student Programme 231
International Postgraduate Research Scholarships (IPRS) 417

FORESTRY ENGINEERING

Any Country
Henrietta Hutton Research Grants 564
Hong Kong Research Grant 564
International Postgraduate Research Scholarship 674
John Moyes Lessells Travel Scholarships 574
Lindbergh Grants 232
Small Research Grants 566

African Nations
Canadian Window on International Development 384
IDRC Doctoral Research Awards 384

Australia
Australian Postgraduate Award 674
Fulbright Postgraduate Scholarship in Science and Engineering 159

Canada
Canadian Window on International Development 384
IDRC Doctoral Research Awards 384

Caribbean Countries
Canadian Window on International Development 384
IDRC Doctoral Research Awards 384

East European Countries
IDRC Doctoral Research Awards 384

Middle East
IDRC Doctoral Research Awards 384

New Zealand
Australian Postgraduate Award 674

South Africa
Canadian Window on International Development 384
IDRC Doctoral Research Awards 384

United States of America
Air Force Summer Faculty Fellowship Program 92
Environmental Public Policy and Conflict Resolution PhD Fellowship 468

HYDRAULIC ENGINEERING

Any Country
Horton (Hydrology) Research Grant 61
International Postgraduate Research Scholarship 674
John Moyes Lessells Travel Scholarships 574
Utilities and Service Industries Training (USIT) 259

African Nations
International Postgraduate Research Scholarships (IPRS) 417

Australia
Fulbright Postgraduate Scholarship in Science and Engineering 159

Canada
CWRA Dillon Consulting Scholarship/Ken Thomson Scholarship 219
Horticultural Research Institute Grants 344
International Postgraduate Research Scholarships (IPRS) 417

Caribbean Countries
International Postgraduate Research Scholarships (IPRS) 417

East European Countries
International Postgraduate Research Scholarships (IPRS) 417

European Union
Panasonic Trust Fellowships 259

Middle East
International Postgraduate Research Scholarships (IPRS) 417

South Africa
International Postgraduate Research Scholarships (IPRS) 417

United Kingdom
International Postgraduate Research Scholarships (IPRS) 417
The Lorch MSc Student Bursary 259
Panasonic Trust Fellowships 259
Utilities and Service Industries Training (USIT) 259

United States of America
DEED (Demonstration of Energy-Efficient Developments) Student Research Grant and Internship 87
Horticultural Research Institute Grants 344
International Postgraduate Research Scholarships (IPRS) 417
Winston Churchill Foundation Scholarship 859

West European Countries
International Postgraduate Research Scholarships (IPRS) 417

INDUSTRIAL ENGINEERING

Any Country
ETH Zurich Excellence Scholarship and Opportunity Award 629
International Postgraduate Research Scholarship 674
John Moyes Lessells Travel Scholarships 574
Johnson's Wax Research Fellowship 443
Royal Commission Industrial Fellowships 563
University of Bristol Postgraduate Scholarships 680

African Nations
International Postgraduate Research Scholarships (IPRS) 417

Australia
Australian Postgraduate Award 674
Fulbright Postgraduate Scholarship in Science and Engineering 159

Canada
International Postgraduate Research Scholarships (IPRS) 417

Caribbean Countries
International Postgraduate Research Scholarships (IPRS) 417

East European Countries
International Postgraduate Research Scholarships (IPRS) 417

Middle East
International Postgraduate Research Scholarships (IPRS) 417

New Zealand
Australian Postgraduate Award 674

South Africa
International Postgraduate Research Scholarships (IPRS) 417

United Kingdom
Frank Knox Fellowships at Harvard University 303
International Postgraduate Research Scholarships (IPRS) 417
Royal Commission Industrial Design Studentships 259
Royal Commission Industrial Fellowships 563

United States of America
Air Force Summer Faculty Fellowship Program 92
Congress Bundestag Youth Exchange for Young Professionals 225

West European Countries

MARINE ENGINEERING AND NAVAL ARCHITECTURE

Any Country

Australia

European Union

United Kingdom

United States of America

MATERIALS ENGINEERING

Any Country

African Nations

Australia

Canada

Caribbean Countries

East European Countries

European Union

Middle East

New Zealand

South Africa

United Kingdom

United States of America

West European Countries

MEASUREMENT/PRECISION ENGINEERING

Any Country

African Nations

Australia

Canada
International Postgraduate Research Scholarships (IPRS) 417

Caribbean Countries
International Postgraduate Research Scholarships (IPRS) 417

East European Countries
International Postgraduate Research Scholarships (IPRS) 417

European Union
EPSRC Studentships 676

Middle East
International Postgraduate Research Scholarships (IPRS) 417

New Zealand
Australian Postgraduate Award 674

South Africa
International Postgraduate Research Scholarships (IPRS) 417

United Kingdom
EPSRC Studentships 676
Frank Knox Fellowships at Harvard University 303
International Postgraduate Research Scholarships (IPRS) 417
Kennedy Scholarships 412

United States of America
Air Force Summer Faculty Fellowship Program 92
DEED (Demonstration of Energy-Efficient Developments) Student Research Grant and Internship 87
International Postgraduate Research Scholarships (IPRS) 417
Winston Churchill Foundation Scholarship 859

West European Countries
International Postgraduate Research Scholarships (IPRS) 417

MINING ENGINEERING

Any Country
Bosworth Smith Trust Fund 373
G Vernon Hobson Bequest 373
International Postgraduate Research Scholarship 674
John Moyes Lessells Travel Scholarships 574
The Tom Seaman Travelling Scholarship 373

African Nations
International Postgraduate Research Scholarships (IPRS) 417

Australia
Australian Postgraduate Award 674
Edgar Pam Fellowship 373
Fulbright Postgraduate Scholarship in Science and Engineering 159

Canada
Edgar Pam Fellowship 373
International Postgraduate Research Scholarships (IPRS) 417

Caribbean Countries
International Postgraduate Research Scholarships (IPRS) 417

East European Countries
International Postgraduate Research Scholarships (IPRS) 417
Synthesys Visiting Fellowship 501

European Union
NERC Research (PhD) Studentships 501
Synthesys Visiting Fellowship 501

Middle East
International Postgraduate Research Scholarships (IPRS) 417

New Zealand
Australian Postgraduate Award 674
Edgar Pam Fellowship 373

South Africa
Edgar Pam Fellowship 373
International Postgraduate Research Scholarships (IPRS) 417

United Kingdom
Edgar Pam Fellowship 373
International Postgraduate Research Scholarships (IPRS) 417
Mining Club Award 373
NERC Research (PhD) Studentships 501
Synthesys Visiting Fellowship 501

United States of America
DEED (Demonstration of Energy-Efficient Developments) Student Research Grant and Internship 87
International Postgraduate Research Scholarships (IPRS) 417
Winston Churchill Foundation Scholarship 859

West European Countries
International Postgraduate Research Scholarships (IPRS) 417
Synthesys Visiting Fellowship 501

NANOTECHNOLOGY

NUCLEAR ENGINEERING

Any Country
Alan F Henry/Paul A Greebler Scholarship 76
ETH Zurich Excellence Scholarship and Opportunity Award 629
John and Muriel Landis Scholarship Awards 77
John Moyes Lessells Travel Scholarships 574

Australia
Fulbright Postgraduate Scholarship in Science and Engineering 159

East European Countries
CERN Technical Student Programme 231

United Kingdom
CERN Technical Student Programme 231

United States of America
Air Force Summer Faculty Fellowship Program 92
DEED (Demonstration of Energy-Efficient Developments) Student Research Grant and Internship 87
Department of Energy Computational Science Graduate Fellowship Program 416
Everitt P Blizard Scholarship 76
James F Schumar Scholarship 77
ONR Summer Faculty Research Program 93
Robert A Dannels Memorial Scholarship 77
Verne R Dapp Memorial Scholarship 77
Walter Meyer Scholarship 78
Winston Churchill Foundation Scholarship 859

West European Countries
CERN Technical Student Programme 231

PETROLEUM AND GAS ENGINEERING

Any Country
ACS PRF Scientific Education (Type SE) Grants 41

ExxonMobil Excellence in Teaching Awards 553
John Moyes Lessells Travel Scholarships 574
Postgraduate Top Up Scholarships in Offshore Engineering and Naval Architecture 819

Australia

Fulbright Postgraduate Scholarship in Science and Engineering 159
Society of Petroleum Engineers Western Australia Scholarships 819

European Union

NERC Research (PhD) Studentships 501

United Kingdom

NERC Research (PhD) Studentships 501

United States of America

Air Force Summer Faculty Fellowship Program 92
DEED (Demonstration of Energy-Efficient Developments) Student Research Grant and Internship 87
Winston Churchill Foundation Scholarship 859

PHYSICAL ENGINEERING

PRODUCTION ENGINEERING

Any Country

ETH Zurich Excellence Scholarship and Opportunity Award 629
John Moyes Lessells Travel Scholarships 574
Singapore-MIT Alliance Graduate Fellowship 500

African Nations

International Postgraduate Research Scholarships (IPRS) 417

Australia

Australian Postgraduate Award 674
Fulbright Postgraduate Scholarship in Science and Engineering 159

Canada

International Postgraduate Research Scholarships (IPRS) 417

Caribbean Countries

International Postgraduate Research Scholarships (IPRS) 417

East European Countries

International Postgraduate Research Scholarships (IPRS) 417

Middle East

International Postgraduate Research Scholarships (IPRS) 417

New Zealand

Australian Postgraduate Award 674
Dick and Mary Earle Scholarship in Technology 632

South Africa

International Postgraduate Research Scholarships (IPRS) 417

United Kingdom

International Postgraduate Research Scholarships (IPRS) 417

United States of America

DEED (Demonstration of Energy-Efficient Developments) Student Research Grant and Internship 87
International Postgraduate Research Scholarships (IPRS) 417
Winston Churchill Foundation Scholarship 859

West European Countries

International Postgraduate Research Scholarships (IPRS) 417

SAFETY ENGINEERING

Any Country

International Postgraduate Research Scholarship 674
John Moyes Lessells Travel Scholarships 574
University of Southampton Postgraduate Studentships 806

African Nations

International Postgraduate Research Scholarships (IPRS) 417

Australia

Australian Postgraduate Award 674
Fulbright Postgraduate Scholarship in Science and Engineering 159

Canada

International Postgraduate Research Scholarships (IPRS) 417
Public Safety Canada Research Fellowship Program in Honour of Stuart Nesbitt White 134
TAC Foundation Scholarships 643

Caribbean Countries

International Postgraduate Research Scholarships (IPRS) 417

East European Countries

CERN Technical Student Programme 231
International Postgraduate Research Scholarships (IPRS) 417

European Union

University of Southampton Engineering Doctorate 806

Middle East

International Postgraduate Research Scholarships (IPRS) 417

New Zealand

Australian Postgraduate Award 674

South Africa

International Postgraduate Research Scholarships (IPRS) 417

United Kingdom

CERN Technical Student Programme 231
International Postgraduate Research Scholarships (IPRS) 417
University of Southampton Engineering Doctorate 806

United States of America

DEED (Demonstration of Energy-Efficient Developments) Student Research Grant and Internship 87
International Postgraduate Research Scholarships (IPRS) 417
Winston Churchill Foundation Scholarship 859

West European Countries

CERN Technical Student Programme 231
International Postgraduate Research Scholarships (IPRS) 417

SOUND ENGINEERING

Any Country

ASA Frederick V. Hunt Postdoctoral Research Fellowship 7
John Moyes Lessells Travel Scholarships 574
University of Southampton Postgraduate Studentships 806
World Universities Network (WUN) International Research Mobility Scheme 806

Australia

Fulbright Postgraduate Scholarship in Science and Engineering 159

European Union

University of Southampton Engineering Doctorate 806

SURVEYING AND MAPPING SCIENCE

FINE AND APPLIED ARTS

GENERAL

African Nations

Australia

Canada

Caribbean Countries

East European Countries

European Union

Middle East

New Zealand

South Africa

United Kingdom

United States of America

West European Countries

ART CRITICISM

ART HISTORY

Any Country

African Nations

Australia

Canada

Caribbean Countries

East European Countries

European Union

AESTHETICS

ART MANAGEMENT

African Nations

Australia

Canada

Caribbean Countries

East European Countries

European Union

Middle East

New Zealand

South Africa

United Kingdom

United States of America

West European Countries

CINEMA AND TELEVISION

Any Country

African Nations

Australia

Canada

Caribbean Countries

European Union

Middle East

New Zealand

South Africa

United Kingdom

United States of America

West European Countries

DANCE

Any Country

Australia

Canada

East European Countries

European Union

Middle East

New Zealand

United Kingdom

United States of America

West European Countries

DESIGN

Any Country

African Nations

Australia

Canada

Caribbean Countries

East European Countries

European Union

Middle East

New Zealand
Australian Postgraduate Award 674

South Africa
International Postgraduate Research Scholarships (IPRS) 417

United Kingdom
Collaborative Doctoral Awards 121
The Costume Society Student Bursary 250
The Costume Society Yarwood Award 251
International Postgraduate Research Scholarships (IPRS) 417
Professional Preparation Master's Scheme 121
Research Preparation Master's Scheme 121

United States of America
Congress Bundestag Youth Exchange for Young Professionals 225
Fulbright Specialist Program 253
The Fulbright-Coventry Award in Design 831
The George and Viola Hoffman Fund 130
International Postgraduate Research Scholarships (IPRS) 417
Virginia Liebeler Biennial Grants for Mature Women (Art) 491
Virginia Liebeler Biennial Grants for Mature Women (Music) 492
Virginia Liebeler Biennial Grants for Mature Women (Writing) 492

West European Countries
International Postgraduate Research Scholarships (IPRS) 417
Janson Johan Helmich Scholarships and Travel Grants 401

DISPLAY AND STAGE DESIGN

Any Country
Banff Centre Scholarship Fund 162
Doctoral Career Developement Scholarship (DCDS) Scheme 4
Ohio Arts Council Individual Excellence Awards 514
Paul D. Fleck Fellowships in the Arts 162

African Nations
International Postgraduate Research Scholarships (IPRS) 417

Canada
International Postgraduate Research Scholarships (IPRS) 417

Caribbean Countries
International Postgraduate Research Scholarships (IPRS) 417

East European Countries
International Postgraduate Research Scholarships (IPRS) 417

European Union
Collaborative Doctoral Awards 121
Professional Preparation Master's Scheme 121
Research Preparation Master's Scheme 121

Middle East
AICF Sharett Scholarship Program 26
International Postgraduate Research Scholarships (IPRS) 417

South Africa
International Postgraduate Research Scholarships (IPRS) 417

United Kingdom
Collaborative Doctoral Awards 121
The Costume Society Patterns of Fashion Award 250
The Costume Society Student Bursary 250
International Postgraduate Research Scholarships (IPRS) 417
Professional Preparation Master's Scheme 121
Research Preparation Master's Scheme 121

United States of America
International Postgraduate Research Scholarships (IPRS) 417

West European Countries
International Postgraduate Research Scholarships (IPRS) 417

FASHION DESIGN

Any Country
CSA Adele Filene Travel Award 251
CSA Stella Blum Student Research Grant 251
CSA Travel Research Grant 251
Ohio Arts Council Individual Excellence Awards 514
Wolfsonian-FIU Fellowship 861

African Nations
International Postgraduate Research Scholarships (IPRS) 417

Canada
International Postgraduate Research Scholarships (IPRS) 417

Caribbean Countries
International Postgraduate Research Scholarships (IPRS) 417

East European Countries
International Postgraduate Research Scholarships (IPRS) 417

European Union
Collaborative Doctoral Awards 121
Professional Preparation Master's Scheme 121
Research Preparation Master's Scheme 121

Middle East
AICF Sharett Scholarship Program 26
International Postgraduate Research Scholarships (IPRS) 417

South Africa
International Postgraduate Research Scholarships (IPRS) 417

United Kingdom
Collaborative Doctoral Awards 121
The Costume Society Museum Placement Award 250
The Costume Society Patterns of Fashion Award 250
The Costume Society Student Bursary 250
The Costume Society Yarwood Award 251
International Postgraduate Research Scholarships (IPRS) 417
MA Fashion Design and Enterprise Scholarship 820
Professional Preparation Master's Scheme 121
Research Preparation Master's Scheme 121

United States of America
International Postgraduate Research Scholarships (IPRS) 417

West European Countries
International Postgraduate Research Scholarships (IPRS) 417

FURNITURE DESIGN

Any Country
Haystack Scholarship 331
Ohio Arts Council Individual Excellence Awards 514
Wolfsonian-FIU Fellowship 861

African Nations
International Postgraduate Research Scholarships (IPRS) 417

Canada
International Postgraduate Research Scholarships (IPRS) 417

Caribbean Countries
International Postgraduate Research Scholarships (IPRS) 417

Canada

International Postgraduate Research Scholarships (IPRS) 417

Caribbean Countries

International Postgraduate Research Scholarships (IPRS) 417

East European Countries

International Postgraduate Research Scholarships (IPRS) 417

European Union

Collaborative Doctoral Awards 121
Professional Preparation Master's Scheme 121
Research Preparation Master's Scheme 121

Middle East

International Postgraduate Research Scholarships (IPRS) 417

South Africa

International Postgraduate Research Scholarships (IPRS) 417

United Kingdom

Collaborative Doctoral Awards 121
International Postgraduate Research Scholarships (IPRS) 417
Professional Preparation Master's Scheme 121
Research Preparation Master's Scheme 121

United States of America

American Academy in Rome Fellowships in Design Art 26
International Postgraduate Research Scholarships (IPRS) 417
Virginia Liebeler Biennial Grants for Mature Women (Art) 491

West European Countries

International Postgraduate Research Scholarships (IPRS) 417

TEXTILE DESIGN

Any Country

CSA Adele Filene Travel Award 251
CSA Stella Blum Student Research Grant 251
CSA Travel Research Grant 251
Hambidge Residency Program Scholarships 329
Haystack Scholarship 331
HWSDA Scholarship Program 329
Ohio Arts Council Individual Excellence Awards 514
School of Textiles and Design Awards 338
Wolfsonian-FIU Fellowship 861

African Nations

International Postgraduate Research Scholarships (IPRS) 417

Canada

International Postgraduate Research Scholarships (IPRS) 417

Caribbean Countries

International Postgraduate Research Scholarships (IPRS) 417

East European Countries

International Postgraduate Research Scholarships (IPRS) 417

European Union

Collaborative Doctoral Awards 121
Professional Preparation Master's Scheme 121
Research Preparation Master's Scheme 121

Middle East

AICF Sharett Scholarship Program 26
International Postgraduate Research Scholarships (IPRS) 417

South Africa

International Postgraduate Research Scholarships (IPRS) 417

United Kingdom

Collaborative Doctoral Awards 121
The Costume Society Patterns of Fashion Award 250
The Costume Society Student Bursary 250
The Costume Society Yarwood Award 251
International Postgraduate Research Scholarships (IPRS) 417
Professional Preparation Master's Scheme 121
Research Preparation Master's Scheme 121

United States of America

International Postgraduate Research Scholarships (IPRS) 417

West European Countries

International Postgraduate Research Scholarships (IPRS) 417

DRAWING AND PAINTING

Any Country

Abbey Harris Mural Fund 3
Akademie Schloss Solitude Fellowships 14
ASCSA Advanced Fellowships 89
Austro-American Association of Boston Stipend 161
Banff Centre Scholarship Fund 162
The Barns-Graham Travel Award 570
Camargo Fellowships 200
The Dawn Jolliffe Botanical Art Bursary 567
Doctoral Career Developement Scholarship (DCDS) Scheme 4
Don Freeman Memorial Grant-in-Aid 617
Fine Arts Work Center in Provincetown Fellowships 297
Franklin Research Grant Program 80
Frederick Douglass Institute Postdoctoral Fellowship 304
Frederick Douglass Institute Predoctoral Dissertation Fellowship 304
Gottlieb Foundation Emergency Assistance Grants 9
Gottlieb Foundation Individual Support Grants 9
Hambidge Residency Program Scholarships 329
Haystack Scholarship 331
International Postgraduate Research Scholarship 674
Jacob Hirsch Fellowship 91
Lewis Walpole Library Fellowship 429
MacDowell Colony Residencies 439
Mackendrick Scholarship 321
MICA Fellowship 444
MICA International Fellowship Award 444
Ohio Arts Council Individual Excellence Awards 514
Paul D. Fleck Fellowships in the Arts 162
Pollock-Krasner Foundation Grant 536
Research Workshops in Arts and Humanities 574
Rhodes University Postdoctoral Fellowship and The Andrew Mellon Postdoctoral Fellowship 549
The RSA John Kinross Scholarships to Florence 570
The Rsa William Littlejohn Award for Excellence and Innovation in Water-Based Media 571
Small Research Grants in Arts and Humanities 575
University of Southampton Postgraduate Studentships 806
World Universities Network (WUN) International Research Mobility Scheme 806

African Nations

Aberystwyth International Excellence Scholarships 4
International Postgraduate Research Scholarships (IPRS) 417

Australia

AAH Humanities Travelling Fellowships 143
Aberystwyth International Excellence Scholarships 4
Australian Postgraduate Award 674
Marten Bequest Travelling Scholarships 645
Portia Geach Memorial Award 645

HANDICRAFTS

MUSIC

African Nations

Australia

Canada

Caribbean Countries

European Union

Middle East

New Zealand

South Africa

United Kingdom

United States of America

West European Countries

CONDUCTING

Any Country

European Union

Middle East

United Kingdom

United States of America

West European Countries

JAZZ AND POPULAR MUSIC

Any Country

European Union

Middle East

South Africa

United Kingdom

United States of America

MUSIC THEORY AND COMPOSITION

Any Country

Australia

Canada

European Union

Middle East

New Zealand

South Africa

United Kingdom

United States of America

RELIGIOUS ART

SCULPTURE

Any Country

African Nations

Australia

Canada

Caribbean Countries

East European Countries

European Union

Middle East

New Zealand

South Africa

United Kingdom

United States of America

West European Countries

THEATRE

Any Country

HOME ECONOMICS

GENERAL

CHILD CARE/CHILD DEVELOPMENT

CLOTHING AND SEWING

CONSUMER STUDIES

HOUSE ARTS AND ENVIRONMENT

HOUSEHOLD MANAGEMENT

NUTRITION

African Nations

Australia

Canada

Caribbean Countries

East European Countries

Middle East

New Zealand

South Africa

United Kingdom

United States of America

West European Countries

LAW

GENERAL

Any Country

African Nations

Australia

AIR AND SPACE LAW

CANON LAW

Any Country

European Union

United Kingdom

United States of America

CIVIL LAW

Any Country

African Nations

Australia

Canada

Caribbean Countries

East European Countries

European Union

Middle East

South Africa

United Kingdom

United States of America

West European Countries

COMMERCIAL LAW

Any Country

African Nations

Australia

Canada

Caribbean Countries

East European Countries

European Union

Middle East

New Zealand

South Africa

United Kingdom

COMPARATIVE LAW

CRIMINAL LAW

West European Countries

PhD Studentships 164

EUROPEAN COMMUNITY LAW

Any Country

Doctoral Career Developement Scholarship (DCDS) Scheme 4
Fernand Braudel Senior Fellowships 294
Franklin Research Grant Program 80
Jean Monnet Fellowships 294
Law & Business in Europe Fellowships 668
Law and Criminology Masters Scholarships 5
Law and Criminology Research Scholarships 5
Max Weber Fellowships 294
Paul H. Nitze School of Advanced International Studies (SAIS) Financial Aid and Fellowships 173
United States Holocaust Memorial Museum Center for Advanced Holocaust Studies Visiting Scholar Programs 649
University of Southampton Postgraduate Studentships 806
World Universities Network (WUN) International Research Mobility Scheme 806

African Nations

Aberystwyth International Excellence Scholarships 4

Australia

Aberystwyth International Excellence Scholarships 4

Canada

Aberystwyth International Excellence Scholarships 4
Harry Bikakis Fellowship 90

Caribbean Countries

Aberystwyth International Excellence Scholarships 4

East European Countries

EUI Postgraduate Scholarships 293

European Union

Collaborative Doctoral Awards 121
EUI Postgraduate Scholarships 293
PhD Studentships 164
Research Preparation Master's Scheme 121
University of Essex Silberrad Scholarships 729

Middle East

Aberystwyth International Excellence Scholarships 4
EUI Postgraduate Scholarships 293

New Zealand

Aberystwyth International Excellence Scholarships 4

South Africa

Aberystwyth International Excellence Scholarships 4

United Kingdom

Collaborative Doctoral Awards 121
EUI Postgraduate Scholarships 293
PhD Studentships 164
Research Preparation Master's Scheme 121
University of Essex Silberrad Scholarships 729

United States of America

Aberystwyth International Excellence Scholarships 4
Harry Bikakis Fellowship 90
Marshall Scholarships 181

West European Countries

EUI Postgraduate Scholarships 293
PhD Studentships 164

HISTORY OF LAW

Any Country

Ahmanson and Getty Postdoctoral Fellowships 646
ASCSA Advanced Fellowships 89
ASCSA Fellowships 89
ASECS (American Society for 18th-Century Studies)/Clark Library Fellowships 646
British Academy Postdoctoral Fellowships 176
Clark Library Short-Term Resident Fellowships 646
Clark Predoctoral Fellowships 646
Clark-Huntington Joint Bibliographical Fellowship 647
CRF European Visiting Research Fellowships 573
David Hume Fellowship 365
Doctoral Career Developement Scholarship (DCDS) Scheme 4
Findel Scholarships and Schneider Scholarships 339
Franklin Research Grant Program 80
Herzog August Library Fellowship 339
Hugh Last and Donald Atkinson Funds Committee Grants 610
International Postgraduate Research Scholarship 674
J Franklin Jameson Fellowship 63
Jacob Hirsch Fellowship 91
Kanner Fellowship In British Studies 647
Law and Criminology Research Scholarships 5
Lewis Walpole Library Fellowship 429
Littleton-Griswold Research Grant 64
M Alison Frantz Fellowship 91
Oscar Broneer Traveling Fellowship 91
Society for the Study of French History Bursaries 613
United States Holocaust Memorial Museum Center for Advanced Holocaust Studies Visiting Scholar Programs 649

Australia

Australian Postgraduate Award 674

Canada

Harry Bikakis Fellowship 90
Oscar Broneer Traveling Fellowship 91
Viscount Bennett Fellowship 203

European Union

Collaborative Doctoral Awards 121
Professional Preparation Master's Scheme 121
Research Preparation Master's Scheme 121

New Zealand

Australian Postgraduate Award 674

United Kingdom

Collaborative Doctoral Awards 121
Hector and Elizabeth Catling Bursary 189
Professional Preparation Master's Scheme 121
Research Preparation Master's Scheme 121

United States of America

ACLS/New York Public Library Fellowship 53
Fulbright Specialist Program 253
Fulbright-Kennan Institute Research Scholarships 411
Harry Bikakis Fellowship 90
Marshall Scholarships 181
NEH Fellowships 91
Oscar Broneer Traveling Fellowship 91

HUMAN RIGHTS

Any Country

ALA Eli M. Oboler Memorial Award 69
The Artellus Scholarships 726
Doctoral Career Developement Scholarship (DCDS) Scheme 4
Fernand Braudel Senior Fellowships 294
Franklin Research Grant Program 80
Giulia Mereu Scholarships 727

African Nations

Australia

Canada

Caribbean Countries

East European Countries

European Union

Middle East

New Zealand

South Africa

United Kingdom

United States of America

West European Countries

INTERNATIONAL LAW

Any Country

African Nations

MASS COMMUNICATION AND INFORMATION SCIENCE

GENERAL

COMMUNICATION ARTS

DOCUMENTATION TECHNIQUES AND ARCHIVING

Any Country

Albert J Beveridge Grant 63
AMIA Kodak Fellowship in Film Preservation 132
ASCSA Advanced Fellowships 89
ASCSA Fellowships 89
Bernadotte E Schmitt Grants 63
The Cloisters Summer Internship for College Students 455
Doctoral Career Developement Scholarship (DCDS) Scheme 4
Eugene Garfield Doctoral Dissertation Fellowship 168
Eugene Garfield Doctoral Dissertation Scholarship 168
International Postgraduate Research Scholarship 674
J Franklin Jameson Fellowship 63
Jacob's Pillow Intern Program 399
Lester J Cappon Fellowship in Documentary Editing 507
Littleton-Griswold Research Grant 64
Research Workshops in Arts and Humanities 574
Small Research Grants in Arts and Humanities 575
Summer Internships for College and Graduate Students 455
Victoria Doctoral Scholarships 836

African Nations

Aberystwyth International Excellence Scholarships 4
International Postgraduate Research Scholarships (IPRS) 417

Australia

Aberystwyth International Excellence Scholarships 4
Practitioner Fellowship 481

Canada

Aberystwyth International Excellence Scholarships 4
International Postgraduate Research Scholarships (IPRS) 417

Caribbean Countries

Aberystwyth International Excellence Scholarships 4
International Postgraduate Research Scholarships (IPRS) 417

East European Countries

International Postgraduate Research Scholarships (IPRS) 417

European Union

CBRL Travel Grant 252
Collaborative Doctoral Awards 121
Professional Preparation Master's Scheme 121
Research Preparation Master's Scheme 121

Middle East

Aberystwyth International Excellence Scholarships 4
International Postgraduate Research Scholarships (IPRS) 417

New Zealand

Aberystwyth International Excellence Scholarships 4

South Africa

Aberystwyth International Excellence Scholarships 4
International Postgraduate Research Scholarships (IPRS) 417

United Kingdom

CBRL Pilot Study Award 252
CBRL Travel Grant 252
Collaborative Doctoral Awards 121
International Postgraduate Research Scholarships (IPRS) 417
Professional Preparation Master's Scheme 121
Research Preparation Master's Scheme 121

United States of America

Aberystwyth International Excellence Scholarships 4
Fulbright Specialist Program 253
International Postgraduate Research Scholarships (IPRS) 417
NEH Fellowships 91

West European Countries

International Postgraduate Research Scholarships (IPRS) 417

JOURNALISM

Any Country

AAS Joyce Tracy Fellowship 28
Arthur F. Burns Fellowship Program 381
CDI Internship 228
David Faddy Scholarships 820
Edward R Murrow Fellowship for Foreign Correspondents 255
Ethel Payne Fellowships 478
Fulbright Institutional Linkages Program 306
Glenn E and Barbara Hodsdon Ullyot Scholarship 236
Herbert Hoover Presidential Library Association Travel Grants 336
International Postgraduate Research Scholarship 674
Jacob's Pillow Intern Program 399
Knight International Journalism Fellowships 381
The McGee Journalism Fellowship in Southern Africa 381
Paul H. Nitze School of Advanced International Studies (SAIS) Financial Aid and Fellowships 173
Postdoctoral Fellowship 236
Pulitzer Center on Crisis Reporting Travel Grants 539
Pulliam Journalism Fellowship 356
Research Workshops in Arts and Humanities 574
Rhodes University Postdoctoral Fellowship and The Andrew Mellon Postdoctoral Fellowship 549
RTDNF Fellowships 542
Small Research Grants in Arts and Humanities 575
Société de Chimie Industrielle (American Section) Fellowship 237
Title VIII-Supported Short-Term Grant 411
United States Holocaust Memorial Museum Center for Advanced Holocaust Studies Visiting Scholar Programs 649
University of Kent Centre for Journalism Scholarships 739
William B. Ruggles Right to Work Journalism Scholarship 489

African Nations

IDRC Doctoral Research Awards 384
International Postgraduate Research Scholarships (IPRS) 417

Australia

Australian Postgraduate Award 674
Edward Wilson Scholarship for Graduate Diploma of Journalism 267

Canada

IDRC Doctoral Research Awards 384
International Postgraduate Research Scholarships (IPRS) 417

Caribbean Countries

IDRC Doctoral Research Awards 384
International Postgraduate Research Scholarships (IPRS) 417

East European Countries

IDRC Doctoral Research Awards 384
International Postgraduate Research Scholarships (IPRS) 417

European Union

Collaborative Doctoral Awards 121
GMF Journalism Program 315
Professional Preparation Master's Scheme 121
Research Preparation Master's Scheme 121

Middle East

IDRC Doctoral Research Awards 384
International Postgraduate Research Scholarships (IPRS) 417

New Zealand

Australian Postgraduate Award 674

South Africa

IDRC Doctoral Research Awards 384

United Kingdom

United States of America

West European Countries

LIBRARY SCIENCE

Any Country

African Nations

Australia

Canada

Caribbean Countries

East European Countries

European Union

Middle East
Aberystwyth International Excellence Scholarships 4
Cunningham Memorial International Fellowship 446
International Postgraduate Research Scholarships (IPRS) 417

New Zealand
Aberystwyth International Excellence Scholarships 4
Cunningham Memorial International Fellowship 446

South Africa
Aberystwyth International Excellence Scholarships 4
Canadian Window on International Development 384
Cunningham Memorial International Fellowship 446
International Postgraduate Research Scholarships (IPRS) 417

United Kingdom
Collaborative Doctoral Awards 121
Cunningham Memorial International Fellowship 446
ESU Travelling Librarian Award 286
International Postgraduate Research Scholarships (IPRS) 417
Professional Preparation Master's Scheme 121
Research Preparation Master's Scheme 121

United States of America
Aberystwyth International Excellence Scholarships 4
ALA Bound to Stay Bound Book Scholarships 68
ALA David H Clift Scholarship 69
ALA Mary V Gaver Scholarship 71
ALA Miriam L Hornback Scholarship 71
ALA NMRT/EBSCO Scholarship 71
ALA Spectrum Initiative Scholarship Program 72
Beard Scholarship 312
The Bound to Stay Bound Books Scholarship 128
CLA Library Research and Development Grants 215
The Frederic G. Melcher Scholarship 128
Fulbright Specialist Program 253
Hubbard Scholarship 312
International Postgraduate Research Scholarships (IPRS) 417
Kennedy Research Grants 408
LITA/LSSI Minority Scholarship in Library and Information Technology 429
LITA/OCLC Minority Scholarship in Library and Information Technology 429
MLA Continuing Education Grants 446
MLA Research, Development and Demonstration Project Award 446
MLA Scholarship 447
MLA Scholarship for Minority Students 447
Thomson Reuters/MLA Doctoral Fellowship 447

West European Countries
Cunningham Memorial International Fellowship 446
International Postgraduate Research Scholarships (IPRS) 417

MASS COMMUNICATION

Any Country
Doctoral Career Developement Scholarship (DCDS) Scheme 4
Jennings Randolph Program for International Peace Dissertation Fellowship 649
Knight International Journalism Fellowships 381
The McGee Journalism Fellowship in Southern Africa 381
United States Holocaust Memorial Museum Center for Advanced Holocaust Studies Visiting Scholar Programs 649

African Nations
Aberystwyth International Excellence Scholarships 4
Canadian Window on International Development 384
International Postgraduate Research Scholarships (IPRS) 417

Australia
AAH Humanities Travelling Fellowships 143
Aberystwyth International Excellence Scholarships 4

Canada
Aberystwyth International Excellence Scholarships 4
BBM Scholarship 202
Canadian Window on International Development 384
International Postgraduate Research Scholarships (IPRS) 417
Jim Bourque Scholarship 114

Caribbean Countries
Aberystwyth International Excellence Scholarships 4
Canadian Window on International Development 384
International Postgraduate Research Scholarships (IPRS) 417

East European Countries
International Postgraduate Research Scholarships (IPRS) 417

Middle East
Aberystwyth International Excellence Scholarships 4
International Postgraduate Research Scholarships (IPRS) 417

New Zealand
Aberystwyth International Excellence Scholarships 4

South Africa
Aberystwyth International Excellence Scholarships 4
Canadian Window on International Development 384
International Postgraduate Research Scholarships (IPRS) 417

United Kingdom
International Postgraduate Research Scholarships (IPRS) 417
UK Fulbright Alistair Cooke Award in Journalism 834

United States of America
Aberystwyth International Excellence Scholarships 4
DEED (Demonstration of Energy-Efficient Developments) Student Research Grant and Internship 87
Environmental Public Policy and Conflict Resolution PhD Fellowship 468
Fulbright Distinguished Chairs Program 252
Fulbright Specialist Program 253
International Postgraduate Research Scholarships (IPRS) 417
IREX Short-Term Travel Grants 391
Robert Bosch Foundation Fellowships 551
Virginia Liebeler Biennial Grants for Mature Women (Writing) 492

West European Countries
International Postgraduate Research Scholarships (IPRS) 417

MEDIA STUDIES

Any Country
AAS Joyce Tracy Fellowship 28
ATM Postgraduate Scholarships 163
Doctoral Career Developement Scholarship (DCDS) Scheme 4
International Postgraduate Research Scholarship 674
Jennings Randolph Program for International Peace Dissertation Fellowship 649
Marshall Memorial Fellowship 315
The McGee Journalism Fellowship in Southern Africa 381
Media, Arts and Design Scholarship 821
Rhodes University Postdoctoral Fellowship and The Andrew Mellon Postdoctoral Fellowship 549
United States Holocaust Memorial Museum Center for Advanced Holocaust Studies Visiting Scholar Programs 649
Victoria Doctoral Scholarships 836

African Nations
Aberystwyth International Excellence Scholarships 4
International Postgraduate Research Scholarships (IPRS) 417

Australia
AAH Humanities Travelling Fellowships 143

MUSEUM MANAGEMENT

MUSEUM STUDIES

PUBLIC RELATIONS AND PUBLICITY

RADIO/TELEVISION BROADCASTING

Middle East

New Zealand

South Africa

United Kingdom

United States of America

West European Countries

RESTORATION OF WORKS OF ART

Any Country

African Nations

Australia

Canada

Caribbean Countries

East European Countries

European Union

Middle East

South Africa

United Kingdom

United States of America

West European Countries

MATHEMATICS AND COMPUTER SCIENCE

GENERAL

Any Country

New Zealand

South Africa

United Kingdom

United States of America

West European Countries

ACTUARIAL SCIENCE

Any Country

African Nations

Australia

Canada

Caribbean Countries

East European Countries

Middle East

South Africa

United Kingdom

United States of America

West European Countries

APPLIED MATHEMATICS

Any Country

African Nations

Australia

Canada

Caribbean Countries

East European Countries

European Union

Middle East

New Zealand

South Africa

United Kingdom

United States of America

West European Countries

ARTIFICIAL INTELLIGENCE

Any Country

African Nations

Australia

Canada

Caribbean Countries

East European Countries

European Union

EPSRC School of Computer Science and Electronic Engineering Research Studentship 726
University of Essex Silberrad Scholarships 729

Middle East

Aberystwyth International Excellence Scholarships 4
International Postgraduate Research Scholarships (IPRS) 417

New Zealand

Aberystwyth International Excellence Scholarships 4
Australian Postgraduate Award 674

South Africa

Aberystwyth International Excellence Scholarships 4
International Postgraduate Research Scholarships (IPRS) 417

United Kingdom

EPSRC School of Computer Science and Electronic Engineering Research Studentship 726
International Postgraduate Research Scholarships (IPRS) 417
University of Essex Silberrad Scholarships 729

United States of America

Aberystwyth International Excellence Scholarships 4
Air Force Summer Faculty Fellowship Program 92
ETS Postdoctoral Fellowships 282
International Postgraduate Research Scholarships (IPRS) 417
Marshall Scholarships 181
Winston Churchill Foundation Scholarship 859

West European Countries

International Postgraduate Research Scholarships (IPRS) 417

COMPUTER SCIENCE

Any Country

Acadia Graduate Awards 7
Adelle and Erwin Tomash Fellowship in the History of Information Processing 232
AUC Graduate Merit Fellowships 107
AUC Laboratory Instruction Graduate Fellowships 108
BP Trust Research Fellowships 573
Bursaries 163
CERN Summer Student Programme 231
Computer Science PhD Studentship 676
Computing Laboratory Scholarship 737
DAGSI Research Fellowships 267
Doctoral Career Developement Scholarship (DCDS) Scheme 4
ESRF Postdoctoral Fellowships 293
ETH Zurich Excellence Scholarship and Opportunity Award 629
ETS Postdoctoral Fellowships 282
ETS Summer Internship Program for Graduate Students 282
Franklin Research Grant Program 80
Graduate Teaching Assistant in Computer Science 747
HFSPO Long-Term Fellowships 387
Hugh Kelly Fellowship 549
Infosys Fellowship for PhD Students 356
Institute for Advanced Study Postdoctoral Residential Fellowships 366
International Postgraduate Research Scholarship 674
Keble College: Ian Palmer Scholarship 772
La Trobe University Postgraduate Research Scholarship 418
Microsoft Fellowship 456
Microsoft Research India PhD Fellowships 456
NCAR Faculty Fellowship Programme 482
NCAR Graduate Student Visitor Programme 482
NCAR Postdoctoral Appointments in the Advanced Study Program 482
New Mexico Information Technology Fellowships 756
Ohio Board of Regents (OBR) Distinguished Doctoral Research Fellowship in Computer Science and Engineering 721
Queen Mary, University of London Research Studentships 540
Rhodes University Postdoctoral Fellowship and The Andrew Mellon Postdoctoral Fellowship 549
Scottish Government Personal Research Fellowships Co-funded by Marie Curie Actions 574
Tomsk Polytechnic University International Scholarship 640
University of Ballarat Postgraduate Research Scholarship 675
University of Bristol Postgraduate Scholarships 680
University of Kent School of Computing Scholarships 741
Victoria Doctoral Scholarships 836

African Nations

Aberystwyth International Excellence Scholarships 4
ESRF Thesis Studentships 293
International Postgraduate Research Scholarships (IPRS) 417

Australia

Aberystwyth International Excellence Scholarships 4
Australian Postgraduate Award 674
ESRF Thesis Studentships 293
NBCF Postdoctoral Fellowship 480

Canada

Aberystwyth International Excellence Scholarships 4
ESRF Thesis Studentships 293
International Postgraduate Research Scholarships (IPRS) 417
Public Safety Canada Research Fellowship Program in Honour of Stuart Nesbitt White 134

Caribbean Countries

Aberystwyth International Excellence Scholarships 4
ESRF Thesis Studentships 293
International Postgraduate Research Scholarships (IPRS) 417

East European Countries

CERN Technical Student Programme 231
ESRF Thesis Studentships 293
International Postgraduate Research Scholarships (IPRS) 417

European Union

EPSRC Doctoral Training Awards for School of Computing 737
EPSRC School of Computer Science and Electronic Engineering Research Studentship 726
Microsoft Research European PhD Scholarship Programme 456
NERC Studentships 164
Paul Ramsay MSc Computer Science Bursary 678
PhD Studentships in Computer Science 748
University of Essex Silberrad Scholarships 729
University of Southampton Engineering Doctorate 806

Middle East

Aberystwyth International Excellence Scholarships 4
ESRF Thesis Studentships 293
International Postgraduate Research Scholarships (IPRS) 417

New Zealand

Aberystwyth International Excellence Scholarships 4
Australian Postgraduate Award 674
ESRF Thesis Studentships 293

South Africa

Aberystwyth International Excellence Scholarships 4
ESRF Thesis Studentships 293
International Postgraduate Research Scholarships (IPRS) 417

United Kingdom

CERN Technical Student Programme 231
EPSRC Doctoral Training Awards for School of Computing 737
EPSRC School of Computer Science and Electronic Engineering Research Studentship 726
ESRF Thesis Studentships 293
Frank Knox Fellowships at Harvard University 303
International Postgraduate Research Scholarships (IPRS) 417
Kennedy Scholarships 412

NERC Studentships 164
Ohio Board of Regents (OBR) Distinguished Doctoral Research Fellowship in Computer Science and Engineering 721
Paul Ramsay MSc Computer Science Bursary 678
PhD Studentships in Computer Science 748
University of Essex Silberrad Scholarships 729
University of Southampton Engineering Doctorate 806

United States of America

Aberystwyth International Excellence Scholarships 4
Air Force Summer Faculty Fellowship Program 92
DEED (Demonstration of Energy-Efficient Developments) Student Research Grant and Internship 87
Department of Energy Computational Science Graduate Fellowship Program 416
ESRF Thesis Studentships 293
ETS Postdoctoral Fellowships 282
ETS Summer Internship Program for Graduate Students 282
Foundation for Science and Disability Student Grant Fund 302
Fulbright Distinguished Chairs Program 252
Fulbright Specialist Program 253
Graduate Assistance in Areas of National Need (GAANN) Award 866
The Graduate Fellowship Award 338
International Postgraduate Research Scholarships (IPRS) 417
Marshall Scholarships 181
NCAR Faculty Fellowship Programme 482
NPSC Fellowship in Physical Sciences 494
Ohio Board of Regents (OBR) Distinguished Doctoral Research Fellowship in Computer Science and Engineering 721
Philip Morris USA Thurgood Marshall Scholarship 639
PhRMAF Postdoctoral Fellowships in Informatics 530
PhRMAF Research Starter Grants in Informatics 531
PhRMAF Sabbatical Fellowships in Informatics 532
Winston Churchill Foundation Scholarship 859

West European Countries

CERN Technical Student Programme 231
ESRF Thesis Studentships 293
International Postgraduate Research Scholarships (IPRS) 417
Microsoft Research European PhD Scholarship Programme 456

STATISTICS

Any Country

Acadia Graduate Awards 7
Ellis R. Ott Scholarship for Applied Statistics and Quality Management 101
ETH Zurich Excellence Scholarship and Opportunity Award 629
ETS Postdoctoral Fellowships 282
Franklin Research Grant Program 80
Hugh Kelly Fellowship 549
International Postgraduate Research Scholarship 674
NCAR Faculty Fellowship Programme 482
NCAR Graduate Student Visitor Programme 482
NCAR Postdoctoral Appointments in the Advanced Study Program 482
Pierre Robillard Award 625
Queen Mary, University of London Research Studentships 540
Rhodes University Postdoctoral Fellowship and The Andrew Mellon Postdoctoral Fellowship 549
Sir James McNeill Foundation Postgraduate Scholarship 465
Statistics: Teaching Assistant Bursaries/ Departmental studentships 793
University of Ballarat Postgraduate Research Scholarship 675
University of Bristol Postgraduate Scholarships 680
University of Kent School of Mathematics, Statistics and Actuarial Science Scholarships 742
Victoria Doctoral Scholarships 836

African Nations

International Postgraduate Research Scholarships (IPRS) 417

Australia

Australian Postgraduate Award 674
NHMRC Early Career Fellowship 486
Pilot Study Grants 481
Practitioner Fellowship 481

Canada

Gertrude M. Cox Scholarship 106
International Postgraduate Research Scholarships (IPRS) 417

Caribbean Countries

International Postgraduate Research Scholarships (IPRS) 417

East European Countries

International Postgraduate Research Scholarships (IPRS) 417

European Union

ESRC 1+3 Awards and +3 Awards 280
Statistics: EPSRC studentships 793
University of Essex Silberrad Scholarships 729
University of Kent School of Mathematics, Statistics and Actuarial Science (SMSAS) 742

Middle East

International Postgraduate Research Scholarships (IPRS) 417

New Zealand

Australian Postgraduate Award 674

South Africa

Allan Gray Senior Scholarship 548
International Postgraduate Research Scholarships (IPRS) 417

United Kingdom

ESRC 1+3 Awards and +3 Awards 280
International Postgraduate Research Scholarships (IPRS) 417
James Watt Research Scholarships 733
Kennedy Scholarships 412
Mr and Mrs David Edward Memorial Award 164
Statistics: EPSRC studentships 793
University of Essex Silberrad Scholarships 729
University of Kent School of Mathematics, Statistics and Actuarial Science (SMSAS) 742

United States of America

Air Force Summer Faculty Fellowship Program 92
ETS Postdoctoral Fellowships 282
ETS Sylvia Taylor Johnson Minority Fellowship in Educational Measurement 282
Fulbright Specialist Program 253
Gertrude M. Cox Scholarship 106
International Postgraduate Research Scholarships (IPRS) 417
Marshall Scholarships 181
NCAR Faculty Fellowship Programme 482
Winston Churchill Foundation Scholarship 859

West European Countries

ESRC 1+3 Awards and +3 Awards 280
International Postgraduate Research Scholarships (IPRS) 417
Mr and Mrs David Edward Memorial Award 164

SYSTEMS ANALYSIS

Any Country

Bursaries 163
Doctoral Career Developement Scholarship (DCDS) Scheme 4
Franklin Research Grant Program 80
HFSPO Long-Term Fellowships 387
Institut Mittag-Leffler Grants 365
International Postgraduate Research Scholarship 674
La Trobe University Postgraduate Research Scholarship 418

African Nations

Australia

Canada

Caribbean Countries

East European Countries

European Union

Middle East

New Zealand

South Africa

United Kingdom

United States of America

West European Countries

MEDICAL SCIENCES

GENERAL

Any Country

African Nations

Australia

Canada

Caribbean Countries

East European Countries

European Union

Middle East

New Zealand

South Africa

United Kingdom

United States of America

West European Countries

ACUPUNCTURE

Any Country

Australia

Canada

BIOMEDICINE

Any Country

African Nations

Australia

Canada

Caribbean Countries

East European Countries

European Union

Middle East

New Zealand

South Africa

United Kingdom

United States of America

West European Countries

FORENSIC MEDICINE AND DENTISTRY

Any Country
HRB Project Grants-General 332

African Nations
International Postgraduate Research Scholarships (IPRS) 417

Australia
Australian Postgraduate Award 674
Biomedical (Dora Lush) and Public Health Postgraduate Scholarships 485
Biomedical Australian (Peter Doherty) Fellowship 485
Career Development Fellowship 486

Canada
CIHR Canadian Graduate Scholarships Doctoral Awards 214
International Postgraduate Research Scholarships (IPRS) 417

Caribbean Countries
International Postgraduate Research Scholarships (IPRS) 417

East European Countries
International Postgraduate Research Scholarships (IPRS) 417

Middle East
International Postgraduate Research Scholarships (IPRS) 417

New Zealand
Australian Postgraduate Award 674

South Africa
International Postgraduate Research Scholarships (IPRS) 417

United Kingdom
Hilda Martindale Trust Exhibitions 340
International Postgraduate Research Scholarships (IPRS) 417

United States of America
International Postgraduate Research Scholarships (IPRS) 417

West European Countries
International Postgraduate Research Scholarships (IPRS) 417

HEALTH ADMINISTRATION

Any Country
AAOHN Professional Development Scholarship 35
AFSP Distinguished Investigator Awards 60
AFSP Pilot Grants 60
AFSP Postdoctoral Research Fellowships 60
AFSP Standard Research Grants 60
AFSP Young Investigator Award 60
All-Ireland Institute for Hospice and Palliative Care Fellowships 331
CIHR Fellowships Program 214
Hastings Center International Visiting Scholars Program 330
HRB Project Grants-General 332
International Postgraduate Research Scholarship 674
IRSST Graduate Studies Scholarships 364
La Trobe University Postgraduate Research Scholarship 418
The Pedro Zamora Public Policy Fellowship 13
Press Ganey Best Place to Practice Award 538
Savoy Foundation Postdoctoral and Clinical Research Fellowships 586
Savoy Foundation Research Grants 586
Savoy Foundation Studentships 587
Sir Allan Sewell Visiting Fellowship 325
University of Bristol Postgraduate Scholarships 680
Women's Health Scholars Awards 2013-2014 515

African Nations
ABCCF Student Grant 113
Hastings Center International Visiting Scholars Program 330
International Postgraduate Research Scholarships (IPRS) 417

Australia
Australian Postgraduate Award 674
The Cancer Council NSW Research Project Grants 220
Early Career Fellowship 428
Harkness Fellowships in Healthcare Policy 244
Hastings Center International Visiting Scholars Program 330
National Collaborative Breast Cancer Research Grant Program 480
NBCF Postdoctoral Fellowship 480
Practitioner Fellowship 481

Canada
CIHR Canadian Graduate Scholarships Doctoral Awards 214
CIHR Doctoral Research Awards 214
International Postgraduate Research Scholarships (IPRS) 417
Métis Centre Fellowship Program 454
PAHO Grants 522
Public Safety Canada Research Fellowship Program in Honour of Stuart Nesbitt White 134

Caribbean Countries
Hastings Center International Visiting Scholars Program 330
International Postgraduate Research Scholarships (IPRS) 417
PAHO Grants 522

East European Countries
Hastings Center International Visiting Scholars Program 330
International Postgraduate Research Scholarships (IPRS) 417

European Union
University of Essex Silberrad Scholarships 729

Middle East
ABCCF Student Grant 113
Hastings Center International Visiting Scholars Program 330
International Postgraduate Research Scholarships (IPRS) 417

New Zealand
Australian Postgraduate Award 674
Harkness Fellowships in Healthcare Policy 244
Hastings Center International Visiting Scholars Program 330

South Africa
Hastings Center International Visiting Scholars Program 330
International Postgraduate Research Scholarships (IPRS) 417

United Kingdom
Harkness Fellowships in Healthcare Policy 244
Hastings Center International Visiting Scholars Program 330
International Postgraduate Research Scholarships (IPRS) 417
University of Essex Silberrad Scholarships 729

United States of America
Congress Bundestag Youth Exchange for Young Professionals 225
International Postgraduate Research Scholarships (IPRS) 417
Lung Health (LH) Research Dissertation Grants 73
Marshall Scholarships 181
NMSS Patient Management Care and Rehabilitation Grants 493
Nurses' Educational Funds Fellowships and Scholarships 513
Packer Fellowships 245
PAHO Grants 522
Pfizer Fellowships in Health Literary/ Clear Health Communication 528
Pfizer Fellowships in Public Health Overview 528
Pfizer Visiting Professorships in Health Literacy 529
PhRMAF Postdoctoral Fellowships in Health Outcomes Research 529
PhRMAF Sabbatical Fellowships in Health Outcomes Research 532
SSRC Abe Fellowship Program 606

West European Countries

Hastings Center International Visiting Scholars Program 330
International Postgraduate Research Scholarships (IPRS) 417

HOMEOPATHY

Any Country

Breast Cancer Campaign PhD Studentships 174
Breast Cancer Campaign Project Grants 174
Breast Cancer Campaign Scientific Fellowships 175
Breast Cancer Campaign Small Pilot Grants 175
HRB Project Grants-General 332
Postgraduate Research Bursaries (Epilepsy Action) 288

Australia

Biomedical (Dora Lush) and Public Health Postgraduate Scholarships 485

Canada

CIHR Canadian Graduate Scholarships Doctoral Awards 214

MEDICAL AUXILIARIES

Any Country

HRB Project Grants-General 332
HRB Translational Research Programmes 332

African Nations

International Postgraduate Research Scholarships (IPRS) 417

Australia

Biomedical (C J Martin) Fellowships 485
Biomedical (Dora Lush) and Public Health Postgraduate Scholarships 485
Career Development Fellowship 486
Clinical (Neil Hamilton Fairley) Fellowship 486
NBCF Postdoctoral Fellowship 480
NHMRC Early Career Fellowship 486
NHMRC Medical and Dental and Public Health Postgraduate Research Scholarships 486
NHMRC/INSERM Exchange Fellowships 486
Practitioner Fellowship 481
Public Health Australian Fellowship 486
Public Health Overseas (Sidney Sax) Fellowships 487
Training Scholarship for Indigenous Health Research 487

Canada

CIHR Canadian Graduate Scholarships Doctoral Awards 214
International Postgraduate Research Scholarships (IPRS) 417

Caribbean Countries

International Postgraduate Research Scholarships (IPRS) 417

East European Countries

International Postgraduate Research Scholarships (IPRS) 417

Middle East

International Postgraduate Research Scholarships (IPRS) 417

New Zealand

Training Scholarship for Indigenous Health Research 487

South Africa

International Postgraduate Research Scholarships (IPRS) 417

United Kingdom

International Postgraduate Research Scholarships (IPRS) 417

United States of America

International Postgraduate Research Scholarships (IPRS) 417
Nurses' Educational Funds Fellowships and Scholarships 513

West European Countries

International Postgraduate Research Scholarships (IPRS) 417

MEDICAL TECHNOLOGY

Any Country

ABMRF/The Foundation for Alcohol Research Project Grant 5
Action Medical Research Project Grants 8
Action Medical Research Training Fellowship 8
Alzheimer's Society Research Grants 25
Breast Cancer Campaign PhD Studentships 174
Breast Cancer Campaign Project Grants 174
Breast Cancer Campaign Scientific Fellowships 175
Breast Cancer Campaign Small Pilot Grants 175
Career Development Program 427
CCFF Fellowships 206
CCFF Special Travel Allowances 207
CCFF Studentships 207
Cooley's Anemia Foundation Research Fellowship 249
Dr Hadwen Trust Research Assistant or Technician 277
Dr Hadwen Trust Research Fellowship 277
Enterprise Fellowships funded by BBSRC 573
ESRF Postdoctoral Fellowships 293
Hastings Center International Visiting Scholars Program 330
HDA Studentship 347
Heart Research UK Novel and Emerging Technologies Grant 333
Heart Research UK Translational Research Project Grants 333
HRB Project Grants-General 332
HRB Translational Research Programmes 332
Meningitis Research Foundation Project Grant 453
MND Research Project Grants 469
Queen Mary, University of London Research Studentships 540
Research Contracts (IAEA) 380
Resuscitation Council Research Fellowships 547
Resuscitation Council Research Grants 547
RP Fighting Blindness Research Grants 189
Savoy Foundation Postdoctoral and Clinical Research Fellowships 586
Savoy Foundation Studentships 587
Translational Research Program 427

African Nations

ESRF Thesis Studentships 293
Hastings Center International Visiting Scholars Program 330

Australia

Biomedical (C J Martin) Fellowships 485
Biomedical (Dora Lush) and Public Health Postgraduate Scholarships 485
Biomedical Australian (Peter Doherty) Fellowship 485
Career Development Fellowship 486
Clinical (Neil Hamilton Fairley) Fellowship 486
Early Career Fellowship 428
Early Career Fellowship 428
ESRF Thesis Studentships 293
Hastings Center International Visiting Scholars Program 330
National Collaborative Breast Cancer Research Grant Program 480
NBCF Postdoctoral Fellowship 480
NHMRC Early Career Fellowship 486
NHMRC Medical and Dental and Public Health Postgraduate Research Scholarships 486
NHMRC/INSERM Exchange Fellowships 486
Practitioner Fellowship 481
Training Scholarship for Indigenous Health Research 487

Canada

ABMRF/The Foundation for Alcohol Research Project Grant 5
CCFF Senior Scientist Research Training Award 206
CIHR Canadian Graduate Scholarships Doctoral Awards 214

African Nations

Australia

Canada

Caribbean Countries

East European Countries

Middle East

New Zealand

South Africa

United Kingdom

United States of America

West European Countries

GASTROENTEROLOGY

Any Country

HAEMATOLOGY

Any Country

Australia

Canada

New Zealand

United Kingdom

United States of America

HEPATHOLOGY

Any Country

Canada

East European Countries

European Union

United Kingdom

United States of America

NEPHROLOGY

Any Country

Australia

Canada

New Zealand

ONCOLOGY

Any Country

African Nations

Australia

Canada

Caribbean Countries

East European Countries

European Union

Middle East

New Zealand

South Africa

United Kingdom

United States of America

West European Countries

OPHTHALMOLOGY

Any Country

ABMRF/The Foundation for Alcohol Research Project Grant 5
Action Medical Research Project Grants 8
AHAF Macular Degeneration Research 62
AHAF National Glaucoma Research 62
CNIB Winston Gordon Award 216
Dorey Bequest & Sir William Lister Travel Awards 560
The E. (Ben) & Mary Hochhausen Access Technology Research Award 216
Ethicon Foundation Fund Travel Award 560
The Fight For Sight Award 561
Fight for Sight Awards 296
HRB Translational Research Programmes 332
Jointly Funded Clinical Research Training Fellowship 447
The Keeler Scholarship 561
Lister Institute Research Prizes 432
LSA Medical Research Grant 438
Ophthalmology Section Travelling Fellowships 577
The Pfizer Ophthalmic Fellowship 561
Ross Purse Doctoral Fellowship 217
University of Bristol Postgraduate Scholarships 680

Canada

ABMRF/The Foundation for Alcohol Research Project Grant 5
CNIB Baker Fellowship Fund 216
Tuck MacPhee Award 217

United States of America

ABMRF/The Foundation for Alcohol Research Project Grant 5
AFAR Research Grants 57
The Glenn/AFAR Breakthroughs in Gerontology Awards 58
Nurses' Educational Funds Fellowships and Scholarships 513
Paul Beeson Career Development Award in Aging Research 58

OTORHINOLARYNGOLOGY

Any Country

Action Medical Research Project Grants 8
American Tinnitus Association Scientific Research Grants 106
HRB Translational Research Programmes 332
Laryngology & Rhinology Section Travel and Equipment Grants 577
Lister Institute Research Prizes 432

Canada

American Otological Society Research Grants 79
American Otological Society Research Training Fellowships 80

United Kingdom

Hilda Martindale Trust Exhibitions 340
Otology Section Norman Gamble Grant 578

United States of America

AFAR Research Grants 57
American Otological Society Research Grants 79
American Otological Society Research Training Fellowships 80
The Glenn/AFAR Breakthroughs in Gerontology Awards 58
Nurses' Educational Funds Fellowships and Scholarships 513
Paul Beeson Career Development Award in Aging Research 58

PAEDIATRICS

Any Country

ABMRF/The Foundation for Alcohol Research Project Grant 5
Action Medical Research Project Grants 8
ASBAH Research Grant 128
Ashok Nathwani Visiting Fellowship 562
Babes in Arms Fellowships 562
BackCare Research Grants 162
FRAXA Grants and Fellowships 303
HRB Translational Research Programmes 332
ICR Studentships 369
LSA Medical Research Grant 438
Meningitis Research Foundation Project Grant 453
Neuroblastoma Society Research Grants 503
Postgraduate Research Bursaries (Epilepsy Action) 288
Queen Mary, University of London Research Studentships 540
RCPCH/VSO Fellowship 562
Research Contracts (IAEA) 380
RESTRACOMP Research Fellowship 344
Resuscitation Council Research Fellowships 547
Resuscitation Council Research Grants 547
Savoy Foundation Postdoctoral and Clinical Research Fellowships 586
Savoy Foundation Research Grants 586
Savoy Foundation Studentships 587
Sir Halley Stewart Trust Grants 600
Translational Research Program 427
University of Bristol Postgraduate Scholarships 680

Australia

Australian Postgraduate Award 674
The Cancer Council NSW Research Project Grants 220

Canada

ABMRF/The Foundation for Alcohol Research Project Grant 5
CTS Research Fellowship Program 219

East European Countries

ECTS Career Establishment Award 290
ECTS Exchange Scholarship Grants 290

European Union

ECTS Career Establishment Award 290
ECTS Exchange Scholarship Grants 290

New Zealand

Australian Postgraduate Award 674

United Kingdom

British Lung Foundation Project Grants 185
ECTS Career Establishment Award 290
ECTS Exchange Scholarship Grants 290
Newlife Foundation-Full and Small Grants Schemes 171
Paediatrics & Child Health Section Trainees Tim David Prize 578
PWSA UK Research Grants 537
REMEDI Research Grants 544
WellChild Pump-Priming Grants 842
WellChild Research Fellowships 842

United States of America

ABMRF/The Foundation for Alcohol Research Project Grant 5
Florence P Kendall Doctoral Scholarships 301
New Investigator Fellowships Training Initiative (NIFTI) 302
Nurses' Educational Funds Fellowships and Scholarships 513
Paul Beeson Career Development Award in Aging Research 58
Promotion of Doctoral Studies (PODS) Scholarships 302
Research Grants (FPT) 302

West European Countries

ECTS Career Establishment Award 290
ECTS Exchange Scholarship Grants 290

PARASITOLOGY

Any Country

Ann Bishop Travelling Award 195
C.A. Wright Memorial Medal 195
Garnham Expeditionary Scholarship 195
HRB Translational Research Programmes 332
Lister Institute Research Prizes 432

RHEUMATOLOGY

Any Country

African Nations

Australia

Canada

Caribbean Countries

East European Countries

European Union

Middle East

New Zealand

South Africa

United Kingdom

United States of America

West European Countries

TROPICAL MEDICINE

Any Country

African Nations

Australia

Canada

Caribbean Countries

Canada

Caribbean Countries

East European Countries

Middle East

New Zealand

South Africa

United Kingdom

United States of America

West European Countries

NURSING

Any Country

African Nations

Australia

Canada

Caribbean Countries

East European Countries

European Union

Middle East

New Zealand

South Africa

United Kingdom

United States of America

West European Countries

OPTOMETRY

Any Country

Australia

Canada

New Zealand

OSTEOPATHY

Any Country

Australia

Canada

United States of America

PHARMACY

Any Country

African Nations

Australia

Canada

Caribbean Countries

East European Countries

Middle East

New Zealand

South Africa

United Kingdom

United States of America

West European Countries

PODIATRY

Any Country

African Nations

Australia

Canada

Caribbean Countries

East European Countries

Middle East

New Zealand

South Africa

United Kingdom

United States of America

West European Countries

PUBLIC HEALTH AND HYGIENE

Any Country

African Nations

Australia

Canada

Caribbean Countries

East European Countries

Middle East

New Zealand

South Africa

United Kingdom

United States of America

West European Countries

DIETETICS

Any Country

West European Countries

SPORTS MEDICINE

Any Country

African Nations

Australia

Canada

Caribbean Countries

East European Countries

Middle East

New Zealand

South Africa

United Kingdom

United States of America

West European Countries

RADIOLOGY

Any Country

African Nations

Australia

Canada

Caribbean Countries

East European Countries

Middle East

New Zealand

South Africa

United Kingdom

United States of America

West European Countries

REHABILITATION AND THERAPY

Any Country

African Nations

Australia

Canada

Caribbean Countries

East European Countries

European Union

Middle East

New Zealand

South Africa

United Kingdom

United States of America

West European Countries

TRADITIONAL EASTERN MEDICINE

Any Country

Australia

TREATMENT TECHNIQUES

Any Country

African Nations

Australia

Canada

Caribbean Countries

East European Countries

Middle East

New Zealand

South Africa

United Kingdom

United States of America

West European Countries

NATURAL SCIENCES

GENERAL

Any Country

African Nations

Australia

Canada

Caribbean Countries

East European Countries

European Union

Middle East

New Zealand

South Africa

United Kingdom

ASTRONOMY AND ASTROPHYSICS

Middle East

New Zealand

South Africa

United Kingdom

United States of America

West European Countries

BIOLOGICAL AND LIFE SCIENCES

Any Country

African Nations

Australia

Canada

Caribbean Countries

East European Countries

European Union

Middle East

New Zealand

South Africa

United Kingdom

United States of America

West European Countries

ANATOMY

Any Country

African Nations

Australia

Canada

Caribbean Countries

East European Countries

European Union

Middle East

New Zealand

South Africa

United Kingdom

United States of America

West European Countries

BIOCHEMISTRY

Any Country

African Nations

Australia

Canada

Caribbean Countries

East European Countries

European Union

Middle East

New Zealand

South Africa

United Kingdom

United States of America

West European Countries

BIOLOGY

Any Country

African Nations

Australia

Canada

Caribbean Countries

East European Countries

European Union

Middle East

New Zealand

South Africa

United Kingdom

United States of America

West European Countries

BIOPHYSICS AND MOLECULAR BIOLOGY

Any Country

African Nations

Australia

Canada

Caribbean Countries

East European Countries

European Union

Middle East

New Zealand

South Africa

United Kingdom

United States of America

West European Countries

BOTANY

Any Country

African Nations

Australia

Canada

Caribbean Countries

East European Countries

European Union

Middle East

New Zealand

South Africa

United Kingdom

United States of America

West European Countries

EMBRYOLOGY AND REPRODUCTION BIOLOGY

GENETICS

African Nations

Australia

Canada

Caribbean Countries

East European Countries

European Union

Middle East

New Zealand

South Africa

United Kingdom

United States of America

West European Countries

HISTOLOGY

Any Country

South Africa

Aberystwyth International Excellence Scholarships 4

United Kingdom

Access to Learning Fund 725
BBSRC Studentship 726
Clinical Immunology & Allergy Section President's Prize 576
FEMS Fellowship 295
Mr and Mrs David Edward Memorial Award 164
NERC Studentships 164
Wellbeing of Women Research Training Fellowship 842

United States of America

Aberystwyth International Excellence Scholarships 4
AFAR Research Grants 57
Air Force Summer Faculty Fellowship Program 92
Eli Lilly and Company-Elanco Research Award 95
The Glenn/AFAR Breakthroughs in Gerontology Awards 58
MARC Faculty Predoctoral Fellowships 490
Marshall Scholarships 181
NFID Advanced Vaccinology Course Travel Grant 484
NIGMS Fellowship Awards for Students With Disabilities 490
NIGMS Postdoctoral Awards 490
NIGMS Research Project Grants (R01) 490
NIH Research Grants 491
Paul Beeson Career Development Award in Aging Research 58
Pfizer Atorvastatin Research Awards Program 528
Research Supplements to Promote Diversity in Health-Related Research 491
Washington University Chancellor's Graduate Fellowship Program 841

West European Countries

FEMS Fellowship 295
Mr and Mrs David Edward Memorial Award 164

LIMNOLOGY

Any Country

Earthwatch Field Research Grants 279
Edmund Niles Huyck Preserve, Inc. Graduate and Postgraduate Grants 281
EMBO Short-Term Fellowships in Molecular Biology 291
ETH Zurich Excellence Scholarship and Opportunity Award 629
Hugh Kelly Fellowship 549
Rhodes University Postdoctoral Fellowship and The Andrew Mellon Postdoctoral Fellowship 549
University of Otago Coursework Master's Scholarship 761
University of Otago Doctoral Scholarships 761
University of Otago International Master's Scholarship 761
Victoria Doctoral Scholarships 836

African Nations

Austrian Academy of Sciences, MSc Course in Limnology and Wetland Ecosystems 160

Australia

University of Otago Master's Scholarships 761

East European Countries

FEMS Fellowship 295

European Union

NERC Research (PhD) Studentships 501

New Zealand

University of Otago Master's Scholarships 761

United Kingdom

FEMS Fellowship 295
Mr and Mrs David Edward Memorial Award 164
NERC Research (PhD) Studentships 501

West European Countries

FEMS Fellowship 295
Mr and Mrs David Edward Memorial Award 164

MARINE BIOLOGY

Any Country

CICOR Postdoctoral Scholar Fellowship in Coastal Oceanography, Climate or Marine Ecosystems 864
Dean John A Knauss Fellowship Program 496
Doctoral Career Developement Scholarship (DCDS) Scheme 4
Earthwatch Field Research Grants 279
EMBO Long-Term Fellowships in Molecular Biology 291
EMBO Short-Term Fellowships in Molecular Biology 291
EMBO Young Investigator 292
ETH Zurich Excellence Scholarship and Opportunity Award 629
Hugh Kelly Fellowship 549
NRC Research Associateships 495
Queen Mary, University of London Research Studentships 540
Research Contracts (IAEA) 380
Rhodes University Postdoctoral Fellowship and The Andrew Mellon Postdoctoral Fellowship 549
SERC Graduate Student Fellowship 603
SERC Postdoctoral Fellowships 603
SERC Predoctoral Fellowships 603
SERC Senior Fellowships 604
Tibor T Polgar Fellowship 345
UFAW Research and Project Awards 655
University of Essex Department of Biological Sciences Studentships 729
University of Otago Coursework Master's Scholarship 761
University of Otago Doctoral Scholarships 761
University of Otago International Master's Scholarship 761
Victoria Doctoral Scholarships 836

African Nations

Aberystwyth International Excellence Scholarships 4
Canadian Window on International Development 384
IDRC Doctoral Research Awards 384
International Masters Scholarships 4

Australia

Aberystwyth International Excellence Scholarships 4
International Masters Scholarships 4
Noel and Kate Monkman Postgraduate Award 400
Our World-Underwater Scholarship Society Rolex Scholarships 521
University of Otago Master's Scholarships 761

Canada

Aberystwyth International Excellence Scholarships 4
Canadian Window on International Development 384
IDRC Doctoral Research Awards 384
International Masters Scholarships 4
Our World-Underwater Scholarship Society Rolex Scholarships 521

Caribbean Countries

Aberystwyth International Excellence Scholarships 4
Canadian Window on International Development 384
IDRC Doctoral Research Awards 384
International Masters Scholarships 4
Our World-Underwater Scholarship Society Rolex Scholarships 521

East European Countries

FEMS Fellowship 295
IDRC Doctoral Research Awards 384
International Masters Scholarships 4
Synthesys Visiting Fellowship 501

European Union

BBSRC Studentship 726

Middle East

New Zealand

South Africa

United Kingdom

United States of America

West European Countries

MICROBIOLOGY

Any Country

African Nations

Australia

Canada

Caribbean Countries

African Nations

Australia

Canada

Caribbean Countries

East European Countries

European Union

Middle East

New Zealand

South Africa

United Kingdom

United States of America

West European Countries

PARASITOLOGY

Any Country

African Nations

Australia

Canada

Caribbean Countries

East European Countries

European Union

Middle East

West European Countries

PHYSIOLOGY

Any Country

Australia

Canada

East European Countries

European Union

New Zealand

South Africa

United Kingdom

United States of America

African Nations

Australia

Canada

Caribbean Countries

East European Countries

European Union

Middle East

New Zealand

South Africa

United Kingdom

United States of America

West European Countries

ZOOLOGY

Any Country

African Nations

Australia

Canada

Caribbean Countries

East European Countries

European Union

Middle East

New Zealand

South Africa

United Kingdom

United States of America

West European Countries

CHEMISTRY

Any Country

ESRF Thesis Studentships 293
Fulbright Specialist Program 253
The Glenn/AFAR Breakthroughs in Gerontology Awards 58
The Graduate Fellowship Award 338
International Postgraduate Research Scholarships (IPRS) 417
Irving Langmuir Award in Chemical Physics 43
MARC Faculty Predoctoral Fellowships 490
McDonnell Center Astronaut Fellowships in the Space Sciences 445
National Energy Technology Laboratory Methane Hydrates Fellowship Program (MHFP) 478
NCAR Faculty Fellowship Programme 482
NIGMS Fellowship Awards for Minority Students 490
NIGMS Fellowship Awards for Students With Disabilities 490
NIGMS Postdoctoral Awards 490
NIGMS Research Project Grants (R01) 490
ONR Summer Faculty Research Program 93
Philip Morris USA Thurgood Marshall Scholarship 639
Research Supplements to Promote Diversity in Health-Related Research 491
Washington University Chancellor's Graduate Fellowship Program 841
Winston Churchill Foundation Scholarship 859
Xerox Technical Minority Scholarship 868

West European Countries

ESRF Thesis Studentships 293
International Postgraduate Research Scholarships (IPRS) 417
Janson Johan Helmich Scholarships and Travel Grants 401
Microsoft Research European PhD Scholarship Programme 456
Mr and Mrs David Edward Memorial Award 164

ANALYTICAL CHEMISTRY

Any Country

Alzheimer's Research Trust, Major Project or Programme 24
Alzheimer's Society Research Grants 25
BP Trust Research Fellowships 573
ETH Zurich Excellence Scholarship and Opportunity Award 629
Hugh Kelly Fellowship 549
JILA Postdoctoral Research Associateship and Visiting Fellowships 404
NRC Research Associateships 495
Queen Mary, University of London Research Studentships 540
Research Contracts (IAEA) 380
Rhodes University Postdoctoral Fellowship and The Andrew Mellon Postdoctoral Fellowship 549
Teagasc Walsh Fellowships 633
Tomsk Polytechnic University International Scholarship 640
University of Bristol Postgraduate Scholarships 680
Utilities and Service Industries Training (USIT) 259
Victoria Doctoral Scholarships 836

African Nations

International Postgraduate Research Scholarships (IPRS) 417

Australia

Fulbright Postgraduate Scholarship in Science and Engineering 159

Canada

International Postgraduate Research Scholarships (IPRS) 417

Caribbean Countries

International Postgraduate Research Scholarships (IPRS) 417

East European Countries

International Postgraduate Research Scholarships (IPRS) 417

European Union

Panasonic Trust Fellowships 259

Middle East

International Postgraduate Research Scholarships (IPRS) 417

South Africa

International Postgraduate Research Scholarships (IPRS) 417

United Kingdom

Hector and Elizabeth Catling Bursary 189
International Postgraduate Research Scholarships (IPRS) 417
The Lorch MSc Student Bursary 259
Mr and Mrs David Edward Memorial Award 164
Panasonic Trust Fellowships 259
Utilities and Service Industries Training (USIT) 259

United States of America

Air Force Summer Faculty Fellowship Program 92
International Postgraduate Research Scholarships (IPRS) 417
ONR Summer Faculty Research Program 93
Washington University Chancellor's Graduate Fellowship Program 841

West European Countries

International Postgraduate Research Scholarships (IPRS) 417
Mr and Mrs David Edward Memorial Award 164

INORGANIC CHEMISTRY

Any Country

Alzheimer's Research Trust, Major Project or Programme 24
Alzheimer's Society Research Grants 25
BP Trust Research Fellowships 573
ETH Zurich Excellence Scholarship and Opportunity Award 629
F. Albert Cotton Award in Synthetic Inorganic Chemistry 42
Hugh Kelly Fellowship 549
JILA Postdoctoral Research Associateship and Visiting Fellowships 404
NRC Research Associateships 495
Queen Mary, University of London Research Studentships 540
Research Contracts (IAEA) 380
Rhodes University Postdoctoral Fellowship and The Andrew Mellon Postdoctoral Fellowship 549
University of Bristol Postgraduate Scholarships 680
University of Kent School of Physical Sciences Scholarships 743
Victoria Doctoral Scholarships 836

African Nations

International Postgraduate Research Scholarships (IPRS) 417

Australia

Fulbright Postgraduate Scholarship in Science and Engineering 159

Canada

International Postgraduate Research Scholarships (IPRS) 417

Caribbean Countries

International Postgraduate Research Scholarships (IPRS) 417

East European Countries

International Postgraduate Research Scholarships (IPRS) 417

Middle East

International Postgraduate Research Scholarships (IPRS) 417

South Africa

International Postgraduate Research Scholarships (IPRS) 417

United Kingdom

Hector and Elizabeth Catling Bursary 189
International Postgraduate Research Scholarships (IPRS) 417
Mr and Mrs David Edward Memorial Award 164
University of Kent School of Physical Sciences Studentships 743

United States of America

Air Force Summer Faculty Fellowship Program 92
International Postgraduate Research Scholarships (IPRS) 417
ONR Summer Faculty Research Program 93
Washington University Chancellor's Graduate Fellowship Program 841

West European Countries

International Postgraduate Research Scholarships (IPRS) 417
Mr and Mrs David Edward Memorial Award 164
University of Kent School of Physical Sciences Studentships 743

ORGANIC CHEMISTRY

Any Country

ACS Award for Creative Work in Synthetic Organic Chemistry 40
ACS Roger Adams Award in Organic Chemistry 41
Alzheimer's Research Trust, Major Project or Programme 24
Alzheimer's Society Research Grants 25
Arthur C. Cope Scholar Awards 42
BP Trust Research Fellowships 573
Ernest Guenther Award in the Chemistry of Natural Products 42
ETH Zurich Excellence Scholarship and Opportunity Award 629
Hugh Kelly Fellowship 549
JILA Postdoctoral Research Associateship and Visiting Fellowships 404
NRC Research Associateships 495
Queen Mary, University of London Research Studentships 540
Research Contracts (IAEA) 380
Rhodes University Postdoctoral Fellowship and The Andrew Mellon Postdoctoral Fellowship 549
Tomsk Polytechnic University International Scholarship 640
University of Bristol Postgraduate Scholarships 680
University of Kent School of Physical Sciences Scholarships 743
Utilities and Service Industries Training (USIT) 259
Victoria Doctoral Scholarships 836

African Nations

International Postgraduate Research Scholarships (IPRS) 417

Australia

Fulbright Postgraduate Scholarship in Science and Engineering 159

Canada

Boehringer Ingelheim Doctoral Research Award 218
Ichikizaki Fund for Young Chemists 218
International Postgraduate Research Scholarships (IPRS) 417

Caribbean Countries

International Postgraduate Research Scholarships (IPRS) 417

East European Countries

International Postgraduate Research Scholarships (IPRS) 417

European Union

Panasonic Trust Fellowships 259

Middle East

International Postgraduate Research Scholarships (IPRS) 417

South Africa

International Postgraduate Research Scholarships (IPRS) 417

United Kingdom

Hickinbottom/Briggs Fellowship 571
International Postgraduate Research Scholarships (IPRS) 417
The Lorch MSc Student Bursary 259
Mr and Mrs David Edward Memorial Award 164
Panasonic Trust Fellowships 259
University of Kent School of Physical Sciences Studentships 743
Utilities and Service Industries Training (USIT) 259

United States of America

Air Force Summer Faculty Fellowship Program 92
Department of Energy Computational Science Graduate Fellowship Program 416
International Postgraduate Research Scholarships (IPRS) 417
ONR Summer Faculty Research Program 93
Washington University Chancellor's Graduate Fellowship Program 841

West European Countries

International Postgraduate Research Scholarships (IPRS) 417
Mr and Mrs David Edward Memorial Award 164
University of Kent School of Physical Sciences Studentships 743

PHYSICAL CHEMISTRY

Any Country

Alzheimer's Research Trust, Major Project or Programme 24
Alzheimer's Society Research Grants 25
BP Trust Research Fellowships 573
CBS Graduate Fellowship Program 203
ETH Zurich Excellence Scholarship and Opportunity Award 629
Hugh Kelly Fellowship 549
JILA Postdoctoral Research Associateship and Visiting Fellowships 404
NRC Research Associateships 495
Peter Debye Award in Physical Chemistry 43
Queen Mary, University of London Research Studentships 540
Rhodes University Postdoctoral Fellowship and The Andrew Mellon Postdoctoral Fellowship 549
Tomsk Polytechnic University International Scholarship 640
University of Bristol Postgraduate Scholarships 680
University of Kent School of Physical Sciences Scholarships 743
Victoria Doctoral Scholarships 836
Welch Scholarship 392

African Nations

International Postgraduate Research Scholarships (IPRS) 417
TWAS-S N Bose National Centre for Basic Sciences Postgraduate Fellowships in Physical Sciences 639

Australia

Fulbright Postgraduate Scholarship in Science and Engineering 159

Canada

CBS Postdoctoral Fellowship (PDF) 203
International Postgraduate Research Scholarships (IPRS) 417

Caribbean Countries

International Postgraduate Research Scholarships (IPRS) 417
TWAS-S N Bose National Centre for Basic Sciences Postgraduate Fellowships in Physical Sciences 639

East European Countries

International Postgraduate Research Scholarships (IPRS) 417

Middle East

International Postgraduate Research Scholarships (IPRS) 417
TWAS-S N Bose National Centre for Basic Sciences Postgraduate Fellowships in Physical Sciences 639

South Africa

International Postgraduate Research Scholarships (IPRS) 417
TWAS-S N Bose National Centre for Basic Sciences Postgraduate Fellowships in Physical Sciences 639

United Kingdom

International Postgraduate Research Scholarships (IPRS) 417
Mr and Mrs David Edward Memorial Award 164
University of Kent School of Physical Sciences Studentships 743

United States of America

West European Countries

EARTH SCIENCES

Any Country

African Nations

Australia

Canada

Caribbean Countries

East European Countries

European Union

Middle East

New Zealand

South Africa

GEOCHEMISTRY

GEOGRAPHY (SCIENTIFIC)

African Nations

Australia

Canada

Caribbean Countries

East European Countries

European Union

Middle East

New Zealand

South Africa

United Kingdom

United States of America

West European Countries

GEOLOGY

Any Country

African Nations

Canadian Window on International Development 384
IDRC Doctoral Research Awards 384
International Postgraduate Research Scholarships (IPRS) 417
Lindemann Trust Fellowships 286
Master Studies in Physical Land Resources Scholarship 382

Australia

Aberystwyth International Excellence Scholarships 4
Australian Postgraduate Award 674
Edgar Pam Fellowship 373
Fulbright Postgraduate Scholarship in Science and Engineering 159
Lindemann Trust Fellowships 286

Canada

Aberystwyth International Excellence Scholarships 4
Canadian Window on International Development 384
Edgar Pam Fellowship 373
GSA Research Grants 310
IDRC Doctoral Research Awards 384
International Postgraduate Research Scholarships (IPRS) 417
Lindemann Trust Fellowships 286

Caribbean Countries

Aberystwyth International Excellence Scholarships 4
Canadian Window on International Development 384
IDRC Doctoral Research Awards 384
International Postgraduate Research Scholarships (IPRS) 417
Master Studies in Physical Land Resources Scholarship 382

East European Countries

IDRC Doctoral Research Awards 384
International Postgraduate Research Scholarships (IPRS) 417
Synthesys Visiting Fellowship 501

European Union

NERC Research (PhD) Studentships 501
Synthesys Visiting Fellowship 501

Middle East

Aberystwyth International Excellence Scholarships 4
IDRC Doctoral Research Awards 384
International Postgraduate Research Scholarships (IPRS) 417
Master Studies in Physical Land Resources Scholarship 382

New Zealand

Aberystwyth International Excellence Scholarships 4
Australian Postgraduate Award 674
Edgar Pam Fellowship 373
Lindemann Trust Fellowships 286

South Africa

Aberystwyth International Excellence Scholarships 4
Canadian Window on International Development 384
Edgar Pam Fellowship 373
IDRC Doctoral Research Awards 384
International Postgraduate Research Scholarships (IPRS) 417
Lindemann Trust Fellowships 286
Master Studies in Physical Land Resources Scholarship 382

United Kingdom

Edgar Pam Fellowship 373
Hector and Elizabeth Catling Bursary 189
International Postgraduate Research Scholarships (IPRS) 417
Lindemann Trust Fellowships 286
Mr and Mrs David Edward Memorial Award 164
NERC Research (PhD) Studentships 501
Synthesys Visiting Fellowship 501

United States of America

Aberystwyth International Excellence Scholarships 4
Air Force Summer Faculty Fellowship Program 92
Environmental Public Policy and Conflict Resolution PhD Fellowship 468
The Graduate Fellowship Award 338
GSA Research Grants 310
International Postgraduate Research Scholarships (IPRS) 417
Norwegian Thanksgiving Fund Scholarship 512
Washington University Chancellor's Graduate Fellowship Program 841

West European Countries

International Postgraduate Research Scholarships (IPRS) 417
Mr and Mrs David Edward Memorial Award 164
Synthesys Visiting Fellowship 501

GEOPHYSICS AND SEISMOLOGY

Any Country

Abdus Salam ICTP Fellowships 4
ACS PRF Scientific Education (Type SE) Grants 41
American Association of Petroleum Geologists Foundation Grants-in-Aid 35
BP Trust Research Fellowships 573
Dublin Institute for Advanced Studies Scholarship in Astronomy, Astrophysics and Geophysics 277
Earthwatch Field Research Grants 279
ETH Zurich Excellence Scholarship and Opportunity Award 629
Hugh Kelly Fellowship 549
JILA Postdoctoral Research Associateship and Visiting Fellowships 404
Research Contracts (IAEA) 380
Rhodes University Postdoctoral Fellowship and The Andrew Mellon Postdoctoral Fellowship 549
SEG Scholarships 618
University of Bristol Postgraduate Scholarships 680
Victoria Doctoral Scholarships 836
WHOI Postdoctoral Scholarship Program 865

African Nations

International Postgraduate Research Scholarships (IPRS) 417
Lindemann Trust Fellowships 286

Australia

Fulbright Postgraduate Scholarship in Science and Engineering 159
Lindemann Trust Fellowships 286

Canada

GSA Research Grants 310
International Postgraduate Research Scholarships (IPRS) 417
Lindemann Trust Fellowships 286

Caribbean Countries

International Postgraduate Research Scholarships (IPRS) 417

East European Countries

International Postgraduate Research Scholarships (IPRS) 417
Synthesys Visiting Fellowship 501

European Union

NERC Research (PhD) Studentships 501
Synthesys Visiting Fellowship 501

Middle East

International Postgraduate Research Scholarships (IPRS) 417

New Zealand

Lindemann Trust Fellowships 286

South Africa

International Postgraduate Research Scholarships (IPRS) 417
Lindemann Trust Fellowships 286

United Kingdom

Hector and Elizabeth Catling Bursary 189

West European Countries

METEOROLOGY

Any Country

African Nations

Caribbean Countries

European Union

Middle East

South Africa

United Kingdom

United States of America

West European Countries

ARCTIC STUDIES

Any Country

European Union

United Kingdom

United States of America

ARID LAND STUDIES

Any Country

European Union

United Kingdom

United States of America

PHYSICS

Any Country

African Nations

Australia

Canada

Caribbean Countries

East European Countries

European Union

Middle East

New Zealand

South Africa

ATOMIC AND MOLECULAR PHYSICS

NUCLEAR PHYSICS

East European Countries

CERN Technical Student Programme 231

European Union

STFC Studentships 679

United Kingdom

CERN Technical Student Programme 231
STFC Studentships 679

United States of America

Air Force Summer Faculty Fellowship Program 92
DEED (Demonstration of Energy-Efficient Developments) Student Research Grant and Internship 87
Department of Energy Computational Science Graduate Fellowship Program 416
Everitt P Blizard Scholarship 76
Walter Meyer Scholarship 78

West European Countries

CERN Technical Student Programme 231

OPTICS

Any Country

Abdus Salam ICTP Fellowships 4
BP Trust Research Fellowships 573
Ernst and Grace Matthaei Research Scholarship 750
ESRF Postdoctoral Fellowships 293
ETH Zurich Excellence Scholarship and Opportunity Award 629
JILA Postdoctoral Research Associateship and Visiting Fellowships 404
Tomsk Polytechnic University International Scholarship 640
University of Kent School of Physical Sciences Scholarships 743
Victoria Doctoral Scholarships 836

African Nations

ESRF Thesis Studentships 293

Australia

ESRF Thesis Studentships 293

Canada

ESRF Thesis Studentships 293

Caribbean Countries

ESRF Thesis Studentships 293

East European Countries

CERN Technical Student Programme 231
ESRF Thesis Studentships 293

Middle East

ESRF Thesis Studentships 293

New Zealand

ESRF Thesis Studentships 293

South Africa

ESRF Thesis Studentships 293

United Kingdom

CERN Technical Student Programme 231
ESRF Thesis Studentships 293
University of Kent School of Physical Sciences Studentships 743

United States of America

Air Force Summer Faculty Fellowship Program 92
Department of Energy Computational Science Graduate Fellowship Program 416
ESRF Thesis Studentships 293
Xerox Technical Minority Scholarship 868

West European Countries

CERN Technical Student Programme 231
ESRF Thesis Studentships 293
University of Kent School of Physical Sciences Studentships 743

SOLID STATE PHYSICS

Any Country

Abdus Salam ICTP Fellowships 4
BP Trust Research Fellowships 573
Doctoral Career Developement Scholarship (DCDS) Scheme 4
ESRF Postdoctoral Fellowships 293
ETH Zurich Excellence Scholarship and Opportunity Award 629
JILA Postdoctoral Research Associateship and Visiting Fellowships 404
Tomsk Polytechnic University International Scholarship 640
University of Kent School of Physical Sciences Scholarships 743
Victoria Doctoral Scholarships 836
Welch Scholarship 392

African Nations

Aberystwyth International Excellence Scholarships 4
ESRF Thesis Studentships 293
TWAS-S N Bose National Centre for Basic Sciences Postgraduate Fellowships in Physical Sciences 639

Australia

Aberystwyth International Excellence Scholarships 4
ESRF Thesis Studentships 293

Canada

Aberystwyth International Excellence Scholarships 4
ESRF Thesis Studentships 293

Caribbean Countries

Aberystwyth International Excellence Scholarships 4
ESRF Thesis Studentships 293
TWAS-S N Bose National Centre for Basic Sciences Postgraduate Fellowships in Physical Sciences 639

East European Countries

CERN Technical Student Programme 231
ESRF Thesis Studentships 293

Middle East

Aberystwyth International Excellence Scholarships 4
ESRF Thesis Studentships 293
TWAS-S N Bose National Centre for Basic Sciences Postgraduate Fellowships in Physical Sciences 639

New Zealand

Aberystwyth International Excellence Scholarships 4
ESRF Thesis Studentships 293

South Africa

Aberystwyth International Excellence Scholarships 4
ESRF Thesis Studentships 293
TWAS-S N Bose National Centre for Basic Sciences Postgraduate Fellowships in Physical Sciences 639

United Kingdom

CERN Technical Student Programme 231
ESRF Thesis Studentships 293
University of Kent School of Physical Sciences Studentships 743

United States of America

Aberystwyth International Excellence Scholarships 4
Air Force Summer Faculty Fellowship Program 92

West European Countries

THERMAL PHYSICS

Any Country

African Nations

Canada

Caribbean Countries

East European Countries

Middle East

South Africa

United Kingdom

United States of America

West European Countries

RECREATION, WELFARE, PROTECTIVE SERVICES

GENERAL

Any Country

African Nations

Australia

Canada

Caribbean Countries

East European Countries

European Union

Middle East

New Zealand

South Africa

United Kingdom

United States of America

West European Countries

African Nations

Australia

Canada

Caribbean Countries

East European Countries

European Union

Middle East

New Zealand

South Africa

United Kingdom

United States of America

West European Countries

ECOLOGY

Any Country

African Nations

Australia

Canada

Caribbean Countries

East European Countries

European Union

Middle East

Aberystwyth International Excellence Scholarships 4
IDRC Doctoral Research Awards 384
IDRC Research Awards 384
International Masters Scholarships 4
International Postgraduate Research Scholarships (IPRS) 417
Research Grant Small 182

New Zealand

Aberystwyth International Excellence Scholarships 4
Australian Postgraduate Award 674
International Masters Scholarships 4

South Africa

Aberystwyth International Excellence Scholarships 4
Canadian Window on International Development 384
IDRC Doctoral Research Awards 384
IDRC Research Awards 384
International Masters Scholarships 4
International Postgraduate Research Scholarships (IPRS) 417
Research Grant Small 182

United Kingdom

Department of Agriculture and Rural Development (DARD) for Northern Ireland 258
International Postgraduate Research Scholarships (IPRS) 417
Mr and Mrs David Edward Memorial Award 164
NERC Research (PhD) Studentships 501
Synthesys Visiting Fellowship 501

United States of America

Aberystwyth International Excellence Scholarships 4
Congress Bundestag Youth Exchange for Young Professionals 225
Environmental Public Policy and Conflict Resolution PhD Fellowship 468
FFGC Scholarship in Ecology 297
Fulbright Specialist Program 253
International Masters Scholarships 4
International Postgraduate Research Scholarships (IPRS) 417
Marshall Scholarships 181
Welder Wildlife Foundation Fellowship 551

West European Countries

International Postgraduate Research Scholarships (IPRS) 417
Mr and Mrs David Edward Memorial Award 164
Synthesys Visiting Fellowship 501

ENVIRONMENTAL MANAGEMENT

Any Country

BIAA Research Scholarship 183
BIAA Study Grants 183
Doctoral Career Developement Scholarship (DCDS) Scheme 4
Earthwatch Field Research Grants 279
EREF Scholarships in Solid Waste Management Research and Education 287
Geography: Andrew Goudie Bursary 769
Geography: Birkett Scholarship at Trinity College 769
Geography: Hitachi Chemical Europe Scholarship with Linacre College 769
Geography: Joan Doll Scholarship 769
Hudson River Graduate Fellowships 345
Hyland R Johns Grant Program 644
Jack Kimmel International Grant Program 644
John Z Duling Grant Program 644
NJWEA Scholarship Award Program 504
Oriel College: Sir Walter Raleigh Scholarship 781
Rhodes University Postdoctoral Fellowship and The Andrew Mellon Postdoctoral Fellowship 549
Scholarship Opportunities Linked to CATIE's Postgraduate Program Including CATIE Scholarship Forming Part of the Scholarship-Loan Program 644
St John's College: Dr Yungtai Hsu Studentship 793
Stanley Smith (UK) Horticultural Trust Awards 623
Tibor T Polgar Fellowship 345
Tomsk Polytechnic University International Scholarship 640
UFAW Animal Welfare Research Training Scholarships 654
UFAW Animal Welfare Student Scholarships 654
UFAW Research and Project Awards 655
UFAW Small Project and Travel Awards 655
University of Calgary Faculty of Law Graduate Scholarship 681
Utilities and Service Industries Training (USIT) 259
Victoria Doctoral Scholarships 836

African Nations

Aberystwyth International Excellence Scholarships 4
Austrian Academy of Sciences, MSc Course in Limnology and Wetland Ecosystems 160
Canadian Window on International Development 384
IDRC Doctoral Research Awards 384
IDRC Research Awards 384
International Masters Scholarships 4
International Postgraduate Research Scholarships (IPRS) 417

Australia

Aberystwyth International Excellence Scholarships 4
Australian Postgraduate Award 674
Dunlop Asia Fellowships 125
Fulbright Postgraduate Scholarship in Science and Engineering 159
International Masters Scholarships 4
Victoria Fellowships 269

Canada

Aberystwyth International Excellence Scholarships 4
Canadian Window on International Development 384
CWRA Dillon Consulting Scholarship/Ken Thomson Scholarship 219
Horticultural Research Institute Grants 344
IDRC Doctoral Research Awards 384
IDRC Research Awards 384
International Masters Scholarships 4
International Postgraduate Research Scholarships (IPRS) 417
OFAF/OFAF Zone 6 Wildlife Research Grant 515

Caribbean Countries

Aberystwyth International Excellence Scholarships 4
Canadian Window on International Development 384
IDRC Doctoral Research Awards 384
IDRC Research Awards 384
International Masters Scholarships 4
International Postgraduate Research Scholarships (IPRS) 417

East European Countries

IDRC Doctoral Research Awards 384
International Masters Scholarships 4
International Postgraduate Research Scholarships (IPRS) 417
Synthesys Visiting Fellowship 501

European Union

Department of Agriculture and Rural Development (DARD) for Northern Ireland 258
NERC Research (PhD) Studentships 501
Panasonic Trust Fellowships 259
Synthesys Visiting Fellowship 501

Middle East

Aberystwyth International Excellence Scholarships 4
IDRC Doctoral Research Awards 384
IDRC Research Awards 384
International Masters Scholarships 4
International Postgraduate Research Scholarships (IPRS) 417

New Zealand

Aberystwyth International Excellence Scholarships 4
Australian Postgraduate Award 674
International Masters Scholarships 4

South Africa

Aberystwyth International Excellence Scholarships 4
Canadian Window on International Development 384
IDRC Doctoral Research Awards 384
IDRC Research Awards 384
International Masters Scholarships 4
International Postgraduate Research Scholarships (IPRS) 417

United Kingdom

Department of Agriculture and Rural Development (DARD) for Northern Ireland 258
Douglas Bomford Trust 258
Environmental Issues Award 258
International Postgraduate Research Scholarships (IPRS) 417
Mr and Mrs David Edward Memorial Award 164
NERC Research (PhD) Studentships 501
Panasonic Trust Fellowships 259
Synthesys Visiting Fellowship 501
Utilities and Service Industries Training (USIT) 259

United States of America

Aberystwyth International Excellence Scholarships 4
Congress Bundestag Youth Exchange for Young Professionals 225
Environmental Public Policy and Conflict Resolution PhD Fellowship 468
Fulbright Specialist Program 253
Horticultural Research Institute Grants 344
International Masters Scholarships 4
International Postgraduate Research Scholarships (IPRS) 417
Welder Wildlife Foundation Fellowship 551

West European Countries

International Postgraduate Research Scholarships (IPRS) 417
Mr and Mrs David Edward Memorial Award 164
Synthesys Visiting Fellowship 501

NATURAL RESOURCES

Any Country

Alberta Law Foundation Graduate Scholarship 680
American Association of Petroleum Geologists Foundation Grants-in-Aid 35
BIAA Research Scholarship 183
BIAA Study Grants 183
Canadian Embassy Faculty Enrichment Program 209
Earthwatch Field Research Grants 279
Edmund Niles Huyck Preserve, Inc. Graduate and Postgraduate Grants 281
Hudson River Graduate Fellowships 345
Hyland R Johns Grant Program 644
Jack Kimmel International Grant Program 644
John Z Duling Grant Program 644
Rhodes University Postdoctoral Fellowship and The Andrew Mellon Postdoctoral Fellowship 549
Scholarship Opportunities Linked to CATIE's Postgraduate Program Including CATIE Scholarship Forming Part of the Scholarship-Loan Program 644
Silverhill Institute of Environmental Research and Conservation Award 598
Tibor T Polgar Fellowship 345
University of Calgary Faculty of Law Graduate Scholarship 681
Utilities and Service Industries Training (USIT) 259
Victoria Doctoral Scholarships 836

African Nations

Austrian Academy of Sciences, MSc Course in Limnology and Wetland Ecosystems 160
Canadian Window on International Development 384
IDRC Doctoral Research Awards 384
IDRC Research Awards 384
International Masters Scholarships 4
International Postgraduate Research Scholarships (IPRS) 417

Australia

Australian Postgraduate Award 674
Fulbright Postgraduate Scholarship in Science and Engineering 159
International Masters Scholarships 4
Victoria Fellowships 269

Canada

Canadian Window on International Development 384
CWRA Dillon Consulting Scholarship/Ken Thomson Scholarship 219
IDRC Doctoral Research Awards 384
IDRC Research Awards 384
International Masters Scholarships 4
International Postgraduate Research Scholarships (IPRS) 417
OFAF/OFAF Zone 6 Wildlife Research Grant 515
OFAH/Oakville and District Rod & Gun Club Conservation Research Grant 515
OFAH/Toronto Sportsmen's Show Sport Fisheries Research Grant 516

Caribbean Countries

Canadian Window on International Development 384
IDRC Doctoral Research Awards 384
IDRC Research Awards 384
International Masters Scholarships 4
International Postgraduate Research Scholarships (IPRS) 417

East European Countries

IDRC Doctoral Research Awards 384
International Masters Scholarships 4
International Postgraduate Research Scholarships (IPRS) 417
Synthesys Visiting Fellowship 501

European Union

Department of Agriculture and Rural Development (DARD) for Northern Ireland 258
NERC Research (PhD) Studentships 501
Synthesys Visiting Fellowship 501

Middle East

IDRC Doctoral Research Awards 384
IDRC Research Awards 384
International Masters Scholarships 4
International Postgraduate Research Scholarships (IPRS) 417

New Zealand

Australian Postgraduate Award 674
International Masters Scholarships 4

South Africa

Canadian Window on International Development 384
IDRC Doctoral Research Awards 384
IDRC Research Awards 384
International Masters Scholarships 4
International Postgraduate Research Scholarships (IPRS) 417

United Kingdom

Department of Agriculture and Rural Development (DARD) for Northern Ireland 258
Douglas Bomford Trust 258
Environmental Issues Award 258
International Postgraduate Research Scholarships (IPRS) 417
Mr and Mrs David Edward Memorial Award 164
NERC Research (PhD) Studentships 501
Synthesys Visiting Fellowship 501
Utilities and Service Industries Training (USIT) 259

United States of America

Campus Ecology Fellowship Program 501
Congress Bundestag Youth Exchange for Young Professionals 225
Environmental Public Policy and Conflict Resolution PhD Fellowship 468
Fulbright Specialist Program 253
International Masters Scholarships 4

International Postgraduate Research Scholarships (IPRS) 417
NOAA Coastal Management Fellowship 510
NOAA Coral Reef Management Fellowship 510
Welder Wildlife Foundation Fellowship 551

West European Countries

International Postgraduate Research Scholarships (IPRS) 417
Mr and Mrs David Edward Memorial Award 164
Synthesys Visiting Fellowship 501

WASTE MANAGEMENT

WILDLIFE AND PEST MANAGEMENT

Any Country

BIAA Research Scholarship 183
BIAA Study Grants 183
Charles L. Dobbins Memorial Scholarship 496
Earthwatch Field Research Grants 279
Hudson River Graduate Fellowships 345
Hyland R Johns Grant Program 644
Jack Kimmel International Grant Program 644
John H. Borden Scholarship 287
John Z Duling Grant Program 644
Rhodes University Postdoctoral Fellowship and The Andrew Mellon Postdoctoral Fellowship 549
Rob and Bessie Welder Wildlife Foundation's Graduate Research Scholarship Program 841
RSPCA Australia Alan White Scholarship for Animal Welfare Research 582
Scholarship Opportunities Linked to CATIE's Postgraduate Program Including CATIE Scholarship Forming Part of the Scholarship-Loan Program 644
Tibor T Polgar Fellowship 345
UFAW Animal Welfare Research Training Scholarships 654
UFAW Animal Welfare Student Scholarships 654
UFAW Research and Project Awards 655
UFAW Small Project and Travel Awards 655
Victoria Doctoral Scholarships 836

African Nations

Canadian Window on International Development 384
IDRC Doctoral Research Awards 384
IDRC Research Awards 384
International Postgraduate Research Scholarships (IPRS) 417

Australia

Australian Postgraduate Award 674
Victoria Fellowships 269

Canada

Canadian Window on International Development 384
IDRC Doctoral Research Awards 384
IDRC Research Awards 384
International Postgraduate Research Scholarships (IPRS) 417
OFAF/OFAF Zone 6 Wildlife Research Grant 515

Caribbean Countries

Canadian Window on International Development 384
IDRC Doctoral Research Awards 384
IDRC Research Awards 384
International Postgraduate Research Scholarships (IPRS) 417

East European Countries

IDRC Doctoral Research Awards 384
International Postgraduate Research Scholarships (IPRS) 417

European Union

NERC Research (PhD) Studentships 501

Middle East

IDRC Doctoral Research Awards 384
IDRC Research Awards 384
International Postgraduate Research Scholarships (IPRS) 417

New Zealand

Australian Postgraduate Award 674

South Africa

Canadian Window on International Development 384
IDRC Doctoral Research Awards 384
IDRC Research Awards 384
International Postgraduate Research Scholarships (IPRS) 417

United Kingdom

International Postgraduate Research Scholarships (IPRS) 417
Mr and Mrs David Edward Memorial Award 164
NERC Research (PhD) Studentships 501

United States of America

Congress Bundestag Youth Exchange for Young Professionals 225
Environmental Public Policy and Conflict Resolution PhD Fellowship 468
Fulbright Specialist Program 253
International Postgraduate Research Scholarships (IPRS) 417
Welder Wildlife Foundation Fellowship 551

West European Countries

International Postgraduate Research Scholarships (IPRS) 417
Mr and Mrs David Edward Memorial Award 164

FIRE PROTECTION SCIENCE

Any Country

Doctoral Career Developement Scholarship (DCDS) Scheme 4

African Nations

Aberystwyth International Excellence Scholarships 4

Australia

Aberystwyth International Excellence Scholarships 4
Australian Postgraduate Award 674

Canada

Aberystwyth International Excellence Scholarships 4

Caribbean Countries

Aberystwyth International Excellence Scholarships 4

Middle East

Aberystwyth International Excellence Scholarships 4

New Zealand

Aberystwyth International Excellence Scholarships 4
Australian Postgraduate Award 674

South Africa

Aberystwyth International Excellence Scholarships 4

United States of America

Aberystwyth International Excellence Scholarships 4

LEISURE STUDIES

African Nations

International Postgraduate Research Scholarships (IPRS) 417
NUFFIC-NFP Fellowships for Master's Degree Programmes 502

Australia
Our World-Underwater Scholarship Society Rolex Scholarships 521

Canada
International Postgraduate Research Scholarships (IPRS) 417
JCCA Graduate Education Scholarship 403
Our World-Underwater Scholarship Society Rolex Scholarships 521

Caribbean Countries
International Postgraduate Research Scholarships (IPRS) 417
Our World-Underwater Scholarship Society Rolex Scholarships 521

East European Countries
International Postgraduate Research Scholarships (IPRS) 417

European Union
Our World-Underwater Scholarship Society Rolex Scholarships 521

Middle East
International Postgraduate Research Scholarships (IPRS) 417
NUFFIC-NFP Fellowships for Master's Degree Programmes 502

New Zealand
Our World-Underwater Scholarship Society Rolex Scholarships 521

South Africa
International Postgraduate Research Scholarships (IPRS) 417
NUFFIC-NFP Fellowships for Master's Degree Programmes 502

United Kingdom
International Postgraduate Research Scholarships (IPRS) 417
Mr and Mrs David Edward Memorial Award 164
Our World-Underwater Scholarship Society Rolex Scholarships 521

United States of America
International Postgraduate Research Scholarships (IPRS) 417
JCCA Graduate Education Scholarship 403
Our World-Underwater Scholarship Society Rolex Scholarships 521

West European Countries
International Postgraduate Research Scholarships (IPRS) 417
Mr and Mrs David Edward Memorial Award 164
Our World-Underwater Scholarship Society Rolex Scholarships 521

MILITARY SCIENCE

Any Country
CDI Internship 228
Doctoral Career Developement Scholarship (DCDS) Scheme 4
HFG Research Program 330
Jennings Randolph Program for International Peace Dissertation Fellowship 649
United States Holocaust Memorial Museum Center for Advanced Holocaust Studies Visiting Scholar Programs 649

African Nations
Aberystwyth International Excellence Scholarships 4

Australia
Aberystwyth International Excellence Scholarships 4

Canada
Aberystwyth International Excellence Scholarships 4

Caribbean Countries
Aberystwyth International Excellence Scholarships 4

Middle East
Aberystwyth International Excellence Scholarships 4

New Zealand
Aberystwyth International Excellence Scholarships 4

South Africa
Aberystwyth International Excellence Scholarships 4

United States of America
Aberystwyth International Excellence Scholarships 4
IREX Short-Term Travel Grants 391

PARKS AND RECREATION

Any Country
Doctoral Career Developement Scholarship (DCDS) Scheme 4
Stanley Smith (UK) Horticultural Trust Awards 623
Victoria Doctoral Scholarships 836

African Nations
Aberystwyth International Excellence Scholarships 4
International Postgraduate Research Scholarships (IPRS) 417

Australia
Aberystwyth International Excellence Scholarships 4
Our World-Underwater Scholarship Society Rolex Scholarships 521

Canada
Aberystwyth International Excellence Scholarships 4
International Postgraduate Research Scholarships (IPRS) 417
JCCA Graduate Education Scholarship 403
Our World-Underwater Scholarship Society Rolex Scholarships 521

Caribbean Countries
Aberystwyth International Excellence Scholarships 4
International Postgraduate Research Scholarships (IPRS) 417
Our World-Underwater Scholarship Society Rolex Scholarships 521

East European Countries
International Postgraduate Research Scholarships (IPRS) 417

European Union
Our World-Underwater Scholarship Society Rolex Scholarships 521

Middle East
Aberystwyth International Excellence Scholarships 4
International Postgraduate Research Scholarships (IPRS) 417

New Zealand
Aberystwyth International Excellence Scholarships 4
Our World-Underwater Scholarship Society Rolex Scholarships 521

South Africa
Aberystwyth International Excellence Scholarships 4
International Postgraduate Research Scholarships (IPRS) 417

United Kingdom
International Postgraduate Research Scholarships (IPRS) 417
Our World-Underwater Scholarship Society Rolex Scholarships 521

United States of America
Aberystwyth International Excellence Scholarships 4
Environmental Public Policy and Conflict Resolution PhD Fellowship 468
International Postgraduate Research Scholarships (IPRS) 417
JCCA Graduate Education Scholarship 403
Our World-Underwater Scholarship Society Rolex Scholarships 521

West European Countries
International Postgraduate Research Scholarships (IPRS) 417
Our World-Underwater Scholarship Society Rolex Scholarships 521

PEACE AND DISARMAMENT

POLICE STUDIES

SOCIAL WELFARE AND SOCIAL WORK

West European Countries

SOCIAL AND COMMUNITY SERVICES

Any Country

African Nations

Australia

Canada

Caribbean Countries

East European Countries

Middle East

New Zealand

South Africa

United Kingdom

United States of America

West European Countries

SPORTS

Any Country

African Nations

Australia

Canada

Caribbean Countries

East European Countries

European Union

Middle East

New Zealand

South Africa

United Kingdom

United States of America

Aberystwyth International Excellence Scholarships 4
Essex/Fulbright Commission Postgraduate Scholarships 727
International Postgraduate Research Scholarships (IPRS) 417
JCCA Graduate Education Scholarship 403
Marshall Scholarships 181

West European Countries

International Postgraduate Research Scholarships (IPRS) 417
Mr and Mrs David Edward Memorial Award 164

SOCIOLOGY OF SPORTS

Any Country

Doctoral Career Developement Scholarship (DCDS) Scheme 4

African Nations

Aberystwyth International Excellence Scholarships 4
International Postgraduate Research Scholarships (IPRS) 417

Australia

Aberystwyth International Excellence Scholarships 4
Australian Postgraduate Award 674
Our World-Underwater Scholarship Society Rolex Scholarships 521

Canada

Aberystwyth International Excellence Scholarships 4
International Postgraduate Research Scholarships (IPRS) 417
JCCA Graduate Education Scholarship 403
Our World-Underwater Scholarship Society Rolex Scholarships 521

Caribbean Countries

Aberystwyth International Excellence Scholarships 4
International Postgraduate Research Scholarships (IPRS) 417
Our World-Underwater Scholarship Society Rolex Scholarships 521

East European Countries

International Postgraduate Research Scholarships (IPRS) 417

European Union

ESRC 1+3 Awards and +3 Awards 280
Our World-Underwater Scholarship Society Rolex Scholarships 521

Middle East

Aberystwyth International Excellence Scholarships 4
International Postgraduate Research Scholarships (IPRS) 417

New Zealand

Aberystwyth International Excellence Scholarships 4
Australian Postgraduate Award 674
Our World-Underwater Scholarship Society Rolex Scholarships 521

South Africa

Aberystwyth International Excellence Scholarships 4
International Postgraduate Research Scholarships (IPRS) 417

United Kingdom

ESRC 1+3 Awards and +3 Awards 280
International Postgraduate Research Scholarships (IPRS) 417
Mr and Mrs David Edward Memorial Award 164
Our World-Underwater Scholarship Society Rolex Scholarships 521

United States of America

Aberystwyth International Excellence Scholarships 4
International Postgraduate Research Scholarships (IPRS) 417
JCCA Graduate Education Scholarship 403
Marshall Scholarships 181
Our World-Underwater Scholarship Society Rolex Scholarships 521

West European Countries

ESRC 1+3 Awards and +3 Awards 280
International Postgraduate Research Scholarships (IPRS) 417
Mr and Mrs David Edward Memorial Award 164
Our World-Underwater Scholarship Society Rolex Scholarships 521

SPORTS MANAGEMENT

Any Country

Harness Track of America Scholarship 329
University of Bristol Postgraduate Scholarships 680

African Nations

International Postgraduate Research Scholarships (IPRS) 417

Australia

Australian Postgraduate Award 674
Dunlop Asia Fellowships 125
Our World-Underwater Scholarship Society Rolex Scholarships 521

Canada

International Postgraduate Research Scholarships (IPRS) 417
JCCA Graduate Education Scholarship 403
Our World-Underwater Scholarship Society Rolex Scholarships 521

Caribbean Countries

International Postgraduate Research Scholarships (IPRS) 417
Our World-Underwater Scholarship Society Rolex Scholarships 521

East European Countries

International Postgraduate Research Scholarships (IPRS) 417

European Union

Our World-Underwater Scholarship Society Rolex Scholarships 521
The University of Glamorgan Sports Scholarships 731

Middle East

International Postgraduate Research Scholarships (IPRS) 417

New Zealand

Australian Postgraduate Award 674
Our World-Underwater Scholarship Society Rolex Scholarships 521

South Africa

International Postgraduate Research Scholarships (IPRS) 417

United Kingdom

International Postgraduate Research Scholarships (IPRS) 417
Mr and Mrs David Edward Memorial Award 164
Our World-Underwater Scholarship Society Rolex Scholarships 521
The University of Glamorgan Sports Scholarships 731

United States of America

Congress Bundestag Youth Exchange for Young Professionals 225
International Postgraduate Research Scholarships (IPRS) 417
JCCA Graduate Education Scholarship 403
Marshall Scholarships 181
Our World-Underwater Scholarship Society Rolex Scholarships 521

West European Countries

International Postgraduate Research Scholarships (IPRS) 417
Mr and Mrs David Edward Memorial Award 164
Our World-Underwater Scholarship Society Rolex Scholarships 521

VOCATIONAL COUNSELLINGRELIGION AND THEOLOGY

GENERAL

Any Country

African Nations

Australia

Canada

Caribbean Countries

East European Countries

European Union

Middle East

New Zealand

South Africa

United Kingdom

CHURCH ADMINISTRATION (PASTORAL WORK)

COMPARATIVE RELIGION

ESOTERIC PRACTICES

HISTORY OF RELIGION

Any Country

Ahmanson and Getty Postdoctoral Fellowships 646
Albert J Beveridge Grant 63
ASCSA Fellowships 89
ASECS (American Society for 18th-Century Studies)/Clark Library Fellowships 646
Bernadotte E Schmitt Grants 63
BIAA Research Scholarship 183
BIAA Study Grants 183
British Academy Small Research Grants 176
Camargo Fellowships 200
Clark Library Short-Term Resident Fellowships 646
Clark-Huntington Joint Bibliographical Fellowship 647
Craig Hugh Smyth Visiting Fellowship 836
Delahaye Memorial Benefaction 815
Earthwatch Field Research Grants 279
Findel Scholarships and Schneider Scholarships 339
Helen McCormack Turner Memorial Scholarship 815
Herbert Hughes Scholarship 815
Herzog August Library Fellowship 339
I Tatti Fellowships 836
Leibniz Institute of European History Fellowships 370
M Alison Frantz Fellowship 91
Mary Radcliffe Scholarship 815
Oscar Broneer Traveling Fellowship 91
Queen Mary, University of London Research Studentships 540
RHYS Curzon-Jones Scholarship 815
Ridley Lewis Bursary 815
SOAS Research Scholarship 588
University of Bristol Postgraduate Scholarships 680
Victoria Doctoral Scholarships 836
W D Llewelyn Memorial Benefaction 816

Australia

AAH Humanities Travelling Fellowships 143

Canada

Oscar Broneer Traveling Fellowship 91

European Union

CBRL Travel Grant 252
Collaborative Doctoral Awards 121
German Historical Institute Doctoral and Postdoctoral Fellowships 314
Professional Preparation Master's Scheme 121
Research Preparation Master's Scheme 121

United Kingdom

Balsdon Fellowship 190
CBRL Travel Grant 252
Collaborative Doctoral Awards 121
Mr and Mrs David Edward Memorial Award 164
Professional Preparation Master's Scheme 121
Research Preparation Master's Scheme 121
Rome Awards 191
Rome Fellowship 192
Rome Scholarships in Ancient, Medieval and Later Italian Studies 192

United States of America

Collaborative Research Grants in the Humanities 54
Fritz Stern Dissertation Prize 314
German Historical Institute Doctoral and Postdoctoral Fellowships 314
Japan Society for the Promotion of Science (JSPS) Fellowship 606
Marshall Scholarships 181
NEH Fellowships 91
Oscar Broneer Traveling Fellowship 91

West European Countries

German Historical Institute Doctoral and Postdoctoral Fellowships 314
Mr and Mrs David Edward Memorial Award 164

HOLY WRITINGS

Any Country

BIAA Research Scholarship 183
BIAA Study Grants 183
Craig Hugh Smyth Visiting Fellowship 836
I Tatti Fellowships 836
SOAS Research Scholarship 588

Australia

AAH Humanities Travelling Fellowships 143

European Union

CBRL Travel Grant 252
Collaborative Doctoral Awards 121
Professional Preparation Master's Scheme 121
Research Preparation Master's Scheme 121

United Kingdom

Balsdon Fellowship 190
CBRL Travel Grant 252
Collaborative Doctoral Awards 121
Professional Preparation Master's Scheme 121
Research Preparation Master's Scheme 121
Rome Awards 191
Rome Fellowship 192
Rome Scholarships in Ancient, Medieval and Later Italian Studies 192

United States of America

ARCE Fellowships 87

RELIGIOUS EDUCATION

Any Country

BIAA Research Scholarship 183
BIAA Study Grants 183
Dempster Fellowship 309
Faculty Studentships 733
Jennings Randolph Program for International Peace Senior Fellowships 650
Louisville Institute Dissertation Fellowship Program 438
University of Kent School of European Culture and Languages Scholarships 742

African Nations

International Postgraduate Research Scholarships (IPRS) 417

Canada

International Postgraduate Research Scholarships (IPRS) 417
Ministry Fellowship 308
North American Doctoral Fellowship 308

Caribbean Countries

International Postgraduate Research Scholarships (IPRS) 417

East European Countries

International Postgraduate Research Scholarships (IPRS) 417

European Union

All Saints Educational Trust Personal Scholarships 21
CBRL Travel Grant 252

Middle East

International Postgraduate Research Scholarships (IPRS) 417

South Africa

International Postgraduate Research Scholarships (IPRS) 417

United Kingdom

All Saints Educational Trust Personal Scholarships 21

United States of America

West European Countries

RELIGIOUS PRACTICE

Any Country

Canada

European Union

United Kingdom

United States of America

West European Countries

RELIGIOUS STUDIES

Any Country

Australia

Canada

European Union

United Kingdom

United States of America

ISLAM

Any Country

Canada

European Union

United Kingdom

United States of America

West European Countries

JUDAIC RELIGIOUS STUDIES

Any Country

Canada

European Union

United Kingdom

United States of America

West European Countries

SOCIOLOGY OF RELIGION

Any Country

Australia

SERVICE TRADES

GENERAL

SOCIAL AND BEHAVIOURAL SCIENCES

GENERAL

African Nations

Australia

Canada

Caribbean Countries

East European Countries

European Union

Middle East

New Zealand

South Africa

United Kingdom

United States of America

West European Countries

ECONOMETRICS

Any Country

African Nations

Australia

Canada

Caribbean Countries

East European Countries

European Union

Middle East

New Zealand

South Africa

United Kingdom

United States of America

West European Countries

ECONOMIC AND FINANCE POLICY

Any Country

African Nations

Australia

Canada

Caribbean Countries

East European Countries

European Union

Middle East

New Zealand

South Africa

United Kingdom

CBRL Travel Grant 252
ESRC 1+3 Awards and +3 Awards 280
EUI Postgraduate Scholarships 293
International Postgraduate Research Scholarships (IPRS) 417
Mr and Mrs David Edward Memorial Award 164
University of Essex Silberrad Scholarships 729

United States of America

Aberystwyth International Excellence Scholarships 4
Fulbright Distinguished Chairs Program 252
Fulbright Specialist Program 253
International Postgraduate Research Scholarships (IPRS) 417
IREX Short-Term Travel Grants 391
Japan Society for the Promotion of Science (JSPS) Fellowship 606
Marshall Scholarships 181
SSRC Abe Fellowship Program 606

West European Countries

ESRC 1+3 Awards and +3 Awards 280
EUI Postgraduate Scholarships 293
International Postgraduate Research Scholarships (IPRS) 417
Mr and Mrs David Edward Memorial Award 164

ECONOMIC HISTORY

Any Country

AIER Summer Fellowship 65
BIAA Study Grants 183
Camargo Fellowships 200
Craig Hugh Smyth Visiting Fellowship 836
CRF European Visiting Research Fellowships 573
David Hume Fellowship 365
Doctoral Career Developement Scholarship (DCDS) Scheme 4
Dumbarton Oaks Fellowships and Junior Fellowships 278
Findel Scholarships and Schneider Scholarships 339
Frederick Douglass Institute Postdoctoral Fellowship 304
Frederick Douglass Institute Predoctoral Dissertation Fellowship 304
German Historical Institute Transatlantic Doctoral Seminar in German History 314
Graduate Dissertation Research Fellowship 457
Harry S Truman Library Institute Dissertation Year Fellowships 330
Herbert Hoover Presidential Library Association Travel Grants 336
Herzog August Library Fellowship 339
I Tatti Fellowships 836
IHS Summer Graduate Research Fellowship 367
Institute for Advanced Study Postdoctoral Residential Fellowships 366
Leibniz Institute of European History Fellowships 370
Lewis Walpole Library Fellowship 429
The Library Company of Philadelphia And The Historical Society of Pennsylvania Visiting Research Fellowships in Colonial and U.S. History and Culture 430
Library Company of Philadelphia Dissertation Fellowships 430
Minda de Gunzberg Graduate Dissertation Completion Fellowship 457
Paul H. Nitze School of Advanced International Studies (SAIS) Financial Aid and Fellowships 173
Postdoctoral Fellowship 236
Queen Mary, University of London Research Studentships 540
Rhodes University Postdoctoral Fellowship and The Andrew Mellon Postdoctoral Fellowship 549
Roosevelt Institute Research Grant 303
Russell Sage Foundation Visiting Scholar Program 582
University of Bristol Postgraduate Scholarships 680
Victoria Doctoral Scholarships 836

African Nations

Aberystwyth International Excellence Scholarships 4
International Postgraduate Research Scholarships (IPRS) 417
St Antony's College: ENI Scholarship 788

Australia

Aberystwyth International Excellence Scholarships 4
Australian Postgraduate Award 674

Canada

Aberystwyth International Excellence Scholarships 4
International Postgraduate Research Scholarships (IPRS) 417

Caribbean Countries

Aberystwyth International Excellence Scholarships 4
International Postgraduate Research Scholarships (IPRS) 417

East European Countries

International Postgraduate Research Scholarships (IPRS) 417

European Union

CBRL Travel Grant 252
ESRC 1+3 Awards and +3 Awards 280
ESRC Studentships 677
German Historical Institute Doctoral and Postdoctoral Fellowships 314
University of Essex Silberrad Scholarships 729

Middle East

Aberystwyth International Excellence Scholarships 4
International Postgraduate Research Scholarships (IPRS) 417

New Zealand

Aberystwyth International Excellence Scholarships 4
Australian Postgraduate Award 674

South Africa

Aberystwyth International Excellence Scholarships 4
International Postgraduate Research Scholarships (IPRS) 417

United Kingdom

Access to Learning Fund 725
Balsdon Fellowship 190
CBRL Travel Grant 252
ESRC 1+3 Awards and +3 Awards 280
ESRC Studentships 677
International Postgraduate Research Scholarships (IPRS) 417
Mr and Mrs David Edward Memorial Award 164
Prix du Québec Award 179
Rome Awards 191
Rome Fellowship 192
Rome Scholarships in Ancient, Medieval and Later Italian Studies 192
University of Essex Silberrad Scholarships 729

United States of America

Aberystwyth International Excellence Scholarships 4
ARCE Fellowships 87
Fritz Stern Dissertation Prize 314
Fulbright Distinguished Chairs Program 252
Fulbright Specialist Program 253
German Historical Institute Doctoral and Postdoctoral Fellowships 314
German Historical Institute Summer Seminar in Germany 314
German Historical Institute Transatlantic Doctoral Seminar in German History 314
International Postgraduate Research Scholarships (IPRS) 417
Japan Society for the Promotion of Science (JSPS) Fellowship 606
Marshall Scholarships 181
SSRC Abe Fellowship Program 606

West European Countries

ESRC 1+3 Awards and +3 Awards 280
German Historical Institute Doctoral and Postdoctoral Fellowships 314
German Historical Institute Transatlantic Doctoral Seminar in German History 314
International Postgraduate Research Scholarships (IPRS) 417
Mr and Mrs David Edward Memorial Award 164

ECONOMICS

Any Country

ABMRF/The Foundation for Alcohol Research Project Grant 5

African Nations

Australia

Canada

Caribbean Countries

East European Countries

European Union

Middle East

New Zealand

South Africa

United Kingdom

Access to Learning Fund 725
Alfa Fellowship Program 225
Balsdon Fellowship 190
CBRL Travel Grant 252
Economics - ESRC Quota Award 766
ESRC 1+3 Awards and +3 Awards 280
ESRC Studentships 677
EUI Postgraduate Scholarships 293
International Postgraduate Research Scholarships (IPRS) 417
Mr and Mrs David Edward Memorial Award 164
Rome Awards 191
Rome Fellowship 192
Rome Scholarships in Ancient, Medieval and Later Italian Studies 192
University of Essex Silberrad Scholarships 729
Utilities and Service Industries Training (USIT) 259

United States of America

Aberystwyth International Excellence Scholarships 4
ABMRF/The Foundation for Alcohol Research Project Grant 5
Alfa Fellowship Program 225
ARCE Fellowships 87
Environmental Public Policy and Conflict Resolution PhD Fellowship 468
Fellowship of the Flemish Community 546
Fulbright Distinguished Chairs Program 252
Fulbright Specialist Program 253
Gilbert Chinard Fellowships 364
Harmon Chadbourn Rorison Fellowship 365
International Postgraduate Research Scholarships (IPRS) 417
IREX Individual Advanced Research Opportunities 391
IREX Short-Term Travel Grants 391
Japan Society for the Promotion of Science (JSPS) Fellowship 606
Kennedy Research Grants 408
Marshall Scholarships 181
NIH Research Grants 491
Robert Bosch Foundation Fellowships 551
Sea Grant/NOAA Fisheries Fellowship 496
SSRC Abe Fellowship Program 606
Washington University Chancellor's Graduate Fellowship Program 841

West European Countries

ESRC 1+3 Awards and +3 Awards 280
EUI Postgraduate Scholarships 293
International Postgraduate Research Scholarships (IPRS) 417
Mr and Mrs David Edward Memorial Award 164

INDUSTRIAL AND PRODUCTION ECONOMICS

Any Country

BIAA Study Grants 183
Doctoral Career Developement Scholarship (DCDS) Scheme 4
Houblon-Norman Fellowships/George Fellowships 344
S S Huebner Foundation for Insurance Education Doctoral Fellowships 583
Sohei Nakayama Memorial Scholarship (Type A, B, C, D and S) 393
Victoria Doctoral Scholarships 836

African Nations

Aberystwyth International Excellence Scholarships 4
International Postgraduate Research Scholarships (IPRS) 417

Australia

Aberystwyth International Excellence Scholarships 4

Canada

Aberystwyth International Excellence Scholarships 4
International Postgraduate Research Scholarships (IPRS) 417
S S Huebner Foundation for Insurance Education Doctoral Fellowships 583

Caribbean Countries

Aberystwyth International Excellence Scholarships 4
International Postgraduate Research Scholarships (IPRS) 417

East European Countries

International Postgraduate Research Scholarships (IPRS) 417

European Union

ESRC 1+3 Awards and +3 Awards 280
University of Essex Silberrad Scholarships 729

Middle East

Aberystwyth International Excellence Scholarships 4
International Postgraduate Research Scholarships (IPRS) 417

New Zealand

Aberystwyth International Excellence Scholarships 4

South Africa

Aberystwyth International Excellence Scholarships 4
International Postgraduate Research Scholarships (IPRS) 417

United Kingdom

Access to Learning Fund 725
ESRC 1+3 Awards and +3 Awards 280
International Postgraduate Research Scholarships (IPRS) 417
Mr and Mrs David Edward Memorial Award 164
University of Essex Silberrad Scholarships 729

United States of America

Aberystwyth International Excellence Scholarships 4
International Postgraduate Research Scholarships (IPRS) 417
Japan Society for the Promotion of Science (JSPS) Fellowship 606
Marshall Scholarships 181
SSRC Abe Fellowship Program 606

West European Countries

ESRC 1+3 Awards and +3 Awards 280
International Postgraduate Research Scholarships (IPRS) 417
Mr and Mrs David Edward Memorial Award 164

INTERNATIONAL ECONOMICS

TAXATION

Any Country

ABMRF/The Foundation for Alcohol Research Project Grant 5
AIER Summer Fellowship 65
Doctoral Career Developement Scholarship (DCDS) Scheme 4
Rhodes University Postdoctoral Fellowship and The Andrew Mellon Postdoctoral Fellowship 549
University of Bristol Postgraduate Scholarships 680
Victoria Doctoral Scholarships 836

African Nations

International Postgraduate Research Scholarships (IPRS) 417

Canada

ABMRF/The Foundation for Alcohol Research Project Grant 5
International Postgraduate Research Scholarships (IPRS) 417

Caribbean Countries

International Postgraduate Research Scholarships (IPRS) 417

East European Countries

International Postgraduate Research Scholarships (IPRS) 417

European Union

University of Essex Silberrad Scholarships 729

Middle East

South Africa

United Kingdom

United States of America

West European Countries

ANCIENT CIVILISATIONS (EGYPTOLOGY, ASSYRIOLOGY)

Any Country

Australia

Canada

European Union

United Kingdom

United States of America

ANTHROPOLOGY

Any Country

World Universities Network (WUN) International Research Mobility Scheme 806

African Nations

British Institute in Eastern Africa Graduate Attachments 184
Canadian Window on International Development 384
IDRC Doctoral Research Awards 384
IDRC Research Awards 384
Primate Conservation Inc. Grants 538

Canada

ABMRF/The Foundation for Alcohol Research Project Grant 5
Canadian Window on International Development 384
IDRC Doctoral Research Awards 384
IDRC Research Awards 384
Mary McNeill Scholarship in Irish Studies 372
Oscar Broneer Traveling Fellowship 91
The Yitzhak Rabin Fellowship Fund for the Advancement of Peace and Tolerance 319

Caribbean Countries

Canadian Window on International Development 384
IDRC Doctoral Research Awards 384
IDRC Research Awards 384

East European Countries

IDRC Doctoral Research Awards 384

European Union

CBRL Travel Grant 252
ESRC 1+3 Awards and +3 Awards 280

Middle East

IDRC Doctoral Research Awards 384
IDRC Research Awards 384

South Africa

Canadian Window on International Development 384
IDRC Doctoral Research Awards 384
IDRC Research Awards 384

United Kingdom

Balsdon Fellowship 190
British Institute in Eastern Africa Graduate Attachments 184
CBRL Travel Grant 252
Emslie Horniman Anthropological Scholarship Fund 556
ESRC 1+3 Awards and +3 Awards 280
Hector and Elizabeth Catling Bursary 189
Prix du Québec Award 179
Rome Awards 191
Rome Fellowship 192
Rome Scholarships in Ancient, Medieval and Later Italian Studies 192
Slawson Awards 566
St Cross College: Graduate Scholarship in Anthropology 790

United States of America

ABMRF/The Foundation for Alcohol Research Project Grant 5
ACC Fellowship Grants Program 125
ARCE Fellowships 87
ASOR W.F. Albright Institute of Archaeological Research/National Endowment of the Humanities Fellowships 92
Fellowships for Intensive Advanced Turkish Language Study in Istanbul, Turkey 88
Fulbright Specialist Program 253
Fulbright-Kennan Institute Research Scholarships 411
IREX Individual Advanced Research Opportunities 391
IREX Short-Term Travel Grants 391
Japan Society for the Promotion of Science (JSPS) Fellowship 606
Mary McNeill Scholarship in Irish Studies 372
NEH ARIT-National Endowment for the Humanities Fellowships for Research in Turkey 88
NEH Fellowships 91
NIH Research Grants 491
Oscar Broneer Traveling Fellowship 91
SSRC Abe Fellowship Program 606
Washington University Chancellor's Graduate Fellowship Program 841

West European Countries

ESRC 1+3 Awards and +3 Awards 280

ETHNOLOGY

Any Country

ABMRF/The Foundation for Alcohol Research Project Grant 5
ASCSA Fellowships 89
Earthwatch Field Research Grants 279
Fondation Fyssen Postdoctoral Study Grants 298
IAS-STS Fellowship Programme 366
Institute of Irish Studies Research Fellowships 372
Jennings Randolph Program for International Peace Senior Fellowships 650
M Alison Frantz Fellowship 91
Phillips Fund for Native American Research 81
Victoria Doctoral Scholarships 836

Canada

ABMRF/The Foundation for Alcohol Research Project Grant 5
Mary McNeill Scholarship in Irish Studies 372

European Union

CBRL Travel Grant 252
ESRC 1+3 Awards and +3 Awards 280

United Kingdom

CBRL Travel Grant 252
ESRC 1+3 Awards and +3 Awards 280

United States of America

ABMRF/The Foundation for Alcohol Research Project Grant 5
ACC Fellowship Grants Program 125
ARCE Fellowships 87
Bicentennial Swedish-American Exchange Fund Travel Grants 249
Japan Society for the Promotion of Science (JSPS) Fellowship 606
Mary McNeill Scholarship in Irish Studies 372
NEH Fellowships 91

West European Countries

ESRC 1+3 Awards and +3 Awards 280

FOLKLORE

Any Country

ASCSA Fellowships 89
Earthwatch Field Research Grants 279
Institute of Irish Studies Research Fellowships 372
M Alison Frantz Fellowship 91
Phillips Fund for Native American Research 81
Victoria Doctoral Scholarships 836

Canada

Mary McNeill Scholarship in Irish Studies 372

European Union

CBRL Travel Grant 252
ESRC 1+3 Awards and +3 Awards 280

United Kingdom

CBRL Travel Grant 252
ESRC 1+3 Awards and +3 Awards 280

United States of America

ACC Fellowship Grants Program 125

African Nations

Australia

Canada

Caribbean Countries

East European Countries

European Union

Middle East

New Zealand

South Africa

United Kingdom

United States of America

African Nations

Australia

Canada

Caribbean Countries

East European Countries

European Union

Middle East

New Zealand

South Africa

United Kingdom

United States of America

West European Countries

HERITAGE PRESERVATION

Any Country

African Nations

Canada

Caribbean Countries

East European Countries

European Union

Middle East

South Africa

United Kingdom

United States of America

West European Countries

INTERNATIONAL RELATIONS

Any Country

African Nations

Australia

Canada

Caribbean Countries

East European Countries

European Union

Middle East

New Zealand

South Africa

United Kingdom

United States of America

West European Countries

POLITICAL SCIENCE AND GOVERNMENT

Any Country

African Nations

Australia

Canada

Caribbean Countries

East European Countries

European Union

Middle East

New Zealand

South Africa

United Kingdom

EXPERIMENTAL PSYCHOLOGY

INDUSTRIAL AND ORGANISATIONAL PSYCHOLOGY

Middle East

International Postgraduate Research Scholarships (IPRS) 417

South Africa

International Postgraduate Research Scholarships (IPRS) 417

United Kingdom

ESRC 1+3 Awards and +3 Awards 280
International Postgraduate Research Scholarships (IPRS) 417
Mr and Mrs David Edward Memorial Award 164

United States of America

ABMRF/The Foundation for Alcohol Research Project Grant 5
International Postgraduate Research Scholarships (IPRS) 417
Marshall Scholarships 181

West European Countries

ESRC 1+3 Awards and +3 Awards 280
International Postgraduate Research Scholarships (IPRS) 417
Mr and Mrs David Edward Memorial Award 164

PERSONALITY PSYCHOLOGY

Any Country

ABMRF/The Foundation for Alcohol Research Project Grant 5
Alzheimer's Society Research Grants 25
BackCare Research Grants 162
Victoria Doctoral Scholarships 836

African Nations

International Postgraduate Research Scholarships (IPRS) 417

Australia

Australian Postgraduate Award 674

Canada

ABMRF/The Foundation for Alcohol Research Project Grant 5
International Postgraduate Research Scholarships (IPRS) 417

Caribbean Countries

International Postgraduate Research Scholarships (IPRS) 417

East European Countries

International Postgraduate Research Scholarships (IPRS) 417

European Union

ESRC 1+3 Awards and +3 Awards 280

Middle East

International Postgraduate Research Scholarships (IPRS) 417

New Zealand

Australian Postgraduate Award 674

South Africa

International Postgraduate Research Scholarships (IPRS) 417

United Kingdom

ESRC 1+3 Awards and +3 Awards 280
International Postgraduate Research Scholarships (IPRS) 417
Mr and Mrs David Edward Memorial Award 164
PWSA UK Research Grants 537

United States of America

ABMRF/The Foundation for Alcohol Research Project Grant 5
International Postgraduate Research Scholarships (IPRS) 417
Marshall Scholarships 181
NIH Research Grants 491

West European Countries

ESRC 1+3 Awards and +3 Awards 280
International Postgraduate Research Scholarships (IPRS) 417
Mr and Mrs David Edward Memorial Award 164

PSYCHOMETRICS

Any Country

ABMRF/The Foundation for Alcohol Research Project Grant 5
Alzheimer's Society Research Grants 25
ETS Summer Internship Program for Graduate Students 282
NARSAD Distinguished Investigator Awards 476
NARSAD Independent Investigator Awards 476
NARSAD Young Investigator Awards 476
Victoria Doctoral Scholarships 836

Canada

ABMRF/The Foundation for Alcohol Research Project Grant 5

United Kingdom

Mr and Mrs David Edward Memorial Award 164

United States of America

ABMRF/The Foundation for Alcohol Research Project Grant 5
ETS Summer Internship Program for Graduate Students 282
Marshall Scholarships 181
NIH Research Grants 491

West European Countries

Mr and Mrs David Edward Memorial Award 164

SOCIAL PSYCHOLOGY

Any Country

ABMRF/The Foundation for Alcohol Research Project Grant 5
Alzheimer's Society Research Grants 25
BackCare Research Grants 162
Breast Cancer Campaign Project Grants 174
Doctoral Career Developement Scholarship (DCDS) Scheme 4
Eileen J Garrett Scholarship 522
Frederick Douglass Institute Postdoctoral Fellowship 304
Frederick Douglass Institute Predoctoral Dissertation Fellowship 304
Harold D. Lasswell Award 342
HFG Research Program 330
John L Stanley Award 342
Joshua Feigenbaum Award 343
Louise Kidder Early Career Award 611
Martinus Nijhoff Award 343
Parapsychology Foundation Grant 522
Postgraduate Research Bursaries (Epilepsy Action) 288
Rhodes University Postdoctoral Fellowship and The Andrew Mellon Postdoctoral Fellowship 549
Shelby Cullom Davis Center Research Projects, Research Fellowships 592
SPSSI Applied Social Issues Internship Program 612
SPSSI Grants-in-Aid Program 612
SPSSI Social Issues Dissertation Award 612
Teagasc Walsh Fellowships 633
Victoria Doctoral Scholarships 836

African Nations

International Postgraduate Research Scholarships (IPRS) 417

Australia

Australian Postgraduate Award 674
Biomedical (Dora Lush) and Public Health Postgraduate Scholarships 485
The Cancer Council NSW Research Project Grants 220
Career Development Fellowship 486
Clinical (Neil Hamilton Fairley) Fellowship 486
CRC Grants 260

Middle East

New Zealand

South Africa

United Kingdom

United States of America

West European Countries

COMPARATIVE SOCIOLOGY

Any Country

Australia

Canada

European Union

New Zealand

United Kingdom

United States of America

West European Countries

FUTUROLOGY

Any Country

European Union

United Kingdom

United States of America

West European Countries

HISTORY OF SOCIETIES

Any Country

Australia

European Union

New Zealand

United Kingdom

United States of America

West European Countries

SOCIAL INSTITUTIONS

Any Country

African Nations

Australia

Canada

Caribbean Countries

East European Countries

European Union

Middle East

New Zealand

South Africa

United Kingdom

United States of America

West European Countries

SOCIAL POLICY

Any Country

African Nations

Australia

Canada

Caribbean Countries

East European Countries

European Union

Middle East

New Zealand

South Africa

United Kingdom

United States of America

West European Countries

SOCIAL PROBLEMS

Any Country

United Kingdom

United States of America

West European Countries

WOMEN'S STUDIES

Any Country

African Nations

Australia

Canada

Caribbean Countries

East European Countries

European Union

Middle East

New Zealand

TRADE, CRAFT AND INDUSTRIAL TECHNIQUES

GENERAL

BUILDING TECHNOLOGIES

ELECTRICAL AND ELECTRONIC EQUIPMENT AND MAINTENANCE

FOOD TECHNOLOGY

Any Country

Baking Industry Scholarship 65
Research Contracts (IAEA) 380
Teagasc Walsh Fellowships 633

Australia

Australian Postgraduate Award 674

New Zealand

Australian Postgraduate Award 674

United States of America

Congress Bundestag Youth Exchange for Young Professionals 225

GRAPHIC ARTS

Any Country

Haystack Scholarship 331
Henry Moore Institute Research Fellowships 335
Lewis Walpole Library Fellowship 429
WSW Hands-On-Art Visiting Artist Project 863
WSW Internships 863

United States of America

Early American Industries Association Research Grants Program 279
Virginia Liebeler Biennial Grants for Mature Women (Art) 491

PRINTING AND PRINTMAKING

Any Country

Artists Fellowships at WSW 863
Artists' Book Residencies at WSW 863
Haystack Scholarship 331
Henry Moore Institute Research Fellowships 335
The Library Company of Philadelphia And The Historical Society of Pennsylvania Visiting Research Fellowships in Colonial and U.S. History and Culture 430
WSW Hands-On-Art Visiting Artist Project 863
WSW Internships 863

East European Countries

FEMS Fellowship 295

United Kingdom

FEMS Fellowship 295

United States of America

Early American Industries Association Research Grants Program 279
Virginia Liebeler Biennial Grants for Mature Women (Art) 491
Virginia Liebeler Biennial Grants for Mature Women (Writing) 492

West European Countries

FEMS Fellowship 295

PUBLISHING AND BOOK TRADE

Any Country

Artists Fellowships at WSW 863
Artists' Book Residencies at WSW 863
Harold White Fellowships 492
Henry Moore Institute Research Fellowships 335
The Library Company of Philadelphia And The Historical Society of Pennsylvania Visiting Research Fellowships in Colonial and U.S. History and Culture 430
WSW Hands-On-Art Visiting Artist Project 863
WSW Internships 863

East European Countries

FEMS Fellowship 295

United Kingdom

FEMS Fellowship 295

United States of America

Early American Industries Association Research Grants Program 279
Virginia Liebeler Biennial Grants for Mature Women (Writing) 492

West European Countries

FEMS Fellowship 295

HEATING, AND REFRIGERATION

Any Country

ASHRAE Grants-in-Aid for Graduate Students 101

United States of America

Congress Bundestag Youth Exchange for Young Professionals 225
DEED (Demonstration of Energy-Efficient Developments) Student Research Grant and Internship 87

LABORATORY TECHNIQUES

Any Country

Los Alamos Graduate Research Assistant Program 436
Research Contracts (IAEA) 380

Australia

Early Career Fellowship 428
Infrastructure Grant 480
National Collaborative Breast Cancer Research Grant Program 480

Canada

Norman and Marion Bright Memorial Award 218

United States of America

Congress Bundestag Youth Exchange for Young Professionals 225
Los Alamos Graduate Research Assistant Program 436

LEATHER TECHNIQUES

African Nations

CSIR (Council of Scientific and Industrial Research)/TWAS Fellowship for Postgraduate Research 636
CSIR (The Council of Scientific and Industrial Research)/TWAS Fellowship for Postdoctoral Research 636

Caribbean Countries

CSIR (Council of Scientific and Industrial Research)/TWAS Fellowship for Postgraduate Research 636
CSIR (The Council of Scientific and Industrial Research)/TWAS Fellowship for Postdoctoral Research 636

Middle East

CSIR (Council of Scientific and Industrial Research)/TWAS Fellowship for Postgraduate Research 636
CSIR (The Council of Scientific and Industrial Research)/TWAS Fellowship for Postdoctoral Research 636

South Africa

CSIR (Council of Scientific and Industrial Research)/TWAS Fellowship for Postgraduate Research 636
CSIR (The Council of Scientific and Industrial Research)/TWAS Fellowship for Postdoctoral Research 636

TRANSPORT AND COMMUNICATIONS

United Kingdom

ESRC 1+3 Awards and +3 Awards 280
International Postgraduate Research Scholarships (IPRS) 417
Kennedy Scholarships 412
Peter Fleming Award 565
RGS-IBG Postgraduate Research Awards 565

United States of America

Congress Bundestag Youth Exchange for Young Professionals 225
Fulbright - Northern Ireland Policy and Goverance Scholar Award 826
Fulbright - Scotland Visiting Professorship at the Glasgow Urban Lab 828
International Postgraduate Research Scholarships (IPRS) 417
North Dakota Indian Scholarship Program 511

West European Countries

ESRC 1+3 Awards and +3 Awards 280
International Postgraduate Research Scholarships (IPRS) 417
Janson Johan Helmich Scholarships and Travel Grants 401

AIR TRANSPORT

Any Country

Royal Aeronautical Society Centennial Scholarship Award 555

African Nations

International Postgraduate Research Scholarships (IPRS) 417

Australia

DOI Women in Freight, Logistics and Marine Management Scholarship 269

Canada

International Postgraduate Research Scholarships (IPRS) 417

Caribbean Countries

International Postgraduate Research Scholarships (IPRS) 417

East European Countries

International Postgraduate Research Scholarships (IPRS) 417

Middle East

International Postgraduate Research Scholarships (IPRS) 417

South Africa

International Postgraduate Research Scholarships (IPRS) 417

United Kingdom

International Postgraduate Research Scholarships (IPRS) 417

United States of America

Congress Bundestag Youth Exchange for Young Professionals 225
International Postgraduate Research Scholarships (IPRS) 417

West European Countries

International Postgraduate Research Scholarships (IPRS) 417

MARINE TRANSPORT AND NAUTICAL SCIENCE

Any Country

De Paepe-Willems Award 389
Ralph Brown Expedition Award 565

Australia

DOI Women in Freight, Logistics and Marine Management Scholarship 269

United States of America

Congress Bundestag Youth Exchange for Young Professionals 225

POSTAL SERVICES

United States of America

Congress Bundestag Youth Exchange for Young Professionals 225

RAILWAY TRANSPORT

African Nations

International Postgraduate Research Scholarships (IPRS) 417

Australia

DOI Women in Freight, Logistics and Marine Management Scholarship 269

Canada

International Postgraduate Research Scholarships (IPRS) 417

Caribbean Countries

International Postgraduate Research Scholarships (IPRS) 417

East European Countries

International Postgraduate Research Scholarships (IPRS) 417

Middle East

International Postgraduate Research Scholarships (IPRS) 417

South Africa

International Postgraduate Research Scholarships (IPRS) 417

United Kingdom

International Postgraduate Research Scholarships (IPRS) 417

United States of America

Congress Bundestag Youth Exchange for Young Professionals 225
International Postgraduate Research Scholarships (IPRS) 417

West European Countries

International Postgraduate Research Scholarships (IPRS) 417

ROAD TRANSPORT

Any Country

Rees Jeffreys Road Fund Bursaries 543
Rees Jeffreys Road Fund Research Grants 543

African Nations

International Postgraduate Research Scholarships (IPRS) 417

Australia

DOI Women in Freight, Logistics and Marine Management Scholarship 269

Canada

International Postgraduate Research Scholarships (IPRS) 417
TAC Foundation Scholarships 643

Caribbean Countries

International Postgraduate Research Scholarships (IPRS) 417

East European Countries

International Postgraduate Research Scholarships (IPRS) 417

Middle East

International Postgraduate Research Scholarships (IPRS) 417

South Africa
International Postgraduate Research Scholarships (IPRS) 417

United Kingdom
International Postgraduate Research Scholarships (IPRS) 417

United States of America
Congress Bundestag Youth Exchange for Young Professionals 225
International Postgraduate Research Scholarships (IPRS) 417

West European Countries
International Postgraduate Research Scholarships (IPRS) 417

TELECOMMUNICATIONS SERVICES

Any Country
Essex Rotary University Travel Grants 727
Sir Eric Berthoud Travel Grant 728

African Nations
International Postgraduate Research Scholarships (IPRS) 417

Canada
International Postgraduate Research Scholarships (IPRS) 417

Caribbean Countries
International Postgraduate Research Scholarships (IPRS) 417

East European Countries
International Postgraduate Research Scholarships (IPRS) 417

Middle East
International Postgraduate Research Scholarships (IPRS) 417

South Africa
International Postgraduate Research Scholarships (IPRS) 417

United Kingdom
Access to Learning Fund 725
International Postgraduate Research Scholarships (IPRS) 417

United States of America
Congress Bundestag Youth Exchange for Young Professionals 225
International Postgraduate Research Scholarships (IPRS) 417

West European Countries
International Postgraduate Research Scholarships (IPRS) 417

TRANSPORT ECONOMICS

Any Country
Rees Jeffreys Road Fund Bursaries 543
Rees Jeffreys Road Fund Research Grants 543

African Nations
International Postgraduate Research Scholarships (IPRS) 417

Australia
DOI Women in Freight, Logistics and Marine Management Scholarship 269

Canada
International Postgraduate Research Scholarships (IPRS) 417
TAC Foundation Scholarships 643

Caribbean Countries
International Postgraduate Research Scholarships (IPRS) 417

East European Countries
International Postgraduate Research Scholarships (IPRS) 417

European Union
ESRC 1 + 3 Awards and + 3 Awards 280

Middle East
International Postgraduate Research Scholarships (IPRS) 417

South Africa
International Postgraduate Research Scholarships (IPRS) 417

United Kingdom
ESRC 1 + 3 Awards and + 3 Awards 280
International Postgraduate Research Scholarships (IPRS) 417

United States of America
Congress Bundestag Youth Exchange for Young Professionals 225
DEED (Demonstration of Energy-Efficient Developments) Student Research Grant and Internship 87
International Postgraduate Research Scholarships (IPRS) 417

West European Countries
ESRC 1 + 3 Awards and + 3 Awards 280
International Postgraduate Research Scholarships (IPRS) 417

TRANSPORT MANAGEMENT

African Nations
International Postgraduate Research Scholarships (IPRS) 417

Australia
DOI Women in Freight, Logistics and Marine Management Scholarship 269

Canada
International Postgraduate Research Scholarships (IPRS) 417
TAC Foundation Scholarships 643

Caribbean Countries
International Postgraduate Research Scholarships (IPRS) 417

East European Countries
International Postgraduate Research Scholarships (IPRS) 417

European Union
ESRC 1 + 3 Awards and + 3 Awards 280

Middle East
International Postgraduate Research Scholarships (IPRS) 417

South Africa
International Postgraduate Research Scholarships (IPRS) 417

United Kingdom
ESRC 1 + 3 Awards and + 3 Awards 280
International Postgraduate Research Scholarships (IPRS) 417

United States of America
Congress Bundestag Youth Exchange for Young Professionals 225
DEED (Demonstration of Energy-Efficient Developments) Student Research Grant and Internship 87
International Postgraduate Research Scholarships (IPRS) 417

West European Countries
ESRC 1 + 3 Awards and + 3 Awards 280
International Postgraduate Research Scholarships (IPRS) 417

INDEX OF AWARDS

INDEX OF DISCONTINUED AWARDS

Alberta Innovates Health Solutions
AIHS Health Research Studentship

British Association for Canadian Studies (BACS)
BACS Travel Awards

British School at Rome (BSR)
Helen Chadwick Fellowship

British Society for Antimicrobial Chemotherapy
Pfizer Anti-Infectives Research Foundation

Broadcast Education Association (BEA)
Broadcast Education Association James S Gilmore Award

Canadian Cancer Society Research Institute (CCSRI)
CCS Equipment Grants for New Investigators
CCS Research Grants for New Investigators
CCS Research Grants to Individuals

Christopher & Dana Reeve Foundation
Reeve Foundation Research Grant

European Calcified Tissue Society
ECTS/Servier Fellowship

Guide Dogs for the Blind Association
Guide Dogs Ophthalmic Research Grant

Institute for Advanced Studies in the Humanities
Andrew W Mellon Foundation East-Central European Fellowships in the Humanities

Memorial Foundation for Jewish Culture
Memorial Foundation for Jewish Culture International Scholarship Programme for Community Service

National Dairy Shrine
Iager Dairy Scholarship

Natural Environment Research Council (NERC)
NERC Advanced Research Fellowships
NERC Postdoctoral Research Fellowships

Royal College of Organists (RCO)
William Robertshaw Exhibition

Royal Geographical Society (with the Institute of British Geographers)
British Airways Travel Bursary
EPSRC Geographical Research Grants
Innovative Geography Teaching Grants
Ray Y. Gildea Jr Award

RSM Erasmus University
RSM Alumni Welcome Scholarship
RSM Erasmus University Dean's Fund Scholarship
RSM Erasmus University Dean's Fund Assistance Award
RSM Erasmus University Gateway Scholarship
RSM Erasmus University MBA Citizenship Scholarship
RSM MBA Corporate Scholarship Programme
RSM-Intermediair EMBA Scholarship
RSM/Far East Scholarship
RSM/Latin America Scholarship

Social Sciences and Humanities Research Council of Canada (SSHRC)
Aid to Research Workshops and Conferences in Canada

Swansea University
The James Callaghan Scholarships

University of Cambridge
Roger Needham Research Studentship

University of East Anglia (UEA)
Charles Pick Fellowship

The Australia Council for the Arts
Aboriginal and Torres Strait Islander Arts Indigenous Arts Worker's Program Grant
Aboriginal and Torres Strait Islander Arts Key Organisations Triennial
Aboriginal and Torres Strait Islander Arts Program
Communities Cultural Development Grants Programme
Communities Cultural Development Grants Residency
Communities Cultural Development Skills and Arts Development
Community Cultural Development Category A
Community Cultural Development Category B
Community Cultural Development Category C
Community Cultural Development Presentation and Promotion
Community Partnerships Residency
Dance Cite International des Arts Residency
Dance Initiative Project Workspace
Dance New Work Creative Development
Dance New Work Presentation
Dance Presentation and Promotion
Dance Program
Dance Skill and Arts Development–Skills Development for Young Emerging Artists
Inter-Arts Fellowship
Inter-Arts Hybrid Art Key Organisations Triennial
Inter-Arts Hybrid Art Program
Inter-Arts Initiative–Artslab
Inter-Arts Initiative-Run_way 2006
Inter-Arts National Residency
Inter-Arts Projects
Inter-Arts Residency
Inter-Arts Residency–International Residency
Literative Key Organisations Triennial
Literature New Work
Literature Presentation and Promotion
Literature Skills and Arts Development
Music Initiative
Music Initiative Sounding Out
Music International Market Development Program International Pathways
Music Key Organisations Triennial
Music National Residency
Music New Work
Music Presentation and Promotion
Music Program
OZCO Community Cultural Development Fellowship
OZCO Dance Grant Initiative: Take Your Partner
OZCO Literature Fellowships
OZCO Literature Grants Initiative: Write in Your Face
OZCO Theatre Fellowship
OZCO Visual Arts Fellowship
Theatre Initiative Dramaturgy Fellowship
Theatre Initiative-Playing the World
Theatre International Residency
Theatre Key Organisations Triennial
Theatre Presentation and Promotion
Visual Arts Emeritus Award
Visual Arts Emeritus Medal
Visual Arts Initiative Maker to Manufacturer to Market
Visual Arts Initiative-Artista Run
Visual Arts Key Organisations Triennial
Visual Arts New Work
Visual Arts Presentation and Promotion

The Jacob Rader Marcus Center of The American Jewish Archives
Ethel Marcus Memorial Fellowship
Marguerite R Jacobs Memorial Award
The Natalie Feld Memorial Fellowship

The University of Leeds
Brotherton Research Scholarship
Chevening-Kulika Charitable Trust - University of Leeds Scholarships
William Wright Smith Scholarship

The University of Waikato
Lee Foundation Grants
MTKS Education Fund International Scholarships

The Wingate Scholarships
Wingate Scholarships

INDEX OF AWARDING ORGANISATIONS

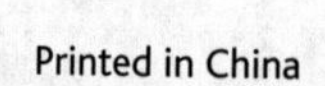
Printed in China